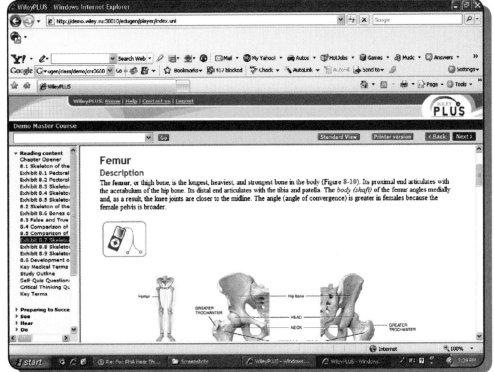

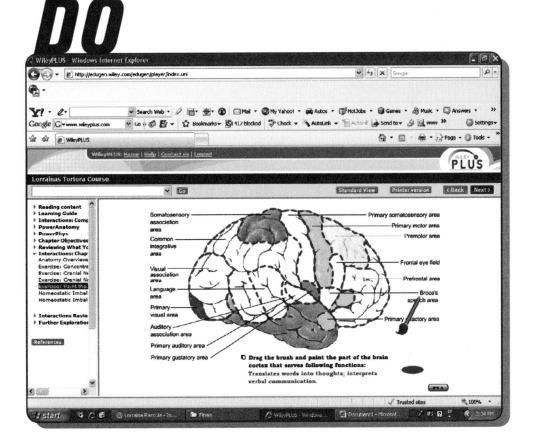

Anatomy and Physiology for the Manual Therapies

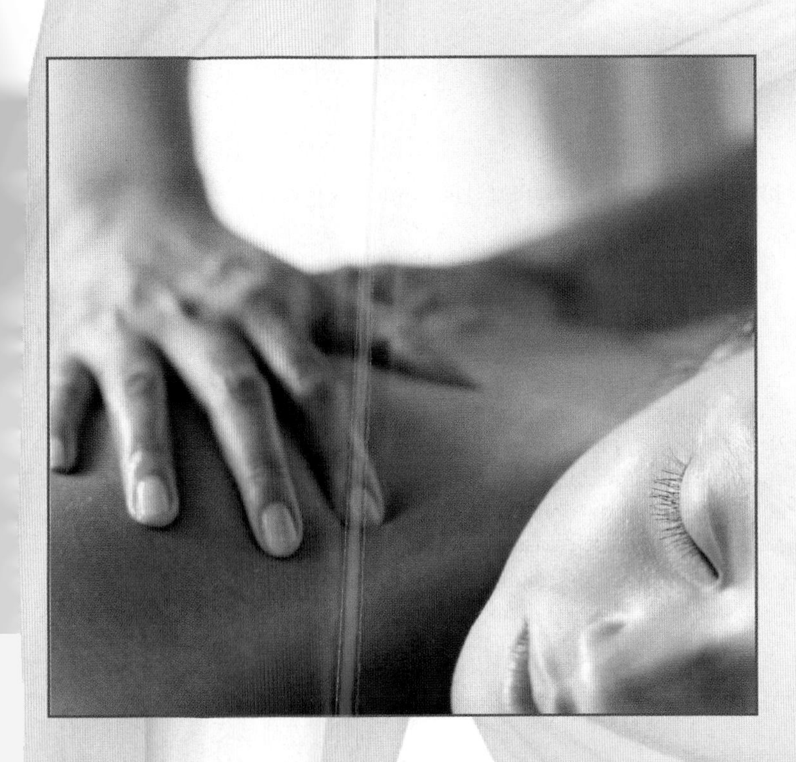

ANDREW J. KUNTZMAN

Sinclair Community College and
SHI Integrative Medical Massage and Traditional Chinese Acupuncture School

GERARD J. TORTORA

Bergen Community College

JOHN WILEY & SONS, INC.

Executive Editor	Bonnie Roesch
Executive Marketing Manager	Clay Stone
Project Editor	Lorraina Raccuia
Production Manager	Lucille Buonocore
Senior Designer	Madelyn Lesure
Senior Illustration Editor	Anna Melhorn
Photo Editor	Hilary Newman
Program Assistant	Lauren Morris
Senior Media Editor	Linda Muriello
Media Project Manager	Bonnie Roth
Outside Production Manager	Suzanne Ingrao
Cover Photo	© Alex Mares-Manton/Getty Images
Cover Illustration	John Gibb of Spiker Studios

This book was typeset by Aptara®, Inc. It was printed and bound by Quad/Graphics Versailles. The cover was printed by Quad/Graphics Versailles.

The paper in this book was manufactured by a mill whose forest management programs include sustained yield harvesting of its timberlands. Sustained yield harvesting principles ensure that the number of trees cut each year does not exceed the amount of new growth.

To order books or for customer service, please call 1-800-CALL WILEY (225-5945).

Library of Congress Cataloging-in-Publication Data

Kuntzman, Andrew J.
Anatomy and physiology for the manual therapies / Andrew J. Kuntzman, Gerald J. Tortora.
p. ; cm.
Includes index.
ISBN 978-0-470-04496-4 (cloth)
1. Human physiology. 2. Human anatomy. 3. Massage therapy 4. Physical therapy. I. Tortora, Gerard J. II. Title.
[DNLM: 1. Anatomy. 2. Physical Therapy Modalities. 3. Physiological Processes. QT 104 K946a 2010]
QP34.5.K86 2010
612—dc22

2008036131

Printed in the United States of America
10 9 8 7 6 5 4 3

ABOUT THE AUTHORS

Andrew J. Kuntzman, Ph.D., Licensed Massage Therapist, teaches anatomy and physiology at Sinclair Community College in Dayton, Ohio. He has taught anatomy and physiology at the medical school, graduate, and undergraduate levels for many years, the majority of this time at Wright State University. Andy has also taught anatomy and physiology, clinical massage, and massage theory to massage therapy students during the past 17 years at SHI Integrative Medical Massage & Traditional Chinese Acupuncture School. He received a bachelor's of science degree in science education, master's in science degree in zoology, and PhD in human anatomy, all from The Ohio State University. He has been honored with numerous teaching awards, including the Award for Innovative Excellence in Teaching, Learning, and Technology, from the Center for Advancement of Teaching and Learning in Jacksonville, FL.

A Licensed Massage Therapist in the state of Ohio, Andy has had his own practice and is the former Chair of the Anatomy and Physiology Curriculum Committee of the Ohio Council of Massage Therapy Schools and an active member of the American Massage Therapy Association.

When Andy is not teaching or working at his massage therapy practice, he enjoys spending time with his wife, four children, and three grandsons. He enjoys watching college football, reading, and gardening.

To my loving wife,
Judy, our four children, Bill,
Andrea, Karen, and Kathy, and three grandsons,
Matt, Andy, and Christopher.
To Karen E. Benedict, LMT, for her invaluable contributions
to the latter phases of this book
A.J.K.

Courtesy of Heidi Chung

Gerard J. Tortora is Professor of Biology and former Biology Coordinator at Bergen Community College in Paramus, New Jersey, where he teaches human anatomy and physiology as well as microbiology. He has taught for 47 years. Jerry received his bachelor's degree in biology from Fairleigh Dickinson University and his master's degree in science education from Montclair State College. He is a member of many professional organizations, including the Human Anatomy and Physiology Society (HAPS), the American Society of Microbiology (ASM), American Association for the Advancement of Science (AAAS), National Education Association (NEA), and the Metropolitan Association of College and University Biologists (MACUB).

Above all, Jerry is devoted to his students and their aspirations. In recognition of this commitment, Jerry was the recipient of MACUB's 1992 President's Memorial Award. In 1996, he received a National Institute for Staff and Organizational Development (NISOD) excellent award from the University of Texas and was selected to represent Bergen Community College in a campaign to increase awareness of the contributions of community colleges to higher education.

Jerry is the author of several best-selling science textbooks and laboratory manuals, a calling that often requires an additional 40 hours per week beyond his teaching responsibilities. Nevertheless, he still makes time for four or five weekly aerobic workouts that include biking and running. He also enjoys attending college basketball and professional hockey games and performances at the Metropolitan Opera House.

To my mother, Angelina M. Tortora.
Her love, guidance, faith, courage,
determination, support, and example continue
to be the cornerstone of my personal and professional life.
G.J.T.

PREFACE

A strong understanding of human anatomy and physiology is an important foundation for students of the manual therapies. Manual therapists are expected to assess and appropriately treat a wide variety of physical problems, so a holistic understanding of human anatomy and physiology is essential. This new textbook *Anatomy and Physiology for the Manual Therapies* offers students who are preparing for careers as massage therapists, physical therapy assistants, exercise therapists, sports medicine, or other careers in the manual therapies, a balanced presentation of anatomy and physiology, unified by the theme of homeostasis.

This unique textbook provides illustrations and text for every system of the body, with a special focus on the musculoskeletal and nervous systems that's tailored to the needs of manual therapists. There are four chapters that focus on the muscular system: Chapter 11: Muscles of the Head and Neck, Chapter 12: Muscles of the Torso, Chapter 13: Muscles of the Upper Limb, and Chapter 14: Muscles of the Lower Limb. These chapters that focus on the muscular system include comprehensive text, clear art that illustrates the skeleton and all major muscle groups including the important suboccipital muscles, art that shows muscle origins and insertions, and applications for manual therapists. In addition, each chapter includes at least one Manual Therapy Application, a boxed feature that relates the chapter's topic directly to the students' future careers.

ORGANIZATION

Anatomy and Physiology for the Manual Therapies begins by providing students with an understanding of the structural and functional levels of the human body, from molecules to organ systems. Chapter 4: Tissues includes description of connective tissues as they relate to the manual therapies. Students then learn the anatomy and physiology of the Integumentary system. A detailed group of chapters follow, offering students a deeper understanding of the skeletal system, joints, and muscles. The nervous system is then covered, with an emphasis on neural communication and maintenance of homeostasis, as well as sensations of pain. The endocrine system, cardiovascular system, and lymphatic systems are covered, giving manual therapy students an understanding of hormonal controls, the heart's anatomy and physiology, blood flow and circulation, as well as the immune system. The next chapters cover the respiratory system, the digestive system, nutrition, and the urinary system, giving students an understanding of how the body uses and processes air and nutrients. The final chapter provides students with an overview of the reproductive systems.

ILLUSTRATED BOOK TOUR

CHAPTER OPENING VIGNETTES

Unique chapter openers get students' attention and connect them with the material that follows. These chapter openers relate the chapter topic to the manual therapies and students' everyday lives.

11 Muscular System: Muscles of the Head and Neck

Manual therapists require an integral knowledge of the muscular system anatomy. Muscles often have injuries in the musculotendinous units or the muscle belly itself. Therefore you need a good understanding of the origin, insertion, and fiber direction for those muscles to effectively treat the key areas. You must also assess patients or clients to determine if the muscle is the primary source of injury or whether it comes from another source such as the nerve. If a nerve is injured it may cause either atrophy of the muscle or spasm, depending on the cause and severity of the injury. Muscles also require large amounts of oxygen to produce ATP for contraction; this in turn requires adequate blood supply. You will study all of these aspects for each of the muscles in the body. Beginning with the head and neck, and continuing with the rest of the body in later chapters, you will cover the major skeletal muscles in the body that provide the basis for manual therapy.

CONTENTS AT A GLANCE

284

MANUAL THERAPY APPLICATIONS

Each chapter includes at least one special box that relates the chapter topic to manual therapists. Some Manual Therapies Applications include Hydrotherapy, Massage of Cancer Patients, Treatment of Whiplash, Back Injuries and Heavy Lifting, and Pregnancy Massage.

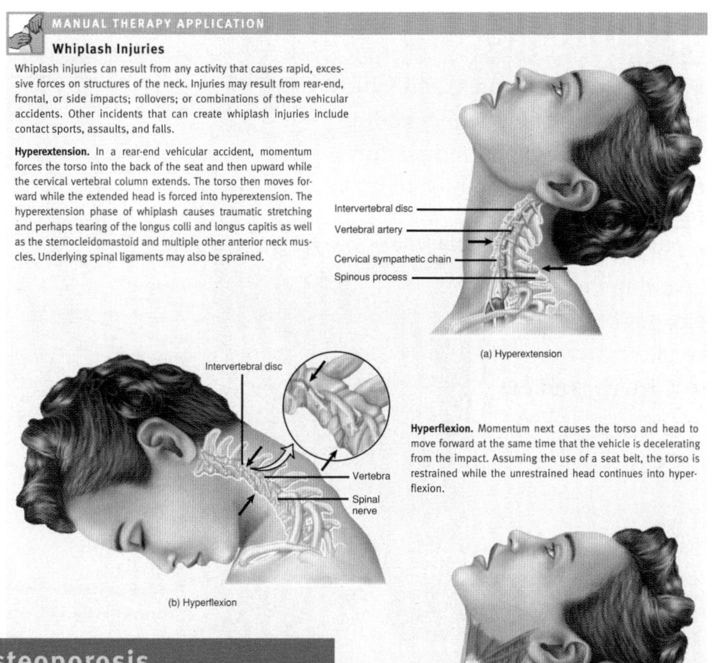

MANUAL THERAPY APPLICATION

Whiplash Injuries

Whiplash injuries can result from any activity that causes rapid, excessive forces on structures of the neck. Injuries may result from rear-end, frontal, or side impacts; rollovers; or combinations of these vehicular accidents. Other incidents that can create whiplash injuries include contact sports, assaults, and falls.

Hyperextension. In a rear-end vehicular accident, momentum forces the torso into the back of the seat and then upward while the cervical vertebral column extends. The torso then moves forward while the extended head is forced into hyperextension. The hyperextension phase of whiplash causes traumatic stretching and perhaps tearing of the longus colli and longus capitis as well as the sternocleidomastoid and multiple other anterior neck muscles. Underlying spinal ligaments may also be sprained.

(a) Hyperextension

Hyperflexion. Momentum next causes the torso and head to move forward at the same time that the vehicle is decelerating from the impact. Assuming the use of a seat belt, the torso is restrained while the unrestrained head continues into hyperflexion.

(b) Hyperflexion

• CLINICAL CONNECTION | Osteoporosis

Osteoporosis (os′-tē-ō-pō-RŌ-sis; *por-* = passageway; *-osis* = condition) is literally a condition of porous bones (Figure 6.12) that affects 10 million people in the United States each year. In addition, about 18 million people have *osteopenia* (low bone mass), which puts them at risk for osteoporosis. The basic problem is that bone resorption outpaces bone deposition. In large part this is due to depletion of calcium from the body—more calcium is lost in urine, feces, and sweat than is absorbed from the diet. Bone mass becomes so depleted that bones fracture, often spontaneously, under the mechanical stresses of everyday living. For example, a hip fracture might result from simply sitting down too quickly. In the United States, osteoporosis results in more than one and a half million fractures a year, mainly in the hip, wrist, and vertebrae. Osteoporosis afflicts the entire skeletal system. In addition to fractures, osteoporosis causes shrinkage of vertebrae, height loss, hunched backs, and bone pain.

The disorder primarily affects middle-aged and elderly people, 80% of them women. Older women suffer from osteoporosis more often than men for two reasons: (1) Women's bones are less massive than men's bones, and (2) production of estrogens in women declines dramatically at menopause, but production of the main androgen, testosterone, wanes gradually and only slightly in older men. Estrogens and testosterone stimulate osteoblast activity and synthesis of bone extracellular matrix. Besides gender, risk factors for developing osteoporosis include

CLINICAL CONNECTIONS

In each chapter, a few Clinical Connections provide students more details on diseases, disorders, and treatments, related to a particular body system. Some examples of topics for Clinical Connections include: Lactose Intolerance, Osteoporosis, Muscular Dystrophy, Phantom Limb Sensation, Regeneration of Heart Cells, Prostate Disorders.

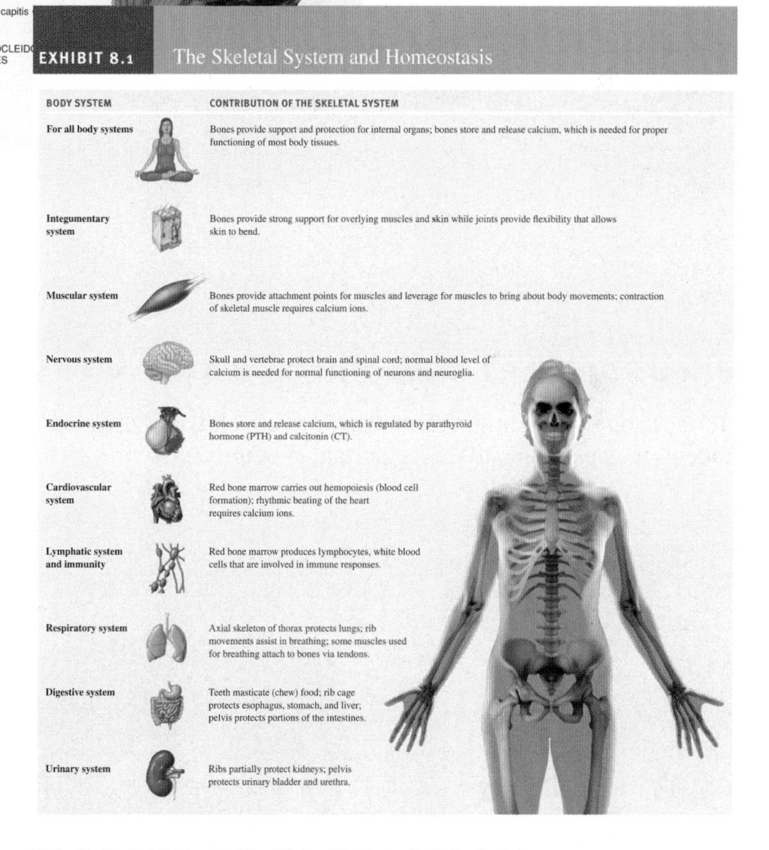

EXHIBIT 8.1 The Skeletal System and Homeostasis

BODY SYSTEM	CONTRIBUTION OF THE SKELETAL SYSTEM
For all body systems	Bones provide support and protection for internal organs; bones store and release calcium, which is needed for proper functioning of most body tissues.
Integumentary system	Bones provide strong support for overlying muscles and skin while joints provide flexibility that allows skin to bend.
Muscular system	Bones provide attachment points for muscles and leverage for muscles to bring about body movements; contraction of skeletal muscle requires calcium ions.
Nervous system	Skull and vertebrae protect brain and spinal cord; normal blood level of calcium is needed for normal functioning of neurons and neuroglia.
Endocrine system	Bones store and release calcium, which is regulated by parathyroid hormone (PTH) and calcitonin (CT).
Cardiovascular system	Red bone marrow carries out hemopoiesis (blood cell formation); rhythmic beating of the heart requires calcium ions.
Lymphatic system and immunity	Red bone marrow produces lymphocytes, white blood cells that are involved in immune responses.
Respiratory system	Axial skeleton of thorax protects lungs; rib movements assist in breathing; some muscles used for breathing attach to bones via tendons.
Digestive system	Teeth masticate (chew) food; rib cage protects esophagus, stomach, and liver; pelvis protects portions of the intestines.
Urinary system	Ribs partially protect kidneys; pelvis protects urinary bladder and urethra.

FOCUS ON HOMEOSTASIS

This feature combines graphic and narrative elements to explain how the system under consideration contributes to the homeostasis of the body as a whole. It is included for the following systems: integumentary, skeletal, muscular, nervous, endocrine, cardiovascular, lymphatic and immune, respiratory, digestive, urinary, and reproductive systems. This holistic approach to the human body will enhance student understanding of the connection between body systems.

THE ILLUSTRATION PROGRAM

Each page is carefully laid out to place related text, figures, and tables near one another, minimizing the need for page turning while reading a topic. Beautiful artwork, carefully chosen photographs and photomicrographs, and unique pedagogical enhancements all combine to make the visual appeal and usefulness of the illustration program in *Anatomy and Physiology for the Manual Therapies* distinctive. You will find exciting three-dimensional illustrations gracing the pages of nearly every chapter in the text. Many of the photomicrographs in this textbook have been provided by Andrew J. Kuntzman.

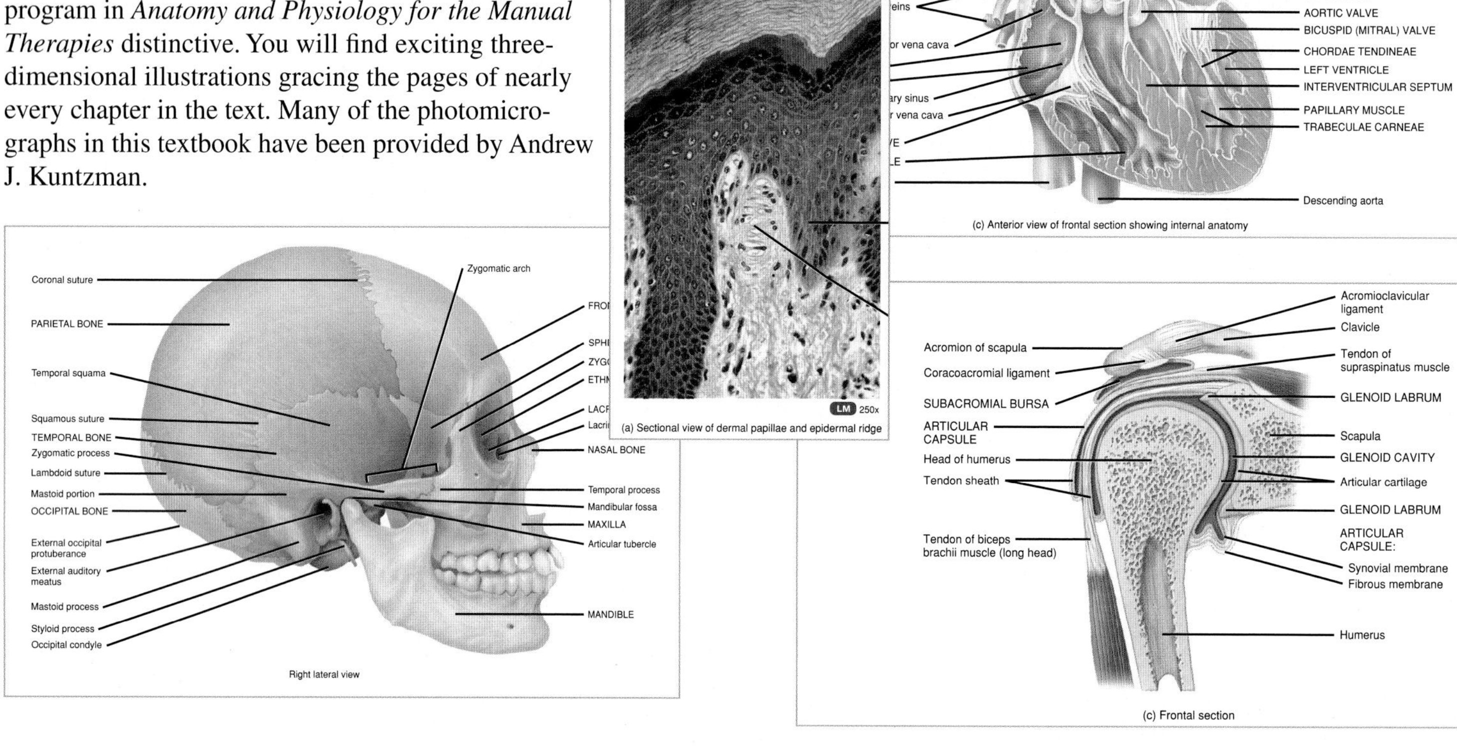

MUSCLE ART AND PHOTOS FOR MANUAL THERAPISTS

Muscles and muscle origins and insertions are clearly illustrated, to give manual therapy students a visual guide to this important system.

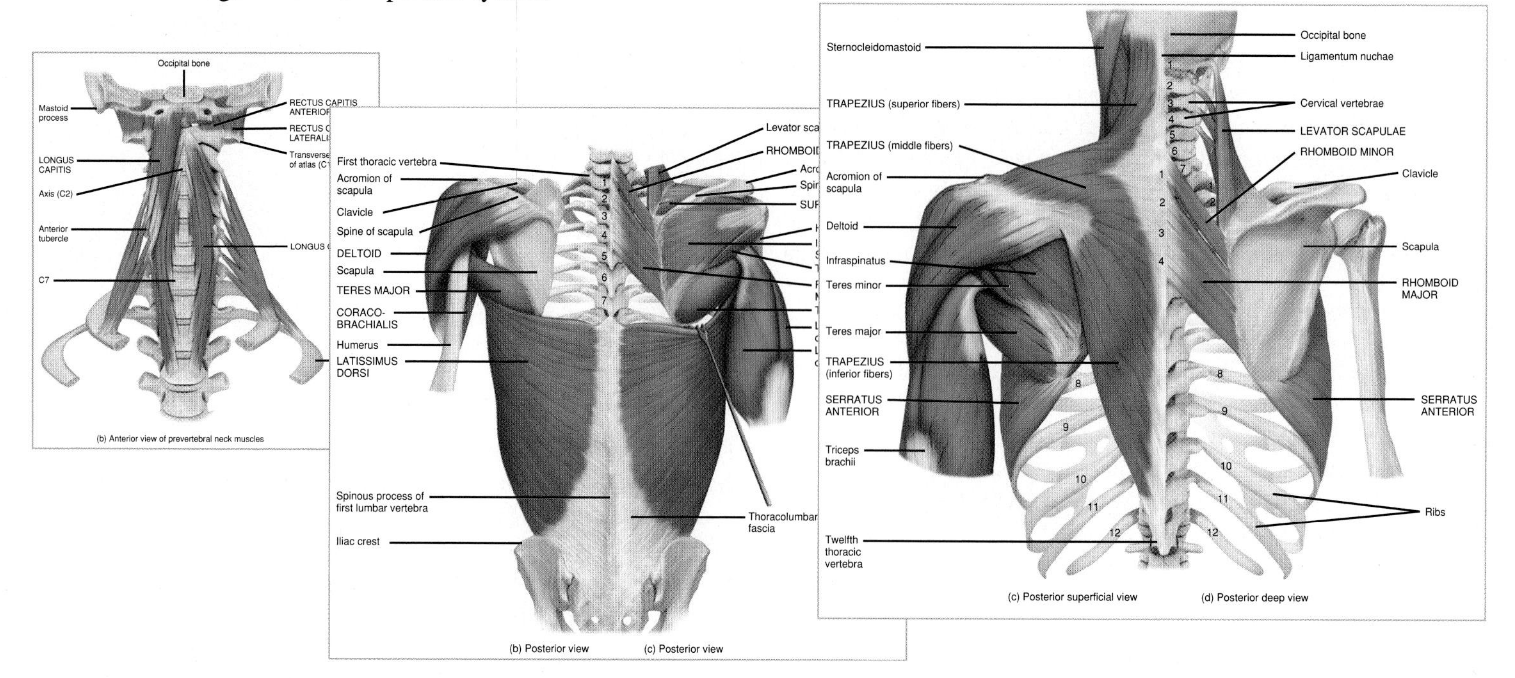

TENSOR FASCIAE LATAE (cut)
SARTORIUS (cut)
RECTUS FEMORIS (cut)
Iliofemoral ligament of hip joint
Inguinal ligament
PECTINEUS (cut)
Pubis
OBTURATOR EXTERNUS
ADDUCTOR LONGUS (cut)
ILIOPSOAS (cut)
PECTINEUS (cut)
ADDUCTOR BREVIS
ADDUCTOR MAGNUS
ADDUCTOR LONGUS (cut)
GRACILIS
Femur
Adductor hiatus
SARTORIUS (cut)
Patella

(b) Anterior deep view

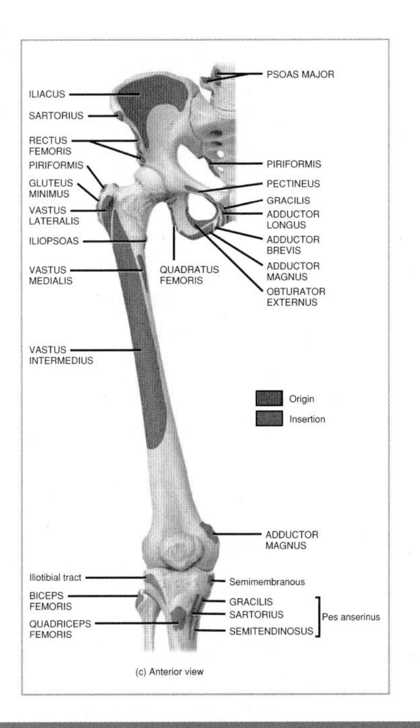

(c) Anterior view

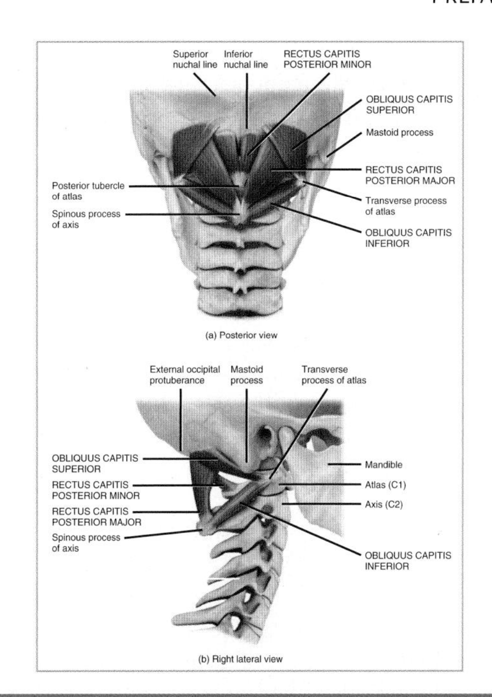

(b) Right lateral view

COMPLETE TEACHING AND LEARNING PACKAGE

This book is accompanied by a host of dynamic resources designed to help you and your students maximize your time and energies. Please contact your Wiley representative for details about these and other resources or visit our website at www.wiley.com/college/kuntzman

AUDIO DOWNLOADS FOR SELECTED ILLUSTRATIONS MP3 downloads, linked to identified illustrations in each chapter give the students the opportunity to hear while they study—as they would in lecture—about the importance and relevance of the structures or concepts that are depicted. These illustrations are identified in each chapter by a distinctive icon.

NEW! REAL ANATOMY. REAL ANATOMY Mark Nielsen and Shawn Miller of the University of Utah, led a team of media and anatomical experts in the creation of this powerful new DVD, ***Real Anatomy***. Their extensive experience in undergraduate anatomy classrooms and cadaver laboratories as well as their passion for the subject matter shine through in this new, user-friendly program with its intuitive interface. The 3-D imaging software allows students to dissect through numerous layers of a real three-dimensional human body to study and learn the anatomical structures of all body systems from multiple perspectives. Histology is viewed via a virtual microscope at varied levels of magnification. Professors can use the program to capture and customize images from a large database of stunning cadaver photographs and clear histology photomicrographs for presentations, quizzing, or testing.

INTERACTIONS: EXPLORING THE FUNCTIONS OF THE HUMAN BODY 3.0 From Tom Lancraft, St. Petersburg College and Frances Frierson, Valencia Community College. Covering all body systems, this dynamic and highly acclaimed program includes anatomical overviews linking form and function, rich animations of complex physiological processes, a variety of creative interactive exercises, concept maps to help students make the connections, and animated clinical case studies. The 3.0 release boasts enhancements based on user feedback, including coverage of ATP, the building blocks of proteins, and dermatomes; a new overview on Special Senses; cardiac muscle; and a revised animation on muscle contraction. Interactions is available in one DVD or in a web-based version, and is fully integrated into *WileyPLUS*.

POWERPHYS by Allen, Harper, Ivlev, and Lancraft. This exceptional resource includes ten self-contained lab modules for exploring physiological principles. Each module contains objectives with illustrated and animated review material, prelab quizzes, prelab reporting, data collection and analysis, and a full lab report with discussion and application questions. Experiments contain randomly generated data, allowing users to experiment multiple times but still arrive at the same conclusions. Available as a stand-alone product, PowerPhys is also bundled with every new copy of the Allen and Harper Laboratory Manual for Anatomy and Physiology, 3rd edition and integrated into *WileyPLUS*.

WileyPLUS is a powerful online tool that provides students and instructors with an integrated suite of teaching and learning resources in one easy-to-use website. With

WileyPLUS, students will come to class better prepared for lectures, get immediate feedback and context-sensitive help on assignments and quizzes, and have access to a full range of interactive learning resources, including a complete online version of their text. A description of some of the resources available to students within *WileyPLUS* appears on the front endpapers of this text. Instructors benefit as well; all the tools and resources to prepare and present dynamic lectures as well as assess student progress and learning are included. New within *WileyPLUS*, ***Quickstart*** is an organizing tool that makes it possible for you to spend less time preparing lectures and grading quizzes and more time teaching and interacting with students. Ask your sales representative to set you up with a test drive, or view a demo online.

VISUAL LIBRARY FOR ANATOMY AND PHYSIOLOGY 4.0 A cross-platform DVD includes all of the illustrations from the textbook in three formats: labeled, unlabeled, and unlabeled with leader lines. The Visual Library also contains many additional illustrations and photographs that could easily enhance lecture or lab. Search for images by chapter or by using keywords.

COMPANION WEBSITES. A dynamic website for students, rich with many activities for review and exploration includes self-quizzes for each chapter, Visual Anatomy review exercises, and weblinks. An access code is bundled with each new text. A dedicated companion website for instructors provides many resources for preparing and presenting lectures. Additionally, this website provides a web version of the Visual Library for Anatomy and Physiology, additional critical thinking questions with answers, an editable test bank, a computerized test bank, transparencies on demand, and clicker questions. These websites can be accessed through www.wiley.com/college/kuntzman.

A BRIEF ATLAS OF THE SKELETON, SURFACE ANATOMY, AND SELECTED MEDICAL IMAGES by Gerard J. Tortora Packaged with every new copy of the text, this atlas of stunning photographs provides a visual reference for both lecture and lab.

LABORATORY MANUAL FOR ANATOMY AND PHYSIOLOGY 3E by Allen and Harper. This newly revised laboratory manual includes multiple activities to enhance student laboratory experience. Illustrations and terminology closely match the text, making this manual the perfect companion. Each copy of the lab manual includes a CD with the ***PowerPhys*** simulation software for the laboratory. *WileyPLUS*, with a wealth of integrated resources including cat, fetal pig, and rat dissection videos, is also available for adoption with this laboratory. The Cat Dissection Laboratory Guide and a Fetal Pig Laboratory Guide, depending upon your dissection needs, are available to package at no additional cost with the main laboratory manual or as stand alone dissection guides.

PHOTOGRAPHIC ATLAS OF THE HUMAN BODY SECOND EDITION by Gerard J. Tortora. This resource is loaded with excellent cadaver photographs and micrographs. The high-quality imagery can be used in the classroom, laboratory, or for studying and review.

ACKNOWLEDGMENTS

We wish to especially thank several academic colleagues for their helpful contributions to this edition. Special thanks to Brandy M. John Hons. B.Sc., RMT, CR, CMAC, owner of A Balanced Body Therapy & Supplies and Massage Therapy Program Coordinator Everest College, for her contributions to the chapter opening vignettes. Her ideas and support throughout the project were extremely helpful. And thanks to Frank Nagy, Ph.D., School of Medicine, Wright State University, for his careful review of portions of the manuscript. Thanks to Sharon L. Barnes, Ph.D., LMT, Executive Director, SHI Integrative Medical Massage and Traditional Chinese Acupuncture School for her support and encouragement throughout the project. We would also like to thank friends and colleagues Ron Bowersock, LMT and Beth L. Wismar, Ph.D., Associate Professor Emeritus, Department of Anatomy, The Ohio State University.

Thanks to the talented group of educators who have prepared the supplementary materials for this text. Iris Berry, Salem State College; Janet K. Copelle, Stautzenberger College; Ron Diana, National Massage Therapy Institute; Edward Johnson, Central Oregon Community College; David Marquart, Tidewater Tech; Mary Ellen Scott, Case Western Reserve University; and Debra Schroeder, American School of Massage.

We are also extremely grateful to our colleagues who have reviewed the manuscript or participated in focus groups and offered numerous suggestions for improvement: Adedayo Adeeko, Sir Sandford Fleming College; Robyn Anthony, Beta Tech Richmond; Michael Arvin, Youngstown College of Massotherapy; Angela Asselin, Wellington College of Remedial Massage Therapies Inc.; Jennette Ball, Onandaga School of Therapeutic Massage; Betty Barker, Spencerian College; Kendra Barnes, ICT Northumberland College; Leigh

Beining, Stautzenberger College; Iris E. Berry, Salem State College; Donald Bisson, Ontario College of Reflexology; Debbie Bomkamp, SHI Massage School; Lisa Boynton, TriOS College; Kelly Brennan, Everest College; Molly Brignall, Ashmead College-Fife; Susan Burgoon, Amarillo College; Janet Carson, Alberta Institute of Massage Therapy; Valerie Chin, Bryan College; Caren C. Clift, Potomac Massage Training Institute; Barbara Cocanour, University of Massachusetts Lowell; Kevin Conley, University of Pittsburgh; Sunny Cooper, CenterPoint Massage & Shiatsu; Dawn Crawford, Tidewater Tech Chesapeake; Julie Dais, Okanagan Valley College of Massage Therapy; Judy Deken, International Professional School of Bodywork; Robin Devine, CCMH-Foothills College of Massage Therapy; Matthew Dodge, Olympic College; Katrina Farber, Massage Institute of Maryland; Brooke Galo, Apollo College; Julie Geyer, Mt. Nittany Institute of Natural Health; Andrea Giaschi, Canadian College of Massage & Hydrotherapy; Lisa Giudici, Onandaga School of Therapeutic Massage; Cindy Goodnetter, A Gathering Place; Ronna Hoglund Hale, TriOS College; John Hokanson, Lincoln College of Technology; Barbara Harwell, Stark State College of Technology; Vicki Hastings, Medical Training College; Janna Kucharski-Howard, New Hampshire Institute for Therapeutic Arts; Cheryl A. Howe, University of Massachusetts Amherst; Nadine Currie Jackson, Atlantic College of Therapeutic Massage; Margaret Jaillet, Mount Wachusett Community College; Bobbie Jennings, CDI College; Edward Johnson, Central Oregon Community College; John Britton Kube, Centura College; Imran Khaliq, Canadian Institute of Traditional Chinese Medicine; Patty J. Kruschke, Tri-City School of Massage; Lazella Lawson, The Master's College; Stephen Leffler, International Professional School of Bodywork; Suzanna Lembeck-Edens, Mt. Nittany Institute of Natural Health; Lori Levisen, Western Career College; Beverly M. Lewis, Career Training Solutions; Jane Marone, University of Illinois-Chicago; David Thomas Marquart, Tidewater Tech; Alex Matthews, Austin Community College; Jose Milan, Auting Community College; Jennifer Minion, Stone Clan Education Center; Anne Morien, Florida School of Massage; Janet Nezon, Canadian College of Massage & Hydrotherapy; Julie Orr, Professional Massage Training Center; Kathleen Paholsky, Schoolcraft College; Janet Pellow, Stautzenberger College; Charlene Penner, International Professional School of Bodywork; Jeffrey Penton, Capps College; Mark Perido, Kansas College of Chinese Medicine; Dianne Polseno, Cortiva Institute; Henry Przybylowicz, The Salter School; Patti Reynolds, Port Townsend School of Massage; Christine E. Sanewsky, Onondaga School of Therapeutic Massage; Mary Ellen Scott, Case Western Reserve University; Janet Self, Port Townsend School of Massage; Prabhat Sharma, Cuyahoga Community College; Jan Sultan, Rolf Institute; Danny Thomas, Canadian Institute of Traditional Chinese Medicine; Jennifer Westacott, Onandaga School of Therapeutic Massage; Rhonda Witherite, Mt. Nittany Institute of Natural Health; Daniel Wood, Ohio College of Massotherapy; Beth Zullinger, Berks Technical Institute.

We wish to thank to James Witte and Prasanthi Pallapu of Auburn University and the Institute for Learning Styles Research for their collaboration with us in developing questions and tools for students to assess, understand, and apply their learning style preferences.

This beautiful textbook would not be possible without the talent and skill of several outstanding medical illustrators. Kevin Sommerville has contributed many illustrations for us over numerous editions. Many new drawings are the work of his talented hands. We so value the long relationship we have with Kevin. John Gibb is responsible for all the outstanding skeletal and muscle illustrations. And we thank the artists of Imagineering Media Services for all they do to enhance the visuals within this text.

A huge thank you goes to the talented publishing professionals at Wiley. In particular, we wish to thank Bonnie Roesch, Executive Editor, who believed in and supported the development of this text. We're grateful to Lorraina Raccuia, Project Editor, for skillfully facilitating the collaboration and day-to-day integration of concepts from the two authors. The development of this text is a testament to her expertise and patience. Thanks to Lauren Morris, Program Assistant, who was always cordial and extremely helpful and Suzanne Ingrao, Production Manager, for her encouragement and skillful work on this project. Thank you to the rest of the Wiley staff for their expertise and diligence: Elizabeth Swain, Senior Production Editor; Hilary Newman, Photo Manager; Anna Melhorn, Senior Illustration Editor; Madelyn Lesure, Senior Designer; Linda Muriello, Senior Media Editor; and Clay Stone, Executive Marketing Manager.

Andrew J. Kuntzman
Biology Department
Sinclair Community College
444 West Third Street
Dayton, OH 45402-1460
andykuntzman@aol.com

Gerard J. Tortora
Department of Science
and Math, S229
Bergen Community College
400 Paramus Road
Paramus, NJ 07652

NOTE TO STUDENTS

You've probably chosen a career in the manual therapies. Your book has a variety of special features that will make your time studying anatomy a more rewarding experience.

As you start to read each section of a chapter, be sure to take note of the **Objectives** at the beginning of the section to help you focus on what is important as you read. At the end of the section, take time to try and answer the **Checkpoint** questions placed there. If you can answer, then you are ready to move on. If you have trouble answering the questions, you may want to re-read the section before continuing.

Studying the figures (illustrations that include artwork and photographs) in this book is as important as reading the text. To get the most out of the visual parts of this book, use the tools we have added to the figures to help you understand the concepts being presented. Start by reading the **Legend,** which explains what the figure is about. Next, study the **Key Concept Statement,** indicated by a "key" icon, which reveals a basic idea portrayed in the figure. Added to many figures you will also find an **Orientation Diagram** to help you understand the perspective from which you are viewing a particular piece of anatomical art. Finally, at the bottom of each figure you will find a **Figure Question,** accompanied by a "question mark" icon. If you try to answer these questions as you go along, they will serve as self-checks to help you understand the material. Often it will be possible to answer a question by examining the figure itself. Other questions will encourage you to integrate the knowledge you've gained by carefully reading the text associated with the figure. Still other questions may prompt you to think critically about the topic at hand or predict a consequence in advance of its description in the text. You will find the answer to each figure question at the end of the chapter in which the figure appears. Selected figures include **Functions** boxes, brief summaries of the functions of the anatomical structure of the system shown (see Figure 3.13 cytoskeleton).

In each chapter you will find that several illustrations are marked with an icon that looks like an MP3 player. This is an indication that a download which narrates and discusses the important elements of that particular illustration is available for your study. You can access these downloads on the student companion website.

Studying physiology requires an understanding of the sequence of processes. Special numbered lists in the narrative that correspond to numbered segments in the illustration facilitate understanding of these physiological processes. This approach is used extensively throughout the book to lend clarity to the flow of *complex* processes.

Learning the complex anatomy and all of the terminology involved for each body system can be a daunting task. For many topics, including the bones, joints, skeletal muscles, surface anatomy, blood vessels, and nerves, we have created special **Exhibits** which organize the material into manageable segments. Each Exhibit consists of an objective, an overview, a tabular

Figure 8.1 **Right pectoral (shoulder) girdle.**

The clavicle is the anterior bone of the pectoral girdle and the scapula is the posterior bone.

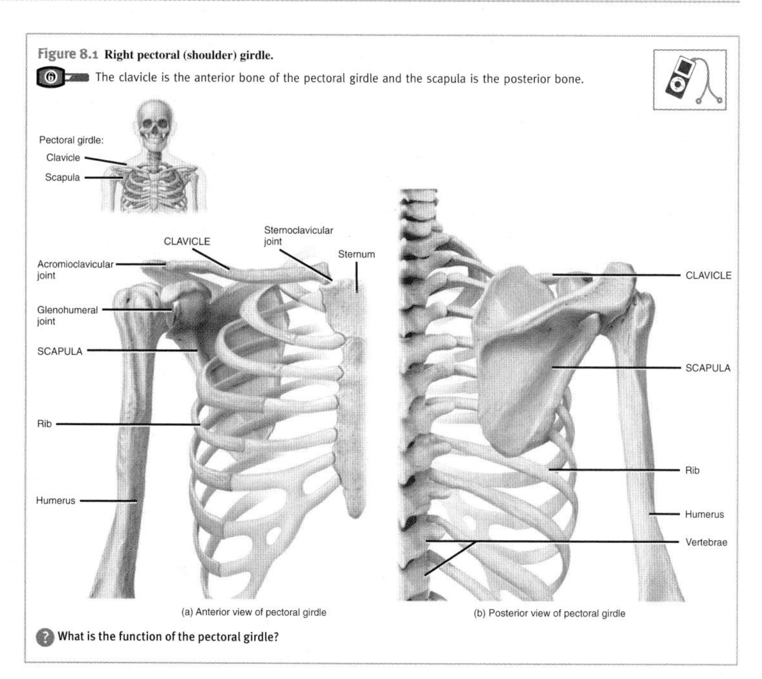

(a) Anterior view of pectoral girdle

(b) Posterior view of pectoral girdle

What is the function of the pectoral girdle?

Figure 15.12 **Signal transmission at a chemical synapse.** Through exocytosis of synaptic vesicles, a presynaptic neuron releases neurotransmitter molecules. After diffusing across the synaptic cleft, the neurotransmitter binds to receptors in the plasma membrane of the postsynaptic neuron and produces a postsynaptic potential.

At a chemical synapse, a presynaptic neuron converts an electrical signal (nerve impulse) into a chemical signal (neurotransmitter release). The postsynaptic neuron then converts the chemical signal back into an electrical signal (postsynaptic potential).

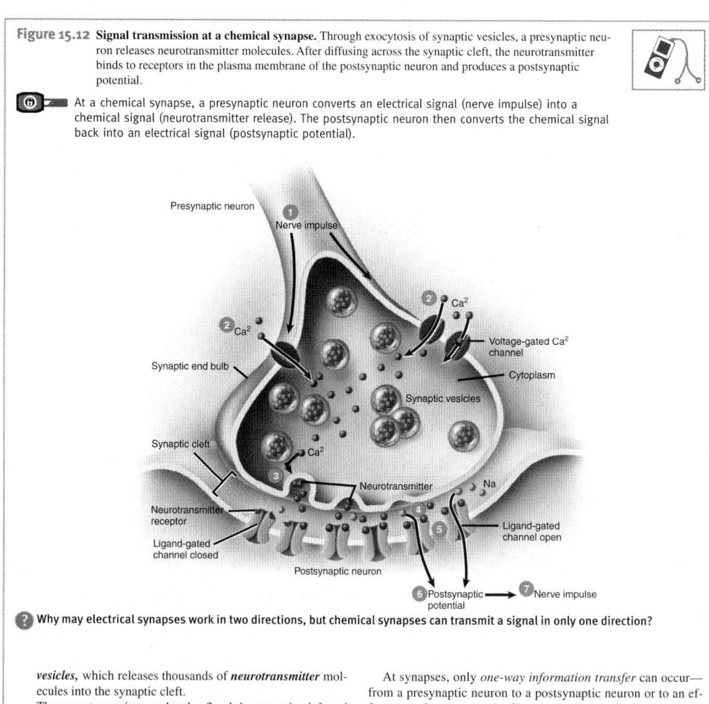

Why may electrical synapses work in two directions, but chemical synapses can transmit a signal in only one direction?

vesicles, which releases thousands of ***neurotransmitter*** molecules into the synaptic cleft.

The neurotransmitter molecules flood the synaptic cleft and bind to ***neurotransmitter receptors*** in the postsynaptic neuron's plasma membrane.

Binding of neurotransmitter molecules opens ion channels, which allows certain ions to flow across the membrane.

As ions flow through the opened channels, the voltage across the membrane changes. Depending on which ions the channels admit, the voltage change may be a depolarization or a hyperpolarization.

If a depolarization occurs in the postsynaptic neuron and reaches threshold, then it triggers one or more nerve impulses.

At synapses, only *one-way information transfer* can occur—from a presynaptic neuron to a postsynaptic neuron or to an effector, such as a muscle fiber or a gland cell. For example, synaptic transmission at a neuromuscular junction (NMJ) proceeds from a somatic motor neuron to a skeletal muscle fiber (but not in the opposite direction). Only synaptic end bulbs of presynaptic neurons can release neurotransmitters, and only the postsynaptic neuron's membrane has the correct receptor proteins to recognize and bind that neurotransmitter. As a result, nerve impulses move along their pathways in one direction within the body. It may be noted that nerve cell conduction in the laboratory may go in both directions when an axon is electrically stimulated in the middle. Within the body, however, synapses cause neuronal signals to move in only one direction.

EXHIBIT 16.2 Brachial Plexus CONTINUED

Figure 16.8 (continued)

Dorsal scapular nerve
Nerve to subclavius
Suprascapular nerve
Superior trunk
Middle trunk
Inferior trunk
Clavicle
Lateral pectoral nerve
Lateral cord
Posterior cord
Axillary nerve
Medial cord
Musculocutaneous nerve
Radial nerve
Median nerve
From C4
C5
C6
C7
C8
T1
Long thoracic
Medial pectoral
Medial brachial cutaneous
Medial antebrachial cutaneous
Scapula
Ulnar nerve
Humerus
Deep branch of radial nerve
Radius
Ulna
Superficial branch of radial nerve
Median nerve
Ulnar nerve
Radial nerve
Superficial branch of ulnar nerve
Digital branch of median nerve
Digital branch of ulnar nerve

(b) Distribution of nerves from the brachial plexus

What five important nerves arise from the brachial plexus?

MANUAL THERAPY APPLICATION

Injuries to the Roots of the Brachial Plexus

Injury to the roots of the brachial plexus (C5–C6) may result from forceful pulling away of the head from the shoulder, as might occur from a heavy fall on the shoulder or excessive stretching of an infant's neck during childbirth. The presentation of this injury is characterized by an upper limb in which the shoulder is adducted, the arm is medially rotated, the elbow is extended, the forearm is pronated, and the wrist is flexed (Figure 16.9a). This condition is called *Erb-Duchenne palsy* or *waiter's tip* position. There is loss of sensation along the lateral side of the arm.

Radial (and axillary) nerve injury can be caused by improperly administered intramuscular injections into the deltoid muscle. The radial nerve may also be injured when a cast is applied too tightly around the mid-humerus. Radial nerve injury is indicated by *wrist drop*, the inability to extend the wrist and fingers (Figure 16.9b). Sensory loss is minimal due to the overlap of sensory innervation by adjacent nerves.

Median nerve injury may result in *median nerve palsy*, which is indicated by numbness, tingling, and pain in the palm and fingers. There is also inability to pronate the forearm and flex the proximal interphalangeal joints of all digits and the distal interphalangeal joints of the second and third digits (Figure 16.9c). In addition, wrist flexion and thumb movements are weak, and are accompanied by adduction of the thumb due to a loss of function of the muscles of the thenar eminence (see Clinical Connection on carpal tunnel syndrome in Chapter 13).

Ulnar nerve injury may result in *ulnar nerve palsy (claw hand)*, which is indicated by an inability to abduct or adduct the fingers, atrophy of the interosseus muscles of the hand, hyperextension of the metacarpophalangeal joints, and flexion of the interphalangeal joints, a condition called claw hand (Figure 16.9d). There is also loss of sensation over the little finger and the medial half of the ring finger.

Long thoracic nerve injury results in paralysis of the serratus anterior muscle. The medial border of the scapula protrudes, giving it the appearance of a wing. When the arm is raised, the vertebral border and inferior angle of the scapula pull away from the thoracic wall and protrude outward, causing the medial border of the scapula to protrude; because the scapula looks like a wing, this condition is called *winged scapula* (Figure 16.9e). The arm cannot be abducted beyond the horizontal position.

CHECKPOINT
Injury of which nerve could cause paralysis of the serratus anterior muscle?

Figure 16.9 **Injuries to the brachial plexus.**

Injuries to the brachial plexus affect the sensations and movements of the upper limbs.

(a) Erb-Duchenne palsy (waiter's tip)
(b) Wrist drop
(c) Median nerve palsy
(d) Ulnar nerve palsy (claw hand)
(e) Winging of right scapula

Injury to which nerve of the brachial plexus affects sensations of the palm and digits?

summary of the relevant anatomy, an associated group of illustrations or photographs, and a checkpoint question.

Some Exhibits also contain relevant Clinical Connections or Manual Therapy Applications. At the end of each chapter are other resources that you will find useful. The **Key Medical Terms** section includes selected terms dealing with both normal and pathological conditions. The **Study Outline** is a concise statement of important topics discussed in the chapter. Page numbers are listed next to key concepts so that you can refer easily to specific passages in the text for clarification or amplification. The **Self-Quiz Questions** are designed to help you evaluate your understanding of the chapter contents. **Critical Thinking Questions** are word problems that allow you to apply the concepts you have studied in the chapter to specific situations. Answers to the Self-Quiz Questions and suggested answers to the Critical Thinking Questions (some of which have no one right answer) appear in an appendix at the end of the book so you can check your progress.

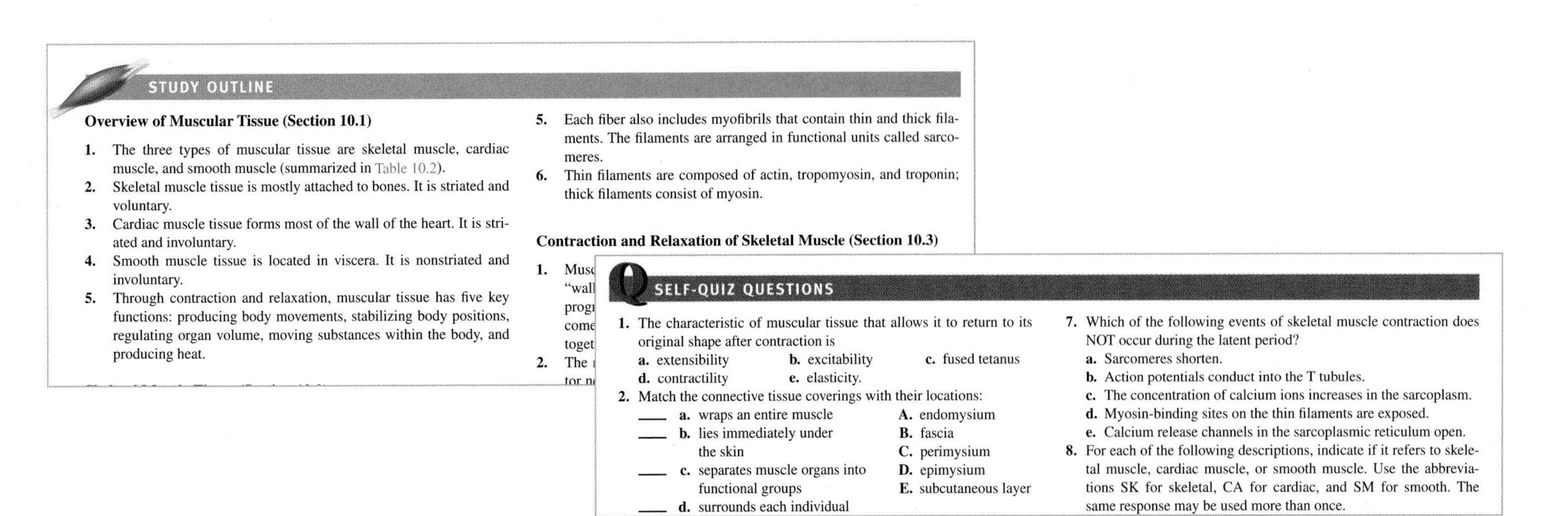

STUDY OUTLINE

Overview of Muscular Tissue (Section 10.1)

1. The three types of muscular tissue are skeletal muscle, cardiac muscle, and smooth muscle (summarized in Table 10.2).
2. Skeletal muscle tissue is mostly attached to bones. It is striated and voluntary.
3. Cardiac muscle tissue forms most of the wall of the heart. It is striated and involuntary.
4. Smooth muscle tissue is located in viscera. It is nonstriated and involuntary.
5. Through contraction and relaxation, muscular tissue has five key functions: producing body movements, stabilizing body positions, regulating organ volume, moving substances within the body, and producing heat.

5. Each fiber also includes myofibrils that contain thin and thick filaments. The filaments are arranged in functional units called sarcomeres.
6. Thin filaments are composed of actin, tropomyosin, and troponin; thick filaments consist of myosin.

Contraction and Relaxation of Skeletal Muscle (Section 10.3)

1. Musc… "wall… progr… come… toget…
2. The … tor n…

SELF-QUIZ QUESTIONS

1. The characteristic of muscular tissue that allows it to return to its original shape after contraction is
   a. extensibility b. excitability c. fused tetanus
   d. contractility e. elasticity.
2. Match the connective tissue coverings with their locations:
   ___ a. wraps an entire muscle — A. endomysium
   ___ b. lies immediately under the skin — B. fascia
   ___ c. separates muscle organs into functional groups — C. perimysium
   ___ d. surrounds each individual — D. epimysium
   E. subcutaneous layer
7. Which of the following events of skeletal muscle contraction does NOT occur during the latent period?
   a. Sarcomeres shorten.
   b. Action potentials conduct into the T tubules.
   c. The concentration of calcium ions increases in the sarcoplasm.
   d. Myosin-binding sites on the thin filaments are exposed.
   e. Calcium release channels in the sarcoplasmic reticulum open.
8. For each of the following descriptions, indicate if it refers to skeletal muscle, cardiac muscle, or smooth muscle. Use the abbreviations SK for skeletal, CA for cardiac, and SM for smooth. The same response may be used more than once.

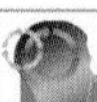

CRITICAL THINKING QUESTIONS

1. Weight-lifter Jamal has been practicing many hours a day, and his muscles have gotten noticeable bigger. He tells you that his muscle cells are "multiplying like crazy and making him get stronger and stronger." Do you believe his explanation? Why or why not?
2. Chicken breasts are composed of "white meat" while chicken legs are composed of "dark meat." The breasts and legs of migrating ducks are dark meat. The breasts of both chickens and ducks are used in flying. How can you explain the differences in the color of the meat (muscles)? How are they adapted for their particular functions?
3. Polio is a disease caused by a virus that can attack the somatic motor neurons in the central nervous system. Individuals who suffer from polio can develop muscle weakness and atrophy. In a certain percentage of cases, the individuals may die due to respiratory paralysis. Re
the symptom

? ANSWERS TO FIGURE QUESTIONS

10.1 In order from the inside toward the outside, the connective tissue layers are endomysium, perimysium, epimysium, fascia, and subcutaneous layer.
10.2 The A band has thick filaments in its center and overlapping thick and thin filaments at each end; the I band has thin filaments.
10.3 Proteins present on the A band are myosin, actin, troponin, and tropomyosin; the I band has actin, troponin, and tropomyosin.
10.4 The I bands disappear. The lengths of the thick and thin filaments do not change during contraction.
10.5 The motor end plate is the region of the sarcolemma near the axon terminal.
10.6 Binding of ATP to the myosin heads detaches them from actin.
10.7 The power stroke occurs during step 6.
10.8 Glycolysis, exchange of phosphate between creatine phosphate and ADP, and glycogen breakdown occur in the cytosol. Oxidation of pyruvic acid, amino acids, and fatty acids (aerobic cellular respiration) occurs in the mitochondria.
10.9 Sarcomeres shorten during the contraction period.
10.10 Fused tetanus occurs when the frequency of stimulation reaches about 90 stimuli per second.
10.11 Holding your head upright without movement involves mainly isometric contractions.
10.12 Pressure by a therapist on the muscle in spasm for 30 seconds or longer usually stretches the muscle and causes the tone of the muscle spindle complex to be reset.
10.13 The walls of hollow organs contain visceral (single-unit) smooth muscle.

At times, you may require extra help to learn specific anatomical features of the various body systems. One way to do this is through the use of **Mnemonics,** aids to help memory. Mnemonics are included throughout the text, some displayed in figures, tables, or exhibits, and some included within the text discussion. We encourage you to use not only the mnemonics provided, but also to create your own to help you learn the multitude of terms involved in your study of human anatomy.

Throughout the text we have included **Pronunciations** and, sometimes, **Word Roots,** for many terms that may be new to you. These appear in parentheses immediately following the new words, and the pronunciations are repeated in the glossary at the back of the book. Look at the words carefully and say them out loud several times. Learning to pronounce a new word will help you remember it and make it a useful part of your medical vocabulary. Take a few minutes now to read the following pronunciation key, so it will be familiar as you encounter new words. The key is repeated at the beginning of the Glossary, page G-1.

PRONUNCIATION KEY

1. The most strongly accented syllable appears in capital letters, for example, bilateral (bī-LAT-er-al) and diagnosis (dī-ag-NŌ-sis).
2. If there is a secondary accent, it is noted by a prime (′), for example, constitution (kon′-sti-TOO-shun) and physiology (fiz′-ē-OL-ō-jē). Any additional secondary accents are also noted by a prime, for example, decarboxylation (dē′-kar-bok′-si-LĀ-shun).
3. Vowels marked by a line above the letter are pronounced with the long sound, as in the following common words:
   ā as in *māke* | *ō* as in *pōle*
   ē as in *bē* | *ū* as in *cūte*
   ī as in *īvy*
4. Vowels not marked by a line above the letter are pronounced with the short sound, as in the following words:
   a as in *above* or *at* | *o* as in *not*
   e as in *bet* | *u* as in *bud*
   i as in *sip*
5. Other vowel sounds are indicated as follows:
   oy as in *oil*
   oo as in *root*
6. Consonant sounds are pronounced as in the following words:
   b as in *bat* | *m* as in *mother*
   ch as in *chair* | *n* as in *no*
   d as in *dog* | *p* as in *pick*
   f as in *father* | *r* as in *rib*
   g as in *get* | *s* as in *so*
   h as in *hat* | *t* as in *tea*
   j as in *jump* | *v* as in *very*
   k as in *can* | *w* as in *welcome*
   ks as in *tax* | *z* as in *zero*
   kw as in *quit* | *zh* as in *lesion*
   l as in *let*

BRIEF CONTENTS

CONTENTS

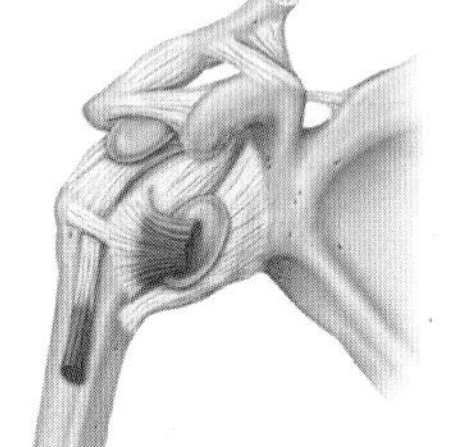

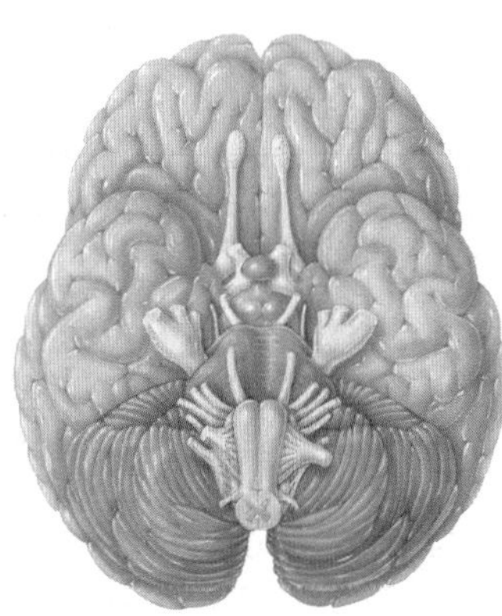

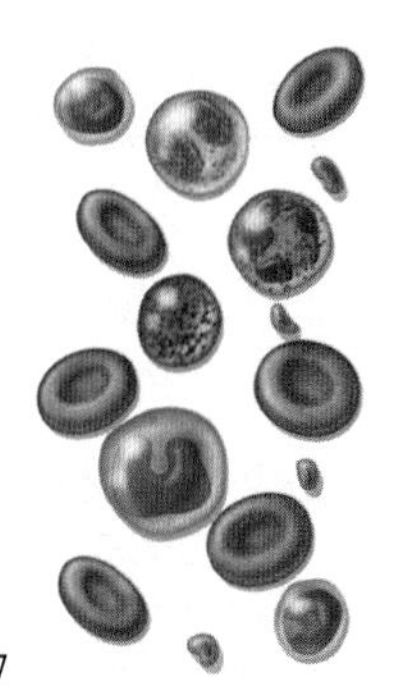

25 THE RESPIRATORY SYSTEM 651

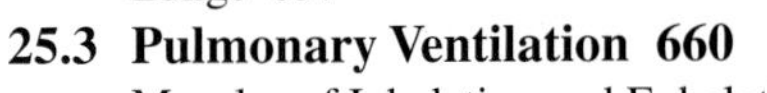

26 THE DIGESTIVE SYSTEM 677

27 NUTRITION AND METABOLISM 707

28 THE URINARY SYSTEM 726

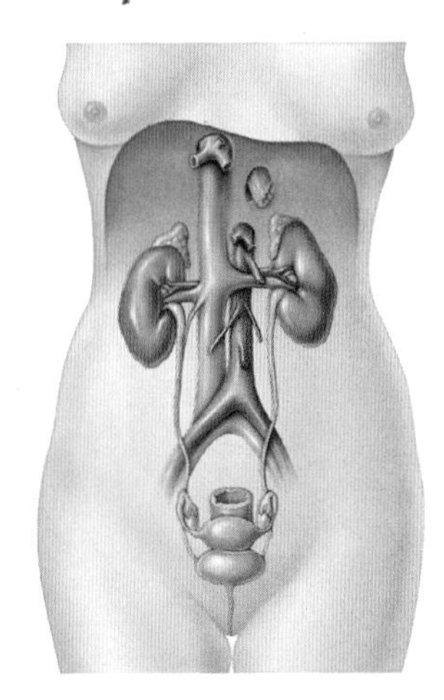

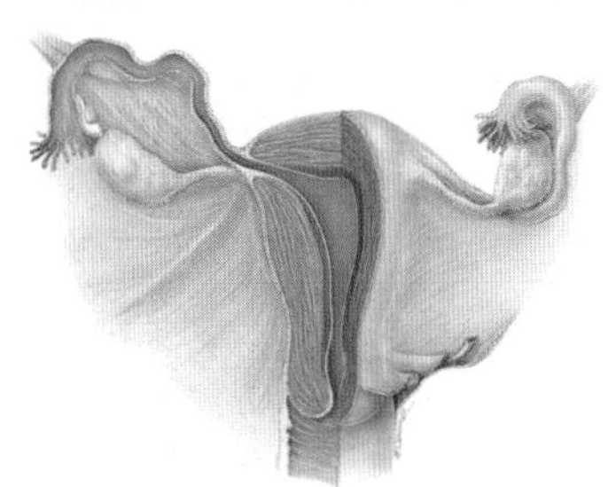

Organization of the Human Body

1

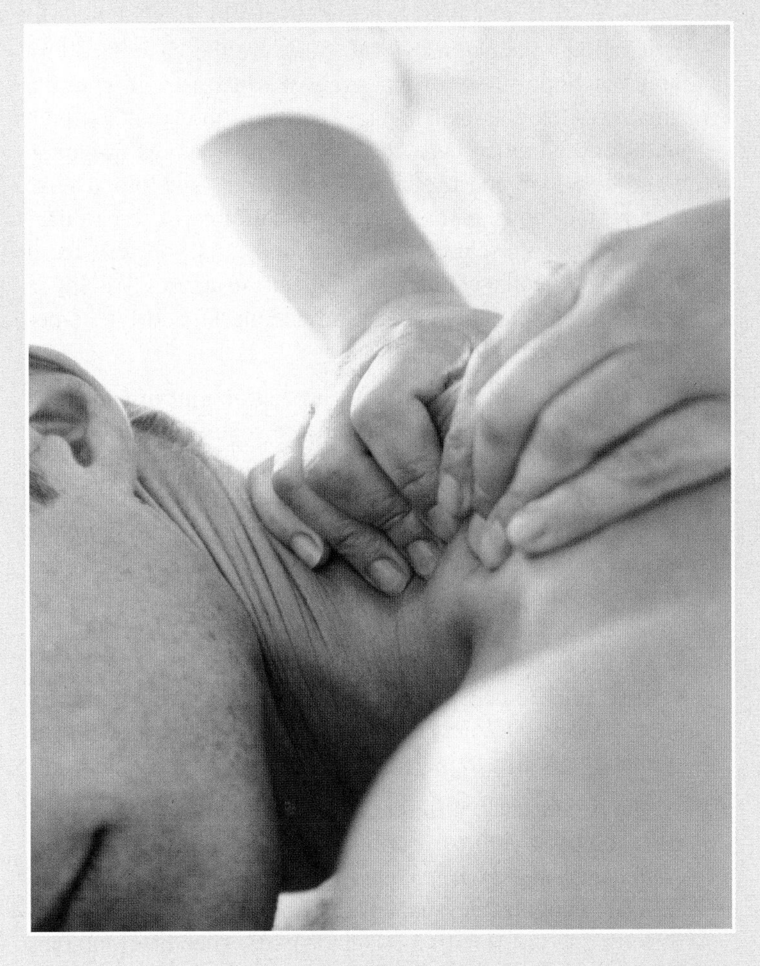

The body is truly amazing. It can function for 100 years or more. It can run a marathon, hug a child, climb a mountain, laugh, sing, and process millions of thoughts. At the same time, the body breathes, digests, maintains its temperature, and heals tiny cuts and scrapes.

Knowledge of anatomy and physiology is the foundation for all health-care providers. You are beginning a fascinating exploration of the human body that will allow you to understand the workings of your own body and the bodies of others. The human body exhibits a stunningly complex design, but it can be easily understood by simply studying it at all levels—from the individual cell to the organism as a whole.

Your goal as a health-care professional will be to assist the body in its own processes in maintaining health and balance, such as healing, to maintain homeostasis. Homeostasis, the way in which the body maintains itself within stable limits, is a major concept considered regularly throughout this book. You will begin your journey through the text by discovering what makes you truly alive and by learning the core terminology that therapists use to discuss the concepts of anatomy, physiology, and pathology with patients, primary health-care providers, and colleagues.

CONTENTS AT A GLANCE

1.1 ANATOMY AND PHYSIOLOGY DEFINED

OBJECTIVE

- Define anatomy and physiology.

The sciences of anatomy and physiology are the foundation for understanding the structures and functions of the human body. ***Anatomy*** (a-NAT-ō-mē; *ana-* = up; *-tomy* = a cutting) is the science of *structure* and the relationships among structures. ***Physiology*** (fiz′-ē-OL-ō-jē; *physio-* = nature; *-logy* = study of) is the science of body *functions*, that is, how the body parts work. The disciplines of anatomy and physiology are described by many as being the foundations of all medical and allied health education. Because function can never be separated completely from structure, we can understand the human body best by studying anatomy and physiology together. We will look at how each structure of the body is designed to carry out a particular function and how the structure of a part often determines the functions it can perform. The bones of the skull, for example, are joined to form a rigid case that protects the brain. The bones of the fingers, by contrast, are more loosely joined, which enables them to perform a variety of movements, such as turning the pages of this book.

CHECKPOINT

1. What is the basic difference between anatomy and physiology?
2. Give your own example of how the structure of a part of the body is related to its function.

1.2 LEVELS OF ORGANIZATION AND BODY SYSTEMS

OBJECTIVES

- Describe the structural organization of the human body.
- Define the body systems and explain how they relate to one another.

The structures of the human body are organized on several levels, similar to the way letters of the alphabet, words, sentences, and paragraphs make up the written language of this text. Listed here, from smallest to largest, are the six levels of organization of the human body: chemical, cellular, tissue, organ, system, and organismal (Figure 1.1).

1. The ***chemical level,*** which can be compared to *letters of the alphabet,* includes ***atoms,*** the smallest units of matter that participate in chemical reactions, and ***molecules,*** two or more atoms joined together. Certain atoms, such as carbon (C), hydrogen (H), oxygen (O), nitrogen (N), calcium (Ca), and others, are essential for maintaining life. Familiar examples of molecules found in the body are DNA (deoxyribonucleic acid), the genetic material passed on from one generation to another; hemoglobin, which carries oxygen in the blood; glucose, commonly known as blood sugar; and vitamins, which are needed for a variety of chemical processes.

2. Molecules combine to form structures at the next level of organization—the ***cellular level. Cells*** are the basic structural and functional units of an organism. Just as *words* are the smallest elements of language, cells are the smallest living units in the human body. Among the many types of cells in your body are muscle cells, nerve cells, and blood cells. Figure 1.1 shows a smooth muscle cell, one of three different kinds of muscle cells in your body. As you will see in Chapter 3, cells contain specialized structures called *organelles,* such as the nucleus, mitochondria, and lysosomes, that perform specific functions.

3. The ***tissue level*** is the next level of structural organization. ***Tissues*** are groups of cells, and the materials surrounding them, that work together to perform a particular function, similar to the way words are put together to form *sentences.* The four basic types of tissues in your body are *epithelial tissue, connective tissue, muscular tissue,* and *nervous tissue.* The similarities and differences among the different types of tissues are the focus of Chapter 4. Note in Figure 1.1 that smooth muscle tissue consists of tightly packed smooth muscle cells together with associated connective tissues, vessels, and nerves. (The supportive structures of this tissue are not illustrated.)

4. At the ***organ level,*** different kinds of tissues join together to form body structures. Similar to the relationship between *sentences* and *paragraphs,* ***organs*** are structures that usually have a recognizable shape, are composed of two or more different types of tissues, and have specific functions. Examples of organs are the stomach, heart, liver, lungs, and brain. Figure 1.1 shows several tissues that make up the stomach. The *serous membrane* is a layer around the outside of the stomach that protects it and reduces friction when the stomach moves and rubs against other organs. Underneath the serous membrane are the *smooth muscle tissue layers,* which contract to churn and mix food and push it on to the next digestive organ, the small intestine. The innermost lining of the stomach is an extremely thin *epithelial tissue layer,* which contributes fluid and chemicals that aid digestion. We see similar patterns of organization throughout most of the body.

5. The next level of structural organization in the body is the ***system level.*** A ***system*** consists of related organs that have a common function (like *chapters* in a book). The example shown in Figure 1.1 is the digestive system, which breaks down and absorbs molecules in food. In the chapters that follow, we will explore the anatomy and physiology of each of the body systems. Table 1.1 starting on page 4 introduces the components and functions of these systems. As you study the body systems, you will discover how they work together to maintain health, protect you from disease, and allow for reproduction of the species.

Figure 1.1 Levels of structural organization in the human body.

The levels of structural organization are chemical, cellular, tissue, organ, system, and organismal.

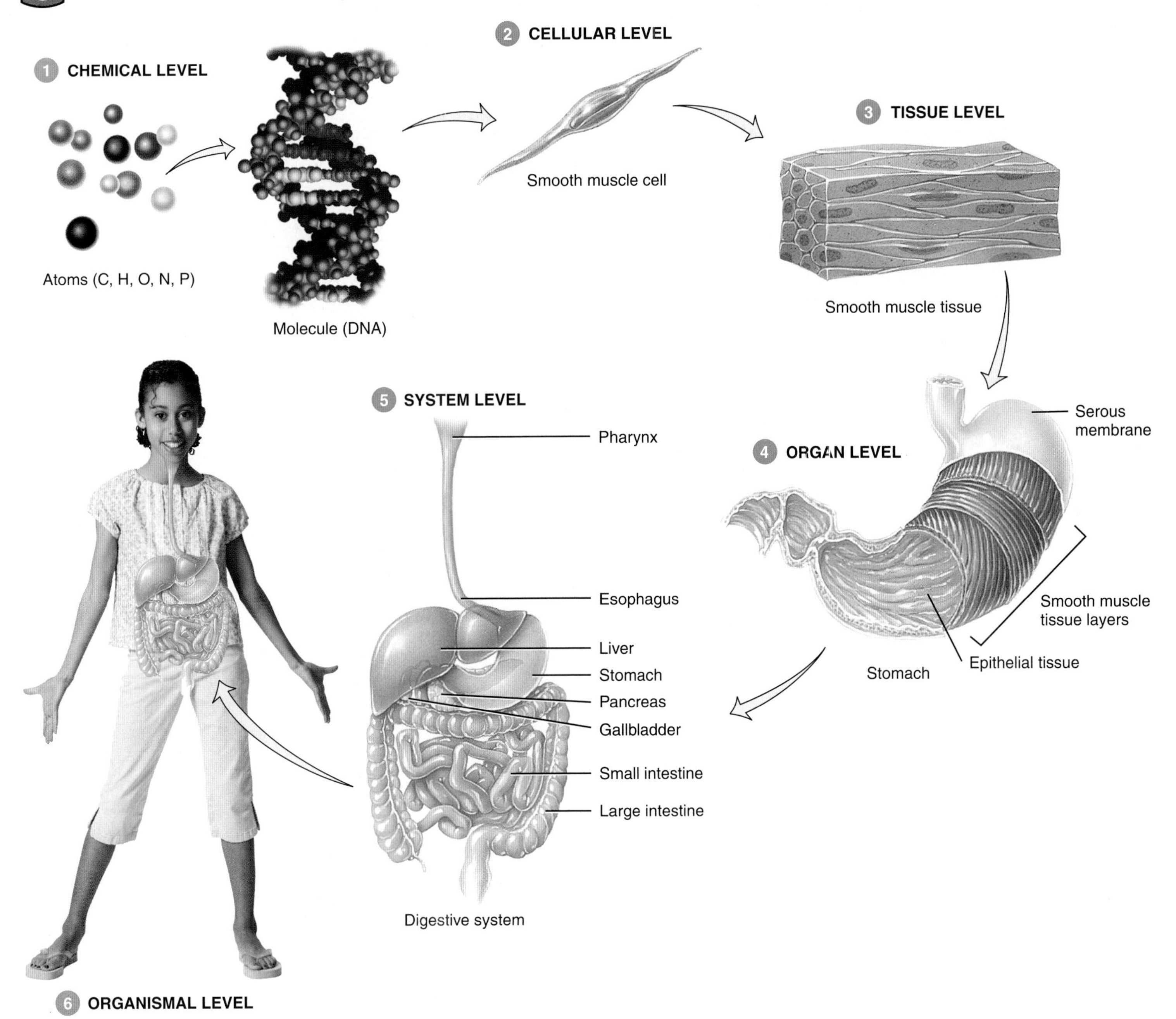

? Which level of structural organization usually has a recognizable shape and is composed of two or more different types of tissues that have a specific function?

6 The ***organismal level*** is the largest level of organization. All the systems of the body combine to make up an ***organism*** (OR-ga-nizm), that is, one human being. An organism can be compared to a *book* in our analogy.

CHECKPOINT

3. Define the following terms: atom, molecule, cell, tissue, organ, system, and organism.
4. Referring to Table 1.1, which body systems help eliminate wastes?

TABLE 1.1

Components and Functions of the Eleven Principal Systems of the Human Body

1. INTEGUMENTARY SYSTEM (CHAPTER 5)

Components: Skin and structures associated with it, such as hair, nails, and sweat and oil glands.

Functions: Helps regulate body temperature; protects the body; eliminates some wastes; helps make vitamin D; detects sensations such as touch, pressure, pain, warmth, and cold.

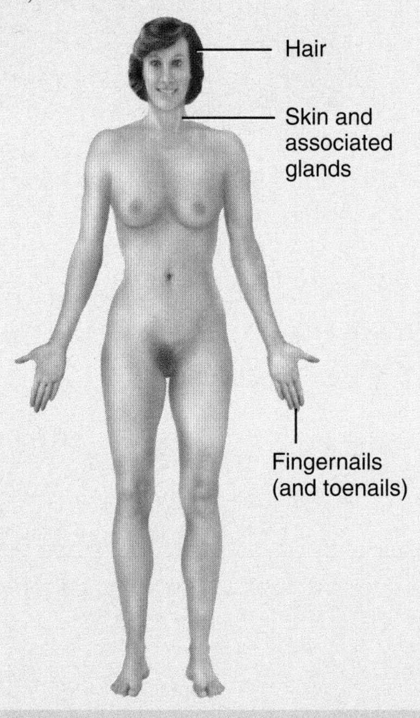

2. SKELETAL SYSTEM (CHAPTERS 6-9)

Components: All the bones and joints of the body and their associated cartilages.

Functions: Supports and protects the body, provides a specific area for muscle attachment, assists with body movements, stores cells that produce blood cells, and stores minerals and lipids (fats).

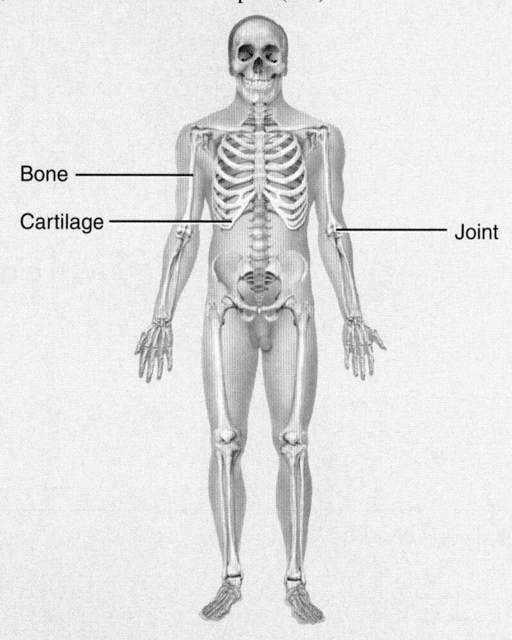

3. MUSCULAR SYSTEM (CHAPTERS 10-14)

Components: Specifically refers to skeletal muscle tissue, which is muscle usually attached to bones (other muscle tissues include smooth and cardiac).

Functions: Participates in bringing about body movements, such as walking, maintains posture, and produces heat.

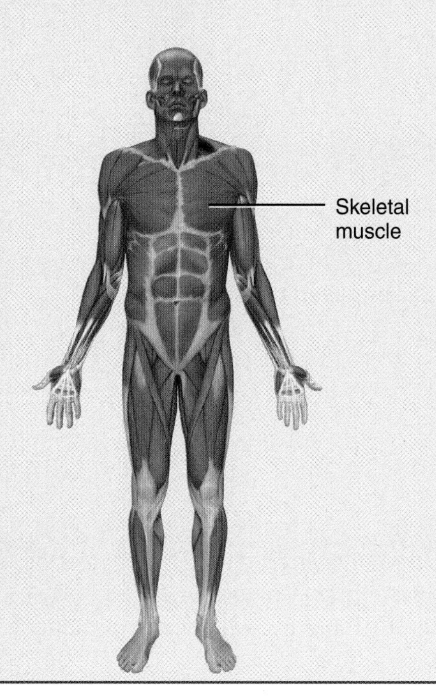

4. NERVOUS SYSTEM (CHAPTERS 15-19)

Components: Brain, spinal cord, nerves, and special sense organs such as the eye and ear.

Functions: Control system that regulates body activities through nerve impulses by detecting changes in the environment, interpreting the changes, and responding to the changes by bringing about muscular contractions or glandular secretions.

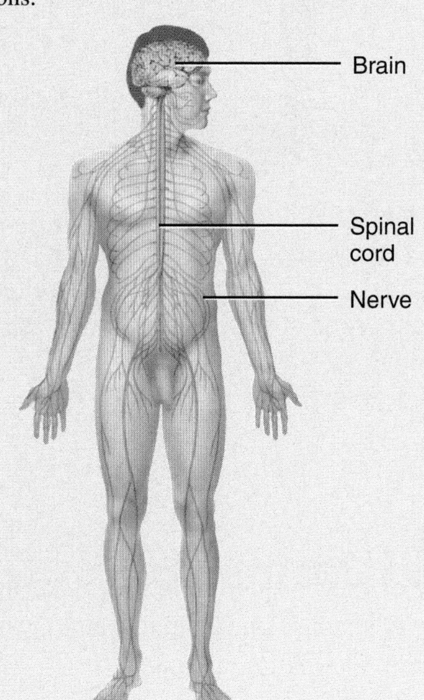

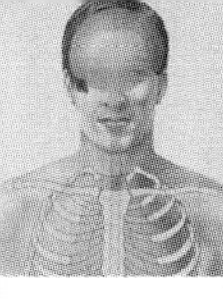

5. ENDOCRINE SYSTEM (CHAPTER 20)

Components: All glands and tissues that produce chemical regulators of body functions, called hormones.

Functions: Control system that regulates body activities through hormones transported by the blood to various target organs.

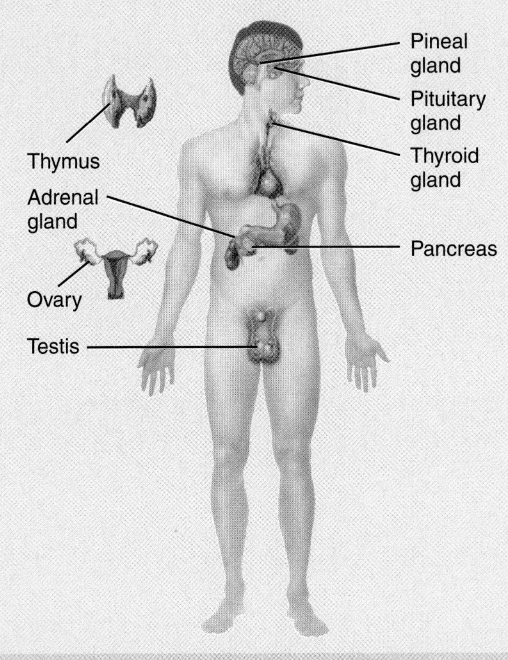

6. CARDIOVASCULAR SYSTEM (CHAPTERS 21-23)

Components: Blood, heart, and blood vessels.

Functions: Heart pumps blood through blood vessels; blood carries oxygen and nutrients to cells and carbon dioxide and wastes away from cells, and helps regulate acidity, temperature, and water content of body fluids; blood components help defend against disease and mend damaged blood vessels.

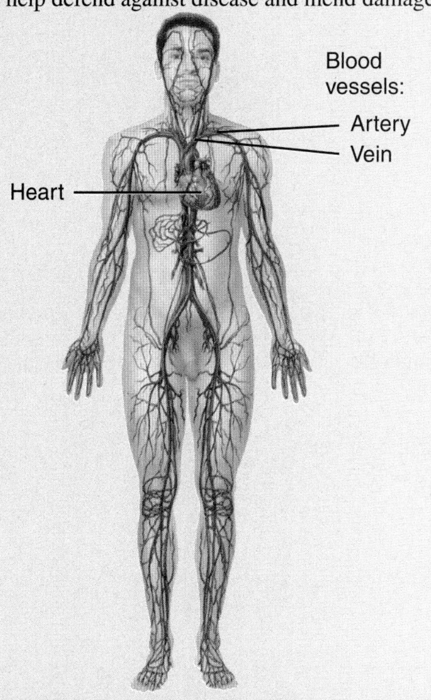

7. LYMPHATIC SYSTEM AND IMMUNITY (CHAPTER 24)

Components: Lymphatic fluid and vessels; spleen, thymus, lymph nodes, and tonsils; cells that carry out immune responses (B cells, T cells, and others).

Functions: Returns proteins and fluid to blood; carries lipids from gastrointestinal tract to blood; contains sites of maturation and proliferation of B cells and T cells that protect against disease-causing microbes.

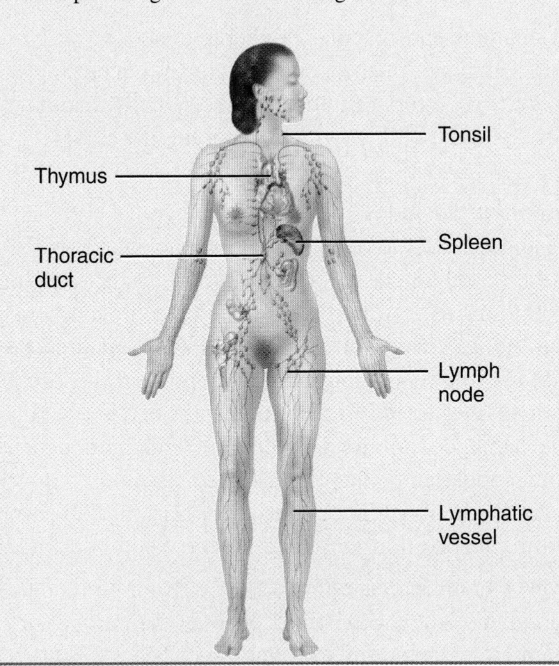

8. RESPIRATORY SYSTEM (CHAPTER 25)

Components: Lungs and air passageways such as the pharynx (throat), larynx (voice box), trachea (windpipe), and bronchial tubes leading into and out of them.

Functions: Transfers oxygen from inhaled air to blood and carbon dioxide from blood to exhaled air; helps regulate acidity of body fluids; air flowing out of lungs through vocal cords produces sounds.

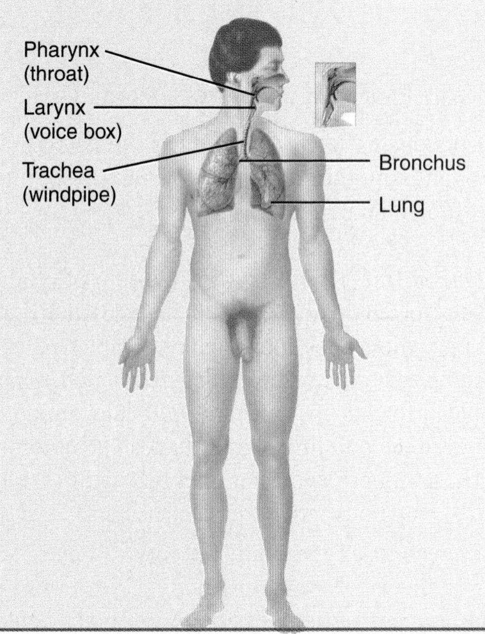

CONTINUES

TABLE 1.1 CONTINUED

Components and Functions of the Eleven Principal Systems of the Human Body

9. DIGESTIVE SYSTEM (CHAPTERS 26-27)

Components: Organs of gastrointestinal tract, including the mouth, pharynx (throat), esophagus, stomach, small and large intestines, rectum, and anus; also includes accessory digestive organs that assist in digestive processes, such as the salivary glands, liver, gallbladder, and pancreas.

Functions: Achieves physical and chemical breakdown of food; absorbs nutrients; eliminates solid wastes.

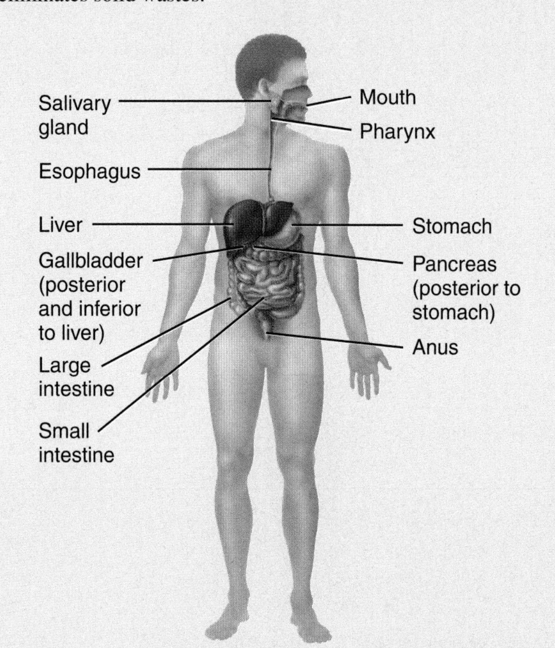

10. URINARY SYSTEM (CHAPTER 28)

Components: Kidneys, ureters, urinary bladder, and urethra.

Functions: Produces, stores, and eliminates urine; eliminates wastes and regulates volume and chemical composition of blood; helps regulate acidity of body fluids; maintains body's mineral balance; helps regulate red blood cell production.

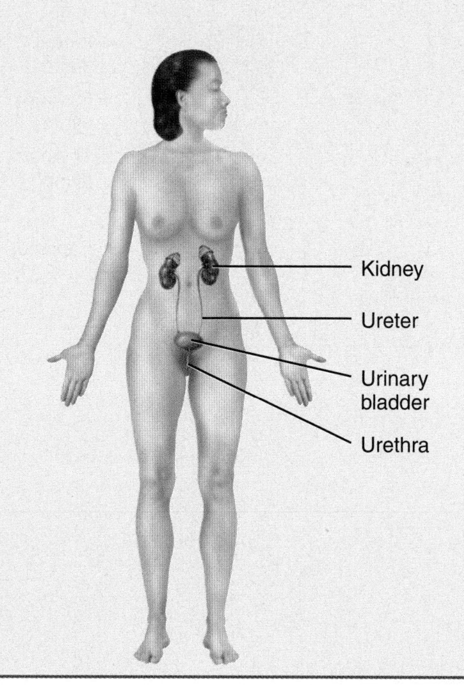

1.3 LIFE PROCESSES

OBJECTIVE

- Define the important life processes of the human body.

All living organisms have certain characteristics that set them apart from nonliving things. The following are six important life processes of the human body.

1. ***Metabolism*** (me-TAB-ō-lizm) is the sum of all the chemical processes that occur in the body. It includes the breakdown of large, complex molecules into smaller, simpler ones as well as the building up of complex molecules from smaller, simpler ones. For example, proteins in food are split into amino acids. The amino acids are the building blocks that can be used to make new proteins. These amino acids can then be used to build new proteins that make up body structures such as muscles and bones.

2. ***Responsiveness*** is the body's ability to detect and react to changes in its internal (inside the body) or external (outside the body) environment. Different cells in the body detect different sorts of changes and respond in characteristic ways. Nerve cells respond to changes in the environment by generating electrical signals, known as nerve impulses. Muscle cells respond to nerve impulses by contracting, which generates force to move body parts.

3. ***Movement*** includes motion of the whole body, individual organs, single cells, and even tiny organelles inside cells. For example, the coordinated action of several muscles and bones enables you to move your body from one place to another by walking or running. After you eat a meal that contains fats, your gallbladder (an organ) contracts and moves bile into the gastrointestinal tract to help in the digestion of fats. When a body tissue is damaged or infected, certain white blood cells move from the bloodstream into the affected tissue to help clean up and repair the area. At the cellular level, various cell components move from one position to another to carry out their functions.

4. ***Growth*** is an increase in body size. It may be due to an increase in (1) the size of existing cells, (2) the number of cells, or (3) the amount of material surrounding cells.

11. REPRODUCTIVE SYSTEMS (CHAPTER 29)

Components: Gonads (testes or ovaries) and associated organs: uterine tubes, uterus, and vagina in females, and epididymis, ductus (vas) deferens, and penis in males. Also, mammary glands in females.

Functions: Gonads produce gametes (sperm or oocytes) that unite to form a new organism and release hormones that regulate reproduction and other body processes; associated organs transport and store gametes. Mammary glands produce milk.

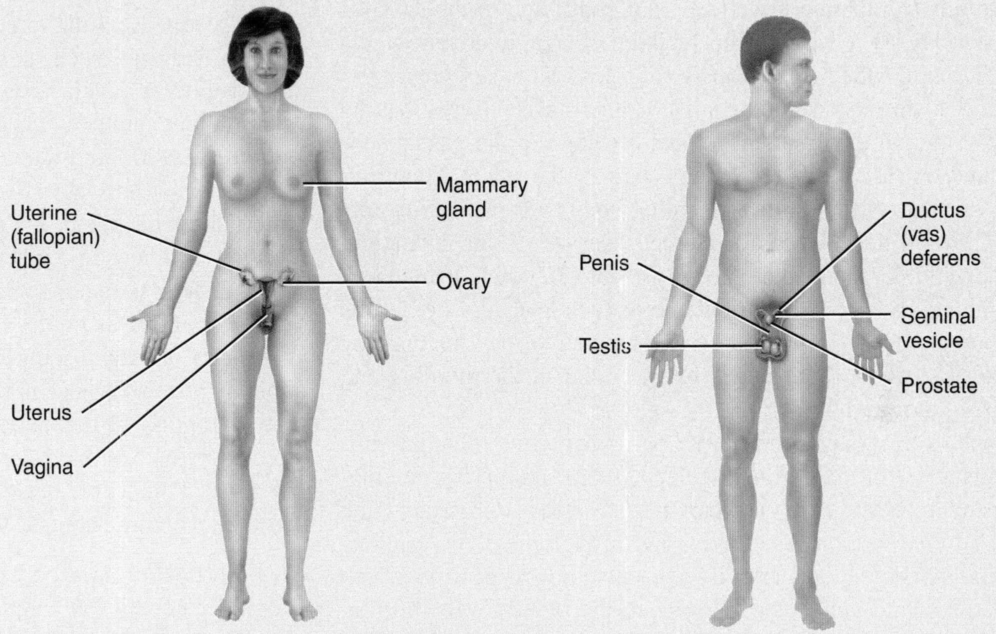

5. ***Differentiation*** (dif′-er-en-shē-Ā-shun) is the process whereby unspecialized cells become specialized cells. Specialized cells differ in structure and function from the unspecialized cells that gave rise to them. For example, specialized red blood cells and several types of white blood cells differentiate from the same unspecialized cells in bone marrow. Similarly, a single fertilized egg cell undergoes tremendous differentiation to develop into a unique individual who is similar to, yet quite different from, either of the parents.

6. ***Reproduction*** (rē-prō-DUK-shun) refers to either (1) the formation of new cells for growth, repair, or replacement or (2) the production of a new individual.

Although not all of these processes occur in cells throughout the body all of the time, when they cease to occur properly cell death may occur. When cell death is extensive and leads to organ failure, the result is death of the organism.

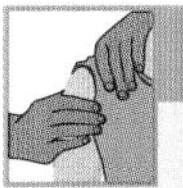

MANUAL THERAPY APPLICATION

Touch

The manual therapies, by definition, involve touching a patient, and the experience of **touch** is unique for each individual. Many variables are involved in determining our experiences of touching and being touched. All cultures, societies, families, and individuals have "rules," mostly unwritten and mostly unconscious, about *if* one should touch, *how* to touch, *whom* it is appropriate to touch, and *where* on the body it is appropriate. Most individuals benefit from touch, but others are offended or even dramatically affected in a negative way by touch. Students of the manual therapies need to assess the issue of touch within themselves as well as that of their patient.

CHECKPOINT

5. What types of movement can occur in the human body?

1.4 HOMEOSTASIS: MAINTAINING LIMITS

OBJECTIVES

- Define homeostasis and explain its importance.
- Describe the components of a feedback system.
- Compare the operation of negative and positive feedback systems.
- Distinguish between symptoms and signs of a disease.

The trillions of cells of the human body need relatively stable conditions to function effectively and contribute to the survival of the body as a whole. The maintenance of relatively stable conditions is called ***homeostasis*** (hō′mē-ō-STĀ-sis; *homeo-* = sameness; *-stasis* = standing still). Homeostasis ensures that the body's internal environment remains steady despite changes inside and outside the body. A large part of the internal environment consists of the fluid surrounding body cells, called ***interstitial fluid*** (in′-ter-STISH-al). Homeostasis keeps the interstitial fluid at a proper temperature of 37° Celsius (98° to 99° Fahrenheit) and maintains adequate nutrient and oxygen levels for body cells to flourish. Note that normal body temperature may fluctuate by 1° Celsius in a 24-hour cycle, being cooler in the morning and higher in the late afternoon.

Each body system contributes to homeostasis in some way. For instance, in the cardiovascular system, alternating contraction and relaxation of the heart propels blood throughout the body's blood vessels. As blood flows through the blood capillaries, the smallest blood vessels, nutrients and oxygen move into interstitial fluid and wastes move into the blood. Cells, in turn, remove nutrients and oxygen from and release their wastes into interstitial fluid. Homeostasis is *dynamic;* that is, it can change over a narrow range that is compatible with maintaining cellular life processes. For example, the level of glucose in the blood is maintained within a narrow range. It normally does not fall too low between meals or rise too high even after eating a high-glucose meal. The brain needs a steady supply of glucose to keep functioning—a low blood glucose level may lead to unconsciousness or even death. A prolonged high blood glucose level, by contrast, can damage blood vessels and cause excessive loss of water in the urine.

Control of Homeostasis: Feedback Systems

Fortunately, every body structure, from cells to systems, has one or more homeostatic devices that work to keep the internal environment within normal limits. The homeostatic mechanisms of the body are mainly under the control of two systems: the nervous system and the endocrine system. The nervous system detects changes from the balanced state and sends messages in the form of ***nerve impulses*** to organs that can counteract the change. For example, when body temperature rises, nerve impulses cause sweat glands to release more sweat, which cools the body as it evaporates. The endocrine system corrects changes by secreting molecules called ***hormones*** (HOR-mōns) into the blood. Hormones affect specific body cells where they cause responses that restore homeostasis. For example, the hormone insulin reduces the blood glucose level when it is too high. Nerve impulses typically cause rapid corrections, whereas hormones usually work more slowly. Hormones work more slowly since they travel via the cardiovascular system and therefore have a wider range of action.

Homeostasis is maintained by means of many feedback systems. A ***feedback system*** or *feedback loop* is a cycle of events in which a condition in the body is continually monitored, evaluated, changed, remonitored, reevaluated, and so on. Each monitored condition, such as body temperature, blood pressure, or blood glucose level, is termed a *controlled condition*. Any disruption that causes a change in a controlled condition is called a ***stimulus.*** Some stimuli come from the external environment, such as intense heat or lack of oxygen. Others originate in the internal environment, for example, a blood glucose level that is too low. Homeostatic imbalances may also occur due to psychological stresses in our social environment—the demands of work and school, for example. In most cases, the disruption of homeostasis is mild and temporary, and the responses of body cells quickly restore balance in the internal environment. In other cases, the disruption of homeostasis may be intense and prolonged, as in poisoning, overexposure to temperature extremes, severe infection, or death. Three basic components make up a feedback system: a receptor, a control center, and an effector (Figure 1.2).

1. A ***receptor*** is a body structure that monitors changes in a controlled condition and sends information called the *input* to a control center. Input is in the form of nerve impulses or chemical signals. Nerve endings in the skin that sense temperature are one of the hundreds of different kinds of receptors in the body.

2. A ***control center*** in the body, for example, the brain, sets the range of values within which a controlled condition should be maintained, evaluates the input it receives from receptors, and generates output commands when they are needed. *Output* is information, in the form of nerve impulses or chemical signals, that is relayed from the control center to an effector.

3. An ***effector*** (e-FEK-tor) is a body structure that receives output from the control center and produces a *response* that changes the controlled condition. Nearly every organ or tissue in the body can behave as an effector. For example, when your body temperature drops sharply, your brain (control center) sends nerve impulses to your skeletal muscles (effectors) that cause you to shiver, which generates heat and raises your temperature.

Feedback systems can produce either negative feedback or positive feedback. If the response reverses a change in a controlled condition, as in the body temperature regulation example, the system is called a negative feedback system. If the response enhances or intensifies a change in the controlled condition, it is referred to as a positive feedback system.

Figure 1.2 Parts of a feedback system. The dashed return arrow on the right side symbolizes negative feedback.

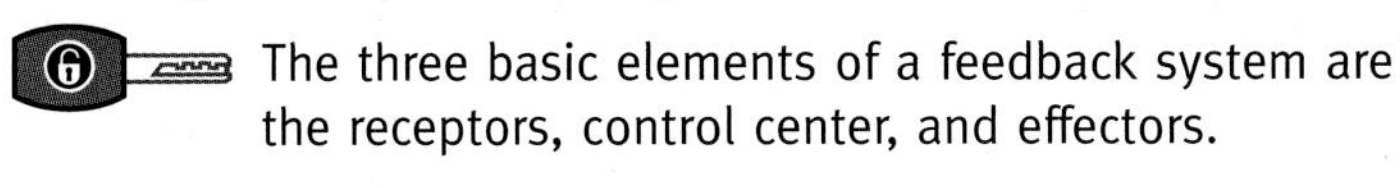
The three basic elements of a feedback system are the receptors, control center, and effectors.

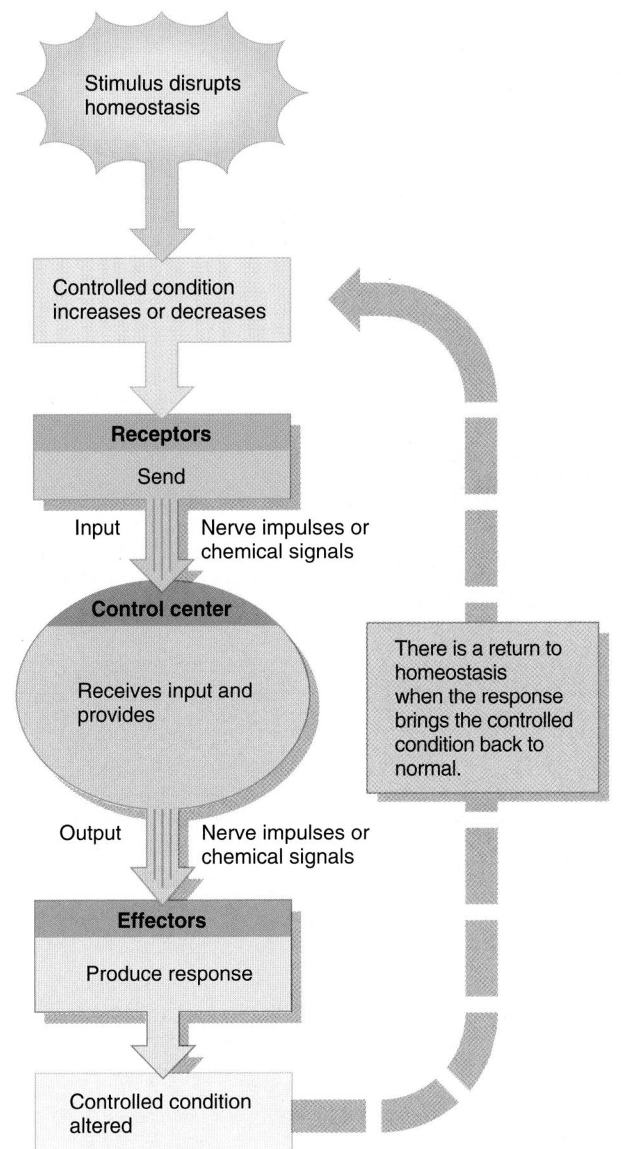

What is the basic difference between negative and positive feedback systems?

Negative Feedback Systems

A ***negative feedback system*** *reverses* a change in a controlled condition. Consider one negative feedback system that helps regulate blood pressure. Blood pressure (BP) is the force exerted by blood as it presses against the walls of blood vessels. When the heart beats faster or harder, BP increases. If a stimulus causes BP (controlled condition) to rise, the following sequence of events occurs (see Figure 1.3). The higher pressure is detected by *baroreceptors,* pressure-sensitive nerve cells located in the walls of certain blood vessels (the receptors). The baroreceptors send nerve impulses (input) to the brain (control center), which interprets the impulses and responds by sending nerve impulses (output) to the heart (the effector). Heart rate decreases, which causes BP to decrease (response). This sequence of events returns the controlled condition—blood pressure—to normal, and homeostasis is restored. This is a negative feedback system because the activity of the effector produces a result, a drop in BP, that reverses the effect of the stimulus. Negative feedback systems tend to regulate conditions in the body that are held fairly stable over long periods, such as blood pressure, blood glucose level, and body temperature.

Positive Feedback Systems

A ***positive feedback system*** enhances a change in a controlled condition. Normal positive feedback systems tend to intensify conditions that don't happen very often, such as childbirth (see Figure 1.4), ovulation, and blood clotting. Because a positive feedback system continually enhances a change in a controlled condition, it must be shut off by some event outside the system. If the action of a positive feedback system isn't stopped, it can "run away" and produce life-threatening changes in the body.

Homeostasis and Good Health

You've seen homeostasis defined as a condition in which the body's internal environment remains relatively stable. The body's ability to maintain homeostasis gives it tremendous healing power and a remarkable resistance to abuse. The physiological processes responsible for maintaining homeostasis are in large part also responsible for your good health.

For most people, lifelong good health is not something that happens effortlessly. Two of the many factors in this balance called health are the environment and your own behavior. Your genetic makeup is also important. Your body's homeostasis is affected by the air you breathe, the food you eat, and even the thoughts you think. The way you live your life can either support or interfere with your body's ability to maintain homeostasis and recover from the inevitable stresses life throws your way.

Let's consider the common cold. You support your natural healing processes when you take care of yourself. Plenty of rest, fluids, fruits, and vegetables allow the immune system to do its job. The cold runs its course, and you are soon back on your feet. If, however, instead of taking care of yourself, you continue

Figure 1.3 Homeostasis of blood pressure by a negative feedback system. Note that the response is fed back into the system and the system continues to lower blood pressure until there is a return to normal blood pressure (homeostasis).

If the response reverses a change in a controlled condition, a system is operating by negative feedback.

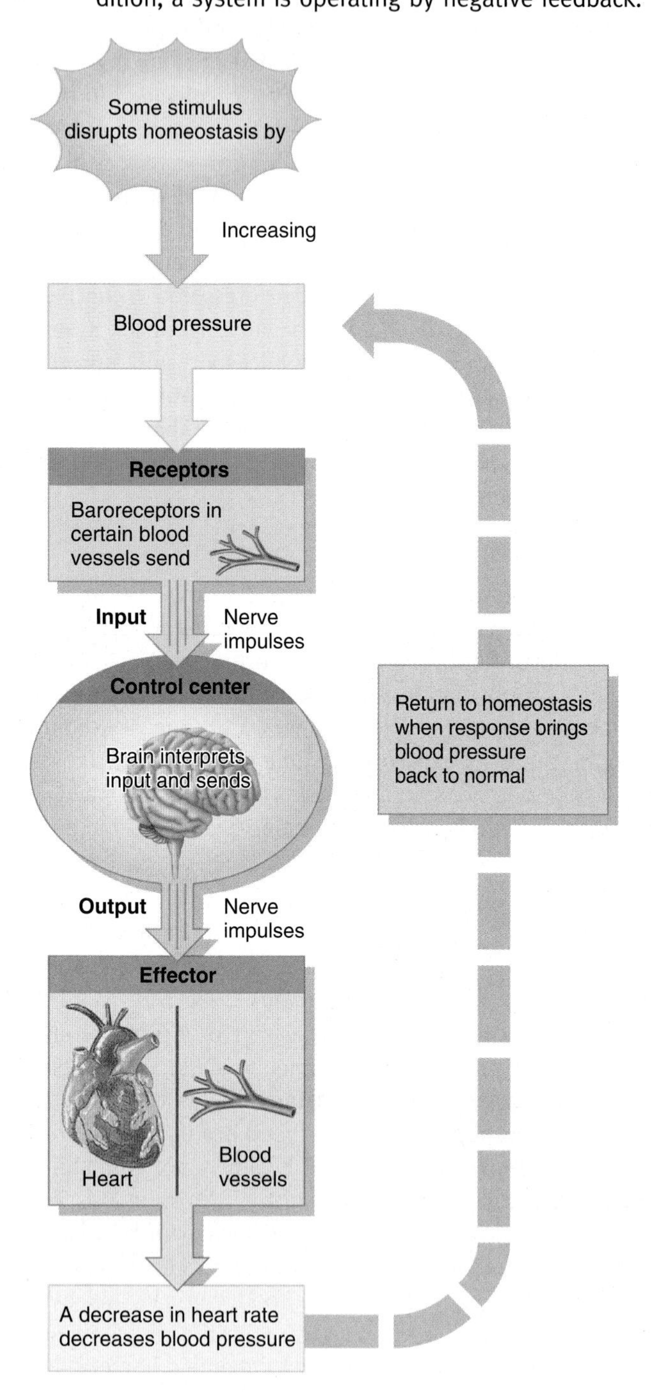

What would happen to the heart rate if some stimulus caused blood pressure to decrease? Would this occur by positive or negative feedback?

Figure 1.4 Positive feedback control of labor contractions during birth of a baby. The solid return arrow symbolizes positive feedback.

If the response enhances or intensifies the stimulus, a system is operating by positive feedback.

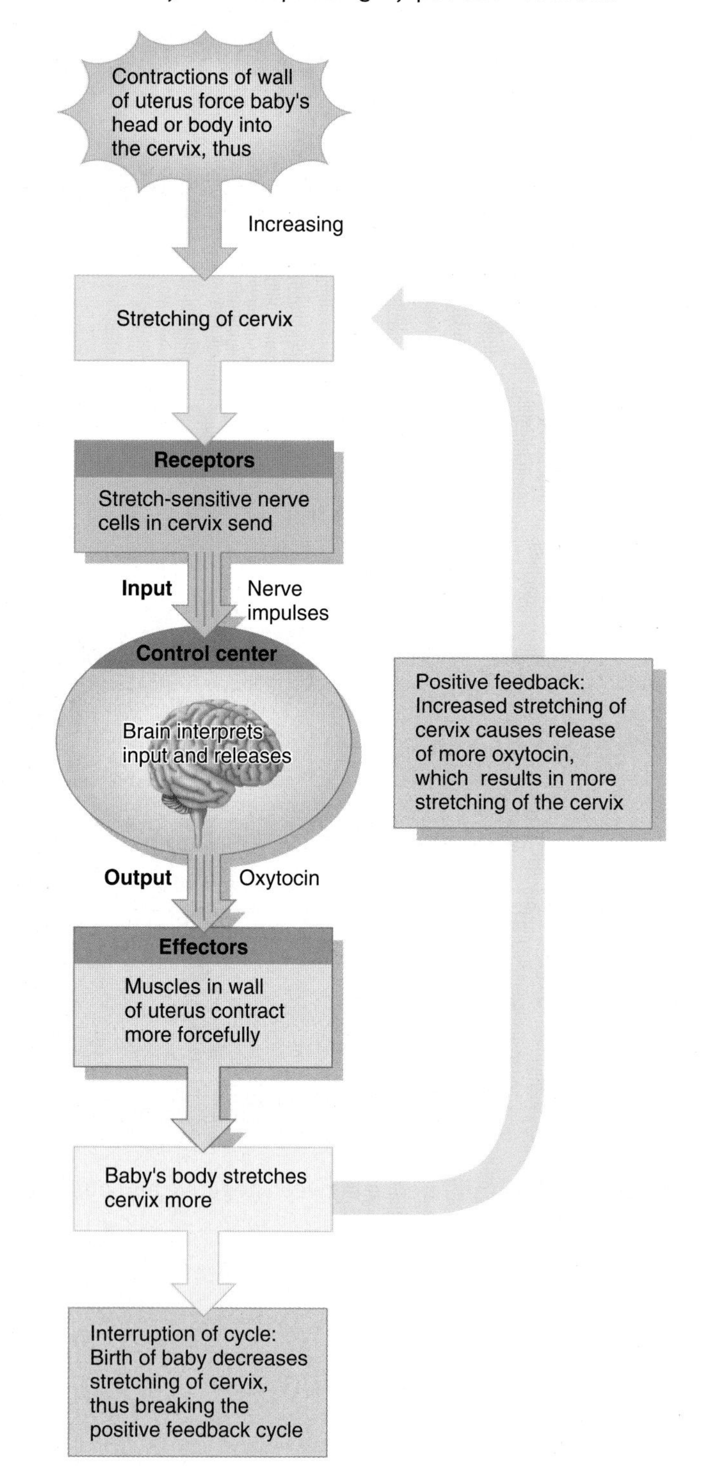

Why do positive feedback systems that are part of a normal physiological response include some mechanism that terminates the system?

to smoke two packs of cigarettes a day, skip meals, and pull several all-nighters studying for an anatomy and physiology exam, you interfere with the immune system's ability to fend off attacking microbes and bring the body back to homeostasis and good health. Other infections take advantage of your weakened state, and soon the cold may have developed into bronchitis or pneumonia.

• CLINICAL CONNECTION | Diagnosis of Disease

Diagnosis (dī′-ag-NŌ-sis; *dia-* = through; *-gnosis* = knowledge) is the identification of a disease or disorder by a physician and is based on a scientific evaluation of the patient's signs and symptoms, medical history, physical examination, and sometimes data from laboratory tests. Taking a *medical history* consists of collecting information about events that might be related to a patient's illness, including the chief complaint, history of present illness, past medical problems, family medical problems, and social history. A *physical examination* is an orderly evaluation of the body and its functions. This process includes *inspection* (looking at or into the body with various instruments for any changes that deviate from normal), *palpation* (pronounced pal-PĀ-shun, feeling body surfaces with the hands), *auscultation* (pronounced aus-cul-TĀ-shun, listening to body sounds, often using a stethoscope), *percussion* (pronounced per-KUSH-un, tapping on body surfaces and listening to the resulting echo), and measuring vital signs (temperature, pulse, respiratory rate, height, weight, and blood pressure). Some common laboratory tests include analyses of blood and urine. •

Many diseases are the result of years of poor health behavior that interferes with the body's natural drive to maintain homeostasis. An obvious example is smoking-related illness. Smoking tobacco exposes sensitive lung tissue to a multitude of chemicals that cause cancer and damage the lung's ability to repair itself. Because diseases such as emphysema and lung cancer are difficult to treat and are very rarely cured, it is much wiser to quit smoking—or never start—than to hope a doctor can "fix" you once you are diagnosed with a lung disease. Developing a lifestyle that works with, rather than against, your body's homeostatic processes helps you maximize your personal potential for optimal health and well-being. The concept of a healthy lifestyle will be discussed in Chapter 27, Nutrition and Metabolism.

Homeostasis and Disease

As long as all of the body's controlled conditions remain within certain narrow limits, body cells function efficiently, homeostasis is maintained, and the body stays healthy. Should one or more components of the body lose their ability to contribute to homeostasis, however, the normal balance among all of the body's processes may be disturbed. If the homeostatic imbalance is moderate, a disorder or disease may occur; if it is severe, death may result.

A ***disorder*** is any abnormality of structure and/or function. ***Disease*** is a more specific term for an illness characterized by a recognizable set of signs and symptoms. ***Signs*** are objective changes that a clinician can observe and measure, such as bleeding, swelling, vomiting, diarrhea, fever, a rash, or paralysis. ***Symptoms*** are subjective changes in body functions that are not apparent to an observer, for example, headache or nausea. Specific diseases alter body structure and function in characteristic ways, usually producing a recognizable cluster of signs and/or symptoms.

MANUAL THERAPY APPLICATION

Assessment

Assessment involves a systematic method of gathering information to make appropriate decisions about *how,* or even *if,* treatment sessions should proceed. Assessment involves taking a history as well as the evaluation of the body using a series of movements or tests to determine if there is an impairment. Tests reveal either a *positive sign,* indicating that the condition is present, or a *negative finding,* indicating that the individual does not have the impairment. The therapist can then determine a clinical impression or refer the patient to a physician for a confirming diagnosis or further testing.

If the assessment and treatment are within the expertise of the manual therapist, a plan of treatment is formally or informally developed for each patient. Different disciplines of manual therapies have various forms of assessment, but the process of assessment is similar for each plan. One form of treatment plan involves development of S.O.A.P. notes:

S—Subjective findings (patient's report of symptoms)
O—Objective findings (therapist's assessment, based on *palpation* and/or visual findings)
A—Action (therapist's choice of techniques and body locations for treatment)
P—Plan of action (therapist's plan for follow-up or additional sessions)

The S.O.A.P. notes can be written appropriately only if the therapist has a commmand of the language of medical terms and knows the anatomy and physiology of the organs under her hands. This chapter introduces medical terms that enable a manual therapist to effectively communicate with other members of the health-care delivery system.

CHECKPOINT

6. What types of disturbances can act as stimuli that initiate a feedback system?
7. How are negative and positive feedback systems similar? How are they different?
8. What is the difference between signs and symptoms of a disease? Give examples of each.

1.5 AGING AND HOMEOSTASIS

OBJECTIVE

- Describe some of the effects of aging.

As you will see later, ***aging*** is a normal process characterized by a progressive decline in the body's ability to restore homeostasis. Aging produces observable changes in structure and function and increases vulnerability to stress and disease. The changes associated with aging are apparent in all body systems. Examples include wrinkled skin, gray hair, loss of bone mass, decreased muscle mass and strength, diminished reflexes, decreased production of some hormones, increased incidence of heart disease, increased susceptibility to infections and cancer, decreased lung capacity, less efficient functioning of the digestive system, decreased kidney function, menopause, and enlarged prostate. These and other effects of aging will be discussed in detail in most chapters.

CHECKPOINT

9. What are some of the signs of aging?

1.6 ANATOMICAL TERMS

OBJECTIVES

- Describe the anatomical position.
- Identify the major regions of the body and relate the common names to the corresponding anatomical terms for various parts of the body.
- Define the directional terms and the anatomical planes and sections used to locate parts of the human body.

The language of anatomy and physiology is very precise. When describing where the wrist is located, is it correct to say, "The wrist is above the fingers"? This description is true if your arms are at your sides. But if you hold your hands up above your head, your fingers would be above your wrists. To prevent this kind of confusion, scientists and health-care professionals refer to one standard anatomical position and use a special vocabulary for relating body parts to one another.

In the study of anatomy, descriptions of any part of the human body assume that the body is in a specific stance called the ***anatomical position*** (an′-a-TOM-i-kal). In the anatomical position, the subject is in the erect position, facing the observer, with the head level and the eyes facing forward. The feet are flat on the floor and directed forward, and the arms are at the sides with the palms turned forward (Figure 1.5). The joints of the body are in extension while in the anatomical position. Two terms describe a reclining body. If the body is lying face down, it is in the *prone* position. If the body is lying face up, it is in the *supine* position.

Names of Body Regions

The human body is divided into several major regions that can be identified externally. These are the head, neck, trunk, upper limbs, and lower limbs (Figure 1.5). The ***head*** consists of the skull and face. The ***skull*** is the part of the head that encloses and protects the brain, and the ***face*** is the front portion of the head that includes the eyes, nose, mouth, forehead, cheeks, and chin. The ***neck*** supports the head and attaches it to the trunk. The ***trunk*** consists of the chest, abdomen, and pelvis. Each ***upper limb (appendage*** or ***extremity)*** is attached to the trunk and consists of the shoulder, armpit, arm (portion of the limb from the shoulder to the elbow), forearm (portion of the limb from the elbow to the wrist), wrist, and hand. Each ***lower limb (appendage*** or ***extremity)*** is also attached to the trunk and consists of the buttock, thigh (portion of the limb from the hip to the knee), leg (portion of the limb from the knee to the ankle), ankle, and foot. The ***groin*** (GROYN) is the area on the front surface of the body, marked by a crease on each side, where the trunk attaches to the thighs.

In Figure 1.5, the corresponding anatomical adjective for each part of the body appears in parentheses next to the common name. For example, if you receive a tetanus shot in your *buttock,* it is a *gluteal* injection; the bulging biceps brachii muscle in your arm is in the *brachial region*. The descriptive form of a body part is based on a Greek or Latin word or "root" for the same part or area. The Latin word for armpit is *axilla* (ak-SIL-a), for example, and thus one of the nerves passing within the armpit is named the axillary nerve. Learning the medical or anatomical term for each region of the body now will save you a lot of frustration in later chapters. You will learn more about the word roots of anatomical and physiological terms as you continue to read this book.

Directional Terms

To locate various body structures, anatomists use specific ***directional terms,*** words that describe the position of one body part relative to another. Several directional terms can be grouped in pairs that have opposite meanings, for example, anterior (front) and posterior (back). Study Exhibit 1.1 and Figure 1.6 to determine, among other things, whether your stomach is superior to your lungs.

Figure 1.5 The anatomical position. The common names and corresponding anatomical terms (in parentheses) are indicated for specific body regions. For example, the head is the cephalic region.

In the anatomical position, the subject stands erectly and facing the observer, with the head level and the eyes facing forward. The feet are flat on the floor and directed forward, and the arms are at the sides with the palms facing forward.

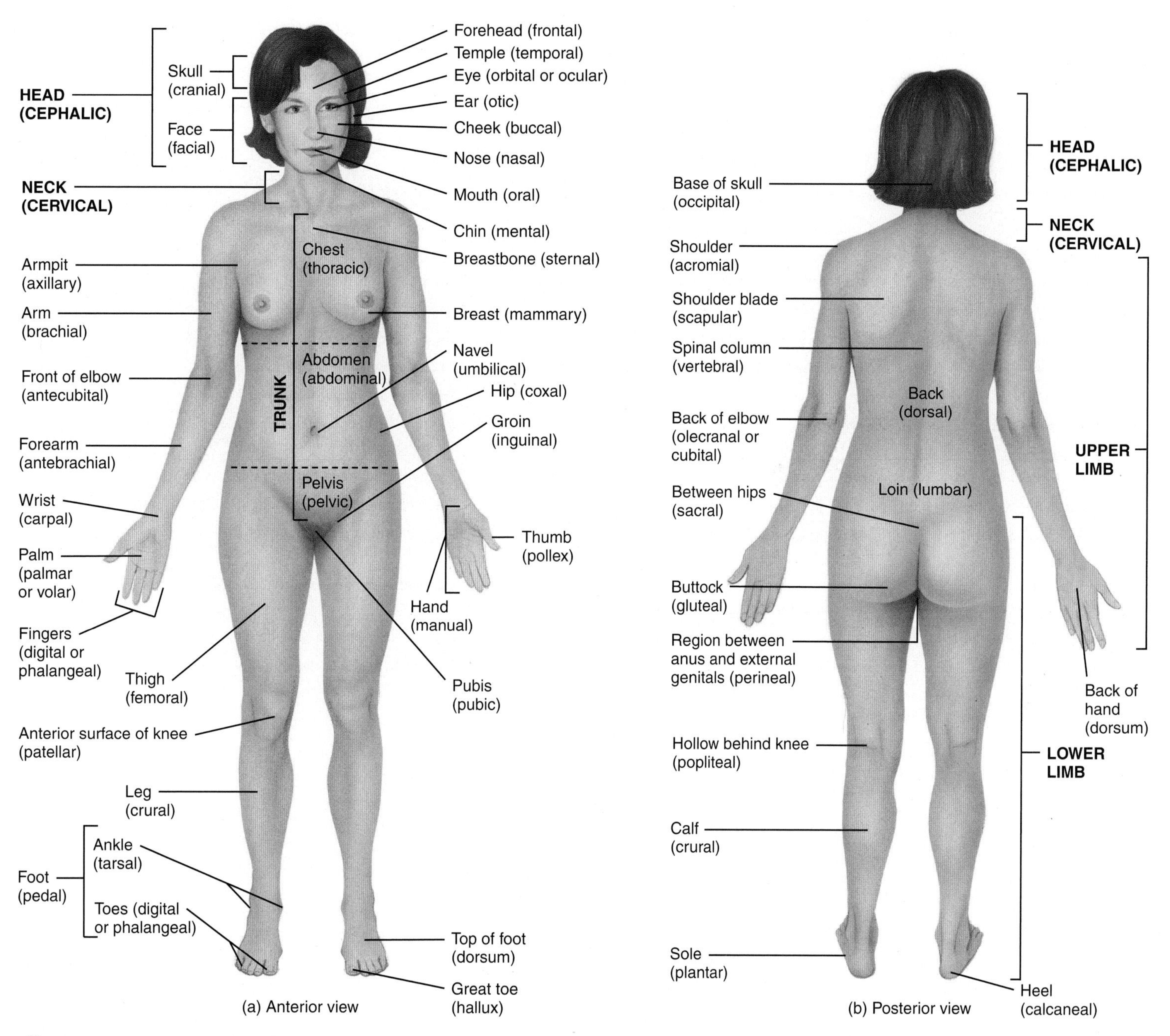

? **Where is a plantar wart located?**

EXHIBIT 1.1 Directional Terms

OBJECTIVE

- Define each directional term used to describe the human body.

Most of the directional terms used to describe the human body can be grouped into pairs that have opposite meanings. For example, *superior* means toward the upper part of the body and *inferior* means toward the lower part of the body. It is important to understand that directional terms have *relative* meanings; they only make sense when used to describe the position of one structure relative to another. For example, your knee is superior to your ankle, even though both are located in the inferior half of the body. Study the directional terms and the example of how each is used. As you read each example, refer to Figure 1.6 to see the location of the structures mentioned.

CHECKPOINT

10. Which directional terms can be used to specify the relationships between (1) the elbow and the shoulder, (2) the left and right shoulders, (3) the sternum and the humerus, and (4) the heart and the diaphragm?

DIRECTIONAL TERM	DEFINITION	EXAMPLE OF USE
Superior (soo′-PĒR-ē-or) (**cephalic** or **cranial**)	Toward the head, or the upper part of a structure.	The heart is superior to the liver.
Inferior (in′-FĒR-ē-or) (**caudal**)	Away from the head, or the lower part of a structure.	The stomach is inferior to the lungs.
Anterior (an-TĒR-ē-or) (**ventral**)	Nearer to or at the front of the body.	The sternum (breastbone) is anterior the heart.
Posterior (pos-TĒR-ē-or) (**dorsal**)	Nearer to or at the back of the body.	The esophagus (food tube) is posterior to the trachea (windpipe).
Medial (MĒ-dē-al)	Nearer to the midline* or midsagittal plane.	The ulna is medial to the radius.
Lateral (LAT-er-al)	Farther from the midline or midsagittal plane.	The lungs are lateral to the heart.
Intermediate (in′-ter-MĒ-dē-at)	Between two structures.	The transverse colon is intermediate between the ascending and descending colons.
Ipsilateral (ip-si-LAT-er-al)	On the same side of the body as another structure.	The gallbladder and ascending colon are ipsilateral.
Contralateral (CON-tra-lat-er-al)	On the opposite side of the body from another structure.	The ascending and descending colons are contralateral.
Proximal (PROK-si-mal)	Nearer to the attachment of a limb to the trunk; nearer the point of origin or the beginning.	The humerus is proximal to the radius.
Distal (DIS-tal)	Farther from the attachment of a limb to the trunk; farther from the point of origin or the beginning.	The phalanges are distal to the carpals.
Superficial (soo′-per-FISH-al)	Toward or on the surface of the body.	The ribs are superficial to the lungs.
Deep (DĒP)	Away from the surface of the body.	The ribs are deep to the skin of the chest and back.

*The ***midline*** is an imaginary vertical line that divides the body into equal right and left sides.

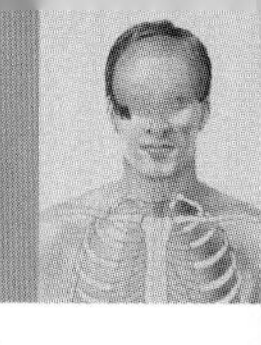

Figure 1.6 Directional terms. Please note that the terms "right" and "left" always refer to that of the patient or client. In all anterior views the right side of the body is located on the left side of the page.

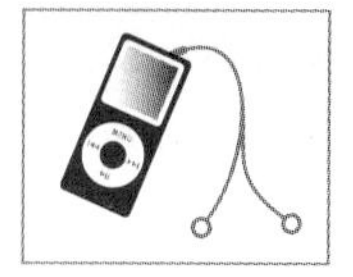

Directional terms precisely locate various parts of the body in relation to one another.

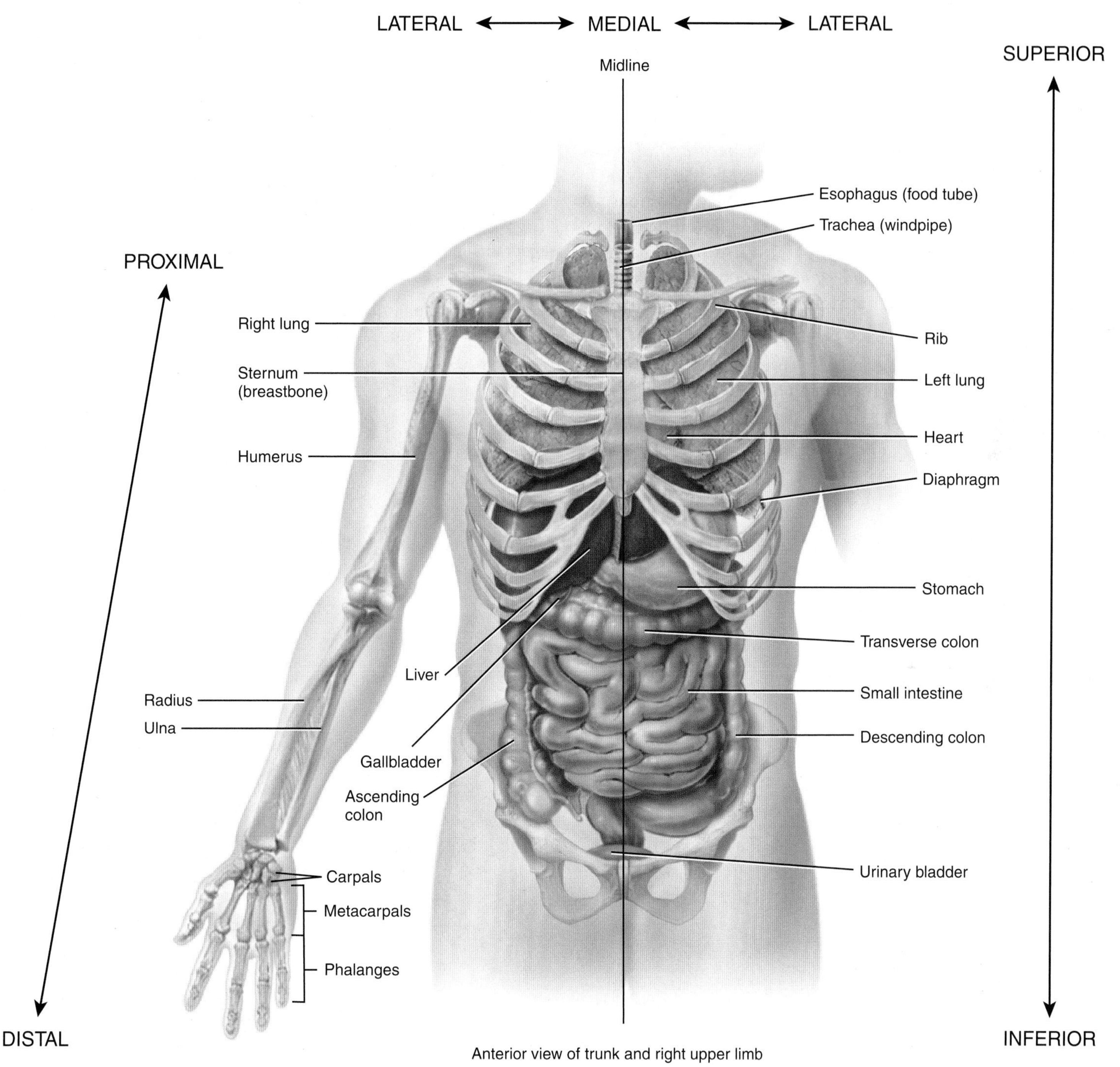

Anterior view of trunk and right upper limb

? Is the radius proximal to the humerus? Is the esophagus anterior to the trachea? Are the ribs superficial to the lungs? Is the urinary bladder medial to the ascending colon? Is the sternum lateral to the descending colon?

Planes and Sections

You will also study parts of the body in four major ***planes,*** that is, imaginary flat surfaces that pass through the body parts (Figure 1.7): sagittal, frontal, transverse, and oblique. A ***sagittal plane*** (SAJ-i-tal; *sagitt-* = arrow) is a vertical plane that divides the body or an organ into right and left sides. More specifically, when such a plane passes through the midline of the body or organ and divides it into *equal* right and left sides, it is called a ***midsagittal (median) plane.*** If the sagittal plane does not pass through the midline but instead divides the body or an organ into *unequal* right and left sides, it is called a ***parasagittal plane*** (*para-* = near). A ***frontal (coronal) plane*** divides the body or an organ into anterior (front) and posterior (back) portions. A ***transverse plane*** divides the body or an organ into superior (upper) and inferior (lower) portions. A transverse plane may also be called a ***cross-sectional*** or ***horizontal plane.*** Sagittal, frontal, and transverse planes are all at right angles to one another. An ***oblique plane*** (ō-BLĒK), by contrast, passes through the body or an organ at an angle between a transverse plane and a sagittal plane or between a transverse plane and a frontal plane.

When you study a body region, you will often view it in section. A ***section*** is a cut of the body or an organ made along one of the planes just described and will be viewed as one flat surface of the three-dimensional structure. It is important to know the plane of the section so you can understand the anatomical relationship of one part to another. Figure 1.8 indicates how three

Figure 1.7 Planes through the human body.

Frontal, transverse, sagittal, and oblique planes divide the body in specific ways.

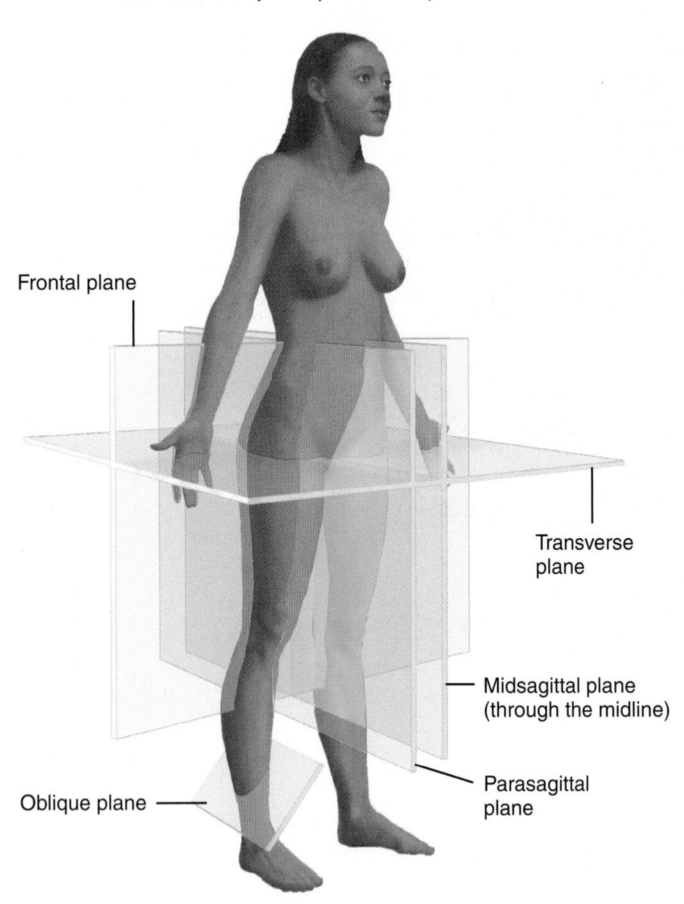

Which plane divides the heart into anterior and posterior portions?

Figure 1.8 Planes and sections through different parts of the brain. The diagrams (left) show the planes and the photographs (right) show the resulting sections. (Note: The "View" arrows in the diagrams indicate the direction from which each section is viewed. This aid is used throughout the book to indicate viewing direction.)

Planes divide the body in various ways to produce sections.

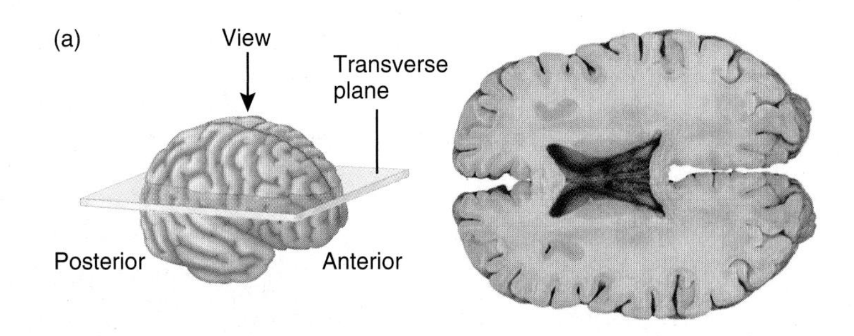

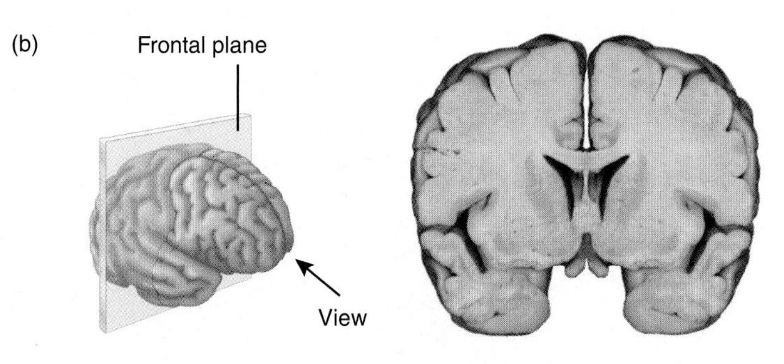

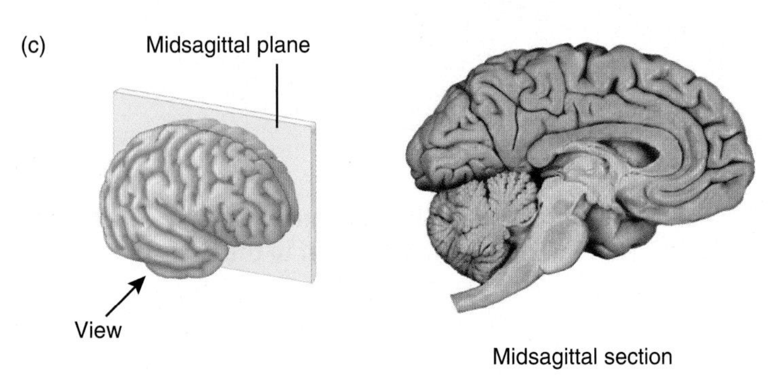

Which plane divides the brain into equal right and left sides?

different sections—a *transverse (cross) section,* a *frontal section,* and a *midsagittal section*—provide different views of the brain.

CHECKPOINT

11. Describe the anatomical position and explain why it is used.
12. Locate each region on your own body, and then identify it by its common name and the corresponding anatomical descriptive form.
13. For each directional term listed in Exhibit 1.1, provide your own example.
14. What are the various planes that may be passed through the body? Explain how each divides the body.

1.7 BODY CAVITIES

OBJECTIVES

- Describe the principal body cavities and the organs they contain.
- Explain why the abdominopelvic cavity is divided into regions and quadrants.

Body cavities are spaces within the body that contain, protect, separate, and support internal organs. Here we discuss several of the larger body cavities (Figure 1.9).

The ***cranial cavity*** (KRĀ-nē-al) is formed by the cranial (skull) bones and contains the brain, and the ***vertebral (spinal)***

Figure 1.9 Body cavities. The dashed lines indicate the border between the abdominal and pelvic cavities.

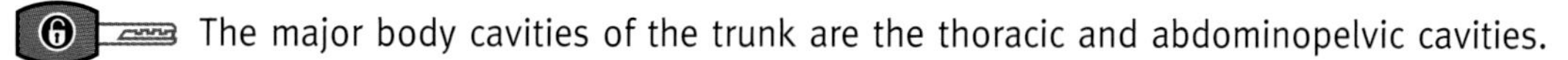

The major body cavities of the trunk are the thoracic and abdominopelvic cavities.

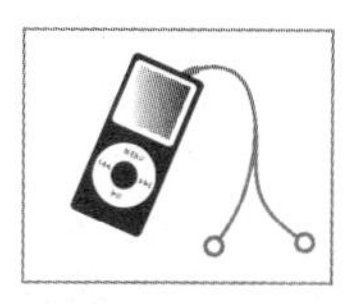

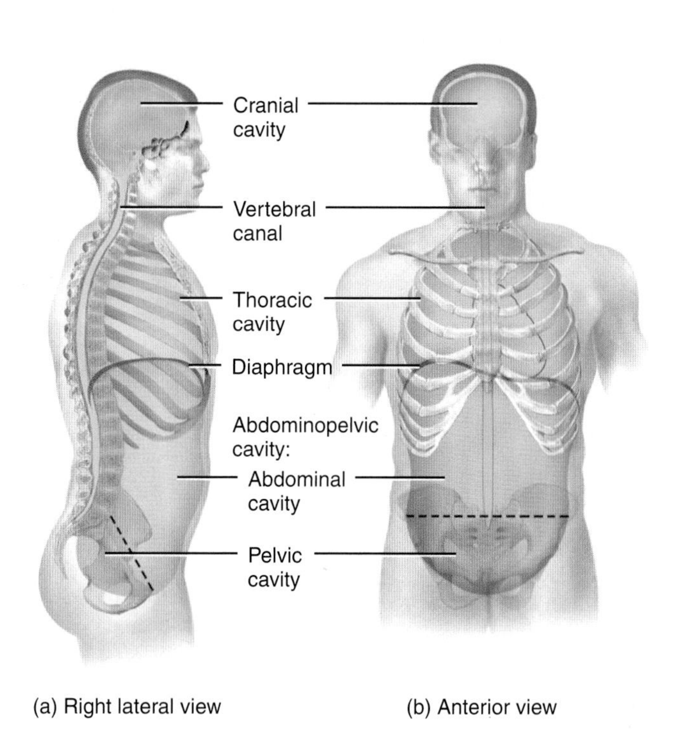

(a) Right lateral view

(b) Anterior view

CAVITY	COMMENTS
Cranial cavity	Formed by cranial bones and contains brain.
Vertebral canal	Formed by vertebral column and contains spinal cord and the beginnings of spinal nerves.
Thoracic cavity*	Chest cavity; contains pleural and pericardial cavities and mediastinum.
Pleural cavity	Each surrounds a lung; the serous membrane of each pleural cavity is the pleura.
Pericardial cavity	Surrounds the heart; the serous membrane of the pericardial cavity is the pericardium.
Mediastinum	Central portion of thoracic cavity between the lungs; extends from sternum to vertebral column and from first rib to diaphragm; contains heart, thymus, esophagus, trachea, and several large blood vessels.
Abdominopelvic cavity	Subdivided into abdominal and pelvic cavities.
Abdominal cavity	Contains stomach, spleen, liver, gallbladder, small intestine, and most of large intestine; the serous membrane of the abdominal cavity is the peritoneum.
Pelvic cavity	Contains urinary bladder, portions of large intestine, and internal organs of reproduction.

* See Figure 1.10 for details of the thoracic cavity.

In which cavities are the following organs located: urinary bladder, stomach, heart, small intestine, lungs, internal female reproductive organs, thymus, spleen, liver? Use the following symbols for your response: T = thoracic cavity, A = abdominal cavity, or P = pelvic cavity.

canal (VER-te-bral) is formed by the bones of the vertebral column (backbone) and contains the spinal cord.

The major body cavities of the trunk are the thoracic and abdominopelvic cavities. The ***thoracic cavity*** (thor-AS-ik; *thorac-* = chest) is the chest cavity. Within the thoracic cavity are three smaller cavities: the ***pericardial cavity*** (per′-i-KAR-dē-al; *peri-* = around; *-cardial* = heart), a fluid-filled space that surrounds the heart, and two ***pleural cavities*** (PLOOR-al; *pleur-* = rib or side), each of which surrounds one lung and contains a small amount of fluid (Figure 1.10).

The central portion of the thoracic cavity is an anatomical region called the ***mediastinum*** (mē-dē-a-STĪ-num; *media-* = middle; *-stinum* = partition). It is between the lungs, extending from the sternum (breastbone) to the vertebral column (backbone), and from the first rib to the diaphragm (see Figure 1.10). The mediastinum contains all thoracic organs except the lungs themselves. Among the structures in the mediastinum are the heart, thymus, esophagus, trachea, and several large blood vessels. The ***diaphragm*** (DĪ-a-fram = partition or wall) is a dome-shaped muscle that powers breathing and separates the thoracic cavity from the abdominopelvic cavity.

The ***abdominopelvic cavity*** (ab-dom′-i-nō-PEL-vik) extends from the diaphragm to the groin. As the name suggests, the abdominopelvic cavity is divided into two portions, although no wall separates them (see Figure 1.9). The upper portion, the ***abdominal cavity*** (ab-DOM-i-nal; *abdomin-* = belly), contains, among other things, the stomach, spleen, liver, gallbladder, small intestine, and most of the large intestine. The lower portion, the ***pelvic cavity*** (PEL-vik; *pelv-* = basin), contains the urinary bladder, portions of the large intestine, and internal organs of the reproductive system. The pelvic cavity is located below the dashed line in Figure 1.9. Organs inside the thoracic and abdominopelvic cavities are called ***viscera*** (VIS-e-ra).

Abdominopelvic Regions and Quadrants

To describe the location of the many abdominal and pelvic organs more precisely, the abdominopelvic cavity may be divided into smaller compartments. In one method, two horizontal and two vertical lines, like a tick-tack-toe grid, partition the cavity into nine ***abdominopelvic regions*** (Figure 1.11). The names of the nine abdominopelvic regions are the *right hypochondriac* (hī′-pō-KON-drē-ak), *epigastric* (ep-i-GAS-trik), *left hypochondriac, right lumbar, umbilical* (um-BIL-i-kal), *left lumbar, right inguinal* (IN-gwi-nal), *pubic* (PŪ-bik), and *left inguinal.* In another method, one horizontal and one vertical line passing through the ***umbilicus*** (um-bi-LĪ-kus; *umbilic-* = navel) or belly button divides the abdominopelvic cavity into ***quadrants*** (KWOD-rantz; *quad-* = one-fourth). The names of the abdominopelvic quadrants are the *right upper quadrant (RUQ), left upper quadrant (LUQ), right lower quadrant (RLQ),* and *left lower quadrant (LLQ).* Whereas the nine-region division is more widely used for anatomical studies, quadrants are more commonly used by clinicians to describe the site of an abdominopelvic pain, mass, or other abnormality.

Figure 1.10 The thoracic cavity. The dashed lines indicate the borders of the mediastinum. Notice that the pericardial cavity surrounds the heart and that the pleural cavities surround the lungs.

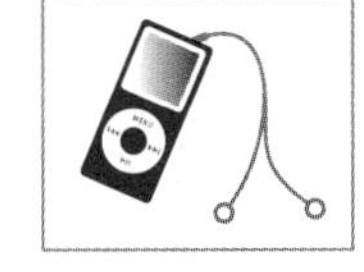

The mediastinum is medial to the lungs; it extends from the sternum to the vertebral column and from the first rib to the diaphragm.

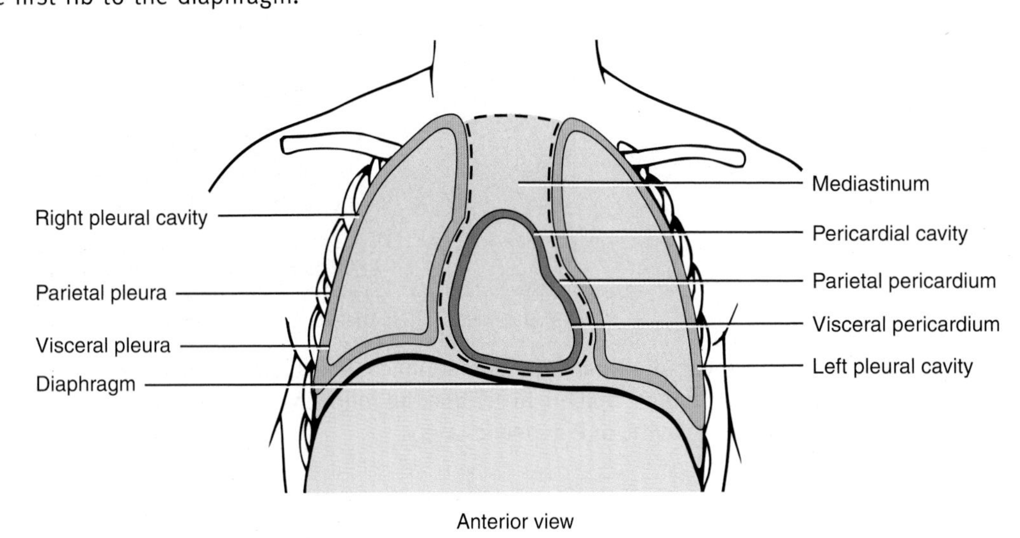

Which of the following structures are contained in the mediastinum: right lung, heart, esophagus, spinal cord, aorta, left pleural cavity?

Figure 1.11 Divisions of the abdominopelvic cavity into regions and quadrants.

The nine-region designation is used for anatomical studies. The quadrant designation is used to locate the site of pain, a mass, or some other abnormality and is typically used by clinicians.

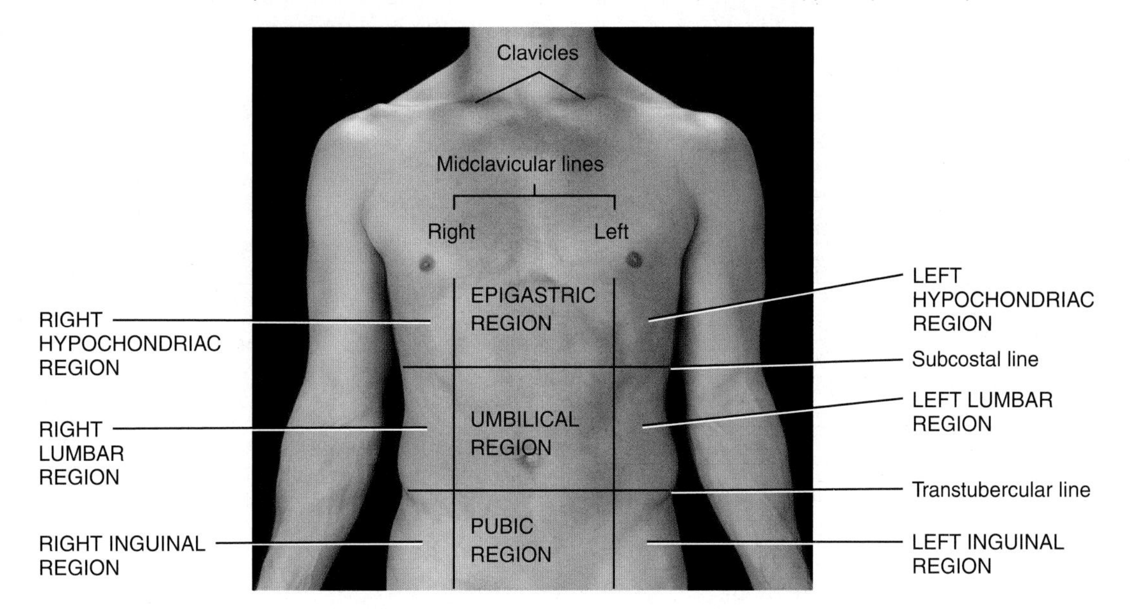

(a) Anterior view showing abdominopelvic regions

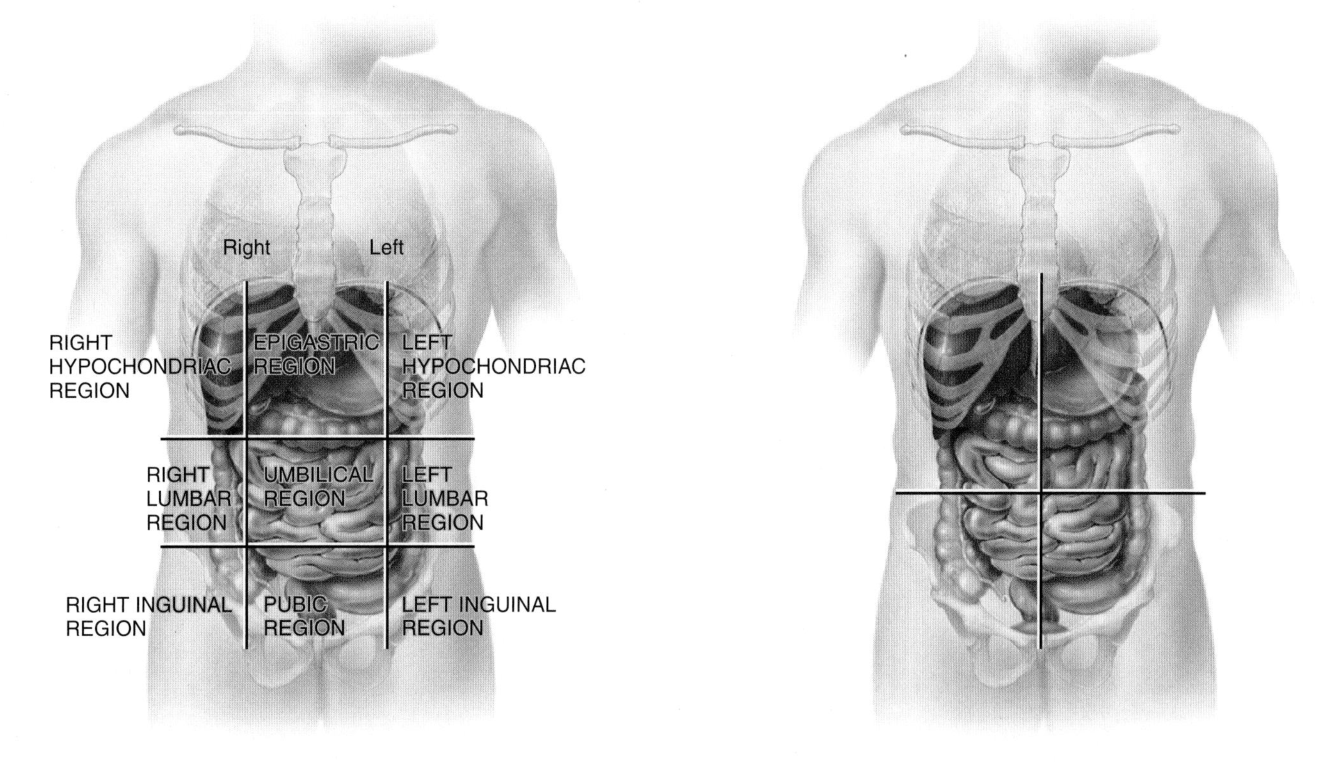

(b) Anterior view showing location of abdominopelvic regions

(c) Anterior superficial view of organs in abdominopelvic quadrants

In which abdominopelvic region is each of the following found: most of the liver, ascending colon, urinary bladder, most of the small intestine? You will have to refer to Figure 1.6 for the names and locations of the organs.

• CLINICAL CONNECTION | Autopsy

An **autopsy** (AW-top-sē; *auto-* = self; *-opsy* = to see) or *necropsy* is a postmortem (after-death) examination of the body and dissection of its internal organs to confirm or determine the cause of death. An autopsy can uncover the existence of diseases not detected during life, determine the extent of injuries, and explain how those injuries may have contributed to a person's death. It also may support the accuracy of diagnostic tests, establish the beneficial and adverse effects of drugs, reveal the effects of environmental influences on the body, provide more information about a disease, assist in the accumulation of statistical data, and educate health-care students. Moreover, an autopsy can reveal conditions that may affect offspring or siblings (such as congenital heart defects). An autopsy may be legally required, such as in the course of a criminal investigation, or to resolve disputes between beneficiaries and insurance companies about the cause of death. •

CHECKPOINT

15. What landmarks separate the various body cavities from one another?
16. Locate the nine abdominopelvic regions and the four abdominopelvic quadrants on yourself, and list some of the organs found in each.

1.8 MEDICAL IMAGING

OBJECTIVE

- Describe the principles and importance of medical imaging procedures in the evaluation of organ functions and the diagnosis of disease.

Medical imaging refers to techniques and procedures used to create images of the human body. Various types of medical imaging allow visualization of structures inside our bodies and are increasingly helpful for precise diagnosis of a wide range of anatomical and physiological disorders. The grandparent of all medical imaging techniques is conventional radiography (x-rays), in medical use since the late 1940s. The newer imaging technologies not only contribute to diagnosis of disease, but they also are advancing our understanding of normal physiology. Table 1.2 describes some commonly used medical imaging techniques. Other imaging methods, such as cardiac catheterization, will be discussed in later chapters.

TABLE 1.2

Common Medical Imaging Procedures

RADIOGRAPHY

Procedure: A single barrage of x-rays passes through the body, producing an image of interior structures on x-ray-sensitive film. The resulting two-dimensional image is a *radiograph* (RĀ-dē-ō-graf′), commonly called an *x-ray*.

Comments: Radiographs are relatively inexpensive, quick, and simple to perform, and usually provide sufficient information for diagnosis. X-rays do not easily pass through dense structures, so bones appear white. Hollow structures, such as the lungs, appear black. Structures of intermediate density, such as skin, fat, and muscle, appear as varying shades of gray. At low doses, x-rays are useful for examining soft tissues such as the breast (*mammography*) and bone density (*bone densitometry*). Radiologic technicians have artificially colored the black-and-white images with color to facilitate visualization by a radiologist and other physicians.

It is necessary to use a substance called a contrast medium to make hollow or fluid-filled structures visible in radiographs. X-rays make structures that contain contrast medium appear white. The medium may be introduced by injection, orally, or rectally depending on the structure to be imaged. Contrast x-rays are used to image blood vessels (*angiography*), the urinary system (*intravenous urography*), and the gastrointestinal tract (*barium contrast x-ray*).

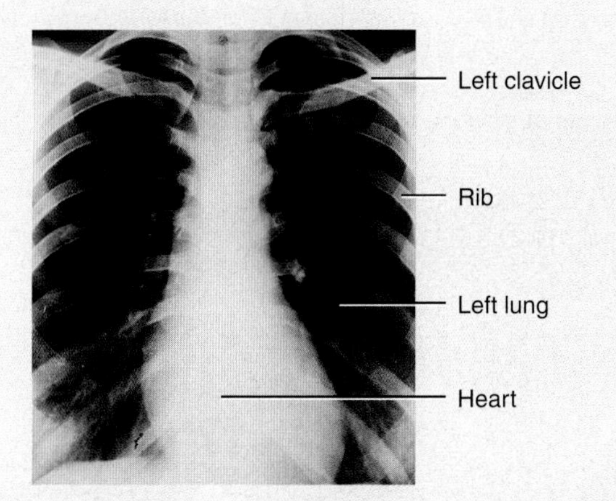

Radiograph of the thorax in anterior view

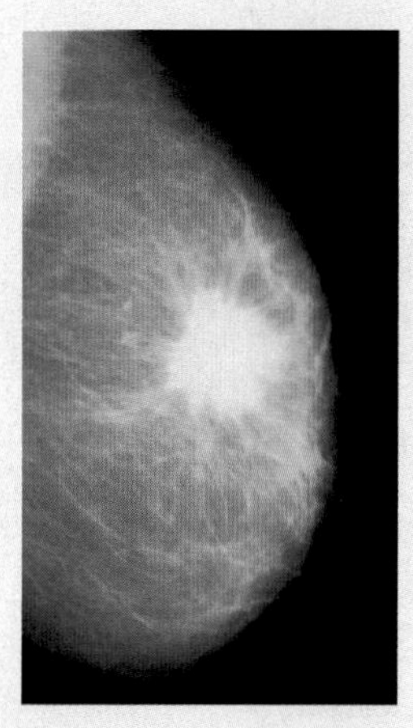

Mammogram of a female breast showing a cancerous tumor (white mass with uneven border)

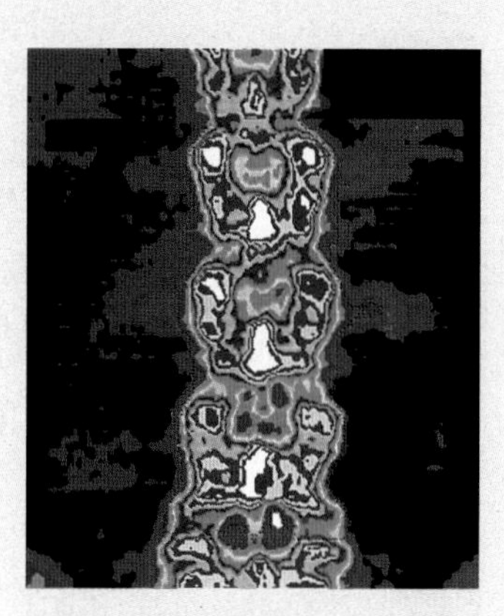

Bone densiometry scan of the lumbar spine in anterior view

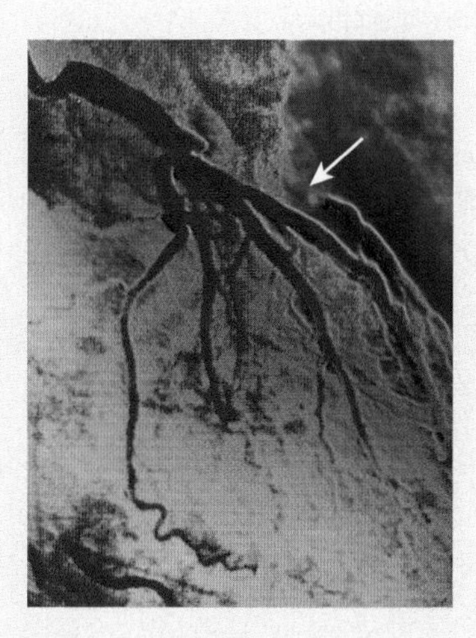

Angiogram of an adult human heart showing a blockage in a coronary artery (arrow)

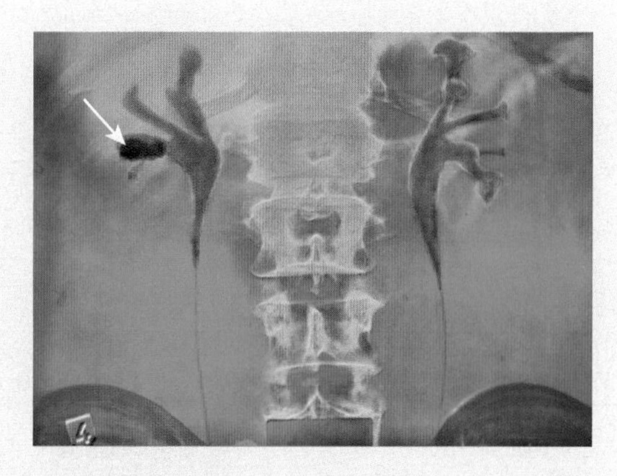

Intravenous urogram showing a kidney stone (arrow) in the right ureter

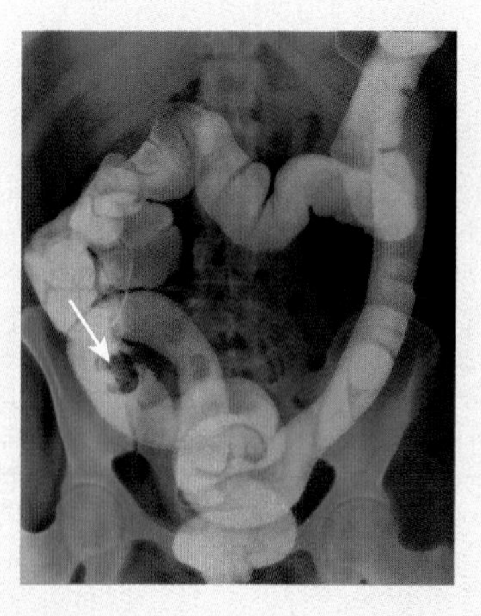

Barium contrast x-ray showing a cancer of the ascending colon (arrow)

CONTINUES

TABLE 1.2 CONTINUED

Common Medical Imaging Procedures

MAGNETIC RESONANCE IMAGING (MRI)

Procedure: The body is exposed to a high-energy magnetic field, which causes protons (small positive particles within atoms, such as hydrogen) in body fluids and tissues to arrange themselves in relation to the field. Then a pulse of radio waves "reads" these ion patterns, and a color-coded image is assembled on a video monitor. The result is a two- or three-dimensional blueprint of cellular chemistry.

Comments: Relatively safe, but can't be used on patients with metal in their bodies. Shows fine details for soft tissues but not for bones. MRI is most useful for differentiating between normal and abnormal tissues. Used to detect tumors and artery-clogging fatty plaques, reveal brain abnormalities, measure blood flow, and detect a variety of musculoskeletal, liver, and kidney disorders.

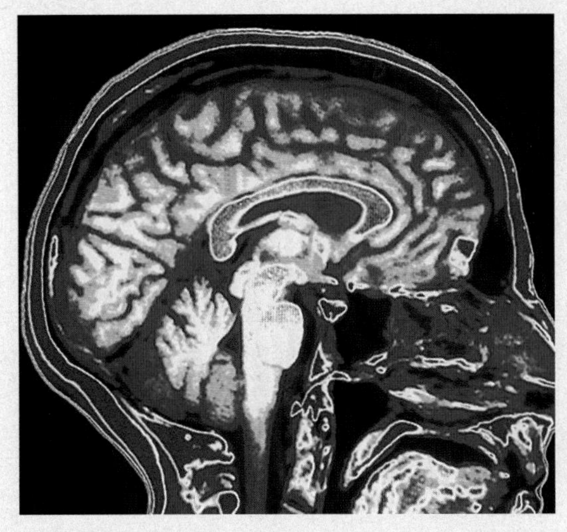

Magnetic resonance image of the brain in sagittal section

COMPUTED TOMOGRAPHY (CT) [FORMERLY CALLED COMPUTERIZED AXIAL TOMOGRAPHY (CAT) SCANNING]

Procedure: Computer-assisted radiography in which an x-ray beam traces an arc at multiple angles around a section of the body. The resulting transverse section of the body, called a *CT scan,* is reproduced on a video monitor.

Comments: Visualizes soft tissues and organs with much more detail than conventional radiographs. Differing tissue densities show up as various shades of gray. Multiple scans can be assembled to build three-dimensional views of structures. Whole-body CT scanning is also used. Typically, such scans actually target the torso. Whole-body CT scanning appears to provide the most benefit in screening for lung cancers, coronary artery disease, and kidney cancers. (Cross-sectional views of anatomical diagrams or CT scans, such as this one, assume that the observer is standing at the feet of the supine patient. The anterior portion of the body is therefore at the top of the illustration; anatomical structures in the left side of the body appear on the right side of the illustration.)

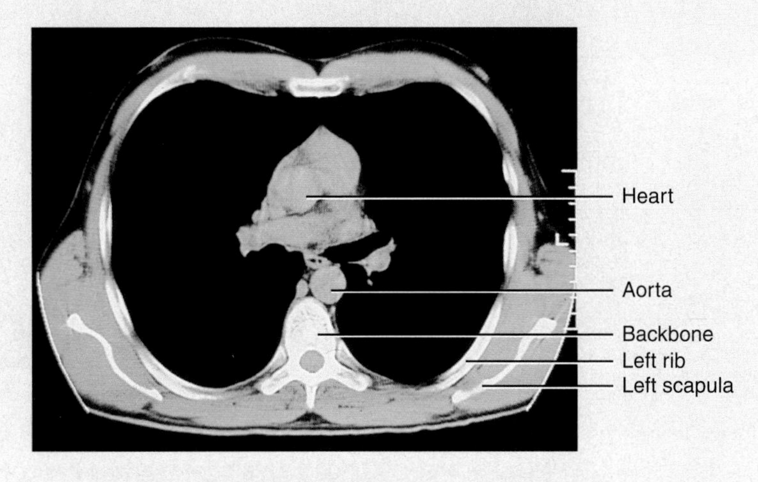

Computed tomography scan of the thorax in inferior view

ULTRASOUND SCANNING

Procedure: High-frequency sound waves produced by a handheld wand reflect off body tissues and are detected by the same instrument. The image, which may be still or moving, is called a *sonogram* (SON-ō-gram) and is reproduced on a video monitor.

Comments: Noninvasive, painless, uses no dyes, and usually considered to be safe. Most commonly used to visualize the fetus during pregnancy. Also used to observe the size, location, and actions of organs and blood flow through blood vessels (*doppler ultrasound*).

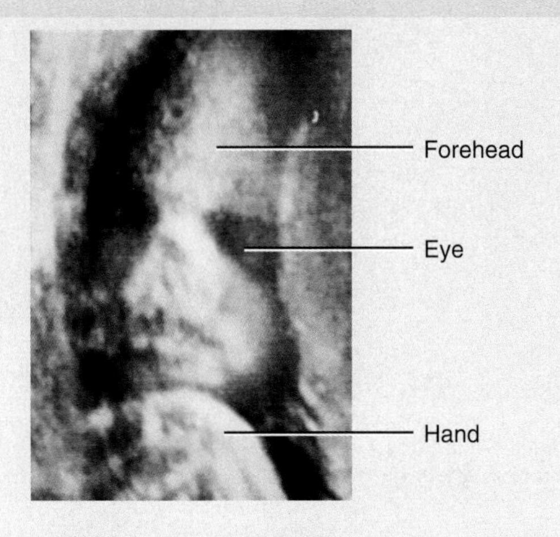

Sonogram of a fetus (Courtesy of Andrew Joseph Tortora and Damaris Soler)

CONTINUES

POSITRON EMISSION TOMOGRAPHY (PET)

Procedure: A substance that emits positrons (positively charged particles) is injected into the body, where tissues take it up. The collision of positrons with negatively charged electrons in body tissues produces gamma rays (similar to x-rays) that are detected by gamma cameras positioned around the subject. A computer receives signals from the gamma cameras and constructs a *PET scan image,* displayed in color on a video monitor. The PET scan shows where the injected substance is being used in the body. In the PET scan image shown here, the black and blue colors indicate minimal activity, whereas the red, orange, yellow, and white colors indicate areas of increasingly greater activity.

Comments: Used to study the physiology of body structures, such as metabolism in the brain or heart.

ANTERIOR

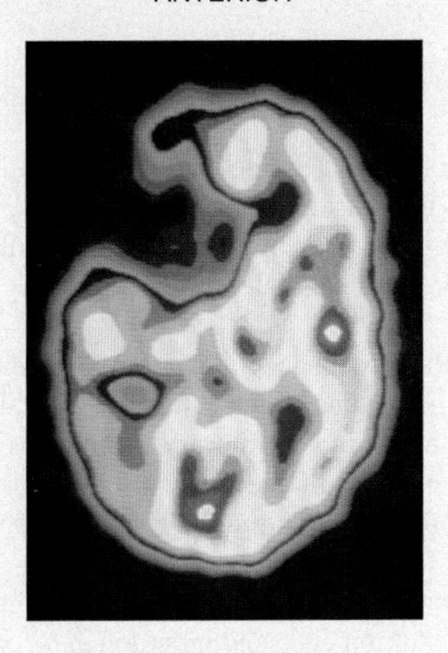

POSTERIOR

Positron emission tomography scan of a transverse section of the brain (darkened area at upper left indicates where a stroke has occurred)

RADIONUCLIDE SCANNING

Procedure: A *radionuclide* (radioactive substance) is introduced intravenously into the body and carried by the blood to the tissue to be imaged. Gamma rays emitted by the radionuclide are detected by a gamma camera outside the subject and fed into a computer. The computer constructs a *radionuclide image* and displays it in color on a video monitor. Areas of intense color take up a lot of the radionuclide and represent high tissue activity; areas of less intense color take up smaller amounts of the radionuclide and represent low tissue activity. *Single-photon-emission computerized tomography (SPECT) scanning* is a specialized type of radionuclide scanning that is especially useful for studying the brain, heart, lungs, and liver.

Comments: Used to study activity of a tissue or organ, such as the heart, thyroid gland, and kidneys.

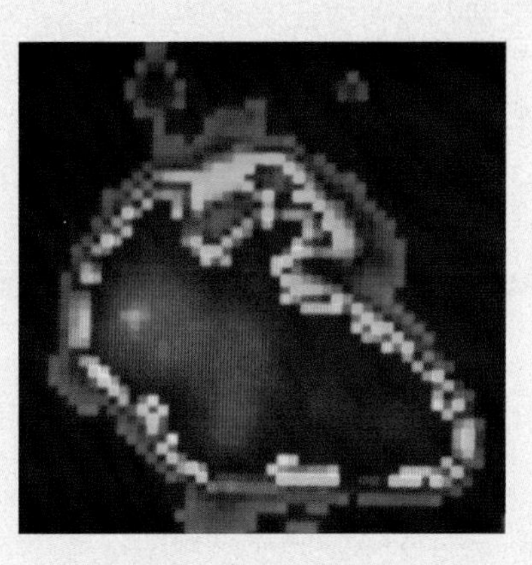

Radionuclide (nuclear) scan of a normal human heart

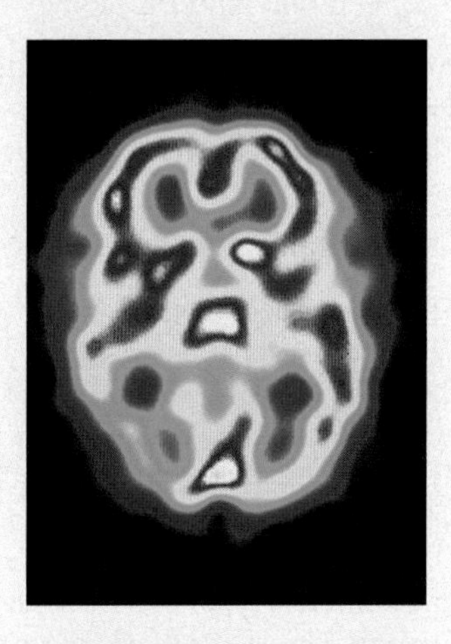

Single-photon-emission computerized tomography (SPECT) scan of a transverse section of the brain (green area at lower left indicates a migraine attack)

ENDOSCOPY

Procedure: The visual examination of the inside of body organs or cavities using a lighted instrument with lenses called an *endoscope.* The image is viewed through an eyepiece on the endoscope or projected onto a monitor.

Comments: Endoscopes are used in a wide variety of procedures; a few examples follow. *Colonoscopy* is used to examine the interior of the colon, which is part of the large intestine. *Laparoscopy* is used to examine the organs within the abdominopelvic cavity. *Arthroscopy* is used to examine the interior of a joint, usually the knee. *Esophagoscopy* and *gastroscopy* are used to examine the interior of the esophagus and stomach, respectively.

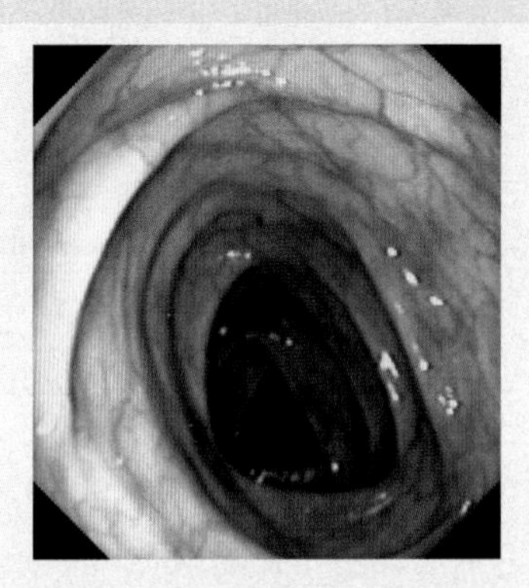

Interior view of the colon as shown by colonoscopy

KEY MEDICAL TERMS ASSOCIATED WITH ORGANIZATION OF THE HUMAN BODY

Most chapters in this text are followed by a glossary of key medical terms that include both normal and pathological conditions. You should familiarize yourself with these terms because they will play an essential role in your medical vocabulary.

Some of these conditions, as well as ones discussed in the text, are referred to as local or systemic. A *local disease* is one that affects one part or a limited area of the body. A *systemic disease* (sis-TEM-ik) affects the entire body or several parts.

Epidemiology (ep′-i-dē-mē-OL-ō-jē; *epi-* = upon; *-demi-* = people) The science that deals with why, when, and where diseases occur and how they are transmitted within a defined human population.

Geriatrics (jer′-ē-AT-riks; *ger-* = old; *-iatrics* = medicine) The science that deals with the medical problems and care of elderly persons.

Pathology (pa-THOL-ō-je; *patho-* = disease) The science that deals with the nature, causes, and development of abnormal conditions and the structural and functional changes that diseases produce.

Pharmacology (far′-ma-KOL-ō-jē; *pharmaco-* = drug) The science that deals with the effects and use of drugs in the treatment of disease.

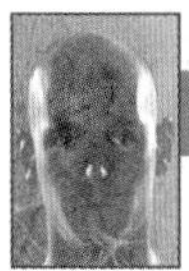

STUDY OUTLINE

Anatomy and Physiology Defined (Section 1.1)

1. Anatomy is the science of structure and the relationships among structures.
2. Physiology is the science of how body structures function.

Levels of Organization and Body Systems (Section 1.2)

1. The human body consists of six levels of organization: chemical, cellular, tissue, organ, system, and organismal.
2. Cells are the basic structural and functional units of an organism and the smallest living units in the human body.
3. Tissues consist of groups of cells and the materials surrounding them that work together to perform a particular function.
4. Organs usually have recognizable shapes, are composed of two or more different types of tissues, and have specific functions.
5. Systems consist of related organs that have a common function.
6. Table 1.1 introduces the eleven systems of the human body: integumentary, skeletal, muscular, nervous, endocrine, cardiovascular, lymphatic, respiratory, digestive, urinary, and reproductive.
7. The human organism is a collection of structurally and functionally integrated systems.
8. Body systems work together to maintain health, protect against disease, and allow for reproduction of the species.

Life Processes (Section 1.3)

1. All living organisms have certain characteristics that set them apart from nonliving things.
2. Among the life processes in humans are metabolism, responsiveness, movement, growth, differentiation, and reproduction.

Homeostasis: Maintaining Limits (Section 1.4)

1. Homeostasis is a condition in which the internal environment of the body remains stable, within certain limits.
2. A large part of the body's internal environment is interstitial fluid, which surrounds all body cells.
3. Homeostasis is regulated by the nervous and endocrine systems acting together or separately. The nervous system detects body changes and sends nerve impulses to maintain homeostasis. The endocrine system regulates homeostasis by secreting hormones.
4. Disruptions of homeostasis come from external and internal stimuli and from psychological stresses. When disruption of homeostasis is mild and temporary, responses of body cells quickly restore balance in the internal environment. If disruption is extreme, the body's attempts to restore homeostasis may fail.
5. A feedback system consists of (1) receptors that monitor changes in a controlled condition and send input to (2) a control center that sets the value at which a controlled condition should be maintained, evaluates the input it receives, and generates output commands when they are needed, and (3) effectors that receive output from the control center and produce a response (effect) that alters the controlled condition.
6. If a response reverses a change in a controlled condition, the system is called a negative feedback system. If a response enhances a change in a controlled condition, the system is referred to as a positive feedback system.
7. One example of negative feedback is the system that regulates blood pressure. If a stimulus causes blood pressure (controlled condition) to rise, baroreceptors (pressure-sensitive nerve cells, the receptors) in blood vessels send impulses (input) to the brain (control center). The brain sends impulses (output) to the heart (effector). As a result, heart rate decreases (response), and blood pressure drops back to normal (restoration of homeostasis).
8. Disruptions of homeostasis—homeostatic imbalances—can lead to disorders, disease, and even death.
9. A disorder is any abnormality of structure and/or function. Disease is a more specific term for an illness with a definite set of signs and symptoms.
10. Symptoms are subjective changes in body functions that are not apparent to an observer, whereas signs are objective changes that can be observed and measured.

Aging and Homeostasis (Section 1.5)

1. Aging produces observable changes in structure and function and increases vulnerability to stress and disease.
2. Changes associated with aging occur in all body systems.

Anatomical Terms (Section 1.6)

1. Descriptions of any region of the body assume the body is in the anatomical position, in which the subject stands erect facing the observer, with the head level and the eyes facing forward, the feet flat on the floor and directed forward, and the arms at the sides, with the palms turned forward.
2. The human body is divided into several major regions: the head, neck, trunk, upper limbs, and lower limbs.
3. Within body regions, specific body parts have common names and corresponding anatomical descriptive forms (adjectives). Examples are chest (thoracic), nose (nasal), and wrist (carpal).
4. Directional terms indicate the relationship of one part of the body to another. Exhibit 1.1 summarizes commonly used directional terms.
5. Planes are imaginary flat surfaces that divide the body or organs into two parts. A midsagittal plane divides the body or an organ into equal right and left sides. A parasagittal plane divides the body or an organ into unequal right and left sides. A frontal plane divides the body or an organ into anterior and posterior portions. A transverse plane divides the body or an organ into superior and inferior portions. An oblique plane passes through the body or an organ at an angle between a transverse plane and a sagittal plane, or between a transverse plane and a frontal plane.
6. Sections result from cuts through body structures. They are named according to the plane on which the cut is made: transverse, frontal, or sagittal.

Body Cavities (Section 1.7)

1. Spaces in the body that contain, protect, separate, and support internal organs are called body cavities.
2. The cranial cavity contains the brain and the vertebral canal contains the spinal cord.
3. The thoracic cavity is subdivided into three smaller cavities: a pericardial cavity, which contains the heart, and two pleural cavities, which each contain a lung.
4. The central portion of the thoracic cavity is the mediastinum. It is located between the lungs and extends from the sternum to the vertebral column and from the first rib to the diaphragm. It contains all thoracic organs except the lungs.
5. The abdominopelvic cavity is separated from the thoracic cavity by the diaphgram and is divided into a superior abdominal cavity and an inferior pelvic cavity.
6. Organs in the thoracic and abdominopelvic cavities are called viscera.
7. Viscera of the abdominal cavity include the stomach, spleen, liver, gallbladder, small intestine, and most of the large intestine.
8. Viscera of the pelvic cavity include the urinary bladder, portions of the large intestine, and internal organs of the reproductive system.
9. To describe the location of organs more easily, the abdominopelvic cavity may be divided into nine abdominopelvic regions by two horizontal and two vertical lines.
10. The names of the nine abdominopelvic regions are right hypochondriac, epigastric, left hypochondriac, right lumbar, umbilical, left lumbar, right inguinal, pubic, and left inguinal.
11. The abdominopelvic cavity may also be divided into quadrants by passing one horizontal and one vertical line through the umbilicus (navel).
12. The names of the abdominopelvic quadrants are right upper quadrant (RUQ), left upper quadrant (LUQ), right lower quadrant (RLQ), and left lower quadrant (LLQ).

Medical Imaging (Section 1.8)

1. Medical imaging refers to techniques and procedures used to create images of the human body. They allow visualization of internal structures to diagnose abnormal anatomy and deviations from normal physiology.
2. Table 1.2 summarizes and illustrates several medical imaging techniques.

SELF-QUIZ QUESTIONS

1. Which of the following best illustrates the concept of increasing levels of organizational complexity?
   a. chemical → tissue → cellular → organ → organismal → system
   b. chemical → cellular → tissue → organ → system → organismal
   c. cellular → chemical → tissue → organismal → organ → system
   d. chemical → cellular → tissue → system → organ → organismal
   e. tissue → cellular → chemical → organ → system → organismal

2. To properly reconnect the disconnected bones of a human skeleton, you would need to have a good understanding of
   a. physiology b. homeostasis
   c. chemistry d. anatomy
   e. feedback systems.

3. Fill in the missing blanks in the following table.

System	Major Organs	Functions
a	b	Regulates body activities by nerve impulses
c	Lymph vessels, spleen, thymus, tonsils, lymph nodes	d
e	f	Supplies oxygen to cells, eliminates carbon dioxide, regulates acid–base balance
Reproductive system	g	h

4. Homeostasis is
   a. the sum of all of the chemical processes in the body
   b. the sign of a disorder or disease
   c. the combination of growth, repair, and energy release that is basic to life
   d. the tendency to maintain constant, favorable internal body conditions
   e. caused by stress.
5. Which of the following is NOT true concerning the life processes?
   a. The pupils of your eyes becoming smaller when exposed to strong light is an example of differentiation.
   b. The ability to walk to your car following class is a result of the life process called movement.
   c. The repair of injured skin would involve the life process of reproduction.
   d. Digesting and absorbing food is an example of metabolism.
   e. Sweating on a hot summer day involves responsiveness.
6. In a negative feedback system,
   a. the controlled condition is never disrupted
   b. there tends to be a "runaway" body response
   c. the change in the controlled condition is reversed
   d. the body part that responds to the output is known as the receptor
   e. the response results in a reinforcement of the original stimulus.
7. The part of a feedback system that receives the input and generates the output command is the
   a. effector **b.** receptor
   c. feedback loop **d.** response
   e. control center.
8. An itch in your axillary region would cause you to scratch
   a. your armpit **b.** the front of your elbow
   c. your neck **d.** the top of your head
   e. your calf.
9. If you were facing a person who is in the correct anatomical position, you could observe the ________ region.
   a. crural **b.** plantar
   c. gluteal **d.** popliteal
   e. scapular
10. Midway through a 5-mile workout, a runner begins to sweat profusely. The sweat glands producing the sweat would be considered which part of a feedback loop?
    a. controlled condition **b.** receptors **c.** stimulus
    d. effectors **e.** control center
11. The right ear is ________ to the right nostril.
    a. intermediate **b.** inferior **c.** lateral
    d. distal **e.** medial
12. Your chin is ________ in relation to your lips.
    a. lateral **b.** superior **c.** deep
    d. posterior **e.** inferior
13. Your skull is ________ in relation to your brain.
    a. intermediate **b.** superior **c.** deep
    d. superficial **e.** proximal
14. A magician is about to separate his assistant's body into superior and inferior portions. The plane through which he will pass his magic wand is the
    a. midsagittal **b.** frontal
    c. transverse **d.** parasagittal
    e. oblique.
15. Which statement is NOT true of body cavities?
    a. The diaphragm separates the thoracic and abdominopelvic cavities.
    b. The organs in the cranial cavity and vertebral canal are called viscera.
    c. The urinary bladder is in the pelvic cavity.
    d. The abdominal cavity is inferior to the thoracic cavity.
    e. The abdominal cavity contains the liver.
16. To find the urinary bladder, you would look in the ________ region.
    a. hypochondriac **b.** umbilical
    c. epigastric **d.** inguinal
    e. pubic
17. Match the following common names and anatomical descriptive adjectives:

____	**a.** axillary	**A.** skull
____	**b.** inguinal	**B.** eye
____	**c.** cervical	**C.** cheek
____	**d.** cranial	**D.** armpit
____	**e.** oral	**E.** arm
____	**f.** brachial	**F.** groin
____	**g.** orbital	**G.** buttock
____	**h.** gluteal	**H.** neck
____	**i.** buccal	**I.** mouth
____	**j.** coxal	**J.** hip

18. Match the following:

____	**a.** contains the urinary bladder and reproductive organs	**A.** cranial cavity
____	**b.** contains the brain	**B.** abdominal cavity
____	**c.** contains the heart	**C.** vertebral canal
____	**d.** region between the lungs, from the breastbone to the backbone	**D.** pelvic cavity
____	**e.** separates the thoracic and abdominal cavities	**E.** pleural cavity
____	**f.** contains a lung	**F.** mediastinum
____	**g.** contains the spinal cord	**G.** diaphragm
____	**h.** contains the stomach and liver	**H.** pericardial cavity

19. Match the following:

____	**a.** transports oxygen, nutrients, and carbon dioxide	**A.** urinary system
____	**b.** breaks down and absorbs food	**B.** digestive system
____	**c.** functions in body movement, posture, and heat production	**C.** endocrine system
____	**d.** regulates body activities through hormones	**D.** integumentary system
____	**e.** supports and protects the body	**E.** muscular system
____	**f.** eliminates wastes and regulates the chemical composition and volume of blood	**F.** skeletal system
____	**g.** protects the body, detects sensations, and helps regulate body temperature	**G.** cardiovascular system

20. Match the following:

____	**a.** observable, measurable change	**A.** systemic
____	**b.** abnormality of function	**B.** symptom
____	**c.** affects the entire body	**C.** sign
____	**d.** subjective changes that aren't easily observed	**D.** disorder

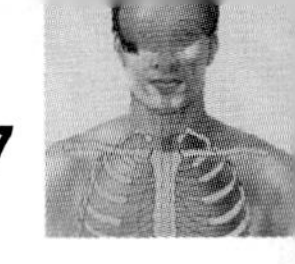

CRITICAL THINKING QUESTIONS

1. You are studying for your first anatomy and physiology exam and want to know which areas of your brain are working hardest as you study. Your classmate suggests that you could have a computed tomography (CT) scan done to assess your brain activity. Would this be the best way to determine brain activity levels?
2. There is much interest in using stem cells to help in the treatment of diseases such as type I diabetes, which is due to a malfunction of some of the normal cells in the pancreas. What would make stem cells useful in disease treatment?
3. On her first anatomy and physiology exam, Heather defined homeostasis as "the condition in which the body approaches room temperature and stays there." Do you agree with Heather's definition?

? ANSWERS TO FIGURE QUESTIONS

1.1 Organs have a recognizable shape and consist of two or more different types of tissues that have a specific function.

1.2 The basic difference between negative and positive feedback systems is that in negative feedback systems, the response reverses a change in a controlled condition, and in positive feedback systems, the response enhances the change in a controlled condition.

1.3 If a stimulus caused blood pressure to decrease, the heart rate would increase due to the operation of this negative feedback system.

1.4 Because positive feedback systems continually intensify or reinforce the original stimulus, some mechanism is needed to end the response before life-threatening changes occur in the body.

1.5 A plantar wart is found on the sole.

1.6 No, the radius is distal to the humerus. No, the esophagus is posterior to the trachea. Yes, the ribs are superficial to the lungs. Yes, the urinary bladder is medial to the ascending colon. No, the sternum is medial to the descending colon.

1.7 The frontal plane divides the heart into anterior and posterior portions.

1.8 The midsagittal plane divides the brain into equal right and left sides.

1.9 Urinary bladder = P, stomach = A, heart = T, small intestine = A, lungs = T, internal female reproductive organs = P, thymus = T, spleen = A, liver = A.

1.10 Some structures in the mediastinum are the heart, esophagus, and aorta.

1.11 The liver is mostly in the epigastric region; the ascending colon is in the right lumbar region; the urinary bladder is in the pubic region; most of the small intestine is in the umblical region.

2 Introductory Chemistry

In this chapter you will learn to appreciate that chemistry is the basis for all components of the body. Everything you eat, drink, and smell is a chemical—water, carbohydrates, salt, proteins, and fats all have some important function in the body. In later chapters you will continue to observe chemistry in action when discussing the workings of many of the body systems such as the muscular system. Your body maintains an intricate balance of chemicals to create stable chemical reactions and functions. Chemistry is the starting point for both structure and function in the body. You'll begin with the basics of atoms, molecules, and compounds and then move on to see how these building blocks work together to create healthy cells, tissues, organs, organ systems, and your total being. You will apply these chemical constructs to the body where they result in the formation of stable tissues. These tissues compartmentalize chemical reactions such as digestion where compounds may be broken down to their simplest form. The opposite actions also occur where the body takes these simple building blocks and builds them into larger, beautifully complex structures such as DNA or various proteins such as actin and myosin. Through these intricate chemicals and chemical reactions, you will see how your body builds tissues from what you ate for breakfast and how it utilizes and recycles certain components that you no longer need.

CONTENTS AT A GLANCE

2.1 INTRODUCTION TO CHEMISTRY

OBJECTIVES

- Define a chemical element, atom, ion, molecule, and compound.
- Explain how chemical bonds form.
- Describe what happens in a chemical reaction and explain why it is important to the human body.

Chemistry (KEM-is-trē) is the science of the structure and interactions of ***matter,*** anything that occupies space and has mass. ***Mass*** is the amount of matter in any living organism or nonliving thing.

Chemical Elements and Atoms

All forms of matter are made up of a limited number of building blocks called ***chemical elements,*** substances that cannot be broken down into a simpler form by ordinary chemical means. At present, scientists recognize 112 different elements. Each element is designated by a ***chemical symbol,*** one or two letters of the element's name in English, Latin, or another language. Examples are H for hydrogen, C for carbon, O for oxygen, N for nitrogen, K for potassium (kalium), Na for sodium (natrium), Fe for iron (ferrum), and Ca for calcium.

Twenty-six different elements normally are present in your body. Just four elements, called the *major elements,* constitute about 96% of the body's mass: oxygen, carbon, hydrogen, and nitrogen. Eight others, the *lesser elements,* contribute 3.8% of the body's mass: calcium (Ca), phosphorus (P), potassium (K), sulfur (S), sodium (Na), chlorine (Cl), magnesium (Mg), and iron (Fe). An additional 14 elements—the *trace elements*—are present in tiny amounts. Together, they account for the remaining 0.2% of the body's mass. Although trace elements are few in number, several have important functions in the body. For example, iodine is needed to make thyroid hormones. The functions of some trace elements are unknown. Table 2.1 lists the main chemical elements of the human body.

TABLE 2.1

Main Chemical Elements in the Body

CHEMICAL ELEMENT (SYMBOL)	% OF TOTAL BODY MASS	SIGNIFICANCE
MAJOR ELEMENTS	**96%**	
Oxygen (O)	65.0	Part of water and many organic (carbon-containing) molecules; used to generate ATP, a molecule used by cells to temporarily store chemical energy.
Carbon (C)	18.5	Forms backbone chains and rings of all organic molecules: carbohydrates, lipids (fats), proteins, nucleic acids (DNA and RNA), and adenosine triphosphate.
Hydrogen (H)	9.5	Constituent of water and most organic molecules; ionized form (H^+) makes body fluids more acidic.
Nitrogen (N)	3.2	Component of all proteins and nucleic acids.
LESSER ELEMENTS	**3.8%**	
Calcium (Ca)		Contributes to hardness of bones and teeth; ionized form (Ca^{2+}) needed for blood clotting, release of hormones, contraction of muscle, and many other processes.
Phosphorus (P)		Component of nucleic acids and ATP; required for normal bone and tooth structure.
Potassium (K)		Ionized form (K^+) is the most plentiful cation (positively charged particle) in intracellular fluid; needed to generate action potentials (impulses).
Sulfur (S)		Component of some vitamins and many proteins.
Sodium (Na)		Ionized form (Na^+) is the most plentiful cation (positively charged particle) in extracellular fluid; essential for maintaining water balance; needed to generate action potentials.
Chlorine (Cl)		Ionized form (Cl^-) is the most plentiful anion (negatively charged particle) in extracellular fluid; essential for maintaining water balance.
Magnesium (Mg)		Ionized form (Mg^{2+}) needed for action of many enzymes, molecules that increase the rate of chemical reactions in organisms.
Iron (Fe)		Ionized forms (Fe^{2+} and Fe^{3+}) are part of hemoglobin (oxygen-carrying protein in red blood cells) and some enzymes (proteins that catalyze chemical reactions in living cells.
TRACE ELEMENTS	**0.2%**	
		Aluminum (Al), boron (B), chromium (Cr), cobalt (Co), copper (Cu), fluorine (F), iodine (I), manganese (Mn), molybdenum (Mo), selenium (Se), silicon (Si), tin (Sn), vanadium (V), and zinc (Zn).

Each element is made up of ***atoms,*** the smallest units of matter that retain the properties and characteristics of the element. A sample of the element carbon, such as pure coal, contains only carbon atoms, and a tank of helium gas contains only helium atoms.

An atom consists of two basic parts: a nucleus and one or more electrons (Figure 2.1). The centrally located ***nucleus*** contains positively charged ***protons*** (p^+) and uncharged (neutral) ***neutrons*** (n^0). Because each proton has one positive charge, the nucleus is positively charged. The ***electrons*** (e^-) are tiny, negatively charged particles that move about in a large space surrounding the nucleus. They do not follow a fixed path or orbit but instead form a negatively charged "cloud" that surrounds the nucleus (Figure 2.1a). The number of electrons in an atom equals the number of protons. Because each electron carries one negative charge, the negatively charged electrons and the positively charged protons balance each other. As a result, each atom is electrically neutral, meaning its total charge is zero.

The number of protons in the nucleus of an atom is called the atom's ***atomic number.*** The atoms of each different kind of element have a different number of protons in the nucleus: A hydrogen atom has 1 proton, a carbon atom has 6 protons, a sodium atom has 11 protons, a chlorine atom has 17 protons, and so on (Figure 2.2). Thus, each type of atom or element has a different atomic number.

Figure 2.1 Two representations of the structure of an atom.
Electrons move about the nucleus, which contains neutrons and protons. (a) In the electron cloud model of an atom, the shading represents the chance of finding an electron in regions outside the nucleus. (b) In the electron shell model, filled circles represent individual electrons, which are grouped into concentric circles according to the shells they occupy. Both models depict a carbon atom, with six protons, six neutrons, and six electrons.

6 An atom is the smallest unit of matter that retains the properties and characteristics of its element.

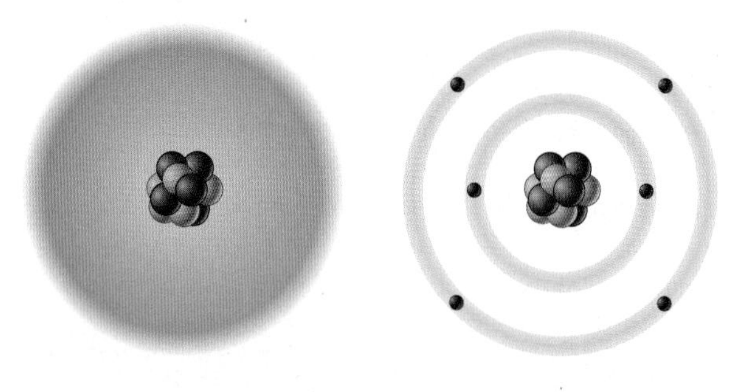

? What is the atomic number of carbon?

Even though their exact positions cannot be predicted, specific groups of electrons are most likely to move about within certain regions around the nucleus. These regions are called ***electron shells,*** which are depicted as two-dimensional circles in Figures 2.1b and 2.2 even though some of their three-dimensional shapes are not spherical. The electron shell nearest the nucleus—the first electron shell—can hold a maximum of 2 electrons. The second electron shell can hold a maximum of 8 electrons, whereas the third typically holds 8 but can hold up to 18 electrons. Higher electron shells (there are as many as seven) can contain many more electrons. The electron shells are filled with electrons in a specific order, beginning with the first shell. Electrons in the outermost shell are the only electrons involved in chemical reactions.

Ions, Molecules, and Compounds

The atoms of each element have a characteristic way of losing, gaining, or sharing their electrons when interacting with other atoms. If an atom either *gives up* or *gains* electrons, it becomes an ***ion*** (Ī-on), an atom that has a positive or negative charge due to unequal numbers of protons and electrons. An ion of an atom is symbolized by writing its chemical symbol followed by the number of its positive (+) or negative (−) charges. For example, Ca^{2+} stands for a calcium ion that has a charge of positive two because it has given up two electrons. Refer to Table 2.1 for the important functions of several ions in the body.

In contrast, when two or more atoms *share* electrons, the resulting combination of atoms is called a ***molecule*** (MOL-e-kyool). A *molecular formula* indicates the number and type of atoms that make up a molecule. A molecule may consist of two or more atoms of the same element, such as an oxygen molecule or a hydrogen molecule, or two or more atoms of different elements, such as a water molecule (Figure 2.3). The molecular formula for a molecule of oxygen is O_2. The subscript 2 indicates there are two atoms of oxygen in the oxygen molecule. In the water molecule, H_2O, one atom of oxygen shares electrons with two atoms of hydrogen. Notice that two hydrogen *molecules* can combine with one oxygen *molecule* to form two water molecules (Figure 2.3).

A ***compound*** is a substance that can be broken down into two or more different elements by ordinary chemical means. Most of the atoms in your body are joined into compounds, for example, water (H_2O). A molecule of oxygen (O_2) is *not* a compound because it consists of atoms of only one element. Note that water has been used as an example of both molecules and compounds. Both definitions are correct since (1) atoms of hydrogen and oxygen share electrons and (2) hydrogen and oxygen are different elements.

A ***free radical*** is an ion or molecule that has an unpaired electron in its outermost shell. (Most of an atom's electrons associate in pairs.) A common example of a free radical is *superoxide*, which is formed by the addition of an electron to an oxygen molecule. Having an unpaired electron makes a free radical unstable and destructive to nearby molecules. Free radicals break apart important body molecules by either giving up their unpaired electron to or taking on an electron from another molecule.

Figure 2.2 **Atomic structures of several atoms that have important roles in the human body.**

 The atoms of different elements have different atomic numbers because they have different numbers of protons.

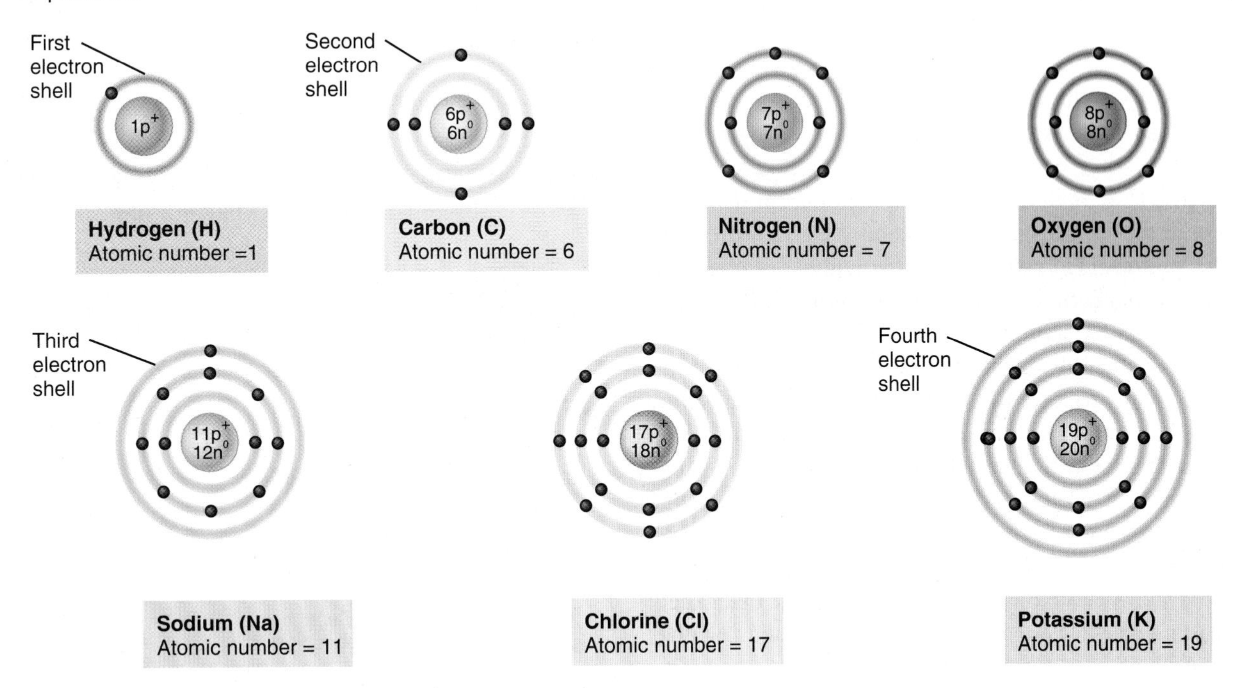

Atomic number = number of protons in an atom

? **Which four of these elements are most abundant in living organisms?**

• CLINICAL CONNECTION | Antioxidants

In our bodies, several processes can generate free radicals. They may result from exposure to ultraviolet radiation in sunlight or to x-rays. Some reactions that occur during normal metabolic processes produce free radicals. Moreover, certain harmful substances, such as carbon tetrachloride (a solvent used in dry cleaning), give rise to free radicals when they participate in metabolic reactions in the body. Among the many disorders and diseases linked to oxygen-derived free radicals are cancer, the buildup of fatty materials in blood vessels (atherosclerosis), Alzheimer disease, emphysema, diabetes mellitus, cataracts, macular degeneration, rheumatoid arthritis, and deterioration associated with aging. Consuming more **antioxidants**—substances that inactivate oxygen-derived free radicals—is thought to slow the pace of damage caused by free radicals. Important dietary antioxidants include selenium, zinc, beta-carotene, and vitamins C and E. Red, blue, or purple fruits and vegetables contain high levels of antioxidants. •

Figure 2.3 **Molecules.**

 A molecule may consist of two or more atoms of the same element or two or more atoms of different elements.

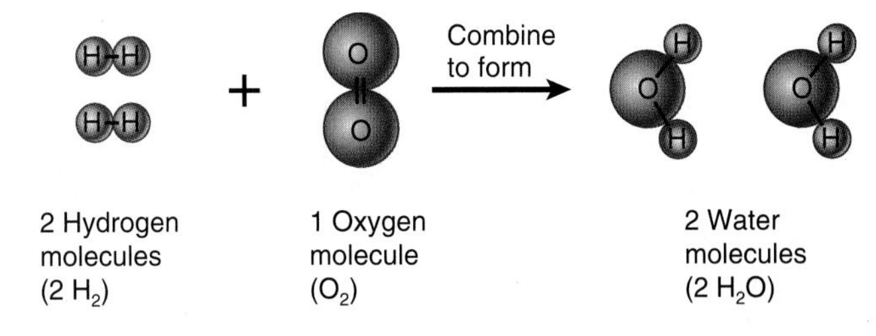

? **Which of the molecules shown here is a compound?**

Chemical Bonds

The forces that bind the atoms of molecules and compounds together, resisting their separation, are ***chemical bonds.*** The chance that an atom will form a chemical bond with another atom depends on the number of electrons in its outermost shell, also called the ***valence shell.*** An atom with an outer shell holding 8 electrons is *chemically stable*, which means it is unlikely to form chemical bonds with other atoms.

The atoms of most biologically important elements do not have 8 electrons in the valence shells. Given the correct conditions, two or more such atoms can interact or bond in ways that produce a chemically stable arrangement of electrons in the outer shell of each atom (*octet rule*). Three general types of chemical bonds are ionic bonds, covalent bonds, and hydrogen bonds.

Ionic Bonds

Positively charged ions and negatively charged ions are attracted to one another. This force of attraction between ions of opposite charges is called an ***ionic bond.*** Consider sodium and chlorine atoms to see how an ionic bond forms (Figure 2.4). Sodium has 1 outer shell electron (Figure 2.4a). If a sodium atom *loses* this electron, it is left with the 8 electrons in its second shell. However, the total number of protons (11) now exceeds the number of electrons (10). As a result, the sodium atom becomes a ***cation*** (KAT-ī-on), a positively charged ion. A sodium ion has a charge of 1+ and is written Na^+. On the other hand, chlorine has 7 outer shell electrons (Figure 2.4b), too many to lose. But if chlorine *accepts* an electron from a neighboring atom, it will have 8 electrons in its third electron shell. When this happens, the total number of electrons (18) exceeds the number of protons (17), and the chlorine atom becomes an ***anion*** (AN-ī-on), a negatively charged ion. The ionic form of chlorine is called a chloride ion. It has a charge of 1− and is written Cl^-. When an atom of sodium donates its sole outer shell electron to an atom of chlorine, the resulting positive and negative charges attract each other to form an ionic bond (Figure 2.4c). The resulting ionic compound is sodium chloride, written NaCl.

Ionic bonds are found in teeth and bones, where they give great strength to the tissue. Most other ions in the body are dissolved in body fluids. An ionic compound that breaks apart into cations and anions when dissolved in water is called an ***electrolyte*** (e-LEK-trō-līt) because the solution can conduct an electrical current. As you will see in later chapters, electrolytes have many important functions. For example, they are critical for controlling water movement within the body, maintaining acid–base balance, and producing nerve impulses.

Covalent Bonds

When a ***covalent bond*** forms, neither of the combining atoms loses or gains electrons. Instead, the atoms form a molecule by *sharing* one, two, or three pairs of their outer shell electrons. Covalent bonds are the most common chemical bonds in the body, and the compounds that result from them form most of the body's structures. Unlike ionic bonds, most covalent bonds do not break apart when the molecule is dissolved in water.

In some covalent bonds, atoms share the electrons equally—one atom does not attract the shared electrons more strongly than the other atom. This is called a *nonpolar covalent bond.* The bonds between two identical atoms always are nonpolar covalent bonds (Figure 2.5a–c). Another example of a nonpolar covalent bond is the single covalent bond that forms between carbon and each atom of hydrogen in a methane molecule (Figure 2.5d).

In a *polar covalent bond,* the sharing of electrons between atoms is unequal—one atom attracts the shared electrons more strongly than the other. A very important example of a polar covalent bond in living systems is the bond between oxygen and hydrogen in a molecule of water (Figure 2.5e). The concepts of nonpolar and polar bonds are critical to the understanding of how substances move through a cell membrane (Chapter 3). Polar covalent bonds in water play an important role in the solubility of substances (see Section 2.2).

Figure 2.4 Ions and ionic bond formation. (a) A sodium atom can attain the stability of 8 electrons in its outermost shell by losing its 1 valence electron; it then becomes a sodium ion, Na^+. (b) A chlorine atom can attain the stability of 8 electrons in its outermost shell by accepting 1 electron; it then becomes a chloride ion, Cl^-. (c) An ionic bond holds Na^+ and Cl^+ together in the ionic compound sodium chloride, NaCl. The electron that is donated or accepted is colored red.

An ionic bond is the force of attraction that holds together oppositely charged ions.

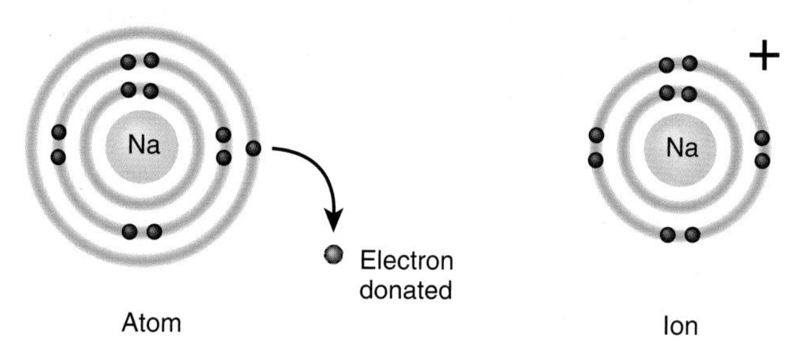

(a) Sodium: 1 valence electron

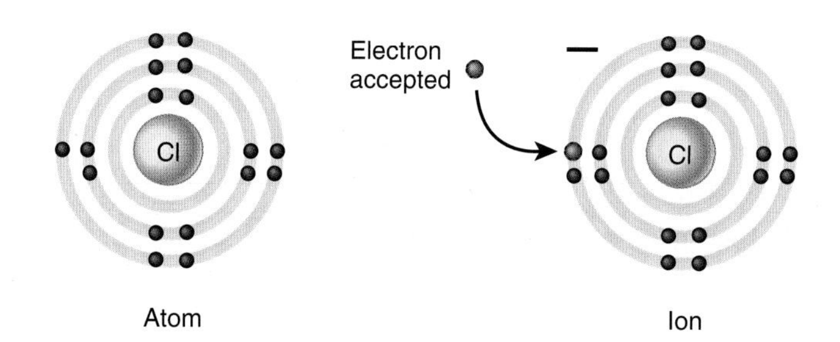

(b) Chlorine: 7 valence electrons

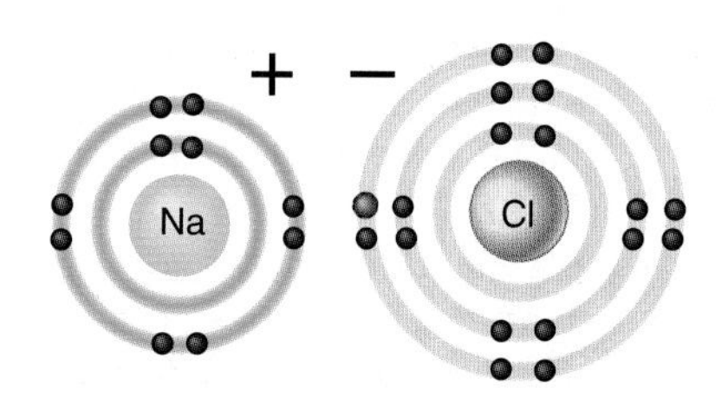

(c) Ionic bond in sodium chloride (NaCl)

Will the element potassium (K) be more likely to form an anion or a cation? Why? (Hint: Look back to Figure 2.2 for the atomic structure of K.)

Hydrogen Bonds

The polar covalent bonds that form between hydrogen atoms and other atoms can give rise to a third type of chemical bond, a hydrogen bond. A ***hydrogen bond*** forms when a hydrogen atom with a partial positive charge (δ^+) attracts the partial negative charge (δ^-)

Figure 2.5 Covalent bond formation. The red electrons are shared equally. To the right are simpler ways to represent these molecules. In a structural formula, each covalent bond is denoted by a straight line between the chemical symbols for two atoms. In a molecular formula, the number of atoms in each molecule is noted by subscripts.

In a covalent bond, two atoms share one, two, or three pairs of electrons in their outer (valence) shells.

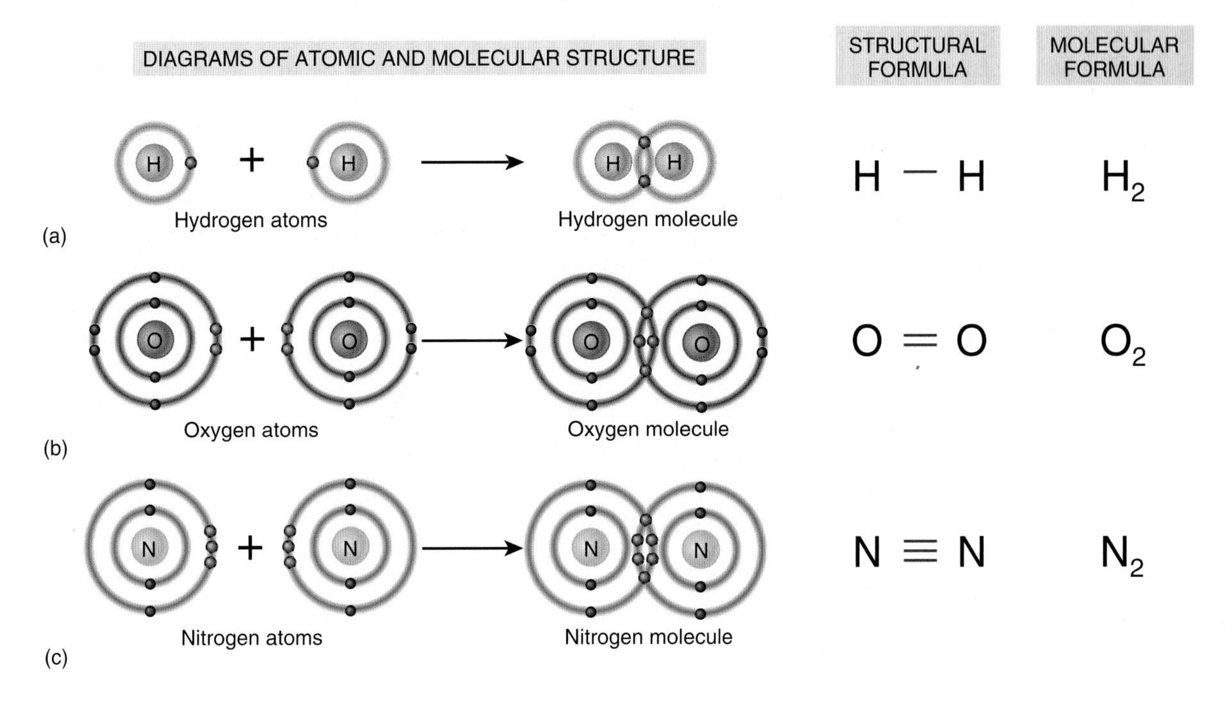

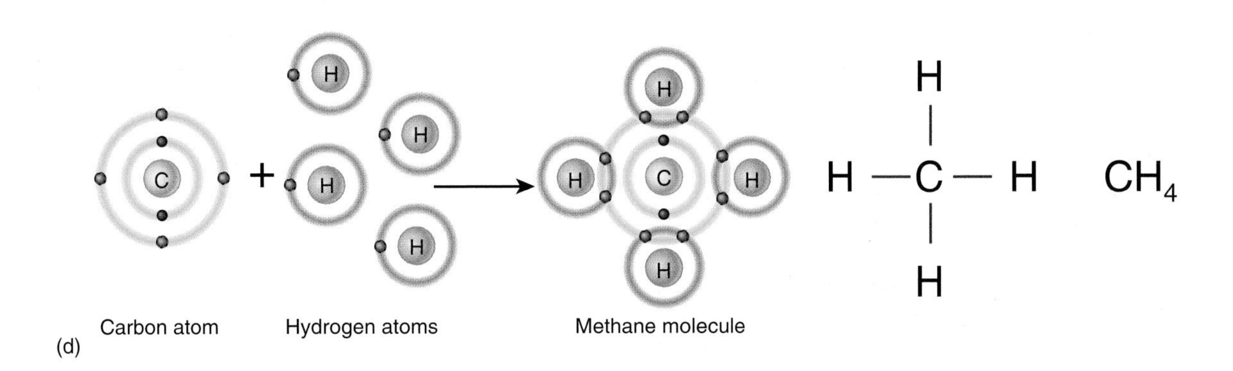

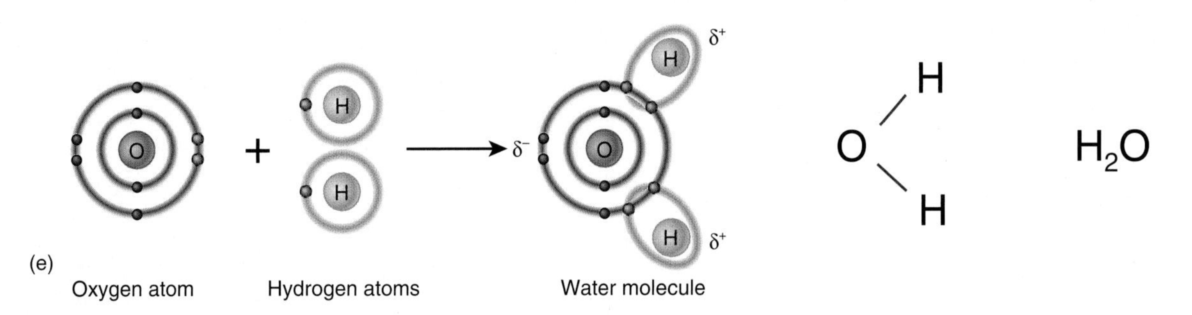

What is the main difference between an ionic bond and a covalent bond?

of neighboring electronegative atoms, most often oxygen or nitrogen. Individual hydrogen bonds are weak, but a large number of hydrogen bonds collectively provide considerable strength and stability for the three-dimensional shape of proteins and nucleic acids. As you will see in Chapter 4, Section 4.4 Connective Tissue, the concept of hydrogen bonding is essential in understanding the bonds between adjacent collagen molecules.

Chemical Reactions

A ***chemical reaction*** occurs when new bonds form and/or old bonds break between atoms. Through chemical reactions, body structures are built and body functions are carried out, processes that involve transfers of energy.

Forms of Energy and Chemical Reactions

Energy (*en-* = in; *-ergy* = work) is the capacity to do work. The two main forms of energy are ***potential energy,*** energy stored by matter due to its *position,* and ***kinetic energy,*** the energy of matter *in motion*. For example, the energy stored in a battery or in a person poised to jump from steps is potential energy. When the battery is used to run a clock or the person jumps, potential energy is converted into kinetic energy. ***Chemical energy*** is a form of potential energy that is stored in the bonds of molecules. In your body, chemical energy in the foods you eat is eventually converted into various forms of kinetic energy, such as mechanical energy, used to walk and talk, and heat energy, used to maintain body temperature. In chemical reactions, breaking old bonds requires an input of energy and forming new bonds releases energy. Because most chemical reactions involve both breaking old bonds and forming new bonds, the *overall reaction* may either release energy or require energy.

Synthesis Reactions

When two or more atoms, ions, or molecules combine to form new and larger molecules, the process is a ***synthesis reaction.*** The word *synthesis* means "to put together." Synthesis reactions can be expressed as follows:

$$\underset{\text{Atom, ion, or molecule A}}{\text{A}} + \underset{\text{Atom, ion, or molecule B}}{\text{B}} \xrightarrow{\text{Combine to form}} \underset{\text{New molecule AB}}{\text{AB}}$$

An example of a synthesis reaction is the synthesis of water from hydrogen and oxygen molecules (see Figure 2.3). The synthesis reactions that occur in your body are collectively referred to as ***anabolism*** (a-NAB-ō-lizm).

Decomposition Reactions

In a ***decomposition reaction,*** a molecule is split apart. The word *decompose* means to break down into smaller parts. Large molecules are split into smaller molecules, ions, or atoms. A decomposition reaction occurs in this way:

$$\underset{\text{Molecule AB}}{\text{AB}} \xrightarrow{\text{Breaks down into}} \underset{\text{Atom, ion, or molecule A}}{\text{A}} + \underset{\text{Atom, ion, or molecule B}}{\text{B}}$$

For example, under the proper conditions, a methane molecule can decompose into one carbon atom and two hydrogen molecules.

The decomposition reactions that occur in your body are collectively referred to as ***catabolism*** (ka-TAB-ō-lizm). The breakdown of large starch molecules into many small glucose molecules during digestion is an example of catabolism.

In general, energy-releasing reactions occur as nutrients, such as glucose, and are broken down via decomposition reactions. Some of the energy released is temporarily stored in a special molecule called ***adenosine triphosphate (ATP)*** (a-DEN-ō-sēn trī-FOS-fāt), which is discussed more fully later in this chapter. The energy transferred to the ATP molecules is then used to drive the energy-requiring synthesis reactions that lead to the building of body structures such as muscles and bones.

Exchange Reactions

Many reactions in the body are ***exchange reactions;*** they consist of both synthesis and decomposition reactions. One type of exchange reaction works like this:

$$\text{AB} + \text{CD} \longrightarrow \text{AD} + \text{BC}$$

The bonds between A and B and between C and D break (decomposition), and new bonds then form (synthesis) between A and D and between B and C. Notice that the atoms or ions in both compounds have "switched partners."

Reversible Reactions

Some chemical reactions proceed in only one direction, as previously indicated by the single arrows. Other chemical reactions may be reversible. ***Reversible reactions*** can go in either direction under different conditions and are indicated by $\rightleftharpoons$, two half-arrows pointing in opposite directions:

$$\text{AB} \underset{\text{Combines to form}}{\overset{\text{Breaks down into}}{\rightleftharpoons}} \text{A} + \text{B}$$

Some reactions are reversible only under special conditions:

$$\text{AB} \underset{\text{Heat}}{\overset{\text{Water}}{\rightleftharpoons}} \text{A} + \text{B}$$

Whatever is written above or below the arrows indicates the condition needed for the reaction to occur. In these reactions, AB breaks down into A and B only when water is added, and A and B react to produce AB only when heat is applied. The sum of all the chemical reactions in the body (including the categories just mentioned) is known as ***metabolism*** (me-TAB-ō-lizm; *metabol* = change). Metabolism and nutrition will be the topics of Chapter 27.

CHECKPOINT

1. Compare the meanings of atomic number, ion, and molecule.
2. What is the significance of the valence (outer) electron shell of an atom?
3. Distinguish among ionic, covalent, and hydrogen bonds.

4. Explain the difference between anabolism and catabolism. Which involves synthesis reactions?

2.2 CHEMICAL COMPOUNDS AND LIFE PROCESSES

OBJECTIVES

- Discuss the functions of water and inorganic acids, bases, and salts.
- Define pH and explain how the body attempts to keep pH within the limits of homeostasis.
- Discuss the functions of carbohydrates, lipids, and proteins.
- Explain the importance of deoxyribonucleic acid (DNA), ribonucleic acid (RNA), and adenosine triphosphate (ATP).

Chemicals in the body can be divided into two main classes of compounds: inorganic and organic. ***Inorganic compounds*** usually lack carbon, are structurally simple, and are held together by ionic or covalent bonds. They include water and many salts, acids, and bases. Two inorganic compounds are exceptions since they contain carbon: carbon dioxide (CO_2) and bicarbonate ion (HCO_3^-). ***Organic compounds,*** by contrast, always contain carbon, usually contain hydrogen, and always have covalent bonds. Examples include carbohydrates, lipids, proteins, nucleic acids, and adenosine triphosphate (ATP). Organic compounds are discussed in more detail in Chapter 27. Large organic molecules called *macromolecules* are formed by covalent bonding of many identical or similar building-block subunits termed *monomers.*

Inorganic Compounds

Water

Water is the most important and most abundant inorganic compound in all living systems, making up 55–60% of body mass in lean adults. With few exceptions, most of the volume of cells and body fluids is water. Several of its properties explain why water is such a vital compound for life.

1. Water is an excellent solvent. A ***solvent*** is a liquid or gas in which some other material, called a ***solute,*** has been dissolved. The combination of solvent plus solute is called a ***solution.*** Water is the solvent that carries nutrients, oxygen, and wastes throughout the body. The versatility of water as a solvent is due to its polar covalent bonds and its "bent" shape (see Figure 2.5e), which allow each water molecule to interact with four or more neighboring ions or molecules.

2. Water participates in chemical reactions. Because water can dissolve so many different substances, it is an ideal medium for chemical reactions. Water also is an active participant in some decomposition and synthesis reactions. During digestion, for example, decomposition reactions break down large nutrient molecules into smaller molecules by the addition of water to the large nutrient molecules. This type of reaction is called ***hydrolysis*** (hī-DROL-i-sis; *-lysis* = to loosen or break apart) (see Figure 2.8). Hydrolysis reactions enable dietary nutrients to be absorbed into the body.

3. Water absorbs and releases heat very slowly. In comparison to most other substances, water can absorb or release a relatively large amount of heat with only a slight change in its own temperature. The large amount of water in the body thus moderates the effect of changes in the environmental temperature, thereby helping maintain the homeostasis of body temperature.

4. Water requires a large amount of heat to change from a liquid to a gas. When the water in sweat evaporates from the skin surface, it takes with it large quantities of heat and provides an excellent cooling mechanism.

5. Water serves as a lubricant. Water is a major part of saliva, mucus, and other lubricating fluids. Lubrication is especially necessary in the thoracic and abdominal cavities, where internal organs touch and slide over one another. It is also needed at joints, where bones, ligaments, and tendons rub against one another.

MANUAL THERAPY APPLICATION

Hydrotherapy

Hydrotherapy refers to the external application of water for therapeutic purposes. Due to the physical properties of water just described, water is well suited to a variety of rehabilitation applications including control of pain and reduction of edema. Water can be utilized as either a superficial heating or cooling agent since it has *high specific heat* and *thermal conductivity.* Assuming that water and air are the same temperature, water holds 4 times more heat energy than the same mass of air and conducts heat 25 times faster than air. Furthermore, water that is not moving transfers heat by *conduction* while moving water transfers heat by *convection,* with the rate of heat transfer increasing as the rate of flow increases. These properties explain the popularity of bathtubs and hot tubs with jets, whirlpools for athletes, and heated swimming pools for therapeutic exercise. *Fomentation,* the application of a hot or warm and moist substance to the body, reduces pain.

Water is also used as a cooling agent and is used therapeutically when the trauma is very recent. Frozen water (ice), again due to the high specific heat and thermal conductivity of water, retains its coldness for a long time. Application of ice to a patient's trauma site causes constriction of blood vessels in the affected area and thus decreased blood flow, swelling, and pain.

Water also has the ability to supply hydrostatic pressure, buoyancy, and resistance. *Hydrostatic pressure* is the pressure applied by water on all surfaces of a person who is immersed at a particular depth. Additionally, a person submerged to the neck in water and floating in an upright position has greater pressure on the lower limbs and lower torso than on the upper part of the body. As you will learn later, this situation facilitates the flow of venous blood and lymph toward the heart. *Buoyancy,* another property of water, is used to decrease compression and stress on weight-bearing joints. *Resistance,* as a function of the viscosity of water, occurs in opposition to the motion of the body, with the resistance also increasing in proportion to the speed of the water or the speed of the patient's activity.

Inorganic Acids, Bases, and Salts

Many inorganic compounds can be classified as acids, bases, or salts. An ***acid*** is a substance that breaks apart or *dissociates* (dis-SŌ-sē-āts′) into one or more *hydrogen ions* (H^+) when it dissolves in water (Figure 2.6a). A ***base,*** by contrast, usually dissociates into one or more *hydroxide ions* (OH^-) when it dissolves in water (Figure 2.6b). A ***salt,*** when dissolved in water, dissociates into cations and anions, neither of which is H^+ or OH^- (Figure 2.6c).

Acids and bases react with one another to form salts. For example, the reaction of hydrochloric acid (HCl) and potassium hydroxide (KOH), a base, produces the salt potassium chloride (KCl), along with water (H_2O). This exchange reaction can be written as follows:

$$\underset{\text{Acid}}{HCl} + \underset{\text{Base}}{KOH} \longrightarrow \underset{\text{Salt}}{KCl} + \underset{\text{Water}}{H_2O}$$

Acid–Base Balance: The Concept of pH

To ensure homeostasis, body fluids must contain balanced quantities of acids and bases. The more hydrogen ions (H^+) dissolved in a solution, the more acidic is the solution; conversely, the more hydroxide ions (OH^-), the more basic (alkaline) is the solution. The chemical reactions that take place in the body are very sensitive to even small changes in the acidity or alkalinity of the body fluids in which they occur. Any departure from the narrow limits of normal H^+ and OH^- concentrations greatly disrupts body functions.

A solution's acidity or alkalinity is expressed on the ***pH scale,*** which extends from 0 to 14 (Figure 2.7). This scale is based on the number of hydrogen ions in a solution. The midpoint of the pH scale is 7, where the numbers of H^+ and OH^- are equal. A solution with a pH of 7, such as pure water, is neutral—neither acidic nor alkaline. A solution that has more H^+ than OH^- is ***acidic*** and has a pH less than 7. A solution that has more OH^- than H^+ is ***basic (alkaline)*** and has a pH greater than 7. A change of one whole number on the pH scale represents a *10-fold* change in the number of H^+. At a pH of 6, there are 10 times more H^+ than at a pH of 7. Put another way, a pH of 6 is 10 times more acidic than a pH of 7, and a pH of 9 is 100 times more alkaline than a pH of 7.

MAINTAINING PH: BUFFER SYSTEMS

Although the pH of various body fluids may differ, the normal limits for each are quite narrow. Table 2.2 shows the pH values for certain body fluids compared with those of common household substances. Homeostatic mechanisms maintain the pH of blood between 7.35 and 7.45, so that it is slightly more basic than pure water. Even though strong acids and bases may be taken into the body or be formed by body cells, the pH of fluids inside and outside cells remains almost constant. One important reason is the presence of ***buffer systems,*** in which chemical compounds called *buffers* convert strong acids or bases into weak acids or bases.

Organic Compounds

Carbohydrates

Carbohydrates are organic compounds and include sugars, glycogen, starches, and cellulose. The elements present in carbohydrates are carbon, hydrogen, and oxygen. In any carbohydrate, the ratio of carbon to hydrogen to oxygen atoms is usually 1:2:1.

Figure 2.6 Acids, bases, and salts. (a) When placed in water, hydrochloric acid (HCl) ionizes into H^+ and Cl^-. (b) When the base potassium hydroxide (KOH) is placed in water, it ionizes into OH^- and K^+. (c) When the salt potassium chloride (KCl) is placed in water, it ionizes into positive and negative ions (K^+ and Cl^-), neither of which is H^+ or OH^-.

Ionization is the separation of inorganic acids, bases, and salts into ions in a solution.

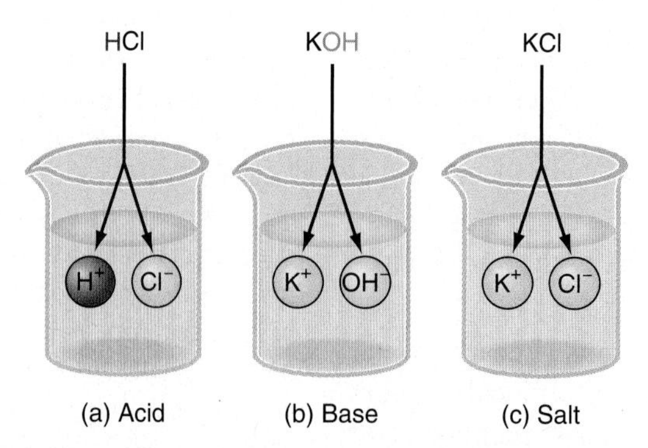

? **The compound $CaCO_3$ (calcium carbonate) dissociates into a calcium ion (Ca^{2+}) and a carbonate ion (CO_3^{2-}). Is it an acid, a base, or a salt? What about H_2SO_4, which dissociates into two H^+ and one SO_4^{2-}?**

Figure 2.7 The pH scale. A pH below 7 indicates an acidic solution, or more H^+ than OH^-. The lower the numerical value of the pH, the more acidic is the solution because the H^+ concentration becomes progressively greater. A pH above 7 indicates a basic (alkaline) solution; that is, there are more OH^- than H^+. The higher the pH, the more basic is the solution.

At pH 7.0 (neutrality), the concentrations of H^+ and OH^- are equal.

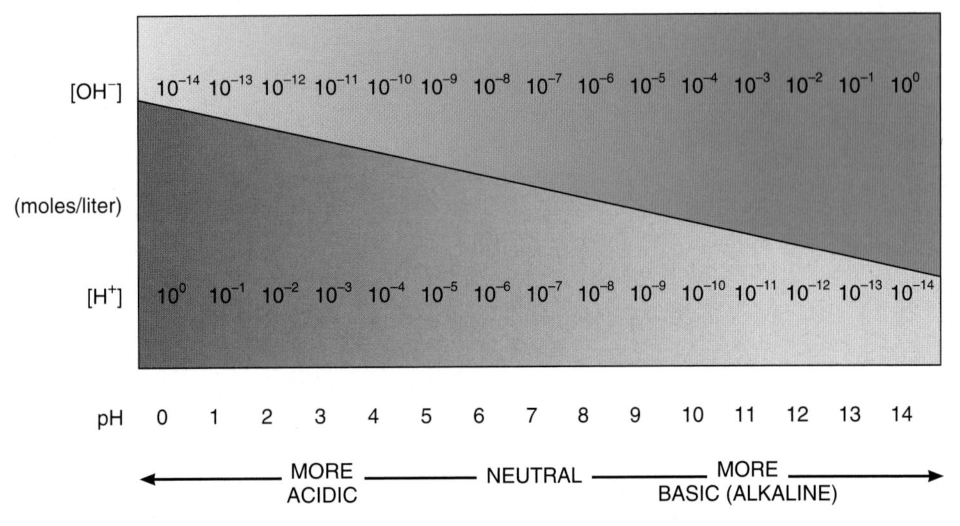

Which pH is more acidic, 6.82 or 6.91? Which pH is closer to neutral, 8.41 or 5.59?

TABLE 2.2

pH Values of Selected Substances

SUBSTANCE*	pH VALUE
• Gastric juice (digestive juice of the stomach)	1.2–3.0
Lemon juice	2.3
Grapefruit juice, vinegar, wine	3.0
Carbonated soft drink	3.0–3.5
Orange juice	3.5
• Vaginal fluid	3.5–4.5
Tomato juice	4.2
Coffee	5.0
• Urine	4.6–8.0
• Saliva	6.35–6.85
Cow's milk	6.8
Distilled (pure) water	7.0
• Blood	7.35–7.45
• Semen (fluid containing sperm)	7.20–7.60
• Cerebrospinal fluid (fluid associated with the nervous system)	7.4
• Pancreatic juice (digestive juice of the pancreas)	7.1–8.2
• Bile (liver secretion that aids fat digestion)	7.6–8.6
Milk of magnesia	10.5
Lye	14.0

• Denotes substances in the human body.

For example, the molecular formula for the small carbohydrate glucose is $C_6H_{12}O_6$. Carbohydrates are divided into three major groups based on size: monosaccharides, disaccharides, and polysaccharides. Monosaccharides and disaccharides are termed ***simple sugars,*** and polysaccharides are also known as ***complex carbohydrates.***

1. ***Monosaccharides*** (mon′-ō-SAK-a-rīds; *mono-* = one; *-sacchar-* = sugar) are the building blocks of carbohydrates. In your body, the principal function of the monosaccharide glucose is to serve as a source of chemical energy for generating the ATP that fuels metabolic reactions. Ribose and deoxyribose are also monosaccharides and are used to make ribonucleic acid (RNA) and deoxyribonucleic acid (DNA), which are described later in this section.

2. ***Disaccharides*** (dī-SAK-a-rīds; *di-* = two) are simple sugars that consist of two monosaccharides joined by a covalent bond. When two monosaccharides combine to form a disaccharide, a molecule of water is lost. When two smaller molecules join to form a larger molecule in a ***dehydration synthesis reaction*** (*de-* = from, down, or out; *-hydra-* = water), a water molecule is formed and removed. As you will see later in the chapter, such reactions occur during synthesis of proteins and other large molecules (see Figure 2.13). For example, the monosaccharides glucose and fructose combine to form the disaccharide sucrose (table sugar) as shown in Figure 2.8. Disaccharides can be split into monosaccharides by adding a molecule of water, a hydrolysis reaction. Sucrose, for example, may be hydrolyzed into its components

Figure 2.8 Dehydration synthesis and hydrolysis of a molecule of sucrose. In the dehydration synthesis reaction (read from left to right), two smaller molecules, glucose and fructose, are joined to form a larger molecule of sucrose. Note the loss of a water molecule. In the hydrolysis reaction (read from right to left), the larger sucrose molecule is broken down into two smaller molecules, glucose and fructose. Here, a molecule of water is added to sucrose for the reaction to occur.

Monosaccharides are the building blocks of carbohydrates.

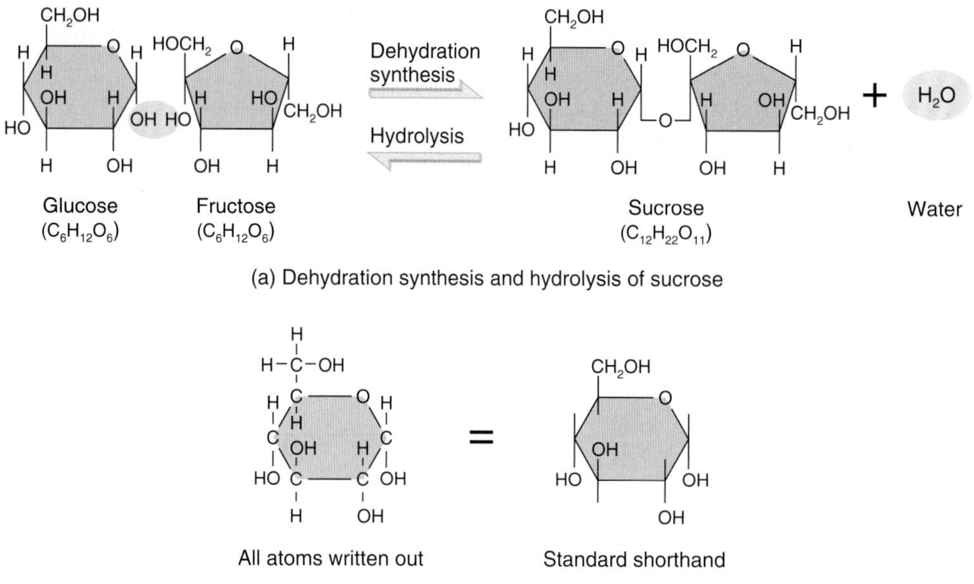

Dehydration synthesis

Hydrolysis

Glucose ($C_6H_{12}O_6$) + Fructose ($C_6H_{12}O_6$) ⇌ Sucrose ($C_{12}H_{22}O_{11}$) + Water (H_2O)

(a) Dehydration synthesis and hydrolysis of sucrose

All atoms written out = Standard shorthand

(b) Alternate chemical structures of organic molecules (shown here is glucose)

How many carbons are there in fructose? In sucrose?

of glucose and fructose by the addition of water (Figure 2.8). Other disaccharides include maltose (glucose + glucose), or malt sugar, and lactose (glucose + galactose), the sugar in milk.

3. ***Polysaccharides*** (pol′-ē-SAK-a-rīds; *poly-* many) are large, complex carbohydrates that contain tens or hundreds of monosaccharides joined through dehydration synthesis reactions. Like disaccharides, polysaccharides can be broken down into monosaccharides through hydrolysis reactions. The main polysaccharide in the human body is *glycogen,* which is made entirely of glucose units joined together in branching chains (Figure 2.9). Glycogen is stored in cells of the liver and in skeletal muscles. If energy demands of the body are high, glycogen is broken down into glucose; when energy demands are low, glucose is built back up into glycogen. *Starches* are also made of glucose units and are polysaccharides made mostly by plants. We digest starches to glucose as another energy source. *Cellulose* is a polysaccharide found in plant cell walls. Although humans cannot digest cellulose, it does provide bulk (roughage or fiber) that helps move feces through the large intestine. Unlike simple sugars, polysaccharides usually are not soluble in water and do not taste sweet.

Figure 2.9 Part of a glycogen molecule, the main polysaccharide in the human body.

Glycogen is made up of glucose units and is the storage form of carbohydrates in the human body.

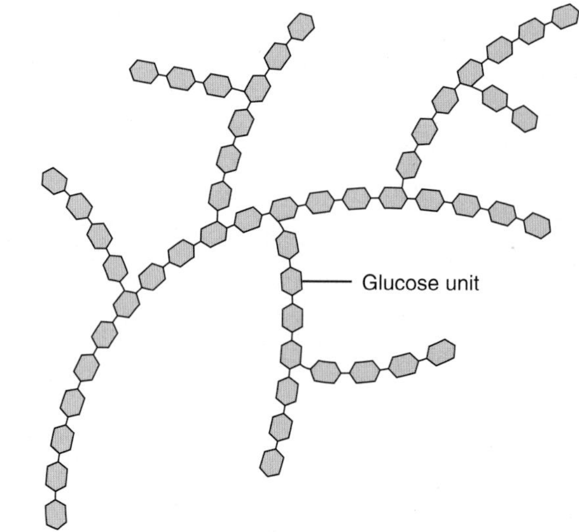

Which body cells store glycogen?

Lipids

Like carbohydrates, ***lipids*** (LIP-ids, *lip-* = fat) contain carbon, hydrogen, and oxygen. Unlike carbohydrates, they do not have a 2:1 ratio of hydrogen to oxygen. The proportion of oxygen atoms in lipids is usually smaller than in carbohydrates, so there are fewer polar covalent bonds. As a result, most lipids are hydrophobic; that is, they are insoluble in water.

The diverse lipid family includes triglycerides (fats and oils), phospholipids (lipids that contain phosphorus), steroids, fatty acids, and fat-soluble vitamins (A, D, E, and K).

The most plentiful lipids in your body and in your diet are the ***triglycerides*** (trī-GLI-cer-īds; *tri-* = three). At room temperature, triglycerides may be either solids (fats) or liquids (oils). They are the body's most highly concentrated form of chemical energy, storing more than twice as much chemical energy per gram as carbohydrates or proteins. Our capacity to store triglycerides in fat tissue, called adipose tissue, for all practical purposes, is unlimited. Excess dietary carbohydrates, proteins, fats, and oils all have the same fate: They are deposited in adipose tissue as triglycerides.

A triglyceride consists of two types of building blocks: a single glycerol molecule and three fatty acid molecules. The three-carbon ***glycerol*** molecule forms the backbone of a triglyceride (Figure 2.10). Three ***fatty acids*** are attached by dehydration synthesis reactions, one to each carbon of the glycerol backbone. The fatty acid chains of a triglyceride may be saturated, monounsaturated, or polyunsaturated. ***Saturated fats*** contain only *single covalent bonds* between fatty acid carbon atoms. Because they do not contain any double bonds between fatty acid carbon atoms, each carbon atom is *saturated with hydrogen atoms* (see palmitic acid and stearic acid in Figure 2.10). Triglycerides with mainly saturated fatty acids are solid at room temperature and occur mostly in meats (especially red meats) and nonskim dairy products (whole milk, cheese, and butter). They also occur in a few tropical plants, such as cocoa, palm, and coconut. ***Monounsaturated fats*** (*mono-* = one) contain fatty acids with *one double covalent bond* between two fatty acid carbon atoms and thus are not completely saturated with hydrogen atoms (see oleic acid in Figure 2.10). Diets that contain large amounts of saturated fats have been associated with disorders such as heart disease and colorectal cancer. Olive oil, peanut oil, canola oil, most nuts, and avocados are rich in triglycerides with monounsaturated fatty acids. Monounsaturated fats are thought to decrease the risk of heart disease. ***Polyunsaturated fats*** (*poly-* = many) contain *more than one double covalent bond* between fatty acid carbon atoms. Corn oil, safflower oil, sunflower oil, soybean oil, and fatty fish (salmon, tuna, and mackerel) contain a high percentage of polyunsaturated fatty acids. Polyunsaturated fats are also believed to decrease the risk of heart disease. However, when products such as margarine and vegetable shortening are made from polyunsaturated fats, compounds called *trans*-fatty acids are produced. *Trans*-fatty acids, like saturated fats, increase the risk of cardiovascular disease.

• CLINICAL CONNECTION | Essential Fatty Acids

A group of fatty acids called **essential fatty acids (EFAs)** are essential to human health. However, they cannot be manufactured by the human body and must be obtained from foods or supplements. Among the more important EFAs are *omega-3 fatty acids, omega-6 fatty acids,* and cis-*fatty acids.*

Figure 2.10 **Triglycerides consist of three fatty acids attached to a glycerol backbone.** The fatty acids vary in length and in the number and location of double bonds between carbon atoms (C=C). Shown here is a triglyceride molecule that contains two saturated fatty acids and one monounsaturated fatty acid.

A triglyceride consists of two types of building blocks: a single glycerol molecule and three fatty acid molecules.

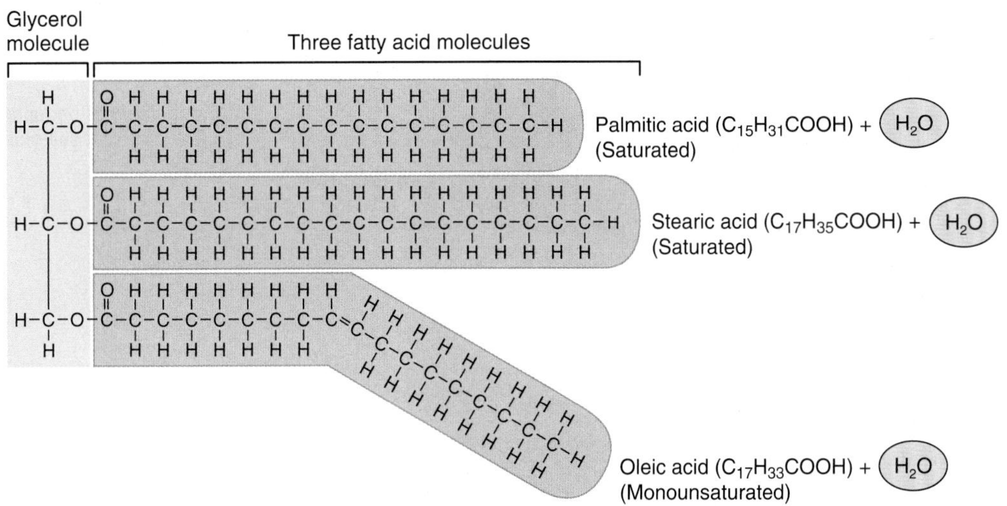

? **How many double bonds are there in a monounsaturated fatty acid?**

Omega-3 and omega-6 fatty acids are polyunsaturated fatty acids that may have a protective effect against heart disease and stroke by lowering total cholesterol, raising HDL (high-density lipoproteins or "good cholesterol") and lowering LDL (low-density lipoproteins or "bad cholesterol"). In addition, they decrease bone loss; reduce symptoms of arthritis due to inflammation; promote wound healing; improve certain skin disorders (psoriasis, eczema, and acne); and improve mental functions. Primary sources of omega-3 fatty acids include flaxseed, fatty fish, oils that have large amounts of polyunsaturated fats, fish oils, and walnuts. Primary sources of omega-6 fatty acids include most processed foods (cereals, breads, white rice), eggs, baked goods, oils with large amounts of polyunsaturated fats, and meats (especially organ meats, such as liver).

Cis-fatty acids are nutritionally beneficial monounsaturated fatty acids that are used by the body to produce hormone-like regulators and cell membranes. However, when *cis*-fatty acids are heated, pressurized, and combined with a catalyst (usually nickel) in a process called *hydrogenation,* they are changed to unhealthy *trans*-fatty acids. Hydrogenation is used by manufacturers to make vegetable oils solid at room temperature and less likely to turn rancid. Hydrogenated or *trans*-fatty acids are common in commercially baked goods (crackers, cakes, and cookies), salty snack foods, some margarines, and fried foods (donuts and french fries). If a product label contains the words "hydrogenated" or "partially hydrogenated," then the product contains *trans*-fatty acids. Among the adverse effects of *trans*-fatty acids are an increase in total cholesterol, a decrease in HDL, an increase in LDL, and an increase in triglycerides. These effects, which can increase the risk of heart disease and other cardiovascular diseases, are similar to those caused by saturated fats. •

Like triglycerides, ***phospholipids*** have a glycerol backbone and two fatty acids attached to the first two carbons (Figure 2.11a). Attached to the third carbon is a phosphate group

Figure 2.11 Phospholipids. (a) In the synthesis of phospholipids, two fatty acids attach to the first two carbons of the glycerol backbone. A phosphate group links a small, charged group to the third carbon in glycerol. In (b), the circle represents the polar head region, and the two wavy lines represent the two nonpolar tails.

Phospholipids are the main lipids in cell membranes.

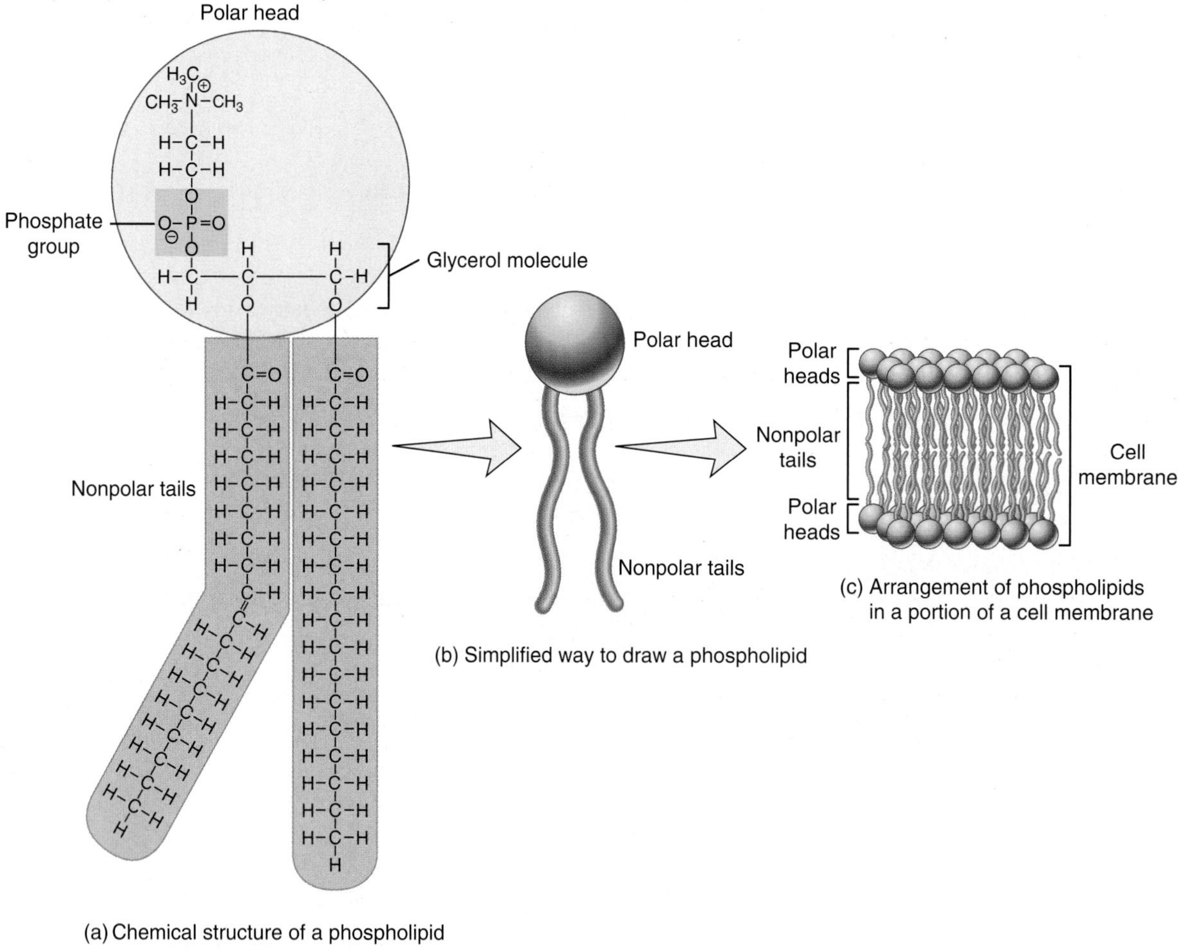

(a) Chemical structure of a phospholipid

(b) Simplified way to draw a phospholipid

(c) Arrangement of phospholipids in a portion of a cell membrane

How does a phospholipid differ from a triglyceride?

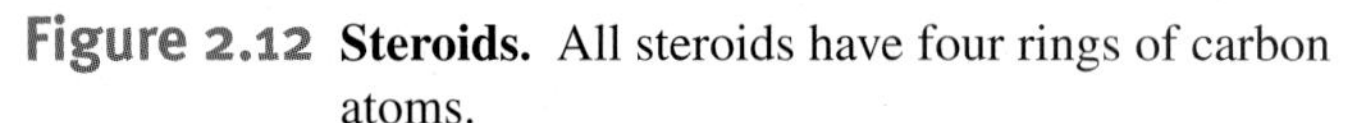

(PO_4^{3-}) that links a small, charged group to the glycerol backbone. Whereas the nonpolar fatty acids form the hydrophobic (*hydro-* = water; *-phobic* = fearing) "tails" of a phospholipid, the polar phosphate group and charged group form the hydrophilic (*-philic* = loving) "head" (Figure 2.11b). Phospholipids line up tails-to-tails in a double row to make up much of the membrane that surrounds each cell (Figure 2.11c).

The structure of **steroids** differs considerably from that of the triglycerides. Steroids have four rings of carbon atoms (colored gold in Figure 2.12). Body cells synthesize other steroids from cholesterol (Figure 2.12a), which has a large nonpolar region consisting of the four rings and a hydrocarbon tail. In the body, the commonly encountered steroids, such as cholesterol, estrogens, testosterone, cortisol, bile salts, and vitamin D, are known as **sterols.** Cholesterol is needed for cell membrane structure; estrogens and testosterone are required for regulating sexual functions; cortisol is necessary for maintaining normal blood sugar levels; bile salts are needed for lipid digestion and absorption; and vitamin D is related to bone growth.

MANUAL THERAPY APPLICATION

Anabolic Steroids: Performance-Enhancing Drugs

Some athletes will do almost anything to win, even at the potential expense of their health. An **anabolic steroid** is a chemical derived from *testosterone,* a steroid hormone. Tissues are broken down by catabolic steroids; in contrast, anabolic steroids build muscle and bone mass by stimulating production of new muscle and bone protein (see Chapters 6 and 10). Use of anabolic steroids thus permits athletes to develop additional musculature within a relatively short time. The additional musculature enables them to train harder and longer, which gives them a competitive edge over their opponents.

Individuals who use anabolic steroids may also seek the services of therapists as another enhancement toward better performance. Therapists should be aware that anabolic steroids have a number of possible *side effects* including liver damage, jaundice, depression, aggression, mood swings, bad breath, development of severe acne on the back and abdomen, nervousness, trembling, and decreased sexual function. When taken by males, there is also a potential for baldness and breast development. When taken by females, anabolic steroids cause male characteristics to develop. Each of the above symptoms could be caused by any number of disorders, but the combination of two or more of these symptoms in a patient may be a sign of steroid abuse.

Proteins

Proteins are large molecules that contain carbon, hydrogen, oxygen, and nitrogen; some proteins also contain sulfur. Much more complex in structure than carbohydrates or lipids, proteins have many roles in the body and are largely responsible for the structure of body cells. For example, proteins termed enzymes speed up particular chemical reactions, other proteins are responsible for contraction of muscles, proteins called antibodies help defend the body against invading microbes, and some hormones are proteins.

Figure 2.12 Steroids. All steroids have four rings of carbon atoms.

Cholesterol is the starting molecule for synthesis of other steroids in the body.

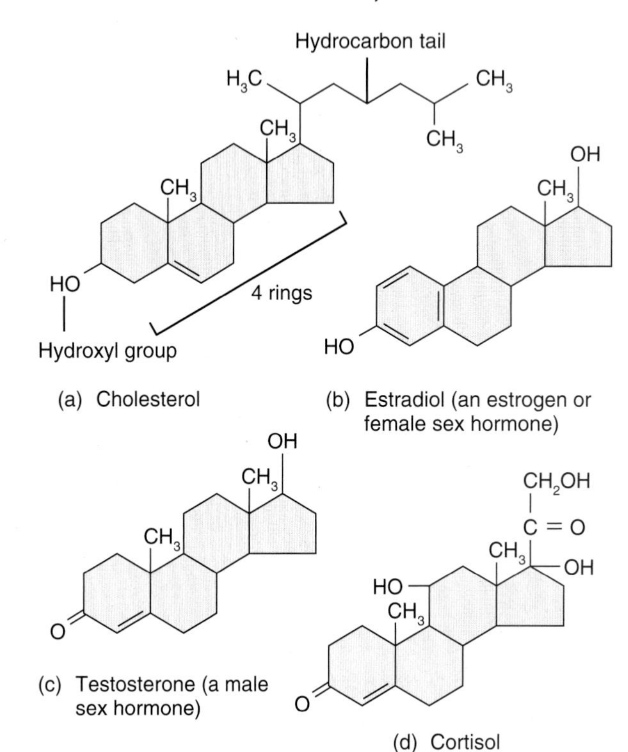

How is the structure of estradiol different from that of testosterone?

Amino acids (a-MĒ-nō) are the building blocks of proteins. All amino acids have an *amino group* (—NH_2) at one end and an acidic *carboxyl group* (—COOH) at the other end. Each of the 20 different amino acids has a different *side chain* (R group) (see Figure 2.13a). The covalent bonds that join amino acids to form more complex molecules are called *peptide bonds* (see Figure 2.13b).

The union of two or more amino acids produces a ***peptide*** (PEP-tīd). When two amino acids combine, the molecule is called a ***dipeptide*** (see Figure 2.13b). Adding another amino acid to a dipeptide produces a ***tripeptide.*** A ***polypeptide*** contains a large number of amino acids. Proteins are polypeptides that contain as few as 50 or as many as 2000 or more amino acids. Because each variation in the number and sequence of amino acids produces a different protein, a great variety of proteins is possible. The situation is similar to using an alphabet of 20 letters to form words. Each letter would be equivalent to an amino acid, and each word would be a different protein.

An alteration in the sequence of amino acids can have serious consequences. For example, a single substitution of an amino

Figure 2.13 Amino acids. (a) In keeping with their name, amino acids have an amino group (shaded blue) and a carboxyl (acid) group (shaded red). The side chain (R group) is shaded gold and is different in each type of amino acid. (b) When two amino acids are chemically united by dehydration synthesis (read from left to right), the resulting covalent bond between them is called a *peptide bond.* The peptide bond is formed at the point where water is lost. Here, the amino acids glycine and alanine are joined to form the dipeptide glycylalanine. Breaking a peptide bond occurs by hydrolysis (read from right to left).

Amino acids are the building blocks of proteins.

Side chain

Amino group

Carboxyl group

(a) Structure of an amino acid

Glycine

Alanine

Dehydration synthesis

Hydrolysis

Peptide bond

Glycylalanine (a dipeptide)

H_2O

Water

(b) Protein formation

How many peptide bonds would there be in a tripeptide?

acid in hemoglobin, a blood protein, can result in a deformed molecule that produces *sickle-cell disease*.

A protein may consist of only one polypeptide or several intertwined polypeptides. A given type of protein has a unique three-dimensional shape because of the ways in which each individual polypeptide twists and folds and associated polypeptides come together. If a protein encounters a hostile environment in which temperature, pH, or ion concentration is significantly altered, it may unravel and lose its characteristic shape. This process is called ***denaturation*** (dē-nā′-chur-Ā-shun). Denatured proteins are no longer functional. A common example of denaturation is seen in frying an egg. In a raw egg the egg-white protein (albumin) is soluble and the egg white appears as a clear, viscous fluid. When heat is applied to the egg, however, the albumin denatures; it changes shape, becomes insoluble, and turns white.

As we have seen, chemical reactions occur only when chemical bonds are made or broken as atoms, ions, or molecules collide with one another. At normal body temperature, such collisions occur too infrequently to maintain life. ***Enzymes*** (EN-zīms) are the living cell's solution to this problem because they speed chemical reactions by increasing the frequency of collisions and by properly orienting the colliding molecules. Substances that can speed chemical reactions without themselves being altered are called ***catalysts*** (KAT-a-lists). In living cells, most enzymes are proteins. The names of enzymes usually end in *-ase*. All enzymes can be grouped according to the types of chemical reactions they catalyze. For example, *oxidases* add oxygen, *kinases* add phosphate, *dehydrogenases* remove hydrogen, *anhydrases* remove water, *ATPases* split ATP, *proteases* break down proteins, and *lipases* break down lipids.

Enzymes catalyze selected reactions with great efficiency and with many built-in controls. Three important properties of enzymes are their specificity, efficiency, and control.

1. **Specificity.** Enzymes are highly specific. Each particular enzyme catalyzes a particular chemical reaction that involves specific ***substrates,*** the molecules on which the enzyme acts, and that gives rise to specific ***products,*** the molecules produced by the reaction. Each of the more than 1000 known enzymes in your body has a characteristic three-dimensional shape with a specific surface configuration that allows it to fit specific substrates.
2. **Efficiency.** Under optimal conditions, enzymes can catalyze reactions at rates that are millions to billions of times more rapid than those of similar reactions occurring without enzymes. A single enzyme molecule can convert substrate molecules to product molecules at rates as high as 600,000 per second.
3. **Control.** Enzymes are subject to a variety of cellular controls. Their rate of synthesis and their concentration at any given time are under the control of a cell's genes. Substances within the cell may either enhance or inhibit activity of a given enzyme. Many enzymes exist in both active and inactive forms. The rate at which the inactive form becomes active or vice versa is determined by the chemical environment inside the cell.

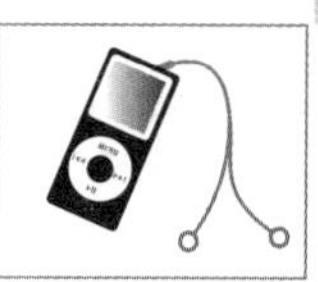

Enzymes decrease the "randomness" of the collisions between molecules. They also help bring the substrates together in the proper orientation so that the reaction can occur. Figure 2.14 depicts how an enzyme works:

1. The substrates make contact with the active site on the surface of the enzyme molecule, forming a temporary intermediate compound called the *enzyme–substrate complex.* In this reaction, the two substrate molecules are sucrose (a disaccharide) and water.
2. The substrate molecules are transformed by the rearrangement of existing atoms, the breakdown of the substrate molecule, or the combination of several substrate molecules into the products of the reaction. Here the products are two monosaccharides: glucose and fructose.
3. After the reaction is completed and the reaction products move away from the enzyme, the unchanged enzyme is free to attach to other substrate molecules.

Figure 2.14 How an enzyme works.

An enzyme speeds up a chemical reaction without being altered or consumed.

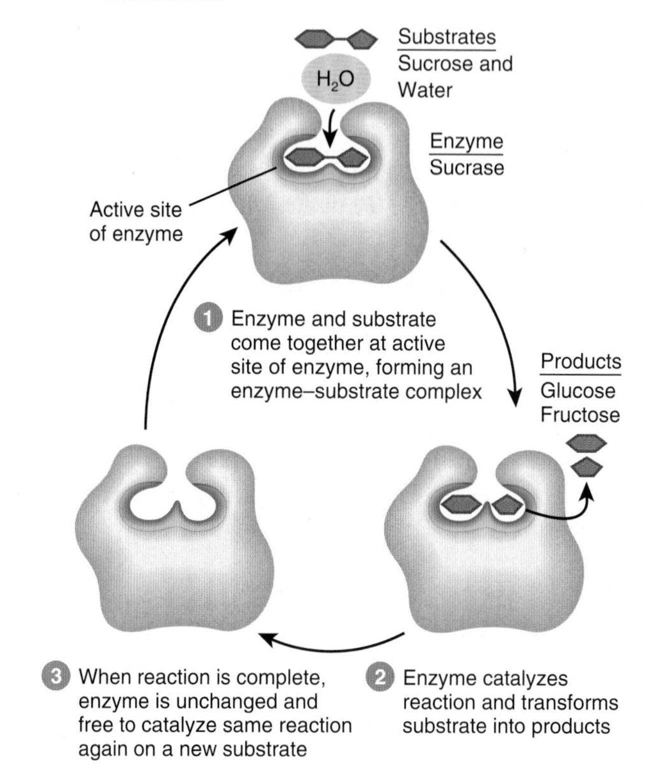

? **What part of an enzyme combines with its substrate?**

• CLINICAL CONNECTION | Lactose Intolerance

Enzyme deficiencies may lead to certain disorders. For example, some people do not produce enough lactase, an enzyme that breaks down the disaccharide lactose into the monosaccharides glucose and galactose. This deficiency causes a condition called **lactose intolerance,** in which undigested lactose retains fluid in the feces, and bacterial fermentation of lactose results in the production of gases. Symptoms of lactose intolerance include diarrhea, gas, bloating, and abdominal cramps after consumption of milk and other dairy products. The severity of symptoms varies from relatively minor to sufficiently serious to require medical attention. Persons with lactose intolerance can take dietary enzyme supplements to aid in the digestion of lactose. •

Nucleic Acids: Deoxyribonucleic Acid (DNA) and Ribonucleic Acid (RNA)

Nucleic acids (noo-KLĒ-ic), so named because they were first discovered in the nuclei of cells, are huge organic molecules that contain carbon, hydrogen, oxygen, nitrogen, and phosphorus. The two kinds of nucleic acids are ***deoxyribonucleic acid (DNA)*** (dē-ok′-sē-rī-bō-noo-KLĒ-ik) and ***ribonucleic acid (RNA).*** A nucleic acid molecule is composed of repeating building blocks called ***nucleotides.*** DNA and RNA will be further discussed in the next chapter.

Adenosine Triphosphate

Adenosine triphosphate (a-DEN-ō-sēn or ***ATP***) is the "energy currency" of living systems (Figure 2.15). ATP transfers the energy liberated in *exergonic catabolic reactions* to power cellular activities that require energy (*endergonic reactions*). Among these cellular activities are muscular contractions, movement of chromosomes during cell division, movement of structures

Figure 2.15 Structures of ATP and ADP. The two phosphate bonds that can be used to transfer energy are indicated in red. Most often energy transfer involves hydrolysis of the terminal phosphate bond of ATP.

ATP transfers chemical energy to power cellular activities.

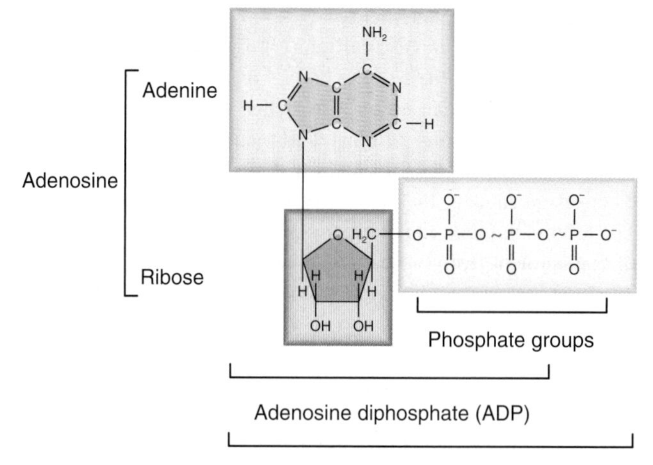

? **What are some cellular activities that depend on energy supplied by ATP?**

within cells, transport of substances across cell membranes, and synthesis of larger molecules from smaller ones. As its name implies, ATP consists of three phosphate groups attached to adenosine, a unit composed of adenine and the five-carbon sugar ribose.

When a water molecule is added to ATP, the third phosphate group (PO_4^{3-}), symbolized by Ⓟ in the following discussion, is removed, and the overall reaction liberates energy. The enzyme that catalyzes the hydrolysis of ATP is called *ATPase.* Removal of the third phosphate group produces a molecule called ***adenosine diphosphate (ADP)*** in the following reaction:

$$\underset{\text{Adenosine triphosphate}}{\text{ATP}} + \underset{\text{Water}}{H_2O} \xrightarrow{\textit{ATPase}} \underset{\text{Adenosine diphosphate}}{\text{ADP}} + \underset{\text{Phosphate group}}{\text{Ⓟ}} + \underset{\text{Energy}}{\text{E}}$$

As noted previously, the energy supplied by the catabolism of ATP into ADP is constantly being used by the cell. As the supply of ATP at any given time is limited, a mechanism exists to replenish it: The enzyme *ATP synthase* catalyzes the addition of a phosphate group to ADP in the following reaction:

$$\underset{\text{Adenosine diphosphate}}{\text{ADP}} + \underset{\text{Phosphate group}}{\text{Ⓟ}} + \underset{\text{Energy}}{\text{E}} \xrightarrow{\textit{ATP synthase}} \underset{\text{Adenosine triphosphate}}{\text{ATP}} + \underset{\text{Water}}{H_2O}$$

Where does the cell get the energy required to produce ATP? The energy needed to attach a phosphate group to ADP is supplied mainly by the catabolism of glucose in a process called ***cellular respiration.*** Cellular respiration has two phases, anaerobic and aerobic:

1. ***Anaerobic phase.*** In a series of reactions that do not require oxygen, glucose is partially broken down by a series of catabolic reactions into pyruvic acid. Each glucose molecule that is converted into a pyruvic acid molecule yields two molecules of ATP.

2. ***Aerobic phase.*** In the presence of oxygen, glucose is completely broken down into carbon dioxide and water. These reactions generate heat and 36 or 38 ATP molecules.

In Chapter 1, you learned that the human body is characterized by various levels of organization; this chapter has just shown you the alphabet of atoms and molecules that is the basis for the language of the body. Now that you have an understanding of the chemistry of the human body, you are ready to form words; in Chapter 3 you will see how atoms and molecules are organized to form structures of cells and perform the activities of cells that contribute to homeostasis.

CHECKPOINT

5. How do inorganic compounds differ from organic compounds?
6. What functions does water perform in the body?
7. Distinguish among saturated, monounsaturated, and polyunsaturated fats.
8. What are the important properties of enzymes?
9. Why are nuclei acids so named?
10. Why is ATP important?

STUDY OUTLINE

Introduction to Chemistry (Section 2.1)

1. Chemistry is the science of the structure and interactions of matter, which is anything that occupies space and has mass. Matter is made up of chemical elements.
2. The elements oxygen (O), carbon (C), hydrogen (H), and nitrogen (N) make up about 96% of the body's mass.
3. Each element is made up of units called atoms, which consist of a nucleus that contains protons and neutrons, and electrons that move about the nucleus in electron shells or clouds. The number of electrons is equal to the number of protons in an atom.
4. The atomic number, the number of protons, distinguishes the atoms of one element from those of another element.
5. An atom that *gives up* or *gains* electrons becomes an ion—an atom that has a positive or negative charge due to having unequal numbers of protons and electrons.
6. A molecule is a substance that consists of two or more chemically combined atoms. The molecular formula indicates the number and type of atoms that make up a molecule.
7. A compound is a substance that can be broken down into two or more different elements by ordinary chemical means.
8. A free radical is a destructive ion or molecule that has an unpaired electron in its outermost shell.
9. Chemical bonds hold the atoms of a molecule together.
10. Electrons in the outermost shell are the parts of an atom that participate in chemical reactions.
11. When outer shell electrons are transferred from one atom to another, the transfer forms ions, whose unlike charges attract each other and form ionic bonds. Positively charged ions are called cations; negatively charged ions are called anions.
12. In a covalent bond, pairs of outer shell electrons are shared between two atoms.
13. Hydrogen bonds are weak bonds between hydrogen and certain other atoms and often occur within large, complex molecules such as proteins and nucleic acids. They add strength and stability and help determine the molecule's three-dimensional shape.
14. Energy is the capacity to do work. Potential energy is energy stored by matter due to its position. Kinetic energy is the energy of matter in motion. Chemical energy is a form of potential energy stored in the bonds of molecules.
15. In a synthesis (anabolic) reaction, two or more atoms, ions, or molecules combine to form a new and larger molecule. In a decomposition (catabolic) reaction, a molecule is split into smaller molecules, ions, or atoms.
16. When nutrients, such as glucose, are broken down via decomposition reactions, some of the energy released is temporarily stored in

adenosine triphosphate (ATP) and then later used to drive energy-requiring synthesis reactions that build body structures, such as muscles and bones.

17. Exchange reactions are combination synthesis and decomposition reactions. Reversible reactions can proceed in both directions under different conditions.

Chemical Compounds and Life Processes (Section 2.2)

1. Inorganic compounds usually are structurally simple and lack carbon. Organic substances always contain carbon, usually contain hydrogen, and always have covalent bonds.
2. Water is the most abundant substance in the body. It is an excellent solvent, participates in chemical reactions, absorbs and releases heat slowly, requires a large amount of heat to change from a liquid to a gas, and serves as a lubricant.
3. Inorganic acids, bases, and salts dissociate into ions in water. An acid ionizes into hydrogen ions (H^+); a base usually ionizes into hydroxide ions (OH^-). A salt ionizes into neither H^+ nor OH^- ions.
4. The pH of body fluids must remain fairly constant for the body to maintain homeostasis. On the pH scale, 7 represents neutrality. Values below 7 indicate acidic solutions, and values above 7 indicate alkaline (basic) solutions.
5. Buffer systems help maintain pH by converting strong acids or bases into weak acids or bases.
6. Carbohydrates include sugars, glycogen, and starches. They may be monosaccharides, disaccharides, or polysaccharides. Carbohydrates provide most of the chemical energy needed to generate ATP. Carbohydrates, and other large, organic molecules, are synthesized via dehydration synthesis reactions, in which a molecule of water is lost. In the reverse process, called hydrolysis, large molecules are broken down into smaller ones upon the addition of water.
7. Lipids are a diverse group of compounds that include triglycerides (fats and oils), phospholipids, and steroids. Triglycerides protect, insulate, provide energy, and are stored in adipose tissue. Phospholipids are important membrane components. Steroids are synthesized from cholesterol.
8. Proteins are constructed from amino acids. They give structure to the body, regulate processes, provide protection, help muscles to contract, transport substances, and serve as enzymes.
9. Enzymes are molecules, usually proteins, that speed up chemical reactions and are subject to a variety of cellular controls.
10. Deoxyribonucleic acid (DNA) and ribonucleic acid (RNA) are nucleic acids consisting of repeating units called nucleotides.
11. Adenosine triphosphate (ATP) is the principal energy-transferring molecule in living systems. When it transfers energy, ATP is decomposed by hydrolysis to adenosine diphosphate (ADP) and Ⓟ. ATP is synthesized from ADP and Ⓟ using primarily the energy supplied by the breakdown of glucose.

SELF-QUIZ QUESTIONS

1. A substance that dissociates in water to form H^+ is called
   a. a base **b.** a salt
   c. a buffer **d.** an acid
   e. a nucleic acid.
2. Ionic bonds are characterized by
   a. sharing electrons between atoms
   b. their ability to form strong, stable bonds
   c. atoms giving away and taking electrons
   d. the type of bonding formed in most organic compounds
   e. an attraction between water molecules.
3. If an atom has two electrons in its second electron shell and its first electron shell is filled, it will most likely
   a. lose two electrons from its second electron shell
   b. lose the electrons from its first electron shell
   c. lose all of the electrons from its first and second electron shells
   d. gain six electrons in its second electron shell
   e. share two electrons in its second electron shell.
4. Matter that cannot be broken down into simpler substances by chemical reactions is known as
   a. a molecule **b.** an antioxidant
   c. a compound **d.** a buffer
   e. a chemical element.
5. Chlorine (Cl) has an atomic number of 17. An atom of chlorine may become a chloride ion (Cl^-) by
   a. losing one electron **b.** losing one neutron
   c. gaining one proton **d.** gaining one electron
   e. gaining two electrons.
6. Which of the following is NOT true?
   a. A substance that separates in water to form some cation other than H^+ and some anion other than OH^- is known as a salt.
   b. A solution that has a pH of 9.4 is acidic.
   c. A solution with a pH of 5 is 100 times more acidic than distilled water, which has a pH of 7.
   d. Buffers help to make the body's pH more stable.
   e. Amino acids are linked by peptide bonds.
7. Which of the following organic compounds are NOT paired with their correct subunits (building blocks)?
   a. glycogen, glucose **b.** proteins, monosaccharides
   c. DNA, nucleotides **d.** lipids, glycerol and fatty acids
   e. ATP, ADP, and P
8. The type of reaction by which a disaccharide is formed from two monosaccharides is known as a
   a. decomposition reaction
   b. hydrolysis reaction
   c. dehydration synthesis reaction
   d. reversible reaction
   e. dissociation reaction.
9. Which of the following is a nucleic acid?
   a. DNA **b.** an enzyme **c.** ADP
   d. glucose **e.** ATP
10. What is the principal energy-transferring molecule in the body?
   a. ADP **b.** RNA **c.** DNA
   d. ATP **e.** NAD
11. Which of the following statements about water is NOT true?
   a. It is involved in many chemical reactions in the body.
   b. It is an important solvent in the human body.
   c. It helps lubricate a variety of structures in the body.
   d. It can absorb a large amount of heat without changing its temperature.
   e. It requires very little heat to change from a liquid to a gas.

12. The difference in H^+ concentration between solutions with a pH of 3 and a pH of 5 is that the solution with the pH of 3 has ________ H^+.
   a. 2 times more b. 5 times more
   c. 10 times more d. 100 times more
   e. 200 times less
13. Which of the following is NOT a true statement about enzyme activity?
   a. Enzymes form a temporary complex with their substrates.
   b. Enzymes are not permanently altered by the chemical reactions they catalyze.
   c. All proteins are enzymes.
   d. Enzymes are considered organic catalysts.
   e. Enzymes are subject to cellular control.
14. An organic compound that consists of C, H, and O and that may be broken down into glycerol and fatty acids is a
   a. triglyceride b. nucleic acid
   c. monosaccharide d. carbohydrate
   e. protein.
15. Why is it important to consume foods that contain antioxidants?
   a. They provide an energy source for the body.
   b. They help inactivate damaging free radicals.
   c. They make up the body's genes.
   d. They act as buffers to help maintain the blood's pH.
   e. They are important solvents in the body.
16. If a protein is exposed to an extremely high temperature, it will
   a. divide b. release energy
   c. become an electrolyte d. form hydrogen bonds
   e. denature.
17. In what form are lipids stored in the adipose (fat) tissue of the body?
   a. triglycerides b. glycogen
   c. cholesterol d. polypeptides
   e. disaccharides
18. Approximately 96% of your body's mass is composed of which of the following elements? Place an X beside each correct answer.

____ calcium	____ iron	____ nitrogen
____ phosphorus	____ sodium	____ chlorine
____ carbon	____ oxygen	____ sulfur
____ hydrogen	____ potassium	____ magnesium

19. Match the following:

____ a. inorganic compound	A. glycogen
____ b. monosaccharide	B. enzyme
____ c. polysaccharide	C. glucose
____ d. component of triglycerides	D. water
____ e. lipase	E. glycerol

CRITICAL THINKING QUESTIONS

1. Your best friend has decided to begin frying his breakfast eggs in margarine instead of butter because he has heard that eating butter is bad for his heart. Has he made a wise choice? Are there other alternatives?
2. A 4-month-old baby is admitted to the hospital with a fever of 102° Fahrenheit (38.9° Celsius). Why is it critical to treat the fever as quickly as possible?
3. During chemistry lab, Maria places sucrose (table sugar) in a glass beaker, adds water, and stirs. As the table sugar disappears, she loudly proclaims that she has chemically broken down the sucrose into fructose and glucose. Is Maria's chemical analysis correct?

? ANSWERS TO FIGURE QUESTIONS

2.1 The atomic number of carbon is 6.

2.2 The four most plentiful elements in living organisms are oxygen, carbon, hydrogen, and nitrogen.

2.3 Water is a compound because it contains atoms of both hydrogen and oxygen.

2.4 K is an electron donor; when it ionizes, it becomes a cation, K^+, because losing 1 electron from the fourth electron shell leaves 8 electrons in the third shell.

2.5 An ionic bond involves the *loss* and *gain* of electrons; a covalent bond involves the *sharing* of pairs of electrons.

2.6 $CaCO_3$ is a salt, and H_2SO_4 is an acid.

2.7 A pH of 6.82 is more acidic than a pH of 6.91. Both pH 8.41 and pH 5.59 are 1.41 pH units from neutral (pH = 7).

2.8 There are 6 carbons in fructose, 12 in sucrose.

2.9 Glycogen is stored in liver and skeletal muscle cells.

2.10 A monounsaturated fatty acid has one double bond.

2.11 A triglyceride has three fatty acid molecules attached to a glycerol backbone, and a phospholipid has two fatty acid tails and a phosphate group attached to a glycerol backbone.

2.12 The only differences between estradiol and testosterone are the number of double bonds and the types of functional groups attached to ring A.

2.13 A tripeptide would have two peptide bonds, each linking two amino acids.

2.14 The enzyme's active site combines with the substrate.

2.15 A few cellular activities that depend on energy supplied by ATP are: muscular contractions, movement of chromosomes, transport of substances across cell membranes, and synthesis reactions.

Cells 3

In this chapter you will learn why it is so important to eat a variety of fruits and vegetables. Some children eat them because their parents won't let them have dessert unless they do. That may have been your motivation when you were a kid, but as an adult you may have learned that a good reason to eat plenty of fruits and vegetables is that these foods contain important compounds, known as phytochemicals (literally, "plant chemicals"), which help to keep your cells healthy. Some phytochemicals block chemicals that can cause damage to your cells and can, in turn, lead to symptoms that you experience with a disease process. Other compounds enhance your body's production of enzymes that render potentially cancer-causing substances harmless. Collectively, the actions of phytochemicals promote healthy cellular function and prevent the types of cellular damage associated with cancer, aging, and heart disease.

CONTENTS AT A GLANCE

3.1 A GENERALIZED VIEW OF THE CELL

OBJECTIVE

- Name and describe the three main parts of a cell.

Figure 3.1 is a generalized view of a cell that shows the main cellular components. Though some body cells lack some cellular structures shown in this diagram, such as secretory vesicles or cilia, many body cells include most of these components. For ease of study, we divide a cell into three main parts: plasma membrane, cytoplasm, and nucleus.

- The ***plasma membrane*** forms a cell's flexible outer surface, separating the cell's internal environment (inside the cell) from its external environment (outside the cell). It regulates the flow of materials into and out of a cell to maintain the appropriate environment for normal cellular activities. The plasma membrane also plays a key role in communication among cells and between cells and their external environment.
- The ***cytoplasm*** (SĪ-tō-plazm; *-plasm* = formed or molded) consists of all the cellular contents between the plasma membrane and the nucleus. Cytoplasm can be divided into two components: cytosol and organelles. ***Cytosol*** (SĪ-tō-sol) is the liquid portion of cytoplasm that consists mostly of water plus dissolved solutes and suspended particles. Within the cytosol are different types of ***organelles*** (or-ga-NELZ = little organs), each of which has a characteristic structure and specific functions.
- The ***nucleus*** (NOO-klē-us = nut kernel) is the largest organelle of a cell. The nucleus acts as the control center for a cell because it contains the genes that control cellular structure and most cellular activities.

Figure 3.1 Generalized view of a body cell.

The cell is the basic, living, structural and functional unit of the body.

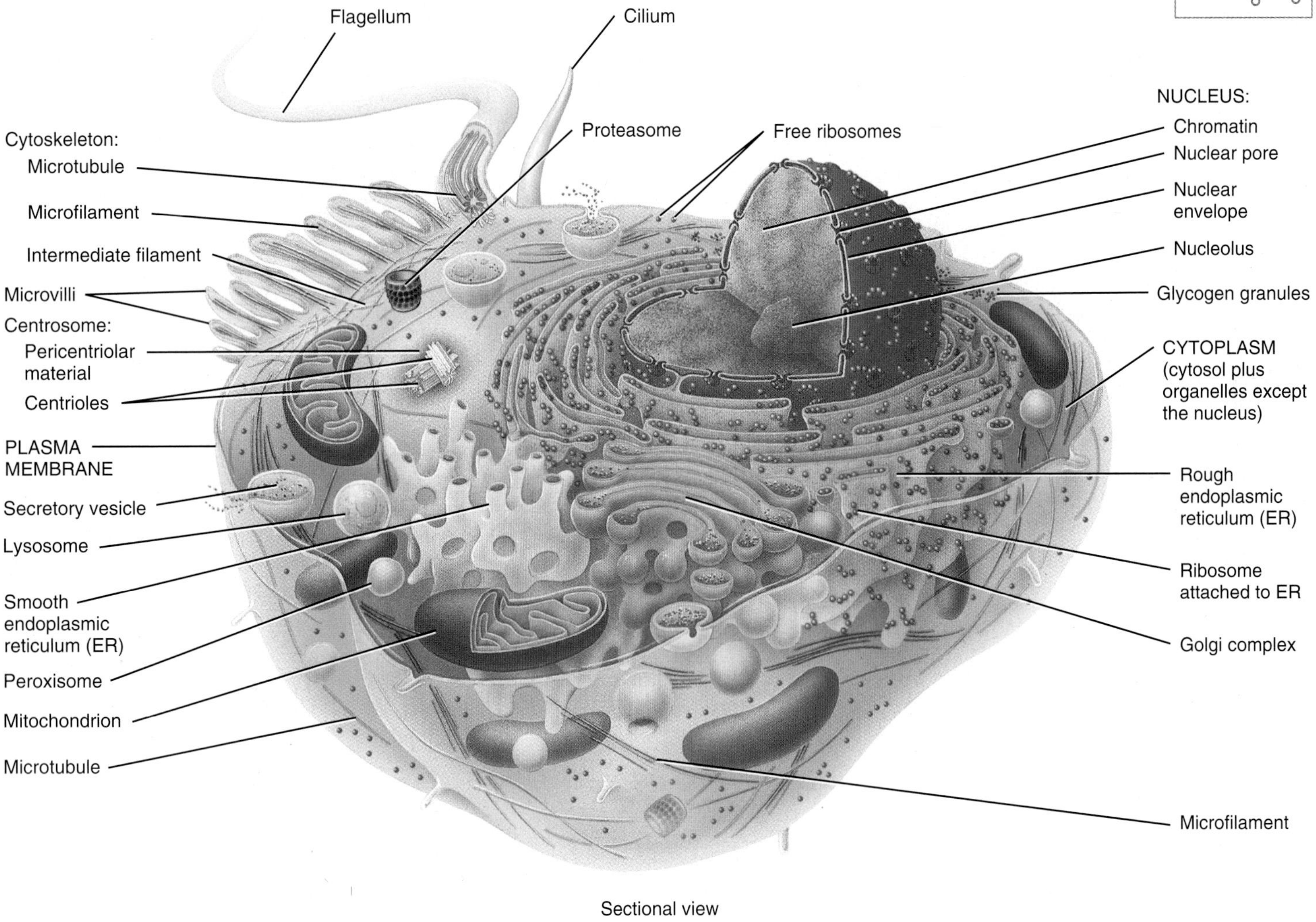

Flagellum
Cilium
Proteasome
Free ribosomes
NUCLEUS:
Chromatin
Nuclear pore
Nuclear envelope
Nucleolus
Glycogen granules
CYTOPLASM (cytosol plus organelles except the nucleus)
Rough endoplasmic reticulum (ER)
Ribosome attached to ER
Golgi complex
Microfilament
Cytoskeleton:
Microtubule
Microfilament
Intermediate filament
Microvilli
Centrosome:
Pericentriolar material
Centrioles
PLASMA MEMBRANE
Secretory vesicle
Lysosome
Smooth endoplasmic reticulum (ER)
Peroxisome
Mitochondrion
Microtubule
Sectional view

What are the three principal parts of a cell?

CHECKPOINT

1. What are the general functions of the three main parts of a cell?

3.2 PLASMA MEMBRANE

OBJECTIVE

- Describe the structure and functions of the plasma membrane.

The ***plasma membrane*** is a flexible yet sturdy barrier that consists mostly of phospholipids (lipids that contain phosphate groups, as studied in Chapter 2) and proteins. Virtually all membrane proteins are ***glycoproteins*** (*glyco-* = carbohydrate), proteins with attached carbohydrate groups. Other molecules present in lesser amounts in the plasma membrane are cholesterol and glycolipids (lipids with attached carbohydrate groups). The basic framework of the plasma membrane is the ***lipid bilayer,*** two back-to-back layers made up of three types of lipid molecules: phospholipids, cholesterol, and glycolipids (Figure 3.2). The glycoproteins in a membrane are of two types—integral and peripheral. ***Integral proteins*** extend into or through the lipid bilayer among the fatty acid tails. ***Peripheral proteins*** are loosely attached to the exterior or interior surface of the membrane. Although many of the proteins can move laterally in the lipid bilayer, each individual protein has a specific orientation with respect to the "inside" and "outside" faces of the membrane.

The plasma membrane allows some substances to move into and out of the cell but restricts the passage of other substances. This property of membranes is called ***selective permeability*** (per′-mē-a-BIL-i-tē). The lipid bilayer part of the membrane is permeable to water and to most lipid-soluble molecules, such as fatty acids, fat-soluble vitamins, steroids, oxygen, and carbon dioxide. The lipid bilayer is *not* permeable to ions or polar molecules, such as glucose and amino acids. These small and medium-sized water-soluble materials may cross the membrane with the assistance of integral proteins. Some integral proteins form ***ion channels*** through which specific ions can move into and out of cells. Other membrane proteins act as ***carriers (transporters),*** which change shape as they move a substance from one side of the membrane to the other. Large molecules such as proteins are unable to pass through the plasma membrane except by transport within vesicles (discussed in Section 3.3).

Generally, the types of lipids in cellular membranes vary only slightly. In contrast, the membranes of different cells and various in-

Figure 3.2 Chemistry and structure of the plasma membrane.

The plasma membrane consists mostly of phospholipids, arranged in a bilayer, and proteins, most of which are glycoproteins. Molecules of cholesterol provide stability of the membrane.

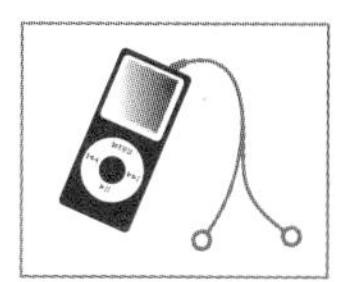

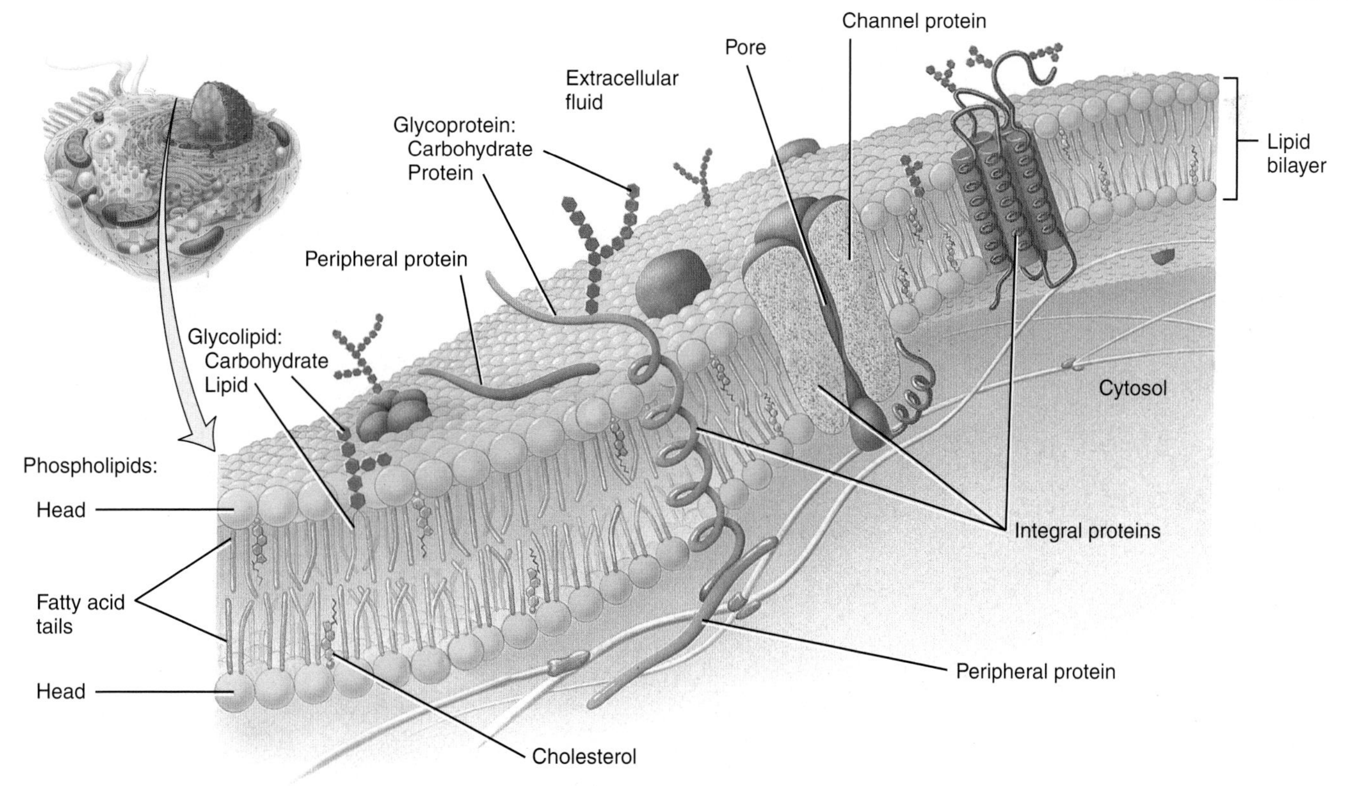

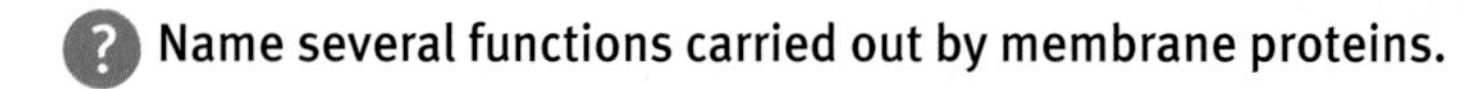

Name several functions carried out by membrane proteins.

tracellular organelles have remarkably different assortments of proteins that determine many of the membrane's functions (Figure 3.3).

- Some integral membrane proteins form ***ion channels,*** *pores* or holes through which specific ions, such as potassium ions (K^+), can flow to get into or out of the cell. Most ion channels are *selective;* they allow only a single type of ion to pass through.
- Other integral proteins act as ***carriers*** (***transporters***) selectively moving a polar substance or ion from one side of the membrane to the other.
- Integral proteins called ***receptors*** serve as cellular recognition sites. Each type of receptor recognizes and binds a specific type of molecule. For instance, insulin receptors bind the hormone insulin. A specific molecule that binds to a receptor is called a ***ligand*** (LĪ-gand; *liga-* = tied) of that receptor.
- Some integral proteins are ***enzymes*** that catalyze specific chemical reactions at the inside or outside surface of the cell.
- Integral proteins may also serve as ***linkers,*** which anchor proteins in the plasma membranes of neighboring cells to one another or to protein filaments inside and outside the cell. Peripheral proteins also serve as enzymes and linkers.
- Membrane glycoproteins and glycolipids often serve as ***cell-identity markers.*** They may enable a cell to recognize other cells of the same kind during tissue formation or to recognize and respond to potentially dangerous foreign cells. The ABO blood type markers are one example of cell-identity markers. When you receive a blood transfusion, the blood type must be compatible with your own.

In addition to the specific functions just mentioned, peripheral proteins help support the plasma membrane, anchor integral proteins, and participate in mechanical activities such as moving materials and organelles within cells, changing cell shape in dividing cells, enabling contraction of muscle cells, and attaching cells to one another.

CHECKPOINT

2. What molecules make up the plasma membrane and what are their functions?
3. What is meant by selective permeability?

3.3 TRANSPORT ACROSS THE PLASMA MEMBRANE

OBJECTIVE

- Describe the processes that transport substances across the plasma membrane.

Movement of materials across its plasma membrane is essential to the life of a cell. Certain substances must move into the cell to support metabolic reactions. Other materials must be moved out because they have been produced by the cell for export or are cellular waste products. Before discussing how materials move into and out of a cell, we need to understand what exactly is being moved as well as the form it needs to take to make its journey.

Figure 3.3 Membrane protein functions.

Membrane proteins largely reflect the functions a cell can perform.

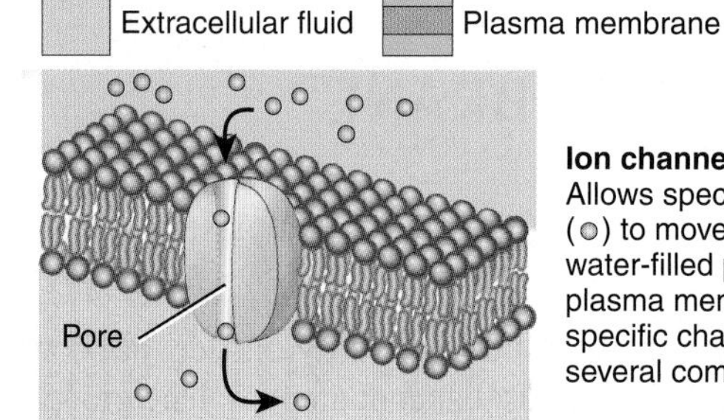

Ion channel (integral)
Allows specific ion (○) to move through water-filled pore. Most plasma membranes include specific channels for several common ions.

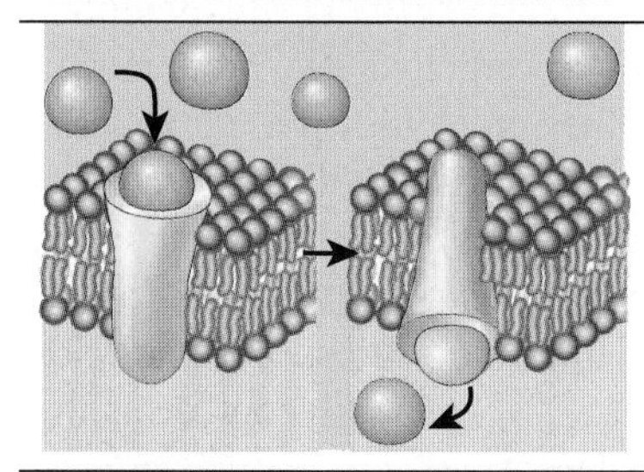

Carrier (integral)
Carries specific substances (○) across membrane by changing shape. For example, amino acids, needed to synthesize new proteins, enter body cells via carriers. Carrier proteins are also known as *transporters.*

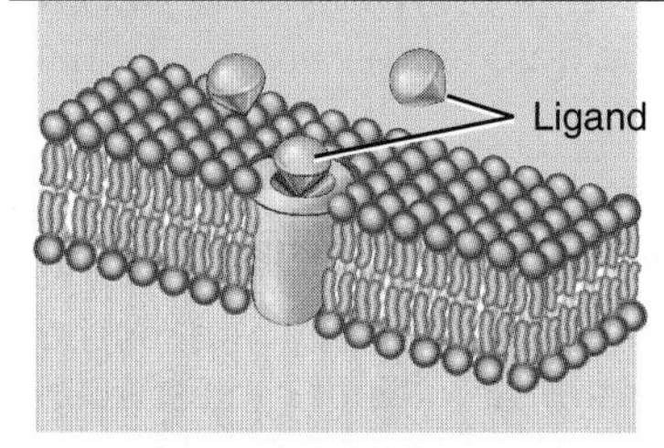

Receptor (integral)
Recognizes specific ligand (▽) and alters cell's function in some way. For example, antidiuretic hormone binds to receptors in the kidneys and changes the water permeability of certain plasma membranes.

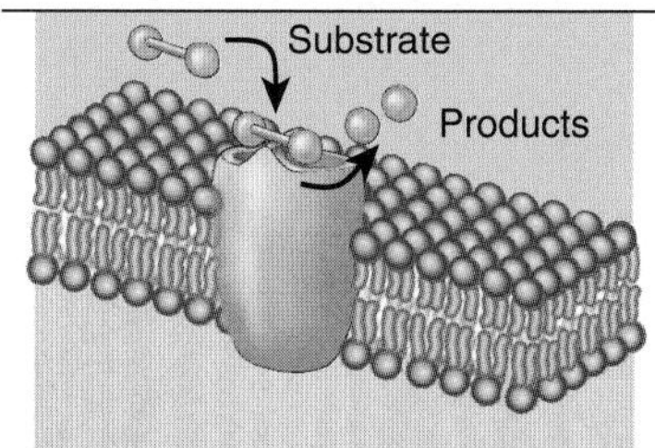

Enzyme (integral and peripheral)
Catalyzes reaction inside or outside cell (depending on which direction the active site faces). For example, lactase protruding from epithelial cells lining your small intestine splits the disaccharide lactose in the milk you drink.

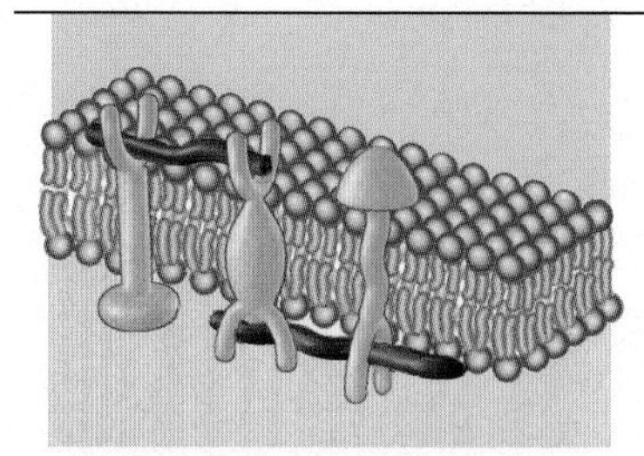

Linker (integral and peripheral)
Anchors filaments inside and outside the plasma membrane, providing structural stability and shape for the cell. May also participate in movement of the cell or link two cells together.

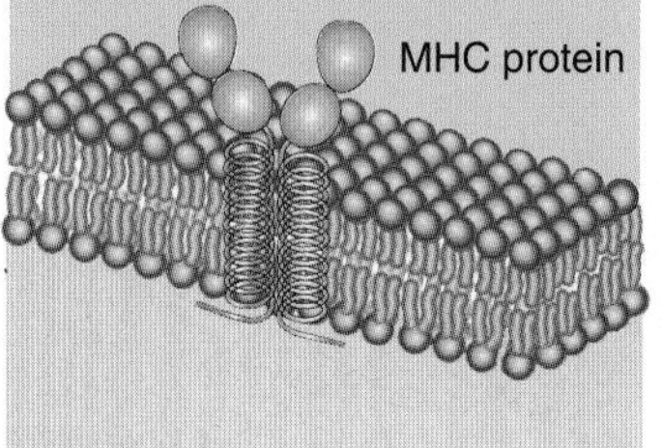

Cell-identity marker (glycoprotein)
Distinguishes your cells from anyone else's (unless you are an identical twin). An important class of such markers are the major histocompatibility (MHC) proteins.

? When stimulating a cell, the hormone insulin first binds to a protein in the plasma membrane. This action best represents which membrane protein function?

About two-thirds of the fluid in your body is contained inside body cells and is called ***intracellular fluid*** or ***ICF*** (in′-tra-SEL-yū-lar; *intra-* = within). ICF is the cytosol of a cell. Fluid outside body cells is called ***extracellular fluid*** or ***ECF*** (*extra-* = outside). As Figure 3.4 illustrates, the ECF in the microscopic spaces between the cells of tissues is ***interstitial fluid*** (in′-ter-STISH-al; *inter-* = between). The ECF in blood vessels is called ***plasma*** (PLAZ-ma), and that in lymphatic vessels is called ***lymph*** (LIMF). The ECF within and around the central nervous system is called ***cerebrospinal fluid (CSF)*** (se-rē′-brō-SPĪ-nal).

Materials dissolved in body fluids include gases, nutrients, ions, and other substances needed to maintain life. Any material dissolved in a fluid is called a ***solute,*** and the fluid in which it is dissolved is the ***solvent.*** Body fluids are dilute ***solutions*** in which a variety of solutes are dissolved in a very familiar solvent, water. The amount of a solute in a solution is its ***concentration.*** A ***concentration gradient*** is a difference in concentration between two different areas, for example, the ICF and ECF. Solutes moving from a high-concentration area (where there are more of them) to a low-concentration area (where there are fewer of them) are said to move *down* or *with* the concentration gradient. Solutes moving from a low-concentration area to a high-concentration area are said to move *up* or *against* the concentration gradient.

Substances move across cellular membranes by passive processes, active transport, and transport in vesicles. ***Passive processes,*** in which a substance moves down its concentration gradient through the membrane, using only its own energy of motion (kinetic energy), include diffusion and osmosis. In ***active processes,*** cellular energy, usually in the form of ATP, is used to "push" the substance through the membrane "uphill" against its concentration gradient. An example is active transport. Another way that some substances may enter and leave cells is an active process in which tiny membrane sacs referred to as ***vesicles*** are used (see Figure 3.18).

Figure 3.4 Body fluids. Intracellular fluid (ICF) is found within cells. Extracellular fluid (ECF) is found outside cells, in blood vessels as plasma, in lymphatic vessels as lymph, between tissue cells as interstitial fluid, and associated with the nervous system as cerebrospinal fluid (not illustrated).

Plasma membranes regulate fluid movements from one compartment to another.

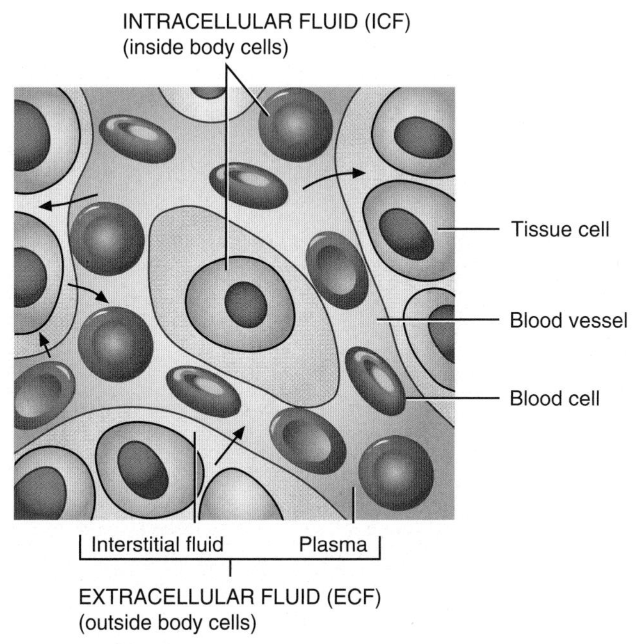

What is another name for intracellular fluid?

Passive Processes

Diffusion

Diffusion (di-FŪ-zhun; *diffus-* = spreading) is the movement of a substance from one place to another due to its kinetic energy. If a particular substance is present in high concentration in one area and in low concentration in another area, more particles of the substance diffuse from the region of high concentration to the region of low concentration than diffuse in the opposite direction. The diffusion of more molecules in one direction than the other is called *net* diffusion. Substances undergoing net diffusion move from a high to a low concentration, or *down their concentration gradient.* After some time, ***equilibrium*** (ē′-kwi-LIB-rē-um) is reached: the substance becomes evenly distributed throughout the solution and the concentration gradient disappears.

Placing a crystal of dye in a water-filled container provides an example of diffusion (Figure 3.5). At the beginning, the color is

Figure 3.5 Principle of diffusion. A crystal of dye placed in a cylinder of water dissolves (beginning), and there is net diffusion from the region of higher dye concentration to regions of lower dye concentration (intermediate). At equilibrium, the dye concentration is uniform throughout the solution.

In diffusion, a substance moves down its concentration gradient.

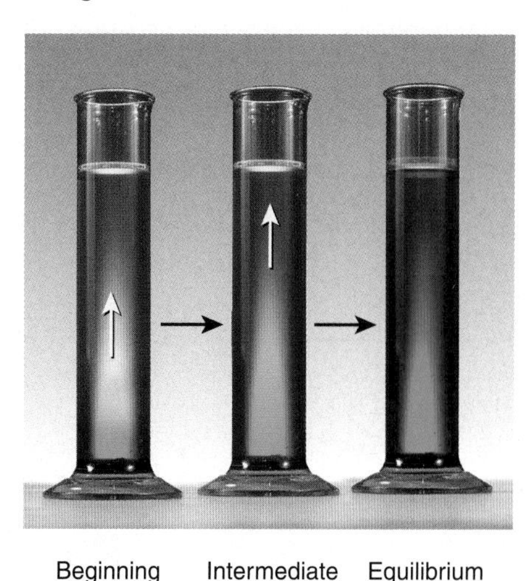

How would having a fever affect body processes that involve diffusion?

most intense just next to the crystal because the crystal is dissolving and the dye concentration is greatest there. At increasing distances, the color is lighter and lighter because the dye concentration is lower and lower. The dye molecules undergo net diffusion, down their concentration gradient, until they are evenly mixed in the water. At equilibrium the solution has a uniform color. In the example of dye diffusion, no membrane was involved. Substances may also diffuse across a membrane, if the membrane is permeable to them.

Now that you have a basic understanding of the nature of diffusion, we consider two types of diffusion: simple diffusion and facilitated diffusion.

In ***simple diffusion,*** substances diffuse across a membrane in one of two ways: lipid-soluble substances diffuse through the lipid bilayer (Figure 3.6). Substances that move across membranes by simple diffusion through the lipid bilayer include lipid-soluble molecules such as oxygen, carbon dioxide, and nitrogen gases; fatty acids, steroids, and fat-soluble vitamins (A, D, E, and K); glycerol; small alcohols; and ammonia. Water and urea (polar molecules) also move through the lipid bilayer. Simple diffusion through the lipid bilayer is important in the exchange of oxygen and carbon dioxide between blood and body cells and between blood and air within the lungs during breathing. It also is the transport method for absorption of lipid-soluble nutrients and release of some wastes from body cells.

Some substances that cannot move through the lipid bilayer do cross the plasma membrane by a passive process called ***facilitated diffusion.*** In this process, the integral membrane protein assists a specific substance to move across the membrane. The membrane protein can be either a membrane channel or a carrier.

In facilitated diffusion involving *ion channels,* ions move down their concentration gradients across the lipid bilayer. Most membrane channels are ion channels, which allow a specific type of ion to move across the membrane through the channel's pore. In typical plasma membranes, the most common ion channels are selective for K^+ (potassium ions) or Cl^- (chloride ions); fewer channels are available for Na^+ (sodium ions) or Ca^{2+} (calcium ions). Many ion channels are gated; that is, a portion of the channel protein acts as a "gate," moving in one direction to open the pore and in another direction to close it (Figure 3.7). When the gates are open, ions diffuse into or out of cells, down their concentration gradient. Gated channels are important for the production of electrical signals by body cells.

In facilitated diffusion involving a *carrier,* the substance binds to a specific carrier on one side of the membrane and is released on the other side after the carrier undergoes a change in shape (Figure 3.8).

Substances that move across plasma membranes by facilitated diffusion include glucose, fructose, galactose, and some vitamins. Glucose enters many body cells by facilitated diffusion as follows (Figure 3.8):

1. Glucose binds to a glucose carrier protein on the outside surface of the membrane.
2. As the carrier undergoes a change in shape, glucose passes through the membrane.
3. The carrier releases glucose on the other side of the membrane.

The selective permeability of the plasma membrane is often regulated to achieve homeostasis. For example, the hormone in-

Figure 3.6 Simple diffusion. Lipid-soluble molecules diffuse through the lipid bilayer.

In simple diffusion there is a net (greater) movement of substances from a region of their higher concentration to a region of their lower concentration.

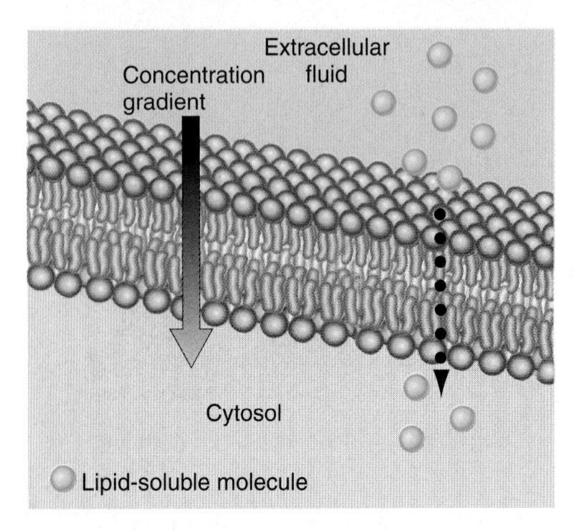

What are some examples of substances that diffuse through the lipid bilayer?

Figure 3.7 Facilitated diffusion of potassium ions (K^+) through a gated K^+ channel. A gated channel is one in which a portion of the channel protein acts as a gate to open or close the channel's pore to the passage of ions.

Ion channels are integral membrane proteins that allow specific small, inorganic ions to pass across the membrane.

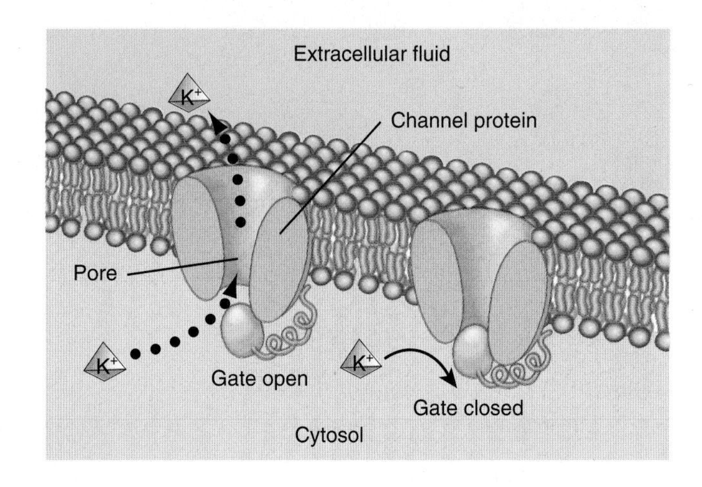

Details of the K^+ channel

Is the concentration of K^+ in body cells higher in the cytosol or in the extracellular fluid?

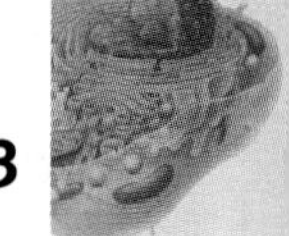

sulin promotes the insertion of more glucose carriers into the plasma membranes of certain cells. Thus, the effect of insulin is to increase entry of glucose into body cells by means of facilitated diffusion.

Osmosis

Osmosis (oz-MŌ-sis) is the net movement of a solvent through a selectively permeable membrane. Like diffusion, it is a passive process. In living systems, the solvent is water, which moves by osmosis across plasma membranes from an area of *higher water concentration* to an area of *lower water concentration.* Another way to understand this idea is to consider the solute concentration: In osmosis, water moves through a selectively permeable membrane from an area of *lower solute concentration* to an area of *higher solute concentration.* During osmosis, water molecules pass through a plasma membrane in two ways: (1) by moving through the lipid bilayer, as previously described, and (2) by moving through integral membrane proteins that function as water channels.

Osmosis occurs only when a membrane is permeable to water but is not permeable to certain solutes. A simple experiment can demonstrate osmosis. Consider a U-shaped tube in which a selectively permeable membrane separates the left and right arms of the tube (Figure 3.9). A volume of pure water is poured into the left arm, and the same volume of a solution containing a solute that cannot pass through the membrane is poured into the right arm (Figure 3.9a). Because the *water* concentration is higher on the left

Figure 3.8 Facilitated diffusion of glucose across a plasma membrane using a carrier. The carrier protein binds to glucose in the extracellular fluid and releases it into the cytosol.

Facilitated diffusion across a membrane involving a carrier is an important mechanism for transporting sugars such as glucose, fructose, and galactose into cells.

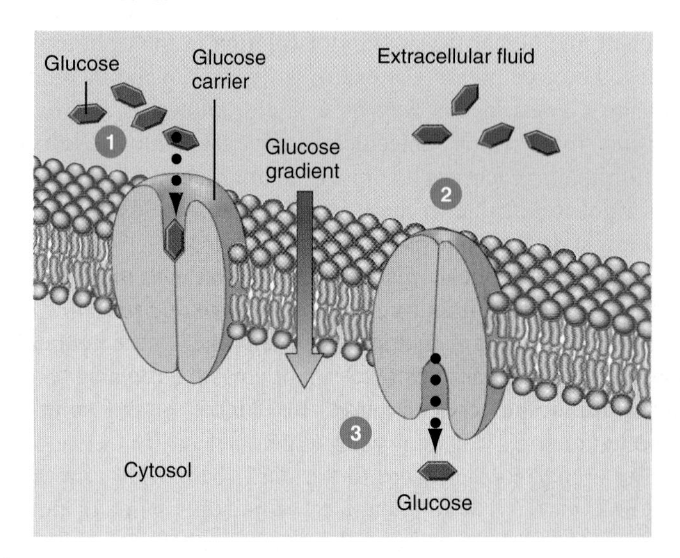

How does insulin alter glucose transport by facilitated diffusion?

Figure 3.9 Principle of osmosis. Water molecules move through the selectively permeable membrane; the solute molecules in the right arm cannot pass through the membrane. (a) As the experiment starts, water molecules move from the left arm into the right arm, down the water concentration gradient. (b) After some time, the volume of water in the left arm has decreased and the volume of solution in the right arm has increased. At equilibrium, there is no net osmosis: hydrostatic pressure forces just as many water molecules to move from right to left as osmosis forces water molecules to move from left to right. (c) If pressure is applied to the solution in the right arm, the starting conditions can be restored. This pressure, which stops osmosis, is equal to the osmotic pressure.

Osmosis in living systems is the movement of water molecules through a selectively permeable membrane.

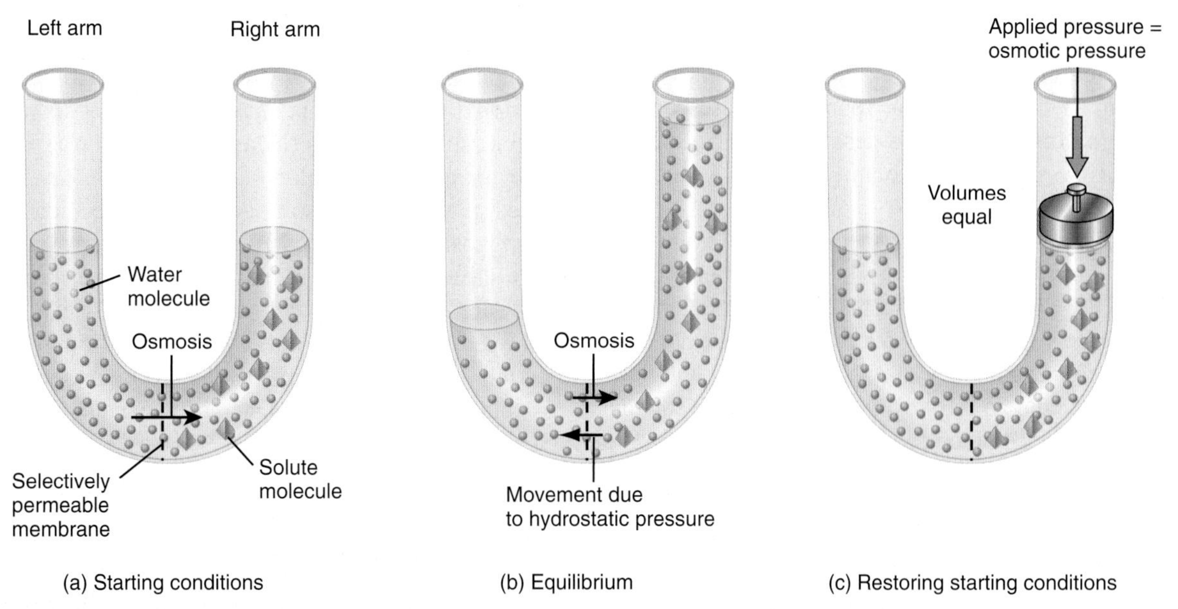

Will the fluid level in the right arm rise until the water concentrations are the same in both arms?

and lower on the right, net movement of water molecules—osmosis—occurs from left to right, so that the water is moving down its concentration gradient. At the same time, the membrane prevents diffusion of the solute from the right arm into the left arm. As a result, the volume of water in the left arm decreases, and the volume of solution in the right arm increases (see Figure 3.9b).

You might think that osmosis would continue until no water remained on the left side, but this is *not* what happens. In this experiment, the higher the column of solution in the right arm becomes, the more pressure it exerts on its side of the membrane. Pressure exerted in this way by a liquid, known as ***hydrostatic pressure,*** forces water molecules to move back into the left arm. Equilibrium is reached when just as many water molecules move from right to left due to the hydrostatic pressure as move from left to right due to osmosis (see Figure 3.9b).

To further complicate matters, the solution with the impermeable solute also exerts a force, called the ***osmotic pressure.*** The osmotic pressure of a solution is proportional to the concentration of the solute particles that cannot cross the membrane—the higher the solute concentration, the higher is the solution's osmotic pressure. Consider what would happen if a piston were used to apply more pressure to the fluid in the right arm of the tube in Figure 3.9. With enough pressure, the volume of fluid in each arm could be restored to the starting volume, and the concentration of solute in the right arm would be the same as it was at the beginning of the experiment (see Figure 3.9c). The amount of pressure needed to restore the starting condition equals the osmotic pressure. So, in our experiment osmotic pressure is the pressure needed to stop the movement of water from the left tube into the right tube. Notice that the osmotic pressure of a solution does not produce the movement of water during osmosis. Rather it is the pressure that would *prevent* such water movement.

Normally, the osmotic pressure of the cytosol is the same as the osmotic pressure of the interstitial fluid outside cells. Because the osmotic pressure on both sides of the plasma membrane (which is selectively permeable) is the same, cell volume remains relatively constant. When body cells are placed in a solution having a different osmotic pressure than cytosol, however, the shape and volume of the cells changes. As water moves by osmosis into or out of the cells, their volume increases or decreases. A solution's ***tonicity*** (*tonic-* = tension) is a measure of the solution's ability to change the volume of cells by altering their water content.

Any solution in which cells maintain their shape and volume is called an ***isotonic solution*** (ī′-sō-TON-ik; *iso-* = same) (Figure 3.10a). This is a solution in which the concentrations of solutes that cannot pass through the plasma membrane are the *same* on both sides. For example, a 0.9% NaCl (sodium chloride, or table salt) solution, called a *normal saline solution,* is isotonic for red blood cells. When red blood cells are bathed in 0.9% NaCl, water molecules enter and exit the cells at the same rate, allowing the red blood cells to maintain their normal shape and volume.

If red blood cells are placed in a ***hypotonic solution*** (hī′-pō-TON-ik; *hypo-* = less than), a solution that has a *lower* concentration of solutes (higher concentration of water) than the cytosol inside the red blood cells (Figure 3.10b), water molecules enter the cells by osmosis faster than they leave. This situation causes the red blood cells to swell and eventually to burst. Rupture of red blood cells is called ***hemolysis*** (hē-MOL-i-sis). A ***hypertonic solution*** (hī-per-TON-ik; *hyper-* = greater than) has a *higher* concentration of solutes (lower concentration of water) than does the cytosol inside red blood cells (Figure 3.10c). When cells are placed in a hypertonic solution, water molecules move out of the cells by osmosis faster than they enter, causing the cells to shrink. Such shrinkage of red blood cells is called ***crenation*** (kre-NĀ-shun). Red blood cells and other body cells may be damaged or destroyed if exposed to either hypertonic or hypotonic solutions. For this reason, most intravenous (IV) solutions, which are infused into the blood of patients via a vein, are isotonic.

Figure 3.10 Principle of osmosis applied to red blood cells (RBCs). (a) In an isotonic solution, there is no water gain or loss by osmosis. (b) In a hypotonic solution, the RBC gains water rapidly by osmosis until it bursts, an event called hemolysis. (c) In a hypertonic solution, the RBC loses water by osmosis so rapidly that it shrinks, a process called crenation. The arrows indicate the direction and degree of water movement into and out of the cells.

An isotonic solution is one in which cells maintain their normal shape and volume.

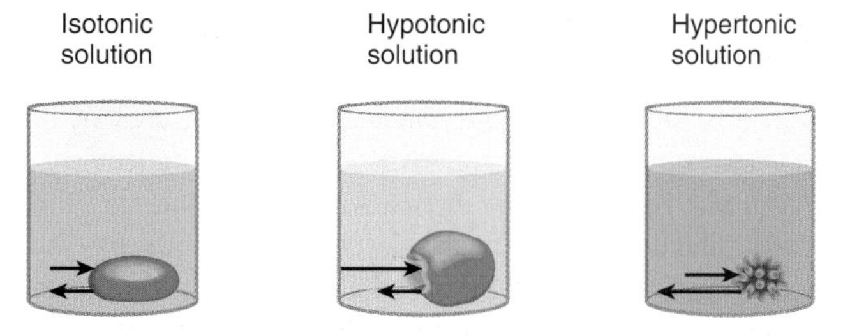

(a) Illustrations showing direction of water movement

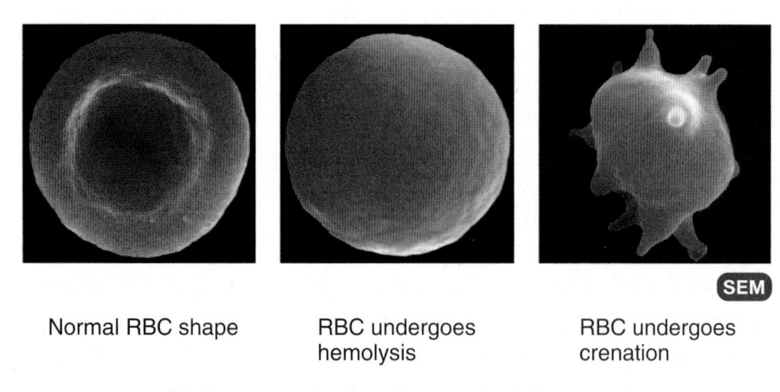

(b) Scanning electron micrographs (all 15,000x)

? Will a 2% solution of NaCl cause hemolysis or crenation of RBCs? Why?

Active Processes

Active Transport

In ***active transport,*** cellular energy is used to transport substances across the membrane against a concentration gradient (from an area of low to an area of high concentration).

Energy derived from splitting ATP changes the shape of a carrier protein, called a ***pump,*** which pumps a substance across a cellular membrane against its concentration gradient. A typical body cell expends about 40% of its ATP on active transport. Drugs that turn off ATP production, such as the poison cyanide, are lethal because they shut down active transport in cells throughout the body. Substances transported across the plasma membrane by active transport are mainly ions, primarily Na^+, K^+, H^+, Ca^{2+}, I^-, and Cl^-.

The most important active transport pump expels sodium ions (Na^+) from cells and brings in potassium ions (K^+). The pump protein also acts as an enzyme to split ATP. Because of the ions it moves, this pump is called the ***sodium–potassium (Na^+/K^+) pump.*** All cells have thousands of sodium–potassium pumps in their plasma membranes. These pumps maintain a low concentration of sodium ions in the cytosol by pumping Na^+ into the extracellular fluid against the Na^+ concentration gradient. At the same time, the pump moves potassium ions into cells against the K^+ concentration gradient. Because K^+ and Na^+ slowly leak back across the plasma membrane down their gradients, the sodium–potassium pumps must operate continually to maintain a low concentration of Na^+ and a high concentration of K^+ in the cytosol. These differing concentrations are crucial for osmotic balance of the two fluids and also for the ability of some cells to generate electrical signals such as action potentials. Figure 3.11 shows how the sodium–potassium pump operates.

1. Three sodium ions (Na^+) in the cytosol bind to the pump protein.
2. Na^+ binding triggers the splitting of ATP into ADP plus a phosphate group ((P)), which also becomes attached to the pump protein. This chemical reaction changes the shape of the pump protein, expelling the three Na^+ into the extracellular fluid. The changed shape of the pump protein then favors binding of two potassium ions (K^+) in the extracellular fluid to the pump protein.
3. The binding of K^+ causes the pump protein to release the phosphate group, which causes the pump protein to return to its original shape.
4. As the pump protein returns to its original shape, it releases the two K^+ into the cytosol. At this point, the pump is ready again to bind Na^+, and the cycle repeats.

Transport in Vesicles

A ***vesicle*** (VES-i-kul) is a small round sac formed by budding off from an existing membrane. Vesicles transport substances from one structure to another within cells, take in substances from extracellular fluid, and release substances into extracellular fluid. Movement of vesicles requires energy supplied by ATP. The two main types of transport in vesicles between a cell and the extracellular fluid that surrounds it are (1) ***endocytosis*** (en′-dō-sī-TŌ-sis; *endo-* = within), in which materials move *into* a cell in a vesicle formed from the plasma membrane, and (2) ***exocytosis*** (eks′-ō-sī-TŌ-sis; *exo-* = out), in which materials move *out of* a cell by the fusion of a vesicle formed inside a cell with the plasma membrane.

ENDOCYTOSIS Substances brought *into* the cell by endocytosis are surrounded by a piece of the plasma membrane, which buds off inside the cell to form a vesicle containing the ingested substances. The two types of endocytosis we consider are phagocytosis and bulk-phase endocytosis.

1. **Phagocytosis.** In ***phagocytosis*** (fag′-ō-sī-TŌ-sis; *phago-* = to eat), large solid particles, such as whole bacteria or viruses or

Figure 3.11 Operation of the sodium–potassium pump. Sodium ions (Na^+) are expelled from the cell, and potassium ions (K^+) are imported into the cell. The pump does not work unless Na^+ and ATP are present in the cytosol and K^+ is present in the extracellular fluid.

The sodium–potassium pump maintains a low intracellular concentration of Na^+.

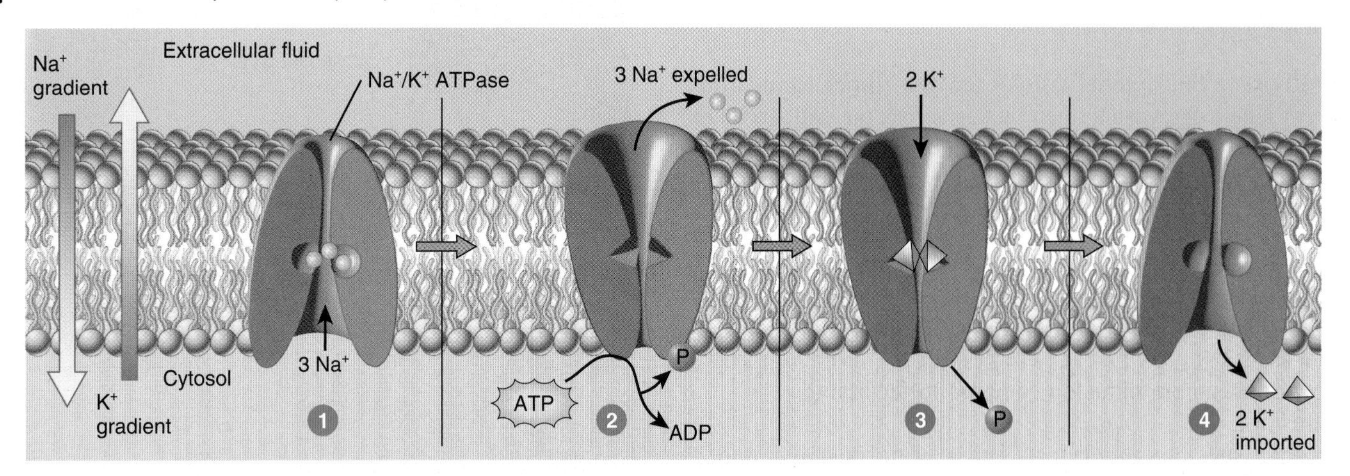

What is the role of ATP in the operation of this pump?

aged or dead cells, are taken in by the cell (Figure 3.12). Phagocytosis begins as the particle binds to a plasma membrane receptor, causing the cell to extend projections of its plasma membrane and cytoplasm, called ***pseudopods*** (SOO-dō-pods; *pseudo-* = false; *-pods* = feet). Two or more pseudopods surround the particle, and portions of their membranes fuse to form a vesicle called a *phagosome* that enters the cytoplasm. The vesicle fuses with one or more lysosomes, and lysosomal enzymes break down the ingested material. In most cases, any undigested materials remain indefinitely in a vesicle called a *residual body.* Phagocytosis occurs only in ***phagocytes,*** cells that are specialized to engulf and destroy bacteria and other foreign substances. Phagocytes include certain types of white blood cells and macrophages that are present in most body tissues. The process of phagocytosis is a vital defense mechanism that helps protect the body from disease.

2. Bulk-Phase Endocytosis (Pinocytosis). In ***bulk-phase endocytosis,*** cells take up vesicles containing tiny droplets of extracellular fluid. The process occurs in most body cells and takes in any and all solutes dissolved in the extracellular fluid. During bulk-phase endocytosis the plasma membrane folds inward and forms a vesicle containing a droplet of extracellular fluid. The vesicle detaches or "pinches off" from the plasma membrane and enters the cytosol. Within the cell, the vesicle fuses with a lysosome, where enzymes degrade the engulfed solutes. The resulting smaller molecules, such as amino acids and fatty acids, leave the lysosome and may be used elsewhere in the cell.

EXOCYTOSIS In contrast with endocytosis, which brings materials into a cell, exocytosis results in *secretion,* the liberation of materials from a cell. All cells carry out exocytosis, but it is especially important in two types of cells: (1) secretory cells that liberate digestive enzymes, hormones, mucus, or other secretions; and (2) nerve cells that release substances called *neurotransmitters* via exocytosis (see Figure 15.12). During exocytosis, membrane-enclosed vesicles called *secretory vesicles* form inside the cell, fuse with the plasma membrane, and release their contents into the extracellular fluid.

Segments of the plasma membrane lost through endocytosis are recovered or recycled by exocytosis. The balance between endocytosis and exocytosis keeps the surface area of a cell's plasma membrane relatively constant.

TRANSCYTOSIS Pinocytotic vesicles are sometimes formed on one side of a cell, the vesicle then moves across the cell, and the contents of the vesicle are released on the other side. Various proteins, such as some antibodies, are transported by transcytosis.

Table 3.1 summarizes the processes by which materials move into and out of cells.

CHECKPOINT

4. What is the key difference between passive and active processes?
5. How does simple diffusion compare to facilitated diffusion?
6. In what ways are endocytosis and exocytosis similar and different?

Figure 3.12 Phagocytosis. Pseudopods surround a particle and the membranes fuse to form a phagosome.

Phagocytosis is a vital defense mechanism that helps protect the body from disease.

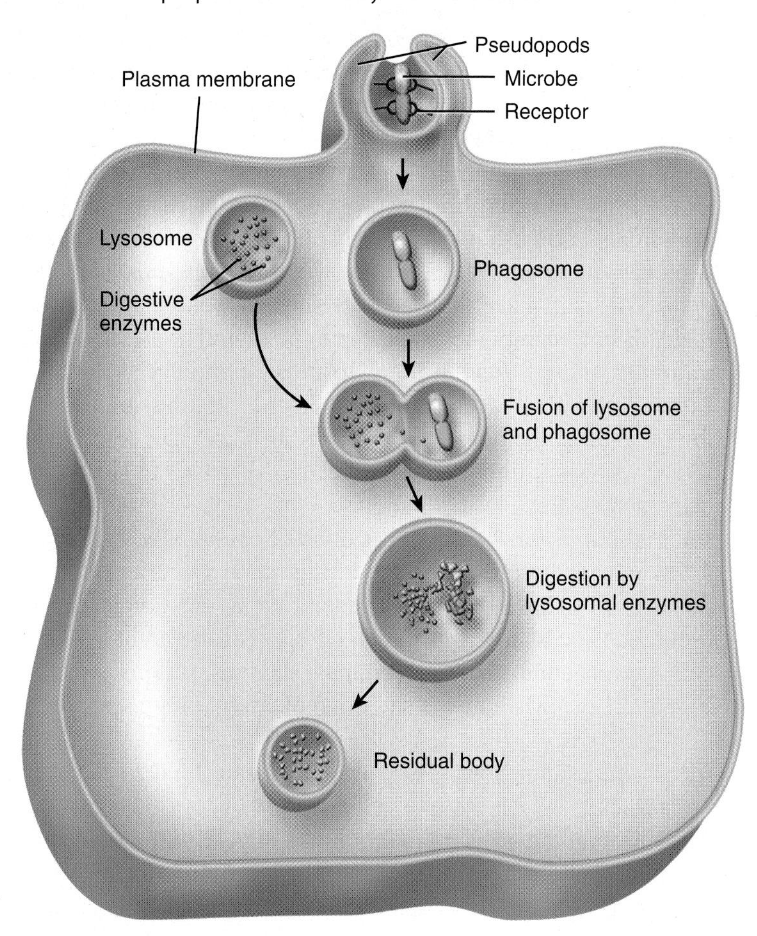

(a) Diagram of the process

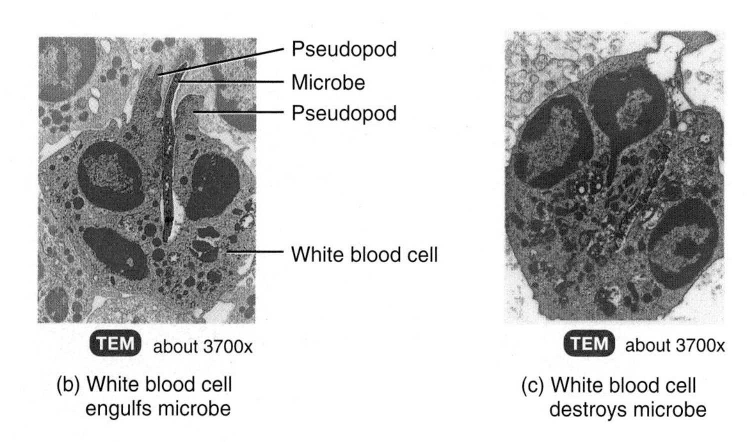

(b) White blood cell engulfs microbe

(c) White blood cell destroys microbe

What triggers pseudopod formation?

TABLE 3.1

Transport of Materials Into and Out of Cells

TRANSPORT PROCESS	DESCRIPTION	SUBSTANCES TRANSPORTED
Passive Processes	Movement of substances down a concentration gradient until equilibrium is reached; do not require cellular energy in the form of ATP.	
Diffusion	Movement of a substance by kinetic energy down a concentration gradient until equilibrium is reached.	
Simple diffusion	Passive movement of a substance through the lipid bilayer of the plasma membrane.	Lipid-soluble molecules: oxygen, carbon dioxide, and nitrogen gases; fatty acids, steroids, and fat-soluble vitamins (A, D, E, K); glycerol, small alcohols; ammonia. Polar molecules: water and urea.
Facilitated diffusion	Passive movement of a substance down its concentration gradient aided by ion channels and carriers.	K^+, Cl^-, Na^+, Ca^{2+}; glucose, fructose, galactose, and some vitamins.
Osmosis	Movement of water molecules across a selectively permeable membrane from an area of higher water concentration to an area of lower water concentration	Water.
Active Processes	Movement of substances against a concentration gradient; requires cellular energy in the form of ATP.	
Active Transport	Transport in which cell expends energy to move a substance across the membrane against its concentration gradient aided by membrane proteins that act as pumps; these integral membrane proteins use energy supplied by ATP.	Na^+, K^+, Ca^{2+}, H^+, I^-, Cl^-, and other ions.
Transport in Vesicles	Movement of substances into or out of a cell in vesicles that bud from the plasma membrane; requires energy supplied by ATP.	
Endocytosis	Movement of substances into a cell in vesicles.	
Phagocytosis	"Cell eating"; movement of a solid particle into a cell after pseudopods engulf it.	Bacteria, viruses, and aged or dead cells.
Bulk-phase endocytosis (Pinocytosis)	"Cell drinking"; movement of extracellular fluid into a cell by infolding of plasma membrane.	Solutes in extracellular fluid.
Exocytosis	Movement of substances out of a cell in secretory vesicles that fuse with the plasma membrane and release their contents into the extracellular fluid.	Neurotransmitters, hormones, and digestive enzymes.
Transcytosis	Movement of a substance through a cell as a result of endocytosis on one side of a cell and exocytosis on the opposite side.	Various proteins, such as some antibodies.

3.4 CYTOPLASM

OBJECTIVE

- Describe the structure and functions of cytoplasm, cytosol, and organelles.

Cytoplasm consists of all the cellular contents between the plasma membrane and the nucleus, and includes both cytosol and organelles.

Cytosol

The ***cytosol (intracellular fluid)*** is the liquid portion of the cytoplasm that surrounds organelles and accounts for about 55% of the total cell volume. Although cytosol varies in composition and consistency from one part of a cell to another, typically it is 75% to 90% water plus various dissolved solutes and suspended particles. Among these are various ions, glucose, amino acids, fatty acids, proteins, lipids, ATP, and waste products. Some cells also contain *lipid droplets* that contain triglycerides and *glycogen granules,* clusters of glycogen molecules. The cytosol is the site of many of the chemical reactions that maintain cell structures and allow cellular growth.

Organelles

Organelles are specialized structures inside cells that have characteristic shapes and specific functions. Each type of organelle is a functional compartment where specific processes take place, and each has its own unique set of enzymes.

The Cytoskeleton

Extending throughout the cytosol, the ***cytoskeleton*** is a network of three different types of protein filaments: microfilaments, intermediate filaments, and microtubules (Figure 3.13).

The thinnest elements of the cytoskeleton are the ***microfilaments*** (mī-krō-FIL-a-ments), which are concentrated at the periphery of a cell and contribute to the cell's strength and shape (Figure 3.13a). Microfilaments have two general functions: providing mechanical support and helping generate movements. They also anchor the cytoskeleton to integral proteins in the plasma membrane and provide support for microscopic, fingerlike projections of the plasma membrane called ***microvilli*** (mī′-krō-VIL-ī; *micro-* = small; *-villi* = tufts of hair; singular is *microvillus*). Because they greatly increase the surface area of the cell, microvilli are abundant on cells involved in absorption, such as the cells that line the small intestine. Some microfilaments extend beyond the plasma membrane and help cells attach to one another or to extracellular materials.

With respect to movement, microfilaments are involved in muscle contraction, cell division, and cell locomotion. Microfilament-assisted movements include the migration of embryonic cells during development, the invasion of tissues by white blood cells to fight infection, and the migration of skin cells during wound healing.

As their name suggests, ***intermediate filaments*** are thicker than microfilaments but thinner than microtubules (Figure 3.13b). They are found in parts of cells subject to tension (such as stretching), help hold organelles such as the nucleus in place, and help attach cells to one another.

The largest of the cytoskeletal components, ***microtubules*** (mī-krō-TOO-būls′) are long, hollow tubes (Figure 3.13c). Microtubules help determine cell shape and function in both the movement of organelles, such as secretory vesicles, within a cell and the migration of chromosomes during cell division. They also are responsible for movements of cilia and flagella.

Centrosome

The ***centrosome*** (SEN-trō-sōm), located near the nucleus, has two components—a pair of centrioles and pericentriolar material (Figure 3.14). The two *centrioles* are cylindrical structures, each of which is composed of nine clusters of three microtubules (a triplet) arranged in a circular pattern. Surrounding the centrioles is the *pericentriolar material* (per′-ē-sen′-trē-Ō-lar), containing hundreds of ring-shaped proteins called *tubulins*. The tubulins are the organizing centers for growth of the mitotic spindle, which plays a critical role in cell division, and for microtubule formation in nondividing cells.

Cilia and Flagella

Microtubules are the main structural and functional components of cilia and flagella, both of which are motile projections of the cell surface. ***Cilia*** (SIL-ē-a; singular is *cilium* = eyelash) are numerous, short, hairlike projections that extend from the surface of the cell (see Figure 3.1). Cilia propel fluids across the surfaces of cells that are firmly anchored in place. The coordinated movement of many cilia on the surface of a cell causes a steady movement of fluid along the cell's surface. Many cells of the respiratory tract, for example, have hundreds of cilia that help sweep foreign particles trapped in mucus away from the lungs. Their movement is paralyzed by nicotine in cigarette smoke. For this reason, smokers may cough often to remove foreign particles from their airways. Cells that line the uterine (fallopian) tubes also have cilia that sweep oocytes (egg cells) toward the uterus.

Figure 3.13 Cytoskeleton.

Extending throughout the cytosol, the cytoskeleton is a network of three kinds of protein filaments: microfilaments, intermediate filaments, and microtubules.

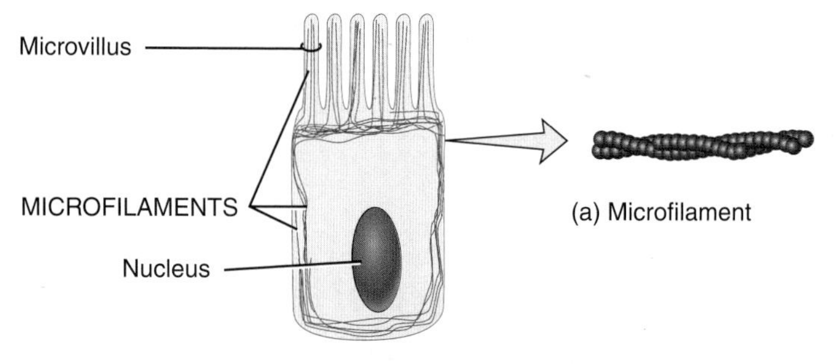

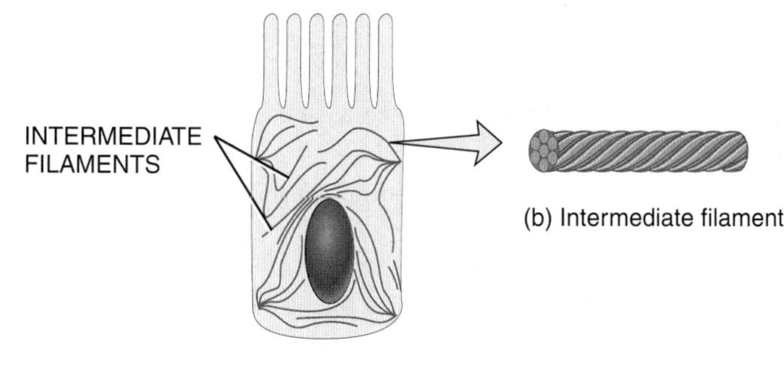

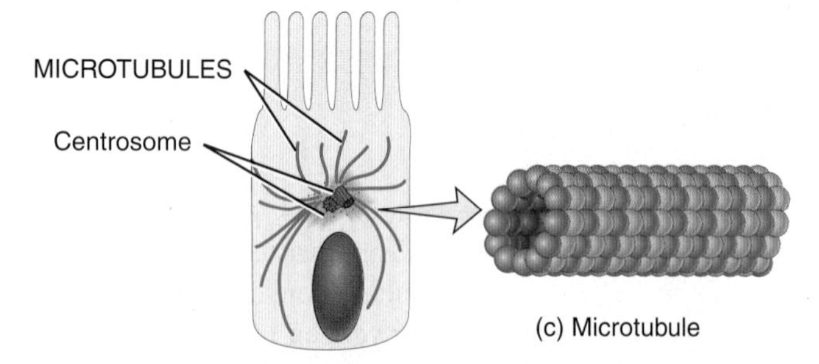

Function

1. Serves as a scaffold that helps to determine a cell's shape and to organize the cellular contents.
2. Aids movement of organelles within the cell, of chromosomes during cell division, and of whole cells such as phagocytes.

? Which cytoskeletal component helps form the structure of centrioles, cilia, and flagella?

Figure 3.14 Centrosome.

Located near the nucleus, the centrosome consists of a pair of centrioles and pericentriolar material.

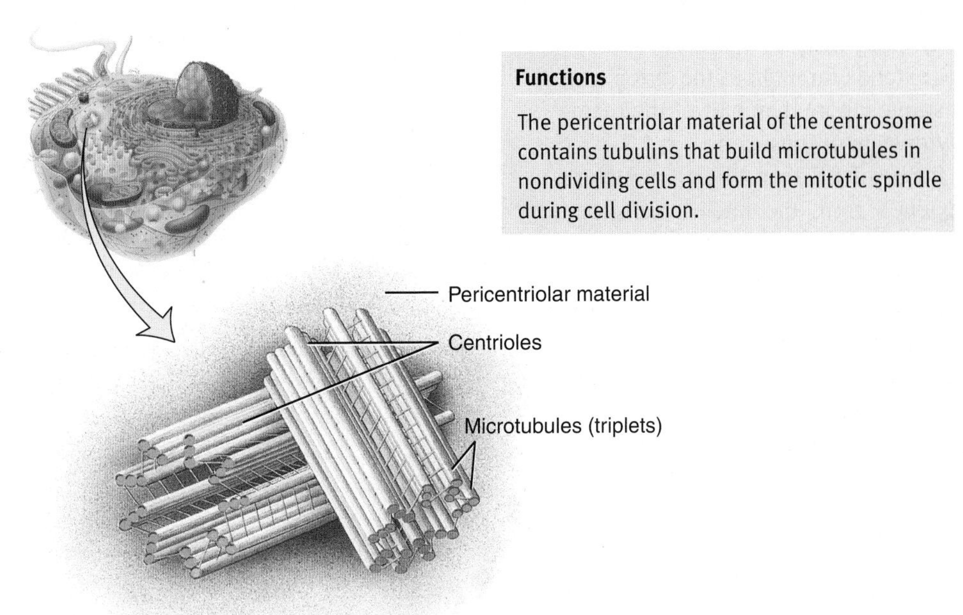

Functions

The pericentriolar material of the centrosome contains tubulins that build microtubules in nondividing cells and form the mitotic spindle during cell division.

? If you observed that a cell did not have a centrosome, what could you predict about its capacity for cell division?

Flagella (fla-JEL-a; singular is *flagellum* = whip) are similar in structure to cilia but are much longer (see Figure 3.1). Flagella usually move an entire cell. The only example of a flagellum in the human body is a sperm cell's tail, which propels the sperm toward its possible union with an oocyte.

Ribosomes

Ribosomes (RĪ-bō-sōms; *-somes* = bodies) are the sites of protein synthesis. Ribosomes are named for their high content of ribonucleic acid (RNA). Besides ribosomal RNA (rRNA), these tiny organelles contain ribosomal proteins. Structurally, a ribosome consists of two subunits, large and small, one about half the size of the other (Figure 3.15). The large and small subunits are made in the nucleolus of the nucleus. Later, they exit the nucleus and are assembled in the cytoplasm where they form a functional ribosome.

Some ribosomes are attached to the outer surface of the nuclear membrane and to an extensively folded membrane called the endoplasmic reticulum. These ribosomes synthesize proteins destined for specific organelles, for insertion in the plasma membrane, or for export from the cell. Other ribosomes are called free ribosomes because they are not attached to other cytoplasmic structures. Free ribosomes synthesize proteins used within the cytosol. Ribosomes are also located within mitochondria, where they synthesize mitochondrial proteins.

Figure 3.15 Ribosomes.

Ribosomes are the sites of protein synthesis.

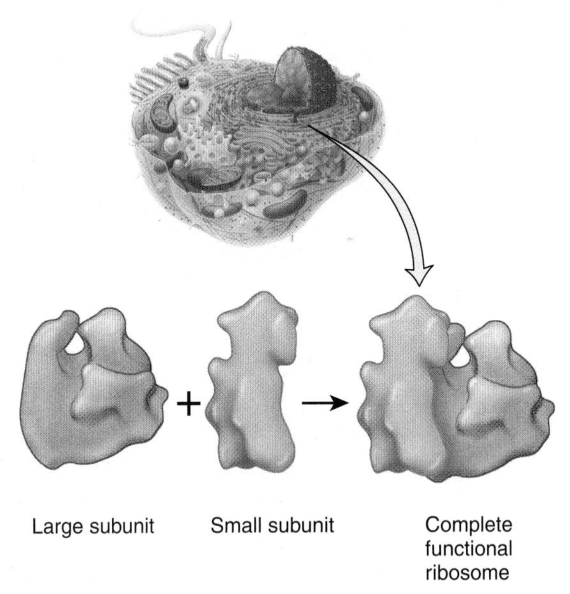

Functions

1. Ribosomes associated with endoplasmic reticulum synthesize proteins destined for insertion in the plasma membrane or secretion from the cell.
2. Free ribosomes synthesize proteins used in the cytosol.
3. Ribosomes within mitochondria synthesize mitochondrial proteins.

? Where are subunits of ribosomes synthesized and assembled?

Endoplasmic Reticulum

The ***endoplasmic reticulum*** (en′-dō-PLAS-mik re-TIK-ū-lum; *-plasmic* = cytoplasm; *reticulum* = network), or ***ER,*** is a network of folded membranes in the form of flattened sacs or tubules (Figure 3.16). The ER extends throughout the cytoplasm and is so extensive that it constitutes more than half of the membranous surfaces within the cytoplasm of most cells.

Cells contain two distinct forms of ER that differ in structure and function. ***Rough ER*** extends from the nuclear envelope (membrane around the nucleus) and appears "rough" because its outer surface is studded with ribosomes. Proteins synthesized by ribosomes attached to rough ER enter the spaces within the ER for processing and sorting. These molecules (glycoproteins and phospholipids) may be incorporated into organelle membranes or the plasma membrane. Thus, rough ER is essentially a "factory" for synthesizing secretory proteins and membrane molecules.

Figure 3.16 Endoplasmic reticulum.

The endoplasmic reticulum is a network of membrane-enclosed sacs or tubules that extend throughout the cytoplasm and connect to the nuclear envelope.

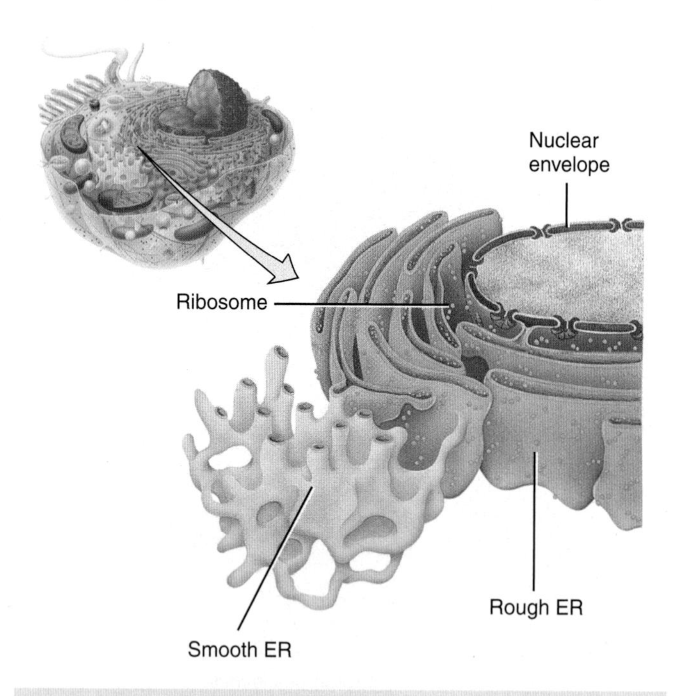

Functions of the Endoplasmic Reticulum

1. Rough ER synthesizes glycoproteins and phospholipids that are transferred into cellular organelles, inserted into the plasma membrane, or secreted during exocytosis.
2. Smooth ER synthesizes fatty acids and steroids, such as estrogens and testosterone; inactivates or detoxifies drugs and other potentially harmful substances; removes the phosphate group from glucose-6-phosphate; and stores and releases calcium ions that trigger contraction in muscle cells.

? What are the structural and functional differences between rough ER and smooth ER?

Smooth ER extends from the rough ER to form a network of membranous tubules (Figure 3.16). As you may already have guessed, smooth ER appears "smooth" because it lacks ribosomes. Smooth ER is where fatty acids and steroids, such as estrogens and testosterone, are synthesized. In liver cells, enzymes of the smooth ER also help release glucose into the bloodstream and inactivate or detoxify a variety of drugs and potentially harmful substances, including alcohol, pesticides, and carcinogens (cancer-causing agents). In muscle cells, calcium ions that are needed for muscle contraction are stored and released from a form of smooth ER called sarcoplasmic reticulum.

• CLINICAL CONNECTION | Smooth ER and Increased Drug Tolerance

One of the functions of smooth ER, as noted earlier, is to detoxify certain drugs. Individuals who repeatedly take such drugs, such as the sedative phenobarbital, develop changes in the smooth ER in their liver cells. Prolonged administration of phenobarbital results in changes in **smooth ER** that result in **increased drug tolerance** to the drug; the same dose no longer produces the same degree of sedation. With repeated exposure to the drug, the amount of smooth ER and its enzymes increases to protect the cell from its toxic effects. As the amounts of smooth ER increases, higher and higher dosages of the drug are needed to achieve the original effect. •

Golgi Complex

After proteins are synthesized on a ribosome attached to rough ER, most are transported to another region of the cell. The first step in the transport pathway is through an organelle called the ***Golgi complex*** (GOL-jē). It consists of 3 to 20 ***cisterns*** (SIS-terns = cavities), flattened membranous sacs with bulging edges, piled on each other like a stack of pita bread (Figure 3.17). Most cells have several Golgi complexes. The Golgi complex is more extensive in cells that secrete proteins.

The main function of the Golgi complex is to modify and package proteins (Figure 3.18). Proteins synthesized by ribosomes on rough ER enter the Golgi complex and are modified to form glycoproteins and lipoproteins. Then, they are sorted and packaged into vesicles. Some of the processed proteins are discharged from the cell by exocytosis. Certain cells of the pancreas release the hormone insulin this way. Other processed proteins become part of the plasma membrane as existing parts of the membrane are lost. Still other processed proteins become incorporated into organelles called lysosomes.

Lysosomes

Lysosomes (LĪ-sō-sōms; *lyso-* = dissolving; *-somes* = bodies) are membrane-enclosed vesicles that may contain as many as 60 different digestive enzymes that can break down a wide variety of molecules once the lysosome fuses with vesicles formed during endocytosis. The lysosomal membrane carries proteins that allow the final products of digestion, such as monosaccharides,

Figure 3.17 Golgi complex.

Most proteins synthesized by ribosomes attached to rough ER pass through the Golgi complex for processing.

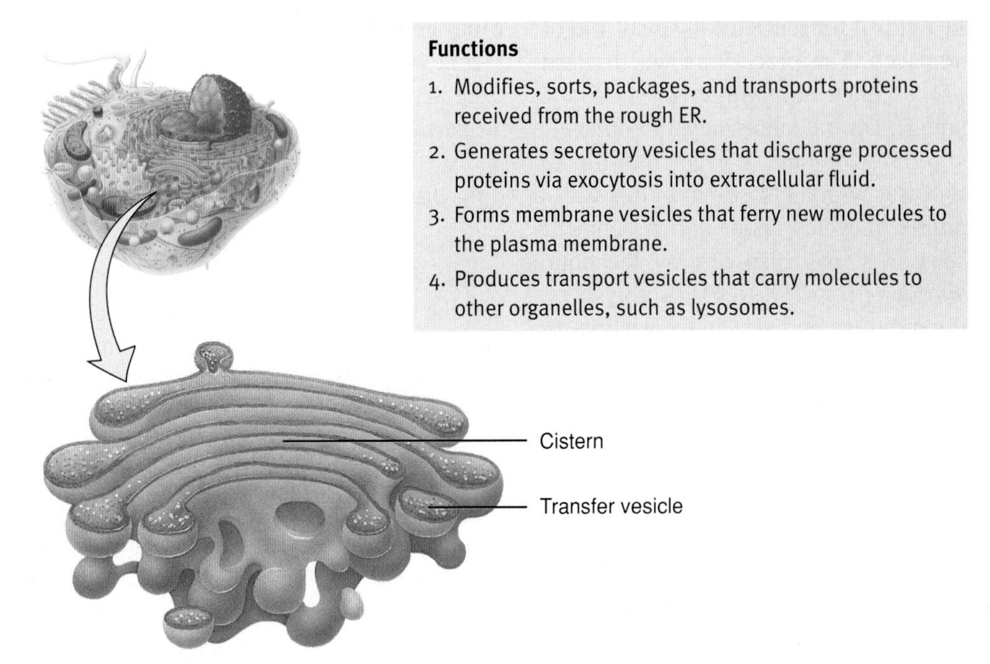

Functions
1. Modifies, sorts, packages, and transports proteins received from the rough ER.
2. Generates secretory vesicles that discharge processed proteins via exocytosis into extracellular fluid.
3. Forms membrane vesicles that ferry new molecules to the plasma membrane.
4. Produces transport vesicles that carry molecules to other organelles, such as lysosomes.
Cistern
Transfer vesicle

What types of body cells are likely to have extensive Golgi complexes?

Figure 3.18 Processing and packaging of proteins by the Golgi complex.

All proteins exported from the cell are processed in the Golgi complex.

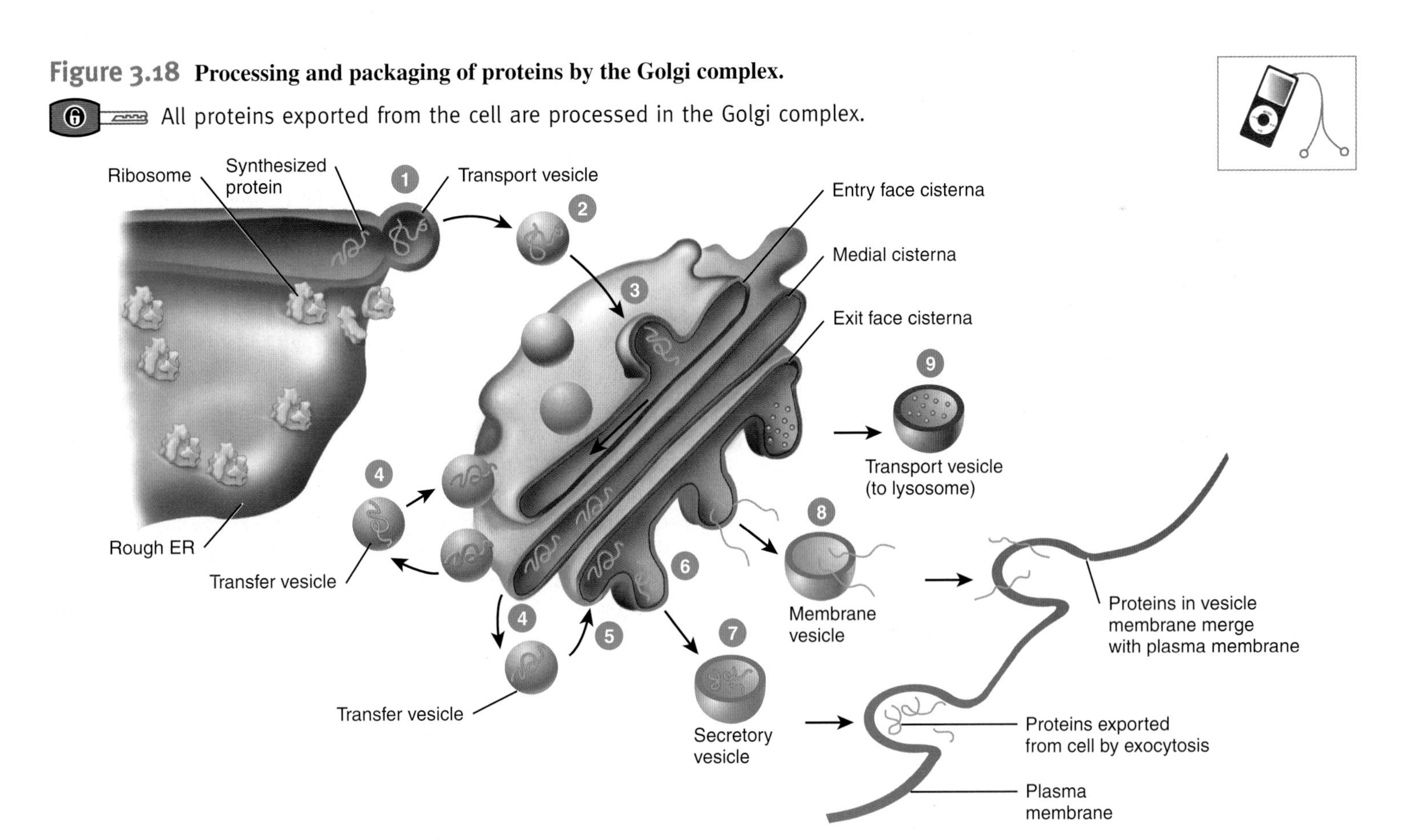

What are the three general destinations for proteins that leave the Golgi complex?

fatty acids, and amino acids, to be transported into the cytosol (Figure 3.19).

Lysosomal enzymes also help recycle worn-out structures. A lysosome can engulf another organelle, digest it, and return the digested components to the cytosol for reuse. In this way, old organelles are continually replaced. The process by which worn-out organelles are digested is called ***autophagy*** (aw-TOF-a-jē; *auto-* = self; *-phagy* = eating). During autophagy, the organelle to be digested is enclosed by a membrane derived from the ER to create a vesicle that then fuses with a lysosome. In this way, a human liver cell, for example, recycles about half its contents every week. Lysosomal enzymes may also destroy the entire cell, a process known as ***autolysis*** (aw-TOL-i-sis). Autolysis occurs in some pathological conditions and also is responsible for the tissue deterioration that occurs just after death.

Peroxisomes

Another group of organelles similar in structure to lysosomes, but smaller, are called ***peroxisomes*** (pe-ROK-si-sōms; *peroxi-* = peroxide; see Figure 3.1). Peroxisomes contain several *oxidases,* which are enzymes that can oxidize (remove hydrogen atoms from) various organic substances. For example, amino acids and fatty acids are oxidized in peroxisomes as part of normal metabolism. In addition, enzymes in peroxisomes oxidize toxic substances. Thus, peroxisomes are very abundant in the liver, where detoxification of alcohol and other damaging substances takes place. A by-product of the oxidation reactions is hydrogen peroxide (H_2O_2), a potentially toxic compound, and associated free radicals such as superoxide. However, peroxisomes also contain an enzyme called *catalase* that decomposes the H_2O_2. Because the generation and degradation of H_2O_2 occurs within the same organelle, peroxisomes protect other parts of the cell from the toxic effects of H_2O_2. Peroxisomes also contain enzymes that destroy superoxide.

Proteasomes

Although lysosomes degrade proteins delivered to them in vesicles, proteins in the cytosol also require disposal at certain times in the life of a cell. Continuous destruction of unneeded, damaged, or faulty proteins is the function of tiny barrel-shaped structures called ***proteasomes*** (PRŌ-tē-a-sōms = protein bodies; see Figure 3.1). A typical body cell contains many thousands of proteasomes, in both the cytosol and the nucleus. Proteasomes are so named because they contain myriad *proteases,* enzymes that cut (catabolize) proteins into small peptides. Once the enzymes of a proteasome have chopped up a protein into smaller chunks, other enzymes break down the peptides into amino acids, which can be recycled into new proteins.

Some diseases may result from failure of proteasomes to degrade abnormal proteins. For example, clumps of misfolded proteins accumulate in brain cells of people with Parkinson disease and Alzheimer disease. Discovering why the proteasomes fail to clear these abnormal proteins is a goal of ongoing research.

Mitochondria

Because they are the site of most ATP production, the "powerhouses" of a cell are its ***mitochondria*** (mī-tō-KON-drē-a; *mito-* = thread; *-chondria* = granules; singular is *mitochondrion*). A cell may have as few as a hundred or as many as several thousand mitochondria, depending on how active the cell is. For example, active cells in muscles, liver, and kidneys use ATP at a high rate and have large numbers of mitochondria. A mitochondrion consists of two membranes, each of which is similar in structure to the plasma membrane (Figure 3.20). The *outer mitochondrial membrane* is smooth, but the *inner mitochondrial membrane* is arranged in a series of folds called ***mitochondrial cristae*** (KRIS-tē; singular is *crista* = ridge). The large central, fluid-filled cavity of a mitochondrion, enclosed by the inner membrane and cristae, is the mitochondrial ***matrix.*** The elaborate folds of the cristae provide an enormous surface area for a series of chemical reactions (Figure 3.20) that provide most of a cell's ATP. Enzymes that catalyze these reactions are located in the matrix and on the cristae. Mitochondria also contain a small number of genes in mitochondrial DNA (mtDNA) and a few ribosomes, enabling them to synthesize some proteins. Additional mitochondria can thus be formed to meet the needs of cells that have become metabolically more active. Unlike nuclear DNA, mtDNA is basically unchanged in each generation and is transmitted only from the mother.

Figure 3.19 Lysosomes.

Lysosomes contain several types of powerful digestive enzymes.

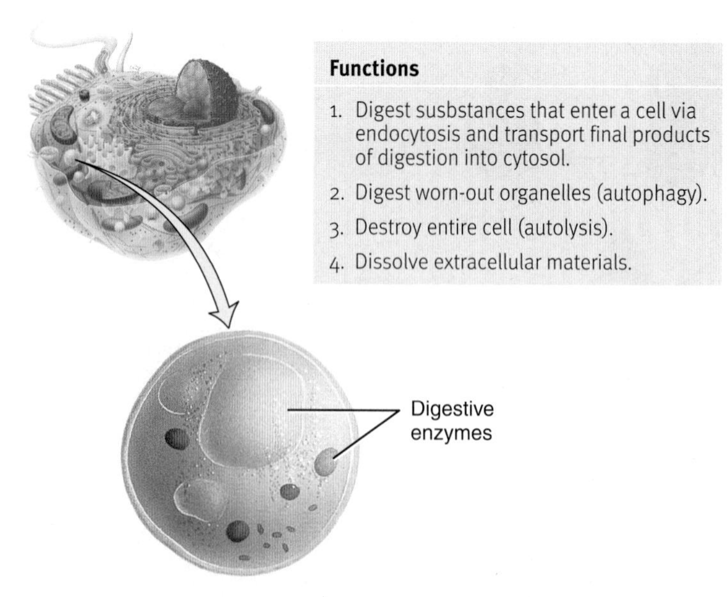

Functions
1. Digest susbstances that enter a cell via endocytosis and transport final products of digestion into cytosol.
2. Digest worn-out organelles (autophagy).
3. Destroy entire cell (autolysis).
4. Dissolve extracellular materials.
Digestive enzymes

What is the name of the process by which worn-out organelles are digested by lysosomes?

Figure 3.20 Mitochondria.

Within mitochondria, chemical reactions generate most of a cell's ATP.

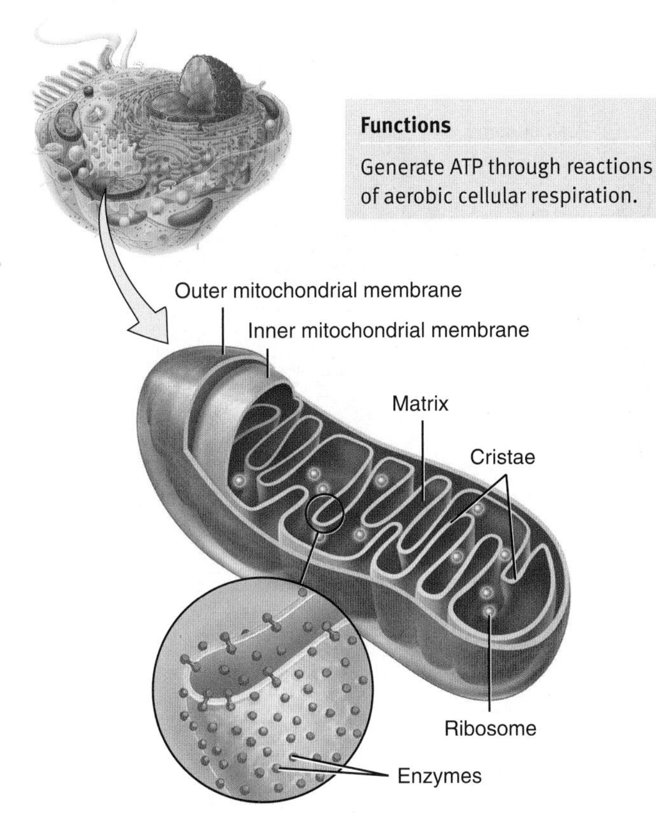

How do the cristae of a mitochondrion contribute to its ATP-producing function?

• CLINICAL CONNECTION Mitochondrial Cytopathies

Mitochondrial cytopathies (sī-TOP-a-thēz; *myo-* = muscle; *-path-* = disease) are rare inherited diseases that include more than 40 different identified genetic features. The common factor among these diseases is that the mitochondria are unable to produce sufficient molecules of ATP. The production of ATP requires hundreds of chemical reactions and each reaction must run perfectly to have a continuous supply of energy. When one or more components of these chemical reactions do not work correctly, there is an energy crisis and the cells cannot function normally. As a result, the incompletely burned food may accumulate as a toxin. Only one organ or system is affected in some patients, whereas in other patients a multitude of organs may be affected. The illness can range in severity from mild to fatal. Mitochondrial diseases affect muscles, cells of the brain, nerves (including visceral nerves), kidneys, heart, pancreas, eyes, or ears. The immune system can be compromised, which can lead to multiple problems. Because of the possibility of so many symptoms, mitochondrial disorders often escape correct diagnosis. In mitochondrial myopathies, muscles become weak and fatigue easily due to inadequate ATP production. One such example is *ophthalmoplegia* (of-thal-mō-PLĒ-jē-a; *ophthalmo-* = eye; *-plegia* = blow or strike), weakness or paralysis of eye muscles. Cyclist Greg Le Mond, a three-time Tour de France champion, had to stop competitive cycling due to a mitochondrial myopathy. •

The organelles just described are critical for the functioning of the cell and collectively for maintaining total body function.

CHECKPOINT

7. What does cytoplasm have that cytosol does not?
8. What is an organelle?
9. Describe the structure and function of ribosomes, the Golgi complex, and mitochondria.

3.5 NUCLEUS

OBJECTIVE

- Describe the structure and functions of the nucleus.

The ***nucleus*** is a spherical or oval structure that usually is the most prominent feature of a cell (see Figure 3.21). Most body cells have a single nucleus, although some, such as mature red blood cells, have none. In contrast, skeletal muscle cells and a few other types of cells have several to many nuclei. A double membrane called the ***nuclear envelope*** separates the nucleus from the cytoplasm. Both layers of the nuclear envelope are lipid bilayers similar to the plasma membrane. The outer membrane of the nuclear envelope is continuous with the rough endoplasmic reticulum and resembles it in structure. Many openings called ***nuclear pores*** pierce the nuclear envelope. Nuclear pores control the movement of substances between the nucleus and the cytoplasm.

Inside the nucleus are one or more spherical bodies called ***nucleoli*** (noo′-KLĒ-ō-lī; singular is *nucleolus*). These clusters of protein, DNA, and RNA are the sites of assembly of ribosomes, which exit the nucleus through the nuclear pores and participate in protein synthesis in the cytoplasm. Cells that synthesize large amounts of protein, such as muscle and liver cells, have more prominent nucleoli.

Also within the nucleus are most of the cell's hereditary units, called ***genes*** (JĒNS), which control cellular structure and direct most cellular activities (see Figure 3.23a). The nuclear genes are arranged along ***chromosomes*** (KRŌ-mō-sōms; *chromo-* = colored). Human somatic (body) cells have 46 chromosomes, 23 inherited from each parent. In a cell that is not dividing, the 46 chromosomes appear as a diffuse, granular mass, which is called ***chromatin*** (KRŌ-ma-tin). The total genetic information carried in a cell or organism is called its ***genome*** (JĒ-nōm).

In the last decade of the twentieth century, the genomes of humans, mice, fruit flies, and more than 50 microbes were sequenced. As a result, research in the field of ***genomics,*** the study of the relationships between the genome and the biological functions of an organism, has flourished. The Human Genome Project began in 1990 as an effort to sequence all of the nearly 3.2 billion nucleotides of our genome and was completed in April 2003. Scientists now know that the total number of genes in the human genome is about 30,000. Information regarding the human genome and how it is affected by the environment seeks to identify and discover the functions of the specific genes that play a role in genetic diseases. Genomic medicine also aims to design new drugs and to provide screening tests to enable physicians

Figure 3.21 Nucleus.

The nucleus contains most of the cell's genes, which are located on chromosomes.

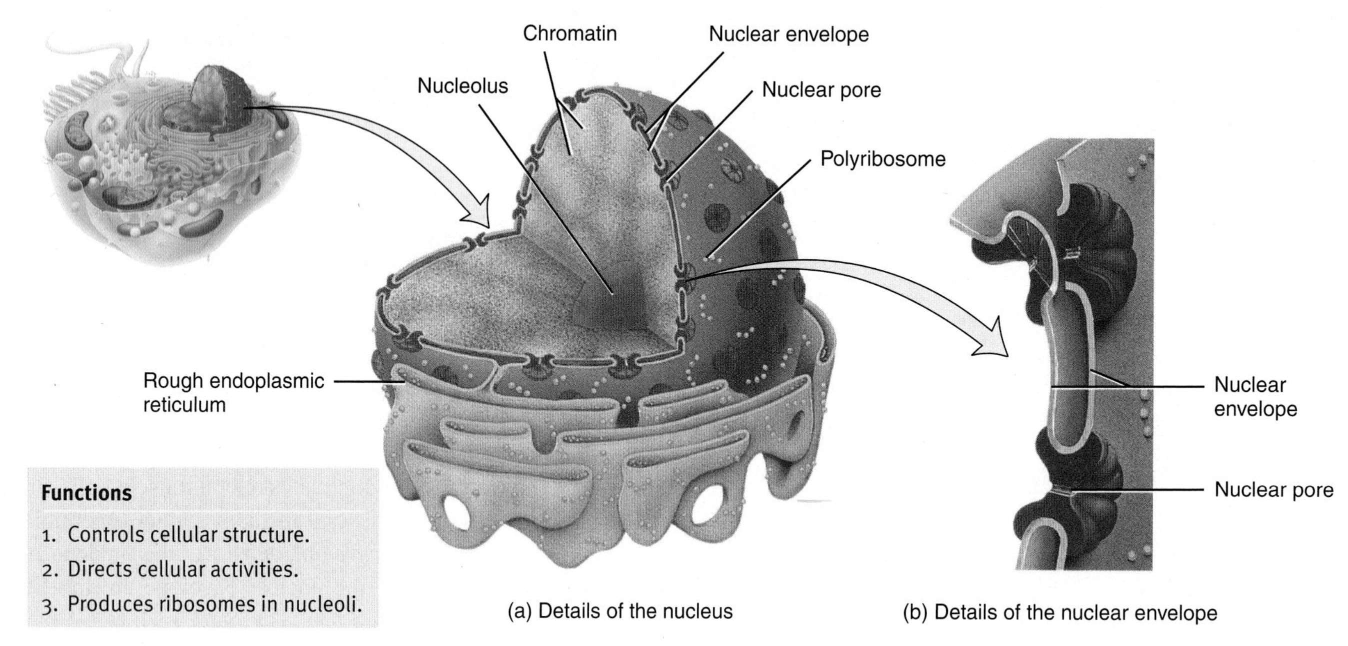

(a) Details of the nucleus

(b) Details of the nuclear envelope

What is chromatin?

to provide more effective counseling and treatment for disorders with significant genetic components such as hypertension (high blood pressure), obesity, diabetes, and cancer.

The main parts of a cell and their functions are summarized in Table 3.2.

CHECKPOINT

10. Why is the nucleus so important in the life of a cell?

3.6 GENE ACTION: PROTEIN SYNTHESIS

OBJECTIVE

- Outline the sequence of events involved in protein synthesis.

Although cells synthesize different substances to maintain homeostasis, much of the cellular machinery is devoted to protein production. Cells constantly synthesize large numbers of diverse proteins. The proteins, in turn, determine the physical and chemical characteristics of cells and, on a larger scale, of organisms.

Genes in DNA provide the instructions for making proteins. To synthesize a protein, the information contained in a specific region of DNA (a gene) is first *transcribed* (copied) to produce a specific molecule of RNA (ribonucleic acid). The RNA then attaches to a ribosome, where the information contained in the RNA is *translated* into a corresponding specific sequence of amino acids to form a new protein molecule (Figure 3.22).

Figure 3.22 Overview of gene expression. Synthesis of a specific protein requires transcription of a gene's DNA into RNA and translation of RNA into a corresponding sequence of amino acids.

Transcription occurs in the nucleus; translation occurs in the cytoplasm.

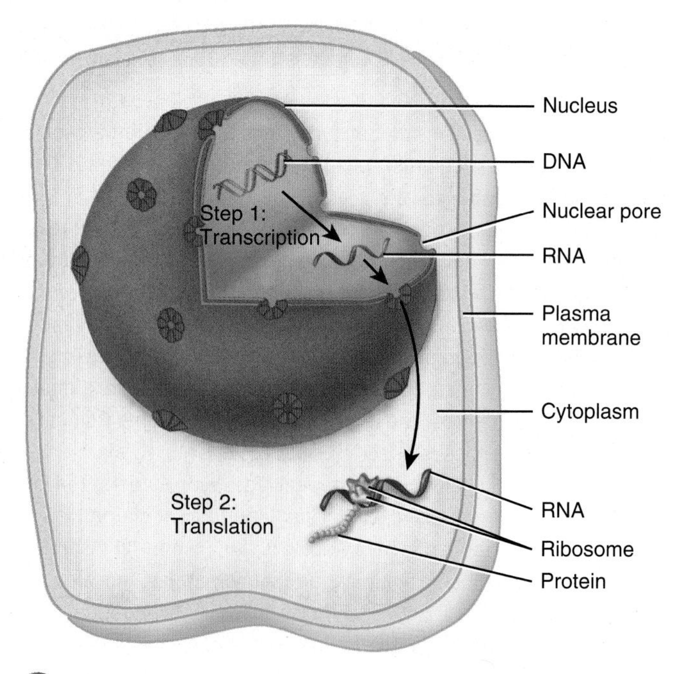

Why are proteins important in the life of a cell?

TABLE 3.2

Cell Parts and Their Functions

PART	DESCRIPTION	FUNCTIONS
PLASMA MEMBRANE	Composed of a lipid bilayer consisting of phospholipids, cholesterol, and glycolipids with various proteins inserted; surrounds cytoplasm.	Protects cellular contents; makes contact with other cells; contains channels, transporters, receptors, enzymes, and cell-identity markers; mediates the entry and exit of substances.
CYTOPLASM	Cellular contents between plasma membrane and nucleus, including cytosol and organelles.	Site of all intracellular activities except those occurring in the nucleus.
Cytosol	Composed of water, solutes, suspended particles, lipid droplets, and glycogen granules.	Liquid in which many of the cell's chemical reactions occur.
Organelles	Specialized cellular structures with characteristic shapes and specific functions.	Each organelle has one or more specific functions.
Cytoskeleton	Network composed of three protein filaments: microfilaments, intermediate filaments, and microtubules.	Maintains shape and general organization of cellular contents; responsible for cell movements.
Centrosome	Paired centrioles plus pericentriolar material.	Pericentriolar material is organizing center for microtubules and mitotic spindle.
Cilia and flagella	Motile cell-surface projections with inner core of microtubules.	Cilia move fluids over a cell's surface; a flagellum moves an entire cell.
Ribosome	Composed of two subunits containing ribosomal RNA and proteins; may be free in cytosol or attached to rough ER.	Protein synthesis.
Endoplasmic reticulum (ER)	Membranous network of folded membranes. Rough ER is studded with ribosomes and is attached to nuclear membrane; smooth ER lacks ribosomes.	Rough ER is site of synthesis of glycoproteins and phospholipids; smooth ER is site of fatty acid and steroid synthesis. Smooth ER also releases glucose into bloodstream, inactivates or detoxifies drugs and potentially harmful substances, and stores and releases calcium ions for muscle contraction.
Golgi complex	A stack of 3–20 flattened membranous sacs called cisterns.	Accepts proteins from rough ER; forms glycoproteins and lipoproteins; stores, packages, and exports proteins.
Lysosome	Vesicle formed from Golgi complex; contains digestive enzymes.	Fuses with and digests contents of vesicles; digests worn-out organelles (autophagy), entire cells (autolysis), and extracellular materials.
Peroxisome	Vesicle containing oxidative enzymes.	Detoxifies harmful substances, such as hydrogen peroxide and associated free radicals.
Proteasome	Tiny barrel-shaped structure that contains proteases, enzymes that cut proteins.	Degrades unneeded, damaged, or faulty proteins by cutting them into small peptides.
Mitochondrion	Consists of outer and inner membranes, cristae, and matrix.	Site of reactions that produce most of a cell's ATP.
NUCLEUS	Consists of nuclear envelope with pores, nucleoli, and chromatin (or chromosomes).	Contains genes, which control cellular structure and direct most cellular activities.

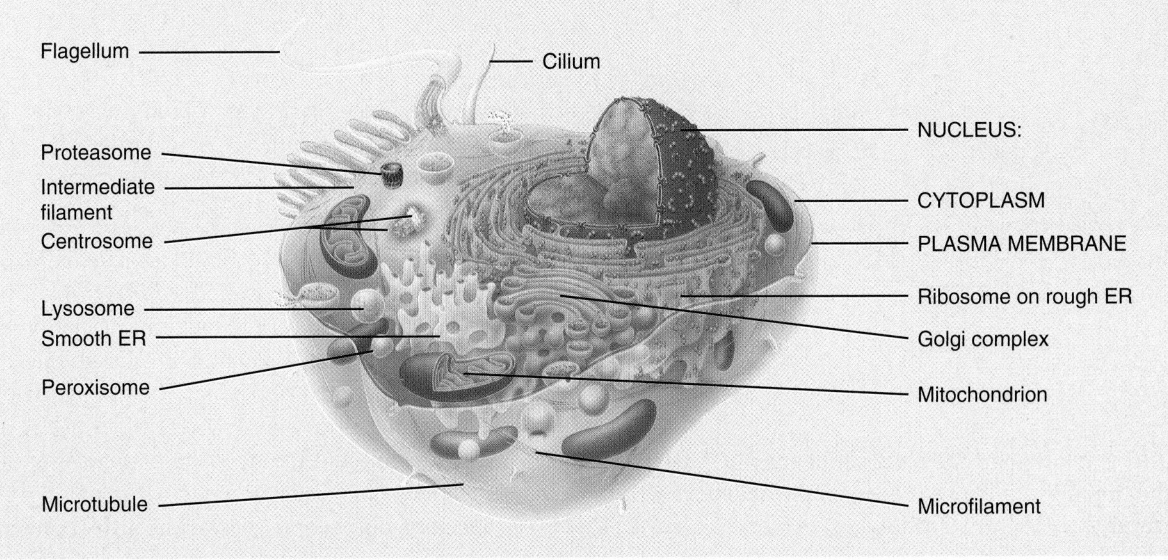

Information is stored in DNA by the sequencing of four ***nucleotides,*** the repeating units of nucleic acids (Figure 3.23). Each sequence of three DNA nucleotides is transcribed as a complementary (corresponding) sequence of three RNA nucleotides. Such a sequence of three successive DNA nucleotides is called a ***base triplet.*** The three successive RNA nucleotides are called a ***codon.*** When translated, a given codon specifies a particular amino acid.

Figure 3.23 Transcription. DNA transcription begins at a promoter and ends at a terminator.

During transcription, the genetic information in DNA is copied to RNA.

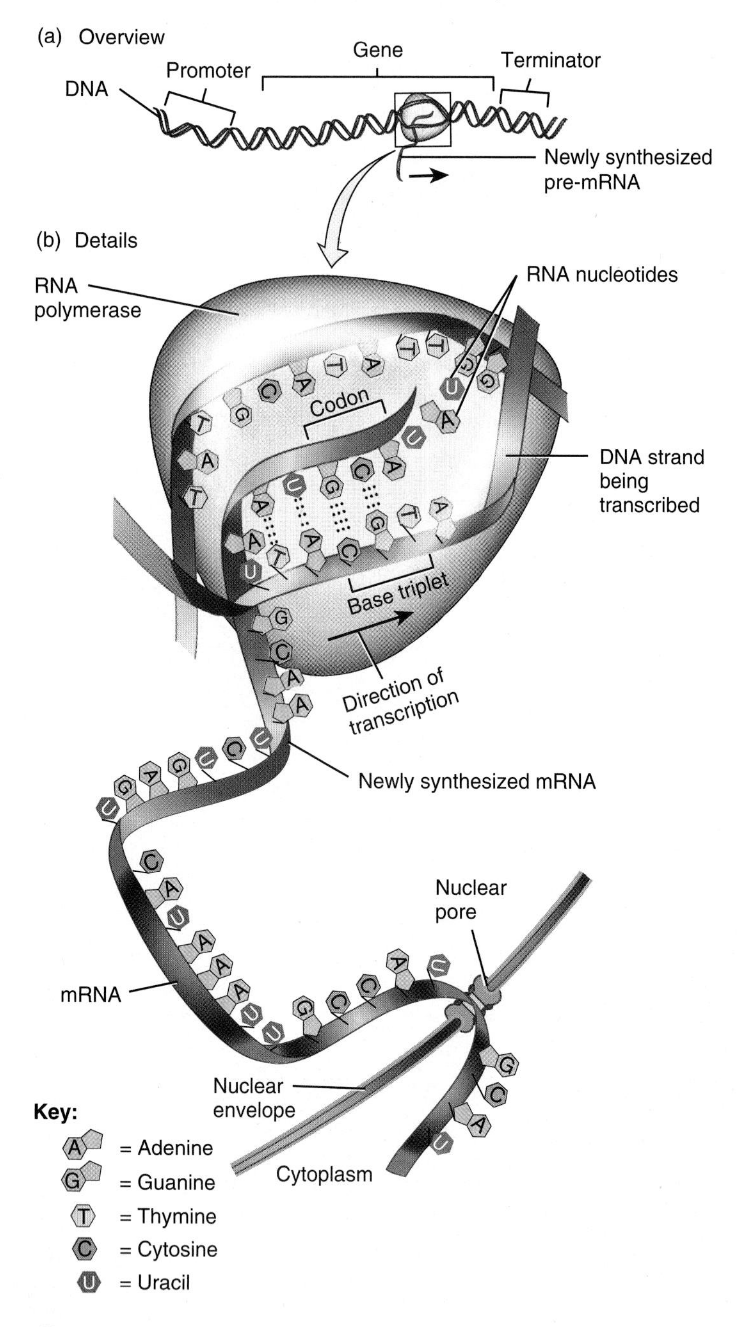

If the DNA template had the base sequence AGCT, what would be the mRNA base sequence, and what enzyme would catalyze DNA transcription?

Transcription

During ***transcription,*** which occurs in the nucleus, the genetic information in DNA base triplets is copied into a complementary sequence of codons in a strand of RNA. Transcription of DNA is catalyzed by the enzyme *RNA polymerase,* which must be instructed where to start the transcription process and where to end it. The segment of DNA where RNA polymerase attaches to it is a special sequence of nucleotides called a ***promoter,*** located near the beginning of a gene (Figure 3.23).

Three kinds of RNA are made from DNA:

- ***Messenger RNA (mRNA)*** directs synthesis of a protein.
- ***Ribosomal RNA (rRNA)*** joins with ribosomal proteins to make ribosomes.
- ***Transfer RNA (tRNA)*** binds to an amino acid and holds it in place on a ribosome until it is incorporated into a protein during translation. Each of the more than 20 different types of tRNA binds to only one of the 20 different amino acids.

During transcription, nucleotides pair in a complementary manner: The nitrogenous base cytosine (C) in DNA dictates the complementary nitrogenous base guanine (G) in the new RNA strand, a G in DNA dictates a C in RNA, a thymine (T) in DNA dictates an adenine (A) in RNA, and an A in DNA dictates a uracil (U) in RNA. As an example, if a segment of DNA had the base sequence ATGCAT, the newly transcribed RNA strand would have the complementary base sequence UACGUA (Figure 3.23).

Transcription of DNA ends at another special nucleotide sequence on DNA called a ***terminator,*** which specifies the end of the gene (Figure 3.23). Upon reaching the terminator, RNA polymerase detaches from the transcribed RNA molecule and the DNA strand. Once synthesized, mRNA, rRNA (in ribosomes), and tRNA leave the nucleus of the cell by passing through a nuclear pore. In the cytoplasm, they participate in the next step in protein synthesis, translation.

Translation

Translation is the process in which mRNA associates with ribosomes and directs synthesis of a protein by converting the sequence of nucleotides in mRNA into a specific sequence of amino acids. Translation occurs in the following way (Figure 3.24):

1. A molecule of mRNA binds to the small ribosomal subunit, and a special tRNA, called *initiator tRNA,* binds to the start codon (AUG) on mRNA, where translation begins.
2. The large ribosomal subunit attaches to the small subunit, creating a functional ribosome. The initiator tRNA fits into position on the ribosome. One end of a tRNA carries a specific amino acid, and the opposite end consists of a triplet of nucleotides called an ***anticodon.*** By pairing between complementary nitrogenous bases, the tRNA anticodon attaches to

Figure 3.24 Protein elongation and termination of protein synthesis during translation.

During protein synthesis the small and large ribosomal subunits join to form a functional ribosome. When the process is complete, they separate.

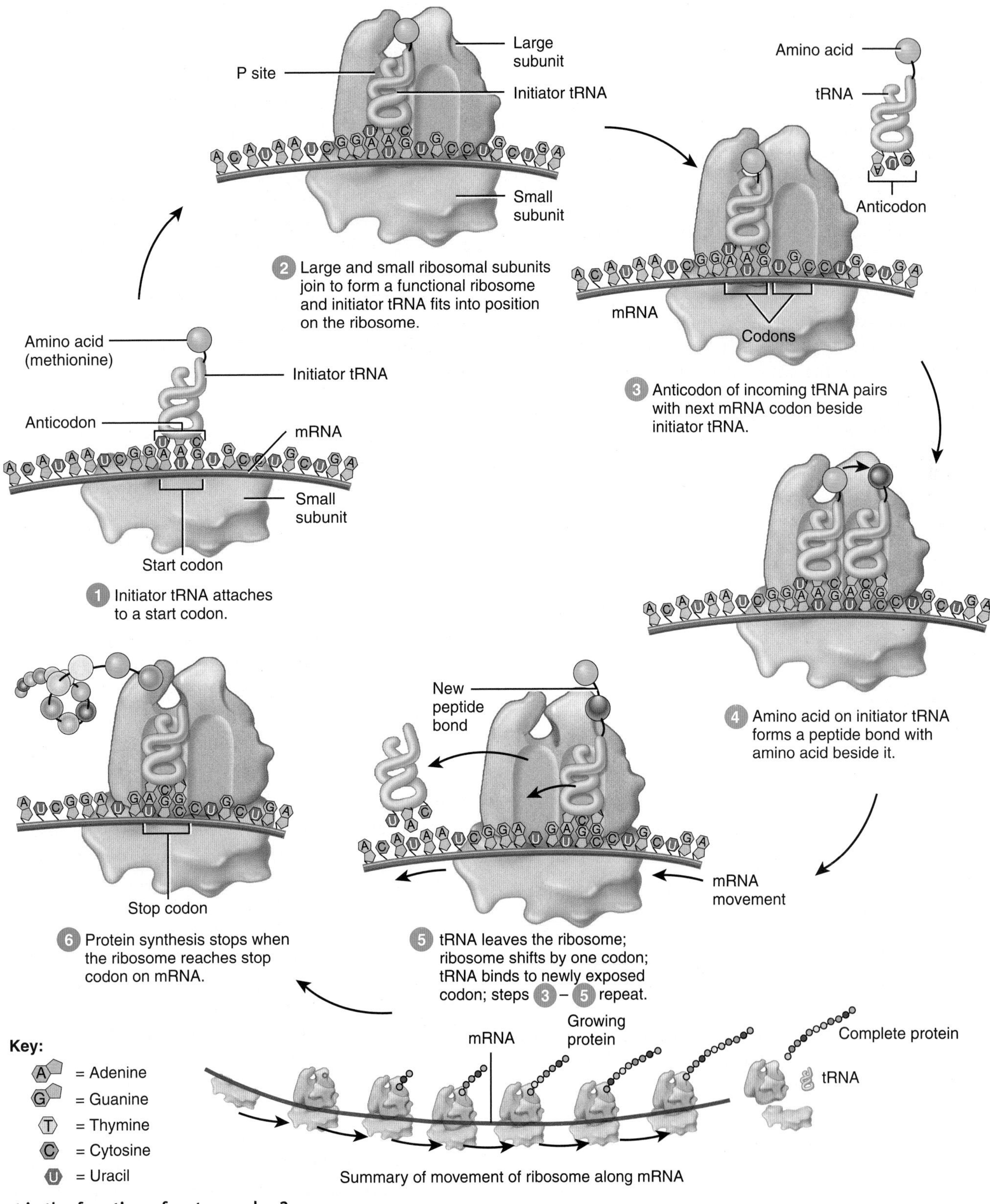

What is the function of a stop codon?

the mRNA codon. For example, if the mRNA codon is AUG, then a tRNA with the anticodon UAC would attach to it.

3. The anticodon of another tRNA with its amino acid attaches to the complementary mRNA codon next to the initiator tRNA.
4. A peptide bond is formed between the amino acids carried by the initiator tRNA and the tRNA next to it.
5. After the peptide bond forms, the empty tRNA detaches from the ribosome, and the ribosome shifts the mRNA strand by one codon. As the tRNA bearing the newly forming protein shifts, another tRNA with its amino acid binds to a newly exposed codon. Steps 3 through 5 repeat again and again as the protein lengthens.
6. Protein synthesis ends when the ribosome reaches a stop codon, at which time the completed protein detaches from the final tRNA. When the tRNA vacates the ribosome, the ribosome splits into its large and small subunits.

Protein synthesis progresses at a rate of about 15 amino acids per second. As the ribosome moves along the mRNA and before it completes synthesis of the whole protein, another ribosome may attach behind it and begin translation of the same mRNA strand. In this way, several ribosomes may be attached to the same mRNA. Such a group of ribosomes is called a ***polyribosome.*** The simultaneous movement of several ribosomes along the same mRNA strand permits a large amount of protein to be produced from each mRNA.

CHECKPOINT

11. Define protein synthesis.
12. Distinguish between transcription and translation.

3.7 SOMATIC CELL DIVISION

OBJECTIVE

- Discuss the stages, events, and significance of somatic cell division.

As most body cells become damaged, diseased, or worn out, they are replaced by ***cell division,*** the process whereby cells reproduce themselves. The two types of cell division are reproductive cell division and somatic cell division. ***Reproductive cell division*** or ***meiosis*** is the process that produces gametes—sperm and oocytes—the cells needed to form the next generation of sexually reproducing organisms. This is described in Chapter 29; here we focus on somatic cell division.

All body cells, except those that produce gametes, are called ***somatic cells*** (sō-MAT-ik; *soma* = body). In ***somatic cell division (mitosis),*** a cell divides into two identical cells. An important part of somatic cell division is replication (duplication) of the DNA sequences that make up genes and chromosomes so that the same genetic material can be passed on to the newly formed cells. After somatic cell division, each newly formed cell has the same number of chromosomes as the original cell. Because somatic cells contain two sets of chromosomes, they are called ***diploid cells*** (DIP-loyd; *dipl-* = double; *-oid* = form). Geneticists use the symbol n to denote the number of different chromosomes in an organism; in humans, $n = 23$. Diploid cells are $2n$. Somatic cell division replaces dead or injured cells and adds new ones for tissue growth. For example, skin cells are continually replaced by somatic cell divisions.

The ***cell cycle*** depicts the sequence of changes that a cell undergoes from the time it forms until it replicates its contents and divides into two cells. In somatic cells, the cell cycle consists of two major periods: interphase, when a cell is not dividing, and the mitotic (M) phase, when a cell is dividing.

Interphase

During ***interphase*** (IN-ter-fāz) the cell replicates its DNA and manufactures additional organelles and cytosolic components, such as centrosomes, prior to cell division. Interphase is a state of high metabolic activity and cell growth.

A microscopic view of a cell during interphase shows a clearly defined nuclear envelope, a nucleolus, and a tangled mass of chromatin (Figure 3.25a). The mitotic phase begins once a cell completes replication of DNA and other activities of interphase.

Mitotic Phase

The ***mitotic phase*** (mī-TOT-ik) of the cell cycle consists of *mitosis,* division of the nucleus, followed by *cytokinesis,* division of the cytoplasm into two cells. The events that take place during mitosis and cytokinesis are plainly visible under a microscope because chromatin condenses into chromosomes.

Nuclear Division: Mitosis

During ***mitosis*** (mī-TŌ-sis; *mitos* = thread), the duplicated chromosomes become exactly segregated, one set into each of two separate nuclei. For clarity, biologists divide the process into four stages: prophase, metaphase, anaphase, and telophase. However, mitosis is a continuous process, with one stage merging imperceptibly into the next.

- ***Prophase.*** During early prophase (PRŌ-fāz), the chromatin fibers condense and shorten into chromosomes that are visible under the light microscope (Figure 3.25b). The condensation process may prevent entangling of the long DNA strands as they move during mitosis. Recall that DNA replication took place during interphase. Thus, each prophase chromosome consists of a pair of identical, double-stranded chromatids. A constricted region of the chromosome, called a ***centromere*** (sen-trō-mer), holds the chromatid pair together. Later in prophase, the pericentriolar material of the two centrosomes starts to form the mitotic spindle, a football-shaped assembly of microtubules (Figure 3.25b). Lengthening of the microtubules between centrosomes pushes the centrosomes to opposite poles (ends) of the cell. Finally, the spindle extends from pole to pole. Then the nucleolus and nuclear envelope break down.
- ***Metaphase.*** During metaphase (MET-a-fāz), the centromeres of the chromatid pairs are aligned along the microtubules of the mitotic spindle at the exact center of the mitotic spindle (Figure 3.25c). This midpoint region is called the ***metaphase plate.***

Figure 3.25 Cell division: mitosis and cytokinesis. Begin the sequence at (a) at the top of the figure and read clockwise to complete the process.

In somatic cell division, a single diploid cell divides to produce two identical diploid cells.

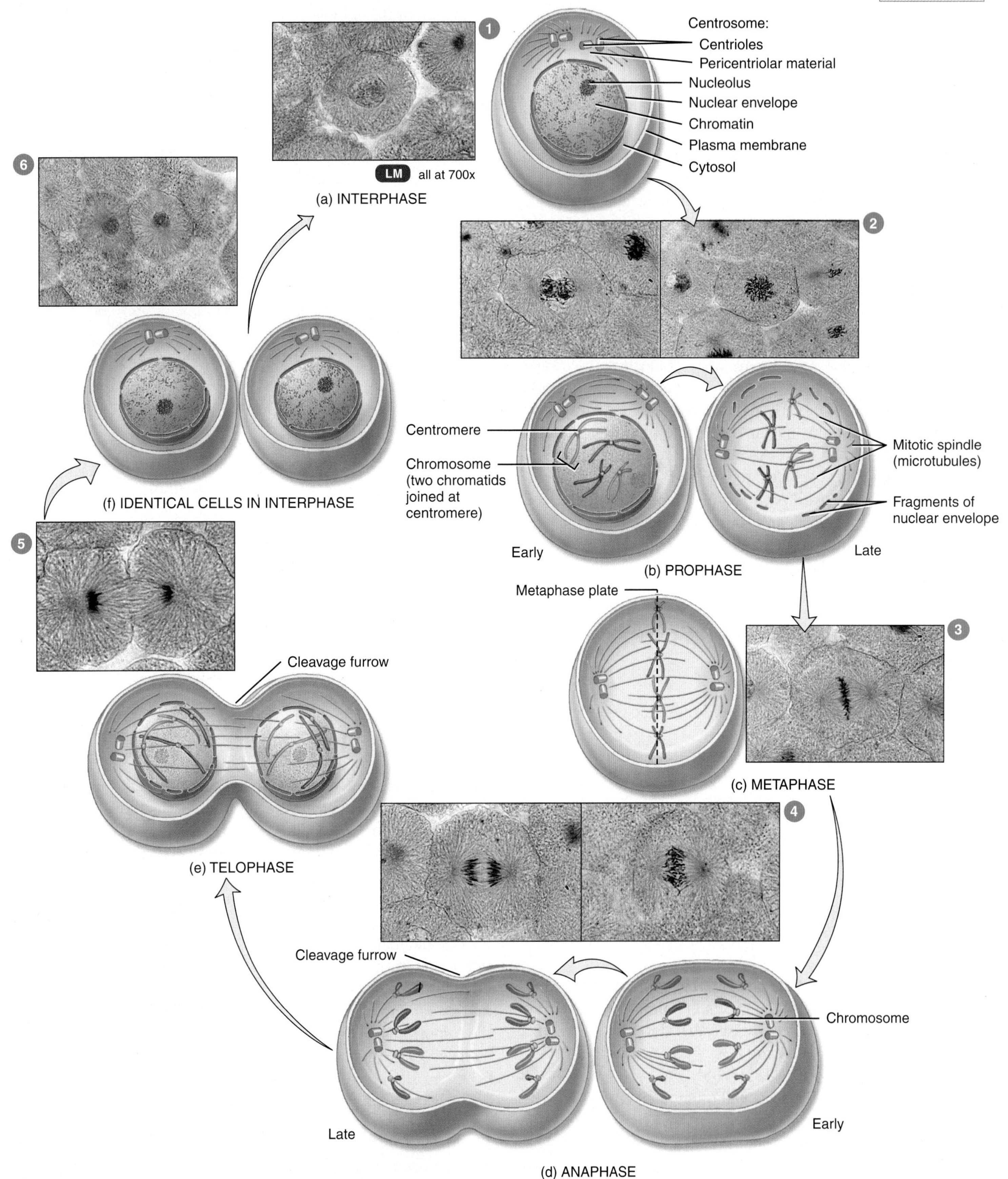

During which phase of mitosis does cytokinesis begin?

- ***Anaphase.*** During anaphase (AN-a-fāz), the centromeres split, separating the two members of each chromatid pair, which move to opposite poles of the cell (see Figure 3.25d). Once separated, the chromatids are called chromosomes. As the chromosomes are pulled by the microtubules of the mitotic spindle during anaphase, they appear V-shaped because the centromeres lead the way and seem to drag the trailing arms of the chromosomes toward the pole.
- ***Telophase.*** The final stage of mitosis, telophase (TEL-ō-fāz), begins after chromosomal movement stops (see Figure 3.25e). The identical sets of chromosomes, now at opposite poles of the cell, uncoil and revert to the threadlike chromatin form. A new nuclear envelope forms around each chromatin mass, nucleoli appear, and eventually the mitotic spindle breaks up.

Cytoplasmic Division: Cytokinesis

Division of a cell's cytoplasm and organelles is called ***cytokinesis*** (sī′-tō-ki-NĒ-sis; *-kinesis* = motion). This process begins late in anaphase or early in telophase with formation of a ***cleavage furrow,*** a slight indentation of the plasma membrane that extends around the center of the cell (see Figure 3.25d,e). Microfilaments in the cleavage furrow pull the plasma membrane progressively inward, constricting the center of the cell like a belt around a waist, and ultimately pinching it in two. After cytokinesis there are two new and separate cells, each with equal portions of cytoplasm and organelles and identical sets of chromosomes. When cytokinesis is complete, interphase begins (see Figure 3.25f).

Table 3.3 summarizes the events of the cell cycle in somatic cell division.

CHECKPOINT

13. Distinguish between somatic and reproductive cell division. Why is each important?
14. What are the major events of each stage of the mitotic phase?

TABLE 3.3

Events of the Somatic Cell Cycle

PHASE	ACTIVITY
Interphase	Period between cell divisions when the chromosomes are not visible under a light microscope. Metabolically active cell duplicates organelles and cytosolic components; replication of DNA and centrosome; cell growth; enzyme and protein synthesis.
Mitotic phase	Parent cell produces identical cells with identical chromosomes; chromosomes visible under light microscope.
Mitosis	Nuclear division; distribution of two sets of chromosomes into separate nuclei.
Prophase	Chromatin fibers condense into paired chromatids; nucleolus and nuclear envelope disappear; each centrosome moves to an opposite pole of the cell.
Metaphase	Centromeres of chromatid pairs line up at metaphase plate.
Anaphase	Centromeres split; identical sets of chromosomes move to opposite poles of cell.
Telophase	Nuclear envelopes and nucleoli reappear; chromosomes resume chromatin form; mitotic spindle disappears.
Cytokinesis	Cytoplasmic division; contractile ring forms cleavage furrow around center of cell, dividing cytoplasm into separate and equal portions.

3.8 CANCER

OBJECTIVE

- Describe the process of carcinogenesis and possible treatments for cancer.

Cancer is a group of diseases characterized by uncontrolled or abnormal cell proliferation. When cells in a part of the body divide without control, the excess tissue that develops is called a ***tumor*** or ***neoplasm*** (NĒ-ō-plazm; *neo-* = new). The study of tumors is called ***oncology*** (on-KOL-ō-jē; *onco-* = swelling or mass). Tumors may be cancerous and often fatal, or they may be harmless. A cancerous neoplasm is called a ***malignant tumor*** or ***malignancy.*** One property of most malignant tumors is their ability to undergo ***metastasis*** (me-TAS-ta-sis), the spread of cancerous cells to other parts of the body. A ***benign tumor*** is a neoplasm that does not metastasize. An example is a wart. Most benign tumors may be surgically removed if they interfere with normal body function or become disfiguring. Some tumors are inoperable and can be fatal.

Growth and Spread of Cancer

Cells of malignant tumors duplicate rapidly and continuously. Cells of the body that have a high rate of cellular division are more at risk for developing cancer. As malignant cells invade surrounding tissues, they trigger ***angiogenesis*** (an′-jē-ō-JEN-e-sis), the growth of new networks of blood vessels. As the cancer grows, it begins to compete with normal tissues for space and nutrients. Eventually, the normal tissue decreases in size and dies. Some malignant cells may detach from the initial (primary) tumor and invade a body cavity or enter the blood or lymph, then circulate to and invade other body tissues, establishing secondary tumors. The pain associated with cancer develops when the tumor presses on nerves or blocks a passageway in an organ so that secretions build up pressure.

Causes of Cancer

Several factors may trigger a normal cell to lose control and become cancerous. One cause is environmental agents: substances in the air we breathe, the water we drink, and the food we eat. A

chemical agent or radiation that produces cancer is called a ***carcinogen*** (car-SIN-ō-jen). Carcinogens induce ***mutations*** (mū-TĀ-shuns), permanent changes in the DNA base sequence of a gene. The World Health Organization estimates that carcinogens are associated with 60–90% of all human cancers. Examples of carcinogens are hydrocarbons found in cigarette tar, radon gas from the earth, and ultraviolet (UV) radiation in sunlight.

Intensive research efforts are now directed toward studying cancer-causing genes, or ***oncogenes*** (ON-kō-jēnz). When inappropriately activated, these genes have the ability to transform a normal cell into a cancerous cell. Most oncogenes derive from normal genes called ***proto-oncogenes*** that regulate growth and development. The proto-oncogene undergoes some change that causes it to either be expressed inappropriately or make its products in excessive amounts or at the wrong time. Some oncogenes cause excessive production of growth factors, chemicals that stimulate cell growth. Others may trigger changes in a cell-surface receptor, causing it to send signals as though it were being activated by a growth factor. As a result, the growth pattern of the cell becomes abnormal.

Some cancers have a viral origin. Viruses are tiny packages of nucleic acids, either RNA or DNA, that can reproduce only while inside the cells they infect. Some viruses, termed ***oncogenic viruses,*** cause cancer by stimulating abnormal proliferation of cells. For instance, the *human papillomavirus (HPV)* causes virtually all cervical cancers in women.

Some studies suggest that certain cancers may be linked to a cell having abnormal numbers of chromosomes. As a result, the cell could potentially have extra copies of oncogenes or too few copies of tumor-suppressor genes, which in either case could lead to uncontrolled cell proliferation. There is also some evidence suggesting that cancer may be caused by normal stem cells that develop into cancerous stem cells capable of forming malignant tumors.

Later in the book, we will discuss the process of inflammation, which is a defensive response to tissue damage. It appears that inflammation contributes to various steps in the development of cancer. Some evidence suggests that chronic inflammation stimulates the proliferation of mutated cells and enhances their survival, promotes angiogenesis, and contributes to invasion and metastasis of cancer cells. There is a clear relationship between certain chronic inflammatory conditions and the transformation of inflamed tissue into a malignant tissue. For example, chronic gastritis (inflammation of the stomach lining) and peptic ulcers may be a causative factor in 60–90% of stomach cancers. Chronic hepatitis (inflammation of the liver) and cirrhosis of the liver are believed to be responsible for about 80% of liver cancers. Colorectal cancer is ten times more likely to occur in patients with chronic inflammatory diseases of the colon, such as ulcerative colitis and Crohn's disease. And the relationship between asbestosis and silicosis, two chronic lung inflammatory conditions, and lung cancer has long been recognized. Chronic inflammation is also an underlying contributor to rheumatoid arthritis, Alzheimer's disease, depression, schizophrenia, cardiovascular disease, and diabetes.

Carcinogenesis: A Multistep Process

Carcinogenesis (kar′-si-nō-JEN-e-sis), the process by which cancer develops, is a multistep process in which as many as 10 distinct mutations may have to accumulate in a cell before it becomes cancerous. In colon cancer, the tumor begins as an area of increased cell proliferation that results from one mutation. This growth then progresses to abnormal, but noncancerous, growths called adenomas. After several more mutations, a carcinoma develops. The fact that so many mutations are needed for a cancer to develop indicates that cell growth is normally controlled with many sets of checks and balances.

MANUAL THERAPY APPLICATION

Massage of Cancer Patients

Appropriate **massage of cancer patients** can ease uncomfortable *symptoms* cancer such as anxiety, pain, fatigue, and nausea. This noninvasive modality may reduce these and other symptoms by 30–50%. Massage therapy will not cure or slow the development of cancer, but it can help to reduce or alleviate some of the side effects associated with invasive treatments and procedures.

Until the past 5 years, all forms of cancer have been considered by most health professionals to be a contraindication for massage. Appropriate massage modalities are, however, now being utilized in oncology units of hospitals, hospice, private practices, and other sites. Positive effects of touch include the following benefits: provide pain relief and reduce the need for pain medication, decrease depression and anxiety, increase circulation including lymph flow, improve sleep, decrease edema and lymphedema, and help prevent bedsores.

Cancer is a frightening disease that affects persons in subtle or profound ways. Massage allows patients to have a period of time away from all the pressures and stress brought about by cancer treatment, family, and work issues. It can enable persons to reconnect with their bodies in a healthful way.

Treatment of Cancer

Many cancers are removed surgically. However, when cancer is widely distributed throughout the body or exists in organs such as the brain whose functioning would be greatly harmed by surgery, other methodologies are used singly or in combination. One of the ways to treat cancer is by *chemotherapy,* the use of anticancer drugs. Some of these drugs stop cell division by inhibiting the formation of the mitotic spindle. Unfortunately, these types of anticancer drugs also kill all types of rapidly dividing cells in the body, thus causing side effects. The use of anti-angiogenesis drugs is at the forefront of cancer research.

Radiation therapy breaks chromosomes, thus blocking cell division. Because cancerous cells divide rapidly, they are more vulnerable to the destructive effects of chemotherapy and radiation therapy than are normal cells. Unfortunately for the patients, hair follicle cells, red bone marrow cells, and cells lining the gastrointestinal tract also are rapidly dividing. Hence, the side effects of chemotherapy and radiation therapy include hair loss due to death of hair follicle cells, vomiting and nausea due to death of cells lining the stomach and intestines, and susceptibility to infection due to slowed production of white blood cells in red bone marrow.

Another potential treatment for cancer that is currently under development is *virotherapy,* the use of viruses to kill cancer cells. The viruses employed in this strategy are designed so that they specifically target cancer cells without affecting the healthy cells of the body. Once inside the body, the viruses bind to cancer cells and then infect them. The cancer cells are eventually killed once the viruses cause cellular lysis (destruction).

CHECKPOINT

15. Describe the significance of angiogenesis in the spread of cancer.
16. List three carcinogens.
17. What is the significance of oncogenic viruses?
18. Describe three general treatments for cancer.

3.9 AGING AND CELLS

OBJECTIVE

- Describe the cellular changes that occur with aging.

Aging is a normal process accompanied by a progressive alteration of the body's homeostatic adaptive responses. It produces observable changes in structure and function of the body and increases vulnerability to environmental stress and disease. The specialized branch of medicine that deals with the medical problems and care of elderly persons is ***geriatrics*** (jer′-ē-AT-riks; *ger-* = old age; *-iatrics* = medicine). ***Gerontology*** (jer′-on-TOL-ō-jē) is the scientific study of the process and problems associated with aging.

Although many millions of new cells are normally produced each minute, several cell types, especially nerves and muscles, have a limited capability to divide. Normal cells grown outside the body divide only a certain number of times and then stop. These observations suggest that cessation of mitosis is a normal, genetically programmed event. According to this view, "aging genes" are part of the genetic blueprint at birth. These genes have an important function in normal cells, but their activities slow over time. They bring about aging by slowing down or halting processes vital to life.

Another aspect of aging involves ***telomeres*** (TĒ-lō-merz), specific DNA sequences found only at the tips of each chromosome. These pieces of DNA protect the tips of chromosomes from erosion and from sticking to one another. However, in most normal body cells each cycle of cell division shortens the telomeres. Eventually, after many cycles of cell division, the telomeres can be completely gone, and even some of the functional chromosomal material may be lost. These observations suggest that erosion of DNA from the tips of our chromosomes contributes greatly to the aging and death of cells. Individuals who experience high levels of stress have significantly shorter telomere length.

Four additional markers of aging are the presence of excess glucose, excess insulin, excess cortisol, and excess free radicals. Glucose, the most abundant sugar in the body, is haphazardly added to proteins inside and outside cells, forming irreversible cross-links between adjacent protein molecules. With advancing age, more cross-links form, which contributes to the stiffening and loss of elasticity that occur in aging tissues.

Free radicals produce oxidative damage in lipids, proteins, or nucleic acids. Some effects are wrinkled skin, stiff joints, and hardened arteries. Naturally occurring enzymes in peroxisomes and in the cytosol normally dispose of free radicals. Certain dietary substances, such as beta-carotene, zinc, vitamin C, vitamin E, and selenium (ACES at a health food store), are antioxidants that inhibit free radical formation.

Some theories of aging explain the process at the cellular level, while others concentrate on regulatory mechanisms operating within the entire organism. For example, the immune system may start to attack the body's own cells. This *autoimmune response* might be caused by changes in certain plasma membrane glycoproteins and glycolipids (cell-identity markers) that cause antibodies to attach to and mark the cell for destruction. As changes in the proteins on the plasma membrane of cells increase, the autoimmune response intensifies, producing the well-known signs of aging.

CHECKPOINT

19. Briefly outline the cellular changes involved in aging.

KEY MEDICAL TERMS ASSOCIATED WITH CELLS

Anaplasia (an′-a-PLĀ-zē-a; *an-* = not; *-plasia* = to shape) The loss of tissue differentiation and function that is characteristic of most malignancies.

Apoptosis (ap′-ōp-TŌ-sis; a falling off, like dead leaves from a tree) An orderly, genetically programmed cell death in which "cell-suicide" genes become activated. Enzymes produced by these genes disrupt the cytoskeleton and nucleus; the cell shrinks and pulls away from neighboring cells; the DNA within the nucleus fragments; and the cytoplasm shrinks, although the plasma membrane remains intact. Phagocytes in the vicinity then ingest the dying cell. Apoptosis removes unneeded cells during development before birth and continues after birth both to regulate the number of cells in a tissue and to eliminate potentially dangerous cells such as cancer cells.

Atrophy (A-trō-fē; *a-* = without; *-trophy* = nourishment) A decrease in the size of cells with subsequent decrease in the size of the affected tissue or organ; wasting away.

Biopsy (BĪ-op-sē; *bio-* = life; *-opsy* = viewing) The removal and microscopic examination of tissue from the living body for diagnosis.

Dysplasia (dis-PLĀ-zē-a; *dys-* = abnormal) Alteration in the size, shape, and organization of cells due to chronic irritation or inflammation; may progress to a neoplasm (tumor formation, usually malignant) or revert to normal if the irritation is removed.

Hyperplasia (hī′-per-PLĀ-zē-a; *hyper-* = over) Increase in the number of cells of a tissue due to an increase in the frequency of cell division.

Hypertrophy (hī-PER-trō-fē) Increase in the size of cells in a tissue without cell division.

Metaplasia (met′-a-PLĀ-zē-a; *meta-* = change) The transformation of one type of cell into another.

Necrosis (ne-KRŌ-sis = death) A pathological type of cell death, resulting from tissue injury, in which many adjacent cells swell, burst, and spill their cytoplasm into the interstitial fluid; the cellular debris usually stimulates an inflammatory response, which does not occur in apoptosis.

Progeny (PROJ-e-nē; *pro-* = forward; *-geny* = production) Offspring or descendants.

Proteomics (prō′-tē-Ō-miks; *proteo-* = protein) The study of the proteome (all of an organism's proteins) in order to identify all the proteins produced; it involves determining how the proteins interact and ascertaining the three-dimensional structure of proteins so that drugs can be designed to alter protein activity to help in the treatment and diagnosis of disease.

Tumor marker A substance introduced into circulation by tumor cells that indicates the presence of a tumor, as well as the specific type. Tumor markers may be used to screen, diagnose, make a prognosis, evaluate a response to treatment, and monitor for recurrence of cancer.

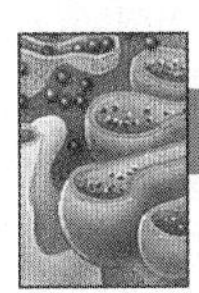

STUDY OUTLINE

A Generalized View of the Cell (Section 3.1)

1. A cell is the basic, living, structural and functional unit of the body.
2. Cell biology is the study of cell structure and function.
3. Figure 3.1 shows a generalized view of a cell that is a composite of many different cells in the body.
4. The principal parts of a cell are the plasma membrane; the cytoplasm, which consists of cytosol and organelles; and the nucleus.

Plasma Membrane (Section 3.2)

1. The plasma membrane surrounds and contains the cytoplasm of a cell; it is composed of lipids and proteins.
2. The lipid bilayer consists of two back-to-back layers of phospholipids, cholesterol, and glycolipids.
3. Integral proteins extend into or through the lipid bilayer, whereas peripheral proteins associate with the inner or outer surface of the membrane.
4. The plasma membrane's selective permeability permits some substances to pass across it more easily than others. The lipid bilayer is permeable to water and to most lipid-soluble molecules. Small- and medium-sized water-soluble materials may cross the membrane with the assistance of integral proteins.
5. Membrane proteins have several functions. Channels and transporters are integral proteins that help specific solutes across the membrane; receptors serve as cellular recognition sites; some membrane proteins are enzymes; and others are cell-identity markers.

Transport Across the Plasma Membrane (Section 3.3)

1. Fluid inside body cells is called intracellular fluid (ICF); fluid outside body cells is extracellular fluid (ECF). The ECF in the microscopic spaces between the cells of tissues is interstitial fluid. The ECF in blood vessels is plasma, that in lymphatic vessels is lymph, and that within and around the brain and spinal cord is cerebrospinal fluid (CSF).
2. Any material dissolved in a fluid is called a solute, and the fluid that dissolves materials is the solvent. Body fluids are dilute solutions in which a variety of solutes are dissolved in the solvent water.
3. The selective permeability of the plasma membrane supports the existence of concentration gradients, which are differences in the concentration of chemicals between one side of the membrane and the other.
4. Materials move through cell membranes by passive processes or by active transport. In passive processes, a substance moves down its concentration gradient across the membrane. In active transport, cellular energy is used to drive the substance "uphill" against its concentration gradient.
5. In transport in vesicles, tiny vesicles either detach from the plasma membrane while bringing materials into the cell or merge with the plasma membrane to release materials from the cell.
6. Diffusion is the movement of substances due to their kinetic energy. In net diffusion, substances move from an area of higher concentration to an area of lower concentration until equilibrium is reached. At equilibrium the concentration is the same throughout the solution.
7. In simple diffusion, lipid-soluble substances move through the lipid bilayer. In facilitated diffusion, substances cross the membrane with the assistance of ion channels or carriers.
8. Osmosis is the movement of water molecules through a selectively permeable membrane from an area of higher to an area of lower water concentration.
9. In an isotonic solution, red blood cells maintain their normal shape; in a hypotonic solution, they gain water and undergo hemolysis; in a hypertonic solution, they lose water and undergo crenation.
10. With the expenditure of cellular energy, usually in the form of ATP, solutes can cross the membrane against their concentration gradient by means of active transport. Actively transported solutes include several ions such as Na^+, K^+, H^+, Ca^{2+}, I^-, and Cl^-.
11. The most important active transport pump is the sodium–potassium pump, which expels Na^+ from cells and brings K^+ in.
12. Transport in vesicles includes both endocytosis (phagocytosis and bulk-phase endocytosis) and exocytosis.
13. Phagocytosis is the ingestion of solid particles. It is an important process used by some white blood cells to destroy bacteria that enter the body. Bulk-phase endocytosis is the ingestion of extracellular fluid.
14. Exocytosis involves movement of secretory or waste products out of a cell by fusion of vesicles with the plasma membrane.

Cytoplasm (Section 3.4)

1. Cytoplasm includes all the cellular contents between the plasma membrane and nucleus; it consists of cytosol and organelles.
2. The liquid portion of cytoplasm is cytosol, composed mostly of water, plus ions, glucose, amino acids, fatty acids, proteins, lipids,

ATP, and waste products; cytosol is the site of many chemical reactions required for a cell's existence.
3. Organelles are specialized cellular structures with characteristic shapes and specific functions.
4. The cytoskeleton is a network of several kinds of protein filaments that extend throughout the cytoplasm; they provide a structural framework for the cell and generate movements. Components of the cytoskeleton are microfilaments, intermediate filaments, and microtubules.
5. The centrosome consists of two centrioles and pericentriolar material. The centrosome serves as a center for organizing microtubules in interphase cells and the mitotic spindle during cell division.
6. Cilia and flagella are motile projections of the cell surface. Cilia move fluid along the cell surface, whereas a flagellum moves an entire cell.
7. Ribosomes, composed of ribosomal RNA and ribosomal proteins, consist of two subunits and are the sites of protein synthesis.
8. Endoplasmic reticulum (ER) is a network of membranes that extends from the nuclear envelope throughout the cytoplasm.
9. Rough ER is studded with ribosomes. Proteins synthesized on the ribosomes enter the ER for processing and sorting. The ER is also where glycoproteins and phospholipids form.
10. Smooth ER lacks ribosomes. It is where fatty acids and steroids are synthesized. Smooth ER also participates in releasing glucose from the liver into the bloodstream, inactivating or detoxifying drugs and other potentially harmful substances, and storing and releasing calcium ions that trigger contraction in muscle cells.
11. The Golgi complex consists of flattened sacs called cisterns that receive proteins synthesized in the rough ER. Within the Golgi cisterns the proteins are modified, sorted, and packaged into vesicles for transport to different destinations. Some processed proteins leave the cell in secretory vesicles, some are incorporated into the plasma membrane, and some enter lysosomes.
12. Lysosomes are membrane-enclosed vesicles that contain digestive enzymes. They function in digestion of worn-out organelles (autophagy) and even in digestion of their own cell (autolysis).
13. Peroxisomes are similar to lysosomes but smaller. They oxidize various organic substances such as amino acids, fatty acids, and toxic substances and, in the process, produce hydrogen peroxide and associated free radicals such as superoxide. The hydrogen peroxide is degraded by an enzyme in peroxisomes called catalase.
14. Proteasomes contain proteases that continually degrade unneeded, damaged, or faulty proteins.
15. Mitochondria consist of a smooth outer membrane, an inner membrane containing folds called cristae, and a fluid-filled cavity called the matrix. They are called "powerhouses" of the cell because they produce most of a cell's ATP.

Nucleus (Section 3.5)

1. The nucleus consists of a double nuclear envelope; nuclear pores, which control the movement of substances between the nucleus and cytoplasm; nucleoli, which produce ribosomes; and genes arranged on chromosomes.
2. Most body cells have a single nucleus; some (red blood cells) have none, whereas others (skeletal muscle cells) have several.
3. Genes control cellular structure and most cellular functions.

Gene Action: Protein Synthesis (Section 3.6)

1. Most of the cellular machinery is devoted to protein synthesis.
2. Cells make proteins by transcribing and translating the genetic information encoded in the sequence of four types of nitrogenous bases in DNA.
3. In transcription, genetic information encoded in the DNA base sequence is copied into a complementary sequence of bases in a strand of messenger RNA (mRNA). Transcription begins on DNA in a region called a promoter.
4. Translation is the process in which mRNA associates with ribosomes and directs synthesis of a protein, converting the nucleotide sequence in mRNA into a specific sequence of amino acids.
5. In translation, mRNA binds to a ribosome, specific amino acids attach to transfer RNA (tRNA), and anticodons of tRNA bind to codons of mRNA, bringing specific amino acids into position on a growing protein.
6. Translation begins at the start codon and terminates at the stop codon.

Somatic Cell Division (Section 3.7)

1. Cell division is the process by which cells reproduce themselves.
2. Cell division that results in an increase in the number of body cells is called somatic cell division; it involves a nuclear division called mitosis plus division of the cytoplasm, called cytokinesis.
3. Cell division that results in the production of sperm and oocytes is called reproductive cell division.
4. The cell cycle is an orderly sequence of events in which a cell duplicates its contents and divides in two. It consists of interphase and a mitotic phase.
5. Before the mitotic phase, the DNA molecules, or chromosomes, replicate themselves so that identical chromosomes can be passed on to the next generation of cells.
6. A cell that is between divisions and is carrying on every life process except division is said to be in interphase.
7. Mitosis is the replication and distribution of two sets of chromosomes into separate and equal nuclei; it consists of prophase, metaphase, anaphase, and telophase.
8. Cytokinesis usually begins late in anaphase and ends in telophase.
9. A cleavage furrow forms and progresses inward, cutting through the cell to form two separate identical cells, each with equal portions of cytoplasm, organelles, and chromosomes.

Cancer (Section 3.8)

1. Cancer is a group of diseases characterized by uncontrolled or abnormal cell proliferation.
2. A cancerous neoplasm is called a malignant tumor or malignancy.
3. Malignant tumors are capable of metastasis, the spread of cancerous cells to other parts of the body.
4. Malignant cells often trigger angiogenesis, the growth of new networks of blood vessels, at the expense of the usual vascularization of healthy tissue.
5. Carcinogens, chemical agents or radiation that cause cancer, induce mutations of DNA base sequences in genes.
6. Cancer-causing genes, oncogenes, may transform normal cells into malignancies.
7. Oncogenic viruses may also stimulate abnormal proliferation of cells.
8. As many as 10 distinct mutations may have to accumulate in a cell before it becomes cancerous.
9. Many cancers are removed surgically.
10. Chemotherapy typically works by stopping cell division; although dividing cancerous cells may be arrested, other fast-growing cells,

such as the lining of the digestive tract and hair cells, are also affected.

11. Radiation therapy breaks chromosomes, thus blocking cell divison of cancerous cells as well as fast-growing cells of normal tissues.

12. Virotherapy is under development and it may affect only cancerous cells, thus sparing normal tissues.

Aging and Cells (Section 3.9)

1. Aging is a normal process accompanied by progressive alteration of the body's homeostatic adaptive responses.

2. Many theories of aging have been proposed, including genetically programmed cessation of cell division, shortening of telomeres, addition of glucose to proteins, buildup of free radicals, and an intensified autoimmune response.

SELF-QUIZ QUESTIONS

1. If the extracellular fluid contains a greater concentration of solutes than the cytosol of the cell, the extracellular fluid is said to be
a. isotonic **b.** hypertonic
c. hypotonic **d.** cytotonic
e. epitonic.

2. The proteins found in the plasma membrane
a. are primarily glycoproteins
b. allow the passage of many substances into the cell
c. allow cells to recognize other cells
d. help anchor cells to each other
e. have all of the above functions.

3. To enter many body cells, glucose must bind to a specific membrane carrier protein, which assists glucose to cross the membrane without using ATP. This type of movement is known as
a. facilitated diffusion **b.** simple diffusion
c. vesicular transport **d.** osmosis
e. active transport.

4. A red blood cell placed in a hypotonic solution undergoes
a. hemolysis **b.** crenation
c. equilibrium **d.** a decrease in osmotic pressure
e. shrinkage.

5. Which of the following normally pass through the plasma membrane only by transport in vesicles?
a. water molecules **b.** sodium ions
c. proteins **d.** oxygen molecules
e. hydrogen ions

6. Which of the following statements is NOT true?
a. A benign tumor is noncancerous.
b. When a cancerous growth presses on nerves, it can cause pain.
c. Angiogenesis is the spread of cancerous cells to other parts of the body.
d. Ultraviolet radiation and radon gas are carcinogens.
e. Cancer is uncontrolled mitosis in abnormal cells.

7. Which of the following processes requires ATP?
a. simple diffusion **b.** active transport
c. osmosis **d.** facilitated diffusion using ion channels
e. facilitated diffusion using carriers

8. Nicotine in cigarette smoke interferes with the ability of cells to rid the breathing passageways of debris. Which organelles are "paralyzed" by nicotine?
a. flagella **b.** ribosomes **c.** microfilaments
d. cilia **e.** lysosomes

9. Many proteins found in the plasma membrane are formed by the ________ and packaged by the ________.
a. ribosomes, Golgi complex **b.** smooth ER, Golgi complex
c. Golgi complex, lysosomes **d.** mitochondria, Golgi complex
e. nucleus, smooth ER

10. Match the following:

____	**a.** cellular movement	**A.** centrosome
____	**b.** selective permeability	**B.** cytoskeleton
____	**c.** protein synthesis	**C.** Golgi complex
____	**d.** lipid synthesis, detoxification	**D.** lysosomes
____	**e.** packages proteins and lipids	**E.** mitochondria
____	**f.** ATP production	**F.** plasma membrane
____	**g.** digest bacteria and worn-out organelles	**G.** ribosomes
		H. smooth ER
____	**h.** forms mitotic spindle	

11. If the smooth endoplasmic reticulum were destroyed, a cell would not be able to
a. form lysosomes **b.** synthesize certain proteins
c. generate energy **d.** phagocytize bacteria
e. synthesize fatty acids and steroids.

12. Water moves into and out of red blood cells through the process of
a. endocytosis **b.** phagocytosis
c. osmosis **d.** active transport
e. facilitated diffusion.

13. A cell undergoing mitosis goes through the following stages in which sequence?
a. interphase, metaphase, prophase, cytokinesis,
b. interphase, prophase, cytokinesis, telophase
c. anaphase, metaphase, prophase, telophase
d. anaphase, metaphase, prophase, cytokinesis
e. prophase, metaphase, anaphase, telophase

14. Transcription involves
a. transferring information from mRNA to tRNA
b. codon binding with anticodons
c. joining amino acids by peptide bonds
d. copying information contained in DNA to mRNA
e. synthesizing the protein on the ribosome.

15. If a DNA strand has a nitrogenous base sequence TACGA, then the sequence of bases on the corresponding mRNA would be
a. ATGCT **b.** AUGCU
c. GUACU **d.** CTGAT
e. AUCUG.

16. Place the following events of protein synthesis in the proper order. **1.** DNA uncoils and mRNA is transcribed. **2.** tRNA with an attached amino acid pairs with mRNA. **3.** mRNA passes from the nucleus into the cytoplasm and attaches to a ribosome. **4.** Protein is formed. **5.** Two amino acids are linked by a peptide bond.
a. 1, 2, 3, 4, 5 **b.** 1, 3, 2, 5, 4
c. 1, 2, 3, 5, 4 **d.** 1, 5, 3, 2, 4
e. 2, 1, 3, 4, 5

17. Match the following descriptions with the phases shown.

____ **a.** nuclear envelope (membrane) and nucleoli reappear
____ **b.** centromeres of the chromatid pairs line up in the center of the mitotic spindle
____ **c.** DNA duplicates
____ **d.** cleavage furrow splits cell into two identical cells
____ **e.** chromosomes move toward opposite poles of cell
____ **f.** chromatids are attached at centromeres; mitotic spindle forms

A. prophase
B. cytokinesis
C. telophase
D. anaphase
E. metaphase
F. interphase

18. If a virus were to enter a cell and destroy its ribosomes, how would the cell be affected?
a. It would be unable to undergo mitosis.
b. It could no longer produce ATP.
c. Movement of the cell would cease.
d. It would undergo autophagy.
e. It would be unable to synthesize proteins.

19. The spread of malignant cells to normal tissues is called
a. angiogenesis **b.** metastasis
c. mitosis **d.** neoplasm
e. proliferation.

20. Breaking chromosomes and thus blocking cell divison of cancerous cells is the goal of
a. chemotherapy **b.** radiation therapy
c. surgery **d.** virotherapy
e. genomics.

CRITICAL THINKING QUESTIONS

1. Mucin is a protein present in saliva and other secretions. When mixed with water, it becomes the slippery substance known as mucus. Trace the route taken by mucin through the cell, from its synthesis to its secretion, listing all the organelles and processes involved.
2. Sam does not consume alcohol, while his brother Sebastian regularly drinks large quantities of alcohol. If we could examine the liver cells of each of these brothers, would we see a difference in smooth ER and peroxisomes? Explain.
3. Marathon runners can become dehydrated due to the extreme physical activity. What types of fluids should they consume in order to rehydrate their cells?

? ANSWERS TO FIGURE QUESTIONS

3.1 The three main parts of a cell are the plasma membrane, cytoplasm, and nucleus.
3.2 Some integral proteins function as channels or transporters to move substances across membranes. Other integral proteins function as receptors. Membrane glycolipids and glycoproteins are involved in cellular recognition.
3.3 The membrane protein that binds to insulin acts as a receptor.
3.4 Another name for intracellular fluid is cytosol.
3.5 Because fever involves an increase in body temperature, the rates of all diffusion processes would increase.
3.6 Oxygen, carbon dioxide, fatty acids, fat-soluble vitamins, and steroids can cross the plasma membrane by simple diffusion through the lipid bilayer.
3.7 The concentration of K^+ is higher in the cytosol of body cells than in extracellular fluids.
3.8 Insulin promotes insertion of glucose carriers in the plasma membrane, which increases cellular glucose uptake by facilitated diffusion.
3.9 The water concentrations can never be the same in the two arms because the left arm contains pure water and the right arm contains a solution that is less than 100% water.
3.10 A 2% solution of NaCl would cause crenation of RBCs because it is hypertonic and promotes water loss and cell shrinkage.
3.11 ATP adds a phosphate group to the pump protein, which changes the pump's three-dimensional shape. ATP transfers energy to power the pump.
3.12 The binding of particles to a plasma membrane receptor triggers pseudopod formation.
3.13 Microtubules help to form centrioles, cilia, and flagella.
3.14 A cell without a centrosome probably would not be able to undergo cell division.
3.15 Large and small ribosomal subunits are synthesized separately in the nucleolus in the nucleus and then come together in the cytoplasm.
3.16 Rough ER has attached ribosomes; smooth ER does not. Rough ER synthesizes proteins that will be used in organelles or plasma membranes or exported from the cell; smooth ER is associated with lipid synthesis and other metabolic reactions.
3.17 Cells that secrete proteins into extracellular fluid have extensive Golgi complexes.
3.18 Some proteins are secreted from the cell by exocytosis, some are incorporated into the plasma membrane, and some occupy transport vesicles that become lysosomes.
3.19 Digestion of worn-out organelles by lysosomes is called autophagy.
3.20 Mitochondrial cristae increase the surface area available for chemical reactions and contain some of the enzymes needed for ATP production.
3.21 Chromatin is a complex of DNA, proteins, and some RNA.
3.22 Proteins determine the physical and chemical characteristics of cells.
3.23 The DNA base sequence AGCT would be transcribed into the RNA base sequence UCGA by RNA polymerase.
3.24 When a ribosome encounters a stop codon at the A site, it releases the completed protein from the final tRNA.
3.25 Cytokinesis usually starts in late anaphase of mitosis.

Tissues

In this chapter you will learn that approximately 100 trillion cells in your body are grouped to form four types of tissues: epithelial, connective, muscular, and nervous. Cells in each tissue are dependent on each other to survive and work together for the benefit of the whole body. *Epithelial tissue* is essentially the "skin" that covers the outside of your body, but it also lines all of your body cavities, organs, and glands. It provides a barrier much like the plasma membrane for each cell; it limits what can exit or enter the body. For example, perspiration from skin helps you release waste products and the body heat that the water holds. *Connective tissue* does exactly what it sounds like; it connects all the parts of the body. In a seemingly unlikely role, connective tissue also separates—one muscle from another, a muscle from a nerve, or an artery from a vein. Like skin, connective tissue can be a barrier to infection. Blood is classified as a connective tissue and provides a way to transmit water and nutrients throughout the body. Connective tissue is of special interest to therapists who treat the fasciae of the body. *Muscular tissue* is the main mover of the body but many are unaware that it is also the body's greatest source of heat; this is the reason you get hot when you exercise or why the muscular contraction you experience when you shiver warms you. The last tissue type, *nervous tissue,* is a regulator of the body, sending and receiving messages to guide the rest of the tissues as to their overall job in the body. Nervous tissue also determines what takes priority when the body is relaxed and healthy or in distress.

CONTENTS AT A GLANCE

4.1 TYPES OF TISSUES

OBJECTIVE

- Name four basic types of tissue that make up the human body and state the characteristics of each.

All of the cells of the body can be classified into about 20 different cell types. These cells are highly organized living units, but they cannot function alone. Nearly all cells work together in groups called tissues. A ***tissue*** is a group of similar cells, usually with a common embryonic origin, that functions together to carry out specialized activities. The 100 trillion or so cells in the body are classified into one of four basic types of tissues based on their structure and functions:

1. ***Epithelial tissue*** (ep′-i-THĒ-lē-al) covers body surfaces; lines body cavities, hollow organs, and ducts (tubes); and forms glands.
2. ***Connective tissue*** protects and supports the body and its organs, binds organs together, stores energy reserves as fat, and provides immunity.
3. ***Muscular tissue*** generates the physical force needed to make body structures move.
4. ***Nervous tissue*** detects changes inside and outside the body and initiates and transmits nerve impulses that coordinate body activities to help maintain homeostasis.

Histology (his-TOL-ō-jē; *hist-* = tissue; *-ology* = study of) is the science that deals with the study of tissues. A ***pathologist*** (pa-THOL-ō-jist; *patho-* = disease) is a physician who specializes in laboratory studies of cells and tissues to help other physicians make accurate diagnoses. One of the principal functions of a pathologist is to examine tissues for any changes that might indicate disease.

In this chapter we discuss all four tissue types but focus on epithelial and connective tissues. Muscular and nervous tissues will be discussed in detail in Chapter 10 and Chapter 15.

CHECKPOINT

1. Define a tissue. What are the four basic types of body tissues?

4.2 CELL JUNCTIONS

OBJECTIVE

- Describe the structure and functions of the five main types of cell junctions.

Most epithelial cells and some muscle and nerve cells are tightly joined into functional units by points of contact between their plasma membranes called ***cell junctions.*** Some cell junctions fuse cells together so tightly that they prevent substances from passing between the cells. Other cell junctions hold cells together so that they don't separate while performing their functions. Still other cell junctions form channels that allow ions and molecules to pass between cells. This permits cells in a tissue to communicate with each other and it also enables nerve or muscle impulses to spread rapidly among cells. Here we consider the five most important types of cell junctions: tight junctions, adherens junctions, desmosomes, hemidesmosomes, and gap junctions (Figure 4.1).

Tight junctions consist of weblike strands of transmembrane proteins that fuse the outer surfaces of adjacent plasma membranes together (Figure 4.1a). Cells of epithelial tissues that line the stomach, intestines, and urinary bladder have many tight junctions to retard the passage of substances between cells and prevent the contents of these organs from leaking into the blood or surrounding tissues.

Adherens junctions (ad-HĒR-ens) contain ***plaque*** (PLAK), a dense layer of proteins on the inside of the plasma membrane that attaches to both membrane proteins and to microfilaments of the cytoskeleton (Figure 4.1b). In epithelial cells, adherens junctions often form extensive zones called ***adhesion belts*** because they encircle the cell similar to the way a belt encircles your waist. Adherens junctions help epithelial surfaces resist separation during various contractile activities, as when food moves through the intestines.

Like adherens junctions, ***desmosomes*** (DEZ-mō-sōms; *desmo-* = band) contain plaque and have transmembrane glycoproteins that extend into the intercellular space between adjacent cell membranes and attach cells to one another (Figure 4.1c). However, unlike adherens junctions, the plaque of desmosomes does not attach to microfilaments. Instead, a desmosome plaque attaches to intermediate filaments, described in Chapter 3. The intermediate filaments extend from desmosomes on one side of the cell across the cytosol to desmosomes on the opposite side of the cell. This structural arrangement contributes to the stability of the cells and tissue. These spot-weld-like junctions are common among the cells that make up the epidermis (the outermost layer of the skin) and among cardiac muscle cells in the heart. Desmosomes prevent epidermal cells from separating under tension and cardiac muscle cells from pulling apart during contraction.

Hemidesmosomes (*hemi-* = half) resemble desmosomes but they do not link adjacent cells. The name arises from the fact that they look like half of a desmosome (Figure 4.1d). On the inside of the plasma membrane, intermediate filaments attach to the plaque. On the outside of the plasma membrane, the transmembrane glycoproteins in the extracellular space attach to the basement membrane (discussed shortly). Thus, hemidesmosomes anchor cells not to each other but to the basement membrane.

At ***gap junctions*** tiny fluid-filled tunnels called ***connexons*** connect neighboring cells (Figure 4.1e). The plasma membranes of gap junctions are not fused together as in tight junctions but are separated by a very narrow intercellular gap. Through the connexons, ions and small molecules can diffuse from the cytosol of one cell to another. The transfer of nutrients, and perhaps wastes, takes place through gap junctions in avascular tissues such as the lens and cornea of the eye. Gap junctions also allow the cells in a tissue to communicate with one another. Gap junctions enable nerve or muscle impulses to spread rapidly

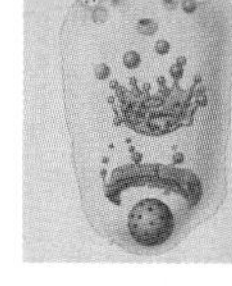

among cells, a process that is crucial for the normal operation of some parts of the nervous system and for the contraction of muscle in the heart, gastrointestinal tract, and uterus.

CHECKPOINT

2. Which type of cell junctions prevent the content of organs from leaking into surrounding tissues?
3. Which types of cell junctions are found in epithelial tissues?

4.3 EPITHELIAL TISSUE

OBJECTIVES

- Discuss the general features of epithelial tissue.
- Describe the structure, location, and function of the various types of epithelial tissues.

Epithelial tissue, or the noun ***epithelium*** (plural is *epithelia*), may be divided into two types: (1) *covering and lining epithe-*

Figure 4.1 Cell junctions.

Most epithelial cells and some muscle and nerve cells contain cell junctions.

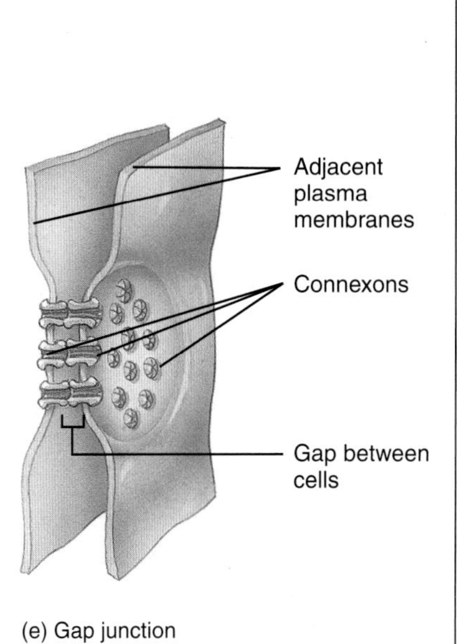

(e) Gap junction

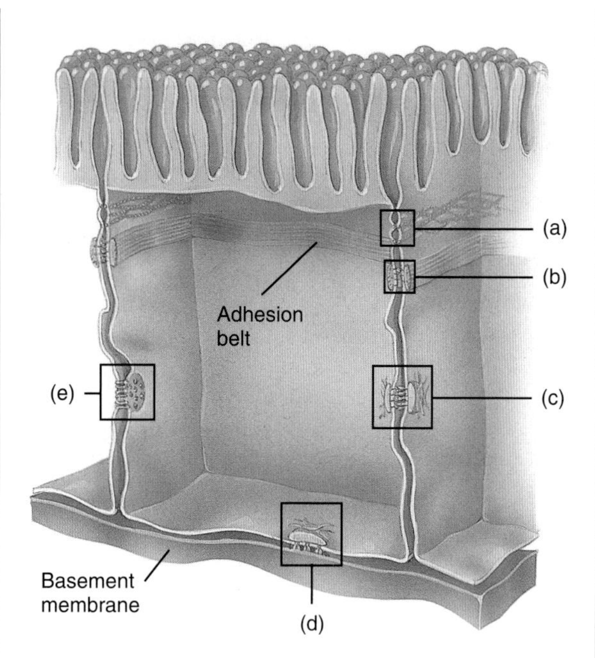

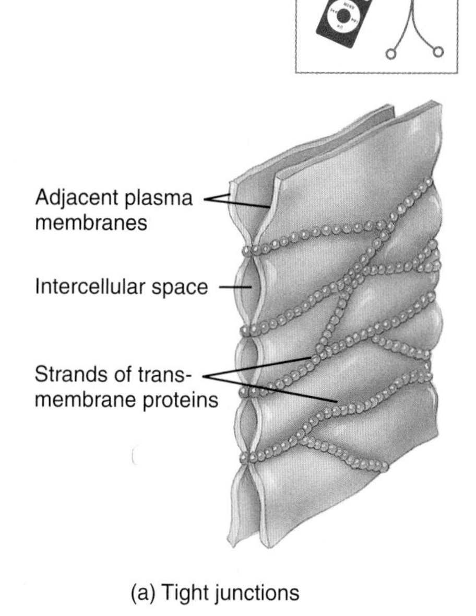

(a) Tight junctions

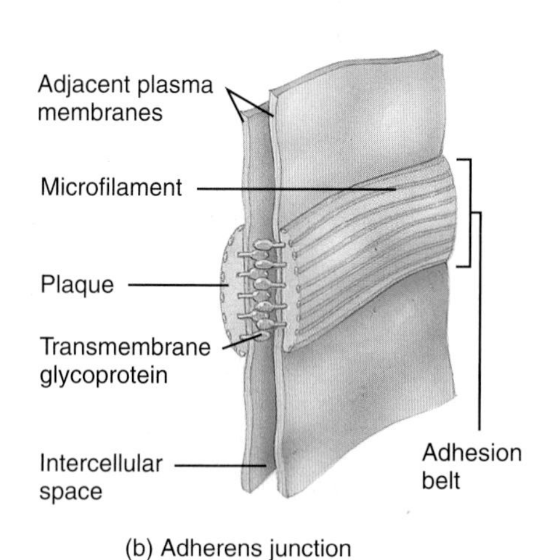

(b) Adherens junction

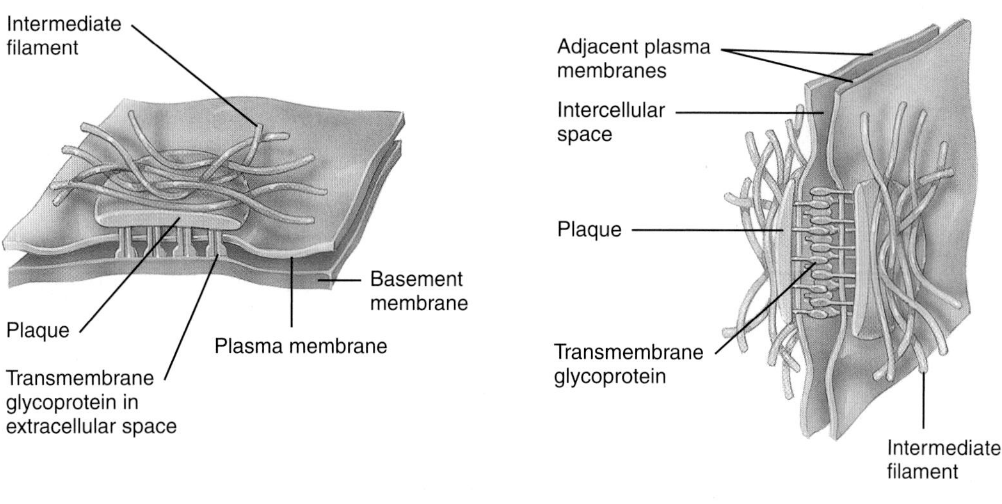

(d) Hemidesmosome

(c) Desmosome

Which type of cell junction functions in communication between adjacent cells?

lium and (2) *glandular epithelium.* As its name suggests, covering and lining epithelium forms the outer covering of the skin and the outer covering of some internal organs. It also lines the inside of body cavities; blood vessels; ducts; and the interiors of the respiratory, digestive, urinary, and reproductive systems. It makes up, along with nervous tissue, the parts of the sense organs for hearing, vision, and touch. Glandular epithelium makes up the secreting portion of glands, such as sweat glands.

General Features of Epithelial Tissue

As you will see shortly, there are many different types of epithelia, each with characteristic structures and functions. However, all of the different types of epithelial tissue also have features in common. General features of epithelial tissue include the following (Figure 4.2):

1. Epithelium consists largely or entirely of closely packed cells with little extracellular material between them, and the cells are arranged in continuous sheets, in either single or multiple layers.

2. Epithelial cells have an *apical* (free) *surface,* which is exposed to a body cavity, lining of an internal organ, or the exterior of the body; *lateral surfaces,* which face adjacent cells on either side; and a *basal surface,* which is attached to a basement membrane. In subsequent discussions, the term *apical layer* refers to the most superficial layer of cells, whereas the term *basal layer* refers to the deepest layer of cells. The ***basement membrane*** is a thin extracellular structure composed of two layers, the basal lamina and the reticular lamina. It is located between the epithelium and the underlying connective tissue layer and helps bind and support the epithelium. The *basal lamina* is derived from epithelial tissues whereas the *reticular lamina* is derived from connective tissues.

3. Epithelia are ***avascular*** (*a-* = without; *-vascular* = blood vessels); that is, they lack blood vessels. The vessels that supply nutrients to and remove wastes from epithelia are located in adjacent connective tissues. The exchange of materials between epithelium and connective tissue occurs by diffusion, as discussed in Chapter 3.

4. Epithelia have a nerve supply.

5. Because epithelium is subject to a certain amount of wear and tear and injury, it has a high capacity for renewal by cell division. (As you learned in Chapter 3, cells that have a high rate of cellular division are more at risk for developing cancer.)

Figure 4.2 Surfaces of epithelial cells and the structure and location of the basement membrane.

The basement membrane is found between epithelium and connective tissue.

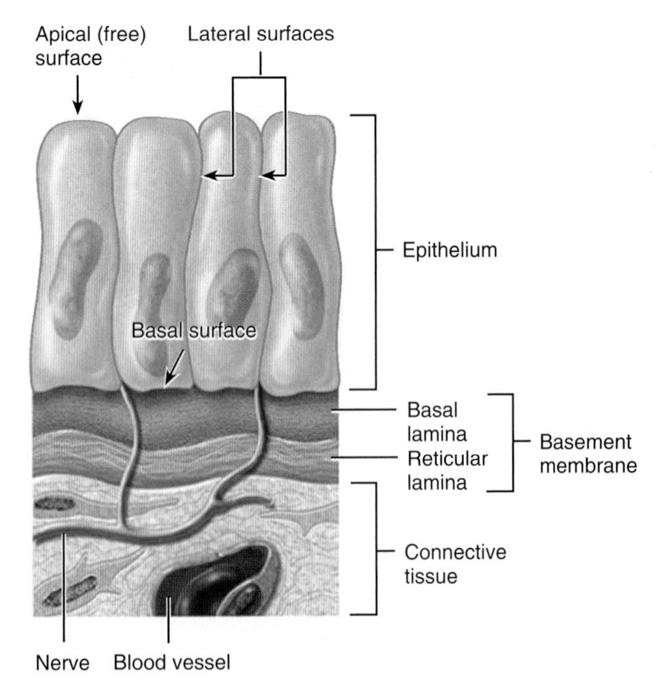

What is the function of the basement membrane?

Covering and Lining Epithelium

Covering and lining epithelium, which covers or lines various parts of the body, is classified according to the arrangement of cells into layers and the shape of the cells (Figure 4.3):

I. ***Arrangement of cells in layers.*** The cells of covering and lining epithelia are arranged in one or more layers depending on the functions the epithelium performs:

   A. *Simple epithelium* is a single layer of cells that functions in diffusion, osmosis, filtration, secretion, and absorption. ***Secretion*** (se-KRĒ-shun) is the production and release of substances such as mucus, sweat, or enzymes. ***Absorption*** (ab-SORP-shun) is the intake of fluids or other substances such as digested food from the intestinal tract.

   B. *Pseudostratified epithelium* (*pseudo-* = false) appears to have multiple layers of cells because the cell nuclei lie at different levels and not all cells reach the apical surface. Cells that do extend to the apical surface may contain cilia; others (goblet cells) secrete mucus. Pseudostratified epithelium is technically classified as a simple epithelium because all its cells rest on the basement membrane.

   C. *Stratified epithelium* (*stratum* = layer) consists of more than one layer of cells that protect underlying tissues in locations where there is considerable wear and tear.

II. ***Cell shapes.***

   A. *Squamous* cells (SKWĀ-mus = flat) are thin, and this allows for the rapid passage of substances through them.

   B. *Cuboidal* cells are nearly as tall as they are wide and are shaped like cubes or hexagons, or may be somewhat pie shaped. They may have microvilli at their apical surface and function in either secretion or absorption.

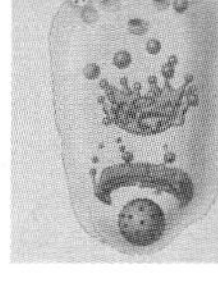

Figure 4.3 Cell shapes and arrangement of layers for covering and lining epithelium.

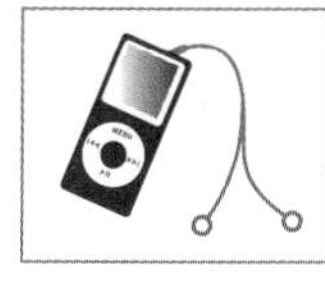

Cell shapes and arrangement of layers are the bases for classifying covering and lining epithelium.

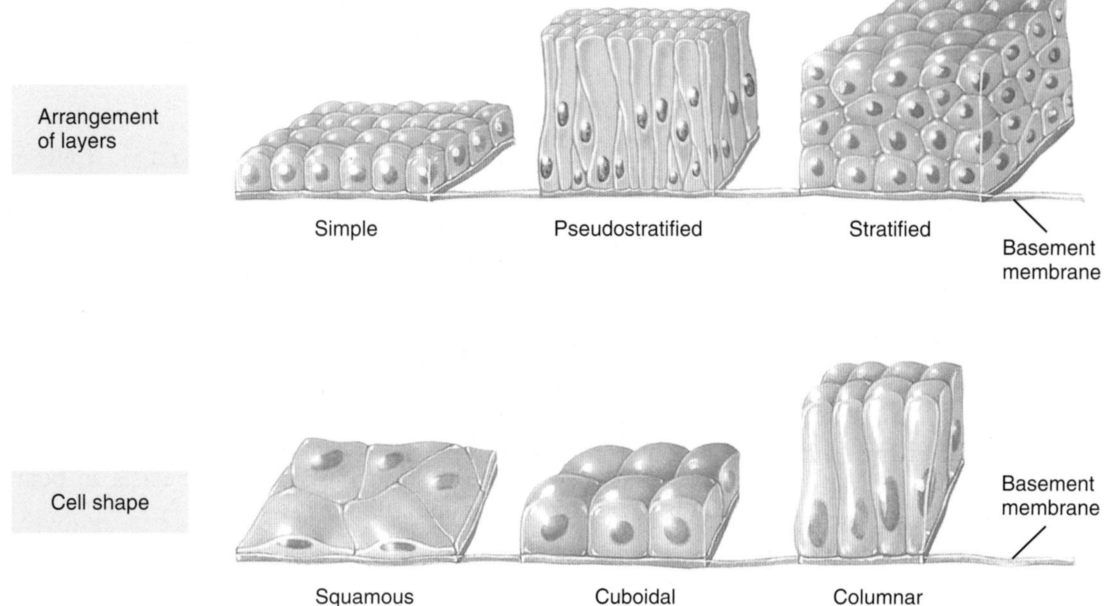

Which cell shape is best adapted for the rapid movement of substances from one cell to another?

C. *Columnar* cells are much taller than they are wide, like columns, and protect underlying tissues. Their apical surfaces may have cilia or microvilli, and they often are specialized for secretion and absorption.

D. *Transitional* cells change shape, from flat to cuboidal and back, as organs such as the urinary bladder stretch (distend) to a larger size and then collapse to a smaller size.

Combining the two characteristics (arrangement of layers and cell shapes), the types of covering and lining epithelia are as follows:

I. Simple epithelium
   A. Simple squamous epithelium
   B. Simple cuboidal epithelium
   C. Simple columnar epithelium (nonciliated and ciliated)
   D. Psuedostratified columnar epithelium (nonciliated and ciliated)

II. Stratified epithelium
   A. Stratified squamous epithelium (keratinized and nonkeratinized)*
   B. Stratified cuboidal epithelium*
   C. Stratified columnar epithelium*
   D. Transitional epithelium

*This classification is based on the shape of the cells at the *apical* surface.

Each of these covering and lining epithelia is described in the following sections and illustrated in Table 4.1. The illustration of each type consists of a photomicrograph, a corresponding diagram, and an inset that identifies a major location of the tissue in the body. Descriptions, certain locations, and functions of the tissues accompany each illustration.

Simple Epithelium

SIMPLE SQUAMOUS EPITHELIUM This tissue consists of a single layer of flat cells that resembles a tiled floor when viewed from its apical surface (Table 4.1A). The nucleus of each cell is a flattened oval or sphere and is centrally located. Simple squamous epithelium is so thin that the flattened nucleus makes a bulge in the cell; there is very little cytoplasm in each cell. Simple squamous epithelium is found in parts of the body where filtration (kidneys) or diffusion (lungs) are priority processes. It is not found in body areas that are subjected to wear and tear.

Owing to specific locations where it is found, simple squamous epithelium that lines the heart, blood vessels, and lymphatic vessels is known as ***endothelium*** (en′-dō-THĒ-lē-um; *endo-* = within; *-thelium* = covering); the type that forms the epithelial layer of serous membranes, such as the peritoneum, pleura, or pericardium is called ***mesothelium*** (mez′-ō-THĒ-lē-um; *meso-* = middle).

SIMPLE CUBOIDAL EPITHELIUM The cuboidal shape of the cells in this tissue (Table 4.1B) is obvious only when the tissue is sectioned and viewed from the side. Cell nuclei are usually round and centrally located. Simple cuboidal epithelium is found in organs such as the thyroid gland and kidneys and performs the functions of secretion and absorption. Note that cells that are strictly cuboidal could not form small tubes; such cuboidal cells are more pie-shaped but they are still nearly as high as they are wide (at the base).

SIMPLE COLUMNAR EPITHELIUM When viewed from the side, the cells of simple columnar epithelium appear like columns with oval nuclei near the base of the cells. Simple columnar epithelium exists in two forms: nonciliated simple columnar epithelium and ciliated simple columnar epithelium. Like cuboidal epithelium, columnar epithelium functions in secretion and absorption. The larger columnar cells, however, contain more organelles and are therefore capable of a higher level of secretion and absorption than are cuboidal cells.

Nonciliated simple columnar epithelium contains absorptive cells and goblet cells (Table 4.1C). ***Absorptive cells*** are columnar epithelial cells with ***microvilli,*** microscopic finger-like projections that increase the surface area of the plasma membrane (see Figure 3.1). Their presence increases the rate of absorption by the absorptive cell. ***Goblet cells*** are modified columnar cells that secrete mucus, a slightly sticky fluid, at their apical surfaces. Before it is released, mucus accumulates in the upper portion of the cell, causing that area to bulge. The whole cell then resembles a goblet or wine glass. Secreted mucus serves as a lubricant for the linings of the digestive, respiratory, reproductive, and most of the urinary tracts. Mucus also helps to trap dust entering the respiratory tract, and it prevents destruction of the stomach lining by acid secreted by the stomach.

Ciliated simple columnar epithelium (Table 4.1D) contains cells with cilia at their apical surface. In a few parts of the upper respiratory tract, ciliated columnar cells are interspersed with goblet cells. Mucus secreted by the goblet cells forms a film over the respiratory surface that traps inhaled foreign particles. The cilia wave in unison and move the mucus and any trapped foreign particles toward the throat, where they can be coughed up and swallowed or spit out. Cilia also help to move oocytes (nearly mature ova) expelled by the ovaries through the uterine tubes into the uterus.

PSEUDOSTRATIFIED COLUMNAR EPITHELIUM As noted earlier, pseudostratified columnar epithelium appears to have several layers because the nuclei of the cells are at various levels (Table 4.1E). Even though all the cells are attached to the basement membrane in a single layer, some cells do not extend to the apical surface. When viewed from the side, these features give the false impression of a multilayered tissue—thus the name pseudostratified epithelium (*pseudo-* = false).

Pseudostratified ciliated columnar epithelium contains cells that extend to the surface and either secrete mucus (goblet cells) or bear cilia. The secreted mucus traps foreign particles and the cilia sweep away mucus for eventual elimination from the body.

Pseudostratified nonciliated columnar epithelium contains cells without cilia and also lacks goblet cells and functions in absorption and protection.

Stratified Epithelium

Stratified epithelium contains two or more layers of cells and therefore is useful for protection of underlying tissues in areas where there is considerable wear and tear. Some cells of stratified epithelia also produce secretions. The name of the specific kind of stratified epithelium depends on the shape of the cells in the apical layer.

STRATIFIED SQUAMOUS EPITHELIUM Cells in the apical layer of this type of epithelium are flat, whereas in the deep layers, cells vary in shape from cuboidal to columnar (Table 4.1F). The basal (deepest) cells continually undergo cell division. As new cells grow, the cells of the basal layer are pushed toward the surface. As they move farther from the deeper layers and from their blood supply in the underlying connective tissue, they become dehydrated, shrunken, and harder. At the apical layer the cells lose their cell junctions and are sloughed off, but they are replaced as new cells continually emerge from the basal cells. Stratified squamous epithelium exists in both keratinized and nonkeratinized forms.

Keratinized stratified squamous epithelium develops a tough layer of keratin in the apical layer and several layers deep to it. *Keratin* is a tough protein that helps protect and waterproof the skin and protect underlying tissues from microbes, heat, and chemicals.

Nonkeratinized stratified squamous epithelium contains living cells as evidenced by nuclei at the free surface. This tissue is found, for example, lining the mouth; it does not contain keratin in the apical layer and remains moist.

Stratified squamous epithelium forms the first line of defense against microbes.

• CLINICAL CONNECTION | Pap Smear

A **Papanicolaou test** (pa-pa-NI-kō-lō), also called a ***Pap test*** or ***Pap smear,*** involves collection and microscopic examination of epithelial cells that have been scraped off the apical layer of a tissue. A very common type of Pap test involves examining the cells from the nonkeratinized stratified squamous epithelium of the vagina and cervix (inferior portion) of the uterus. This type of Pap test is performed mainly to detect early changes in the cells of the female reproductive system that may indicate a precancerous condition or cancer. In performing a Pap smear, a physician collects cells, which are then smeared on a microscope slide. The slides are then sent to a laboratory for analysis. Pap tests should be started within three years of the onset of sexual activity, or age 21, whichever comes first. Annual screening is recommended for females ages 21–30 and every 2–3 years for females age 30 or older following three consecutive negative Pap tests. •

STRATIFIED CUBOIDAL EPITHELIUM This fairly rare type of epithelium sometimes consists of more than two layers of cells (Table 4.1G). Cells in the apical layer are cuboidal. Its function is mainly protective; in some locations it also functions in secretion and absorption.

TABLE 4.1

Epithelial Tissues: Covering and Lining Epithelium

SIMPLE EPITHELIUM

A. Simple squamous epithelium

Description: Single layer of flat cells; centrally located nucleus.

Location: Lines heart, blood vessels, lymphatic vessels, air sacs of lungs, glomerular (Bowman's) capsule of kidneys, and inner surface of the tympanic membrance (eardrum); forms epithelial layer of serous membranes (mesothelium), such as the peritoneum.

Function: Filtration, diffusion, osmosis, and secretion in serous membranes.

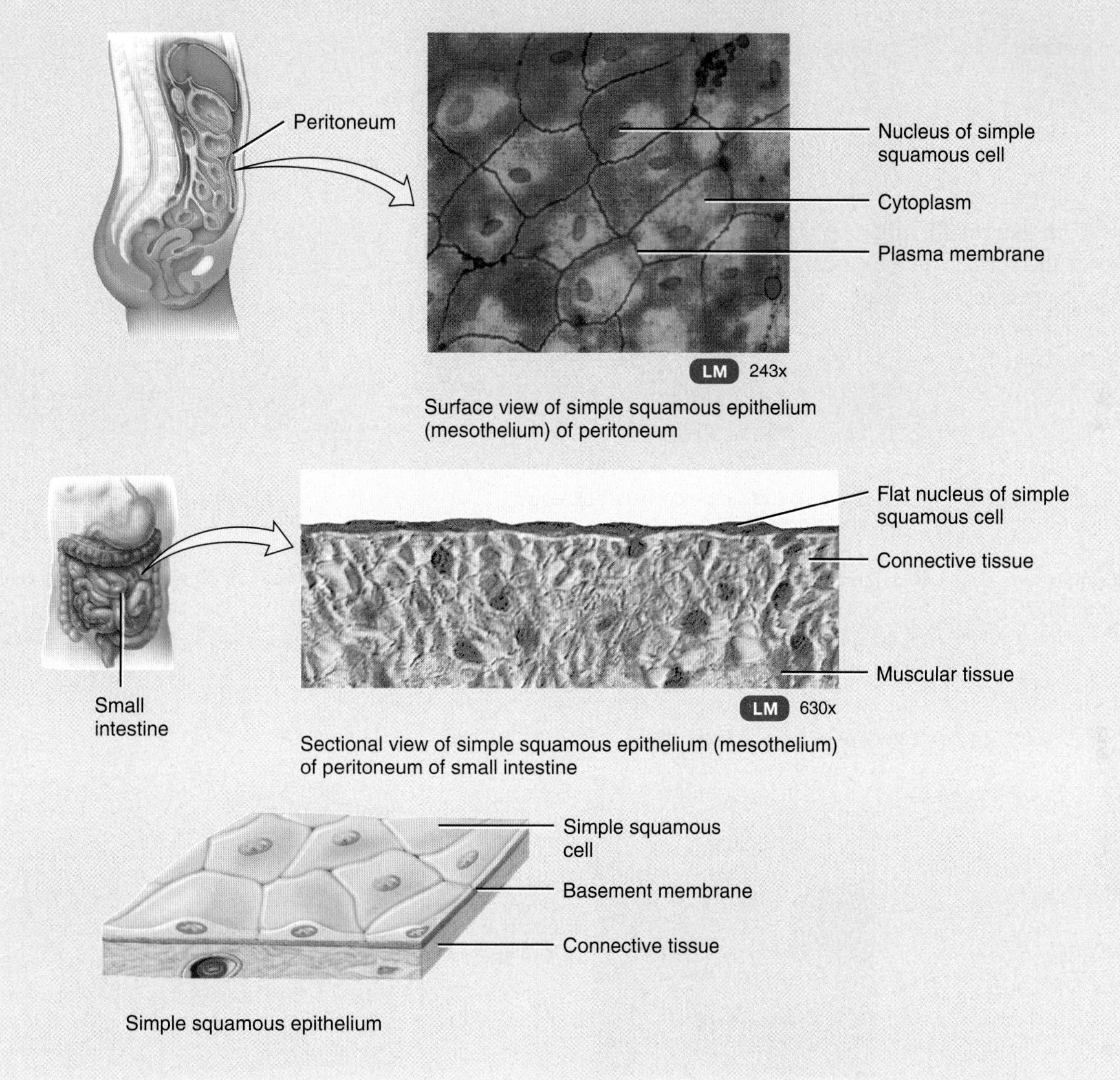

Surface view of simple squamous epithelium (mesothelium) of peritoneum

Sectional view of simple squamous epithelium (mesothelium) of peritoneum of small intestine

Simple squamous epithelium

CONTINUES

Stratified Columnar Epithelium This type of tissue also is uncommon. Usually the layer or layers near the basement membrane consist of shortened, irregularly shaped cells; only the apical layer of cells is columnar in form (Table 4.1H). This type of epithelium functions in protection and secretion.

Transitional Epithelium This type of stratified epithelium is variable in appearance, depending on whether the organ it lines is unstretched or stretched. It is, therefore, in a somewhat constant state of transition (change). In its unstretched state (Table 4.1I), transitional epithelium looks similar to stratified cuboidal epithelium, except that the cells in the apical layer tend to be large and rounded (often described as being dome shaped). As the cells are stretched, they become flatter, giving the appearance of stratified squamous epithelium. Because of its elasticity, transitional epithelium lines hollow structures that are subjected to expansion from within, such as the urinary bladder. It allows organs to stretch to hold a variable amount of fluid without rupturing.

TABLE 4.1 CONTINUED

Epithelial Tissues: Covering and Lining Epithelium

SIMPLE EPITHELIUM

B. Simple cuboidal epithelium

Description: Single layer of cube-shaped cells; centrally located nucleus.

Location: Lines kidney tubules and smaller ducts of many glands, and makes up the secreting portion of some glands such as the thyroid gland, covers surface of ovary, lines anterior surface of capsule of the lens of the eye, forms the pigmented epithelium at the posterior surface of the eye.

Function: Secretion and absorption.

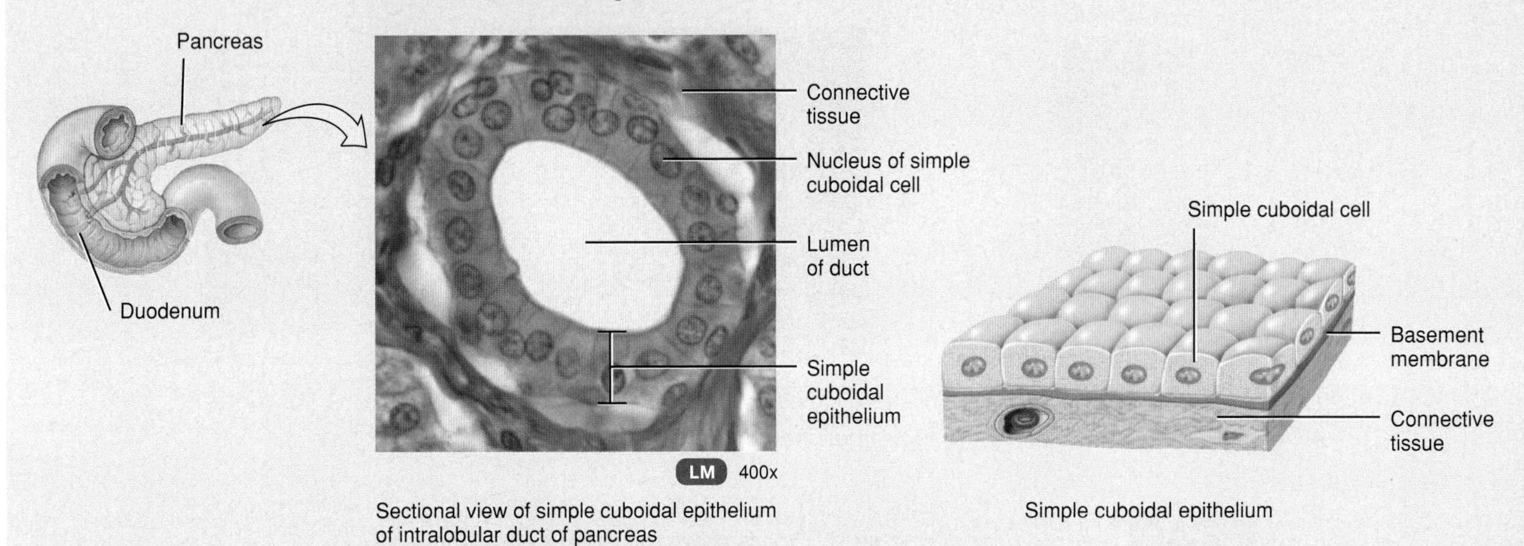

Sectional view of simple cuboidal epithelium of intralobular duct of pancreas

Simple cuboidal epithelium

C. Nonciliated simple columnar epithelium

Description: Single layer of nonciliated column-like cells with nuclei near bases of cells; contains goblet cells and cells with microvilli in some locations.

Location: Lines most of the gastrointestinal tract (from the stomach to the anus), ducts of many glands, and gallbladder.

Function: Secretion and absorption.

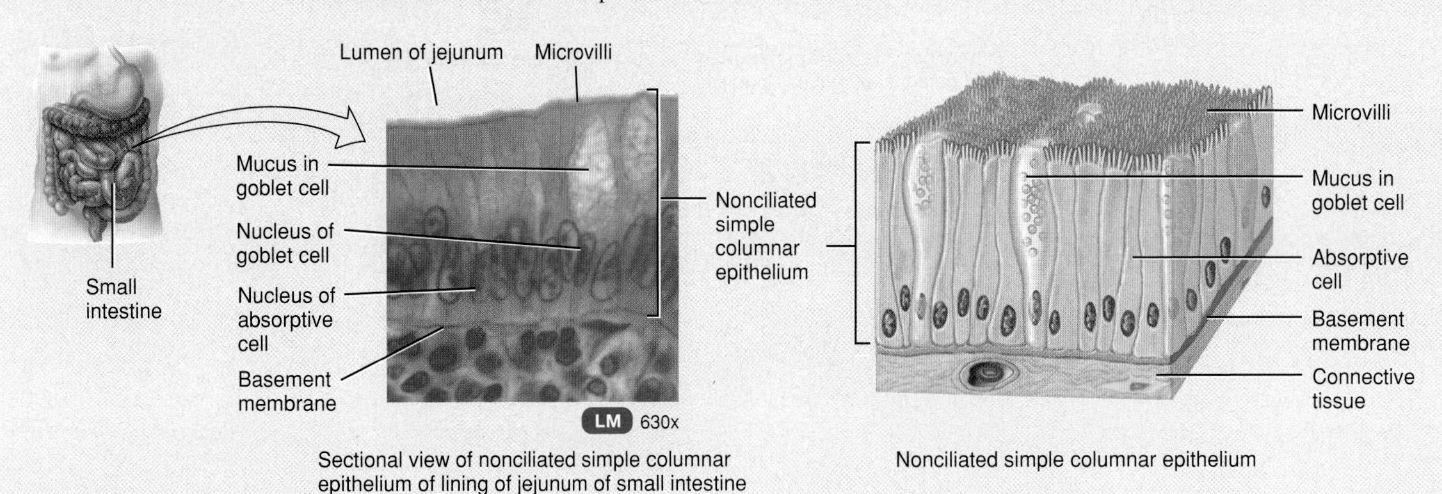

Sectional view of nonciliated simple columnar epithelium of lining of jejunum of small intestine

Nonciliated simple columnar epithelium

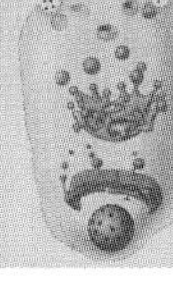

D. Ciliated simple columnar epithelium

Description: Single layer of ciliated column-like cells with nuclei near bases; contains goblet cells in some locations.

Location: Lines a few portions of upper respiratory tract, uterine (fallopian) tubes, uterus, some paranasal sinuses, and central canal of spinal cord.

Function: Moves mucus and other substances by ciliary action.

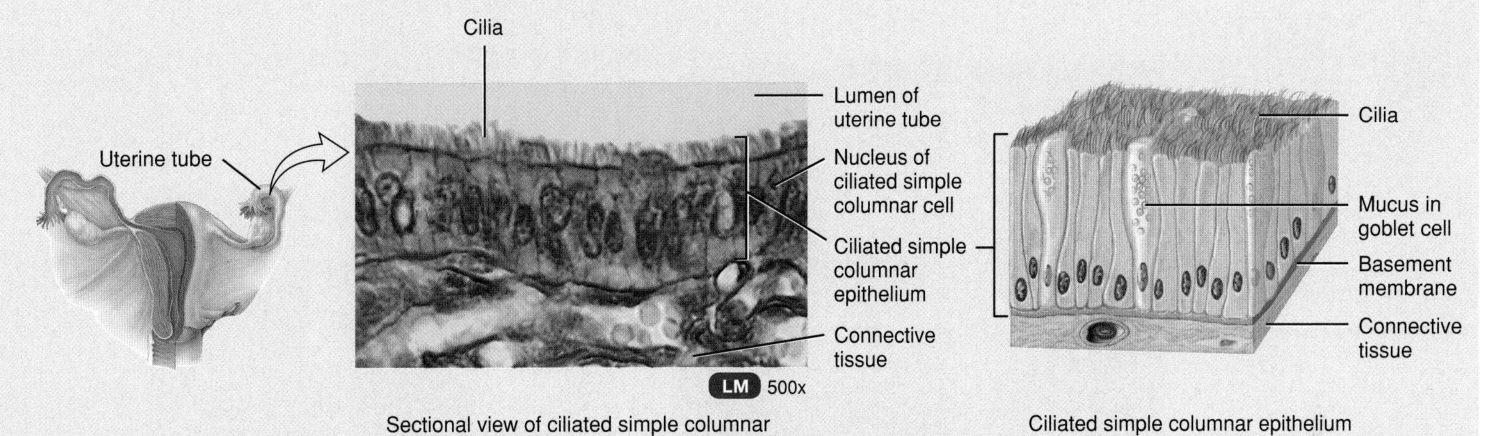

Sectional view of ciliated simple columnar epithelium of uterine tube

Ciliated simple columnar epithelium

E. Pseudostratified columnar epithelium

Description: Not a true stratified tissue; nuclei of cells are at different levels; all cells are attached to basement membrane, but not all reach the apical surface.

Location: Psuedostratified ciliated columnar epithelium lines the airways of most of upper respiratory tract; pseudostratified nonciliated columnar epithelium lines larger ducts of many glands, epididymis, and part of male urethra.

Function: Secretion and movement of mucus by ciliary action (ciliated); absorption and protection (nonciliated).

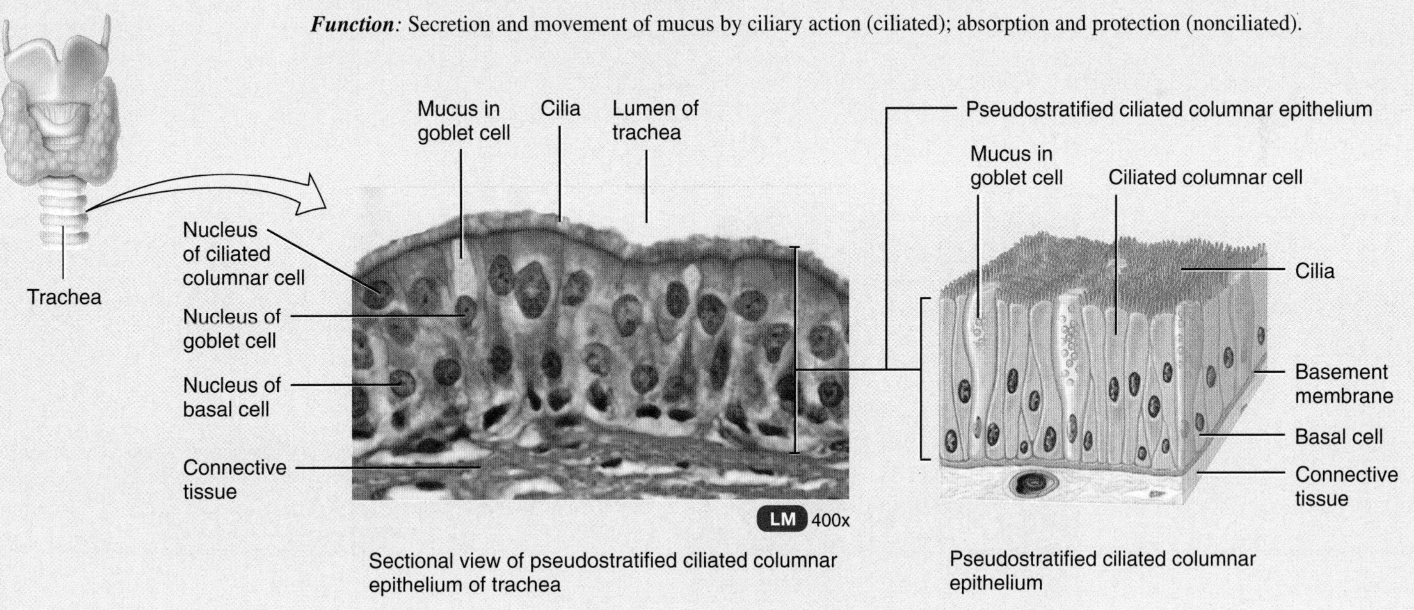

Sectional view of pseudostratified ciliated columnar epithelium of trachea

Pseudostratified ciliated columnar epithelium

CONTINUES

TABLE 4.1 CONTINUED

Epithelial Tissues: Covering and Lining Epithelium

STRATIFIED EPITHELIUM

F. Stratified squamous epithelium

Description: Several layers of cells; cuboidal to columnar shape in deep layers; squamous cells form the apical layer and several layers deep to it; cells from the basal layer replace surface cells as they are lost.

Location: Keratinized variety forms superficial layer of skin; nonkeratinized variety lines wet surfaces, such as lining of the mouth, esophagus, part of epiglottis, part of pharynx, and vagina, and covers the tongue.

Function: Protection.

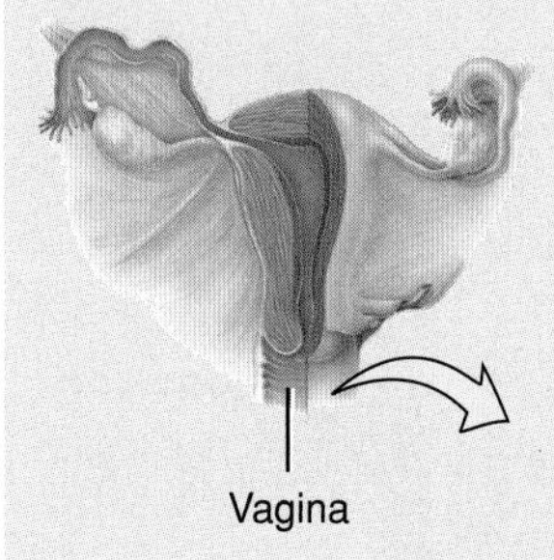

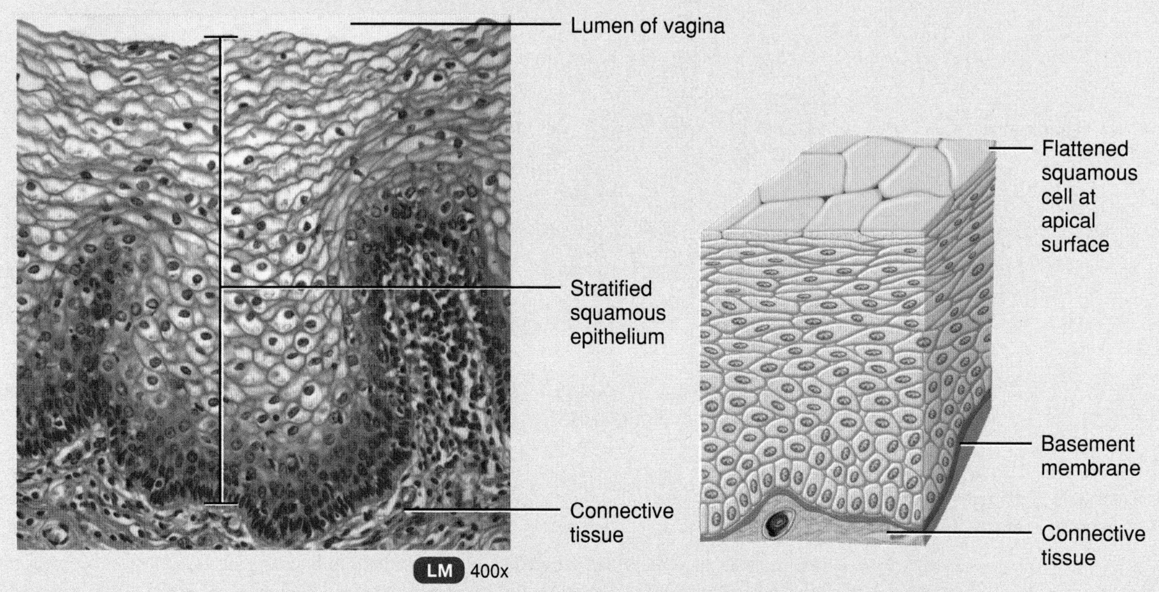

Sectional view of stratified squamous epithelium of vagina

Stratified squamous epithelium

G. Stratified cuboidal epithelium

Description: Two or more layers of cells in which cells in the apical layer are cube-shaped.

Location: Ducts of adult sweat glands and esophageal glands and part of male urethra.

Function: Protection and limited secretion and absorption.

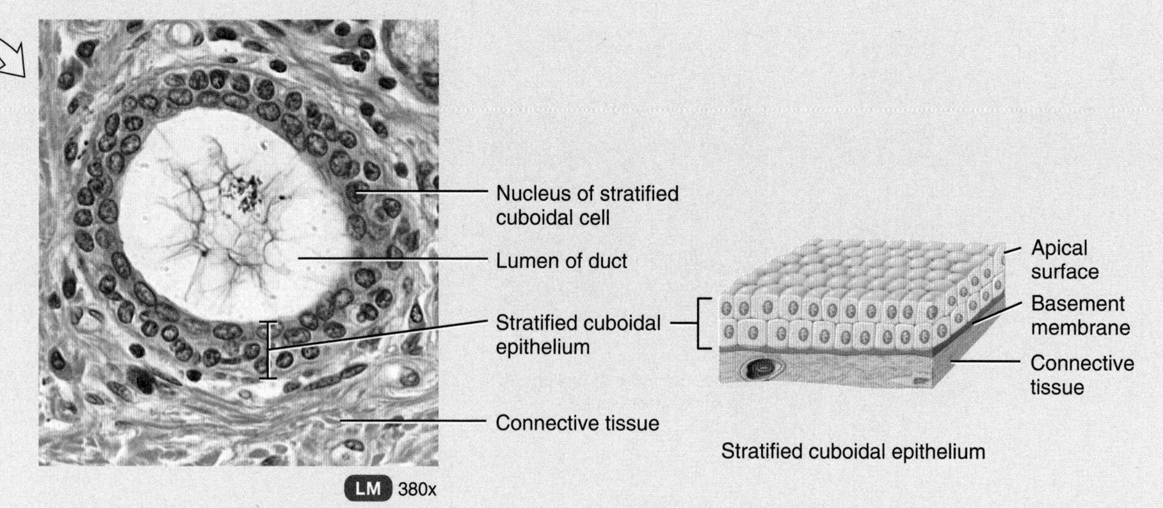

Sectional view of stratified cuboidal epithelium of the duct of an esophageal gland

Stratified cuboidal epithelium

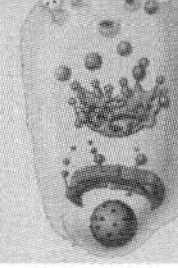

H. Stratified columnar epithelium

Description: Several layers of irregularly shaped cells; only the apical layer has columnar cells.

Location: Lines part of urethra, large excretory ducts of some glands such as esophageal glands, small areas in anal mucous membrane, and a part of the conjunctiva of the eye.

Function: Protection and secretion.

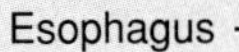

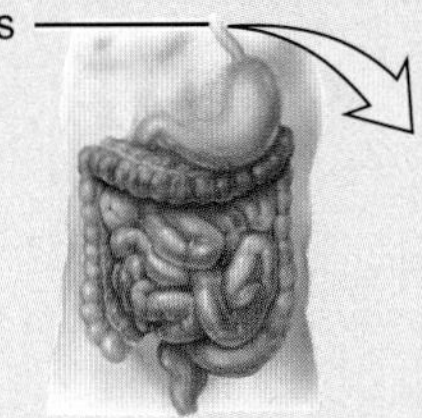

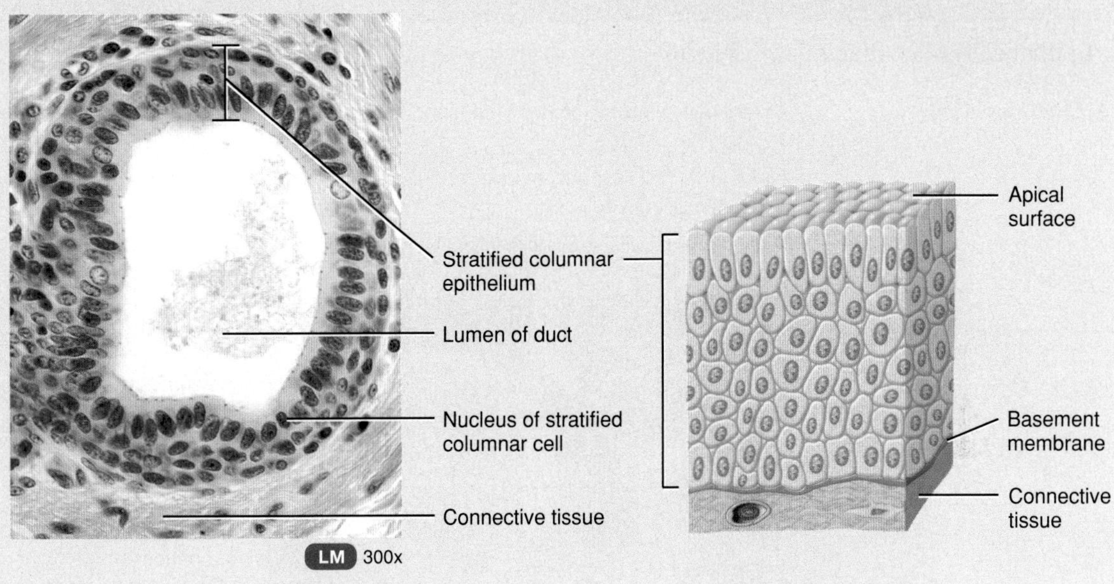

Sectional view of stratified columnar epithelium of the duct of an esophageal gland

Stratified columnar epithelium

I. Transitional epithelium

Description: Appearance is variable (transitional); shape of cells in apical layer ranges from squamous (when stretched) to cuboidal (when relaxed).

Location: Lines urinary bladder and portions of ureters and urethra.

Function: Permits distention.

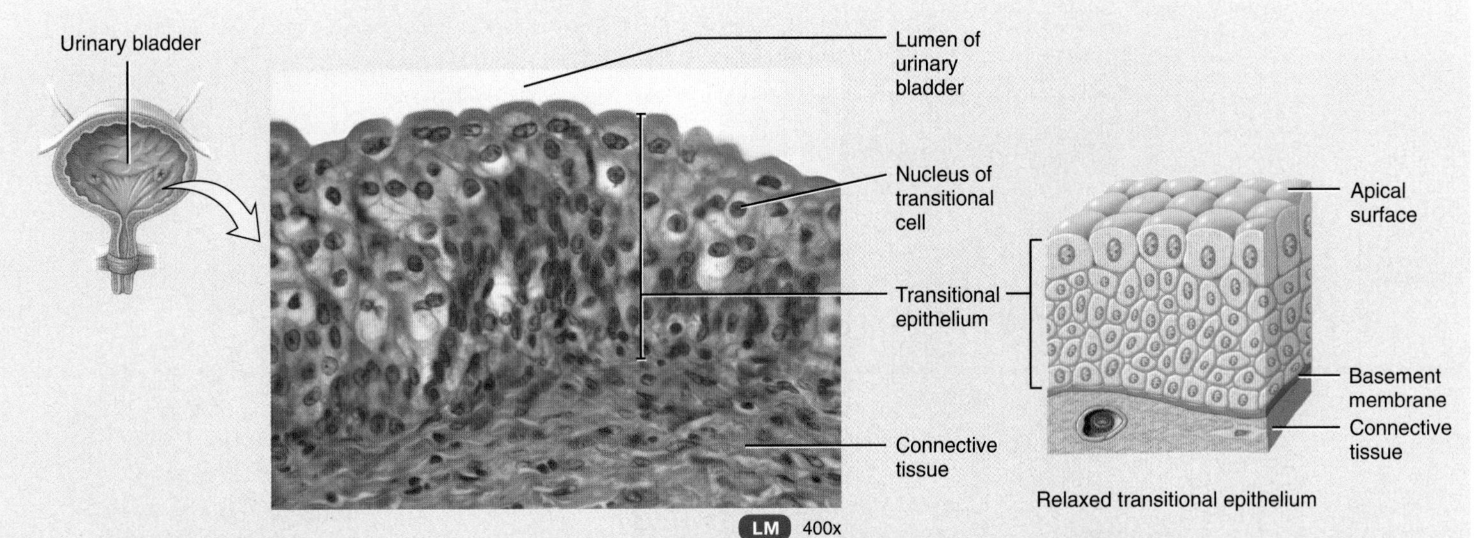

Relaxed transitional epithelium

Sectional view of transitional epithelium of urinary bladder in relaxed state

Glandular Epithelium

The function of glandular epithelium, secretion, is accomplished by glandular cells that often lie in clusters deep to the covering and lining epithelium. A ***gland*** may consist of a single cell or a group of cells that secrete substances into ducts (tubes), onto a surface, or into the blood. All glands of the body are classified as either endocrine or exocrine.

The secretions of ***endocrine glands*** (EN-dō-krin; *endo-* = within; *-crine* = secretion) (Table 4.2A) enter the interstitial fluid and then diffuse directly into the bloodstream without flowing through a duct. These secretions, called *hormones,* regulate many metabolic and physiological activities to maintain homeostasis. The pituitary, thyroid, and adrenal glands are examples of endocrine glands. Endocrine glands will be described in detail in Chapter 20.

Table 4.2

Epithelial Tissue: Glandular Epithelium

A. Endocrine glands

Description: Secretory products (hormones) diffuse into blood after passing through interstitial fluid.

Location: Examples include pituitary gland at base of brain, pineal gland in brain, thyroid and parathyroid glands near larynx (voice box), adrenal glands superior to kidneys, pancreas near stomach, ovaries in pelvic cavity, testes in scrotum, and thymus in thoracic cavity.

Function: Produce hormones that regulate various body activities.

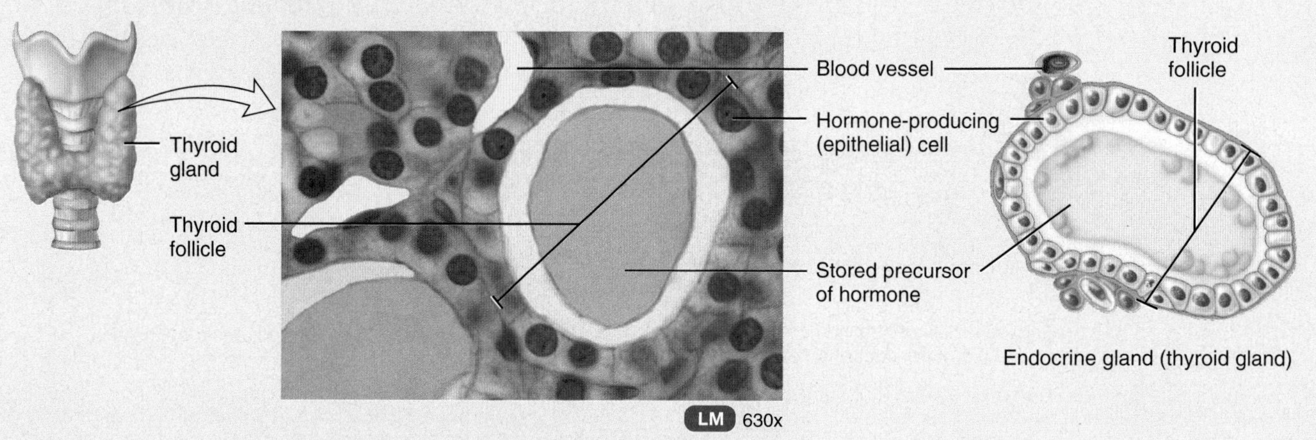

Sectional view of endocrine gland (thyroid gland)

Endocrine gland (thyroid gland)

B. Exocrine glands

Description: Secretory products released into ducts.

Location: Sweat, oil, and earwax glands of the skin; digestive glands such as salivary glands, which secrete into mouth cavity, and pancreas, which secretes into the small intestine.

Function: Produce substances such as sweat, oil, earwax, saliva, or digestive enzymes.

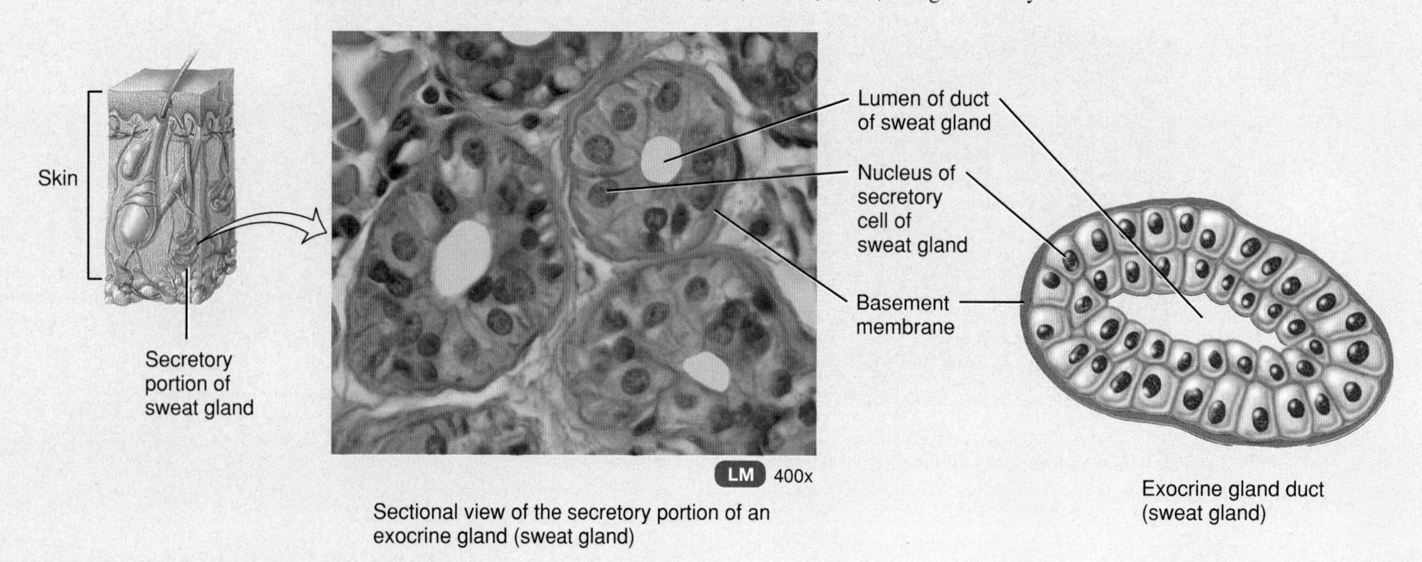

Sectional view of the secretory portion of an exocrine gland (sweat gland)

Exocrine gland duct (sweat gland)

Exocrine glands (EK-sō-krin; *exo-* = outside; Table 4.2B) secrete their products into ducts that empty onto the surface of a covering and lining epithelium such as the skin surface or the lumen of a hollow organ. The secretions of exocrine glands include mucus, sweat, oil, earwax, saliva, and digestive enzymes. Examples of exocrine glands include sudoriferous (sweat) glands, which produce sweat to help lower body temperature, and salivary glands, which secrete saliva. Saliva contains mucus and digestive enzymes among other substances. As you will learn later in the text, some glands of the body, such as the pancreas, ovaries, and testes, are mixed glands that contain both endocrine and exocrine tissue.

Structural Classification of Exocrine Glands

Exocrine glands are classified as unicellular or multicellular. As the name implies, ***unicellular glands*** are single-celled. Goblet cells are important unicellular exocrine glands that secrete mucus directly onto the apical surface of a lining epithelium. Most glands are ***multicellular glands,*** composed of many cells that form a distinctive microscopic structure or macroscopic organ. Examples include sudoriferous, sebaceous (oil), and salivary glands.

Functional Classification of Exocrine Glands

The functional classification of exocrine glands is based on how their secretions are released. Secretions of ***merocrine glands*** (MER-ō-krin; *mero-* = a part), also known as ***eccrine*** (EK-rin) ***glands,*** are synthesized on ribosomes attached to rough ER; processed, sorted, and packaged by the Golgi complex; and released from the cell in secretory vesicles via exocytosis (Figure 4.4a). Most exocrine glands of the body are merocrine glands. Examples include the salivary glands and pancreas. ***Apocrine glands*** (AP-ō-krin; *apo-* = from) accumulate their secretory product at the apical surface of the secreting cell. Then, that portion of the cell pinches off from the rest of the cell to release the secretion (Figure 4.4b). The remaining part of the cell repairs itself and repeats the process. The cells of ***holocrine glands*** (HŌ-lō-krin; *holo-* = entire) accumulate a secretory product in their cytosol. As the secretory cell matures, it ruptures and becomes the secretory product (Figure 4.4c). The sloughed off cell is replaced by a new cell. One example of a holocrine gland is a sebaceous gland of the skin.

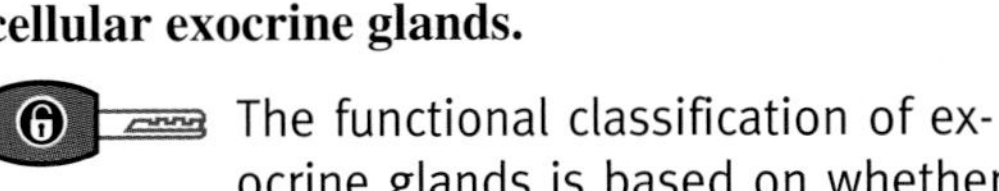

Figure 4.4 Functional classification of multicellular exocrine glands.

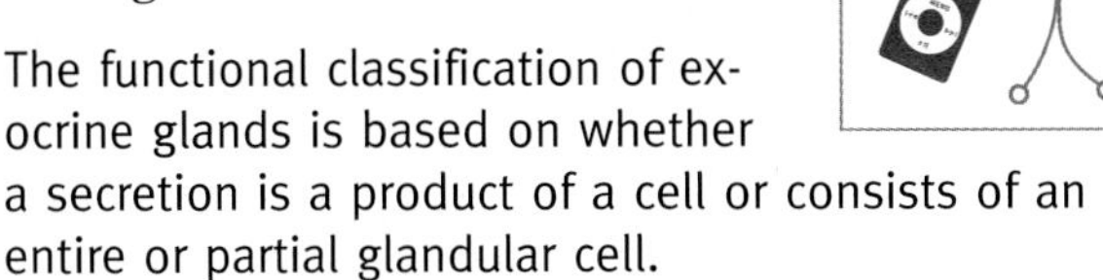

The functional classification of exocrine glands is based on whether a secretion is a product of a cell or consists of an entire or partial glandular cell.

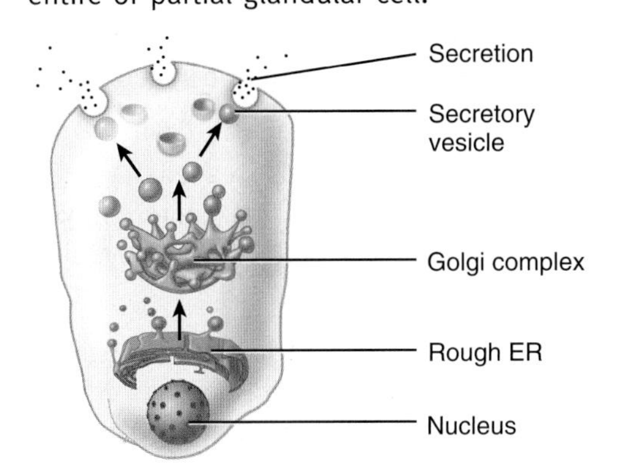

(a) Merocrine (eccrine) secretion

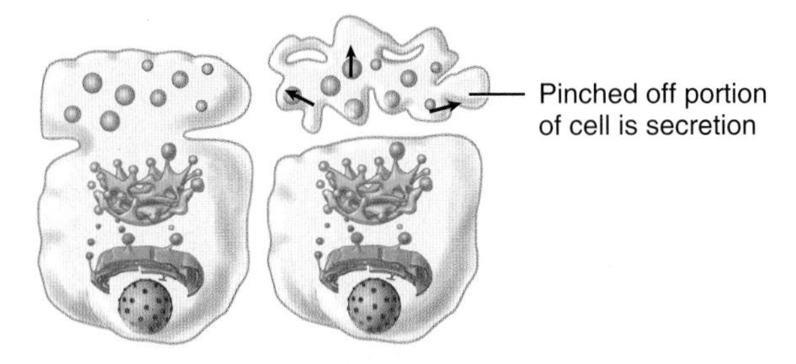

(b) Apocrine secretion

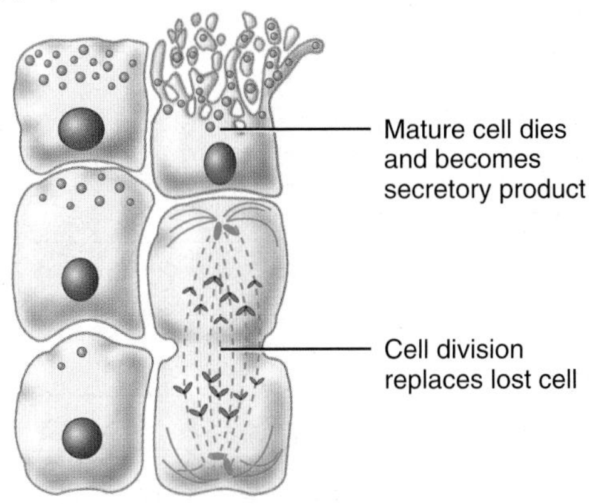

(c) Holocrine secretion

What class of glands are sebaceous (oil) glands? Salivary glands?

CHECKPOINT

4. What characteristics are common to all epithelial tissues?
5. Describe the various cell shapes and layering arrangements of epithelium.
6. Explain how the structure of the following kinds of epithelium is related to the functions of each: simple squamous, simple cuboidal, simple columnar (nonciliated and ciliated), pseudostratified columnar (nonciliated and ciliated), stratified squamous (keratinized and nonkeratinized), stratified cuboidal, stratified columnar, and transitional.

4.4 CONNECTIVE TISSUE

OBJECTIVES

- Discuss the general features of connective tissue.
- Describe the structure, location, and function of the various types of connective tissue.

Connective tissue is the most abundant and widely distributed tissue in the body. In its various forms, connective tissue has a variety of functions. It binds together, supports, and strengthens

ens other body tissues; protects and insulates internal organs; compartmentalizes structures such as skeletal muscles; is the major transport system within the body (blood is a fluid connective tissue); is the major site of stored energy reserves (adipose, or fat, tissue); and is the main site of immune responses.

General Features of Connective Tissue

Connective tissue consists of two basic elements: cells and extracellular matrix. A connective tissue's ***extracellular matrix*** (MĀ-triks) is the material between its widely spaced cells. The extracellular matrix consists of protein fibers and ground substance, the material between the cells and the fibers. The extracellular matrix is usually secreted by the connective tissue cells and determines the tissue's qualities. For instance, in cartilage, the extracellular matrix is firm but pliable. The extracellular matrix of bone, by contrast, is hard and not pliable.

In contrast to epithelia, connective tissues do not usually occur on body surfaces. Also, unlike epithelia, connective tissues usually are highly vascular; that is, they have a rich blood supply. Cartilage is an exception and it is avascular. Tendons and ligaments have a scanty blood supply. Except for cartilage, connective tissues, like epithelia, are supplied with nerves.

Connective Tissue Cells

The types of connective tissue cells vary according to the type of tissue and include the following (Figure 4.5):

1. ***Fibroblasts*** (FĪ-brō-blasts; *fibro-* = fibers) are large, flat cells with branching processes. They are present in nearly all connective tissues, and usually are the most numerous cell type. Fibroblasts migrate through the connective tissue, secreting the various fibers and ground substance of the extracellular matrix.

2. ***Macrophages*** (MAK-rō-fā-jez; *macro-* = large; *-phages* = eaters) develop from monocytes, a type of white blood cell. Macrophages have an irregular shape with short branching projections and are capable of engulfing bacteria and cellular debris by phagocytosis.

3. ***Plasma cells*** are small cells that develop from a type of white blood cell called a B lymphocyte. Plasma cells secrete antibodies, proteins that attack or neutralize foreign substances in the body. Thus, plasma cells are an important part of the body's immune response.

4. ***Mast cells*** are abundant alongside the blood vessels that supply connective tissue. They produce histamine, a chemical that dilates small blood vessels as part of the inflammatory response, the body's reaction to injury or infection. Mast cells can also kill bacteria.

5. ***Adipocytes*** (A-di-pō-sīts), also called fat cells or adipose cells, are connective tissue cells that store triglycerides (fats). They are found below the skin and around organs such as the heart and kidneys.

White blood cells are not normally found in significant numbers in connective tissues. However, in response to certain conditions, white blood cells can leave blood and enter connective tissues. For example, *neutrophils* gather at sites of infection and *eosinophils* migrate to sites of parasitic invasion and allergic responses.

Figure 4.5 Representative cells and fibers present in connective tissues.

6 Fibroblasts are usually the most numerous connective tissue cells.

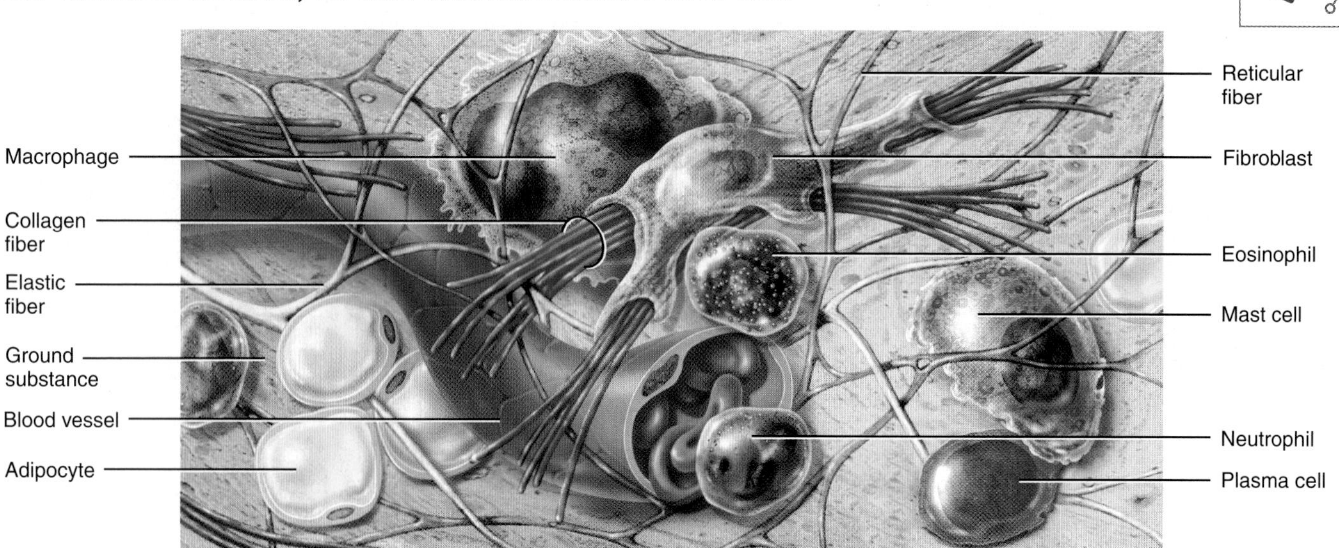

? What is the function of fibroblasts?

Connective Tissue Extracellular Matrix

Each type of connective tissue has unique properties, based on the specific extracellular materials between the cells. The extracellular matrix consists of a fluid, gel, or solid ground substance plus various protein fibers.

Ground Substance

Ground substance, the component of a connective tissue between the cells and fibers, supports cells, binds them together, and provides a medium through which substances are exchanged between the blood and cells. The ground substance plays an active role in how tissues develop, migrate, proliferate, and change shape, and in how they carry out their metabolic functions.

Ground substance contains water and an assortment of large organic molecules, many of which are complex combinations of polysaccharides and proteins. For example, the polysaccharide ***hyaluronic acid*** (hī′-a-loo-RON-ik) is a viscous, slippery substance that binds cells together, lubricates joints, and helps maintain the shape of the eyeballs. It also appears to play a role in helping phagocytes migrate through connective tissue during development and wound repair. White blood cells, sperm cells, and some bacteria produce *hyaluronidase,* an enzyme that breaks apart hyaluronic acid and causes the ground substance of connective tissue to become watery. The ability to produce hyaluronidase enables white blood cells to move through connective tissues to reach sites of infection and enables sperm cells to penetrate the ovum during fertilization. It also accounts for how bacteria spread through connective tissues.

Another ground substance is the polysaccharide ***chondroitin sulfate*** (kon-DROY-tin), which provides support and adhesiveness in connective tissues in bone, cartilage, skin, and blood vessels. ***Glucosamine*** is a protein, polysaccharide molecule.

In recent years, chondroitin sulfate and glucosamine have been used as nutritional supplements either alone or in combination to promote and maintain the structure and function of joint cartilage, to provide pain relief from osteoarthritis, and to reduce joint inflammation. Although these supplements have benefited some individuals with moderate to severe osteoarthritis, the benefit is minimal in lesser cases. More research is needed to determine how they act and why they help some people and not others.

Fibers

Fibers in the extracellular matrix strengthen and support connective tissues. Three types of fibers are embedded in the matrix between the cells: collagen fibers, elastic fibers, and reticular fibers.

Collagen fibers (KOL-a-jen; *colla-* = glue) are very strong and resist pulling forces, but they are not stiff, which promotes tissue flexibility. These fibers often occur in bundles lying parallel to one another (Figure 4.5). The bundle arrangement affords great strength. Chemically, collagen fibers consist of the protein *tropocollagen.* Tropocollagen molecules form *collagen fibrils* and then, through hydrogen bonding, fibrils are organized into larger *collagen fibers.* Collagen, in its multiple states of organization, is the most abundant protein in your body, representing about 25% of total protein. Collagen fibers are found in most types of connective tissues, including bone, cartilage, tendons, and ligaments.

Manual therapists are aware that many people "hold their stress" in a specific area of the body. This area may be chronically tender or may be the first area to become tender following a change—an increase or decrease—in the level of physical activity. These areas contain increased numbers of collagen fibers. *Hydrogen bonding,* as you learned in Chapter 2, is responsible for maintaining the three-dimensional integrity of many proteins; although any one hydrogen bond is weak, the extraordinarily large number of them in a collagen fiber, and between collagen fibers, makes collagen-containing tissues exceptionally strong. The analogy of a rope is appropriate since the individual components of a rope are exceedingly thin and weak filaments, yet the large number of them twisted together produces a structure with great strength. Hydrogen bonding is thus responsible for the strength of a collagen *molecule* as well as collagen *fibers.* However, with either greatly increased or decreased activity in the muscle tissue of a specific area of the body, molecules bind to each other and collagen fibers increase. The connective tissues become much more dense and thus reduce the range of motion (ROM) of the area.

Elastic fibers, which are smaller in diameter than collagen fibers, branch and join together to form a network within a tissue (Figure 4.5). An elastic fiber consists of molecules of a protein called *elastin* surrounded by a glycoprotein named *fibrillin,* which is essential to the stability of an elastic fiber. Elastic fibers are strong but can be stretched up to one-and-a-half times their relaxed length without breaking. Equally important, elastic fibers have the ability to return to their original shape after being stretched, a property called *elasticity.* Elastic fibers are plentiful in skin, blood vessel walls, and lung tissue.

• CLINICAL CONNECTION | Marfan syndrome

Marfan syndrome (MAR-fan) is an inherited disorder caused by a defective fibrillin gene. The result is abnormal development of elastic fibers. Tissues rich in elastic fibers are malformed or weakened. Structures affected most seriously are the connective tissue covering layer of bones (periosteum), the ligament that suspends the lens of the eye, and the walls of the large arteries. People with Marfan syndrome tend to be tall and have disproportionately long arms, legs, fingers, and toes (such as some basketball players). A common symptom is blurred vision caused by displacement of the lens of the eye. The most life-threatening complication of Marfan syndrome is weakening of the aorta (the main artery that emerges from the heart), which can suddenly burst. The death rate of basketball players at a young age is disproportionately high. •

Reticular fibers (*reticul-* = net), consisting of *collagen fibrils* and a coating of glycoprotein, provide support in the walls of blood vessels and form branching networks around fat cells, nerve fibers, and skeletal and smooth muscle cells. Produced by fibroblasts, they are much thinner than collagen fibers. Like collagen fibers, reticular fibers provide support and strength and also form the ***stroma*** (STRŌ-ma = bed or covering) or supporting framework of many soft organs, such as the spleen and lymph nodes. These fibers also help form the basement membrane (see Figure 4.2).

Classification of Connective Tissues

Because of the diversity of cells and extracellular matrix and the differences in their relative proportions, the classification of connective tissues is not always clear-cut. We offer the following scheme:

I. Embryonic connective tissue
  A. Mesenchyme
  B. Mucous connective tissue
II. Mature connective tissue
  A. Loose connective tissue
    1. Areolar connective tissue
    2. Adipose tissue
    3. Reticular connective tissue
  B. Dense connective tissue
    1. Dense regular connective tissue
    2. Dense irregular connective tissue
    3. Elastic connective tissue
  C. Cartilage
    1. Hyaline cartilage
    2. Fibrocartilage
    3. Elastic cartilage
  D. Bone tissue
  E. Liquid connective tissue
    1. Blood tissue
    2. Lymph

Note that our classification scheme has two major subclasses of connective tissues: embryonic and mature. ***Embryonic connective tissue*** is present primarily in the *embryo,* the developing human from fertilization through the first two months of pregnancy, and in the *fetus,* the developing human from the third month of pregnancy to birth.

One example of embryonic connective tissue is ***mesenchyme*** (MEZ-en-kīm), the tissue from which all other connective tissues eventually arise (Table 4.3A). Mesenchyme is composed of irregularly shaped cells, a semifluid ground substance, and delicate reticular fibers. In spite of its classification as an embryonic tissue, mesenchymal cells are present throughout life and, as described later in this chapter, are the stem cells from which all connective tissues continue to form.

Another kind of embryonic tissue is ***mucous connective tissue (Wharton's jelly),*** found mainly in the umbilical cord of the fetus. Mucous connective tissue is a form of mesenchyme that contains widely scattered fibroblasts, a more viscous, jelly-like ground substance, and collagen fibers (Table 4.3B).

The second major subclass of connective tissue, ***mature connective tissue,*** is present in the newborn. Its cells arise from mesenchyme. In the next section we explore the numerous types of mature connective tissue.

Types of Mature Connective Tissue

Loose Connective Tissue

The fibers in ***loose connective tissue*** are loosely arranged among the many cells. The types of loose connective tissue are areolar connective tissue, adipose tissue, and reticular connective tissue.

Areolar Connective Tissue One of the most widely distributed connective tissues in the body is ***areolar connective tissue*** (a-RĒ-ō-lar; *areol-* = a small space). It contains several kinds of cells, including fibroblasts, macrophages, plasma cells, mast cells, adipocytes, and a few white blood cells (Table 4.4A). All three types of fibers—collagen, elastic, and reticular—are arranged randomly throughout the tissue. Areolar connective tissue has been called the "packing material" of the body since it is found in and around nearly every structure of the body. Combined with adipose tissue, areolar connective tissue forms the *subcutaneous layer,* the layer of tissue that attaches the skin to underlying tissues and organs.

Adipose Tissue Adipose tissue is a loose connective tissue in which the cells, called ***adipocytes*** (*adipo-* = fat), are specialized for storage of triglycerides (fats) (Table 4.4B). Because the cell fills with a single, large triglyceride droplet, the cytoplasm and nucleus are pushed to the periphery of the cell. Adipose tissue is found wherever areolar connective tissue is located. Adipose tissue is a good insulator and can therefore reduce heat loss through the skin. It is a major energy reserve and generally supports and protects various organs. As the amount of adipose tissue increases with weight gain, new blood vessels form. Thus, an obese person has many more miles of blood vessels than does a lean person, a situation that can cause high blood pressure.

Reticular Connective Tissue Reticular connective tissue consists of fine interlacing reticular fibers and reticular cells, cells that are connected to each other and form a network (Table 4.4C). Reticular connective tissue forms the stroma (supporting framework) of certain organs, helps bind together smooth muscle cells, and filters worn-out blood cells and bacteria.

Dense Connective Tissue

Dense connective tissue contains more numerous, thicker, and denser fibers (more closely packed) but fewer cells than loose connective tissue. There are three types: dense regular connective tissue, dense irregular connective tissue, and elastic connective tissue.

Dense Regular Connective Tissue In this tissue, bundles of collagen fibers are arranged *regularly* in parallel patterns that provide the tissue with great strength (Table 4.4D). The tissue structure withstands pulling along the axis of the fibers. Fibroblasts, which produce the fibers and ground substance, appear in rows between the fibers. The tissue is silvery white and tough, yet somewhat pliable. Examples are tendons and most ligaments. The predominant collagen fibers are not living (since

TABLE 4.3

Embryonic Connective Tissues

A. Mesenchyme

Description: Consists of irregularly shaped mesenchymal cells embedded in a semifluid ground substance that contains reticular fibers.

Location: Under skin and along developing bones of embryo; some mesenchymal cells are found in adult connective tissue, especially along blood vessels.

Function: Forms all other types of connective tissue.

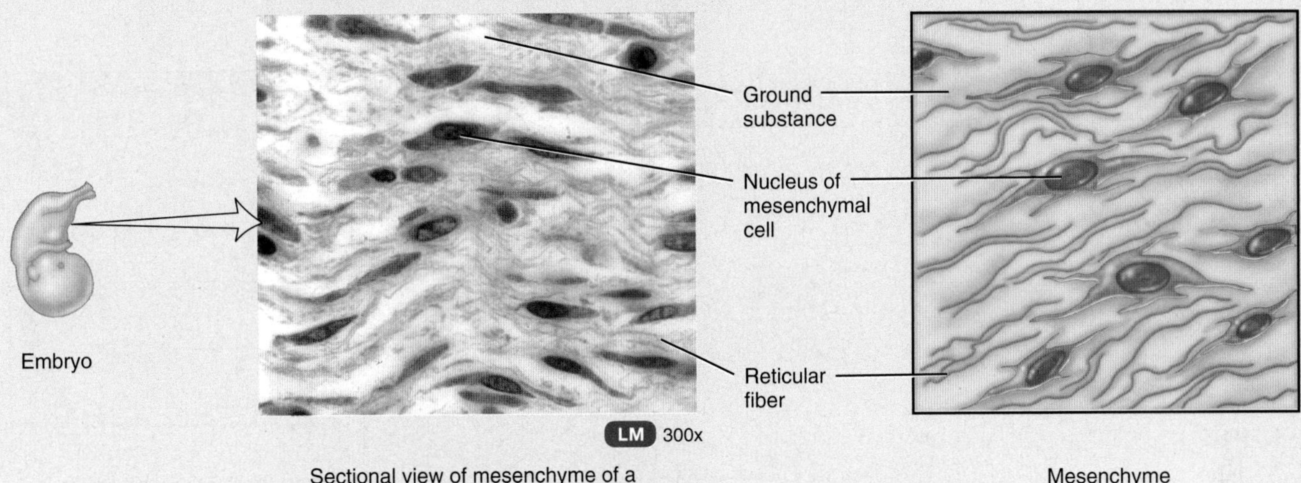

Sectional view of mesenchyme of a developing embryo

Mesenchyme

B. Mucous connective tissue

Description: Consists of widely scattered fibroblasts embedded in a viscous, jelly-like ground substance that contains fine collagen fibers.

Location: Umbilical cord of fetus.

Function: Support.

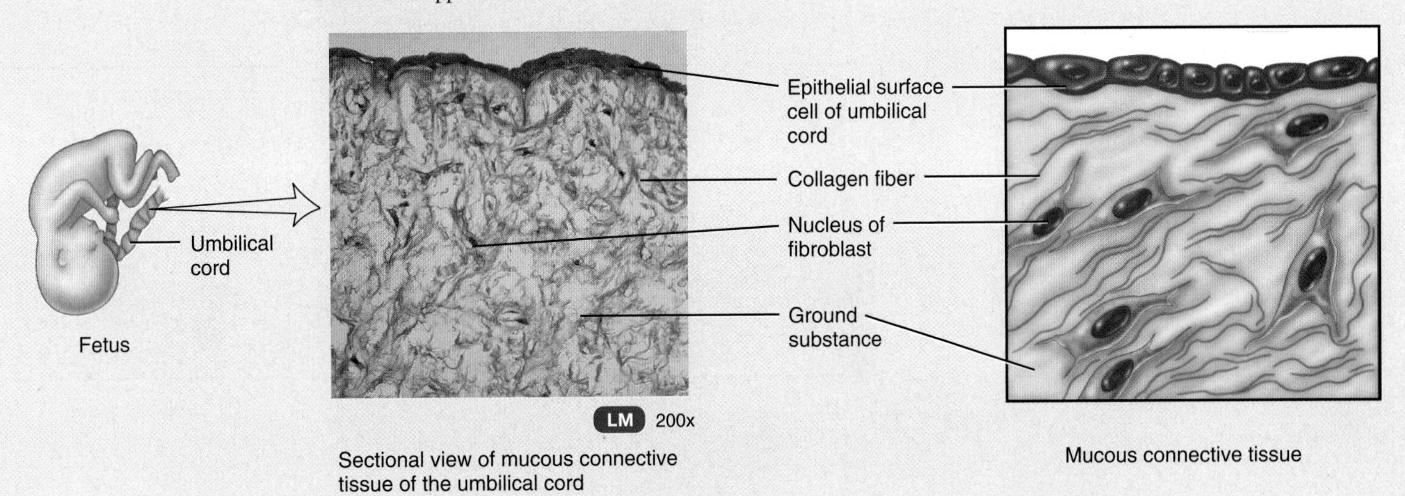

Sectional view of mucous connective tissue of the umbilical cord

Mucous connective tissue

they were secreted by fibroblasts); damaged tendons and ligaments are therefore very slow to heal.

Dense Irregular Connective Tissue This tissue contains collagen fibers that are packed more closely together than in loose connective tissue and are usually *irregularly* arranged (Table 4.4E). It is found in parts of the body where pulling forces are exerted in various directions. The tissue usually occurs in sheets, such as in the dermis of the skin, which underlies the epidermis. Heart valves, the perichondrium (the membrane surrounding cartilage), and the periosteum (the covering around bone) are examples of dense irregular connective tissues.

Elastic Connective Tissue Branching elastic fibers predominate in elastic connective tissue (Table 4.4F), giving the unstained tissue a yellowish color. Fibroblasts are present in the

TABLE 4.4

Mature Connective Tissues

LOOSE CONNECTIVE TISSUE*

A. Areolar connective tissue

Description: Consists of fibers (collagen, elastic, and reticular) and several kinds of cells (fibroblasts, macrophages, plasma cells, adipocytes, and mast cells) embedded in a semifluid ground substance.

Location: Subcutaneous layer deep to skin; papillary (superficial) region of dermis of skin; connective tissue layer of mucous membranes; and around blood vessels, nerves, and body organs.

Function: Strength, elasticity, and support.

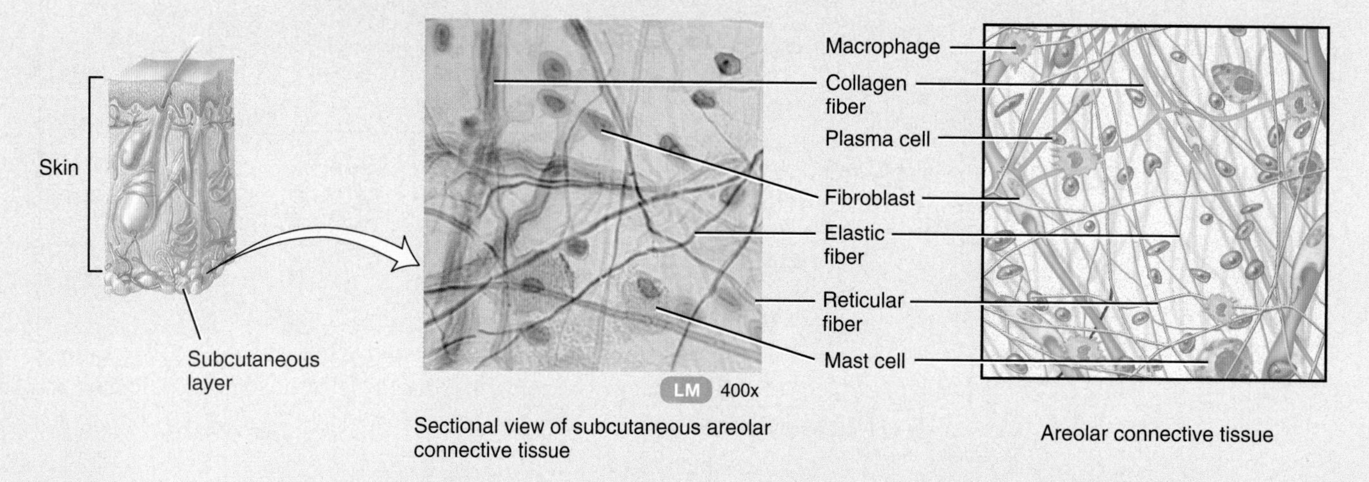

Sectional view of subcutaneous areolar connective tissue

Areolar connective tissue

B. Adipose tissue

Description: Consists of adipocytes, cells specialized to store triglycerides (fats) as a large centrally located droplet; nucleus and cytoplasm are peripherally located.

Location: Subcutaneous layer deep to skin, around heart and kidneys, yellow bone marrow, and padding around joints and behind eyeball in eye socket.

Function: Reduces heat loss through skin, serves as an energy reserve, supports, and protects.

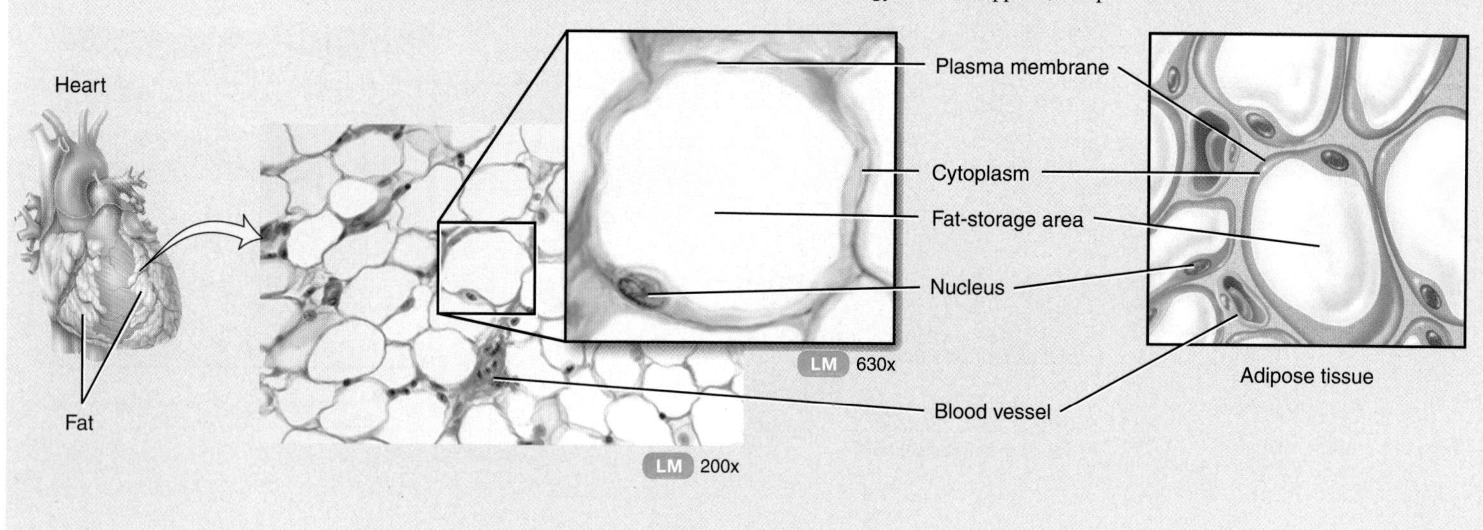

Adipose tissue

*Note: Some of the listed structures will be studied in later chapters and are placed here for reference.

spaces between the fibers. Elastic connective tissue is quite strong and can recoil to its original shape after being stretched. Elasticity is important to the normal functioning of lung tissue, which recoils as you exhale, and elastic arteries, whose recoil between heartbeats helps maintain blood flow.

Cartilage

Cartilage (KAR-ti-lij) consists of a dense network of collagen fibers or elastic fibers firmly embedded in chondroitin sulfate, a rubbery component of the ground substance. Cartilage can endure considerably more stress than loose and dense connective

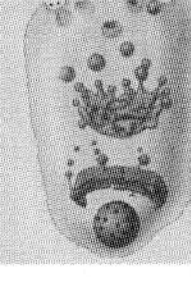

LOOSE CONNECTIVE TISSUE*

C. Reticular connective tissue

Description: A network of interlacing reticular fibers and reticular cells.

Location: Stroma (supporting framework) of liver, spleen, lymph nodes; red bone marrow, which gives rise to blood cells; reticular lamina of the basement membrane; and around blood vessels and muscles.

Function: Forms stroma of certain organs; binds together smooth muscle tissue cells; filters and removes worn-out blood cells in the spleen and microbes in lymph nodes.

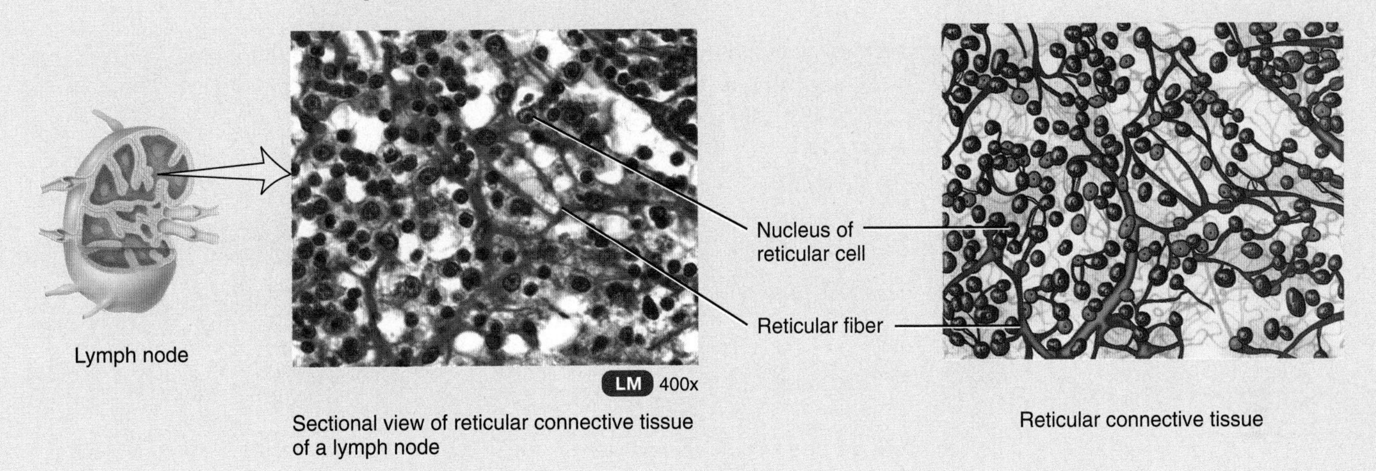

Sectional view of reticular connective tissue of a lymph node

Reticular connective tissue

DENSE CONNECTIVE TISSUE

D. Dense regular connective tissue

Description: Extracellular matrix in a gross specimen looks shiny white; consists mainly of collagen fibers arranged in parallel bundles; fibroblasts present in rows between bundles.

Location: Forms tendons (attach muscle to bone), most ligaments (attach bone to bone), and aponeuroses (sheetlike tendons that attach muscle to muscle or muscle to bone).

Function: Provides strong attachment between various structures.

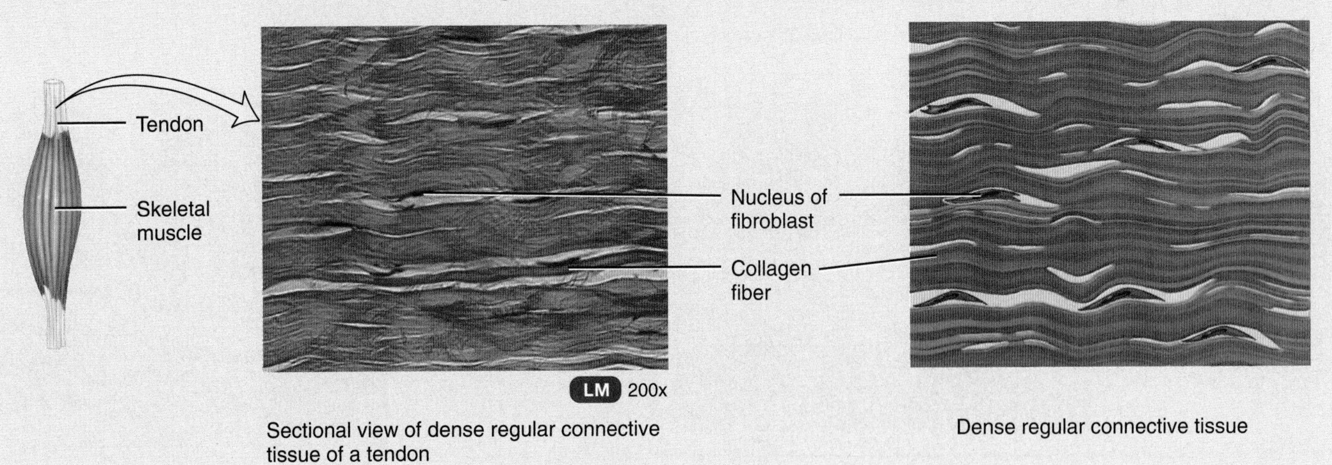

Sectional view of dense regular connective tissue of a tendon

Dense regular connective tissue

CONTINUES

tissues. Whereas the strength of cartilage is due to its collagen fibers, its resilience (ability to assume its original shape after deformation) is due to chondroitin sulfate.

The cells of mature cartilage, called ***chondrocytes*** (KON-drō-sīts; *chondro-* = cartilage), occur singly or in groups within spaces in the extracellular matrix called ***lacunae*** (la-KOO-nē = little lakes; singular is *lacuna*). The surface of most cartilage is surrounded by a membrane of dense irregular connective tissue called the ***perichondrium*** (per′-i-KON-drē-um; *peri-* = around). Unlike other connective tissues, cartilage has no blood vessels or

TABLE 4.4 CONTINUED

Mature Connective Tissues

DENSE CONNECTIVE TISSUE

E. Dense irregular connective tissue

Description: Consists predominantly of randomly arranged collagen fibers and a few fibroblasts.

Location: Fasciae (tissues beneath skin and around muscles and other organs), reticular (deeper) region of dermis of skin, periosteum of bone, perichondrium of cartilage, joint capsules, membrane capsules around various organs (kidneys, liver, testes, lymph nodes), pericardium of the heart, and heart valves.

Function: Provides strength.

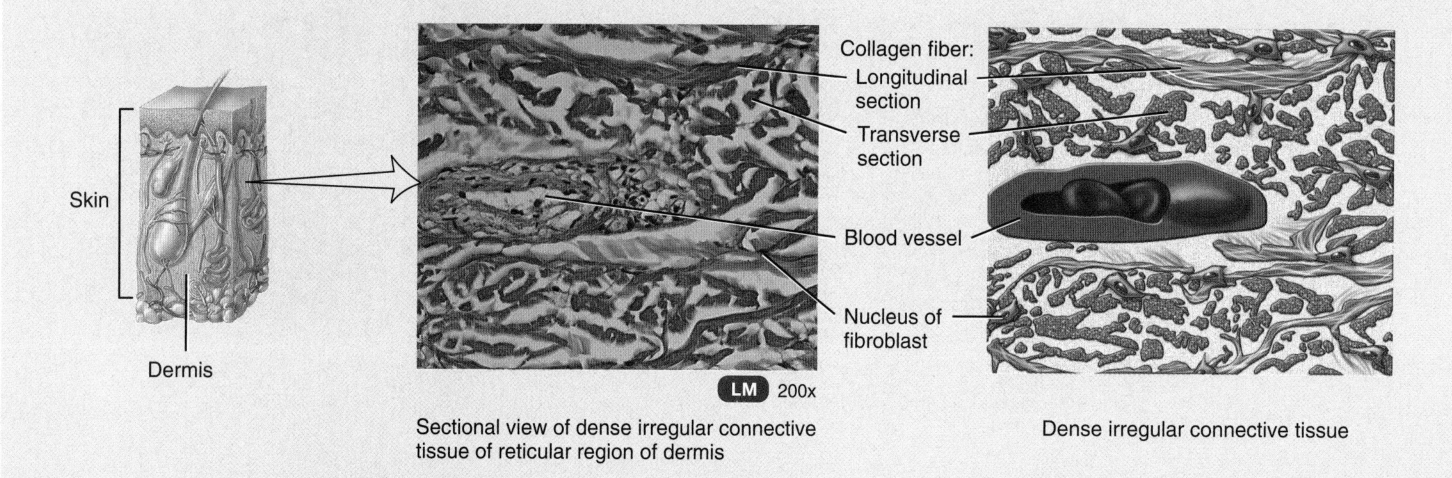

Sectional view of dense irregular connective tissue of reticular region of dermis

Dense irregular connective tissue

F. Elastic connective tissue

Description: Consists predominantly of elastic fibers; fibroblasts are present in spaces between fibers.

Location: Lung tissue, walls of elastic arteries, trachea, bronchial tubes, true vocal cords, suspensory ligament of penis, and ligaments between vertebrae.

Function: Allows stretching of various organs.

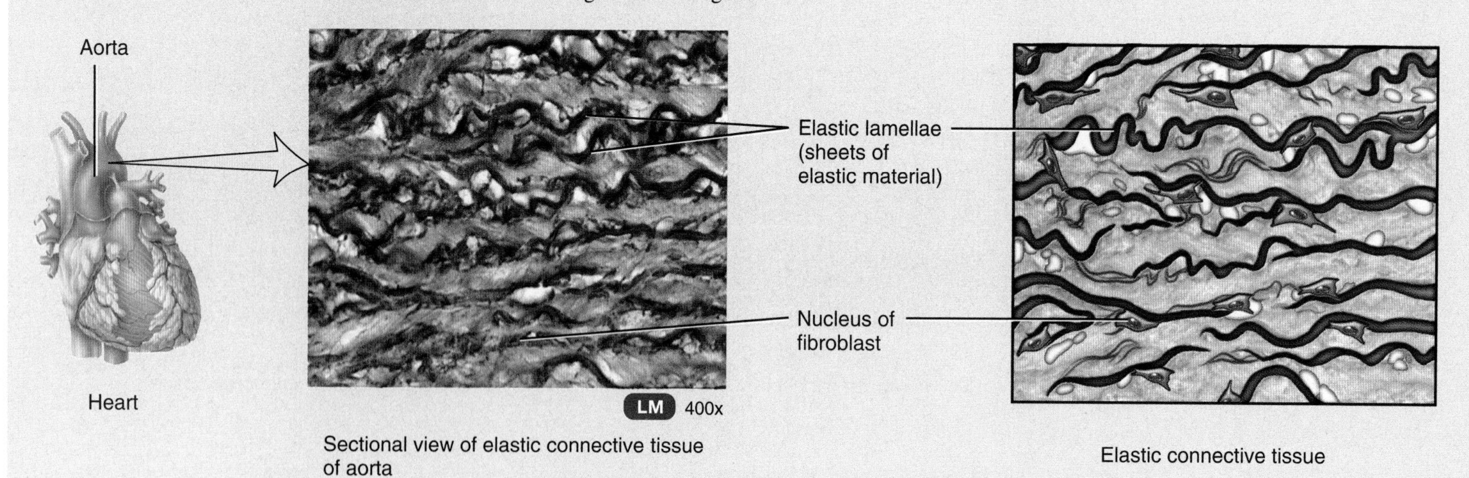

Sectional view of elastic connective tissue of aorta

Elastic connective tissue

nerves, except in the perichondrium. Since cartilage has no blood supply, it heals poorly following an injury. The three types of cartilage are hyaline cartilage, fibrocartilage, and elastic cartilage.

Hyaline Cartilage This type of cartilage contains a resilient gel as its ground substance and appears in the body as a bluish-white, shiny substance. The fine collagen fibers are not visible

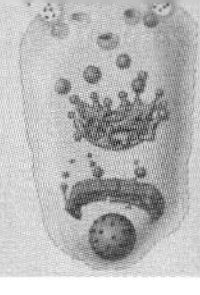

CARTILAGE

G. Hyaline cartilage

Description: Consists of a bluish-white, shiny ground substance with fine collagen fibers and many chondrocytes; most abundant type of cartilage.

Location: Ends of long bones, anterior ends of ribs, nose, parts of larynx, trachea, bronchi, bronchial tubes, and embryonic and fetal skeleton.

Function: Provides smooth surfaces for movement at joints, as well as flexibility and support.

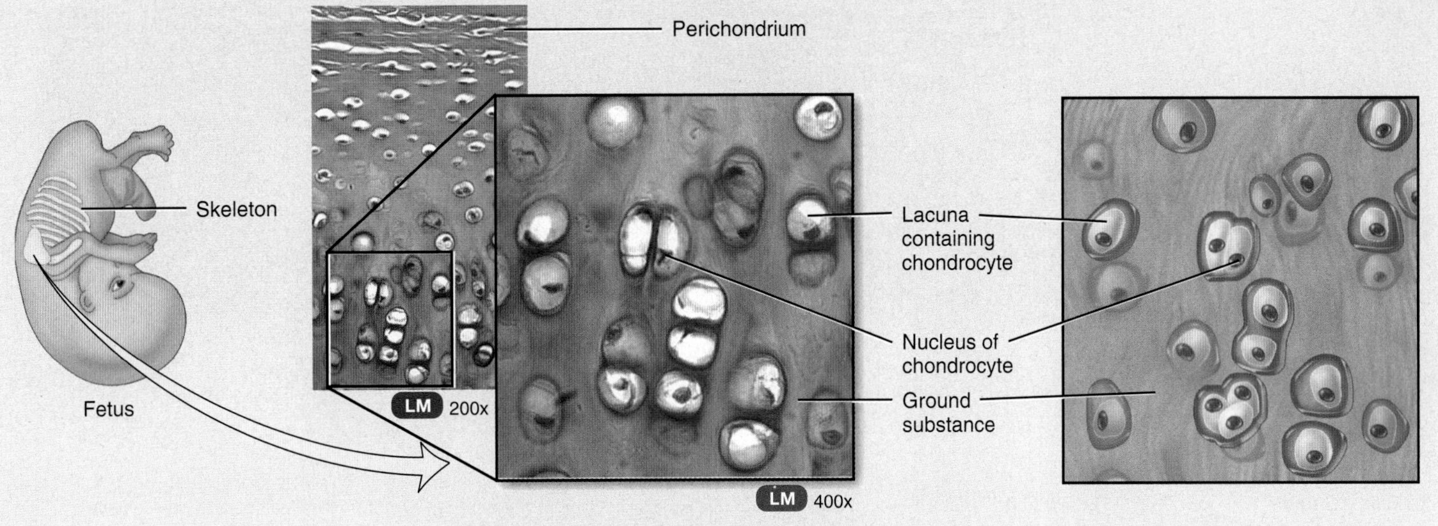

Sectional view of hyaline cartilage of a developing fetal bone and details of several chondrocytes

Hyaline cartilage

H. Fibrocartilage

Description: Consists of chondrocytes scattered among bundles of collagen fibers within the extracellular matrix.

Location: Pubic symphysis (point where hip bones join anteriorly), intervertebral discs (discs between vertebrae), menisci (cartilage pads) of knee, and portions of tendons that insert into cartilage.

Function: Support and joining structures together.

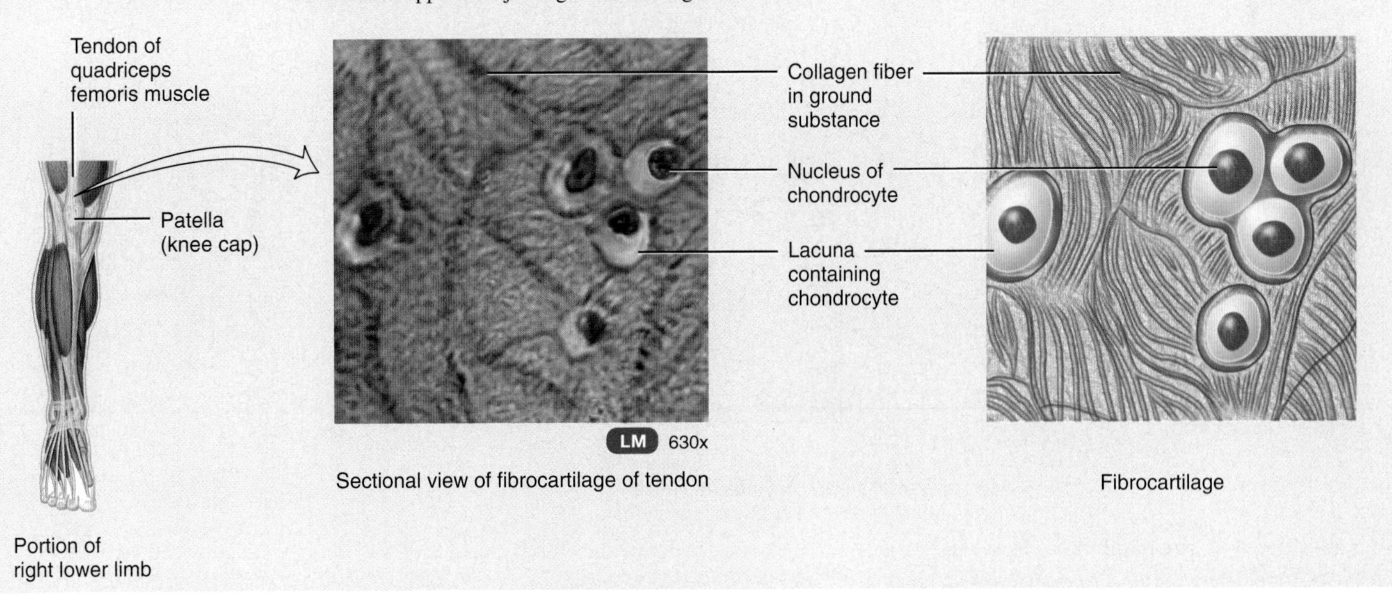

Sectional view of fibrocartilage of tendon

Fibrocartilage

Portion of right lower limb

CONTINUES

with ordinary staining techniques, and prominent chondrocytes are found in lacunae (Table 4.4G). Most hyaline cartilage is surrounded by a perichondrium. The exceptions are the articular cartilage in joints and at the epiphyseal plates, the regions where bones lengthen as a person grows. Hyaline cartilage is the most abundant cartilage in the body. It affords flexibility and support

TABLE 4.4 CONTINUED

Mature Connective Tissues

CARTILAGE

I. Elastic cartilage

Description: Consists of chondrocytes located in a threadlike network of elastic fibers within the matrix.

Location: Lid on top of larynx (epiglottis), part of external ear (auricle), and auditory (eustachian) tubes.

Function: Gives support and maintains shape.

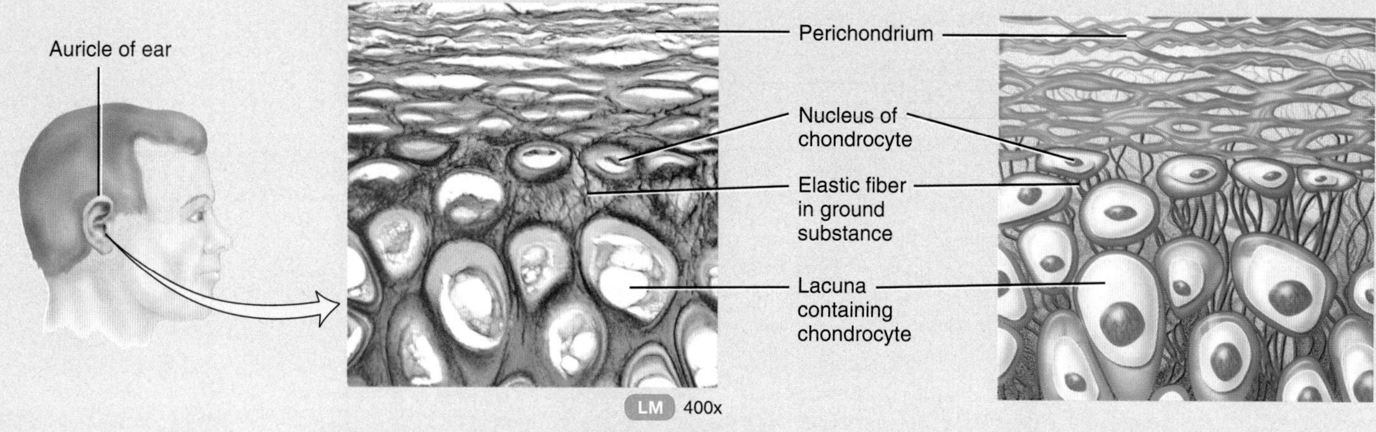

Sectional view of elastic cartilage of auricle of ear

Elastic cartilage

BONE TISSUE

J. Compact bone

Description: Compact bone tissue consists of osteons (haversian systems) that contain lamellae, lacunae, osteocytes, canaliculi, and central (haversian) canals. By contrast, spongy bone tissue (see Figure 6.3) consists of thin columns called trabeculae; spaces between trabeculae are filled with red bone marrow.

Location: Both compact and spongy bone tissue make up the various parts of bones of the body.

Function: Support, protection, storage; houses blood-forming tissue; serves as levers that act with muscle tissue to enable movement.

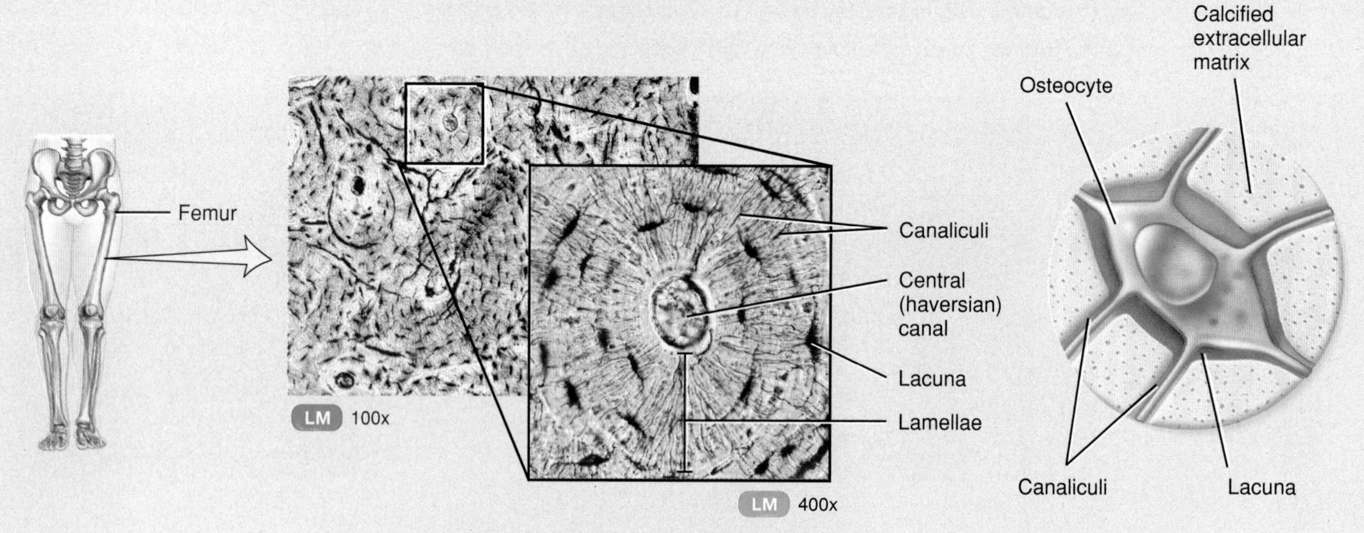

Sectional view of several osteons (haversian systems) of femur (thigh bone) and details of one osteon

Details of an osteocyte

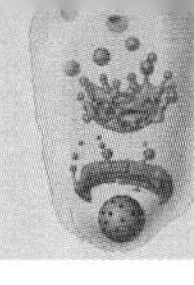

LIQUID CONNECTIVE TISSUE

K. Blood

Description: Consists of blood plasma and formed elements: red blood cells (erythrocytes), white blood cells (leukocytes), and platelets (thrombocytes).

Location: Within blood vessels (arteries, arterioles, capillaries, venules, and veins) and within the chambers of the heart.

Function: Red blood cells transport oxygen and some carbon dioxide; white blood cells carry on phagocytosis and are involved in allergic reactions and immune system responses; platelets are essential for the clotting of blood.

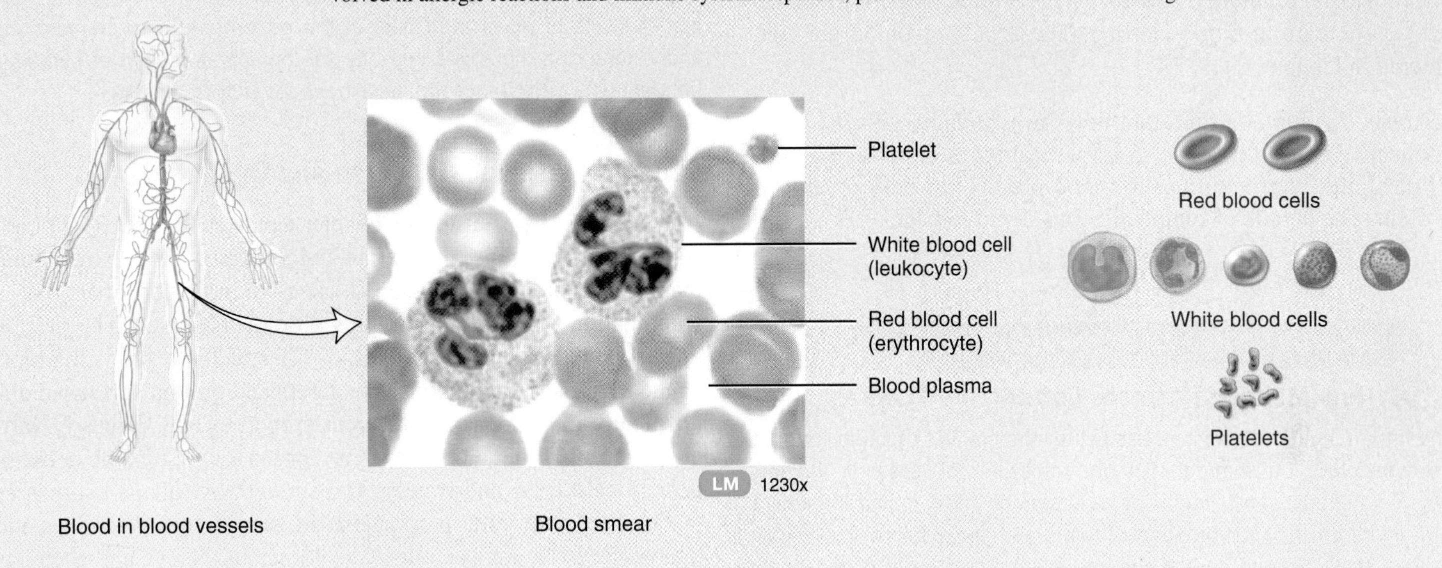

Blood in blood vessels

Blood smear

L. Lymph

Description: Consists of extracellular matrix similar to blood plasma but with much less protein; several cell types are present.

Location: Within lymphatic vessels and tissues.

Function: Transports excess interstitial fluid back to venous blood; transports lipids from the intestines; essential for immunity.

and, at joints, reduces friction and absorbs shock. Hyaline cartilage is the weakest of the three types of cartilage.

FIBROCARTILAGE Chondrocytes are scattered among clearly visible bundles of collagen fibers within the extracellular matrix of this type of cartilage (Table 4.4H). Fibrocartilage lacks a perichondrium. This tissue combines strength and rigidity and is the strongest of the three types of cartilage. One location of fibrocartilage is in the discs between vertebrae (backbones).

ELASTIC CARTILAGE In elastic cartilage, chondrocytes are located within a threadlike network of elastic fibers within the extracellular matrix (Table 4.4I). A perichondrium is present. Elastic cartilage provides strength and elasticity and maintains the shape of certain structures, such as the external ear.

Bone Tissue

Bones are organs composed of several different connective tissues, including ***bone*** or *osseous tissue* (OS-ē-us). Bone tissue has several functions. It supports soft tissues, protects delicate structures, and works with skeletal muscles to generate movement. Bone stores calcium and phosphorus; stores red bone marrow, which produces blood cells; and houses yellow bone marrow, a storage site for triglycerides. The details of bone tissue are presented in Chapter 6.

• CLINICAL CONNECTION Tissue Engineering

The technology of **tissue engineering** allows scientists to grow new tissues in the laboratory to replace damaged tissues in the body. Tissue engineers have already developed laboratory-grown versions of skin and cartilage. In the procedure, scaffolding beds of biodegradable synthetic materials or collagen are used as substrates that permit body cells such as skin cells or cartilage cells to be cultured. As the cells divide and assemble, the scaffolding degrades, and the new, permanent tissue is then implanted in the patient. Other structures being developed by tissue engineers include bones, tendons, heart valves, bone marrow, and intestines. Work is also underway to develop insulin-producing cells for diabetics, dopamine-producing cells for Parkinson disease patients, and even entire livers and kidneys. •

Liquid Connective Tissue

BLOOD TISSUE ***Blood tissue*** (or simply ***blood***) is a connective tissue with a liquid extracellular matrix called ***blood plasma,*** a pale yellow fluid that consists mostly of water with a wide variety of dissolved substances: nutrients, wastes, enzymes, hormones, respiratory gases, and ions. Suspended in the plasma are red blood cells, white blood cells, and platelets (Table 4.4K). ***Red blood cells*** transport oxygen to body cells and help remove carbon dioxide from them. ***White blood cells*** are involved in phagocytosis, immunity, and allergic reactions. ***Platelets*** (PLĀT-lets) participate in blood clotting. The details of blood are considered in Chapter 21.

LYMPH ***Lymph*** is a fluid that flows in lymphatic vessels. It is a connective tissue that consists of several types of cells in a clear extracellular matrix similar to blood plasma but with much less protein. The details of lymph are considered in Chapter 24.

MANUAL THERAPY APPLICATION

How Massage Affects Connective Tissues

The importance of connective tissues for the manual therapies cannot be overstated. The manual therapy professions of massage therapy, physical therapy, sports medicine, and athletic training, just to name a few, focus on the manipulation of joints and soft tissues. Connective tissues surround and spread through every other tissue of the body. Furthermore, this type of tissue "connects" organs with one another, forming a continuous network throughout the body, and is in very close proximity to nearly every cell in the body. This network provides insight into the seemingly inexplicable occasion when, for example, body work on the right foot causes the patient to perceive a change in the right shoulder. Work on connective tissues is subtle and requires that the therapist be "in tune" with what he or she is trying to accomplish.

Massage affects connective tissues in serveral ways. Reduced range of motion, as a result of the process of hydrogen bonding between molecules and fibers, is evident in advancing age. As one example, the connective tissue (fascia) between the scapula and rib cage typically thickens as we age. Assuming no trauma to the area, and assuming the lack of aggressive exercise in the sedentary lifestyle of many persons, progression of hydrogen bonding of collagen may be evident to a manual therapist. In a 20-year-old patient, therapists can slide their hands under the inferior border of the scapula a distance of approximately 7.6 cm (3 in.). With each additional 10 years of age, this distance commonly decreases by approximately 1.3 cm (0.5 in.). An elderly person, again assuming a typical lifestyle of decreasing activity with advancing age, often has a scapula that is essentially "cemented" to the rib cage by connective tissue. The shoulder at this point has an extremely limited range of motion.

Similarly, the fasciae that surrounds muscles or muscle groups (as will be discussed in Chapter 10) commonly thickens with advancing age. Particularly in the limbs, these fascial tubes can severely limit activity of the enveloped muscle. The shuffling gait, greatly reduced stride, and reduction in range of motion of certain elderly persons can usually be attributed to the process just described. As multiple connective tissues surrounding muscle cells, muscle organs, and groups of muscles thicken, the nourishment for muscle cells declines and weakened muscles are now working within "concrete-like" collars. Intervention at any point in the process by manual therapists can increase fluidity of the tissues through thixotropy (thik-SUT-rō-pē; *thixis* = touching; *tropē* = turning), the phenomenon by which certain gels become more fluid when warmed and shaken and more gel-like upon standing. More physical activity by the patient, passive movement by a manual therapist, or warming of the tissue through kneading and other modalities are all effective ways to increase the fluidity of the tissues. Owing to the younger age and higher levels of physical activity of the patients, athletic trainers and sports medicine therapists typically see this phenomenon of hydrogen bonding only with tissues that have been severely traumatized.

Connective Tissues: Diverse and Dynamic

As you learned earlier in this chapter, connective tissue (CT) consists of various cells and extracellular matrix. The extracellular matrix may contain collagen, elastic, or reticular fibers—or a combination of these fibers—plus ground substance. The protein fiber collagen makes up the bulk of CT structures and is the most abundant protein in the body. The fibers and ground substance differ greatly in CT; CT has different properties and shapes in various organs within the body. CT can form a tough, flexible network or may be diffuse and watery. The strength of collagen surpasses that of steel wire (after proportions are considered) in tendons and ligaments. CT is transparent in the cornea of the eye and accounts for the toughness of leather, the viscosity of gelatin, and the cohesiveness of glue. When a chemical called hyaline is added to the ground substance it becomes a form of cartilage; similarly, CT becomes bone when mineral salts become incorporated.

The various types of connective tissues are continuously changing in a growing person. The process slows in adulthood; nevertheless, connective tissues are in a state of flux at some level throughout a lifetime. The CT in any given location in the body may change dramatically throughout the life cycle, from conception through death.

The ground substance of CT is not to be confused with interstitial fluid. Interstitial fluid consists of blood plasma, nutrients, and hormones that have, through the process of diffusion, moved out of the bloodstream. The ground substance of CT, on the other hand, is produced by fibroblasts. The chemistry of ground substance varies with the location where it is produced. Protein combines with carbohydrates to form molecules called mucopolysaccharides. When the mucopolysaccharides are small, the resulting ground substance is somewhat watery. When the mucopolysaccharides are larger and more numerous, the ground substance is more viscous and resembles gelatin. The substances are, however, dynamic and can periodically change between the fluid and gel states in healthy tissue.

Embedded within the ground substance of CT are the fibers, mostly collagen fibers in many tissues. Collagen fibers are predominant in tendons, ligaments, aponeuroses, the dermis of the skin, bone, blood vessels, the framework (stroma) of virtually all visceral organs, and fasciae. Collagen fibers provide all of these structures with shape, resiliency, flexibility, strength, and structural integrity. Strength and flexibility may appear to be contradictory properties, but think of a collagen fiber as a miniature rope. Collagen fibers are strong and flexible.

Connective tissue fibers, manufactured by fibroblasts, are not living matter. Fibroblasts retain the unique property of being able to move from one part of the body to another and then producing the fibers and ground substance that is appropriate to the tissue at any given time. Fibroblasts are key in wound healing. Scar tissue is composed of new fibers, primarily collagen, produced by fibroblasts that have migrated to the area. The process of forming scar tissue (fibrosis) occurs when dead or damaged cells are replaced by the connective tissue stroma of that organ.

The abilities of connective tissues to fluctuate with the tissue conditions tends to decline with a decrease in physical activities and with advanced aging. As noted earlier, ground substance becomes more fluid when it is warmed and stirred up and more gel-like when it sits without being disturbed (thixotropy). Poor cellular nutrition and sedentary habits weaken all connective tissues, stiffen them, and accelerate their biological aging. This thixotropic effect provides one of the foundations of manual therapies: Skilled hands can warm and mix connective tissues. The ground substance then becomes more fluid and thus the exchange of nutrients and wastes is enhanced between blood and cells.

Collagen is formed by fibroblasts. Amino acids are linked in a specific sequence, as you learned in Chapter 2, and then form a triple-helix molecule. These individual collagen molecules, now called *tropocollagen,* are moved into the ground substance that surrounds the fibroblast. The anatomical and physiological conditions dictate the fate of each molecule of tropocollagen. As discussed earlier, the tropocollagen molecules may follow any number of pathways to help form the collagenous component of areolar, dense regular, and dense irregular connective tissues, cartilage, or bone. The tropocollagen molecules typically organize into collagen fibrils that then organize into collagen fibers. Again, the bonding at each level after the peptide linking of the amino acids involves hydrogen bonds.

CHECKPOINT

7. What are the features of the cells, ground substance, and fibers that make up connective tissue?
8. How are the structures of the following connective tissues related to their functions: areolar connective tissue, adipose tissue, reticular connective tissue, dense regular connective tissue, dense irregular connective tissue, elastic connective tissue, hyaline cartilage, fibrocartilage, elastic cartilage, bone tissue, blood tissue, and lymph?

4.5 MEMBRANES

OBJECTIVES

- Define a membrane.
- Describe the classification of membranes.

Membranes are flat sheets of pliable tissue that cover or line a part of the body. The combination of an epithelial layer and an underlying connective tissue layer constitutes an ***epithelial membrane.*** The principal epithelial membranes of the body are mucous membranes; serous membranes; and the cutaneous membrane, or skin. (Skin will be discussed in detail in Chapter 5 and is not discussed here.) Another kind of membrane, a synovial membrane, lines joints and contains connective tissue but no epithelium.

Mucous Membranes

A ***mucous membrane*** or ***mucosa*** (mū-KŌ-sa) lines a body cavity that opens directly to the exterior. Mucous membranes line the entire digestive, respiratory, and reproductive systems and much of the urinary system (see Figure 4.6a). The epithelial layer of a mucous membrane secretes mucus, which prevents the cavities from drying out. It also traps particles in the respiratory passageways, lubricates and absorbs food as it moves through the gastrointestinal tract, and secretes digestive enzymes. The connective tissue layer (areolar connective tissue) helps bind the epithelium to the underlying structures. It also provides the epithelium with oxygen and nutrients and removes wastes via its blood vessels.

Serous Membranes

A ***serous membrane*** (SĒR-us = watery) lines a body cavity that does not open directly to the exterior, and it also covers the organs that lie within the cavity. Serous membranes consist of two parts: a parietal layer and a visceral layer (see Figure 4.6b). The ***parietal layer*** (pa-RĪ-e-tal; *pariet-* = wall) is the part attached to the cavity wall, and the ***visceral layer*** (*viscer-* = body organ) is the part that covers and attaches to the organs inside these cavities. Each layer consists of areolar connective tissue sandwiched between two sheets of *mesothelium.* Mesothelium is a simple squamous epithelium. It secretes ***serous fluid,*** a watery lubricating fluid that allows organs to glide easily over one another or to slide against the walls of cavities.

Recall from Chapter 1 that the serous membrane lining the thoracic cavity and covering the lungs is the ***pleura*** (PLOO-ra). The serous membrane lining the heart cavity and covering the heart is the ***pericardium*** (per-i-KAR-dē-um). The serous membrane lining the abdominal cavity and covering the abdominal organs is the ***peritoneum*** (per-i-tō-NĒ-um).

Synovial Membranes

Synovial membranes (sin-Ō-vē-al) line the cavities of most joints. They are composed of areolar connective tissue and adipose

Figure 4.6 Membranes.

A membrane is a flat sheet of pliable tissues that covers or lines a part of the body.

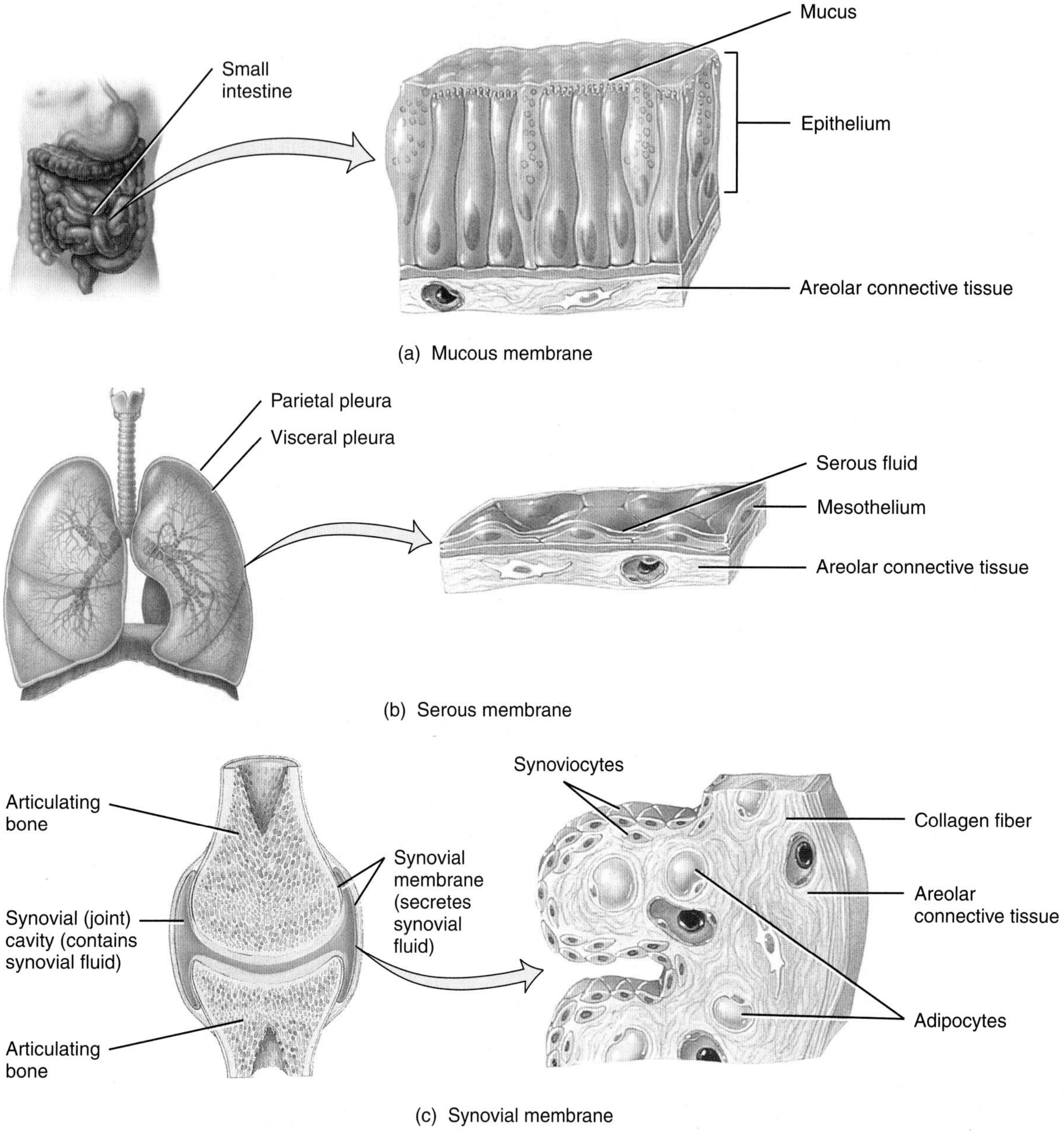

(a) Mucous membrane

(b) Serous membrane

(c) Synovial membrane

What is an epithelial membrane?

pose tissue with collagen fibers; they do not have an epithelial layer. Synovial membranes contain cells (synoviocytes) that secrete ***synovial fluid*** (Figure 4.6c). This fluid lubricates the ends of bones as they move at joints, nourishes the cartilage covering the bones, and removes microbes and debris from the joint cavity.

CHECKPOINT

9. Define the following kinds of membranes: mucous, serous, cutaneous, and synovial.
10. Where is each type of membrane located in the body? What are their functions?

4.6 MUSCULAR TISSUE

OBJECTIVES

- Describe the functions of muscular tissue.
- Contrast the locations of the three types of muscular tissue.

Muscular tissue consists of elongated cells called *muscle fibers* that are highly specialized to generate force. As a result of this characteristic, muscular tissue produces motion, maintains posture, and generates heat. It also offers protection. Based on its location

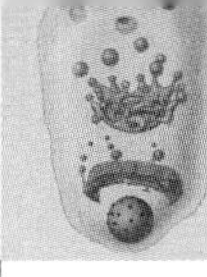

TABLE 4.5

Muscular Tissues

A. Skeletal muscle tissue

Description: Long, cylindrical, striated fibers with many peripherally located nuclei; voluntary control.

Location: Usually attached to bones by tendons.

Function: Motion, posture, heat production, and protection.

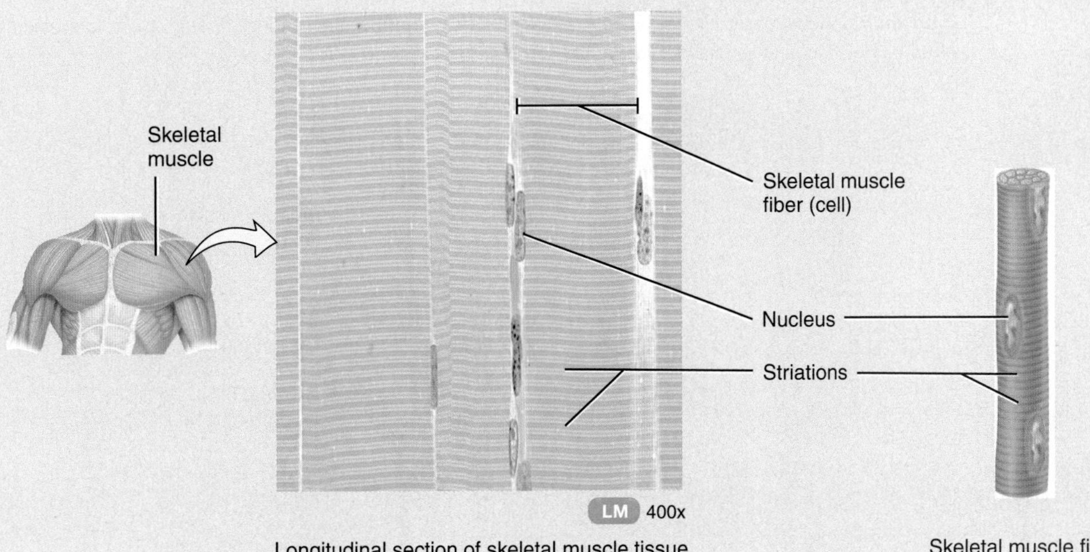

Longitudinal section of skeletal muscle tissue

Skeletal muscle fiber

B. Cardiac muscle tissue

Description: Branched striated fibers with one or two centrally located nuclei; contains intercalated discs; involuntary control.

Location: Heart wall.

Function: Pumps blood to all parts of the body.

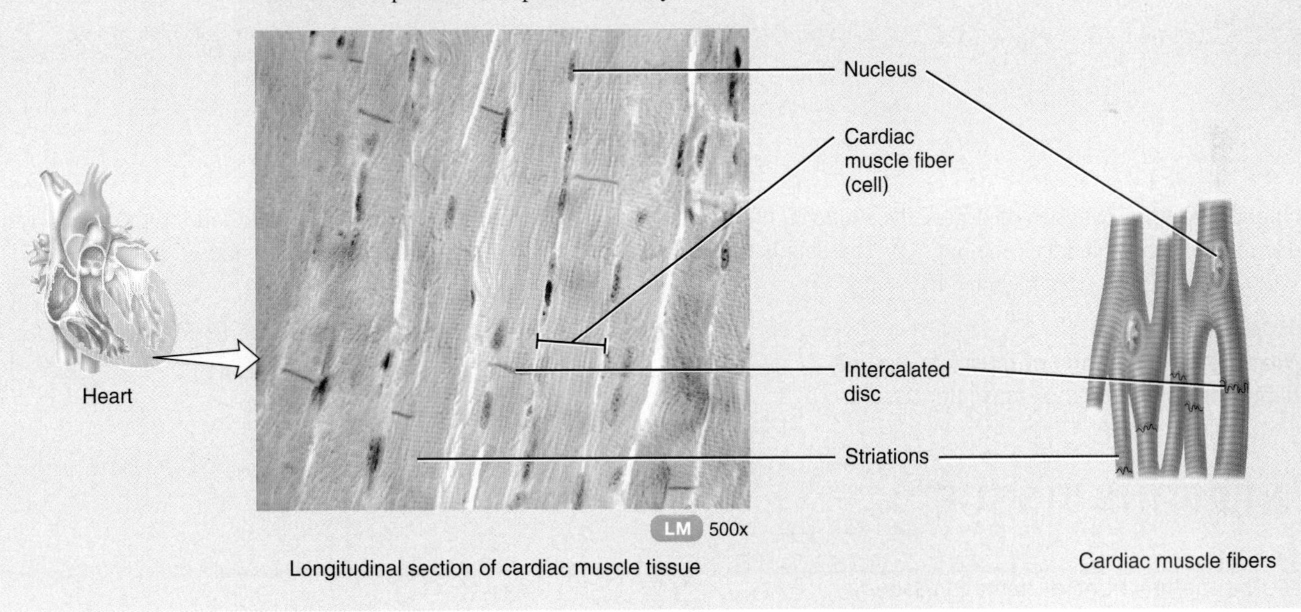

Longitudinal section of cardiac muscle tissue

Cardiac muscle fibers

CONTINUES

and certain structural and functional characteristics, muscular tissue is classified into three types: skeletal, cardiac, and smooth. ***Skeletal muscle tissue*** is named for its location—it is usually attached to the bones of the skeleton (Table 4.5A). ***Cardiac muscle tissue*** forms the bulk of the wall of the heart (Table 4.5B). ***Smooth muscle tissue*** is located in the walls of hollow internal structures

TABLE 4.5 CONTINUED

Muscular Tissues

C. Smooth muscle tissue

Description: Spindle-shaped (thickest in middle and tapering at both ends), nonstriated fibers with one centrally located nucleus; involuntary control.

Location: Iris of the eyes, walls of hollow internal structures such as blood vessels, airways to the lungs, stomach, intestines, gallbladder, urinary bladder, and uterus.

Function: Motion (constriction of blood vessels and airways, propulsion of foods through gastrointestinal tract, contraction of urinary bladder and gallbladder).

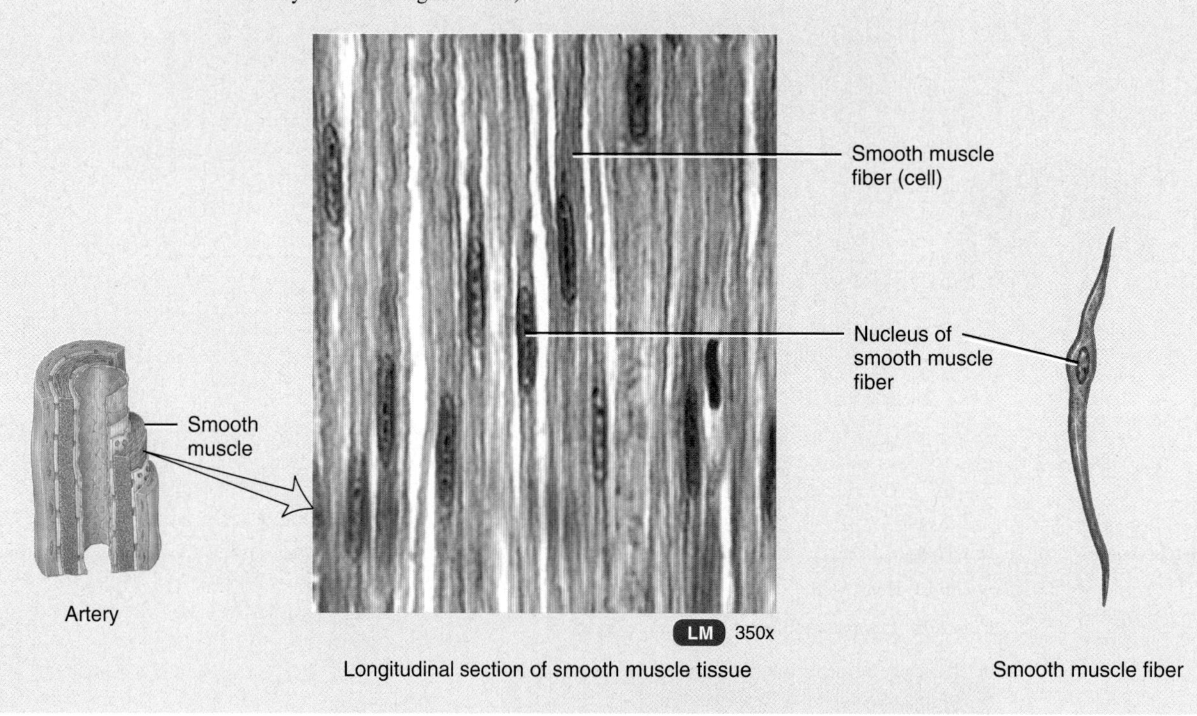

Longitudinal section of smooth muscle tissue

Smooth muscle fiber

such as blood vessels, airways to the lungs, the stomach, intestines, gallbladder, and urinary bladder (Table 4.5C). The details of muscular tissue will be presented in Chapter 10.

CHECKPOINT

11. What are the functions of muscular tissue?
12. Name the three types of muscular tissue.

4.7 NERVOUS TISSUE

OBJECTIVE

- Describe the functions of nervous tissue.

Despite the awesome complexity of the nervous system, it consists of only two principal types of cells: neurons and neuroglia (Table 4.6). ***Neurons*** (NOO-rons; *neur-* = nerve, nerve tissue, nervous system) or ***nerve cells,*** are sensitive to various stimuli. They convert stimuli into nerve impulses (action potentials) and conduct these impulses to other neurons, to muscle fibers, or to glands. ***Neuroglia*** (noo-RŌG-lē-a; *-glia* = glue) do not generate or conduct nerve impulses, but they do have many other important supportive functions. The detailed structure and function of neurons and neuroglia will be considered in Chapter 15.

CHECKPOINT

13. How do neurons differ from neuroglia?

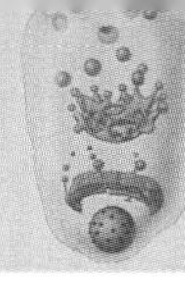

TABLE 4.6

Nervous Tissue

Description: Consists of neurons (nerve cells) and neuroglia. Neurons consist of a cell body and processes extending from the cell body (multiple dendrites and a single axon). Neuroglia do not generate or conduct nerve impulses but have other important supporting functions.

Location: Nervous system.

Function: Exhibits sensitivity to various types of stimuli, converts stimuli into nerve impulses (action potentials), and conducts nerve impulses to other neurons, muscle fibers, or glands.

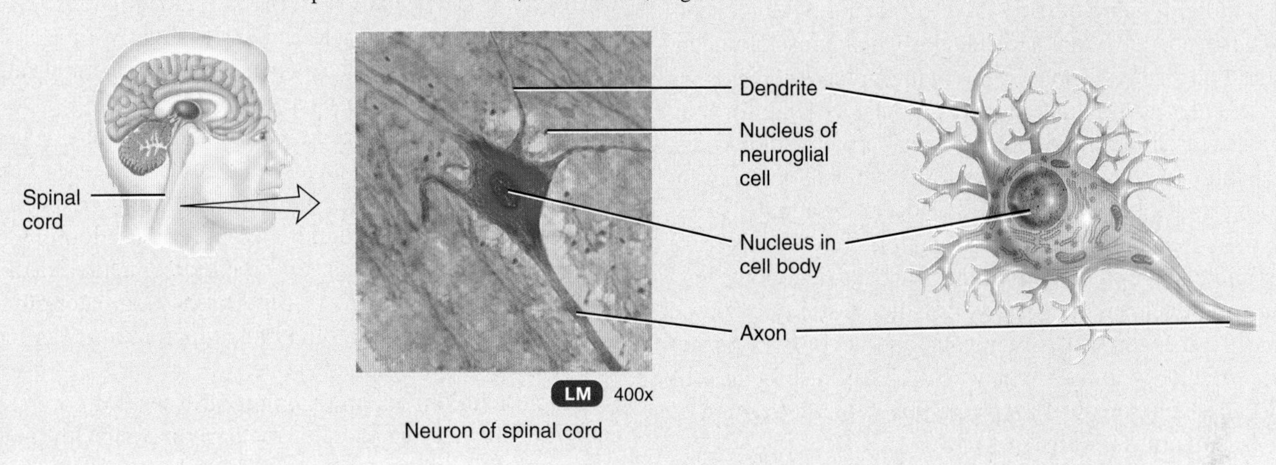

Neuron of spinal cord

4.8 TISSUE REPAIR: RESTORING HOMEOSTASIS

OBJECTIVE

- Describe the role of tissue repair in restoring homeostasis.

Tissue repair is the process that replaces worn-out, damaged, or dead cells. New cells originate by cell division from the ***stroma,*** the supporting connective tissue, or from the ***parenchyma*** (par-EN-ki-ma), cells that constitute the functioning part of the tissue or organ. In adults, each of the four basic tissue types (epithelial, connective, muscular, and nervous) has a different capacity for replenishing parenchymal cells lost by damage, disease, or other processes.

Epithelial cells, which endure considerable wear and tear (and even injury) in some locations, have a continuous capacity for renewal. In some cases, immature, undifferentiated cells called ***stem cells*** divide to replace lost or damaged cells. For example, stem cells reside in protected locations in the epithelia of the skin and gastrointestinal tract to replenish cells sloughed from the apical layer.

Some connective tissues contain mesenchyme and thus also have a continuous capacity for renewal. One example is bone, which has an ample blood supply. Other connective tissues such as cartilage can replenish cells less readily in part because of a smaller blood supply.

Muscular tissue has a relatively poor capacity for renewal of lost cells. Cardiac muscle fibers can be produced from stem cells under special conditions. Skeletal muscle tissue does not divide rapidly enough to replace extensively damaged muscle fibers. Smooth muscle fibers can proliferate to some extent, but they do so much more slowly than the cells of epithelial or connective tissues.

Nervous tissue has the poorest capacity for renewal. Although experiments have revealed the presence of some stem cells in the brain, they normally do not undergo mitosis to replace damaged neurons.

If parenchymal cells accomplish the repair, ***tissue regeneration*** is possible, and a near-perfect reconstruction of the injured tissue may occur. However, if fibroblasts of the stroma are active in the repair, the replacement tissue is a new connective tissue. The fibroblasts synthesize collagen and other matrix materials that aggregate to form scar tissue, a process known as ***fibrosis.*** Because scar tissue is not specialized to perform the functions of the parenchymal tissue, the original function of the tissue or organ is impaired.

• CLINICAL CONNECTION | Adhesions

Scar tissue can form **adhesions** (ad-HĒ-shuns), abnormal joining of tissues. Adhesions commonly form in the abdomen around a site of previous inflammation such as an inflamed appendix or gallbladder. More importantly for manual therapists, adhesions may form following joint injury, surgery in any part of the body, or prolonged immobilization. Although adhesions do not always cause problems, they can decrease tissue flexibility, cause obstruction, and make a subsequent operation more difficult. Manual therapists are often challenged to reduce troublesome adhesions; multiple techniques can be utilized for this purpose. Sometimes surgery is needed to remove adhesions. •

CHECKPOINT

14. How do the four basic types of tissue compare with respect to capacity for repair?

4.9 AGING AND TISSUES

OBJECTIVE

- Describe the effects of aging on tissues.

Generally, tissues heal faster and leave less obvious scars in the young than in the aged. In fact, surgery performed on fetuses leaves no scars. The younger body is generally in a better nutritional state, its tissues have a better blood supply, and its cells have a higher metabolic rate. Thus, cells can synthesize needed materials and divide more quickly. The extracellular components of tissues also change with age. Glucose, the most abundant sugar in the body, plays a role in the aging process. Glucose is haphazardly added to proteins inside and outside cells, forming irreversible cross-links between adjacent protein molecules. With advancing age, more cross-links form, which contributes to the stiffening and loss of elasticity that occur in aging tissues. Collagen fibers, responsible for the strength of tendons, increase in number and change in quality with aging. Elastin, another extracellular component, is responsible for the elasticity of blood vessels and skin. It thickens, fragments, and acquires a greater affinity for calcium with age—changes that may also be associated with the development of *atherosclerosis,* the deposition of fatty materials in arterial walls.

Excess Adiposity

As people age, skeletal muscles do atrophy, or shrink, whereas adipose tissue tends to grow. Muscular tissue is specialized for its contractile function and does not "turn into" fat tissue, which is specialized for energy storage. Although some of these changes in body composition appear to be an inevitable part of the aging process, scientists believe much of the loss of muscular tissue can be prevented with exercise training.

Adiposity becomes too much of a good thing when it leads to health problems, which can include hypertension (high blood pressure), poor blood sugar regulation (including type 2 diabetes), heart disease, certain cancers, gallstones, arthritis, and backaches. In general, the risk of health problems associated with excess adipose tissue increases in a dose-dependent fashion: the greater the excess weight, the greater is the risk. People with a great deal of excess fat have a much higher risk for health problems than people who are only slightly too fat. But the health effects of excess fat depend on several important factors besides quantity of adipose tissue.

People who carry extra fat on the torso are at greater risk for hypertension, type 2 diabetes, and artery disease than people whose extra fat resides in the hips and thighs. Especially risky is excess fat stored around the viscera (abdominal organs). Adipocytes in this area appear to be more "metabolically active" than those under the skin. Visceral fat has a greater effect on the levels of blood sugar and blood fat, which in turn can lead to the health problems mentioned above.

People with a moderate amount of excess adipose tissue who eat a healthy diet and exercise regularly have health risks similar to those of their leaner peers. This observation suggests that some of the health risks seen in overweight people may be caused by poor health habits (such as too much food, too much alcohol, or too little exercise) rather than by the presence of excess adipose tissue.

CHECKPOINT

15. What types of health problems are caused by adiposity?

KEY MEDICAL TERMS ASSOCIATED WITH TISSUES

Tissue rejection An immune response of the body directed at foreign proteins in a transplanted tissue or organ; immunosuppressive drugs, such as cyclosporine, have enabled clinicians to largely overcome tissue rejection in heart-, kidney-, and liver-transplant patients.

Tissue transplantation The replacement of a diseased or injured tissue or organ; the most successful transplants involve use of a person's own tissues or those from an identical twin.

Xenotransplantation (zen′-ō-trans′-plan-TĀ-shun; *xeno-* = strange, foreign) The replacement of a diseased or injured tissue or organ with cells or tissues from an animal. Only a few cases of successful xenotransplantation exist to date.

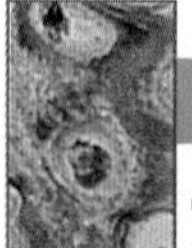

STUDY OUTLINE

Types of Tissues (Section 4.1)

1. A tissue is a group of similar cells that usually has a similar embryological origin and is specialized for a particular function.
2. The various tissues of the body are classified into four basic types: epithelial, connective, muscular, and nervous.

Cell Junctions (Section 4.2)

1. Cell junctions are points of contact between adjacent plasma membranes.
2. Tight juctions form fluid tight seals between cells; adherens junctions, desmosomes, and hemidesmosomes anchor cells to one an-

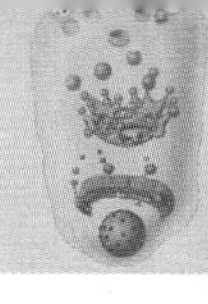

other or to the basement membrane; and gap juctions permit electrical and chemical signals to pass between cells.

Epithelial Tissue (Section 4.3)

1. The general types of epithelia include covering and lining epithelium and glandular epithelium.
2. Some general characteristics of epithelium are as follows: It consists mostly of cells with little extracellular material, is arranged in sheets, is attached to connective tissue by a basement membrane, is avascular (no blood vessels), has a nerve supply, and can replace itself.
3. Epithelial layers can be simple (one layer) or stratified (several layers). The cell shapes may be squamous (flat), cuboidal (cube-like), columnar (rectangular), or transitional (variable).
4. Simple squamous epithelium consists of a single layer of flat cells (Table 4.1A). It is found in parts of the body where filtration or diffusion are priority processes. One type, endothelium, lines the heart and blood vessels. Another type, mesothelium, forms the serous membranes that line the thoracic and abdominal cavities and cover the organs within them.
5. Simple cuboidal epithelium consists of a single layer of cube-shaped cells that function in secretion and absorption (Table 4.1B). It is found covering the ovaries, in the kidneys and eyes, and lining some glandular ducts.
6. Nonciliated simple columnar epithelium consists of a single layer of nonciliated rectangular cells (Table 4.1C). It lines most of the gastrointestinal tract. Specialized cells containing microvilli perform absorption. Goblet cells secrete mucus.
7. Ciliated simple columnar epithelium consists of a single layer of ciliated rectangular cells (Table 4.1D). It is found in a few portions of the upper respiratory tract, where it moves foreign particles trapped in mucus out of the respiratory tract.
8. Pseudostratified columnar epithelium has only one layer but gives the appearance of many (Table 4.1E). The ciliated variety moves mucus by ciliary action; the nonciliated variety protects and absorbs.
9. Stratified squamous epithelium consists of several layers of cells; cells in the apical layer and several layers deep to it are flat (Table 4.1F). It is protective. A nonkeratinized variety lines the mouth; a keratinized variety forms the epidermis, the most superficial layer of the skin.
10. Stratified cuboidal epithelium consists of several layers of cells; cells in the apical layer are cube-shaped (Table 4.1G). It is found in adult sweat glands and a portion of the male urethra. It functions in protection and limited secretion and absorption.
11. Stratified columnar epithelium consists of several layers of cells; cells in the apical layer are column-shaped (Table 4.1H). It is found in a portion of the male urethra and large excretory ducts of some glands where it functions in protection and secretion.
12. Transitional epithelium consists of several layers of cells whose appearance varies with the degree of stretching (Table 4.1I). It lines the urinary bladder.
13. A gland is a single cell or a group of epithelial cells adapted for secretion.
14. Endocrine glands secrete hormones into interstitial fluid and then the blood (Table 4.2A).
15. Exocrine glands (mucous, sweat, oil, and digestive glands) secrete into ducts or directly onto a free surface (Table 4.2B).

Connective Tissue (Section 4.4)

1. Connective tissue is the most abundant body tissue.
2. Connective tissue consists of cells and an extracellular matrix of ground substance and fibers; it has abundant extracellular matrix with relatively few cells. It does not usually occur on free surfaces, has a nerve supply (except for cartilage), and is highly vascular (except for cartilage, tendons, and ligaments).
3. Cells in connective tissue include fibroblasts (secrete matrix), macrophages (perform phagocytosis), mast cells (produce histamine), and adipocytes (store fat).
4. The ground substance and fibers make up the extracellular matrix.
5. The ground substance supports and binds cells together, provides a medium for the exchange of materials, and is active in influencing cell functions.
6. The fibers in the extracellular matrix provide strength and support and are of three types: (a) Collagen fibers (composed of collagen) are found in large amounts in bone, tendons, and ligaments; (b) elastic fibers (composed of elastin, fibrillin, and other glycoproteins) are found in skin, blood vessel walls, and lungs; and (c) reticular fibers (composed of collagen and glycoprotein) are found around fat cells, nerve fibers, and skeletal and smooth muscle cells.
7. Connective tissue is subdivided into loose connective tissue, dense connective tissue, cartilage, bone tissue, blood tissue, and lymph.
8. Loose connective tissue includes areolar connective tissue, adipose tissue, and reticular connective tissue.
9. Areolar connective tissue consists of the three types of fibers, several cells, and a semifluid ground substance (Table 4.4A). It is found in the subcutaneous layer; in mucous membranes; and around blood vessels, nerves, and body organs.
10. Adipose tissue consists of adipocytes, which store triglycerides (Table 4.4B). It is found in the subcutaneous layer, around organs, and in the yellow bone marrow.
11. Reticular connective tissue consists of reticular fibers and reticular cells and is found in the liver, spleen, and lymph nodes (Table 4.4C).
12. Dense connective tissue includes dense regular connective tissue, dense irregular connective tissue, and elastic connective tissue.
13. Dense regular connective tissue consists of parallel bundles of collagen fibers and fibroblasts (Table 4.4D). It forms tendons, most ligaments, and aponeuroses.
14. Dense irregular connective tissue consists of usually randomly arranged collagen fibers and a few fibroblasts (Table 4.4E). It is found in fasciae, the dermis of skin, and membrane capsules around organs.
15. Elastic connective tissue consists of branching elastic fibers and fibroblasts (Table 4.4F). It is found in the walls of large arteries, lungs, trachea, and bronchial tubes.
16. Cartilage contains chondrocytes and has a rubbery matrix (chondroitin sulfate) containing collagen and elastic fibers.
17. Hyaline cartilage is found in the embryonic skeleton, at the ends of bones, in the nose, and in respiratory structures (Table 4.4G). It is flexible, allows movement, and provides support.
18. Fibrocartilage is found in the pubic symphysis, intervertebral discs, and menisci (cartilage pads) of the knee joint (Table 4.4H).
19. Elastic cartilage maintains the shape of organs such as the epiglottis of the larynx, auditory (eustachian) tubes, and external ear (Table 4.4I).
20. Bone or osseous tissue supports, protects, helps provide movement, stores minerals, and houses blood-forming tissue (Table 4.4J).

21. Blood tissue is liquid connective tissue that consists of blood plasma and formed elements—red blood cells, white blood cells, and platelets. Its cells transport oxygen and carbon dioxide, carry on phagocytosis, participate in allergic reactions, provide immunity, and bring about blood clotting (Table 4.4K).
22. Lymph, the extracellular fluid that flows in lymphatic vessels, is also a liquid connective tissue. It is a clear fluid similar to blood plasma but with less protein (Table 4.4L).

Connective Tissue: Diverse and Dynamic

1. All tissues of the body are enveloped by and contain connective tissue (CT) and thus CT forms a continuous network that runs throughout the body.
2. Connective tissue is highly variable in its composition. It can form a tough, flexible network or may be diffuse and watery.
3. The CT can vary within an organ. As the organ changes during growth or disease, as extreme examples, its CT can change dramatically.
4. The ground substance of CT, produced by fibroblasts, may periodically change between the fluid and gel states, even in healthy tissue.
5. Collagen fibers predominate in tendons, ligaments, aponeuroses, the dermis of the skin, bone, blood vessels, the framework (stroma) of virtually all visceral organs, and deep fasciae.
6. Collagen fibers are responsible for the shape, resiliency, flexibility, strength, and structural integrity of organs.
7. Fibroblasts have the unique property of being able to move from one part of the body to another and then producing the fibers and ground substance that is appropriate to the tissue at any given time. The process of wound healing often involves the production of scar tissue that is formed when fibroblasts migrate to the area and produce collagen. This process is called fibrosis.
8. The abilities of CT to fluctuate with the tissue conditions tends to decline with a decrease in physical activities and with advanced aging.
9. Thixotropy is a phenomenon whereby ground substance becomes more fluid when it is warmed and mixed up and more gel-like when it sits without being disturbed. Manual therapists may be able to reverse this process in their patients or clients.
10. Poor cellular nutrition and sedentary habits weaken all connective tissues, stiffen them, and accelerate their biological aging.
11. Hydrogen bonding is responsible for the great strength of a collagen molecule as well as the collagen fibers.
12. The formation of collagen fibers by fibroblasts is as follows: Amino acids are linked in a specific sequence, and a triple-helix molecule is formed called tropocollagen. Depending on the physiological state of the tissue, tropocollagen may organize into collagen fibrils that then organize into collagen fibers. At every level just described, hydrogen bonding is responsible.
13. The range of motion (ROM) of any part of the musculoskeletal system decreases with the increase in hydrogen bonding that normally accompanies the aging process.
14. Fasciae that surround muscles or muscle groups thicken with advancing age. As this process continues, the nourishment for cells declines and the weakened muscles are now surrounded by "concrete-like" collars that reduce ROM and reduce nourishment for the cells.

Membranes (Section 4.5)

1. An epithelial membrane consists of an epithelial layer overlying a connective tissue layer. Examples are mucous, serous, and synovial membranes (Figure 4.7).
2. Mucous membranes line cavities that open to the exterior, such as the gastrointestinal tract.
3. Serous membranes line closed cavities (pleura, pericardium, peritoneum) and cover the organs in the cavities. These membranes consist of parietal and visceral layers.
4. Synovial membranes line joint cavities and do not have an epithelial layer.

Muscular Tissue (Section 4.6)

1. Muscular tissue consists of cells (fibers) that are specialized for contraction. It provides motion, maintenance of posture, heat production, and protection.
2. Skeletal muscle tissue is attached to bones (Table 4.5A), cardiac muscle tissue forms most of the heart wall (Table 4.5B), and smooth muscle tissue is found in the walls of hollow internal structures (blood vessels and viscera) (Table 4.5C).

Nervous Tissue (Section 4.7)

1. The nervous system is composed of neurons (nerve cells) and neuroglia (protective and supporting cells) (Table 4.6).
2. Neurons are sensitive to stimuli, convert stimuli into nerve impulses, and conduct nerve impulses.

Tissue Repair: Restoring Homeostasis (Section 4.8)

1. Tissue repair is the replacement of worn-out, damaged, or dead cells by healthy ones.
2. Stem cells may divide to replace lost or damaged cells.

Aging and Tissues (Section 4.9)

1. Tissues heal faster and leave less obvious scars in the young than in the aged; surgery performed on fetuses leaves no scars.
2. The extracellular components of tissues, such as collagen and elastic fibers, also change with age.

SELF-QUIZ QUESTIONS

1. Epithelial tissue functions in
   a. conducting nerve impulses
   b. storing fat
   c. covering and lining the body and its parts
   d. movement
   e. storing minerals.

2. Epithelial tissue is classified according to
   a. its location
   b. its function
   c. the composition of the matrix
   d. the shape and arrangement of its cells
   e. whether it is under voluntary or involuntary control.

3. Mucous membranes are
  a. composed of three layers
  b. found in body cavities that open to the body's exterior
  c. located at the ends of bones
  d. found lining the thoracic cavity
  e. capable of producing synovial fluid.

4. Which of the following is NOT a type of connective tissue?
  a. blood **b.** adipose
  c. reticular **d.** cuboidal
  e. cartilage

5. Which of the following is true concerning connective tissue?
  a. Except for cartilage, tendons, and ligaments, connective tissue has a rich blood supply.
  b. Connective tissue is classified according to cell shape and arrangement.
  c. The cells of connective tissue are generally closely joined.
  d. Loose connective tissue consists of many fibers arranged in a regular pattern.
  e. The fibers in connective tissue are composed of lipids.

6. Match the following tissue types with their descriptions.

____ **a.** fat storage	**A.** simple cuboidal epithelium
____ **b.** waterproofs the skin	**B.** simple squamous epithelium
____ **c.** forms the stroma (framework) of many organs	**C.** adipose
	D. fibrocartilage
____ **d.** composes the intervertebral discs	**E.** reticular connective
	F. smooth muscle
____ **e.** stores red bone marrow, protects, supports	**G.** keratinized stratified squamous epithelium
____ **f.** nonstriated, involuntary	**H.** bone
____ **g.** found in lungs, involved in diffusion	
____ **h.** found in kidney tubules, involved in absorption	

7. If you were going to design a hollow organ that needed to expand and have stretchability, which of the following epithelial and connective tissues might you use?
  a. transitional epithelium and elastic connective tissue
  b. stratified columnar epithelium and adipose tissue
  c. simple columnar epithelium and dense regular connective tissue
  d. simple squamous epithelium and hyaline cartilage
  e. transitional epithelium and reticular connective tissue

8. Which of the following statements is NOT true concerning epithelial tissue?
  a. The cells of epithelial tissue are closely packed.
  b. The basal layer of cells rests on a basement membrane.
  c. Epithelial tissue has a nerve supply.
  d. Epithelial tissue undergoes rapid rates of cell division.
  e. Epithelial tissue is well supplied with blood vessels.

9. Where would you find smooth muscle tissue?
  a. the heart **b.** attached to the bones
  c. in joints **d.** the discs between the vertebrae
  e. in the walls of hollow organs

10. A connective tissue with a liquid matrix is
  a. elastic cartilage **b.** blood
  c. areolar **d.** reticular
  e. osseous.

11. The interior of your nose is lined with
  a. a mucous membrane
  b. smooth muscle tissue
  c. a synovial membrane
  d. keratinized stratified squamous epithelium
  e. a serous membrane.

12. The four main types of tissue are
  a. epithelial, embryonic, blood, nervous
  b. blood, connective, muscular, nervous
  c. connective, epithelial, muscular, nervous
  d. stratified, muscular, striated, nervous
  e. epithelial, connective, muscular, membranous.

13. Which of the following materials would NOT be found in the extracellular matrix of connective tissue?
  a. collagen fibers **b.** elastic fibers
  c. keratin **d.** reticular fibers
  e. hyaluronic acid

14. Which connective tissue cells secrete antibodies?
  a. mast cells **b.** adipocytes
  c. macrophages **d.** plasma cells
  e. chondrocytes

15. Modified columnar epithelial cells that secrete mucus are ________ cells.
  a. microvilli **b.** keratinized
  c. mast **d.** fibroblast
  e. goblet

16. Which tissue is characterized by branching cells connected to each other by intercalated discs?
  a. skeletal muscle **b.** nervous
  c. bone **d.** cardiac muscle
  e. smooth muscle

17. Stratified squamous epithelium functions in
  a. protection **b.** contraction
  c. absorption **d.** stretching
  e. transport.

18. What tissue type is found in tendons?
  a. dense irregular connective tissue
  b. elastic connective tissue
  c. dense regular connective tissue
  d. pseudostratified epithelium
  e. areolar tissue

19. In what tissue type would you find central (haversian) canals, lacunae, and osteocytes?
  a. bone **b.** hyaline cartilage
  c. fibrocartilage **d.** dense irregular connective tissue
  e. elastic cartilage

20. Which of the following statements is true concerning glandular tissue?
  a. Endocrine glands are composed of connective tissue, whereas exocrine glands are composed of modified epithelium.
  b. Endocrine gland secretions diffuse into interstitial fluid and then into the bloodstream, whereas exocrine gland secretions enter ducts.
  c. A sweat gland is an example of an endocrine gland.
  d. Endocrine glands contain ducts, whereas exocrine glands do not.
  e. Exocrine glands produce substances known as hormones.

CRITICAL THINKING QUESTIONS

1. Many manual therapists utilize both heat and pressure modalities during a portion of their treatment sessions. One technique involves gentle pressure on a heated bag that is positioned on a patient, thus utilizing heat and pressure simultaneously. What is the primary goal of therapists who practice in this manner?
2. Imagine that you live 50 years in the future, and you can custom-design a human to suit the environment. Your assignment is to customize the human's tissue so that the individual can survive on a large planet with gravity, a cold, dry climate, and a thin atmosphere. What adaptations would you incorporate into the structure and/or amount of tissues, and why?
3. You've been on a "bread-and-water" diet for 3 weeks and have noticed that a cut on your shin won't heal and bleeds easily. Why?

? ANSWERS TO FIGURE QUESTIONS

4.1 Gap junctions allow cellular communication via passage of electrical and chemical signals between adjacent cells.

4.2 The basement membrane provides a physical support for an epithelium.

4.3 Substances would move most rapidly through squamous cells because they are so thin.

4.4 Sebaceous (oil) glands are holocrine glands, and salivary glands are merocrine glands.

4.5 Fibroblasts secrete the fibers and ground substance of the extracellular matrix.

4.6 An epithelial membrane is a membrane that consists of an epithelial tissue layer and an underlying layer of connective tissue.

The Integumentary System 5

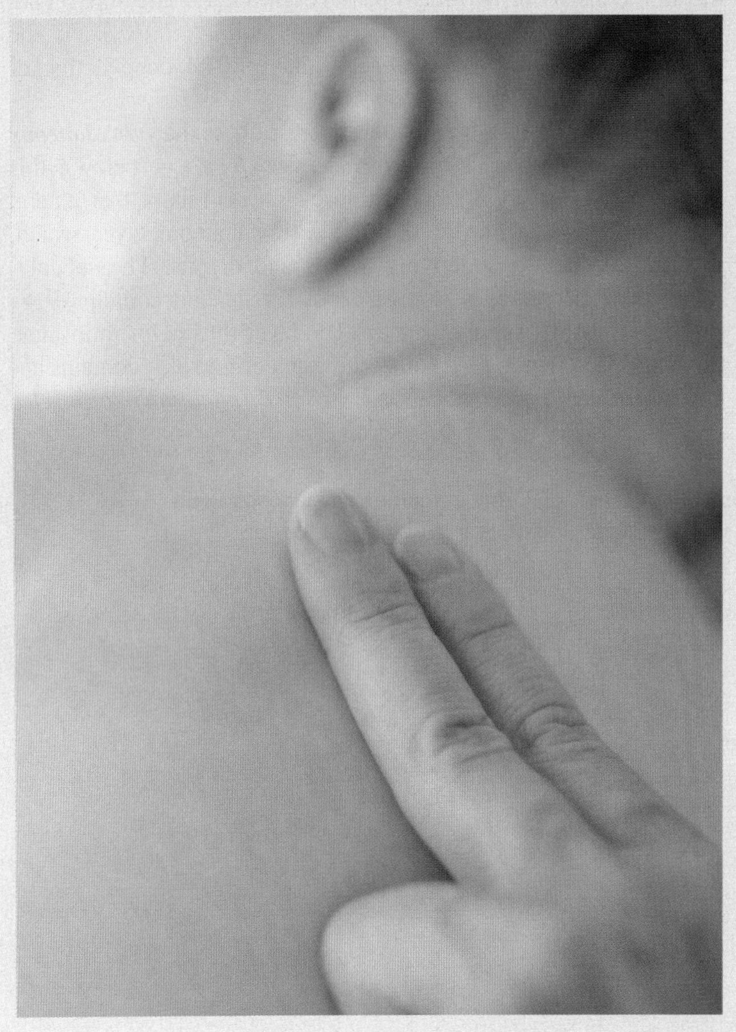

Although some persons say that the eyes are the windows to the soul, you could also say that the skin is the window to the body's health. A change in the integumentary system can be viewed from a homeostatic perspective. If the body is not functioning as well as it should, you often see symptoms arise in the skin. As a health-care professional you will begin assessment by first observing the body's surface, including the skin, for subtle or dramatic changes in texture, tone, temperature, and tenderness (the four T's) and color. Essentially, you are observing signs of tissue health or disease. These observations help you decide whether a patient can be treated or needs to be referred to a physician. Treatment may be based on whether an area of the patient 's body is a local contraindication (an area to avoid in treatment), such as a small cut or wart, or a general contraindication (the person should not be treated at all), such as a patient with a fever.

The skin is full of sensory nerve endings that make your body aware of the sensations about you. In fact, the first sense that is fully developed in babies is "the sense of touch." Studies show that massaging an infant and stimulating the nervous receptors in the skin actually aids growth of the body by up to 25%. As such, massage therapy is often used as one form of treatment for premature babies.

CONTENTS AT A GLANCE

5.1 STRUCTURE OF THE SKIN

OBJECTIVES

- Describe the layers of the epidermis and the cells that compose them.
- Compare the composition of the papillary and reticular regions of the dermis.
- Explain the basis for different skin colors.

The ***integumentary system*** (in-teg-ū-MEN-tar-ē; *in* = inward; *tegere* = to cover) is made up of organs and tissues such as skin, hair, oil and sweat glands, nails, and sensory receptors. The ***skin*** or ***cutaneous membrane*** (kū-TĀ-nē-us), which covers the external surface of the body, is the largest organ of the body in both surface area and weight. In adults, the skin covers an area of about 22 square feet and weighs 8–11 lb. It ranges in thickness from 0.5 mm (0.02 in.) on the eyelids to 4.0 mm (0.16 in.) on the heels. For your reference, a penny is 0.16 in. thick. Structurally, the skin consists of two main parts (Figure 5.1). The superficial, thinner portion, which is composed of *epithelial tissue,* is the ***epidermis*** (ep′-i-DERM-is; *epi-* = above). The deeper, thicker *connective tissue* part is the ***dermis.***

Deep to the dermis, but not part of the skin, is the ***subcutaneous (subQ) layer.*** Also called the ***hypodermis*** (*hypo-* = below), this layer consists of areolar and adipose tissues. Fibers that extend from the dermis anchor the skin to the subcutaneous layer, which, in turn, attaches to underlying tissues and organs. The subcutaneous layer serves as a storage depot for fat and contains large blood vessels that supply the skin. The layer thus helps to insulate the body. This region (and sometimes the dermis) also contains encapsulated nerve endings called ***lamellated (pacinian) corpuscles***

Figure 5.1 Components of the integumentary system. The skin consists of a superficial, thin epidermis and a deep, thicker dermis. Deep to the skin is the subcutaneous layer, which attaches the dermis to underlying organs and tissues.

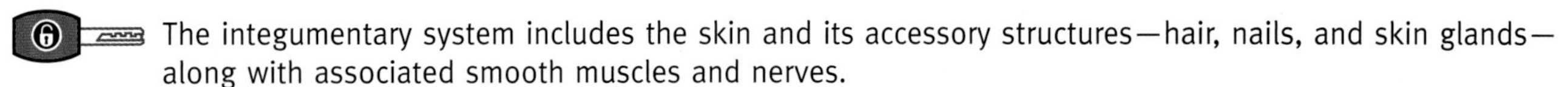

The integumentary system includes the skin and its accessory structures—hair, nails, and skin glands—along with associated smooth muscles and nerves.

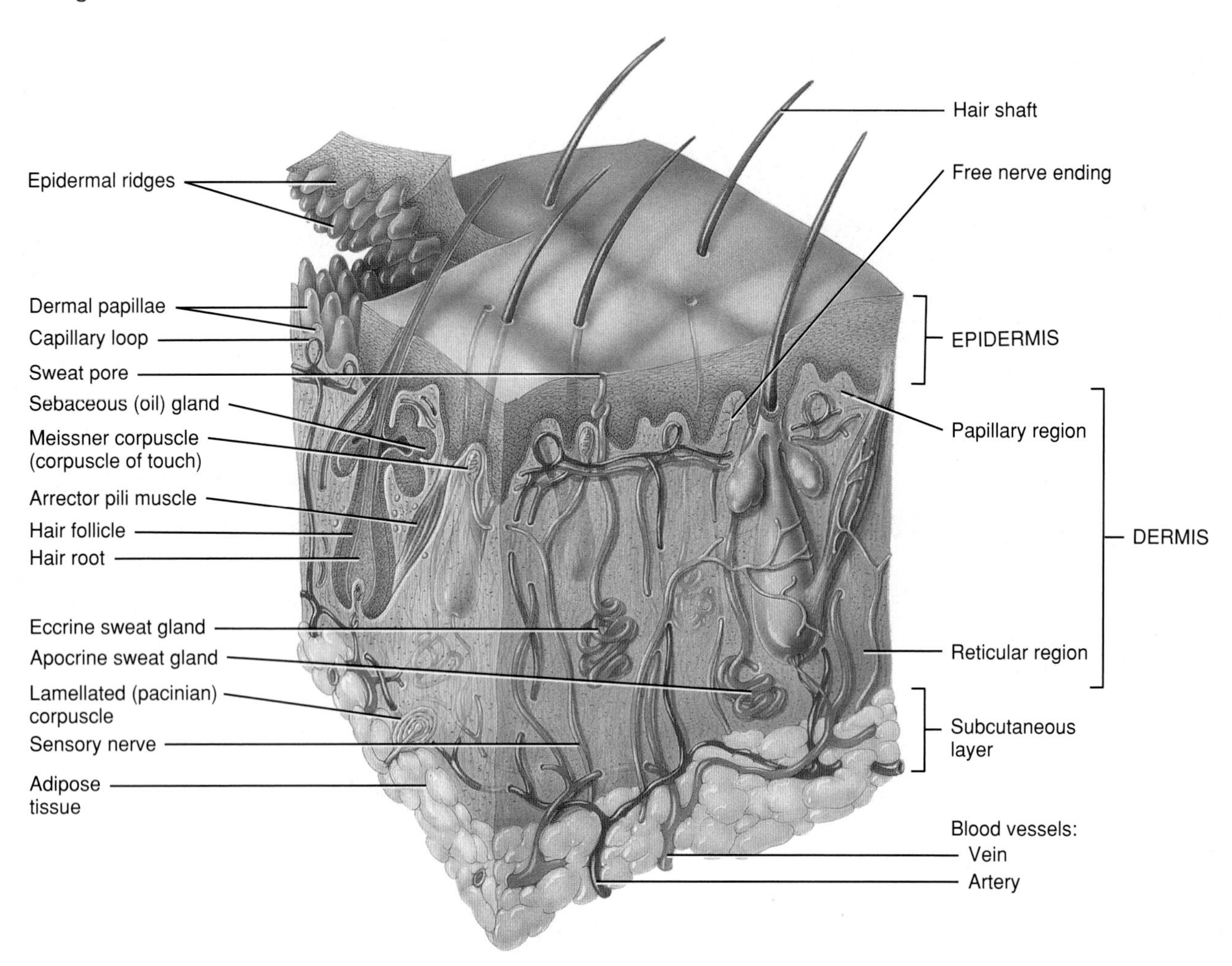

(a) Sectional view of skin and subcutaneous layer

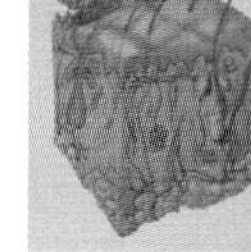

(pa-SIN-ē-an) that are sensitive to pressure (see Figure 5.1). Note that some authorities consider the hypodermis to be the third layer of skin. ***Dermatology*** (der′-ma-TOL-ō-jē; *dermato-* = skin; *-logy* = study of) is the medical specialty that deals with the diagnosis and treatment of integumentary system disorders.

Epidermis

The ***epidermis*** is composed of keratinized stratified squamous epithelium. As you learned in Chapter 4, this tissue is avascular. The average epidermis is 0.1 mm (0.004 in.) in thickness; since blisters are formed when fluid separates the skin near the epidermal–dermal junction, the thinness of the epidermis can be realized by noting the exceedingly thin covering of a blister. The epidermis contains four principal types of cells: keratinocytes, melanocytes, Langerhans cells, and Merkel cells (see Figure 5.2). About 90% of epidermal cells are ***keratinocytes*** (ker-a-TIN-ō-sīts; *keratino-* = hornlike; *-cytes* = cells), which are arranged in four or five layers and produce the protein ***keratin*** (KER-a-tin) (Figure 5.2c). Recall from Chapter 4 that keratin is a tough, fibrous protein that helps protect the skin and underlying tissues from abrasion, heat, microbes, and chemicals. Keratinocytes also produce lamellar granules, which release a water-repellent sealant that decreases water entry and loss and inhibits the entry of foreign materials.

About 8% of the epidermal cells are ***melanocytes*** (MEL-a-nō-sīts; *melano-* = black), which develop from the ectoderm of a developing embryo and produce the pigment melanin (see Figure 5.2d). Their long, slender projections extend between the keratinocytes and transfer melanin granules to them. ***Melanin*** (MEL-a-nin) is a yellow-red or brown-black pigment that contributes to skin color and absorbs damaging ultraviolet

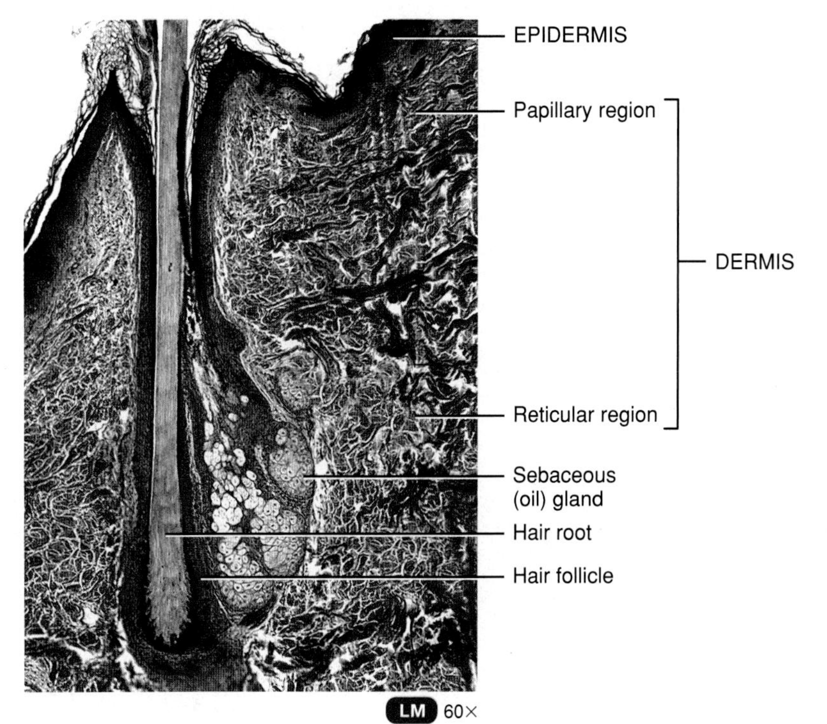

(b) Sectional view of skin

Functions

1. Regulates body temperature.
2. Stores blood.
3. Protects body from external environment.
4. Detects cutaneous sensations.
5. Excretes and absorbs substances.
6. Synthesizes vitamin D.

? What types of tissues make up the epidermis and the dermis?

(UV) light. Once inside keratinocytes, the melanin granules cluster to form a protective veil over the nucleus, on the side toward the skin surface. In this way, they shield the nuclear DNA from damage by UV light. Although their melanin granules effectively protect keratinocytes, melanocytes themselves are particularly susceptible to damage by UV light.

Langerhans cells (LANG-er-hans) arise from red bone marrow and migrate to the epidermis (Figure 5.2e), where they con-

Figure 5.2 Types of cells in the epidermis. Besides keratinocytes, the epidermis contains melanocytes, which produce the pigment melanin; Langerhans cells, which participate in immune responses; and Merkel cells, which function in the sensation of touch.

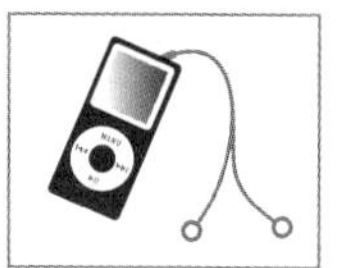

Most of the epidermis consists of keratinocytes, which produce the protein keratin (protects underlying tissues) and lamellar granules (contains a waterproof sealant). The epidermis consists of keratinized stratified squamous epithelium.

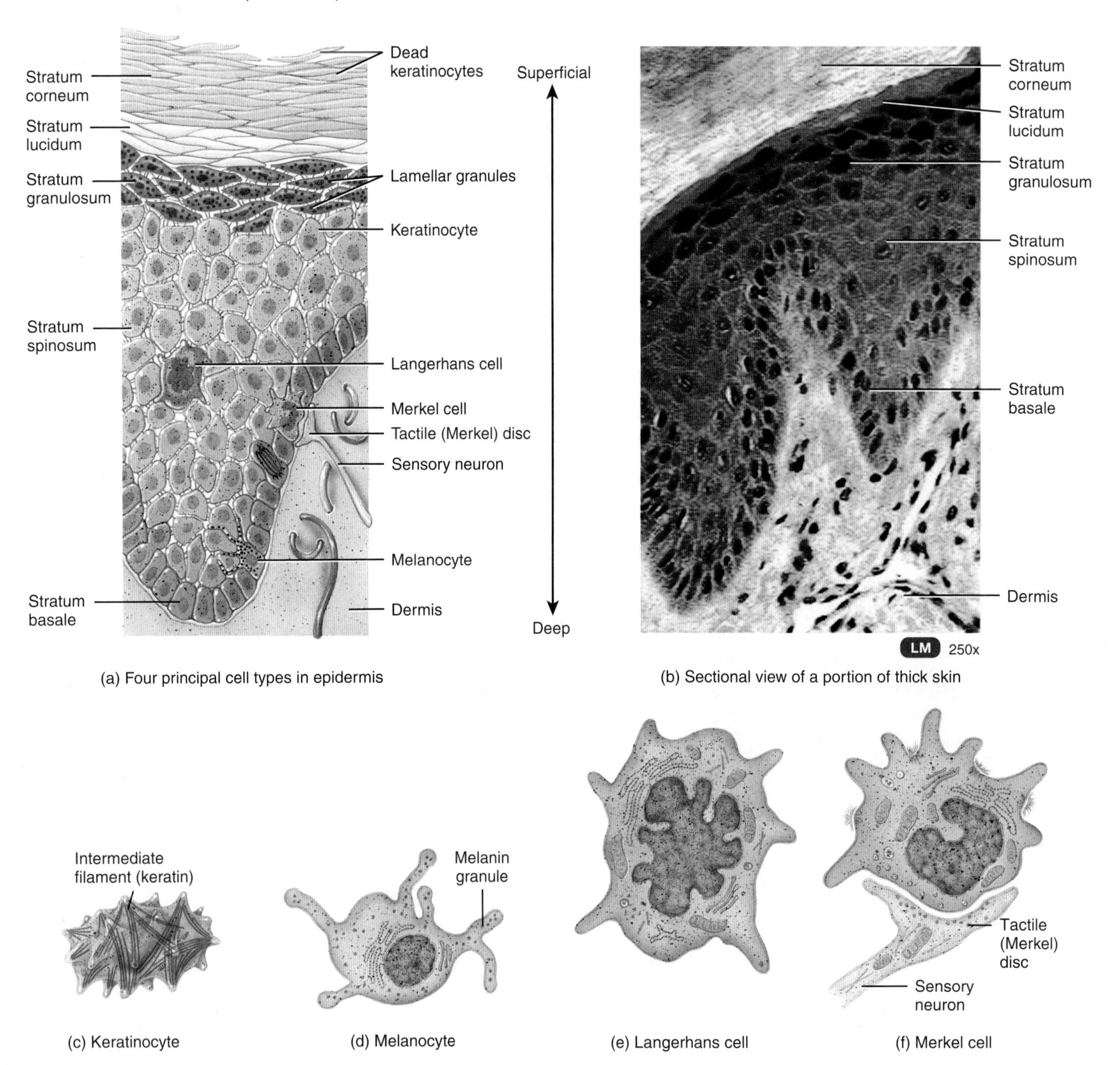

(a) Four principal cell types in epidermis

(b) Sectional view of a portion of thick skin

(c) Keratinocyte

(d) Melanocyte

(e) Langerhans cell

(f) Merkel cell

Which epidermal layer includes stem cells that continually undergo cell division?

stitute a small fraction of the epidermal cells. They participate in immune responses mounted against microbes that invade the skin, and are easily damaged by UV light.

Merkel cells (MER-kel) are the least numerous of the epidermal cells. They are located in the deepest layer of the epidermis, where they contact the flattened process of a sensory neuron (nerve cell), a structure called a ***tactile (Merkel) disc*** (see Figure 5.2f). Merkel cells and tactile discs detect touch sensations.

Several distinct layers of keratinocytes in various stages of development form the epidermis (see Figure 5.2a,b). In most regions of the body the epidermis has four strata or layers: stratum basale (germinativum), stratum spinosum, stratum granulosum, and a thin stratum corneum. This is called ***thin skin.*** Where exposure to friction is greatest, such as in the fingertips, palms, and soles, the epidermis has five layers: stratum basale, stratum spinosum, stratum granulosum, stratum lucidum, and a thick stratum corneum. This is called ***thick skin.*** The details of thin and thick skin are discussed later in the chapter.

Stratum Basale

The deepest layer of the epidermis is the ***stratum basale*** (ba-SA-lē; *basal-* = base), composed of a single row of cuboidal or columnar keratinocytes. Some cells in this layer are *stem cells* that undergo cell division to continually produce new keratinocytes. The nuclei of keratinocytes in the stratum basale are large, and their cytoplasm contains many ribosomes, a small Golgi complex, a few mitochondria, and some rough endoplasmic reticulum. The cytoskeleton within keratinocytes of the stratum basale includes scattered intermediate filaments, called *tonofilaments.* The tonofilaments are composed of a protein that will form keratin in more superficial epidermal layers. Tonofilaments attach to desmosomes, which bind cells of the stratum basale to each other and to the cells of the adjacent stratum spinosum, and to hemidesmosomes, which bind the keratinocytes to the basement membrane between the epidermis and the dermis. Melanocytes and Merkel cells with their associated tactile discs are scattered among the keratinocytes of the basal layer. The stratum basale is also known as the ***stratum germinativum*** (jer′-mi-na-TĒ-vum; *germ-* = sprout) to indicate its role in forming new cells.

• CLINICAL CONNECTION | Skin Grafts

New skin cannot regenerate if an injury destroys a large area of the stratum basale and its stem cells. Skin wounds of this magnitude require skin grafts in order to heal. A **skin graft** involves covering the wound with a patch of healthy skin taken from a donor site. To avoid tissue rejection, the transplanted skin is usually taken from the same individual (*autograft*) or an identical twin (*isograft*). If skin damage is so extensive that an autograft would cause harm, a self-donation procedure called *autologous skin transplantation* (aw-TOL-ō-gus) may be used. In this procedure, performed most often for severely burned patients, small amounts of an individual's epidermis are removed and the keratinocytes are cultured in the laboratory to produce thin sheets of skin. The new skin is transplanted back to the patient so that it covers the burn wound and generates a permanent skin. Also available as skin grafts for wound coverage are products (Apligraft and Transite) grown in the laboratory from the foreskins of circumcised infants. •

Stratum Spinosum

Superficial to the stratum basale is the ***stratum spinosum*** (spi-NŌ-sum; *spinos-* = thornlike), where 8 to 10 layers of many-sided keratinocytes fit closely together. This layer provides strength and flexibility to the skin. These keratinocytes have the same organelles as cells of the stratum basale. When cells of the stratum spinosum are prepared for microscopic examination, they shrink and pull apart so that they seem to be covered with thornlike spines (see Figure 5.2a), although they appear rounded and larger in living tissue. Each spiny projection in a prepared tissue section is a point where bundles of tonofilaments are inserting into a desmosome, tightly joining the cells to one another. This arrangement provides both strength and flexibility to the skin. Langerhans cells and projections of melanocytes are also present in this layer.

Stratum Granulosum

At about the middle of the epidermis, the ***stratum granulosum*** (gran-ū-LŌ-sum; *granulos-* = little grains) consists of three to five layers of flattened keratinocytes that are undergoing apoptosis. (Recall from Chapter 3 that apoptosis is an orderly, genetically programmed cell death in which the nucleus fragments before the cells die.) The nuclei and other organelles of these cells begin to degenerate, and tonofilaments become more apparent. A distinctive feature of cells in this layer is the presence of darkly staining granules of a protein called ***keratohyalin*** (ker′-a-tō-HĪ-a-lin), which converts the tonofilaments into keratin. Also present in the keratinocytes are membrane-enclosed ***lamellar granules,*** which release a lipid-rich secretion. This secretion fills the spaces between cells of the stratum granulosum, stratum lucidum, and stratum corneum. The lipid-rich secretion acts as a water-repellent sealant, retarding loss and entry of water and entry of foreign materials. As their nuclei break down during apoptosis, the keratinocytes of the stratum granulosum can no longer carry on vital metabolic reactions, and they die. Thus, the stratum granulosum marks the transition between the deeper, metabolically active strata and the dead cells of the more superficial strata.

Stratum Lucidum

The ***stratum lucidum*** (LOO-si-dum; *lucid-* = clear) is present only in the thick skin of areas such as the fingertips, palms, and soles. It consists of three to five layers of flattened clear, dead keratinocytes that contain large amounts of keratin and thickened plasma membranes. Note that the appearance of this layer varies with the stain that was used. Sometimes the layer appears clear, thus the name stratum lucidum, but many newer histological stains cause this layer to be the darkest of the five (see Figure 5.2b).

Stratum Corneum

The ***stratum corneum*** (KOR-nē-um; *corne-* = horn or horny) commonly consists of 25 to 30 layers of flattened dead keratinocytes. These cells are continuously shed and replaced by cells from the deeper strata. The interior of the cells contains mostly keratin. Between the cells are lipids from lamellar granules that help make this layer an effective water-repellent barrier. Its multiple layers of dead cells also help to protect deeper layers from injury and microbial invasion. Constant exposure of skin to friction stimulates the formation of a *callus,* an abnormal thickening of the stratum corneum.

Keratinization and Growth of the Epidermis

Newly formed cells in the stratum basale are slowly pushed to the surface. As the cells move from one epidermal layer to the next, they accumulate more and more keratin, a process called ***keratinization*** (ker′-a-tin-i-ZĀ-shun). Then they undergo apoptosis (cell suicide). Eventually the keratinized cells slough off and are replaced by underlying cells that, in turn, become keratinized. The whole process by which cells form in the stratum basale, rise to the surface, become keratinized, and slough off takes about 4 weeks. The rate of cell division in the stratum basale increases when the outer layers of the epidermis are stripped away, as occurs in abrasions and burns. The mechanisms that regulate this remarkable growth are not well understood, but hormone-like proteins such as ***epidermal growth factor (EGF)*** play a role. An excessive amount of keratinized cells shed from the skin of the scalp is called ***dandruff.***

Table 5.1 summarizes the distinctive features of the epidermal strata.

TABLE 5.1

Summary of Epidermal Strata

STRATUM	DESCRIPTION
Basale	Deepest layer, composed of a single row of cuboidal or columnar keratinocytes that contain scattered tonofilaments (intermediate filaments); stem cells undergo cell division to produce new keratinocytes; melanocytes, Langerhans cells, and Merkel cells associated with tactile discs are scattered among the keratinocytes.
Spinosum	Eight to ten rows of many-sided keratinocytes with bundles of tonofilaments; includes projections of melanocytes and Langerhans cells.
Granulosum	Three to five rows of flattened keratinocytes, in which organelles are beginning to degenerate; cells contain the protein keratohyalin, which converts tonofilaments into keratin, and lamellar granules, which release a lipid-rich, water-repellent secretion.
Lucidum	Present only in skin of fingertips, palms, and soles; consists of three to five rows of clear, flat, dead keratinocytes with large amounts of keratin.
Corneum	Twenty-five to thirty rows of dead, flat keratinocytes that contain mostly keratin.

Dermis

The second, deeper part of the skin, the ***dermis,*** is composed mainly of connective tissue. Blood vessels, nerves, glands, and hair follicles are embedded in dermal tissue. Based on its tissue structure, the dermis can be divided into a papillary region and a reticular region.

The ***papillary region*** makes up about one-fifth of the thickness of the total layer (see Figure 5.1). It consists of areolar connective tissue containing fine elastic fibers. Its surface area is greatly increased by small, finger-like structures that project into the undersurface of the epidermis called ***dermal papillae*** (pa-PIL-ē = nipples). These nipple-shaped structures project into the epidermis and some contain ***capillary loops*** (blood capillaries). Some dermal papillae also contain tactile receptors called ***corpuscles of touch*** or ***Meissner corpuscles*** (MĪS-ner) (Figure 5.3), encapsulated nerve endings that are sensitive to touch, and ***free nerve endings,*** dendrites that lack any apparent structural specialization. Different free nerve endings initiate signals that give rise to sensations of warmth, coolness, pain, tickling, and itching.

The ***reticular region*** (*reticul-* = netlike), which is attached to the subcutaneous layer, consists of dense irregular connective tissue containing fibroblasts, bundles of collagen, and some coarse elastic fibers. The collagen fibers in the reticular region interlace in a netlike manner. A few adipose cells, hair follicles, nerves, sebaceous (oil) glands (found only in thin skin), and sudoriferous (sweat) glands occupy the spaces between fibers.

The combination of collagen and elastic fibers in the reticular region provides the skin with strength, ***extensibility*** (ek-sten′-si-BIL-i-tē) (ability to stretch), and ***elasticity*** (e-las-TIS-i-tē) (ability to return to original shape after stretching). The extensibility of skin can be readily seen around joints where furrows are present. Note especially the dorsum of the extended wrist. Gross examination of the skin reveals small diamond-shaped areas; histological examination of the same areas demonstrates shallow furrows that permit expansion of the skin as the wrist is flexed (see thin skin in Figure 5.7b). Extreme stretching in pregnancy and obesity may produce small tears in the dermis, causing ***striae*** (STRĪ-ē = streaks), or stretch marks, visible as red or silvery white streaks on the skin surface.

The surfaces of the palms, fingers, soles, and toes have a series of ridges and grooves. They appear either as straight lines or as a pattern of loops and whorls, as on the tips of the digits. These ***epidermal ridges*** develop during the third month of fetal development as downward projections of the epidermis into the dermis between the dermal papillae of the papillary region (Figure 5.3a). These epidermal ridges, especially in the fingers and toes, increase the surface area of the epidermis and thus increase the grip of the hand or foot by increasing friction. Because the ducts of sweat glands open between the epidermal ridges as sweat pores,

Figure 5.3 Interface of the epidermis and the dermis. Dermal papillae and epidermal ridges complement one another and form an interface that minimizes the possible separation of the dermis and epidermis. A corpuscle of touch is shown in a dermal papilla in (a).

Epithelial tissues of the epidermal layer and the connective tissues of the dermal layer are closely adherent.

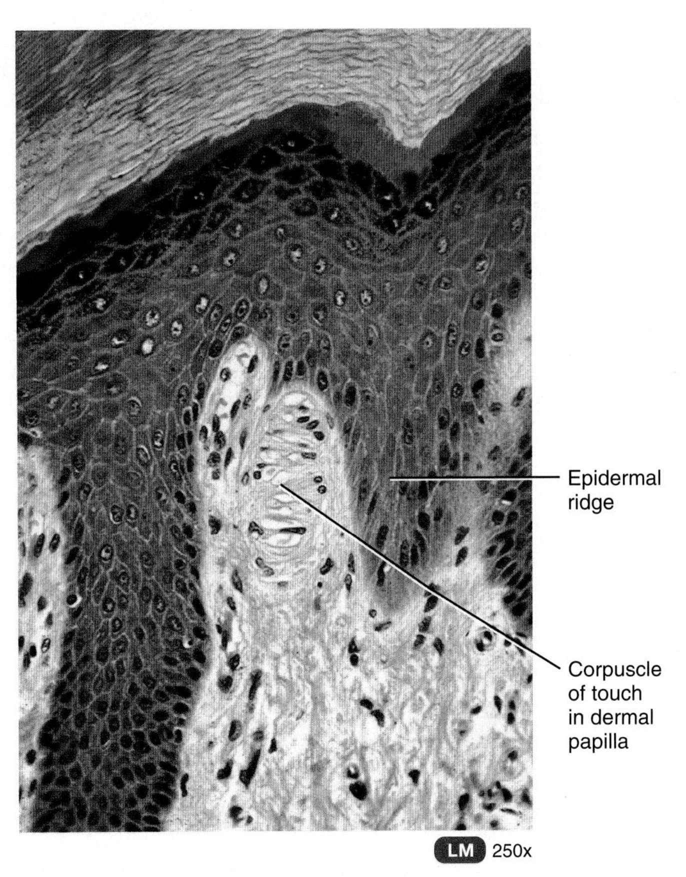

(a) Sectional view of dermal papillae and epidermal ridge

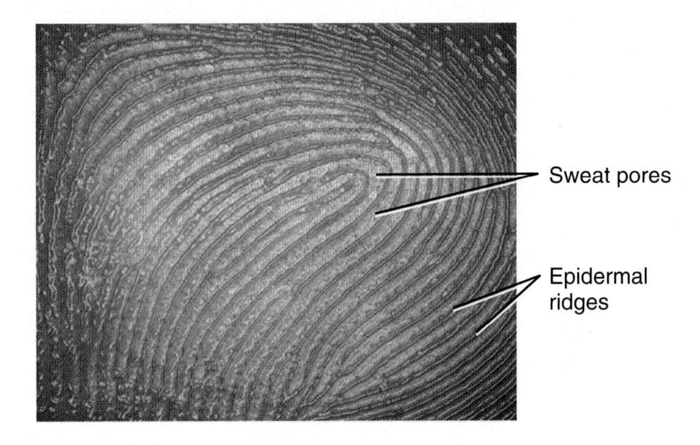

(b) Epidermal ridges and sweat pores

What are the functions of epidermal ridges?

the sweat and ridges form ***fingerprints*** (or ***footprints***) upon touching a smooth object (Figure 5.3b). The epidermal ridge pattern is genetically determined and is unique for each individual. Normally, the ridge pattern does not change during life, except to enlarge, and thus can serve as the basis for identification. The study of the pattern of epidermal ridges is called ***dermatoglyphics*** (der′-ma-tō-GLIF-iks; *glyphe* = carved work).

Table 5.2 summarizes the structural features of the papillary and reticular regions of the dermis.

TABLE 5.2

Summary of Papillary and Reticular Regions of the Dermis

REGION	DESCRIPTION
Papillary	The superficial portion of the dermis (about one-fifth); consists of areolar connective tissue with elastic fibers; contains dermal papillae that house capillaries, corpuscles of touch, and free nerve endings.
Reticular	The deeper portion of the dermis (about four-fifths); consists of dense irregular connective tissue with bundles of collagen and some coarse elastic fibers. Spaces between fibers contain some adipose cells, hair follicles, nerves, sebaceous glands, and sudoriferous glands.

Skin Color

Melanin, hemoglobin, and carotene are three pigments that impart a wide variety of colors to skin. The amount of ***melanin*** causes the skin's color to vary from pale yellow to reddish-brown. The difference between the two forms of melanin, *pheomelanin* (yellow to red) and *eumelanin* (brown to black), is most apparent in the hair. Melanocytes, the melanin-producing cells, are most plentiful in the epidermis of the penis, nipples of the breasts, area just around the nipples (areolae), face, and limbs. They are also present in mucous membranes. Because the *number* of melanocytes is about the same in all people, differences in skin color are due mainly to the *amount of pigment* the melanocytes produce and transfer to keratinocytes. In some people, melanin accumulates in patches called *freckles.* In the elderly, *age (liver) spots* may develop. These flat blemishes look like freckles and range in color from light brown to black. Like freckles, age spots are accumulations of melanin. A round, flat, or raised area that represents a benign localized overgrowth of melanocytes and usually develops in childhood or adolescence is called a ***nevus*** (NĒ-vus), or a ***mole*** (see Figure 5.9a). Melanocytes synthesize melanin from the amino acid *tyrosine* in the presence of an enzyme called *tyrosinase.* Synthesis occurs in an organelle called a ***melanosome.*** Exposure to UV light increases the enzymatic activity within melanosomes and thus increases melanin production. Both the amount and darkness of melanin increase upon UV exposure, which gives the skin a tanned appearance and helps protect the body against further UV radiation. Melanin absorbs UV radiation, prevents damage to DNA in epidermal cells, and neutralizes free radicals that form in the skin following damage by UV radiation. Thus, within limits, melanin serves a protective function. As you will see later, however, repeatedly exposing the skin to UV light may cause skin cancer. Persons who wish, however, to keep a suntan need to expose their skin repeatedly because the tan is lost when the melanin-containing keratinocytes are shed from the stratum corneum.

Dark-skinned individuals have large amounts of melanin in the epidermis. Consequently, the epidermis has a dark pigmentation and skin color ranges from yellow to red to tan to black. Light-skinned individuals have little melanin in the epidermis. Thus, the epidermis appears translucent and skin color ranges from pink to red depending on the amount and oxygen content of the blood moving through capillaries in the dermis. The red color is due to ***hemoglobin,*** the oxygen-carrying pigment in red blood cells.

Carotene (KAR-ō-tēn; *carot-* = carrot) is a yellow-orange pigment that gives egg yolks and carrots their color. This precursor of vitamin A, which is used to synthesize pigments needed for vision, accumulates in the stratum corneum and fatty areas of the dermis and subcutaneous layer in response to excessive dietary intake. In fact, so much carotene may be deposited in the skin after eating large amounts of carotene-rich foods that the skin color actually turns orange, which is especially apparent in light-skinned individuals. Decreasing carotene intake eliminates the problem.

Albinism (AL-bin-izm; *albin-* = white) is the inherited inability of an individual to produce melanin. Most ***albinos*** (al-BĪ-nōs), people affected by albinism, have melanocytes that are unable to synthesize tyrosinase. Melanin is missing from their hair, eyes, and skin and as a result the hair and skin are colorless or whitish; the eyes appear red from the rich blood supply to the eye while the colorless iris permits this redness to be evident.

In another condition, called ***vitiligo*** (vit-i-LĪ-gō), the partial or complete loss of melanocytes from patches of skin produces irregular white spots of varying size. The loss of melanocytes may be related to an immune system malfunction in which antibodies attack the melanocytes.

• CLINICAL CONNECTION Skin and Mucous Membrane Color as a Diagnostic Clue

The **color of skin and mucous membranes can provide clues for diagnosing** certain conditions. When blood is not picking up an adequate amount of oxygen from the lungs, as in someone who has stopped breathing, the mucous membranes, nail beds, and skin appear bluish or ***cyanotic*** (sī-a-NOT-ik; *cyan-* = blue). ***Jaundice*** (JON-dis; *jaund-* = yellow) is due to a buildup of the yellow pigment bilirubin in the skin. This condition gives a yellowish appearance to the skin and the whites of the eyes, and usually indicates liver disease. ***Erythema*** (er-e-THĒ-ma; *eryth-* = red), redness of the skin, is caused by engorgement of capillaries in the dermis with blood due to skin injury, exposure to heat, infection, inflammation, or allergic reactions. ***Pallor*** (PAL-or), also known as paleness or blanching of the skin, may occur in conditions such as shock and anemia. All skin color changes are observed most readily in people with lighter-colored skin and may be more difficult to discern in people with darker skin. However, examination of the nail beds and gums can provide some information about circulation in individuals with darker skin. •

CHECKPOINT

1. What structures are included in the integumentary system?
2. What are the main differences between the epidermis and dermis of the skin?
3. Compare the composition of the papillary and reticular regions of the dermis.
4. What are the three pigments in the skin and how do they contribute to skin color?

5.2 ACCESSORY STRUCTURES OF THE SKIN

OBJECTIVE

- Describe the structure and functions of hair, skin glands, and nails.

Accessory structures of the skin—hair, skin glands, and nails—develop from the embryonic epidermis. They have a host of important functions. For example, hair and nails protect the body, and sweat glands help regulate body temperature.

Hair

Hairs, or *pili* (PĪ-lī), are present on most skin surfaces except the palms, palmar surfaces of the fingers, the soles, and plantar surfaces of the feet. In adults, hair usually is most heavily distributed across the scalp, in the eyebrows, in the axillae (armpits), and around the external genitalia. Genetic and hormonal influences largely determine the thickness and the pattern of distribution of hairs.

Although the protection it offers is limited, hair on the head guards the scalp from injury and the sun's rays. It also decreases heat loss from the scalp. Eyebrows and eyelashes protect the eyes from foreign particles, as does hair in the nostrils and in the external ear canal. Touch receptors (hair root plexuses) associated with hair follicles are activated whenever a hair is moved even slightly. Thus, hairs also function in sensing light touch.

Each hair is composed of columns of dead, keratinized epidermal cells bonded together by extracellular proteins. The ***shaft*** is the superficial portion of the hair, which projects above the surface of the skin (see Figure 5.4a). The ***root*** is the portion of the hair deep to the shaft that penetrates into the dermis, and sometimes into the subcutaneous layer. The shaft and root of the hair both consist of three concentric layers of cells: medulla, cortex, and cuticle of the hair (see Figure 5.4c,d). The inner *medulla,* which may be lacking in thinner hair, is composed of two or three rows of irregularly shaped cells. The middle *cortex* forms the major part of the shaft and consists of elongated cells. The *cuticle of the hair,* the outermost layer, consists of a single layer of thin, flat cells that are the most heavily keratinized. Cuticle cells on the shaft are arranged like shingles on the side of a house, with their free edges pointing toward the end of the hair (see Figure 5.4b).

Surrounding the root of the hair is the ***hair follicle*** (FOL-li-kul), which is made up of an external root sheath and an internal root sheath, together referred to as an ***epithelial root sheath*** (see Figure 5.4c,d). The *external root sheath* is a downward continuation of the epidermis. The *internal root sheath* is produced by the matrix (described shortly) and forms a cellular tubular sheath of epithelium between the external root sheath and the hair. The dense dermis surrounding the hair follicle is called the ***dermal root sheath.*** The base of each hair follicle is an onion-shaped structure, the ***bulb*** (see Figure 5.4c). This structure houses a nipple-shaped indentation, the ***papilla of the hair,*** which contains areolar connective tissue and many blood vessels that nourish the growing hair follicle. The bulb also contains a germinal layer of cells called the ***hair matrix.*** The hair matrix cells arise from the stratum basale, the site of cell division. Hence, hair matrix cells are responsible for the growth of existing hairs, and they produce new hairs when old hairs are shed. This replacement process occurs within the same follicle. Hair matrix cells also give rise to the cells of the internal root sheath.

Sebaceous (oil) glands (discussed shortly) and a bundle of smooth muscle cells are also associated with hairs (see Figure 5.4a). The smooth muscle is the ***arrector pili*** (a-REK-tor PĪ-lē; *arrect-* = to raise). It extends from the superficial dermis of the skin to the dermal root sheath around the side of the hair follicle. In its normal position, hair emerges at an angle to the surface of the skin. Under physiological or emotional stress, such as cold or fright, autonomic nerve endings stimulate the arrector pili muscles to contract, which pull the hair shafts perpendicular to the skin surface. This action causes "goose bumps" or "gooseflesh" because the skin around the shaft forms slight elevations.

Surrounding each hair follicle are dendrites of neurons, forming a ***hair root plexus*** (PLEK-sus) that is sensitive to touch (see Figure 5.4a). The hair root plexus generates nerve impulses if the hair shaft is moved.

Each hair follicle goes through a growth cycle, which consists of a growth stage and a resting stage. During the ***growth stage,*** cells of the matrix differentiate, keratinize, and die. As new cells are added at the base of the hair root, the hair grows longer. In time, the growth of the hair stops and the ***resting stage*** begins. After the resting stage, a new growth cycle begins. The old hair root falls out or is pushed out of the hair follicle, and a new hair begins to grow in its place. Individual scalp hairs grow for 2 to 6 years and rest for about 3 months. This variation in growth accounts for the short time that it takes for bangs or sideburns to appear ragged. At any time, about 85% of scalp hairs are in the growth stage. Visible hair is dead, but until the hair is pushed out of its follicle by a new hair, portions of its root within the scalp are alive.

Normal hair loss in the adult scalp is about 70–100 hairs per day. Both the rate of growth and the replacement cycle may be altered by illness, radiation therapy, chemotherapy, age, genetics, gender, and severe emotional stress. Barbers and beauticians (estheticians) sometimes will note a change in the health of a long-time customer before the change is evident to most other persons. Rapid weight-loss diets that severely restrict calories or protein increase hair loss. The rate of shedding also increases for three to four months after childbirth. ***Alopecia*** (al′-ō-PĒ-shē-a), the partial or complete lack of hair, may result from genetic factors, aging, endocrine disorders, chemotherapy, or skin disease.

Hair Color

The color of hair is due primarily to the amount and type of melanin in its keratinized cells. Melanin is synthesized by melanocytes scattered in the matrix of the bulb and passes into cells of the cortex and medulla of the hair (see Figure 5.4c). Dark-colored hair contains mostly brown to black melanin; blond and red hair contains yellow to red melanin. Hair becomes gray because of a progressive decline in melanin production. White hair results from the lack of melanin and the accumulation of air bubbles in the shaft.

Hair and Hormones

At puberty, when the testes begin secreting significant quantities of androgens (masculinizing sex hormones), males develop the typical male pattern of hair growth, including a beard and a hairy chest. In females at puberty, the ovaries and the adrenal glands produce small quantities of androgens, which promote hair growth in the axillae and pubic region. Occasionally, a tumor of the adrenal glands, testes, or ovaries produces an excessive amount of androgens. The result in females or prepubertal

Figure 5.4 Hair.

Hairs are growths of epidermal derivatives composed of dead, keratinized epidermal cells.

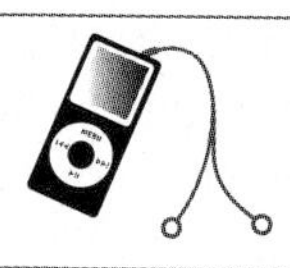

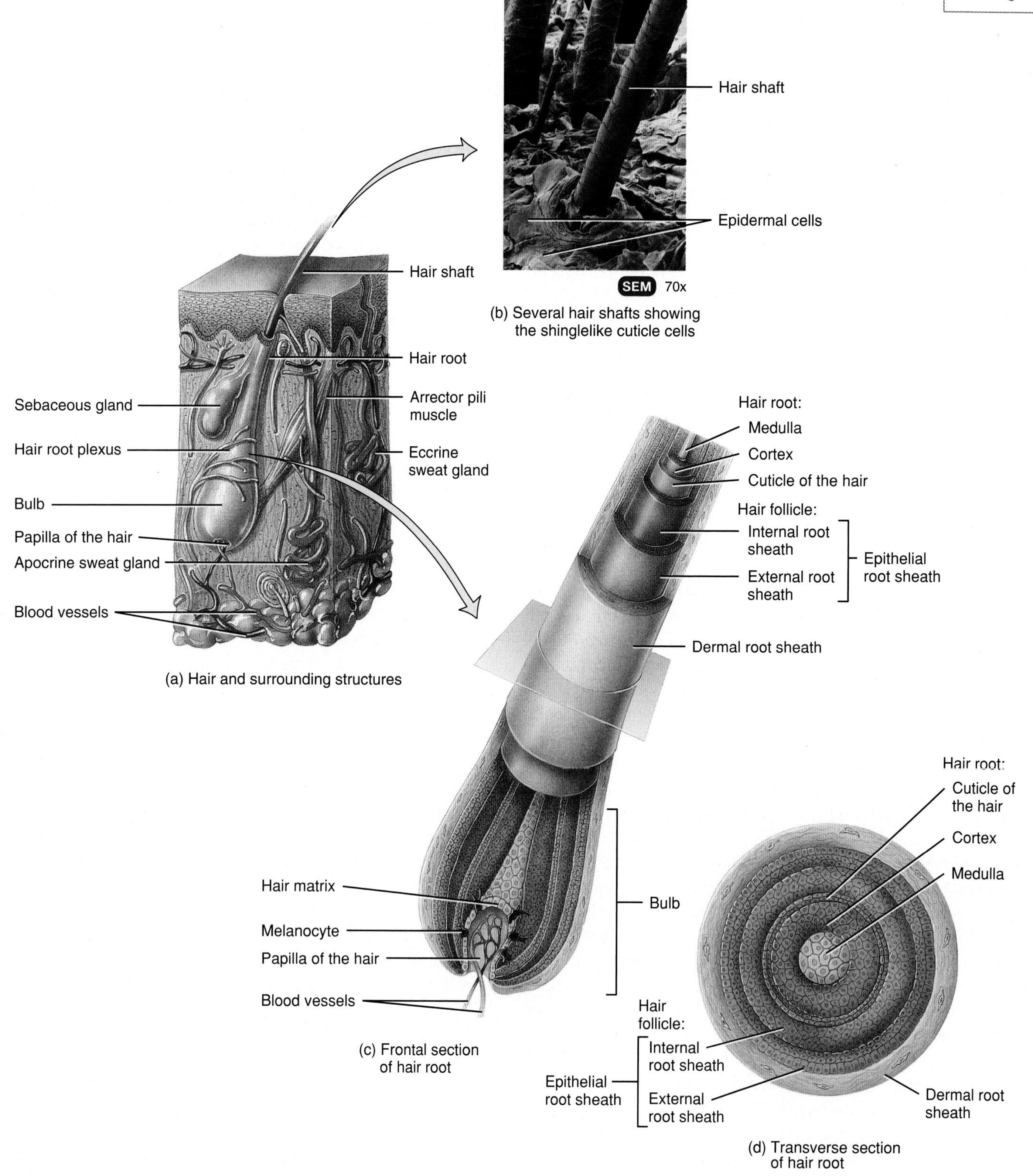

(a) Hair and surrounding structures

(b) Several hair shafts showing the shinglelike cuticle cells

(c) Frontal section of hair root

(d) Transverse section of hair root

Which part of a hair produces a new hair by cell division?

males is ***hirsutism*** (HER-soo-tizm; *hirsut-* = shaggy), a condition of excessive body hair.

Surprisingly, androgens also must be present for the occurrence of the most common form of baldness, ***androgenic alopecia*** or ***male-pattern baldness.*** In genetically predisposed adults, androgens inhibit hair growth. In men, hair loss usually begins with a receding hairline followed by hair loss in the temples and crown. Women are more likely to have thinning of hair on top of the head. The first drug approved for enhancing scalp hair growth was minoxidil (Rogaine®). It causes vasodilation (widening of blood vessels), thus increasing circulation. In about a third of the people who try it, minoxidil improves hair growth, causing scalp follicles to enlarge and lengthening the growth cycle. For many, however, the hair growth is meager. Minoxidil does not help people who already are bald.

Skin Glands

Recall from Chapter 4 that glands are epithelial cells that secrete a substance. Several kinds of exocrine glands are associated with the skin: sebaceous (oil) glands, sudoriferous (sweat) glands, and ceruminous glands. Mammary glands, which are specialized sudoriferous glands that secrete milk, are discussed in Chapter 29 along with the female reproductive system.

Sebaceous Glands

Sebaceous glands (se-BĀ-shus; *sebace-* = greasy) or ***oil glands*** are simple, branched acinar (rounded) glands. With few exceptions, they are connected to hair follicles (Figure 5.5) where the secreting portion of a sebaceous gland lies in the dermis and usually opens into the neck of a hair follicle. In

Figure 5.5 Histology of skin glands.

Sebaceous glands secrete their product directly into the space surrounding a hair follicle. Eccrine and apocrine sweat glands secrete their product into ducts.

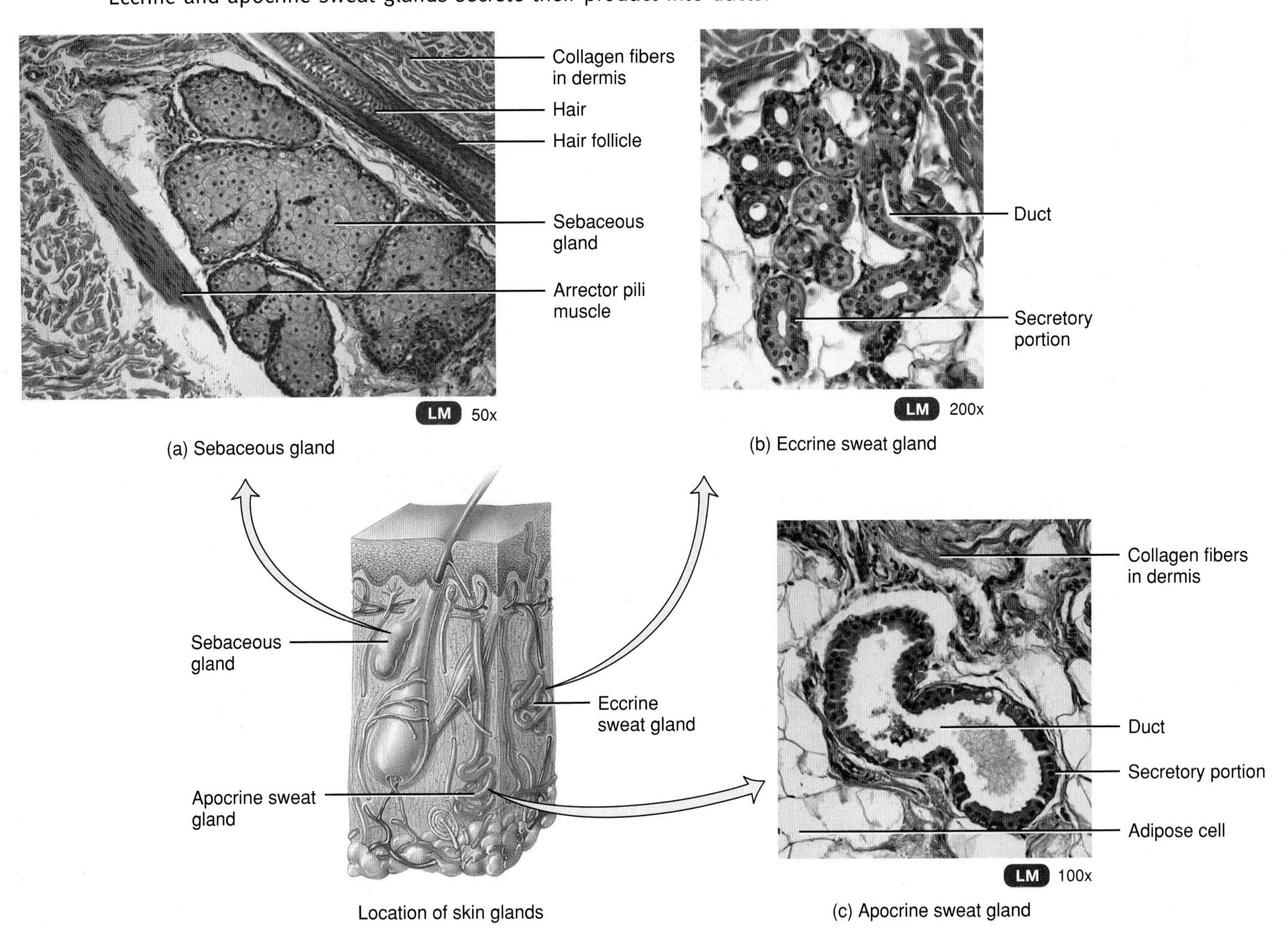

What is the main function of eccrine sweat glands?

some locations, however, such as the lips, glans penis, labia minora, and tarsal glands of the eyelids, sebaceous glands open directly onto the skin. Absent in the palms and soles, sebaceous glands are small in most areas of the trunk and limbs but large in the skin of the breasts, face, neck, and superior chest.

Sebaceous glands (see Figure 5.5a) secrete an oily substance called ***sebum*** (SĒ-bum), a mixture of triglycerides, cholesterol, proteins, and inorganic salts. Sebum coats the surface of hairs and helps keep them from drying and becoming brittle. Sebum also prevents excessive evaporation of water from the skin, keeps the skin soft and pliable, and inhibits the growth of certain bacteria. Clogged sebaceous glands may cause acne.

Sudoriferous Glands

There are three to four million ***sweat glands,*** or ***sudoriferous glands*** (soo′-dor-IF-er-us; *sudori-* = sweat; *-ferous* = bearing), in the body. The cells of these glands release sweat, or perspiration, onto the skin surface through pores or into hair follicles. Sweat glands are divided into two main types, eccrine and apocrine, based on their structure, location, and type of secretion.

Eccrine sweat glands (EK-rin = secreting outwardly), also known as ***merocrine sweat glands,*** are much more common than apocrine sweat glands (see Figure 5.5b). They are distributed throughout the skin of most regions of the body, especially in the skin of the forehead, palms, and soles. Eccrine sweat glands are not present, however, in the margins of the lips, nail beds of the fingers and toes, glans penis, glans clitoris, labia minora, and eardrums. The secretory portion of eccrine sweat glands is located mostly in the deep dermis (sometimes in the upper subcutaneous layer). The excretory duct projects through the dermis and epidermis and ends as a pore at the surface of the epidermis.

The sweat produced by eccrine sweat glands (about 600 mL per day) consists of water, ions (mostly Na^+ and Cl^-), urea, uric acid, ammonia, amino acids, glucose, and lactic acid along with other waste products. The main function of eccrine sweat glands is to help regulate body temperature through evaporation. As sweat evaporates, large quantities of heat energy leave the body surface. Sweat that evaporates from the skin before it is perceived as moisture is termed ***insensible perspiration.*** Sweat that is excreted in larger amounts and that is seen as moisture on the skin is called ***sensible perspiration.*** The product of eccrine sweat glands does not have an odor, at least not until it combines with bacteria on the skin.

Apocrine sweat glands (AP-ō-krin; *apo-* = separated from) are found mainly in the skin of the axilla (armpit), groin, areolae (pigmented areas around the nipples) of the breasts, bearded regions of the face in postpubescent males, and the clitoris and labia minora in postpubescent females. These glands were once thought to release their secretions in an apocrine manner (see Figure 5.5c)—by pinching off a portion of the cell. We now know, however, that their secretion is via exocytosis, which is characteristic of merocrine glands (see Figure 4.4a). Nevertheless, the term *apocrine* is still used. The product of apocrine glands does have an odor. The secretory portion of these sweat glands is located mostly in the subcutaneous layer, and the excretory duct opens into hair follicles (see Figure 5.5). Their secretory product is slightly viscous compared to eccrine secretions and contains the same components as eccrine sweat plus lipids and proteins. Eccrine sweat glands start to function soon after birth, but apocrine sweat glands do not begin to function until puberty. Apocrine sweat glands are stimulated during emotional stress and sexual excitement; these secretions are commonly known as a "cold sweat."

Table 5.3 presents a comparison of eccrine and apocrine sweat glands.

Ceruminous Glands

Modified sweat glands in the external ear, called ***ceruminous glands*** (se-RŪ-mi-nus; *cer-* = wax), produce a yellowish, waxy secretion. The secretory portions of ceruminous glands lie in the subcutaneous layer, deep to sebaceous glands. Their excretory ducts open either directly onto the surface of the external auditory canal (ear canal) or into ducts of sebaceous glands. The combined secretion of the ceruminous and sebaceous glands is called ***cerumen*** (se-ROO-men), or ***earwax.*** Cerumen, together with hairs in the external auditory canal, provides a sticky barrier that impedes the entrance of foreign bodies.

Nails

Nails are plates of tightly packed, hard, dead, keratinized epidermal cells that form a clear, solid covering over the dorsal surfaces of the distal portions of the digits. Each nail consists of a nail body, a free edge, and a nail root (Figure 5.6). The ***nail body***

TABLE 5.3

Comparison of Eccrine and Apocrine Sweat Glands

FEATURE	ECCRINE SWEAT GLANDS	APOCRINE SWEAT GLANDS
Distribution	Throughout skin of most regions of the body, especially in skin of forehead, palms, and soles.	Skin of the axilla, groin, areolae, bearded regions of the face, clitoris, and labia minora.
Location of secretory portion	Mostly in deep dermis.	Mostly in subcutaneous layer.
Termination of excretory duct	Surface of epidermis.	Hair follicle.
Secretion	Less viscous; consists of water, ions (Na^+, Cl^-), urea, uric acid, ammonia, amino acids, glucose, and lactic acid.	More viscous; consists of the same components as eccrine sweat glands plus lipids and proteins.
Functions	Regulation of body temperature and waste removal	Stimulated during emotional stress and sexual excitement.
Onset of function	Soon after birth.	Puberty.
Odor	No.	Yes.

Figure 5.6 Nails. Shown is a fingernail.

Nail cells arise by transformation of superficial cells of the nail matrix.

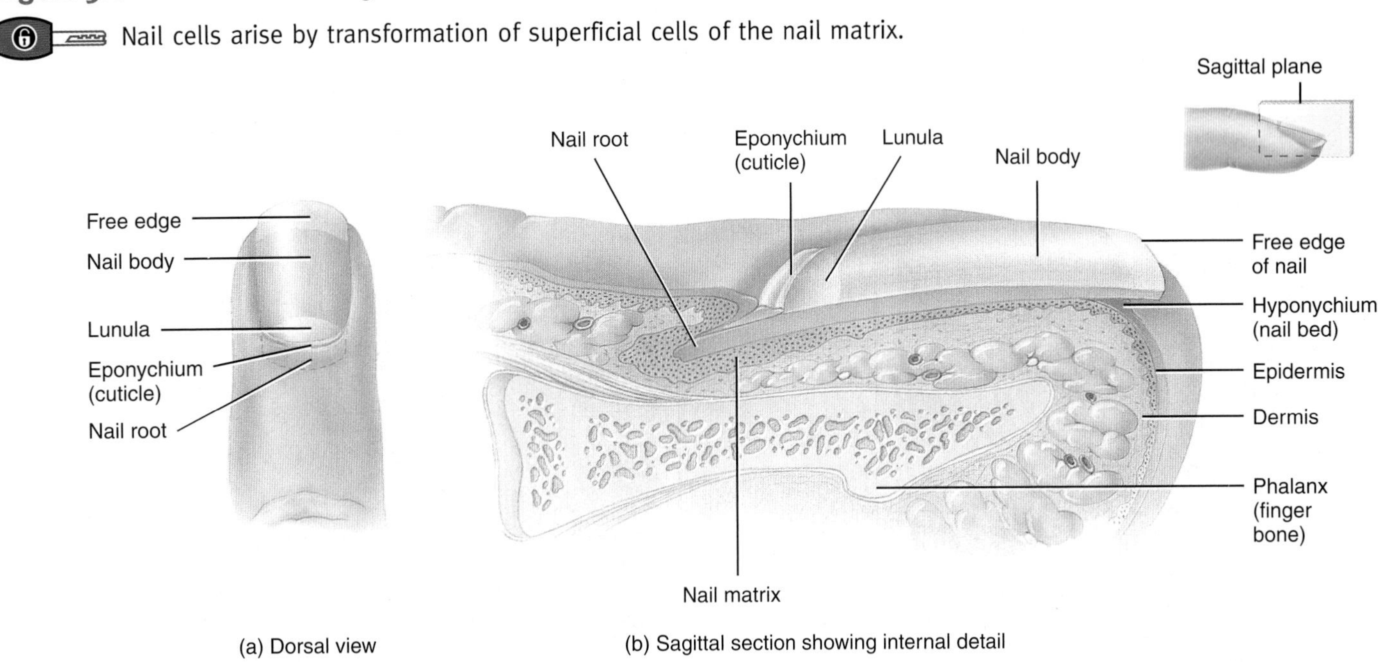

(a) Dorsal view

(b) Sagittal section showing internal detail

Why are nails so hard?

is the visible portion of the nail, the ***free edge*** is the part that may extend past the distal end of the digit, and the ***nail root*** is the portion that is buried in a fold of skin. Below the nail body is a region of epithelium and a deeper layer of dermis. Most of the nail body appears pink because of blood flowing through the capillaries in the underlying dermis. The free edge is white because there are no underlying capillaries. The whitish, crescent-shaped area of the proximal end of the nail body is called the ***lunula*** (LOO-noo-la = little moon). It appears whitish because the vascular tissue underneath does not show through due to a thickened region of epithelium in the area. Beneath the free edge is a thickened region of stratum corneum called the ***hyponychium*** (hī′-pō-NIK-ē-um; *hypo-* = below; *-onych* = nail), which secures the nail to the fingertip. The ***eponychium*** (ep′-ō-NIK-ē-um; *ep-* = above) or ***cuticle*** is a narrow band of epidermis that extends from and adheres to the margin (lateral border) of the nail wall. It occupies the proximal border of the nail and consists of stratum corneum.

The proximal portion of the epithelium deep to the nail root is the ***nail matrix,*** where superficial cells divide by mitosis to produce new nail cells. Nail growth occurs by the transformation of superficial cells of the matrix into nail cells. The growth rate of nails is determined by the rate of mitosis in matrix cells, which is influenced by factors such as a person's age, health, and nutritional status. Nail growth also varies according to the season, the time of day, and environmental temperature. The average growth in the length of fingernails is about 1 mm (0.04 in.) per week. The growth rate is somewhat slower in toenails.

Functionally, nails help us grasp and manipulate small objects in various ways, provide protection against trauma to the ends of the digits, and allow us to scratch various parts of the body.

CHECKPOINT

5. Describe the structure of a hair. What produces "goose bumps"?
6. Contrast the locations and functions of sebaceous (oil) glands, sudoriferous (sweat) glands, and ceruminous glands.
7. Describe the principal parts of a nail.

5.3 TYPES OF SKIN

OBJECTIVE

- Compare structural and functional differences in thin and thick skin.

Although the skin over the entire body is similar in structure, there are quite a few local variations related to thickness of the epidermis, strength, flexibility, degree of keratinization, distribution and type of hair, density and types of glands, pigmentation, vascularity (blood supply), and innervation (nerve supply). Two major types of skin are recognized on the basis of the thickness of the epidermal layer: ***thin (hairy) skin*** and ***thick (hairless) skin*** (see Figure 5.7). Table 5.4 presents a comparison of the features of thin and thick skin.

CHECKPOINT

8. What criteria are used to distinguish thin and thick skin?

Figure 5.7 Types of skin.

The terms thick and thin skin refer to the relative thickness of the epidermis.

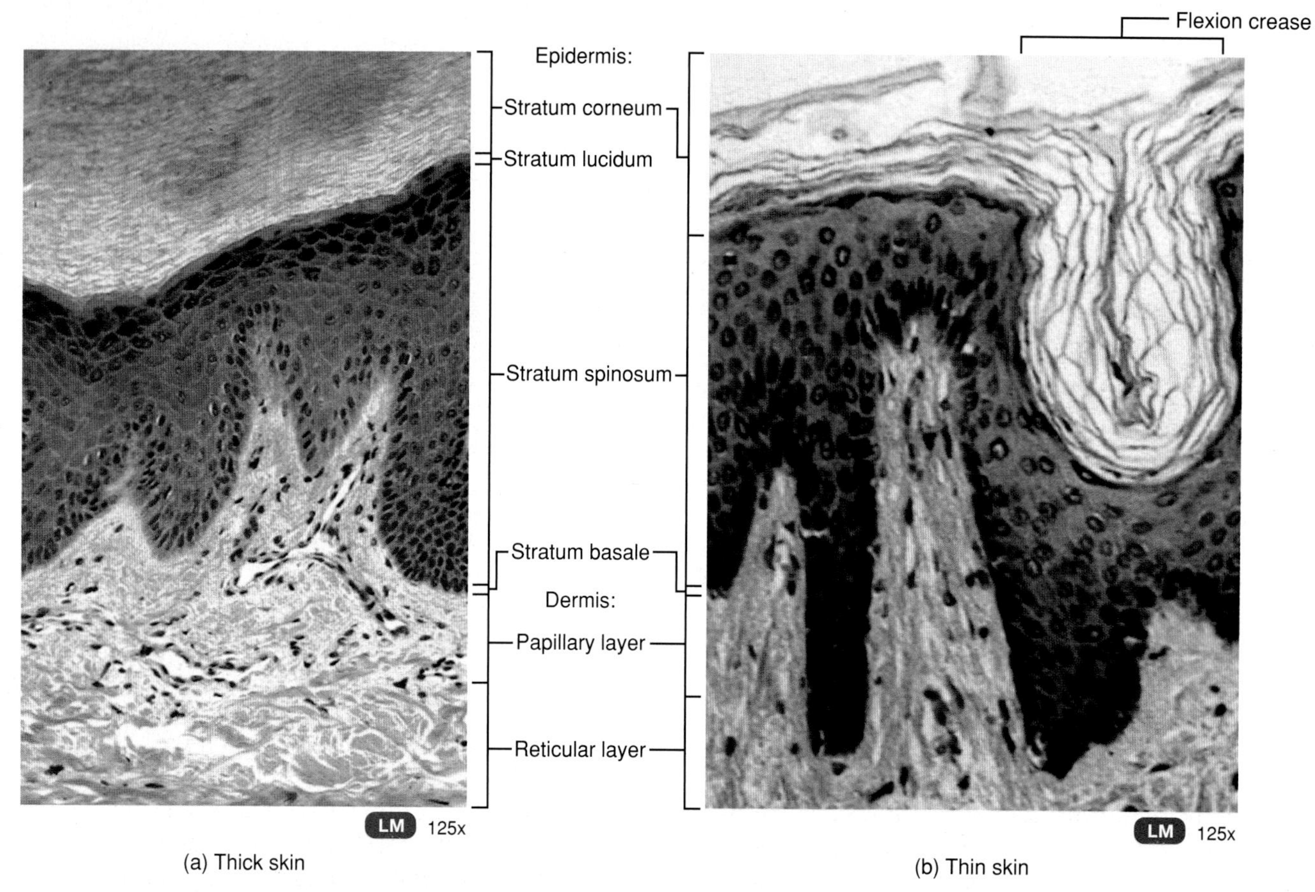

(a) Thick skin (b) Thin skin

What layer of the epidermis is present in thick skin but is lacking in thin skin?

TABLE 5.4

Comparison of Thin and Thick Skin

FEATURE	THICK SKIN	THIN SKIN
Distribution	Palms, palmar surface of digits, and soles.	All parts of the body except palms and palmar surface of digits, and soles.
Epidermal thickness	0.6–4.5 mm (0.024–0.18 in.).	0.10–0.15 mm (0.004–0.006 in.).
Epidermal strata	Thick strata lucidum, spinosum, and corneum.	Stratum lucidum essentially lacking; thinner strata spinosum and corneum.
Epidermal ridges	Present due to well-developed and more numerous dermal papillae.	Lacking due to poorly developed and fewer dermal papillae.
Hair follicles and arrector pili muscles	Absent.	Present.
Sebaceous glands	Absent.	Present.
Sudoriferous glands	More numerous.	Fewer.
Sensory receptors	Denser.	Sparser.

5.4 FUNCTIONS OF THE SKIN

OBJECTIVE

- Describe how the skin contributes to the regulation of body temperature, protection, sensation, excretion and absorption, and synthesis of vitamin D.

Following are the major functions of the skin:

1. **Body temperature regulation.** The skin contributes to the homeostatic regulation of body temperature by liberating sweat at its surface and by adjusting the flow of blood in the dermis (discussed in detail in Chapter 20).

2. **Blood reservoir.** The dermis has an extensive network of blood vessels that stores 8–10% of the total blood in a resting adult.

3. **Protection.** Keratin in the skin protects underlying tissues from microbes, abrasion, heat, and chemicals, and the tightly interlocked keratinocytes resist invasion by microbes. Lipids released by lamellar granules inhibit evaporation of water from the skin surface, thus protecting the body from dehydration. Oily sebum prevents hairs from drying out and contains bactericidal chemicals that kill surface bacteria. The acidic pH of perspira-

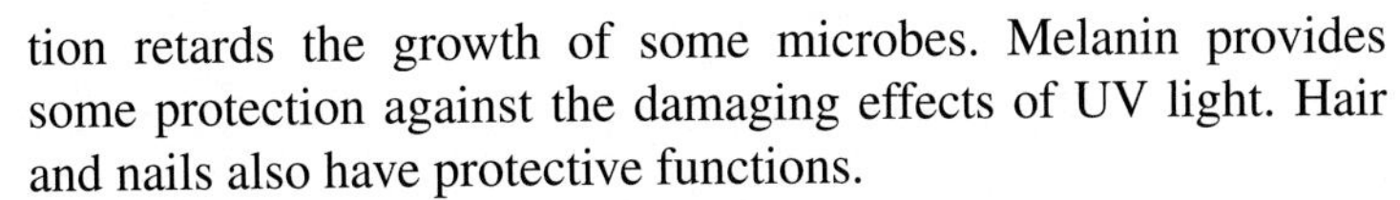

tion retards the growth of some microbes. Melanin provides some protection against the damaging effects of UV light. Hair and nails also have protective functions.

4. Cutaneous sensations. ***Cutaneous sensations*** are those that arise in the skin. These include tactile sensations—touch, pressure, vibration, and tickling—as well as thermal sensations such as warmth and coolness. Another cutaneous sensation, pain, usually is an indication of impending or actual tissue damage. Chapter 15 provides more details on the topic of cutaneous sensations.

5. Excretion and absorption. The skin normally has a small role in *excretion*, the elimination of substances from the body, and *absorption*, the passage of materials from the external environment into body cells.

6. Synthesis of vitamin D. Exposure of the skin to ultraviolet radiation activates vitamin D precursors. Ultimately vitamin D is converted to its active form, a hormone called calcitriol, that aids in the absorption of calcium and phosphorus from the gastrointestinal tract into the blood. People who avoid sun exposure and individuals who live in colder, northern climates may experience vitamin D deficiency if it is not included in their diet or as supplements.

CHECKPOINT

9. In what two ways does the skin help regulate body temperature?
10. In what ways does the skin serve as a protective barrier?
11. What sensations arise from stimulation of neurons in the skin?

5.5 MAINTAINING HOMEOSTASIS: SKIN WOUND HEALING

OBJECTIVES

- Describe the processes of epidermal and dermal wound healing.
- Describe the formation of scars of the skin.

Skin damage sets in motion a sequence of events that repairs the skin to its normal (or near-normal) structure and function. Two kinds of wound-healing processes can occur, depending on the depth of the injury. Epidermal wound healing occurs following wounds that affect only the epidermis; deep wound healing occurs following wounds that penetrate the dermis.

Epidermal Wound Healing

Even though the central portion of an epidermal wound may extend to the dermis, the edges of the wound usually involve only slight damage to superficial epidermal cells. Common types of epidermal wounds include abrasions, in which a portion of skin has been scraped away, and minor burns.

In response to an epidermal injury, basal cells of the epidermis surrounding the wound break contact with the basement membrane. The cells then enlarge and migrate across the wound (see Figure 5.8a). The cells appear to migrate as a sheet until advancing cells from opposite sides of the wound meet. When epidermal cells encounter one another, they stop migrating due to a cellular response called ***contact inhibition.*** Migration of the epidermal cells stops completely when each is finally in contact with other epidermal cells on all sides.

As the basal epidermal cells migrate, a hormone called *epidermal growth factor* stimulates basal stem cells to divide and replace the ones that have moved into the wound. The relocated basal epidermal cells divide to build new strata, thus thickening the new epidermis (see Figure 5.8b).

Dermal Wound Healing

Dermal wound healing occurs when an injury extends to the dermis and subcutaneous layer. Because multiple tissue layers must be repaired, the healing process is more complex than in epidermal wound healing. In addition, because scar tissue is formed in the dermis, the healed tissue loses some of its normal function. Dermal wound healing occurs in four phases: an inflammatory phase, a migratory phase, a proliferative phase, and a maturation phase. During the ***inflammatory phase,*** a blood clot forms in the wound and loosely unites the wound edges (see Figure 5.8c). As its name implies, this phase of dermal wound healing involves ***inflammation*** (in′-fla-MĀ-shun), a vascular and cellular response that helps eliminate microbes, foreign material, and dying tissue in preparation for repair. The vasodilation and increased permeability of blood vessels associated with inflammation enhance delivery of helpful cells. These include phagocytic white blood cells called neutrophils; monocytes, which develop into macrophages that phagocytize microbes; and mesenchymal cells, which develop into fibroblasts.

The three phases that follow do the work of repairing the wound. In the ***migratory phase,*** the clot becomes a scab, and epithelial cells migrate beneath the scab to bridge the wound. Fibroblasts migrate along fibrin threads and begin synthesizing scar tissue (collagen fibers and glycoproteins), and damaged blood vessels begin to regrow. During this phase, the tissue filling the wound is called ***granulation tissue.*** The ***proliferative phase*** is characterized by extensive growth of epithelial cells beneath the scab, deposition by fibroblasts of collagen fibers in random patterns, and continued growth of blood vessels. Finally, during the ***maturation phase,*** the scab sloughs off once the epidermis has been restored to normal thickness. Collagen fibers become more organized, fibroblasts decrease in number, and blood vessels are restored to normal (see Figure 5.8d).

Deep wound healing may result in the formation of a scar. A scar on the surface is caused when the wound gapes sufficiently that the epidermal cells do not migrate and meet. Fibroblasts may fill the entire deep wound area with connective tissue. In the beginning, a recent scar is redder than the surrounding tissue because the blood vessels of the dermis penetrate the tissue (see Figure 5.8e). In time, however, the blood vessels within the scar atrophy and the scar tissue is now typically lighter in color than

Figure 5.8 Skin wound healing.

In an epidermal wound, the injury is restricted to the epidermis; in a dermal wound, the injury extends deep into the dermis; in a gaping wound, a scar may form.

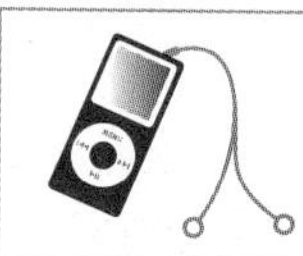

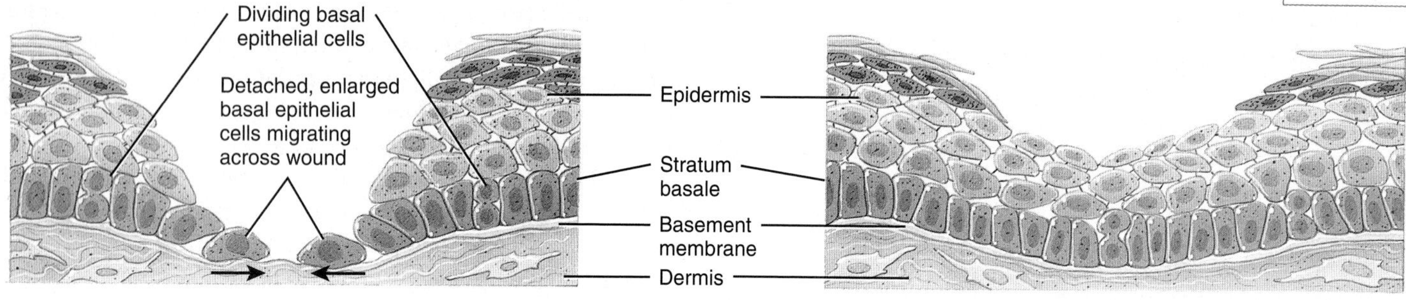

(a) Division of basal epithelial cells and migration across wound

(b) Thickening of epidermis

Epidermal wound healing

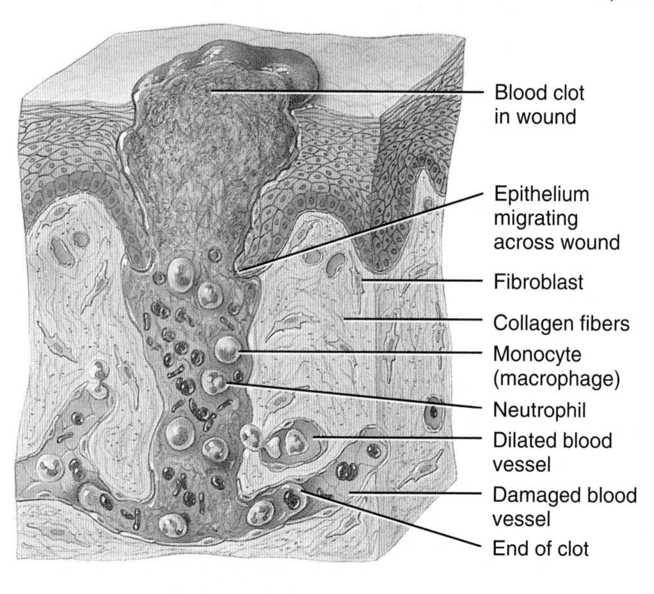

(c) Inflammatory phase

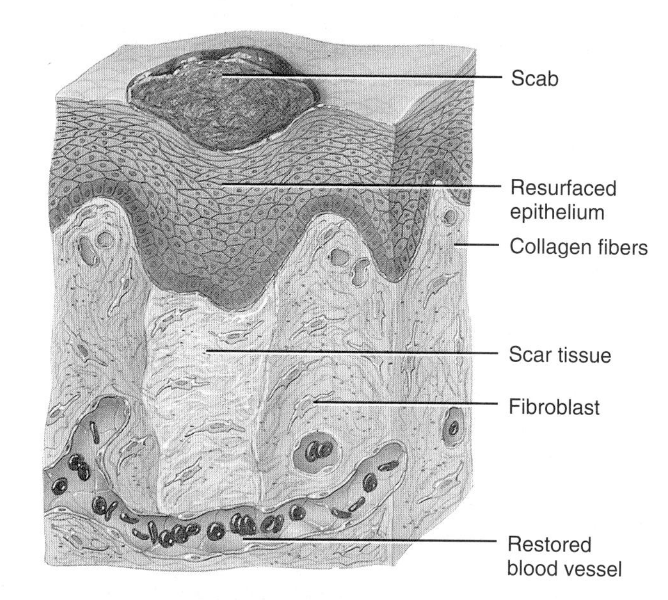

(d) Maturation phase

Dermal wound healing

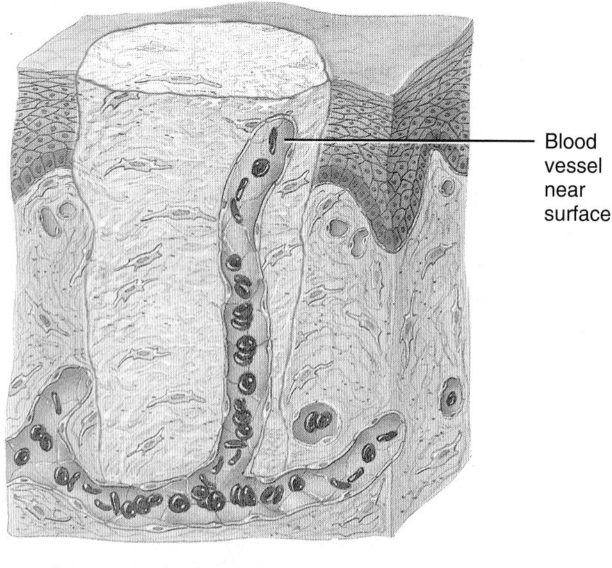

(e) Recent scar

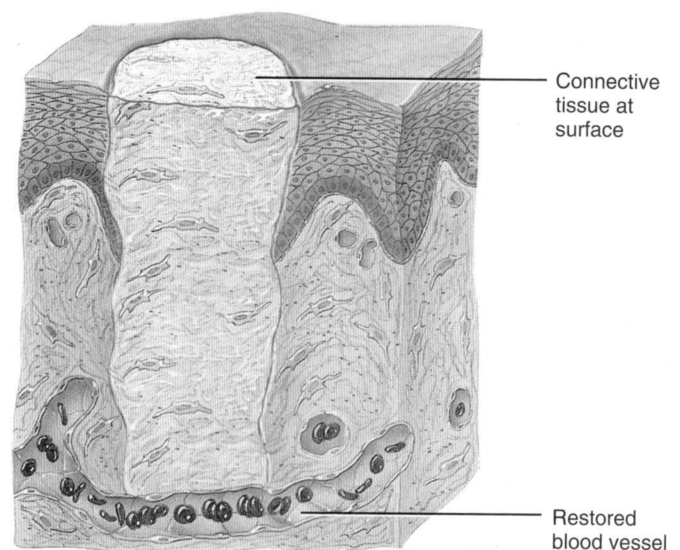

(f) Old scar

Scar tissue formation

Would you expect an epidermal wound to bleed? Why or why not?

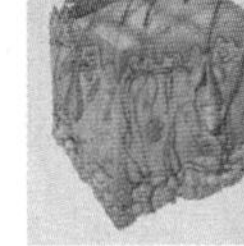

the surrounding tissue (see Figure 5.8f). Remember that the coloration of the skin is largely due to melanocytes, cells that are rarely present in the dermis. Furthermore, if the initial wound (lesion) is located in a hairy area, the scar will be hairless since hair is an epidermal derivative.

The process of scar tissue formation is called ***fibrosis.*** Sometimes, so much scar tissue is formed during deep wound healing that a raised scar—one that is elevated above the normal epidermal surface—results. If such a scar remains within the boundaries of the original wound, it is a ***hypertrophic scar.*** If it extends beyond the boundaries into normal surrounding tissues, it is a ***keloid scar.*** Scar tissue differs from normal skin in that its collagen fibers are more densely arranged, it has decreased elasticity, it has fewer blood vessels, and it may or may not contain the same number of hairs, skin glands, or sensory structures as undamaged skin. Scars are usually lighter in color than normal skin due to the lack of melanocytes in the dermis, the arrangement of collagen fibers, and the scarcity of blood vessels.

MANUAL THERAPY APPLICATION

Physiological Effects of Massage

Massage of the skin, whether by stroking (effleurage) or deep kneading (petrissage), has several **physiological effects** and is greatly beneficial. Massage increases blood and lymph flow and this increased flow increases nutrition as well as improves immune system response in the superficial tissues. Increased blood flow is evidenced by hyperemia (reddening) of the skin after massage. Centripetal (toward the heart) friction of the skin also increases sudoriferous (sweat) gland activity. The ability to regulate body temperature and to eliminate waste products is thus enhanced; friction dissipates excess heat by as much as 95%, both by radiation (bringing blood to the surface of skin) and by the evaporation of sweat. Sebaceous (oil) gland activity also increases with massage. Increased sebaceous gland activity helps to maintain the elasticity of the skin, keeps hair supple, and inhibits the growth of some bacteria. Centripetal friction helps maintain the elasticity of the subcutaneous layer, the layer that attaches skin to underlying structures. Movement of the skin may be inhibited by fascial adhesions or scar tissue. Centripetal friction moves skin over underlying structures and thus aids the breakdown of scar tissue in that layer. Centrifugal friction, movement of the therapist's hands away from the heart, is also a very effective treatment.

CHECKPOINT

12. Cells of which layer are most responsible for epidermal wound healing?
13. Describe the four phases of dermal wound healing.
14. Describe the basis of forming the general color of a new scar and an older scar.

5.6 SKIN CONDITIONS IMPORTANT TO THERAPISTS

OBJECTIVES

- Describe three types of skin cancer.
- List five risk factors for skin cancer.
- Describe the four levels of a skin burn.

Skin Cancer

Excessive exposure to the sun has caused virtually all of the one million cases of ***skin cancer*** diagnosed annually in the United States. There are three common forms of skin cancer. ***Basal cell carcinomas*** account for about 78% of all skin cancers. The tumors arise from cells in the stratum basale of the epidermis and rarely metastasize. Although usually harmless, the untreated tumor may eventually grow and compromise other tissues. ***Squamous cell carcinomas,*** which account for about 20% of all skin cancers, arise from keratinocytes of the epidermis, and they have a variable tendency to metastasize. Most arise from preexisting lesions of damaged tissue on sun-exposed skin. Basal and squamous cell carcinomas are together known as *nonmelanoma skin cancer.* They are 50% more common in males than in females. ***Actinic keratosis*** is a precancerous condition that can lead to squamous cell carcinoma in 10–20% of persons who suffer from the crusty lesions that do not heal normally.

Malignant melanomas arise from melanocytes and account for about 2% of all skin cancers. The estimated lifetime risk of developing melanoma is 1 in 75, double the risk only 20 years ago. In part, this increase is due to depletion of the ozone layer, which absorbs some UV light high in the atmosphere. But the main reason for the increase is that more people are spending more time in the sun and in tanning beds. Malignant melanomas metastasize rapidly and can kill a person within months of diagnosis.

The key to successful treatment of malignant melanoma is early detection. The early warning signs of malignant melanoma are identified by the acronym ABCD (Figure 5.9b). *A* is for *asymmetry;* malignant melanomas tend to lack symmetry. *B* is for *border;* malignant

Figure 5.9 Comparison of a normal nevus (mole) and a malignant melanoma.

Excessive exposure to the sun accounts for almost all cases of skin cancer.

(a) Normal nevus (mole)

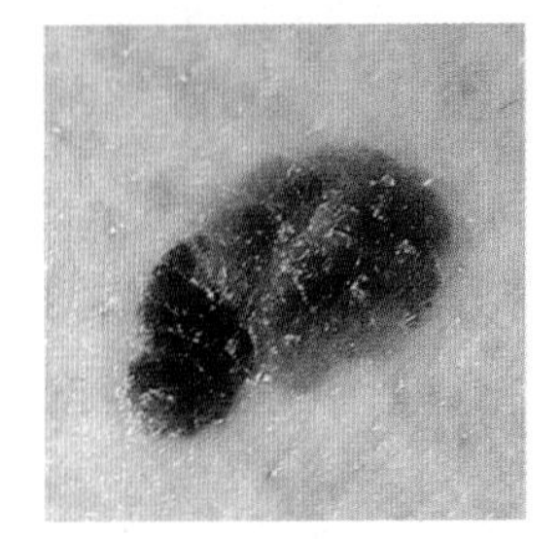

(b) Malignant melanoma

Which is the most common type of skin cancer?

melanomas have irregular—notched, indented, scalloped, or indistinct—borders. *C* is for *color;* malignant melanomas have uneven coloration and may contain several colors. *D* is for *diameter;* ordinary moles typically are smaller than 6 mm (0.25 in.), about the size of a pencil eraser. Once a malignant melanoma has the characteristics of A, B, and C, it is usually larger than 6 mm.

Among the risk factors for skin cancer are the following:

1. ***Skin type.*** Individuals with light-colored skin who never tan but always burn are at high risk.
2. ***Sun exposure.*** People who live in areas with many days of sunlight per year and at high altitudes (where UV light is more intense) have a higher risk of developing skin cancer. Likewise, people who engage in outdoor occupations and those who have suffered three or more severe sunburns have a higher risk.
3. ***Family history.*** Skin cancer rates are higher in some families than in others.
4. ***Age.*** Older people are more prone to skin cancer owing to longer total exposure to sunlight.
5. ***Immunological status.*** Immunosuppressed individuals have a higher incidence of skin cancer.

Burns

A ***burn*** is tissue damage caused by excessive heat, electricity, radioactivity, or corrosive chemicals that denature (destroy) the proteins in the skin cells. A rug burn or a rope burn, as examples, can cause the same symptoms. Burns destroy some of the skin's important contributions to homeostasis—protection against microbial invasion and dehydration, and regulation of body temperature.

Burns are graded according to their severity. A *first-degree burn* involves only the epidermis (Figure 5.10a). It is characterized by mild pain and erythema (redness) but no blisters. Skin functions remain intact. A *second-degree burn* destroys the epidermis and part of the dermis (Figure 5.10b). Some skin functions are lost. In a second-degree burn, redness, blister formation, edema, and pain result. In a blister the epidermis separates from the dermis due to the accumulation of tissue fluid between them. Associated structures, such as hair follicles, sebaceous glands, and sweat glands, usually are not injured. If there is no infection, second-degree burns heal without skin grafting in about 3 to 4 weeks, but scarring may result. First- and second-degree burns are collectively referred to as *partial-thickness burns.*

Figure 5.10 Burns.

A burn is tissue damage caused by agents that destroy the proteins in skin cells.

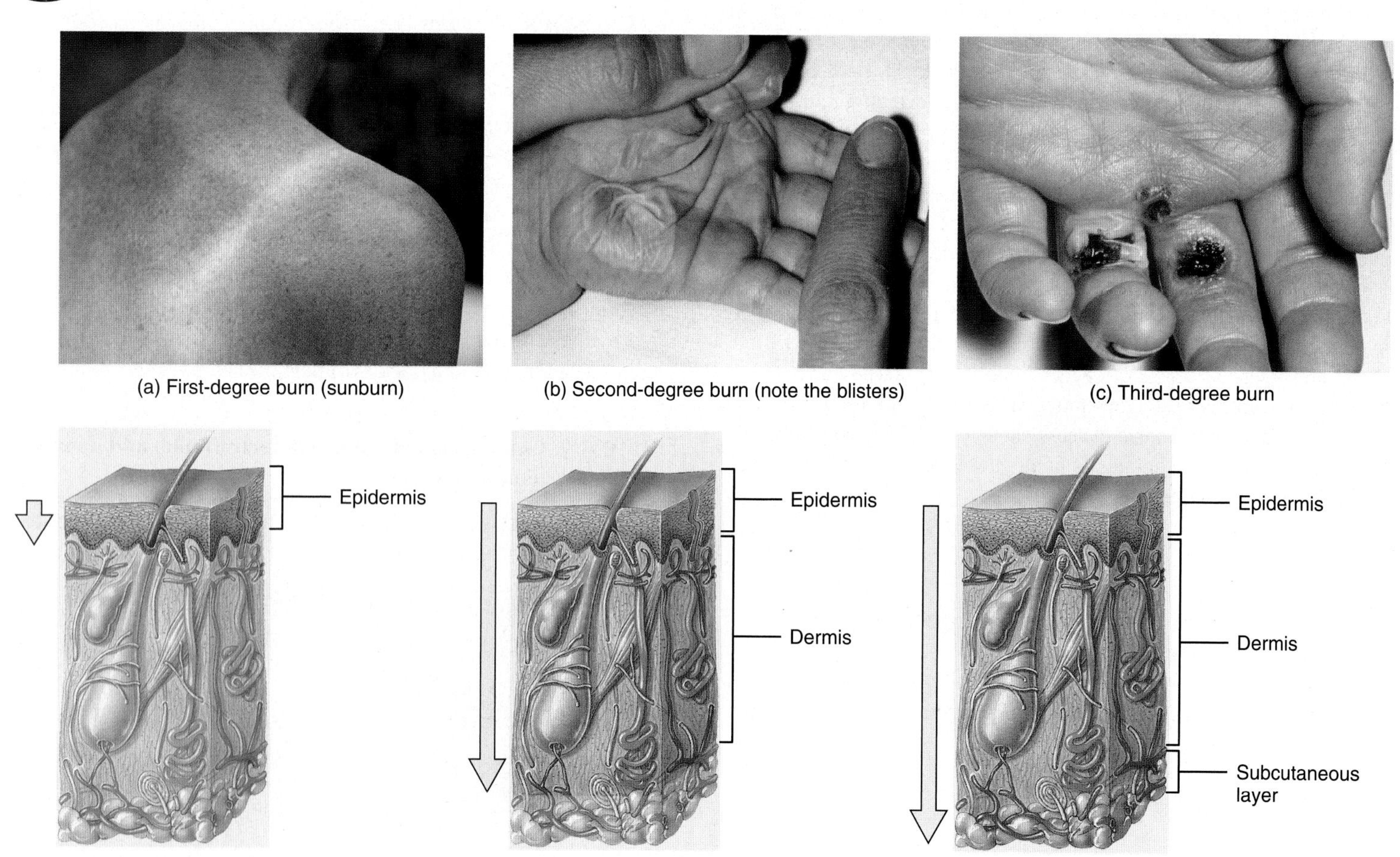

What factors determine the seriousness of a burn?

A *third-degree burn* or *full-thickness burn* destroys the epidermis, dermis, and subcutaneous layer (Figure 5.10c). Most skin functions are lost. Such burns vary in appearance from marble-white to mahogany colored to charred, dry wounds. There is marked edema, and the burned region is numb because sensory nerve endings have been destroyed. Regeneration occurs slowly, and much granulation tissue forms before being covered by epithelium. Skin grafting may be required to promote healing and to minimize scarring.

A *fourth-degree burn* (not illustrated) destroys not only skin but also tissues that are deep to the skin such as fascia, muscles, and bones.

The seriousness of a burn is determined by its depth and extent of area involved (Figure 5.11), as well as the person's age and general health. According to the American Burn Association's classification of burn injury, a *major burn* includes third-degree burns over 10% of body surface area; or second-degree burns over 25% of body surface area; or any third-degree burns on the face, hands, feet, or *perineum* (per′-i-NĒ-um, which includes the anal and urogenital regions). When the burn area exceeds 70%, more than half the victims die.

Figure 5.11 Rule-of-nines method for determining the extent of a burn. The percentages are the approximate proportions of the body surface area.

The rule of nines is a quick rule for estimating the surface area affected by a burn in an adult.

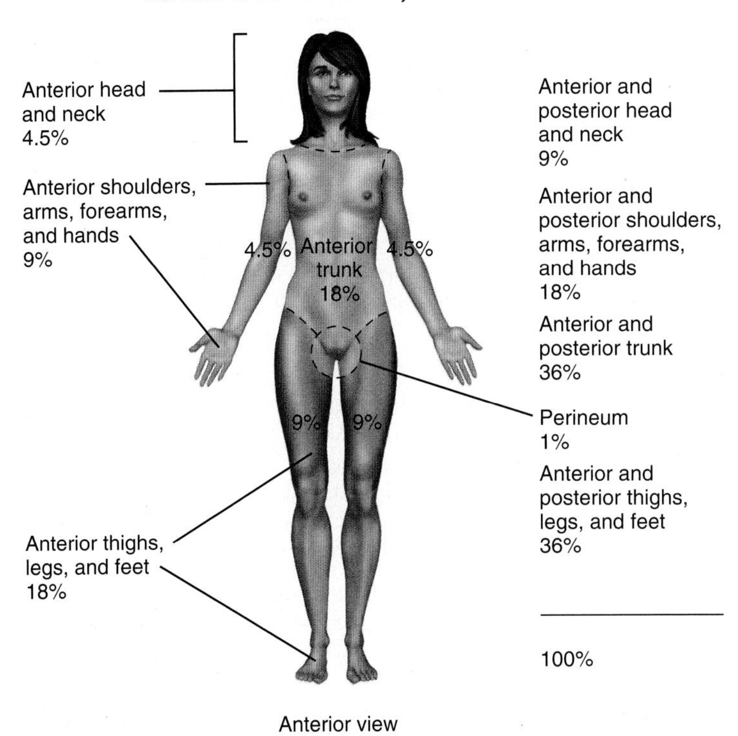

What percentage of the body would be burned if only the anterior trunk and anterior left upper limb were involved?

Pressure Ulcers

Pressure ulcers, also known as *decubitus ulcers* (dē-KŪ-bi-tus) or *bedsores,* are caused by a constant deficiency of blood flow to tissues (Figure 5.12). Typically the affected tissue overlies a bony projection that has been subjected to prolonged pressure against an object such as a bed, cast, or splint. If the pressure is relieved in a few hours, redness occurs but no lasting tissue damage results. Blistering of the affected area may indicate superficial damage; a reddish-blue discoloration may indicate deep tissue damage. Prolonged pressure causes tissue ulceration. Small breaks in the epidermis become infected, and the sensitive subcutaneous layer and deeper tissues are damaged. Eventually, the tissue dies. Pressure ulcers occur most often in bedridden patients. The biggest danger for someone with pressure ulcers is the possibility of infection, which can be life threatening for someone who is bedridden and already immune suppressed. With proper care, pressure ulcers are preventable, but they can develop very quickly in patients who are very old or very ill.

MANUAL THERAPY APPLICATION

Assessing Skin Lesions

As was stated on the first page of this chapter, it is important for manual therapists to be able to **assess skin lesions** to determine whether or not a patient with skin lesions should be referred to a physician. Skin lesions may result from an injury or from a pathological change in skin tissues.

Skin conditions are beyond the intended scope of this work. Students are highly encouraged to seek additional assistance from a pathology textbook since patients may present with impetigo, eczema, psoriasis, fungal infections, or a multitude of other skin conditions.

CHECKPOINT

15. What are the risk factors for developing skin cancer?
16. Describe the classification of burns by severity.

Figure 5.12 Pressure ulcers.

A pressure ulcer is a shedding of epithelium caused by a constant deficiency of blood flow to tissues.

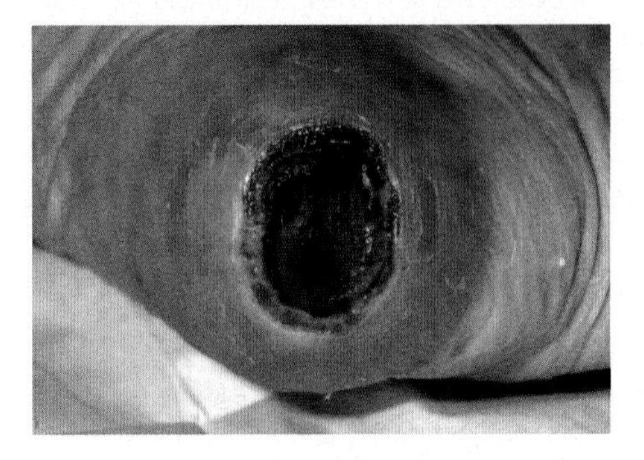

Pressure ulcer on heel

What parts of the body are usually affected by pressure ulcers?

5.7 AGING AND THE INTEGUMENTARY SYSTEM

OBJECTIVE

- Describe the effects of aging on the integumentary system.

Most infants and children encounter relatively few problems with the skin as it ages. With the arrival of adolescence, however, some teens develop acne (inflammation of sebaceous glands). The pronounced effects of skin aging do not become noticeable until people reach their late forties. Most of the age-related changes occur in the proteins in the dermis. Collagen fibers in the dermis begin to decrease in number, stiffen, break apart, and disorganize into a shapeless, matted tangle. Elastic fibers lose some of their elasticity, thicken into clumps, and fray, an effect that is greatly accelerated in the skin of smokers. Fibroblasts, which produce both collagen and elastic fibers, decrease in number. As a result, the skin forms the characteristic crevices and furrows known as *wrinkles*.

With further aging, Langerhans cells dwindle in number and macrophages become less-efficient phagocytes, thus decreasing the skin's immune responsiveness. Moreover, the decreased size of sebaceous glands leads to dry and broken skin that is more susceptible to infection. Production of sweat diminishes, which probably contributes to the increased incidence of heat stroke in the elderly. There is a decrease in the number of functioning melanocytes, resulting in gray hair and atypical skin pigmentation. An increase in the size of some melanocytes produces pigmented blotching (age spots). Walls of blood vessels in the dermis become thicker and less permeable, and subcutaneous adipose tissue is lost. Aged skin (especially the dermis) is thinner than young skin, and the migration of epidermal cells from the basal layer to the epidermal surface slows considerably. With the onset of old age, skin heals poorly and becomes more susceptible to pathological conditions such as skin cancer and pressure sores. Growth of nails and hair slows during the second and third decades of life. The nails also may become more brittle with age, often due to dehydration or repeated use of cuticle remover or nail polish. Several cosmetic anti-aging treatments are available to diminish the effects of aging or sun-damaged skin.

The many ways skin contributes to homeostasis of other body systems are listed in Exhibit 5.1.

CHECKPOINT

17. Which portion of the skin is involved in most age-related changes? Give several examples.

KEY MEDICAL TERMS ASSOCIATED WITH THE INTEGUMENTARY SYSTEM

Abrasion (a-BRĀ-shun; *ab-* = away; *-rasion* = scraped) An area where skin has been scraped away.

Blister A collection of serous fluid within the epidermis or between the epidermis and dermis, due to short-term but severe friction. The term ***bulla*** (BUL-a) refers to a large blister.

Callus (KAL-lus = hard skin) An area of hardened and thickened skin that is usually seen in palms and soles, due to persistent pressure and friction.

Cold sore A lesion, usually in oral mucous membranes, caused by Type 1 herpes simplex virus (HSV) transmitted by oral or respiratory routes. The virus remains dormant until triggered by factors such as ultraviolet light, hormonal changes, and emotional stress. Also called a ***fever blister.***

Comedo (KOM-ē-dō; comedo = to eat up) A collection of sebaceous material and dead cells in the hair follicle and excretory duct of the sebaceous (oil) gland. Usually found over the face, chest, and back, and more commonly during adolescence. Also called a **black-head**.

Contact dermatitis (der-ma-TĪ-tis; *dermat-* = skin; *-itis* = inflammation of) Inflammation of the skin characterized by redness, itching, and swelling and caused by exposure of the skin to chemicals that bring about an allergic reaction, such as poison ivy toxin.

Corn (KORN) A painful conical thickening of the stratum corneum of the epidermis found principally over toe joints and between the toes, often caused by friction or pressure. Corns may be hard or soft, depending on their location. Hard corns are usually found over toe joints, and soft corns are usually found between the fourth and fifth toes.

Cyst (SIST; *cyst* = sac containing fluid) A sac with a distinct connective tissue wall, containing a fluid or other material.

Eczema (EK-ze-ma; *ekzeo-* = to boil over) An inflammation of the skin characterized by patches of red, blistering, dry, extremely itchy skin. It occurs mostly in skin creases in the wrists, backs of the knees, and fronts of the elbows. It typically begins in infancy and many children outgrow the condition. The cause is unknown but is linked to genetics and allergies.

Erysipelas (er′-i-SIP-e-las) A streptococcal infection of the skin that may, if not properly treated, become systemic and involve the lymphatic and circulatory systems. A cardinal sign of erysipelas is a very sharp margin between the red and tender involved skin and the uninvolved skin.

Frostbite Local destruction of skin and subcutaneous tissue on exposed surfaces as a result of extreme cold. In mild cases, the skin is blue and swollen and there is slight pain. In severe cases there is considerable swelling, some bleeding, no pain, and blistering. If untreated, gangrene may develop. Frostbite is treated by rapid rewarming.

Hemangioma (he-man′-jē-Ō-ma; *hem-* = blood; *-angi-* = blood vessel; *-oma* = tumor) Commonly called a *birthmark*, a hemangioma is a localized benign tumor of the skin and subcutaneous layer that results from an abnormal increase in blood vessels. One type is a ***port-wine stain,*** a flat, pink, red, or purple lesion present at birth, usually at the nape of the neck. Certain hemangiomas last a lifetime; others gradually fade and may disappear.

Hives Reddened elevated patches of skin that are often itchy. Most commonly caused by infections, physical trauma, medications, emotional stress, food additives, and certain food allergies. Also called ***urticaria*** (ūr-ti-KAR-ē-a).

Impetigo (im′-pe-TĪ-gō) An extremely contagious bacterial infection that is prevalent in children. It commonly begins around the mouth, nose, eyes, and ears, the areas that are commonly scratched and thus vulnerable to the invasion by either staphylococcus or streptococcus bacteria.

EXHIBIT 5.1 The Integumentary System and Homeostasis

BODY SYSTEM		CONTRIBUTION OF THE INTEGUMENTARY SYSTEM
For all body systems		Skin and hair provide barriers that protect all internal organs from damaging agents in the external environment; sweat glands and skin blood vessels regulate body temperature, needed for proper functioning of other body systems.
Skeletal system	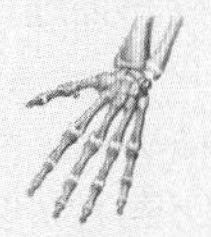	Skin helps activate vitamin D, needed for proper absorption of dietary calcium and phosphorus to build and maintain bones.
Muscular system		Skin helps provide calcium ions, needed for muscle contraction.
Nervous system	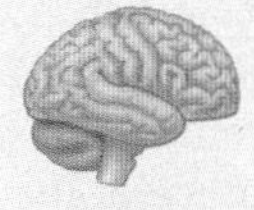	Nerve endings in skin and subcutaneous tissue provide input to the brain for touch, pressure, thermal, and pain sensations. From an embryological point of view, some experts consider the skin to be an extension of the brain and spinal cord.
Endocrine system	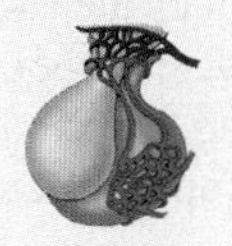	Keratinocytes in skin help activate vitamin D to calcitriol, a hormone that aids absorption of dietary calcium and phosphorus.
Cardiovascular system	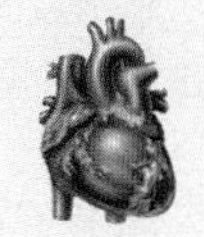	Local chemical changes in dermis cause widening and narrowing of skin blood vessels, which help adjust blood flow to the skin.
Lymphatic system and immunity	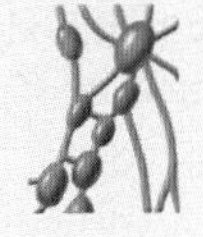	Skin is "first line of defense" in immunity, providing mechanical barriers and chemical secretions that discourage penetration and growth of microbes; Langerhans cells in epidermis participate in immune responses by recognizing and processing foreign antigens; macrophages in the dermis phagocytize microbes that penetrate the skin surface.
Respiratory system	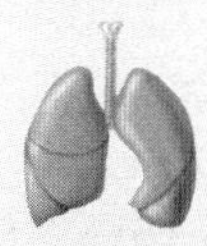	Hairs in nose filter dust particles from inhaled air; stimulation of pain nerve endings in skin may alter breathing rate.
Digestive system	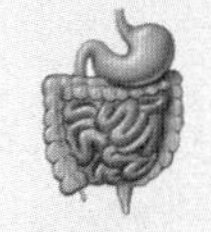	Skin helps activate vitamin D to the hormone calcitriol, which promotes absorption of dietary calcium and phosphorus in the small intestine.
Urinary system	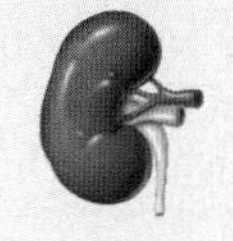	Kidney cells receive partially activated vitamin D hormone from skin and convert it to calcitriol; some waste products are excreted from body in sweat, contributing to excretion by urinary system.
Reproductive system	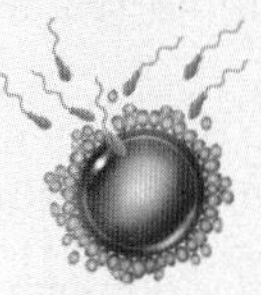	Nerve endings in skin and subcutaneous tissue respond to erotic stimuli, thereby contributing to sexual pleasure; suckling of a baby stimulates nerve endings in skin, leading to milk ejection; mammary glands (modified sweat glands) produce milk; skin stretches during pregnancy as fetus enlarges.

Keratosis (ker′-a-TŌ-sis; *kera-* = horn) Formation of a hardened growth of epidermal tissue, such as a *solar keratosis,* a premalignant lesion of the sun-exposed skin of the face and hands.

Lice Contagious arthropods that include two basic forms. ***Head lice*** are tiny, jumping arthropods that suck blood from the scalp. They lay eggs, called nits, and their saliva causes itching that may lead to complications. ***Pubic lice*** are tiny arthropods that do not jump and they look like miniature crabs.

Papule (PAP-ūl; *papula* = pimple) A small, round skin elevation less than 1 cm in diameter. One example is a pimple.

Pruritus (proo-RĪ-tus; *pruri-* = to itch) Itching, one of the most common dermatological disorders. It may be caused by skin disorders (infections), systemic disorders (cancer, kidney failure), psychogenic factors (emotional stress), or allergic reactions.

Sepsis (SEP-sis) A widespread and dangerous bacterial infection. It is the leading cause of death in severely burned patients.

Tinea corporis A fungal infection characterized by scaling, itching, and sometimes painful lesions that may appear on any part of the body, also known as ***ringworm.*** Fungi thrive in warm, moist places such as skin folds of the groin, where it is known as ***tinea cruris (jock itch)*** or between the toes where it is called ***tinea pedis (Athlete's foot).***

Wart Mass produced by uncontrolled growth of epithelial skin cells; caused by a papillomavirus. Most warts are noncancerous.

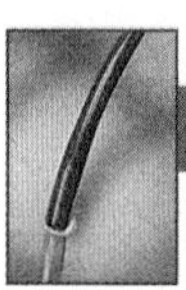

STUDY OUTLINE

Structure of the Skin (Section 5.1)

1. The integumentary system consists of organs such as the skin, blood vessels, and nerves and its accessory structures such as hair, nails, and skin glands.
2. The skin is the largest organ of the body in surface area and weight. The principal parts of the skin are the epidermis (superficial) and dermis (deep).
3. The subcutaneous layer (hypodermis) is deep to the dermis and not part of the skin. It anchors the dermis to underlying tissues and organs, and it contains lamellated (pacinian) corpuscles.
4. The types of cells in the epidermis are keratinocytes, melanocytes, Langerhans cells, and Merkel cells.
5. The epidermal layers, from deep to superficial, are the stratum basale (undergoes cell division and produces all other layers), stratum spinosum (provides strength and flexibility), stratum granulosum (contains keratin and lamellar granules), stratum lucidum (in thick skin only), and stratum corneum (sloughs off as dead skin) (see Table 5.1).
6. The dermis consists of papillary and reticular regions. The papillary region is composed of areolar connective tissue containing fine elastic fibers, dermal papillae, and Meissner corpuscles. The reticular region is composed of dense irregular connective tissue containing interlaced collagen and coarse elastic fibers, adipose tissue, hair follicles, nerves, sebaceous (oil) glands, and ducts of sudoriferous (sweat) glands.
7. Epidermal ridges provide the basis for fingerprints and footprints.
8. The color of skin is due to melanin, carotene, and hemoglobin.

Accessory Structures of the Skin (Section 5.2)

1. Accessory structures of the skin—hair, skin glands, and nails—develop from the embryonic epidermis.
2. A hair consists of a shaft, most of which is superficial to the surface, a root that penetrates the dermis and sometimes the subcutaneous layer, and a hair follicle.
3. Associated with each hair follicle is a sebaceous (oil) gland, an arrector pili muscle, and a hair root plexus.
4. New hairs develop from division of matrix cells in the bulb; hair replacement and growth occur in a cyclic pattern consisting of alternating growth and resting stages.
5. Hairs offer a limited amount of protection—from the sun, heat loss, and entry of foreign particles into the eyes, nose, and ears. They also function in sensing light touch.
6. Sebaceous (oil) glands are usually connected to hair follicles; they are absent from the palms and soles. Sebaceous glands produce sebum, which moistens hairs and waterproofs the skin. Clogged sebaceous glands may produce acne.
7. There are two types of sudoriferous (sweat) glands: eccrine and apocrine. Eccrine sweat glands have an extensive distribution; their ducts terminate at pores at the surface of the epidermis. Apocrine sweat glands are limited to the skin of the axillae, groin, and areolae; their ducts open into hair follicles. They begin functioning at puberty and are stimulated during emotional stress and sexual excitement. Mammary glands are specialized sudoriferous glands that secrete milk.
8. Ceruminous glands are modified sudoriferous glands that secrete cerumen (ear wax). They are found in the external auditory canal (ear canal).
9. Nails are hard, keratinized epidermal cells over the dorsal surfaces of the distal portions of the digits.
10. The principal parts of a nail are the nail body, free edge, nail root, lunula, eponychium (cuticle), and matrix. Cell division of the matrix cells produces new nails.

Types of Skin (Section 5.3)

1. Thin skin covers all parts of the body except for the palms, palmar surfaces of the digits, and the soles.
2. Thick skin covers the palms, palmar surfaces of the digits, and soles.

Functions of the Skin (Section 5.4)

1. Skin functions include body temperature regulation, blood storage, protection, sensation, excretion and absorption, and synthesis of vitamin D.
2. The skin participates in thermoregulation by liberating sweat at its surface and by adjusting the flow of blood in the dermis.
3. The skin provides physical, chemical, and biological barriers that help protect the body.
4. Cutaneous sensations include tactile sensations, thermal sensations, and pain.

Maintaining Homeostasis: Skin Wound Healing (Section 5.5)

1. In an epidermal wound, the central portion of the wound usually extends down to the dermis; the wound edges involve only superficial damage to the epidermal cells.
2. Epidermal wounds are repaired by enlargement and migration of basal cells, contact inhibition, and division of migrating and stationary basal cells.

3. During the inflammatory phase of dermal wound healing, a blood clot unites the wound edges, epithelial cells migrate across the wound, vasodilation and increased permeability of blood vessels enhance delivery of phagocytes, and mesenchymal cells develop into fibroblasts.
4. During the migratory phase, fibroblasts migrate along fibrin threads and begin synthesizing collagen fibers and glycoproteins.
5. During the proliferative phase, epithelial cells grow extensively.
6. During the maturation phase, the scab sloughs off, the epidermis is restored to normal thickness, collagen fibers become more organized, fibroblasts begin to disappear, and blood vessels are restored to normal. Scar tissue may differ from normal skin in elevation, elasticity, color, and complement of accessory structures.
7. The process of scar tissue formation is called fibrosis.

Skin Conditions Important to Therapists (Section 5.6)

1. Basal cell carcinomas are the most common type of skin cancer and they are the least harmful.
2. Squamous cell carcinomas may or may not metastasize.
3. Malignant melanomas are the most rare and the most harmful type of skin cancer.
4. Risk factors for developing skin cancer include skin type, sun exposure, family history, age, and immunological status.
5. First-degree burns involve only the epidermis.
6. Second-degree burns destroy the epidermis and part of the dermis.
7. Third-degree burns destroy the epidermis, dermis, and hypodermis.
8. Fourth-degree burns destroy the skin, hypodermis, and deeper tissues such as fascia, muscles, and bones.
9. Pressure ulcers (decubitus ulcers) are caused by a constant deficiency of blood flow to a specific area.

Aging and the Integumentary System (Section 5.7)

1. Most effects of aging begin to occur when people reach their late forties.
2. Among the effects of aging are wrinkling of skin, loss of subcutaneous adipose tissue, atrophy of sebaceous glands, and decrease in the number of melanocytes and Langerhans cells.

SELF-QUIZ QUESTIONS

Fill in the blanks in the following statements.

1. The epidermal layer that is found in thick skin but not in thin skin is the ________ .
2. The most common sweat glands that release a watery secretion are ________ sweat glands; modified sweat glands in the ear are ________ glands; sweat glands located in the axillae, groin, areolae, and beards of males and that release a thick, lipid-rich secretion are ________ sweat glands.

Indicate whether the following statements are true or false.

3. An individual with a dark skin color has more melanocytes than a fair-skinned person.
4. In order to permanently prevent growth of an unwanted hair, you must destroy the hair matrix.

Choose the one best answer to the following questions.

5. The layer of the epidermis that contains stem cells undergoing mitosis is the
   a. stratum corneum **b.** stratum lucidum **c.** stratum basale **d.** stratum spinosum **e.** stratum granulosum.
6. The substance that helps promote mitosis in epidermal skin cells is
   a. keratohyalin **b.** melanin **c.** carotene **d.** collagen **e.** epidermal growth factor.
7. Which of the following is *not* a function of skin?
   a. calcium production **b.** vitamin D synthesis **c.** protection **d.** excretion of wastes **e.** temperature regulation
8. To expose underlying tissues in the bottom of the foot, a foot surgeon must first cut through the skin. Place the following layers in the order that the scalpel would cut: (1) stratum lucidum, (2) stratum corneum, (3) stratum basale, (4) stratum granulosum, (5) stratum spinosum.
   a. 3, 5, 4, 1, 2 **b.** 2, 1, 5, 4, 3 **c.** 2, 1, 4, 5, 3 **d.** 1, 3, 5, 4, 2 **e.** 3, 4, 5, 1, 2
9. Aging of the skin can result in
   a. an increase in collagen and elastic fibers
   b. a decrease in the activity of sebaceous glands
   c. a thickening of the skin
   d. an increased blood flow to the skin
   e. an increase in toenail growth.
10. Which of the following is *not* true?
    a. Albinism is an inherited inability of melanocytes to produce melanin.
    b. Striae occur when the dermis is overstretched to the point of tearing.
    c. In order to prevent excessive scarring, physicians should bring edges of wide wounds close together.
    d. The papillary layer of the dermis is directly responsible for fingerprints.
    e. Much of the body's fat is located in the dermis of the skin.
11. A patient is brought into the emergency room suffering from a burn. The patient does not feel any pain at the burn site. Using a gentle pull on a hair, the examining physician can remove entire hair follicles from the patient's arm. This patient is suffering from what type of burn?
    a. third degree **b.** second degree **c.** first degree **d.** partial thickness **e.** localized
12. Which of the following statements are true? (1) Nails are composed of tightly packed, hard, keratinized cells of the epidermis that form a clear, solid covering over the dorsal surface of the terminal end of digits. (2) The free edge of the nail is white due to the absence of underlying capillaries. (3) Nails help us grasp and manipulate small objects. (4) Nails protect the ends of digits from trauma. (5) Nail color is due to a combination of melanin and carotene.
    a. 1, 2, and 3 **b.** 1, 3, and 4 **c.** 1, 2, 3, and 4 **d.** 2, 3, and 4 **e.** 1, 3, and 5

13. Match the following:

____ **a.** produce the protein that helps protect the skin and underlying tissues from light, heat, microbes, and many chemicals
____ **b.** produce a pigment that contributes to skin color and absorbs ultraviolet light
____ **c.** cells that arise from red bone marrow, migrate to the epidermis, and participate in immune responses
____ **d.** cells thought to function in the sensation of touch
____ **e.** located in the dermis, they function in the sensations of warmth, coolness, pain, itching, and tickling
____ **f.** smooth muscles associated with the hair follicles; when contracted, they pull the hair shafts perpendicular to the skin's surface
____ **g.** an abnormal thickening of the epidermis
____ **h.** release a lipid-rich secretion that functions as a water-repellent sealant in the stratum granulosum
____ **i.** pressure-sensitive cells found mostly in the subcutaneous layer
____ **j.** associated with hair follicles, these secrete an oily substance that helps prevent hair from becoming brittle, prevents evaporation of water from the skin's surface, and inhibits the growth of certain bacteria

A. Merkel cells
B. callus
C. keratinocytes
D. Langerhans cells
E. melanocytes
F. free nerve endings
G. sebaceous glands
H. lamellar granules
I. lamellated (pacinian) corpuscles
J. arrector pili

14. Match the following:

____ **a.** deep region of the dermis composed primarily of dense irregular connective tissue
____ **b.** composed of keratinized stratified squamous epithelial tissue
____ **c.** not considered part of the skin, it contains areolar and adipose tissues and blood vessels; attaches skin to underlying tissues and organs
____ **d.** superficial region of the dermis; composed of areolar connective tissue

A. subcutaneous layer (hypodermis)
B. papillary region
C. reticular region
D. epidermis

15. Match the following and place the phases of deep wound healing in the correct order:

____ **a.** epithelial cells migrate under scab to bridge the wound; formation of granulation tissue
____ **b.** sloughing of scab; reorganization of collagen fibers; blood vessels return to normal
____ **c.** vasodilation and increased permeability of blood vessels to deliver cells involved in phagocytosis; clot formation
____ **d.** extensive growth of epithelial cells beneath scab; random deposition of collagen fibers; continued growth of blood vessels
____ **e.** Correct order of phases **A, B, C, D.**

A. proliferative phase
B. inflammatory phase
C. maturation phase
D. migratory phase

CRITICAL THINKING QUESTIONS

1. Three-year-old Michael was having his first haircut. As the barber started to snip his hair, Michael cried, "Stop! You're killing it!" He then pulled his own hair, yelling, "Ouch! See! It's alive!" Is Michael right about his hair?
2. Michael's twin sister Michelle scraped her knee at the playgroup. She told her mother that she wanted "new skin that doesn't leak." Her mother promised that new skin would soon appear under the bandage. How does new skin grow?
3. Andrew is training for the Megaman triathlon. After hours in running shoes and damp locker rooms, his feet are a mess! He has calluses, warts, and athlete's foot. What are the causes of his misery?
4. Fifteen-year-old Jeremy has a bad case of "blackheads." According to his Aunt Frieda, Jeremy's skin problems are from too much late-night TV, frozen pizza, and cheddar popcorn. Explain the real cause of blackheads to Aunt Frieda.

? ANSWERS TO FIGURE QUESTIONS

5.1 The epidermis is composed of epithelial tissue, whereas the dermis is made up of connective tissue.
5.2 The stratum basale is the layer of the epidermis that contains stem cells.
5.3 Epidermal ridges increase the grip of the hand or foot by increasing friction and form fingerprints or toeprints.
5.4 New hairs are produced by cell division in the hair matrix.
5.5 The main function of eccrine sweat glands is to help regulate body temperature through evaporation.
5.6 Nails are hard because they are composed of tightly packed, keratinized epidermal cells.
5.7 The stratum lucidum is present in thick skin but is lacking in thin skin.
5.8 Since the epidermis is avascular, an epidermal wound would not produce any bleeding.
5.9 Basal cell carcinoma is the most common type of skin cancer.
5.10 The seriousness of a burn is determined by the depth and extent of the area involved, and also the individual's age and general health.
5.11 About 22.5% of the body would be involved [4.5% (anterior arm) + 18% (anterior trunk)].
5.12 Pressure ulcers typically develop in tissues that overlie bony projections subjected to pressure, such as the shoulders, hips, buttocks, heels, and ankles.

Bone Tissue 6

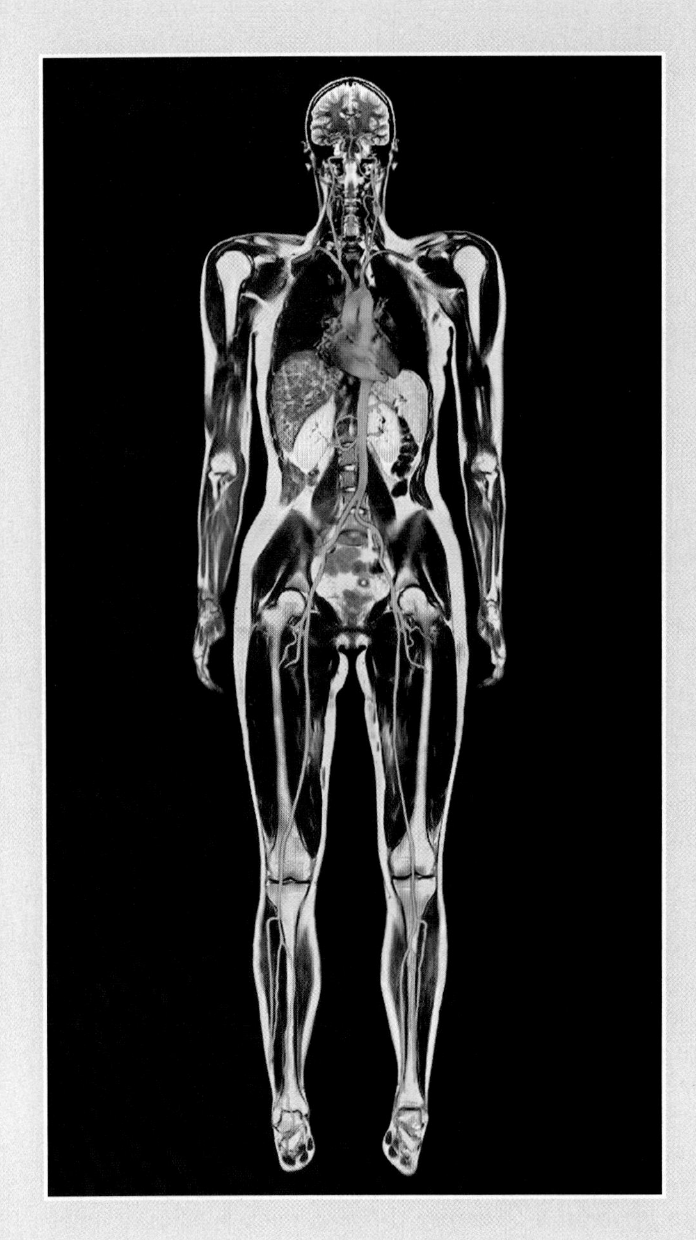

You know from the song "the hip bone's connected to the leg bone" that bones connect with other bones. Bones are attached to most skeletal muscles by tendons and enable movement of the body. They are also a storehouse for minerals such as calcium. The bones remodel themselves on a daily basis to provide useable minerals and to replace the ones they have given up to the rest of the body for use elsewhere (i.e. muscle contraction). Thus normal adult bone is continuously repairing and remodeling its minerals and is strong enough for the strenuous physical demands you place on it every day. Bone also provides a storehouse for triglycerides (fat) and is the main producer of blood cells in adults (in the red bone marrow). These important functions are often overlooked when you think of bones as only a source of support that prevents you from falling into a pile of mush on the floor or only for protection of organs from physical harm. In light of these varied functions, bone is one of the best examples of homeostasis where the different systems work together for the betterment of the whole body.

When you see a skeleton you may think of bone as nonliving, but it is amazing how truly alive bone is. It is full of cells that work together in a communal setting sharing food and removing waste from a dense, compact housing structure.

CONTENTS AT A GLANCE

6.1 FUNCTIONS OF BONE AND THE SKELETAL SYSTEM

OBJECTIVE

- Describe the six main functions of the skeletal system.

In this chapter we will help you understand how bones form and age, and how weight-bearing exercise can improve bone density and strength. The study of bone structure and the treatment of bone disorders is called ***osteology*** (os-tē-OL-ō-jē; *osteo-* = bone; *-logy* = study of).

Knowledge of the musculoskeletal system is paramount for manual therapists. Your knowledge of bone at the cellular level will better enable you to understand the complexities of the teenage athlete who is growing in both size and muscle strength. On the other end of the age spectrum, knowledge of bone at the cellular level will better enable you to understand the complexities of the elderly who are losing muscle strength and thus bone composition.

Bone tissue makes up about 18% of the weight of the human body. The skeletal system performs several basic functions:

1. ***Support.*** The skeleton serves as the structural framework for the body by supporting soft tissues and providing attachment points for the tendons of most skeletal muscles.
2. ***Protection.*** The skeleton protects the most important internal organs from injury. For example, cranial bones protect the brain, vertebrae (backbones) protect the spinal cord, and the rib cage protects the heart and lungs.
3. ***Assistance in movement.*** Most skeletal muscles attach to bones; when they contract, they pull on bones to produce movement. This function is discussed in detail in Chapter 10.
4. ***Mineral homeostasis.*** Bone tissue stores several minerals, especially calcium and phosphorus, which contribute to the strength of bone. On demand, bone releases minerals into the blood to maintain critical mineral balances (homeostasis) and to distribute the minerals to other parts of the body.
5. ***Blood cell production.*** Within certain bones, a connective tissue called ***red bone marrow*** produces red blood cells, white blood cells, and platelets, a process called ***hemopoiesis*** (hēm-ō-poy-Ē-sis; *hemo-* = blood; *-poiesis* = making) or ***hematopoiesis***. Red bone marrow consists of developing blood cells, adipocytes, fibroblasts, and macrophages within a network of reticular fibers. It is present in developing bones of the fetus and in some adult bones, such as the pelvis, ribs, breastbone, vertebrae (backbones), skull, and ends of the bones of the arm and thigh.
6. ***Triglyceride storage. Yellow bone marrow*** consists mainly of adipose cells, which store triglycerides. The stored triglycerides are a potential chemical energy reserve.

CHECKPOINT

1. What are the primary functions of the skeletal system?
2. How do red and yellow bone marrow differ in composition and function?

6.2 STRUCTURE OF BONE

OBJECTIVE

- Describe the parts of a long bone.

We will now examine the structure of bone at the macroscopic level. Macroscopic bone structure may be analyzed by considering the parts of a long bone, such as the humerus (the arm bone) shown in Figure 6.1a. A *long bone* is one that has greater length than width. A typical long bone consists of the following parts:

1. The ***diaphysis*** (dī-AF-i-sis = growing between) is the bone's shaft or body—the long, cylindrical, main portion of the bone.
2. The ***epiphyses*** (e-PIF-i-sēz = growing over; singular is *epiphysis*) are the distal and proximal ends of the bone.
3. The ***metaphyses*** (me-TAF-i-sēz; *meta-* = between; singular is *metaphysis*) are the regions in a mature bone where the diaphysis joins the epiphyses. In a growing bone, each metaphysis includes an ***epiphyseal (growth) plate*** (ep′-i-FIZ-ē-al), a layer of hyaline cartilage that allows the diaphysis of the bone to grow in length (described later in the chapter). When a bone ceases to grow in length at about ages 18–21, the cartilage in the epiphyseal plate is replaced by bone; the resulting bony structure is known as the ***epiphyseal line.***
4. The ***articular cartilage*** is a thin layer of hyaline cartilage covering the part of the epiphysis where the bone forms an articulation (joint) with another bone. Articular cartilage reduces friction and absorbs shock at freely movable joints.
5. The ***periosteum*** (per′-ē-OS-tē-um; *peri-* = around) is a tough sheath of dense irregular connective tissue, and its associated blood vessels, that surrounds the bone surface wherever it is not covered by articular cartilage. The bone-forming cells of the periosteum enable bone to grow in thickness, but not in length. The periosteum also protects the bone, assists in fracture repair, helps nourish bone tissue, and serves as an attachment point for ligaments and tendons. It is attached to the underlying bone through ***perforating (Sharpey's) fibers,*** thick bundles of collagen fibers that extend from the periosteum into the extracellular bone matrix. It is also rich in nerve fibers; damage to the nerves of the periosteum accounts for much of the pain associated with fractures.
6. The ***medullary cavity*** (MED-ū-lar′-ē; *medulla-* = marrow, pith) or ***marrow cavity*** is the space within the diaphysis that contains fatty yellow bone marrow in adults.
7. The ***endosteum*** (end-OS-tē-um; *endo-* = within) is a thin membrane that lines the medullary cavity. This vascular layer contains a single layer of bone-forming cells and a small amount of connective tissue.

CHECKPOINT

3. Diagram the parts of a long bone, and list the functions of each part.

Figure 6.1 **Parts of a long bone.** The spongy bone tissue of the epiphyses and metaphyses contains red bone marrow, whereas the medullary cavity of the diaphysis contains yellow bone marrow (in adults).

A long bone is covered by articular cartilage at its proximal and distal epiphyses and by periosteum around the diaphysis.

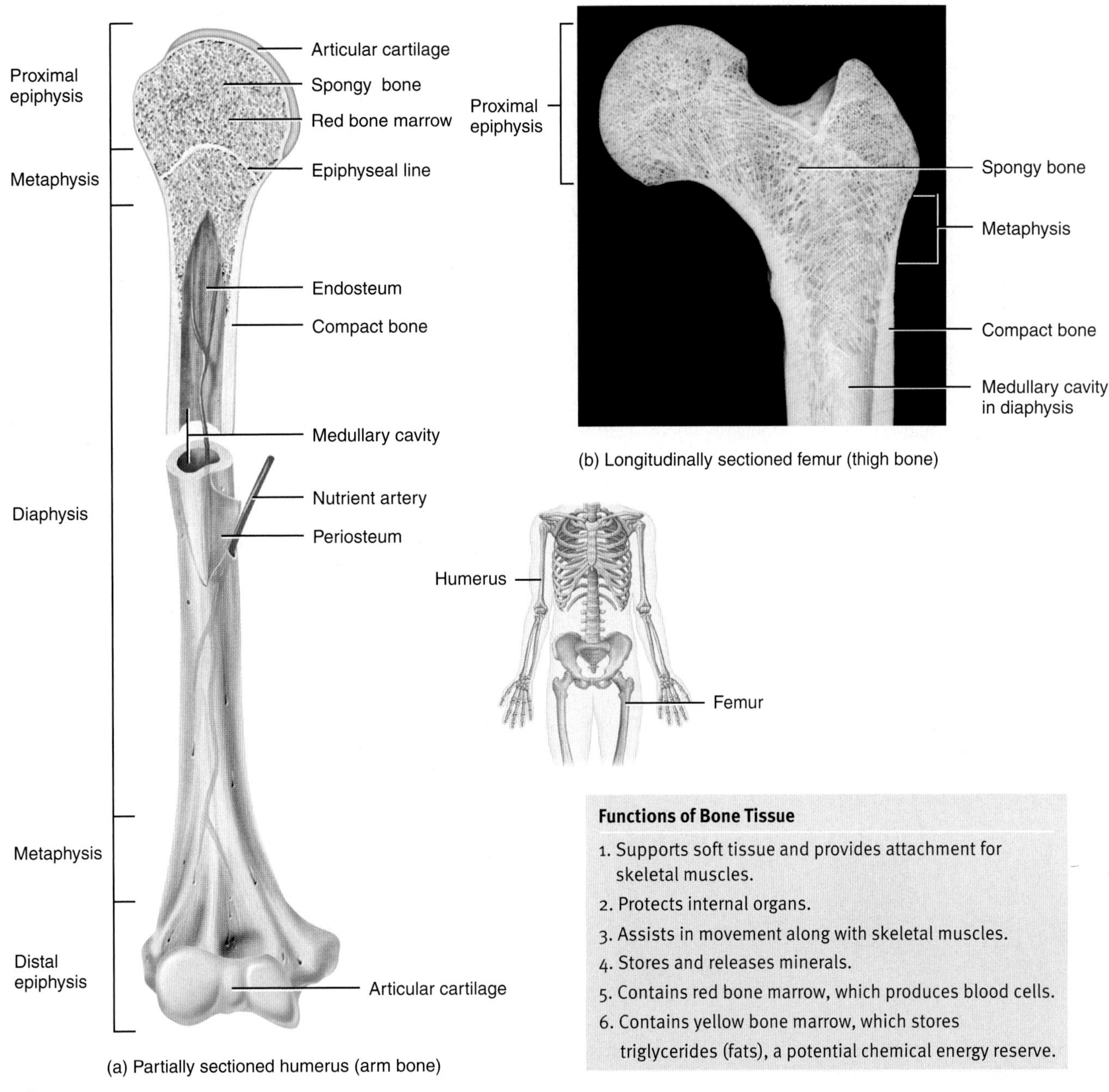

(a) Partially sectioned humerus (arm bone)

(b) Longitudinally sectioned femur (thigh bone)

Functions of Bone Tissue

1. Supports soft tissue and provides attachment for skeletal muscles.
2. Protects internal organs.
3. Assists in movement along with skeletal muscles.
4. Stores and releases minerals.
5. Contains red bone marrow, which produces blood cells.
6. Contains yellow bone marrow, which stores triglycerides (fats), a potential chemical energy reserve.

What is the functional significance of the periosteum?

6.3 HISTOLOGY OF BONE TISSUE

OBJECTIVE

- Describe the histological features of bone tissue.

We will now examine the structure of bone at the microscopic level. Like other connective tissues, ***bone***, or ***osseous tissue*** (OS-ē-us), contains an abundant extracellular matrix that surrounds widely separated cells. The extracellular matrix is about 25% water, 25% collagen fibers, and 50% crystallized mineral salts. The most abundant mineral salt is calcium phosphate [$Ca_3(PO_4)_2$]. It combines with another mineral salt, calcium hydroxide [$Ca(OH)_2$], to form crystals of ***hydroxyapatite*** (hī-drok′-sē-AP-a-tīt). As the crystals form, they combine with still other mineral salts, such as calcium carbonate ($CaCO_3$), and ions such as magnesium, fluoride, potassium, and sulfate. As these mineral salts are deposited in the framework formed by the collagen fibers of the extracellular matrix, they crystallize and the tissue hardens. This process of ***calcification*** (kal′-si-fi-KĀ-shun) is initiated by bone-building cells called osteoblasts.

It was once thought that calcification simply occurred when enough mineral salts were present to form crystals. We now know that the process requires the presence of collagen fibers. Mineral salts first begin to crystallize in the microscopic spaces between collagen fibers. After the spaces are filled, mineral crystals accumulate around the collagen fibers.

Although a bone's *hardness* depends on the crystallized inorganic mineral salts (hydroxyapatite), a bone's *flexibility* depends on its collagen fibers. Like reinforcing metal rods in concrete, collagen fibers and other organic molecules provide *tensile strength,* resistance to being stretched or torn apart. Soaking a bone in an acidic solution, such as vinegar, dissolves its mineral salts, causing the bone to become rubbery and flexible. As you will see shortly, when the need for particular minerals arises or as part of bone formation or breakdown, bone cells called osteoclasts secrete enzymes and acids that break down both the mineral salts and the collagen fibers of bone extracellular matrix.

Four types of cells are present in bone tissue: osteogenic cells, osteoblasts, osteocytes, and osteoclasts (Figure 6.2).

1. ***Osteogenic cells*** (os′-tē-ō-JEN-ik; *-genic* = producing) are unspecialized stem cells derived from mesenchyme, the tissue from which almost all connective tissues are originally formed. They are the only bone cells to undergo cell division; the resulting cells develop into osteoblasts. Osteogenic cells are found along the inner portion of the periosteum, in the endosteum, and in the canals within bone that contain blood vessels.

2. ***Osteoblasts*** (OS-tē-ō-blasts′; *-blasts* = buds or sprouts) are bone-building cells. They synthesize and secrete collagen fibers and other organic components needed to build the extracellular matrix of bone tissue, and they initiate calcification (described shortly). As osteoblasts surround themselves with extracellular

Figure 6.2 Types of cells in bone tissue.

Osteogenic cells undergo cell division and develop into osteoblasts, which secrete bone extracellular matrix and become osteocytes when they are surrounded by the matrix. Osteoclasts develop from multiple monocytes.

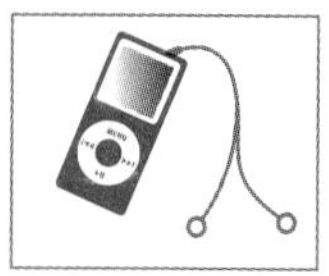

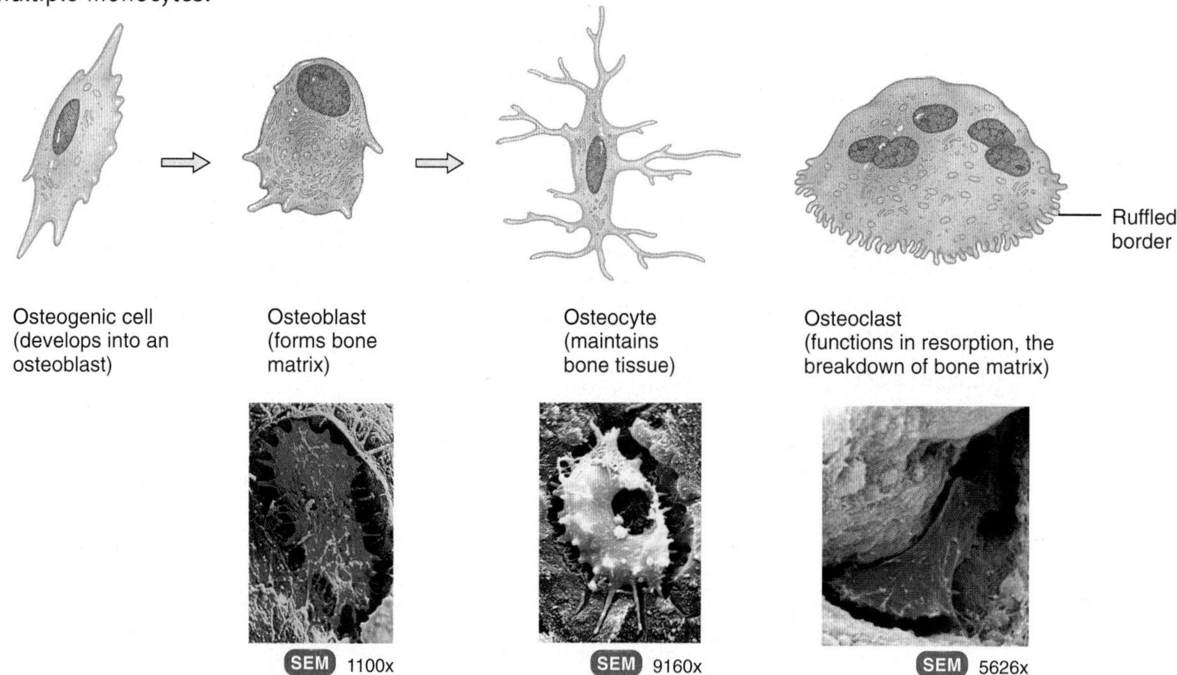

? Why is bone resorption important?

matrix, they become trapped in their own secretions and become osteocytes. (Note: *Blasts* in bone or any other connective tissue secrete the extracellular matrix.)

3. ***Osteocytes*** (OS-tē-ō-sīts′; *-cytes* = cells), mature bone cells, are the main cells in bone tissue and maintain its daily metabolism, such as the exchange of nutrients and wastes with the blood. Like osteoblasts, osteocytes do not undergo cell division. (Note: *Cytes* in bone, or any other tissue, maintain the tissue.)

4. ***Osteoclasts*** (OS-tē-ō-clasts′; *-clast* = break) are huge cells derived from the fusion of as many as 50 monocytes (a type of white blood cell) and are concentrated in the endosteum. On the side of the cell that faces the bone surface, the plasma membrane of the osteoclasts is deeply folded into a *ruffled border.* Here the cell releases powerful lysosomal enzymes and acids that digest the protein and mineral components of the underlying extracellular bone matrix. This breakdown of bone extracellular matrix, termed ***resorption*** (rē-SORP-shun), is part of the normal development, growth, maintenance, and repair of bone. (Note: *Clasts* in bone break down extracellular matrix.) As you will see later, in response to certain hormones, osteoclasts help regulate blood calcium level (see Section 6.7). They are also target cells for drug therapy used to treat osteoporosis (see Section 6.8).

Bone is not completely solid but has many small spaces between its cells and extracellular matrix components. Some spaces serve as channels for blood vessels that supply bone cells with nutrients. Other spaces act as storage areas for red bone marrow. Depending on the size and distribution of the spaces, the regions of a bone may be categorized as either compact or spongy (see Figure 6.1). Overall, about 80% of the skeleton is compact bone and 20% is spongy bone.

Compact Bone Tissue

Compact (dense) bone tissue contains few spaces (see Figure 6.3a) and is the strongest form of bone tissue. It is found beneath the periosteum of all bones and makes up the bulk of the diaphyses of long bones. Compact bone tissue provides protection and support and resists the stresses produced by weight and movement.

Blood vessels, lymphatic vessels, and nerves from the periosteum penetrate compact bone through transverse ***perforating*** or ***Volkmann's canals*** (FŌLK-mans). The vessels and nerves of the perforating canals connect with those of the medullary cavity, periosteum, and ***central*** or ***haversian canals*** (ha-VER-shun). The central canals run longitudinally through the bone. Around the central canals are ***concentric lamellae*** (la-MEL-ē)—rings of calcified extracellular matrix much like the rings of a tree trunk. Between the lamellae are small spaces called ***lacunae*** (la-KOO-nē = little lakes; singular is *lacuna*), which contain osteocytes. Radiating in all directions from the lacunae are tiny ***canaliculi*** (kan′-a-LIK-ū-lī = small channels) filled with extracellular fluid. Inside the canaliculi are slender finger-like processes of osteocytes (see inset at the right in Figure 6.3a). Neighboring osteocytes communicate via gap junctions. The canaliculi connect lacunae with one another and with the central canals, forming an intricate, miniature system of interconnected canals throughout the bone. This system provides many routes for nutrients and oxygen to reach the osteocytes and for the removal of wastes.

The components of compact bone tissue are arranged into repeating structural units called ***osteons*** (OS-tē-ons) ***(haversian systems)*** (see Figure 6.3a). Each osteon is a tubelike cylinder that consists of a central (haversian) canal with its concentrically arranged lamellae, lacunae, osteocytes, and canaliculi. Osteons in compact bone tissue are aligned in the same direction along lines of stress. In the shaft, for example, they are parallel to the long axis of the bone. As a result, the shaft of a long bone resists bending or fracturing even when considerable force is applied from either end. The osteons of a long bone can be compared to a stack of logs; each log is made up of rings of hard material, and it requires considerable force to fracture them all. The lines of stress in a bone change as a baby learns to walk. This also occurs in response to repeated strenuous physical activity, fractures, or physical deformity. Thus, the organization of osteons is not static but changes over time in response to the physical demands placed on the skeleton. This is known as *Wolff's law.*

The areas between osteons contain ***interstitial lamellae*** (in′-ter-STISH-al), which also have lacunae with osteocytes and canaliculi. Interstitial lamellae are fragments of older osteons that have been partially destroyed during bone rebuilding or growth. Lamellae that encircle the bone just beneath the periosteum or encircle the medullary cavity are called ***circumferential lamellae.***

Spongy (Cancellous) Bone Tissue

In contrast to compact bone tissue, ***spongy (cancellous) bone tissue*** does not contain osteons. Despite what the name seems to imply, the term "spongy" does not refer to the texture of the bone, only its appearance (see Figure 6.3b). Spongy bone consists of lamellae arranged in an irregular lattice of thin columns called ***trabeculae*** (tra-BEK-ū-lē = little beams; singular is *trabecula*). The macroscopic spaces between the trabeculae help make bones lighter and can sometimes be filled with red bone marrow. Within each trabecula are lacunae that contain osteocytes. Canaliculi radiate outward from the lacunae. Because the osteocytes of spongy bone are located on the superficial surfaces of trabeculae, they receive nourishment directly from the blood circulating through the medullary cavities.

Spongy bone tissue makes up most of the bone tissue of short, flat, and irregularly shaped bones. It also forms most of the epiphyses of long bones and a narrow rim around the medullary cavity of the diaphysis of long bones.

At first glance, the structure of the osteons of compact bone tissue appears to be highly organized, and the trabeculae of spongy bone tissue appear to be randomly arranged. However, the trabeculae of spongy bone tissue are precisely oriented along lines of stress, a characteristic that helps bones resist stresses and transfer force without breaking. Spongy bone tissue tends to be located where bones are not heavily stressed or where stresses are applied from many directions.

Figure 6.3 Histology of compact and spongy bone. (a) Sections through the diaphysis of a long bone, from the surrounding periosteum on the right, to compact bone in the middle, to spongy bone and the medullary cavity on the left. (b and c) Trabeculae of spongy bone are illustrated at the macroscopic and microscopic levels.

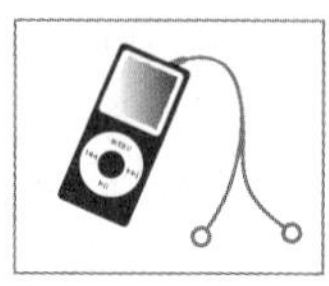

Osteocytes lie in lacunae arranged in concentric circles around a central (haversian) canal in compact bone and in irregularly arranged lacunae in the trabeculae of spongy bone.

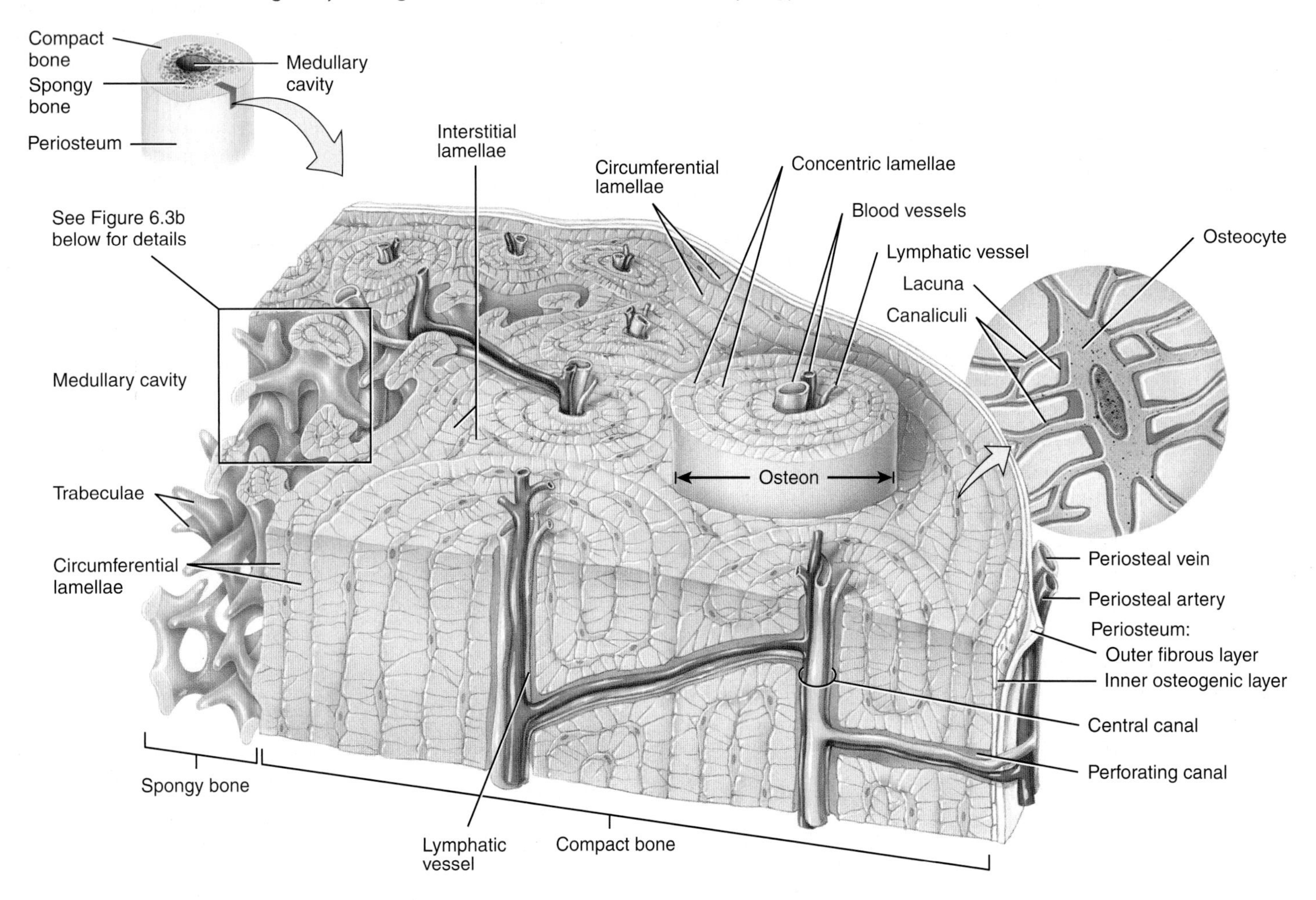

(a) Osteons (haversian systems) in compact bone and trabeculae in spongy bone

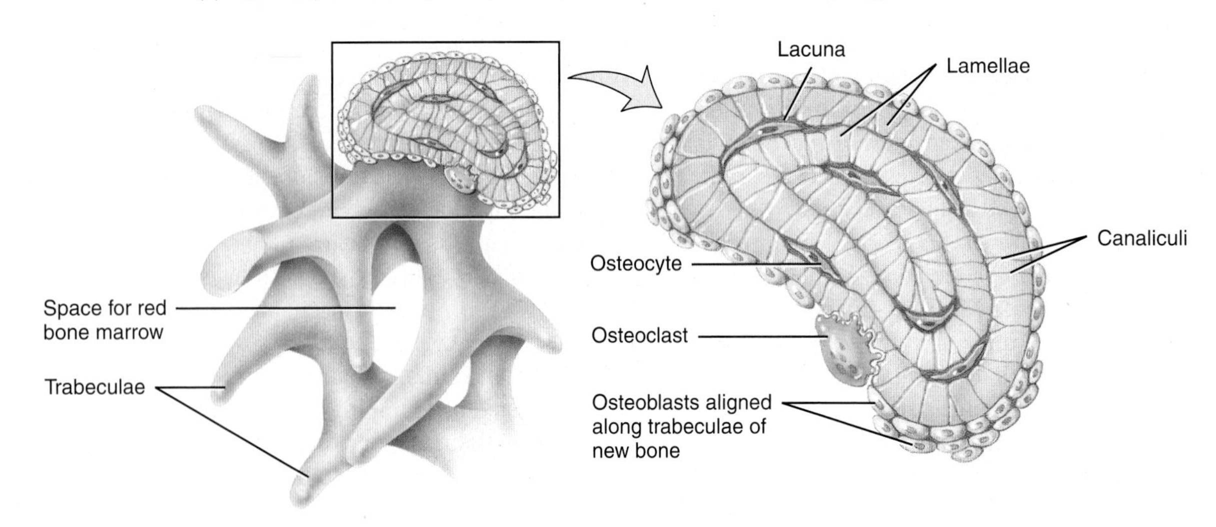

(b) Enlarged aspect of spongy bone trabeculae

(c) Details of a section of a trabecula

? As people age, some central (haversian) canals may become blocked. What effect would this have on the surrounding osteocytes?

Spongy bone tissue is different from compact bone tissue in two respects. First, spongy bone tissue is light, which reduces the overall weight of a bone so that it moves more readily when pulled by a skeletal muscle. Second, the trabeculae of spongy bone tissue support and protect the red bone marrow. The spongy bone tissue in the hip bones, ribs, breastbone, vertebrae, and the ends of certain long bones is where red bone marrow is stored and, thus, where hemopoiesis (blood cell production) occurs in adults.

• CLINICAL CONNECTION | Bone Scan

A **bone scan** is a diagnostic procedure that takes advantage of the fact that bone is living tissue. A small amount of a radioactive tracer compound that is readily absorbed by bone is injected intravenously. The degree of uptake of the tracer is related to the amount of blood flow to the bone. A scanning device (gamma camera) measures the radiation emitted from the bones, and the information is translated into a photograph that can be read like an x-ray on a monitor. Normal bone tissue is identified by a consistent gray color throughout because of its uniform uptake of the radioactive tracer. Darker or lighter areas may indicate bone abnormalities. Darker areas, called "hot spots," are areas of increased metabolism that absorb more of the radioactive tracer due to increased blood flow. Hot spots may indicate bone cancer, abnormal healing of fractures, or abnormal bone growth. Lighter areas, called "cold spots," are areas of decreased metabolism that absorb less of the radioactive tracer due to decreased blood flow. Cold spots may indicate problems such as degenerative bone disease, decalcified bone, fractures, bone infections, Paget's disease, and rheumatoid arthritis. A bone scan detects abnormalities 3 to 6 months sooner than standard x-ray procedures and exposes the patient to less radiation. A bone scan is the standard test for bone density screening, which is particularly important in screening females for osteoporosis. •

CHECKPOINT

4. Why is bone considered a connective tissue?
5. Describe the four types of cells in bone tissue.
6. What is the composition of the matrix of bone tissue?
7. Distinguish between spongy and compact bone tissue in terms of microscopic appearance, location, and function.

6.4 BLOOD AND NERVE SUPPLY OF BONE

OBJECTIVE

- Describe the blood and nerve supply of bone.

Bone is richly supplied with blood. Blood vessels, which are especially abundant in portions of bone containing red bone marrow, pass into bones from the periosteum. We will consider the blood supply of a long bone such as the mature tibia (shinbone) shown in Figure 6.4.

Periosteal arteries (per-ē-OS-tē-al) accompanied by nerves enter the diaphysis through many perforating (Volkmann's) canals and supply the periosteum and outer part of the compact bone (see Figure 6.3a). Near the center of the diaphysis, a large ***nutrient artery*** passes through a hole in compact bone called the ***nutrient foramen.*** On entering the medullary cavity, the nutrient artery divides into proximal and distal branches that supply both the inner part of compact bone tissue of the diaphysis and the spongy bone tissue and red marrow as far as the epiphyseal plates (or lines). Some bones, like the tibia, have only one nutrient artery; others like the femur (thighbone) have several. The ends of long bones are supplied by the metaphyseal and epiphyseal arteries, which arise from arteries that supply the associated joint (Figure 6.4). The ***metaphyseal arteries*** (met-a-FIZ-ē-al) enter the metaphyses of a long bone and, together with the nutrient artery, supply the red bone marrow and bone tissue of the metaphyses. The ***epiphyseal arteries*** (ep′-i-FIZ-ē-al) enter the epiphyses of a long bone and supply the red bone marrow and bone tissue of the epiphyses.

Veins that carry blood away from long bones are evident in three places: (1) One or two ***nutrient veins*** accompany the nutrient artery in the diaphysis; (2) numerous ***epiphyseal veins*** and ***metaphyseal veins*** exit with their respective arteries in the epiphyses and metaphyses; and (3) many small ***periosteal veins*** exit with their respective arteries in the periosteum (Figure 6.4).

Figure 6.4 Blood supply of a mature long bone, the tibia (shinbone).

Bone is richly supplied with blood vessels.

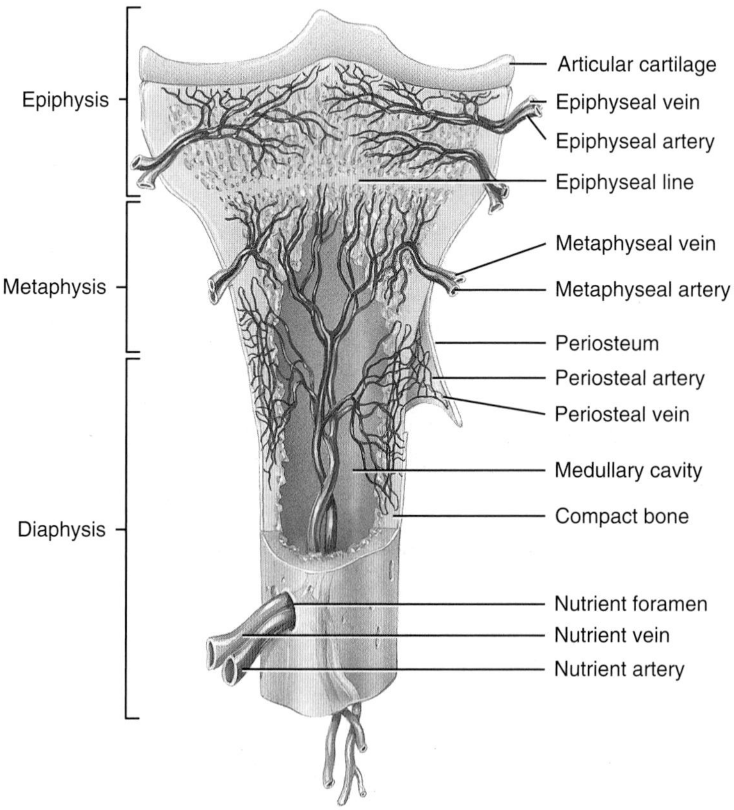

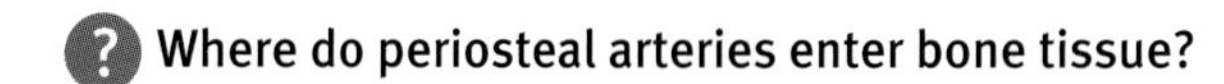

? Where do periosteal arteries enter bone tissue?

Nerves accompany the blood vessels that supply bones. The periosteum is rich in sensory nerves, some of which carry pain sensations. These nerves are especially sensitive to tearing or tension, which explains the severe pain resulting from a fracture or a bone tumor. For the same reason there is some pain associated with a bone marrow needle biopsy. In this procedure, a needle is inserted into the middle of the bone to withdraw a sample of red bone marrow to examine it for conditions such as leukemias, metastatic neoplasms, lymphoma, Hodgkin's disease, and aplastic anemia. As the needle penetrates the periosteum, pain is felt. Once it passes through, there is little pain.

CHECKPOINT

8. Explain the location and roles of the nutrient arteries, nutrient foramina, metaphyseal arteries, epiphyseal arteries, and periosteal arteries.

6.5 BONE FORMATION

OBJECTIVE

- Describe the steps involved in intramembranous and endochondral ossification.

The process by which bone forms is called ***ossification*** (os′-i-fi-KĀ-shun; *ossi-* = bone; *-fication* = making) or ***osteogenesis*** (os′-tē-ō-JEN-e-sis). The "skeleton" of a human embryo is composed of loose mesenchymal cells, which are shaped like bones and are the sites where ossification occurs. These "bones" provide the template for subsequent ossification, which begins during the sixth week of embryonic development and follows one of two patterns. Recall that ***mesenchyme*** is a connective tissue found mostly in an embryo and is the tissue from which most other connective tissues develop.

The two methods of bone formation, which both involve the replacement of a preexisting connective tissue with bone, do not lead to differences in the structure of mature bones, but are simply different methods of bone development. In the first type of ossification, called ***intramembranous ossification*** (in′-tra-MEM-bra-nus; *intra-* = within; *-membran-* = membrane), bone forms directly within mesenchyme arranged in sheetlike layers that resemble membranes. In the second type, ***endochondral ossification*** (en′-dō-KON-dral; *endo-* = within; *-chondral* = cartilage), bone forms within hyaline cartilage that develops from mesenchyme.

Intramembranous Ossification

Intramembranous ossification is the simpler of the two methods of bone formation. The flat bones of the skull, mandible (lower jawbone), and part of the clavicle (collar bone) are formed in this way. Also, the "soft spots" that help the fetal skull pass through the birth canal later harden as they undergo intramembranous ossification, which occurs as follows (Figure 6.5):

1. ***Development of the ossification center.*** At the site where the bone will develop, specific chemical messages cause the mesenchymal cells to cluster together and differentiate, first into osteogenic cells and then into osteoblasts. The site of such a cluster is called an ***ossification center.*** Osteoblasts secrete the organic extracellular matrix of bone until they are surrounded by it.
2. ***Calcification.*** Next, the secretion of extracellular matrix stops and the cells, now called osteocytes, lie in lacunae and extend their narrow cytoplasmic processes into canaliculi that radiate in all directions. Within a few days, calcium and other mineral salts are deposited and the extracellular matrix hardens or calcifies (calcification).
3. ***Formation of trabeculae.*** As the bone extracellular matrix forms, it develops into trabeculae that fuse with one another to form spongy bone. Blood vessels grow into the spaces between the trabeculae. Connective tissue that is associated with the blood vessels in the trabeculae differentiates into red bone marrow.
4. ***Development of the periosteum.*** In conjuction with the formation of trabeculae, mesenchyme condenses at the periphery of the bone and develops into the periosteum. Eventually, a thin layer of compact bone replaces the surface layers of the spongy bone, but spongy bone remains in the center. Much of the newly formed bone is remodeled (destroyed and reformed) as the bone is transformed into its adult size and shape.

Endochondral Ossification

The replacement of cartilage by bone is called ***endochondral ossification***. Although most bones of the body are formed in this way, the process is best observed in a long bone. It proceeds as follows (see Figure 6.6):

1. ***Development of the cartilage model.*** At the site where the bone is going to form, specific chemical messages cause the mesenchymal cells to crowd together in the shape of the future bone, and then develop into chondroblasts. The chondroblasts secrete cartilage extracellular matrix, producing a ***cartilage model*** consisting of hyaline cartilage. A membrane called the ***perichondrium*** (per-i-KON-drē-um) develops around the cartilage model.
2. ***Growth of the cartilage model.*** Once chondroblasts become deeply buried in the cartilage extracellular matrix, they are called chondrocytes. The cartilage model grows in length by continual cell division of chondrocytes accompanied by further secretion of the cartilage extracellular matrix. This type of growth is termed ***interstitial growth*** and results in an increase in length. In contrast, growth of the cartilage in thickness is due mainly to the addition of more extracellular matrix material to the periphery of the model by new chondroblasts that develop from the perichondrium. This growth pattern, resulting from extracellular matrix deposited on the cartilage surface, is called ***appositional growth*** (a-pō-ZISH-i-nal).

Figure 6.5 **Intramembranous ossification.** Illustrations ① and ② show a smaller field of vision at higher magnification than illustrations ③ and ④. Refer to this figure as you read the corresponding numbered paragraphs in the text.

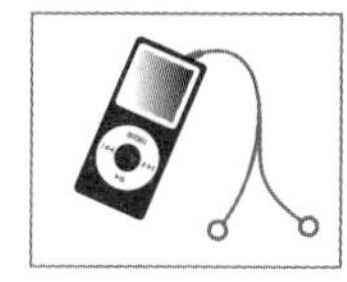

Intramembranous ossification involves the formation of bone within mesenchyme arranged in sheet-like layers that resemble membranes.

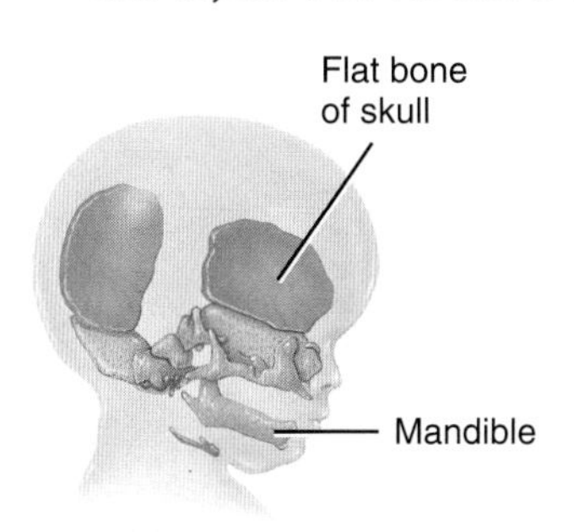

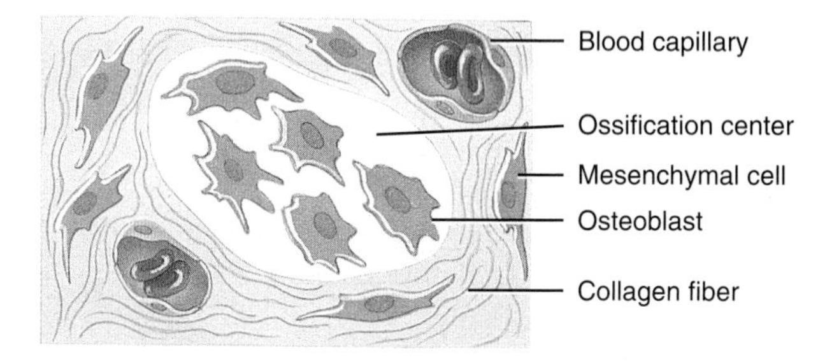

① Development of ossification center: osteoblasts secrete organic extracellular matrix

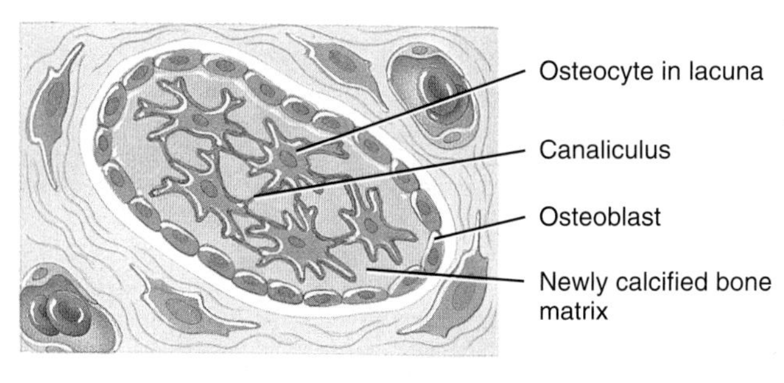

② Calcification: calcium and other mineral salts are deposited and extracellular matrix calcifies (hardens)

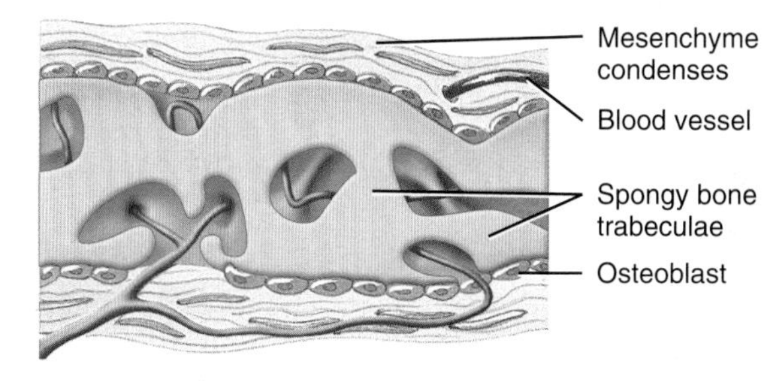

③ Formation of trabeculae: extracellular matrix develops into trabeculae that fuse to form spongy bone

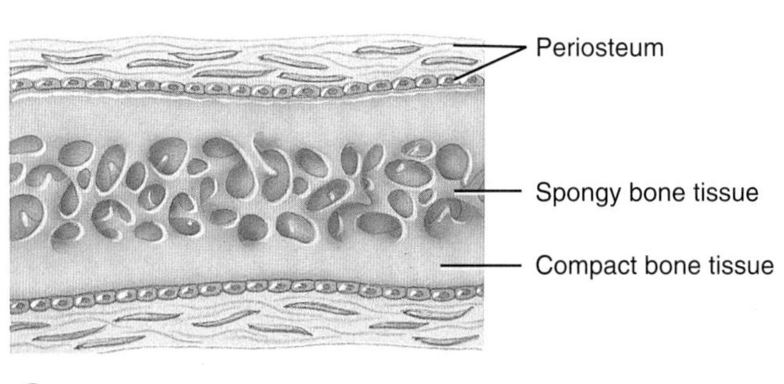

④ Development of the periosteum: mesenchyme at the periphery of the bone develops into the periosteum

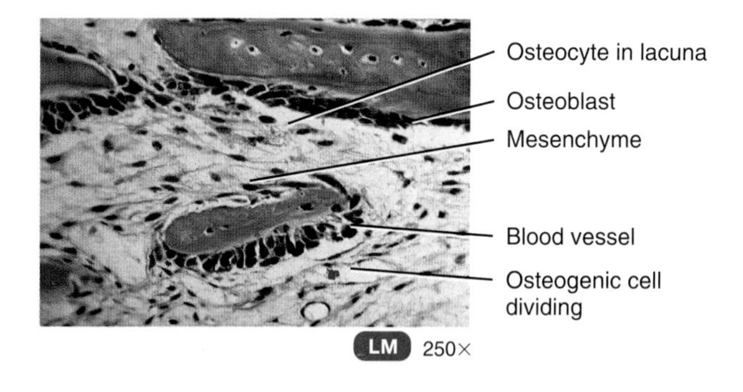

? Which bones of the body develop by intramembranous ossification?

As the cartilage model continues to grow, chondrocytes in its mid-region hypertrophy (increase in size) and the surrounding cartilage extracellular matrix begins to calcify. Other chondrocytes within the calcifying cartilage die because nutrients can no longer diffuse quickly enough through the extracellular matrix. As these chondrocytes die, lacunae form and eventually merge into small cavities.

③ ***Development of the primary ossification center.*** Primary ossification proceeds *inward* from the external surface of the bone. A nutrient artery penetrates the perichondrium and the calcifying cartilage model through a nutrient foramen in the mid-region of the cartilage model, stimulating osteogenic cells in the perichondrium to differentiate into osteoblasts. Once the perichondrium starts to form bone, it is known as

Figure 6.6 Endochondral ossification.

During endochondral ossification, bone gradually replaces a cartilage model.

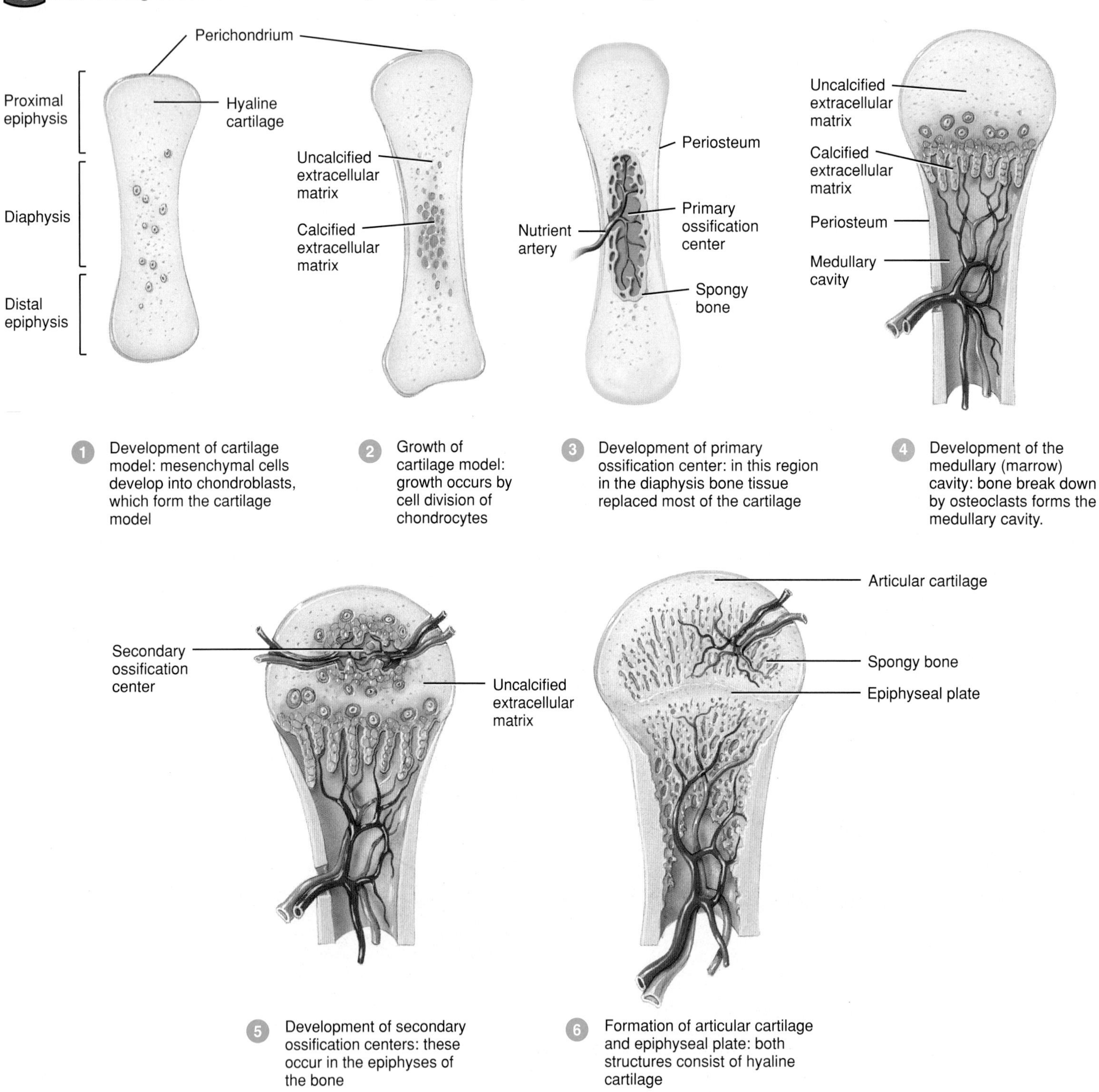

① Development of cartilage model: mesenchymal cells develop into chondroblasts, which form the cartilage model

② Growth of cartilage model: growth occurs by cell division of chondrocytes

③ Development of primary ossification center: in this region in the diaphysis bone tissue replaced most of the cartilage

④ Development of the medullary (marrow) cavity: bone break down by osteoclasts forms the medullary cavity.

⑤ Development of secondary ossification centers: these occur in the epiphyses of the bone

⑥ Formation of articular cartilage and epiphyseal plate: both structures consist of hyaline cartilage

? Where in the cartilage model do secondory ossification centers develop during endochondral ossification?

the ***periosteum.*** Near the middle of the model, periosteal capillaries grow into the disintegrating calcified cartilage, inducing growth of a ***primary ossification center,*** a region where bone tissue will replace most of the cartilage. Osteoblasts then begin to deposit bone extracellular matrix over the remnants of calcified cartilage, forming spongy

bone trabeculae. Primary ossification spreads toward both ends of the cartilage model.

4. ***Development of the medullary (marrow) cavity.*** As the primary ossification center grows toward the ends of the bone, osteoclasts break down some of the newly formed spongy bone trabeculae. This activity leaves a cavity, the medullary (marrow) cavity, in the diaphysis (shaft). Eventually, most of the wall of the diaphysis is replaced by compact bone.

5. ***Development of the secondary ossification centers.*** When branches of the epiphyseal artery enter the epiphyses, ***secondary ossification centers*** develop, usually around the time of birth. Bone formation is similar to that in primary ossification centers. One difference, however, is that spongy bone remains in the interior of the epiphyses (no medullary cavities are formed there). In contrast to primary ossification, secondary ossification proceeds *outward* from the center of the epiphysis toward the outer surface of the bone.

6. ***Formation of articular cartilage and the epiphyseal plate.*** The hyaline cartilage that covers the epiphyses becomes the articular cartilage. Prior to adulthood, hyaline cartilage remains between the diaphysis and epiphysis as the ***epiphyseal (growth) plate,*** which is responsible for the lengthwise growth of long bones.

CHECKPOINT

9. What are the major events of intramembranous ossification and endochondral ossification and how are they different?

6.6 BONE GROWTH

OBJECTIVE

- Describe how bones grow in length and thickness.

During childhood, bones throughout the body grow in thickness by ***appositional growth,*** and long bones lengthen by the addition of bone material on the diaphyseal side of the epiphyseal plate by ***interstitial growth.***

Growth in Length

To understand how a bone grows in length (interstitial growth), you need to know some of the details of the structure of the epiphyseal plate (Figure 6.7). The ***epiphyseal (growth) plate*** (ep-i-FIZ-ē-al) is a layer of hyaline cartilage in the metaphysis of a growing bone that consists of four zones (Figure 6.7b):

1. ***Zone of resting cartilage.*** This layer is nearest the epiphysis and consists of small, scattered chondrocytes. The term "resting" is used because the cells do not function in bone growth. Rather, they anchor the epiphyseal plate to the epiphysis of the bone.

2. ***Zone of proliferating cartilage.*** Slightly larger chondrocytes in this zone are arranged like stacks of coins. These chondrocytes divide to replace those that die at the diaphyseal side of the epiphyseal plate.

3. ***Zone of hypertrophic cartilage*** (hī-per-TRŌ-fik). This layer consists of large, maturing chondrocytes arranged in columns.

Figure 6.7 The epiphyseal (growth) plate is a layer of hyaline cartilage in the metaphysis of a growing bone. The epiphyseal plate appears as a dark band between whiter calcified areas in the radiograph shown in (a).

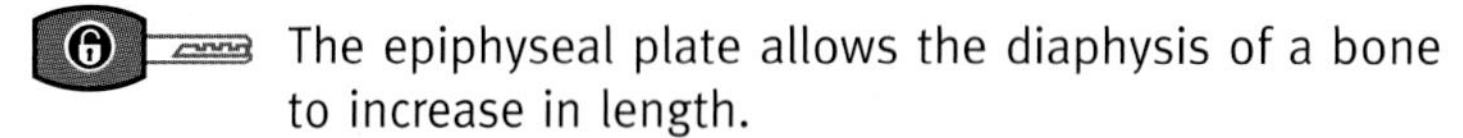

The epiphyseal plate allows the diaphysis of a bone to increase in length.

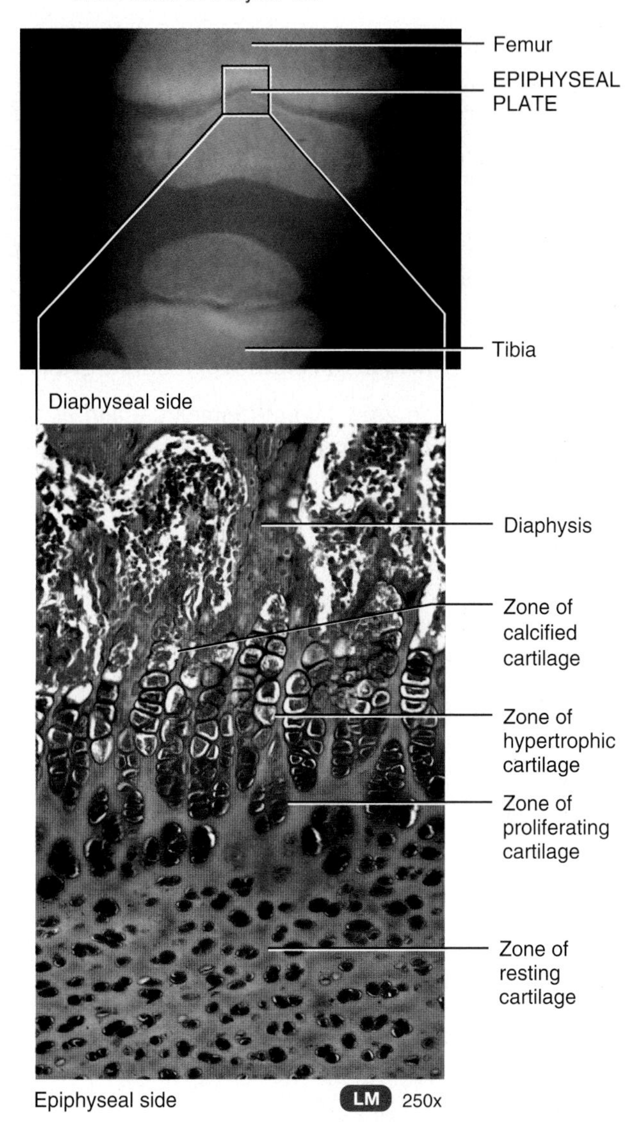

(b) Histology of the epiphyseal plate

What activities of the epiphyseal plate account for the lengthwise growth of the diaphysis?

4. ***Zone of calcified cartilage.*** The final zone of the epiphyseal plate is only a few cells thick and consists mostly of chondrocytes that are dead because the extracellular matrix around them has calcified. Osteoclasts dissolve the calcified cartilage, and osteoblasts and capillaries from the diaphysis invade the area. The osteoblasts lay down bone extracellular matrix, replacing the calcified cartilage. As a result, the zone of calcified cartilage becomes "new diaphysis" that is firmly cemented to the rest of the diaphysis of the bone.

The activity at the epiphyseal plate is the only way that the diaphysis can increase in length. As a bone grows, new chondrocytes are formed on the epiphyseal side of the plate, while old chondrocytes on the diaphyseal side of the plate are replaced by bone. In this way the thickness of the epiphyseal plate remains relatively constant, but the bone on the diaphyseal side increases in length.

At about age 18 in females and 21 in males, the epiphyseal plates close; the epiphyseal cartilage cells stop dividing, and bone replaces all the cartilage. The epiphyseal plate fades, leaving a bony structure called the ***epiphyseal line.*** The appearance of the epiphyseal line signifies that the bone has stopped growing in length. The clavicle is the last bone to stop growing. If a bone fracture damages the epiphyseal plate, the fractured bone may be shorter than normal once adult stature is reached. This is because damage to cartilage accelerates closure of the epiphyseal plate, thus inhibiting lengthwise growth of the bone.

Growth in Thickness

Unlike cartilage, which can thicken by both interstitial and appositional growth, bone can grow in thickness (diameter) only by ***appositional growth*** (Figure 6.8):

1. At the bone surface, cells in the periosteum differentiate into osteoblasts, which secrete collagen fibers and other organic molecules that form bone extracellular matrix. The os-

Figure 6.8 Bone growth in thickness: appositional growth.

Cartilage can grow by both interstitial and appositional growth, but bone can grow in diameter only by appositional growth.

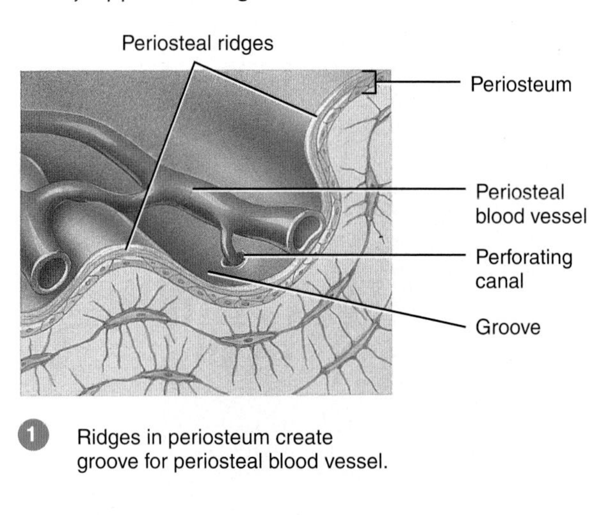

1 Ridges in periosteum create groove for periosteal blood vessel.

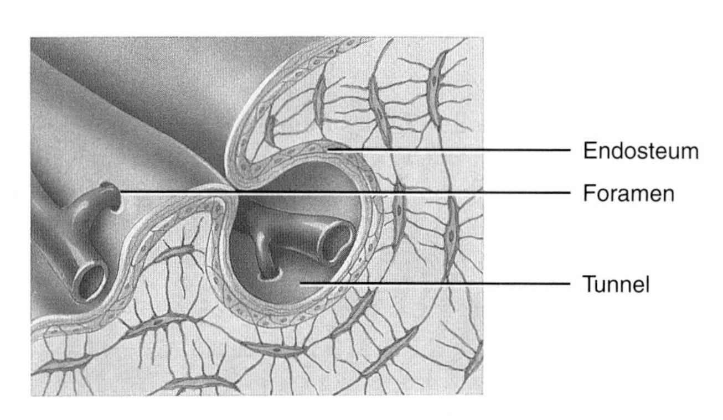

2 Periosteal ridges fuse, forming an endosteum-lined tunnel.

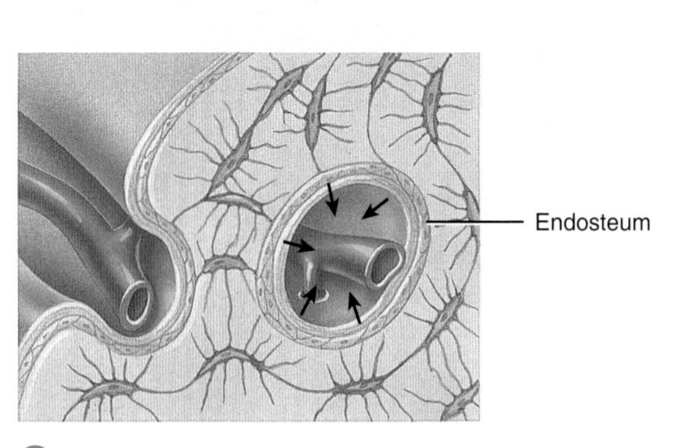

3 Osteoblasts in endosteum build new concentric lamellae inward toward center of tunnel, forming a new osteon.

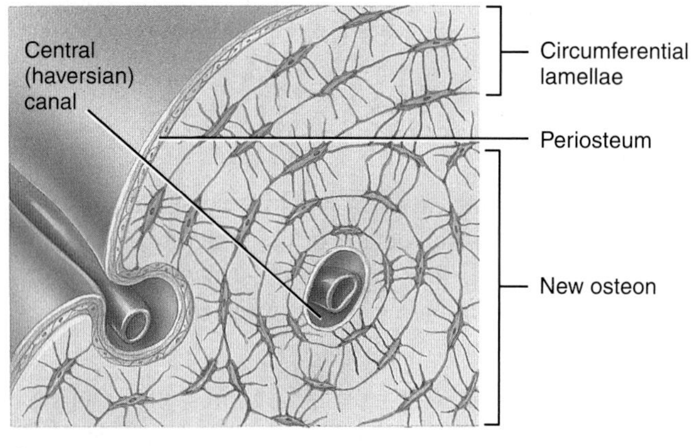

4 Bone grows outward as osteoblasts in periosteum build new circumferential lamellae. Osteon formation repeats as new periosteal ridges fold over blood vessels.

? How does the medullary cavity enlarge during growth in thickness?

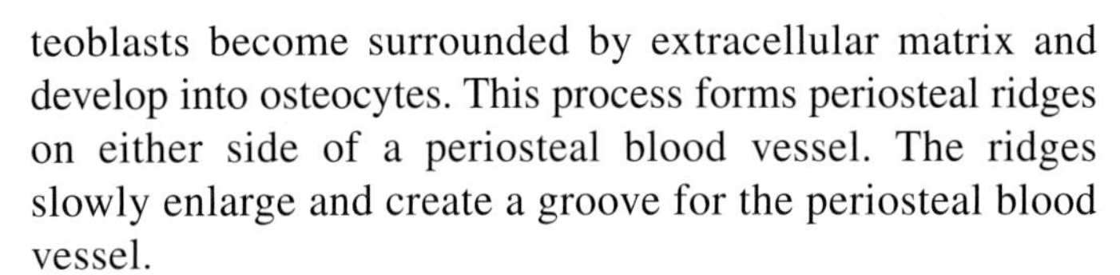

teoblasts become surrounded by extracellular matrix and develop into osteocytes. This process forms periosteal ridges on either side of a periosteal blood vessel. The ridges slowly enlarge and create a groove for the periosteal blood vessel.

2. Eventually, the ridges fold together and fuse, and the groove becomes a tunnel that encloses the blood vessel. The former periosteum now becomes the endosteum that lines the tunnel.
3. Osteoblasts in the endosteum deposit bone extracellular matrix, forming new concentric lamellae. The formation of additional concentric lamellae proceeds inward toward the periosteal blood vessel. In this way, the tunnel fills in, and a new osteon is created.
4. As an osteon is forming, osteoblasts under the periosteum deposit new outer circumferential lamellae, further increasing the thickness of the bone. As additional periosteal blood vessels become enclosed as in step 2, the growth process continues.

As new bone tissue is deposited on the outer surface of bone, the bone tissue lining the medullary cavity is destroyed by osteoclasts in the endosteum. In this way, the medullary cavity enlarges as the bone increases in thickness.

CHECKPOINT

10. Describe the zones of the epiphyseal plate and their functions.
11. Explain how bone grows in length and in thickness.
12. What is the significance of the epiphyseal line?

6.7 BONES AND HOMEOSTASIS

OBJECTIVES

- Describe the processes involved in bone remodeling.
- Describe the sequence of events in repair of a fracture.
- Describe the role of bone in calcium homeostasis.

Bone Remodeling

Like skin, bone forms before birth but continually renews itself thereafter. ***Bone remodeling*** is the ongoing replacement of old bone tissue by new bone tissue. It involves ***bone resorption,*** the removal of minerals and collagen fibers from bone by osteoclasts, and ***bone deposition,*** the addition of minerals and collagen fibers to bone by osteoblasts. Thus, bone resorption results in the destruction of bone extracellular matrix, whereas bone deposition results in the formation of bone extracellular matrix. At any given time, about 5% of the total bone mass in the body is being remodeled. The renewal rate for compact bone tissue is about 4% per year and for spongy bone tissue it is about 20% per year. Remodeling also takes place at different rates in different regions of the body. The distal portion of the thighbone (femur) is replaced about every four months. In contrast, bone in certain areas of the shaft of the femur will not be replaced completely during an individual's life. Even after bones have reached their adult shapes and sizes, old bone is continually destroyed and new bone is formed in its place. Remodeling also removes injured bone, replacing it with new bone tissue. Remodeling may be triggered by factors such as exercise, sedentary lifestyle, and changes in diet.

Remodeling has several other benefits. Because the strength of bone is related to the degree to which it is stressed, if newly formed bone is subjected to heavy loads, it will grow thicker and therefore be stronger than the old bone. Also, the shape of a bone can be altered for proper support based on the stress patterns experienced during the remodeling process. Finally, new bone is more resistant to fracture than old bone.

• CLINICAL CONNECTION | Remodeling and Orthodontics

Orthodontics (or-thō-DON-tiks) is the branch of dentistry concerned with the prevention and correction of poorly aligned teeth. The movement of teeth by braces places a stress on the bone that forms the sockets that anchor the teeth. In response to this artificial stress, osteoclasts and osteoblasts remodel the sockets so that the teeth align properly. •

During the process of bone resorption, an osteoclast attaches tightly to the bone surface at the endosteum or periosteum and forms a leakproof seal at the edges of its ruffled border (see Figure 6.2). Then it releases protein-digesting lysosomal enzymes and several acids into the sealed pocket. The enzymes digest collagen fibers and other organic substances while the acids dissolve the bone minerals. Working together, several osteoclasts carve out a small tunnel in the old bone. The degraded bone proteins and extracellular matrix minerals, mainly calcium and phosphorus, enter an osteoclast by endocytosis, cross the cell in vesicles, and undergo exocytosis on the side opposite the ruffled border. Now in the interstitial fluid, the products of bone resorption diffuse into nearby blood capillaries. Once a small area of bone has been resorbed, osteoclasts depart and osteoblasts move in to rebuild the bone in that area.

A delicate balance exists between the actions of osteoclasts and osteoblasts. Should too much new tissue be formed, the bones become abnormally thick and heavy. If too much mineral material is deposited in the bone, the surplus may form thick bumps, called *spurs,* on the bone that interfere with movement at joints. Excessive loss of calcium or tissue weakens the bones, and they may break, as occurs in osteoporosis, or they may become too flexible, as in rickets and osteomalacia. Abnormal acceleration of the remodeling process results in a condition called *Paget's disease*, in which the newly formed bone, especially that of the pelvis, limbs, lower vertebrae, and skull, becomes hard and brittle and fractures easily.

• CLINICAL CONNECTION | Rickets and Osteomalacia

Rickets and **osteomalacia** (os-tē-ō-ma-LĀ-shē-a; *malacia* = softness) are two forms of the same disease that result from inadequate calcification of the extracellular bone matrix, usually caused by a vitamin D dificiency. Rickets is a disease of children in which the growing bones become "soft" or rubbery and are easily deformed. Because new bone formed at the epiphyseal (growth) plates fails to ossify, bowed legs and deformities of the skull, rib cage, and pelvis are common. Osteomalacia is the adult counterpart of rickets, sometimes called *adult rickets*. New bone formed during remodeling fails to calcify, and the person experiences varying degrees of pain and tenderness in bones, especially the hip and legs. Bone fractures also result from minor trauma. Prevention and treatment for rickets and osteomalacia consists of the administration of adequate vitamin D. •

Factors Affecting Bone Growth and Bone Remodeling

Normal bone metabolism—growth in the young and bone remodeling in the adult—depends on several factors. These include adequate dietary intake of minerals and vitamins, as well as sufficient levels of several hormones.

1. ***Minerals.*** Large amounts of calcium and phosphorus are needed while bones are growing, as are smaller amounts of fluoride, magnesium, iron, and manganese. These minerals are also necessary during bone remodeling.

2. ***Vitamins.*** Vitamin C is needed for synthesis of collagen, the main bone protein, and also for differentiation of osteoblasts into osteocytes. Vitamins K and B_{12} also are needed for protein synthesis, whereas vitamin A stimulates activity of osteoblasts.

3. ***Hormones.*** During childhood, the hormones most important to bone growth are the insulin-like growth factors (IGFs), which are produced by the liver and bone tissue. IGFs stimulate osteoblasts, promote cell division at the epiphyseal plate and in the periosteum, and enhance synthesis of the proteins needed to build new bone. IGFs are produced in response to the secretion of human growth hormone (hGH) from the anterior lobe of the pituitary gland (see Section 20.3). Thyroid hormones (T_3 and T_4) from the thyroid gland also promote bone growth by stimulating osteoblasts.

At puberty, the secretion of hormones known as sex hormones causes a dramatic effect on bone growth. The ***sex hormones*** include estrogens (produced by the ovaries) and androgens such as testosterone (produced by the testes). Although females have much higher levels of estrogens and males have higher levels of androgens, females also have low levels of androgens and males have low levels of estrogens. The adrenal glands of both sexes produce androgens, and other tissues, such as adipose tissue, can convert androgens to estrogens. These hormones are responsible for increased osteoblast activity and synthesis of bone extracellular matrix and the sudden "growth spurt" that occurs during the teenage years. Estrogens also promote changes in the skeleton that are typical of females, such as widening of the pelvis. Ultimately sex hormones, especially estrogens in both sexes, shut down growth at epiphyseal plates, causing elongation of the bones to cease. Lengthwise growth of bones typically ends earlier in females than in males due to their higher levels of estrogens.

During adulthood, sex hormones contribute to bone remodeling by slowing resorption of old bone and promoting deposition of new bone. One way that estrogens slow resorption is by promoting apoptosis (programmed death) of osteoclasts. As you will see shortly, parathyroid hormone, calcitriol (the active form of vitamin D), and calcitonin are other hormones that can affect bone remodeling.

4. ***Weight-bearing exercises.*** Moderate-weighted exercises maintain sufficient stress on the bones to increase and maintain bone density.

• CLINICAL CONNECTION | Hormonal Abnormalities That Affect Height

Hormonal abnormalities can affect bone growth and can cause a person to be abnormally tall or short (see Figure 20.3 in Section 20.3). Oversecretion of hGH during childhood produces ***giantism***, in which a person becomes much taller and heavier than normal. Undersecretion of hGH produces ***pituitary dwarfism*** (pi-TOO-i-tār-ē), in which a person has short stature. (A dwarf has a normal-sized head and torso but small limbs; a midget has a proportioned head, torso, and limbs.) Because estrogens terminate growth at the epiphyseal (growth) plates, both men and women who lack estrogens or receptors for estrogens grow taller than normal. Oversecretion of hGH during adulthood is called ***acromegaly*** (ak′-rō-MEG-a-lē). Although hGH cannot produce further lengthening of the long bones because the epiphyseal (growth) plates are already closed, the bones of the hands, feet, and jaws thicken and other tissues enlarge. In addition, the eyelids, lips, tongue, and nose enlarge, and the skin thickens and develops furrows, especially on the forehead and soles. hGH has been used to induce growth in short-statured children. •

Fracture and Repair of Bone

A ***fracture*** (FRAK-choor) is any break in a bone. Fractures are named according to their severity, the shape or position of the fracture line, or even the physician who first described them. Some fractures are described by combining several of the terms below. Among the common types of fractures are the following (Figure 6.9):

- **Open (compound) fracture:** The broken ends of the bone protrude through the skin (Figure 6.9a).
- **Closed (simple) fracture:** Does not break the skin.
- **Comminuted fracture** (KOM-i-noo-ted; *com-* = together; -*minuted* = crumbled): The bone splinters at the site of impact, and smaller bone fragments lie between the two main fragments (Figure 6.9b). This is the most difficult fracture to treat.
- **Greenstick fracture:** A partial fracture in which one side of the bone is broken and the other side bends; occurs only in

Figure 6.9 Types of bone fractures. Illustrations are shown on the left and radiographs are shown on the right.

A fracture is any break in a bone.

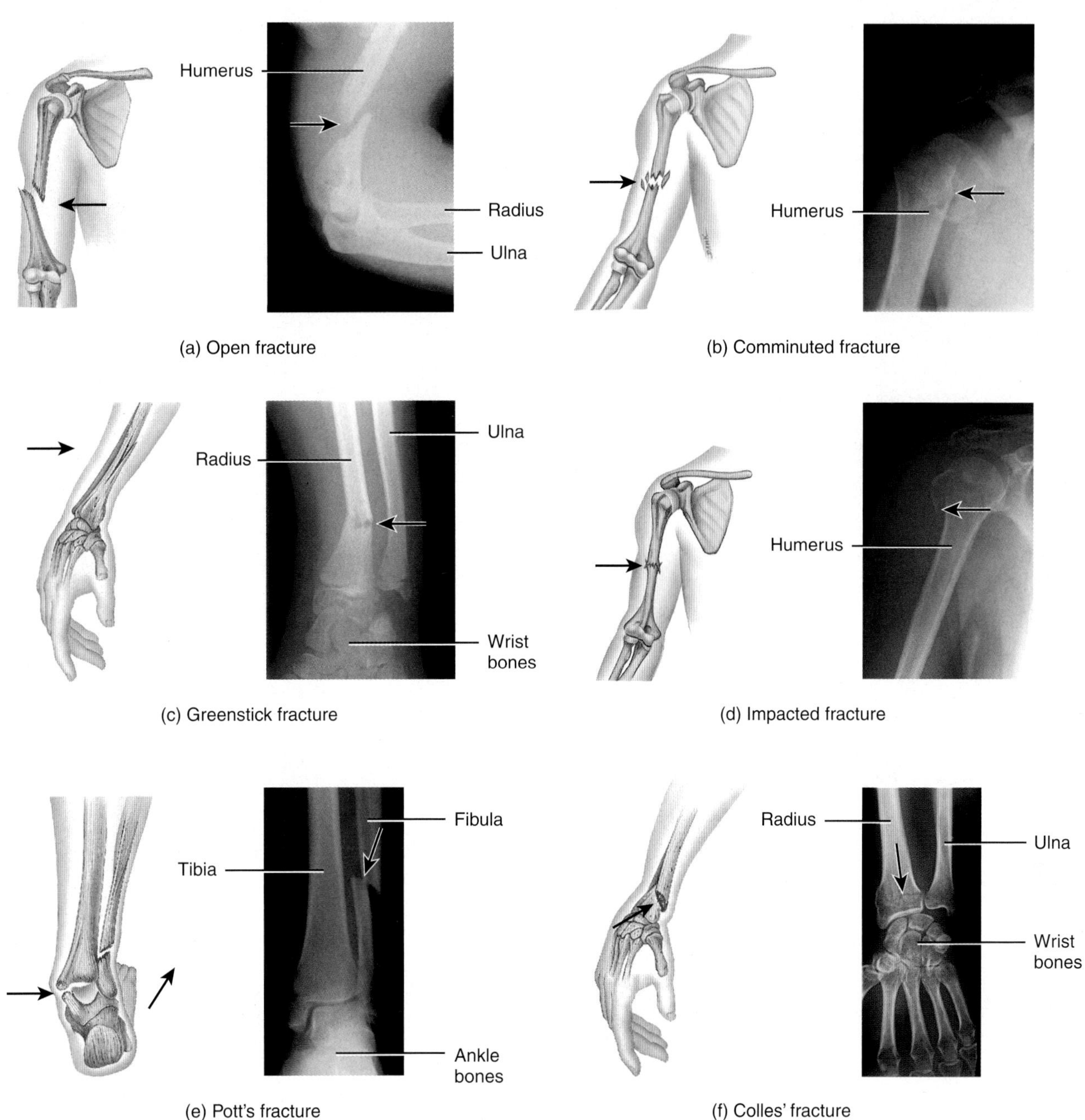

What is the difference between an open fracture and a closed fracture?

children, whose bones are not yet fully ossified and contain more organic material than inorganic material (Figure 6.9c).

- **Impacted fracture:** One end of the fractured bone is forcefully driven into the interior of the other (Figure 6.9d).
- **Pott's fracture:** A fracture of the distal end of the lateral leg bone (fibula), with serious injury of the distal tibial articulation (Figure 6.9e).
- **Colles' fracture** (KOL-ez): A fracture of the distal end of the lateral forearm bone (radius) in which the distal fragment is displaced posteriorly (Figure 6.9f).

In some cases, a bone may fracture without visibly breaking. A ***stress fracture*** is a series of microscopic fissures in bone that forms without any evidence of injury to other tissues. In healthy adults, stress fractures result from repeated, strenuous activities such as running, jumping, or aerobic dancing. Stress fractures also result from disease processes that disrupt normal bone calcification, such as osteoporosis (discussed in Section 6.8). About 25% of stress fractures involve the tibia. Although standard x-ray images often fail to reveal the presence of stress fractures, they show up clearly in a bone scan.

The repair of a bone fracture involves the following steps (Figure 6.10):

1. ***Formation of fracture hematoma.*** Blood vessels crossing the fracture line are broken. As blood leaks from the torn ends of the vessels, it forms a clot around the site of the

Figure 6.10 Steps in repair of a bone fracture.

Bone heals more rapidly than cartilage because its blood supply is more plentiful.

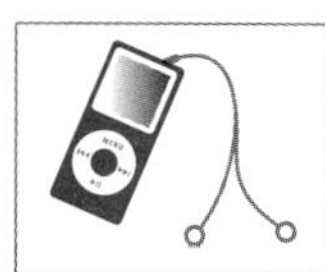

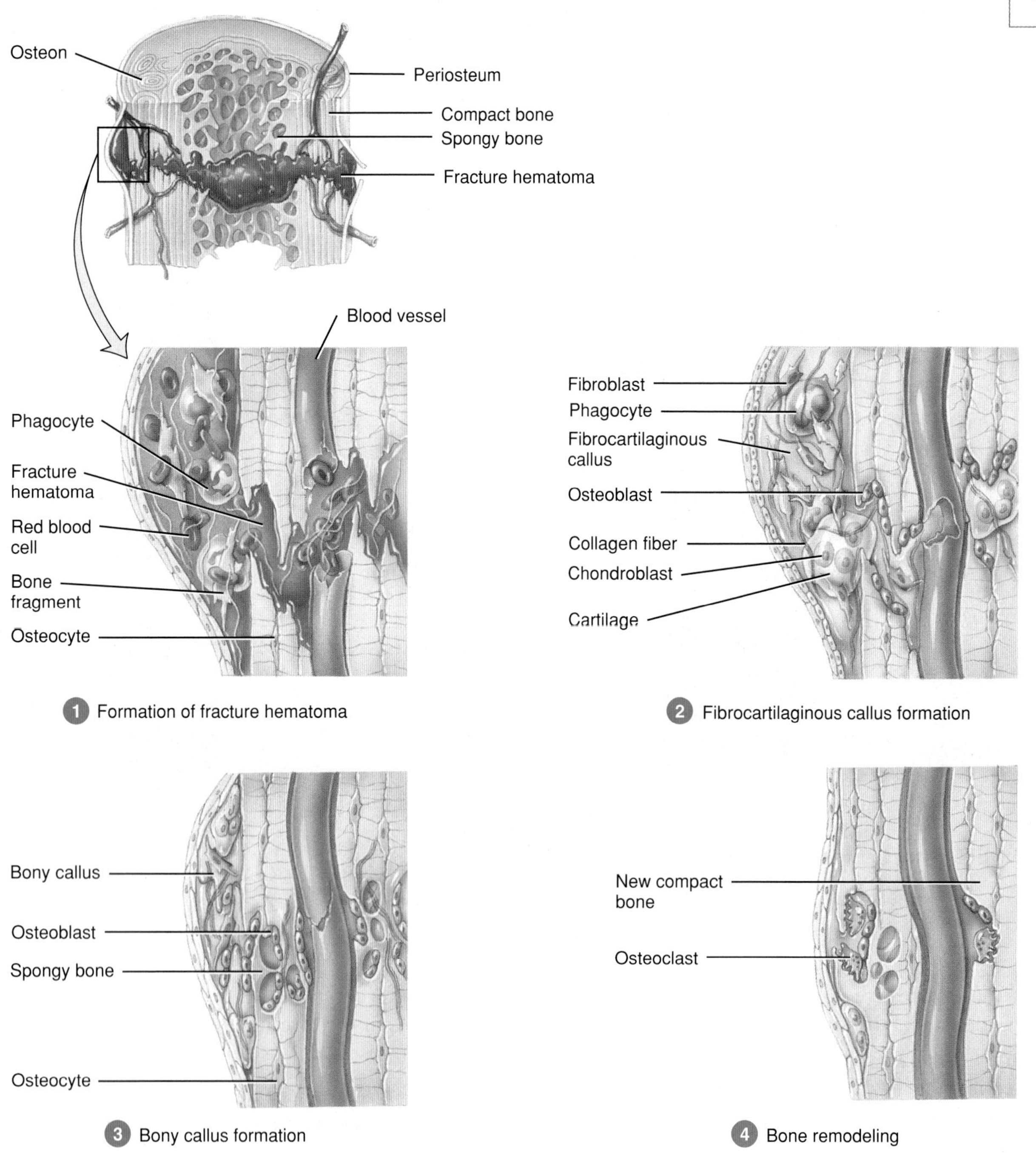

Why does it sometimes take months for a fracture to heal?

fracture. This clot, called a ***fracture hematoma*** (hē′-ma-TŌ-ma; *hemat-* = blood; *-oma* = tumor), usually forms 6 to 8 hours after the injury. Circulation of blood stops where the fracture hematoma forms and nearby bone cells die. Swelling and inflammation occur in response to dead bone cells, producing additional cellular debris. Phagocytes (neutrophils and macrophages) and osteoclasts begin to remove the dead or damaged tissue in and around the fracture hematoma. This stage may last up to several weeks.

2. ***Fibrocartilaginous callus formation.*** Fibroblasts from the periosteum invade the fracture site and produce collagen fibers. In addition, cells from the periosteum develop into chondroblasts and begin to produce fibrocartilage in this region. These events lead to the development of a ***fibrocartilaginous callus*** (fī-brō-kar-ti-LAJ-i-nus), a mass of repair tissue consisting of collagen fibers and cartilage that bridges the broken ends of the bone. Formation of the fibrocartilaginous callus takes about 3 weeks.
3. ***Bony callus formation.*** In areas closer to well-vascularized healthy bone tissue, osteogenic cells develop into osteoblasts, which begin to produce spongy bone trabeculae. The trabeculae join living and dead portions of the original bone fragments. In time, the fibrocartilage is converted to spongy bone, and the callus is then referred to as a ***bony callus.*** The bony callus lasts about 3 to 4 months.
4. ***Bone remodeling.*** The final phase of fracture repair is ***bone remodeling*** of the callus. Dead portions of the original fragments of broken bone are gradually resorbed by osteoclasts. Compact bone replaces spongy bone around the periphery of the fracture. Sometimes, the repair process is so thorough that the fracture line is undetectable, even in a radiograph (x-ray). However, a thickened area on the surface of the bone remains as evidence of a healed fracture, and eventually a healed bone may be stronger than it was before the break.

Although bone has a generous blood supply, healing sometimes takes months. The calcium and phosphorus needed to strengthen and harden new bone are deposited only gradually, and bone cells generally grow and reproduce slowly. The temporary disruption in their blood supply also helps explain the slowness of healing of severely fractured bones.

MANUAL THERAPY APPLICATION

Treatments for Fractures

Treatments for fractures vary according to age, type of fracture, and the bone involved. The ultimate goals of fracture treatment are *realignment* of bone fragments, *immobilization* to maintain realignment, and *restoration* of function. For bones to unite properly, fractured ends must be brought into alignment, a process called *reduction.* In *closed reduction,* fractured ends of a bone are brought into alignment by manual manipulation, and the skin remains intact. In *open reduction,* fractured ends of a bone are brought into alignment by a surgical procedure in which internal fixation devices such as screws, plates, pins, rods, and wires are used. Following reduction, a fractured bone may be kept immobilized by a cast, sling, splint, elastic bandage, external fixation device, or a combination of these devices. Bone regeneration may be increased by appropriate electrical *stimulation* and exercise as the fracture heals. Derivative massage, massage of a specific part of the body to indirectly influence a related body part, will increase circulation and decrease the healing time of a fractured bone.

Bone's Role in Calcium Homeostasis

Bone is the body's major calcium reservoir, storing 99% of total body calcium. One way to maintain the level of calcium in the blood is to control the rates of calcium resorption from bone into blood and of calcium deposition from blood into bone. Both nerve and muscle cells depend on a stable level of calcium ions (Ca^{2+}) in extracellular fluid to function properly. Blood clotting also requires Ca^{2+}. Also, many enzymes require Ca^{2+} as a cofactor (an additional substance needed for an enzymatic reaction to occur). For this reason, the blood plasma level of Ca^{2+} is very closely regulated between 9 and 11 mg/100 mL. Even small changes in Ca^{2+} concentration outside this range may prove fatal—the heart may stop (cardiac arrest) if the concentration goes too high, or breathing may cease (respiratory arrest) if the level falls too low. The role of bone in calcium homeostasis is to help "buffer" the blood Ca^{2+} level, releasing Ca^{2+} into blood plasma (using osteoclasts) when the level decreases, and absorbing Ca^{2+} (using osteoblasts) when the level rises.

Exchange of Ca^{2+} is regulated by hormones, the most important of which is ***parathyroid hormone (PTH)*** secreted by the parathyroid glands. This hormone increases blood Ca^{2+} level. PTH secretion operates via a negative feedback system (see Figure 6.11). If some stimulus causes the blood Ca^{2+} level to decrease, parathyroid gland cells (receptors) detect this change and increase their production of a molecule known as cyclic adenosine monophosphate (cyclic AMP). The gene for PTH within the nucleus of a parathyroid gland cell (the control center) detects the intracellular increase in cyclic AMP (the input). As a result, PTH synthesis speeds up, and more PTH (the output) is released into the blood. The presence of higher levels of PTH increases the number and activity of osteoclasts (effectors), which step up the pace of bone resorption. The resulting release of Ca^{2+} from bone into blood returns the blood Ca^{2+} level to normal.

PTH also acts on the kidneys (effectors) to decrease loss of Ca^{2+} in the urine, so more is retained in the blood. PTH stimulates formation of ***calcitriol*** (the active form of vitamin D), a hormone that promotes absorption of calcium from foods in the gastrointestinal tract into the blood. Both of these actions also help elevate blood Ca^{2+} level.

Another hormone works to decrease blood Ca^{2+} level. When blood Ca^{2+} rises above normal, *parafollicular cells* in the thyroid gland secrete ***calcitonin (CT)*** (kal-si-TŌ-nin). CT inhibits activity

Figure 6.11 Negative feedback system for the regulation of blood calcium (Ca^{2+}) concentration. PTH = parathyroid hormone.

Release of calcium from bone matrix and retention of calcium by the kidneys are the two main ways that blood calcium level can be increased.

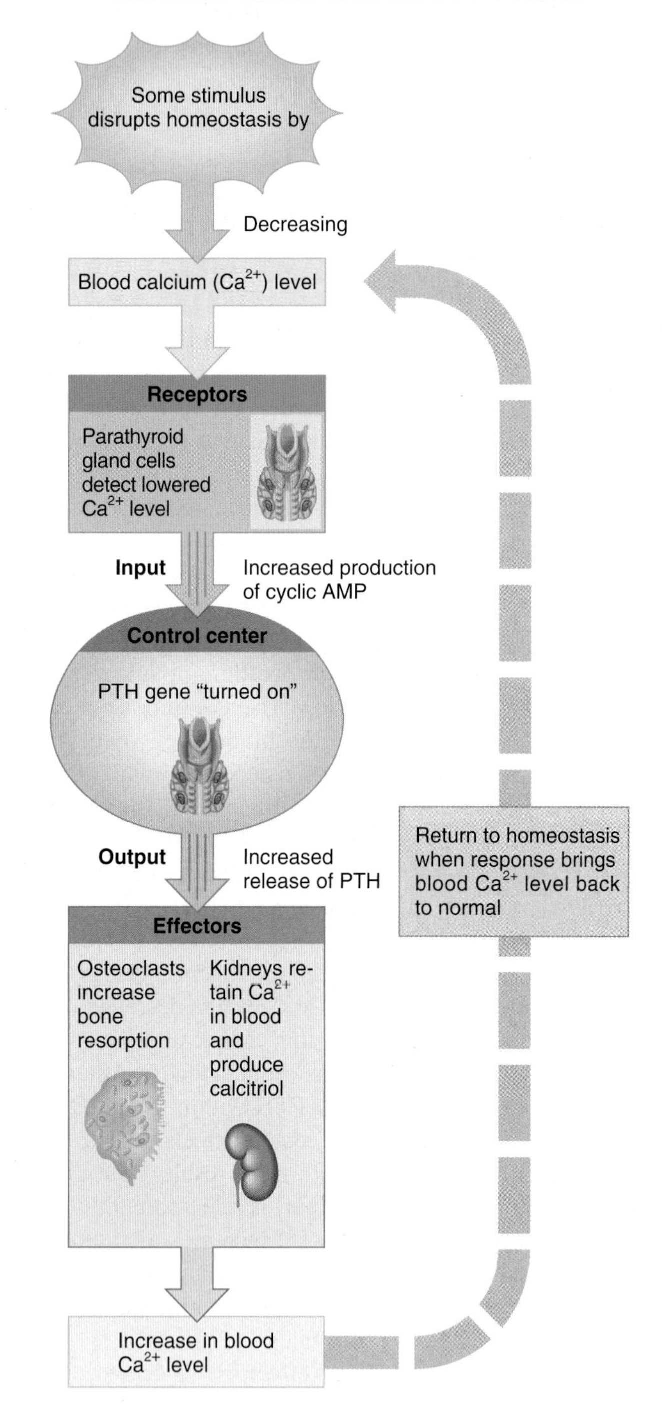

? What body functions depend on proper levels of Ca^{2+}?

of osteoclasts, speeds blood Ca^{2+} uptake by bone, and accelerates Ca^{2+} deposition into bones. The net result is that CT promotes bone formation and decreases blood Ca^{2+} level. Despite these effects, the role of CT in normal calcium homeostasis is uncertain because it can be completely absent without causing symptoms. Nevertheless, calcitonin harvested from salmon (Miacalcin®) is an effective drug for treating osteoporosis because it slows bone resorption.

Figure 20.10 in Section 20.5 summarizes the roles of parathyroid hormone, calcitriol, and calcitonin in regulation of blood Ca^{2+} level.

MANUAL THERAPY APPLICATION

Effects of Exercise on Bone

The **effects of exercise on bone** can be seen in various ways. Within limits, bone tissue has the ability to alter its strength in response to changes in *mechanical stress*. When placed under mechanical stress, bone tissue becomes stronger through increased deposition of mineral salts and production of collagen fibers by osteoblasts. Without mechanical stress, bone does not remodel normally because bone resorption occurs more quickly than bone formation.

The main mechanical stresses on bone are those that result from the *pull of skeletal muscles* and the *pull of gravity*. If a person is bedridden or has a fractured bone in a cast, the strength of the unstressed bones diminishes because of the loss of bone minerals and decreased numbers of collagen fibers. Astronauts subjected to the microgravity of space also lose bone mass. In both cases, bone loss can be dramatic—as much as 1% per week. Bones of athletes, which are repetitively and highly stressed, become notably thicker and stronger than those of nonathletes. Weight-bearing activities, such as walking or moderate weight lifting, help build and retain bone mass. Adolescents and young adults should engage in regular weight-bearing exercise to help build total mass prior to its inevitable reduction with aging. Even elderly people can strengthen their bones by engaging in weight-bearing exercise.

Therapists will adjust their procedures to account for decreased density of bone in certain patients. Deep procedures would be contraindicated in patients taking certain drugs, including chemotherapy treatments, as well as in the very young and the very old.

CHECKPOINT

13. Define bone remodeling, and describe the roles of osteoblasts and osteoclasts in the process.
14. Define a fracture and outline the four steps involved in fracture repair.
15. What are some of the important functions of calcium in the body?

6.8 AGING AND BONE TISSUE

OBJECTIVE

- Describe the effects of aging on bone tissue.

From birth through adolescence, more bone tissue is produced than is lost during bone remodeling. In young adults the rates of bone deposition and resorption are about the same. As the level of sex hormones diminishes during middle age, especially in women after menopause, a decrease in bone mass occurs because bone resorption by osteoclasts outpaces bone deposition by osteoblasts. In old age, loss of bone through resorption occurs more rapidly than bone gain. Because women's bones generally are smaller and less massive than men's bones to begin with, loss of bone mass in old age typically has a greater adverse effect in females. These factors contribute to the higher incidence of osteoporosis in females.

There are two principal effects of aging on bone tissue: loss of bone mass and brittleness. Loss of bone mass results from ***demineralization*** (dē-min′-er-al-i-ZĀ-shun) that includes the loss of calcium and other minerals from bone extracellular matrix. This loss usually begins after age 30 in females, accelerates greatly around age 45 as levels of estrogens decrease, and continues until as much as 30% of the calcium in bones is lost by age 70. Once bone loss begins in females, about 8% of bone mass is lost every 10 years. In males, calcium loss typically does not begin until after age 60, and about 3% of bone mass is lost every 10 years. The loss of calcium from bones is one of the problems in osteoporosis (described shortly).

The second principal effect of aging on the skeletal system, ***brittleness,*** results from a decreased rate of protein synthesis. Recall that the organic part of bone extracellular matrix, mainly collagen fibers, gives bone its tensile strength. The loss of tensile strength causes the bones to become very brittle and susceptible to fracture. In some elderly people, collagen fiber synthesis slows, in part, due to diminished production of human growth hormone. In addition to increasing the susceptibility to fractures, loss of bone mass also leads to deformity, pain, loss of height, and loss of teeth.

Figure 6.12 **Comparison of spongy bone tissue from (a) a normal young adult and (b) a person with osteoporosis.** Notice the weakened trabeculae in (b). Compact bone tissue is similarly affected by osteoporosis.

In osteoporosis, bone resorption outpaces bone deposition, so bone mass decreases.

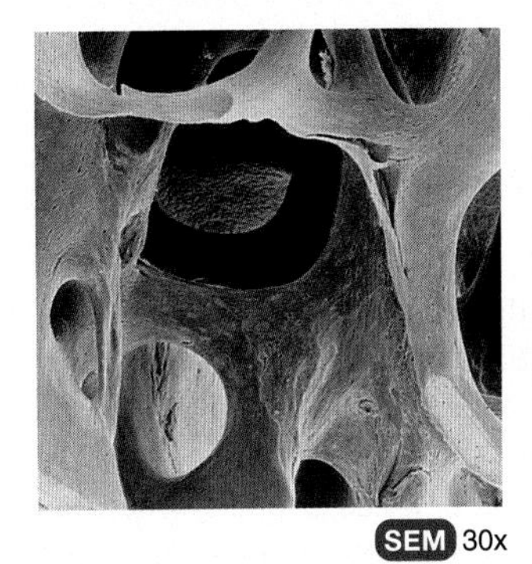

(a) Normal bone

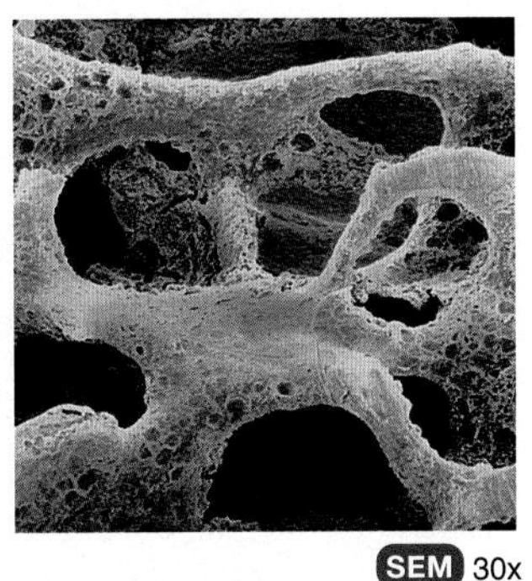

(b) Osteoporotic bone

If you wanted to develop a drug to lessen the effects of osteoporosis, would you look for a chemical that inhibits the activity of osteoblasts or that of osteoclasts?

• CLINICAL CONNECTION Osteoporosis

Osteoporosis (os′-tē-ō-pō-RŌ-sis; *por-* = passageway; *-osis* = condition) is literally a condition of porous bones (Figure 6.12) that affects 10 million people in the United States each year. In addition, about 18 million people have *osteopenia* (low bone mass), which puts them at risk for osteoporosis. The basic problem is that bone resorption outpaces bone deposition. In large part this is due to depletion of calcium from the body—more calcium is lost in urine, feces, and sweat than is absorbed from the diet. Bone mass becomes so depleted that bones fracture, often spontaneously, under the mechanical stresses of everyday living. For example, a hip fracture might result from simply sitting down too quickly. In the United States, osteoporosis results in more than one and a half million fractures a year, mainly in the hip, wrist, and vertebrae. Osteoporosis afflicts the entire skeletal system. In addition to fractures, osteoporosis causes shrinkage of vertebrae, height loss, hunched backs, and bone pain.

The disorder primarily affects middle-aged and elderly people, 80% of them women. Older women suffer from osteoporosis more often than men for two reasons: (1) Women's bones are less massive than men's bones, and (2) production of estrogens in women declines dramatically at menopause, but production of the main androgen, testosterone, wanes gradually and only slightly in older men. Estrogens and testosterone stimulate osteoblast activity and synthesis of bone extracellular matrix. Besides gender, risk factors for developing osteoporosis include a family history of the disease, European or Asian ancestry, thin or small body build, an inactive lifestyle, cigarette smoking, a diet low in calcium and vitamin D, more than two alcoholic drinks a day, and the use of certain medications.

Osteoporosis is diagnosed by taking a family history and undergoing a *bone mineral density* (BMD) test. BMD tests are performed like x-rays and measure bone density. They can also be used to confirm a diagnosis of osteoporosis, determine the rate of bone loss, and monitor the effects of treatment.

Treatment options for osteoporosis are varied. With regard to nutrition, a diet high in calcium is important to reduce the risk of fractures. Vitamin D is necessary for the body to utilize calcium. In terms of exercise, regularly performing weight-bearing exercises has been shown to maintain and build bone mass. Among these exercises are walking, jogging, hiking, climbing stairs, playing tennis, and dancing. Resistance exercises, such as weight lifting, build bone strength and muscle mass.

Medications used to treat osteoporosis are generally of two types: (1) *Antiresorptive drugs* that slow down the progression of bone loss and (2) *bone-building drugs* that promote increasing bone mass. Among the antiresorptive drugs are (1) *bisphosphonates*, which inhibit osteoclasts (Fosamax®, Actonel®, Boniva®, and calcitonin); (2) selective estrogen receptor modulators, which mimic the effects of estrogens without unwanted side effects (Raloxifene®, Evista®); and (3) estrogen replacement therapy (ERT), which replaces estrogens lost during and after menopause (Premarin®), and hormone replacement therapy (HRT), which replaces estrogens and progesterone lost during and after menopause (Prempro®). HRT also helps maintain and increase bone mass, and women on HRT have an increased risk of heart disease, breast cancer, stroke, blood clots, and dementia. Among the bone-forming drugs is parathyroid hormone (PTH), which stimulates osteoblasts to produce new bone (Fortes®). Others are under development. •

CHECKPOINT

16. What is demineralization, and how does it affect the functioning of bone?
17. What changes occur in the organic part of bone extracellular matrix with aging?

KEY MEDICAL TERMS ASSOCIATED WITH BONE TISSUE

Osteogenic sarcoma (os′-tē-ō-JEN-ik sar-KŌ-ma; *sarcoma* = connective tissue tumor) Bone cancer that primarily affects osteoblasts and occurs most often in teenagers during their growth spurt; the most common sites are the metaphyses of the thighbone (femur), shinbone (tibia), and arm bone (humerus). Metastases occur most often in lungs; treatment consists of multidrug chemotherapy and removal of the malignant growth, or amputation of the limb.

Osteomyelitis (os′-tē-ō-mī-e-LĪ-tis) An infection of bone characterized by high fever, sweating, chills, pain, nausea, pus formation, edema, and warmth over the affected bone and rigid overlying muscles. It is often caused by bacteria, usually *Staphylococcus aureus*. The bacteria may reach the bone from outside the body (through open fractures, penetrating wounds, or orthopedic surgical procedures); from other sites of infection in the body (abscessed teeth, burn infections, urinary tract infections, or upper respiratory infections) via the blood; and from adjacent soft tissue infections (as occurs in diabetes mellitus). The patient would be under medical supervision if this condition were present. The therapist may be required to wear masks and/or gloves if the patient is immunosuppressed or wait until the infection has abated before performing treatment.

Osteopenia (os′-tē-ō-PĒ-nē-a; *penia* = poverty) Reduced bone mass due to a decrease in the rate of bone synthesis to a level too low to compensate for normal bone resorption; any decrease in bone mass below normal. An example is osteoporosis.

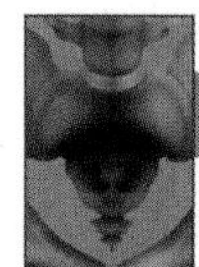

STUDY OUTLINE

Functions of Bone and the Skeletal System (Section 6.1)

1. A bone is made up of several different tissues: bone or osseous tissue, cartilage, dense connective tissues, epithelium, adipose tissue, and nervous tissue.
2. The entire framework of bones and their cartilages constitutes the skeletal system.
3. The skeletal system functions in support, protection, movement, mineral homeostasis, blood cell production, and triglyceride storage.

Structure of Bone (Section 6.2)

1. Parts of a typical long bone are the diaphysis (shaft), proximal and distal epiphyses (ends), metaphyses, articular cartilage, periosteum, medullary (marrow) cavity, and endosteum.

Histology of Bone Tissue (Section 6.3)

1. Bone tissue consists of widely separated cells surrounded by large amounts of extracellular matrix.
2. The four principal types of cells in bone tissue are osteogenic cells, osteoblasts (bone-building cells), osteocytes (maintain daily activity of bone), and osteoclasts (bone-destroying cells).
3. The extracellular matrix of bone contains abundant mineral salts (mostly hydroxyapatite) and collagen fibers.
4. Compact bone tissue consists of osteons (haversian systems) with little space between them.
5. Compact bone tissue lies over spongy bone tissue in the epiphyses and makes up most of the bone tissue of the diaphysis. Functionally, compact bone tissue is the strongest form of bone and protects, supports, and resists stress.
6. Spongy bone tissue does not contain osteons. It consists of trabeculae surrounding many red bone marrow–filled spaces.
7. Spongy bone tissue forms most of the structure of short, flat, and irregular bones, and the interior of the epiphyses in long bones. Functionally, spongy bone tissue trabeculae offer resistance along lines of stress, support and protect red bone marrow, and make bones lighter for easier movement.

Blood and Nerve Supply of Bone (Section 6.4)

1. Long bones are supplied by periosteal, nutrient, metaphyseal and epiphyseal arteries; veins accompany the arteries.
2. Nerves accompany blood vessels in bone; the periosteum is rich in sensory neurons.

Bone Formation (Section 6.5)

1. Bone forms by a process called ossification (osteogenesis), which begins when mesenchymal cells become transformed into osteogenic cells. These undergo cell division and give rise to cells that differentiate into osteoblasts, osteoclasts, and osteocytes.
2. Ossification begins during the sixth week of embryonic life. The two types of ossification, intramembranous and endochondral, involve the replacement of a preexisting connective tissue with bone.
3. Intramembranous ossification refers to bone formation directly within mesenchymal tissue arranged in sheetlike layers that resemble membranes.

4. Endochondral ossification refers to bone formation within hyaline cartilage that develops from mesenchyme. The primary ossification center of a long bone is in the diaphysis. Cartilage degenerates, leaving cavities that merge to form the medullary cavity. Osteoblasts lay down bone. Next, ossification occurs in the epiphyses, where bone replaces cartilage, except for the epiphyseal plate.

Bone Growth (Section 6.6)

1. The epiphyseal (growth) plate consists of four zones: zone of resting cartilage, zone of proliferating cartilage, zone of hypertrophic cartilage, and zone of calcified cartilage.
2. Because of the cell division in the epiphyseal plate, the diaphysis of a bone increases in length during interstitial growth.
3. Bone grows in thickness or diameter due to the addition of new bone tissue by periosteal osteoblasts around the outer surface of the bone (appositional growth).

Bones and Homeostasis (Section 6.7)

1. Bone remodeling is an ongoing process in which osteoclasts carve out small tunnels in old bone tissue and then osteoblasts rebuild it.
2. In bone resorption, osteoclasts release enzymes and acids that degrade collagen fibers and dissolve mineral salts.
3. Dietary minerals (especially calcium and phosphorus) and vitamins (C, K, and B_{12}) are needed for bone growth and maintenance. Insulin-like growth factors (IGFs), human growth hormone, thyroid hormones, estrogens, and androgens stimulate bone growth.
4. Sex hormones slow resorption of old bone and promote new bone deposition.
5. A fracture is any break in a bone.
6. Fracture repair involves formation of a fracture hematoma, a fibrocartilaginous callus, and a bony callus, and bone remodeling.
7. Types of fractures include closed (simple), open (compound), comminuted, greenstick, impacted, stress, Pott's, and Colles'.
8. Bone is the major reservoir for calcium in the body.
9. Parathyroid hormone (PTH) secreted by the parathyroid gland increases blood Ca^{2+} level, whereas calcitonin (CT) from the thyroid gland has the potential to decrease blood Ca^{2+} level. Vitamin D (in its active form calcitriol) enhances absorption of calcium and phosphate and thus raises the blood levels of these substances.

Aging and Bone Tissue (Section 6.8)

1. The principal effect of aging is demineralization, a loss of calcium from bones, which is due to reduced osteoblast activity.
2. Another effect is decreased production of extracellular matrix proteins (mostly collagen fibers), which makes bones more brittle and thus more susceptible to fracture.

SELF-QUIZ QUESTIONS

Fill in the blanks in the following statements.

1. Bone growth in length is called ________ growth, and bone growth in diameter (thickness) is called ________ growth.
2. The crystallized inorganic mineral salts in bone contribute to bone's ________, whereas the collagen fibers and other organic molecules provide bone with ________.

Indicate whether the following statements are true or false.

3. Bone resorption involves increased activity of osteoclasts.
4. The formation of bone from cartilage is known as endochondral ossification.
5. The growth of bone is controlled primarily by hormones.

Choose the one best answer to the following questions.

6. Place in order the steps involved in intramembranous ossification. (1) Bony matrices fuse to form trabeculae. (2) Clusters of osteoblasts form a center of ossification that secretes the organic extracellular matrix. (3) Spongy bone is replaced with compact bone on the bone's surface. (4) Periosteum develops on the bone's periphery. (5) The extracellular matrix hardens by deposition of calcium and mineral salts.
   a. 2, 4, 5, 1, 3 **b.** 4, 3, 5, 1, 2 **c.** 1, 2, 5, 4, 3 **d.** 2, 5, 1, 4, 3 **e.** 5, 1, 3, 4, 2

7. Place in order the steps involved in endochondral ossification. (1) Nutrient artery invades the perichondrium. (2) Osteoclasts create a marrow cavity. (3) Chondrocytes enlarge and calcify. (4) Secondary ossification centers appear at epiphyses. (5) Osteoblasts become active in the primary ossification center.
   a. 3, 1, 5, 2, 4 **b.** 3, 1, 5, 4, 2 **c.** 1, 3, 5, 2, 4 **d.** 1, 2, 3, 5, 4 **e.** 2, 5, 4, 3, 1

8. Spongy bone differs from compact bone because spongy bone
   a. is composed of numerous osteons (haversian systems)
   b. is found primarily in the diaphyses of long bones, and compact bone is found primarily in the epiphyses of long bones
   c. contains osteons all aligned in the same direction along lines of stress
   d. does not contain osteocytes contained in lacunae
   e. is composed of trabeculae that are oriented along lines of stress

9. A primary effect that weight-bearing exercise has on bones is to
   a. provide oxygen for bone development
   b. increase the demineralization of bone;
   c. maintain and increase bone mass
   d. stimulate the release of sex hormones for bone growth
   e. utilize the stored triglycerides from the yellow bone marrow.

10. Place in order the steps involved in the repair of a bone fracture. (1) Osteoblast production of trabeculae and bony callus formation; (2) formation of a hematoma at the site of fracture; (3) resorption of remaining bone fragments and remodeling of bone; (4) migration of fibroblasts to the fracture site; (5) bridging of broken ends of bones by a fibrocartilaginous callus.

a. 2, 4, 5, 1, 3 **b.** 2, 5, 4, 1, 3 **c.** 1, 2, 5, 4, 3
d. 2, 5, 1, 3, 4 **e.** 5, 2, 4, 1, 3

11. Match the following:

____ **a.** space within the shaft of the bone that contains yellow bone marrow
____ **b.** triglyceride storage tissue
____ **c.** hemopoietic tissue
____ **d.** thin layer of hyaline cartilage covering the ends of bones where they form a joint
____ **e.** distal and proximal ends of bones
____ **f.** the long, cylindrical main portion of the bone; the shaft
____ **g.** in a growing bone, the region that contains the epiphyseal plate
____ **h.** the tough membrane that surrounds the bone surface wherever cartilage is not present
____ **i.** a layer of hyaline cartilage in the area between the shaft and end of a growing bone
____ **j.** membrane lining the medullary cavity
____ **k.** a remnant of the active epiphyseal plate; a sign that the bone has stopped growing in length
____ **l.** bundles of collagen fibers that attach periosteum to bone

A. articular cartilage
B. endosteum
C. medullary cavity
D. diaphysis
E. epiphyses
F. metaphysis
G. periosteum
H. red bone marrow
I. yellow bone marrow
J. perforating (Sharpey's) fibers
K. epiphyseal line
L. epiphyseal plate

12. Match the following:

____ **a.** decreases blood calcium levels by accelerating calcium deposition in bones and inhibiting osteoclasts
____ **b.** required for collagen synthesis
____ **c.** during childhood, promote growth at epiphyseal plate; production stimulated by human growth hormone
____ **d.** involved in bone growth by increasing osteoblast activity; cause long bones to stop growing in length
____ **e.** required for protein synthesis
____ **f.** active form of vitamin D; raises blood calcium levels by increasing absorption of calcium from digestive tract
____ **g.** raises blood calcium levels by increasing bone resorption

A. PTH
B. CT
C. calcitriol
D. insulin-like growth factors
E. sex hormones
F. vitamin C
G. vitamin K

13. Match the following:

____ **a.** small spaces between lamellae that contain osteocytes
____ **b.** perforating canals that penetrate compact bone; carry blood vessels, lymphatic vessels, and nerves from the periosteum
____ **c.** areas between osteons; fragments of old osteons
____ **d.** cells that secrete the components required to build bone
____ **e.** microscopic unit of compact bone tissue
____ **f.** interconnected, tiny canals filled with extracellular fluid; connect lacunae to each other and to the central canal
____ **g.** canals that extend longitudinally through the bone and connect blood vessels and nerves to the osteocytes
____ **h.** large cells derived from monocytes and involved in bone resorption
____ **i.** irregular lattice of thin columns of bone found in spongy bone tissue
____ **j.** rings of hard calcified matrix found just beneath the periosteum and in the medullary cavity
____ **k.** mature cells that maintain the daily metabolism of bone
____ **l.** an opening in the shaft of the bone allowing an artery to pass into the bone
____ **m.** unspecialized stem cells derived from mesenchyme

A. osteogenic cells
B. osteocytes
C. osteon (haversian system)
D. Volkmann's canals
E. circumferential lamellae
F. osteoblasts
G. trabeculae
H. interstitial lamellae
I. canaliculi
J. osteoclasts
K. nutrient foramen
L. lacunae
M. haversian (central) canals

14. Match the following:

____ **a.** column-like layer of maturing chondrocytes
____ **b.** layer of small, scattered chondrocytes anchoring the epiphyseal plate to the bone
____ **c.** layer of actively dividing chondrocytes
____ **d.** region of dead chondrocytes

A. zone of hypertrophic cartilage
B. zone of calcified cartilage
C. zone of proliferating cartilage
D. zone of resting cartilage

15. Match the following:

____ **a.** a broken bone in which one end of the fractured bone is driven into the other end
____ **b.** a condition of porous bones characterized by decreased bone mass and increased susceptibility to fractures
____ **c.** splintered bone, with smaller fragments lying between main fragments
____ **d.** a broken bone that does not break through the skin
____ **e.** a partial break in a bone in which one side of the bone is broken and the other side bends
____ **f.** a broken bone that protrudes through the skin
____ **g.** microscopic bone breaks resulting from inability to withstand repeated stressful impact
____ **h.** oversecretion of hGH during childhood
____ **i.** condition characterized by failure of new bone formed by remodeling to calcify in adults
____ **j.** an infection of bone

A. closed (simple) fracture
B. open (compound) fracture
C. impacted fracture
D. greenstick fracture
E. stress fracture
F. comminuted fracture
G. osteoporosis
H. osteomalacia
I. giantism
J. osteomyelitis

CRITICAL THINKING QUESTIONS

1. Taryn is a high school senior who is undergoing a strenuous running regimen for several hours a day in order to qualify for her state high school track meet. Lately she has experienced intense pain in her right leg that is hindering her workouts. Her physician performs an examination of her right leg. The doctor doesn't notice any outward evidence of injury; he then orders a bone scan. What does her doctor suspect the problem is?
2. While playing basketball, 9-year-old Marcus fell and broke his left arm. The arm was placed in a cast and appeared to heal normally. As an adult, Marcus was puzzled because it seemed that his right arm is longer than his left arm. He measured both arms and he was correct—his right arm *is* longer! How would you explain to Marcus what happened?
3. Astronauts in space exercise as part of their daily routine, yet they still have problems with bone weakness after prolonged stays in space. Why does this happen?

? ANSWERS TO FIGURE QUESTIONS

6.1 The periosteum is essential for growth in bone thickness, bone repair, and bone nutrition. It also serves as a point of attachment for ligaments and tendons.

6.2 Bone resorption is necessary for the development, growth, maintenance, and repair of bone.

6.3 The central (haversian) canals are the main blood supply to the osteocytes of an osteon (haversian system), so their blockage would lead to death of the osteocytes.

6.4 Periosteal arteries enter bone tissue through perforations (perforating or Volkmann's canals).

6.5 Flat bones of the skull, mandible (lower jawbone), and part of the clavicle develop by intramembranous ossification.

6.6 Secondary ossification centers develop in the regions of the cartilage model that will give rise to the epiphyses.

6.7 The lengthwise growth of the diaphysis is caused by cell divisions in the zone of proliferating cartilage and maturation of the cells in the zone of hypertrophic cartilage.

6.8 The medullary cavity enlarges by activity of the osteoclasts in the endosteum.

6.9 In an open fracture the ends of the bone break through the skin; in a closed fracture they do not.

6.10 Healing of bone fractures can take months because calcium and phosphorus deposition is a slow process, and bone cells generally grow and reproduce slowly.

6.11 Heartbeat, respiration, nerve cell functioning, enzyme functioning, and blood clotting all depend on proper levels of calcium.

6.12 A drug that inhibits the activity of osteoclasts might lessen the effects of osteoporosis.

The Skeletal System: The Axial Skeleton

7

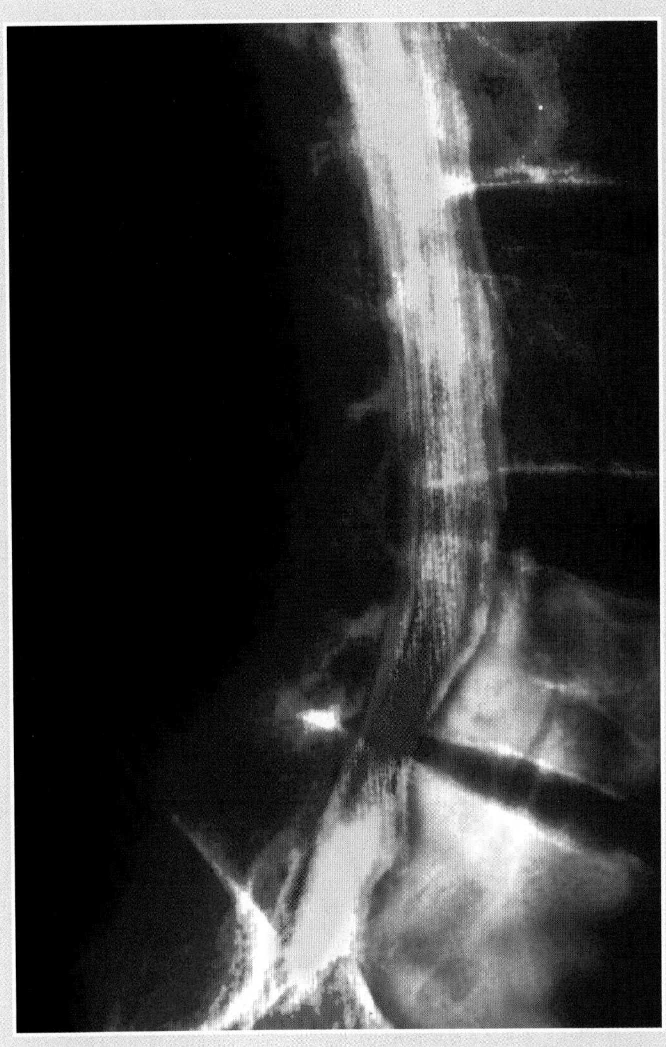

In this chapter you will learn that the axial skeleton, all of the skeleton except for those bones associated with the four limbs, forms the central axis of the body. Literally, you could say that everything in the body revolves around this central line (axis) of the body. The 26 bones of the vertebral column are truly amazing! They provide the support required for an upright posture and yet enable the flexibility required to bend over and tie your shoes. The way you balance against gravity when you walk depends largely on this axis. When the body is in good alignment, you usually do not notice because the proper alignment of the bones allows for strong but flexible muscles creating a balanced posture. The bones and muscles work together to create a stable core position.

As a manual therapist, you will find this chapter and following chapters to contain valuable information for your chosen career. Because the skeletal system forms the framework of the body, a familiarity with the names, shapes, and positions of individual bones and the landmarks on them will help you locate muscle attachments, organ locations, and abnormal conditions in the body. Palpation of bones is an important component of assessment and treatment for many conditions addressed by manual therapists.

CONTENTS AT A GLANCE

7.1 DIVISIONS OF THE SKELETAL SYSTEM

OBJECTIVE

- Describe how the skeleton is divided into axial and appendicular divisions.

The adult human skeleton consists of 206 named bones. The skeletons of infants and children have more than 206 bones because some of their bones, such as the hip bones and certain bones of the backbone, fuse later in life. There are two principal divisions in the adult skeleton (Figure 7.1): the ***axial skeleton*** and the ***appendicular skeleton***. The longitudinal ***axis,*** or center, of the human body is a straight line that runs through the body's center of gravity. This imaginary line extends through the head and down to the space between the feet. The axial skeleton consists of the bones that lie around the axis: skull bones, auditory ossicles (ear bones), hyoid bone, ribs, breastbone, and bones of the backbone. Although the auditory ossicles are not considered part of the axial or appendicular skeleton, but rather a separate group of bones, they are placed with the axial skeleton for convenience. The appendicular skeleton contains the bones of the ***upper*** and ***lower limbs (extremities),*** plus the bones called ***girdles*** that connect the limbs to the axial skeleton. Table 7.1 presents the standard grouping of the 80 bones of the axial skeleton and the 126 bones of the appendicular skeleton.

CHECKPOINT

1. Which bones make up the axial and appendicular divisions of the skeleton?

Table 7.1

The Bones of the Adult Skeletal System

DIVISION OF THE SKELETON	STRUCTURE	NUMBER OF BONES
Axial Skeleton		
	Skull	
	Cranium	8
	Face	14
	Hyoid	1
	Auditory ossicles	6
	Vertebral column	26
	Thorax	
	Sternum	1
	Ribs	24
		Subtotal = 80
Appendicular Skeleton		
	Pectoral (shoulder) girdles	
	Clavicle	2
	Scapula	2
	Upper limbs	
	Humerus	2
	Ulna	2
	Radius	2
	Carpals	16
	Metacarpals	10
	Phalanges	28
	Pelvic (hip) girdle	
	Hip, pelvic, or coxal bone	2
	Lower limbs	
	Femur	2
	Patella	2
	Fibula	2
	Tibia	2
	Tarsals	14
	Metatarsals	10
	Phalanges	28
		Subtotal = 126
	Total in an adult skeleton	**= 206**

7.2 TYPES OF BONES

OBJECTIVE

- Classify bones based on their shape or location.

Almost all the bones of the body can be classified into five principal types on the basis of shape: long, short, flat, irregular, and sesamoid (see Figure 7.2). As you learned in Chapter 6, ***long bones*** have greater length than width and consist of a shaft and a variable number of extremities (ends). They are slightly curved for strength. A curved bone absorbs the stress of the body's weight at several different points so that the stress is evenly distributed. If such bones were straight, the weight of the body would be unevenly distributed and the bone would more easily fracture. Long bones consist mostly of *compact bone tissue,* which is dense and has few spaces, but they also contain considerable amounts of *spongy bone tissue,* which has larger spaces (see Section 6.3). Long bones include those in the thigh (femur), leg (tibia and fibula), toes (phalanges), arm (humerus), forearm (ulna and radius), and fingers (phalanges). Please note that phalanges of the toes and fingers are physically short bones but meet all the criteria previously listed and are therefore classified as long bones.

Short bones are somewhat cube-shaped and nearly equal in length and width. They consist of spongy bone except at the surface, where there is a thin layer of compact bone. Examples of short bones are the wrist or carpal bones (except for the pisiform, which is a sesamoid bone) and the ankle or tarsal bones (except for the calcaneus, which is an irregular bone).

Flat bones are generally thin and composed of two nearly parallel plates of compact bone enclosing a layer of spongy bone. Flat bones afford considerable protection and provide extensive areas for muscle attachment. Flat bones include the cra-

Figure 7.1 Divisions of the skeletal system.

The adult human skeleton consists of 206 bones grouped in axial and appendicular divisions.

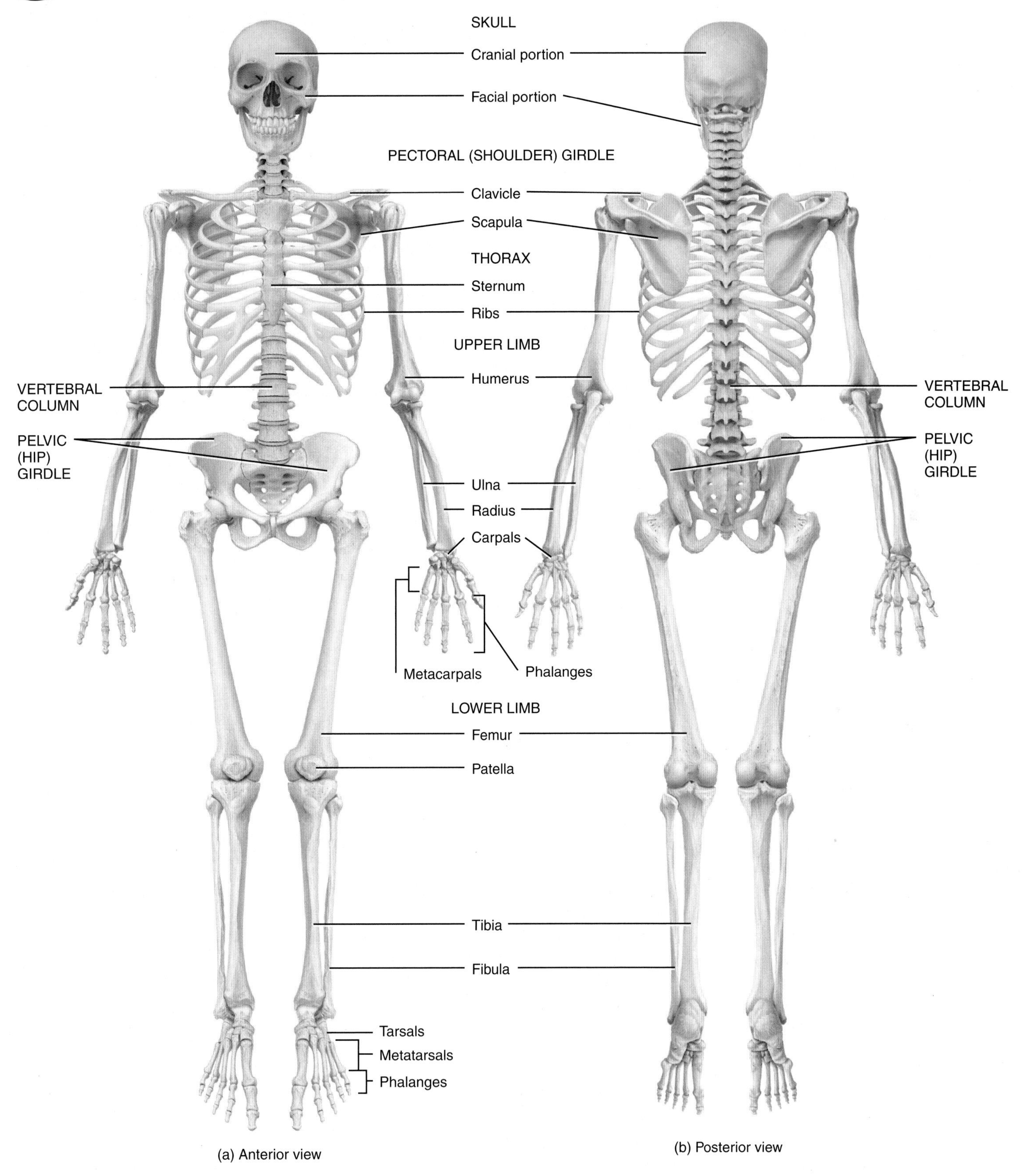

Which of the following structures are part of the axial skeleton, and which are part of the appendicular skeleton? Skull, clavicle, vertebral column, shoulder girdle, humerus, pelvic girdle, and femur.

Figure 7.2 Types of bones based on shape.

The shapes of bones largely determine their functions.

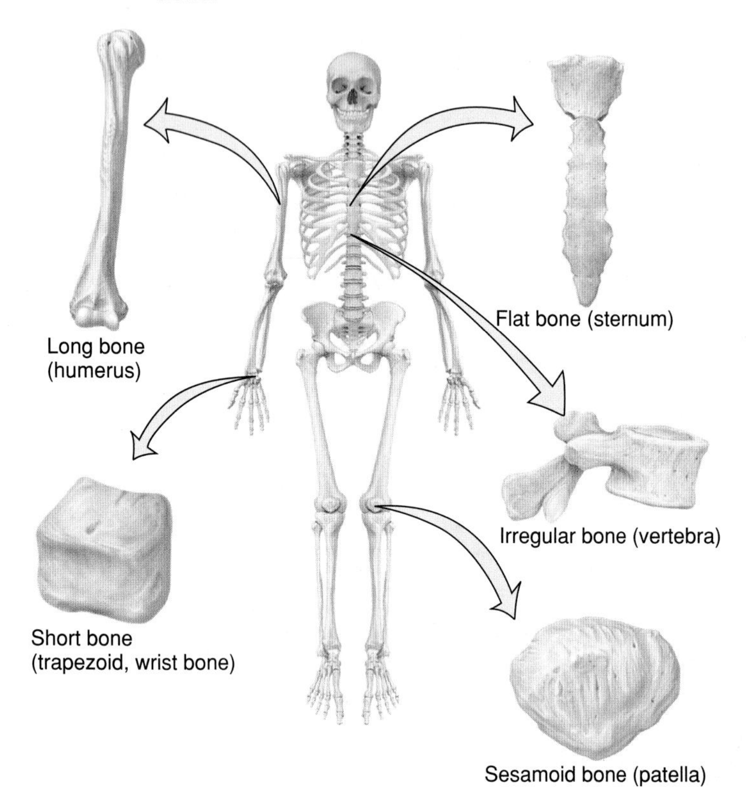

? Which type of bone primarily provides protection and a large surface area for muscle attachment?

nial bones, which protect the brain; the breastbone (sternum) and ribs, which protect organs in the thorax; and the shoulder blades (scapulae), which also protect the lungs.

Irregular bones have complex shapes and cannot be grouped into any of the three categories just described. They also vary in the amount of spongy and compact bone present. Such bones include the backbones (vertebrae), certain facial bones, and the aforementioned heel bone (calcaneus).

Sesamoid bones (SES-a-moyd = shaped like a sesame seed) develop in certain tendons where there is considerable friction, tension, or physical stress. They are not always completely ossified and measure only a few millimeters in diameter except for the two patellae (kneecaps), the largest of the sesamoid bones. Sesamoid bones vary in number from person to person except for the patellae, which are located in the quadriceps femoris tendon (see Section 14.1) and are normally present in all individuals. Functionally, sesamoid bones protect tendons from excessive wear and tear, and they often change the direction of pull of a tendon, which improves the mechanical advantage at a joint.

In the upper limbs, sesamoid bones usually occur only in the joints of the palmar surface of the hands. Two frequently encountered sesamoid bones are in the tendons of the adductor pollicis and flexor pollicis brevis muscles at the metacarpophalangeal joint of the thumb (see Section 8.2). In the lower limbs, aside from the patellae, two sesamoid bones normally occur on the plantar surface of the foot in the tendons of the flexor hallucis brevis muscle at the metatarsophalangeal joint of the great (big) toe (see Figure 8.16).

An additional type of bone is not included in this classification by shape, but instead is classified by location. ***Sutural bones*** (SOŌ-chur-al; *sutura* = seam), also known as ***Wormian bones,*** are small bones located within the sutures (joints) of certain cranial bones (see Figure 7.6). The number of sutural bones varies greatly from person to person.

During active growth of the skeleton, red bone marrow is progressively replaced by yellow bone marrow in most of the long bones. In adults, red bone marrow is restricted to flat bones such as the ribs, sternum (breastbone), and skull; irregular bones such as vertebrae (backbones) and hip bones; long bones such as the proximal epiphyses of the femur (thigh bone) and humerus (arm bone); and some short bones.

CHECKPOINT

2. Give examples of long, short, flat, and irregular bones.

7.3 BONE SURFACE MARKINGS

OBJECTIVE

- Describe the principal surface markings on bones and the functions of each.

Bones have characteristic ***surface markings,*** structural features adapted for specific functions. Most are not present at birth but develop later in response to certain forces and are most prominent during adult life. In response to tension on a bone surface where tendons, ligaments, aponeuroses, and fasciae pull on the periosteum of bone, new bone is deposited, resulting in raised or roughened areas. Conversely, compression on a bone surface results in a depression.

There are two major types of surface markings: (1) *depressions and openings,* which form joints or allow the passage of soft tissues (such as blood vessels and nerves), and (2) *processes, projections or outgrowths,* which either help form joints or serve as attachment points for connective tissue (such as ligaments and tendons). Table 7.2 describes the various surface markings and provides examples of each.

CHECKPOINT

3. List and describe several bone surface markings and give an example of each.

7.4 SKULL

OBJECTIVES

- Name the cranial and facial bones and indicate whether they are paired or single.
- Describe the following special features of the skull: sutures, paranasal sinuses, and fontanels.

The ***skull,*** which contains 22 bones, rests on the superior end of the vertebral column (backbone). It includes two sets of bones: cranial bones and facial bones (Table 7.3). The ***cranial bones***

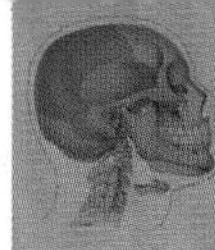

Table 7.2

Bone Surface Markings

MARKING	DESCRIPTION	EXAMPLE
DEPRESSIONS AND OPENINGS: SITES ALLOWING THE PASSAGE OF SOFT TISSUE (NERVES, BLOOD VESSELS, LIGAMENTS, TENDONS) OR FORMATION OF JOINTS		
Fissure (FISH-ur)	Narrow slit between adjacent parts of bones through which blood vessels or nerves pass.	Superior orbital fissure of the sphenoid bone (Figure 7.12).
Foramen (fō-RĀ-men = hole; plural is *foramina*)	Opening through which blood vessels, nerves, or ligaments pass.	Optic foramen of the sphenoid bone (Figure 7.12).
Fossa (FOS-a = trench; plural is *fossae*, FOS-ē)	Shallow depression.	Coronoid fossa of the humerus (Figure 8.5a).
Sulcus (SUL-kus = groove; plural is *sulci*, SUL-sī)	Furrow along a bone surface that accommodates a blood vessel, nerve, or tendon.	Intertubercular sulcus of the humerus (Figure 8.5a).
Canal (ka-NAL)	A tubular structure through a bone.	Hypoglossal canal of the occipital bone (Figure 7.8a)
Meatus (mē-Ā-tus = passageway; plural is *meati*, mē-Ā-tī)	Tubelike opening into a canal.	External auditory meatus of the temporal bone (Figure 7.4).
PROCESSES: PROJECTIONS OR OUTGROWTHS ON BONE THAT FORM JOINTS OR ATTACHMENT POINTS FOR CONNECTIVE TISSUE, SUCH AS LIGAMENTS AND TENDONS		
Processes that form joints		
Condyle (KON-dīl; *condylus* = knuckle)	Large, round protuberance at the end of a bone.	Lateral condyle of the femur (Figure 8.13).
Facet (FAS-et or fa-SET)	Smooth, flat articular surface.	Facet of superior articular process of vertebra (Figure 7.18a).
Head	Rounded articular projection supported on the neck (constricted portion) of a bone.	Head of the femur (Figure 8.13).
Processes that form attachment points for connective tissue		
Crest	Prominent ridge or elongated projection.	Iliac crest of the hip bone (Figure 8.10b).
Epicondyle (*epi-* = above)	Projection above a condyle.	Medial epicondyle of the femur (Figure 8.13).
Line (*linea*)	Long, narrow ridge or border (less prominent than a crest).	Linea aspera of the femur (Figure 8.13b).
Spinous process (SPĪ-nus)	Sharp, slender projection.	Spinous process of a vertebra (Figure 7.18).
Trochanter (trō-KAN-ter)	Very large projection.	Greater trochanter of the femur (Figure 8.13).
Tubercle (TOO-ber-kul; *tuber-* = knob)	Small, rounded projection.	Greater tubercle of the humerus (Figure 8.5).
Tuberosity	Large, rounded, usually roughened projection.	Ischial tuberosity of the hip bone (Figure 8.10b).

(*crani-* = brain case) form the cranial cavity, which encloses and protects the brain. The eight cranial bones are the frontal bone, two parietal bones, two temporal bones, the occipital bone, the sphenoid bone, and the ethmoid bone. Fourteen ***facial bones*** form the face: two nasal bones, two maxillae (maxillary bones), two zygomatic bones, the mandible, two lacrimal bones, two palatine bones, two inferior nasal conchae, and the vomer. Figures 7.3 through 7.12 illustrate these bones from different viewing directions.

General Features and Functions

Besides forming the large cranial cavity, the skull also forms several smaller cavities, including the nasal cavity and orbits (eye sockets), which open to the exterior. Certain skull bones also contain cavities called paranasal sinuses that are lined with mucous membranes and open into the nasal cavity. Also within the skull are small cavities that house the structures involved in hearing and equilibrium.

Other than the auditory ossicles, which are involved in hearing and are located within the temporal bones (Section 19.7), the

Table 7.3

Summary of Bones of the Adult Skull

CRANIAL BONES	FACIAL BONES
Frontal (1)*	Nasal (2)
Parietal (2)	Maxillae (2)
Temporal (2)	Zygomatic (2)
Occipital (1)	Mandible (1)
Sphenoid (1)	Lacrimal (2)
Ethmoid (1)	Palatine (2)
	Inferior nasal conchae (2)
	Vomer (1)

*The numbers in parentheses indicate how many of each bone are present.

mandible is the only movable bone of the skull. Most of the skull bones are held together by slightly movable joints called sutures, which are especially noticeable on the outer surface of the skull.

The skull has numerous surface markings, such as foramina and fissures through which blood vessels and nerves pass. You will learn the names of important skull bone surface markings as the various bones are described.

The cranial bones have other functions in addition to protecting the brain. Their inner surfaces attach to membranes (meninges) that stabilize the positions of the brain, blood vessels, and nerves. The outer surfaces of cranial bones provide large areas of attachment for muscles that move various parts of the head. The bones also provide attachment for some muscles that are involved in producing facial expressions. Besides forming the framework of the face, the facial bones protect and provide support for the entrances to the digestive and respiratory systems. Together, the cranial and facial bones protect and support the delicate special sense organs for vision, taste, smell, hearing, and equilibrium (balance).

Cranial Bones

Frontal Bone

The ***frontal bone*** forms the forehead (the anterior part of the cranium), the roofs of the *orbits* (eye sockets), and most of the anterior part of the cranial floor (Figure 7.3). Soon after birth

Figure 7.3 Skull (anterior view).

6 The skull consists of cranial bones and facial bones.

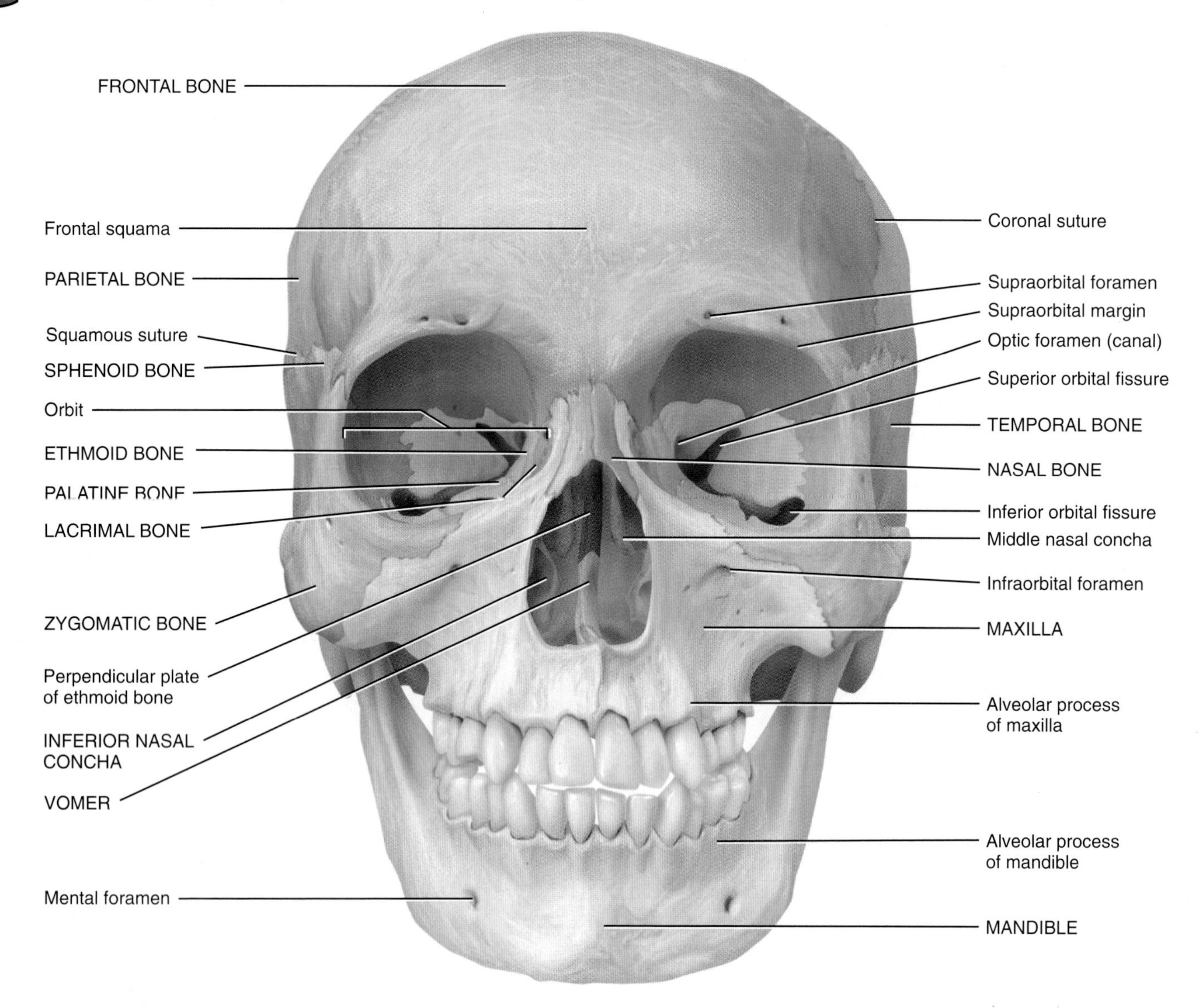

Which of the bones shown here are cranial bones?

the left and right sides of the frontal bone are united by a suture called the *frontal (metopic) suture,* which usually disappears by age 6–8.

If you examine the anterior view of the skull in Figure 7.3, you will note the *frontal squama,* a scalelike plate of bone that forms the forehead. It gradually slopes inferiorly from the coronal suture on top of the skull, then angles abruptly and becomes almost vertical. Superior to the orbits the frontal bone thickens, forming the *supraorbital margin* (*supra-* = above; *-orbital* = wheel rut).

From the supraorbital margin the frontal bone extends posteriorly to form the roof of the orbit and part of the floor of the cranial cavity. Within the supraorbital margin, slightly medial to its midpoint, is a hole called the *supraorbital foramen* through which the supraorbital nerve and artery pass. Sometimes this foramen is incomplete and is called the *supraorbital notch.* The *frontal sinuses* lie deep to the frontal squama.

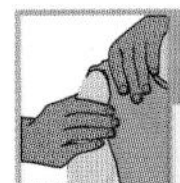

MANUAL THERAPY APPLICATION

Treatment for Frontal Headaches

Many persons who suffer from **frontal headaches** will profit from applying pressure, or having pressure applied by a therapist, to the *supraorbital foramen (notch).* Some patients have two foramina, some have two notches, and some have one of each. Compression of the nerves and arteries that pass through this foramen for 30 to 45 seconds commonly alleviates a frontal headache.

Parietal Bones

The two ***parietal bones*** (pa-RĪ-e-tal; *pariet-* = wall) form the greater portion of the sides and roof of the cranial cavity (Figure 7.4). The internal surfaces of the parietal bones contain many protrusions and depressions that accommodate the blood

Figure 7.4 Skull (lateral view).

The zygomatic arch is formed by the zygomatic process of the temporal bone and the temporal process of the zygomatic bone.

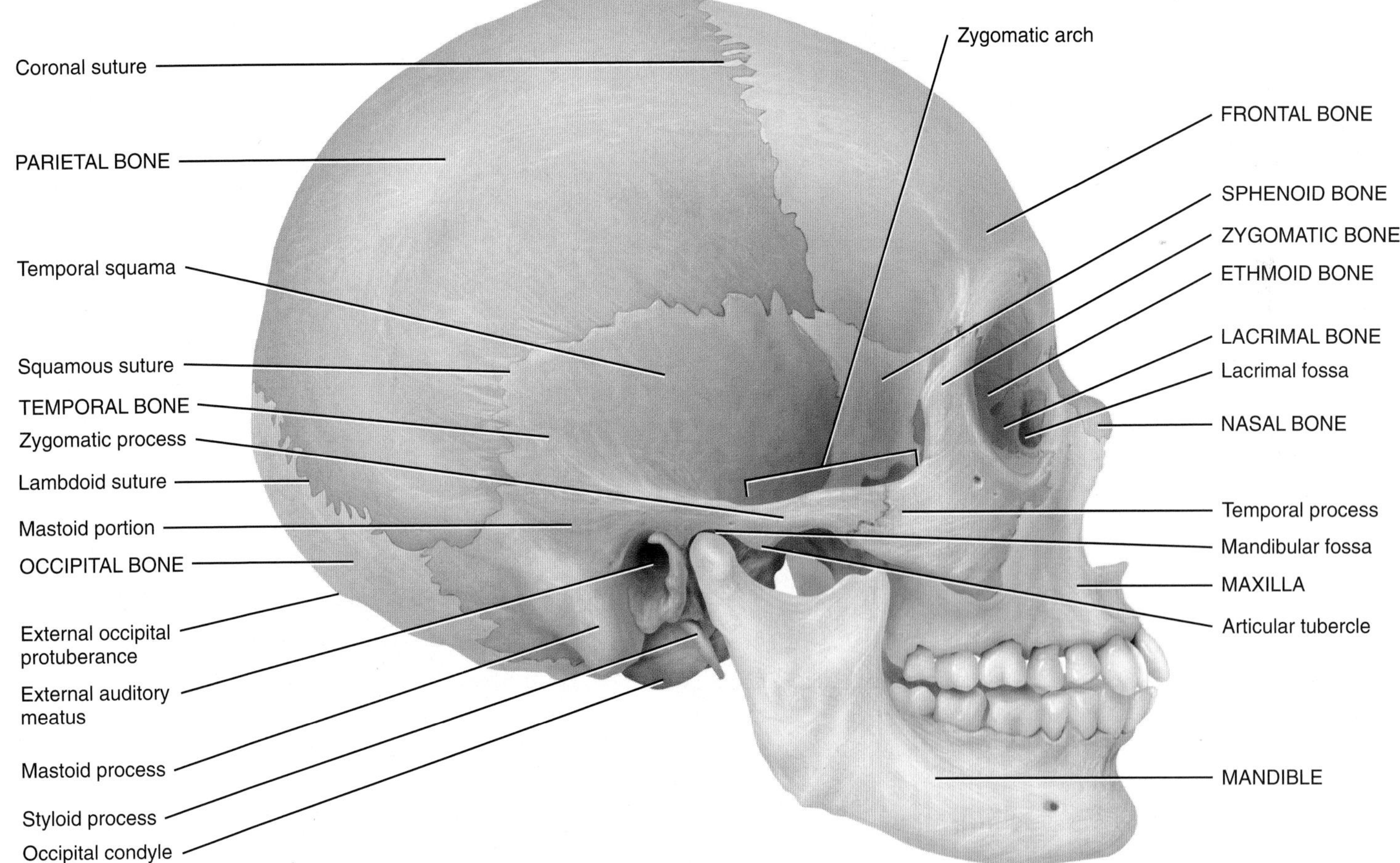

What major bones are joined by (1) the squamous suture, (2) the lambdoid suture, and (3) the coronal suture?

vessels supplying the dura mater, the superficial membrane (meninx) covering the brain (see Figure 7.5).

Temporal Bones

The two ***temporal bones*** (*tempor-* = temple) form the inferior lateral aspects of the cranium and part of the cranial floor. In the lateral view of the skull (see Figure 7.4), note the *temporal squama,* the thin, flat portion of the temporal bone that forms the anterior and superior part of the temple (the region of the cranium around the ear). Projecting from the inferior portion of the temporal squama is the *zygomatic process,* which articulates (forms a joint) with the temporal process of the zygomatic (cheek) bone. Together, the zygomatic process of the temporal bone and the temporal process of the zygomatic bone form the *zygomatic arch.*

On the inferior posterior surface of the zygomatic process of the temporal bone is a socket called the *mandibular fossa.* Anterior to the mandibular fossa is a rounded elevation, the *articular tubercle* (see Figure 7.4). The mandibular fossa and articular tubercle articulate with the mandible (lower jawbone) to form the *temporomandibular joint (TMJ).*

Located posteriorly on the temporal bone is the *mastoid portion* (*mastoid* = breast-shaped) (see Figure 7.4). The mastoid portion is located posterior to the *external auditory meatus* (*meatus* = passageway), or ear canal, which directs sound waves into the ear. In the adult, this portion of the bone contains several *mastoid "air cells"* (deep to this illustration). These tiny air-filled compartments are separated from the brain by thin bony partitions. ***Mastoiditis*** (inflammation of the mastoid air cells) can spread an infection to the middle ear and to the brain.

The *mastoid process* is a rounded projection of the mastoid portion of the temporal bone posterior to the external auditory meatus. It serves as a point of attachment for several neck

Figure 7.5 Skull (sagittal section). Although the hyoid bone is not part of the skull, it is included in the illustration for reference.

The cranial bones are the frontal, parietal, temporal, occipital, sphenoid, and ethmoid bones. The facial bones are the nasal bone, maxillae, zygomatic bones, lacrimal bones, palatine bones, mandible, and vomer.

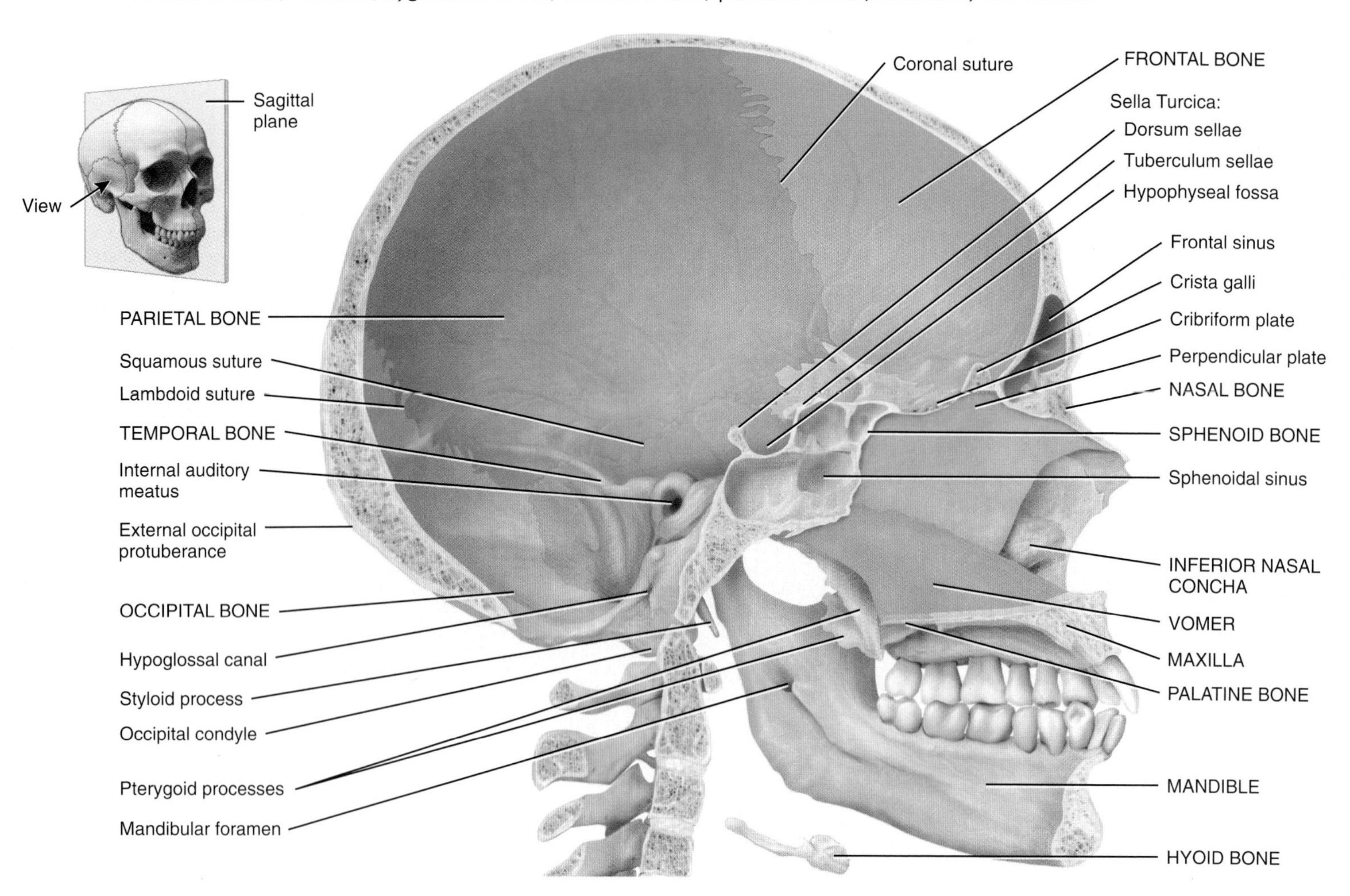

Medial view of sagittal section

With which bones does the temporal bone articulate?

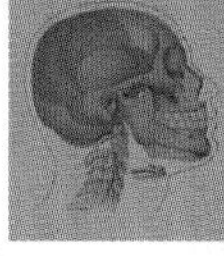

muscles (see Figure 7.4). The *internal auditory meatus* (Figure 7.5) is the opening through which the vestibulocochlear (VIII) and a small trunk of the facial (VII) cranial nerves pass. The *styloid process* (*styl-* = stake or pole) projects inferiorly from the inferior surface of the temporal bone and serves as a point of attachment for muscles and ligaments of the tongue and neck (see Figure 7.4). Between the styloid process and the mastoid process is the *stylomastoid foramen,* through which the main trunk of the facial (VII) nerve and stylomastoid artery pass (see Figure 7.7).

At the floor of the cranial cavity (see Figure 7.8a) is the *petrous portion* (*petrous* = rock) of the temporal bone. This portion is triangular and located at the base of the skull between the sphenoid and occipital bones. The petrous portion houses the internal ear and the middle ear, structures involved in hearing and equilibrium. It also contains the *carotid foramen,* through which the carotid artery passes (see Figure 7.7). Posterior to the carotid foramen and anterior to the occipital bone is the *jugular foramen,* a passageway for the jugular vein.

Occipital Bone

The **occipital bone** (ok-SIP-i-tal; *occipit-* = back of head) forms the posterior part and most of the base of the cranium (Figure 7.6; also see Figure 7.4). The *foramen magnum* (= large hole) is in the inferior part of the bone. Within this foramen, the medulla oblongata (inferior part of the brain) connects with the spinal cord. The vertebral and spinal arteries also pass through this foramen. The *occipital condyles* are two oval processes with convex surfaces, one on either side of the foramen magnum

Figure 7.6 Skull (posterior view).

The occipital bone forms most of the posterior and inferior portions of the cranium.

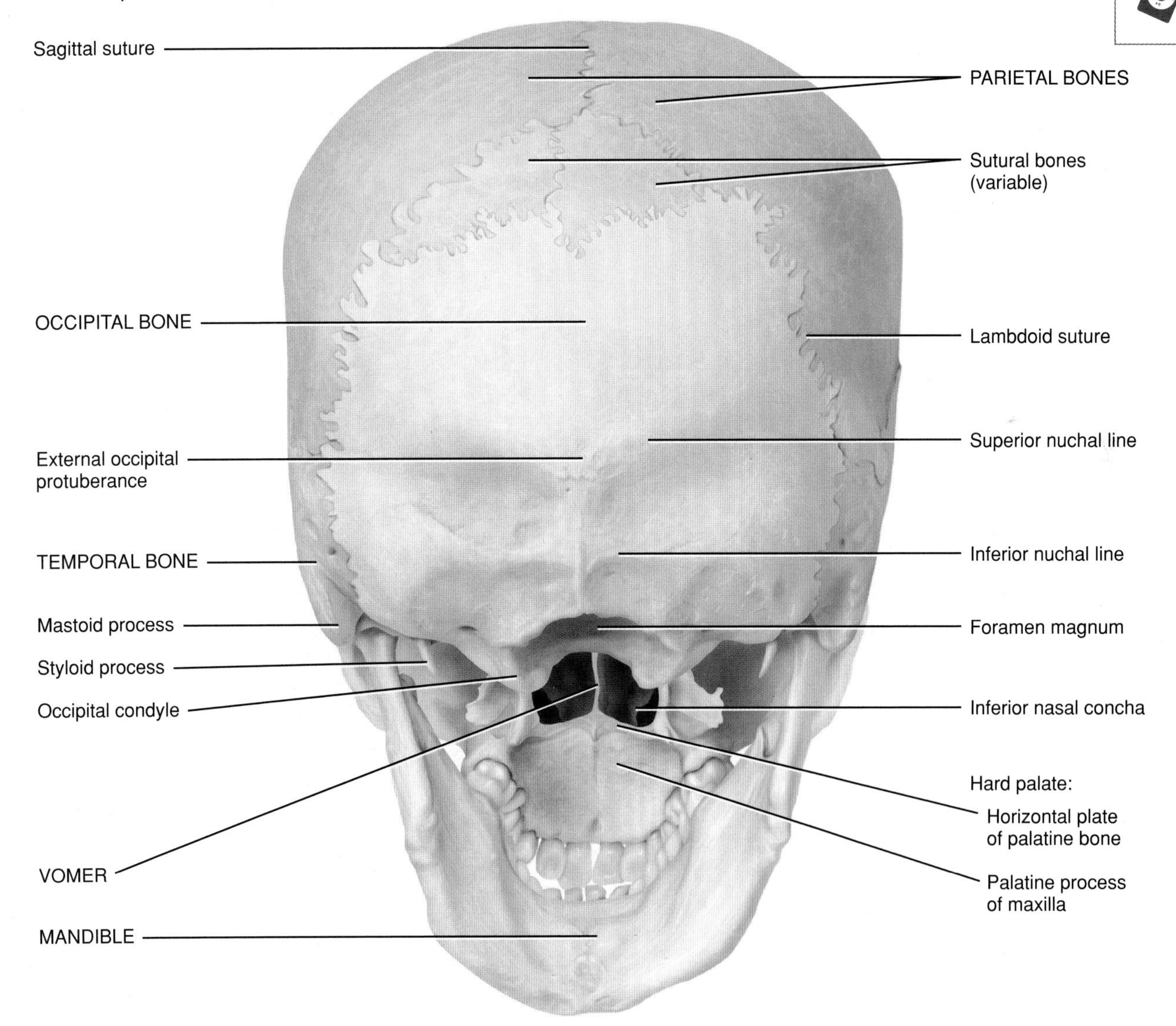

Which bones form the posterior, lateral portion of the cranium?

(Figure 7.7). They articulate with depressions on the first cervical vertebra (atlas) to form the *atlanto-occipital joints.* Superior to each occipital condyle on the inferior surface of the skull is the *hypoglossal canal* (*hypo-* = under; *-glossal* = tongue), through which the hypoglossal (XII) nerve and a branch of the ascending pharyngeal artery pass (see Figures 7.5 and 7.8a).

The *external occipital protuberance* is a prominent midline projection on the posterior surface of the bone just superior and posterior to the foramen magnum. You may be able to feel this structure as a definite bump on the back of your head, just above your neck (see Figure 7.6). A large fibrous, elastic ligament, the *ligamentum nuchae* (*nucha-* = nape of neck), which helps support the head, extends from the external occipital protuberance to the seventh cervical vertebra. Extending laterally from the protuberance are two curved lines, the *superior nuchal lines,* and below these are two *inferior nuchal lines,* which are areas of muscle attachment (Figure 7.7). It is possible to view the parts of the occipital bone, as well as surrounding structures, in the inferior view of the skull in Figure 7.7.

Sphenoid Bone

The ***sphenoid bone*** (SFĒ-noyd = wedge-shaped) lies anterior to the middle part of the base of the skull (Figures 7.7 and 7.8). This bone is called the keystone of the cranial floor because it articulates with all the other cranial bones, holding them together. Viewing the floor of the cranium superiorly (Figure 7.8a), note the sphenoid articulations: anteriorly with the frontal and

Figure 7.7 Skull (inferior view). The mandible (lower jawbone) has been removed.

The occipital condyles of the occipital bone articulate with the first cervical vertebra to form the atlanto-occipital joints.

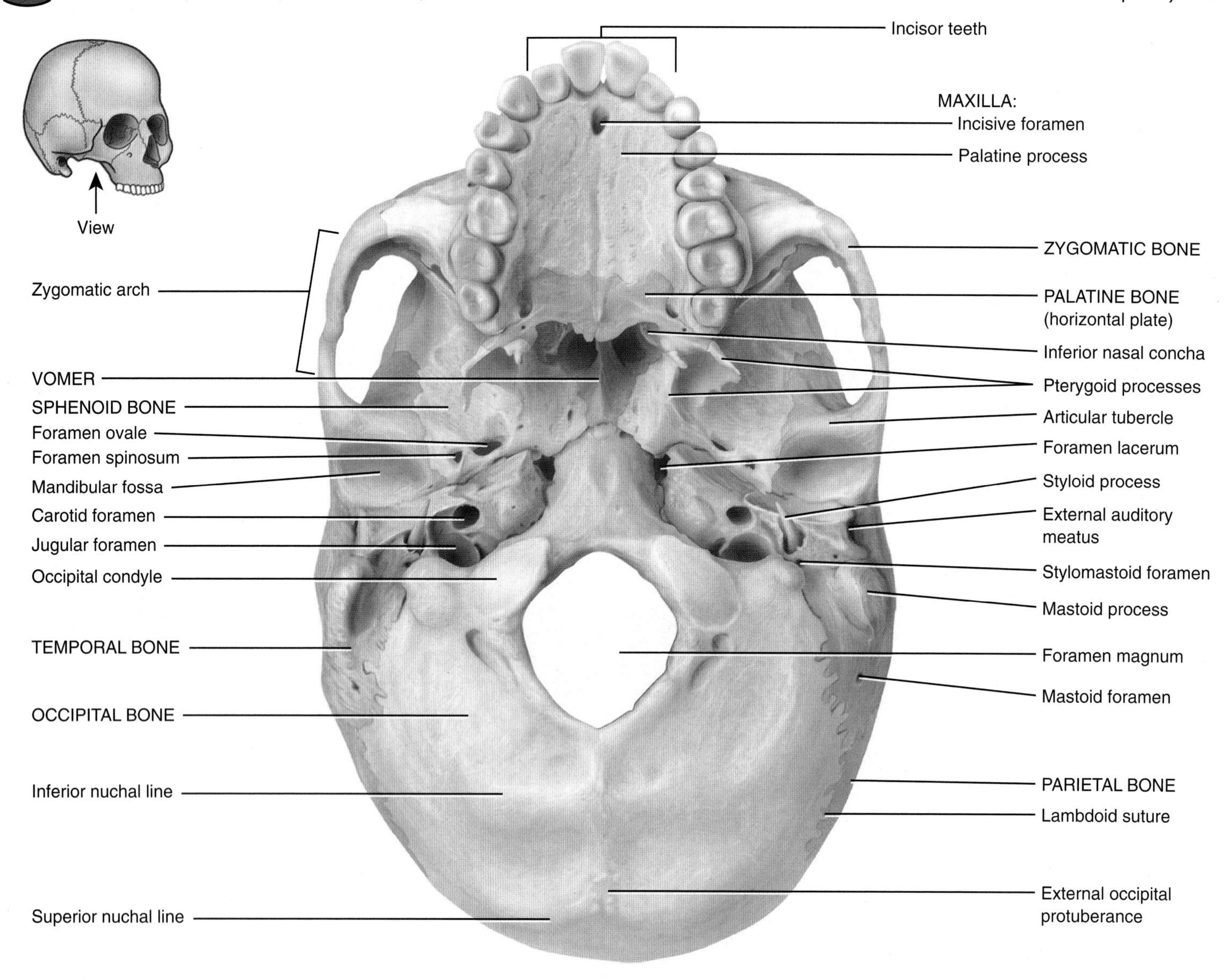

What parts of the nervous system join together within the foramen magnum?

Figure 7.8 **Sphenoid bone.**

The sphenoid bone is called the keystone of the cranial floor because it articulates with all other cranial bones, helping to hold them together.

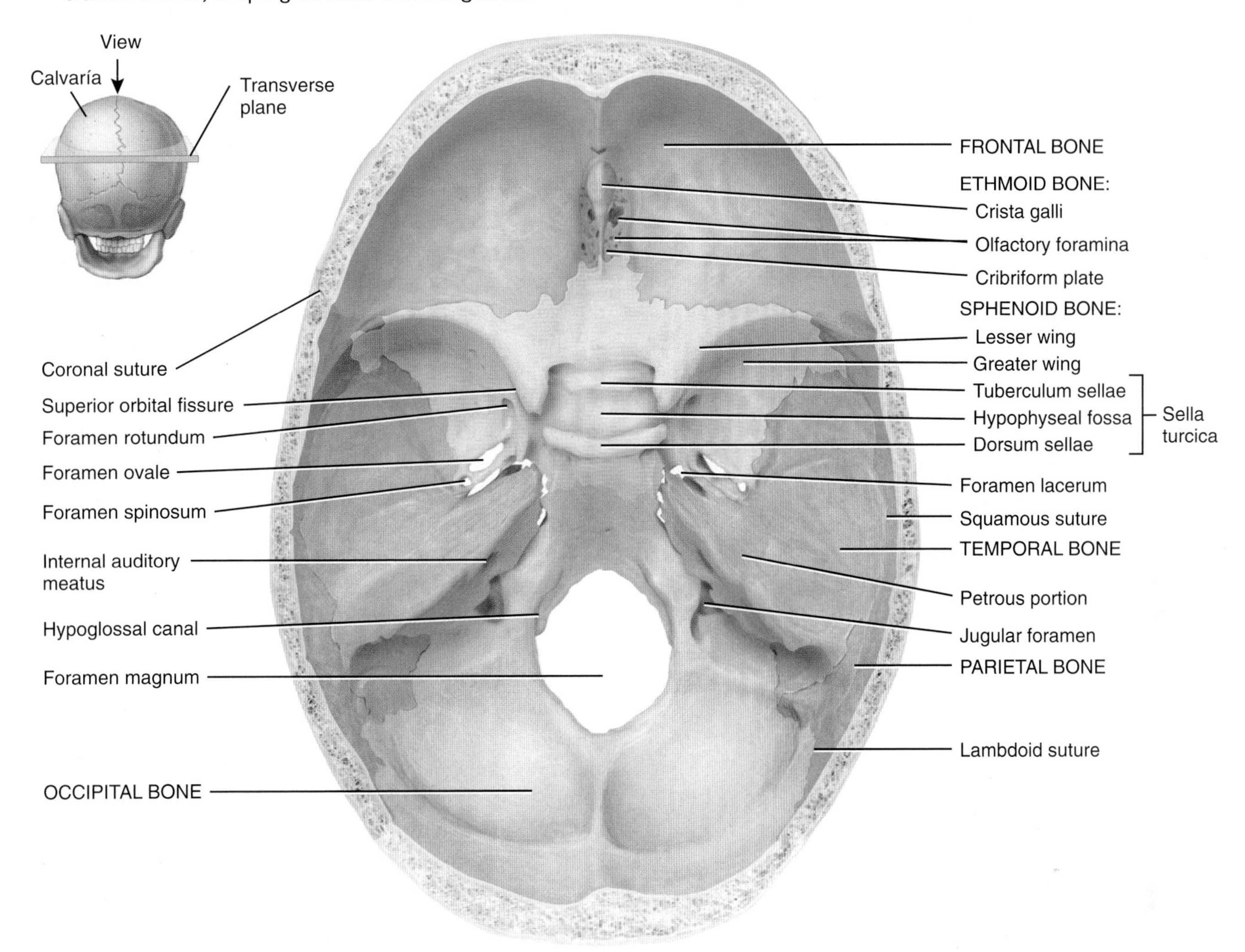

(a) Superior view of sphenoid bone in floor of cranium

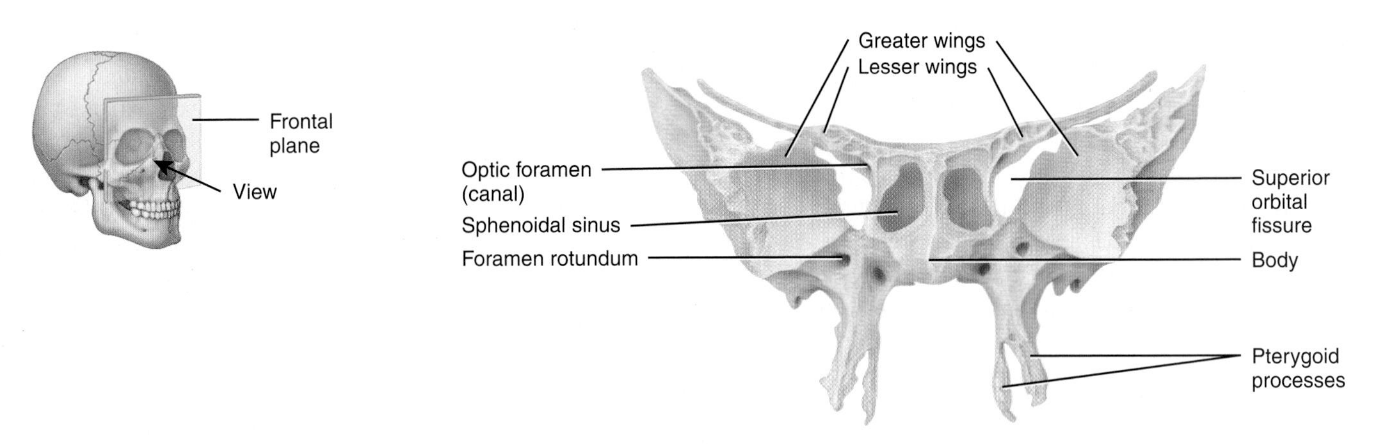

(b) Anterior view of sphenoid bone

Name the bones that articulate with the sphenoid bone, starting with the crista galli of the ethmoid bone and going in a clockwise direction.

ethmoid bones, laterally with the temporal and parietal bones, superiorly with the parietal bones, and posteriorly with the occipital bone. It lies posterior and slightly superior to the nasal cavity and forms part of the floor, sidewalls, and rear wall of the orbit (see Figure 7.12). The superior portion of the skull, indicated above the transverse plane, is termed the calvaria and has been removed to show the sphenoid.

The shape of the sphenoid resembles a bat with outstretched wings (see Figure 7.8b). The *body* of the sphenoid is the cubelike medial portion between the ethmoid and occipital bones. It contains *sphenoidal sinuses,* which drain into the nasal cavity (see Figure 7.13). The *sella turcica* (SEL-a TUR-si-ka; *sella* = saddle; *turcica* = Turkish) is a bony saddle-shaped structure on the superior surface of the body of the sphenoid (see Figure 7.8a). The anterior part of the sella turcica, which forms the horn of the saddle, is a ridge called the *tuberculum sellae.* The seat of the saddle is a depression, the *hypophyseal fossa* (hī-pō-FIZ-ē-al), which contains the pituitary gland. The posterior part of the sella turcica, which forms the back of the saddle, is another ridge called the *dorsum sellae.*

The *greater wings* of the sphenoid project laterally from the body, forming the anterolateral floor of the cranium. The greater wings also form part of the lateral wall of the skull just anterior to the temporal bone and can be viewed externally (see Figure 7.4). The *lesser wings,* which are smaller than the greater wings, form a ridge of bone anterior and superior to the greater wings. They form part of the floor of the cranium and the posterior part of the orbit of the eye.

Between the body and lesser wing just anterior to the sella turcica is the *optic foramen* (*optic* = eye), through which the optic (II) nerve and ophthalmic artery pass. Lateral to the body between the greater and lesser wings is a somewhat triangular slit called the *superior orbital fissure.* This fissure may also be seen in the anterior view of the orbit in Figure 7.12.

In Figures 7.7 and 7.8b you can see the *lateral* and *medial pterygoid processes* (TER-i-goyd = winglike) extending from the inferior part of the sphenoid bone. These structures project inferiorly from the points where the body and greater wings unite and form the lateral posterior region of the nasal cavity. Some of the muscles that move the mandible attach to the pterygoid processes. At the base of the lateral pterygoid process in the greater wing is the *foramen ovale* (= oval), an opening for the mandibular branch of the trigeminal (V) nerve. Another foramen, the *foramen spinosum* (= resembling a spine), lies at the posterior angle of the sphenoid and transmits the middle meningeal blood vessels. The *foramen lacerum* (= lacerated) is bounded anteriorly by the sphenoid bone and medially by the sphenoid and occipital bones. Although it is an empty space in a prepared skull, the foramen is covered in part by a layer of fibrocartilage in living subjects. It transmits a branch of the ascending pharyngeal artery. Another foramen associated with the sphenoid bone is the *foramen rotundum* (= round) located at the junction of the anterior and medial parts of the sphenoid bone. Through it passes the maxillary branch of the trigeminal (V) nerve.

Ethmoid Bone

The ***ethmoid bone*** (ETH-moyd = like a sieve) is a light, spongelike bone located on the midline in the anterior part of the cranial floor medial to the orbits (Figure 7.9). It is anterior to the sphenoid bone and posterior to the nasal bones. The ethmoid bone forms (1) part of the anterior portion of the cranial floor; (2) the medial wall of the orbits; (3) the superior portion of the nasal septum, a partition that divides the nasal cavity into right and left sides; and (4) most of the superior sidewalls of the nasal cavity. The ethmoid bone is a major superior supporting structure of the nasal cavity.

The *lateral masses* of the ethmoid bone compose most of the wall between the nasal cavity and the orbits. They contain 3 to 18 air spaces (cells). The ethmoidal cells together form the *ethmoidal sinuses* (see Figure 7.13). The *perpendicular plate* forms the superior portion of the nasal septum (see Figure 7.11). The *cribriform plate* (*cribri-* = sieve) lies in the anterior floor of the cranium and forms the roof of the nasal cavity. The cribriform plate contains the *olfactory foramina* (*olfact-* = smell), through which the many portions of the olfactory (I) nerve pass. Projecting superiorly from the cribriform plate is a sharp triangular process called the *crista galli* (*crista* = crest; *galli* = cock). This structure serves as a point of attachment for the membranes (meninges) that cover the brain.

The paired lateral masses of the ethmoid bone contain two thin, scroll-shaped projections lateral to the nasal septum. These are called the *superior nasal concha* (KONG-ka = shell) or *turbinate* and the *middle nasal concha* or *turbinate.* The plural form is *conchae* (KONG-kē). A third pair of conchae, the inferior nasal conchae, are separate bones (discussed shortly). The conchae increase the vascular and mucous membrane surface area in the nasal cavities, which warms and moistens inhaled air before passing into the lungs. The conchae also cause inhaled air to swirl; the result is that many inhaled particles strike and become trapped in the mucus that lines the nasal passageways. The action of the conchae helps cleanse inhaled air before it passes into the rest of the respiratory tract. The superior nasal conchae also participate in the sense of smell.

Facial Bones

The shape of the face changes dramatically during the first two years after birth. The brain and cranial bones expand, the teeth form and erupt, and the paranasal sinuses increase in size. Growth of the face ceases at about 16 years of age. The 14 facial bones include two nasal bones, two maxillae (maxillary bones), two zygomatic bones, the mandible, two lacrimal bones, two palatine bones, two inferior nasal conchae, and the vomer.

Nasal Bones

The paired ***nasal bones*** meet at the midline (see Figure 7.3) and form part of the bridge of the nose. The major structural portion of the nose consists of cartilage.

Maxillae

The paired ***maxillae*** (mak-SIL-ē = jawbones; singular is *maxilla*) unite to form the upper jawbone. They articulate with every bone of the face except the mandible, or lower jawbone (see Figures 7.3, 7.4, and 7.7). The maxillae (maxillary bones)

Figure 7.9 Ethmoid bone

The ethmoid bone forms part of the anterior portion of the cranial floor, the medial wall of the orbits, the superior portions of the nasal septum, and most of the sidewalls of the nasal cavity.

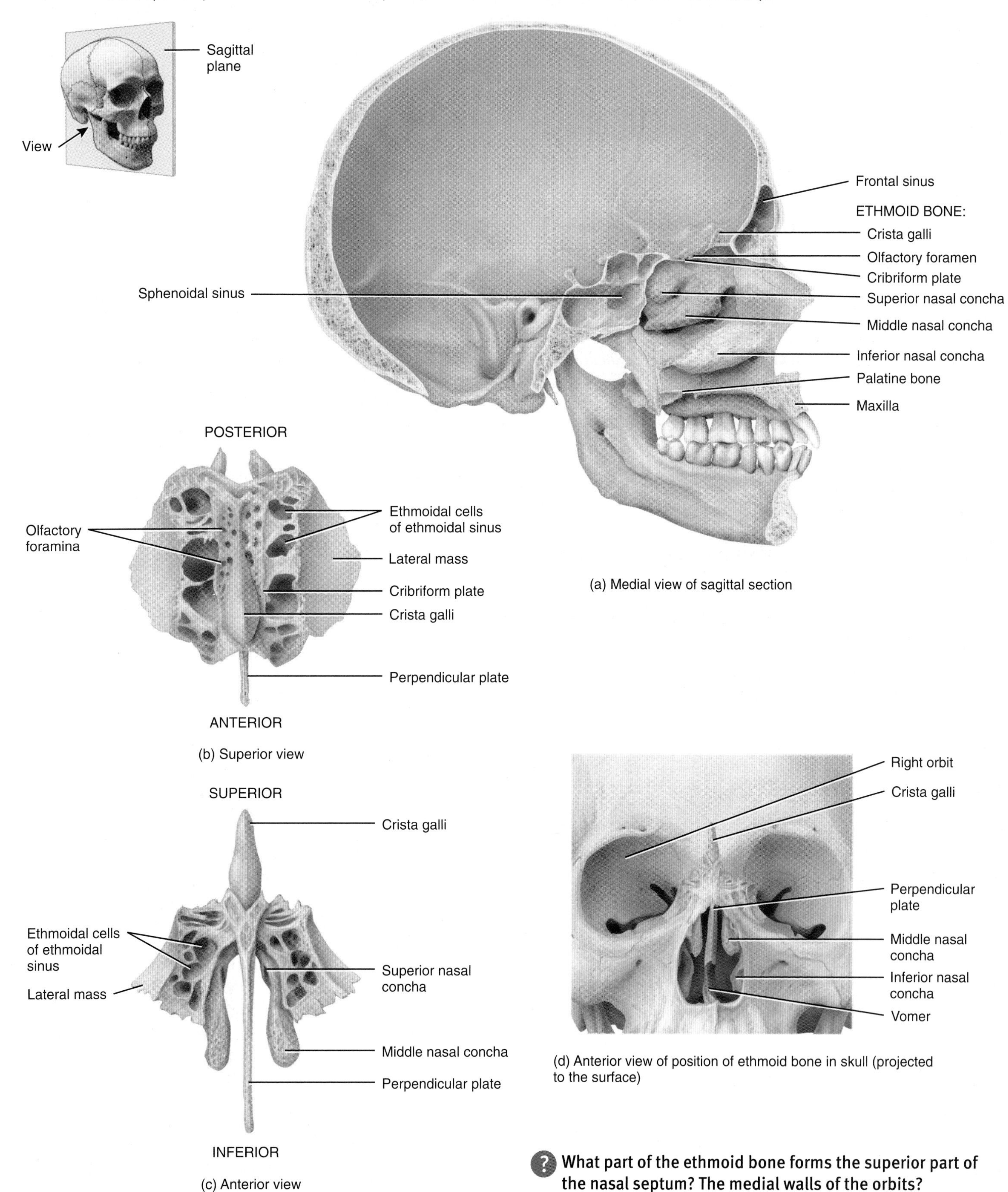

(a) Medial view of sagittal section

(b) Superior view

(c) Anterior view

(d) Anterior view of position of ethmoid bone in skull (projected to the surface)

What part of the ethmoid bone forms the superior part of the nasal septum? The medial walls of the orbits?

form part of the floors of the orbits, part of the lateral walls and floor of the nasal cavity, and most of the hard palate. The *hard palate* is a bony partition composed of the palatine processes of the maxillae and horizontal plates of the palatine bones that forms the roof of the mouth (see Figure 7.7).

Each maxilla contains a large *maxillary sinus* that empties into the nasal cavity (see Figure 7.13). The *alveolar process* (al-VĒ-ō-lar; *alveol-* = small cavity) is an arch that contains the *alveoli* (sockets) for the maxillary (upper) teeth. The *palatine process* is a horizontal projection of the maxilla that forms the anterior three-quarters of the hard palate. The union and fusion of the two maxillary bones is normally completed before birth.

The *infraorbital foramen* (*infra-* = below; *orbital* = orbit), which can be seen in the anterior view of the skull in Figure 7.3, is an opening in the maxilla inferior to the orbit. Through it passes the infraorbital nerve and blood vessels and a branch of the maxillary division of the trigeminal (V) nerve. Another prominent foramen in the maxilla is the *incisive foramen* (= incisor teeth) just posterior to the incisor teeth (see Figure 7.7). It transmits branches of the greater palatine blood vessels and nasopalatine nerve. A final structure associated with the maxilla and sphenoid bone is the *inferior orbital fissure,* which is located between the greater wing of the sphenoid and the maxilla (see Figure 7.12).

• CLINICAL CONNECTION | Cleft Palate and Cleft Lip

Usually the palatine processes of the maxillary bones unite during weeks 10 to 12 of embryonic development. Failure to do so can result in one type of **cleft palate.** The condition may also involve incomplete fusion of the horizontal plates of the palatine bones (see Figure 7.7). Another form of this condition, called **cleft lip,** involves a split in the upper lip. Cleft lip and cleft palate often occur together. Depending on the extent and position of the cleft, speech and swallowing may be affected. In addition, children with cleft palate tend to have many ear infections that can lead to hearing loss. Facial and oral surgeons recommend closure of cleft lip during the first few weeks following birth, and surgical results are excellent. Repair of cleft palate typically is done between 12 and 18 months of age, ideally before the child begins to talk. Speech therapy may be needed, because the palate is important for pronouncing consonants, and orthodontic therapy may be needed to align the teeth. Again, results are usually excellent. Folic acid (one of the B vitamins) supplementation during pregnancy decreases the incidence of cleft palate and cleft lip. •

Zygomatic Bones

The two ***zygomatic bones*** (*zygo-* = yokelike), commonly called cheekbones, form the prominences of the cheeks (see Figure 7.3) and part of the lateral wall and floor of each orbit (see Figure 7.12). They articulate with the frontal, maxilla, sphenoid, and temporal bones.

The *temporal process* of the zygomatic bone projects posteriorly and articulates with the zygomatic process of the temporal bone to form the *zygomatic arch* (see Figure 7.4).

Lacrimal Bones

The paired ***lacrimal bones*** (LAK-ri-mal; *lacrim-* = teardrops) are thin and roughly resemble a fingernail in size and shape (see Figures 7.3, 7.4, and 7.12). These bones, the smallest bones of the face, are posterior and lateral to the nasal bones and form a part of the medial wall of each orbit. The lacrimal bones each contain a *lacrimal fossa,* a vertical tunnel formed with the maxilla that houses the *lacrimal sac,* a structure that gathers tears and passes them into the nasal cavity (see Figure 7.12).

Palatine Bones

The two L-shaped ***palatine bones*** (PAL-a-tīn) form the posterior portion of the hard palate, part of the floor and lateral wall of the nasal cavity, and a small portion of the floors of the orbits (see Figures 7.7 and 7.11). The posterior portion of the hard palate, which separates the nasal cavity from the oral cavity, is formed by the *horizontal plates* of the palatine bones (see Figures 7.6 and 7.7).

Inferior Nasal Conchae

The two ***inferior nasal conchae*** are inferior to the middle nasal conchae of the ethmoid bone (see Figures 7.3 and 7.9a). These scroll-like bones form a part of the inferior lateral wall of the nasal cavity and project into the nasal cavity. The inferior nasal conchae are separate bones; they are not part of the ethmoid bone. All three pairs of nasal conchae help swirl and filter air before it passes into the lungs. However, only the superior nasal conchae of the ethmoid bone are involved in the sense of smell.

Vomer

The ***vomer*** (VŌ-mer = plowshare) is a roughly triangular bone on the floor of the nasal cavity that articulates superiorly with the perpendicular plate of the ethmoid bone and inferiorly with both the maxillae and palatine bones along the midline (see Figures 7.3, 7.7, and 7.11). It is one of the components of the nasal septum.

Mandible

The ***mandible*** (*mand-* = to chew), or lower jawbone, is the largest, strongest facial bone (Figure 7.10). It is the only movable skull bone (other than the auditory ossicles). In the lateral view, you can see that the mandible consists of a curved, horizontal portion, the *body,* and two perpendicular portions, the *rami* (RĀ-mī- = branches). The *angle* of the mandible is the area where each *ramus* (singular form) meets the body. Each ramus has a posterior *condylar process* (KON-di-lar) that articulates with the mandibular fossa and articular tubercle of the temporal bone (see Figure 7.4) to form the ***temporomandibular joint (TMJ).*** It also has an anterior *coronoid process* (KOR-ō-noyd) to which the temporalis muscle attaches. The depression between the coronoid and condylar processes is called the *mandibular notch.* The *alveolar process* is an arch containing the *alveoli* (sockets) for the mandibular (lower) teeth.

The *mental foramen* (*ment-* = chin) is approximately inferior to the second premolar tooth. It is near this foramen that dentists

Figure 7.10 Mandible.

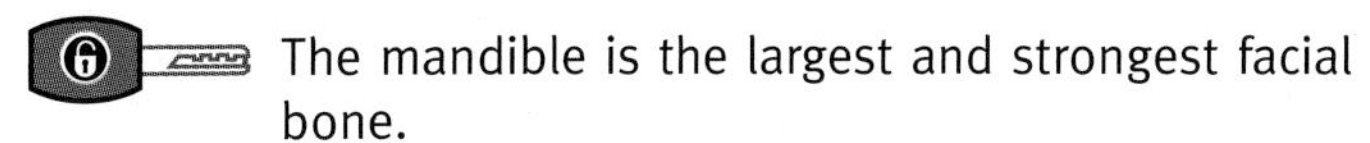

The mandible is the largest and strongest facial bone.

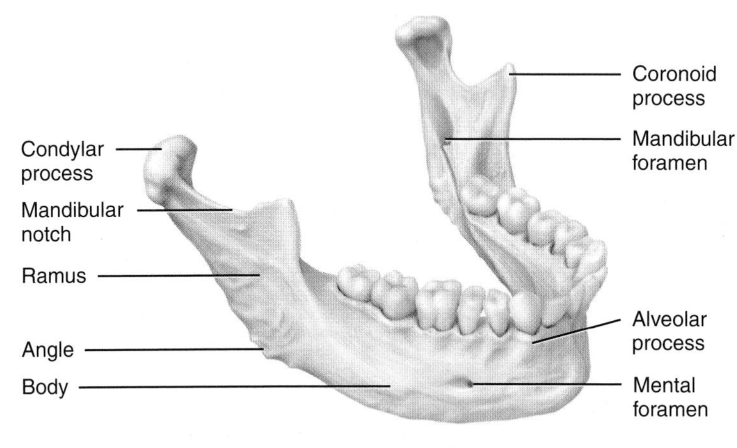

What is the distinctive functional feature of the mandible among all the skull bones?

reach the mental nerve when injecting anesthetics. Another foramen associated with the mandible is the *mandibular foramen* on the medial surface of each ramus, another site often used by dentists to inject anesthetics. The mandibular foramen is the beginning of the *mandibular canal,* which runs obliquely in the ramus and anteriorly to the body. Through the canal pass the inferior alveolar nerves and blood vessels, which are distributed to the mandibular teeth.

• CLINICAL CONNECTION | Temporomandibular Joint Syndrome

One problem associated with the temporomandibular joint is **temporomandibular joint (TMJ) syndrome.** It is characterized by dull pain around the ear, tenderness of the jaw muscles, a clicking or popping noise when opening or closing the mouth, limited or abnormal opening of the mouth, headache, tooth sensitivity, and abnormal wearing of the teeth. TMJ syndrome can be caused by improperly aligned teeth, grinding or clenching the teeth, trauma to the head and neck, or arthritis. Treatment may involve applying moist heat or ice, eating soft foods, taking pain relievers such as aspirin, muscle retraining, adjusting or reshaping the teeth, massage, orthodontic treatment, or surgery. •

Nasal Septum

The inside of the nose, called the nasal cavity, is divided into right and left sides by a vertical partition called the ***nasal septum*** (SEP-tum). The three components of the nasal septum are the vomer, septal cartilage, and the perpendicular plate of the ethmoid bone (Figure 7.11). The anterior border of the vomer articulates with the septal cartilage, which is hyaline cartilage, to form the anterior portion of the septum. The superior border of the vomer articulates with the perpendicular plate of the ethmoid bone to form the remainder of the nasal septum. The inferior border of the vomer articulates with the maxillae and the palatine bones. The term "broken nose," in most cases, refers to damage to the septal cartilage rather than the nasal bones themselves.

Figure 7.11 Nasal septum.

The structures that form the nasal septum are the perpendicular plate of the ethmoid bone, the vomer, and septal cartilage.

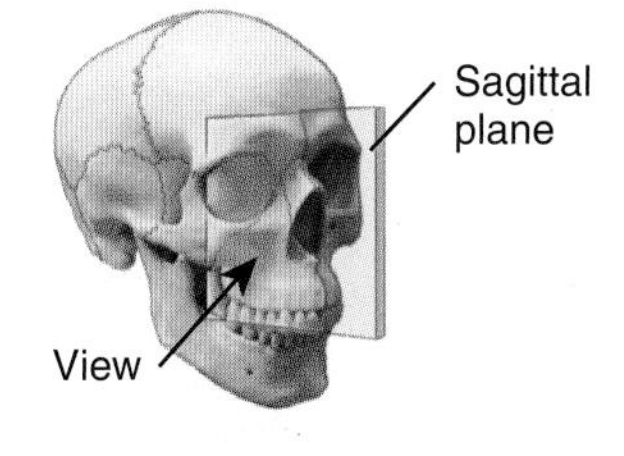

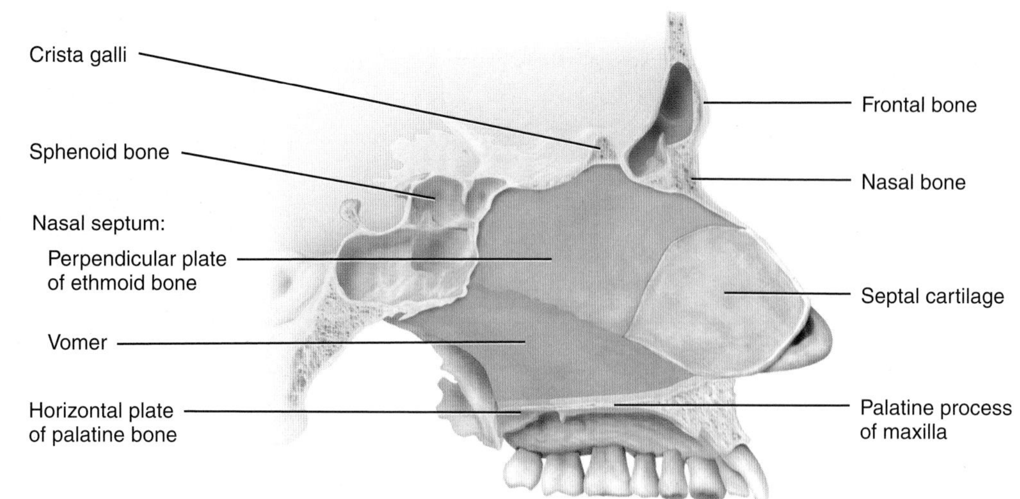

What is the function of the nasal septum?

Orbits

Seven bones of the skull join to form each ***orbit*** (OR-bit) (eye socket), which contains the eyeball and associated structures (Figure 7.12). The three cranial bones of the orbit are the frontal, sphenoid, and ethmoid; the four facial bones are the palatine, zygomatic, lacrimal, and maxilla. Each pyramid-shaped orbit has four regions that converge posteriorly:

1. Parts of the frontal and sphenoid bones comprise the *roof* of the orbit.

2. Parts of the zygomatic and sphenoid bones form the *lateral wall* of the orbit.

3. Parts of the maxilla, zygomatic, and palatine bones make up the *floor* of the orbit.

4. Parts of the maxilla, lacrimal, ethmoid, and sphenoid bones form the *medial wall* of the orbit.

Associated with each orbit are five openings:

1. The *optic foramen* is at the junction of the roof and medial wall.

2. The *superior orbital fissure* is at the superior lateral angle of the apex.

3. The *inferior orbital fissure* is at the junction of the lateral wall and floor.

4. The *supraorbital foramen* is on the medial side of the supraorbital margin of the frontal bone.

5. The *lacrimal fossa* is in the lacrimal bone.

Foramina

We mentioned most of the ***foramina*** (openings for blood vessels, nerves, or ligaments) of the skull in the descriptions of the cranial and facial bones that they penetrate. As preparation for studying other systems of the body, especially the nervous and cardiovascular systems, these foramina and the structures passing through them are listed in Table 7.4. For your convenience and for future reference, the foramina are listed alphabetically.

Unique Features of the Skull

The skull exhibits several unique features not seen in other bones of the body. These include sutures, paranasal sinuses, and fontanels.

Sutures

A ***suture*** (SOO-chur = seam) is an immovable or slightly movable joint in an adult that is found only between skull bones and that holds most skull bones together. Sutures in the skulls of infants and children often are more movable. The names of many sutures reflect the bones they unite. For example, the frontozygomatic suture is between the frontal bone and the zygomatic bone. Similarly, the sphenoparietal suture is between the sphenoid bone and the parietal bone. In other cases, however, the names of sutures are not so obvious. Of the many sutures found in the skull, we will identify only four prominent ones:

1. The ***coronal suture*** (kō-RŌ-nal; *coron-* = crown) unites the frontal bone and both parietal bones (see Figure 7.4).

Figure 7.12 Details of the orbit (eye socket).

The orbit is a pyramid-shaped structure that contains the eyeball and associated structures.

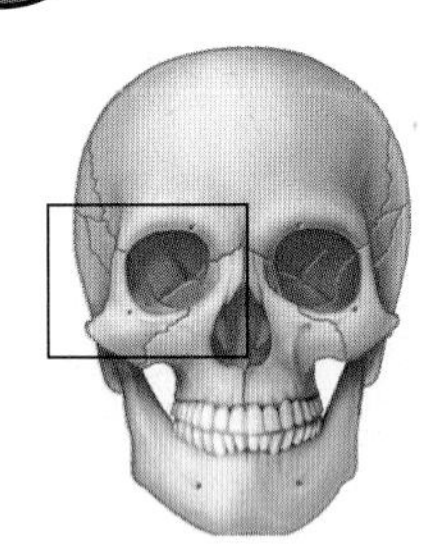

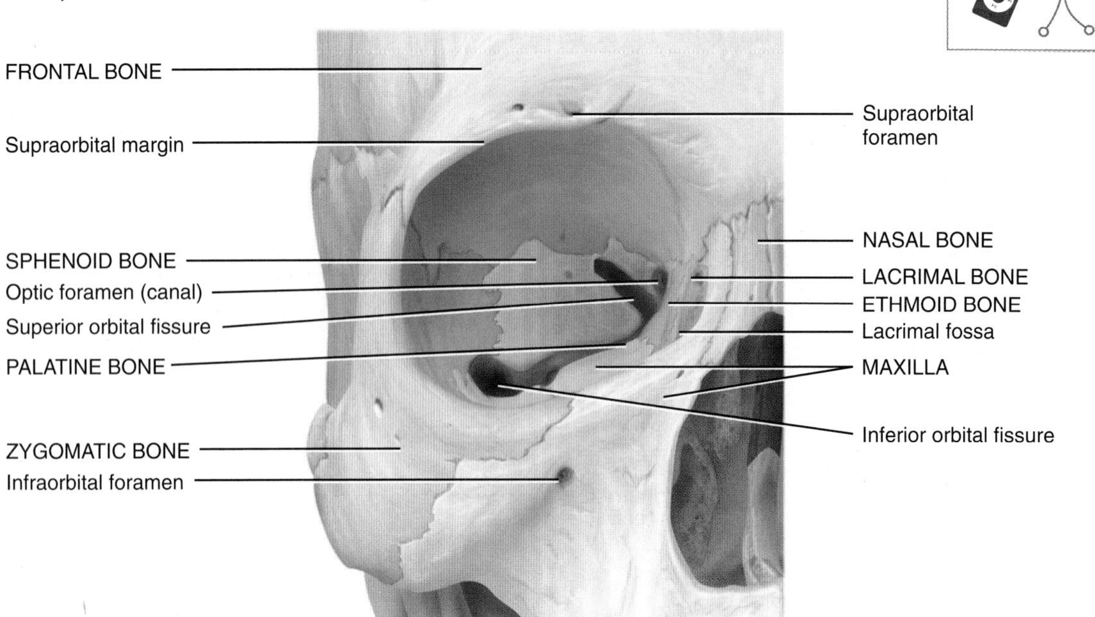

Anterior view showing the bones of the right orbit

Which seven bones form the orbit?

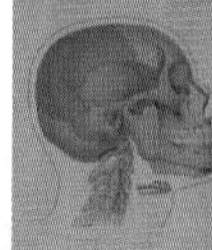

Table 7.4

Principal Foramina of the Skull

FORAMEN	LOCATION	STRUCTURES PASSING THROUGH*
Carotid (relating to carotid artery in neck)	Petrous portion of temporal bone (Figure 7.7).	Internal carotid artery and sympathetic nerves for eyes.
Hypoglossal (*hypo-* = under; *glossus* = tongue)	Superior to base of occipital condyles (Figure 7.8a).	Cranial nerve XII (hypoglossal) and branch of ascending pharyngeal artery.
Infraorbital (*infra-* = below)	Inferior to orbit in maxilla (Figure 7.12).	Infraorbital nerve and blood vessels and a branch of the maxillary division of cranial nerve V (trigeminal).
Jugular (*jugul-* = the throat)	Posterior to carotid canal between petrous portion of temporal bone and occipital bone (Figure 7.8a).	Internal jugular vein, cranial nerves IX (glossopharyngeal), X (vagus), and XI (accessory).
Lacerum (= lacerated)	Bounded anteriorly by sphenoid bone, posteriorly by petrous portion of temporal bone, and medially by sphenoid and occipital bones (Figure 7.8a).	Branch of ascending pharyngeal artery.
Magnum (= large)	Occipital bone (Figure 7.7).	Medulla oblongata and its membranes (meninges), cranial nerve XI (accessory), and vertebral and spinal arteries.
Mandibular (*mand-* = to chew)	Medial surface of ramus of mandible (Figure 7.10).	Inferior alveolar nerve and blood vessels.
Mental (*ment-* = chin)	Inferior to second premolar tooth in mandible (Figure 7.10).	Mental nerve and vessels.
Olfactory (*olfact-* = to smell)	Cribriform plate of ethmoid bone (Figure 7.8a).	Cranial nerve I (olfactory).
Optic (= eye)	Between superior and inferior portions of small wing of sphenoid bone (Figure 7.12).	Cranial nerve II (optic) and ophthalmic artery.
Ovale (= oval)	Greater wing of sphenoid bone (Figure 7.8a).	Mandibular branch of cranial nerve V (trigeminal).
Rotundum (= round)	Junction of anterior and medial parts of sphenoid bone (Figure 7.8).	Maxillary branch of cranial nerve V (trigeminal).
Stylomastoid (*stylo-* = stake or pole)	Between styloid and mastoid processes of temporal bone (Figure 7.7).	Cranial nerve VII (facial) and stylomastoid artery.
Supraorbital (*supra-* = above)	Supraorbital margin of orbit in frontal bone (Figure 7.12).	Supraorbital nerve and artery.

*The cranial nerves listed here are described in Table 17.4.

2. The ***sagittal suture*** (SAJ-i-tal; *sagitt-* = arrow) unites the two parietal bones on the superior midline of the skull (see Figure 7.6). The sagittal suture is so named because in the infant, before the bones of the skull are firmly united, the suture and the fontanels (soft spots) associated with it resemble an arrow.

3. The ***lambdoid suture*** (LAM-doyd) unites the two parietal bones to the occipital bone. This suture is so named because of its resemblance to the Greek letter lambda (Λ), as can be seen in Figure 7.6. Sutural bones may occur within the sagittal and lambdoid sutures.

4. The ***squamous sutures*** (SKWĀ-mus; *squam-* = flat) unite the parietal and temporal bones on the lateral aspects of the skull (see Figure 7.4).

Paranasal Sinuses

The **paranasal sinuses** (par′-a-NĀ-zal SĪ-nus-ez; *para-* = beside) are cavities in certain cranial and facial bones near the nasal cavity (see Figure 7.13). They are most evident in a sagittal section of the skull. The paranasal sinuses are lined with mucous membranes that are continuous with the lining of the nasal cavity. Skull bones containing the paranasal sinuses are the frontal, sphenoid, ethmoid, and maxillary. Besides producing mucus, the paranasal sinuses serve as resonating (echo) chambers for sound as we speak or sing.

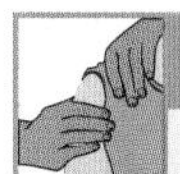

MANUAL THERAPY APPLICATION

Sinusitis

Secretions produced by the mucous membranes of the paranasal sinuses drain into the nasal cavity. An inflammation of the membranes due to an allergic reaction or infection is called **sinusitis.** If the membranes swell enough to block drainage into the nasal cavity, fluid pressure builds up in the paranasal sinuses, and a sinus headache results. A severely deviated nasal septum or nasal polyps, growths that can be removed surgically, may also cause chronic sinusitis. *Reflexology treatments* involving a therapist squeezing each fingertip for 30 seconds will cause the blocked sinuses to open in approximately two of three patients.

Figure 7.13 Paranasal sinuses (projected to the surface).

Paranasal sinuses are mucous membrane–lined spaces in the frontal, sphenoid, ethmoid, and maxillary bones that connect to the nasal cavity.

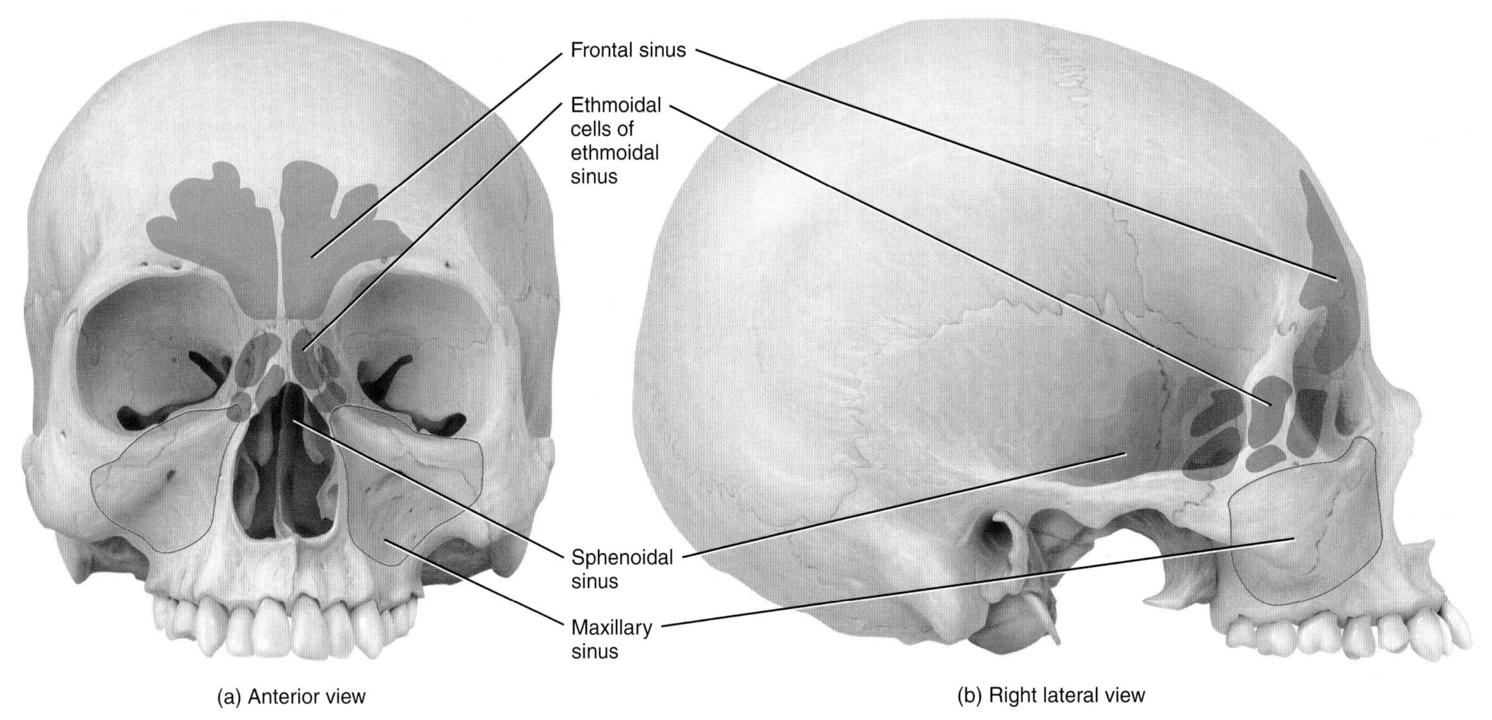

What are the functions of the paranasal sinuses?

Fontanels

The skeleton of a newly formed embryo consists of cartilage or mesenchyme arranged in sheetlike layers that resemble membranes shaped like bones. Gradually, ossification occurs—bone replaces the cartilage and mesenchyme. At birth, mesenchyme-filled spaces called ***fontanels*** (fon-ta-NELZ = little fountains) are present between the cranial bones (Figure 7.14). Commonly called "soft spots," fontanels are areas of unossified mesenchyme. Eventually, they will be replaced with bone by intramembranous ossification and become sutures. Functionally, the fontanels provide some flexibility to the fetal skull. They allow the skull to change shape as it passes through the birth canal and permit rapid growth of the brain during infancy. Although an infant may have many fontanels at birth, the form and location of six are fairly constant:

1. The unpaired ***anterior fontanel,*** located at the midline between the two parietal bones and the frontal bone, is roughly diamond-shaped and is the largest fontanel. It usually closes 18 to 24 months after birth.

2. The unpaired ***posterior fontanel*** is located at the midline between the two parietal bones and the occipital bone. Because it is much smaller than the anterior fontanel, it generally closes about 2 months after birth.

3. The paired ***anterolateral fontanels,*** located laterally between the frontal, parietal, temporal, and sphenoid bones, are

Figure 7.14 Fontanels at birth.

Fontanels are mesenchyme-filled spaces between cranial bones that are present at birth.

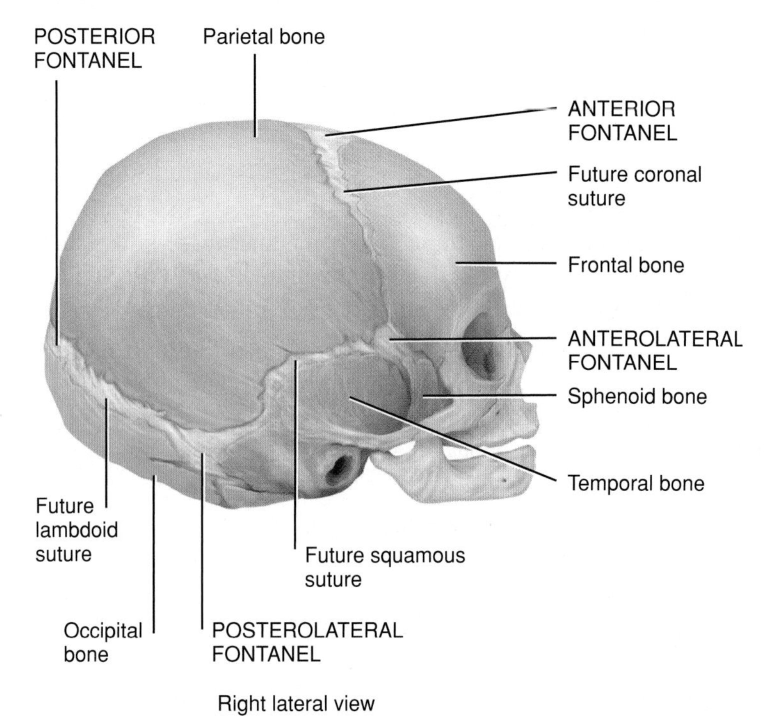

Which fontanel is bordered by four different skull bones?

small and irregular in shape. Normally, they close about 3 months after birth.

4. The paired ***posterolateral fontanels,*** located laterally between the parietal, occipital, and temporal bones, are irregularly shaped. They begin to close 1 to 2 months after birth, but closure is generally not complete until 12 months.

The amount of closure in fontanels helps a physician gauge the degree of brain development. In addition, the anterior fontanel serves as a landmark for withdrawal of blood for analysis from the superior sagittal sinus (a large vein on the midline surface of the brain).

CHECKPOINT

4. Describe the general features of the skull.
5. What bones constitute the orbit?
6. What structures make up the nasal septum?
7. Define the following: foramen, suture, paranasal sinus, and fontanel.

7.5 HYOID BONE

OBJECTIVE

- Describe the relationship of the hyoid bone to the skull.

The single ***hyoid bone*** (= U-shaped) is a unique component of the axial skeleton because it does not articulate with any other bone. Rather, it is suspended from the styloid processes of the temporal bones by ligaments and muscles (see Figure 11.8). Located in the anterior neck between the mandible and larynx (Figure 7.15a), the hyoid bone supports the tongue, providing attachment sites for some tongue muscles and for muscles of the neck and pharynx. The hyoid bone consists of a horizontal *body* and paired projections called the *lesser horns* and the *greater horns* (Figure 7.15b,c). Muscles and ligaments attach to these paired projections.

The hyoid bone, as well as cartilages of the larynx and trachea, are often fractured during strangulation. As a result, they are carefully examined at autopsy when strangulation is suspected.

CHECKPOINT

8. What are the functions of the hyoid bone?

Figure 7.15 Hyoid bone.

The hyoid bone supports the tongue, providing attachment sites for muscles of the tongue, neck, and pharynx.

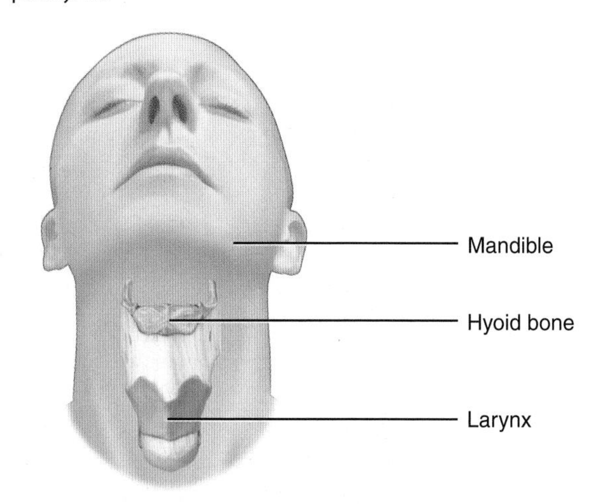

(a) Position of hyoid

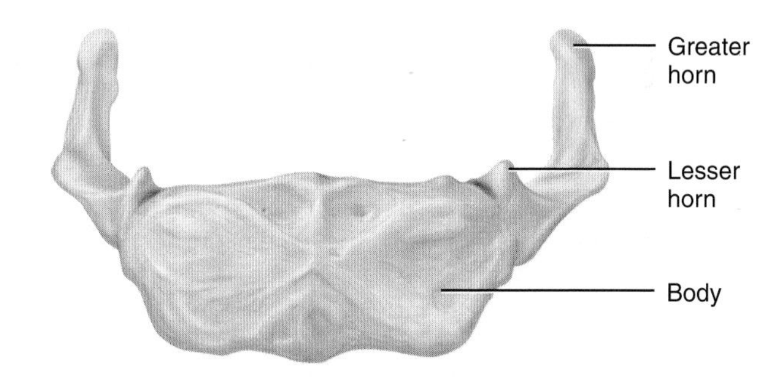

(b) Anterior view

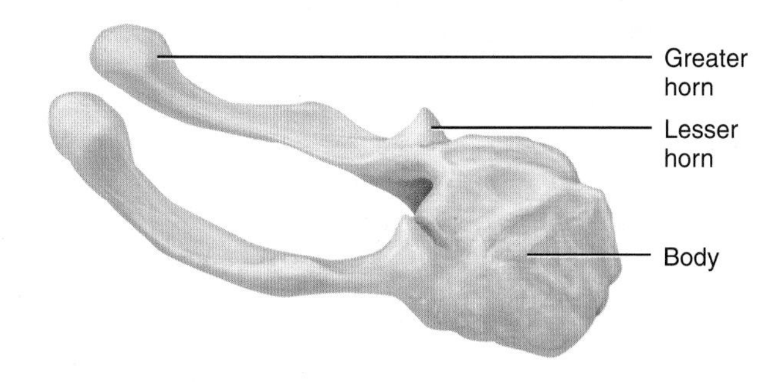

(c) Right lateral view

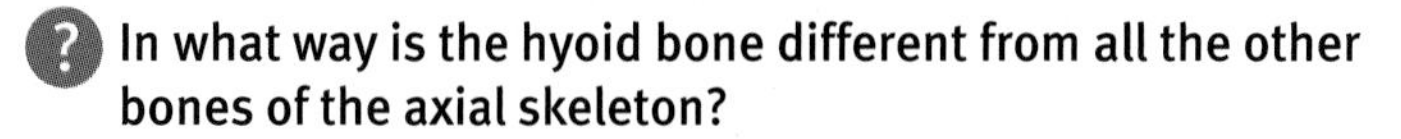

In what way is the hyoid bone different from all the other bones of the axial skeleton?

7.6 VERTEBRAL COLUMN

OBJECTIVE

- Identify the regions and normal curves of the vertebral column and describe its structural and functional features.

The ***vertebral column,*** also called the *spine, spinal column,* or *backbone,* makes up about two-fifths of the total height of the body and is composed of a series of bones called ***vertebrae*** (VER-te-brā; singular is *vertebra*). The vertebral column consists of bone and connective tissue; the spinal cord that it surrounds and protects consists of nervous tissue. The length of the column is about 71 cm (28 in.) in an average adult male and about 61 cm (24 in.) in an average adult female. The vertebral column functions as a strong, flexible rod with elements that can move forward, move backward, move sideways, and rotate. It encloses and protects the spinal cord, supports the head, and serves as a point of attachment for the ribs, pelvic girdle, and muscles of the back.

Figure 7.16 Vertebral column. The numbers in parentheses in (a) indicate the number of vertebrae in each region. In (d), the relative size of the disc has been enlarged for emphasis.

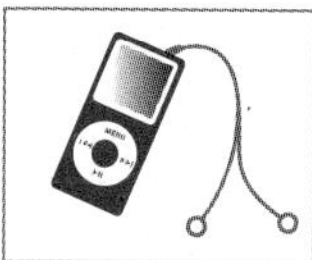

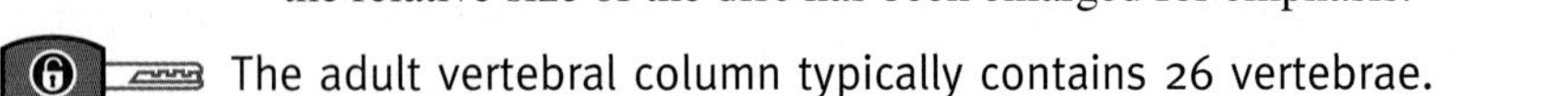
The adult vertebral column typically contains 26 vertebrae.

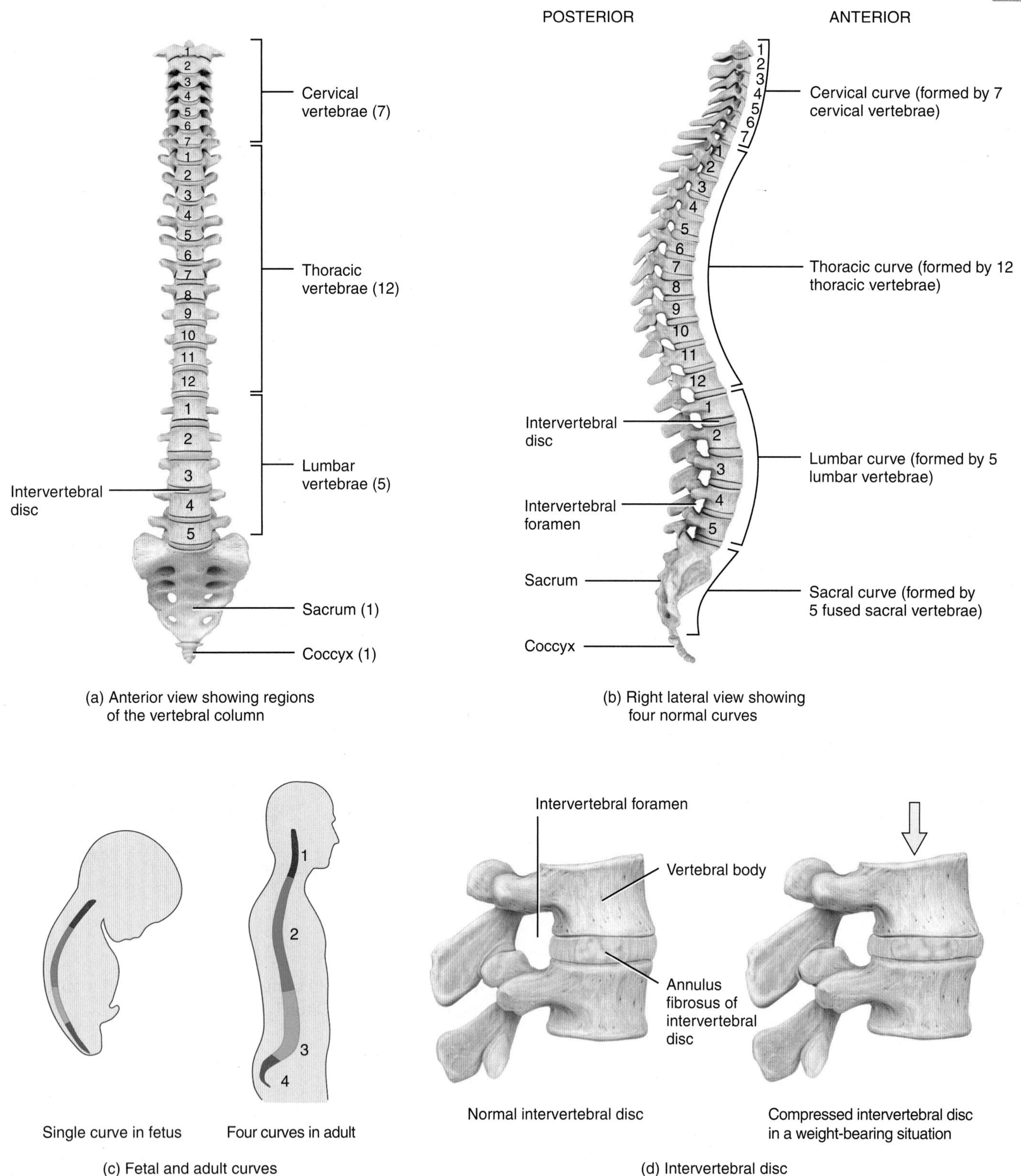

(a) Anterior view showing regions of the vertebral column

(b) Right lateral view showing four normal curves

(c) Fetal and adult curves

(d) Intervertebral disc

Which curves of the adult vertebral column are concave (relative to the anterior side of the body)?

The total number of vertebrae during early development is 33. Then, several vertebrae in the sacral and coccygeal regions fuse. As a result, the adult vertebral column, also called the spinal column, typically contains 26 vertebrae (see Figure 7.16a). These are distributed as follows:

- 7 ***cervical vertebrae*** (*cervic-* = neck) are in the neck region.
- 12 ***thoracic vertebrae*** (*thorax* = chest) are posterior to the thoracic cavity.
- 5 ***lumbar vertebrae*** (*lumb-* = loin) support the lower back.
- 1 ***sacrum*** (SĀ-krum = sacred bone) consists of five fused ***sacral vertebrae.***
- 1 ***coccyx*** (KOK-siks = cuckoo, because the shape resembles the bill of a cuckoo bird), consists of four fused ***coccygeal vertebrae*** (kok-SIJ-ē-al).

Normal Curves of the Vertebral Column

When viewed from the side, the adult vertebral column shows four slight bends called ***normal curves*** (Figure 7.16b). Relative to the front of the body, the *cervical* and *lumbar curves* are convex (bulging out), whereas the *thoracic* and *sacral curves* are concave (cupping in). The curves of the vertebral column increase its strength, help maintain balance in the upright position, absorb shocks during walking, and help protect the vertebrae from fracture.

In the fetus, there is only a single anteriorly concave curve (Figure 7.16c). At about the third month after birth, when an infant begins to hold its head erect, the cervical curve develops. Later, when the child sits up, stands, and walks, the lumbar curve develops. The thoracic and sacral curves are called *primary curves* because they form first during fetal development. The cervical and lumbar curves are known as *secondary curves* because they begin to form later, several months after birth. All curves are fully developed by age 10. However, secondary curves may be progressively lost in old age.

Various conditions may exaggerate the normal curves of the vertebral column, or the column may acquire a lateral bend, resulting in ***abnormal curves*** of the vertebral column. Three such abnormal curves—kyphosis, lordosis, and scoliosis—are described in Section 7.8.

Intervertebral Discs

Between the bodies of adjacent vertebrae from the second cervical vertebra to the sacrum are ***intervertebral discs*** (in′-ter-VER-te-bral; *inter* = between) (Figure 7.16d). Each disc has an outer fibrous ring consisting of fibrocartilage called the *annulus fibrosus* (*annulus* = ringlike) and an inner soft, pulpy, highly elastic substance called the *nucleus pulposus* (*pulposus* = pulplike). The discs form strong joints, permit various movements of the vertebral column, and absorb vertical shock. Under compression, they flatten and broaden; with age, the nucleus pulposus hardens and becomes less elastic. Narrowing of the discs and compression of vertebrae results in a decrease in height with age.

MANUAL THERAPY APPLICATION

Herniated (Slipped) Disc

Like many other processes in the body, disc problems can be of varying degrees. The disc may be slightly bulged or radically bulged, or the disc may rupture and is called a **herniated (slipped) disc.** Persons may have a problem with one disc or multiple discs. Patients with disc problems commonly seek the services of manual therapists. Intervertebral discs function as shock absorbers and are constantly being compressed. If the anterior and posterior ligaments attached to the discs become injured or weakened, pressure developed in the nucleus pulposus may be great enough to rupture the surrounding annulus fibrosus (Figure 7.17). This condition is called a herniated (slipped) disc. Because the lumbar region bears much of the weight of the body, and is the region of the most flexing and bending, herniated discs most often occur in the lumbar area.

Frequently, with a herniated disc, the nucleus pulposus herniates (protrudes) posteriorly. This movement exerts pressure on the spinal nerves, causing acute pain. If the roots of the sciatic nerve, which (with some additional stops on the way) passes from the spinal cord to the foot, are compressed, the pain radiates down the posterior thigh, through the calf, and occasionally into the foot. If pressure is exerted on the spinal cord itself, some of its neurons may be destroyed. Treatment options include bed rest, medications for pain, physical therapy and exercises, lengthening of paravertebral muscles by a manual therapist, and traction. A person with a herniated disc may also undergo a *laminectomy,* a procedure in which parts of the laminae of the vertebra and parts of the intervertebral disc are removed to relieve pressure on nerves.

Figure 7.17 Herniated (slipped) disc.

6 Most often the nucleus pulposus herniates posteriorly.

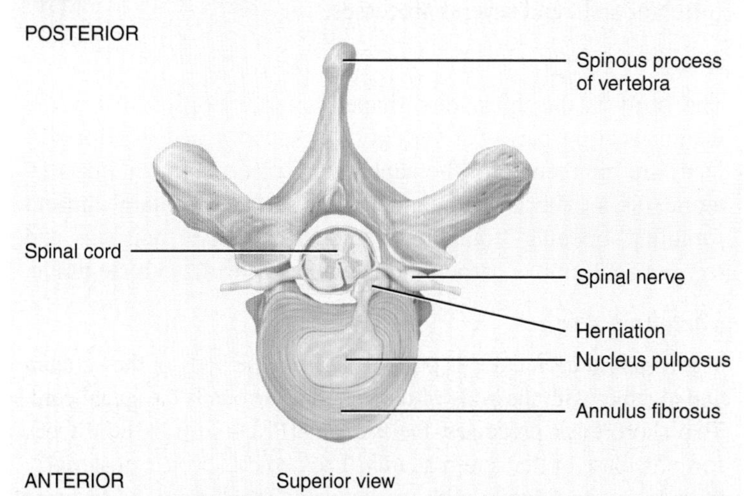

Why do most herniated discs occur in the lumbar region?

Parts of a Typical Vertebra

Even though vertebrae in different regions of the spinal column vary in size, shape, and detail, they are similar enough that we can discuss the structures (and the functions) of a typical

Figure 7.18 Structure of a typical vertebra, as illustrated by a thoracic vertebra. In (b), only one spinal nerve has been included, and it has been extended beyond the intervertebral foramen for clarity.

A vertebra consists of a body, a vertebral arch, and several processes.

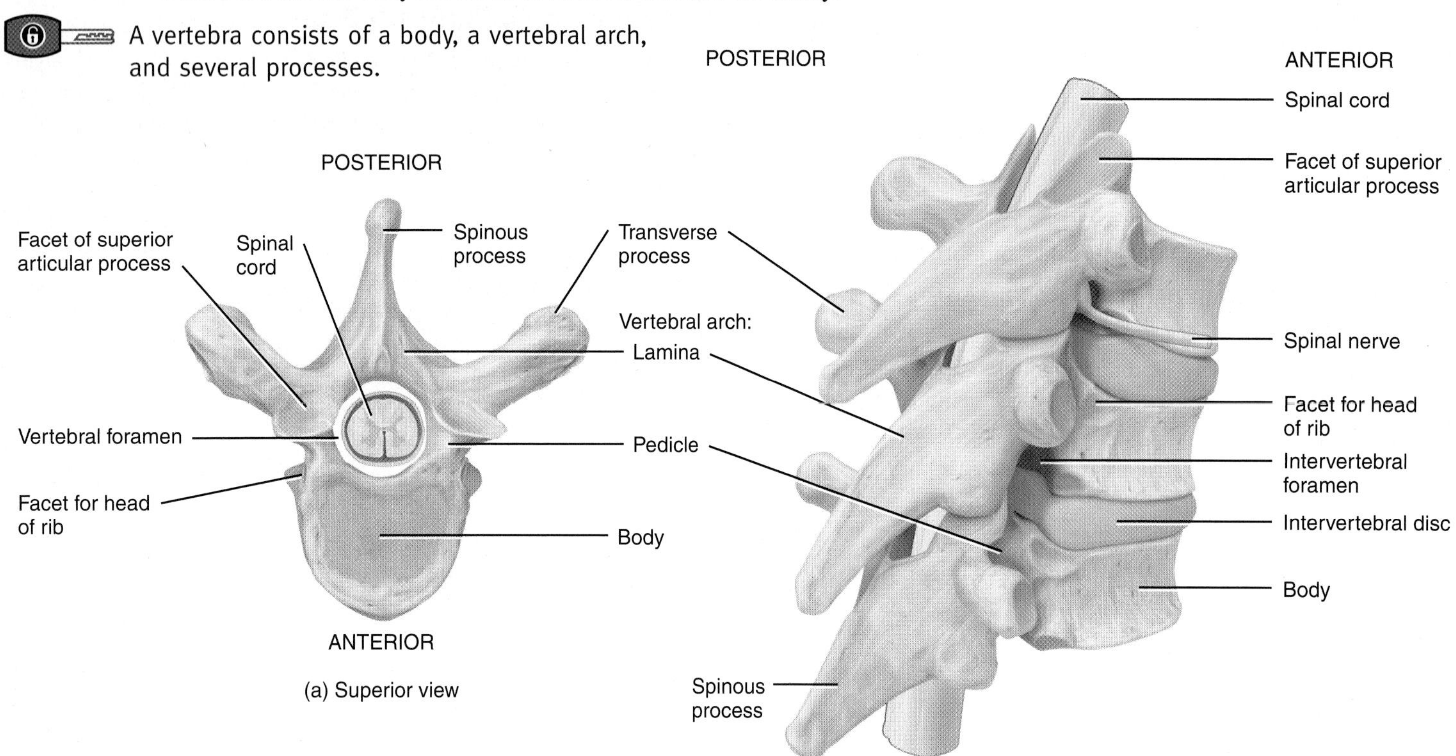

What are the functions of the vertebral and intervertebral foramina?

vertebra (Figure 7.18). Vertebrae typically consist of a body, a vertebral arch, and several processes.

Body

The **body** is the thick, disc-shaped anterior portion that is the weight-bearing part of a vertebra. Its superior and inferior surfaces are roughened for the attachment of cartilaginous intervertebral discs. The anterior and lateral surfaces contain nutrient foramina, openings for blood vessels that deliver nutrients and oxygen and remove carbon dioxide and wastes from bone tissue.

Vertebral Arch

The ***vertebral arch*** extends posteriorly from the body of the vertebra and together with the body of the vertebra surrounds the spinal cord. Two short, thick processes, the *pedicles* (PED-i-kuls = little feet), form the base of the vertebral arch. The pedicles project posteriorly from the body to unite with the laminae. The *laminae* (LAM-i-nē = thin layers) are the flat parts that join to form the posterior portion of the vertebral arch. The *vertebral foramen* lies between the vertebral arch and body and contains the spinal cord, adipose tissue, areolar connective tissue, and blood vessels. Collectively, the vertebral foramina of all vertebrae form the *vertebral (spinal) canal.* The pedicles exhibit superior and inferior indentations called *vertebral notches.* When the vertebral notches are stacked on top of one another, they form an opening between adjoining vertebrae on both sides of the column. Each opening, called an *intervertebral foramen,* permits the passage of a single spinal nerve.

Processes

Seven ***processes*** arise from the vertebral arch. At the point where a lamina and pedicle join, *transverse processes* extend laterally, one on each side. A single *spinous process (spine)* projects posteriorly from the junction of the laminae. These three processes serve as points of attachment for muscles. The remaining four processes form joints with other vertebrae above or below. The two *superior articular processes* of a vertebra articulate (form joints) with the two inferior articular processes of the vertebra immediately superior to them. In turn, the two *inferior articular processes* of that vertebra articulate with the two superior articular processes of the vertebra immediately inferior to them, and so on.

The articulating surfaces of the articular processes are referred to as *facets* (= little faces), and are covered with hyaline cartilage. The articulations formed between the bodies and articular facets of successive vertebrae are called *intervertebral joints.*

Regions of the Vertebral Column

We turn now to the five regions of the vertebral column, beginning superiorly and moving inferiorly. Note that vertebrae in each region are numbered in sequence, from superior to inferior.

Cervical Region

The bodies of ***cervical vertebrae*** (C1–C7) are smaller than those of thoracic vertebrae (Figure 7.19d). The vertebral arches, however, are larger. All cervical vertebrae have three foramina: one

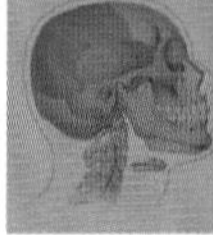

Figure 7.19 **Cervical vertebrae.**

The cervical vertebrae are found in the neck region.

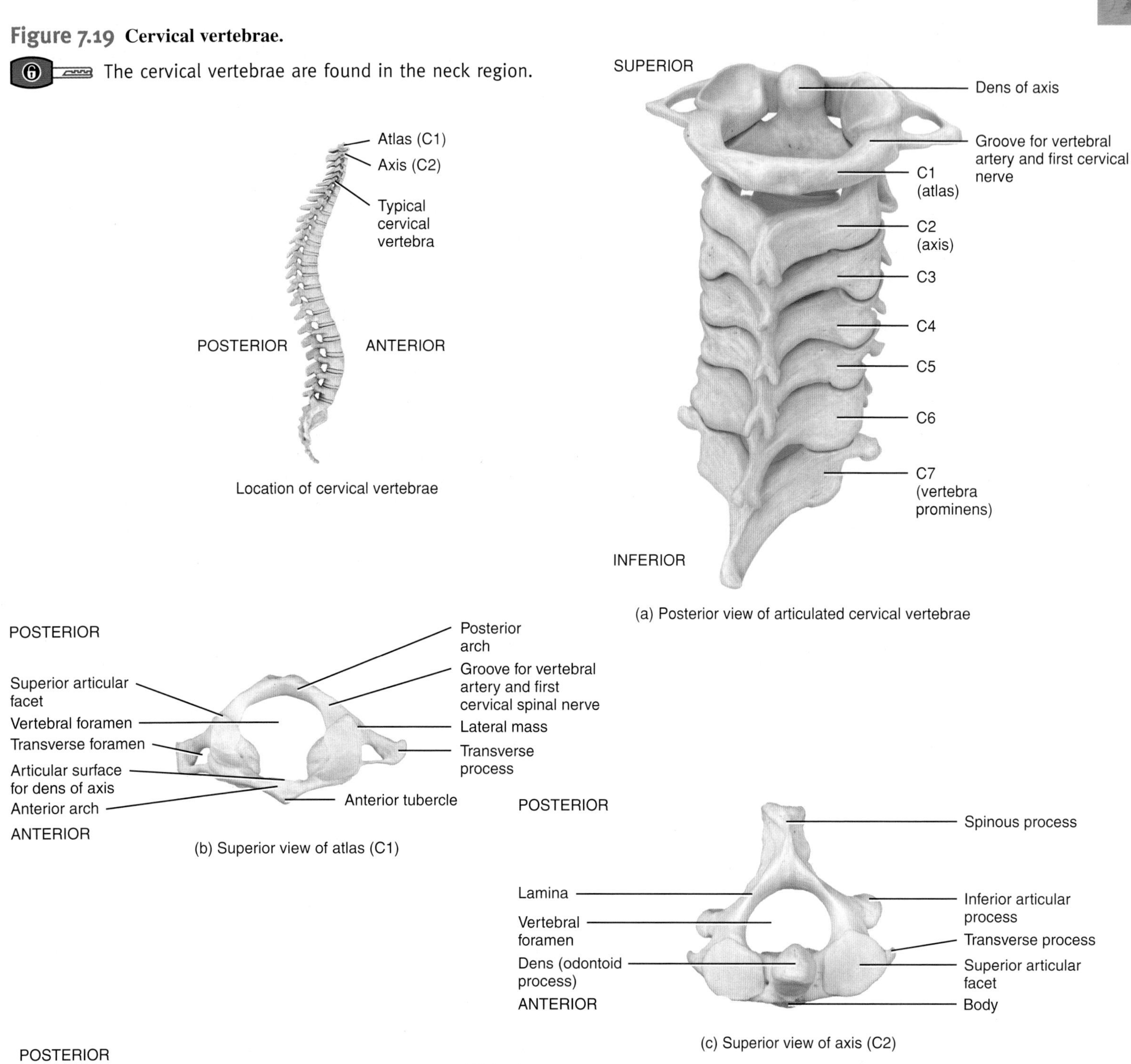

Which vertebrae permit you to move your head to signify "no"?

vertebral foramen and two transverse foramina (Figure 7.19d). The vertebral foramina of cervical vertebrae are the largest in the spinal column because they house the cervical enlargement of the spinal cord. Each cervical transverse process contains a *transverse foramen* through which the vertebral artery and its accompanying vein and nerve fibers pass. The spinous processes of C2 through C6 are often *bifid,* that is, split into two parts (Figure 7.19a,d).

The first two cervical vertebrae differ considerably from the others. Named after the mythological Atlas, who supported the world on his shoulders, the first cervical vertebra (C1), the ***atlas,*** supports the head (see Figure 7.19a,b). The atlas is a ring of bone with *anterior* and *posterior arches* and large *lateral masses.* It lacks a body and a spinous process. The superior surfaces of the lateral masses, called *superior articular facets,* are concave and articulate with the occipital condyles of the occipital bone to form the paired *atlanto-occipital joints.* These articulations permit the movement seen when moving the head to signify "yes." The inferior surfaces of the lateral masses, the *inferior articular facets,* articulate with the second cervical vertebra. The transverse processes and transverse foramina of the atlas are quite large.

The second cervical vertebra (C2), called the ***axis*** (see Figure 7.19a,c), does have a body. A peglike process called the *dens* (= tooth) or *odontoid process* projects superiorly through the anterior portion of the vertebral foramen of the atlas. Embryologically, the dens is the body of the atlas that has been shifted onto the axis. The dens makes a pivot on which the atlas and head rotate. This arrangement permits side-to-side movement of the head, as when you move your head to signify "no." The articulation formed between the anterior arch of the atlas and dens of the axis, and between their articular facets, is called the *atlanto-axial joint.* In some instances of trauma, the dens of the axis may be driven into the medulla oblongata of the brain. This type of injury is the usual cause of death from whiplash injuries.

The third through sixth cervical vertebrae (C3–C6), represented by the vertebra in Figure 7.19d, correspond to the structural pattern of the typical cervical vertebra previously described. The seventh cervical vertebra (C7), called the *vertebra prominens,* is somewhat different (see Figure 7.19a). It has a single large spinous process that may be seen and felt at the base of the neck.

Thoracic Region

Thoracic vertebrae T1–T12 (Figure 7.20) are considerably larger and stronger than cervical vertebrae. In addition, the spinous processes on T1 and T2 are long, laterally flattened, and directed inferiorly. In contrast, the spinous processes on T11 and T12 are shorter, broader, and directed more posteriorly. Compared to cervical vertebrae, thoracic vertebrae also have longer and larger transverse processes.

The most distinguishing feature of thoracic vertebrae is that they articulate with the ribs. Except for T11 and T12, the transverse processes have facets for articulating with the *tubercles* of

Figure 7.20 Thoracic vertebrae.

The thoracic vertebrae are found in the chest region and articulate with the ribs.

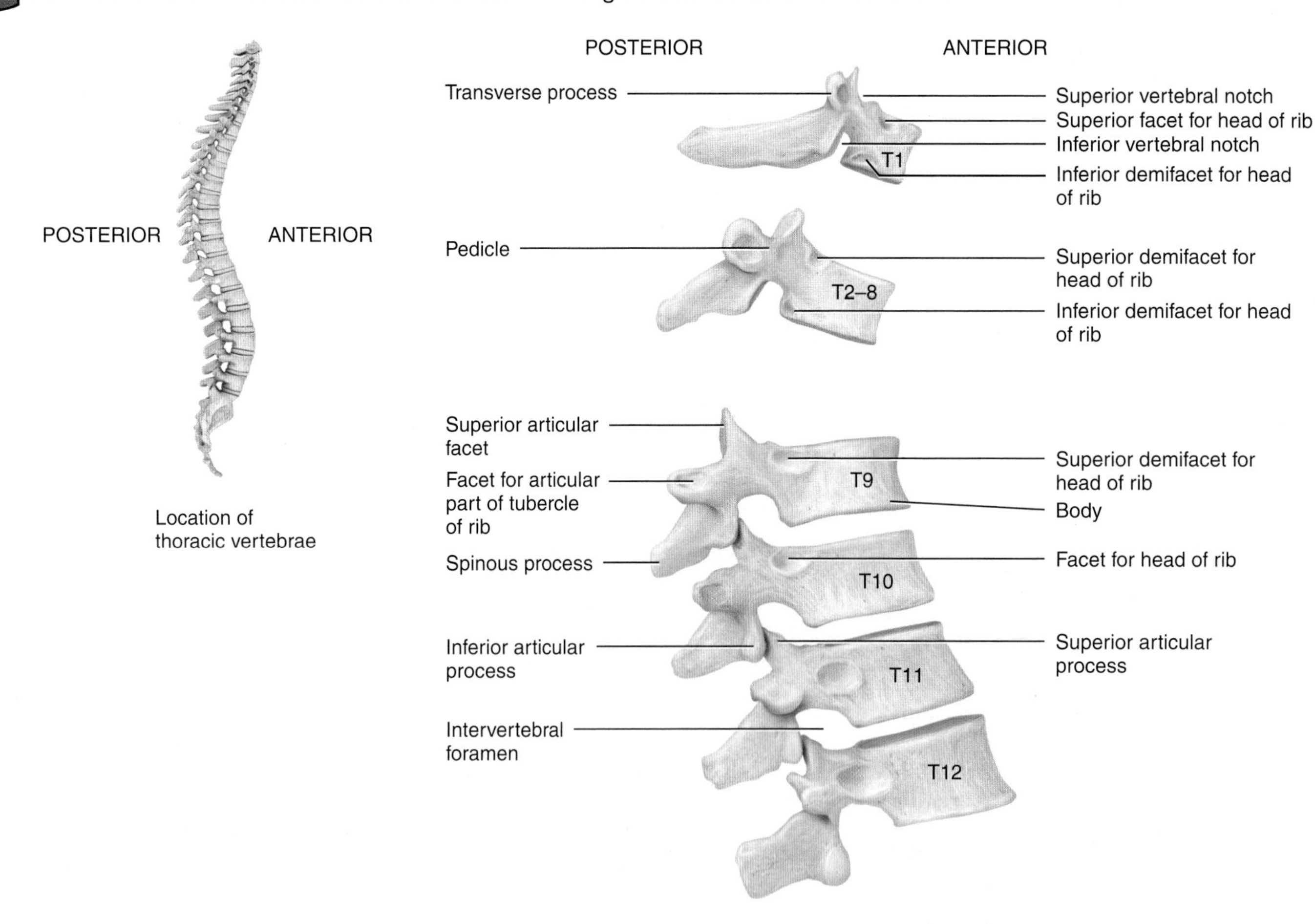

(a) Right lateral view of several articulated thoracic vertebrae

the ribs. The bodies of thoracic vertebrae also have either facets or demifacets (half-facets) for articulation with the *heads* of the ribs. The articulations between the thoracic vertebrae and ribs are called *vertebrocostal joints.* As you can see in Figure 7.20, T1 has a superior facet and an inferior demifacet, one on each side of the vertebral body. T2–T8 have a superior and inferior demifacet, one on each side of the vertebral body. T9 has a superior demifacet on each side of the vertebral body, and T10–T12 have a superior facet on each side of the vertebral body. Movements of the thoracic region are limited by thin intervertebral discs and by the attachment of the ribs to the sternum.

Lumbar Region

The ***lumbar vertebrae*** (L1–L5) are the largest and strongest in the vertebral column (see Figure 7.21) because the amount of body weight supported by the vertebrae increases toward the inferior end of the backbone. Their various projections are short and thick. The superior articular processes are directed medially instead of superiorly, and the inferior articular processes are directed laterally instead of inferiorly. The spinous processes are quadrilateral in shape, thick and broad, and project nearly straight posteriorly. The spinous processes are well adapted for the attachment of the large back muscles.

A summary of the major structural differences among cervical, thoracic, and lumbar vertebrae is presented in Table 7.5.

Sacrum

The ***sacrum*** is a triangular bone formed by the union of five sacral vertebrae (S1–S5), indicated in Figure 7.22a. The sacral vertebrae begin to fuse in individuals between 16 and 18 years of age, a process usually completed by age 30. The sacrum serves as a strong foundation for the pelvic girdle. It is positioned at the posterior portion of the pelvic cavity where its lateral surfaces fuse to the two hip bones. The female sacrum is shorter, wider, and more curved between S2 and S3 than the male sacrum (see Table 8.1).

The concave anterior side of the sacrum faces the pelvic cavity. It is smooth and contains four *transverse lines (ridges)* that mark the joining of the sacral vertebral bodies (see Figure 7.22a). At the ends of these lines are four pairs of *anterior sacral foramina.* The lateral portion of the superior surface of the sacrum contains a smooth surface called the *sacral ala* (= wing), which is formed by the fused transverse processes of the first sacral vertebra (S1).

The convex, posterior surface of the sacrum contains a *median sacral crest,* which is the fused spinous processes of the upper sacral vertebrae; two *lateral sacral crests,* which are the fused transverse processes of the sacral vertebrae; and four pairs of *posterior sacral foramina* (see Figure 7.22b). These foramina connect with the anterior sacral foramina to allow passage of nerves and blood vessels. The *sacral canal* is a continuation of

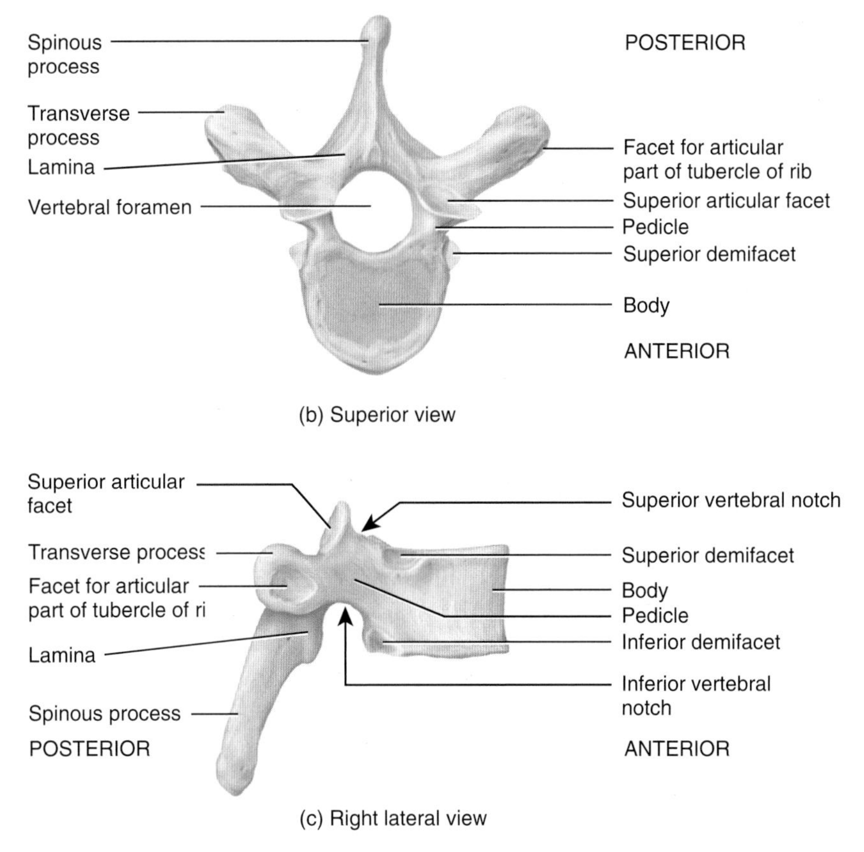

(b) Superior view

(c) Right lateral view

Which parts of the thoracic vertebrae articulate with the ribs?

Figure 7.21 Lumbar vertebrae.

Lumbar vertebrae are found in the lower back.

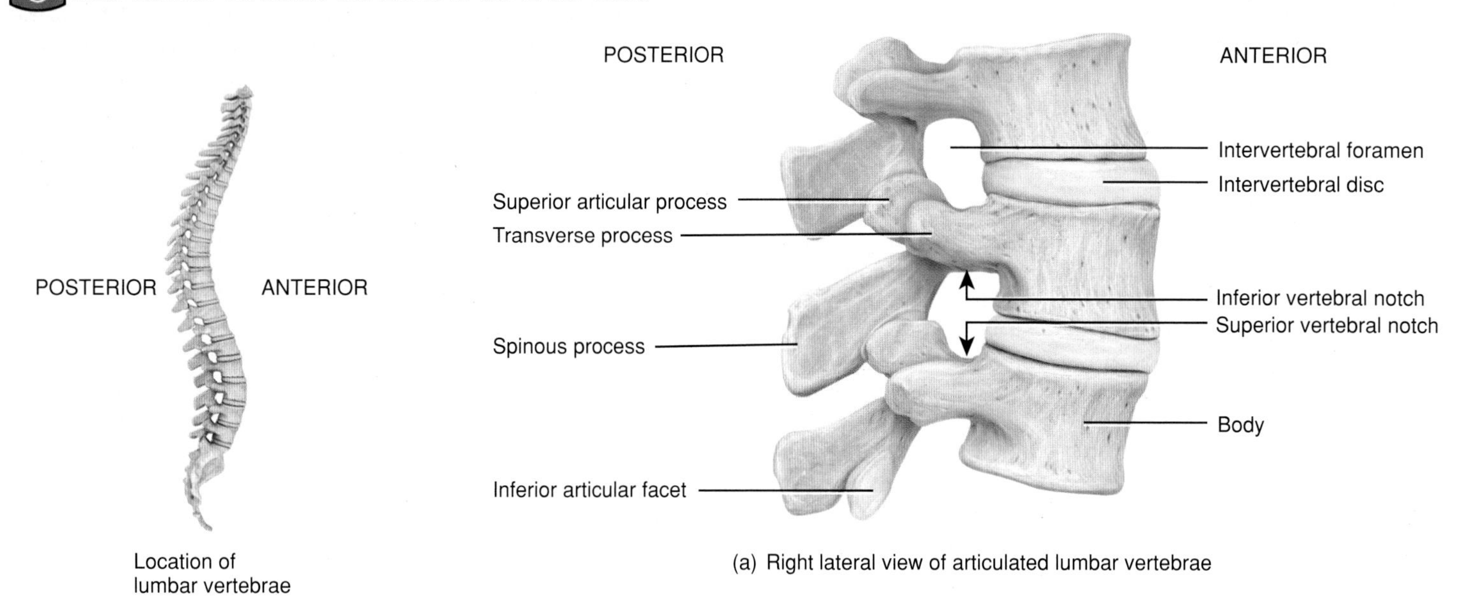

Location of lumbar vertebrae

(a) Right lateral view of articulated lumbar vertebrae

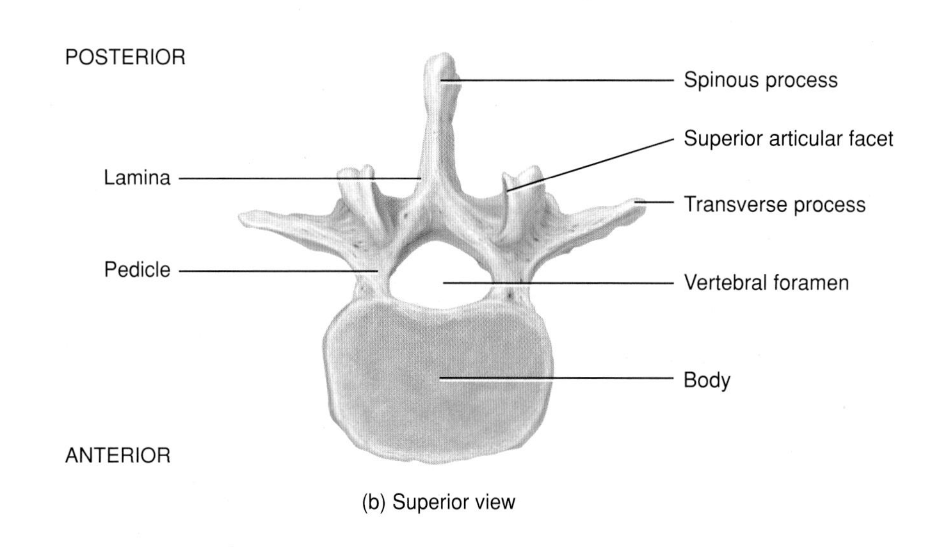

(b) Superior view

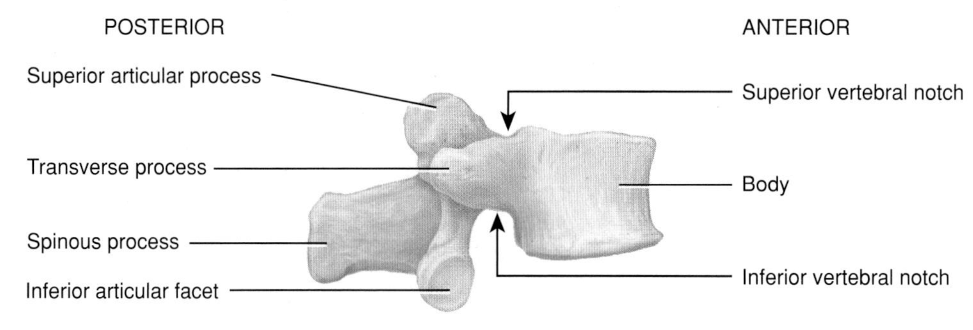

(c) Right lateral view

Why are the lumbar vertebrae the largest and strongest in the vertebral column?

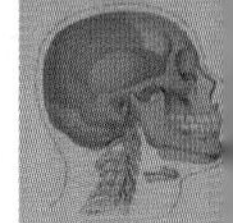

Table 7.5

Comparison of Major Structural Features of Cervical, Thoracic, and Lumbar Vertebrae

CHARACTERISTIC	CERVICAL	THORACIC	LUMBAR
Overall structure			
Body	Small.	Larger.	Largest.
Foramina	One vertebral and two transverse.	One vertebral.	One vertebral.
Spinous processes	Slender and often bifid (C2–C6).	Long and fairly thick (most project inferiorly).	Short and blunt (project posteriorly rather than inferiorly).
Transverse processes	Small.	Fairly large.	Large and blunt.
Articular facets for ribs	Absent.	Present.	Absent.
Direction of articular facets			
Superior	Posterosuperior.	Posterolateral.	Medial.
Inferior	Anteroinferior.	Anteromedial.	Lateral.
Size of intervertebral discs	Thick relative to size of vertebral bodies.	Thin relative to size of vertebral bodies.	Thickest.

Figure 7.22 Sacrum and coccyx.

The sacrum is formed by the union of five sacral vertebrae, and the coccyx is formed by the union of usually four coccygeal vertebrae.

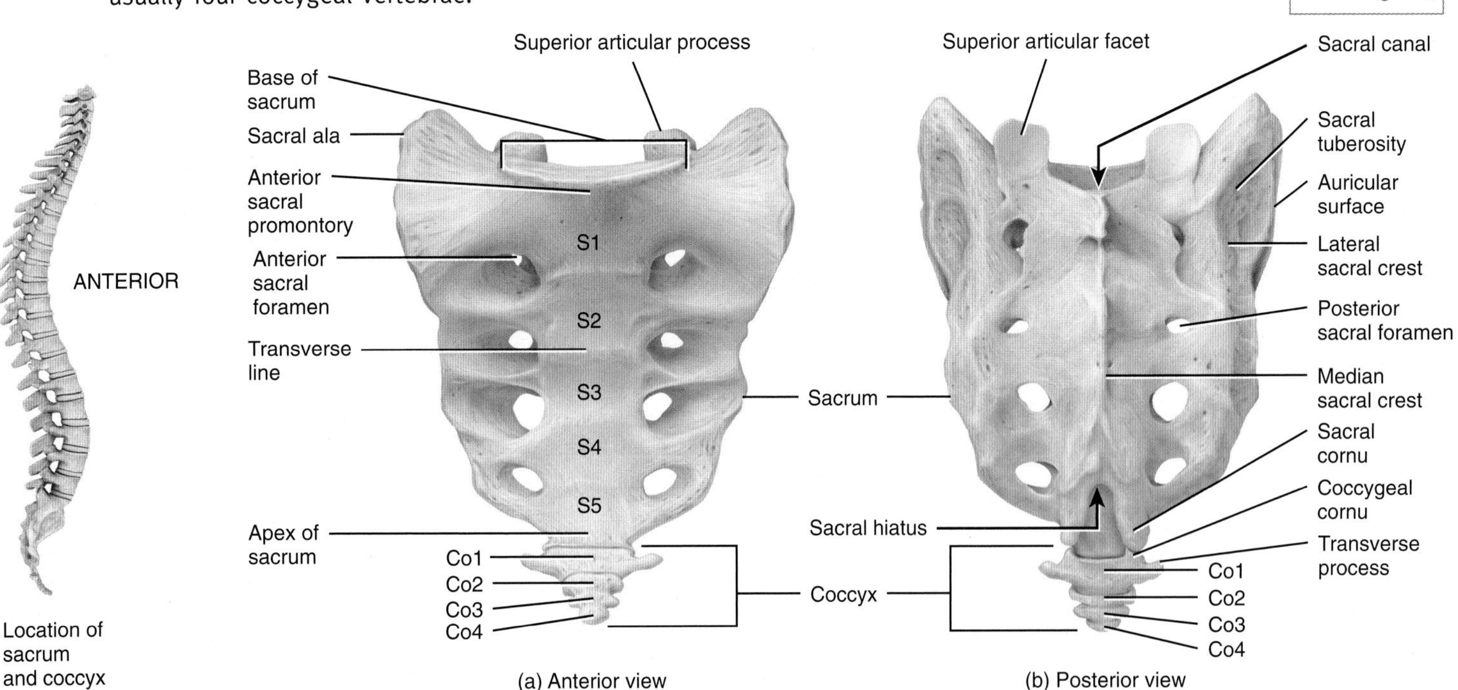

How many foramina pierce the sacrum, and what is their function?

the vertebral canal. The laminae of the fifth sacral vertebra, and sometimes the fourth, fail to meet. This leaves an inferior entrance to the vertebral canal called the *sacral hiatus* (hī-Ā-tus = opening). On either side of the sacral hiatus are the *sacral cornua* (KOR-noo-a; *cornu* = horn), the inferior articular processes of the fifth sacral vertebra. They are connected by ligaments to the coccyx.

The narrow inferior portion of the sacrum is known as the *apex.* The broad superior portion of the sacrum is called the *base.* The anteriorly projecting border of the base, called the *sacral promontory* (PROM-on-tō-rē), is one of the points used for measurements of the pelvis. On both lateral surfaces the sacrum has a large ear-shaped *auricular surface* that articulates with the ilium of each hip bone to form the *sacroiliac joint.* Posterior to the auricular surface is a roughened surface, the *sacral tuberosity,* that contains depressions for the attachment of ligaments. The sacral tuberosity is another surface of the sacrum that unites with the hip bones to form the sacroiliac joints. The *superior articular processes* of the sacrum articulate with the inferior articular processes of the fifth lumbar vertebra, and the base of the sacrum articulates with the body of the fifth lumbar vertebra, to form the *lumbosacral joint.*

Coccyx

The ***coccyx*** is also triangular in shape and is usually formed by the fusion of four coccygeal vertebrae, indicated in Figure 7.22 as Co1–Co4. The coccygeal vertebrae fuse when a person is between 20 and 30 years of age. The dorsal surface of the body of the coccyx contains two long *coccygeal cornua* that are connected by ligaments to the sacral cornua. The coccygeal cornua are the pedicles and superior articular processes of the first coccygeal vertebra. On the lateral surfaces of the coccyx are a series of *transverse processes,* the first pair being the largest. The coccyx articulates superiorly with the apex of the sacrum. In females, the coccyx is less curved anteriorly; in males, it is more curved anteriorly (see Table 8.1).

CHECKPOINT

9. What are the functions of the vertebral column?
10. When do the secondary vertebral curves develop?
11. What are the principal distinguishing characteristics of the bones of the various regions of the vertebral column?

7.7 THORAX

OBJECTIVE

- Identify the bones of the thorax.

The term ***thorax*** (THŌ-raks) refers to the entire chest. The skeletal part of the thorax, the ***thoracic cage,*** is a bony enclosure formed by the sternum, costal cartilages, ribs, and the bodies of the thoracic vertebrae (Figure 7.23). The thoracic cage is narrower at its superior end and broader at its inferior end and is flattened from front to back. It encloses and protects the organs in the thoracic and superior abdominal cavities and provides support for the bones of the shoulder girdle and upper limbs.

Sternum

The ***sternum,*** or breastbone, is a flat, narrow bone located in the center of the anterior thoracic wall that measures about 15 cm (6 in.) in length and consists of three parts (Figure 7.23). The superior part is the ***manubrium*** (ma-NOO-brē-um = handle-like); the middle and largest part is the ***body;*** and the inferior, smallest part is the ***xiphoid process*** (ZĪ-foyd = sword-shaped). The segments of the sternum typically fuse by age 25 and the points of fusion are marked by transverse ridges.

The junction of the manubrium and body forms the *sternal angle.* The manubrium has a depression on its superior surface, the *suprasternal notch.* Lateral to the suprasternal notch are *clavicular notches* that articulate with the medial ends of the clavicles to form the *sternoclavicular joints.* The manubrium also articulates with the costal cartilages of the first rib and part of the second rib to form the *sternocostal joints.*

The body of the sternum articulates directly or indirectly with the costal cartilages of the part of the second rib and the third through tenth ribs. Note that the second rib articulates with both the manubrium and the body at the sternal angle. The xiphoid process consists of hyaline cartilage during infancy and childhood and usually does not ossify completely until about age 40. No ribs are attached to it, but the xiphoid process provides attachment for some abdominal muscles. Incorrect positioning of the hands of a rescuer during cardiopulmonary resuscitation (CPR) may fracture the xiphoid process (whether cartilaginous or bony), driving it into internal organs. During thoracic surgery, the sternum may be split along the midline and the halves spread apart to allow surgeons access to structures in the thoracic cavity such as the thymus, heart, and great vessels of the heart. After the surgery, the halves of the sternum are held together with wire sutures.

Ribs

Twelve pairs of ***ribs*** give structural support to the sides of the thoracic cavity (Figure 7.23b). The ribs increase in length from the first through seventh, then decrease in length to the twelfth rib. Each articulates posteriorly with its corresponding thoracic vertebra.

The first through seventh pairs of ribs have a direct anterior attachment to the sternum by a strip of hyaline cartilage called *costal cartilage* (*cost-* = rib). The costal cartilages contribute to the elasticity of the thoracic cage and prevent various blows to the chest from fracturing the sternum and/or ribs. The ribs that have costal cartilages and attach directly to the sternum are called *true (vertebrosternal) ribs.* The remaining five pairs of ribs are termed *false ribs* because their costal cartilages either attach indirectly to the sternum or do not attach to the sternum at all. The cartilages of the eighth, ninth, and tenth pairs of ribs

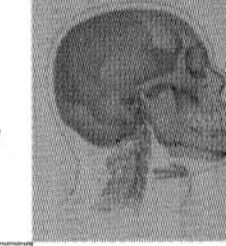

Figure 7.23 Skeleton of the thorax.

The bones of the thorax enclose and protect organs in the thoracic cavity and upper abdominal cavity.

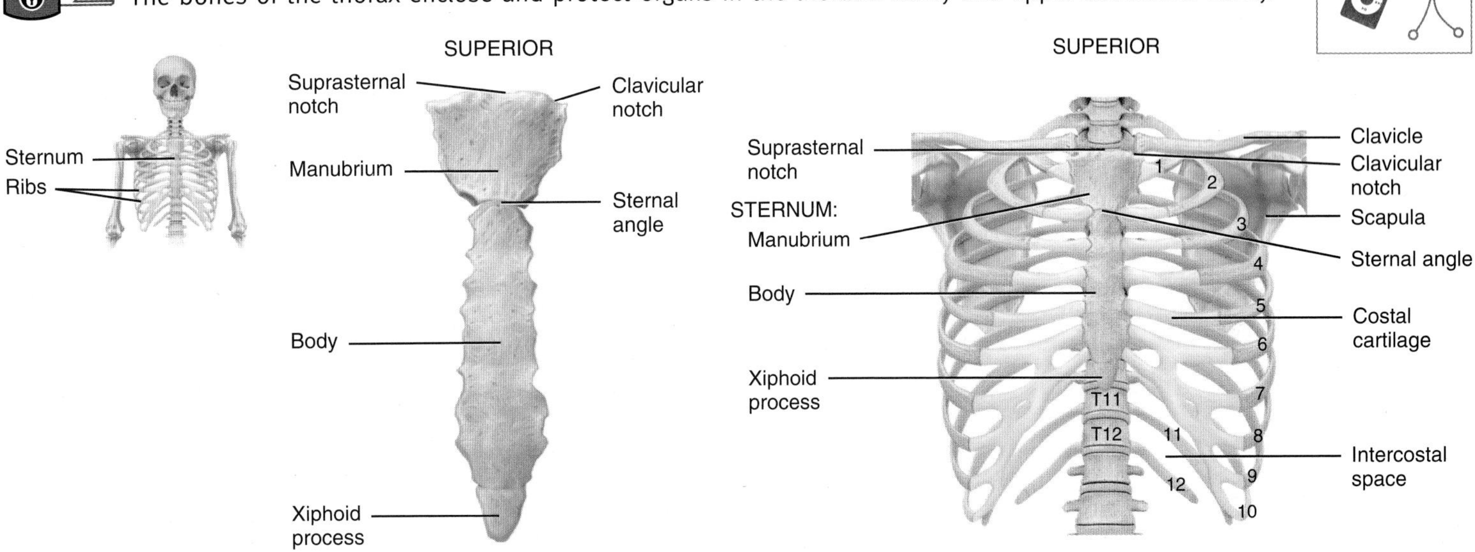

(a) Anterior view of sternum

(b) Anterior view of skeleton of thorax

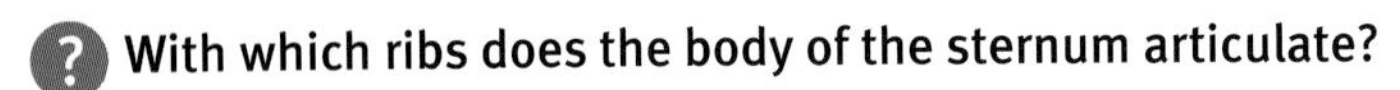

With which ribs does the body of the sternum articulate?

attach to one another and then to the cartilages of the seventh pair of ribs. These false ribs are called *vertebrochondral ribs.* The eleventh and twelfth pairs of ribs are false ribs designated as *floating (vertebral) ribs* because their anterior ends do not attach to the sternum at all. These ribs attach only posteriorly to the thoracic vertebrae. Inflammation of one or more costal cartilages, called *costochondritis,* is characterized by local tenderness and pain in the anterior chest wall that may radiate. The symptoms mime the chest pain associated with a heart attack (angina pectoris).

Figure 7.24a shows the parts of a typical (third through ninth) rib. The *head* is a projection at the posterior end of the rib. The facets of the head fit into either a facet on the body of one vertebra or into the demifacets of two adjoining vertebrae to form *vertebrocostal joints.* The *neck* is a constricted portion just lateral to the head. A knoblike structure on the posterior surface where the neck joins the body is called a *tubercle* (TOO-ber-kul). The *nonarticular part* of the tubercle attaches to a ligament (lateral costotransverse ligament) that attaches the transverse process of a vertebra to the nonarticular part of the corresponding rib. The *articular part* of the tubercle articulates with the facet of a transverse process of the inferior of the two vertebrae to which the head of the rib is connected (see Figure 7.24c). These articulations also form vertebrocostal joints. The *body (shaft)* is the main part of the rib. A short distance beyond the tubercle, an abrupt change in the curvature of the shaft occurs. This point is called the *costal angle.* The inner surface of the rib has a *costal groove* that protects blood vessels and a small nerve.

In summary, the posterior portion of the rib is connected to a thoracic vertebra by its head and the articular part of a tubercle. The facet of the head fits into a facet on the body of one vertebra or into the demifacets of two adjoining vertebrae. The articular part of the tubercle articulates with the facet of the transverse process of the inferior vertebra.

If you examine Figure 7.23b, you will notice that the first rib is the shortest, broadest, and most sharply curved. The first rib is an important landmark because of its close relationship to the nerves of the *brachial plexus* (the entire nerve supply of the shoulder and upper limb), two major blood vessels, the subclavian artery and vein, and two skeletal muscles, the anterior and middle scalene muscles. The superior surface of the first rib has two shallow grooves, one for the subclavian vein and one for the subclavian artery and inferior trunk of the brachial plexus. The second rib is thinner, less curved, and considerably longer than the first. Unlike the paired facets of the typical ribs, the tenth rib has a single articular facet on its head. The eleventh and twelfth ribs also have single articular facets on their heads, but no necks, tubercles, or costal angles.

Spaces between ribs, called *intercostal spaces,* are occupied by intercostal muscles, blood vessels, and nerves. Surgical access to the lungs or other structures in the thoracic cavity is commonly obtained through an intercostal space. Special rib retractors are used to create a wide separation between ribs. The costal cartilages are sufficiently elastic in younger individuals to permit considerable bending without breaking.

Structures passing between the thoracic cavity and the neck pass through an opening called the ***superior thoracic aperture (thoracic outlet).*** Among these structures are the trachea, esophagus, nerves, and blood vessels that supply and drain the head, neck, and upper limbs. The aperture is bordered by the first thoracic vertebra

Figure 7.24 The structure of ribs. Each rib has a head, a neck, and a body.

Each rib articulates posteriorly with its corresponding thoracic vertebra.

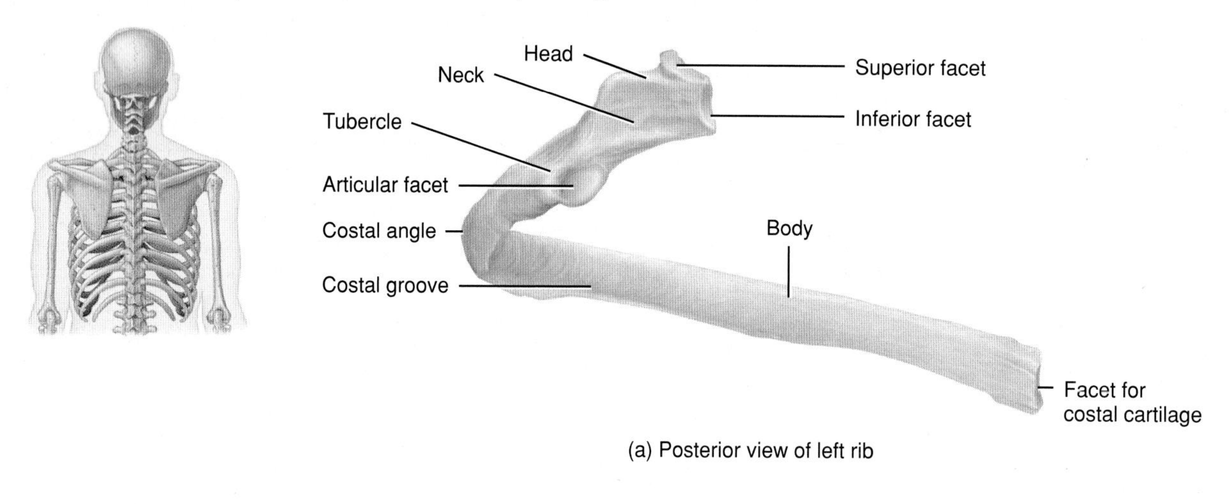

(a) Posterior view of left rib

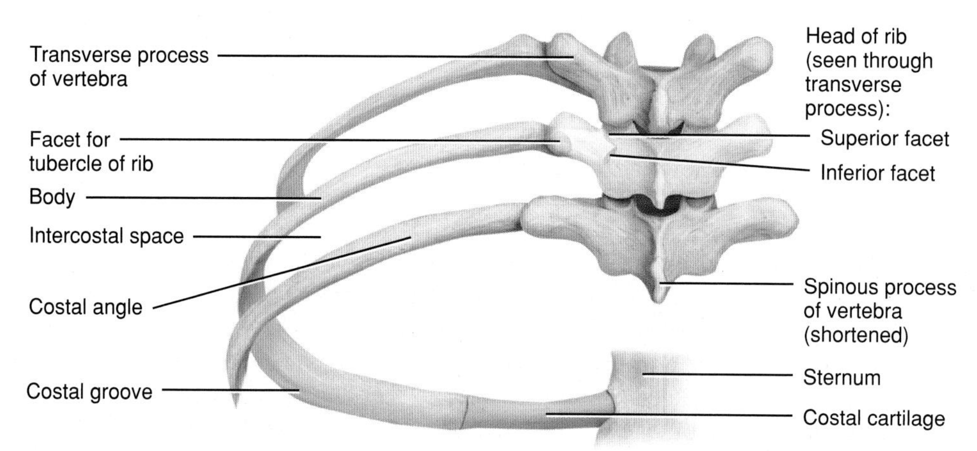

(b) Posterior view of left ribs articulated with thoracic vertebrae and sternum

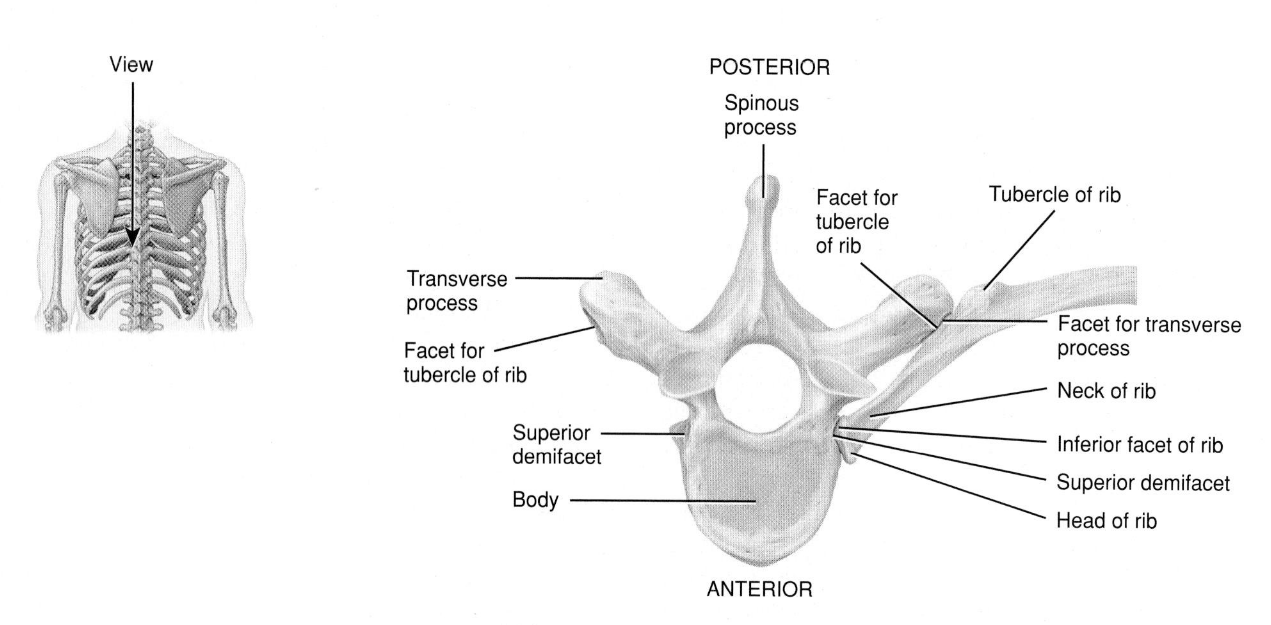

(c) Superior view of left rib articulated with thoracic vertebra

How does a rib articulate with a thoracic vertebra?

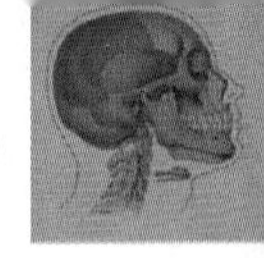

(posteriorly), the first pair of ribs and their cartilages, and the superior border of the manubrium of the sternum. Structures passing between the thoracic cavity and abdominal cavity pass through the ***inferior thoracic aperture.*** Through this large opening, which is closed by the diaphragm, pass structures such as the esophagus, nerves, and large blood vessels. This aperture is bordered by the twelfth thoracic vertebra (posteriorly), the eleventh and twelfth pair of ribs, the costal cartilages of ribs 7 through 10, and the joint between the body and xiphoid process of the sternum (anteriorly).

MANUAL THERAPY APPLICATION

Rib Fractures, Dislocations, and Separations

Rib fractures are the most common chest injuries, and they usually result from direct blows, most often from impact with a steering wheel, falls, or crushing injuries to the chest. Ribs tend to break at the point where the greatest force is applied, but they may also break at their weakest point—the site of greatest curvature, which is just anterior to the costal angle. The middle ribs are most commonly fractured. In some cases, fractured ribs may puncture the heart, great vessels of the heart, lungs, trachea, bronchi, esophagus, spleen, liver, and kidneys. Rib fractures are usually quite painful.

Dislocated ribs, which are common in body contact sports, involve displacement of a costal cartilage from the sternum, with resulting pain, especially during deep inhalations. **Separated ribs** involve displacement of a rib and its costal cartilage; as a result, a rib may move superiorly, overriding the rib above and causing severe pain. Treatment of dislocated and separated ribs usually requires the services of a chiropractor or other health professional. Once the affected structures are aligned, a manual therapist may be asked to assist the healing process for the patient. As stated previously, derivative massage is helpful for healing of any fracture; it is also helpful with the healing of soft tissues involved in dislocations and separations.

CHECKPOINT

12. What bones form the skeleton of the thorax?
13. What are the functions of the bones of the thorax?
14. How are ribs classified?

7.8 DISORDERS OF THE AXIAL SKELETON

OBJECTIVE

- Describe three types of abnormal curves in the vertebral column, spina bifida, and fractures of the vertebral column.

Abnormal Curves of the Vertebral Column

Various conditions may exaggerate the normal curves of the vertebral column, or the column may acquire a lateral bend, resulting in ***abnormal curves*** of the vertebral column.

Scoliosis (skō-lē-Ō-sis; *scolio-* = crooked) is a lateral bending of the vertebral column, usually in the thoracic or lumbar region (Figure 7.25a). This is the most common of the abnormal curves. The causes of scoliosis are many and varied. In the most common type, *idiopathic scoliosis*, the cause is unknown. It may result from *congenitally* (present at birth) malformed vertebrae, chronic sciatica, paralysis of muscles on one side of the vertebral column, shortened muscles on one side of the vertebral column, physical trauma or stress, poor posture, or one leg being shorter than the other. A large percentage of patients with scoliosis develop the problem between the ages of 8 and 18 and girls are more commonly affected than boys.

Kyphosis (kī-FŌ-sis; *kyphos-* = hump; also known as ***humpback*** or ***hunchback***) is an exaggeration of the thoracic curve of the vertebral column (Figure 7.25b). In tuberculosis of the spine, vertebral bodies may partially collapse, causing an acute

Figure 7.25 Abnormal curves of the vertebral column.

 An abnormal curve is the result of an exaggerated normal curve.

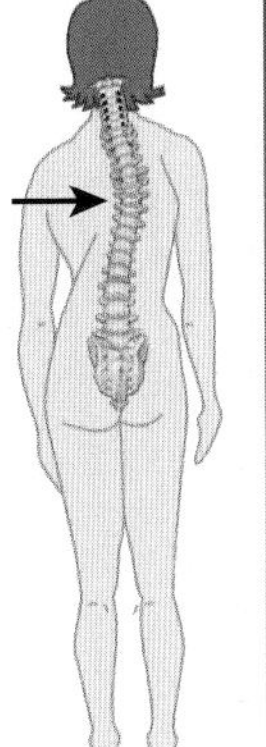

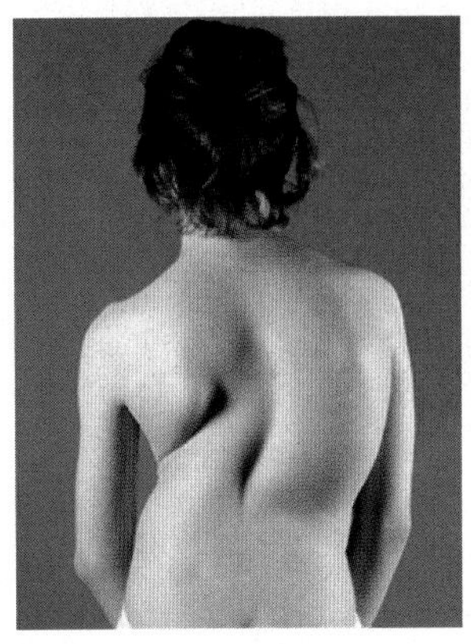

(a) Scoliosis

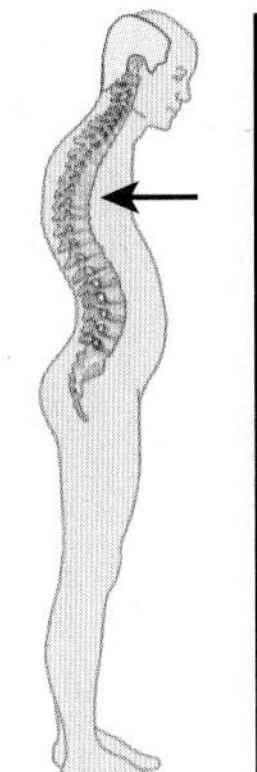

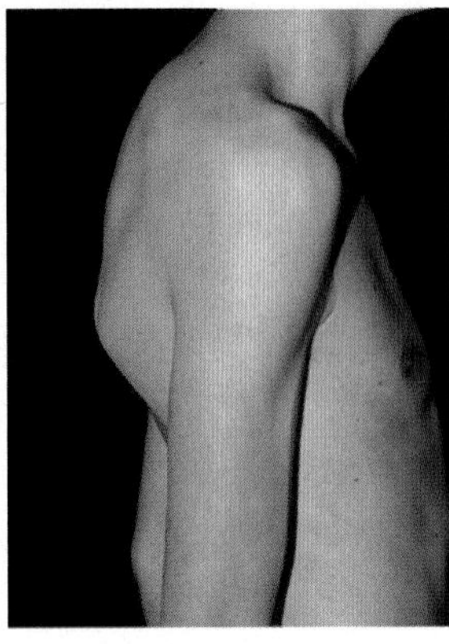

(b) Kyphosis

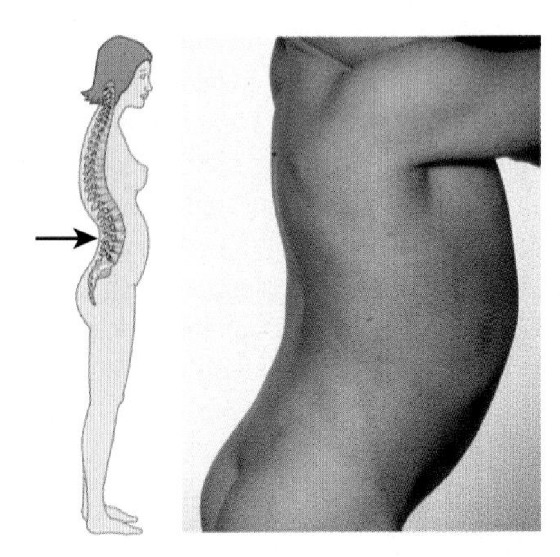

(c) Lordosis

 **Which abnormal curve is common in women with advanced osteroporosis?**

angular bending of the vertebral column. In the elderly, degeneration of the intervertebral discs leads to kyphosis. Kyphosis may also be caused by rickets and poor posture. It is also common in females with advanced osteoporosis. The term *round-shouldered* is an expression for mild kyphosis.

Lordosis (lor-DŌ-sis; *lord-* = bent backward), sometimes called ***hollow back*** or ***sway back,*** is an exaggeration of the lumbar curve of the vertebral column (Figure 7.25c). It may result from increased weight of the abdomen as in pregnancy, or extreme obesity, poor posture, rickets, osteoporosis, or tuberculosis of the spine.

MANUAL THERAPY APPLICATION

Treatment of Scoliosis

Scoliosis varies from mild to severe and, in addition to the lateral curvature, usually results in varying levels of pain as spinal nerves are compressed. More severe cases may require surgery to correct or reduce the problems; such surgeries may involve alignment and fusion of the affected vertebrae and/or installation of metal rods to stabilize the vertebral column. Moderate cases of scoliosis may be treated with a full-torso brace. Such devices may be worn for a number of years and must be replaced periodically to accommodate the growth of the person, usually a teenager. Braces are commonly prescribed to prevent further curvature rather than to straighten the existing condition. Manual therapy may serve the role as an adjunct treatment to these more severe cases of scoliosis. As stated repeatedly, manual therapy has the potential to reduce stress, reduce pain, increase blood flow, and thus facilitate healing.

Scoliosis may also be caused by shortened muscles on one side of the vertebral column, sciatica, or a "shortened leg." Appropriate manual therapy is often successful in the treatment of these conditions and therefore will positively affect the alignment of the curvature. Some cases of sciatica may be treated, for example, by release of a chronically shortened piriformis muscle that may be impinging on the nerve. When a "short leg" is caused by imbalance of the hips, manual therapy may permit the realignment of the pelvis. When scoliosis occurs as a result of shortened muscles on one side, manual therapy may reduce the curvature and reduce the pain normally associated with the abnormal curvature. Muscles of the back are described in Chapter 12. Scoliosis may present as a C-shaped bend, but commonly presents as an S-shaped curve. A compensatory curve may develop to maintain the appearance that the body is in alignment. Covered by clothing, the patient may thus have a seemingly normal vertebral alignment but is shorter in stature.

Chronically shortened muscles on one side of the vertebral column cause the curvature of the vertebral column to point toward the contralateral (opposite) side. Musculature and connective tissues would thus be worked by the therapist on the concave side of the curvature. As described in Chapter 4, fasciae and other connective tissues associated with muscle (Chapter 10) will also be shortened. Deep myofascial work involving appropriate areas of the back will thus be required for successful treatment.

Spina Bifida

Spina bifida (SPĪ-na BIF-i-da) is a congenital defect of the vertebral column in which laminae of L5 and/or S1 fail to develop normally and unite at the midline. The least serious form is called *spina bifida oculta.* It occurs in L5 or S1 and produces no symptoms. The only evidence of its presence is a small dimple with a tuft of hair in the overlying skin. Several types of spina bifida involve protrusion of the meninges (membranes) and/or spinal cord through the defect in the laminae and are collectively termed *spina bifida cystica* because of the presence of a cystlike sac protruding from the backbone (Figure 7.26). If the sac contains the meninges from the spinal cord and cerebrospinal fluid, the condition is called *spina bifida with meningocele* (me-NING-gō-sēl). If the spinal cord and/or its nerve roots are in the sac, the condition is called *spina bifida with meningomyelocele* (me-ning-gō-MĪ-e-lō-sēl). The larger the cyst and the number of neural structures it contains, the more serious the neurological problems. In severe cases, there may be partial or complete paralysis, partial or complete loss of urinary bladder and bowel control, and the absence of reflexes. An increased risk of spina bifida is associated with low levels of a B vitamin called folic acid during pregnancy. Spina bifida may be diagnosed prenatally by a test of the mother's blood for a substance produced by the fetus called alphafetoprotein, by sonography, or by amniocentesis (withdrawal of amniotic fluid for analysis).

Figure 7.26 Spina bifida.

Spina bifida is caused by a failure of laminae to unite at the midline.

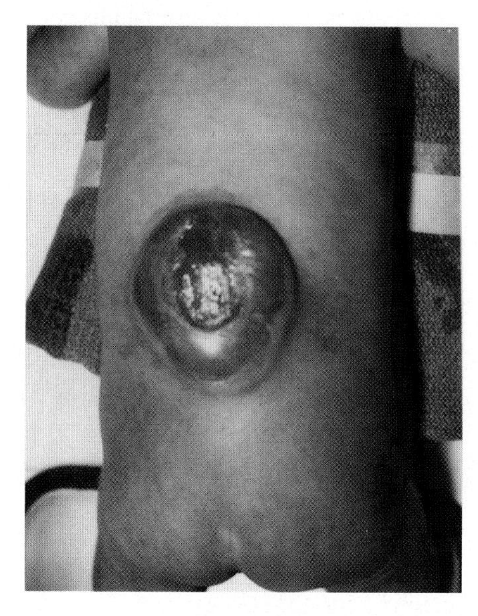

? Deficiency of which B vitamin is linked to spina bifida?

• CLINICAL CONNECTION | Fractures of the Vertebral Column

Fractures of the vertebral column often involve C1, C2, C4–T7, and T12–L2. Cervical or lumbar fractures usually result from a flexion–compression type of injury such as might be sustained in landing on the feet or buttocks after a fall or having a weight fall on the shoulders. Cervical vertebrae may be fractured or dislodged by a fall on the head with acute flexion of the neck, as might happen on diving into shallow water or being thrown from a horse. Dislocation may result from the sudden forward-then-backward jerk ("whiplash") that may occur in an automobile crash. Spinal cord or spinal nerve damage may occur as a result of fractures of the vertebral column. •

CHECKPOINT

15. How do scoliosis, hyphosis, and lordosis differ?
16. Distinguish among the types of spina bifida
17. How are fractures of the vertebral column caused?

KEY MEDICAL TERMS ASSOCIATED WITH AXIAL SKELETON

Craniostenosis (krā-nē-ō-sten-Ō-sis; *cranio-* = skull; *-stenosis* = narrowing) Premature closure of one or more cranial sutures during the first 18 to 20 months of life, resulting in a distorted skull. Premature closure of the sagittal suture produces a long, narrow skull, whereas premature closure of the coronal suture results in a broad skull. Premature closure of all sutures restricts brain growth and development; surgery is necessary to prevent brain damage.

Craniotomy (krā-nē-OT-ō-mē; *cranio-* = skull; *-tome* = cutting) Surgical procedure in which part of the cranium is removed. It may be performed to remove a blood clot, a brain tumor, or a sample of brain tissue for biopsy.

Laminectomy (lam′-i-NEK-tō-mē; *lamina-* = layer) Surgical procedure to remove a vertebral lamina. It may be performed to access the vertebral canal and relieve the symptoms of a herniated disc.

Lumbar spine stenosis (*sten-* = narrowed) Narrowing of the spinal canal in the lumbar part of the vertebral column, due to hypertrophy of surrounding bone or soft tissues. It may be caused by arthritic changes in the intervertebral discs and is a common cause of back and leg pain.

Spinal fusion (FŪ-zhun) Surgical procedure in which two or more vertebrae of the vertebral column are stabilized with a bone graft or synthetic device. It may be performed to treat a fracture of a vertebra or following removal of a herniated disc.

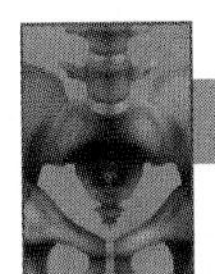

STUDY OUTLINE

Division of the Skeletal System (Section 7.1)

1. The axial skeleton consists of bones arranged along the longitudinal axis. The parts of the axial skeleton are the skull, auditory ossicles (ear bones), hyoid bone, vertebral column, sternum, and ribs.
2. The parts of the appendicular skeleton are the pectoral (shoulder) girdles, bones of the upper limbs, pelvic (hip) girdle, and bones of the lower limbs.

Types of Bones (Section 7.2)

1. On the basis of shape, bones are classified as long, short, flat, irregular, or sesamoid. Sesamoid bones develop in tendons or ligaments.
2. Sutural bones are found within the sutures of certain cranial bones.

Bone Surface Markings (Section 7.3)

1. Surface markings are structural features visible on the surfaces of bones.
2. Each marking—whether a depression, an opening, or a process—is structured for a specific function, such as joint formation, muscle attachment, or passage of nerves and blood vessels (see Table 7.2).

Skull (Section 7.4)

1. The 22 bones of the skull include cranial bones and facial bones.
2. The eight cranial bones include the frontal, parietal (2), temporal (2), occipital, sphenoid, and ethmoid.
3. The 14 facial bones are the nasal (2), maxillae (2), zygomatic (2), lacrimal (2), palatine (2), inferior nasal conchae (2), vomer, and mandible.
4. The nasal septum consists of the vomer, perpendicular plate of the ethmoid, and septal cartilage. The nasal septum divides the nasal cavity into left and right sides.
5. Seven skull bones form each of the orbits (eye sockets).
6. The foramina of the skull bones provide passages for nerves and blood vessels (Table 7.4).

7. Sutures are immovable or slightly movable joints that connect most bones of the skull. Examples are the coronal, sagittal, lambdoid, and squamous sutures.
8. Paranasal sinuses are cavities in bones of the skull lined with mucous membranes that are continuous with the nasal cavity. The frontal, sphenoid, and ethmoid bones and the maxillae contain paranasal sinuses.
9. Fontanels are mesenchyme-filled spaces between the cranial bones of fetuses and infants. The major fontanels are the anterior, posterior, anterolaterals (2), and posterolaterals (2). After birth, the fontanels fill in with bone and become sutures.

Hyoid Bone (Section 7.5)

1. The hyoid bone is a U-shaped bone that does not articulate with any other bone.
2. It supports the tongue and provides attachment for some tongue muscles and for some muscles of the throat and neck.

Vertebral Column (Section 7.6)

1. The vertebral column, sternum, and ribs constitute the skeleton of the body's trunk.
2. The 26 bones of the adult vertebral column are the cervical vertebrae (7), the thoracic vertebrae (12), the lumbar vertebrae (5), the sacrum (5 fused vertebrae), and the coccyx (usually 4 fused vertebrae).
3. The adult vertebral column contains four normal curves (cervical, thoracic, lumbar, and sacral) that give strength, support, and balance.
4. The vertebrae are similar in structure, each usually consisting of a body, vertebral arch, and seven processes. Vertebrae in the different regions of the column vary in size, shape, and detail.

Thorax (Section 7.7)

1. The thoracic skeleton consists of the sternum, ribs and costal cartilages, and thoracic vertebrae.
2. The thoracic cage protects vital organs in the chest area and upper abdomen.

Disorders of the Axial Skeleton (Section 7.8)

1. Scoliosis is a lateral bending of the vertebral column, kyphosis is an exaggeration of the thoracic curve, and lordosis is an exaggeration of the lumbar curve.
2. Spina bifida is a congenital defect of the vertebral column in which the laminae of L5 and/or S1 fail to develop normally at the midline.
3. Fractures of the vertebral column often involve C1, C2, C4–T7, and T12–L2.

SELF-QUIZ QUESTIONS

Choose the one best answer to the following questions.

1. A foramen is
   a. a cavity within a bone
   b. a depression
   c. a hole for blood vessels and nerves
   d. a ridge
   e. a site for muscle attachment.
2. Normal skulls have two of each of the following *except*
   a. parietal bones
   b. nasal bones
   c. temporal bones
   d. sphenoid bones
   e. maxillae.
3. Tears pass into the nasal cavity through a tunnel formed in part by the
   a. nasal bone
   b. lacrimal bone
   c. vomer
   d. both a and b are correct
   e. a, b, and c are all correct.
4. Which of the following bones is correctly matched with the process?
   a. zygomatic bone–temporal process
   b. maxilla–pterygoid process
   c. vomer–mandibular process
   d. temporal–occipital process
   e. sphenoid–coronoid process
5. All of the following articulate with the maxilla except the
   a. nasal bone
   b. frontal bone
   c. palatine bone
   d. mandible
   e. zygomatic bone.
6. Which of the following is *not* a component of the orbit?
   a. temporal
   b. ethmoid
   c. zygomatic
   d. maxilla
   e. lacrimal
7. The nasal septum is formed by parts of the
   a. ethmoid bone
   b. nasal bone
   c. vomer
   d. palatine bone
   e. Both a and c are correct.
8. Which of the following does *not* contain a paranasal sinus?
   a. zygomatic
   b. sphenoid
   c. ethmoid
   d. frontal
   e. maxilla
9. The skeleton of the thorax
   a. is formed by 12 pairs of ribs and costal cartilages, the sternum, and 12 thoracic vertebrae
   b. protects the internal chest organs, as well as the liver
   c. is narrower at its superior end
   d. aids in supporting the bones of the shoulder girdle
   e. is described by all of the above.
10. The meninges (coverings) of the brain attach to the crista galli, which is part of the
    a. frontal bone
    b. temporal bone
    c. sphenoid bone
    d. parietal bone
    e. ethmoid bone.

Complete the following.

11. The first cervical vertebra is the ________, and the second cervical vertebra is the ________.
12. According to shape classification, phalanges are ________ bones, carpals are ________ bones, and ribs are ________ bones.
13. Bones located in sutures are called sutural or ________ bones.
14. The only bone of the axial skeleton that does not articulate with any other bone is the ________.

Are the following statements true or false?

15. The opening in the occipital bone through which the medulla oblongata connects with the spinal cord is the foramen magnum.
16. The sagittal suture joins the parietal bones.
17. Sesamoid bones protect tendons from wear and tear and may improve the mechanical advantage at a joint by altering the direction of pull of a tendon.
18. The sacral promontory projects anteriorly from the apex of the sacrum.
19. The tubercle of a rib articulates with demifacets on the bodies of adjacent vertebrae.

Matching

20. Match the following (answers may be used more than once):

____	**a.** mandibular fossa	**A.** zygomatic
____	**b.** optic foramen	**B.** frontal
____	**c.** superior nasal concha	**C.** axis
____	**d.** superior and inferior nuchal lines	**D.** ethmoid
____	**e.** horizontal plate	**E.** maxilla
____	**f.** pterygoid process	**F.** temporal
____	**g.** coronoid process	**G.** sternum
____	**h.** mastoid process	**H.** occipital
____	**i.** temporal process	**I.** palatine
____	**j.** supraorbital margin	**J.** mandible
____	**k.** olfactory foramina	**K.** sphenoid
____	**l.** sella turcica	
____	**m.** mental foramen	
____	**n.** external auditory meatus	
____	**o.** perpendicular plate	
____	**p.** palatine process	
____	**q.** xiphoid process	
____	**r.** dens (odontoid process)	

CRITICAL THINKING QUESTIONS

1. Jimmy is in a car accident. He can't open his mouth and has been told that he suffers from the following: black eye, broken nose, broken cheek, broken upper jaw, damaged eye socket, and punctured lung. Describe *exactly* what structures have been affected by his car accident.

2. A new mother brings her newborn infant home and has been told by her well-meaning friend not to wash the baby's hair for several months because the water and soap could "get through that soft area in the top of the head and cause brain damage." Explain to her why this is not true.

? ANSWERS TO FIGURE QUESTIONS

7.1 The skull and vertebral column are part of the axial skeleton. The clavicle, shoulder girdle, humerus, pelvic girdle, and femur are part of the appendicular skeleton.

7.2 Flat bones protect underlying organs and provide a large surface area for muscle attachment.

7.3 The frontal, parietal, sphenoid, ethmoid, and temporal bones are cranial bones.

7.4 The parietal and temporal bones are joined by the squamous suture, the parietal and occipital bones are joined by the lambdoid suture, and the parietal and frontal bones are joined by the coronal suture.

7.5 The temporal bone articulates with the parietal, sphenoid, zygomatic, and occipital bones.

7.6 The parietal bones form the posterior, lateral portion of the cranium.

7.7 The medulla oblongata of the brain connects with the spinal cord in the foramen magnum.

7.8 From the crista galli of the ethmoid bone, the sphenoid articulates with the frontal, parietal, temporal, occipital, temporal, parietal, and frontal bones, ending again at the crista galli of the ethmoid bone.

7.9 The perpendicular plate of the ethmoid bone forms the superior part of the nasal septum, and the lateral masses compose most of the medial walls of the orbits.

7.10 The mandible is the only movable skull bone, other than the auditory ossicles.

7.11 The nasal septum divides the nasal cavity into right and left sides.

7.12 Bones forming the orbit are the frontal, sphenoid, zygomatic, maxilla, lacrimal, ethmoid, and palatine. The palatine bone is not visible in the figure.

7.13 The paranasal sinuses produce mucus and serve as resonating chambers for vocalization.

7.14 The anterolateral fontanel is bordered by four different skull bones, the frontal, parietal, temporal, and sphenoid.

7.15 The hyoid bone does not articulate with any other bone.

7.16 The thoracic and sacral curves of the vertebral column are concave relative to the anterior of the body.

7.17 Most herniated discs occur in the lumbar region because it bears most of the body weight and most flexing and bending occur there compared to other regions of the spine.

7.18 The vertebral foramina enclose the spinal cord; the intervertebral foramina provide spaces through which spinal nerves exit the vertebral column.

7.19 The atlas moving on the axis permits movement of the head to signify "no."

7.20 The facets and demifacets on the bodies of the thoracic vertebrae articulate with the heads of the ribs, and the facets on the transverse processes of these vertebrae articulate with the tubercles of the ribs.

7.21 The lumbar vertebrae are the largest and strongest in the body because the amount of weight supported by vertebrae increases toward the inferior end of the vertebral column.

7.22 There are four pairs of sacral foramina, for a total of eight. Each anterior sacral foramen joins a posterior sacral foramen at the intervertebral foramen. Nerves and blood vessels pass through these tunnels in the bone.

7.23 The body of the sternum articulates directly or indirectly with ribs 2–10.

7.24 The facet on the head of a rib fits into a facet on the body of a vertebra, and the articular part of the tubercle of a rib articulates with the facet of the transverse process of a vertebra.

7.25 Kyphosis is common in women with advanced osteoporosis.

7.26 Deficiency of folic acid is associated with spina bifida.

The Skeletal System: The Appendicular Skeleton

8

The appendicular skeleton includes the bones that make up the upper and lower limbs as well as the bones of the two girdles that attach the limbs to the axial skeleton. The bones of the girdles, and their corresponding limbs, are connected with one another and with skeletal muscles, permitting you to do things such as walk, lift, grasp, use a computer, and dance.

Bones also play an important role in providing a storage site for calcium in the body, which not only keeps the framework strong but also is important in muscle contraction and thought transmission which you will see in other chapters. As you determined in the previous chapter, palpation of bones of the appendicular skeleton is an important component of the work accomplished by manual therapists.

CONTENTS AT A GLANCE

8.1 PECTORAL (SHOULDER) GIRDLE

OBJECTIVE

- Identify the bones of the pectoral (shoulder) girdle and their principal markings.

The ***pectoral*** (PEK-tō-ral) or ***shoulder girdles*** attach the bones of the upper limbs to the axial skeleton (Figure 8.1). Each of the two pectoral girdles consists of a clavicle and a scapula. The clavicle is the anterior bone and articulates with the manubrium of the sternum at the *sternoclavicular joint.* The scapula is the posterior bone and articulates with the clavicle at the *acromioclavicular joint* and with the humerus at the *glenohumeral (shoulder) joint.* The pectoral girdles do not articulate with the vertebral column and are held in position instead by complex muscle attachments.

Clavicle

Each slender, S-shaped ***clavicle*** (KLAV-i-kul = key), or *collarbone,* lies horizontally across the anterior part of the thorax superior to the first rib (Figure 8.1). The medial half of the clavicle is convex anteriorly, whereas the lateral half is concave anteriorly (Figure 8.2). The medial end of the clavicle, called the *sternal end,* is rounded in cross section and articulates with the manubrium of the sternum to form the *sternoclavicular joint.* The broad, flat, lateral end, the *acromial end* (a-KRŌ-mē-al), articulates with the acromion of the scapula. This joint is called the *acromioclavicular joint* (see Figure 8.1). The *conoid tubercle* (KŌ-noyd = conelike)

Figure 8.1 Right pectoral (shoulder) girdle.

The clavicle is the anterior bone of the pectoral girdle and the scapula is the posterior bone.

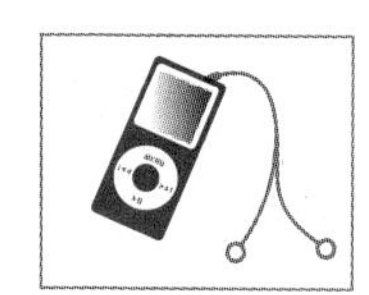

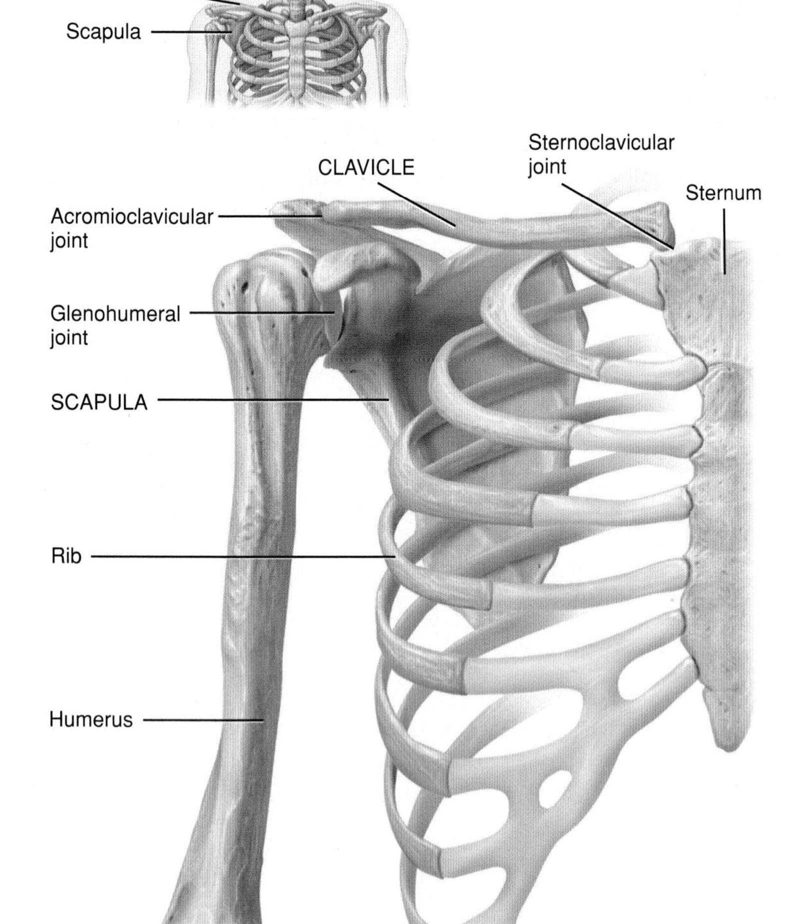

(a) Anterior view of pectoral girdle

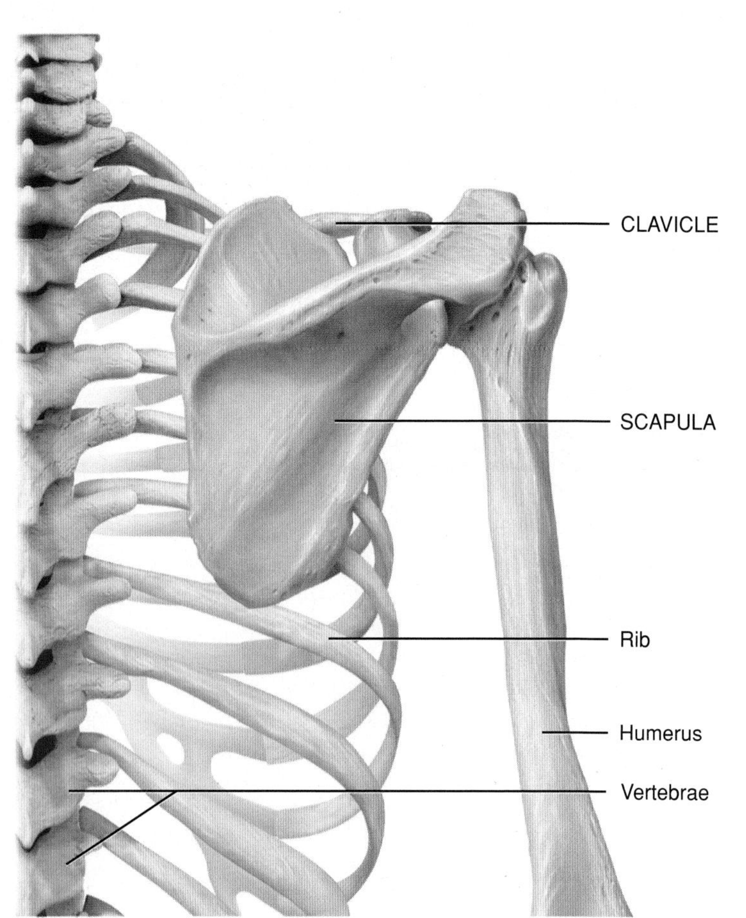

(b) Posterior view of pectoral girdle

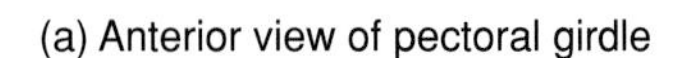

What is the function of the pectoral girdle?

on the inferior surface of the lateral end of the bone is a point of attachment for the conoid ligament, which attaches the clavicle and the scapula. As its name implies, the *impression for the costoclavicular ligament* on the inferior surface of the sternal end is a point of attachment for the costoclavicular ligament.

• CLINICAL CONNECTION | Fractured Clavicle

The clavicle transmits mechanical force from the upper limb to the trunk. If the force transmitted to the clavicle is excessive, as in falling on one's outstretched arm, a **fractured clavicle** may result. The clavicle is one of the most frequently broken bones in the body. The clavicular mid-region is the most frequent fracture site because the weakest point of the clavicle is at the junction of the clavicle's two curves. Even in the absence of a fracture, compression of the clavicle as a result of automobile accidents involving the use of shoulder harness seatbelts often causes damage to the median nerve, which lies between the clavicle and the second rib. A fractured clavicle is usually treated with a sling to keep the arm from moving outward. Falling on an outstretched upper limb can also fracture the bones of the hand, forearm, and arm. •

Scapula

Each ***scapula*** (SCAP-ū-la; plural is *scapulae*), or *shoulder blade,* is a large, triangular, flat bone situated in the superior part of the posterior thorax between the levels of the second and seventh ribs (see Figure 8.1). A prominent ridge called the *spine* runs diagonally across the posterior surface of the flattened, triangular *body* of the scapula (see Figure 8.3b). The lateral end of the spine projects as a flattened, expanded process called the *acromion* (a-KRŌ-mē-on; *acrom-* = topmost), easily felt as the high point of the shoulder. Tailors measure the length of the upper limb from the acromion. The acromion articulates with the acromial end of the clavicle to form the *acromioclavicular joint.* Inferior to the acromion is a shallow depression, the *glenoid cavity,* that accepts the head of the humerus (arm bone) to form the *glenohumeral joint* (see Figure 8.1).

The thin edge of the scapula closer to the vertebral column is called the *medial (vertebral) border* (see Figure 8.3a–b). The medial borders of the scapulae lie about 5 cm (2 in.) from the vertebral column. The thick edge of the scapula closer to the arm is called the *lateral (axillary) border.* The medial and lateral borders join at the *inferior angle.* The superior edge of the scapula, called the *superior border,* joins the vertebral border at the *superior angle.* The *scapular notch* is a prominent indentation along the superior border through which the suprascapular nerve passes.

At the lateral end of the superior border of the scapula is a projection of the anterior surface called the *coracoid process* (KOR-a-koyd = like a crow's beak), to which the tendons of muscles attach. Superior and inferior to the posteriorly located spine are

Figure 8.2 Right clavicle.

The clavicle articulates medially with the manubrium of the sternum and laterally with the acromion of the scapula.

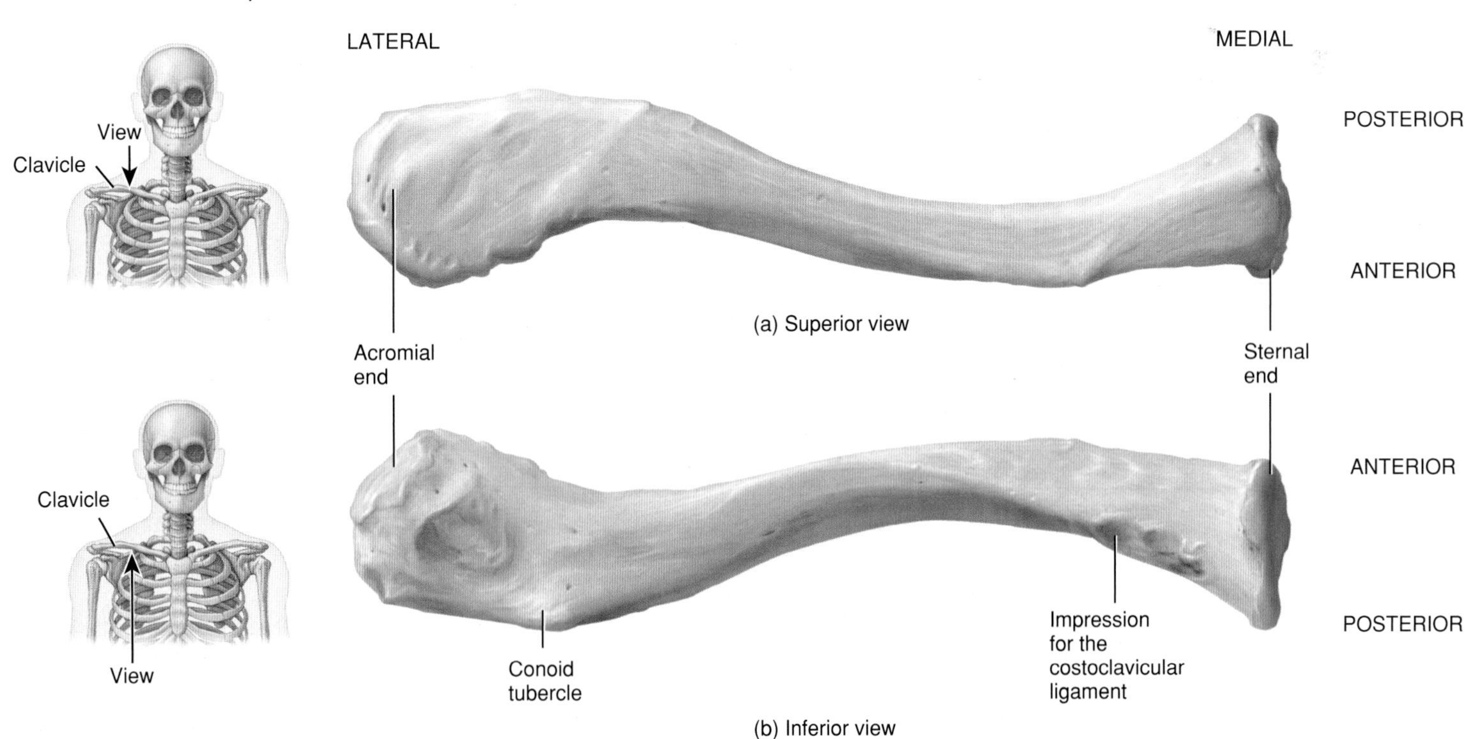

Which part of the clavicle is its weakest point?

Figure 8.3 Right scapula (shoulder blade).

The glenoid cavity of the scapula articulates with the head of the humerus to form the shoulder joint.

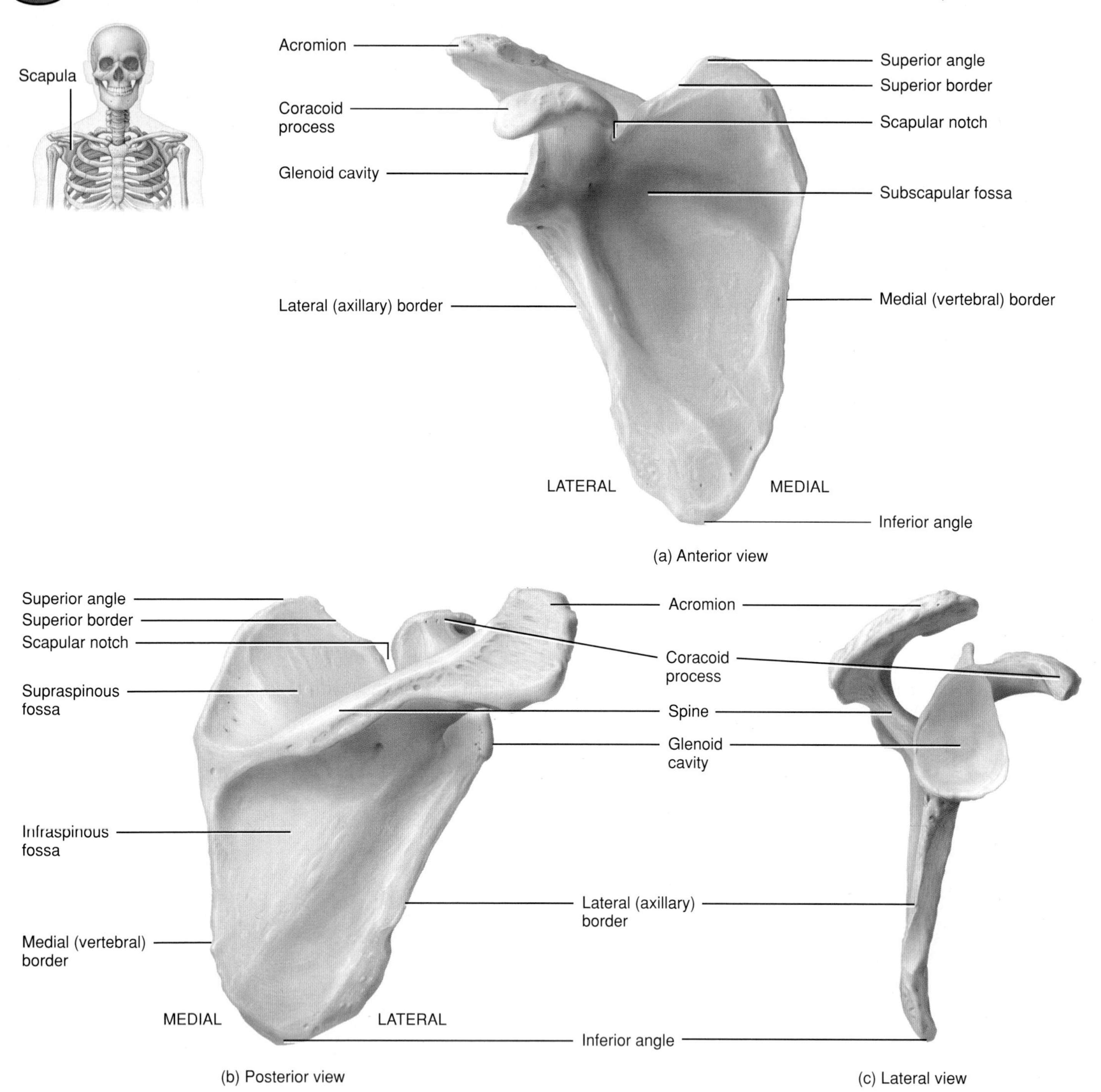

Which part of the scapula forms the high point of the shoulder?

two fossae: the *supraspinous fossa* (sū-pra-SPĪ-nus) and the *infraspinous fossa* (in-fra-SPĪ-nus), respectively. Both serve as surfaces of attachment for the tendons of the supraspinatus and infraspinatus muscles of the shoulder. On the anterior surface is a slightly hollowed-out area called the *subscapular fossa,* also a surface of attachment for the tendons of shoulder muscles.

CHECKPOINT

1. Which bones or parts of bones of the pectoral girdle form the sternoclavicular, acromioclavicular, and glenohumeral joints?

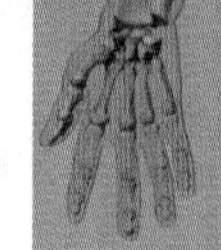

8.2 UPPER LIMB (EXTREMITY)

OBJECTIVES

- Identify the bones of the upper limb and their principal markings.
- Describe the joints between the upper limb bones.

Each ***upper limb (extremity)*** has 30 bones: the humerus in the arm; the ulna and radius in the forearm; and the 8 carpals in the carpus (wrist), the 5 metacarpals in the metacarpus (palm), and the 14 phalanges (bones of the digits) in the hand (Figure 8.4).

Humerus

The ***humerus*** (HŪ-mer-us), or arm bone, is the longest and largest bone of the upper limb (see Figure 8.5). It articulates proximally with the scapula and distally at the elbow with both the ulna and the radius.

The proximal end of the humerus features a rounded *head* that articulates with the glenoid cavity of the scapula to form the *glenohumeral joint.* Distal to the head is the *anatomical neck,* which is visible as an oblique and roughened groove. The *greater tubercle* is a lateral projection distal to the anatomical neck. It is the most laterally palpable bony landmark of the shoulder region. The *lesser tubercle* projects anteriorly. Between both tubercles runs an *intertubercular sulcus (bicipital groove).* The *surgical neck* is a constriction in the humerus just distal to the tubercles, where the head tapers to the shaft; it is so named because fractures often occur here.

The *body (shaft)* of the humerus is roughly cylindrical at its proximal end, but it gradually becomes triangular until it is flattened and broad at its distal end. Laterally, at the middle portion of the shaft, there is a roughened, V-shaped area called the *deltoid tuberosity.* This area serves as a point of attachment for the tendons of the deltoid muscle.

Several prominent features are evident at the distal end of the humerus. The *capitulum* (ka-PIT-ū-lum; *capit-* = head) is a rounded knob on the lateral aspect of the bone that articulates with the head of the radius. The *radial fossa* is an anterior depression that receives the head of the radius when the forearm is flexed (bent). The *trochlea* (TRŌK-lē-a), located medial to the capitulum, is a spool-shaped surface that articulates with the ulna. The *coronoid fossa* (KOR-ō-noyd = crown-shaped) is an anterior depression that receives the coronoid process of the ulna when the forearm is flexed. The *olecranon fossa* (ō-LEK-ra-non = elbow) is a posterior depression that receives the olecranon of the ulna when the forearm is extended (straightened). The *medial epicondyle* and *lateral epicondyle* are rough projections on either side of the distal end of the humerus to which the tendons of most muscles of the forearm are attached. The ulnar nerve may be easily palpated by rolling a finger over the skin surface above the posterior surface of the medial epicondyle.

Figure 8.4 Right upper limb.

Each upper limb consists of a humerus, ulna, radius, carpals, metacarpals, and phalanges.

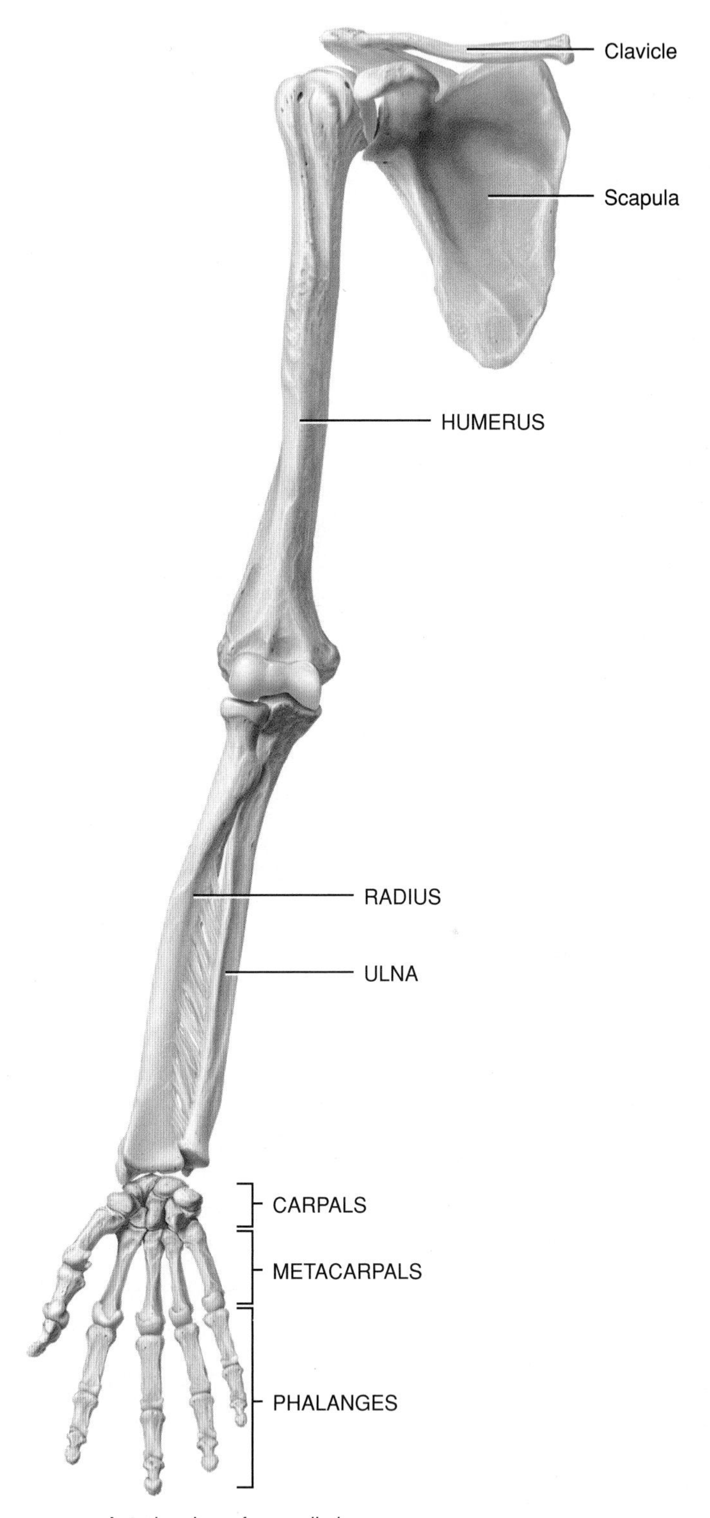

Anterior view of upper limb

How many bones make up each upper limb?

Figure 8.5 Right humerus in relation to the scapula, ulna, and radius.

The humerus is the longest and largest bone of the upper limb.

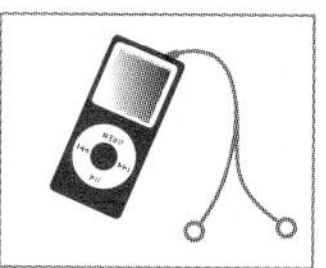

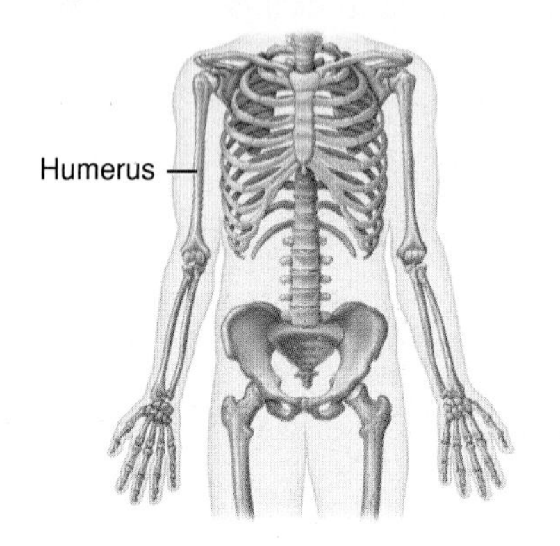

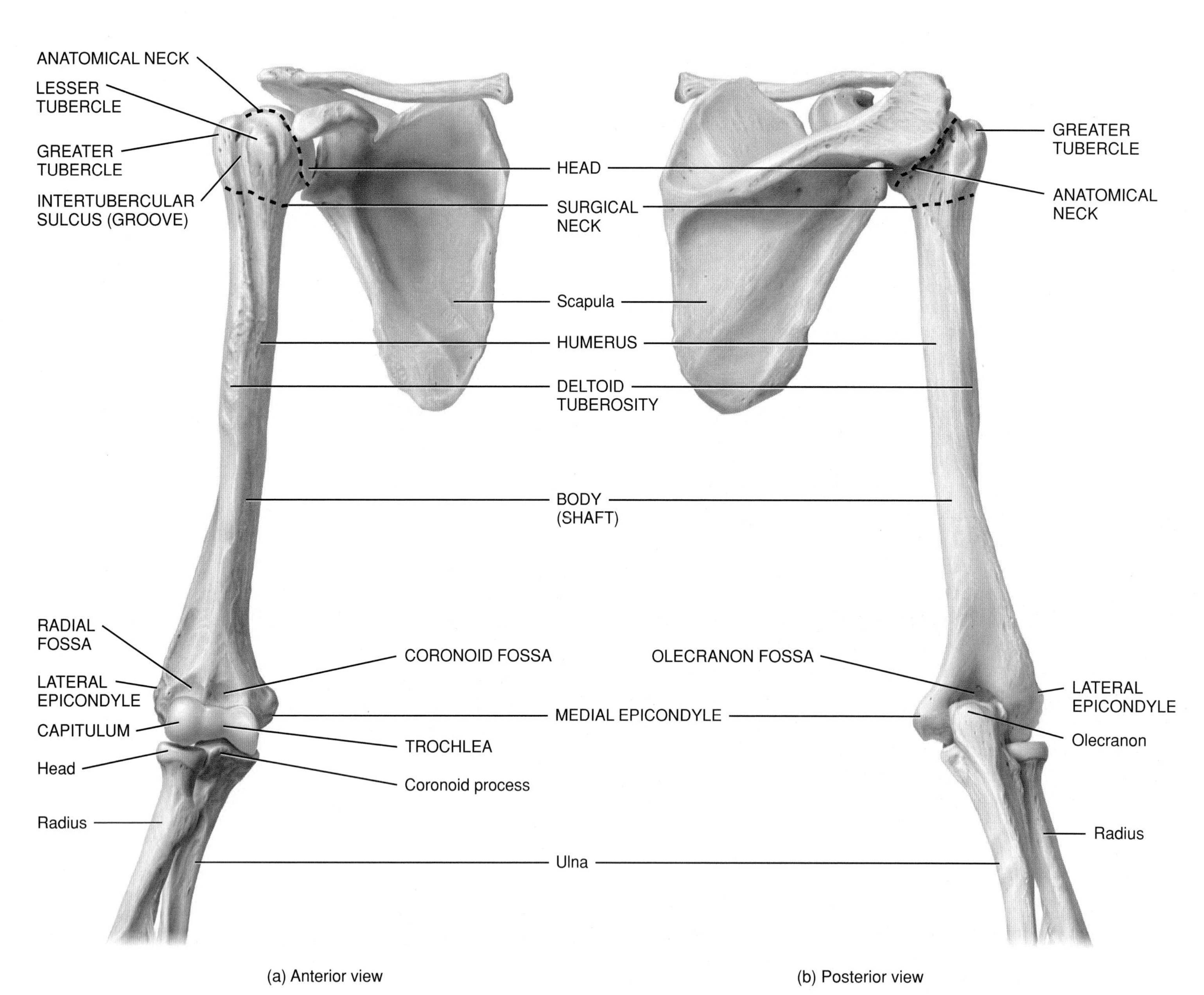

(a) Anterior view

(b) Posterior view

Which parts of the humerus articulate with the radius at the elbow? With the ulna at the elbow?

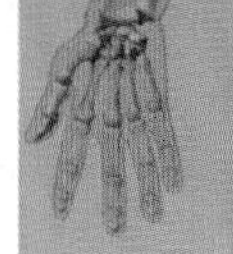

Ulna and Radius

The ***ulna*** is located on the medial aspect (the little-finger side) of the forearm and is longer than the radius (Figure 8.6). It is sometimes convenient to use an aid to help remember information that may be unfamiliar. Such an aid is called a *mnemonic device* (nē-MON-ik = memory). One such mnemonic to help you remember the location of the ulna in relation to the hand is "*p.u.*" (the *p*inky is on the *u*lna side). At the proximal end of the ulna (Figure 8.6b) is the *olecranon,* which forms the prominence of the elbow. The *coronoid process* (Figure 8.6a) is an anterior projection that, together with the olecranon, receives the trochlea of the humerus. The *trochlear notch* is a large curved area between the olecranon

Figure 8.6 Right ulna and radius in relation to the humerus and carpals.

In the forearm, the longer ulna is on the medial side, whereas the shorter radius is on the lateral side.

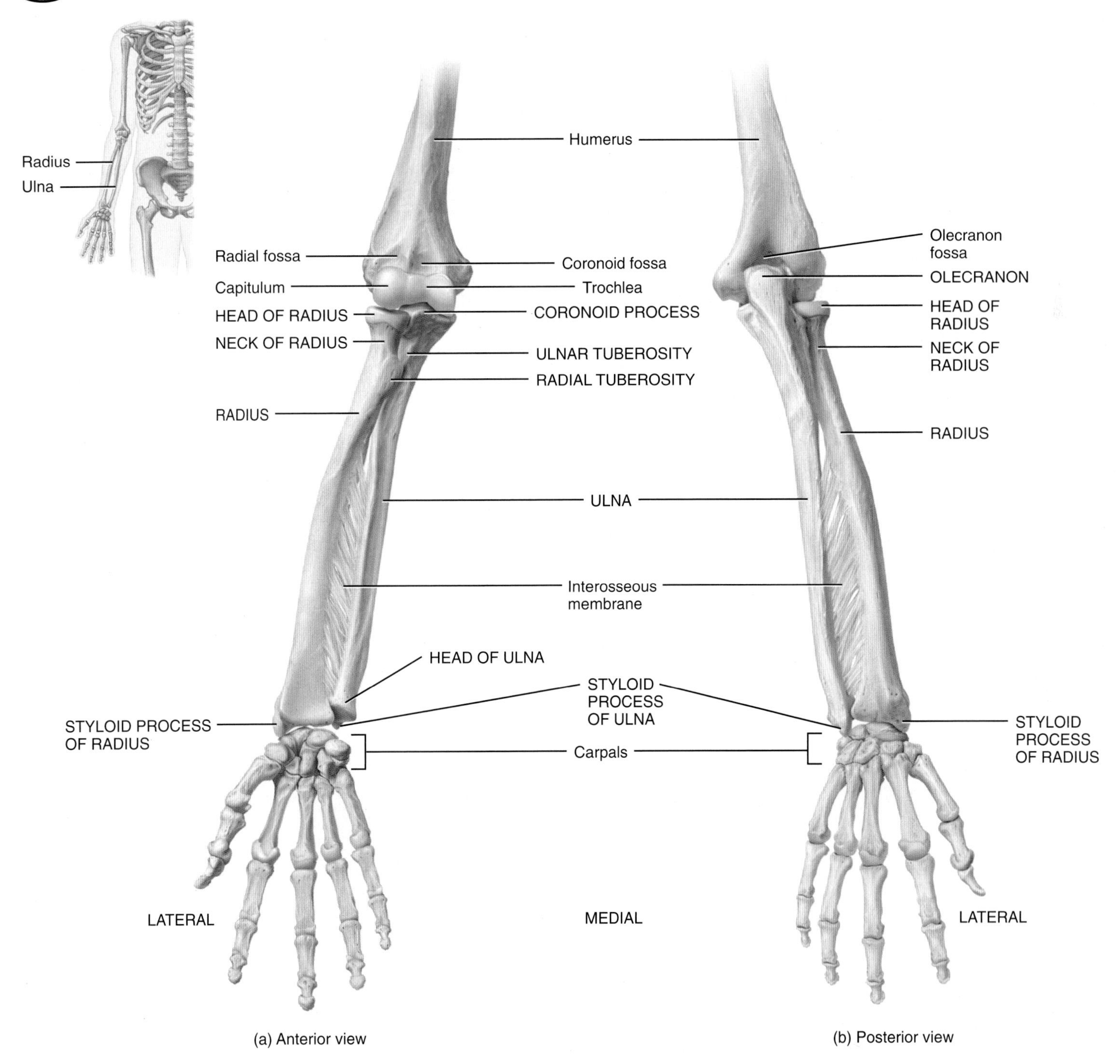

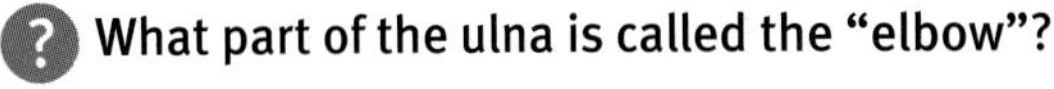
What part of the ulna is called the "elbow"?

and coronoid process that forms part of the elbow joint (see Figure 8.7b). On the lateral side of the coronoid process is a depression, the *radial notch,* which receives the head of the radius. Just inferior to the coronoid process is the *ulnar tuberosity.* The distal end of the ulna consists of a *head;* a *styloid process* is located on the posterior side of the ulna's distal end (the head).

The ***radius*** is located on the lateral aspect (thumb side) of the forearm (see Figure 8.6). The proximal end of the radius has a disc-shaped *head* that articulates with the capitulum of the humerus and the radial notch of the ulna. Inferior to the head is the constricted *neck.* A roughened area inferior to the neck on the medial side, called the *radial tuberosity,* is a point of attachment for the tendons of the biceps brachii muscle. The shaft of the radius widens distally to form a *styloid process* on the lateral side. Fracture of the distal end of the radius is the most common fracture in adults older than 50 years.

The ulna and radius articulate with the humerus at the *elbow joint.* The articulation occurs in two places: where the head of the radius articulates with the capitulum of the humerus and where the trochlear notch of the ulna receives the trochlea of the humerus (see Figures 8.6 and 8.7).

The ulna and the radius connect with one another at three sites (Figure 8.7a–c). First, a broad, flat, fibrous connective tissue called the *interosseous membrane* (in′-ter-OS-ē-us; *inter-* = between, *-osseous* = bone) joins the shafts of the two bones. This membrane also provides a site of attachment for some tendons of deep skeletal muscles of the forearm. The ulna and radius articulate at their proximal and distal ends. Proximally, the head of the radius articulates with the ulna's *radial notch,* a depression that is lateral and inferior to the trochlear notch (Figure 8.7b). This articulation is the *proximal radioulnar joint.* Distally, the head of the ulna articulates with the *ulnar notch* of the radius (Figure 8.7c). This articulation is the *distal radioulnar joint.* Finally, the distal end of the radius articulates with three bones of the wrist—the lunate, the scaphoid, and the triquetrum—to form the *radiocarpal (wrist) joint.*

Carpals, Metacarpals, and Phalanges

The ***carpus*** (wrist) is the proximal region of the hand and consists of eight small bones, the ***carpals,*** joined to one another by ligaments (Figure 8.8). Articulations between carpal bones are called *intercarpal joints.* The carpals are arranged in two transverse rows of four bones each. Their names reflect their shapes. The carpals in the proximal row, from lateral to medial, are the ***scaphoid*** (SKAF-oyd = boatlike), ***lunate*** (LOO-nāt = moon-shaped), ***triquetrum*** (trī-KWĒ-trum = three-cornered), and ***pisiform*** (PĪS-i-form = pea-shaped). The carpals in the distal row, from lateral to medial, are the ***trapezium*** (tra-PĒ-zē-um = four-sided figure with no two sides parallel), ***trapezoid*** (TRAP-e-zoyd = four-sided figure with two sides parallel), ***capitate*** (KAP-i-tāt = head-shaped), and ***hamate*** (HAM-āt = hooked).

The capitate is the largest carpal bone; its rounded projection, the head, articulates with the lunate. The hamate is named for a large hook-shaped projection on its anterior surface. In about 70% of carpal fractures, only the scaphoid is broken. This is because the force of a fall on an outstretched hand is transmitted from the capitate through the scaphoid to the radius (Figure 8.8b).

The concave space formed by the pisiform and hamate (on the ulnar side), the scaphoid and trapezium (on the radial side), plus the *flexor retinaculum* (fibrous bands of deep fascia) is the ***carpal tunnel.*** The long flexor tendons of the digits and thumb and the median nerve pass through the carpal tunnel. Narrowing of the carpal tunnel may give rise to a condition called carpal tunnel syndrome (described in Exhibit 13.4).

Figure 8.7 Articulations formed by the right ulna and radius. (a) Elbow joint. (b) Joint surfaces at proximal end of the ulna. (c) Joint surfaces at distal ends of radius and ulna. The ulna and radius are also attached by the interosseous membrane.

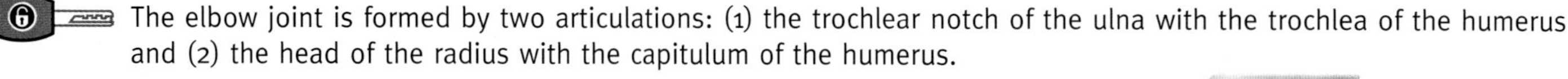

The elbow joint is formed by two articulations: (1) the trochlear notch of the ulna with the trochlea of the humerus and (2) the head of the radius with the capitulum of the humerus.

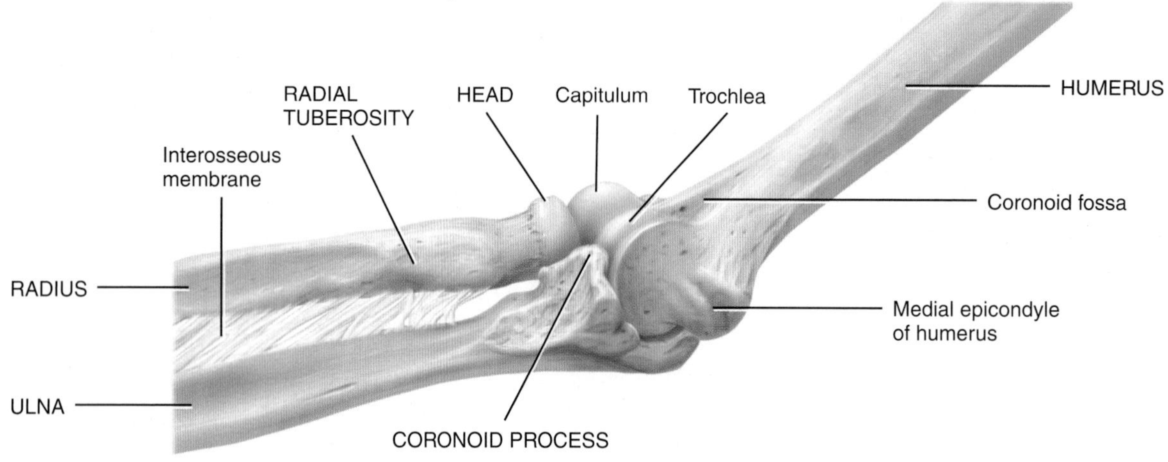

(a) Medial view in relation to humerus

A mnemonic for learning the names of the carpal bones is also shown in Figure 8.8. The first letters of the carpal bones from lateral to medial (proximal row, then distal row) correspond to the first letter of each word in the mnemonic. Students are encouraged to develop their own mnemonics that contain personalized people, events, etc., that you will be better able to remember.

The ***metacarpus*** (*meta-* = beyond), or palm, is the intermediate region of the hand and consists of five bones called ***metacarpals***. Each metacarpal bone consists of a proximal *base,* an intermediate *shaft,* and a distal *head* (see Figure 8.8b). The metacarpal bones are numbered I to V (or 1–5), starting with the thumb, from lateral to medial. The bases articulate with the distal row of carpal bones to form the *carpometacarpal joints.* The heads articulate with the proximal phalanges to form the *metacarpophalangeal joints.* The heads of the metacarpals are commonly called "knuckles" and are readily visible in a clenched fist.

The ***phalanges*** (fa-LAN-jēz; *phalan-* = a battle line), or bones of the digits, make up the distal region of the hand. There are 14 phalanges in the five digits of each hand and, like the metacarpals, the digits are numbered I to V (or 1–5), beginning with the thumb, from lateral to medial. A single bone of a digit is referred to as a *phalanx* (FĀ-lanks). Each phalanx consists of a proximal *base,* an intermediate *shaft,* and a distal *head.* The thumb (*pollex*) has two phalanges, and there are three phalanges in each of the other four digits. In order from the thumb, these other four digits are commonly referred to as the index finger, middle finger, ring finger, and little finger. The first row of phalanges, the *proximal row,* articulates with the metacarpal bones and second row of phalanges. The second row of phalanges, the *middle row,* articulates with the proximal row and the third row. The third row of phalanges, the *distal row,* articulates with the middle row. The thumb has no middle phalanx. Joints between phalanges are called *interphalangeal joints.*

CHECKPOINT

2. Name the bones that form the upper limb, from proximal to distal.
3. Describe the joints of the upper limb.

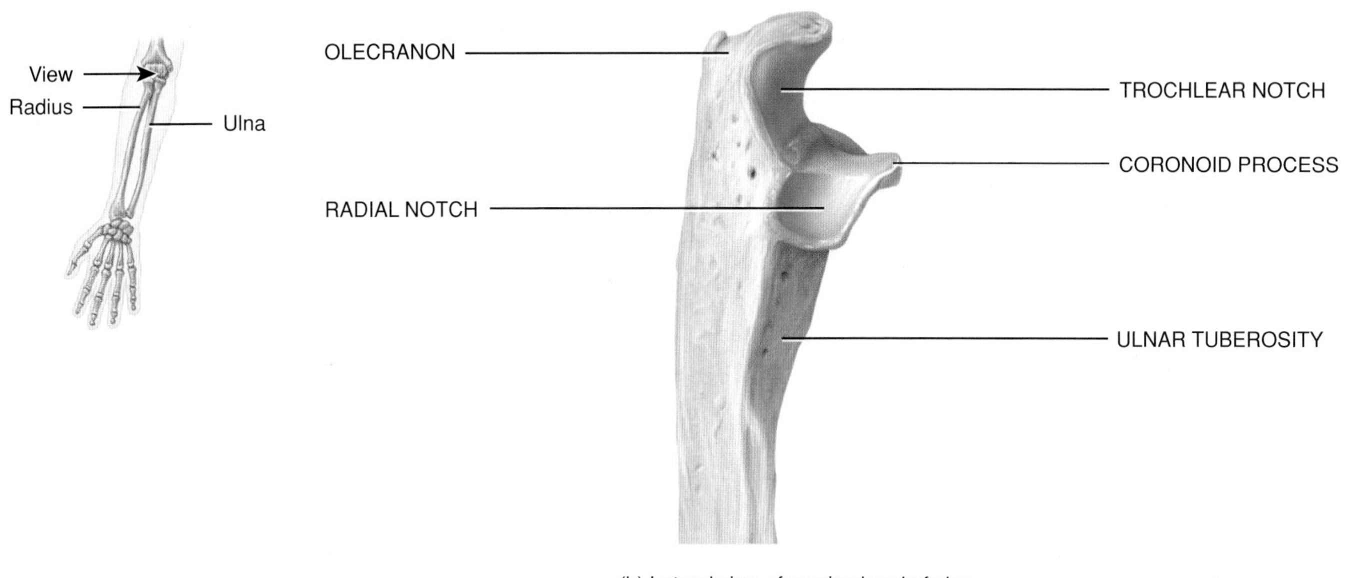

(b) Lateral view of proximal end of ulna

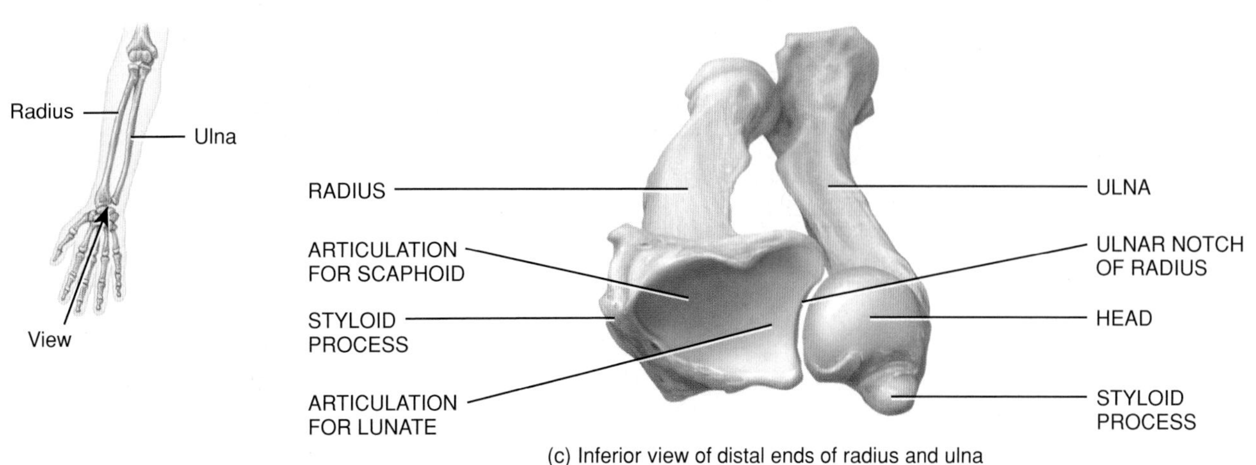

(c) Inferior view of distal ends of radius and ulna

How many points of attachment are there between the radius and ulna?

Figure 8.8 **Right wrist and hand in relation to the ulna and radius.**

The skeleton of the hand consists of the proximal carpals, the intermediate metacarpals, and the distal phalanges.

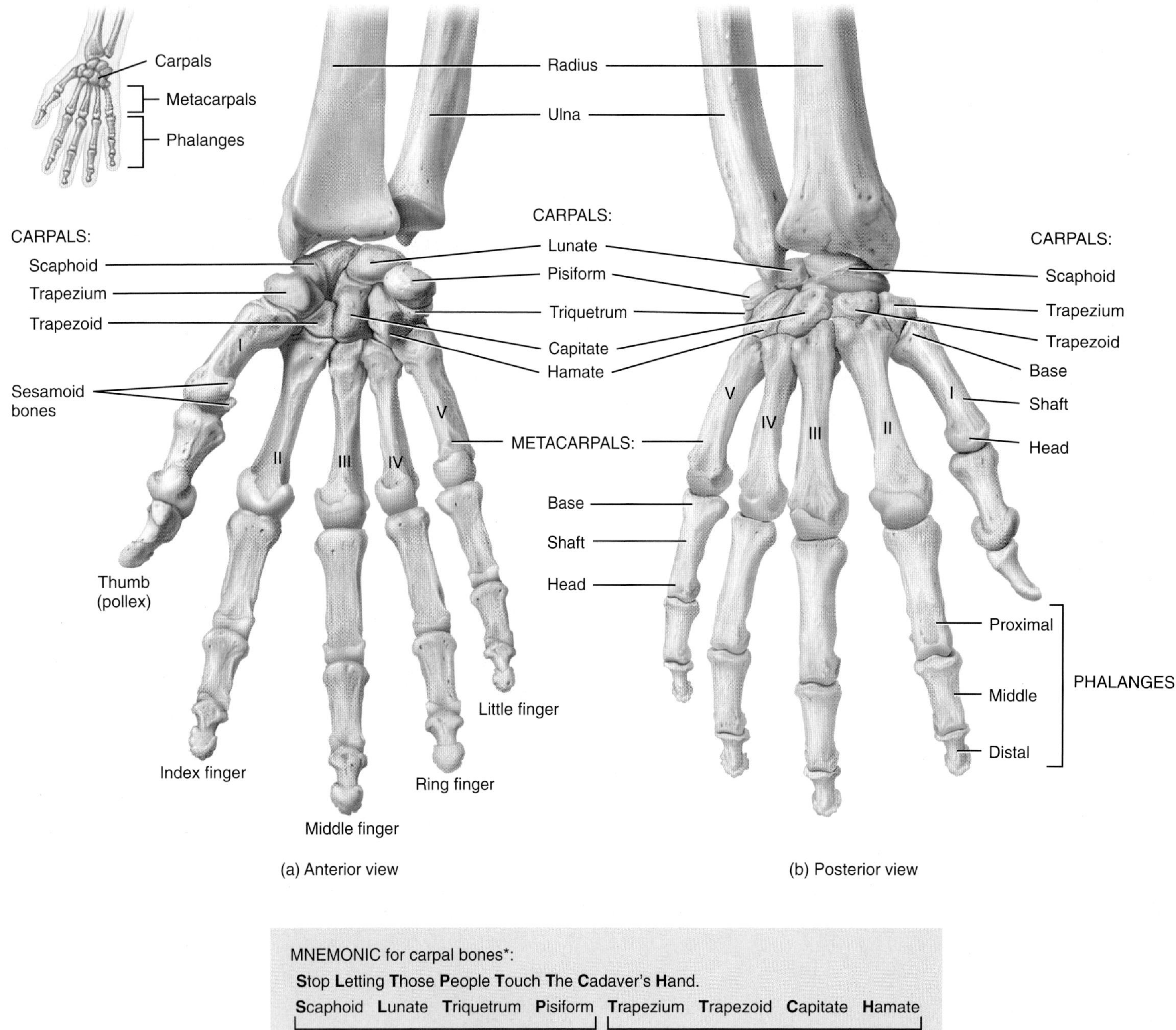

(a) Anterior view

(b) Posterior view

MNEMONIC for carpal bones*:

Stop **L**etting **T**hose **P**eople **T**ouch **T**he **C**adaver's **H**and.

Scaphoid	**L**unate	**T**riquetrum	**P**isiform	**T**rapezium	**T**rapezoid	**C**apitate	**H**amate
Proximal row				Distal row			
Lateral → Medial				Lateral → Medial			

* *Edward Tanner, University of Alabama, School of Medicine*

Which is the most frequently fractured wrist bone?

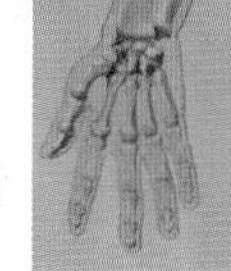

8.3 PELVIC (HIP) GIRDLE

OBJECTIVES

- Identify the bones of the pelvic girdle and their principal markings.
- Describe the division of the pelvic girdle into false and true pelves.

The ***pelvic (hip) girdle*** consists of the two ***hip bones,*** also called ***coxal bones*** (KOK-sal; *cox-* = hip) or ***os coxa*** (Figure 8.9). The hip bones unite anteriorly at a joint called the ***pubic symphysis*** (PŪ-bik SIM-fi-sis). They unite posteriorly with the sacrum at the two *sacroiliac joints.* The complete ring composed of the hip bones, pubic symphysis, and sacrum forms a deep, basin-like structure called the ***bony pelvis*** (*pelv-* = basin). The plural of pelvis is *pelves* (PEL-vēz) or *pelvises.* Functionally, the bony pelvis provides a strong and stable support for the vertebral column and pelvic organs. The pelvic girdle of the bony pelvis also accepts the bones of the lower limbs, connecting them to the axial skeleton.

Each of the two hip bones of a newborn consists of three bones separated by cartilage: a superior *ilium,* an inferior and anterior *pubis,* and an inferior and posterior *ischium.* By approximately age 23, the three separate bones fuse together (see Figure 8.10a). Although the hip bones function as single bones, anatomists and allied health professionals commonly discuss them as though they still consisted of three bones.

Figure 8.9 Bony pelvis. Shown here is the female bony pelvis.

The hip bones are united anteriorly at the pubic symphysis and posteriorly at the sacrum to form the bony pelvis.

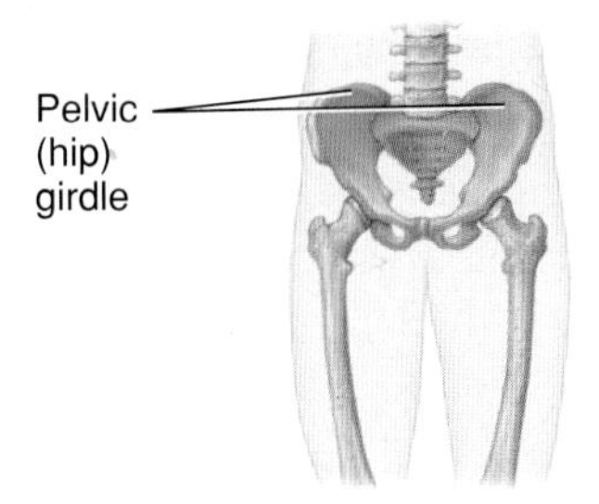

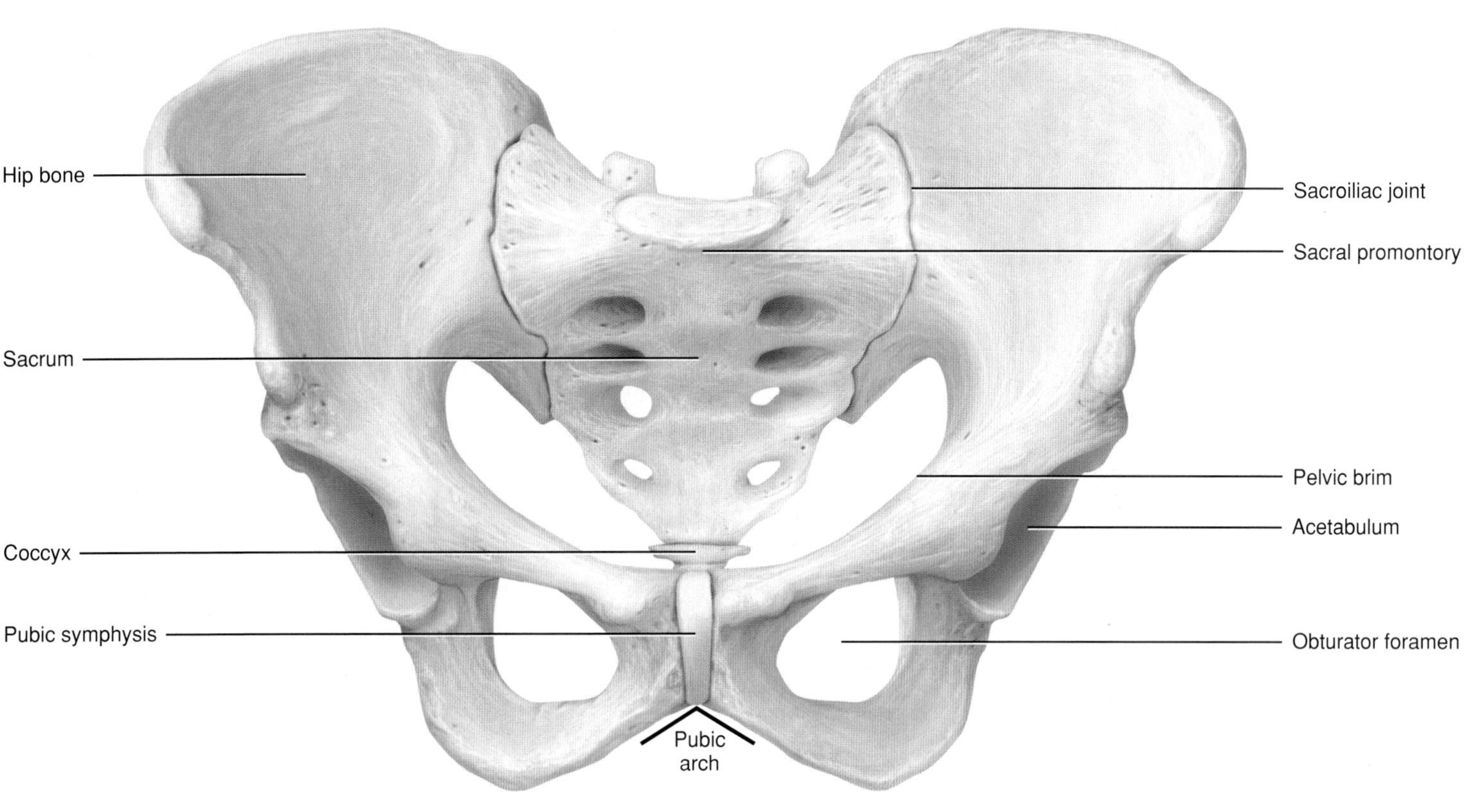

Anterosuperior view of pelvic girdle

What are the functions of the bony pelvis?

Ilium

The ***ilium*** (IL-ē-um = flank) is the largest of the three components of the hip bone (Figure 8.10). A superior *ala* (= wing) and an inferior *body,* which enters into the formation of the *acetabulum,* which is the socket for the head of the femur, comprise the ilium. Its superior border, the *iliac crest,* ends anteriorly in a blunt *anterior superior iliac spine* (commonly known as the ASIS). Below this spine is the *anterior inferior iliac spine* (AIIS). Posteriorly, the iliac crest ends in a sharp *posterior superior iliac spine* (PSIS). Below this spine is the *posterior inferior iliac spine* (PIIS). The anterior superior iliac spine is an exceedingly important anatomical landmark for manual therapists. The spines serve as points of attachment for the tendons of the muscles of the trunk, hip, and thighs. Below the posterior inferior iliac spine is the *greater sciatic notch* (sī-AT-ik), through which the sciatic nerve, the longest nerve in the body, passes. Bruising of the anterior superior iliac spine and associated soft tissues, such as occurs in body contact sports such as football, is called a ***hip pointer.***

The medial surface of the ilium contains the *iliac fossa,* a concavity where the iliacus muscle attaches. Posterior to this fossa are the *iliac tuberosity,* a point of attachment for the sacroiliac ligament, and the *auricular surface* (*auric-* = ear-shaped), which articulates with the sacrum to form the *sacroiliac joint* (see Figures 8.9 and 8.10c). Projecting anteriorly and inferiorly from the auricular surface is a ridge called the *arcuate line* (AR-kū-āt; *arc-* = bow).

The other conspicuous markings of the ilium are three arched lines on its lateral surface called the *posterior gluteal line* (*glut-* = buttock), the *anterior gluteal line,* and the *inferior gluteal line.* The tendons of the gluteal muscles attach to the ilium between these lines.

Ischium

The ***ischium*** (IS-kē-um = hip), the inferior, posterior portion of the hip bone (Figure 8.10), is composed of a superior *body* and an inferior *ramus* (*ram-* = branch; plural is *rami*). The ramus joins the pubis. Features of the ischium include the prominent *ischial spine,* a *lesser sciatic notch* below the spine, and a large, rough, and thickened *ischial tuberosity.* When sitting upright on a hard surface, this prominent tuberosity is just deep to the skin and commonly begins hurting after a relatively short time. Together,

Figure 8.10 Right hip bone. The lines of fusion of the ilium, ischium, and pubis depicted in (a) are not always visible in an adult.

The acetabulum is the socket formed where the three parts of the hip bone converge.

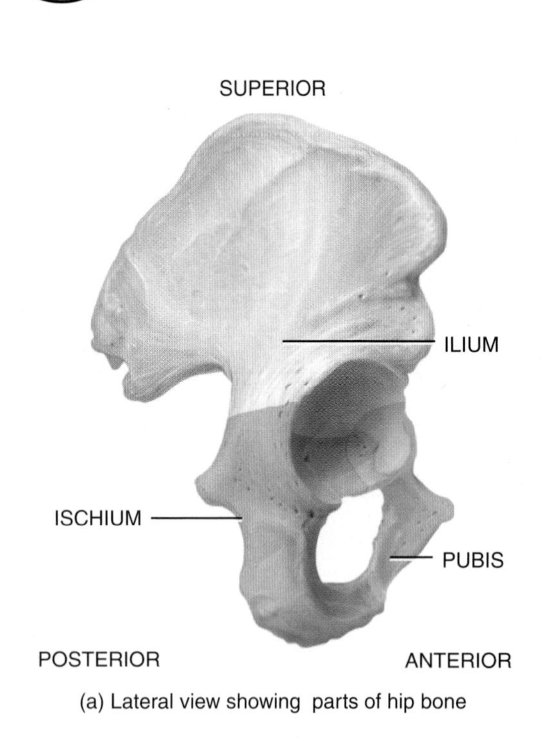

(a) Lateral view showing parts of hip bone

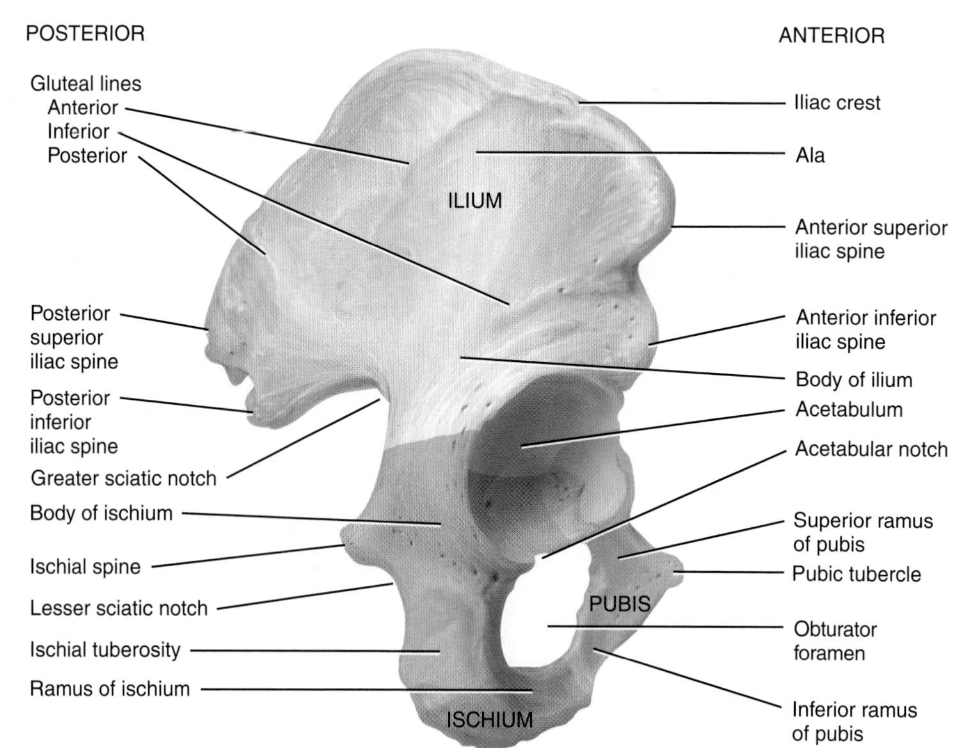

(b) Detailed lateral view

the ramus and the pubis surround the *obturator foramen* (OB-too-rā-tōr; *obtur-* = closed up), the largest foramen in the skeleton. The foramen is so named because, even though blood vessels and nerves pass through it, it is nearly completely closed by the fibrous *obturator membrane.*

Pubis

The ***pubis*** or ***os pubis,*** meaning pubic bone, is the anterior and inferior part of the hip bone (Figure 8.10). A *superior ramus,* an *inferior ramus,* and a *body* between the rami comprise the pubis. The anterior border of the body is the *pubic crest,* and at its lateral end is a projection called the *pubic tubercle.* This tubercle is the beginning of a raised line, the *pectineal line* (pek-TIN-ē-al), which extends superiorly and laterally along the superior ramus to merge with the arcuate line of the ilium. These lines, as you will see shortly, are important landmarks for distinguishing the superior and inferior portions of the bony pelvis.

The *pubic symphysis* is the joint between the two hip bones (see Figure 8.9). It consists of a disc of fibrocartilage. Inferior to this joint, the inferior rami of the two pubic bones converge to form the *pubic arch.* In the later stages of pregnancy, the hormone relaxin (produced by the ovaries and placenta) increases the flexibility of the pubic symphysis to ease delivery of the baby. Weakening of the joint, together with an already compromised center of gravity due to an enlarged uterus, also alters the gait during pregnancy.

The *acetabulum* (as-e-TAB-ū-lum = vinegar cup) is a deep fossa formed by the ilium, ischium, and pubis. It functions as the socket that accepts the rounded head of the femur. Together, the acetabulum and the femoral head form the *hip (coxal) joint.* On the inferior side of the acetabulum is a deep indentation, the *acetabular notch.* It forms a foramen through which blood vessels and nerves pass, and it serves as a point of attachment for ligaments of the femur (e.g., the ligament of the head of the femur).

False and True Pelves

The bony pelvis is divided into superior and inferior portions by a boundary called the *pelvic brim* (see Figure 8.11a). You can trace the pelvic brim by following the landmarks around parts of the hip bones to form the outline of an oblique plane. Beginning

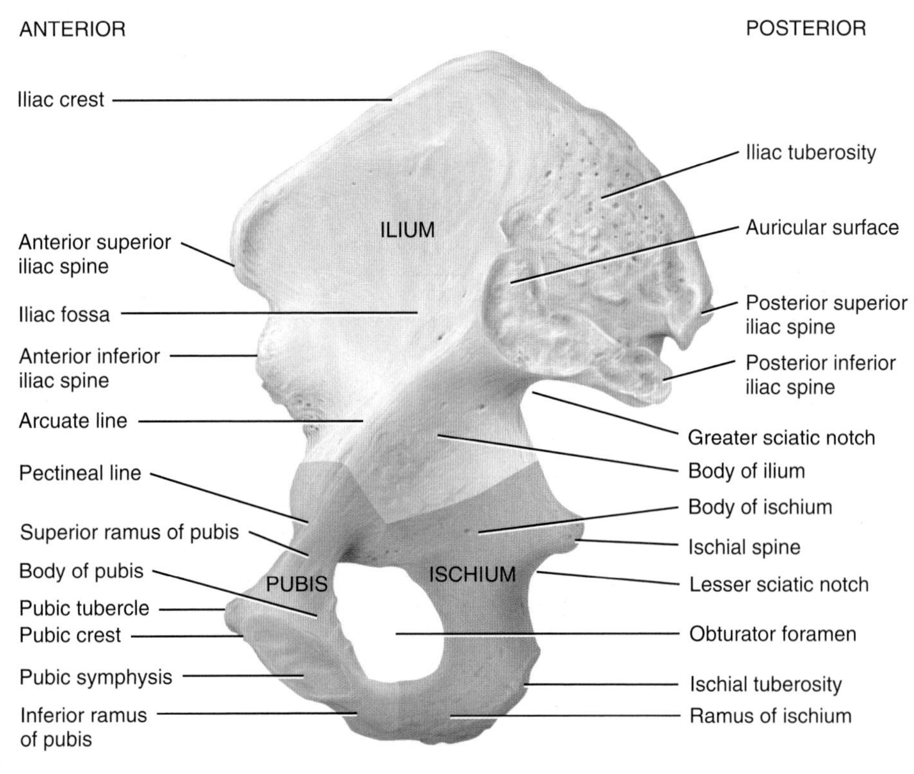

(c) Detailed medial view

Which part of the hip bone articulates with the femur? With the sacrum?

Figure 8.11 **True and false pelves.** Shown here is the female pelvis. For simplicity, in part (a) the landmarks of the pelvic brim are shown only on the left side of the body, and the outline of the pelvic brim is shown only on the right side. The entire pelvic brim is shown in Table 8.1.

The true and false pelves are separated by the pelvic brim.

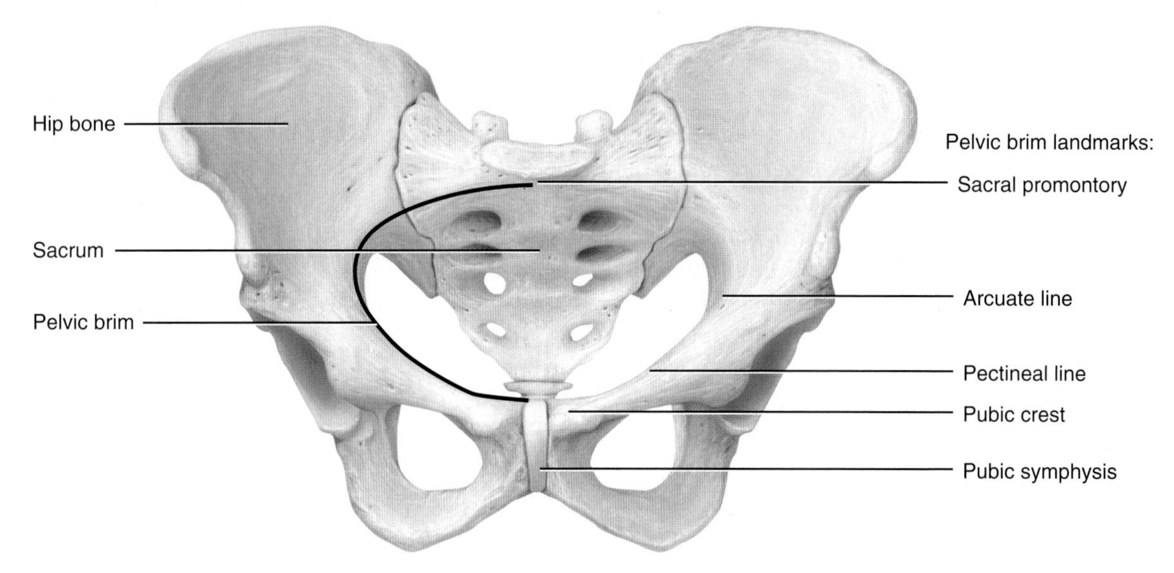

(a) Anterosuperior view of pelvic girdle

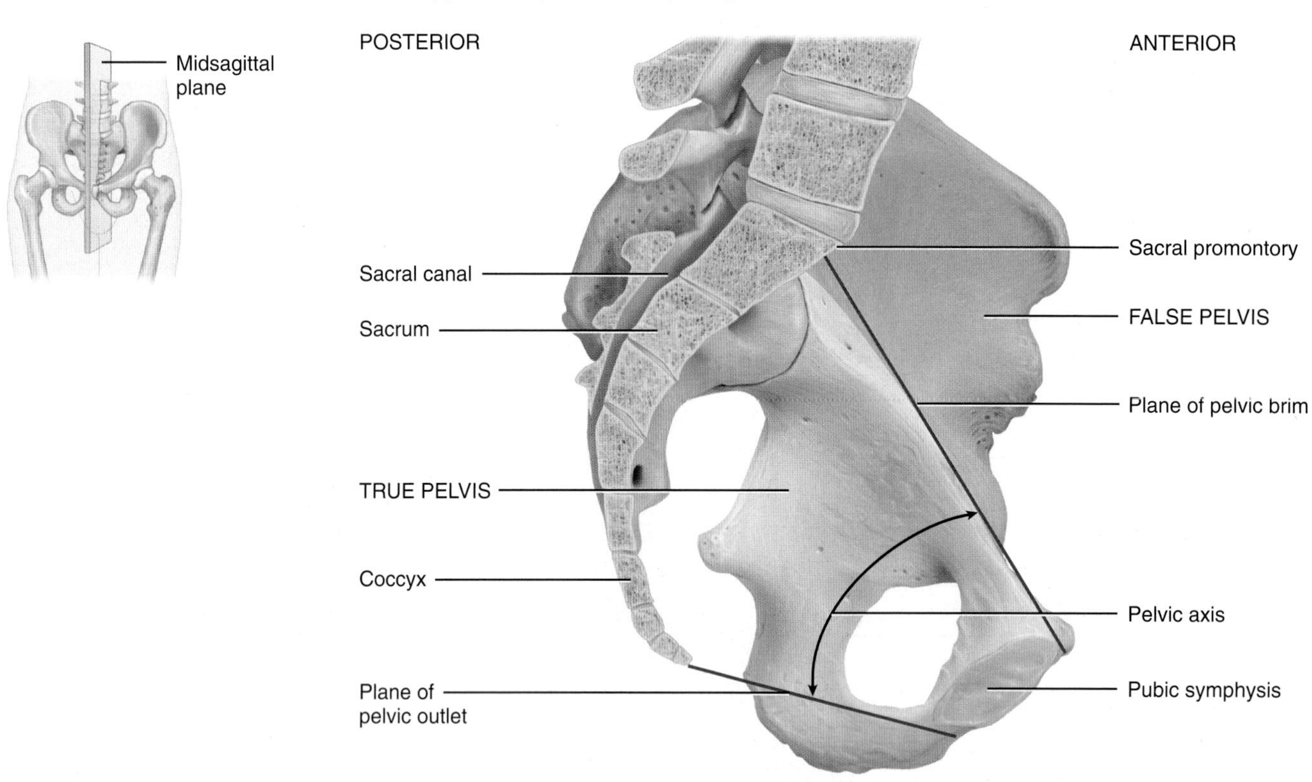

(b) Midsagittal section indicating locations of true and false pelves

What is the significance of the pelvic axis?

posteriorly at the *sacral promontory* of the sacrum, trace laterally and inferiorly along the *arcuate lines* of the ilium. Continue inferiorly along the *pectineal lines* of the pubis. Finally, trace anteriorly to the superior portion of the *pubic symphysis.* Together, these points form an oblique plane that is higher in the back than in the front. The circumference of this plane is the pelvic brim.

The portion of the bony pelvis superior to the pelvic brim is the ***false (greater) pelvis*** (Figure 8.11b). It is bordered by the lumbar vertebrae posteriorly, the superior portions of the hip bones laterally, and the abdominal wall anteriorly. The space enclosed by the false pelvis is part of the abdomen; it does not contain pelvic organs, except for the urinary bladder (when it is full) and the uterus during pregnancy.

The portion of the bony pelvis inferior to the pelvic brim is the ***true (lesser) pelvis*** (Figure 8.11b). It is bounded by the sacrum and coccyx posteriorly, inferior portions of the ilium and ischium laterally, and the pubic bones anteriorly. The true pelvis surrounds the pelvic cavity (see Figure 1.9). The superior opening of the true pelvis, bordered by the pelvic brim, is called the *pelvic inlet;* the inferior opening of the true pelvis is the *pelvic outlet.* The *pelvic axis* is an imaginary line that curves through the true pelvis from the central point of the plane of the pelvic inlet to the central point of the plane of the pelvic outlet. During childbirth the pelvic axis is the route taken by the baby's head as it descends through the pelvis.

CHECKPOINT

4. Describe the distinguishing characteristics of the individual bones of the pelvic girdle.
5. Distinguish between the false and true pelves.

8.4 COMPARISON OF FEMALE AND MALE PELVES

OBJECTIVE

- Compare the principal structural differences between female and male pelves.

In the following discussion it is assumed that the male and female are comparable in age and physical stature. Generally, the bones of a male are larger and heavier than those of a female and have larger surface markings. Gender-related differences in the features of bones are readily apparent when comparing the female and male pelves. Most of the structural differences in the pelves are adaptations to the requirements of pregnancy and childbirth. The female's pelvis is wider and shallower than the male's. Consequently, there is more space in the true pelvis of the female, especially in the pelvic inlet and pelvic outlet, which accommodate the passage of the infant at birth. Other significant structural differences between pelves of females and males are listed and illustrated in Table 8.1.

CHECKPOINT

6. Why are structural differences between female and male pelves important?

Table 8.1

Comparison of Female and Male Pelves

POINT OF COMPARISON	FEMALE	MALE
General structure	Light and thin.	Heavy and thick.
False (greater) pelvis	Shallow.	Deep.
Pelvic brim (inlet)	Larger and more oval.	Smaller and heart-shaped.
Acetabulum	Small and faces anteriorly.	Large and faces laterally.
Obturator foramen	Oval.	Round.
Pubic arch	Greater than 90° angle.	Less than 90° angle.

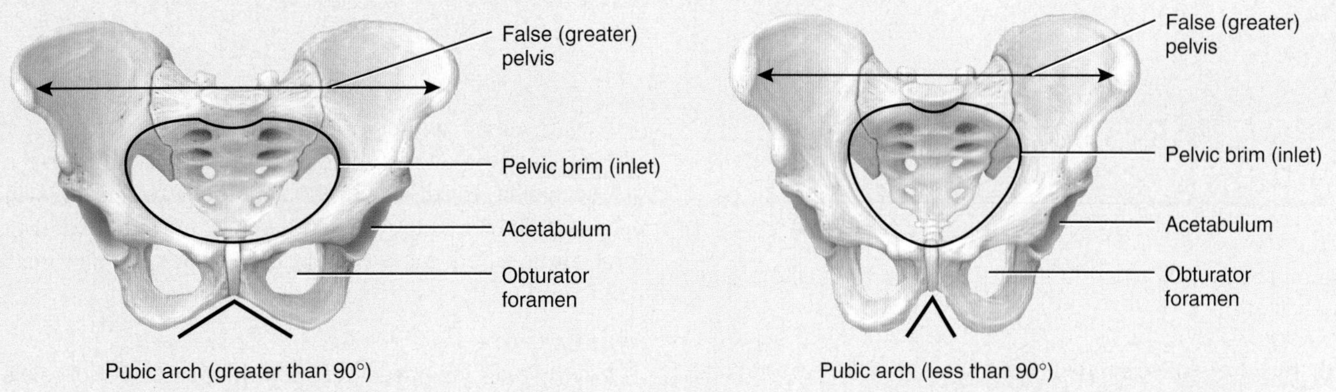

Anterior views

CONTINUES

TABLE 8.1 CONTINUED

POINT OF COMPARISON	FEMALE	MALE
Illiac crest	Less curved.	More curved.
Ilium	Less vertical.	More vertical.
Greater sciatic notch	Wide.	Narrow.
Coccyx	More movable and more curved anteriorly.	Less movable and less curved anteriorly.
Sacrum	Shorter, wider (see anterior views), and less curved anteriorly.	Longer, narrower (see anterior views), and more curved anteriorly.
Pelvic outlet	Wider.	Narrower.
Ischial tuberosity	Shorter, farther apart, and more medially projecting.	Longer, closer together, and more laterally projecting.

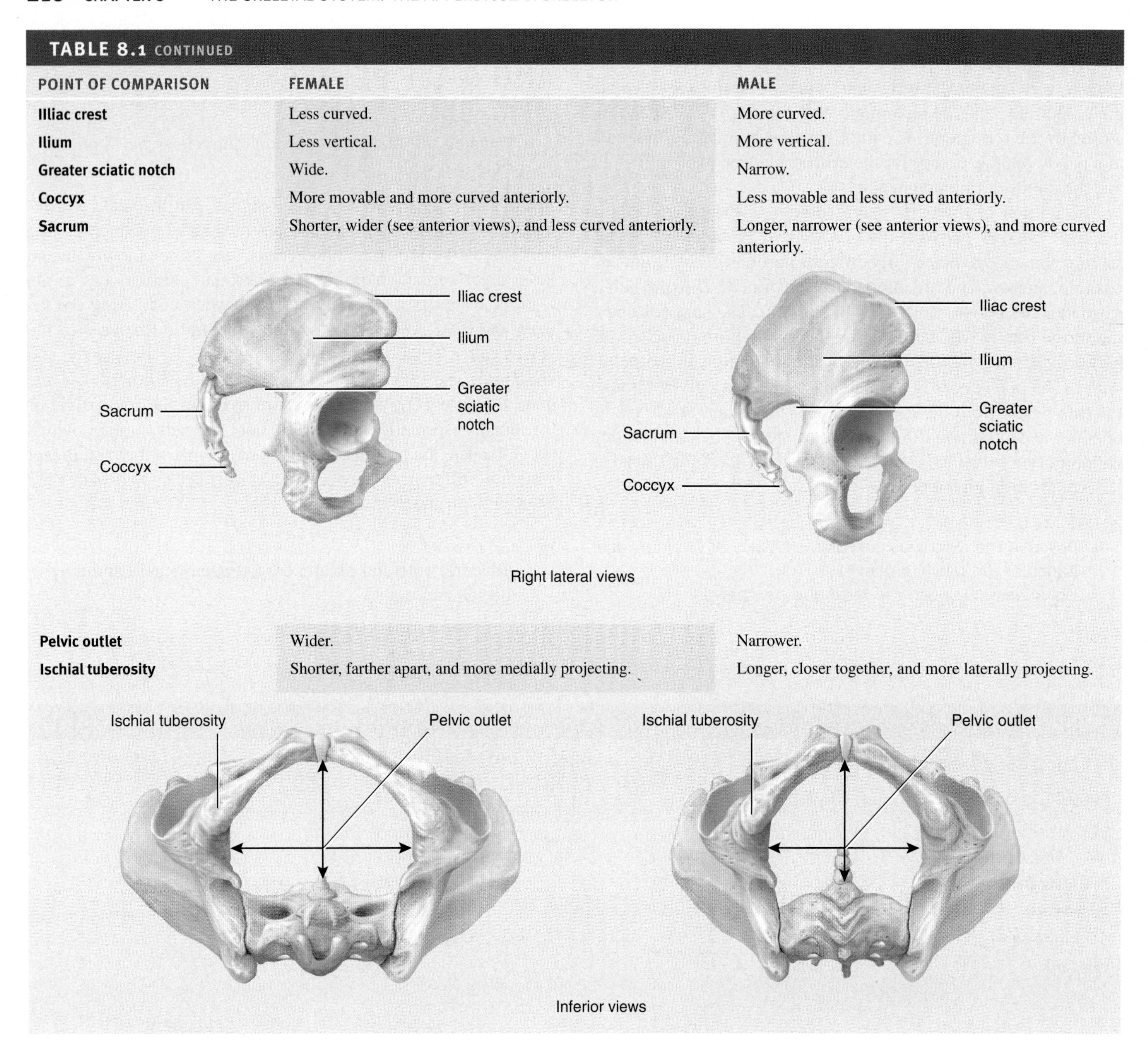

Right lateral views

Inferior views

8.5 COMPARISON OF PECTORAL AND PELVIC GIRDLES

OBJECTIVE

- Compare the principal difference between the pectoral and pelvic girdles.

Now that we have have studied the structures of the pectoral and pelvic girdles, we can note some of their significant differences. The pectoral girdle does not directly articulate with the vertebral column, but the pelvic girdle does so via the sacroiliac joint. The sockets (glenoid cavities) for the upper limbs in the pectoral girdle are shallow and maximize movement, in contrast to the sockets (acetabula) for the lower limbs in the pelvic girdle, which are deep and allow less movement. Overall, the structure of the pectoral girdle offers more mobility than strength, and that of the pelvic girdle offers more strength than mobility.

CHECKPOINT

7. Why are differences between the pectoral and pelvic girdle important?
8. Why is it important that the pelvic girdle articulate with the axial skeleton?

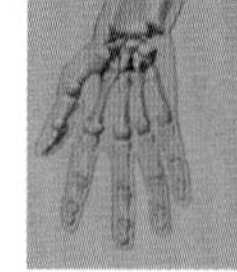

8.6 LOWER LIMB (EXTREMITY)

OBJECTIVE

- Identify the bones of the lower limb and their principal markings.

Each ***lower limb (extremity)*** has 30 bones: the femur in the thigh; the patella (kneecap); the tibia and fibula in the leg; and the 7 tarsals in the tarsus (ankle), the 5 metatarsals in the metatarsus, and the 14 phalanges (bones of the digits) in the foot (Figure 8.12).

Femur

The ***femur***, or thigh bone, is the longest, heaviest, and strongest bone in the body (see Figure 8.13). Its proximal end articulates with the acetabulum of the hip bone. Its distal end articulates with the tibia and patella (Figure 8.12). The *body (shaft)* of the femur angles medially and, as a result, the knee joints are closer to the midline. The angle is greater in females than in males because the female pelvis is broader.

The proximal end of the femur consists of a rounded *head* that articulates with the acetabulum of the hip bone to form the *hip (coxal) joint.* The head contains a small centered depression (pit) called the *fovea capitis* (FŌ-vē-a CAP-i-tis; *fovea* = pit; *capitis* = of the head). The ligament of the head of the femur connects the fovea capitis (see Figure 8.13c) of the femur to the acetabulum of the hip bone. The *neck* of the femur is a constricted region distal to the head. The *greater trochanter* (trō-KAN-ter) and *lesser trochanter* are projections that serve as points of attachment for the tendons of some of the thigh and buttock muscles. The greater trochanter is the prominence felt and seen anterior to the hollow on the side of the hip. It is a landmark commonly used to locate the site for intramuscular injections into the lateral surface of the thigh. The lesser trochanter is inferior and medial to the greater trochanter. Between the anterior surfaces of the trochanters is a narrow *intertrochanteric line* (see Figure 8.13a). A ridge called the *intertrochanteric crest* appears between the posterior surfaces of the trochanters (see Figure 8.13b).

Inferior to the intertrochanteric crest on the posterior surface of the body of the femur is a vertical ridge called the *gluteal tuberosity.* It blends into another vertical ridge called the *linea aspera* (LIN-ē-a AS-per-a; *asper* = rough). Both ridges serve as attachment points for the tendons of several thigh muscles.

The distal end of the femur expands to include the *medial condyle* and the *lateral condyle.* These articulate with the medial and lateral condyles of the tibia. Superior to the condyles are the *medial epicondyle* and the *lateral epicondyle,* to which ligaments of the knee joint attach. A depressed area between the condyles on the posterior surface is called the *intercondylar fossa* (in-ter-KON-di-lar). The *patellar surface* is located between the condyles on the anterior surface.

Figure 8.12 Right lower limb.

Each lower limb consists of a femur, patella (kneecap), tibia, fibula, tarsals (ankle bones), metatarsals, and phalanges (bones of the digits).

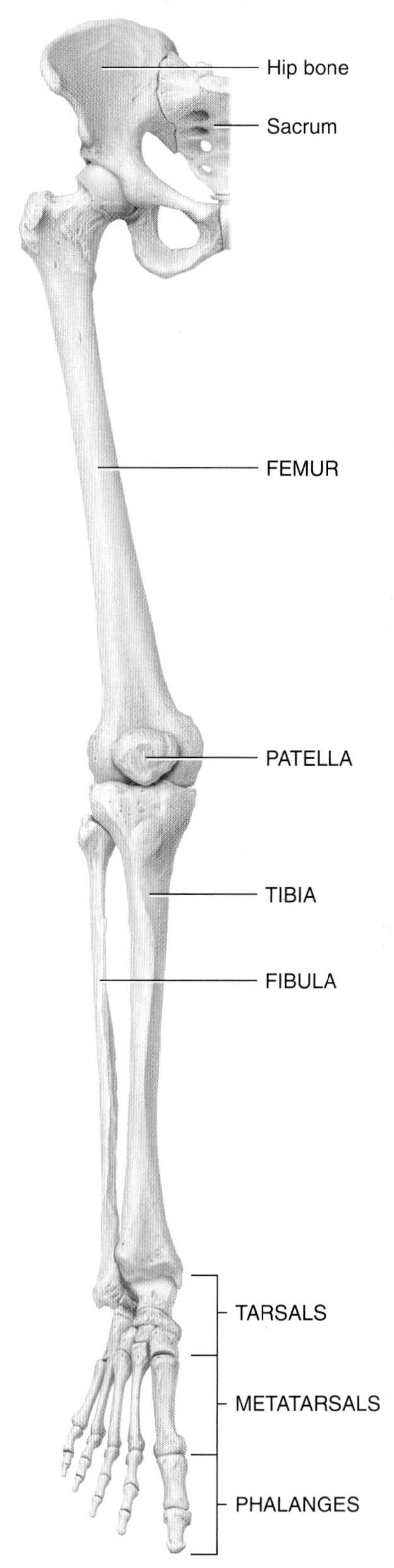

Anterior view of lower limb

How many bones make up each lower limb?

Figure 8.13 Right femur in relation to the hip bone, patella, tibia, and fibula.

The acetabulum of the hip bone and head of the femur articulate to form the hip joint.

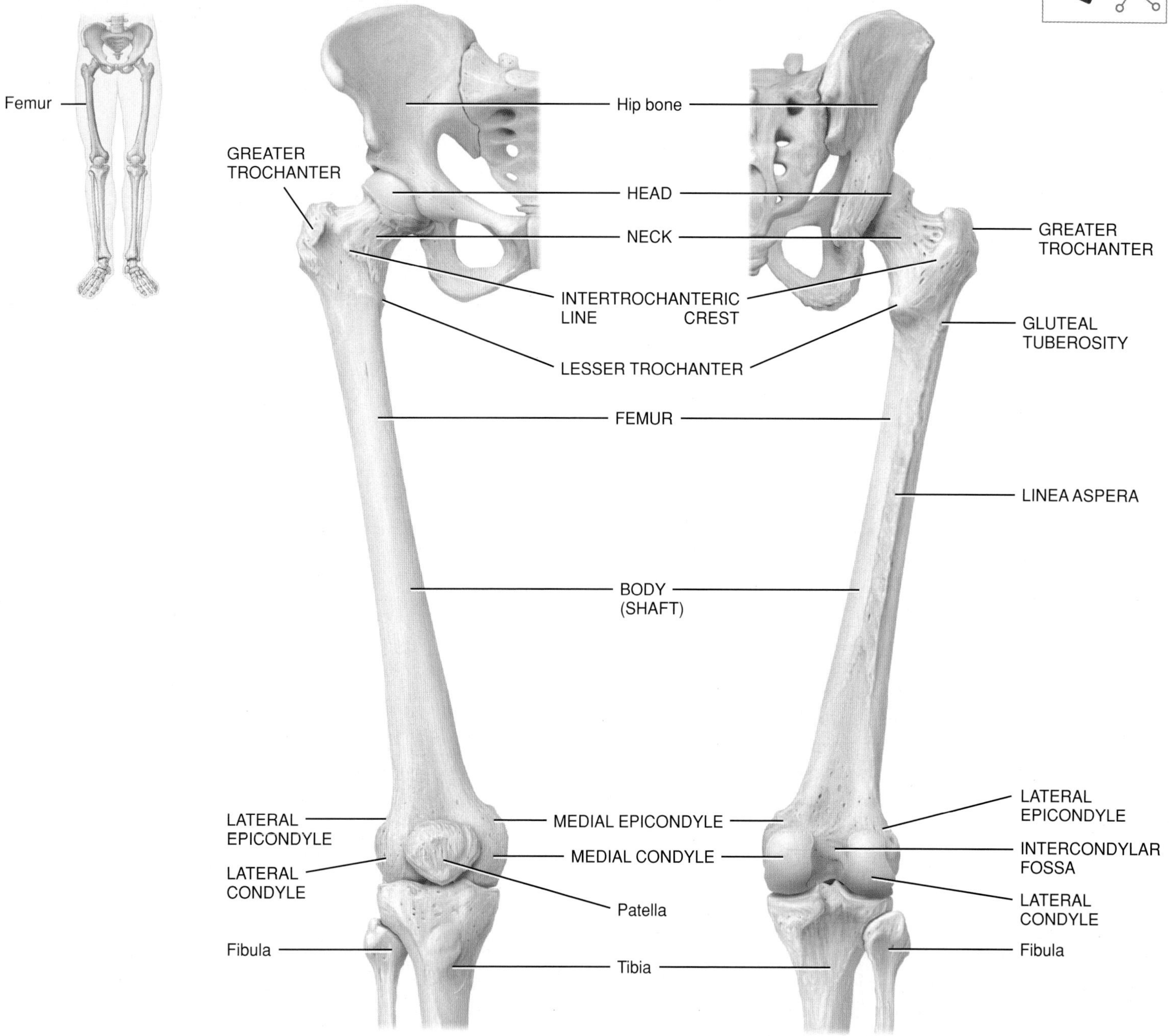

(a) Anterior view

(b) Posterior view

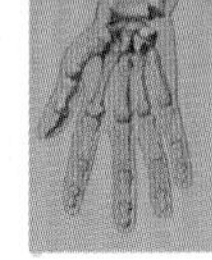

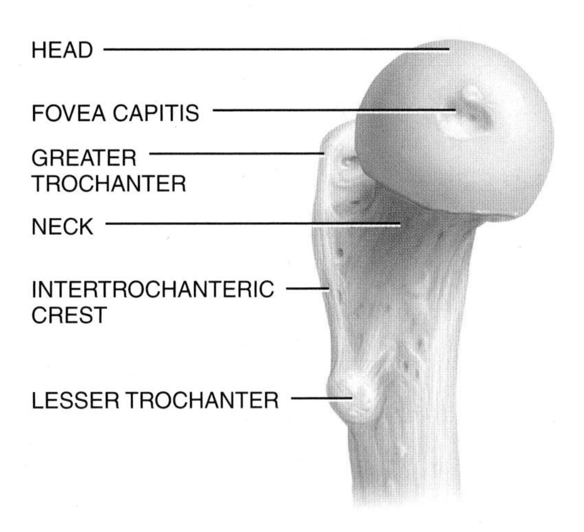

(c) Medial view of proximal end of femur

Why is the angle of convergence of the femurs greater in females than males?

Patella

The ***patella*** (= little dish), or kneecap, is a small, triangular bone located anterior to the knee joint (Figure 8.14). It is a sesamoid bone that develops in the tendon of the quadriceps femoris muscle. The broad superior end of the patella is called the *base.* The pointed inferior end is the *apex.* The posterior surface contains two *articular facets,* one for the medial condyle of the femur and the other for the lateral condyle of the femur. The patellar ligament attaches the patella to the tibial tuberosity (see Figure 14.1). The *patellofemoral joint,* between the posterior surface of the patella and the patellar surface of the femur, is the intermediate component of the *tibiofemoral (knee) joint.* The functions of the patella are to increase the leverage of the tendon of the quadriceps femoris muscle (see Figure 14.1), to maintain the position of the tendon when the knee is bent (flexed), and to protect the anterior aspect of the knee joint.

MANUAL THERAPY APPLICATION

Patellofemoral Stress Syndrome

Patellofemoral stress syndrome ("runner's knee") is one of the most common problems runners experience. During normal flexion and extension of the knee, the patella tracks (glides) superiorly and inferiorly in the groove between the femoral condyles. In patellofemoral stress syndrome, normal tracking does not occur; instead, the patella tracks laterally as well as superiorly and inferiorly, and increased pressure on the joint causes aching or tenderness around or deep to the patella. Pain typically occurs after a person has been sitting for a while, especially after exercise. It is worsened by squatting or walking down stairs. One cause of runner's knee is the constant walking, running, or jogging on the same side of the road. Because roads or streets that do not have sidewalks and storm sewers (called high-crown roads) slope on the sides, the knee that is closer to the center of the road endures greater mechanical stress because it does not fully extend during a stride. Other predisposing factors include having knock-knees, running on hills, and running long distances.

Figure 8.14 Right patella.

The patella articulates with the lateral and medial condyles of the femur.

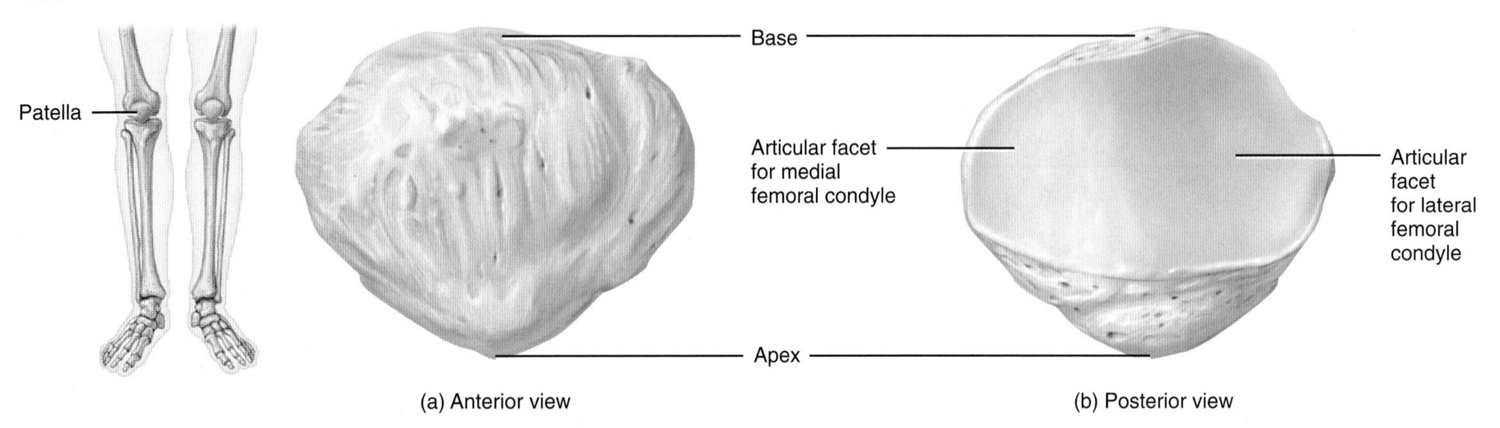

(a) Anterior view

(b) Posterior view

The patella is classified as which type of bone? Why?

Tibia and Fibula

The ***tibia,*** or shin bone, is the larger, medial, weight-bearing bone of the leg (Figure 8.15). The tibia articulates at its proximal end with the femur and fibula and at its distal end with the fibula and the talus bone of the ankle. The tibia and fibula, like the ulna and radius, are connected by an interosseous membrane.

The proximal end of the tibia is expanded into a *lateral condyle* and a *medial condyle.* These articulate with the condyles of the femur to form the lateral and medial *tibiofemoral (knee) joints.* The inferior surface of the lateral condyle articulates with the head of the fibula. The slightly concave condyles are separated by an upward projection called the *intercondylar eminence*

Figure 8.15 Right tibia and fibula in relation to the femur, patella, and talus.

The tibia articulates with the femur and fibula proximally, and with the fibula and talus distally.

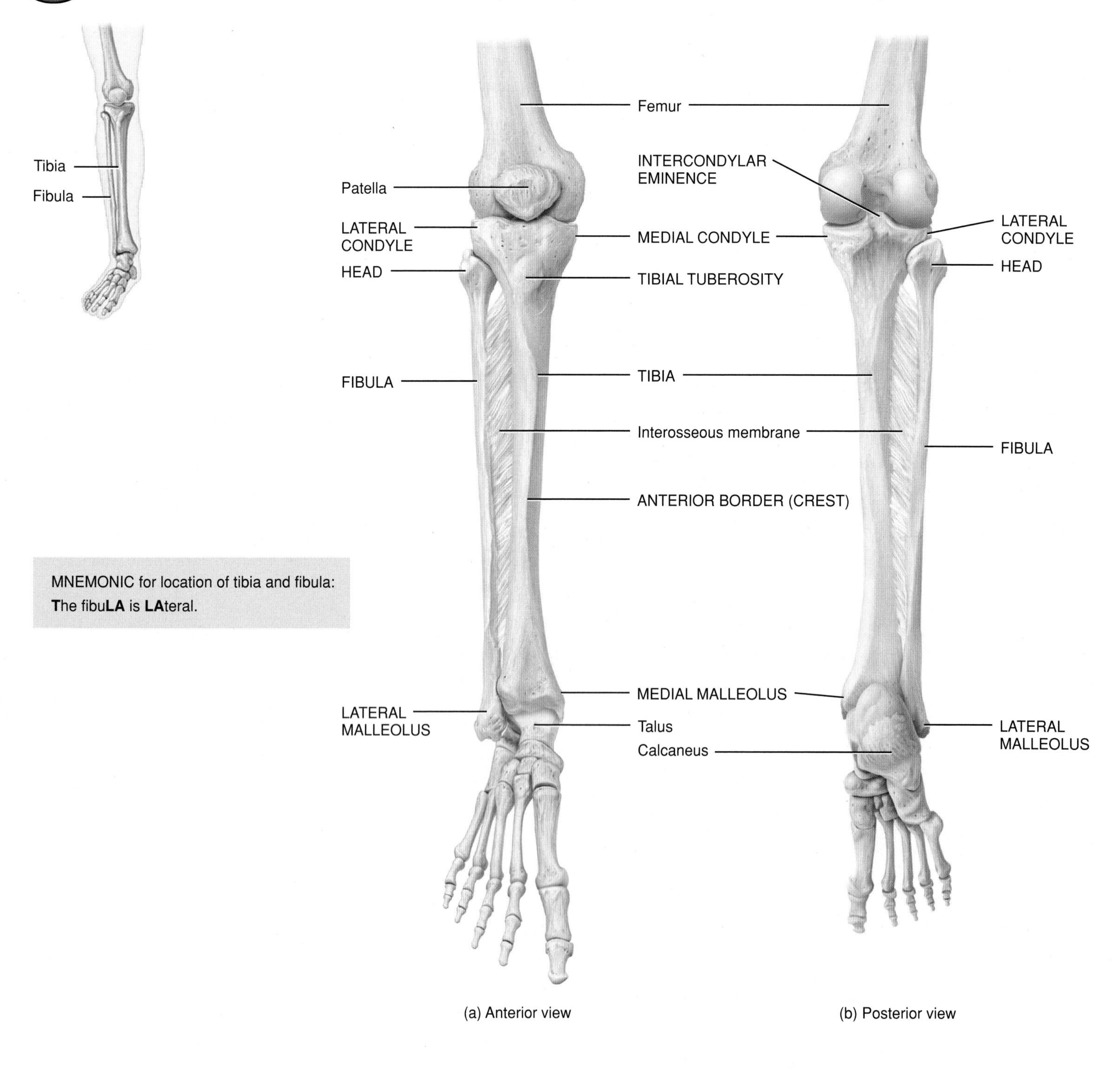

(Figure 8.15b). The *tibial tuberosity* on the anterior surface is a point of attachment for the patellar ligament. Inferior to and continuous with the tibial tuberosity is a sharp ridge that can be felt below the skin and is known as the *anterior border (crest),* also known as the shin.

The medial surface of the distal end of the tibia forms the *medial malleolus* (mal-LĒ-ō-lus = hammer). This structure articulates with the talus of the ankle and forms the prominence that can be felt on the medial surface of the ankle. The *fibular notch* (Figure 8.15c) articulates with the distal end of the fibula to form the *distal tibiofibular joint.* Of all the long bones of the body, the tibia is the most frequently fractured and is also the most frequent site of an open (compound) fracture.

The ***fibula*** is parallel and lateral to the tibia, but is considerably smaller. The *head* of the fibula, the proximal end, articulates with the inferior surface of the lateral condyle of the tibia below the level of the knee joint (Figure 8.15) to form the *proximal tibiofibular joint.* The distal end has a projection called the *lateral malleolus* that articulates with the talus of the ankle. This forms the prominence on the lateral surface of the ankle. As noted previously, the fibula also articulates with the tibia at the fibular notch.

• CLINICAL CONNECTION | Cartilage Implants

As indicated in Chapters 4 and 6, cartilages do not have a direct blood supply. Damaged cartilages, therefore, do not heal and are usually surgically replaced with materials that have characteristics of strength and flexibility similar to cartilage. **Cartilage implants,** however, may deteriorate in time due to mechanical forces; natural chemicals within the body may also cause the implant to deteriorate over time by corrosion. The metals include a particular type of stainless steel that resists corrosion, titanium, titanium alloys (a mixture of metals), tantalum, and cobalt–chromium alloys. Some of these metals are strong, others are flexible, and others have both properties. Choice of the metal by the surgeon depends on the location of the implant and on the level of activity of the patient. The opposing surface is commonly covered with a special form of polyethylene or a ceramic material. Both are very durable and when a metal implant moves on the polyethylene or ceramic implant, the contact is smooth and the amount of wear is minimal. •

Tarsals, Metatarsals, and Phalanges

The ***tarsus*** (ankle) is the proximal region of the foot and consists of seven ***tarsals*** (see Figure 8.16). They include the ***talus*** (TĀ-lus = ankle bone) and ***calcaneus*** (kal-KĀ-nē-us = heel), located in the posterior part of the foot. The calcaneus is the largest and strongest tarsal bone. The anterior tarsal bones are the ***navicular*** (= like a little boat), three ***cuneiform bones*** (= wedge-shaped) called the ***third (lateral), second (intermediate),*** and ***first (medial) cuneiforms,*** and the ***cuboid*** (= cube-shaped). Joints between tarsal bones are called *intertarsal joints.* The talus, the most superior tarsal bone, is the only bone of the foot that articulates with the fibula and tibia. It articulates on one side with the medial malleolus of the tibia and on the other side with the lateral malleolus of the fibula. These articulations form the *talocrural (ankle) joint.* During walking, the talus transmits about half the weight of the body to the calcaneus. The remainder is transmitted to the other tarsal bones.

The ***metatarsus*** is the intermediate region of the foot and consists of five ***metatarsal bones*** numbered I to V (or 1–5) from the medial to lateral position (see Figure 8.16). Like the metacarpals of the palm, each metatarsal consists of a proximal *base,* an intermediate *shaft,* and a distal *head.* The metatarsals articulate proximally with the first, second, and third cuneiform bones and with the cuboid to form the *tarsometatarsal joints.* Distally, they articulate with the proximal row of phalanges to form the

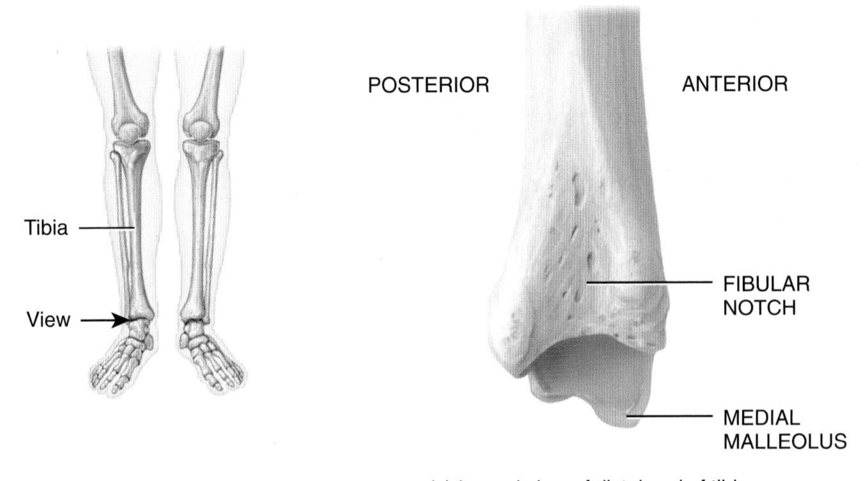

(c) Lateral view of distal end of tibia

? Which leg bone bears the weight of the body?

Figure 8.16 Right foot.

The skeleton of the foot consists of the proximal tarsals, the intermediate metatarsals, and the distal phalanges.

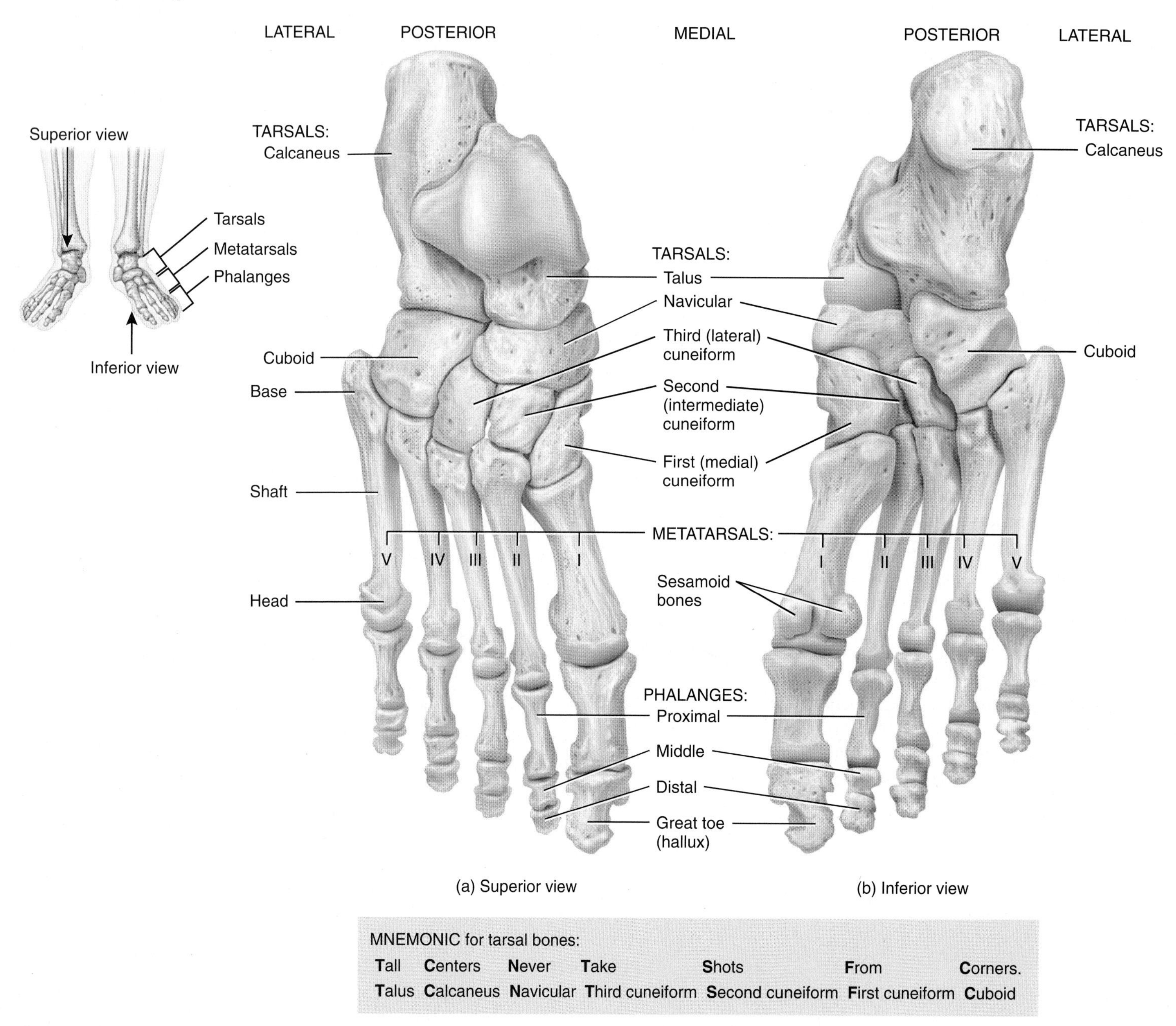

Which tarsal bone articulates with the tibia and fibula?

metatarsophalangeal joints. The first metatarsal is thicker than the others because it bears more weight.

The ***phalanges*** comprise the distal component of the foot and resemble those of the hand both in number and arrangement. The toes are numbered I to V (or 1–5) beginning with the great toe, from medial to lateral (Figure 8.16). Each *phalanx* (singular) consists of a proximal *base,* an intermediate *shaft,* and a distal *head.* The great, or big toe (*hallux*), has two large, heavy phalanges called proximal and distal phalanges. The other four toes each have three phalanges—proximal, middle, and distal. Joints between phalanges of the foot, like those of the hand, are called *interphalangeal joints.*

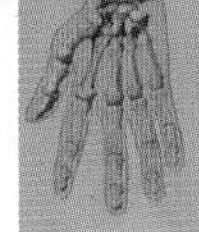

Arches of the Foot

The bones of the foot are arranged in two ***arches*** (Figure 8.17). The arches enable the foot to support the weight of the body, provide an ideal distribution of body weight over the soft and hard tissues of the foot, and provide leverage when walking. The arches are not rigid; they yield as weight is applied and spring back when the weight is lifted, thus helping to absorb shocks. Usually, the arches are fully developed by the time children reach age 12 or 13.

The ***longitudinal arch*** has two parts, both of which consist of tarsal and metatarsal bones arranged to form an arch from the anterior to the posterior part of the foot. The *medial part* of the longitudinal arch originates at the calcaneus. It rises to the talus and descends through the navicular, the three cuneiforms, and the heads of the three medial metatarsals. The *lateral part* of the longitudinal arch also begins at the calcaneus. It rises at the cuboid and descends to the heads of the two lateral metatarsals.

The ***transverse arch*** is found between the medial and lateral aspects of the foot and is formed by the navicular, three cuneiforms, and the bases of the five metatarsals.

As noted earlier, one function of the arches is to distribute body weight over the soft and hard tissues of the body. Normally, the ball of the foot carries about 40% of the weight and the heel carries about 60%. The ball of the foot is the padded portion of the sole superficial to the heads of the metatarsals. When a person wears high-heeled shoes, however, the distribution of weight changes so that the ball of the foot may carry up to 80% and the heel only 20%. As a result, the fat pads at the ball of the foot are damaged, joint pain develops, and structural changes in bones may occur.

• CLINICAL CONNECTION | Flatfoot and Clawfoot

The bones composing the arches are held in position by ligaments and tendons. If these ligaments and tendons are weakened, the height of the medial longitudinal arch may decrease or "fall." The result is **flatfoot,** the causes of which include excessive weight, postural abnormalities, weakened supporting tissues, and genetic predisposition. A custom-designed arch support often is prescribed to treat flatfoot.

Clawfoot is a condition in which the medial longitudinal arch is abnormally elevated. It is often caused by muscle deformities, such as may occur in diabetics whose neurological lesions lead to atrophy of muscles of the foot. •

To appreciate the skeletal system's contributions to homeostasis of other body systems, examine *Focus on Homeostasis: The Skeletal System.*

CHECKPOINT

9. Name the bones that form the lower limb, from proximal to distal.
10. Describe the joints of the lower limbs.
11. What are the functions of the arches of the foot?

Figure 8.17 Arches of the right foot.

Arches help the foot support and distribute the weight of the body and provide leverage during walking.

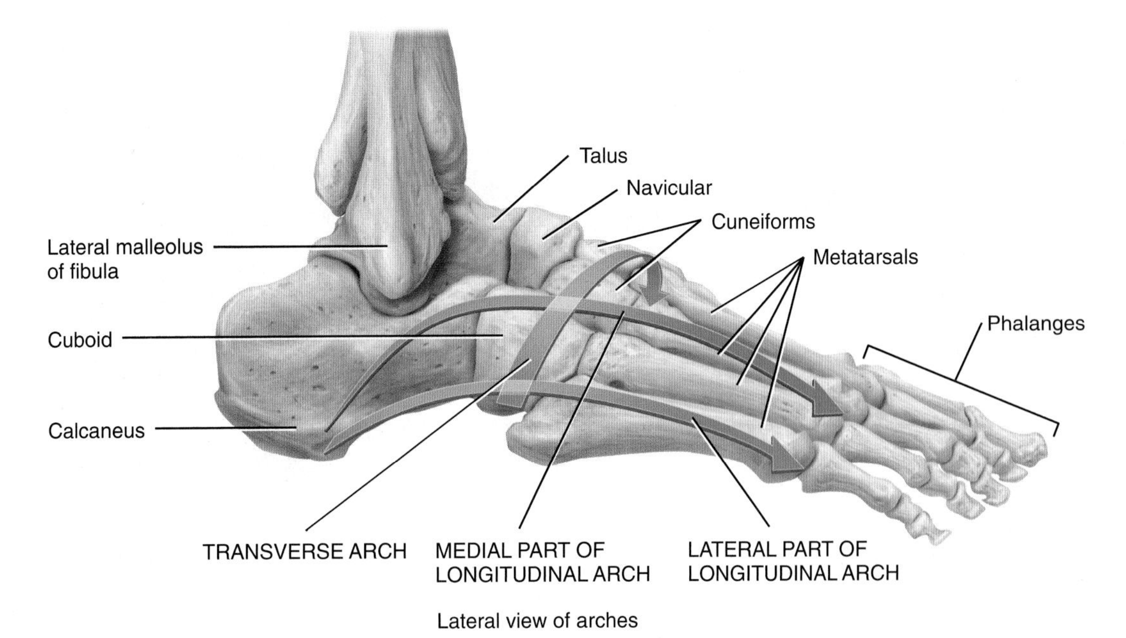

Lateral view of arches

What structural feature of the arches allows them to absorb shocks?

EXHIBIT 8.1 The Skeletal System and Homeostasis

BODY SYSTEM		CONTRIBUTION OF THE SKELETAL SYSTEM
For all body systems		Bones provide support and protection for internal organs; bones store and release calcium, which is needed for proper functioning of most body tissues.
Integumentary system	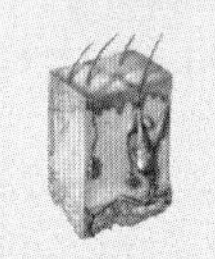	Bones provide strong support for overlying muscles and skin while joints provide flexibility that allows skin to bend.
Muscular system		Bones provide attachment points for muscles and leverage for muscles to bring about body movements; contraction of skeletal muscle requires calcium ions.
Nervous system	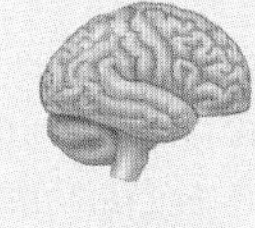	Skull and vertebrae protect brain and spinal cord; normal blood level of calcium is needed for normal functioning of neurons and neuroglia.
Endocrine system	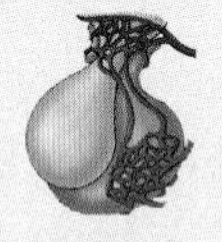	Bones store and release calcium, which is regulated by parathyroid hormone (PTH) and calcitonin (CT).
Cardiovascular system	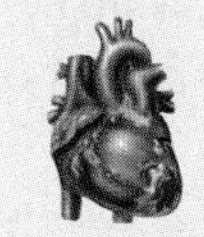	Red bone marrow carries out hemopoiesis (blood cell formation); rhythmic beating of the heart requires calcium ions.
Lymphatic system and immunity		Red bone marrow produces lymphocytes, white blood cells that are involved in immune responses.
Respiratory system	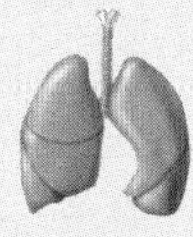	Axial skeleton of thorax protects lungs; rib movements assist in breathing; some muscles used for breathing attach to bones via tendons.
Digestive system	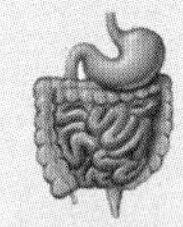	Teeth masticate (chew) food; rib cage protects esophagus, stomach, and liver; pelvis protects portions of the intestines.
Urinary system	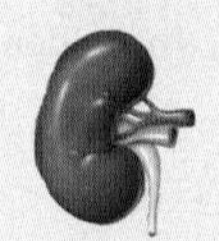	Ribs partially protect kidneys; pelvis protects urinary bladder and urethra.
Reproductive systems	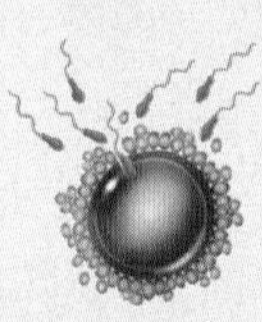	Pelvis protects ovaries, uterine (fallopian) tubes, and uterus in females and part of ductus (vas) deferens and accessory glands in males; bones are an important source of calcium needed for milk synthesis during lactation.

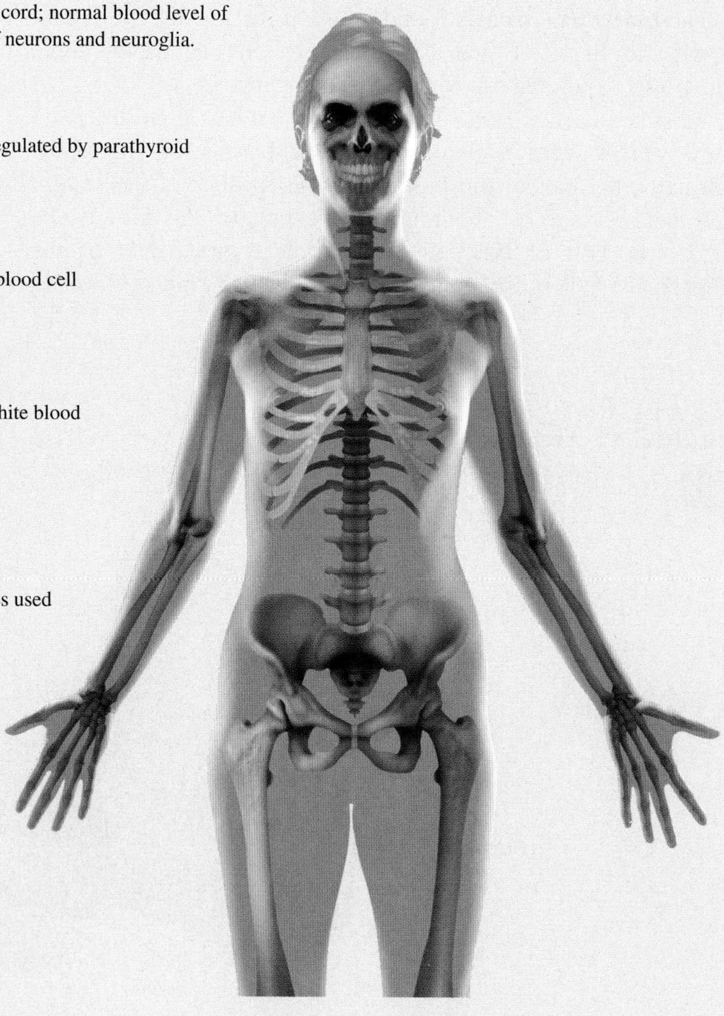

KEY MEDICAL TERMS ASSOCIATED WITH APPENDICULAR SKELETON

Clubfoot or ***talipes equinovarus*** (*-pes* = foot; *equino-* = horse) An inherited deformity in which the foot is twisted inferiorly and medially, and the angle of the arch is increased; occurs in 1 of every 1000 births. Treatment consists of manipulating the arch to a normal curvature by casts or adhesive tape, usually soon after birth. Corrective shoes or surgery may also be required.

Genu valgum (JĒ-noo VAL-gum; *genu* = knee; *valgum* = bent outward) A deformity in which the knees are abnormally close together and the space between the ankles is increased due to a lateral angulation of the tibia. Also called ***knock-knee***.

Genu varum (JĒ-noo VAR-um; *varum* = bent toward the midline) A deformity in which the knees are abnormally separated and the lower limbs are bowed laterally. Also called ***bowleg***.

Hallux valgus (HAL-uks VAL-gus; *hallux* = great toe) Angulation of the great toe away from the midline of the body, typically caused by wearing tightly fitting shoes. Involves lateral deviation of the proximal phalanx of the great toe and medial displacement of metatarsal I. Also called a ***bunion***.

STUDY OUTLINE

Pectoral (Shoulder) Girdle (Section 8.1)

1. Each pectoral (shoulder) girdle consists of a clavicle and scapula.
2. Each pectoral girdle attaches an upper limb to the axial skeleton.

Upper Limb (Extremity) (Section 8.2)

1. Each of the two upper limbs (extremities) contains 30 bones.
2. The bones of each upper limb include the humerus, the ulna, the radius, the carpals, the metacarpals, and the phalanges.

Pelvic (Hip) Girdle (Section 8.3)

1. The pelvic (hip) girdle consists of two hip bones.
2. Each hip bone consists of three fused bones: the ilium, pubis, and ischium.
3. The hip bones, sacrum, and pubic symphysis form the bony pelvis. It supports the vertebral column and pelvic viscera and attaches the lower limbs to the axial skeleton.
4. The true pelvis is separated from the false pelvis by the pelvic brim.

Comparison of Female and Male Pelves (Section 8.4)

1. Bones of males are generally larger and heavier than bones of females, with more prominent markings for muscle attachment.
2. The female pelvis is adapted for pregnancy and childbirth. Gender-related differences in pelvic structure are listed and illustrated in Table 8.1.

Comparison of Pectoral and Pelvic Girdles (Section 8.5)

1. The pectoral girdles do not directly articulate with the vertebral column; the pelvic girdle does.
2. The glenoid cavities of the scapulae are shallow and maximize movement; the acetabula of the hip bones are deep and allow less movement.

Lower Limb (Extremity) (Section 8.6)

1. Each of the two lower limbs (extremities) contains 30 bones.
2. The bones of each lower limb include the femur, the patella, the tibia, the fibula, the tarsals, the metatarsals, and the phalanges.
3. The bones of the foot are arranged in two arches, the longitudinal arch and the transverse arch, to provide support and leverage.

SELF-QUIZ QUESTIONS

Choose the one best answer to the following questions.

1. The olecranon is at the proximal end of the
   a. humerus **b.** radius **c.** ulna
   d. tibia **e.** scapula.
2. The tibia articulates distally with the
   a. femur **b.** fibula **c.** talus
   d. cuboid **e.** both b and c are correct.
3. Which structures are on the posterior surface of the upper limb?
   a. radial fossa and radial notch
   b. trochlea and capitulum
   c. coronoid process and coronoid fossa
   d. olecranon process and olecranon fossa
   e. lesser tubercle and intertubercular sulcus
4. The bones of the pectoral girdle
   a. articulate with the sternum anteriorly and the vertebrae posteriorly
   b. include both clavicles, both scapulae, and the manubrium of the sternum
   c. are considered to be part of the axial skeleton
   d. articulate with the head of the humerus, forming the shoulder joint
   e. are described by none of the above
5. The anatomical name for the socket into which the humerus fits is the
   a. acetabulum **b.** coronoid fossa **c.** glenoid cavity
   d. supraspinous fossa **e.** iliac fossa.
6. All of the following are tarsal bones *except* the
   a. cuboid **b.** triquetrum **c.** navicular
   d. first cuneiform **e.** third cuneiform.
7. Which of the following articulates with the sacrum?
   a. ilium **b.** ischium
   c. pubis **d.** both a and b are correct
   e. all three bones articulate with the sacrum

8. The greater trochanter is a large bony prominence located
   a. on the proximal part of the humerus
   b. on the proximal part of the femur
   c. near the tibial tuberosity
   d. on the ilium
   e. on the posterior surface of the scapula.
9. Which of the following statements regarding the male pelvis is *not* true?
   a. The bones are heavier and thicker than in the female.
   b. The male pelvis is narrow and deep.
   c. The coccyx is less curved anteriorly.
   d. The pelvic outlet is narrower than in the female.
   e. None of the above statements is true.
10. Which of the following is not a carpal bone?
   a. hamate b. cuboid c. pisiform
   d. trapezium e. scaphoid.

Complete the following.

11. The greater sciatic notch is an indentation seen on the _____.
12. The sesamoid bone that forms in the tendon of the quadriceps femoris muscle is the _____.
13. The lesser trochanter is on the _____ surface of the _____.
14. The portion of the pelvis superior to the pelvic brim is the _____ pelvis.
15. The _____ of the _____ fits into a depression called the acetabulum.

Are the following statements true or false?

16. The metatarsals are proximal to the tarsals.
17. The hamate is a carpal bone with a hook-shaped projection on its anterior surface.
18. The acromial extremity is located on the lateral end of the clavicle.
19. The capitulum of the humerus articulates with the styloid process of the radius.

Matching

20. Match the following bony landmarks and bones. Answers may be used more than once.

____	a. fibular notch	A. radius
____	b. acromion	B. femur
____	c. coronoid process	C. tibia
____	d. lateral malleolus	D. humerus
____	e. deltoid tuberosity	E. scapula
____	f. radial tuberosity	F. ischium
____	g. anterior inferior iliac spine	G. ulna
____	h. linea aspera	H. fibula
____	i. ischial tuberosity	I. ilium
____	j. coracoid process	J. clavicle
____	k. medial malleolus	K. pubis
____	l. conoid tubercle	
____	m. glenoid cavity	
____	n. coronoid fossa	
____	o. obturator foramen	
____	p. gluteal tuberosity	
____	q. trochlear notch	

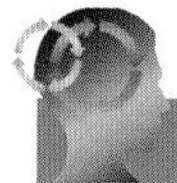

CRITICAL THINKING QUESTIONS

1. Mr. Smith's dog Rover dug up a complete set of human bones in the woods near his house. After examining the scene, the local police collected the bones and transported them to the coroner's office for identification. Later, Mr. Smith read in the newspaper that the bones belonged to an elderly female. How was this determined?
2. A proud dad holds his 5-month-old baby girl upright on her feet while supporting her under her arms. He states that she can never be a dancer because her feet are too flat. Is this true? Why or why not?
3. The local newspaper reported that Farmer White caught his hand in a piece of machinery last Tuesday. He lost the lateral two fingers of his left hand. His daughter, who is taking high school science, reports that Farmer White has three remaining phalanges. Is she correct, or does she need a refresher course in anatomy? Support your answer.

? ANSWERS TO FIGURE QUESTIONS

8.1 The pectoral girdles attach the upper limbs to the axial skeleton.
8.2 The weakest part of the clavicle is its midregion at the junction of the two curves.
8.3 The acromion of the scapula forms the high point of the shoulder.
8.4 Each upper limb has 30 bones.
8.5 The radius articulates at the elbow with the capitulum and radial fossa of the humerus. The ulna articulates at the elbow with the trochlea, coronoid fossa, and olecranon fossa of the humerus.
8.6 The olecranon is the "elbow" part of the ulna.
8.7 There are three points of attachment between the radius and ulna. The radius and ulna form the proximal and distal radioulnar joints. Their shafts are also connected by the interosseous membrane.
8.8 The scaphoid is the most frequently fractured wrist bone.
8.9 The bony pelvis attaches the lower limbs to the axial skeleton and supports the vertebral column and pelvic viscera.
8.10 The femur articulates with the acetabulum of the hip bone; the sacrum articulates with the auricular surface of the hip bone.
8.11 The pelvic axis is the course taken by a baby's head as it descends through the pelvis during childbirth.
8.12 Each lower limb has 30 bones.
8.13 The angle of convergence of the femurs is greater in females than males because the female pelvis is broader.
8.14 The patella is classified as a sesamoid bone because it develops in a tendon (the tendon of the quadriceps femoris muscle of the thigh).
8.15 The tibia is the weight-bearing bone of the leg.
8.16 The talus is the only tarsal bone that articulates with the tibia and the fibula.
8.17 Because the arches are not rigid, they yield when weight is applied and spring back when weight is lifted, allowing them to absorb the shock of walking.

Joints

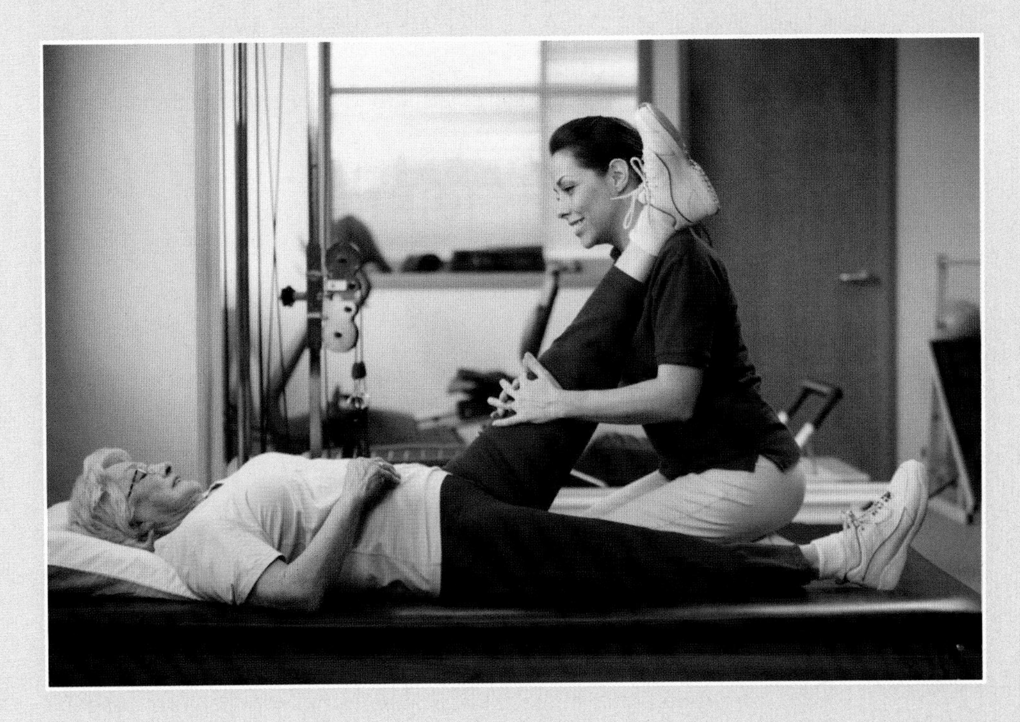

Joint diseases account for more than half of all chronic conditions in the elderly. People are living longer and conditions like arthritis and osteoporosis have a great impact on quality of life. For many years, people believed that exercise accelerated joint degeneration. Scientists now believe that a sedentary lifestyle leads to loss of strength in muscles, tendons, ligaments, and other joint structures, which makes movement even more painful and difficult. When muscles and joints atrophy, the resulting weakness makes joints less stable and more vulnerable to injury. Thirty minutes of appropriate activity a day helps to strengthen joint structures and delay the progress of arthritis. Appropriate activities include walking, stretching exercises, cycling, and swimming. Keeping joints healthy and more flexible can improve fitness, functional status, and quality of life.

CONTENTS AT A GLANCE

9.1 JOINT CLASSIFICATIONS

OBJECTIVES

- Describe how the structure of a joint determines its function.
- Describe the structural and functional classes of joints.

Bones are too rigid to bend without being damaged. Fortunately, flexible connective tissues form joints that hold bones together while still permitting some degree of movement. A ***joint***, also called an ***articulation*** (ar-tik-ū-LĀ-shun) or ***arthrosis*** (ar-THRŌ-sis), is a point of contact between two bones, between bone and cartilage, or between bone and teeth. When we say one bone *articulates* with another bone, we mean that the bones form a joint. Because most movements of the body occur at joints, you can appreciate their importance. Imagine how a cast over your knee joint makes walking difficult, or how a splint on a finger limits your ability to manipulate small objects.

The scientific study of joints is termed ***arthrology*** (ar-THROL-ō-jē; *arthr-* = joint; *-logy* = study of). The study of motion of the joints of the human body is called ***kinesiology*** (ki-nē-sē′-OL-ō-jē; *kinesi-* = movement). Note that this definition of kinesiology is not to be confused with the definition utilized by numerous alternative health-care practitioners who use the term kinesiology to describe a process of selecting specific supplements for patients.

Joints are classified structurally based on their anatomical characteristics and functionally based on the type of movement they permit. The structural classification of joints is based on (1) the presence or absence of a space between the articulating bones, called a synovial cavity, and (2) the type of connective tissue that binds the bones together. Structurally, joints are classified as one of the following types:

- **Fibrous joints** (FĪ-brus): There is no synovial cavity and the bones are held together by a solid mass of dense irregular connective tissue that is rich in collagen fibers.
- **Cartilaginous joints** (kar-ti-LAJ-i-nus): There is no synovial cavity and the bones are held together by cartilage.
- **Synovial joints** (sī-NŌ-vē-al; *syn-* = together): There is a synovial cavity and the bones are united by the dense irregular connective tissue that is part of the articular capsule, and often by accessory ligaments.

The functional classification of joints relates to the degree of movement they permit. Functionally, joints are classified as one of the following types:

- **Synarthrosis** (sin′-ar-THRŌ-sis): An immovable joint. The plural is *synarthroses.*
- **Amphiarthrosis** (am′-fē-ar-THRŌ-sis; *amphi-* = on both sides) A joint with slight movement. The plural is *amphiarthroses.*
- **Diarthrosis** (dī-ar-THRŌ-sis = movable joint): A freely movable joint. The plural is *diarthroses.* All diarthroses are synovial joints. They have a variety of shapes and permit several different types of movements (discussed later).

The following sections present the joints of the body according to their structural classifications. As we examine the structure of each type of joint, we will also describe its functions.

CHECKPOINT

1. How are joints classified structurally and functionally?

9.2 FIBROUS JOINTS

OBJECTIVE

- Describe the structure and functions of the three types of fibrous joints.

As previously noted, ***fibrous joints*** lack a synovial cavity, and the articulating bones are held together by a solid mass of dense irregular connective tissue. These joints, which permit slight movement or no movement, include sutures, syndesmoses, and interosseous membranes.

Sutures

A ***suture*** (SOO-chur; *sutur-* = seam) is a fibrous joint composed of a thin layer of dense irregular connective tissue. Such joints are found only in the skull. Examples include the coronal suture between the parietal and frontal bones (Figure 9.1a) and the squamosal suture between the parietal and temporal bones (Figure 9.1b). The irregular, interlocking edges of sutures give them added strength and decrease their chance of fracturing. A suture is classified functionally as a synarthrosis (immovable) or an amphiarthrosis (slightly movable). In older individuals, several sutures are immovable, but in infants and children they are slightly movable.

One type of suture, although present during growth of the skull, is replaced by bone in the adult. Such a suture is called a ***synostosis*** (sin′-os-TŌ-sis; *os-* = bone), or bony joint—a joint in which there is a complete fusion of bone across the suture line. For example, the frontal bone grows in halves that join together. Usually the halves are completely fused by age 6 and the suture becomes obscure. If the suture persists beyond age 6, it is called a ***frontal (metopic) suture*** (me-TŌ-pik; *metopon* = forehead). A synostosis is classified functionally as a synarthrosis (immovable).

MANUAL THERAPY APPLICATION

Movement at Sutures

It is possible to have a slight degree of **movement at sutures.** The jigsaw puzzle appearance of sutures makes it difficult to envision movement between adjacent bones of the skull (see Figure 9.1a). As illustrated in Figure 9.1b, however, adjacent skull bones can experience a shearing motion. This low level of movement at the sutures contributes to the compliance, a measure of the ease with which a structure is deformed, and the elasticity of the skull. Bone misalignment as a result of shearing movement may lead to headaches and other issues involving flow of the underlying cerebrospinal fluid. *Craniosacral therapy* is based on the assumption that gentle correction by a therapist's manipulation may realign skull bones at the suture lines.

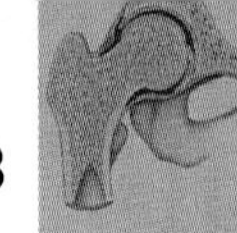

Figure 9.1 Fibrous joints.

At a fibrous joint the bones are held together by dense irregular connective tissue.

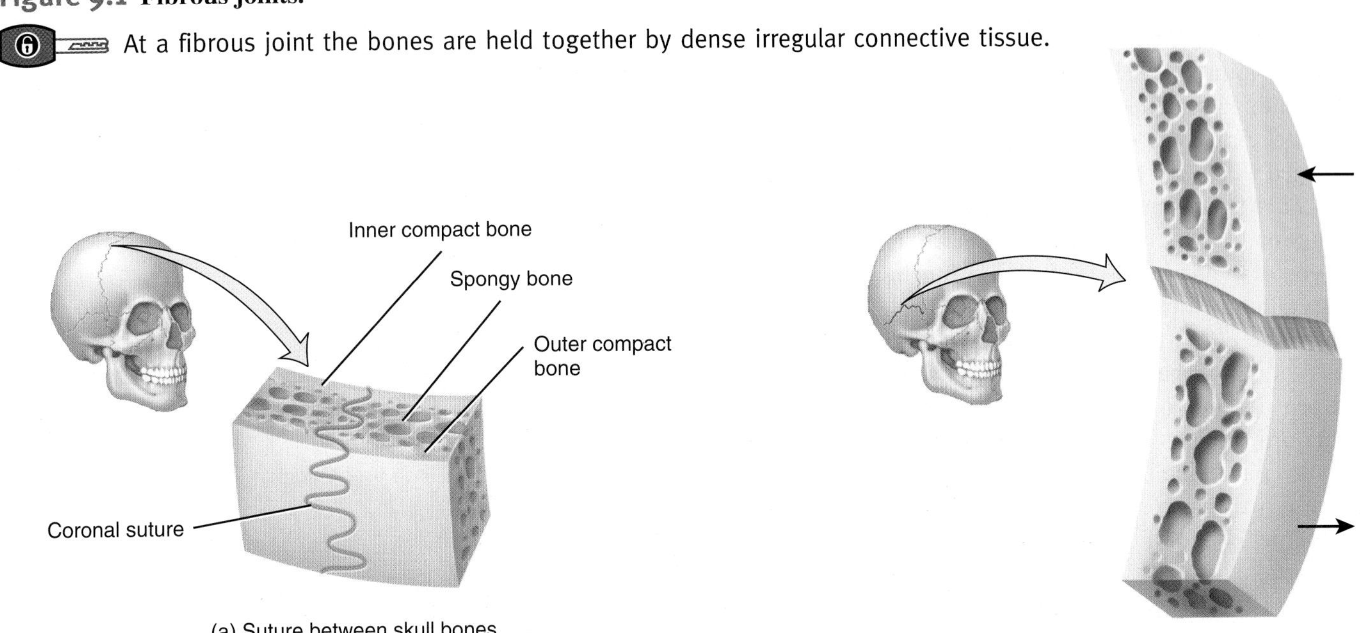

(a) Suture between skull bones

(b) Slight movement at suture

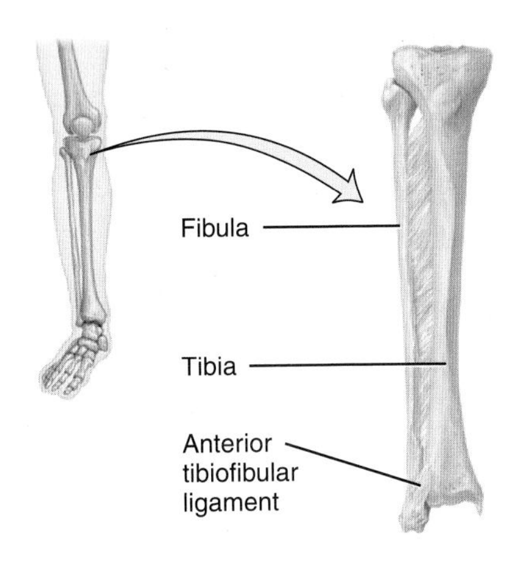

Syndesmosis between tibia and fibula at distal talofibular joint

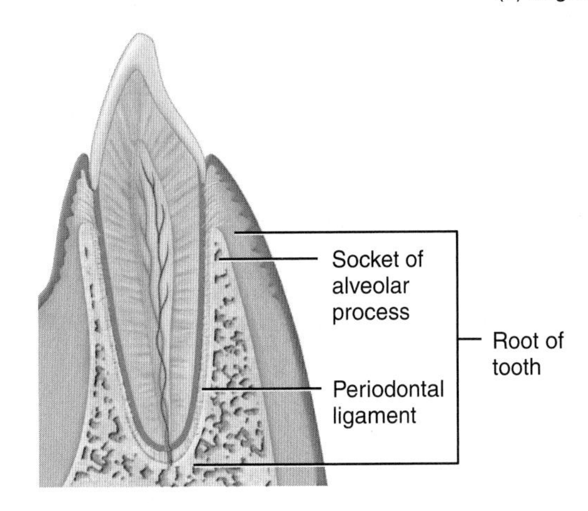

Syndesmosis between tooth and socket of alveolar process (gomphosis)

(c) Syndesmosis

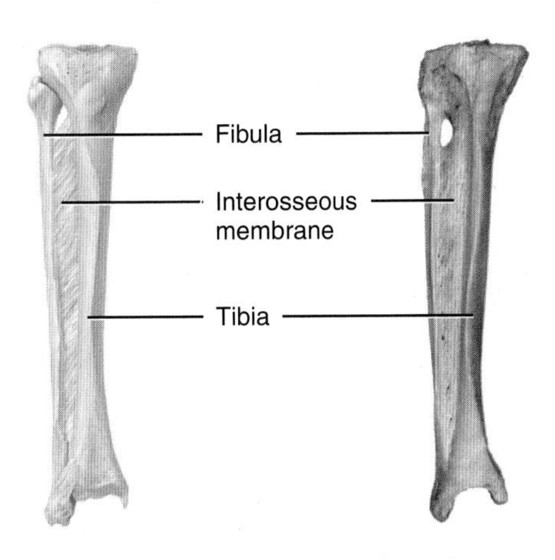

(d) Interosseous membrane between diaphyses of tibia and fibula

Functionally, why are sutures classified as amphiarthroses and synarthroses?

Syndesmoses

A ***syndesmosis*** (sin′-dez-MŌ-sis; *syndesmo-* = band or ligament) is a fibrous joint in which there is a greater distance between the articulating bones and more dense irregular connective tissue than in a suture. The dense irregular connective tissue is typically arranged as a bundle (ligament) (see Figure 9.1c, left). One example of a syndesmosis is the distal tibiofibular joint, where the anterior tibiofibular ligament connects the tibia and fibula. This type of syndesmosis is classified functionally as an amphiarthrosis (slightly movable).

Another example of a syndesmosis is a ***gomphosis*** (gom-FŌ-sis; *gompho-* = a bolt or nail) or *dentoalveolar joint* in which a cone-shaped peg fits into a socket. The only examples of gomphoses are the articulations between the roots of the teeth and their sockets (alveoli) in the maxillae and mandible (Figure 9.1c, right). The dense irregular connective tissue between a tooth and its socket is the periodontal ligament (membrane). A gomphosis is classified functionally as a synarthrosis (immovable). Through the use of wires and screws or other appliances, orthodontists are able to realign teeth. Inflammation and degeneration of the gums, periodontal ligament, and bone is called *periodontal disease.*

Interosseous Membranes

The final category of fibrous joint is the ***interosseous membrane***, a substantial sheet of dense irregular connective tissue that binds neighboring long bones and permits slight movement (amphiarthrosis). There are two principal interosseous membrane joints in the human body. One occurs between the radius and ulna in the forearm (see Figure 8.6a,b in Section 8.2) and the other occurs between the tibia and fibula in the leg (see Figure 9.1d).

CHECKPOINT

2. Which fibrous joints are synarthroses? Which are amphiarthroses?

Figure 9.2 Cartilaginous joints.

At a cartilaginous joint the bones are held together by cartilage.

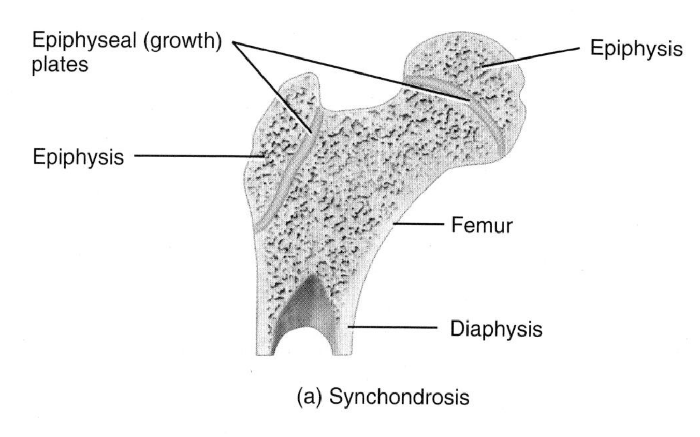

(a) Synchondrosis

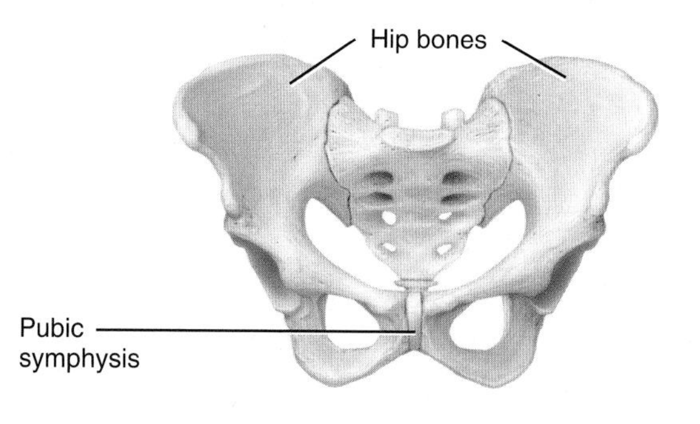

(b) Symphysis

What is the structural difference between a synchondrosis and a symphysis?

9.3 CARTILAGINOUS JOINTS

OBJECTIVE

- Describe the structure and functions of the two types of cartilaginous joints.

Like a fibrous joint, a ***cartilaginous joint*** lacks a synovial cavity and allows slight movement or no movement. Here the articulating bones are tightly connected, either by hyaline cartilage or by fibrocartilage (see Table 4.4H). The two types of cartilaginous joints are synchondroses and symphyses.

Synchondroses

A ***synchondrosis*** (sin′-kon-DRŌ-sis; *chondro-* = cartilage) is a cartilaginous joint in which the connecting material is hyaline cartilage. An example of a synchondrosis is the epiphyseal (growth) plate that connects the epiphysis and diaphysis of a growing bone (Figure 9.2a). A photomicrograph of the epiphyseal plate is shown in Figure 6.8. Functionally, a synchondrosis is a synarthrosis (immovable). When bone elongation ceases, bone replaces the hyaline cartilage, and the synchondrosis becomes a synostosis, a bony joint. Another example of a synchondrosis is the joint between the first rib and the manubrium of the sternum, which also ossifies during adult life and becomes an immovable synostosis (see Figure 7.23).

Symphyses

A ***symphysis*** (SIM-fi-sis = growing together) is a cartilaginous joint in which the ends of the articulating bones are covered with hyaline cartilage, but the bones are connected by a broad, flat disc of fibrocartilage. All symphyses occur in the midline of the body. The pubic symphysis between the anterior surfaces of the hip bones is one example of a symphysis (Figure 9.2b). This type of joint is also found at the junction of the manubrium and body of the sternum (see Figure 7.23 in Section 7.7) and at the intervertebral joints between the bodies of vertebrae (see Figure 7.21a). A portion of the intervertebral disc is made

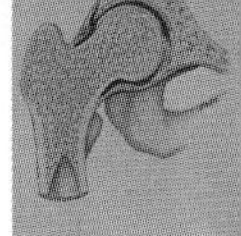

up of fibrocartilage. A symphysis is classified as an amphiarthrosis (slightly movable).

CHECKPOINT

3. Which cartilaginous joints are synarthroses? Which are amphiarthroses?

9.4 SYNOVIAL JOINTS

OBJECTIVE

- Describe the structure of synovial joints.

Structure of Synovial Joints

Synovial joints (si-NŌ-vē-al) have certain characteristics that distinguish them from other joints. The unique characteristic of a synovial joint is the presence of a space called a ***synovial (joint) cavity*** between the articulating bones (Figure 9.3). The synovial cavity allows a joint to be freely movable. Hence, all synovial joints are classified functionally as diarthroses. The bones at a synovial joint are covered by a layer of hyaline cartilage called ***articular cartilage***. The cartilage covers the articulating surfaces of the bones with a smooth, slippery surface but does not bind them together. Articular cartilage reduces friction between bones in the joint during movement and helps to absorb shock.

Articular Capsule

A sleevelike ***articular capsule*** surrounds a synovial joint, encloses the synovial cavity, and unites the articulating bones. The articular capsule is composed of two layers, an outer fibrous membrane and an inner synovial membrane (Figure 9.3). The ***fibrous membrane*** usually consists of dense, irregular connective tissue (mostly collagen fibers) that attaches to the periosteum of the articulating bones. The flexibility of the fibrous membrane permits considerable movement at a joint while its great tensile strength (resistance to stretching) helps prevent the bones from dislocating. The fibers of some fibrous membranes are arranged in parallel bundles that are highly adapted for resisting strains. Such fiber bundles, called ***ligaments*** (*liga-* = bound or tied), are often designated by individual names. The strength of ligaments is one of the principal mechanical factors that holds bones close together in a synovial joint. The inner layer of the articular capsule, the ***synovial membrane***, is composed of areolar connective tissue with elastic fibers. At many synovial joints, the synovial membrane also includes accumulations of adipose tissue called ***articular fat pads***. An example is the *infrapatellar fat pad* in the knee (see Figure 9.15c).

Figure 9.3 Structure of a typical synovial joint. Note the two layers of the articular capsule—the fibrous membrane and the synovial membrane. Synovial fluid lubricates the joint cavity which is located between the synovial membrane and the articular cartilage.

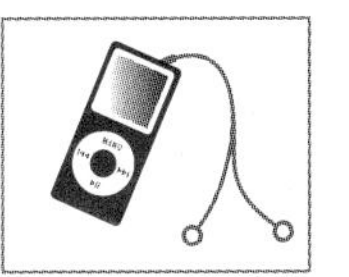

The distinguishing feature of a synovial joint is the synovial cavity between the articulating bones.

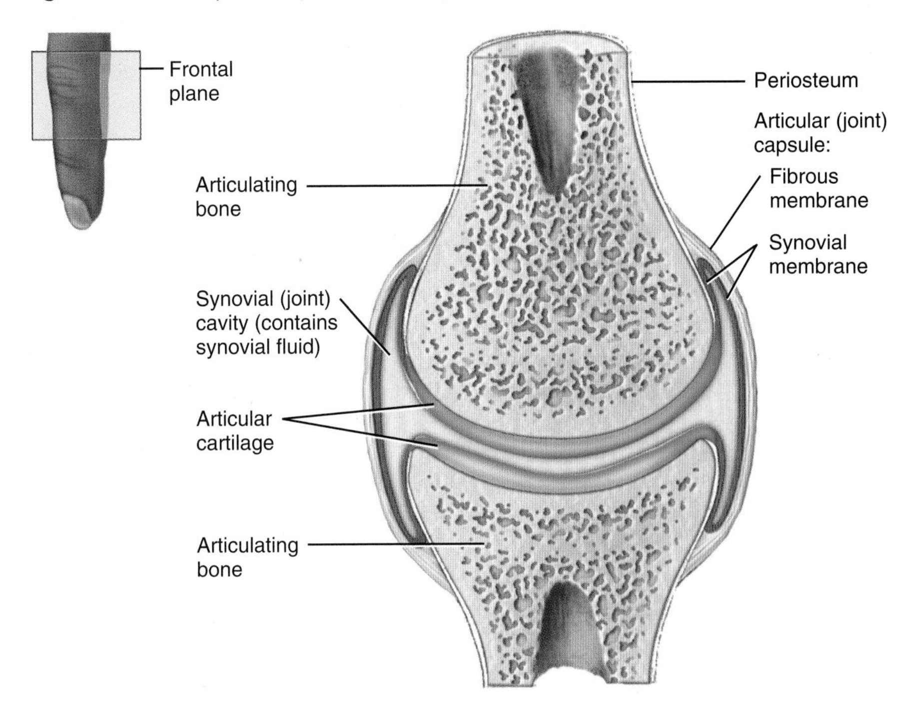

What is the functional classification of synovial joints?

Synovial Fluid

The synovial membrane secretes ***synovial fluid*** (*ov-* = egg), which forms a thin film over the surfaces within the articular capsule. This viscous, clear or pale yellow fluid was named for its similarity in appearance and consistency to the uncooked egg white (albumin) of a chicken. Synovial fluid consists of hyaluronic acid, secreted by fibroblast-like cells in the synovial membrane, and interstitial fluid filtered from blood plasma. Its functions include reducing friction by lubricating the joint, absorbing shocks, and supplying oxygen and nutrients to and removing carbon dioxide and metabolic wastes from the chondrocytes within articular cartilage. (Recall that cartilage is an avascular tissue, so it does not have blood vessels to perform the latter function.) Synovial fluid also contains phagocytic cells that remove microbes and the debris that results from normal wear and tear in the joint. When a synovial joint is immobile for a time (perhaps the person is bedridden or has had a cast applied) the fluid becomes quite viscous (gel-like). As joint movement increases (actively or passively), the fluid becomes less viscous. One of the benefits of warming up before exercise is that it stimulates the production and secretion of synovial fluid; more fluid means less stress on the joints during exercise.

We are all familiar with the cracking sounds heard as certain joints move, or the popping sounds that arise when a person pulls on the fingers to crack their knuckles. According to one theory, when the synovial cavity expands, the pressure of the synovial fluid decreases, creating a partial vacuum. The suction draws carbon dioxide and oxygen out of blood vessels in the synovial membrane, forming bubbles in the fluid. When the bubbles burst, the cracking or popping sound is heard.

• CLINICAL CONNECTION | Aspiration of Synovial Fluid

As a result of various injuries or diseases, there may be an excessive buildup of synovial fluid in a joint cavity, resulting in pain and decreased mobility. In order to relieve the pressure and ease the pain, **aspiration of synovial fluid** (as′-pi-RĀ-shun) may be necessary. In this procedure, a needle is inserted into the joint cavity and the fluid is withdrawn into a syringe; the fluid is then analyzed for diagnostic purposes. For example, the fluid may contain bacteria, which confirms a diagnosis of infection, or it may contain urea crystals, which confirms a diagnosis of gout. It is also possible to inject medication into a joint cavity; anti-inflammatory drugs such as cortisol are also used to decrease joint swelling caused by retention of fluid from the blood that entered the joint. •

Accessory Ligaments and Articular Discs

Many synovial joints also contain ***accessory ligaments*** called extracapsular ligaments and intracapsular ligaments. *Extracapsular ligaments* lie outside the articular capsule. Examples are the fibular and tibial collateral ligaments of the knee joint (see Figure 9.15d,f). *Intracapsular ligaments* occur within the articular capsule but are excluded from the synovial cavity by folds of the synovial membrane. Examples are the anterior and posterior cruciate ligaments of the knee joint (see Figure 9.15d-f).

Inside some synovial joints, such as the knee, pads of fibrocartilage lie between the articular surfaces of the bones and are attached to the fibrous capsule. These pads are called ***articular discs*** or ***menisci*** (me-NIS-sī or me-NIS-kī; the singular is *meniscus*). Figure 9.15d,f depicts the lateral and medial menisci in the knee joint. The discs usually subdivide the synovial cavity into two separate spaces. This separation can allow separate movements to occur in each space. As you will see later, separate movements also occur in the respective compartments of the temporomandibular joint (TMJ) (see Exhibit 9.1). By modifying the shape of the joint surfaces of the articulating bones, articular discs allow two bones of different shapes to fit together more tightly. Articular discs also help to maintain the stability of the joint and direct the flow of synovial fluid to the areas of greatest friction.

MANUAL THERAPY APPLICATION

Torn Cartilage and Arthroscopy

The tearing of articular menisci in the knee, commonly called **torn cartilage,** occurs often among athletes. Such damaged cartilage will begin to wear and may precipitate arthritis unless surgically removed (meniscectomy). Surgical repair of the torn cartilage may be performed by **arthroscopy** (ar-THROS-kō-pē). This minimally invasive procedure involves examination of the interior of a joint, usually the knee, with an *arthroscope,* a lighted pencil-thin instrument used for visualization. Arthroscopy is used to determine the nature and extent of damage following knee injury and to monitor the progression of disease and the effects of therapy. In addition, the insertion of surgical instruments along with the arthroscope enables a physician to remove torn cartilage and repair damaged cruciate ligaments in the knee; to obtain tissue samples for analysis; and to perform surgery on other joints, such as the shoulder, elbow, ankle, and wrist. Therapists can enhance healing through the use of techniques that increase blood flow to the area. Therapeutic procedures can also reduce the incidence and severity of adhesions that commonly develop with this type of medical intervention.

Patients who have had a cartilage transplant are typically referred to physical therapy after the surgery. Physical therapists can help the person to improve the strength of the quadriceps and hamstring muscles even when the knee joint is immobilized. Manual therapists can work on the surgical areas during healing to increase the effectiveness of the healing process and to prevent adhesions.

Nerve and Blood Supply

The nerves that supply a joint are the same as those that supply the skeletal muscles that move the joint. Synovial joints contain many nerve endings that are distributed to the articular capsule and associated ligaments. Some of the nerve endings convey information about pain from the joint to the spinal cord and brain for processing. Other nerve endings are responsive to the degree of movement and stretch at a joint. This information is also re-

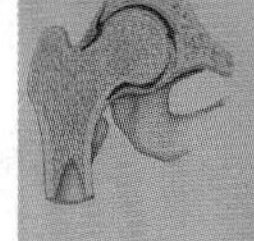

layed to the spinal cord and brain, which may respond by sending impulses through different nerves to the muscles to adjust body movements.

Although many of the components of synovial joints are avascular, arteries in the vicinity send out numerous branches that penetrate the ligaments and articular capsule to deliver oxygen and nutrients. Veins remove carbon dioxide and wastes from the joints. The arterial branches from several different arteries typically join together around a joint before penetrating the articular capsule. The chondrocytes of articular cartilage of a synovial joint receive oxygen and nutrients from synovial fluid derived from blood, whereas all other joint tissues are supplied directly by arteries. Carbon dioxide and wastes pass from chondrocytes of articular cartilage into synovial fluid and then into veins; carbon dioxide and wastes from all other joint structures pass directly into veins.

Bursae and Tendon Sheaths

The various movements of the body create friction between moving parts. Saclike structures called ***bursae*** (BER-sē = purses; singular is *bursa*) are strategically situated to alleviate friction in some joints, such as the shoulder and knee joints (see Figures 9.12 and 9.15c). Bursae are not strictly parts of synovial joints, but they do resemble joint capsules because their walls consist of connective tissue lined by a synovial membrane. They are also filled with a small amount of fluid similar to synovial fluid. Bursae are located between the skin and bone, tendons and bones, muscles and bones, and ligaments and bones. The fluid-filled bursal sacs cushion the movement of these body parts over one another. An acute or chronic inflammation of a bursa is called ***bursitis*** (bur-SĪ-tis). It is usually caused by irritation from repeated, excessive exertion of a joint. The condition may also be caused by trauma, an acute or chronic infection (including syphilis and tuberculosis), or rheumatoid arthritis (described in Section 9.7). Symptoms include pain, swelling, tenderness, and limited movement. Treatment may include oral anti-inflammatory agents and injections of cortisol-like steroids.

Structures called tendon sheaths also reduce friction at joints. ***Tendon sheaths*** are tubelike bursae that wrap around tendons which experience considerable friction. This occurs where tendons pass through synovial cavities, such as the tendon of the biceps brachii muscle at the shoulder joint (see Figure 9.12c). Tendon sheaths are also found at the wrist and ankle, where many tendons come together in a confined space, and in the fingers and toes, where there is a great deal of movement.

Types of Synovial Joints

Although all synovial joints are similar in structure, the shapes of the articulating surfaces vary and thus various types of movements are possible. Accordingly, synovial joints are divided into six subtypes: planar, hinge, pivot, condyloid, saddle, and ball-and-socket.

Planar Joints

The articulating surfaces of bones in a ***planar joint*** are flat or slightly curved (see Figure 9.4a). Planar joints primarily permit side-to-side and back-and-forth gliding movements (described shortly). Many planar joints are *biaxial* because they permit movement around two axes. An *axis* is a straight line around which a rotating (revolving) body moves. Some examples of planar joints are the intercarpal joints (between carpal bones at the wrist); intertarsal joints (between tarsal bones at the ankle); sternoclavicular joints (between the manubrium of the sternum and the clavicle); acromioclavicular joints (between the acromion of the scapula and the clavicle); sternocostal joints (between the sternum and ends of the costal cartilages at the tips of the second through seventh pairs of ribs); and vertebrocostal joints (between the heads and tubercles of ribs and transverse processes of thoracic vertebrae). X-ray films made during wrist and ankle movements reveal some rotation of the small carpal and tarsal bones in addition to their predominant gliding movements.

Hinge Joints

In a ***hinge joint***, the convex surface of one bone fits into the concave surface of another bone (see Figure 9.4b). As the name implies, hinge joints produce an angular, opening-and-closing motion like that of a hinged door. In most joint movements, one bone remains in a fixed position while the other moves around an axis. Hinge joints are said to be *monaxial (uniaxial)* because they typically allow motion around a single axis. Examples of hinge joints are the knee, elbow, ankle, and interphalangeal joints (between the phalanges of the fingers and toes).

Pivot Joints

In a ***pivot joint***, the rounded or pointed surface of one bone articulates with a ring formed partly by another bone and partly by a ligament (see Figure 9.4c). A pivot joint is monaxial because it allows rotation only around its own longitudinal axis. Examples of pivot joints are the atlanto-axial joint, in which the atlas rotates around the axis and permits the head to turn from side-to-side as in signifying "no" (see Figure 9.9a), and the radioulnar joints that enable the palms to turn anteriorly and posteriorly (see Figure 9.10h).

Condyloid Joints

In a ***condyloid joint*** (KON-di-loyd; *condyl-* = knuckle), the convex oval-shaped projection of one bone fits into the oval-shaped depression of another bone (see Figure 9.4d). A condyloid joint is *biaxial* because the movement it permits is around two axes (flexion-extension and abduction-adduction). Notice that you can move your index finger both up-and-down and from side-to-side. Examples are the radiocarpal (wrist) and the metacarpophalangeal joints (between the metacarpals and phalanges) of the second through fifth digits.

Saddle Joints

In a ***saddle joint***, the articular surface of one bone is saddle-shaped, and the articular surface of the other bone fits into the

Figure 9.4 Subtypes of synovial joints. For each subtype, a drawing of the actual joint and a simplified diagram are shown.

Synovial joints are classified into subtypes on the basis of the shapes of the articulating bone surfaces.

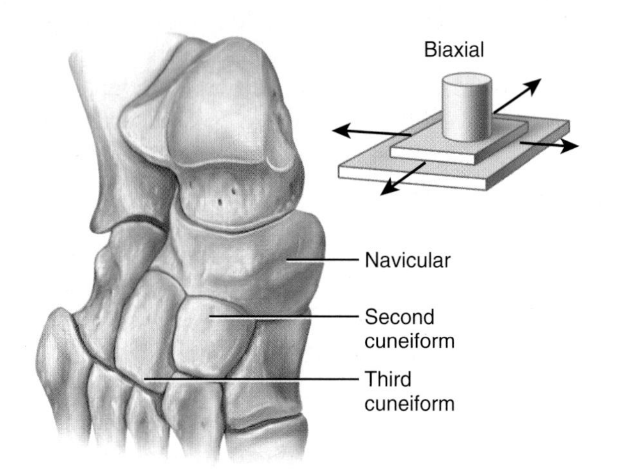

(a) Planar joint between navicular and second and third cuneiforms of tarsus (ankle) in foot

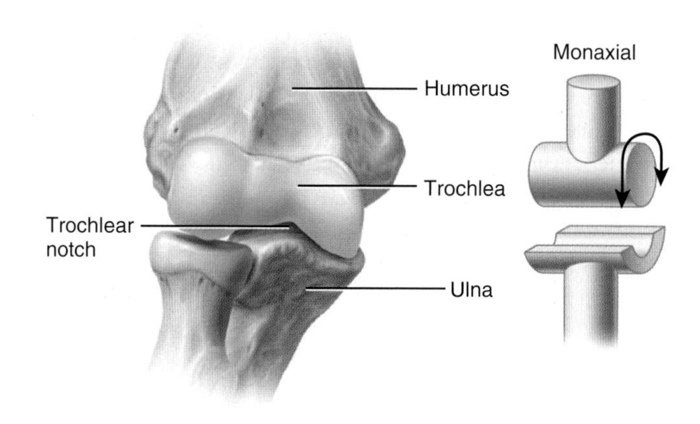

(b) Hinge joint between trochlea of humerus and trochlear notch of ulna at the elbow

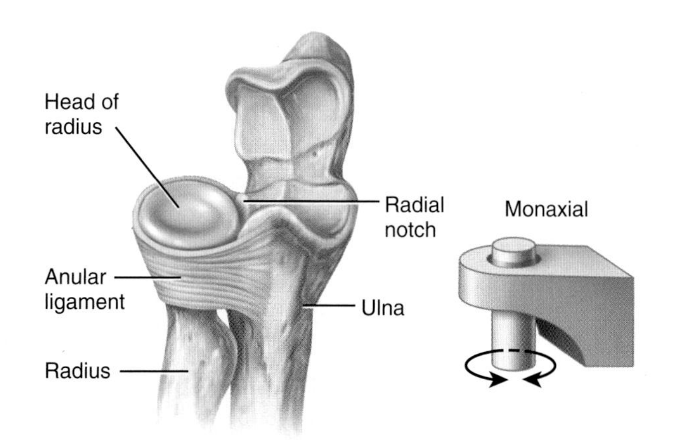

(c) Pivot joint between head of radius and radial notch of ulna

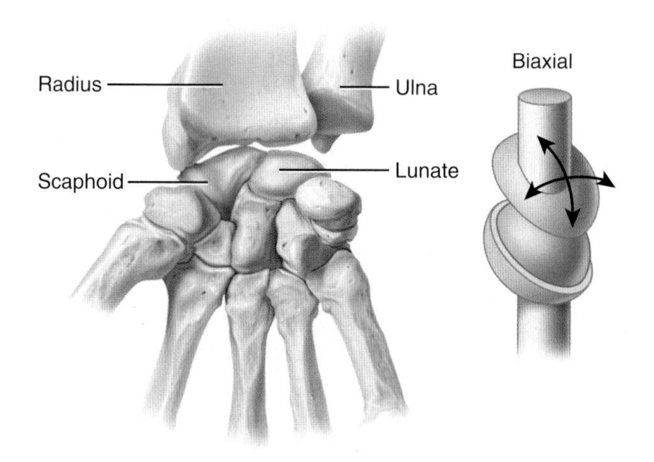

(d) Condyloid joint between radius and scaphoid and lunate bones of carpus (wrist)

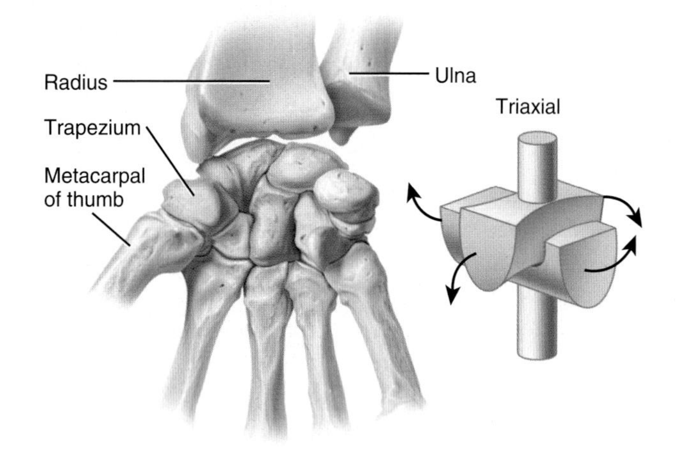

(e) Saddle joint between trapezium of carpus (wrist) and metacarpal of thumb

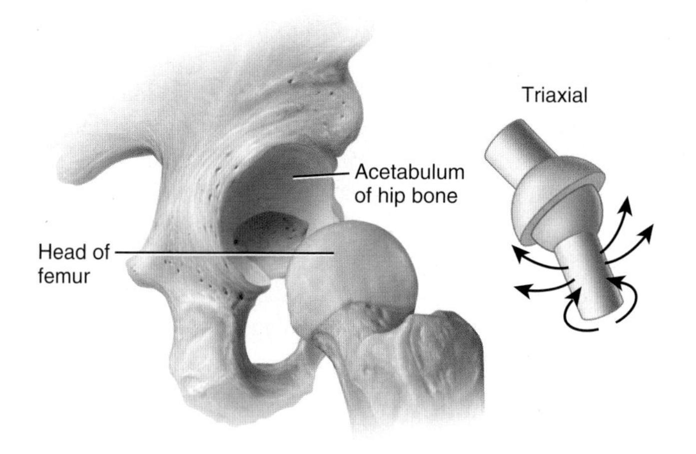

(f) Ball-and-socket joint between head of femur and acetabulum of hip bone

Which of the joints here are biaxial?

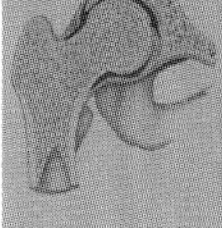

"saddle" as a sitting rider would (Figure 9.4e). A saddle joint is a modified condyloid joint in which the movement is somewhat freer. Saddle joints are *triaxial,* permitting movements around three axes (flexion-extension, abduction-adduction, and rotation). An example of a saddle joint is the carpometacarpal joint between the trapezium of the carpus and metacarpal of the thumb.

Ball-and-Socket Joints

A ***ball-and-socket joint*** consists of the ball-like surface of one bone fitting into a cuplike depression of another bone (Figure 9.4f). Such joints are *triaxial* and permit movements around three axes (flexion-extension, abduction-adduction, and rotation). Examples of functional ball-and-socket joints are the shoulder and hip joints. At the shoulder joint the head of the humerus fits into the glenoid cavity of the scapula. At the hip joint the head of the femur fits into the acetabulum of the hip bone.

Table 9.1 summarizes the structural and functional categories of joints.

TABLE 9.1

Summary of Structural and Functional Classification of Joints

STRUCTURAL CLASSIFICATION	DESCRIPTION	FUNCTIONAL CLASSIFICATION	EXAMPLE
FIBROUS No Synovial Cavity; Articulating Bones Held Together by Dense Irregular Connective Tissue.			
Suture	Articulating bones united by a thin layer of dense irregular connective tissue, found between bones of the skull. With age, some sutures are replaced by a synostosis, in which separate cranial bones fuse into a single bone.	Synarthrosis (immovable) and amphiarthrosis (slightly movable).	Coronal suture.
Syndesmosis	Articulating bones united by more dense irregular connective tissue than a suture, usually a ligament.	Amphiarthrosis (slightly movable) and synarthrosis (immovable).	Distal tibiofibular joint and gomphosis.
Interosseous membrane	Articulating bones united by a substantial sheet of dense irregular connective tissue.	Amphiarthrosis (slightly movable).	Between the radius and ulna and between the tibia and fibula.
CARTILAGINOUS No Synovial Cavity; Articulating Bones United by Hyaline Cartilage or Fibrocartilage.			
Synochondrosis	Connecting material is hyaline cartilage; becomes a synostosis when bone elongation ceases.	Synarthrosis.	Epiphyseal (growth) plate between the diaphysis and epiphysis of a long bone.
Symphysis	Connecting material is a broad, flat disc of fibrocartilage.	Amphiarthrosis.	Pubic symphysis and intervertebral joints.
SYNOVIAL Characterized by a Synovial Cavity, Articular Cartilage, and an Articular (Joint) Capsule; May Contain Accessory Ligaments, Articular Discs, and Bursae.			
Planar	Articulated surfaces are flat or slightly curved.	Many biaxial diarthroses (freely movable): back-and-forth and side-to-side movements.	Intercarpal, intertarsal, sternocostal (between sternum and the second to seventh pairs of ribs), and vertebrocostal joints.
Hinge	Convex surface fits into a concave surface.	Monaxial (uniaxial) diarthrosis: flexion–extension.	Knee (modified hinge), elbow, ankle, and interphalangeal joints.
Pivot	Rounded or pointed surface fits into a ring formed partly by bone and partly by a ligament.	Monaxial diarthrosis: rotation.	Atlanto-axial and radioulnar joints.
Condyloid	Oval-shaped projection fits into an oval-shaped depression.	Biaxial diarthrosis: flexion–extension, abduction–adduction.	Radiocarpal and metacarpophalangeal joints.
Saddle	Articular surface of one bone is saddle-shaped and the articular surface of the other bone "sits" in the saddle.	Triaxial diarthrosis: flexion–extension, abduction–adduction, and rotation.	Carpometacarpal joint between trapezium and thumb.
Ball-and-socket	Ball-like surface fits into a cuplike depression.	Triaxial diarthrosis: flexion–extension, abduction–adduction, and rotation.	Shoulder and hip joints.

• CLINICAL CONNECTION | Double-Jointedness

A **double-jointed** person does not have extra joints. Individuals who are double-jointed have greater flexibility in their articular capsules and ligaments; the resulting increase in range of motion allows them to entertain fellow partygoers with activities such as touching their thumbs to their wrists and putting their ankles or elbows behind their necks. Unfortunately, such flexible joints are less structurally stable and are more easily dislocated. •

CHECKPOINT

4. How does the structure of synovial joints classify them as diarthroses?
5. What are the functions of articular cartilage, the articular capsule, synovial fluid, articular discs, and bursae?
6. Where in the body can each subtype of synovial joint be found?

9.5 TYPES OF MOVEMENTS AT SYNOVIAL JOINTS

OBJECTIVE

- Describe the types of movements that can occur at synovial joints.

Anatomists, physical therapists, kinesiologists (professionals who treat injuries or disorders through active or passive movements), and other health-care practitioners use specific terminology to designate movements that can occur at synovial joints. These precise terms may indicate the form of motion, the direction of movement, or the relationship of one body part to another during movement. Movements at synovial joints are grouped into four main categories: (1) gliding, (2) angular movements, (3) rotation, and (4) special movements. The last category includes movements that occur only at certain joints.

Gliding

Gliding is a simple movement in which relatively flat bone surfaces move back and forth and from side to side with respect to one another (Figure 9.5). There is no significant alteration of the angle between the bones. Gliding movements are limited in range due to the structure of the articular capsule and associated ligaments and bones. Gliding occurs at planar joints (see Figure 9.4a).

Angular Movements

In ***angular movements***, there is an increase or a decrease in the angle between articulating bones. The principal angular movements are flexion, extension, lateral extension, hyperextension, abduction, adduction, and circumduction. These movements are always discussed with respect to the body in the anatomical position.

Flexion, Extension, Lateral Flexion, and Hyperextension

Flexion and extension are opposite movements. In ***flexion*** (FLEK-shun; *flex-* = to bend) there is a decrease in the angle between articulating bones; in ***extension*** (eks-TEN-shun; *exten-* = to stretch out) there is an increase in the angle between articulating bones, often to restore a part of the body to the anatomical position after it has been flexed (Figure 9.6). Both movements usually occur along the sagittal plane. Hinge, pivot, condyloid, saddle, and ball-and-socket joints all permit flexion and extension (see Figure 9.4b–f).

All of the following are examples of flexion:

- Bending the head toward the chest at the atlanto-occipital joint between the atlas (the first vertebra) and the occipital bone of the skull, and at the cervical intervertebral joints between the cervical vertebrae (Figure 9.6a).
- Bending the trunk forward at the intervertebral joints.
- Moving the humerus forward at the shoulder joint, as in swinging the arms forward while walking (Figure 9.6b).
- Moving the forearm toward the arm at the elbow joint between the humerus, ulna, and radius (Figure 9.6c).
- Moving the palm toward the forearm at the wrist or radiocarpal joint between the radius and carpals (Figure 9.6d).
- Bending of the digits of the hands or feet at the interphalangeal joints between phalanges.
- Moving the femur forward at the hip joint between the femur and hip bone, as in walking (Figure 9.6e).
- Moving the leg toward the thigh at the knee or tibiofemoral joint between the tibia, femur, and patella, as occurs when bending the knee (Figure 9.6f).

Figure 9.5 Gliding movements at synovial joints.

Gliding motion consists of side-to-side and back-and-forth movements.

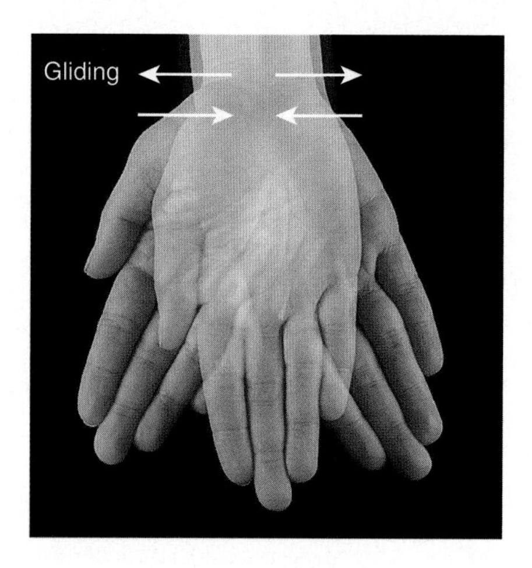

Intercarpal joints

What are two examples of joints that permit gliding movements?

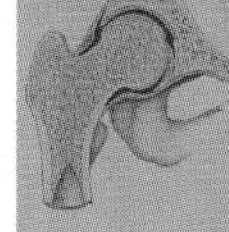

Figure 9.6 Angular movements at synovial joints—flexion, extension, hyperextension, and lateral flexion.

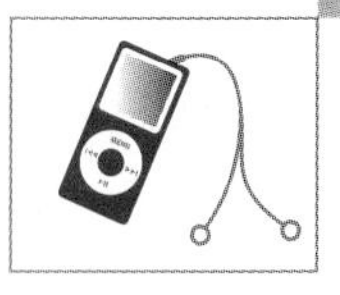

In angular movements, there is an increase or decrease in the angle between articulating bones.

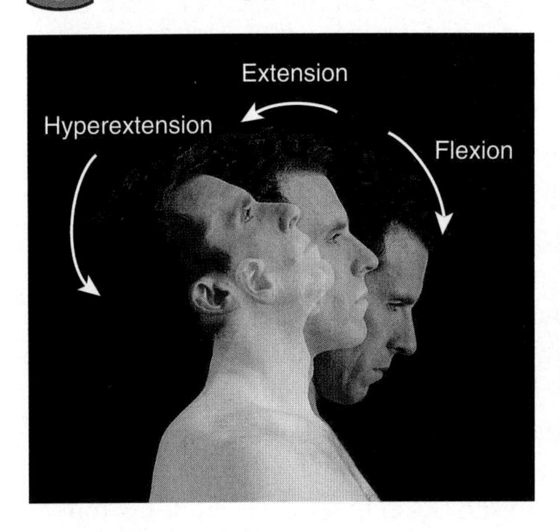

(a) Atlanto-occipital and cervical intervertebral joints

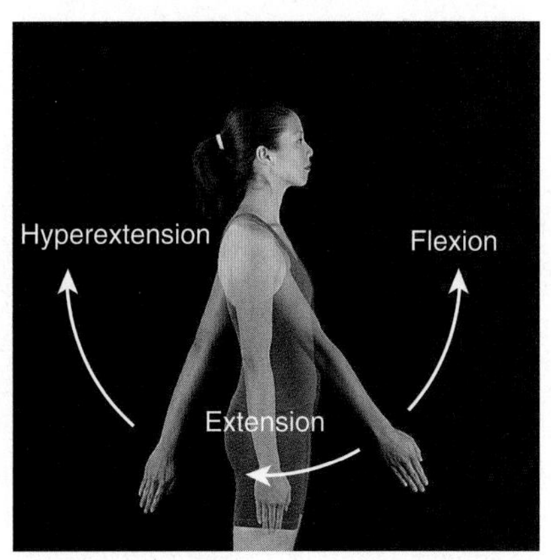

(b) Shoulder joint

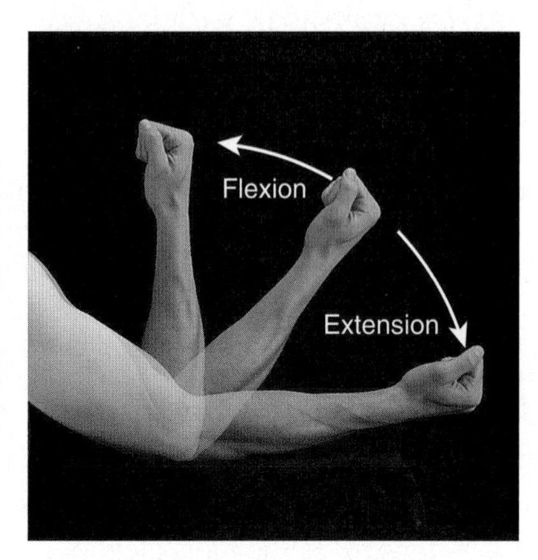

(c) Elbow joint

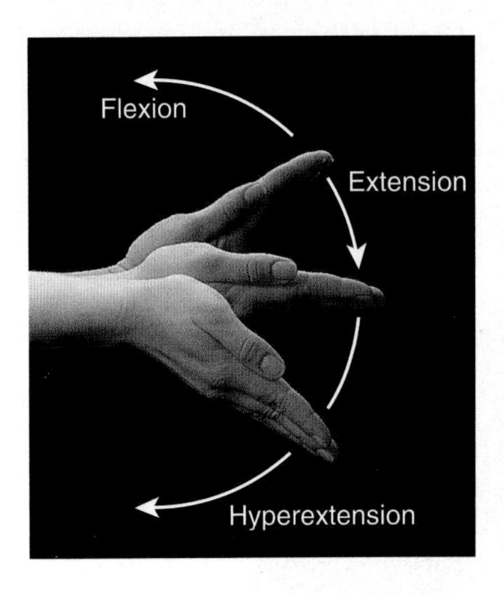

(d) Wrist joint

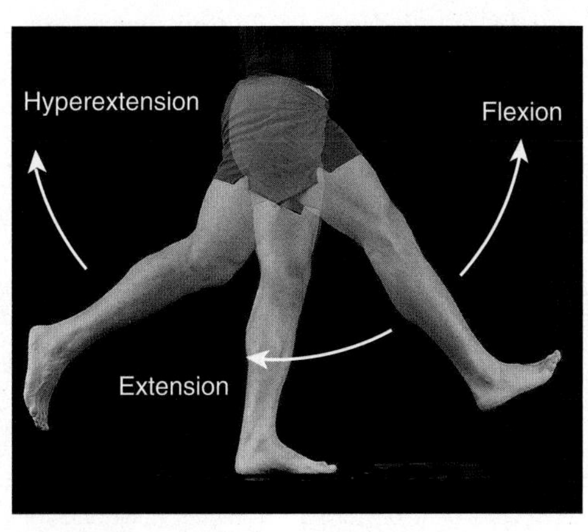

(e) Hip joint

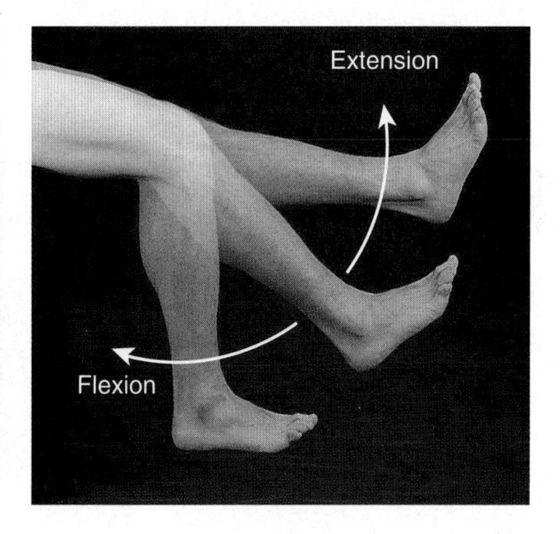

(f) Knee joint

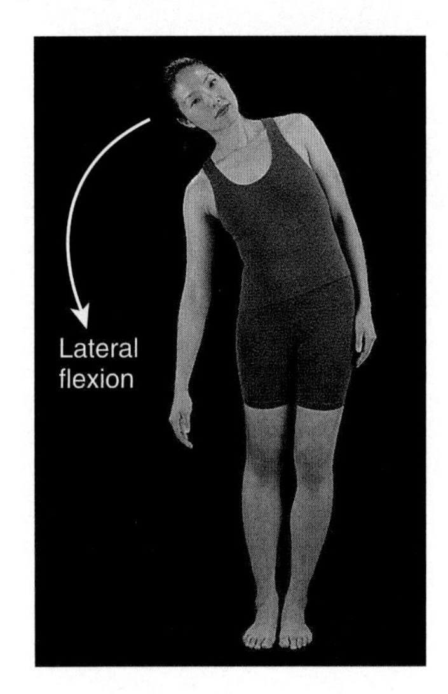

(g) Intervertebral joints

? **What are two examples of flexion that do not occur along the sagittal plane?**

Although flexion and extension usually occur along the sagittal plane, there are a few exceptions. For example, flexion of the thumb involves movement of the thumb medially across the palm at the carpometacarpal joint between the trapezium and metacarpal of the thumb, as in touching the thumb to the opposite side of the palm (see Figure 13.11g). Another example is movement of the trunk sideways to the right or left at the waist as in a side bend. This movement, which occurs along the frontal plane and involves the intervertebral joints, is called ***lateral flexion*** (Figure 9.6g). Movement back to the anatomical position from a laterally flexed position is called ***reduction***.

Continuation of extension beyond the anatomical position is called ***hyperextension*** (hī′-per-ek-STEN-shun; *hyper-* = beyond or excessive). Examples of hyperextension include:

- Bending the head backward at the atlanto-occipital and cervical intervertebral joints (Figure 9.6a), as in looking at the ceiling.
- Bending the trunk backward at the intervertebral joints, as in a back bend.

- Moving the humerus backward at the shoulder joint, as in swinging the arms backward while walking (see Figure 9.6b).
- Moving the palm backward at the wrist joint (see Figure 9.6d).
- Moving the femur posteriorly at the hip joint, as in walking (see Figure 9.6e).

Hyperextension of hinge joints, such as the elbow, interphalangeal, and knee joints, is usually prevented by factors such as the arrangement of ligaments and the anatomical alignment of the bones.

Abduction, Adduction, and Circumduction

Abduction (ab-DUK-shun; *ab-* = away; *-duct-* = to lead) is the movement of a bone away from the midline, whereas ***adduction*** (ad-DUK-shun; *ad-* = toward) is the movement of a bone toward the midline. Both movements usually occur along the frontal plane. Condyloid, saddle, and ball-and-socket joints permit abduction and adduction. Examples of abduction include moving the humerus laterally at the shoulder joint, moving the palm laterally at the wrist joint, and moving the femur laterally at the hip joint (Figure 9.7a–c). The movement that returns each of these body parts to the anatomical position is adduction (Figure 9.7a–c).

With respect to the digits, the midline of the body is not used as a point of reference for abduction and adduction. In abduction of the fingers (but not the thumb), an imaginary line is drawn through the longitudinal axis of the middle (longest) finger, and the fingers move away (spread out) from the middle finger (Figure 9.7d). In abduction of the thumb, the thumb moves away from the palm in the sagittal plane (see Figure 13.11d). Abduction of the toes is relative to an imaginary line drawn through the second toe. Adduction of the fingers and toes involves returning them to the anatomical position. Adduction of the thumb moves the thumb toward the palm in the sagittal plane (see Figure 13.11d).

Circumduction (ser-kum-DUK-shun; *circ-* = circle) is movement of the distal end of a body part in a circle (Figure 9.8). Circumduction occurs as a result of a continuous sequence of flexion, abduction, extension, and adduction. It does not occur

Figure 9.7 Angular movements at synovial joints—abduction and adduction.

Abduction and adduction usually occur along the frontal plane.

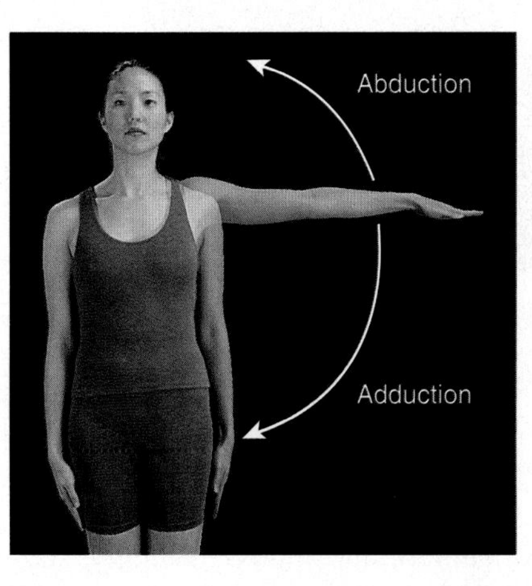

(a) Shoulder joint

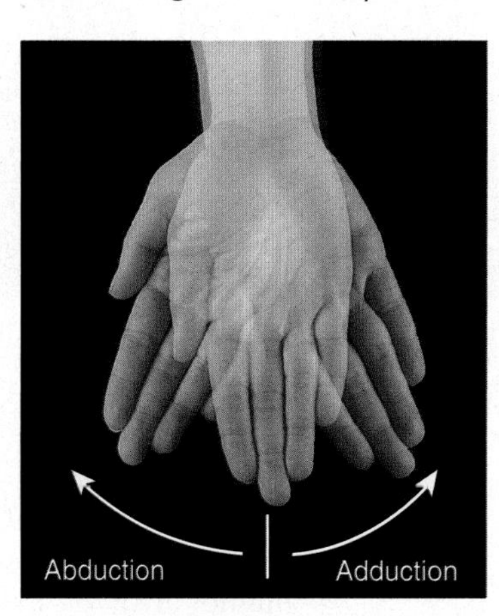

(b) Wrist joint

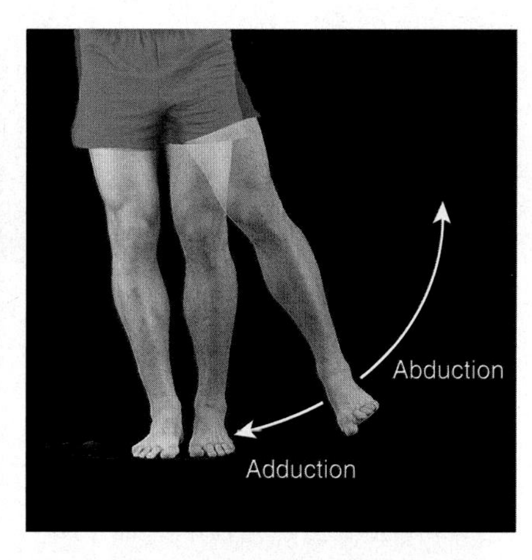

(c) Hip joint

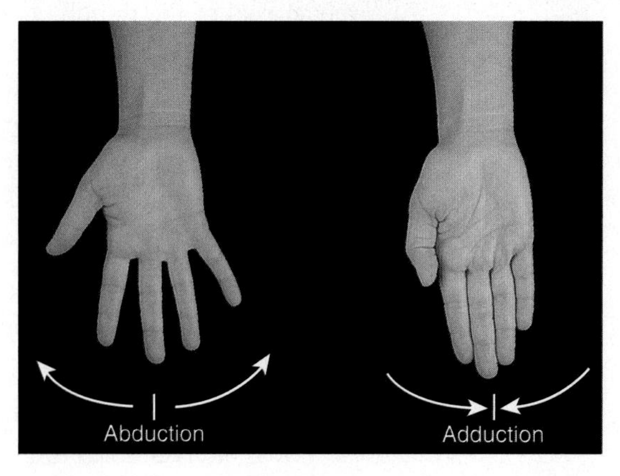

(d) Metacarpophalangeal joints of the fingers (not the thumb)

Is considering adduction as "adding your limb to your trunk" an effective learning device?

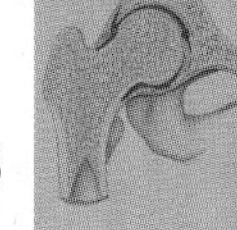

Figure 9.8 Angular movements at synovial joints—circumduction.

Circumduction is the movement of the distal end of a body part in a circle.

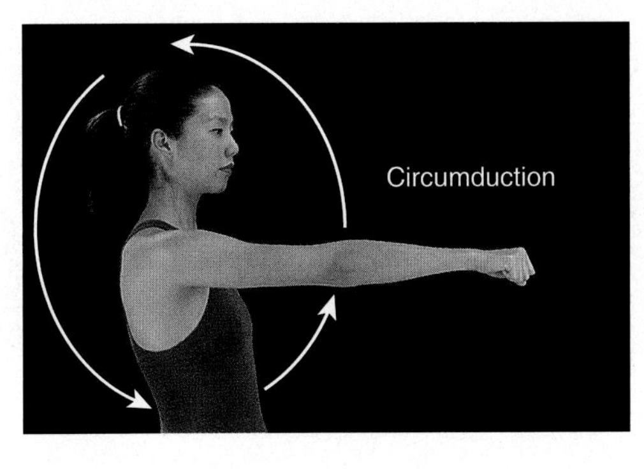

(a) Shoulder joint

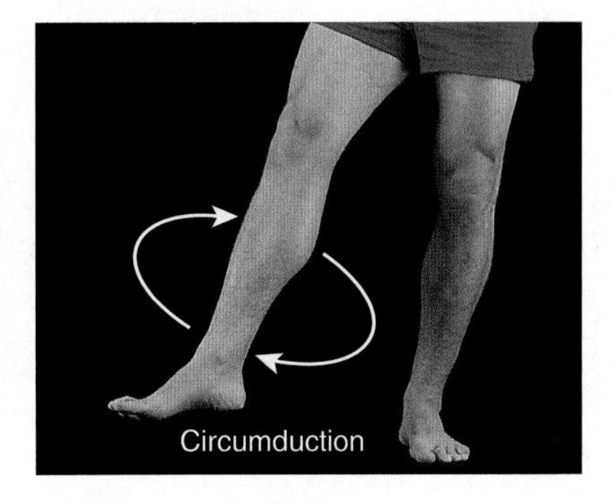

(b) Hip joint

Which movements in continuous sequence produce circumduction?

along a separate plane of movement. Examples of circumduction are moving the humerus in a circle at the shoulder joint (Figure 9.8a), moving the hand in a circle at the wrist joint, moving the thumb in a circle at the carpometacarpal joint, moving the fingers in a circle at the metacarpophalangeal joints (between the metacarpals and phalanges), and moving the femur in a circle at the hip joint (Figure 9.8b). Although the shoulder and hip joints permit circumduction, it is more limited in the hip joints than the shoulder joints due to the tension on certain ligaments and muscles in the hip joints and the depth of the acetabulum in the hip joint (see Exhibits 9.2 and 9.4).

Rotation

In ***rotation*** (rō-TĀ-shun; *rota-* = revolve), a bone revolves around its own longitudinal axis. Pivot and ball-and-socket joints permit rotation. One example is turning the head from side to side at the atlanto-axial joint, as in signifying "no" (Figure 9.9a). Another is turning the trunk from side to side at the intervertebral joints while keeping the hips and lower limbs in the anatomical position. In the limbs, rotation is defined relative to the midline, and specific qualifying terms are used. If the anterior surface of a bone of the limb is turned toward the midline, the movement is called *medial (internal) rotation.* You can medially rotate the humerus at the shoulder joint as follows: Starting in the anatomical position, flex your elbow and then draw your palm across the chest (Figure 9.9b). Medial rotation of the forearm at the radioulnar joints (between the radius and ulna) involves turning the palm medially from the anatomical position (see Figure 9.10h). You can medially rotate the femur at the hip joint as follows: Lie on your back, bend your knee, and then move your leg and foot laterally from the midline. Although you are moving your leg and foot laterally, the femur is rotating medially (Figure 9.9c). Medial rotation of the leg at the knee joint can be produced by sitting on

Figure 9.9 Rotation at synovial joints.

In rotation, a bone revolves around its own longitudinal axis.

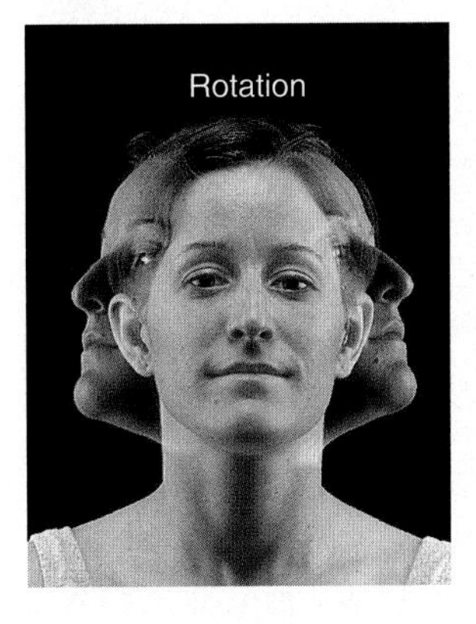

(a) Atlanto-axial joint

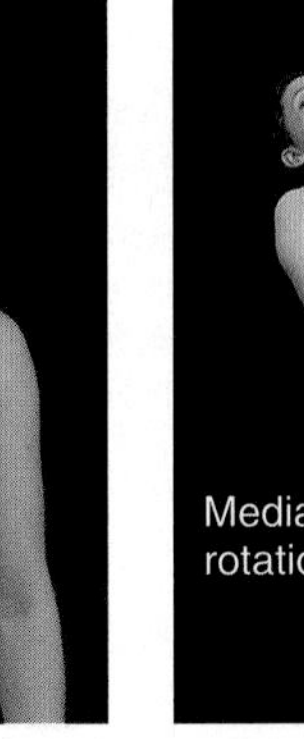

(b) Shoulder joint

(c) Hip joint

How do medial and lateral rotation differ?

Figure 9.10 Special movements at synovial joints.

Special movements occur only at certain synovial joints.

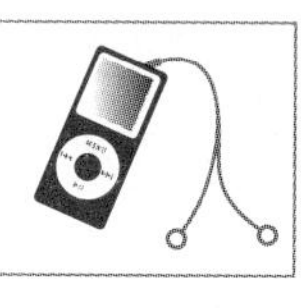

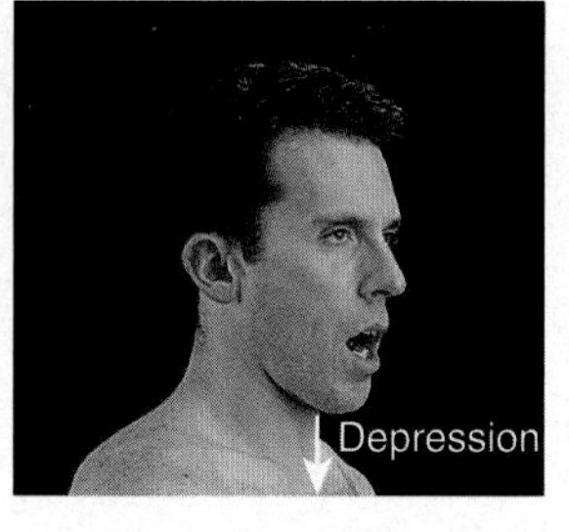

(a) Temporomandibular joint (b)

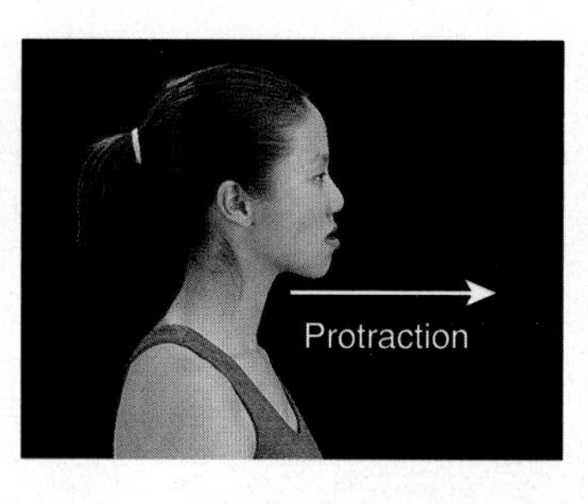

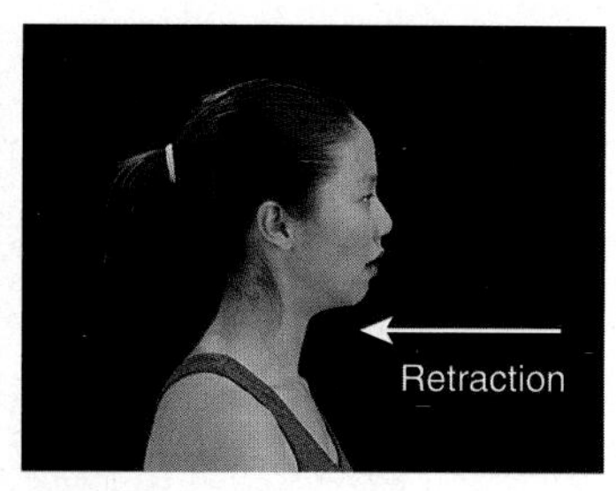

(c) Temporomandibular joint (d)

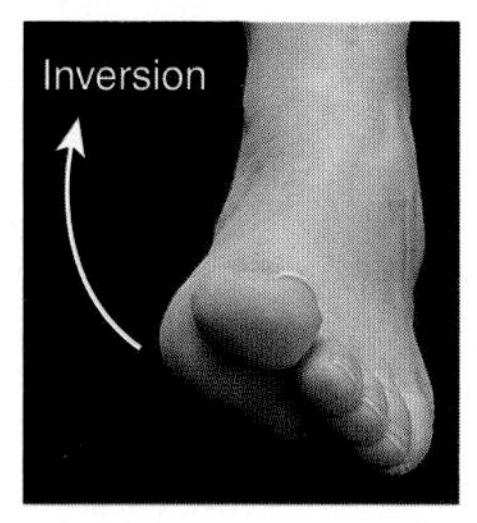

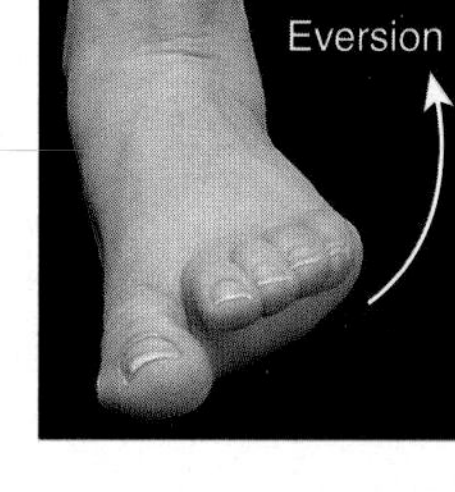

(e) Intertarsal joints (f)

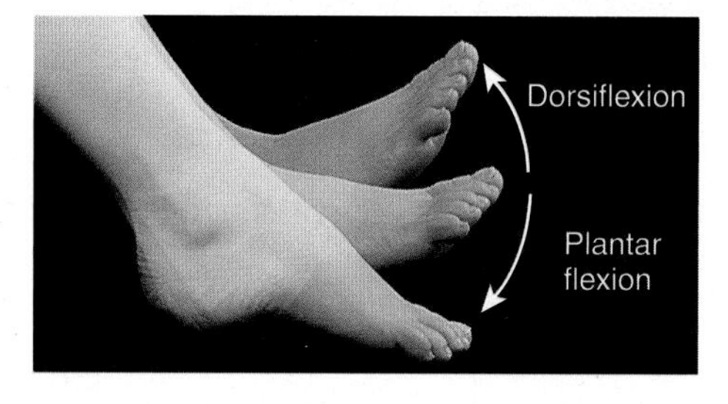

(g) Ankle joint

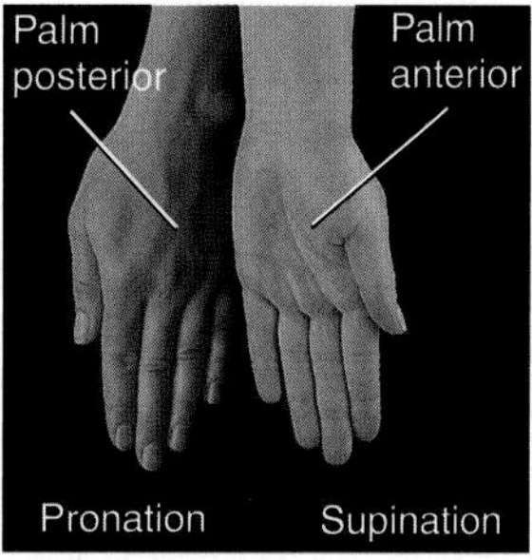

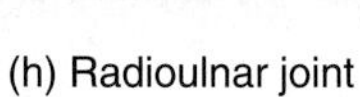

(h) Radioulnar joint

(i) Carpometacarpal joint

What movement occurs when the thumb moves across the palm to touch the ring finger?

a chair, bending your knee, raising your lower limb off the floor, and turning your toes medially. If the anterior surface of the bone of a limb is turned away from the midline, the movement is called *lateral (external) rotation* (see Figure 9.9b,c).

Special Movements

As noted previously, special movements occur only at certain joints. They include elevation, depression, protraction, retraction, inversion, eversion, dorsiflexion, plantar flexion, supination, pronation, and opposition (Figure 9.10):

- ***Elevation*** (el-e-VĀ-shun = to lift up) is an upward movement of a part of the body, such as closing the mouth at the temporomandibular joint (between the mandible and temporal bone) to elevate the mandible (Figure 9.10a) or shrugging the shoulder at the acromioclavicular joint to elevate the scapula. Other bones that may be elevated (or depressed) include the hyoid, clavicle, and ribs.
- ***Depression*** (dē-PRESH-un = to press down) is a downward movement of a part of the body, such as opening the mouth to depress the mandible (Figure 9.10b) or returning shrugged shoulders to the anatomical position to depress the scapulae.
- ***Protraction*** (prō-TRAK-shun = to draw forth) is a movement of a part of the body anteriorly in the transverse plane. You can protract your mandible at the temporomandibular joint by thrusting it outward (Figure 9.10c) or protract your

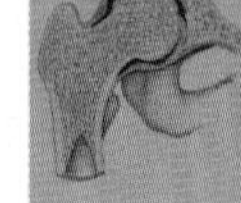

clavicles at the acromioclavicular and sternoclavicular joints by crossing your arms.

- ***Retraction*** (rē-TRAK-shun = to draw back) is a movement of a protracted part of the body back to the anatomical position (Figure 9.10d).
- ***Inversion*** (in-VER-zhun = to turn inward) is movement of the soles of the feet medially at the intertarsal joints (between the tarsals) (Figure 9.10e). Physical therapists and other health-care providers also refer to inversion of the feet as *supination.*
- ***Eversion*** (ē-VER-zhun = to turn outward) is a movement of the soles laterally at the intertarsal joints (Figure 9.10f). Some professionals also refer to eversion of the feet as *pronation.*
- ***Dorsiflexion*** (dor-si-FLEK-shun) refers to bending of the foot at the ankle or talocrural joint (between the tibia, fibula, and talus) in the direction of the dorsum (superior surface) (Figure 9.10g). Dorsiflexion occurs when you stand on your heels.
- ***Plantar flexion*** (PLAN-tar) involves bending of the foot at the ankle joint in the direction of the plantar or inferior surface (Figure 9.10g), as when you elevate your body by standing on your toes.
- ***Supination*** (soo-pi-NĀ-shun) is a movement of the forearm at the proximal and distal radioulnar joints in which the palm is turned anteriorly (Figure 9.10h). This position of the palms is one of the defining features of the anatomical position.
- ***Pronation*** (prō-NĀ-shun) is a movement of the forearm at the proximal and distal radioulnar joints in which the distal end of the radius crosses over the distal end of the ulna and the palm is turned posteriorly (Figure 9.10h).
- ***Opposition*** (op-ō-ZISH-un) is the movement of the thumb at the carpometacarpal joint (between the trapezium and metacarpal of the thumb) in which the thumb moves across the palm to touch the tips of the fingers on the same hand (Figure 9.10i). This is the distinctive digital movement that gives humans and other primates the ability to grasp and manipulate objects very precisely.

A summary of the movements that occur at synovial joints is presented in Table 9.2.

CHECKPOINT

7. Alone or with a partner, demonstrate each movement listed in Table 9.2.

TABLE 9.2

Summary of Movements at Synovial Joints

MOVEMENT	DESCRIPTION
Gliding	Movement of relatively flat bone surfaces back-and-forth and side-to-side over one another; little change in the angle between bones.
Angular	Increase or decrease in the angle between bones.
Flexion	Decrease in the angle between articulating bones, usually in the sagittal plane.
Lateral flexion	Movement of the trunk in the frontal plane.
Extension	Increase in the angle between articulating bones, usually in the sagittal plane.
Hyperextension	Extension beyond the anatomical position.
Abduction	Movement of a bone away from the midline, usually in the frontal plane.
Adduction	Movement of a bone toward the midline, usually in the frontal plane.
Circumduction	Flexion, abduction, extension, and adduction in succession, in which the distal end of a body part moves in a circle.
Rotation	Movement of a bone around its longitudinal axis; in the limbs, it may be medial (toward midline) or lateral (away from midline).
Special	Occurs at specific joints.
Elevation	Superior movement of a body part.
Depression	Inferior movement of a body part.
Protraction	Anterior movement of a body part in the transverse plane.
Retraction	Posterior movement of a body part in the transverse plane.
Inversion	Medial movement of the sole.
Eversion	Lateral movement of the sole.
Dorsiflexion	Bending the foot in the direction of the dorsum (superior surface).
Plantar flexion	Bending the foot in the direction of the plantar surface (sole).
Supination	Movement of the forearm that turns the palm anteriorly.
Pronation	Movement of the forearm that turns the palm posteriorly.
Opposition	Movement of the thumb across the palm to touch fingertips on the same hand.

9.6 SELECTED JOINTS OF THE BODY

In Chapters 7 and 8 we discussed the major bones and their markings. In this chapter we have examined how joints are classified according to their structure and function, and we have introduced the movements that occur at joints. Table 9.3 (selected joints of the axial skeleton) and Table 9.4 (selected joints of the appendicular skeleton) will help you integrate the information you have learned in all three chapters. These tables list some of the major joints of the body according to their articular components (the bones that enter into their formation), their structural and functional classification, and the type(s) of movement that occurs at each joint.

Next we examine in detail six selected joints of the body in a series of exhibits. Each exhibit considers a specific synovial joint and contains (1) a definition—a description of the type of joint and the bones that form the joint; (2) the anatomical components—a description of the major connecting ligaments, articular disc (if present), articular capsule, and other distinguishing features of the joint; and (3) the joint's possible movements. Each exhibit also refers you to a figure that illustrates the joint. The joints described are the temporomandibular joint (TMJ), shoulder (humeroscapular or glenohumeral) joint, elbow joint, hip (coxal) joint, knee (tibiofemoral) joint, and ankle (talocrural) joint. Because these joints are described in detail in Exhibits 9.1 through 9.6, they are not included in Tables 9.3 and 9.4.

TABLE 9.3

Selected Joints of the Axial Skeleton

JOINT	ARTICULAR COMPONENTS	CLASSIFICATION	MOVEMENTS
Suture	Between skull bones.	*Structural:* fibrous. *Functional:* slightly movable or immovable.	Slight.
Atlanto-occipital	Between superior articular facets of atlas and occipital condyles of occipital bone.	*Structural:* synovial (condyloid). *Functional*: freely movable.	Flexion and extension of head and slight lateral flexion of head to either side.
Atlanto-axial	(1) Between dens of axis and anterior arch of atlas and (2) between lateral masses of atlas and axis.	*Structural:* synovial (pivot) between dens and anterior arch, and synovial (planar) between lateral masses. *Functional*: freely movable.	Rotation of head.
Intervertebral	(1) Between vertebral bodies and (2) between vertebral arches.	*Structural:* cartilaginous (symphysis) between vertebral bodies, and synovial (planar) between vertebral arches. *Functional:* slightly movable between vertebral bodies, and freely movable between vertebral arches.	Flexion, extension, lateral flexion, and rotation of vertebral column.
Vertebrocostal	(1) Between facets of heads of ribs and facets of bodies of adjacent thoracic vertebrae and intervertebral discs between them and (2) between articular part of tubercles of ribs and facets of transverse processes of thoracic vertebrae.	*Structural:* synovial (planar). *Functional:* freely movable.	Slight gliding.
Sternocostal	Between sternum and first seven pairs of ribs.	*Structural:* cartilaginous (synchondrosis) between sternum and first pair of ribs, and synovial (planar) between sternum and second through seventh pairs of ribs. *Functional:* immovable between sternum and first pair of ribs, and freely movable between sternum and second through seventh pairs of ribs.	None between sternum and first pair of ribs; slight gliding between sternum and second through seventh pairs of ribs.
Lumbosacral	(1) Between body of fifth lumbar vertebra and base of sacrum and (2) between inferior articular facets of fifth lumbar vertebra and superior articular facets of first vertebra of sacrum.	*Structural:* cartilaginous (symphysis) between body and base, and synovial (planar) between articular facets. *Functional:* slightly movable between body and base, and freely movable between articular facets.	Flexion, extension, lateral flexion, and rotation of vertebral column

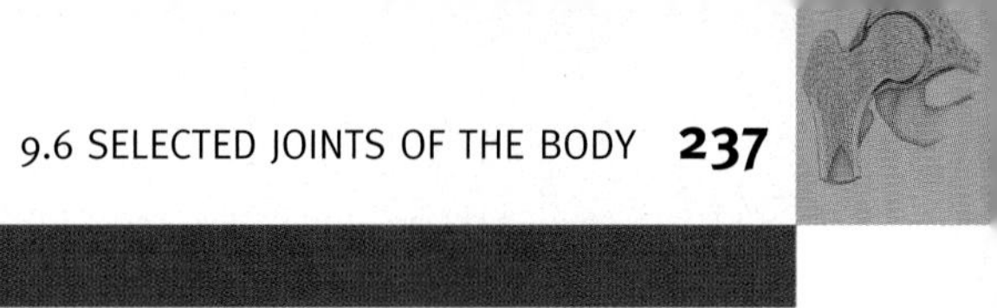

TABLE 9.4

Selected Joints of the Appendicular Skeleton

JOINT	ARTICULAR COMPONENTS	CLASSIFICATION	MOVEMENTS
Sternoclavicular	Between sternal end of clavicle, manubrium of sternum, and first costal cartilage.	*Structural:* synovial (planar and pivot). *Functional*: freely movable.	Gliding, with limited movements in nearly every direction.
Acromioclavicular	Between acromion of scapula and acromial end of clavicle.	*Structural:* synovial (planar). *Functional*: freely movable.	Gliding and rotation of scapula on clavicle.
Radioulnar	Proximal radioulnar joint between head of radius and radial notch of ulna; distal radioulnar joint between ulnar notch of radius and head of ulna.	*Structural:* synovial (pivot). *Functional*: freely movable.	Rotation of forearm.
Wrist (radiocarpal)	Between distal end of radius and scaphoid, lunate, and triquetrum of carpus.	*Structural:* synovial (condyloid). *Functional*: freely movable.	Flexion, extension, abduction, adduction, circumduction, and slight hyperextension of wrist.
Intercarpal	Between proximal row of carpal bones, distal row of carpal bones, and between both rows of carpal bones (midcarpal joints).	*Structural:* synovial (planar), except for hamate, scaphoid, and lunate (midcarpal) joint, which is synovial (saddle). *Functional*: freely movable.	Gliding plus flexion, extension, abduction, adduction, and slight rotation at midcarpal joints.
Carpometacarpal	Carpometacarpal joint of thumb between trapezium of carpus and first metacarpal; carpometacarpal joints of remaining digits formed between carpus and second through fifth metacarpals.	*Structural:* synovial (saddle) at thumb and synovial (planar) at remaining digits. *Functional*: freely movable.	Flexion, extension, abduction, adduction, and circumduction at thumb, and gliding at remaining digits.
Metacarpophalangeal and metatarsophalangeal	Between heads of metacarpals (or metatarsals) and bases of proximal phalanges.	*Structural:* synovial (condyloid). *Functional*: freely movable.	Flexion, extension, abduction, adduction, and circumduction of phalanges.
Interphalangeal	Between heads of phalanges and bases of more distal phalanges.	*Structural:* synovial (hinge). *Functional*: freely movable.	Flexion and extension of phalanges.
Sacroiliac	Between auricular surfaces of sacrum and ilia of hip bones.	*Structural:* synovial (planar). *Functional*: freely movable.	Slight gliding (even more so during pregnancy).
Pubic symphysis	Between anterior surfaces of hip bones.	*Structural:* cartilaginous (symphysis). *Functional*: slightly movable.	Slight movements (even more so during pregnancy).
Tibiofibular	Proximal tibiofibular joint between lateral condyle of tibia and head of fibula; distal tibiofibular joint between distal end of fibula and fibular notch of tibia.	*Structural:* synovial (planar) at proximal joint, and fibrous (syndesmosis) at distal joint. *Functional*: freely movable at proximal joint, and slightly movable at distal joint.	Slight gliding at proximal joint, and slight rotation of fibula during dorsiflexion of foot.
Intertarsal	Subtalar joint between talus and calcaneus of tarsus; talocalcaneonavicular joint between talus and calcaneus and navicular of tarsus; calcaneocuboid joint between calcaneus and cuboid of tarsus.	*Structural*: synovial (planar) at subtalar and calcaneocuboid joints, and synovial (saddle) at talocalcaneonavicular joint. *Functional*: freely movable.	Inversion and eversion of foot.
Tarsometatarsal	Between three cuneiforms of tarsus and bases of five metatarsal bones.	*Structural:* synovial (planar). *Functional*: freely movable.	Slight gliding.

OBJECTIVE

- Describe the anatomical components of the temporomandibular joint and explain the movements that can occur at this joint.

Definition

The ***temporomandibular joint (TMJ)*** (Figure 9.11) is a combined hinge and planar joint formed by the condylar process of the mandible and the mandibular fossa and articular tubercle of the temporal bone. The temporomandibular joint is the only freely movable joint between skull bones; all other skull joints are sutures and therefore immovable or slightly movable.

Anatomical Components

1. ***Articular disc (meniscus).*** Fibrocartilage disc that separates the joint cavity into superior and inferior compartments, each with a synovial membrane (Figure 9.11c).
2. ***Articular capsule.*** Thin, fairly loose envelope around the circumference of the joint (Figure 9.11a).
3. ***Lateral ligament.*** Two short bands on the lateral surface of the articular capsule that extend inferiorly and posteriorly from the inferior border and tubercle of the zygomatic process of the temporal bone to the lateral and posterior aspect of the neck of the mandible. The lateral ligament is covered by the parotid gland and helps strengthen the joint laterally and prevent displacement of the mandible (Figure 9.11a).
4. ***Sphenomandibular ligament.*** Thin band that extends inferiorly and anteriorly from the spine of the sphenoid bone to the ramus of the mandible (Figure 9.11b). It does not contribute significantly to the strength of the joint.
5. ***Stylomandibular ligament.*** Thickened band of deep cervical fascia that extends from the styloid process of the temporal bone to the inferior and posterior border of the ramus of the mandible. This ligament separates the parotid gland from the submandibular gland and inhibits movement of the mandible at the TMJ (Figure 9.11a,b).

Movements

In the temporomandibular joint, only the mandible moves because the maxilla is firmly anchored to other bones of the skull by sutures. Accordingly, the mandible may function in depression (jaw opening) and elevation (jaw closing), which occurs in the inferior compartment, and protraction, retraction, lateral displacement, and slight rotation, which occur in the superior compartment (see Figure 9.10a–d).

• CLINICAL CONNECTION | Dislocated Mandible

A *dislocation* (dis′-lō-KĀ-shun; *dis-* = apart) or *luxation* (luks-Ā-shun; *luxatio* = dislocation) is the displacement of a bone from a joint with tearing of ligaments, tendons, and articular capsules. It is usually caused by a blow or fall, although unusual physical effort may be a factor. For example, if the condylar processes of the mandible pass anterior to the articular tubercles when you yawn or take a large bite, a **dislocated mandible** (anterior displacement) may occur. When the mandible is displaced in this manner, the mouth remains wide open and the person is unable to close it. This may be corrected by pressing the thumbs downward on the lower molar teeth and pushing the mandible backward. Other causes of a dislocated mandible include a lateral blow to the chin when the mouth is open and a fracture of the mandible. •

CHECKPOINT

8. What distinguishes the temporomandibular joint from the other joints of the skull?

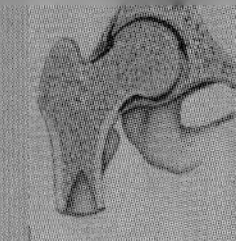

Figure 9.11 Temporomandibular joint (TMJ).

The TMJ is the only freely movable joint between skull bones.

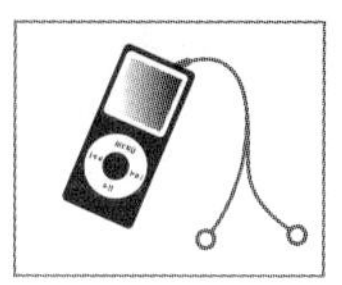

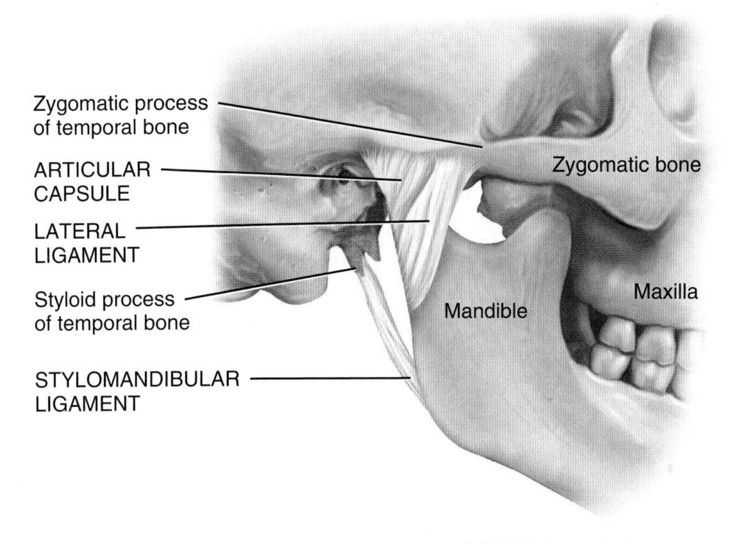

(a) Right lateral view

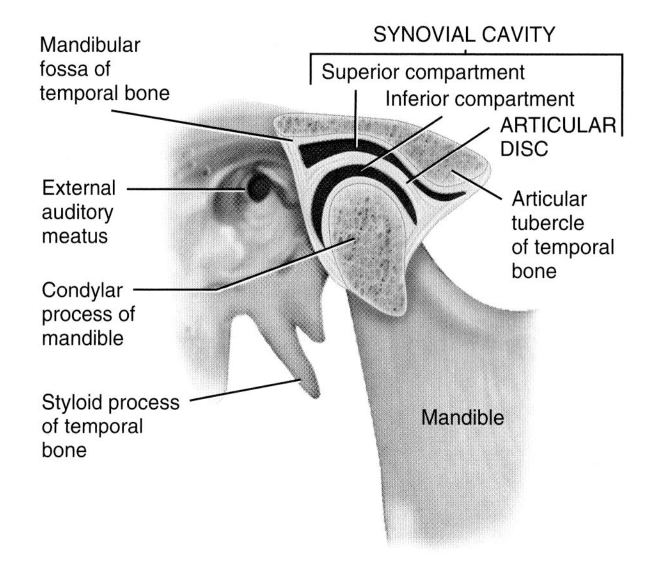

(c) Sagittal section viewed from right

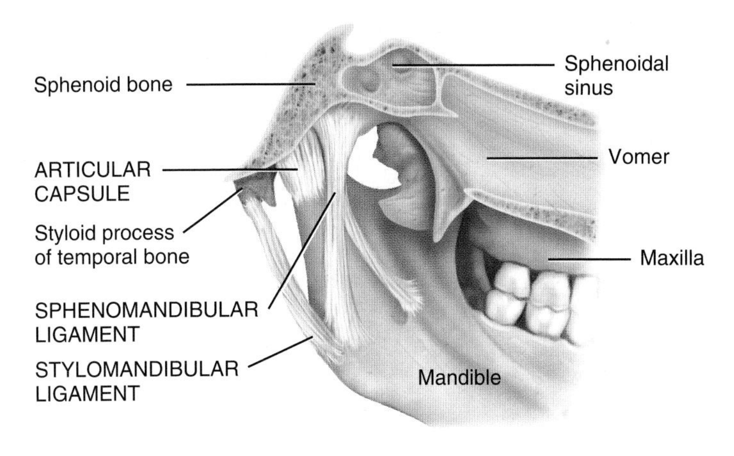

(b) Left medial view

Which ligament prevents displacement of the mandible?

OBJECTIVE

- Describe the anatomical components of the shoulder joint and the movements that can occur at this joint.

Definition

The ***shoulder joint*** (Figure 9.12) is a ball-and-socket joint formed by the head of the humerus and the glenoid cavity of the scapula. It also is referred to as the *humeroscapular* or *glenohumeral joint.*

Anatomical Components

1. ***Articular capsule.*** Thin, loose sac that completely envelops the joint and extends from the glenoid cavity to the anatomical neck of the humerus. The inferior part of the capsule is its weakest area (Figure 9.12a,c,d).
2. ***Coracohumeral ligament.*** Strong, broad ligament that strengthens the superior part of the articular capsule and extends from the coracoid process of the scapula to the greater tubercle of the humerus (Figure 9.12a). The ligament strengthens the superior part of the articular capsule, and reinforces the anterior aspect of the articular capsule.
3. ***Glenohumeral ligaments.*** Three thickenings of the articular capsule over the anterior surface of the joint. They extend from the glenoid cavity to the lesser tubercle and anatomical neck of the humerus. These ligaments are often indistinct or absent and provide only minimal strength (Figure 9.12a,b). This ligament plays a role in joint stabilization when the humerus approaches or exceeds its limits of motion.
4. ***Transverse humeral ligament.*** Narrow sheet extending from the greater tubercle to the lesser tubercle of the humerus (Figure 9.12a). The ligament functions as a retinaculum for the long tendon of the biceps brachii muscle.
5. ***Glenoid labrum.*** Narrow rim of fibrocartilage around the edge of the glenoid cavity. It slightly deepens and enlarges the glenoid cavity (Figure 9.12b,c,d).
6. ***Bursae.*** Four bursae (see Section 9.4) are associated with the shoulder joint. They are the *subscapular bursa* (Figure 9.12a), *subdeltoid bursa, subacromial bursa* (Figure 9.12a–c), and *subcoracoid bursa.*

Figure 9.12 Right shoulder (humeroscapular or glenohumeral) joint.

Most of the stability of the shoulder joints results from the arrangement of the rotator cuff muscles.

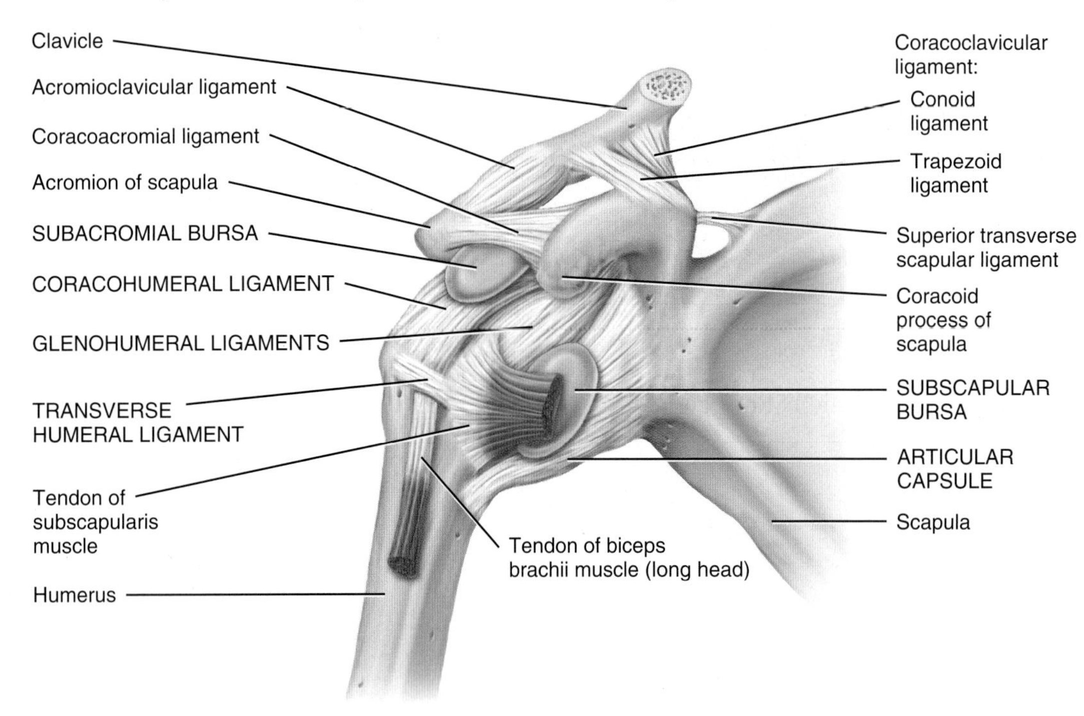

(a) Anterior view

Movements

The shoulder joint allows flexion, extension, abduction, adduction, medial rotation, lateral rotation, and circumduction of the arm (see Figures 9.6–9.9). It has more freedom of movement than any other joint of the body. This freedom results from the looseness of the articular capsule and shallowness of the glenoid cavity in relation to the large size of the head of the humerus.

Although the ligaments of the shoulder joint strengthen it to some extent, most of the strength results from the muscles that surround the joint, especially the *rotator cuff muscles.* These muscles (supraspinatus, infraspinatus, teres minor, and subscapularis) join the scapula to the humerus (see also Figure 13.4d,e). The tendons of the rotator cuff muscles encircle the joint (except for the inferior portion) and fuse with the articular capsule. The rotator cuff muscles work as a group to hold the head of the humerus in the glenoid cavity.

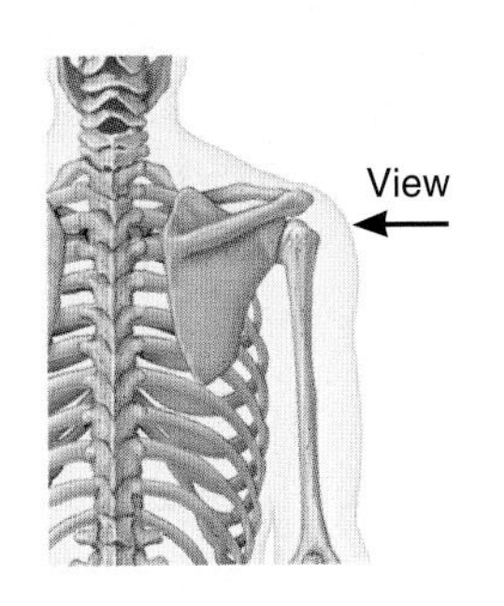

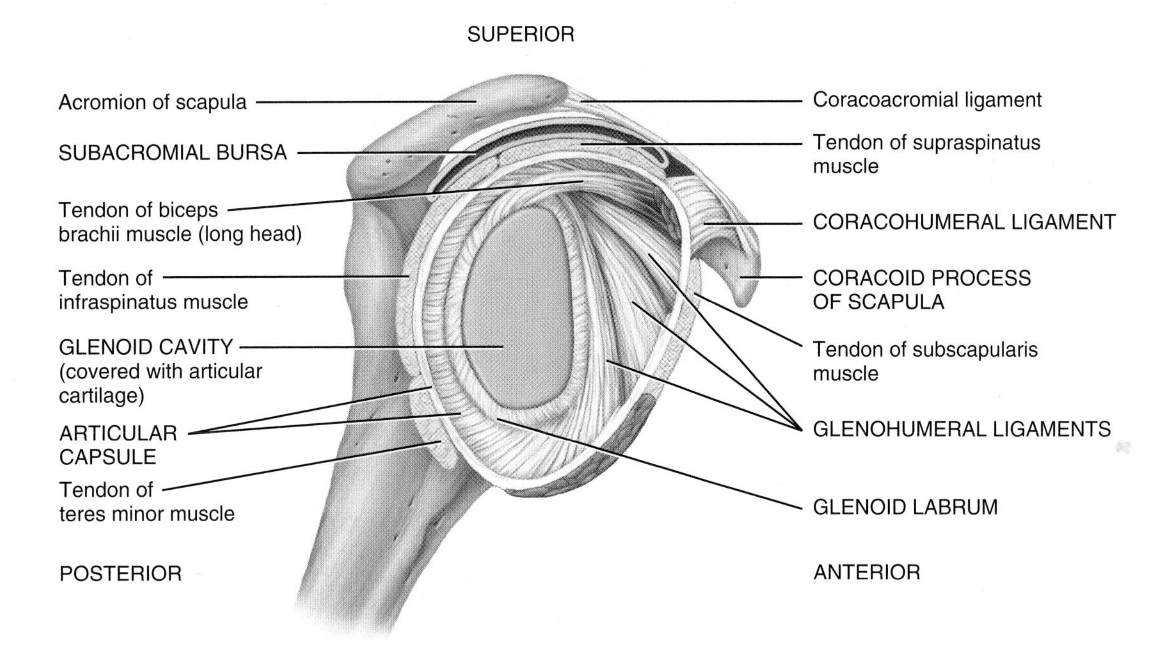

(b) Lateral view (opened with synovial membrane removed)

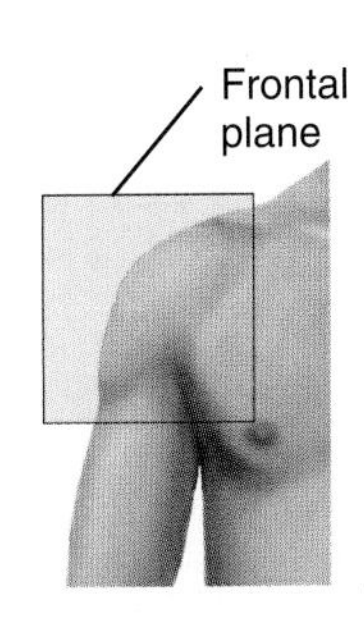

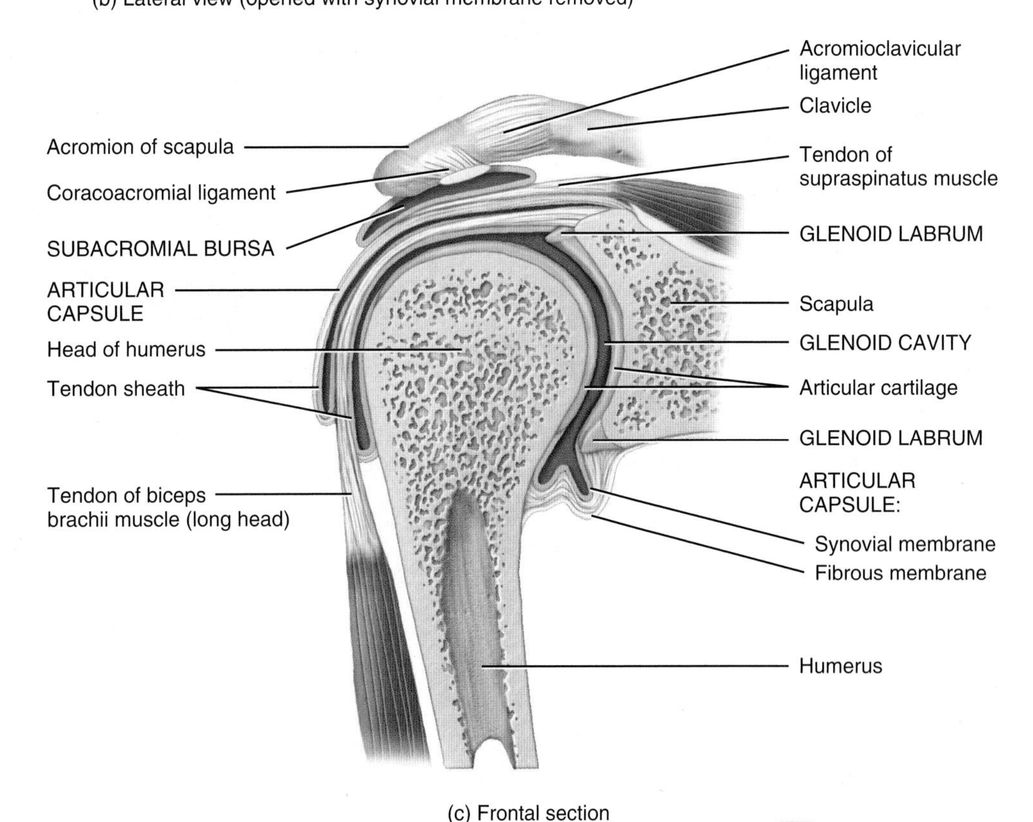

(c) Frontal section

EXHIBIT 9.2 CONTINUES

• CLINICAL APPLICATION | Shoulder Joint Injuries

Torn Glenoid Labrum The fibrocartilaginous labrum may tear away from the glenoid cavity. This causes the joint to catch or feel like it's slipping out of place. The shoulder may indeed become dislocated as a result. A torn labrum is reattached to the glenoid surgically with anchors and sutures. The repaired joint is more stable.

Arthritis of the Shoulder Joint Damage to the joint cartilage may result from aging or use. Wear and tear may also lead to loose pieces of bone or bone spurs in the joint. During surgery, bone spurs and/or rough parts are removed and smoothed. Any loose pieces of bone are removed from the joint space. Bone may also be scraped or shaved to promote the growth of new cartilage.

Dislocated and Separated Shoulder The joint most commonly dislocated in adults is the shoulder joint because its socket is quite shallow and the bones are held together mainly by supporting muscles. Usually in a **dislocated shoulder,** the head of the humerus becomes displaced inferiorly, where the articular capsule is least protected. Dislocations of the mandible, elbow, fingers, knee, or hip are less common.

A **separated shoulder** refers to an injury of the acromioclavicular joint, a joint formed by the acromion of the scapula and the acromial end of the clavicle. This condition is usually the result of forceful trauma to the joint, as when the shoulder strikes the ground in a fall.

Treating Shoulder Injuries *Shoulder arthroscopy* allows a physician to see and work inside the shoulder joint through small incisions. During surgery the arthroscope sends live video images from inside the joint to a monitor. These images permit the physician to diagnose and treat many shoulder problems. Because arthroscopy utilizes smaller incisions, recovery is usually shorter and less painful than recovery after open surgery. •

CHECKPOINT

9. Which tendons at the shoulder joint of a baseball pitcher are most likely to be torn due to excessive circumduction?

Figure 9.12 (continued)

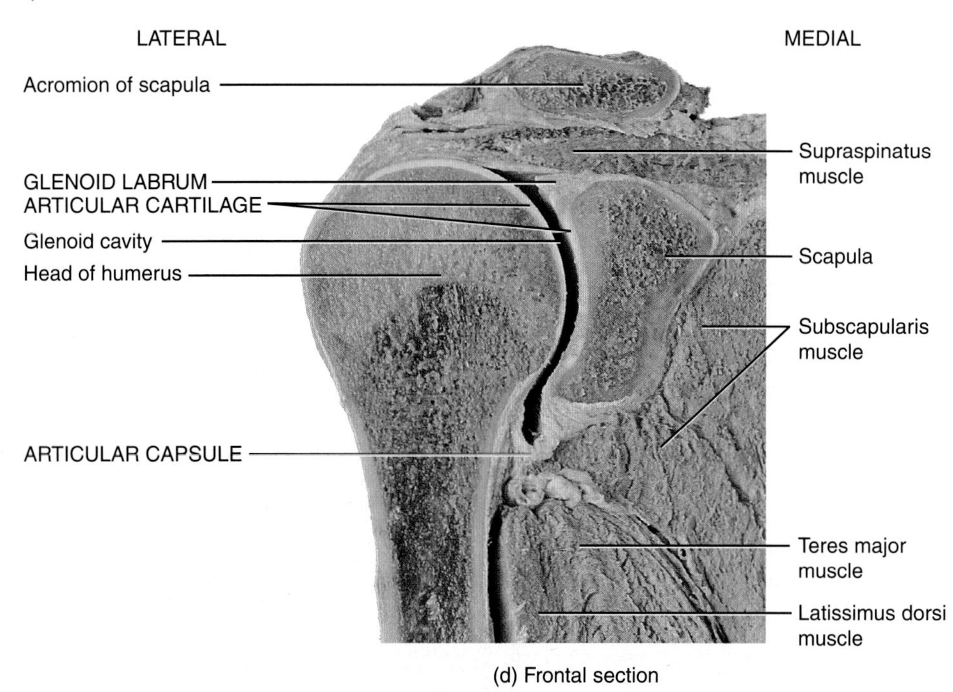

(d) Frontal section

? Why does the shoulder joint have more freedom of movement than any other joint of the body?

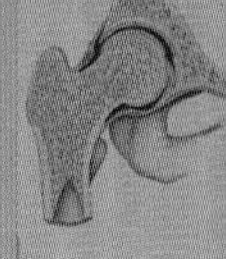

EXHIBIT 9.3 Elbow Joint

OBJECTIVE

- Describe the anatomical components of the elbow joint and the movements that can occur at this joint.

Definition

The ***elbow joint*** (Figure 9.13) is a hinge joint formed by the trochlea and capitulum of the humerus, the trochlear notch of the ulna, and the head of the radius.

Anatomical Components

1. ***Articular capsule.*** The anterior part of the articular capsule covers the anterior part of the elbow joint, from the radial and coronoid fossae of the humerus to the coronoid process of the ulna and the anular ligament of the radius. The posterior part extends from the capitulum, olecranon fossa, and lateral epicondyle of the humerus to the anular ligament of the radius, the olecranon of the ulna, and the ulna posterior to the radial notch (Figure 9.13).
2. ***Ulnar collateral ligament.*** Thick, triangular ligament that extends from the medial epicondyle of the humerus to the coronoid process and olecranon of the ulna (Figure 9.13a). Part of the ligament deepens the socket for the trochlea of the humerus.
3. ***Radial collateral ligament.*** Strong, triangular ligament that extends from the lateral epicondyle of the humerus to the anular ligament of the radius and the radial notch of the ulna (Figure 9.13b). The anular ligament of the radius holds the head of the radius in the radial notch of the ulna.

Movements

The elbow joint allows flexion and extension of the forearm (see Figure 9.6c).

• CLINICAL APPLICATION — Dislocation of the Radial Head

A **dislocation of the radial head** is the most common upper limb dislocation in children. In this injury, the head of the radius slides past or ruptures the radial anular ligament, a ligament that forms a collar around the head of the radius at the proximal radioulnar joint. Dislocation is most apt to occur when a strong pull is applied to the forearm while it is extended and supinated, for instance, while swinging a child around with outstretched arms. •

CHECKPOINT

10. At the elbow joint, which ligaments connect (a) the humerus and the ulna, and (b) the humerus and the radius?

Figure 9.13 Right elbow joint.

The elbow joint is formed by parts of three bones: humerus, ulna, and radius.

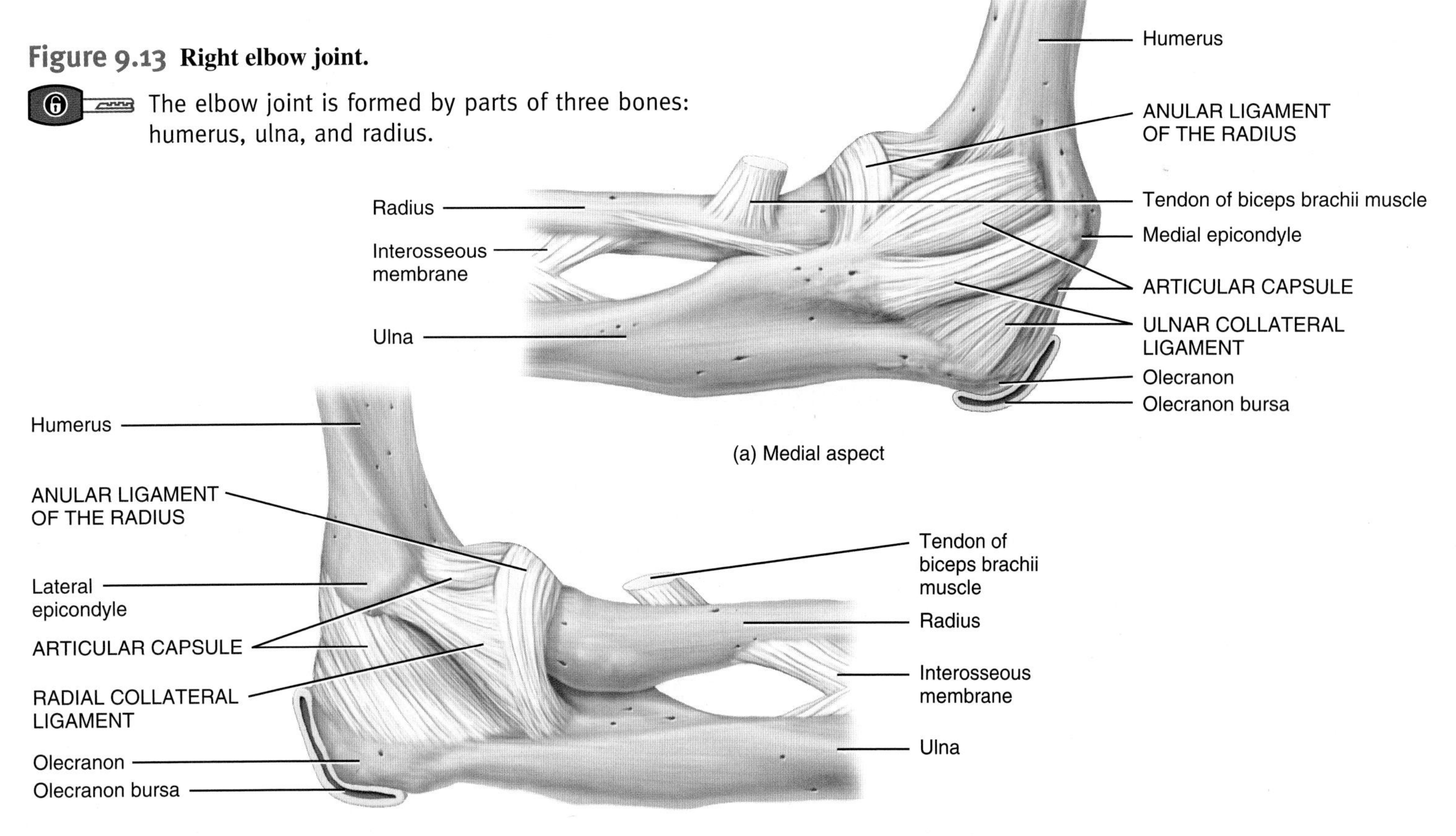

Which movements are possible at a hinge joint?

EXHIBIT 9.4 Hip Joint

OBJECTIVE

- Describe the anatomical components of the hip joint and the movements that can occur at this joint.

Definition

The ***hip joint*** *(coxal joint)* is a ball-and-socket joint formed by the head of the femur and the acetabulum of the hip bone (Figure 9.14).

Anatomical Components

1. ***Articular capsule.*** Very dense and strong capsule that extends from the rim of the acetabulum to the neck of the femur (Figure 9.14c). One of the strongest structures of the body, the capsule consists of circular and longitudinal fibers. The circular fibers, called the *zona orbicularis,* form a collar around the neck of the femur. Accessory ligaments known as the iliofemoral ligament, pubofemoral ligament, and ischiofemoral ligament reinforce the longitudinal fibers.
2. ***Iliofemoral ligament.*** Thickened portion of the articular capsule that extends from the anterior inferior iliac spine of the hip bone to the intertrochanteric line of the femur (Figure 9.14a,b). This ligament is said to be the strongest ligament in the body and prevents hyperextension of the femur at the hip joint.
3. ***Pubofemoral ligament.*** Thickened portion of the articular capsule that extends from the pubic part of the rim of the acetabulum to the neck of the femur (Figure 9.14a). The ligament prevents overabduction of the femur at the hip joint and strengthens the articular capsule.
4. ***Ischiofemoral ligament.*** Thickened portion of the articular capsule that extends from the ischial wall of the acetabulum to the neck of the femur (Figure 9.14b). This ligament slackens during adduction and tenses during abduction and strengthens the articular capsule.
5. ***Ligament of the head of the femur.*** Flat, triangular band (primarily a synovial fold) that extends from the fossa of the acetabulum to the fovea capitis of the head of the femur (Figure 9.14c). The ligament contains a small artery to the head of the femur.
6. ***Acetabular labrum.*** Fibrocartilage rim attached to the margin of the acetabulum that enhances the depth of the acetabulum. Because the diameter of the acetabular rim is smaller than that of the head of the femur, dislocation of the femur is rare (Figure 9.14c).
7. ***Transverse ligament of the acetabulum.*** Strong ligament that crosses over the acetabular notch. It supports part of the acetabular labrum and is connected with the ligament of the head of the femur and the articular capsule (Figure 9.14c).

Movements

The hip joint allows flexion, extension, abduction, adduction, circumduction, medial rotation, and lateral rotation of the thigh (see Figures 9.6–9.9). The extreme stability of the hip joint is related to the very strong articular capsule and its accessory ligaments, the manner in which the femur fits into the acetabulum, and the muscles surrounding the joint. Although the shoulder and hip joints are both ball-and-socket joints, the hip joints do not have as wide a range of motion. Flexion is limited by the anterior surface of the thigh coming into contact with the anterior abdominal wall when the knee is flexed and by tension of the hamstring muscles when the knee is extended. Extension is limited by tension of the iliofemoral, pubofemoral, and ischiofemoral ligaments. Abduction is limited by the tension of the pubofemoral ligament, and

Figure 9.14 Right hip (coxal) joint.

The articular capsule of the hip joint is one of the strongest structures in the body.

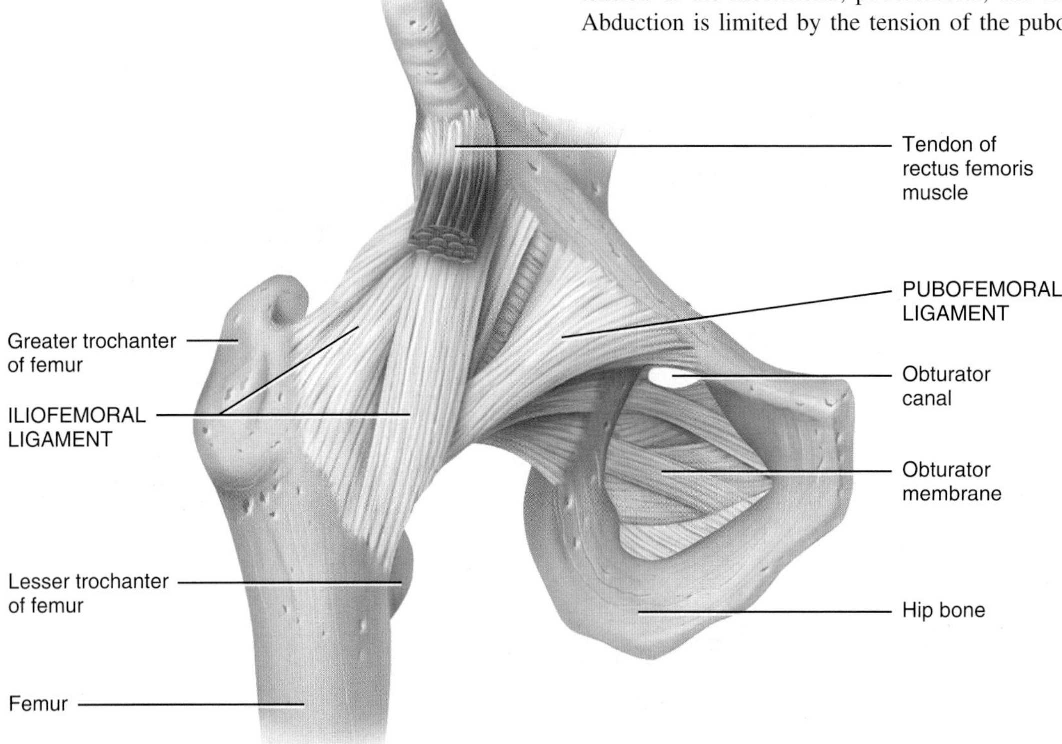

(a) Anterior view

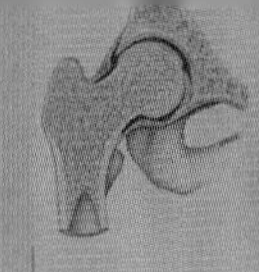

adduction is limited by contact with the opposite limb and tension in the ligament of the head of the femur. Medial rotation is limited by the tension in the ischiofemoral ligament, and lateral rotation is limited by tension in the iliofemoral and pubofemoral ligaments.

CHECKPOINT

11. What factors limit the degree of flexion and abduction at the hip joint?

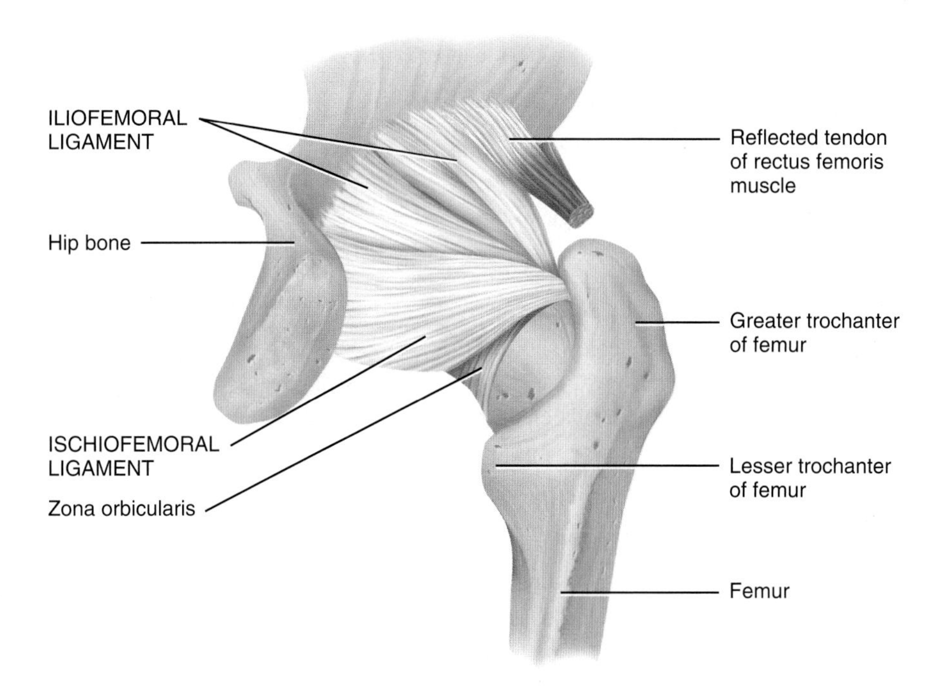

(b) Posterior view

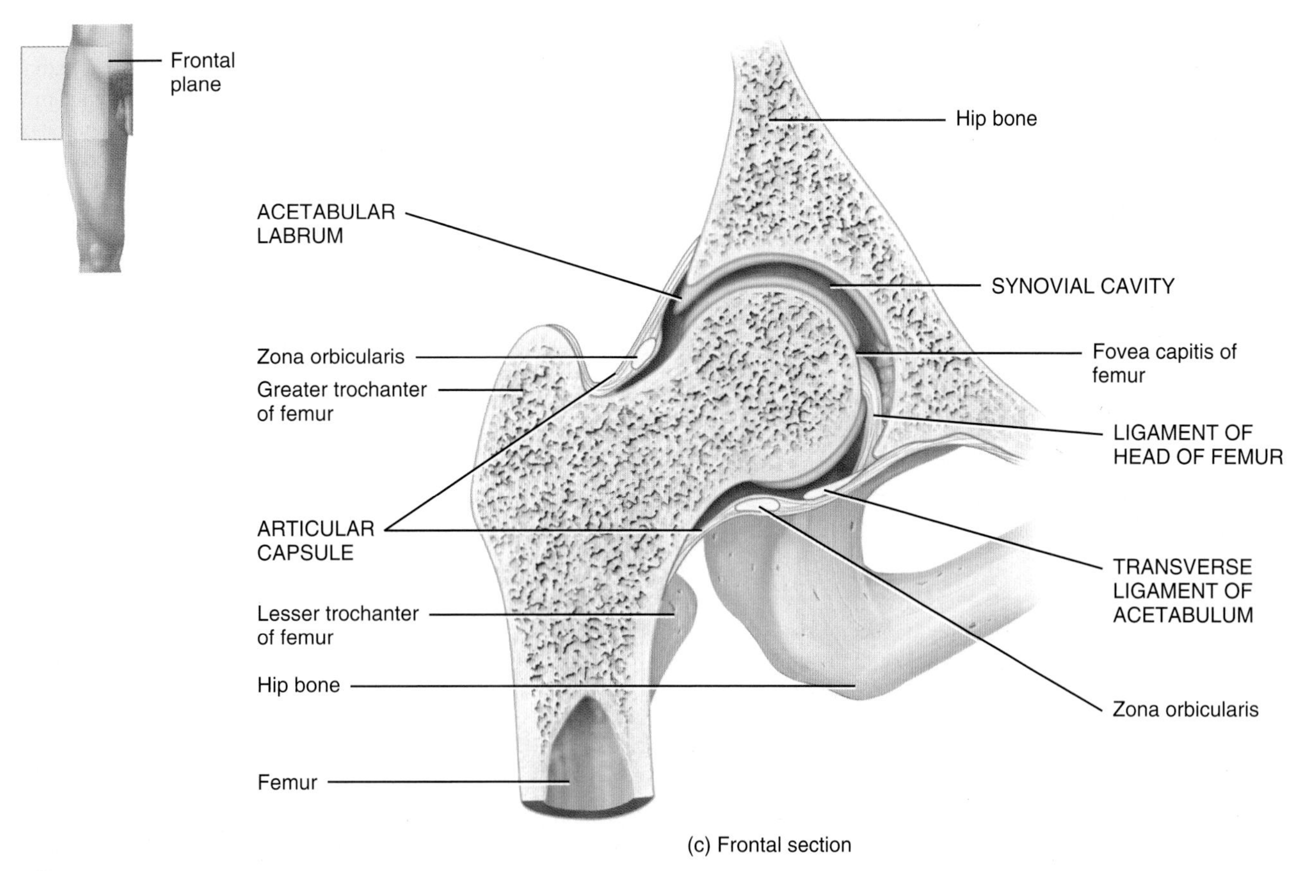

(c) Frontal section

? Which ligaments limit the degree of extension that is possible at the hip joint?

EXHIBIT 9.5 Knee Joint

OBJECTIVE

- Describe the main anatomical components of the knee joint and explain the movements that can occur at this joint.

Definition

The ***knee joint*** *(tibiofemoral joint)* is the largest and most complex joint of the body (Figure 9.15), actually a modified hinge joint consisting of three joints within a single synovial cavity:

1. Laterally is a tibiofemoral joint, between the lateral condyle of the femur, lateral meniscus, and lateral condyle of the tibia, which is weight bearing.
2. Medially is a second tibiofemoral joint, between the medial condyle of the femur, medial meniscus, and medial condyle of the tibia.
3. An intermediate patellofemoral joint, between the patella and the patellar surface of the femur.

Anatomical Components

1. ***Articular capsule.*** No complete, independent capsule unites the bones. The ligamentous sheath surrounding the joint consists mostly of muscle tendons or their expansions (Figure 9.15a,b). There are, however, some capsular fibers connecting the articulating bones.
2. ***Medial and lateral patellar retinacula.*** Fused tendons of insertion of the quadriceps femoris muscle and the fascia lata (deep fascia of thigh) that strengthen the anterior surface of the joint (Figure 9.15a).
3. ***Patellar ligament.*** Continuation of the common tendon of insertion of the quadriceps femoris muscle that extends from the patella to the tibial tuberosity. This ligament also strengthens the anterior surface of the joint. The posterior surface of the ligament is separated from the synovial membrane of the joint by an ***infrapatellar fat pad*** (Figure 9.15a,c).
4. ***Oblique popliteal ligament.*** Broad, flat ligament that extends from the intercondylar fossa of the femur to the head of the tibia (Figure 9.15b). The tendon of the semimembranosus muscle is superficial to this ligament and passes from the medial condyle of the tibia to the lateral condyle of the femur. The ligament and tendon strengthen the posterior surface of the joint.
5. ***Arcuate popliteal ligament.*** Extends from the lateral condyle of the femur to the styloid process of the head of the fibula. It strengthens the lower lateral part of the posterior surface of the joint (Figure 9.15b).
6. ***Tibial collateral ligament.*** Broad, flat ligament on the medial surface of the joint that extends from the medial condyle of the femur to the medial condyle of the tibia (Figure 9.15a,b,d,f). Tendons of the sartorius, gracilis, and semitendinosus muscles, all of which strengthen the medial aspect of the joint, cross the ligament. Because the tibial collateral ligament is firmly attached to the medial meniscus, tearing of the ligament frequently results in tearing of the meniscus and damage to the anterior cruciate ligament, described under 8a.
7. ***Fibular collateral ligament.*** Strong, rounded ligament on the lateral surface of the joint that extends from the lateral condyle of the femur to the lateral side of the head of the fibula (Figure 9.15a,b,d,f). It strengthens the lateral aspect of the joint. The ligament is covered by the tendon of the biceps femoris muscle. The tendon of the popliteal muscle is deep to the ligament.
8. ***Intracapsular ligaments.*** Ligaments within the capsule that connect the tibia and femur. The anterior and posterior cruciate ligaments (KROO-shē-āt = shaped like a cross) are named based on their origins relative to the intercondylar area of the tibia. Following their originations, they cross on their way to their destinations on the femur.
   a. *Anterior cruciate ligament (ACL).* Extends posteriorly and laterally from a point *anterior* to the intercondylar area of the tibia to the posterior part of the medial surface of the lateral condyle of the femur (Figure 9.15d,f). The ACL limits hyperextension of the knee and prevents the anterior sliding of the tibia on the femur. This ligament is stretched or torn in about 70% of all serious knee injuries.
   b. *Posterior cruciate ligament (PCL).* Extends anteriorly and medially from a depression on the *posterior* intercondylar area of the tibia and lateral meniscus to the anterior part of the lateral surface of the medial condyle of the femur (Figure 9.15d,f). The PCL prevents the posterior sliding of the tibia on the femur, especially when the knee is flexed. This is very important when walking down stairs or a steep incline.
9. ***Articular discs (menisci).*** Two fibrocartilage discs between the tibial and femoral condyles that help compensate for the irregular shapes of the bones and circulate synovial fluid.
   a. *Medial meniscus.* Semicircular piece of fibrocartilage (C-shaped). Its anterior end is attached to the anterior intercondylar fossa of the tibia, anterior to the anterior cruciate ligament. Its posterior end is attached to the posterior intercondylar fossa of the tibia between the attachments of the posterior cruciate ligament and lateral meniscus (Figure 9.15d–f).
   b. *Lateral meniscus.* Nearly circular piece of fibrocartilage (approaches an incomplete O in shape). Its anterior end is attached anterior to the intercondylar eminence of the tibia, and lateral and posterior to the anterior cruciate ligament. Its posterior end is attached posterior to the intercondylar eminence of the tibia, and anterior to the posterior end of the medial meniscus (Figure 9.15d–f). The medial and lateral menisci are connected to each other by the *transverse ligament* (Figure 9.15d,e) and to the margins of the head of tibia by the *coronary ligaments* (not illustrated).
10. The more important ***bursae*** of the knee include the following:
    a. *Prepatellar bursa* between the patella and skin (Figure 9.15c).
    b. *Infrapatellar bursa* between superior part of the tibia and patellar ligament (Figure 9.15a,c).
    c. *Suprapatellar bursa* between inferior part of femur and deep surface of quadriceps femoris muscle (Figure 9.15a,c).

Movements

The knee joint allows flexion, extension, slight medial rotation, and lateral rotation of the leg in the flexed position (see Figures 9.6f and 9.9c).

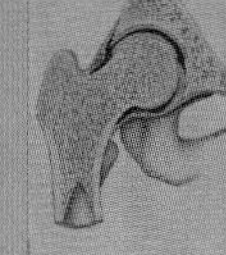

Figure 9.15 Right knee (tibiofemoral) joint.

The knee joint is the largest and most complex joint in the body.

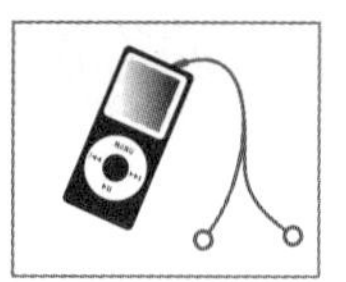

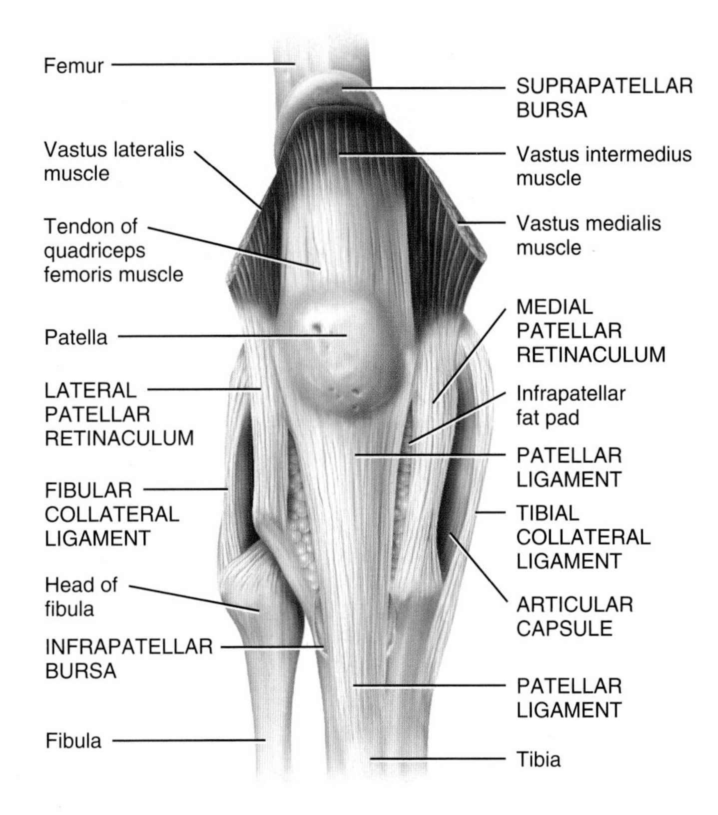

(a) Anterior superficial view

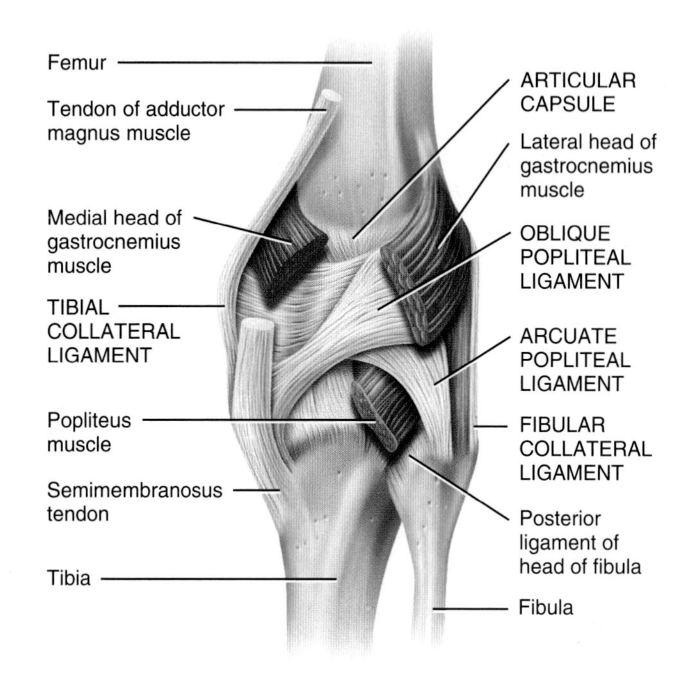

(b) Posterior deep view

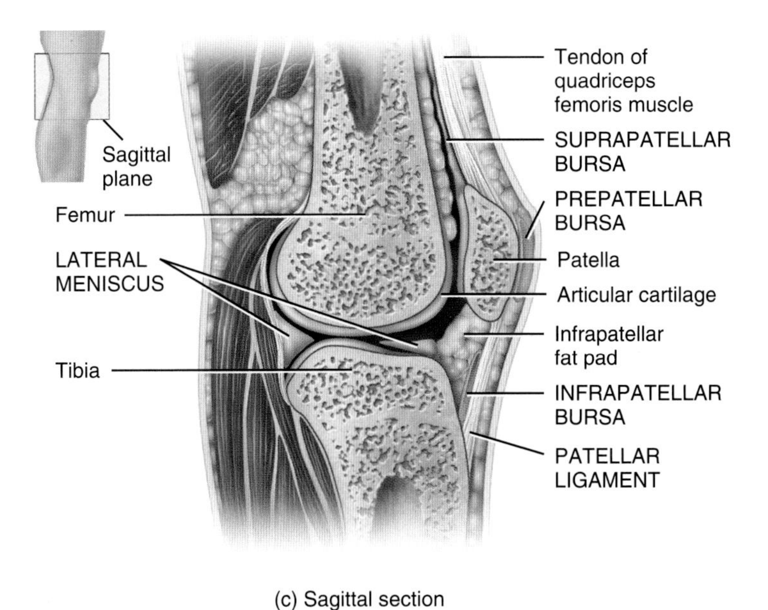

(c) Sagittal section

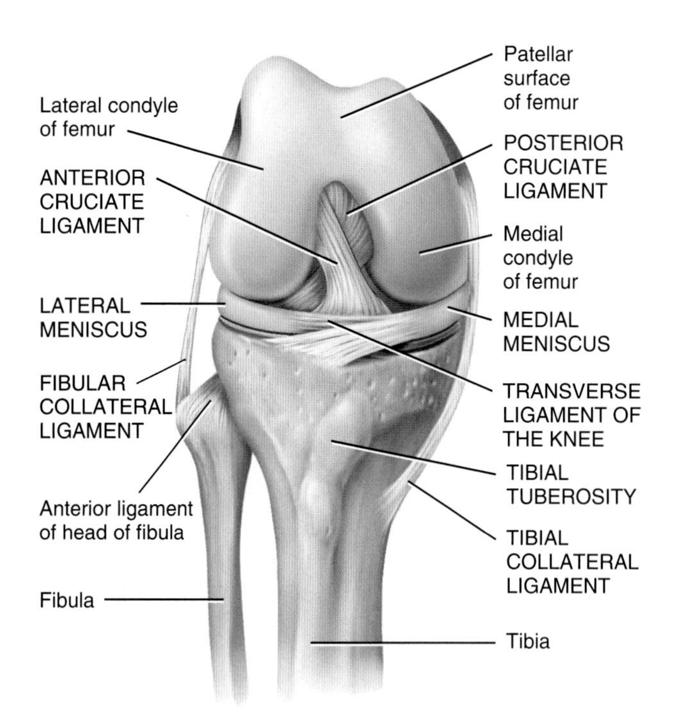

(d) Anterior deep view of flexed knee

EXHIBIT 9.5 CONTINUES

• CLINICAL APPLICATION Knee Injuries

The knee joint is the joint most vulnerable to damage because it is a mobile, weight-bearing joint and its stability depends almost entirely on its associated ligaments and muscles. Further, there is no correspondence of the articulating bones. Following are several kinds of **knee injuries.** A **swollen knee** may occur immediately or hours after an injury. The initial swelling is due to escape of blood from damaged blood vessels adjacent to areas involving rupture of the anterior cruciate ligament, damage to synovial membranes, torn menisci, fractures, or collateral ligament sprains. Delayed swelling is due to excessive production of synovial fluid, a condition commonly referred to as "water on the knee." A common type of knee injury in football is **rupture of the tibial collateral ligaments,** often associated with tearing of the anterior cruciate ligament and medial meniscus (torn cartilage). Usually, a hard blow to the lateral side of the knee while the foot is fixed on the ground causes the damage. A **dislocated knee** refers to the displacement of the tibia relative to the femur. The most common type is dislocation anteriorly, resulting from hyperextension of the knee. A frequent consequence of a dislocated knee is damage to the popliteal artery. •

CHECKPOINT

12. What are the opposing functions of the anterior and posterior cruciate ligaments?

Figure 9.15 (continued)

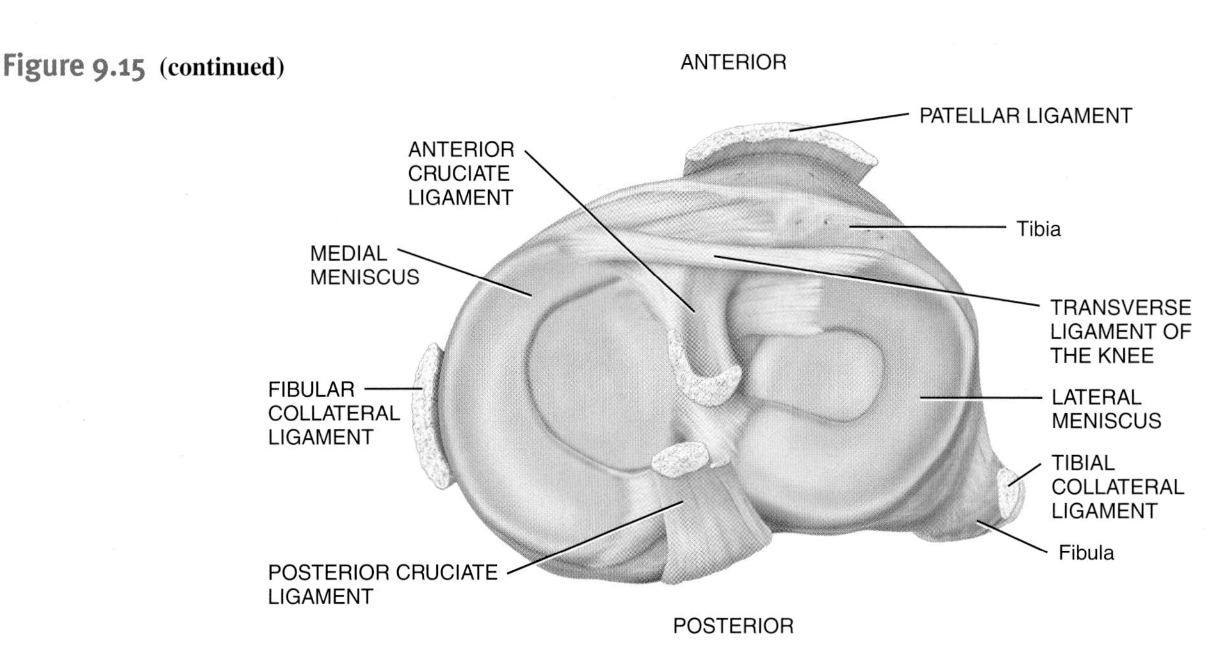

(e) Superior view of menisci

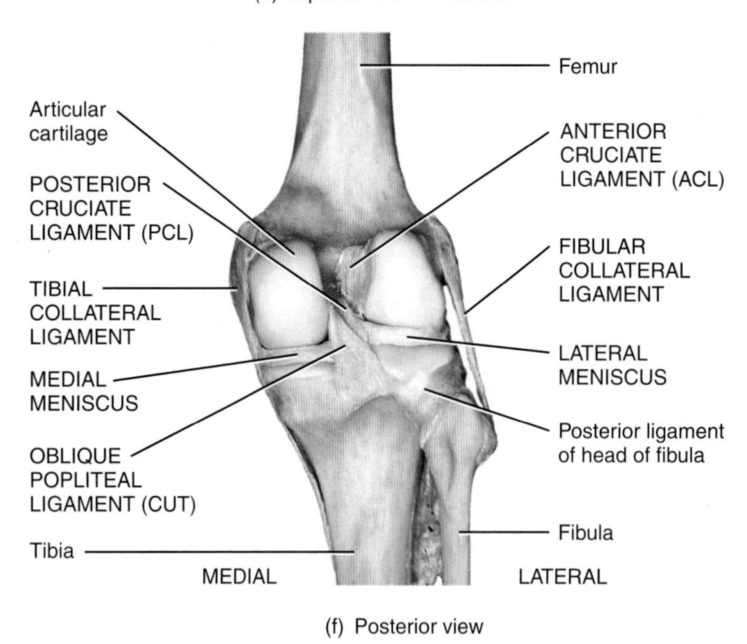

(f) Posterior view

? What movement occurs at the knee joint when the quadriceps femoris (anterior thigh) muscles contract?

EXHIBIT 9.6 Ankle Joint

OBJECTIVE

- Describe the anatomical components of the ankle joint and explain the movements that can occur at this joint.

Definition

The ***ankle joint*** *(talocrural joint)* is a hinge joint formed by (1) the distal end of the tibia and its medial malleolus with the talus and (2) the lateral malleolus of the fibula with the talus (Figure 9.16). It is a strong and stable joint due to the shapes of the articulating bones, the strength of its ligaments, and the tendons that surround it.

Anatomically, the ankle is the region that extends from the distal region of the leg to the proximal region of the foot and contains the ankle joint. In this transition region, there is a change in orientation from a vertical position of the bones and muscles and associated structures in the leg to a horizontal position of the structures in the foot. As a result, there is a turning anteriorly of the tendons, blood vessels, and nerves in the ankle as they enter the foot.

As the structures pass from the leg into the foot at the ankle from a vertical to a horizontal orientation, they are anchored by thickenings of deep fascia called ***retinacula*** (ret-i-NAK-yoo-la; *retineo* = to hold back). Two principal retinacula are the superior and inferior extensor retinacula.

Anatomical Components

1. ***Articular capsule.*** Completely surrounds the joint and is attached superiorly to the tibia and fibula and inferiorly to the talus. The capsule is thin (and weak) anteriorly and posteriorly to permit dorsiflexion and plantar flexion.

Figure 9.16 Right ankle (talocrural) joint.

The strength and stability of the ankle joint are due to the shapes of the articulating bone, the strength of its ligaments, and the tendons that surround it.

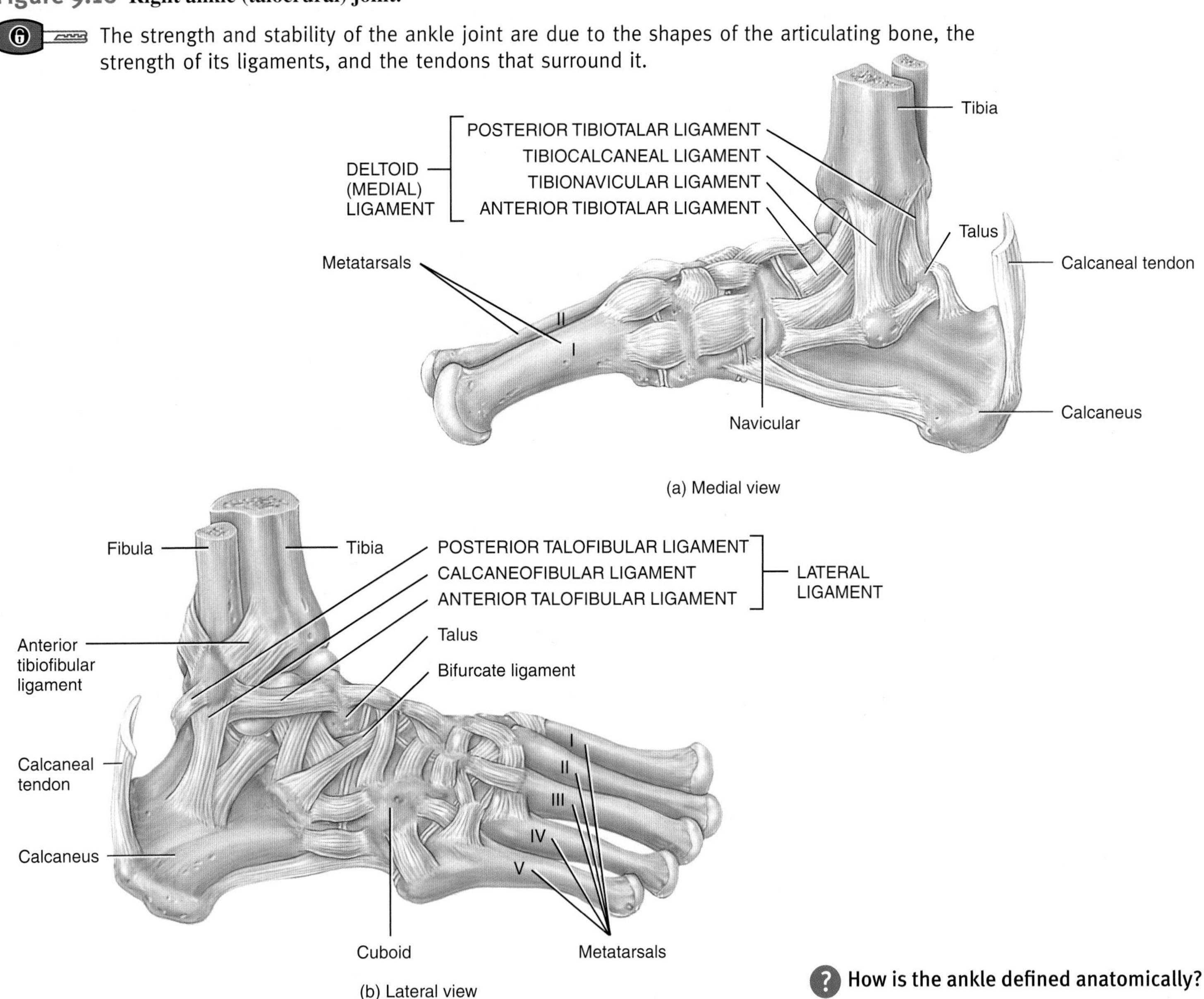

How is the ankle defined anatomically?

EXHIBIT 9.6 CONTINUES

2. ***Deltoid (medial) ligament.*** Strong, flat, triangular ligament that extends from the medial malleolus to the talus, navicular, and calcaneus of the tarsus. It is divisible into superficial and deep parts. The superficial components from anterior to posterior are the *tibionavicular ligament, tibiocalcaneal ligament,* and *posterior tibiotalar ligament.* The deep component is the *anterior tibiotalar ligament* (see Figure 9.16a). The deltoid ligament strengthens the medial aspect of the ankle joint.
3. ***Lateral ligament.*** Not as strong as the deltoid ligament. It extends from the lateral malleolus to the talus and calcaneus and is divisible into three components: *anterior talofibular ligament, posterior talofibular ligament,* and *calcaneofibular ligament* (see Figure 9.16b). The lateral ligament strengthens the lateral aspect of the ankle joint.

Movements

The ankle joint permits dorsiflexion and plantar flexion (see Figure 9.10g).

CHECKPOINT

13. What are retinacula? What is their function?

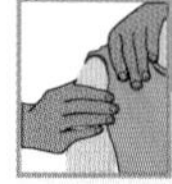

MANUAL THERAPY APPLICATION

Sprain and Strain

A **sprain** is the forcible wrenching or twisting of a joint that stretches or tears its *ligaments* but does not dislocate bones. It occurs when ligaments are stressed beyond normal capacity. Sprains also may damage surrounding blood vessels, muscles, tendons, or nerves. Severe sprains may be so painful that the joint cannot be moved. There is considerable swelling, which results from hemorrhage of ruptured blood vessels. The lateral ankle joint is most often sprained; the lower back is another common location of sprains.

A **strain** is a stretched or partially torn *muscle.* It often occurs when a muscle contracts suddenly and powerfully—for example, in leg muscles of sprinters when they accelerate quickly. Sprains and strains may occur separately or even together in more traumatic injuries such as whiplash.

Initially the injury should be treated with ***RICE:*** rest, ice, compression, and elevation. RICE therapy may be used on muscle strains, ligament sprains, joint inflammation, suspected fractures, and bruises. Many injuries result from overusing muscles and joints and these overuse injuries should not be ignored.

The four components of RICE therapy are:

- ***Rest*** the injured area to avoid further damage to the tissues. Stop the activity immediately. Avoid exercise or other activities that cause pain or swelling to the injured area. Rest is needed for repair, and exercising before an injury has healed may increase the probability of re-injury.
- ***Ice*** the injured area as soon as possible. Applying ice slows blood flow to the area, reduces swelling, and relieves pain. Ice works effectively when applied for 20 minutes, off for 40 minutes, back on for 20 minutes, etc. The ice may be shielded from direct contact with the skin by adding any form of cloth. Moderate to severe injuries should be iced intermittently for approximately 3 days.
- ***Compression*** by wrap or bandage helps to reduce swelling. Care must be taken to compress the injured area but not to block blood flow. If the compression wrap becomes uncomfortable, it should be loosened until blood flow is restored.
- ***Elevation*** of the injured area above the level of the heart, when possible, will reduce potential swelling.

When the injury is no longer painful, commonly 4 or 5 days after the trauma, the process of healing will be enhanced through the use of multiple techniques used by manual therapists that increase blood flow to the area.

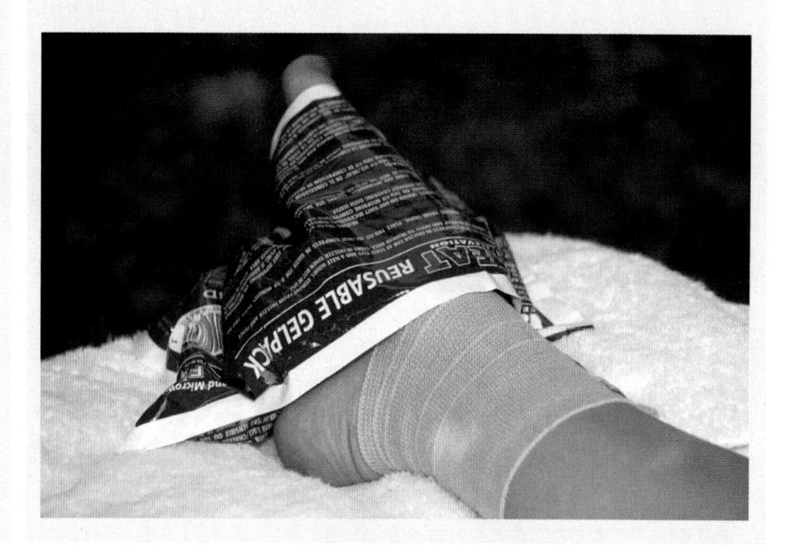

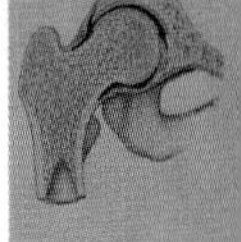

9.7 FACTORS AFFECTING CONTACT AND RANGE OF MOTION AT SYNOVIAL JOINTS

OBJECTIVE

- Describe six factors that influence the type of movement and range of motion possible at a synovial joint.

The articular surfaces of synovial joints contact one another and determine the type and range of motion that is possible. ***Range of motion (ROM)*** refers to the range, measured in degrees of a circle, through which the bones of a joint can be moved. The following factors contribute to keeping the articular surfaces in contact and affect range of motion:

1. ***Structure or shape of the articulating bones.*** The structure or shape of the articulating bones determines how closely they can fit together. The articular surfaces of some bones have a complementary relationship with one another. This spatial relationship is very obvious at the hip joint, where the head of the femur articulates with the acetabulum of the hip bone. An interlocking fit allows rotational movement.

2. ***Strength and tension (tautness) of the joint ligaments.*** The different components of a fibrous capsule are tense or taut only when the joint is in certain positions. Tense ligaments not only restrict the range of motion but also direct the movement of the articulating bones with respect to each other. In the knee joint, for example, the anterior cruciate ligament is taut and the posterior cruciate ligament is loose when the knee is straightened, and the reverse occurs when the knee is bent.

3. ***Arrangement and tension of the muscles.*** Muscle tension reinforces the restraint placed on a joint by its ligaments, and thus restricts movement. A good example of the effect of muscle tension on a joint is seen at the hip joint. When the thigh is fixed with the knee extended, the movement is restricted by the tension of the hamstring muscles on the posterior surface of the thigh. But if the knee is flexed, the tension on the hamstring muscles is lessened, and the thigh can be raised farther.

4. ***Contact of soft parts.*** The point at which one body surface contacts another may limit mobility. For example, if you bend your arm at the elbow, it can move no farther after the anterior surface of the forearm meets with and presses against the biceps brachii muscle of the arm. Joint movement may also be restricted by the presence of adipose tissue.

5. ***Hormones.*** Joint flexibility may also be affected by hormones. For example, relaxin, a hormone produced by the placenta and ovaries, increases the flexibility of the fibrocartilage of the pubic symphysis and loosens the ligaments between the sacrum, hip bone, and coccyx toward the end of pregnancy. These changes permit expansion of the pelvic outlet, which assists in delivery of the baby.

6. ***Disuse.*** Movement at a joint may be restricted if a joint has not been used for an extended period. For example, if an elbow joint is immobilized by a cast, range of motion at the joint may be limited for a time after the cast is removed. Disuse may also result in decreased amounts of synovial fluid, diminished flexibility of ligaments and tendons, and ***muscular atrophy,*** a reduction in size or wasting of a muscle.

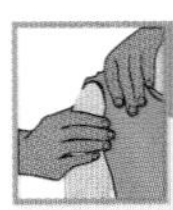

MANUAL THERAPY APPLICATION

Improving Joint Movements through Manual Therapies

When a patient is experiencing problems with the movements of daily living, exercises specific to a particular joint may be suggested by manual therapists. When joint movement is even more restricted, manual therapists may facilitate either passive or resistive joint movements. Passive joint movements occur during massage, while resistive joint movements occur during exercise. **Improving joint movements through manual therapies** results in:

1. Increased flexibility, range of motion (ROM), and overall dexterity.
2. Increased circulation and lymph flow through joints with ensuing increased nutrition to the muscles, ligaments, and tendons associated with the joint.
3. Decreased swelling and resultant decreased pain commonly associated with connective tissue disorders such as sprains, rheumatoid arthritis, and fibromyalgia.
4. Increased muscle strength (size and tone).

• CLINICAL CONNECTION | Rheumatism and Arthritis

Rheumatism (ROO-ma-tizm) is any painful disorder of the supporting structures of the body—bones, ligaments, tendons, or muscles—that is not caused by infection or injury. **Arthritis** is a form of rheumatism in which the joints are swollen, stiff, and painful. It afflicts about 45 million people in the United States.

Rheumatoid arthritis (RA) is an autoimmune disease in which the immune system of the body attacks its own tissues—in this case, its own cartilage and joint linings. RA is characterized by inflammation of the joint, which causes swelling, pain, and loss of function. Usually, this form of arthritis occurs bilaterally: If one wrist is affected, the other is also likely to be affected, although often not to the same degree. The primary symptom of RA is inflammation of the synovial membrane. If untreated, the membrane thickens, and synovial fluid accumulates. The resulting pressure causes pain and tenderness. The membrane then produces an abnormal granulation tissue, called *pannus,* that adheres to the surface of the articular cartilage and sometimes erodes the cartilage completely. When the cartilage is destroyed, fibrous tissue joins the exposed bone ends. The fibrous tissue ossifies and fuses the joint so that it becomes immovable—the ultimate crippling effect of rheumatoid arthritis. The growth of the granulation tissue causes the distortion of the fingers that characterizes hands of RA sufferers.

Massage to the joints of an RA patient is contraindicated as the condition can spread to other areas of the body.

Osteoarthritis (OA) (os′-tē-ō-ar-THRĪ-tis) is a degenerative joint disease in which joint cartilage is gradually lost. It results from a combination of aging, irritation of the joints, and wear and abrasion. Commonly known as "wear-and-tear" arthritis, osteoarthritis is the leading cause of disability in older persons. Osteoarthritis is a progressive disorder of synovial joints, particularly weight-bearing joints. Articular cartilage deteriorates and new bone forms in the subchondral areas and at the margins of the joint. The cartilage slowly degenerates, and as the bone ends become exposed, spurs (small bumps) of new osseous tissue are deposited on them. These spurs decrease the space of the joint cavity and restrict joint movement. Unlike rheumatoid arthritis, osteoarthritis affects mainly the articular cartilage, although the synovial membrane often becomes inflamed late in the disease. A major distinction between osteoarthritis and rheumatoid arthritis is that osteoarthritis first afflicts the larger joints (knees, hips), whereas rheumatoid arthritis first strikes smaller joints. RA is usually symmetrical whereas OA affects the joints used the most.

Massage to the muscles that pass over the affected joints is indicated. Therapists must take into consideration other conditions such as osteoporosis when applying increased pressure to bones and joints. •

CHECKPOINT

14. How do the strength and tension of ligaments determine range of motion?

9.8 ARTHROPLASTY

OBJECTIVE

- Explain the procedures involved in arthroplasty, and describe how a total hip replacement is performed.

Joints that have been severely damaged by a disease such as arthritis, or by an injury, may be replaced surgically with artificial joints in a procedure referred to as ***arthroplasty*** (AR-thrō-plas′-tē; *arthr-* = joint; *-plasty* = plastic repair of). Although most joints in the body can undergo arthroplasty, the ones most commonly replaced are the hips, knees, and shoulders. During the procedure, the ends of the damaged bones are removed and the artificial components are fixed in place. Artificial joint components are made of metal, ceramic, or plastic. The goals of arthroplasty are to relieve pain and increase range of motion.

Thousands of partial hip replacements, involving only the femur, are performed annually. In a total hip replacement, the procedure involves both the acetabulum and head of the femur (Figure 9.17). The damaged portions of the acetabulum and head of the femur are replaced by prefabricated prostheses (artificial devices). The acetabulum is shaped to accept the new socket, the head of the femur is removed, and the center of the bone is shaped to fit the femoral component. The acetabular component consists of polyethylene, whereas the femoral component is composed of cobalt–chromium, titanium alloys, or stainless steel. These materials are designed to withstand a high degree of stress. Once the appropriate acetabular and femoral components are selected, they are attached to the healthy portion of bone with acrylic cement, which forms an interlocking mechanical bond. Researchers are continually seeking to improve the strength of the cement and devise ways to stimulate bone growth around the implanted area.

CHECKPOINT

15. Which joints of the body most commonly undergo arthroplasty?

Figure 9.17 Total hip replacement.

In a total hip replacement, damaged portions of the acetabulum and head of the femur are replaced by prostheses.

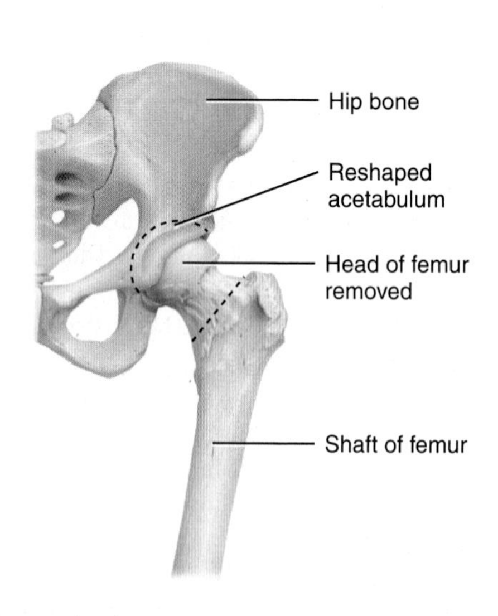

(a) Preparation for total hip replacement

Artificial acetabulum
Artificial femoral head
Artificial metal shaft

(b) Components of an artificial hip joint

Hip bone
Artificial acetabulum
Artificial femoral head
Artificial metal shaft
Shaft of femur

(c) Radiograph of an artificial hip joint

What is the purpose of arthroplasty?

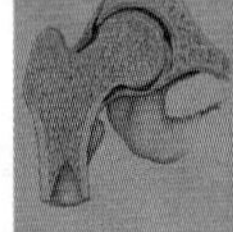

9.9 AGING AND JOINTS

OBJECTIVE

- Explain the effects of aging on joints.

Aging usually results in decreased production of synovial fluid in joints. In addition, the articular cartilage becomes thinner with age, and ligaments shorten and lose some of their flexibility. The effects of aging on joints, which vary considerably from one person to another, are influenced by genetic factors and by wear and tear. Although degenerative changes in joints may begin in those as young as 20 years of age, most changes do not occur until much later. By age 80, almost everyone develops some type of degeneration in the knees, elbows, hips, and shoulders. It is also common for elderly individuals to develop degenerative changes in the vertebral column, resulting in a hunched-over posture and pressure on nerve roots. One type of arthritis, called osteoarthritis, is at least partially age-related. Nearly everyone over age 70 has evidence of some osteoarthritic changes. Stretching and aerobic exercises that attempt to maintain full range of motion are very important in minimizing the effects of aging. They help to maintain the effective functioning of ligaments, tendons, muscles, synovial fluid, and articular cartilage.

• CLINICAL CONNECTION | Ankylosing Spondylitis

Ankylosing spondylitis (ang′-ki-LŌ-sing spon′-di-LĪ-tis; *ankyle* = stiff; *spondyl* = vertebra) is an inflammatory disease that affects joints between vertebrae (intervertebral) and between the sacrum and hip bone (sacroiliac joint). The cause is unknown. The disease is more common in males and has its onset between the ages of 20 and 40 years. It is characterized by pain and stiffness in the hips and lower back that progress upward along the backbone. Inflammation can lead to *ankylosis* (severe or complete loss of movement at a joint) and *kyphosis* (hunchback). Treatment consists of anti-inflammatory drugs, heat, massage, and supervised exercise. •

CHECKPOINT

16. Which joints show evidence of degeneration in nearly all individuals as aging progresses?

KEY MEDICAL TERMS ASSOCIATED WITH JOINTS

Arthralgia (ar-THRAL-jē-a; *arthr-* = joint; *-algia* = pain) Pain in a joint.

Bursectomy (bur-SEK-tō-mē; *-ectomy* = removal of) Removal of a bursa.

Chondritis (kon-DRĪ-tis; *chondr-* = cartilage) Inflammation of cartilage.

Subluxation (sub-luks-Ā-shun) A partial or incomplete dislocation.

Synovitis (sin′-ō-VĪ-tis) Inflammation of a synovial membrane in a joint.

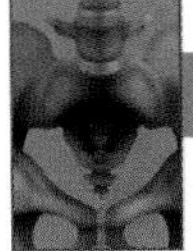

STUDY OUTLINE

Joint Classifications (Section 9.1)

1. A joint (articulation or arthrosis) is a point of contact between two bones, between bone and cartilage, or between bone and teeth.
2. Structural classification is based on the presence or absence of a synovial cavity and the type of connecting tissue. Structurally, joints are classified as fibrous, cartilaginous, or synovial.
3. Functional classification of joints is based on the degree of movement permitted. Joints may be synarthroses (immovable), amphiarthroses (slightly movable), or diarthroses (freely movable).

Fibrous Joints (Section 9.2)

1. The bones of fibrous joints are held together by dense irregular connective tissue.
2. These joints include slightly movable or immovable sutures (found between skull bones), slightly movable syndesmoses (such as the distal tibiofibular joint), immovable gomphoses (roots of teeth in the sockets in the mandible and maxilla), and slightly movable interosseous membranes (found between the radius and ulna and tibia and fibula).

Cartilaginous Joints (Section 9.3)

1. The bones of cartilaginous joints are held together by cartilage.
2. These joints include immovable synchondroses united by hyaline cartilage (epiphyseal plates between diaphyses and epiphyses) and slightly movable symphyses united by fibrocartilage (pubic symphysis).

Synovial Joints (Section 9.4)

1. Synovial joints contain a space between bones called the synovial (joint) cavity. All synovial joints are freely movable.
2. Other characteristics of synovial joints are the presence of articular cartilage and an articular (joint) capsule, made up of a fibrous capsule and a synovial membrane.
3. The synovial membrane secretes synovial fluid, which forms a thin, viscous film over the surfaces within the articular capsule.
4. Many synovial joints also contain accessory ligaments (extracapsular and intracapsular) and articular discs (menisci).
5. Synovial joints contain an extensive nerve and blood supply. The nerves convey information about pain, joint movements, and the degree of stretch at a joint. Blood vessels penetrate the articular capsule and ligaments.

6. Bursae are saclike structures, similar in structure to joint capsules, that alleviate friction in joints such as the shoulder and knee joints.
7. Tendon sheaths are tubelike bursae that wrap around tendons where there is considerable friction.
8. Types of synovial joints include planar, hinge, pivot, condyloid, saddle, and ball-and-socket.
9. In a planar joint the articulating surfaces are flat, and the bones glide back-and-forth and side-to-side (many are biaxial); examples are joints between carpals and tarsals.
10. In a hinge joint, the convex surface of one bone fits into the concave surface of another, and motion is angular around one axis (monaxial); examples are the elbow, knee, and ankle joints.
11. In a pivot joint, a round or pointed surface of one bone fits into a ring formed by another bone and a ligament, and movement is rotational (monaxial); examples are the atlanto-axial and radioulnar joints.
12. In a condyloid joint, an oval projection of one bone fits into an oval cavity of another, and motion is angular around two axes (biaxial); examples include the wrist joint and metacarpophalangeal joints of the second through fifth digits.
13. In a saddle joint, the articular surface of one bone is shaped like a saddle and the other bone fits into the saddle like a sitting rider; motion is angular around three axes (triaxial). An example is the carpometacarpal joint between the trapezium and the metacarpal of the thumb.
14. In a ball-and-socket joint, the ball-shaped surface of one bone fits into the cuplike depression of another; motion is angular around three axes (triaxial). Examples include the shoulder and hip joints.
15. Table 9.1 summarizes the structural and functional categories of joints.

Types of Movements at Synovial Joints (Section 9.5)

1. In a gliding movement, the nearly flat surfaces of bones move back and forth and from side to side. Gliding movements occur at planar joints.
2. In angular movements, a change in the angle between bones occurs. Examples are flexion–extension, lateral flexion, hyperextension, and abduction–adduction. Circumduction refers to flexion, abduction, extension, and adduction in succession. Angular movements occur at hinge, condyloid, saddle, and ball-and-socket joints.
3. In rotation, a bone moves around its own longitudinal axis. Rotation can occur at pivot and ball-and-socket joints.
4. Special movements occur at specific synovial joints. Examples are elevation–depression, protraction–retraction, inversion–eversion, dorsiflexion–plantar flexion, supination–pronation, and opposition.
5. Table 9.2 summarizes the various types of movements at synovial joints.

Selected Joints of the Body (Section 9.6)

1. A summary of selected joints of the body, including articular components, structural and functional classifications, and movements, is presented in Tables 9.3 and 9.4.
2. The temporomandibular joint (TMJ) is between the condyle of the mandible and mandibular fossa and articular tubercle of the temporal bone (Exhibit 9.1).
3. The shoulder (humeroscapular or glenohumeral) joint is between the head of the humerus and glenoid cavity of the scapula (Exhibit 9.2).
4. The elbow joint is between the trochlea of the humerus, the trochlear notch of the ulna, and the head of the radius (Exhibit 9.3).
5. The hip (coxal) joint is between the head of the femur and acetabulum of the hip bone (Exhibit 9.4).
6. The knee (tibiofemoral) joint is between the patella and patellar surface of the femur; the lateral condyle of the femur, the lateral meniscus, and the lateral condyle of the tibia; and the medial condyle of the femur, the medial meniscus, and the medial condyle of the tibia (Exhibit 9.5).
7. The ankle (talocrural) joint is formed by the distal end of the tibia and its medial malleolus with the talus and the lateral malleolus of the fibula (Exhibit 9.6).

Factors Affecting Contact and Range of Motion at Synovial Joints (Section 9.7)

1. The ways that articular surfaces of synovial joints contact one another determines the type of movement possible.
2. Factors that contribute to keeping the surfaces in contact and affect range of motion are structure or shape of the articulating bones, strength and tension of the ligaments, arrangement and tension of the muscles, apposition of soft parts, hormones, and disuse.

Arthroplasty (Section 9.8)

1. Arthroplasty refers to the surgical replacement of joints.
2. The most commonly replaced joints are the hips, knees, and shoulders.

Aging and Joints (Section 9.9)

1. With aging, a decrease in synovial fluid, thinning of articular cartilage, and decreased flexibility of ligaments occur.
2. Most individuals experience some degeneration in the knee, elbow, hip, and shoulder joints due to the aging process.

SELF-QUIZ QUESTIONS

Choose the one best answer to the following questions.

1. Choose the pair of terms that is most closely associated or matched.
   a. wrist joint–pronation **b.** intertarsal joints–inversion
   c. elbow joint–hyperextension **d.** ankle joint–eversion
   e. interphalangeal joint–circumduction
2. Which of the following is a ball-and-socket joint?
   a. temporomandibular joint **b.** knee joint
   c. shoulder joint **d.** elbow joint
   e. both b and c are correct

3. Which of the following structures is not associated with the knee joint?
   a. glenoid labrum
   b. patellar ligament
   c. infrapatellar bursa
   d. cruciate ligaments
   e. tibial collateral ligament
4. The lambdoid suture is an example of a(n)
   a. synarthrosis
   b. amphiarthrosis
   c. diarthrosis
   d. fibrous joint
   e. both b and d are correct.
5. Pads of fibrocartilage that extend into the space between articulating bones in the knee are called
   a. ligaments
   b. bursae
   c. articular capsules
   d. menisci
   e. gomphoses.
6. The joint at which the atlas rotates around the axis is an example of a
   a. hinge joint
   b. ball-and-socket joint
   c. gliding joint
   d. saddle joint
   e. pivot joint.
7. The inner layer of the articular capsule is the
   a. synovial membrane
   b. fibrous membrane
   c. bursa
   d. meniscus
   e. articular cartilage.
8. A broad flat disc of fibrocartilage joins the bones at the
   a. knee joint
   b. shoulder joint
   c. pubic symphysis
   d. coronal suture
   e. atlanto-occipital joint.

Complete the following:

9. The epiphyseal plate is an example of the structural joint classification known as a ________ because ________ joins the epiphysis and diaphysis of the growing bone.
10. When the fibers of the fibrous membrane of an articular capsule are arranged in parallel bundles, the structure is called a(n) ________.
11. The temporomandibular joint is functionally classified as a(n) ________.
12. The ends of bones at synovial joints are covered with a protective layer of ________.
13. Saclike structures that reduce friction between body parts at joints are called ________.

Are the following statements true or false?

14. Synovial fluid becomes more viscous as joint activity increases.
15. Tipping the head backward at the atlanto-occipital joint is an example of hyperextension.
16. Raising the arm to point straight ahead is an example of abduction.
17. The nerves that supply a joint are different from those that supply the skeletal muscles that move that joint.
18. A shoulder dislocation is an injury to the acromioclavicular joint.

Matching

19. Match the following:

____ a. distal tibiofibular joint	A. synovial, saddle
____ b. pubic symphysis	B. gomphosis
____ c. coronal suture	C. syndesmosis
____ d. tooth in alveolar socket	D. synovial, pivot
____ e. atlanto-axial joint	E. cartilaginous (fibrocartilage)
____ f. intercarpal joints	F. cartilaginous (hyaline cartilage)
____ g. elbow joint	G. fibrous joint, amphiarthrosis
____ h. carpometacarpal joint of thumb	H. synovial, planar
____ i. epiphyseal plate	I. synovial, hinge

20. Match the following:

____ a. movement of a body part anteriorly, horizontally to the ground	A. pronation
____ b. horizontal movement of an anteriorly projected body part back into the anatomical position	B. plantar flexion
____ c. movement of the soles medially at the intertarsal joints	C. eversion
____ d. movement of the soles laterally	D. retraction
____ e. action that occurs when you stand on your toes	E. inversion
____ f. position of foot when the heel is on the floor and rest of the foot is raised	F. protraction
____ g. movement of the forearm to turn the palm anteriorly	G. dorsiflexion
____ h. movement of the forearm to turn the palm posteriorly	H. supination

CRITICAL THINKING QUESTIONS

1. Katie loves pretending that she's a human cannonball. As she jumps off the diving board, she assumes the proper position before she pounds into the water: head and thighs tucked against her chest; back rounded; arms pressed against her sides while her forearms, crossed in front of her shins, hold her legs tightly folded against her chest. Use the proper anatomical terms to describe the position of Katie's back, head, and free limbs.
2. During football practice, Jeremiah was tackled and twisted his lower leg. There was a sharp pain, followed immediately by swelling of the knee joint. The pain and swelling worsened throughout the remainder of the afternoon until Jeremiah could barely walk. The coach told Jeremiah to see a doctor who might want to "drain the water off his knee." What was the coach referring to and what specifically do you think happened to Jeremiah's knee joint to cause these symptoms?
3. After lunch, during a particularly long and dull class video, Antonio became sleepy and yawned. To his dismay, he was then unable to close his mouth. Explain what happened and what should be done to correct this problem.

? ANSWERS TO FIGURE QUESTIONS

9.1 Functionally, sutures are classified as synarthroses because they are immovable; and amphiarthroses because they are slightly movable.
9.2 The structural difference between a synchondrosis and a symphysis is the type of cartilage that holds the joint together: hyaline cartilage in a synchondrosis and fibrocartilage in a symphysis.
9.3 Functionally, synovial joints are diarthroses, freely movable joints.
9.4 Many planar and condyloid joints are biaxial joints.
9.5 Gliding movements occur at intercarpal joints and at intertarsal joints.
9.6 Two examples of flexion that do not occur along the sagittal plane are flexion of the thumb and lateral flexion of the trunk.
9.7 Yes. When you adduct your arm or leg, you bring it closer to the midline of the body, thus "adding" it to the trunk.
9.8 Circumduction involves flexion, abduction, extension, and adduction in continuous sequence.
9.9 The anterior surface of a bone or limb rotates toward the midline in medial rotation, and away from the midline in lateral rotation.
9.10 Moving the thumb across the palm to touch any of the four fingers is called opposition.
9.11 The lateral ligament prevents displacement of the mandible.
9.12 The shoulder joint is the most freely movable joint in the body because of the looseness of its articular capsule and the shallowness of the glenoid cavity in relation to the size of the head of the humerus.
9.13 A hinge joint permits flexion and extension.
9.14 Tension in three ligaments—iliofemoral, pubofemoral, and ischiofemoral—limits the degree of extension at the hip joint.
9.15 Contraction of the quadriceps femoris muscle causes extension at the knee joint.
9.16 Anatomically, the ankle is the region between the distal aspect of the leg and the proximal aspect of the foot.
9.17 The purpose of arthroplasty is to relieve joint pain and permit greater range of motion.

Muscular Tissue

10

When you think about muscle, you may instantly think of movements such as walking or throwing a ball. *Skeletal muscles* pull on the bones to produce these movements. Also, approximately 85% of your body heat comes from skeletal muscle contraction. However, *smooth muscles* are responsible for the movement of substances within your body, such as the food you eat, and *cardiac muscle* forms the bulk of the heart. Not only does skeletal muscle keep you moving and warm, it keeps you stable by contracting certain muscles to maintain posture in the body. Muscle imbalance in the body can lead to many musculoskeletal injuries, especially when repetitive movements are involved. Such musculoskeletal injuries often cause patients to seek the services of manual therapists. You will examine the three muscle types and the mechanism by which skeletal muscle contracts in this chapter.

CONTENTS AT A GLANCE

10.1 OVERVIEW OF MUSCULAR TISSUE

OBJECTIVE

- Describe the types, functions, and properties of muscular tissue.

Muscle spasms play a very important role in the practices of manual therapists. The assessment and successful treatment of muscle spasms require the structural and functional knowledge of muscles at the cellular level. Muscle spasms often occur after an injury, but many conditions described in this chapter may cause the spasms. If not resolved, the spasms may become chronic. Patients seldom remember the cause, if due to any injury; they only know that an area of the body hurts 24/7. Spasms commonly present as a tightness that can be palpated by the therapist.

As you will learn in this chapter, individual muscle cells typically take turns contracting, even when the load on a muscle is very heavy. A muscle spasm involves muscle cells that are always "on," a condition that is foreign to the normal working of muscle cells. Overworked muscles send messages to the brain where the stimulus is perceived as pain. The pain may be at any level, from mild to debilitating. The level of pain may, but not always, dictate the type of professional that the patient seeks for assistance. Medical (therapeutic) massage therapists often resolve pain by techniques explained in this chapter. Physical therapists may resolve pain with appropriate massage, heat, cold, electrical stimulation, or ultrasound. Chiropractors may resolve pain by adjusting misaligned bony structures as well as by any of the techniques previously mentioned. Acupuncturists are often quite successful in resolving pain. Physicians may treat severe pain with trigger point injections. Appropriate exercise will likely be advised by all of these practitioners. Regardless of the technique, pain will be resolved only when the affected muscle cells cease to contract continuously.

You have just studied the skeletal system and joints. Movements such as throwing a ball, biking, and walking require an interaction among bones, joints, and muscles which together form an integrated system called the ***musculoskeletal system.*** To understand how muscles produce different movements, you will learn in Chapters 11–14 where specific muscles attach on individual bones and the types of joints acted on by the contracting muscles. The primary focus of this chapter is skeletal muscle tissue.

The scientific study of muscles is known as ***myology*** (mī-OL-ō-jē; *my-* = muscle; *-logy* = study of). The branch of medical science concerned with the prevention or correction of disorders of the musculoskeletal system is called ***orthopedics*** (or′-thō-PĒ-diks; *ortho-* = correct; *-pedi-* = child). (Although this is a correct spelling of the term, the American Academy of Orthopaedic Surgeons has officially adopted the spelling of *orthopaedics.*)

Types of Muscular Tissue

Depending on the percentage of body fat, gender, and exercise regimen, muscular tissue constitutes about 40% to 50% of the total body weight and is composed of highly specialized cells. Recall from Chapter 4 that the three types of muscular tissue are skeletal, cardiac, and smooth. As the name suggests, most ***skeletal muscle tissue*** is attached to bones and moves parts of the skeleton. It is ***striated;*** that is, *striations,* or alternating light and dark protein bands, are visible under a microscope. Because skeletal muscle can typically be made to contract and relax by conscious control, it is ***voluntary.*** Due to the presence of a small number of cells that can undergo cell division, skeletal muscle has a limited capacity for regeneration.

Cardiac muscle tissue, found only in the heart, forms the bulk of the heart wall. The heart pumps blood through blood vessels to all parts of the body. Like skeletal muscle tissue, cardiac muscle tissue is ***striated.*** Unlike skeletal muscle tissue, however, it is ***involuntary:*** Its contractions are not under conscious control. Cardiac muscle can regenerate under certain conditions. This will be explained in Chapter 22.

Smooth muscle tissue is located in the walls of hollow internal structures, such as blood vessels, airways, the stomach, and the intestines. It participates in internal processes such as digestion and the regulation of blood pressure. Smooth muscle is ***nonstriated*** (lacks striations) and ***involuntary*** (not under conscious control). Although smooth muscle tissue has considerable capacity to regenerate when compared with other muscle tissues, this capacity is limited when compared to other types of tissues, for example, epithelium.

Functions of Muscular Tissue

Through sustained contraction (rare) or alternating contraction and relaxation, muscular tissue has five key functions: producing body movements, stabilizing body positions, regulating organ volume, moving substances within the body, and generating heat.

1. **Producing body movements.** Body movements such as walking, running, writing, or nodding the head rely on the integrated functioning of skeletal muscles, bones, and joints.

2. **Stabilizing body positions.** Skeletal muscle contractions stabilize joints and help maintain body positions, such as standing or sitting. Postural muscles contract continuously when a person is awake; for example, sustained contractions of your neck muscles hold your head upright.

3. **Regulating organ volume.** Sustained contractions of ring-like bands of smooth muscles called *sphincters* prevent outflow of the contents of a hollow organ. Temporary storage of food in the stomach or urine in the urinary bladder is possible because smooth muscle sphincters close off the outlets of these organs.

4. **Moving substances within the body.** Cardiac muscle contractions pump blood through the body's blood vessels.

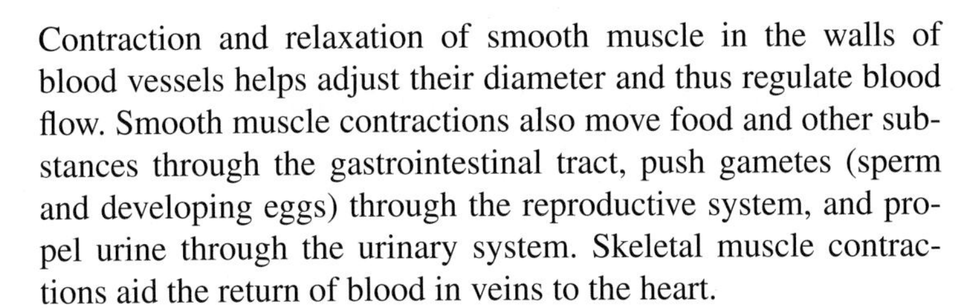

Contraction and relaxation of smooth muscle in the walls of blood vessels helps adjust their diameter and thus regulate blood flow. Smooth muscle contractions also move food and other substances through the gastrointestinal tract, push gametes (sperm and developing eggs) through the reproductive system, and propel urine through the urinary system. Skeletal muscle contractions aid the return of blood in veins to the heart.

5. **Producing heat.** As muscular tissue contracts, it produces heat. Much of the heat released by muscles is used to maintain normal body temperature. Involuntary contractions of skeletal muscle, known as shivering, can help warm the body by greatly increasing the rate of heat production.

Properties of Muscular Tissue

Muscular tissue has four special properties that enable it to function and contribute to homeostasis:

1. ***Electrical excitability*** (ek-sīt′-a-BIL-i-tē), a property of both muscle and nerve cells, is the ability to respond to certain stimuli by producing electrical signals called ***action potentials (impulses).*** Action potentials can travel along a cell's plasma membrane due to the presence of specific ion channels. For muscle cells, two main types of stimuli trigger action potentials. One is autorhythmic *electrical signals* arising in the muscular tissue itself, as in the heart's pacemaker. The other is *chemical stimuli,* such as neurotransmitters released by neurons, hormones distributed by the blood, or even local changes in pH.

2. ***Contractility*** (kon′-trak-TIL-i-tē) is the ability of muscular tissue to contract forcefully when stimulated by an action potential. When a skeletal muscle contracts, it generates tension (force of contraction) while pulling on its attachment points. In some muscle contractions, the muscle develops tension (force of contraction) but does not shorten. An example is holding this book in an outstretched hand. In other muscle contractions, the tension generated is great enough to overcome the load (resistance) of the object to be moved so the muscle shortens and movement occurs. An example is lifting this book off a table.

3. ***Extensibility*** (ek-sten′-si-BIL-i-tē) is the ability of muscular tissue to stretch without being damaged. Extensibility allows a muscle to contract forcefully even if it is already stretched. Normally, smooth muscle is subject to the greatest amount of stretching. For example, each time your stomach fills with food, the muscle in its wall is stretched. Cardiac muscle also is stretched each time the heart fills with blood.

4. ***Elasticity*** (e-las-TIS-i-tē) is the ability of muscular tissue to return to its original length and shape after contraction or extension.

CHECKPOINT

1. What features distinguish the three types of muscular tissue?
2. What are the general functions of muscular tissue?
3. What are the four properties of muscular tissue?

10.2 SKELETAL MUSCLE TISSUE

OBJECTIVES

- Explain the relation of connective tissue components, blood vessels, and nerves to skeletal muscles.
- Describe the histology of a skeletal muscle fiber.

Each skeletal muscle is a separate organ composed of hundreds to thousands of cells, which are called ***muscle fibers*** because of their elongated shapes. Use of the term "fiber" to describe a muscle cell should not be confused with connective tissue fibers, which are nonliving products of fibroblasts, or with nerve fibers, which are the axons and dendrites of neurons. A host of connective tissues surround muscle fibers and whole muscles, and blood vessels and nerves penetrate the muscles.

Connective Tissue Components

Connective tissue surrounds and protects muscular tissue. The ***subcutaneous layer,*** or ***hypodermis,*** which separates muscle from skin (see Figure 5.1), is composed of areolar connective tissue and adipose tissue. It provides a pathway for nerves, blood vessels, and lymphatic vessels to enter and exit muscles. The adipose tissue of the subcutaneous layer stores most of the body's triglycerides, serves as an insulating layer that reduces heat loss, and protects muscles from physical trauma. **Fascia** (FASH-ē-a = bandage) is a dense sheet or broad band of irregular connective tissue that lines the body wall and limbs and supports and surrounds muscles and other organs of the body. As you will see, fascia holds muscles with similar functions together (see Figure 11.21). Fascia allows free movement of muscles, carries nerves, blood vessels, and lymphatic vessels, and fills spaces between muscles.

Three layers of connective tissue extend from the fascia to protect and strengthen skeletal muscle (see Figure 10.1). The outermost layer, encircling the entire muscle, is the ***epimysium*** (ep-i-MĪZ-ē-um; *epi-* = upon). ***Perimysium*** (per′-i-MĪZ-ē-um; *peri-* = around) surrounds bundles of 10 to 100 or more muscle fibers called ***fascicles*** (FAS-i-kuls = little bundles). Finally, ***endomysium*** (en′-dō-MĪZ-ē-um; *endo-* = within) wraps each individual muscle fiber. Epimysium, perimysium, and endomysium extend beyond the muscle belly as a ***tendon,*** a cord of dense regular connective tissue composed of parallel bundles of collagen fibers, or an ***aponeurosis*** (ap′-ō-noo-RŌ-sis), a flat sheet of collagen fibers. Their function is to attach a muscle to a bone. Connective tissue fibers of tendons and aponeuroses blend with the connective tissue fibers of the periosteum (see Figure 10.1).

Figure 10.1 Organization of skeletal muscle and its connective tissue coverings.

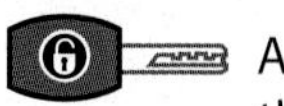

A skeletal muscle consists of individual muscle fibers (cells) bundled into fasciculi and surrounded by three connective tissue layers.

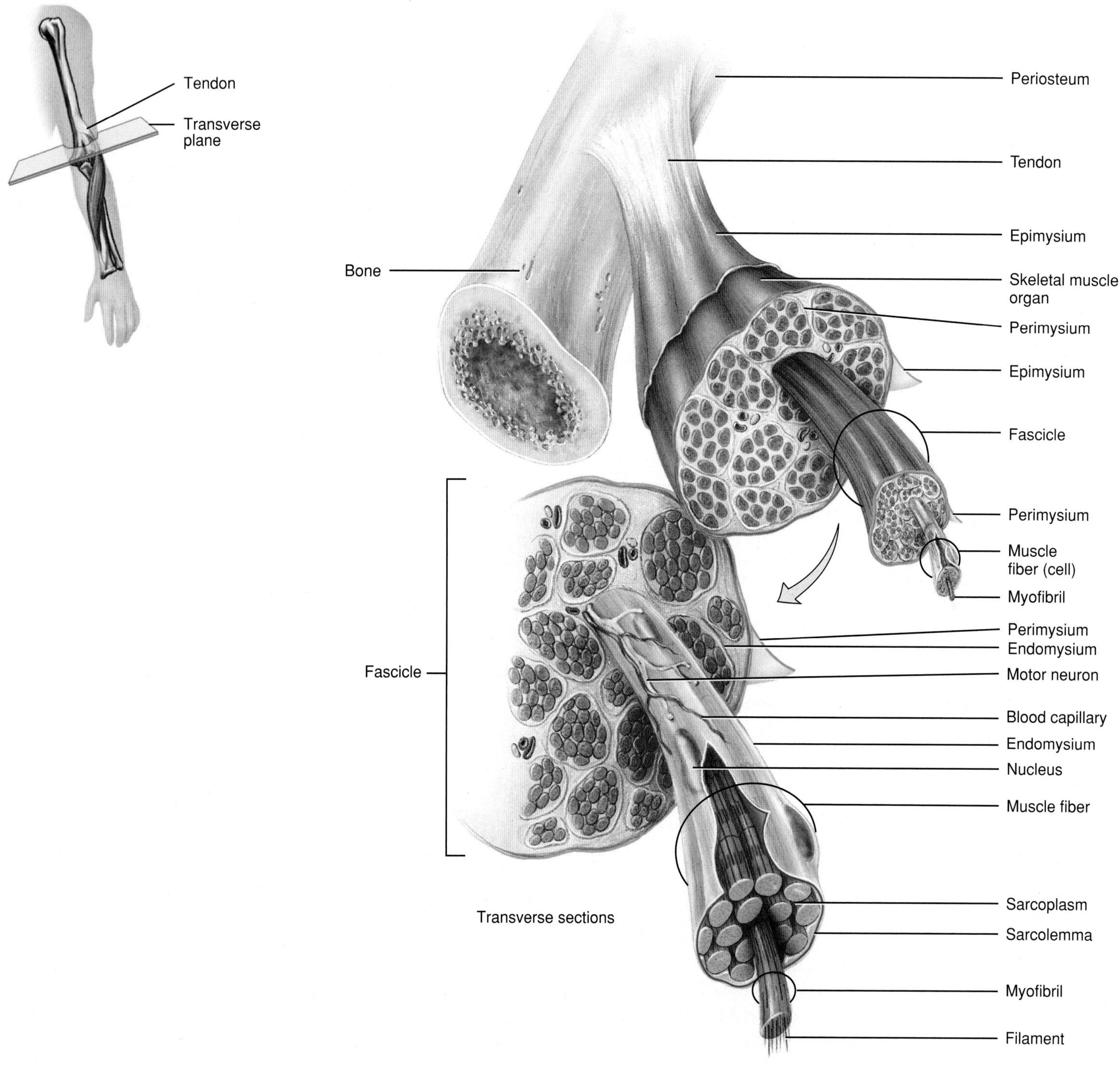

Components of a skeletal muscle

Starting with the connective tissue that surrounds an individual muscle fiber (cell) and working toward the outside, list the connective tissue layers in order.

• CLINICAL CONNECTION Fibromyalgia

Fibromyalgia (*algia* = painful condition) is a painful, nonarticular rheumatic disorder that usually appears between the ages of 25 and 50. An estimated 3 million people in the United States suffer from fibromyalgia, which is 15 times more common in women than in men. It should be noted that the term fibromyalgia is considered by some clinicians to represent a more generalized set of symptoms since there are nearly 40 different medical names for this condition. The disorder affects the fibrous connective tissue components of muscles, tendons, and ligaments. A striking sign is pain that results from gentle pressure at specific "tender points." Even without pressure, there is pain, tenderness, and stiffness of muscles, tendons, and surrounding soft tissues. Persons with severe fibromyalgia may be disabled and unable to be employed. Besides muscle pain, those with fibromyalgia report severe fatigue, poor sleep, headaches, depression, and inability to carry out their daily activities. Often, a gentle aerobic fitness program is beneficial in reducing the pain. Medications that affect neurogenic conditions may also be helpful in treating fibromyalgia. •

Nerve and Blood Supply

Skeletal muscles are well supplied with nerves and blood vessels (Figure 10.1), both of which are directly related to contraction, the chief characteristic of muscle. Muscle contraction also requires plentiful ATP and therefore large amounts of nutrients and oxygen for ATP synthesis. Moreover, the waste products of these ATP-producing reactions must be eliminated. Thus, prolonged muscle action depends on a rich blood supply to deliver nutrients and oxygen and to remove wastes.

Generally, an artery and one or two veins accompany each nerve that penetrates a skeletal muscle organ. Within the endomysium, microscopic blood vessels called capillaries are distributed so that each muscle fiber is in close proximity with one or more capillaries. Each skeletal muscle fiber also makes "contact" with the terminal portion of a neuron. The section on neuromuscular junctions will illustrate that the two fibers don't quite touch.

Histology

Microscopic examination of a skeletal muscle reveals that it consists of thousands of elongated, cylindrical ***muscle fibers*** arranged parallel to one another (Figure 10.2a). Each muscle fiber is covered by a plasma membrane called the ***sarcolemma*** (sar′-kō-LEM-ma; *sarco-* = flesh; *-lemma* = sheath). ***Transverse (T) tubules*** tunnel in from the surface toward the center of each muscle fiber. Multiple nuclei lie at the periphery of the fiber, under the sarcolemma. The muscle fiber's cytoplasm, called ***sarcoplasm*** (SAR-kō-plazm), contains many mitochondria that produce large amounts of ATP. Extending throughout the sarcoplasm is ***sarcoplasmic reticulum*** (sar′-kō-PLAZ-mik re-TIK-ū-lum) or ***SR,*** a network of fluid-filled membrane-enclosed tubules (similar to smooth endoplasmic reticulum) that stores calcium ions required for muscle contraction. Also in the sarcoplasm are numerous molecules of ***myoglobin*** (mī′-ō-GLŌ-bin), a reddish pigment similar to hemoglobin in blood. In addition to the characteristic color it lends to skeletal muscle, myoglobin stores oxygen until it is needed by mitochondria to generate ATP.

• CLINICAL CONNECTION Muscular Dystrophy

The term **muscular dystrophy** refers to a group of inherited muscle-destroying diseases that cause progressive degeneration of skeletal muscle fibers. The most common form of muscular dystrophy is *DMD* or *Duchenne muscular dystrophy* (doo-SHĀN). Because the mutated gene is on the X chromosome, which males have only one of, DMD strikes boys almost exclusively. Worldwide, about 1 in every 3500 male babies are born with DMD each year. The disorder usually becomes apparent between the ages of 2 and 5, when parents notice the child falls often and has difficulty running, jumping, and hopping. By age 12 most boys with DMD are unable to walk. Respiratory or cardiac failure usually causes death between the ages of 20 and 30.

In DMD, the gene that codes for the protein dystrophin is mutated and little or no dystrophin is present (dystrophin provides structural reinforcement for the sarcolemma of skeletal muscle fibers). Without the reinforcing effect of dystrophin, the sarcolemma easily tears during muscle contraction. Because their plasma membranes are damaged, muscle fibers slowly rupture and die. •

Extending along the entire length of the muscle fiber are cylindrical structures called ***myofibrils*** (mī′-ō-FĪ-brils). Each myofibril, in turn, consists of two types of protein filaments called ***thin filaments*** and ***thick filaments*** (Figure 10.2b), which do not extend the entire length of a muscle fiber. Filaments overlap in specific patterns and form compartments called ***sarcomeres*** (SAR-kō-mērs; *-mere* = part), the basic functional units of striated muscle fibers (Figure 10.2b,c). Sarcomeres are separated from one another by zigzagging zones of dense protein material called ***Z discs.*** Within each sarcomere a darker area, called the ***A band,*** extends the entire length of the thick filaments. At the center of each A band is a narrow ***H zone,*** which contains only the thick filaments. At both ends of the A band, thick and thin filaments overlap. A lighter-colored area to either side of the A band, called the ***I band,*** contains the rest of the thin filaments but no thick filaments. Each I band extends into two sarcomeres, divided in half by a Z disc (Figure 10.2c). The alternating darker A bands and lighter I bands give the muscle fiber its striated appearance. Supporting proteins that hold the thick filament together at the center of the H zone form the ***M line,*** so-named because it is in the middle of the sarcomere.

Thick filaments are composed of protein molecules called ***myosin*** (MĪ-ō-sin). This molecule is shaped like two golf clubs twisted together (see Figure 10.3a). The *myosin tails* (golf club handles) are arranged parallel to each other, forming the shaft of the thick filament. The heads of the golf clubs pro-

Figure 10.2 Organization of skeletal muscle from gross to molecular levels.

The structural organization of a skeletal muscle from macroscopic to microscopic is as follows: skeletal muscle, fascicle (bundle of muscle fibers), muscle fiber, myofibril, and thin and thick filaments.

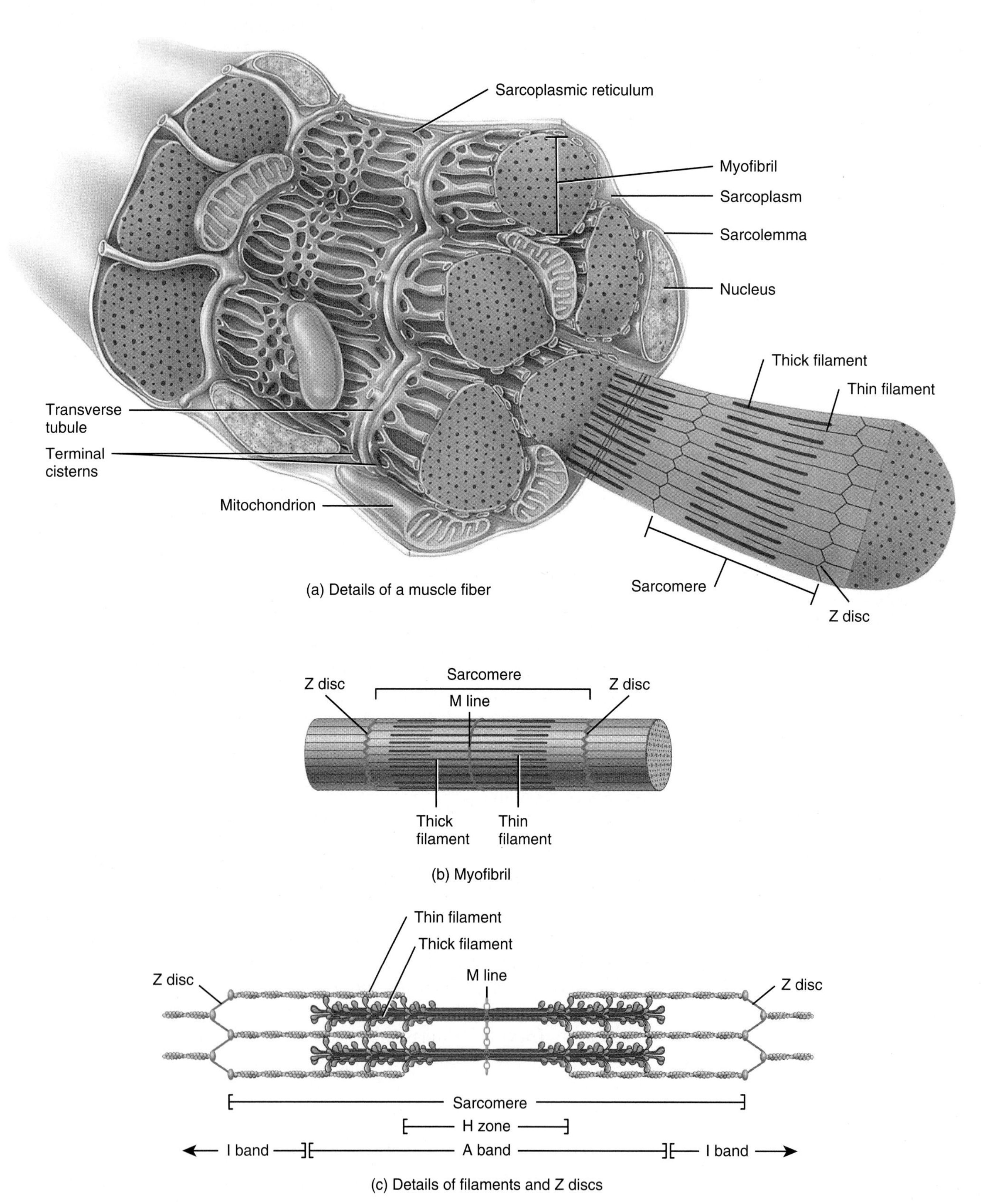

Which filaments are part of the A band and I band?

Figure 10.3 Detailed structure of filaments. (a) About 300 myosin molecules compose a thick filament. The myosin tails all point toward the center of the sarcomere. (b) Thin filaments contain actin, troponin, and tropomyosin.

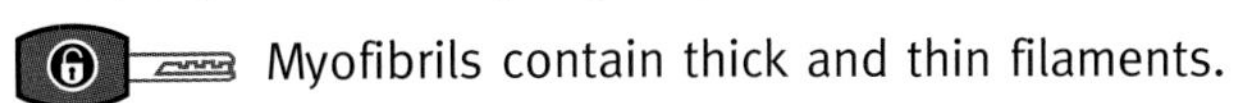

Myofibrils contain thick and thin filaments.

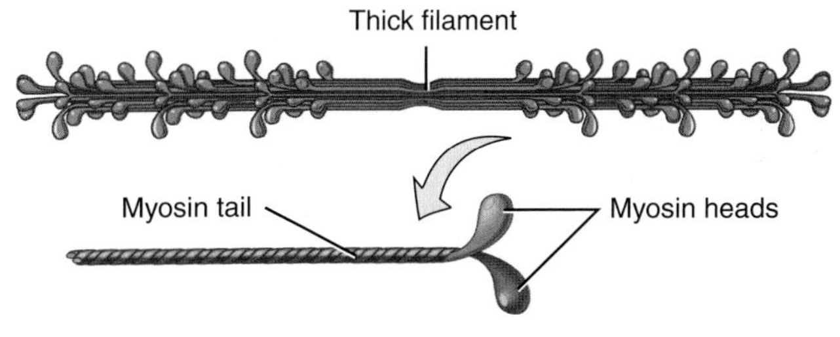

(a) One thick filament and a myosin molecule

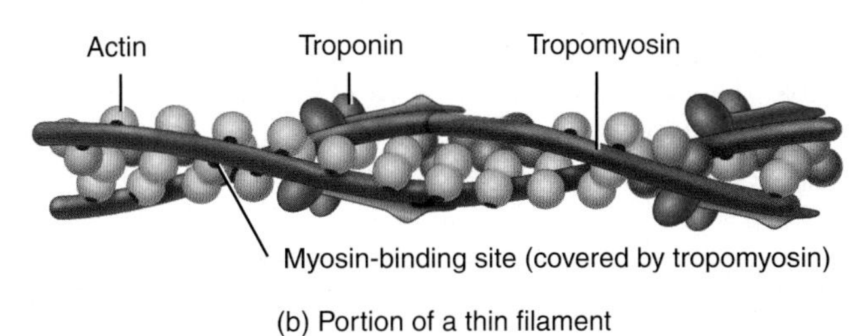

(b) Portion of a thin filament

What proteins are present in the A band and in the I band?

ject outward from the surface of the shaft. These projecting heads are referred to as *myosin heads.*

Thin filaments are anchored to the Z discs. Their main component is the protein molecule called ***actin*** (AK-tin). Individual actin molecules join to form an actin filament that is twisted into a helix (Figure 10.3b). Each actin molecule contains a *myosin-binding site,* where a myosin head can attach. The thin filaments contain two other protein molecules, ***tropomyosin*** and ***troponin.*** In a relaxed muscle, myosin is blocked from binding to actin because strands of tropomyosin cover the myosin-binding sites on the actin molecules. The tropomyosin strands, in turn, are held in place by troponin molecules. You will soon learn that when calcium ions (Ca^{2+}) bind to troponin, it undergoes a change in shape; this change moves tropomyosin away from myosin-binding sites on actin and muscle contraction subsequently begins as myosin binds to actin.

MANUAL THERAPY APPLICATION

Muscular Atrophy and Hypertrophy

Muscular atrophy (A-trō-fē; *a-* = without, *-trophy* = nourishment) is a wasting away of muscles. Individual muscle fibers decrease in size because of progressive loss of myofibrils. The atrophy that occurs if muscles are not used is termed *disuse atrophy.* Bedridden individuals and people with casts experience disuse atrophy because the number of nerve impulses to inactive muscle is greatly reduced. If the nerve supply to a muscle is disrupted or cut, the muscle undergoes *denervation atrophy.* In about 6 months to 2 years, the muscle will be one-quarter of its original size, and many of the muscle fibers will be replaced by fibrous connective tissue. The transition to connective tissue, when complete, cannot be reversed.

Muscular hypertrophy (hī-PER-trō-fē; *hyper-* = above or excessive) is an increase in muscle fiber diameter owing to the production of more myofibrils, mitochondria, sarcoplasmic reticulum, and other cytoplasmic structures. It results from very forceful, repetitive muscular activity, such as strength training. Because hypertrophied muscles contain more myofibrils, they are capable of contractions that are more forceful.

CHECKPOINT

4. What type of connective tissue coverings are associated with skeletal muscle?
5. Why is a rich blood supply important for muscle contraction?
6. What is a sarcomere? What does a sarcomere contain?

10.3 CONTRACTION AND RELAXATION OF SKELETAL MUSCLE

OBJECTIVE

- Explain how skeletal muscle fibers contract and relax.

Sliding Filament Mechanism

During muscle contraction, myosin heads of the thick filaments pull on the thin filaments, causing the thin filaments to slide toward the center of a sarcomere (see Figure 10.4a,b). As the thin filaments slide, the I bands and H zones become narrower (see Figure 10.4b) and eventually disappear altogether when the muscle is maximally contracted (see Figure 10.4c).

The thin filaments slide past the thick filaments because the myosin heads move like the oars of a boat, pulling on the actin molecules of the thin filaments. Although the sarcomere shortens because of the increased overlap of thin and thick filaments, the lengths of the thin and thick filaments do not change. The sliding of filaments and shortening of sarcomeres in turn cause the shortening of the muscle fibers. This process, the ***sliding filament mechanism*** of muscle contraction, occurs only when the level of calcium ions (Ca^{2+}) is high enough and ATP is available, for reasons you will see shortly.

Neuromuscular Junction

Before a skeletal muscle fiber can contract, it must be stimulated by an electrical signal called a ***muscle action potential*** delivered by its neuron called a ***motor neuron.*** A single somatic motor neuron along with all the muscle fibers it stimulates is called a ***motor unit.*** Stimulation of one somatic motor neuron causes all the muscle fibers in that motor unit to contract at the same time. Muscles that control small, precise movements, such as the mus-

Figure 10.4 Sliding filament mechanism of muscle contraction, as it occurs in two adjacent sarcomeres.

During muscle contraction, thin filaments move inward toward the M line of each sarcomere.

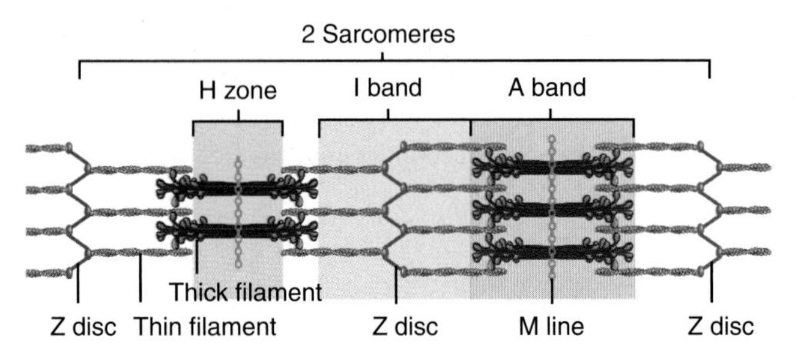

(a) Relaxed muscle

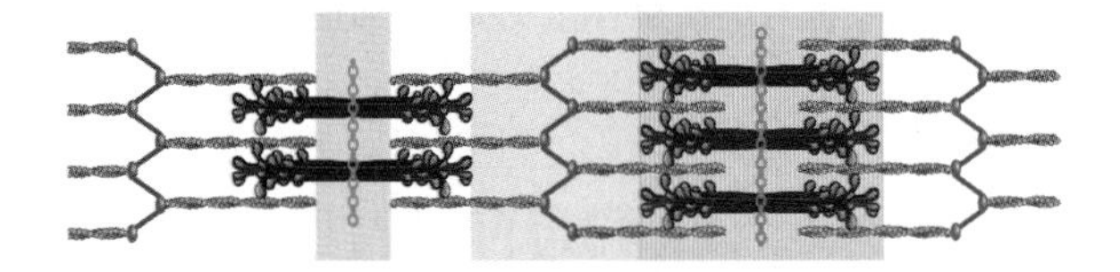

(b) Partially contracted muscle

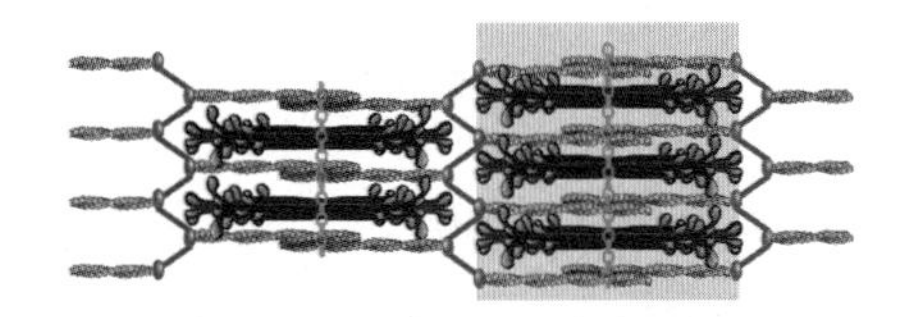

(c) Maximally contracted muscle

What happens to the I bands as muscle contracts? Do the lengths of the thick and thin filaments change during contraction?

cles that move the eyes, have 10 to 20 muscle fibers per motor unit. Muscles of the body that are responsible for large, powerful movements, such as the biceps brachii in the arm and the gastrocnemius in the leg, may have as many as 2000 muscle fibers in some motor units.

As the *axon* (long process) of a motor neuron enters a skeletal muscle, it divides into branches called *axon terminals* that approach—but do not touch—the sarcolemma of a muscle fiber (Figure 10.5a,b). The ends of the axon terminals enlarge into swellings known as *synaptic end bulbs,* which contain *synaptic vesicles* filled with a chemical *neurotransmitter*. The region of the sarcolemma near the axon terminal is called the ***motor end plate.*** The space between the axon terminal and sarcolemma is the ***synaptic cleft*** (sin-AP-tik). The synapse formed between the axon terminals of a motor neuron and the motor end plate of a muscle fiber is known as the ***neuromuscular junction (NMJ)*** (noo-rō-MUS-kū-lar). At the NMJ, a somatic motor neuron excites a skeletal muscle fiber in the following way (Figure 10.5c):

1. **Release of acetylcholine.** Arrival of the nerve impulse at the synaptic end bulbs triggers release of the neurotransmitter ***acetylcholine (ACh)*** (as′-ē-til-KŌ-lēn). ACh then floods the synaptic cleft between the motor neuron and the motor end plate.
2. **Activation of ACh receptors.** Binding of ACh to its receptor in the motor end plate opens ion channels that allow small cations, especially sodium ions (Na^+), to flow across the membrane.
3. **Generation of muscle action potential.** The inflow of Na^+ (down its concentration gradient) generates a muscle action potential. The muscle action potential then travels along the sarcolemma and through the T tubules. Each nerve impulse normally elicits one muscle action potential. If another nerve impulse releases more acetylcholine, then steps 2 and 3 repeat. See Chapter 15 for the details of nerve impulse generation.
4. **Breakdown of ACh.** The effect of ACh lasts only briefly because the neurotransmitter is rapidly broken down in the synaptic cleft by an enzyme called ***acetylcholinesterase (AChE)*** (as′-ē-til-kō′-lin-ES-ter-ās).

• CLINICAL CONNECTION | Myasthenia Gravis

Myasthenia gravis (mī-as-THĒ-nē-a GRAV-is) is an autoimmune disease that causes chronic, progressive damage of the neuromuscular junction. In people with myasthenia gravis, the immune system inappropriately produces antibodies that bind to and block some ACh receptors, thereby decreasing the number of functional ACh receptors at the motor end plates of skeletal muscles (Figure 10.5). As the disease progresses, more ACh receptors are lost. Thus, muscles become increasingly weaker, fatigue more easily, and may eventually cease to function.

Myasthenia gravis occurs in about 1 in 10,000 people and is more common in women. The muscles of the face and neck are most often affected. Initial symptoms include weakness of the eye muscles, which may produce double vision, and difficulty in swallowing. Later, the person has difficulty chewing and talking. Eventually the muscles of the limbs may become involved. Death may result from paralysis of the respiratory muscles, but often the disorder does not progress to that stage. •

Physiology of Contraction

Both Ca^{2+} and energy, in the form of ATP, are needed for muscle contraction. When a muscle fiber is relaxed (not contracting), there is a low concentration of Ca^{2+} in the sarcoplasm because the membrane of the sarcoplasmic reticulum contains Ca^{2+} active transport pumps that continually transport Ca^{2+} from the sarcoplasm into the sarcoplasmic reticulum (see Figure 10.7, 7). However, when a muscle action potential travels along the sarcolemma and into the transverse tubule system, Ca^{2+}-release channels open (see Figure 10.7, 4), allowing Ca^{2+} to escape into the sarcoplasm. The Ca^{2+} binds to troponin molecules in the thin filaments, causing the troponin to change shape. This change in shape moves troponin away from the myosin-binding sites on actin (see Figure 10.7, 5). Once

Figure 10.5 Structure of the neuromuscular junction (NMJ), the synapse between a motor neuron and a skeletal muscle fiber.

A neuromuscular junction includes the axon terminal of a motor neuron, containing synaptic vesicles filled with acetylcholine, plus the motor end plate of a muscle fiber.

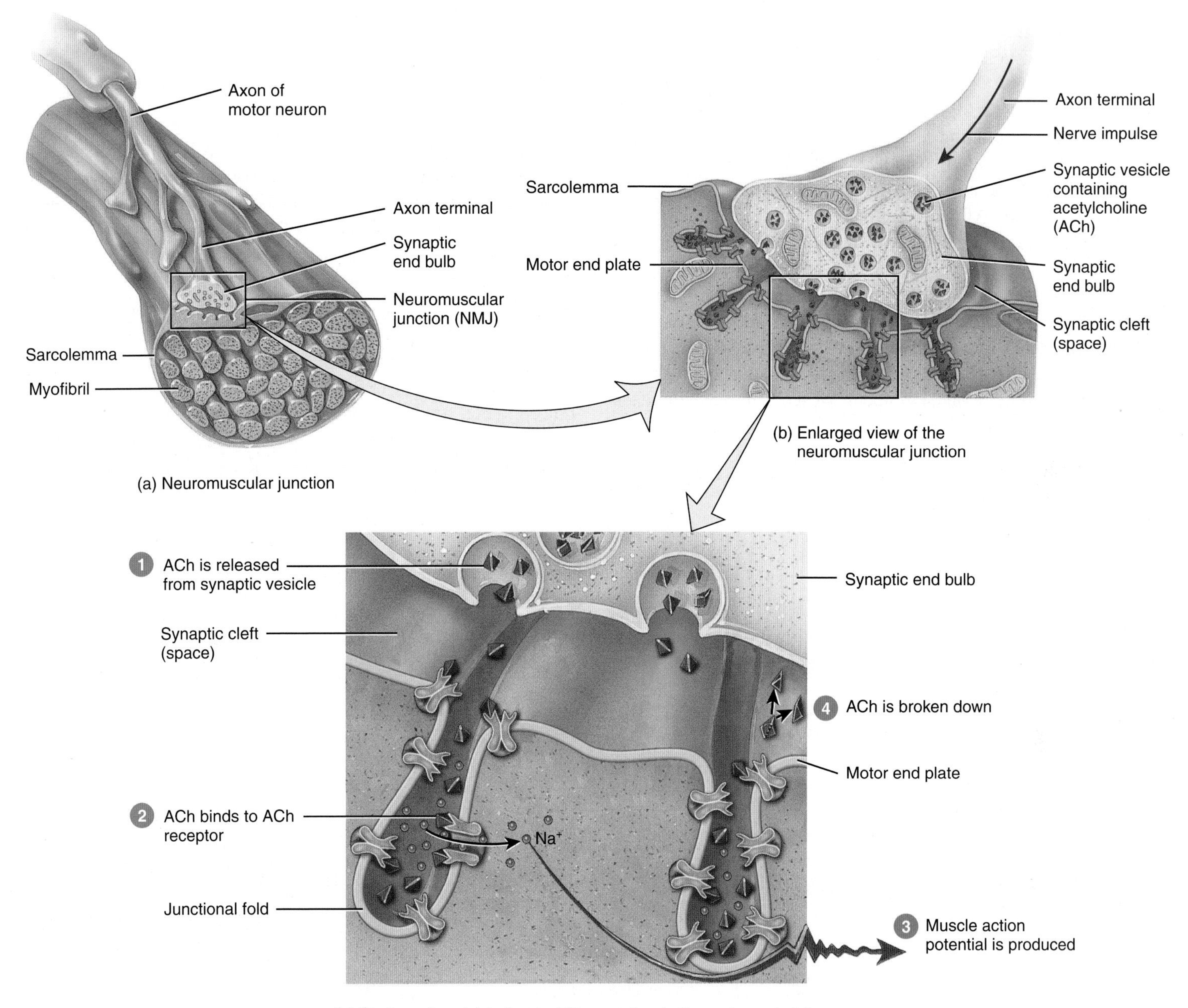

(c) Binding of acetylcholine to ACh receptors in the motor end plate

What is the motor end plate?

the myosin-binding sites are uncovered, the ***contraction cycle***—the repeating sequence of events that causes the filaments to slide—begins, as shown in Figure 10.6:

1. **Splitting ATP.** The myosin heads contain ATPase, an enzyme that splits ATP into ADP (adenosine diphosphate) and Ⓟ (a phosphate group). This splitting reaction transfers energy to the myosin head, although ADP and Ⓟ remain attached to it.
2. **Forming cross-bridges.** The energized myosin heads attach to the myosin-binding sites on actin, and release the phosphate groups. When myosin heads attach to actin during contraction, they are referred to as ***cross-bridges*** (Figure 10.6).
3. **Power stroke.** After the cross-bridges form, the ***power stroke*** occurs. During the power stroke, the cross-bridge rotates or swivels and releases the ADP. The force produced as hundreds of cross-bridges swivel slides the thin filament past the thick filament toward the center of the sarcomere.
4. **Binding ATP and detaching.** At the end of the power stroke, the cross-bridges remain firmly attached to actin. When they bind another molecule of ATP, the myosin heads detach from actin.

As the myosin ATPase again splits ATP, the myosin head is reoriented and energized, ready to combine with another myosin-binding site farther along the thin filament. The contraction cycle repeats as long as ATP and Ca^{2+} are available in the sarcoplasm. At any one instant, some of the myosin heads are attached to actin, forming cross-bridges and generating force, and other myosin heads are detached from actin and getting ready to bind again. During a maximal contraction, the sarcomere can shorten by as much as half its resting length.

Relaxation

Two changes permit a muscle fiber to relax after it has contracted. First, the neurotransmitter acetylcholine is rapidly broken down by the enzyme acetylcholinesterase (AChE). When nerve action potentials cease, release of ACh stops, and AChE rapidly breaks down the ACh already present in the synaptic cleft. This ends the generation of muscle action potentials, and the Ca^{2+} release channels in the sarcoplasmic reticulum membrane close.

Second, calcium ions are rapidly transported from the sarcoplasm into the sarcoplasmic reticulum. As the level of Ca^{2+} in the sarcoplasm falls, tropomyosin slides back over the myosin-binding sites on actin. Once the myosin-binding sites are covered, the thin filaments slip back to their relaxed positions. Figure 10.7 summarizes the events of contraction and relaxation in a muscle fiber.

Figure 10.6 The contraction cycle. Sarcomeres shorten through repeated cycles in which the myosin heads (cross-bridges) attach to actin, rotate, and detach.

During the power stroke of contraction, cross-bridges rotate and move the thin filaments past the thick filaments toward the center of the sarcomere.

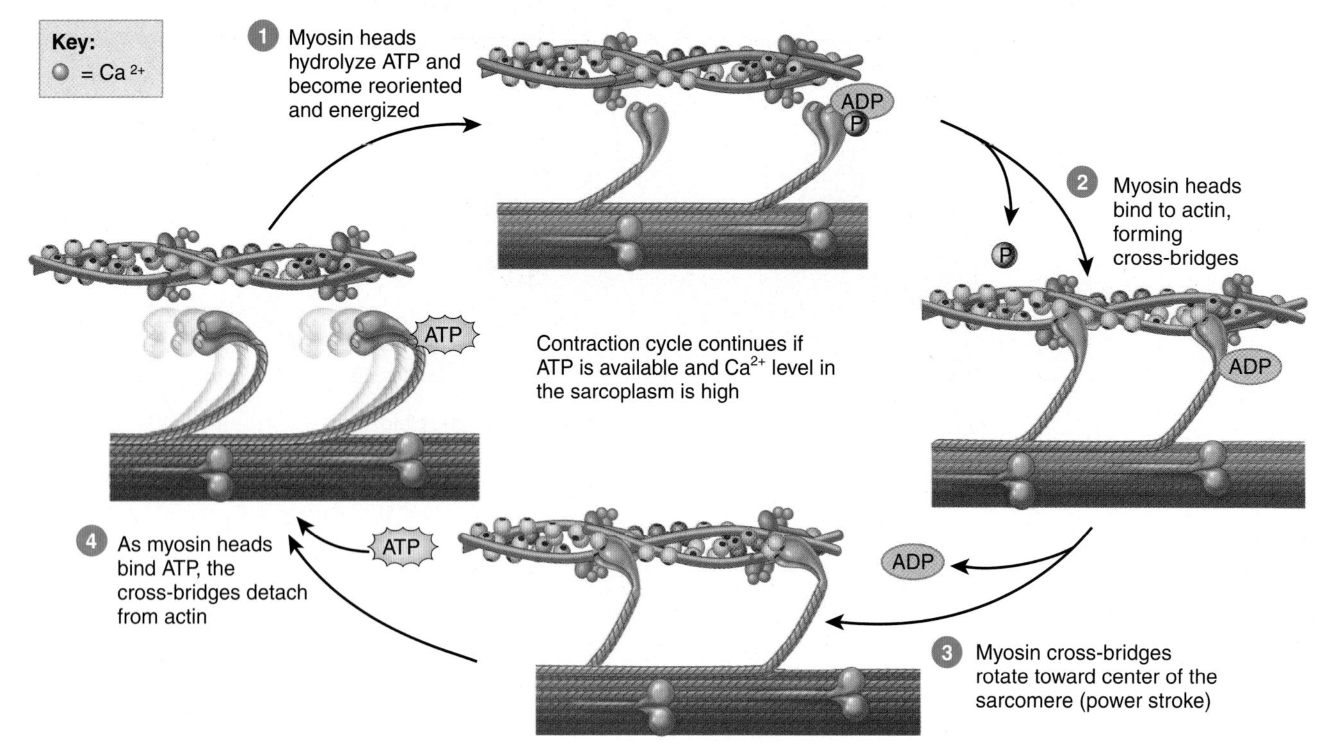

? What causes cross-bridges to detach from actin?

Figure 10.7 Summary of the events of contraction and relaxation in a skeletal muscle fiber.

Acetylcholine released at the neuromuscular junction triggers a muscle action potential, which leads to muscle contraction.

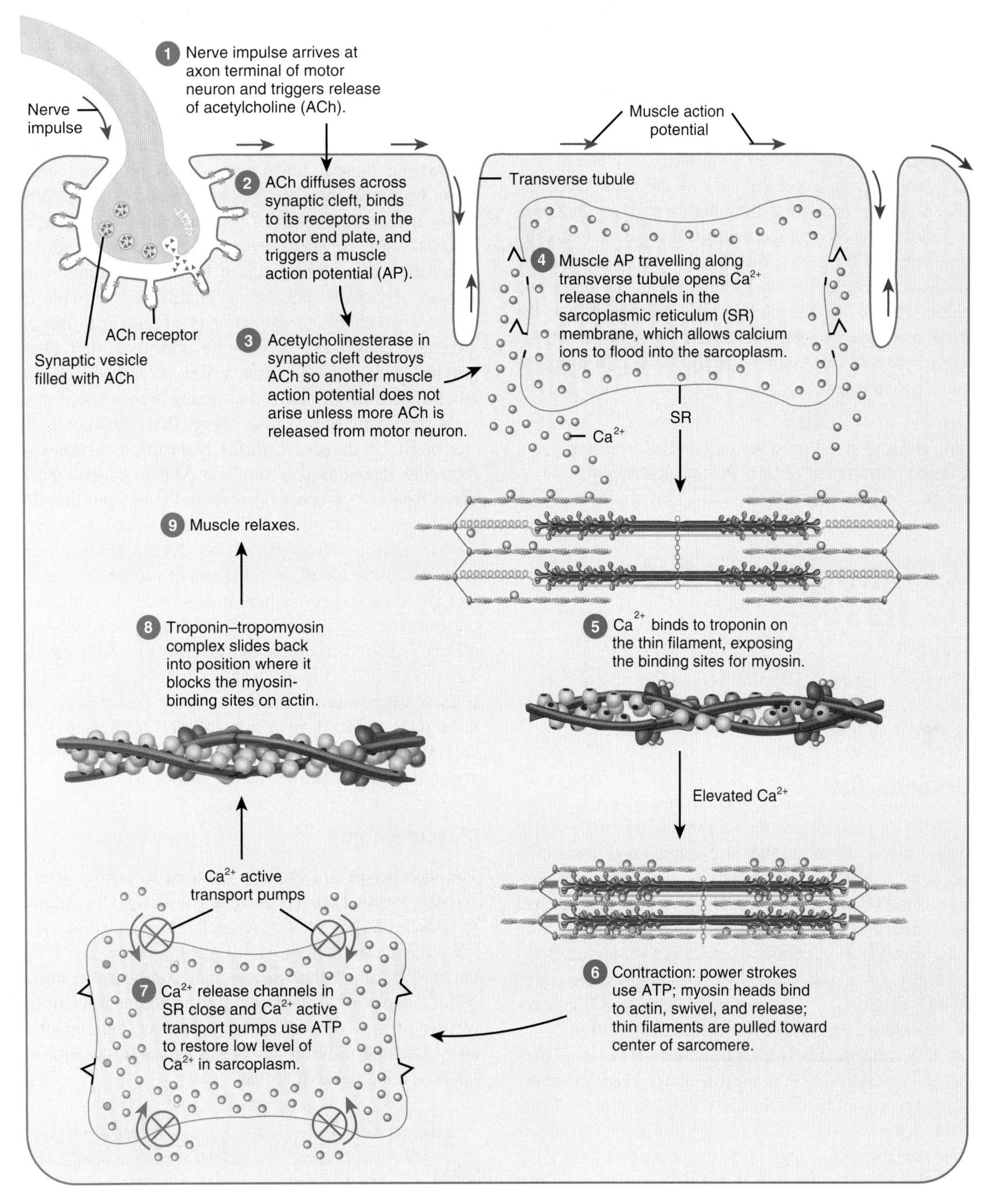

The power stroke occurs during which numbered step in this figure?

Muscle Tone

Even when a whole muscle is not contracting, a small number of its motor units are involuntarily activated to produce a sustained contraction of their muscle fibers. This process results in ***muscle tone*** (*tonos* = tension). Remember that skeletal muscle is traditionally classified as voluntary; the concept of muscle tone is an exception. To sustain muscle tone, small groups of motor units are alternately active and inactive in a constantly shifting pattern. Muscle tone keeps skeletal muscles firm, but it does not result in a contraction strong enough to produce movement. For example, the tone of muscles in the back of the neck keeps the head upright and prevents it from slumping forward on the chest. Recall that skeletal muscle contracts only after it is activated by acetylcholine released by nerve impulses in its motor neurons. Hence, muscle tone is established by neurons in the brain and spinal cord that excite the muscle's motor neurons. When the motor neurons serving a skeletal muscle are damaged or cut, the muscle becomes ***flaccid*** (FLA-sid = flabby), a state of limpness in which muscle tone is lost.

CHECKPOINT

7. Explain how a skeletal muscle contracts and relaxes.
8. What is the importance of the neuromuscular junction?

10.4 METABOLISM OF SKELETAL MUSCLE TISSUE

OBJECTIVES

- Describe the sources of ATP and oxygen for muscle contraction.
- Define muscle fatigue and list its possible causes.

Energy for Contraction

Unlike most cells of the body, skeletal muscle fibers often switch between virtual inactivity, when they are relaxed and using only a modest amount of ATP, and great activity, when they are contracting and using ATP at a rapid pace. However, the ATP present inside muscle fibers is enough to power maximal contraction for only a few seconds. If strenuous exercise is to continue, additional ATP must be synthesized. Muscle fibers have three sources for ATP production: (1) creatine phosphate, (2) anaerobic cellular respiration, and (3) aerobic cellular respiration.

While at rest, muscle fibers produce more ATP than they need. Some of the excess ATP is used to make ***creatine phosphate,*** an energy-rich molecule that is unique to muscle fibers (Figure 10.8a). One of ATP's high-energy phosphate groups is transferred to creatine, forming creatine phosphate and ADP (adenosine diphosphate). ***Creatine*** is a small, amino acid–like molecule that is synthesized in the liver, kidneys, and pancreas and derived from certain foods (milk, red meat, fish), then transported to muscle fibers. While muscle is contracting, the high-energy phosphate group can be transferred from creatine phosphate back to ADP, quickly forming new ATP molecules. Together, creatine phosphate and ATP provide enough energy for muscles to contract maximally for about 15 seconds. This energy is sufficient for short bursts of intense activity, for example, running a 100-meter dash.

When intense muscle activity continues past the 15-second mark, the supply of creatine phosphate is depleted. The next source of ATP is *glycolysis,* a series of cytosolic reactions that produces two molecules of ATP by breaking down a glucose molecule to pyruvic acid. Glucose passes easily from the blood into contracting muscle fibers and also is produced within muscle fibers by breakdown of glycogen (Figure 10.8b). When oxygen levels are low as a result of vigorous muscle activity, most of the pyruvic acid is converted to *lactic acid,* a process called ***anaerobic cellular respiration*** because it occurs without using oxygen. Anaerobic cellular respiration can provide enough energy for about 30 to 40 seconds of maximal muscle activity. Together, conversion of creatine phosphate and glycolysis can provide enough ATP to run a 400-meter race (essentially once around the track that surrounds many high school football fields).

Muscle activity that lasts longer than 30 seconds depends increasingly on ***aerobic cellular respiration,*** a series of oxygen-requiring reactions that produce ATP in mitochondria. Muscle fibers have two sources of oxygen: (1) oxygen that diffuses into them from the blood and (2) oxygen released by myoglobin in the sarcoplasm. ***Myoglobin*** is an oxygen-binding protein found only in muscle fibers. It binds oxygen when oxygen is plentiful and releases oxygen when it is scarce. If enough oxygen is present, pyruvic acid enters the mitochondria, where it is completely oxidized in reactions that generate ATP, carbon dioxide, water, and heat (Figure 10.8c). In comparison with anaerobic cellular respiration, aerobic cellular respiration yields much more ATP, about 36 molecules of ATP from each glucose molecule. In activities that last more than 10 minutes, aerobic cellular respiration provides most of the needed ATP.

Muscle Fatigue

The inability of a muscle to contract forcefully after prolonged activity is called ***muscle fatigue*** (fa-TĒG). One important factor in muscle fatigue is lowered release of calcium ions from the sarcoplasmic reticulum, resulting in a decline of Ca^{2+} levels in the sarcoplasm. Other factors that contribute to muscle fatigue include depletion of creatine phosphate, insufficient oxygen, depletion of glycogen and other nutrients, buildup of lactic acid and ADP, and failure of nerve impulses in the motor neuron to release enough acetylcholine.

MANUAL THERAPY APPLICATION

Lactic Acid Fuels Intensive Exercise

It has been long believed that muscles build up lactic acid as a waste product during heavy exercise and that it causes unwanted muscular aches and pains that may last several days. However, it is now known

Figure 10.8 Production of ATP for muscle contraction. (a) Creatine phosphate, formed from ATP while the muscle is relaxed, transfers a high-energy phosphate group to ADP, forming ATP, during muscle contraction. (b) Breakdown of muscle glycogen into glucose and production of pyruvic acid from glucose via glycolysis produce both ATP and lactic acid. Because no oxygen is needed, this is an anaerobic pathway. (c) Within mitochondria, pyruvic acid, fatty acids, and amino acids are used to produce ATP via aerobic cellular respiration, an oxygen-requiring set of reactions.

During a long-term event, such as a marathon race, most ATP is produced aerobically.

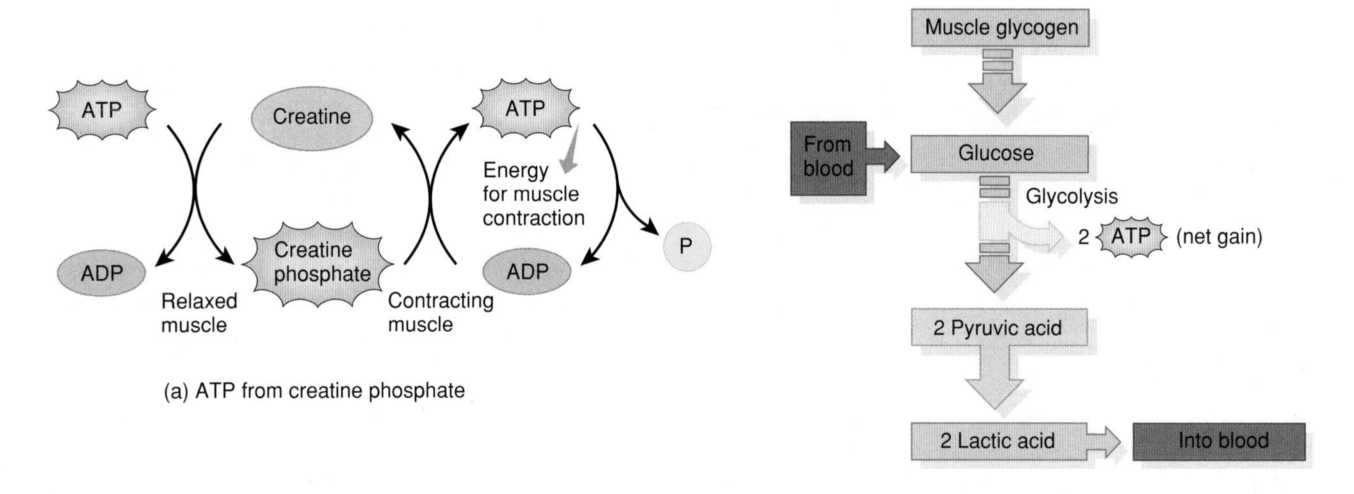

(b) ATP from anaerobic cellular respiration

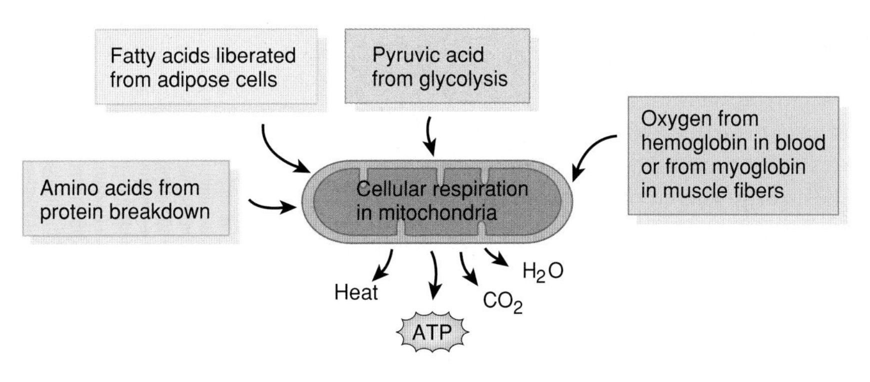

(c) ATP from aerobic cellular respiration

Where inside a skeletal muscle fiber are the events shown here occurring?

that **lactic acid fuels intensive exercise** and is an important *fuel* for muscles, not merely a by-product of hard work; furthermore, the heart can use lactic acid for fuel under intense exercise. Lactic acid produced in skeletal muscle cells enters the bloodstream; it can be converted into glucose by the liver and released back into the bloodstream. It may then be used by the heart or by skeletal muscles to sustain a high level of exercise.

When exercising at a low intensity level the body fuels its muscles with glucose; with higher intensity exercise and reduced oxygen available to cells, lactic acid accumulates faster than it is removed. A signal of its presence is a *burning* sensation in muscles. When the intensity is increased, skeletal muscles require more fuel. As the fuel requirements of muscles escalate to meet the increased demand, glucose stored as glycogen in muscles is not sufficient. The mitochondria provide the additional power from lactic acid that muscle cells manufacture and that is converted back to glucose.

During heavy exercise, we may assume that uncomfortable muscle fatigue and burning, caused by the accumulation of lactic acid, are a signal to slow down or stop. Rather than let the discomfort stall exercise, it should be viewed as an indicator that something positive is occurring, which makes it possible to continue training. The burning sensation that accompanies accumulation of lactic acid is usually gone soon after a strenuous exercise session. The process of removing the excess lactic acid and resultant soreness will be enhanced by a vigorous massage.

Even before actual muscle fatigue occurs, a person may have feelings of tiredness and the desire to cease activity; this response, called *central fatigue,* is caused by changes in the central nervous system (brain and spinal cord). Although its exact mechanism is unknown, it may be a protective mechanism to stop a person from exercising before muscles become damaged. Perhaps you have experienced a long day of physical exertion followed by a telephone call in the evening with a friend asking you to do something with them. Your response may be, "Sorry, I'm just too tired." A few minutes later the telephone may ring again and another person asks you to do something that is much more exciting. Your response may be, "Give me 30 minutes to change clothes!" The brain has modified the feeling of central fatigue in this case and has overriden muscular fatigue.

Oxygen Consumption after Exercise

During prolonged periods of muscle contraction, increases in breathing and blood flow enhance oxygen delivery to muscular tissue. After muscle contraction has stopped, heavy breathing continues for a time, and oxygen consumption remains above the resting level. The term ***oxygen debt*** refers to the added oxygen, over and above the oxygen consumed at rest, that is taken into the body after exercise. This extra oxygen is used to "pay back" or restore metabolic conditions to the resting level in three ways: (1) to convert lactic acid back into glycogen stores in the liver, (2) to resynthesize creatine phosphate and ATP, and (3) to replace the oxygen removed from myoglobin.

The metabolic changes that occur *during exercise,* however, account for only some of the extra oxygen used *after exercise.* Only a small amount of resynthesis of glycogen occurs from lactic acid. Instead, glycogen stores are replenished much later from dietary carbohydrates. Much of the lactic acid that remains after exercise is converted back to pyruvic acid and used for ATP production via aerobic cellular respiration. Ongoing changes after exercise also boost oxygen use. First, the elevated body temperature after strenuous exercise increases the rate of chemical reactions throughout the body. Faster reactions use ATP more rapidly, and more oxygen is needed to produce ATP. Second, the heart and muscles used in breathing are still working harder than they were at rest, and thus they consume more ATP. Third, tissue repair processes are occurring at an increased pace. For these reasons, ***recovery oxygen uptake*** is a better term than oxygen debt for the elevated use of oxygen after exercise.

CHECKPOINT

9. What are the sources of ATP for muscle fibers?
10. What factors contribute to muscle fatigue?
11. Why is the term *recovery oxygen uptake* more accurate than *oxygen debt*?

10.5 CONTROL OF MUSCLE TENSION

OBJECTIVES

- Explain the three phases of a twitch contraction.
- Describe how the frequency of stimulation and motor unit recruitment affect muscle tension.
- Compare the three types of skeletal muscle fibers.
- Distinguish between isotonic and isometric contractions.

The contraction that results from a single muscle action potential, a muscle twitch, has significantly smaller force than the maximum force or tension that the fiber is capable of producing. The total tension that a *single* muscle fiber can produce depends mainly on the rate at which nerve impulses arrive at its neuromuscular junction. The number of impulses per second is the *frequency of stimulation.* When considering the contraction of a *whole* muscle, the total tension it can produce depends on the number of muscle fibers that are contracting in unison.

Twitch Contraction

A ***twitch contraction*** is a brief contraction of all the muscle fibers in a motor unit in response to a single action potential in its motor neuron. Figure 10.9 shows a recording of a muscle contraction, called a ***myogram*** (MĪ-ō-gram). Twitches are a phenomenon of the laboratory; they do not occur in the body. Note that a brief delay, called the *latent period,* occurs between application of the stimulus (time zero on the graph) and the beginning of contraction. During the latent period, the muscle action potential sweeps over the sarcolemma and calcium ions are released from the sarcoplasmic reticulum. During the second phase, the *contraction period* (upward tracing), repetitive power strokes are occurring, generating tension or force of contraction. In the third phase, the *relaxation period* (downward tracing),

Figure 10.9 Myogram of a twitch contraction. The arrow indicates the time at which the stimulus occurred.

A myogram is a record of a muscle contraction.

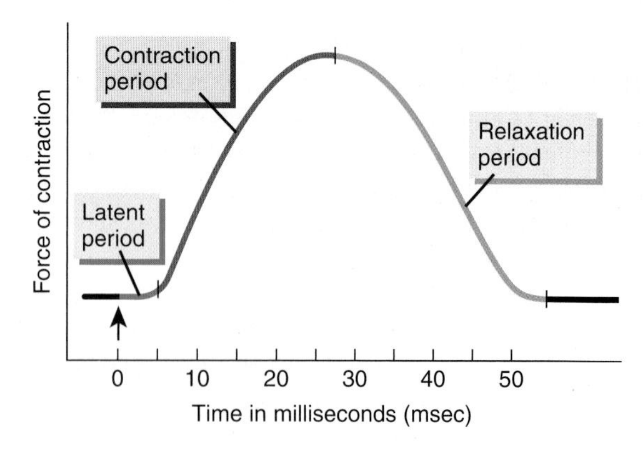

During which period do sarcomeres shorten?

power strokes cease because the level of Ca^{2+} in the sarcoplasm is decreasing to the resting level. (Recall that calcium ions are actively transported back into the sarcoplasmic reticulum.)

Frequency of Stimulation

If a second stimulus arrives before a muscle fiber has completely relaxed, the second contraction will be stronger than the first because the second contraction begins when the fiber is at a higher level of tension (Figure 10.10a,b). This phenomenon, in which stimuli arriving one after the other before a muscle fiber has completely relaxed and causes larger contractions, is called ***wave summation.*** When a skeletal muscle fiber is stimulated at a rate of 20 to 30 times per second, it can only partially relax between stimuli. The result is a sustained but wavering contraction called ***unfused (incomplete) tetanus*** (*tetan-* = rigid, tense; Figure 10.10c). When a skeletal muscle fiber is stimulated at a higher rate of 80 to 100 times per second, it does not relax at all. The result is ***fused (complete) tetanus,*** a sustained contraction in which individual twiches cannot be detected (Figure 10.10d).

Motor Unit Recruitment

The process by which the number of contracting motor units is increased is called ***motor unit recruitment.*** Normally, the various motor neurons to a whole muscle fire *asynchronously* (at different times): While some motor units are contracting, others are relaxed. This pattern of motor unit activity delays muscle fatigue by allowing alternately contracting motor units to relieve one another, so that the contraction can be sustained for long periods.

Recruitment is one factor responsible for producing smooth movements rather than a series of jerky movements. Precise movements are brought about by small changes in muscle contraction. Typically, the muscles that produce precise movements are composed of small motor units. In this way, when a motor unit is recruited or turned off, only slight changes occur in muscle tension. On the other hand, large motor units are active where large tension is needed and precision is less important.

Types of Skeletal Muscle Fibers

Skeletal muscles contain three types of muscle fibers, which are present in varying proportions in different muscles of the body: The fiber types are (1) slow oxidative fibers, (2) fast oxidative–glycolytic fibers, and (3) fast glycolytic fibers.

Slow oxidative (SO) fibers or *red fibers* are small in diameter and appear dark red because they contain a large amount of myoglobin. Because they have many large mitochondria, SO fibers generate ATP mainly by aerobic cellular respiration, which is why they are called oxidative fibers. These fibers are

Figure 10.10 **Myograms showing the effects of different frequencies of stimulation.** (a) Single twitch. (b) When a second stimulus occurs before the muscle has relaxed, wave summation occurs, and the second contraction is stronger than the first. (The dashed line indicates the force of contraction expected in a single twitch.) (c) In unfused tetanus, the curve looks jagged due to partial relaxation of the muscle between stimuli. (d) In fused tetanus, the contraction force is steady and sustained.

Due to wave summation, the tension produced during a sustained contraction is greater than during a single twitch.

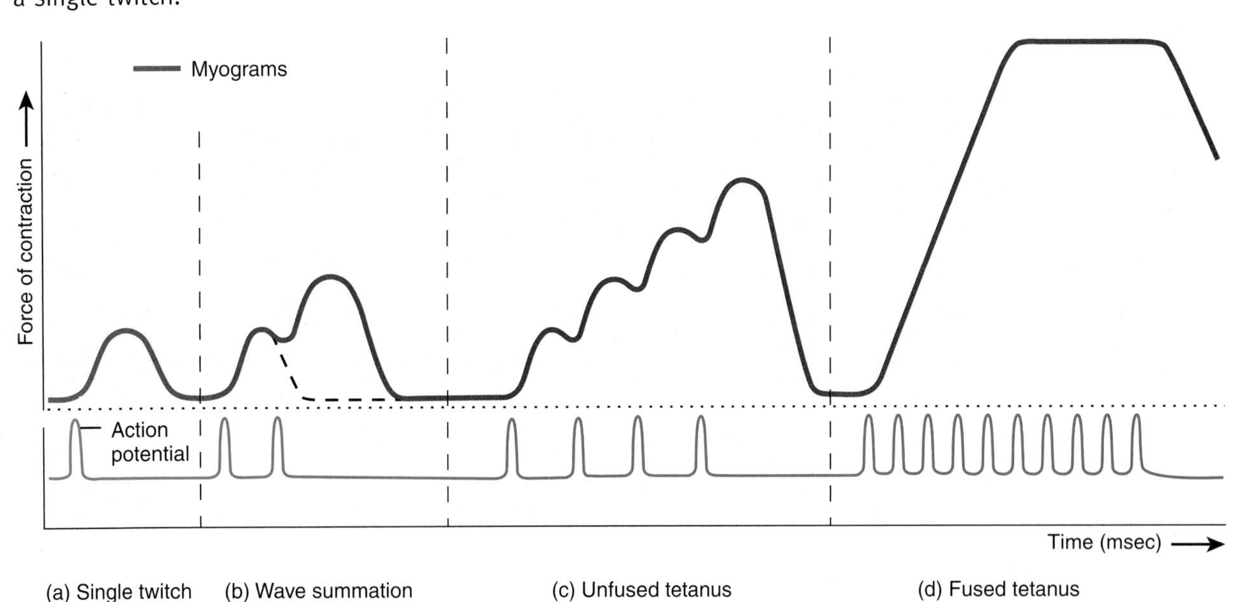

What frequency of stimulation is needed to produce fused tetanus?

said to be "slow" because the contraction cycle proceeds at a slower pace than in "fast" fibers. SO fibers are very resistant to fatigue and are capable of prolonged, sustained contractions.

Fast oxidative–glycolytic (FOG) fibers are intermediate in diameter between the other two types. Like slow oxidative fibers, they contain a large amount of myoglobin, and thus appear dark red. FOG fibers can generate considerable ATP by aerobic cellular respiration, which gives them a moderately high resistance to fatigue. Because their glycogen content is high, they also generate ATP by anaerobic glycolysis. These fibers are "fast" because they contract and relax more quickly than SO fibers.

Fast glycolytic (FG) fibers or *white fibers* are largest in diameter, contain the most myofibrils, and generate the most powerful and most rapid contractions. They have a low myoglobin content and few mitochondria. FG fibers contain large amounts of glycogen and generate ATP mainly by anaerobic glycolysis. They are used for intense movements of short duration, but they fatigue quickly. Strength-training programs that engage a person in activities requiring great strength for short times produce increases in the size, strength, and glycogen content of FG fibers.

Most skeletal muscle organs are a mixture of all three types of skeletal muscle fibers, about half of which are SO fibers. The proportions vary somewhat, depending on the action of the muscle, the person's training program, and genetic factors. For example, the continually active postural muscles of the neck, back, and legs have a high proportion of SO fibers. Muscles of the shoulders and arms, in contrast, are not constantly active but are used intermittently and briefly to produce large amounts of tension, such as in lifting and throwing. These muscles have a high proportion of FG fibers. Leg muscles, which not only support the body but are also used for walking and running, have large numbers of both SO and FOG fibers.

Even though most skeletal muscles are a mixture of all three types of skeletal muscle fibers, the skeletal muscle fibers of any given motor unit are all of the same type. The different motor units in a muscle are recruited in a specific order, depending on need. For example, if weak contractions suffice to perform a task, only SO motor units are activated. If more force is needed, the motor units of FOG fibers are also recruited. Finally, if maximal force is required, motor units of FG fibers are also called into action.

Table 10.1 compares the characteristics of these three types of skeletal muscle fibers.

Isometric and Isotonic Contractions

Muscle contractions may be either isotonic or isometric. In an ***isotonic contraction*** (*iso-* = equal; *-tonic* = tension), the tension (force of contraction) in the muscle remains almost constant while the muscle changes its length. Isotonic contractions are used for body movements and for moving objects. The two types of isotonic contractions are concentric and eccentric. In a *concentric isotonic contraction,* if the tension generated is great enough to overcome the resistance of the object to be moved, the muscle shortens and pulls on another structure, such as a tendon, to produce movement and to reduce the angle at a joint. Picking a book up off a table involves concentric isotonic contractions of the biceps brachii muscle in the arm (Figure 10.11a).

Figure 10.11 Comparison between isotonic (concentric and eccentric) and isometric contractions. Parts (a) and (b) show isotonic contractions of the biceps brachii muscle in the arm; part (c) illustrates isometric contractions of the shoulder and arm muscles.

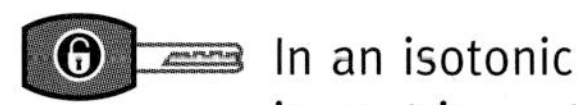

In an isotonic contraction, tension remains constant as muscle length decreases or increases; in an isometric contraction, tension increases greatly without a change in muscle length.

(a) Concentric contraction while picking up a book

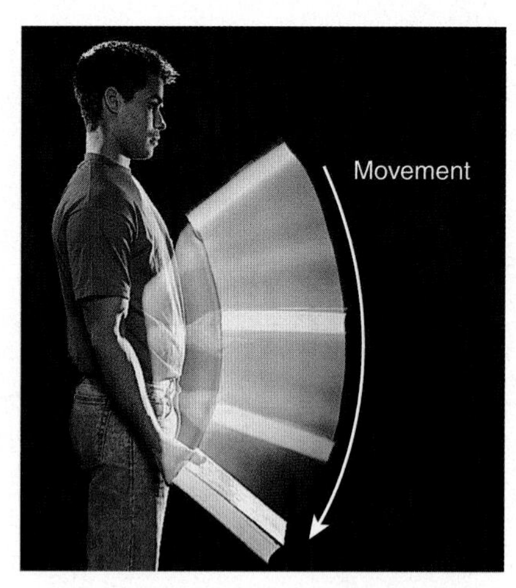

(b) Eccentric contraction while lowering a book

(c) Isometric contraction while holding a book steady

? What type of contraction occurs in your neck muscles while you are walking?

By contrast, as you lower the book to place it back on the table, the previously shortened biceps lengthens in a controlled manner while it continues to contract. When the length of a muscle increases during a contraction, the contraction is an *eccentric isotonic contraction* (Figure 10.11b). During an eccentric contraction, the tension exerted by the myosin crossbridges resists movement of a load (the book, in this case) and slows the lengthening process.

In an ***isometric contraction*** (*metro* = measure or length), the tension generated is not enough to exceed the resistance of the object to be moved and the muscle does not change its length (Figure 10.11c). Isometric contractions occur when you try to lift a box but the box does not move because it is too heavy. Isometric contractions are important for maintaining posture and supporting objects in a fixed position.

CHECKPOINT

12. Define the following terms: myogram, twitch contraction, wave summation, unfused tetanus, and fused tetanus.
13. What characteristics distinguish the three types of skeletal muscle fibers?
14. Provide examples of isometric and isotonic contractions.

10.6 MUSCLE SPASMS

OBJECTIVES

- Describe the possible sources of muscle spasms.
- Describe abnormal contractions of skeletal muscle.
- Describe the significance of gamma motor neurons. Describe multiple treatments for muscle spasms.

Table 10.1

Characteristics of the Three Types of Skeletal Muscle Fibers

Slow oxidative fiber
Fast glycolytic fiber
Fast oxidative-glycolytic fiber

LM 440x

Transverse section of three types of skeletal muscle fibers

	SLOW OXIDATIVE (SO) FIBERS	FAST OXIDATIVE–GLYCOLYTIC (FOG) FIBERS	FAST GLYCOLYTIC (FG) FIBERS
STRUCTURAL CHARACTERISTICS			
Fiber diameter	Smallest.	Intermediate.	Largest.
Myoglobin content	Large amount.	Large amount.	Small amount.
Mitochondria	Many.	Many.	Few.
Capillaries	Many.	Many.	Few.
Color	Red.	Red to pink.	White (pale).
FUNCTIONAL CHARACTERISTICS			
Capacity for generating ATP and method used	High capacity, by aerobic (oxygen-requiring) cellular respiration.	Intermediate capacity, by both aerobic (oxygen-requiring) cellular respiration and anaerobic (does not require oxygen) cellular respiration (glycolysis).	Low capacity, by anaerobic cellular respiration (glycolysis).
Rate of ATP hydrolysis by myosin ATPase	Slow.	Fast.	Fast.
Contraction velocity	Slow.	Fast.	Fast.
Fatigue resistance	High.	Intermediate.	Low.
Creatine kinase	Lower amount.	Intermediate amount.	Highest amount.
Glycogen stores	Low.	Intermediate.	High.
Order of recruitment	First.	Second.	Third.
Location where fibers are abundant	Postural muscles such as those of the neck.	Lower limb muscles.	Upper limb muscles.
Primary functions of fibers	Maintaining posture and aerobic endurance activities.	Walking, sprinting.	Rapid, intense movements of short duration.

Muscle spasms have many names, many causes, and many potential treatments. Spasms are involuntary and often painful contractions of the muscles. Very often only one or two portions of a muscle organ are in actual spasm. Muscle spasms are complex and are not fully understood. Knowledge associated with spasms is critical, however, to the practice of all manual therapies.

Muscle spasms are sometimes also called cramps, charley horses, knots, trigger points, or areas of hypertonicity. Muscle spasms commonly occur when a muscle is injured or overused. Some spasms occur when the nerve to the muscle is irritated, such as when a herniated disk presses on a nerve. Abnormal curvature of the spine (see Section 7.6 Vertebral Column) and other bony alignment problems can also cause chronic spasms. Other potential causes of spasm include stress (*physiological stress* such as from cancer or a broken leg or *emotional stress* from relationships, financial problems, deadlines, or a host of other stressors), dehydration, alcoholism, medications, reduced levels of calcium or magnesium in the body, other nutritional imbalances, pregnancy, hypothyroidism, and kidney failure.

A muscle in spasm may feel tight, hard, or bulging (some practitioners use the term "a rope" to describe an area of spasm) and the pain can be mild to severe and even debilitating. A mild to moderate spasm may not pull the two muscle attachments closer together. Chronic spasms can, however, shorten a muscle to the point that one or both of the bony attachments may be moved. For example, as you will learn in Chapter 12, the muscle called the quadratus lumborum is attached to the twelfth rib and to the crest of the ilium. Chronic contraction (spasm) of this muscle can elevate the pelvis on one side to the point that one leg may be "shorter" than the other. The muscle is thus also known as the "hip hiker."

• CLINICAL CONNECTION | Abnormal Contractions of Skeletal Muscle

One kind of **abnormal contraction of a skeletal muscle** is a **spasm,** a sudden involuntary contraction of a single muscle in a large group of muscles. A painful spasmodic contraction is known as a **cramp.** A ***tic*** is a spasmodic twitching made involuntarily by muscles that are ordinarily under voluntary control. Twitching of the eyelid and facial muscles are examples of tics. A **tremor** is a rhythmic, involuntary, purposeless contraction that produces a quivering or shaking movement. A **fasciculation** (fa-sik-ū-LĀ-shun) is an involuntary, brief twitch of an entire motor unit that is visible under the skin; it occurs irregularly and is not associated with movement of the affected muscle. Fasciculations may be seen in multiple sclerosis (see Clinical Connection in Chapter 15) or in amyotrophic lateral sclerosis (Lou Gehrig's disease). A **fibrillation** is a spontaneous contraction of a single muscle fiber that is not visible under the skin but can be recorded by electromyography. Fibrillations may signal destruction of motor neurons. •

Figure 10.12a illustrates a typical skeletal muscle organ attached to a bone. Blood vessels and nerves commonly enter and leave near the middle of the muscle belly. Figure 10.12c illustrates the smaller blood vessels within the muscle organ. Figure 10.12b shows a muscle spindle and a tendon organ on a skeletal muscle fiber and demonstrates that there are four types of neurons within the nerve; two types of neurons are motor and two types are sensory. Neurons will be studied in Chapter 15 and muscle spindles and tendon organs will be further studied in Chapter 19.

Muscle spindles are the ***proprioceptors*** (PRŌ-prē-ō-sep′-tors) in skeletal muscles that monitor changes in the length of skeletal muscles. By adjusting how vigorously a muscle spindle responds to contraction or stretching of a skeletal muscle, the brain sets an overall level of ***muscle tone,*** the small degree of contraction that is present while the muscle is at rest.

Each of the many *muscle spindles* in each muscle organ consists of several slowly adapting sensory nerve endings that wrap around 3 to 10 specialized muscle fibers, called *intrafusal muscle fibers* (in′-tra-FŪ-sal = within a spindle) (see Figure 16.16). A *muscle spindle capsule* composed of connective tissue encloses the sensory nerve endings and intrafusal fibers and anchors the spindle to the endomysium and perimysium (Figure 10.12b). Muscle spindles are interspersed among most skeletal muscle fibers and aligned parallel to them. The main function of muscle spindles is to measure *muscle length*—how much a muscle is being contracted or stretched. The resulting nerve impulses propagate into the central nervous system. Information from muscle spindles arrives quickly at somatic sensory areas of the cerebral cortex, which allows conscious perception of limb positions and movements. At the same time, impulses from muscle spindles pass to the cerebellum, where the input is used to coordinate muscle contractions.

In addition to their sensory nerve endings near the middle of intrafusal fibers, muscle spindles contain somatic motor neurons called *gamma motor neurons.* These somatic motor neurons terminate near both ends of the intrafusal fibers and adjust the tension in each muscle spindle to variations in the length of the muscle organ. For example, when a muscle shortens, gamma motor neurons stimulate the ends of the intrafusal fibers to contract slightly. This keeps the intrafusal fibers taut and maintains the sensitivity of the muscle spindle to stretching of the muscle. As the frequency of impulses in its gamma motor neuron increases, a muscle spindle becomes more sensitive to stretching of its mid-region.

Surrounding muscle spindles are ordinary skeletal muscle fibers, called *extrafusal muscle fibers* (*extrafusal* = outside a spindle), which are supplied by large-diameter A fibers called *alpha motor neurons* (to be studied in Chapter 15). The cell bodies of both gamma and alpha motor neurons are located in the spinal cord (or in the brain stem for muscles in the head). During the stretch reflex, impulses in muscle spindle sensory axons propagate into the spinal cord and brain stem and activate alpha motor neurons that connect to extrafusal muscle fibers in the same muscle. In this way, activation of its muscle spindles causes contraction of a skeletal muscle, which relieves the stretching.

Tendon organs, which are also proprioceptors, are located at the junction of a tendon and a muscle. By initiating tendon reflexes, tendon organs protect tendons and their associated muscles from damage due to excessive tension. (When a muscle contracts, it may exert a force that pulls the points of attachment of the muscle at either end toward each other.

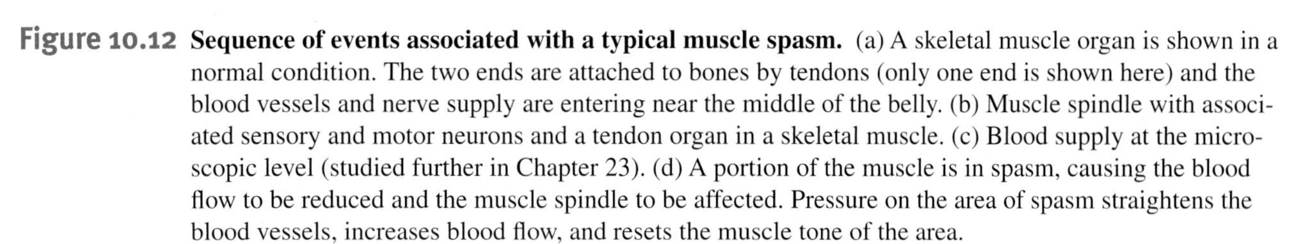
Figure 10.12 Sequence of events associated with a typical muscle spasm. (a) A skeletal muscle organ is shown in a normal condition. The two ends are attached to bones by tendons (only one end is shown here) and the blood vessels and nerve supply are entering near the middle of the belly. (b) Muscle spindle with associated sensory and motor neurons and a tendon organ in a skeletal muscle. (c) Blood supply at the microscopic level (studied further in Chapter 23). (d) A portion of the muscle is in spasm, causing the blood flow to be reduced and the muscle spindle to be affected. Pressure on the area of spasm straightens the blood vessels, increases blood flow, and resets the muscle tone of the area.

Massage can eliminate or reduce the severity of muscle spasms.

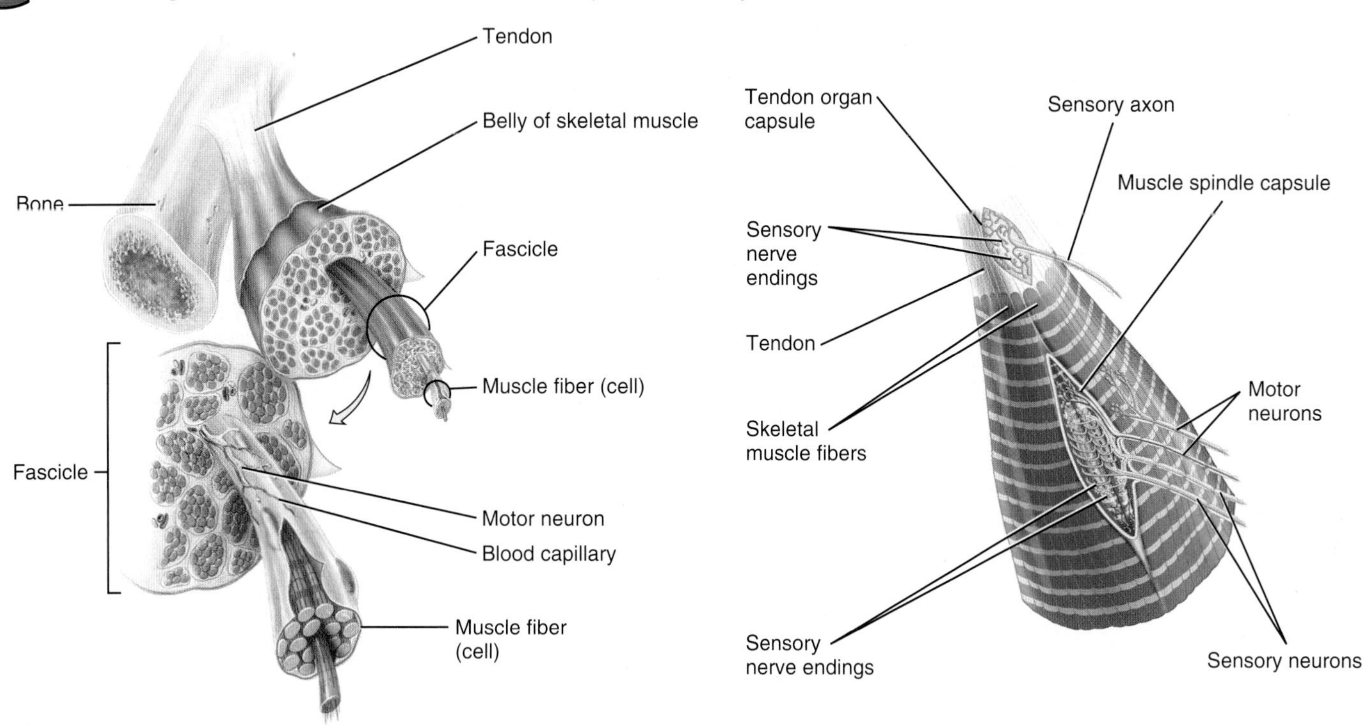

(a) Blood and nerve supply of a skeletal muscle fiber

(b) Muscle spindle and associated neurons and tendon organ in a skeletal muscle

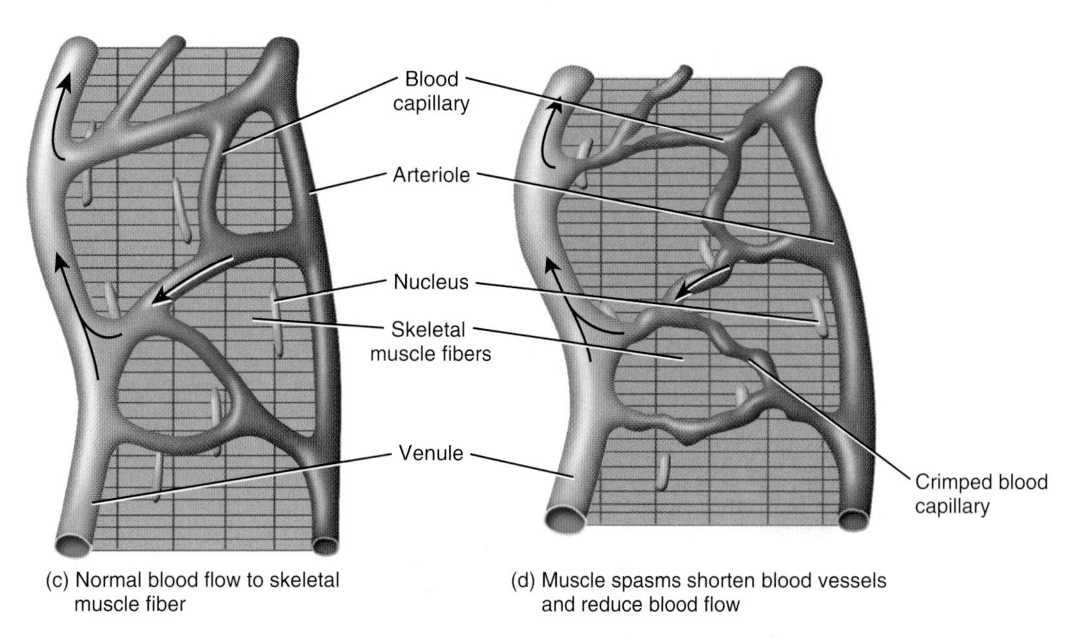

(c) Normal blood flow to skeletal muscle fiber

(d) Muscle spasms shorten blood vessels and reduce blood flow

Can you name a way in which the blood supply can be reestablished in a muscle that is in spasm?

This force is the muscle tension.) Each *tendon organ* consists of a thin *tendon organ capsule* composed of connective tissue that encloses a few tendon fascicles (bundles of collagen fibers). Penetrating the capsule are one or more *sensory nerve endings* that entwine among and around the collagen fibers of the tendon. When tension is applied to a muscle, the tendon organs generate nerve impulses that propagate into the CNS, providing information about changes in muscle tension. Tendon reflexes decrease muscle tension by causing muscle relaxation.

As noted previously, the area of hypertonicity (active spasm) is often confined to one or two portions of the muscle. A number of physiological and anatomical events are occurring within the area of spasm. (1) Shortening of the muscle belly causes the blood vessels in the area to become more convoluted or crimped, and the flow of blood to the area and distal to the area of spasm is reduced (see Figure 10.12d). (2) As noted in Section 10.5, normally the various motor neurons to a whole muscle fire *asynchronously*. While some motor units are contracting, others are relaxed. This pattern of motor unit activity delays muscle fatigue by allowing alternately contracting motor units to relieve one another, so that the contraction can be sustained for long periods. During a muscle spasm, however, the motor units in the area fire much more frequently, perhaps even continuously. Muscles are not designed for this level of continuous work. (3) Sensory neurons to the limbic system and other parts of the brain set a new, but incorrect, muscle tone and the gamma motor neurons then send impulses that cause the intrafusal fibers to stay contracted. (4) Lactic acid and other metabolic substances are produced at an accelerated pace within the area of spasm. Since the blood flow to the area has been significantly reduced, the lactic acid accumulates locally. (5) Nociceptors, sensory neurons for pain, respond to the lactic acid and waste products of muscle metabolism and send messages to the cerebral cortex where the neuronal impulses are interpreted as a burning sensation or pain in the area of hypertonicity. (6) At this point many persons with muscle spasms will seek the services of a manual therapist or other health practitioner.

In addition to the muscle spasms just described, myofascial dysfunctions can often be found in antagonistic (opposing) muscles or muscle groups. When a patient is unable to completely extend (straighten) the elbow, there are a few approaches that manual therapists can use to treat this condition. Combining palpation, medical history, symptomatology, and knowledge of anatomy and physiology, an assessment is made. The patient may complain of pain in the region of either antagonistic muscle. Assessment may indicate that one muscle is in spasm while the antagonistic muscle is weakened or lengthened. The palpatory feel to the therapist of a lengthened muscle is one of the tissue moving beyond the norm. Such a situation (in any joint) leaves the patient vulnerable to strains, sprains, and other soft tissue injuries.

One manual therapy school of thought would say that if the shortened muscle is lengthened by manual therapy techniques, the opposite muscle will tend to strengthen and balance over time. This is often the position of massage therapists and others. Another school of thought would say that the weakened or lax muscle must be strengthened by manual therapy techniques and then the shortened muscle will lengthen over time. This is often the position of physical therapists and others. An ideal treatment would combine the two protocols: lengthening of the shortened muscle by a clinical massage therapist and strengthening of the weakened muscle by a physical therapist.

MANUAL THERAPY APPLICATION

Treatment of Muscle Spasms

As noted previously, the causes of muscle spasms are many and varied. Other health professionals may be required to treat muscle spasms that are secondary to more complex diseases. The following discussion focuses on **treatment of muscle spasms** that are *not* secondary to other issues.

Several methods are utilized to treat muscle spasms, and choosing among them is the responsibility of the practitioner. (1) Exercise is a very important treatment and it may be the primary source of treatment or may supplement other treatments. (2) Slow stretching often brings relief. Spasms are, however, often localized within a muscle and stretching of the entire muscle may not be entirely effective. (3) Physical therapists and chiropractors may apply cold, heat, ultrasound, or electrical stimulation. (4) Chiropractors may realign bones if a misalignment is the primary cause of a spasm. (5) Physicians may prescribe pain medication, muscle relaxants, or trigger point injections; the analgesic injection may relieve pain immediately. (6) Acupuncturists may increase the level of endorphins in the region of spasm by inserting thin needles and thus also reset muscle tone to normal. (7) Clinical massage therapists also can very often successfully treat muscle spasms. Therapeutic massage can be slightly uncomfortable when it is being done, but the patient usually improves immediately or within 36 hours of treatment. The primary effect of all successful treatments is to interrupt cyclical neuronal circuits that involve *gamma motor neurons* and therefore *reset* muscle tone to a "normal" level. This concept will be discussed further in Chapter 19.

Figure 10.12c illustrates the normal blood flow and nerve supply to a skeletal muscle cell. A spasm causes the muscle to shorten, and one of the effects is to crimp the blood vessels and thus reduce blood flow to the area. A second effect of a spasm is to change the neuronal signals to and from the muscle spindle. One treatment of the spasm by a manual therapist is to apply pressure from a finger or a massage tool. This pressure causes a stretch of skin and its underlying connective tissues, a stretch of multiple forms of connective tissues in and around muscle, and then stretching of muscle fibers. If the pressure in the area of a spasm is held for 30 seconds or more, vessels often straighten, blood flow is reestablished, and metabolic waste products are able to move out of the area into the venous flow. Most importantly, however, the proprioceptors of muscle spindles are "reset" and gamma and alpha motor neurons are no longer stimulated to shorten the muscle in the area of the spasm.

CHECKPOINT

15. List possible sources of muscle spasms.
16. List the names given to abnormal contractions of skeletal muscle.

17. What is the significance of the four types of neurons present in each nerve connected to a muscle?
18. Describe manual therapy treatments for muscle spasms.

10.7 EXERCISE AND SKELETAL MUSCLE TISSUE

OBJECTIVE

- Describe the effects of exercise on skeletal muscle tissue.

The relative ratio of fast glycolytic (FG) and slow oxidative (SO) fibers in each muscle is genetically determined and helps account for individual differences in physical performance. For example, people with a higher proportion of FG fibers (see Table 10.1) often excel in activities that require periods of intense activity, such as weight lifting or sprinting. People with higher percentages of SO fibers are better at activities that require endurance, such as long-distance running.

Although the total number of skeletal muscle fibers usually does not increase, the characteristics of those present can change to some extent. Various types of exercises can induce changes in the fibers in a skeletal muscle. Endurance-type (aerobic) exercises, such as running or swimming, cause a gradual transformation of some FG fibers into fast oxidative–glycolytic (FOG) fibers. The transformed muscle fibers show slight increases in diameter, number of mitochondria, blood supply, and strength. Endurance exercises also result in cardiovascular and respiratory changes that cause skeletal muscles to receive better supplies of oxygen and nutrients but do not increase muscle mass. By contrast, exercises that require great strength for short periods produce an increase in the size and strength of FG fibers. The increase in size is due to increased synthesis of thick and thin filaments. The overall result is muscle enlargement (hypertrophy), as evidenced by the bulging muscles of body builders.

A certain degree of elasticity is an important attribute of skeletal muscles and their connective tissue attachments. Greater elasticity contributes to a greater degree of *flexibility,* increasing range of motion of a joint. When a relaxed muscle is physically stretched, its ability to lengthen is limited by connective tissue structures, such as fasciae. Regular stretching gradually lengthens these structures, but the process occurs very slowly. To see an improvement in flexibility, stretching exercises must be performed regularly—daily, if possible—for many weeks.

Effective Stretching

Tissues stretch best when slow, gentle force is applied at elevated tissue temperatures. An external source of heat, such as hot packs or ultrasound, may also be used. But 10 or more minutes of muscular contraction is also a good way to raise muscle temperature. Exercise heats muscle more deeply and thoroughly. That's where the term "warm-up" comes from. It's important to warm up *before* stretching, not vice versa. Stretching cold muscles does not increase flexibility and may cause injury.

Strength Training

Strength training exercise results not only in stronger muscles, but in many other health benefits as well. Strength training helps to increase bone strength, increasing the deposition of bone minerals in young adults and helping to prevent, or at least slow, their loss in later life. By increasing muscle mass, strength training raises resting metabolic rate, the amount of energy expended at rest, so a person can eat more food without gaining weight. Strength training helps to prevent back injury and injury from participation in sports and other physical activities. Psychological benefits include reductions in feelings of stress and fatigue. As exercise tolerance increases from repeated training, it takes increasingly longer before lactic acid is produced in the muscle, so there is a reduced probability of muscle spasms.

CHECKPOINT

19. Explain how the characteristics of skeletal muscle fibers may change with exercise.

10.8 CARDIAC MUSCLE TISSUE

OBJECTIVE

- Describe the structure and function of cardiac muscle tissue.

Most of the heart consists of ***cardiac muscle tissue.*** Like skeletal muscle, cardiac muscle is also *striated,* but its action is *involuntary:* Its alternating cycles of contraction and relaxation are not consciously controlled. Cardiac muscle fibers often are branched; are shorter in length and larger in diameter than skeletal muscle fibers; and have a single, centrally located nucleus (see Figure 22.2d). Cardiac muscle fibers interconnect with one another by irregular transverse thickenings of the sarcolemma called ***intercalated discs*** (in-TER-ka-lāt-ed = to insert between). The intercalated discs hold the fibers together and contain *gap junctions*, studied in Chapter 4, that allow muscle action potentials to spread quickly from one cardiac muscle fiber to another. Cardiac muscle tissue has an endomysium and perimysium, but lacks an epimysium.

A major difference between skeletal muscle and cardiac muscle is the source of stimulation. We have seen that skeletal muscle tissue contracts only when stimulated by acetylcholine released by a nerve impulse in a motor neuron. In contrast, the heart beats because some of the cardiac muscle fibers act as a pacemaker to initiate each cardiac contraction. The built-in or intrinsic rhythm of heart contractions is called ***autorhythmicity*** (aw′-tō-rith-MIS-i-tē). Several hormones and neurotransmitters can increase or decrease heart rate by speeding or slowing the heart's pacemaker.

Under normal resting conditions, cardiac muscle tissue contracts and relaxes an average of about 75 times a minute. Thus, cardiac muscle tissue requires a constant supply of oxygen and nutrients. The mitochondria in cardiac muscle fibers are larger and more numerous than in skeletal muscle fibers and produce most of the needed ATP via aerobic cellular respiration. As stated previously, cardiac muscle fibers can use lactic acid, released by skeletal muscle fibers during exercise, to make ATP.

CHECKPOINT

20. What are the major structural and functional differences between cardiac and skeletal muscle tissue?

10.9 SMOOTH MUSCLE TISSUE

OBJECTIVE

- Describe the structure and function of smooth muscle tissue.

Smooth muscle tissue is found in the skin, many internal organs, and blood vessels. Like cardiac muscle, smooth muscle is *involuntary.* Smooth muscle fibers are considerably smaller in length and diameter than skeletal muscle fibers and are tapered at both ends. Within each fiber is a single, oval, centrally located nucleus (Figure 10.13). In addition to thick and thin filaments, smooth muscle fibers also contain *intermediate filaments.* Because the various filaments have no regular pattern of overlap, smooth muscle fibers lack alternating dark and light bands and thus appear *nonstriated,* or smooth.

In smooth muscle fibers, the thin filaments attach to structures called *dense bodies,* which are functionally similar to Z discs in striated muscle fibers. Some dense bodies are dispersed throughout the sarcoplasm; others are attached to the sarcolemma. Bundles of intermediate filaments also attach to dense bodies and stretch from one dense body to another. During contraction, the sliding filament mechanism involving thick and thin filaments generates tension that is transmitted to intermediate filaments. These, in turn, pull on the dense bodies attached to the sarcolemma, causing a lengthwise shortening of the muscle fiber.

There are two kinds of smooth muscle tissue, visceral and multiunit. The more common type is ***visceral (single-unit) smooth muscle tissue*** (Figure 10.13a). It is found in sheets that wrap around to form part of the walls of small arteries and veins and hollow viscera such as the stomach, intestines, uterus, and urinary bladder. The fibers in visceral muscle tissue are tightly bound together in a continuous network. Like cardiac muscle, visceral smooth muscle is autorhythmic. Because the fibers connect to one another by gap junctions, muscle action potentials spread throughout the network. When a neurotransmitter, hormone, or autorhythmic signal stimulates one fiber, the muscle action potential spreads to neighboring fibers, which then contract in unison, as a single unit.

The second kind of smooth muscle tissue, ***multiunit smooth muscle tissue*** (Figure 10.13b), consists of individual fibers, each with its own motor nerve endings. Unlike stimulation of a single visceral muscle fiber, which causes contraction of many adjacent fibers, stimulation of a single multiunit smooth muscle fiber causes contraction of that fiber only. Multiunit smooth muscle

Figure 10.13 Histology of smooth muscle tissue.

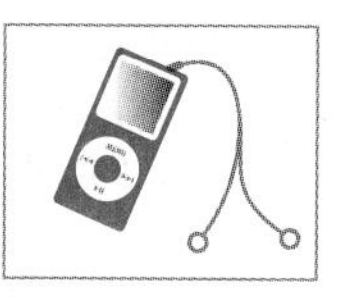

Smooth muscle lacks striations—it looks "smooth"—because the thick and thin filaments and intermediate filaments are irregularly arranged.

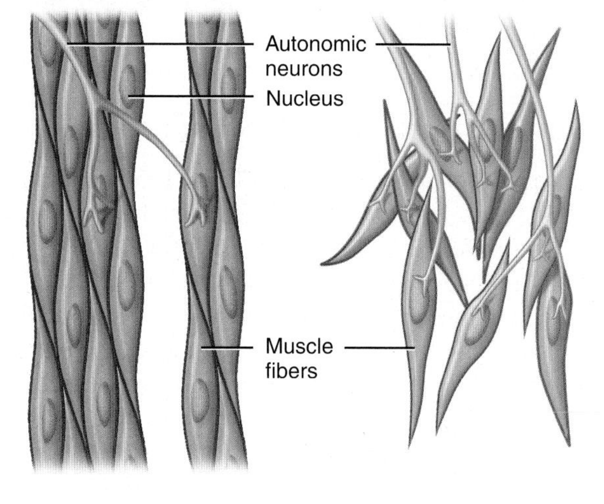

Which type of smooth muscle is found in the walls of hollow organs?

tissue is found in the walls of large arteries, in large airways to the lungs, in the arrector pili muscles attached to hair follicles, and in the internal eye muscles.

Compared with contraction in a skeletal muscle fiber, contraction in a smooth muscle fiber starts more slowly and lasts much longer. Calcium ions enter smooth muscle fibers slowly and also move slowly out of the muscle fiber when excitation declines, which delays relaxation. The prolonged presence of Ca^{2+} in the cytosol provides for ***smooth muscle tone,*** a state of continued partial contraction. Smooth muscle tissue can thus sustain long-term tone, which is important in the walls of blood vessels and in the walls of organs that maintain pressure on their contents. Finally, smooth muscle can both shorten and stretch to a greater extent than other muscle types. Stretchiness permits smooth muscle in the wall of hollow organs such as the uterus, stomach, intestines, and urinary bladder to expand as their contents enlarge, while still retaining the ability to contract.

Most smooth muscle fibers contract or relax in response to nerve impulses from the autonomic (involuntary) nervous system. In addition, many smooth muscle fibers contract or relax in response to stretching; hormones; or local factors such as changes in pH, oxygen and carbon dioxide levels, temperature, and ion concentrations. For example, the hormone epinephrine, released by the adrenal medulla, causes relaxation of smooth muscle in the airways and in some blood vessel walls.

Table 10.2 presents a summary of the major characteristics of the three types of muscular tissue.

Table 10.2

Summary of the Principal Features of Muscular Tissue

CHARACTERISTICS	SKELETAL MUSCLE	CARDIAC MUSCLE	SMOOTH MUSCLE
Cell appearance and features	Long cylindrical fiber with many peripherally located nuclei; striated; unbranched.	Branched cylindrical fiber, usually with one centrally located nucleus; intercalated discs join neighboring fibers; striated.	Fiber is thickest in the middle, tapered at each end, and has one centrally located nucleus; not striated.
Location	Primarily attached to bones by tendons.	Heart.	Walls of hollow viscera, airways, blood vessels, iris and ciliary body of the eye, arrector pili of hair follicles.
Fiber diameter	Very large (10–100 μm)*.	Large (10–20 μm).	Small (3–8 μm).
Fiber length	Very large (100 μm–30 cm = 12 inches).	Small (50–100 μm).	Intermediate (30–200 μm).
Sarcomeres	Yes.	Yes.	No.
Transverse tubules	Yes, aligned with each A–I band junction.	Yes, aligned with each Z disc.	No.
Speed of contraction	Fast.	Moderate.	Slow.
Nervous control	Voluntary.	Involuntary.	Involuntary.
Capacity for regeneration	Limited.	Limited.	Considerable compared with other muscle tissues, but limited compared with tissues such as epithelium.

*1 micrometer (μm) = 1/25,000 of an inch.

CHECKPOINT

21. How do visceral and multiunit smooth muscle differ?
22. What are the major structural and functional differences between smooth and skeletal muscle tissue?

10.10 AGING AND MUSCULAR TISSUE

OBJECTIVE

- Explain the effects of aging on skeletal muscle.

Between the ages of 30 and 50, humans undergo a slow, progressive loss of skeletal muscle mass that is replaced largely by nonfunctional fibrous connective tissue and adipose tissue. An estimated 10% of muscle mass is lost during these years. In part, this decline may be due to decreased levels of physical activity. Accompanying the loss of muscle mass is a decrease in maximal strength, a slowing of muscle reflexes, and a loss of flexibility. In some muscles, a selective loss of muscle fibers of a given type may occur. With aging, the relative number of slow oxidative (SO) fibers appears to increase. This could be due either to atrophy of the other fiber types or their conversion into slow oxidative fibers. Aerobic activities and strength training programs are effective in older people and can slow or even reverse the age-associated decline in muscular performance. Another 40% of muscle is typically lost between the ages of 50 and 80. Loss of muscle strength is usually not perceived by persons until they reach the age of 60 to 65. At that point it is most common for muscles of the lower limbs to weaken before those of the upper limbs. Thus the independence of the elderly may be affected when it becomes difficult to climb stairs or get up from a seated position.

Assuming that there is not a chronic medical condition for which exercise is contraindicated, exercise has been shown to be effective at any age. It is not necessary for the elderly to buy expensive equipment or join a health club to improve their muscular strength. For example, two cans of soup make great weights for maintaining or increasing the strength of the upper limbs. When soup cans become too easy, grab two large cans of kidney beans. Lower limbs can initially be strengthened by walking. When possible, the walking should be fun and for some persons that could occur in a nearby park when the weather permits or in an enclosed mall during inclement weather. Persons of all ages should probably park their car at the farthest corner of a mall parking lot rather than cruising for 15 minutes to find a parking space close to the stores. Within reason, packages carried back to the car after shopping will strengthen both upper and lower limb musculature. If accessible, plan on climbing one flight of steps at least twice a day. Those persons who are just beginning to sense that they are losing muscular strength would do well to hide the remote to the television set. Getting out of the chair, walking across the room, and stooping to push the buttons several times during the evening will help delay the inevitable loss of muscular strength as the aging process continues.

CHECKPOINT

23. Why does muscle strength decrease with aging?

KEY MEDICAL TERMS ASSOCIATED WITH MUSCULAR TISSUE

Electromyography or ***EMG*** (e-lek′-trō-mī-OG-ra-fē; *electro-* = electricity; *-myo-* = muscle; *-graphy* = to write) The recording and study of electrical changes that occur in muscular tissue.

Hypertonia (hī-per-TŌ-nē-a; *hyper-* = above) Increased muscle tone, characterized by increased muscle stiffness and sometimes associated with a change in normal reflexes.

Hypotonia (hī-pō-TŌ-nē-a; *hypo-* = below) Decreased or lost muscle tone.

Muscle strain Tearing of a muscle because of forceful impact, accompanied by bleeding and severe pain. Also known as a *charley horse* or pulled muscle. It often occurs in contact sports and typically affects the quadriceps femoris muscle on the anterior surface of the thigh.

Myalgia (mī-AL-jē-a; *-algia* = painful condition) Pain in or associated with muscles.

Myoma (mī-Ō-ma; *-oma* = tumor) A tumor consisting of muscular tissue.

Myomalacia (mī′-ō-ma-LĀ-shē-a; *-malacia* = soft) Pathological softening of muscle tissue.

Myositis (mī′-ō-SĪ-tis; *-itis* = inflammation of) Inflammation of muscle fibers (cells).

Myotonia (mī′-ō-TŌ-nē-a; *-tonia* = tension) Increased muscular excitability and contractility, with decreased power of relaxation; tonic spasm of the muscle.

STUDY OUTLINE

Overview of Muscular Tissue (Section 10.1)

1. The three types of muscular tissue are skeletal muscle, cardiac muscle, and smooth muscle (summarized in Table 10.2).
2. Skeletal muscle tissue is mostly attached to bones. It is striated and voluntary.
3. Cardiac muscle tissue forms most of the wall of the heart. It is striated and involuntary.
4. Smooth muscle tissue is located in viscera. It is nonstriated and involuntary.
5. Through contraction and relaxation, muscular tissue has five key functions: producing body movements, stabilizing body positions, regulating organ volume, moving substances within the body, and producing heat.

Skeletal Muscle Tissue (Section 10.2)

1. Connective tissue coverings associated with skeletal muscle extend from fascia and include the epimysium, covering an entire muscle; perimysium, covering fascicles; and endomysium, covering individual muscle fibers.
2. Tendons are extensions of connective tissue beyond muscle fibers that attach the muscle to bone.
3. Skeletal muscles are well supplied with nerves and blood vessels, which provide nutrients and oxygen for contraction.
4. Skeletal muscle consists of muscle fibers (cells) covered by a sarcolemma that features tunnel-like extensions, the transverse tubules. The fibers contain sarcoplasm, multiple nuclei, many mitochondria, myoglobin, and sarcoplasmic reticulum.
5. Each fiber also includes myofibrils that contain thin and thick filaments. The filaments are arranged in functional units called sarcomeres.
6. Thin filaments are composed of actin, tropomyosin, and troponin; thick filaments consist of myosin.

Contraction and Relaxation of Skeletal Muscle (Section 10.3)

1. Muscle contraction occurs when myosin heads attach to and "walk" along the thin filaments at both ends of a sarcomere, progressively pulling the thin filaments toward the center of a sarcomere. As the thin filaments slide inward, the Z discs come closer together, and the sarcomere shortens.
2. The neuromuscular junction (NMJ) is the synapse between a motor neuron and a skeletal muscle fiber. The NMJ includes the axon terminals and synaptic end bulbs of a somatic motor neuron plus the adjacent motor end plate of the muscle fiber sarcolemma.
3. A somatic motor neuron and all of the muscle fibers it stimulates form a motor unit. A single motor unit may include as few as 10 or as many as 3000 muscle fibers.
4. When a nerve impulse reaches the synaptic end bulbs of a motor neuron, it triggers the release of acetylcholine (ACh) from synaptic vesicles. ACh diffuses across the synaptic cleft and binds to ACh receptors, initiating a muscle action potential. Acetylcholinesterase then quickly destroys ACh.
5. An increase in the level of Ca^{2+} in the sarcoplasm, caused by the muscle action potential, starts the contraction cycle; a decrease in the level of Ca^{2+} turns off the contraction cycle.

6. The contraction cycle is the repeating sequence of events that causes sliding of the filaments: (1) myosin ATPase splits ATP and becomes energized, (2) the myosin head attaches to actin forming a cross-bridge, (3) the cross-bridge generates force as it swivels or rotates toward the center of the sarcomere (power stroke), and (4) binding of ATP to myosin detaches myosin from actin. The myosin head again splits ATP, returns to its original position, and binds to a new site on actin as the cycle continues.
7. Ca^{2+} active transport pumps continually remove Ca^{2+} from the sarcoplasm into the sarcoplasmic reticulum (SR). When the level of Ca^{2+} in the sarcoplasm decreases, the troponin–tropomyosin complexes slide back over and cover the myosin-binding sites, and the muscle fiber relaxes.
8. Continual involuntary activation of a small number of motor units produces muscle tone, which is essential for maintaining posture.

Metabolism of Skeletal Muscle Tissue (Section 10.4)

1. Muscle fibers have three sources for ATP production: creatine phosphate, anaerobic cellular respiration, and aerobic cellular respiration.
2. The transfer of a high-energy phosphate group from creatine phosphate to ADP forms new ATP molecules. Together, creatine phosphate and ATP provide enough energy for muscles to contract maximally for about 15 seconds.
3. Glucose is converted to pyruvic acid in the reactions of glycolysis, which yield two ATPs without using oxygen. These anaerobic reactions can provide enough ATP for about 30 to 40 seconds of maximal muscle activity.
4. Muscular activity that lasts longer than half a minute depends on aerobic cellular respiration, mitochondrial reactions that require oxygen to produce ATP. Aerobic cellular respiration yields about 36 molecules of ATP from each glucose molecule.
5. The inability of a muscle to contract forcefully after prolonged activity is muscle fatigue.
6. Elevated oxygen use after exercise is called recovery oxygen uptake.

Control of Muscle Tension (Section 10.5)

1. A twitch contraction is a brief contraction of all the muscle fibers in a motor unit in response to a single action potential.
2. A record of a contraction is called a myogram. It consists of a latent period, a contraction period, and a relaxation period.
3. Wave summation is the increased strength of a contraction that occurs when a second stimulus arrives before the muscle has completely relaxed after a previous stimulus.
4. Repeated stimuli can produce unfused tetanus, a sustained muscle contraction with partial relaxation between stimuli; more rapidly repeating stimuli will produce fused tetanus, a sustained contraction without partial relaxation between stimuli.
5. Motor unit recruitment is the process of increasing the number of active motor units.
6. On the basis of their structure and function, skeletal muscle fibers are classified as slow oxidative (SO), fast oxidative–glycolytic (FOG), and fast glycolytic (FG) fibers.
7. Most skeletal muscles contain a mixture of all three fiber types; their proportions vary with the typical action of the muscle.
8. The motor units of a muscle are recruited in the following order: first SO fibers, then FOG fibers, and finally FG fibers.
9. In an isometric contraction, there is no change in the length of a muscle, but the muscle develops considerable tension. In an isotonic contraction, there is a change in the length of a muscle, but no change in its tension.

Muscle Spasms (Section 10.6)

1. Muscle spasms have many names, many causes, and many potential treatments.
2. Muscle spasms commonly occur when a muscle is injured or overused.
3. Muscle spindles are the proprioceptors in skeletal muscles that monitor changes in the length of skeletal muscles.
4. Information from the muscle spindles is utilized by the brain to set a level of muscle tone for each muscle organ.
5. Muscle tone is transmitted through gamma motor neurons to the spindles.
6. In the area of the spasm, blood flow is reduced and all motor units are contracting very frequently, perhaps continuously.
7. Metabolic waste products of muscular contraction accumulate in the area of the spasm and nociceptors transmit signals that the brain interprets as pain.
8. Although many treatment modalities are available, many manual therapists utilize pressure to straighten the blood vessels and thus restore the flow of blood to normal. At the same time, the pressure causes proprioceptors to set the degree of muscle tone back to "normal."

Exercise and Skeletal Muscle Tissue (Section 10.7)

1. Various types of exercises can induce changes in the fibers in a skeletal muscle. Endurance-type (aerobic) exercises cause a gradual transformation of some fast glycolytic (FG) fibers into fast oxidative–glycolytic (FOG) fibers.
2. Exercises that require great strength for short periods produce an increase in the size and strength of fast glycolytic (FG) fibers. The increase in size is due to increased synthesis of thick and thin filaments.

Cardiac Muscle Tissue (Section 10.8)

1. Cardiac muscle tissue, which is striated and involuntary, is found only in the heart.
2. Each cardiac muscle fiber usually contains a single centrally located nucleus and exhibits branching.
3. Cardiac muscle fibers are connected by means of intercalated discs, which hold the muscle fibers together and allow muscle action potentials to quickly spread from one cardiac muscle fiber to another.
4. Cardiac muscle tissue contracts when stimulated by its own autorhythmic fibers. Due to its continuous, rhythmic activity, cardiac muscle depends greatly on aerobic cellular respiration to generate ATP.

Smooth Muscle Tissue (Section 10.9)

1. Smooth muscle tissue is nonstriated and involuntary.
2. In addition to thin and thick filaments, smooth muscle fibers contain intermediate filaments and dense bodies.
3. Visceral (single-unit) smooth muscle is found in the walls of hollow viscera and of small blood vessels. Many visceral fibers form a network that contracts in unison.

4. Multiunit smooth muscle is found in large blood vessels, large airways to the lungs, arrector pili muscles, and the eye. The fibers contract independently rather than in unison.
5. The duration of contraction and relaxation is longer in smooth muscle than in skeletal muscle.
6. Smooth muscle fibers can be stretched considerably and still retain the ability to contract.
7. Smooth muscle fibers contract in response to nerve impulses, stretching, hormones, and local factors.

Aging and Muscular Tissue (Section 10.10)

1. Beginning at about 30 years of age, there is a slow, progressive loss of skeletal muscle, which is replaced by fibrous connective tissue and fat.
2. Aging also results in a decrease in muscle strength, slower muscle reflexes, and loss of flexibility.
3. Exercise of all forms helps to delay the loss of muscle strength.

SELF-QUIZ QUESTIONS

1. The characteristic of muscular tissue that allows it to return to its original shape after contraction is
a. extensibility **b.** excitability **c.** fused tetanus
d. contractility **e.** elasticity.

2. Match the connective tissue coverings with their locations:
____ **a.** wraps an entire muscle
____ **b.** lies immediately under the skin
____ **c.** separates muscle organs into functional groups
____ **d.** surrounds each individual muscle fiber
____ **e.** divides muscle fibers into fascicles

A. endomysium
B. fascia
C. perimysium
D. epimysium
E. subcutaneous layer

3. Which of the following statements about skeletal muscle tissue is NOT true?
a. Skeletal muscle requires a large blood supply.
b. Skeletal muscle fibers have many mitochondria.
c. The arrangement of thick and thin filaments produces the striations in skeletal muscle tissue.
d. Skeletal muscle fibers contain gap junctions that help conduct action potentials from one fiber to another.
e. A skeletal muscle fiber has many nuclei.

4. Match the following:
____ **a.** network of tubules that stores calcium
____ **b.** pigment that stores oxygen
____ **c.** composed of myosin
____ **d.** composed of actin, tropomyosin, and troponin
____ **e.** tunnel-like extensions of sarcolemma

A. thick filaments
B. transverse tubules
C. sarcoplasmic reticulum
D. myoglobin
E. thin filaments

5. The sarcolemma is the equivalent of the
a. cytoplasm **b.** nucleus **c.** plasma membrane
d. endoplasmic reticulum **e.** mitochondrion.

6. You begin an intensive weight lifting plan because you want to enter a weight lifting contest. During the activity of weight lifting, your skeletal muscles will obtain energy (ATP) primarily through
a. anaerobic cellular respiration
b. the complete breakdown of pyruvic acid in the mitochondria
c. hyperplasia
d. hypertrophy
e. aerobic cellular respiration.

7. Which of the following events of skeletal muscle contraction does NOT occur during the latent period?
a. Sarcomeres shorten.
b. Action potentials conduct into the T tubules.
c. The concentration of calcium ions increases in the sarcoplasm.
d. Myosin-binding sites on the thin filaments are exposed.
e. Calcium release channels in the sarcoplasmic reticulum open.

8. For each of the following descriptions, indicate if it refers to skeletal muscle, cardiac muscle, or smooth muscle. Use the abbreviations SK for skeletal, CA for cardiac, and SM for smooth. The same response may be used more than once.
____ **a.** involuntary
____ **b.** multinucleated
____ **c.** striated
____ **d.** contain intercalated discs
____ **e.** elongated, cylindrical cells
____ **f.** voluntary
____ **g.** cells that taper at both ends
____ **h.** nonstriated
____ **i.** muscle fibers contract individually
____ **j.** autorhythmic

9. When ATP in the sarcoplasm is exhausted, the muscle must rely on _____ to quickly produce more ATP from ADP for contraction.
a. acetylcholine **b.** creatine phosphate
c. lactic acid **d.** pyruvic acid
e. acetylcholinesterase

10. A motor unit consists of
a. a transverse tubule and its associated sarcomeres
b. a motor neuron and all of the muscle fibers it stimulates
c. a muscle and all of its motor neurons
d. all of the filaments encased within a sarcomere
e. the motor end plate and the transverse tubules.

11. Thick filaments
a. include actin, troponin, and tropomyosin
b. compose the I band
c. stretch the entire length of a sarcomere
d. have binding sites for Ca^{2+}
e. have myosin heads (cross-bridges) used for the power stroke.

12. The substance that prevents the continuous stimulation of a muscle fiber is
a. Ca^{2+} **b.** acetylcholinesterase
c. ATP **d.** acetylcholine
e. troponin–tropomyosin.

13. Which of the following is NOT associated with muscle fatigue?
 a. depletion of creatine phosphate
 b. lack of oxygen
 c. decrease in Ca^{2+} levels in the sarcoplasm
 d. decrease in lactic acid levels
 e. lack of glycogen

14. All of the following may result in an increase in muscle size EXCEPT
 a. denervation atrophy b. weight training
 c. human growth hormone d. testosterone
 e. isotonic contraction.

15. Arrange the following in the correct order for skeletal muscle fiber contraction.
 1. Sarcoplasmic reticulum releases Ca^{2+}.
 2. Ca^{2+} combines with troponin.
 3. Acetylcholine is released from the axon terminal.
 4. Action potential travels into transverse tubules.
 5. Energized myosin heads (cross-bridges) attach to actin.
 6. Thin filaments slide toward the center of the sarcomere.
 a. 3, 4, 1, 2, 5, 6 b. 4, 3, 2, 1, 5, 6
 c. 1, 2, 3, 4, 5, 6 d. 4, 1, 3, 5, 2, 6
 e. 3, 1, 4, 5, 2, 6

16. Your instructor asks you to pick up a box of books and carry them to the library in another building. You try to pick up the box, but the box is too heavy to move. Which of the following types of muscle contractions would you be utilizing?
 a. hypertonic b. isotonic only c. spastic
 d. isometric only e. isometric and isotonic

17. Match the following:
 ____ a. extend from the thick filaments
 ____ b. contain myosin-binding site
 ____ c. dense area that separates sarcomeres
 ____ d. contain acetylcholine
 ____ e. striated zone of the sarcomere composed of thick and thin filaments
 ____ f. space between axon terminal and the sarcolemma
 ____ g. striated zone of the sarcomere composed of thin filaments only
 ____ h. region of sarcolemma near the adjoining axon terminal

 A. I band
 B. synaptic vesicles
 C. myosin heads
 D. Z discs
 E. motor end plate
 F. actin molecules
 G. A band
 H. synaptic cleft

CRITICAL THINKING QUESTIONS

1. Weight-lifter Jamal has been practicing many hours a day, and his muscles have gotten noticeably bigger. He tells you that his muscle cells are "multiplying like crazy and making him get stronger and stronger." Do you believe his explanation? Why or why not?
2. Chicken breasts are composed of "white meat" while chicken legs are composed of "dark meat." The breasts and legs of migrating ducks are dark meat. The breasts of both chickens and ducks are used in flying. How can you explain the differences in the color of the meat (muscles)? How are they adapted for their particular functions?
3. Polio is a disease caused by a virus that can attack the somatic motor neurons in the central nervous system. Individuals who suffer from polio can develop muscle weakness and atrophy. In a certain percentage of cases, the individuals may die due to respiratory paralysis. Relate your knowledge of how muscle fibers function to the symptoms exhibited by infected individuals.

? ANSWERS TO FIGURE QUESTIONS

10.1 In order from the inside toward the outside, the connective tissue layers are endomysium, perimysium, epimysium, fascia, and subcutaneous layer.

10.2 The A band has thick filaments in its center and overlapping thick and thin filaments at each end; the I band has thin filaments.

10.3 Proteins present on the A band are myosin, actin, troponin, and tropomyosin; the I band has actin, troponin, and tropomyosin.

10.4 The I bands disappear. The lengths of the thick and thin filaments do not change during contraction.

10.5 The motor end plate is the region of the sarcolemma near the axon terminal.

10.6 Binding of ATP to the myosin heads detaches them from actin.

10.7 The power stroke occurs during step 6.

10.8 Glycolysis, exchange of phosphate between creatine phosphate and ADP, and glycogen breakdown occur in the cytosol. Oxidation of pyruvic acid, amino acids, and fatty acids (aerobic cellular respiration) occurs in the mitochondria.

10.9 Sarcomeres shorten during the contraction period.

10.10 Fused tetanus occurs when the frequency of stimulation reaches about 90 stimuli per second.

10.11 Holding your head upright without movement involves mainly isometric contractions.

10.12 Pressure by a therapist on the muscle in spasm for 30 seconds or longer usually stretches the muscle and causes the tone of the muscle spindle complex to be reset.

10.13 The walls of hollow organs contain visceral (single-unit) smooth muscle.

11 The Muscular System: The Muscles of the Head and Neck

Manual therapists require an integral knowledge of the muscular system anatomy. Muscles often have injuries in the musculotendinous units or the muscle belly itself. Therefore you need a good understanding of the origin, insertion, and fiber direction for those muscles to effectively treat the key areas. You must also assess patients or clients to determine if the muscle is the primary source of injury or whether it comes from another source such as the nerve. If a nerve is injured it may cause either atrophy of the muscle or spasm, depending on the cause and severity of the injury. Muscles also require large amounts of oxygen to produce ATP for contraction; this in turn requires adequate blood supply. You will study all of these aspects for each of the muscles in the body. Beginning with the head and neck, and continuing with the rest of the body in later chapters, you will cover the major skeletal muscles in the body that provide the basis for manual therapy.

CONTENTS AT A GLANCE

11.1 HOW SKELETAL MUSCLES PRODUCE MOVEMENTS

OBJECTIVES

- Describe the relationship between bones and skeletal muscles in producing body movements.
- Define lever and fulcrum, and compare the three types of levers based on location of the fulcrum, effort, and load.
- Identify the types of fascicle arrangements in a skeletal muscle, and relate the arrangements to strength of contraction and range of motion.
- Explain how the prime mover, antagonist, synergist, and fixator in a muscle group work together to produce movement.

Knowledge of skeletal muscle structure and function is the basic foundation for health professionals, such as those in the manual therapies, who work with patients whose normal patterns of movement and physical mobility have been disrupted by physical trauma, levels of exercise, surgery, or muscular paralysis. This chapter introduces many of the major muscles in the body. The focus of this chapter, however, is on muscles of the head and neck. The subsequent three chapters will focus on the torso, upper limbs, and the lower limbs. Developing a working knowledge of these key aspects of skeletal muscle anatomy will enable you to understand how normal movements occur.

The muscular system comprises approximately 700 individual, voluntarily controlled muscle organs of your body. Most muscles are found on both the right and left sides of the body (which essentially halves the total number of muscles to be learned). We'll identify the attachment sites, actions, and innervation—the nerve or nerves that stimulate it to contract—of each muscle described. Almost all of the muscles that make up the muscular system, such as the biceps brachii muscle, include skeletal muscle tissue, connective tissue, blood vessels, and nerves. The function of most muscles is to produce movements of body parts. Certain muscles also function to stabilize bones so that other skeletal muscles can execute a movement more effectively.

Skeletal muscle function may be abnormal due to disease or damage of any of the components of a motor unit: motor neurons, neuromuscular junctions, or muscle fibers. The term ***neuromuscular disease*** encompasses problems at all three sites; the term ***myopathy*** (mī-OP-a-thē; *-pathy* = disease) signifies a disease or disorder of the skeletal muscle tissue itself.

Muscle Attachment Sites: Origin and Insertion

As we have already mentioned, not all skeletal muscles produce movements. Skeletal muscles that do produce movements do so by exerting force on tendons, which in turn pull on bones or other structures (such as skin or other muscles). Most muscles cross at least one joint and are usually attached to articulating bones that form the joint (see Figure 11.1a).

When a skeletal muscle contracts, it usually pulls one of the articulating bones toward the other. The two articulating bones typically do not move equally in response to contraction. One bone remains stationary or near its original position, either because other muscles stabilize that bone by contracting and pulling it in the opposite direction, or because its structure makes it less movable. Ordinarily, the attachment of a muscle's tendon to the stationary bone is called the ***origin*** (OR-i-jin); the attachment of the muscle's other tendon to the movable bone is called the ***insertion*** (in-SER-shun). A good analogy is a spring on a screen door. In this example, the part of the spring attached to the door frame is the origin; the part attached to the door represents the insertion. A useful rule of thumb is that the origin is usually proximal and the insertion distal, especially in the limbs; the insertion is usually pulled toward the origin. The fleshy portion of the muscle between the tendons is called the ***belly*** (*gaster*). The ***actions*** of a muscle are the main movements that occur when the muscle contracts. In our door spring example, the action would be the closing of the door. Certain muscles are also capable of ***reverse muscle action (RMA):*** During specific movements of the body the actions are reversed and therefore the positions of the origin and insertion of a specific muscle are switched.

Muscles that move a body part often do not cover the moving part. Figure 11.1b shows that although one of the functions of the biceps brachii muscle is to move the forearm, the belly of the muscle lies over the arm, not over the forearm. You will also see that muscles that cross two joints, such as the rectus femoris and sartorius of the thigh, have more complex actions than muscles that cross only one joint.

Lever Systems and Leverage

In producing movement, bones act as levers, and joints function as the fulcrums of these levers. A ***lever*** is a rigid structure that can move around a fixed point called a ***fulcrum,*** Ⓕ. A lever is acted on at two different points by two different forces: the ***effort*** (E), which causes movement, and the ***load*** Ⓛ or ***resistance,*** which opposes movement. The effort is the force exerted by muscular contraction, whereas the load is typically the weight of the body part that is moved. Motion occurs when the effort applied to the bone at the insertion exceeds the load. Consider the biceps brachii muscle flexing the forearm at the elbow as an object is lifted (see Figure 11.1b). When the forearm is raised, the elbow is the fulcrum. The weight of the forearm plus the weight of the object in the hand is the load. The force of contraction of the biceps brachii muscle pulling the forearm up is the effort.

Levers produce trade-offs between effort, speed, and range of motion. A lever operates at a *mechanical advantage*—has ***leverage***—when a smaller effort can move a heavier load. Here the trade-off is that the effort must move a greater distance (must have a longer range of motion) and faster than the load. Recall from Chapter 9 that range of motion (ROM) refers to the range, measured in degrees of a circle, through which the bones of a joint can be moved. The lever formed by the mandible at the

Figure 11.1 Relationship of skeletal muscles to bones. (a) Muscles are attached to bones by tendons known as the origin and the insertion. (b) Skeletal muscles produce movements by pulling on bones. Bones serve as levers, and joints act as fulcrums for the levers. Here the lever–fulcrum principle is illustrated by the movement of the forearm. Note where the load (resistance) and effort are applied in this example.

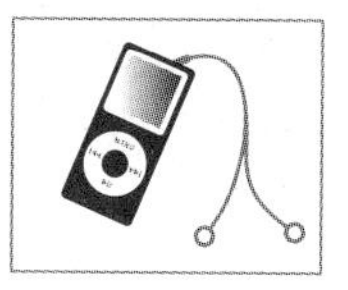

In the limbs, the origin of a muscle is usually proximal and the insertion is usually distal.

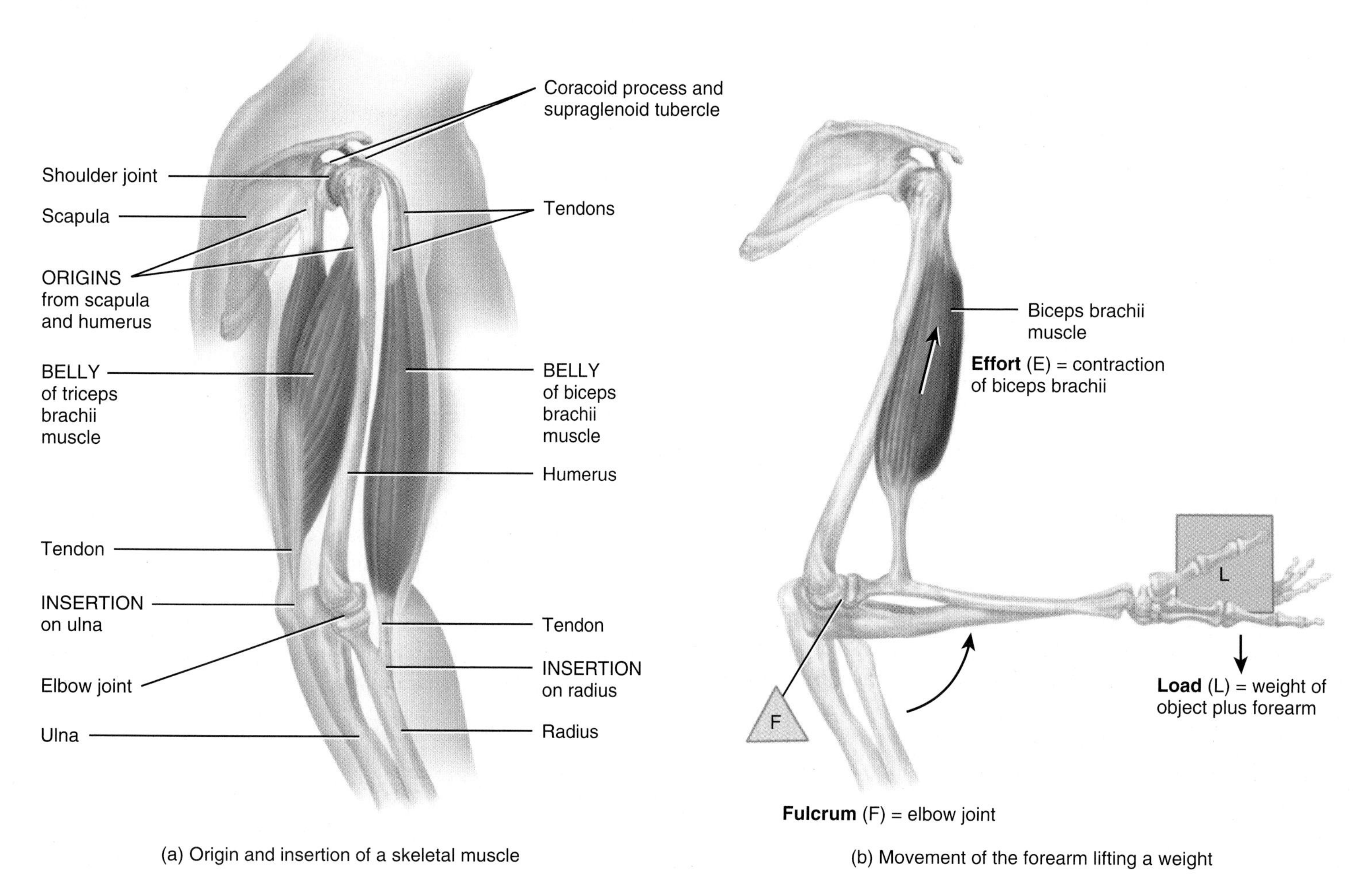

(a) Origin and insertion of a skeletal muscle

(b) Movement of the forearm lifting a weight

Where are the bellies of the muscles that extend the forearm located?

temporomandibular joints (fulcrums) and the effort provided by contraction of the jaw muscles produce a high mechanical advantage that crushes food. In contrast, a lever operates at a *mechanical disadvantage* when a larger effort moves a lighter load. In this case the trade-off is that the effort must move more slowly and for a shorter distance than the load. The lever formed by the humerus at the shoulder joint (fulcrum) and the effort provided by the back and shoulder muscles produces a mechanical "disadvantage" that enables a major-league pitcher to hurl a baseball at nearly 100 miles per hour.

The positions of the effort, load, and fulcrum on the lever determine whether the lever operates at a mechanical advantage or disadvantage. When the load is close to the fulcrum and the effort is applied farther away, the lever operates at a mechanical advantage. When you chew food, the load (the food) is positioned close to the fulcrums (your temporomandibular joints) while your jaw muscles exert effort farther away from the joints. By contrast, when the effort is applied close to the fulcrum and the load is farther away, the lever operates at a mechanical disadvantage. When a pitcher throws a baseball, the back and shoulder muscles apply intense effort very close to the fulcrum (the shoulder joint) while the lighter load (the ball) is propelled at the far end of the lever (the arm bone).

Levers are categorized into three types according to the positions of the fulcrum, the effort, and the load:

1. The fulcrum is between the effort and the load in ***first-class levers*** (Figure 11.2a). (Think E*F*L; the *fulcrum* is in the middle.) Note that on one side of the body, the order might be E*F*L but on the opposite side it might be L*F*E. Although the sequence is reversed, the *F* is still in the middle. Scissors and seesaws are examples of first-class levers. A first-class lever can produce either

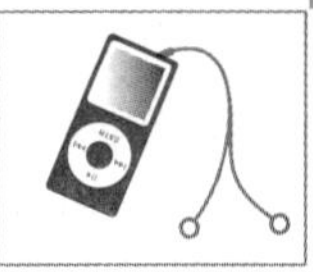

Figure 11.2 Types of levers.

Levers are divided into three types based on the placement of the fulcrum, effort, and load (resistance).

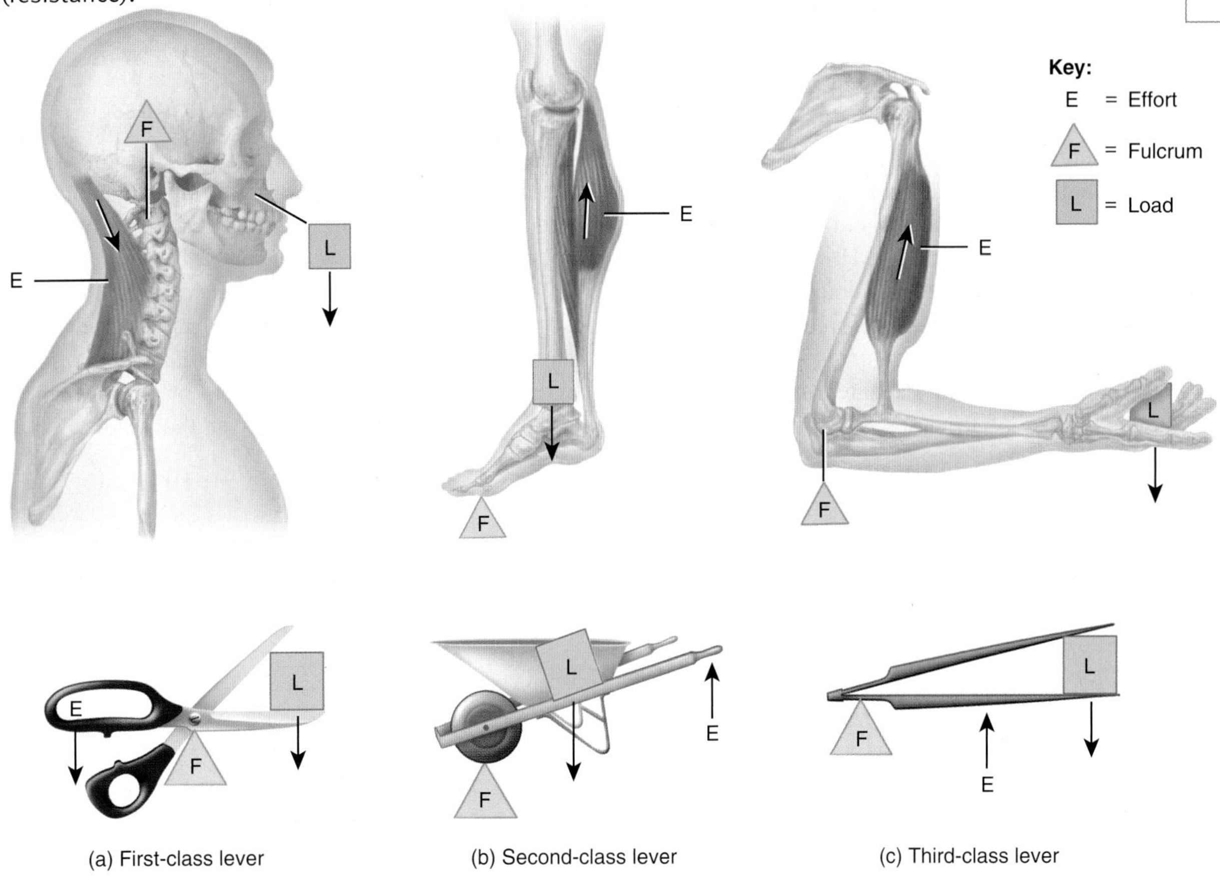

Which type of lever produces the most force?

a mechanical advantage or disadvantage depending on whether the effort or the load is closer to the fulcrum. (Think of an adult and a child on a seesaw.) As we've seen in the preceding examples, if the effort (child) is farther from the fulcrum than the load (adult), a heavy load can be moved, but not very far or fast. If the effort is closer to the fulcrum than the load, only a lighter load can be moved, but it moves far and fast. There are few first-class levers in the body. One example is the lever formed by the head resting on the vertebral column (Figure 11.2a). When the head is raised, the contraction of the posterior neck muscles provides the effort (E), the joint between the atlas and the occipital bone (atlanto-occipital joint) forms the fulcrum (F), and the weight of the anterior portion of the skull is the load (L).

2. The load is between the fulcrum and the effort in ***second-class levers*** (Figure 11.2b). (Think FLE; the *load* is in the middle.) They operate like a wheelbarrow. Second-class levers always produce a mechanical advantage because the load is always closer to the fulcrum than the effort. This arrangement sacrifices speed and range of motion for force; this type of lever produces the most force. One example is raising the body on the toes. The body is the load (L), the ball of the foot is the fulcrum (F), and the contraction of the calf muscle to pull the heel upward is the effort (E).

3. The effort is between the fulcrum and the load in ***third-class levers*** (Figure 11.2c). (Think FEL; *effort* is in the middle.) These levers operate like a pair of forceps (tweezers) and are the most common levers in the body. Third-class levers always produce a mechanical disadvantage because the effort is always closer to the fulcrum than the load. In the body, this arrangement favors speed and range of motion over force. The elbow joint, the biceps brachii muscle, and the bones of the arm and forearm are one example of a third-class lever (Figure 11.2c). As we have seen, in flexing the forearm at the elbow, the elbow joint is the fulcrum (F), the contraction of the biceps brachii muscle provides the effort (E), and the weight of the hand and forearm is the load (L). Note in Figure 11.2c that it is the position of the *tendon* that attaches to the lever that is important; the position of the muscle belly is not relevant when discussing levers. Another example of the action of a third-class lever is adduction of the thigh, in which the hip joint is the fulcrum, the contraction of the adductor muscles is the effort, and the thigh is the load.

Effects of Fascicle Arrangement

Recall from Chapter 10 that the skeletal muscle fibers (cells) within a muscle are arranged in bundles known as ***fascicles.*** Within a fascicle, all muscle fibers are parallel to one another. The fascicles, however, may form one of five patterns with respect to the tendons: parallel, fusiform (shaped like a cigar), circular, triangular, or pennate (shaped like a feather) (Table 11.1).

Fascicular arrangement affects a muscle's power and range of motion. As a muscle fiber contracts, it shortens to about 50% of its resting length. Thus, the longer the fibers in a muscle, the greater the range of motion it can produce. However, the power of a muscle depends not on length but on its total cross-sectional area, because a short fiber can contract as forcefully as a long one. Fascicular arrangement often represents a compromise between power and range of motion. Pennate muscles, for instance, have a large number of fascicles distributed over their tendons, giving them greater power but a smaller range of motion. Parallel muscles, in contrast, have comparatively few fascicles that extend the length of the muscle, so they have a greater range of motion but less power.

Coordination within Muscle Groups

Movements of the body often occur as the result of several skeletal muscles acting as a group rather than acting alone. Most skeletal muscles are arranged in opposing (antagonistic) pairs at joints—that is, flexors/extensors, abductors/adductors, and so on. Within opposing pairs, one muscle, called the ***prime mover,*** also known as the ***agonist*** (= leader), contracts to cause an action while the other muscle, the ***antagonist*** (*anti-* = against), stretches and yields to the effects of the prime mover. In the process of flexing the forearm at the elbow, for instance, the

TABLE 11.1

Arrangement of Fascicles

PARALLEL

Fascicles parallel to longitudinal axis of muscle; terminate at either end in flat tendons.

Example: Sternohyoid muscle (see Figure 11.8).

FUSIFORM

Fascicles nearly parallel to longitudinal axis of muscle; terminate in flat tendons; muscle tapers toward tendons, where diameter is less than at belly.

Example: Digastric muscle (see Figure 11.8).

CIRCULAR

Fascicles in concentric circular arrangements form sphincter muscles that enclose an orifice (opening).

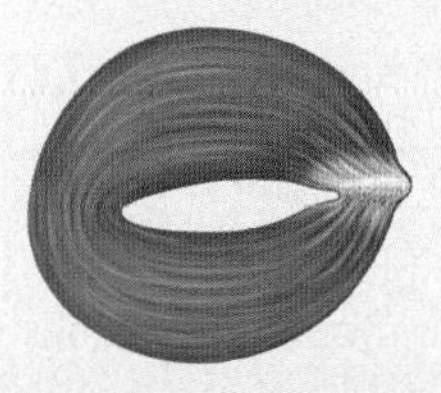

Example: Orbicularis oculi muscle (see Figure 11.4).

TRIANGULAR

Fascicles spread over broad area converge at thick central tendon; gives muscle a triangular appearance.

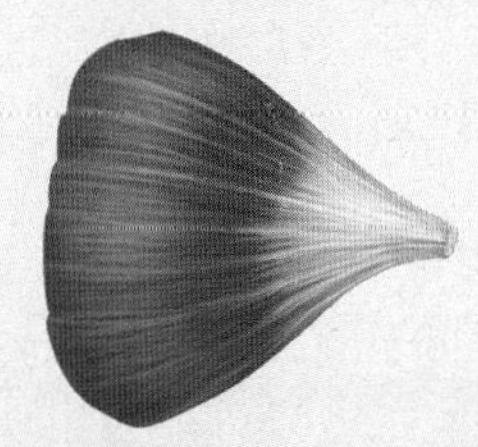

Example: Pectoralis major muscle (see Figure 11.3a).

PENNATE

Short fascicles in relation to total muscle length; tendon extends nearly entire length of muscle.

Unipennate

Fascicles are arranged on only one side of tendon.

Example: Extensor digitorum longus muscle (see Figure 14.6g).

Bipennate

Fascicles are arranged on both sides of centrally positioned tendons.

Example: Rectus femoris muscle (see Figure 14.4b).

Multipennate

Fascicles attach obliquely from many directions to several tendons.

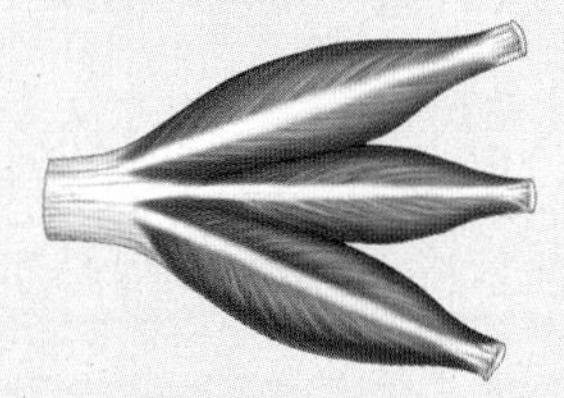

Example: Deltoid muscle (see Figure 11.3).

biceps brachii is the prime mover, and the triceps brachii is the antagonist (see Figure 11.1a). The antagonist and prime mover are usually located on opposite sides of the bone or joint, as is the case in this example.

With an opposing pair of muscles, the roles of the prime mover and antagonist can switch for different movements. For example, while extending the forearm at the elbow (i.e., lowering the load shown in Figure 11.1), the triceps brachii becomes the prime mover, and the biceps brachii is the antagonist. If a prime mover and its antagonist contract at the same time with equal force, there will be no movement. Such activity is critical for a gymnast supporting the weight of his body on the rings without movement of the upper limbs.

Muscles (muscle organs) sometimes assist or aid other muscles and are called ***synergists*** (SIN-er-gists; *syn-* = together; *-ergon* = work). Synergists are usually located close to the prime mover. The biceps brachii muscle, for example, spans both the shoulder and elbow joints. In flexion of the arm at the shoulder, the coracobrachialis muscle is a synergist to the biceps brachii muscle (Figure 13.4a), whereas in flexion of the forearm at the elbow, the brachialis muscle is a synergist to the biceps brachii muscle.

Some muscles in a group also act as ***fixators***, which stabilize the origin of the prime mover so that the prime mover can act more efficiently. Fixators steady the proximal end of a limb while movements occur at the distal end. For example, the scapula in the pectoral (shoulder) girdle is a freely movable bone that serves as the origin for several muscles that move the arm. When the arm muscles contract, the scapula must be held steady. As another example, in abduction of the arm, the deltoid muscle serves as the prime mover, whereas fixators (pectoralis minor, trapezius, subclavius, serratus anterior muscles, and others) hold the scapula firmly against the posterior of the chest (see Figure 13.1). The insertion of the deltoid muscle pulls on the humerus to abduct the arm. As you will learn in Chapter 13, the supraspinatus muscle is a synergist in the action of abduction of the arm. Under different conditions—that is, for different movements—and at different times, many muscles may act alternatively as prime movers, antagonists, synergists, or fixators.

In the limbs, a ***compartment*** is a group of skeletal muscles, along with their associated blood vessels and nerves, that have a common function. In the upper limbs, for example, flexor compartment muscles are anterior, whereas extensor compartment muscles are typically posterior. Recognizing this compartmentalization, learning of muscle positions, actions, and innervations will require significantly less effort if they are learned by compartment, rather than individually. Muscles of a given compartment commonly have the same major action, same innervation, and same blood supply.

CHECKPOINT

1. Using the terms origin, insertion, and belly in your discussion, describe how skeletal muscles produce body movements by pulling on bones.
2. Describe the three types of levers, and give an example of a first- and third-class lever found in the body.
3. Describe the various arrangements of fascicles.
4. Define the roles of the prime mover (agonist), antagonist, synergist, and fixator in producing various movements of the upper limb.

11.2 HOW SKELETAL MUSCLES ARE NAMED

OBJECTIVE

- Explain seven features used in naming skeletal muscles.

The names of most of the skeletal muscles contain combinations of the word roots of their distinctive features. You can remember the names of muscles by learning the terms that refer to muscle features, such as the pattern of the muscle's fascicles; the size, shape, action, number of origins, and location of the muscle; and the sites of origin and insertion of the muscle. Familiarity with the names of the muscles will give you clues to their features. Study Table 11.2 to become familiar with the terms used in muscle names.

CHECKPOINT

5. Select 10 muscles in Figure 11.3 and identify the features on which their names are based. (Hint: *Use the prefix, suffix, and root of each muscle's name as a guide.*)

11.3 PRINCIPAL SKELETAL MUSCLES OF THE HEAD AND NECK

Exhibits and figures in this and the next three chapters will assist you in learning the names of the principal skeletal muscles in the body. The muscles in the exhibits are divided into groups according to the part of the body on which they act. As you study groups of muscles in the exhibits of these four chapters on muscle, refer to Figure 11.3 to see how each group is related to the others.

The exhibits contain the following elements:

- ***Objective.*** This statement describes what you should learn from the exhibit.
- ***Overview.*** These paragraphs provide a general introduction to the muscles under consideration and emphasize how the muscles are organized within various regions. The discussion also highlights any distinguishing features of the muscles.
- ***Muscle names.*** The word roots indicate how most of the muscles are named. As noted previously, once you have mastered the naming of the muscles, you can more easily understand their actions.
- ***Origins, insertions, and actions.*** You are also given the origin, insertion, actions, and reverse muscle actions (if any) of each muscle.

TABLE 11.2

Characteristics Used to Name Muscles

NAME	MEANING	EXAMPLE	FIGURE
DIRECTION: ORIENTATION OF MUSCLE FASCICLES RELATIVE TO THE BODY'S MIDLINE			
Rectus	Parallel to midline.	Rectus abdominis.	12.2
Transverse	Perpendicular to midline.	Transversus abdominis.	12.2
Oblique	Diagonal to midline.	External oblique.	12.2
SIZE: RELATIVE SIZE OF THE MUSCLE			
Maximus	Largest.	Gluteus maximus.	14.1d
Minimus	Smallest.	Gluteus minimus.	14.1e
Longus	Long.	Adductor longus.	14.1b
Brevis	Short.	Adductor brevis.	14.1b
Latissimus	Widest.	Latissimus dorsi.	11.3b
Longissimus	Longest.	Longissimus capitis.	12.7a
Magnus	Large.	Adductor magnus.	14.1b
Major	Larger.	Pectoralis major.	11.3a
Minor	Smaller.	Pectoralis minor.	13.1a
Vastus	Huge.	Vastus lateralis.	14.4b
SHAPE: RELATIVE SHAPE OF THE MUSCLE			
Deltoid	Triangular.	Deltoid.	11.3a,b
Trapezius	Trapezoid.	Trapezius.	11.3b
Serratus	Saw-toothed.	Serratus anterior.	13.1b
Rhomboid	Diamond-shaped.	Rhomboid major.	13.1d
Orbicularis	Circular.	Orbicularis oculi.	11.4a
Pectinate	Comblike.	Pectineus.	14.1a
Piriformis	Pear-shaped.	Piriformis.	14.1e
Platys	Flat.	Platysma.	11.3a
Quadratus	Square, four-sided.	Quadratus femoris.	14.1e
Gracilis	Slender.	Gracilis.	14.4a
ACTION: PRINCIPAL ACTION OF THE MUSCLE			
Flexor	Decreases a joint angle.	Flexor carpi radialis.	13.8a
Extensor	Increases a joint angle.	Extensor carpi ulnaris.	13.8d
Abductor	Moves a bone away from the midline.	Abductor pollicis longus.	13.8e
Adductor	Moves a bone closer to the midline.	Adductor longus.	14.4a
Levator	Raises or elevates a body part.	Levator scapulae.	13.1a
Depressor	Lowers or depresses a body part.	Depressor labii inferioris.	11.4a
Supinator	Turns palm anteriorly.	Supinator.	13.6h
Pronator	Turns palm posteriorly.	Pronator teres.	13.6i
Sphincter	Decreases the size of an opening.	External anal sphincter.	12.6
Tensor	Makes a body part rigid.	Tensor fasciae latae.	14.1a
Rotator	Rotates a bone around its longitudinal axis.	Rotatore.	12.7a
NUMBER OF ORIGINS: NUMBER OF TENDONS OF ORIGIN			
Biceps	Two origins.	Biceps brachii.	13.6a
Triceps	Three origins.	Triceps brachii.	13.6b
Quadriceps	Four origins.	Quadriceps femoris.	14.1a
LOCATION: STRUCTURE NEAR WHICH A MUSCLE IS FOUND			
Example: Temporalis, a muscle near the temporal bone.			11.4d
ORIGIN AND INSERTION: SITES WHERE MUSCLE ORIGINATES AND INSERTS			
Example: Sternocleidomastoid, originating on the sternum and clavicle and inserting on the mastoid process of the temporal bone.			11.11a

- ***Innervation.*** This section lists the nerve or nerves that cause contraction of each muscle. In general, cranial nerves, which arise from the lower parts of the brain, serve muscles in the head region. Spinal nerves, which arise from the spinal cord within the vertebral column, innervate muscles in the rest of the body. Cranial nerves are designated by both a name and a Roman numeral—for example, the facial (VII) nerve. Spinal nerves are numbered in groups according to the part of the spinal cord from which they arise: C = cervical (neck region), T = thoracic (chest region), L = lumbar (lower back region), and S = sacral (buttocks region). An example is T1, the first thoracic spinal nerve.
- ***Relating muscles to movements.*** These exercises will help you organize the muscles in the body region under consideration according to the actions they produce.
- ***Figures.*** The figures in the exhibits may present superficial and deep, anterior and posterior, or medial and lateral views to show each muscle's position as clearly as possible. Please note that the muscle names in all capital letters are specifically referred to in the tables of the exhibit.
- ***Surface Anatomy.*** In this chapter we will begin to take a closer look at ***surface anatomy.*** Surface anatomy is the study of the anatomical landmarks on the exterior of the body. A knowledge of surface anatomy will help you to not only identify structures on the body's exterior, but also locate the positions of various internal structures.

The study of surface anatomy involves two related, although distinct, activities: visualization and palpation. ***Visualization*** involves looking in a very selective and purposeful manner at a specific part of the body. ***Palpation*** (pal-PĀ-shun) means using the sense of touch to determine the location of an internal part of the body through the skin. Like visualization, palpation is performed selectively and purposefully, and it supplements information already gained through other methods, including visualization. Because of the varying depth of different structures from the surface of the body and due to differences in the thickness of the subcutaneous layer over different parts of the body between females (thicker) and males (thinner), palpations may range from light to moderate to deep.

Knowledge of surface anatomy has many applications, both anatomically and clinically. From an anatomical viewpoint, a knowledge of surface anatomy can provide valuable information regarding the location of structures such as bones, muscles, blood vessels, nerves, lymph nodes, and internal organs. Clinically, a knowledge of surface anatomy is the basis for conducting a physical examination and performing certain diagnostic tests. Health-care professionals use their understanding of surface anatomy to learn where to take the pulse, measure blood pressure, draw blood, stop bleeding, insert needles and tubes, make surgical incisions, reduce (straighten) fractured bones, and listen to sounds made by the heart, lungs, and intestines. Clinicians also draw on a knowledge of surface anatomy to assess the status of lymph nodes and to identify the presence of tumors or other unusual masses in the body.

Because living bodies are best suited for studying surface anatomy, you will find it helpful to visualize and/or palpate the various structures described on your own body as you study each region. Following a brief introduction to a region of the body, you will be presented with a list of the prominent structures to locate. This directed approach will help to organize your learning efforts. Labeled photographs illustrate most of the structures listed in each region.

Principal skeletal muscles of the head and neck are presented in Exhibits 11.1 through 11.6.

Figure 11.3 Principal superficial skeletal muscles.

Most movements require several skeletal muscles acting in groups rather than individually. Note the large areas of connective tissue that are illustrated in white.

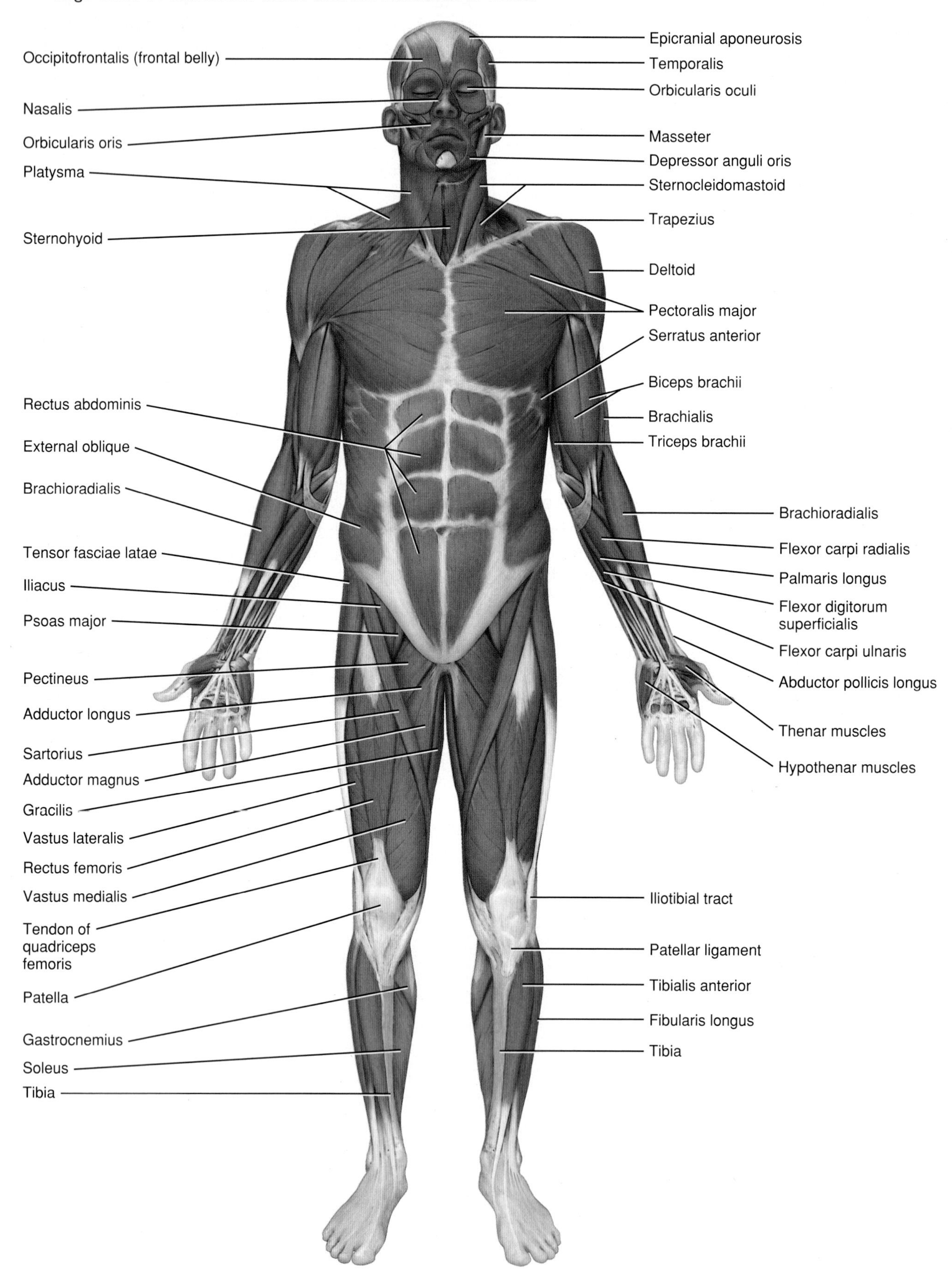

(a) Anterior view

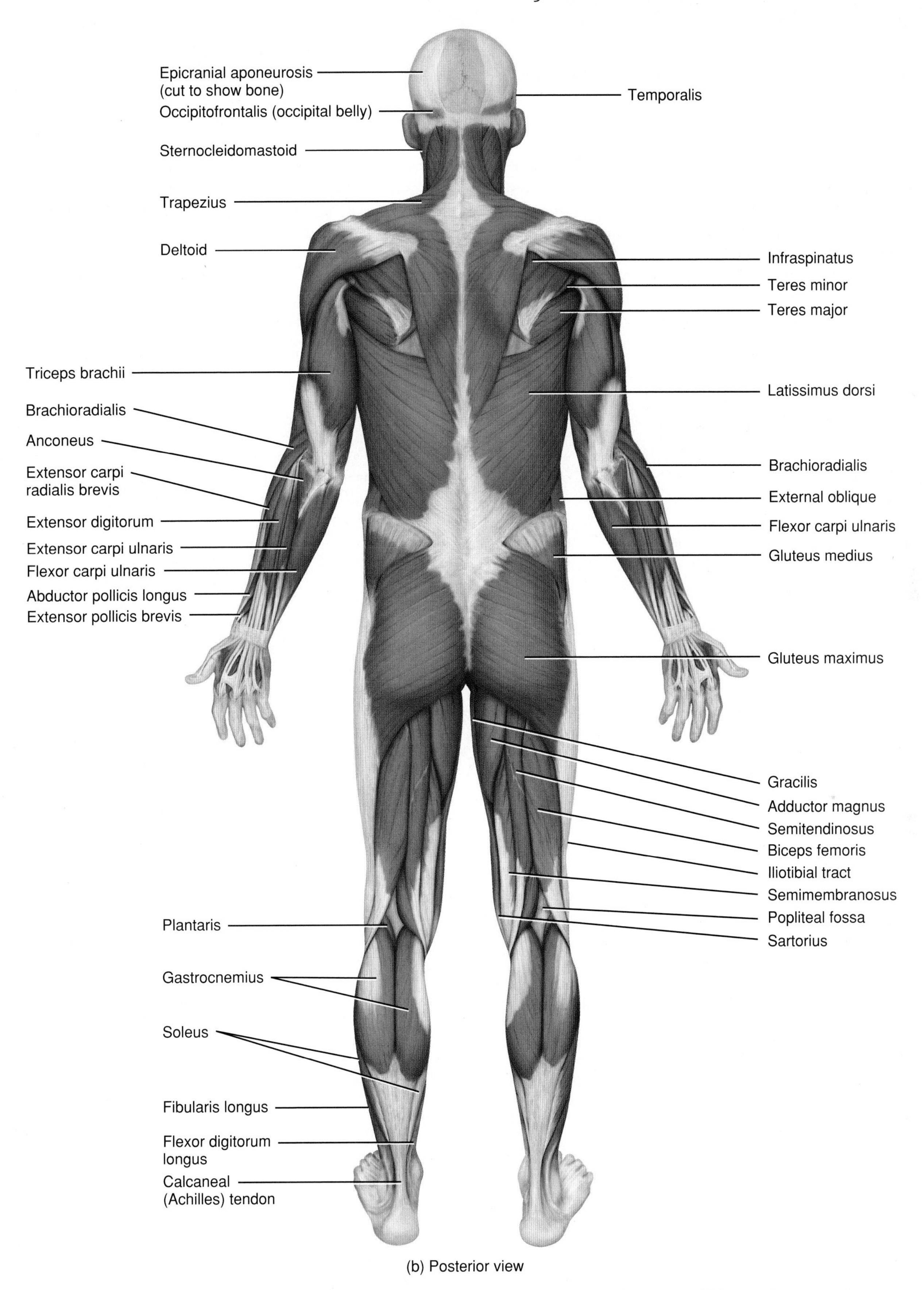

(b) Posterior view

Give an example of a muscle named for each of the following characteristics: direction of fibers, shape, action, size, origin and insertion, location, and number of tendons of origin.

EXHIBIT 11.1 Muscles of Facial Expression

OBJECTIVES

- Describe the origin, insertion, and innervation of the muscles of facial expression.
- Describe major surface features of the face.

The ***head*** (*cephalic region,* or *caput*) contains the brain and sense organs (eyes, ears, nose, and tongue) and is divided into the cranium and face. The ***cranium*** (*skull*, or *brain case*) is that portion of the head that surrounds and protects the brain; the ***face*** is the anterior portion of the head.

The muscles of facial expression (see Figure 11.4) provide us with the ability to express a wide variety of emotions, including displeasure, surprise, fear, and happiness. The muscles themselves lie within the layers of superficial fascia (connective tissue beneath the skin). They usually originate in the fascia or bones of the skull, but they are unusual muscles in that they insert into skin or other muscles rather than bones. Because of their insertions, the muscles of facial expression move the skin rather than a joint when they contract.

The ***occipitofrontalis*** is an unusual muscle in this group because it is made up of two parts: an anterior part called the ***frontal belly (frontalis),*** which is superficial to the frontal bone, and a posterior part called the ***occipital belly (occipitalis),*** which is superficial to the occipital bone. The two muscular portions are held together by a strong aponeurosis (sheetlike

MUSCLE	ORIGIN	INSERTION	ACTION	INNERVATION
SCALP MUSCLES				
Occipitofrontalis (ok-sip′-i-tō-frun-TĀ-lis)				
Frontal belly (frontalis)	Epicranial aponeurosis (galea aponeurotica).	Skin superior to supraorbital margin.	Draws scalp anteriorly, as in frowning. RMA: Raises eyebrows and wrinkles skin of forehead horizontally as in a look of surprise.	Facial (VII) nerve.
Occipital belly (occipitalis) (*occipit-* = back of the head)	Occipital bone and mastoid process of temporal bone.	Epicranial aponeurosis.	Draws scalp posteriorly.	Facial (VII) nerve.
MOUTH MUSCLES				
Orbicularis oris (or-bi′-kū-LAR-is OR-is; *orb-* = circular; *oris* = of the mouth)	Muscle fibers surrounding opening of mouth.	Skin at corner of mouth.	Closes and protrudes lips, as in kissing; compresses lips against teeth; and shapes lips during speech.	Facial (VII) nerve.
Zygomaticus major (zī-gō-MA-ti-kus; *zygomatic* = cheek bone; *major* = greater)	Zygomatic bone.	Skin at angle of mouth and orbicularis oris.	Draws angle of mouth superiorly and laterally, as in smiling.	Facial (VII) nerve.
Zygomaticus minor (*minor* = lesser)	Zygomatic bone.	Upper lip.	Elevates (raises) upper lip, exposing maxillary (upper) teeth.	Facial (VII) nerve.
Nasalis	Maxilla.	Skin over cartilaginous portion of nose.	Dilates external nares.	Facial (VII) nerve.
Levator labii superioris (le-VĀ-tor LĀ-bē-ī soo-per′-ē-OR-is; *levator* = raises or elevates; *labii* = lip; *superioris* = upper)	Superior to infraorbital foramen of maxilla.	Skin at angle of mouth and orbicularis oris.	Elevates (raises) upper lip.	Facial (VII) nerve.
Depressor labii inferioris (de-PRE-sor LĀ-bē-ī; *depressor* = depresses or lowers; *inferioris* = lower)	Mandible.	Skin of lower lip.	Depresses (lowers) lower lip.	Facial (VII) nerve.

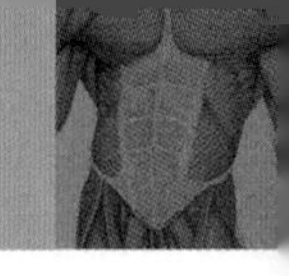

tendon), the ***epicranial aponeurosis*** (ep-i-KRĀ-nē-al ap-ō-noo-RŌ-sis), also known as ***galea aponeurotica*** (GĀ-lē-a ap-ō-noo-RO-ti-ka), which covers the superior and lateral surfaces of the skull. As will be discussed in the next three chapters, muscles on the plantar surface of the foot; posterior leg, thigh, and hip; back; and posterior neck all have fascial connections. Tension (tightness) of one or more muscles in any of these locations has the potential to affect the epicranial aponeurosis of the occipitofrontalis and, therefore, may cause headaches and other symptoms.

The ***buccinator*** muscle forms the major muscular portion of the cheek. The buccinator muscle is so named because it compresses the cheeks (*bucc-* = cheek) during blowing—for example, when a musician plays a wind instrument such as a trumpet. It also functions in whistling, blowing, and sucking, and assists in chewing.

Among the noteworthy muscles of facial expression are those surrounding the orifices (openings) of the head, such as the eyes, nose, and mouth. These muscles function as *sphincters* (SFINGK-ters), which close the orifices, and *dilators,* which open the orifices. For example, the ***orbicularis oculi*** muscle closes the eye, and the ***levator palpebrae superioris*** muscle opens it. Other muscles cause the lips to change orientation, for example, the ***depressor labii inferioris*** and ***zygomaticus major***. All muscles listed in this group are innervated by the same nerve, the ***facial nerve,*** also known as cranial nerve VII.

MUSCLE	ORIGIN	INSERTION	ACTION	INNERVATION
MOUTH MUSCLES (continued)				
Depressor anguli oris (*angul* = angle or corner)	Mandible.	Angle of mouth.	Draws angle of mouth laterally and inferiorly, as in opening mouth.	Facial (VII) nerve.
Levator anguli oris	Inferior to infraorbital foramen.	Skin of lower lip and orbicularis oris.	Draws angle of mouth laterally and superiorly.	Facial (VII) nerve.
Buccinator (BUK-si-nā′-tor; *bucc-* = cheek)	Alveolar processes of maxilla and mandible near the molar teeth.	Orbicularis oris.	Presses cheeks against teeth and lips, as in whistling, blowing, and sucking; draws corner of mouth laterally, and assists in mastication (chewing) by keeping food between the teeth (and not between teeth and cheeks).	Facial (VII) nerve.
Risorius (ri-ZOR-ē-us; *risor* = laughter)	Fascia over parotid (salivary) gland.	Skin at angle of mouth.	Draws angle of mouth laterally, as in grimacing.	Facial (VII) nerve.
Mentalis (men-TĀ-lis; *ment-* = chin)	Mandible near the midline.	Skin of inferior chin.	Elevates and protrudes lower lip and pulls skin of chin up, as in pouting.	Facial (VII) nerve.
NECK MUSCLE				
Platysma (pla-TIZ-ma; *platy-* = flat, broad)	Fascia over deltoid and pectoralis major muscles.	Mandible, muscle around angle of mouth, and skin of lower face.	Draws outer part of lower lip inferiorly and posteriorly as in pouting; depresses mandible.	Facial (VII) nerve.
ORBIT AND EYEBROW MUSCLES				
Orbicularis oculi (or-bi′-kū-LAR-is OK-ū-lī; *oculi* = of the eye)	Medial wall of orbit.	Circular path around orbit.	Closes eye.	Facial (VII) nerve.
Corrugator supercilii (KOR-a-gā′-tor soo′-per-SI-lē-ī; *corrugat* = wrinkle; *supercilii* = of the eyebrow)	Medial end of superciliary arch of frontal bone.	Skin of eyebrow.	Draws eyebrow inferiorly and medially and wrinkles skin of forehead vertically as in frowning.	Facial (VII) nerve.

CONTINUES

Figure 11.4 **Muscles of the face and neck.** (a) This anterior superficial view would be seen when the skin and subcutaneous layer of the face are removed. (b) This anterior deep view would be seen after removal of all the superficial muscles. (c) Origins (red) and insertions (blue) of muscles of the face are illustrated on the anterior skull. Note that the image includes many more origins than insertions since muscles of facial expression insert into skin or other muscles rather than bone. (d) Right lateral superficial view; the skin and superficial fascia have been removed. (e) Origins and insertions of muscles of the face are illustrated on the right lateral skull.

When they contract, muscles of facial expression move the skin rather than a joint. Muscles of facial expression, muscles that move the mandible, and muscles that move the head on the vertebral column are closely intertwined.

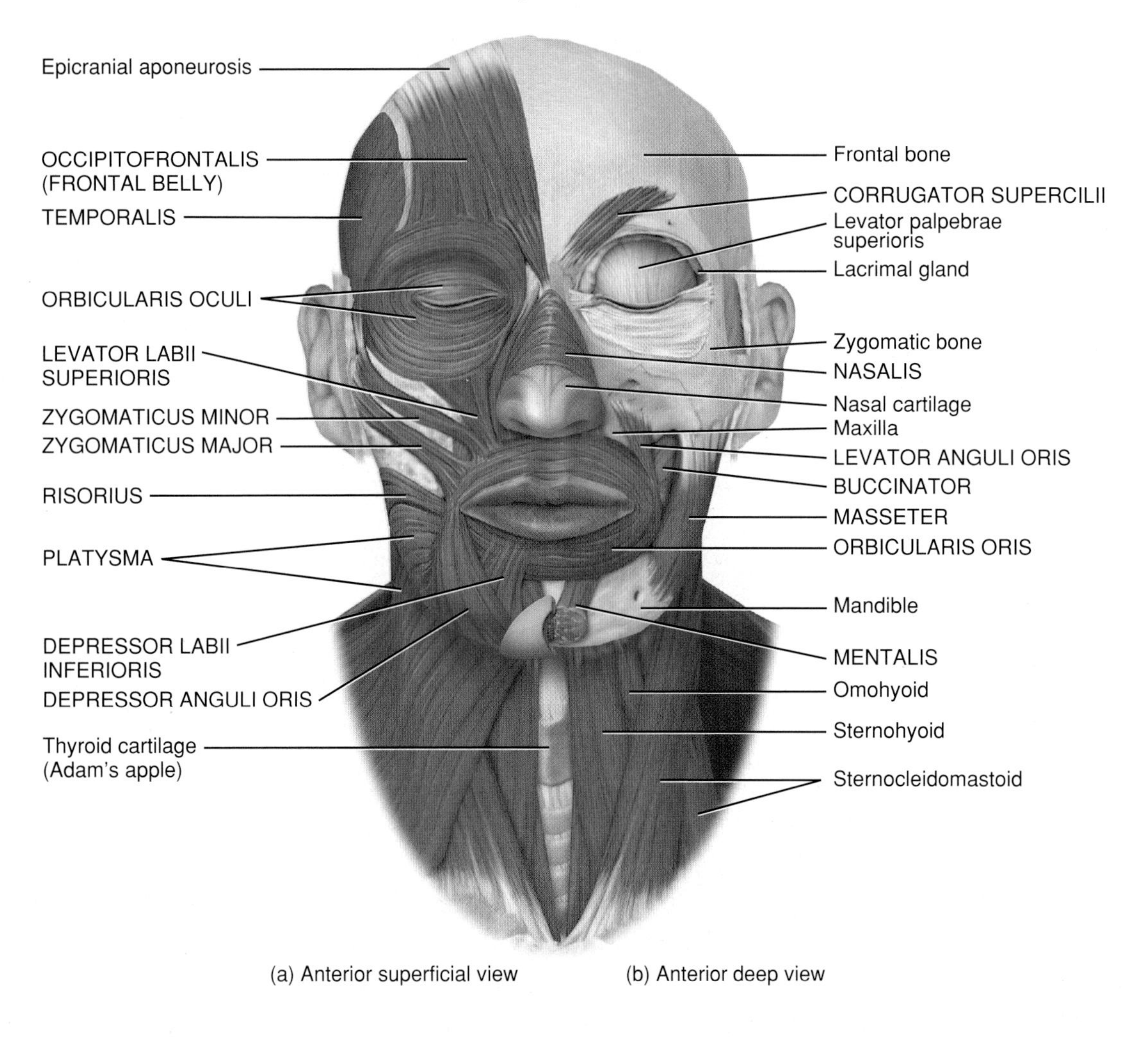

(a) Anterior superficial view (b) Anterior deep view

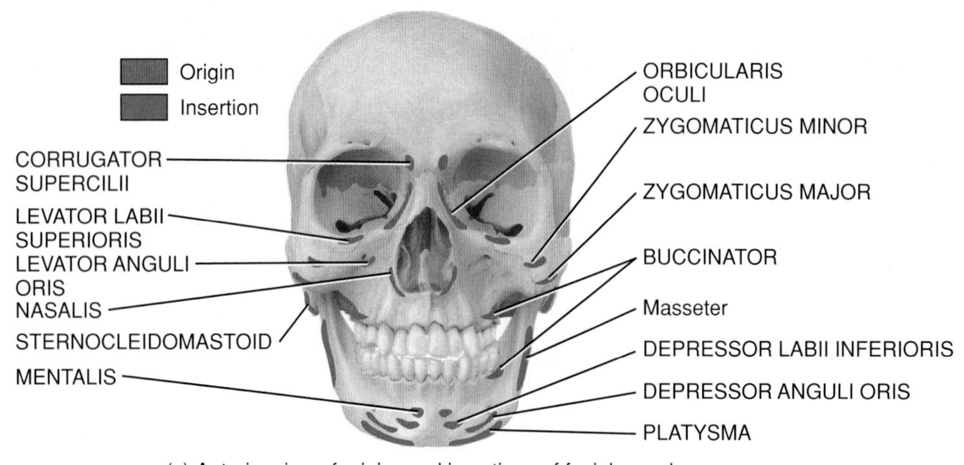

(c) Anterior view of origins and insertions of facial muscles

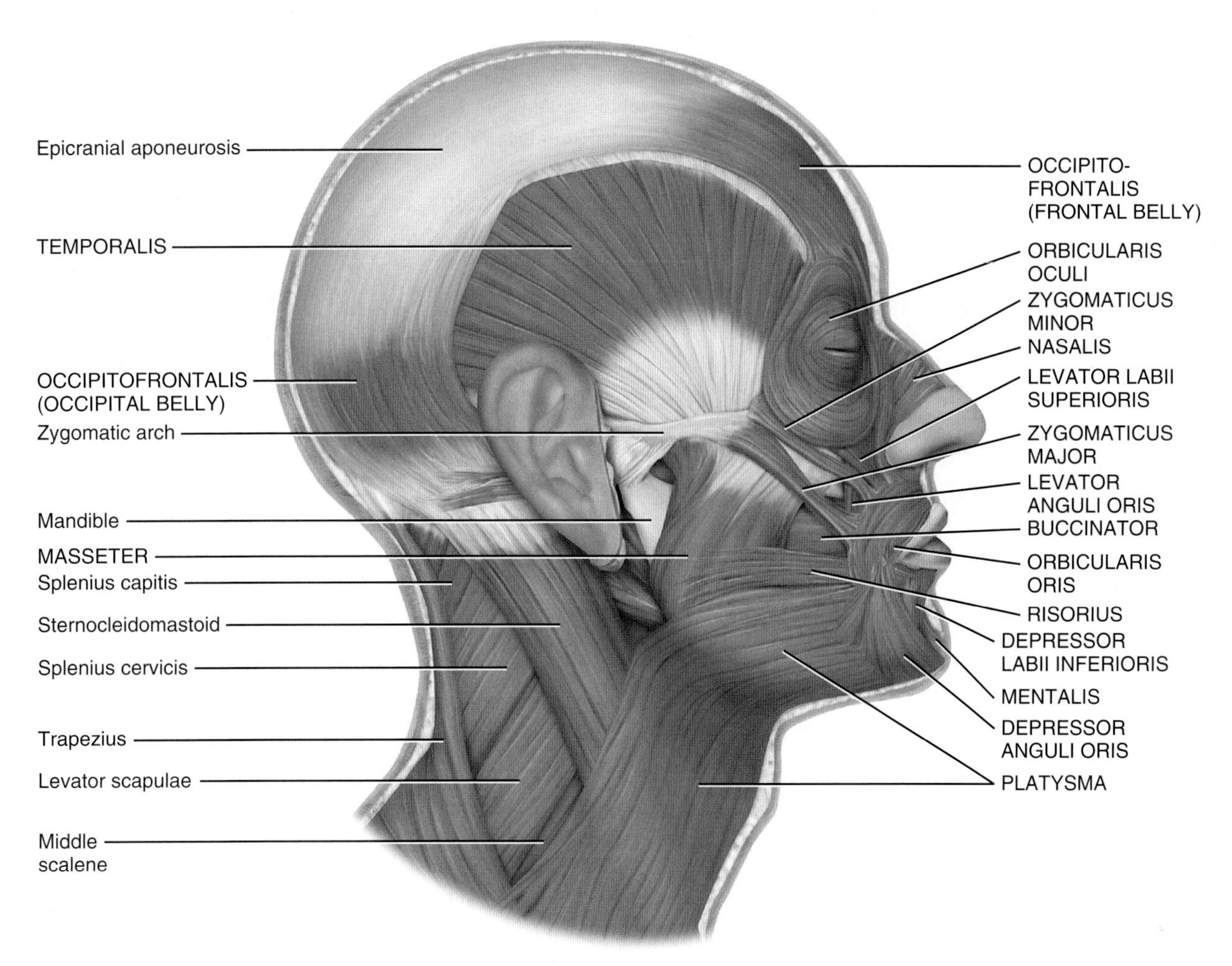

(d) Right lateral superficial view

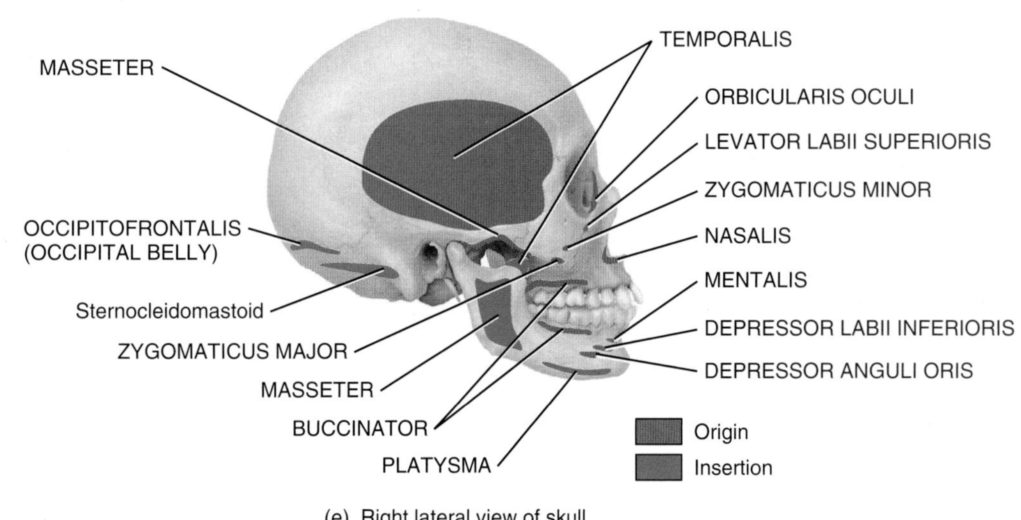

(e) Right lateral view of skull

Which muscles of facial expression cause frowning, smiling, pouting, and squinting?

CONTINUES

Surface Anatomy of the Head

Several bony structures of the skull (Figure 11.5) can be detected through palpation, including the following:

- **Sagittal suture.** Move the fingers from side to side over the superior aspect of the scalp in order to palpate this suture (see Figure 7.6).
- **Coronal and lambdoid sutures.** Use the same side-to-side technique to palpate these sutures located on the frontal and occipital regions of the skull (see Figure 7.4).
- **External occipital protuberance.** This is the most prominent bony landmark on the occipital region of the skull (see Figure 7.6).
- **Orbit.** The entire circumference of the orbit can be palpated. Deep to the eyebrow, the superior aspect of the orbit, you can palpate the **supraorbital margin** of the frontal bone.
- **Nasal bones.** These can be palpated in the nasal region between the orbits on either side of the midline. If you wear glasses, the bridge of the glasses sits on these bones.
- **Mandible.** The ***ramus*** (vertical portion), ***body*** (horizontal portion), and ***angle*** (area where the ramus meets the body) of the mandible can be easily palpated at the mental and buccal regions of the head.
- **Zygomatic arch and zygomatic bones.** These can be palpated in the zygomatic region (see Figure 7.3).
- **Mastoid process.** This is the prominent bony landmark situated posterior to the ear that is easily palpated in the auricular region (see Figure 7.4).

Muscles of Facial Expression

Several muscles of facial expression can be palpated while they are contracting:

- **Occipitofrontalis.** By raising and lowering the eyebrows, it is possible to palpate the frontal belly and the occipital belly in the frontal and occipital regions, respectively, as they alternately contract and the scalp moves forward and backward.
- **Orbicularis oculi.** By closing the eyes and placing the fingers on the eyelids, the muscle can be palpated in the orbital region by tightly squeezing the eyes shut.
- **Corrugator supercilii.** By frowning and drawing the eyebrows toward each other, this muscle can be felt above the nose near the medial end of the eyebrow.
- **Zygomaticus major.** By smiling, this muscle can be palpated between the corner of the mouth and zygomatic bone.
- **Depressor labii inferioris.** This muscle can be palpated in the mental region between the lower lip and chin when moving the lower lip inferiorly to expose the lower teeth.
- **Orbicularis oris.** By closing the lips tightly, this muscle can be palpated in the oral region around the margin of the lips.

Muscles of Mastication

Two muscles of mastication (see Exhibit 11.3) can be palpated while they are contracting:

- **Temporalis.** By alternately clenching the teeth and then opening the mouth, it is possible to palpate this muscle in the temporal region just superior to the zygomatic arch.
- **Masseter.** Again, by alternately clenching the teeth and then opening the mouth, this muscle can be palpated over the ramus of the mandible.

Relating Muscles to Movements

Arrange the muscles in this exhibit into two groups: (1) those that act on the mouth and (2) those that act on the eyes.

• CLINICAL CONNECTION | Bell's Palsy

Bell's palsy, also known as **facial paralysis,** is a unilateral paralysis of the muscles of facial expression. It is due to damage or disease of the facial (VII) nerve. Possible causes include inflammation of the facial nerve due to an ear infection, ear surgery that damages the facial nerve, or infection by the herpes simplex virus. The paralysis causes the entire side of the face to droop in severe cases. The person cannot wrinkle the forehead, close the eye, or pucker the lips on the affected side. Drooling and difficulty in swallowing also occur. Eighty percent of patients recover completely within a few weeks to a few months. For others, paralysis is permanent. The symptoms of Bell's palsy mimic those of a stroke. •

CHECKPOINT

6. Why do the muscles of facial expression move the skin rather than a joint?

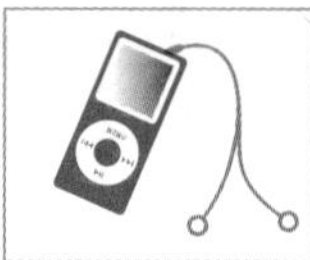

Figure 11.5 Surface anatomy of the head.

Several muscles can be palpated while they are contracting.

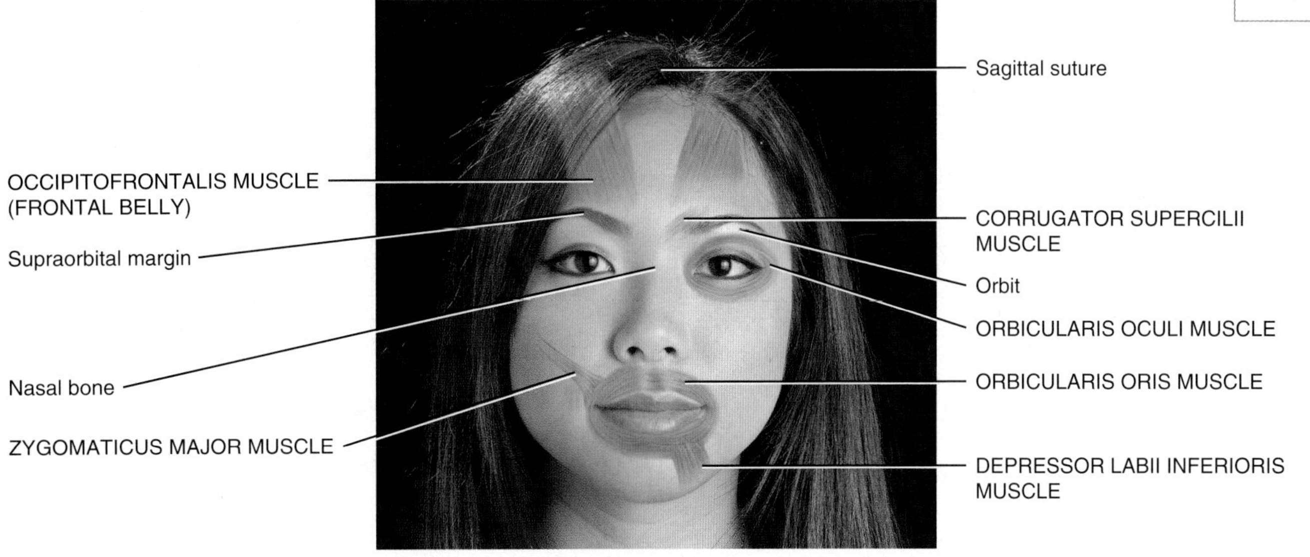

(a) Anterior view

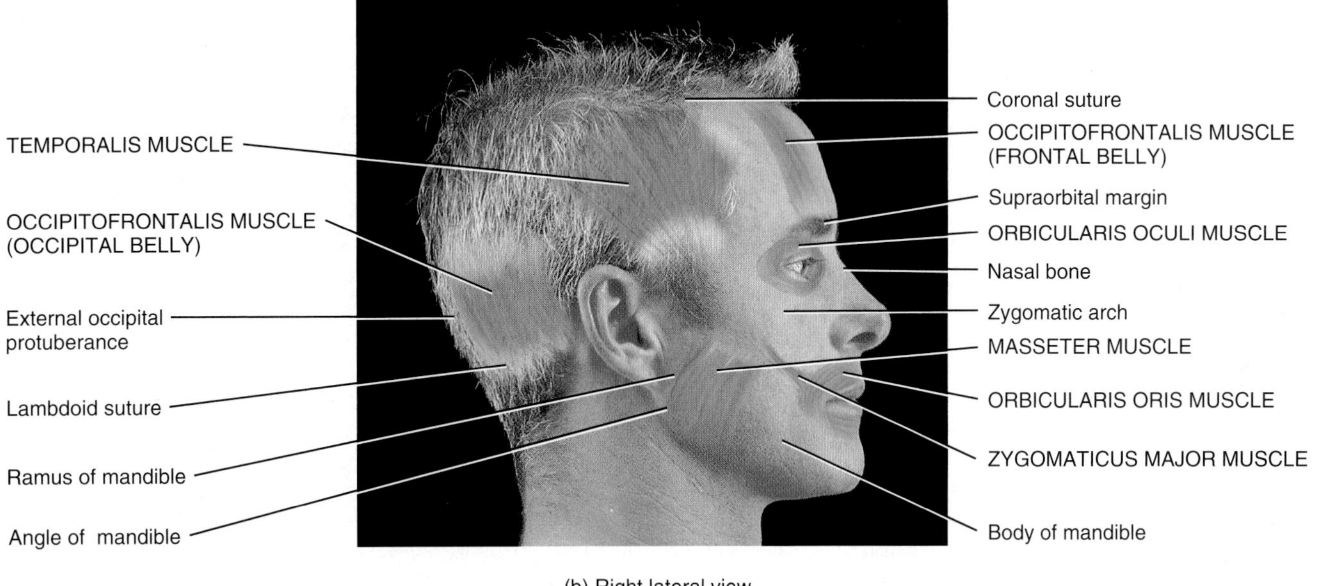

(b) Right lateral view

? When palpating the sagittal suture, what sheetlike tendon is also palpated deep to the skin?

EXHIBIT 11.2 Muscles That Move the Eyeballs and Upper Eyelids (Extrinsic Eye Muscles)

OBJECTIVE

- Describe the origin, insertion, action, and innervation of the extrinsic eye muscles.

Muscles that move the eyeballs are called ***extrinsic eye muscles*** because they originate outside the eyeballs (in the orbit) and insert on its outer surface, the sclera (Figure 11.6). The extrinsic eye muscles are some of the fastest contracting and most precisely controlled skeletal muscles in the body.

Three pairs of extrinsic eye muscles control movements of the eyeballs: (1) superior and inferior recti, (2) lateral and medial recti, and (3) superior and inferior oblique. The four recti muscles (superior, inferior, lateral, and medial) arise from a common tendinous ring at the back of the orbit and insert into the sclera of the eyeball. The names of the four recti muscles clearly imply their actions. When the ***superior rectus*** muscle contracts, the eye looks up; as the ***inferior rectus*** muscle contracts, the eye looks down. Contraction of the ***lateral rectus*** muscle makes the eye move laterally, and when the ***medial rectus*** muscle contracts, the eye moves medially.

The oblique muscles—superior and inferior—rotate the eyeball on its axis. The ***superior oblique*** muscle originates posteriorly near the tendinous ring, passes anteriorly, and ends in a round tendon that extends through a pulley-like loop called the *trochlea* (= pulley) where it turns and inserts on the superior, lateral eyeball. When the superior oblique muscle contracts, the eyeball moves inferiorly and laterally. The ***inferior oblique*** muscle originates on the maxilla at the anterior, medial floor of the orbit. It then passes posteriorly and laterally and inserts on the posteriolateral aspects of the eyeball. Because of this arrangement, the inferior oblique muscle moves the eyeball superiorly and laterally.

The ***levator palpebrae superioris,*** unlike the recti and oblique muscles, does not move the eyeballs. Rather, it raises the upper eyelids, that is, opens the eyes. It is therefore an antagonist to the orbicularis oculi, which closes the eye.

Relating Muscles to Movements

Arrange the muscles in this exhibit according to their actions on the eyeballs: (1) elevation, (2) depression, (3) abduction, (4) adduction, (5) medial rotation, and (6) lateral rotation. The same muscle may be mentioned more than once.

CHECKPOINT

7. Which muscles contract and relax in each eye as you gaze to your left without moving your head?

MUSCLE	ORIGIN	INSERTION	ACTION	INNERVATION
Superior rectus (*rectus* = fascicles parallel to midline)	Common tendinous ring (attached to orbit around optic foramen).	Superior and central part of eyeball.	Moves eyeball superiorly (elevation) and medially (adduction), and rotates it medially.	Oculomotor (III) nerve.
Inferior rectus	Same as above.	Inferior and central part of eyeball.	Moves eyeball inferiorly (depression) and medially (adduction), and rotates it medially.	Oculomotor (III) nerve.
Lateral rectus	Same as above.	Lateral side of eyeball.	Moves eyeball laterally (abduction).	Abducens (VI) nerve.
Medial rectus	Same as above.	Medial side of eyeball.	Moves eyeball medially (adduction).	Oculomotor (III) nerve.
Superior oblique (*oblique* = fascicles diagonal to midline)	Sphenoid bone, superior and medial to the common tendinous ring in the orbit.	Eyeball between superior and lateral recti. The muscle inserts into the superior and lateral surfaces of the eyeball via a tendon that passes through the trochlea.	Moves eyeball inferiorly (depression) and laterally (abduction), and rotates it medially.	Trochlear (IV) nerve.
Inferior oblique	Maxilla in floor of orbit.	Eyeball between inferior and lateral recti.	Moves eyeball superiorly (elevation) and laterally (abduction) and rotates it laterally.	Oculomotor (III) nerve.
Levator palpebrae superioris (le-VĀ-tor PAL-pe-brē soo-per′-ē-OR-is; *palpebrae* = eyelids)	Roof of orbit near optic foramen.	Skin and tarsal plate of upper eyelid.	Elevates upper eyelid (opens eye).	Oculomotor (III) nerve.

Figure 11.6 **Extrinsic muscles of the eyeball.**

The extrinsic muscles of the eyeball are among the fastest contracting and most precisely controlled skeletal muscles in the body.

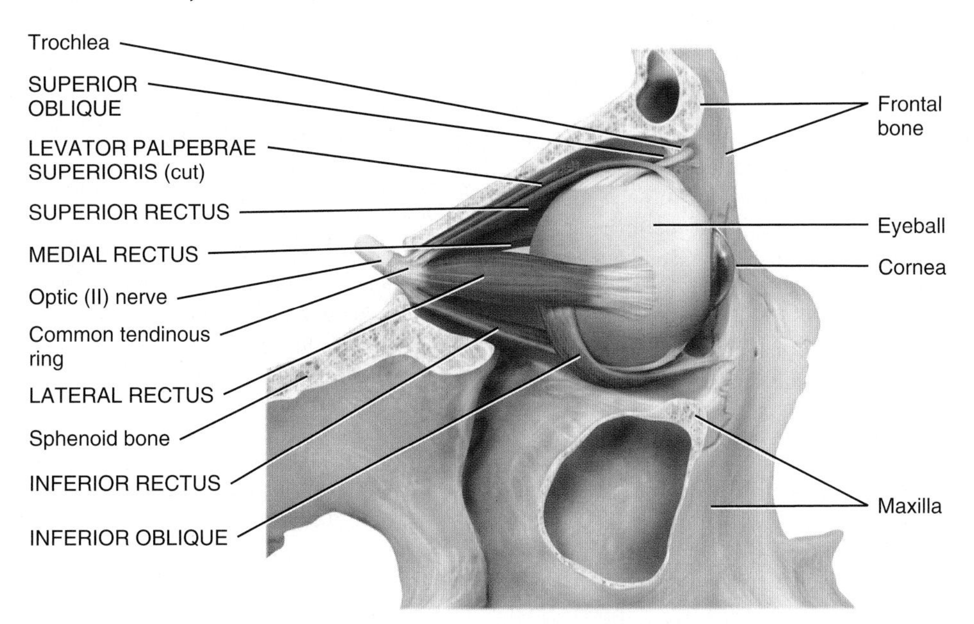

(a) Lateral view of right eyeball

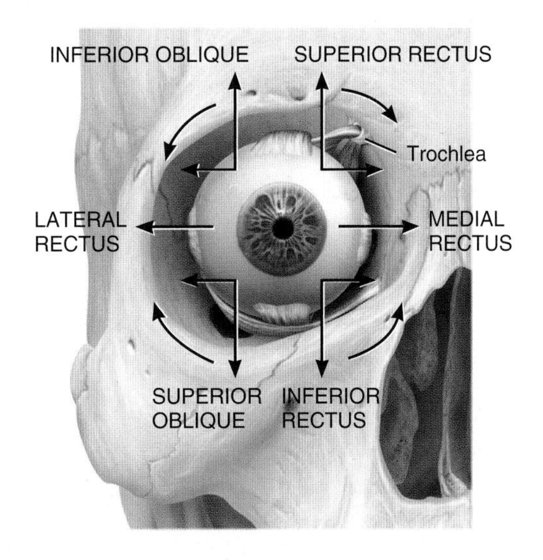

(b) Movements of right eyeball in response to contraction of extrinsic muscles

How does the inferior oblique muscle move the eyeball superiorly and laterally?

EXHIBIT 11.3 Muscles That Move the Mandible and Assist in Mastication (Chewing) and Speech

OBJECTIVE

- Describe the origin, insertion, action, and innervation of the muscles that move the mandible.

The muscles that move the mandible (lower jaw bone) at the *temporomandibular joint* (*TMJ*) are known as the muscles of *mastication* (chewing) (Figure 11.7). These muscles also assist in speech. Of the four pairs of muscles involved in mastication, three are powerful closers of the jaw and account for the strength of the bite: ***masseter, temporalis,*** and ***medial pterygoid.*** Of these, the masseter is the strongest muscle of mastication. The ***medial*** and ***lateral pterygoid*** muscles assist in mastication by moving the mandible from side to side to help grind food. Additionally, these two muscles protract the mandible (thrust it forward). The masseter muscle has been removed in Figure 11.7 to illustrate the deeper pterygoid muscles; the masseter can be seen in Figure 11.4d. Note the enormous bulk of the temporalis and masseter muscles in Figure 11.4d compared to the smaller mass of the two pterygoid muscles.

Note: A mnemonic for muscles of mastication is **T**eeny **M**ice **M**ake **P**etite **L**ittle **P**rints = **T**emporalis, **M**asseter, **M**edial **P**terygoid, and **L**ateral **P**terygoid.

Relating Muscles to Movements

Arrange the muscles in this exhibit according to their actions on the mandible: (1) elevation, (2) depression, (3) retraction, (4) protraction, and (5) side-to-side movement. The same muscle may be mentioned more than once.

• CLINICAL CONNECTION | Gravity and the Mandible

As just noted, three of the four muscles of mastication close the mandible and only the lateral pterygoid opens the mouth. The force of **gravity on the mandible** offsets this imbalance. When the masseter, temporalis, and medial pterygoid muscles relax, the mandible drops. Now you know why the mouth of many persons, particularly the elderly, is open while the person is asleep in a chair. In contrast, astronauts in zero gravity must work hard to open their mouths. •

CHECKPOINT

8. What would happen if you lost tone in the masseter and temporalis muscles?

MUSCLE	ORIGIN	INSERTION	ACTION	INNERVATION
Masseter (MA-se-ter = a chewer) (see Figure 11.4d)	Maxilla and zygomatic arch.	Angle and ramus of mandible.	Elevates mandible, as in closing mouth.	Mandibular division of trigeminal (V) nerve.
Temporalis (tem′-pō-RĀ-lis; *tempor-* = time or temples)	Temporal bone.	Coronoid process and ramus of mandible.	Elevates and retracts mandible.	Mandibular division of trigeminal (V) nerve.
Medial pterygoid (TER-i-goyd; *medial* = closer to midline; *pterygoid* = like a wing)	Medial surface of lateral portion of pterygoid process of sphenoid bone; maxilla.	Angle and ramus of mandible.	Elevates and protracts (protrudes) mandible and moves mandible from side to side.	Mandibular division of trigeminal (V) nerve.
Lateral pterygoid (*lateral* = farther from midline)	Greater wing and lateral surface of lateral portion of pterygoid process of sphenoid bone.	Condyle of mandible; temporomandibular joint (TMJ).	Protracts mandible, depresses mandible as in opening mouth, and moves mandible from side to side.	Mandibular division of trigeminal (V) nerve.

Figure 11.7 Muscles that move the mandible and assist in mastication (chewing) and speech.

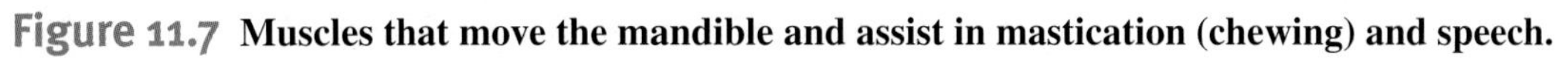

The muscles that move the mandible are also known as muscles of mastication.

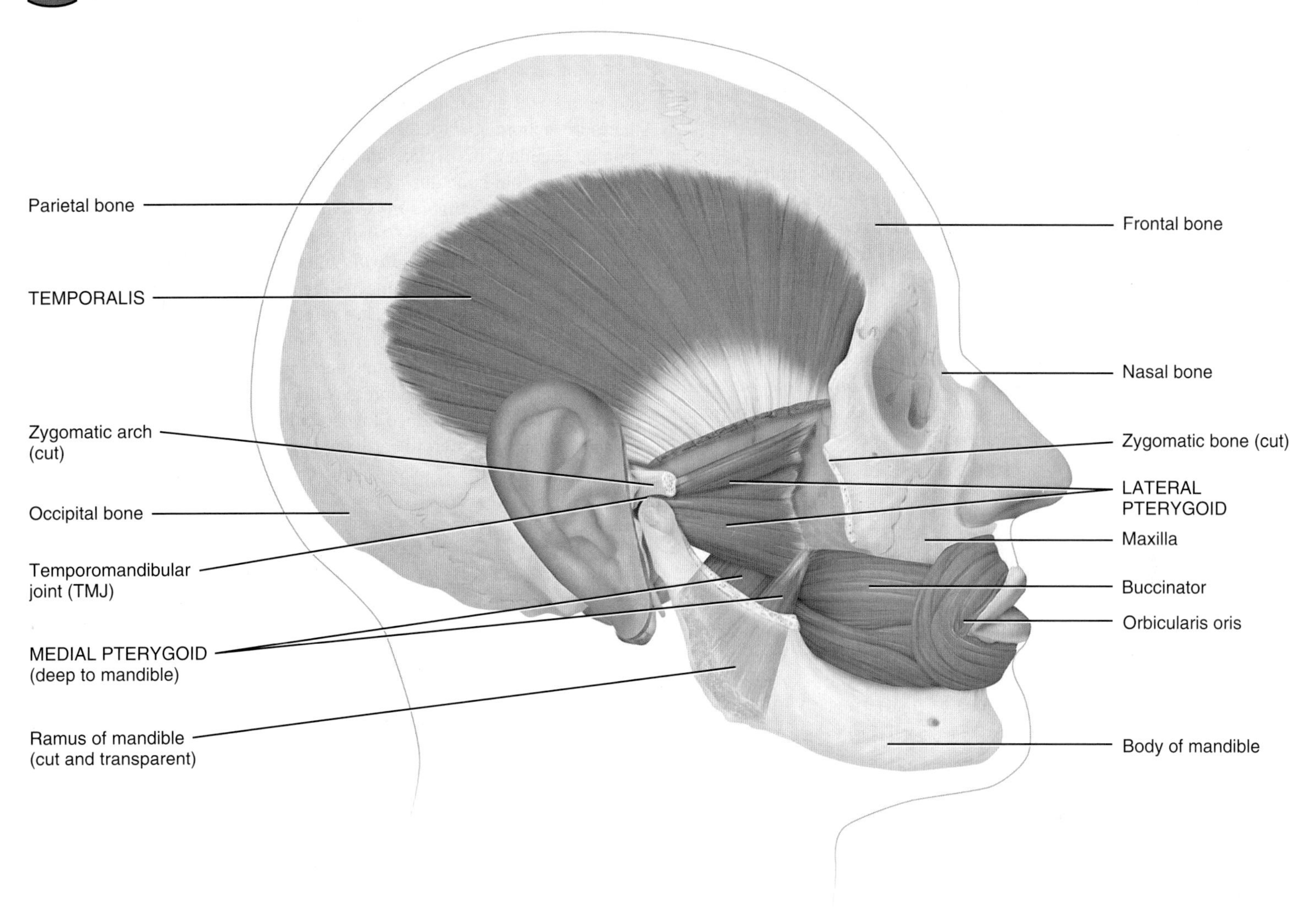

Right lateral superficial view

Which is the strongest muscle of mastication?

EXHIBIT 11.4 Muscles of the Anterior Neck That Assist in Deglutition (Swallowing) and Speech

OBJECTIVE

- Describe the origin, insertion, action, and innervation of the muscles of the anterior neck.
- Describe the muscles responsible for elevating and depressing the "Adam's apple" when swallowing.

The neck (*cervical*) is the region of the body that connects the head to the trunk and is divided into an *anterior cervical region*, two *lateral cervical regions*, and a *posterior cervical region* (*nucha*).

Muscles of the neck are particularly complex since multiple muscles cross multiple joints and thus have multiple actions. Muscle bellies located in the neck may attach to the head (to either cranial and facial bones), to cervical or thoracic vertebrae, to the shoulder, or to the thorax. Discussion of muscles of the neck may thus be found in Chapters 11, 12, and 13 of this text.

Two groups of muscles are associated with the anterior aspect of the neck (Figure 11.8): the ***suprahyoid muscles,*** so called because they are located superior to the hyoid bone, and the ***infrahyoid muscles,*** named for their position inferior to the hyoid bone. Both groups of muscles stabilize the hyoid bone, allowing it to serve as a firm base on which the tongue can move.

As a group, the ***suprahyoid muscles*** elevate the hyoid bone, floor of the oral cavity, and tongue during swallowing. As its name suggests, the ***digastric*** muscle (*di-* = two) has two bellies, anterior and posterior, united by an intermediate tendon that is held in position by a fibrous loop. This muscle elevates the hyoid bone and larynx (voice box) during swallowing and speech. In a *reverse muscle action (RMA)*, when the hyoid is stabilized, the digastric depresses the mandible and is therefore synergistic to the lateral pterygoid in the opening of the mouth. Together, the ***stylohyoid, mylohyoid,*** and ***geniohyoid*** muscles elevate the hyoid bone during deglutition (swallowing).

Most of the ***infrahyoid muscles*** depress the hyoid bone, and some move the thyroid cartilage (Adam's apple) of the larynx during swallowing and speech. The ***omohyoid*** muscle, like the digastric muscle, is composed of two bellies connected by an intermediate tendon. In this case, however, the two bellies are referred to as *superior* and *inferior,* rather than anterior and posterior. Together, the omohyoid, ***sternohyoid,*** and ***thyrohyoid*** muscles depress the hyoid bone. In addition, the ***sternothyroid*** muscle depresses the thyroid cartilage (Adam's apple) of the larynx to produce low sounds; the RMA of the thyrohyoid muscle elevates the thyroid cartilage to produce high sounds.

MUSCLE	ORIGIN	INSERTION	ACTION	INNERVATION
SUPRAHYOID MUSCLES				
Digastric (dī′-GAS-trik; *di-* = two; *-gastr-* = belly)	Anterior belly from inner side of inferior border of mandible near the chin; posterior belly from mastoid process of temporal bone.	Body of hyoid bone via an intermediate tendon.	Elevates hyoid bone. RMA: Depresses mandible, as in opening the mouth.	Anterior belly: mandibular division of trigeminal (V) nerve. Posterior belly: facial (VII) nerve.
Stylohyoid (stī′-lō-HI-oid; *stylo* = stake or pole, styloid process of temporal bone; *hyo* = U-shaped, pertaining to hyoid bone)	Styloid process of temporal bone.	Body of hyoid bone.	Elevates hyoid bone and draws it posteriorly.	Facial (VII) nerve.
Mylohyoid (mī′-lō-HĪ-oid; *mylo-* = mill)	Inner surface of mandible.	Body of hyoid bone.	Elevates hyoid bone and floor of mouth. RMA: Depresses mandible.	Mandibular division of trigeminal (V) nerve.
Geniohyoid (jē′-nē-ō-HĪ-oid; *genio-* = chin) (not illustrated)	Inner surface of mandible.	Body of hyoid bone.	Elevates hyoid bone; draws hyoid bone and tongue anteriorly. RMA: Depresses mandible.	C1.
INFRAHYOID MUSCLES				
Omohyoid (ō-mō-HĪ-oid; *omo-* = relationship to the shoulder)	Superior border of scapula and superior transverse ligament.	Body of hyoid bone.	Depresses hyoid bone.	Branches of spinal nerves C1–C3.
Sternohyoid (ster′-nō-HĪ-oid; *sternoī* = sternum)	Medial end of clavicle and manubrium of sternum.	Body of hyoid bone.	Depresses hyoid bone.	Branches of spinal nerves C1–C3.
Sternothyroid (ster′-nō-THĪ-roid; *thyro* = thyroid gland)	Manubrium of sternum.	Thyroid cartilage of larynx.	Depresses thyroid cartilage of larynx.	Branches of spinal nerves C1–C3.
Thyrohyoid (thī-rō-HĪ-oid)	Thyroid cartilage of larynx.	Greater horn of hyoid bone.	Depresses hyoid bone. RMA: Elevates thyroid cartilage.	Branches of spinal nerves C1–C2 and hypoglossal (XII) nerve.

Figure 11.8 Muscles of the floor of the oral cavity, the anterior neck, and infrahyoid muscles.

The suprahyoid muscles elevate the hyoid bone, the floor of the oral cavity, and the tongue during swallowing. Infrahyoid muscles are antagonistic to the suprahyoid muscles and therefore help stabilize the hyoid bone.

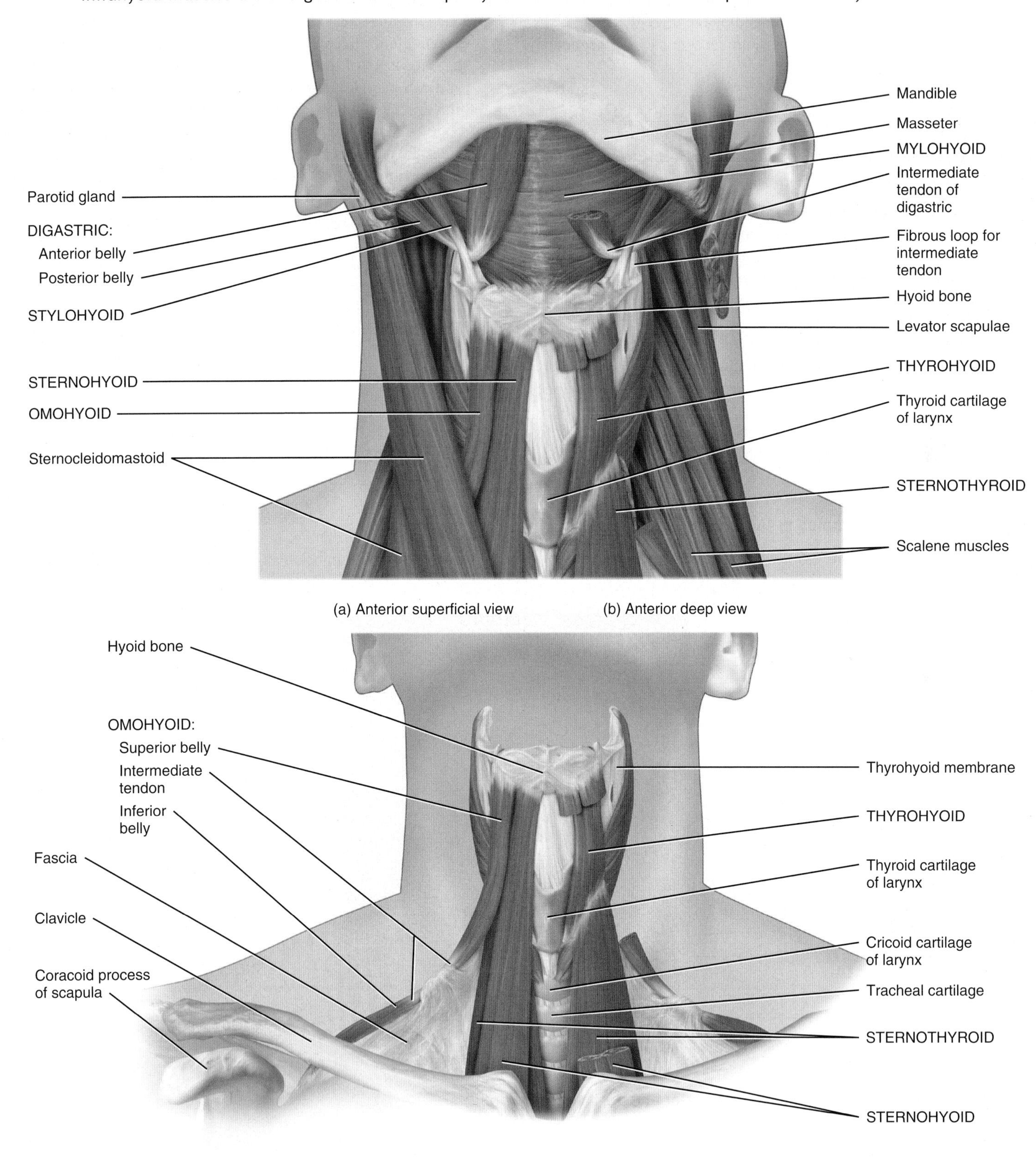

What is the combined action of the suprahyoid and infrahyoid muscles?

CONTINUES

Relating Muscles to Movements

Arrange the muscles in this exhibit according to the following actions on the hyoid bone: (1) elevating it, (2) drawing it anteriorly, (3) drawing it posteriorly, and (4) depressing it; and on the thyroid cartilage: (1) elevating it and (2) depressing it. The same muscle may be mentioned more than once.

Surface Anatomy of the Neck

The following are the major surface features of the neck, most of which are illustrated in Figure 11.9.

- **Thyroid cartilage (Adam's apple).** The largest of the cartilages that compose the larynx (voice box). It is the most prominent structure in the midline of the anterior cervical region. The common carotid artery *bifurcates* (divides) at the level of the superior border of the thyroid cartilage to form the internal and external carotid arteries.
- **Hyoid bone.** Located just superior to the thyroid cartilage. It is the first structure palpated in the midline inferior to the chin. It is easily palpated laterally as you move posterior from its midline body onto the greater horn (see Figure 7.15).
- **Cricoid cartilage.** A laryngeal cartilage located just inferior to the thyroid cartilage, it attaches the larynx to the trachea (windpipe). After you pass your finger over the cricoid cartilage moving inferiorly, your fingertip sinks in. The cricoid cartilage is used as a landmark for performing a tracheotomy.
- **Thyroid gland.** Two-lobed gland just inferior to the larynx with one lobe on either side.
- **Sternocleidomastoid muscle.** Forms the major portion of the lateral aspect of the neck. If you rotate your head to either side, you can palpate the muscle from its origins on the sternum and clavicle to its insertion on the mastoid process of the temporal bone. As discussed in Exhibit 11.6, the sternocleidomastoid divides the neck into anterior and posterior triangles.
- **Subclavian artery.** Located just lateral to the inferior portion of the sternocleidomastoid muscle. Pressure on this artery can stop bleeding in the upper limb because this vessel supplies blood to the entire limb.
- **External jugular vein.** Located superficial to the sternocleidomastoid muscle, this vessel is readily seen if you are angry or if your collar is too tight.
- **Trapezius muscle.** Extends inferiorly and laterally from the base of the skull and occupies a portion of the lateral cervical region. A "stiff neck" is frequently associated with inflammation of this muscle.

CHECKPOINT

9. Why do the two bellies of the digastric muscle have different innervations?

Figure 11.9 Surface anatomy of the neck.

The anatomical subdivisions of the neck are the anterior cervical region, lateral cervical regions, and posterior cervical region.

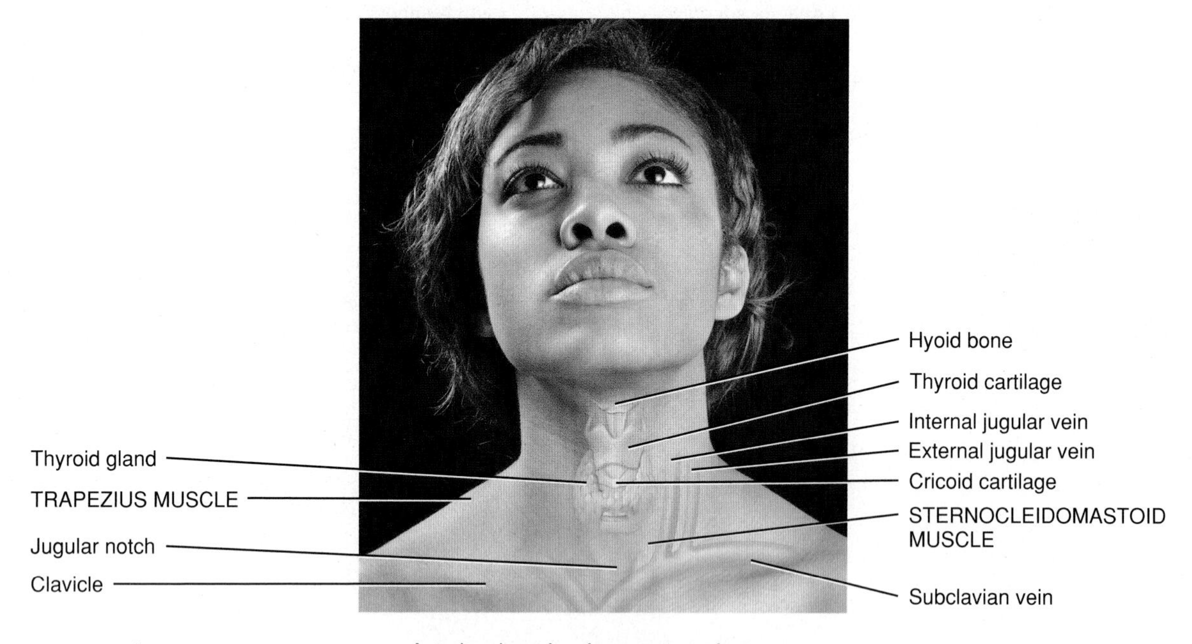

Anterior view of surface anatomy features

Which muscle divides the neck into anterior and posterior triangles?

EXHIBIT 11.5 Muscles of the Anterior Neck That Assist in Elevating the Ribs or Flexing the Neck and Head

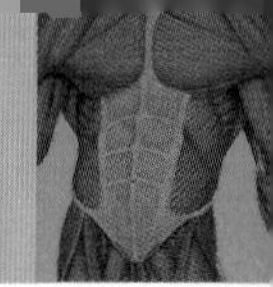

OBJECTIVE

- Describe the origin, insertion, action, and innervation of the muscles of the neck that assist in elevating the ribs or flexing the neck and head.

The three scalene muscles originate on transverse processes of the cervical vertebrae and thus portions of their bellies are within the lateral aspect of the neck. The insertions, however, are on anterior portions of the ribs. Therefore, the scalene muscles are discussed in this exhibit.

Within the ***scalene*** group (see Figure 11.10), the ***anterior scalene*** muscle is anterior to the middle scalene muscle. The ***middle scalene*** muscle is intermediate in placement and is the longest and largest of the scalene muscles. The ***posterior scalene*** muscle is posterior to the middle scalene muscle and is the smallest of the scalene muscles. These muscles flex, laterally flex, and rotate the head and assist in deep inhalation. The precise location of the scalene muscles is important to manual therapists since spasm of the scalenes may mimic the symptoms of carpal tunnel syndrome. This topic will be studied further in Chapter 16.

Prevertebral muscles of the anterior neck include ***longus colli, longus capitis, rectus capitis anterior,*** and ***rectus capitis lateralis*** (see Figure 11.10b). As the name of the group implies, the bellies of most of these muscles lie directly on or near the anterior surface of the vertebral column. When contracting bilaterally, all four prevertebral muscles cause flexion of the head or the cervical spine. This prevertebral group is antagonistic to the powerful extensors of the posterior neck (to be studied in the next chapter). The bony attachments of the neck are shown in Figure 11.10c.

MUSCLE	ORIGIN	INSERTION	ACTION	INNERVATION
SCALENES (SKĀ-LĒNZ)				
Anterior scalene (SKĀ-lēn; *anterior* = front; *scalene* = uneven)	Transverse processes of C3–C6.	First rib.	Acting together, right and left anterior scalene and middle scalene muscles elevate first ribs during deep inhalation. RMA: Flex cervical vertebrae; acting singly, laterally flex and slightly rotate cervical vertebrae.	Cervical spinal nerves.
Middle scalene	Transverse processes of C2–C7.	First rib.		Cervical spinal nerves.
Posterior scalene	Transverse processes of C4–C6.	Second rib.	Acting together, right and left posterior scalene elevate second ribs during deep inhalation. RMA: Flex cervical vertebrae and acting singly, laterally flex and slightly rotate cervical vertebrae.	Cervical spinal nerves.
PREVERTEBRAL MUSCLES				
Longus capitis (LON-gus KAP-i-tus)	Transverse processes of C3–C6.	Base of occipital bone anterior to foramen magnum.	Acting together, flexes the neck and limits hyperextension; acting singly, rotates head to same side.	Cervical spinal nerves.
Longus colli (LON-gus KŌ-lī)	Anterior vertebral surfaces C3–T3.	Transverse processes of upper cervical vertebrae.	Acting together, flexes the neck and limits hyperextension; acting singly, rotates neck to same side.	Cervical spinal nerves.
Rectus capitis anterior (REK-tus)	Lateral surface of atlas (C1).	Base of occipital bone anterior to foramen magnum.	Acting together, flexion at atlanto-occipital (AO) joint; acting singly, lateral flexion at AO joint.	C1.
Rectus capitis lateralis	Transverse processes of C1.	Base of occipital bone lateral to foramen magnum.	Acting together, flexion at AO joint; acting singly, lateral flexion at AO joint.	C1.

CONTINUES

Figure 11.10 Anterior muscles of the neck. (a) Three scalene muscles. (b) Prevertebral neck muscles. (c) Bony attachments of the neck with origins shown in red and insertions shown in blue; each muscle is illustrated on only one side.

Many small muscles are responsible for the stability and flexibility of the head and neck.

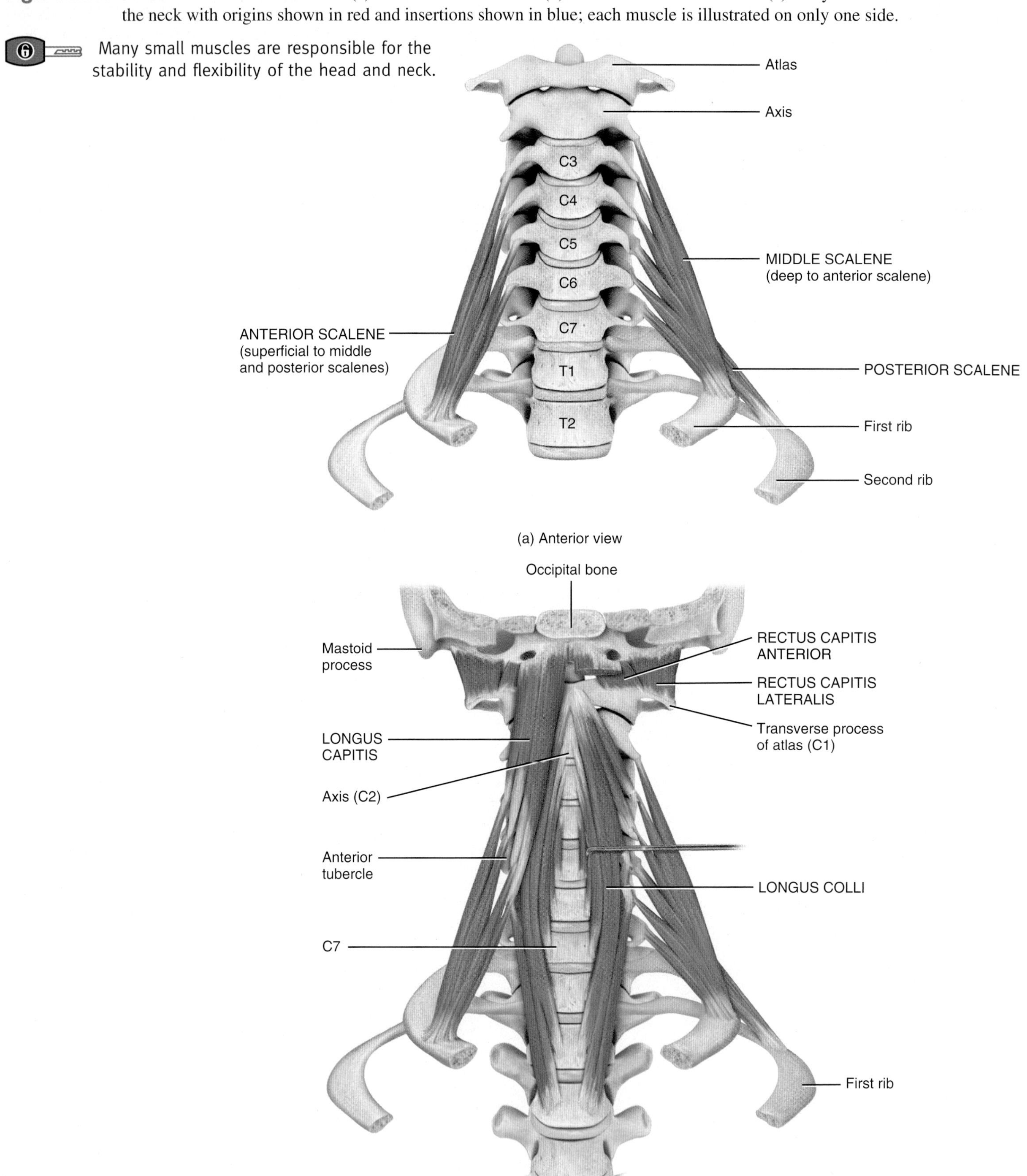

(a) Anterior view

(b) Anterior view of prevertebral neck muscles

Relating Muscles to Movements

Arrange the muscles in this exhibit according to the following actions on the head and neck: (1) flexion, (2) lateral flexion, (3) extension, (4) rotation to side opposite the contracting muscle, and (5) rotation to the same side as the contracting muscle. The same muscle may be used more than once.

CHECKPOINT

10. Which muscles of the anterior neck assist in elevating the ribs during deep inhalation?

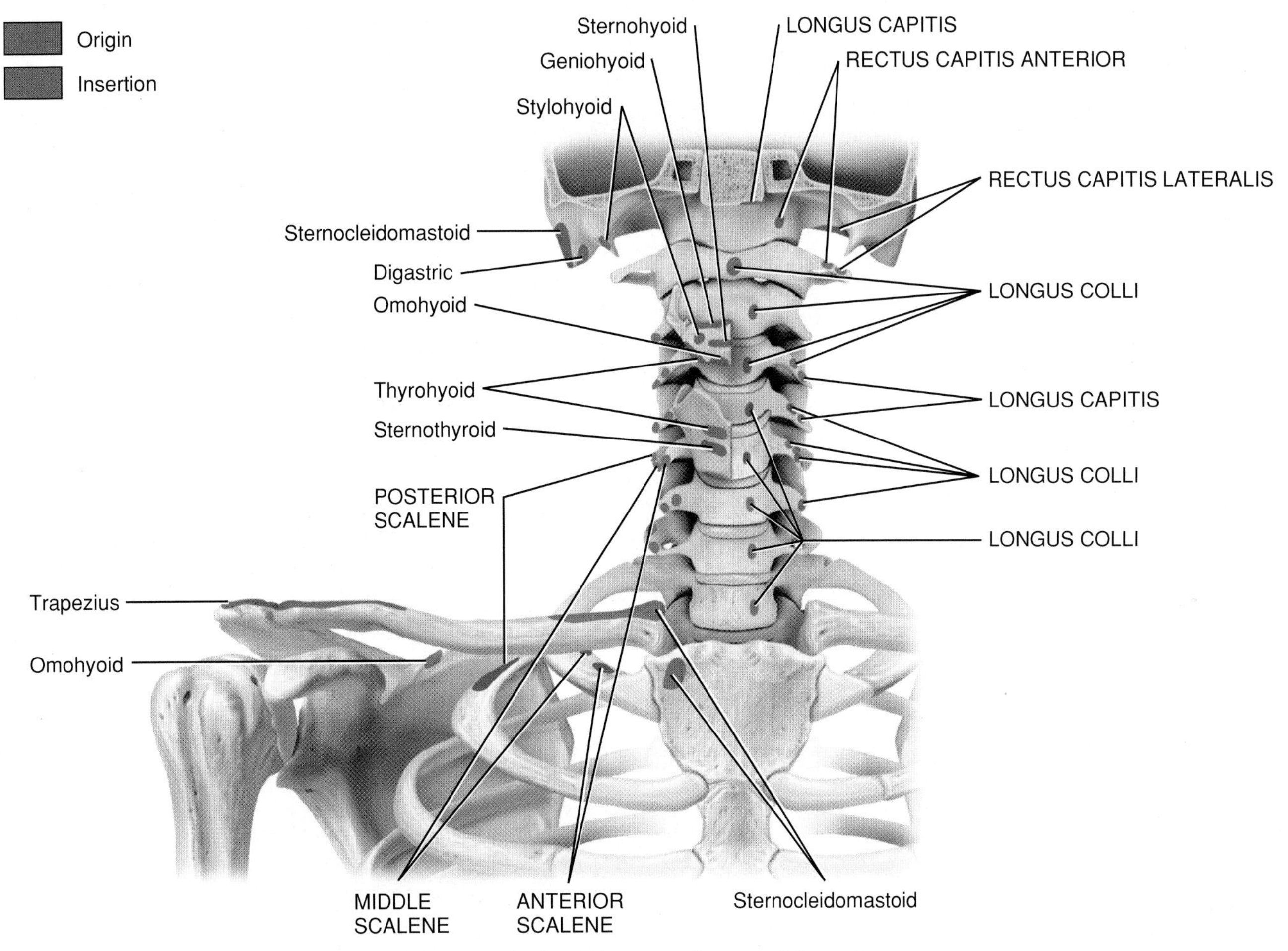

(c) Anterior view of bony attachments of the neck; each muscle is illustrated on one side only

? Anterior muscles of the neck lie anterior to what structure?

EXHIBIT 11.6 Muscles of the Lateral Neck That Move the Head

OBJECTIVE

- Describe the origin, insertion, action, and innervation of the muscles of the lateral neck that move the head.

The ***sternocleidomastoid (SCM)*** is one of the largest and most important muscles of the neck. Portions of the muscle are located in the anterior neck, whereas others are located in the posterior neck. The bellies of the paired SCM muscles thus pass through the right and left lateral aspects of the neck.

The sternocleidomastoid muscle is an important landmark that divides the neck into two major triangles: anterior and posterior. The triangles are important because of the structures that lie within their boundaries. For physicians, especially surgeons, the further subdivision of the two major triangles is clinically very significant. The ***anterior triangle*** is bordered superiorly by the mandible, inferiorly by the sternum, medially by the cervical midline, and laterally by the anterior border of the sternocleidomastoid muscle (Figure 11.11a). The ***posterior triangle*** is bordered inferiorly by the clavicle, anteriorly by the posterior border of the sternocleidomastoid muscle, and posteriorly by the anterior border of the trapezius muscle.

The SCM is superficial and easily palpable in the anterior and lateral aspects of the neck. As will be seen in Chapter 23, the external jugular vein is superficial to the middle portion of the SCM. Deep to the SCM are mutiple muscles, some discussed previously and some to be discussed in the next chapter. The carotid sinus of the common carotid artery (see Chapter 23) lies directly medial to the SCM. Pressure on the carotid sinus, whether from the hands of a therapist or from the increase in diameter of the SCM in spasm, can lower blood pressure. Each SCM consists of two bellies (Figure 11.11a); they are more evident near the anterior attachments. The separation of the two bellies is variable and thus more evident in some persons than in others. The two bellies insert as the ***sternal head*** and the ***clavicular head*** of the SCM. The bellies also function differently; muscular spasm in the two bellies cause somewhat different symptoms. Remember from the last chapter that each muscle spindle involves two sensory neurons and two motor neurons (see Figure 10.12). Activation of the sensory neurons due to muscle spasm is relayed to the brain and may be interpreted as referred pain (see Figure 19.2).

The three scalene muscles are located in the anterolateral neck, deep to the sternocleidomastoid. They were presented in Exhibit 11.5.

Relating Muscles to Movements

Arrange the three scalene muscles (Exhibit 11.5) and the sternocleidomastoid muscles according to the following actions on the head and neck: (1) flexion, (2) lateral flexion, and (3) extension.

CHECKPOINT

11. What are the reverse muscle actions of these four muscles?

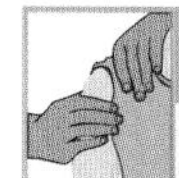

MANUAL THERAPY APPLICATION

Sternocleidomastoid (SCM) as an Example of Referred Trigger Points

Referred pain from the ***clavicular division*** of the SCM may be perceived as pain deep to and around the ear and therefore the SCM may account for the "unexplained" earache. Proprioceptors of the SCM also play a large role in providing information to the inner ear, thus accounting for possible symptoms of dizziness, vertigo, and nausea. Referred pain may also involve the forehead.

Pain from the ***sternal division*** of the SCM may be referred to the superior cranial portion (calvaria) of the skull (see Figure 7.8a). This pain is clearly perceived to be superficial and in the scalp at the top of the head. Since this area is superficial to only the epicranial aponeurosis (see Figure 11.4), some therapists do not correlate pain in this area with muscular spasm, specifically the SCM. Pain referred from the muscle may also produce symptoms of sore throat, toothache, temporomandibular joint (TMJ) syndrome, pain deep within the orbits, and excessive lacrimation or tear production. This example illustrates that **trigger point** assessment and therapy by a manual therapist sometimes enable patients to have a reduction in pain when other traditional treatments have been unable to effectively treat the cause of the pain.

MUSCLE	ORIGIN	INSERTION	ACTION	INNERVATION
Sternocleidomastoid (ster′-nō-klī-dō-MAS-toyd; *sterno-* = breastbone; *cleido-* = clavicle; *mastoid* = mastoid process of temporal bone)	Sternal head: manubrium of sternum; clavicular head: medial third of clavicle.	Mastoid process of temporal bone and lateral half of superior nuchal line of occipital bone.	Acting together (bilaterally), flex cervical portion of vertebral column, extend head at AO joint; acting singly (unilaterally), laterally flex the neck and head to the same side and rotate the head to the side opposite the contracting muscle. RMA: Elevate the sternum during forced inhalation.	Accessory (XI) nerve, C2, and C3.
Anterior scalene (SKĀ-lēn; *anterior* = front; *scalene* = uneven)	Transverse processes of C3–C6.	First rib.	Acting together, right and left anterior scalene and middle scalene muscles elevate first ribs during deep inhalation. RMA: Flex cervical vertebrae; acting singly, laterally flex and slightly rotate cervical vertebrae.	Cervical spinal nerves.
Middle scalene	Transverse processes of C2–C7.	First rib.	Acting together, right and left posterior scalene elevate second ribs during deep inhalation. RMA: Flex cervical vertebrae and acting singly, laterally flex and slightly rotate cervical vertebrae.	Cervical spinal nerves.
Posterior scalene	Transverse processes of C4–C6.	Second rib.		Cervical spinal nerves.

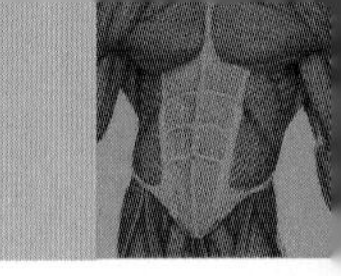

Figure 11.11 Triangles of the neck. (a) The sternocleidomastoid muscle divides the neck into two principal triangles: anterior and posterior. (b) Muscle attachment sites on the skull.

The anatomical triangles of the neck are important landmarks because of the structures that lie within their boundaries.

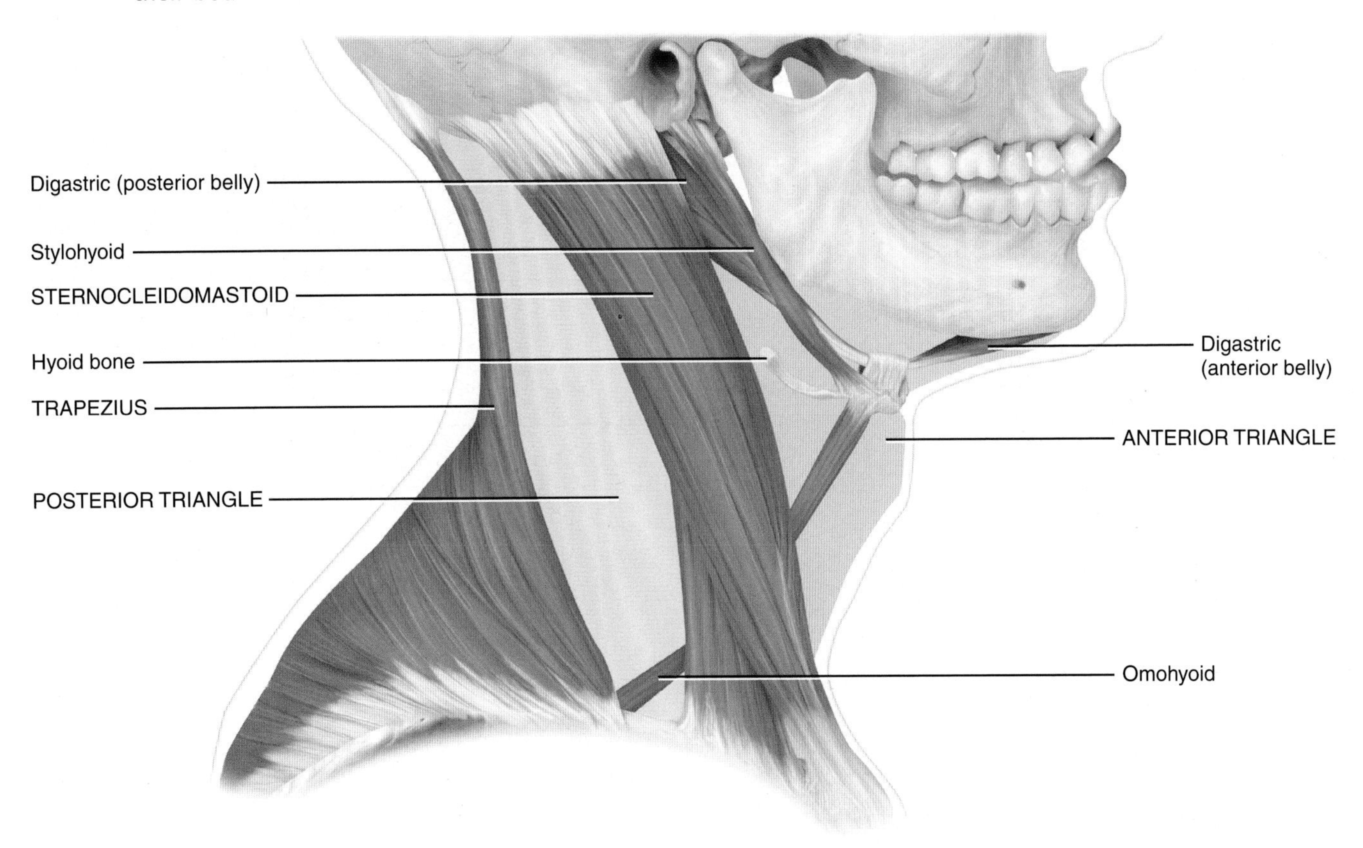

(a) Right lateral view of triangles of neck

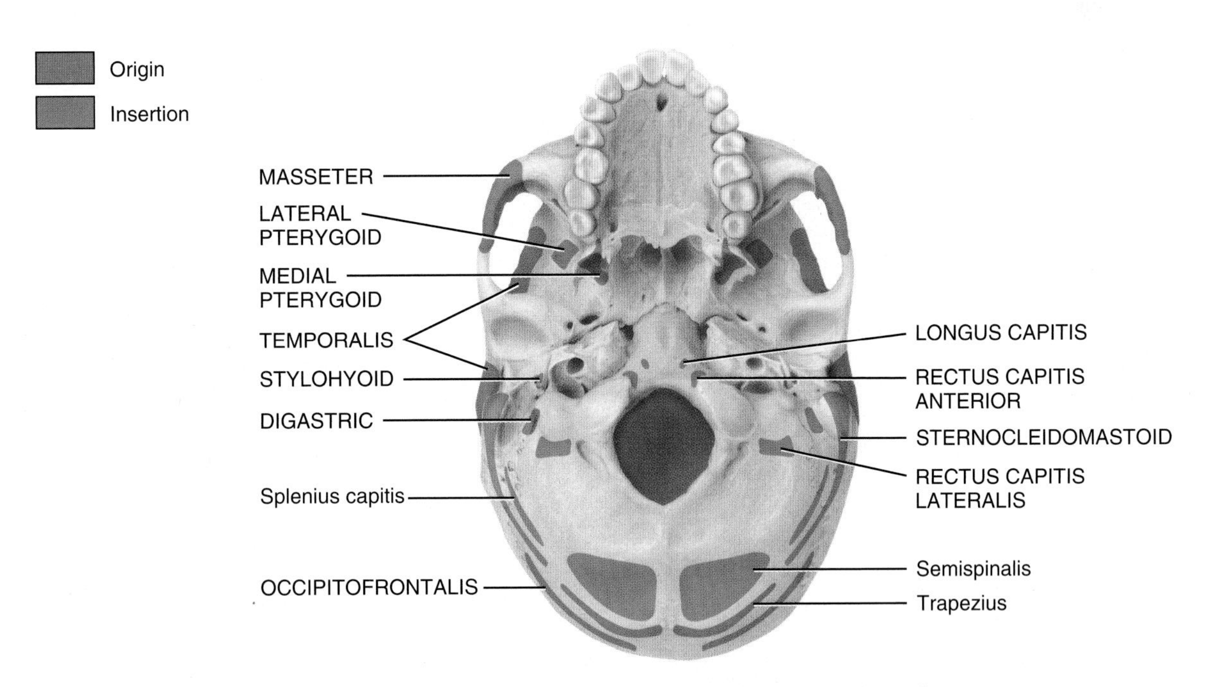

(b) Inferior view of muscle attachment sites on the skull

Why are triangles important?

CONTINUES

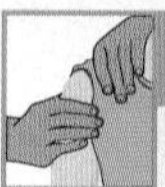

MANUAL THERAPY APPLICATION

Whiplash Injuries

Whiplash injuries can result from any activity that causes rapid, excessive forces on structures of the neck. Injuries may result from rear-end, frontal, or side impacts; rollovers; or combinations of these vehicular accidents. Other incidents that can create whiplash injuries include contact sports, assaults, and falls.

Hyperextension. In a rear-end vehicular accident, momentum forces the torso into the back of the seat and then upward while the cervical vertebral column extends. The torso then moves forward while the extended head is forced into hyperextension. The hyperextension phase of whiplash causes traumatic stretching and perhaps tearing of the longus colli and longus capitis as well as the sternocleidomastoid and multiple other anterior neck muscles. Underlying spinal ligaments may also be sprained.

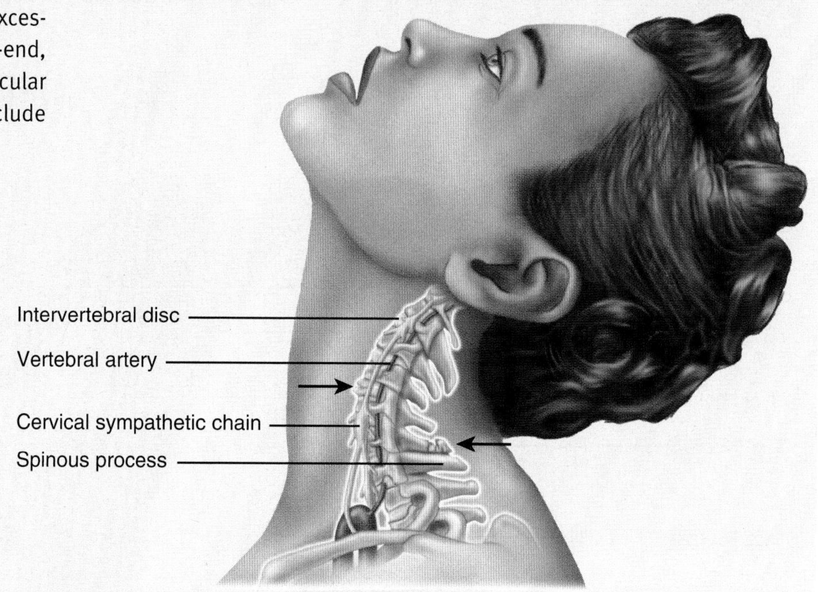

(a) Hyperextension

Intervertebral disc
Vertebra
Spinal nerve

(b) Hyperflexion

Hyperflexion. Momentum next causes the torso and head to move forward at the same time that the vehicle is decelerating from the impact. Assuming the use of a seat belt, the torso is restrained while the unrestrained head continues into hyperflexion.

Damage to muscles. Posterior cervical muscles commonly affected during hyperextension include the trapezius, splenius capitis, semispinalis capitis, levator scapulae, and the suboccipital muscles. Anterior muscles commonly affected during hyperflexion are the longus colli, sternocleidomastoid, and the scalene muscles. Any or all of these muscles may be stretched, strained, or torn. Blood vessels and resultant blood flow to these muscles may also be compromised to varying degrees.

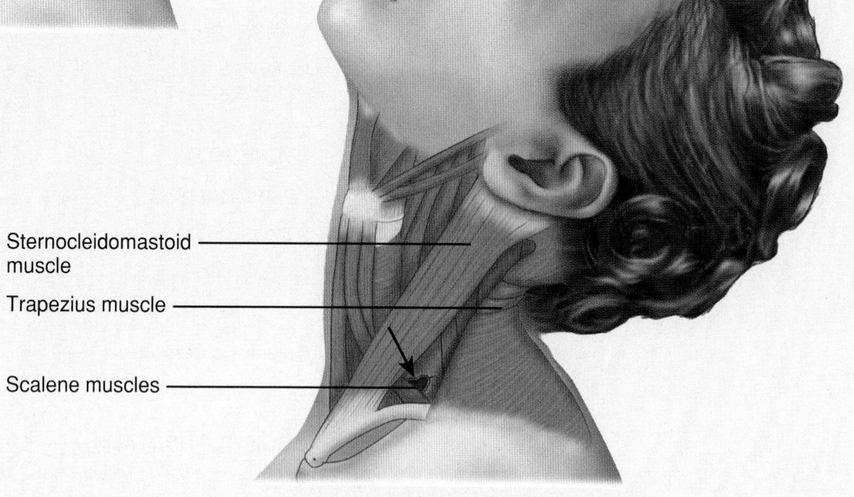

(c) Damage to muscle

Damage to ligaments. Multiple short, thin ligaments may be stretched, sprained, torn, or may produce fractures of the bone to which they are attached.

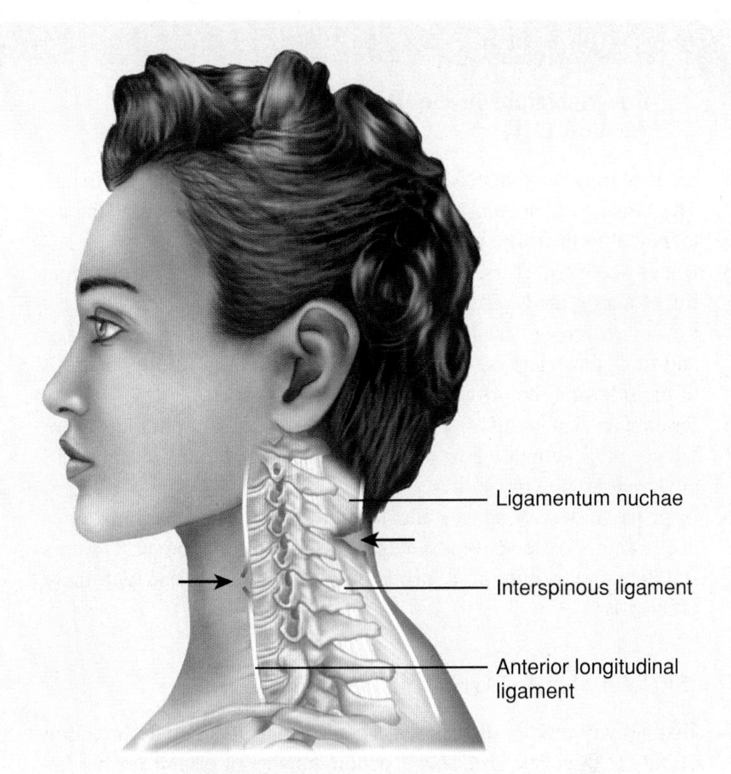

(d) Damage to ligaments

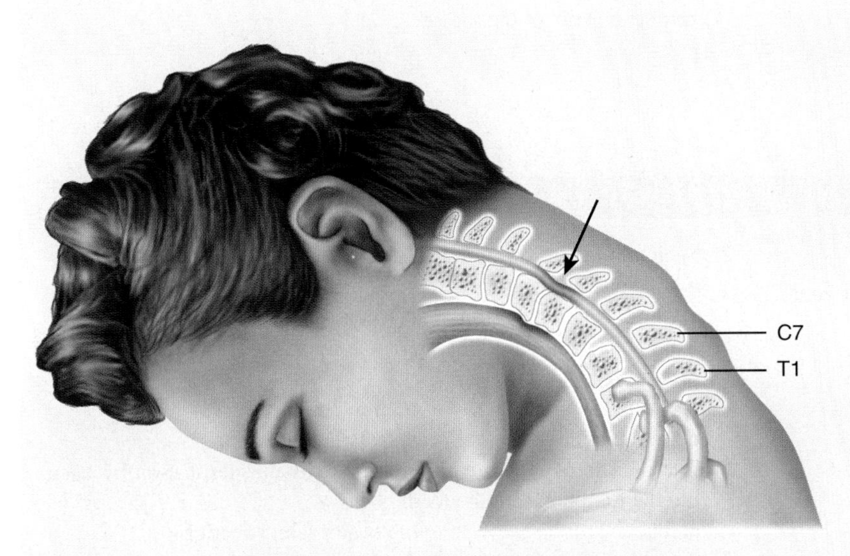

(e) Damage to bone and nervous tissue

Damage to bone and nervous tissue C7 and T1 are most vulnerable to compression fractures. Vertebral spines may be fractured and even torn from the vertebral bodies. Superior articular processes may be fractured, and intervertebral discs may be herniated or develop clefts.

Entrapment of the spinal cord within the vertebral canal is always a possible consequence of whiplash injuries. Herniation of intervertebral discs, the presence of bone fragments within the narrow vertebral canal, and/or the development of scar tissue around the spinal cord and/or spinal nerves can compromise the activities of nervous tissue. In severe whiplash injuries such damage to nervous tissue could be fatal. Less severe whiplash injuries may lead to pain, loss of sensation, or paralysis.

Pain may begin immediately or may not be present until weeks after the trauma. Left untreated or inappropriately treated, whiplash injuries can persist for months or years. As discussed in Chapter 10, spasm may be relieved by direct pressure on the shortened muscle bellies. Healing of ligament sprains in the cervical area will be facilitated by the increased blood flow associated with massage. Access to the longus colli and the longus capitis muscles is difficult, since they are deep to the larynx, trachea, esophagus, major blood vessels, and multiple visceral organs. For this reason, treatment techniques for the prevertebral muscles may be one of the last techniques taught to students of the manual therapies or may be offered as a postgraduate seminar.

STUDY OUTLINE

How Skeletal Muscles Produce Movements (Section 11.1)

1. Skeletal muscles that produce movement do so by pulling on bones.
2. The attachment to the more stationary bone is the origin; the attachment to the more movable bone is the insertion.
3. Bones serve as levers, and joints serve as fulcrums. Two different forces act on the lever: load (resistance) and effort.
4. Levers are categorized into three types—first-class, second-class, and third-class levers (most common)—according to the positions of the fulcrum, the effort, and the load on the lever.
5. Fascicular arrangements include parallel, fusiform, circular, triangular, and pennate. Fascicular arrangement affects a muscle's power and range of motion.
6. A prime mover produces the desired action; an antagonist produces an opposite action. Synergists assist a prime mover. Fixators stabilize the origin of a prime mover so that it can act more efficiently.

How Skeletal Muscles Are Named (Section 11.2)

1. Distinctive features of different skeletal muscles include direction of muscle fascicles; size, shape, action, number of origins (or heads), and location of the muscle; and sites of origin and insertion of the muscle.
2. Most skeletal muscles are named based on combinations of these features.

Principal Skeletal Muscles of the Head and Neck (Section 11.3)

1. Muscles of facial expression move the skin rather than a joint when they contract, and they permit us to express a wide variety of emotions.
2. The extrinsic muscles that move the eyeballs are among the fastest contracting and most precisely controlled skeletal muscles in the body. They permit us to elevate, depress, abduct, adduct, and medially and laterally rotate the eyeballs.
3. Muscles that move the mandible are also known as the muscles of mastication because they are involved in chewing.
4. Muscles of the floor of the anterior neck, called suprahyoid muscles, are located above the hyoid bone. They elevate the hyoid bone, oral cavity, and tongue during swallowing.
5. Infrahyoid muscles depress the hyoid bone or the larynx. The suprahyoid and infrahyoid muscles work antagonistically to maintain the proper position for the hyoid bone according to the desired action.
6. The bellies of the sternocleidomastoid and the three scalene muscles are located in the lateral neck.
7. Prevertebral muscles flex the neck and are antagonistic to the powerful extensors of the the neck.

SELF-QUIZ QUESTIONS

Choose the one best answer to the following questions.

1. Which of these statements is FALSE when you hyperextend your head as if to look at the sky?
   a. The weight of the face and jaw serves as the load (L).
   b. The posterior neck muscles provide the E (effort).
   c. The F (fulcrum) is the atlanto-occipital joint.
   d. This is an example of a second-class lever.
2. Choose the muscle(s) that depress(es) the mandible:
   a. lateral pterygoid b. digastric
   c. medial pterygoid d. both a and b are correct.
3. Which of the following suprahyoid muscles elevates the hyoid bone but does not depress the mandible?
   a. digastric b. geniohyoid
   c. mylohoid d. stylohoid
4. The term muscular system refers to ________ muscle.
   a. cardiac b. skeletal
   c. smooth d. all of the above
5. Parallel muscles have comparatively few muscle fasciculi that extend the length of the muscle organ and thus they have a greater ________ but less ________.
   a. power; range of motion b. range of motion; power
6. Which of the following eye muscles does NOT originate on or near the common tendinous ring?
   a. inferior oblique b. inferior rectus
   c. lateral rectus d. medial rectus
7. Which of the following muscles does not originate on the mandible?
   a. buccinator b. depressor anguli oris
   c. levator labii superioris d. mentalis
8. Contraction of the right sternocleidomastoid muscle causes the head to ________.
   a. extend b. flex
   c. rotate to the left d. rotate to the right
9. The muscle of facial expression that aids in mastication by keeping the food between the teeth is the
   a. buccinator b. levator labii superioris
   c. orbicularis oris d. risorius.
10. Which of the following muscles does not insert into the orbicularis oris?
   a. buccinator b. levator labii superioris
   c. mentalis d. zygomaticus major

Fill in the blanks in the following statements.

11. A muscle that contracts to cause the desired action is called the ________. Muscles that assist or cooperate with the muscle that causes the desired action are known as ________.
12. The attachment of a muscle's tendon to the stationary bone is called the ________, and the attachment of the muscle's other tendon to the movable bone is called the ________.
13. In an anatomical lever system, ________ act as levers, ________ act as fulcrums, and the effort is provided by ________.

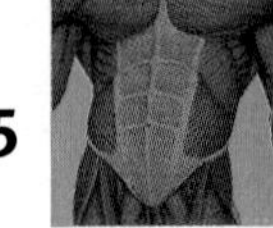

14. The sternocleidomastoid muscles insert on the ________ of the temporal bone.
15. A circular muscle that decreases the size of an opening, such as the mouth, is known as a ________.

Indicate whether the following statement is true or false.

16. The action of the biceps brachii on the forearm at the elbow joint illustrates the structure and action of a first-class lever system.

Matching.

17. Match the following muscles with the appropriate nerves:

____	**a.** muscles of facial expression	**A.** Occulomotor (III) nerve
____	**b.** muscles that move the mandible	**B.** Trigeminal (V) nerve
____	**c.** levator palpebrae superioris	**C.** Facial (VII) nerve
____	**d.** digastric	

CRITICAL THINKING QUESTIONS

1. During a facelift, the cosmetic surgeon accidentally severs the facial nerve on the right side of the face. What are some of the effects this would have on the patient, and what muscles are involved?
2. Anthropology graduate students are learning about "bite force" in Neanderthals. Since this is a significant feature of chewing, it is critical to understand muscle attachments onto the mandible, and specifically what actions these muscles produce. Distinguish the attachment sites and actions of the four muscles of mastication.

? ANSWERS TO FIGURE QUESTIONS

11.1 The belly of the muscle that extends the forearm, the triceps brachii, is located posterior to the humerus.

11.2 Second-class levers produce the most force.

11.3 For muscles named after their various characteristics, here are possible correct responses (for others, see Table 11.2): direction of fibers: external oblique; shape: deltoid; action: extensor digitorum; size: gluteus maximus: origin and insertion: sternocleidomastoid; location: tibialis anterior; number of tendons of origin: biceps brachii.

11.4 The corrugator supercilii muscle is involved in frowning; the zygomaticus major muscle contracts when you smile; the platysma muscles contribute to pouting; the orbicularis oculi muscle contributes to squinting.

11.5 When palpating the sagittal suture, the sheetlike epicranial aponeurosis (galea aponeurotica) is also palpated deep to the skin.

11.6 The inferior oblique muscle moves the eyeball superiorly and laterally because it originates at the anteromedial aspect of the floor of the orbit and inserts on the posterolateral aspect of the eyeball.

11.7 The masseter is the strongest muscle of mastication (see Figure 11.4d).

11.8 The suprahyoid and infrahyoid muscles stabilize the hyoid bone to assist in tongue movements.

11.9 The sternocleidomastoid muscle divides the neck into anterior and posterior triangles.

11.10 The anterior muscles lie deep within the neck and are anterior to the vertebral column.

11.11 The triangles in the neck formed by the sternocleidomastoid muscles are important anatomically and surgically because of the structures that lie within their boundaries.

12 The Muscular System: The Muscles of the Torso

As mentioned in previous chapters, the bones and muscles work together to create a stable center axis for the body. Muscles of the torso have been equated to the "core" muscles: those that attach to the spine and to the pelvic–hip complex. A current trend in exercise is to focus on strengthening the core of the body, particularly the abdominal muscles. Without strong core muscles, you cannot maintain balance, flexibility, and stabilization when using large muscles for gait or bending to pick up items. Having a well-aligned core allows the appropriate space for anatomical compartments such as the thoracic cavity that houses the lungs and heart. You will often find it difficult to breathe in cases where the body is misaligned due to muscle imbalance since you lose some of this anatomical space. Later in this chapter, you will also study breathing, which is of great importance to relax the body and provide adequate amounts of oxygen during activity. Exercise programs like pilates and yoga use those two basic themes—core stabilization and breathing—as the basis for all stretching and strengthening exercises.

CONTENTS AT A GLANCE

OBJECTIVES

- Describe the origin, insertion, action, and innervation of the muscles that act on the abdominal wall.
- Describe major surface features of the abdomen and pelvis.

In this chapter we will examine the muscles required for core stabilization of the body and for breathing, beginning with the muscles that act on the abdominal wall. The anterolateral abdominal wall is composed of skin, fascia, and four major pairs of muscles: the external oblique, internal oblique, transversus abdominis, and rectus abdominis. The first three muscles named are arranged from superficial to deep (Figures 12.1, 12.2, and 12.3). The ***external oblique*** is the superficial muscle. Its fascicles extend inferiorly and medially. The ***internal oblique*** is the intermediate flat muscle. Its fascicles extend at right angles to those of the external oblique. The ***transversus abdominis*** is the deep muscle, with most of its fascicles directed transversely around the abdominal wall. Together, the external oblique, internal oblique, and transversus abdominis form three layers of muscle around the abdomen.

The ***rectus abdominis*** (see Figures 12.1, 12.2, 12.3, and 12.4) is a long muscle that extends the entire length of the anterior abdominal wall, originating at the pubic crest and pubic symphysis and inserting on the cartilages of ribs 5–7 and the xiphoid process of the sternum. The anterior surface of the muscle is interrupted by three or more transverse fibrous bands of tissue called ***tendinous intersections,*** believed to be remnants of embryological development.

In each layer of the anterolateral abdominal wall, the muscle fascicles extend in a different direction (see Figure 12.2). This is a structural arrangement that affords considerable protection for the abdominal viscera, especially when the muscles have good tone. As a group, the muscles of the anterolateral abdominal wall help contain and protect the abdominal viscera; flex, laterally flex, and rotate the vertebral column at the intervertebral joints; compress the abdomen during forced exhalation;

Figure 12.1 Muscles of the male anterolateral abdominal wall.

Muscles of the torso help protect the abdominal viscera, move the vertebral column, and assist breathing and abdominal functions.

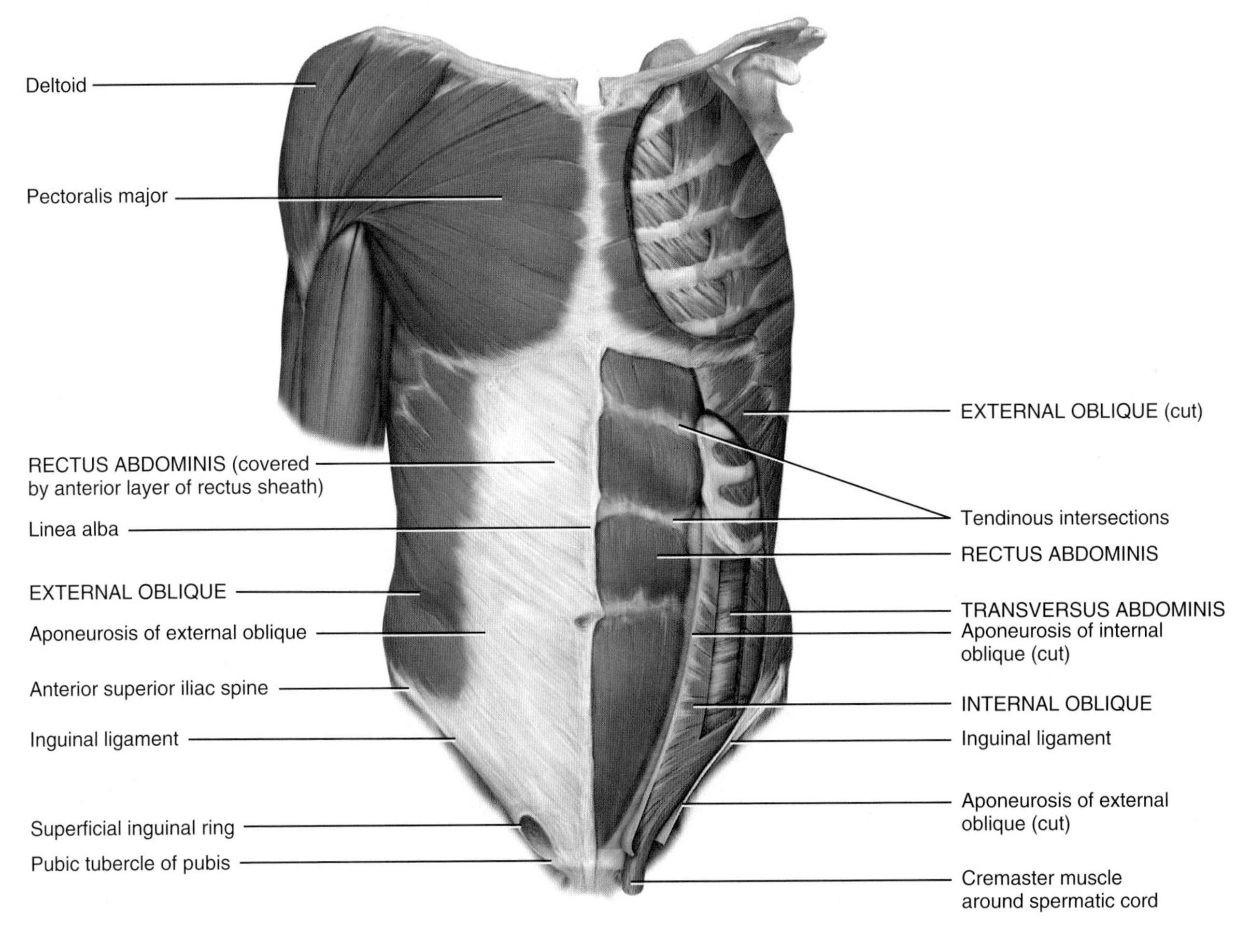

Which abdominal muscles laterally flex and rotate the vertebral column?

CONTINUES

Figure 12.2 Schematic of muscles of the anterolateral abdominal wall.

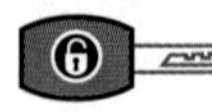

Muscle fibers of the anterolateral abdominal wall run in four different directions.

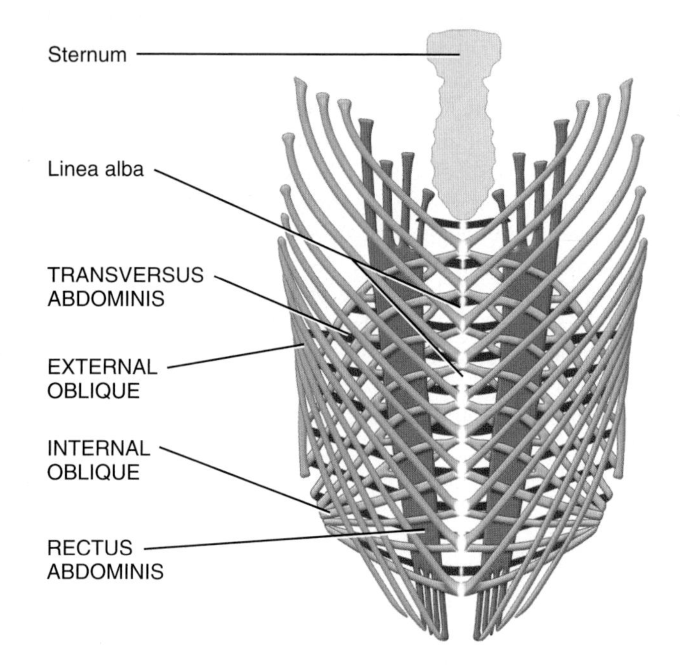

Schematic of anterolateral abdominal wall

What actions are aided by the compression of the anterolateral abdominal wall?

and produce the force required for defecation, urination, or childbirth. Activity of the abdominal muscles also affects the flow of venous blood through the abdomen as a result of intra-abdominal pressure changes.

The aponeuroses (sheathlike tendons) of the external oblique, internal oblique, and transversus abdominis muscles form the ***rectus sheaths,*** which enclose the rectus abdominis muscles (Figure 12.3). The sheaths meet at the midline to form the ***linea alba.*** In the latter stages of pregnancy, the linea alba stretches to increase the distance between the rectus abdominis muscles. The inferior free border of the external oblique aponeurosis plus some collagen fibers forms the ***inguinal ligament*** (IN-gwi-nal), which runs from the anterior superior iliac spine to the pubic tubercle (see Figure 12.1). Just superior to the medial end of the inguinal ligament is a triangular slit in the aponeurosis referred to as the ***superficial inguinal ring*** (see Figure 12.1a), the outer opening of the ***inguinal canal.*** The inguinal canal contains the spermatic cord and ilioinguinal nerve in males, and the round ligament of the uterus and ilioinguinal nerve in females; these structures will be studied in Chapter 29. The cremaster muscle encircles the spermatic cord (see Figure 12.1b).

A ***hernia*** (HER-nē-a) is a protrusion of an organ through a structure that normally contains it, which creates a lump that can be seen or felt through the skin's surface. The inguinal region is a weak area in the abdominal wall. It is often the site of an ***inguinal hernia,*** a rupture or separation of a portion of the inguinal area of the abdominal wall resulting in the protrusion of a part of the small intestine. Inguinal hernia is much more common in males than in females because the inguinal canals in males are larger to accommodate the spermatic cord and ilioinguinal nerve. Treatment of hernias most often involves surgery. The organ that protrudes is "tucked" back into the abdominal cavity and the defect in the abdominal muscles is repaired. In addition, a mesh is often applied to reinforce the area of weakness.

The posterior abdominal wall is formed by the lumbar vertebrae, parts of the ilia of the hip bones, psoas major, and iliacus muscles (described in Exhibit 14.1), and the ***quadratus lumborum*** muscle (see Figures 12.5b and 12.5d). The anterolateral abdominal wall can contract and distend; the posterior abdominal wall is bulky and stable by comparison.

Figure 12.3 Rectus sheaths and linea alba.

The rectus sheath and the linea alba are composed of aponeuroses of three abdominal muscles.

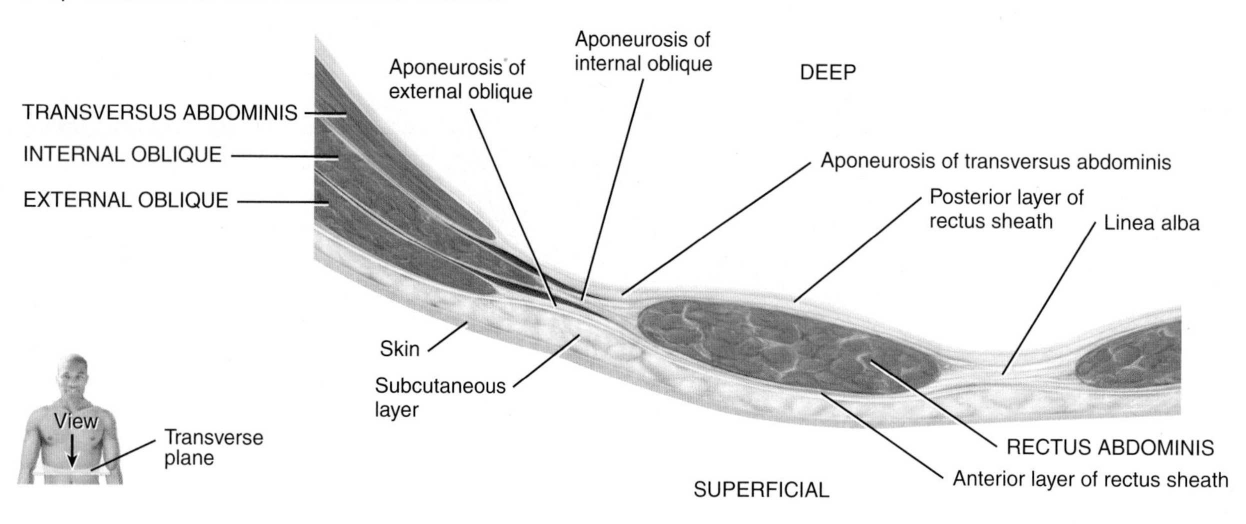

Superior view of transverse section of anterior abdominal wall superior to umbilicus (navel)

Which of the anterolateral abdominal muscles contribute to the formation of the posterior layer of the rectus sheath?

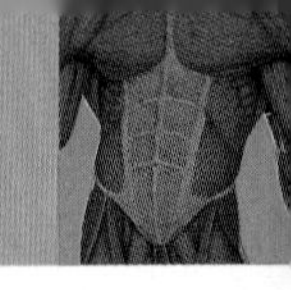

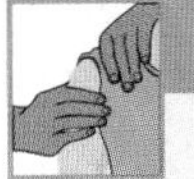

MANUAL THERAPY APPLICATION

The Quadratus Lumborum and "Short Leg" Syndrome

The **quadratus lumborum** muscle is the deepest muscle of the lumbar region of the back; its anterior surface is covered by the peritoneum of the abdomen. The muscle is therefore more difficult to access for the manual therapist. The origin is on the posterior third of the iliac crest (see Figure 12.5b) and the iliolumbar ligament. The insertion is on the inferior border of the twelfth rib and the transverse processes of L1–L4. The functions of quadratus lumborum are complex. Acting bilaterally, the muscles pull the twelfth ribs inferiorly during forced exhalation, such as during a cough. The muscles also stabilize the twelfth ribs to prevent their elevation during deep inhalation. The muscles also extend the lumbar portion of the vertebral column and thus are synergistic to the muscles of the back (see Exhibit 12.4) and antagonistic to the anterolateral muscles described previously. Acting unilaterally, the quadratus lumborum laterally flexes the vertebral column and is thus synergistic with the external and internal abdominal oblique muscles. During reverse muscle action (RMA), when positions of the origins and insertions are switched, the quadratus lumborum elevates the pelvis. Unilateral chronic contraction (spasm) of the quadratus lumborum muscle will elevate the pelvis on the same side, thus giving the muscle its nickname, the "hip hiker." When the pelvis is elevated on one side it causes the lower limb of that side to elevate, thus leading to a functional **"short leg" syndrome,** also known as leg length discrepancy (LLD). A misaligned pelvis, due to spasm of the quadratus lumborum, can compromise the normal action of virtually every muscle associated with upright posture and walking. Many muscles, from the plantar aspect of the foot to the epicranial aponeurosis, may change their tone and function to compensate for the misaligned pelvis. The resultant low back pain, sometimes known as lumbago, can be debilitating. When not recognized as a chronic spasm of the quadratus lumborum, the pain can be misdiagnosed as a disc problem or as a failed back surgery. Pain can be referred to (be perceived in) the sacroiliac joint, the deep buttock, or the area of the greater trochanter. Assessment of this functional problem may be difficult. One approach is for the therapist to gently tug the patient's ankles inferiorly to straighten the patient's hips on the table and then to visually evaluate the relative positions of the two medial malleoli. Another way to evaluate LLD is to have the patient stand in the anatomical position: The foot of a shorter leg will be closer to the midline of the body than the other foot. Note that unilateral contraction of the iliopsoas can also produce a "short leg."

MUSCLE	ORIGIN	INSERTION	ACTION	INNERVATION
Rectus abdominis (REK-tus ab-DOM-in-is; *rectus* = fascicles parallel to midline: *abdominis* = abdomen)	Pubic crest and pubic symphysis.	Cartilage of ribs 5–7 and xiphoid process.	Flexes vertebral column, especially lumbar portion, and compresses abdomen to aid in defecation, urination, forced exhalation, and childbirth. RMA: Flexes pelvis on the vertebral column.	Thoracic spinal nerves T7–T12.
External oblique (ō-BLĒK; *external* = closer to surface; *oblique* = fascicles diagonal to midline)	Ribs 5–12.	Iliac crest and linea alba.	Acting together (bilaterally), compress abdomen and flex vertebral column; acting singly (unilaterally), laterally flex vertebral column, especially lumbar portion, and rotate vertebral column.	Thoracic spinal nerves T7–T12 and the iliohypogastric nerve.
Internal oblique (ō-BLĒK; *internal* = farther from surface)	Iliac crest, inguinal ligament, and thoracolumbar fascia.	Cartilage of ribs 7–10 and linea alba.	Acting together, compress abdomen and flex vertebral column; acting singly, laterally flex vertebral column, especially lumbar portion, and rotate vertebral column.	Thoracic spinal nerves T8–T12, the iliohypogastric nerve, and ilioinguinal nerve.
Transversus abdominis (tranz-VER-sus ab-DOM-in-is; *transverse* = fascicles perpendicular to midline)	Iliac crest, inguinal ligament, lumbar fascia, and cartilages of ribs 5–10.	Xiphoid process, linea alba, and pubis.	Compresses abdomen.	Thoracic spinal nerves T8–T12, iliohypogastric nerve, and ilioinguinal nerve.
Quadratus lumborum (kwod-RĀ-tus lum-BŌR-um: *quad-* = four; *lumbo-* = lumbar region) (See Figure 12.5b)	Iliac crest and iliolumbar ligament.	Inferior border of rib 12 and L1–L4.	Acting together, pull twelfth ribs inferiorly during forced exhalation, fix twelfth ribs to prevent their elevation during deep inhalation, and help extend lumbar portion of vertebral column; acting singly, laterally flex vertebral column, especially lumbar portion. RMA: Elevates hip bone, commonly on one side.	Thoracic spinal nerve T12 and lumbar spinal nerves L1–L3 or L1–L4.

Figure 12.4 Surface features of the anterolateral abdomen and pelvis.

The linea alba is a frequent site for an abdominal incision because cutting through it severs no muscles and few blood vessels or nerves.

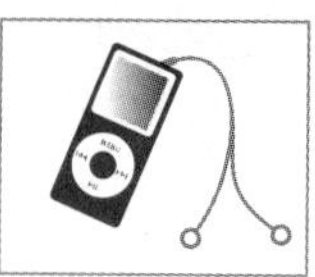

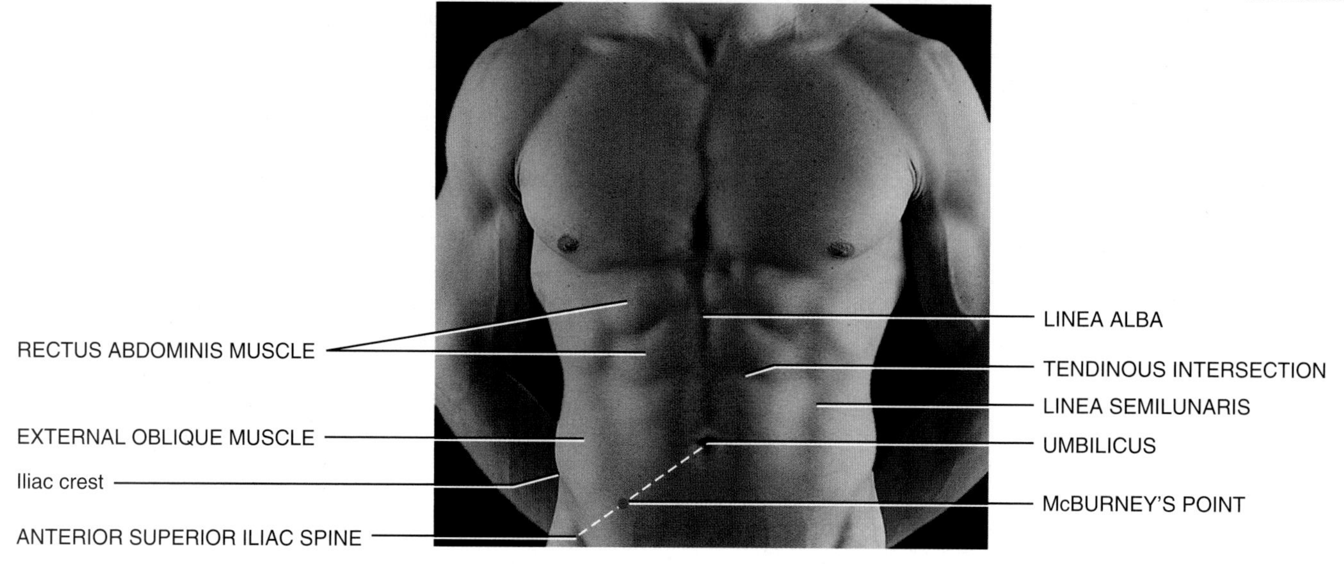

(a) Anterior view of abdomen

Surface Features of the Abdomen and Pelvis

Following are some of the prominent surface anatomy features of the *abdomen* and *pelvis* (Figure 12.4):

- **Umbilicus.** Also called the *navel,* the umbilicus marks the site of attachment of the umbilical cord to the fetus. It is level with the intervertebral disc between the bodies of vertebrae L3 and L4.
- **External oblique muscle.** Located inferior to the serratus anterior muscle (see Figure 12.1). The aponeurosis of the muscle on its inferior border is the ***inguinal ligament,*** a structure along which hernias frequently occur.
- **Rectus abdominis muscles.** Located just lateral to the midline of the abdomen. They can be seen by raising the shoulders while in the supine position without using the arms.
- **Linea alba.** Flat, tendinous raphe (RĀ-fē = the line of union to two contiguous, bilaterally symmetric structures) forming a furrow along the midline between the rectus abdominis muscles. The furrow extends from the xiphoid process to the pubic symphysis. It is broad superior to the umbilicus and narrow inferior to it. The linea alba is a frequently selected site for abdominal surgery because an incision through it severs no muscles and only a few blood vessels and nerves.

• CLINICAL CONNECTION McBurney's Point

A clinically important site located two-thirds of the way down an imaginary line drawn between the umbilicus and anterior superior iliac spine is **McBurney's point**, an important landmark related to the appendix. When the appendix must be removed, in an operation called an appendectomy, an oblique incision is made through McBurney's point. Moreover, pressure of the finger on McBurney's point produces pain in acute *appendicitis,* inflammation of the appendix, thus aiding in diagnosis. •

• CLINICAL CONNECTION The Six-Pack

Rectus abdominis is segmented by fibrous bands that run transversely or obliquely across the rectus abdominis muscle called *tendinous intersections.* These intersections illustrate the development of the muscle from several embryological structures. There are usually three intersections, one at the level of the umbilicus, one near the xiphoid process, and one midway between the other two. A fourth intersection is sometimes found below the level of the umbilicus. These tendinous intersections are fused with the anterior wall of the rectus sheath but have no connections to the posterior abdominal wall. Muscular persons may possess easily demonstrated intersections as the result of exercise and the ensuing hypertrophy of the rectus muscle. Hypertrophy of the muscle tissue, of course, has no effect on the connective tissue of the intersections. Body builders focus on the development of the **"six-pack"** effect of the abdomen. Small percentages of the population have a variant of the intersections and are able to develop an "eight-pack." •

- **Linea semilunaris.** The lateral edge of the rectus abdominis muscle (Figure 12.4) can be seen as a line that crosses the costal margin at the top of the ninth costal cartilage.
- **Iliac crest.** Superior margin of the ilium of the hip bone. It forms the outline of the superior border of the buttock. When you rest your hands on your hips, they rest on the iliac crests. A horizontal line drawn across the highest point of each iliac crest is called the ***supracristal line,*** which intersects the spinous process of the fourth lumbar vertebra. This vertebra is a landmark for performing a spinal tap (see Exhibit 16.4).

CHECKPOINT

1. Which muscles do you contract when you "suck in your gut," thereby compressing the anterior abdominal wall?

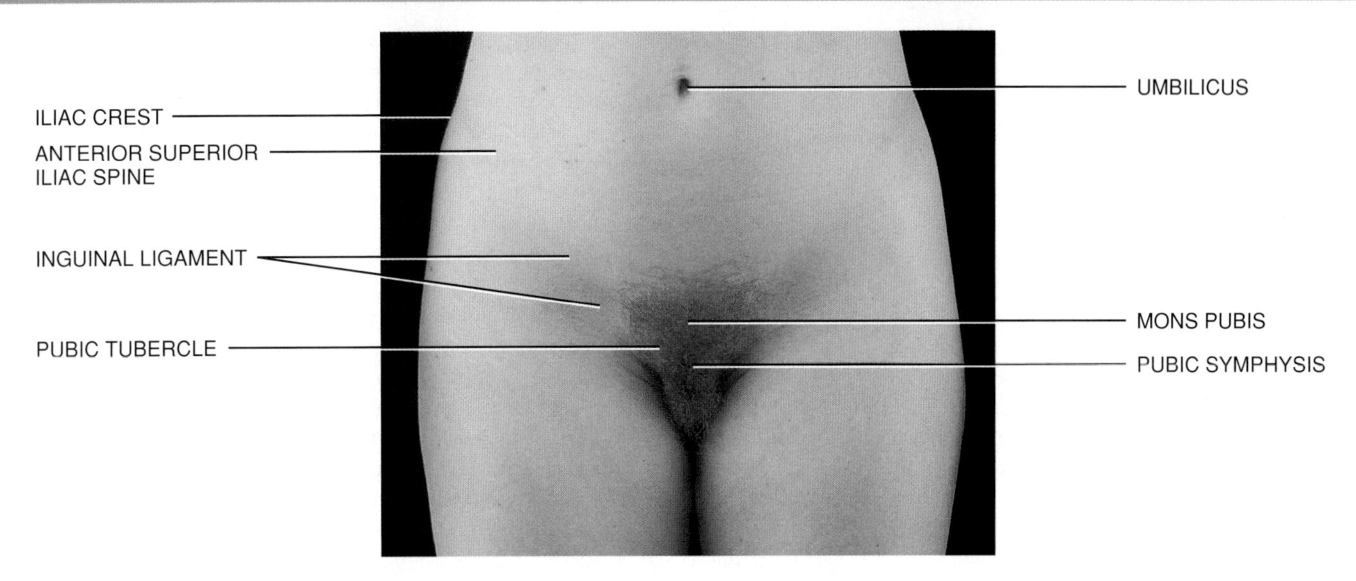

(b) Anterior view of pelvis

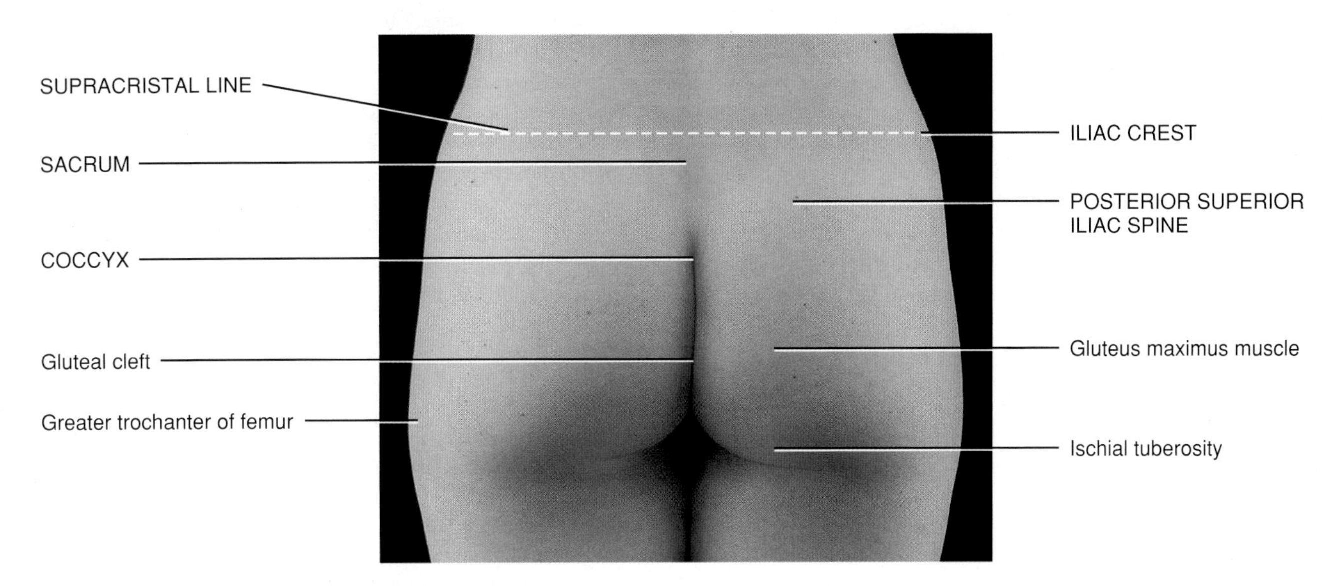

(c) Posterior view of pelvis

What is the clinical significance of McBurney's point?

- **Anterior superior iliac spine.** The anterior end of the iliac crest that lies at the superior lateral end of the fold of the groin.
- **Posterior superior iliac spine.** The posterior end of the iliac crest, usually indicated by a dimple in the skin that coincides with the middle of the sacroiliac joint, where the hip bone attaches to the sacrum.
- **Pubic tubercle.** Projection on the superior border of the pubis of the hip bone. Attached to it is the medial end of the ***inguinal ligament,*** the inferior free edge of the aponeurosis of the external oblique muscle that forms the ***inguinal canal.*** The lateral end of the ligament is attached to the anterior superior iliac spine. The spermatic cord in males and the round ligament of the uterus in females pass through the inguinal canal.
- **Pubic symphysis.** Anterior joint of the hip bones; palpated as a firm resistance in the midline at the inferior portion of the anterior abdominal wall.
- **Mons pubis.** An elevation of adipose tissue covered by skin and pubic hair that is anterior to the pubic symphysis.
- **Sacrum.** The fused spinous processes of the sacrum, called the ***median sacral crest,*** can be palpated beneath the skin superior to the ***gluteal cleft,*** a depression along the midline that separates the buttocks (and is part of the discussion on the buttocks in Section 14.1).
- **Coccyx.** The inferior surface of the tip of the coccyx can be palpated in the gluteal cleft, about 2.5 cm (1 in.) posterior to the anus.

Relating Muscles to Movements

Arrange the muscles in this exhibit according to the following actions on the vertebral column: (1) flexion, (2) lateral flexion, (3) extension, and (4) rotation. The same muscle may be mentioned more than once.

EXHIBIT 12.2 Muscles of the Thorax Used in Breathing

OBJECTIVES

- Describe the origin, insertion, action, and innervation of the muscles used in breathing.
- Describe the major structures that pass through openings in the diaphragm.

The muscles described here alter the size of the thoracic cavity so that breathing can occur (Figure 12.5). Inhalation (breathing in) occurs when the thoracic cavity increases in size, and exhalation (breathing out) occurs when the thoracic cavity decreases in size.

The dome-shaped ***diaphragm*** separates the thoracic and abdominal cavities. The diaphragm has a convex superior surface that forms the floor of the thoracic cavity (Figure 12.5b,c) and a concave, inferior surface that forms the roof of the abdominal cavity (Figure 12.5b,d). The muscular portion of the diaphragm is around the periphery of the muscle. The fibers of the muscular portion converge and insert into the *central tendon*, a strong aponeurosis located near the center of the muscle (Figure 12.5b,c,d). The central tendon fuses with the inferior surface of the pericardium (covering of the heart) and the pleurae (coverings of the lungs).

The diaphragm has three major openings through which the aorta, esophagus, and inferior vena cava pass between the thorax and abdomen (Figure 12.5c,d). These structures and the openings include the aorta, along with the thoracic duct and azygous vein, which pass through the ***aortic hiatus;*** the esophagus with accompanying vagus (X) nerves, which pass through the ***esophageal hiatus;*** and the inferior vena cava, which passes through the ***caval opening (foramen for the vena cava).*** In a condition called a hiatus hernia, a portion of the stomach protrudes superiorly through the esophageal hiatus.

The diaphragm is the most important muscle that powers breathing. During contraction, it depresses into a flatter shape, increasing the thoracic cavity volume, which results in inhalation. As the diaphragm relaxes, it elevates back to the dome shape, decreasing the thoracic cavity volume to produce exhalation. Movements of the diaphragm also help return blood from the abdomen to the heart. Together, the diaphragm and anterolateral abdominal muscles can be voluntarily contracted to help increase pressure in the abdomen to evacuate the pelvic contents during defecation, urination, and childbirth. This mechanism is further assisted when you take a deep breath and close the larynx. The trapped air in the respiratory system prevents the diaphragm from elevating. The increase in intra-abdominal pressure helps support the vertebral column and also helps prevent flexion during weight lifting. This greatly assists the back muscles in lifting a heavy weight.

Other muscles involved in breathing are called intercostal muscles (Figure 12.5a,b,c). They span the intercostal spaces, the spaces between ribs. The 11 pairs of ***external intercostal*** muscles occupy the superficial layer, and their fibers run obliquely and anteriorly from the rib above to the rib below. They elevate the ribs during inhalation to help expand the thoracic cavity. The 11 pairs of ***internal intercostal*** muscles lie deep to the external intercostals. The fibers of these muscles are at right angles to fibers of the external intercostal muscles and run obliquely and posteriorly from the rib below to the rib above. They draw adjacent ribs together during forced exhalation to help decrease the size of the thoracic cavity. The deepest muscle layer is made up of ***innermost intercostal*** muscles. They are poorly developed in most persons and run in the same direction as the internal intercostals (Figure 12.5b,c).

Note: A mnemonic for the action of the intercostal muscles is singing "Old MacDonald had a farm, **E, I, E, I, O**" = **E**xternal **I**ntercostals **E**levate during **I**nspiration, **O**h!"

MUSCLE	ORIGIN	INSERTION	ACTION	INNERVATION
Diaphragm (DĪ-a-fram; *dia-* = across; *-phragm* = wall)	Xiphoid process of the sternum, costal cartilages and portions of ribs 6–12, lumbar vertebrae and their intervertebral discs.	Central tendon.	Contraction of the diaphragm causes it to flatten and increases the vertical dimension of the thoracic cavity, resulting in inhalation; relaxation of the diaphragm causes it to move superiorly and decreases the vertical dimension of the thoracic cavity, resulting in exhalation.	Phrenic nerve, which contains axons from cervical spinal nerves C3–C5.
External intercostals (in′-ter-KOS-tals; *external* = closer to surface; *inter-* = between; *costa* = rib)	Inferior border of rib above.	Superior border of rib below.	Contraction elevates the ribs and increases the anteroposterior and lateral dimensions of the thoracic cavity, resulting in inhalation; relaxation depresses the ribs and decreases the anteroposterior and lateral dimensions of the thoracic cavity, resulting in exhalation.	Thoracic spinal nerves T2–T12.
Internal intercostals (*internal* = farther from surface)	Superior border of rib below.	Inferior border of rib above.	Contraction draws adjacent ribs together to further decrease the anteroposterior and lateral dimensions of the thoracic cavity during forced exhalation.	Thoracic spinal nerves T2–T12.
Innermost intercostals	Superior border of rib below.	Inferior border of rib above.	Action is the same as for internal intercostals; formerly considered to be a deep layer of the internal intercostals.	Thoracic spinal nerves T2–T12.

As you will see in Chapter 25, the diaphragm and external intercostal muscles are used during quiet inhalation and exhalation. However, during deep, forceful inhalation, as occurs during exercise or playing a wind instrument, the sternocleidomastoid, scalene, and pectoralis minor muscles are also used. During deep, forceful exhalation, the external oblique, internal oblique, transversus abdominis, rectus abdominis, and internal intercostals are also active.

CHECKPOINT

2. Which muscles do you contract during normal inhalation and exhalation?
3. Which muscles do you contract during forced inhalation and forced exhalation?

Figure 12.5 Muscles of the thorax used in breathing, as seen in a male. Muscles used in breathing alter the size of the thoracic cavity, thus enabling ventilation and assisting in venous return of blood to the heart.

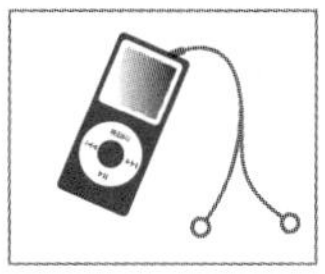

Openings in the diaphragm permit the passage of the aorta, esophagus, and inferior vena cava.

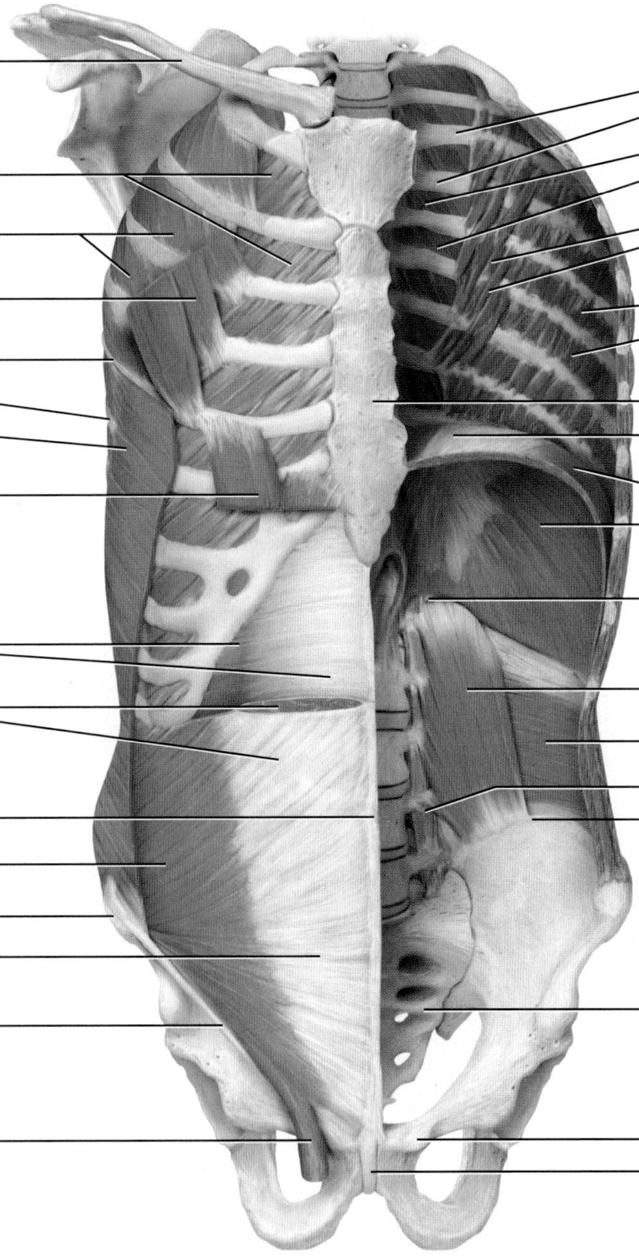

(a) Anterior superficial view (b) Anterior deep view

CONTINUES

Figure 12.5 (continued)

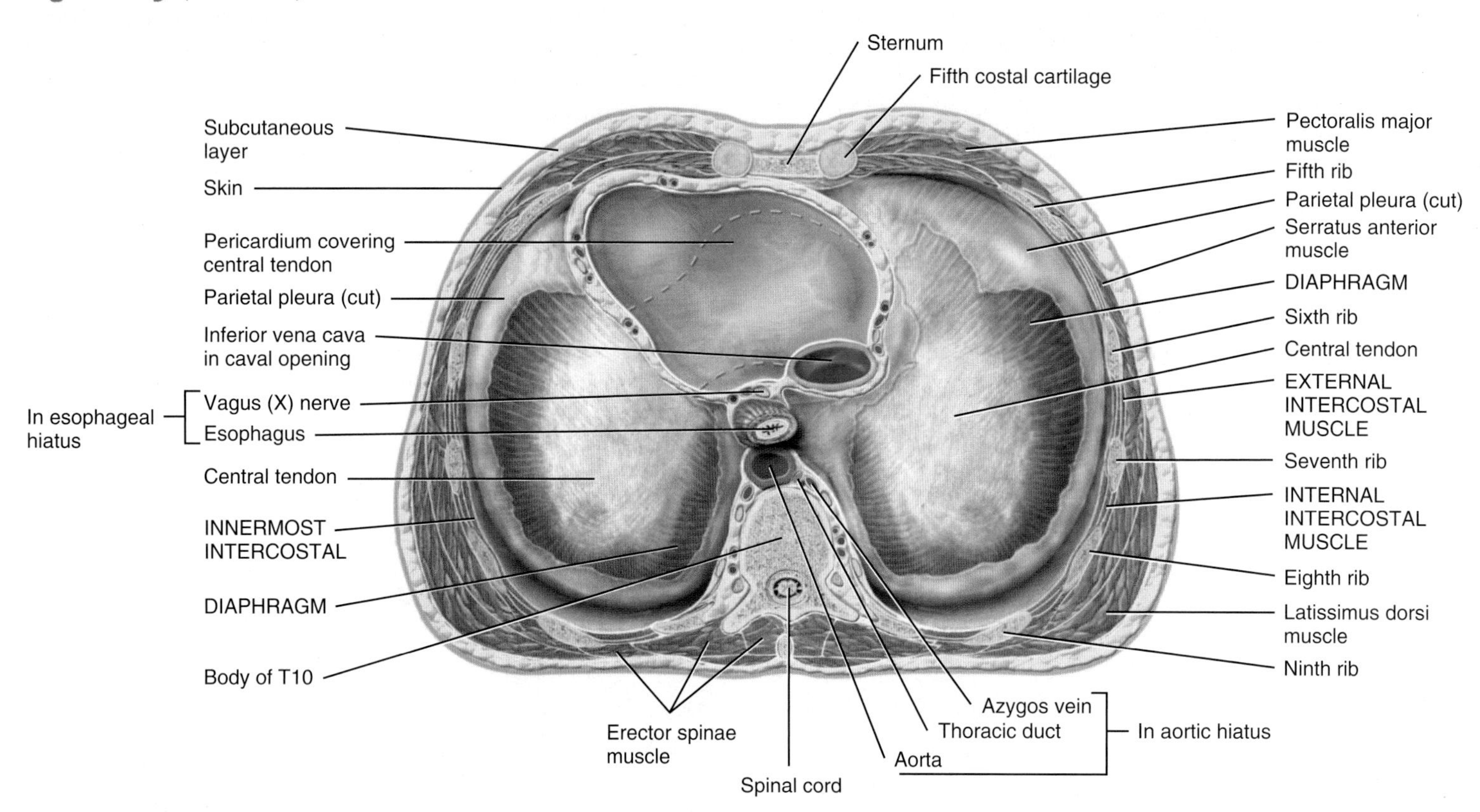

(c) Superior view of diaphragm

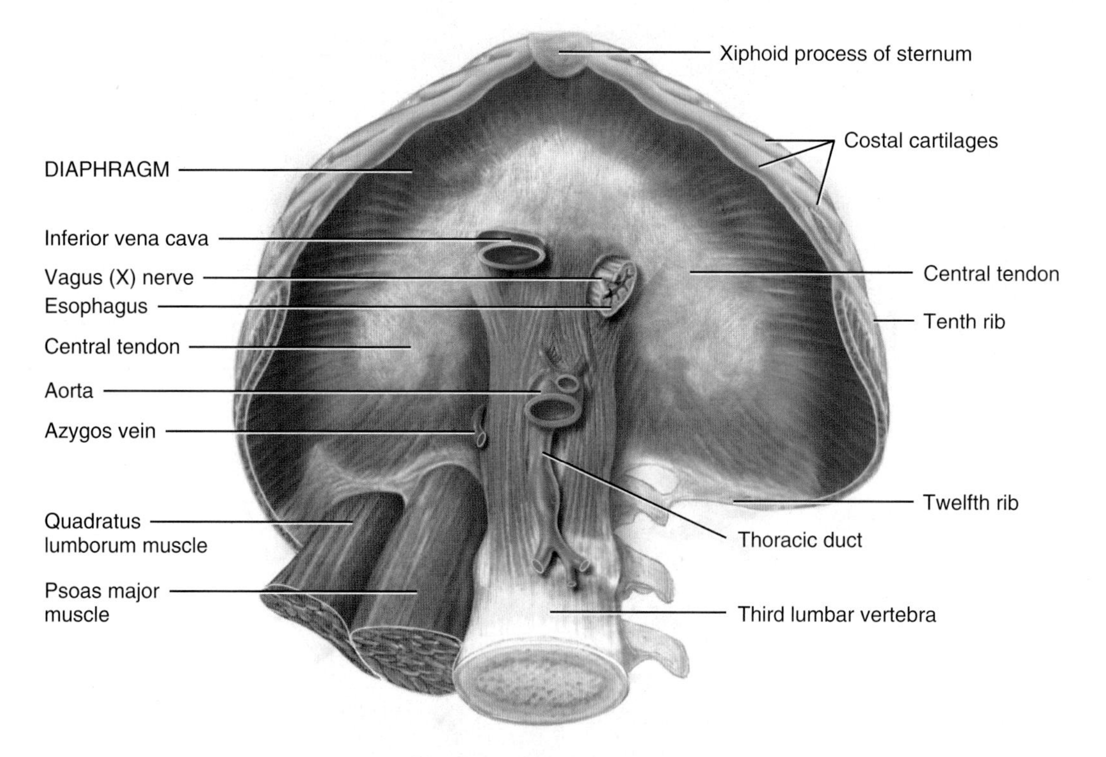

(d) Inferior view of diaphragm

? Which muscle associated with breathing is innervated by the phrenic nerve?

EXHIBIT 12.3 Muscles of the Pelvic Diaphragm and Perineum That Support the Pelvic Viscera

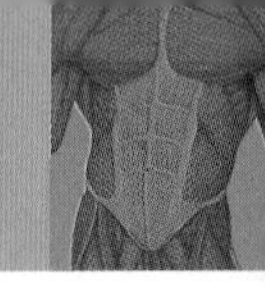

OBJECTIVE

- Describe the origin, insertion, action, and innervation of the muscles of the pelvic diaphragm and the perineum.

The muscles of the pelvic floor are the levator ani and ischiococcygeus. Together with the fascia covering their internal and external surfaces these muscles are referred to as the ***pelvic diaphragm,*** which stretches from the pubis anteriorly to the coccyx posteriorly, and from one lateral wall of the pelvis to the other. This arrangement gives the pelvic diaphragm the appearance of a funnel suspended from its attachments. The pelvic diaphragm separates the pelvic cavity above from the perineum below. The anal canal and urethra pierce the pelvic diaphragm in both sexes, and the vagina also goes through it in females.

The two components of the ***levator ani*** muscle are the ***pubococcygeus*** and ***iliococcygeus.*** Figure 12.6a shows these muscles in the female and Figure 12.6b illustrates them in the male. The levator ani is the largest and most important muscle of the pelvic floor. It supports the pelvic viscera and resists the inferior thrust that accompanies increases in intra-abdominal pressure during functions such as forced exhalation, coughing, vomiting, urination, and defecation. The muscle also functions as a sphincter at the anorectal junction, urethra, and vagina. In addition to assisting the levator ani, the ***ischiococcygeus*** pulls the coccyx anteriorly after it has been pushed posteriorly during defecation or childbirth.

The ***perineum*** is the region of the trunk inferior to the pelvic diaphragm. It is a diamond-shaped area that extends from the pubic symphysis anteriorly, to the coccyx posteriorly, and to the ischial tuberosities laterally. The female and the male perineums may be compared in Figure 12.6. A transverse line drawn between the ischial tuberosities divides the perineum into an anterior ***urogenital triangle*** that contains the external genitals and a posterior ***anal triangle*** that contains the anus. Several perineal muscles insert into the perineal body of the perineum. Clinically, the perineum is very important to physicians who care for women during pregnancy and treat disorders related to the female genital tract, urogenital organs, and the anorectal region.

The muscles of the perineum are arranged in two layers; ***superficial*** and ***deep.*** The muscles of the superficial layer are the ***superficial transverse perineal,*** the ***bulbospongiosus,*** and the ***ischiocavernosus*** (Figure 12.6). The deep muscles of the male perineum are the ***deep transverse perineal*** and ***external urethral sphincter*** (Figure 12.6b). The deep muscles of the female perineum are the ***compressor urethrae, sphincter urethrovaginalis,*** and ***external urethral sphincter*** (Figure 12.6a). The deep muscles of the perineum assist in urination and ejaculation in males and facilitate closing of the vagina in females. The ***external anal sphincter*** closely adheres to the skin around the margin of the anus and keeps the anal canal and anus closed except during defecation.

• CLINICAL CONNECTION | Injury of Levator Ani and Urinary Stress Incontinence

During childbirth, the levator ani muscle supports the head of the fetus, and the muscle may be injured during a difficult childbirth or traumatized during an *episiotomy* (a cut made with surgical scissors to prevent or direct tearing of the perineum during the birth of a baby). The consequence of such injury may be **urinary stress incontinence,** that is, the leakage of urine whenever intra-abdominal pressure is increased—for example, during coughing. One way to treat urinary stress incontinence is to strengthen and tighten the muscles that support the pelvic viscera. This is accomplished by *Kegel exercises,* the alternate contraction and relaxation of muscles of the pelvic floor. To find the correct muscles, the person imagines that she is urinating and then contracts the muscles as if stopping in midstream. The muscles should be held for a count of three, then relaxed for a count of three. This should be done 5–10 times each hour—sitting, standing, and lying down. Kegel exercises are also encouraged during pregnancy to strengthen the muscles for delivery. •

Relating Muscles to Movements

Arrange the muscles in this exhibit according to the following actions: (1) supporting and maintaining the position of the pelvic viscera; (2) resisting an increase in intra-abdominal pressure; (3) constriction of the anus, urethra, and vagina; (4) expulsion of urine and semen, and (5) erection of the clitoris and penis. The same muscle may be mentioned more than once.

CHECKPOINT

4. Which muscles are strengthened by Kegel exercises?
5. What are the borders and contents of the urogenital triangle and the anal triangle?

CONTINUES

Muscles of the Pelvic Diaphragm and Perineum That Support the Pelvic Viscera CONTINUED

MUSCLE	ORIGIN	INSERTION	ACTION	INNERVATION
PELVIC DIAPHRAGM MUSCLES				
Levator ani (le-VĀ-tor Ā-nē; *levator* = raises; *ani* = anus)	This muscle is divisible into two parts, the pubococcygeus and iliococcygeus.			
Pubococcygeus (pū′-bō-kok-SIJ-ē-us; *pubo-* = pubis; *-coccygeus* = coccyx)	Pubis and ischial spine.	Coccyx, urethra, anal canal, perineal body of perineum (wedge-shaped mass of fibrous tissue in center of perineum), and anococcygeal ligament (narrow fibrous band that extends from anus to coccyx).	Supports and maintains position of pelvic viscera; resists increase in intra-abdominal pressure during forced exhalation, coughing, vomiting, urination, and defecation; constricts anus, urethra, and vagina.	Sacral spinal nerves S2–S4.
Iliococcygeus (il′-ē-ō-kok-SIJ-ē-us; *ilio-* = ilium)	Ischial spine.	Coccyx.	As above.	Sacral spinal nerves S2–S4.
Ischiococcygeus (is′-kē-ō-kok-SIJ-ē-us; *ischio-* = hip)	Ischial spine.	Lower sacrum and upper coccyx.	Supports and maintains position of pelvic viscera; resists increase in intra-abdominal pressure during forced exhalation, coughing, vomiting, urination, and defecation; pulls coccyx anteriorly following defecation or childbirth.	Sacral spinal nerves S4–S5.
SUPERFICIAL PERINEAL MUSCLES				
Superficial transverse perineal (per-i-NĒ-al; *superficial* = closer to surface; *transverse* = across; *perineus* = perineum)	Ischial tuberosity.	Perineal body of perineum.	Stabilizes perineal body of perineum.	Pudendal nerve of sacral plexus.
Bulbospongiosus (bul′-bō-spon′-jē-Ō-sus; *bulb-* = bulb; *-spongio-* = sponge)	Perineal body of perineum.	Perineal membrane of deep muscles of perineum, corpus spongiosum of penis, and deep fascia on dorsum of penis in male; pubic arch and root and dorsum of clitoris in female.	Helps expel urine during urination, helps propel semen along urethra, assists in erection of the penis in male; constricts vaginal orifice and assists in erection of clitoris in female.	Pudendal nerve of sacral plexus.
Ischiocavernosus (is′-kē-ō-ka′-ver-NŌ-sus; *ischio-* = the hip)	Ischial tuberosity and ischial and pubic rami.	Corpora cavernosa of penis in male and clitoris in female and pubic symphysis.	Maintains erection of penis in male and clitoris in female by decreasing venous drainage.	Pudendal nerve of sacral plexus.
DEEP PERINEAL MUSCLES				
Deep transverse perineal (per-i-NĒ-al; *deep* = farther from surface)	Ischial ramus.	Perineal body of perineum.	Helps expel last drops of urine and semen in male.	Pudendal nerve of sacral plexus.
External urethral sphincter (ū-RĒ-thral SFINGK-ter)	Ischial and pubic rami.	Median raphe in male and vaginal wall in female.	Helps expel last drops of urine and semen in male and urine in female.	Sacral spinal nerve S4 and pudendal nerve.
Compressor urethrae (ū-RĒ-thrē)	Ischiopubic ramus.	Blends with partner on other side anterior to urethra.	Serves as an accessory sphincter of the urethra.	Pudendal nerve of sacral plexus.
Sphincter urethrovaginalis (ū-RĒ-thrō-vaj-i-NAL-is)	Perineal body.	Blends with partner on other side anterior to urethra.	Serves as an accessory sphincter of the urethra and facilitates closing of the vagina.	Pudendal nerve of sacral plexus.
External anal sphincter (Ā-nal)	Anococcygeal ligament.	Perineal body of perineum.	Keeps anal canal and anus closed.	Sacral spinal nerve S4 and pudendal nerve.

Figure 12.6 Muscles of the pelvic floor, as seen in the female and male perineum.

Muscles of the pelvic floor and perineum support the pelvic viscera, function as sphincters, and assist in urination, erection of the penis and clitoris, ejaculation, and defecation.

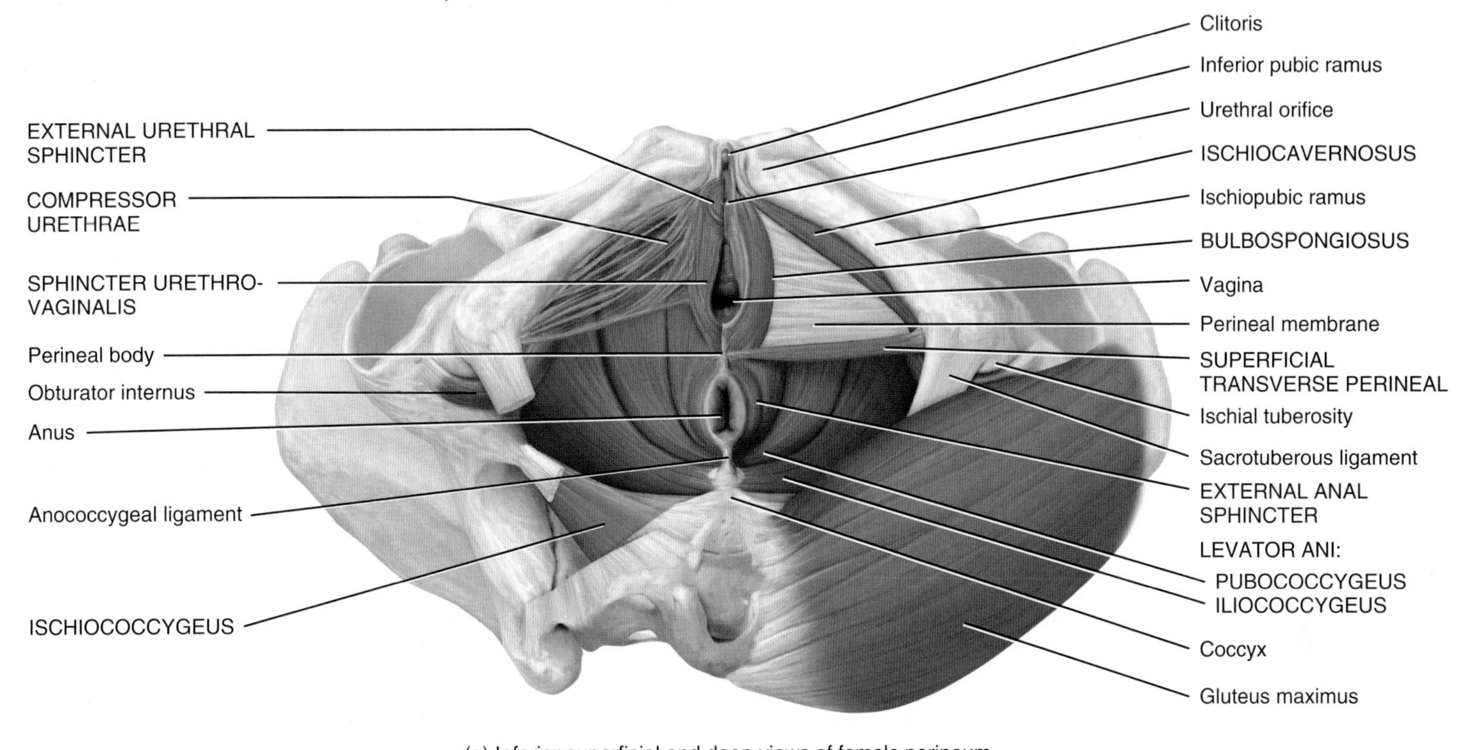

(a) Inferior superficial and deep views of female perineum

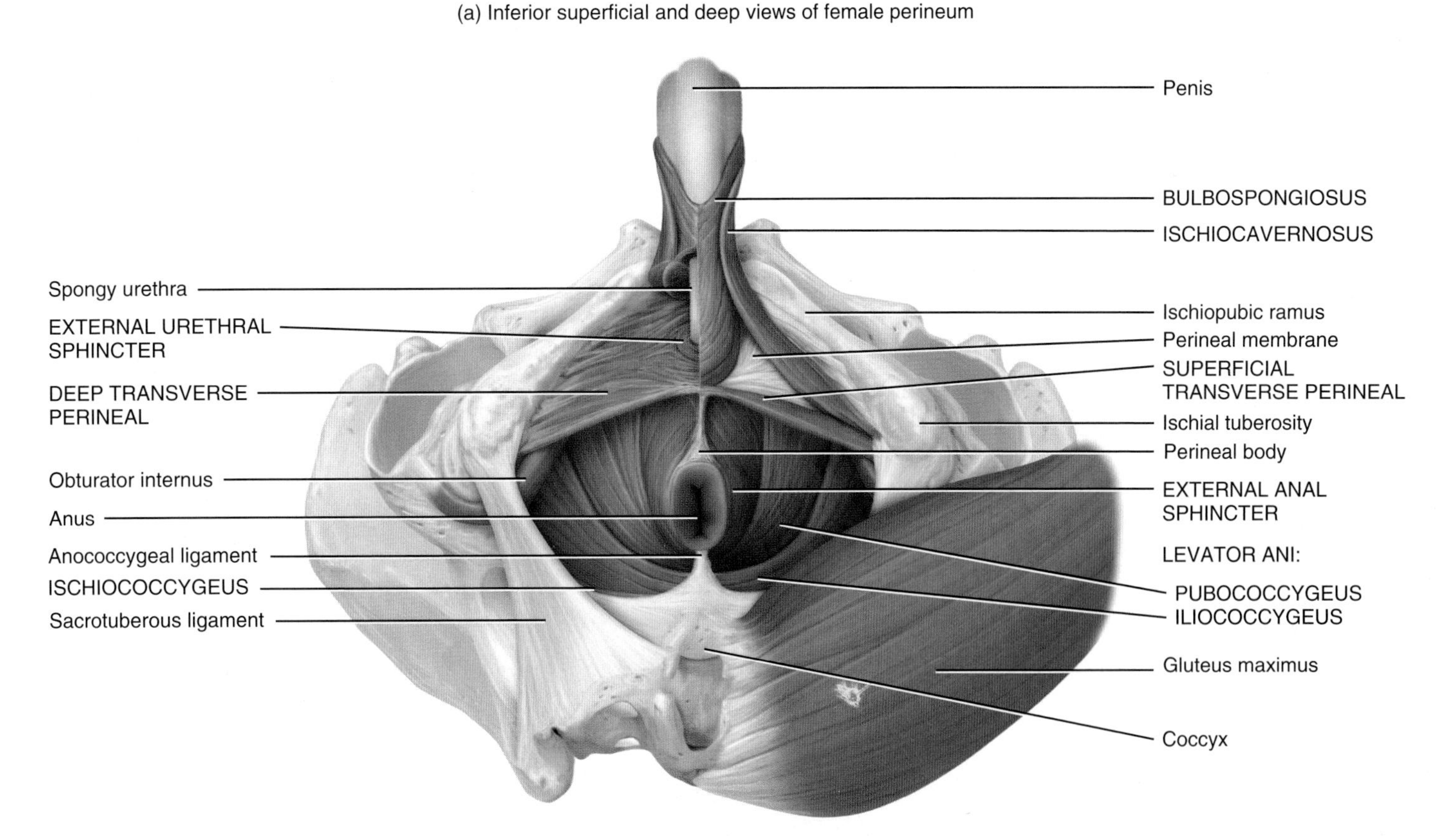

(b) Inferior superficial and deep views of male perineum

What are the borders of the pelvic diaphragm?

EXHIBIT 12.4

Muscles of the Neck and Back That Act on the Posterior Head, Posterior Neck, Back, and Vertebral Column

OBJECTIVE

- Describe the origin, insertion, action, and innervation of the muscles that act on the posterior head, posterior neck, and the vertebral column.

Balance and movement of the head on the vertebral column involve the action of many muscles. Muscles of the anterior neck were described in Chapter 11. As stated in that chapter, muscles of the posterior neck are, in most cases, continuations of back muscles and are discussed in this chapter. Study of the neck and back muscles is further complicated by the presence of two very large superficial muscles of the back, the trapezius and the latissimus dorsi (see Figures 13.1c and 13.4b), whose actions are primarily associated with the upper limbs and will be discussed in Chapter 13. The trapezius, however, is a very important muscle of extension of both the head and neck.

The muscles that move the vertebral column (Figure 12.7) are quite complex because they have multiple origins and insertions and there is considerable overlap among them. The ***splenius*** muscles are attached

MUSCLE	ORIGIN	INSERTION	ACTION	INNERVATION
SPLENIUS (SPLĒ-nē-us)				
Splenius capitis (KAP-i-tis, *splenium* = bandage; *capit-* = head)	Ligamentum nuchae and spinous processes of C7–T4.	Occipital bone and mastoid process of temporal bone.	Acting together (bilaterally), extend head; acting singly (unilaterally), laterally flex and/or rotate head to same side as contracting muscle.	Middle cervical spinal nerves.
Splenius cervicis (SER-vi-kis; *cervic-* = neck)	Spinous processes of T3–T6.	Transverse processes of C1–C2 or C1–C4.	Acting together; extend head; acting singly, laterally flex and/or rotate head to same side as contracting muscle.	Inferior cervical spinal nerves.
ERECTOR SPINAE (e-REK-tor SPĪ-nē) Consists of iliocostalis muscles (lateral), longissimus muscles (intermediate), and spinalis muscles (medial).				
ILIOCOSTALIS GROUP (LATERAL)				
Iliocostalis cervicis (il′-ē-ō-kos-TĀL-is; *ilio-* = ilium; *-costa-* = rib).	Ribs 1–6.	Transverse processes of C4–C6.	Acting together, muscles of each region (cervical, thoracic, and lumbar) extend and maintain erect posture of vertebral column of their respective regions; acting singly, laterally flex vertebral column of their respective regions to the same side as the contracting muscle.	Cervical and thoracic spinal nerves.
Iliocostalis thoracis (thō-RĀ-sis; *thorac-* = chest)	Ribs 7–12.	Ribs 1–6.		Thoracic spinal nerves.
Iliocostalis lumborum (lum-BOR-um; *lumbo-* = lumbar region)	Iliac crest.	Ribs 7–12.		Lumbar spinal nerves.
LONGISSIMUS GROUP (INTERMEDIATE)				
Longissimus capitis (lon-JIS-i-mus = longest)	Articular processes of C4–C7 and transverse processes of T1–T4.	Mastoid process of temporal bone.	Acting together, both longissimus capitis muscles extend head; acting singly, rotate head to same side as contracting muscle.	Middle and inferior cervical spinal nerves.
Longissimus cervicis	Transverse processes of T4–T5.	Transverse processes of C2–C6.	Acting together longissimus cervicis and both longissimus thoracis muscles extend vertebral column of their respective regions; acting singly, laterally flex vertebral column of their respective regions.	Cervical and superior thoracic, spinal nerves.
Longissimus thoracis	Transverse processes of lumbar vertebrae.	Transverse processes of all thoracic and superior lumbar vertebrae and ribs 9 and 10.		Thoracic and lumbar spinal nerves.
SPINALIS GROUP (MEDIAL)				
Spinalis capitis (spi-NĀ-lis; *spinal-* = vertebral column)	Often absent or very small; arises with semispinalis capitis.	Occipital bone (with semispinalis).	Acting together, muscles of each region (cervical, thoracic, and lumbar) extend vertebral column of their respective regions.	Cervical and superior thoracic spinal nerves.
Spinalis cervicis	Ligamentum nuchae and spinous process of C7.	Spinous process of axis.		Inferior cervical and thoracic spinal nerves.
Spinalis thoracis	Spinous processes of T10–L2.	Spinous processes of superior thoracic vertebrae.		Thoracic spinal nerves.

MUSCLE	ORIGIN	INSERTION	ACTION	INNERVATION
TRANSVERSOSPINALES (trans-ver-sō-spi-NĀ-lēz)				
Semispinalis capitis (sem′-ē-spī-NĀ-lis; *semi-* = partially or one-half)	Articular processes of C4–C6 and transverse processes of C7–T7.	Occipital bone.	Acting together, extend head; acting singly, rotate head to side opposite contracting muscle.	Cervical and thoracic spinal nerves.
Semispinalis cervicis	Transverse processes of T1–T5.	Spinous processes of C1–C5.	Acting together, both semispinalis cervicis and both semispinalis thoracis muscles extend vertebral column of their respective regions; acting singly, rotate head to side opposite contracting muscle.	Cervical and thoracic spinal nerves.
Semispinalis thoracis	Transverse processes of T6–T10.	Spinous processes of C6–T4.		Thoracic spinal nerves.
Multifidus (mul-TIF-i-dus: *multi-* = many; *-fid-* = segmented)	Sacrum, ilium, transverse processes of lumbar, thoracic, and C4–C7.	Spinous process of a more superior vertebra.	Acting together, extend vertebral column; acting singly, weakly laterally flex vertebral column and weakly rotate vertebral column to side opposite contracting muscle.	Cervical, thoracic, and lumbar spinal nerves.
Rotatores (rō-ta-TŌ-rēz; singular is **rotatore**; *rotatore* = to rotate)	Transverse processes of all vertebrae.	Spinous process of vertebra superior to the one of origin.	Acting together, weakly extend vertebral column; acting singly, weakly rotate vertebral column to side opposite contracting muscle.	Cervical, thoracic, and lumbar spinal nerves.
SEGMENTAL (seg-MEN-tal)				
Interspinales (in-ter-spī-NĀ-lēz; *inter-* = between)	Superior surface of all spinous processes.	Inferior surface of spinous process of vertebra superior to the one of origin.	Acting together, weakly extend vertebral column; acting singly, stabilize vertebral column during movement.	Cervical, thoracic, and lumbar spinal nerves.
Intertransversarii (in′-ter-trans-vers-AR-ē-ī; singular is **intertransversarius**)	Transverse processes of all vertebrae.	Transverse process of vertebra superior to the one of origin.	Acting together, extend vertebral column; acting singly, weakly laterally flex vertebral column and stabilize it during movements.	Cervical, thoracic, and lumbar spinal nerves.
SUBOCCIPITAL (CRANIOVERTEBRAL JOINT) MUSCLES				
Rectus capitis posterior major	Spinous process of the axis (C2).	Middle third of inferior nuchal line.	Acting together, extend head; acting singly, rotate head to same side.	Dorsal ramus of C1 (suboccipital nerve).
Rectus capitis posterior minor	Posterior tubercle of the atlas (C1).	Medial third of inferior nuchal line.	Extend or retract head.	Dorsal ramus of C1 (suboccipital nerve).
Obliquus (ob-LĪ-kwus) **capitis inferior**	Spinous process of the axis (C2).	Transverse process of atlas (C1).	Rotate head to the same side.	Dorsal ramus of C1 (suboccipital nerve).
Obliquus capitis superior	Transverse process of the atlas (C1).	Above the insertion of rectus capitis posterior major.	Acting together, extend or tilt head; acting singly, rotate and/or laterally flex head to same side.	Dorsal ramus of C1 (suboccipital nerve).

CONTINUES

Figure 12.7 Muscles that move the vertebral column and head. Muscles that move the vertebral column and head are quite complex because they have multiple origins and insertions and because there is considerable overlap among them.

The erector spinae group (iliocostalis, longissimus, and spinalis muscles) is the largest muscular mass of the body and is the chief extensor of the vertebral column. Note that the trapezius and occipitofrontalis muscles have been removed in this figure.

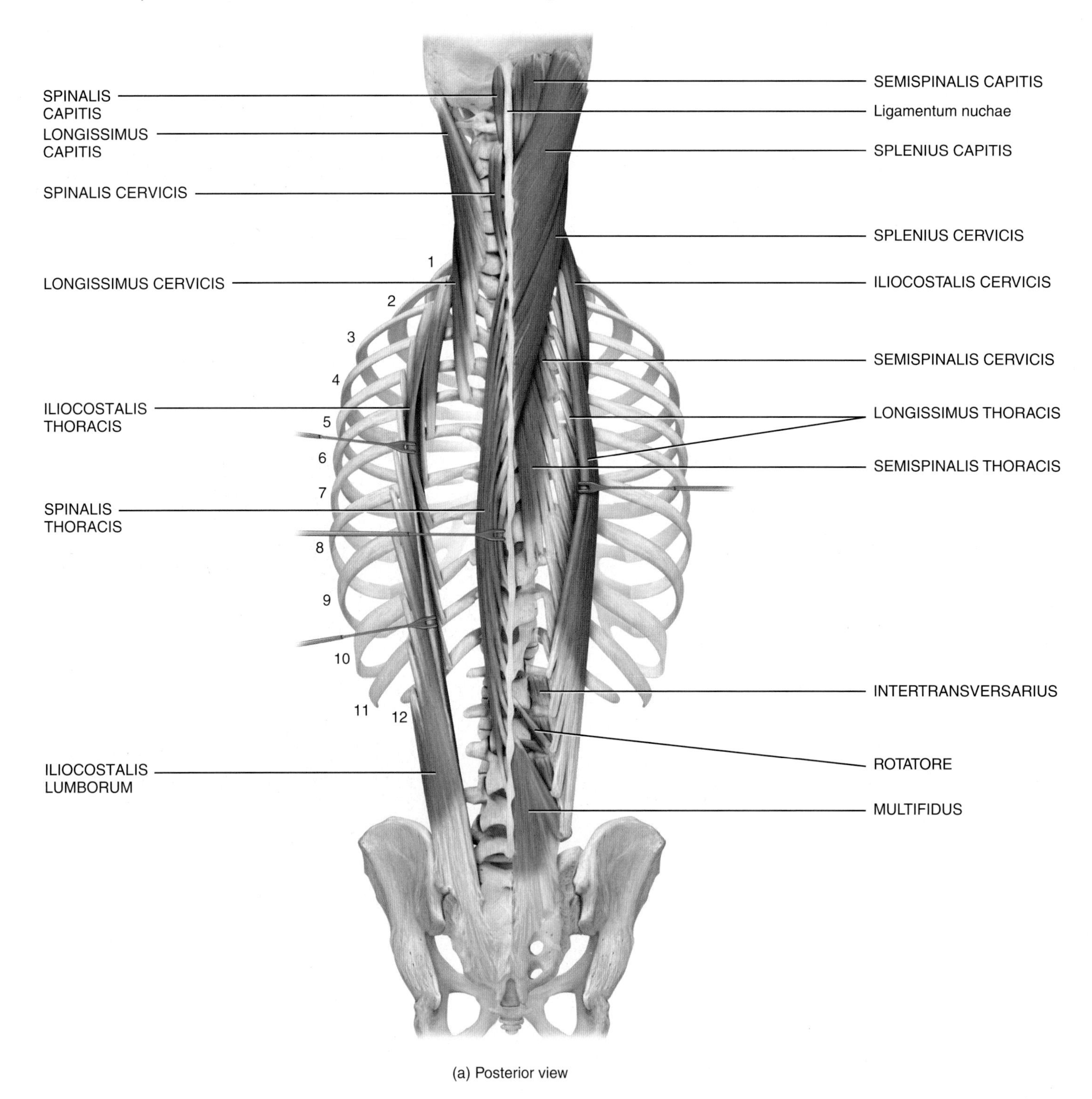

(a) Posterior view

to the sides and back of the neck (Figure 12.7a,b). The two muscles in this group are named on the basis of their insertions: ***splenius capitis*** (head region) and ***splenius cervicis*** (cervical region). They extend, laterally flex, and rotate the head at the neck.

The ***erector spinae*** (sacrospinalis) is the largest muscle mass of the back, forming a prominent bulge on either side of the vertebral column. It is the chief extensor of the vertebral column. It is also important in controlling flexion, lateral flexion, and rotation of the vertebral column and in maintaining the lumbar curve. The erector spinae consists of three groups of muscles: ***iliocostalis group, longissimus group,*** and ***spinalis group.*** The ***iliocostalis group*** consists of three muscles: the ***iliocostalis cervicis, iliocostalis thoracis*** (thoracic region), and ***iliocostalis lumborum*** (lumbar region). The longissimus group resembles a herringbone and consists of three muscles: ***longissimus capitis, longissimus cervicis,*** and ***longissimus thoracis.*** The iliocostalis group is the most lateral, the longissimus group is intermediate, and the spinalis group is the most medial. The spinalis group also consists of three muscles: the ***spinalis capitis, spinalis cervicis,*** and ***spinalis thoracis.***

The ***transversospinales,*** so named because their fibers run from the transverse processes to the spinous processes of the vertebrae, extend, laterally flex, and rotate the vertebral column, and rotate the head. The ***segmental*** muscles unite the spinous and transverse processes of consecutive vertebrae, and function primarily in stabilizing the vertebral column during its movements (Figure 12.7c). Note in Exhibit 12.1 that the rectus abdominis, external oblique, internal oblique, and quadratus lumborum muscles also play roles in moving the vertebral column.

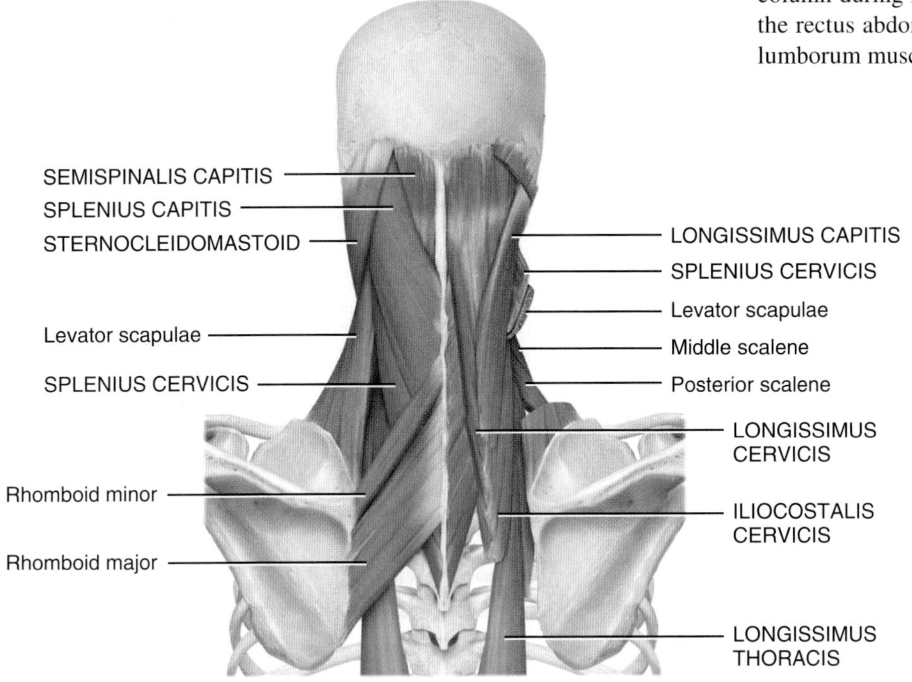

(b) Posterior superficial view (left) and deep view (right)

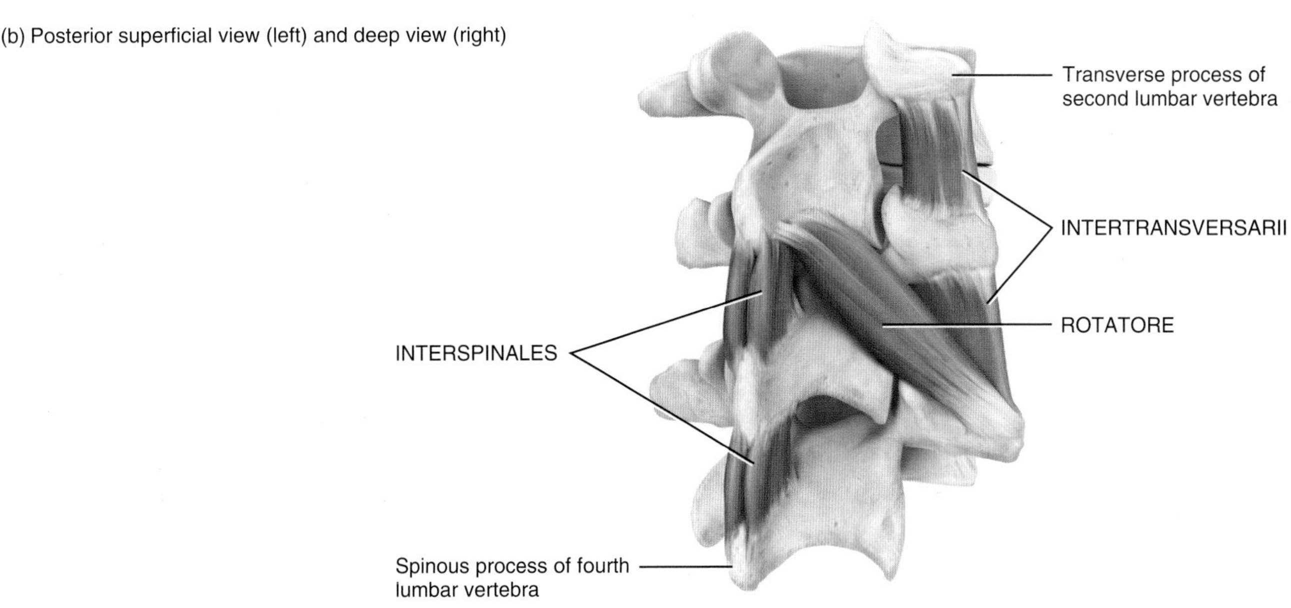

(c) Posterolateral view

What muscles originate at the midline and extend laterally and upward to their insertion?

CONTINUES

MANUAL THERAPY APPLICATION

Back Injuries and Heavy Lifting

Next to headaches, medical experts note that back problems are the most common medical complaint that causes people to seek treatment. The four factors associated with increased risk of **back injuries** are the amount of force, repetition, posture, and stress applied to the vertebral column. Poor physical condition, poor posture, lack of exercise, and excessive body weight contribute to the number and severity of sprains and strains. Back pain caused by a muscle strain or ligament sprain will normally heal within a short time and may never cause further problems. If ligaments and muscles are weak, however, discs in the lower back can become weakened and may herniate (rupture) with excessive lifting or a sudden fall. After years of back abuse, or with aging, the discs may simply wear out and cause chronic pain. Degeneration of the spine due to aging is often misdiagnosed as a sprain or strain. Although spinal degeneration is an inevitable part of aging, close collaboration between the primary physician, the spine specialist, and manual therapists can produce effective treatment.

Performing full flexion at the waist, as in touching your toes, overstretches the erector spinae muscles. Muscles that are overstretched cannot contract effectively since the zone of overlap in a sarcomere shortens and fewer cross-bridges make contact with thin filaments (see Figure 10.6). Straightening up (extension) from such a position is therefore initiated by the hamstring muscles on the back of the thigh and the gluteus maximus muscles of the buttocks. The erector spinae muscles join in as the degree of flexion decreases. Improperly lifting a heavy weight, however, can strain the erector spinae muscles. The result can be painful muscle spasms, tearing of tendons and ligaments of the lower back, and herniating of intervertebral discs. The lumbar muscles are adapted for maintaining posture, not for lifting. This is why it is important to bend at the knees and use the powerful extensor muscles of the thighs and buttocks while lifting a heavy load.

Suboccipital Muscles

Four small paired muscles specifically move the head on the vertebral column and are known collectively as the suboccipital (craniovertebral joint) muscles (Figure 12.8a,b). The muscles are ***rectus capitis posterior major, rectus capitis posterior minor, obliquus capitis superior,*** and ***obliquus capitis inferior.*** Collectively the four pairs of suboccipital muscles provide stability and control fine movements of the head and atlas. Actions include the delicate control of nodding, lateral bending, retraction, and rotation of the head. The vertebral artery passes among the suboccipital muscles on its way from transverse foramina of the vertebral column to the foramen magnum of the skull. The suboccipital muscles are a common source of headache.

Relating Muscles to Movements

Arrange the muscles in this exhibit according to the following actions on the head at the atlanto-occipital and intervertebral joints: (1) extension, (2) lateral flexion, (3) rotation to the same side as contracting muscle, and (4) rotation to opposite side as contracting muscle; and to the following actions on the vertebral column at the intervertebral joints: (1) flexion, (2) extension, (3) lateral flexion, (4) rotation, and (5) stabilization. The same muscle may be mentioned more than once.

CHECKPOINT

6. What is the largest muscle group of the back?

Figure 12.8 Suboccipital (craniovertebral joint) muscles that move the head on the vertebral column.

The four small paired suboccipital muscles extend and rotate the head.

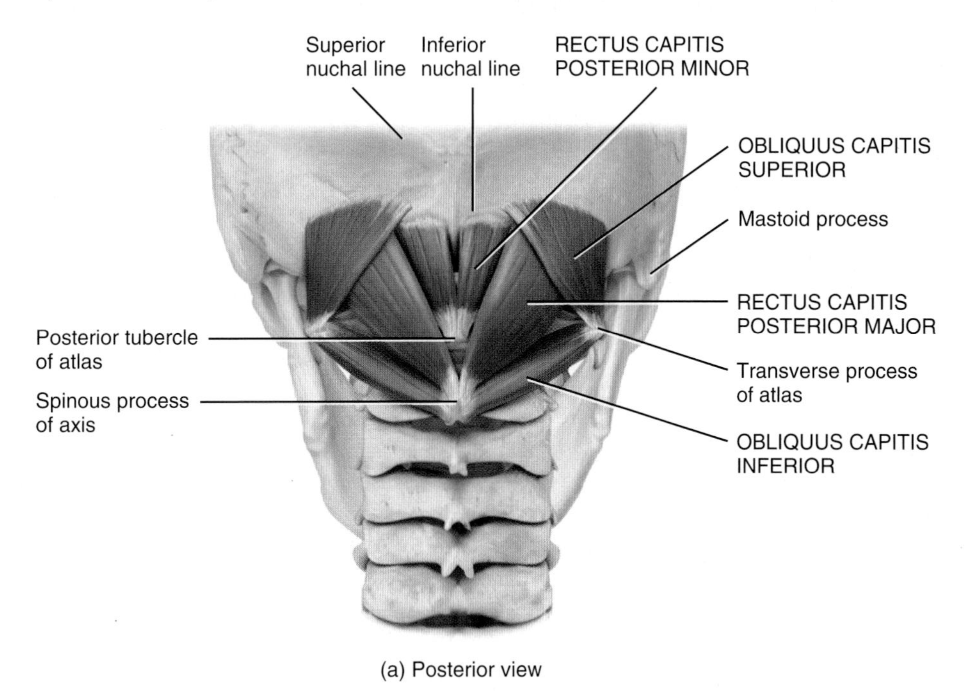

(a) Posterior view

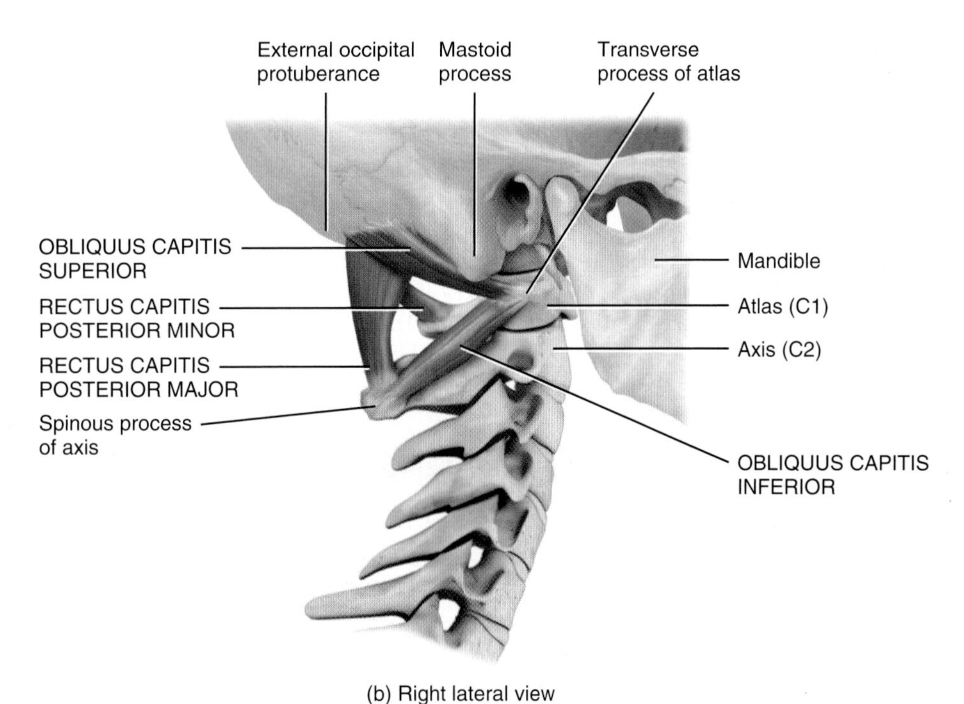

(b) Right lateral view

Which suboccipital muscle also laterally flexes the head to the same side?

STUDY OUTLINE

Muscles of the abdomen that act on the abdominal wall (Exhibit 12.1)

1. Muscles of the abdomen that act on the abdominal wall help contain and protect the abdominal viscera, move the vertebral column, compress the abdomen, and produce the force required for defecation, urination, vomiting, and childbirth.
2. The anterolateral abdominal wall contains four pairs of muscles that are required for core stabilization of the body.
3. In each layer of the anterolateral abdominal wall, the muscle fascicles extend in a different direction. This structural arrangement affords considerable variation in the movements of the abdomen.
4. The more stable posterior wall is formed by lumbar vertebrae, parts of the ilia of the hip bones, the psoas major, iliacus, and quadratus lumborum muscles.

Muscles of the thorax used in breathing (Exhibit 12.2)

1. Four sets of muscles alter the size of the thoracic cavity so that breathing can occur.
2. Inhalation occurs when the thoracic cavity increases in size and exhalation occurs when the thoracic cavity decreases in size.
3. The dome-shaped diaphragm separates the thoracic and abdominal cavities. Contraction of this muscle causes movement of the muscle inferiorly and thus the volume of the thoracic cavity is increased during inhalation. Relaxation of the diaphragm causes it to move superiorly and the reduced volume of the thoracic cavity leads to exhalation.
4. The superior surface of the diaphragm fuses with the inferior surfaces of the pericardium (covering the heart) and the pleurae (coverings of the lungs).
5. The diaphragm has three major openings through which blood vessels and nerves pass between the thorax and abdomen.
6. The 11 pairs of external intercostal muscles elevate the ribs during inhalation to help expand the thoracic cavity.
7. The 11 pairs of internal intercostal muscles draw adjacent ribs together during forced exhalation to help decrease the size of the thoracic cavity.
8. The innermost intercostal muscles, poorly developed in most persons, run in the same direction as the internal intercostals.

Muscles of the pelvic diaphragm and perineum that support the pelvic viscera (Exhibit 12.3)

1. Muscles of the pelvic floor, together with the fasciae covering their internal and external surfaces, form the funnel-shaped floor of the pelvic cavity.
2. These muscles support the pelvic viscera and resist the inferior thrust that accompanies increased intra-abdominal pressure during such functions such as forced exhalation, coughing, vomiting, urination, and defecation.
3. Muscles of the perineum are located inferior to the pelvic diaphragm.
4. Muscles of the deep layer of the perineum assist in urination and ejaculation in males and urination and closing of the vagina in females.
5. The external anal sphincter of the perineum keeps the anal canal and anus closed except during defection.

Muscles of the neck and back that act on the posterior head, posterior neck, back, and vertebral column (Exhibit 12.4)

1. Muscles that move the vertebral column are quite complex because they have multiple origins and insertions and because there is considerable overlap.
2. The erector spinae (sacrospinalis) muscles are the chief extensors of the vertebral column, but they also function in lateral flexion, rotation, and in controlling flexion.
3. Groups of the erector spinae muscles serve the lumbar, thoracic, and cervical portions of the vertebral column.
4. The suboccipital muscles insert on the skull or atlas and are responsible for stability and control of fine movements of the head and atlas.

SELF-QUIZ QUESTIONS

Complete the following:

1. The diaphragm inserts onto the ________ tendon.
2. The diamond-shaped area that extends from the pubic symphysis anteriorly, to the coccyx posteriorly, and to the ischial tuberosities laterally is the ________.
3. A circular muscle that decreases the size of an opening, such as the anus, is known as a ________.

Choose the one best answer to the following questions.

4. During exhalation the diaphragm
   a. contracts and flattens **b.** relaxes and flattens
   c. contracts and forms a dome **d.** relaxes and forms a dome
   e. does not contract or relax.
5. The splenius muscles arise from the midline and extend ________ and ________ to their insertions.
   a. laterally; inferiorly **b.** laterally; superiorly
   c. medially; inferiorly **d.** medially; superiorly
6. The inferior free border of the ________ muscle, plus some collagen fibers, form the inguinal ligament.
   a. external oblique **b.** quadratus lumborum
   c. rectus abdominis **d.** transversus abdominis
7. The diaphragm has ________ major openings through which various structures pass between the abdomen and thorax.
   a. one **b.** two **c.** three **d.** four
8. Tendinous intersections are found in the ________ muscle.
   a. external oblique **b.** quadratus lumborum
   c. rectus abdominis **d.** transversus abdominis
9. Most of the attachments of the ________ muscle group are on ribs.
   a. iliocostalis **b.** longissimus
   c. rotatores **d.** spinalis
10. Which of the following muscles does NOT compress the abdomen and thus aid in defecation and urination?
   a. external oblique **b.** internal oblique
   c. quadratus lumborum **d.** rectus abdominis
   e. transversus abdominis

11. The splenius muscles acting bilaterally ________ the head and acting unilaterally rotate the head to the ________ side.
    a. extend; opposite b. extend; same
    c. flex; opposite d. flex; same
12. Most of the attachments of the ________ muscle group are on transverse processes.
    a. iliocostalis b. interspinalis
    c. longissimus d. spinalis
13. Which of the following structures is NOT an insertion of the transversus abdominis?
    a. linea alba b. pubis
    c. twelfth rib d. xiphoid process
14. During deep inspiration, the diaphragm contracts and pulls the central tendon ________.
    a. anteriorly b. inferiorly
    c. medially d. superiorly
15. The fascicles of the interspinales and intertransversarii muscles are directed ________ from the points of origin.
    a. laterally
    b. parallel to the axis of the vertebral column
    c. medially
    d. at right angles to each other
16. The sacrum is usually pulled anteriorly following defecation or childbirth by
    a. coccygeus b. iliococcygeus
    c. pubococcygeus d. all of the above.
17. Sometimes known as the "hip hiker," the ________ is an abdominal muscle located on the posterior and deep surface of the abdomen.
    a. external oblique b. longissimus
    c. quadratus lumborum d. rectus abdominis
18. Which of the following muscles does NOT originate on the iliac crest?
    a. external oblique b. internal oblique
    c. quadratus lumborum d. transversus abdominis
19. Obliquus capitis inferior originates on the ________ of the axis and inserts on ________ of the atlas.
    a. transverse processes; spinous processes
    b. transverse processes; transverse processes
    c. spinous processes; spinous processes
    d. spinous processes; transverse processes
20. The internal intercostals function during
    a. normal exhalation b. forced exhalation
    c. normal inhalation d. forced inhalation.

CRITICAL THINKING QUESTIONS

1. While taking the bus to the supermarket, 11-year-old Desmond informs his mother that he has to "go to the bathroom" (urinate). His mother tells him he must "hold it" until they arrive at the store. What muscles must remain contracted in order for him to prevent urination?
2. Wyman has been doing a lot of heavy lifting lately. He has suddenly noticed a bulge on the anterior aspect of his torso down near his groin. What has probably happened to him? Do you think he needs to see a doctor? Why or why not?

? ANSWERS TO FIGURE QUESTIONS

12.1 Acting singly, the external oblique and internal oblique muscles laterally flex and rotate the vertebral column.

12.2 Defecation, urination, and childbirth are aided by compression of the anterolateral abdominal wall.

12.3 The posterior layer of the rectus sheath is formed by the aponeuroses of the transversus abdominis and the internal oblique muscles.

12.4 Pain produced by the pressure of a finger on McBurney's point indicates acute appendicitis.

12.5 The diaphragm is innervated by the phrenic nerve.

12.6 The borders of the pelvic diaphragm are the pubic symphysis anteriorly, the coccyx posteriorly, and the walls of the pelvis laterally.

12.7 The splenius muscles arise from the midline and extend laterally and superiorly to their insertion.

12.8 The obliquus capitis superior also laterally flexes the head to the same side.

13 The Muscular System: The Muscles of the Upper Limb

Muscles of the upper limb (extremity) are amazingly diverse in form and function. Some span large areas while others are quite small. The actions of these muscles are complex because many of them cross more than one joint and contraction can move one bone or the other or perhaps both bones at once. Therefore, many muscles may work together to stabilize one bone to provide a large variety of individual movements or combined movements. The shoulder is *flexible,* being attached by a bony girdle and supported mainly by muscle. This allows the arm and hand a huge range of motion required for activities such as throwing a baseball. This to some degree compromises the strength of the shoulder (versus other joints such as the hip). However, the shoulder is still quite *strong* and provides the stability required for lifting heavy objects while moving at the same time.

CONTENTS AT A GLANCE

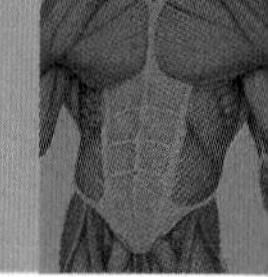

EXHIBIT 13.1 Muscles of the Thorax That Move the Pectoral Girdle

OBJECTIVE

- Describe the origin, insertion, action, and innervation of the muscles that move the pectoral girdle.

Muscles of the upper limb are arranged in diverse groups; muscles are superficial, deep, or very deep. Four muscles, the pectoralis major, deltoid, trapezius, and latissimus dorsi muscles, are not only superficial but also have a large surface area and dominate the superficial musculature of the shoulder region. The concept of ***reverse muscle action (RMA),*** described previously in Section 11.1, is well illustrated in particular muscles of the upper limb and will be described in this chapter.

The main action of the muscles that move the pectoral girdle is to stabilize the scapula so it can function as a steady origin for most of the muscles that move the humerus. Because scapular movements usually accompany humeral movements in the same direction, the muscles also move the scapula to increase the range of motion of the humerus. For example, it would not be possible to raise the arm above the head if the scapula did not move with the humerus. During abduction, the scapula follows the humerus by rotating upward.

Muscles that move the pectoral girdle can be classified into two groups based on their location in the thorax: ***anterior*** and ***posterior thoracic muscles*** (Figure 13.1). The anterior thoracic muscles are the subclavius, pectoralis minor, and serratus anterior. The ***subclavius*** is a small, cylindrical muscle under the clavicle that extends from the first rib to the clavicle. It steadies the clavicle during movements of the pectoral girdle. It also helps hold the only bony articulation of the upper limb with the axial skeleton (the sternoclavicular joint) when, for example, hanging from a bar.

The ***pectoralis minor*** is a thin, flat, triangular muscle that is deep to the pectoralis major (Figure 13.1a). This muscle causes, among other actions, protraction of the scapula. The scapulae of persons who spend a lot of time with their arms in front of them, such as pianists, factory workers, and those who use computers, may develop chronically contracted pectoralis minor muscles. As described previously (see Section 10.6,

MUSCLE	ORIGIN	INSERTION	ACTION	INNERVATION
ANTERIOR THORACIC MUSCLES				
Subclavius (sub-KLĀ-vē-us; *sub-* = under; *clavius* = clavical)	First rib.	Clavicle.	Depresses and moves clavicle anteriorly (protraction) and helps stabilize pectoral girdle.	Subclavian nerve.
Pectoralis minor (pek′-tō-RĀ-lis; *pector-* = the breast, chest, thorax; *minor* = lesser)	Usually ribs 3–5.	Coracoid process of scapula.	Abducts scapula and rotates it downward. RMA: Elevates third through fifth ribs during forced inhalation when scapula is fixed.	Medial pectoral nerve.
Serratus anterior (ser-Ā-tus; *serratus* = saw-toothed; *anterior* = front)	Usually ribs 1–8.	Vertebral border and inferior angle of anterior surface of scapula.	Abducts scapula and rotates it upward. RMA: Elevates ribs when scapula is stabilized; known as "boxer's muscle" because it is important in horizontal arm movements such as punching and pushing.	Long thoracic nerve.
POSTERIOR THORACIC MUSCLES				
Trapezius (tra-PĒ-zē-us = trapezoid-shaped)	Superior nuchal line of occipital bone, ligamentum nuchae, and spines of C7–T12.	Clavicle and acromion and spine of scapula.	Superior fibers elevate scapula; middle fibers adduct scapula; inferior fibers depress scapula; superior and inferior fibers together rotate scapula upward; stabilizes scapula. RMA: Superior fibers can help extend head.	Accessory (XI) nerve and cervical spinal nerves C3–C5.
Levator scapulae (le-VĀ-tor SKA-pū-lē; *levator* = raises; *scapulae* = of the scapula)	Transverse processes of C1–C4.	Superior vertebral border of scapula.	Elevates scapula and rotates it downward.	Dorsal scapular nerve and cervical spinal nerves C3–C5.
Rhomboid major (rom-BOYD = rhomboid or diamond-shaped)	Spines of T2–T5.	Vertebral border of scapula inferior to spine.	Elevates and adducts scapula and rotates it downward; stabilizes scapula.	Dorsal scapular nerve.
Rhomboid minor	Spines of C7–T1.	Vertebral border of scapula near spine.	Elevates and adducts scapula and rotates it downward; stabilizes scapula.	Dorsal scapular nerve.

CONTINUES

Muscle Spasms), contracted muscles become shorter and wider. Moreover, the brachial plexus runs between the pectoralis minor and the rib cage. Chronic contraction of this muscle can thus compress nerves and emulate the symptoms of carpal tunnel syndrome (see Manual Therapy Applications in Exhibit 13.4). Besides its role in movements of the scapula, the pectoralis minor muscle assists in forced inhalation.

The ***serratus anterior*** is a large, flat, fan-shaped muscle between the ribs and scapula. It is so named because of the saw-toothed appearance of its origins on the ribs. This muscle can be highly developed in body builders and athletes (Figure 13.1a–d). It is an antagonist of the rhomboids and is responsible for abduction of the scapula. A large portion of the belly is deep to the anterior scapula. The muscle is thus riding over the rib cage. The production of hydrogen bonds during the aging process, as discussed in Section 4.4, often causes the serratus anterior of elderly persons to become "cemented" between the rib cage and the scapula. Range of motion of the scapula is thus often greatly reduced as the aging process continues. The lateral and inferior portion of the breast lies superficial to the serratus anterior muscle.

The posterior thoracic muscles are the trapezius, levator scapulae, rhomboid major, and rhomboid minor.

The ***trapezius*** is a large, flat, triangular sheet of muscle extending from the skull and vertebral column medially to the pectoral girdle laterally. It is the most superficial back muscle and covers the posterior neck region and superior portion of the trunk. The two trapezius muscles form a trapezoid (diamond-shaped quadrangle), hence its name. Discounting body builders and professional athletes, the trapezius muscle of most persons is less than 0.25 in. thick. The three sets of fibers

Figure 13.1 Muscles of the thorax that move the pectoral girdle.

Muscles that move the pectoral girdle originate on the axial skeleton and insert on the clavicle or scapula.

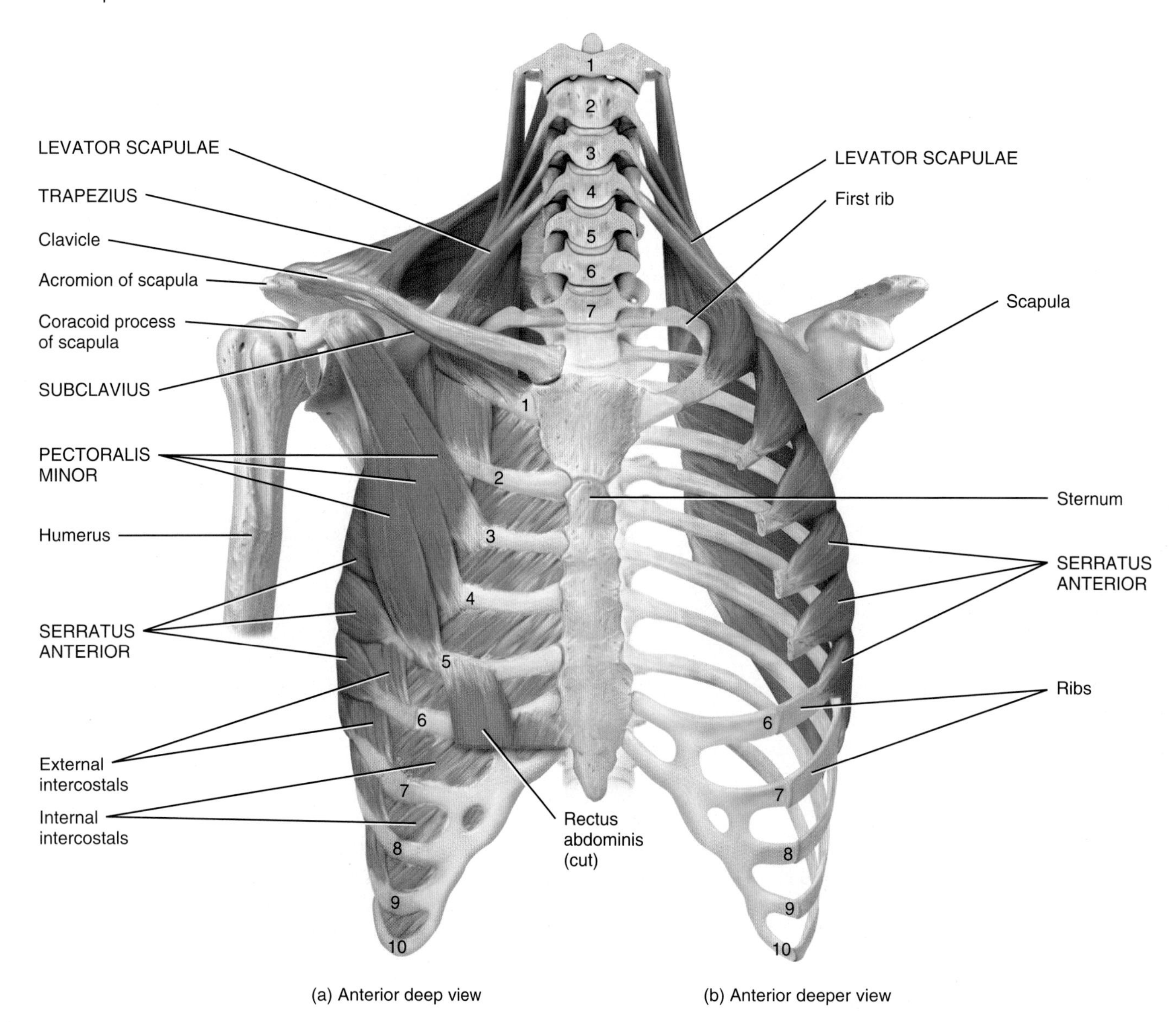

(a) Anterior deep view

(b) Anterior deeper view

(superior, middle, and inferior) enable this muscle to cause multiple actions (Figure 13.1c). The superior fibers extend the head and neck. The weight of the head (about 12 pounds) is functionally doubled with each inch that the head flexes from the neutral position (directly over the atlas). For example, for the person who has a head-forward posture of 2 in. from neutral, the trapezius and smaller muscles of the posterior neck need to contract as though the head weighed 36 pounds. The trapezius is thus overworked in such persons and becomes painful. The head-forward position is usually a habit that the patient can correct with practice.

The ***levator scapulae*** is a narrow, elongated muscle in the posterior portion of the neck. It is deep to the sternocleidomastoid and trapezoid muscles. This muscle contains a twist in the belly (Figure 13.1d). The twist inverts the superior and inferior fibers as they approach the insertion and increases the leverage of each fiber. The fibers become untwisted and stretched as a bone is moved away from the anatomical or neutral position. A muscle that is twisted has the ability to contract with greater force. The insertion is on and near the superior angle of the scapula and pain of the muscle is usually perceived here. The origin, however, goes as high as the atlas, and therefore the belly is longer than one might think. As its name suggests, one of its actions is to elevate the scapula. Its reverse muscle action (RMA), when the origin and insertion are switched, is to pull posteriorly on (extend) the neck, particularly when the person has a head-forward posture, as demonstrated in Figure 13.2. Normal alignment of the body includes positioning of the external auditory meatus directly superior to the acromion process of the scapula. As discussed in Chapter 10, chronic shortening of any muscle can reset the tone of the muscle spindles, and the muscle can

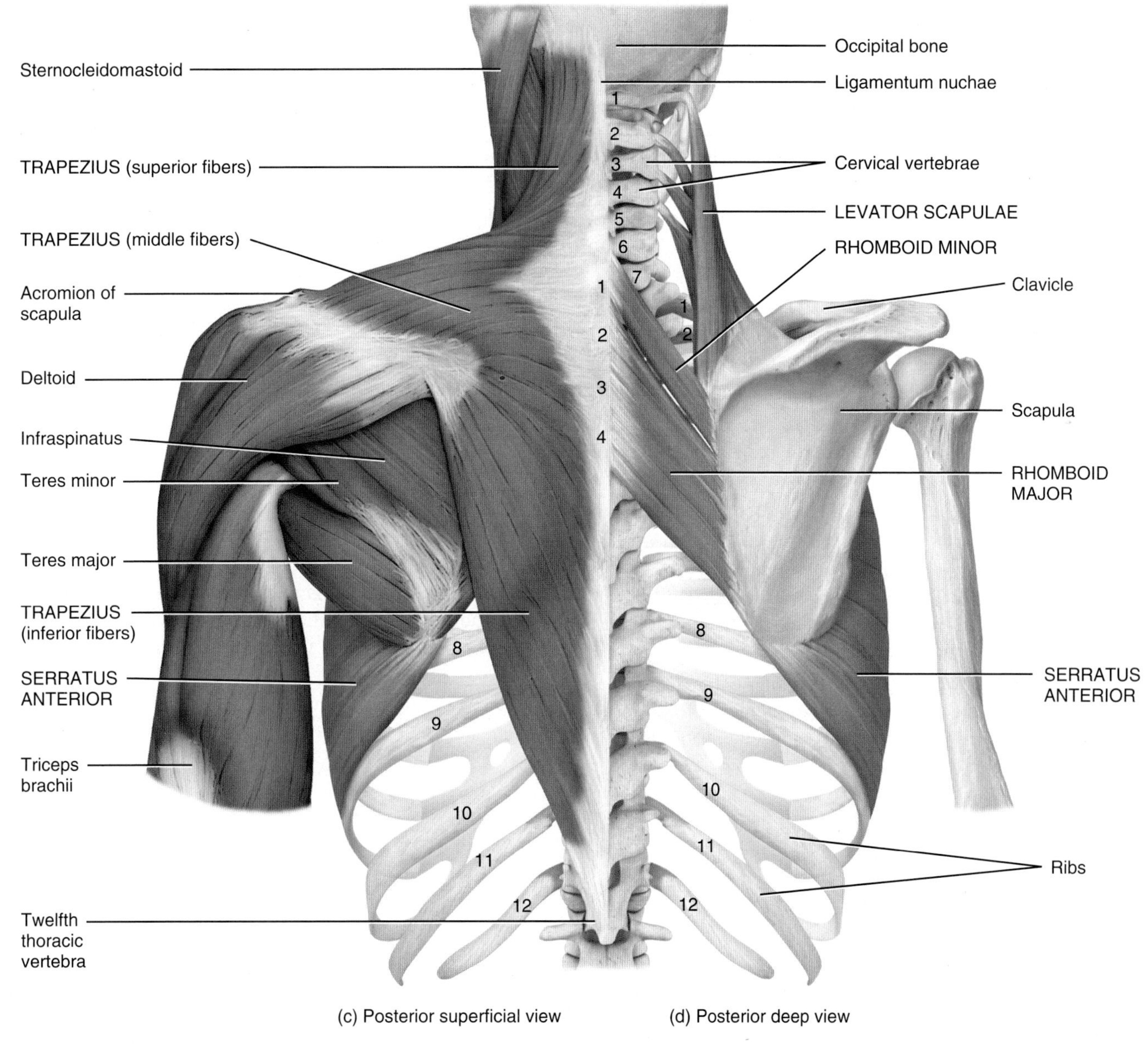

(c) Posterior superficial view (d) Posterior deep view

What is the main action of the muscles that move the pectoral girdle?

CONTINUES

therefore develop a compromised blood supply and become painful. Persons who wish to have both hands free and hold a telephone cradle by elevating the shoulder are also likely to develop a chronically shortened levator scapulae muscle and thus experience pain near the superior angle of the scapula and an inability to fully turn their neck when checking the blind spot while driving.

The ***rhomboid major*** and ***rhomboid minor*** lie deep to the trapezius and are not always distinct from each other (see Figure 13.1d). They appear as parallel bands that pass inferiorly and laterally from the vertebrae to the scapula. Their names are based on their shape, that is, a rhomboid (an oblique parallelogram). The rhomboid major is about two times wider than the rhomboid minor. The two muscles are often identified by their attachments. The muscles lie deep to the trapezius and superficial to the erector spinae. The rhomboids and the trapezius are functionally the only muscles holding the upper limb to the posterior axial skeleton. The rhomboids are usually very thin and are easily overworked by lifting or repetitive movements of the upper limbs; thus pain and a burning sensation between the shoulder blades is common for many people. Both muscles are used when forcibly lowering the raised upper limbs, as in driving a stake with a sledgehammer. Emotional stress can also cause spasm of the rhomboids in many people. Manual therapists who develop pain in the rhomboids sometimes back up to a corner of a refrigerator or a wall and lean themselves backward such that the corner presses into the rhomboids and helps to relieve the pain.

Movements of the Scapula

To understand the actions of muscles that move the scapula, it is first helpful to review the various movements of the scapula:

- **Elevation:** Superior movement of the scapula, such as shrugging the shoulders or lifting a weight over the head.
- **Depression:** Inferior movement of the scapula, as in pulling down on a rope attached to a pulley.
- **Abduction (protraction):** Movement of the scapula laterally and anteriorly, as in doing a push-up or punching.
- **Adduction (retraction):** Movement of the scapula medially and posteriorly, as in pulling the oars in a rowboat.
- **Upward rotation:** Movement of the inferior angle of the scapula laterally so that the glenoid cavity is moved upward. This movement is required to move the humerus past the horizontal as in raising the arms in a jumping jack.
- **Downward rotation:** Movement of the inferior angle of the scapula medially so that the glenoid cavity is moved downward. This movement is seen when a gymnast on parallel bars supports the weight of the body on the hands.

Relating Muscles to Movements

Arrange the muscles in this exhibit according to the following actions on the scapula: (1) depression, (2) elevation, (3) abduction, (4) adduction, (5) upward rotation, and (6) downward rotation. The same muscle may be mentioned more than once.

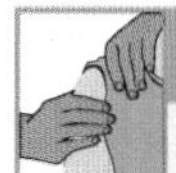

MANUAL THERAPY APPLICATION

Structural and Functional Analysis

Structural analysis includes visual and palpatory assessment of a patient while he is sitting, standing, or lying without movement. For example, a manual therapist might position herself behind the patient who is standing and observe the heights of the two scapulae. If the heights are different, the therapist would attempt to determine what would contribute to a height discrepancy. Pain in the shoulder region or the back may result from such misalignments of the skeleton.

Functional analysis includes palpating the structure during motion to assess the functioning of the involved musculature and joints. For example, a manual therapist might stand behind the patient who is standing and ask her to place her hands (palms out) on her lower back. The manual therapist would observe the movement of the scapulae for symmetry of motion. If one scapula is "winged out," meaning that the medial edge of the scapula has moved posteriorly and is no longer near the rib cage, the therapist might conclude that the action may have been the result of a whiplash injury in the past. The serratus anterior muscle helps hold the scapula against the rib cage and is innervated by the long thoracic nerve that arises from spinal nerves C5–C7. A whiplash injury involving this part of the cervical region may compromise this nerve and reduce the strength of the serratus anterior muscle, thus causing the winged appearance of the scapula. Such an injury would be confirmed by, or added to, the written history of the patient. It should be noted that during the intake history, many patients inadvertently forget to mention such injuries, which could have occurred years ago.

CHECKPOINT

1. What muscles in this exhibit are used to raise your shoulders, lower your shoulders, join your hands behind your back, and join your hands in front of your chest?

Muscles of the Thorax and Shoulder That Move the Humerus

OBJECTIVE

- Describe the origin, insertion, action, and innervation of the muscles that move the humerus.
- Describe the surface features of the shoulder, axilla, and back.

Surface Features of the Shoulder

The *shoulder,* or *acromial region*, is located at the lateral aspect of the clavicle, where the clavicle joins the scapula and the scapula joins the humerus. The region presents several conspicuous surface features (Figure 13.2).

- **Acromioclavicular joint.** A slight elevaion at the lateral end of the clavicle. It is the joint between the acromion of the scapula and the clavicle.
- **Acromion.** The expanded lateral end of the spine of the scapula that forms the top of the shoulder. It can be palpated about 2.5 cm (1 in.) distal to the acromioclavicular joint.
- **Humerus.** The **greater tubercle** of the humerus may be palpated on the superior aspect of the shoulder. It is the most laterally palpable bony structure.

Surface Features of the Armpit

The armpit region, or ***axilla,*** is a pyramid-shaped area at the junction of the arm and the chest that enables blood vessels and nerves to pass between the neck and the free upper limbs (Figure 13.3).

- **Apex.** The apex of the axilla is surrounded by the clavicle, scapula, and first rib.
- **Base.** The base of the axilla is formed by the concave skin and fascia that extends from the arm to the chest wall. It contains hair, and deep to the base the axillary lymph nodes can be palpated.
- **Anterior wall.** The anterior wall of the axilla is composed mainly of the pectoralis major muscle (anterior exillary fold).
- **Posterior wall.** The posterior wall of the axilla is formed mainly by the teres major and latissimus dorsi muscles (posterior axillary fold).
- **Medial wall.** The medial wall of the axilla is formed by ribs 1–4 and their corresponding intercostal muscles, plus the overlying serratus anterior muscle.
- **Lateral wall.** Finally, the lateral wall of the axilla is formed by the coracobrachialis and biceps brachii muscles and the superior portion of the shaft of the humerus. Passing through the axilla are the axillary artery and vein, branches of the brachial plexus, and axillary

Figure 13.2 Surface features of the shoulder.

The deltoid muscle gives the shoulder its rounded prominence.

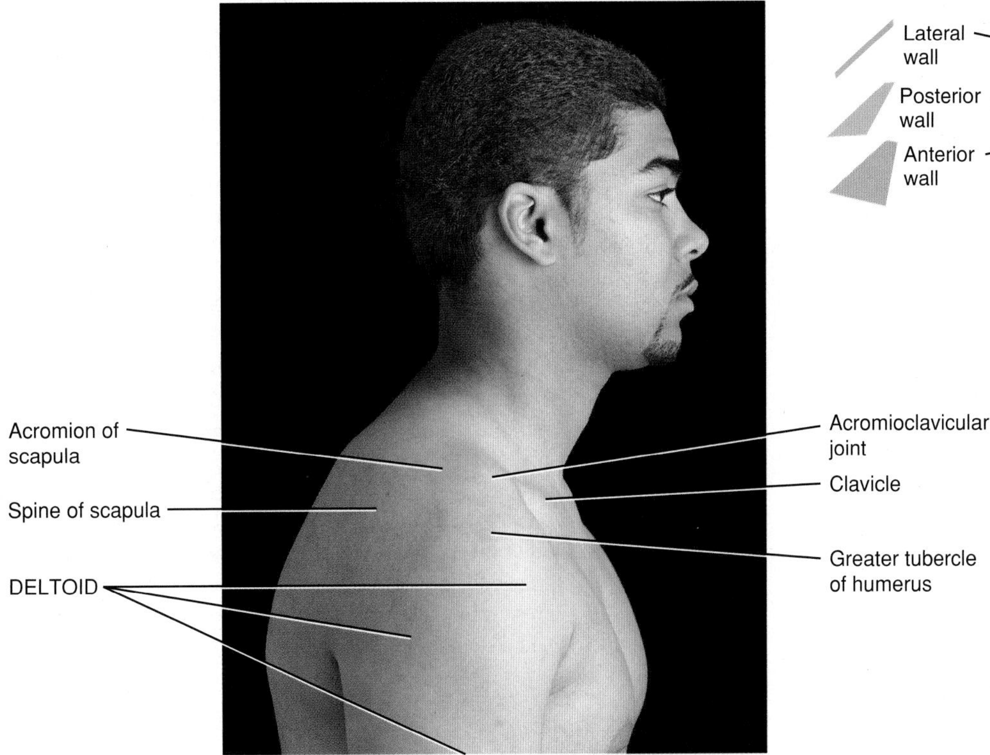

Right Lateral View

Which structure forms the top of the shoulder?

Figure 13.3 Axilla schematic.

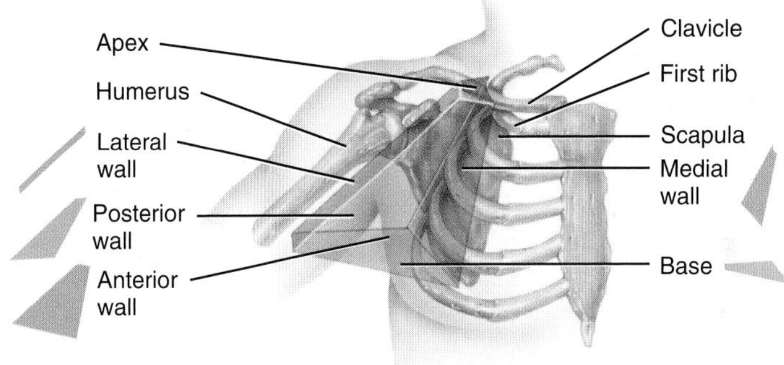

Location and parts of the axilla

What structures comprise the medial and lateral walls of the axilla?

CONTINUES

lymph nodes. All these structures are surrounded by a considerable amount of axillary fat.

Of the nine muscles that cross the shoulder joint, all except the pectoralis major and latissimus dorsi originate on the scapula. The pectoralis major and latissimus dorsi thus are called ***axial muscles*** because they originate on the axial skeleton. The remaining seven muscles, the ***scapular muscles,*** arise from the scapula (see Exhibit 13.1).

Of the two axial muscles that move the humerus, the ***pectoralis major*** is a large, thick, fan-shaped muscle that covers the superior part of the thorax and forms the anterior fold of the axilla. When this muscle and the latissimus dorsi are well developed, the axilla is deepened. By placing the thumb in the axilla of the patient, manual therapists are able to place the fingers on the belly of the muscle and thus massage both surfaces simultaneously. The pectoralis major has a twist in it that improves its contraction strength; the clavicular head inserts more distally and the sternocostal head inserts more proximally.

The ***latissimus dorsi*** is a broad, triangular muscle located on the inferior part of the back (Figure 13.4b,c). The muscle forms most of the posterior wall of the axilla. The reverse muscle action (RMA) of the latissimus dorsi enables the spine and torso to be elevated. It is commonly called the "swimmer's muscle" because its many actions are used while swimming; consequently, many competitive swimmers have well-developed "lats." These movements are seen in a paraplegic when transferring from a wheelchair. Similarly, if the arm is stabilized, as when hanging from a bar, latissimus dorsi will assist in extension of the spine. Like the pectoralis major and levator scapulae muscles, latissimus dorsi has a twist in it near the insertion that increases its contraction effectiveness.

Among the scapular muscles, the ***deltoid*** is a thick, powerful shoulder muscle that covers the shoulder joint and forms the rounded contour of the shoulder. This muscle is a frequent site of intramuscular injections. As you study the deltoid, note that its fascicles originate from three different points and that each group of fascicles moves the humerus differently. These points of insertion are the clavicle, acromion, and spine of the scapula (Figure 13.4b); they are the same three points as the insertions of the trapezius. The muscle has three sets of fibers (anterior, middle, and posterior) that enable it to function as three distinct muscles that are used in flexion, abduction, rotation, or extension of the humerus; see the photo of the arm that shows the deltoid in Figure 13.2.

The ***subscapularis*** is a large triangular muscle that fills the subscapular fossa of the scapula and forms a small part in the apex of the posterior wall of the axilla (Figure 13.3). The ***supraspinatus,*** a rounded muscle named for its location in the supraspinous fossa of the scapula, lies deep to the trapezius and has a belly that is about the size of your thumb (Figure 13.4c–e). The tendon of insertion slides back and forth beneath the acromion; inflammation can thus cause swelling and pain. See the discussion on impingement in the Clinical Connection in this exhibit. The supraspinatus is responsible for the first 15° of abduction of the humerus; the middle fibers of the deltoid do not cause the beginning of abduction. The supraspinatus has therefore been called the "suitcase muscle" since it is essentially acting alone to keep a suitcase from rubbing your leg as you carry it. No wonder travel through a large airport is so exhausting and painful for your shoulder.

The ***infraspinatus*** is a triangular muscle, also named for its location in the infraspinous fossa of the scapula. A portion of the muscle is superficial and other portions are deep to the trapezius and to the deltoid (see Figure 13.4c,e). Thick fascial layers cause the infraspinatus to feel more dense than surrounding muscles on palpation. This muscle often develops trigger points (knots) in the muscle or adheres to the glenohumeral joint capsule and results in a condition called *adhesive capsulitis* which limits movement of the arm dramatically.

The ***teres major*** is a thick, flattened muscle inferior to the teres minor that also helps form part of the posterior wall of the axilla. It is also a synergist of the latissimus dorsi (Figure 13.4b,c). The two muscles have been called the "handcuff muscles" since their combined actions are to bring the arms into position behind the back. The teres major rotates the arm medially and the teres minor rotates it laterally.

The ***teres minor*** is a small, cylindrical, elongated muscle, located between the teres major and the infraspinatus muscles (Figure 13.4c,e). Its belly lies parallel to the inferior edge of the infraspinatus and is sometimes indistinguishable from the infraspinatus.

The ***coracobrachialis*** is an elongated, narrow muscle in the arm, located in the lateral wall of the axilla along with the biceps brachii (Figure 13.4a). Its point of origin, the coracoid process of the scapula, in many people is tender on palpation. Since three muscles are attached here, tenderness at the site implies a problem with one or more of three muscles: the coracobrachialis, pectoralis minor, or biceps brachii muscles.

Four deep muscles of the shoulder—supraspinatus, infraspinatus, teres minor, and subscapularis—strengthen and stabilize the shoulder joint. These muscles join the scapula to the humerus. Their flat tendons fuse together to form the ***rotator (musculotendinous) cuff,*** a nearly complete circle of tendons around the shoulder joint, like the cuff on a shirtsleeve. The four rotator cuff muscles are often described as the "SITS" muscles and this could serve as a mnemonic for remembering the names.

The supraspinatus muscle is especially subject to wear and tear because of its location between the head of the humerus and acromion of the scapula, which compress its tendon during shoulder movements, especially abduction of the arm. This is further aggravated by poor posture with slouched shoulders and medially rotated shoulders that also increase compression of the supraspinatus tendon.

The subscapularis is the only rotator cuff muscle that attaches to the lesser tubercle of the humerus; the other three attach to the greater tubercle. Sandwiched between the serratus anterior and subscapular fossa (on the anterior scapula), the subscapularis is difficult to access in many patients (Figure 13.4a,d). When the therapist places the fingers of one hand on the inferior vertebral border of the scapula, the other hand can pull the patient's shoulder over the fingers. This technique can be used to stretch some of the fibers of the trapezius, the rhomboids, the serratus anterior, and the subscapularis simultaneously. As described in Section 4.4, hydrogen bonds within connective tissues commonly increase in number during the aging process and, with the lack of aerobic exercise, reduce the flexibility of the scapula.

Figure 13.4 **Muscles of the thorax and shoulder that move the humerus.**

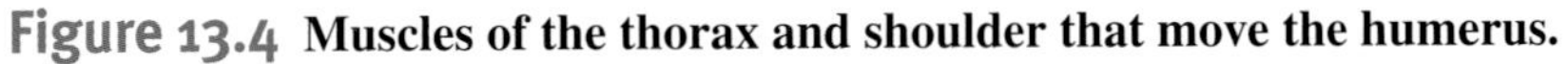
The strength and stability of the shoulder joint are provided by the tendons that form the rotator cuff.

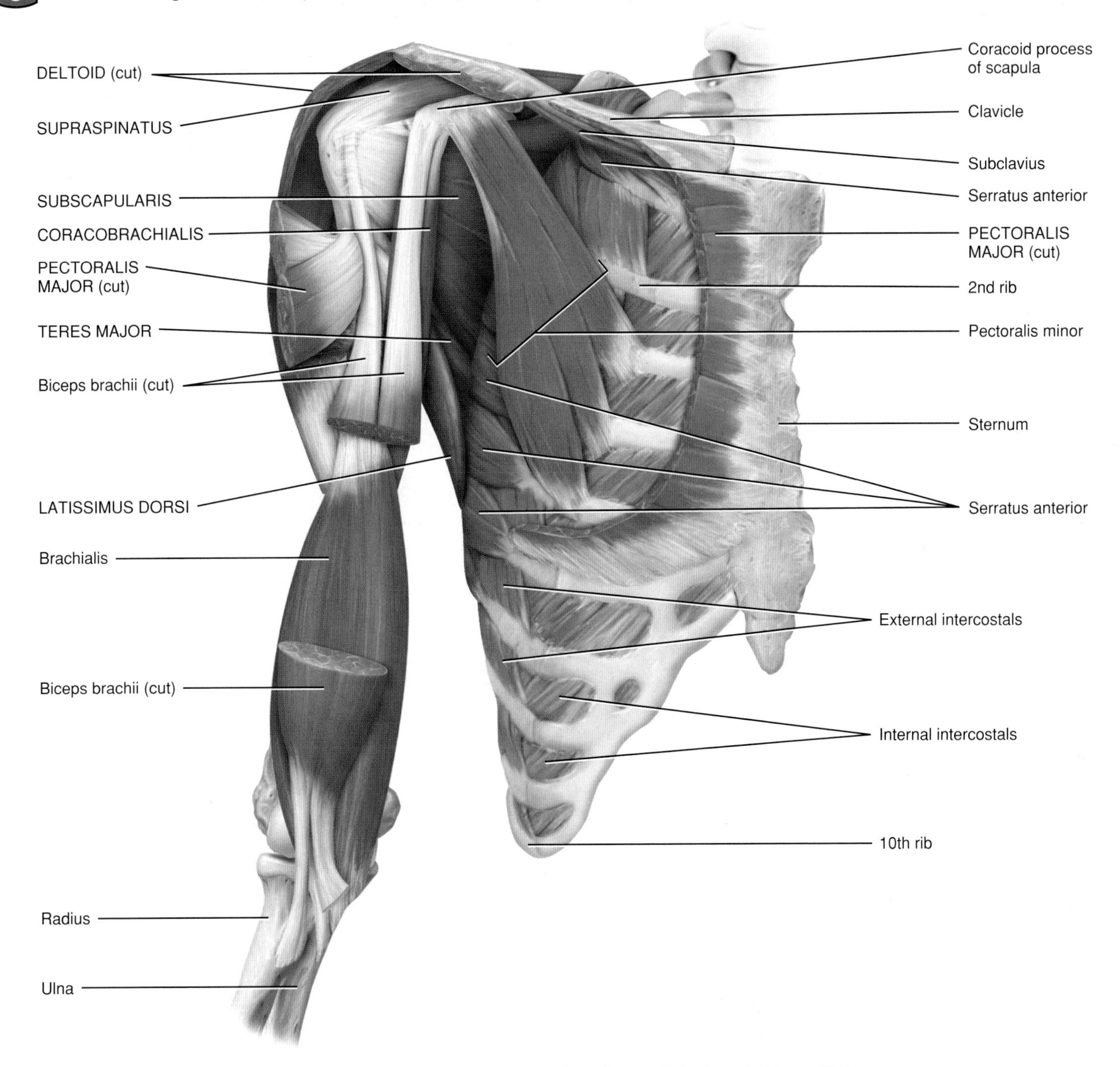

(a) Anterior deep view (the intact pectoralis major muscle is shown in Figure 12.1)

CONTINUES

Figure 13.4 (continued)

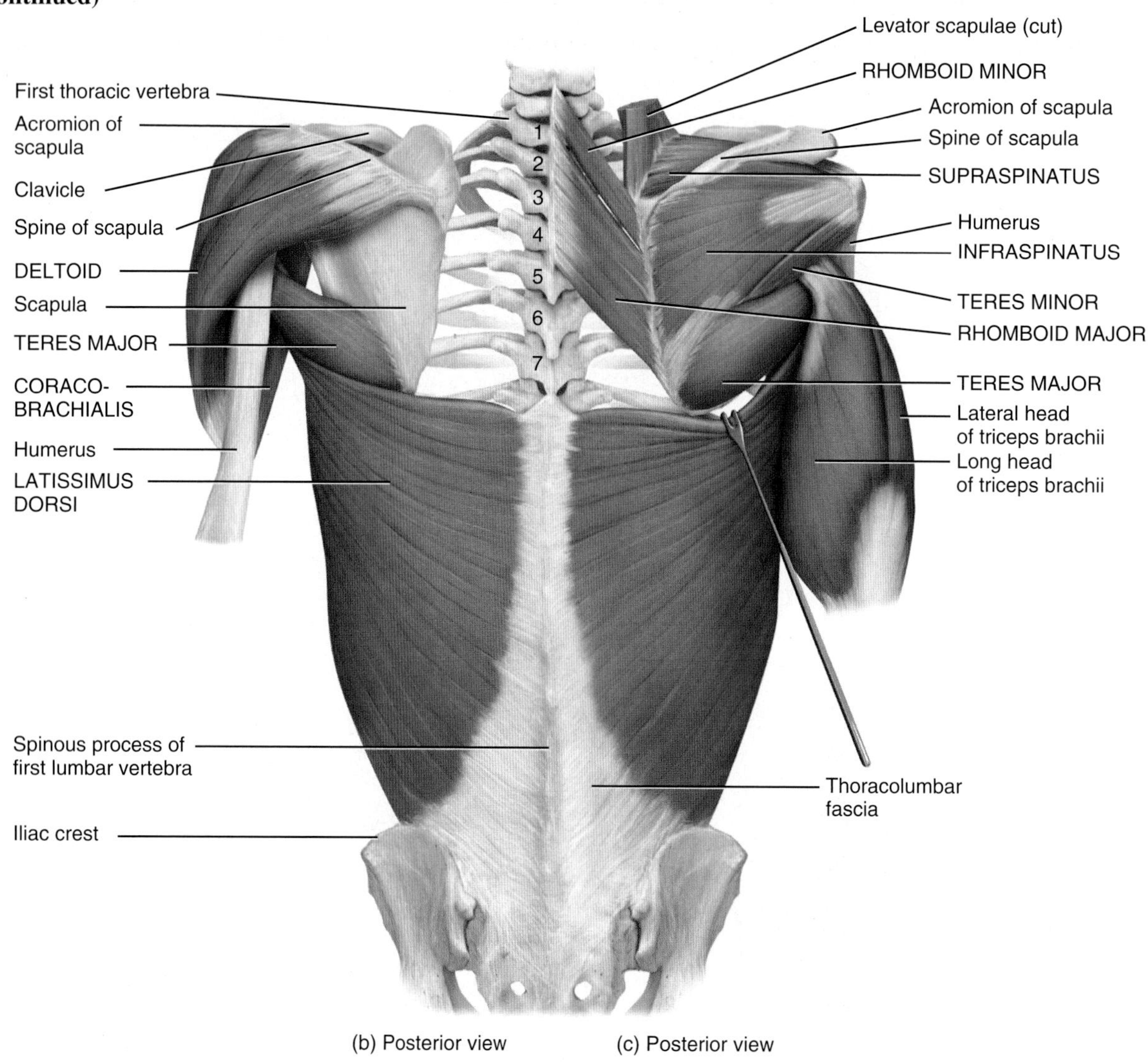

(b) Posterior view (c) Posterior view

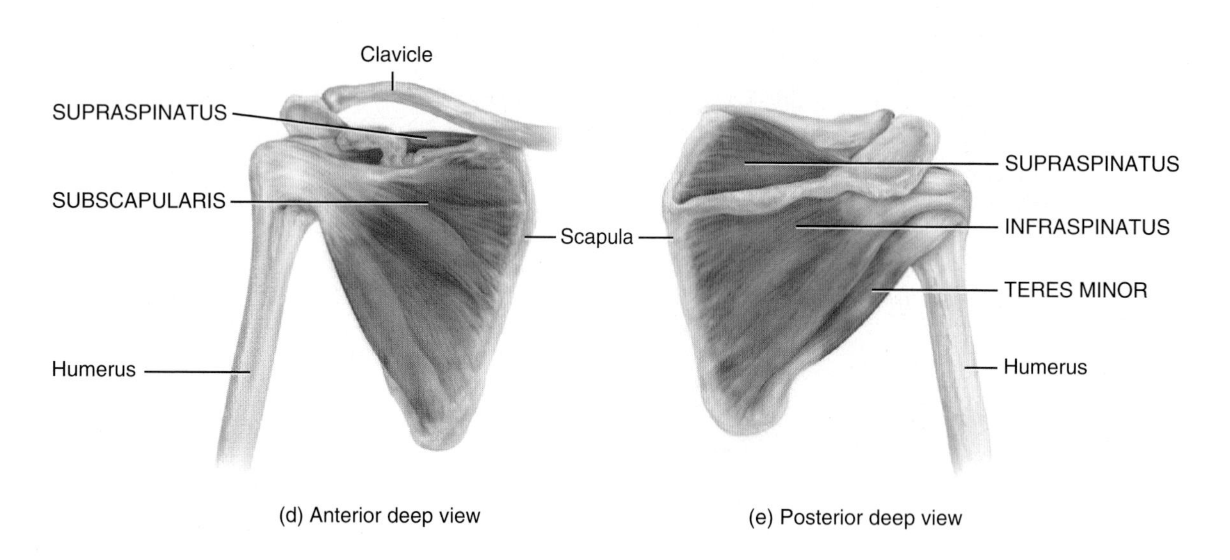

(d) Anterior deep view (e) Posterior deep view

? **Which of the rotator cuff muscles inserts on the anterior humerus?**

• CLINICAL CONNECTION | Rotator Cuff Injuries and Impingement Syndrome

Rotator cuff injury is common among baseball pitchers, volleyball players, racket sports players, and swimmers due to shoulder movements that involve vigorous circumduction. It also occurs as a result of wear and tear, trauma, and repetitive motions in certain occupations, such as painters or those required to place items on a shelf above the head. Although muscle bellies may be damaged, the tendon of one or more of the four muscles is usually partially or completely torn. Most often, there is tearing of the supraspinatus muscle tendon of the rotator cuff. This tendon is especially predisposed to wear-and-tear because of its location between the head of the humerus and acromion of the scapula, which compresses the tendon during shoulder movements.

One of the most common causes of shoulder pain and dysfunction in athletes is known as **impingement syndrome.** The repetitive movement of the arm over the head may put athletes at risk. Impingement syndrome may also be caused by a direct blow or stretch injury. Continual pinching of the supraspinatus tendon as a result of overhead motions causes it to become inflamed and results in pain. If movement is continued despite the pain, the tendon may degenerate near the attachment to the humerus and ultimately may tear away from the bone (rotator cuff injury). Treatment consists of resting the injured tendons, strengthening the shoulder through exercise, massage therapy, and finally surgery if the injury is particularly severe. During surgery, an inflamed bursa may be removed, bone may be trimmed, and/or the coracoacromial ligament may be detached. Torn rotator cuff tendons may be trimmed and then reattached with sutures, anchors, or surgical tacks. These steps make more space, thus relieving pressure and allowing the arm to move freely. •

MUSCLE	ORIGIN	INSERTION	ACTION	INNERVATION
AXIAL MUSCLES THAT MOVE THE HUMERUS				
Pectoralis major (pek′-tō-RĀ-lis; *pector-* = chest; *major* = larger) (see also Figure 12.1)	Clavicle (clavicular head), sternum, and costal cartilages of ribs 2–6 (sternocostal head).	Greater tubercle and lateral lip of the intertubercular sulcus of humerus.	As a whole, adducts and medially rotates arm at shoulder joint; clavicular head flexes arm, and sternocostal head extends the flexed arm to side of trunk.	Medial and lateral pectoral nerves.
Latissimus dorsi (la-TIS-i-mus DOR-sī; *latissimus* = widest; *dorsi* = of the back)	Spines of T7–L5, lumbar vertebrae, crests of sacrum and ilium, ribs 9–12 via thoracolumbar fascia.	Medial lip of intertubercular sulcus of humerus.	Extends, adducts, and medially rotates arm at shoulder joint; draws arm inferiorly and posteriorly.	Thoracodorsal nerve.
SCAPULAR MUSCLES THAT MOVE THE HUMERUS				
Deltoid (DEL-toyd = triangularly shaped)	Acromial extremity of clavicle (anterior fibers), acromion of scapula (lateral fibers), and spine of scapula (posterior fibers).	Deltoid tuberosity of humerus.	Lateral fibers abduct arm at shoulder joint; anterior fibers flex and medially rotate arm at shoulder joint; posterior fibers extend and laterally rotate arm at shoulder joint.	Axillary nerve.
Teres major (TE-rēz; *teres* = long and round)	Inferior angle of scapula.	Medial lip of intertubercular sulcus of humerus.	Extends arm at shoulder joint and assists in adduction and medial rotation of arm at shoulder joint.	Lower subscapular nerve.
Coracobrachialis (kor′-a-kō-brā-kē-Ā-lis; *coraco-* = coracoid process of the scapula; *brachi-* = arm)	Coracoid process of scapula.	Middle of medial surface of shaft of humerus.	Flexes and adducts arm at shoulder joint.	Musculocutaneous nerve.
Supraspinatus (soo-pra-spī-NĀ-tus; *supra-* = above; *spina-* = spine of the scapula)	Supraspinous fossa of scapula.	Greater tubercle of humerus.	Assists deltoid muscle in abducting arm at shoulder joint.	Suprascapular nerve.
Infraspinatus (in′-fra-spī-NĀ-tus; *infra-* = below)	Infraspinous fossa of scapula.	Greater tubercle of humerus.	Laterally rotates arm at shoulder joint.	Suprascapular nerve.
Teres minor	Inferior lateral border of scapula.	Greater tubercle of humerus.	Laterally rotates and extends arm at shoulder joint.	Axillary nerve.
Subscapularis (sub-scap′-ū-LĀ-ris; *sub-* = below; *scapularis* = scapula)	Subscapular fossa of scapula.	Lesser tubercle of humerus.	Medially rotates arm at shoulder joint.	Upper and lower subscapular nerve.

CONTINUES

Surface Features of the Back

In addition to the muscles that we've discussed in detail, there are several other superficial bones and muscles that form the surface features of the ***back*** (Figure 13.5).

- **Vertebral spines.** The spinous processes of vertebrae, especially the thoracic and lumbar vertebrae, are quite prominent when the vertebral column is flexed.
- **Scapulae.** These easily identifiable surface landmarks on the back lie between ribs 2 and 7. In fact, it is also possible to palpate some ribs on the back. Depending on how lean a person is, it might be possible to palpate various parts of the scapula, such as the **vertebral border, axillary border, inferior angle, spine,** and **acromion.** The spinous process of T3 is at about the same level as the spine of the scapula, and the spinous process of T7 is approximately opposite the inferior angle of the scapula.
- **Erector spinae (sacrospinalis) muscle.** Located on either side of the vertebral column between the skull and iliac crests.
- **Posterior axillary fold.** Formed by the latissimus dorsi and teres major muscles, the posterior axillary fold can be palpated between the fingers and thumb at the posterior aspect of the axilla (armpit region); forms the posterior wall of the axilla.
- **Triangle of auscultation** (aw-skul-TĀ-shun; *ausculto-* = listening). A triangular region of the back just medial to the inferior part of the scapula, where the rib cage is not covered by superficial muscles. It is bounded by the latissimus dorsi and trapezius muscles and vertebral border of the scapula. The triangle of auscultation is a landmark of clinical significance because in this area respiratory sounds can be heard clearly through a stethoscope pressed against the skin. If a patient folds the arms across the chest and bends forward, the lung sounds can be heard clearly in the intercostal space between ribs 6 and 7.

Relating Muscles to Movements

Arrange the muscles in this exhibit according to the following actions on the humerus at the shoulder joint: (1) flexion, (2) extension, (3) abduction, (4) adduction, (5) medial rotation, and (6) lateral rotation. The same muscle may be mentioned more than once.

CHECKPOINT

2. Why are the two muscles that cross the shoulder joint called axial muscles, and the seven others called scapular muscles?

Figure 13.5 **Surface features of the back.**

The posterior boundary of the axilla, the posterior axillary fold, is formed mainly by the latissimus dorsi and teres major muscles.

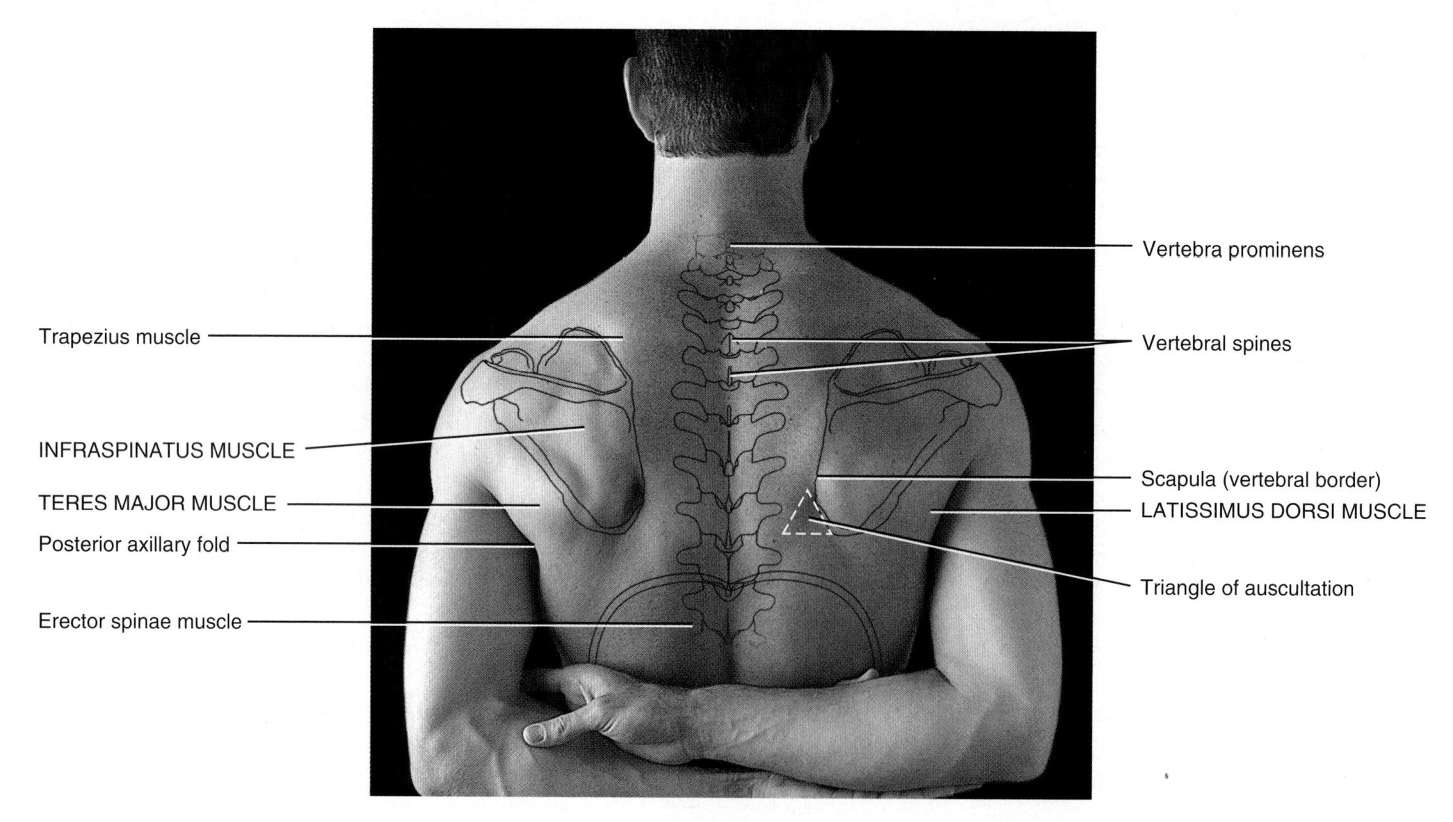

What is the clinical significance of the triangle of auscultation?

EXHIBIT 13.3 Muscles of the Arm That Move the Radius and Ulna

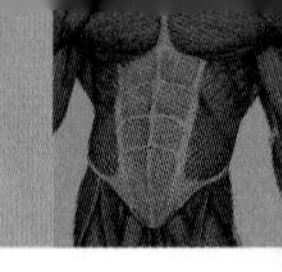

OBJECTIVE

- Describe the origin, insertion, action, and innervation of the muscles that move the radius and ulna.

Most of the muscles that move the radius and ulna cause flexion and extension at the elbow, which is a hinge joint. The biceps brachii, brachialis, and brachioradialis muscles are the flexor muscles. The extensor muscles are the triceps brachii and the anconeus (Figure 13.6a,b).

The ***biceps brachii*** is the large muscle located on the anterior surface of the arm (Figure 13.6a). As indicated by its name, it has two heads of origin (long and short), both from the scapula. The muscle spans both the shoulder and elbow joints. In addition to its role in flexing the forearm at the elbow joint, it also supinates the forearm at the radioulnar joints and flexes the arm at the shoulder joint. The tendon of the distal muscle attaches to the radius and also becomes aponeurotic (meaning that the cross-sectional view changes from round to a flat, sheetlike aponeurosis) on the surface of the muscles of the forearm. The bicipital aponeurosis covers and helps protect the median nerve and the brachial artery. This aponeurotic sheet of fascia also tends to support a vein (see Chapter 23) by forming a sturdy platform, and the connective tissue fibers of the aponeurosis keep the vein from rolling. This site is thus a favorite of health-care professionals who wish to insert a needle into a vein for either withdrawal of blood or injection of medications. Since the biceps brachii is the primary supinator of the forearm, the belly is shorter and thicker when the forearm is supinated. Even children, when showing off their muscles,

MUSCLE	ORIGIN	INSERTION	ACTION	INNERVATION
ANTERIOR (FLEXOR) COMPARTMENT OF THE ARM				
Biceps brachii (BĪ-ceps BRĀ-kē-ī; *biceps* = two heads of origin; *brachii* = arm)	Long head: tubercle above glenoid cavity of scapula (supraglenoid tubercle); short head: coracoid process of scapula.	Radial tuberosity of radius and bicipital aponeurosis.*	Flexes forearm at elbow joint, supinates forearm at radioulnar joints, and flexes arm at shoulder joint.	Musculocutaneous nerve.
Brachialis (brā-kē-Ā-lis)	Distal, anterior surface of humerus.	Ulnar tuberosity and coronoid process of ulna.	Flexes forearm at elbow joint.	Musculocutaneous and radial nerves.
Brachioradialis (brā′-kē-ō-rā-dē-Ā-lis; *radi-* = radius)	Lateral border of distal end of humerus.	Superior to styloid process of radius.	Flexes forearm at elbow joint: supinates and pronates forearm at radioulnar joints to neutral position.	Radial nerve.
POSTERIOR (EXTENSOR) COMPARTMENT OF THE ARM				
Triceps brachii (TRĪ-ceps = three heads of origin)	Long head: infraglenoid tubercle, a projection inferior to glenoid cavity of scapula; lateral head: lateral and posterior surface of humerus superior to radial groove; medial head: entire posterior surface of humerus inferior to a groove for the radial nerve.	Olecranon of ulna.	Extends forearm at elbow joint and extends arm at shoulder joint.	Radial nerve.
Anconeus (an-KŌ-nē-us = the elbow)	Lateral epicondyle of humerus.	Olecranon and superior portion of shaft of ulna.	Extends forearm at elbow joint.	Radial nerve.
FOREARM PRONATORS				
Pronator teres (PRŌ-nā-tor TE-rēz; *pronator* = turns palm posteriorly; *teres* = round and long)	Medial epicondyle of humerus and coronoid process of ulna.	Midlateral surface of radius.	Pronates forearm at proximal radioulnar joint and weakly flexes forearm at elbow joint.	Median nerve.
Pronator quadratus (kwod-RĀ-tus = square, four-sided)	Distal portion of shaft of ulna.	Distal portion of shaft of raidus	Pronates forearm at distal radioulnar joint.	Median nerve.
FOREARM SUPINATOR				
Supinator (SOO-pi-nā-tor = turns palm anteriorly)	Lateral epicondyle of humerus and ridge near radial notch of ulna (supinator crest).	Lateral surface of proximal one-third of radius.	Supinates forearm at both radioulnar joints.	Deep radial nerve.

*The **bicipital aponeurosis** is a broad aponeurosis from the tendon of insertion of the biceps brachii muscle that descends medially across the brachial artery and fuses with deep fascia over the forearm flexor muscles. It also helps to protect the median nerve and brachial artery.

CONTINUES

usually use this pose. Look at your biceps brachii muscle when the shoulder and elbow joints are flexed and when the forearm is pronated and then supinated.

The ***brachialis*** is deep to the biceps brachii muscle. It is the most powerful flexor of the forearm at the elbow joint. For this reason, it is called the "workhorse" of the elbow flexors. Its thick belly is wider than that of biceps brachii (see Figure 13.6a,f). Sandwiched between the biceps brachii and triceps brachii, the lateral edge of the brachialis is usually distinguishable on palpation of the lateral arm.

The ***brachioradialis*** is located in the proximal half of the forearm (see Figures 13.6g, 13.8a,d). Its belly is superficial and separates the bellies of the extensor muscles from those of the flexor muscles of the forearm. It flexes the forearm at the elbow joint, especially when a quick movement is required or when a weight is lifted slowly during flexion of the forearm, as well as in supination and pronation of the forearm. It is therefore well developed in mechanics or others who spend a lot of time using a screwdriver or a wrench.

The ***triceps brachii*** is the large muscle located on the posterior surface of the arm (Figure 13.6b). It is the only muscle belly of the posterior compartment of the arm. This muscle is the more powerful of the extensors of the forearm at the elbow joint and is an antagonist to the biceps brachii. As its name implies, it has three heads of origin, one from the scapula (long head) and two from the humerus (lateral and medial heads). The long head crosses the shoulder joint between teres major and teres minor; the other heads do not. The ***anconeus*** is a small muscle located in the posterior compartment of the forearm that assists the triceps brachii in extending the forearm at the elbow joint (see Figure 13.6b,h).

Some muscles that move the radius and ulna are involved in pronation and supination at the radioulnar joints. The pronators, as suggested by their names, are the pronator teres and pronator quadratus muscles. ***Pronator teres*** (see Figures 13.6i, 13.8a) is located in the proximal forearm between the brachioradialis and the forearm flexor muscles. In this region of the upper limb, it is the only muscle with obliquely running fibers. If this muscle impinges on the median nerve that runs deep to it, the patient may experience *pronator teres syndrome*. When these patients are asked to make a fist, they are able to flex only the fourth and fifth digits.

Figure 13.6 Muscles of the arm that move the radius and ulna.

The anterior arm muscles flex the forearm, and the posterior arm muscles extend it.

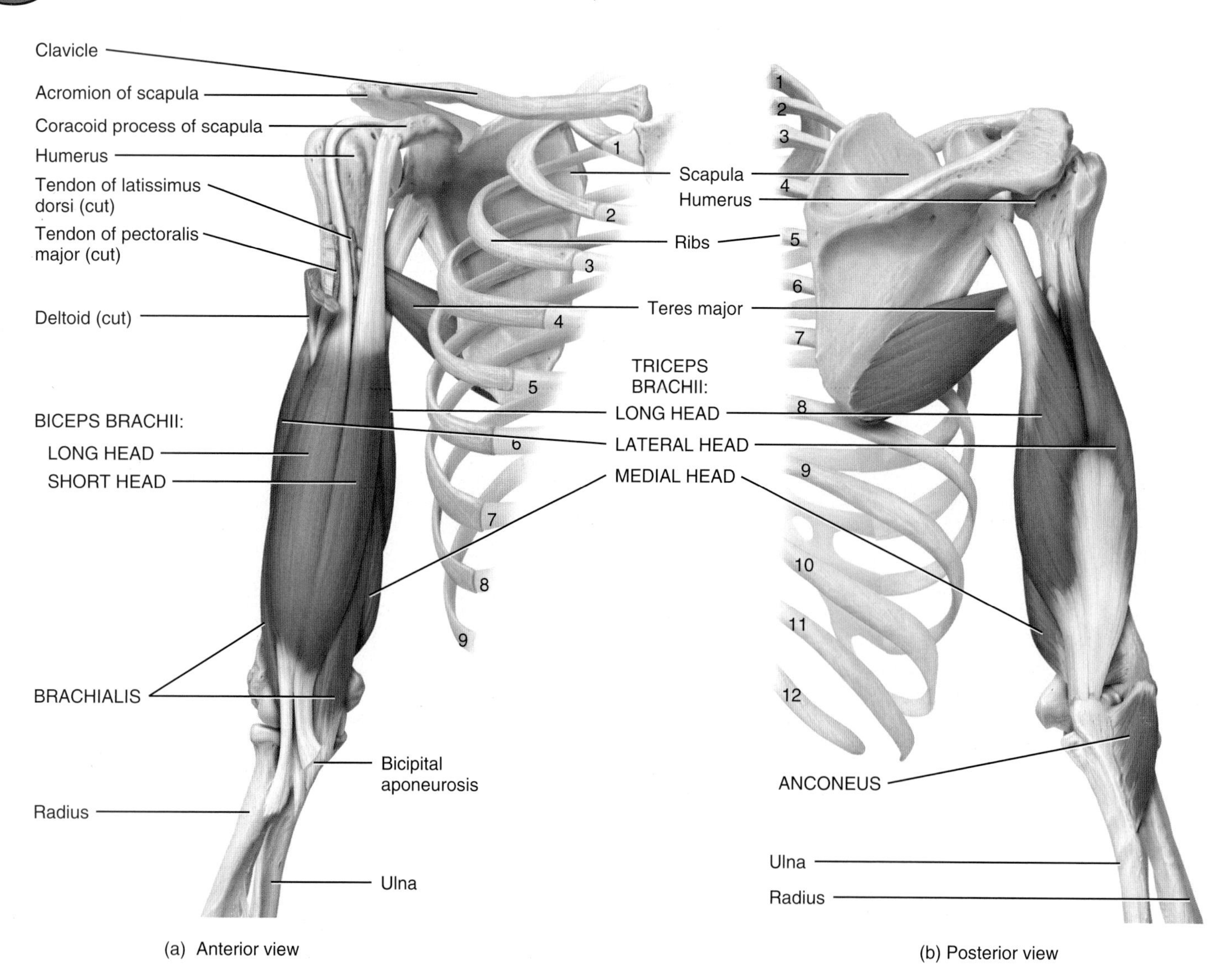

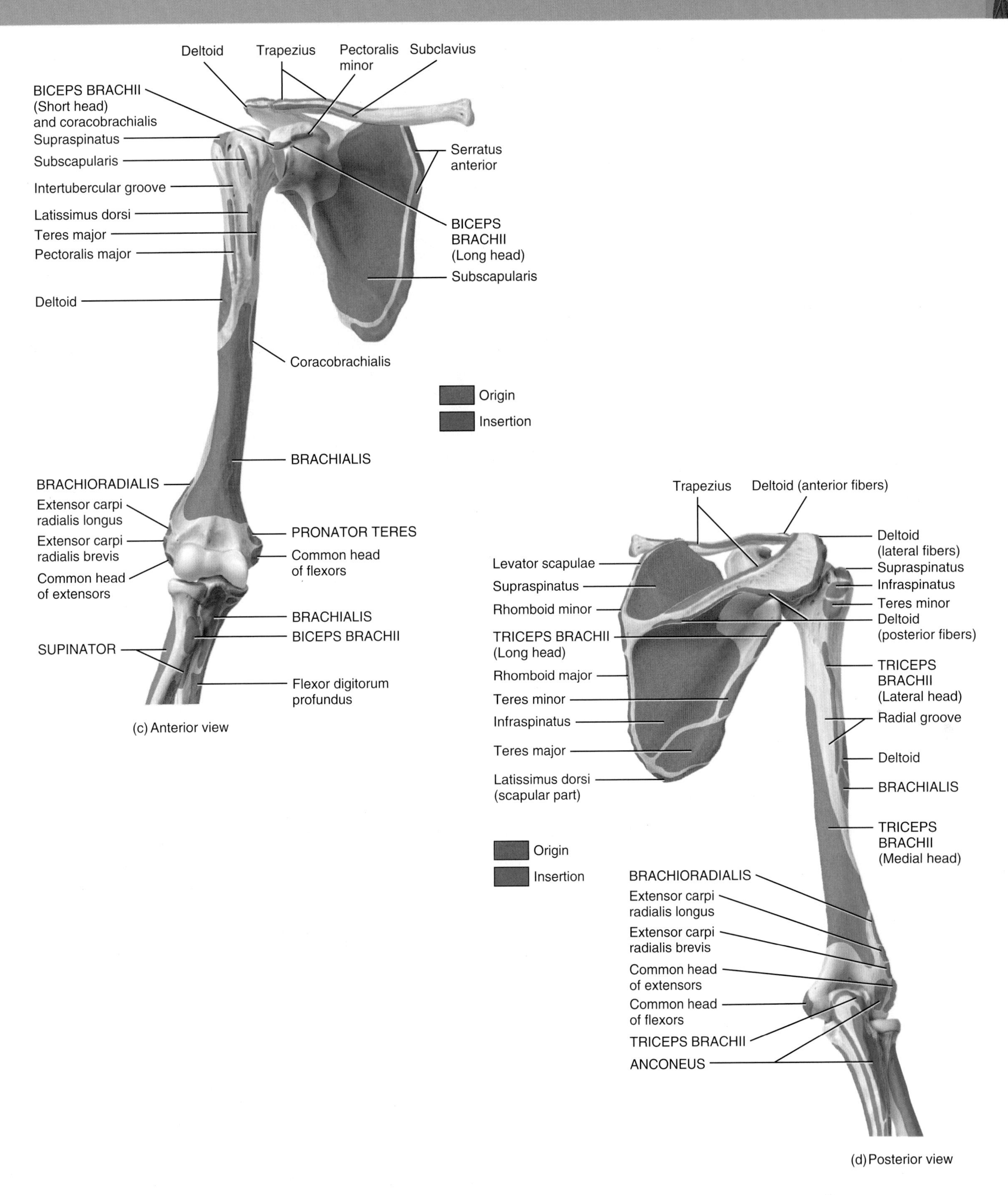

(c) Anterior view

(d) Posterior view

CONTINUES

Figure 13.6 (continued)

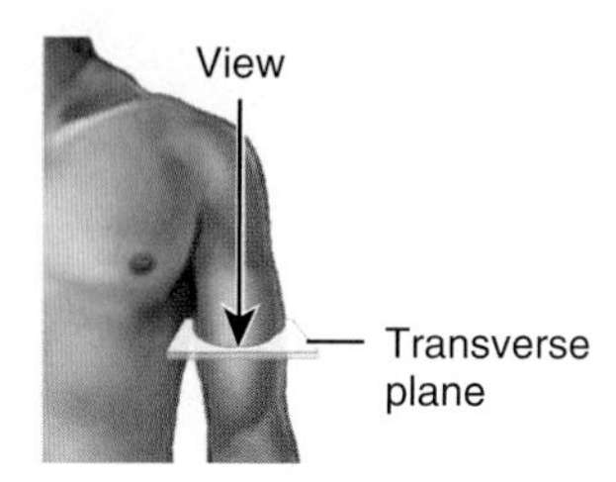

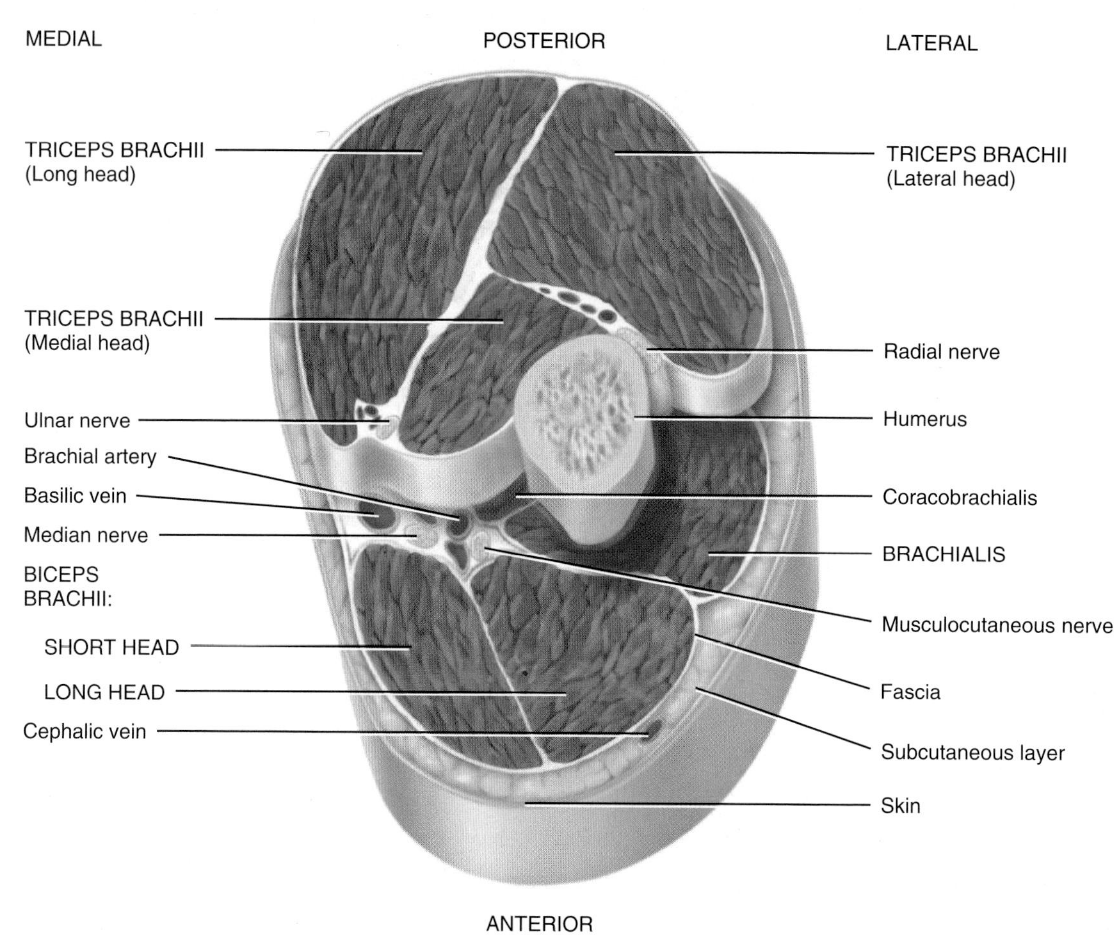

(e) Superior view of transverse section of arm

The ***pronator quadratus*** is flat and four-sided (*quadratus*); its fibers run transversely, at right angles to the radius and ulna. It is located in the distal forearm and is synergistic to the pronator teres. Both of the pronator muscles cause the radius to cross the ulna in the act of pronation of the forearm.

The supinator of the forearm is aptly named the ***supinator*** muscle (see Figures 13.6h, 13.8b,c). Its thin muscle belly lies lateral to the elbow joint and is deep to the forearm extensor muscles. The supinator is antagonistic to both of the pronator muscles and pulls the radius into the anatomical position from pronation. You use the powerful action of the supinator when you twist a corkscrew or turn a screw with a screwdriver.

In the limbs, functionally related skeletal muscles and their associated blood vessels and nerves are grouped together by fascia into regions called ***compartments*** (Figure 13.6e). In the arm, the biceps brachii, brachialis, and coracobrachialis muscles make up the ***anterior (flexor) compartment.*** The triceps brachii muscle forms the ***posterior (extensor) compartment.***

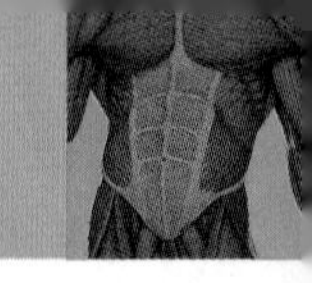

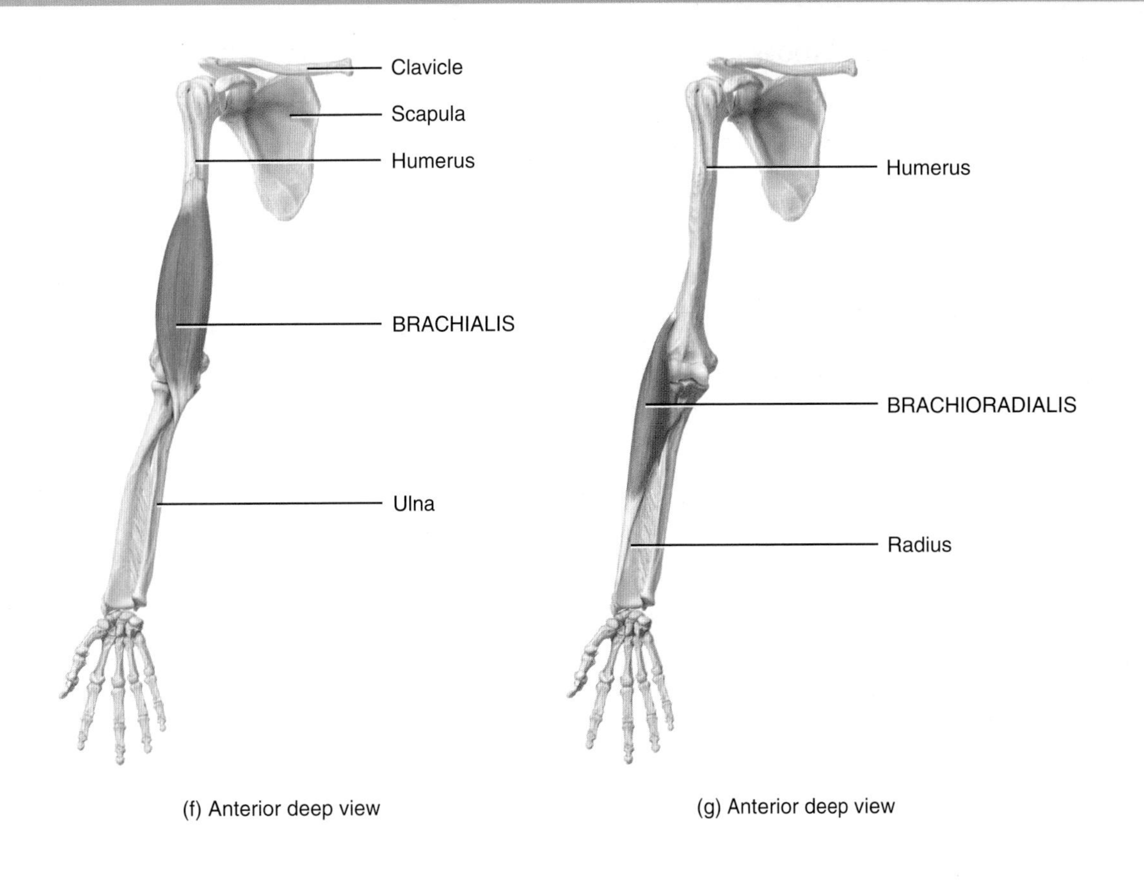

(f) Anterior deep view

(g) Anterior deep view

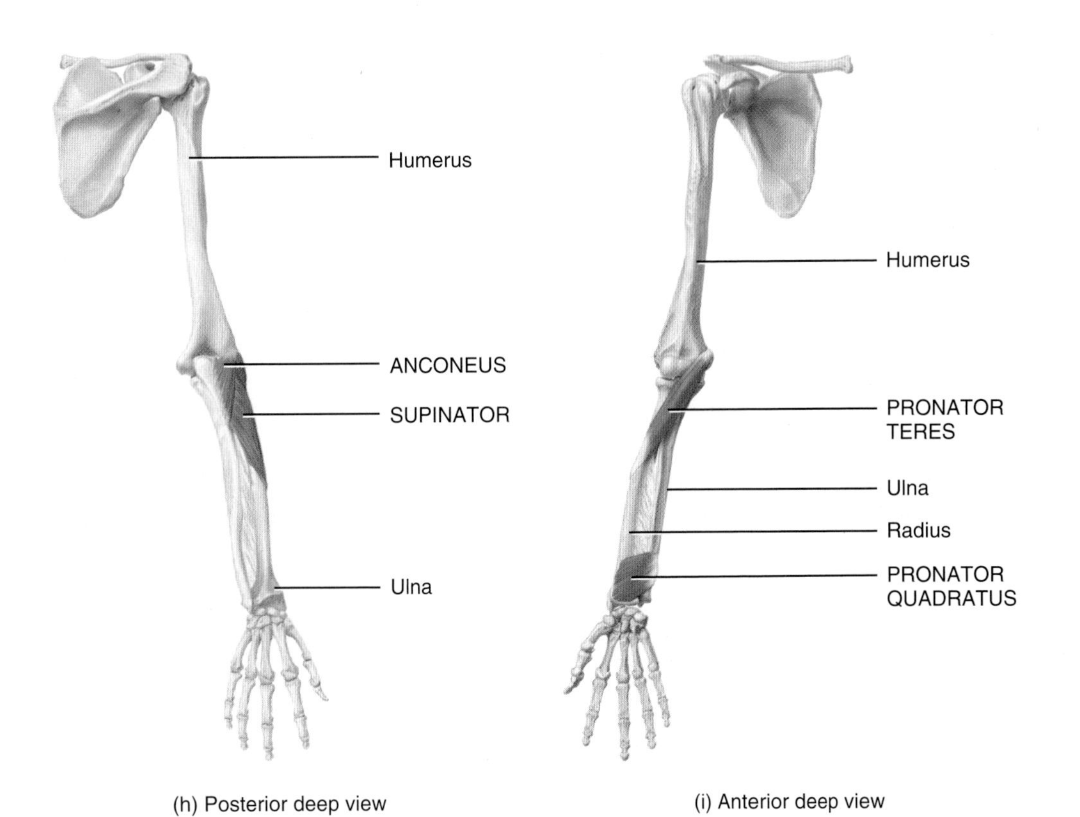

(h) Posterior deep view

(i) Anterior deep view

? Which muscles are the most powerful flexor and the most powerful extensor of the forearm?

CONTINUES

Surface Features of the Arm and Elbow

The ***arm,*** or *brachium*, is the region between the shoulder and elbow. The ***elbow,*** or *cubitus*, is the region where the arm and forearm join. The arm and elbow present several surface anatomy features (Figure 13.7).

- **Humerus.** This arm bone may be palpated along its entire length, especially near the elbow (see descriptions of the medial and lateral epicondyles that follow).
- **Biceps brachii muscle.** Forms the bulk of the anterior surface of the arm. On the medial side of the muscle is a groove that contains the **brachial artery.**
- **Triceps brachii muscle.** Forms the bulk of the posterior surface of the arm.
- **Medial epicondyle.** Medial projection of the humerus near the elbow.
- **Lateral epicondyle.** Lateral projection of the humerus near the elbow.
- **Olecranon.** Projection of the proximal end of the ulna between and slightly superior to the epicondyles when the forearm is extended; it forms the elbow.
- **Ulnar nerve.** Can be palpated in a groove posterior to the medial epicondyle. The "funny bone" is the region where the ulnar nerve rests against the medial epicondyle. Hitting the nerve at this point produces a sharp pain along the medial side of the forearm that almost everyone would agree isn't very funny.
- **Cubital fossa.** Triangular space in the anterior region of the elbow bounded proximally by an imaginary line between the humeral epicondyles, laterally by the medial border of the brachioradialis muscle, and medially by the lateral border of the pronator teres muscle; contains the tendon of the biceps brachii muscle, the median cubital vein, brachial artery and its terminal branches (radial and ulnar arteries), and parts of the median and radial nerves.
- **Median cubital vein.** Crosses the cubital fossa obliquely and connects the lateral cephalic with the medial basilic veins of the arm. The median cubital vein is frequently used to withdraw blood from a vein for diagnostic purposes or to introduce substances into blood, such as medications, contrast media for radiographic procedures, nutrients, and blood cells and/or plasma for transfusion.
- **Brachial artery.** Continuation of the axillary artery that passes posterior to the coracobrachialis muscle and then medial to the biceps brachii muscle. It enters the middle of the cubital fossa and passes deep to the bicipital aponeurosis, which separates it from the median cubital vein. **Blood pressure** is usually measured in the brachial artery, when the cuff of a *sphygmomanometer* (blood pressure instrument) is wrapped around the arm and a stethoscope is placed over the brachial artery in the cubital fossa. Pulse can also be detected in the artery in the cubital fossa. However, blood pressure can be measured at any artery where you can obstruct blood flow. This becomes important in case the brachial artery cannot be utilized. In such situations, the radial or popliteal arteries might be used to obtain a blood pressure reading.
- **Bicipital aponeurosis.** An aponeurosis that inserts the biceps brachii muscle into the deep fascia in the medial aspect of the forearm. It can be felt when the muscle contracts.

MANUAL THERAPY APPLICATION

Tenosynovitis

Tenosynovitis (ten′-ō-sin-ō-VĪ-tis) is an inflammation of the tendons, tendon sheaths, and synovial membranes surrounding certain joints. The tendons most often affected are at the wrists, shoulders, elbows, finger joints, ankles, and feet. The affected sheaths sometimes become visibly swollen because of fluid accumulation. Tenderness and pain are frequently associated with movement of the body part. The condition often follows trauma, strain, excessive exercise, or other stressors. For example, tenosynovitis of the dorsum of the foot may also be caused by tying shoelaces too tightly. Gymnasts are prone to developing the condition as a result of chronic, repetitive, and maximum hyperextension at the wrists. Other repetitive movements involving activities such as typing, haircutting, carpentry, and assembly line work can also result in tenosynovitis. Manual therapy procedures can reduce swelling by moving venous blood through the affected area and can also enhance arterial blood flow that will facilitate healing.

Relating Muscles to Movements

Arrange the muscles in this exhibit according to the following actions on the elbow joint: (1) flexion and (2) extension; the following actions on the forearm at the radioulnar joints: (1) supination and (2) pronation; and the following actions on the humerus at the shoulder joint: (1) flexion and (2) extension. The same muscle may be mentioned more than once.

CHECKPOINT

3. Flex your arm. Which group of muscles is contracting? Which group of muscles must relax so that you can flex your arm?

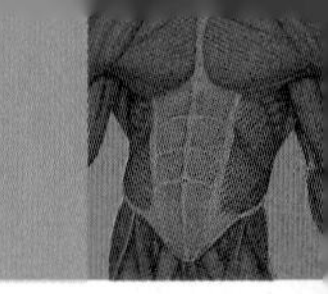

Figure 13.7 Surface features of the axilla, arm, and elbow.

The axilla becomes deeper when the musculature forming the walls is more developed.

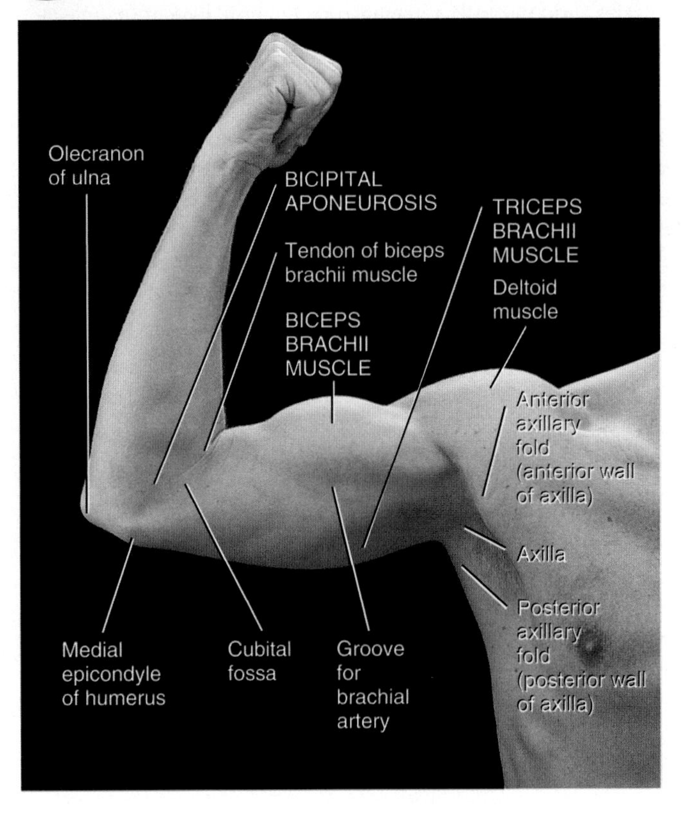

(a) Medial view of arm

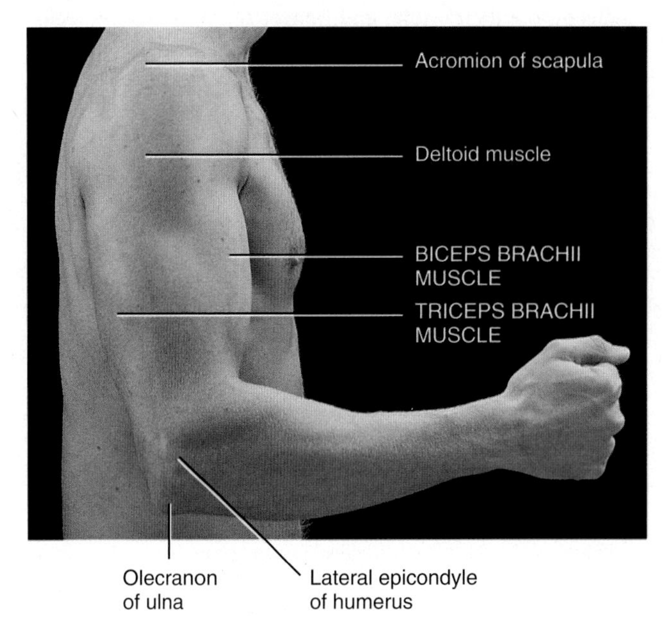

(b) Right lateral view of arm

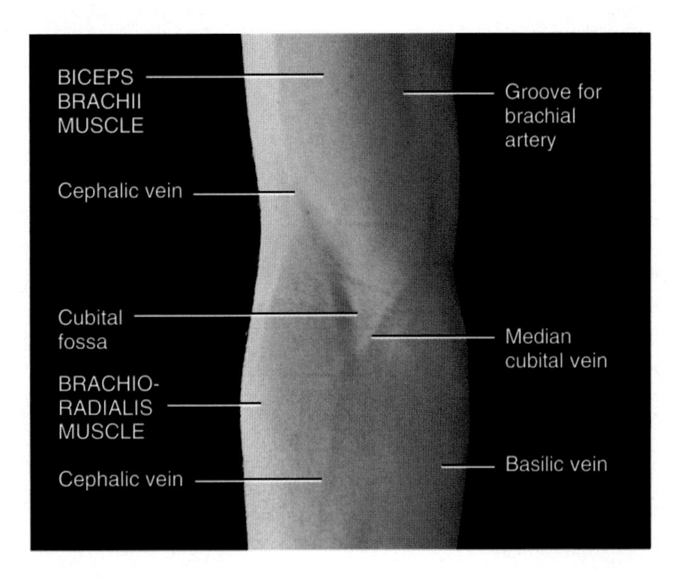

(c) Anterior view of cubital fossa

What muscles form the anterior and posterior axillary folds?

EXHIBIT 13.4 Muscles of the Forearm That Move the Wrist, Hand, and Digits

OBJECTIVE

- Describe the origin, insertion, action, and innervation of the muscles that move the wrist, hand, thumb, and fingers.

Muscles of the forearm that move the wrist, hand, thumb, and fingers are many and varied (Figure 13.8a–e). Those in this group that act on the digits are known as ***extrinsic muscles of the hand*** (*ex-* = outside) because they originate *outside* the hand and insert within it. As you will see, the names for the muscles that move the wrist, hand, and digits give some indication of their origin, insertion, or action. Based on location and function, the muscles of the forearm are divided into two groups: (1) anterior compartment muscles and (2) posterior compartment muscles. The ***anterior (flexor) compartment*** muscles of the forearm originate on the humerus, typically insert on the carpals, metacarpals, and phalanges, and function as flexors. The bellies of these muscles form the bulk of the anterior forearm. One of the muscles in the superficial anterior compartment, the ***palmaris longus*** muscle, is missing in about 10% of individuals (usually in the left forearm) and is commonly used for tendon repair. The ***posterior (extensor) compartment*** muscles of the forearm originate on the humerus, insert on the metacarpals and phalanges, and function as extensors. Within each compartment, the muscles are grouped as superficial or deep.

The ***superficial anterior compartment*** muscles are arranged in the following order from lateral to medial: ***flexor carpi radialis, palmaris***

MUSCLE	ORIGIN	INSERTION	ACTION	INNERVATION
SUPERFICIAL ANTERIOR (FLEXOR) COMPARTMENT OF THE FOREARM				
Flexor carpi radialis (FLEK-sor KAR-pē-rā′-dē-Ā-lis; *flexor* = decreases angle at joint; *carpi* = of the wrist; *radi-* = radius)	Medial epicondyle of humerus.	Second and third metacarpals.	Flexes and abducts hand (radial deviation) at wrist joint.	Median nerve.
Palmaris longus (pal-MA-ris LON-gus; *palma-* = palm; *longus* = long)	Medial epicondyle of humerus.	Palmar aponeurosis (fascia in center of palm).	Weakly flexes hand at wrist joint.	Median nerve.
Flexor carpi ulnaris (ul-NAR-is = of the ulna)	Medical epicondyle of humerus and superior posterior border of ulna.	Pisiform, hamate, and base of fifth metacarpal.	Flexes and adducts hand (ulnar deviation) at wrist joint.	Ulnar nerve
Flexor digitorum superficialis (di-ji-TOR-um soo′-per-fish′-ē-Ā-lis; *digit* = finger or toe; *superficialis* = closer to surface)	Medial epicondyle of humerus, coronoid process of ulna, and a ridge along lateral margin or anterior surface (anterior oblique line) of radius.	Middle phalanx of each finger.*	Flexes middle phalanx of each finger at proximal interphalangeal joint, proximal phalanx of each finger at metacarpophalangeal joint, and hand at wrist joint.	Median nerve.
DEEP ANTERIOR (FLEXOR) COMPARTMENT OF THE FOREARM				
Flexor pollicis longus (POL-li-sis = of the thumb)	Anterior surface of radius and interosseous membrane (sheet of fibrous tissue that holds shafts of ulna and radius together).	Base of distal phalanx of thumb.	Flexes distal phalanx of thumb at interphalangeal joint.	Median nerve.
Flexor digitorum profundus (prō-FUN-dus = deep)	Anterior medial surface of body of ulna.	Base of distal phalanx of each finger.	Flexes distal and middle phalanges of each finger at interphalangeal joints, proximal phalanx of each finger at metacarpophalangeal joint, and hand at wrist joint.	Median and ulnar nerves.
SUPERFICIAL POSTERIOR (EXTENSOR) COMPARTMENT OF THE FOREARM				
Extensor carpi radialis longus (eks-TEN-sor = increases angle at joint)	Lateral supracondylar ridge of humerus.	Second metacarpal.	Extends and abducts hand at wrist joint (radial deviation).	Radial nerve.
Extensor carpi radialis brevis (BREV-is = short)	Lateral epicondyle of humerus.	Third metacarpal.	Extends and abducts hand at wrist joints (ulnar deviation).	Radial nerve.

*Reminder: The thumb or pollex is the first digit and has two phalanges: proximal and distal. The remaining digits, the fingers, are numbered II–V (2–5), and each has three phalanges: proximal, middle, and distal.

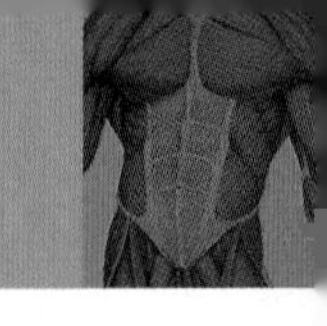

longus, and ***flexor carpi ulnaris*** (the ulnar nerve and artery are just lateral to the tendon of this muscle at the wrist). The ***flexor digitorum superficialis*** muscle is deep to the other three muscles and is the largest superficial muscle in the forearm. These muscles make up the fleshy mass that is deep to the hairless skin of the anterior forearm. The common origin is on the medial epicondyle. The palmaris longus muscle inserts into the thickened palmar aponeurosis. The tendons of the flexor digitorum superficialis and flexor digitorum profundus muscles pass through the carpal tunnel; the tendon of palmaris longus lies superficial to the flexor retinaculum.

The ***deep anterior compartment*** muscles are arranged in the following order from lateral to medial: ***flexor pollicis longus*** (the only flexor of the distal phalanx of the thumb) and ***flexor digitorum profundus*** (ends in four tendons that insert into the distal phalanges of the fingers).

The ***superficial posterior compartment*** muscles are arranged in the following order from lateral to medial: ***extensor carpi radialis longus, extensor carpi radialis brevis, extensor digitorum*** (occupies most of the posterior surface of the forearm and divides into four tendons that insert into the middle and distal phalanges of the fingers), ***extensor digiti minimi*** (a slender muscle usually connected to the extensor digitorum), and the ***extensor carpi ulnaris.***

The five superficial extensor muscles of the forearm all originate on or near the lateral epicondyle of the humerus. The bellies of these su-

MUSCLE	ORIGIN	INSERTION	ACTION	INNERVATION
SUPERFICIAL POSTERIOR (EXTENSOR) COMPARTMENT OF THE FOREARM (CONTINUED)				
Extensor digitorum	Lateral epicondyle of humerus.	Distal and middle phalanges of each finger.	Extends distal and middle phalanges of each finger at interphalangeal joints, proximal phalanx of each finger at metacarpophalangeal joint, and hand at wrist joint.	Radial nerve.
Extensor digiti minimi (DIJ-i-tē MIN-i-mē; *digit* = finger or toe; *minimi* = smallest)	Lateral epicondyle of humerus.	Tendon of extensor digitorum on fifth phalanx.	Extends proximal phalanx of little finger at metacarpophalangeal joint and hand at wrist joint.	Deep radial nerve.
Extensor carpi ulnaris	Lateral epicondyle of humerus and posterior border of ulna.	Fifth metacarpal.	Extends and adducts hand at wrist joint (ulnar deviation).	Deep radial nerve.
DEEP POSTERIOR (EXTENSOR) COMPARTMENT OF THE FOREARM				
Abductor pollicis longus (ab -DUK-tor = moves part away from midline)	Posterior surface of middle of radius and ulna and interosseous membrane.	First metacarpal.	Abducts and extends thumb at carpometacarpal joint and abducts hand at wrist joint.	Deep radial nerve.
Extensor pollicis brevis	Posterior surface of middle of radius and interosseous membrane.	Base of proximal phalanx of thumb.	Extends proximal phalanx of thumb at metacarpophalangeal joint, first metacarpal of thumb at carpometacarpal joint, and hand at wrist joint.	Deep radial nerve.
Extensor pollicis longus	Posterior surface of middle of ulna and interosseous membrane.	Base of distal phalanx of thumb.	Extends distal phalanx of thumb at interphalangeal joint, first metacarpal of thumb at carpometacarpal joint, and abducts hand at wrist joint.	Deep radial nerve.
Extensor indicis (IN-di-kis = index)	Posterior surface of ulna and interosseous membrane.	Tendon of extensor digitorum of index finger.	Extends distal and middle phalanges of index finger at interphalangeal joints, proximal phalanx of index finger at metacarpophalangeal joint, and hand at wrist joint.	Deep radial nerve.

CONTINUES

perficial muscles lie deep to the hairy skin of the posterior forearm. The tendons of extensor digitorum are easily seen beneath the skin on the dorsum of the hand.

The ***deep posterior compartment*** muscles are arranged in the following order from lateral to medial: ***abductor pollicis longus, extensor pollicis brevis, extensor pollicis longus,*** and ***extensor indicis.***

The tendons of the muscles of the forearm that attach to the wrist or continue into the hand, along with blood vessels and nerves, are held close to bones by strong fasciae. The tendons are also surrounded by tendon sheaths. At the wrist, the deep fascia is thickened into fibrous bands called ***retinacula*** (*retinacul* = a holdfast). The ***flexor retinaculum*** is located over the palmar surface of the carpal bones. The long flexor tendons of the digits and wrist and the median nerve pass deep to the flexor retinaculum (see Figure 13.9). The ***extensor retinaculum*** is located over the dorsal surface of the carpal bones. The extensor tendons of the wrist and digits pass deep to it.

Figure 13.8 Muscles of the forearm that move the wrist, hand, and digits.

The anterior compartment muscles function as flexors, adductors, and abductors. The posterior compartment muscles function as extensors, adductors, and abductors.

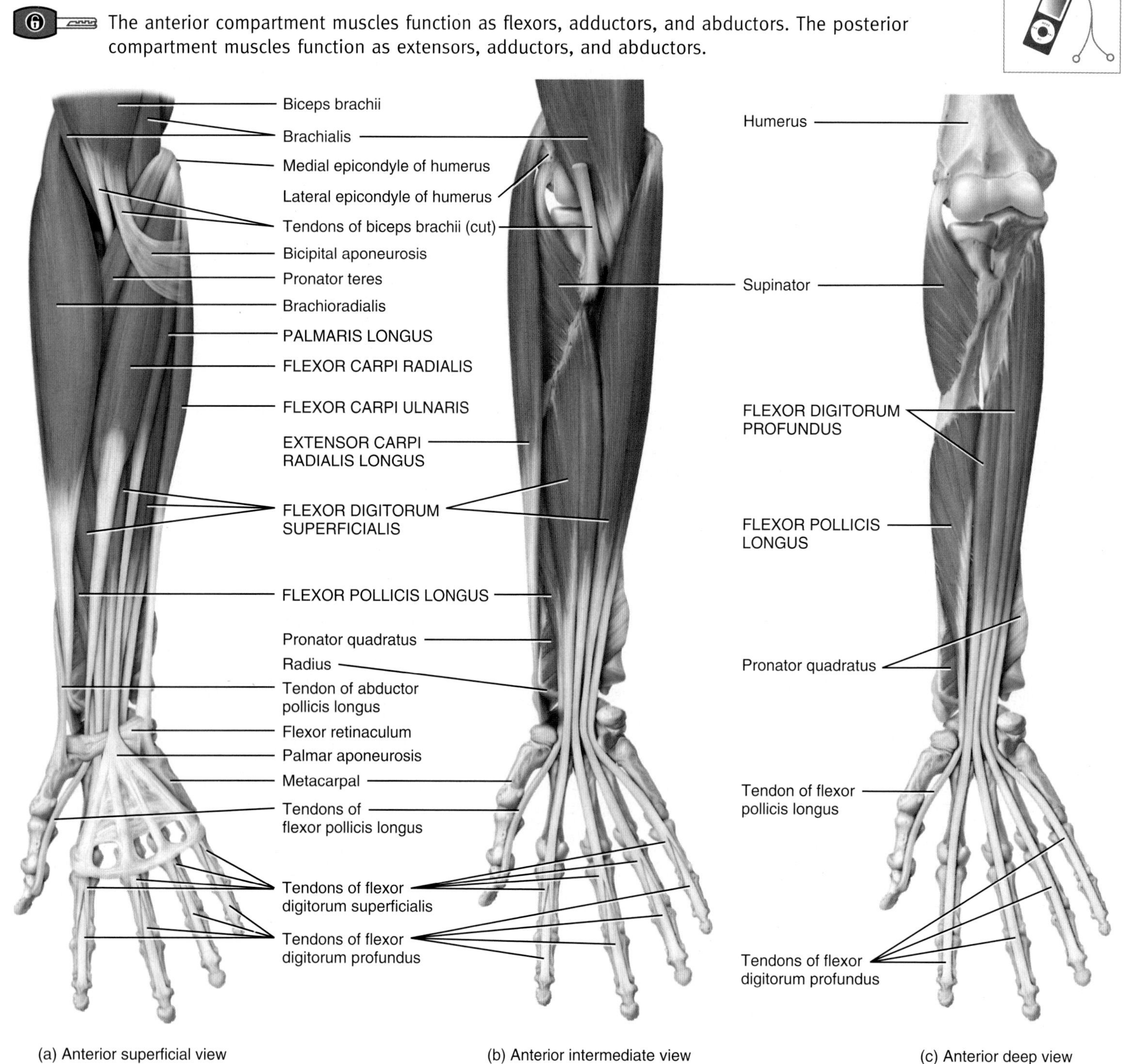

(a) Anterior superficial view

(b) Anterior intermediate view

(c) Anterior deep view

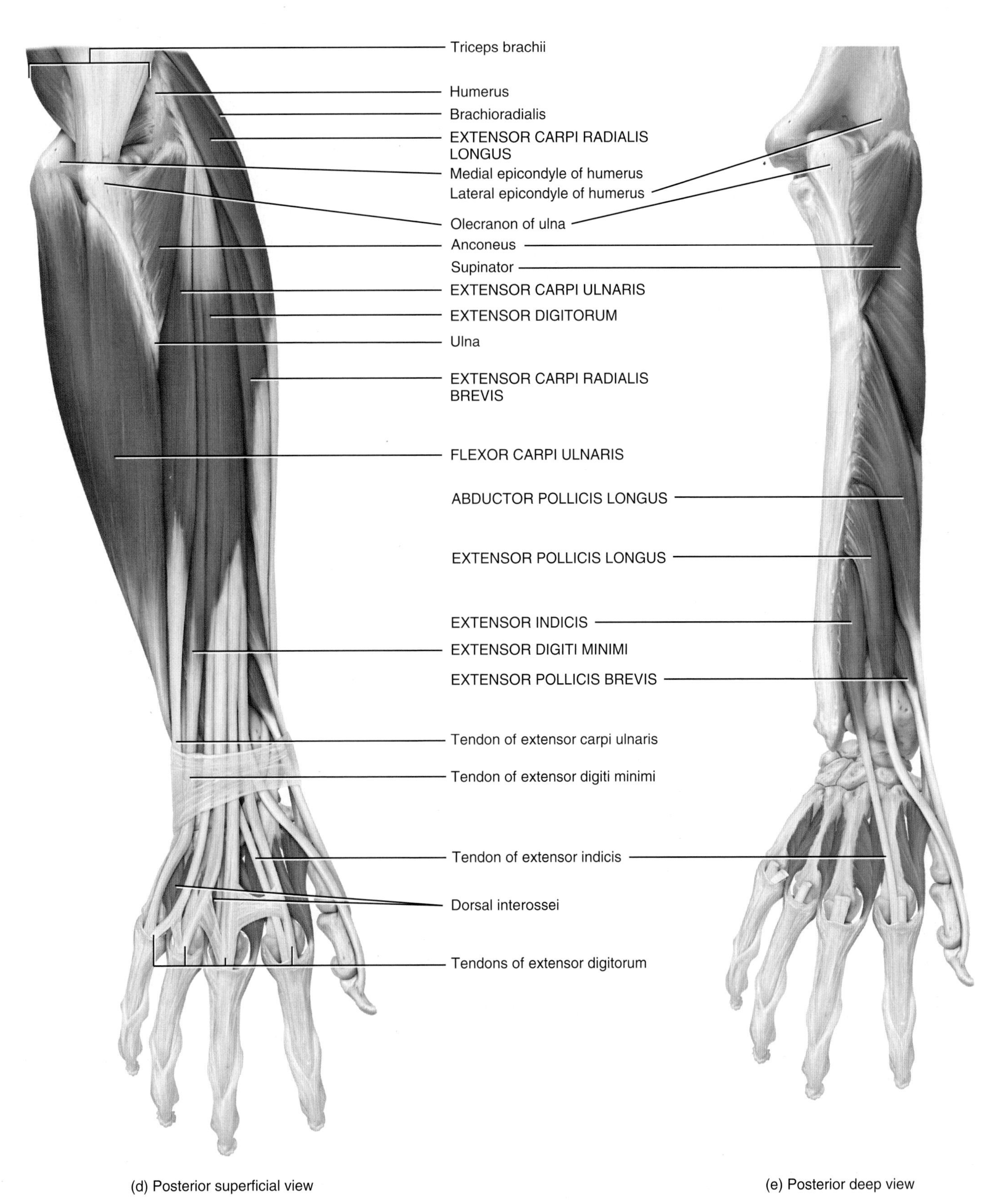

(d) Posterior superficial view

(e) Posterior deep view

CONTINUES

Figure 13.8 (continued)

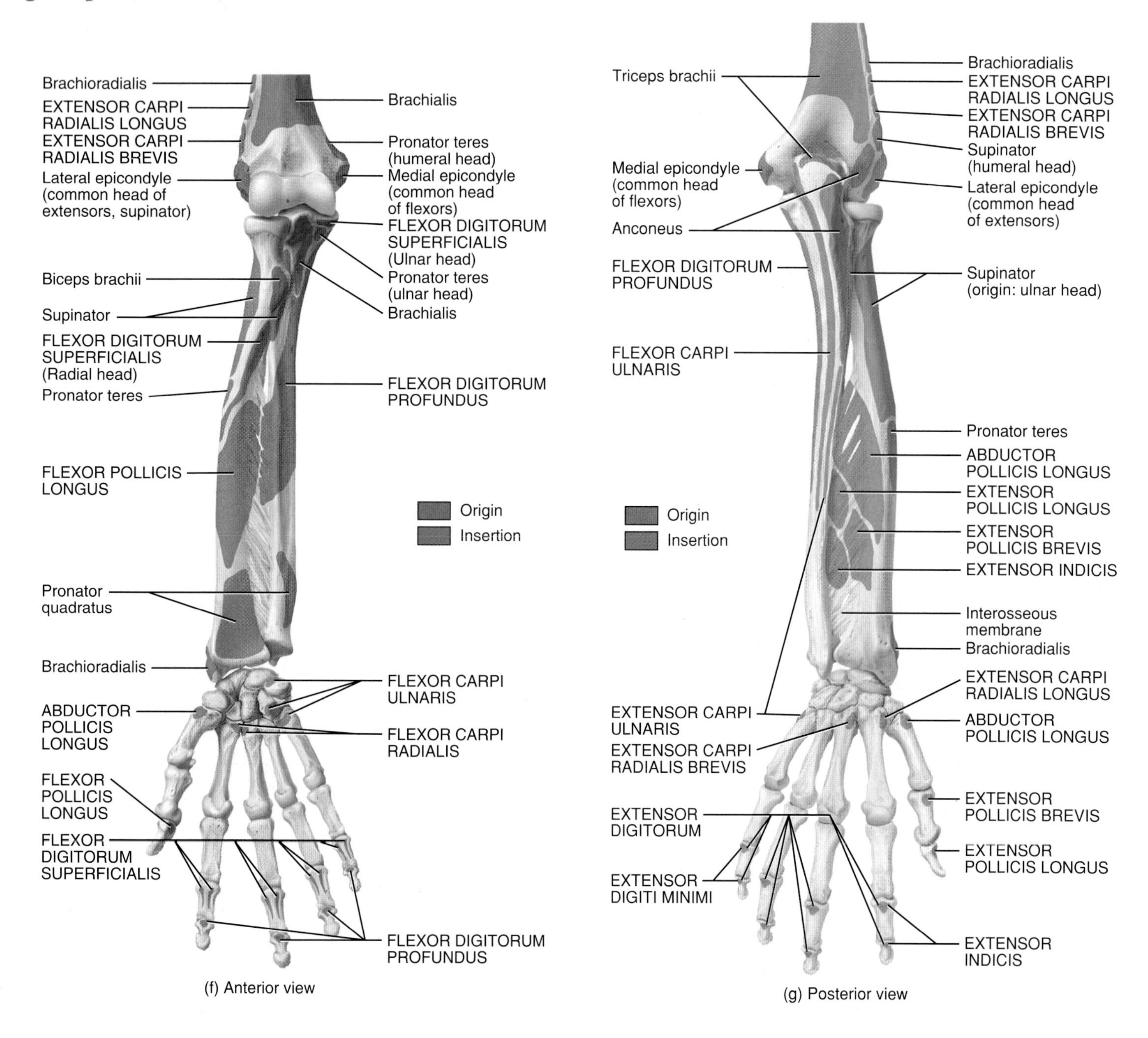

(f) Anterior view

(g) Posterior view

Which of the flexor muscles does not pass deep to the flexor retinaculum?

MANUAL THERAPY APPLICATION

Repetitive Strain or Motion Injuries

Repetitive strain or motion injuries (RSIs) include a large number of conditions resulting from overuse of equipment, poor posture, poor body mechanics, or activity that requires repeated movements; for example, various conditions of assembly line workers. Examples of overuse of equipment include overuse of a computer, hammer, guitar, or piano, to name a few.

Tennis elbow can be caused by strain of the extensor muscles of the forearm. It is so named because a series of forceful backhand strokes in tennis can cause pain of the lateral elbow joint (*lateral epicondylitis*). The extensor muscles strain or sprain, resulting in pain. Tennis elbow can be caused by a sudden trauma or repetitive actions in many types of daily activities.

Little-league elbow typically develops as a result of a heavy pitching schedule and/or a schedule that involves throwing curve balls, especially among youngsters. In this disorder, the elbow may enlarge, fragment, or separate.

Golfer's elbow can be caused by strain of the flexor muscles, especially flexor carpi radialis, as a result of repetitive movements such as swinging a golf club. Strain can, however, be caused by many actions. Pianists, violinists, movers, weight lifters, bikers, and those who use computers are among those who may develop pain near the medial epicondyle (*medial epicondylitis*).

Carpal tunnel pain is caused by compression of the median nerve. The carpal tunnel is a narrow passageway formed anteriorly by the flexor retinaculum and posteriorly by the carpal bones. Through this tunnel pass the median nerve, the most superficial structure, and the long flexor tendons for the digits (Figure 13.9). Structures within the carpal tunnel, especially the median nerve, are vulnerable to compression, and the resulting condition is known as **carpal tunnel syndrome**. The person may experience numbness, tingling, or pain of the wrist and hand. Compression within the tunnel usually results from inflamed and thickened tendon sheaths of flexor tendons, fluid retention, excessive exercise, infection, trauma, and/or repetitive activities that involve flexion of the wrist such as keyboarding, cutting hair, and playing a piano.

Treatment may be progressive if the problem worsens. Initial treatment may include aspirin or ibuprofen (both are anti-inflammatory drugs). Treatment may progress to an injection of cortisone into the carpal tunnel. Persons may be asked to keep the wrist straight to minimize movement of the inflamed tendon sheaths; some type of splint or brace may be prescribed. Continued pain may necessitate surgery to cut (release) the transverse carpal ligament and thus relieve the compression of the nerve. It should be noted that "carpal tunnel pain" can also be caused by compression of the median nerve in two areas of the shoulder. When this occurs, carpal tunnel surgery will not alleviate the pain. Furthermore, scar tissue formed after the surgery may exacerbate the problem. Nerve compression in the shoulder area is discussed in Chapter 16.

Compression of the median nerve can also occur between the anterior and middle scalenes or deep to the pectoralis minor. Pain in the wrist or hand is perceived by the patient that is identical to the pain of true carpal tunnel syndrome. Massage of the scalenes and pectoralis minor can usually lengthen those muscles and thereby reduce impingement on the median nerve. By lengthening these muscles, a manual therapist can usually determine within minutes whether the pain of the wrist and hand may be a function of compression of the median nerve in the neck or axilla or compression of the median nerve within the carpal tunnel.

Figure 13.9 Transverse (cross) section of wrist.

 The structure of the carpal tunnel is illustrated in relation to the thenar and hypothenar muscles.

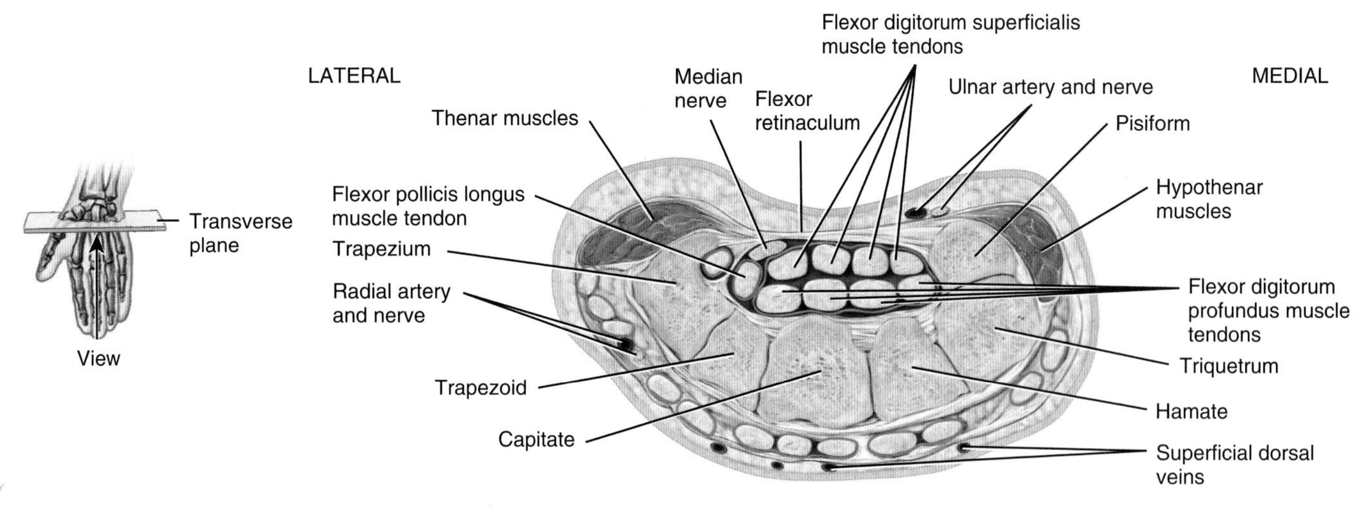

 **What nerve may become compressed within the carpal tunnel?**

CONTINUES

Surface Features of the Forearm and Wrist

The ***forearm*** or *antebrachium*, is the region between the elbow and wrist. The ***wrist,*** or carpus, is between the forearm and palm. Following are some prominent surface anatomy features of the forearm and wrist (Figure 13.10).

- **Ulna.** The medial bone of the forearm. It can be palpated along its entire length from the olecranon (see Figure 8.6a,b in Section 8.2) to the **styloid process**, a projection on the distal end of the bone at the medial (little finger) side of the wrist. The **head of the ulna** is a conspicuous enlargement just proximal to the styloid process.
- **Radius.** The distal half of the radius can be palpated proximal to the thumb side of the hand. The proximal half is covered by muscles. The **styloid process** of the radius is a projection on the distal end of the bone at the lateral (thumb) side of the wrist.
- **Flexor carpi radialis muscle.** The tendon of this muscle is on the lateral side of the forearm about 1 cm medial to the styloid process of the radius.
- **Palmaris longus muscle.** The tendon of this muscle is medial to the flexor carpi radialis tendon and can be seen quite prominently if the wrist is slightly flexed and the base of the thumb and little finger are drawn together. About 15 to 20% of individuals do not have this muscle in at least one arm.
- **Flexor digitorum superficialis muscle.** The tendon of this muscle is medial to the palmaris longus tendon and can be palpated by flexing the fingers at the metacarpophalangeal and proximal interphalangeal joints.
- **Flexor carpi ulnaris muscle.** The tendon of this muscle is on the medial aspect of the forearm.
- **Radial artery.** Located on the lateral aspect of the wrist between the flexor carpi radialis tendon and styloid process of the radius. It is frequestly used to take a pulse.
- **Pisiform bone.** Medial bone of the proximal row of carpals that can be palpated as a projection distal and anterior to the styloid process of the ulna.
- **"Anatomical snuffbox."** A triangular depression between **tendons of the extensor pollicis brevis** and **extensor pollicis longus muscles**. It derives its name from a habit of previous centuries in which a person would take a pinch of snuff (powdered tobacco or scented powder) and place it in the depression before sniffing it into the nose. The styloid process of the radius, the base of the first metacarpal, trapezium, scaphoid, and deep branch of the radial artery can all be palpated in the depression.
- **Wrist creases.** Three more or less constant lines on the anterior aspect of the wrist (named *proximal*, *middle*, and *distal*) where the skin is firmly attached to underlying deep fascia.

Relating Muscles to Movements

Arrange the muscles in this exhibit according to the following actions on the wrist joint: (1) flexion, (2) extension, (3) abduction, and (4) adduction; the following actions on the fingers at the metacarpophalangeal joints: (1) flexion and (2) extension; the following actions on the fingers at the interphalangeal joints: (1) flexion and (2) extension; the following actions on the thumb at the carpometacarpal, metacarpophalangeal, and interphalangeal joints: (1) extension and (2) abduction; and the following action on the thumb at the interphalangeal joint: flexion. The same muscle may be mentioned more than once.

CHECKPOINT

4. Which muscles and actions of the wrist, hand, and digits are used when writing?

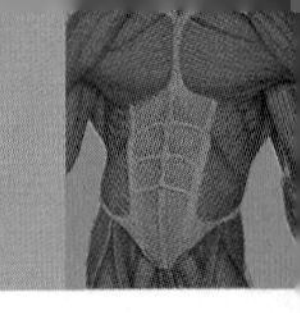

Figure 13.10 Surface features of the forearm and wrist.

Muscles of the forearm are most easily identified by locating their tendons near the wrist and tracing them proximally.

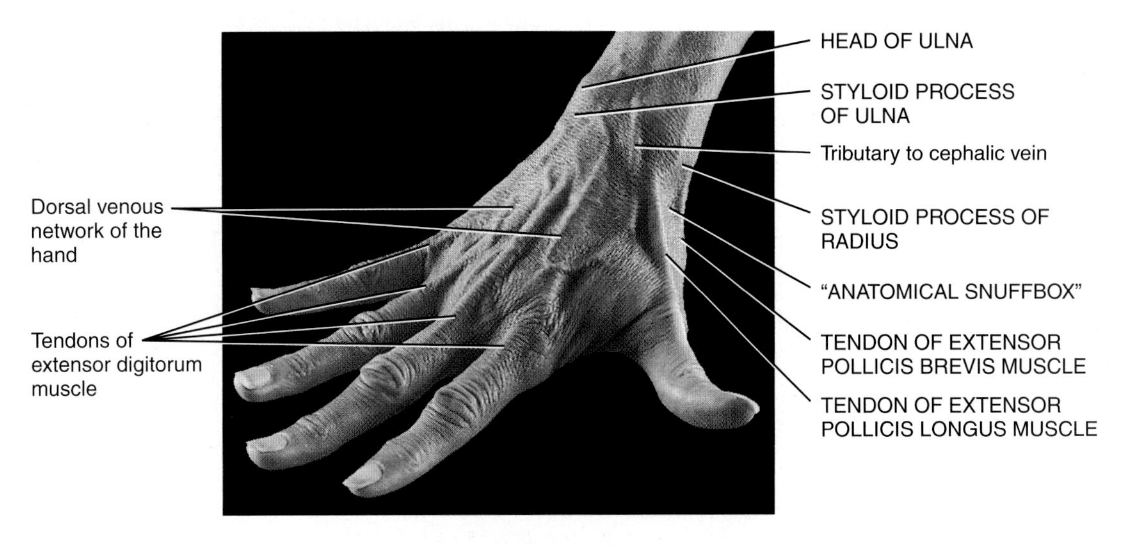

(a) Dorsum of wrist

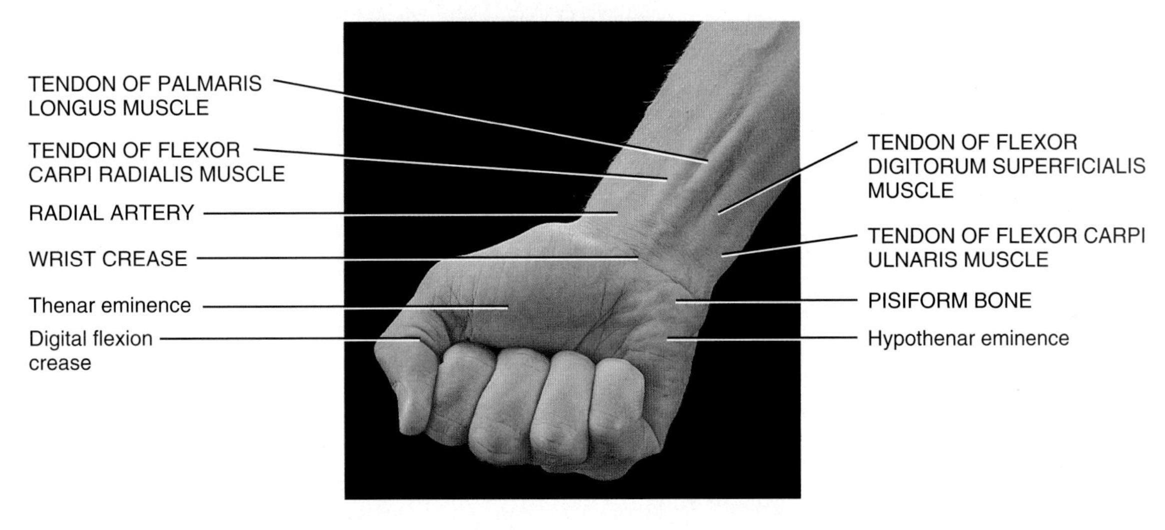

(b) Anterior aspect of wrist

What tendons form the boundaries of the "anatomical snuffbox"?

EXHIBIT 13.5 Muscles of the Palm That Move the Digits—Intrinsic Muscles of the Hand

OBJECTIVE

- Describe the origin, insertion, action, and innervation of the intrinsic muscles of the hand.

Several of the muscles discussed in Exhibit 13.4 move the digits in various ways and are known as extrinsic muscles of the hand. They produce the powerful but crude movements of the digits. The ***intrinsic muscles of the hand*** (Figure 13.11) in the palm produce the weak but intricate and precise movements of the digits that characterize the human hand. The muscles in this group are so named because their origins and insertions are *within* the hand.

The intrinsic muscles of the hand are divided into three groups: (1) ***thenar,*** (2) ***hypothenar,*** and (3) ***intermediate.*** The thenar muscles include the abductor pollicis brevis, opponens pollicis, and flexor pollicis brevis. The ***abductor pollicis brevis*** is a thin, short, relatively broad superficial muscle on the lateral side of the thenar eminence. The ***opponens pollicis*** is a small, triangular muscle that is deep to the abductor pollicis brevis muscle. The ***flexor pollicis brevis*** is a short,

MUSCLE	ORIGIN	INSERTION	ACTION	INNERVATION
THENAR (LATERAL ASPECT OF PALM)				
Abductor pollicis brevis (ab-DUK-tor POL-li-sis BREV-is; *abductor* = moves part away from middle; *pollic-* = the thumb; *brevis* = short)	Flexor retinaculum, scaphoid, and trapezium.	Lateral side of proximal phalanx of thumb.	Abducts thumb at carpometacarpal joint.	Median nerve.
Opponens pollicis (op-PŌ-nenz = opposes)	Flexor retinaculum and trapezium.	Lateral side of first metacarpal (thumb).	Moves thumb across palm to meet any finger (opposition) at the carpometacarpal joint.	Median nerve.
Flexor pollicis brevis (FLEK-sor = decreases angle at joint)	Flexor retinaculum, trapezium, capitate, and trapezoid.	Lateral side of proximal phalanx of thumb.	Flexes thumb at carpometacarpal and metacarpophalangeal joints.	Median and ulnar nerves.
Adductor pollicis (ad-DUK-tor = moves part toward midline)	Oblique head: capitate and second and third metacarpals; transverse head: third metacarpal.	Medial side of proximal phalanx of thumb by a tendon containing a sesamoid bone.	Adducts thumb at carpometacarpal and metacarpophalangeal joints.	Ulnar nerve.
HYPOTHENAR (MEDIAL ASPECT OF PALM)				
Abductor digiti minimi (DIJ-ī-tē MIN-i-mē; *digit* = finger or toe; *minimi* = little)	Pisiform and tendon of flexor carpi ulnaris.	Medial side of proximal phalanx of little finger.	Abducts and flexes little finger at metacarpophalangeal joint.	Ulnar nerve.
Flexor digiti minimi brevis	Flexor retinaculum and hamate.	Medial side of proximal phalanx of little finger.	Flexes little finger at carpometacarpal and metacarpophalangeal joints.	Ulnar nerve.
Opponens digiti minimi	Flexor retinaculum and hamate.	Medial side of fifth metacarpal (little finger).	Moves little finger across palm to meet thumb (opposition) at the carpometacarpal joint.	Ulnar nerve.
INTERMEDIATE (MIDPALMAR)				
Lumbricals (LUM-bri-kals; *lumbric-* = earthworm) (four muscles)	Lateral sides of tendons and flexor digitorum profundus of each finger.	Lateral sides of tendons of extensor digitorum on proximal phalanges of each finger.	Flex each finger at metacarpophalangeal joints and extend each finger at interphalangeal joints.	Median and ulnar nerves.
Palmar interossei (PAL-mar in′-ter-OS-ē-ī; *palmar* = palm; *inter-* = between: *-ossei* = bones) (three muscles)	Sides of shafts of metacarpals of three digits.	Sides of bases of proximal phalanges of three digits.	Adduct three fingers at metacarpophalangeal joints, flex three fingers at metacarpophalangeal joints, and extend three fingers at interphalangeal joints.	Ulnar nerve.
Dorsal interossei (DOR-sal in′-ter-OS-ē-ī; *dorsal* = back surface) (four muscles).	Adjacent sides of metacarpals.	Proximal phalanx of each finger.	Abduct fingers 2–4 at metacarpophalangeal joints, flex fingers 2–4 at metacarpophalangeal joints, and extend each finger at interphalangeal joints.	Ulnar nerve.

Figure 13.11 Muscles of the palm that move the digits—intrinsic muscles of the hand.

The intrinsic muscles of the hand produce the intricate and precise movements of the digits that characterize the human hand.

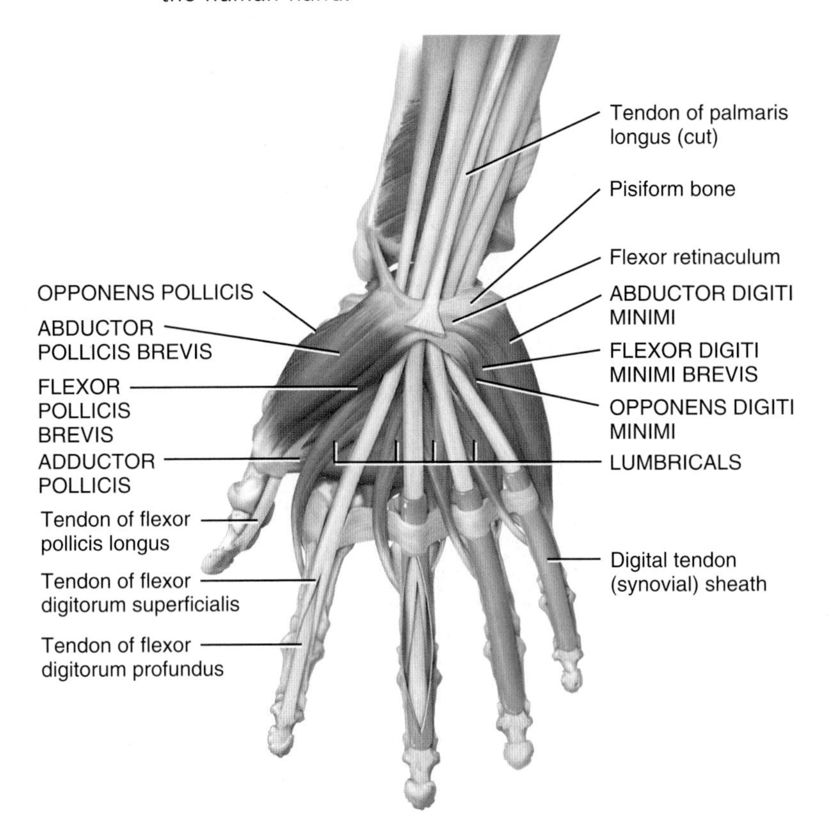

(a) Anterior superficial view

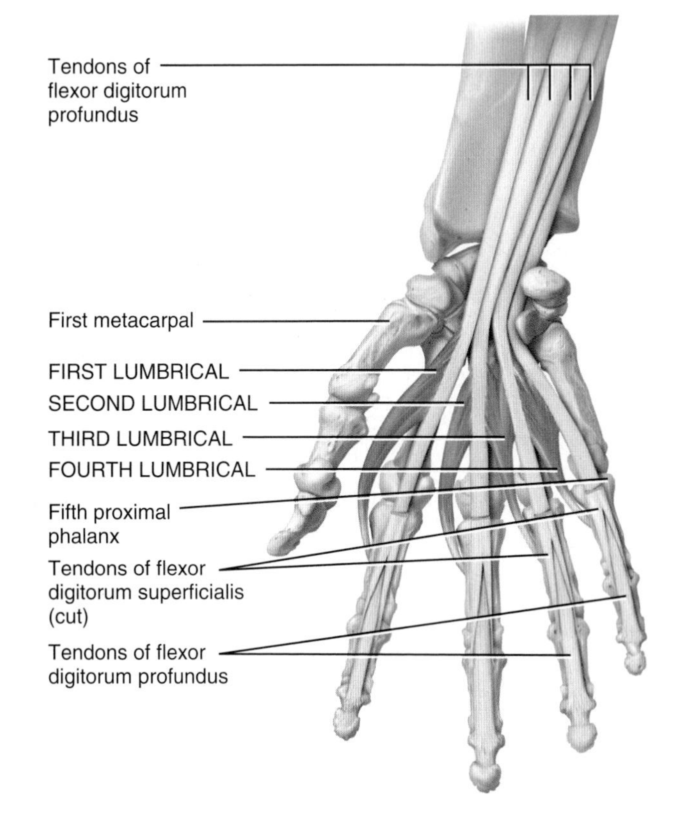

(b) Anterior intermediate view showing lumbricals

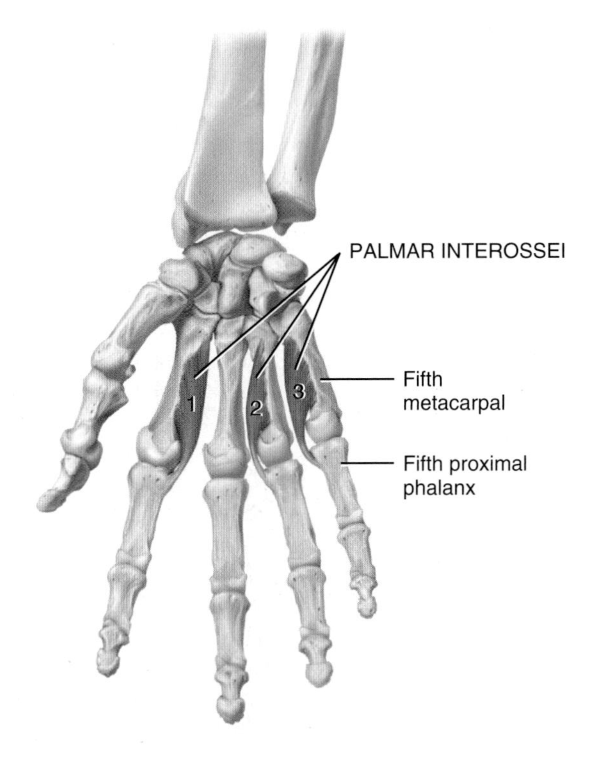

(c) Anterior deep view of palmar interossei

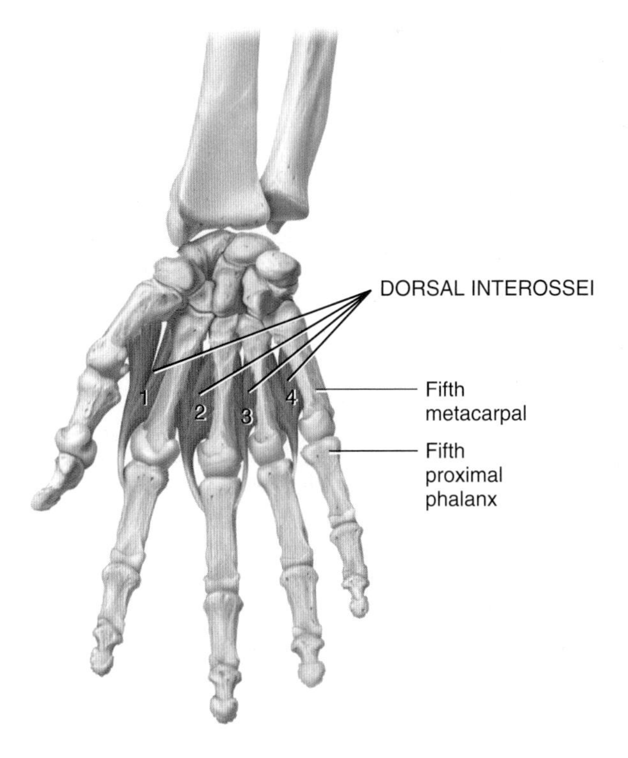

(d) Anterior deep view of dorsal interossei

CONTINUES

Figure 13.11 (continued)

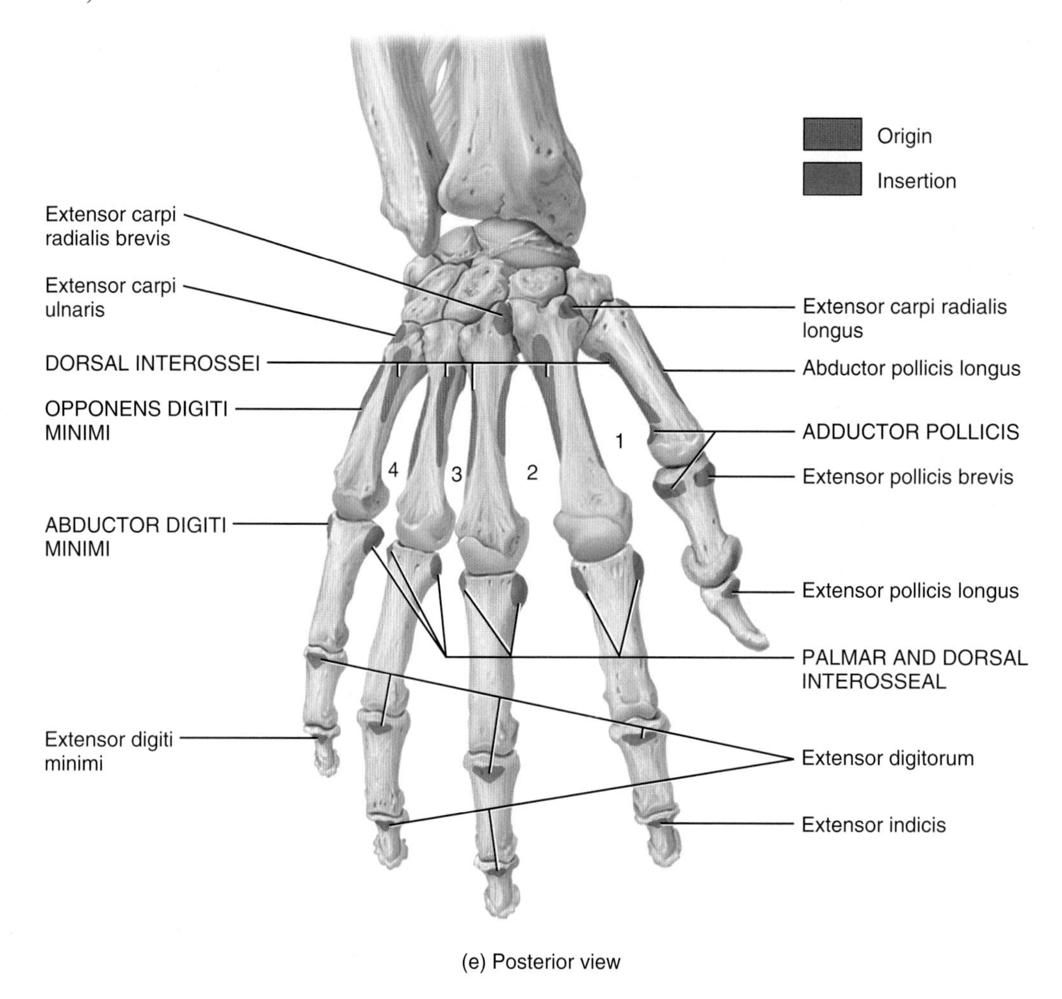

(e) Posterior view

wide muscle that is medial to the abductor pollicis brevis muscle. The three thenar muscles plus the adductor pollicis form the ***thenar eminence,*** the lateral rounded contour on the palm that is also called the *ball of the thumb.* The ***adductor pollicis*** also acts on the thumb. The muscle is fan-shaped and has two heads (oblique and transverse) separated by a gap through which the radial artery passes.

The three hypothenar muscles act on the little finger and form the ***hypothenar eminence,*** the medial rounded contour on the palm that is also called the *ball of the little finger.* The hypothenar muscles are the abductor digiti minimi, flexor digiti minimi brevis, and opponens digiti minimi. The ***abductor digiti minimi*** is a short, wide muscle and is the most superficial of the hypothenar muscles. It is a powerful muscle that plays an important role in grasping an object with outspread fingers. The ***flexor digiti minimi brevis*** muscle is also short and wide and is lateral to the abductor digiti minimi muscle. The ***opponens digiti minimi*** muscle is triangular and deep to the other two hypothenar muscles.

The 11 intermediate (midpalmar) muscles act on all the digits. The intermediate muscles include the lumbricals, palmar interossei, and dorsal interossei. The ***lumbricals*** (= worm-shaped), as their name indicates, are squiggly. They originate from and insert into the tendons of other muscles (flexor digitorum profundus and extensor digitorum). The ***palmar interossei*** are the smaller and more anterior of the interossei muscles. The ***dorsal interossei*** are the posterior interossei muscles. Both sets of interossei muscles are located between the metacarpals and are important in abduction, adduction, flexion, and extension of the fingers, and involving movements in skilled activities such as writing, typing, and playing a piano.

The functional importance of the hand is readily apparent when you consider that certain hand injuries can result in permanent disability. Most of the dexterity of the hand depends on movements of the thumb. The general activities of the hand are free motion, power grip (forcible movement of the fingers and thumb against the palm, as in squeezing), precision handling (a change in position of a handled object that requires exact control of finger and thumb positions, as in winding a watch or threading a needle), and pinch (compression between the thumb and index finger or between the thumb and first two fingers).

Movements of the thumb are very important in the precise activities of the hand, and they are defined in different planes from comparable movements of other digits because the thumb is positioned at a right

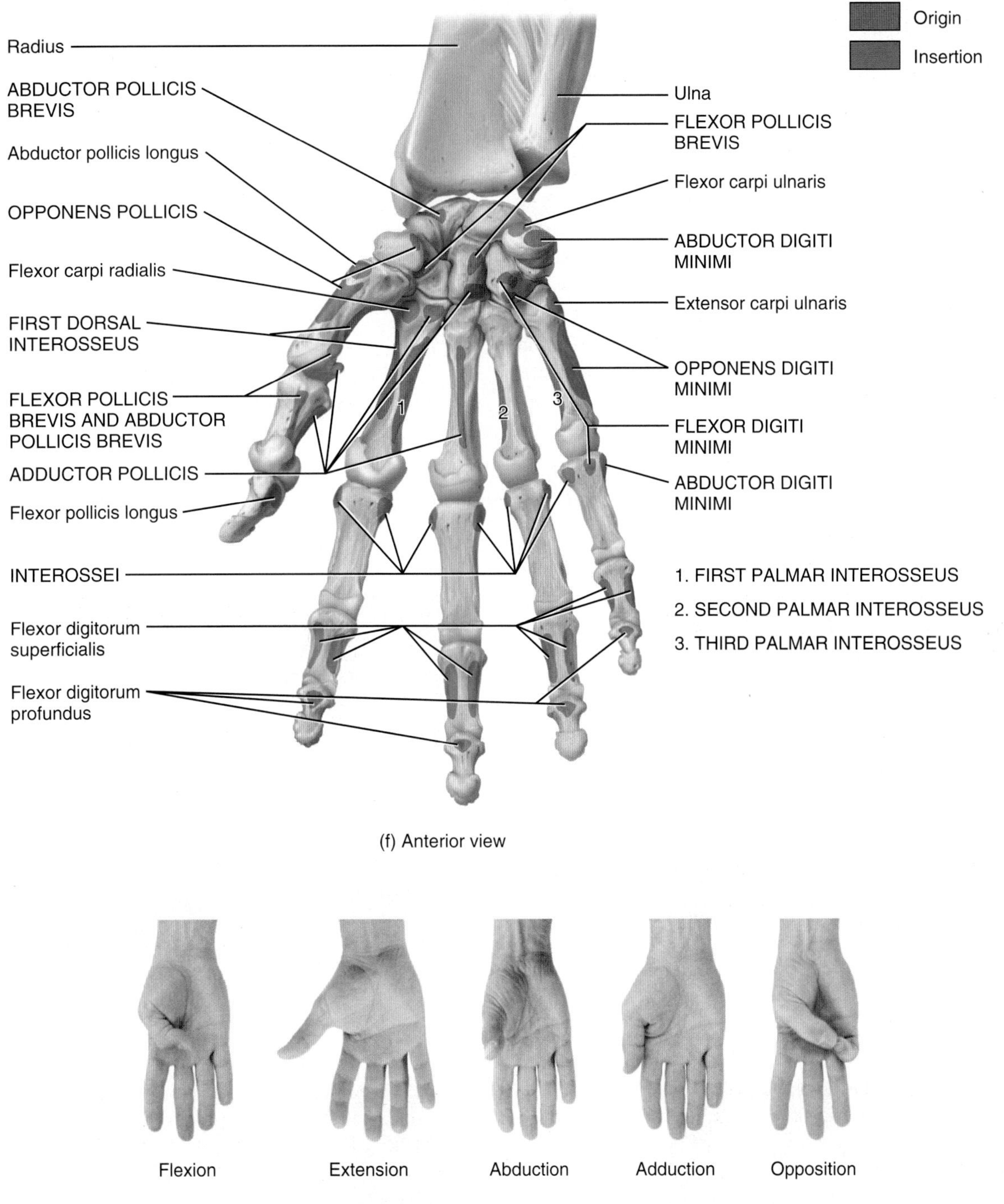

(f) Anterior view

Flexion Extension Abduction Adduction Opposition

(g) Movements of the thumb

Muscles of the thenar eminence act on which digit?

angle to the other digits. The five principal movements of the thumb are illustrated in Figure 13.11g and include *flexion* (movement of the thumb medially across the palm), *extension* (movement of the thumb laterally away from the palm), *abduction* (movement of the thumb in an anteroposterior plane away from the palm), *adduction* (movement of the thumb in an anteroposterior plane toward the palm), and *opposition* (movement of the thumb across the palm so that the tip of the thumb meets the tip of a finger). Opposition is the single most distinctive digital movement that gives humans and other primates the ability to grasp and manipulate objects precisely.

CONTINUES

Surface Features of the Hand

The ***hand,*** or *manus,* is the region from the wrist to the termination of the upper limb; it has several conspicuous surface features (Figure 13.12).

- **Knuckles.** Commonly refers to the dorsal aspect of the heads of metacarpals 2–5 (or II–V), but also includes the dorsal aspects of the metacarpophalangeal and interphalangeal joints.
- **Dorsal venous network of the hand (dorsal venous arch).** Superficial veins on the dorsum of the hand that drain blood into the cephalic vein. It can be displayed by compressing the blood vessels at the wrist for a few moments as the hand is opened and closed.
- **Tendon of extensor digiti minimi muscle.** This can be seen on the dorsum of the hand in line with the phalanx of the little finger.
- **Tendons of extensor digitorum muscle.** These can be seen on the dorsum of the hand in line with the phalanges of the ring, middle, and index fingers.

Relating Muscles to Movements

Arrange the muscles in the exhibit according to the following actions on the thumb at the carpometacarpal and metacarpophalangeal joints: (1) abduction, (2) adduction, (3) flexion, and (4) opposition; and the following actions on the fingers at the metacarpophalangeal and interphalageal joints: (1) abduction (2) adduction, (3) flexion, and (4) extension. The same muscle may be mentioned more than once.

CHECKPOINT

5. How do the actions of the extrinsic and intrinsic muscles of the hand differ?

Figure 13.12 Surface features of the hand.

Several tendons on the dorsum of the hand can be identified by their alignment with the phalanges of the digits.

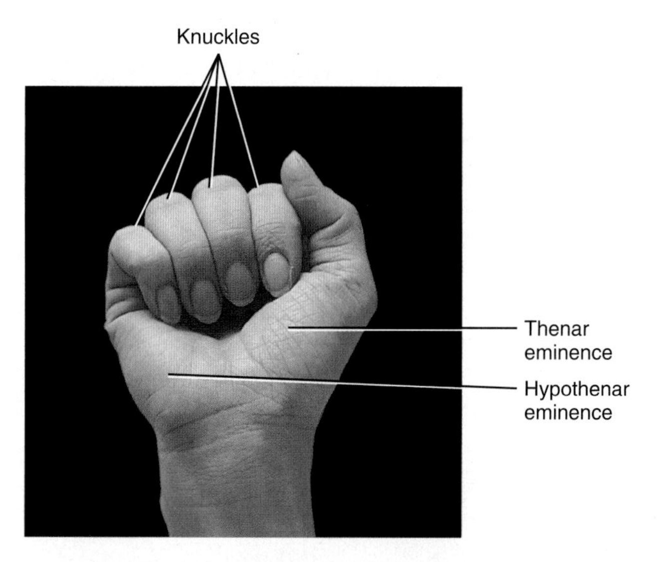

(a) Palmar and dorsal view

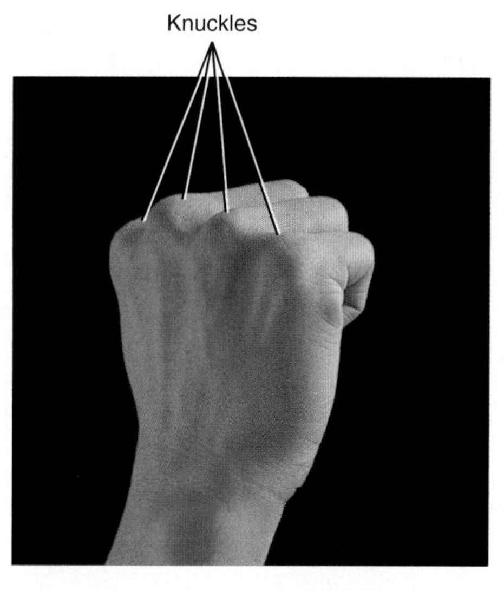

(b) Dorsal view

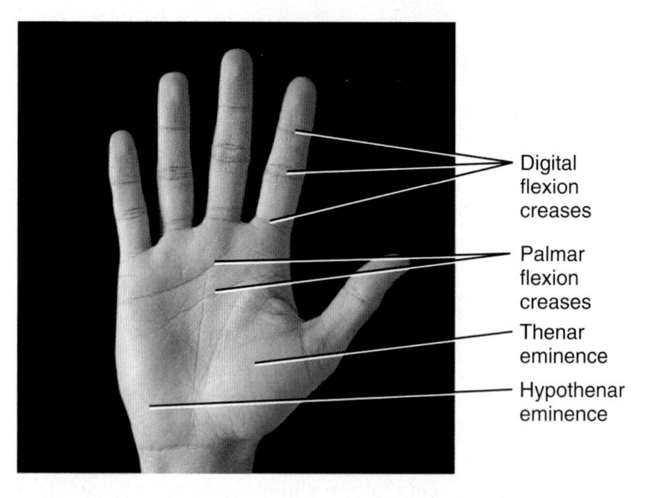

(c) Palmar view

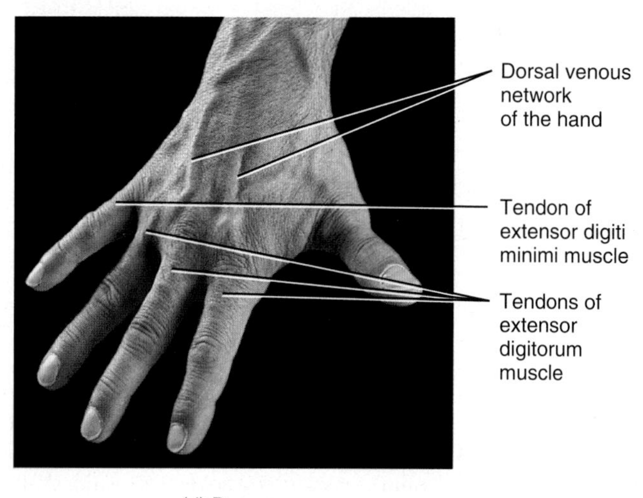

(d) Dorsum

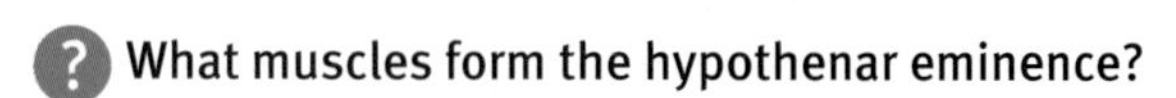

What muscles form the hypothenar eminence?

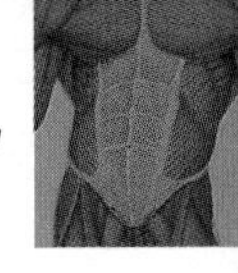

STUDY OUTLINE

Muscles of the thorax that move the pectoral girdle (Exhibit 13.1)

1. Muscles that move the pectoral girdle stabilize the scapula so it can function as a stable point of origin for most of the muscles that move the humerus.
2. Anterior thoracic muscles that move the pectoral girdle are the subclavius, pectoralis minor, and serratus anterior.
3. Posterior thoracic muscles that move the pectoral girdle are the trapezius, levator scapulae, rhomboid major, and rhomboid minor.
4. The serratus anterior is an antagonist of the rhomboids and is responsible for abduction of the scapula.
5. The trapezius is the most superficial muscle of the back with three sets of fibers (superior, middle, and inferior) that enable this muscle to cause multiple actions.

Muscles of the thorax and shoulder that move the humerus (Exhibit 13.2)

1. Muscles that move the humerus originate for the most part on the scapula; the remaining muscles originate on the axial skeleton.
2. The two axial muscles that move the humerus, the pectoralis major and the latissimus dorsi, form the anterior and posterior axillary folds, respectively, and a twist in each muscle near the insertion increases the contraction strength.
3. The prominence of the shoulder is formed by the deltoid muscle, a frequent site of intramuscular injection.
4. The rotator (musculotendinous) cuff muscles, which include the supraspinatus, infraspinatus, teres minor, and subscapularis, strengthen and stabilize the shoulder joint.

Muscles of the arm that move the radius and ulna (Exhibit 13.3)

1. Muscles that move the radius and ulna are involved in flexion and extension at the elbow joint and are organized into flexor and extensor compartments.
2. The biceps brachii muscle spans both the shoulder and elbow joints with two heads of origin (long and short), both from the scapula, and an insertion on the radius and bicipital aponeurosis.
3. The triceps brachii, the only muscle belly located in the posterior compartment of the arm, is an antagonist to the biceps brachii.
4. The axilla is a pyramid-shaped area at the junction of the arm and chest that contains blood vessels and nerves that pass between the neck and upper limbs.

Muscles of the forearm that move the wrist, hand, and digits (Exhibit 13.4)

1. Muscles that move the wrist, hand, and digits are many and varied; those muscles that act on the digits are called extrinsic muscles of the hand.
2. Based on location and function, the muscles of the forearm are divided into two groups: (1) anterior (flexor) compartment muscles and (2) posterior (extensor) compartment muscles.
3. The superficial anterior muscles of the forearm have a common origin on the medial epicondyle of the humerus, insert on the carpals, metacarpals, and phalanges, and function as flexors.
4. The superficial posterior muscles of the forearm have a common origin on the lateral epicondyle of the humerus, insert on the metacarpals and phalanges, and function as extensors.
5. Carpal tunnel syndrome is caused by compression of the median nerve within the narrow passageway formed anteriorly by the flexor retinaculum and posteriorly by carpal bones.
6. Several muscles of the forearm are more easily identifiable by their tendons.

Muscles of the palm that move the digits—intrinsic muscles of the hand (Exhibit 13.5)

1. The intrinsic muscles of the hand are important in skilled activities and provide humans with the ability to grasp and manipulate objects precisely.
2. The thenar muscles act on the thumb and form the lateral aspect of the palm (thenar eminence).
3. The hypothenar muscles act on the little finger and form the medial aspect of the palm (hypothenar eminence).
4. The hand contains the extensor tendons on the dorsum, and the thenar and hypothenar eminences on the palm.
5. The intermediate (midpalmar) muscles act on the digits and include the lumbricals, palmar interossei, and dorsal interossei.

SELF-QUIZ QUESTIONS

1. Impingement syndrome, common among athletes, is usually caused by the frequent pinching of
   a. coracobrachialis **b.** deltoid
   c. subscapularis **d.** supraspinatus.
2. Pronator quadratus originates on the ________ and inserts on the ________.
   a. radius; radius **b.** radius; ulna
   c. ulna; ulna **d.** ulna; radius
3. Which of the following muscles is NOT innervated by the median nerve?
   a. flexor carpi radialis **b.** flexor digitorum superficialis
   c. flexor carpi ulnaris **d.** palmaris longus
4. Which muscle is commonly called the "swimmer's muscle"?
   a. deltoid **b.** latissimus dorsi
   c. pectoralis major **d.** supraspinatus
5. Which of the following muscles does not adduct the scapula?
   a. rhomboid major **b.** rhomboid minor
   c. serratus anterior **d.** trapezius
6. The anterior fibers of the deltoid ________ and ________ rotate the arm at the shoulder.
   a. extend; laterally **b.** extend; medially
   c. flex; laterally **d.** flex; medially

7. The muscle that originates on the medial epicondyle of the humerus and inserts on second and third metacarpals is the
   a. flexor carpi radialis
   b. flexor digitorum superficialis
   c. flexor carpi ulnaris
   d. palmaris longus.
8. The origin of the levator scapulae is
   a. C1–C4 **b.** C4–C8
   c. T1–T4 **d.** superior vertebral border of the scapula.
9. Which muscle is NOT innervated, at least partially, by the musculocutaneous nerve?
   a. brachialis **b.** biceps brachii
   c. brachioradialis **d.** coracobrachialis
10. Which of the following is not an innervation of the trapezius?
   a. accessory (cranial nerve XI)
   b. C2
   c. C3
   d. C4
11. The innervation of the triceps brachii is the ________ nerve.
   a. median **b.** radial
   c. musculocutaneous **d.** two of the above
12. Which muscle is NOT part of the rotator cuff?
   a. infraspinatus **b.** subscapularis
   c. teres major **d.** teres minor
13. Which of the following muscles does NOT originate on the scapula?
   a. infraspinatus **b.** subscapularis
   c. pectoralis major **d.** supraspinatus
14. Which of the following muscles inserts on the radial tuberosity and bicipital aponeurosis?
   a. biceps brachii **b.** brachioradialis
   c. brachialis **d.** coracobrachialis
15. The origin of the rhomboid minor is
   a. C7–T1 **b.** T2–T5
   c. C7–T5 **d.** C6–T6
16. Which muscle inserts on the middle of the medial surface of the shaft of the humerus?
   a. coracobrachialis **b.** latissimus dorsi
   c. supraspinatus **d.** teres minor
17. Which muscle usually originates on the third, fourth, and fifth ribs and inserts on the coracoid process?
   a. coracobrachialis **b.** pectoralis minor
   c. serratus anterior **d.** subclavius
18. Carpal tunnel syndrome may be caused by compression of the ________ nerve.
   a. median **b.** radial
   c. musculocutaneous **d.** ulnar
19. The brachialis inserts on the
   a. humerus **b.** radius
   c. ulna **d.** two of the above.
20. The tendon of palmaris longus lies
   a. deep to extensor retinaculum
   b. superficial to extensor retinaculum
   c. deep to flexor retinaculum
   d. superficial to flexor retinaculum.

CRITICAL THINKING QUESTIONS

1. George is an automobile mechanic and he presents with a "sore elbow." His therapist assesses that the pain is localized on both the lateral and medial aspects of the proximal forearm. Explain.
2. Minor league pitcher José has been throwing a hundred pitches a day in order to perfect his curve ball. Lately he has experienced pain in his pitching arm. The doctor diagnosed a torn rotator cuff. José was confused because he thought his cuffs were only found on shirtsleeves, not inside his shoulder. Explain to José what the doctor means and how the injury could affect his arm movement.

? ANSWERS TO FIGURE QUESTIONS

13.1 The main action of the muscles that move the pectoral girdle is to stabilize the scapula to assist in movements of the humerus.

13.2 The acromion of the scapula forms the top of the shoulder.

13.3 The medial wall of the axilla is formed by ribs 1-4, their corresponding intercostal muscles, and the serratus anterior. The lateral wall is formed primarily by the superior portion of the shaft of the humerus, the coracobrachialis, and the biceps brachii.

13.4 The subscapularis is the only rotator cuff muscle that inserts on the lesser tubercle located on the anterior humerus.

13.5 With the aid of a stethoscope, respiratory sounds can be heard clearly in the triangle of auscultation.

13.6 The brachialis is the most powerful forearm flexor; the triceps brachii is the most powerful forearm extensor.

13.7 The anterior axillary fold is formed by the pectoralis major muscle; the posterior axillary fold is formed by the latissimus dorsi and the teres major muscles.

13.8 The palmaris longus is the only flexor muscle of the anterior compartment of the forearm that does not pass deep to the flexor retinaculum.

13.9 The median nerve may become compressed within the carpal tunnel.

13.10 The "anatomical snuffbox" is formed by the tendons of the extensor pollicis brevis and extensor pollicis longus muscles.

13.11 Muscles of the thenar eminence act on the thumb.

13.12 The hypothenar eminence is formed by the abductor digiti minimi and opponens digiti minimi muscles.

The Muscular System: The Muscles of the Lower Limb

14

You will now begin looking at the attachment of the lower body muscles to the bones. Reviewing the bones of the lower limb would be helpful before beginning this chapter since muscles often derive their name from their origin or insertion, action, or shape. The main action of the lower body muscles is to provide for ambulation (walking). As therapists, you will often assess patients by viewing their gait (the way they walk) as soon as you see them. You may evaluate the movement of the hip muscles that maintain a balanced posture. You may observe the anterior thigh to see if the hip is flexed and knee straightened to swing the leg out. You may watch the anterior leg muscles to see if the foot lifts high enough to clear the floor. Then as the person steps we watch to see the posterior thigh muscles bend the knee and extend the hip and the lower posterior compartment push the foot down to start the whole process again. The intrinsic foot muscles then provide a strong, flexible support for the enormous pressure per square inch when landing on the foot bones. When assessing a person's gait you will need to be aware of the innervation of nerves to the muscles to determine if the disturbance in gait is due to the muscle or nerve or damage to both. For example, weakness in the gluteus medius muscle would give you a swagger like a runway model. Next time you are sitting in the mall, take a look at the way people walk. You will see them in a new muscular way.

CONTENTS AT A GLANCE

14.1 INTRODUCTION TO THE MUSCLES OF THE LOWER LIMB (EXTREMITY)

OBJECTIVES

- Describe the functional differences between muscles of the upper and lower limbs.
- Describe the value of studying limb muscles by compartment.

As you will see, muscles of the lower limbs are larger and more powerful than those of the upper limbs because of differences in *function*. Whereas upper limb muscles are characterized by versatility of movement and manipulation, lower limb muscles function in stability, locomotion (ambulation), and maintenance of posture. In addition, muscles of the lower limbs often cross two joints and act equally on both.

The hip muscles are massive and serve as powerful movers and stabilizers of the femur. These muscles counteract the loads imposed by supporting the whole body weight on two limbs and by maintaining balance and stability during walking. Although not indicated in the actions column of the table accompanying Exhibit 14.1, an important function of nearly all hip muscles is *stabilization* of the hip with relation to either the vertebral column or the femur.

The most powerful muscles of the lower limb lie in the *posterior* hip, the *anterior* thigh, and the *posterior* leg. Muscles in these locations are primarily responsible for standing from a seated position.

Muscles of the thigh and leg are divided by fasciae into *compartments*. The thigh has three compartments: anterior, posterior, and medial. The leg also has three compartments: anterior, posterior, and lateral. Experience has shown that the task is less daunting when approached from the standpoint of learning muscles by compartment. You will note that the primary actions of most of the muscles in any given compartment are nearly the same. The muscle innervations within any compartment are also nearly the same. Furthermore, when blood vessels are studied in Chapter 23, you will find that the same major blood vessel services all components of a compartment.

The intrinsic muscles of the foot originate and insert within the foot. These muscles are responsible for fine-tuning the actions caused by the foot muscles that are found in the leg. The intrinsic muscles are also responsible for the abduction and adduction of the toes.

Students typically have access to textbook images, skeletons, models, charts, and video images while they are initially studying the skeletal system. A few students have the rare opportunity to see a cadaver presentation. In practice, however, manual therapists will be *observing* and *palpating* through the skin. A working knowledge of surface anatomy is therefore crucial to successful assessment and treatment.

Finally, you will learn the structures of the skeletal system as fragmented portions of the whole. You will independently learn the name, origin, insertion, actions, and innervation of each muscle. Hopefully, you will also learn, when applicable, the information for an entire compartment as a whole. There is nothing wrong with the approach of learning fragmented information; the human body, however, is not fragmented. Every portion of the body is interconnected by *connective tissue*, a continuum of cells, fibers, networks of fibers, fascial sheets, and layers of fascial sheets. Recognizing this *continuum* will enable you to better understand that activity of the patient in one area may have an influence on other portions of the body or perhaps the entire body.

• CLINICAL CONNECTION | Compartment Syndrome

As noted earlier in this chapter, skeletal muscles in the limbs are organized into functional units called *compartments*. In a disorder called **compartment syndrome,** some external or internal pressure constricts the structures within a compartment, resulting in damaged blood vessels and subsequent reduction of the blood supply (ischemia) to the structures within the compartment. Symptoms include pain, burning, pressure, pale skin, and paralysis. Common causes of compartment syndrome include crushing and penetrating injuries, contusion (damage to subcutaneous tissues without the skin being broken), muscle strain (overstretching of a muscle), or an improperly fitted cast. The pressure increase in the compartment can have serious consequences, such as hemorrhage, tissue injury, and edema (buildup of interstitial fluid). Because fasciae that enclose the compartments are very strong, accumulated blood and interstitial fluid cannot escape, and the increased pressure can literally choke off the blood flow and deprive nearby muscles and nerves of oxygen. One treatment option is *fasciotomy* (fash-ē-OT-ō-mē), a surgical procedure in which muscle fascia is cut to relieve the pressure. Without intervention, nerves can suffer damage, and muscles can develop scar tissue that results in permanent shortening of the muscles, a condition called *contracture*. If left untreated, tissues may die and the limb may no longer function. Once the syndrome has reached this stage, amputation may be the only treatment option. •

CHECKPOINT

1. What are the functional differences between the muscles of the upper and lower limbs?
2. What is the value of studying muscles of the limbs by compartment?
3. What tissue is responsible for connecting skeletal muscles from widely separated locations in the body?

Muscles of the Gluteal Region That Move the Femur

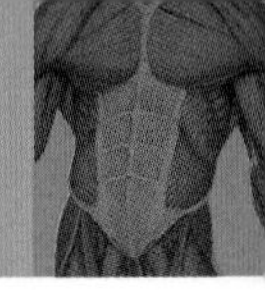

OBJECTIVE

- Describe the origin, insertion, action, and innervation of the muscles that move the femur.
- Describe the major muscles that can be seen and palpated in the gluteal region.

The majority of muscles that move the femur originate on the pelvic girdle and insert on the femur (Figure 14.1). The ***psoas major, psoas minor,*** and ***iliacus*** muscles share a common insertion (lesser trochanter of the femur) and are collectively known as the ***iliopsoas*** (il′-ē-ō-SŌ-as) muscle.

The ***psoas minor,*** not described elsewhere in this book, is present in only half of the population. When present, it originates on T12 and L1 and inserts into the fascia just superior to the inguinal ligament. The iliopsoas is the most powerful flexor of the thigh and therefore is important for walking, running, and standing. The iliopsoas is very susceptible to pathological shortening, especially in individuals with immobilization conditions and those persons with a sedentary lifestyle. Pathological shortening of the iliopsoas causes a limitation of hip extension, increased anterior tilt of the pelvis, and other problems with posture. When walking, persons with contracture of the iliopsoas are bent at the waist and have a short gait. Due to reverse muscle action (RMA) these same persons are unable to raise the torso from the supine position without using their arms. The sacroiliac joints are compromised, as well as lumbar intervertebral joints. With time, the lumbar bodies may also undergo degenerative changes.

There are three gluteal muscles: gluteus maximus, gluteus medius, and gluteus minimus (see Figures 14.1d,e and 14.3). The ***gluteus maximus*** is the largest and heaviest of the three muscles and is one of the largest muscles in the body. This large, superficial muscle of the buttock region is quadrilateral in form. It is coarsely fasciculated with the fascicles directed inferiorly and laterally. It is a powerful extensor of the thigh or, in its RMA, a powerful extensor of the torso. One of its insertions, into the iliotibial tract, helps hold (lock) the knee in extension. This action is critical for standing and also for walking when the center of gravity is passing over the extended knee as each step is taken. The gluteus maximus is innervated solely by the inferior gluteal nerve, a branch of the sacral plexus. The ***gluteus medius*** is mostly deep to the gluteus maximus and is a powerful abductor of the femur at the hip joint. It is a common site for an intramuscular injection. The ***gluteus minimus*** is the smallest of the gluteal muscles and lies deep to the gluteus medius. Its insertion, like that of the gluteus medius, is on the greater trochanter and therefore the actions of the two muscles are essentially the same. Both muscles abduct and medially rotate the femur, which are important actions in walking. When one foot is raised, these two muscles hold the pelvic bone of the opposite side to the greater trochanter and prevent the collapse of the pelvis on the unsupported side. At the same time, the rotary action helps swing the pelvis forward with each step. It should be noted that one of the actions of the gluteus maximus is *lateral* rotation while one action of the gluteus medius and minimus muscles is *medial* rotation.

The ***tensor fasciae latae*** muscle is a fusiform muscle located on the lateral surface of the thigh. This muscle lies between two layers of the *fascia lata,* a layer of deep fascia composed of dense connective tissue that encircles the entire thigh. The fascia lata is well developed laterally where, together with the tendons of the tensor fasciae latae and gluteus maximus muscles, it forms a structure called the ***iliotibial tract (band).*** This structure is best known clinically as the ***IT band*** and inserts into the lateral condyle of the tibia. The origin of the tensor fasciae latae is on the iliac crest and anterior superior iliac spine (ASIS) and its tendon of insertion blends with the fibers of the iliotibial tract. The IT band, perhaps three fingers wide, has such strength that it commonly flattens the lateral side of the otherwise rounded thigh (as seen in transverse section in Figure 14.2). The belly of the tensor fasciae latae lies between the gluteus maximus and sartorius muscles. It keeps the iliotibial tract taut and thus helps maintain the extended knee in the erect posture. Any patient who stands or walks extensively may present with pain at the site of the attachments for the tensor fasciae latae—ASIS and iliotibial tract (band). The IT band has been called the "hornet's nest" due to the frequency and severity of tenderness. Such tenderness signals chronic shortening of the gluteus maximus, tensor fasciae latae, or both muscles.

The ***Piriformis, superior Gemellus, Obturator internus, inferior Gemellus, Obturator externus,*** and ***Quadratus femoris*** muscles are all deep to the gluteus maximus muscle and function as lateral rotators of the femur at the hip joint. These muscles have essentially the same insertion, on or near the greater trochanter of the femur, and they all have essentially the same actions—to laterally rotate and abduct the femur. From superior to inferior, the six muscles can be remembered with the mnemonic ***P****iece* ***G****oods* ***O****ften* ***G****et into* ***O****ld* ***Q****uilts.*

The ***piriformis*** is a very active muscle when moving the toes to the lateral side of the heel or when the feet are fixed and the body rotates, as might be done, for example, when standing in front of the sink and washing dishes or when working on an assembly line. The piriformis is also activated to limit vigorous and/or rapid medial rotation. As with any other muscle that is overworked, the piriformis may become chronically contracted. As described previously in Chapter 10, a contracted muscle becomes not only shorter but fatter. As seen in Figure 14.1e, the sciatic nerve also exits the pelvis through the greater sciatic notch and is thus in contact with the piriformis muscle. Occasionally a portion of the sciatic nerve actually pierces the belly of the piriformis. Pressure on the sciatic nerve by the contracted piriformis muscle may cause pain of multiple levels and multiple locations. Note in Exhibit 16.4 that the ***sciatic nerve*** is actually two nerves, the tibial and the common fibular nerves, that are wrapped with a common sheath of connective tissue. Pressure on the sciatic nerve by the piriformis may affect either nerve or both. Sciatic pain may thus be localized to the hip, it may travel down the posterior thigh, it may involve the leg, or it may cause numbness, tingling, or pain all the way to the toes. Sciatic pain may, of course, also be caused by herniated (slipped) discs or degenerating discs in the lower lumbar area. More commonly, however, the pain of sciatica is due to a pathological contracture of the piriformis muscle due to overuse. When this occurs, the sacrum is pulled posteriorly by the muscle and may become somewhat locked in position. Also, the lower muscle fibers may produce a strong shearing force on the sacroiliac (SI) joint and cause the ipsilateral inferior auricular surface of the ilium to move anteriorly while the sacrum and coccyx are pulled posteriorly. By causing relaxation of the piriformis muscle, a manual therapist may reduce or eliminate sciatic pain, decrease ipsilateral SI pain, and cause the sacrum of the patient to become more flexible.

CONTINUES

MUSCLE	ORIGIN	INSERTION	ACTION	INNERVATION
Iliopsoas				
Psoas major (SŌ-as; *psoa* = a muscle of the loin)	Transverse processes and bodies of lumbar vertebrae.	With iliacus into lesser trochanter of femur.	Psoas major and iliacus muscles acting together flex thigh at hip joint, rotate thigh laterally. RMA: Flex trunk on the hip as in sitting up from the supine position.	Lumbar spinal nerves.
Iliacus (il′-ē-A-cus; *iliac-* = ilium)	Iliac fossa and sacrum.	With psoas major into lesser trochanter of femur.		Femoral nerve.
Gluteus maximus (GLOO-tē-us MAK-si-mus; *glute-* = rump or buttock; *maximus* = largest)	Iliac crest, sacrum, coccyx, sacrotuberous ligament and thoracolumbar aponeurosis.	Iliotibial tract of fascia lata and superior and lateral part of linea aspera (gluteal tuberosity).	Extends thigh at hip joint and laterally rotates thigh; helps lock knee in extension. RMA: Extends torso.	Inferior gluteal nerve.
Gluteus medius (MĒ-de-us = middle)	Posterior ilium.	Greater trochanter of femur.	Abducts thigh at hip joint and medially rotates thigh.	Superior gluteal nerve.
Gluteus minimus (MIN-i-mus = smallest)	Posterior ilium.	Greater trochanter of femur.	Abducts thigh at hip joint and medially rotates thigh.	Superior gluteal nerve.
Tensor fasciae latae (TEN-sor FA-shē-ē LĀ-tē; *tensor* = makes tense; *fasciae* = of the band; *lat-* = wide)	Iliac crest and anterior superior iliac spine (ASIS).	Lateral condyle of tibia by way of the iliotibial tract.	Flexes and abducts thigh at hip joint; helps lock knee in extension.	Superior gluteal nerve.
Piriformis (pir-i-FOR-mis = pear- shaped)	Anterior sacrum.	Superior border of greater trochanter of femur.	Laterally rotates and abducts thigh at hip joint.	Sacral spinal nerves.
Obturator internus (OB-too-rā′-tor in-TER-nus; *obturator* = obturator foramen; *intern-* = inside)	Inner surface of obturator membrane and its bony margins.	Medial surface of greater trochanter of femur.	Laterally rotates and abducts thigh at hip joint.	Nerve to obturator internus.
Obturator externus (ex-TER-nus = outside)	Outer surface of obturator membrane and its bony margins.	Deep depression inferior to greater trochanter (trochanteric fossa) of femur.	Laterally rotates and abducts thight at hip joint.	Obturator nerve.
Superior gemullus (jem-EL-lus; *superior* = above; *gemall-* = twins)	Ischial spine.	Medial surface of greater trochanter of femur.	Laterally rotates and abducts thigh at hip joint.	Nerve to obturator internus.
Inferior gemellus (*inferior* = below)	Ischial tuberosity.	Medial surface of greater trochanter of femur.	Laterally rotates and abducts thigh at hip joint.	Nerve to quadratus femoris.
Quadratus femoris (kwod-RĀ-tus FEM-or-is; *quad* = square, four-sided; *femoris* = femur)	Ischial tuberosity.	Intertrochanteric crest on posterior femur.	Laterally rotates and stabilizes hip joint.	Nerve to quadratus femoris.
Adductor longus (ad-DUK-tor LONG-us; *adductor* = moves part closer to midline; *longus* = long)	Pubic crest and pubic symphysis.	Middle third of linea aspera of femur.	Adducts and flexes thigh at hip joint and rotates thigh.* RMA: Extends thigh.	Obturator nerve.
Adductor brevis (BREV-is = short)	Inferior ramus of pubis.	Superior half of linea aspera of femur.	Adducts and flexes thigh at hip joint and rotates thigh.* RMA: Extends thigh.	Obturator nerve.
Adductor magnus (MAG-nus = large)	Inferior ramus of pubis and ischium to ischial tuberosity.	Distal linea aspera and medial epicondyle of femur.	Adducts thigh at hip joint and rotates thigh; anterior part flexes thigh at hip joint, and posterior part extends thigh at hip joint.*	Obturator and sciatic nerves.
Pectineus (pek-TIN-ē-us; *pectin* = a comb)	Superior ramus of pubis.	Pectineal line of femur, between lesser trochanter and linea aspera.	Flexes and adducts thigh at hip joint.	Femoral nerve.

*All adductors are unique muscles that cross the thigh joint obliquely from an anterior origin to a posterior insertion. As a result they laterally rotate the hip joint when the foot is off the ground, but medially rotate the hip joint when the foot is on the ground.

Figure 14.1 Muscles of the pelvic region that move the femur.

Most muscles that move the femur originate on the pelvic girdle and insert on the femur.

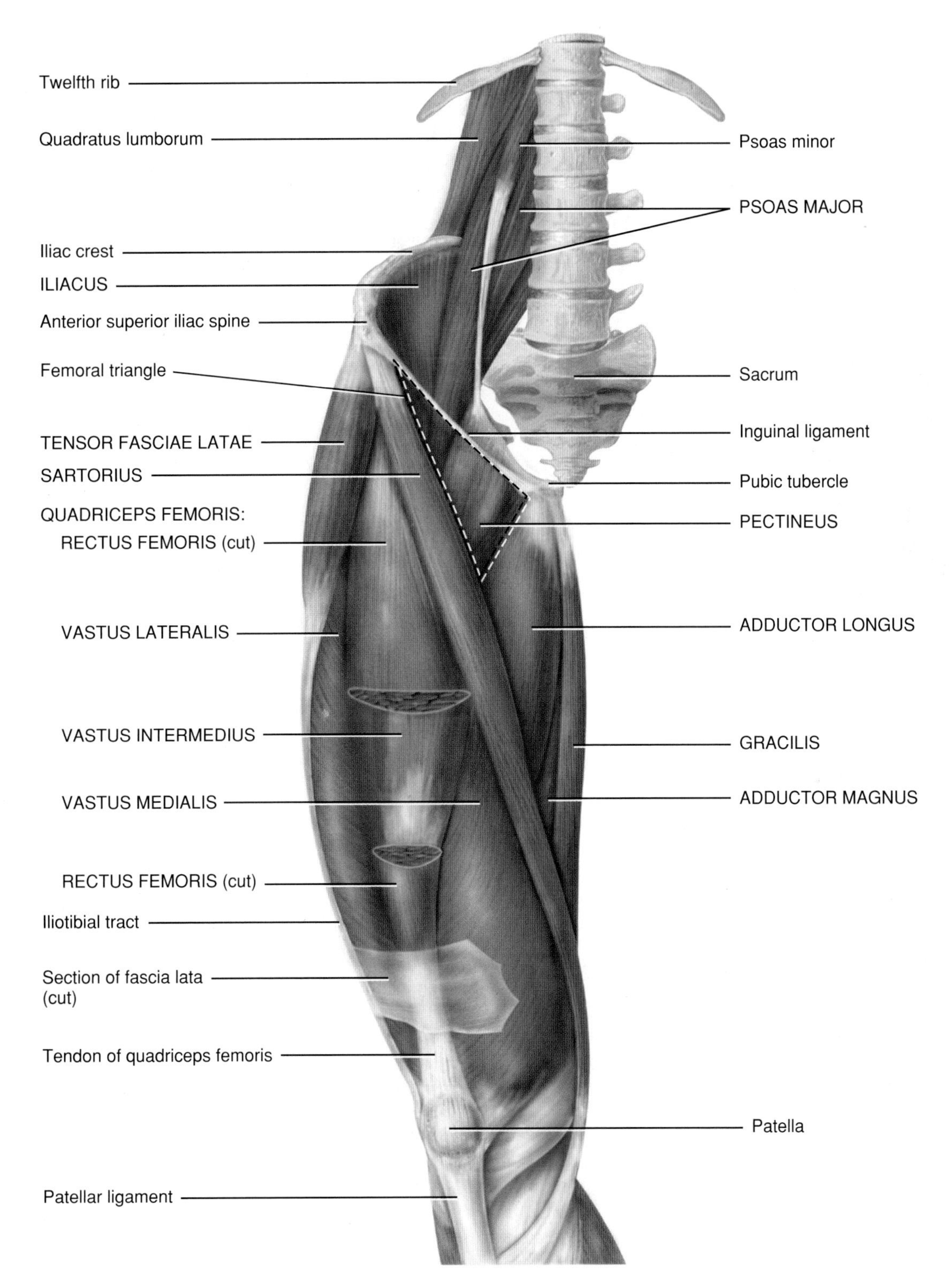

(a) Anterior superficial view (the femoral triangle is indicated by a dashed line)

CONTINUES

Figure 14.1 (continued)

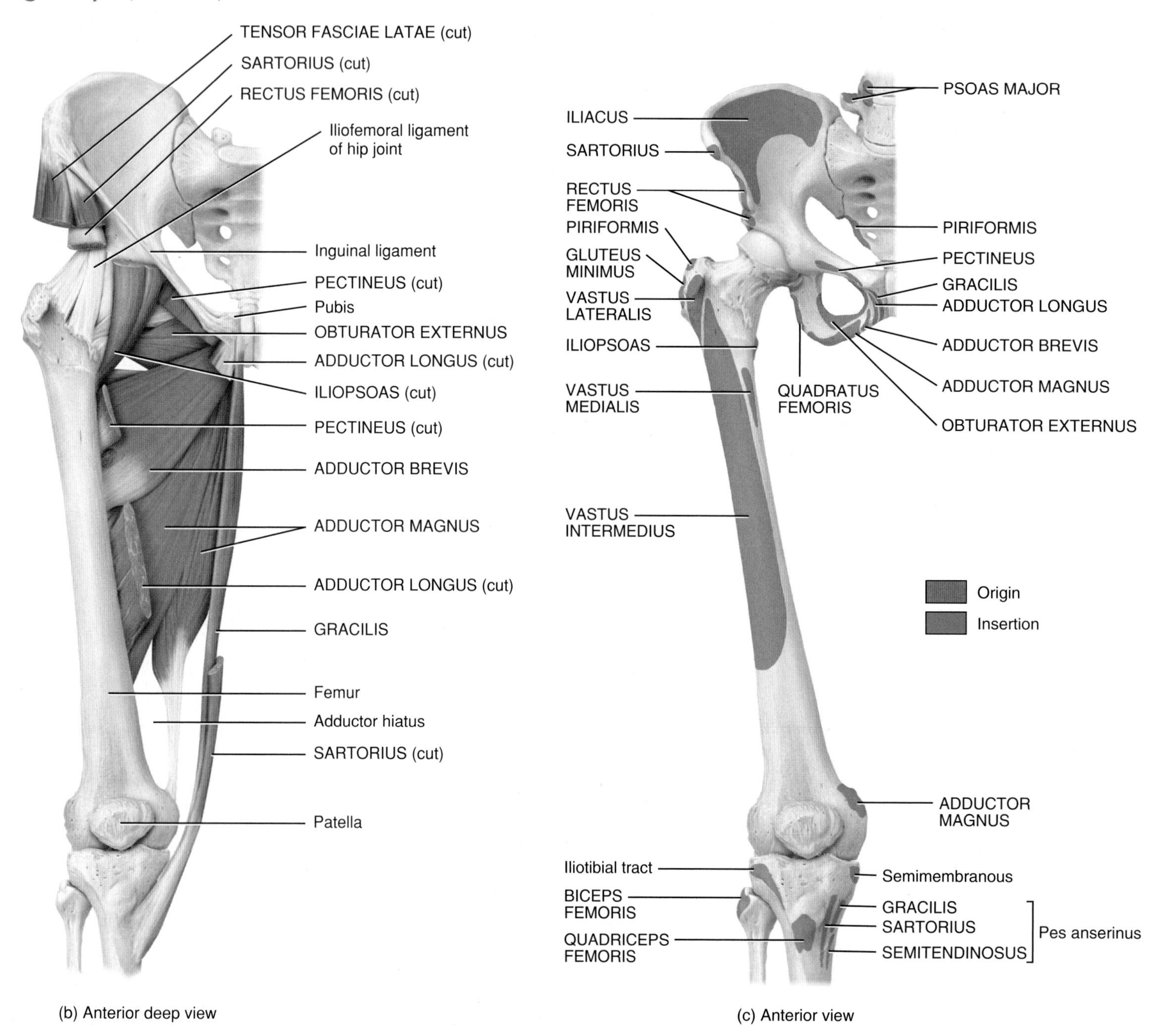

(b) Anterior deep view

(c) Anterior view

Three muscles on the medial aspect of the thigh are the adductor longus, adductor brevis, and adductor magnus. They originate on the pubic bone and insert on the femur. All three muscles adduct, flex, and rotate the femur at the hip joint. These muscles are responsible for adduction of the thigh, but they can also function as either flexors or extensors. The ***adductor magnus*** muscle, as the name implies, is a massive muscle. Some of the fibers are anterior to the plane of the femoral head and therefore assist in flexion. Other fibers are posterior to the plane of the femoral head and assist in extension. The distal tendons of insertion are split just superior to the medial epicondyle of the femur; the space formed by the split is called the ***adductor hiatus*** (Figure 14.1b). As you will see in Chapter 23, the femoral artery courses through the anterior thigh but passes through the adductor hiatus to the posterior compartment. At this point the femoral artery becomes the popliteal artery. The ***adductor longus*** and ***adductor brevis*** muscles assist in the flexion of the thigh from neutral (the anatomical position) to nearly 80°, but they assist in extension when the insertion in the linea aspera becomes superior to the origin on the pubis. Most activities, however, do not require flexion of more than 80° of the thigh at the hip. The ***pectineus*** muscle also adducts and flexes the femur at the hip joint.

Technically, the adductor muscles and pectineus muscles are components of the medial compartment of the thigh and could be included in Exhibit 14.2. However, they are discussed here because they act on the femur.

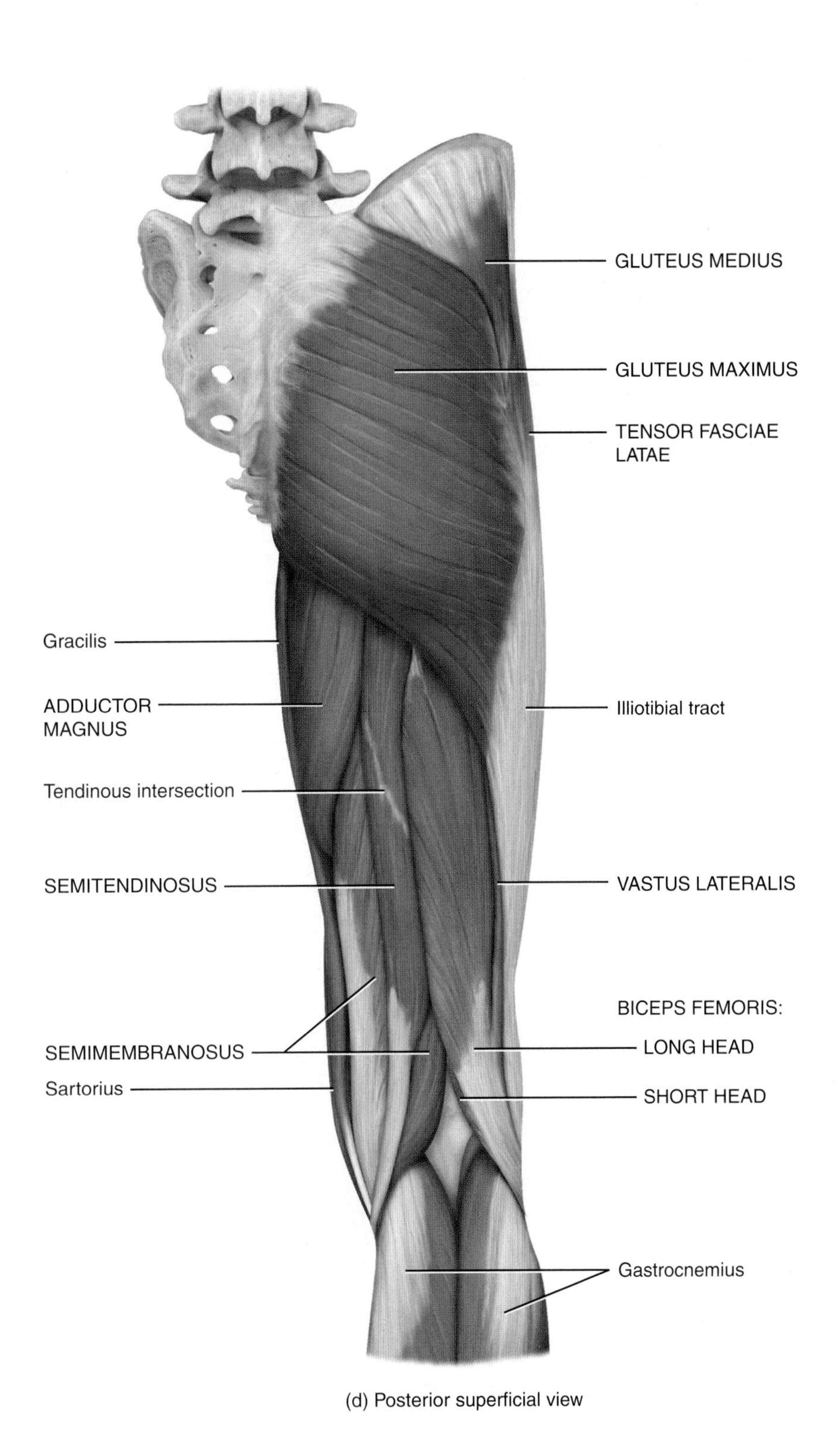

(d) Posterior superficial view

CONTINUES

Figure 14.1 (continued)

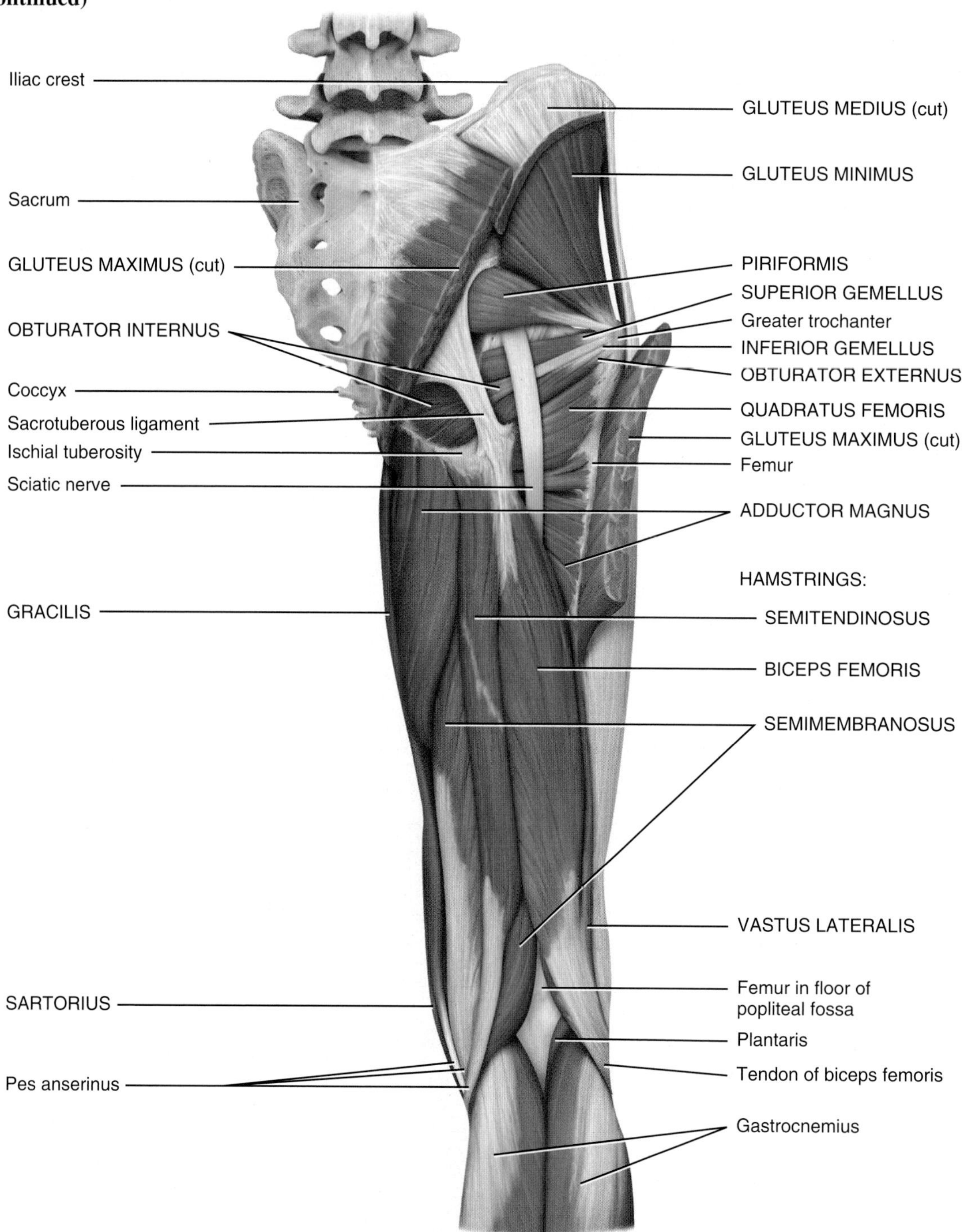

(e) Posterior superficial and deep view

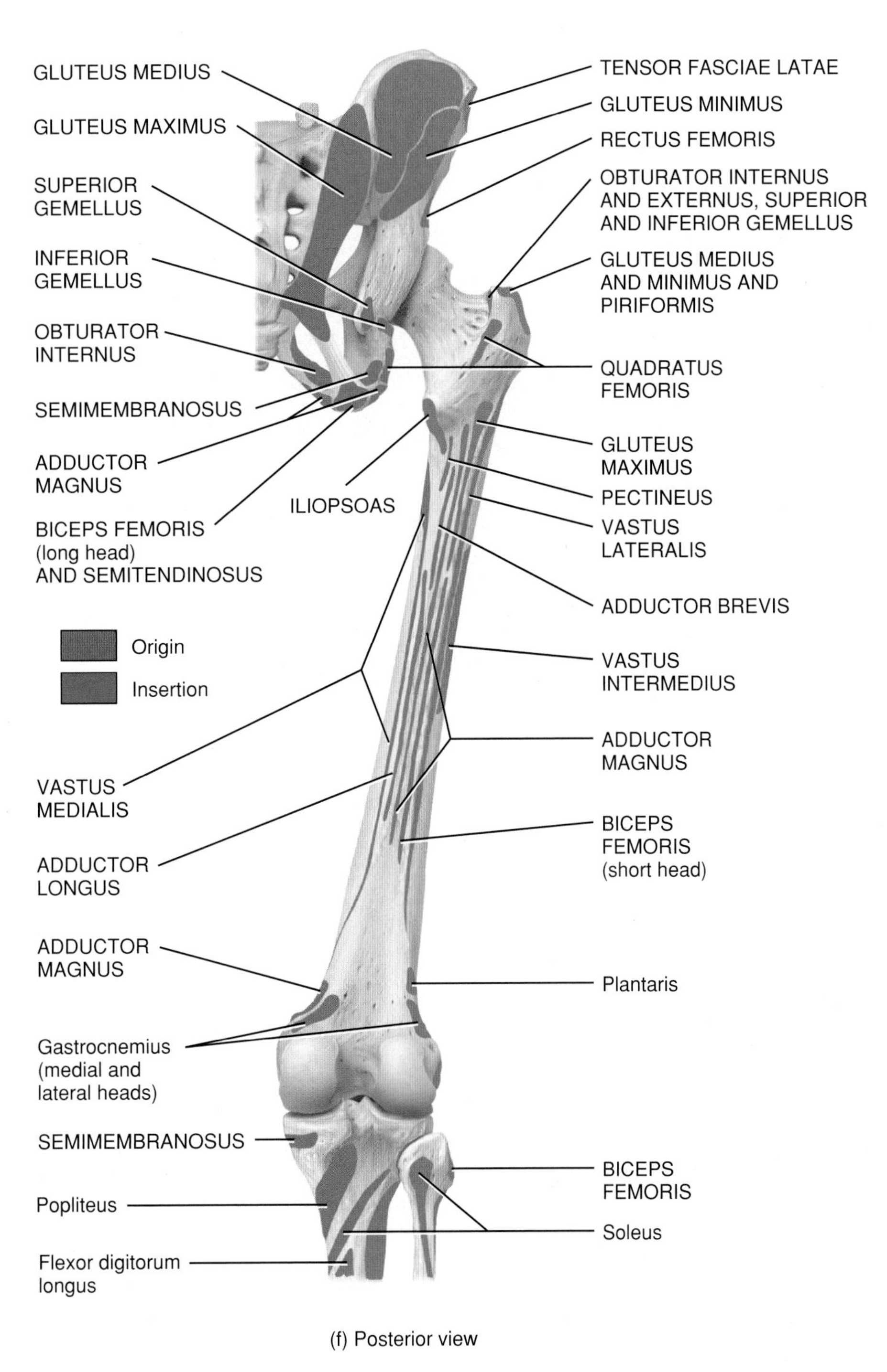

(f) Posterior view

What are the principal differences between the muscles of the upper and lower limbs?

CONTINUES

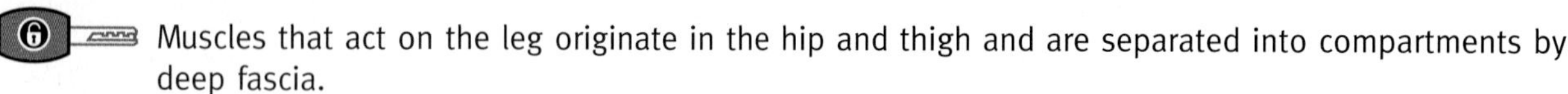

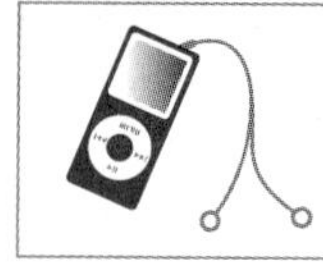

Figure 14.2 Muscles of the gluteal region that move the femur and tibia and fibula.

Muscles that act on the leg originate in the hip and thigh and are separated into compartments by deep fascia.

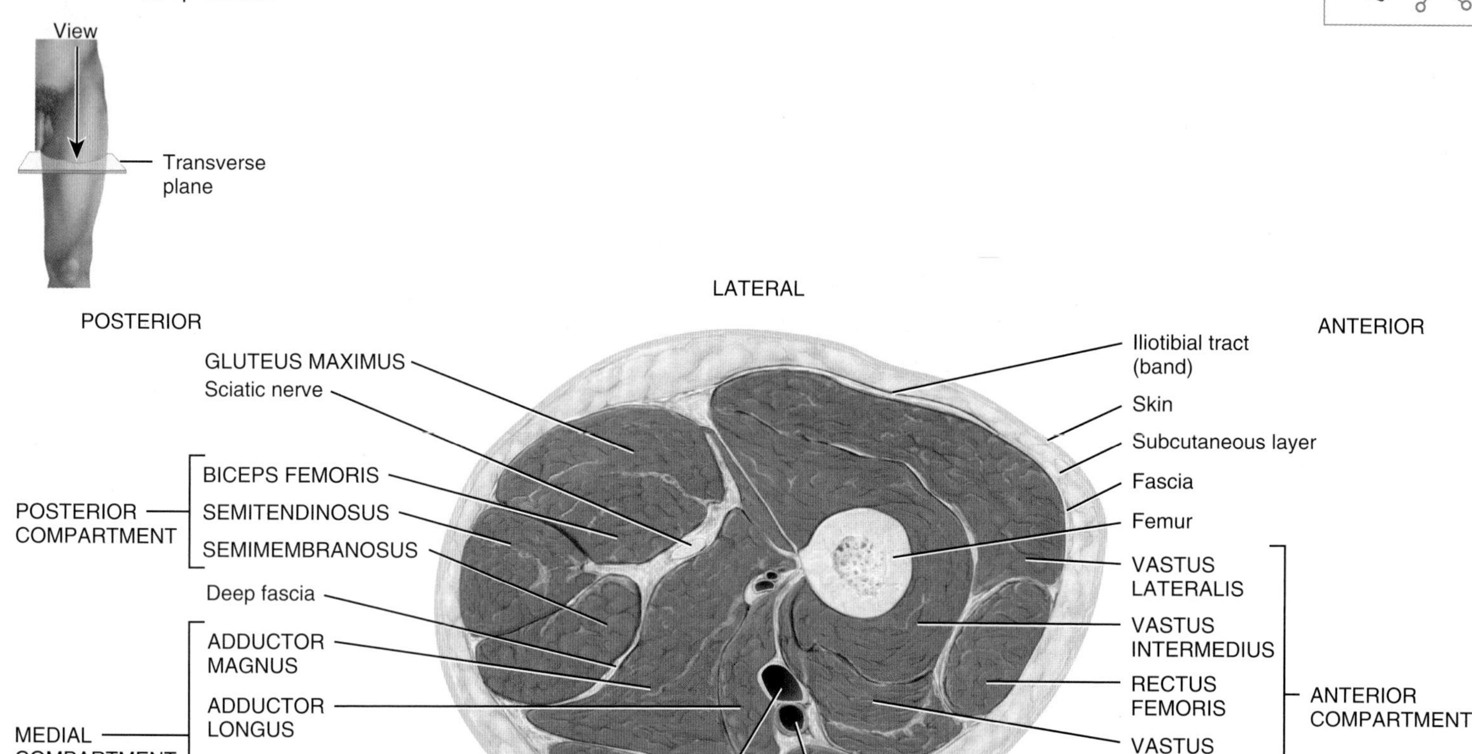

Superior view of transverse section of thigh

What five muscles are found in the medial compartment?

At the junction between the trunk and lower limb is a space called the ***femoral triangle,*** an area that is important for its contents. The base is formed superiorly by the inguinal ligament, medially by the lateral border of the adductor longus muscle, and laterally by the medial border of the sartorius muscle. The apex is formed by the crossing of the adductor longus by the sartorius muscle (see Figure 14.1a). A mnemonic for the femoral triangle is *NAVEL.* Starting laterally the structures include the femoral *N*erve and its branches, the femoral *A*rtery and several of its branches, the femoral *V*ein and its proximal tributaries, an "*E*mpty space," and the deep inguinal *L*ymphatic vessels and nodes. The 'empty space' is actually the femoral canal and is not really empty because it contains the lymphatics and a small amount of fat. The pectineus forms most of the floor of the triangle. The femoral artery is easily accessible within the triangle and is the site for insertion of catheters that may extend into the aorta and ultimately into the coronary vessels of the heart. Such catheters are utilized during cardiac catheterization, coronary angiography, and other procedures involving the heart. Inguinal hernias frequently appear in this area.

The pectineus forms the medial floor of the femoral triangle. Chronic contraction of the pectineus muscle produces pain deep in the groin. As described in Section 11.1, a muscle with a short lever arm is designed for power rather than speed. The belly lies between the iliopsoas and the adductor longus and, not surprisingly, the actions of the pectineus are flexion and adduction of the femur.

Surface Features of the Buttock

The ***buttock,*** or *gluteal region,* is formed mainly by the gluteus maximus muscle. The outline of the superior border of the buttock is formed by the iliac crest. Following are some surface features of the buttock (Figure 14.3).

- **Gluteus maximus muscle.** Forms the major portion of the prominence of the buttock.
- **Gluteus medius muscle.** Superior, anterior, and lateral to the gluteus maxmus muscle. A common site for an intramuscular injection.

- **Gluteal cleft.** Depression along the midline that separates the left and right buttocks.
- **Gluteal fold.** Inferior limit of the buttock that roughly corresponds to the inferior margin of the gluteus maximus muscle.
- **Ischial tuberosity.** Just superior to the medial side of the gluteal fold, the tuberosity bears the weight of the body when seated.
- **Greater trochanter.** A projection of the proximal end of the femur on the lateral side of the thigh. It is about 20 cm (8 in.) inferior to the highest point of the iliac crest.
- **Posterior superior iliac spine (PSIS).** A sharp spine on the posterior aspect of the iliac crest.

MANUAL THERAPY APPLICATION

Groin Pull (Strain) and RICE Therapy

The five major muscles—adductor magnus, adductor longus, adductor brevis, pectineus, and gracilis—of the inner thigh function to move the lower limbs medially. This muscle group is important in activities such as sprinting, hurdling, and horseback riding. A rupture or tear of one or more of these muscles can cause a **groin pull (strain).**

Groin pulls (strains) most often occur during sprinting or twisting, or from kicking a solid, perhaps stationary object. Symptoms of a groin pull (strain) may be sudden or may not surface until the day after the injury, and include sharp pain in the inguinal region, swelling, bruising, or inability to contract the muscles.

Most strains and other sports injuries should be treated initially with ***RICE*** therapy, which stands for ***R***est, ***I***ce, ***C***ompression, and ***E***levation. Immediately apply ice, and rest and elevate the injured part. Then apply an elastic bandage, if possible, to compress the injured tissue. Continue using RICE for 2 to 3 days, and resist the temptation to apply heat, which may worsen the swelling. In consultation with a physician, follow-up treatment may include alternating moist heat and ice to enhance blood flow in the injured area. Sometimes it is helpful to take over-the-counter nonsteroidal anti-inflammatory drugs (NSAIDs) or to have local injections of corticosteroids. During the recovery period, it is important to keep active, using an alternative fitness program that does not worsen the original injury. Finally, careful exercise is needed to rehabilitate the injured area, perhaps under the care of a physical therapist. All such activities on the part of the therapist should be determined for the patient in consultation with a physician.

Relating Muscles to Movements

Arrange the muscles in this exhibit according to the following actions on the thigh at the hip joint: (1) flexion, (2) extension, (3) abduction, (4) adduction, (5) medial rotation, and (6) lateral rotation. The same muscle may be mentioned more than once.

CHECKPOINT

4. What is the origin of most muscles that move the femur?

Figure 14.3 Surface features of the buttock.

6 The gluteus medius muscle is a frequent site for an intramuscular injection.

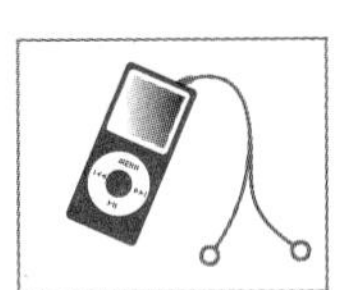

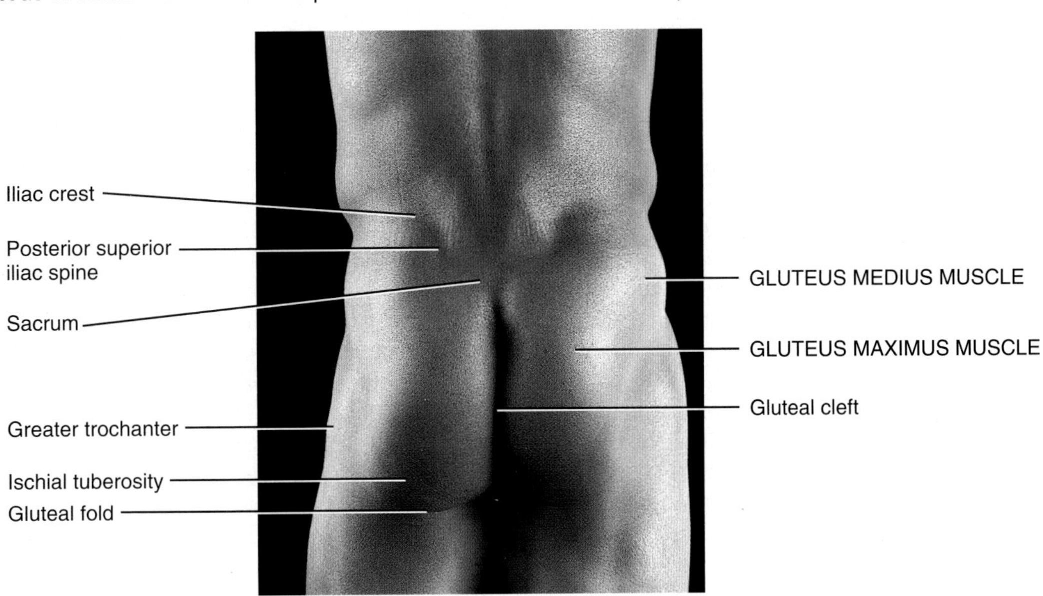

Posterior view of pelvis

What is the direction of the fascicles of the gluteus maximus muscle?

Muscles of the Thigh That Move the Femur, Tibia, and Fibula

OBJECTIVE

- Describe the origin, insertion, action, and innervation of the muscles that act on the femur, tibia, and fibula.
- Describe the major muscles that can be seen and palpated in the thigh and knee regions.

Deep fasciae separates the muscles that act on the femur and tibia and fibula into medial, anterior, and posterior compartments. The muscles of the ***medial (adductor) compartment of the thigh*** adduct the femur at the hip joint (Figure 14.4a). (See also the adductor magnus, adductor longus, adductor brevis, and pectineus, which are components of the medial compartment, in Figures 14.1 and 14.2.) The ***gracilis,*** the other muscle in the medial compartment, not only adducts the thigh, but also flexes the leg at the knee joint. For this reason, it is discussed here. The gracilis is a long, straplike muscle on the medial aspect of the thigh and knee (Figure 14.1a–e). The tendon of insertion is the intermediate structure of the three tendons that compose the *pes anserinus,* as described in the Manual Therapy Application in this exhibit.

The muscles of the ***anterior (extensor) compartment of the thigh*** extend the leg (and flex the thigh). This compartment contains the quadriceps femoris and sartorius muscles (Figure 14.4b). The ***quadriceps femoris*** muscle is the largest muscle in the body, covering most of the anterior surface and sides of the thigh, and it is the great extensor muscle of the leg. The muscle is actually a composite muscle, usually described as four separate muscles: (1) ***rectus femoris,*** on the anterior aspect of the thigh; (2) ***vastus lateralis,*** on the lateral aspect of the thigh; (3) ***vastus medialis,*** on the medial aspect of the thigh; and (4) ***vastus intermedius,*** located deep to the rectus femoris between the vastus lateralis and vastus medialis. The rectus femoris originates on the anterior inferior iliac spine (AIIS) and thus crosses two joints; it joins the iliopsoas and pectineus muscles in their role as flexors of the femur at the hip. Chronic contraction of any portion of this large muscle causes the patella to track abnormally in the patellar groove and thus causes weakness and pain in the knee area. Weakness of the muscle, particularly the vastus medialis and vastus lateralis, also causes abnormal tracking. Physicians or physical therapists can prescribe specific exercises that selectively strengthen just one of the four muscles of the quadriceps.

The common tendon for the four muscles is known as the ***quadriceps tendon,*** which inserts into the patella. The tendon continues below the patella as the ***patellar ligament,*** which attaches to the tibial tuberosity. Elevating the tendon from the condyles improves the angle of pull on the tibial tuberosity. *Patellar tendonitis,* inflammation of the quadriceps tendon, is often reported by a patient having run up an incline prior to the onset of symptoms. Any issue involving abnormal movement of the patella is known as *patellofemoral dysfunction* (see Manual Therapy Application in Section 8.6).

The ***sartorius,*** the longest single muscle in the body, is a narrow muscle that forms a band across the thigh from the ilium of the hip bone to the medial side of the tibia. The various movements it produces (flexion of the leg at the knee joint and flexion, abduction, and lateral rotation at the hip joint) help effect the cross-legged sitting position in which the heel of one limb is placed on the knee of the opposite limb. It originates on the ASIS, crosses the proximal superficial thigh obliquely, descends almost vertically in the distal thigh, crosses the medial condyle of the femur, and inserts into the pes anserinus. The sartorius has *tendinous intersections* similar to that of the rectus abdominis. Many individual muscle fibers are thus shortened. Recall from Section 11.1 the effect of fiber length on movement. The sartorius assists other muscles of the hip in its multiple actions—flexion of the femur at the hip, flexion of the leg at the knee, and lateral rotation of the femur. It is known as the tailor's muscle because tailors often assume a cross-legged sitting position. (Because the major action of the sartorius muscle is to move the thigh rather than the leg, it could have been included in Exhibit 14.1.) Both sartorius muscles are extremely active in sports such as basketball, while the left sartorius muscle is, for example, active during a right-handed tennis serve.

The muscles of the ***posterior (flexor) compartment of the thigh*** flex the leg (and extend the thigh). This compartment is composed of three muscles collectively called the ***hamstrings:*** (1) biceps femoris, (2) semitendinosus, and (3) semimembranosus (Figure 14.4c). The hamstrings extend from a common origin on the ischial tuberosity. The hamstrings are so named because their tendons are long and stringlike in the popliteal area. Because the hamstrings span two joints (hip and knee), the hamstrings extend the hip or flex the knee or both. In walking, as the foot leaves the ground to a take a step forward, the hamstrings initially contract and take the weight of the partially flexed leg. As soon as hip flexion begins, the hamstrings relax and permit knee extension in the advancing limb. It should be noted that when sitting erectly on a hard surface, such as bleachers at a ball game, the majority of the body weight is resting on the tendons of the hamstrings. After a relatively short period of sitting, the hamstring tendons may become painful. A cushion helps distribute the weight of the body over a greater area and the hamstring tendons are thus not as painful.

The ***biceps femoris*** has two heads; the *long head* originates with the tendon of the semitendinosus on the ischial tuberosity and also from the sacrotuberous ligament. The *short head* arises from the linea aspera. The two heads are innervated by two nerves: The tibial portion of the sciatic often sends two branches into the long head, and the short head is innervated by a branch of the common fibular nerve.

The ***semitendinosus,*** as just stated, arises with the tendon of the long head of the biceps femoris from the ischial tuberosity. The rounded tendon of insertion forms the medial border of the popliteal fossa before joining the *pes anserinus*. Since it shares its tendon of origin with the long head of the biceps femoris, it is not surprising that the muscles share the same innervation. In fact, the same branch of the tibial nerve that innervates the biceps femoris often continues directly to the semitendinosus. The semitendinosus, like the sartorius, contains tendinous intersections (see Figure 14.1d) that shorten the fiber length.

The ***semimembranosus*** originates from the ischial tuberosity by a long flat tendon that lies deep to the proximal half of the muscle and adjacent to the tendon of adductor magnus.

The ***popliteal fossa*** (see Figure 14.1e) is a diamond-shaped space on the posterior aspect of the knee bordered laterally by the tendons of the biceps femoris muscle and medially by the tendons of the semitendinosus and semimembranosus muscles.

MUSCLE	ORIGIN	INSERTION	ACTION	INNERVATION
MEDIAL (ADDUCTOR) COMPARTMENT OF THE THIGH				
Adductor magnus (ad-DUK-tor MAG-nus)	See Exhibit 14.1			
Adductor longus (LONG-us)	See Exhibit 14.1			
Adductor brevis (BREV-is)	See Exhibit 14.1			
Pectineus (pek-TIN-ē-us)	See Exhibit 14.1			
Gracilis (GRAS-i-lis; *gracilis* = slender)	Body and inferior ramus of pubis.	Proximal part of medial surface of body of tibia (pes anserinus).	Adducts thigh at hip joint, medially rotates thigh, and flexes leg at knee joint.	Obturator nerve.
ANTERIOR (EXTENSOR) COMPARTMENT OF THE THIGH				
Quadriceps femoris (KWOD-ri-ceps FEM-or-is: *quadriceps* = four heads of origin; *femoris* = femur)		Patella via quadriceps tendon and then tibial tuberosity via patellar ligament.	All four heads extend leg at knee joint; rectus femoris muscle acting alone also flexes thigh at hip joint.	Femoral nerve.
Rectus femoris (REK-tus FEM-or-is; *rectus* = fascicles parallel to midline)	Anterior inferior iliac spine (AIIS).			
Vastus lateralis (VAS-tus lat′-e-RĀ-lis; *vast* = huge; *lateralis* = lateral)	Greater trochanter and linea aspera of femur.			
Vastus medialis (mē-dē-Ā-lis = medial)	Linea aspera of femur.			
Vastus intermedius (in′-ter-MĒ-dē-us = middle)	Anterior and lateral surfaces of body of femur.			
Sartorius (sar-TOR-ē-us; *sartor* = tailor; longest muscle in body)	Anterior superior iliac spine (ASIS).	Proximal part of medial surface of body of tibia (pes anserinus).	Weakly flexes leg at knee joint; weakly flexes, abducts, and laterally rotates thigh at hip joint.	Femoral nerve.
POSTERIOR (FLEXOR) COMPARTMENT OF THE THIGH				
Hamstrings A collective designation for three separate muscles.				
Biceps femoris (BĪ-ceps = two heads of origin)	Long head arises from ischial tuberosity and sacrotuberous ligament; short head arises from linea aspera of femur.	Head of fibula.	Flexes leg at knee joint and extends thigh at hip joint.	Tibial and common fibular (peroneal) nerves from the sciatic nerve.
Semitendinosus (sem′-ē-ten-di-NŌ-sus; *semi-* = half; *-tendo* = tendon)	Ischial tuberosity.	Proximal part of medial surface of shaft of tibia (pes anserinus).	Flexes leg at knee joint and extends thigh at hip joint.	Tibial nerve from the sciatic nerve.
Semimembranosus (sem′-ē-mem-bra-NŌ-sus; *-membran-* = membrane)	Ischial tuberosity.	Medial condyle of tibia.	Flexes leg at knee joint and extends thigh at hip joint.	Tibial nerve from the sciatic nerve.

CONTINUES

Figure 14.4 Muscles of the thigh that act on the femur, tibia, and fibula.

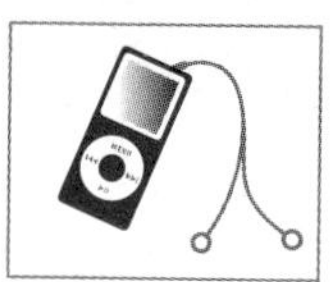

Muscles that act on the leg originate in the hip and thigh and are separated into compartments by deep fascia.

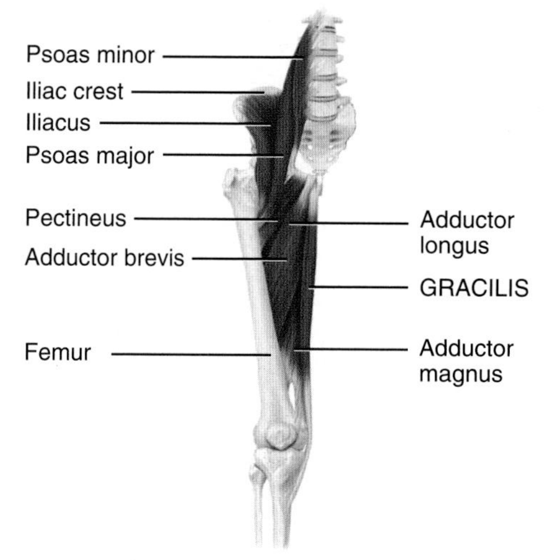

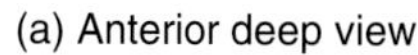
(a) Anterior deep view

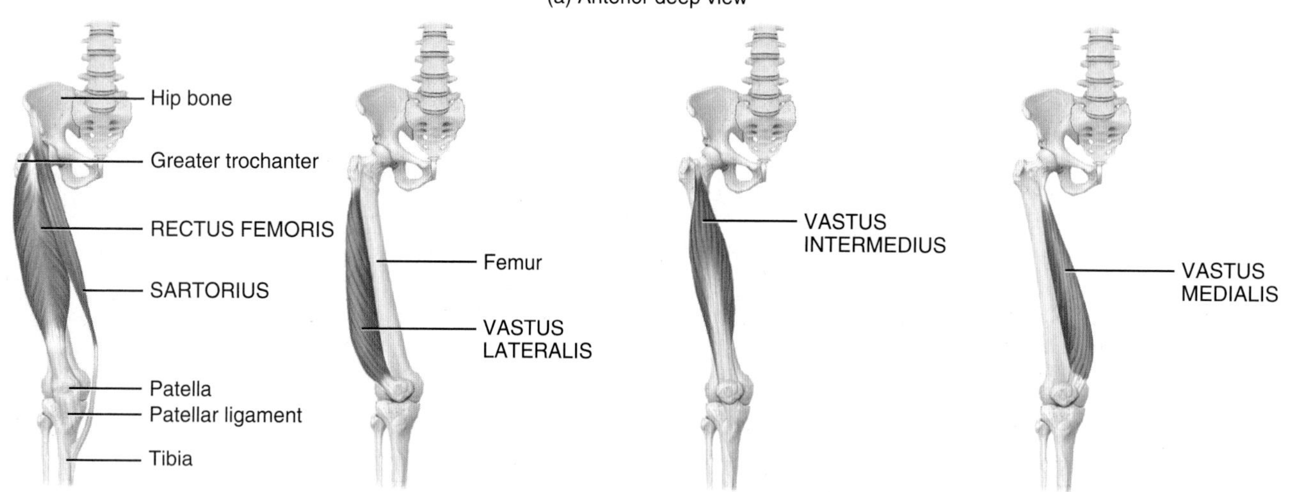

(b) Anterior views

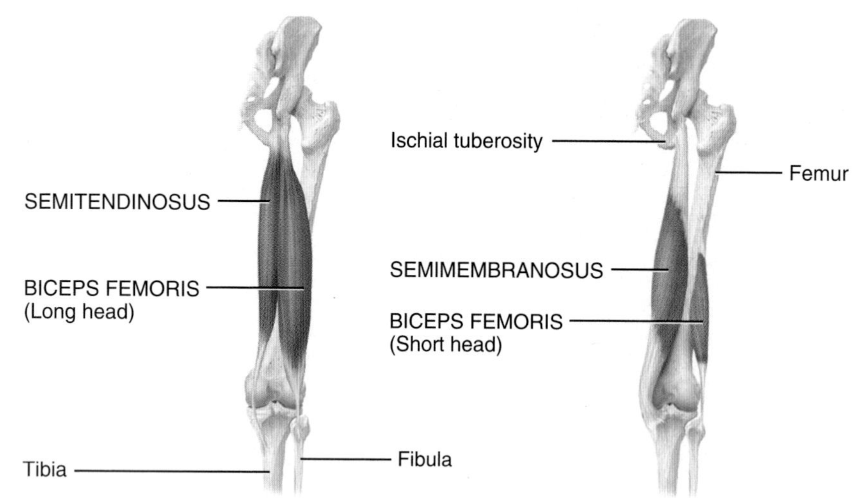

(c) Posterior deep views

Which of these muscles cross two joints?

Surface Features of the Thigh and Knee

The ***thigh,*** or *femoral region,* is the region from the hip to the knee. The ***knee,*** or *genu,* is the region where the thigh and leg join. Several muscles are clearly visible in the thigh. Following are several surface features of the thigh and knee (Figure 14.5).

- **Sartorius muscle.** Superficial anterior muscle that can be traced from the lateral aspect of the thigh to the medial aspect of the knee.
- **Quadriceps femoris muscle.** Three of the four components of the muscle can be seen: **rectus femoris** at the midpoint of the anterior aspect of the thigh; **vastus medialis** at the anteromedial aspect of the thigh; and **vastus lateralis** at the anterolateral aspect of the thigh (see Figure 14.4b).
- **Adductor longus muscle.** Located at the superior aspect of the medial thigh. It is the most anterior of the three adductor muscles (see Figure 14.4a).
- **Femoral triangle.** A space at the proximal end of the thigh formed by the inguinal ligament superiorly, the sartorius muscle laterally, and the adductor longus muscle medially.
- **Hamstring muscles.** Superficial, posterior thigh muscles located below the gluteal folds. The tendons of the hamstring muscles can be palpated laterally and medially on the posterior aspect of the knee.

The knee presents several distinguishing surface features.

- **Patella.** Also called the *kneecap,* this large sesamoid bone is located within the tendon of the quadriceps femoris muscle on the anterior surface of the knee along the midline.
- **Patellar ligament.** Continuation of the quadriceps femoris tendon inferior to the patella. Infrapatellar fat pads cushion it on both sides.
- **Medial condyle of femur.** Medial projection on the distal end of the femur.
- **Medial condyle of tibia.** Medial projection on the proximal end of the tibia.
- **Lateral condyle of femur.** Lateral projection on the distal end of the femur.
- **Lateral condyle of tibia.** Lateral projection on the proximal end of the tibia. All four condyles can be palpated just inferior to the patella on either side of the patellar ligament.

Figure 14.5 Surface features of the thigh and knee.

The quadriceps femoris and hamstrings form the bulk of the musculature of the thigh.

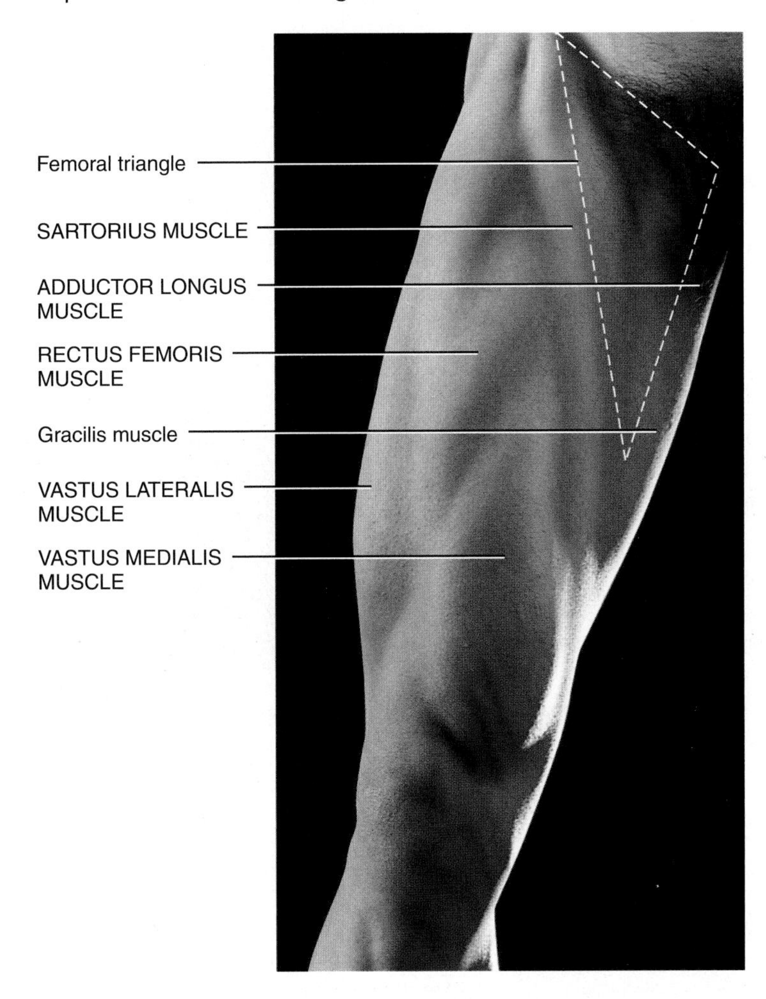

(a) Anterior view of thigh

CONTINUES

Figure 14.5 (continued)

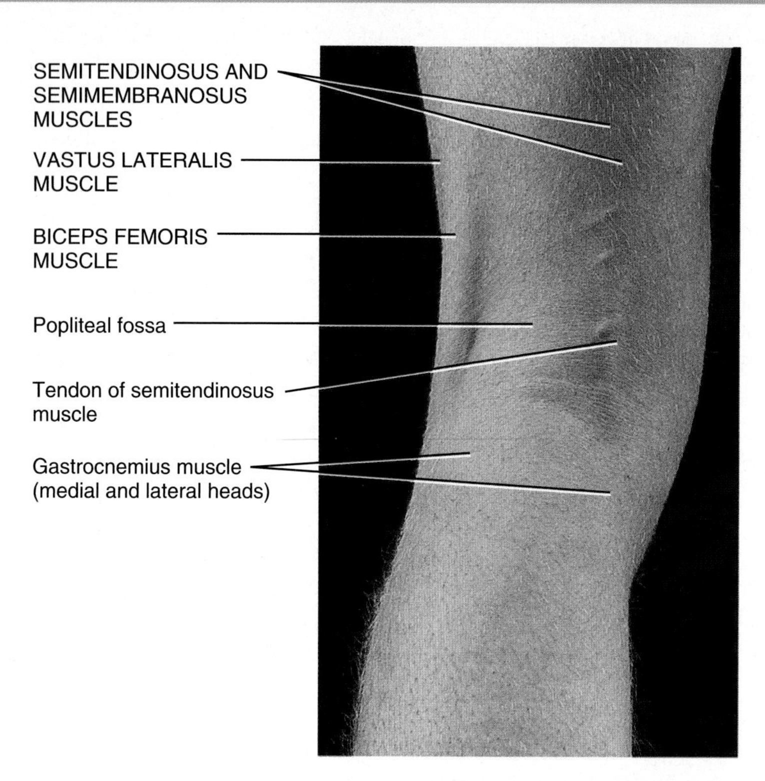

(b) Posterior view of popliteal fossa

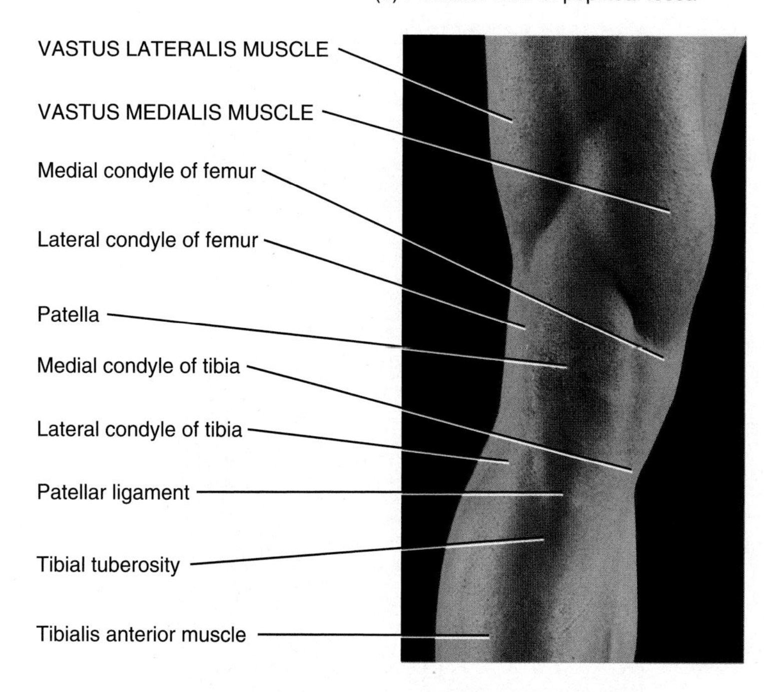

(c) Anterior view of knee

Which muscles form the borders of the popliteal fossa?

- **Popliteal fossa** (pop-LIT-ē-al). A diamond-shaped area on the posterior aspect of the knee that is clearly visible when the knee is flexed. The fossa is bordered superolaterally by the biceps femoris muscle, superomedially by the semimembranosus and semitendinosus muscles, and inferolaterally and inferomedially by the lateral and medial heads of the gastrocnemius muscle, respectively. The **head of the fibula** can be easily palpated on the lateral side of the popliteal fossa. The fossa also contains the popliteal artery and vein. It is sometimes possible to detect a pulse in the popliteal artery. The presence of this superficial arterial blood supply dictates that deep pressure not be applied by therapists. Some techniques encourage therapists to work on the posterior leg, jump over the popliteal fossa, and continue on the posterior thigh.

• CLINICAL CONNECTION | Pulled Hamstrings

A strain or partial tear of the proximal hamstring muscles is referred to as **pulled hamstrings** or **hamstring strains.** Like pulled groins (see Exhibit 14.1), they are common sports injuries in individuals who run very hard and/or are required to perform quick starts and stops. Sometimes the violent muscular exertion required to perform a feat tears away a part of the tendinous origins of the hamstrings, especially the biceps femoris, from the ischial tuberosity. This is usually accompanied by a contusion (bruising), tearing of some of the muscle fibers, and rupture of blood vessels, producing a hematoma (collection of blood) and sharp pain. Adequate training with good balance between the quadriceps femoris and hamstrings and stretching exercises before running or competing are important in preventing this injury. •

Relating Muscles to Movements

Arrange the muscles in this exhibit according to the following actions on the thigh at the hip joint: (1) abduction, (2) adduction, (3) lateral rotation, (4) flexion, and (5) extension; and according to the following actions on the leg at the knee joint: (1) flexion and (2) extension. The same muscle may be mentioned more than once.

CHECKPOINT

5. Which muscles are part of the medial, anterior, and posterior compartments of the thigh?

MANUAL THERAPY APPLICATION

Pes Anserinus

Pes anserinus (pes an-ser-ĪN-us; *pes* = footlike; *anserinus* = goose), also known as the *superficial pes anserinus,* is a site of attachment for three muscles: sartorius, gracilis, and semitendinosus. It is located just deep to the skin at the medial border of the tibial tuberosity (see Figure 14.1e). The tendons of the three muscles attach in series and reminded early anatomists of the webbing of a goose's foot; therefore, the structure was named *pes anserinus*. It is a spot that is tender in many persons on palpation by a manual therapist. Tenderness of pes anserinus signals possible spasm, increased tension, unbalanced strength, or excessive physical strain or activity in one or more of the three named muscle bellies. It should be noted that the three muscles are located in three different compartments of the thigh, have three different nerves, and attach to three outlying points of the pelvis. This is important because it shows that the pelvis is being stabilized from several points of origin, which is a testament to the significance of the pes anserinus. The **sartorius** muscle is in the anterior compartment, is innervated by the femoral nerve, and arises from the most lateral point of the pelvis, the anterior superior iliac spine (ASIS). The **gracilis** muscle is in the medial compartment, is innervated by the obturator nerve, and arises from the body and inferior ramus near the pubic symphysis, the most medial point of the pelvis. The **semitendinosus** muscle is in the posterior compartment, is innervated by the tibial nerve from the sciatic nerve, and arises from the ischial tuberosity, the most posterior point of the pelvis. This arrangement of the three muscle bellies resembles an inverted tripod; they help to stabilize the pelvis as well as reinforce the medial knee.

Muscles of the Leg That Move the Foot and Toes

OBJECTIVE

- Describe the origin, insertion, action, and innervation of the muscles that move the foot and toes.

Muscles that move the foot and toes are located in the leg (Figure 14.6). The muscles of the leg, like those of the thigh, are divided by deep fascia into three compartments: anterior, lateral, and posterior. The ***anterior compartment of the leg*** consists of muscles that dorsiflex the foot. In a situation analogous to the wrist, the tendons of the muscles of the anterior compartment are held firmly to the ankle by thickenings of deep fasciae called the ***superior extensor retinaculum*** (*transverse ligament of the ankle*) and ***inferior extensor retinaculum*** (*cruciate ligament of the ankle*).

The extensor muscles lie lateral to the tibia and anterior to the fibula. These extensor muscles extend the toes, meaning that they straighten or pull the toes upward. The same action occurs at the ankle joint, but pulling the entire foot superiorly at the ankle is difficult to call extension when the ankle joint is bending. Therefore, the terms ***dorsiflexion*** and ***plantar flexion*** are used to describe the hinge action at the ankle joint (see Section 9.5). There are essentially five extensor muscles. Three are the ***tibialis anterior, extensor digitorum longus,*** and ***extensor hallucis longus*** muscles. The inferior and lateral part of the extensor digitorum longus is separated slightly and is known as ***fibularis (peroneus) tertius.*** The ***extensor digitorum brevis,*** confined to the dorsum of the foot, is also an extensor of the toes and is included in this group of muscles. The ***deep fibular (peroneal) nerve*** enters the anterior compartment through the superior portion of the interosseus membrane. It supplies all five of these muscles. The tendons of all but the extensor digitorum brevis are held in position by superior and inferior extensor retinacula. These retinacula redirect the force of the tendons and keep the tendons from bow-stringing or slipping out of place. The tendons of these muscles are enclosed within synovial sheaths or tubular bursae. These structures may become inflamed and swollen and cause tenosynovitis (see Exhibit 13.3).

Within the anterior compartment the extensor muscles lie lateral to the tibia and anterior to the fibula (Figure 14.6g). The tibialis anterior is a long, thick muscle against the lateral surface of the tibia, where it is easy to palpate. The extensor hallucis longus is a thin muscle between and partly deep to the tibialis anterior and extensor digitorum longus muscles. This featherlike muscle is lateral to the tibialis anterior muscle, where it can also be palpated easily. The fibularis (peroneus) tertius muscle is embryologically part of the extensor digitorum longus, with which it shares a common origin.

The ***lateral (fibular) compartment of the leg*** contains two muscles that plantar flex and evert the foot: the ***fibularis (peroneus) longus*** and ***fibularis (peroneus) brevis.*** The bellies of both of these muscles insert into the foot via tendons that pass posterior to the lateral malleolus. The lateral malleolus acts as a pulley and redirects the force of these two muscles. The fibularis brevis attaches to the base of the fifth metatarsal. The fibularis longus moves from lateral to medial by lying in a groove of the plantar cuboid. It inserts into the base of the plantar surface of the first metatarsal. The two muscles plantar flex and evert the foot. The importance of eversion and inversion of the foot can be seen when walking over an uneven surface. These muscles will maintain the weight of the body for a short time. When the brain is unaware of uneven surfaces or the severity of the unevenness, the muscles responsible for eversion and inversion do not work appropriately and walking on the uneven terrain may result in a sprained ankle.

The ***posterior compartment of the leg*** consists of muscles in superficial and deep groups (Figure 14.6d,e). The superficial muscles share a common tendon of insertion, the ***calcaneal (Achilles) tendon,*** the strongest tendon of the body. It inserts into the calcaneal bone of the ankle. In many people this tendon is approximately the size of their index finger, and a healthy tendon can sustain a weight of nearly a ton. Rupture of such a strong structure is usually the result of repetitive chronic loads such as might occur in athletes. The blood supply of the tendon may be compromised by chronic or repetitive trauma, although each event may be minor. Compromised blood supply to an organ (see Section 4.4) may result in degeneration and gradual weakening of this tendon. The whiplike sound of a rupturing tendon is loud enough that spectators behind home plate in a baseball stadium may hear the snap of a ruptured tendon of the center fielder who is attempting to catch a ball.

The superficial and most of the deep muscles plantar flex the foot at the ankle joint. The superficial muscles of the posterior compartment are the gastrocnemius, soleus, and plantaris—the so-called calf muscles. The large size of these muscles is directly related to the characteristic upright stance of humans. The ***gastrocnemius*** is the most superficial muscle and forms the prominence of the calf. This muscle has two bellies that originate on the medial and lateral epicondyles of the femur. The two bellies fuse, and at midcalf the muscle becomes a wide and flat tendon. The muscle fibers of the gastrocnemius are so short that the muscle is unable to flex the knee and plantar flex the foot at the same time. Recall from Chapter 10 that short muscle fibers have a small range of motion. As is stated in the tabular portion of this exhibit, the gastrocnemius can accomplish both movements—plantar flexion of the foot and flexion of the knee—but never at the same time. The ***soleus,*** which lies deep to the gastrocnemius, is broad and flat (Figure 14.6g). It derives its name from its resemblance to a flat fish (sole). The soleus muscle is the more powerful of the two muscles and is solely responsible for plantar flexion of the foot when the knee is bent. The ***plantaris*** is a small muscle that may be absent; conversely, sometimes there are two of them in each leg. Its belly is superior to the lateral head of gastrocnemius, and its tendon, sometimes called the "freshman nerve," is very long and thin.

Muscles of the deep posterior compartment include the popliteus, a weak flexor of the knee, and three plantar flexors of the foot and toes, the tibialis posterior, flexor hallucis longus, and flexor digitorum longus (Figure 14.6g). The ***popliteus*** is a triangular muscle that forms the floor of the popliteal fossa. The ***tibialis posterior*** is the deepest muscle in the posterior compartment. It lies between the flexor digitorum longus and flexor hallucis longus muscles. The tibialis posterior helps maintain the longitudinal and transverse arches of the foot, and this action helps explain possible spasms and tightness of the muscle as a result of repetitive use or abuse of the feet. Spasms of the tibialis posterior can give rise to pain that is called shin splints. The ***flexor digitorum longus*** is smaller than the ***flexor hallucis longus,*** even though the former flexes four toes and the latter flexes only the great toe at the interphalangeal joint. The flexor hallucis longus lies on the fibular side of the leg. Due to the strength of this muscle it is sometimes called the "push-off" muscle since it is responsible for much of the spring of the step. Tendons of these muscles course around the posterior and inferior surfaces of the medial malleolus of the tibia.

MUSCLE	ORIGIN	INSERTION	ACTION	INNERVATION
ANTERIOR COMPARTMENT OF THE LEG				
Tibialis anterior (tib′-ē-Ā-lis an-TĒR-ē-or; *tibialis* = tibia; *anterior* = front)	Lateral condyle and body of tibia and interosseous membrane (sheet of fibrous tissue that holds shafts of tibia and fibula together).	First metatarsal and first (medial) cuneiform.	Dorsiflexes foot at ankle joint and inverts (supinates) foot at intertarsal joints.	Deep fibular (peroneal) nerve.
Extensor hallucis longus (eks-TEN-sor HAL-ū-sis LON-gus; *extensor* = increases angle at joint; *halluc-* = hallux or great toe; *longus* = long)	Anterior surface of middle third of fibula and interosseous membrane.	Distal phalanx of great toe.	Dorsiflexes foot at ankle joint and extends proximal phalanx of great toe at metatarsophalangeal joint.	Deep fibular (peroneal) nerve.
Extensor digitorum longus (di′-ji-TOR-um LON-gus; digit- = finger or toe)	Lateral condyle of tibia, anterior surface of fibula, and interosseous membrane.	Middle and distal phalanges of toes 2–5.*	Dorsiflexes foot at ankle joint and extends distal and middle phalanges of each toe at interphalangeal joints and proximal phalanx of each toe at metatarsophalangeal joint.	Deep fibular (peroneal) nerve.
Fibularis (peroneus) tertius (fib-ū-LĀ-ris; per′-Ō-nē-us; *peron-* = fibula; TER-shus; *tertius* = third)	Distal third of fibula and interosseous membrane.	Base of fifth metatarsal.	Dorsiflexes foot at ankle joint and everts (pronates) foot at intertarsal joints.	Deep fibular (peroneal) nerve.
LATERAL (FIBULAR) COMPARTMENT OF THE LEG				
Fibularis (peroneus) longus.	Head and body of fibula.	First metatarsal and first cuneiform.	Plantar flexes foot at ankle joint and everts (pronates) foot at intertarsal joints.	Superficial fibular (peroneal) nerve.
Fibularis (peroneus) brevis (BREV-is = short).	Distal half of body of fibula.	Base of fifth metatarsal.	Plantar flexes foot at ankle joint and everts (pronates) foot at intertarsal joints.	Superficial fibular (peroneal) nerve.
SUPERFICIAL POSTERIOR COMPARTMENT OF THE LEG				
Gastrocnemius (gas′-trok-NĒ-mē-us; *gastro-* = belly; *cnem* = leg).	Lateral and medial condyles of femur and capsule of knee.	Calcaneus by way of calcaneal (Achilles) tendon.	Plantar flexes foot at ankle joint and flexes leg at knee joint.	Tibial nerve.
Soleus (SŌ-lē-us; *sole* = a type of flat fish).	Head of fibula and medial border of tibia.	Calcaneus by way of calcaneal (Achilles) tendon.	Plantar flexes foot at ankle joint.	Tibial nerve.
Plantaris (plan-TĀR-is = sole).	Lateral epicondyle of femur.	Calcaneus by way of calcaneal (Achilles) tendon.	Plantar flexes foot at ankle joint and flexes leg at knee joint.	Tibial nerve.
DEEP POSTERIOR COMPARTMENT OF THE LEG				
Popliteus (pop-LIT-ē-us = the back of the knee)	Lateral condyle of femur.	Proximal tibia.	Flexes leg at knee joint and medially rotates tibia to unlock the extended knee.	Tibial nerve.
Tibialis posterior (*posterior* = back).	Proximal tibia, fibula, and interosseous membrane.	Second, third, and fourth metatarsals; navicular; and all three cuneiform.	Plantar flexes foot at ankle joint and inverts (supinates) foot at intertarsal joints.	Tibial nerve.
Flexor digitorum longus (FLEK-sor = decreases angle at joint).	Middle third of posterior surface of tibia.	Distal phalanges of toes 2–5.	Plantar flexes foot at ankle joint; flexes distal and middle phalanges of each toe at interphalangeal joints and proximal phalanx of each toe at metatarsophalangeal joint.	Tibial nerve.
Flexor hallucis longus.	Inferior two-thirds of posterior portion of fibula.	Distal phalanx of great toe.	Plantar flexes foot at ankle joint; Flexes distal phalanx of great toe at interphalangeal joint and proximal phalanx of great toe at metatarsophalangeal joint.	Tibial nerve.

*Reminder:The great toe or hallux is the first toe and has two phalanges: proximal and distal. The remaining toes are numbered II–V (2–5), and each has three phalanges: proximal, middle, and distal.

CONTINUES

Figure 14.6 Muscles of the leg that move the foot and toes.

The superficial muscles of the posterior compartment share a common tendon of insertion, the calcaneal (Achilles) tendon, that inserts into the calcaneal bone of the ankle.

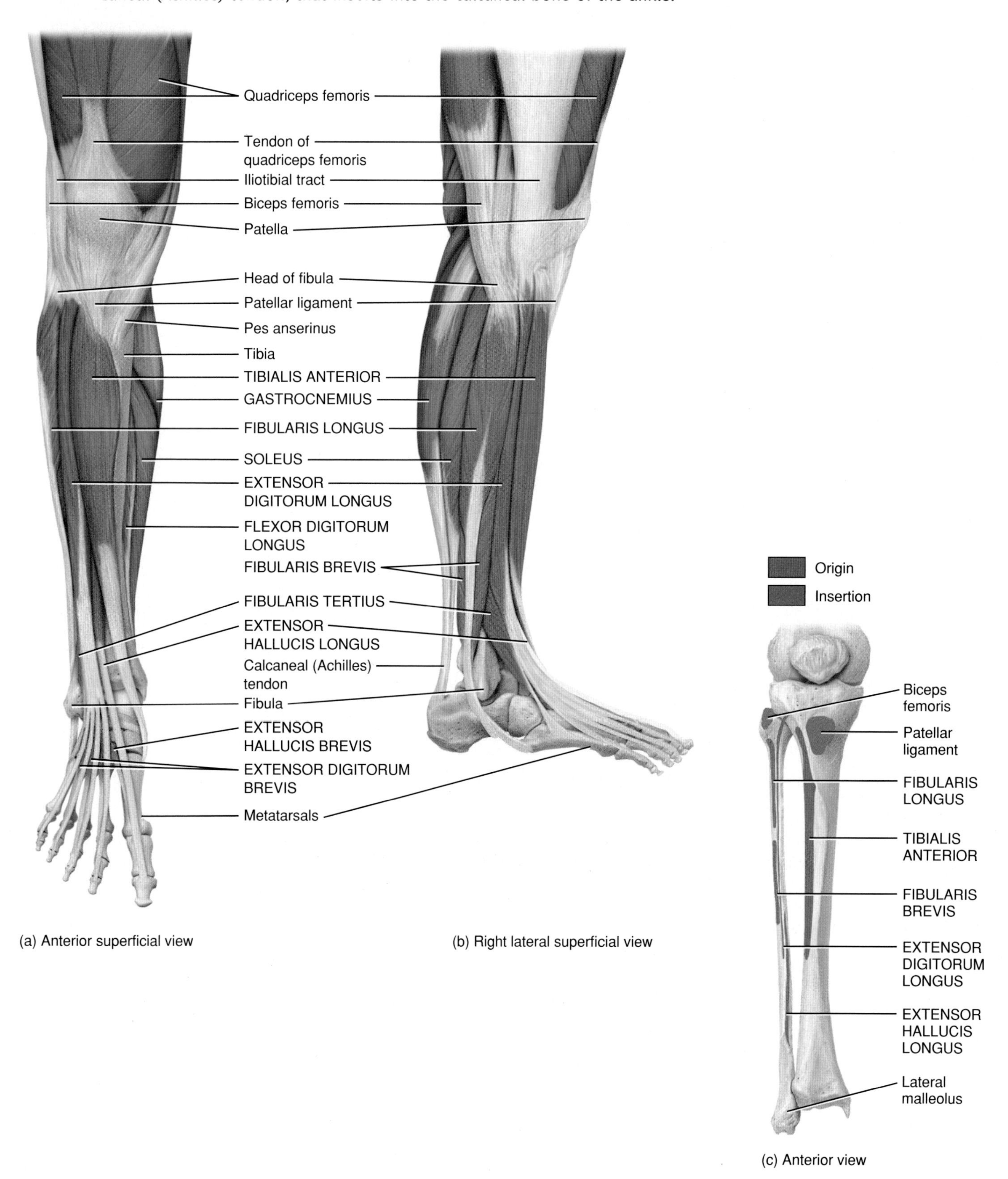

(a) Anterior superficial view

(b) Right lateral superficial view

(c) Anterior view

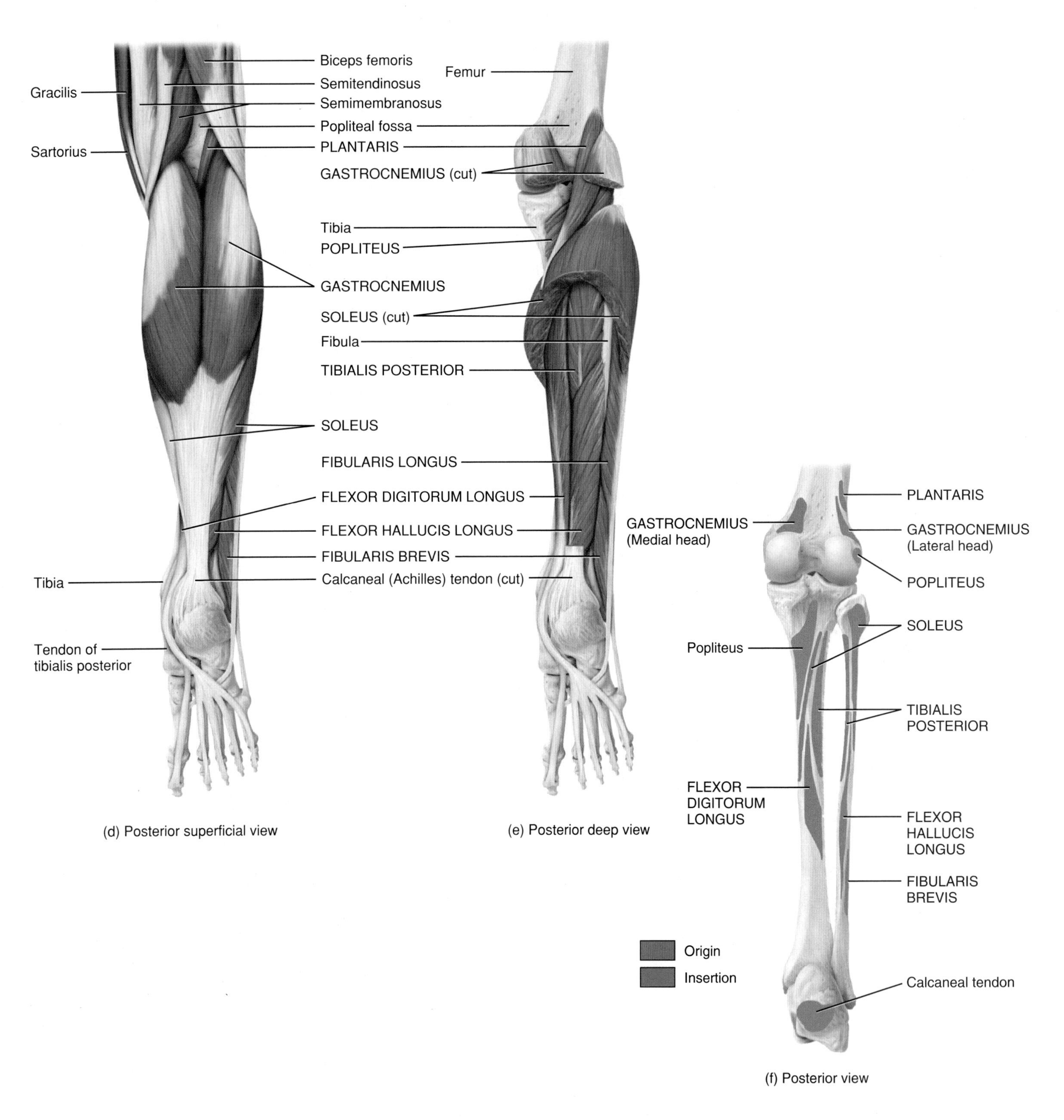

(d) Posterior superficial view

(e) Posterior deep view

(f) Posterior view

CONTINUES

Figure 14.6 (continued)

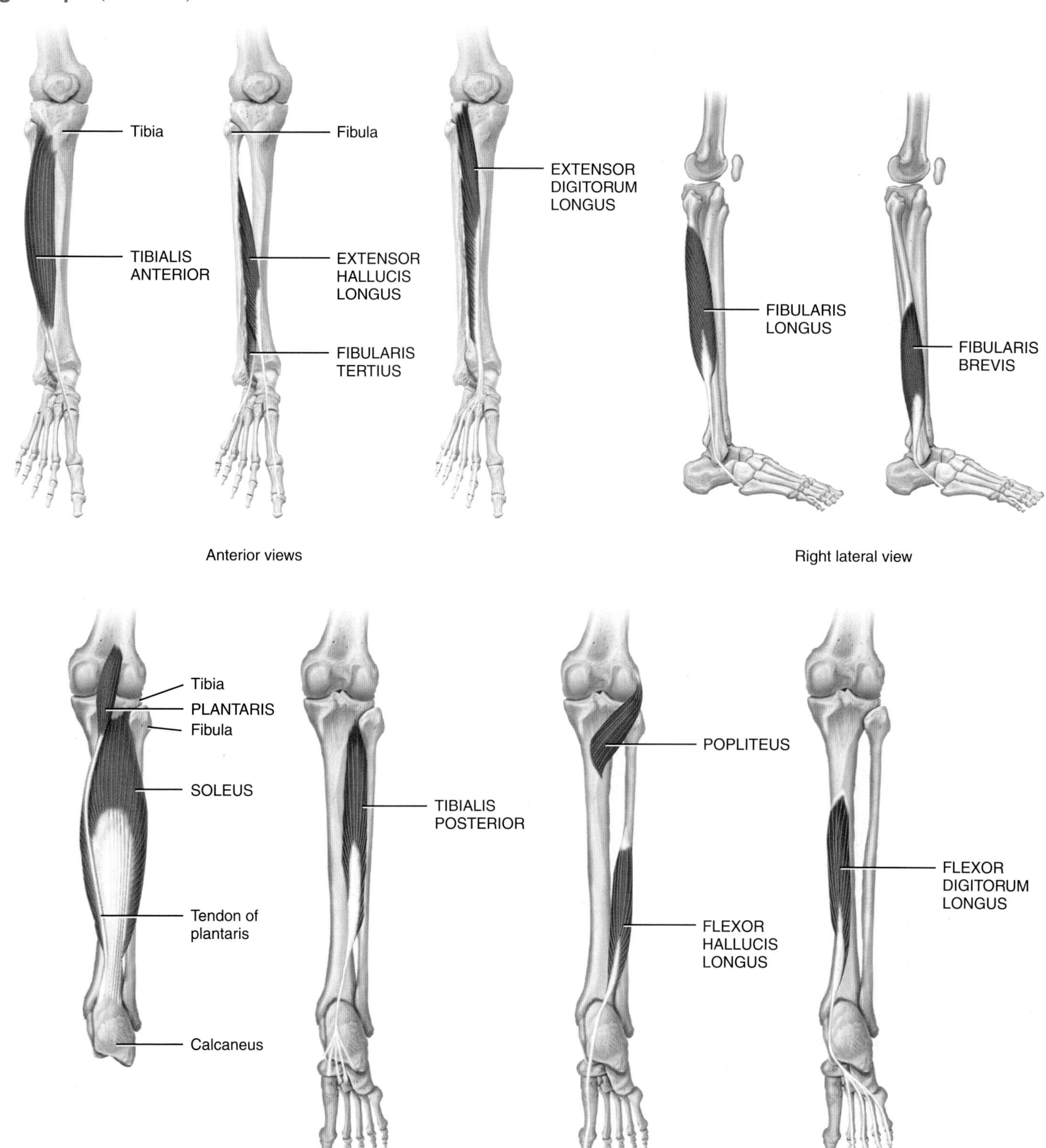

(g) Anterior, right lateral, and posterior views

What structures firmly hold the tendons of the anterior compartment muscles to the ankle?

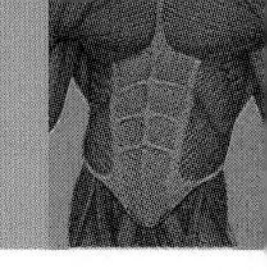

Surface Features of the Leg, Ankle, and Foot

The *leg,* or *crus,* is the region between the knee and ankle. The ***ankle,*** or *tarsus,* is between the leg and foot. The ***foot*** is the region from the ankle to the termination of the free lower limb. Following are several surface anatomy features of the leg, ankle, and foot (Figure 14.7).

- **Tibial tuberosity.** Bony prominence on the superior, anterior surface of the tibia into which the patellar ligament inserts.
- **Tibialis anterior muscle.** Lies against the lateral surface of the tibia, where it is easy to palpate, particularly when the foot is dorsiflexed.
- **Tibia.** The medial surface and anterior border (shin) of the tibia are subcutaneous and can be palpated throughout the length of the bone.
- **Fibularis (peroneus) longus muscle.** A superficial lateral muscle that overlies the fibula.
- **Gastrocnemius muscle.** Forms the bulk of the mid-portion and superior portion of the posterior aspect of the leg. The medial and lateral bellies can be seen clearly in a person standing on tiptoe.
- **Soleus muscle.** Located deep to the gastrocnemius muscle.
- **Calcaneal (Achilles) tendon.** Prominent tendon of the gastrocnemius and soleus muscles on the posterior aspect of the ankle; inserts into the **calcaneus** (heel) bone of the foot.
- **Lateral malleolus of fibula.** Projection of the distal end of the fibula that forms the lateral prominence of the ankle. The head of the fibula, at the proximal end of the bone, lies at the same level as the tibial tuberosity.
- **Medial malleolus of tibia.** Projection of the distal end of the tibia that forms the medial prominence of the ankle.
- **Dorsal venous arch.** Superficial veins on the dorsum of the foot that unite to form the small and great saphenous veins.
- **Tendons of extensor digitorum longus muscle.** Visible in line with phalanges (2–5).
- **Tendon of extensor hallucis longus muscle.** Visible in line with phalanx I (great toe). Pulsations in the dorsalis pedis artery may be felt in most people just lateral to this tendon where the blood vessel passes over the navicular and cuneiform bones of the tarsus.

Figure 14.7 Surface features of the leg, ankle, and foot.

6 The “ankle bones” are the distal ends of the tibia and the fibula.

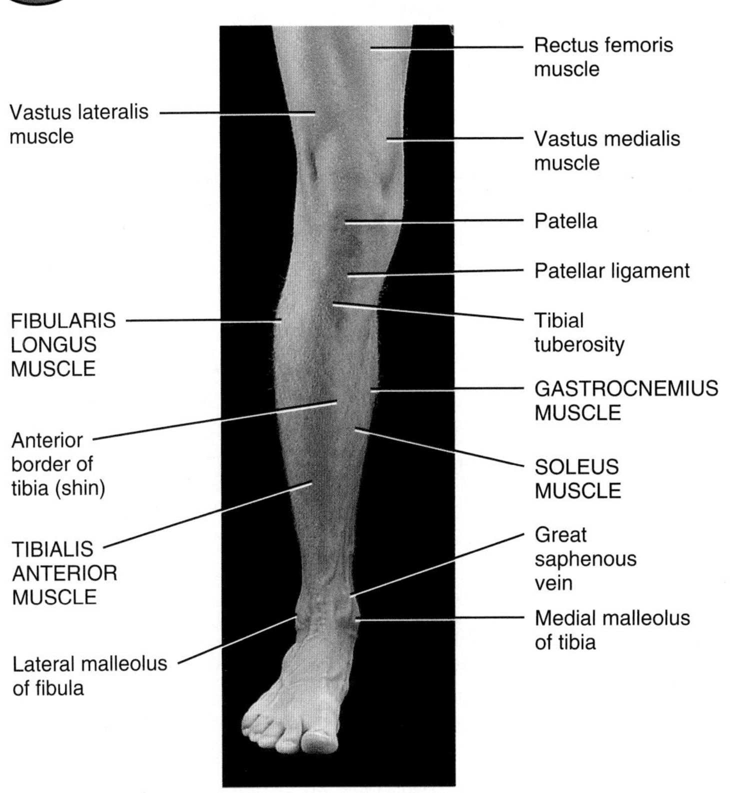

(a) Anterior view of leg, ankle, and foot

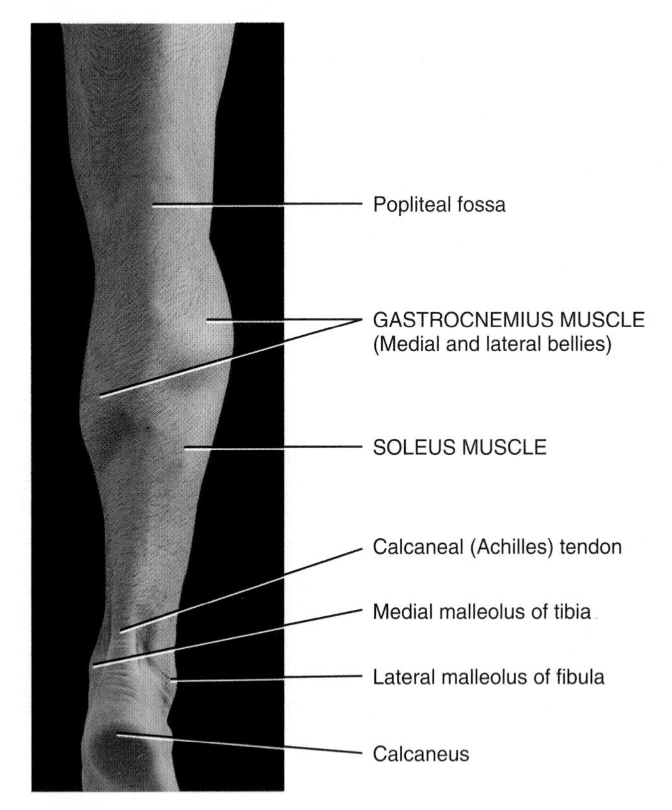

(b) Posterior view of leg and ankle

CONTINUES

Figure 14.7 (continued)

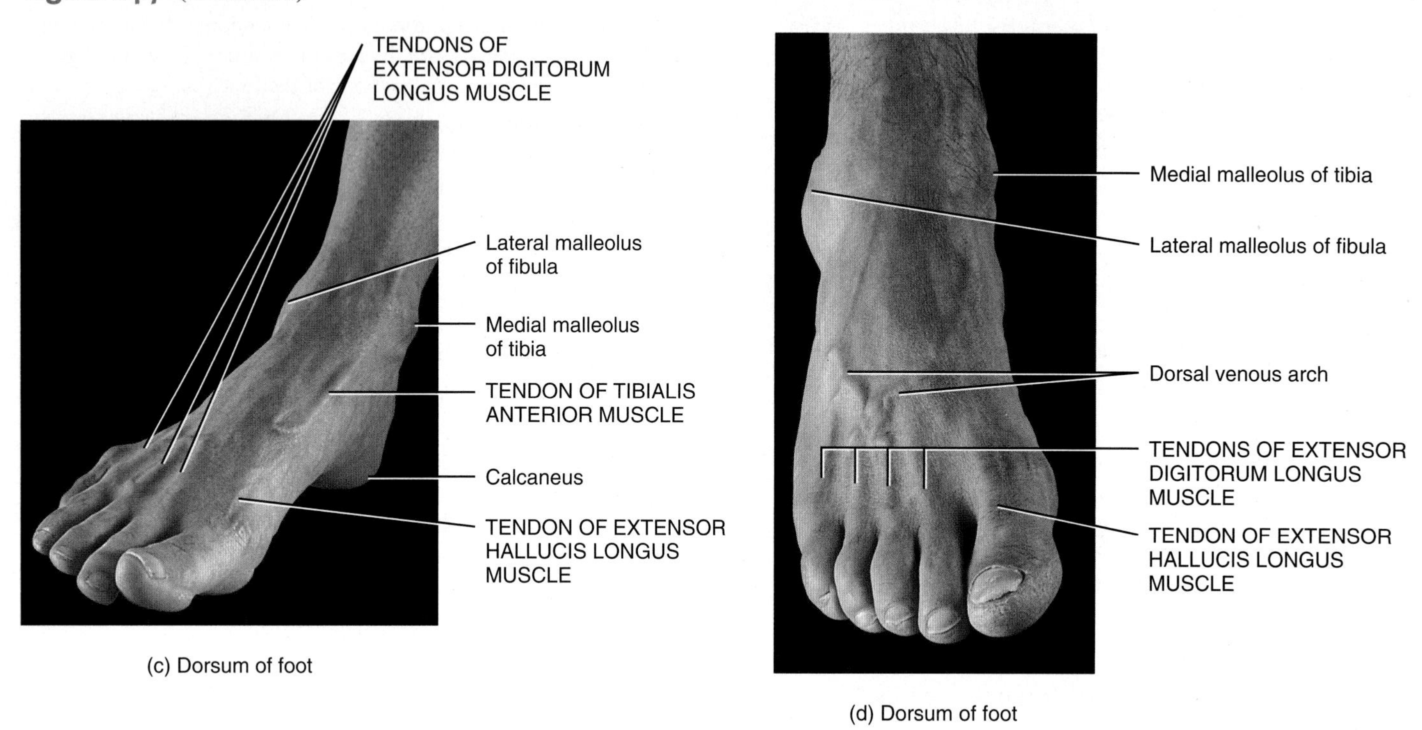

(c) Dorsum of foot

(d) Dorsum of foot

? The tendon of which muscle is most prominent in the medial aspect of the dorsum of the foot during dorsiflexion?

MANUAL THERAPY APPLICATION

Running-Related Injuries and Shin Splints

Many individuals who jog or run sustain some type of **running-related injury**. Although such injuries may be minor, some can be quite serious. Untreated or inappropriately treated minor injuries may become chronic. Among runners, common sites of injury include the ankle, knee, calcaneal (Achilles) tendon, hip, groin, leg, foot, and back. Of these, the knee often is the most severely injured area.

Running-related injuries are frequently related to faulty training techniques. Poorly constructed or worn-out running shoes can also contribute to injury, as can any biomechanical problem aggravated by running; such biomechanical problems include fallen arches, diagnosed by a physician or podiatrist, or arches that collapse and overextend while running. Be aware that your patient may purchase new running shoes only when the old ones get dirty or begin to look bad; patients seldom check their shoes for signs of wear or deterioration of the materials.

Shin splint syndrome, or simply **shin splints,** refers to pain or soreness along the tibia, specifically the medial, distal two-thirds. It may be caused by tendinitis of the anterior compartment muscles, especially the tibialis anterior muscle, inflammation of the periosteum (periostitis) around the tibia, or stress fractures of the tibia. It is also very common for the tibialis posterior muscle to be in spasm or be inflamed, perhaps as a result of the collapsing and extending of the arch of the foot during movement. The belly of the tibialis posterior muscle is deep to the gastrocnemius and the soleus muscles and therefore difficult to access. One edge of the belly, however, is accessible to the therapist on the entire medial side of the tibia. The insertion of the tibialis posterior is, however, directly accessible to the therapist. The tendinitis may occur in a trained athlete but usually occurs when poorly conditioned persons run on hard or banked surfaces with poorly supportive running shoes. The condition may also occur with vigorous activity of the legs following a period of relative inactivity or running in cold weather without proper warm-up. The muscles in the anterior compartment (mainly the tibialis anterior) can be strengthened to balance the stronger muscles of the posterior compartment.

Relating Muscles to Movements

Arrange the muscles in this exhibit according to the following actions on the foot at the ankle joint: (1) dorsiflexion and (2) plantar flexion; according to the following actions on the foot at the intertarsal joints: (1) inversion and (2) eversion; and according to the following actions on the toes at the metatarsophalangeal and interphalangeal joints: (1) flexion and (2) extension. The same muscle may be mentioned more than once.

CHECKPOINT

6. What are the superior extensor retinaculum and inferior extensor retinaculum?

EXHIBIT 14.4 Intrinsic Muscles of the Foot That Move the Toes

OBJECTIVE

- Describe the origin, insertion, action, and innervation of the intrinsic muscles of the foot.

The muscles in this exhibit are termed ***intrinsic muscles of the foot*** because they originate and insert *within* the foot (Figure 14.8). The muscles of the hand are specialized for precise and intricate movements, but those of the foot are limited to support and locomotion. The deep fascia of the foot forms the ***plantar aponeurosis (fascia)*** that extends from the calcaneus bone to the phalanges of the toes. The aponeurosis supports the longitudinal arch of the foot and encloses the flexor tendons of the foot.

The intrinsic muscles of the foot are divided into two groups: ***dorsal*** and ***plantar.*** There is only one dorsal muscle, the ***extensor digitorum brevis,*** a four-part muscle deep to the tendons of the extensor digitorum longus muscle, which extends toes 2–5 at the metatarsophalangeal joints.

The plantar muscles are arranged in four layers (Figure 14.8a–e). The most superficial layer is called the first layer. Three muscles are in the first layer. The ***abductor hallucis,*** which lies along the medial border of the sole, is comparable to the abductor pollicis brevis in the hand and abducts the great toe at the metatarsophalangeal joint. The ***flexor digitorum brevis,*** which lies in the middle of the sole, flexes toes 2–5 at the interphalangeal and metatarsophalangeal joints. The ***abductor***

Figure 14.8 Intrinsic muscles of the foot that move the toes.

Whereas the muscles of the hand are specialized for precise and intricate movements, those of the foot are limited to support and movement.

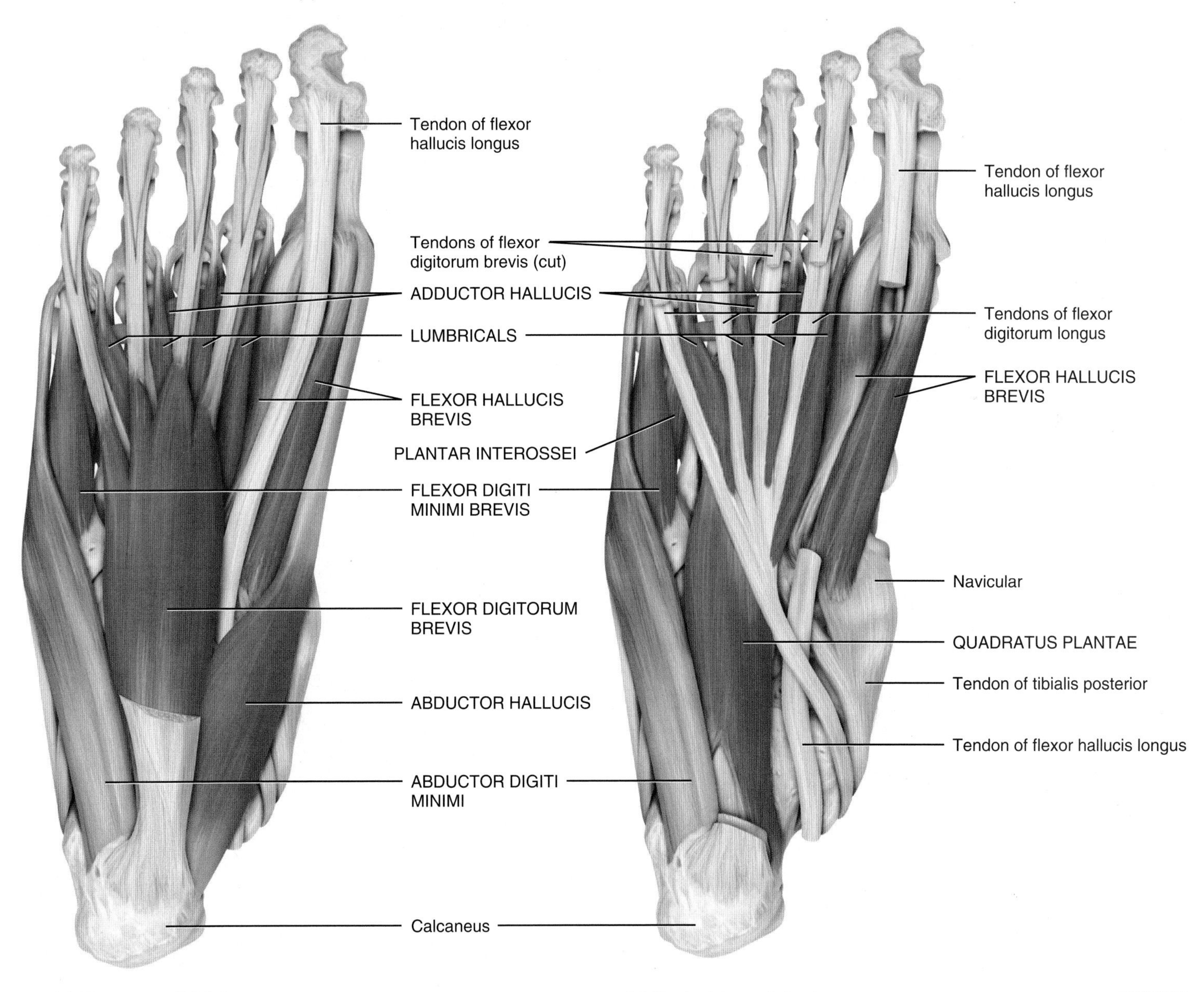

CONTINUES

Figure 14.8 (continued)

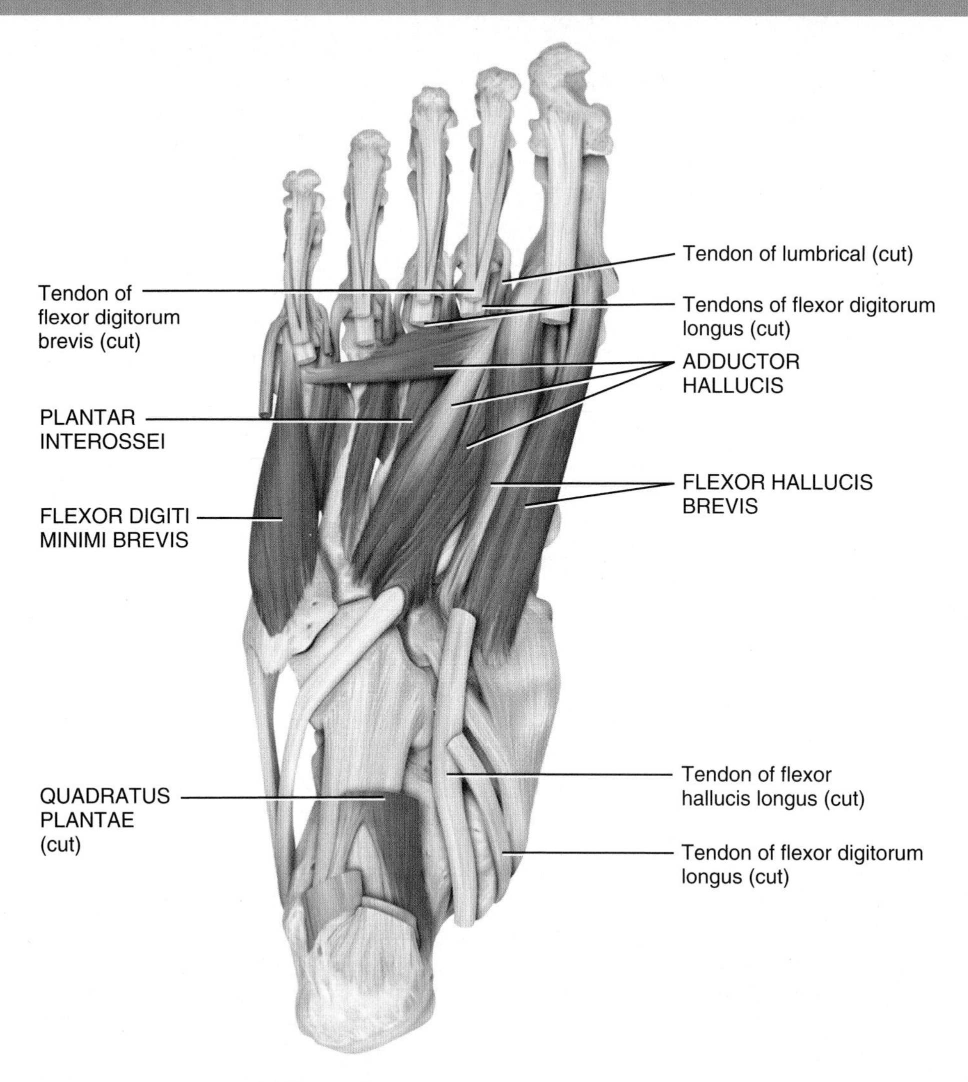

(c) Plantar deeper view

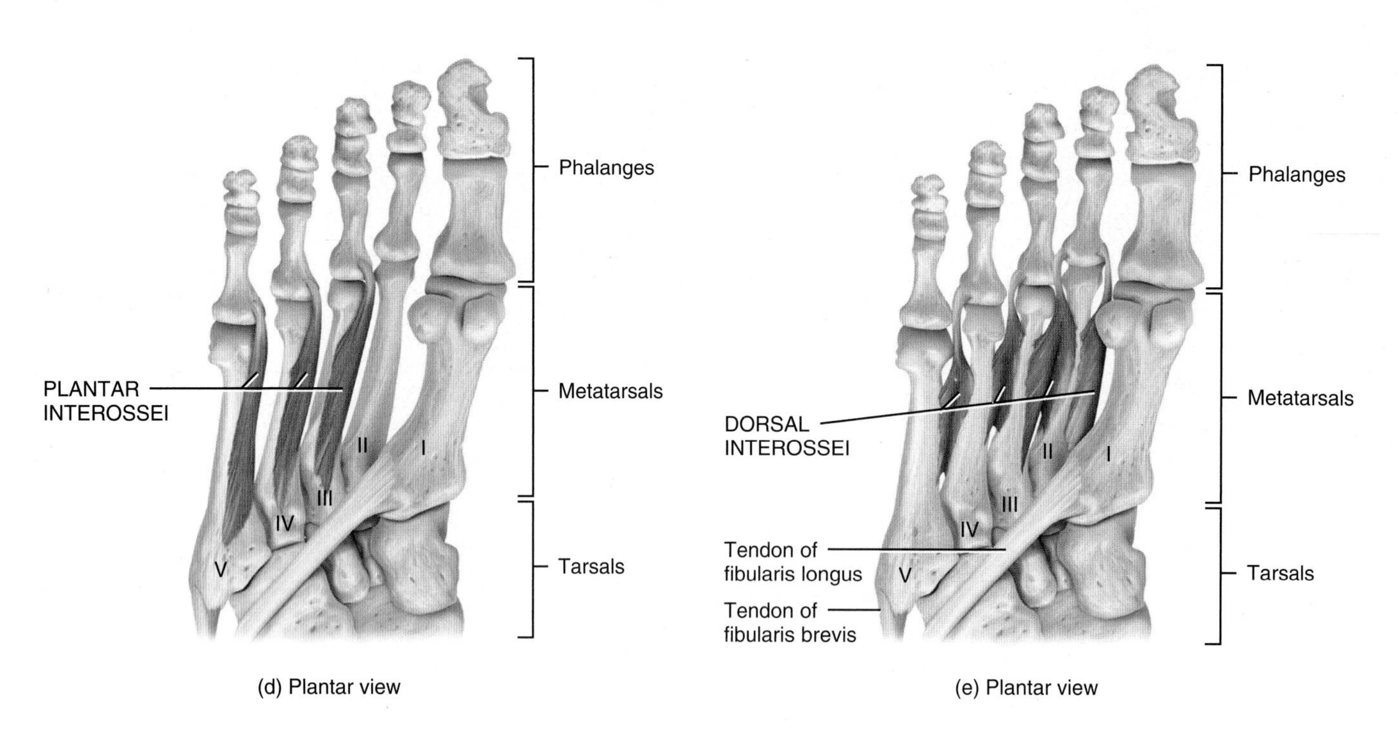

(d) Plantar view

(e) Plantar view

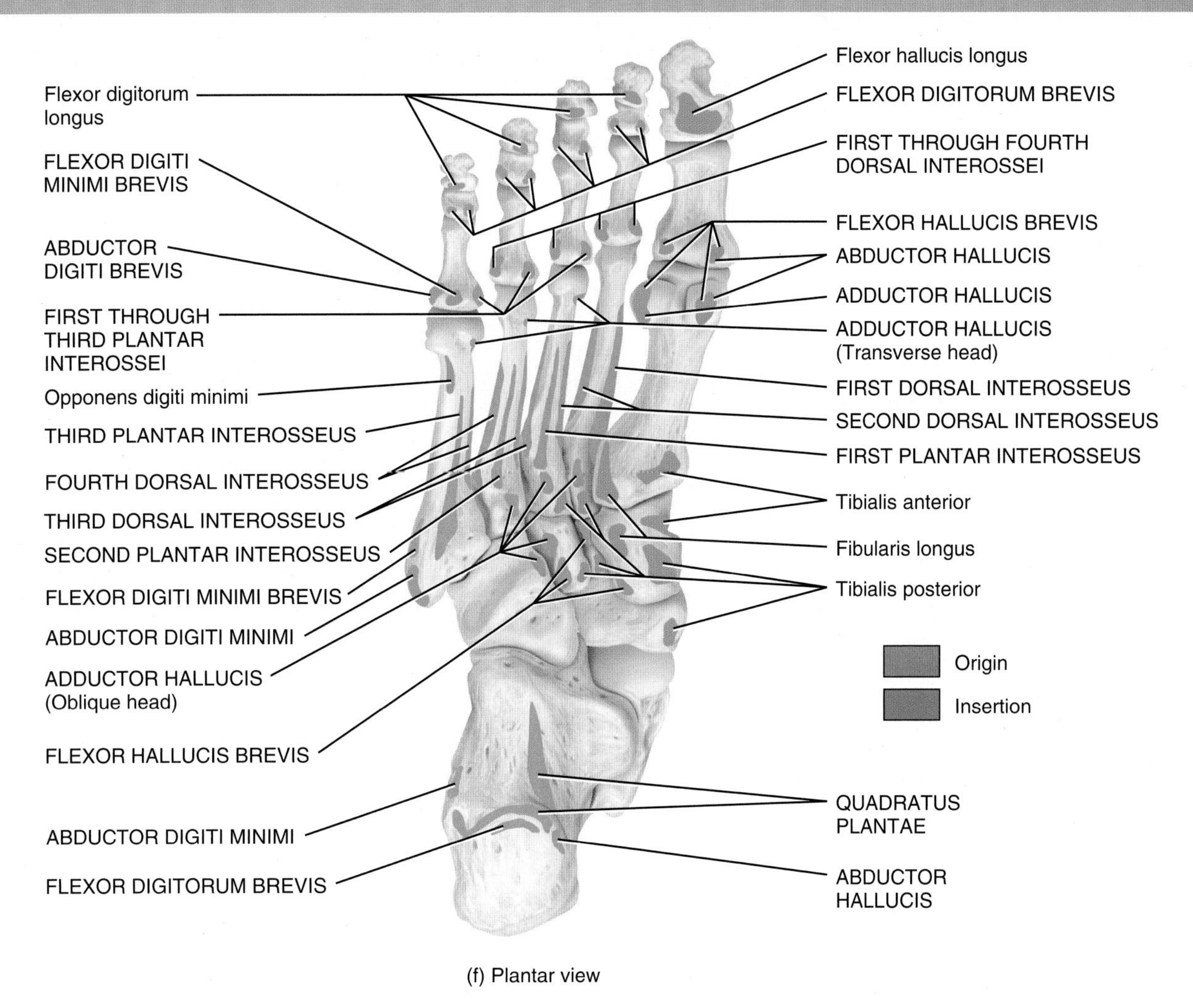

(f) Plantar view

What is the only muscle belly located on the dorsal surface of the foot?

digiti minimi, which lies along the lateral border of the sole, is comparable to the same muscle in the hand and abducts the little toe.

The second layer consists of the ***quadratus plantae,*** a rectangular muscle that arises by two heads and flexes toes 2–5 at the metatarsophalangeal joints, and the ***lumbricals,*** four small muscles that are similar to the lumbricals in the hands. They flex the proximal phalanges and extend the distal phalanges of toes 2–5.

Three muscles make up the third layer. The ***flexor hallucis brevis,*** which lies adjacent to the plantar surface of the metatarsal of the great toe, is comparable to the same muscle in the hand and flexes the great toe. The ***adductor hallucis,*** which has an oblique and transverse head like the adductor pollicis in the hand, adducts the great toe. The ***flexor digiti minimi brevis,*** which lies superficial to the metatarsal of the little toe, is comparable to the same muscle in the hand and flexes the little toe.

The fourth layer is the deepest and consists of two muscle groups. The ***dorsal interossei*** are four muscles that abduct toes 2–4, flex the proximal phalanges, and extend the distal phalanges. The three ***plantar interossei*** adduct toes 3–5, flex the proximal phalanges, and extend the distal phalanges. The interossei of the feet are similar to those of the hand. However, their actions are relative to the midline of the second digit rather than the third digit as in the hand.

CONTINUES

EXHIBIT 14.4 Intrinsic Muscles of the Foot That Move the Toes CONTINUED

MUSCLE	ORIGIN	INSERTION	ACTION	INNERVATION
DORSAL				
Extensor digitorum brevis (eks-TEN-sor di-ji-TOR-um BREV-is; *extensor* = increases angle at joint; *digit* = finger or toe; *brevis* = short (see Figure 14.6a, b).	Calcaneus and inferior extensor retinaculum.	Tendons of extensor digitorum longus on toes 2–4 and proximal phalanx of great toe.*	Extensor hallucis brevis* extends great toe at metatarsophalangeal joint and extensor digitorum brevis extends toes 2–4 at interphalangeal joints.	Deep fibular (peroneal) nerve.
PLANTAR				
FIRST LAYER (MOST SUPERFICIAL)				
Abductor hallucis (ab-DUK-tor HAL-ū-cis; *abductor* = moves part away from midline; *hallucis* = hallux or great toe).	Calcaneus, plantar aponeurosis, and flexor retinaculum.	Medial side of proximal phalanx of great toe with the tendon of the flexor hallucis brevis.	Abducts and flexes great toe at metatarsophalangeal joint.	Medial plantar nerve.
Flexor digitorum brevis (FLEK-sor = decreases angle at joint).	Calcaneus, plantar aponeurosis, and flexor retinaculum.	Sides of middle phalanx of toes 2–5.	Flexes toes 2–5 at proximal interphalangeal and metatarsophalangeal joints.	Medial plantar nerve.
Abductor digiti minimi (DIJ-i-tē MIN-i-mē; *minimi* = little)	Calcaneus and plantar aponeurosis.	Lateral side of proximai phalanx of little toe with the tendon of the flexor digiti minimi brevis.	Abducts and flexes little toe at metatarsophalangeal joint.	Lateral plantar nerve.
SECOND LAYER				
Quadratus plantae (kwod-RĀ-tus PLAN-tē; *quad* = square four-sided; *planta* = the sole)	Calcaneus.	Tendons of flexor digitorum longus.	Assists flexor digitorum longus to only flex toes 2–5 at interphalangeal and metatararsophalangeal joints.	Lateral plantar nerve.
Lumbricals (LUM-bri-kals; *lumbric* = earthworm).	Tendons of flexor digitorum longus.	Tendons of extensor digitorum longus on proximal phalanges of toes 2–5.	Extend toes 2–5 at interphalangeal joints and flex toes 2–5 at metatarsophalangeal joints.	Medial and lateral plantar nerves.
THIRD LAYER				
Flexor hallucis brevis	Cuboid and third (lateral) cuneiform.	Medial and lateral sides of proximal phalanx of great toe via a tendon containing a sesamoid bone.	Flexes great toe at metatarsophalangeal joint.	Medial plantar nerve.
Adductor hallucis (ad-DUK-tor = moves part closer to midline)	Metatarsals 2–4, ligaments of 3–5 metatarsophalangeal joints, and tendon of fibularis (peroneus) longus.	Lateral side of proximal phalanx of great toe.	Adducts and flexes great toe at metatarsophalangeal joint.	Lateral plantar nerve.
Flexor digiti minimi brevis	Metatarsal 5 and tendon of fibularis (peroneus) longus.	Lateral side of proximal phalanx of little toe.	Flexes little toe at metatarsophalangeal joint.	Lateral plantar nerve.
FOURTH LAYER (DEEPEST)				
Dorsal interossei (DOR-sal in-ter-OS-ē-ī) (not illustrated)	Adjacent side of all metatarsals.	Proximal phalanges: both sides of toe 2 and lateral side of toes 3 and 4.	Abducts and flexes toes 2–4 at metatarsophalangeal joints and extends toes at interphalangeal joints.	Lateral plantar nerve.
Plantar interossei (PLAN-tar).	Metatarsals 3–5.	Medial side of proximal phalanges of toes 3–5.	Adducts and flexes proximal metatarsophalangeal joints and extends toes at interphalangeal joints.	Lateral plantar nerve.

*The tendon that inserts into the proximal phalanx of the great toe, together with its belly, is often described as a separate muscle, the extensor hallucis brevis:

MANUAL THERAPY APPLICATION

Plantar Fasciitis

Plantar fasciitis (fas-ē-Ī-tis) or **painful heel syndrome** is an inflammatory reaction due to chronic irritation of the plantar aponeurosis (fascia) at its origin on the calcaneus (heel bone). The aponeurosis becomes less elastic with age. This condition is also related to weight-bearing activities (walking, jogging, lifting heavy objects), improperly constructed or fitting shoes, excess weight (puts pressure on the feet), and poor biomechanics (flat feet, high arches, and abnormalities in gait may cause uneven distribution of weight on the feet). Plantar fasciitis is the most common cause of heel pain in runners and arises in response to the repeated impact of running. Treatments include ice, deep heat, stretching exercises, as well as deep kneading procedures by a therapist, prosthetics (such as shoe inserts or heel lifts), steroid injections, and surgery.

Relating Muscles to Movements

Arrange the muscles in this exhibit according to the following actions on the great toe at the metatarsophalangeal joint: (1) flexion, (2) extension, (3) abduction, and (4) adduction; and according to the following actions on toes 2–5 at the metatarsophalangeal and interphalangeal joints: (1) flexion, (2) extension, (3) abduction, and (4) adduction. The same muscle may be mentioned more than once.

CHECKPOINT

7. How do the intrinsic muscles of the hand and foot differ in function?

14.2 MUSCLE INTERACTIONS

OBJECTIVES

- Describe the "connectedness" of skeletal muscles throughout the body.
- Describe the properties of connective tissue fascia that contribute to the "connectedness" of muscles.

The Interconnectedness of the Whole Body

Our knowledge of anatomy and physiology is constantly changing. Most people recognize that our understanding of physiological mechanisms is expanding exponentially as new technology permits study at lower and smaller levels. For example, there are researchers who spend their careers studying the mechanisms involved in moving ions through channels in cell membranes. A common perception of anatomy, however, is that our knowledge of the subject is static. For example, some people believe that muscles and bones have not changed, we continue to have two lungs, and the heart is still in the same place and still has four chambers. For the most part, muscles continue to be studied in isolation, as has been presented in this and the previous three chapters. The very term anatomy, as defined in Section 1.1, literally means "to cut up." In the study of muscles, the goal of most anatomists, medical students, and others seeking knowledge of the human muscular system has been to remove the skin and fascia to expose the "important" structures like muscles, major blood vessels, and named nerves.

Studies of fascia and other connective tissue components of the body, however, have demonstrated that the human body is more than a series of parts. The parts contain and are enveloped within voluminous connective tissues. Connective tissues form an anatomical network throughout the body. Although not yet fully accepted by the medical community, there are studies suggesting that connective tissue functions as a signaling system responsive to mechanical movement and that it pervade the entire body. Fasciae also contain free and encapsulated nerve endings that suggest a proprioceptive function. The science of acupuncture, over 2,000 years old, is built on the premise that a network of meridians, present within connective tissue membranes, function to interconnect all parts of the body.

A well-known property of connective tissue is its plasticity in response to different levels of mechanical stresses. These changes may take place over the course of days, weeks, or months, after a change in posture or activity. As discussed in Section 10.2, connective tissues are known to change their composition of ground substance as well as the orientation and density of the fibers in response to the existing cellular environment.

Studies indicate that fasciae contain cells known as myofibroblasts that have contractile properties similar to that of smooth muscles. Fascial structures, therefore, may constitute a contractile organ that pervades the entire body. Such activity would account for chronic tissue contractures as well as smooth muscle–like contractions that may occur within minutes, hours, or days of a strenuous trauma such as lifting a very heavy object. Every practicing manual therapist has heard from a patient that they could feel something, for example, near the right shoulder when the therapist was working on the right calf. This seemingly inexplicable phenomenon is a function of the connective tissue network that envelopes the entire body. A critical component of the network is the sacrotuberous ligament. It connects the sacral fascia and the erector spinae with the biceps femoris. In his work, *Anatomy Trains*, Thomas W. Myers demonstrates multiple lines of interconnected muscles and fascia. As one of many examples, Myers illustrates how the plantar foot is connected to the anterior skull through specific muscle fascia and the periosteum of multiple bony landmarks and is organized along a specific line of pull. He also states that either chronic contraction or weakness of muscles or fascia at any point along this demonstrable route can affect any other portion of the muscular pathway. This concept explains how a problem in the lower limb, for example, could express itself as a frontal headache.

As this new concept becomes generally accepted, it will become more evident that the human body is a whole, not just a collection of parts. In this age of medical specialization, fragmentation is the practice. Practicing manual therapists already know, at some level, the connectedness of the human body.

Posture and Interactions between Muscles

Posture is affected by the interplay of multiple muscles of the back, abdomen, pelvis, and thigh. Figure 14.9 illustrates three postures. The pelvis is tilted anteriorly in a normal posture when the major muscles indicated are not actively contracting but are in a state of normal muscle tone (Figure 14.9a). A posture similar to the anatomical position causes the pelvis to tilt such that the anterior superior iliac spine (ASIS) and the posterior superior iliac spines (PSIS) are level (Figure 14.9b). Note that elevating the chest causes the abdominal muscles to become active and elevate the anterior pelvis; at the same time the gluteal muscles and the hamstrings are contracting actively and are pulling the posterior portion of the pelvis inferiorly. A slumped posture may be the result of generalized depression or may be due to poorly conditioned abdominal muscles. The posture is characterized by an excessive anterior pelvic tilt and accentuated lordotic curvature. The result may be chronic and cause pathological shortening of back muscles, the iliopsoas, and muscles of the anterior and medial compartments of the thigh (Figure 14.9c).

Figure 14.9 Interactions between muscles of the back, abdomen, pelvis, and thigh.

⑥ Posture is affected by the interplay of multiple muscles.

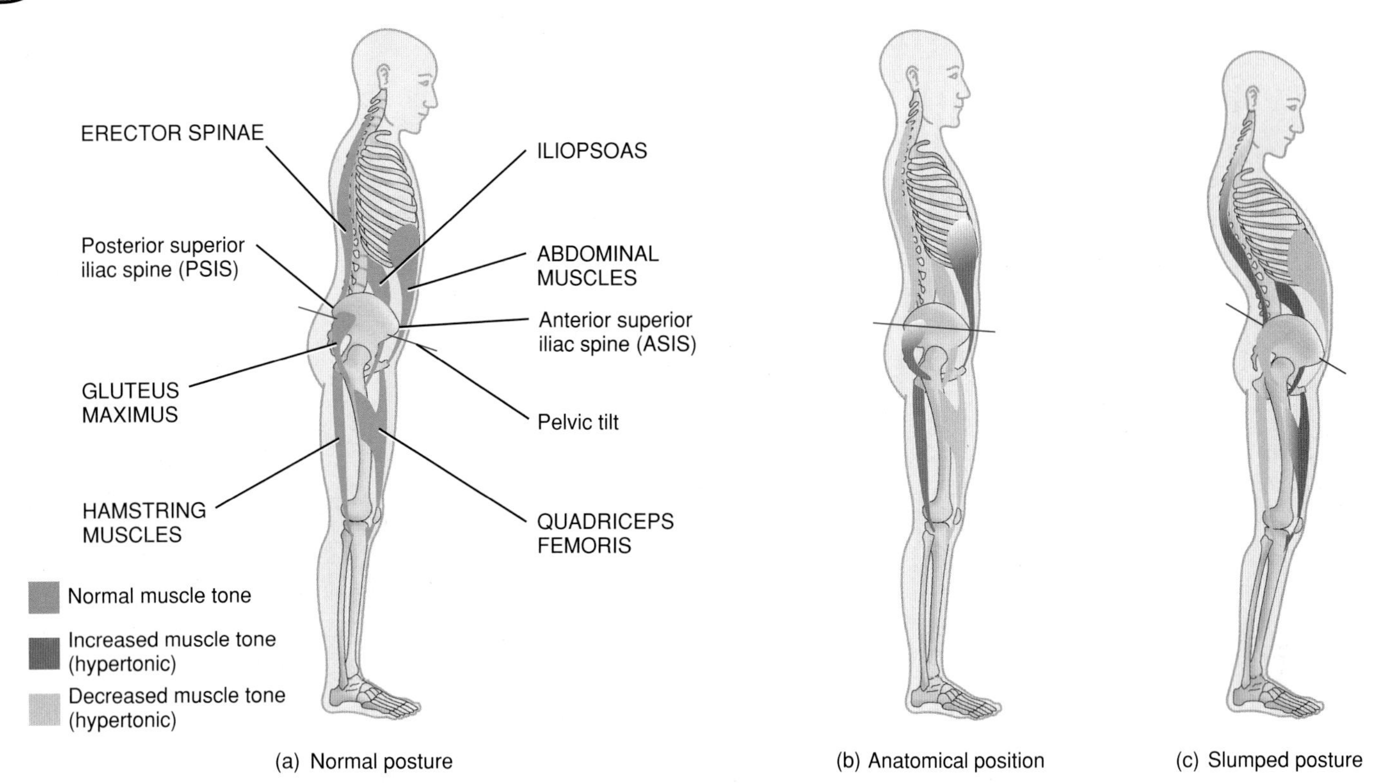

(a) Normal posture (b) Anatomical position (c) Slumped posture

? Which posterior muscles are the most contracted when a person is standing in a slumped posture?

CHECKPOINT

8. Describe connective tissue structures such as fascia, periosteum, epimysium, tendons, and ligaments that may form continuous pathways throughout the body.

EXHIBIT 14.5 Contributions of the Muscular System to Homeostasis

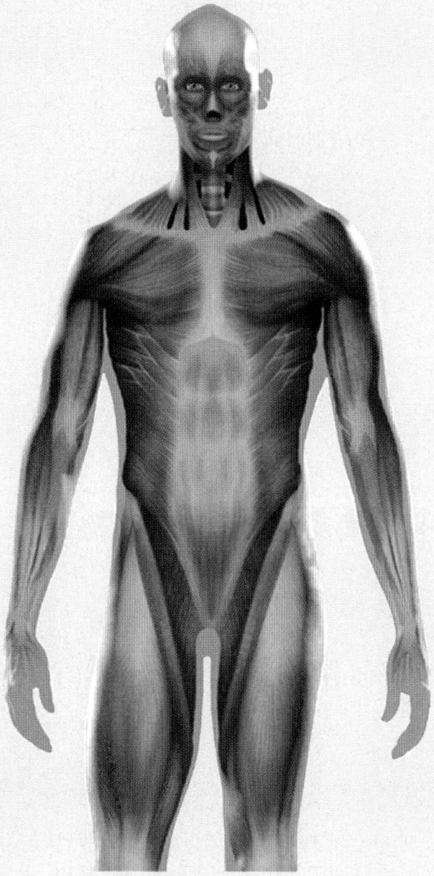

BODY SYSTEM		CONTRIBUTION TO HOMEOSTASIS
For all body systems		The muscular system and muscle tissues produce body movements, stabilize body positions, move substances within the body, and produce heat that helps maintain normal body temperature.
Integumentary system	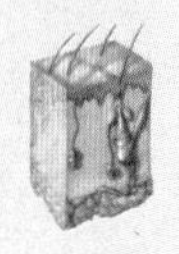	Pull of skeletal muscles on attachments to skin of face causes facial expressions; muscular exercise increases skin blood flow.
Skeletal system	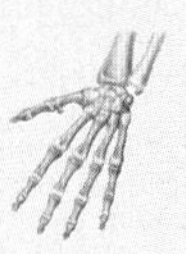	Skeletal muscle causes movement of body parts by pulling on attachments to bones; skeletal muscle provides stability for bones and joints.
Nervous system	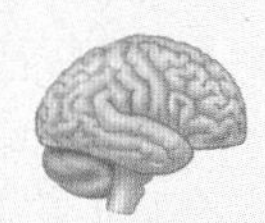	Smooth, cardiac, and skeletal muscles carry out commands for the nervous system; shivering—involuntary contraction of skeletal muscles that is regulated by the brain—generates heat to raise body temperature.
Endocrine system		Regular activity of skeletal muscles (exercise) improves action and signaling mechanisms of some hormones, such as insulin; muscles protect some endocrine glands.
Cardiovascular system		Cardiac muscle powers pumping action of heart; contraction and relaxation of smooth muscle in blood vessel walls help adjust the amount of blood flowing through various body tissues; contraction of skeletal muscles in the legs assists return of blood to the heart; regular exercise causes cardiac hypertrophy (enlargement) and increases heart's pumping efficiency; lactic acid produced by active skeletal muscles may be used for ATP production by the heart.
Lymphatic system and immunity		Skeletal muscles protect some lymph nodes and lymphatic vessels and promote flow of lymph inside lymphatic vessels; exercise may increase or decrease some immune responses.
Respiratory system	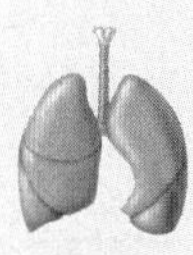	Skeletal muscles involved with breathing cause air to flow into and out of the lungs; smooth muscle fibers adjust size of airways; vibrations in skeletal muscles of larynx control air flowing past vocal cords, regulating voice production; coughing and sneezing, due to skeletal muscle contractions, help clear airways; regular exercise improves efficiency of breathing.
Digestive system	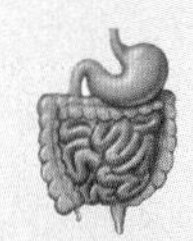	Skeletal muscles protect and support organs in the abdominal cavity; alternating contraction and relaxation of skeletal muscles power chewing and initiate swallowing; smooth muscle sphincters control volume of organs of the gastrointestinal (GI) tract; smooth muscles in walls of GI tract mix and move its contents through the tract.
Urinary system	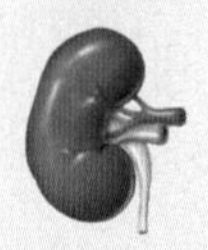	Skeletal and smooth muscle sphincters and smooth muscle in wall of urinary bladder control whether urine is stored in the urinary bladder or voided (urination).
Reproductive systems	 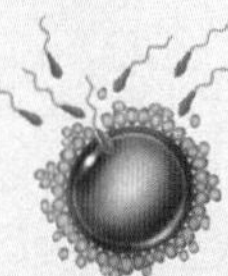	Skeletal and smooth muscle contractions eject semen from male; smooth muscle contractions propel oocyte along uterine tube, help regulate flow of menstrual blood from uterus, and force baby from uterus during childbirth; during intercourse, skeletal muscle contractions are associated with orgasm and pleasurable sensations in both sexes.

STUDY OUTLINE

Introduction to the muscles of the lower limb (extremity) (Section 14.1)

1. Muscles that move the femur originate for the most part on the pelvic girdle and insert on the femur; these muscles are larger and more powerful than comparable muscles in the upper limb.
2. The most powerful muscles of the lower limb lie in the posterior hip, the anterior thigh, and the posterior leg.
3. The thigh is divided by fasciae into anterior, posterior, and medial compartments.
4. The primary actions and muscle innervations within any compartment are nearly the same.

Muscles of the gluteal region that move the femur (Exhibit 14.1)

1. The iliopsoas and the gluteus maximus are powerful flexor and extensor muscles of the hip.
2. A group of six muscles, known collectively as lateral rotators of the hip, lie deep to the gluteus maximus.
3. Four muscles in the medial aspect of the thigh are responsible for adduction of the lower limb.
4. The piriformis muscle, one of the lateral rotators, can impinge on the sciatic nerve and thus cause pain in a portion or most of the lower limb.

Muscles of the thigh that move the femur, tibia, and fibula (Exhibit 14.2)

1. Muscles that move the femur, tibia, and fibula are separated into medial, anterior, and posterior compartments.
2. Muscles of the medial compartment of the thigh adduct the femur at the hip joint.
3. Muscles of the anterior compartment of the thigh extend the leg at the knee; one muscle also flexes the thigh at the hip.
4. Muscles of the posterior compartment of the thigh flex the leg at the knee; they may also extend the thigh at the hip.

Muscles of the leg that move the foot and toes (Exhibit 14.3)

1. Muscles that move the foot and toes are divided into anterior, lateral, and posterior compartments.
2. Muscles of the anterior compartment of the leg dorsiflex the foot and extend the toes.
3. Muscles of the lateral compartment of the leg plantar flex and evert the foot.
4. Muscles of the posterior compartment of the leg plantar flex the foot at the ankle.

Intrinsic muscles of the foot that move the toes (Exhibit 14.4)

1. Intrinsic muscles of the foot, unlike those of the hand, are limited to the functions of support and locomotion.
2. All intrinsic muscles originate and insert within the foot.
3. Collectively the muscles assist the larger muscles located in the leg in the actions of flexion and extension of the toes.
4. Intrinsic muscles are responsible for abduction and adduction of the toes.

Muscle interactions (Section 14.2)

1. Recent studies of fascia and other connective tissue components of the body have demonstrated that the human body is more than a series of parts.
2. Connective tissues form an anatomical network throughout the body.
3. Connective tissues illustrate plasticity in response to different levels of mechanical stress and thus can change their composition, orientation, and density of fibers in response to the existing cellular environment.

SELF-QUIZ QUESTIONS

Choose the one best answer to the following questions.

1. Which of the following muscles does *not* flex the thigh?
   a. rectus femoris **b.** gracilis
   c. sartorius **d.** tensor fascia latae.
2. The iliotibial tract is composed of the tendon of the gluteus maximus muscle, the deep fascia that encircles the thigh, and the tendon of which of the following muscles?
   a. iliacus **b.** gluteus minimus
   c. tensor fascia latae **d.** adductor longus
3. Which muscle does NOT originate on the pubic symphysis, pubic crest, or pubic rami?
   a. adductor magnus **b.** pectineus
   c. gracilis **d.** vastus medialis
4. Which muscle inserts on or near the linea aspera?
   a. adductor magnus **b.** gluteus maximus
   c. pectineus **d.** all of the above
5. The longest single muscle in the body is the
   a. adductor longus **b.** longissimus thoracis
   c. peroneus longus **d.** sartorius.
6. Which of the muscles is (are) NOT lateral rotators of the hip?
   a. gemellus inferior and superior
   b. gluteus medius and minimus
   c. obturator externus and internus
   d. quadratus femoris
7. Piriformis originates on the _____ surface of the sacrum.
   a. anterior **b.** lateral
   c. medial **d.** posterior
8. The only muscle of the posterior compartment of the leg that does NOT plantar flex the foot is the
   a. gastrocnemius **b.** flexor digitorum longus
   c. plantaris **d.** popliteus.
9. Plantar fasciitis may be related to aging and _____.
   a. excess weight **b.** improper shoes
   c. poor biomechanics **d.** weight-bearing activities
   e. all of the above
10. Tibialis anterior _____ and _____ the foot.
    a. dorsiflexes; everts **b.** dorsiflexes; inverts
    c. plantar flexes; everts **d.** plantar flexes; inverts

11. Which of the following is NOT a major action of the iliopsoas?
a. flexion of the thigh at the hip
b. flexion of the trunk on the hip
c. lateral rotation of the thigh
d. medial rotation of the thigh

12. Quadriceps femoris inserts on the
a. anterior inferior iliac spine
b. linea aspera
c. tendon of quadriceps femoris
d. tibial tuberosity.

13. Fibularis (peroneus) longus _____ and _____ the foot.
a. dorsiflexes; everts **b.** dorsiflexes; inverts
c. plantar flexes; everts **d.** plantar flexes; inverts

14. The (superficial) pes anserinus does NOT include the
a. biceps femoris **b.** gracilis
c. sartorius **d.** semitendinosus.

15. Muscles of the _____ compartment of the leg are innervated by the tibial nerve.
a. anterior **b.** lateral
c. medial **d.** posterior

16. Flexion of four toes is caused by the
a. extensor digitorum longus
b. flexor digitorum longus
c. flexor hallucis longus
d. tibialis posterior.

17. Which of the following muscles is NOT innervated by the superior gluteal nerve?
a. gluteus maximus **b.** gluteus medius
c. gluteus minimus **d.** tensor fasciae latae

18. The hamstrings are _____ of the thigh and _____ of the leg.
a. extensors; extensors **b.** extensors; flexors
c. flexors; extensors **d.** flexors; flexors

19. Development of scar tissue that results in permanent shortening of skeletal muscles is a condition known as
a. contracture **b.** fibrillation
c. fibrosis **d.** spasm.

20. Which muscle can either flex or extend the thigh at the hip?
a. adductor magnus **b.** gluteus maximus
c. quadratus femoris **d.** tensor fasciae latae

CRITICAL THINKING QUESTIONS

1. A group of elderly retired men meet every Tuesday morning for breakfast in a local restaurant. Last week the discussion, and ensuing arguments, centered on sciatica. Charlie said that his pain went down to the knee while Sam said that his pain went all the way down into the foot. Rufus had pain in his hip and upper thigh. All three men had been diagnosed by physicians with sciatica. If you had been present at this breakfast discussion, how could you have explained the big differences in pain with all three men having the same diagnosis.

2. Lou is a serious bicycle rider on the weekend. During the week he spends time before work in the gym and his goal is to strengthen his leg muscles. A friend noticed the progressive development of the musculature of Lou's calves and complimented him on his "nice gastrocs." What muscle is more likely to have been developed but the result was to push the gastrocnemius more posteriorly?

? ANSWERS TO FIGURE QUESTIONS

14.1 Upper limb muscles exhibit diversity of movement; lower limb muscles function in stability, locomotion, and maintenance of posture. In addition, lower limb muscles usually cross two joints and act equally on both.

14.2 Muscles located in the medial compartment of the thigh include the adductor magnus, adductor longus, adductor brevis, pectineus, and gracilis.

14.3 The fascicles of the gluteus maximus are directed inferiorly and laterally from the iliac crest.

14.4 The gracilis, rectus femoris, sartorius, and hamstring muscles (the biceps femoris, semitendinosus, and semimembranosus) cross both the hip and knee joints.

14.5 The popliteal fossa is bordered superolaterally by the biceps femoris, superomedially by the semimembranosus and semitendinosus, and inferolaterally and inferomedially by the lateral and medial heads of the gastrocnemius.

14.6 The superior and inferior extensor retinacula firmly hold the tendons of the anterior compartment muscles to the ankle.

14.7 The tendon of the tibialis anterior is most prominent during dorsiflexion.

14.8 The only muscle belly located on the dorsal surface of the foot is the extensor digitorum brevis.

14.9 The lumbar portion of the erector spinae muscles are the most contracted in a person in a slumped position.

Nervous Tissue 15

Nervous tissue is one of the four main tissue types. It acts together with the endocrine system (Chapter 20) to regulate homeostasis in the body. The nervous system has many similarities with the endocrine system; they share control of the activities of the body to keep it within optimal limits. However, the nervous system is extremely fast acting but shorter lived in action than the endocrine (hormonal) system. Think of how quickly you reflexively move when you accidentally put your hand on a hot stove or step on a tack. The nervous system uses a series of electrochemical signals to receive information from the receptors of the body in the peripheral nervous system (PNS) regions and sends them to the central nervous system (CNS), the brain and spinal cord, to coordinate our actions. A new message is then sent to an effector organ or muscle to take action. This whole process of sending information from receptor to coordinator to reactor takes only a fraction of a second. That would not sound so amazing, if not for the fact that this is happening at millions of places in the body at once. Nervous tissue monitors every body activity such as breathing, digestion, and the beating of your heart. You do not even need to actively think about these things since they are done for you automatically (or autonomically) without your conscious thought.

CONTENTS AT A GLANCE

15.1 OVERVIEW OF THE NERVOUS SYSTEM

OBJECTIVES

- List the structures and basic functions of the nervous system.
- Describe the organization of the nervous system.

Even in this age of technology and computers, no computer built today can rival the complexity of the human nervous system. The nervous system is a network of billions of interconnected nerve cells (neurons) that receive stimuli, coordinate this sensory information, and cause the body to respond appropriately. The individual neurons transmit messages by means of a complicated electrochemical process. With a mass of only 3% of the total body weight, the nervous system is one of the smallest yet most complex of the 11 body systems. The two main subdivisions of the nervous system are the ***central nervous system (CNS),*** which consists of the brain and spinal cord, and the ***peripheral*** (pe-RIF-er-al) ***nervous system (PNS),*** which includes all nervous tissue outside the CNS.

Together, the nervous and endocrine systems share responsibility for regulating homeostasis, which means keeping the body's internal environment within normal limits. The objective is the same but the two systems achieve that objective differently. The nervous system regulates body activities by responding rapidly using nerve impulses (action potentials); the endocrine system responds more slowly, though no less effectively, by releasing hormones. Furthermore, the effects of nervous system activation are very short lived while the effects of endocrine system activation may last for minutes or hours. Certain of the manual therapies may utilize routine sensory, motor, and reflex tests to assess the role of the nervous system in maintaining homeostasis.

The nervous system is also responsible for our perceptions, behaviors, and memories, and initiates all voluntary movements. Because the nervous system is quite complex, we will consider different aspects of its structure and function in several related chapters. This chapter focuses on the organization of the nervous system and the properties of the cells that make up nervous tissue—neurons (nerve cells) and neuroglia (different cells that support the activities of neurons). In the chapters that follow, we will examine the structure and functions of the spinal cord and spinal nerves (Chapter 16) and of the brain and cranial nerves (Chapter 17). Then we will discuss the autonomic nervous system, the part of the nervous system that operates without voluntary control (Chapter 18). Next, we examine the somatic senses—touch, pressure, warmth, cold, pain, and others—and the sensory and motor pathways to understand how nerve impulses pass into the spinal cord and brain or from the spinal cord and brain to muscles and glands (Chapter 19). Our exploration of the nervous system concludes with a discussion of the special senses: smell, taste, vision, hearing, and equilibrium (Chapter 19). Since the endocrine system is sometimes called the neuroendocrine system, references will also be made to the nervous system in our discussion of the endocrine system (Chapter 20).

Structures of the Nervous System

The nervous system is an intricate, highly organized network of billions of neurons and even more neuroglia. The structures that make up the nervous system include the brain, cranial nerves and their branches, the spinal cord, spinal nerves and their branches, ganglia, enteric plexi, and sensory receptors (Figure 15.1).

A ***nerve*** is a bundle of hundreds to thousands of axons plus associated connective tissue and blood vessels that lie outside the brain and spinal cord. Nerves follow a defined path and serve specific regions of the body. The skull encloses the ***brain,*** which contains about 100 billion neurons. Twelve pairs (right and left) of ***cranial nerves,*** numbered I through XII, emerge from the base of the brain. For example, cranial nerve I carries signals for the sense of smell from the nose to the brain.

The ***spinal cord*** connects to the brain, contains about 100 million neurons, and is encircled by the bones of the vertebral column. Emerging from the spinal cord are 31 pairs of ***spinal nerves,*** each serving a specific region on the right or left side of the body. ***Ganglia*** (GANG-lē-a = swelling or knot; singular is *ganglion*) contain cell bodies of neurons, are located outside the brain and spinal cord, and are closely associated with cranial and spinal nerves.

The walls of organs of the gastrointestinal tract contain extensive networks of neurons, called ***enteric plexuses,*** that help regulate the digestive system (Figure 15.1). ***Sensory receptors*** are dendrites of sensory neurons (such as sensory receptors in the skin) or separate, specialized cells that monitor changes in the internal or external environment (such as photoreceptors in the retina of the eye).

The branch of medical science that deals with the normal functioning and disorders of the nervous system is ***neurology*** (noo-ROL-ō-jē; *neuro-* = nerve or nervous system; *-logy* = study of). A ***neurologist*** is a physician who specializes in the diagnosis and treatment of disorders of the nervous system.

Functions of the Nervous System

Individual neurons, just described, carry incoming signals, or communicate with an array of neurons, or carry signals to effectors that produce an action. The nervous system thus carries out a complex assortment of tasks, such as sensing smells, producing speech, remembering past events, providing signals that control body movements, and regulating the operation of internal organs. These diverse activities are grouped into three basic functions: sensory, integrative, and motor.

- ***Sensory function.*** The sensory receptors *detect* many different types of stimuli, both within your body, such as an increase in blood temperature, and outside your body, such as a touch on your arm. ***Sensory*** or ***afferent neurons*** (AF-erent NOOR-onz; *af-* = toward; *-ferrent* = carried) carry this

Figure 15.1 Major structures of the nervous system.

The nervous system includes the brain, cranial nerves, spinal cord, spinal nerves, ganglia, enteric plexuses, and sensory receptors.

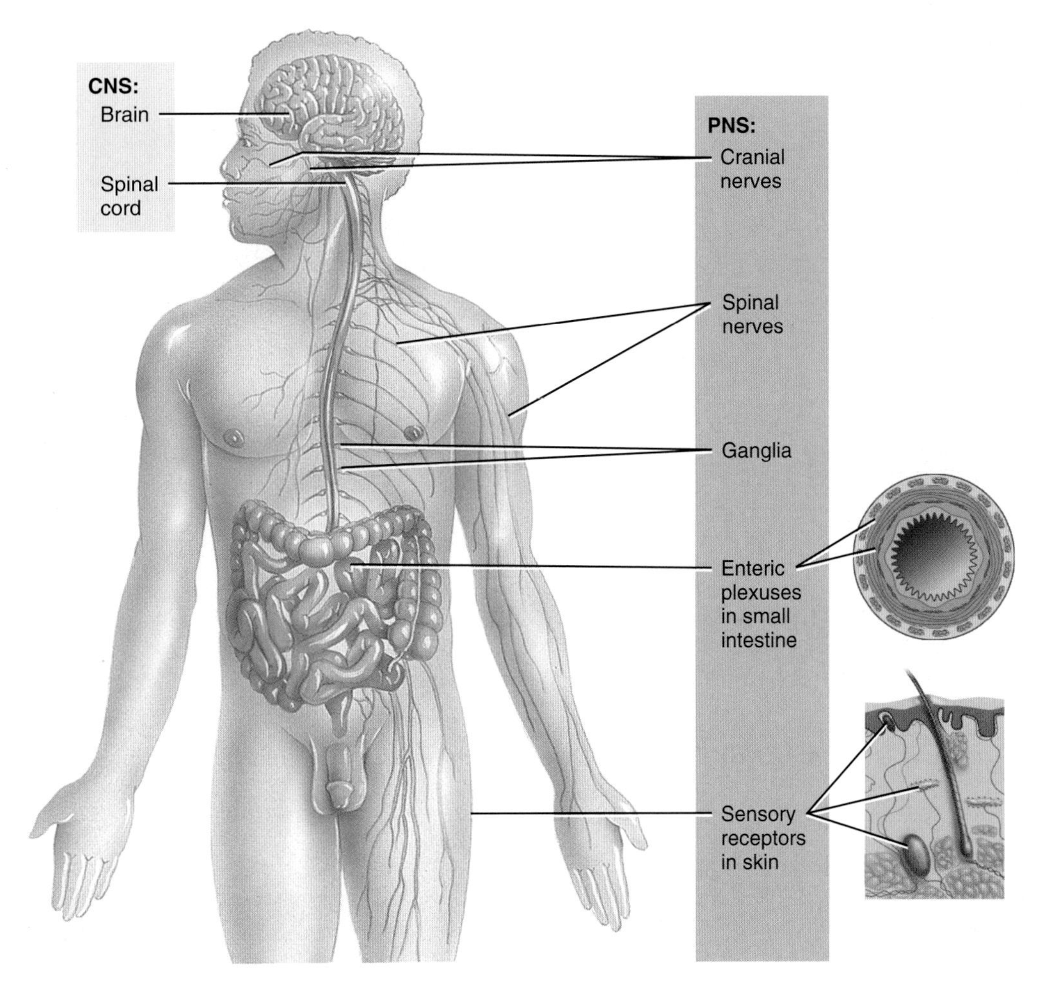

What is the total number of cranial and spinal nerves in the human body?

sensory information *into* the brain and spinal cord through cranial and spinal nerves.

- ***Integrative function.*** The nervous system *integrates* (processes) sensory information by analyzing and storing some of it and by making decisions for appropriate responses. An important integrative function is ***perception,*** the conscious awareness of sensory stimuli. Perception occurs in the brain. Many of the neurons that participate in integration are ***interneurons (association neurons),*** whose axons extend for only a short distance and contact nearby neurons in the brain or spinal cord. Interneurons comprise the vast majority of neurons in the body.
- ***Motor function.*** Once a sensory stimulus is received, the nervous system may elicit an appropriate motor response such as muscular contraction or glandular secretion. The neurons that serve this function are ***motor (efferent) neurons*** (EF-er-ent; *ef-* = away from). Motor neurons carry information *from* the brain toward the spinal cord or out of the brain and spinal cord to ***effectors*** (muscles and glands) through cranial and spinal nerves. Stimulation of the effectors by motor neurons causes muscles to contract and glands to secrete.

Organization of the Nervous System

The CNS integrates and correlates many different kinds of incoming sensory information. The CNS is also the source of thoughts, emotions, and memories. Most nerve impulses that stimulate muscles to contract and glands to secrete originate in the CNS. Structural components of the PNS are cranial nerves and their branches, spinal nerves and their branches, ganglia, and sensory receptors. The PNS is further subdivided into a ***somatic nervous system (SNS)*** (*somat-* = body), an ***autonomic nervous system (ANS)*** (*auto-* = self; *-nomic* =

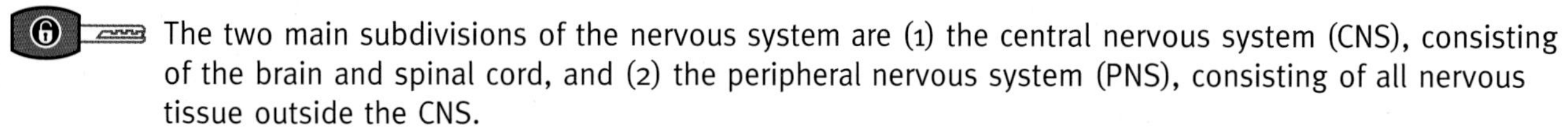

Figure 15.2 Organization of the nervous system. Subdivisions of the PNS are the somatic nervous system (SNS), the autonomic nervous system (ANS), and the enteric nervous system (ENS).

The two main subdivisions of the nervous system are (1) the central nervous system (CNS), consisting of the brain and spinal cord, and (2) the peripheral nervous system (PNS), consisting of all nervous tissue outside the CNS.

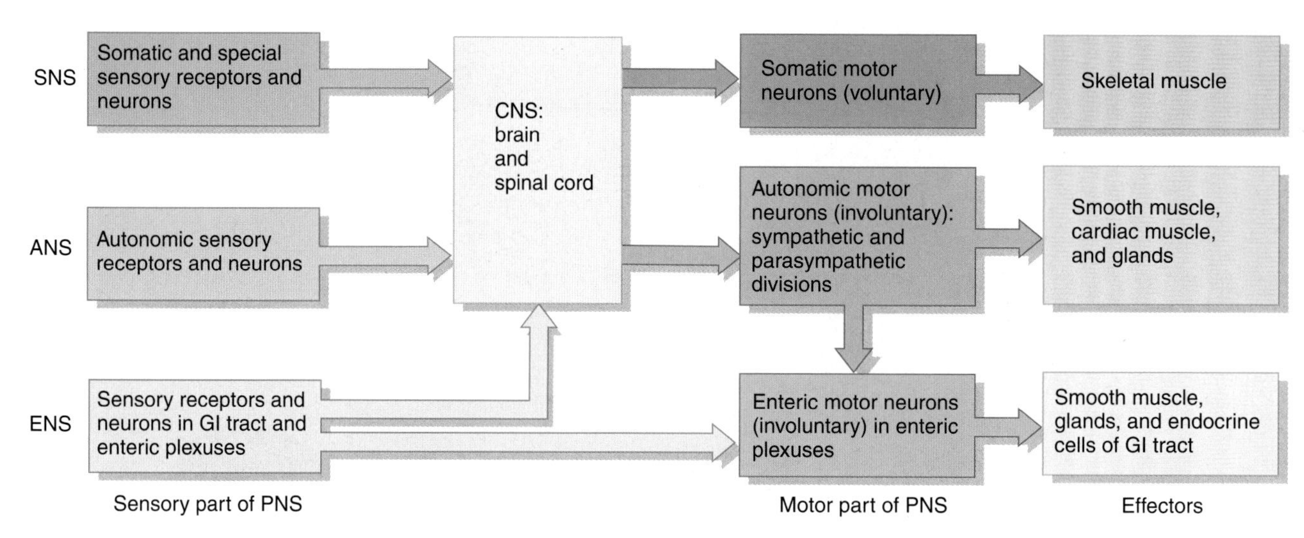

Which types of neurons carry input to the CNS and output from the CNS?

law), and an ***enteric nervous system (ENS)*** (*enter-* = intestines) (Figure 15.2).

The somatic nervous system consists of (1) sensory neurons that convey information from somatic receptors in the head, body wall, viscera, and limbs and from receptors for the special senses of vision, hearing, taste, and smell to the CNS and (2) motor neurons that conduct impulses from the CNS to *skeletal muscles* only. Because these motor responses can be consciously controlled, the action of this part of the PNS is *voluntary*.

The ANS (the focus of Chapter 18) consists of motor neurons that conduct nerve impulses from the CNS to *smooth muscle, cardiac muscle*, and *glands*. Because its motor responses are not normally under conscious control, the action of the ANS is *involuntary*. The ANS consists of two divisions, *sympathetic* and *parasympathetic*. With a few exceptions, effectors are innervated by both divisions, and the two divisions usually have opposing actions. For example, sympathetic neurons speed the heartbeat, and parasympathetic neurons slow it down. In general, the sympathetic division helps support exercise and/or emergency actions, so-called fight-or-flight responses, and the parasympathetic division takes care of "rest-and-digest" activities.

The enteric system is the "brain of the gut," and its operation is involuntary. Its neurons extend most of the length of the gastrointestinal (GI) tract. Sensory neurons of the enteric nervous system monitor chemical changes within the GI tract and the stretching of its walls. Enteric motor neurons govern contraction of GI tract smooth muscle, secretions of the GI tract organs, such as acid secretion by the stomach, and activity of GI tract endocrine cells.

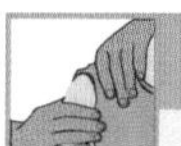

MANUAL THERAPY APPLICATION

Physiological Effects of Appropriate Massage on Nervous Tissue

Our consideration of nervous tissue encompasses five chapters (Chapters 15–19) and this presentation includes the physiological effects of massage on certain structures that will be studied later. Most comments, however, apply to all nervous tissue. The following description is not complete, but includes many of the widely accepted **physiological effects of appropriate massage on nervous tissue:**

1. Depending on the techniques utilized, massage can be either stimulative or sedative to nervous tissue as well as other tissues associated with it, for example, muscles.
2. Massage releases or reduces emotional stress.
3. General massage tends to quiet the sympathetic division of the autonomic nervous system (see Chapter 18), that portion of the nervous system that responds to fight-or-flight situations.
4. Massage enhances the development and growth of nervous tissue, especially in newborn children (see Chapter 29).
5. Massage affects exteroceptors, interoceptors, and proprioceptors which, through reflexes, affect a large number of internal organs.
6. Massage increases the production and release of a number of neurotransmitters and other substances from nervous tissue that facilitate homeostasis.

CHECKPOINT

1. What are the components of the CNS and PNS?
2. What kinds of problems would result from damage of sensory neurons, interneurons, and motor neurons?
3. What are the components and functions of the somatic, autonomic, and enteric nervous systems? Which subdivisions have involuntary actions?

15.2 HISTOLOGY OF NERVOUS TISSUE

OBJECTIVES

- Contrast the histological characteristics and functions of neurons and neuroglia.
- Distinguish between gray matter and white matter.

Nervous tissue consists of two types of cells: neurons and neuroglia. ***Neurons*** (nerve cells) are the basic information-processing units of the nervous system and are specialized for nerve impulse (action potential) conduction. They provide most of the unique functions of the nervous system, such as sensing, thinking, remembering, controlling muscle activity, and regulating glandular secretions. ***Neuroglia*** (noo-RŌG-lē-a; *glia* = glue) support, nourish, and protect the neurons and maintain homeostasis in the interstitial fluid that bathes neurons.

Neurons

Neurons usually have three parts: (1) a cell body, (2) dendrites, and (3) an axon (Figure 15.3). The ***cell body*** contains a nucleus surrounded by cytoplasm that includes typical organelles such as rough endoplasmic reticulum and ribosomes (known as Nissl bodies in neurons), lysosomes, mitochondria, and a Golgi complex (Figure 15.3). Most cellular molecules needed for a neuron's operation are synthesized in the cell body.

Two kinds of processes (extensions) emerge from the cell body of a neuron: multiple dendrites and a single axon. The cell body and the ***dendrites*** (= little trees) are the receiving or input parts of a neuron. Usually, dendrites are short, tapering, and highly branched, forming a tree-shaped array of processes that emerge from the cell body. The second type of process, the ***axon***, conducts nerve impulses toward another neuron, a muscle fiber, or a gland cell. An axon is a long, thin, cylindrical projection that often joins the cell body at a cone-shaped elevation called the ***axon hillock*** (= small hill). The part of the axon closest to the axon hillock is the ***initial segment.*** In most neurons nerve impulses arise at the junction of the axon hillock and the initial segment, an area called the ***trigger zone,*** from which they travel along the axon to their destination. An axon contains mitochondria, microtubules, and neurofibrils. Because rough endoplasmic reticulum is not present, protein synthesis does not occur in the axon. The cytoplasm of an axon, called ***axoplasm,*** is surrounded by a plasma membrane known as the ***axolemma*** (*lemma* = sheath or husk). Some axons have side branches called ***axon collaterals.*** The axon and axon collaterals end by dividing into many fine processes called ***axon terminals*** (Figure 15.3).

It should be noted that axons and dendrites are often collectively referred to as *nerve fibers*. Remember from Chapter 4 that connective tissue *fibers*, such as collagen fibers, are nonliving structures made by cells. In Chapter 10 you learned that a muscle *fiber* was another name for the entire muscle cell. Be aware of the different uses of the same word.

The site where two neurons or a neuron and an effector cell can communicate is termed a ***synapse.*** The tips of most axon terminals swell into ***synaptic end bulbs.*** These bulb-shaped structures contain *synaptic vesicles*, tiny sacs that store chemicals called *neurotransmitters*. The neurotransmitter molecules released from synaptic vesicles are the means of communication at a synapse.

Because some substances synthesized or recycled in the neuron cell body are needed in the axon or at the axon terminals, two types of transport systems carry materials from the cell body to the axon terminals and back. The slower system, which moves materials about 1–5 mm per day, is called ***slow axonal transport.*** It conveys axoplasm in one direction only—from the cell body toward the axon terminals. Slow axonal transport supplies new axoplasm to developing or regenerating axons and replenishes axoplasm in growing and mature axons.

Fast axonal transport, which is capable of moving materials a distance of 200–400 mm per day, uses proteins that function as "motors" to move materials in both directions—away from and toward the cell body—along the surfaces of microtubules. Fast axonal transport moves various organelles and materials that form the membranes of the axolemma, synaptic end bulbs, and synaptic vesicles. Some materials transported back to the cell body are degraded or recycled; others influence neuronal growth.

Structural Diversity in Neurons

Neurons display great diversity in size and shape. For example, their cell bodies range in diameter from 5 micrometers (μm) (slightly smaller than a red blood cell) up to 135 μm (barely large enough to see with the unaided eye). The pattern of dendritic branching is varied and distinctive for neurons in different parts of the nervous system. A few small neurons lack an axon, and many others have very short axons. The longest axons are almost as long as a person is tall, extending from the toes to the lowest part of the brain.

Both functional and structural features are used to classify the various neurons in the body. Recall that neurons are classified as sensory neurons, interneurons, or motor neurons based on function. Structurally, neurons are classified according to the number of processes extending from the cell body (Figure 15.4 on page 409).

1. ***Multipolar neurons*** usually have several dendrites and one axon (see also Figure 15.3). Most neurons in the brain and spinal cord are of this type.
2. ***Bipolar neurons*** have one main dendrite and one axon. They are found only in the retina of the eye, in the inner ear, and in the olfactory (*olfact-* = to smell) area of the brain (Figure 15.4b).

Figure 15.3 Structure of a typical multipolar neuron. Arrows indicate the direction of information flow: dendrites → cell body → axon → axon terminals → synaptic end bulbs. The break indicates that the axon is longer than shown

The basic parts of a typical neuron are several dendrites, a cell body, and an axon.

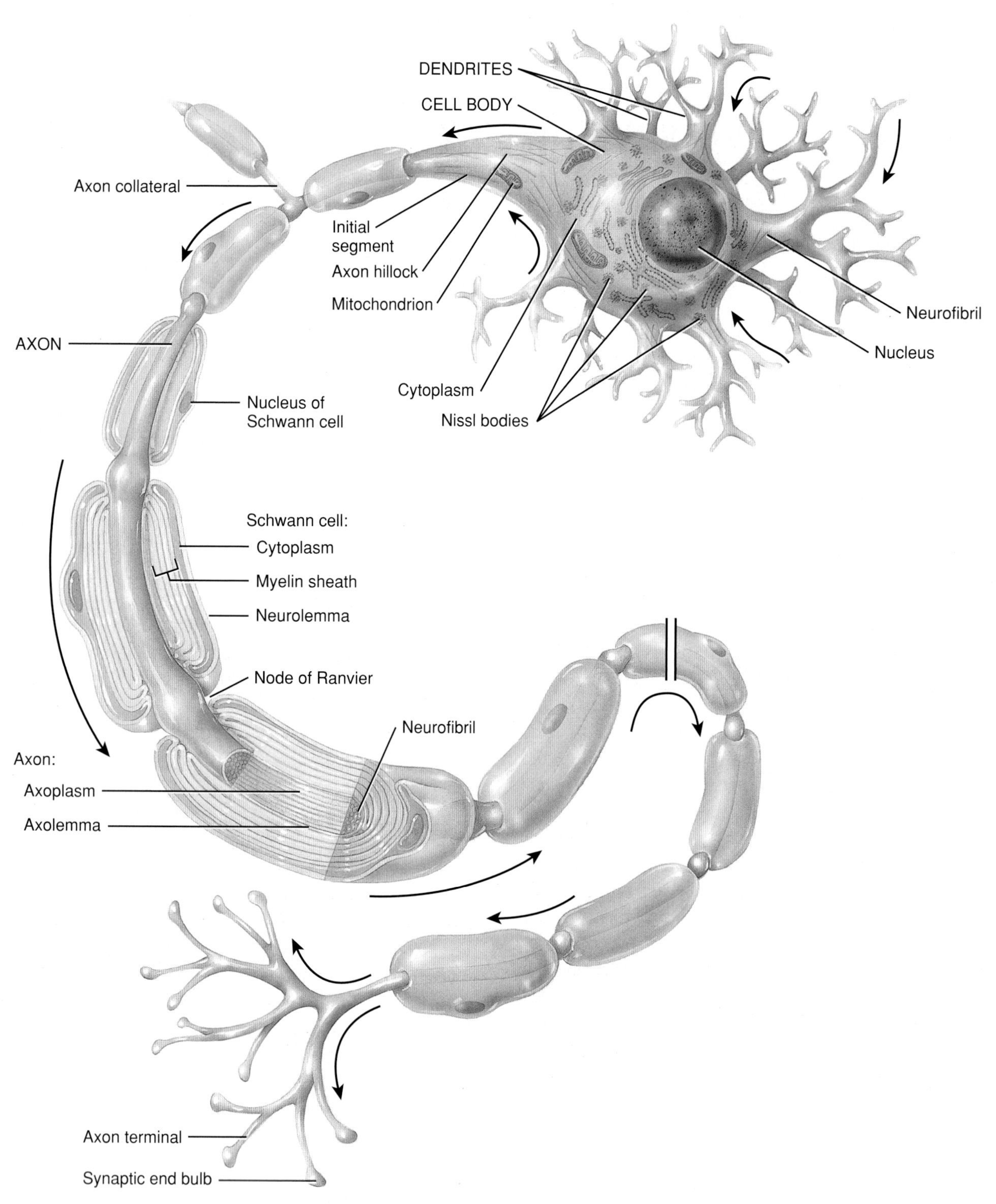

What roles do the dendrites, cell body, and axon play in communication of signals?

3. ***Unipolar neurons*** are sensory neurons that begin in the embryo as bipolar neurons. During development, the axon and dendrite fuse into a single process that divides into two branches a short distance from the cell body. Both branches have the characteristic structure and function of an axon. They are long, cylindrical processes that propagate action potentials. However, the

Figure 15.4 Structural classification of neurons. Breaks indicate that axons are longer than shown; and arrows indicate the direction of information flow.

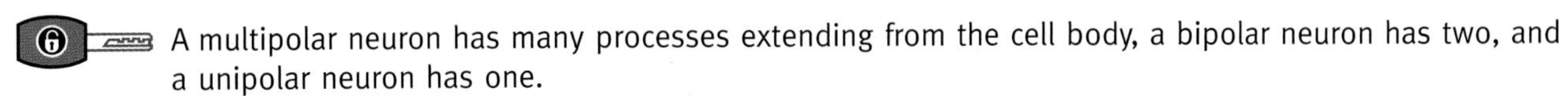

A multipolar neuron has many processes extending from the cell body, a bipolar neuron has two, and a unipolar neuron has one.

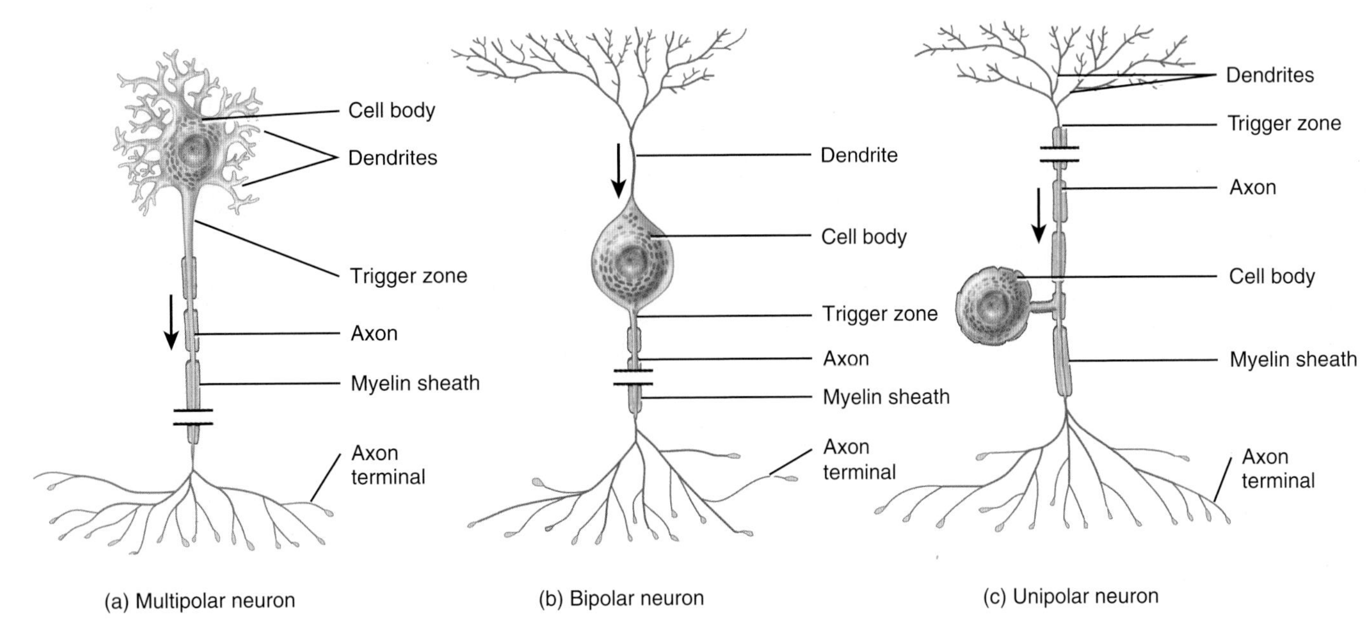

What process occurs at a trigger zone?

axon branch that extends into the periphery has dendrites at its distal tip, whereas the axon branch that extends into the CNS ends in synaptic end bulbs. The dendrites monitor a sensory stimulus such as touch or stretching. The trigger zone for nerve impulses in a unipolar neuron is at the junction of the dendrites and axon (Figure 15.4c). The impulses then propagate toward the synaptic end bulbs. The cell bodies of most unipolar neurons are located in the ganglia of spinal and cranial nerves.

Some neurons are named for the histologist who first described them or for an aspect of their shape or appearance; examples include ***Purkinje cells*** (pur-KIN-jē) in the cerebellum (Figure 15.5a) and ***pyramidal cells*** (pi-RAM-i-dal), found in the cerebral cortex of the brain, which have pyramid-shaped cell bodies (Figure 15.5b). Often, a neuron can be identified by a distinctive pattern of dendritic branching. Note in both of these cells that multiple dendrites have the potential of receiving multiple incoming signals that may or may not lead to a nerve impulse over the single axon.

Myelination

The axons of most neurons are surrounded by a ***myelin sheath,*** a many-layered covering composed of lipid and protein (Figure 15.6). Like insulation covering an electrical wire, the myelin sheath insulates the axon of a neuron and increases the speed of nerve impulse conduction. Two types of neuroglia produce myelin sheaths: Schwann cells (in the PNS) and oligodendrocytes (in the CNS). Each Schwann cell wraps about 1

Figure 15.5 Two examples of CNS neurons. Arrows indicate the direction of information flow.

The dendritic branching pattern often is distinctive for a particular type of neuron.

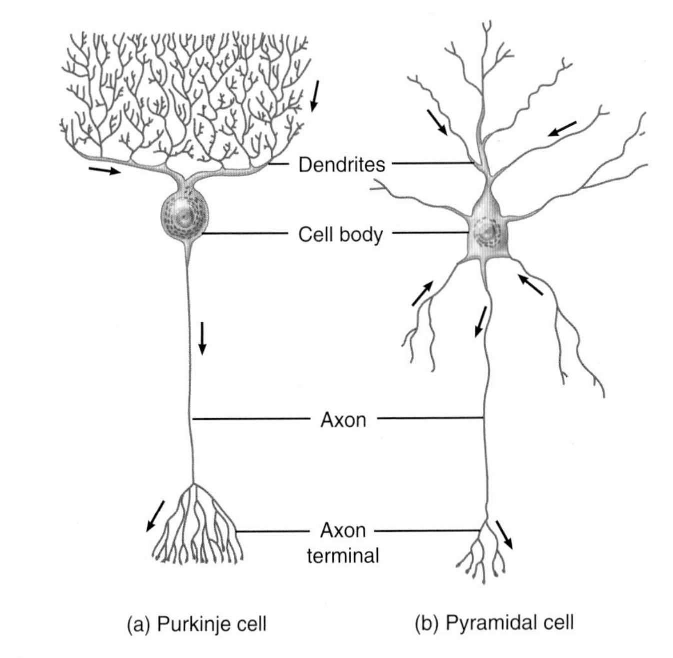

Where do pyramidal cells get their name?

Figure 15.6 Myelinated and unmyelinated axons. Notice that one layer of Schwann cell plasma membrane nearly surrounds each unmyelinated axon.

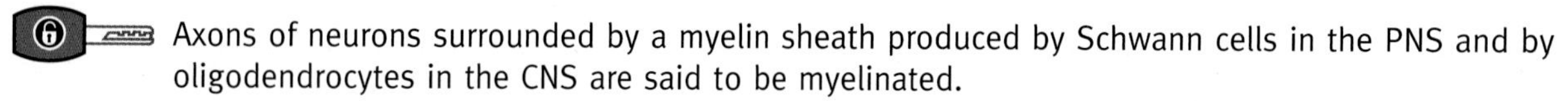

Axons of neurons surrounded by a myelin sheath produced by Schwann cells in the PNS and by oligodendrocytes in the CNS are said to be myelinated.

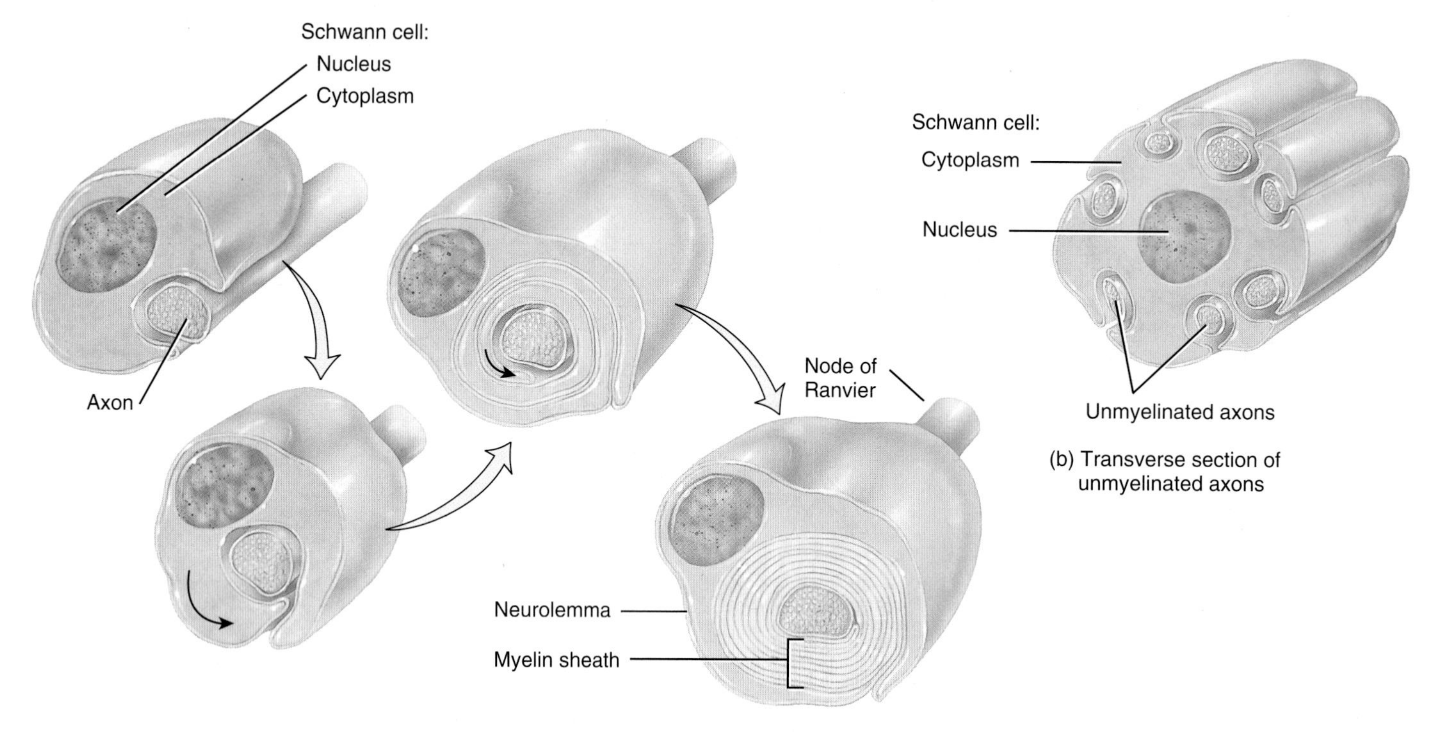

(a) Transverse sections of stages in the formation of a myelin sheath

(b) Transverse section of unmyelinated axons

What is the functional advantage of myelination?

millimeter (1 mm = 0.04 in.) of a single axon's length by spiraling many times around the axon (Figure 15.6a). Eventually, multiple layers of glial plasma membrane surround the axon, with the Schwann cell's cytoplasm and nucleus forming the outermost layer. The inner portion, consisting of up to 100 layers of Schwann cell membrane, is the myelin sheath. The outer nucleated cytoplasmic layer of the Schwann cell, which encloses the myelin sheath, is the ***neurolemma (sheath of Schwann).*** A neurolemma is found only around axons in the PNS. When an axon is injured, the neurolemma aids regeneration by forming a regeneration tube that guides and stimulates regrowth of the axon. Gaps in the myelin sheath, called ***nodes of Ranvier*** (RON-vē-ā), appear at intervals along the axon (see also Figure 15.3). Each Schwann cell (neurolemmocyte) wraps one axon segment between two nodes.

In the CNS, an oligodendrocyte myelinates parts of several axons. Each oligodendrocyte puts forth about 15 broad, flat processes that spiral around CNS axons, forming multiple myelin sheaths (see Table 15.2). A neurolemma is not present, however, because the oligodendrocyte cell body and nucleus do not envelop the axon. Nodes of Ranvier are present, but they are fewer in number. Axons in the CNS display little regrowth after injury. This is thought to be due, in part, to the absence of a neurolemma, and in part to an inhibitory influence exerted by the oligodendrocytes on axon regrowth. Axons with a myelin sheath are said to be ***myelinated,*** and those without are said to be ***unmyelinated.***

The amount of myelin increases from birth to maturity, and its presence greatly increases the speed of nerve impulse conduction. By the time a baby starts to talk, most myelin sheaths are partially formed, but myelination continues into the teenage years. An infant's responses to stimuli are neither so rapid nor so

• CLINICAL CONNECTION Multiple Sclerosis

Multiple sclerosis (MS) is a disease that causes progressive destruction of myelin sheaths of neurons in the CNS. It afflicts about 2 million people worldwide and affects females twice as often as males. MS is an autoimmune disease—the body's own immune system spearheads the attack. The condition's name describes the anatomical pathology: In *multiple* regions, the myelin sheaths deteriorate to *scleroses,* which are hardened scars or plaques. The destruction of myelin sheaths slows and then short-circuits the conduction of nerve impulses.

The most common form of the condition is *relapsing–remitting MS,* which usually appears in early adulthood. The first symptoms may include a feeling of heaviness or weakness in the muscles, abnormal sensations, or double vision. An attack is followed by a period of remission during which the symptoms temporarily disappear. One attack follows another over the years. The result is a progressive loss of function interspersed with remission periods, during which symptoms abate. •

coordinated as those of an older child or a young adult, in part because myelination is still in progress during infancy.

Gray and White Matter

In a freshly dissected section of the brain or spinal cord, some regions look white and glistening, and others appear gray. The ***white matter*** of nervous tissue consists primarily of myelinated axons of many neurons. The whitish color of myelin, the phospholipid bilayer, gives white matter its name. The ***gray matter*** of nervous tissue contains neuronal cell bodies, dendrites, unmyelinated axons, axon terminals, and neuroglia. It looks grayish, rather than white, because the cellular organelles impart a gray color and there is little or no myelin in these areas. Blood vessels are present in both white and gray matter.

In the spinal cord, the outer white matter surrounds an inner core of gray matter shaped like a butterfly or the letter H (Figure 15.7a). In the brain, a thin shell of gray matter (cortex) covers the surface of the largest parts of the brain, the cerebrum and cerebellum (see Figures 17.2 and 17.3). When used to describe nervous tissue, a ***nucleus*** is a cluster of neuronal cell bodies within the CNS. Many nuclei of gray matter lie deep within the brain. Much of the CNS white matter consists of ***tracts,*** which are bundles of axons in the CNS that extend for some distance up or down the spinal cord or connect parts of the brain with each other and with the spinal cord (Table 15.1).

TABLE 15.1

Summary of Terminology

	COLLECTION OF NERVE CELL BODIES	COLLECTION OF NERVE FIBERS
Central nervous system	Nucleus	Tract
Peripheral nervous system	Ganglion	Nerve

Neuroglia

Neuroglia or *glial cells* constitute about half the volume of the CNS. Their name derives from the idea of early histologists that they were the "glue" that held nervous tissue together. We now know that neuroglia are not merely passive bystanders but rather active participants in the operation of nervous tissue. Generally, neuroglia are smaller than neurons, and they are 5 to 50 times more numerous. In contrast to neurons, glia do not generate or conduct nerve impulses, and they can multiply and divide in the mature nervous system. In cases of injury or disease, neuroglia multiply to fill in the spaces formerly occupied by neurons. Brain tumors derived from glia, called ***gliomas,*** often are highly malignant and grow rapidly. Of the six types of neuroglia, ***astrocytes, oligodendrocytes, microglia,*** and ***ependymal cells*** are found only in the CNS. The remaining two types—***Schwann cells*** and ***satellite cells***—are present in the PNS. Table 15.2 shows the appearance of neuroglia and lists their functions.

CHECKPOINT

4. What are the functions of the dendrites, cell body, axon, and synaptic end bulbs of a neuron?
5. Which cells produce myelin in nervous tissue, and what is the function of a myelin sheath?
6. What are the functions of neuroglia?

Figure 15.7 Distribution of gray and white matter in the spinal cord and brain.

White matter primarily consists of myelinated axons of many neurons. Gray matter consists of neuron cell bodies, dendrites, axon terminals, unmyelinated axons, and neuroglia.

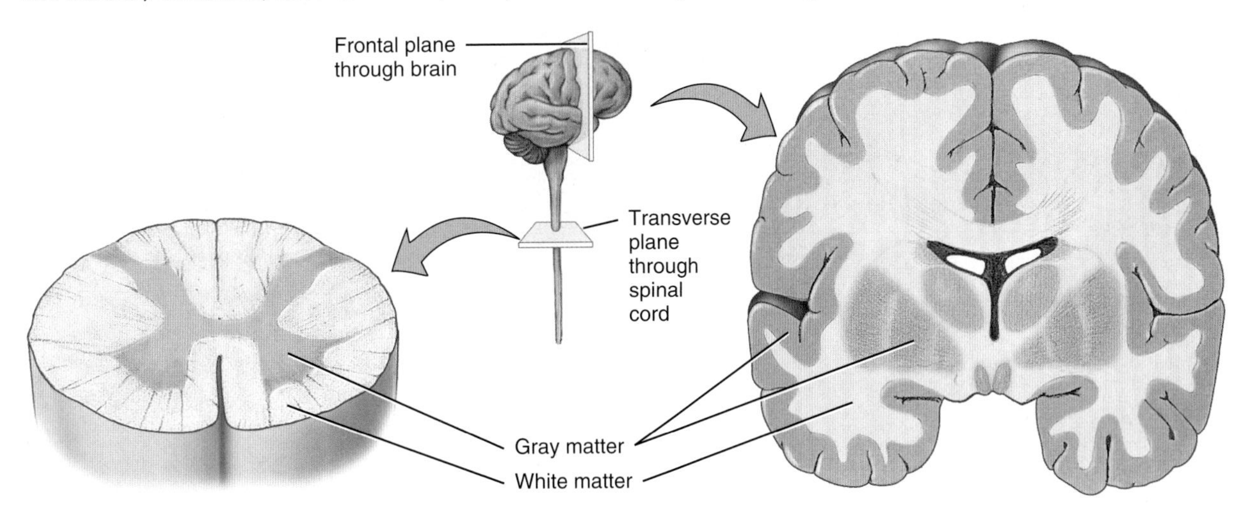

What is responsible for the white appearance of white matter?

TABLE 15.2

Neuroglia in the CNS and PNS

TYPE OF NEUROGLIAL CELL	FUNCTIONS	TYPE OF NEUROGLIAL CELL	FUNCTIONS
CENTRAL NERVOUS SYSTEM			
Astrocytes (AS-trō-sīts; *astro-* = star; *-cyte* = cell).	Support neurons; help protect neurons from harmful substances; help maintain proper chemical environment for generation of nerve impulses; assist with growth and migration of neurons during brain development; play a role in learning and memory; help form the blood–brain barrier.	**Oligodendrocytes** (OL-i-gō-den′-drō-sīts; *oligo-* = few; *-dendro-* = tree)	Produce and maintain myelin sheath around several adjacent axons of CNS neurons.
Microglia (mī-KROG-lē-a; *micro-* = small)	Protect CNS cells from some diseases by engulfing invading microbes; migrate to areas of injured nerve tissue where they clear away debris of dead cells.	**Ependymal cells** (ep-EN-di-mal; *epen-* = above; *-dym-* = garment)	Line ventricles of the brain (cavities filled with cerebrospinal fluid) and central canal of the spinal cord; form cerebrospinal fluid and assist in its circulation.
PERIPHERAL NERVOUS SYSTEM			
Schwann cells (SCHWON)	Produce and maintain myelin sheath around a single axon of a PNS neuron; participate in regeneration of PNS axons.	**Satellite cells** (SAT-i-līt).	Support neurons in PNS ganglia and regulate exchange of materials between neurons and interstitial fluid.

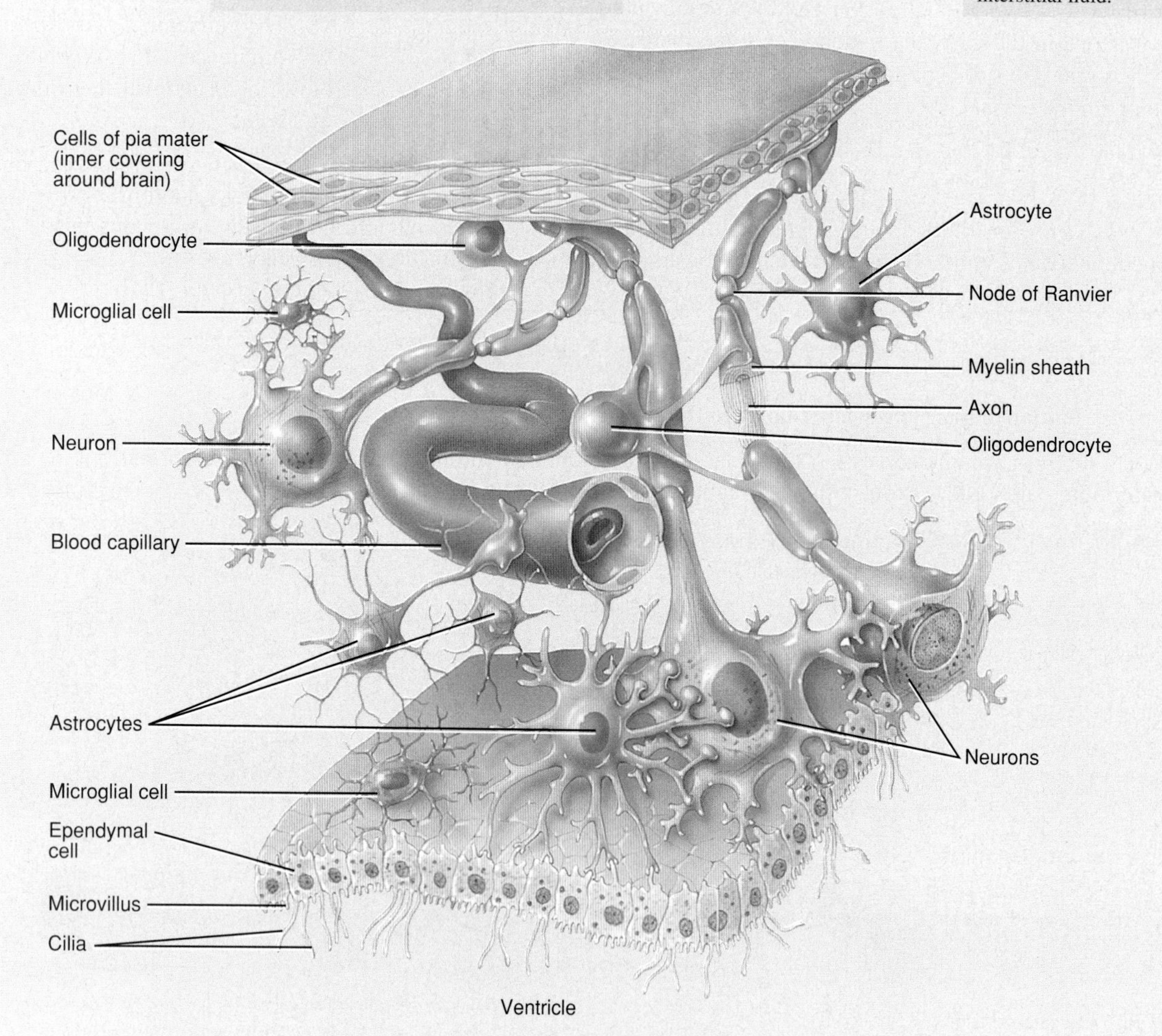

15.3 ELECTRICAL SIGNALS IN NEURONS

OBJECTIVE

- Describe how a nerve impulse is generated and conducted.

Like muscle fibers, neurons are electrically excitable. They communicate with one another using two types of electrical signals: (1) *Graded potentials* are used only for short-distance communication. (2) *Action potentials* allow communication over both short and long distances within the body. Recall that an action potential in a muscle fiber is called a *muscle action potential.* When an action potential occurs in a neuron (nerve cell), it is called a *nerve action potential (nerve impulse).* To understand the functions of graded potentials and action potentials, consider how the nervous system allows you to feel the smooth surface of a pen that you have picked up from a table (Figure 15.8):

1. As you touch the pen, a graded potential develops in a sensory receptor in the skin of the fingers.
2. The graded potential triggers the axon of the sensory neuron to form a nerve action potential, which travels along the axon into the CNS and ultimately causes the release of neurotransmitter at a synapse with an interneuron.
3. The neurotransmitter stimulates the interneuron to form a graded potential in its dendrites and cell body.
4. In response to the graded potential, the axon of the interneuron forms a nerve action potential. The nerve action potential travels along the axon, which results in the release of neurotransmitter at the next synapse with another interneuron.
5. This process of neurotransmitter release at a synapse followed by the formation of a graded potential and then a nerve action potential occurs over and over as interneurons in higher parts of the brain (such as the thalamus and cerebral cortex) are activated. Once interneurons are activated in the ***cerebral cortex,*** the outer part of the brain, perception occurs and you are able to feel the smooth surface of the pen touch your fingers. As you will learn in Chapter 17, ***perception,*** the conscious awareness of a sensation, is primarily a function of the cerebral cortex.

Suppose that you want to use the pen to write a letter. The nervous system would respond in the following way (Figure 15.8):

6. A stimulus in the brain causes a graded potential to form in the dendrites and cell body of an ***upper motor neuron,*** a type of motor neuron that synapses with a lower motor neuron farther down in the CNS to contract a skeletal muscle. The graded potential subsequently causes a nerve action potential to occur in the axon of the upper motor neuron, followed by neurotransmitter release.
7. The neurotransmitter generates a graded potential in a ***lower motor neuron,*** a type of motor neuron that directly supplies skeletal muscle fibers. The graded potential triggers the formation of a nerve action potential and then release of neurotransmitter at neuromuscular junctions formed with skeletal muscle fibers that control movements of the fingers.
8. The neurotransmitter stimulates the formation of muscle action potentials in these muscle fibers. The muscle action potentials cause the muscle fibers of the fingers to contract, which allows you to write with the pen.

Action Potentials

Neurons communicate with one another by means of ***nerve action potentials,*** also called ***nerve impulses.*** Recall from Chapter 10 that a muscle fiber contracts in response to a muscle action potential. The generation of action potentials in both muscle fibers and neurons depends on two basic features of the plasma membrane: the existence of a resting membrane potential and the presence of specific types of ion channels. Many body cells exhibit a ***membrane potential,*** a difference in the amount of electrical charge on the inside of the plasma membrane as compared to the outside. The membrane potential is like voltage stored in a battery. A cell that has a membrane potential is said to be ***polarized.*** When muscle fibers and neurons are "at rest" (not conducting action potentials), the voltage across the plasma membrane is termed the ***resting membrane potential.***

If you connect the positive and negative terminals of a battery with a piece of metal (look in the battery compartment of your portable radio), an *electrical current* carried by electrons flows from the battery, allowing you to listen to your favorite music. In living tissues, the flow of *ions* (rather than electrons) constitutes electrical currents. The main sites where ions can flow across the membrane are through the pores of various types of ion channels.

Graded potentials and action potentials occur because the membranes of neurons contain many different kinds of ion channels that open or close in response to specific stimuli. Because the lipid bilayer of the plasma membrane is a good electrical insulator, the main paths for current to flow across the membrane are through the ion channels.

Ion Channels

When open, ion channels allow specific ions to diffuse across the plasma membrane from where the ions are more concentrated to where they are less concentrated. Similarly, positively charged ions will move toward a negatively charged area, and negatively charged ions will move toward a positively charged area. As ions diffuse across a plasma membrane to equalize differences in charge or concentration, the result is a flow of current that can change the membrane potential.

There are several types of ion channels. ***Leakage channels*** allow a small but steady stream of ions to leak across the membrane following the electrochemical gradient. Because plasma membranes typically have many more potassium ion (K^+) leakage channels than sodium ion (Na^+) leakage channels, the membrane's permeability to K^+ is much higher than its permeability to Na^+. ***Gated channels,*** in contrast, open and close on command in response to chemical or mechanical stimuli (see Figure 3.7 in Section 3.3 Transport Across the Plasma Membrane). ***Voltage-gated channels,*** channels that open in response to a change in membrane potential, are used to generate and conduct action potentials.

Figure 15.8 Overview of nervous system functions.

Graded potentials and nerve and muscle action potentials are involved in the relay of sensory stimuli, integrative functions such as perception, and motor activities.

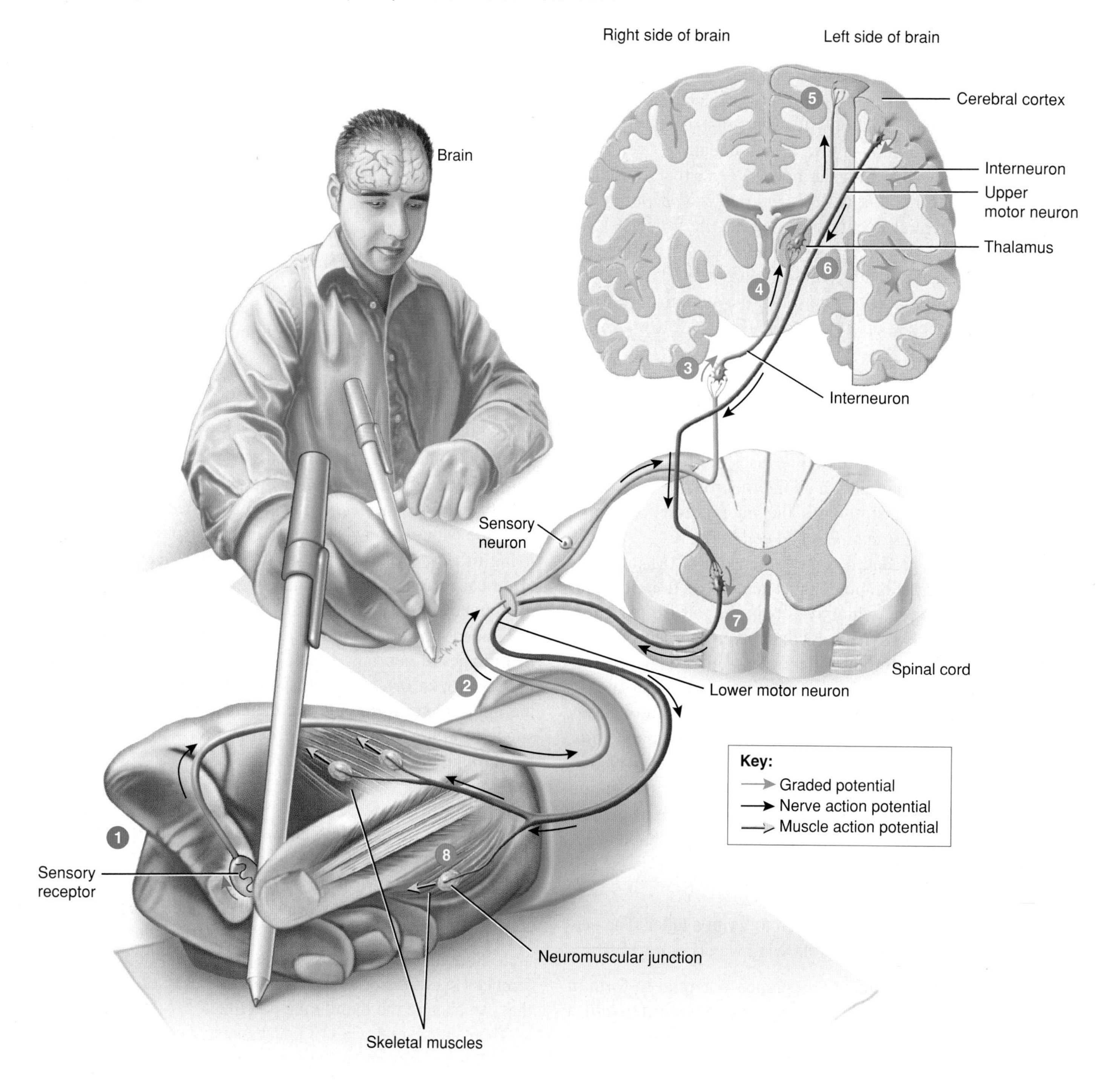

In which region of the brain does perception primarily occur?

Resting Membrane Potential

In a resting neuron, the outside surface of the plasma membrane has a relatively positive charge and the inside surface has a negative charge. The separation of positive and negative electrical charges is a form of potential energy, which can be measured in volts. For example, two 1.5-volt batteries can power a portable CD player. Voltages produced by cells typically are much smaller and are measured in millivolts (1 millivolt = 1 mV = 1/1000 volt). In neurons, the resting membrane potential is about –70 mV. The minus sign indicates that the inside of the membrane is negative relative to the outside.

The resting membrane potential arises from the unequal distributions of various ions in cytosol and interstitial fluid (Figure 15.9). Interstitial fluid is rich in sodium ions (Na^+) and chloride ions (Cl^-). Inside cells, the main positively charged ions in the cytosol are potassium ions (K^+), and the two dominant negatively charged ions are phosphates attached to organic molecules, such as the three phosphates in ATP (adenosine triphosphate), and amino acids in proteins. Because the concentration of K^+ is higher in cytosol and since plasma membranes have many K^+ leakage channels, potassium ions diffuse down their concentration gradient—out of cells into the interstitial fluid. As more and more positive potassium ions exit, the inside of the membrane becomes increasingly negative, and the outside of the membrane becomes increasingly positive. Another factor contributes to the negativity on the inner cell membrane. Most negatively charged ions inside the cell are not free to leave. They cannot follow the K^+ out of the cell as they are attached to large proteins or other large molecules.

Membrane permeability to Na^+ is very low because there are only a few sodium leakage channels. Nevertheless, sodium ions do slowly diffuse inward, down their concentration gradient. Left unchecked, such inward leakage of Na^+ would eventually dissipate the resting membrane potential. The small inward Na^+ leak and outward K^+ leak are offset by the action of the sodium–potassium pumps (see Figure 3.11), which help maintain the resting membrane potential by pumping out Na^+ as fast as it leaks in. At the same time, the sodium–potassium pumps bring K^+ back into the cell.

Generation of Action Potentials

An ***action potential (AP)*** is a sequence of rapidly occurring events that decrease and reverse the membrane potential and eventually restore it to the resting state. The ability of muscle fibers and neurons to convert stimuli into action potentials is called ***electrical excitability***. A ***stimulus*** is anything in the cell's environment that can change the resting membrane potential. When a stimulus causes the membrane

Figure 15.9 The distribution of ions that produces the resting membrane potential.

The resting membrane potential is determined by three major factors: (1) unequal distribution of ions in the interstitial fluid and cytosol; (2) inability of most anions to leave the cell; and (3) the Na^+/K^+ pump.

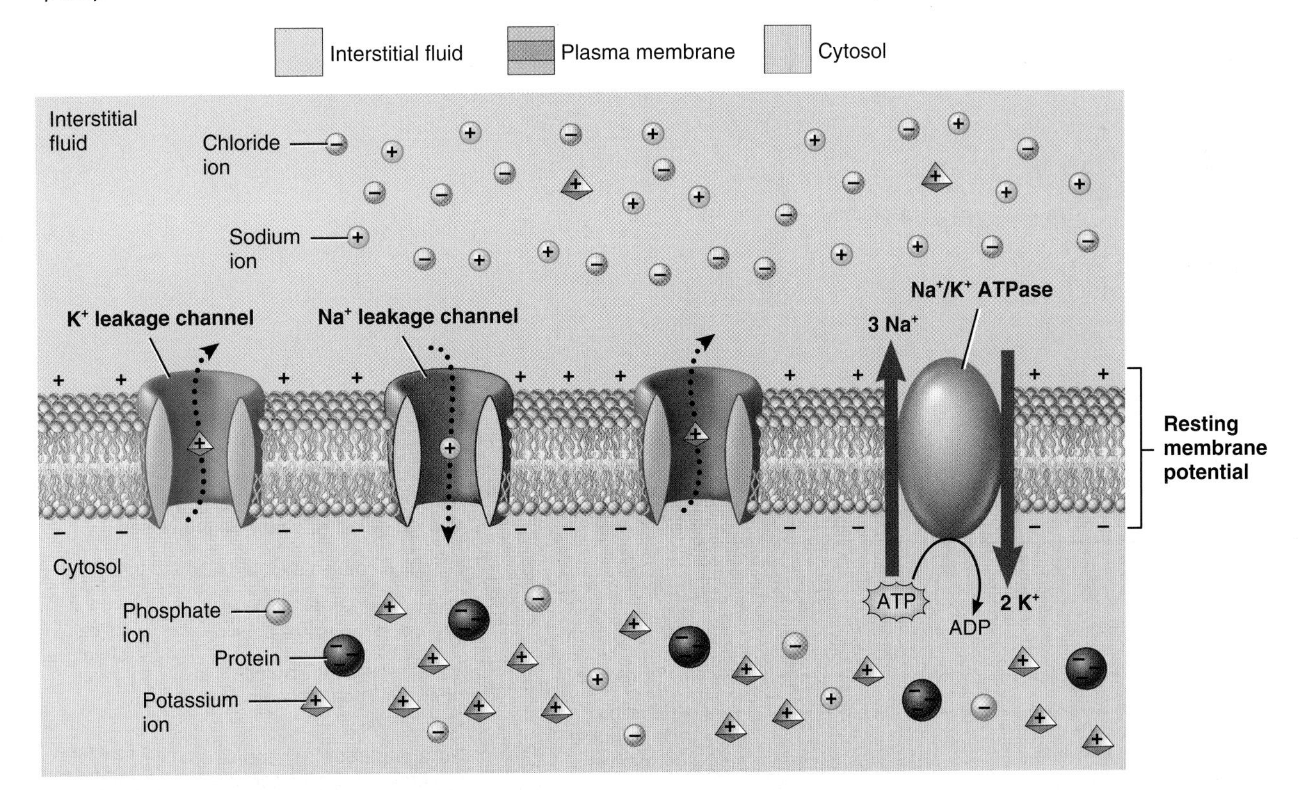

Suppose that the plasma membrane of a neuron has more Na^+ leakage channels than K^+ leakage channels. What effect would this have on the resting membrane potential?

to depolarize to a critical level, called ***threshold*** (typically about –55 mV), then an action potential arises. An action potential has two main phases: a depolarizing phase and a repolarizing phase (Figure 15.10). During ***depolarization*** the negative membrane potential becomes less negative, reaches zero, and then becomes positive. During ***repolarization,*** the membrane potential is restored to its resting state of −70 mV. In neurons, the depolarizing and repolarizing phases of an action potential typically last about one millisecond (1/1000 sec).

During an action potential, depolarization to threshold briefly opens two types of voltage-gated ion channels. In neurons, these channels are present mainly in the plasma membrane of the axon and axon terminals. First, a threshold depolarization opens voltage-gated Na^+ channels. As these channels open, about 20,000 sodium ions rush into the cell, resulting in depolarization. The inflow of Na^+ causes the membrane potential to pass 0 mV and finally reach +30 mV. Second, the threshold depolarization also opens voltage-gated K^+ channels. The voltage-gated K^+ channels open more slowly, so their opening occurs at about the same time that the voltage-gated Na^+ channels are automatically closing. As the K^+ channels open, potassium ions flow out of the cell, producing the repolarizing phase.

While the voltage-gated K^+ channels are open, outflow of K^+ may be large enough to cause an ***after-hyperpolarizing phase*** of the action potential (Figure 15.10). During hyperpolarization, the membrane potential becomes even *more negative* than the resting level. Finally, as K^+ channels close, the membrane potential returns to the resting level of −70 mV.

Action potentials arise according to the ***all-or-none principle.*** As long as a stimulus is strong enough to cause depolarization to threshold, the voltage-gated Na^+ and K^+ channels open, and an action potential occurs. A much stronger stimulus cannot cause a larger action potential because the size of an action potential is always the same. A weak stimulus that fails to cause a threshold-level depolarization does not elicit an action potential. For a brief time after an action potential begins, a muscle fiber or neuron cannot conduct another action potential. This time is called the ***refractory period.*** During the ***absolute refractory period,*** even a very strong stimulus cannot initiate a second action potential. The ***relative refractory period*** is the period of time during which a scecond action potential can be initiated, but only by a larger-than-normal stimulus.

Figure 15.10 Action potential (AP) or impulse. When a stimulus depolarizes the membrane to threshold (−55 mV), an AP is generated. The action potential arises at the trigger zone (here, at the junction of the axon hillock and the initial segment) and then propagates along the axon to the axon terminals. The green-colored regions of the neuron in the inset indicate the parts that typically have voltage-gated Na^+ and K^+ channels (axon plasma membrane and axon terminals).

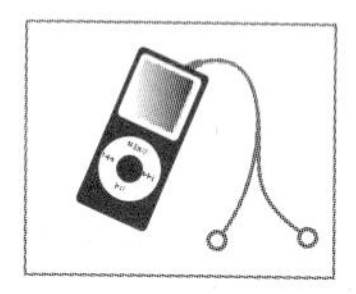

An action potential consists of a depolarizing phase and a repolarizing phase, which may be followed by an after-hyperpolarizing phase.

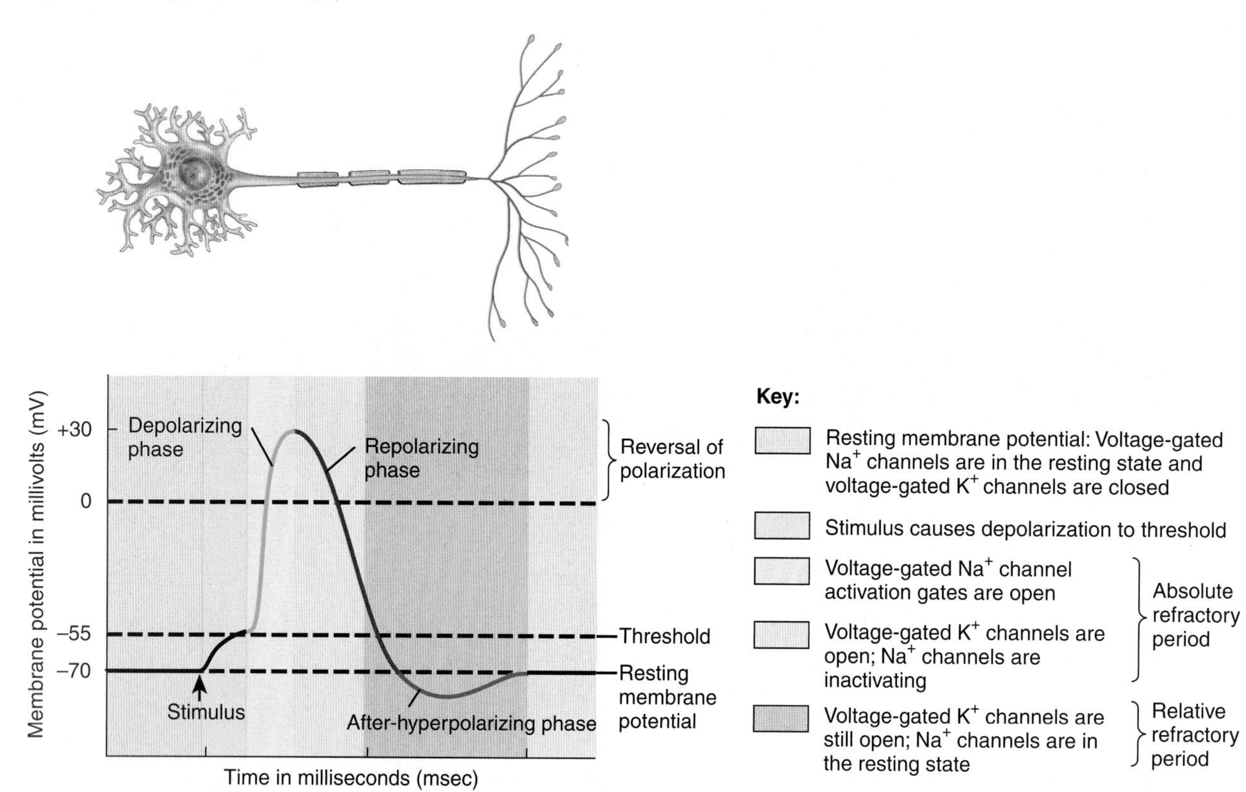

Which channels are open during the depolarizing phase? During the repolarizing phase?

• CLINICAL CONNECTION | Epilepsy

Epilepsy is a disorder characterized by short, recurrent, periodic attacks of motor, sensory, or psychological malfunction, although it almost never affects intelligence. The attacks, called *epileptic seizures*, afflict about 1% of the world's population. They are initiated by abnormal, synchronous electrical discharges from millions of neurons in the brain. As a result, lights, noises, or smells may be sensed when the eyes, ears, and nose have not been stimulated. In addition, the skeletal muscles of a person having a seizure may contract involuntarily. *Partial seizures* begin in a small focus on one side of the brain and produce milder symptoms; *generalized seizures* involve larger areas on both sides of the brain and loss of consciousness.

Epilepsy has many causes, including brain damage at birth (the most common cause); metabolic disturbances such as insufficient glucose or oxygen in the blood; infections; toxins; loss of blood or low blood pressure; head injuries; and tumors and abscesses of the brain. However, most epileptic seizures have no demonstrable cause. •

Conduction of Nerve Impulses

To communicate information from one part of the body to another, nerve impulses must travel from where they arise, usually at the trigger zone of the axon to the axon terminals. This type of impulse movement, which operates by positive feedback, is called ***conduction*** or ***propagation*** (Figure 15.11). Depolarization to threshold at the axon hillock opens voltage-gated Na^+ channels and the resulting inflow of sodium ions depolarizes the adjacent membrane to threshold, which opens even more voltage-gated Na^+ channels, a positive feedback effect. Thus, a nerve impulse self-conducts along the axon plasma membrane. This situation is similar to pushing on the first domino in a long row: when the push on the first domino is strong enough, that domino falls against the second domino, and eventually the entire row topples.

The type of action potential conduction that occurs in unmyelinated axons (and muscle fibers) is called ***continuous conduction.*** In this case, each adjacent segment of the plasma

Figure 15.11 Conduction (propagation) of a nerve impulse after it arises at the trigger zone. Dotted lines indicate ionic current flow. The insets show the path of current flow. (a) In continuous conduction along an unmyelinated axon, ionic currents flow across each adjacent segment of the membrane. (b) In saltatory conduction along a myelinated axon, the action potential (nerve impulse) at the first node generates ionic currents in the cytosol and interstitial fluid that open voltage-gated Na^+ channels at the second node, and so on at each subsequent node.

Unmyelinated axons exhibit continuous conduction; myelinated axons exhibit saltatory conduction.

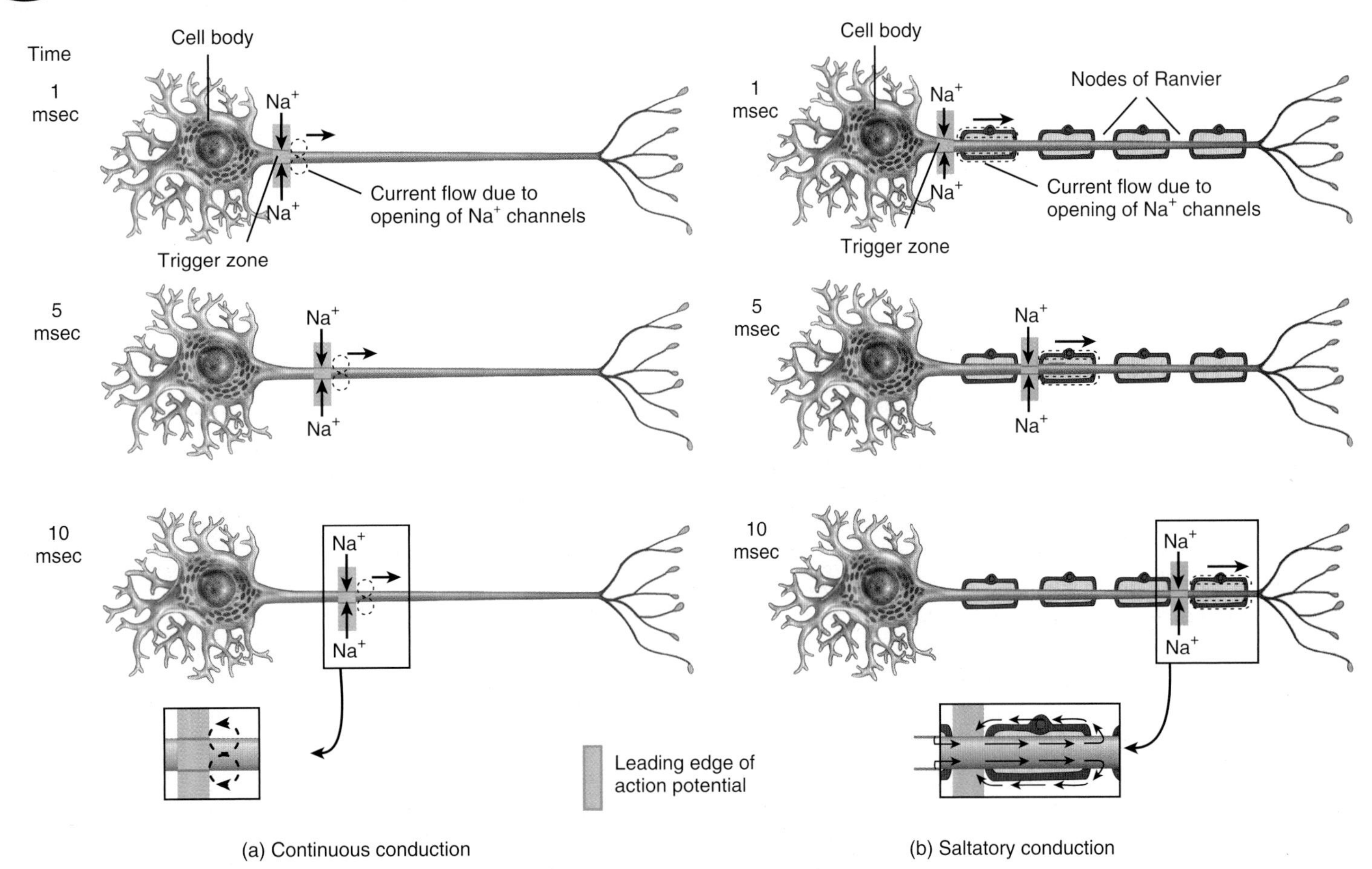

What factors determine the speed of propagation of an action potential?

membrane depolarizes to threshold and generates an action potential that depolarizes the next portion of the membrane (see Figure 15.11a). Note that the impulse has traveled only a relatively short distance after 10 milliseconds (10 msec).

In myelinated axons, conduction is somewhat different. The voltage-gated Na^+ and K^+ channels are located primarily at the nodes of Ranvier, the gaps in the myelin sheath. When a nerve impulse conducts along a myelinated axon, current carried by Na^+ and K^+ flows through the interstitial fluid surrounding the myelin sheath and through the cytosol from one node to the next (see Figure 15.11b). The nerve impulse at the first node generates ionic currents that open voltage-gated Na^+ channels at the second node and trigger a nerve impulse there. Then the nerve impulse from the second node generates an ionic current that opens voltage-gated Na^+ channels at the third node, and so on. Each node depolarizes and then repolarizes. Note that the impulse has traveled much farther along the myelinated axon in Figure 15.11b in the same time interval. Because current flows across the membrane only at the nodes, the impulse appears to leap from node to node as each nodal area depolarizes to threshold. This type of impulse conduction is called ***saltatory conduction*** (SAL-ta-tō-rē; *saltat-* = leaping).

The diameter of the axon and the presence or absence of a myelin sheath are the most important factors that determine the speed of nerve impulse conduction. Axons with large diameters conduct impulses faster than those with small diameters. Also, myelinated axons conduct impulses faster than do unmyelinated axons.

Effect of Axon Diameter

Larger-diameter axons propagate impulses faster than smaller ones due to their larger surface areas. All of the largest-diameter axons, called ***A fibers,*** are myelinated. A fibers have a brief absolute refractory period and conduct impulses at speeds of 12 to 130 m/sec (27–290 mi/hr). The axons of sensory neurons that propagate impulses associated with touch, pressure, position of joints, and some thermal sensations are A fibers, as are the axons of motor neurons that conduct impulses to skeletal muscles.

B fibers are axons with intermediate diameters. Like A fibers, B fibers are myelinated and exhibit saltatory conduction at speeds up to 15 m/sec (34 mi/hr). B fibers have a somewhat longer absolute refractory period than A fibers. B fibers conduct sensory nerve impulses from the viscera to the brain and spinal cord. They also constitute all the axons of the autonomic motor neurons that extend from the brain and spinal cord to the ANS relay stations called autonomic ganglia.

C fibers are the smallest diameter axons and all are unmyelinated. Nerve impulse propagation along a C fiber ranges from 0.5 to 2 m/sec (1–4 mi/hr). C fibers exhibit the longest absolute refractory periods. These unmyelinated axons conduct some sensory impulses for pain, touch, pressure, heat, and cold from the skin, and pain impulses from the viscera. Autonomic motor fibers that extend from autonomic ganglia to stimulate the heart, smooth muscle, and glands are C fibers. Examples of motor functions of B and C fibers are constricting and dilating the pupils, increasing and decreasing the heart rate, and contracting and relaxing the urinary bladder.

Axons with the largest diameters are all myelinated and therefore capable of saltatory conduction. The smallest diameter axons are unmyelinated, so their conduction is continuous. Axons conduct impulses at higher speeds when warmed and at lower speeds when cooled. Pain resulting from tissue injury such as that caused by a minor burn can be reduced by the application of ice because cooling slows conduction of nerve impulses along the axons of pain-sensitive neurons.

CHECKPOINT

7. What are the meanings of the terms resting membrane potential, depolarization, repolarization, nerve impulse, and refractory period?
8. How is saltatory conduction different from continuous conduction?

15.4 SYNAPTIC TRANSMISSION

OBJECTIVE

- Explain the events of synaptic transmission and the type of neurotransmitters used.

Now that you know how action potentials arise and conduct along the axon of an individual neuron, we can explore how neurons communicate with one another. At synapses, neurons communicate with other neurons or with effectors by a series of events known as ***synaptic transmission.*** In Chapter 10 we examined the events occurring at the neuromuscular junction, the synapse between a somatic motor neuron and a skeletal muscle fiber (see Figure 10.5). Synapses between neurons operate in a similar way. The neuron sending the signal is called the ***presynaptic neuron*** (*pre-* = before), and the neuron receiving the message is called the ***postsynaptic neuron*** (*post-* = after).

Events at a Synapse

Although the presynaptic and postsynaptic neurons are in close proximity at a synapse, their plasma membranes do not touch. They are separated by the ***synaptic cleft,*** a tiny space filled with interstitial fluid. Because nerve impulses cannot conduct across the synaptic cleft, an alternate, indirect form of communication occurs across this space. A typical synapse operates as follows (Figure 15.12):

1. A nerve impulse arrives at a synaptic end bulb of a presynaptic axon.
2. The depolarizing phase of the nerve impulse opens *voltage-gated Ca^{2+} channels,* which are present in the membrane of synaptic end bulbs. Because calcium ions are more concentrated in the interstitial fluid, Ca^{2+} flows into the synaptic end bulb through the opened channels.
3. An increase in the concentration of Ca^{2+} inside the synaptic end bulb triggers exocytosis of some of the ***synaptic***

Figure 15.12 Signal transmission at a chemical synapse. Through exocytosis of synaptic vesicles, a presynaptic neuron releases neurotransmitter molecules. After diffusing across the synaptic cleft, the neurotransmitter binds to receptors in the plasma membrane of the postsynaptic neuron and produces a postsynaptic potential.

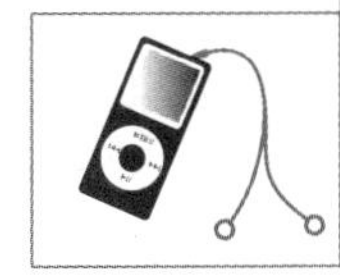

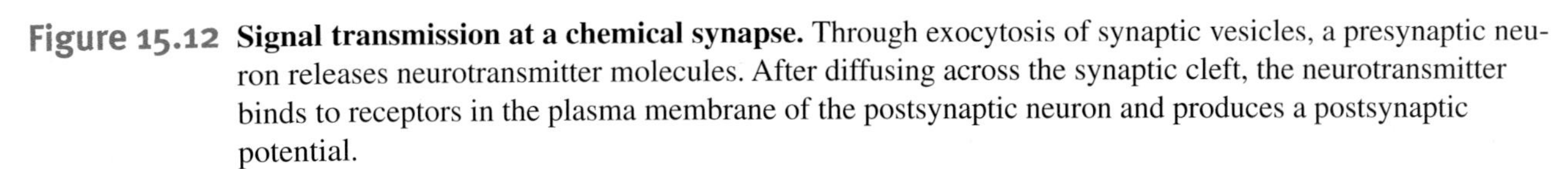

At a chemical synapse, a presynaptic neuron converts an electrical signal (nerve impulse) into a chemical signal (neurotransmitter release). The postsynaptic neuron then converts the chemical signal back into an electrical signal (postsynaptic potential).

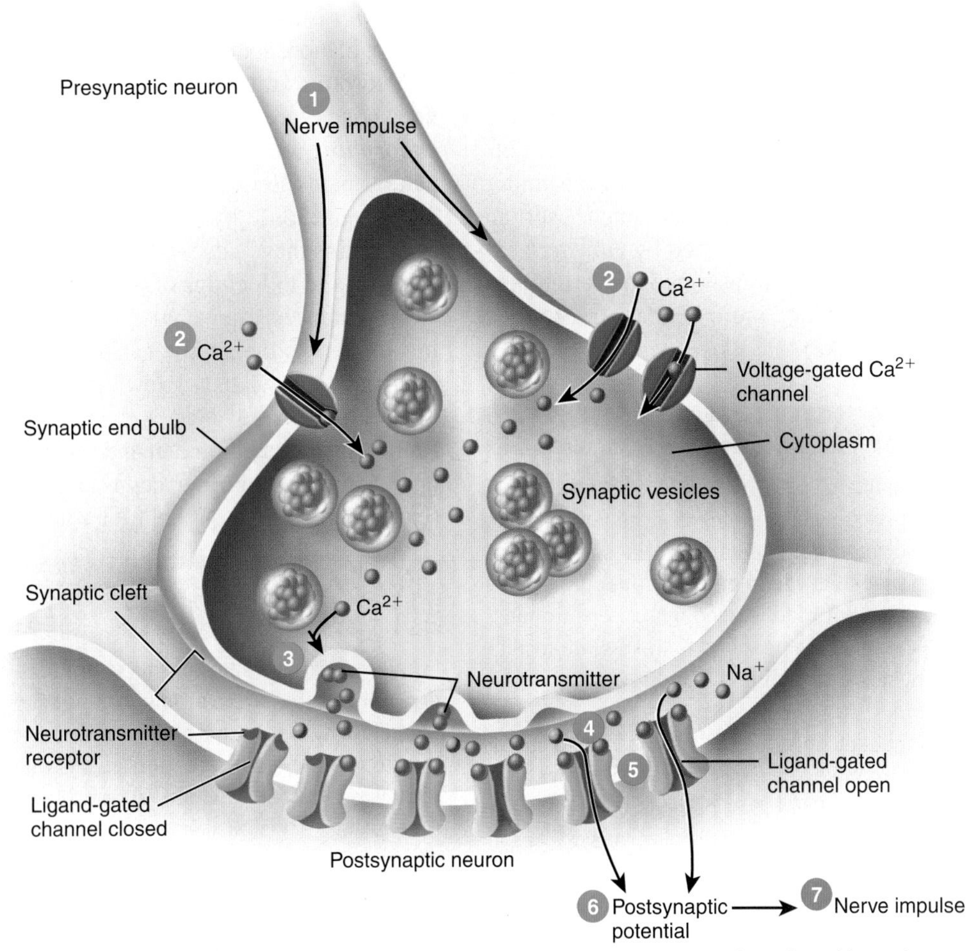

? Why may electrical synapses work in two directions, but chemical synapses can transmit a signal in only one direction?

vesicles, which releases thousands of ***neurotransmitter*** molecules into the synaptic cleft.

4. The neurotransmitter molecules flood the synaptic cleft and bind to ***neurotransmitter receptors*** in the postsynaptic neuron's plasma membrane.
5. Binding of neurotransmitter molecules opens ion channels, which allows certain ions to flow across the membrane.
6. As ions flow through the opened channels, the voltage across the membrane changes. Depending on which ions the channels admit, the voltage change may be a depolarization or a hyperpolarization.
7. If a depolarization occurs in the postsynaptic neuron and reaches threshold, then it triggers one or more nerve impulses.

At synapses, only *one-way information transfer* can occur—from a presynaptic neuron to a postsynaptic neuron or to an effector, such as a muscle fiber or a gland cell. For example, synaptic transmission at a neuromuscular junction (NMJ) proceeds from a somatic motor neuron to a skeletal muscle fiber (but not in the opposite direction). Only synaptic end bulbs of presynaptic neurons can release neurotransmitters, and only the postsynaptic neuron's membrane has the correct receptor proteins to recognize and bind that neurotransmitter. As a result, nerve impulses move along their pathways in one direction within the body. It may be noted that nerve cell conduction in the laboratory may go in both directions when an axon is electrically stimulated in the middle. Within the body, however, synapses cause neuronal signals to move in only one direction.

When a postsynaptic neuron depolarizes, the effect is excitatory: if threshold is reached, one or more nerve impulses are initiated. By contrast, hyperpolarization has an inhibitory effect on the postsynaptic neuron: As the membrane potential moves farther away from threshold, nerve impulses are less likely to arise. A typical neuron in the CNS receives input from 1,000 to 10,000 synapses. Some of this input is excitatory and some inhibitory. The sum of all the excitatory and inhibitory effects at any given time determines whether one or more impulses will occur in the postsynaptic neuron.

A neurotransmitter affects the postsynaptic neuron, muscle fiber, or gland cell as long as it remains bound to its receptors. Thus, removal of the neurotransmitter is essential for normal synaptic function. Neurotransmitter is removed in one of three ways: (1) Some of the released neurotransmitter molecules diffuse away from the synaptic cleft, and once a neurotransmitter molecule is out of reach of its receptors it can no longer exert an effect; (2) some neurotransmitters are destroyed by enzymes; (3) many neurotransmitters are actively transported back into the neuron that released them (reuptake); and others are transported into neighboring neuroglia (uptake).

• CLINICAL CONNECTION | Depression

Depression is a disorder characterized by a mixture of psychological and physical symptoms, and is marked by changes in nervous system function. Depression is associated with imbalances in some of the chemicals that transmit messages between nerve cells (neurotransmitters). Sometimes the body does not produce enough of a neurotransmitter. Other times the nerve cells do not respond to the neurotransmitter as they should. One of the neurotransmitters that plays an important role in depression is *serotonin.*

Several therapeutically important drugs selectively block reuptake of specific neurotransmitters. For example, the drug fluoxetine (Prozac®) is a *selective serotonin reuptake inhibitor (SSRI).* By blocking reuptake of serotonin, Prozac prolongs the activity of this neurotransmitter at synapses in the brain and thus delays the onset of subsequent synaptic transmissions. SSRIs provide relief for those suffering from some forms of depression. •

Neurotransmitters

About 125 substances are either known or suspected neurotransmitters. Most neurotransmitters are synthesized and loaded into synaptic vesicles in the synaptic end bulbs, close to their site of release. One of the best-studied neurotransmitters is ***acetylcholine (ACh),*** which is released by many PNS neurons and by some CNS neurons. ACh is an excitatory neurotransmitter at some synapses, such as the neuromuscular junction. It is also known to be an inhibitory neurotransmitter at other synapses. For example, parasympathetic neurons slow heart rate by releasing ACh at inhibitory synapses.

Several amino acids are neurotransmitters in the CNS. ***Glutamate*** and ***aspartate*** have powerful excitatory effects. Two other amino acids, ***gamma-aminobutyric*** (GAM-ma am-i-nō-bū-TIR-ik) ***acid (GABA)*** and ***glycine,*** are important inhibitory neurotransmitters. Antianxiety drugs such as diazepam (Valium®) enhance the action of GABA.

Some neurotransmitters are modified amino acids. These include norepinephrine, dopamine, and serotonin. ***Norepinephrine (NE)*** plays roles in arousal (awakening from deep sleep), dreaming, and regulating mood. Brain neurons containing the neurotransmitter ***dopamine (DA)*** are active during emotional responses, addictive behaviors, and pleasurable experiences. In addition, dopamine-releasing neurons help regulate skeletal muscle tone and some aspects of movement due to contraction of skeletal muscles. One form of schizophrenia is due to accumulation of excess dopamine. ***Serotonin*** is thought to be involved in sensory perception, temperature regulation, control of mood, appetite, and the onset of sleep.

Neurotransmitters consisting of amino acids linked by peptide bonds are called ***neuropeptides*** (noor-ō-PEP-tīds). The ***endorphins*** (en-DOR-fins) are neuropeptides that are the body's natural painkillers. Acupuncture may produce analgesia (loss of pain sensation) by increasing the release of endorphins. Endorphins have also been linked to improved memory and learning and to feelings of pleasure or euphoria.

An important newcomer to the ranks of recognized neurotransmitters is the simple gas ***nitric oxide (NO),*** which is different from all previously known neurotransmitters because it is not synthesized in advance and packaged into synaptic vesicles. Rather, it is formed on demand, diffuses out of cells that produce it and into neighboring cells, and acts immediately. Some research suggests that NO plays a role in learning and memory.

• CLINICAL CONNECTION | Neurotransmitters and Food

Everyone who has enjoyed the soothing relaxation of a good meal has experienced the effect of food on mood. Neurons manufacture neurotransmitters from chemicals that come from food, so you could say that the story of the food–mood link begins with digestion. Here is how **neurotransmitters and food** are related. Many neurotransmitters are made from amino acids, which are the basic building blocks of proteins. Amino acids are made available when your body digests the protein in the food you eat. For example, the neurotransmitter serotonin is made from the amino acid tryptophan, and both dopamine and norepinephrine are synthesized from the amino acid tyrosine.

Regulation of neurotransmitter levels in the brain is quite complicated and depends not only on the availability of amino acid (and other) precursors, but also on competition of these precursors for entry into the brain. Serotonin leads to feelings of relaxation and sleepiness.

Although serotonin is manufactured from the amino acid tryptophan, high-protein foods do not lead to higher levels of tryptophan in the blood or brain. After a high-protein meal, tryptophan must compete with more than 20 other amino acids for entry into the central nervous system, so its concentration in the brain remains relatively low. On the other hand, consumption of carbohydrate-rich foods, such as bread, pasta, potatoes, or sweets, is associated with an increase in the synthesis and release of serotonin in the brain. The result: Carbohydrates help us feel relaxed and sleepy. •

CHECKPOINT

9. How are neurotransmitters removed after they are released from synaptic vesicles?

15.5 REGENERATION AND REPAIR OF NERVOUS TISSUE

OBJECTIVES

- Define plasticity and neurogenesis.
- Describe the events involved in damage and repair of peripheral nerves.

Throughout your life, your nervous system exhibits ***plasticity,*** the capability to change based on experience. At the level of individual neurons, the changes that can occur include the sprouting of new dendrites, synthesis of new proteins, and changes in synaptic contacts with other neurons. Undoubtedly, both chemical and electrical signals drive the changes that occur. Despite plasticity, however, mammalian neurons have very limited powers of ***regeneration,*** the capability to replicate or repair themselves. In the PNS, damage to dendrites and myelinated axons may be repaired if the cell body remains intact and if the Schwann cells that produce myelination remain active. In the CNS, little or no repair of damage to neurons occurs. Even when the cell body remains intact, a severed axon in the CNS cannot be repaired or regrown.

Neurogenesis in the CNS

Neurogenesis—the birth of new neurons from undifferentiated stem cells—occurs regularly in some animals. For example, new neurons appear and disappear every year in some songbirds. Until relatively recently, the dogma in humans and other primates was "no new neurons" in the adult brain. Then, in 1992, Canadian researchers published their unexpected finding that ***epidermal growth factor (EGF)*** stimulated cells taken from the brains of adult mice to proliferate into both neurons and astrocytes. Previously, EGF was known to trigger mitosis in a variety of nonneuronal cells and to promote wound healing and tissue regeneration. In 1998 scientists discovered that significant numbers of new neurons do arise in the adult human hippocampus, an area of the brain that is crucial for learning.

The nearly complete lack of neurogenesis in other regions of the brain and spinal cord seems to result from two factors: (1) inhibitory influences from neuroglia, particularly oligodendrocytes, and (2) absence of growth-stimulating cues that were present during fetal development. Axons in the CNS are myelinated by oligodendrocytes that do not form neurolemmas (sheaths of Schwann). In addition, CNS myelin is one of the factors inhibiting regeneration of neurons. Perhaps this same mechanism stops axonal growth once a target region has been reached during development. Also, after axonal damage, nearby astrocytes proliferate rapidly, forming a type of scar tissue that acts as a physical barrier to regeneration. Thus, injury of the brain or spinal cord usually is permanent. Ongoing research seeks ways to improve the environment for existing spinal cord axons to bridge the injury gap. Scientists also are trying to find ways to stimulate dormant stem cells to replace neurons lost through damage or disease and to develop tissue-cultured neurons that can be used for transplantation purposes.

Damage and Repair in the PNS

Axons and dendrites that are associated with a neurolemma may undergo repair if (1) the cell body is intact, (2) the Schwann cells (neurolemmocytes) are functional, and (3) scar tissue formation does not occur too rapidly (Figure 15.13). Most nerves in the PNS consist of processes that are covered with a neurolemma.

As occurs with most other systems in the body, varying degrees of damage may occur in a nerve of the PNS. The mildest form of damage that produces clinical deficits is called *neurapraxia* (noor-a-PRAK-sē-a); there is a loss of nerve conduction, but the axon does not degenerate and recovery is complete. More severe damage results in degeneration of the axon distal to the site of the lesion and is called *axonotmesis* (ak′-son-ot-MĒ-sis); the connective tissue coverings are left intact and Wallerian degeneration of axons occurs (described next). The most severe damage to a nerve, in which the associated connective tissues are also damaged, is called *neurotmesis* (noo-rot-MĒ-sis); recovery of nerve function is highly unlikely.

A person who experiences neurapraxia or axonotmesis of a nerve in an upper limb, for example, has a good chance of regaining nerve function. When there is damage to an axon, changes usually occur both in the cell body of the affected neuron and in the portion of the axon distal to the site of injury. Changes may also occur in the portion of the axon proximal to the site of injury.

About 24 to 48 hours after injury to a process of a normal peripheral neuron (see Figure 15.13b), the ***Nissl bodies*** break up into fine granular masses. This alteration is called ***chromatolysis*** (krō′-ma-TOL-i-sis; *chromato-* = color; *-lysis* = destruction). By the third to fifth day, the part of the axon distal to the damaged region becomes slightly swollen and then breaks up into fragments; the myelin sheath also deteriorates (see Figure 15.13b). Even though the axon and myelin sheath degenerate, the neurolemma remains. Degeneration of the distal portion of the axon and myelin sheath is called ***Wallerian degeneration.***

Following chromatolysis, signs of recovery in the cell body become evident. Macrophages phagocytize the debris. Synthesis of RNA and protein accelerates, which favors rebuilding or ***regeneration*** of the axon. The Schwann cells on either side of the injured site multiply by mitosis, grow toward each other, and may form a ***regeneration tube*** across the injured area (see Figure 15.13c). The tube guides growth of a new axon from the proximal area across the injured area into the distal area previously occupied by the original axon. However, new axons cannot grow if the gap at the site of injury is too large or if the gap becomes filled with collagen fibers.

During the first few days following damage, buds of regenerating axons begin to invade the tube formed by the Schwann cells (see Figure 15.13b). Axons from the proximal area grow at a rate of about 1.5 mm (0.06 in.) per day across the area of damage, find their way into the distal regeneration tubes, and grow toward the distally located receptors and effectors. Thus, some sensory and motor connections are reestablished and some functions restored. In time, the Schwann cells form a new myelin sheath.

Figure 15.13 Damage and repair of a neuron in the PNS.

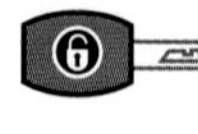

Myelinated axons in the peripheral nervous system may be repaired if the cell body remains intact and if Schwann cells (neurolemmocytes) remain active.

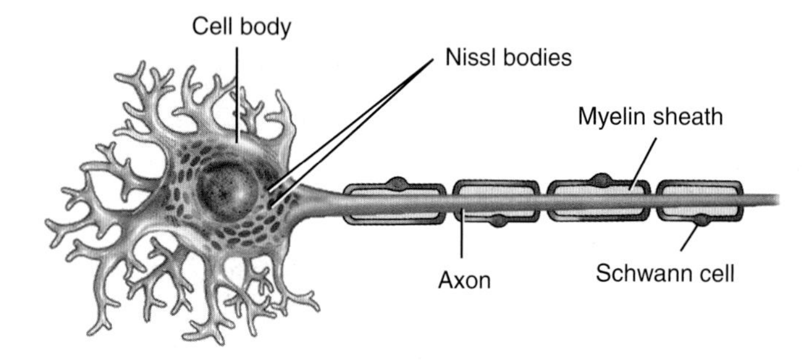

(a) Normal neuron

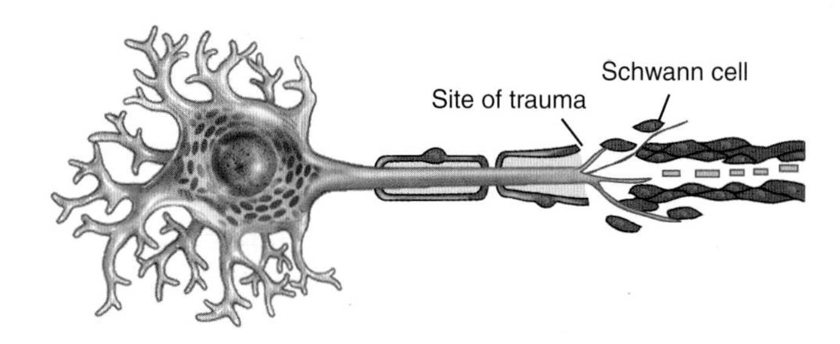

(b) Chromatolysis and Wallerian degeneration

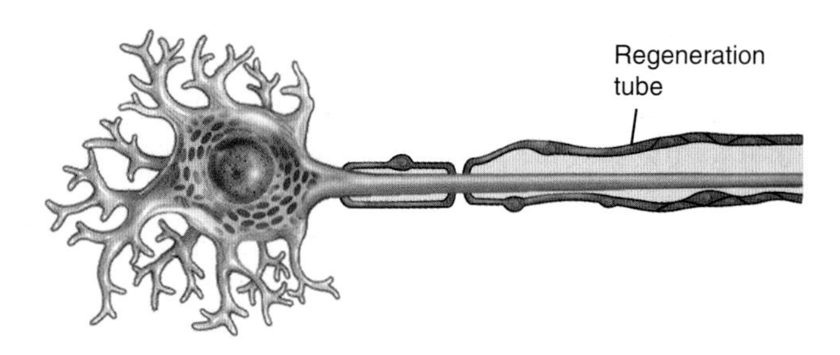

(c) Regeneration

 **What is the role of the neurolemma in regeneration?**

MANUAL THERAPY APPLICATION

Repair of Damaged Nerves

In the next chapter you will learn that some neurons travel from the lower spinal cord to the great toe. The overall growth rate of 1.5 mm per day is approximately 2 inches per month. Assuming that trauma of a patient occurred in a peripheral nerve, but near the spinal cord, it could take over 2 years for the **repair of damaged nerves** as evidenced by the return of sensation and function of the great toe. In another scenario, if a nerve (a bundle of neurons, some of which are sensory and others are motor) is completely severed, elastic fibers around the nerve cause the two ends to be retracted. When this occurs, the two ends must be connected surgically. Although a surgeon will attempt to align the two cut ends by aligning the blood vessels that are servicing the outside of the nerve, it is rare that the alignment of neurons is exactly correct. As long as the nerve cell bodies are intact and scar tissue does not block the process, the neurons will regenerate and axonal growth into neurolemmal tunnels (regeneration tubes) will take place. The axons may, however, grow through different tunnels. Assuming that regeneration is complete, the brain may send messages down the "wrong" motor neurons and therefore the actions of the person may be inappropriate. Similarly, sensory neurons growing through different neurolemmal tunnels will result in inaccurate perceptions in the brain. Physical therapy and other modalities may be required to retrain the brain so that the appropriate actions and perceptions will occur. As described previously, manual therapy can be of value to the patient by maximizing the flow of nutrients into the areas of healing (regeneration).

CHECKPOINT

10. What factors contribute to a lack of neurogenesis in most parts of the brain?
11. What is the function of the regeneration tube in repair of neurons?

KEY MEDICAL TERMS ASSOCIATED WITH NERVOUS TISSUE

Demyelination (GĒ-mī-e-li-NĀ-shun) Loss or destruction of myelin sheaths around axons in the CNS or PNS.

Guillain-Barré syndrome (GBS) (GĒ-an ba-RĀ) A demyelinating disorder in which macrophages remove myelin from PNS axons. It is a common cause of sudden paralysis and may result from the immune system's response to a bacterial infection. Most patients recover completely or partially, but about 15% remain paralyzed.

Neuroma (noo-ROM-a) A tumor consisting of nervous tissue. A mature nerve cell (neuron) cannot reproduce and therefore cannot form a tumor. Neuroglial cells are responsible for tumors of the nervous system.

Neuropathy (noo-ROP-a-thē; *neuro-* = a nerve; *-pathy* = disease) Any disorder that affects the nervous system, but particularly a disorder of a cranial or spinal nerve.

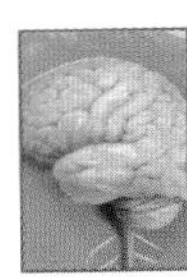

STUDY OUTLINE

Overview of the Nervous System (Section 15.1)

1. Components of the nervous system include the brain, 12 pairs of cranial nerves and their branches, the spinal cord, 31 pairs of spinal nerves and their branches, sensory receptors, ganglia, and enteric plexuses.
2. Three basic functions of the nervous system are detecting stimuli (sensory function); analyzing, integrating, and storing sensory information (integrative function); and responding to integrative decisions (motor function).
3. Sensory (afferent) neurons provide input to the CNS; motor (efferent) neurons carry output from the CNS to effectors.
4. The two main subsystems of the nervous system are (1) the central nervous system (CNS), the brain and spinal cord, and (2) the peripheral nervous system (PNS), all nervous tissues outside the brain and spinal cord.
5. The PNS also is subdivided into the somatic nervous system (SNS), autonomic nervous system (ANS), and enteric nervous system (ENS).
6. The SNS consists of (1) sensory neurons that conduct impulses from somatic and special sense receptors to the CNS, and (2) motor neurons from the CNS to skeletal muscles.
7. The ANS contains two divisions, sympathetic and parasympathetic, that convey impulses from the CNS to smooth muscle tissue, cardiac muscle tissue, and glands.
8. The ENS consists of neurons in two enteric plexuses that extend the length of the gastrointestinal (GI) tract; it monitors sensory changes and controls operation of the GI tract.

Histology of Nervous Tissue (Section 15.2)

1. Nervous tissue consists of two types of cells: neurons and neuroglia. Neurons are specialized for nerve impulse conduction and provide most of the unique functions of the nervous system, such as sensing, thinking, remembering, controlling muscle activity, and regulating glandular secretions. Neuroglia support, nourish, and protect the neurons and maintain homeostasis in the interstitial fluid that bathes neurons.
2. Most neurons have three parts. The dendrites are the main receiving or input region. Integration occurs in the cell body. The output component typically is a single axon, which conducts nerve impulses toward another neuron, a muscle fiber, or a gland cell.
3. Two types of neuroglia produce myelin sheaths: oligodendrocytes myelinate axons in the CNS and Schwann cells myelinate axons in the PNS.
4. White matter primarily contains myelinated axons; gray matter contains neuronal cell bodies, dendrites, axon terminals, unmyelinated axons, and neuroglia.
5. In the spinal cord, gray matter forms an H-shaped inner core that is surrounded by white matter. In the brain, a thin, superficial shell of gray matter covers the cerebrum and cerebellum. Nuclei of gray matter also lie deep within the brain.
6. Neuroglia include astrocytes, oligodendrocytes, microglia, ependymal cells, Schwann cells, and satellite cells (see Table 15.2).

Electrical Signals in Neurons (Section 15.3)

1. Neurons communicate with one another using nerve action potentials, also called nerve impulses.
2. Generation of action potentials depends on the existence of a resting membrane potential and the presence of voltage-gated channels for Na^+ and K^+.
3. A typical value for the resting membrane potential (difference in electrical charge across the plasma membrane) is -70 mV (inside negative). A cell that exhibits a membrane potential is polarized.
4. The resting membrane potential arises due to an unequal distribution of ions on either side of the plasma membrane and a higher membrane permeability to K^+ than to Na^+. The level of K^+ is higher inside and the level of Na^+ is higher outside, a situation that is maintained by sodium–potassium pumps (Na^+/K^+ ATPases).
5. The ability of muscle fibers and neurons to respond to a stimulus and convert it into action potentials is called excitability.
6. During an action potential, voltage-gated Na^+ and K^+ channels open in sequence. Opening of voltage-gated Na^+ channels results in depolarization, the loss and then reversal of membrane polarization (from -70 mV to $+30$ mV). Then, opening of voltage-gated K^+ channels allows repolarization, recovery of the membrane potential to the resting level.
7. According to the all-or-none principle, if a stimulus is strong enough to generate an action potential, the impulse generated is of a constant size.
8. During the refractory period, another action potential cannot be generated.
9. Nerve impulse conduction that occurs as a step-by-step process along an unmyelinated axon is called continuous conduction. In saltatory conduction, a nerve impulse "leaps" from one node of Ranvier to the next along a myelinated axon.
10. Axons with larger diameters conduct impulses faster than those with smaller diameters; myelinated axons conduct impulses faster than unmyelinated axons.

Synaptic Transmission (Section 15.4)

1. Neurons communicate with other neurons and with effectors at synapses in a series of events known as synaptic transmission.
2. At a synapse, neurotransmitters are released from a presynaptic neuron into the synaptic cleft and then bind to receptors on the postsynaptic neuron's plasma membrane.
3. An excitatory neurotransmitter depolarizes the postsynaptic neuron's membrane, brings the membrane potential closer to threshold, and increases the chance that one or more action potentials will arise. An inhibitory neurotransmitter hyperpolarizes the membrane of the postsynaptic neuron, thereby inhibiting action potential generation.
4. Neurotransmitters are removed in three ways: diffusion, enzymatic destruction, and reuptake by neurons or neuroglia.
5. Important neurotransmitters include acetylcholine (ACh), glutamate, aspartate, gamma-amino butyric acid (GABA), glycine, norepinephrine (NE), dopamine (DA), serotonin, neuropeptides, and nitric oxide (NO).

Regeneration and Repair of Nervous Tissue (Section 15.5)

1. The nervous system exhibits plasticity (the capability to change based on experience), but it has very limited powers of regeneration (the capability to replicate or repair damaged neurons).

2. Neurogenesis, the birth of new neurons from undifferentiated stem cells, is normally very limited. Repair of damaged axons does not occur in most regions of the CNS.

3. Axons and dendrites that are associated with the neurolemma in the PNS may undergo repair if the cell body is intact, the Schwann cells are functional, and scar tissue formation does not occur too rapidly.

SELF-QUIZ QUESTIONS

1. Which of the following is NOT correctly matched?
   a. central nervous system: composed of the brain and spinal cord
   b. somatic nervous system: includes motor neurons to skeletal muscles
   c. sympathetic nervous system: includes motor neurons to skeletal, smooth, and cardiac muscles
   d. peripheral nervous system: includes cranial and spinal nerves
   e. autonomic nervous system: includes parasympathetic and sympathetic divisions

2. The portion of the nervous system that regulates the gastrointestinal (GI) tract is the
   a. somatic nervous system **b.** sympathetic division
   c. integrative division **d.** central nervous system
   e. enteric nervous system.

3. Damage to dendrites would interfere with a neuron's ability to
   a. receive input **b.** make proteins
   c. release neurotransmitters **d.** form myelin
   e. conduct nerve impulses to another neuron.

4. The type of cell that produces myelin sheaths around axons in the CNS is the
   a. astrocyte **b.** myelinocyte
   c. Schwann cell **d.** oligodendrocyte
   e. microglia.

5. A bundle of axons in the CNS is
   a. a tract **b.** a nucleus
   c. a mixed nerve **d.** a ganglion
   e. an enteric plexus.

6. Which of the following is NOT true concerning the repair of nervous tissue?
   a. If the cell body is not damaged, neurons in the PNS may be able to repair themselves.
   b. In the CNS, myelin inhibits neuronal regeneration.
   c. Injury to the CNS is usually permanent.
   d. Active Schwann cells contribute to the repair process in the PNS.
   e. A regeneration tube forms across the injured area of a PNS neuron that undergoes repair.

7. In a resting neuron
   a. there is a high concentration of K^+ outside the cell
   b. negatively charged ions move freely through the plasma membrane
   c. the sodium–potassium pumps help maintain the low concentration of Na^+ inside the cell
   d. the outside surface of the plasma membrane has a negative charge
   e. the plasma membrane is highly permeable to Na^+.

8. The depolarizing phase of a nerve impulse is caused by a
   a. rush of Na^+ into the neuron
   b. rush of Na^+ out of the neuron
   c. rush of K^+ into the neuron
   d. rush of K^+ out of the neuron
   e. pumping of K^+ into the neuron.

9. Saltatory conduction occurs
   a. in unmyelinated axons **b.** at the nodes of Ranvier
   c. in the smallest-diameter axons **d.** in skeletal muscle fibers
   e. in cardiac muscle fibers.

10. Place the following events in the correct order of occurrence:
    1. Voltage-gated Na^+ channels open and permit Na^+ to rush inside the neuron.
    2. The Na^+/K^+ pump restores the ions to their original sites.
    3. A stimulus of threshold strength is applied to the neuron.
    4. The membrane polarization changes from negative (−55 mV) to positive (+30 mV).
    5. Voltage-gated K^+ channels open, and K^+ flows out of the neurons.
    a. 4, 1, 2, 3, 5 **b.** 4, 3, 1, 2, 5
    c. 3, 1, 4, 2, 5 **d.** 5, 3, 1, 4, 2
    e. 3, 1, 4, 5, 2

11. If a stimulus is strong enough to generate an action potential in a neuron, the impulse generated is of a constant size. A stronger stimulus cannot generate a larger impulse. This is known as
    a. the principle of polarization–depolarization
    b. saltatory conduction
    c. the all-or-none principle
    d. the principle of reflex action
    e. the absolute refractory period.

12. The speed of nerve impulse conduction is increased by
    a. cold **b.** a very strong stimulus
    c. small diameter of the axon **d.** myelination
    e. astrocytes.

13. For a signal to be transmitted by means of a chemical synapse from a presynaptic neuron to a postsynaptic neuron,
    a. the presynaptic neuron must be touching the postsynaptic neuron
    b. the postsynaptic neuron must contain neurotransmitter receptors
    c. there must be gap junctions present between the two neurons
    d. the postsynaptic neuron needs to release neurotransmitters from its synaptic vesicles
    e. the neurons must be myelinated.

14. What would happen at the postsynaptic neuron if the total inhibitory effects of the neurotransmitters were greater than the total excitatory effects?
    a. A nerve impulse would be generated.
    b. It would be easier to generate a nerve impulse when the next stimulus was received.
    c. The nerve impulse would be rerouted to another neuron.
    d. No nerve impulse would be generated.
    e. The neurotransmitter would be broken down more quickly.

15. Match the following neurotransmitters with their descriptions.
    ____ **a.** inhibitory amino acid in the CNS
    ____ **b.** a gaseous neurotransmitter that is not packaged into synaptic vesicles
    ____ **c.** excitatory amino acid in the CNS
    ____ **d.** body's natural painkillers
    ____ **e.** helps regulate mood
    ____ **f.** neurotransmitter that activates skeletal muscle fibers

    A. serotonin
    B. acetylcholine
    C. endorphins
    D. GABA
    E. nitric oxide
    F. glutamate

16. Match the following.

____ **a.** the portion of a neuron containing the nucleus	**A.** synaptic end bulb
____ **b.** rounded structure at the distal end of an axon terminal	**B.** motor neuron
____ **c.** highly branched, input part of a neuron	**C.** sensory neuron
____ **d.** sac in which neurotransmitter is stored	**D.** dendrite
____ **e.** neuron located entirely within the CNS	**E.** interneuron
____ **f.** long, cylindrical process that conducts impulses toward another neuron	**F.** nucleus
____ **g.** produces myelin sheath in PNS	**G.** myelin sheath
____ **h.** unmyelinated gap in the myelin sheath	**H.** Schwann cell
____ **i.** substance that increases the speed of nerve impulse conduction	**I.** cell body
____ **j.** neuron that conveys information from a receptor to the CNS	**J.** node of Ranvier
____ **k.** neuron that conveys information from the CNS to an effector	**K.** ganglion
____ **l.** bundle of many axons in the PNS	**L.** nerve
____ **m.** bundle of many axons in the CNS	**M.** neurotransmitter
____ **n.** group of cell bodies in the PNS	**N.** tract
____ **o.** group of cell bodies in the CNS	**O.** synaptic vesicle
____ **p.** substance used for communication at chemical synapses	**P.** axon

CRITICAL THINKING QUESTIONS

1. The buzzing of the alarm clock woke Carrie. She stretched, yawned, and started to salivate as she smelled the brewing coffee. She could feel her stomach rumble. List the divisions of the nervous system that are involved in each of these actions.
2. Baby Ming is learning to crawl. He also likes to pull himself onto window sills, gnawing on the painted wood of his century-old home as he looks out the windows. Lately his mother, an anatomy and physiology student, has noticed some odd behavior and took Ming to the pediatrician. Blood work determined that Ming had a high level of lead in his blood, ingested from the old leaded paint on the window sill. The doctor indicated that lead poisoning is a type of demyelination disorder. Why should Ming's mother be concerned?

? ANSWERS TO THE FIGURE QUESTIONS

15.1 The total number of cranial and spinal nerves in the human body is $(12 \times 2) + (31 \times 2) = 86$.

15.2 Sensory or afferent neurons carry input to the CNS. Motor or efferent neurons carry output from the CNS.

15.3 Dendrites and the cell body receive input; the axon conducts nerve impulses (action potentials) and transmits the message to another neuron or effector cell by releasing a neurotransmitter at its synaptic end bulbs.

15.4 In most neurons, nerve impulses arise at the "trigger zone."

15.5 The cell body of a pyramidal cell is shaped like a pyramid.

15.6 Myelination increases the speed of nerve impulse conduction.

15.7 Myelin makes white matter look shiny and white.

15.8 Perception primarily occurs in the cerebral cortex.

15.9 More Na^+ ions would leak into the cell and fewer K^+ ions would leak out of the cell, which would make the resting membrane potential more inside-positive.

15.10 Voltage-gated Na^+ channels are open during the depolarizing phase, and voltage-gated K^+ channels are open during the repolarizing phase.

15.11 The diameter of an axon, presence or absence of a myelin sheath, and temperature determine the speed of propagation of an action potential.

15.12 In some electrical synapses (gap junctions), ions may flow equally well in either direction, so either neuron may be the presynaptic one. At a chemical synapse, one neuron releases neurotransmitters and the other neuron has receptors that bind this chemical. Thus, the signal can proceed in only one direction.

15.13 The neurolemma provides a regeneration tube that guides regrowth of a severed axon.

16 The Spinal Cord and Spinal Nerves

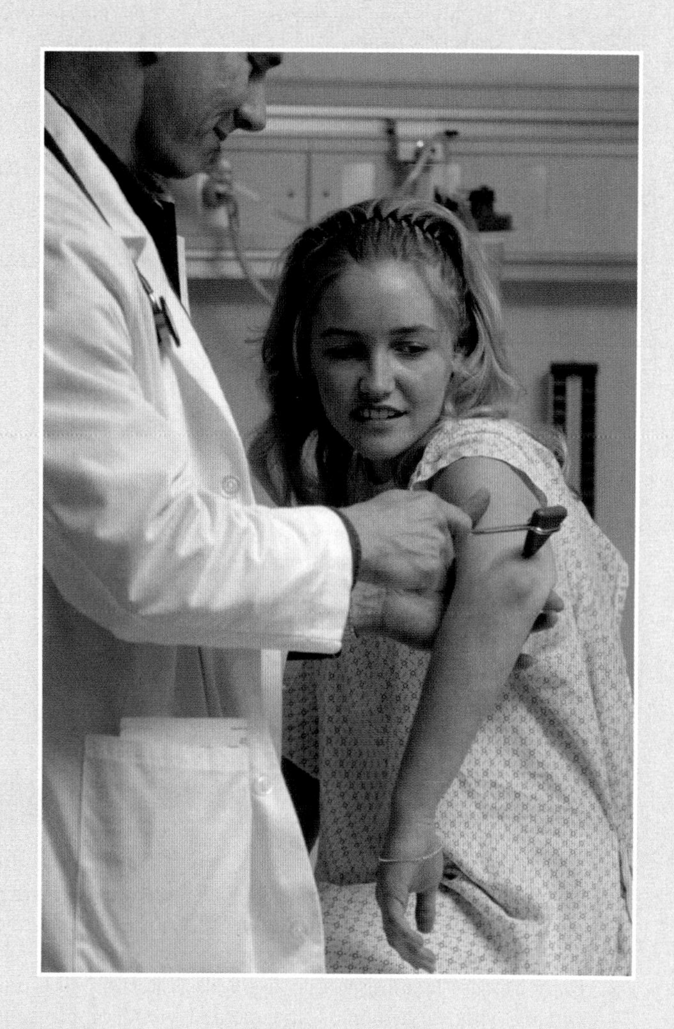

Early anatomists made a distinction between the brain and spinal cord. Today we know that the brain and spinal cord are really just one large, interconnected group of nervous tissues known as the central nervous system (CNS). Since the nervous system is so complex, it is more convenient to study the individual parts rather than the whole. However, it is important to think of the nervous system as one complex mass of interconnected neurons. Function or dysfunction of any part may affect many seemingly independent neuronal structures.

The spinal cord contains a series of "pathways" that relay sensory information along fibers to the processing centers and then react by sending information along different fibers for motor function. The spinal cord allows us to make quick responses, such as pulling the foot away quickly when we step on a sharp tack. We lift our foot before we have a chance to think. This is an example of a spinal cord reflex—a quick, automatic response to certain kinds of stimuli that involves neurons only in the spinal nerves and spinal cord. Reflexes are simply preprogrammed reactions to strong stimuli such as pain, touch, temperature, or pressure. An example is when a physician strikes near your elbow with a reflex hammer and an extensor reflex pathway causes your upper limb to straighten.

CONTENTS AT A GLANCE

16.1 SPINAL CORD ANATOMY

OBJECTIVE

- Describe the protective structures and the gross anatomical features of the spinal cord.

The spinal cord is continuous with the medulla oblongata of the brain. Both portions of the CNS contain gray and white matter for specialized processing of information.

Protective Structures

Two types of connective tissue coverings—bony vertebrae and tough, connective tissue meninges—plus a cushion of cerebrospinal fluid (produced in the brain) surround and protect the delicate nervous tissue of the spinal cord.

Vertebral Column

The spinal cord is located within the vertebral canal. As you learned in Chapter 7, the vertebral foramina of all the vertebrae, stacked one on top of the other, form the vertebral canal. The surrounding vertebrae provide a sturdy shelter for the enclosed spinal cord (see Figure 16.6). The vertebral ligaments, meninges, and cerebrospinal fluid provide additional protection.

Meninges

The ***meninges*** (me-NIN-jēz) form three connective tissue coverings that encircle the spinal cord and brain. The ***spinal meninges*** surround the spinal cord (Figure 16.1) and are continuous with the ***cranial meninges,*** which encircle the brain (shown in Figure 17.2). The most superficial of the three spinal meninges, the ***dura mater*** (DOO-ra MĀ-ter = tough mother), is composed of dense, irregular connective tissue. It forms a sac from the level of the foramen magnum in the occipital bone, where it is continuous with the dura mater of the brain, to the second sacral vertebra (S2). The spinal cord is also protected by a cushion of fat and connective tissue located in the ***epidural space,*** a space between the dura mater and the wall of the vertebral canal (see Figure 16.6).

The middle ***meninx*** (MĒ-ninks; singular form of *meninges*) is an avascular covering called the ***arachnoid mater*** (a-RAK-noyd; *arachn-* = spider; *-oid* = similar to) because of its spidersweb arrangement of delicate collagen fibers and some elastic fibers. It is deep to the dura mater and is continuous with the arachnoid mater of the brain. Between the dura mater and the arachnoid mater is a thin ***subdural space,*** which contains interstitial fluid.

The innermost meninx is the ***pia mater*** (PĪ-a MĀ-ter; *pia* = delicate), a thin transparent connective tissue layer that adheres to the surface of the spinal cord and brain. It consists of interlacing bundles of collagen fibers and some fine elastic fibers. Within the pia mater are many blood vessels that supply oxygen and nutrients to the spinal cord. Between the arachnoid mater and the pia mater is the ***subarachnoid space,*** which contains cerebrospinal fluid (CSF).

All three spinal meninges cover the spinal nerve roots up to the point where they exit the spinal column through the intervertebral foramina. As you will see later in the chapter, spinal nerve roots are structures that connect spinal nerves to the spinal cord. Triangular-shaped membranous extensions of the pia mater suspend the spinal cord in the middle of its dural sheath. These extensions, called ***denticulate ligaments*** (den-TIK-ū-lāt = small tooth), are thickenings of the pia mater. They project laterally and fuse with the arachnoid mater and inner surface of the dura mater between the anterior and posterior nerve roots of spinal nerves on either side (Figures 16.1 and see Figure 16.6). Extending along the length of the spinal cord, the denticulate ligaments protect the spinal cord against sudden displacement that could result in shock.

External Anatomy of the Spinal Cord

The ***spinal cord,*** although roughly cylindrical, is flattened slightly in its anterior–posterior dimension. In adults, it extends from the medulla oblongata, the inferior part of the brain, to the inferior border of the first lumbar vertebra (L1) or the superior border of the second lumbar vertebra (L2) (see Figure 16.2). In newborn infants, it extends to the third or fourth lumbar vertebra. During early childhood, both the spinal cord and the vertebral

Figure 16.1 Gross anatomy of the spinal cord.

Meninges are connective tissue coverings that surround the spinal cord and brain.

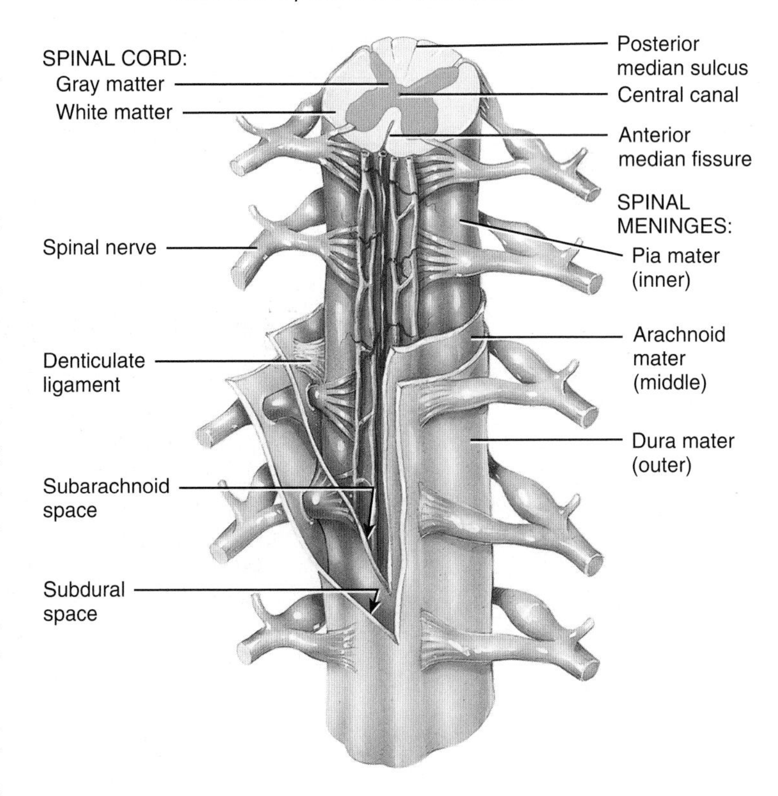

Which layer of the meninges forms the denticulate ligaments?

Figure 16.2 External anatomy of the spinal cord and spinal nerves.

The spinal cord extends from the medulla oblongata of the brain to the superior border of the second lumbar vertebra.

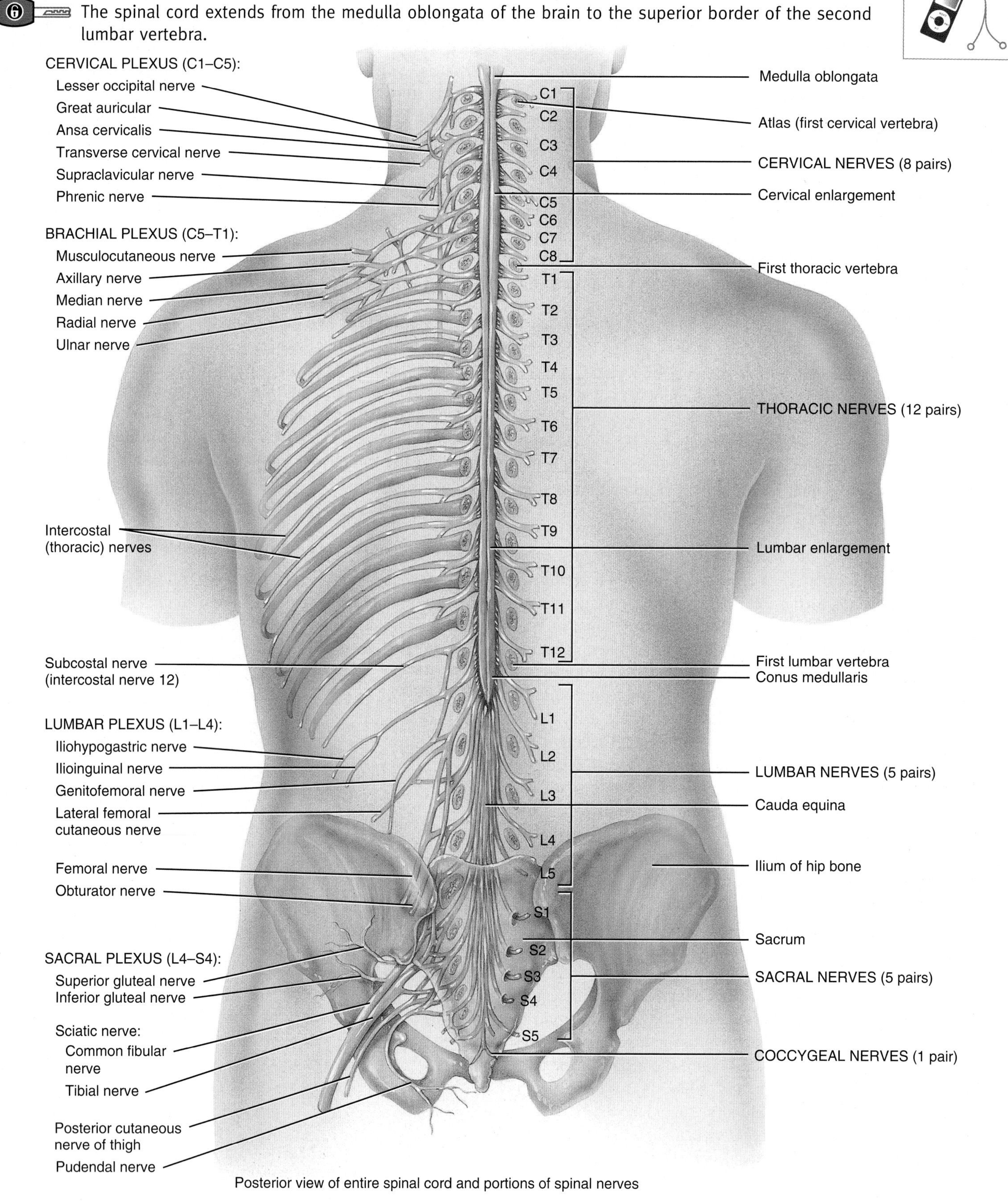

Posterior view of entire spinal cord and portions of spinal nerves

Which spinal nerves do not form a plexus?

column grow longer as part of overall body growth. Elongation of the spinal cord stops around age 4 or 5, but growth of the vertebral column continues. Thus, the spinal cord does not extend the entire length of the adult vertebral column. The length of the adult spinal cord ranges from 42 to 45 cm (16–18 in.). Its diameter is about 2 cm (0.75 in.) in the midthoracic region, somewhat larger in the lower cervical and midlumbar regions, and smallest at the inferior tip.

When the spinal cord is viewed externally, two conspicuous enlargements can be seen. The superior enlargement, the ***cervical enlargement,*** extends from the fourth cervical vertebra (C4) to the first thoracic vertebra (T1). Nerves to and from the upper limbs arise from the cervical enlargement. The inferior enlargement, called the ***lumbar enlargement,*** extends from the ninth to the twelfth thoracic vertebra (T9–T12). Nerves to and from the lower limbs arise from the lumbar enlargement.

Inferior to the lumbar enlargement, the spinal cord terminates as a tapering, conical structure called the ***conus medullaris*** (KŌ-nus med-ū-LAR-is; *conus* = cone), which ends at the level of the intervertebral disc between the first and second lumbar vertebrae in adults. Arising from the conus medullaris is the ***filum terminale*** (FĪ-lum ter-mi-NAL-ē = terminal filament), an extension of the pia mater that extends inferiorly and anchors the spinal cord to the coccyx.

Because the spinal cord is shorter than the vertebral column, nerves that arise from the lumbar, sacral, and coccygeal regions of the spinal cord do not leave the vertebral column at the same level they exit the cord. The roots of these spinal nerves angle inferiorly in the vertebral cavity from the end of the spinal cord like wisps of hair. Appropriately, the roots of these nerves are collectively named the ***cauda equina*** (KAW-da ē-KWĪ-na), meaning "horse's tail" (Figure 16.2).

Internal Anatomy of the Spinal Cord

Two grooves penetrate the white matter of the spinal cord and divide it into right and left sides (Figure 16.3). The ***anterior median fissure*** is a deep, wide groove on the anterior (ventral) side. The ***posterior median sulcus*** is a shallower, narrow groove on the posterior (dorsal) side. The gray matter of the spinal cord is shaped like the letter H or a butterfly and is surrounded by white matter. The gray matter consists of dendrites and cell bodies of neurons, unmyelinated axons, and neuroglia. The white matter consists primarily of bundles of myelinated axons of neurons.

Figure 16.3 Internal anatomy of the spinal cord: the organization of gray matter and white matter. For this and other illustrations of transverse sections of the spinal cord, circles represent cell bodies and dendrites, lines represent axons, and Y-shaped forks represent axon terminals. Blue, red, and green arrows indicate the direction of nerve impluse propagation.

The posterior gray horn contains axons of sensory neurons and cell bodies of interneurons; the lateral gray horn contains cell bodies of autonomic neurons; and the anterior gray horn contains cell bodies of somatic motor neurons.

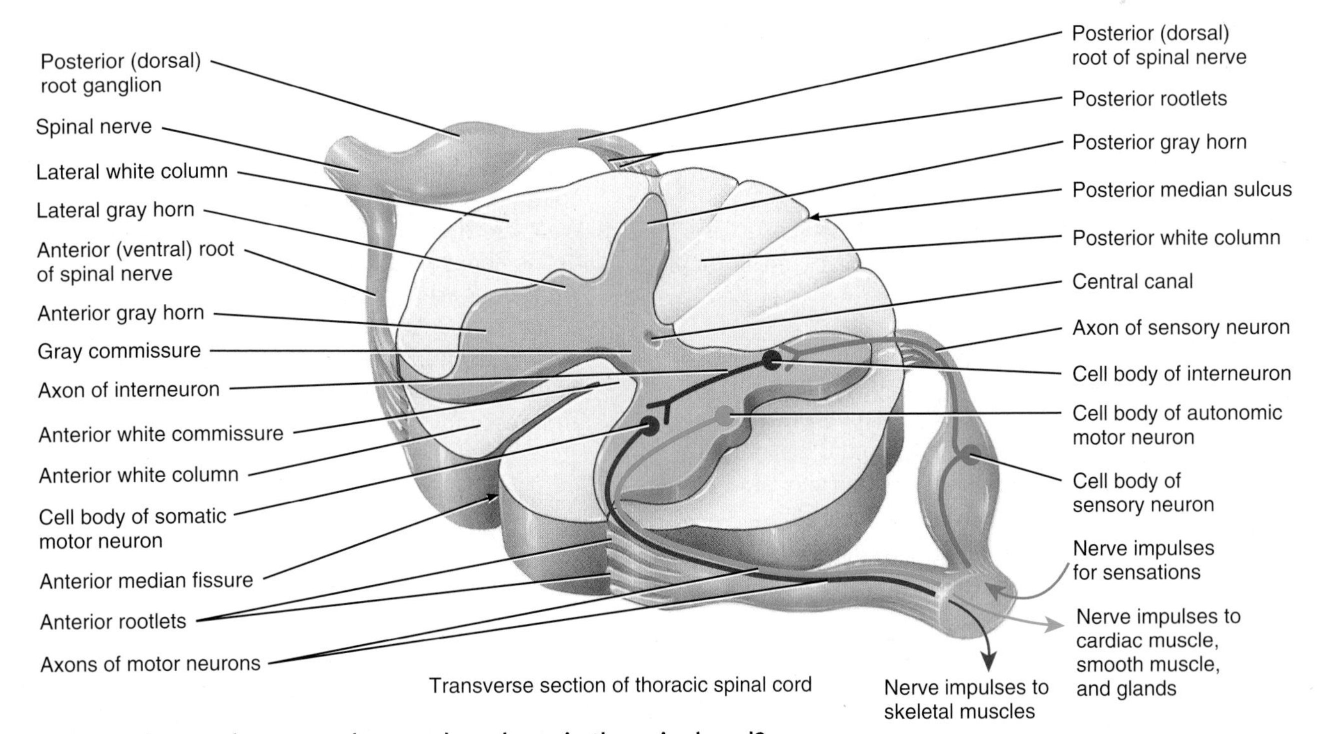

What is the difference between a horn and a column in the spinal cord?

The ***gray commissure*** (KOM-mi-shur) forms the crossbar of the H. In the center of the gray commissure is a small space called the ***central canal;*** it extends the entire length of the spinal cord and is filled with cerebrospinal fluid. At its superior end, the central canal is continuous with the fourth ventricle (a space that contains cerebrospinal fluid) in the medulla oblongata of the brain. Anterior to the gray commissure is the ***anterior (ventral) white commissure,*** which connects the white matter of the right and left sides of the spinal cord.

In the gray matter of the spinal cord and brain, clusters of neuronal cell bodies form functional groups called ***nuclei.*** *Sensory nuclei* receive input from sensory receptors via sensory neurons, and *motor nuclei* provide output to effector tissues via motor neurons. The gray matter on each side of the spinal cord is subdivided into regions called ***horns.*** The ***anterior (ventral) gray horns*** contain *somatic motor nuclei,* which provide nerve impulses for contraction of skeletal muscles. The ***posterior (dorsal) gray horns*** contain *somatic* and *autonomic sensory nuclei.* Between the anterior and posterior gray horns are the ***lateral gray horns,*** which are present only in the thoracic, upper lumbar, and sacral segments of the spinal cord. The lateral horns contain *autonomic motor nuclei* that regulate the activity of smooth muscle, cardiac muscle, and glands.

The white matter, like the gray matter, is organized into regions. The anterior and posterior gray horns divide the white matter on each side into three broad areas called ***columns:*** (1) ***anterior (ventral) white columns,*** (2) ***posterior (dorsal) white columns,*** and (3) ***lateral white columns.*** Each column, in turn, contains distinct bundles of axons having a common origin or destination and carrying similar information. These bundles, which may extend long distances up or down the spinal cord, are called ***tracts.*** Tracts are bundles of axons in the CNS; recall that nerves are bundles of axons in the peripheral nervous system (PNS). ***Sensory (ascending) tracts*** consist of axons that conduct nerve impulses toward the brain. Tracts consisting of axons that carry nerve impulses from the brain are called ***motor (descending) tracts.*** Sensory and motor tracts of the spinal cord are continuous with sensory and motor tracts in the brain.

The internal organization of the spinal cord allows sensory input and motor output to be processed by the spinal cord in the following way (Figure 16.4):

1. Sensory receptors detect a sensory stimulus.
2. Sensory neurons convey this sensory input in the form of nerve impulses along their axons, which extend from sensory receptors into the spinal nerve and then into the dorsal root. From the dorsal root, axons of sensory neurons may proceed along three possible paths (see steps 3, 4, and 5).
3. Axons of sensory neurons may extend into the white matter of the spinal cord and ascend to the brain as part of a sensory tract.
4. Axons of sensory neurons may enter the dorsal gray horn and synapse with interneurons whose axons extend into the white matter of the spinal cord and then ascend to the brain as part of a sensory tract.
5. Axons of sensory neurons may enter the dorsal gray horn and synapse with interneurons that in turn synapse with somatic motor neurons that are involved in spinal reflex pathways. Spinal cord reflexes are described in more detail later in this chapter.
6. Motor output from the spinal cord to skeletal muscles involves somatic motor neurons of the ventral gray horn. Many somatic motor neurons are regulated by the brain. Axons from higher brain centers form motor tracts that descend from the brain into the white matter of the spinal cord. There they synapse with the somatic motor neurons either directly or indirectly by first synapsing with interneurons that in turn synapse with somatic motor neurons.
7. When activated, somatic motor neurons convey motor output in the form of nerve impulses along their axons, which sequentially pass through the ventral gray horn and ventral root to enter the spinal nerve. From the spinal nerve, axons of somatic motor neurons extend to skeletal muscles of the body.
8. Motor output from the spinal cord to cardiac muscle, smooth muscle, and glands involves autonomic motor neurons of the lateral gray horn. When activated, autonomic motor neurons convey motor output in the form of nerve impulses along their axons, which sequentially pass through the lateral gray horn, ventral gray horn, and ventral root to enter the spinal nerve.
9. From the spinal nerve, axons of autonomic motor neurons from the spinal cord synapse with another group of autonomic motor neurons located in the PNS. The axons of this second group of autonomic motor neurons in turn synapse with cardiac muscle, smooth muscle, and glands. You will learn more about autonomic motor neurons when the autonomic nervous system is described in Chapter 18.

CHECKPOINT

1. Where are the spinal meninges located? Where are the epidural, subdural, and subarachnoid spaces located?
2. What are the cervical and lumbar enlargements?
3. Define conus medullaris, filum terminale, and cauda equina.
4. What does each of the following terms mean? Gray commissure, central canal, anterior gray horn, lateral gray horn, posterior gray horn, anterior white column, lateral white column, posterior white column, ascending tract, and descending tract.

16.2 SPINAL NERVES

OBJECTIVES

- Describe the components, connective tissue coverings, and branching of a spinal nerve.
- Define a plexus, and identify the distribution of nerves of the cervical, brachial, lumbar, and sacral plexuses.
- Describe the clinical significance of dermatomes.

Figure 16.4 Processing of sensory input and motor output by the spinal cord.

Sensory input is conveyed from sensory receptors to the posterior gray horns of the spinal cord, whereas motor output is conveyed from the anterior and lateral gray horns of the spinal cord to effectors (muscles and glands).

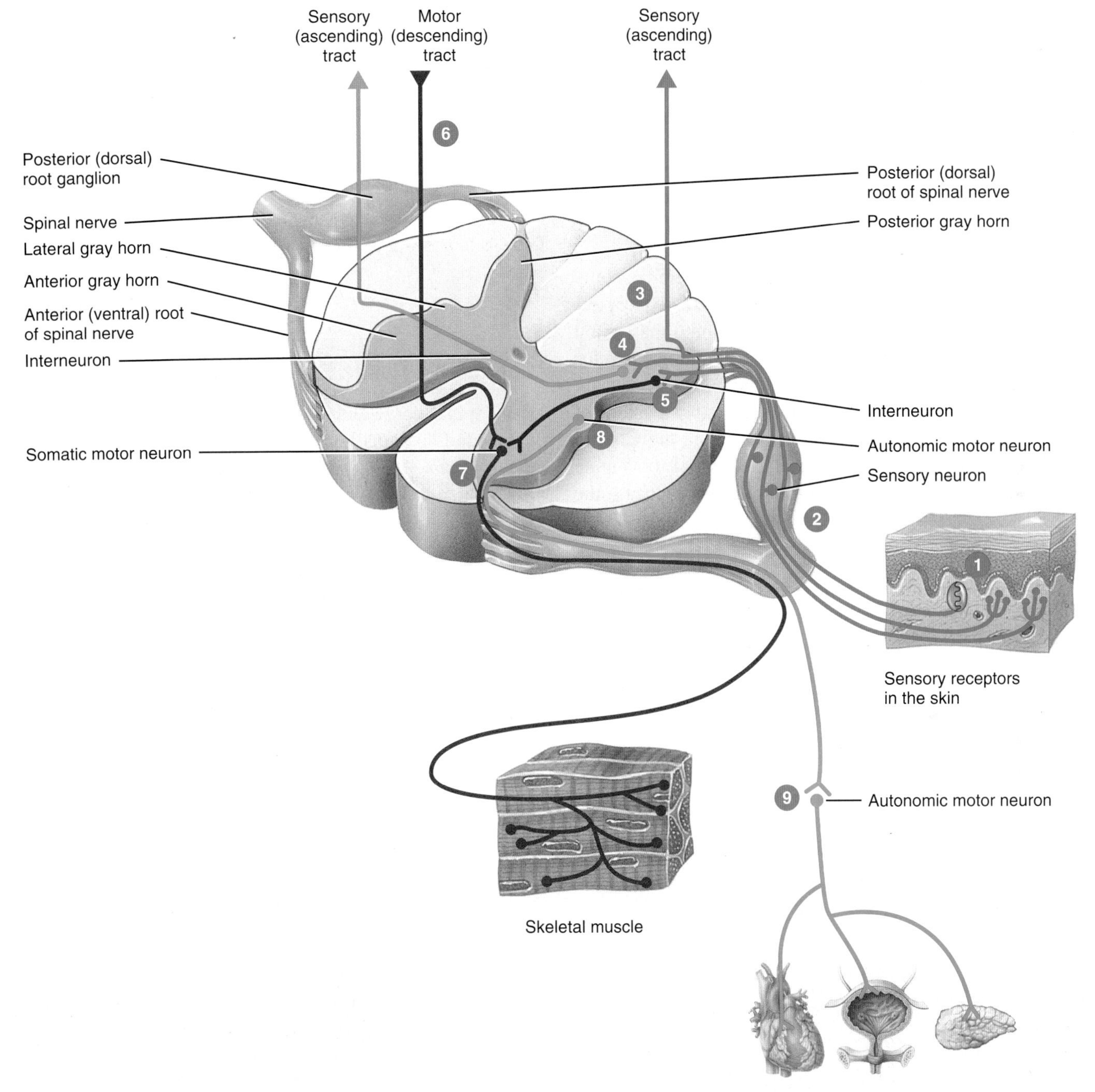

? Lateral gray horns are found in which segments of the spinal cord?

Spinal nerves, part of the peripheral nervous system (PNS), are the paths of communication between the spinal cord and the nerves supplying specific regions of the body. Spinal cord organization appears to be segmented because the 31 pairs of spinal nerves emerge at regular intervals from intervertebral foramina (see Figure 16.2). Indeed, each pair of spinal nerves is said to arise from a *spinal segment.* Within the spinal cord there is no obvious segmentation but, for convenience, the naming of spinal nerves is based on the segment in which they are located. There are 8 pairs of *cervical nerves* represented as

C1–C8, 12 pairs of *thoracic nerves* (T1–T12), 5 pairs of *lumbar nerves* (L1–L5), 5 pairs of *sacral nerves* (S1–S5), and 1 pair of *coccygeal nerves* (Co1) for a total of 31 pairs (see Figure 16.2). The first cervical pair emerges between the atlas (first cervical vertebra) and the occipital bone. All other spinal nerves emerge from the vertebral column through the intervertebral foramina between adjoining vertebrae. Not all spinal cord segments are aligned with their corresponding vertebrae. Recall that the spinal cord ends near the level of the superior border of the second lumbar vertebra, and that the roots of the lumbar, sacral, and coccygeal nerves descend at an angle to reach their respective foramina before emerging from the vertebral column. This arrangement constitutes the cauda equina (see Figure 16.2).

Two bundles of axons, called ***roots,*** connect each spinal nerve to a segment of the cord by a series of small rootlets (see Figure 16.3). The ***posterior (dorsal) root*** and rootlets contain only sensory axons, which conduct nerve impulses from sensory receptors in the skin, muscles, and internal organs into the central nervous system. Each posterior root has a swelling, the ***posterior (dorsal) root ganglion,*** which contains the cell bodies of sensory neurons. The ***anterior (ventral) root*** and rootlets contain axons of motor neurons, which conduct nerve impulses from the CNS to effectors (muscles and glands). The dorsal and ventral roots unite to form a spinal nerve at the intervertebral foramen. Because the dorsal root contains sensory axons and the ventral root contains motor axons, a spinal nerve is classified as a *mixed nerve*.

• CLINICAL CONNECTION | Spinal Nerve Root Damage

As you have just learned, spinal nerve roots exit from the vertebral canal through intervertebral foramina. The most common cause of **spinal nerve root damage** is a herniated intervertebral disc. Damage to vertebrae as a result of osteoporosis, osteoarthritis, cancer, or trauma can also damage spinal nerve roots. Symptoms of spinal nerve root damage include pain, muscle weakness, and loss of feeling. Rest, manual therapy, pain medications, and epidural injections are the most widely used conservative treatments. It is recommended that 6 to 12 weeks of conservative therapy be attempted first. If the pain continues, is intense, or is impairing normal functioning, surgery is often the next step. •

Connective Tissue Coverings of Spinal Nerves

Each spinal nerve and cranial nerve consists of many individual axons and contains layers of protective connective tissue coverings (Figure 16.5). Individual axons within a nerve, whether myelinated or unmyelinated, are wrapped in ***endoneurium*** (en′-

Figure 16.5 Organization and connective tissue coverings of a spinal nerve. [Part (b) from Richard G. Kessel and Randy H. Kardon, *Tissues and Organs: A Text-Atlas of Scanning Electron Microscopy.* Copyright © 1979 by W. H. Freeman and Company. Reprinted by permission.]

Three layers of connective tissue wrappings protect axons: Endoneurium surrounds individual axons, perineurium surrounds bundles of axons (fascicles), and epineurium surrounds an entire nerve.

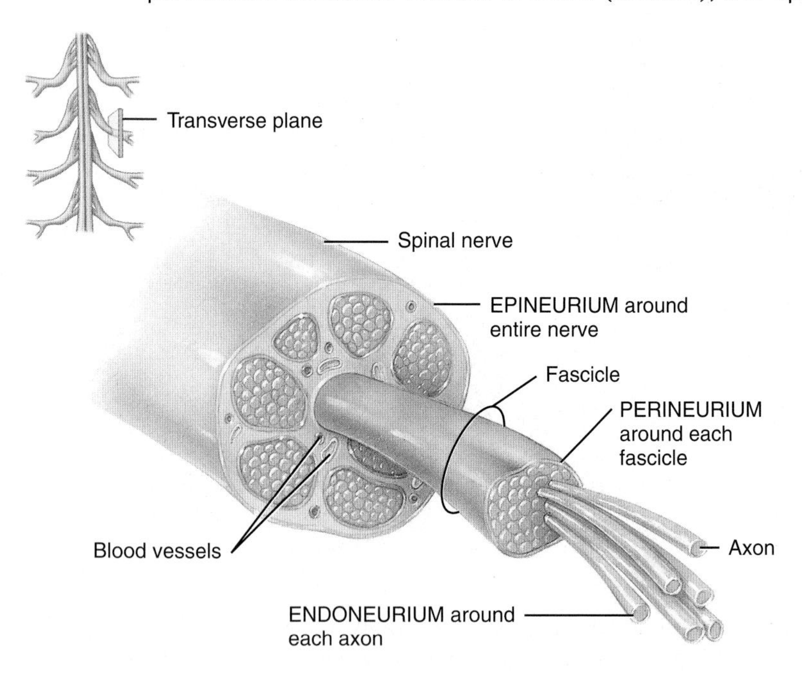

(a) Transverse section showing the coverings of a spinal nerve

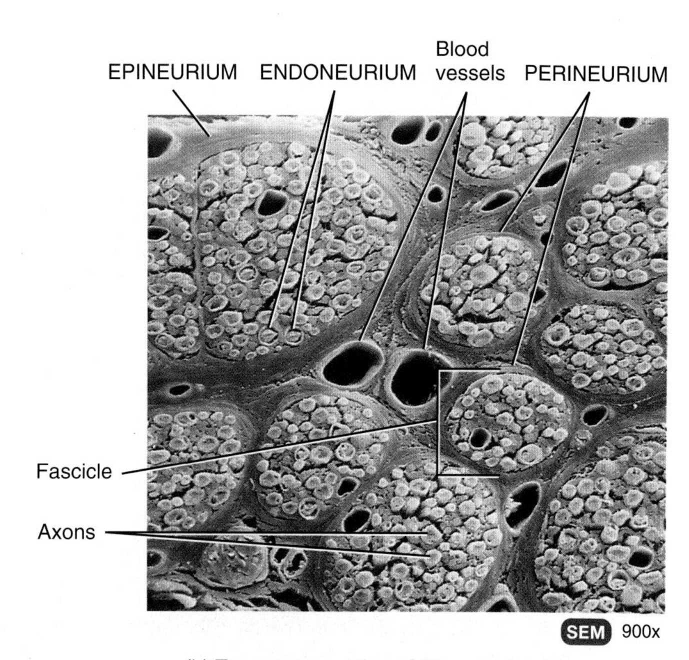

(b) Transverse section of 12 nerve fascicles

? **Why are all spinal nerves classified as mixed nerves?**

dō-NOO-rē-um; *endo-* = within or inner; *-neurium* = nerve), the innermost layer. Groups of axons with their endoneurium are arranged in bundles called ***fascicles,*** each of which is wrapped in ***perineurium*** (per′-i-NOO-rē-um; *peri-* = around), the middle layer. The outermost covering over the entire nerve is the ***epineurium*** (ep′-i-NOO-rē-um; *epi-* = over). The dura mater of the spinal meninges fuses with the epineurium as the nerve passes through the intervertebral foramen. Note the presence of many blood vessels, which nourish nerves, within all three layers of connective tissue (Figure 16.5b). You may recall from Chapter 10 that the connective tissue coverings of skeletal muscles—endomysium, perimysium, and epimysium—are similar in organization to those of nerves.

Distribution of Spinal Nerves

Branches

A short distance after passing through its intervertebral foramen, a spinal nerve divides into several branches (Figure 16.6). These branches are known as ***rami*** (RĀ-mī = branches). The ***posterior (dorsal) ramus*** (RĀ-mus; singular form) serves the deep muscles and skin of the dorsal surface of the trunk. The ***anterior (ventral) ramus*** serves the muscles and structures of the upper and lower limbs and the skin of the lateral and ventral surfaces of the trunk. In addition to posterior and anterior rami, spinal nerves also give off a ***meningeal branch.*** This branch reenters the vertebral cavity through the intervertebral foramen and supplies the vertebrae, vertebral ligaments, blood vessels of the spinal cord, and meninges. Other branches of a spinal nerve are the ***rami communicantes*** (kō-mū-ni-KAN-tēz), components of the autonomic nervous system that will be discussed in Chapter 18.

Plexuses

Axons from the anterior rami of spinal nerves, except for thoracic nerves T2–T12, do not innervate the body structures directly. Instead, they form networks on both the left and right sides of the body by joining with various numbers of axons from anterior rami of adjacent nerves. Such a network of axons is called a ***plexus*** (= braid or network). The principal plexuses

Figure 16.6 Branches of a typical spinal nerve, shown in transverse section through the thoracic portion of the spinal cord.

 The branches of a spinal nerve are the posterior ramus, the anterior ramus, the meningeal branch, and the rami communicantes.

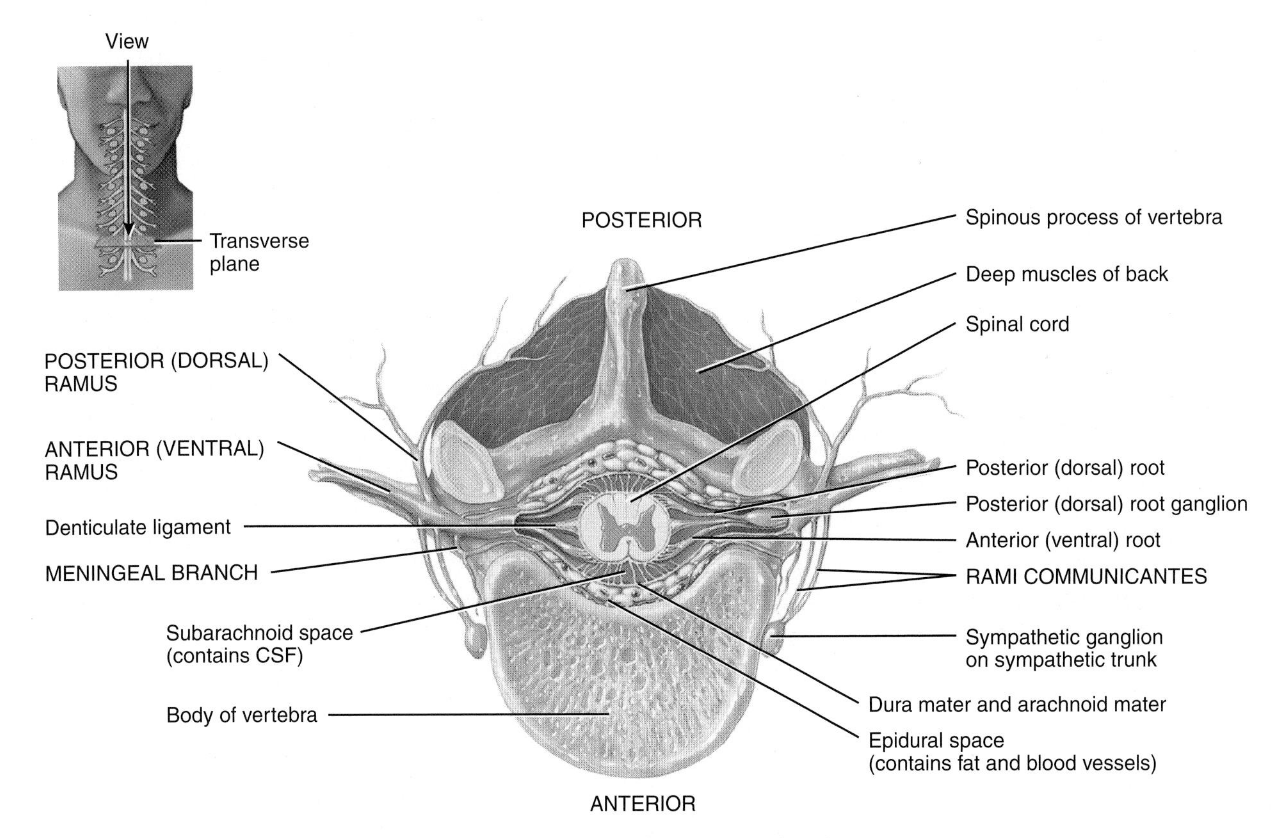

Transverse section of thoracic spinal cord

? Which branches of spinal nerves serve the upper and lower limbs?

are the cervical, brachial, lumbar, and sacral. A smaller ***coccygeal plexus*** is also present. Refer to Figure 16.2 to see their relationships to one another. Emerging from the plexi are nerves bearing names that are often descriptive of the general regions they serve or the course they take. Each of the nerves, in turn, may have several branches named for the specific structures they innervate.

Exhibits 16.1–16.4 summarize the principal plexi. The anterior rami of spinal nerves T2–T12 are called intercostal nerves and are discussed next.

EXHIBIT 16.1 Cervical Plexus

OBJECTIVE

- Describe the origin and distribution of the cervical plexus.

The ***cervical plexus*** (SER-vi-kul) is formed by the roots (ventral rami) of the first four cervical nerves (C1–C4), with contributions from C5 (Figure 16.7). There is one on each side of the neck alongside the first four cervical vertebrae.

The cervical plexus supplies the skin and muscles of the head, neck, and superior part of the shoulders and chest. The phrenic nerve arises from the cervical plexus and supplies motor fibers to the diaphragm. Branches of the cervical plexus also run parallel to two cranial nerves, the accessory (XI) nerve and hypoglossal (XII) nerve.

Injuries to the Cervical Nerves

Complete transection of the spinal cord above the origin of the phrenic nerves (C3, C4, and C5) causes respiratory arrest. Breathing stops because the phrenic nerves no longer send nerve impulses to the diaphragm.

NERVE	ORIGIN	DISTRIBUTION
SUPERFICIAL (SENSORY) BRANCHES		
Lesser occipital	C2	Skin of scalp posterior and superior to ear.
Great auricular (aw-RIK-ū-lar)	C2–C3	Skin anterior, inferior, and over ear, and over parotid glands.
Transverse cervical	C2–C3	Skin over anterior aspect of neck.
Supraclavicular	C3–C4	Skin over superior portion of chest and shoulder.
DEEP (LARGELY MOTOR) BRANCHES		
Ansa cervicalis (AN-sa ser-vi-KAL-is)		This nerve divides into superior and inferior roots.
Superior root	C1	Infrahyoid and geniohyoid muscles of neck.
Inferior root	C2–C3	Infrahyoid muscles of neck.
Phrenic (FREN-ik)	C3–C5	Diaphragm.
Segmental branches	C1–C5	Prevertebral (deep) muscles of neck, levator scapulae, and middle scalene muscles.

CHECKPOINT

5. Which nerve that arises from the cervical plexus causes contraction of the diaphragm?

Intercostal Nerves

The anterior rami of spinal nerves T2–T12 are not part of the plexi and are known as ***intercostal (thoracic) nerves.*** These nerves directly connect to the structures they supply in the intercostal spaces. After leaving its intervertebral foramen, the anterior ramus of nerve T2 innervates the intercostal muscles of the second intercostal space and supplies the skin of the axilla and posteromedial aspect of the arm. Nerves T3–T6 extend along the costal grooves of the ribs and then to the intercostal muscles and skin of the anterior and lateral chest wall. Nerves T7–T12 supply the intercostal muscles and abdominal muscles, and the overlying skin. The posterior rami of the intercostal nerves supply the deep back muscles and skin of the posterior aspect of the thorax.

Figure 16.7 Cervical plexus in anterior view.

The cervical plexus supplies the skin and muscles of the head, neck, superior portion of the shoulders and chest, and diaphragm.

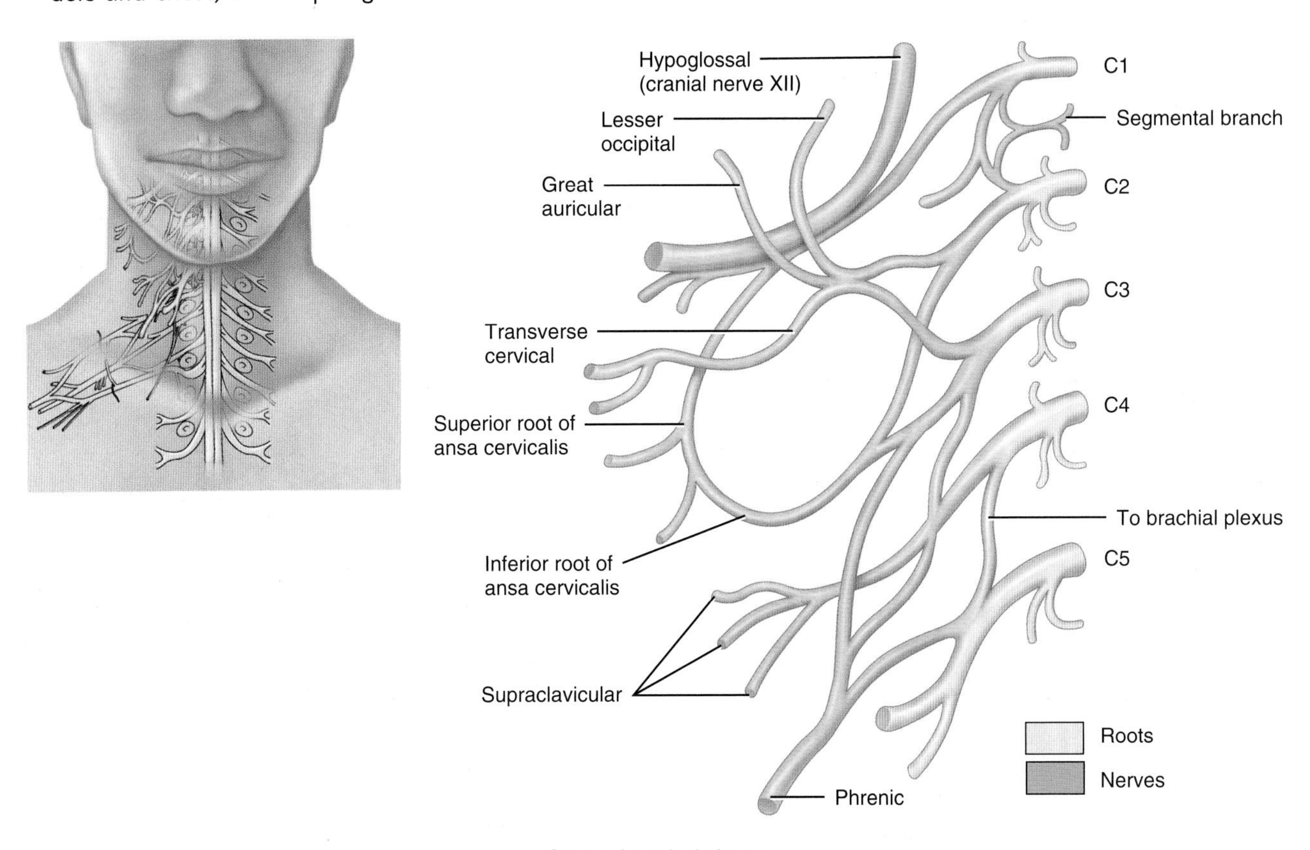

Why does complete severing of the spinal cord at level C2 cause respiratory arrest?

EXHIBIT 16.2 Brachial Plexus

OBJECTIVE

- Describe the origin, distribution, and effects of damage to the brachial plexus.

The roots (ventral rami) of spinal nerves C5–C8 and T1 form the ***brachial plexus*** (BRĀ-kē-al), which extends inferiorly and laterally on either side of the last four cervical and first thoracic vertebrae (Figure 16.8a). It passes between the anterior and middle scalene muscles and above the first rib posterior to the clavicle; the plexus goes deep to the pectoralis minor muscle and then enters the axilla.

Since the brachial plexus is so complex, an explanation of its various parts is helpful. As with the cervical and other plexi, the ***roots*** are the ventral rami of the spinal nerves. The roots of several spinal nerves unite to form ***trunks*** in the inferior part of the neck. These are the *superior, middle,* and *inferior trunks.* Posterior to the clavicles, the trunks divide into ***divisions***, called the *anterior* and *posterior divisions.* In the axillae, the divisions unite to form ***cords*** called the *lateral, medial,* and *posterior cords.* The cords are named for their relationship to the axillary artery, a large artery that supplies blood to the upper limb. The principal ***nerves*** of the brachial plexus branch from the cords.

The brachial plexus provides the entire nerve supply of the shoulders and upper limbs (see Figure 16.8b). Five important nerves arise from the brachial plexus: (1) ***axillary*** supplies the deltoid and teres minor muscles, (2) ***musculocutaneous*** supplies the flexors of the arm, (3) ***radial*** supplies the muscles on the posterior aspect of the arm and forearm, (4) ***median*** supplies most of the muscles of the anterior forearm and some of the muscles of the hand, and (5) ***ulnar*** supplies the anteromedial muscles of the forearm and most of the muscles of the hand.

MANUAL THERAPY APPLICATION

Thoracic Outlet Syndrome

Compression of the brachial plexus on one or more of its nerves is sometimes known as **thoracic outlet syndrome.** The subclavian artery and subclavian vein may also be compressed. The compression may result from spasm of the scalene or pectoralis minor muscles, the presence of a cervical rib (an embryological anomaly), or misaligned ribs. The patient may experience pain, numbness, weakness, or tingling in the upper limb, across the upper thoracic area, and over the scapula on the affected side. The symptoms of thoracic outlet syndrome are exaggerated during physical or emotional stress because the added stress increases the contraction of the involved muscles.

NERVE	ORIGIN	DISTRIBUTION
MOTOR ONLY		
Dorsal scapular (SKAP-ū-lar)	C5	Levator scapulae, rhomboid major, and rhomboid minor muscles.
Suprascapular	C5–C6	Supraspinatus and infraspinatus muscles.
Upper subscapular	C5–C6	Subscapularis muscle.
Lower subscapular	C5–C6	Subscapularis and teres major muscles.
Nerve to subclavius (sub-KLĀ-vē-us)	C5–C6	Subclavius muscle.
Long thoracic (thor-RAS-ik)	C5–C7	Serratus anterior muscle.
Lateral pectoral (PEK-to-ral)	C5–C7	Pectoralis major muscle.
Musculocutaneous (mus′-kū-lō-kū-TĀN-ē-us)	C5–C7	Coracobrachialis, biceps brachii, and brachialis muscles.
Thoracodorsal (tho-RĀ-kō-dor-sal)	C6–C8	Latissimus dorsi muscle.
Medial pectoral	C8–T1	Pectoralis major and pectoralis minor muscles.
MIXED		
Axillary (AK-si-lar-ē)	C5–C6	Deltoid and teres minor muscles; skin over deltoid and superior posterior aspect of arm.
Median	C5–T1	Flexors of forearm, except flexor carpi ulnaris and some muscles of the hand (lateral palm); skin of lateral two-thirds of palm of hand and fingers.
Radial	C5–T1	Triceps brachii, anconeus, and extensor muscles of forearm; skin of posterior arm and forearm, lateral two-thirds of dorsum of hand, and fingers over proximal and middle phalanges.
Ulnar	C8–T1	Flexor carpi ulnaris, flexor digitorum profundus, and most muscles of the hand; skin of medial side of hand, little finger, and medial half of ring finger.
SENSORY ONLY		
Medial cutaneous nerve of arm (kū′-TĀ-nē-us)	C8–T1	Skin of medial and posterior aspects of distal third of arm.
Medial cutaneous nerve of forearm	C8–T1	Skin of medial and posterior aspects of forearm.

Figure 16.8 Brachial plexus in anterior view.

The brachial plexus supplies the shoulders and upper limbs.

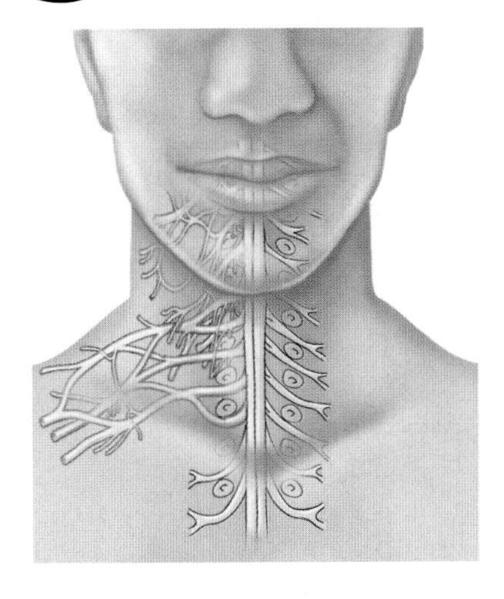

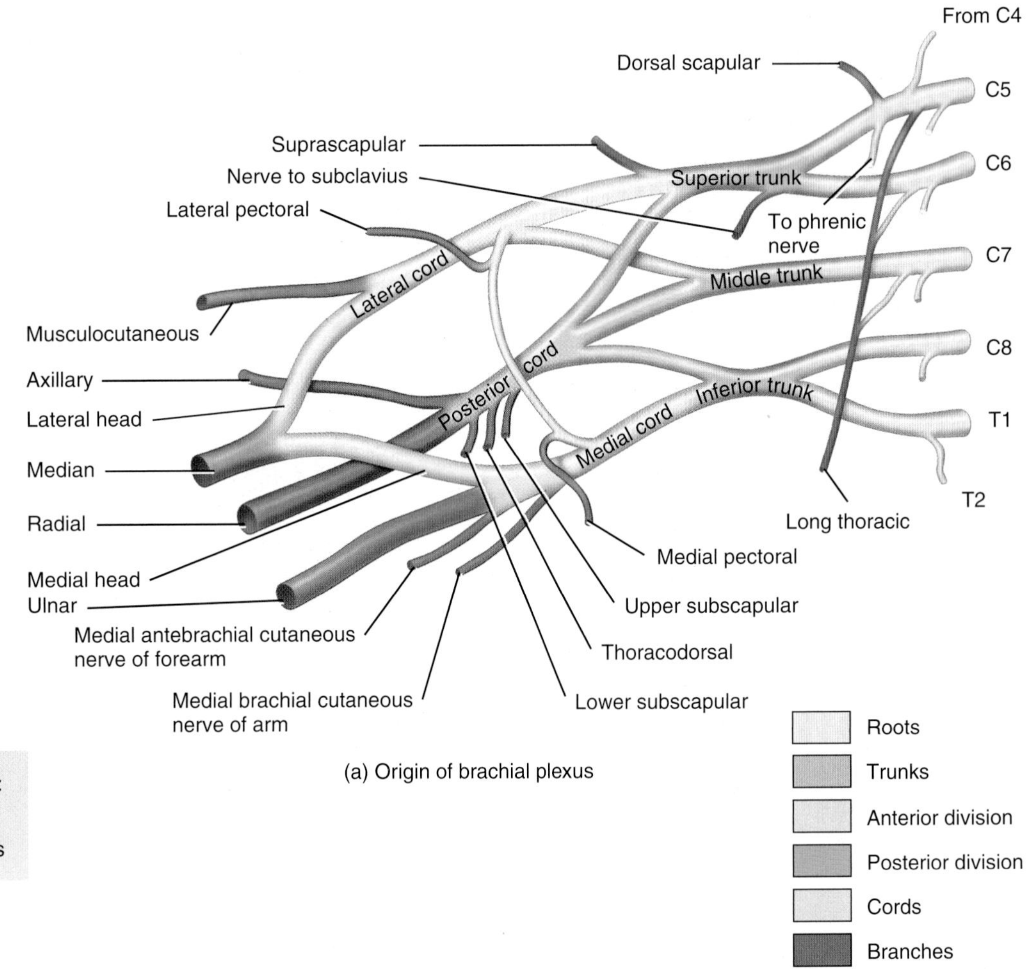

(a) Origin of brachial plexus

MNEMONIC for subunits of the brachial plexus:
Risk **T**akers **D**on't **C**autiously **B**ehave.
Roots, **T**runks, **D**ivisions, **C**ords, **B**ranches

CONTINUES

Figure 16.8 (continued)

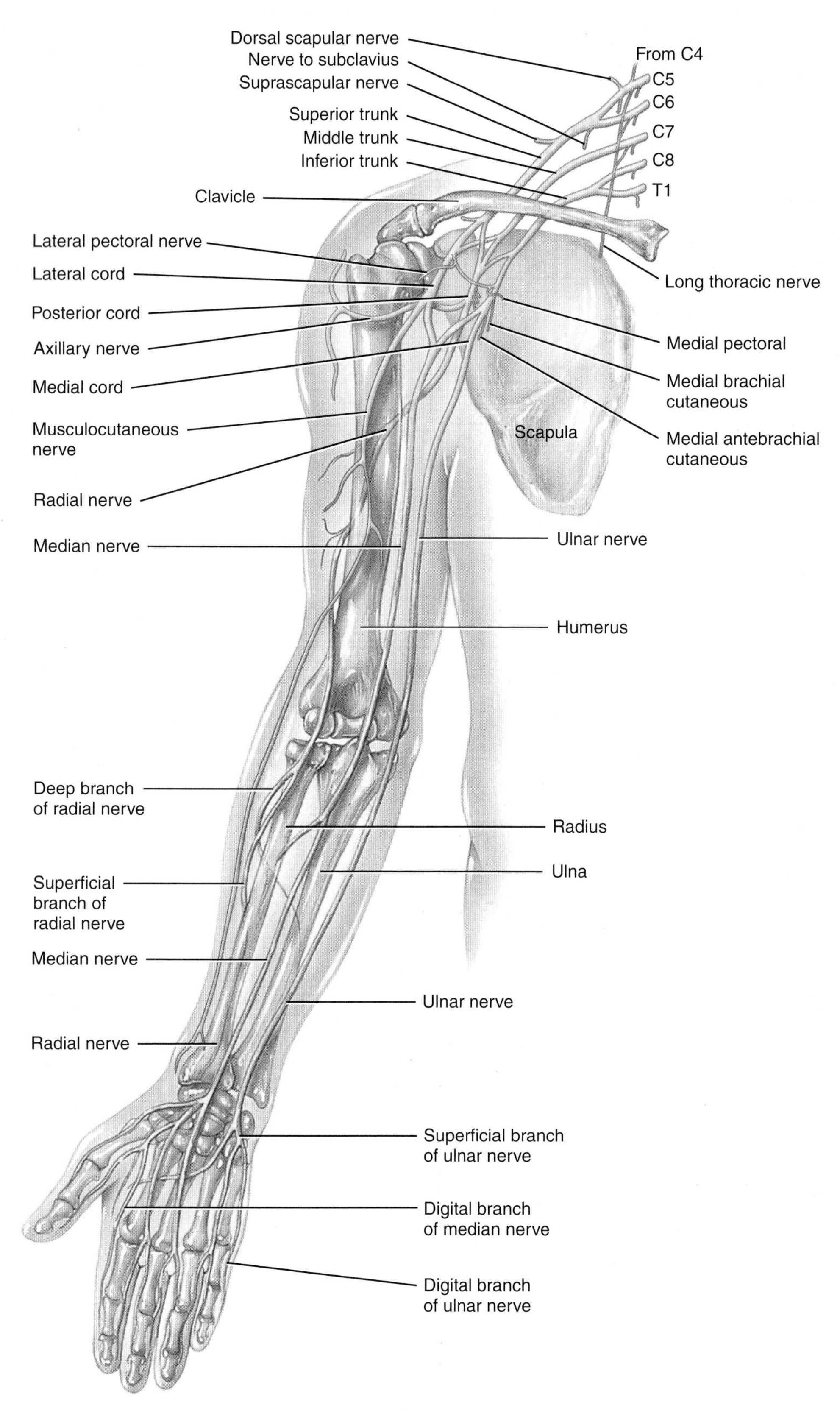

(b) Distribution of nerves from the brachial plexus

? What five important nerves arise from the brachial plexus?

MANUAL THERAPY APPLICATION

Injuries to the Roots of the Brachial Plexus

Injury to the roots of the brachial plexus (C5–C6) may result from forceful pulling away of the head from the shoulder, as might occur from a heavy fall on the shoulder or excessive stretching of an infant's neck during childbirth. The presentation of this injury is characterized by an upper limb in which the shoulder is adducted, the arm is medially rotated, the elbow is extended, the forearm is pronated, and the wrist is flexed (Figure 16.9a). This condition is called *Erb-Duchenne palsy* or *waiter's tip* position. There is loss of sensation along the lateral side of the arm.

Radial (and axillary) nerve injury can be caused by improperly administered intramuscular injections into the deltoid muscle. The radial nerve may also be injured when a cast is applied too tightly around the mid-humerus. Radial nerve injury is indicated by *wrist drop*, the inability to extend the wrist and fingers (Figure 16.9b). Sensory loss is minimal due to the overlap of sensory innervation by adjacent nerves.

Median nerve injury may result in *median nerve palsy*, which is indicated by numbness, tingling, and pain in the palm and fingers. There is also inability to pronate the forearm and flex the proximal interphalangeal joints of all digits and the distal interphalangeal joints of the second and third digits (Figure 16.9c). In addition, wrist flexion and thumb movements are weak, and are accompanied by adduction of the thumb due to a loss of function of the muscles of the thenar eminence (see Clinical Connection on carpal tunnel syndrome in Chapter 13).

Ulnar nerve injury may result in *ulnar nerve palsy (claw hand)*, which is indicated by an inability to abduct or adduct the fingers, atrophy of the interosseus muscles of the hand, hyperextension of the metacarpophalangeal joints, and flexion of the interphalangeal joints, a condition called claw hand (Figure 16.9d). There is also loss of sensation over the little finger and the medial half of the ring finger.

Long thoracic nerve injury results in paralysis of the serratus anterior muscle. The medial border of the scapula protrudes, giving it the appearance of a wing. When the arm is raised, the vertebral border and inferior angle of the scapula pull away from the thoracic wall and protrude outward, causing the medial border of the scapula to protrude; because the scapula looks like a wing, this condition is called *winged scapula* (Figure 16.9e). The arm cannot be abducted beyond the horizontal position.

CHECKPOINT

6. Injury of which nerve could cause paralysis of the serratus anterior muscle?

Figure 16.9 Injuries to the brachial plexus.

6 Injuries to the brachial plexus affect the sensations and movements of the upper limbs.

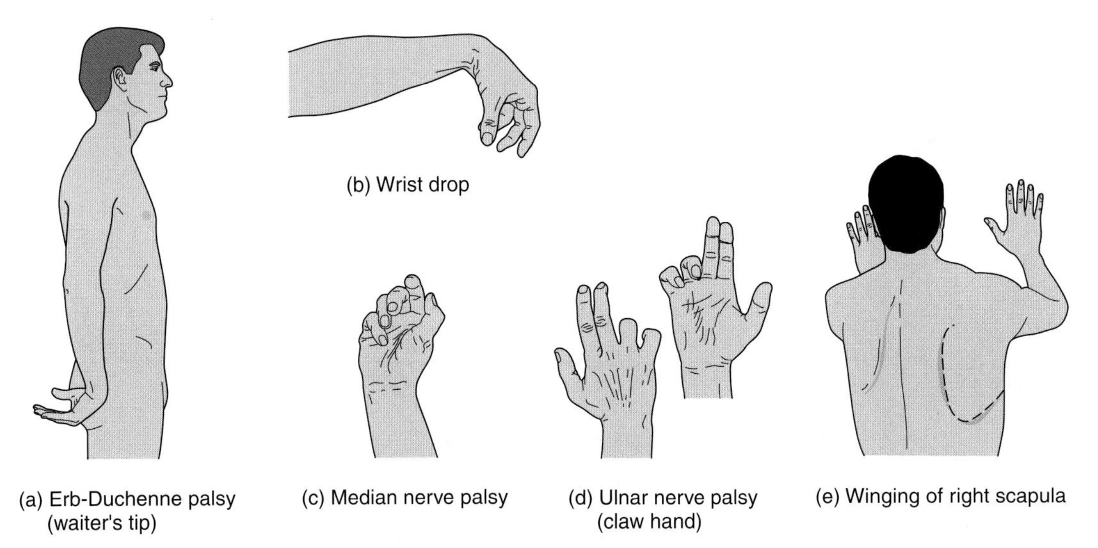

? Injury to which nerve of the brachial plexus affects sensations of the palm and digits?

EXHIBIT 16.3 Lumbar Plexus

OBJECTIVE

- Describe the origin and distribution of the lumbar plexus.

The roots (ventral rami) of spinal nerves L1–L4 form the ***lumbar plexus*** (LUM-bar) (Figure 16.10). Unlike the brachial plexus, there is no intricate intermingling of fibers in the lumbar plexus. On either side of the first four lumbar vertebrae, the lumbar plexus passes obliquely outward, posterior to the psoas major muscle and anterior to the quadratus lumborum muscle. It then gives rise to its peripheral nerves.

The lumbar plexus supplies the anterolateral abdominal wall, external genitals, and part of the lower limbs.

NERVE	ORIGIN	DISTRIBUTION
Iliohypogastric (il′-ē-ō-hī-pō-GAS-trik)	L1	Muscles of anterolateral abdominal wall; skin of inferior abdomen and buttock.
Ilioinguinal (il′-ē-ō-IN-gwi-nal)	L1	Muscles of anterolateral abdominal wall; skin of superior medial aspect of thigh, root of penis and scrotum in male, and labia majora and mons pubis in female.
Genitofemoral (jen′-i-to-FEM-or-al)	L1–L2	Cremaster muscle; skin over middle anterior surface of thigh, scrotum in male, and labia majora in female.
Lateral cutaneous nerve of thigh	L2–L3	Skin over lateral, anterior, and posterior aspects of thigh.
Femoral	L2–L4	The largest nerve arising from the lumbar plexus. It is distributed to the flexor muscles of thigh and extensor muscles of leg; and skin over anterior and medial aspect of thigh and medial side of leg and foot.
Obturator (OB-too-rā-tor)	L2–L4	Adductor muscles of thigh; skin over medial aspect of thigh.

CHECKPOINT

What structures are supplied by the lumbar plexus?

Figure 16.10 Lumbar plexus in anterior view.

The lumbar plexus supplies the anterolateral abdominal wall, external genitals, and part of the lower limbs.

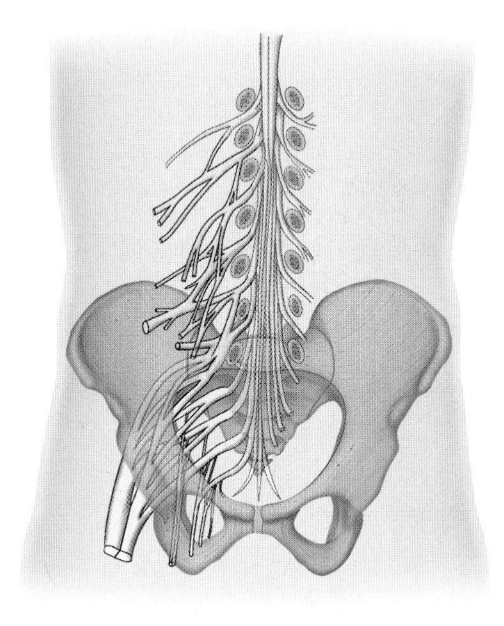

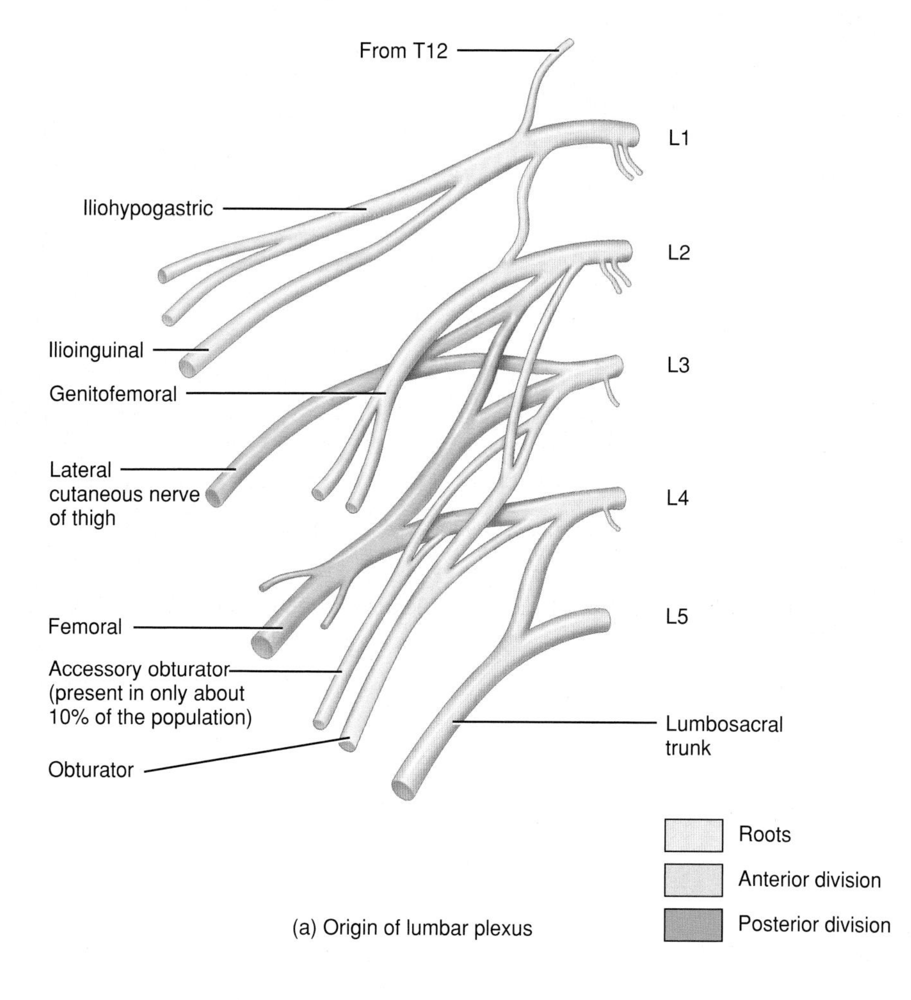

(a) Origin of lumbar plexus

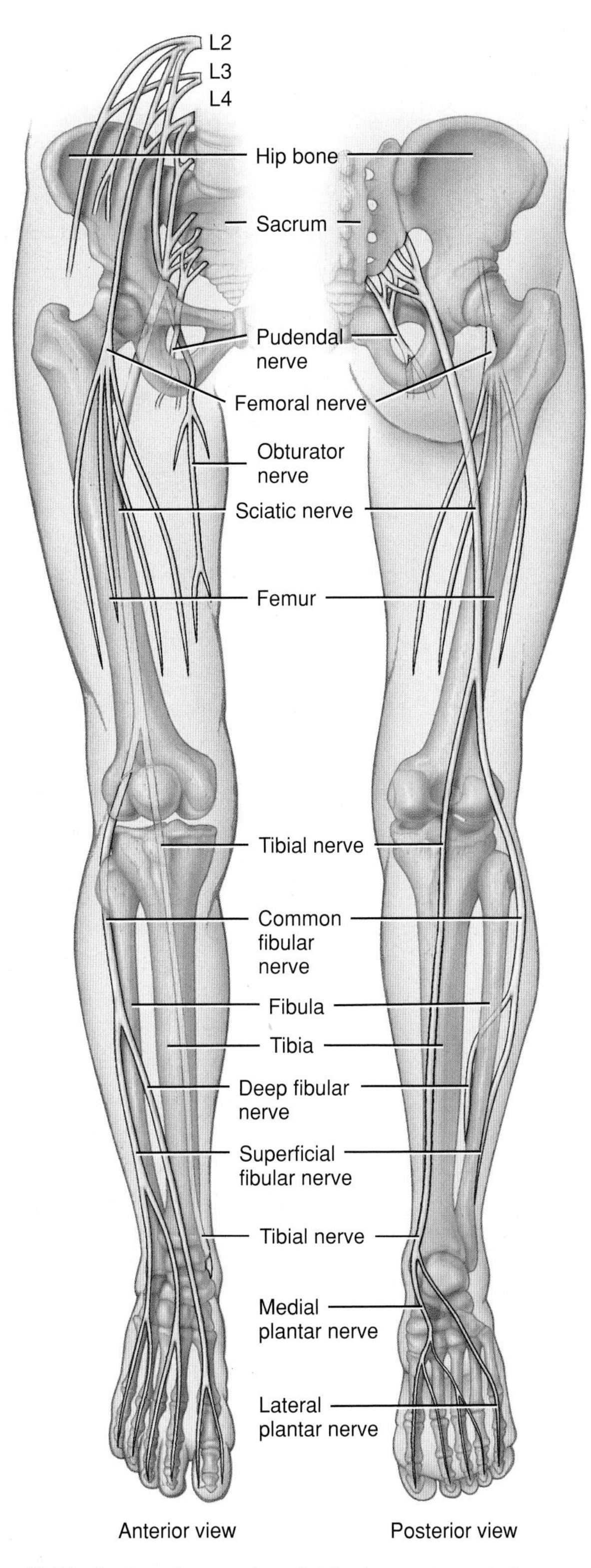

(b) Distribution of nerves from the lumbar and sacral plexuses

? What are the origins of the lumbar plexus?

EXHIBIT 16.4 Sacral and Coccygeal Plexuses

OBJECTIVE

- Describe the origin and distribution of the sacral and coccygeal plexuses.

The roots (ventral rami) of spinal nerves L4–L5 and S1–S4 form the ***sacral plexus*** (SĀ-kral) (Figure 16.11). This plexus is situated largely anterior to the sacrum. The sacral plexus supplies the buttocks, perineum, and lower limbs. The largest nerve in the body—the sciatic nerve—arises from the sacral plexus.

The roots (ventral rami) of spinal nerves S4–S5 and the coccygeal nerves form a small ***coccygeal plexus,*** which supplies a small area of skin in the coccygeal region.

CHECKPOINT

7. Injury of which nerve causes footdrop?

NERVE	ORIGIN	DISTRIBUTION
Superior gluteal (GLOO-tē-al)	L4–L5 and S1	Gluteus minimus, gluteus medius, and tensor fasciae latae muscles.
Inferior gluteal	L5–S2	Gluteus maximus muscle.
Nerve to piriformis (pir-i-FORM-is)	S1–S2	Piriformis muscle.
Nerve to quadratus femoris (quad-RĀ-tus FEM-or-is) and **inferior gemellus**(jem-EL-us)	L4–L5 and S1	Quadratus femoris and inferior gemellus muscles.
Nerve to obturator internus (OB-too-rā′-tor in-TER-nus) and **superior gemellus**	L5–S2	Obturator internus and superior gemellus muscles.
Perforating cutaneous (kū′-TĀ-ne-us)	S2–S3	Skin over inferior medial aspect of buttock.
Posterior cutaneous nerve of thigh	S1–S3	Skin over anal region, inferior lateral aspect of buttock, superior posterior aspect of thigh, superior part of calf, scrotum in male, and labia majora in female.
Pudendal (pū′-DEN-dal)	S2–S4	Muscles of perineum; skin of penis and scrotum in male and clitoris, labia majora, labia minora, and vagina in female.
Sciatic (sī-AT-ik)	L4–S3	The longest nerve in the body. It is actually two nerves—tibial and common fibular—bound together by a common sheath of connective tissue. It splits into its two divisions, usually at the knee. (See below for distributions.) As the sciatic nerve descends through the thigh, it sends branches to hamstring muscles and the adductor magnus.
Tibial (TIB-ē-al)	L4–S3	Gastrocnemius, plantaris, soleus, popliteus, tibialis posterior, flexor digitorum longus, and flexor hallucis longus muscles. Branches of tibial nerve in foot are medial plantar nerve and lateral plantar nerve.
Medial plantar (PLAN-tar) (see Figure 16.10b)		Abductor hallucis, flexor digitorum brevis, and flexor hallucis brevis muscles; skin over medial two-thirds of plantar surface of foot.
Lateral plantar (see Figure 16.10b)		Remaining muscles of foot not supplied by medial plantar nerve; skin over lateral third of plantar surface of foot.
Common fibular (FIB-ū-lar)	L4–S2	Divides into a superficial fibular and a deep fibular branch.
Superficial fibular		Fibularis longus and fibularis brevis muscles; skin over distal third of anterior aspect of leg and dorsum of foot.
Deep fibular		Tibialis anterior, extensor hallucis longus, fibularis tertius, extensor digitorum longus, and extensor digitorum brevis muscles; skin on adjacent sides of great and second toes.

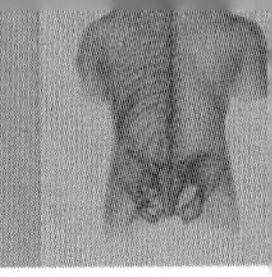

Figure 16.11 Sacral and coccygeal plexuses in anterior view. The distribution of the nerves of the sacral plexus is shown in Figure 16.10b.

The sacral plexus supplies the buttocks, perineum, and lower limbs.

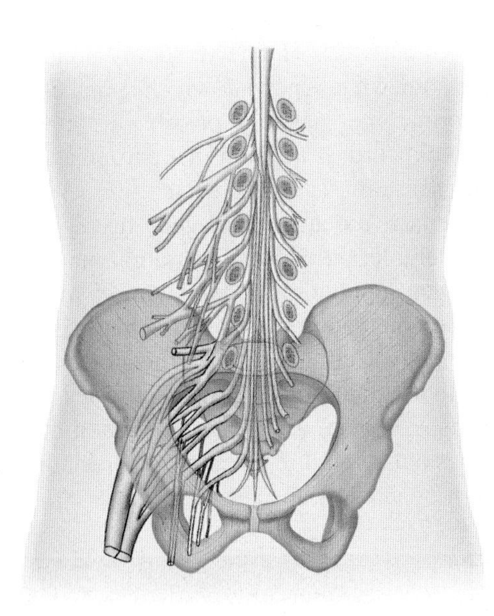

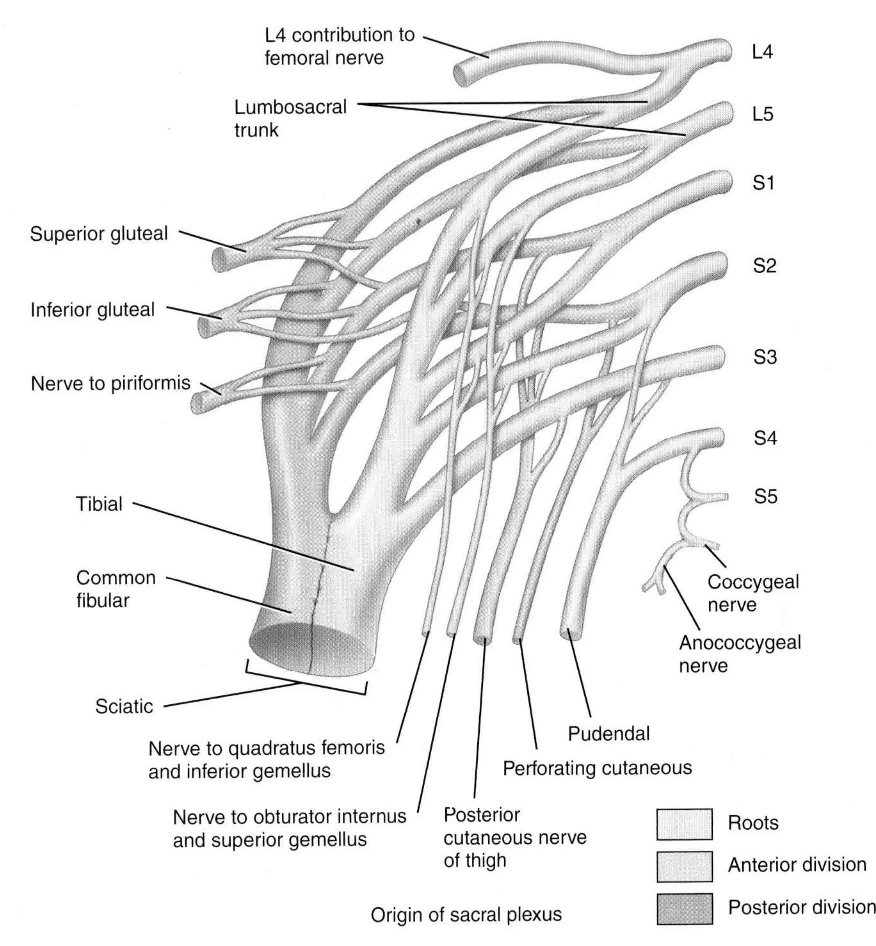

Origin of sacral plexus

What are the origins of the sacral plexus?

• CLINICAL CONNECTION | Spinal Tap

In a **spinal tap (lumbar puncture),** a local anesthetic is given, and a long needle is inserted into the subarachnoid space. During this procedure, the patient lies on the side with the vertebral column flexed as in assuming the fetal position. Flexion of the vertebral column increases the distance between the spinous processes of the vertebrae, which allows easy access to the subarachnoid space. As you will soon learn, the spinal cord ends around the second lumbar vertebra (L2); however, the spinal meninges extend to the second sacral vertebra (S2). Between vertebrae L2 and S2 the spinal meninges are present, but the spinal cord is absent. Consequently, a spinal tap is normally performed in adults between vertebrae L3 and L4 or L4 and L5 because this region provides safe access to the subarachnoid space without the risk of damaging the spinal cord. (A line drawn across the highest points of the iliac crests, called the *supracristal line,* passes through the spinous process of the fourth lumbar vertebra.) A spinal tap is used to withdraw cerebrospinal fluid (CSF) for diagnostic purposes; to introduce antibiotics, contrast media for myelography, or anesthetics; to administer chemotherapy; to measure CSF pressure; and/or to evaluate the effects of treatment for diseases such as meningitis. •

MANUAL THERAPY APPLICATION

Sciatic Nerve Injury

The most common form of back pain is caused by compression or irritation of the sciatic nerve, the longest nerve in the human body. It is actually two nerves—tibial and common fibular—bound together by a common sheath of connective tissue. It splits into its two divisions, usually at the knee. **Sciatic nerve injury** results in *sciatica* (sī-AT-i-ka), pain that may extend from the buttock down the posterior and lateral aspect of the leg and the lateral aspect of the foot. The sciatic nerve may be injured because of a herniated (slipped) disc, dislocated hip, osteoarthritis of the lumbosacral spine, pathological shortening of the lateral rotator muscles of the thigh (especially piriformis), pressure from the uterus during pregnancy, inflammation, irritation, or an improperly administered gluteal intramuscular injection. In addition, sitting on a wallet or other object for a long period of time can also compress the nerve and induce pain. The muscular causes of sciatic nerve pain are discussed in Chapter 14.

In many sciatic nerve injuries, the common fibular portion is the most affected, frequently from fractures of the fibula or by pressure from casts or splints over the thigh or leg. Damage to the common fibular nerve causes the foot to be plantar flexed, a condition called *foot drop,* and inverted, a condition called *equinovarus.* There is also loss of function along the anterolateral aspects of the leg and dorsum of the foot and toes. Injury to the tibial portion of the sciatic nerve results in dorsiflexion of the foot plus eversion, a condition called *calcaneovalgus.* Loss of sensation on the sole also occurs. Treatments for sciatica are similar to those outlined earlier for a herniated (slipped) disc—rest, pain medications, exercises, ice or heat, and massage. The topics of nerve injuries and the effects on muscles cannot be well differentiated.

A very common cause of sciatic pain is spasm of the piriformis muscle. Remember that spasm of a muscle causes the muscle belly to shorten and to become thicker. The sciatic nerve exits the bony sacrum and usually lies deep to the piriformis; in a small percentage of the population the sciatic nerve actually pierces the belly of the piriformis. The level of compression of the sciatic nerve by the piriformis is highly variable and the pain experienced, as stated earlier, is highly variable. The piriformis is a lateral rotator of the thigh. Overuse of the muscle can occur with repetitive lateral rotation of the thigh (and subsequent lateral movement of the foot) as in dancing. The lateral rotators also can be overused by planting the feet solidly on the floor and then rotating the torso, as in some assembly-line work. The piriformis is very deep and its location may be difficult for the beginning student to find. As seen in Figure 14.1d, the piriformis is located along a line between the middle of the sacrum and the greater trochanter. Access to this deep muscle is possible only after softening the gluteal muscles. Whereas most manual therapy treatments involve moving the therapist's hands along the length of a muscle belly, in the case of the piriformis the therapist can locate and then deeply plant her thumb into the piriformis. Passive movement of the patient's flexed leg causes the belly of the piriformis to slide beneath the stationary thumb of the therapist.

Dermatomes

The skin over the entire body is supplied by somatic sensory neurons that carry nerve impulses from the skin into the spinal cord and brain. Each spinal nerve, except for C1, contains sensory neurons that serve a specific, predictable segment of the body. One of the cranial nerves, the trigeminal (V) nerve, serves most of the skin of the face and scalp. The area of the skin that provides sensory input to the CNS via one pair of spinal nerves or the trigeminal (V) nerve is called a ***dermatome*** (*derma-* = skin; *-tome* = thin segment) (Figure 16.12). The nerve supply in adjacent dermatomes overlaps somewhat. Knowing which spinal cord segments supply each dermatome makes it possible to locate damaged regions of the spinal cord. If the skin in a particular region is stimulated but the sensation is not perceived, the nerves supplying that dermatome are probably damaged. In regions where the overlap is considerable, little loss of sensation may result if only one of the nerves supplying the dermatome is damaged. Information about the innervation patterns of spinal nerves can also be used therapeutically. Cutting posterior roots or infusing local anesthetics can block pain either permanently or transiently. Because dermatomes overlap, deliberate production of a region of complete anesthesia may require that at least three adjacent spinal nerves be cut or blocked by an anesthetic drug.

• CLINICAL CONNECTION | Shingles

Shingles is an acute infection of the peripheral nervous system caused by herpes zoster (HER-pēz ZOS-ter), the virus that also causes chickenpox. After a person recovers from chickenpox, the virus retreats to a posterior root ganglion. If the virus is reactivated, the immune system usually prevents it from spreading. From time to time, however, the reactivated virus overcomes a weakened immune system, leaves the ganglion, and travels down sensory neurons of the skin by fast axonal transport (see Section 16.3). The result is pain, discoloration of the skin, and a characteristic line of skin blisters. The line of blisters marks the distribution (dermatome) of the particular cutaneous sensory nerve belonging to the infected posterior root ganglion. •

CHECKPOINT

8. How are spinal nerves named and numbered? Why are all spinal nerves classified as mixed nerves?
9. Compare the names of the connective tissue coverings of spinal nerves and muscle organs. What is found within a fascicle of a nerve and of a muscle organ?
10. How do spinal nerves connect to the spinal cord?
11. Which regions of the body are supplied by plexuses and by intercostal nerves?

16.3 SPINAL CORD PHYSIOLOGY

OBJECTIVES

- Describe the functions of the major sensory and motor tracts of the spinal cord.
- Describe the functional components of a reflex arc and the ways reflexes maintain homeostasis.

The spinal cord has two principal functions in maintaining homeostasis: nerve impulse propagation and integration of infor-

Figure 16.12 Distribution of dermatomes.

A dermatome is an area of skin that provides sensory input to the CNS via the posterior roots of one pair of spinal nerves or via the trigeminal (V) cranial nerve.

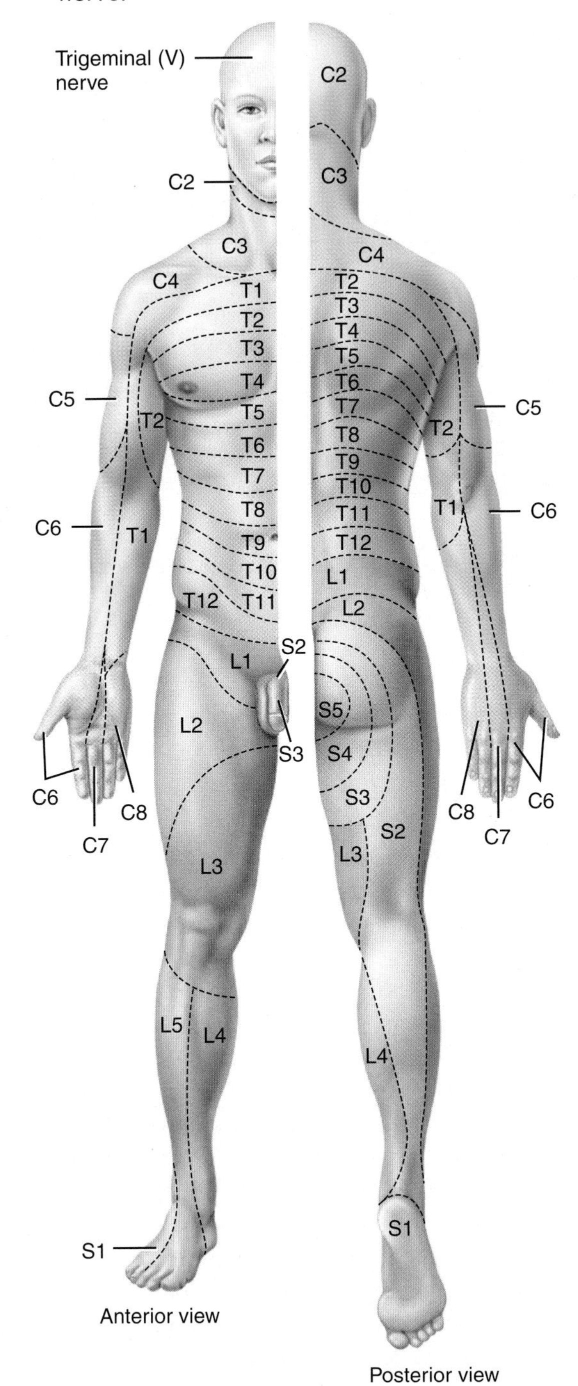

Which is the only spinal nerve that does not have a corresponding dermatome?

mation. The *white matter tracts* in the spinal cord are highways for nerve impulse propagation. Sensory input travels along these tracts toward the brain, and motor output travels from the brain along these tracts toward skeletal muscles and other effector tissues. The *gray matter* of the spinal cord receives and integrates incoming and outgoing information.

Sensory and Motor Tracts

As noted previously, one of the ways the spinal cord promotes ***homeostasis*** is by conducting nerve impulses along tracts. Often, the name of a tract indicates its position in the white matter and where it begins and ends. For example, the anterior spinothalamic tract is located in the *anterior* white column; it begins in the *spinal cord* and ends in the *thalamus* (a region of the brain). Notice that the location of the axon terminals comes last in the name. This regularity in naming allows you to determine the direction of information flow along any tract named according to this convention. Because the anterior spinothalamic tract conveys nerve impulses from the spinal cord toward the brain, it is a sensory (ascending) tract. Figure 16.13 highlights the major sensory and motor tracts in the spinal cord.

Somatic sensory pathways relay information from the somatic sensory receptors to the primary somatosensory area in the cerebral cortex and to the cerebellum (see Section 17.1 The Brain). The pathways to the cerebral cortex consist of thousands of sets of three neurons: a first-order neuron, a second-order neuron, and a third-order neuron.

1. ***First-order neurons*** conduct impulses from somatic receptors into the brain stem or spinal cord. From the face, mouth, teeth, and eyes, somatic sensory impulses propagate along *cranial nerves* into the brain stem. From the neck, trunk, limbs, and posterior aspect of the head, somatic sensory impulses propagate along *spinal nerves* into the spinal cord.
2. ***Second-order neurons*** conduct impulses from the brain stem and spinal cord to the thalamus. Axons of second-order neurons *decussate* (cross over to the opposite side) in the brain stem or spinal cord before ascending to the thalamus. Thus, all somatic sensory information from one side of the body reaches the thalamus on the opposite side.
3. ***Third-order neurons*** conduct impulses from the thalamus to the primary somatosensory area of the cortex on the same side.

Somatic sensory impulses ascend to the cerebral cortex via two main pathways: (1) the posterior column–medial lemniscus pathway, and (2) the anterolateral spinothalamic pathways.

Nerve impulses for touch, pressure, vibration, and conscious proprioception (awareness of the positions of body parts) from the limbs, truck, neck, and posterior head ascend to the cerebral cortex along the ***posterior column–medial lemniscus pathway*** (see Figure 16.14a). The name of the pathway comes from the names of two white-matter tracts that convey the impulses: the

Figure 16.13 Locations of selected sensory and motor tracts, shown in transverse section of the spinal cord. Sensory tracts are indicated on one half and motor tracts on the other half of the cord, but in fact all tracts are present on both sides.

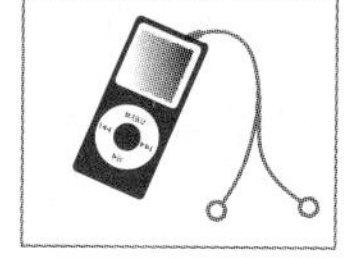

The name of the tract often indicates its location in the white matter and where it begins and ends.

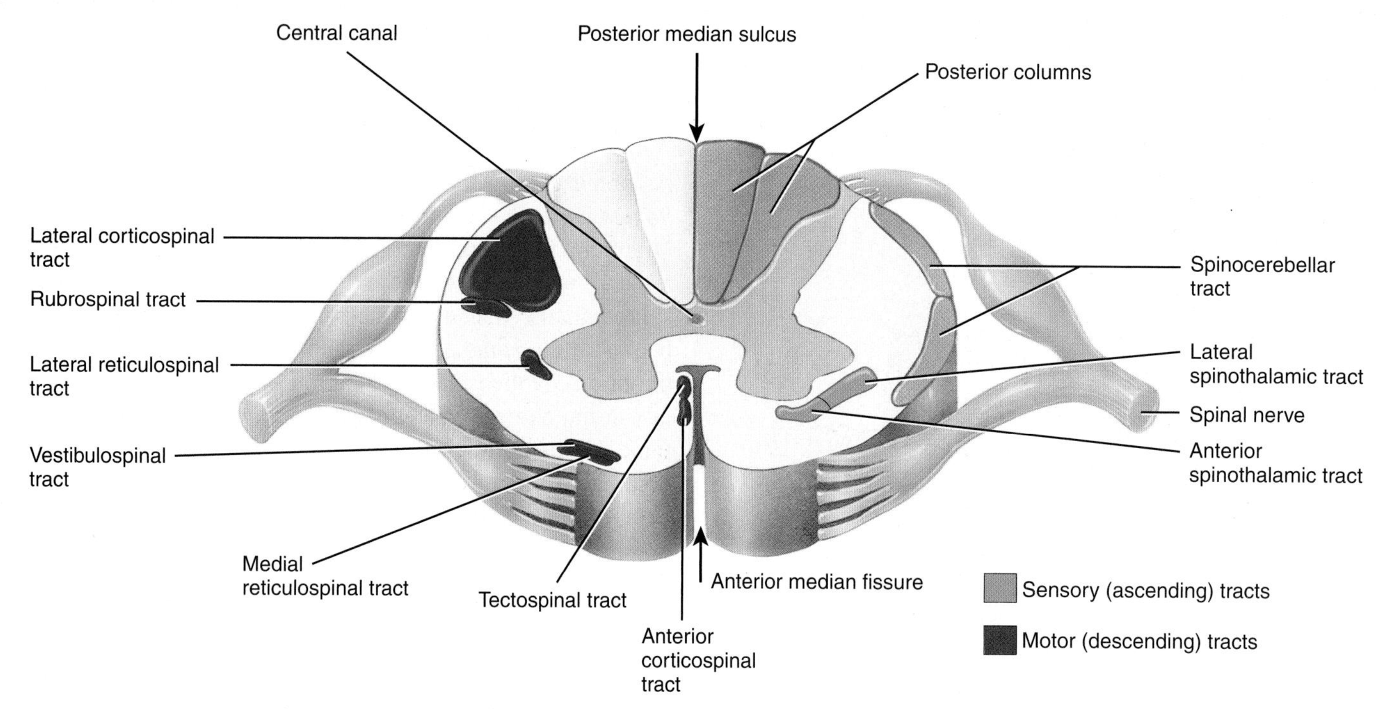

? Based on its name, list the origin and destination of the spinothalamic tract. Is this a sensory or a motor nerve?

posterior column of the spinal cord and the medial lemniscus of the brain stem (see Figure 17.5).

Nerve impulses for pain, temperature, itch, and tickle from the limbs, trunk, neck, and posterior head ascend to the cerebral cortex along the ***anterolateral*** or ***spinothalamic pathway*** (spī-nō-tha-LAM-ik). This pathway begins in two spinal cord tracts, the ***lateral*** and ***anterior spinothalamic tracts*** (Figure 16.14b).

The spinocerebellar tracts are the major routes that proprioceptive impulses take to reach the cerebellum. Although they are not consciously perceived, sensory impulses conveyed to the cerebellum along these pathways are critical for posture, balance, and coordination of movements.

The sensory systems keep the CNS informed of changes in the external and internal environments. The sensory information is integrated (processed) by interneurons in the spinal cord and brain. Responses to the integrative decisions (muscular contractions of all three types of muscles and glandular secretions) are brought about by motor activities.

Neurons in the brain and spinal cord coordinate all voluntary and involuntary movements. All ***somatic motor pathways*** involve at least two motor neurons (see Figure 16.15). The cell bodies of ***upper motor neurons*** are in the higher integration centers of the CNS. The axons of ***lower motor neurons*** extend out of the brain stem to stimulate skeletal muscles in the head and out of the spinal cord to stimulate skeletal muscles in the limbs and trunk.

The cerebral cortex, the outer part of the brain, plays a major role in controlling precise voluntary muscular movements. Other brain regions provide important integration for regulation of automatic movements, such as arm swinging during walking. Motor output to skeletal muscles travels down the spinal cord in two types of descending pathways: direct and indirect. The ***direct motor pathways*** in the spinal cord include the *lateral corticospinal* and *anterior corticospinal tracts.* They convey nerve impulses that originate in the cerebral cortex and are destined to cause precise, *voluntary* movements of skeletal muscles. ***Indirect motor pathways*** located in the spinal cord include the *rubrospinal, reticulospinal, tectospinal,* and *vestibulospinal tracts.* They convey nerve impulses from the brain stem and other parts of the brain that govern *automatic movements* and help coordinate body movements with visual stimuli. Indirect pathways also maintain skeletal muscle tone, maintain contraction of postural muscles, and play a major role in equilibrium by regulating muscle tone in response to movements of the head.

Figure 16.14 Somatic sensory pathways. (a) In the posterior column–medial lemniscus pathway, the first-order neuron in the pathway ascends to the medulla oblongata via the posterior column (white matter located on the posterior side of the spinal cord). In the medulla, it synapses with a second-order neuron, which then extends through the medial lemniscus to the thalamus on the opposite side. The third-order neuron extends from the thalamus to the cerebral cortex. (b) In the anterolateral pathway, the first-order neuron synapses with a second-order neuron in the spinal cord gray matter. The second-order neuron extends to the thalamus on the opposite side, and the third-order neuron extends from the thalamus to the cerebral cortex.

Nerve impulses for somatic sensations conduct to the primary somatosensory area (postcentral gyrus) of the cerebral cortex.

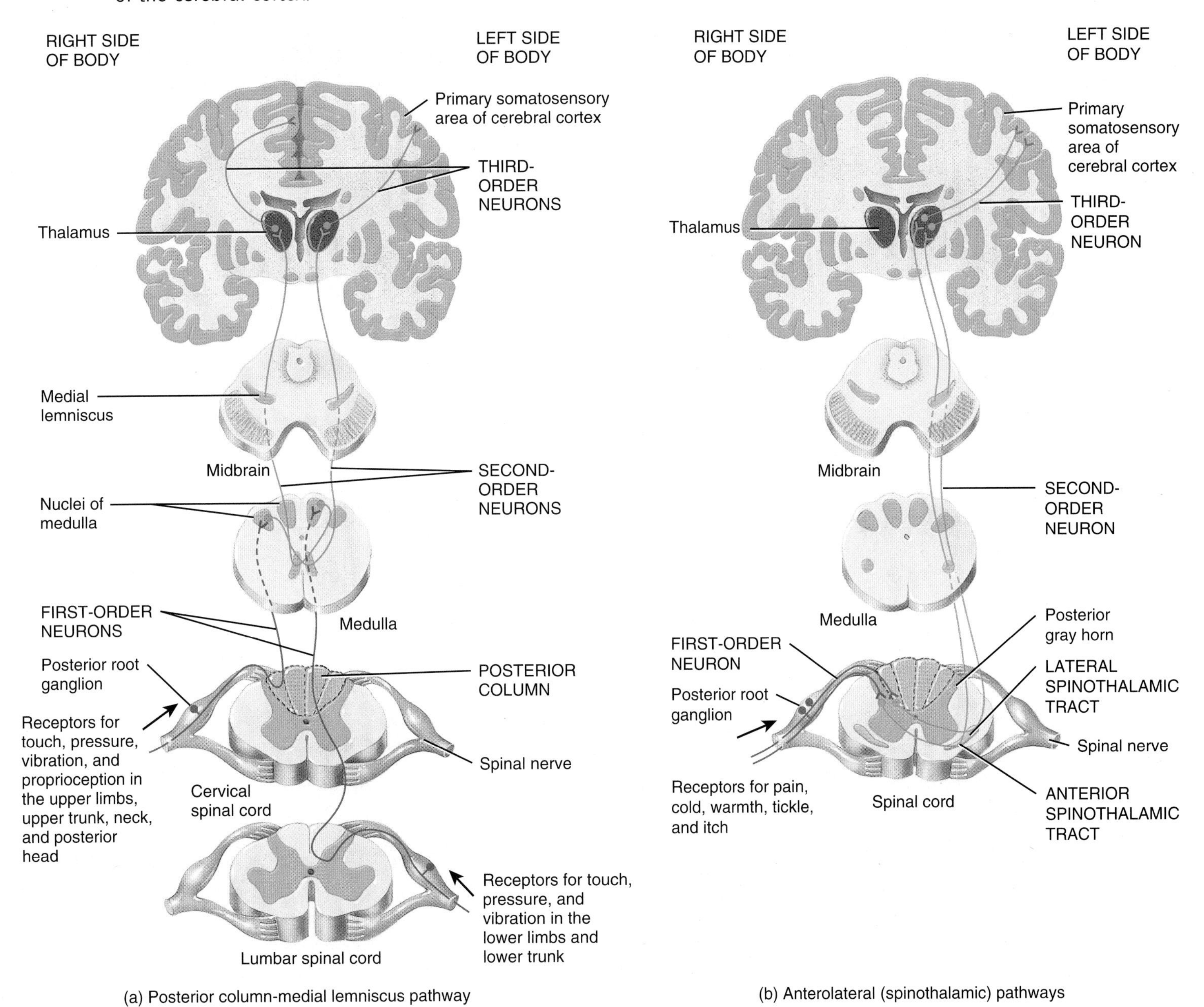

Which somatic sensations could be lost due to damage of the spinothalamic tracts?

Figure 16.15 Somatic motor pathways. Shown here are the two most direct pathways whereby signals initiated by the primary motor area in one hemisphere control skeletal muscles on the opposite side of the body.

Lower motor neurons stimulate skeletal muscles to produce movements.

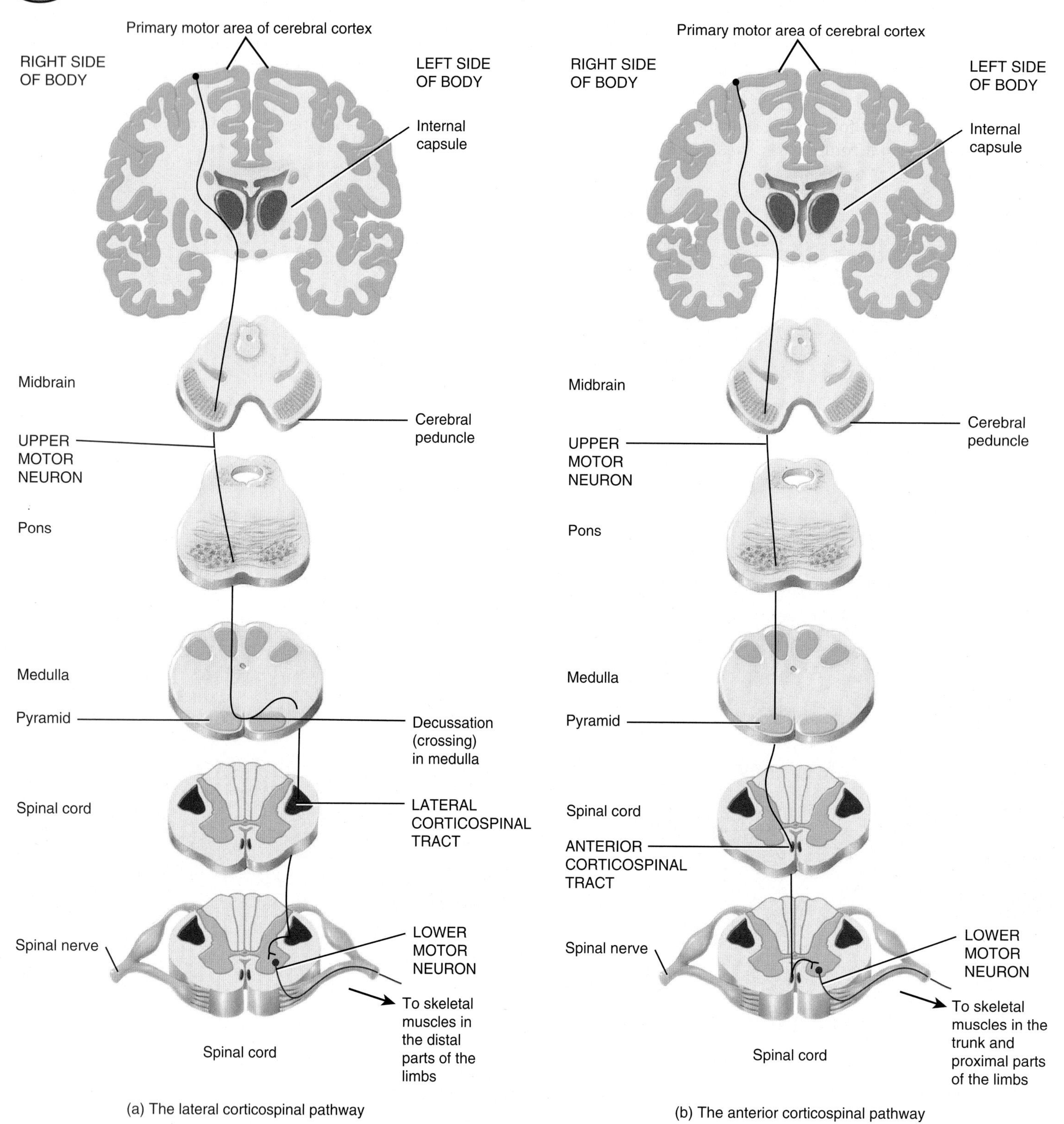

(a) The lateral corticospinal pathway

(b) The anterior corticospinal pathway

What two spinal cord tracts conduct impulses along axons of upper motor neurons?

MANUAL THERAPY APPLICATION

Working with Patients with Paralysis

Damage or disease of lower motor neurons produces **flaccid paralysis** of muscles on the same (ipsilateral) side of the body: The muscles lack voluntary control and reflexes, muscle tone is decreased or lost, and the muscle remains flaccid (limp). Injury or disease of upper motor neurons causes **spastic paralysis** of muscles on the opposite (contralateral) side of the body. In this condition muscle tone is increased, reflexes are exaggerated, and pathological reflexes appear.

Manual therapists should be aware that patients with spinal cord injury and therefore confined to wheelchairs have varying degrees of spastic paralysis. Patients with moderate spasticity are usually prescribed medications that reduce the severity. Patients with severe spasticity commonly have their lower limbs strapped to the wheelchair so that the uncontrolled movements of the limbs do not cause bruising, fracture, or other trauma. When a patient is on your table, only the slightest environmental stimulus may cause uncontrolled spastic movements. Application of hot or cold lubricants; effleurage, pettrisage, or other techniques; and warm or cool temperature of the room are a few stimuli that will induce spastic movements. If you are dressing the patient after treatment, tying the shoe laces too tightly is another example of a stimulus that will be problematic for the patient with severe spasticity.

Muscle Spindles

The next section on spinal reflexes refers to specialized receptors in skeletal muscle tissue known as muscle spindles and tendon organs. Muscle spindles are the proprioceptors in skeletal muscles that monitor changes in the length of skeletal muscles and participate in stretch reflexes (see Figure 16.18). By adjusting how vigorously a muscle spindle responds to stretching of a skeletal muscle, the brain sets an overall level of ***muscle tone,*** the small degree of contraction that is present while the muscle is at rest.

Each ***muscle spindle*** consists of several slowly adapting sensory nerve endings that wrap around 3 to 10 specialized muscle fibers, called ***intrafusal fibers*** (*intrafusal* = within a spindle). A connective tissue capsule encloses the sensory nerve endings and intrafusal fibers and anchors the spindle to the endomysium and perimysium (see Figure 16.16). Muscle spindles are interspersed among most skeletal muscle fibers and aligned parallel to them. In muscles that produce finely controlled movements, such as those of the fingers or eyes, muscle spindles are plentiful. Muscles involved in coarser but more forceful movements, like the quadriceps femoris and hamstring muscles of the thigh, have fewer muscle spindles. The only skeletal muscles that lack spindles are the tiny muscles of the middle ear.

The main function of muscle spindles is to measure *muscle length*—how much a muscle is being stretched. Either sudden or prolonged stretching of the central areas of the intrafusal muscle fibers stimulates the sensory nerve endings. The resulting nerve impulses propagate into the CNS. Information from muscle spindles arrives quickly at the somatic sensory areas of the cerebral cortex, which allows conscious perception of limb positions and movements. At the same time, impulses from muscle spindles pass to the cerebellum, where the input is used to coordinate muscle contractions.

In addition to their sensory nerve endings near the middle of intrafusal fibers, muscle spindles contain motor neurons called ***gamma motor neurons.*** These motor neurons terminate near both ends of the intrafusal fibers and adjust the tension in a muscle spindle to variations in the length of the muscle. For example, when a muscle shortens, gamma motor neurons stimulate the ends of the intrafusal fibers to contract slightly. This keeps the intrafusal fibers taut and maintains the sensitivity of the muscle spindle to stretching of the muscle. As the frequency of impulses in its gamma motor neuron increases, a muscle spindle becomes more sensitive to stretching of its mid-region.

Surrounding muscle spindles are ordinary skeletal muscle fibers, called ***extrafusal muscle fibers*** (*extrafusal* = outside a spindle), which are supplied by large-diameter A fibers called ***alpha motor neurons.*** The cell bodies of both gamma and alpha motor neurons are located in the anterior gray horn of the spinal cord (or in the brain stem for muscles in the head). During the stretch reflex (described later), impulses in muscle spindle sensory axons propagate into the spinal cord and brain stem and activate alpha motor neurons that connect to extrafusal muscle fibers in the same muscle. In this way, activation of its muscle spindles causes contraction of a skeletal muscle, which relieves the stretching.

Tendon Organs

Tendon organs are located at the junction of a tendon and a muscle. By initiating tendon reflexes (described later), tendon organs protect tendons and their associated muscles from damage due to excessive tension (see Figure 16.19). When a muscle contracts, it exerts a force that pulls the points of attachment of the muscle at either end toward each other. This force is the muscle tension. Each ***tendon organ*** (see Figure 16.16) consists of a thin capsule of connective tissue that encloses a few tendon fascicles (bundles of collagen fibers). Penetrating the capsule are one or more sensory nerve endings that entwine among and around the collagen fibers of the tendon. When tension is applied to a muscle, the tendon organs generate nerve impulses that propagate into the CNS, providing information about changes in muscle tension. Tendon reflexes decrease muscle tension by causing muscle relaxation.

Reflexes and Reflex Arcs

The second way the spinal cord promotes ***homeostasis*** is by serving as an integrating center for some reflexes. A ***reflex*** is a fast, automatic, unplanned sequence of actions that occurs in response to a particular stimulus. Some reflexes are inborn, such as pulling your hand away from a hot surface before you even feel that it is hot. Other reflexes are learned or acquired. For instance, you learn many reflexes while acquiring driving expertise. Slamming on the brakes in an emergency is one example. When integration takes place in the spinal cord gray matter, the

Figure 16.16 **Two types of proprioceptors: a muscle spindle and a tendon organ.** In muscle spindles, which monitor changes in skeletal muscle length, sensory nerve endings wrap around the central portion of intrafusal muscle fibers. In tendon organs, which monitor the force of muscle contraction, sensory nerve endings are activated by increasing tension on a tendon. If you examine Figure 16.18, you can see the relationship of a muscle spindle to the spinal cord as a component of a stretch reflex. In Figure 16.19, you can see the relationship of a tendon organ to the spinal cord as a component of a tendon reflex.

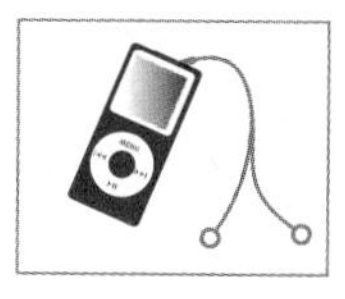

Proprioceptors provide information about body position and movement.

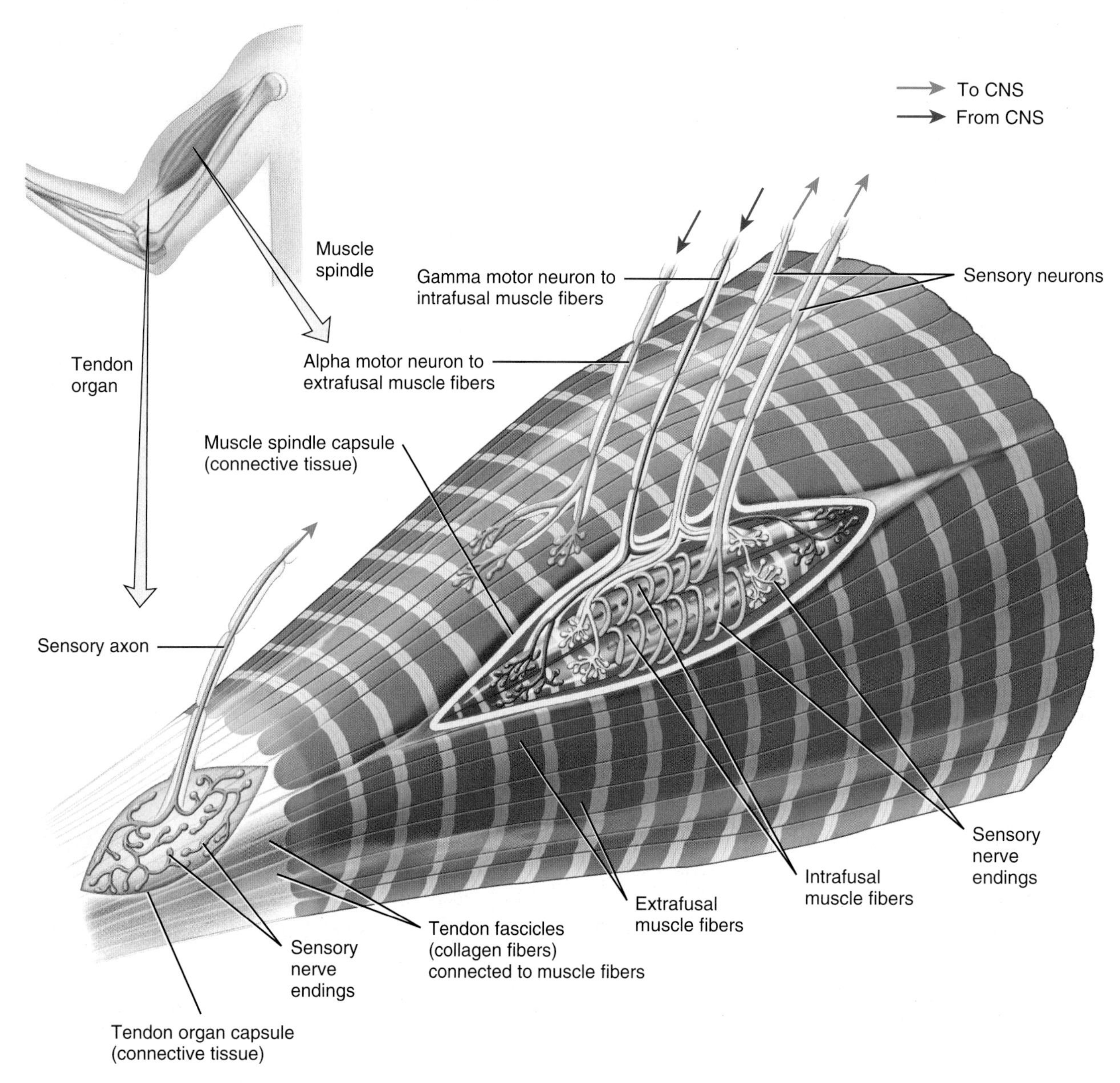

How is a muscle spindle activated?

reflex is a ***spinal reflex.*** An example is the familiar patellar reflex (knee jerk). If integration occurs in the brain stem rather than the spinal cord, the reflex is called a ***cranial reflex.*** An example is the tracking movements of your eyes as you read this sentence. You are probably most aware of ***somatic reflexes,*** which involve contraction of skeletal muscles. Equally important, however, are the ***autonomic (visceral) reflexes,*** which generally are not consciously perceived. They involve responses of smooth muscle, cardiac muscle, and glands. As you will see in Chapter 18, body functions such as heart rate, digestion, urination, and defecation are controlled by the autonomic nervous system through autonomic reflexes.

Nerve impulses propagating into, through, and out of the CNS follow specific pathways, depending on the kind of information, its origin, and its destination. The pathway followed by nerve impulses that produce a reflex is a ***reflex arc*** **(*reflex circuit*).** A reflex arc includes the following five functional components (Figure 16.17):

1. **Sensory receptor.** The distal end of a sensory neuron (dendrite) or an associated sensory structure serves as a sensory receptor. It responds to a specific ***stimulus***—a change in the internal or external environment—by producing a graded potential called a generator (or receptor) potential (see Section 15.3). If a generator potential reaches the threshold level of depolarization, it will trigger one or more nerve impulses in the sensory neuron.
2. **Sensory neuron.** The nerve impulses propagate from the sensory receptor along the axon of the sensory neuron to the axon terminals, which are located in the gray matter of the spinal cord or brain stem.
3. **Integrating center.** One or more regions of gray matter within the CNS act as an integrating center. In the simplest type of reflex, the integrating center is a single synapse between a sensory neuron and a motor neuron. A reflex pathway having only one synapse in the CNS is termed a ***monosynaptic reflex arc*** (*mono-* = one). More often, the integrating center consists of one or more interneurons, which may relay impulses to other interneurons as well as to a motor neuron. A ***polysynaptic reflex arc*** (*poly-* = many) involves more than two types of neurons and more than one CNS synapse.
4. **Motor neuron.** Impulses triggered by the integrating center propagate out of the CNS along a motor neuron to the part of the body that will respond.
5. **Effector.** The part of the body that responds to the motor nerve impulse, such as a muscle or gland, is the effector. Its action is called a reflex. If the effector is skeletal muscle, the reflex is a ***somatic reflex.*** If the effector is smooth muscle, cardiac muscle, or a gland, the reflex is an ***autonomic (visceral) reflex.***

Stretch Reflex

A ***stretch reflex*** causes contraction of a skeletal muscle (the effector) in response to stretching of the muscle. This type of reflex occurs via a monosynaptic reflex arc. The reflex can occur by activation of a single sensory neuron that forms one synapse in the CNS with a single motor neuron. Stretch reflexes can be elicited by tapping on tendons attached to muscles at the elbow, wrist, knee, and ankle joints.

Figure 16.17 General components of a reflex arc. The arrows show the direction of nerve impulse propagation.

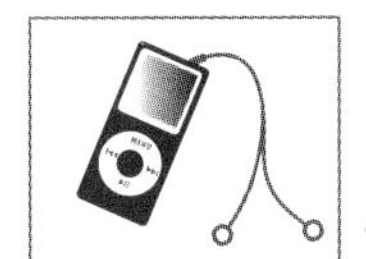

A reflex is a fast, predictable sequence of involuntary actions that occur in response to certain changes in the environment.

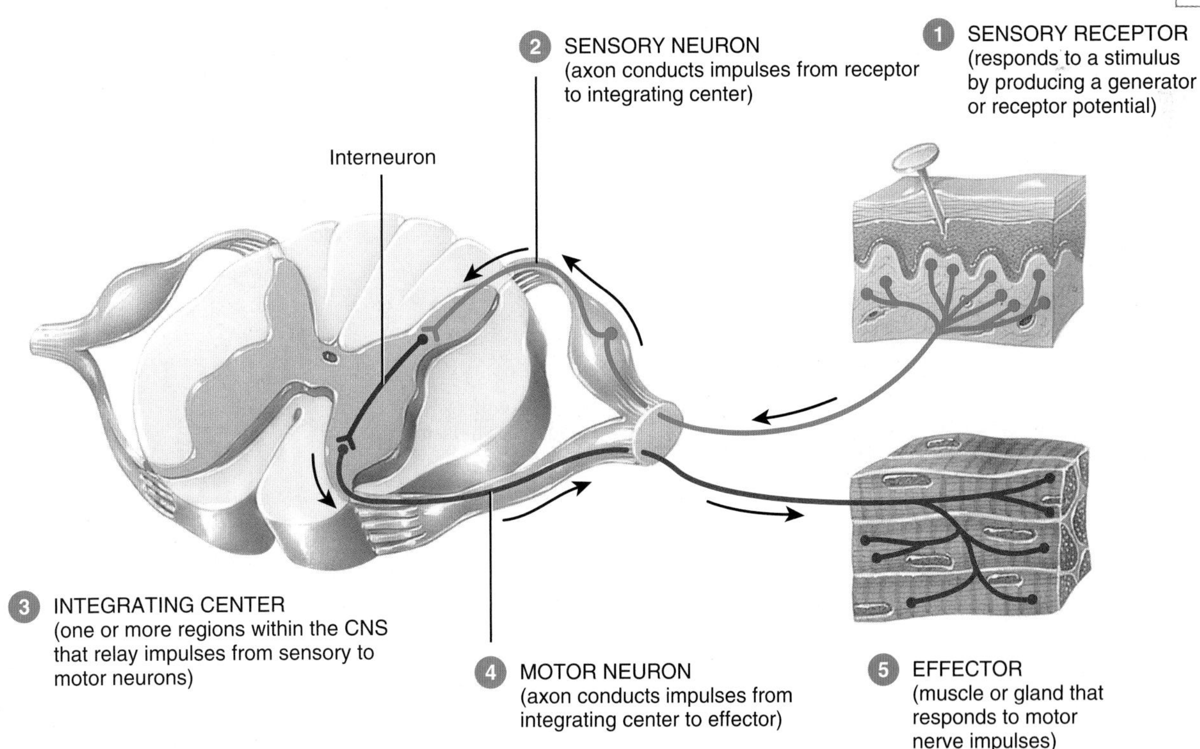

What initiates a nerve impulse in a sensory nerve? Which branch of the nervous system includes all integrating centers for reflexes?

A stretch reflex operates as follows (Figure 16.18):

1. Slight stretching of a muscle stimulates sensory receptors in the ***muscle spindles.*** The spindles monitor changes in the length of the muscle.
2. In response to being stretched, a muscle spindle generates one or more nerve impulses that propagate along a somatic sensory neuron through the posterior root of the spinal nerve and into the spinal cord.
3. In the spinal cord (integrating center), the sensory neuron makes an excitatory synapse with and thereby activates a motor neuron in the anterior gray horn.
4. If the excitation is strong enough, one or more nerve impulses arise in the motor neuron and propagate along its axon, which extends from the spinal cord into the anterior root and through peripheral nerves to the stimulated muscle. The axon terminals of the motor neuron form neuromuscular junctions (NMJs) with skeletal muscle fibers of the stretched muscle.
5. Acetylcholine released by nerve impulses at the NMJs triggers one or more muscle action potentials in the stretched muscle (effector), and the muscle contracts. Thus, muscle stretch is followed by muscle contraction, which relieves the stretching.

Figure 16.18 Stretch reflex. This monosynaptic reflex arc has only one synapse in the CNS—between a single sensory neuron and a single motor neuron. A polysynaptic reflex arc to antagonistic muscles that includes two synapses in the CNS and one interneuron is also illustrated. Plus signs (+) indicate excitatory synapses; the minus sign (−) indicates an inhibitory synapse.

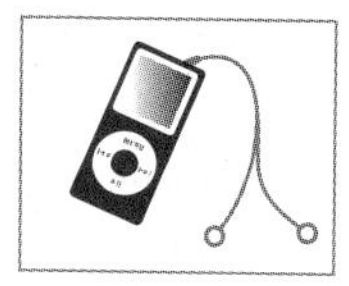

The stretch reflex causes contraction of a muscle that has been stretched.

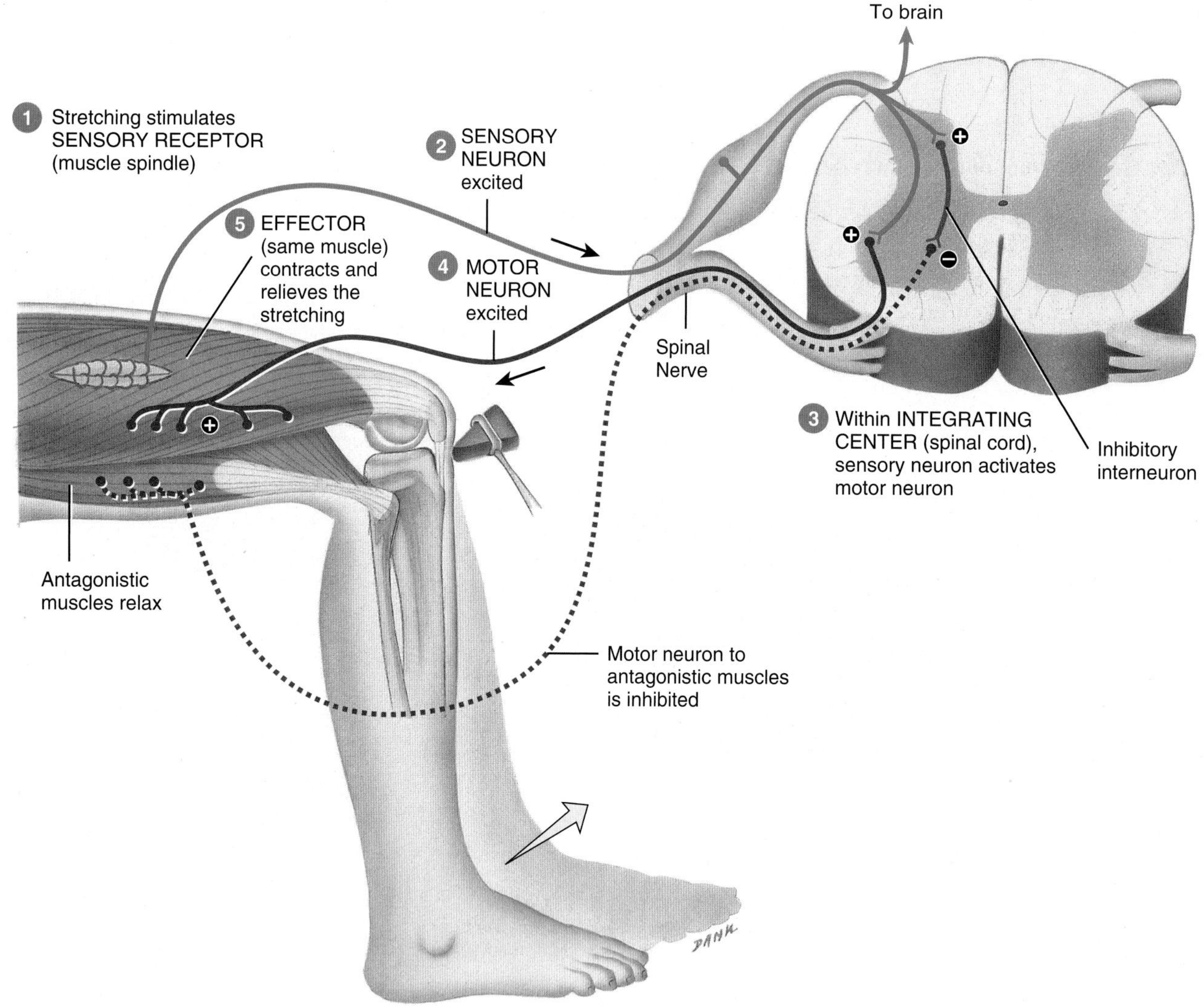

What makes this an ipsilateral reflex?

In the reflex arc just described, sensory nerve impulses enter the spinal cord on the same side from which motor nerve impulses leave it. This arrangement is called an ***ipsilateral reflex*** (ip′-si-LAT-er-al = same side). All monosynaptic reflexes are ipsilateral.

Although the stretch reflex pathway itself is monosynaptic (two neurons and one synapse), a polysynaptic reflex arc to the antagonistic muscles operates at the same time. This arc involves three neurons and two synapses. An axon collateral (branch) from the muscle spindle sensory neuron also synapses with an inhibitory interneuron in the integrating center. In turn, the interneuron synapses with and inhibits a motor neuron that normally excites the antagonistic muscles (Figure 16.18). Thus, when the stretched muscle contracts during a stretch reflex, antagonistic muscles that oppose the contraction relax. This type of arrangement, in which the components of a neural circuit simultaneously cause contraction of one muscle and relaxation of its antagonists, is termed ***reciprocal inhibition.*** Reciprocal inhibition prevents conflict between opposing muscles and is vital in coordinating body movements.

Axon collaterals of the muscle spindle sensory neuron also relay nerve impulses to the brain over specific ascending pathways. In this way, the brain receives input about the state of stretch or contraction of skeletal muscles, enabling it to coordinate muscular movements. The nerve impulses that pass to the brain also allow conscious awareness that the reflex has occurred.

The stretch reflex can also help maintain posture. For example, if a standing person begins to lean forward, the gastrocnemius and other calf muscles are stretched. Consequently, stretch reflexes are initiated in these muscles, which cause them to contract and reestablish the body's upright posture. Similar types of stretch reflexes occur in the muscles of the shin when a standing person begins to lean backward.

Tendon Reflex

The stretch reflex, just described, operates as a feedback mechanism to control muscle *length* by causing muscle contraction. In contrast, the ***tendon reflex*** operates as a feedback mechanism to control muscle *tension* by causing muscle relaxation before muscle force becomes so great that tendons might be torn. Although the tendon reflex is less sensitive than the stretch reflex, it can override the stretch reflex when tension is great, making you drop a very heavy weight, for example. Like the stretch reflex, the tendon reflex is ipsilateral. The sensory receptors for this reflex are ***tendon (Golgi tendon) organs*** which lie within a tendon near its junction with a muscle. In contrast to muscle spindles, which are sensitive to changes in muscle length, tendon organs detect and respond to changes in muscle tension that are caused by passive stretch or muscular contraction.

A tendon reflex operates as follows (see Figure 16.19):

1. As the tension applied to a tendon increases, the tendon organ (sensory receptor) is stimulated (depolarized to threshold).
2. Nerve impulses arise and propagate into the spinal cord along a sensory neuron.
3. Within the spinal cord (integrating center), the sensory neuron activates an inhibitory interneuron that synapses with a motor neuron.
4. The inhibitory neurotransmitter inhibits (hyperpolarizes) the motor neuron, which then generates fewer nerve impulses.
5. The muscle relaxes and relieves excess tension.

Thus, as tension on the tendon organ increases, the frequency of inhibitory impulses increases; inhibition of the motor neurons to the muscle developing excess tension (effector) causes relaxation of the muscle. In this way, the tendon reflex protects the tendon and muscle from damage due to excessive tension.

Note in Figure 16.19 that the sensory neuron from the tendon organ also synapses with an excitatory interneuron in the spinal cord. The excitatory interneuron, in turn, synapses with motor neurons controlling antagonistic muscles. Thus, while the tendon reflex brings about relaxation of the muscle attached to the tendon organ, it also triggers contraction of antagonists. Here we have another example of reciprocal inhibition. The sensory neuron also relays nerve impulses to the brain by way of sensory tracts, thus informing the brain about the state of muscle tension throughout the body.

Flexor Reflex

Another reflex involving a polysynaptic reflex arc results when, for instance, you step on a tack. In response to such a painful stimulus, you immediately withdraw your leg. This reflex, called the ***flexor*** or ***withdrawal reflex,*** operates as follows (see Figure 16.20):

1. Stepping on a tack stimulates the dendrites (sensory receptor) of a pain-sensitive neuron.
2. This sensory neuron then generates nerve impulses, which propagate into the spinal cord.
3. Within the spinal cord (integrating center), the sensory neuron activates interneurons that extend to several spinal cord segments.
4. The interneurons activate motor neurons in several spinal cord segments. As a result, the motor neurons generate nerve impulses, which propagate toward the axon terminals.
5. Acetylcholine released by the motor neurons causes the flexor muscles in the thigh (effectors) to contract, producing withdrawal of the leg. This reflex is protective because contraction of flexor muscles moves a limb away from the source of a possibly damaging stimulus.

The flexor reflex, like the stretch reflex, is ipsilateral—the incoming and outgoing impulses propagate into and out of the same side of the spinal cord. The flexor reflex also illustrates another feature of polysynaptic reflex arcs. Moving your entire lower or upper limb away from a painful stimulus involves contraction of more than one muscle group. Hence, several

Figure 16.19 **Tendon reflex.** This reflex arc is polysynaptic—more than one CNS synapse and more than two different neurons are involved in the pathway. The sensory neuron synapses with two interneurons. An inhibitory interneuron causes relaxation of the effector, and a stimulatory interneuron causes contraction of the antagonistic muscle. Plus signs (+) indicate excitatory synapses; the minus sign (-) indicates an inhibitory synapse.

The tendon reflex causes relaxation of the muscle attached to the stimulated tendon organ.

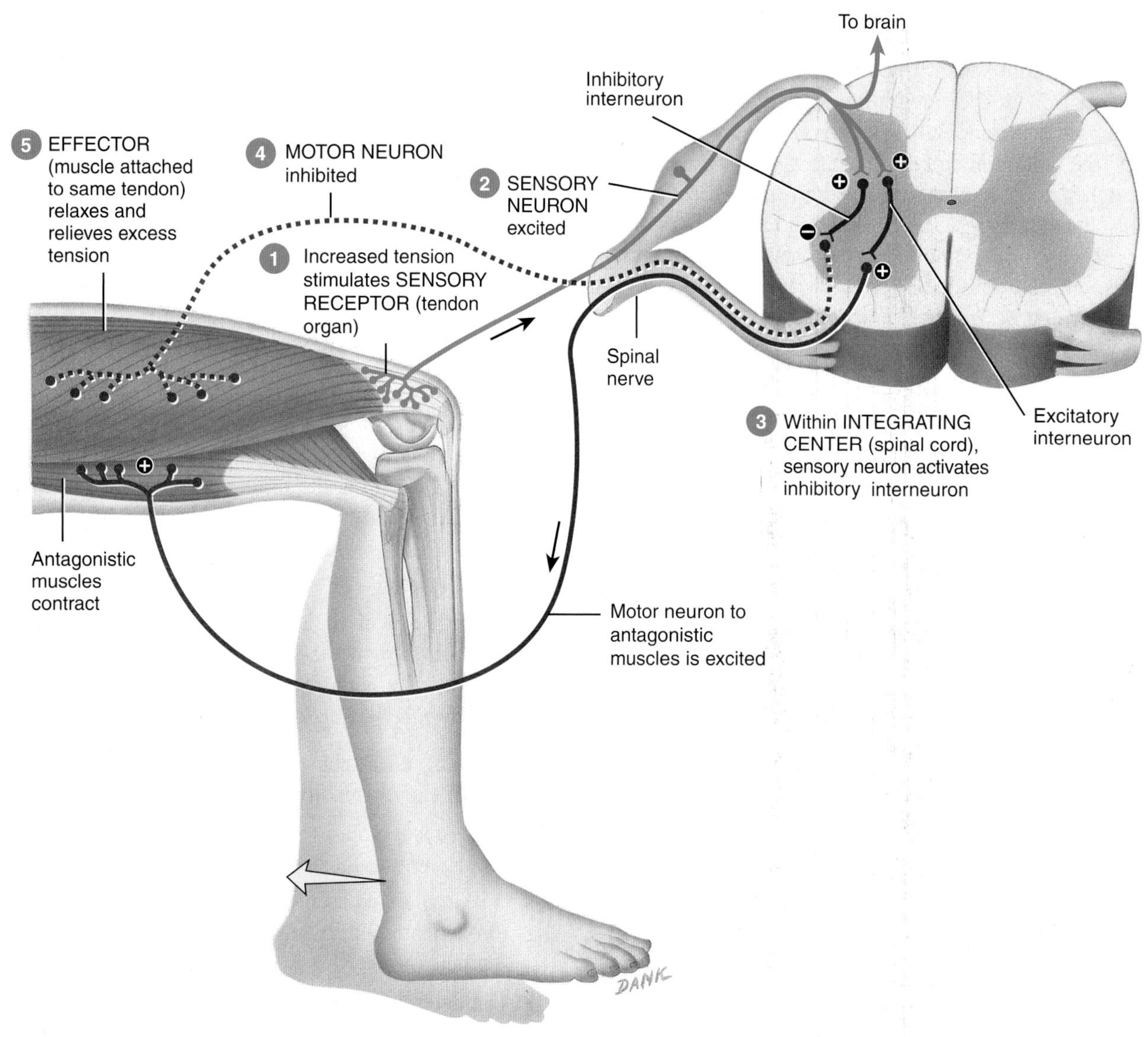

What makes this reciprocal inhibition?

motor neurons must simultaneously convey impulses to several limb muscles. Because nerve impulses from one sensory neuron ascend and descend in the spinal cord and activate interneurons in several segments of the spinal cord, this type of reflex is called an ***intersegmental reflex arc*** (*inter-* = between). Through intersegmental reflex arcs, a single sensory neuron can activate several motor neurons, thereby stimulating more than one effector. The monosynaptic stretch reflex, in contrast, involves muscles receiving nerve impulses from one spinal cord segment only.

Figure 16.20 **Flexor (withdrawal) reflex.** This reflex arc is polysynaptic and ipsilateral. Plus signs (+) indicate excitatory synapses.

The flexor reflex causes withdrawal of a part of the body in response to a painful stimulus.

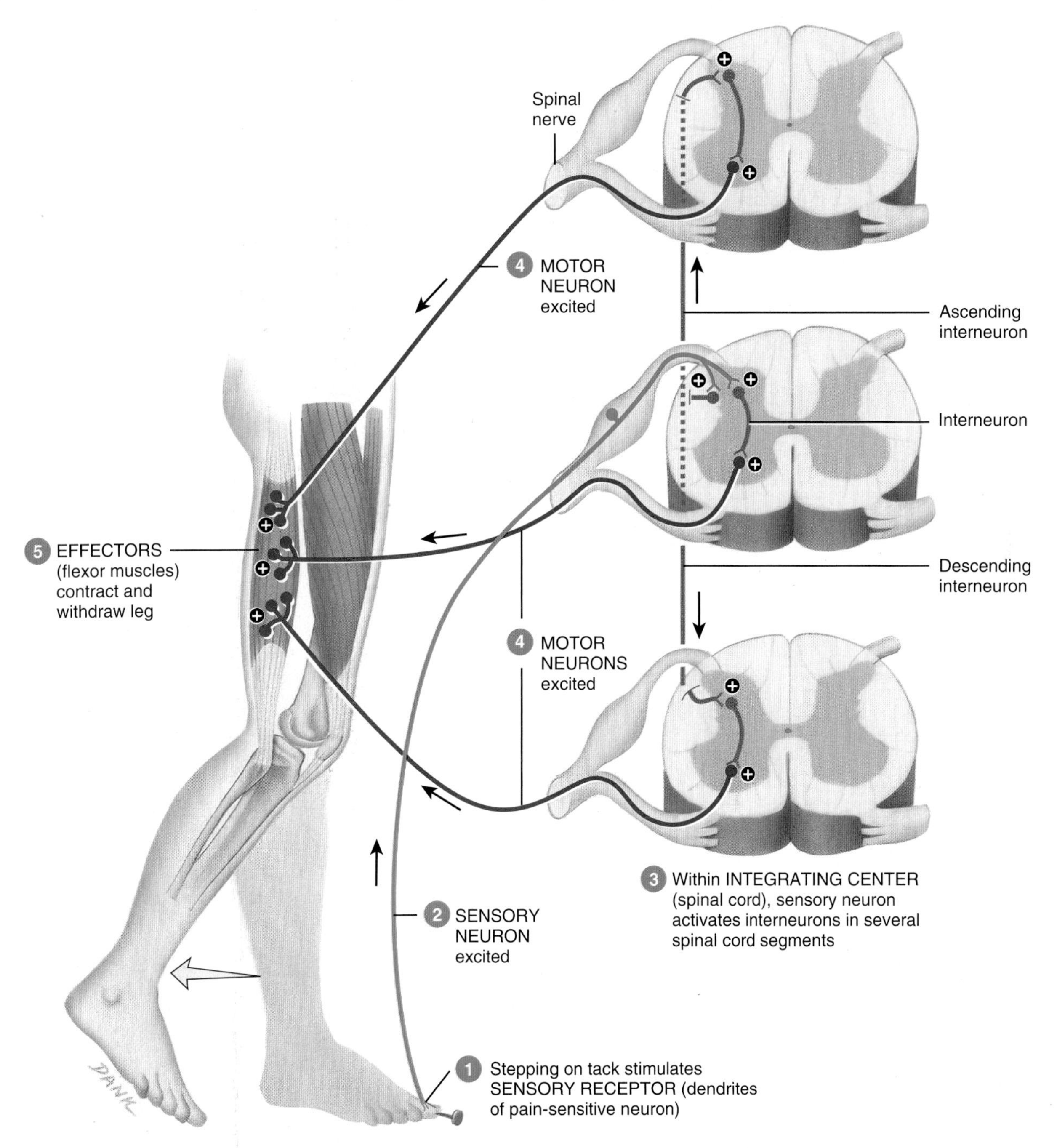

? Why is the flexor reflex classified as an intersegmental reflex arc?

Crossed Extensor Reflex

Something else may happen when you step on a tack: You may start to lose your balance as your body weight shifts to the other foot. Besides initiating the flexor reflex that causes you to withdraw the limb, the pain impulses from stepping on the tack also initiate a ***crossed extensor reflex*** to help you maintain your balance; it operates as follows (Figure 16.21):

1. Stepping on a tack stimulates the sensory receptor of a pain-sensitive neuron in the right foot.

Figure 16.21 Crossed extensor reflex. The flexor reflex arc is shown (at left) to enable comparison with the crossed extensor reflex arc. Plus signs (+) indicate excitatory synapses.

A crossed extensor reflex causes contraction of muscles that extend joints in the limb opposite a painful stimulus.

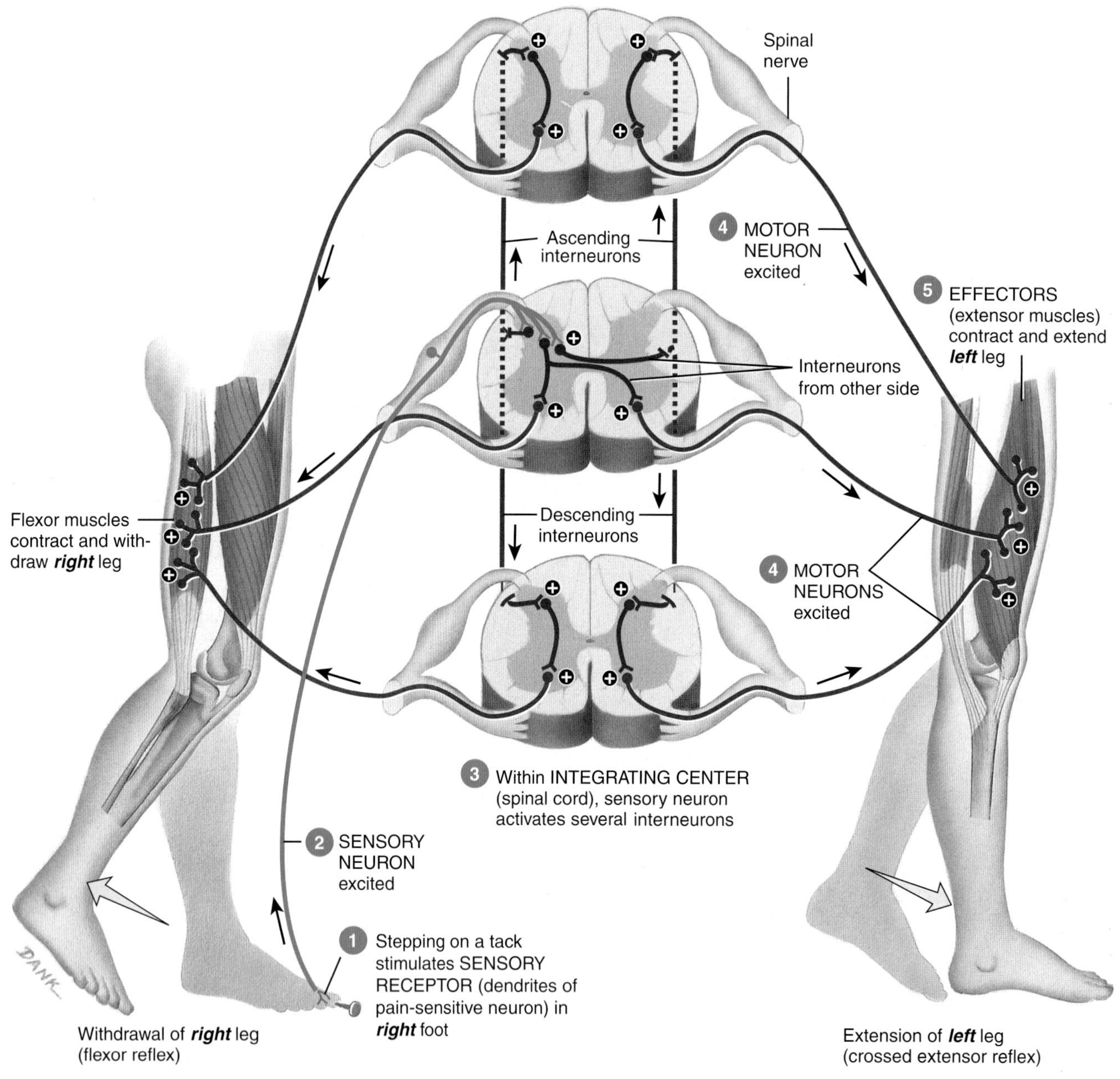

Why is the crossed extensor reflex classified as a contralateral reflex arc?

2. This sensory neuron then generates nerve impulses, which propagate into the spinal cord.
3. Within the spinal cord (integrating center), the sensory neuron activates several interneurons that synapse with motor neurons on the left side of the spinal cord in several spinal cord segments. Thus, incoming pain signals cross to the opposite side through interneurons at that level, and at several levels above and below the point of entry into the spinal cord.
4. The interneurons excite motor neurons in several spinal cord segments that innervate extensor muscles. The motor neurons, in turn, generate more nerve impulses, which propagate toward the axon terminals.
5. Acetylcholine released by the motor neurons causes extensor muscles in the thigh (effectors) of the unstimulated left limb to contract, producing extension of the left leg. In this way, weight can be placed on the foot that must now support the entire body. A comparable reflex occurs with painful stimulation of the left lower limb or either upper limb.

Unlike the flexor reflex, which is an ipsilateral reflex, the crossed extensor reflex involves a ***contralateral reflex arc*** (kon′-tra-LAT-er-al = opposite side): Sensory impulses enter one side of the spinal cord and motor impulses exit on the opposite side. Thus, a crossed extensor reflex synchronizes the extension of the contralateral limb with the withdrawal (flexion) of the stimulated limb. Reciprocal innervation also occurs in both the flexor reflex and the crossed extensor reflex. In the flexor reflex, when the flexor muscles of a painfully stimulated lower limb are contracting, the extensor muscles of the same limb are relaxing to some degree. If both sets of muscles contracted at the same time, the two sets of muscles would pull on the bones in opposite directions, which might immobilize the limb. Because of reciprocal innervation, one set of muscles contracts while the other relaxes.

Reflexes and Diagnosis

Reflexes are often used for diagnosing disorders or assessing the integrity of the nervous system and locating injured tissue. If a reflex ceases to function or functions abnormally, the physician may suspect that the damage lies somewhere along a particular conduction pathway. Many somatic reflexes can be tested simply by tapping or stroking the body. Among the somatic reflexes of clinical significance are the following:

- **Patellar reflex (knee jerk).** This stretch reflex involves extension of the leg at the knee joint by contraction of the quadriceps femoris muscle in response to tapping the patellar ligament (see Figure 16.18). This reflex is blocked by damage to the sensory or motor nerves supplying the muscle or to the integrating centers in the second, third, or fourth lumbar segments of the spinal cord. It is often absent in people with chronic diabetes mellitus or neurosyphilis, both of which cause degeneration of nerves. It is exaggerated in disease or injury involving certain motor tracts descending from the higher centers of the brain to the spinal cord.
- **Achilles reflex (ankle jerk).** This stretch reflex involves extension (plantar flexion) of the foot by contraction of the gastrocnemius and soleus muscles in response to tapping the calcaneal (Achilles) tendon. Absence of the Achilles reflex indicates damage to the nerves supplying the posterior leg muscles or to neurons in the lumbosacral region of the spinal cord. This reflex may also disappear in people with chronic diabetes, neurosyphilis, alcoholism, and subarachnoid hemorrhages. An exaggerated Achilles reflex indicates cervical cord compression or a lesion of the motor tracts of the first or second sacral segments of the cord.
- **Babinski sign.** This reflex results from gentle stroking of the lateral outer margin of the sole. The great toe dorsiflexes, with or without a lateral fanning of the other toes. This phenomenon normally occurs in children under 1½ years of age and is due to incomplete myelination of fibers in the corticospinal tract (see Section 15.2). A positive Babinski sign after age 1½ is abnormal and indicates an interruption of the corticospinal tract as the result of a lesion of the tract, usually in the upper portion. The normal response after age 1½ is the *plantar flexion reflex,* or *negative Babinski*—a curling under of all the toes.
- **Abdominal reflex.** This reflex involves contraction of the muscles that compress the abdominal wall in response to stroking the side of the abdomen. The response is an abdominal muscle contraction that causes the umbilicus to move in the direction of the stimulus. Absence of this reflex is associated with lesions of the corticospinal tracts. It may also be absent because of lesions of the peripheral nerves, lesions of integrating centers in the thoracic part of the cord, or multiple sclerosis.

Most autonomic reflexes are not practical diagnostic tools because it is difficult to stimulate visceral effectors, which are deep inside the body. An exception is the ***pupillary light reflex,*** in which the pupils of both eyes decrease in diameter when either eye is exposed to light. Because the reflex arc includes synapses in lower parts of the brain, the absence of a normal pupillary light reflex may indicate brain damage or injury.

CHECKPOINT

12. Which spinal cord tracts are ascending tracts? Which are descending tracts?
13. Compare and contrast the posterior column–medial lemniscus pathways in terms of function.
14. How are somatic and autonomic reflexes similar and different?
15. Describe the mechanism and function of a stretch reflex, tendon reflex, flexor (withdrawal) reflex, and crossed extensor reflex.
16. What does each of the following terms mean in relation to reflex arcs? Monosynaptic, ipsilateral, polysynaptic, intersegmental, contralateral, and reciprocal inhibition.

16.4 TRAUMATIC INJURIES OF THE SPINAL CORD

OBJECTIVES

- Define the terms used to describe paralysis of one, two, or four limbs.
- Describe the symptoms associated with complete transection and hemisection of the spinal cord.
- Describe spinal shock.

Most spinal cord injuries are due to trauma as a result of factors such as automobile accidents, falls, contact sports, diving, or acts of violence. The effects of the injury depend on the extent of direct trauma to the spinal cord or compression of the cord by fractured or displaced vertebrae or blood clots. Although any segment of the spinal cord may be involved, most common sites of injury are in the cervical, lower thoracic, and upper lumbar regions. Depending on the location and extent of spinal cord damage, paralysis may occur: ***monoplegia*** (*mono-* = one; *-plegia* = blow or strike) is paralysis of one limb only; ***diplegia*** (*di-* = two) is paralysis of both upper limbs or both lower limbs; ***paraplegia*** (*para-* = beyond) is paralysis of both lower limbs; ***hemiplegia*** (*hemi-* = half) is paralysis of the upper limb, trunk, and lower limb on one side of the body; and ***quadriplegia*** (*quad-* = four) is paralysis of all four limbs.

Complete transection (tran-SEK-shun; *trans-* = across; *-section* = a cut) of the spinal cord means that the cord is severed from one side to the other, thus cutting all sensory and motor tracts. It results in a loss of all sensations and voluntary movement *below* the level of the transection. A person will have permanent loss of all sensations in dermatomes below the injury because ascending nerve impulses cannot propagate past the transection to reach the brain. At the same time, voluntary muscle contractions will be lost below the transection because nerve impulses descending from the brain also cannot pass. The extent of paralysis of skeletal muscles depends on the level of injury.

The following list outlines which muscle functions may be *retained* at progressively lower levels of spinal cord transection.

- C1–C3: No function maintained from the neck down; ventilator needed for breathing.
- C4–C5: Diaphragm, which allows breathing.
- C6–C7: Some arm and chest muscles, which allows feeding, some dressing, and propelling wheelchair.
- T1–T3: Intact arm function.
- T4–T9: Control of trunk above the umbilicus.
- T10–L1: Most thigh muscles, which allows walking with long leg braces.
- L1–L2: Most leg muscles, which allows walking with short leg braces.

Hemisection is a partial transection of the cord on either the right or left side. Following complete transection, and to varying degrees after hemisection, spinal shock occurs. ***Spinal shock*** is an immediate response to spinal cord injury characterized by temporary ***areflexia*** (ā′-rē-FLEX-sē-a), loss of reflex function. The areflexia occurs in parts of the body served by spinal nerves below the level of the injury. Signs of acute spinal shock include slow heart rate, low blood pressure, flaccid paralysis of skeletal muscles, loss of somatic sensations, and urinary bladder dysfunction. Spinal shock may begin within 1 hour after injury and may last from several minutes to several months, after which reflex activity gradually returns.

In many cases of traumatic injury of the spinal cord, the patient may have an improved outcome if an anti-inflammatory corticosteroid drug called methylprednisolone is given within 8 hours of the injury. This is because the degree of neurological deficit is greatest immediately following traumatic injury as a result of edema (collection of fluid within tissues) as the immune system responds to injury.

CHECKPOINT

17. Define the symptoms associated with monoplegia, diplegia, hemiplegia, and quadriplegia.
18. Define areflexia. Describe the symptoms of spinal shock.

KEY MEDICAL TERMS ASSOCIATED WITH THE SPINAL CORD AND SPINAL NERVES

Epidural block Injection of an anesthetic drug into the epidural space, the space between the dura mater and the vertebral column, in order to cause a temporary loss of sensation. Such injections in the lower lumbar region are used to control pain during childbirth.

Meningitis (men-in-JĪ-tis; *-itis* = inflammation) Inflammation of the meninges due to an infection, usually caused by a bacterium or virus. Symptoms include fever, headache, stiff neck, vomiting, confusion, lethargy, and drowsiness. Bacterial meningitis is much more serious and is treated with antibiotics. Viral meningitis has no specific treatment. Bacterial meningitis may be fatal if not treated promptly; viral meningitis usually resolves on its own in 1–2 weeks. A vaccine is available to help protect against some types of bacterial meningitis.

Myelitis (mī-e-LĪ-tis; *myel-* = spinal cord) Inflammation of the spinal cord.

Myelography (mī-e-LOG-ra-fē; *myel-* = marrow, *-graph* = to write) A procedure in which a CT scan or x-ray image of the spinal cord is taken after injection of a radiopaque dye (contrast medium) to diagnose abnormalities such as tumors and herniated intervertebral discs. MRI has largely replaced myelography because the former shows greater detail, is safer, and is simpler.

Nerve block Loss of sensation in a region due to injection of a local anesthetic; an example is local dental anesthesia.

Neuralgia (noo-RAL-jē-a; *neur-* = nerve; *-algia* = pain) Attacks of pain along the entire course or a branch of a sensory nerve.

Neuritis Inflammation of one or several nerves that may result from irritation to the nerve produced by direct blows, bone fractures, contusions, or penetrating injuries. Additional causes include infections, vitamin deficiency (usually thiamine), and poisons such as carbon monoxide, carbon tetrachloride, heavy metals, and some drugs.

Paresthesia (par-es-THĒ-zē-a; *par-* = departure from normal; *-esthesia* = sensation) An abnormal sensation such as burning, pricking, tickling, or tingling resulting from a disorder of a sensory nerve.

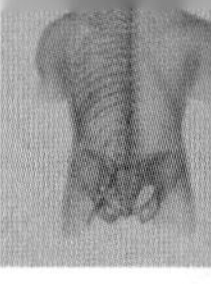

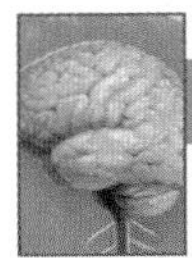

STUDY OUTLINE

Spinal Cord Anatomy (Section 16.1)

1. The spinal cord is protected by the vertebral column, the meninges, cerebrospinal fluid, and denticulate ligaments.
2. The three meninges are coverings that run continuously around the spinal cord and brain. They are the dura mater, arachnoid mater, and pia mater.
3. The spinal cord begins as a continuation of the medulla oblongata and ends at about the second lumbar vertebra (L2) in an adult.
4. The spinal cord contains cervical and lumbar enlargements that serve as points of origin for nerves to the limbs.
5. The tapered inferior portion of the spinal cord is the conus medullaris, from which arise the filum terminale and cauda equina.
6. Spinal nerves connect to each segment of the spinal cord by two roots. The posterior or dorsal root contains sensory axons, and the anterior or ventral root contains motor neuron axons.
7. The anterior median fissure and the posterior median sulcus partially divide the spinal cord into right and left sides.
8. The gray matter in the spinal cord is divided into horns, and the white matter into columns. In the center of the spinal cord is the central canal, which runs the length of the spinal cord.
9. Parts of the spinal cord observed in transverse section are the gray commissure; central canal; anterior, posterior, and lateral gray horns; and anterior, posterior, and lateral white columns, which contain ascending and descending tracts. Each part has specific functions.
10. The spinal cord conveys sensory and motor information by way of ascending and descending tracts, respectively.

Spinal Nerves (Section 16.2)

1. The 31 pairs of spinal nerves are named and numbered according to the region and level of the spinal cord from which they emerge.
2. There are 8 pairs of cervical, 12 pairs of thoracic, 5 pairs of lumbar, 5 pairs of sacral, and 1 pair of coccygeal nerves.
3. Spinal nerves typically are connected with the spinal cord by a posterior root and an anterior root. All spinal nerves contain both sensory and motor axons (are mixed nerves).
4. Three connective tissue coverings associated with spinal nerves are the endoneurium, perineurium, and epineurium.
5. Branches of a spinal nerve include the posterior ramus, anterior ramus, meningeal branch, and rami communicantes.
6. The anterior rami of spinal nerves, except for T2–T12, form networks of nerves called plexi.
7. Emerging from the plexi are nerves bearing names that typically describe the general regions they supply or the route they follow.
8. Nerves of the cervical plexus supply the skin and muscles of the head, neck, and upper part of the shoulders; they run parallel with some cranial nerves and innervate the diaphragm.
9. Nerves of the brachial plexus supply the upper limbs and several neck and shoulder muscles.
10. Nerves of the lumbar plexus supply the anterolateral abdominal wall, external genitals, and part of the lower limbs.
11. Nerves of the sacral plexus supply the buttocks, perineum, and part of the lower limbs.
12. Nerves of the coccygeal plexus supply the skin of the coccygeal region.
13. Anterior rami of nerves T2–T12 do not form plexi and are called intercostal (thoracic) nerves. They are distributed directly to the structures they supply in intercostal spaces.
14. Sensory neurons within spinal nerves and the trigeminal (V) cranial nerve serve specific, constant segments of the skin called dermatomes.
15. Knowledge of dermatomes helps a physician determine which segment of the spinal cord or which spinal nerve is damaged.

Spinal Cord Physiology (Section 16.3)

1. The white matter tracts in the spinal cord are highways for nerve impulse propagation. Along these tracts, sensory input travels toward the brain, and motor output travels from the brain toward skeletal muscles and other effector tissues.
2. Sensory input travels along two main pathways in the white matter of the spinal cord: the posterior column–medical lemniscus pathway and the anterolateral (spinothalamic) pathways.
3. Motor output travels along two main routes in the white matter of the spinal cord: direct pathways and indirect pathways.
4. A second major function of the spinal cord is to serve as an integrating center for spinal reflexes. This integration occurs in the gray matter.
5. Muscle spindles monitor changes in the length of skeletal muscles and participate in stretch reflexes.
6. Tendon organs monitor changes in the tension of skeletal muscles.
7. A reflex is a fast, predictable sequence of involuntary actions, such as muscle contractions or glandular secretions, which occurs in response to certain changes in the environment.
8. Reflexes may be spinal or cranial and somatic or autonomic (visceral).
9. The components of a reflex arc are sensory receptor, sensory neuron, integrating center, motor neuron, and effector.
10. Somatic spinal reflexes include the stretch reflex, the tendon reflex, the flexor (withdrawal) reflex, and the crossed extensor reflex; all exhibit reciprocal inhibition.
11. A two-neuron or monosynaptic reflex arc consists of one sensory neuron and one motor neuron. A stretch reflex, such as the patellar reflex, is an example.
12. The stretch reflex is ipsilateral and is important in maintaining muscle tone.
13. A polysynaptic reflex arc contains sensory neurons, interneurons, and motor neurons. The tendon reflex, flexor (withdrawal) reflex, and crossed extensor reflex are examples.
14. The tendon reflex is ipsilateral and prevents damage to muscles and tendons when muscle force becomes too extreme. The flexor reflex is ipsilateral and moves a limb away from the source of a painful stimulus. The crossed extensor reflex extends the limb contralateral to a painfully stimulated limb, allowing the weight of the body to shift when a supporting limb is withdrawn.
15. Several important somatic reflexes are used to diagnose various disorders. These include the patellar reflex, Achilles reflex, Babinski sign, and abdominal reflex.

Traumatic Injuries of the Spinal Cord (Section 16.4)

1. Types of paralysis are dependent on the location and extent of spinal nerve damage.

2. Complete transection of the spinal cord results in a loss of sensation and voluntary movements below the level of transection.

3. Spinal shock is an immediate response to spinal cord injury characterized by temporary areflexia, a loss of reflex function.

SELF-QUIZ QUESTIONS

Fill in the blanks in the following statements.

1. Because they contain both sensory and motor axons, spinal nerves are considered to be _____ nerves.
2. The five components of a reflex arc, in order from the beginning to the end, are (1) _____, (2) _____, (3) _____, (4) _____, and (5) _____.

Indicate whether the following statements are true or false.

3. Gray matter of the spinal cord contains somatic motor and sensory nuclei, and autonomic motor and sensory nuclei, and functions to receive and integrate both incoming and outgoing information.
4. The epidural space is located between the wall of the vertebral canal and the pia mater.

Choose the one best answer to the following questions.

5. Which of the following is NOT true? (1) Dermatomes are areas of the body that are stimulated by motor neurons exiting specific spinal nerves. (2) The stretch reflex helps to maintain muscle tone. (3) The Achilles reflex is an example of a stretch reflex. (4) The abdominal reflex is used to diagnose problems with autonomic reflexes. (5) Spinal nerves T2–T12 do not enter into the formation of a plexus.
   a. 1, 2, and 4 **b.** 2 and 5 **c.** 1 and 4
   d. 1, 3, and 5 **e.** 1, 3, and 4
6. While identifying and labeling cadaver muscles, your lab partner accidentally pokes your finger with a pin. Place the following steps in the correct order from beginning to end of your body's response. (1) Impulses travel through anterior (ventral) root of spinal nerve(s). (2) Sensory neuron relays impulse to spinal cord. (3) Motor impulses reach muscles, causing withdrawal of the affected limb. (4) Integrating centers interpret sensory impulses, and then generate motor impulses. (5) Sensory receptor activated by stimulus. (6) Impulse travels through posterior (dorsal) root of spinal nerve.
   a. 5, 3, 6, 4, 1, 2 **b.** 5, 2, 1, 4, 6, 3
   c. 5, 2, 6, 4, 1, 3 **d.** 3, 5, 1, 2, 4, 6
   e. 2, 1, 5, 4, 6, 3
7. The connective tissue surrounding each individual axon is
   a. endoneurium **b.** epineurium
   c. perineurium **d.** fascicle
   e. arachnoid mater.
8. The tracts of the posterior column are involved in (1) proprioception, (2) touch, (3) pain, (4) temperature, (5) pressure, and (6) vibration.
   a. 1, 2, 4, and 5 **b.** 2, 4, 6, and 7
   c. 1, 2, 6, and 7 **d.** 3, 4, 5, 6, and 7
   e. 1, 2, 5, and 6
9. Which of the following is a motor tract?
   a. spinocerebellar **b.** spinothalamic
   c. corticospinal **d.** posterior column
10. Cutting the posterior root of a spinal nerve would
    a. interfere with the circulation of cerebrospinal fluid
    b. impair motor control of skeletal muscles
    c. interfere with the ability of the brain to transmit motor impulses
    d. impair motor control of organs
    e. interfere with the flow of sensory impulses.
11. Which of the following statements is *false*?
    a. The two main spinal cord sensory paths are the spinothalamic and anterior columns.
    b. The spinothalamic tracts convey impulses for sensing pain, temperature, touch, and deep pressure.
    c. Direct pathways convey nerve impulses destined to cause precise, voluntary movements of skeletal muscles.
    d. Indirect pathways convey nerve impulses that program automatic movements, help coordinate body movements with visual stimuli, maintain skeletal muscle tone and posture, and contribute to equilibrium.
    e. The direct pathways are motor pathways.
12. Which of the following are *true*? (1) The anterior (ventral) gray horns contain cell bodies of neurons that cause skeletal muscle contraction. (2) The gray commissure connects the white matter of the right and left sides of the spinal cord. (3) Cell bodies of autonomic motor neurons are located in the lateral gray horns. (4) Sensory (ascending) tracts conduct motor impulses down the spinal cord. (5) Gray matter in the spinal cord consists of cell bodies of neurons, neuroglia, unmyelinated axons, and dendrites of interneurons and motor neurons.
    a. 1, 2, 3, and 5 **b.** 2 and 4 **c.** 2, 3, 4, and 5
    d. 1, 3, and 5 **e.** 1, 2, 3, and 4.

Matching

13. Match the following (some answers may be used more than once):

____ **a.** a reflex resulting in the contraction of a skeletal muscle when it is stretched	**A.** stretch reflex
____ **b.** receptor that monitors changes in muscle length	**B.** tendon reflex
____ **c.** a balance-maintaining reflex	**C.** flexor (withdrawal) reflex
____ **d.** operates as a feedback mechanism to control muscle tension by causing muscle relaxation when muscle force becomes too extreme	**D.** crossed extensor reflex
____ **e.** reflex arc that consists of one sensory and one motor neuron	**E.** intersegmental reflex arc
____ **f.** acts as a feedback mechanism to control muscle length by causing muscle contraction	**F.** contralateral reflex arc
____ **g.** sensory impulses enter on one side of the spinal cord and motor impulses exit on the opposite side	**G.** ipsilateral reflex arc
____ **h.** occurs when sensory nerve impulse travels up and down the spinal cord, thereby activating several motor neurons and more than one effector	**H.** muscle spindle
	I. tendon (Golgi tendon) organ
	J. reciprocal innervation
	K. monosynaptic reflex
	L. polysynaptic reflex

____ **i.** polysynaptic reflex initiated in response to a painful stimulus
____ **j.** receptor that monitors changes in muscle tension
____ **k.** maintains proper muscle tone
____ **l.** reflex pathway that contains sensory neurons, interneurons, and motor neurons
____ **m.** motor nerve impulses exit the spinal cord on the same side that sensory impulses entered the spinal cord
____ **n.** protects the tendon and muscle from damage due to excessive tension
____ **o.** a neural circuit that coordinates body movements by causing contraction of one muscle and relaxation of antagonistic muscles or relaxation of a muscle and contraction of the antagonists

14. Match the following (answers may be used more than once).

____ **a.** provides the entire nerve supply of the shoulders and upper limbs
____ **b.** provides the nerve supply of the skin and muscles of the head, neck, and superior part of the shoulders and chest
____ **c.** provides the nerve supply of the anterolateral abdominal wall, external genitals, and part of the lower limbs
____ **d.** supplies the buttocks, perineum, and lower limbs
____ **e.** formed by the anterior rami of C1–C4 with some contribution by C5
____ **f.** formed by anterior rami of S4–S5 and coccygeal nerves
____ **g.** formed by the anterior rami of L1–L4
____ **h.** formed by the anterior rami of C5–C8 and T1
____ **i.** formed by the anterior rami of L4–L5 and S1–S4
____ **j.** phrenic nerve arises from this plexus
____ **k.** median nerve arises from this plexus
____ **l.** sciatic nerve arises from this plexus
____ **m.** femoral nerve arises from this plexus
____ **n.** supplies a small area of skin in coccygeal region
____ **o.** injury to this plexus can affect breathing

A. cervical plexus
B. brachial plexus
C. lumbar plexus
D. sacral plexus
E. coccygeal plexus

15. Match the following:

____ **a.** the joining together of the anterior rami of adjacent nerves
____ **b.** spinal nerve branches that serve the deep muscles and skin of the posterior surface of the trunk
____ **c.** spinal nerve branches that serve the muscles and structures of the upper and lower limbs and the lateral and ventral trunk
____ **d.** area of the spinal cord from which nerves to and from the upper limbs arise
____ **e.** area of the spinal cord from which nerves to and from the lower limbs arise
____ **f.** the roots form the nerves that arise from the inferior part of the spinal cord but do not leave the vertebral column at the same level as they exit the cord
____ **g.** contains motor neuron axons and conducts impulses from the spinal cord to the peripheral organs and cells
____ **h.** avascular covering of spinal cord composed of delicate collagen fibers and some elastic fibers
____ **i.** contains sensory neuron axons and conducts impulses from the peripheral receptors into the spinal cord
____ **j.** superficial spinal cord covering of dense, irregular connective tissue
____ **k.** an extension of the pia mater that anchors the spinal cord to the coccyx
____ **l.** extending the length of the spinal cord, these pia mater thickenings fuse with the arachnoid mater and dura mater and help to protect the spinal cord from shock and sudden displacement
____ **m.** thin transparent connective tissue composed of interlacing bundles of collagen fibers and some elastic fibers adhering to the spinal cord's surface
____ **n.** space within the spinal cord filled with cerebrospinal fluid
____ **o.** spinal nerve branch that supplies vertebrae, vertebral ligaments, blood vessels of the spinal cord, and meninges

A. cervical enlargement
B. lumbar enlargement
C. central canal
D. denticulate ligaments
E. cauda equina
F. meningeal branch
G. pia mater
H. arachnoid mater
I. dura mater
J. posterior (dorsal) root
K. anterior (ventral) root
L. posterior (dorsal) ramus
M. anterior (ventral) ramus
N. plexus
O. filum terminale

CRITICAL THINKING QUESTIONS

1. Karen had pain and weakness of the right upper limb, across the upper thoracic area, and over the right scapula. Her right hand was also cold, even though it was August when she consulted a therapist. Karen reported during her history that the pain had gotten worse right after her cat had died. Why would the therapist assess that Karen may have thoracic outlet syndrome? What group of nerves were probably involved? Why might the death of Karen's cat have contributed to her symptoms?
2. Bill was a driver for an interstate trucking company. He made an appointment with Judy, a manual therapist, for pain in his left hip, thigh, and leg. Judy noticed that Bill seemingly carried a large wallet in his left hip pocket. What was Judy's first thought about Bill's pain, even before he was on the table?
3. After a few days of using her new crutches, Kathy's arms and hands felt tingly and numb. The physical therapist said Kathy had a case of "crutch palsy" from improper use of her crutches. Kathy had been leaning her armpits on the crutches while hobbling along. What caused the numbness in her arms and hands?
4. Andrea flicked on the light when she heard her husband's yell. Lou was bouncing on his left foot while holding his right foot in his hand. A pin was sticking out of the bottom of his foot. Explain Lou's response to stepping on the pin.
5. Evalina's severe headaches and other symptoms were suggestive of meningitis, so her physician ordered a spinal tap. List the structures that the needle will pierce from the most superficial to the deepest. Why would the physician order a test in the spinal region to check a problem in Evalina's head?
6. Sunil has developed an infection that is destroying cells in the anterior gray horns in the lower cervical region of the spinal cord. What kinds of symptoms would you expect to occur?
7. Allyson is in a car accident and suffers spinal cord compression in the lower spinal cord. Although she is in pain, she cannot distinguish when the doctor is touching her calf or her toes and she is having trouble telling how her lower limbs are positioned. What part of the spinal cord has been affected by the accident?

? ANSWERS TO FIGURE QUESTIONS

16.1 Denticulate ligaments are formed by extensions of the pia mater.

16.2 Spinal nerves T2–T12 do not form a plexus.

16.3 A horn is an area of gray matter, and a column is a region of white matter in the spinal cord.

16.4 Lateral gray horns are found in the thoracic, upper lumbar, and sacral segments of the spinal cord.

16.5 All spinal nerves are classified as mixed (with both sensory and motor components) because their posterior roots contain sensory axons and their anterior roots contain motor axons.

16.6 The anterior rami serve the upper and lower limbs.

16.7 Severing the spinal cord at level C2 causes respiratory arrest because it prevents descending nerve impulses from reaching the phrenic nerve, which stimulates contraction of the diaphragm, the main muscle needed for breathing.

16.8 The axillary, musculocutaneous, radial, median, and ulnar nerves are five important nerves that arise from the brachial plexus.

16.9 Injury to the median nerve affects sensations of the palm and digits.

16.10 The lumbar plexus originates from the roots of the spinal nerves L1-L4.

16.11 The origins of the sacral plexus are the anterior rami of spinal nerves L4–L5 and S1–S4.

16.12 The only spinal nerve without a corresponding dermatome is C1.

16.13 The spinothalamic tract originates in the spinal cord and ends in the thalamus. The prefix (spino-) indicates the origin and the suffix of the nerve (-thalamic) indicates its destination; the nerve is therefore an ascending sensory tract.

16.14 The spinothalamic tracts carry the sensations of pain, cold, warmth, tickle, and itch.

16.15 The lateral corticospinal and the anterior corticospinal tracts conduct impulses along axons of upper motor neurons.

16.16 Muscle spindles are activated when the central areas of the intrafusal fibers are stretched.

16.17 A sensory receptor produces a generator potential, which triggers a nerve impulse if the generator potential reaches threshold. Reflex integrating centers are in the CNS.

16.18 In an ipsilateral reflex, the sensory and motor neurons are on the same side of the spinal cord.

16.19 Reciprocal inhibition is a type of arrangement of a neural circuit involving simultaneous contraction of one muscle and relaxation of its antagonist.

16.20 The flexor reflex is intersegmental because impulses go out over motor neurons located in several spinal nerves, each arising from a different segment of the spinal cord.

16.21 The crossed extensor reflex is a contralateral reflex arc because the motor impulses leave the spinal cord on the side opposite the entry of sensory impulses.

The Brain and Cranial Nerves

17

In this chapter we cover the major components of the brain, which are varied in function but can work together very quickly. For example, the limbic portion of our brain often links our nose and our memory, hence the happy emotions that could remind us of Grandma when we smell cookies or the potentially calming effect of lavender as it decreases our heart rate. The brain is the center of intellect, and the interactions between areas of the brain can affect our thoughts, emotions, memories, and behavior, often without our even being aware of it. Among its multiple functions, the brain directs our behavior toward others. With ideas that excite, artistry that dazzles, or rhetoric that mesmerizes, a person's thoughts and actions may influence and shape the lives of many others.

Your brain contributes to homeostasis by receiving sensory input, integrating new and stored information, making decisions, and causing motor reactions. Essentially, the brain is the control center for sending information to the body and its many systems. This chapter explores how the brain is protected and nourished, what functions occur in the major regions of the brain, and how the spinal cord and the 12 pairs of cranial nerves connect with the brain to form the control center of the human body. It is important that therapists recognize that multiple inputs create the total patient treatment experience, inputs such as type of touch, smell of a candle, sounds of relaxing music, the professionalism of the therapist, and the temperature of the room, to name some of the obvious inputs. The brain then puts all of this information together and, among other things, the patient decides whether or not to schedule another appointment.

CONTENTS AT A GLANCE

17.1 THE BRAIN

OBJECTIVES

- Discuss how the brain is protected and supplied with blood.
- Name the major parts of the brain and explain the function of each part.

We now consider the major parts of the brain, how the brain is protected, and how it is related to the spinal cord and cranial nerves.

Major Parts and Protective Coverings

The ***brain*** is one of the largest organs of the body, consisting of about 100 billion neurons and 10–50 trillion neuroglia with a mass of about 1300 g (almost 3 lb). The four major parts are the brain stem, diencephalon, cerebrum, and cerebellum (Figure 17.1). The ***brain stem*** is continuous with the spinal cord and consists of the medulla oblongata, pons, and midbrain. Above the brain stem is the ***diencephalon*** (dī′-en-SEF-a-lon; *di-* = through; *-encephalon* = brain), consisting mostly of the thalamus, hypothalamus, and epithalamus. Supported on the diencephalon and brain stem and forming the bulk of the brain is the ***cerebrum*** (se-RĒ-brum = brain). The surface of the cerebrum is composed of a thin layer of gray matter, the *cerebral cortex* (*cortex* = rind or bark), beneath which lies the cerebral white matter. Posterior to the brain stem is the ***cerebellum*** (ser′-e-BEL-um = little brain).

The brain is protected by the cranium and cranial meninges. The ***cranial meninges*** have the same names as the spinal meninges: the outermost ***dura mater,*** middle ***arachnoid mater,*** and innermost ***pia mater*** (see Figure 17.2a on page 466). Three extensions of the dura mater separate parts of the brain. (1) The ***falx cerebri*** (FALKS SER-ē-brē; *falx* = sickle-shaped) separates the two hemispheres (sides) of the cerebrum. (2) The ***falx cerebelli*** (cer-e-BEL-ī) separates the two hemispheres of the cerebellum. (3) The ***tentorium cerebelli*** (ten-TŌ-rē-um = tent) separates the cerebrum from the cerebellum (see Figure 17.2a,b).

Brain Blood Supply and the Blood–Brain Barrier

Although the brain constitutes only about 2% of total body weight, it requires about 20% of the body's oxygen supply. If blood flow to the brain stops, even briefly, unconsciousness may result. Brain neurons that are totally deprived of oxygen for 4 minutes or longer may be permanently injured. Blood supplying the brain also contains glucose, the main source of energy for brain cells. Because virtually no glucose is stored in the brain, the supply of glucose also must be continuous. If blood entering the brain has a low level of glucose, mental confusion, dizziness, convulsions, and loss of consciousness may occur.

• CLINICAL CONNECTION | Cerebrovascular Accident and Transient Ischemic Attack

The most common brain disorder is a **cerebrovascular accident (CVA),** also called a **stroke** or **brain attack.** CVAs affect approximately 500,000 people a year in the United States and represent the third leading cause of death, behind heart attacks and cancer. A CVA is characterized by abrupt onset of persisting symptoms, such as paralysis or loss of sensation, that arise from destruction of brain tissue. Common causes of CVAs are hemorrhage from a blood vessel in the pia mater or brain, blood clots, and formation of cholesterol-containing atherosclerotic plaques that block brain blood flow. The risk factors implicated in CVAs are high blood pressure, high blood cholesterol, heart disease, narrowed carotid arteries, transient ischemic attacks (discussed next), diabetes, smoking, obesity, and excessive alcohol intake.

A **transient ischemic attack (TIA)** is an episode of temporary cerebral dysfunction caused by impaired blood flow to part of the brain. Symptoms include dizziness, weakness, numbness, or paralysis in a limb or in one side of the body; drooping of one side of the face; headache; slurred speech or difficulty understanding speech; and a partial loss of vision or double vision. Sometimes nausea or vomiting also occur. The onset of symptoms is sudden and reaches maximum intensity almost immediately. A TIA usually persists for 5 to 10 minutes and only rarely lasts as long as 24 hours. It leaves no persistent neurological deficits. The causes of TIAs include blood clots, atherosclerosis, and certain blood disorders. •

The existence of a ***blood–brain barrier (BBB)*** protects brain cells from certain harmful substances and pathogens by preventing passage of many substances from blood into brain tissue. This barrier consists of very tightly sealed blood capillaries (microscopic blood vessels) plus an abundance of astrocytes that surround the blood capillaries. However, lipid-soluble substances such as oxygen, carbon dioxide, alcohol, most anesthetic agents, and essential oils used in aromatherapy, easily cross the blood–brain barrier. Trauma, certain toxins, and inflammation can cause a breakdown of the blood–brain barrier. A few areas of the brain do not contain the BBB, including the hypothalamus, pituitary gland, and pineal gland.

Cerebrospinal Fluid

The spinal cord and brain are further protected against chemical and physical injury by ***cerebrospinal fluid (CSF).*** CSF, a blood derivative, is a clear, colorless liquid that carries oxygen, glucose, and other needed chemicals from the blood to neurons and neuroglia and removes wastes and toxic substances produced by brain and spinal cord cells. CSF circulates through the subarachnoid space (between the arachnoid mater and pia mater), around the brain and spinal cord, and through cavities in the brain known as ***ventricles*** (VEN-tri-kuls = little cavities). There are

Figure 17.1 Brain. The pituitary gland is discussed together with the endocrine system in Chapter 20.

The four major parts of the brain are the brain stem, cerebellum, diencephalon, and cerebrum.

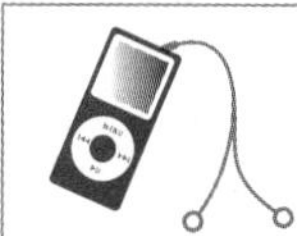

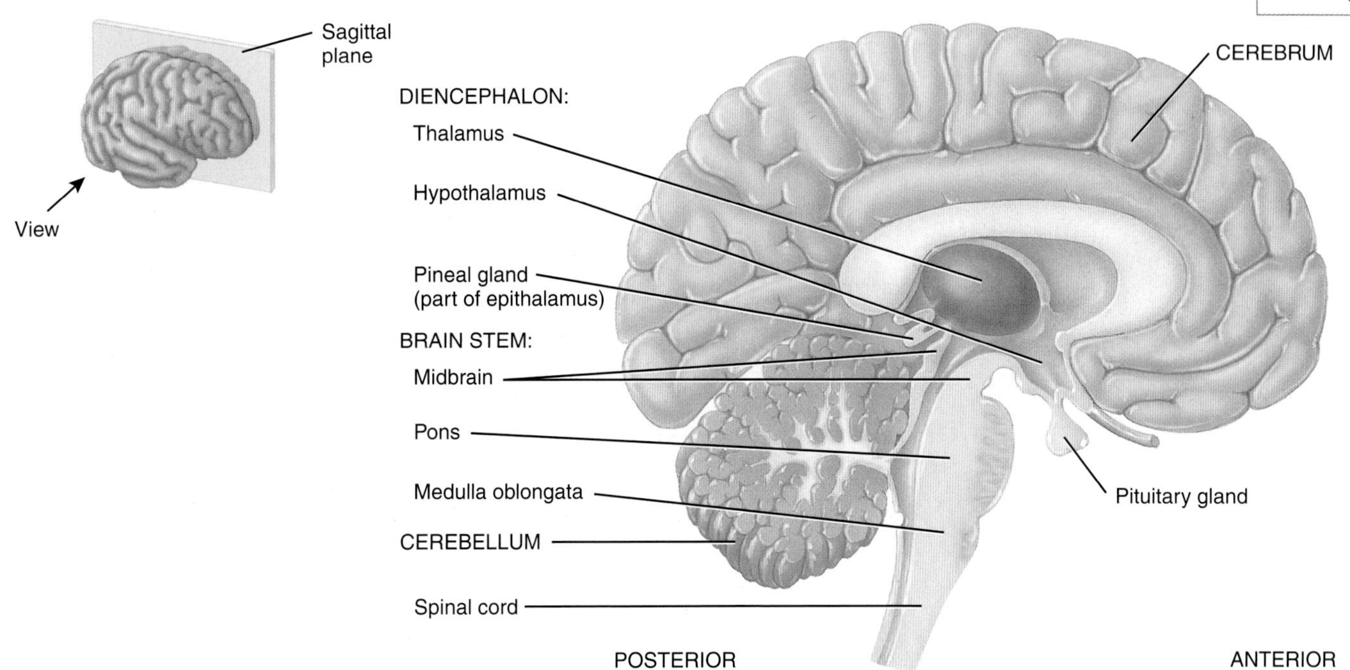

(a) Medial view of sagittal section

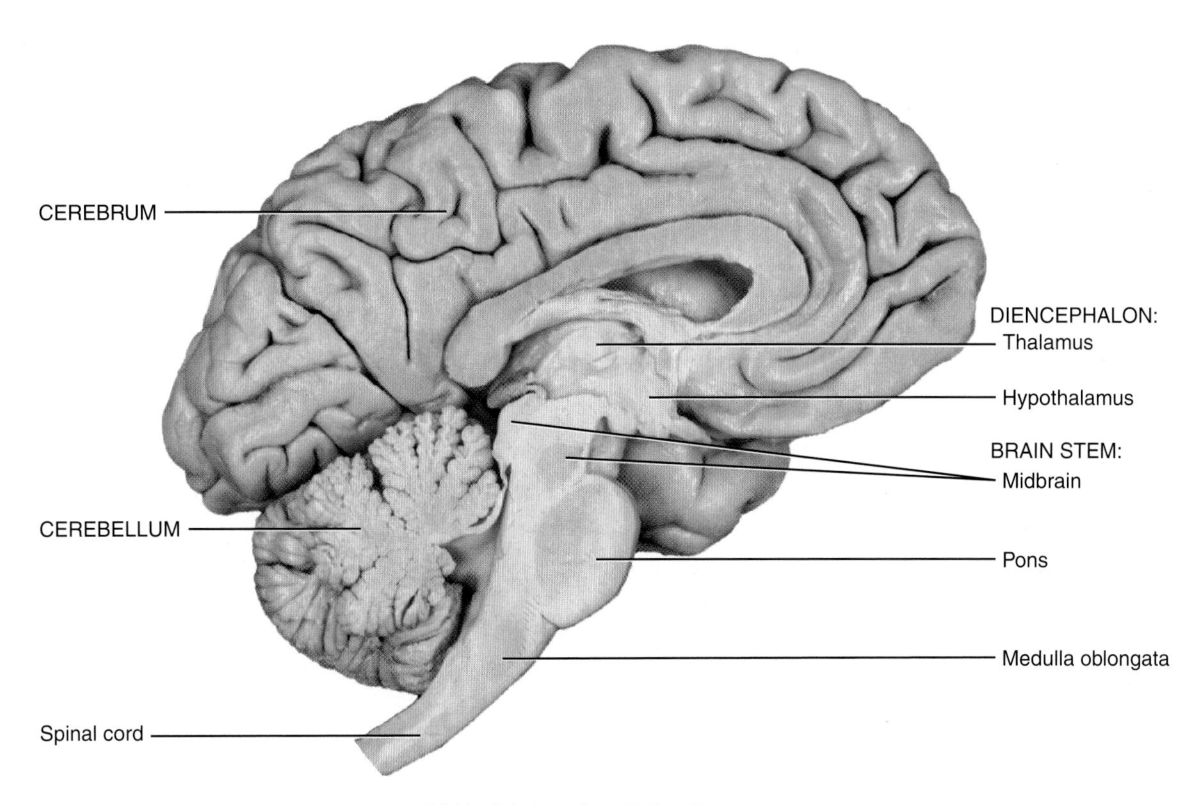

(b) Medial view of sagittal section

Which part of the brain is the largest? Second largest?

Figure 17.2 The protective coverings of the brain.

Cranial bones and the cranial meninges protect the brain.

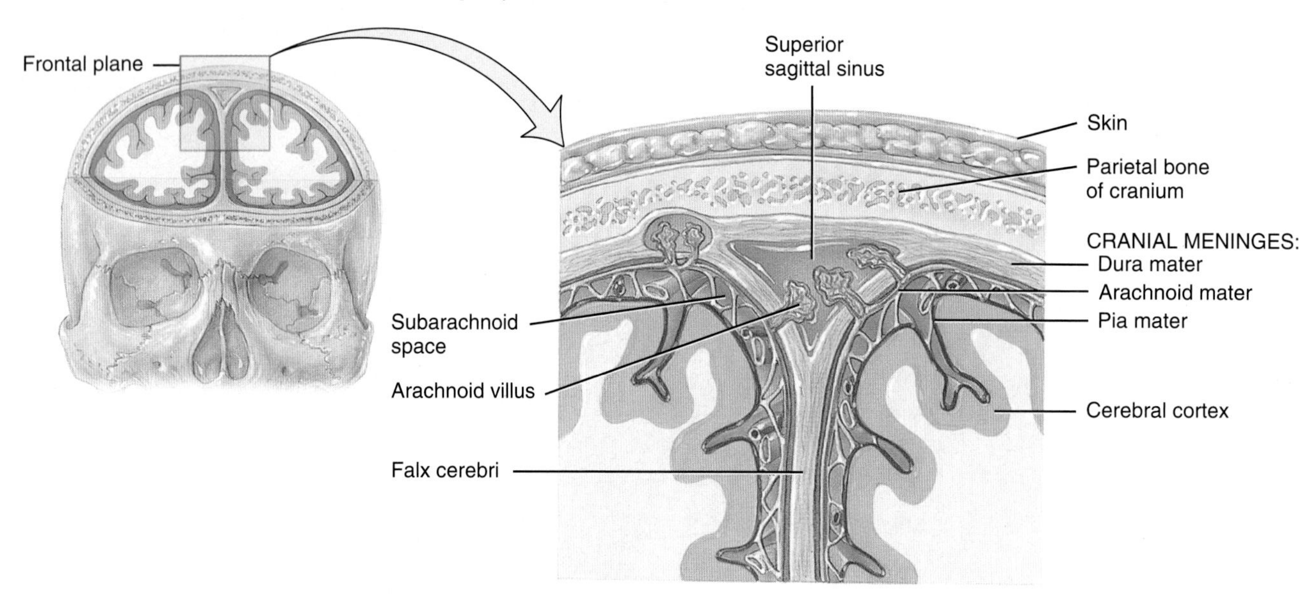

(a) Anterior view of frontal section through skull showing the cranial meninges

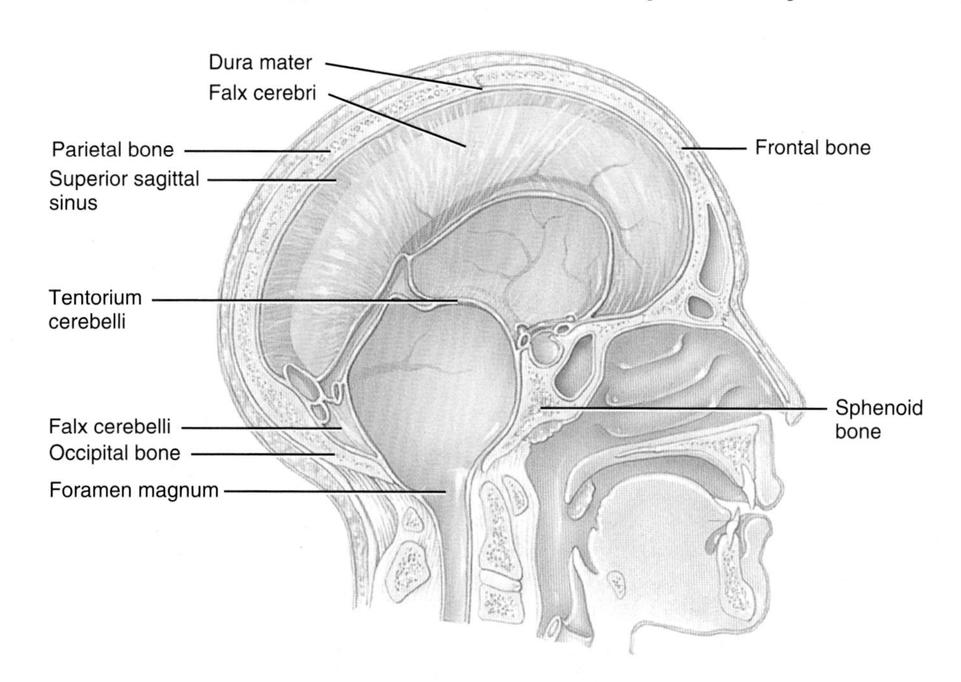

(b) Sagittal section of extensions of the dura mater

What are the three layers of the cranial meninges, from superficial to deep?

four ventricles: two ***lateral ventricles,*** one ***third ventricle,*** and one ***fourth ventricle*** (Figure 17.3). The ***aqueduct of the midbrain (cerebral aqueduct)*** connects the third and fourth ventricles. Other openings and tubes connect the ventricles with one another, with the central canal of the spinal cord, and with the subarachnoid space.

The sites of CSF production are the ***choroid plexuses*** (KŌ-royd = membrane-like), which are specialized networks of capillaries in the walls of the ventricles (Figure 17.3). Covering the capillaries of the choroid plexus are ependymal cells, which form cerebrospinal fluid from blood plasma by filtration and secretion. From the fourth ventricle, CSF flows into the central canal of the spinal cord and into the subarachnoid space around the surface of the brain and spinal cord. CSF is thus within and around the outside of the central nervous system (CNS); this column of fluid serves as a shock absorber that reduces trauma to the brain and spinal cord. CSF is gradually reabsorbed into the blood through ***arachnoid villi,***

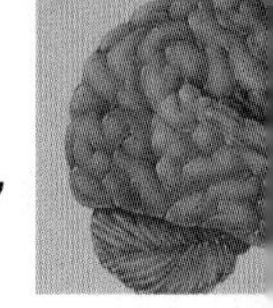

Figure 17.3 Ventricles of the brain and flow of CSF.

Cerebrospinal fluid (CSF) protects the brain and spinal cord and delivers nutrients from the blood to the brain and spinal cord; CSF also removes wastes from the brain and spinal cord to the blood.

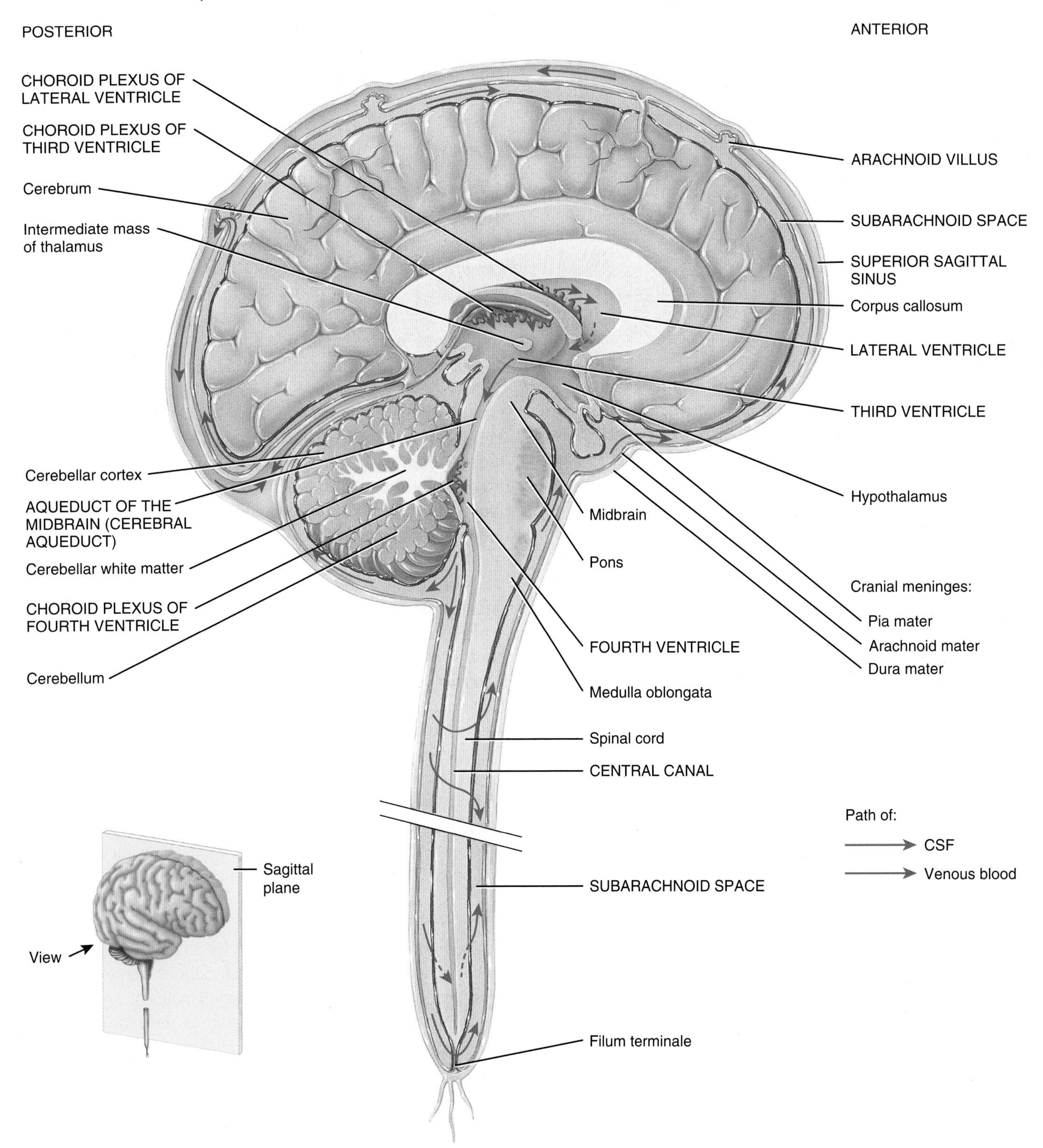

Sagittal section of brain and spinal cord

Where is cerebrospinal fluid produced and where is it reabsorbed?

which are finger-like extensions of the arachnoid mater (see Figures 17.2a and 17.3). The CSF drains primarily into a vein called the ***superior sagittal sinus*** (see Figures 17.2a,b and 17.3). Normally, the volume of CSF remains constant at 80 to 150 mL (3 to 5 oz) because it is absorbed as rapidly as it is formed.

Brain Stem

The brain stem is the part of the brain between the spinal cord and the diencephalon. It consists of three regions: (1) the medulla oblongata, (2) pons, and (3) midbrain. Extending through the brain stem is the reticular formation, a region where gray and white matter are intermingled.

Medulla Oblongata

The ***medulla oblongata*** (me-DOOL-la ob′-long-GA-ta), or simply ***medulla,*** is a continuation of the spinal cord (Figure 17.4). It forms the inferior part of the brain stem. Within the medulla's white matter are all sensory (ascending) and motor (descending) tracts extending between the spinal cord and other parts of the brain.

The medulla also contains several nuclei, which are masses of gray matter where neurons form synapses with one another. Two major nuclei are the ***cardiovascular center,*** which regulates the rate and force of the heartbeat and the diameter of blood vessels, and the ***medullary rhythmicity area,*** which adjusts the basic rhythm of breathing. Nuclei associated with sensations of touch, pressure vibration, and conscious proprioception (awareness of the position of body parts) are located in the posterior part of the medulla. Many ascending sensory axons form synapses in these nuclei. Other nuclei in the medulla oblongata control reflexes for swallowing, vomiting, coughing, hiccupping, and sneezing. Finally, the medulla contains nuclei associated with five pairs of cranial nerves (Figure 17.4): vestibulocochlear (VIII) nerves, glossopharyngeal (IX) nerves, vagus (X) nerves, accessory (XI) nerves, and hypoglossal (XII) nerves.

• CLINICAL CONNECTION | Injury to the Medulla

Given the many vital activities controlled by the medulla, it is not surprising that **injury to the medulla** from a hard blow to the back of the head or upper neck can be fatal. The medulla can also be damaged, even fatally damaged, by a blow such as an uppercut from a boxer when the skull is moved violently on the vertebral column and the dens of the axis impinges on the medulla. Damage to the medullary rhythmicity area is particularly serious and can rapidly lead to death. Symptoms of nonfatal injury to the medulla may include paralysis and loss of sensation on the opposite side of the body, and irregularities in breathing or heart rhythm. Alcohol overdose also suppresses the medullary rhythmicity area and may result in death. •

Pons

The ***pons*** (= bridge) is above the medulla and anterior to the cerebellum (Figure 17.4). Like the medulla, the pons consists of both nuclei and tracts. As its name implies, the pons is a bridge that connects parts of the brain with one another. These connections are bundles of axons. Some axons of the pons connect the right and left sides of the cerebellum. Others are part of ascending sensory tracts and descending motor tracts. Several nuclei in the pons are the sites where signals for voluntary skeletal movements that originate in the cerebral cortex are relayed into the cerebellum. Other nuclei in the pons help control breathing. The pons also contains nuclei associated with the following four pairs of cranial nerves (Figure 17.4): trigeminal (V) nerves, abducens (VI) nerves, facial (VII) nerves, and vestibulocochlear (VIII) nerves.

Midbrain

The ***midbrain*** connects the pons to the diencephalon. The anterior part of the midbrain consists of a pair of large tracts called ***cerebral peduncles*** (pe-DUNG-kuls or PĒ-dung-kuls = little feet) (see Figures 17.4 and 17.5). They contain axons of motor neurons that conduct nerve impulses from the cerebrum to the spinal cord, medulla, and pons.

Nuclei of the midbrain include the ***substantia nigra*** (sub-STAN-shē-a = substance; NĪ-gra = black), which is large and darkly pigmented. Loss of these neurons is associated with Parkinson disease. Also present are the ***red nuclei,*** which look reddish due to their rich blood supply and an iron-containing pigment in their neuronal cell bodies. Axons from the cerebellum and cerebral cortex form synapses in the red nuclei, which function with the cerebellum to coordinate muscular movements. Other nuclei in the midbrain are associated with two pairs of cranial nerves (see Figures 17.4 and 17.5): oculomotor (III) nerves and trochlear (IV) nerves.

The midbrain also contains nuclei that appear as four rounded bumps on the posterior surface. The two superior bumps are the ***superior colliculi*** (ko-LIK-ū-lī = little hills; singular is *colliculus*) (see Figure 17.5). Several reflex arcs pass through the superior colliculi: tracking and scanning movements of the eyes and reflexes that govern movements of the eyes, head, and neck in response to visual stimuli. The two ***inferior colliculi*** are part of the auditory pathway, relaying impulses from the receptors for hearing in the ear to the thalamus. Both the superior and inferior colliculi are reflex centers for the startle reflex, sudden movements of the head and body that occur when you are surprised by a loud noise.

Reticular Formation

In addition to the well-defined nuclei already described, much of the brain stem consists of small clusters of neuronal cell bodies (gray matter) intermingled with small bundles of myelinated axons (white matter). This region is known as the ***reticular formation*** (*ret-* = net) due to its netlike arrangement of white matter and gray matter (see Figure 17.5). Neurons within the reticular formation have both ascending (sensory) and descending (motor) functions.

The ascending part of the reticular formation is called the ***reticular activating system (RAS),*** which consists of sensory axons that project to the cerebral cortex. When the RAS is stimulated, many nerve impulses pass upward to widespread areas of the cerebral cortex. Activity of the RAS is often equated with the term ***attention span.*** The attention span of most adults is approximately 20 minutes. You may therefore note that a person wishing to keep your attention for more than 20 minutes, such as a cleric or a teacher, may tell a joke or do something else that will grab your attention for another 20-minute span. Another action of the RAS results in a state of wakefulness called ***consciousness.*** The RAS helps maintain consciousness and is active during awakening from sleep. Inactivation of the RAS produces ***sleep,*** a state of partial unconsciousness from which an individual can be aroused. The reticular formation's main descending function is to help regulate muscle tone, which is the slight degree of involuntary contraction in normal resting skeletal muscles.

Diencephalon

The ***diencephalon*** extends from the brain stem to the cerebrum and surrounds the third ventricle. Major regions of the diencephalon include the thalamus, hypothalamus, and epithalamus (see Figure 17.1).

Figure 17.4 Inferior aspect of the brain, showing the brain stem and cranial nerves.

The brain stem consists of the medulla oblongata, pons, and midbrain.

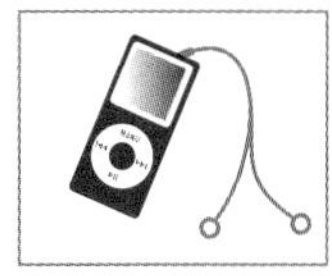

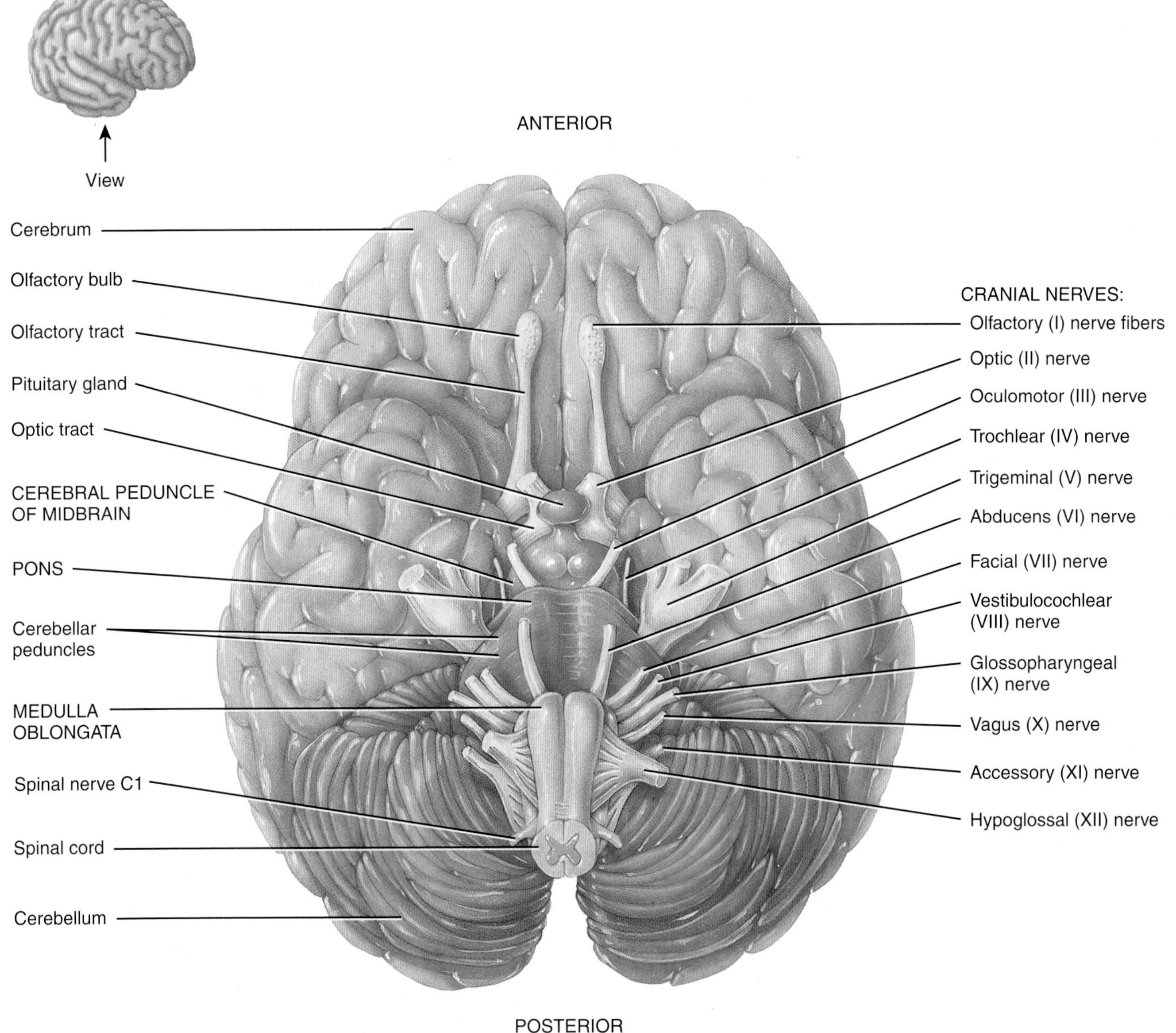

Inferior aspect of brain

Which part of the brain contains the cerebellar peduncles? The cerebral peduncles?

Figure 17.5 **Midbrain.**

The midbrain connects the pons to the diencephalon.

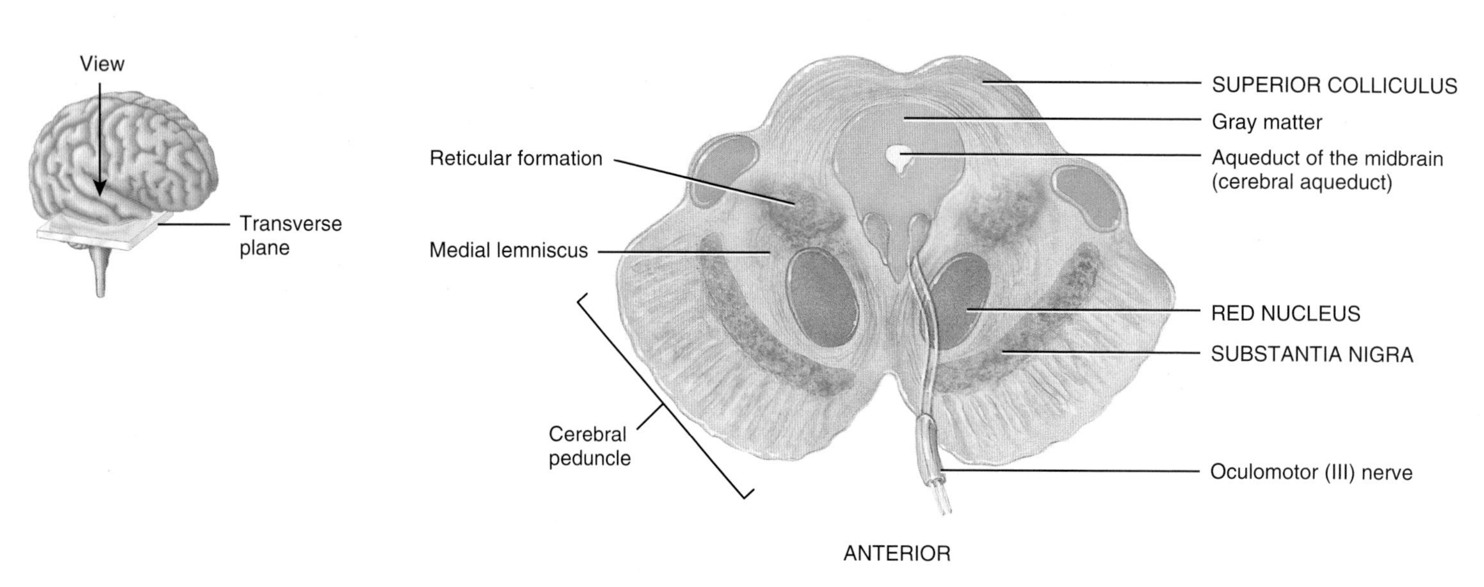

Transverse section of midbrain

What functions are carried out by the superior colliculi?

Thalamus

The ***thalamus*** (THAL-a-mus = inner chamber) consists of paired oval masses of gray matter, organized into nuclei, with interspersed tracts of white matter (see Figures 17.1 and 17.7). The right and left portions of the thalamus are connected by the ***intermediate mass*** (see Figure 17.3), a structure that spans the third ventricle. Nuclei of the thalamus are important relay stations for sensory impulses that are conducted to the cerebral cortex from the spinal cord, brain stem, cerebellum, and other parts of the cerebrum. The term relay station indicates that there are synapses present in the thalamus. The second-order neuron of every sensation of the body, except for smell, synapses in the thalamus (see Figure 16.14). As was discussed in Chapter 15, not all electrical impulses in a neuron are transmitted to the next neuron. The thalamus thus acts as a filter that permits "important" signals to pass to the cerebrum but filters out the "unimportant" signals that bombard our CNS on a continuing basis. Of course, what is important information to one person may not be considered important to the next.

The thalamus contributes to motor functions by transmitting information from the cerebellum and basal ganglia to motor areas of the cerebral cortex. It also relays nerve impulses between different areas of the cerebrum. The thalamus contributes to the maintenance of consciousness.

Hypothalamus

The ***hypothalamus*** (*hypo-* = under) is the small portion of the diencephalon that lies below the thalamus and above the pituitary gland (see Figures 17.1 and 17.7). Although its size is small, the hypothalamus controls many important body activities, most of them related to homeostasis. The chief functions of the hypothalamus are as follows:

1. **Control of the ANS.** The hypothalamus controls and integrates activities of the autonomic nervous system (ANS), which regulates contraction of smooth and cardiac muscles and the secretions of many glands. Through the ANS, the hypothalamus helps to regulate activities such as heart rate, movement of food through the gastrointestinal tract, and contraction of the urinary bladder.
2. **Control of the pituitary gland and production of hormones.** The hypothalamus controls the release of several hormones from the pituitary gland and thus serves as a primary connection between the nervous system and endocrine system. The hypothalamus also produces two hormones that are stored in the pituitary gland prior to their release; these will be discussed in Chapter 20.
3. **Regulation of emotional and behavioral patterns.** Together with the limbic system (described shortly), the hypothalamus regulates feelings of rage, aggression, pain, and pleasure, and the behavioral patterns related to sexual arousal.
4. **Regulation of eating and drinking.** The hypothalamus regulates eating behavior via two antagonistic nuclei. There are exceptions, but many of us are able to wear the same-sized clothing several years later, implying that the two antagonistic centers for eating are in homeostasis even though we may have eaten over a thousand meals (plus snacks) per year. The hypothalamus similarly contains two nuclei in its ***thirst center.*** When certain cells in the hypothalamus are stimulated by rising osmotic pressure of

the interstitial fluid, they cause the sensation of thirst. The intake of water by drinking restores the osmotic pressure to normal, removing the stimulation and relieving the thirst.

5. Control of body temperature. If the temperature of blood flowing through the hypothalamus is above normal, the hypothalamus directs the autonomic nervous system to stimulate activities that promote heat loss. If, however, blood temperature is below normal, the hypothalamus generates impulses that promote heat production and retention. The hypothalamus is one of the few structures of the brain that does not contain a blood-brain barrier and thus it is better structured to monitor the temperature of blood.

6. Regulation of circadian rhythms and states of consciousness. The hypothalamus establishes patterns of awakening and sleep that occur on a circadian (daily) schedule.

Epithalamus

The ***pineal gland*** (PĪN-ē-al = pinecone-like), located in the ***epithalamus,*** is about the size of a small pea and protrudes from the posterior midline of the third ventricle (see Figure 17.1a). Because the pineal gland secretes the hormone *melatonin,* it is part of the endocrine system and will be mentioned again in Chapter 20. Melatonin promotes sleepiness and contributes to the setting of the body's biological clock.

Cerebellum

The ***cerebellum*** consists of two ***cerebellar hemispheres,*** which are located posterior to the medulla and pons and below the cerebrum (see Figures 17.1, 17.3, and 17.4). The surface of the cerebellum, called the ***cerebellar cortex,*** consists of gray matter. Beneath the cortex is ***white matter*** that resembles the branches of a tree (see Figure 17.3). Deep within the white matter are masses of gray matter, the ***cerebellar nuclei.*** The cerebellum attaches to the brain stem by bundles of axons called ***cerebellar peduncles*** (see Figure 17.4).

The cerebellum compares intended movements programmed by the cerebral cortex with what is actually happening. It constantly receives sensory impulses from muscles, tendons, joints, equilibrium receptors, and visual receptors. The cerebellum helps to smooth and coordinate complex sequences of skeletal muscle contractions. It regulates posture and balance and is essential for all skilled motor activities, from catching a baseball to dancing.

• CLINICAL CONNECTION | Ataxia

Damage to the cerebellum through trauma or disease disrupts muscle coordination, a condition called **ataxia** (a-TAK-sē-a; *a-* = without; *-taxia* = order). Blindfolded people with ataxia cannot touch the tip of their nose with a finger because they cannot coordinate movement with their sense of where a body part is located. Another sign of ataxia is a changed speech pattern due to uncoordinated speech muscles. Cerebellar damage may also result in staggering or abnormal walking movements. People who consume too much alcohol show signs of ataxia because alcohol inhibits activity of the cerebellum. •

Cerebrum

The ***cerebrum*** consists of the ***cerebral cortex*** (an outer rim of gray matter), an internal region of cerebral white matter, and gray matter nuclei deep within the white matter (Figure 17.6). The cerebrum provides us with the ability to read, write, and speak; to make calculations and compose music; to remember the past and plan for the future; and to create. During embryonic development, when there is a rapid increase in brain size, the gray matter of the cerebral cortex enlarges much faster than the underlying white matter. As a result, the cerebral cortex rolls and folds upon itself to fit into the cranial cavity. The folds are called ***gyri*** (JĪ-rī = circles; singular is *gyrus*) or ***convolutions*** (Figure 17.6). The deep grooves between folds are ***fissures;*** the shallow grooves are ***sulci*** (SUL-sī = groove; singular is *sulcus,* SUL-kus). The ***transverse fissure*** separates the cerebrum and the cerebellum. The ***longitudinal fissure*** separates the cerebrum into right and left halves called ***cerebral hemispheres.*** The two hemispheres are connected internally by the ***corpus callosum*** (kal-LŌ-sum; *corpus* = body; *callosum* = hard), a broad band of white matter containing axons that extend between the two hemispheres (see Figures 17.3 and 17.7).

Each cerebral hemisphere has four lobes that are named after the bones that cover them: ***frontal lobe, parietal lobe, temporal lobe,*** and ***occipital lobe*** (see Figure 17.6). The ***central sulcus*** separates the frontal and parietal lobes. A major gyrus, the ***precentral gyrus,*** is located immediately anterior to the central sulcus. The precentral gyrus contains the primary motor area of the cerebral cortex. The ***postcentral gyrus,*** located immediately posterior to the central sulcus, contains the primary somatosensory area of the cerebral cortex, which is discussed shortly. The ***lateral cerebral sulcus*** separates the frontal lobe from the temporal lobe. The ***parieto-occipital sulcus*** separates the parietal lobe from the occipital lobe. A fifth part of the cerebrum, the ***insula,*** cannot be seen at the surface of the brain because it lies within the lateral cerebral sulcus, deep to the parietal, frontal, and temporal lobes (see Figures 17.6 and 17.7).

The ***cerebral white matter*** consists of axons, the *association tracts,* that transmit impulses between gyri in the same hemisphere, *commisural tracts* that connect gyri in one cerebral hemisphere to the corresponding gyri in the opposite cerebral hemisphere via the ***corpus callosum*** (see Figures 17.3 and 17.7); and *projection tracts* that pass through the ***internal capsule*** (see Figure 17.7b) and connect the cerebrum to other parts of the brain and spinal cord.

Deep within each cerebral hemisphere are three nuclei (masses of gray matter) that are collectively termed the ***basal ganglia*** (see Figure 17.7). (Recall that "ganglion" usually means a collection of neuronal cell bodies *outside* the CNS. The name here is the one exception to that general rule.) They are the *globus pallidus* (*globus* = ball; *pallidus* = pale), the *putamen* (pū-TĀ-men = shell), and the *caudate nucleus* (*caud-* = tail). A major function of the basal ganglia is to help initiate and terminate movements. They also help regulate the muscle tone required for specific body movements and control subconscious contractions of skeletal muscles, such as automatic arm swings while walking.

Figure 17.6 Cerebrum. The inset in (a) indicates the differences among a gyrus, a sulcus, and a fissure. Because the insula cannot be seen externally, it has been projected to the surface in (b).

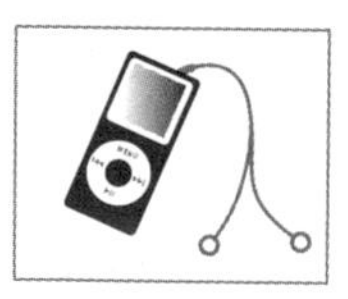

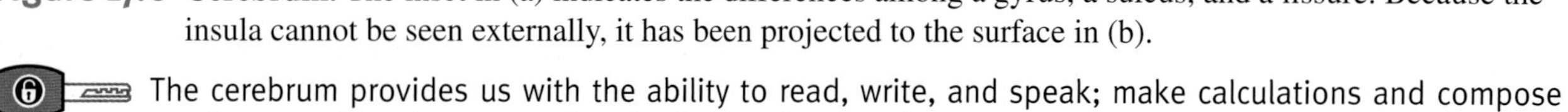

The cerebrum provides us with the ability to read, write, and speak; make calculations and compose music; remember the past and make future plans; and create.

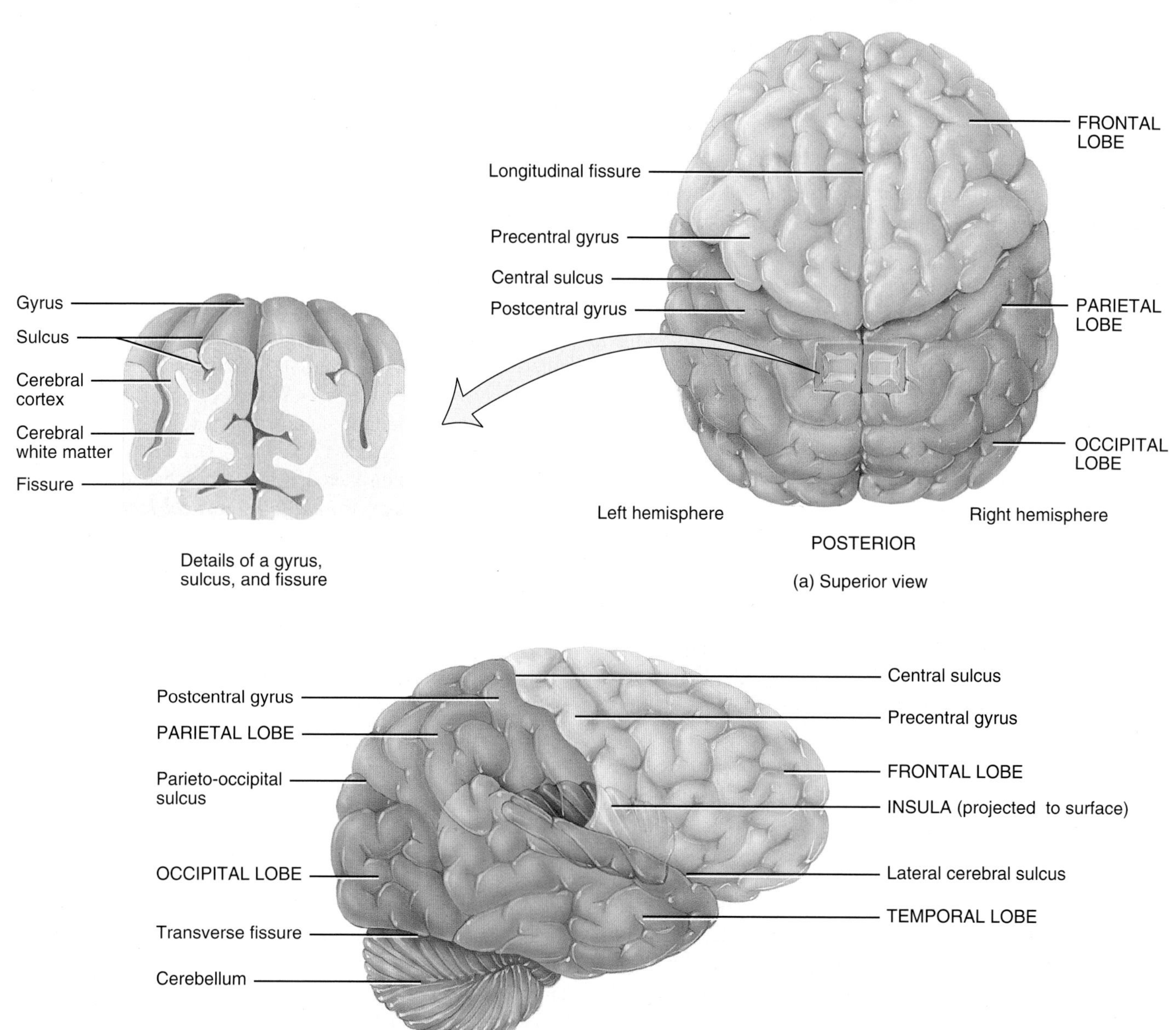

Details of a gyrus, sulcus, and fissure

(a) Superior view

(b) Right lateral view

What are the names of the four lobes of the cerebrum? What is the fifth part of the cerebrum called?

MANUAL THERAPY APPLICATION

Parkinson Disease

Parkinson disease (PD) is a progressive disorder of the CNS that typically affects its victims around age 60. Neurons that extend from the substantia nigra to the putamen and caudate nucleus, where they release the neurotransmitter dopamine (DA), degenerate in PD. The cause of PD is unknown, but toxic environmental chemicals, such as pesticides, herbicides, and carbon monoxide, are suspected contributing agents. Only 5% of PD patients have a family history of the disease.

Figure 17.7 Basal ganglia. In (a) the basal ganglia have been projected to the surface and are shown in purple; in (b) they are also shown in purple.

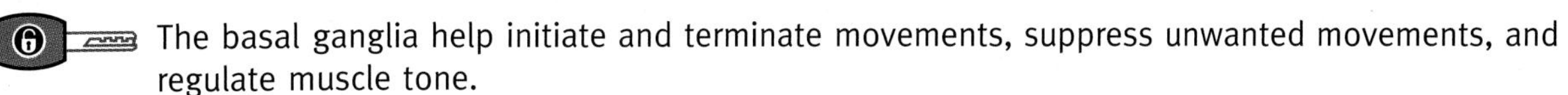

The basal ganglia help initiate and terminate movements, suppress unwanted movements, and regulate muscle tone.

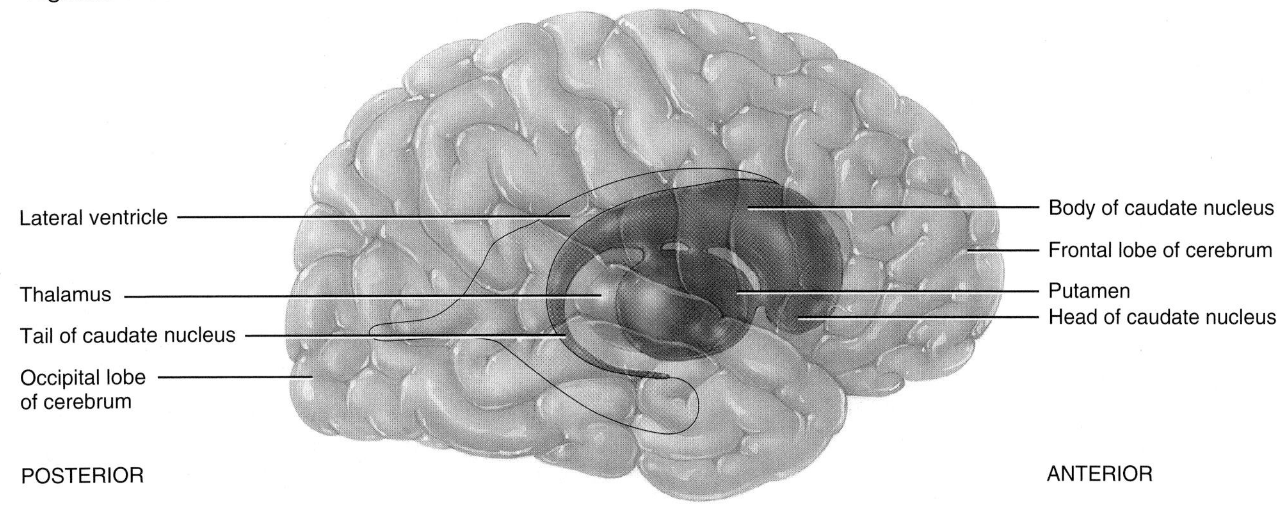

(a) Lateral view of right side of brain

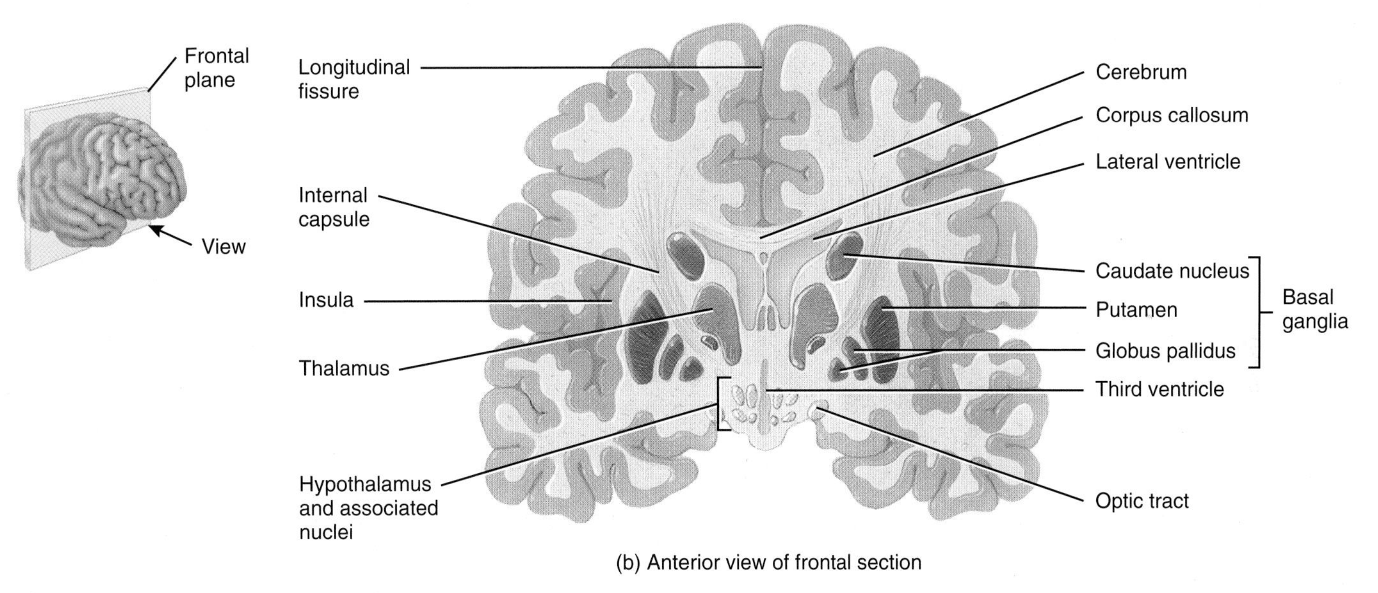

(b) Anterior view of frontal section

Where are the basal ganglia located relative to the thalamus?

In PD patients, involuntary skeletal muscle contractions often interfere with voluntary movement. For instance, the muscles of the upper limb may alternately contract and relax, causing the hand to shake. This shaking, called **tremor,** is the most common symptom of PD. Also, muscle tone may increase greatly, causing rigidity of the involved body part. Rigidity of the facial muscles gives the face a masklike appearance. The expression is characterized by a wide-eyed, unblinking stare and a slightly open mouth with uncontrolled drooling.

Motor performance is also impaired by **bradykinesia** (*brady-* = slow), slowness of movements. Activities such as shaving, cutting food, and buttoning a shirt take longer and become increasingly more difficult as the disease progresses. Muscular movements also exhibit **hypokinesia** (*hypo-* = under), decreasing range of motion. For example, words are written smaller, letters are poorly formed, and eventually handwriting becomes illegible. Often, walking is impaired; steps become shorter and shuffling, and arm swing diminishes. Even speech may be affected.

Studies of patients with progressive and degenerative diseases and disorders, such as Parkinson disease, spinal cord injury, and multiple sclerosis, indicate that massage may be beneficial in increasing the quality of sleep, improving daily functioning, and decreasing levels of stress hormones.

Limbic System

Encircling the upper part of the brain stem and the corpus callosum is a complex ring of structures on the inner border of the cerebrum and floor of the diencephalon that constitutes the ***limbic system*** (*limbic* = border) (Figure 17.8). The limbic system is sometimes called the "emotional brain" because it plays a primary role in a range of emotions, including pain, pleasure, docility, affection, and anger. Although behavior is a function of the entire nervous system, the limbic system controls most of its involuntary aspects related to survival. Animal experiments suggest that it has a major role in controlling the overall pattern of behavior. Together with parts of the cerebrum, the limbic system also functions in memory; damage to the limbic system causes memory impairment. One portion of the limbic system, the *hippocampus*, is seemingly unique for the central nervous system; the hippocampus has cells that have been reported to be capable of mitosis. Thus this portion of the brain that is responsible for some aspects of memory may develop new neurons, even in the elderly.

Figure 17.8 The limbic system. The components of the limbic system are shaded green and have been projected to the surface.

The limbic system, a complex group of structures surrounding the thalamus, governs emotional aspects of behavior.

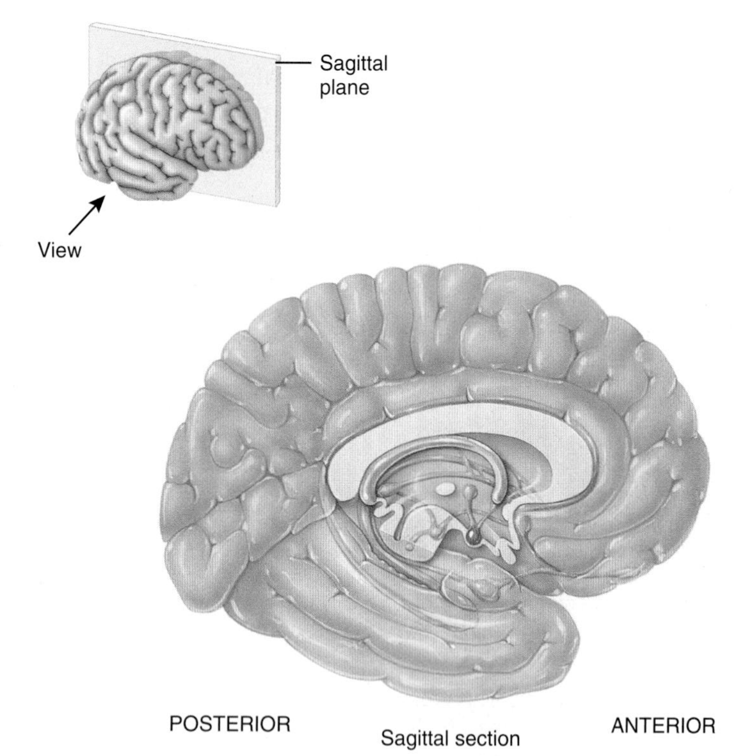

Where in the brain is the limbic system located?

Functional Areas of the Cerebral Cortex

Specific types of sensory, motor, and integrative signals are processed in certain regions of the cerebral cortex (Figure 17.9). Generally, ***sensory areas*** receive sensory information and are involved in ***perception,*** the conscious awareness of a sensation; ***motor areas*** initiate movements; and ***association areas*** deal with more complex integrative functions such as memory, emotions, reasoning, will, judgment, personality traits, and intelligence.

SENSORY AREAS Sensory input to the cerebral cortex flows mainly to the posterior half of the cerebral hemispheres, to regions behind the central sulci. In the cerebral cortex, primary sensory areas receive sensory information that has been relayed from peripheral sensory receptors through lower regions of the brain.

The ***primary somatosensory area*** (sō′-mat-ō-SEN-sō-rē) is posterior to the central sulcus of each cerebral hemisphere in the postcentral gyrus of the parietal lobe (Figure 17.9). It receives nerve impulses for touch, proprioception (joint and muscle position), pain, itching, tickle, and temperature and is involved in the perception of these sensations. The primary somatosensory area allows you to pinpoint where sensations originate, so that you know exactly where on your body to swat that mosquito. The *primary visual area,* located in the occipital lobe, receives visual information and is involved in visual perception. The *primary auditory area,* located in the temporal lobe, receives information for sound and is involved in auditory perception. The *primary gustatory area,* located at the base of the postcentral gyrus, receives impulses for taste and is involved in gustatory perception. The *primary olfactory area,* located on the medial aspect of the temporal lobe (and thus not visible in Figure 17.9), receives impulses for smell and is involved in olfactory perception.

MOTOR AREAS Motor output from the cerebral cortex flows mainly from the anterior part of each hemisphere. Among the most important motor areas are the primary motor area and Broca's speech area (Figure 17.9). The ***primary motor area*** is located in the precentral gyrus of the frontal lobe in each hemisphere. Each region in the primary motor area controls voluntary contractions of specific muscles on the opposite side of the body. ***Broca's speech area*** (BRŌ-kaz) is located in the frontal lobe close to the lateral cerebral sulcus. Speaking and understanding language are complex activities that involve several sensory, association, and motor areas of the cortex. In 97% of the population, these language areas are localized in the *left* hemisphere. Neural connections between Broca's speech area, the premotor area, and primary motor area activate muscles needed for speaking and breathing muscles. The *premotor area,* immediately anterior to the primary motor area, generates nerve impulses that cause a specific group of muscles to contract in a specific sequence, for example, to write a word.

Figure 17.9 **Functional areas of the cerebrum.** Broca's speech area and Wernicke's area are in the left cerebral hemisphere of most people; they are shown here to indicate their relative locations.

Particular areas of the cerebral cortex process sensory, motor, and integrative signals.

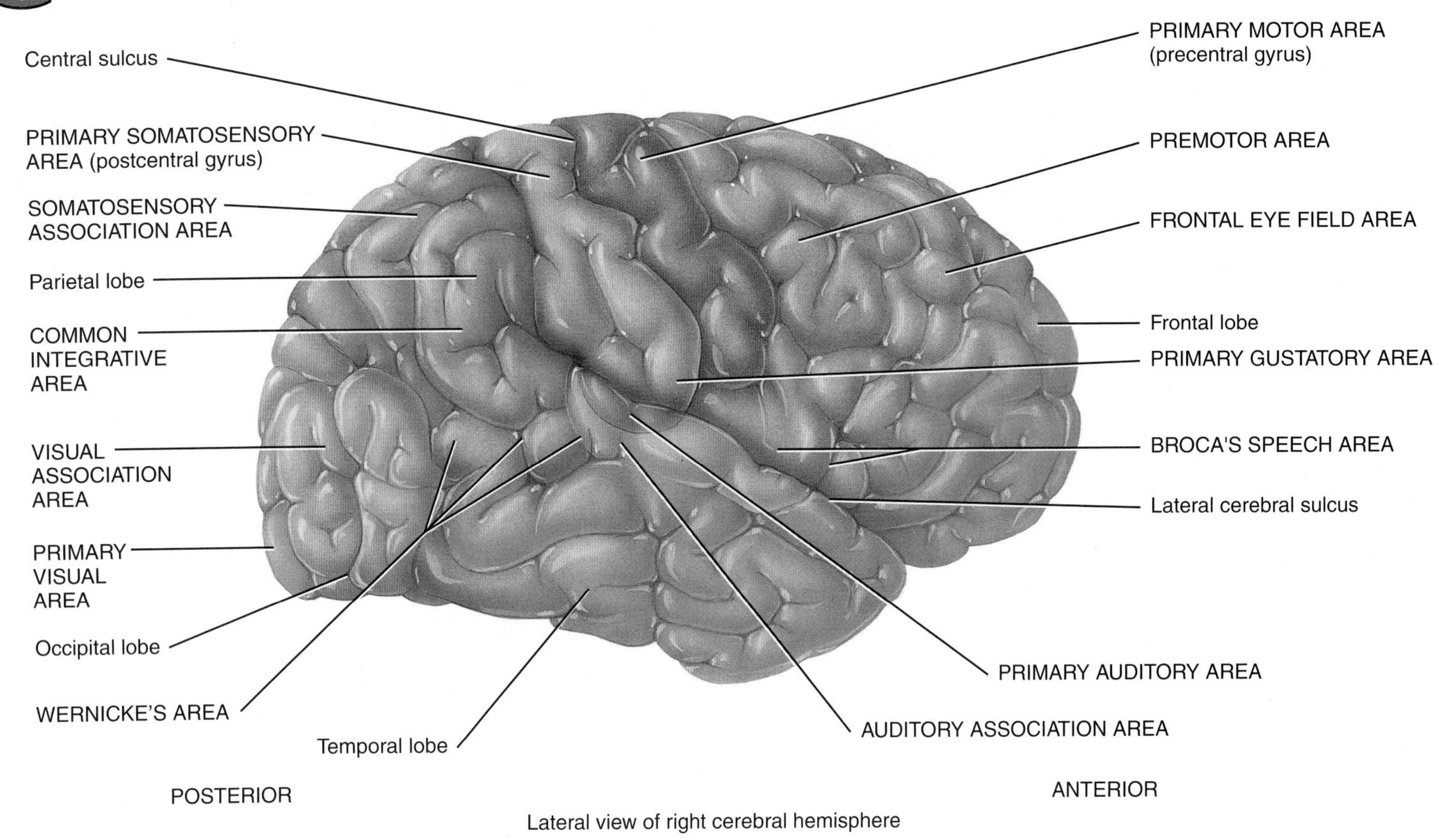

Lateral view of right cerebral hemisphere

Which part of the cerebrum localizes exactly where somatic sensations occur?

Association Areas The association areas of the cerebrum consist of some motor and sensory areas, plus large areas on the lateral surfaces of the occipital, parietal, and temporal lobes and on the frontal lobes anterior to the motor areas. Tracts connect association areas to one another. The ***somatosensory association area,*** just posterior to the primary somatosensory area, integrates and interprets somatic sensations such as the exact shape and texture of an object. Another role of the somatosensory association area is the storage of memories of past sensory experiences, enabling you to compare current sensations with previous experiences. For example, the somatosensory association area usually allows you to recognize objects such as a pencil, a tennis ball, and a paperclip simply by touching them. Certain persons, however, have never experienced a tennis ball and therefore would not be able to identify one by touching it. The ***visual association area,*** located in the occipital lobe, relates present and past visual experiences and is essential for recognizing and evaluating what is seen. The ***auditory association area,*** located below the primary auditory area in the cortex of the temporal lobe, allows you to recognize a particular sound as speech, music, or noise.

Wernicke's area, a broad region in the *left* temporal and parietal lobes, interprets the meaning of speech by recognizing spoken words. It is active as you translate words into thoughts. The regions in the *right* hemisphere that correspond to Broca's and Wernicke's areas in the left hemisphere also contribute to verbal communication by adding emotional content, for instance, anger or joy, to spoken words. The ***common integrative area*** receives, interprets, and responds appropriately to nerve impulses from the somatosensory, visual, and auditory association areas, and from the primary gustatory area, primary olfactory area, the thalamus, and parts of the brain stem. The ***frontal eye field area,*** located in the cortex of the frontal lobe, controls voluntary scanning movements of the eyes, such as those that occur while you are reading this sentence. Since the frontal eye field must be trained, experts suggest that parents read to their children while on their lap and that the parent follow the reading with a fingertip. The child will thus learn that the normal pattern for the English language, for example, is left to right and then drop one line. The multiple sensory, motor, and association areas are interconnected and form the basis of *study skills.*

• CLINICAL CONNECTION | Aphasia

Injury to language areas of the cerebral cortex results in **aphasia** (a-FĀ-zē-a; *a-* = without; *-phasia* = speech), an inability to use or comprehend words. Damage to Broca's speech area results in *nonfluent aphasia,* an inability to properly form words. People with nonfluent aphasia know what they wish to say but cannot properly speak the words. Damage to Wernicke's area, the common integrative area, or the auditory association area results in *fluent aphasia,* characterized by faulty understanding of spoken or written words. A person experiencing this type of aphasia may produce strings of words that have no meaning ("word salad"). For example, someone with fluent aphasia might say, "I rang car porch dinner light river pencil." •

Hemispheric Lateralization

Although the brain is quite symmetrical, there are subtle anatomical differences between the two hemispheres. The hemispheres are also functionally different in some ways, with each specializing in certain functions. This functional asymmetry is termed ***hemispheric lateralization.***

As you have seen, the left hemisphere receives sensory signals from and controls the right side of the body, and the right hemisphere receives sensory signals from and controls the left side of the body. In addition, the left hemisphere is more important for spoken and written language, numerical and scientific skills, ability to use and understand sign language, and reasoning in most people. Patients with damage in the left hemisphere, for example, often have difficulty speaking. The right hemisphere is more important for musical and artistic awareness; creativity; spatial and pattern perception; recognition of faces and emotional content of language; and for generating mental images of sight, sound, touch, taste, and smell. Most individuals have a combination of lateralization; for example, some perons are 90% left brained, others are 70% right brained, and a few about 50–50. Individuals who have a more balanced lateralization are less likely to experience the devastating effects of a stroke on their dominant side.

Memory

Without memory, we would repeat mistakes and be unable to learn. Similarly, we would not be able to repeat our successes or accomplishments, except by chance. ***Memory*** is the process by which information acquired through learning is stored and retrieved. For an experience to become part of memory, it must produce structural and functional changes in the brain. The parts of the brain known to be involved with memory include the association areas of the frontal, parietal, occipital, and temporal lobes; parts of the limbic system, including the hippocampus; and the diencephalon. Recall that the hippocampus is the one area of the brain that is seemingly able to produce new neurons. Memories for motor skills, such as how to serve a tennis ball, are stored in the basal ganglia and cerebellum as well as in the cerebral cortex.

• CLINICAL CONNECTION | Alzheimer Disease

Alzheimer disease (AD) (ALTZ-hī-mer) is a disabling senile dementia, the loss of reasoning and ability to care for oneself, that afflicts about 11% of the population over age 65. In the United States, about 4 million people suffer from AD. Claiming over 100,000 lives a year, AD is the fourth leading cause of death among the elderly, after heart disease, cancer, and stroke. The cause of most AD cases is still unknown, but evidence suggests it is due to a combination of genetic factors, environmental or lifestyle factors, and the aging process. An environmental risk factor for developing AD is a history of head injury. A similar dementia occurs in boxers, probably caused by repeated blows to the head.

Individuals with AD initially have trouble remembering recent events. They then become confused and forgetful, often repeating questions or getting lost while traveling to familiar places. Disorientation grows, memories of past events disappear, and episodes of paranoia, hallucination, or violent changes in mood may occur. As their minds continue to deteriorate, they lose their ability to read, write, talk, eat, or walk. The disease culminates in dementia. A person with AD usually dies of some complication that afflicts bedridden patients, such as pneumonia. •

Electroencephalogram (EEG)

At any instant, brain neurons are generating millions of nerve impulses. Taken together, these electrical signals are called ***brain waves.*** Brain waves generated by neurons close to the brain surface, mainly neurons in the cerebral cortex, can be detected by metal electrodes placed on the forehead and scalp. A record of such waves is called an ***electroencephalogram*** (e-lek′-trō-en-SEF-a-lō-gram) or ***EEG.*** Electroencephalograms are useful for studying normal brain functions, such as changes that occur during sleep. Neurologists also use them to diagnose a variety of brain disorders, such as epilepsy, tumors, metabolic abnormalities, sites of trauma, and degenerative diseases.

Table 17.1 summarizes the principal parts of the brain and their functions.

CHECKPOINT

1. What is the significance of the blood–brain barrier?
2. What structures are the sites of CSF production, and where are they located?
3. Where are the medulla, pons, and midbrain located relative to one another?
4. What functions are governed by nuclei in the brain stem?
5. What are two important functions of the reticular formation?
6. Why is the hypothalamus considered part of both the nervous system and the endocrine system?
7. What are the functions of the cerebellum and basal ganglia?
8. Where are the primary somatosensory area and primary motor area located in the brain? What are their functions?
9. What areas of the cerebral cortex are needed for normal language abilities?
10. Compare and contrast the posterior column–medial lemniscus pathway and the spinothalamic pathways.

TABLE 17.1

Summary of Functions of Principal Parts of the Brain

PART	FUNCTION
Brain Stem 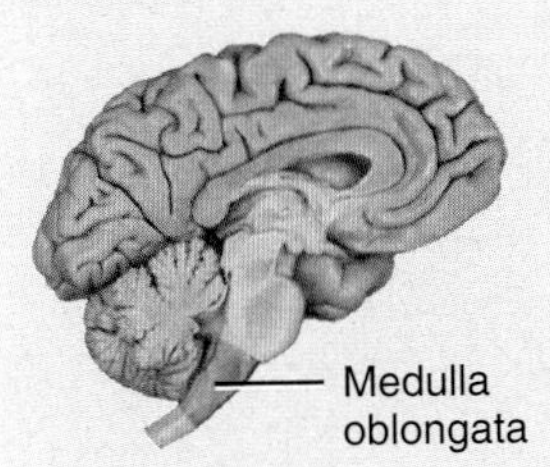	*Medulla oblongata:* Contains sensory (ascending) tracts and motor (descending) tracts. Reticular formation (also in pons, midbrain, and diencephalon) functions in consciousness and arousal. Vital centers regulate heartbeat and breathing (together with pons). Other centers coordinate swallowing, vomiting, coughing, sneezing, and hiccupping. Contains nuclei of origin for cranial nerves VIII, IX, X, XI, and XII.
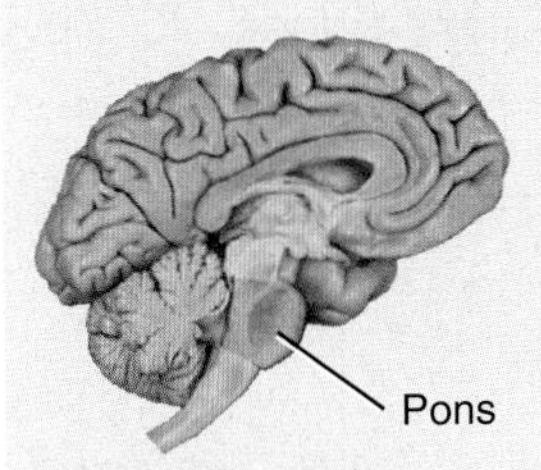	*Pons:* Contains sensory tracts and motor tracts. Contains nuclei of origin for cranial nerves V, VI, VII, and VIII. Together with the medulla, helps control breathing.
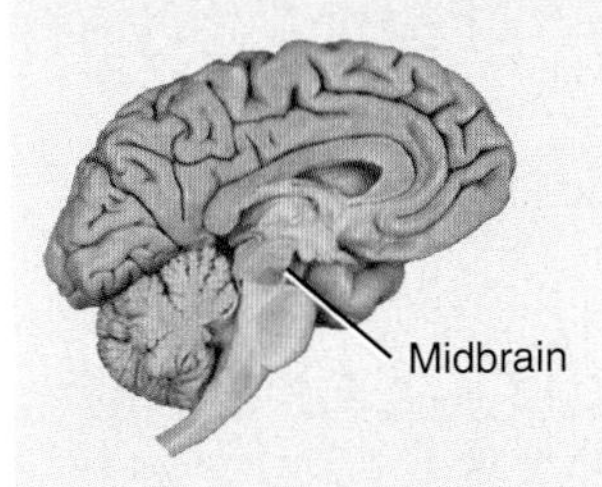	*Midbrain:* Contains sensory tracts and motor tracts. Superior colliculi; coordinate movements of head, eyes, and trunk in response to visual stimuli. Inferior colliculi coordinate movement of head, eyes, and trunk in response to auditory stimuli. Most of substantia nigra and red nucleus contribute to control of movement. Contains nuclei of origin for cranial nerves III and IV.
Diencephalon 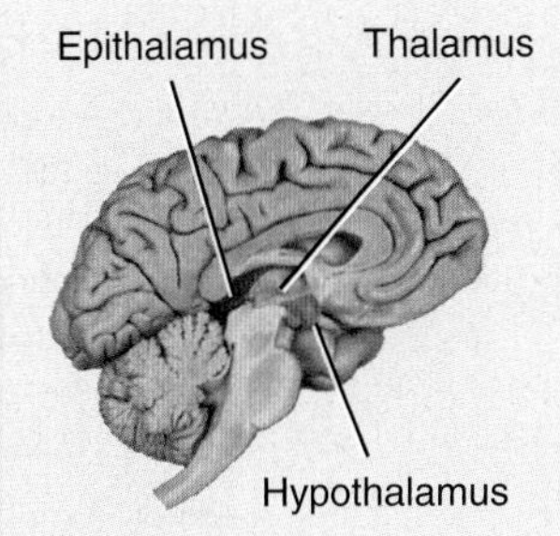	*Thalamus:* Relays almost all sensory input impulses to the cerebral cortex. Contributes to motor functions by transmitting information from the cerebellum and basal ganglia to motor areas of the cerebral cortex. Plays a role in maintaining consciousness. *Hypothalamus:* Controls and integrates activities of the autonomic nervous system and pituitary gland. Regulates emotional and behavioral patterns and circadian rhythms. Controls body temperature and regulates eating and drinking behavior. Helps maintain the waking state and establishes patterns of sleep. *Epithalamus:* Contains the pineal gland, which secretes the hormone melatonin.
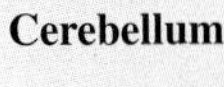 **Cerebellum** 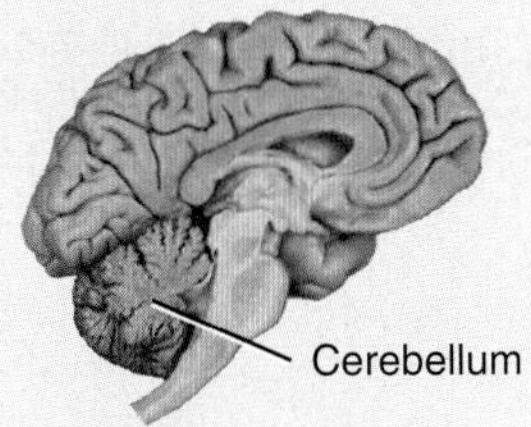	Compares intended movements with what is actually happening to smooth and coordinate complex, skilled movements. Regulates posture and balance.
 **Cerebrum** 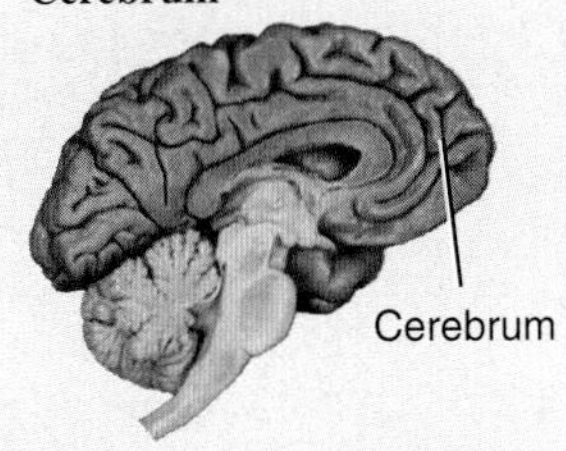	Sensory areas are involved in the perception of sensory information; motor areas control the execution of voluntary movements, and association areas deal with more complex integrative functions such as memory, personality traits, and intelligence. Basal ganglia help initiate and terminate movements, suppress unwanted movements, and regulate muscle tone. Limbic system functions in emotional aspects of behavior related to survival.

17.2 CRANIAL NERVES

OBJECTIVE

- Identify the 12 pairs of cranial nerves by name and number and give the primary functions of each.

The 12 pairs of ***cranial nerves,*** like spinal nerves, are part of the peripheral nervous system. The cranial nerves are designated with roman numerals and with names (see Figure 17.4). The roman numerals indicate the order (anterior to posterior) in which the nerves emerge from the brain. The names indicate the distribution or function.

Cranial nerves emerge from the nose (cranial nerve I), the eyes (cranial nerve II), the inner ear (cranial nerve VIII), the brain stem (cranial nerves III–XII), and the spinal cord (cranial nerve XI). Three cranial nerves (cranial nerves I, II, and VIII) contain only sensory axons and thus are *sensory nerves.* Five cranial nerves (III, IV, VI, XI, and XII) contain only motor axons of motor neurons as they leave the brain stem and are called *motor nerves.* The rest of the cranial nerves (V, VII, IX, and X) are *mixed nerves* because they contain axons of both sensory and motor neurons. Cranial nerves III, VII, IX, and X include both somatic and autonomic motor axons. The somatic axons stimulate skeletal muscles; the autonomic axons, which are part of the parasympathetic division, go to glands, smooth muscle, and cardiac muscle.

Table 17.2 lists the cranial nerves, along with their components (sensory or mixed) and functions.

TABLE 17.2

Summary of Cranial Nerves (see Figure 17.4)

NUMBER	NAME*	COMPONENTS	FUNCTION
I	**Olfactory nerve** (ol-FAK-tō-rē; *olfact-* = to smell)	***Sensory:*** Axons in the lining of the nose.	Smell.
II	**Optic nerve** (OP-tik; *opti-* = eye, vision)	***Sensory:*** Axons from the retina of the eye.	Vision.
III	**Oculomotor nerve** (ok′-ū-lō-MŌ-tor; *oculo-* = eye; *-motor* = mover)	***Motor:*** Axons of somatic motor neurons that innervate muscles of upper eyelid and four muscles that move the eyeballs (superior rectus, medial rectus, inferior rectus, and inferior oblique) plus axons of parasympathetic neurons that pass to two sets of smooth muscles—the ciliary muscle of the eyeball and the circular muscles of the iris.	Movement of upper eyelid and eyeball; alters shape of lens for near vision and constricts pupil.
IV	**Trochlear nerve** (TRŌK-lē-ar; *trochle-* = a pulley)	***Motor:*** Axons of somatic motor neurons that innervate the superior oblique muscles.	Movement of the eyeball.
V	**Trigeminal nerve** (trī-JEM-i-nal = triple, for its three branches)	***Mixed***	
		Sensory part: Consists of three branches: the *ophthalmic nerve* contains axons from the scalp and forehead skin; the *maxillary nerve* contains axons from the lower eyelid, nose, upper teeth, upper lip, and pharynx; and the *mandibular nerve* contains axons from the tongue, lower teeth, and the lower side of the face.	Touch, pain, and temperature sensations and proprioception (muscle sense).
		Motor part: Axons of somatic motor neurons that innervate muscles used in chewing.	Chewing.
VI	**Abducens nerve** (ab-DOO-senz; *ab-* = away; *-ducens* = to lead)	***Motor:*** Axons of somatic motor neurons that innervate the lateral rectus muscles.	Movement of eyeball.
VII	**Facial nerve** (FĀ-shal = face)	***Mixed***	
		Sensory part: Axons from taste buds on anterior tongue and axons from proprioceptors in muscles of face and scalp.	Taste, and proprioception (muscle sense); touch, pain, temperature sensations.
		Motor part: Axons of somatic motor neurons that innervate facial, scalp, and neck muscles plus parasympathetic axons that stimulate lacrimal (tear) glands and salivary glands.	Facial expressions; secretion of tears and saliva.
VIII	**Vestibulocochlear nerve** (ves-tib-ū-lō-KŌK-lē-ar; *vestibulo-* = small cavity; *-cochlear* = a spiral, snail-like)	***Sensory***	
		Vestibular branch: Axons from semicircular canals, saccule, and utricle (organs of equilibrium).	Equilibrium.
		Cochlear branch: Axons from spiral organ (organ of Corti).	Hearing.
IX	**Glossopharyngeal nerve** (glos′-ō-fa-RIN-jē-al; *glosso-* = tongue; *-pharyngeal* = throat)	***Mixed***	
		Sensory part: Axons from taste buds and somatic sensory receptors on posterior part of tongue, from proprioceptors in some swallowing muscles, and from stretch receptors in carotid sinus and chemoreceptors in carotid body.	Taste and somatic sensations (touch, pain, and temperature) from tongue; proprioception (muscle sense) in some swallowing muscles; monitoring blood pressure; monitoring oxygen and carbon dioxide in blood for regulation of breathing.
		Motor part: Axons of somatic motor neurons that innervate swallowing muscles of throat plus parasympathetic axons that stimulate one of the salivary glands.	Swallowing; speech; secretion of saliva.

TABLE 17.2 CONTINUED

NUMBER	NAME*	COMPONENTS	FUNCTION
X	**Vagus nerve** (VĀ-gus; *vagus* = vagrant or wandering)	***Mixed*** ***Sensory part:*** Axons from taste buds in pharynx (throat) and epiglottis; proprioceptors in muscles of neck and throat, from stretch receptors and chemoreceptors in carotid sinus and carotid body, from chemoreceptors in aortic body, and from visceral sensory receptors in most organs of the thoracic and abdominal cavities.	Taste and somatic sensations (touch, pain, and temperature) from pharynx and epiglottis; monitoring of blood pressure; monitoring of oxygen and carbon dioxide in blood for regulation of breathing; sensations from visceral organs in thorax and abdomen.
		Motor part: Axons of somatic motor neurons that innervate skeletal muscles of the throat and neck plus parasympathetic axons that supply smooth muscle in the airways, esophagus, stomach, small intestine, most of the large intestine, and gallbladder; cardiac muscle in the heart; and glands of the gastrointestinal tract.	Swallowing, coughing, and voice production; smooth muscle contraction and relaxation in organs of the gastrointestinal tract; slowing of the heart rate; secretion of digestive fluids.
XI	**Accessory nerve** (ak-SES-ō-re = assisting)	***Motor:*** Axons of somatic motor neurons that innervate the sternocleidomastoid and trapezius muscles.	Movements of head and shoulders.
XII	**Hypoglossal nerve** (hī′-pō-GLOS-al; *hypo-* = below; *-glossal* = tongue)	***Motor:*** Axons of somatic motor neurons that innervate muscles of tongue.	Movement of tongue during speech and swallowing.

*A mnemonic that can be used to remember the names of the nerves is "**O**h, **O**h, **O**h, **t**o **t**ouch **a**nd **f**eel **v**ery **g**reen **v**egetables—**AH**!" Each boldfaced letter corresponds to the first letter of a pair of cranial nerves.

CHECKPOINT

11. What is the difference between a mixed cranial nerve and a sensory cranial nerve?

17.3 AGING AND THE NERVOUS SYSTEM

OBJECTIVE

- Describe the effects of aging on the nervous system.

The brain grows rapidly during the first few years of life. Growth is due mainly to an increase in the size of neurons already present, the proliferation and growth of neuroglia, the development of dendritic branches and synaptic contacts, and continuing myelination of axons. From early adulthood onward, brain mass declines. By the time a person reaches age 80, the brain weighs about 7% less than it did in young adulthood. Although the number of neurons present does not decrease very much, the number of synaptic contacts declines. Associated with the decrease in brain mass is a decreased capacity for sending nerve impulses to and from the brain. As a result, processing of information diminishes. Conduction velocity decreases, voluntary motor movements slow down, and reflex times increase.

CHECKPOINT

12. How is brain mass related to age?

KEY MEDICAL TERMS ASSOCIATED WITH THE BRAIN

Agnosia (ag-NŌ-zē-a; *a-* = without; *-gnosia* = knowledge) Inability to recognize the significance of sensory stimuli such as sounds, sights, smells, tastes, and touch.

Apraxia (a-PRAK-sē-a; *-praxia* = coordinated) Inability to carry out purposeful movements in the absence of paralysis.

Consciousness (KON-shus-nes) A state of wakefulness in which an individual is fully alert, aware, and oriented, partly as a result of feedback between the cerebral cortex and reticular activating system.

Delirium (dē-LIR-ē-um = off the track) A transient disorder of abnormal cognition and disordered attention accompanied by disturbances of the sleep–wake cycle and psychomotor behavior (hyperactivity or hypoactivity of movements and speech). Also called **acute confusional state (ACS).**

Dementia (de-MEN-shē-a; *de-* = away from; *-mentia* = mind) Permanent or progressive general loss of intellectual abilities, including impairment of memory, judgment, abstract thinking, and changes in personality.

Encephalitis (en′-sef-a-LĪ-tis) An acute inflammation of the brain caused by either a direct attack by any of several viruses or an allergic reaction to any of the many viruses that are normally harmless to the central nervous system. If the virus affects the spinal cord as well, the condition is called encephalomyelitis.

Encephalopathy (en-sef′-a-LOP-a-thē; *encephalo* = brain; *-pathos* = disease) Any disorder of the brain.

Lethargy (LETH-ar-jē) A condition of functional sluggishness.

Microcephaly (mī-krō-SEF-a-lē; *micro-* = small; *-cephal* = head) A congenital condition that involves the development of a small brain and skull and frequently results in mental retardation.

Reye's syndrome (RĪZ) Occurs after a viral infection, particularly chickenpox or influenza, most often in children or teens who have taken aspirin; characterized by vomiting and brain dysfunction (disorientation, lethargy, and personality changes) that may progress to coma and death.

Stupor (STOO-por) Unresponsiveness from which a patient can be aroused only briefly and only by vigorous and repeated stimulation.

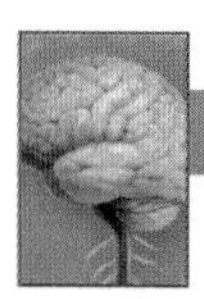

STUDY OUTLINE

The Brain (Section 17.1)

1. The major parts of the brain are the brain stem, diencephalon, cerebellum, and cerebrum (see Table 17.1). The brain stem consists of the medulla oblongata, pons, and midbrain. The diencephalon consists of the thalamus, hypothalamus, and pineal gland.
2. The brain is well supplied with oxygen and nutrients. Any interruption of the oxygen supply to the brain can weaken or permanently damage brain cells. Glucose deficiency may produce dizziness, convulsions, and unconsciousness.
3. The blood–brain barrier (BBB) limits the passage of certain material from the blood into the brain.
4. The brain is protected by cranial bones, meninges, and cerebrospinal fluid.
5. The cranial meninges are continuous with the spinal meninges and are named dura mater, arachnoid mater, and pia mater.
6. Cerebrospinal fluid is formed in the choroid plexuses and circulates continually through the subarachnoid space, ventricles, and central canal.
7. Cerebrospinal fluid protects by serving as a shock absorber. It also delivers nutritive substances from the blood and removes wastes.
8. The medulla oblongata, or medulla, is continuous with the upper part of the spinal cord. It contains regions for regulating heart rate, diameter of blood vessels, breathing, swallowing, coughing, vomiting, sneezing, and hiccupping. Portions of cranial nerves VIII–XII originate in the medulla.
9. The pons links parts of the brain with one another; it relays impulses for voluntary skeletal movements from the cerebral cortex to the cerebellum, and it contains two regions that control breathing. Cranial nerves V–VII and part of VIII are associated with the pons.
10. The midbrain connects the pons to the diencephalon. It conveys motor impulses from the cerebrum to the medulla, pons, and spinal cord, sends sensory impulses from the medulla to the thalamus, and mediates auditory and visual reflexes. It also contains cranial nerves III and IV.
11. The reticular formation is a netlike arrangement of gray and white matter extending throughout the brain stem that alerts the cerebral cortex to incoming sensory signals and helps regulate muscle tone and maintain consciousness.
12. The thalamus contains nuclei that serve as relay stations for sensory impulses to the cerebral cortex. It also contributes to motor functions by transmitting information from the cerebellum and basal ganglia to motor areas of the cerebral cortex.
13. The hypothalamus is inferior to the thalamus. It controls the autonomic nervous system, secretes hormones, functions in rage and aggression, governs body temperature, regulates food and fluid intake, and establishes circadian rhythms.
14. The cerebellum occupies the inferior and posterior aspects of the cranial cavity. It attaches to the brain stem by cerebellar peduncles. It coordinates movements and helps maintain posture and balance.
15. The cerebrum is the largest part of the brain. Its cortex contains gyri (convolutions), fissures, and sulci. The cerebral lobes are frontal, parietal, temporal, and occipital.
16. The white matter is deep to the cortex and consists of myelinated and unmyelinated axons extending to other CNS regions.
17. The basal ganglia are several groups of nuclei in each cerebral hemisphere. They help control automatic movements of skeletal muscles and help regulate muscle tone.
18. The limbic system encircles the upper part of the brain stem and the corpus callosum. It functions in emotional aspects of behavior and memory.
19. The sensory areas of the cerebral cortex receive and perceive sensory information. The motor areas govern muscular movement. The association areas are concerned with emotional and intellectual processes.
20. Subtle anatomical differences exist between the two cerebral hemispheres, and each has some unique functions.
21. Memory, the ability to store and recall thoughts, involves persistent changes in the brain.
22. Brain waves that are generated by the cerebral cortex are recorded as an electroencephalogram (EEG), which may be used to diagnose epilepsy, abnormalities, and tumors.

Cranial Nerves (Section 17.2)

1. Twelve pairs of cranial nerves emerge from the brain.
2. Like spinal nerves, cranial nerves are part of the peripheral nervous system (PNS). See Table 17.2 for the names, components, and functions of each of the cranial nerves.

Aging and the Nervous System (Section 17.3)

1. The brain grows rapidly during the first few years of life.
2. Age-related effects involve loss of brain mass and decreased capacity for sending nerve impulses.

SELF-QUIZ QUESTIONS

1. The diencephalon is composed of the
   a. medulla, pons, and hypothalamus
   b. midbrain, hypothalamus, and thalamus
   c. cerebellum and midbrain
   d. medulla, pons, and midbrain
   e. hypothalamus, thalamus, and pineal gland.
2. Which of the following statements about the blood supply to the brain is NOT true?
   a. The brain needs a constant supply of glucose delivered by the blood.
   b. The structure of the brain capillaries allows selective passage of certain materials from the blood into the brain.
   c. The glucose brought to the brain can be stored for future use.
   d. Brain neurons that are totally deprived of oxygen for 4 minutes or more may be permanently injured.
   e. The brain requires about 20% of the body's oxygen supply.
3. After a car accident, Joe exhibits severe dizziness, difficulty in walking, and slurred speech. He may have damaged his
   a. cerebellum b. pons
   c. reticular activating system d. fifth cranial nerve
   e. midbrain.
4. Which of the following is NOT a function of cerebrospinal fluid?
   a. protection b. circulation
   c. conduction of nerve impulses d. nutrition
   e. shock absorption
5. Which part of the brain contains the centers that control the heart rate and breathing rhythm?
   a. medulla b. midbrain
   c. cerebellum d. thalamus
   e. pons
6. The part of the brain that serves as a link between the nervous and endocrine systems is the
   a. reticular formation b. hypothalamus
   c. pons d. brain stem
   e. cerebellum.
7. Which of the following is NOT a function of the hypothalamus?
   a. regulates food intake
   b. controls body temperature
   c. regulates feelings of rage and aggression
   d. helps establish sleep patterns
   e. allows crude interpretation of pain and pressure
8. The part(s) of the brain concerned with memory, reasoning, judgment, and intelligence is (are) the
   a. sensory areas b. limbic system
   c. motor areas d. cerebellum
   e. association areas.
9. A broad band of white matter that connects the two cerebral hemispheres is the
   a. corpus callosum b. gyrus
   c. insula d. ascending tract
   e. basal ganglia
10. The ringing of your alarm clock in the morning wakes you up by stimulating the
    a. thalamus b. reticular activating system
    c. Broca's area d. basal ganglia
    e. spinal cord.
11. Match the following functions to the primary lobe in which they are located:

____	a. contains primary visual area that allows interpretation of shape and color	A. frontal lobe
____	b. receives impulses for smell	B. parietal lobe
____	c. contains primary motor area that controls muscle movement	C. occipital lobe
____	d. receives sensory impulses for touch, pain, and temperature	D. temporal lobe

12. When entering a restaurant, you are bombarded with many different sensory stimuli. The part of the brain that combines all of those sensory inputs so that you can respond appropriately is the
    a. somatosensory association area
    b. common integrative area
    c. premotor area
    d. Wernicke's area
    e. hypothalamus
13. Which cranial nerves contain only sensory fibers?
    a. olfactory, optic, and vestibulocochlear
    b. optic and oculomotor
    c. optic and trochlear
    d. optic and olfactory
    e. vagus and facial
14. Which two of the following cranial nerves are NOT involved in controlling movement of the eyeball?
    a. oculomotor b. trochlear
    c. facial d. abducens
    e. trigeminal
15. Match the following:

____	a. organization of white matter in the spinal cord	A. longitudinal fissure
____	b. absorb cerebrospinal fluid	B. sulci
____	c. small elevations in the cerebral cortex	C. ventricles
____	d. separates the cerebrum into right and left halves	D. anterior median fissure
____	e. brain cavities where CSF circulates	E. columns
____	f. shallow grooves in the cerebral cortex	F. arachnoid villi
____	g. wide groove in the spinal cord	G. gyri

CRITICAL THINKING QUESTIONS

1. An elderly relative suffered a CVA (stroke) and now has difficulty moving her right arm, and she also has speech problems. What areas of the brain were damaged by the stroke?

2. Nicki has recently had a viral infection and now she cannot move the muscles on the right side of her face. In addition, she is experiencing a loss of taste and a dry mouth and she cannot close her right eye. What cranial nerve has been affected by the viral infection?

? ANSWERS TO FIGURE QUESTIONS

17.1 The largest part of the brain is the cerebrum; the second largest is the cerebellum.

17.2 From superficial to deep, the three cranial meninges are the dura mater, arachnoid mater, and pia mater.

17.3 Cerebrospinal fluid is produced in the choroid plexus of each of the four ventricles and is reabsorbed into the blood by the arachnoid villi that project into the superior sagittal sinus.

17.4 The cerebellar peduncles are located in the pons; the cerebral peduncles are located in the midbrain.

17.5 The superior colliculi are centers that govern movements of the eyes, head, and neck in response to visual stimuli.

17.6 The four lobes of the cerebrum are the frontal lobe, parietal lobe, temporal lobe, and occipital lobe; the fifth part is the insula.

17.7 The basal ganglia are lateral, superior, and inferior to the thalamus.

17.8 The limbic system encircles the superior part of the brain stem and the corpus callosum.

17.9 Somatic sensations are localized in the primary somatosensory area within the postcentral gyrus of the cerebrum.

The Autonomic Nervous System 18

The autonomic nervous system (ANS) acts without our conscious thought. It regulates contraction of cardiac muscle and smooth muscle, and glandular secretion in our bodies. The derivation is *auto-* = self and *-nomic* = law; in other words, the ANS was originally believed to be self-governing. The medulla oblongata and hypothalamus of the brain are now known to be the main integrative centers for the ANS.

The ANS is divided into the sympathetic and parasympathetic divisions. The sympathetic division is very important when we are under stressful situations such as a medical emergency. It allows us to react quickly and gives us a surge of energy to deal with emergency situations. The sympathetic division mediates what is often called the fight-or-flight reaction. The parasympathetic division is known as the rest-and-digest division since its main functions are to heal and to aid digestion. The two divisions work in conjunction so when one division is turned on the other division is normally turned down. This is why, when we are under stress for a long time, we may develop high blood pressure. Every night, sleep forces a person into the parasympathetic state, and therefore we are able to repair tissues and to maintain homeostasis of the body. A goal for manual therapists may be to facilitate a patient's movement toward the parasympathetic state so that his body may better heal itself and decrease stress.

CONTENTS AT A GLANCE

18.1 INTRODUCTION TO THE AUTONOMIC NERVOUS SYSTEM

The part of the nervous system that regulates smooth muscle, cardiac muscle, and certain glands is the ***autonomic nervous system (ANS).*** Recall that together the ANS and somatic nervous system compose the peripheral nervous system; see Figure 15.2. In this chapter, we compare the structural and functional features of the somatic and autonomic nervous systems. Then we discuss the anatomy of the motor portion of the ANS and compare the organization and actions of its two major branches, the sympathetic and parasympathetic divisions.

The fight-or-flight response of the sympathetic nervous system is very helpful when you encounter a snarling dog or need to escape from a burning building. But when the emergency is over, your parasympathetic nervous system needs time to help your body relax and recover. What happens when stress builds up, and no recovery occurs? When your days are filled with negative stress and an overactivated sympathetic nervous system, stress-related health problems may develop. Chronic, unrelenting, overwhelming stress interferes with the body's ability to maintain homeostasis and health. Learning relaxation and stress reduction skills can reduce the harmful effects of stress on the body.

18.2 COMPARISON OF SOMATIC AND AUTONOMIC NERVOUS SYSTEMS

OBJECTIVE

- Compare the main structural and functional differences between the somatic and autonomic parts of the nervous system.

As you learned in Chapter 15, the somatic nervous system includes both sensory and motor neurons. The sensory neurons convey input from receptors for the special senses (vision, hearing, taste, smell, and equilibrium, which will be described in Chapter 19) and from receptors for somatic senses (pain, temperature, touch, and proprioceptive sensations). All these sensations normally are consciously perceived. In turn, somatic motor neurons synapse with skeletal muscle—the effector tissue of the somatic nervous system—and produce conscious, voluntary movements. When a somatic motor neuron stimulates a skeletal muscle, the muscle contracts. If somatic motor neurons cease to stimulate a muscle, the result is a paralyzed, limp muscle that has no muscle tone. In addition, even though we are generally not conscious of breathing, the muscles that generate breathing movements are skeletal muscles controlled by somatic motor neurons. If the respiratory motor neurons become inactive, breathing stops.

The input to the ANS also comes from ***sensory neurons.*** Many of these neurons are associated with sensory receptors that monitor internal conditions, such as blood CO_2 level or the degree of stretching in the walls of internal organs or blood vessels. When the viscera are functioning properly, these sensory signals usually are not consciously perceived.

Autonomic motor neurons regulate ongoing activities in their effector tissues, which are cardiac muscle, smooth muscle, and glands, by both excitation and inhibition. Unlike skeletal muscle, these tissues often function to some extent even if their nerve supply is damaged. The heart continues to beat, for instance, when it is removed for transplantation into another person. Examples of autonomic responses are changes in the diameter of the pupil, dilation and constriction of blood vessels, and changes in the rate and force of the heartbeat. Because most autonomic responses cannot be consciously altered or suppressed to any great degree, they are the basis for polygraph ("lie detector") tests. However, practitioners of yoga or other techniques of meditation and those who employ biofeedback methods may learn how to modulate ANS activities. For example, they may be able to voluntarily decrease their heart rate or blood pressure.

Figure 18.1 compares somatic and autonomic motor neurons. The axon of a somatic motor neuron extends all the way from the central nervous system (CNS) to the skeletal muscle fibers that it stimulates (Figure 18.1a). By contrast, autonomic motor pathways consist of sets of *two* motor neurons (Figure 18.1b). The first neuron, called the ***preganglionic neuron,*** has its cell body in the CNS, either in the lateral gray horn of the spinal cord or in a nucleus of the brain stem. Its axon extends from the CNS as part of a cranial or spinal nerve to an ***autonomic ganglion,*** where it synapses with the second neuron. (Recall that a ganglion is a collection of neuronal cell bodies usually outside the CNS.) The second neuron, the ***postganglionic neuron,*** lies entirely outside the CNS in the peripheral nervous system (PNS). Its cell body is located in an autonomic ganglion, and its axon extends from the ganglion to the effector (smooth muscle, cardiac muscle, or a gland). The effect of the postganglionic neuron on the effector may be either excitation (causing contraction of smooth or cardiac muscle or increasing secretions of glands) or inhibition (causing relaxation of smooth or cardiac muscle or decreasing secretions of glands). In contrast, a single somatic motor neuron extends from the CNS and always excites its effector (causing contraction of skeletal muscle) (Figure 18.1a). Another difference between autonomic and somatic motor neurons is that all somatic motor neurons release acetylcholine (ACh) as their neurotransmitter. Some autonomic motor neurons release ACh; others release norepinephrine (NE).

The output (motor) part of the ANS has two main branches: the ***sympathetic division*** and the ***parasympathetic division.*** Many organs have ***dual innervation;*** that is, they receive impulses from both sympathetic and parasympathetic neurons. In general, nerve impulses from one division stimulate the organ to increase its activity (excitation), whereas impulses from the other division decrease the organ's activity (inhibition). For example, an increased rate of nerve impulses from the sympathetic division increases heart rate, and an increased rate of nerve impulses from the parasympathetic division decreases heart rate.

Figure 18.1 Motor neuron pathways in the (a) somatic nervous system and (b) autonomic nervous system (ANS). Note that somatic motor neurons release acetylcholine (ACh); autonomic motor neurons release either ACh or norepinephrine (NE).

Somatic nervous system stimulation always excites its effectors (skeletal muscle fibers); stimulation by the autonomic nervous system either excites or inhibits visceral effectors.

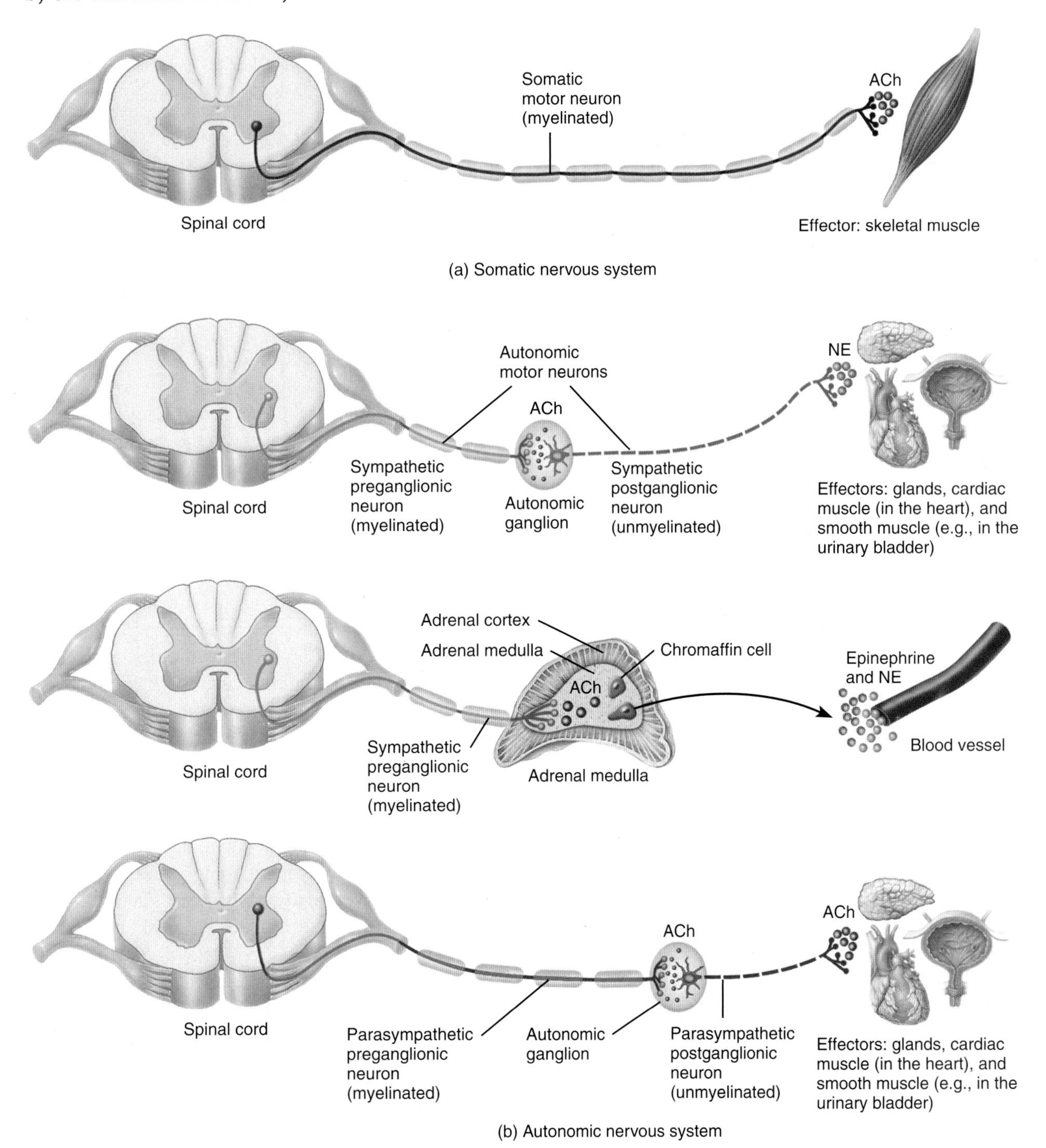

What does dual innervation mean?

TABLE 18.1

Comparison of Somatic and Autonomic Nervous Systems

PROPERTY	SOMATIC	AUTONOMIC
Effectors	Skeletal muscles.	Cardiac muscle, smooth muscle, and glands.
Type of control	Mainly voluntary.	Mainly involuntary.
Neural pathway	One motor neuron extends from CNS and synapses directly with a skeletal muscle fiber.	One motor neuron extends from the CNS and synapses with another motor neuron in a ganglion; the second motor neuron synapses with an autonomic effector.
Neurotransmitter	Acetylcholine.	Acetylcholine or norepinephrine.
Action of neurotransmitter on effector	Always excitatory (causing contraction of skeletal muscle).	May be excitatory (causing contraction of smooth muscle, increased heart rate, increased force of heart contraction, or increased secretions from glands) or inhibitory (causing relaxation of smooth muscle, decreased heart rate, or decreased secretions from glands).

Table 18.1 summarizes the similarities and differences between the somatic and autonomic nervous systems.

CHECKPOINT

1. Why is the autonomic nervous system so named?
2. What are the main input and output components of the autonomic nervous system?

18.3 STRUCTURE OF THE AUTONOMIC NERVOUS SYSTEM

OBJECTIVE

- Identify the structural features of the autonomic nervous system.

We now examine the location and distribution of preganglionic neurons, ganglia, and postganglionic neurons and how they relate to the activities of the autonomic nervous system.

Organization of the Sympathetic Division

The ***sympathetic division*** of the ANS is also called the *thoracolumbar division* (thōr′-a-kō-LUM-bar) because the outflow of sympathetic nerve impulses comes from the thoracic and lumbar segments of the spinal cord (Figure 18.2). The sympathetic preganglionic neurons have their cell bodies in the 12 thoracic and the first 2 (and sometimes 3) lumbar segments of the spinal cord. The preganglionic axons emerge from the spinal cord through the anterior root of a spinal nerve along with axons of somatic motor neurons. After exiting the cord, the sympathetic preganglionic axons extend to a sympathetic ganglion.

In the sympathetic ganglia, sympathetic preganglionic neurons synapse with postganglionic neurons. Because the sympathetic trunk ganglia are near the spinal cord, most sympathetic preganglionic axons are short. ***Sympathetic trunk ganglia,*** also known as ***vertebral chain ganglia*** and ***paravertebral ganglia,*** lie in two vertical rows, one on either side of the vertebral column (Figure 18.2). Most postganglionic axons emerging from sympathetic trunk ganglia supply organs above the diaphragm. Other sympathetic ganglia, the ***prevertebral ganglia,*** also known as ***collateral ganglia,*** lie anterior to the vertebral column and close to the large abdominal arteries. These include the *celiac ganglion* (SĒ-lē-ak), the *superior mesenteric ganglion,* the *inferior mesenteric ganglion,* the *aorticorenal ganglion,* and the *renal ganglion.* In general, postganglionic axons emerging from the prevertebral ganglia innervate organs below the diaphragm.

The cervical portion of each sympathetic trunk is located in the neck and is subdivided into superior, middle, and inferior ganglia (Figure 18.2). Postganglionic neurons leaving the ***superior cervical ganglion*** serve the head and heart. They are distributed to sweat glands, smooth muscles of the eye, blood vessels of the face, lacrimal glands, nasal mucosa, the heart, and the submandibular, sublingual, and parotid salivary glands. Postganglionic neurons leaving the ***middle cervical ganglion*** and the ***inferior cervical ganglion*** innervate the heart.

In the thoracic region, postganglionic axons from the sympathetic trunk serve the heart, lungs, and bronchi. Some axons from thoracic levels also supply sweat glands, blood vessels, and smooth muscles of hair follicles in the skin. In the abdomen, axons of postganglionic neurons leaving the prevertebral ganglia follow the course of various arteries to abdominal and pelvic autonomic effectors.

A single sympathetic preganglionic axon has many branches and may synapse with 20 or more postganglionic neurons. Thus, nerve impulses that arise in a single preganglionic neuron may activate many different postganglionic neurons that in turn synapse with several autonomic effectors. This pattern helps explain why sympathetic responses can affect organs throughout the body almost simultaneously.

Some sympathetic preganglionic axons pass through the sympathetic trunk without terminating in it. Beyond the trunk, they form nerves known as ***splanchnic nerves*** (SPLANK-nik; Figure 18.2 and see Figure 18.4), which extend to and terminate in the outlying prevertebral ganglia. Some sympathetic preganglionic

Figure 18.2 Structure of the sympathetic division of the autonomic nervous system. Solid lines represent preganglionic axons; dashed lines represent postganglionic axons. Although the innervated structures are shown for one side of the body for diagrammatic purposes, the sympathetic division actually innervates tissues and organs on both sides.

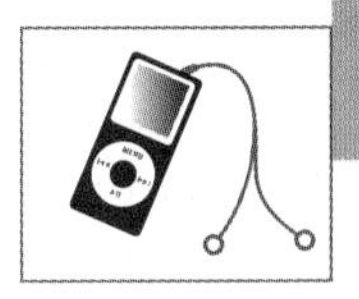

Cell bodies of sympathetic preganglionic neurons are located in the lateral horn gray matter in the 12 thoracic and first 2 (and sometimes 3) lumbar segments of the spinal cord.

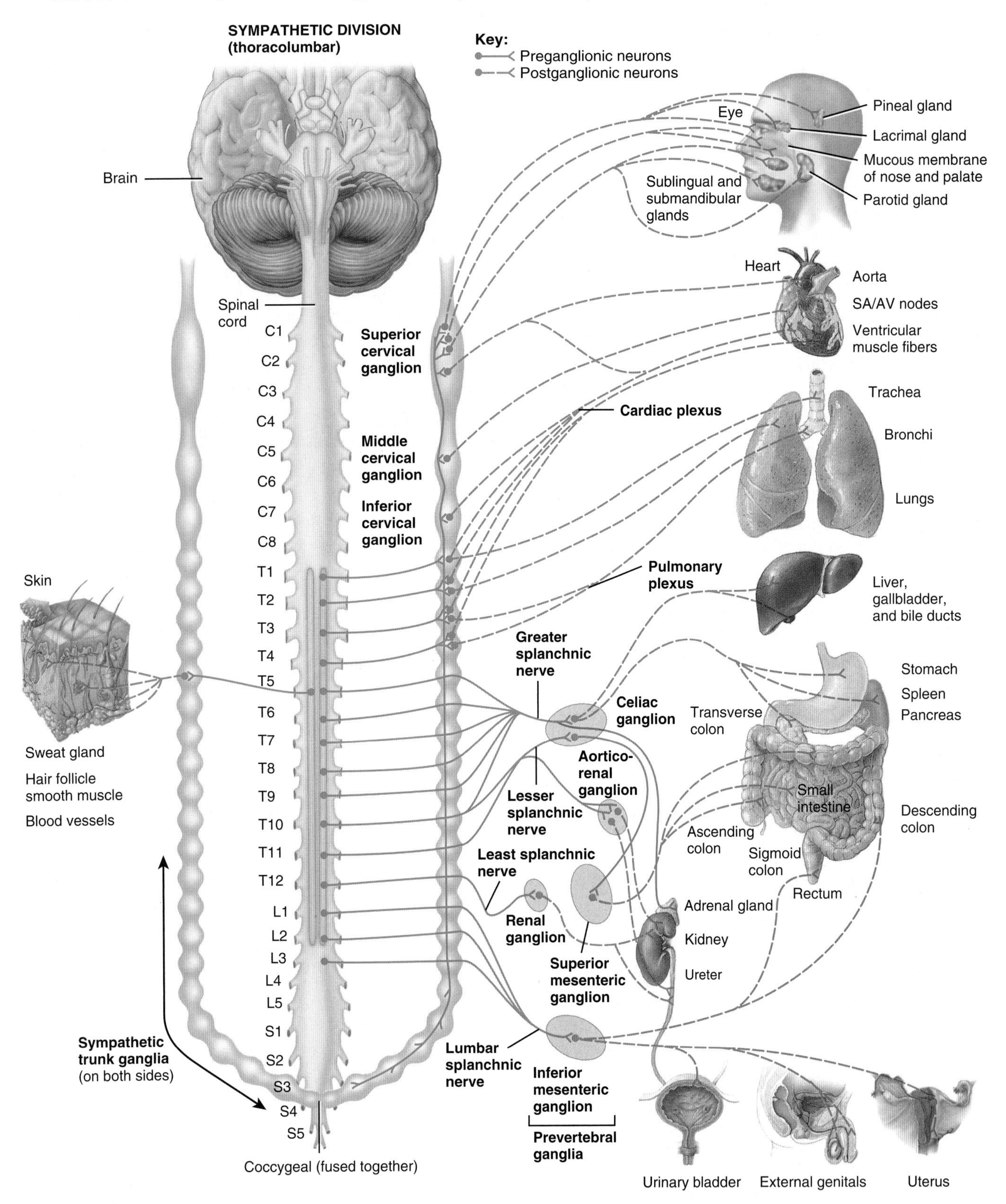

Which division, sympathetic or parasympathetic, has longer preganglionic axons? Why?

neurons that enter the sympathetic trunk ascend to the superior cervical ganglion, where they synapse with postganglionic neurons. The axons of some of these postganglionic neurons leave the sympathetic trunk by forming cephalic periarterial nerves, nerves that extend to the head by wrapping around and following the course of various arteries (such as the carotid arteries) that pass from the neck to the head (see Figure 18.4). Additional splanchnic nerves are located within the abdominopelvic cavity (see Figures 18.2 and 18.3).

Once the axon of a preganglionic neuron of the sympathetic division enters a sympathetic trunk ganglion, it may follow one of four paths (Figure 18.3):

1. An axon may synapse with postganglionic neurons in the sympathetic trunk ganglion that it first reaches.
2. An axon may ascend or descend to a higher or lower sympathetic trunk ganglion before synapsing with postganglionic neurons. The axons of incoming sympathetic preganglionic neurons that pass up or down the sympathetic trunk collectively form the ***sympathetic chains,*** the fibers on which the ganglia are strung.
3. An axon may continue, without synapsing, through the sympathetic trunk ganglion to end at a prevertebral ganglion and synapse with postganglionic neurons there.
4. An axon may also pass, without synapsing, through the sympathetic trunk ganglion and a prevertebral ganglion and then extend to cells in the adrenal medullae called chromaffin cells that are functionally similar to sympathetic postganglionic neurons.

After exiting through the intervertebral foramina, the myelinated preganglionic sympathetic axons enter a short pathway called a *white ramus* before passing to the nearest sympathetic trunk ganglion on the same side (Figure 18.3). Collectively, the white rami are called the ***white rami communicantes*** (kō-mū-ni-KAN-tēz; singular is *ramus communicans*). Thus, white rami communicantes are structures containing sympathetic preganglionic axons that connect the anterior ramus of the spinal nerve with the ganglia of the sympathetic trunk. The "white" in their name indicates that they contain myelinated axons. Only the thoracic and first two or three lumbar nerves have white rami communicantes. The axons of some postganglionic neurons leave the sympathetic trunk by entering a short pathway called a ***gray ramus*** and merge with the anterior ramus of a spinal nerve to supply visceral effectors such as sweat glands, smooth muscle in blood vessels, and arrector pili muscles of hair follicles. Therefore, ***gray rami communicantes*** are structures containing sympathetic postganglionic axons that connect the ganglia of the sympathetic trunk to spinal nerves. The "gray" in their name indicates that they contain unmyelinated axons. Gray rami communicantes outnumber the white rami because there is a gray ramus leading to each of the 31 pairs of spinal nerves.

The sympathetic division of the ANS also includes part of the adrenal glands (see Figure 18.2). The inner part of the adrenal gland, the ***adrenal medulla*** (me-DUL-a), develops from the same embryonic tissue as the sympathetic ganglia, and its cells are similar to sympathetic postganglionic neurons. Rather than extending to another organ, however, these cells release hormones into the blood. Upon stimulation by sympathetic preganglionic neurons, cells of the adrenal medulla release a mixture of hormones—about 80% ***epinephrine*** and 20% ***norepinephrine.*** These hormones circulate throughout the body and intensify responses elicited by sympathetic postganglionic neurons.

Figure 18.3 Types of connections between ganglia and postganglionic neurons in the sympathetic division of the ANS.
Numbers correspond to descriptions in the text. Also illustrated are gray and white rami communicantes.

Sympathetic ganglia lie in two chains on either side of the vertebral column (see Figure 16.6) and near large abdominal arteries that are anterior to the vertebral column.

• CLINICAL CONNECTION | Autonomic Dysreflexia

Autonomic dysreflexia is an exaggerated response of the sympathetic division of the ANS that occurs in about 85% of individuals with spinal cord injury at or above the level of T6. The condition occurs due to interruption of the control of ANS neurons by higher centers. When certain sensory impulses, such as those resulting from stretching of a full urinary bladder, are unable to ascend the spinal cord, mass stimulation of the sympathetic nerves below the level of injury occurs. Among the effects of increased sympathetic activity is severe vasoconstriction, which elevates blood pressure. In response, the cardiovascular center in the medulla oblongata (1) increases parasympathetic output via the vagus nerve, which decreases heart rate, and (2) decreases sympathetic output, which causes dilation of blood vessels above the level of the injury. Autonomic dysreflexia is characterized by a pounding headache; severe high blood pressure (hypertension); flushed, warm skin with profuse sweating above the injury level; pale, cold, and dry skin below the injury level; and anxiety. It is an emergency condition that requires immediate intervention. If untreated, autonomic dysreflexia can cause seizures, stroke, or heart attack. •

Organization of the Parasympathetic Division

The ***parasympathetic division*** is also called the *craniosacral division* (krā′-nē-ō-SĀ-kral) because the outflow of parasympathetic nerve impulses comes from certain cranial nerve nuclei and sacral segments of the spinal cord. The cell bodies of parasympathetic preganglionic neurons are located in the nuclei of four cranial nerves (III, VII, IX, and X) in the brain stem and in the second through fourth sacral segments of the spinal cord (S2, S3, and S4) (see Figure 18.4). Parasympathetic preganglionic axons emerge from the CNS as part of a cranial nerve or as part of the anterior root of a spinal nerve. Axons of the vagus (X)

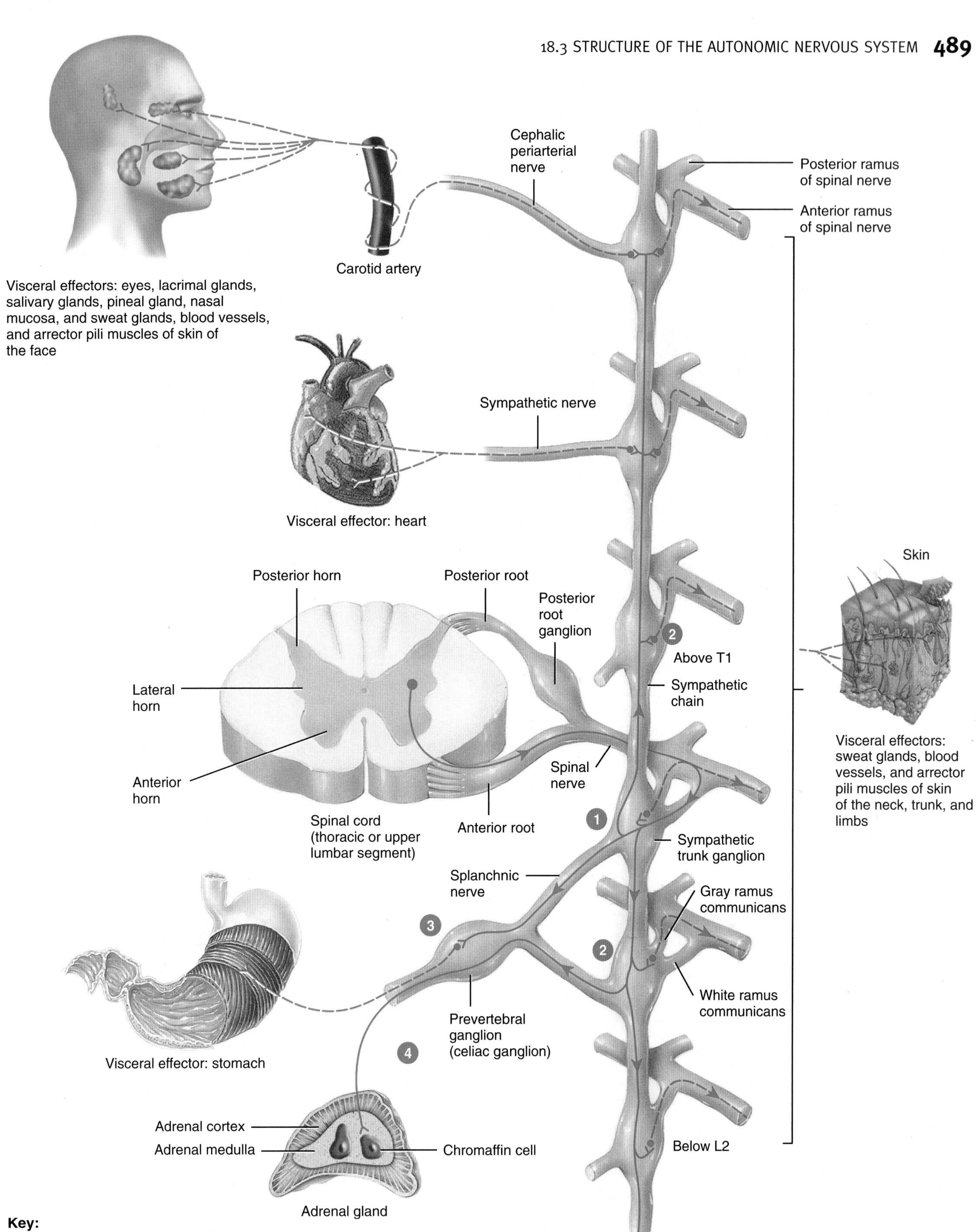

Which organs are supplied by most postganglionic axons of sympathetic trunk ganglia?

Figure 18.4 Structure of the parasympathetic division of the autonomic nervous system. Solid lines represent preganglionic axons; dashed lines represent postganglionic axons. Although the innervated structures are shown for only one side of the body for diagrammatic purposes, the parasympathetic division actually innervates tissues and organs on both sides.

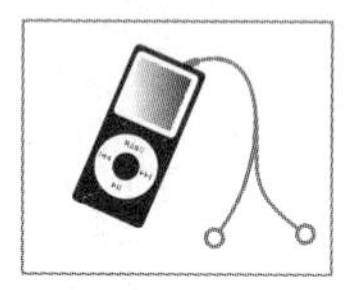

Cell bodies of parasympathetic preganglionic neurons are located in brain stem nuclei and in the lateral gray matter in the second through fourth sacral segments of the spinal cord.

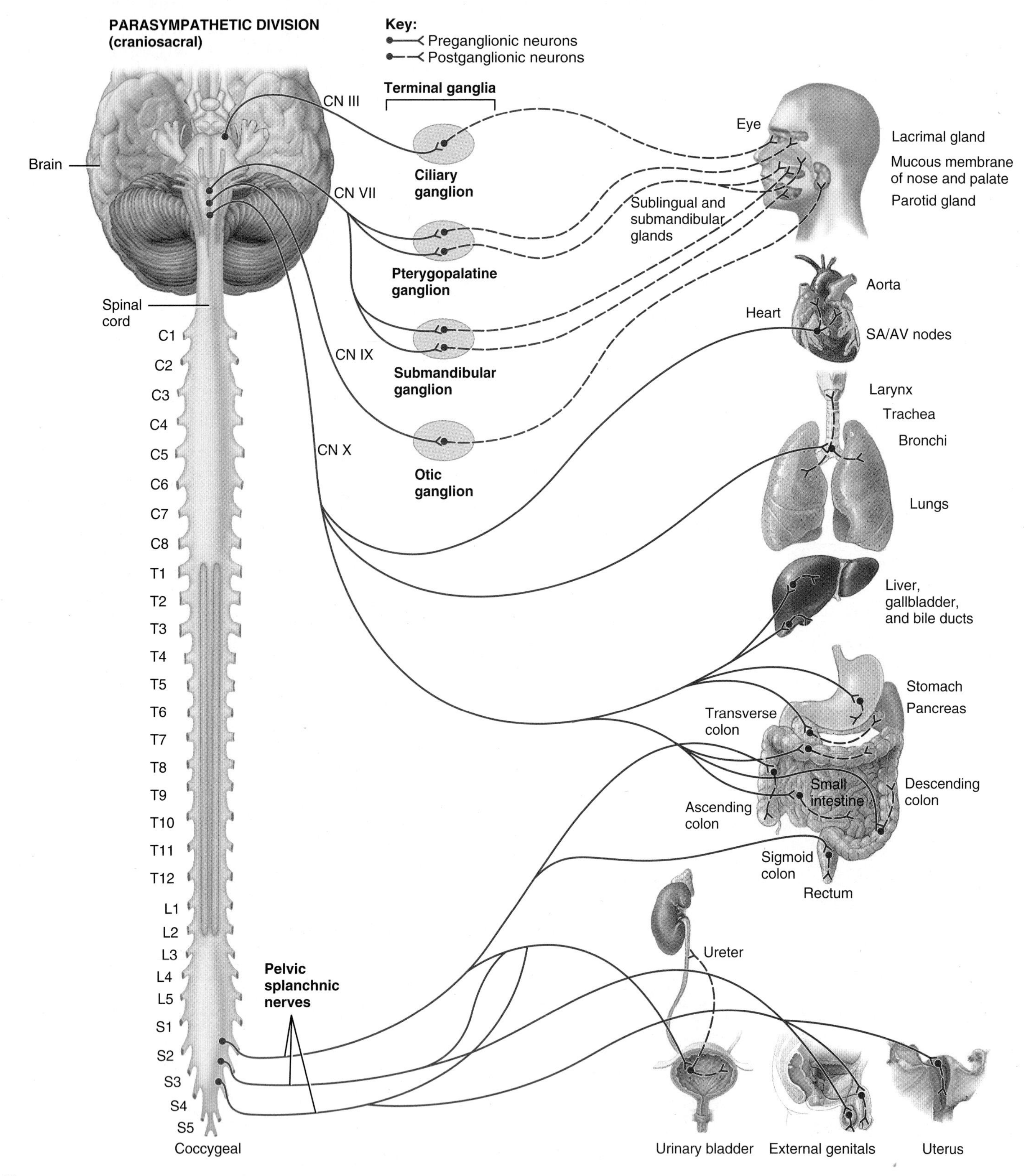

Which ganglia are associated with the parasympathetic division? Sympathetic division?

nerve carry nearly 80% of the total parasympathetic outflow. In the thorax, axons of the vagus nerve extend to ganglia in the heart and the airways of the lungs. In the abdomen, axons of the vagus nerve extend to ganglia in the liver, stomach, pancreas, small intestine, and part of the large intestine. Parasympathetic preganglionic axons exit the sacral spinal cord in the anterior roots of the second through fourth sacral nerves. The axons then extend to ganglia in the walls of the colon, ureters, urinary bladder, and reproductive organs.

Preganglionic axons of the parasympathetic division synapse with postganglionic neurons in ***terminal ganglia,*** which are located close to or actually within the wall of the innervated organ. Terminal ganglia in the head receive preganglionic axons from the oculomotor (III), facial (VII), or glossopharyngeal (IX) cranial nerves and supply structures in the head (see Figure 18.4). Axons in the vagus (X) nerve extend to many terminal ganglia in the thorax and abdomen. Because the axons of parasympathetic preganglionic neurons extend from the brain stem or sacral spinal cord to a terminal ganglion in an innervated organ, they are longer than most of the axons of sympathetic preganglionic neurons (compare Figures 18.2 and 18.4).

In contrast to the preganglionic axons, most parasympathetic postganglionic axons are very short because the terminal ganglia lie in the walls of their autonomic effectors. In the ganglion, the preganglionic neuron usually synapses with only four or five postganglionic neurons, all of which supply the same effector. Thus, parasympathetic responses are localized to a single effector.

CHECKPOINT

3. Describe the locations of sympathetic trunk ganglia, prevertebral ganglia, and terminal ganglia. Which types of autonomic neurons synapse in each type of ganglion?
4. How can the sympathetic division produce simultaneous effects throughout the body, when parasympathetic effects typically are localized to specific organs?

18.4 FUNCTIONS OF THE AUTONOMIC NERVOUS SYSTEM

OBJECTIVE

- Describe the functions of the sympathetic and parasympathetic divisions of the autonomic nervous system.

ANS Neurotransmitters

Neurotransmitters are chemical substances released by neurons at synapses. Autonomic neurons release neurotransmitters at synapses between neurons (preganglionic to postganglionic) and at synapses with autonomic effectors. Some ANS neurons release acetylcholine; others release norepinephrine (see Figure 18.1).

ANS neurons that release *acetylcholine* include (1) all sympathetic and parasympathetic preganglionic neurons, (2) all parasympathetic postganglionic neurons, and (3) a few sympathetic postganglionic neurons. Because acetylcholine is quickly inactivated by the enzyme *acetylcholinesterase (AChE),* parasympathetic effects are short-lived and localized.

Most sympathetic postganglionic neurons release the neurotransmitter *norepinephrine (NE).* Because norepinephrine is inactivated much more slowly than acetylcholine and because the adrenal medulla also releases epinephrine and norepinephrine into the bloodstream, the effects of activation of the sympathetic division are longer lasting and more widespread than those of the parasympathetic division. For instance, your heart continues to pound for several minutes after a near miss at a busy intersection due to the long-lasting effects of the sympathetic division.

Activities of the ANS

As noted earlier, many body organs receive instructions from both divisions of the ANS, which typically work in opposition to one another. The balance between sympathetic and parasympathetic activity or "tone" is regulated by the hypothalamus. Typically, the hypothalamus turns up sympathetic tone at the same time it turns down parasympathetic tone, and vice versa. A few structures receive only sympathetic innervation—sweat glands, arrector pili muscles attached to hair follicles in the skin, the kidneys, the spleen, most blood vessels, and the adrenal medullae (see Figure 18.2). In these structures there is no opposition from the parasympathetic division. Still, an increase in sympathetic tone has one effect, and a decrease in sympathetic tone produces the opposite effect.

Sympathetic Activities

During physical or emotional stress, high sympathetic tone favors body functions that can support vigorous physical activity and rapid production of ATP. At the same time, the sympathetic division reduces body functions that favor the storage of energy. Besides physical exertion, a variety of emotions—such as fear, embarrassment, or rage—stimulate the sympathetic division. Visualizing body changes that occur during "E situations" (exercise, emergency, excitement, embarrassment) will help you remember most of the sympathetic responses. Activation of the sympathetic division and release of hormones by the adrenal medullae result in a series of physiological responses collectively called the ***fight-or-flight response,*** in which the following occur:

1. The pupils of the eyes dilate.
2. Heart rate, force of heart contraction, and blood pressure increase.
3. The airways dilate, allowing faster movement of air into and out of the lungs.
4. The blood vessels that supply nonessential organs such as the kidneys and gastrointestinal tract constrict, which reduces blood flow through these tissues. The result is a slowing of urine formation and digestive activities, which are not essential during exercise.
5. Blood vessels that supply organs involved in exercise or fighting off danger—skeletal muscles, cardiac muscle, liver, and adipose tissue—dilate, which allows greater blood flow through these tissues.

6. Liver cells break down glycogen to glucose, and adipose cells break down triglycerides to fatty acids and glycerol, providing molecules that can be used by body cells for ATP production.

7. Release of glucose by the liver increases the blood glucose level.

8. Processes that are not essential for meeting the stressful situation are inhibited. For example, muscular movements of the gastrointestinal tract and digestive secretions decrease or even stop.

MANUAL THERAPY APPLICATION

Mind–Body Exercise: An Antidote to Stress

When we think of exercise, we usually think of toning up our muscles and maybe our hearts. When other people think of exercise, their focus is on toning up neural input from the parasympathetic division of the autonomic nervous system. As described in this chapter, activation of the parasympathetic division helps restore homeostasis in many systems and is associated with feelings of relaxation.

Mind–Body Harmony

Mind–body exercise refers to exercise systems such as tai chi, hatha yoga, and many forms of the martial arts that couple muscular activity with an internally directed focus. These exercise systems are directed at improving the mind as well as the body. Their internally directed focus usually includes an awareness of breathing, energy, and other physical sensations. Practitioners often refer to this internal awareness as "mindfulness," meaning that the exerciser is open to physical and emotional sensations with an understanding, nonjudgmental attitude. A mindful attitude is typical of many kinds of meditation and relaxation practices. For example, when practicing a yoga pose, you would think something like "Deep, steady breathing; relax into the pose; shoulders pulling back, neck lengthening," rather than "That person next to me sure is flexible; I'm really a failure at this stuff." Of course, in real life such external thoughts do sneak in, but practitioners can redirect their attention back to a more neutral style.

Mind–Body Benefits

People practicing mind–body activities reap benefits from both the physical and mental activity. These practices increase muscular strength and flexibility, posture, balance, and coordination, and if performed vigorously, they can even improve cardiovascular health and endurance to some extent. In addition, the stress relief provided by the activity extends into both physical and psychological realms. Feelings of mental relaxation and emotional well-being translate into better resting blood pressure, a healthier immune system, and more relaxed muscles. Less stress can also mean an improvement in health habits. Those therapists or patients who practice mind-body exercise often improve their eating habits and reduce harmful behaviors such as cigarette smoking.

Parasympathetic Activities

In contrast to the fight-or-flight activities of the sympathetic division, the parasympathetic division enhances ***rest-and-digest*** activities. Parasympathetic responses support body functions that conserve and restore body energy during times of rest and recovery. In the quiet intervals between periods of exercise, parasympathetic impulses to the digestive glands and the smooth muscle of the gastrointestinal tract predominate over sympathetic impulses. This allows energy-supplying food to be digested and absorbed. At the same time, parasympathetic responses reduce body functions that support physical activity.

The acronym *SLUDD* can be helpful in remembering five parasympathetic responses. It stands for salivation (S), lacrimation (L), urination (U), digestion (D), and defecation (D). Mainly the parasympathetic division stimulates all of these activities. Besides the increasing SLUDD responses, other important parasympathetic responses are "three decreases": decreased heart rate, decreased diameter of airways, and decreased diameter (constriction) of the pupils.

Table 18.2 lists the responses of glands, cardiac muscle, and smooth muscle to stimulation by the sympathetic and parasympathetic divisions of the ANS.

• CLINICAL CONNECTION | Raynaud Phenomenon

In **Raynaud phenomenon** (rā-NŌ), the fingers and toes become ischemic (lack blood) after exposure to cold or with emotional stress. The condition is due to excessive sympathetic stimulation of smooth muscle in the arterioles of the fingers and toes. When the arterioles constrict in response to sympathetic stimulation, blood flow is greatly diminished. Symptoms are colorful—red, white, and blue. Fingers and toes may look white due to blockage of blood flow or look blue (cyanotic) due to deoxygenated blood in capillaries. With rewarming after cold exposure, the arterioles may dilate, causing the fingers and toes to look red. The disorder is most common in young women and occurs more often in cold climates. •

CHECKPOINT

5. What are some examples of the opposite effects of the sympathetic and parasympathetic divisions of the autonomic nervous system?
6. What happens during the fight-or-flight response?
7. Why is the parasympathetic division of the ANS considered the rest-and-digest division?

18.5 INTEGRATION AND CONTROL OF AUTONOMIC FUNCTIONS

OBJECTIVES

- Describe the components of an autonomic reflex.
- Explain the relationship of the hypothalamus to the ANS.

Autonomic Reflexes

Autonomic reflexes are responses that occur when nerve impulses pass through an autonomic reflex arc. These reflexes play a key role in regulating controlled conditions in the body, such as *blood pressure,* by adjusting heart rate, force of ventricular contraction, and blood vessel diameter; *digestion,* by adjusting

TABLE 18.2

Functions of the Autonomic Nervous System

EFFECTOR	EFFECT OF SYMPATHETIC STIMULATION	EFFECT OF PARASYMPATHETIC STIMULATION
GLANDS		
Sweat	Increased sweating.	No known effect.
Lacrimal (tear)	Slight secretion of tears.	Secretion of tears.
Adrenal medulla	Secretion of epinephrine and norepinephrine.	No known effect.
Pancreas	Inhibition of secretion of digestive enzymes and insulin (hormone that lowers blood glucose level); secretion of glucagon (hormone that raises blood glucose level).	Secretion of digestive enzymes and insulin.
Posterior pituitary	Secretion of antidiuretic hormone (ADH).	No known effect.
Liver*	Breakdown of glycogen into glucose, synthesis of new glucose, and release of glucose into the blood; decreases bile secretion.	Promotes synthesis of glycogen; increases bile secretion.
Adipose tissue*	Breakdown of triglycerides and release of fatty acids into blood.	No known effect.
CARDIAC MUSCLE		
Heart	Increased heart rate and increased force of contraction.	Decreased heart rate and decreased force of contraction.
SMOOTH MUSCLE		
Radial muscle of iris of eye	Dilation of the pupil.	No known effect.
Circular muscle of iris of eye	No known effect.	Constriction of the pupil.
Ciliary muscle of eye	Relaxation to adjust shape of lens for distant vision.	Contraction to adjust shape of lens for close vision.
Gallbladder and ducts	Relaxation to facilitate storage of bile in the gallbladder.	Contraction, enabling release of bile into the small intestine.
Stomach and intestines	Decreased motility (movement); contraction of sphincters.	Increased motility; relaxation of sphincters.
Lungs (smooth muscle of bronchi)	Widening of the airways (bronchodilation).	Narrowing of the airways (bronchoconstriction).
Urinary bladder	Relaxation of muscular wall; contraction of internal sphincter.	Contraction of muscular wall; relaxation of internal sphincter.
Spleen	Contraction and discharge of stored blood into general circulation.	No known effect.
Smooth muscle of hair follicles	Contraction that results in erection of hairs, producing goose bumps.	No known effect.
Uterus	Inhibits contraction in nonpregnant women; stimulates contraction in pregnant women.	Minimal effect.
Sex organs	In men, causes ejaculation of semen.	Vasodilation; erection of clitoris (women) and penis (men).
Salivary glands (arterioles)	Decreases secretion of saliva.	Stimulates secretion of saliva.
Gastric glands and intestinal glands (arterioles)	Inhibits secretion.	Promotes secretion.
Kidney (arterioles)	Decreases production of urine.	No known effect.
Skeletal muscle (arterioles)	Vasodilation in most, which increases blood flow.	No known effect.
Heart (coronary arterioles)	Vasodilation in most, which increases blood flow.	Causes slight constriction, which decreases blood flow.

*Listed with glands because they release substances into the blood.

the motility (movement) and muscle tone of the gastrointestinal tract; and *defecation* and *urination,* by regulating the opening and closing of sphincters.

The components of an autonomic reflex arc are as follows:

- **Receptor.** Like the receptor in a somatic reflex arc (see Figure 16.17), the receptor in an autonomic reflex arc is the distal end of a sensory neuron, which responds to a stimulus and produces a change that will ultimately trigger nerve impulses. Autonomic sensory receptors are mostly associated with interoceptors.
- **Sensory neuron.** Conducts nerve impulses from receptors to the CNS.

- **Integrating center.** Interneurons within the CNS relay signals from sensory neurons to motor neurons. The main integrating centers for most autonomic reflexes are located in the hypothalamus and brain stem. Some autonomic reflexes, such as those for urination and defecation, have integrating centers in the spinal cord.
- **Motor neurons.** Nerve impulses triggered by the integrating center propagate out of the CNS along motor neurons to an effector. In an autonomic reflex arc, two motor neurons connect the CNS to an effector: The preganglionic neuron conducts motor impulses from the CNS to an autonomic ganglion, and the postganglionic neuron conducts motor impulses from an autonomic ganglion to an effector (see Figure 18.1).
- **Effector.** In an autonomic reflex arc, the effectors are smooth muscle, cardiac muscle, and glands, and the reflex is called an autonomic reflex.

Autonomic Control by Higher Centers

Normally, we are not aware of muscular contractions of our digestive organs, our heartbeat, changes in the diameter of our blood vessels, and pupil dilation and constriction because the integrating centers for these autonomic responses are in the spinal cord or the lower regions of the brain. Somatic or autonomic sensory neurons deliver input to these centers, and autonomic motor neurons provide output that adjusts activity in the visceral effector, usually without our conscious perception.

The hypothalamus is the major control and integration center of the ANS. The hypothalamus receives sensory input related to visceral functions, olfaction (smell), and gustation (taste), as well as changes in temperature, osmolarity, and levels of various substances in blood. It also receives input relating to emotions from the limbic system. Output from the hypothalamus influences autonomic centers both in the brain stem (such as the cardiovascular, salivation, swallowing, and vomiting centers) and the spinal cord (such as the defecation and urination reflex centers in the sacral spinal cord).

Anatomically, the hypothalamus is connected to both the sympathetic and parasympathetic divisions of the ANS by axons of neurons with dendrites and cell bodies in various hypothalamic nuclei. The axons form tracts from the hypothalamus to sympathetic and parasympathetic nuclei in the brain stem and spinal cord through relays in the reticular formation. The posterior and lateral parts of the hypothalamus control the sympathetic division. Stimulation of these areas produces an increase in heart rate and force of contraction, a rise in blood pressure due to constriction of blood vessels, an increase in body temperature, dilation of the pupils, and inhibition of the gastrointestinal tract. In contrast, the anterior and medial parts of the hypothalamus control the parasympathetic division. Stimulation of these areas results in a decrease in heart rate, lowering of blood pressure, constriction of the pupils, and increased secretion and motility of the gastrointestinal tract.

CHECKPOINT

8. Give three examples of controlled conditions in the body that are kept in homeostatic balance by autonomic reflexes.
9. How does an autonomic reflex arc differ from a somatic reflex arc?

KEY MEDICAL TERMS ASSOCIATED WITH THE AUTONOMIC NERVOUS SYSTEM

Autonomic nerve neuropathy (noo-ROP-a-thē) If a neuropathy (specifically a disorder of a cranial or spinal nerve) affects one or more autonomic nerves, there can be multiple effects on the autonomic nervous system that interfere with reflexes. These include fainting and low blood pressure when standing (orthostatic hypotension) due to decreased sympathetic control of the cardiovascular system, constipation, urinary incontinence, and impotence. This type of neuropathy is often caused by long-term diabetes mellitus and is known as ***diabetic retinopathy.***

Biofeedback A technique in which an individual is provided with information regarding an autonomic response such as heart rate, blood pressure, or skin temperature. Various electronic monitoring devices provide visual or auditory signals about the autonomic responses. By concentrating on positive thoughts, individuals learn to alter autonomic responses. For example, biofeedback has been used to decrease heart rate and blood pressure and increase skin temperature in order to decrease the severity of migraine headaches.

Dysautonomia (dis-aw-tō-NŌ-mē-a; *dys-* = difficult; *-autonomia* = self-governing) An inherited disorder in which the autonomic nervous system functions abnormally, resulting in reduced tear gland secretions, poor vasomotor control, motor incoordination, skin blotching, absence of pain sensation, difficulty in swallowing, hyporeflexia, excessive vomiting, and emotional instability.

Hyperhidrosis (hi′-per-hī-DRŌ-sis; *hyper-* = above or too much; *-hidrosis* = sweat; *-osis* = condition) Excessive or profuse sweating due to intense stimulation of sweat glands.

Mass reflex In cases of severe spinal cord injury above the level of the sixth thoracic vertebra, stimulation of the skin or overfilling of a visceral organ (such as the urinary bladder or colon) below the level of the injury results in intense activation of autonomic and somatic output from the spinal cord as reflex activity returns. The exaggerated response occurs because there is no inhibitory input from the brain. The mass reflex consists of flexor spasms of the lower limbs, evacuation of the urinary bladder and colon, and profuse sweating below the level of the lesion.

Megacolon (*mega-* = big) An abnormally large colon. In congenital megacolon, parasympathetic nerves to the distal segment of the colon do not develop properly. Loss of motor function in the segment causes massive dilation of the normal proximal colon. The condition results in extreme constipation, abdominal distension, and, occasionally, vomiting. Surgical removal of the affected segment of the colon corrects the disorder.

Reflex sympathetic dystrophy (RSD) A syndrome that includes spontaneous pain, painful hypersensitivity to stimuli such as light touch, and excessive coldness and sweating in the involved body part.

The disorder frequently involves the forearms, hands, knees, and feet. It appears that activation of the sympathetic division of the autonomic nervous system due to traumatized nociceptors as a result of trauma or surgery on bones or joints is involved. Treatment consists of anesthetics and physical therapy. Clinical studies also suggest that the drug baclofen can be used to reduce pain and restore normal function to the affected body part. Also called ***complex regional pain syndrome type 1.***

Vagotomy (vā-GOT-ō-mē; *-tome* = incision) Cutting the vagus (X) nerve. It is frequently done to decrease the production of hydrochloric acid in persons with ulcers.

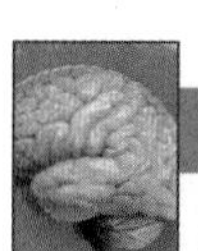

STUDY OUTLINE

Comparison of Somatic and Autonomic Nervous Systems (Section 18.2)

1. The part of the nervous system that regulates smooth muscle, cardiac muscle, and certain glands is the autonomic nervous system (ANS). The ANS usually operates without conscious control from the cerebral cortex, but other brain regions, mainly the hypothalamus and brain stem, regulate it.
2. The axons of somatic motor neurons extend from the CNS and synapse directly with an effector (skeletal muscle). Autonomic motor pathways consist of two motor neurons. The axon of the first motor neuron extends from the CNS and synapses in a ganglion with the second motor neuron; the second neuron synapses with an effector (smooth muscle, cardiac muscle, or a gland).
3. The output (motor) portion of the ANS has two divisions: sympathetic and parasympathetic. Most body organs receive dual innervation; usually one ANS division causes excitation and the other causes inhibition.
4. Somatic motor neurons release acetylcholine (ACh), and autonomic motor neurons release either acetylcholine or norepinephrine (NE).
5. Somatic nervous system effectors are skeletal muscles; ANS effectors include cardiac muscle, smooth muscle, and glands.
6. Table 18.1 compares the somatic and autonomic nervous systems.

Structure of the Autonomic Nervous System (Section 18.3)

1. The sympathetic division of the ANS is also called the thoracolumbar division because the outflow of sympathetic nerve impulses comes from the thoracic and lumbar segments of the spinal cord. Cell bodies of sympathetic preganglionic neurons are in the 12 thoracic and the first 2 (and sometimes 3) lumbar segments of the spinal cord.
2. Sympathetic ganglia are classified as sympathetic trunk ganglia (lateral to the vertebral column) or prevertebral ganglia (anterior to the vertebral column).
3. A single sympathetic preganglionic axon may synapse with 20 or more postganglionic neurons. Sympathetic responses can affect organs throughout the body almost simultaneously.
4. The parasympathetic division is also called the craniosacral division because the outflow of parasympathetic nerve impulses comes from cranial nerve nuclei and from sacral segments of the spinal cord. The cell bodies of parasympathetic preganglionic neurons are located in the nuclei of cranial nerves III, VII, IX, and X in the brain stem and in three sacral segments of the spinal cord (S2, S3, and S4).
5. Parasympathetic ganglia are called terminal ganglia and are located near or within autonomic effectors. Parasympathetic terminal ganglia are close to or in the walls of their autonomic effectors, so most parasympathetic postganglionic axons are very short. In the ganglion, the preganglionic neuron usually synapses with only four or five postganglionic neurons, all of which supply the same effector. Thus, parasympathetic responses are localized to a single effector.

Functions of the Autonomic Nervous System (Section 18.4)

1. Some ANS neurons release acetylcholine, and others release norepinephrine; the result is excitation in some cases and inhibition in others.
2. ANS neurons that release acetylcholine include (1) all sympathetic and parasympathetic preganglionic neurons, (2) all parasympathetic postganglionic neurons, and (3) a few sympathetic postganglionic neurons.
3. Most sympathetic postganglionic neurons release the neurotransmitter norepinephrine (NE). The effects of NE are longer lasting and more widespread than those of acetylcholine.
4. Activation of the sympathetic division causes widespread responses and is referred to as the fight-or-flight response. Activation of the parasympathetic division produces more restricted responses that typically are concerned with rest-and-digest activities.
5. Table 18.2 summarizes the main functions of the sympathetic and parasympathetic divisions of the ANS.

Integration and Control of Autonomic Functions (Section 18.5)

1. An autonomic reflex adjusts the activities of smooth muscle, cardiac muscle, and glands.
2. An autonomic reflex arc consists of a receptor, a sensory neuron, an integrating center, two autonomic motor neurons, and a visceral effector.
3. The hypothalamus is the major control and integration center of the ANS. It is connected to both the sympathetic and the parasympathetic divisions.

SELF-QUIZ QUESTIONS

1. In comparing the somatic nervous system with the autonomic nervous system, which of the following statements is true?
   a. The autonomic nervous system controls involuntary movements in skeletal muscle.
   b. The somatic nervous system controls voluntary activity in glands and smooth muscle.
   c. The autonomic nervous system controls involuntary activity in cardiac muscle, smooth muscle, and glands.
   d. The autonomic nervous system produces voluntary activity in smooth muscle and glands.
   e. The somatic nervous system controls involuntary movements in smooth muscle, cardiac muscle, and glands.

2. Neurons in the autonomic nervous system include
   a. two motor neurons and one ganglion
   b. one motor neuron and two ganglia
   c. two motor neurons and two ganglia
   d. one motor and one sensory neuron, and no ganglia
   e. one motor and one sensory neuron, and one ganglion.
3. Which statement is NOT true?
   a. Most sympathetic postganglionic neurons release norepinephrine.
   b. Parasympathetic preganglionic neurons release acetylcholine.
   c. Sympathetic effects are more localized and short-lived than parasympathetic effects.
   d. The effects from norepinephrine tend to be long-lasting.
   e. Branches of a single postganglionic neuron in the sympathetic division extend to many organs.
4. Which of the following pairs is mismatched?
   a. acetylcholine, parasympathetic nervous system
   b. fight-or-flight, sympathetic nervous system
   c. conserves body energy, parasympathetic nervous system
   d. rest-and-digest, parasympathetic nervous system
   e. norepinephrine, parasympathetic nervous system
5. Which of the following statements is NOT true concerning the autonomic nervous system?
   a. Most autonomic responses cannot be consciously controlled.
   b. In general, if the sympathetic division increases the activity in a specific organ, then the parasympathetic division decreases the activity of that organ.
   c. Sensory receptors monitor internal body conditions.
   d. Sensory neurons include pre- and postganglionic neurons.
   e. Most visceral effectors receive dual innervation.
6. Which part of the central nervous system contains centers that regulate the autonomic nervous system?
   a. hypothalamus b. cerebellum
   c. spinal cord d. basal ganglia
   e. thalamus
7. Place the following structures in the correct order as they relate to an autonomic nervous system response from receipt of the stimulus to response:
   1. visceral effector
   2. centers in the CNS
   3. autonomic ganglion
   4. receptor and autonomic sensory neuron
   5. preganglionic neuron
   6. postganglionic neuron
   a. 4, 5, 2, 3, 6, 1 b. 5, 6, 2, 3, 1, 4
   c. 1, 6, 3, 5, 2, 4 d. 4, 2, 5, 3, 6, 1
   e. 2, 4, 5, 6, 3, 1
8. Which of the following activities would NOT be monitored by autonomic sensory neurons?
   a. carbon dioxide levels in the blood
   b. hearing and equilibrium
   c. blood pressure
   d. stretching of the walls of visceral organs
   e. nausea from damaged viscera
9. The autonomic ganglia associated with the parasympathetic division are the
   a. trunk ganglia b. prevertebral ganglia
   c. posterior root ganglia d. terminal ganglia
   e. basal ganglia
10. Which of these statements about the parasympathetic division of the autonomic nervous system is NOT true? The parasympathetic division
    a. arises from the cranial nerves in the brain stem and sacral spinal cord segments
    b. is concerned with conserving and restoring energy
    c. uses acetylcholine as its neurotransmitter
    d. has ganglia near or within visceral effectors
    e. initiates responses in preganglionic neurons that synapse with 20 or more postganglionic neurons.
11. Which nerve carries most of the parasympathetic output from the brain?
    a. spinal b. vagus
    c. oculomotor d. facial
    e. glossopharyngeal
12. Which of the following would NOT be affected by the autonomic nervous system?
    a. heart b. intestines
    c. urinary bladder d. skeletal muscle
    e. reproductive organs
13. Which of the following neurons release norepinephrine?
    a. somatic motor neurons
    b. sympathetic postganglionic neurons
    c. sympathetic preganglionic neurons
    d. parasympathetic postganglionic neurons
    e. parasympathetic preganglionic neurons
14. Match the following:
    ____ a. cluster of cell bodies outside the CNS
    ____ b. cell body is in ganglion; unmyelinated axon extends to effector
    ____ c. cell body lies inside the CNS; myelinated axon extends to ganglion
    ____ d. their postganglionic axons innervate organs below the diaphragm
    ____ e. their postganglionic axons supply organs above the diaphragm
    ____ f. contain the cell bodies and dendrites of parasympathetic postganglionic neurons

    A. sympathetic trunk ganglia
    B. prevertebral ganglia
    C. ganglion
    D. terminal ganglia
    E. preganglionic neuron
    F. postganglionic neuron
15. For each of the following, place a P if it refers to increased activity of the parasympathetic division or an S if it refers to increased activity of the sympathetic division.
    ____ a. dilates pupils
    ____ b. decreases heart rate
    ____ c. causes bronchoconstriction
    ____ d. stimulates breakdown of triglycerides
    ____ e. inhibits secretion of digestive enzymes and insulin
    ____ f. stimulates the gastrointestinal tract
    ____ g. occurs during exercise
    ____ h. causes release of glucose from the liver
    ____ i. dilates blood vessels to cardiac muscle

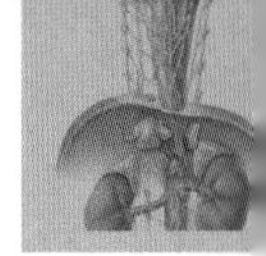

CRITICAL THINKING QUESTIONS

1. It's Thanksgiving and you've just eaten a huge turkey dinner with all the trimmings. Now you're going to watch the big game on TV, if you can make it to the couch! Which division of the nervous system will be handling your body's postdinner activities? Give examples of some organs and the effects on their functions.
2. Anthony wanted a toy on the top of the bookcase, so he climbed up the shelves. His mother ran in when she heard the crash and lifted the heavy bookcase with one arm while pulling her son out with the other. Later that day, she could not lift the bookcase back into position by herself. How do you explain the temporary "supermom" effect?
3. Taylor was watching a scary late-night horror movie when she heard a door slam and a cat's yowl. The hair rose on her arms and she was covered with goose bumps. Trace the pathway taken by the impulses from her CNS to her arms.
4. In the novel *The Hitchhiker's Guide to the Galaxy*, the character Zaphod Beebleborox has two heads and therefore two brains. Is this what is meant by dual innervation? Explain.

? ANSWERS TO FIGURE QUESTIONS

18.1 Dual innervation means that a body organ receives neural innervation from both sympathetic and parasympathetic neurons of the ANS.

18.2 Most parasympathetic preganglionic axons are longer than most sympathetic preganglionic axons; this is because most parasympathetic ganglia are in the walls of visceral organs, but most sympathetic ganglia are close to the spinal cord in the sympathetic trunk.

18.3 Most postganglionic axons of sympathetic trunk ganglia supply organs above the diaphragm.

18.4 Terminal ganglia are associated with the parasympathetic division; sympathetic trunk and prevertebral ganglia are associated with the sympathetic division.

19 Somatic and Special Senses

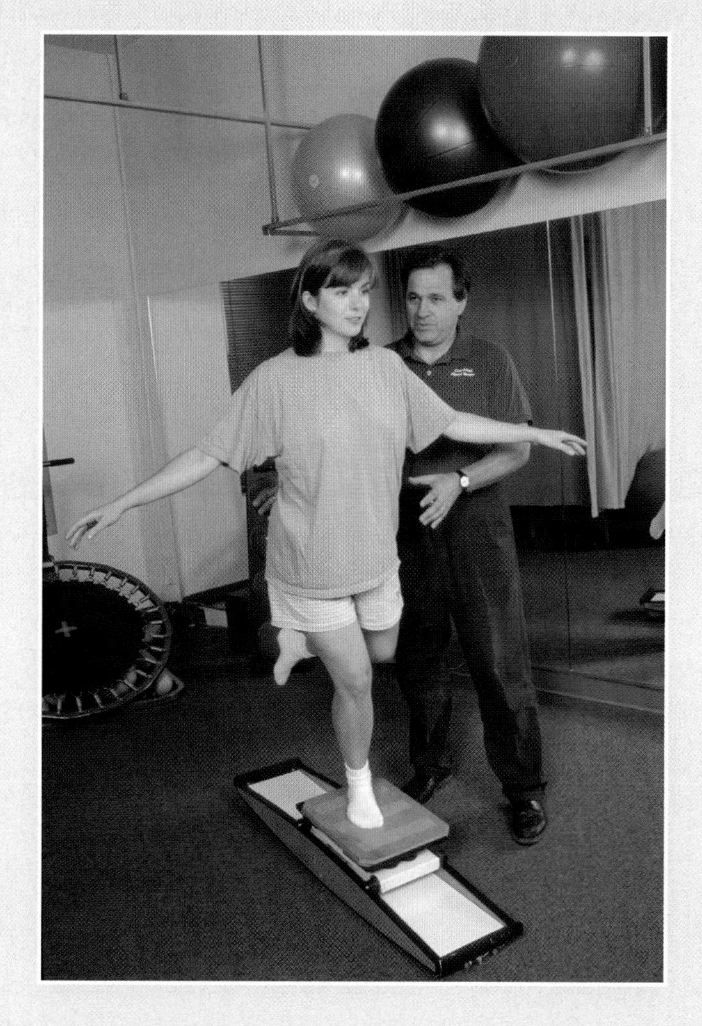

As children we learned the five senses: touch, smell, taste, sight, and hearing. Touch is the first to fully develop in an individual. Slowly, we begin to smell and taste as we eat and drink, and we develop unique likes and dislikes. We begin to develop a good sense of sight, bringing the world from hazy black and white into vivid color. Lastly, we clearly distinguish sound so that we can understand complex sentences.

We often consider balance a sixth sensation. Receptors for balance are activated in the ear when we are moving and in the cerebellum of the brain when we are not moving. Receptors in our skin, other body organs, and virtually all areas of the body are specialized to detect different information that travels along the sensory nerves to the brain and spinal cord to be interpreted. We need to train ourselves to recognize our responses and our sensations, as children and perhaps again later in life after injury, to detect what is a "good" sensation or an uncomfortable one. Balance retraining is often done after injuries like sprains to retrain the tissues of the area to support the actions performed. The acute trauma can become chronic if the area is not retrained. A wobble board is a rehabilitation device often used to treat ankle injuries and other defects with our internal balancing systems.

CONTENTS AT A GLANCE

19.1 OVERVIEW OF SENSATIONS

OBJECTIVE

- Define a sensation and describe the conditions needed for a sensation to occur.

Most of us are aware of sensory input to the central nervous system (CNS) from structures associated with smell, taste, vision, hearing, and equilibrium (balance). These five senses are known as the ***special senses.*** The other senses are termed ***general senses*** and include both somatic senses and visceral senses. ***Somatic senses*** (*somat-* = of the body) include tactile sensations (touch, pressure, and vibration); thermal sensations (warm and cold); pain sensations; and proprioceptive sensations (joint and muscle position and movements of the limbs and head). Each of these senses is considered to be a sensory modality. ***Visceral senses*** provide information about conditions within internal organs. Each unique type of sensation—such as touch, pain, vision, or hearing—is called a ***sensory modality*** (mō-DAL-i-tē).

Definition of Sensation

Sensation is the conscious or subconscious awareness of changes in the external or internal environment. For a sensation to occur, four conditions must be satisfied:

1. A *stimulus,* or change in the environment, capable of activating certain sensory neurons, must occur. A stimulus that activates a sensory receptor may be in the form of light, heat, pressure, mechanical energy, or chemical energy.

2. A *sensory receptor* must convert the stimulus to an electrical signal, which ultimately produces one or more nerve impulses if it is large enough. Most sensory systems are spontaneously active, and thus the stimuli increase or decrease that activity.

3. The nerve impulses must be *conducted* along a neural pathway from the sensory receptor to the brain.

4. A region of the brain must receive and *integrate* the nerve impulses into a sensation.

Characteristics of Sensations

As you learned in Chapter 15, ***perception*** is the conscious awareness and interpretation of sensations and is primarily a function of the cerebral cortex. You seem to see with your eyes, hear with your ears, and feel pain in an injured part of your body. This is because sensory nerve impulses from each part of the body arrive in a specific region of the cerebral cortex, which interprets the sensation as coming from the stimulated sensory receptors. A given sensory neuron carries information for one type of sensation only. Neurons relaying impulses for touch, for example, do not also conduct impulses for pain. The specialization of sensory neurons enables environmental energy to be converted into electrical energy such that light energy is ultimately perceived by the brain as sight and mechanical vibrations are ultimately perceived as sounds.

A characteristic of most sensory receptors is ***adaptation,*** a decrease in the strength of a sensation during a prolonged stimulus. Adaptation is caused in part by a decrease in the responsiveness of sensory receptors. As a result of adaptation, the perception of a sensation may fade or disappear even though the stimulus persists. For example, when you first step into a hot shower, the water may feel very hot, but soon the sensation decreases to one of comfortable warmth even though the stimulus (the high temperature of the water) does not change. Receptors vary in how quickly they adapt. Receptors associated with pressure, touch, and smell adapt rapidly. Slowly adapting receptors monitor stimuli associated with pain, body position, and the chemical composition of the blood.

Types of Sensory Receptors

Both structural and functional characteristics of sensory receptors can be used to group them into different classes (Table 19.1). Structurally, the simplest are ***free nerve endings,*** which are bare dendrites lacking any structural specializations at their ends that can be seen under a light microscope. Receptors for pain, temperature, tickle, itch, and some touch sensations are free nerve endings. Receptors for other somatic and visceral sensations, such as pressure and vibration, and some touch sensations have ***encapsulated nerve endings.*** Their dendrites are enclosed in a connective tissue capsule with a distinctive microscopic structure. Still other sensory receptors consist of specialized, ***separate cells*** that synapse with sensory neurons, for example, hair cells in the inner ear.

Another way to group sensory receptors is functionally—according to the type of stimulus they detect. Most stimuli are in the form of mechanical energy, such as sound waves or pressure changes; electromagnetic energy, such as light or heat; or chemical energy, such as in a molecule of glucose.

- ***Mechanoreceptors*** are sensitive to mechanical stimuli such as the deformation, stretching, or bending of cells. Mechanoreceptors provide sensations of touch, pressure, vibration, proprioception, hearing, and equilibrium. They also monitor the stretching of blood vessels and internal organs.
- ***Thermoreceptors*** detect changes in temperature.
- ***Nociceptors*** respond to painful stimuli resulting from physical or chemical damage to tissue.
- ***Photoreceptors*** detect light that strikes the retina of the eye.
- ***Chemoreceptors*** detect chemicals in the mouth (taste), nose (smell), and body fluids.
- ***Osmoreceptors*** detect the osmotic pressure of body fluids.

CHECKPOINT

1. What is a sensory modality?
2. How is sensation different from perception?
3. What is the difference between rapidly adapting and slowly adapting receptors?

TABLE 19.1

Classifcation of Sensory Receptors

BASIS OF CLASSIFICATION	DESCRIPTION
STRUCTURE	
Free nerve endings	Bare dendrites are associated with pain, thermal, tickle, itch, and some touch sensations.
Encapsulated nerve endings	Dendrites enclosed in a connective tissue capsule for pressure, vibration, and some touch sensations.
Separate cells	Receptor cells synapse with first-order sensory neurons; located in the retina of the eye (photoreceptors), inner ear (hair cells), and taste buds of the tongue (gustatory receptor cells).
FUNCTION	
Mechanoreceptors	Detect mechanical pressure; provide sensations of touch, pressure, vibration, proprioception, and hearing and equilibrium; also monitor stretching of blood vessels and internal organs.
Thermoreceptors	Detect changes in temperature.
Nociceptors	Respond to painful stimuli resulting from physical or chemical damage to tissue.
Photoreceptors	Detect light that strikes the retina of the eye.
Chemoreceptors	Detect chemicals in mouth (taste), nose (smell), and body fluids.
Osmoreceptors	Sense the osmotic pressure of body fluids.

19.2 SOMATIC SENSES

OBJECTIVES

- Describe the location and function of the receptors for tactile, thermal, and pain sensations.
- Identify the receptors for proprioception and describe their functions.

Somatic sensations arise from stimulation of sensory receptors in the skin, mucous membranes, muscles, tendons, and joints. The sensory receptors for somatic sensations are distributed unevenly. Some parts of the body surface are densely populated with receptors, and other parts contain only a few. The areas with the largest numbers of sensory receptors are the tip of the tongue, the lips, and the fingertips.

Tactile Sensations

The ***tactile sensations*** (TAK-tīl; *tact-* = touch) are touch, pressure, vibration, itch, and tickle. Itch and tickle sensations are detected by free nerve endings. All other tactile sensations are detected by a variety of encapsulated mechanoreceptors (see Table 19.2). Tactile receptors in the skin or subcutaneous layer include corpuscles of touch, hair root plexuses, type I and II cutaneous mechanoreceptors, lamellated corpuscles, and free nerve endings (Figure 19.1).

Touch

Sensations of ***touch*** generally result from stimulation of tactile receptors in the skin or subcutaneous layer. There are two types of rapidly adapting touch receptors. ***Corpuscles of touch,*** also known as ***Meissner corpuscles*** (MĪS-ner), are located in the dermal papillae of hairless skin. Each corpuscle is an egg-shaped mass of dendrites enclosed by a capsule of connective tissue. They are abundant in the fingertips, hands, eyelids, tip of the tongue, lips, nipples, soles, clitoris, and tip of the penis. ***Hair root plexuses*** consist of free nerve endings wrapped around hair follicles in hairy skin. Hair root plexuses detect movements on the surface of the skin that disturb hairs. For example, an insect landing on a hair causes movement of the hair shaft that stimulates the free nerve endings.

There are also two types of slowly adapting touch receptors. ***Type I cutaneous mechanoreceptors,*** also known as ***Merkel discs,*** are saucer-shaped, flattened free nerve endings that make contact with Merkel cells of the stratum basale; they are plentiful in the fingertips, hands, lips, and external genitalia. ***Type II cutaneous mechanoreceptors*** (***Ruffini corpuscles***) are elongated, encapsulated receptors located deep in the dermis, and in ligaments and tendons as well. Present in the hands and abundant on the soles, they are most sensitive to stretching that occurs as digits or limbs are moved.

Pressure and Vibration

Pressure is a sustained sensation that is felt over a larger area and occurs in deeper tissues than touch. Receptors that contribute to sensations of pressure include corpuscles of touch, type I mechanoreceptors, and lamellated corpuscles. ***Lamellated*** (***pacinian***) ***corpuscles*** (pa-SIN-ē-an) are large oval structures composed of a multilayered connective tissue capsule that encloses a nerve ending (Figure 19.1). Like corpuscles of touch, lamellated corpuscles adapt rapidly. They are widely distributed in the body: in the dermis and subcutaneous layer; in tissues that underlie mucous and serous membranes; around joints, tendons, and muscles; in the periosteum; and in the mammary glands, external genitalia, and certain viscera, such as the pancreas and urinary bladder.

Sensations of ***vibration*** result from rapidly repetitive sensory signals from tactile receptors. The receptors for vibration sensations are corpuscles of touch and lamellated corpuscles.

Figure 19.1 Structure and location of sensory receptors in the skin and subcutaneous layer.

The somatic sensations of touch, pressure, vibration, warmth, cold, and pain arise from sensory receptors in the skin, subcutaneous layer, and mucous membranes.

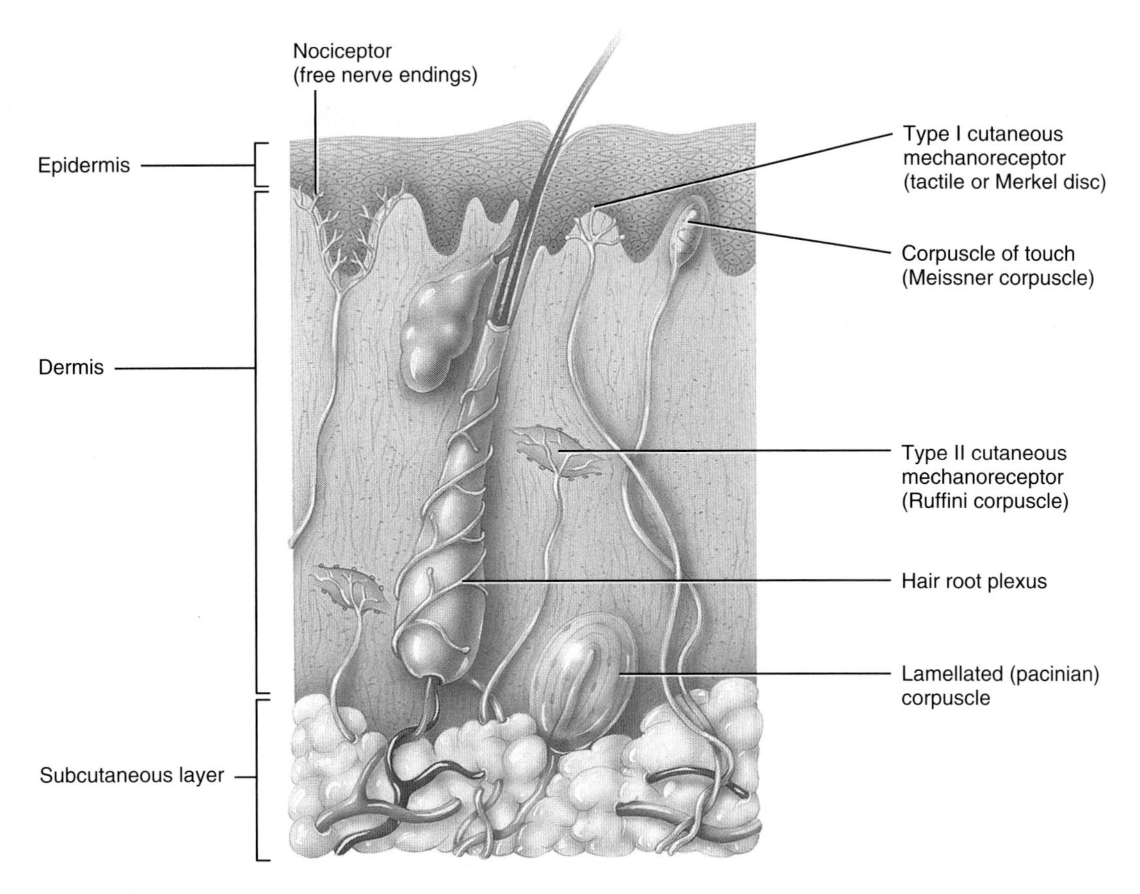

Which sensations can arise when free nerve endings are stimulated?

Corpuscles of touch can detect lower-frequency vibrations; lamellated corpuscles detect higher-frequency vibrations.

Itch and Tickle

The ***itch*** sensation results from stimulation of free nerve endings by certain chemicals, such as bradykinin, often as a result of a local inflammatory response. Receptors for the ***tickle*** sensation are thought to be free nerve endings and lamellated corpuscles. This intriguing sensation typically arises only when someone else touches you, not when you touch yourself. The explanation of this puzzle seems to lie in the nerve impulses that conduct to and from the cerebellum when you are moving your fingers and touching yourself that don't occur when someone else is tickling you.

• CLINICAL CONNECTION | Phantom Limb Sensation

Patients who have had a limb amputated may still experience sensations such as itching, pressure, tingling, or pain as if the limb were still there. This phenomenon is called **phantom limb sensation.** One explanation for phantom limb sensations is that the cerebral cortex interprets impulses arising in the proximal portions of sensory neurons that previously carried impulses from the limb as coming from the nonexistent (phantom) limb. In another explanation for phantom limb sensations, neurons in the brain that previously received sensory impulses from the missing limb are thought to be still active, giving rise to false sensory perceptions.

Some potentially life-saving amputations are required soon after a trauma. However, when an amputation is known in advance, it has been demonstrated that admitting the patient to the hospital 2 days prior to surgery and anesthetizing the portion of the body to be amputated causes a dramatic drop in the incidence of phantom limb sensation. •

Thermal Sensations

Thermoreceptors are free nerve endings. Two distinct ***thermal sensations***—coldness and warmth—are mediated by different receptors. Temperatures between 10° and 40°C (50–105°F) activate *cold receptors,* which are located in the epidermis. *Warm receptors* are located in the dermis and are activated by temperatures between 32° and 48°C (90–118°F). Cold and warm receptors both adapt rapidly at the onset of a stimulus but continue to generate nerve impulses more slowly throughout a prolonged stimulus. Temperatures below

10°C and above 48°C stimulate mainly nociceptors, rather than thermoreceptors, producing painful sensations. Note that the temperature ranges overlap; 105°F activates cold receptors only when the environmental temperature (or a fever) is in excess of 105°F.

Pain Sensations

The sensory receptors for pain, called ***nociceptors*** (nō′-sē-SEP-tors; *noci-* = harmful), are free nerve endings (see Figure 19.1). Nociceptors are found in practically every tissue of the body except the brain, and they respond to several types of stimuli. Excessive stimulation of sensory receptors, excessive stretching of a structure, prolonged muscular contractions, inadequate blood flow to an organ, or the presence of certain chemical substances can all produce the sensation of pain. Pain may persist even after a pain-producing stimulus is removed because pain-causing chemicals linger and because nociceptors exhibit very little adaptation. The lack of adaptation of nociceptors serves a protective function. If there were adaptation to painful stimuli, irreparable tissue damage could result.

There are two types of pain: fast and slow. The perception of ***fast pain*** occurs very rapidly, usually within 0.1 second after a stimulus is applied. This type of pain is also known as acute, sharp, or pricking pain. The pain felt from a needle puncture or knife cut to the skin are examples of fast pain. Fast pain is not felt in deeper (visceral) tissues of the body. The perception of ***slow pain*** begins a second or more after a stimulus is applied. It then gradually increases in intensity over a period of several seconds or minutes. This type of pain, which may be excruciating, is also referred to as chronic, burning, aching, or throbbing pain. Slow pain can occur both in the skin and in deeper tissues or internal organs. An example is the pain associated with a toothache.

• CLINICAL CONNECTION Analgesia

Some pain sensations occur out of proportion to minor damage or persist chronically for no obvious reason. In such cases, **analgesia** (*an-* = without; *-algesia* = pain) or pain relief is needed. Analgesic drugs such as aspirin and ibuprofen (for example, Advil®) block formation of some chemicals that stimulate nociceptors. Local anesthetics, such as Novocaine®, provide short-term pain relief by blocking conduction of nerve impulses. Morphine and other opiate drugs alter the quality of pain perception in the brain; pain is still sensed, but it is no longer perceived as so unpleasant. •

Fast pain is very precisely localized to the stimulated area. For example, if someone pricks you with a pin, you know exactly which part of your body was stimulated. Somatic slow pain is well localized but more diffuse (involves large areas); it usually appears to come from a larger area of the skin. In many instances of visceral pain, the pain is felt in or just deep to the skin that overlies the stimulated organ, or in a surface area far from the stimulated organ. This phenomenon is called ***referred pain*** (Figure 19.2). In general, the visceral organ involved and the

Figure 19.2 Distribution of referred pain. The colored parts of the diagrams indicate skin areas to which visceral pain is referred.

Nociceptors are present in almost every tissue of the body.

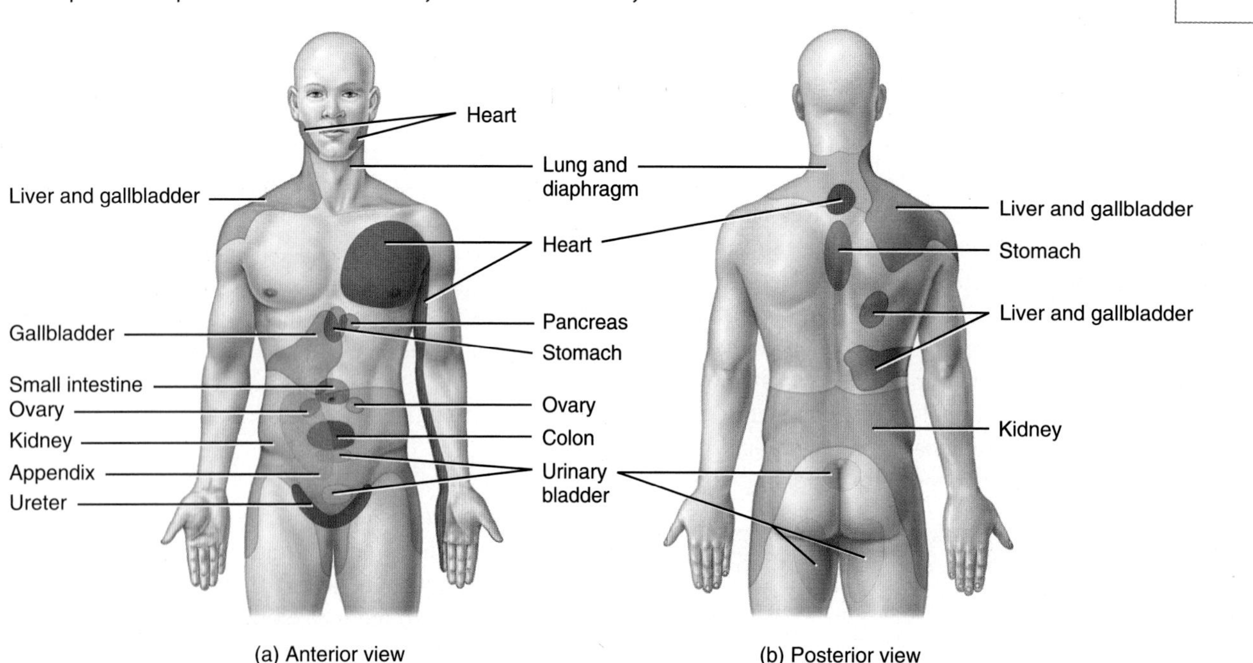

? Which visceral organ has the broadest area for referred pain?

area in which the pain is referred are served by the same segment of the spinal cord. For example, sensory neurons from the heart, the skin over the heart, and the skin along the medial aspect of the left arm enter spinal cord segments T1 to T5. Thus, the pain of a heart attack typically is felt in the skin over the heart and along the left arm. It should be noted, however, that some people experience pain in the right arm, left shoulder, between the shoulder blades, or mandible when they are having angina pain or a heart attack.

Pain Management

Pain that persists for longer than 2 or 3 months despite appropriate treatment is known as ***chronic pain.*** The most common forms of chronic pain are low back pain and headache. Cancer, arthritis, fibromyalgia, and many other disorders are associated with chronic pain. People experiencing chronic pain often experience chronic frustration as they are sent from one specialist to another in search of a diagnosis. The goal of pain management programs, developed to help people with chronic pain, is to decrease pain as much as possible, and then help patients learn to cope with whatever pain remains. Because no single treatment works for everyone, pain management programs typically offer a wide variety of treatments from surgery and nerve blocks to acupuncture and exercise therapy. Following are some of the therapies that complement medical and surgical treatment for the management of chronic pain.

COUNSELING Pain used to be regarded as a purely physical response to physical injury. Psychological factors are now understood to serve as important mediators in the perception of pain. Feelings such as fear and anxiety strengthen the pain perceptions. Pain may be used to avoid certain situations, or to gain attention. Depression and associated symptoms such as sleep disturbances can contribute to chronic pain. Psychological counseling techniques can help people with chronic pain confront issues that may be worsening their pain.

RELAXATION AND MEDITATION Relaxation and meditation techniques may reduce pain by decreasing anxiety and giving people a sense of personal control. Some of these techniques include massage, deep breathing, visualization of positive images, and muscular relaxation. Others encourage people to become more aware of thoughts and situations that increase or decrease pain or provide a mental distraction from the sensations of pain.

EXERCISE People with chronic pain tend to avoid movement because it hurts. Inactivity causes muscles and joint structures to atrophy, which may eventually cause the pain to worsen. Regular exercise and improved fitness help to relieve pain. Why? Exercise stimulates the production of endorphins, chemicals produced by the body to relieve pain. It improves self-confidence, can serve as a distraction from pain, and improves sleep quality, which is often a problem for people with chronic pain. As described in Chapter 10, exercise also has the potential to decrease the viscosity of ground substance of the extracellular matrix of connective tissues. This action enables adjacent tissues to receive more nourishment and better enables these tissues to rid themselves of waste products.

MANUAL THERAPY APPLICATION

Reducing Pain through Massage

Appropriate massage can reduce both acute (fast) and chronic (slow) pain of certain conditions. Acute pain, due to tense muscles and the accompanying ischemia to muscles, may be relieved by appropriate massage techniques, as described in Chapter 10. Therapeutic massage may reduce chronic pain through a neural-gating mechanism in which sensory stimulation from the massage, carried over A and B fibers, floods the brain and blocks the receipt of pain over C fibers. The fiber types were described in Chapter 15. Appropriate massage also induces the release of endorphins (described previously). It should be noted that the terms appropriate massage, therapeutic massage, clinical massage, professional massage, and medical massage are not to be confused with relaxation massage as is often received on cruise ships, in spas, and in certain hair salons.

Proprioceptive Sensations

Proprioceptive sensations (prō-prē-ō-SEP-tive; *proprio-* = one's own) allow us to know where our head and limbs are located and how they are moving even if we are not looking at them, so that we can walk, type, or dress without using our eyes. ***Kinesthesia*** (kin′-es-THĒ-zē-a; *kin-* = motion; *-esthesia* = perception) is the perception of body movements. Proprioceptive sensations arise in receptors termed ***proprioceptors.*** Proprioceptors are located in skeletal muscles (muscle spindles), in tendons (tendon organs) (see Figure 16.16), in and around synovial joints (joint kinesthetic receptors), and in the inner ear (hair cells). Those proprioceptors embedded in muscles, tendons, and synovial joints inform us of the degree to which muscles are contracted, the amount of tension on tendons, and the positions of joints. Hair cells of the inner ear monitor the orientation of the head relative to the ground and head position during movements. Proprioceptive sensations also allow us to estimate the weight of objects and determine the muscular effort necessary to perform a task. For example, as you pick up a bag you quickly realize whether it contains popcorn or books, and you then exert the correct amount of effort needed to lift it.

Nerve impulses for conscious proprioception pass along sensory tracts in the spinal cord and brain stem and are relayed to the primary somatosensory area (postcentral gyrus) in the parietal lobe of the cerebral cortex (see Figure 17.9). Proprioceptive impulses also pass to the cerebellum, where they contribute to the cerebellum's role in coordinating skilled movements. Because proprioceptors adapt slowly and only slightly, the brain continually receives nerve impulses related to the position of different body parts and makes adjustments to ensure coordination.

Table 19.2 summarizes the receptors for somatic sensations.

CHECKPOINT

4. Which somatic sensory receptors are encapsulated?
5. Which somatic sensory receptors mediate touch sensations?
6. How does fast pain differ from slow pain?
7. What is referred pain, and how is it useful in diagnosing internal disorders?

19.3 SPECIAL SENSES AND OLFACTION: SENSE OF SMELL

OBJECTIVE

- Describe the receptors for olfaction and the olfactory pathway to the brain.

Receptors for the special senses—smell, taste, sight, hearing, and equilibrium—are housed in complex sensory organs such as the eyes and ears. Like the general senses, the special senses allow us to detect changes in our environment.All special senses, except for the eye, are the concern of ***otorhinolaryngology*** (ō′-tō-rī′-nō-lar′-in-GOL-ō-jē; *oto-* = ear; *-rhino-* = nose; *-laryngo-* = larynx), the science that deals with the ears, nose, and throat and their disorders.

The nose contains 10–100 million receptors for the sense of smell, or ***olfaction*** (ol-FAK-shun; *olfact-* = smell). Because some nerve impulses for smell and taste propagate to the limbic system, certain odors and tastes can evoke strong emotional responses or a flood of memories.

Structure of the Olfactory Epithelium

The ***olfactory epithelium*** occupies the superior portion of the nasal cavity (Figure 19.3a) and consists of three types of cells: olfactory receptors, supporting cells, and basal stem cells (Figure 19.3b). ***Olfactory receptors*** are the first-order neurons of the olfactory pathway. Several cilia called ***olfactory hairs***

TABLE 19.2

Summary of Receptors for Somatic Sensations

RECEPTOR TYPE	RECEPTOR'S STRUCTURE AND LOCATION	SENSATIONS
TACTILE RECEPTORS		
Corpuscles of touch (Meissner corpuscles)	Capsule surrounds mass of dendrites in dermal papillae of hairless skin.	Fine touch, pressure, and slow vibrations.
Hair root plexuses	Free nerve endings wrapped around hair follicles in skin.	Touch.
Type I cutaneous mechanoreceptors (tactile or Merkel discs)	Saucer-shaped free nerve endings make contact with Merkel cells in epidermis.	Touch and pressure.
Type II cutaneous mechanoreceptors (Ruffini corpuscles)	Elongated capsule surrounds dendrites deep in dermis and in ligaments and tendons.	Stretching of skin.
Lamellated (pacinian) corpuscles	Oval, layered capsule surrounds dendrites; present in dermis and subcutaneous layer, submucosal tissues, joints, periosteum, and some viscera.	Pressure, fast vibrations, and tickling.
Itch and tickle receptors	Free nerve endings and lamellated corpuscles in skin and mucous membranes.	Itching and tickling.
THERMORECEPTORS		
Warm receptors and cold receptors	Free nerve endings in skin and mucous membranes of mouth, vagina, and anus.	Warmth or cold.
PAIN RECEPTORS		
Nociceptors	Free nerve endings in every tissue of the body except the brain.	Pain.
PROPRIOCEPTORS		
Muscle spindles	Sensory nerve endings wrap around central area of encapsulated intrafusal muscle fibers within most skeletal muscles.	Muscle length.
Tendon organs	Capsule encloses collagen fibers and sensory nerve endings at junction of tendon and muscles.	Muscle tension.
Joint kinesthetic receptors	Lamellated corpuscles, Ruffini corpuscles, tendon organs, and free nerve endings.	Joint position and movement.

project from a knob-shaped tip on each olfactory receptor. The olfactory hairs are the parts of the olfactory receptor that respond to inhaled chemicals. Chemicals that have an odor and can therefore stimulate the olfactory hairs are called ***odorants.*** The axons of olfactory receptors extend from the olfactory epithelium to the olfactory bulb. ***Supporting cells*** are columnar epithelial cells of the mucous membrane lining the nose. They provide physical support, nourishment, and electrical insulation for the olfactory receptors, and they help detoxify chemicals that come in contact with the olfactory epithelium. ***Basal cells*** are stem cells located between the bases of the supporting cells and continually undergo cell division to produce new olfactory receptors, which live for only a month or so before being replaced. This process is remarkable because olfactory receptors are neurons and, in general, mature neurons are not replaced. ***Olfactory glands*** produce mucus that moistens the surface of the olfactory epithelium and serves as a solvent for inhaled odorants.

Stimulation of Olfactory Receptors

Many attempts have been made to distinguish among and classify "primary" sensations of smell. Our ability to recognize about 10,000 different odors probably depends on patterns of activity in the brain that arise from activation of many different combinations of olfactory receptors. Olfactory receptors react to odorant molecules by producing an electrical signal that triggers one or more nerve impulses. Adaptation (decreasing sensitivity) to odors occurs rapidly.

The Olfactory Pathway

On each side of the nose, about 40 bundles of the slender, unmyelinated axons of olfactory receptors extend through about 20 holes in the cribriform plate of the ethmoid bone (Figure 19.3b). These bundles of axons collectively form the right and left ***olfactory***

Figure 19.3 Olfactory epithelium and olfactory receptors. (a) Location of olfactory epithelium in the nasal cavity. (b) Anatomy of olfactory receptors, consisting of first-order neurons whose axons extend through the cribriform plate and terminate in the olfactory bulb.

6 The olfactory epithelium consists of olfactory receptors, supporting cells, and basal cells.

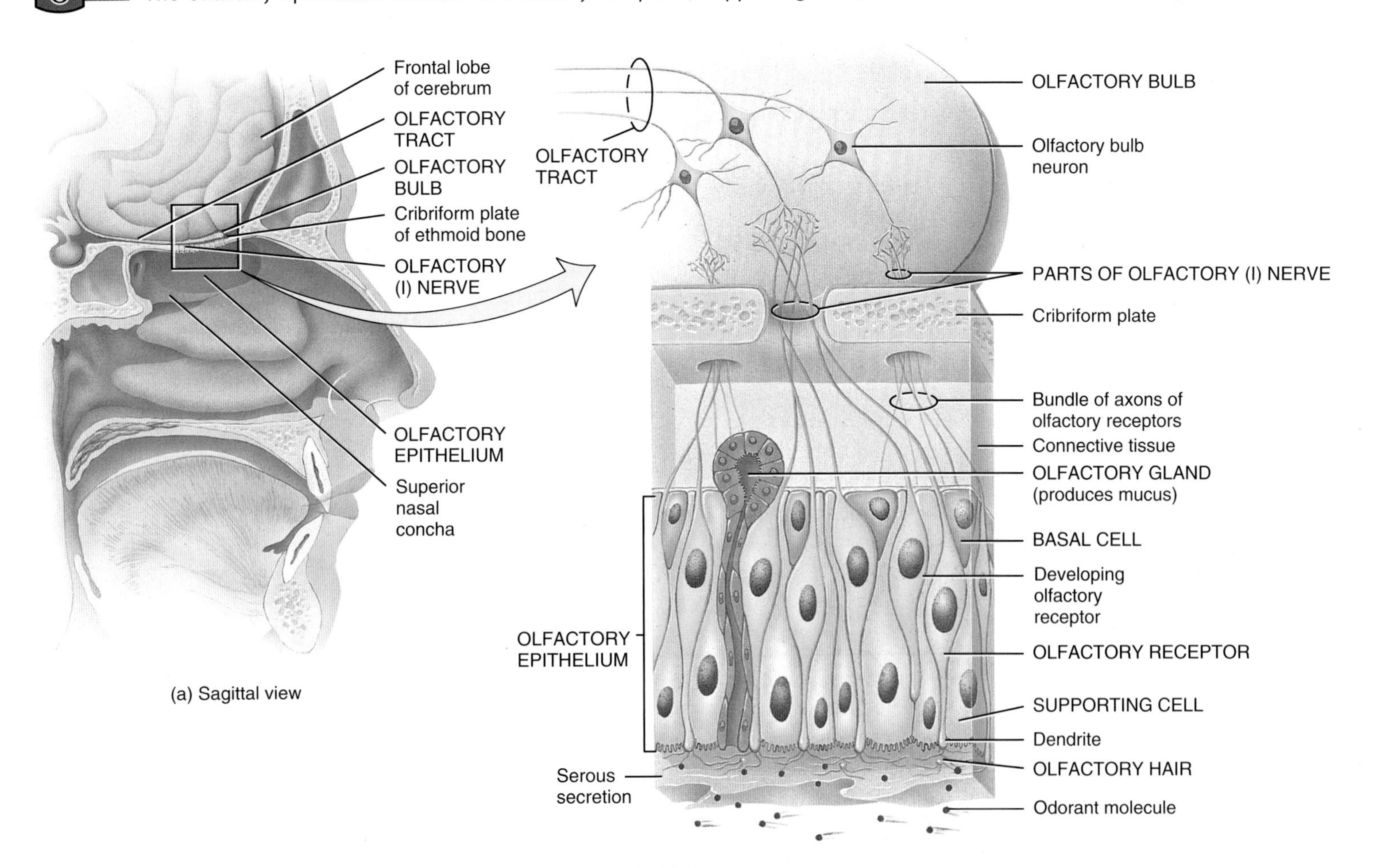

? **What is the life span of an olfactory receptor?**

(I) nerves. The olfactory nerves terminate in the brain in paired masses of gray matter called the ***olfactory bulbs,*** which are located below the frontal lobes of the cerebrum. Within the olfactory bulbs, the axon terminals of olfactory receptors—the first-order neurons—form synapses with the dendrites and cell bodies of second-order neurons in the olfactory pathway.

The axons of the neurons extending from the olfactory bulb form the ***olfactory tract.*** Some of the axons of the olfactory tract project to the *primary olfactory area* in the temporal lobe of the cerebral cortex (see Section 17.1), where conscious awareness of smell begins. Other axons of the olfactory tract project to the limbic system and hypothalamus; these connections account for emotional and memory-evoked responses to odors. Examples include sexual excitement upon smelling a certain perfume or nausea upon smelling a food that once made you violently ill.

CHECKPOINT

8. What functions are carried out by the three types of cells in the olfactory epithelium?
9. Define the following terms: olfactory nerve, olfactory bulb, and olfactory tract.

19.4 GUSTATION: SENSE OF TASTE

OBJECTIVE

- Describe the receptors for gustation and the gustatory pathway to the brain.

Taste or ***gustation*** (gus-TĀ-shun; *gust-* = taste) is much simpler than olfaction because only five primary tastes can be distinguished: *sour, sweet, bitter, salty,* and *umami* (ū-MAM-ē). The umami taste is described as "meaty" or "savory." All other flavors, such as chocolate, pepper, and coffee, are combinations of the five primary tastes, plus the accompanying olfactory and tactile (touch) sensations. Odors from food can pass upward from the mouth into the nasal cavity, where they stimulate olfactory receptors. Because olfaction is much more sensitive than taste, a given concentration of a food substance may stimulate the olfactory system thousands of times more strongly than it stimulates the gustatory system. When you have a cold or are suffering from allergies and cannot taste your food, it is actually olfaction that is blocked, not taste.

Structure of Taste Buds

The receptors for taste sensations are located in the ***taste buds*** (Figure 19.4). Most of the nearly 10,000 taste buds of a young adult are on the tongue, but some are also found on the roof of the mouth, pharynx (throat), and epiglottis (cartilage lid over the voice box). The number of taste buds declines with age. Taste buds are found in elevations on the tongue called ***papillae*** (pa-PIL-ē; singular is *papilla*), which provide a rough texture to the upper surface of the tongue (Figure 19.4a,b). ***Vallate papillae*** (VAL-āt = wall-like) form an inverted V-shaped row at the back of the tongue. ***Fungiform papillae*** (FUN-ji-form = mushroom-like) are small, red, mushroom-shaped elevations scattered over the entire surface of the tongue. In addition, the entire surface of the tongue has ***filiform papillae*** (FIL-i-form = threadlike), which contain touch receptors but no taste buds.

Each ***taste bud*** is an oval body consisting of three types of epithelial cells: supporting cells, gustatory receptor cells, and basal cells (Figure 19.4c). The ***supporting cells*** surround about 50 ***gustatory receptor cells.*** A single, long ***gustatory hair*** projects from each gustatory receptor cell to the external surface through the ***taste pore,*** an opening in the taste bud. ***Basal cells*** are stem cells that produce supporting cells, which then develop into gustatory receptor cells that have a life span of about 10 days. The gustatory receptor cells are separate receptor cells. They do not have an axon (like olfactory receptors) but rather synapse with dendrites of the first-order sensory neurons of the gustatory pathway.

Stimulation of Gustatory Receptors

Chemicals that stimulate gustatory receptor cells are known as ***tastants.*** Once a tastant is dissolved in saliva, it can enter taste pores and make contact with the plasma membrane of the gustatory hairs. The result is an electrical signal that stimulates release of neurotransmitter molecules from the gustatory receptor cell. Nerve impulses are triggered when these neurotransmitter molecules bind to their receptors on the dendrites of the first-order sensory neuron. The dendrites branch profusely and contact many gustatory receptors in several taste buds. Individual gustatory receptor cells may respond to more than one of the five primary tastes. Complete adaptation (loss of sensitivity) to a specific taste can occur in 1 to 5 minutes of continuous stimulation.

If all tastants cause release of neurotransmitter from many gustatory receptor cells, why do foods taste different? The answer to this question is thought to lie in the patterns of nerve impulses in groups of first-order taste neurons that synapse with the gustatory receptor cells. Different tastes arise from activation of different groups of taste neurons. In addition, although each individual gustatory receptor cell responds to more than one of the five primary tastes, it may respond more strongly to some tastants than to others.

The Gustatory Pathway

Three cranial nerves contain axons of first-order gustatory neurons that innervate the taste buds. The facial (VII) nerve and glossopharyngeal (IX) nerve serve the tongue; the vagus (X) nerve serves the throat and epiglottis. From taste buds, impulses propagate along these cranial nerves to the medulla oblongata. From the medulla, some axons carrying taste signals project to the limbic system and the hypothalamus, and others project to the thalamus. Taste signals that project from the thalamus to the *primary gustatory area* in the parietal lobe of the cerebral cortex (see Figure 17.9) give rise to the conscious perception of taste.

Figure 19.4 **The relationship of gustatory receptor cells in taste buds to tongue papillae.**

Gustatory receptor cells are located in taste buds.

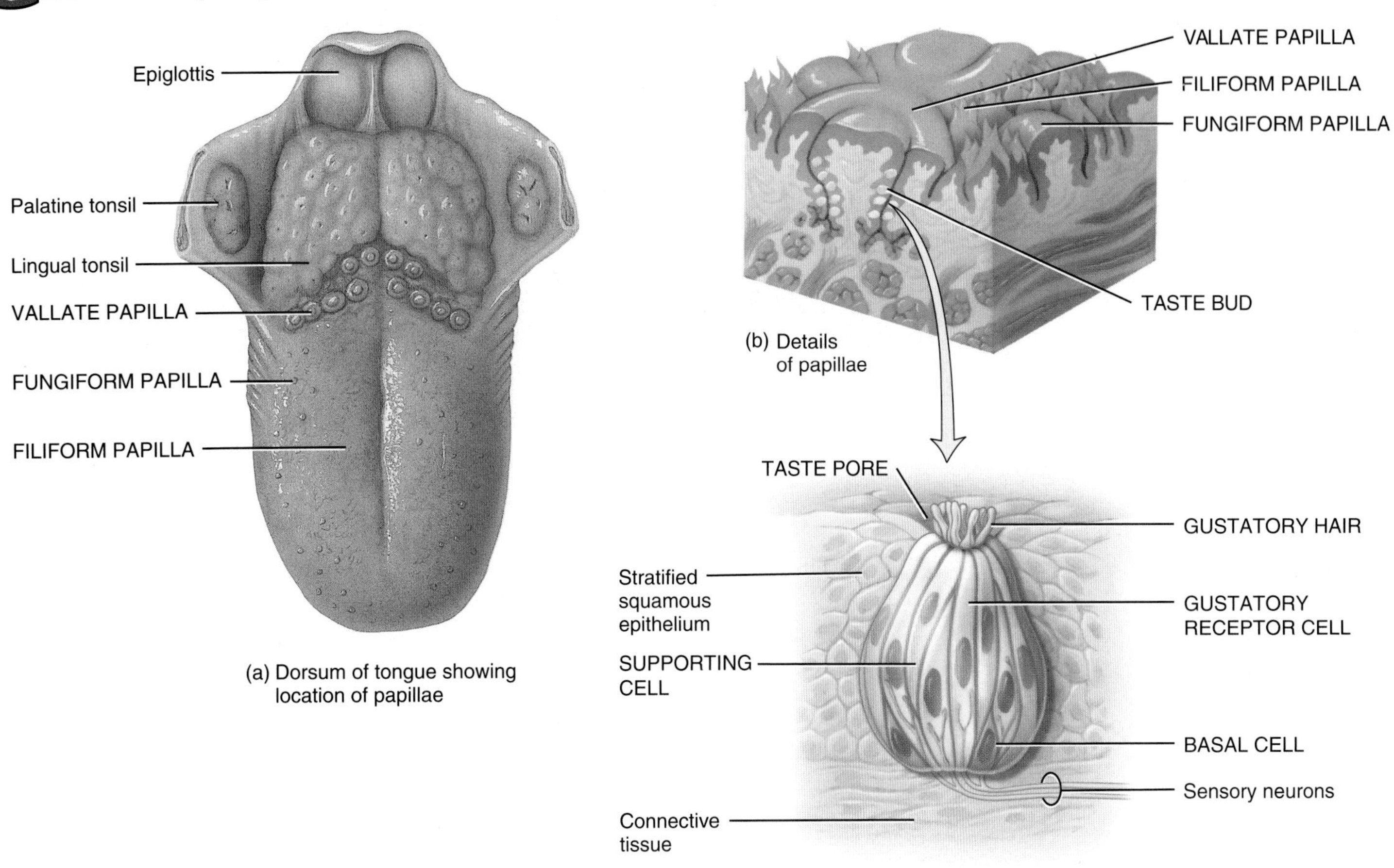

What role do basal cells play in taste buds.

CHECKPOINT

10. How do olfactory receptors and gustatory receptor cells differ in structure and function?
11. Compare the olfactory and gustatory pathways.

19.5 VISION

OBJECTIVES

- Describe the accessory structures of the eye, the layers of the eyeball, the lens, the interior of the eyeball, image formation, and binocular vision.
- Describe the receptors for vision and the visual pathway to the brain.

More than half the sensory receptors in the human body are located in the eyes, and a large part of the cerebral cortex is devoted to processing visual information. In this section, we examine the accessory structures of the eye, the eyeball itself, the formation of visual images, the physiology of vision, and the visual pathway from the eye to the brain. ***Ophthalmology*** (of′-thal-MOL-ō-jē; *ophthalmo-* = eye; *-logy* = study of) is the science that deals with the eye and its disorders.

Accessory Structures of the Eye

The ***accessory structures*** of the eye are the eyebrows, eyelashes, eyelids, extrinsic muscles that move the eyeballs, and lacrimal (tear-producing) apparatus. The ***eyebrows*** and ***eyelashes*** help protect the eyeballs from foreign objects, perspiration, and direct rays of the sun (see Figure 19.5a). The upper and lower ***eyelids*** (***palpabrae***) shade the eyes during sleep, protect the eyes from excessive light and foreign objects, and spread lubricating secretions over the eyeballs (by blinking). ***Meibomian*** (***tarsal***) ***glands*** (mī-BŌ-mē-an) in each eyelid produce a fluid that helps keep the eyelids from adhering to each other. Six extrinsic eye muscles cooperate to move each eyeball right, left, up, down, and diagonally: the *superior rectus, inferior rectus, lateral rectus, medial rectus, superior oblique,* and *inferior oblique* (Chapter 11). Neurons in the brain stem and cerebellum coordinate and synchronize the movements of the eyes.

The ***lacrimal apparatus*** (*lacrima* = tear) is a group of glands, ducts, canals, and sacs that produce and drain ***lacrimal fluid*** or ***tears*** (see Figure 19.5b). The right and left ***lacrimal glands*** are each about the size and shape of an almond. They secrete tears through the ***lacrimal ducts*** onto the surface of the upper eyelid. Tears then pass over the surface of the eyeball toward the nose to

Figure 19.5 Accessory structures of the eye.

Accessory structures of the eye include the eyebrows, eyelashes, eyelids, extrinsic eye muscles, and the lacrimal apparatus.

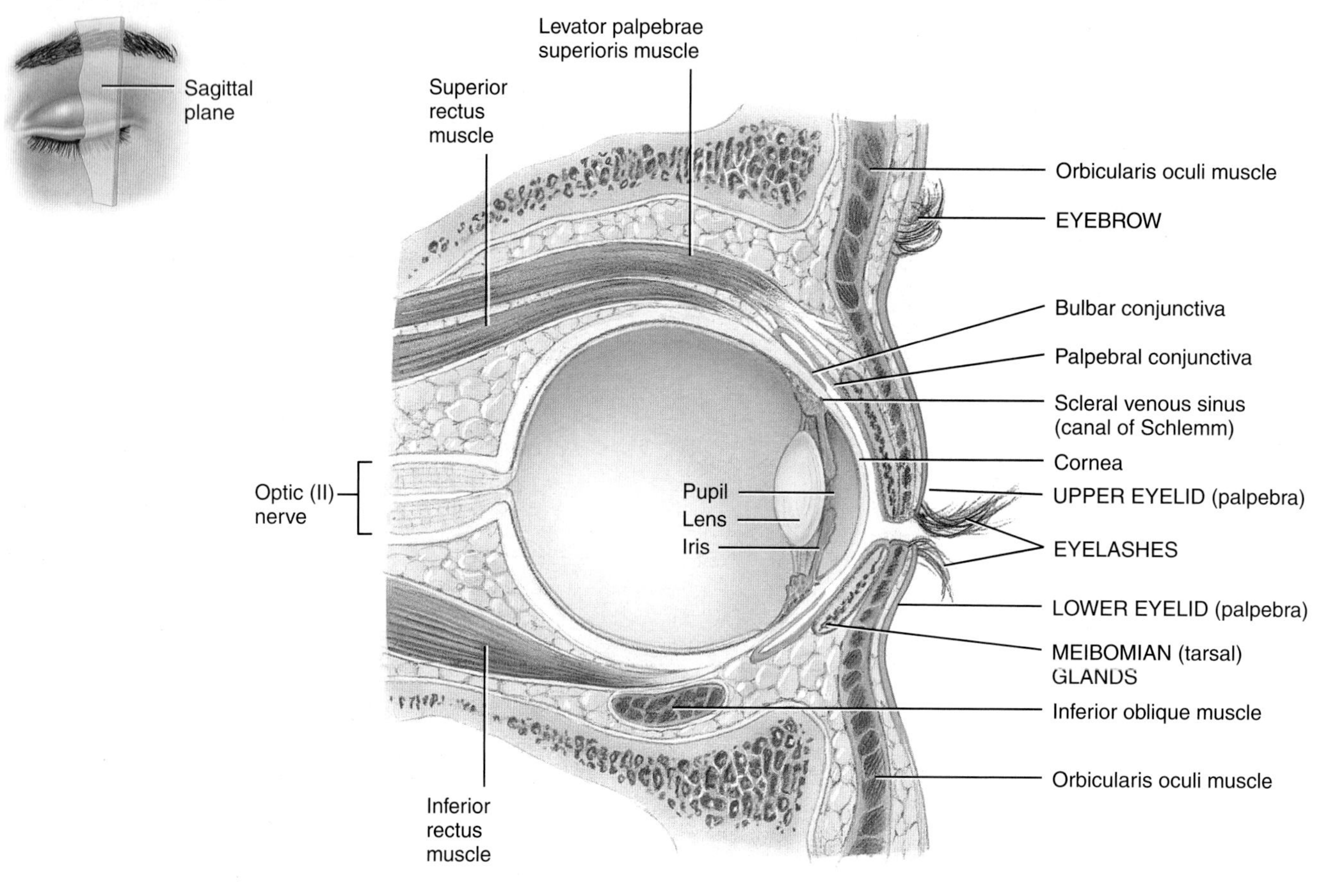

(a) Sagittal section of eye and its accessory structures

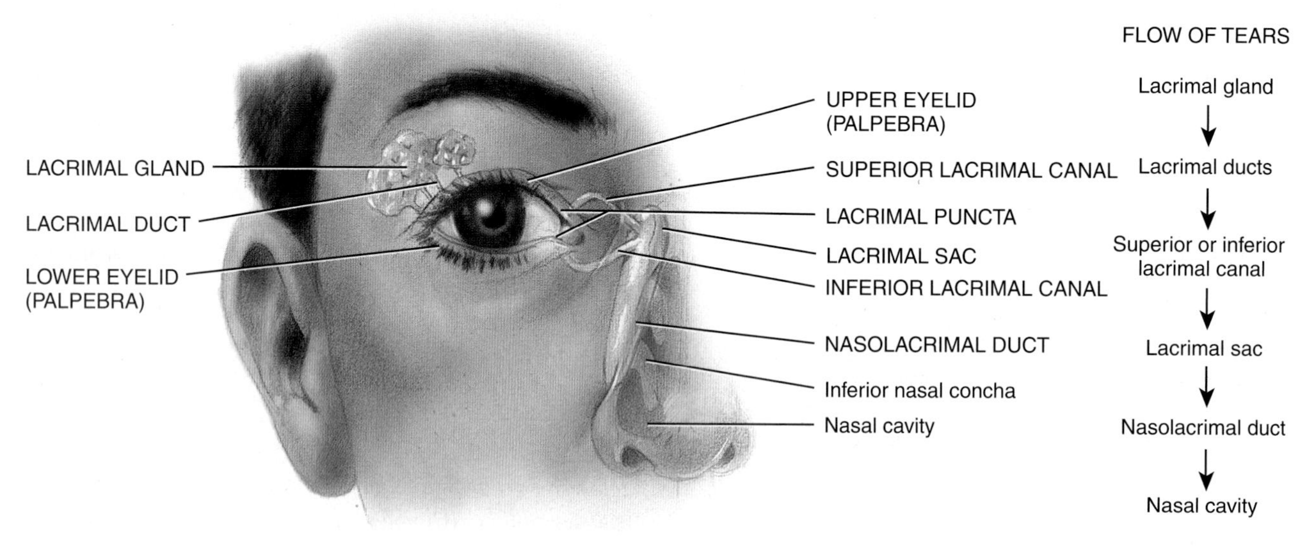

(b) Anterior view of the lacrimal apparatus

What is lacrimal fluid, and what are its functions?

enter two small openings called ***lacrimal puncta.*** From here the tears pass into two ***lacrimal canals,*** which lead into the ***lacrimal sac*** and then into the ***nasolacrimal duct,*** which allow the tears to drain into the nasal cavity.

Tears are a watery solution containing salts, some mucus, and a bacteria-killing enzyme called ***lysozyme.*** Tears clean, lubricate, and moisten the portion of the eyeball exposed to the air to prevent it from drying. Normally, tears are cleared away by evaporation or by passing into the nasal cavity as fast as they are produced. If, however, an irritating substance makes contact with the eye, the lacrimal glands are stimulated to oversecrete and tears accumulate. This protective mechanism dilutes and washes away the irritant. Only humans express emotions, both happiness and sadness, by ***crying.*** In response to parasympathetic stimulation, the lacrimal glands produce excessive tears that may spill over the edges of the eyelids and even fill the nasal cavity with fluid. This is how crying produces a runny nose.

Layers of the Eyeball

The adult ***eyeball*** measures about 2.5 cm (1 in.) in diameter and is divided into three layers: fibrous tunic, vascular tunic, and retina.

Fibrous Tunic

The ***fibrous tunic*** is the outer coat of the eyeball. It consists of an anterior cornea and a posterior sclera. The ***cornea*** (KOR-nē-a) is a transparent fibrous coat that covers the colored iris (Figure 19.6). Because it is curved, the cornea helps focus light rays onto the retina. The ***sclera*** (SKLE-ra = hard), the "white" of the eye, is a coat of dense connective tissue that covers all of the entire eyeball except the cornea. The sclera gives shape to the

Figure 19.6 Structure of the eyeball.

The wall of the eyeball consists of three layers: the fibrous tunic, the vascular tunic, and the retina.

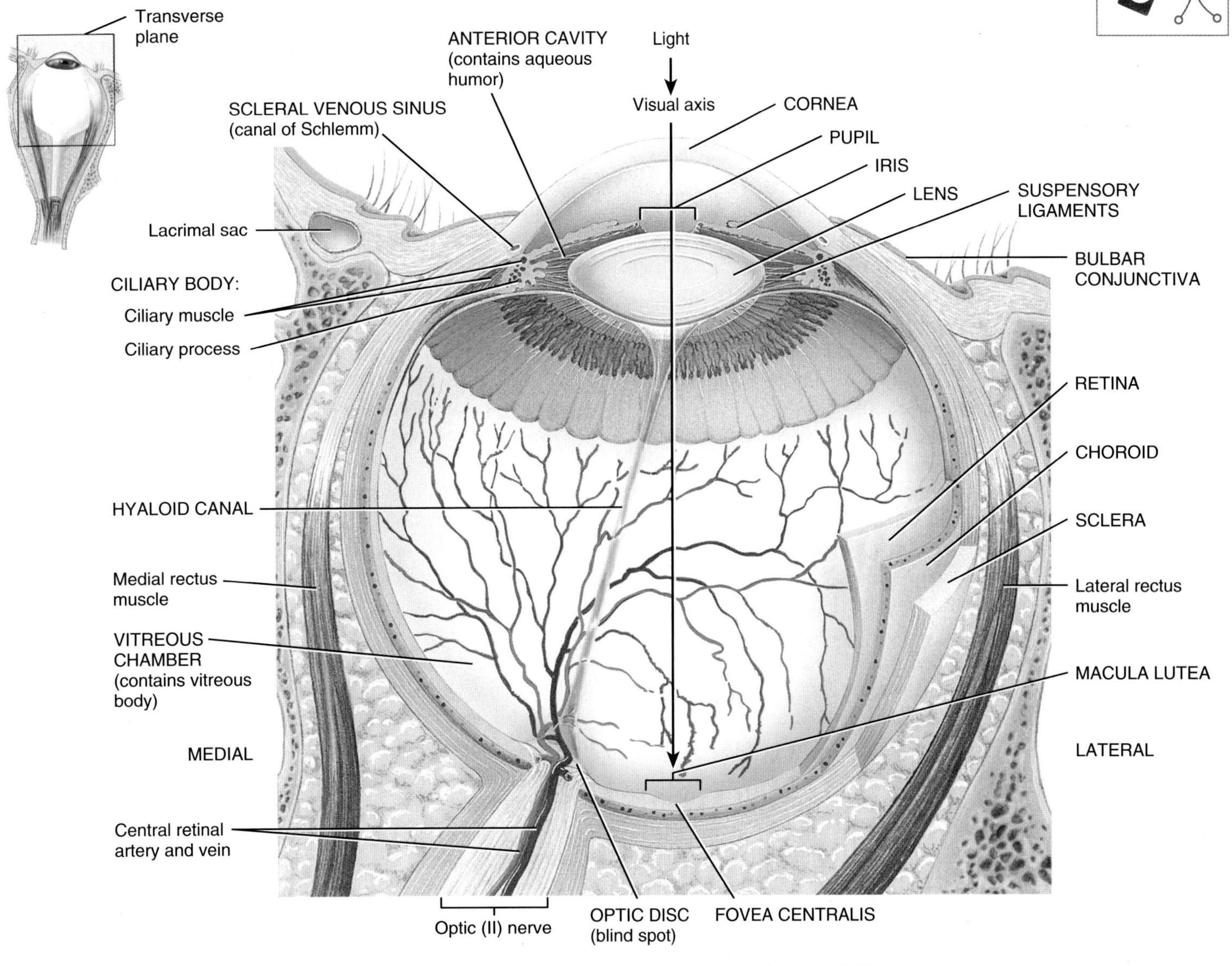

Transverse plane
ANTERIOR CAVITY (contains aqueous humor)
Light
Visual axis
SCLERAL VENOUS SINUS (canal of Schlemm)
CORNEA
PUPIL
IRIS
LENS
SUSPENSORY LIGAMENTS
Lacrimal sac
BULBAR CONJUNCTIVA
CILIARY BODY:
Ciliary muscle
Ciliary process
RETINA
CHOROID
HYALOID CANAL
SCLERA
Medial rectus muscle
Lateral rectus muscle
VITREOUS CHAMBER (contains vitreous body)
MACULA LUTEA
MEDIAL
LATERAL
Central retinal artery and vein
Optic (II) nerve
OPTIC DISC (blind spot)
FOVEA CENTRALIS

Superior view of transverse section of right eyeball

What are the components of the fibrous tunic and vascular tunic?

eyeball, makes it more rigid, and protects its inner parts. An epithelial layer called the ***bulbar conjunctiva*** (kon′-junk-TĪ-va) covers the sclera but not the cornea, and the ***palpebral conjunctiva*** lines the inner surface of the eyelids (see Figure 19.5a).

Vascular Tunic

The ***vascular tunic*** (***uvea***) is the middle layer of the eyeball and is composed of the choroid, ciliary body, and iris. The ***choroid*** (KŌ-royd) is a thin membrane that lines most of the internal surface of the sclera. It contains many blood vessels that help nourish the retina. The choroid also contains melanocytes that produce the pigment melanin, which causes this layer to appear dark brown in color. Melanin in the choroid absorbs stray light rays, which prevents reflection and scattering of light within the eyeball. As a result, the image cast on the retina by the cornea and lens remains sharp and clear.

At the front of the eye, the choroid becomes the ***ciliary body*** (SIL-ē-ar′-ē). The ciliary body consists of the *ciliary processes,* folds on the inner surface of the ciliary body whose capillaries secrete a fluid called aqueous humor, and the *ciliary muscle,* a smooth muscle that alters the shape of the lens for viewing objects up close or at a distance. The ***lens,*** a transparent structure that focuses light rays onto the retina, is constructed of many layers of elastic protein fibers. *Suspensory ligaments* (*zonular fibers*) attach the lens to the ciliary muscle and hold the lens in position.

A common cause of blindness is a loss of transparency of the lens known as a ***cataract*** (KAT-a-rakt). The lens becomes cloudy (less transparent) due to changes in the structure of the lens proteins. Cataracts often occur with aging but may also be caused by injury, excessive exposure to ultraviolet rays, certain medications (such as long-term use of steroids), or complications of other diseases (for example, diabetes). People who smoke also have increased risk of developing cataracts. Fortunately, sight can usually be restored by surgical removal of the old lens and implantation of an artificial one.

The ***iris*** (= colored circle) is the colored part of the eyeball. It includes both circular and radial smooth muscle fibers. The hole in the center of the iris, through which light enters the eyeball, is the ***pupil*** (Figure 19.7). The smooth muscle of the iris regulates the amount of light passing through the lens. When the eye is stimulated by bright light, the parasympathetic division of the autonomic nervous system (ANS) causes contraction of the circular muscles of the iris, which decreases the size of the pupil (constriction). When the eye must adjust to dim light, the sympathetic division of the ANS causes the radial muscles to contract, which increases the size of the pupil (dilation).

Figure 19.7 Responses of the pupil to light of varying brightness.

Contraction of the circular muscles causes constriction of the pupil; contraction of the radial muscles causes dilation of the pupil.

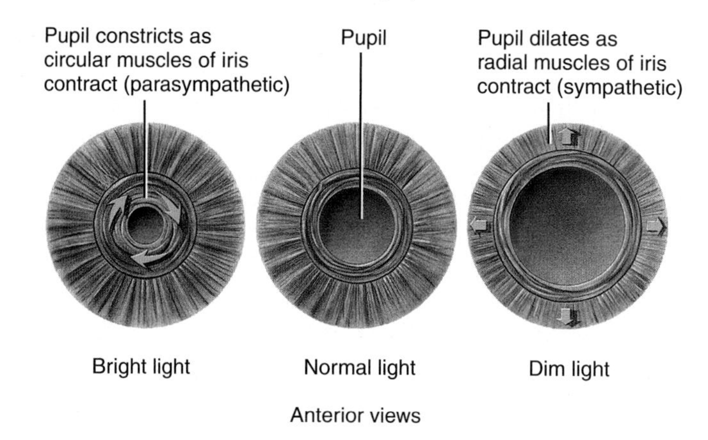

? Which division of the autonomic nervous system causes pupillary constriction? Which causes pupillary dilation?

Retina

The third and inner coat of the eyeball, the ***retina,*** lines the posterior three-quarters of the eyeball and is the beginning of the visual pathway (Figure 19.8). Two layers make up the retina: the neural layer and the pigmented layer. The *neural layer* of the retina is a multilayered outgrowth of the brain. Three distinct layers of retinal neurons—the ***photoreceptor layer,*** the ***bipolar cell layer,*** and the ***ganglion cell layer***—are separated by two zones, the outer and inner synaptic layers, where synaptic contacts are made. Note that light passes through the ganglion and bipolar cell layers and both synaptic layers before it reaches the photoreceptor layer.

The *pigmented layer* of the retina is a sheet of melanin-containing epithelial cells located between the choroid and the neural part of the retina. The melanin in the pigmented layer of the retina, like in the choroid, also helps to absorb stray light rays. ***Photoreceptors*** are specialized cells that begin the process by which light rays are ultimately converted to nerve impulses. There are two types of photoreceptors: rods and cones. ***Rods*** allow us to see shades of gray in dim light, such as moonlight. Brighter lights stimulate the ***cones,*** giving rise to highly acute color vision. Three types of cones are present in the retina: (1) *blue cones,* which are sensitive to blue light; (2) *green cones,* which are sensitive to green light; and (3) *red cones,* which are sensitive to red light. Color vision results from the stimulation of various combinations of these three types of cones. Just as an artist can obtain almost any color by mixing them on a palette, the cones can code for different colors by differential stimulation. There are about 6 million cones and 120 million rods. Cones are most densely concentrated in the ***fovea centralis,*** a small depression in the center of the ***macula lutea*** (MAK-ū-la LOO-tē-a), or yellow spot, in the exact center of the retina (see Figure 19.6). The fovea centralis is the area of highest *visual acuity* or *resolution* (sharpness of vision) because of its high concentration of cones. The main reason that you move your head and eyes while looking at something, such as the words of this sentence, is to place images of interest on your fovea. Rods are absent from the fovea centralis and macula lutea and increase in numbers toward the periphery of the retina.

From photoreceptors, information flows through the outer synaptic layer to the bipolar cells of the bipolar cell layer, and then from bipolar cells through the inner synaptic layer to the

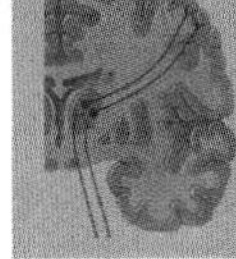

Figure 19.8 Microscopic structure of the retina. The downward blue arrow at left indicates the direction of the signals passing through the neural layer of the retina. Eventually, nerve impulses arise in ganglion cells and propagate along their axons, which make up the optic (II) nerve.

In the retina, visual signals pass from photoreceptors to bipolar cells to ganglion cells.

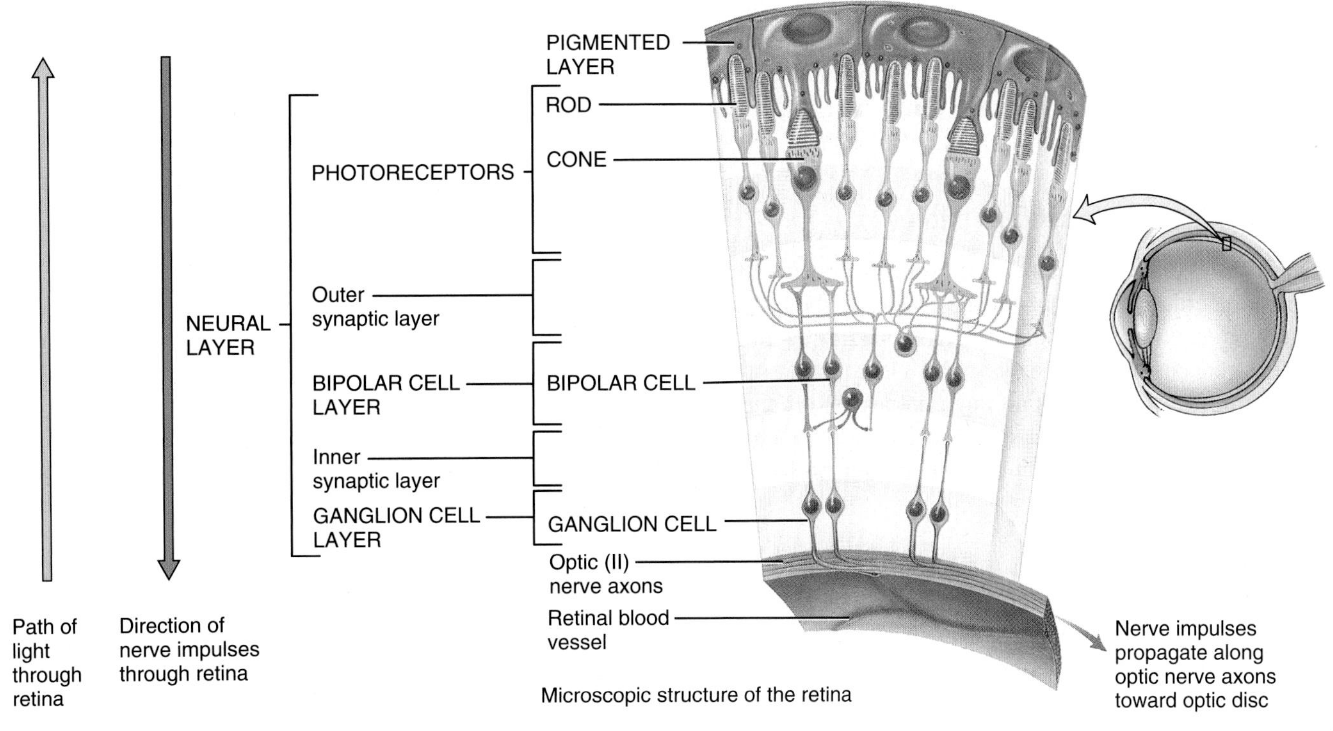

What are the two types of photoreceptors, and how do their functions differ?

ganglion cells of the ganglion cell layer (Figure 19.8). Between 6 and 600 rods synapse with a single bipolar cell in the outer synaptic layer; a cone usually synapses with just one bipolar cell. The convergence of many rods onto a single bipolar cell increases the light sensitivity of rod vision but slightly blurs the image that is perceived. Cone vision, although less sensitive, has higher acuity because of the one-to-one synapses between cones and their bipolar cells. The axons of the ganglion cells extend posteriorly to a small area of the retina called the ***optic disc*** (***blind spot***), where they all exit as the optic (II) nerve (see Figure 19.6). Because the optic disc contains no rods or cones, we cannot see an image that strikes the blind spot. Normally, you are not aware of having a blind spot, but you can easily demonstrate its presence. Cover your left eye and gaze directly at the cross below. Then increase or decrease the distance between the book and your eye. At some point, the square will disappear as its image falls on the blind spot.

+ ■

Interior of the Eyeball

The lens divides the interior of the eyeball into two cavities, the anterior cavity and the vitreous chamber. The ***anterior cavity*** (see Figure 19.6) lies anterior to the lens and is filled with ***aqueous humor*** (ĀK-wē-us HŪ-mer; *aqua* = water), a watery fluid similar to cerebrospinal fluid. Blood capillaries of the ciliary processes of the ciliary body secrete aqueous humor into the anterior cavity. It then drains into the ***scleral venous sinus*** (***canal of Schlemm***), an opening where the sclera and cornea meet (see Figures 19.5 and 19.6), and reenters the blood. The aqueous humor helps maintain the shape of the eye and nourishes the lens and cornea, neither of which has blood vessels. Normally, aqueous humor is completely replaced about every 90 minutes.

In ***glaucoma*** (glaw-KŌ-ma), the most common cause of blindness in the United States, a buildup of aqueous humor within the anterior cavity causes an abnormally high intraocular pressure. Persistent pressure results in a progression from mild visual impairment to irreversible destruction of the retina, damage to the optic nerve, and blindness. Because glaucoma is painless, and because the other eye initially compensates to a large extent for the loss of vision, a person may experience considerable retinal damage and loss of vision before the condition is diagnosed.

Behind the lens is a larger posterior cavity of the eyeball, the ***vitreous chamber*** (see Figure 19.6). It contains a clear, jelly-like substance called the ***vitreous body,*** which forms during embryonic life and consists mostly of water plus collagen fibers and hyaluronic acid. The vitreous body helps prevent the eyeball from collapsing and holds the retina flush against the choroid.

An inconspicuous ***hyaloid canal,*** an embryological remnant, passes through the vitreous body.

The pressure in the eye, called ***intraocular pressure,*** is produced mainly by the aqueous humor with a smaller contribution from the vitreous body. Intraocular pressure maintains the shape of the eyeball and keeps the retina smoothly pressed against the choroid so the retina is well nourished and forms clear images. Normal intraocular pressure (about 16 mm Hg) is maintained by a balance between production and drainage of the aqueous humor. Table 19.3 summarizes the structures of the eyeball.

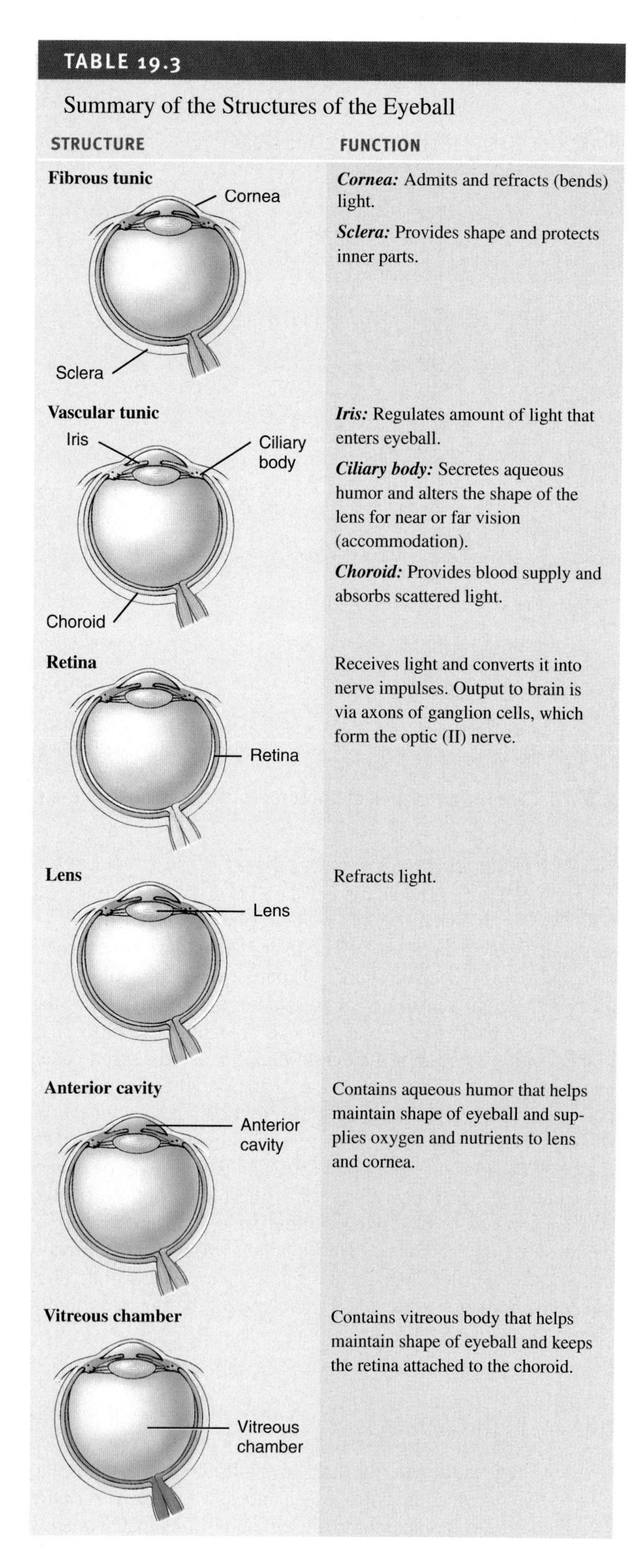

TABLE 19.3

Summary of the Structures of the Eyeball

STRUCTURE	FUNCTION
Fibrous tunic	***Cornea:*** Admits and refracts (bends) light. ***Sclera:*** Provides shape and protects inner parts.
Vascular tunic	***Iris:*** Regulates amount of light that enters eyeball. ***Ciliary body:*** Secretes aqueous humor and alters the shape of the lens for near or far vision (accommodation). ***Choroid:*** Provides blood supply and absorbs scattered light.
Retina	Receives light and converts it into nerve impulses. Output to brain is via axons of ganglion cells, which form the optic (II) nerve.
Lens	Refracts light.
Anterior cavity	Contains aqueous humor that helps maintain shape of eyeball and supplies oxygen and nutrients to lens and cornea.
Vitreous chamber	Contains vitreous body that helps maintain shape of eyeball and keeps the retina attached to the choroid.

Image Formation and Binocular Vision

In some ways the eye is like a camera: Its optical elements focus an image of some object on a light-sensitive "film"—the retina—while ensuring the correct amount of light makes the proper "exposure." To understand how the eye forms clear images of objects on the retina, we must examine three processes: (1) the refraction or bending of light by the lens and cornea, (2) the change in shape of the lens, and (3) constriction or narrowing of the pupil.

Refraction of Light Rays

When light rays traveling through a transparent substance (such as air) pass into a second transparent substance with a different density (such as water), they bend at the junction between the two substances. This bending is called ***refraction*** (Figure 19.9a). About 75% of the total refraction of light occurs at the cornea. Then, the lens of the eye further refracts the light rays so that they come into exact focus on the retina. Images focused on the retina are inverted (upside-down) (Figure 19.9b,c). They also undergo right-to-left reversal; that is, light from the right side of an object strikes the left side of the retina, and vice versa. The reason the world does not look inverted and reversed is that the brain "learns" early in life to coordinate visual images with the orientations of objects. The brain stores the inverted and reversed images we acquire when we first reach for and touch objects and interprets those visual images as being correctly oriented in space.

When an object is more than 6 meters (20 ft) away from the viewer, the light rays reflected from the object are nearly parallel to one another, and the curvatures of the cornea and lens exactly focus the image on the retina (Figure 19.9b). However, light rays from objects closer than 6 meters are divergent rather than parallel (Figure 19.9c). The rays must be refracted more if they are to be focused on the retina. This additional refraction is accomplished by changes in the shape of the lens.

Accommodation

A surface that curves outward, like the surface of a ball, is said to be *convex*. The convex surface of a lens refracts incoming light rays toward each other, so that they eventually intersect. The lens of the eye is convex on both its anterior and posterior surfaces, and its ability to refract light increases as its curvature becomes greater. When the eye is focusing on a close object, the lens becomes more convex and refracts the light rays more. This increase in the curvature of the lens for near vision is called ***accommodation*** (Figure 19.9c).

When you are viewing distant objects, the ciliary muscle of the ciliary body is relaxed and the lens is fairly flat because it is stretched in all directions by taut suspensory ligaments (zonular fibers). When you view a close object, the ciliary muscle contracts, which pulls the ciliary process and choroid forward toward the lens. This action releases tension on the lens, allowing it to become rounder (more convex), which increases its focusing power and causes greater convergence of the light rays.

The normal eye, known as an ***emmetropic eye*** (em′-e-TROP-ik), can sufficiently refract light rays from an object 6 m (20 ft) away so that a clear image is focused on the retina (Figure 19.10a). Many people, however, lack this ability because of refraction abnormalities. Among these abnormalities is ***myopia*** (mī-Ō-pē-a), or nearsightedness, which occurs when the eyeball is too long relative to the focusing power of the cornea and lens. Myopic individuals can see nearby objects clearly, but not distant objects. In ***hyperopia*** (hī′-per-Ō-pē-a) or farsightedness, also known as ***hypermetropia*** (hī′-per-me-TRŌ-pē-a), the eyeball length is

Figure 19.9 Refraction of light rays. (a) Refraction is the bending of light rays at the junction of two transparent substances with different densities. (b) The cornea and lens refract light rays from distant objects so the image is focused on the retina. (c) In accommodation, the lens becomes more spherical, which increases the refraction of light.

Images focused on the retina are inverted and left-to-right reversed.

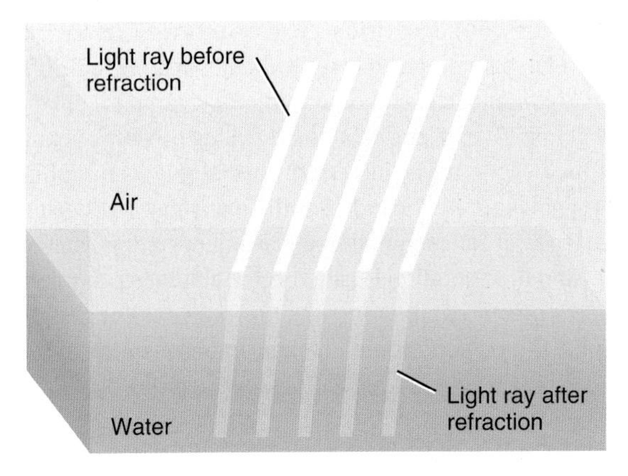

(a) Refraction of light rays

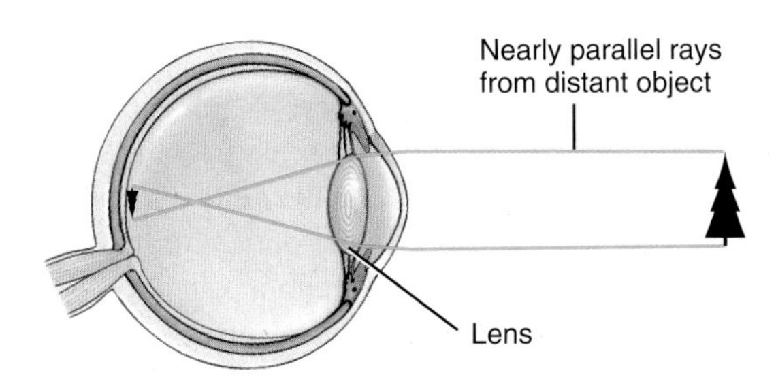

(b) Viewing distant object

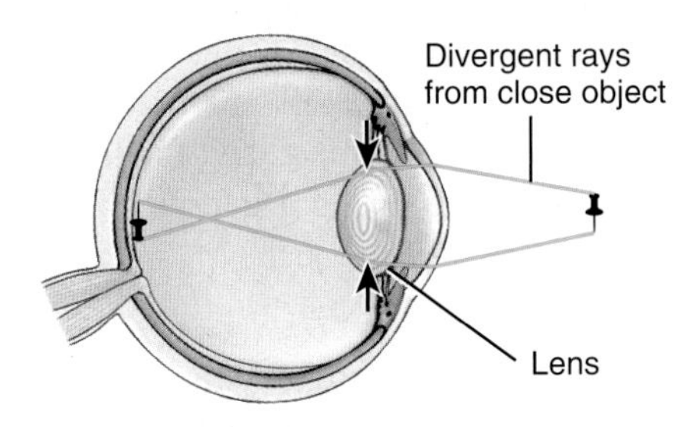

(c) Accommodation

What changes occur during accommodation for near vision?

Figure 19.10 Normal and abnormal refraction in the eyeball. (a) In the normal (emmetropic) eye, light rays from an object are bent sufficiently by the cornea and lens to focus on the fovea centralis. (b) In the nearsighted (myopic) eye, the image is focused in front of the retina. The condition may result from an elongated eyeball or thickened lens. (c) Correction of myopia is by use of a concave lens that diverges entering light rays so that they come into focus directly on the retina. (d) In the farsighted (hyperopic) eye, the image is focused behind the retina. The condition results from a shortened eyeball or a thin lens. (e) Correction of hyperopia is by a convex lens that causes entering light rays to converge so they focus directly on the retina.

In uncorrected myopia (nearsightedness), only close objects can be seen clearly; in uncorrected hyperopia (farsightedness), only distant objects can be seen clearly.

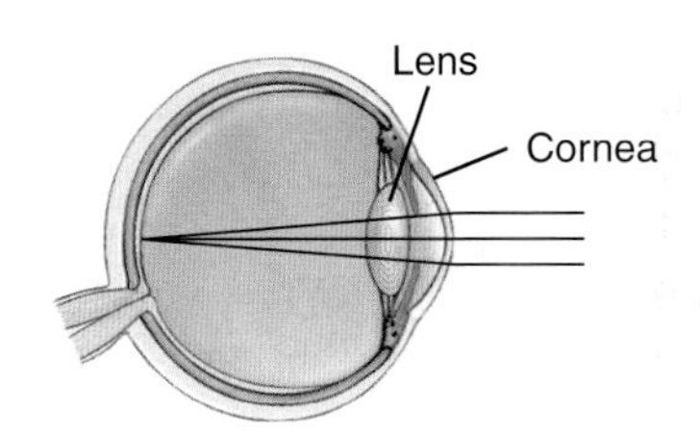

(a) Normal (emmetropic) eye

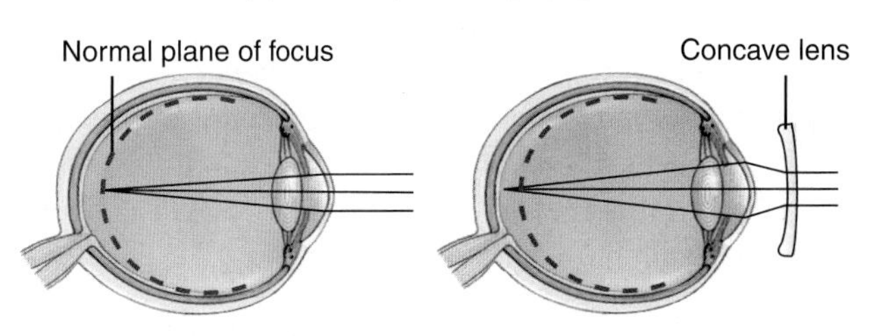

(b) Nearsighted (myopic) eye, uncorrected

(c) Nearsighted (myopic) eye, corrected

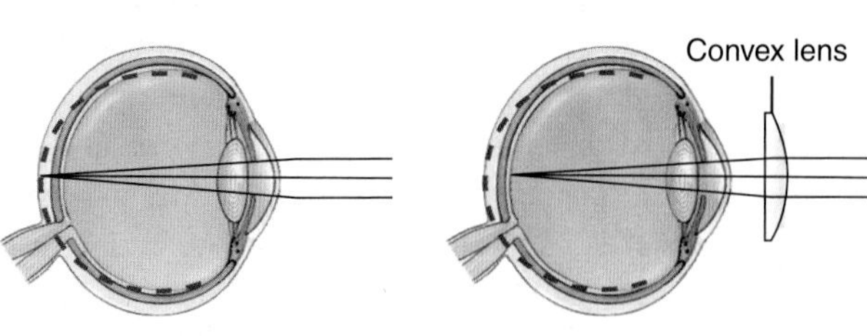

(d) Farsighted (hyperopic) eye, uncorrected

(e) Farsighted (hyperopic) eye, corrected

What is presbyopia?

short relative to the focusing power of the cornea and lens. Hyperopic individuals can see distant objects clearly, but not nearby objects. Figure 19.10b–e illustrates these conditions and shows how they are corrected. Another refraction abnormality is ***astigmatism*** (a-STIG-ma-tizm), in which either the cornea or the lens has an irregular curvature.

With aging, the lens loses some of its elasticity, so its ability to accommodate decreases. At about age 40, people who have not previously worn glasses begin to require them for close vision, such as reading. Many people state that their "arms are too short" since a longer focal distance is required for reading. This condition is called ***presbyopia*** (prez′-bē-Ō-pē-a; *presby-* = old; *-opia* = pertaining to the eye or vision). It should be noted, however, that some people develop presbyopia long before the age of 40. Presbyopia can be corrected through the use of bifocal or transitional (no line) lenses.

Constriction of the Pupil

Constriction of the pupil is a narrowing of the diameter of the hole through which light enters the eye due to contraction of the circular muscles of the iris. This autonomic reflex occurs simultaneously with accommodation and prevents light rays from entering the eye through the periphery of the lens. Light rays entering at the periphery of the lens would not be brought to focus on the retina and would result in blurred vision. The pupil, as noted earlier, also constricts in bright light to limit the amount of light that strikes the retina.

Convergence

In humans, both eyes focus on only one set of objects, a characteristic called ***binocular vision.*** This feature of our visual system allows the perception of depth and an appreciation of the three-dimensional nature of objects. When you stare straight ahead at a distant object, the incoming light rays are aimed directly at the pupils of both eyes and are refracted to comparable spots on the two retinas. As you move closer to the object, your eyes must rotate toward the nose if the light rays from the object are to strike comparable points on both retinas. ***Convergence*** is the name for this automatic movement of the two eyeballs toward the midline, which is caused by the coordinated action of the extrinsic eye muscles. The nearer the object, the greater is the convergence needed to maintain binocular vision.

Stimulation of Photoreceptors

After an image is formed on the retina by refraction, accommodation, constriction of the pupil, and convergence, light rays must be converted into neural signals. The initial step in this process is the absorption of light rays by the rods and cones of the retina. To understand how absorption occurs, it is necessary to understand the role of photopigments.

A ***photopigment*** (***visual pigment***) is a substance that can absorb light and undergo a change in structure. The photopigment in rods is called ***rhodopsin*** (*rhodo-* = rose; *-opsin* = related to vision) and is composed of a protein called *opsin* and a derivative of vitamin A called *retinal.* Any amount of light in a darkened room causes some rhodopsin molecules to split into opsin and retinal and initiate a series of chemical changes in the rods. When the light level is dim, opsin and retinal recombine into rhodopsin as fast as rhodopsin is split apart. Rods usually are nonfunctional in daylight, however, because rhodopsin is split apart faster than it can be reformed. After going from bright sunlight into a dark room, it takes about 40 minutes before the rods function maximally.

Cones function in bright light and provide color vision. As in rods, absorption of light rays causes breakdown of photopigment molecules. The photopigments in cones also contain retinal, but there are three different opsin proteins—one in each of the three types of cones. The cone photopigments reform much more quickly than the rod photopigment. When going from the dark into the light, it takes the cones only seconds to function maximally.

• CLINICAL CONNECTION Night Blindness and Color Blindness

The complete loss of cone vision causes a person to become legally blind. In contrast, a person who loses rod vision mainly has difficulty seeing in dim light and thus should not, for example, drive at night. Prolonged vitamin A deficiency and the resulting below-normal amount of rhodopsin may cause **night blindness** (**nyctalopia**), an inability to see well at low light levels. An individual with an absence or deficiency of one of the three types of cones from the retina cannot distinguish some colors from others and the condition is called **color-blindness.** In the most common type, *red–green color blindness,* either red cones or green cones are missing. Thus, the person cannot distinguish between red and green. •

The Visual Pathway

After stimulation by light, the rods and cones trigger electrical signals in bipolar cells. Bipolar cells transmit both excitatory and inhibitory signals to ganglion cells. The ganglion cells become depolarized and generate nerve impulses. The axons of the ganglion cells exit the eyeball as the ***optic*** (***II***) ***nerve*** (Figure 19.11) and extend posteriorly to the ***optic chiasm*** (KĪ-azm = a crossover, as in the letter X). In the optic chiasm, about half of the axons from each eye cross to the opposite side of the brain. After passing through the optic chiasm, the axons, now part of the ***optic tract,*** terminate in the thalamus. Here they synapse with neurons whose axons project to the primary visual areas in the occipital lobes of the cerebral cortex (see Figure 17.9). Because of crossing at the optic chiasm, the right side of the brain receives signals from both eyes for interpretation of visual sensations from the left side of an object, and the left side of the brain receives signals from both eyes for interpretation of visual sensations from the right side of an object.

CHECKPOINT

12. List and describe the accessory structures of the eye and the structural components of the eyeball.
13. Describe the processing of visual signals in the retina and the neural pathway for vision.

Figure 19.11 **Visual pathway.**

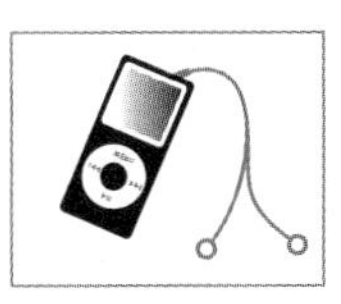

At the optic chiasm, half of the retinal ganglion cell axons from each eye cross to the opposite side of the brain.

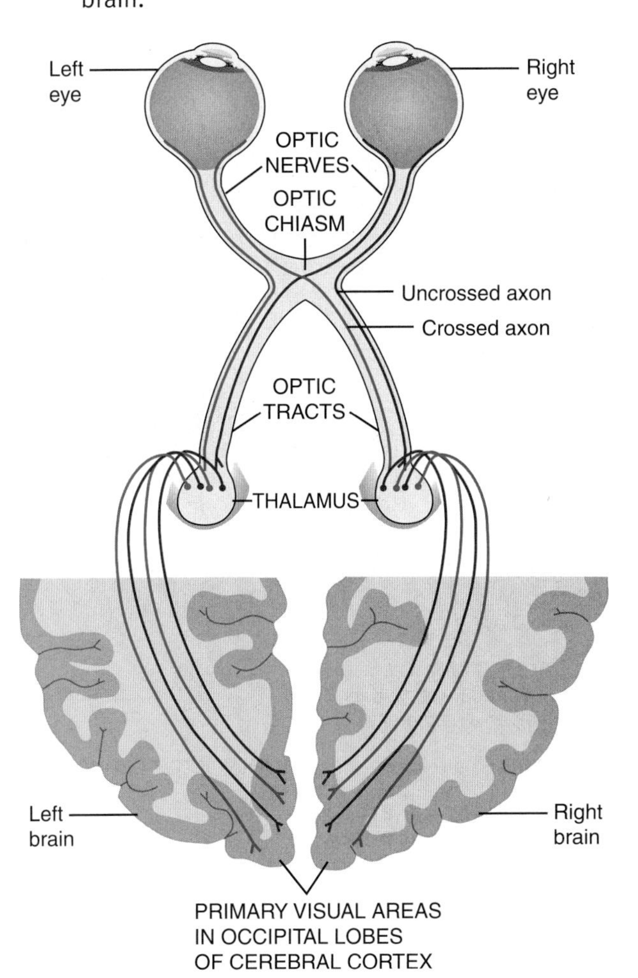

What is the correct order of structures that carry nerve impulses from the retina to the occipital lobe?

14. What types of cells make up the neural layer and the pigmented layer of the retina?
15. How do photopigments respond to light and recover in darkness?

19.6 HEARING AND EQUILIBRIUM

OBJECTIVES

- Describe the structures of the external (outer), middle, and internal (inner) ear.
- Describe the receptors for hearing and equilibrium and their pathways to the brain.

The ear is a marvelously sensitive structure. Its sensory receptors can convert sound vibrations into electrical signals 1000 times faster than photoreceptors can respond to light. Beside receptors for sound waves, the ear also contains receptors for equilibrium (balance).

Anatomy of the Ear

The ear is divided into three main regions: (1) the external (outer) ear, which collects sound waves and channels them inward; (2) the middle ear, which conveys sound vibrations to the oval window; and (3) the internal (inner) ear, which houses the receptors for hearing and equilibrium.

External (Outer) Ear

The ***external*** **(*outer*)** ***ear*** collects sound waves and passes them inward (see Figure 19.12). It consists of an auricle, external auditory canal, and eardrum. The ***auricle*** **(*pinna*)**, the part of the ear that you can see, is a skin-covered flap of elastic cartilage shaped like the flared end of a trumpet (see Figure 19.15). It plays a small role in collecting sound waves and directing them toward the ***external auditory canal*** **(*meatus*)** (*audit-* = hearing), a curved tube that extends from the auricle and directs sound waves toward the eardrum. The canal contains a few hairs and ***ceruminous glands*** (se-RŪ-mi-nus; *cer-* = wax), which secrete ***cerumen*** (se-RŪ-men) (earwax). The hairs and cerumen help prevent foreign objects from entering the ear. The ***tympanic membrane*** (tim-PAN-ik; *tympan-* = a drum), also called the ***eardrum,*** is a thin, semitransparent partition between the external auditory canal and the middle ear. Sound waves cause the tympanic membrane to vibrate. Tearing of the tympanic membrane, due to trauma or infection, is called a ***perforated eardrum.***

Middle Ear

The ***middle ear*** is a small, air-filled cavity between the tympanic membrane and internal ear (see Figure 19.12). An opening in the anterior wall of the middle ear leads directly into the ***auditory tube,*** commonly known as the ***eustachian tube,*** which connects the middle ear with the upper part of the throat. When the auditory tube is open, air pressure can equalize on both sides of the tympanic membrane. Otherwise, abrupt changes in air pressure on one side of the tympanic membrane might cause it to rupture. During swallowing and yawning, the tube opens, which explains why yawning or chewing gum can sometimes help equalize the pressure changes that occur while flying in an airplane.

Extending across the middle ear and attached to it by means of ligaments are three tiny bones called ***auditory ossicles*** (OS-si-kuls) that are named for their shapes: the ***malleus*** (MAL-ē-us), ***incus*** (ING-kus), and ***stapes*** (STĀ-pēz), commonly called the hammer, anvil, and stirrup (see Figure 19.12). Equally tiny skeletal muscles control the amount of movement of these bones to prevent damage by excessively loud noises. The stapes fits into a small opening in the thin bony partition between the middle and internal ear called the ***oval window,*** where the inner ear begins. Directly below the oval window is another opening, the ***round window,*** which is enclosed by a membrane called the ***secondary tympanic membrane.***

Figure 19.12 Anatomy of the ear.

The ear has three principal regions: the external (outer) ear, the middle ear, and the internal (inner) ear.

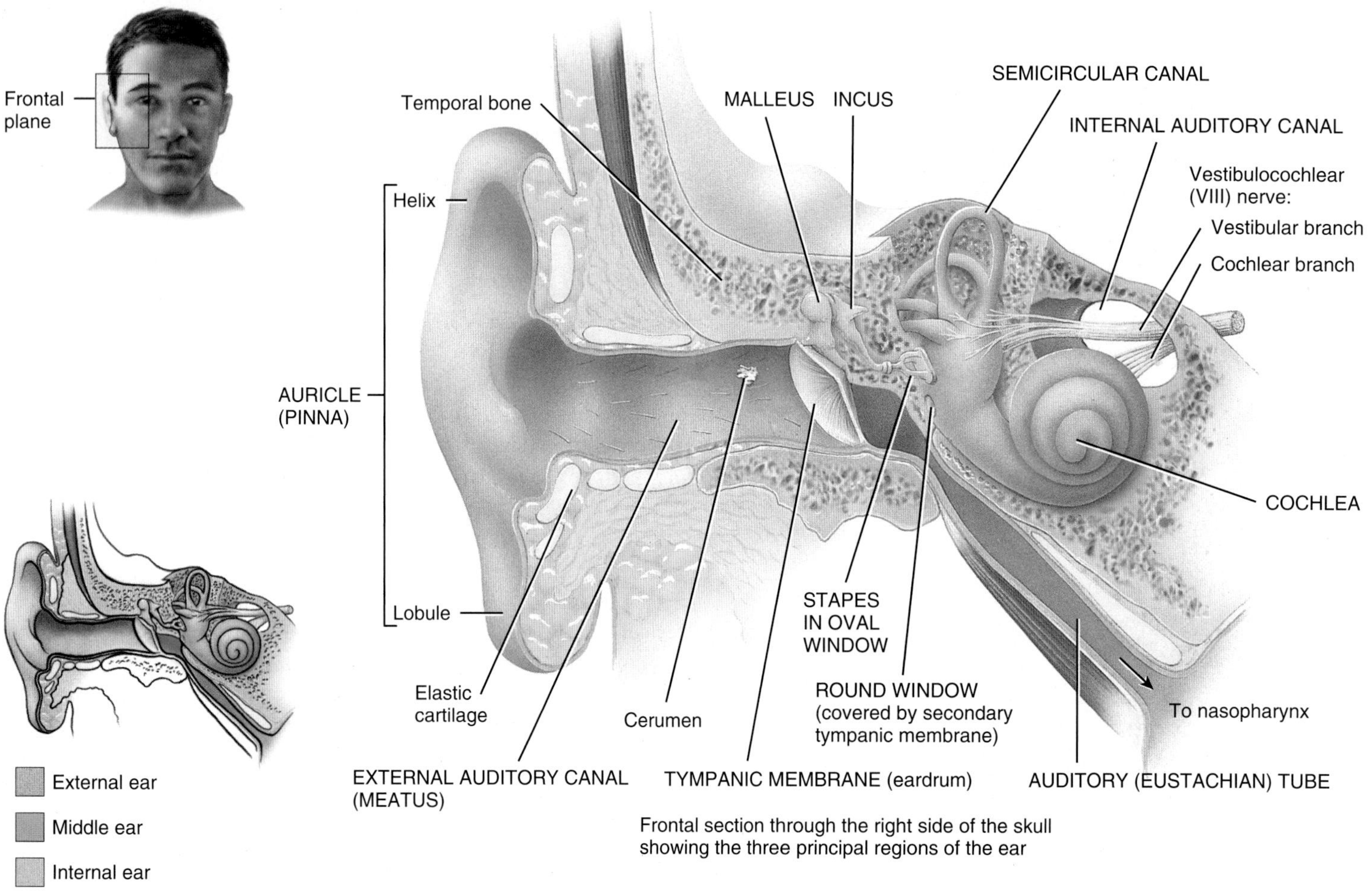

Frontal plane

Temporal bone, MALLEUS, INCUS, SEMICIRCULAR CANAL, INTERNAL AUDITORY CANAL, Vestibulocochlear (VIII) nerve: Vestibular branch, Cochlear branch, Helix, AURICLE (PINNA), Lobule, COCHLEA, STAPES IN OVAL WINDOW, ROUND WINDOW (covered by secondary tympanic membrane), Elastic cartilage, Cerumen, To nasopharynx, EXTERNAL AUDITORY CANAL (MEATUS), TYMPANIC MEMBRANE (eardrum), AUDITORY (EUSTACHIAN) TUBE

External ear
Middle ear
Internal ear

Frontal section through the right side of the skull showing the three principal regions of the ear

To which structure of the external ear does the malleus of the middle ear attach?

Otitis media is an acute infection of the middle ear caused primarily by bacteria and associated with infections of the nose and throat. Symptoms include pain; malaise (discomfort or uneasiness); fever; and a reddening and outward bulging of the tympanic membrane, which may rupture unless prompt treatment is received (this may involve draining pus from the middle ear). Bacteria from the nasopharynx passing into the auditory tube is the primary cause of all middle ear infections. Children are more susceptible than adults to middle ear infections because their auditory tubes are almost horizontal, which decreases drainage.

Internal (Inner) Ear

The ***internal (inner) ear*** is divided into the outer bony labyrinth and inner membranous labyrinth (Figure 19.13a). The ***bony labyrinth*** (LAB-i-rinth) is a series of cavities in the temporal bone, including the cochlea, vestibule, and semicircular canals. The cochlea is the sense organ for hearing, and the vestibule and semicircular canals are the sense organs for equilibrium and balance. The bony labyrinth contains a fluid called *perilymph.* This fluid surrounds the inner ***membranous labyrinth,*** a series of sacs and tubes with the same general shape as the bony labyrinth. The membranous labyrinth contains a fluid called *endolymph.*

The ***vestibule*** (VES-ti-būl) is the oval-shaped middle part of the bony labyrinth. The membranous labyrinth in the vestibule consists of two sacs called the ***utricle*** (Ū-tri-kul = little bag) and ***saccule*** (SAK-ūl = little sac). Behind the vestibule are the three bony ***semicircular canals.*** The anterior and posterior semicircular canals are both vertical, and the lateral canal is horizontal. One end of each canal enlarges into a swelling called the ***ampulla*** (am-PUL-la = little jar). The portions of the membranous labyrinth that lie inside the bony semicircular canals are called the ***semicircular ducts,*** which connect with the utricle of the vestibule (Figure 19.13a).

A transverse section through the ***cochlea*** (KOK-lē-a = snail's shell), a bony spiral canal that resembles a snail's shell, shows that it is divided into three channels: cochlear duct, scala vestibuli, and scala tympani. The *cochlear duct* is a continuation of the membranous labyrinth into the cochlea; it is filled with endolymph. The channel above the cochlear duct is the *scala vestibuli,* which ends

Figure 19.13 Semicircular canals, vestibule, and cochlea of the right ear. Note that the cochlea makes nearly three complete turns. The outer, cream-colored area is part of the bony labyrinth; the inner, pink-colored area is the membranous labyrinth.

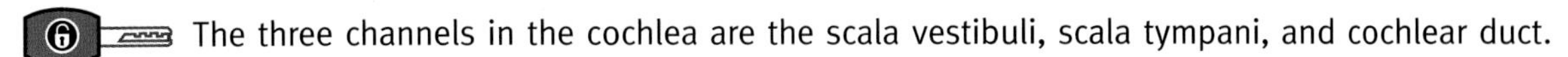

The three channels in the cochlea are the scala vestibuli, scala tympani, and cochlear duct.

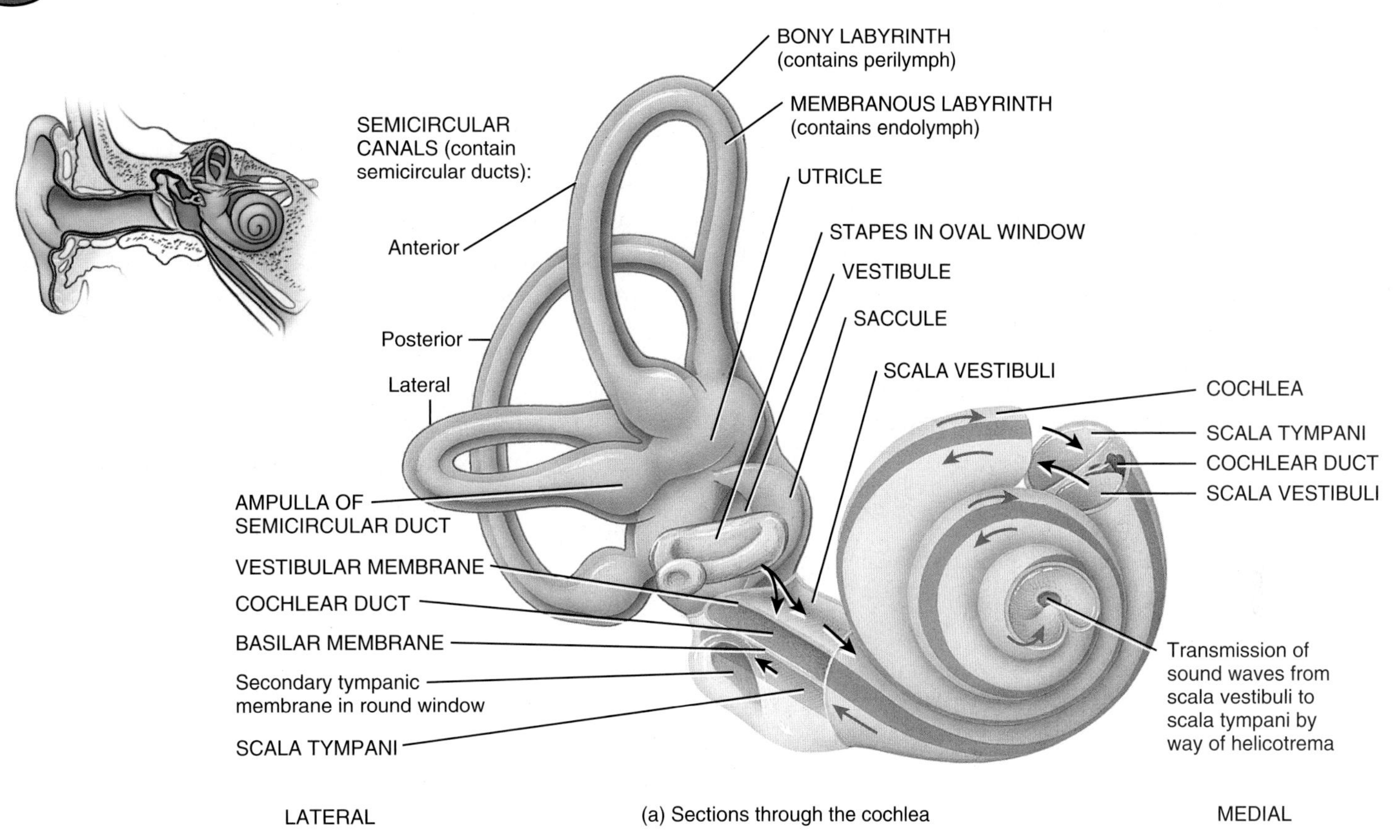

(a) Sections through the cochlea

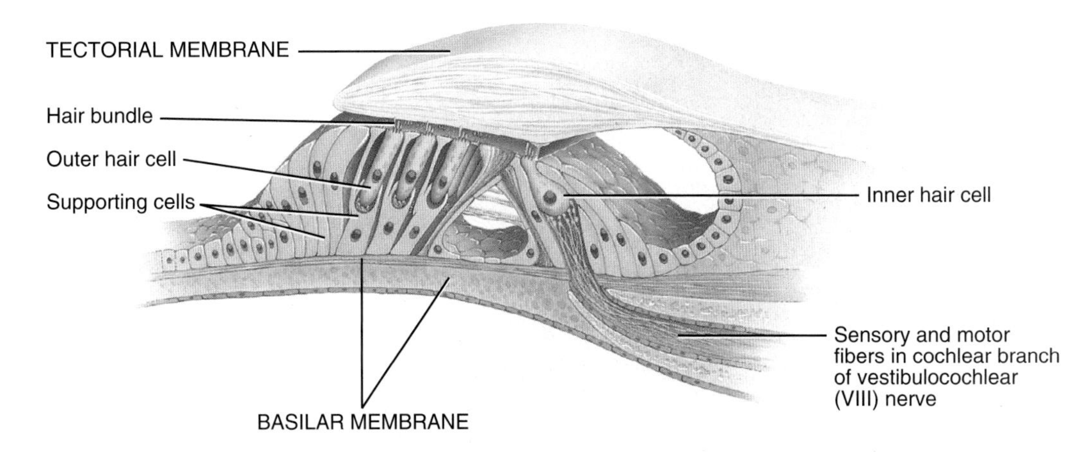

(b) Enlargement of spiral organ (organ of Corti)

What are the three subdivisions of the bony labyrinth?

at the oval window. The channel below the cochlear duct is the *scala tympani,* which ends at the round window. Both the scala vestibuli and scala tympani are part of the bony labyrinth of the cochlea and are filled with perilymph. The scala vestibuli and scala tympani are completely separated, except for an opening at the apex of the cochlea called the *helicotrema.* Between the cochlear duct and the scala vestibuli is the *vestibular membrane*. Between the cochlear duct and scala tympani is the *basilar membrane*.

Resting on the basilar membrane is the ***spiral organ*** (***organ of Corti***), the organ of hearing (Figure 19.13b). The spiral organ consists of *supporting cells* and *hair cells.* The hair cells, the receptors for auditory sensations, have long processes at their free ends that extend into the endolymph of the cochlear duct. The hair cells form synapses with sensory and motor neurons in the cochlear branch of the vestibulocochlear (VIII) nerve. The *tectorial membrane,* a flexible gelatinous membrane, covers the hair cells.

Physiology of Hearing

The events involved in stimulation of hair cells by sound waves are as follows (Figure 19.14):

1. The auricle directs sound waves into the external auditory canal.
2. Sound waves striking the tympanic membrane cause it to vibrate. The distance and speed of its movement depend on the intensity and frequency of the sound waves. More intense (louder) sounds produce larger vibrations. The tympanic membrane vibrates slowly in response to low-frequency (low-pitched) sounds and rapidly in response to high-frequency (high-pitched) sounds.
3. The central area of the tympanic membrane connects to the malleus, which also starts to vibrate. The vibration is transmitted from the malleus to the incus and then to the stapes.
4. As the stapes moves back and forth, it pushes the oval window in and out.
5. The movement of the oval window sets up fluid pressure waves in the perilymph of the cochlea. As the oval window bulges inward, it pushes on the perilymph of the scala vestibuli.
6. The fluid pressure waves are transmitted from the scala vestibuli to the scala tympani and eventually to the membrane covering the round window, causing it to bulge outward into the middle ear. (See 9 in the figure.)
7. As the pressure waves deform the walls of the scala vestibuli and scala tympani, they also push the vestibular membrane back and forth, creating pressure waves in the endolymph inside the cochlear duct.
8. The pressure waves in the endolymph cause the basilar membrane to vibrate, which moves the hair cells of the spiral organ against the tectorial membrane. Bending of their hairs stimulates the hair cells to release neurotransmitter molecules at synapses with sensory neurons that are part of the vestibulocochlear (VIII) nerve (see Figure 19.13b).

Figure 19.14 Physiology of hearing shown in the right ear. The numbers correspond to the events listed in the text. The cochlea has been uncoiled to more easily visualize the transmission of sound waves and their distortion of the vestibular and basilar membranes of the cochlear duct.

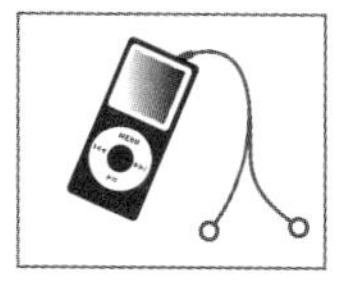

6 Hair cells of the spiral organ (organ of Corti) convert a mechanical vibration into an electrical signal.

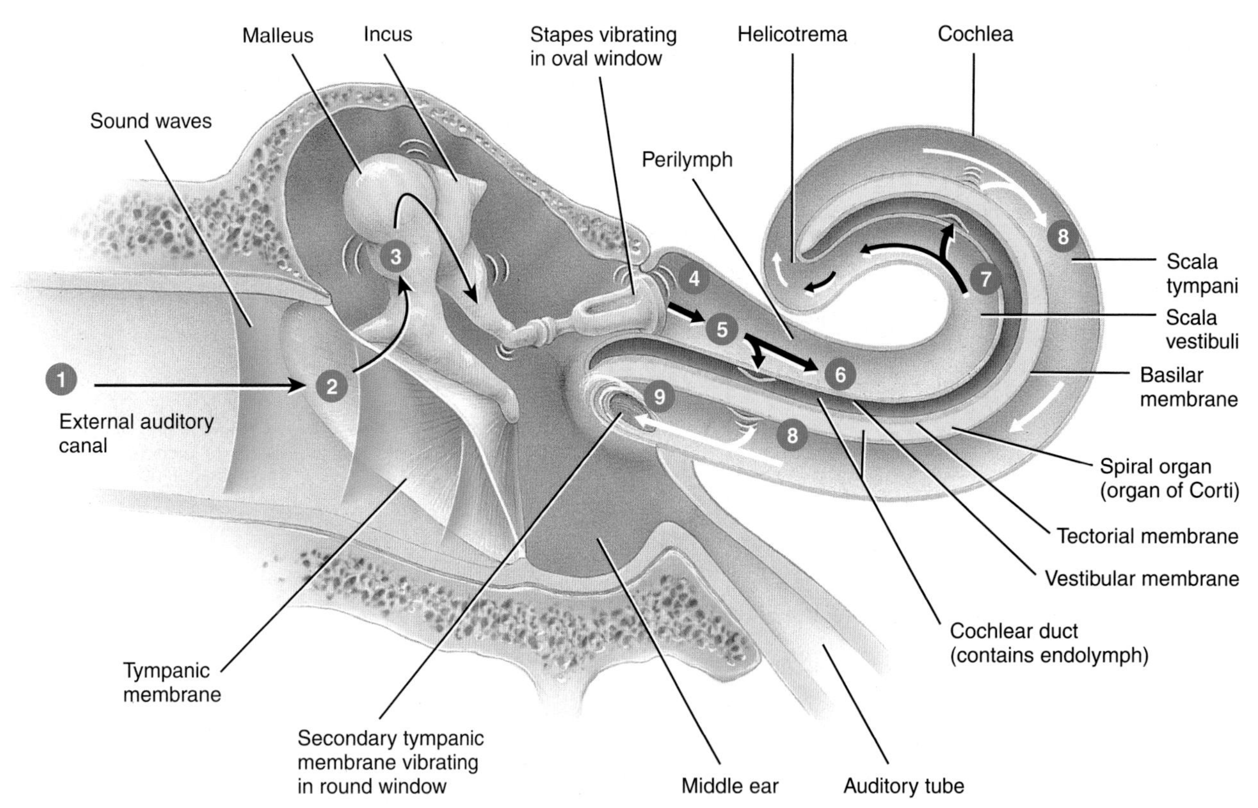

? Which part of the basilar membrane vibrates most vigorously in response to high-frequency (high-pitched) sounds?

Then, the sensory neurons generate nerve impulses that conduct along the vestibulocochlear (VIII) nerve.

Sound waves of various frequencies cause certain regions of the basilar membrane to vibrate more intensely than other regions. Each segment of the basilar membrane is "tuned" for a particular pitch. Because the membrane is narrower and stiffer at the base of the cochlea (closer to the oval window), high-frequency (high-pitched) sounds induce maximal vibrations in this region. Toward the helicotrema, the basilar membrane is wider and more flexible; low-frequency (low-pitched) sounds cause maximal vibration of the basilar membrane there. Loudness is determined by the intensity of sound waves. High-intensity sound waves cause larger vibrations of the basilar membrane, which leads to a higher frequency of nerve impulses reaching the brain. Louder sounds also may stimulate a larger number of hair cells.

Auditory Pathway

Sensory neurons in the cochlear branch of each vestibulocochlear (VIII) nerve terminate in the medulla oblongata on the same side of the brain. From the medulla, axons ascend to the midbrain, then to the thalamus, and finally to the primary auditory area in the temporal lobe (see Figure 17.9). Because many auditory axons cross to the opposite side, the right and left primary auditory areas receive nerve impulses from both ears.

Deafness

Deafness is significant or total hearing loss. *Sensorineural deafness* is caused by either impairment of hair cells in the cochlea or damage of the cochlear branch of the vestibulocochlear nerve. This type of deafness may be caused by atherosclerosis, which reduces blood supply to the ears; repeated exposure to loud noise, which destroys hair cells of the spiral organ; or certain drugs such as aspirin and streptomycin. *Conduction deafness* is caused by impairment of the external and middle ear mechanisms for transmitting sounds to the cochlea. It may be caused by otosclerosis, the deposition of new bone around the oval window; impacted cerumen; injury to the tympanic membrane; or aging, which often results in thickening of the tympanic membrane and stiffening of the joints of the auditory ossicles.

Physiology of Equilibrium

Now that you have learned about the anatomy of the internal ear structures for equilibrium, we briefly cover the physiology of balance, or how you are able to stay on your feet after tripping over your roommate's shoes. There are two types of ***equilibrium*** (balance). One kind, called ***static equilibrium,*** refers to the maintenance of the position of the body (mainly the head) relative to the force of gravity. Body movements that stimulate the receptors for static equilibrium include tilting the head and *linear* acceleration or deceleration, such as when the body is being moved in an elevator or in a car that is speeding up or slowing down. The second kind, ***dynamic equilibrium,*** is the maintenance of body position (mainly the head) in response to sudden movements such as *rotational* acceleration or deceleration. Collectively, the receptor organs for equilibrium within the saccule, utricle, and membranous semicircular ducts are called the ***vestibular apparatus*** (ves-TIB-ū-lar).

The walls of both the utricle and the saccule contain a small, thickened region called a ***macula*** (MAK-ū-la; *macula* = spot), which is a receptor for static equilibrium. The two maculae (plural) provide sensory information on the position of the head in space and help maintain appropriate posture and balance.

The three membranous semicircular ducts lie at right angles to one another in three planes (see Figure 19.13a). The positioning permits detection of linear and rotational acceleration or deceleration.

Equilibrium Pathways

Most of the vestibular branch axons of the vestibulocochlear (VIII) nerve enter the brain stem and then extend to the medulla or the cerebellum, where they synapse with the next neurons in the equilibrium pathways. From the medulla, some axons conduct nerve impulses along the cranial nerves that control eye movements and head and neck movements. Other axons form a spinal cord tract that conveys impulses for regulation of muscle tone in response to head movements. Various pathways among the medulla, cerebellum, and cerebrum enable the cerebellum to play a key role in maintaining equilibrium. The cerebellum continuously receives sensory information from the utricle and saccule. In response, the cerebellum makes adjustments to the signals going from the motor cortex to specific skeletal muscles to maintain equilibrium.

Table 19.4 summarizes the structures of the ear related to hearing and equilibrium.

Reflexology via the Ear

Reflexology is a type of alternative medicine utilized by some manual therapists. It is based on manipulating key points of the body to balance the natural energy flow. Although similar to acupuncture, reflexology originated in Europe rather than Asia and was brought to the United States in the early 1900s. Many reflexologists prefer working on the feet or hands of a patient, but the ear is also used. Reflexologists believe that pressing on a particular part of the auricle will cause a reflex reaction in a corresponding body part.

One form of reflexology utilizes specific locations on the external ear, such as the tragus, concha, helix, antihelix, and antitragus (see Figure 19.15). Here's a brief description of the surface anatomy of the ear:

- **Auricle** (AW-ri-kul). Shell-shaped portion of the external ear; also called the ***pinna.*** It funnels sound waves into the external auditory canal and plays an important role in helping to *localize* sound (determine its location). Just posterior to the auricle, the mastoid process of the temporal bone can be palpated.

TABLE 19.4

Summary of Structures of the Ear

REGIONS OF THE EAR AND KEY STRUCTURES	FUNCTIONS
External (Outer) Ear 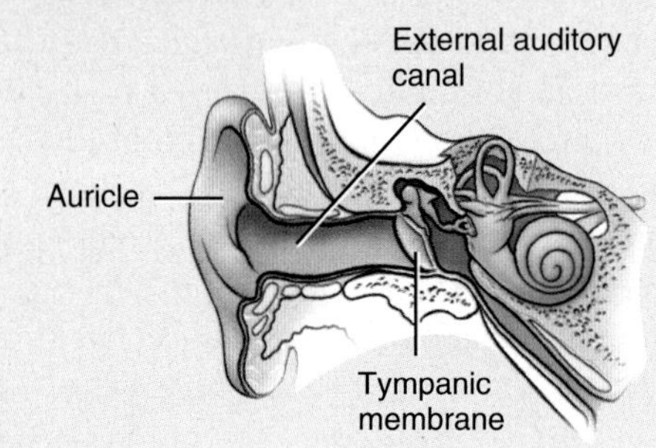 	***Auricle (pinna):*** Collects sound waves. ***External auditory canal (meatus):*** Directs sound waves to the tympanic membrane. ***Tympanic membrane (eardrum):*** Sound waves cause it to vibrate, which, in turn, causes the malleus to vibrate.
Middle Ear 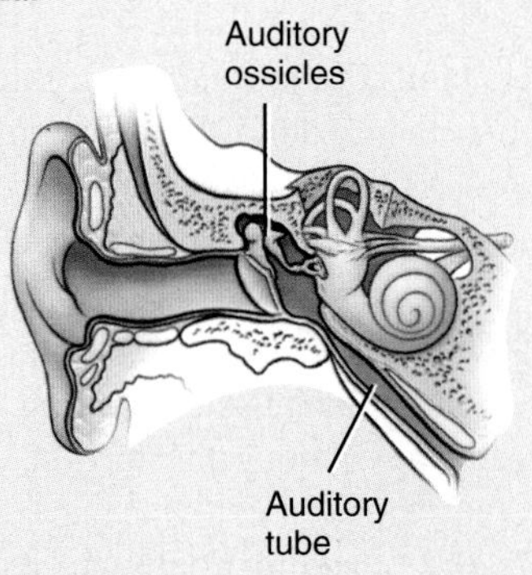 	***Auditory ossicles:*** Transmit and amplify vibrations from tympanic membrane to oval window. ***Auditory (eustachian) tube:*** Equalizes air pressure on both sides of the tympanic membrane.
Internal (Inner) Ear 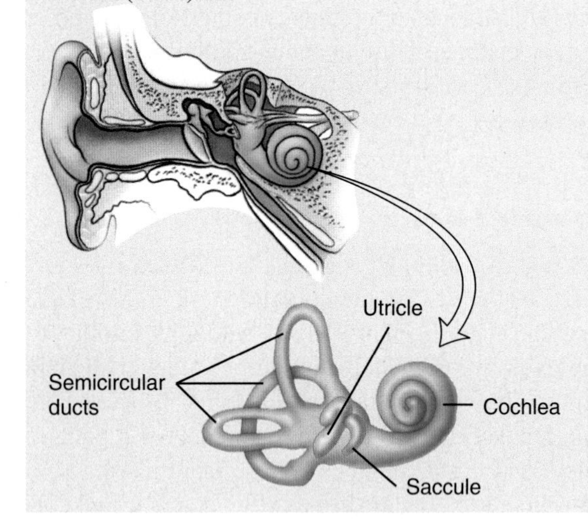 	***Cochlea:*** Contains a series of fluids, channels, and membranes that transmit vibrations to the spiral organ (organ of Corti), the organ of hearing; hair cells in the spiral organ trigger nerve impulses in the cochlear branch of the vestibulocochlear (VIII) nerve. ***Vestibular apparatus:*** Includes semicircular ducts, utricle, and saccule, which generate nerve impulses that propagate along the vestibular branch of the vestibulocochlear (VIII) nerve. ***Semicircular ducts:*** Contain receptors primarily for dynamic equilibrium (maintenance of body position, mainly the head, in response to rotational acceleration). ***Utricle and Saccule:*** Contain receptors primarily for static equilibrium (maintenance of body position, mainly the head, relative to the force of gravity).

- **Tragus** (TRĀ-gus = goat; when hair grows on the tragus it is thought to resemble the beard on a goat's chin). Cartilaginous projection anterior to the external auditory canal. Anterior and inferior to the tragus, it is possible to feel the **temporomandibular joint (TMJ).** As you open and close your jaw, you can feel the movement of the mandibular condyle.
- **Antitragus.** Cartilaginous projection opposite the tragus.
- **Concha** (KON-ka). Hollow of the auricle.
- **Helix.** Superior and posterior free margin of the auricle.
- **Antihelix.** Semicircular ridge superior and posterior to the tragus.
- **Triangular fossa.** Depression in the superior portion of the antihelix.
- **Lobule.** Inferior portion of the auricle; it does not contain cartilage. Commonly referred to as the earlobe.
- **External auditory canal (meatus).** Canal about 3 cm (1 in.) long extending from the external ear to the tympanic

Figure 19.15 Surface anatomy of the right ear.

The most conspicuous surface feature of the ear is the auricle.

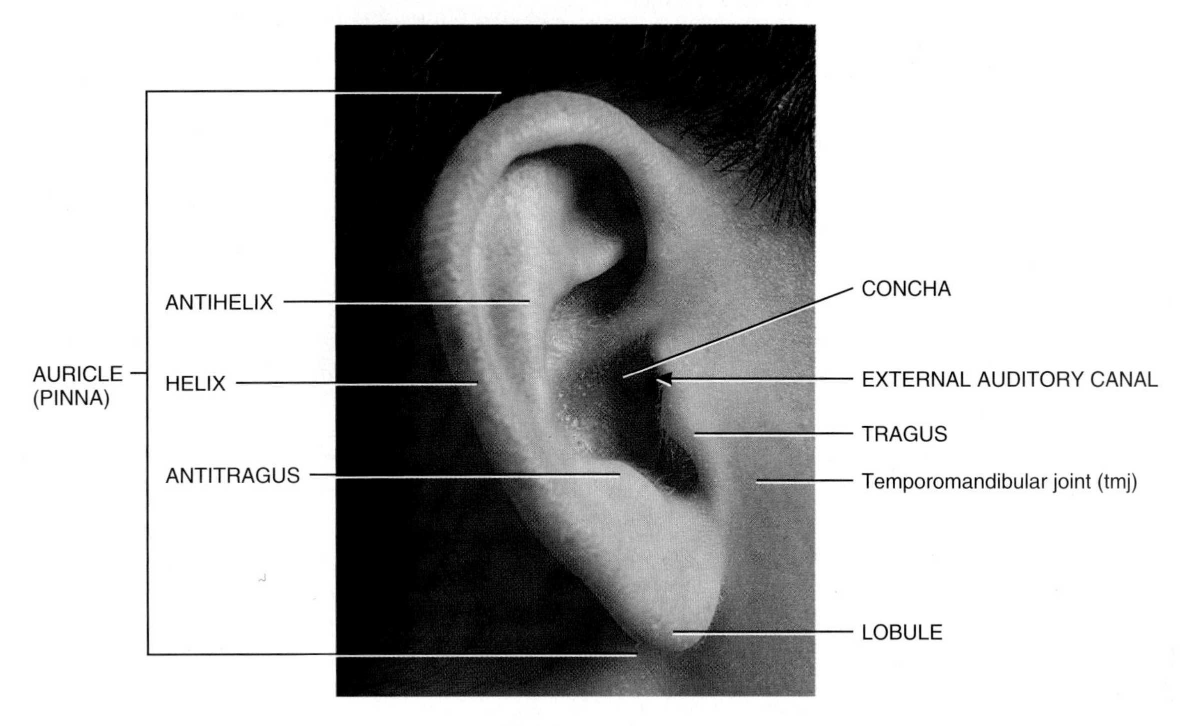

Is the tragus or the antitragus more anterior?

membrane (eardrum). It contains ceruminous glands that secrete cerumen (earwax). The **condylar process** of the mandible can be palpated by placing your little finger in the canal and opening and closing your mouth.

CHECKPOINT

16. How are sound waves transmitted from the auricle to the spiral organ of Corti?
17. How do hair cells in the cochlea convert mechanical vibrations into electrical impulses?
18. What is the pathway for auditory impulses from the cochlea to the cerebral cortex?
19. What is the contribution of sensations generated by the vestibular apparatus to the cerebellum?

19.7 AGING AND THE SPECIAL SENSES

OBJECTIVE

- Describe the age-related changes that occur in the eyes and ears.

Most people do not experience any problems with the senses of smell and taste until about age 50. This loss is due to a gradual loss of olfactory receptors and gustatory receptor cells coupled with their slower rate of replacement as we age.

Several age-related changes occur in the eyes. As noted earlier, the lens loses some of its elasticity and thus cannot change shape as easily, resulting in presbyopia. Cataracts (loss of transparency of the lenses) also occur with aging. In old age, the sclera ("white" of the eye) becomes thick and rigid and develops a yellowish or brownish coloration due to many years of exposure to ultraviolet light, wind, and dust. The sclera may also develop random splotches of pigment, especially in people with dark complexions. The iris fades or develops irregular pigment. The muscles that regulate the size of the pupil weaken with age and the pupils become smaller, react more slowly to light, and dilate more slowly in the dark. For these reasons, elderly people find that objects are not as bright, their eyes may adjust more slowly when going outdoors, and they have problems going from brightly lit to darkly lit places. Some diseases of the retina are more likely to occur in old age, including age-related macular disease and detached retina. A disorder called glaucoma develops in the eyes of aging people as a result of the buildup of aqueous humor. The number of mucous cells in the conjunctiva and tear production may decrease with age, resulting in dry eyes. The eyelids lose their elasticity, becoming baggy and wrinkled. The amount of fat around the orbits may decrease, causing the eyeballs to sink into the orbits. Finally, as we age the sharpness of vision decreases, color and depth perception are reduced, and vitreal floaters increase. Vitreal floaters give the appearance of moving spots before the eyes.

By about age 60, around 25% of individuals experience a noticeable hearing loss, especially for higher pitched sounds. The age-associated progressive loss of hearing in both ears is called ***presbycusis*** (pres′-bī-KŪ-sis; *presby-* = old; *-acou-* = hearing; *-sis* = condition). It may be related to damaged and lost hair cells in the spiral organ or degeneration of the nerve pathway for hearing. Tinnitus (ringing, roaring, or clicking in the ears) and dysfunction of the vestibular apparatus also occur more frequently in the elderly.

CHECKPOINT

20. What are some age-related issues of the special senses?

Now that our exploration of the nervous system and sensations is completed, you can appreciate the many ways that the nervous system contributes to homeostasis of other body systems (Exhibit 19.1). Next, in Chapter 20, we will see how the hormones released by the endocrine system also help maintain homeostasis of many body processes.

KEY MEDICAL TERMS ASSOCIATED WITH SOMATIC AND SPECIAL SENSES

Age-related macular disease (AMD) Degeneration of the macula lutea of the retina in persons 50 years of age and older.

Anosmia (an-OZ-mē-a; *a-* = without; *-osmi-* = smell, odor) Total lack of the sense of smell.

Cochlear implant A device that translates sounds into electrical signals that can be interpreted by the brain. It is especially useful for people with deafness caused by damage to hair cells in the cochlea.

Conjunctivitis (pinkeye) An inflammation of the conjunctiva; when caused by bacteria such as pneumococci, staphylococci, or *Hemophilus influenzae,* it is very contagious and more common in children. May also be caused by irritants, such as dust, smoke, or pollutants in the air, in which case it is not contagious.

Detached retina Detachment of the neural portion of the retina from the pigment epithelium due to trauma, disease, or age-related degeneration. The result is distorted vision and blindness.

Impacted cerumen Affects the transmission of sound waves through the external auditory canal (meatus). Impacted cerumen may cause the speech of infants to be affected and may account for much of the loss of hearing in the elderly.

LASIK (laser-assisted in-situ keratomileusis) Surgery with a laser to correct the curvature of the cornea for conditions such as nearsightedness, farsightedness, and astigmatism.

Ménière's disease An increased amount of endolymph; causes fluctuating hearing loss, tinnitus, and vertigo.

Nystagmus (nis-TAG-mus = nodding or drowsy) A rapid involuntary movement of the eyeballs, possibly caused by a disease of the central nervous system. It is associated with conditions that cause vertigo.

Otalgia (ō-TAL-jē-a; *ot-* = ear; *-algia* = pain) Earache.

Retinoblastoma (ret-i-nō-blas-TŌ-ma; *-oma* = tumor) A tumor arising from immature retinal cells; it accounts for 2% of childhood cancers.

Scotoma (skō-TŌ-ma = darkness) An area of reduced or lost vision in the visual field.

Strabismus (stra-BIZ-mus) An imbalance in the extrinsic eye muscles that causes a misalignment of one eye so that its line of vision is not parallel with that of the other eye (cross-eyes) and both eyes are not pointed at the same object at the same time; the condition produces a squint.

Trachoma (tra-KŌ-ma) A serious form of conjunctivitis and the greatest single cause of blindness in the world. It is caused by the bacterium *Chlamydia trachomatis*. The disease produces an excessive growth of subconjunctival tissue and invasion of blood vessels into the cornea, which progresses until the entire cornea is opaque, causing blindness.

Vertigo (VER-ti-gō = dizziness) A sensation of spinning or movement in which the world seems to revolve or the person seems to revolve in space.

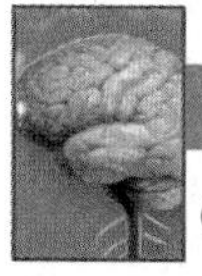

STUDY OUTLINE

Overview of Sensations (Section 19.1)

1. Sensation is the conscious or subconscious awareness of external and internal stimuli.
2. Two general classes of senses are (1) general senses, which include somatic senses and visceral senses, and (2) special senses, which include smell, taste, vision, hearing, and equilibrium (balance).
3. The conditions for a sensation to occur are reception of a stimulus by a sensory receptor, conversion of the stimulus into one or more nerve impulses, conduction of the impulses to the brain, and integration of the impulses by a region of the brain.
4. Sensory impulses from each part of the body arrive in specific regions of the cerebral cortex.
5. Adaptation is a decrease in sensation during a prolonged stimulus. Some receptors are rapidly adapting; others are slowly adapting.
6. Receptors can be classified structurally by their microscopic features as free nerve endings, encapsulated nerve endings, or separate cells. Functionally, receptors are classified by the type of stimulus they detect as mechanoreceptors, thermoreceptors, nociceptors, photoreceptors, osmoreceptors, and chemoreceptors.

Somatic Senses (Section 19.2)

1. Somatic sensations include tactile sensations (touch, pressure, vibration, itch, and tickle), thermal sensations (heat and cold), pain sensations, and proprioceptive sensations (joint and muscle position sense and movements of the limbs). Receptors for these sensations are located in the skin, mucous membranes, muscles, tendons, and joints.

EXHIBIT 19.1 Contributions of the Nervous System to Homeostasis

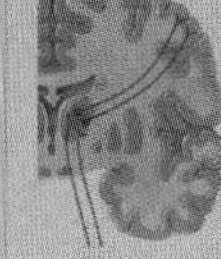

BODY SYSTEM		CONTRIBUTION TO HOMEOSTASIS
For all body systems		Together with hormones from the endocrine system, nerve impulses provide communication and regulation of most body tissues.
Integumentary system		Sympathetic nerves of the autonomic nervous system (ANS) control contraction of smooth muscles attached to hair follicles and secretion of perspiration from sweat glands.
Skeletal system	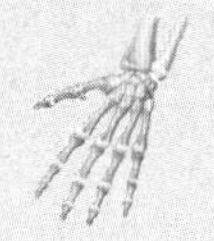	Nociceptors (pain receptors) in bone tissue warn of bone trauma or damage.
Muscular system		Somatic motor neurons receive instructions from motor areas of the brain and stimulate contraction of skeletal muscles to bring about body movements. The basal ganglia and reticular formation set the level of muscle tone. The cerebellum coordinates skilled movements.
Endocrine system	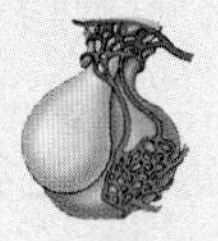	The hypothalamus regulates secretion of hormones from the anterior and posterior pituitary. The ANS regulates secretion of hormones from the adrenal medulla and pancreas.
Cardiovascular system	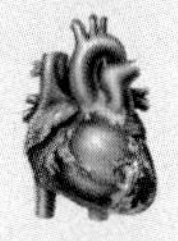	The cardiovascular center in the medulla oblongata provides nerve impulses to the ANS that govern heart rate and the forcefulness of the heartbeat. Nerve impulses from the ANS also regulate blood pressure and blood flow through blood vessels.
Lymphatic system and immunity		Certain neurotransmitters help regulate immune responses. Activity in the nervous system may increase or decrease immune responses.
Respiratory system	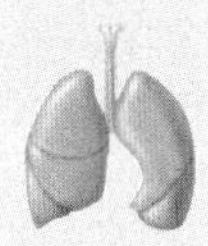	Respiratory areas in the brain stem control breathing rate and depth. The ANS helps regulate the diameter of airways.
Digestive system	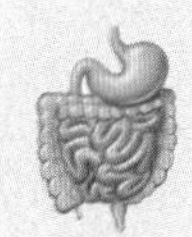	The ANS and enteric nervous system (ENS) help regulate digestion. The parasympathetic division of the ANS stimulates many digestive processes.
Urinary system	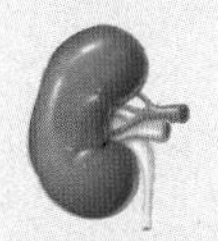	The ANS helps regulate blood flow to kidneys, thereby influencing the rate of urine formation; brain and spinal cord centers govern emptying of urinary bladder.
Reproductive systems	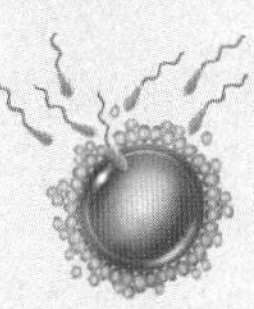	The hypothalamus and limbic system govern a variety of sexual behaviors; the ANS brings about erection of the penis in males and the clitoris in females and ejaculation of semen in males. The hypothalamus regulates release of anterior pituitary hormones that control the gonads (ovaries and testes). Nerve impulses elicited by touch stimuli from a suckling infant cause the release of oxytocin and milk ejection in nursing mothers.

2. Receptors for touch include corpuscles of touch (Meissner corpuscles), hair root plexuses, type I cutaneous mechanoreceptors (Merkel discs), and type II cutaneous mechanoreceptors (Ruffini corpuscles). Receptors for pressure and vibration are corpuscles of touch, type I mechanoreceptors, and lamellated (pacinian) corpuscles. Tickle and itch sensations result from stimulation of free nerve endings and lamellated corpuscles.
3. Thermoreceptors, free nerve endings in the epidermis and dermis, adapt to continuous stimulation.
4. Nociceptors are free nerve endings that are located in nearly every body tissue; they provide pain sensations.
5. Proprioceptors inform us of the degree to which muscles are contracted, the amount of tension present in tendons, the positions of joints, and the orientation of the head.

Olfaction: Sense of Smell (Section 19.4)

1. The olfactory epithelium in the upper portion of the nasal cavity contains olfactory receptors, supporting cells, and basal stem cells.
2. Individual olfactory receptors respond to hundreds of different odorant molecules by producing an electrical signal that triggers one or more nerve impulses. Adaptation (decreasing sensitivity) to odors occurs rapidly.
3. Axons of olfactory receptors form the olfactory nerves, which convey nerve impulses to the olfactory bulbs. From there, impulses conduct via the olfactory tracts to the limbic system, hypothalamus, and cerebral cortex (temporal lobe).

Gustation: Sense of Taste (Section 19.5)

1. The five primary tastes are salty, sweet, sour, bitter, and umami.
2. The receptors for gustation, the gustatory receptor cells, are located in taste buds.
3. To be tasted, substances must be dissolved in saliva.
4. Gustatory receptor cells trigger impulses in cranial nerves VII (facial), IX (glossopharyngeal), and X (vagus). Impulses for taste conduct to the medulla oblongata, limbic system, hypothalamus, thalamus, and the primary gustatory area in the parietal lobe of the cerebral cortex.

Vision (Section 19.6)

1. Accessory structures of the eyes include the eyebrows, eyelids, eyelashes, the lacrimal apparatus (which produces and drains tears), and extrinsic eye muscles (which move the eyes).
2. The eyeball has three layers: (1) fibrous tunic (sclera and cornea), (2) vascular tunic (choroid, ciliary body, and iris), and (3) retina.
3. The retina consists of a neural layer (photoreceptor layer, bipolar cell layer, and ganglion cell layer) and a pigmented layer (a sheet of melanin-containing epithelial cells).
4. The anterior cavity contains aqueous humor; the vitreous chamber contains the vitreous body.
5. Image formation on the retina involves refraction of light rays by the cornea and lens, which focus an inverted image on the fovea centralis of the retina.
6. For viewing close objects, the lens increases its curvature (accommodation), and the pupil constricts to prevent light rays from entering the eye through the periphery of the lens.
7. Improper refraction may result from myopia (nearsightedness), hypermetropia (farsightedness), or astigmatism (irregular curvature of the cornea or lens).
8. Movement of the eyeballs toward the nose to view an object is called convergence.
9. The first step in vision is the absorption of light rays by photopigments in rods and cones (photoreceptors). Stimulation of the rods and cones then activates bipolar cells, which in turn activate the ganglion cells.
10. Nerve impulses arise in ganglion cells and conduct along the optic nerve, through the optic chiasm and optic tract, to the thalamus. From the thalamus, neurons extend to the primary visual area in the occipital lobe of the cerebral cortex.

Hearing and Equilibrium (Section 19.7)

1. The external (outer) ear consists of the auricle, external auditory canal (meatus), and tympanic membrane (eardrum).
2. The middle ear consists of the auditory (eustachian) tube, auditory ossicles, oval window, and round window.
3. The internal (inner) ear consists of the bony labyrinth and membranous labyrinth. The internal ear contains the spiral organ (organ of Corti), the organ of hearing.
4. Sound waves enter the external auditory canal, strike the tympanic membrane, pass through the ossicles, strike the oval window, set up pressure waves in the perilymph, strike the vestibular membrane and scala tympani, increase pressure in the endolymph, vibrate the basilar membrane, and stimulate hair cells in the spiral organ.
5. Hair cells release neurotransmitter molecules that can initiate nerve impulses in sensory neurons.
6. Sensory neurons in the cochlear branch of the vestibulocochlear (VIII) nerve terminate in the medulla oblongata. Auditory signals then pass to the midbrain, thalamus, and temporal lobes.
7. Static equilibrium is the orientation of the body relative to the pull of gravity. The maculae of the utricle and saccule are the receptor organs of static equilibrium.
8. Dynamic equilibrium is the maintenance of body position in response to rotation, acceleration, and deceleration. Receptors in the semicircular ducts are the receptor organs of dynamic equilibrium.
9. Most vestibular branch axons of the vestibulocochlear (VIII) nerve enter the brain stem and terminate in the medulla and pons; other axons extend to the cerebellum.

Aging and Special Senses (Section 19.8)

1. Most people do not experience problems with the senses of smell and taste until about age 50.
2. Among the age-related changes to the eyes are presbyopia, cataracts, difficulty adjusting to light, macular disease, glaucoma, dry eyes, and decreased sharpness of vision.
3. With age there is a progressive loss of hearing and tinnitus occurs more frequently.

SELF-QUIZ QUESTIONS

1. You enter a sauna and it feels awfully hot, but soon the temperature feels comfortably warm. What have you experienced?
a. damage to your thermoreceptors
b. sensory adaptation
c. a change in the temperature of the sauna
d. inactivation of your thermoreceptors
e. damage to the parietal lobe

2. The lacrimal glands produce ________ which drain(s) into the ________.
a. tears; anterior cavity
b. tears; nasal cavity
c. aqueous humor; anterior chamber
d. aqueous humor; anterior cavity
e. aqueous humor; scleral venous sinus

3. The spiral organ (organ of Corti)
a. contains hair cells
b. is responsible for equilibrium
c. is filled with perilymph
d. is another name for the auditory (eustachian) tube
e. contains photoreceptors.

4. Equilibrium and the activities of muscles and joints are monitored by
a. olfactory receptors
b. nociceptors
c. tactile receptors
d. proprioceptors
e. thermoreceptors.

5. In the retina, cone photoreceptors
a. are more numerous than rods
b. contain the photopigment rhodopsin
c. are more sensitive to low light level than are rods
d. reform their photopigments more slowly than do rods
e. provide higher acuity vision than do rods.

6. Which of the following is NOT required for a sensation to occur?
a. the presence of a stimulus
b. a receptor specialized to detect a stimulus
c. the presence of slowly adapting receptors
d. a sensory neuron to conduct impulses
e. a region of the brain for integration of the nerve impulse

7. Match each receptor with its function.

____ **a.** color vision	**A.** lamellated (pacinian) corpuscle
____ **b.** taste	**B.** type I cutaneous mechanoreceptor
____ **c.** smell	**C.** rod photoreceptor
____ **d.** vision in dim light	**D.** nociceptor
____ **e.** stretch in a muscle	**E.** gustatory receptor cell
____ **f.** pressure	**F.** olfactory receptor
____ **g.** touch	**G.** muscle spindle
____ **h.** detection of pain	**H.** cone photoreceptors

8. For taste to occur
a. the mouth must be dry
b. the chemical must be in contact with the basal cells
c. filiform papillae must be stimulated
d. the limbic system needs to be activated
e. the gustatory hair must be stimulated by the dissolved chemical.

9. Which of the following characteristics of taste is NOT true?
a. Olfaction can affect taste.
b. Three cranial nerves conduct the impulses for taste to the brain.
c. Taste adaptation occurs quickly.
d. Humans can recognize about 10 primary tastes.
e. Taste receptors are located in taste buds on the tongue, on the roof of the mouth, in the throat, and in the epiglottis.

10. You are seated at your desk and drop your pencil. As you lean over to retrieve it, what is occurring in your internal (inner) ear?
a. Receptors in the utricle and saccule are responding to changes in static equilibrium.
b. The hair cells in the cochlea are responding to changes in dynamic equilibrium.
c. Receptors of each semicircular duct are responding to changes in dynamic equilibrium.
d. The cochlear branch of the vestibulocochlear (VIII) nerve begins to transmit nerve impulses to the brain.
e. The auditory (eustachian) tube makes adjustments for varying air pressures.

11. Kinesthesia is the
a. perception of body movements
b. ability to identify an object by feeling it
c. sensation of weightlessness that occurs in outer space
d. decrease in sensitivity of receptors to a prolonged stimulus
e. movement of body parts in a rhythmic manner.

12. Which of the following is NOT true about nociceptors?
a. They respond to stimuli that may cause tissue damage.
b. They consist of free nerve endings.
c. They can be activated by excessive stimuli from other sensations.
d. They are found in virtually every body tissue except the brain.
e. They adapt very rapidly.

13. Which of the following is NOT a function of tears?
a. moisten the eye
b. wash away eye irritants
c. destroy certain bacteria
d. lubricate the eye
e. provide nutrients to the cornea

14. Transmission of vibrations (sound waves) from the tympanic membrane to the oval window is accomplished by
a. neurons
b. the tectorial membrane
c. the auditory ossicles
d. the endolymph
e. the auditory (eustachian) tube.

15. Match the following:

____ **a.** focuses light rays onto the retina	**A.** sclera
____ **b.** regulates the amount of light entering the eye	**B.** choroid
____ **c.** contains aqueous humor	**C.** lacrimal gland
____ **d.** contains blood vessels that help nourish the retina	**D.** lens
____ **e.** produces tears	**E.** retina
____ **f.** dense connective tissue that provides shape to the eye	**F.** iris
____ **g.** contains photoreceptors	**G.** anterior cavity

16. Which of the following structures refracts light rays entering the eye?
a. cornea
b. sclera
c. pupil
d. retina
e. conjunctiva

17. Your 45-year-old neighbor has begun to have difficulty reading the morning newspaper. You explain that this condition is known as ________ and is due to ________.
a. myopia, inability of his eyes to properly focus light on his retinas
b. night blindness, a vitamin A deficiency
c. binocular vision, the eyes focusing on two different objects
d. astigmatism, an irregularity in the curvature of the lens
e. presbyopia, the loss of elasticity in the lens

18. Damage to cells in the fovea centralis would interfere with
a. dynamic equilibrium
b. accommodation
c. visual acuity
d. ability to see in dim light
e. intraocular pressure.

19. Place the following events concerning the visual pathway in the correct order:
1. Nerve impulses exit the eye via the optic nerve.
2. Optic tract axons terminate in the thalamus.
3. Light reaches the retina.
4. Rods and cones are stimulated.
5. Synapses occur in the thalamus and continue to the primary visual area in the occipital lobe.
6. Ganglion cells generate nerve impulses.
a. 4, 1, 2, 5, 6, 3 **b.** 5, 4, 1, 3, 2, 6 **c.** 3, 4, 6, 1, 5, 2
d. 3, 4, 6, 1, 2, 5 **e.** 3, 4, 5, 6, 1, 2

20. Place the following events of the auditory pathway in the correct order:
1. Hair cells in the spiral organ bend as they rub against the tectorial membrane.
2. Movement in the oval window begins movement in the perilymph.
3. Nerve impulses exit the ear via the vestibulocochlear (VIII) nerve.
4. The eardrum and auditory ossicles transmit vibrations from sound waves.
5. Pressure waves from the perilymph cause bulging of the round window and formation of pressure waves in the endolymph.
a. 4, 2, 5, 1, 3 **b.** 4, 5, 2, 3, 1 **c.** 5, 3, 2, 4, 1
d. 3, 4, 5, 1, 2 **e.** 2, 4, 1, 5, 3

CRITICAL THINKING QUESTIONS

1. A therapist detected that a muscle spasm in her patient's levator scapulae muscle had just diminshed. The patient reported that the pain had reduced, but was still present. Why did the pain linger?
2. As you help your neighbor put drops in her 6-year-old daughter's eyes, the daughter states, "That medicine tastes bad." How do you explain to the neighbor how her daughter can "taste" the eyedrops?
3. The shift nurse gives ailing 80-year-old Gertrude a menu so she can choose her morning breakfast. Gertrude complains that she is having trouble reading the menu and asks the nurse to read it to her. As the nurse begins to read, Gertrude loudly asks her to "speak up and turn off the buzzing." What does the nurse know about aging and the special senses that help to explain Gertrude's comments?

? ANSWERS TO FIGURE QUESTIONS

19.1 Pain, thermal sensations, and tickle and itch arise with activations of different free nerve endings.
19.2 The kidneys have the broadest area for referred pain.
19.3 An olfactory receptor has a life span of about 1 month.
19.4 Basal cells develop into gustatory receptor cells.
19.5 Lacrimal fluid, or tears, is a watery solution containing salts, some mucus, and lysozyme that protects, cleans, lubricates, and moistens the eyeball.
19.6 The fibrous tunic consists of the cornea and sclera; the vascular tunic consists of the choroid, ciliary body, and iris.
19.7 The parasympathetic division of the ANS causes pupillary constriction; the sympathetic division causes pupillary dilation.
19.8 The two types of photoreceptors are rods and cones; rods provide black-and-white vision in dim light, and cones provide high visual acuity and color vision in bright light.
19.9 During accommodation the ciliary muscle contracts, causing the suspensory ligaments (zonular fibers) to slacken. The lens then becomes more convex, increasing its focusing power.
19.10 Presbyopia is the loss of lens elasticity that typically occurs with aging.
19.11 Structures carrying visual impulses from the retina to the occipital lobe are as follows: axons of ganglion cells → optic (II) nerve → optic chiasm → optic tract → thalamus → primary visual area in occipital lobe of the cerebral cortex.
19.12 The malleus of the middle ear is attached to the tympanic membrane (eardrum), which is part of the external ear.
19.13 The three subdivisions of the bony labyrinth are the semicircular canals, vestibule, and cochlea.
19.14 The region of the basilar membrane close to the oval and round windows vibrates most vigorously in response to high-frequency sounds.
19.15 The tragus of the ear is more anterior than the antitragus.

The Endocrine System 20

The endocrine (hormonal) system is one of the two control systems in the body, the other being the nervous system. Most glands in the body release regulatory substances into the interstitial fluid (the fluid that surrounds cells) that then enter the bloodstream. The blood carries the hormones to their specific target tissues. The interaction between these hormones and the metabolism of the body is usually controlled by negative feedback. As you learned in Chapter 1, the lack or excess of a hormone usually creates a response by negative feedback that returns the body to its optimal state. Hormones may be stimulating (causing a reaction) or inhibiting (preventing or limiting a previous response of another hormone). Hormones can create a variety of long-lived reactions in the body that regulate growth, sexual development, control of reproductive cycles, lactation, labor, pigmentation of the skin, water conservation, speed of metabolism, sleep cycles, and reduction of stress. When we are under stress, our body produces cortisol from the adrenal gland. A side effect of constant cortisol secretion is conversion of sugar to fat in the body. Therefore, we may gain weight when we are stressed out. This is another reason to treat stress in its early stages rather than let it get out of control.

CONTENTS AT A GLANCE

20.1 ENDOCRINE GLANDS

OBJECTIVE

- List the components of the endocrine system.

The ***endocrine system*** consists of several endocrine glands plus many hormone-secreting cells in organs that have functions besides secreting hormones. In contrast to the nervous system, which controls body activities through the release of neurotransmitters at synapses, the endocrine system releases hormones into interstitial fluid (fluid that surrounds cells) and then into the bloodstream. The circulating blood delivers hormones to virtually all cells throughout the body, and cells that recognize a particular hormone will respond. The nervous system and endocrine system often work together. For example, certain parts of the nervous system stimulate or inhibit the release of hormones by the endocrine system. Typically, the endocrine system acts more slowly than the nervous system, which often produces an effect within a fraction of a second. Moreover, the effects of hormones linger until they are cleared from the blood. The liver inactivates some hormones, and the kidneys excrete others in the urine.

As you learned in Chapter 4, two types of glands are present in the body: exocrine glands and endocrine glands. ***Exocrine glands*** secrete their products into *ducts* that carry the secretions into a body cavity, into the lumen of an organ, or onto the outer surface of the body. Sweat glands, oil glands, mucous glands, and digestive glands are examples of exocrine glands. The cells of ***endocrine glands,*** by contrast, secrete their products (hormones) into interstitial fluid; then the hormones diffuse into blood capillaries, and blood carries them throughout the body.

Disorders of the endocrine system often involve either ***hyposecretion*** (*hypo-* = too little or under), inadequate release of a hormone, or ***hypersecretion*** (*hyper-* = too much or above), excessive release of a hormone. In other cases, the problem is faulty hormone receptors or an inadequate number of receptors.

The endocrine glands include the pituitary, thyroid, parathyroid, adrenal, and pineal glands (Figure 20.1). In addition, several organs and tissues are not exclusively classified as endocrine glands but contain cells that secrete hormones. These include the hypothalamus, thymus, pancreas, ovaries, testes, kidneys, stomach, liver, small intestine, skin, heart, adipose tissue, and placenta. ***Endocrinology*** (en′-dō-kri-NOL-ō-jē; *endo-* = within; *-crino-* = to secrete; *-logy* = study of) is the scientific and medical specialty concerned with hormonal secretions and the diagnosis and treatment of disorders of the endocrine system.

CHECKPOINT

1. Why are organs such as the kidneys, stomach, heart, and skin considered part of the endocrine system?

20.2 HORMONE ACTION

OBJECTIVES

- Define target cells and describe the role of hormone receptors.
- Describe the two general mechanisms of action of hormones.

Target Cells and Hormone Receptors

Although a given hormone travels throughout the body in the blood, it affects only specific ***target cells.*** Hormones, like neurotransmitters, influence their target cells by chemically binding to specific protein ***receptors.*** Only the target cells for a given hormone have receptors that bind and recognize the hormone. For example, thyroid-stimulating hormone (TSH) binds to receptors on cells of the thyroid gland, but it does not bind to cells of the ovaries because ovarian cells do not have TSH receptors. Generally, a target cell has 2,000 to 100,000 receptors for a particular hormone.

Chemistry of Hormones

Chemically, some hormones are soluble in lipids (fats) and others are soluble in water. The ***lipid-soluble hormones*** include steroid hormones, thyroid hormones, and nitric oxide. *Steroid hormones* are made from cholesterol. The two *thyroid hormones* (T_3 and T_4) are made by attaching iodine atoms to the amino acid tyrosine. The gas *nitric oxide* (*NO*) is synthesized from the amino acid arginine.

Most of the ***water-soluble hormones*** are made from amino acids. For instance, the amino acid tyrosine is modified to form the hormones *epinephrine* and *norepinephrine* (which are also neurotransmitters). Other water-soluble hormones consist of short chains of amino acids (peptide hormones), such as *antidiuretic hormone (ADH)* and *oxytocin,* or longer chains of amino acids (protein hormones), for instance, *insulin* and *human growth hormone (hGH).*

Mechanisms of Hormone Action

The response to a hormone depends on both the hormone and the target cell. Various target cells respond differently to the same hormone. Insulin, for example, stimulates the synthesis of glycogen in liver cells but the synthesis of triglycerides in adipose cells. To exert an effect, a hormone first must "announce its arrival" to a target cell by binding to its receptors. The receptors for lipid-soluble hormones are located inside target cells, and the receptors for water-soluble hormones are part of the plasma membrane of target cells.

Control of Hormone Secretions

The release of most hormones occurs in short bursts, with little or no secretion between bursts. When stimulated, an endocrine gland releases its hormone in more frequent bursts, increasing the concentration of the hormone in the blood. In the absence of stimulation, the blood level of the hormone decreases as the hormone is inactivated or excreted. Regulation of secretion normally prevents overproduction or underproduction of any given hormone.

Hormone secretion is regulated by (1) signals from the nervous system, (2) chemical changes in the blood, and (3) other hormones. For example, nerve impulses to the adrenal medullae

Figure 20.1 Location of many endocrine glands. Also shown are other organs that contain endocrine cells and associated structures.

Endocrine glands secrete hormones, which circulating blood delivers to target tissues.

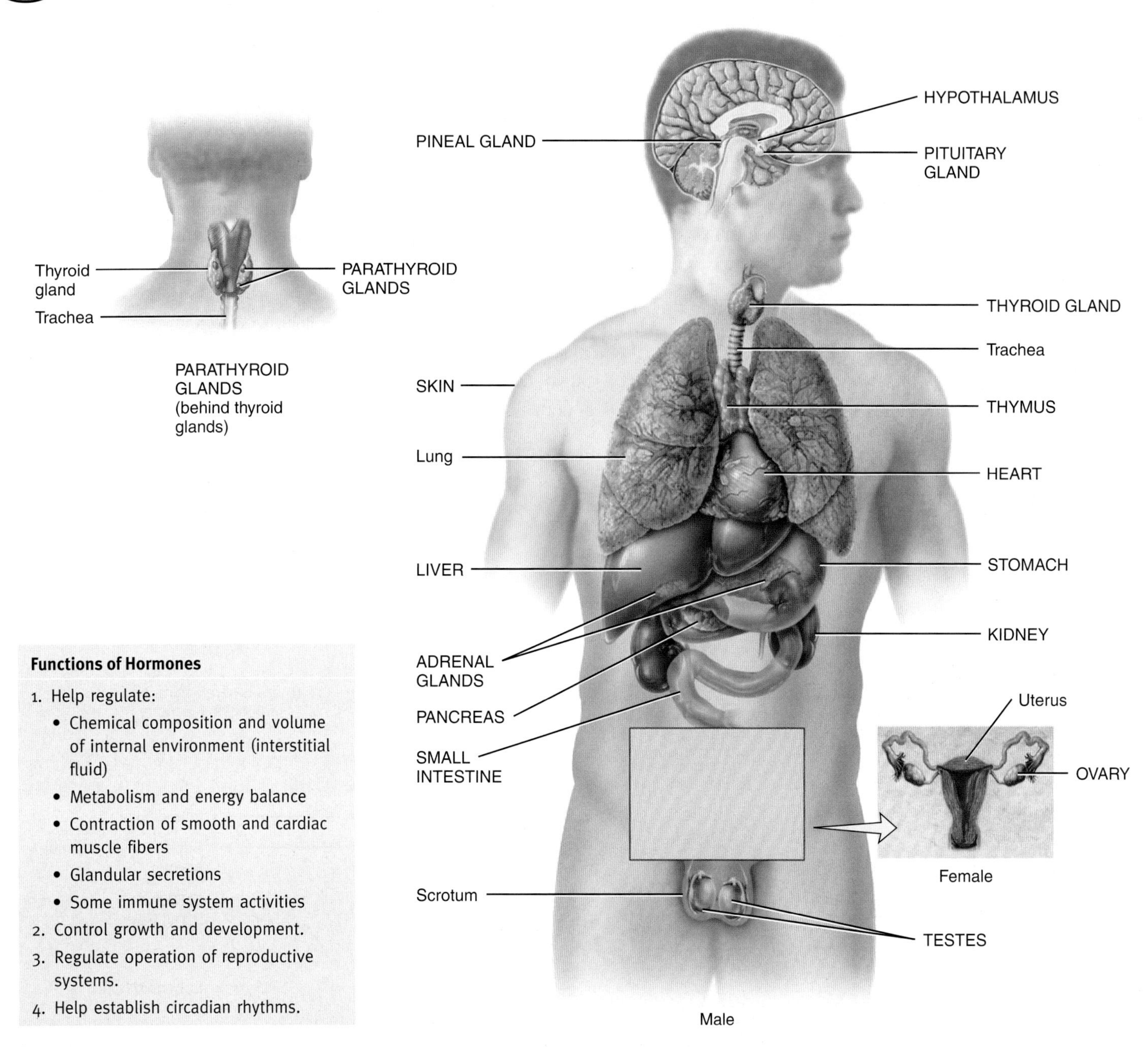

What is the basic difference between endocrine glands and exocrine glands?

regulate the release of epinephrine and norepinephrine; blood Ca^{2+} level regulates the secretion of parathyroid hormone; and a hormone from the anterior pituitary (adrenocorticotropic hormone, ACTH) stimulates the release of cortisol by the adrenal cortex. Most systems that regulate secretion of hormones work by negative feedback, but a few operate by positive feedback. For example, during childbirth, the hormone oxytocin stimulates contractions of the uterus, and uterine contractions, in turn, stimulate more oxytocin release, a positive feedback effect.

CHECKPOINT

2. Chemically, what types of molecules are hormones?
3. What are the general ways in which blood hormone levels are regulated?

20.3 HYPOTHALAMUS AND PITUITARY GLAND

OBJECTIVES

- Describe the locations of and relationship between the hypothalamus and the pituitary gland.
- Describe the functions of each hormone secreted by the pituitary gland.

For many years, the ***pituitary gland*** (pi-TOO-i-tār-ē) or ***hypophysis*** (hī-POF-i-sis) was considered the "master" endocrine gland because it secretes several hormones that control other endocrine glands. We now know that the pituitary gland itself has a master—the ***hypothalamus.*** This small region of the brain is the major link between the nervous and endocrine systems. Cells in the hypothalamus synthesize and secrete at least nine hormones, and the anterior pituitary gland secretes and synthesizes seven. Together, these hormones play important roles in the regulation of virtually all aspects of growth, development, metabolism, and homeostasis.

The pituitary gland is about the size of a large pea and has two lobes: a larger ***anterior pituitary*** (***adenohypophysis***) and a smaller ***posterior pituitary*** (***neurohypophysis***) (Figure 20.2). Both lobes of the pituitary gland rest in the ***hypophyseal fossa*** (hī′-pō-FIZ-ē-al), a cup-shaped depression in the sella turcica (SEL-a TUR-si-ka; *sella* = saddle; *turcica* = Turkish) of the sphenoid bone. A stalklike structure, the ***infundibulum,*** attaches

Figure 20.2 Hypothalamus and pituitary gland, and their blood supply. Releasing and inhibiting hormones synthesized by hypothalamic neurosecretory cells are transported within axons and released at the axon terminals (b). The hormones diffuse into capillaries of the hypothalamus and are carried by the hypophyseal portal veins to the anterior pituitary.

Hypothalamic hormones are an important link between the nervous and endocrine systems.

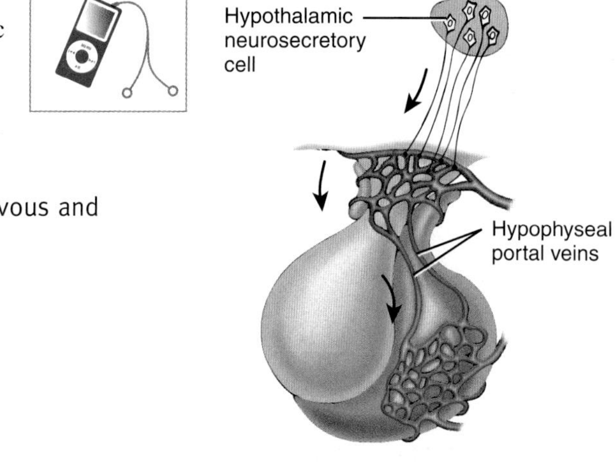

(b) Path of releasing and inhibiting hormones

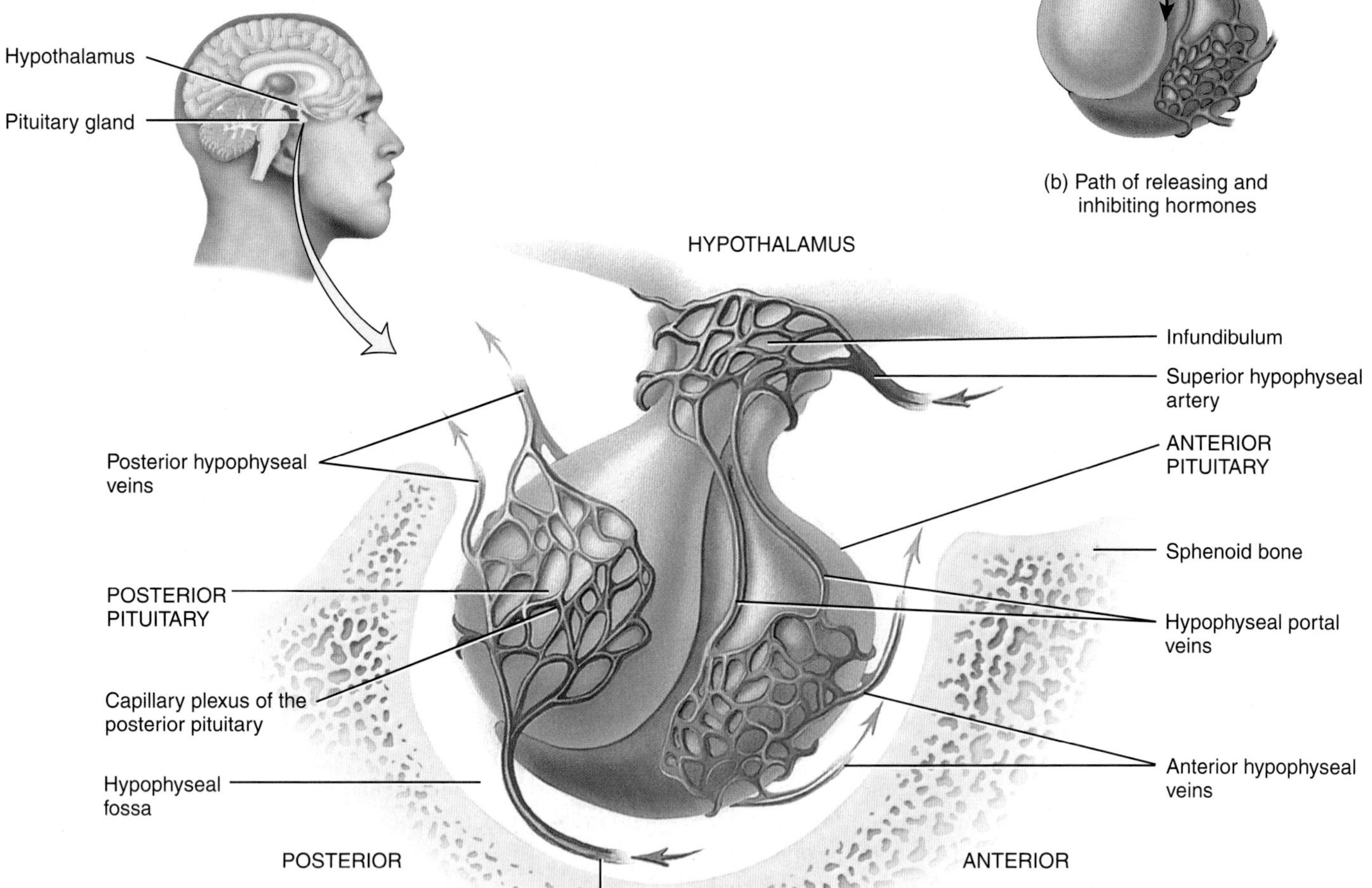

(a) Relationship of the hypothalamus to the pituitary gland

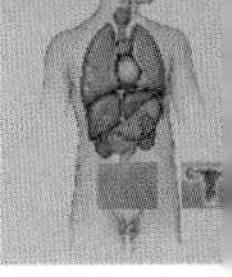

the pituitary gland to the hypothalamus. Hypothalamic hormones that release or inhibit hormones reach the anterior pituitary through a series of blood vessels. The *superior hypophyseal arteries* bring blood into the hypothalamus. Then it passes through a capillary network into *hypophyseal portal veins*. From here, it passes through another capillary network and finally exits the anterior pituitary through the *anterior hypophyseal veins*.

In contrast to the anterior lobe, the posterior lobe has a neural connection with the hypothalamus (see Figure 20.4). The posterior pituitary is actually part of the brain. It contains axons of hypothalamic neurons called *neurosecretory cells*. Hormones made by these cells pass down the axons to a capillary network in the posterior lobe. The hormones are released at the axon terminals into the capillaries and then exit the posterior lobe through the *posterior hypophyseal veins*.

Anterior Pituitary Hormones

The anterior pituitary synthesizes and secretes hormones that regulate a wide range of bodily activities, from growth to reproduction. Secretion of anterior pituitary hormones is stimulated by ***releasing hormones*** and suppressed by ***inhibiting hormones***, both produced by neurosecretory cells of the hypothalamus. The hypophyseal portal veins deliver the hypothalamic releasing and inhibiting hormones from the hypothalamus to the anterior pituitary (Figure 20.2). This direct route allows the releasing and inhibiting hormones to act quickly on cells of the anterior pituitary before the hormones are diluted or destroyed in the general circulation. Those anterior pituitary hormones that act on other endocrine glands are called ***tropic hormones*** (TRŌ-pik) or ***tropins***.

Human Growth Hormone and Insulin-like Growth Factors

***Human growth hormone* (*hGH*)**, also known as *somatotropin* and *somatotropic hormone (STH)*, is the most abundant anterior pituitary hormone. The main function of hGH is to promote synthesis and secretion of small protein hormones called *insulin-like growth factors* (*IGFs*) or *somatomedins*. IGFs are so named because some of their actions are similar to those of insulin. In response to hGH, cells in the liver, skeletal muscles, cartilage, bones, and other tissues secrete IGFs, which may either enter the bloodstream or act locally. IGFs stimulate protein synthesis, help maintain muscle and bone mass, and promote healing of injuries and tissue repair. They also enhance breakdown of triglycerides (fats), which releases fatty acids into the blood, and breakdown of liver glycogen, which releases glucose into the blood. Cells throughout the body can use the released fatty acids and glucose for the production of ATP.

The anterior pituitary releases hGH in bursts that occur every few hours, especially during sleep. Two hypothalamic hormones primarily control secretion of hGH: *growth hormone–releasing hormone* (*GHRH*) promotes secretion of human growth hormone, and *growth hormone–inhibiting hormone* (*GHIH*) suppresses it. Blood glucose level is a major regulator of GHRH and GHIH secretion. Low blood glucose level (hypoglycemia) stimulates the hypothalamus to secrete GHRH. By means of negative feedback, an increase in blood glucose concentration above the normal level (hyperglycemia) inhibits release of GHRH. By contrast, hyperglycemia stimulates the hypothalamus to secrete GHIH and hypoglycemia inhibits release of GHIH.

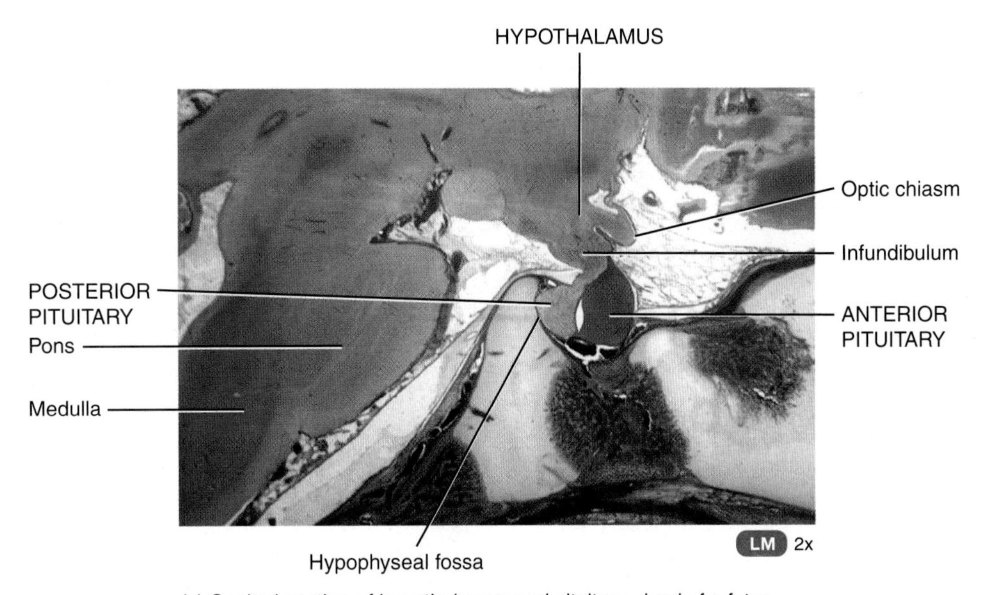

(c) Sagittal section of hypothalamus and pituitary gland of a fetus

What is the functional importance of the hypophyseal portal veins?

• CLINICAL CONNECTION | Pituitary Gland Disorders

Several disorders of the anterior pituitary involve human growth hormone (hGH). Undersecretion of hGH during the growth years slows bone growth, and the epiphyseal plates close before normal height is reached. This condition is called **pituitary dwarfism.** Other organs of the body also fail to grow, and the body proportions are childlike. Whereas a *dwarf* has a normal sized head and torso but small limbs, a *midget* has a proportioned head, torso, and limbs. Oversecretion of hGH during childhood results in **giantism (gigantism),** an abnormal increase in the length of long bones. The person grows to be very tall, but body proportions are about normal. Figure 20.3a shows identical twins; one brother developed giantism due to a pituitary tumor. Oversecretion of hGH during adulthood is called **acromegaly** (ak′-rō-MEG-a-lē). Although hGH cannot produce further lengthening of the long bones because the epiphyseal plates are already closed, the bones of the hands, feet, cheeks, and jaws thicken and other tissues enlarge (Figure 20.3b). •

Thyroid-Stimulating Hormone

Thyroid-stimulating hormone (TSH) stimulates the synthesis and secretion of thyroid hormones by the thyroid gland. *Thyrotropin-releasing hormone* (*TRH*) from the hypothalamus controls TSH secretion. Release of TRH, in turn, depends on blood levels of thyroid hormones, which inhibit secretion of TRH via negative feedback. There is no thyrotropin-inhibiting hormone.

Follicle-Stimulating Hormone and Luteinizing Hormone

In females, the ovaries are the targets for ***follicle-stimulating hormone (FSH)*** and ***luteinizing hormone (LH).*** Each month FSH initiates the development of several ovarian follicles and LH triggers ovulation (described in Chapter 29). After ovulation, LH stimulates formation of the corpus luteum in the ovary and the secretion of progesterone (a female sex hormone) by the corpus luteum. FSH and LH also stimulate follicular cells to secrete estrogens. In males, FSH stimulates sperm production in the testes, and LH stimulates the testes to secrete testosterone. *Gonadotropin-releasing hormone* (*GnRH*) from the hypothalamus stimulates release of FSH and LH. The release of GnRH, FSH, and LH is suppressed by estrogens in females and by testosterone in males through negative feedback systems. There is no gonadotropin-inhibiting hormone.

Prolactin

Prolactin (PRL), together with other hormones, initiates and maintains milk production by the mammary glands. Ejection of milk from the mammary glands depends on the hormone ***oxytocin,*** which is released from the posterior pituitary (discussed shortly). The function of prolactin is unknown in males, but prolactin hypersecretion causes erectile dysfunction (impotence, the inability to have an erection of the penis). In females, *prolactin-inhibiting hormone* (*PIH*) suppresses release of prolactin most of the time. Each month, just before menstruation begins, the secretion of PIH diminishes and the blood level of prolactin rises, but not enough to stimulate milk production. As

Figure 20.3 Hypersecretion of human growth hormone (hGH) may cause giantism or acromegaly.

Giantism occurs during the growth years, prior to closure of the epiphyseal plates, and acromegaly occurs during adulthood.

(a) A 22-year old man with pituitary giantism shown beside his identical twin

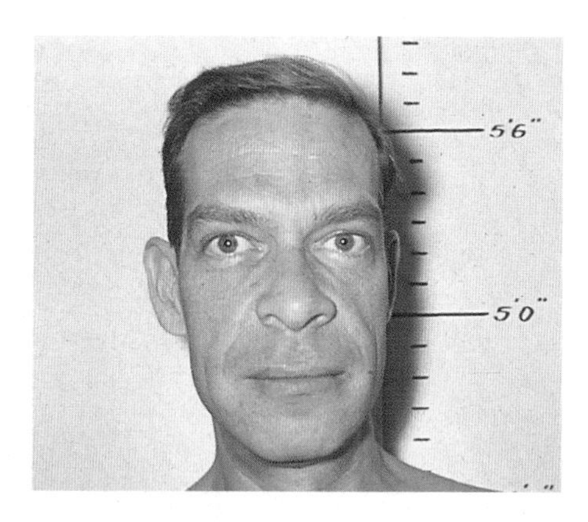

(b) Acromegaly (excess hGH during adulthood)

? What type of bone growth causes thickening of bones in a person with acromegaly?

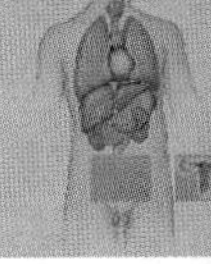

the menstrual cycle begins anew, PIH is again secreted and the prolactin level drops. During pregnancy, very high levels of estrogens promote secretion of *prolactin-releasing hormone* (*PRH*), which in turn stimulates release of prolactin.

Adrenocorticotropic Hormone

Adrenocorticotropic hormone **(*ACTH*)** or ***corticotropin*** controls the production and secretion by the cortex (outer portion) of the adrenal glands of hormones called glucocorticoids. *Corticotropin-releasing hormone (CRH)* from the hypothalamus stimulates secretion of ACTH. Stress-related stimuli, such as low blood glucose or physical trauma, and interleukin-1, a substance produced by macrophages, also stimulate release of ACTH. Glucocorticoids cause negative feedback inhibition of both CRH and ACTH release.

Melanocyte-Stimulating Hormone

There is little circulating ***melanocyte-stimulating hormone*** **(*MSH*)** in humans. Although an excessive amount of MSH causes darkening of the skin, the function of normal levels of MSH is unknown. The presence of MSH receptors in the brain suggests it may influence brain activity such as appetite and sexual behavior. Excessive *corticotropin-releasing hormone (CRH)* can stimulate MSH release, and *dopamine* inhibits MSH release.

ACTH and MSH are both derived from a precursor polypeptide in the anterior pituitary called *proopiomelanocortin* (*POMC*) (prō-Ō-pē-ō-mel′-a-nō-kor′-tin). From this large molecule as many as 10 biologically active peptides with diverse cellular functions can be formed. Other important products of POMC include beta-endorphin and met-enkephalin; both are opioid peptides with widespread pain-reducing actions in the brain.

Posterior Pituitary Hormones

The ***posterior pituitary*** contains the axons and axon terminals of more than 10,000 neurosecretory cells whose cell bodies are in the hypothalamus (Figure 20.4). Although the posterior pituitary does not *synthesize* hormones, it does *store* and *release* two hormones. In the hypothalamus, the hormones ***oxytocin*** (ok′-sē-TŌ-sin; *oxytoc-* = quick birth) and ***antidiuretic hormone*** **(*ADH*)** are synthesized and packaged into secretory vesicles within the cell bodies of different neurosecretory cells. The vesicles then move down the axons to the

Figure 20.4 Axons of hypothalamic neurosecretory cells extend from the hypothalamus to the posterior pituitary. Hormone molecules synthesized in the cell body of a neurosecretory cell are packaged into secretory vesicles that move down to the axon terminals. Nerve impulses trigger release of the hormone.

Oxytocin and antidiuretic hormone are synthesized in the hypothalamus and released into the capillary plexus of the posterior pituitary.

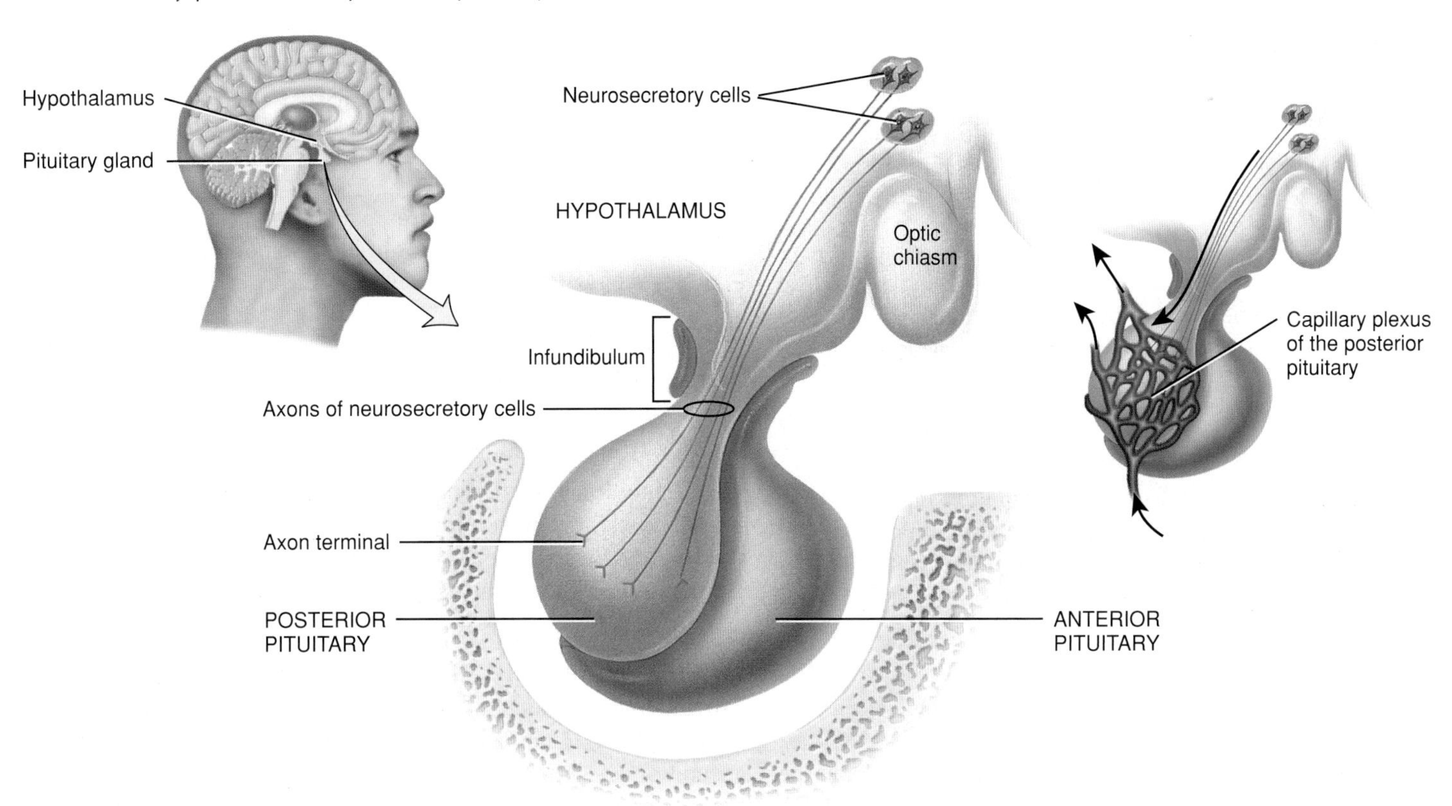

Which lobe of the pituitary gland does not synthesize the hormones it releases? Where are its hormones produced?

axon terminals in the posterior pituitary. Nerve impulses that arrive at the axon terminals trigger release of these hormones into the capillaries of the posterior pituitary.

Oxytocin

During and after delivery of a baby, ***oxytocin*** has two target tissues: the mother's uterus and breasts. During delivery, oxytocin enhances contraction of smooth muscle cells in the wall of the uterus (positive feedback); after delivery, it stimulates milk ejection ("letdown") from the mammary glands in response to the mechanical stimulus provided by a suckling infant. Together, milk production and ejection constitute ***lactation.*** The function of oxytocin in males and in nonpregnant females is not clear. Experiments with animals have suggested actions within the brain that foster parental caretaking behavior toward young offspring. Oxytocin also may be partly responsible for the feelings of sexual pleasure during and after intercourse.

Antidiuretic Hormone

An ***antidiuretic*** (an′-ti-dī-ū-RET-ik; *anti-* = against; *-dia-* = throughout; *-ouresis* = urination) is a substance that decreases urine production. ***Antidiuretic hormone* (*ADH*)** causes the kidneys to retain more water, thus decreasing urine volume. In the absence of ADH, urine output increases more than 10-fold, from the normal 1–2 liters to about 20 liters a day. ADH also decreases the water lost through sweating and causes constriction of arterioles. This hormone's other name, ***vasopressin*** (*vaso-* = vessel; *-pressin* = pressing or constricting), reflects its effect on increasing blood pressure.

The amount of ADH secreted varies with blood osmotic pressure and blood volume. Blood osmotic pressure is proportional to the concentration of solutes in the blood plasma. When body water is lost faster than it is taken in, a condition termed ***dehydration,*** the blood volume falls and blood osmotic pressure rises. Figure 20.5 shows regulation of ADH secretion and the actions of ADH on its target tissues.

1. High blood osmotic pressure—due to dehydration or a drop in blood volume because of hemorrhage, diarrhea, or excessive sweating—stimulates ***osmoreceptors,*** neurons in the hypothalamus that monitor blood osmotic pressure.
2. Osmoreceptors activate the hypothalamic neurosecretory cells that synthesize and release ADH.
3. When neurosecretory cells receive excitatory input from the osmoreceptors, they generate nerve impulses that cause the release of ADH in the posterior pituitary. The ADH then diffuses into blood capillaries of the posterior pituitary.
4. The blood carries ADH to three target tissues: the kidneys, sweat glands, and smooth muscle in blood vessel walls. The kidneys respond by retaining more water, which decreases urine output. Secretory activity of sweat glands decreases, which lowers the rate of water loss by perspiration from the skin. Smooth muscle in the walls of arterioles (small arteries) contracts in response to high levels of ADH, which constricts (narrows) the lumen of these blood vessels and increases blood pressure.
5. Low blood osmotic pressure inhibits the osmoreceptors.
6. Inhibition of osmoreceptors reduces or stops ADH secretion. The kidneys then retain less water by forming a larger volume of urine, secretory activity of sweat glands increases, and arterioles dilate. The blood volume and osmotic pressure of body fluids return to normal.

Figure 20.5 Regulation of secretion and actions of antidiuretic hormone (ADH).

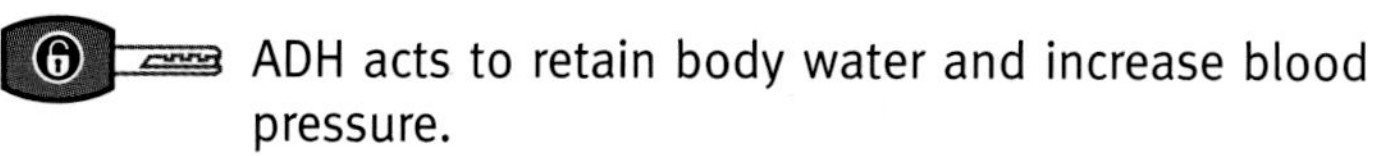

ADH acts to retain body water and increase blood pressure.

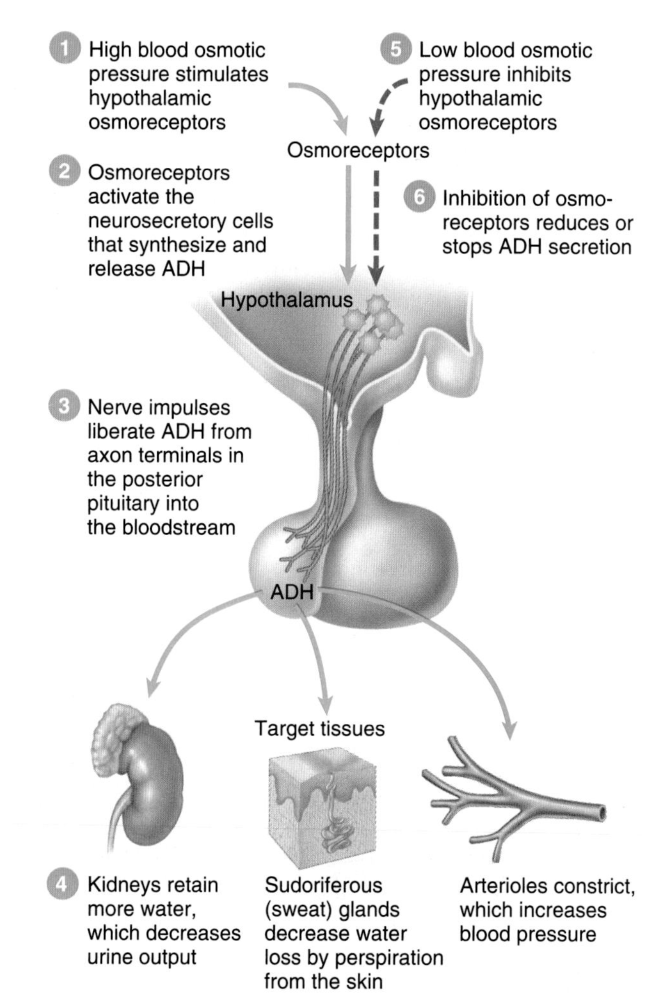

If you drank a liter of water, what effect would this have on the osmotic pressure of your blood, and how would the level of ADH change in your blood?

Secretion of ADH can also be altered in other ways. Pain, stress, trauma, anxiety, acetylcholine, nicotine, and drugs such as morphine, tranquilizers, and some anesthetics stimulate ADH secretion. Alcohol inhibits ADH secretion, thereby increasing urine output. The resulting dehydration may cause both the thirst and the headache typical of a hangover.

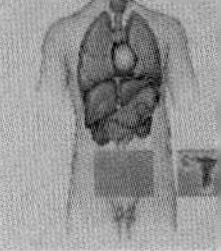

The most common abnormality of the posterior pituitary is ***diabetes insipidus*** (dī-a-BĒ-tēs in-SIP-i-dus; *diabetes* = overflow; *insipidus* = tasteless). This disorder is due to defects in ADH receptors or an inability to secrete ADH. Usually the disorder is caused by a brain tumor, head trauma, or brain surgery that damages the posterior pituitary or the hypothalamus. A common symptom is excretion of large volumes of urine, with resulting dehydration and thirst. Because so much water is lost in the urine, a person with diabetes insipidus may die of dehydration if deprived of water for only a day or so.

Table 20.1 lists the pituitary gland hormones and summarizes their actions.

CHECKPOINT

4. In what respect is the pituitary gland actually two glands?
5. How do hypothalamic releasing and inhibiting hormones influence secretions of anterior pituitary hormones?

20.4 THYROID GLAND

OBJECTIVE

- Describe the location, hormones, and functions of the thyroid gland.

The butterfly-shaped ***thyroid gland*** is located just below the larynx (voice box). It is composed of right and left *lateral lobes*, one on either side of the trachea, that are connected by an *isthmus* (IS-mus = a narrow passage) anterior to the trachea (see Figure 20.6a).

Microscopic spherical sacs called ***thyroid follicles*** (see Figure 20.6b) make up most of the thyroid gland. The wall of each thyroid follicle consists primarily of cells called *follicular cells*, which produce two hormones: ***thyroxine*** (thī-ROK-sēn), also called ***T_4*** because it contains four atoms of iodine, and ***triiodothyronine*** (trī-ī′-ō-dō-THĪ-rō-nēn) (***T_3***), which contains three atoms of iodine. T_3 and T_4 are also known as ***thyroid hormones***. The central cavity of each thyroid follicle contains stored thyroid hormones. As T_4 circulates in the blood and enters cells throughout the body, most of it is converted to T_3 by removal of one iodine atom.

A smaller number of cells called *parafollicular cells* lie between the follicles (see Figure 20.6b). They produce the hormone ***calcitonin*** (discussed shortly).

Actions of Thyroid Hormones

Because most body cells have receptors for thyroid hormones, T_3 and T_4 exert their effects throughout the body.

Thyroid hormones increase *basal metabolic rate* (*BMR*), the rate of oxygen consumption under standard or basal conditions (awake, at rest, and fasting). The BMR rises due to increased synthesis and use of ATP. As cells use more oxygen to produce the ATP, more heat is given off, and body temperature rises. In this way, thyroid hormones play an important role in the maintenance of normal body temperature. The thyroid hormones also stimulate protein synthesis, increase the use of glucose and fatty acids for ATP production, increase the breakdown of triglycerides (one

TABLE 20.1

Summary of Pituitary Gland Hormones and Their Actions

HORMONE	ACTIONS
ANTERIOR PITUITARY HORMONES	
Human growth hormone (hGH)	Stimulates liver, muscle, cartilage, bone, and other tissues to synthesize and secrete insulin-like growth factors (IGFs). IGFs promote growth of body cells, protein synthesis, tissue repair, breakdown of triglycerides, and elevation of blood glucose level.
Thyroid-stimulating hormone (TSH)	Stimulates synthesis and secretion of thyroid hormones by the thyroid gland.
Follicle-stimulating hormone (FSH)	In females, initiates development of oocytes and induces secretion of estrogens by the ovaries. In males, stimulates testes to produce sperm.
Luteinizing hormone (LH)	In females, stimulates secretion of estrogens and progesterone, ovulation, and formation of corpus luteum. In males, stimulates testes to produce testosterone.
Prolactin (PRL)	In females, stimulates milk production by the mammary glands.
Adrenocorticotropic hormone (ACTH), also known as **corticotropin**	Stimulates secretion of glucocorticoids (mainly cortisol) by the adrenal cortex.
Melanocyte-stimulating hormone (MSH)	Exact role in humans is unknown but may influence brain activity. When present in excess, can cause darkening of skin.
POSTERIOR PITUITARY HORMONES	
Oxytocin	Stimulates contraction of smooth muscle cells of uterus during childbirth. Stimulates milk ejection from the mammary glands.
Antidiuretic hormone (ADH), also known as vasopressin	Conserves body water by decreasing urine output. Decreases water loss through sweating. Raises blood pressure by constricting (narrowing) arterioles.

Figure 20.6 Location and histology of the thyroid gland.

Thyroid hormones regulate (1) oxygen use and basal metabolic rate, (2) cellular metabolism, and (3) growth and development.

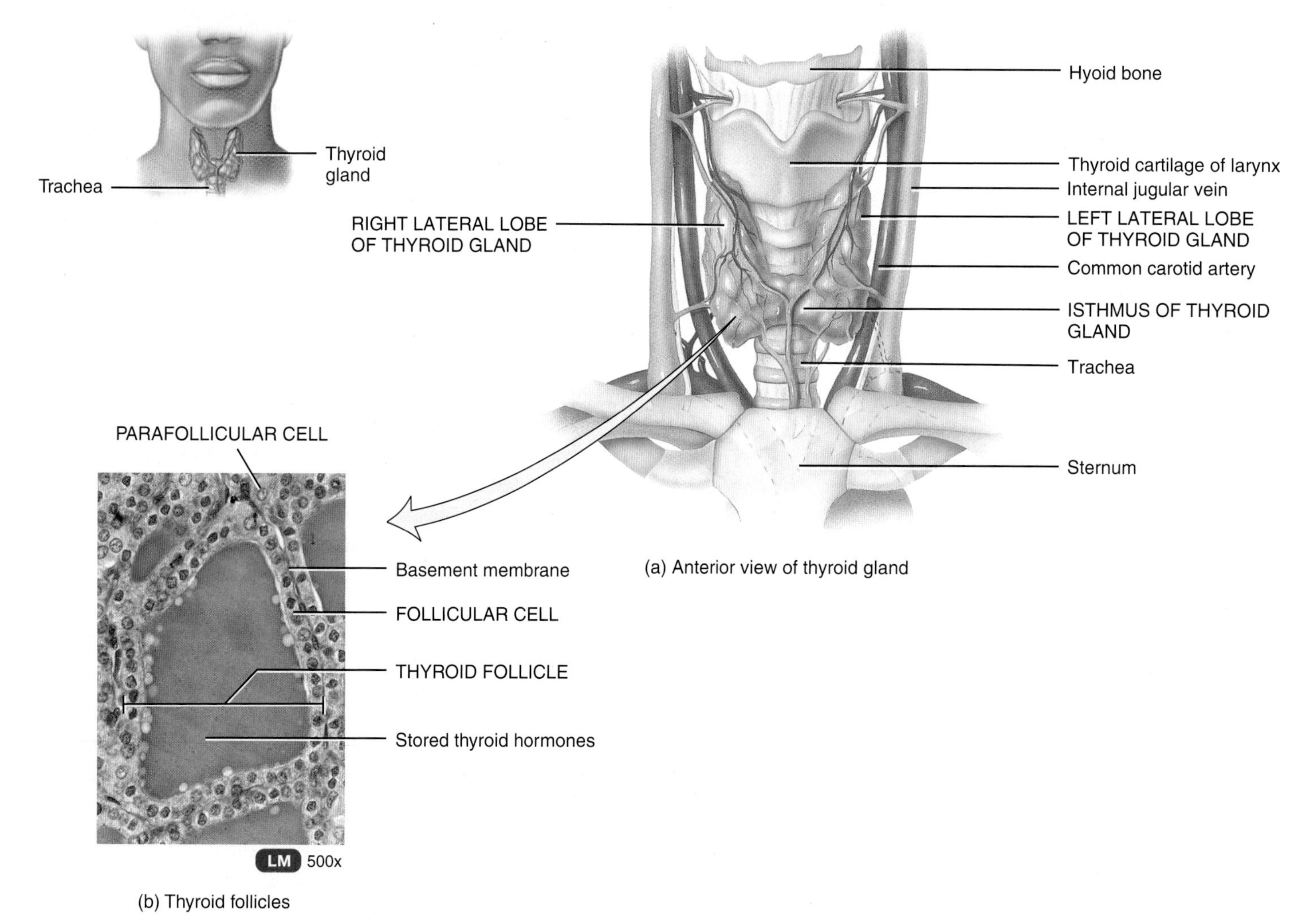

(a) Anterior view of thyroid gland

(b) Thyroid follicles

Which cells secrete T_3 and T_4? Which cells secrete calcitonin? Which of these hormones are also called thyroid hormones?

type of lipolysis), and enhance cholesterol excretion, thus reducing blood cholesterol level. Together with human growth hormone and insulin, thyroid hormones stimulate body growth, particularly the growth of the nervous and skeletal systems.

Control of Thyroid Hormone Secretion

Thyrotropin-releasing hormone (TRH) from the hypothalamus and thyroid-stimulating hormone (TSH) from the anterior pituitary stimulate synthesis and release of thyroid hormones, as shown in Figure 20.7:

1. Low blood levels of thyroid hormones or low metabolic rate stimulate the hypothalamus to secrete TRH.
2. TRH is carried to the anterior pituitary, where it stimulates secretion of TSH.
3. TSH stimulates thyroid follicular cell activity, including thyroid hormone synthesis and secretion, and growth of the follicular cells.
4. The thyroid follicular cells release thyroid hormones into the blood until the metabolic rate returns to normal.
5. An elevated level of thyroid hormones inhibits release of TRH and TSH (negative feedback).

Conditions that increase ATP demand—a cold environment, low blood glucose, high altitude, and pregnancy—also increase secretion of the thyroid hormones.

Figure 20.7 Regulation of secretion and actions of thyroid hormones. TRH = thyrotropin-releasing hormone, TSH = thyroid-stimulating hormone, T_3 = triiodothyronine, and T_4 = thyroxine.

TSH promotes release of thyroid hormones (T_3 and T_4) by the thyroid gland.

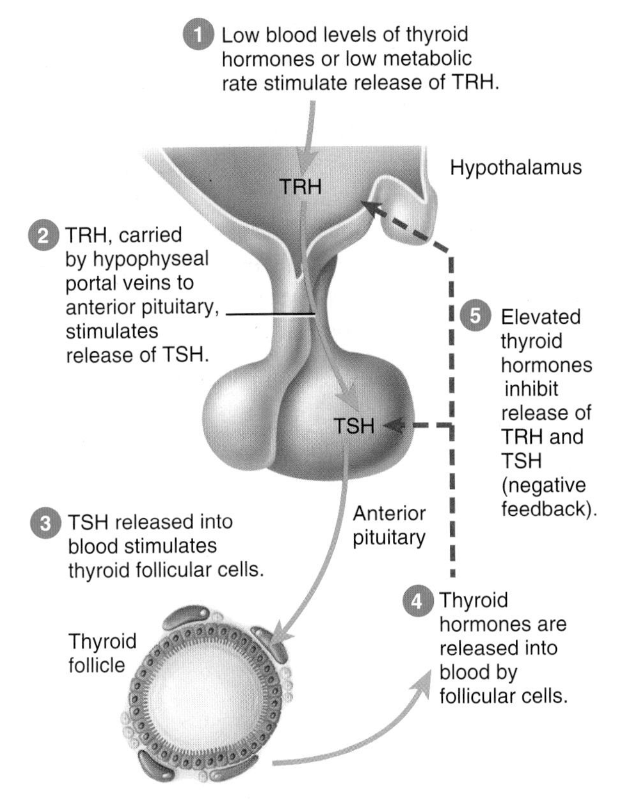

Actions of Thyroid Hormones:

- Increase basal metabolic rate
- Increase body temperature
- Stimulate protein synthesis
- Increase use of glucose and fatty acids for ATP production
- Stimulate lipolysis
- Regulate development and growth of nervous tissue and bones

? How could an iodine-deficient diet lead to goiter, which is an enlargement of the thyroid gland?

• CLINICAL CONNECTION | Thyroid Gland Disorders

Thyroid gland disorders affect all major body systems and are among the most common endocrine disorders. **Congenital hypothyroidism,** hyposecretion of thyroid hormones that is present at birth, has devastating consequences if not treated promptly. Previously termed *cretinism,* this condition causes severe mental retardation. At birth, the baby typically is normal because lipid-soluble maternal thyroid hormones crossed the placenta during pregnancy and allowed normal development. Most states require testing of all newborns to ensure adequate thyroid function. If congenital hypothyroidism exists, oral thyroid hormone treatment must be started soon after birth and continued for life. Hypothyroidism during the adult years produces **myxedema** (miks-e-DĒ-ma), which occurs about five times more often in females than in males. A hallmark of this disorder is edema (accumulation of interstitial fluid) that causes the facial tissues to swell and look puffy. A person with myxedema has a slow heart rate, low body temperature, sensitivity to cold, dry hair and skin, muscular weakness, general lethargy, and a tendency to gain weight easily.

Excess secretion of thyroid hormones is known as *hyperthyroidism.* Symptoms of hyperthyroidism include increased heart rate and more forceful heartbeats, increased blood pressure, and increased nervousness. The most common form of hyperthyroidism is **Graves' disease,** which also occurs much more often in females than in males, usually before age 40. Graves' disease is an autoimmune disorder in which the person produces antibodies that mimic the action of thyroid-stimulating hormone (TSH). The antibodies continually stimulate the thyroid gland to grow and produce thyroid hormones. Thus, the thyroid gland may enlarge to two to three times its normal size, a condition called **goiter** (GOY-ter; *guttur* = throat) (Figure 20.8a). Goiter also occurs in other thyroid diseases and if dietary intake of iodine is inadequate. Graves' patients often have a peculiar edema behind the eyes, called *exophthalmos* (ek′-sof-THAL-mos), which causes the eyes to protrude (Figure 20.8b). •

Figure 20.8 Disorders of the Thyroid Gland.

The thyroid gland may enlarge, a condition called goiter (a), and edema may occur behind the eyes, called exophthalmos (b).

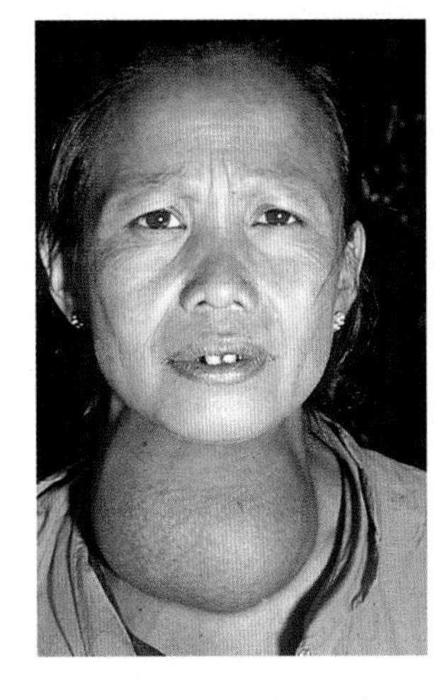

(a) Goiter (enlargement of thyroid gland)

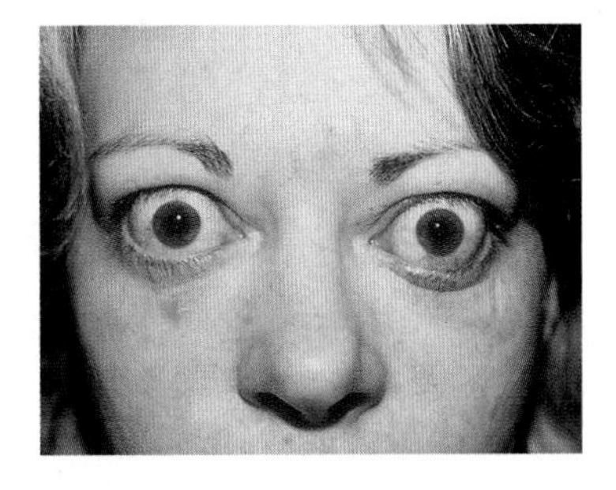

(b) Exophthalmos (excess thyroid hormones, as in Graves' disease)

? What conditions may result from hypothyroidism?

Calcitonin

The hormone produced by the parafollicular cells of the thyroid gland is ***calcitonin (CT)*** (kal-si-TŌ-nin). Calcitonin can decrease the level of calcium in the blood by inhibiting the action of osteoclasts, the cells that break down bone. The secretion of calcitonin is controlled by a negative feedback system (see Figure 20.10).

When its blood level is high, calcitonin lowers the amount of blood calcium and phosphates by inhibiting bone resorption (breakdown of bone extracellular matrix) by osteoclasts and by accelerating uptake of calcium and phosphates into bone extracellular matrix. Miacalcin®, a calcitonin extract derived from salmon that is 10 times more potent than human calcitonin, is prescribed to treat ***osteoporosis,*** a disorder in which the pace of bone breakdown exceeds the pace of bone rebuilding.

CHECKPOINT

6. How is the secretion of T_3 and T_4 regulated?
7. What are the actions of the thyroid hormones and calcitonin?

20.5 PARATHYROID GLANDS

OBJECTIVE

- Describe the location, hormones, and functions of the parathyroid glands.

The ***parathyroid glands*** (*para-* = beside) are small, round masses of glandular tissue that are partially embedded in the posterior surface of the thyroid gland (Figure 20.9). Usually, one superior and one inferior parathyroid gland are attached to each thyroid lobe. Within the parathyroid glands are secretory cells called ***chief cells*** that release ***parathyroid hormone (PTH).***

PTH is the major regulator of the levels of calcium (Ca^{2+}), magnesium (Mg^{2+}), and phosphate (HPO_4^{2-}) ions in the blood. PTH increases the number and activity of osteoclasts, which break down bone extracellular matrix and release Ca^{2+} and HPO_4^{2-} into the blood. PTH also produces three changes in the kidneys. First, it slows the rate at which Ca^{2+} and Mg^{2+} are lost from blood into the urine. Second, it increases loss of HPO_4^{2-} from blood to urine. Because more is lost in the urine than is gained from the bones, PTH decreases blood HPO_4^{2-} level and increases blood Ca^{2+} and Mg^{2+} levels. Third, PTH promotes formation of the hormone ***calcitriol,*** the active form of vitamin D. Calcitriol acts on the gastrointestinal tract to increase the rate of Ca^{2+}, Mg^{2+}, and HPO_4^{2-} absorption from foods into the blood.

The blood calcium level directly controls the secretion of calcitonin and parathyroid hormone via negative feedback, and the two hormones have opposite effects on blood Ca^{2+} level (Figure 20.10).

1. A higher-than-normal level of calcium ions (Ca^{2+}) in the blood stimulates parafollicular cells of the thyroid gland to release more calcitonin.
2. Calcitonin (CT) inhibits the activity of osteoclasts, thereby decreasing blood Ca^{2+} level.
3. A lower-than-normal level of Ca^{2+} in the blood stimulates

Figure 20.9 Location of the parathyroid glands.

The parathyroid glands, normally four in number, are embedded in the posterior surface of the thyroid gland.

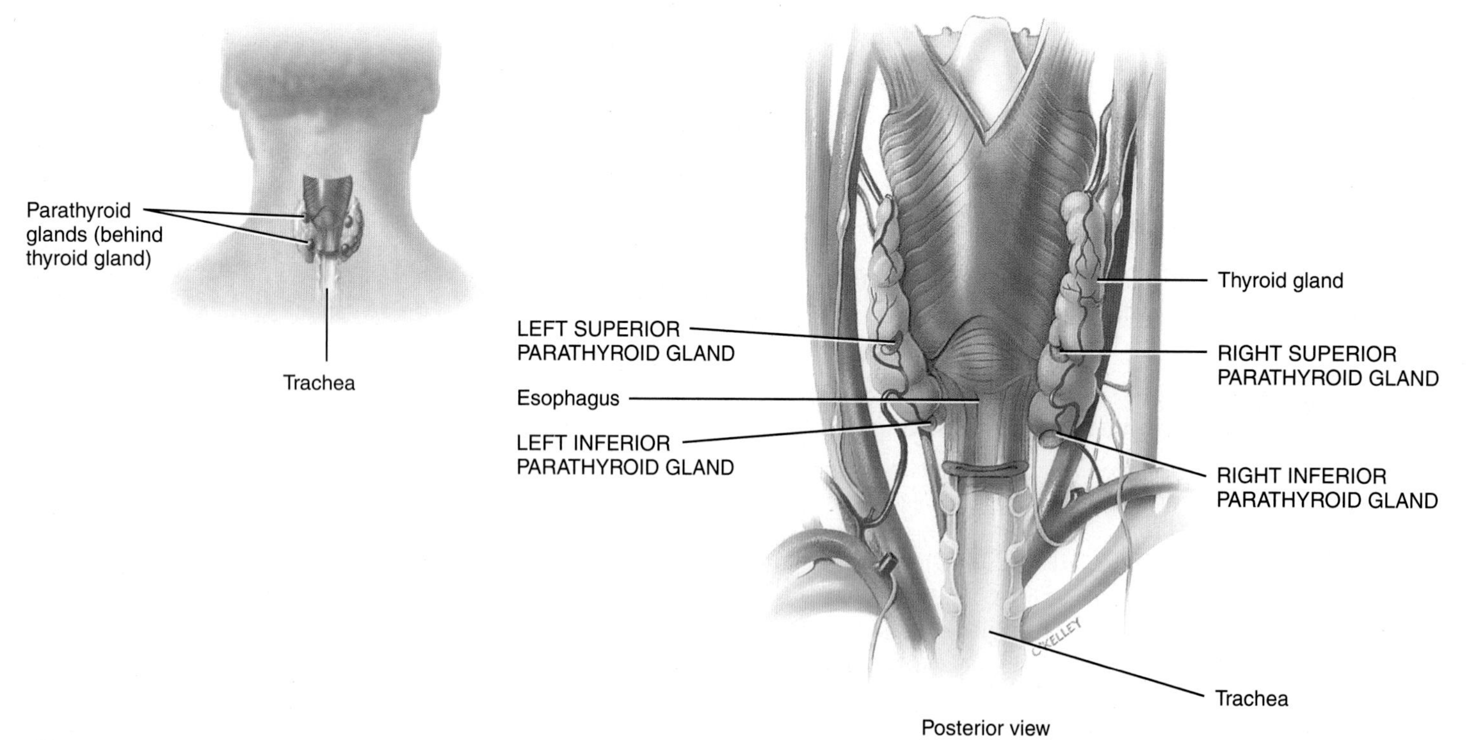

What effect does parathyroid hormone have on osteoclasts?

chief cells of the parathyroid gland to release more parathyroid hormone (PTH).

4. PTH increases the number and activity of osteoclasts, which break down bone and release Ca^{2+} into the blood. PTH also slows loss of Ca^{2+} in the urine. Both actions of PTH raise the blood level of Ca^{2+}.
5. PTH also stimulates the kidneys to synthesize and release calcitriol, the active form of vitamin D.
6. Calcitriol stimulates increased absorption of Ca^{2+} from foods in the gastrointestinal tract, which helps increase the blood level of Ca^{2+}.

• CLINICAL CONNECTION | Parathyroid Gland Disorders

Hypoparathyroidism—too little parathyroid hormone—leads to a deficiency of Ca^{2+}, which causes neurons and muscle fibers to depolarize and produce action potentials spontaneously. This leads to twitches, spasms, and **tetany** (maintained contraction) of skeletal muscle. The leading cause of hypoparathyroidism is accidental damage to the parathyroid glands or to their blood supply during surgery to remove the thyroid gland. •

CHECKPOINT

8. How is secretion of PTH regulated?
9. In what ways are the actions of PTH and calcitriol similar and different?

20.6 PANCREATIC ISLETS

OBJECTIVE

- Describe the location, hormones, and functions of the pancreatic islets.

The ***pancreas*** (*pan-* = all; *-creas* = flesh) is a flattened organ located in the curve of the duodenum, the first part of the small intestine (see Figure 20.11a). It has both endocrine functions, discussed shortly, and exocrine functions, discussed in Chapter 26. The endocrine part of the pancreas consists of clusters of cells called ***pancreatic islets*** or ***islets of Langerhans*** (LANG-er-hanz). Some of the islet cells, the ***alpha cells,*** secrete the hormone ***glucagon*** (GLOO-ka-gon), and other islet cells, the ***beta cells,*** secrete ***insulin*** (IN-soo-lin). The islets also contain abundant blood capillaries and are surrounded by cells that form the exocrine part of the pancreas (see Figure 20.11b).

Figure 20.10 The roles of calcitonin (green arrows), parathyroid hormone (blue arrows), and calcitriol (orange arrows) in calcium homeostasis.

With respect to regulation of blood Ca^{2+} level, calcitonin and PTH are antagonists.

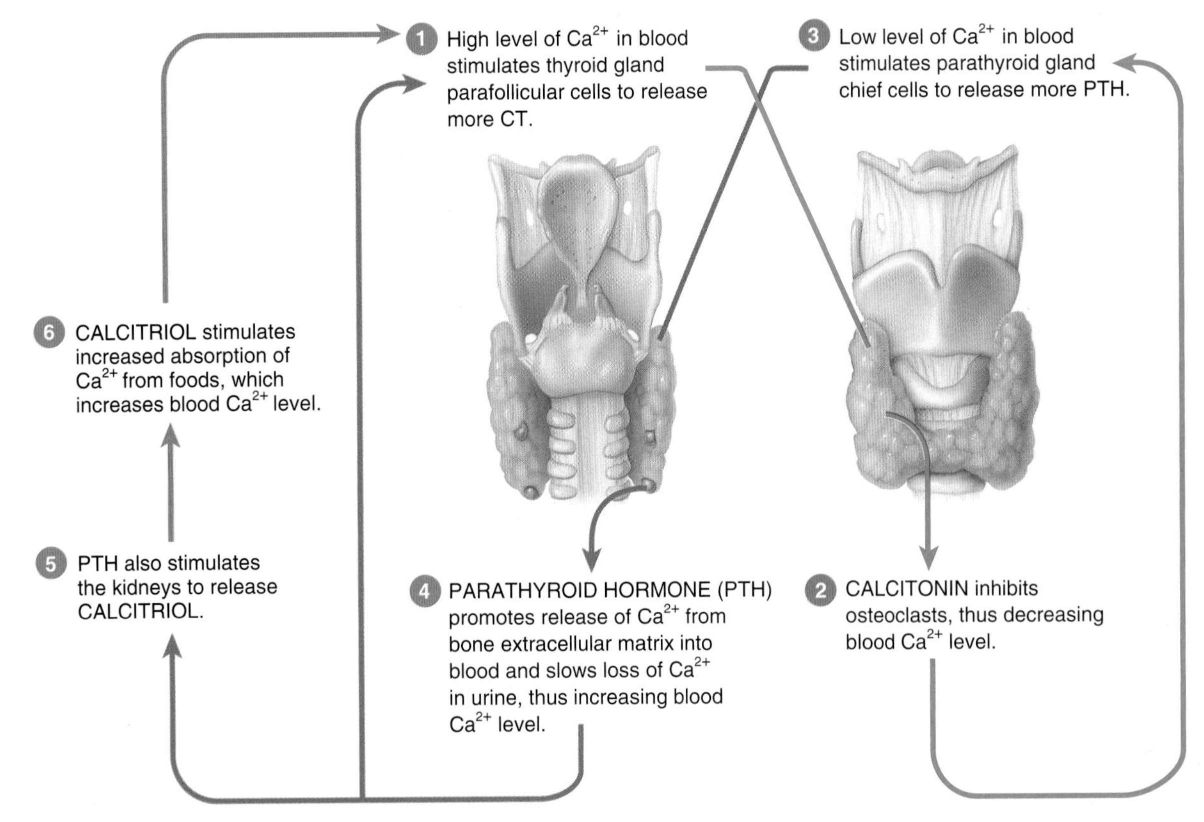

1 High level of Ca2+ in blood stimulates thyroid gland parafollicular cells to release more CT.
2 CALCITONIN inhibits osteoclasts, thus decreasing blood Ca2+ level.
3 Low level of Ca2+ in blood stimulates parathyroid gland chief cells to release more PTH.
4 PARATHYROID HORMONE (PTH) promotes release of Ca2+ from bone extracellular matrix into blood and slows loss of Ca2+ in urine, thus increasing blood Ca2+ level.
5 PTH also stimulates the kidneys to release CALCITRIOL.
6 CALCITRIOL stimulates increased absorption of Ca2+ from foods, which increases blood Ca2+ level.

What are the primary target tissues for PTH, CT, and calcitriol?

Figure 20.11 Location and histology of the pancreas.

6 Pancreatic hormones regulate blood glucose level.

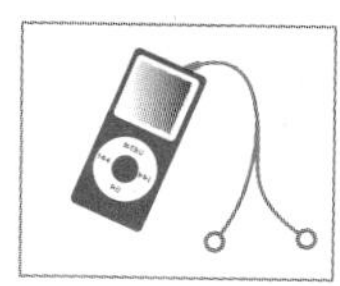

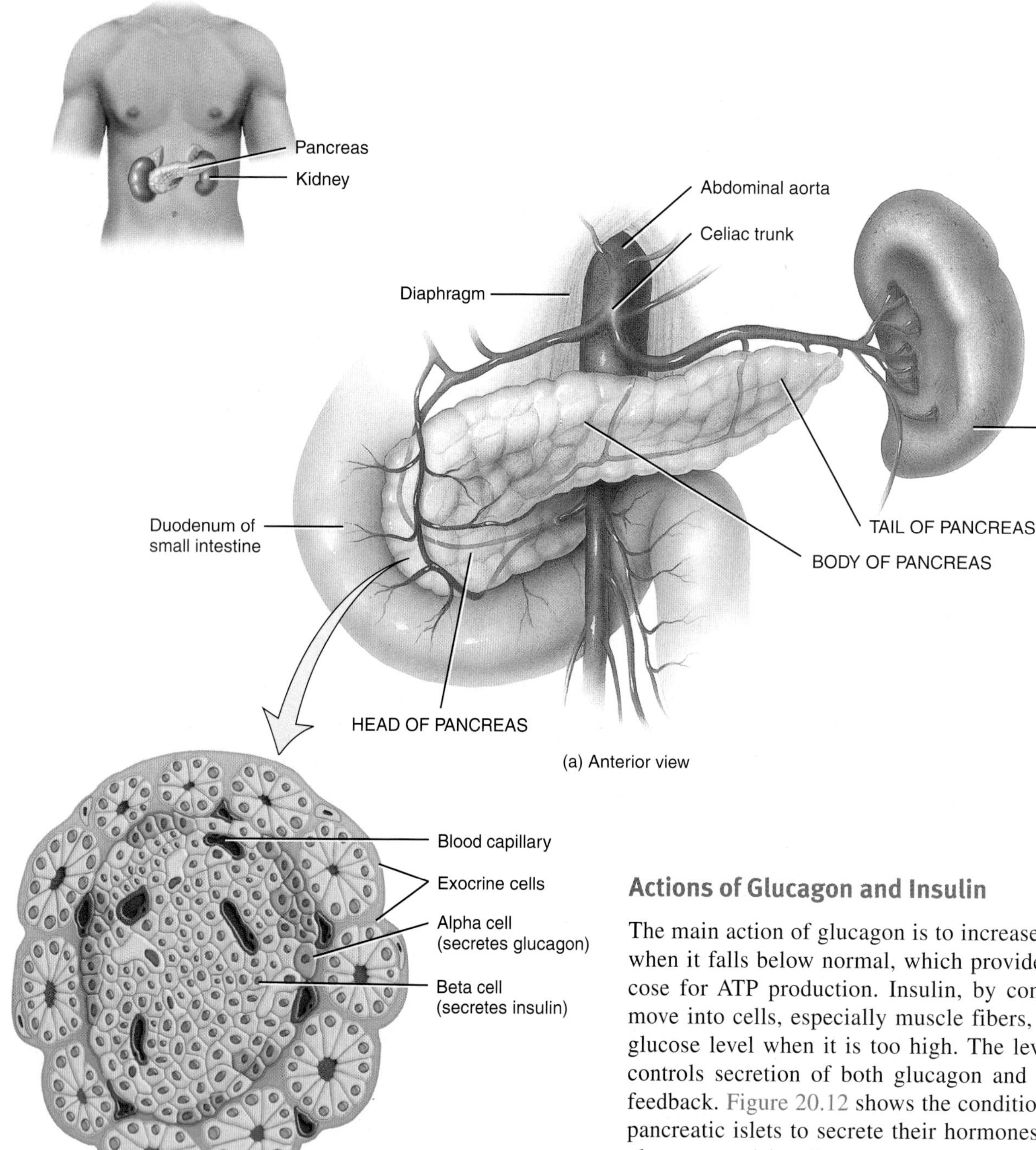

(a) Anterior view

(b) Pancreatic islet and surrounding acini

? Is the pancreas an exocrine gland or an endocrine gland?

Actions of Glucagon and Insulin

The main action of glucagon is to increase blood glucose level when it falls below normal, which provides neurons with glucose for ATP production. Insulin, by contrast, helps glucose move into cells, especially muscle fibers, which lowers blood glucose level when it is too high. The level of blood glucose controls secretion of both glucagon and insulin via negative feedback. Figure 20.12 shows the conditions that stimulate the pancreatic islets to secrete their hormones, the ways in which glucagon and insulin produce their effects on blood glucose level, and the negative feedback control of hormone secretion.

1. Low blood glucose level (hypoglycemia) stimulates secretion of glucagon.
2. Glucagon acts on liver cells (hepatocytes) to promote breakdown of glycogen into glucose and formation of glucose from lactic acid and certain amino acids.
3. As a result, the liver releases glucose into the blood more rapidly, and blood glucose level rises.
4. If blood glucose continues to rise, high blood glucose level

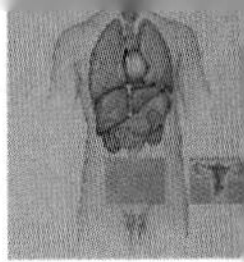

Figure 20.12 Negative feedback regulation of the secretion of glucagon (blue arrows) and insulin (orange arrows).

Low blood glucose stimulates secretion of glucagon; high blood glucose stimulates secretion of insulin.

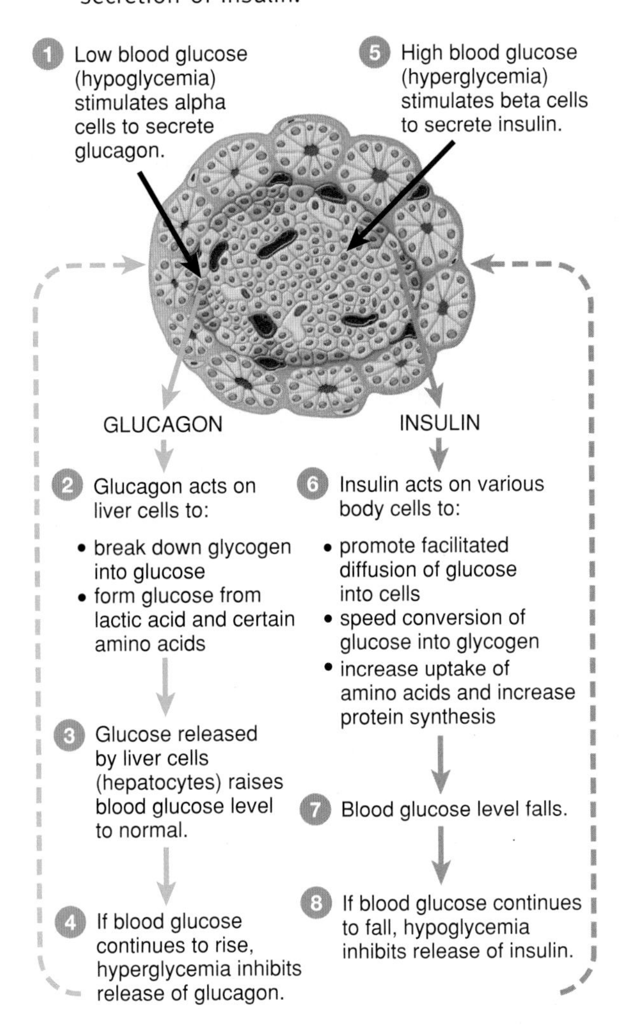

1 Low blood glucose (hypoglycemia) stimulates alpha cells to secrete glucagon.
5 High blood glucose (hyperglycemia) stimulates beta cells to secrete insulin.
GLUCAGON
INSULIN
2 Glucagon acts on liver cells to:
• break down glycogen into glucose
• form glucose from lactic acid and certain amino acids
6 Insulin acts on various body cells to:
• promote facilitated diffusion of glucose into cells
• speed conversion of glucose into glycogen
• increase uptake of amino acids and increase protein synthesis
3 Glucose released by liver cells (hepatocytes) raises blood glucose level to normal.
7 Blood glucose level falls.
4 If blood glucose continues to rise, hyperglycemia inhibits release of glucagon.
8 If blood glucose continues to fall, hypoglycemia inhibits release of insulin.

? Why is glucagon sometimes called an "anti-insulin" hormone?

(hyperglycemia) inhibits release of glucagon by alpha cells (negative feedback).

5. At the same time, high blood glucose level stimulates secretion of insulin.
6. Insulin acts on various cells in the body to promote facilitated diffusion of glucose into cells; to speed conversion of glucose into glycogen; to increase uptake of amino acids by cells; and to increase protein synthesis.
7. As a result, blood glucose level falls.
8. If blood glucose level drops below normal, low blood glucose inhibits release of insulin by beta cells (negative feedback).

In addition to affecting glucose metabolism, insulin promotes the uptake of amino acids into body cells and increases the synthesis of proteins and fatty acids within cells. Therefore, insulin is an important hormone when tissues are developing, growing, or being repaired.

Release of insulin and glucagon is also regulated by the autonomic nervous system (ANS). The parasympathetic division of the ANS stimulates secretion of insulin, for instance, during digestion and absorption of a meal. The sympathetic division of the ANS, by contrast, stimulates secretion of glucagon, as happens during exercise.

• CLINICAL CONNECTION | Pancreatic Islet Disorders

The most common endocrine disorder is **diabetes mellitus** (MEL-i-tus; *melli-* = honey sweetened), caused by an inability either to produce or to use insulin. Diabetes mellitus is the fourth leading cause of death by disease in the United States, primarily because of its damage to the cardiovascular system. Because insulin is unavailable to aid the movement of glucose into body cells, blood glucose level is high and glucose "spills" into the urine (glucosuria). Hallmarks of diabetes mellitus are the three "polys"; *polyuria,* excessive urine production due to an inability of the kidneys to reabsorb water; *polydipsia,* excessive thirst; and *polyphagia,* excessive eating.

Both genetic and environmental factors contribute to onset of the two types of diabetes mellitus—type 1 and type 2—but the exact mechanisms are still unknown. In **type 1 diabetes,** insulin level is low because the person's immune system destroys the pancreatic beta cells. Most commonly, type 1 diabetes develops in people younger than age 20, though it persists throughout life. By the time symptoms arise, 80–90% of the islet beta cells have been destroyed.

Because insulin is not present to aid the entry of glucose into body cells, most cells use fatty acids to produce ATP. Stores of triglycerides in adipose tissue are broken down to fatty acids and glycerol. The by-products of fatty acid breakdown—organic acids called *ketones* or *ketone bodies*—accumulate. Buildup of ketones causes blood pH to fall, a condition known as *ketoacidosis.* Unless treated quickly, ketoacidosis can cause death.

Type 2 diabetes is much more common than type 1. It most often occurs in people who are over 35 and overweight. The high glucose levels in the blood often can be controlled by diet, exercise, and weight loss. Sometimes, an antidiabetic drug such as *glyburide* (Diabeta®) is used to stimulate secretion of insulin by pancreatic beta cells. Although some type 2 diabetics need insulin, many have a sufficient amount (or even a surplus) of insulin in the blood. For these people, diabetes arises not from a shortage of insulin but because target cells become less sensitive to it. Elevated insulin levels stimulate the sympathetic nervous system, which increases blood pressure. Smoking, alcohol consumption, poor diet, and a sedentary lifestyle, predispose a person to the development of type 2 diabetes.

Hyperinsulinism most often results when a diabetic injects too much insulin. The main symptom is *hypoglycemia,* decreased blood glucose level, which occurs because the excess insulin stimulates too much uptake of glucose by body cells. When blood glucose falls, neurons are deprived of the steady supply of glucose they need to function effectively. Severe hypoglycemia leads to mental disorientation, convulsions, unconsciousness, and shock and is termed *insulin shock.* Death can occur quickly unless blood glucose is restored to normal levels. •

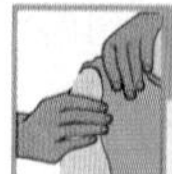

MANUAL THERAPY APPLICATION

Effects of Massage on the Diabetic Patient

Type 1 and type 2 diabetes mellitus have different pathologies, but patients with either type have the same symptom—elevated blood glucose. Treatment of diabetes therefore involves maintenance of blood glucose levels within healthy physiological limits. A physician will diagnose and treat diabetics; manual therapy can make the job easier for the physician and reduce the severity of many symptomatic ramifications of diabetes.

One of the **effects of massage on the diabetic patient** is to improve circulation which, in theory, improves delivery of insulin from the pancreas to tissues and improves uptake of insulin at the cellular level. Diabetes may cause thickening of connective tissues throughout the body, but especially in fascia associated with muscles. As discussed in Chapter 4, massage therapy can reduce or reverse the processes associated with myofascial thickening. With severely diabetic patients, the anxiety of monitoring their sugar levels on a constant basis, of making timely injections with the proper dosage of insulin, and of living with a debilitating disease can cause significant stress. As stated previously, massage shifts the body toward the parasympathetic mode, and there is a reduction of stress hormones and perhaps a release of endorphins.

Diabetes is complicated and there are many parameters to be considered by the patient and the therapist. Although levels vary greatly for each patient, it has been reported that massage for an hour typically drops blood sugar levels by 20 to 40 points. Depending on the blood sugar levels when beginning the therapy, such a decrease may or may not be considered significant. It may be important for some diabetic patients to monitor their glucose levels before they get off the table and then take appropriate steps to correct for this change. Diabetics who are receiving manual therapy should be aware of the symptoms associated with hypoglycemia, which include irritability, rapid heartbeat, excessive sweating, change in personality, unconsciousness, and even death. Therapists who treat diabetics are well advised to become familiar with these symptoms.

CHECKPOINT

10. What are the functions of insulin?
11. How are blood levels of glucagon and insulin controlled?

20.7 ADRENAL GLANDS

OBJECTIVE

- Describe the location, hormones, and functions of the adrenal glands.

There are two ***adrenal (suprarenal) glands,*** one lying atop each kidney (Figure 20.13a). Each adrenal gland has regions that produce different hormones: the outer ***adrenal cortex,*** which makes up 85% of the gland, and the inner ***adrenal medulla.*** A connective tissue capsule covers the gland.

Figure 20.13 Location and histology of the adrenal (suprarenal) glands.

The adrenal cortex secretes steroid hormones that are essential for life; the adrenal medulla secretes epinephrine and norepinephrine.

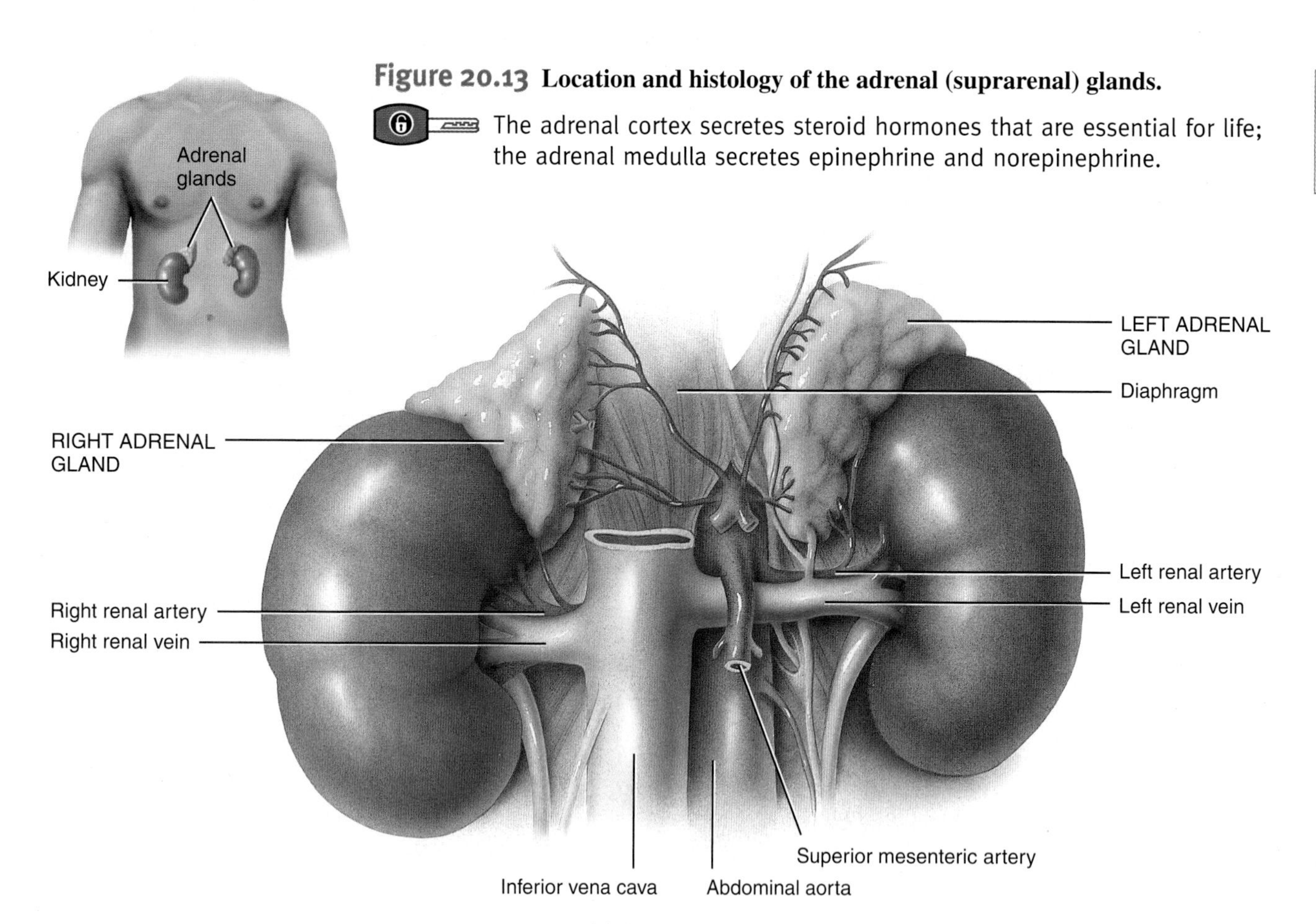

(a) Anterior view

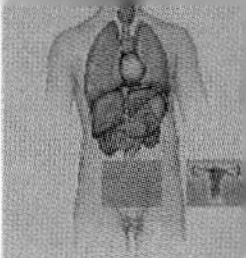

Adrenal Cortex Hormones

The adrenal cortex consists of three zones (Figure 20.13c), each of which synthesizes and secretes different steroid hormones that are essential for life. The outer zone releases hormones called mineralocorticoids because they affect mineral homeostasis. The middle zone releases hormones called glucocorticoids because they affect glucose homeostasis. The inner zone releases androgens (steroid hormones that have masculinizing effects).

Mineralocorticoids

Aldosterone (al-DO-ster-ōn) is the major ***mineralocorticoid*** (min′-er-al-ō-KOR-ti-koyd). It regulates homeostasis of two mineral ions, namely, sodium ions (Na^+) and potassium ions (K^+). Aldosterone increases reabsorption of Na^+ from the urine into the blood, and it stimulates excretion of K^+ into the urine. It also helps adjust blood pressure and blood volume, and promotes excretion of H^+ in the urine. Such removal of acids from the body can help prevent acidosis (blood pH below 7.35).

Secretion of aldosterone occurs as part of the *renin–angiotensin–aldosterone pathway* (RĒ-nin an′-jē-ō-TEN-sin) (see Figure 20.14). Conditions that initiate this pathway include dehydration, Na^+ deficiency, or hemorrhage, which decrease blood volume and blood pressure. Lowered blood pressure stimulates the kidneys to secrete the enzyme *renin,* which promotes a reaction in the blood that forms *angiotensin I.* As blood flows through the lungs, another enzyme called *angiotensin-converting enzyme* (*ACE*) converts inactive angiotensin I into the active hormone *angiotensin II.* Angiotensin II stimulates the adrenal cortex to secrete aldosterone. Aldosterone, in turn, acts on the kidneys to promote the return of Na^+ and water to the blood, and elimination of K^+ in the urine. As more water returns to the blood (and less is lost in the urine), blood volume increases. As blood volume increases, blood pressure increases to normal.

Glucocorticoids

The most abundant ***glucocorticoid*** (gloo′-kō-KOR-ti-koyd; *gluco-* = sugar; *-cortic-* = bark, shell) is ***cortisol.*** Cortisol and other glucocorticoids have the following actions:

- Protein breakdown. Glucocorticoids increase the rate of protein breakdown, mainly in muscle fibers, and thus increase the liberation of amino acids into the bloodstream. The amino acids may be used by body cells for synthesis of new proteins or for ATP production.
- Glucose formation. Upon stimulation by glucocorticoids, liver cells may convert certain amino acids or lactic acid to glucose, which neurons and other cells can use for ATP production.
- Breakdown of triglycerides. Glucocorticoids stimulate the breakdown of triglycerides in adipose tissue. The fatty acids thus released into the blood can be used for ATP production by many body cells.
- Anti-inflammatory effects. Glucocorticoids inhibit white blood cells that participate in inflammatory responses. They

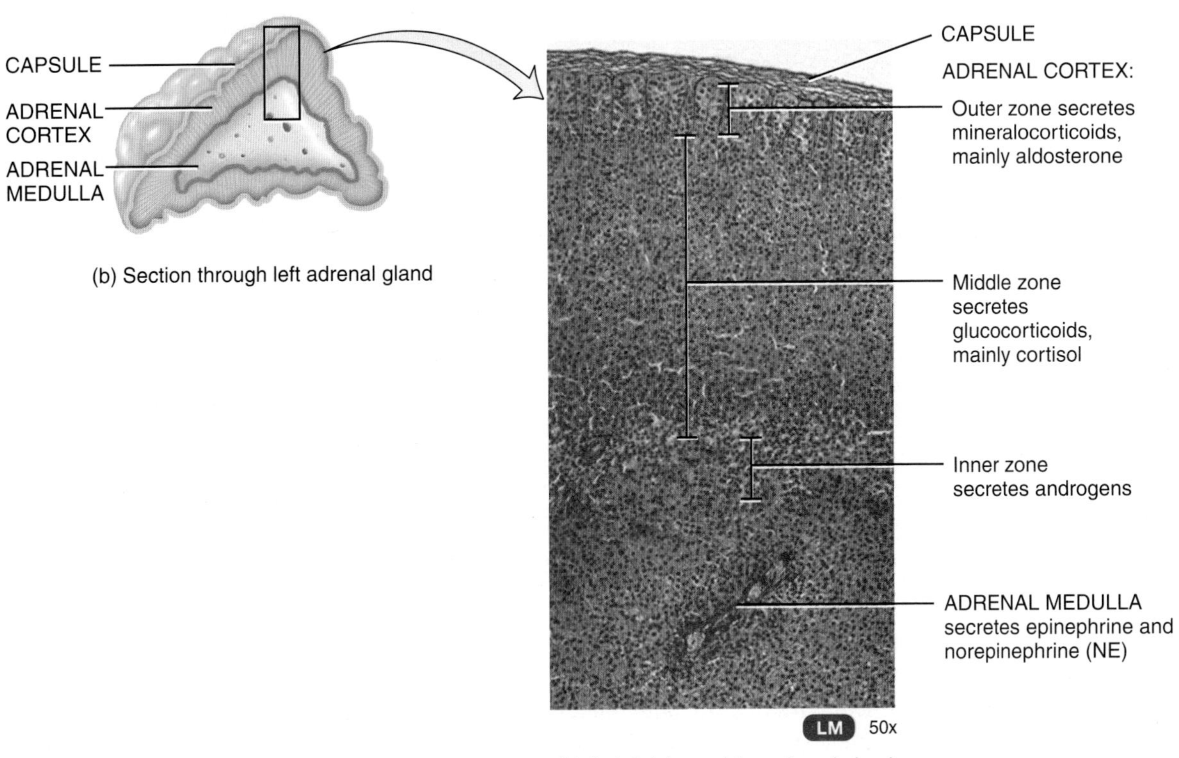

(b) Section through left adrenal gland

(c) Subdivisions of the adrenal gland

What hormones are secreted by the three zones of the adrenal cortex?

Figure 20.14 The renin–angiotensin–aldosterone pathway.

Aldosterone helps regulate blood volume, blood pressure, and levels of Na^+ and K^+ in the blood.

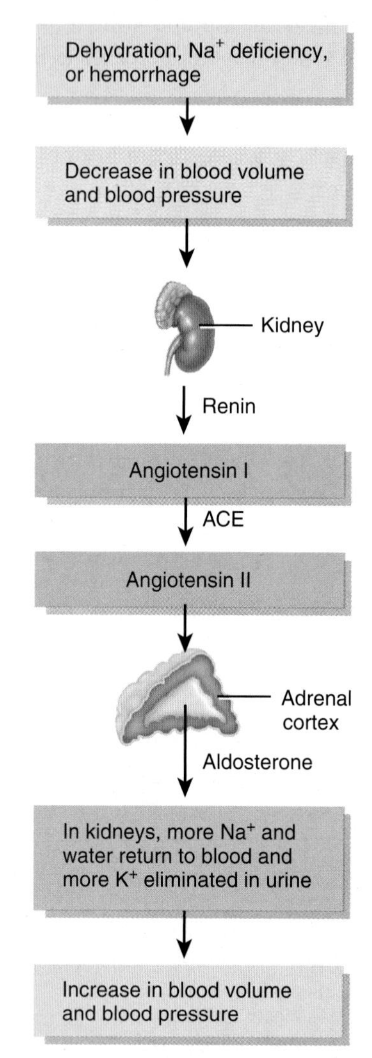

Could a drug that *blocks* the action of angiotensin–converting enzyme (ACE) be used to raise or lower blood pressure?

are often used in the treatment of chronic inflammatory disorders such as rheumatoid arthritis. Unfortunately, glucocorticoids also retard tissue repair, which slows wound healing.

- Depression of immune responses. High doses of glucocorticoids depress immune responses. For this reason, glucocorticoids are prescribed for organ transplant recipients to decrease the risk of tissue rejection by the immune system.

The control of secretion of cortisol (and other glucocorticoids) occurs by negative feedback. A low blood level of cortisol stimulates neurosecretory cells in the hypothalamus to secrete corticotropin-releasing hormone (CRH). The hypophyseal portal veins carry CRH to the anterior pituitary, where it stimulates release of ACTH (adrenocorticotropic hormone). ACTH, in turn, stimulates cells of the adrenal cortex to secrete cortisol. As the level of cortisol rises, it exerts negative feedback inhibition both on the anterior pituitary to reduce release of ACTH and on the hypothalamus to reduce release of CRH.

• CLINICAL CONNECTION | Adrenal Gland Disorders

Hypersecretion of cortisol by the adrenal cortex produces **Cushing's syndrome.** The condition is characterized by the breakdown of muscle proteins and redistribution of body fat, resulting in spindly arms and legs accompanied by a rounded "moon face" (Figure 20.15), "buffalo hump" on the back, and pendulous (hanging) abdomen. The elevated level of cortisol causes hyperglycemia, osteoporosis, weakness, hypertension, increased susceptibility to infection, decreased resistance to stress, and mood swings. Hyposecretion of glucocorticoids and aldosterone causes **Addison's disease.** Symptoms include mental lethargy, anorexia, nausea and vomiting, weight loss, hypoglycemia, and muscular weakness. Loss of aldosterone leads to elevated K^+ and decreased Na^+ in the blood; low blood pressure; dehydration; and decreased cardiac output, cardiac arrhythmias, and even cardiac arrest. The skin may have a "bronzed" appearance that often is mistaken for a suntan. Such was true in the case of President John F. Kennedy, whose Addison's disease was known to only a few while he was alive. •

Figure 20.15 Cushing's syndrome may be caused by hypersecretion of cortisol.

One symptom of Cushing's syndrome is a rounded "moon face."

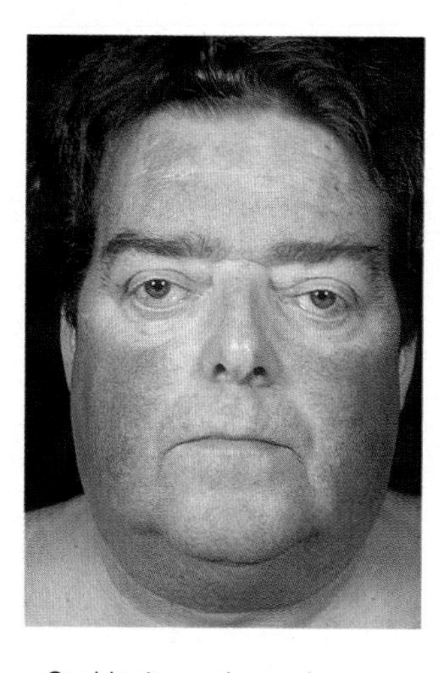

Cushing's syndrome (excess glucocorticoids)

What other physical characteristics may accompany Cushing's syndrome?

Androgens

In both males and females, the adrenal cortex secretes small amounts of weak ***androgens.*** After puberty in males, androgens are also released in much greater quantity by the testes. Thus, the amount of androgens secreted by the adrenal gland in males is usually so low that their effects are insignificant. In females, however, adrenal androgens play important roles: They contribute to libido (sex drive) and are converted into estrogens (feminizing sex steroids) by other body tissues. After menopause, when ovarian secretion of estrogens ceases, all female estrogens come from conversion of adrenal androgens. Adrenal androgens also stimulate growth of axillary (armpit) and pubic hair in boys and girls and contribute to the growth spurt before puberty. Although control of adrenal androgen secretion is not fully understood, the main hormone that stimulates its secretion is ACTH.

• CLINICAL CONNECTION | Congenital Adrenal Hyperplasia

Congenital adrenal hyperplasia (CAH) is a group of genetic disorders in which one or more enzymes needed for the production of cortisol or aldosterone, or both, are absent. Because the cortisol level is low, secretion of ACTH by the anterior pituitary is high due to lack of negative feedback inhibition. ACTH, in turn, stimulates growth and secretory activity of the adrenal cortex. As a result, both adrenal glands are enlarged. However, certain steps leading to synthesis of cortisol are blocked. Thus, precursor molecules build up, and some of these are weak androgens that can be converted to testosterone. The result is *virilism,* or masculinization. In a female, virile characteristics include growth of a beard, development of a much deeper voice and a masculine distribution of body hair, growth of the clitoris so it may resemble a penis, atrophy of the breasts, and increased muscularity that produces a masculine physique. In males, virilism causes the same characteristics as in females, plus rapid development of the male sexual organs and emergence of male sexual desires. •

Adrenal Medulla Hormones

The innermost region of each adrenal gland, the adrenal medulla, consists of sympathetic postganglionic cells of the autonomic nervous system (ANS) that are specialized to secrete hormones. The two main hormones of the adrenal medullae are ***epinephrine*** and ***norepinephrine (NE),*** also called ***adrenaline*** and ***noradrenaline.***

In stressful situations and during exercise, impulses from the hypothalamus stimulate sympathetic preganglionic neurons, which in turn stimulate the cells of the adrenal medullae to secrete epinephrine and norepinephrine. These two hormones greatly augment the fight-or-flight response. By increasing heart rate and force of contraction, epinephrine and norepinephrine increase the pumping output of the heart, which increases blood pressure. They also increase blood flow to the heart, liver, skeletal muscles, and adipose tissue; dilate airways to the lungs; and increase blood levels of glucose and fatty acids. Like the glucocorticoids of the adrenal cortex, epinephrine and norepinephrine also help the body resist stress.

Usually benign tumors of the adrenal medulla, called ***pheochromocytomas*** (fē′-ō-krō′-mō-sī-TŌ-mas; *pheo-* = dusky; *-chromo-* = color; *-cyto-* = cell), cause oversecretion of epinephrine and norepinephrine. The result is a prolonged version of the fight-or-flight response: rapid heart rate, headache, high blood pressure, high levels of glucose in blood and urine, an elevated basal metabolic rate (BMR), flushed face, nervousness, sweating, and decreased gastrointestinal motility.

CHECKPOINT

12. How do the adrenal cortex and adrenal medulla compare with regard to their location and histology?
13. How is secretion of adrenal cortex hormones regulated?

20.8 OVARIES AND TESTES

OBJECTIVE

- Describe the location, hormones, and functions of the ovaries and testes.

Gonads are the organs that produce gametes—sperm in males and oocytes in females. The female gonads, the ***ovaries,*** are paired oval bodies located in the pelvic cavity (see Figure 20.1). They produce the female sex hormones ***estrogens*** and ***progesterone.*** Along with FSH and LH from the anterior pituitary, the female sex hormones regulate the menstrual cycle, maintain pregnancy, and prepare the mammary glands for lactation. They also help establish and maintain the female body shape.

The ovaries also produce ***inhibin,*** a protein hormone that inhibits secretion of follicle-stimulating hormone (FSH). During pregnancy, the ovaries and placenta produce a peptide hormone, called ***relaxin,*** that increases the flexibility of the pubic symphysis during pregnancy and helps dilate the uterine cervix during labor and delivery. These actions enlarge the birth canal, which helps ease the baby's passage.

The male gonads, the ***testes,*** are oval glands that lie in the scrotum (see Figure 20.1). They produce ***testosterone,*** the primary androgen or male sex hormone. Testosterone stimulates descent of the testes before birth, regulates production of sperm, and stimulates the development and maintenance of masculine characteristics such as beard growth and deepening of the voice. The testes also produce ***inhibin,*** which inhibits secretion of FSH. The detailed structure of the ovaries and testes and the specific roles of sex hormones will be discussed in Chapter 29.

CHECKPOINT

14. Why are the ovaries and testes included among the endocrine glands?

20.9 PINEAL GLAND

OBJECTIVE

- Describe the location, hormone, and functions of the pineal gland.

The ***pineal gland*** (PIN-ē-al = pinecone shape) is a small endocrine gland attached to the roof of the third ventricle of the brain at the midline (see Figures 20.1 and 17.1). One hormone secreted by the pineal gland is ***melatonin,*** which contributes to setting the body's biological clock. More melatonin is released in darkness and during sleep; less melatonin is liberated in strong sunlight. In animals that breed during specific seasons, melatonin inhibits reproductive functions. Whether melatonin influences human reproductive function, however, is still unclear. Melatonin levels are higher in children and decline with age into adulthood, but there is no evidence that changes in melatonin secretion correlate with the onset of puberty and sexual maturation.

MANUAL THERAPY APPLICATION

Seasonal Affective Disorder

Seasonal affective disorder (SAD) is a type of depression that afflicts some people during the winter months, when day length is short. It is thought to be due, in part, to overproduction of melatonin. Bright light therapy—repeated exposure to artificial light—can provide relief. A few manual therapists are offering bright light therapy to their patients as an optional service.

CHECKPOINT

15. What is the relationship between melatonin secretion and sleep?

20.10 OTHER HORMONES

OBJECTIVE

- List the hormones secreted by cells in tissues and organs other than endocrine glands, and describe their functions.

Hormones from Other Endocrine Cells

Cells in organs other than those usually classified as endocrine glands have an endocrine function and secrete hormones. Table 20.2 summarizes these hormones and their actions.

Prostaglandins and Leukotrienes

Two families of molecules derived from fatty acids, the ***prostaglandins*** (pros′-ta-GLAN-dins), or ***PGs,*** and the ***leukotrienes*** (loo-kō-TRĪ-ēns), or ***LTs,*** act locally as hormones in most tissues of the body. Virtually all body cells except red blood cells release these local hormones in response to chemical and mechanical stimuli. Because the PGs and LTs act close to their sites of release, they appear in only tiny quantities in the blood.

Leukotrienes stimulate movement of white blood cells and mediate inflammation. The prostaglandins alter smooth muscle contraction, glandular secretions, blood flow, reproductive processes, platelet function, respiration, nerve impulse transmis-

TABLE 20.2

Summary of Hormones Produced by Other Organs and Tissues that Contain Endocrine Cells

SOURCE AND HORMONE	ACTIONS
THYMUS	
Thymosin	Promotes the maturation of T cells (a type of white blood cell that destroys microbes and foreign substances) and may retard the aging process.
GASTROINTESTINAL TRACT	
Gastrin	Promotes secretion of gastric juice and increases movements of the stomach.
Glucose-dependent insulinotropic peptide (GIP)	Stimulates release of insulin by pancreatic beta cells.
Secretin	Stimulates secretion of pancreatic juice and bile.
Cholecystokinin (CCK)	Stimulates secretion of pancreatic juice, regulates release of bile from the gallbladder, and brings about a feeling of fullness after eating.
KIDNEY	
Erythropoietin (EPO)	Increases rate of red blood cell production.
HEART	
Atrial natriuretic peptide (ANP)	Decreases blood pressure.
ADIPOSE TISSUE	
Leptin	Suppresses appetite and may increase the activity of FSH and LH.
PLACENTA	
Human chorionic gonadotropin (hCG)	Stimulates the ovary to continue production of estrogens and progesterone during pregnancy.

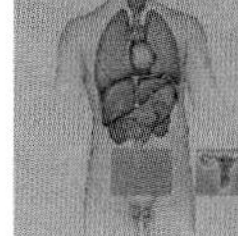

sion, fat metabolism, and immune responses. PGs also have roles in inflammation, promoting fever, and intensifying pain.

CHECKPOINT

16. What hormones are secreted by the gastrointestinal tract, placenta, kidneys, skin, adipose tissue, and heart?
17. What are some functions of prostaglandins and leukotrienes?

20.11 THE STRESS RESPONSE

OBJECTIVE

- Describe how the body responds to stress.

It is impossible to remove all stress from our everyday lives. Any stimulus that produces a stress response is called a *stressor*. A stressor may be almost any disturbance—heat or cold, environmental poisons, toxins given off by bacteria, heavy bleeding from a wound or surgery, or a strong emotional reaction. Stressors may be pleasant or unpleasant, and they vary among people and even within the same person at different times. When homeostatic mechanisms are successful in counteracting stress, the internal environment remains within normal physiological limits. If stress is extreme, unusual, or long lasting, it elicits the ***stress response,*** a sequence of bodily changes that can progress through three stages: (1) an initial fight-or-flight response, (2) a slower resistance reaction, and eventually (3) exhaustion.

The ***fight-or-flight response,*** initiated by nerve impulses from the hypothalamus to the sympathetic division of the autonomic nervous system (ANS), including the adrenal medullae, quickly mobilizes the body's resources for immediate physical activity. It brings huge amounts of glucose and oxygen to the organs that are most active in warding off danger: the brain, which must become highly alert; the skeletal muscles, which may have to fight off an attacker or flee; and the heart, which must work vigorously to pump enough blood to the brain and muscles. Reduction of blood flow to the kidneys, however, promotes the release of renin, which sets into motion the renin–angiotensin–aldosterone pathway (see Figure 20.14). Aldosterone causes the kidneys to retain Na^+, which leads to water retention and elevated blood pressure. Water retention also helps preserve body fluid volume in the case of severe bleeding.

The second stage in the stress response is the ***resistance reaction.*** Unlike the short-lived fight-or-flight response, which is initiated by nerve impulses from the hypothalamus, the resistance reaction is initiated in large part by hypothalamic-releasing hormones and is a longer-lasting response. The hormones involved are corticotropin-releasing hormone (CRH), growth hormone-releasing hormone (GHRH), and thyrotropin-releasing hormone (TRH).

CRH stimulates the anterior pituitary to secrete ACTH, which in turn stimulates the adrenal cortex to release more cortisol. Cortisol then stimulates release of glucose by liver cells, breakdown of triglycerides into fatty acids, and catabolism of proteins into amino acids. Tissues throughout the body can use the resulting glucose, fatty acids, and amino acids to produce ATP or to repair damaged cells. Cortisol also reduces inflammation. A second hypothalamic-releasing hormone, GHRH, causes the anterior pituitary to secrete human growth hormone (hGH). Acting via insulin-like growth factors, hGH stimulates breakdown of triglycerides and glycogen. A third hypothalamic-releasing hormone, TRH, stimulates the anterior pituitary to secrete thyroid-stimulating hormone (TSH). TSH promotes secretion of thyroid hormones, which stimulate the increased use of glucose for ATP production. The combined actions of hGH and TSH thereby supply additional ATP for metabolically active cells.

The resistance stage helps the body continue fighting a stressor long after the fight-or-flight response dissipates. Generally, it is successful in seeing us through a stressful episode, and our bodies then return to normal. Occasionally, however, the resistance stage fails to combat the stressor: The resources of the body may eventually become so depleted that they cannot sustain the resistance stage, and ***exhaustion*** ensues. Prolonged exposure to high levels of cortisol and other hormones involved in the resistance reaction causes wasting of muscles, suppression of the immune system, ulceration of the gastrointestinal tract, and failure of pancreatic beta cells. In addition, pathological changes may occur because resistance reactions persist after the stressor has been removed.

Although the exact role of stress in human diseases is not known, it is clear that stress can temporarily inhibit certain components of the immune system. Stress-related disorders include gastritis, ulcerative colitis, irritable bowel syndrome, hypertension, asthma, rheumatoid arthritis, migraine headaches, anxiety, and depression. People under stress also are at a greater risk of developing a chronic disease or dying prematurely.

• CLINICAL CONNECTION | Posttraumatic Stress Disorder

Posttraumatic stress disorder (PTSD) may develop in someone who has experienced, witnessed, or learned about a physically or psychologically distressing event. The immediate cause of PTSD appears to be the specific stressors associated with the events. Among the stressors are terrorism, hostage taking, imprisonment, serious accidents, torture, sexual or physical abuse, violent crimes, and natural disasters. In the United States, PTSD affects 10% of females and 5% of males. Symptoms of PTSD include reliving the event through nightmares or flashbacks; loss of interest and lack of motivation; poor concentration; irritability; and insomnia. •

CHECKPOINT

18. What is the role of the hypothalamus during stress?
19. How are stress and immunity related?

20.12 AGING AND THE ENDOCRINE SYSTEM

OBJECTIVE

- Describe the effects of aging on the endocrine system.

Although some endocrine glands shrink as we get older, their performance may or may not be compromised. Production of human growth hormone by the anterior pituitary decreases, which is one cause of muscle atrophy as aging proceeds. The thyroid gland often decreases its output of thyroid hormones with age, causing a decrease in metabolic rate, an increase in body fat, and hypothyroidism, which is seen more often in older people. Because there is less negative feedback (lower levels of thyroid hormones), the level of thyroid-stimulating hormone increases with age.

With aging, the blood level of PTH rises, perhaps due to inadequate dietary intake of calcium. In a study of older women who took 2400 mg/day of supplemental calcium, blood levels of PTH were as low as those in younger women. Both calcitriol and calcitonin levels are lower in older persons. Together, the rise in PTH and the fall in calcitonin heighten the age-related decrease in bone mass that leads to osteoporosis and increased risk of fractures.

The adrenal glands contain increasingly more fibrous tissue and produce less cortisol and aldosterone with advancing age. However, production of epinephrine and norepinephrine remains normal. The pancreas releases insulin more slowly with age, and receptor sensitivity to glucose declines. As a result, blood glucose levels in older people increase faster and return to normal more slowly than in younger individuals.

The thymus is largest in infancy. After puberty, its size begins to decrease, and thymic tissue is replaced by adipose and areolar connective tissue. In older adults, the thymus has atrophied significantly. However, it still produces new T cells for immune responses.

The ovaries decrease in size with advancing age, and they no longer respond to FSH and LH. The resultant decreased output of estrogens leads to conditions such as osteoporosis, high blood cholesterol, and atherosclerosis. FSH and LH levels are high due to less negative feedback inhibition of estrogens. Although testosterone production by the testes decreases with age, the effects are not usually apparent until very old age, and many elderly males can still produce active sperm in normal numbers.

CHECKPOINT

20. Which hormone is related to the muscle atrophy that occurs with aging?

To appreciate the many ways the endocrine system contributes to the homeostasis of other body systems, examine *Contributions of the Endocrine System to Homeostasis* (Exhibit 20.1).

KEY MEDICAL TERMS ASSOCIATED WITH THE ENDOCRINE SYSTEM

Gynecomastia (gī-ne′-kō-MAS-tē-a; *gyneco-* = woman; *-mast-* = breast) Excessive development of mammary glands in a male. Sometimes a tumor of the adrenal gland may secrete sufficient amounts of estrogen to cause the condition.

Hirsutism (HER-soo-tizm; *hirsut-* = shaggy) Presence of excessive bodily and facial hair in a male pattern, especially in women; may be due to excess androgen production caused by tumors or drugs.

Thyroid crisis (storm) A severe state of hyperthyroidism that can be life-threatening. It is characterized by high body temperature, rapid heart rate, high blood pressure, gastrointestinal symptoms (abdominal pain, vomiting, diarrhea), agitation, tremors, confusion, seizures, and possibly coma.

Virilizing adenoma (*aden-* = gland; *-oma* = tumor) Tumor of the adrenal gland that liberates excessive androgens, causing virilism (masculinization) in females. Occasionally, adrenal tumor cells liberate estrogens to the extent that a male patient develops gynecomastia. Such a tumor is called a *feminizing adenoma*.

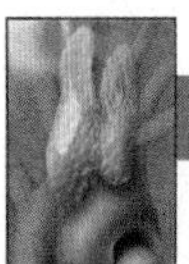

STUDY OUTLINE

Endocrine Glands (Section 20.1)

1. The nervous system controls homeostasis through the release of neurotransmitters; the endocrine system uses hormones.
2. The nervous system causes all three types of muscles to contract and many glands to secrete; the endocrine system affects virtually all body tissues.
3. Exocrine glands (sweat, oil, mucous, digestive) secrete their products through ducts into body cavities or onto body surfaces.
4. Endocrine glands secrete hormones into interstitial fluid. Then, the hormones diffuse into the blood.
5. The endocrine system consists of endocrine glands and several organs that contain endocrine tissue.

Hormone Action (Section 20.2)

1. Hormones affect only specific target cells that have the proper receptors to bind a given hormone.
2. Chemically, hormones are either lipid-soluble (steroids, thyroid hormones, and nitric oxide) or water-soluble (modified amino acids, peptides, and proteins).
3. Hormone secretion is controlled by signals from the nervous system, chemical changes in the blood, and other hormones.

Hypothalamus and Pituitary Gland (Section 20.3)

1. The pituitary gland is attached to the hypothalamus and consists of two lobes: the anterior pituitary and the posterior pituitary.
2. Hormones of the pituitary gland are controlled by inhibiting and releasing hormones produced by the hypothalamus. The hypophyseal portal veins carry hypothalamic releasing and inhibiting hormones from the hypothalamus to the anterior pituitary.
3. The anterior pituitary synthesizes and secretes human growth hormone (hGH), thyroid-stimulating hormone (TSH), follicle-stimulating hormone (FSH), luteinizing hormone (LH), prolactin

Contributions of the Endocrine System to Homeostasis

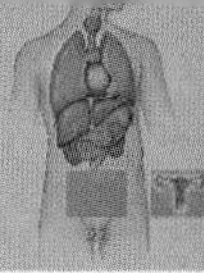

BODY SYSTEM — **CONTRIBUTION OF THE ENDOCRINE SYSTEM**

For all body systems

Together with the nervous system, hormones of the endocrine system regulate the activity and growth of target cells throughout the body. Several hormones regulate metabolism, uptake of glucose, and molecules used for ATP production by body cells.

Integumentary system

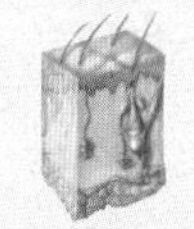

Androgens stimulate the growth of axillary and pubic hair and activation of sebaceous glands. Melanocyte-stimulating hormone (MSH) can cause darkening of the skin.

Skeletal system

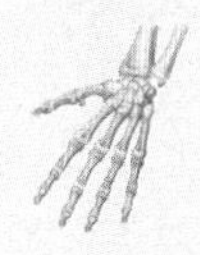

Human growth hormone (hGH) and insulin-like growth factors (IGFs) stimulate bone growth. Estrogens cause closure of epiphyseal plates at the end of puberty and help maintain bone mass in adults. Parathyroid hormone (PTH) promotes the release of calcium and other minerals from bone extracellular matrix into the blood. Thyroid hormones are needed for normal development and growth of the skeleton.

Muscular system

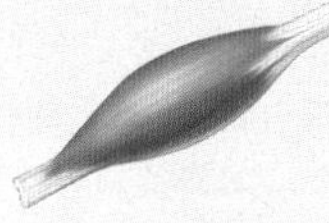

Epinephrine and norepinephrine help increase blood flow to exercising muscles. PTH maintains the proper level of Ca^{2+} in blood and interstitial fluid, which is needed for muscle contraction. Glucagon, insulin, and other hormones regulate metabolism in muscle fibers. IGFs, thyroid hormones, and insulin stimulate protein synthesis and thereby help maintain muscle mass.

Nervous system

Several hormones, especially thyroid hormones, insulin, and IGFs, influence growth and development of the nervous system.

Cardiovascular system

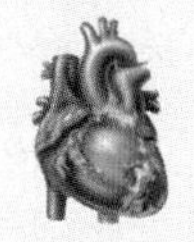

Erythropoietin (EPO) promotes the production of red blood cells. Aldosterone and antidiuretic hormone (ADH) increase blood volume. Epinephrine and norepinephrine increase the heart's rate and force of contraction. Several hormones elevate blood pressure during exercise and other stresses.

Lymphatic system and immunity

Glucocorticoids such as cortisol depress inflammation and immune responses. Hormones from the thymus promote maturation of T cells, a type of white blood cell that participates in immune responses.

Respiratory system

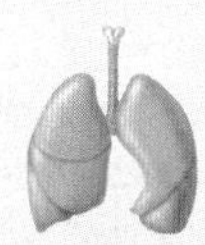

Epinephrine and norepinephrine dilate (widen) the airways during exercise and other stresses. Erythropoietin regulates the amount of oxygen carried in the blood by adjusting the number of red blood cells.

Digestive system

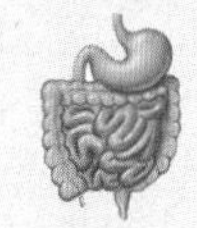

Epinephrine and norepinephrine depress activity of the digestive system. Gastrin, cholecystokinin, secretin, and glucose-dependent insulinotropic peptide (GIP) help regulate digestion. Calcitriol promotes absorption of dietary calcium. Leptin suppresses appetite.

Urinary system

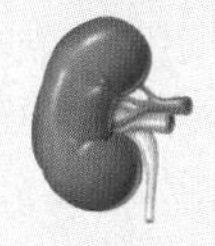

ADH, aldosterone, and atrial natriuretic peptide (ANP) adjust the rate of loss of water and ions in the urine, thereby regulating blood volume and ion levels in the blood.

Reproductive systems

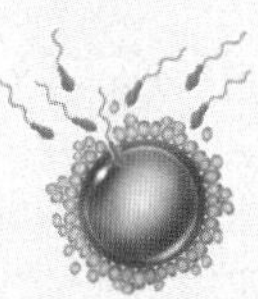

Hypothalamic-releasing and -inhibiting hormones, follicle-stimulating hormone (FSH), and luteinizing hormone (LH) regulate the development, growth, and secretions of the gonads (ovaries and testes). Estrogens and testosterone contribute to the development of oocytes and sperm and stimulate the development of sexual characteristics. Prolactin promotes milk production in the mammary glands. Oxytocin causes contraction of the uterus and ejection of milk from the mammary glands.

(PRL), adrenocorticotropic hormone (ACTH), and melanocyte-stimulating hormone (MSH).
4. Human growth hormone (hGH) stimulates body growth through insulin-like growth factors (IGFs) and is controlled by growth hormone–releasing hormone (GHRH) and growth hormone–inhibiting hormone (GHIH).
5. TSH regulates thyroid gland activities and is controlled by thyrotropin-releasing hormone (TRH).
6. FSH and LH regulate activities of the gonads—ovaries and testes—and are controlled by gonadotropin-releasing hormone (GnRH).
7. PRL helps stimulate milk production. Prolactin-inhibiting hormone (PIH) suppresses release of prolactin. Prolactin-releasing hormone (PRH) stimulates a rise in prolactin level during pregnancy.
8. ACTH regulates activities of the adrenal cortex and is controlled by corticotropin-releasing hormone (CRH).
9. The posterior pituitary contains axon terminals of neurosecretory cells whose cell bodies are in the hypothalamus.
10. Hormones made in the hypothalamus and released by the posterior pituitary include oxytocin, which stimulates contraction of the uterus and ejection of milk from the breasts, and antidiuretic hormone (ADH), which stimulates water reabsorption by the kidneys and constriction of arterioles.
11. Oxytocin secretion is stimulated by uterine stretching and by suckling during nursing; ADH secretion is controlled by the osmotic pressure of the blood and blood volume.
12. Table 20.1 summarizes the hormones of the anterior and posterior pituitary.

Thyroid Gland (Section 20.4)

1. The thyroid gland is located below the larynx.
2. It consists of thyroid follicles composed of follicular cells, which secrete the thyroid hormones thyroxine (T_4) and triiodothyronine (T_3), and parafollicular cells, which secrete calcitonin.
3. Thyroid hormones regulate oxygen use and metabolic rate, cellular metabolism, and growth and development. Secretion is controlled by TRH from the hypothalamus and thyroid-stimulating hormone (TSH) from the anterior pituitary.
4. Calcitonin (CT) can lower the blood level of calcium; its secretion is controlled by the level of calcium in the blood.

Parathyroid Glands (Section 20.5)

1. The parathyroid glands are embedded on the posterior surfaces of the thyroid.
2. Parathyroid hormone (PTH) regulates the homeostasis of calcium, magnesium, and phosphate by increasing blood calcium and magnesium levels and decreasing blood phosphate level. PTH secretion is controlled by the level of calcium in the blood.

Pancreatic Islets (Section 20.6)

1. The pancreas lies in the curve of the duodenum. It has both endocrine and exocrine functions.
2. The endocrine portion consists of pancreatic islets or islets of Langerhans, which are made up of alpha and beta cells.
3. Alpha cells secrete glucagon, and beta cells secrete insulin.
4. Glucagon increases blood glucose level, and insulin decreases blood glucose level. Secretion of both hormones is controlled by the level of glucose in the blood.

Adrenal Glands (Section 20.7)

1. The adrenal glands are located above the kidneys. They consist of an outer cortex and inner medulla.
2. The adrenal cortex is divided into three zones: The outer zone secretes mineralocorticoids; the middle zone secretes glucocorticoids; and the inner zone secretes androgens.
3. Mineralocorticoids (mainly aldosterone) increase sodium and water reabsorption and decrease potassium reabsorption. Secretion is controlled by the renin–angiotensin–aldosterone pathway.
4. Glucocorticoids (mainly cortisol) promote normal metabolism, help resist stress, and decrease inflammation. Secretion is controlled by ACTH.
5. Androgens secreted by the adrenal cortex stimulate growth of axillary and pubic hair, aid the prepubertal growth spurt, and contribute to libido.
6. The adrenal medullae secrete epinephrine and norepinephrine (NE), which are released under stress.

Ovaries and Testes (Section 20.8)

1. The ovaries are located in the pelvic cavity and produce estrogens, progesterone, and inhibin. These sex hormones regulate the menstrual cycle, maintain pregnancy, and prepare the mammary glands for lactation. They also help establish and maintain the female body shape.
2. The testes lie inside the scrotum and produce testosterone and inhibin. Testosterone regulates production of sperm and stimulates the development and maintenance of masculine characteristics such as beard growth and deepening of the voice.

Pineal Gland (Section 20.9)

1. The pineal gland, attached to the roof of the third ventricle in the brain, secretes melatonin, which contributes to setting the body's biological clock.

Other Hormones (Section 20.10)

1. Body tissues other than those normally classified as endocrine glands contain endocrine tissue and secrete hormones. These include the gastrointestinal tract, placenta, kidneys, and heart. (See Table 20.2.)
2. Prostaglandins and leukotrienes act locally in most body tissues.

The Stress Response (Section 20.11)

1. Stressors include surgical operations, poisons, infections, fever, and strong emotional responses.
2. If stress is extreme, it triggers the stress response, which occurs in three stages: the fight-or-flight response, resistance reaction, and exhaustion.
3. The fight-or-flight response is initiated by nerve impulses from the hypothalamus to the sympathetic division of the autonomic nervous system and the adrenal medullae. This response rapidly increases circulation and promotes ATP production.
4. The resistance reaction is initiated by releasing hormones secreted by the hypothalamus. Resistance reactions are longer lasting and accelerate breakdown reactions to provide ATP for counteracting stress.
5. Exhaustion results from depletion of body resources during the resistance stage.

6. Stress may trigger certain diseases by inhibiting the immune system.

Aging and the Endocrine System (Section 20.12)

1. Although some endocrine glands shrink as we get older, their performance may or may not be compromised.
2. Production of human growth hormone, thyroid hormones, cortisol, aldosterone, and estrogens decrease with advancing age.
3. With aging, the blood levels of TSH, LH, FSH, and PTH rise.
4. The pancreas releases insulin more slowly with age, and receptor sensitivity to glucose declines.
5. After puberty, thymus size begins to decrease, and thymic tissue is replaced by adipose and areolar connective tissue.

SELF-QUIZ QUESTIONS

1. Which of the following is NOT true concerning hormones?
   a. Responses to hormones are generally slower and longer lasting than the responses stimulated by the nervous system.
   b. Hormones are generally controlled by negative feedback systems.
   c. The hypothalamus inhibits the release of some hormones.
   d. Most hormones are released steadily throughout the day.
   e. Hormone secretion is determined by the body's need to maintain homeostasis.
2. Which of the following statements is NOT true?
   a. The secretion of hormones by the anterior pituitary is controlled by hypothalamic-releasing or -inhibiting hormones.
   b. The pituitary gland is attached to the hypothalamus by the infundibulum.
   c. Hypophyseal portal veins connect the posterior pituitary to the hypothalamus.
   d. The anterior pituitary constitutes the majority of the pituitary gland.
   e. The posterior pituitary releases hormones produced by neurosecretory cells of the hypothalamus.
3. The hormone that promotes milk release from the mammary glands and that stimulates the uterus to contract is
   a. oxytocin **b.** prolactin **c.** relaxin
   d. calcitonin **e.** follicle-stimulating hormone.
4. The gland that prepares the body to react to stress by releasing epinephrine is the
   a. posterior pituitary **b.** anterior pituitary **c.** pineal
   d. adrenal **e.** pancreas.
5. To help prevent rejection, organ transplant patients could be given
   a. glucocorticoids **b.** calcitonin
   c. mineralocorticoids **d.** vasopressin
   e. melanocyte-stimulating hormone.
6. A female who is sluggish, gaining weight, and has a low body temperature may be having problems with her
   a. pancreas **b.** parathyroid glands
   c. adrenal medullae **d.** ovaries
   e. thyroid gland.
7. Destruction of the alpha cells of the pancreas might result in
   a. hypoglycemia **b.** seasonal affective disorder
   c. acromegaly **d.** hyperglycemia
   e. decreased urine output.
8. Which of the following is NOT true concerning human growth hormone (hGH) and insulin-like growth factors?
   a. They stimulate protein synthesis.
   b. They have one primary target tissue in the body.
   c. They stimulate skeletal muscle growth.
   d. Hyposecretion in childhood results in dwarfism.
   e. Hypoglycemia can stimulate the release of hGH from the pituitary gland.
9. Follicle-stimulating hormone (FSH) acts on ________ and luteinizing hormone (LH) acts on ________.
   a. the ovaries, the testes
   b. the testes, the ovaries
   c. the ovaries and testes, the ovaries and testes
   d. the ovaries, the mammary glands
   e. the ovaries and uterus, the testes.
10. An injection of adrenocorticotropic hormone (ACTH) would
    a. stimulate the ovaries
    b. influence thyroid gland activity
    c. stimulate the release of cortisol
    d. cause uterine contractions
    e. decrease urine output.
11. Which of the following is NOT true concerning glucocorticoids?
    a. They help to control electrolyte balance.
    b. They help provide resistance to stress.
    c. They help promote normal metabolism.
    d. They are anti-inflammatory hormones.
    e. They provide the body with energy.
12. Mineralocorticoids
    a. help prevent the loss of potassium from the body
    b. are secreted based on the renin–angiotensin–aldosterone pathway
    c. increase the rate of sodium loss in the urine
    d. are involved in lowering the body's blood pressure
    e. increase water loss from the body by increasing urine production.
13. A lack of iodine in the diet affects the production of which hormone?
    a. calcitonin **b.** parathyroid hormone
    c. aldosterone **d.** thyroxine
    e. glucagon.
14. Which of the following hormones with opposite effects are correctly paired?
    a. parathyroid hormone, thyroid hormones
    b. parathyroid hormone, calcitonin
    c. oxytocin, glucocorticoids
    d. aldosterone, oxytocin
    e. thyroid hormones, thymosin.
15. The hormone that functions as part of a positive feedback cycle is
    a. cortisol **b.** testosterone
    c. oxytocin **d.** insulin
    e. thyroxine.
16. In a dehydrated person, you would expect to see an increased release of
    a. parathyroid hormone **b.** aldosterone
    c. insulin **d.** melatonin
    e. inhibin.

17. Match the following:

____ **a.** produces thyroid hormones
____ **b.** secretes insulin
____ **c.** releases hormones into capillaries of the posterior pituitary
____ **d.** stores oxytocin
____ **e.** secretes glucagon
____ **f.** produces calcitonin
____ **g.** secretes steroid hormones

A. posterior pituitary
B. adrenal cortex
C. follicular cell
D. alpha cell
E. parafollicular cell
F. beta cell
G. axon from neurosecretory cell

18. Match the following:

____ **a.** diabetes insipidus
____ **b.** diabetes mellitus
____ **c.** myxedema
____ **d.** Cushing's syndrome
____ **e.** Addison's disease
____ **f.** tetany

A. hypersecretion of glucocorticoids
B. hyposecretion of antidiuretic hormone
C. hyposecretion of insulin
D. hyposecretion of parathyroid hormone
E. hyposecretion of thyroid hormone
F. hyposecretion of glucocorticoids

19. For each of the following, indicate at which stage they would occur as part of the stress response. Use F to indicate fight-or-flight response, R to indicate resistance reaction, and E to indicate exhaustion.

____ **a.** initiated by hypothalamic-releasing hormones
____ **b.** initiated by the sympathetic division of the autonomic nervous system
____ **c.** immediately prepares the body for action
____ **d.** increases cortisol release
____ **e.** short-lived response
____ **f.** body resources become depleted
____ **g.** increased release of many hormones that ensure a continued ATP supply
____ **h.** failure of pancreatic beta cells
____ **i.** nonessential body functions inhibited

CRITICAL THINKING QUESTIONS

1. Patrick was diagnosed with diabetes mellitus on his eighth birthday. His 65-year-old aunt was just diagnosed with diabetes also. Patrick is having a hard time understanding why he needs injections, while his aunt controls her blood sugar with diet and oral medication. Why is his aunt's treatment different from his?

2. Melatonin has been suggested as a possible aid for sleeping problems due to jet lag and rotating work schedules (shift work). It may also be involved in seasonal affective disorder (SAD). Explain how melatonin may affect sleeping.

3. Brian is in a 50-mile bike-a-thon on a hot summer day. He's breathing dust at the back of the pack, he's sweating profusely, and now he's lost his water bottle. Brian is not having a good time. How will his hormones respond to decreased intake of water and the stress of the situation?

? ANSWERS TO FIGURE QUESTIONS

20.1 Secretions of endocrine glands diffuse into interstitial fluid and then into the blood; exocrine secretions flow into ducts that lead into body cavities or to the body surface.

20.2 The hypophyseal portal veins connect capillaries in the hypothalamus, where hypothalamic-releasing and -inhibiting hormones are secreted, to the anterior pituitary, where the hormones act.

20.3 Bones grow in thickness by appositional growth (growth at the outer surface).

20.4 The posterior pituitary releases hormones synthesized in the hypothalamus.

20.5 Absorption of a liter of water in the intestines would decrease the osmotic pressure of your blood plasma, turning off secretion of ADH and decreasing the ADH level in your blood.

20.6 Follicular cells secrete T_3 and T_4, also known as thyroid hormones. Parafollicular cells secrete calcitonin.

20.7 Lack of iodine in the diet leads to diminished production of T_3 and T_4, causing an increased release of TSH that results in growth (enlargement) of the thyroid gland (goiter).

20.8 Congenital hypothyroidism, previously termed cretinism, may occur in an infant and myxedema may occur in an adult.

20.9 PTH increases the number and activity of osteoclasts.

20.10 Target tissues for PTH are bones and kidneys; target tissue for CT is bone; target tissue for calcitriol is the GI tract.

20.11 The pancreas is both an exocrine and an endocrine gland.

20.12 Glucagon is considered an "anti-insulin" hormone because it has several effects that are opposite to those of insulin.

20.13 The outer zone of the adrenal cortex secretes mineralocorticoids, the middle zone secretes glucocorticoids, and the inner zone secretes adrenal androgens.

20.14 Because drugs that block ACE lower blood pressure, they are used to treat high blood pressure (hypertension).

20.15 Persons with Cushing's syndrome may have spindly limbs, a "buffalo hump" on the back, and a pendulous (hanging) abdomen.

The Cardiovascular System: The Blood

21

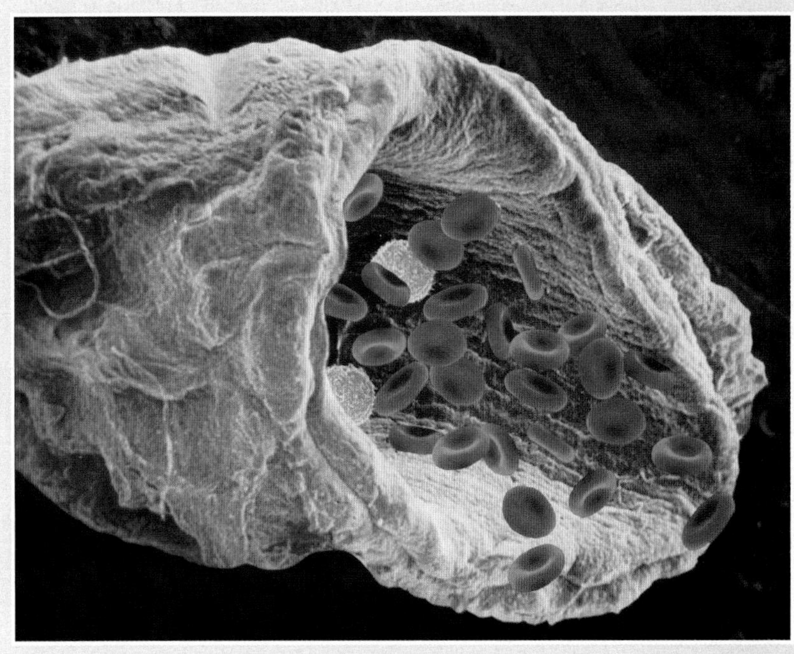

Blood is the main transporter in the body. It transports oxygen, nutrients, immune cells, clotting agents, and wastes. It is also essential to circulate fluids to maintain water balance. In fact, 91.5% of the blood is water; the remainder consists of red blood cells, white blood cells, platelets, electrolytes, proteins, and other molecules being transported. Red blood cells carry oxygen and a portion of the carbon dioxide. White blood cells are immune cells that keep us free of many invaders. Platelets help clot our blood. Owing to all of these properties, blood becomes a wonderful medium for biochemical tests. We can do many tests on blood to see if we are healthy, for example, checking nutrient levels. We can also detect infection by looking at the circulating number and types of white blood cells. Blood tests are therefore a relatively inexpensive first step in either detecting or ruling out disease or injury.

The **cardiovascular system** (*cardio-* = heart; *vascular* = blood or blood vessels) consists of three interrelated components: blood, the heart, and blood vessels. The focus of this chapter is blood; the next two chapters will cover the heart and blood vessels, respectively. Functionally, the cardiovascular system transports substances to and from body cells. To perform its function, blood must circulate throughout the body. The heart serves as the pump for circulation, and blood vessels carry blood from the heart to body cells and from body cells back to the heart. The branch of science concerned with the study of blood, blood-forming tissues, and disorders associated with them is **hematology** (hēm-a-TOL-ō-jē; *hemo* or *hemato-* = blood; *-logy* = study of).

CONTENTS AT A GLANCE

21.1 FUNCTIONS OF BLOOD

OBJECTIVE

- List and describe the functions of blood.

Blood is a liquid connective tissue that consists of cells surrounded by extracellular matrix. Blood has three general functions: transportation, regulation, and protection.

1. **Transportation.** Blood transports oxygen from the lungs to cells throughout the body and carbon dioxide (a waste product of cellular respiration; see Chapter 25) from the cells to the lungs. It also carries nutrients from the gastrointestinal tract to body cells, heat and waste products away from cells, and hormones from endocrine glands to other body cells.
2. **Regulation.** Blood helps regulate the pH of body fluids. The heat-absorbing and coolant properties of the water in blood plasma (see Section 2.2) and its variable rate of flow through the skin help adjust body temperature. Blood osmotic pressure also influences the water content of cells.
3. **Protection.** Blood clots (becomes gel-like) in response to an injury, which protects against its excessive loss from the cardiovascular system. In addition, white blood cells protect against disease by carrying on phagocytosis and producing proteins called antibodies. Blood contains additional proteins, called interferons and complement, that also help protect against disease.

CHECKPOINT

1. Name several substances transported by blood.
2. How is blood protective?

21.2 COMPONENTS OF WHOLE BLOOD

OBJECTIVE

- Discuss the formation, components, and functions of whole blood.

Blood is denser and more viscous (thicker) than water. The temperature of blood is about 38°C (100.4°F). Its pH is slightly alkaline, ranging from 7.35 to 7.45. Blood constitutes about 8% of the total body weight. The blood volume is 5 to 6 liters (1.5 gal) in an average-sized adult male and 4 to 5 liters (1.2 gal) in an average-sized adult female. The difference in volume is due to differences in average body size.

Whole blood is composed of two portions: (1) ***blood plasma,*** a liquid extracellular matrix that contains dissolved substances, and (2) ***formed elements,*** which are cells and cell fragments. If a sample of blood is centrifuged (spun at high speed) in a small glass tube, the cells (which are more dense) sink to the bottom of the tube and the lighter-weight blood plasma (which is less dense) forms a layer on top (Figure 21.1a). Blood is about 45% formed elements and 55% plasma. Normally, more than 99% of the formed elements are red blood cells (RBCs). The percentage of total blood volume occupied by red blood cells is termed the ***hematocrit*** (hē-MAT-ō-krit). Pale, colorless white blood cells (WBCs) and platelets occupy less than 1% of total blood volume. They form a very thin layer, called the buffy coat, between the packed RBCs and blood plasma in centrifuged blood. Figure 21.1b shows the composition of blood plasma and the numbers of the various types of formed elements in blood.

Blood Plasma

When the formed elements are removed from blood, a straw-colored liquid called ***blood plasma*** (or simply ***plasma***) remains. Plasma is about 91.5% water, 7% proteins, and 1.5% solutes other than proteins. Proteins in the blood, the plasma proteins, are synthesized mainly by the liver. The most plentiful plasma proteins are the ***albumins,*** which account for about 54% of all plasma proteins. Among other functions, albumins help maintain proper blood osmotic pressure, which is an important factor in the exchange of fluids across capillary walls. ***Globulins,*** which compose 38% of plasma proteins, include ***antibodies,*** defensive proteins produced during certain immune responses. ***Fibrinogen*** makes up about 7% of plasma proteins and is a key protein in the formation of blood clots. Other solutes in plasma include electrolytes, nutrients, gases, regulatory substances such as enzymes and hormones, vitamins, and waste products; this long list accounts for approximately 1.5% of blood plasma.

Formed Elements

The ***formed elements*** of the blood are the following (see Figure 21.2):

I. Red blood cells
II. White blood cells
   A. Granular leukocytes (contain conspicuous granules that are visible under a light microscope after staining)
      1. Eosinophils
      2. Basophils
      3. Neutrophils
   B. Agranular leukocytes (no granules are visible under a light microscope after staining)
      1. Monocytes
      2. T and B lymphocytes
      3. Natural killer cells
III. Platelets

Formation of Blood Cells

The process by which the formed elements of blood develop is called ***hemopoiesis*** (hē-mō-poy-Ē-sis; *-poieis* = making). Before birth, hemopoiesis first occurs in the yolk sac of an embryo and later in the liver, spleen, thymus, and lymph nodes of a fetus. In the last three months before birth, red bone marrow

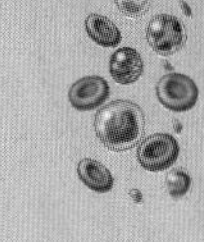

Figure 21.1 Components of blood in a normal adult.

Blood is a connective tissue that consists of blood plasma (liquid) plus formed elements (red blood cells, white blood cells, and platelets).

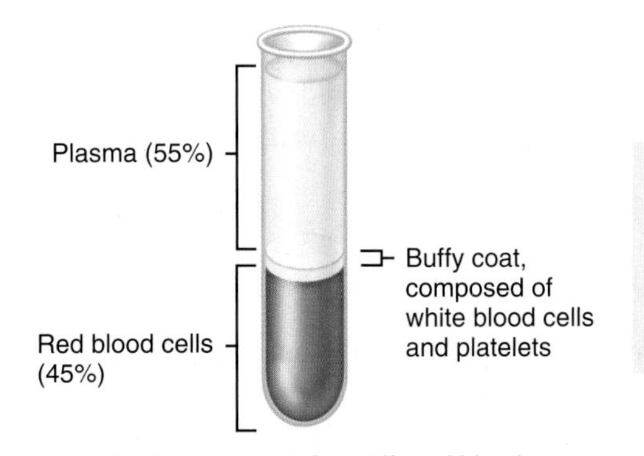

(a) Appearance of centrifuged blood

Functions of Blood

1. Transports oxygen, carbon dioxide, nutrients, hormones, heat, and wastes.
2. Regulates pH, body temperature, and water content of cells.
3. Protects against blood loss through clotting, and against disease through phagocytic white blood cells and antibodies.

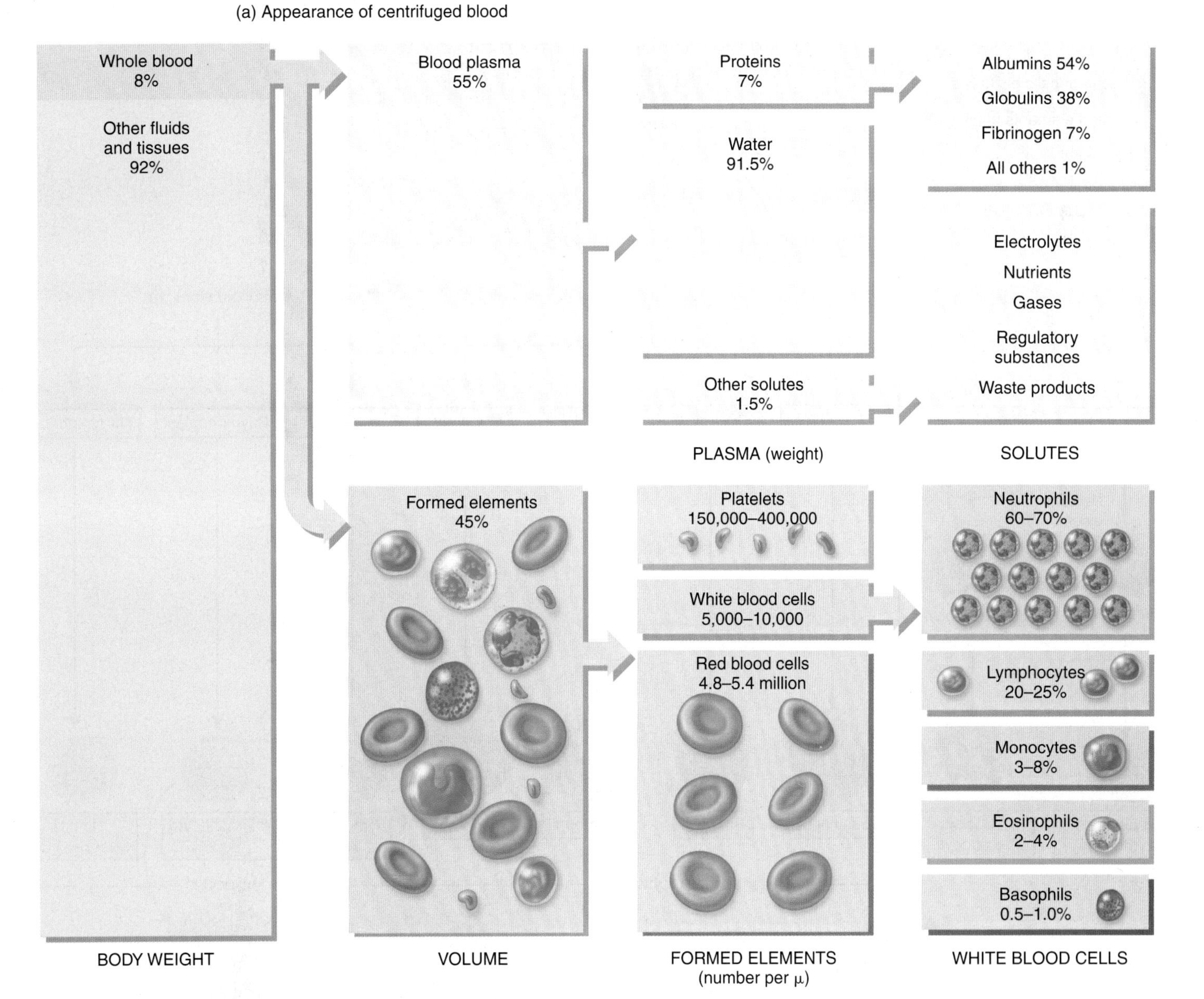

(b) Components of blood

What is the approximate volume of blood in your body?

becomes the primary site of hemopoiesis and continues as the source of blood cells after birth and throughout life.

Red bone marrow is a highly vascularized connective tissue located in the microscopic spaces between trabeculae of spongy bone tissue (see Chapter 6). It is present chiefly in bones of the axial skeleton, pectoral and pelvic girdles, and the proximal epiphyses of the humerus and femur. About 0.05–0.1% of red bone marrow cells are cells called ***pluripotent stem cells*** (ploo-RIP-ō-tent; *pluri-* = several). Pluripotent stem cells are cells that have the capacity to develop into many different types of cells (Figure 21.2).

In response to stimulation by specific hormones, pluripotent stem cells generate two other types of stem cells that have the capacity to develop into fewer types of cells: *myeloid stem cells* and *lymphoid stem cells* (Figure 21.2). Myeloid stem cells begin their development in red bone marrow and differentiate into several types of cells from which red blood cells, platelets,

Figure 21.2 Origin, development, and structure of blood cells. Some of the generations of some cell lines have been omitted.

Blood cell production, called hemopoiesis, occurs mainly in red bone marrow after birth.

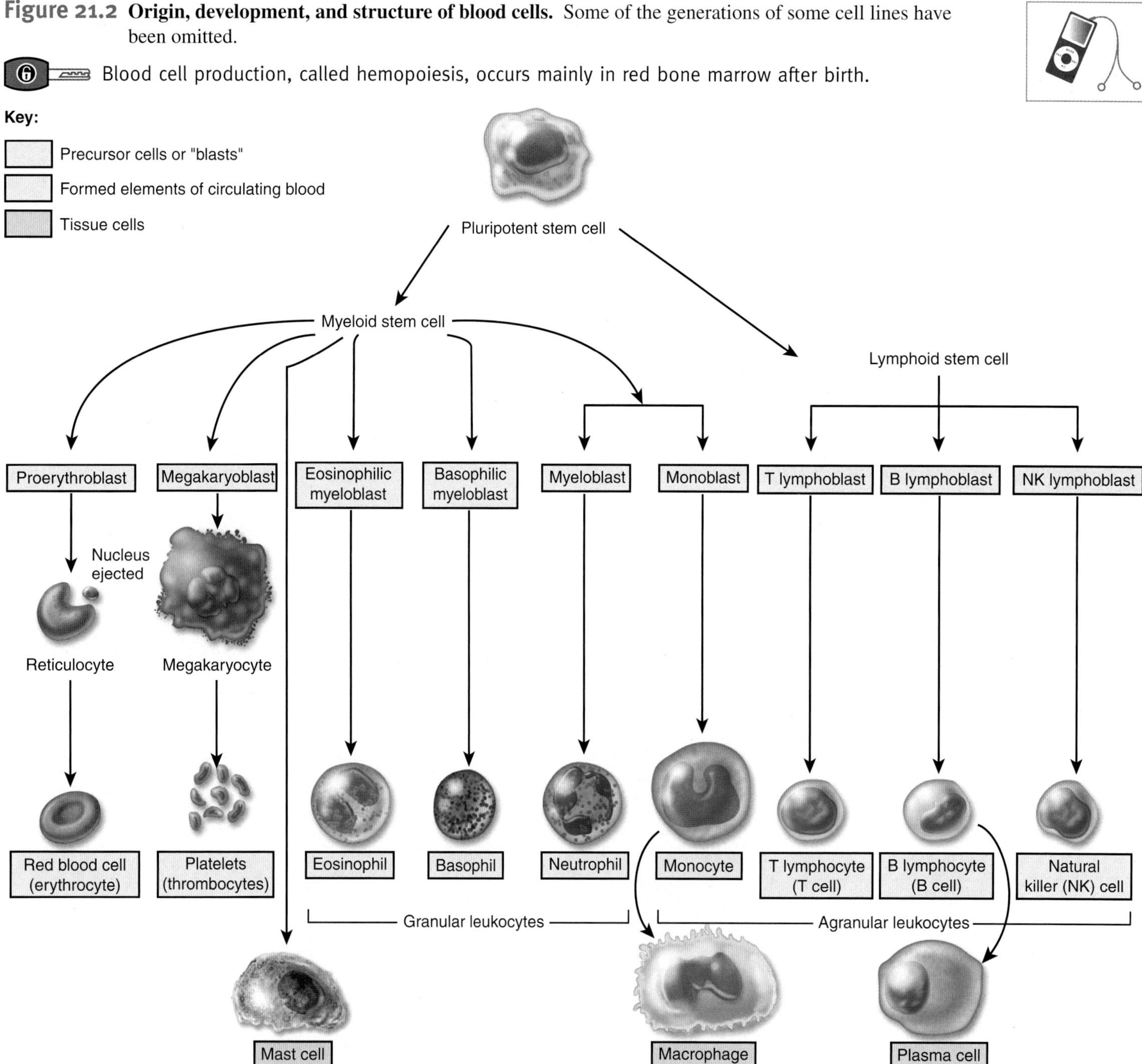

Do all agranular leukocytes develop from lymphoid stem cells?

eosinophils, basophils, neutrophils, and monocytes develop. Lymphoid stem cells begin their development in red bone marrow but complete it in lymphatic tissues. They differentiate into cells from which the T and B lymphocytes and natural killer (NK) cells develop. Intermediate cells between the two types of stem cells and cells found in the bloodstream are termed *precursor cells* or *"blasts"* (gray boxes in Figure 21.2).

• CLINICAL CONNECTION Bone Marrow Transplant

A **bone marrow transplant** is the replacement of cancerous or abnormal red bone marrow with healthy red bone marrow in order to establish normal blood cell counts. The defective red bone marrow is destroyed by high doses of chemotherapy and whole body radiation just before the transplant takes place. These treatments kill the cancer cells and destroy the patient's immune system in order to decrease the chance of transplant rejection. The red bone marrow from a donor is usually removed from the hip bone under general anesthesia with a syringe and is then injected into the recipient's vein, much like a blood transfusion. The injected marrow migrates to the recipient's red bone marrow cavities, and the stem cells in the marrow multiply. If all goes well, the recipient's red bone marrow is replaced entirely by healthy, noncancerous cells.

Bone marrow transplants have been used to treat aplastic anemia, certain types of leukemia, severe combined immunodeficiency disease (SCID), Hodgkin's disease, non-Hodgkin's lymphoma, multiple myeloma, thalassemia, sickle-cell disease, breast cancer, ovarian cancer, testicular cancer, and hemolytic anemia. However, there are some drawbacks (negative side effects). Since the recipient's white blood cells have been completely destroyed by chemotherapy and radiation, the patient is extremely vulnerable to infection. (It takes about 2–3 weeks for transplanted bone marrow to produce enough white blood cells to protect against infection.) In addition, transplanted red bone marrow may produce T lymphocytes that attack the recipient's tissues. Another drawback is that patients must take immunosuppressive drugs for life. Because these drugs reduce the level of immune system activity, they increase the risk of infection. •

Red Blood Cells

RBC STRUCTURE ***Red blood cells (RBCs)*** or ***erythrocytes*** (e-RITH-rō-sīts; *erythro-* = red; *-cyte* = cell) contain the oxygen-carrying protein ***hemoglobin,*** which is a pigment that gives whole blood its red color. Hemoglobin also transports about 23% of the carbon dioxide in the blood. A healthy adult male has about 5.4 million red blood cells per microliter (μL) of blood, and a healthy adult female has about 4.8 million. (One drop of blood is about 50 μL.) Again, this difference in the genders reflects differences in the average body size. To maintain normal numbers of RBCs, new mature cells must enter the circulation at the astonishing rate of at least 2 million per second, a pace that balances the equally high rate of RBC destruction. RBCs are biconcave (concave on both sides) discs averaging about 8 micrometers (μm)* in diameter. Mature RBCs lack a nucleus and other organelles and can neither reproduce nor carry on extensive metabolic activities. However, all of their internal space is available for oxygen and carbon dioxide transport. Essentially, RBCs consist of a selectively permeable plasma membrane, cytosol, and hemoglobin. Since a biconcave disc has a much greater surface area for its volume (compared to a sphere or a cube), this shape provides a large surface area for the diffusion of gas molecules into and out of a RBC.

*One micrometer = 1 one-millionth of a meter (10^{-6}) or 1/25,000 of an inch.

• CLINICAL CONNECTION Anemia

Anemia is a condition in which the oxygen-carrying capacity of blood is reduced. Many types of anemia exist; all are characterized by reduced numbers of RBCs or a decreased amount of hemoglobin in the blood. The person feels fatigued and is intolerant of cold, both of which are related to lack of oxygen needed for ATP and heat production. Also, the skin appears pale, due to the low content of red-colored hemoglobin circulating in skin blood vessels. Among the most important types of anemia are the following:

- *Iron-deficiency anemia,* the most prevalent kind of anemia, is caused by inadequate absorption of iron, excessive loss of iron, or insufficient intake of iron. Women are at greater risk for iron-deficiency anemia due to monthly menstrual blood loss.
- *Pernicious anemia* is caused by insufficient hemopoiesis resulting from an inability of the stomach to produce intrinsic factor (needed for absorption of dietary vitamin B_{12}).
- *Hemorrhagic anemia* is due to an excessive loss of RBCs through bleeding resulting from large wounds, stomach ulcers, or especially heavy menstruation.
- In *hemolytic anemia,* RBC plasma membranes rupture prematurely. The condition may result from inherited defects or from outside agents such as parasites, toxins, or antibodies from incompatible transfused blood.
- *Thalassemia* (thal-a-SĒ-mē-a) is a group of hereditary hemolytic anemias in which there is an abnormality in one or more of the four polypeptide chains of the hemoglobin molecule. Thalassemia occurs primarily in populations from countries bordering the Mediterranean Sea.
- *Aplastic anemia* results from destruction of the red bone marrow caused by toxins, gamma radiation, and certain medications that inhibit enzymes needed for hemopoiesis. •

RBC PRODUCTION The formation of all blood cells is called hemopoiesis; the formation of just RBCs is termed ***erythropoiesis*** (e-rith′-rō-poy-Ē-sis). Near the end of erythropoiesis, an RBC precursor ejects its nucleus and becomes a ***reticulocyte*** (re-TIK-ū-lō-sīt; see Figure 21.2). Loss of the nucleus causes the center of the cell to indent, producing the RBC's distinctive biconcave shape. Reticulocytes, which are about 34% hemoglobin and retain some mitochondria, ribosomes, and endoplasmic reticulum, pass from red bone marrow into the bloodstream. Reticulocytes usually develop into mature RBCs within 1 to 2 days after their release from bone marrow.

Normally, erythropoiesis and destruction of RBCs proceed at the same pace. If the oxygen-carrying capacity of the blood falls because erythropoiesis is not keeping up with RBC destruction, RBC production increases (see Figure 21.3). The controlled condition

Figure 21.3 Negative feedback regulation of erythropoiesis (red blood cell formation). Lower oxygen content of air at higher altitudes, anemia, and circulatory problems may reduce oxygen delivery to the body.

The main stimulus for erythropoiesis is hypoxia, a decrease in the oxygen-carrying capacity of the blood.

Some stimulus disrupts homeostasis by
Decreasing
Oxygen delivery to kidneys (and other tissues)
Receptors
Kidney cells detect low oxygen level
Input
Increased erythropoietin secreted into blood
Control center
Proerythroblasts in red bone marrow mature more quickly into reticulocytes
Return to homeostasis when oxygen delivery to kidneys increases to normal
Output
More reticulocytes enter circulating blood
Effectors
Larger number of RBCs in circulation
Increased oxygen delivery to tissues

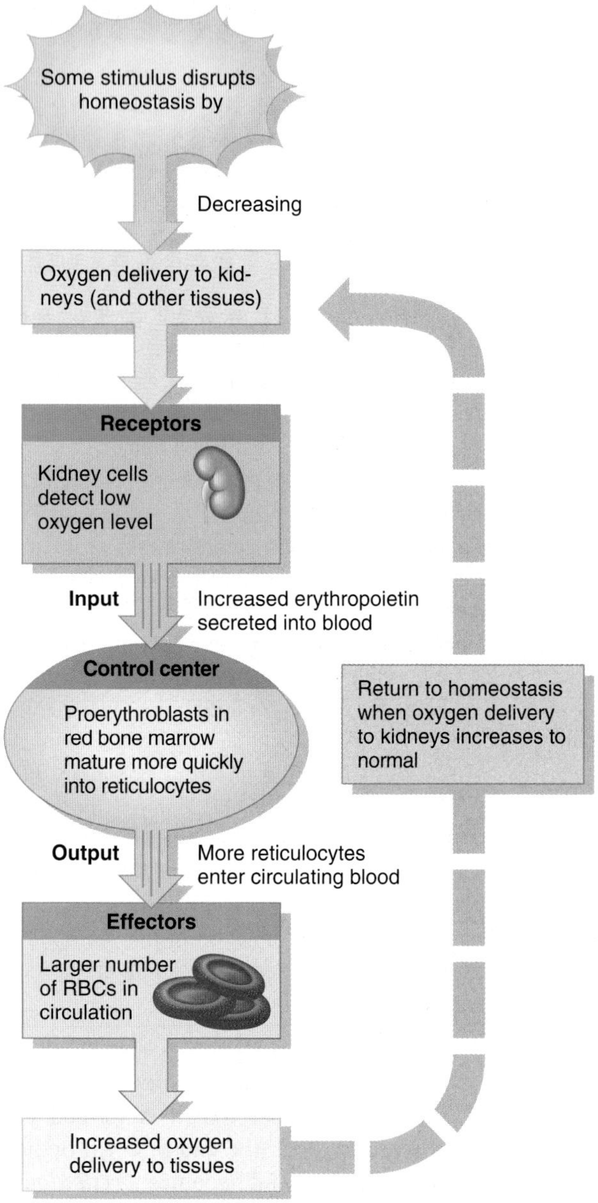

How might your hematocrit change if you moved from a town at sea level to a high mountain village?

in this particular negative feedback loop is the amount of oxygen delivered to the kidneys (and thus to body tissues in general). ***Hypoxia*** (hī-POKS-ē-a), a deficiency of oxygen, stimulates increased release of ***erythropoietin*** (e-rith′-rō-POY-e-tin), or EPO, a hormone made by the kidneys. EPO circulates through the blood to the red bone marrow, where it stimulates erythropoiesis. The larger the number of RBCs in the blood, the higher is the oxygen delivery to the tissues. A person with prolonged hypoxia may develop a life-threatening condition called ***cyanosis*** (sī-a-NŌ-sis), characterized by a bluish-purple skin coloration most easily seen in the nails and mucous membranes. Oxygen delivery may fall due to anemia or circulatory problems that reduce blood flow to tissues. Premature newborns often exhibit anemia, due in part to inadequate production of erythropoietin. During the first weeks after birth, the liver, not the kidneys, produces most EPO. Because the liver is less sensitive than the kidneys to hypoxia, newborns have a smaller EPO response to anemia than do adults. In addition, in infants, fetal hemoglobin is converted into adult hemoglobin; since fetal hemoglobin carries up to 30% more oxygen, the loss of fetal hemoglobin makes the anemia worse.

A test that measures the rate of erythropoiesis is called a ***reticulocyte count.*** This and several other tests related to red blood cells are explained in Table 21.1.

• CLINICAL CONNECTION | Blood Doping

Delivery of oxygen to muscles is a limiting factor in muscular feats. As a result, increasing the oxygen-carrying capacity of the blood enhances athletic performance, especially in endurance events. Because RBCs transport oxygen, athletes have tried several means of increasing their RBC count, known as **blood doping** or artificially induced polycythemia (an abnormally high number of RBCs), to gain a competitive edge. Athletes have enhanced their RBC production by injecting Epoetin alfa (Procrit® or Epogen®), a drug that is used to treat anemia by stimulating the production of RBCs by red bone marrow. Practices that increase the number of RBCs are dangerous because they raise the viscosity of the blood, which increases the resistance to blood flow and makes the blood more difficult for the heart to pump. Increased viscosity also contributes to high blood pressure and increased risk of stroke. During the 1980s, at least 15 competitive cyclists died from heart attacks or strokes linked to suspected use of Epoetin alfa. Although the International Olympics Committee bans Epoetin alfa use, enforcement is difficult because the drug is identical to naturally occurring erythropoietin (EPO).

So-called *natural blood doping* is seemingly the key to the success of marathon runners from Kenya. The average altitude throughout Kenya's highlands is about 2000 meters (6600 feet) above sea level; other areas of Kenya are even higher. Altitude training greatly improves fitness, endurance, and performance. At these higher altitudes, the body increases the production of red blood cells, which means that exercise greatly oxygenates the blood. When these runners compete in Boston, as one example, the altitude is just above sea level, and the bodies of these runners contain more erythrocytes than do the bodies of persons who trained at lower altitudes. A number of training camps have been established in Kenya that now attract endurance athletes from all over the world. •

RBC LIFE CYCLE Red blood cells live only about 120 days because of wear and tear on their plasma membranes as they squeeze through blood capillaries, some of which are smaller than RBCs. Worn-out red blood cells are removed from circulation as follows (Figure 21.4).

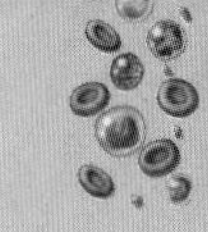

Figure 21.4 Formation and destruction of red blood cells, and the recycling of hemoglobin components.

The rate of RBC formation by red bone marrow equals the rate of RBC destruction by macrophages.

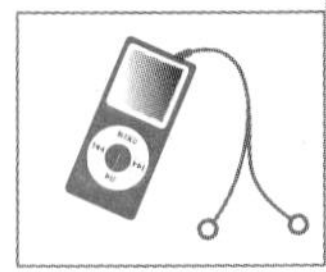

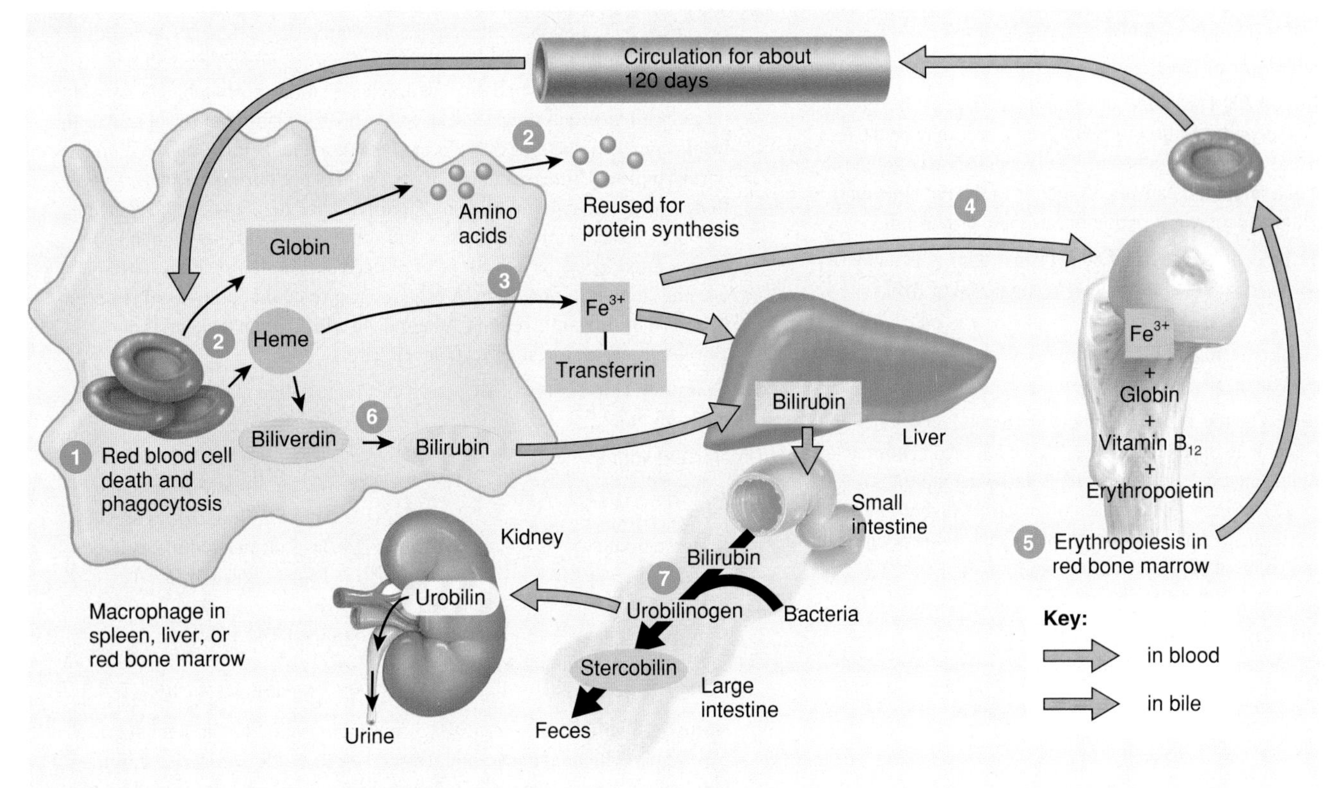

What is the function of transferrin?

1. Macrophages in the spleen, liver, and red bone marrow phagocytize ruptured and worn-out red blood cells, splitting apart the heme and globin portions of hemoglobin.
2. The protein globin is broken down into amino acids, which can be reused by body cells to synthesize other proteins.
3. Iron removed from the heme portion associates with the plasma protein ***transferrin*** (trans-FER-in; *trans-* = across; *-ferr-* = iron), which acts as a transporter.
4. The iron–transferrin complex is then carried to red bone marrow, where RBC precursor cells use it in hemoglobin synthesis. Iron is needed for the heme portion of the hemoglobin molecule, and amino acids are needed for the globin portion. Vitamin B_{12} is also needed for synthesis from hemoglobin. (The lining of the stomach must produce a protein called *intrinsic factor* for absorption of dietary vitamin B_{12} from the GI tract into the blood.)
5. Erythropoiesis in red bone marrow results in the production of red blood cells, which enter the circulation.
6. When iron is removed from heme, the non-iron portion of heme is converted to ***biliverdin*** (bil-i-VER-din), a green pigment, and then into ***bilirubin*** (bil-ē-ROO-bin), a yellow-orange pigment. Bilirubin enters the blood and is transported to the liver. Within the liver, bilirubin is secreted by liver cells into bile, which passes into the small intestine and then into the large intestine.
7. In the large intestine, bacteria convert bilirubin into ***urobilinogen*** (ūr-ō-bī-LIN-ō-jen). Some urobilinogen is absorbed back into the blood, converted to a yellow pigment called ***urobilin*** (ūr-ō-BĪ-lin), and excreted in urine. Most urobilinogen is eliminated in feces in the form of a brown pigment called ***stercobilin*** (ster-kō-BĪ-lin), which gives feces its characteristic color.

Because free iron ions bind to and damage molecules in cells or in the blood, transferrin acts as a protective "protein escort" during transport of iron ions. As a result, plasma contains virtually no free iron.

White Blood Cells

WBC STRUCTURE AND TYPES Unlike red blood cells, ***white blood cells (WBCs)*** or ***leukocytes*** (LOO-kō-sīts; *leuko-* = white) have nuclei and a full complement of other organelles but they do not contain hemoglobin. WBCs are classified as either

TABLE 21.1

Common Medical Tests Involving Blood

A. Reticulocyte count (indicates the rate of erythropoiesis)

Normal value: 0.5% to 1.5%

Abnormal values: A high reticulocyte count might indicate the presence of bleeding or *hemolysis* (rupture of erythrocytes), or it may be the response of someone who is iron deficient. Low reticulocyte count in the presence of anemia might indicate a malfunction of the red bone marrow, owing to a nutritional deficiency, pernicious anemia, or leukemia.

B. Hematocrit (the percentage of red blood cells in blood). A hematocrit of 40 means that 40% of the volume of blood is composed of RBCs.

Normal values:

Females: 38 to 46 (average 42)
Males: 40 to 54 (average 47)

Abnormal values: The test is used to diagnose anemia, polycythemia (an increased percentage of red blood cells above 55), and abnormal states of hydration. Anemia may vary from mild (hematocrit of 35) to severe (hematocrit of less than 15). Athletes often have a higher-than-average hematocrit, and the average hematocrit of persons living at high altitude is greater than that of persons living at sea level.

C. Differential white blood cell count (the percentage of each type of white blood cells in a sample of 100 WBCs)

Normal values:

Type of WBC	Percentage
Neutrophils	60–70
Eosinophils	2–4
Basophils	0.5–1
Lymphocytes	20–25
Monocytes	3–8

Abnormal values: A high neutrophil count might result from bacterial infections, burns, stress, or inflammation; a low neutrophil count might be caused by radiation, certain drugs, or vitamin B_{12} deficiency. A high eosinophil count could indicate allergic reactions, parasitic infections, autoimmune disease, or adrenal insufficiency; a low eosinophil count could be caused by certain drugs, stress, or Cushing's syndrome. Basophils could be elevated in some types of allergic responses, leukemias, cancers, and hyperthyroidism; decreases in basophils could occur during pregnancy, ovulation, stress, and hyperthyroidism. High lymphocyte counts could indicate viral infections, immune diseases, and some leukemias; low lymphocyte counts might occur as a result of prolonged severe illness, high steroid levels, and immunosuppression. A high monocyte count could result from certain viral or fungal infections, tuberculosis (TB), some leukemias, and chronic diseases; low monocyte levels rarely occur.

D. Complete blood count (CBC) (provides information about the formed elements in blood)*

Normal values:

RBC Count	About 5.4 million/μL in males About 4.8 million/μL in females
Hemoglobin	14–18 g/dL in adult males 12–16 g/dL in adult females
Hematocrit	See B
WBC count	5,000–10,000 cells/μL
Differential white blood count	See C
Platelet count	150,000–400,00 per μL

Abnormal values: Increased RBC count, hemoglobin, and hematocrit occur in polycythemia, congenital heart disease, and hypoxia; decreased RBC count, hemoglobin, and hematocrit occur in hemorrhage and certain types of anemia. Increased WBC counts may indicate acute or chronic infections, trauma, leukemia, or stress (see also C). Decreased WBC counts could indicate anemia and viral infections (see also C). High platelet counts may indicate cancer, trauma, or cirrhosis. Low platelet counts could indicate anemia, allergic conditions, or hemorrhage.

*Not all components of a CBC have been included.

• CLINICAL CONNECTION | Sickle-Cell Disease

The RBCs of a person with **sickle-cell disease (SCD)** contain Hb-S, an abnormal kind of hemoglobin. When Hb-S gives up oxygen to the interstitial fluid, it forms long, stiff, rodlike structures that bend the erythrocyte into a sickle shape (Figure 21.5). The sickled cells rupture easily. Even though the loss of RBCs stimulates erythropoiesis, it cannot keep pace with hemolysis; hemolytic anemia is the result. Prolonged oxygen reduction may eventually cause extensive tissue damage. •

Figure 21.5 Red blood cells from a person with sickle-cell disease.

The red blood cells of a person with sickle-cell disease contain an abnormal type of hemoglobin called Hb-S.

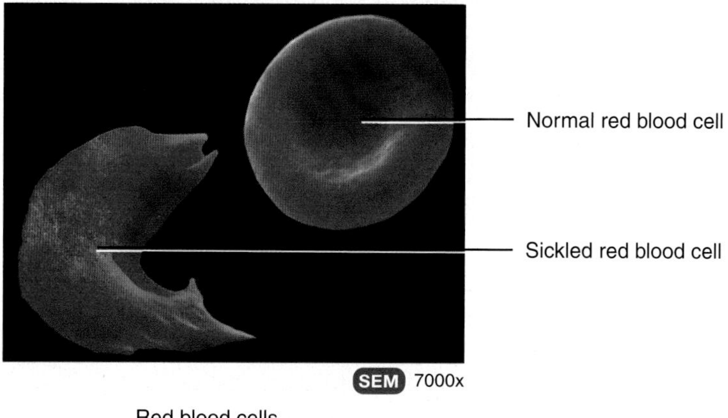

? **What are some symptoms of sickle-cell anemia?**

granular or agranular, depending on whether they contain chemical-filled cytoplasmic granules (vesicles) that are made visible by staining when viewed through a light microscope (Figure 21.6). The granular leukocytes include ***eosinophils*** (ē-ō-SIN-ō-fils), ***basophils*** (BĀ-sō-fils), and ***neutrophils*** (NOO-trō-fils). The agranular leukocytes include ***monocytes*** (MON-ō-sīts′), ***lymphocytes,*** and ***natural killer (NK) cells.*** (See Table 21.2 for the sizes and microscopic characteristics of WBCs.)

Figure 21.6 Types of white blood cells.

The shapes of their nuclei and the staining properties of their cytoplasmic granules distinguish white blood cells from one another.

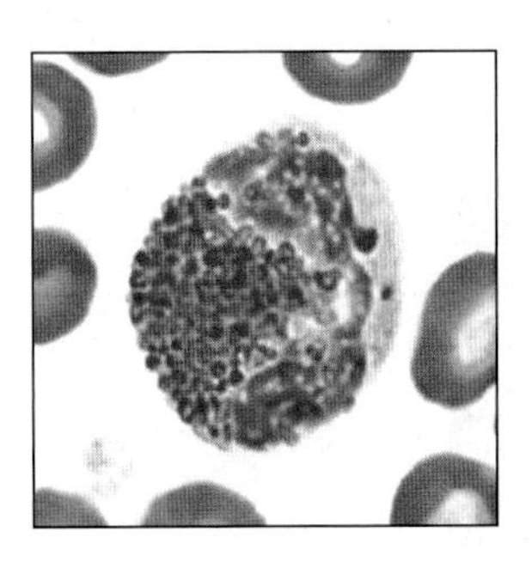

(a) Eosinophil

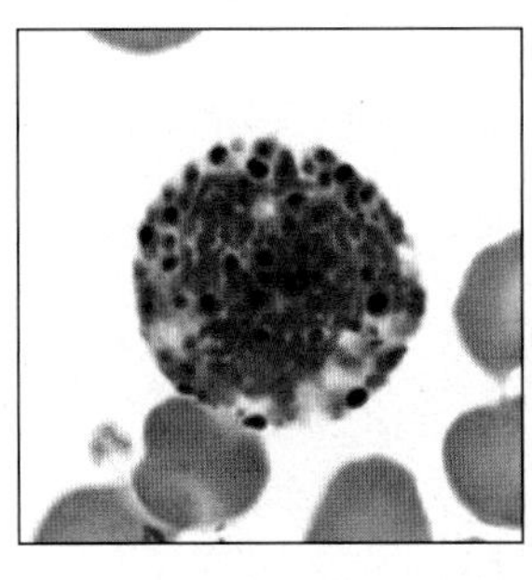

(b) Basophil

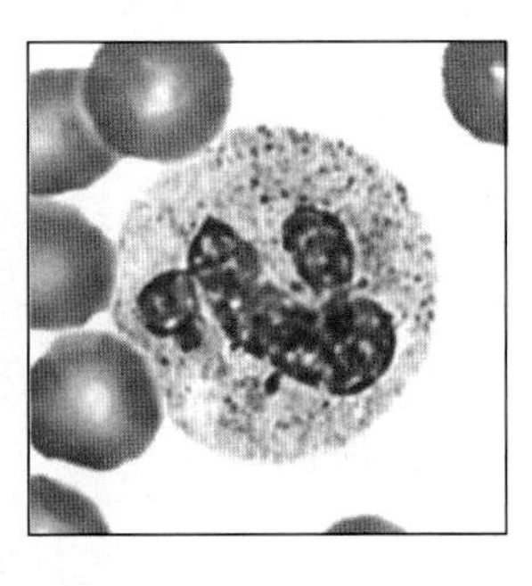

(c) Neutrophil

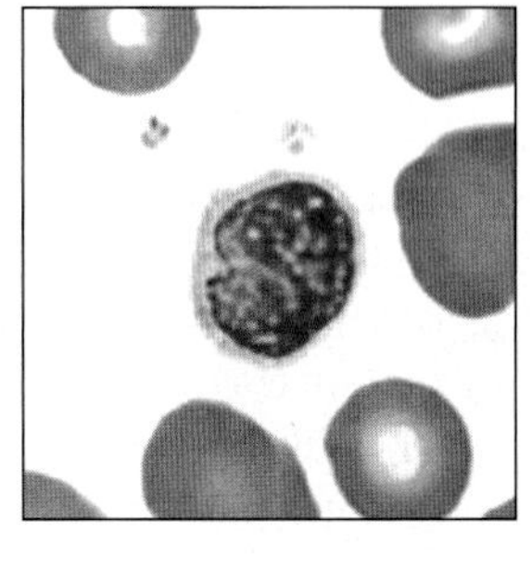

(d) Lymphocyte

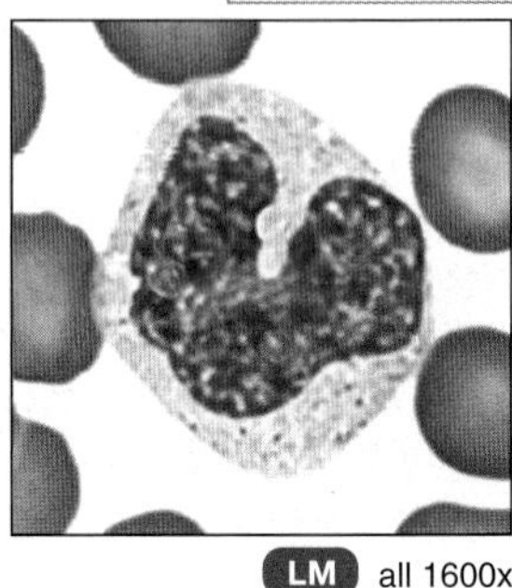

(e) Monocyte

LM all 1600x

? Which WBCs are called granular leukocytes? Why?

WBC Functions The skin and mucous membranes of the body are continuously exposed to microbes (microscopic organisms), such as bacteria, some of which are capable of invading deeper tissues and causing disease. Once microbes enter the body, some WBCs combat them by ***phagocytosis,*** and others produce antibodies. Neutrophils respond first to bacterial invasion, carrying on phagocytosis and releasing enzymes such as lysozyme that destroy certain bacteria. Monocytes take longer to reach the site of infection than neutrophils, but they eventually arrive in larger numbers. Monocytes that migrate into infected tissues develop into cells called ***wandering macrophages*** (*macro-* = large; *-phages* = eaters) (see Figure 21.2), which can phagocytize many more microbes than neutrophils. They also clean up cellular debris following an infection.

Eosinophils leave the capillaries and enter interstitial fluid. They release enzymes that combat inflammation in allergic reactions. Eosinophils also phagocytize antigen–antibody complexes (described in Chapter 24) and are effective against certain parasitic worms. A high eosinophil count often indicates an allergic condition or a parasitic infection.

Basophils are also involved in inflammatory and allergic reactions. They leave capillaries, enter tissues, and can liberate heparin, histamine, and serotonin. These substances intensify the inflammatory reaction.

Three types of lymphocytes—T cells, B cells, and natural killer (NK) cells—are the major combatants in immune responses, which are described in detail in Chapter 24. T cells attack viruses, fungi, transplanted cells, cancer cells, and some bacteria. B cells develop into *plasma cells* (see Figure 21.2), which produce antibodies that help destroy bacteria and inactivate their toxins. Natural killer cells attack a wide variety of infectious microbes and certain spontaneously arising tumor cells.

White blood cells and other nucleated body cells have proteins, called ***major histocompatibility (MHC) antigens,*** protruding from their plasma membrane into the extracellular fluid. These "cell identity markers" are unique for each person (except identical twins). Although RBCs (which do not possess nuclei) possess blood group antigens, they lack the MHC antigens. Rejection of an incompatible tissue transplant is due, in part, to differences in donor and recipient MHC antigens. The MHC antigens are used to type tissues to identify compatible donors and recipients and thus reduce the chance of tissue rejection.

WBC Life Span Red blood cells outnumber white blood cells about 700 to 1. There are normally about 5,000 to 10,000 WBCs per microliter of blood. Bacteria have continuous access to the body through the mouth, nose, and pores of the skin. Furthermore, many cells, especially those of epithelial tissue, age and die daily, and their remains must be removed. However, a WBC can phagocytize only a certain amount of material before it interferes with the WBC's own metabolic activities. Thus, the life span of most WBCs is only a few days. During a period of infection, many WBCs live only a few hours. However, some T and B cells remain in the body for years.

Leukocytosis (loo-kō-sī-TŌ-sis), an increase in the number of WBCs, is a normal, protective response to stresses such as invading microbes, strenuous exercise, anesthesia, and surgery. Leukocytosis usually indicates an inflammation or infection. Because each type of white blood cell plays a different role, determining the percentage of each type in the blood assists in diagnosing the condition. This test, called a ***differential white blood cell count,*** measures the number of each kind of white cell in a sample of 100 white blood cells (Table 21.1). An abnormally low level of white blood cells (below 5,000 cells/μL), called ***leukopenia*** (loo-kō-PĒ-nē-a), is never beneficial; it may be caused by exposure to radiation, shock, and certain chemotherapeutic agents.

WBC Production Leukocytes develop in red bone marrow. As shown in Figure 21.2, monocytes and granular leukocytes develop from a myeloid stem cell. T and B lymphocytes and natural killer cells develop from a lymphoid stem cell.

• CLINICAL CONNECTION | Leukemia

The term **leukemia** (loo-KĒ-mē-a; *leuko-* = white) refers to a group of red bone marrow cancers in which abnormal white blood cells multiply uncontrollably. The accumulation of the cancerous white blood cells in red bone marrow interferes with the production of red blood cells, white blood cells, and platelets. As a result, the oxygen-carrying capacity of the blood is reduced, an individual is more susceptible to infection, and blood clotting is abnormal. In most leukemias, the cancerous white blood cells spread to the lymph nodes, liver, and spleen, causing them to enlarge. All leukemias produce the usual symptoms of anemia (fatigue, intolerance to cold, and pale skin). In addition, weight loss, fever, night sweats, excessive bleeding, and recurrent infections may also occur. •

Platelets

Thrombopoietin (throm′-bō-POY-e-tin), a hormone produced in the liver, stimulates pluripotent stem cells to differentiate into cells that produce platelets (see Figure 21.2). Some myeloid stem cells develop into cells called *megakaryoblasts,* which in turn transform into *megakaryocytes,* huge cells that splinter into 2,000–3,000 fragments in the red bone marrow and then enter the bloodstream. Each fragment, enclosed by a piece of the megakaryocyte cell membrane, is a ***platelet (thrombocyte).*** Between 150,000 and 400,000 platelets are present in each microliter of blood. Platelets have an irregular disc shape, have a diameter of 2–4 μm, and exhibit many vesicles but no nucleus (Figure 21.7). When blood vessels are damaged, platelets help stop blood loss by forming a platelet plug. Their vesicles also contain chemicals that promote blood clotting (both processes are described shortly). After their short life span of 5–9 days, platelets are removed by macrophages in the spleen and liver.

MANUAL THERAPY APPLICATION

Effects of Massage on Blood

The **effects of massage on blood** are indirect through the manipulation of soft tissue throughout the body. As described in Chapter 4, massage of connective tissues increases the fluidity (decreases the viscosity) of ground substance extracellular matrix. Massage can also increase the hemoglobin and red blood cell counts as well as the platelet count. The oxygen-carrying capacity of blood is thus increased for a limited time after general massage.

Education of the patient, as stated previously, is an important role for manual therapists. The media contains regular presentations about the benefits of physical exercise. When a patient asks the therapist about the issue, the following may be helpful information. Assuming that the patient remains hydrated (by drinking water), regular physical activity increases plasma volume. An increase in plasma volume means that the blood is more dilute, or "thinner," with a low percentage of red blood cells and less fibrinogen, and consequently a reduced risk of blood clotting. Several studies have shown that vigorous exercise also reduces platelet stickiness (see Section 21.3 for platelet release reaction) and enhances fibrolytic activity. These effects may help to explain why active people are at lower risk for heart disease and stroke. A sedentary lifestyle, by contrast, leads to increased clotting risk: Blood thickens as plasma volume decreases. Sedentary people have stickier platelets, which together with higher levels of fibrinogen are more likely to form clots.

Although we separate the cardiovascular system into four chapters (Chapters 21–24) to facilitate learning the material, the effects of massage on the entire cardiovascular system (blood, heart, blood vessels, and lymphatics) will eventually become evident. Massage can significantly enhance the overall nutrition of tissues throughout the body.

Figure 21.7 Scanning electron micrograph and photomicrograph of the formed elements of blood.

The formed elements of blood are red blood cells (RBCs), white blood cells (WBCs), and platelets.

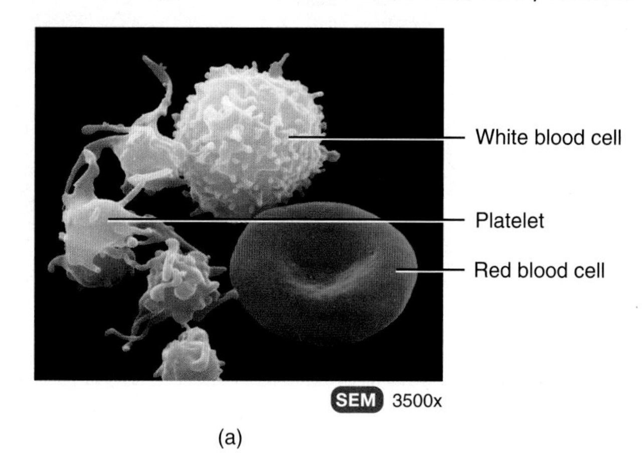

(a)

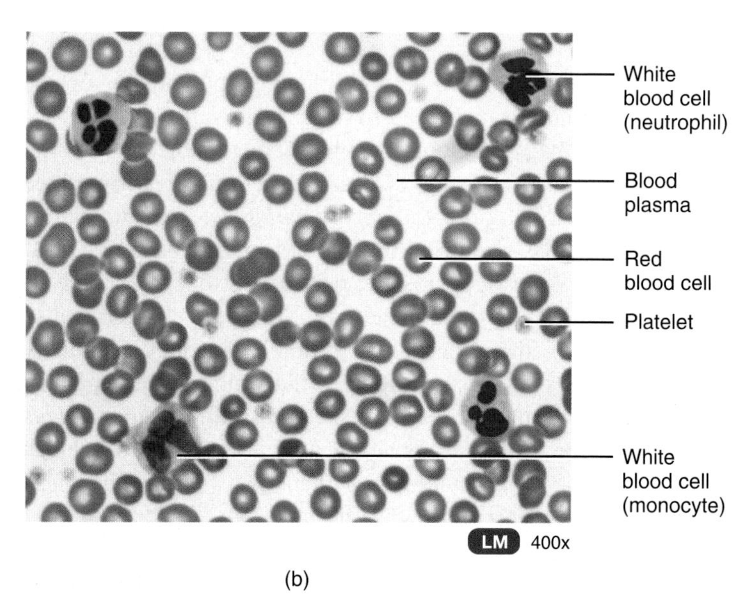

(b)

? Which formed elements of the blood are cell fragments?

Table 21.2 presents a summary of the formed elements in blood.

TABLE 21.2

Summary of Formed Elements in Blood

NAME AND APPEARANCE	NUMBER	CHARACTERISTICS*	FUNCTIONS
Red Blood Cells (RBCs) or Erythrocytes	4.8 million/μL in females; 5.4 million/μL in males	7–8 μm diameter, biconcave discs, without nuclei; live for about 120 days.	Hemoglobin within RBCs transports most of the oxygen and part of the carbon dioxide in the blood.
White Blood Cells (WBCs) or Leukocytes	5,000–10,000 cells/μL	Most live for a few hours to a few days.†	Combat pathogens and other foreign substances that enter the body.
Granular leukocytes			
Neutrophils	60–70% of all WBCs	10–12 μm diameter; nucleus has 2–5 lobes connected by thin strands of chromatin; cytoplasm has very fine, pale lilac granules.	Phagocytosis. Destruction of bacteria with lysozyme.
Eosinophils	2–4% of all WBCs	10–12 μm diameter; nucleus usually has 2 lobes connected by a thick strand of chromatin; large, red-orange granules fill the cytoplasm.	Combat the effects of histamine in allergic reactions, phagocytize antigen–antibody complexes, and destroy certain parasitic worms.
Basophils	0.5–1% of all WBCs	8–10 μm diameter; nucleus has 2 lobes; large cytoplasmic granules appear deep blue-purple.	Liberate heparin, histamine, and serotonin in allergic reactions that intensify the overall inflammatory response.
Agranular leukocytes			
Lymphocytes (T cells, B cells, and natural killer cells)	20–25% of all WBCs	Small lymphocytes are 6–9 μm in diameter; large lymphocytes are 10–14 μm in diameter; nucleus is round or slightly indented; cytoplasm forms a rim around the nucleus that looks sky blue; the larger the cell, the more cytoplasm is visible.	Mediate immune responses, including antigen–antibody reactions. B cells develop into plasma cells, which secrete antibodies. T cells attack invading viruses, cancer cells, and transplanted tissue cells. Natural killer cells attack a wide variety of infectious microbes and certain spontaneously arising tumor cells.
Monocytes	3–8% of all WBCs	12–20 μm diameter; nucleus is kidney shaped or horseshoe shaped; cytoplasm is blue-gray and has foamy appearance.	Phagocytosis (after transforming into fixed or wandering macrophages).
Platelets (Thrombocytes)	150,000–400,000/μL	2–4 μm diameter cell fragments that live for 5–9 days; contain many vesicles but no nucleus.	Form platelet plug in hemostasis; release chemicals that promote vascular spasm and blood clotting.

*Colors are those seen when using Wright's stain.

†Some lymphocytes, called T and B memory cells, can live for many years.

CHECKPOINT

3. Briefly outline the process of hemopoiesis.
4. What is erythropoiesis? How does erythropoiesis affect the hematocrit? What factors speed up and slow down erythropoiesis?
5. What functions do eosinophils, basophils, neutrophils, monocytes, T cells, B cells, and natural killer cells perform?
6. How are leukocytosis and leukopenia different? What is a differential white blood cell count?

21.3 HEMOSTASIS

OBJECTIVE

- Describe the various mechanisms that prevent blood loss.

Hemostasis (hē-mō-STĀ-sis; *-stasis* = standing still) is a sequence of responses that stops bleeding when blood vessels are injured. (Be sure not to confuse the two words *hemostasis* and *homeostasis.*) The hemostatic response must be quick, localized to the region of damage, and carefully controlled. When successful, hemostasis prevents ***hemorrhage*** (HEM-o-rij; *-rhage* = burst forth), the loss of a large amount of blood from the vessels. Hemostasis can prevent hemorrhage from smaller blood vessels, but extensive hemorrhage from larger vessels usually requires medical intervention.

Three mechanisms can reduce loss of blood from blood vessels: (1) vascular spasm, (2) platelet plug formation, and (3) blood clotting (coagulation).

Vascular Spasm

When a blood vessel is damaged, the smooth muscle in its wall contracts immediately, a response called a ***vascular spasm.*** Vascular spasm reduces blood loss for several minutes to several hours, during which time the other hemostatic mechanisms begin to operate. The spasm is probably caused by damage to the smooth muscle and by reflexes initiated by pain receptors. As platelets accumulate at the damaged site, they release chemicals that enhance vasoconstriction (narrowing of a blood vessel), thus maintaining the vascular spasm.

Platelet Plug Formation

When platelets come into contact with parts of a damaged blood vessel, their characteristics change drastically and they quickly come together to form a platelet plug that helps fill the gap in the injured blood vessel wall.

Platelet plug formation occurs as follows (Figure 21.8):

1. Initially, platelets contact and stick to parts of a damaged blood vessel, such as collagen fibers of the connective tissue underlying the damaged endothelial cells. This process is called ***platelet adhesion.***
2. Due to adhesion, the platelets become activated, and their characteristics change dramatically. They extend many projections that enable them to contact and interact with one another, and they begin to liberate chemicals. This phase is called the ***platelet release reaction.*** The chemicals activate nearby platelets and sustain the vascular spasm, which decreases blood flow through the injured vessel.
3. The release of platelet chemicals makes other platelets in the area sticky, and the stickiness of the newly recruited and activated platelets causes them to adhere to the originally activated platelets. This gathering of platelets is called ***platelet aggregation.*** Eventually, the accumulation and attachment of large numbers of platelets form a mass called a ***platelet plug.***

A platelet plug is very effective in preventing blood loss in a small vessel. Although initially the platelet plug is loose, it becomes quite tight when reinforced by fibrin threads formed during clotting (Figure 21.9). A platelet plug can stop blood loss completely if the hole in a blood vessel is not too large.

Figure 21.8 Platelet plug formation.

A platelet plug can stop blood loss completely if the hole in a blood vessel is small enough.

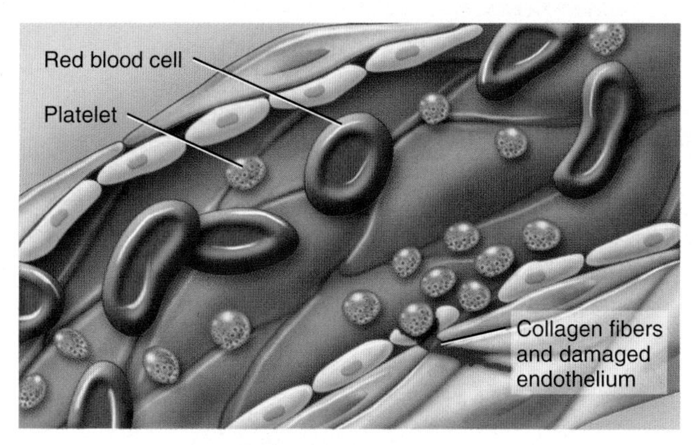

1 Platelet adhesion

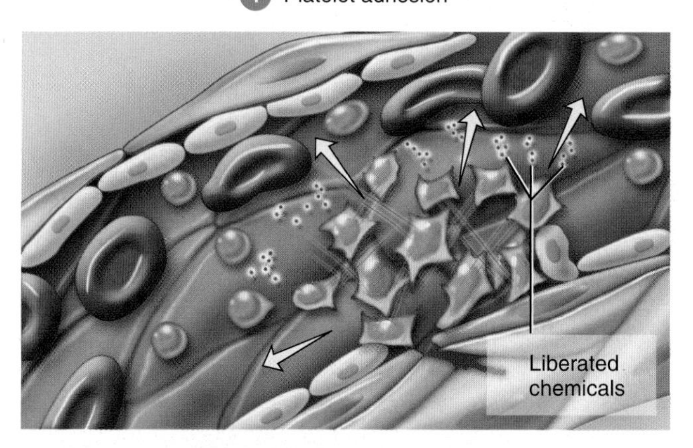

2 Platelet release reaction

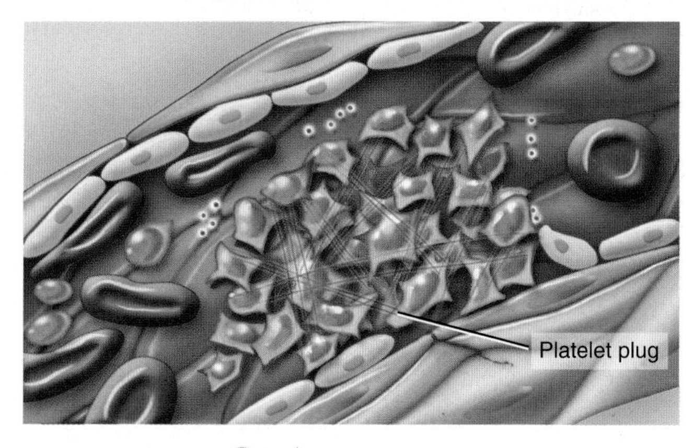

3 Platelet aggregation

? Along with platelet plug formation, which two mechanisms contribute to hemostasis?

Blood Clotting

Normally, blood remains in its liquid form as long as it stays within its vessels. If it is withdrawn from the body, however, it thickens and forms a gel. Eventually, the gel separates from the liquid. The straw-colored liquid, called ***serum,*** is simply plasma minus the clotting proteins. The gel is called a ***clot*** and consists of a network of insoluble protein fibers called ***fibrin*** in which the formed elements of blood are trapped (Figure 21.9).

The process of clot formation, called ***clotting (coagulation),*** is a series of chemical reactions that culminates in the formation of fibrin threads. If blood clots too easily, the result can be

Figure 21.9 The blood clotting cascade (simplified).

During blood clotting, the clotting factors activate each other, resulting in a cascade of reactions that includes positive feedback cycles.

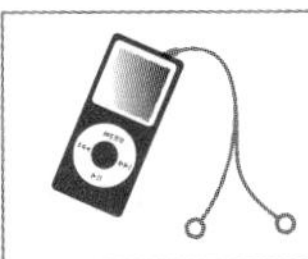

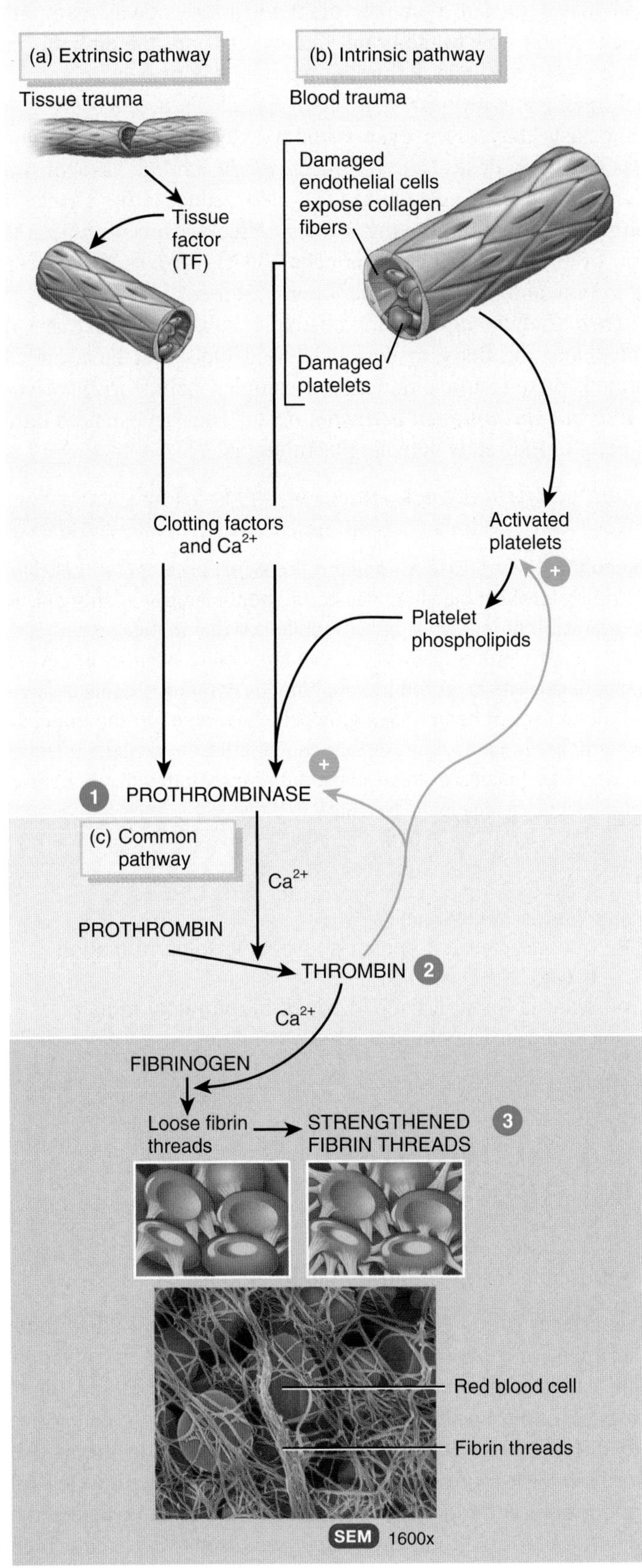

What is the outcome of the first stage of blood clotting?

thrombosis, clotting in an undamaged blood vessel. If the blood takes too long to clot, hemorrhage can result.

Clotting is a complex process in which various chemicals known as ***clotting factors*** activate each other. Clotting (coagulation) factors include calcium ions (Ca^{2+}), several enzymes that are made by liver cells and released into the blood, and various molecules associated with platelets or released by damaged tissues. Normal clotting depends on adequate levels of vitamin K in the body. Although vitamin K is not involved in actual clot formation, it is required for the synthesis of four clotting factors.

Clotting occurs in three stages (Figure 21.9):

1. ***Prothrombinase*** is formed.
2. Prothrombinase converts ***prothrombin*** (a plasma protein formed by the liver with the help of vitamin K) into the enzyme ***thrombin.***
3. Thrombin converts soluble ***fibrinogen*** (another plasma protein formed by the liver) into insoluble fibrin. Fibrin forms the threads of the clot. (Cigarette smoke contains substances that interfere with fibrin formation.)

Prothrombinase can be formed in two ways, by either the extrinsic or the intrinsic pathway of blood clotting. The ***extrinsic pathway*** of blood clotting occurs rapidly, within seconds. It is so named because damaged tissue cells release a tissue protein called ***tissue factor (TF)*** into the blood from outside (extrinsic to) blood vessels (Figure 21.9a). Following several additional reactions that require calcium ions (Ca^{2+}) and several clotting factors (not named here), tissue factor is eventually converted into prothrombinase. This completes the extrinsic pathway.

The ***intrinsic pathway*** of blood clotting (Figure 21.9b) is more complex than the extrinsic pathway, and it occurs more slowly, usually requiring several minutes. The intrinsic pathway is so named because its activators are either in direct contact with blood or contained within (intrinsic to) the blood. If endothelial cells lining the blood vessels become roughened or damaged, blood can come in contact with collagen fibers in the adjacent connective tissue. Such contact activates clotting factors. In addition, trauma to endothelial cells activates platelets, causing them to release phospholipids that can also activate certain clotting factors. After several additional reactions that require Ca^{2+} and several clotting factors, prothrombinase is formed. Once formed via the common pathway, thrombin activates more platelets, resulting in the release of more platelet phospholipids, an example of a positive feedback cycle. Both the extrinsic and intrinsic pathways can be activated at the same time since blood vessel and surrounding tissue damage usually occur simultaneously.

Clot formation occurs locally; it does not extend beyond the wound site into the general circulation. One reason for this is that fibrin has the ability to absorb and inactivate up to nearly 90% of the thrombin formed from prothrombin. This helps stop the spread of thrombin into the blood and thus inhibits clotting except at the wound.

Clot Retraction and Blood Vessel Repair

Once a clot is formed, it plugs the ruptured area of the blood vessel and thus stops blood loss. ***Clot retraction*** is the consolidation or tightening of the fibrin clot. The fibrin threads attached to the damaged surfaces of the blood vessel gradually contract as platelets pull on them. As the clot retracts, it pulls the edges of the damaged vessel closer together, decreasing the risk of further damage. Permanent repair of the blood vessel can then take place. In time, fibroblasts form connective tissue in the ruptured area, and new endothelial cells repair the vessel lining.

Hemostatic Control Mechanisms

Many times a day little clots start to form, often at a site of minor roughness inside a blood vessel. Usually, small, inappropriate clots dissolve in a process called ***fibrinolysis*** (fī-bri-NOL-i-sis). When a clot is formed, an inactive plasma enzyme called ***plasminogen*** is incorporated into the clot. Both body tissues and blood contain substances that can activate plasminogen to ***plasmin,*** an active plasma enzyme. Once plasmin is formed, it can dissolve the clot by digesting fibrin threads. Plasmin also dissolves clots at sites of damage once the damage is repaired. Among the substances that activate plasminogen are thrombin and tissue plasminogen activator (tPA), which is normally found in many body tissues and is liberated into the blood after a vascular injury.

• CLINICAL CONNECTION | Anticoagulants

Patients who are at increased risk of forming blood clots may receive **anticoagulants** (an′-tē-kō-AG-ū-lant), drugs that delay, suppress, or prevent blood clotting. Examples are heparin and warfarin. Heparin, an anticoagulant that is produced by mast cells and basophils, inhibits the conversion of prothrombin to thrombin, thereby preventing blood clot formation. Heparin extracted from animal tissues is often used to prevent clotting during hemodialysis and after open heart surgery. Coumadin® (warfarin) acts as an antagonist to vitamin K and thus blocks synthesis of four clotting factors. To prevent clotting in donated blood, blood banks and laboratories often add a substance that removes Ca^{2+}, for example, CPD (citrate phosphate dextrose). •

Clotting in Blood Vessels

Despite fibrinolysis and the action of anticoagulants, blood clots sometimes form within blood vessels. The endothelial surfaces of a blood vessel may be roughened as a result of ***atherosclerosis*** (accumulation of fatty substances on arterial walls), trauma, or infection. These conditions also make the platelets that are attracted to the rough spots more sticky. Clots may also form in blood vessels when blood flows too slowly, allowing clotting factors to accumulate in high enough concentrations to initiate a clot.

Clotting in an unbroken blood vessel is called ***thrombosis*** (*thromb-* = clot; *-osis* = a condition of). The clot itself, called a ***thrombus,*** may dissolve spontaneously. If it remains intact, however, the thrombus may become dislodged and be swept away in the blood. A blood clot, bubble of air, fat from broken bones, or a piece of debris transported by the bloodstream is called an ***embolus*** (*em-* = in; *-bolus* = a mass; plural is *emboli*). Because emboli often form in veins, where blood flow is slower, the most common site for the embolus to become lodged is in the lungs, a condition called ***pulmonary embolism.*** Massive emboli in the lungs may result in right ventricular failure and death in a few minutes or hours. An embolus that breaks away from an arterial wall may lodge in a smaller diameter artery downstream. If it blocks blood flow to the brain, kidney, or heart, the embolus can cause a stroke, kidney failure, or heart attack, respectively.

In patients with heart and blood vessel disease, the events of hemostasis may occur even without external injury to a blood vessel. At low doses (i.e., 81 mg), ***aspirin*** inhibits vasoconstriction and platelet aggregation. It also reduces the chance of thrombus formation. Owing to these effects, aspirin reduces the risk of transient ischemic attacks (TIA), strokes, myocardial infarction, and blockage of peripheral arteries.

Thrombolytic agents are chemical substances that are injected into the body to dissolve blood clots that have already formed and to restore circulation. Examples include ***streptokinase*** and ***tissue plasminogen activator (tPA).*** These substances either directly or indirectly activate plasminogen.

• CLINICAL CONNECTION | Hemophilia

Hemophilia (hē-mō-FIL-ē-a; *-philia* = loving) is an inherited deficiency of clotting in which bleeding may occur spontaneously or after only minor trauma. Different types of hemophilia are due to deficiencies of different blood clotting factors and exhibit varying degrees of severity. Hemophilia is characterized by spontaneous or traumatic subcutaneous and intramuscular hemorrhaging, nosebleeds, blood in the urine, and hemorrhages in joints that produce pain and tissue damage. Treatment involves transfusions of fresh plasma or concentrates of the deficient clotting factor to relieve the tendency to bleed. •

CHECKPOINT

7. What is hemostasis?
8. How do vascular spasm and platelet plug formation occur?
9. What is fibrinolysis? Why does blood rarely remain clotted inside blood vessels?

21.4 BLOOD GROUPS AND BLOOD TYPES

OBJECTIVE

- Describe the ABO and Rh blood groups.

The surfaces of red blood cells contain a genetically determined assortment of ***antigens*** composed of glycolipids and glycoproteins called ***agglutinogens*** (ag-loo-TIN-ō-jenz). Based on the presence or absence of various antigens, blood is categorized into different ***blood groups.*** Within a given blood group there may be two or more different ***blood types.*** There are at least 24 blood groups and more than 100 antigens that can be detected on the surface of red blood cells. Here we discuss two major blood groups: ABO and Rh.

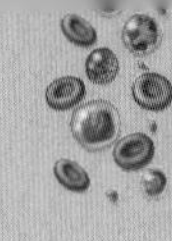

ABO Blood Group

The ***ABO blood group*** is based on two antigens called A and B (Figure 21.10). People whose RBCs display only antigen A have type A blood. Those who have only antigen B are type B. Individuals who have both A and B antigens are type AB, and those who have neither antigen A nor B are type O. In about 80% of the population, soluble antigens of the ABO type appear in saliva and other body fluids, in which case blood type can be identified from a sample of saliva. The incidence of ABO blood types varies among different population groups, as indicated in Table 21.3.

In addition to antigens on RBCs, blood plasma usually contains ***antibodies*** or ***agglutinins*** (a-GLOO-ti-nins) that react with the A or B antigens if the two are mixed. These are the ***anti-A antibody,*** which reacts with antigen A, and the ***anti-B antibody,*** which reacts with antigen B. The antibodies present in each of the four ABO blood types are also shown in Figure 21.10. You do not have antibodies that react with your own antigens, but you do have antibodies for any antigens that your RBCs lack. For example, if you have type A blood, it means that you have A antigens on the surfaces of your RBCs, but anti-B antibodies in your blood plasma. If you had anti-A antibodies in your blood plasma, they would attack your RBCs.

Rh Blood Group

The Rh blood group is so named because the Rh antigen, ***Rh factor,*** was first found in the blood of the rhesus monkey. People whose RBCs have the Rh antigen are designated Rh^+ (Rh positive); those who lack the Rh antigen are designated Rh^- (Rh negative). The percentages of Rh^+ and Rh^- individuals in various populations are shown in Table 21.3. Under normal circumstances, plasma does not contain anti-Rh antibodies. If an Rh^- person receives an Rh^+ blood transfusion, however, the immune system starts to make anti-Rh antibodies that do remain in the blood.

TABLE 21.3

Blood Types in the United States

	BLOOD TYPE (PERCENTAGE)				
POPULATION GROUP	O	A	B	AB	Rh^+
European-American	45	40	11	4	85
African-American	49	27	20	4	95
Korean-American	32	28	30	10	100
Japanese-American	31	38	21	10	100
Chinese-American	42	27	25	6	100
Native American	79	16	4	1	100

• CLINICAL CONNECTION | Hemolytic Disease of the Newborn

Hemolytic disease of the newborn (HDN), also known as **erythroblastosis fetalis,** is a problem that results from Rh incompatibility between a mother and her fetus. Normally, no direct contact occurs between maternal and fetal blood while a woman is pregnant. However, if a small amount of Rh^+ blood leaks from the fetus through the placenta into the bloodstream of an Rh^- mother, her body starts to make anti-Rh antibodies. Because the greatest possibility of fetal blood transfer occurs at delivery, the first-born baby typically is not affected. HDN is prevented by giving all Rh^- women an injection of anti-Rh antibodies called anti-Rh gamma globulin (RhoGAM) soon after every delivery, miscarriage, or abortion. These antibodies destroy any Rh antigens that are present so the mother doesn't produce her own antibodies to them. In the case of an Rh^+ mother, there are no complications, because she cannot make anti-Rh antibodies. •

Figure 21.10 Antigens and antibodies involved in the ABO blood grouping system.

Your plasma does not contain antibodies that could react with the antigens on your red blood cells.

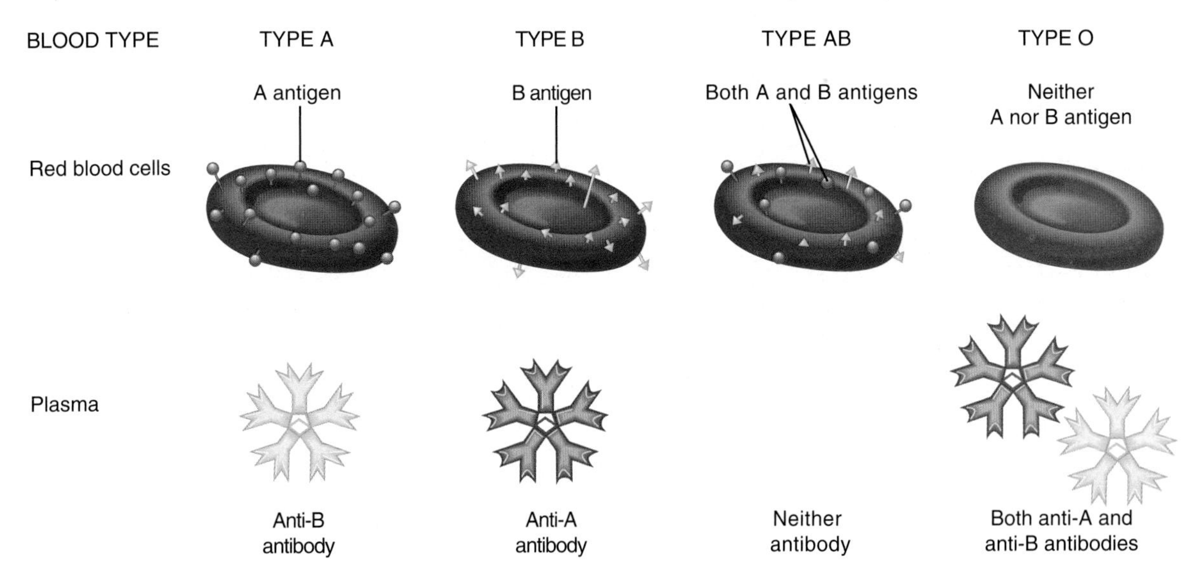

Which antibodies are present in type O blood?

Transfusions

Despite the differences in RBC antigens, blood (the "gift of life" according to the American Red Cross) is the most easily shared of human tissues, saving many thousands of lives every year through transfusions. A ***transfusion*** (trans-FŪ-zhun) is the transfer of whole blood or blood components (red blood cells only or plasma only) into the bloodstream. Most often a transfusion is given to alleviate anemia or when blood volume is low, for example, after a severe hemorrhage. Blood contains antigens and antibodies other than those associated with the ABO and Rh systems, and they can cause transfusion problems. Thus, blood should always be carefully matched before transfusion.

CHECKPOINT

10. What is the basis for distinguishing the various blood groups?
11. What precautions must be taken before giving a blood transfusion?

We will next direct our attention to the heart, the second major component of the cardiovascular system.

KEY MEDICAL TERMS ASSOCIATED WITH BLOOD

Autologous preoperative transfusion (aw-TOL-o-gus trans-FŪ-zhun; *auto-* = self) Donating one's own blood in preparation for surgery; can be done up to 6 weeks before elective surgery. Also called ***predonation.***

Blood bank A facility that collects and stores a supply of blood for future use by the donor or others. Because blood banks have now assumed additional and diverse functions (immunohematology reference work, continuing medical education, bone and tissue storage, and clinical consultation), they are more appropriately referred to as *centers of transfusion medicine.*

Cyanosis (sī-a-NŌ-sis; *cyano-* = blue) Slightly bluish/dark-purple skin discoloration, most easily seen in the nail beds and mucous membranes, due to an increased quantity of reduced hemoglobin (hemoglobin not combined with oxygen) in systemic blood.

Hemochromatosis (hē-mō-krō-ma-TŌ-sis; *-chroma-* = color) Disorder of iron metabolism characterized by excess deposits of iron in tissues (especially the liver, heart, pituitary gland, gonads, and pancreas) that result in discoloration (bronzing) of the skin, cirrhosis, diabetes mellitus, and bone and joint abnormalities.

Jaundice (*jaund-* = yellow) An abnormal yellowish discoloration of the sclerae of the eyes, skin, and mucous membranes due to excess bilirubin (yellow-orange pigment) in the blood that is produced when the heme pigment in aged red blood cells is broken down.

Phlebotomist (fle-BOT-ō-mist; *phlebo-* = vein; *-tom-* = cut) A technician who specializes in withdrawing blood.

Polycythemia (pol′-ē-sī-THĒ-mē-a) An abnormal increase in the number of red blood cells, in which the hematocrit is above 55, the upper limit of normal.

Septicemia (sep-ti-SĒ-mē-a; *septic-* = decay; *-emia* = condition of blood) An accumulation of toxins or disease-causing bacteria in the blood. Also called ***blood poisoning.***

Thrombocytopenia (throm′-bō-sī′-tō-PĒ-nē-a; *-penia* = poverty) Very low platelet count that results in a tendency to bleed from capillaries.

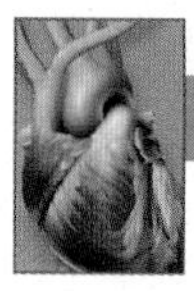

STUDY OUTLINE

Functions of Blood (Section 21.1)

1. Blood transports oxygen, carbon dioxide, nutrients, wastes, and hormones.
2. It helps to regulate pH, body temperature, and water content of cells.
3. It prevents blood loss through clotting and combats microbes and toxins through the action of certain phagocytic white blood cells or specialized plasma proteins.

Components of Whole Blood (Section 21.2)

1. Physical characteristics of whole blood include a viscosity greater than that of water, a temperature of 38°C (100.4°F), and a pH range between 7.35 and 7.45.
2. Blood constitutes about 8% of body weight in an adult.
3. Blood consists of 55% plasma and 45% formed elements.
4. The formed elements in blood include red blood cells (erythrocytes), white blood cells (leukocytes), and platelets. Hematocrit is the percentage of red blood cells in whole blood.
5. Plasma contains 91.5% water, 7% proteins, and 1.5% solutes other than proteins.
6. Principal solutes include proteins (albumins, globulins, fibrinogen), nutrients, hormones, respiratory gases, electrolytes, and waste products.
7. Hemopoiesis, the formation of blood cells from pluripotent stem cells, occurs in red bone marrow.
8. Red blood cells (RBCs) are biconcave discs without nuclei that contain hemoglobin.
9. The function of the hemoglobin in red blood cells is to transport oxygen and a portion of carbon dioxide.
10. Red blood cells live about 120 days. A healthy male has about 5.4 million RBCs/μL of blood and a healthy female has about 4.8 million RBCs/μL.
11. RBC formation, called erythropoiesis, occurs in adult red bone marrow. It is stimulated by hypoxia, which stimulates release of erythropoietin (EPO) by the kidneys.
12. A reticulocyte count is a diagnostic test that indicates the rate of erythropoiesis.
13. After phagocytosis of aged red blood cells by macrophages, hemoglobin is recycled.
14. White blood cells (WBCs) are nucleated cells. The two principal types are granular leukocytes (eosinophils, basophils, neutrophils) and agranular leukocytes (monocytes and lymphocytes).
15. The general function of WBCs is to combat inflammation and infection. Neutrophils and macrophages (which develop from monocytes) do so through phagocytosis.
16. Eosinophils combat inflammation in allergic reactions, phagocytize antigen–antibody complexes, and combat parasitic worms; basophils liberate heparin, histamine, and serotonin in allergic reactions that intensify the inflammatory response.

17. T cells (lymphocytes) are effective against viruses, fungi, and cancer cells. B cells (lymphocytes) are effective against bacteria and other toxins. Natural killer (NK) cells attack microbes and tumor cells.
18. White blood cells usually live for only a few hours or a few days. Normal blood contains 5,000 to 10,000 WBCs/μL.
19. Platelets are irregular disc-shaped cell fragments without nuclei.
20. Platelets are formed from megakaryocytes and take part in hemostasis by forming a platelet plug.
21. Normal blood contains 150,000 to 400,000 platelets/μL.

Hemostasis (Section 21.3)

1. Hemostasis, the stoppage of bleeding, involves vascular spasm, platelet plug formation, and blood clotting.
2. In vascular spasm, the smooth muscle of a blood vessel wall contracts.
3. Platelet plug formation is the aggregation of platelets to stop bleeding.
4. A clot is a network of insoluble protein fibers (fibrin) in which formed elements of blood are trapped. The chemicals involved in clotting are known as clotting factors.
5. Blood clotting involves a series of reactions that may be divided into three stages: formation of prothrombinase by either the extrinsic or intrinsic pathway, conversion of prothrombin into thrombin, and conversion of soluble fibrinogen into insoluble fibrin.
6. Normal coagulation involves clot retraction (tightening of the clot) and fibrinolysis (dissolution of the clot).
7. Anticoagulants (for example, heparin) prevent clotting.
8. Clotting in an unbroken blood vessel is called thrombosis. A thrombus (clot) that moves from its site of origin is called an embolus.

Blood Groups and Blood Types (Section 21.4)

1. In the ABO system, the antigens on RBCs, called A and B, determine blood type. Plasma contains antibodies termed anti-A and anti-B antibodies.
2. In the Rh system, individuals whose erythrocytes have Rh antigens are designated as Rh^+. Those who lack the antigen are Rh^-.

SELF-QUIZ QUESTIONS

1. A hematocrit is
   a. used to measure the quantity of the five types of white blood cells
   b. essential for determining a person's blood type
   c. the percentage of red blood cells in whole blood
   d. also known as a platelet count
   e. involved in blood clotting.
2. Match the following:

   ____ a. involved in certain immune responses
   ____ b. develop into mature red blood cells
   ____ c. required for vitamin B_{12}
   ____ d. most abundant plasma protein
   ____ e. blood after formed elements are removed
   ____ f. plasma without clotting proteins
   ____ g. needed for blood clotting

   A. albumin
   B. fibrinogen
   C. intrinsic factor
   D. antibodies
   E. plasma
   F. serum
   G. reticulocytes
3. In adults, erythropoiesis takes place in
   a. the liver
   b. yellow bone marrow
   c. red bone marrow
   d. lymphatic tissue
   e. the kidneys.
4. Which of the following pigments contributes to the yellow color in urine?
   a. hemoglobin
   b. stercobilin
   c. biliverdin
   d. urobilin
   e. bilirubin
5. Which of the following statements is NOT true about red blood cells?
   a. The production of red blood cells is known as erythropoiesis.
   b. Red blood cells originate from pluripotent stem cells.
   c. Hypoxia increases the production of red blood cells.
   d. The liver takes part in the destruction and recycling of red blood cell components.
   e. Red blood cells have a lobed nucleus and granular cytoplasm.
6. A primary function of red blood cells is to
   a. maintain blood volume
   b. help blood clot
   c. provide immunity against some diseases
   d. clean up debris following infection
   e. deliver oxygen to the cells of the body.
7. If a differential white blood cell count indicated higher than normal numbers of basophils, what may be occurring in the body?
   a. chronic infection
   b. allergic reaction
   c. leukopenia
   d. initial response to invading bacteria
   e. hemostasis
8. In a person with blood type A, the antibodies that would normally be present in the plasma are
   a. anti-A antibody
   b. anti-B antibody
   c. both anti-A and anti-B antibodies
   d. neither anti-A nor anti-B
   e. anti-O antibodies.
9. Hemolytic disease of the newborn (HDN) may occur in the fetus of a second pregnancy if
   a. the mother is Rh^+ and the baby is Rh^-
   b. the mother is Rh^+ and the baby is Rh^+
   c. the mother is Rh^- and the baby is Rh^-
   d. the mother is Rh^- and the baby is Rh^+
   e. the father is Rh^- and the mother is Rh^+.
10. Place the following steps of hemostasis in the correct order.
    1. clot retraction
    2. prothrombinase formed
    3. fibrinolysis by plasmin
    4. vascular spasm
    5. conversion of prothrombin into thrombin
    6. platelet plug formation
    7. conversion of fibrinogen into fibrin

    a. 4, 6, 2, 5, 7, 1, 3
    b. 5, 4, 7, 6, 2, 3, 1
    c. 2, 5, 6, 7, 1, 4, 3
    d. 4, 6, 5, 2, 7, 1, 3
    e. 4, 2, 6, 5, 3, 7, 1

11. Which of the following is NOT a normal component of blood plasma?
a. albumins **b.** fibrinogen
c. hemoglobin **d.** globulins
e. water

12. How does aspirin prevent thrombosis?
a. It inhibits platelet aggregation.
b. It interferes with Ca^{2+} absorption.
c. It inhibits the conversion of prothrombin to thrombin.
d. It acts as an enzyme to dissolve the thrombus.
e. It prevents the accumulation of fatty substances on blood vessel walls.

13. Match the following:
____ **a.** become wandering macrophages **A.** neutrophils
____ **b.** respond first to bacterial invasion **B.** eosinophils
____ **c.** are involved in allergic reactions **C.** basophils
____ **d.** destroy antigen–antibody complexes; combat inflammation **D.** monocytes

14. Hemostasis is
a. maintenance of a steady state in the body
b. an abnormal increase in leukocytes
c. an anticoagulant produced by some leukocytes
d. a series of events that stop bleeding
e. excess loss of blood

15. Which of the following are mismatched?
a. white blood cell count below 5,000 cells/μL, leukopenia
b. red blood cell count of 250,000 cells/μL, normal adult male
c. white blood cell count above 10,000 cells/μL, leukocytosis
d. platelet count of 300,000 cells/μL, normal adult
e. pH 7.4, normal blood

16. An individual with type A blood has ________ in the plasma membranes of red blood cells.
a. antigen A
b. antigen B
c. major histocompatibility antigen A
d. antigen A and antigen Rh
e. antigen B and antigen Rh

17. Mrs. Smith arrives at a health clinic with her ill daughter Beth. It is suspected that Beth has recently developed a bacterial infection. It is likely that Beth's leukocyte count will be ________ cells/μL of blood, a condition known as ________. A differential white blood cell count shows an abnormally high percentage of ________.
a. 20,000, leukopenia, neutrophils
b. 5,000, leukocytosis, monocytes
c. 7,000, leukocytosis, basophils
d. 2,000, leukopenia, platelets
e. 20,000, leukocytosis, neutrophils

18. Clot retraction
a. draws torn edges of the damaged vessel closer together
b. dissolves clots
c. is also known as the intrinsic pathway
d. involves the formation of fibrin from fibrinogen
e. helps prevent the formation of an embolus.

19. Persons with blood type AB are sometimes referred to as universal recipients because their blood
a. lacks A and B antigens
b. lacks anti-A and anti-B antibodies
c. possesses type O antigens and anti-O antibodies
d. has natural immunity to disease
e. contains A and B antigens.

20. A thrombus that is being transported by the bloodstream is called
a. a plasma protein **b.** a platelet
c. an embolus **d.** a wandering macrophage
e. a reticulocyte.

CRITICAL THINKING QUESTIONS

1. Shilpa has recently been on broad-spectrum antibiotics for a recurrent urinary bladder infection. While slicing vegetables, she cut herself and had difficulty stopping the bleeding. How could the antibiotics have played a role in her bleeding?

2. Mrs. Brown is in kidney failure. Her recent blood tests indicated a hematocrit of 22. Why is her hematocrit low? What can she be given to raise her hematocrit?

3. Thomas has hepatitis, which is disrupting his liver functions. What kinds of symptoms would he be experiencing based on the role(s) of the liver related to blood?

? ANSWERS TO FIGURE QUESTIONS

21.1 Blood volume is about 8% of your body mass, roughly 5–6 liters in males and 4–5 liters in females.

21.2 No, monocytes are classified as agranular leukocytes, but they develop from myeloid stem cells.

21.3 Once you moved to high altitude, your hematocrit would increase due to increased secretion of erythropoietin.

21.4 Transferrin is a plasma protein that transports iron in the blood.

21.5 Some symptoms of sickle-cell disease are anemia, mild jaundice, joint pain, shortness of breath, rapid heart rate, abdominal pain, fever, and fatigue.

21.6 Eosinophils, basophils, and neutrophils are called granular leukocytes because all have cytoplasmic granules that are visible through a light microscope when stained.

21.7 Platelets are cell fragments.

21.8 Along with platelet plug formation, vascular spasm and blood clotting contribute to hemostasis.

21.9 The outcome of the first stage of clotting is the formation of prothrombinase.

21.10 Type O blood contains both anti-A and anti-B antibodies.

The Cardiovascular System: The Heart

22

The heart is the pump of the cardiovascular system and it circulates the blood by two main circuits. The pulmonary circuit (the lungs) picks up oxygen and removes carbon dioxide; the systemic circuit delivers oxygen and nutrients to the rest of the body and picks up waste products. The heart is a highly coordinated organ with its own internal firing system for contraction called the sinoatrial node (the pacemaker). This allows for efficient contraction of both sides of the heart at the same time and delivery of blood to both circuits simultaneously. Both sides of the heart work in unison, bringing the blood from the lungs and the body to the upper two chambers in the heart (the atria) and then to the lower two chambers (the ventricles) and then out to the lungs and the rest of the body systems again. The heart's two sides have similar physiological actions. Anatomically, the heart has different amounts of muscle in the walls of the ventricles since the left ventricle must push the blood harder to reach all areas of the body (which are farther away than the nearby lungs). Disorders of the heart often manifest themselves as poor circulation in the extremities or as pain from the heart muscle when sufficient oxygen is unavailable. We will see in further chapters that the blood, heart, blood vessels, and lungs work cooperatively in the delivery of oxygen to the tissues.

CONTENTS AT A GLANCE

22.1 STRUCTURE AND ORGANIZATION OF THE HEART

OBJECTIVES

- Describe the location of the heart and the structure and functions of the pericardium.
- Describe the layers of the heart wall and the chambers of the heart.
- Identify the major blood vessels that enter and exit the heart.
- Describe the structure and functions of the valves of the heart.

Location and Coverings of the Heart

The ***heart*** is situated between the two lungs in the thoracic cavity, with about two-thirds of its mass lying to the left of the body's midline (Figure 22.1b). Your heart is about the size of your closed fist. The pointed end, the ***apex,*** is formed by the tip of the left ventricle, a lower chamber of the heart, and rests on the diaphragm. The ***superior border (base)*** of the heart is formed by the atria (upper chambers of the heart), mostly the left atrium, into which the four pulmonary veins open, and a portion of the right atrium that receives the superior and inferior vena cavae (see Figure 22.3a,b). The superior border lies opposite the

Figure 22.1 Position of the heart and associated blood vessels in the mediastinum (dashed outline) in the thoracic cavity.

The heart is located between the lungs, with two-thirds of its mass to the left of the midline.

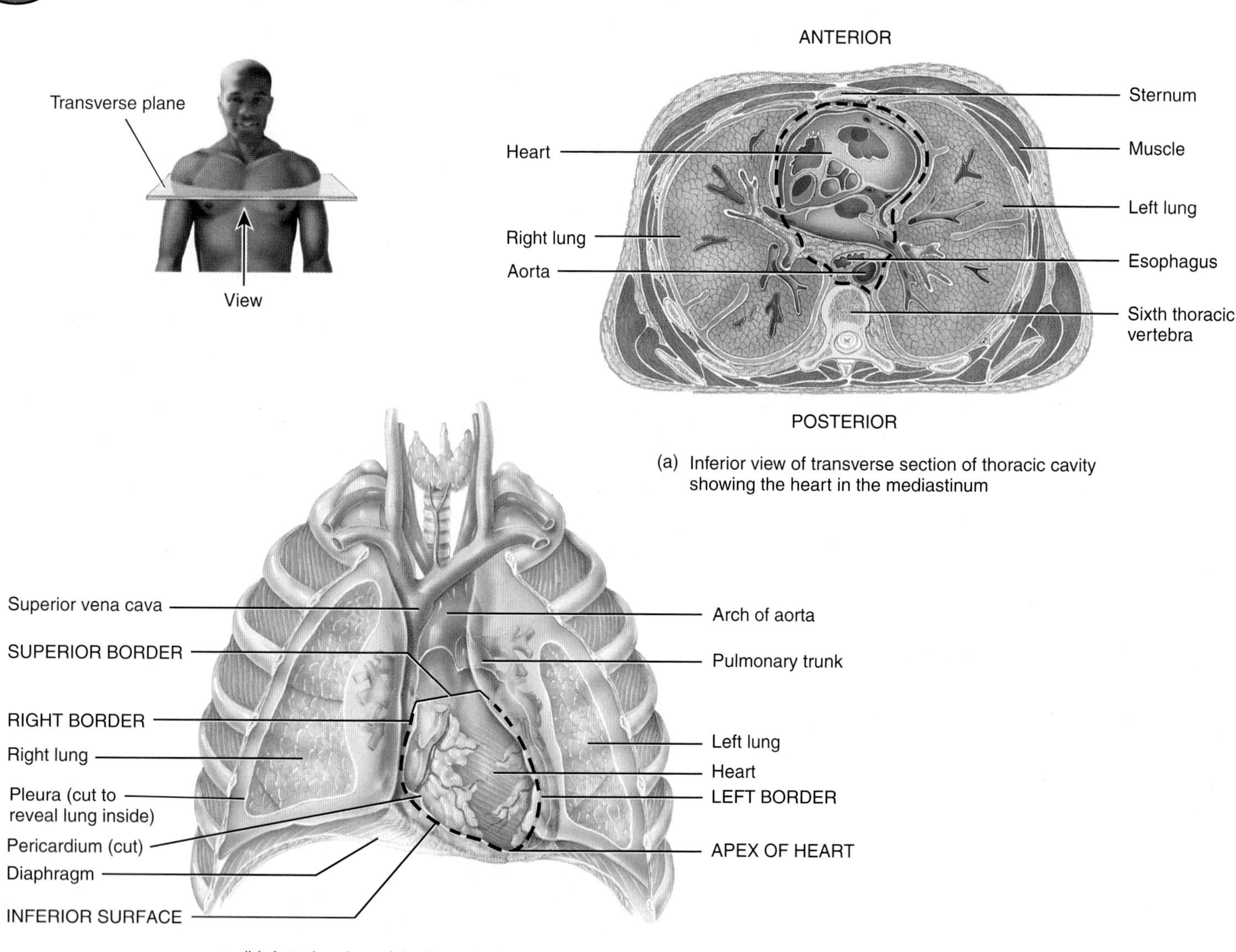

(a) Inferior view of transverse section of thoracic cavity showing the heart in the mediastinum

(b) Anterior view of the heart in the thoracic cavity

What is the mediastinum?

apex. The ***right*** and ***left borders*** of the heart face the right and left lungs. The ***inferior surface*** is the part of the heart between the right border and the apex and rests mostly on the diaphragm. The heart is located within the ***mediastinum,*** an anatomical region that includes all the visceral organs within the thoracic cavity except for the lungs (Figure 22.1a).

The membrane that surrounds and protects the heart and holds it in place is the ***pericardium*** (*peri-* = around; *-cardio-* = heart). It consists of two parts: the *fibrous pericardium* and the *serous pericardium* (Figure 22.2a,b). The outer ***fibrous pericardium*** is a tough, inelastic, dense irregular connective tissue. It prevents overstretching of the heart, provides protection, and anchors the heart in place. The fibrous pericardium near the apex of the heart is partially fused to the membrane that covers the diaphragm and therefore movement of the diaphragm, as in deep breathing, facilitates the movement of blood by the heart.

The inner ***serous pericardium*** is a thinner, more delicate membrane that forms a double layer around the heart. The outer ***parietal layer*** of the serous pericardium is fused to the fibrous pericardium, and the inner ***visceral layer*** of the serous pericardium, also called the ***epicardium*** (*epi-* = on top of), adheres tightly to the surface of the heart. Between the parietal and

Figure 22.2 Pericardium and the heart wall.

The pericardium is a sac that surrounds and protects the heart.

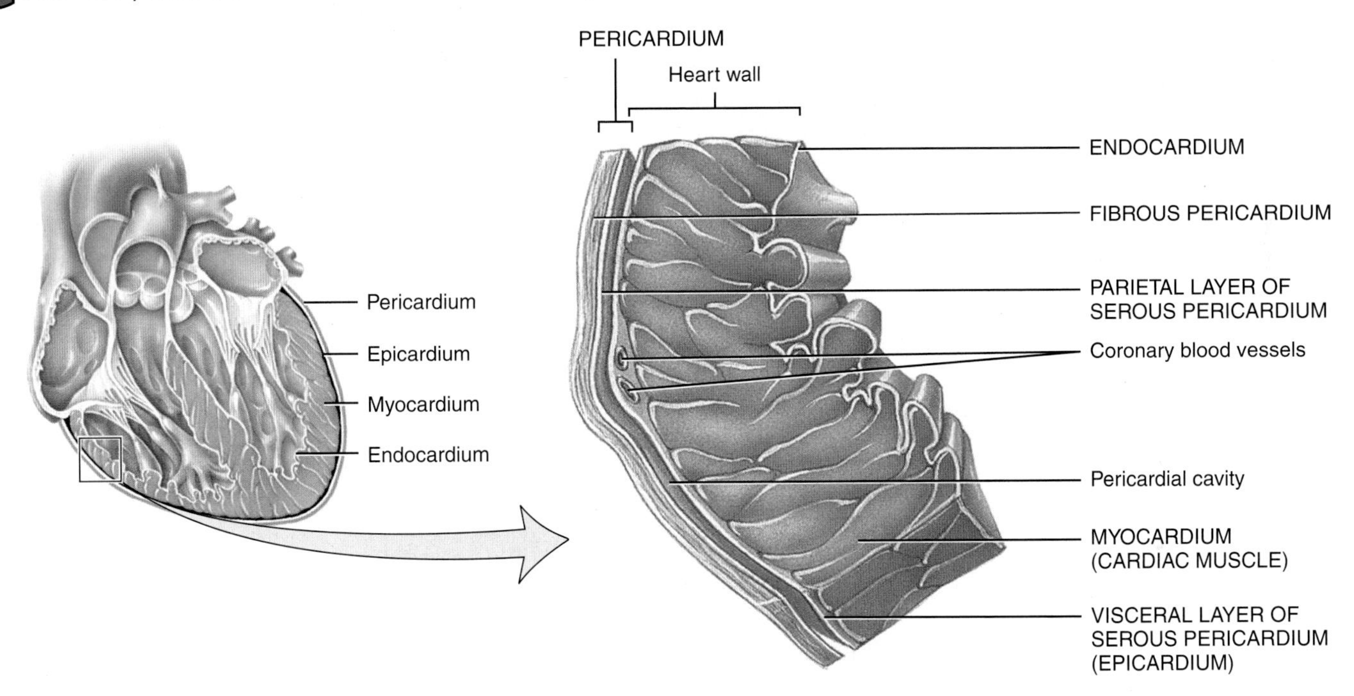

(a) Portion of pericardium and right ventricular heart wall showing the divisions of the pericardium and layers of the heart wall

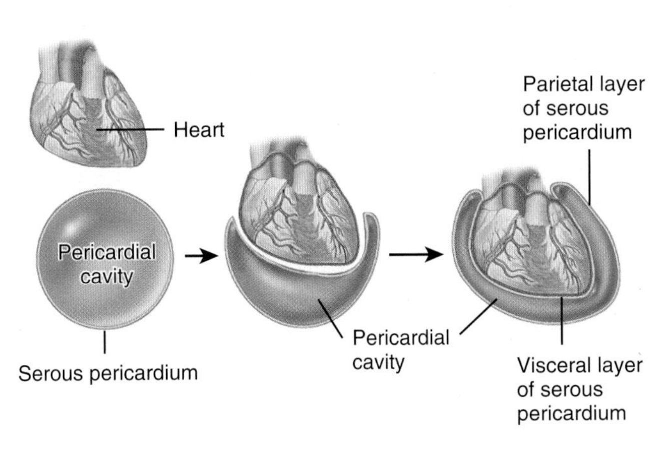

(b) Simplified relationship of the serous pericardium to the heart

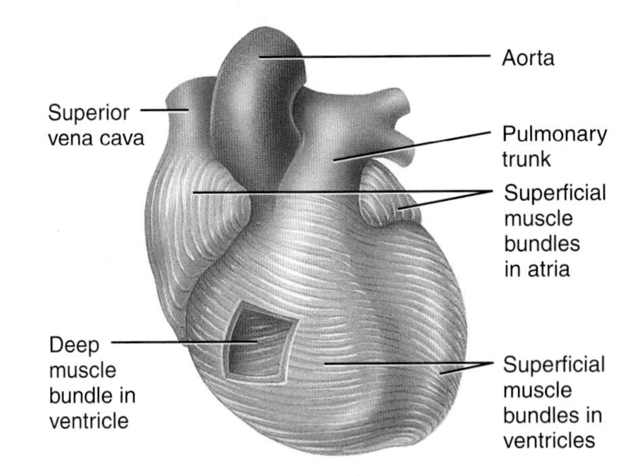

(c) Cardiac muscle bundles of the myocardium

CONTINUES

Figure 22.2 (continued)

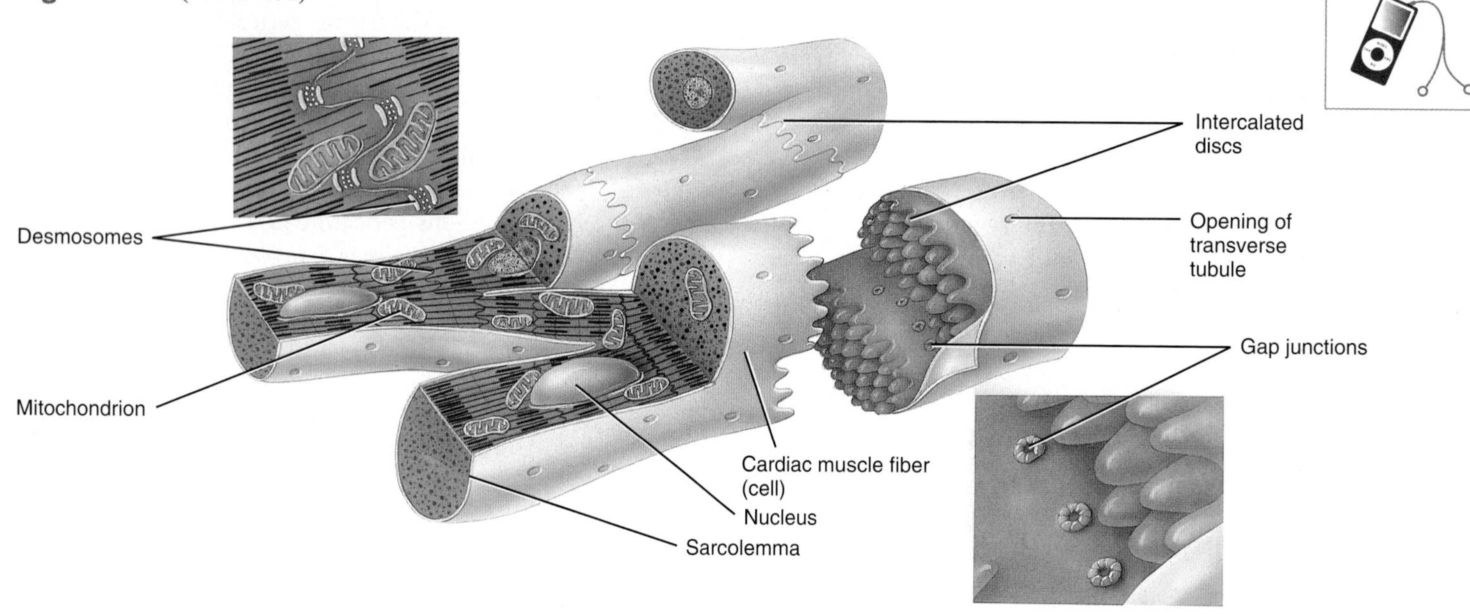

(d) Cardiac muscle fibers

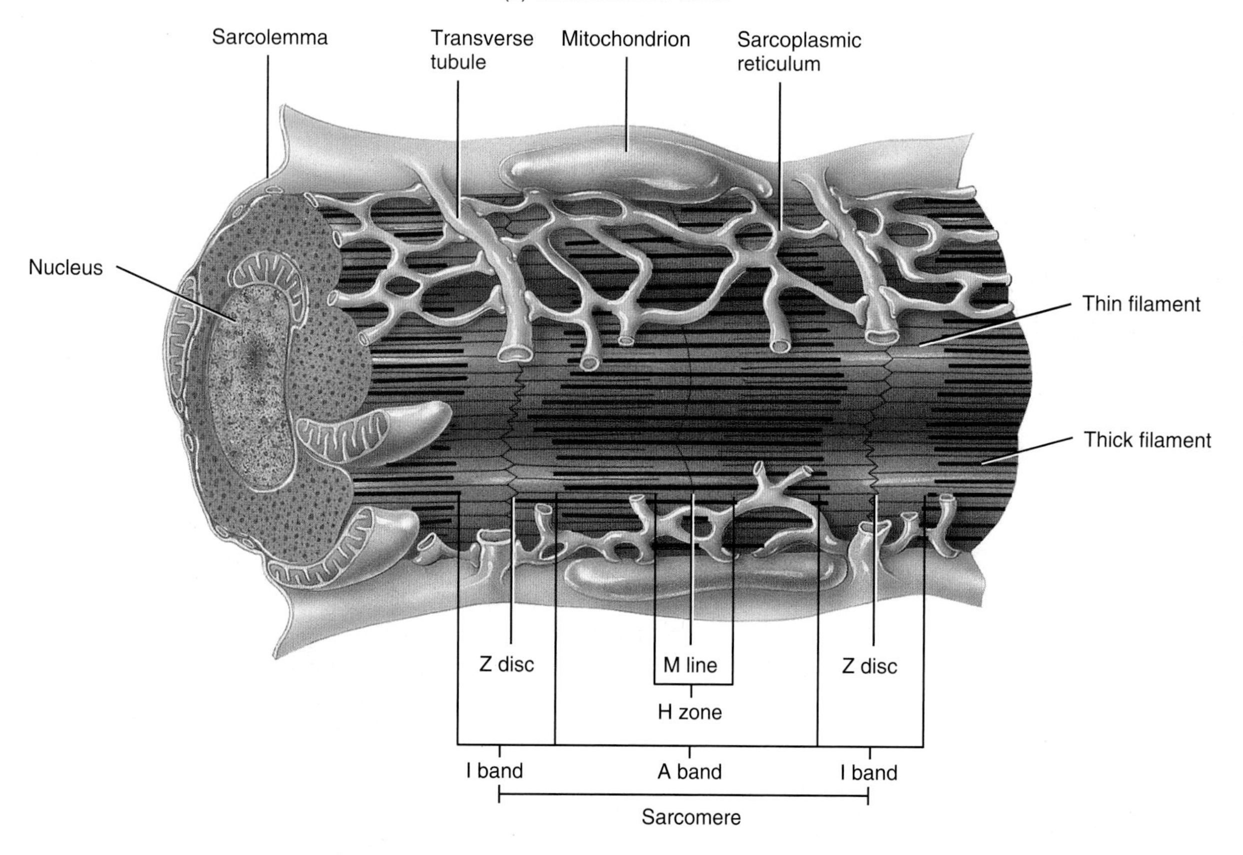

(e) Arrangement of components in a cardiac muscle fiber

? What are the functions of intercalated discs in cardiac muscle fibers?

visceral layers of the serous pericardium is a thin film of lubricating fluid. This fluid, known as ***pericardial fluid,*** reduces friction between the membranes as the heart moves. The ***pericardial cavity*** is the space that contains the pericardial fluid.

• CLINICAL CONNECTION Pedicarditis

Inflammation of the pericardium is called **pericarditis** (pĕr-ī-kar-DĪ-tis). In one form of this condition, there is a buildup of pericardial fluid. If a great deal of fluid accumulates, this is a life-threatening condition because the fluid compresses the heart, a condition called *cardiac tamponade* (tam-pon-ĀD). As a result of the compression, ventricular filling is decreased, cardiac output is reduced, venous return to the heart is diminished, blood pressure falls, and breathing is difficult. •

Heart Wall

The wall of the heart (see Figure 22.2a) is composed of three layers: *epicardium* (external layer), *myocardium* (middle layer), and *endocardium* (inner layer). The ***epicardium,*** which is also known as the visceral layer of serous pericardium, is the thin, transparent outer layer of the wall. It is composed of mesothelium and connective tissue (see Figure 22.2b).

The ***myocardium*** (*myo-* = muscle) consists of cardiac muscle tissue, which constitutes the bulk of the heart. This tissue is found only in the heart and is specialized in structure and function. The myocardium is responsible for the pumping action of the heart. Cardiac muscle fibers (cells) are involuntary, striated, and branched, and the tissue is arranged in interlacing bundles of fibers (see Figure 22.2c). The pattern of thick and thin filaments is similar to that found in skeletal muscle (Chapter 10). In skeletal muscle, however, the transverse tubules are located over the zone of overlap whereas in cardiac muscle the sarcoplasmic reticulum does not contain terminal cisternae and the transverse tubules are located over the Z discs (Figure 22.2e).

Cardiac muscle fibers form two separate networks—one atrial and one ventricular. Each cardiac muscle fiber connects with other fibers in the networks by thickenings of the sarcolemma (plasma membrane) called ***intercalated discs.*** Within the discs are ***gap junctions*** that allow action potentials to conduct from one cardiac muscle fiber to the next (Figure 22.2d). The intercalated discs also link cardiac muscle fibers to one another so they do not pull apart. Each network contracts as a functional unit, so the atria contract separately from the ventricles. In response to a single action potential, cardiac muscle fibers develop a prolonged contraction, 10–15 times longer than the contraction observed in skeletal muscle fibers. Also, the refractory period of a cardiac fiber lasts longer than the contraction itself. Thus, another contraction of cardiac muscle cannot begin until relaxation is well underway. For this reason, tetanus (maintained contraction) cannot occur in cardiac muscle tissue.

The ***endocardium*** (*endo-* = within) is a thin layer of simple squamous epithelium that lines the inside of the myocardium and covers the valves of the heart and the tendons attached to the valves. It is continuous with the epithelial lining of the large blood vessels.

• CLINICAL CONNECTION Regeneration of Heart Cells

The heart of a heart attack survivor often has regions of *infarcted* (dead) cardiac muscle tissue that typically are replaced over time with noncontractile fibrous scar tissue. Our inability to repair damage from a heart attack has been attributed to a lack of stem cells in cardiac muscle and to the absence of mitosis in mature cardiac muscle fibers. One study, however, provides evidence for significant **regeneration of heart cells.** Evidently, stem cells can migrate from the blood into the heart and differentiate into functional muscle and endothelial cells. The hope is that researchers can learn how to "turn on" such regeneration of heart cells to treat people with heart failure or *cardiomyopathy* (diseased heart). •

Chambers of the Heart

The heart contains four chambers (see Figure 22.3). The two upper chambers are the ***atria*** (entry halls or chambers), and the two lower chambers are the ***ventricles*** (little bellies). Between the right atrium and left atrium is a thin partition called the ***interatrial septum*** (*inter-* = between; *septum* = a dividing wall or partition). A prominent feature of this septum is an oval depression called the ***fossa ovalis.*** It is the remnant of the ***foramen ovale,*** an opening in the fetal heart that directs blood from the right to left atrium in order to bypass the nonfunctioning fetal lungs. The foramen ovale normally closes soon after birth. An ***interventricular septum*** separates the right ventricle from the left ventricle (see Figure 22.3c). On the anterior surface of each atrium is a wrinkled pouchlike structure called an ***auricle*** (OR-i-kul; *auri-* = ear), so named because of its resemblance to a dog's ear. Each auricle slightly increases the capacity of an atrium so it can hold a greater volume of blood.

Also on the surface of the heart are a series of grooves, called ***sulci*** (SUL-sī), that contain coronary blood vessels and a variable amount of fat. Each *sulcus* (SUL-kus) marks the external boundary between two chambers of the heart. The deep ***coronary sulcus*** (*coron-* = resembling a crown) encircles most of the heart and marks the boundary between the superior atria and inferior ventricles. The ***anterior interventricular sulcus*** is a shallow groove on the anterior surface of the heart that marks the boundary between the right and left ventricles (see Figure 22.3a). This sulcus continues around to the posterior surface of the heart as the ***posterior interventricular sulcus,*** which marks the boundary between the ventricles on the posterior aspect of the heart (see Figure 22.3a,b).

The thickness of the myocardium of the chambers varies according to the amount of work each chamber has to perform. The walls of the atria are thin compared to those of the ventricles because the atria need only enough cardiac muscle tissue to deliver blood into the ventricles (see Figure 22.3c). The right ventricle pumps blood only to the lungs (pulmonary circulation); the left ventricle pumps blood to all other parts of the body (systemic circulation). The left ventricle must work harder than the right ventricle to maintain the same rate of blood flow, so the muscular wall of the left ventricle is considerably thicker than

Figure 22.3 Structure of the heart: surface features and internal anatomy. Throughout this book, illustrations of blood vessels that carry oxygenated blood (which looks bright red) are colored red, whereas those that carry deoxygenated blood (which looks dark red) are colored blue.

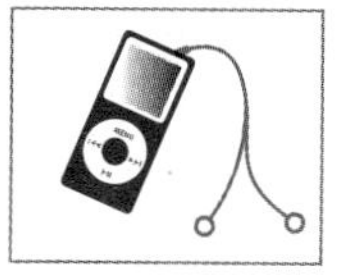

Sulci are grooves that contain blood vessels and fat and mark the external boundaries between the various chambers.

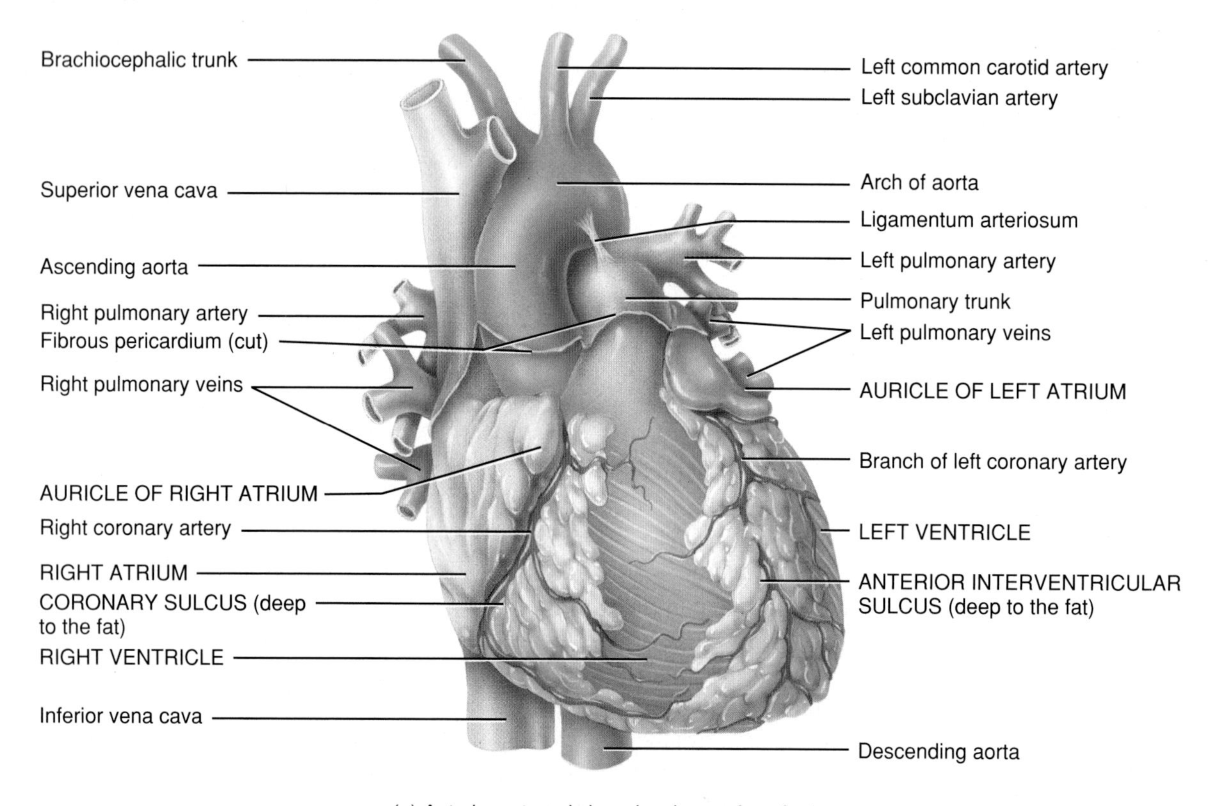

(a) Anterior external view showing surface features

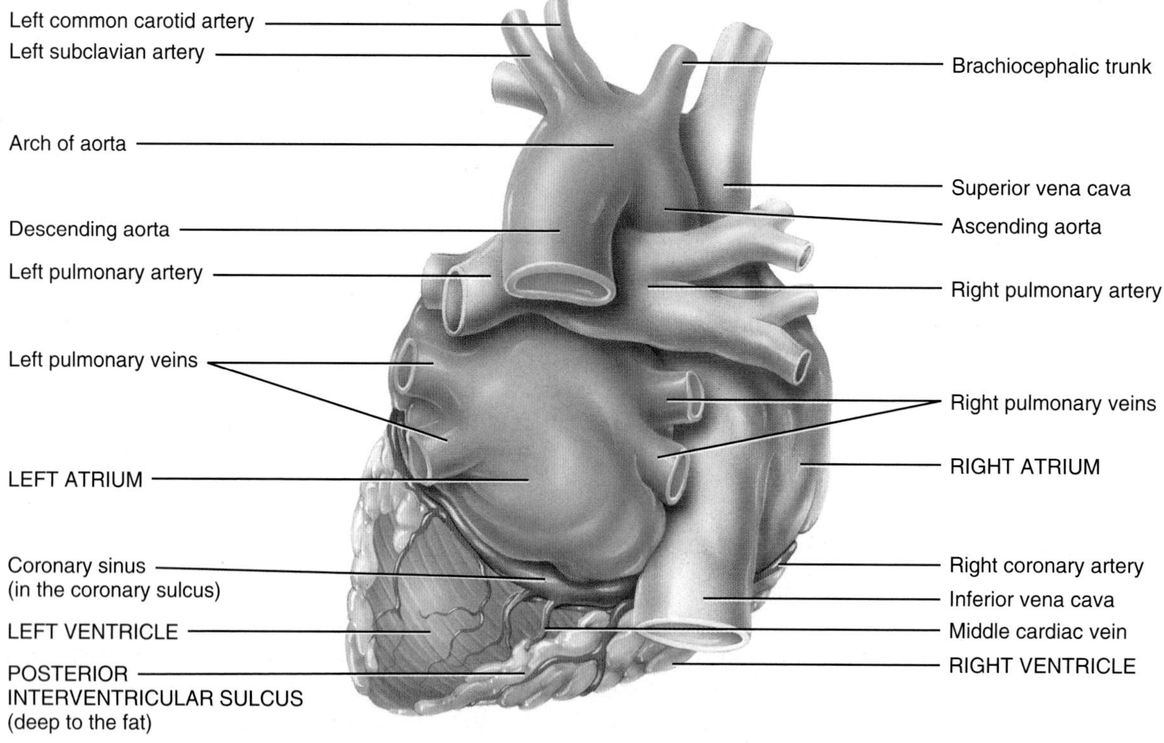

(b) Posterior external view showing surface features

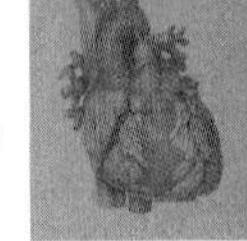

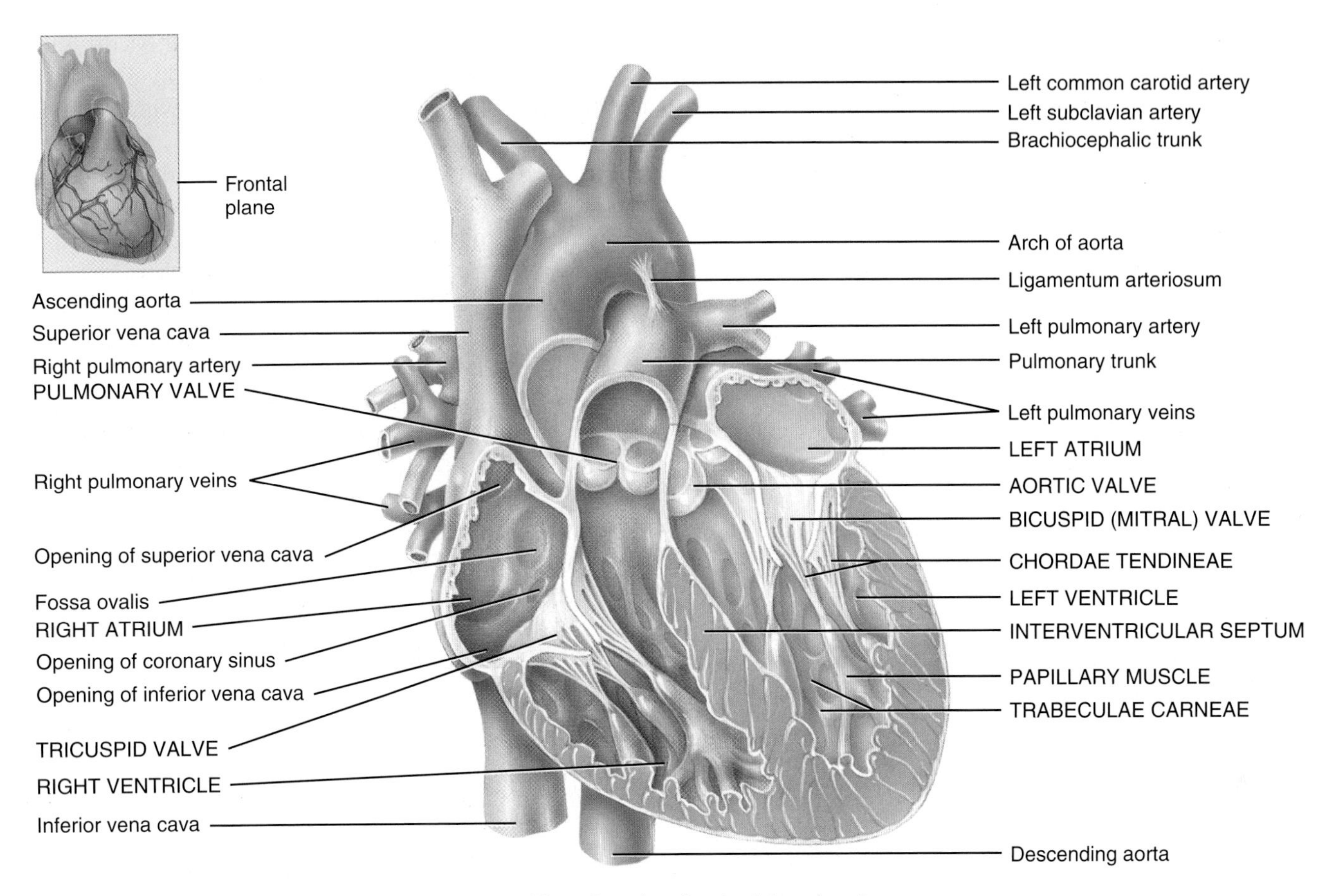

(c) Anterior view of frontal section showing internal anatomy

? **The coronary sulcus forms an external boundary between which chambers of the heart?**

the wall of the right ventricle to overcome the greater pressure (Figure 22.3c).

Great Vessels of the Heart

The right atrium receives deoxygenated blood (oxygen-poor blood that has given up some of its oxygen to cells) through three ***veins,*** blood vessels that return blood to the heart. The ***superior vena cava*** (VĒ-na CĀ-va; *vena* = vein; *cava* = hollow, a cave) brings blood mainly from parts of the body above the heart; the ***inferior vena cava*** brings blood mostly from parts of the body below the heart; and the ***coronary sinus*** drains blood from most of the vessels supplying the wall of the heart (Figure 22.3b,c). The right atrium then delivers the deoxygenated blood into the right ventricle, which pumps it into the ***pulmonary trunk.*** The pulmonary trunk divides into a ***right*** and ***left pulmonary artery,*** each of which carries deoxygenated blood to the corresponding lung. ***Arteries*** are blood vessels that carry blood away from the heart. In the lungs, the deoxygenated blood unloads carbon dioxide and picks up oxygen. This ***oxygenated blood*** (oxygen-rich blood that has picked up oxygen as it flows through the lungs) then enters the left atrium via four ***pulmonary veins.*** The blood then passes into the left ventricle, which pumps the blood into the ***ascending aorta.*** From here the oxygenated blood is carried to all parts of the body.

Between the pulmonary trunk and arch of the aorta is a structure called the ***ligamentum arteriosum*** (Figure 22.3a,c). It is the remnant of the ***ductus arteriosus,*** a blood vessel in fetal circulation that allows most blood to bypass the nonfunctional fetal lungs (see Figure 23.17).

Valves of the Heart

As each chamber of the heart contracts, it pushes a volume of blood into a ventricle or out of the heart into an artery. To prevent the blood from flowing backward, the heart has four valves composed of dense connective tissue covered by endothelium. These valves open and close in response to pressure changes as the heart contracts and relaxes.

As their names imply, ***atrioventricular (AV) valves*** lie between the atria and ventricles (Figure 22.3c). The atrioventricular valve between the right atrium and right ventricle is called the ***tricuspid valve*** because it consists of three cusps (leaflets). The pointed ends of the cusps project into the ventricle. Tendon-like cords, called ***chordae tendineae*** (KOR-dē ten-DI-nē-ē; *chord-* = cord; *tend-* = tendon), connect the pointed ends to

papillary muscles (*papill-* = nipple), cardiac muscle projections located on the inner surface of the ventricles. The chordae tendineae prevent the valve cusps from pushing up into the atria when the ventricles contract and are aligned to allow the valve leaflets to tightly close the valve (Figure 22.4a,b).

The atrioventricular valve between the left atrium and left ventricle is called the ***bicuspid (mitral) valve.*** It has two cusps that work in the same way as the cusps of the tricuspid valve. For blood to pass from an atrium to a ventricle, an atrioventricular valve must open.

The opening and closing of the valves are due to pressure differences across the valves. When there is a pressure gradient the valve opens, and when the valve opens the blood moves. When blood moves from an atrium to a ventricle, the valve is pushed open, the papillary muscles relax, and the chordae tendineae slacken (Figure 22.4a). When a ventricle contracts, the pressure of the ventricular blood drives the cusps upward until their edges meet and close the opening (Figure 22.4b). At the same time, contraction of the papillary muscles and tightening of the chordae tendineae help prevent the cusps from swinging upward into the atrium.

Near the origin of the pulmonary trunk and aorta are ***semilunar valves*** called the ***pulmonary valve*** and the ***aortic valve*** that prevent blood from flowing back into the heart (see Figure 22.3c). The pulmonary valve lies in the opening where the pulmonary trunk leaves the right ventricle. The aortic valve is situated at the opening between the left ventricle and the aorta. Each valve consists of three semilunar (half-moon-shaped) cusps that attach to the artery wall. Like the atrioventricular valves, the semilunar valves permit blood to flow in one direction only—in this case, from the ventricles into the arteries. When the ventricles contract, pressure builds up within them. The semilunar valves open when pressure in the ventricles exceeds the pressure in the arteries, permitting ejection of blood from the ventricles into the pul-

Figure 22.4 Atrioventricular (AV) valves, semilunar valves, and the fibrous skeleton of the heart. The bicuspid and tricuspid valves operate in a similar manner.

Heart valves open and close in response to pressure changes as the heart contracts and relaxes.

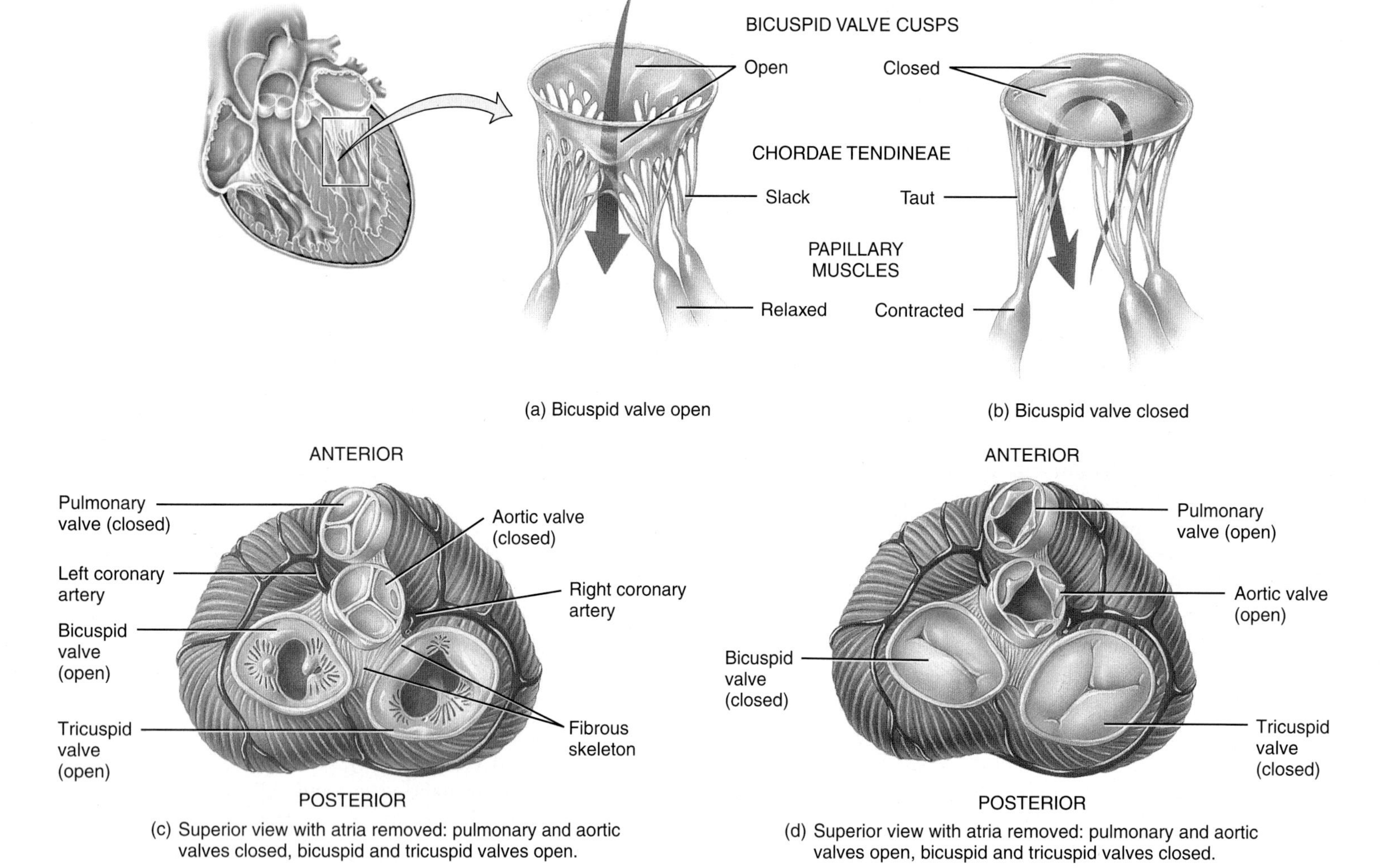

(c) Superior view with atria removed: pulmonary and aortic valves closed, bicuspid and tricuspid valves open.

(d) Superior view with atria removed: pulmonary and aortic valves open, bicuspid and tricuspid valves closed.

What is the function of heart valves?

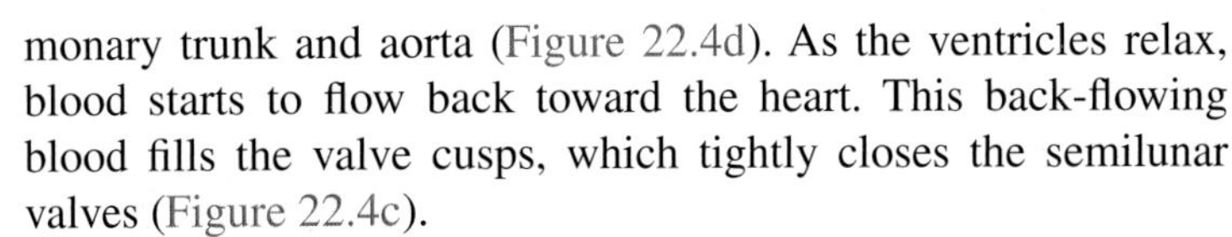

monary trunk and aorta (Figure 22.4d). As the ventricles relax, blood starts to flow back toward the heart. This back-flowing blood fills the valve cusps, which tightly closes the semilunar valves (Figure 22.4c).

In addition to cardiac muscle tissue, the heart wall also contains dense connective tissue that forms the ***fibrous skeleton of the heart*** (Figure 22.4c,d). Essentially, the fibrous skeleton consists of four dense connective tissue rings that surround the valves of the heart, fuse with one another, and merge with the interventricular septum. In addition to forming a structural foundation for the heart valves, the fibrous skeleton prevents overstretching of the valves as blood passes through them. It also serves as an attachment for bundles of cardiac muscle fibers. During atrial contraction, the two atria are pulled inferiorly toward the fibrous skeleton; during ventricular contraction, the fibrous skeleton stabilizes the lower chambers as they contract. The connective tissue of the fibrous skeleton of the heart also acts as an electrical insulator between the atria and ventricles.

• CLINICAL CONNECTION | Heart Valve Disorders

When heart valves operate normally, they open fully and close completely at the proper times. A narrowing of a heart valve opening that restricts blood flow is known as **stenosis** (ste-NŌ-sis = a narrowing); failure of a valve to close completely is termed **insufficiency (incompetence).** In **mitral stenosis,** scar formation or a congenital defect causes narrowing of the mitral valve. One cause of mitral insufficiency, in which there is a backflow of blood from the left ventricle into the left atrium, is **mitral valve prolapse (MVP).** In MVP, one or both cusps of the mitral valve protrude into the left atrium during ventricular contraction. Mitral valve prolapse is one of the most common valvular disorders, affecting as much as 30% of the population. It is more prevalent in women than in men. As with most other disorders of the body, there are varying degrees of disorder of the mitral valve and in many cases mitral valve prolapse does not pose a serious threat. In **aortic stenosis,** the aortic valve is narrowed, and in **aortic insufficiency,** there is backflow of blood from the aorta into the left ventricle.

If a heart valve cannot be repaired surgically, then the valve must be replaced. Tissue (biologic) valves may be provided by human donors or pigs; sometimes mechanical (artificial) valves made of plastic or metal are used. The aortic valve is the most commonly replaced heart valve.

Certain infectious diseases can damage or destroy the heart valves. One example is **rheumatic fever,** an acute systemic inflammatory disease that usually occurs after a streptococcal infection of the throat. The bacteria trigger an immune response in which antibodies produced to destroy the bacteria instead attack and inflame the connective tissues in joints, heart valves, and other organs. Even though rheumatic fever may weaken the entire heart wall, most often it damages the mitral and aortic valves. •

CHECKPOINT

1. Identify the location of the heart.
2. Describe the various layers of the pericardium and the heart wall.
3. How do atria and ventricles differ in structure and function?

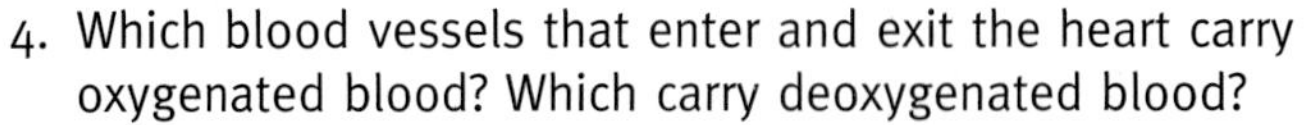

4. Which blood vessels that enter and exit the heart carry oxygenated blood? Which carry deoxygenated blood?
5. In correct sequence, which heart chambers, heart valves, and blood vessels would a drop of blood encounter from the time it flows out of the right atrium until it reaches the aorta?

22.2 BLOOD FLOW AND BLOOD SUPPLY OF THE HEART

OBJECTIVES

- Explain how blood flows through the heart.
- Describe the clinical importance of the blood supply of the heart.

Blood Flow through the Heart

Blood flows through the heart from areas of higher blood pressure to areas of lower blood pressure. As the walls of the atria contract, the pressure of the blood within them increases. This increased blood pressure forces blood through the open AV valves, allowing atrial blood to flow through the AV valves into the ventricles.

After the atria are finished contracting, the walls of the ventricles contract, increasing ventricular blood pressure and pushing blood through the semilunar valves into the pulmonary trunk and aorta. At the same time, the pressure gradient and the chordae tendineae close the AV valves, preventing backflow of ventricular blood into the atria. Figure 22.5 summarizes the flow of blood through the heart.

Blood Supply of the Heart

Nutrients are not able to diffuse quickly enough from blood in the chambers of the heart to supply all the layers of cells that make up the heart wall. For this reason, the myocardium has its own network of blood vessels, the ***coronary*** (*coron-* = crown) or ***cardiac circulation.*** The ***coronary arteries*** branch from the ascending aorta and encircle the heart like a crown encircles the head (see Figure 22.6a). While the heart is contracting, little blood flows in the coronary arteries because they are squeezed shut. When the heart relaxes, however, the high pressure of blood in the aorta propels blood through the coronary arteries, into capillaries, and then into ***coronary (cardiac) veins*** (see Figure 22.6b).

Two coronary arteries, the right and left coronary arteries, branch from the ascending aorta and supply oxygenated blood to the myocardium (see Figure 22.6a). The ***left coronary artery*** passes inferior to the left auricle and divides into the anterior interventricular and circumflex branches. The ***anterior interventricular branch*** or *left anterior descending (LAD) artery* is in the anterior interventricular sulcus and supplies oxygenated blood to the walls of both ventricles. The ***circumflex branch*** lies in the coronary sulcus and distributes oxygenated blood to the walls of the left ventricle and left atrium.

Figure 22.5 Systemic and pulmonary circulations.

The left side of the heart pumps oxygenated blood into the systemic circulation to all tissues of the body except the air sacs (alveoli) of the lungs. The right side of the heart pumps deoxygenated blood into the pulmonary circulation to the air sacs (alveoli) of the lungs.

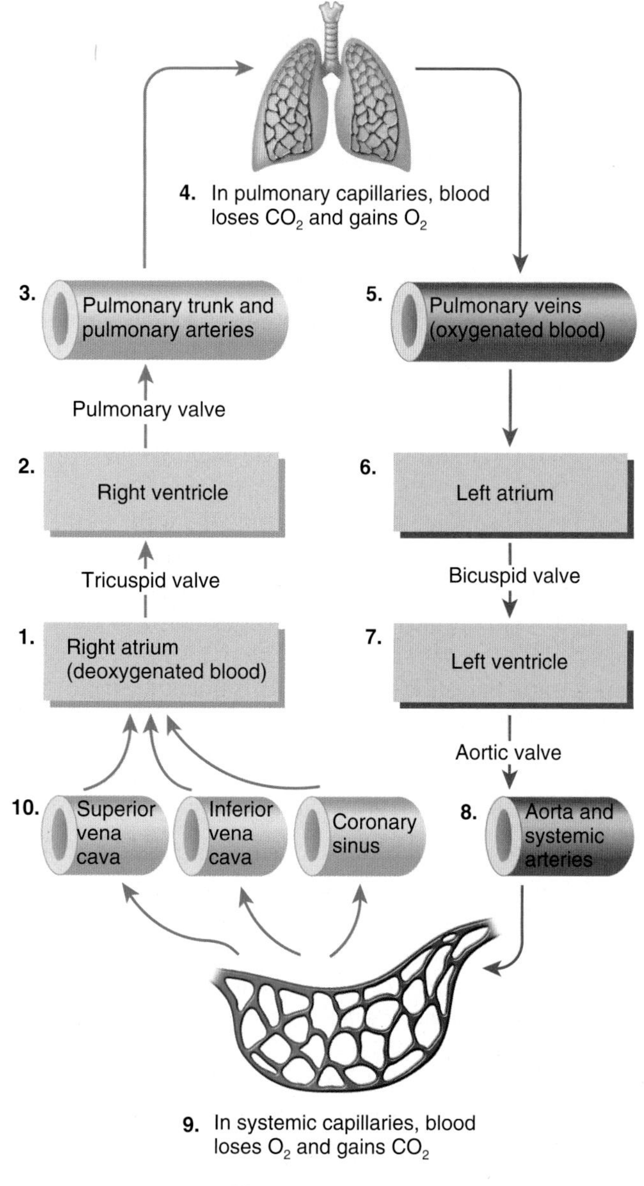

Diagram of blood flow

Which numbers constitute the pulmonary circulation? Which constitute the systemic circulation?

The ***right coronary artery*** supplies small branches (*atrial branches*) to the right atrium. It continues inferior to the right auricle and ultimately divides into the posterior interventricular and marginal branches. The ***posterior interventricular branch*** follows the posterior interventricular sulcus and supplies the walls of the two ventricles with oxygenated blood. The ***marginal branch*** transports oxygenated blood to the myocardium of the right ventricle.

Most parts of the body receive blood from branches of more than one artery, and where two or more arteries supply the same region, they usually connect. These connections, called ***anastomoses*** (a-nas′-tō-MŌ-sēs), provide alternate routes for blood to reach a particular organ or tissue. The myocardium contains many anastomoses that connect branches of a given coronary artery or extend between branches of different coronary arteries. They provide detours for arterial blood if a main route becomes obstructed. Thus, heart muscle may receive sufficient oxygen even if one of its coronary arteries is partially blocked.

Most of the deoxygenated blood, which carries carbon dioxide and wastes, is ultimately collected by a large vein on the posterior surface of the heart, the ***coronary sinus*** (Figure 22.6b), which empties into the right atrium.

Myocardial Ischemia and Infarction

Partial obstruction of blood flow in the coronary arteries may cause ***myocardial ischemia*** (is-KĒ-mē-a; *ische-* = to obstruct; *-emia* = in the blood), a condition of reduced blood flow to the myocardium. Usually, ischemia causes ***hypoxia*** (reduced oxygen supply), which may weaken cells without killing them. ***Angina pectoris*** (an-JĪ-na or AN-ji-na PEK-tō-ris), which literally means "strangled chest," is a severe pain that usually accompanies myocardial ischemia. Typically, sufferers describe it as a tightness or squeezing sensation, as though the chest were in a vise. The pain associated with angina pectoris is often referred to the neck, chin, or down the left arm to the elbow. ***Silent myocardial ischemia,*** an ischemic episode without pain, is particularly dangerous because the person has no forewarning of an impending heart attack.

A complete obstruction to blood flow in a coronary artery may result in a ***myocardial infarction*** (in-FARK-shun), or ***MI,*** commonly called a ***heart attack. Infarction*** means the death of an area of tissue because of interrupted blood supply. Because the heart tissue distal to the obstruction dies and is replaced by noncontractile scar tissue, the heart muscle loses some of its strength. Depending on the size and location of the infarcted (dead) area, an infarction may disrupt the conduction system of the heart and cause sudden death by triggering ventricular fibrillation. Treatment for a myocardial infarction may involve injection of a thrombolytic (clot-dissolving) agent such as streptokinase or tPA, plus heparin (an anticoagulant), or performing coronary angioplasty or coronary artery bypass grafting. Fortunately, heart muscle can remain alive in a resting person if it receives as little as 10–15% of its normal blood supply.

CHECKPOINT

6. Describe the main force that causes blood to flow through the heart.
7. Why is it that blood flowing through the chambers within the heart cannot supply sufficient oxygen or remove enough carbon dioxide from the myocardium?

Figure 22.6 The coronary circulation. The views are drawn as if the heart were transparent to reveal blood vessels on the posterior aspect.

The right and left coronary arteries deliver blood to the heart; the coronary veins drain blood from the heart into the coronary sinus.

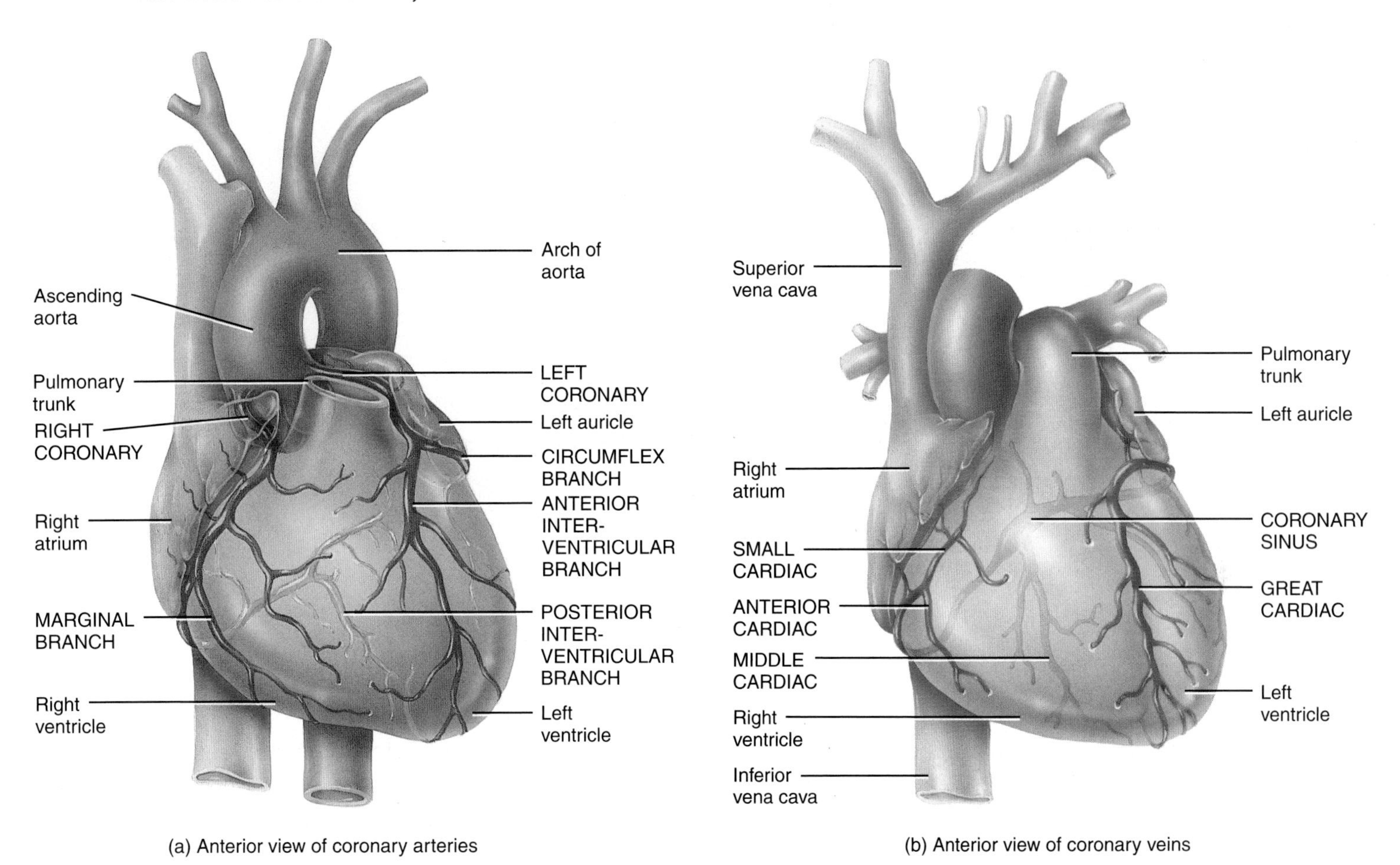

? Which coronary blood vessel delivers oxygenated blood to the walls of the left atrium and left ventricle?

22.3 CONDUCTION SYSTEM OF THE HEART

OBJECTIVE

- Explain how each heartbeat is initiated and maintained.

About 1% of the cardiac muscle fibers are different from all other cardiac fibers because they can generate action potentials over and over and do so in a rhythmical pattern. They continue to stimulate a heart to beat even after it is removed from the body—for example, to be transplanted into another person—and all of its nerves have been cut. The nerves regulate the heart rate, but do not determine it. These cells have two important functions: They act as a ***pacemaker,*** setting the rhythm for the entire heart, and they form the ***conduction system,*** the route for action potentials throughout the heart muscle. The conduction system is thus composed of modified cardiac muscle cells, not nervous tissue. The conduction system ensures that cardiac chambers are stimulated to contract in a coordinated manner, which makes the heart an effective pump. Cardiac action potentials pass through the following components of the conduction system (see Figure 22.7):

1. Normally, cardiac excitation begins in the ***sinoatrial (SA) node,*** located in the right atrial wall just inferior to the opening of the superior vena cava. An action potential spontaneously arises in the SA node and then conducts throughout both atria via gap junctions in the intercalated discs of atrial fibers (see Figure 22.7). Following the action potential, the two atria finish contracting at the same time.
2. By conducting along atrial muscle fibers, the action potential also reaches the ***atrioventricular (AV) node,*** located in the interatrial septum, just anterior to the opening of the coronary sinus. At the AV node, the action potential slows considerably, providing time for the atria to empty their blood into the ventricles.

Figure 22.7 **Conduction system of the heart.** The SA node, located in the right atrial wall, is the heart's pacemaker, initiating cardiac action potentials that cause contraction of the heart's chambers. The arrows indicate the flow of action potentials through the atria. The route of action potentials through the numbered components of the conduction system is described in the text.

The conduction system ensures that cardiac chambers contract in a coordinated manner.

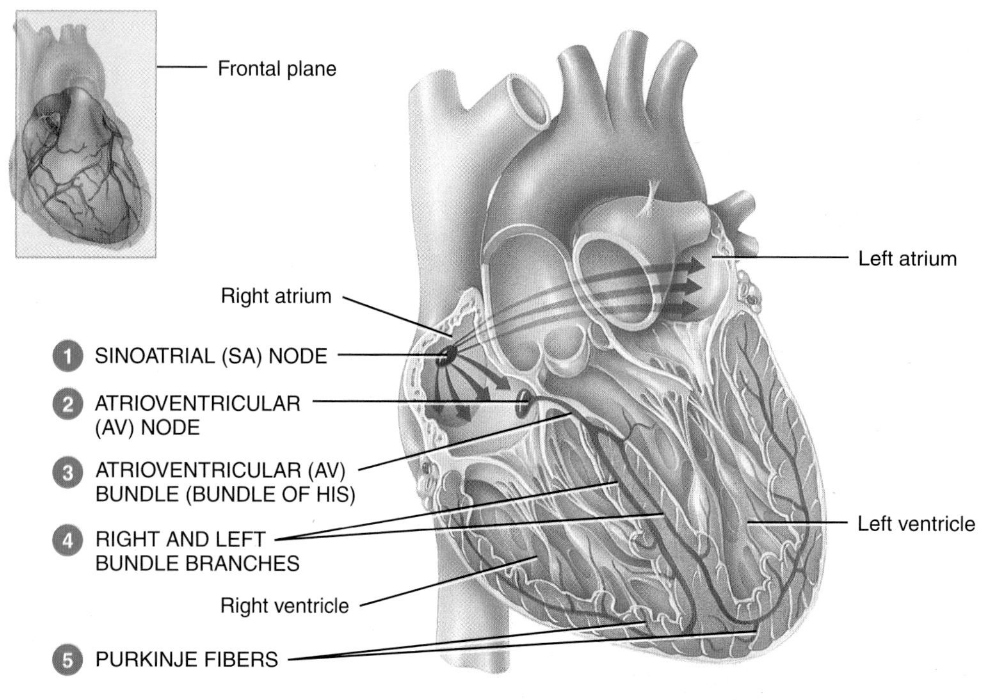

Which component of the conduction system provides the only route for action potentials to conduct between the atria and the ventricles?

3. From the AV node, the action potential enters the ***atrioventricular (AV) bundle*** (also known as the ***bundle of His***), in the interventricular septum. Owing to the position and composition of the fibrous skeleton of the heart, the AV bundle is the only site where action potentials can normally conduct from the atria to the ventricles.
4. After conducting along the AV bundle, the action potential then enters both the ***right*** and ***left bundle branches*** that course through the interventricular septum toward the apex of the heart.
5. Finally, large-diameter ***Purkinje fibers*** (pur-KIN-jē) rapidly conduct the action potential, first to the apex and then upward to the rest of the ventricular myocardium. A fraction of a second after the atria contract, the ventricles simultaneously contract.

The SA node initiates action potentials about 100 times per minute, faster than any other region of the conducting system. Thus, the SA node sets the rhythm for contraction of the heart—it is the ***pacemaker*** of the heart. Various hormones and neurotransmitters can speed or slow pacing of the heart by SA node fibers. In a person at rest, for example, acetylcholine (ACh) released by the parasympathetic division of the autonomic nervous system (ANS) typically slows SA node pacing to about 75 action potentials per minute, causing 75 heartbeats per minute. ACh is only one of many factors in the regulation of heart rate. If the SA node becomes diseased or damaged, the slower AV node fibers can become the pacemaker. With pacing by the AV node, however, heart rate is slower, only 40 to 60 beats/min. If the activity of both nodes is suppressed, the heartbeat may still be maintained by the AV bundle, a bundle branch, or Purkinje fibers. These fibers generate action potentials very slowly, about 20 to 35 times per minute. At such a low heart rate, blood flow to the brain is inadequate.

• CLINICAL CONNECTION | Artificial Pacemakers

When the heart rate is too low, normal heart rhythm can be restored and maintained by surgically implanting an **artificial pacemaker,** a device that sends out small electrical currents to stimulate the heart to contract. A pacemaker consists of a battery and impulse generator and is usually implanted beneath the skin just inferior to the clavicle. The pacemaker is connected to one or two flexible wires (leads) that are threaded through the superior vena cava and then passed into the right atrium and right ventricle. Many of the new models of pacemakers, called *activity-adjusted pacemakers,* contain sensors and automatically speed up the heartbeat during exercise. •

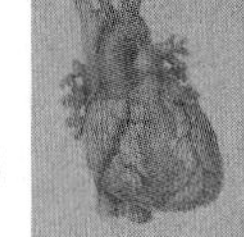

CHECKPOINT

8. Describe the path of an action potential through the conduction system.

22.4 ELECTROCARDIOGRAM

OBJECTIVE

- Describe the meaning and diagnostic value of an electrocardiogram.

Conduction of action potentials through the heart generates electrical currents that can be picked up by electrodes placed on the skin. A recording of the electrical changes that accompany the heartbeat is called an ***electrocardiogram*** (e-lek-trō-KAR-dē-ō-gram), which is abbreviated as either ***ECG*** or ***EKG*** (the Dutch spelling).

Three clearly recognizable waves accompany each heartbeat, which is a series of mechanical actions; the electrical activity of these three waves precedes the heartbeat. The first, called the ***P wave,*** is a small upward deflection on the ECG (Figure 22.8); it represents *atrial depolarization,* the depolarizing phase of the cardiac action potential as it spreads from the SA node throughout both atria. Depolarization causes contraction. Thus, a fraction of a second after the P wave begins, the atria contract. The second wave, called the ***QRS complex,*** begins as a downward deflection (Q); continues as a large, upright, triangular wave (R); and ends as a downward wave (S). The QRS complex represents the onset of *ventricular depolarization,* as the cardiac action potential spreads through the ventricles. Shortly after the QRS complex begins, the ventricles start to contract. The third wave is the ***T wave,*** a dome-shaped upward deflection that indicates *ventricular repolarization* and occurs just before the ventricles start to relax. Repolarization of the atria is not usually evident in an ECG because it is masked by the larger QRS complex; the two electrical events happening at the same time contributes to the larger amplitude of the QRS complex.

Variations in the size and duration of the waves of an ECG are useful in diagnosing abnormal cardiac rhythms and conduction patterns and in following the course of recovery from a heart attack. An ECG can also reveal the presence of a living fetus.

Figure 22.8 Normal electrocardiogram (ECG) of a single heartbeat. P wave = atrial depolarization; QRS complex = onset of ventricular depolarization plus atrial repolarization; T wave = ventricular repolarization.

An electrocardiogram is a recording of the electrical activity that initiates each heartbeat.

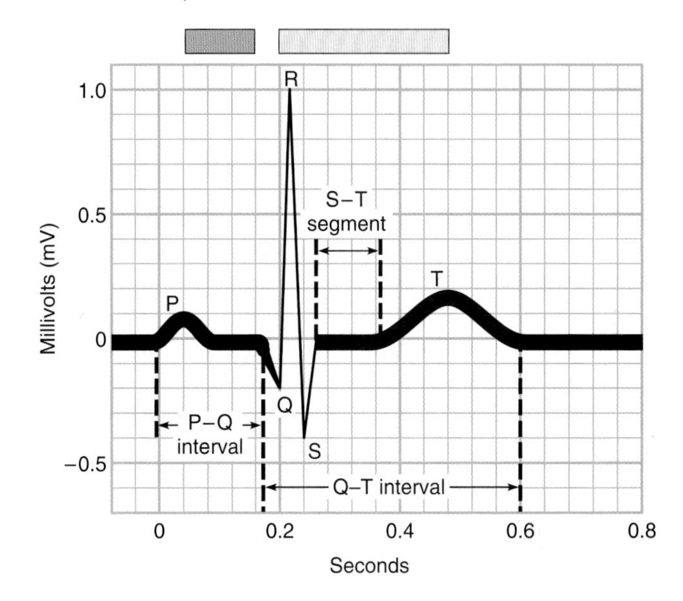

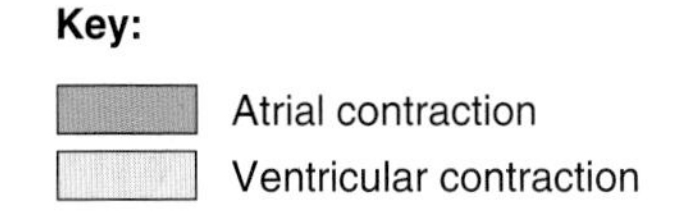

What event occurs in response to atrial depolarization?

Arrhythmias

The usual rhythm of heartbeats, established by the SA node, is called ***normal sinus rhythm.*** The term ***arrhythmia*** (a-RITH-mē-a; *a-* = without) or ***dysrhythmia*** refers to an abnormal rhythm as a result of a defect in the conduction system of the heart. The heart may beat irregularly, too fast, or too slowly. Symptoms include chest pain, shortness of breath, light-headedness, dizziness, and fainting. Arrhythmias may be caused by factors that stimulate the heart, such as stress, caffeine, alcohol, nicotine, cocaine, and certain drugs that contain caffeine or other stimulants. Arrhythmias may also be caused by a congenital defect, coronary artery disease, myocardial infarction, hypertension, defective heart valves, rheumatic heart disease, hyperthyroidism, and potassium deficiency.

One serious arrhythmia is called a ***heart block.*** The most common heart block occurs in the atrioventricular node, which conducts impulses from the atria to the ventricles. This disturbance is called *atrioventricular (AV) block.* In ***atrial flutter,*** the atrial rhythm averages between 240 and 360 beats per minute. The condition is essentially rapid atrial contractions accompanied by AV block. ***Atrial fibrillation*** is an uncoordinated contraction of the atrial muscles. When the muscle fibrillates, the muscle fibers of the atrium quiver individually instead of contracting together, canceling out the pumping of the atrium. ***Ventricular fibrillation (VF)*** is characterized by uncoordinated haphazard ventricular muscle contractions. Ventricular ejection ceases, and circulatory failure and death occur.

A condition known as ***Wolff-Parkinson-White (WPW)*** syndrome can lead to extremely rapid heart rates. This arrhythmia can produce symptoms such as palpitations, light-headedness, or even loss of consciousness. In rare instances, WPW syndrome can lead to cardiac arrest. In WPW syndrome, an extra electrical pathway exists between the atria and the ventricles. This extra pathway may at times encourage a rapid rhythm. Instead of allowing the next heartbeat to begin at the SA node, the extra pathway can "pick up" an electrical impulse in the ventricles and send it abnormally back upward to the atria. When this happens, the impulse begins to travel abnormally in a rapid, circular manner, causing a rapid heart rate.

CHECKPOINT

9. What is the significance of the P wave, QRS complex, and T wave?

22.5 THE CARDIAC CYCLE

OBJECTIVE

- Describe the phases of the cardiac cycle.

A single ***cardiac cycle*** includes all the events associated with one heartbeat. In a normal cardiac cycle, the two atria contract while the two ventricles relax; then, while the two ventricles contract, the two atria relax. The term ***systole*** (SIS-tō-lē = contraction) refers to the phase of contraction; ***diastole*** (dī-AS-tō-lē = dilation or expansion) refers to the phase of relaxation. A cardiac cycle consists of systole and diastole of both atria plus systole and diastole of both ventricles.

Pressure and Volume Changes during the Cardiac Cycle

In each cardiac cycle, the atria and ventricles alternately contract and relax, forcing blood from areas of higher pressure to areas of lower pressure. As a chamber of the heart contracts, blood pressure within it increases. Figure 22.9 shows the relation between the heart's electrical signals (ECG) and changes in atrial pressure, ventricular pressure, aortic pressure, and ventricular volume during the cardiac cycle. The pressures given in Figure 22.9 apply to the left side of the heart; pressures on the right side are considerably lower. Each ventricle, however, expels the same volume of blood per beat, and the same pattern exists for both pumping chambers. When heart rate is 75 beats/min, a cardiac cycle lasts 0.8 sec. To examine and correlate the events taking place during a cardiac cycle, we begin with atrial systole.

Atrial Systole

During ***atrial systole,*** which lasts about 0.1 sec, the atria are contracting. At the same time, the ventricles are relaxed.

1. Depolarization of the SA node causes atrial depolarization, marked by the P wave in the ECG.
2. Atrial depolarization causes atrial systole. As the atria contract, they exert pressure on the blood within, which forces blood through the open AV valves into the ventricles.
3. Atrial systole contributes a final 25 mL of blood to the volume already in each ventricle (about 105 mL). The end of atrial systole is also the end of ventricular diastole (relaxation). Thus, each ventricle contains about 130 mL at the end of its relaxation period (diastole). This blood volume is called the ***end-diastolic volume (EDV).***
4. The QRS complex in the ECG marks the onset of ventricular depolarization.

Ventricular Systole

During ***ventricular systole,*** which lasts about 0.3 sec, the ventricles are contracting. At the same time, the atria are relaxed, in ***atrial diastole.***

5. Ventricular depolarization causes ventricular systole. As ventricular systole begins, pressure rises inside the ventricles and pushes blood up against the atrioventricular (AV) valves, forcing them shut. For about 0.05 sec, both the SL (semilunar) and AV valves are closed. This is the period of ***isovolumetric contraction*** (*iso-* = same). During this interval, cardiac muscle fibers are contracting and exerting force but are not yet shortening. Thus, the muscle contraction is isometric (same length). Moreover, because all four valves are closed, ventricular volume remains the same (isovolumic).
6. Continued contraction of the ventricles causes pressure inside the chambers to rise sharply. When left ventricular pressure surpasses aortic pressure at about 80 millimeters of mercury (mmHg) and right ventricular pressure rises above the pressure in the pulmonary trunk (about 20 mmHg), both SL valves open. At this point, ejection of blood from the heart begins. The period when the SL valves are open is ***ventricular ejection*** and lasts for about 0.25 sec. The pressure in the left ventricle continues to rise to about 120 mmHg, whereas the pressure in the right ventricle climbs to about 25–30 mmHg.
7. The left ventricle ejects about 70 mL of blood into the aorta and the right ventricle ejects the same volume of blood into the pulmonary trunk. The volume remaining in each ventricle at the end of systole, about 60 mL, is the ***end-systolic volume (ESV). Stroke volume (SV),*** the volume ejected per beat from each ventricle, equals end-diastolic volume minus end-systolic volume: SV = EDV − ESV. At rest, the stroke volume is about 130 mL − 60 mL = 70 mL (a little more than 2 oz).
8. The T wave in the ECG marks the onset of ventricular repolarization.

Relaxation Period

During the ***relaxation period,*** which lasts about 0.4 sec, the atria and the ventricles are both relaxed. As the heart beats faster and faster, the relaxation period becomes shorter and shorter, whereas the durations of atrial systole and ventricular systole shorten only slightly.

9. Ventricular repolarization causes ***ventricular diastole.*** As the ventricles relax, pressure within the chambers falls, and blood in the aorta and pulmonary trunk begins to flow backward toward the regions of lower pressure in the ventricles. Backflowing blood catches in the valve cusps and closes the SL valves. The aortic valve closes at a pressure of about 100 mmHg. Rebound of blood off the closed cusps of the aortic valve produces the *dicrotic wave* on the aortic pressure curve. After the SL valves close, there is a brief interval when ventricular blood volume does not change because all four valves are closed. This is the period of ***isovolumetric relaxation.***

Figure 22.9 Cardiac cycle. (a) ECG. (b) Changes in left atrial pressure (green line), left ventricular pressure (blue line), and aortic pressure (red line) as they relate to the opening and closing of heart valves. (c) Heart sounds. (d) Changes in left ventricular volume. (e) Phases of the cardiac cycle.

⑥ A cardiac cycle is composed of all the events associated with one heartbeat.

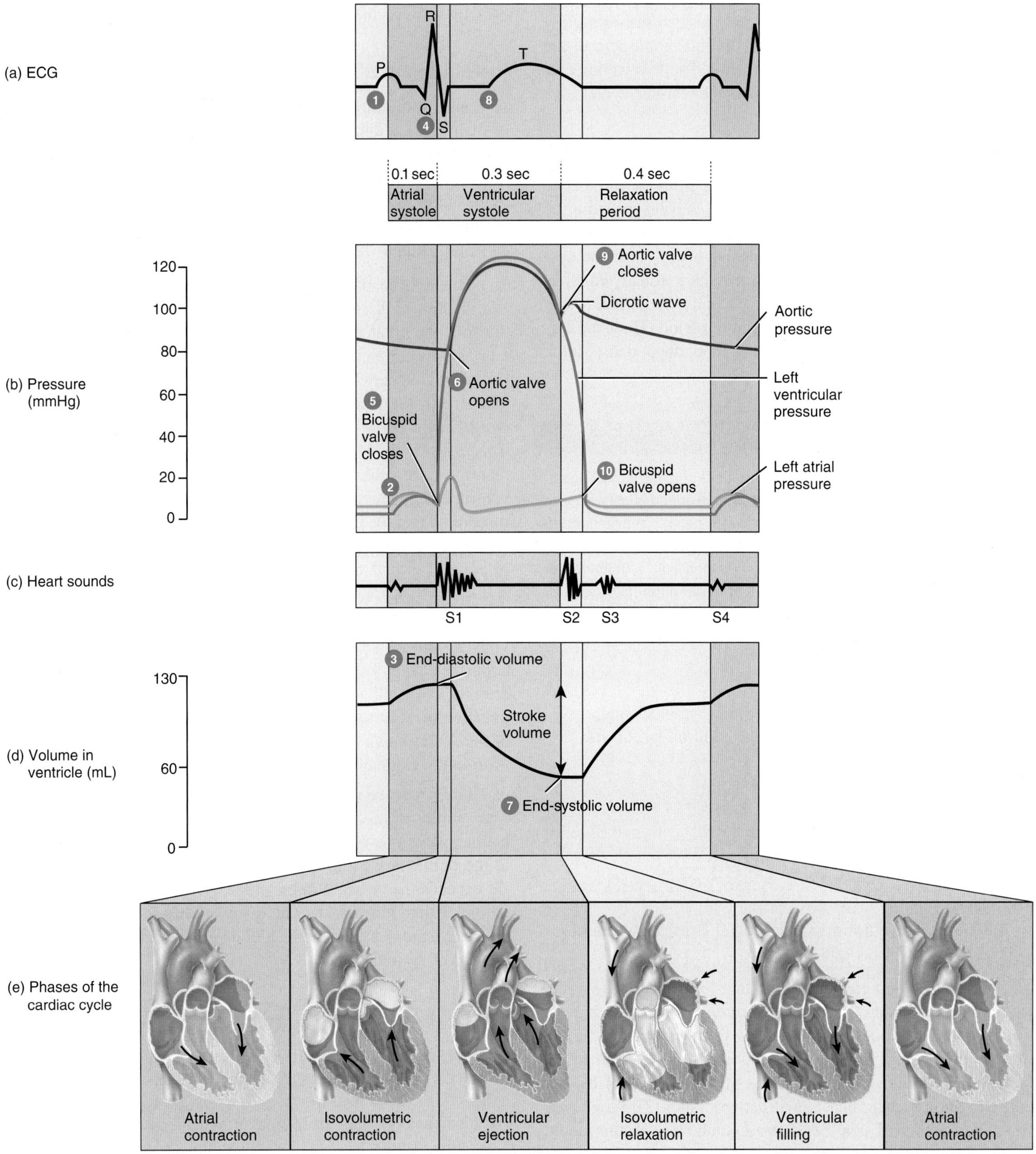

? How much blood remains in each ventricle at the end of ventricular diastole in a resting person? What is this volume called?

(10) As the ventricles continue to relax, the pressure falls quickly. When ventricular pressure drops below atrial pressure, the AV valves open, and ***ventricular filling*** begins. The major part of ventricular filling occurs just after the AV valves open. Blood that has been flowing into and building up in the atria during ventricular systole then rushes rapidly into the ventricles. At the end of the relaxation period, the ventricles are about three-quarters full. The P wave appears in the ECG, signaling the start of another cardiac cycle.

Heart Sounds

The sound of the heartbeat comes primarily from the turbulence in blood flow created by the closure of the valves, not from the contraction of the heart muscle. The first sound, ***lubb,*** is a long, booming sound from the AV valves closing after ventricular systole begins. The second sound, a short, sharp sound, ***dupp,*** is from the semilunar valves closing at the end of ventricular systole. There is a pause during the relaxation period. Thus, the cardiac cycle is heard as lubb, dupp, pause; lubb, dupp, pause; lubb, dupp, pause.

• CLINICAL CONNECTION | Heart Murmurs

Heart sounds provide valuable information about the mechanical operation of the heart. A **heart murmur** is an abnormal sound consisting of a clicking, rushing, or gurgling noise that is heard before, between, or after the normal heart sounds, or that may mask the normal heart sounds. Heart murmurs in children are extremely common and usually do not represent a health condition. These types of heart murmurs often subside or disappear with growth. Although some heart murmurs in adults are innocent, most often a murmur indicates a valve disorder. •

CHECKPOINT

10. Explain the events that occur during atrial systole, ventricular systole, and the relaxation period of the cardiac cycle.
11. What causes the heart sounds?

22.6 CARDIAC OUTPUT

OBJECTIVE

- Define cardiac output, explain how it is calculated, and describe how it is regulated.

The volume of blood ejected per minute from the left ventricle into the aorta is called the ***cardiac output (CO).*** (Note that the same amount of blood is also ejected from the right ventricle into the pulmonary trunk.) Cardiac output is determined by (1) ***stroke volume (SV),*** the amount of blood ejected by the left ventricle during each beat (contraction), and (2) ***heart rate (HR),*** the number of heartbeats per minute. In a resting adult, stroke volume averages 70 mL, and heart rate is about 75 beats per minute. Thus the average cardiac output in a resting adult is

$$\begin{aligned} \text{Cardiac output} &= \text{stroke volume} \times \text{heart rate} \\ &= 70 \text{ mL/beat} \times 75 \text{ beats/min} \\ &= 5250 \text{ mL/min or } 5.25 \text{ liters/min} \end{aligned}$$

Factors that increase stroke volume or heart rate, such as exercise, increase cardiac output.

Regulation of Stroke Volume

Although some blood is always left in the ventricles at the end of their contraction, a healthy heart pumps out the blood that has entered its chambers during the previous diastole.

The more blood that returns to the heart during diastole, the more blood that is ejected during the next systole. Three factors regulate stroke volume and ensure that the left and right ventricles pump equal volumes of blood:

1. The degree of stretch in the heart before it contracts. Within limits, the more the heart is stretched as it fills during diastole, the greater the force of contraction during systole, a relationship known as the ***Frank–Starling law of the heart.*** The situation is somewhat like stretching a rubber band: The more you stretch the heart, the more forcefully it contracts. In other words, within physiological limits, the heart pumps all the blood it receives. If the left side of the heart pumps a little more blood than the right side, a larger volume of blood returns to the right ventricle. On the next beat the right ventricle contracts more forcefully, and the two sides are again in balance.

2. The forcefulness of contraction of ventricular muscle fibers. Even at a constant degree of stretch, the heart can contract more or less forcefully when certain substances are present. Stimulation of the sympathetic division of the autonomic nervous system (ANS), hormones such as epinephrine and norepinephrine, increased Ca^{2+} level in the intracellular fluid, and the drug digitalis all increase the force of contraction of cardiac muscle fibers. In contrast, inhibition of the sympathetic division of the ANS, anoxia, acidosis, some anesthetics, and increased K^+ level in the extracellular fluid decrease contraction force.

3. The pressure required to eject blood from the ventricles. The semilunar valves open, and ejection of blood from the heart begins, when pressure in the right ventricle exceeds the pressure in the pulmonary trunk and when the pressure in the left ventricle exceeds the pressure in the aorta. When the required pressure is higher than normal, the valves open later than normal, stroke volume decreases, and more blood remains in the ventricles at the end of systole.

• CLINICAL CONNECTION | Congestive Heart Failure

In **congestive heart failure (CHF),** the heart is a failing pump. It pumps blood less and less effectively, leaving more blood in the ventricles at the end of each cycle. The result is a positive feedback cycle: Less-effective pumping leads to even lower pumping capability. Often, one side of the heart starts to fail before the other. If the left ventricle fails first, it can't pump out all the blood it receives, and blood backs up in the lungs. The result is *pulmonary edema,* fluid accumulation in the lungs that can lead to suffocation. If the right ventricle fails first, blood backs up in the systemic blood vessels. In this case, the resulting *peripheral edema* is usually most noticeable as swelling in the feet and ankles. Common causes of CHF are coronary artery disease (described next), long-term high blood pressure, myocardial infarctions, alcoholism, and valve disorders. •

Coronary Artery Disease

***Coronary artery disease* (*CAD*)** is a serious medical problem that affects about 7 million people and causes nearly 750,000 deaths in the United States each year. CAD is defined as the effects of the accumulation of atherosclerotic plaques (described shortly) in coronary arteries that lead to a reduction in blood flow to the myocardium. Some individuals have no signs or symptoms, others experience *angina pectoris* (chest pain), and still others suffer a heart attack.

People who possess combinations of certain risk factors are more likely to develop CAD. ***Risk factors*** include smoking, high blood pressure, diabetes, high cholesterol levels, obesity, "type A" personality, sedentary lifestyle, a family history of CAD, age, and gender. Smoking is undoubtedly the number-one risk factor in all CAD-associated diseases, roughly doubling the risk of morbidity and mortality.

A number of other risk factors (all modifiable) have also been identified as significant predictors of CAD. ***C-reactive proteins* (*CRPs*)** are proteins produced by the liver or present in blood in an inactive form that are converted to an active form during inflammation. CRPs may play a direct role in the development of atherosclerosis by promoting the uptake of LDLs by macrophages. ***Lipoprotein(a)*** is an LDL-like particle that binds to endothelial cells, macrophages, and blood platelets, may promote the proliferation of smooth muscle fibers, and inhibits the breakdown of blood clots. ***Fibrinogen*** is a glycoprotein involved in blood clotting that may help regulate cellular proliferation, vasoconstriction, and platelet aggregation. ***Homocysteine*** is an amino acid that may induce blood vessel damage by promoting platelet aggregation and smooth muscle fiber proliferation.

Atherosclerosis (ath′-er-ō-skler-Ō-sis) is a progressive disease characterized by the formation in the walls of large- and medium-sized arteries of lesions called ***atherosclerotic plaques*** (Figure 22.10). Inflammation, a defensive response of the body to tissue damage, plays a key role in the development of atherosclerotic plaques. Because most atherosclerotic plaques expand away from the bloodstream rather than into it, blood can flow through an artery with relative ease, often for decades. There are typically no symptoms until the vessel is at least 85% occluded.

TREATMENT OF CAD Treatment options for CAD include ***drugs*** (antihypertensives, nitroglycerine, beta blockers, cholesterol-lowering drugs, and clot-dissolving agents) and various surgical and nonsurgical procedures designed to increase the blood supply to the heart.

***Coronary artery bypass grafting* (*CABG*)** is a surgical procedure in which a blood vessel from another part of the body is attached ("grafted") to a coronary artery to bypass an area of blockage. A piece of the grafted blood vessel is sutured between the aorta and the unblocked portion of the coronary artery (see Figure 22.11a).

Another procedure used to treat CAD is ***percutaneous transluminal coronary angioplasty* (*PTCA*)** (*percutaneous* = through the skin; *trans-* = across; *lumen* = an opening or channel in a tube; *angio-* = blood vessel; *-plasty* = to mold or to shape). In this procedure, a balloon catheter is inserted into an artery of an arm or thigh and gently guided into a coronary artery (see Figure 22.11b). While dye is released, angiograms (x-rays of blood vessels) are taken to locate the plaques. Next, the catheter is advanced to the point of obstruction, and a balloon-like device is

Figure 22.10 Photomicrographs of transverse sections of (a) a normal artery and (b) an artery partially obstructed by an atherosclerotic plaque.

⑥ Inflammation plays a key role in the development of atherosclerotic plaques.

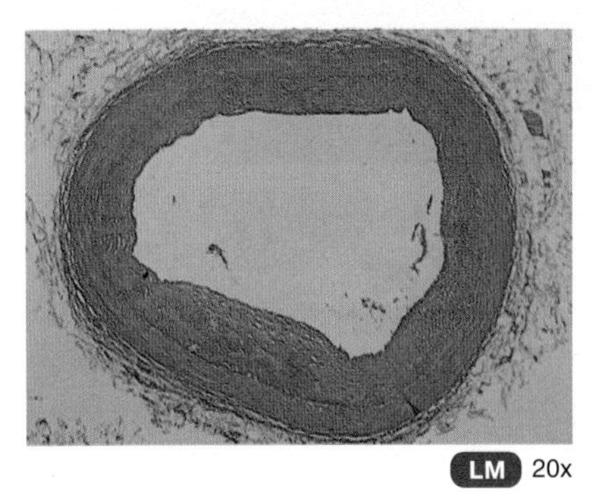

(a) Normal artery

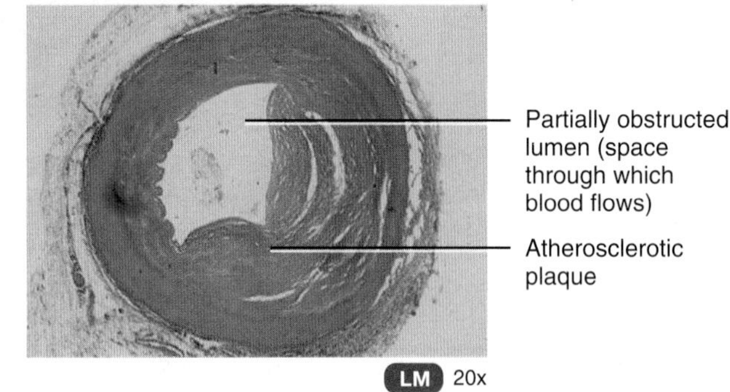

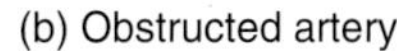

(b) Obstructed artery

? What substances are part of an atherosclerotic plaque?

Figure 22.11 Procedures for reestablishing blood flow in occluded coronary arteries.

Treatment options for CAD include drugs and various surgical procedures.

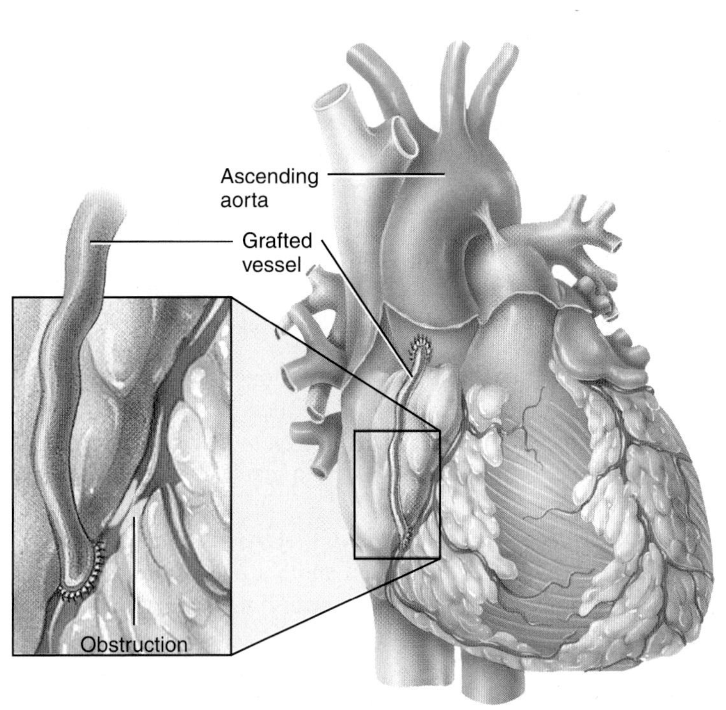

(a) Coronary artery bypass grafting (CABG)

Balloon
Atherosclerotic plaque
Narrowed lumen of artery
Coronary artery

Balloon catheter with uninflated balloon is threaded to obstructed area in artery

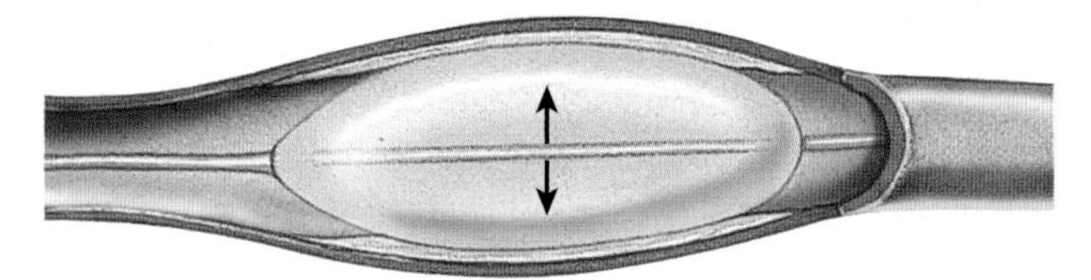

When balloon is inflated, it stretches arterial wall and squashes atherosclerotic plaque

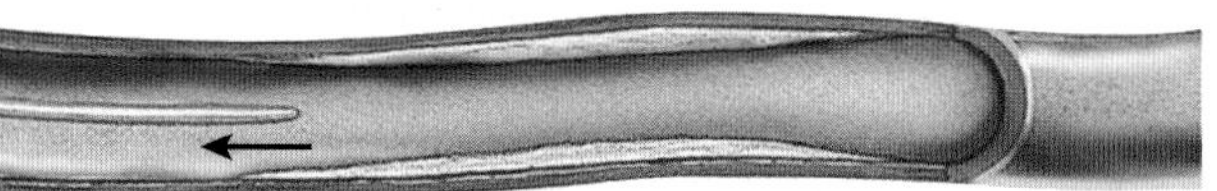

After lumen is widened, balloon is deflated and catheter is withdrawn

(b) Percutaneous transluminal coronary angioplasty (PTCA)

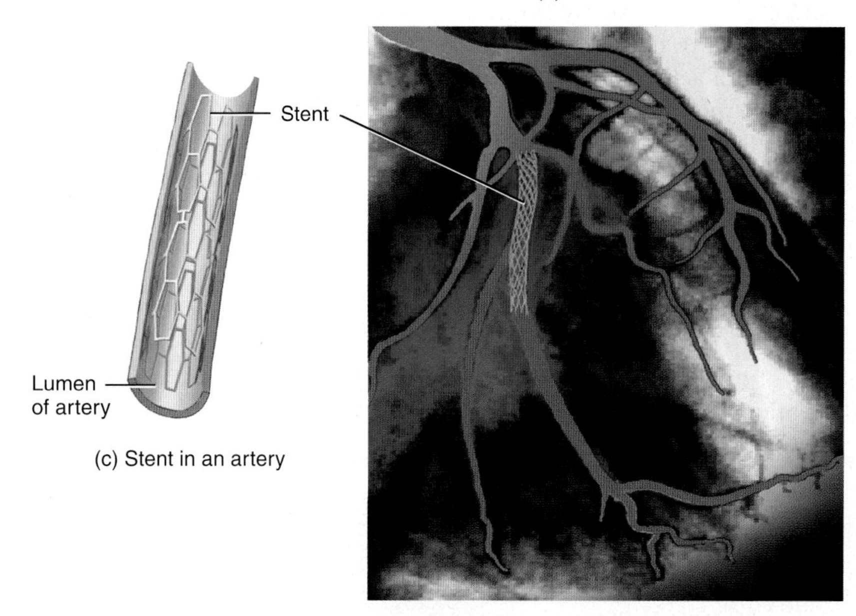

(c) Stent in an artery

(d) Angiogram showing a stent in a coronary artery

Which diagnostic procedure for CAD is used to visualize coronary blood vessels?

inflated with air to squash the plaque against the blood vessel wall. Because 30–50% of PTCA-opened arteries fail due to restenosis (renarrowing) within 6 months after the procedure is done, a stent may be inserted via a catheter. A ***stent*** is a metallic, fine wire tube that is permanently placed in an artery to keep the artery patent (open), permitting blood to circulate (Figure 22.11c,d).

Regulation of Heart Rate

Adjustments to the heart rate are important in the short-term control of cardiac output and blood pressure. If left to itself, the sinoatrial node would set a constant heart rate of about 100 beats/min. However, tissues require different volumes of blood flow under different conditions. During exercise, for example, cardiac output rises to supply working tissues with increased amounts of oxygen and nutrients. The most important factors in the regulation of heart rate are the autonomic nervous system and the hormones epinephrine and norepinephrine, released by the adrenal glands.

Autonomic Regulation of Heart Rate

The nervous system regulation of the heart originates in the ***cardiovascular (CV) center*** in the medulla of the brain stem. This region of the brain stem receives input from a variety of sensory receptors and from higher brain centers, such as the limbic system and cerebral cortex. The cardiovascular center then directs appropriate output by increasing or decreasing the frequency of nerve impulses sent out to both the sympathetic and parasympathetic branches of the ANS (Figure 22.12).

Arising from the CV center are sympathetic neurons that ultimately reach the heart via ***cardiac accelerator nerves.*** They innervate the conduction system, atria, and ventricles. The norepinephrine released by cardiac accelerator nerves increases the heart rate. Also arising from the CV center are parasympathetic neurons that reach the heart via the ***vagus (X) nerves.*** These parasympathetic neurons extend to the conduction system and atria. The neurotransmitter they release—acetylcholine (ACh)—decreases the heart rate by slowing the pacemaking activity of the SA node.

Several types of sensory receptors provide input to the cardiovascular center. For example, ***baroreceptors*** (*baro-* = pressure), neurons sensitive to blood pressure changes, are strategically located in the arch of the aorta and carotid arteries (arteries in the neck that supply blood to the brain). If there is an increase in blood pressure, the baroreceptors send nerve impulses along sensory neurons that are part of the glossopharyngeal (IX) and vagus (X) nerves to the CV center (Figure 22.12). The cardiovascular center responds by putting out more nerve impulses along the parasympathetic (motor) neurons, also part of the vagus (X) nerves, and by decreasing cardiac accelerator output. The resulting decrease in heart rate lowers cardiac output and thus lowers blood pressure. If blood pressure falls, baroreceptors do not stimulate the cardiovascular center. As a result of this lack of stimulation, heart rate increases, cardiac output increases, and

Figure 22.12 Autonomic nervous system regulation of heart rate.

The cardiovascular center in the medulla oblongata controls both sympathetic and parasympathetic nerves that innervate the heart.

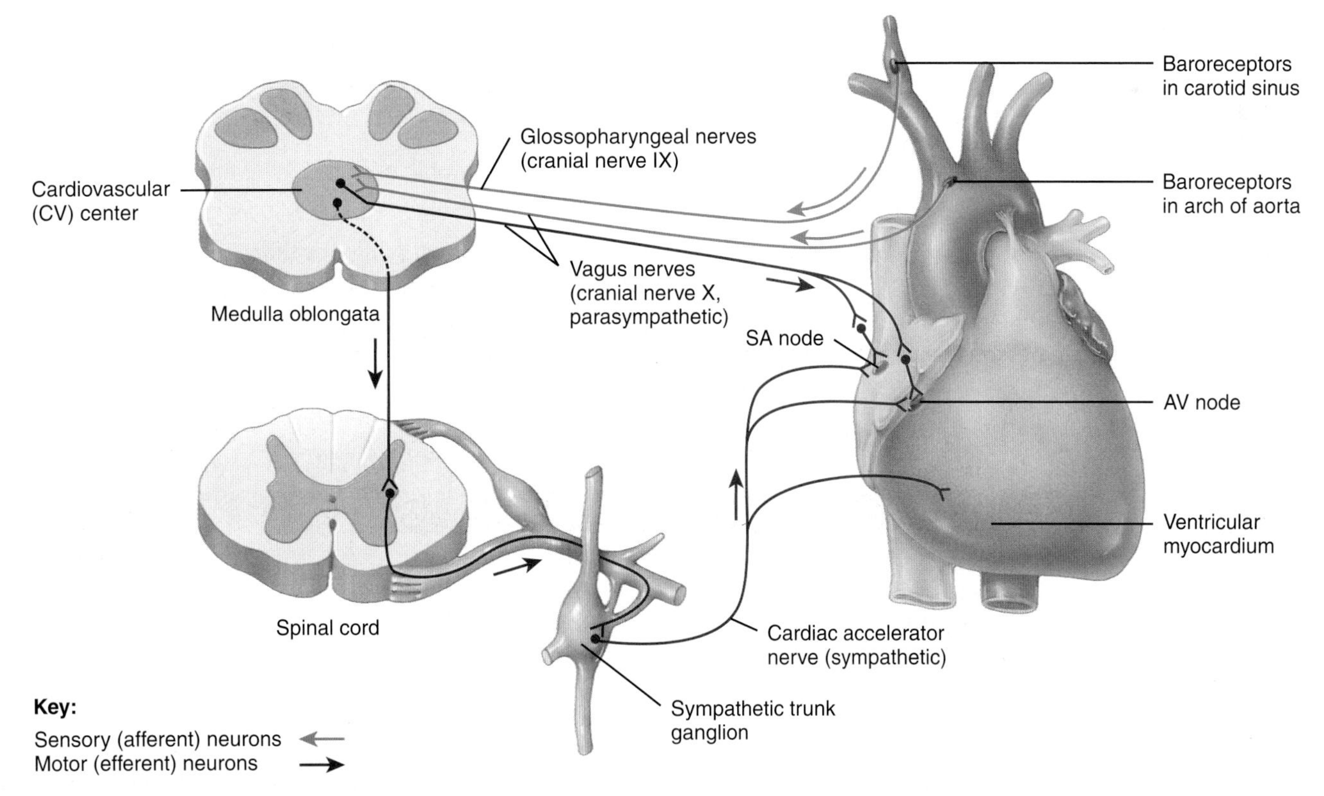

Which cranial nerves conduct impulses to the cardiovascular center from baroreceptors in the carotid sinuses and the arch of the aorta?

blood pressure increases to the normal level. ***Chemoreceptors,*** neurons sensitive to chemical changes in the blood, detect changes in blood levels of chemicals such as O_2, CO_2, and H^+. Their relationship to the cardiovascular center is considered in Chapter 23 with regard to blood pressure.

Chemical Regulation of Heart Rate

Certain chemicals influence both the basic physiology of cardiac muscle and its rate of contraction. Chemicals with major effects on the heart fall into one of two categories:

1. **Hormones.** Epinephrine and norepinephrine (from the adrenal medullae) enhance the heart's pumping effectiveness by increasing both heart rate and contraction force. Exercise, stress, and excitement cause the adrenal medullae to release more hormones. Thyroid hormones also increase heart rate and force of contraction. One sign of *hyperthyroidism* (excessive levels of thyroid hormones) is *tachycardia* (elevated resting heart rate).
2. **Ions.** Elevated blood levels of K^+ or Na^+ decrease heart rate and contraction force. A moderate increase in extracellular and intracellular Ca^{2+} level increases heart rate and contraction force.

Other Factors in Heart Rate Regulation

Age, gender, physical fitness, and body temperature also influence the resting heart rate. A newborn baby is likely to have a resting heart rate over 120 beats per minute; the rate then declines throughout childhood to the adult level of 75 beats per minute. Adult females generally have slightly higher resting heart rates than adult males, although regular exercise tends to bring resting heart rate down in both sexes. As adults age, their heart rates may increase.

Increased body temperature, such as occurs during fever or strenuous exercise, increases heart rate by causing the SA node to discharge more rapidly. Decreased body temperature decreases heart rate and force of contraction. During surgical repair of certain heart abnormalities, it is helpful to slow a patient's heart rate by deliberately cooling the body.

CHECKPOINT

12. Describe how stroke volume is regulated.
13. How does the autonomic nervous system help regulate heart rate?

22.7 EXERCISE AND THE HEART

OBJECTIVE

- Explain the relationship between exercise and the heart.

Regardless of the current level, a person's cardiovascular fitness can be improved at any age with regular exercise. Some types of exercise are more effective than others for improving the health of the cardiovascular system. ***Aerobics,*** any activity that works large body muscles for at least 20 minutes, elevates cardiac output and accelerates metabolic rate. Three to five such sessions a week are usually recommended for improving the health of the cardiovascular system. Brisk walking, running, bicycling, cross-country skiing, and swimming are examples of aerobic activities.

Sustained exercise increases the oxygen demand of the muscles. Whether the demand is met depends mainly on the adequacy of cardiac output and proper functioning of the respiratory system. After several weeks of training, a healthy person increases maximal cardiac output, thereby increasing the maximal rate of oxygen delivery to the tissues. Oxygen delivery also rises because skeletal muscles develop more capillary networks in response to long-term training.

During strenuous activity, a well-trained athlete can achieve a cardiac output double that of a sedentary person, in part because training causes hypertrophy (enlargement) of the heart. Even though the heart of a well-trained athlete is larger, ***resting cardiac output*** is about the same as in a healthy untrained person, because stroke volume is increased while heart rate is decreased. The resting heart rate of a trained athlete often is only 40–60 beats per minute (***resting bradycardia***). Regular exercise also helps to reduce blood pressure, anxiety, and depression; control weight; and increase the body's ability to dissolve blood clots by increasing fibrinolytic activity.

CHECKPOINT

14. What is aerobic exercise? Why are aerobic exercises beneficial?

The heart is the blood pump for the cardiovascular system, but it is the blood vessels that distribute blood to all parts of the body and collect blood from them. In the next chapter we will see how blood vessels accomplish this.

KEY MEDICAL TERMS ASSOCIATED WITH THE HEART

Angiocardiography (an-jē-ō-kar-dē-OG-ra-fē; *angio-* = vessel; *-cardio-* = heart) X-ray examination of the heart and great blood vessels after injection of a radiopaque dye into the bloodstream.

Cardiac arrest (KAR-dē-ak a-REST) A clinical term meaning cessation of an effective heartbeat. The heart may be completely stopped or may be in ventricular fibrillation.

Cardiac catheterization (kath-e-ter-i-ZĀ-shun) Procedure that is used to visualize the heart's coronary arteries, chambers, valves, and great vessels. It may also be used to measure pressure in the heart and blood vessels; to assess cardiac output; and to measure the flow of blood through the heart and blood vessels, the oxygen content of blood, and the status of the heart valves and conduction system. The basic procedure involves inserting a catheter into a peripheral vein (for right heart catheterization) or artery (for left heart catheterization) and guiding it under fluoroscopy (x-ray observation).

Cardiac rehabilitation (rē-ha-bil-i-TĀ-shun) A supervised program of progressive exercise, psychological support, education, and training to enable a patient to resume normal activities following a myocardial infarction.

Cardiomegaly (kar-dē-ō-MEG-a-lē; *mega-* = large) Heart enlargement.

Cardiopulmonary resuscitation (kar-dē-ō-PUL-mō-ner-ē re-sus-i-TĀ-shun) (***CPR***) The artificial establishment of normal or near-normal respiration and circulation. The ***ABCs*** of cardiopulmonary resuscitation are ***Airway, Breathing,*** and ***Circulation,*** meaning the rescuer must establish an airway, provide artificial ventilation if breathing has stopped, and reestablish circulation if there is inadequate cardiac action.

Cor pulmonale (***CP***) (kor pul-mōn-AL-ē; *cor* = heart; *pulmon-* = lung) Right ventricular hypertrophy caused by hypertension (high blood pressure) in the pulmonary circulation.

Palpitation (pal-pi-TĀ-shun) A fluttering of the heart or abnormal rate or rhythm of the heart.

Paroxysmal tachycardia (par-ok-SIZ-mal tak-e-KAR-dē-a) A period of rapid heartbeats that begins and ends suddenly.

Rheumatic fever (roo-MAT-ik) An acute systemic inflammatory disease that usually occurs after a streptococcal infection of the throat. The bacteria trigger an immune response in which antibodies that are produced to destroy the bacteria attack and inflame the connective tissues in joints, heart valves, and other organs. Even though rheumatic fever may weaken the entire heart wall, most often it damages the bicuspid (mitral) and aortic valves.

Sudden cardiac death The unexpected cessation of circulation and breathing due to an underlying heart disease such as ischemia, myocardial infarction, or a disturbance in cardiac rhythm.

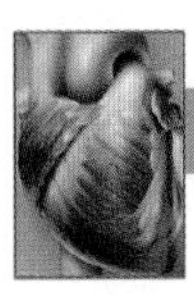

STUDY OUTLINE

Structure and Organization of the Heart (Section 22.1)

1. The heart is situated between the lungs, with about two-thirds of its mass to the left of the midline.
2. The pericardium consists of an outer fibrous layer and an inner serous pericardium.
3. The serous pericardium is composed of a parietal layer and a visceral layer.
4. Between the parietal and visceral layers of the serous pericardium is the pericardial cavity, a space filled with pericardial fluid that reduces friction between the two membranes.
5. The wall of the heart has three layers: epicardium, myocardium, and endocardium.
6. The chambers include two upper atria and two lower ventricles.
7. The blood flows through the heart from the superior and inferior venae cavae and the coronary sinus to the right atrium, through the tricuspid valve to the right ventricle, and through the pulmonary trunk to the lungs.
8. From the lungs, blood flows through the pulmonary veins into the left atrium, through the bicuspid valve to the left ventricle, and out through the aorta.
9. Four valves prevent the backflow of blood in the heart.
10. Atrioventricular (AV) valves, between the atria and their ventricles, are the tricuspid valve on the right side of the heart and the bicuspid (mitral) valve on the left.
11. The atrioventricular valves, chordae tendineae, and their papillary muscles stop blood from flowing back into the atria.
12. Each of the two arteries that leave the heart has a semilunar valve.

Blood Flow and Blood Supply of the Heart (Section 22.2)

1. Blood flows through the heart from areas of higher pressure to areas of lower pressure.
2. The pressure is related to the size and volume of a chamber.
3. The movement of blood through the heart is controlled by the opening and closing of the valves and the contraction and relaxation of the myocardium.
4. Coronary circulation delivers oxygenated blood to the myocardium and removes carbon dioxide from it.
5. Deoxygenated blood returns to the right atrium via the coronary sinus.
6. Malfunctions of this system can result in angina pectoris or myocardial infarction (MI).

Conduction System of the Heart (Section 22.3)

1. The conduction system consists of specialized cardiac muscle tissue that generates and distributes action potentials.
2. Components of this system are the sinoatrial (SA) node (pacemaker), atrioventricular (AV) node, atrioventricular (AV) bundle (bundle of His), bundle branches, and Purkinje fibers.

Electrocardiogram (Section 22.4)

1. The record of electrical changes during each cardiac cycle is referred to as an electrocardiogram (ECG).
2. A normal ECG consists of a P wave (depolarization of atria), QRS complex (onset of ventricular depolarization), and T wave (ventricular repolarization).
3. The ECG is used to diagnose abnormal cardiac rhythms and conduction patterns.

The Cardiac Cycle (Section 22.5)

1. A cardiac cycle consists of systole (contraction) and diastole (relaxation) of the chambers of the heart.
2. The phases of the cardiac cycle are (a) the relaxation period, (b) atrial systole, and (c) ventricular systole.
3. A complete cardiac cycle takes 0.8 sec at an average heartbeat of 75 beats per minute.
4. The first heart sound (lubb) represents the closing of the atrioventricular valves. The second sound (dupp) represents the closing of semilunar valves.

Cardiac Output (Section 22.6)

1. Cardiac output (CO) is the amount of blood ejected by the left ventricle into the aorta each minute: CO = stroke volume × beats per minute.
2. Stroke volume (SV) is the amount of blood ejected by a ventricle during ventricular systole. It is related to stretch on the heart before it contracts, forcefulness of contraction of ventricular muscle fibers, and the amount of pressure required to eject blood from the ventricles.
3. Nervous control of the cardiovascular system originates in the cardiovascular center in the medulla oblongata.
4. Sympathetic impulses increase heart rate and force of contraction; parasympathetic impulses decrease heart rate.

5. Heart rate is affected by hormones (epinephrine, norepinephrine, thyroid hormones), ions (Na^+, K^+, Ca^{2+}), age, gender, physical fitness, and body temperature.

Exercise and the Heart (Section 22.7)

1. Sustained exercise increases oxygen demand on muscles.
2. Among the benefits of aerobic exercise are increased maximal cardiac output, decreased blood pressure, weight control, and increased ability to dissolve clots.

SELF-QUIZ QUESTIONS

1. Match the following:

a. valve between the left atrium and left ventricle
b. valve between the right atrium and right ventricle
c. chamber that pumps blood to the lungs
d. chamber that pumps blood into the aorta
e. chamber that receives oxygenated blood from lungs
f. chamber that receives deoxygenated blood from body
g. valve between the left ventricle and aorta
h. valve between the right ventricle and pulmonary trunk

A. aortic valve
B. right atrium
C. left atrium
D. bicuspid (mitral) valve
E. pulmonary valve
F. right ventricle
G. left ventricle
H. tricuspid

2. Which of the following statements describes the pericardium?
a. It is a layer of nervous tissue.
b. It lines the inside of the myocardium.
c. It is continuous with the epithelial lining of the large blood vessels.
d. It is responsible for the contraction of the heart.
e. It is a membrane that surrounds and protects the heart.

3. Which blood vessel primarily delivers deoxygenated blood from parts of the body above the heart?
a. pulmonary vein **b.** thoracic aorta
c. pulmonary artery **d.** inferior vena cava
e. superior vena cava

4. An embolus originating in the coronary sinus would first enter the
a. right atrium **b.** pulmonary veins
c. left atrium **d.** right ventricle
e. aorta.

5. The chordae tendineae and papillary muscles of the heart
a. are responsible for connecting cardiac muscle fibers for the spread of action potentials
b. can develop self-excitability and stimulate contraction
c. help prevent the atrioventricular valves from protruding into the atria when the ventricles contract
d. help anchor and protect the heart
e. form the cusps (flaps) of the heart valves.

6. Which chamber of the heart has the thickest layer of myocardium?
a. right ventricle **b.** right atrium
c. left ventricle **d.** left atrium
e. coronary sinus

7. The normal pacemaker of the heart is the
a. sinoatrial (SA) node **b.** atrioventricular (AV) node
c. Purkinje fibers **d.** atrioventricular (AV) bundle
e. right bundle branch.

8. In normal heart action,
a. the right atrium and ventricle contract, followed by the contraction of the left atrium and ventricle
b. the order of contraction is right atrium, then right ventricle, then left atrium, then left ventricle
c. the two atria contract together, and then the two ventricles contract together
d. the right atrium and left ventricle contract, followed by the contraction of the left atrium and right ventricle
e. all four chambers of the heart contract and then relax simultaneously.

9. Heart sounds are produced by
a. contraction of the myocardium
b. closure of the heart valves
c. the flow of blood in the coronary arteries
d. the flow of blood in the ventricles
e. the transmission of action potentials through the conduction system.

10. Heart rate and strength of contraction are controlled by the cardiovascular center, which is located in the
a. cerebrum **b.** pons
c. right atrium **d.** medulla oblongata
e. atrioventricular node.

11. The portion of the ECG that corresponds to atrial depolarization is the
a. R peak
b. space between the T wave and P wave
c. T wave
d. P wave
e. QRS complex.

12. The opening of the semilunar valves is due to the pressure in the
a. ventricles exceeding the pressure in the aorta and pulmonary trunk
b. ventricles exceeding the pressure in the atria
c. atria exceeding the pressure in the ventricles
d. atria exceeding the pressure in the aorta and pulmonary trunk
e. aorta and pulmonary trunk exceeding the pressure in the ventricles.

13. On the anterior surface of each atrium is a wrinkled pouchlike structure called the
a. anterior interventricular sulcus
b. coronary sulcus
c. auricle
d. interatrial septum
e. posterior interventricular sulcus.

14. The Frank–Starling law of the heart
a. is important in maintaining equal blood output from both ventricles
b. is used in reference to the force of contraction of the atria
c. results in a decreased heart rate
d. causes blood to accumulate in the lungs
e. is related to the stretching of the cardiac muscle cells in the atria.

15. Which of the following sequences best represents the pathway of an action potential through the heart's conduction system?
1. sinoatrial (SA) node
2. Purkinje fibers
3. atrioventricular (AV) bundle

4. atrioventricular (AV) node
5. right and left bundle branches
   - **a.** 1, 4, 3, 2, 5
   - **b.** 4, 1, 3, 5, 2
   - **c.** 3, 4, 1, 2, 5
   - **d.** 1, 4, 3, 5, 2
   - **e.** 2, 5, 3, 4, 1

16. Which of the following is NOT true concerning ventricular filling during the cardiac cycle?
- **a.** The atrioventricular (AV) valves are open.
- **b.** The ventricles fill to 75% of their capacity before the atria contract.
- **c.** The remaining 25% of the ventricular blood is forced into the ventricles when the atria contract.
- **d.** The semilunar valves are open.
- **e.** Ventricular filling begins when the ventricular pressure drops below the atrial pressure, causing the AV valves to open.

17. Cardiac output
- **a.** equals stroke volume (SV) × blood pressure (BP)
- **b.** equals stroke volume (SV) × heart rate (HR)
- **c.** is calculated using the formula for the Frank–Starling law of the heart
- **d.** is about 70 mL in the average adult male
- **e.** equals blood pressure (BP) × heart rate (HR)

18. Most heart problems are due to
- **a.** old age
- **b.** leakages at the valves
- **c.** problems in the coronary circulation
- **d.** the failure of the conduction system
- **e.** infections in the heart coverings

19. Using the situations that follow, indicate if the heart rate would speed up (A) or slow down (B).
- ____ **a.** sympathetic stimulation of the sinoatrial (SA) node
- ____ **b.** decrease in blood pressure
- ____ **c.** fever
- ____ **d.** parasympathetic stimulation of the heart's conduction system
- ____ **e.** release of epinephrine
- ____ **f.** elevated K^+ level
- ____ **g.** release of acetylcholine
- ____ **h.** strenuous exercise
- ____ **i.** stimulation by the vagus (X) nerve
- ____ **j.** fear, anger, stress
- ____ **k.** cooling the body
- ____ **l.** hypoxia
- ____ **m.** excessive thyroid hormones

20. Match the following
- ____ **a.** may cause a heart murmur
- ____ **b.** heart compression
- ____ **c.** inflammation of heart covering
- ____ **d.** heart chamber contraction
- ____ **e.** chest pain from ischemia
- ____ **f.** heart attack
- ____ **g.** heart chamber

- **A.** pericarditis
- **B.** mitral valve prolapse
- **C.** myocardial infarction
- **D.** angina pectoris
- **E.** diastole
- **F.** systole
- **G.** cardiac tamponade

CRITICAL THINKING QUESTIONS

1. Your uncle had an artificial pacemaker inserted after his last bout with heart trouble. What is the function of a pacemaker? For which heart structure does the pacemaker substitute?
2. Andre was strolling across a four-lane highway when a car suddenly appeared out of nowhere. As he finished sprinting across the road, he felt his heart racing. Trace the route of the signal from his brain to his heart.
3. Christopher, a member of the college's cross-country ski team, volunteered to have his heart function evaluated by the exercise physiology class. His resting pulse rate was 40 beats per minute. Assuming that he has an average cardiac output (CO), determine Christopher's stroke volume. Next, Christopher rode an exercise bike until his heart rate had risen to 60 beats per minute. Assuming that his SV stayed constant, calculate Christopher's CO during this moderate exercise.
4. Rosa's Great-aunt Frieda likes to say that she has complaining feet and a mumbling heart. Aunt Frieda's physician uses the terms "edema" and murmur." Explain Aunt Frieda's medical condition.

? ANSWERS TO FIGURE QUESTIONS

22.1 The mediastinum is the mass of tissue that extends from the sternum to the vertebral column between the lungs.

22.2 The intercalated discs hold the cardiac muscle fibers together and enable action potentials to propagate from one muscle fiber to another.

22.3 The coronary sulcus forms a boundary between the atria and ventricles.

22.4 Heart valves prevent the backflow of blood.

22.5 Numbers 2 (right ventricle) through 6 (left atrium) depict the pulmonary circulation, whereas numbers 7 (left ventricle) through 1 (right atrium) depict the systemic circulation.

22.6 The circumflex artery delivers oxygenated blood to the left atrium and left ventricle.

22.7 The only electrical connection between the atria and the ventricles is the atrioventricular (AV) bundle.

22.8 Atrial depolarization causes contraction of the atria.

22.9 The amount of blood in each ventricle at the end of ventricular diastole—called the end-diastolic volume—is about 130 mL in a resting person.

22.10 Fatty substances, cholesterol, and smooth muscle fibers make up atherosclerotic plaques.

22.11 Coronary angiography is used to visualize many blood vessels.

22.12 Impulses to the cardiovascular center pass from baroreceptors in the carotid sinuses via the glossopharyngeal (IX) nerves and from baroreceptors in the arch of the aorta via the vagus (X) nerves.

23 The Cardiovascular System: Blood Vessels and Circulation

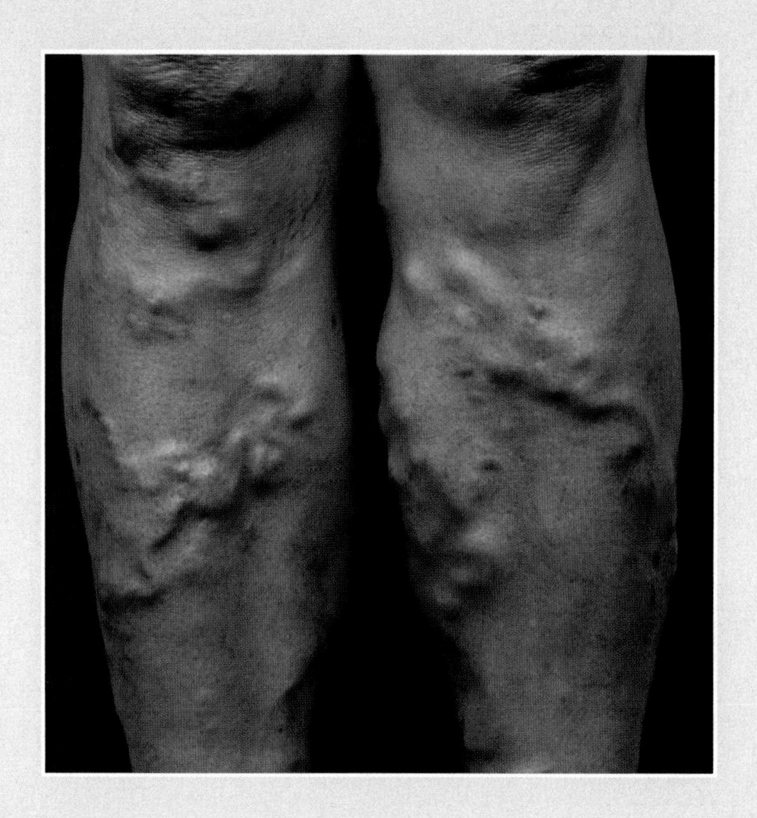

Physical activity helps to protect the cardiovascular system in many ways. It improves blood cholesterol levels and blood sugar regulation. People who exercise regularly have lower rates of inflammation, and less inflammation suggests lower risk of artery disease. The cardiovascular system contributes to the homeostasis of other body systems by transporting and distributing blood throughout the body to deliver materials such as oxygen, nutrients, and hormones and to carry away wastes. This transport is accomplished by blood vessels, which form closed circulatory routes for blood to travel from the heart to body organs and back again.

Arteries progressively branch and form microscopic blood vessels called capillaries that form an efficient exchange surface for oxygen, nutrients, and the returning waste. Capillaries then slowly enlarge to form veins, which are under less pressure since they are farther from the heart. Veins rely mostly on the action of actively contracting muscles in the body to push blood up the limbs against gravity. In massage, the direction of most Swedish strokes is toward the heart to further assist skeletal muscles in their "milking" action. When you stand for a long time, and the muscles of the legs are not contracting and assisting the veins, the veins may become weak and eventually may develop varicose veins.

CONTENTS AT A GLANCE

23.1 BLOOD VESSEL STRUCTURE AND FUNCTION

OBJECTIVES

- Compare the structure and function of the different types of blood vessels.
- Describe how substances enter and leave the blood in capillaries.
- Explain how venous blood returns to the heart.

There are five major types of blood vessels: arteries, arterioles, capillaries, venules, and veins. ***Arteries*** (AR-ter-ēz) carry blood away from the heart to body tissues. Two large arteries—the aorta and the pulmonary trunk—emerge from the heart and branch out into medium-sized arteries that serve various regions of the body. These medium-sized arteries then divide into small arteries, which, in turn, divide into still smaller arteries called ***arterioles*** (ar-TER-ē-ōls). Arterioles within a tissue or organ branch into numerous microscopic vessels called ***capillaries*** (KAP-i-lar′-ēz). Groups of capillaries within a tissue reunite to form small veins called ***venules*** (VEN-ūls). These, in turn, merge to form progressively larger vessels called veins. ***Veins*** (VĀNZ) are the blood vessels that convey blood from the tissues back to the heart.

At any one time, systemic veins and venules contain about 64% of the total volume of blood in the system, systemic arteries and arterioles about 13%, systemic capillaries about 7%, pulmonary blood vessels about 9%, and the heart chambers about 7%. Because veins contain so much of the blood, certain veins function as ***blood reservoirs.*** The main blood reservoirs are the veins of the abdominal organs (especially the liver and spleen) and the skin. Blood can be diverted quickly from these reservoirs to other parts of the body, for example, to skeletal muscles to support increased muscular activity.

Arteries and Arterioles

The walls of arteries have three layers of tissue surrounding a hollow space, the ***lumen,*** through which the blood flows (Figure 23.1a). The inner layer is composed of ***endothelium,*** a type of simple squamous epithelium; a basement membrane; and an elastic tissue called the internal elastic lamina. The middle layer consists of smooth muscle and elastic tissue. The outer layer is composed mainly of elastic and collagen fibers.

Sympathetic fibers of the autonomic nervous system innervate vascular smooth muscle. An increase in sympathetic stimulation typically causes the smooth muscle to contract, squeezing the vessel wall and narrowing the lumen. Such a decrease in the diameter of the lumen of a blood vessel is called ***vasoconstriction.*** In contrast, when sympathetic stimulation decreases, or in the presence of certain chemicals (such as nitric oxide or lactic acid), smooth muscle fibers relax. The resulting increase in lumen diameter is called ***vasodilation.*** Additionally, when an artery or arteriole is damaged, its smooth muscle contracts, producing vascular spasm of the vessel. Such a vasospasm limits blood flow through the damaged vessel and helps reduce blood loss if the vessel is small.

The largest-diameter arteries contain a high proportion of elastic fibers in their middle layer, and their walls are relatively thin in proportion to their overall diameter. Such arteries are called ***elastic arteries.*** These arteries help propel blood onward while the ventricles are relaxing. As blood is ejected from the heart into elastic arteries, their highly elastic walls stretch, accommodating the surge of blood. Then, while the ventricles are relaxing, the elastic fibers in the artery walls recoil, which forces blood onward toward the smaller arteries. Examples include the aorta and the brachiocephalic, common carotid, subclavian, vertebral, pulmonary, and common iliac arteries. Medium-sized arteries, on the other hand, contain more smooth muscle and fewer elastic fibers than elastic arteries. Such arteries are called ***muscular arteries*** and are capable of greater vasoconstriction and vasodilation to adjust the rate of blood flow. Examples include the brachial artery (arm) and radial artery (forearm).

An ***arteriole*** (small artery) is a very small, almost microscopic, artery that delivers blood to capillaries. The smallest arterioles consist of little more than a layer of endothelium covered by a few smooth muscle fibers (see Figure 23.2a). Arterioles play a key role in regulating blood flow from arteries into capillaries. During vasoconstriction, blood flow from the arterioles to the capillaries is restricted; during vasodilation, the flow is significantly increased. A change in diameter of arterioles can also significantly alter blood pressure; vasodilation decreases blood pressure and vasoconstriction increases blood pressure.

• CLINICAL CONNECTION | Aneurysm

An **aneurysm** (AN-ū-rizm) is a thin, weakened section of the wall of an artery or a vein that bulges outward, forming a balloon-like sac. Common causes are atherosclerosis, syphilis, congenital blood vessel defects, and trauma. If untreated, the aneurysm enlarges and the blood vessel wall becomes so thin that it bursts. The result is massive hemorrhage along with shock, severe pain, stroke, or death. •

Capillaries

Capillaries (*capillar-* = hairlike) are microscopic vessels that connect arterioles to venules (Figure 23.1c). Capillaries are present near almost every body cell, and they are known as exchange vessels because they permit the exchange of nutrients and wastes between the body's cells and the blood. The number of capillaries varies with the metabolic activity of the tissue they serve. Body tissues with high metabolic requirements, such as muscles, the liver, the kidneys, and the nervous system, have extensive capillary networks. Tissues with lower metabolic requirements, such as tendons and ligaments, contain fewer capillaries. A few tissues—all covering and lining epithelia, the cornea and lens of the eye, and cartilage—lack capillaries completely.

Figure 23.1 Comparative structure of blood vessels. The relative size of the capillary in (c) is enlarged for emphasis. Note the valve in the vein.

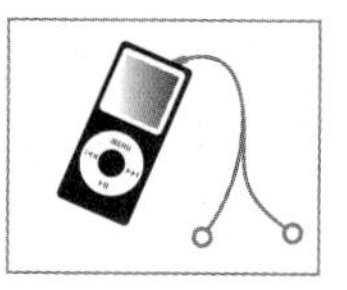

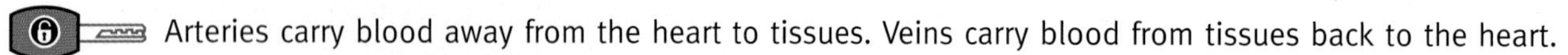

Arteries carry blood away from the heart to tissues. Veins carry blood from tissues back to the heart.

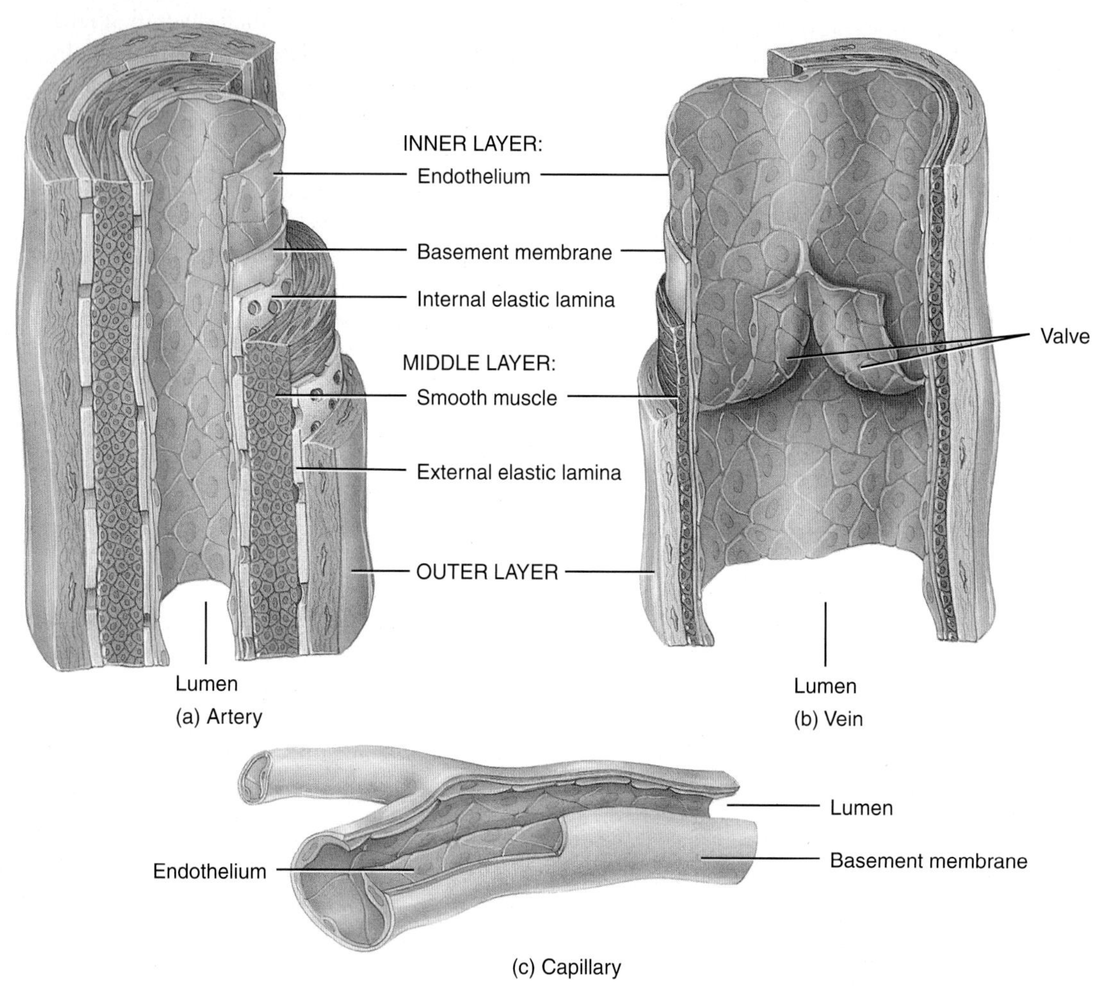

Which vessel—the femoral artery or the femoral vein—has a thicker wall? Which has a wider lumen?

Structure of Capillaries

A capillary consists of a layer of endothelium that is surrounded by basement membrane (Figure 23.1c). Because capillary walls are very thin, many substances easily pass through them to reach tissue cells from the blood or to enter the blood from interstitial fluid. The walls of all other blood vessels are too thick to permit the exchange of substances between blood and interstitial fluid. Depending on how tightly their endothelial cells are joined, different types of capillaries have varying degrees of permeability and are found in different areas of the body.

In some regions, capillaries link arterioles to venules directly. In other places, they form extensive branching networks called ***capillary beds*** (Figure 23.2). Blood flows through only a small part of a tissue's capillary network when metabolic needs are low. But when a tissue becomes active, the entire capillary network fills with blood. The flow of blood in capillaries is regulated by smooth muscle fibers in arteriole walls and by ***precapillary sphincters,*** rings of smooth muscle at the point where capillaries branch from arterioles (Figure 23.2a,b). When precapillary sphincters relax, more blood flows into the connected capillaries; when precapillary sphincters contract, less blood flows through their capillaries and flows through ***thoroughfare channels*** (Figure 23.2b).

Capillary Exchange

Because of the small diameter of capillaries, blood flows more slowly through them than through larger blood vessels. The slow flow aids the prime mission of the entire cardiovascular system: to keep blood flowing through capillaries so that ***capillary exchange***—the movement of substances into and out of capillaries—can occur.

Capillary blood pressure, the pressure of blood against the walls of capillaries, "pushes" fluid out of capillaries into interstitial fluid. An opposing pressure, termed ***blood colloid osmotic pressure,*** "pulls" fluid into capillaries. (Recall that osmotic pressure is the pressure of a fluid due to its solute concentration. The

Figure 23.2 Arteriole, capillaries, and venule. Precapillary sphincters regulate the flow of blood through capillary beds.

Arterioles regulate blood flow into capillaries, where nutrients, gases, and wastes are exchanged between blood and interstitial fluid.

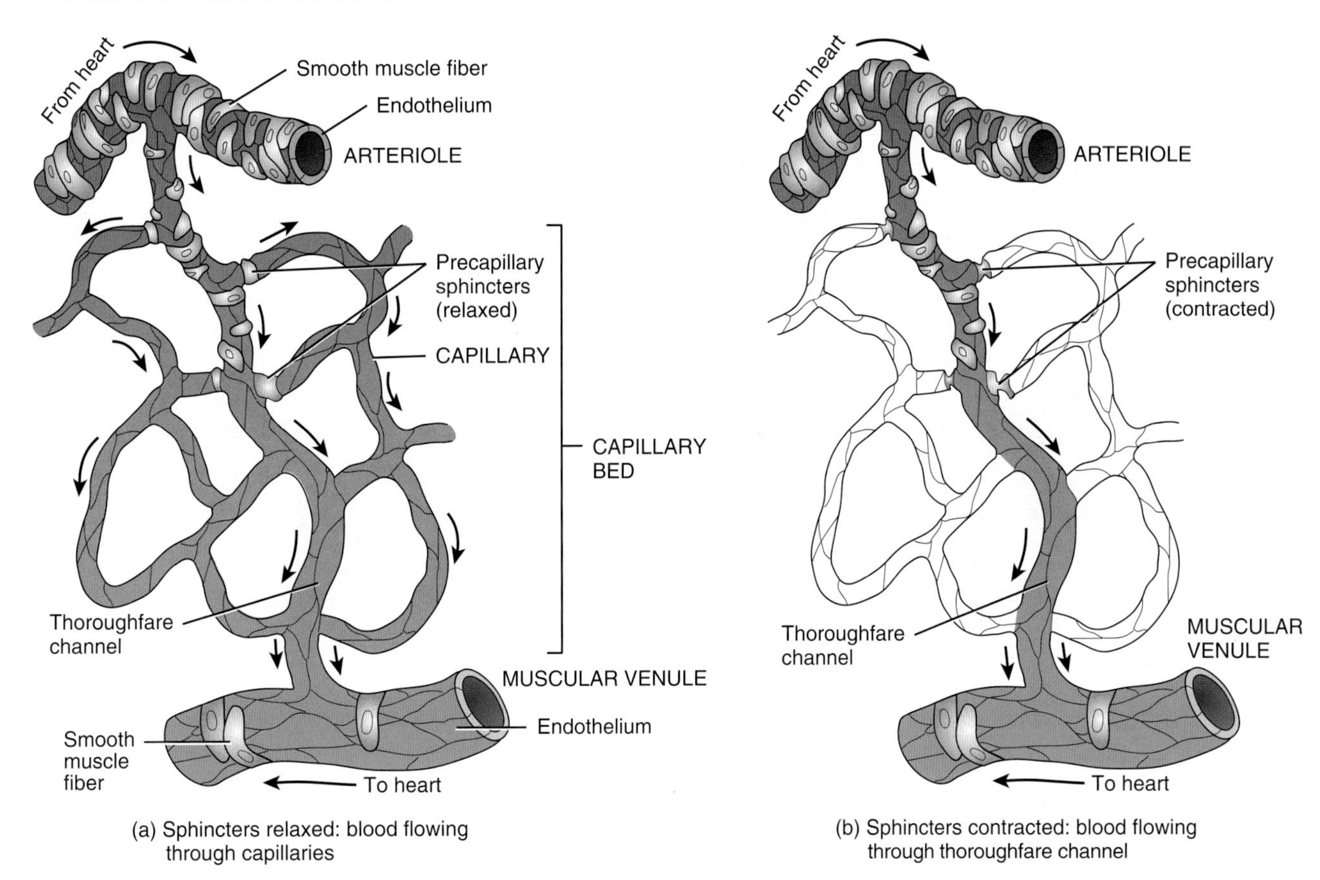

Why do metabolically active tissues have extensive capillary networks?

higher the solute concentration, the greater is the osmotic pressure.) Most solutes are present in nearly equal concentrations in blood and interstitial fluid, but the presence of proteins in plasma and their virtual absence in interstitial fluid gives blood the higher osmotic pressure. Blood colloid osmotic pressure is osmotic pressure due mainly to plasma proteins.

Capillary blood pressure is higher than blood colloid osmotic pressure for about the first half of the length of a typical capillary. Thus, water and solutes flow out of the blood capillary into the surrounding interstitial fluid, a movement called ***filtration*** (see Figure 23.3). Because capillary blood pressure decreases progressively as blood flows along a capillary, at about the capillary's midpoint, blood pressure drops below blood colloid osmotic pressure. Then, water and solutes move from interstitial fluid into the blood capillary, a process termed ***reabsorption.*** Normally, about 85% of the filtered fluid is reabsorbed. The excess filtered fluid and the few plasma proteins that do escape enter lymphatic capillaries and eventually are returned by the lymphatic system (see Chapter 24) to the cardiovascular system.

Localized changes in each capillary network can regulate vasodilation and vasoconstriction. When vasodilators are released by tissue cells, they cause dilation of nearby arterioles and relaxation of precapillary sphincters. Then, blood flow into the capillary networks increases and O_2 delivery to the tissue rises. Vasoconstrictors have the opposite effect. The ability of a tissue to automatically adjust its blood flow to match its metabolic demands is called ***autoregulation.***

Venules and Veins

When several capillaries unite, they form venules. Venules receive blood from capillaries and empty blood into veins, which return blood to the heart.

Structure of Venules and Veins

Venules (= little veins) are similar in structure to arterioles; their walls are thinner near the capillary end and thicker as they progress toward the heart. ***Veins*** are structurally similar to arteries, but their middle and inner layers are thinner (see Figure 23.1b).

Figure 23.3 Capillary exchange.

Capillary blood pressure pushes fluid out of capillaries (filtration); blood colloid osmotic pressure pulls fluid into capillaries (reabsorption).

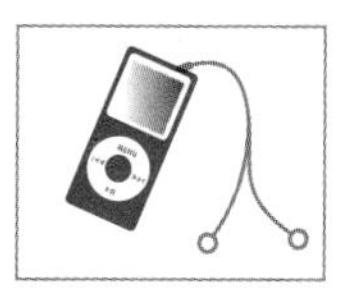

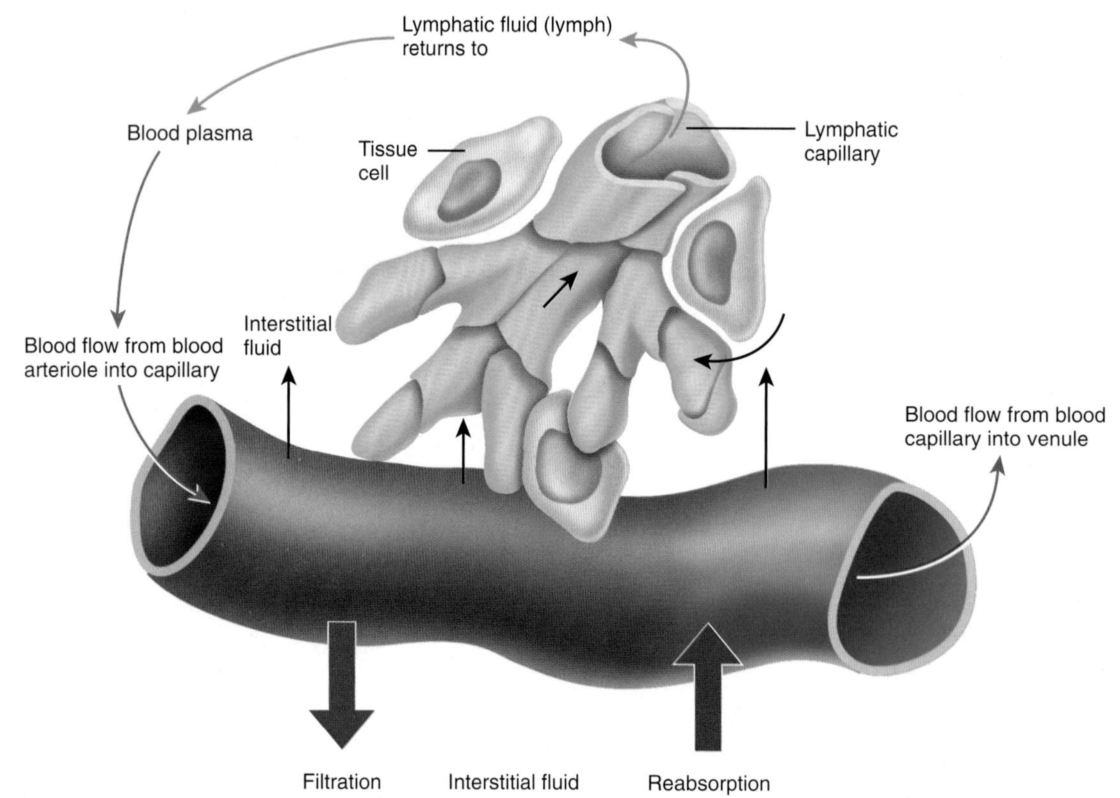

What happens to excess filtered fluid and proteins that are not reabsorbed?

The outer layer of veins is the thickest layer. The lumen of a vein is wider than that of a corresponding artery.

In some veins, the inner layer folds inward to form ***valves*** that prevent the backflow of blood. In people with weak venous valves, gravity forces blood backward through the valve. This increases venous blood pressure, which pushes the vein's wall outward. After repeated overloading, the walls lose their elasticity and become stretched and flabby, a condition called ***varicose veins.***

By the time blood leaves the capillaries and moves into veins, it has lost a great deal of pressure. This can be observed in the blood leaving a cut vessel: Blood flows from a cut vein slowly and evenly, whereas it gushes out of a cut artery in rapid spurts. When a blood sample is needed, it is usually collected from a vein because pressure is low in veins and also because many of them are close to the skin surface.

Venous Return

Venous return, the volume of blood flowing back to the heart through systemic veins, typically occurs due to pressure generated in three ways: (1) contractions of the heart, (2) the skeletal muscle pump, and (3) the respiratory pump. Massage or other manual techniques also facilitate the movement of venous blood. Blood pressure is generated by contraction of the heart's ventricles and is measured in millimeters of mercury, abbreviated mmHg. The pressure difference from venules (averaging about 16 mmHg) to the right atrium (0 mmHg), although small, normally is sufficient to cause venous return to the heart. When you stand, the pressure pushing blood up the veins in your lower limbs is barely enough to overcome the force of gravity pushing it back down.

The ***skeletal muscle pump*** operates as follows (Figure 23.4):

1. While standing at rest, both the venous valve closer to the heart and the one farther from the heart in this part of the leg are open, and blood flows upward toward the heart.
2. Contraction of leg muscles, such as when you stand on tiptoes or take a step, compresses the vein. The compression pushes blood through the valve closer to the heart, an action called *milking*. At the same time, the valve farther from the heart in the uncompressed segment of the vein closes as some blood is pushed against it. People who are immobilized through injury or disease lack these contractions of leg muscles. As a result, their venous return is slower and they may develop serious circulation problems. Again, regularly scheduled massage or other manual techniques on a patient

Figure 23.4 Action of the skeletal muscle pump in returning blood to the heart. Steps are described in the text.

Milking refers to skeletal muscle contractions that drive venous blood toward the heart.

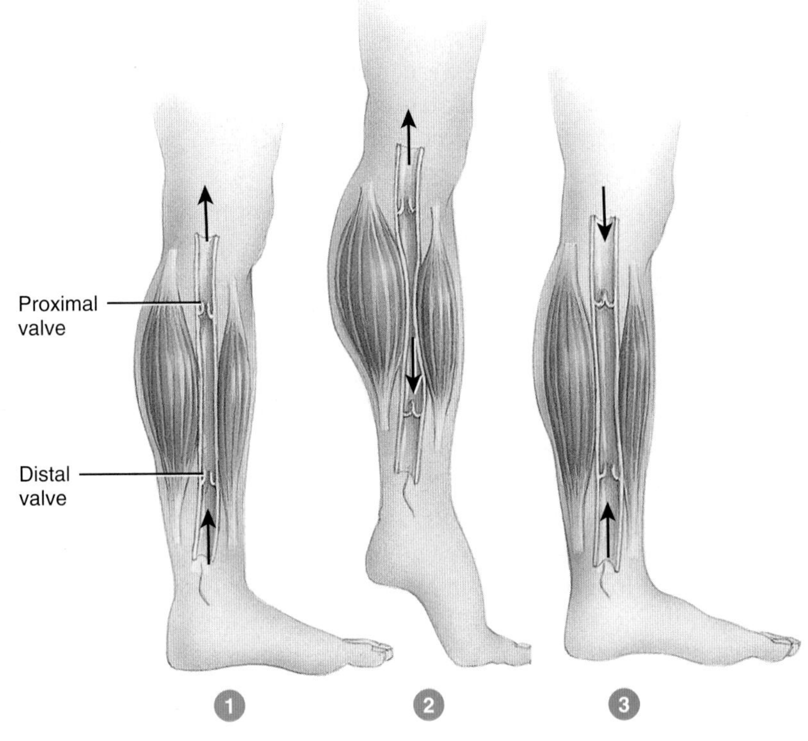

Aside from cardiac contractions, which mechanisms act as pumps to boost venous return?

or client may be sufficient to minimize possible circulation problems.

3. Just after muscle relaxation, pressure falls in the previously compressed section of the vein, which causes the valve closer to the heart to close. The valve farther from the heart now opens because blood pressure in the foot is higher than in the leg, and the vein fills with blood from the foot. Movement of any venous blood toward the heart causes the distal vessels to become filled with more highly oxygenated blood than was previously present around the tissues.

The ***respiratory pump*** is also based on alternating compression and decompression of veins. During inhalation (breathing in) the diaphragm moves downward, which causes a decrease in pressure in the thoracic cavity and an increase in pressure in the abdominal cavity. As a result, abdominal veins are compressed, and a greater volume of blood moves from the compressed abdominal veins into the decompressed thoracic veins and then into the right atrium. When the pressures reverse during exhalation (breathing out), the valves in the veins prevent backflow of blood from the thoracic veins to the abdominal veins.

• CLINICAL CONNECTION | Shock

Shock is a failure of the cardiovascular system to deliver enough O_2 and nutrients to meet cellular metabolic needs. The causes of shock are many and varied, but all are characterized by inadequate blood flow to body tissues. Common causes of shock include loss of body fluids, as occurs in hemorrhage, dehydration, burns, excessive vomiting, diarrhea, or sweating. If shock persists, cells and organs become damaged, and cells may die unless proper treatment begins quickly.

Although the symptoms of shock vary with the severity of the condition, the following are commonly observed: systolic blood pressure lower than 90 mmHg; rapid resting heart rate due to sympathetic stimulation and increased blood levels of epinephrine and norepinephrine; weak, rapid pulse due to reduced cardiac output and fast heart rate; cool, pale skin due to vasoconstriction of skin blood vessels; sweating due to sympathetic stimulation; reduced urine formation and output due to increased levels of aldosterone and antidiuretic hormone (ADH); altered mental state due to reduced oxygen supply to the brain; thirst due to loss of extracellular fluid; and nausea due to impaired circulation to digestive organs. •

CHECKPOINT

1. How do arteries, capillaries, and veins differ in function?
2. Distinguish between filtration and absorption.
3. What factors contribute to blood flow back to the heart?

23.2 BLOOD FLOW THROUGH BLOOD VESSELS

OBJECTIVES

- Define blood pressure and describe how it varies throughout the systemic circulation.
- Identify the factors that affect blood pressure and vascular resistance.
- Describe how blood pressure and blood flow are regulated.

We saw in Chapter 22 that cardiac output (CO) depends on stroke volume and heart rate. Two other factors influencing cardiac output and the proportion of blood that flows through specific circulatory routes are blood pressure and vascular resistance.

Blood Pressure

As you have just learned, blood flows from regions of higher pressure to regions of lower pressure; the greater the pressure difference, the greater is the blood flow. Contraction of the ventricles generates ***blood pressure (BP)***, the pressure exerted by blood on the walls of a blood vessel. BP is highest in the aorta and large systemic arteries, where in a resting, young adult, it rises to about 110 mmHg during *systole* (contraction) and drops to about 70 mmHg during *diastole* (relaxation). Blood pressure falls progressively as the distance from the left ventricle increases

Figure 23.5 Blood pressure changes as blood flows through the systemic circulation. The dashed line is the mean (average) pressure in the aorta, arteries, and arterioles.

 Blood pressure falls progressively as blood flows from systemic arteries through capillaries and back to the right atrium.

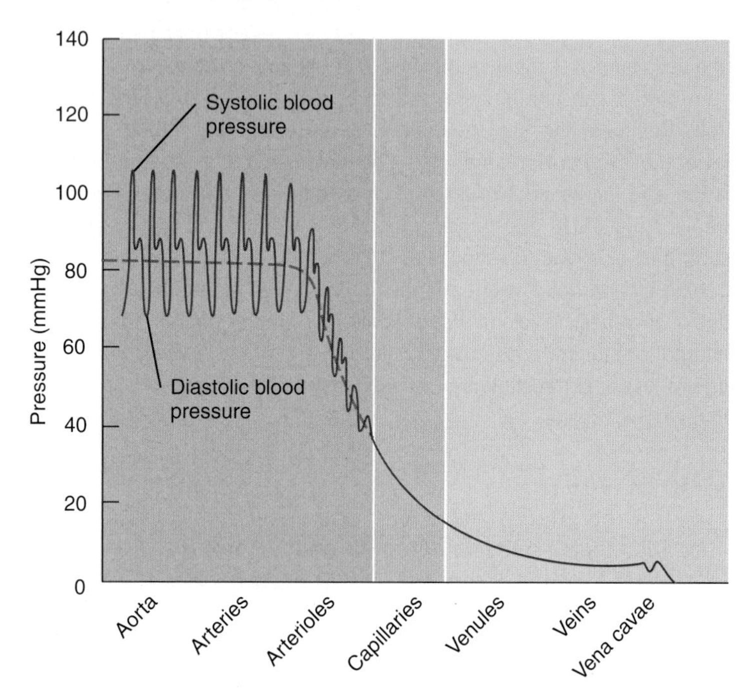

Is the mean blood pressure in the aorta closer to systolic or to diastolic pressure?

(Figure 23.5), to about 35 mmHg as blood passes into systemic capillaries. At the venous end of capillaries, blood pressure drops to about 16 mmHg. Blood pressure continues to drop as blood enters systemic venules and then veins, and it reaches 0 mmHg as blood returns to the right atrium.

Blood pressure depends in part on the total volume of blood in the cardiovascular system. The normal volume of blood in an adult is about 5 liters (5.3 qt). Any decrease in this volume, as from hemorrhage, decreases the amount of blood that is circulated through the arteries. A modest decrease can be compensated by homeostatic mechanisms that help maintain blood pressure, but if the decrease in blood volume is greater than 10% of total blood volume, blood pressure drops, with potentially life-threatening results. Conversely, anything that increases blood volume, such as water retention in the body, tends to increase blood pressure. Hence, the use of salt is contraindicated for cardiac patients.

Resistance

Vascular resistance is the opposition to blood flow due to friction between blood and the walls of blood vessels. An increase in vascular resistance increases blood pressure; a decrease in vascular resistance has the opposite effect. Vascular resistance depends on (1) size of the blood vessel lumen, (2) blood viscosity, and (3) total blood vessel length.

1. **Size of the lumen.** The smaller the lumen of a blood vessel, the greater is its resistance to blood flow. Vasoconstriction narrows the lumen, and vasodilation widens it. Normally, moment-to-moment fluctuations in blood flow through a given tissue are due to vasoconstriction and vasodilation of the tissue's arterioles. As arterioles dilate, resistance decreases, and blood pressure falls. As arterioles constrict, resistance increases, and blood pressure rises.
2. **Blood viscosity.** The *viscosity* (thickness) of blood depends mostly on the ratio of red blood cells to plasma (fluid) volume, and to a smaller extent on the concentration of proteins in plasma. The higher the blood's viscosity, the higher is the resistance. Any condition that increases the viscosity of blood, such as dehydration or *polycythemia* (an unusually high number of red blood cells), thus increases blood pressure. A depletion of plasma proteins or red blood cells, as a result of anemia or hemorrhage, decreases viscosity and thus decreases blood pressure.
3. **Total blood vessel length.** Resistance to blood flow increases when the total length of all blood vessels in the body increases. The longer the blood vessel, the greater is the contact between the vessel wall and the blood. The greater the contact between the vessel wall and the blood, the greater is the friction. An estimated 650 km (about 400 miles) of additional blood vessels develop for each extra kilogram (2.2 lb) of fat, one reason why overweight individuals may have higher blood pressure.

Regulation of Blood Pressure and Blood Flow

Several interconnected negative feedback systems control blood pressure and blood flow by adjusting heart rate, stroke volume, vascular resistance, and blood volume. Some systems allow rapid adjustments to cope with sudden changes, such as the drop in blood pressure in the brain that occurs when you stand up; others provide long-term regulation. The body may also require adjustments to the distribution of blood flow. During exercise, for example, a greater percentage of blood flow is diverted to skeletal muscles.

Role of the Cardiovascular Center

In Chapter 22 we noted how the ***cardiovascular (CV) center*** in the medulla oblongata helps regulate heart rate and stroke volume. The CV center also controls the neural and hormonal negative feedback systems that regulate blood pressure and blood flow to specific tissues.

INPUT The cardiovascular center receives input from higher brain regions: the cerebral cortex, limbic system, and hypothalamus (Figure 23.6). For example, even before you start to run a race, your heart rate may increase due to nerve impulses conveyed from the limbic system to the CV center. If your body temperature rises during a race, the hypothalamus sends nerve

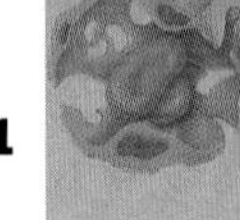

Figure 23.6 The cardiovascular (CV) center. Located in the medulla oblongata, the CV center receives input from higher brain centers, proprioceptors, baroreceptors, and chemoreceptors. It provides output to both the sympathetic and parasympathetic divisions of the autonomic nervous system.

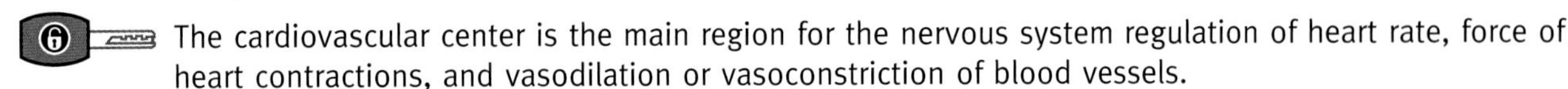

The cardiovascular center is the main region for the nervous system regulation of heart rate, force of heart contractions, and vasodilation or vasoconstriction of blood vessels.

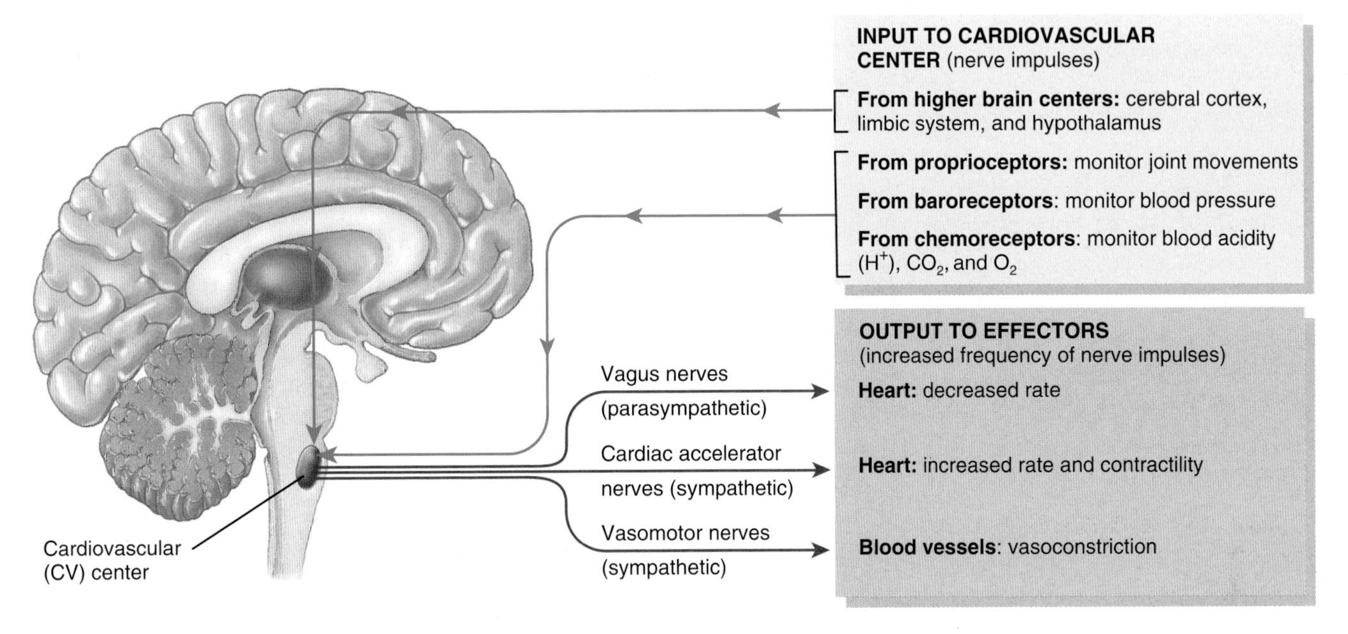

What types of effector tissues are regulated by the cardiovascular center?

impulses to the CV center. The resulting vasodilation of skin blood vessels allows heat to dissipate more rapidly from the surface of the skin.

The CV center also receives input from three main types of sensory receptors: proprioceptors, baroreceptors, and chemoreceptors. ***Proprioceptors,*** which monitor movements of joints and muscles, provide input to the cardiovascular center during physical activity, such as playing tennis, and cause the rapid increase in heart rate at the beginning of exercise.

Baroreceptors (pressure receptors) are located in the aorta, internal carotid arteries (arteries in the neck that supply blood to the brain), and other large arteries in the neck and chest. They send impulses continuously to the cardiovascular center to help regulate blood pressure. If blood pressure falls, the baroreceptors are stretched less, and they send nerve impulses at a slower rate to the cardiovascular center (see Figure 23.7). In response, the cardiovascular center decreases parasympathetic stimulation of the heart and increases sympathetic stimulation of the heart. As the heart beats faster and more forcefully, and as vascular resistance increases, blood pressure increases to the normal level.

By contrast, when an increase in blood pressure is detected, the baroreceptors send impulses at a faster rate. The cardiovascular center responds by increasing parasympathetic stimulation and decreasing sympathetic stimulation. The resulting decreases in heart rate and force of contraction lower cardiac output, and vasodilation lowers vascular resistance. Decreased cardiac output and decreased vascular resistance both lower blood pressure.

MANUAL THERAPY APPLICATION

Arising Slowly From a Massage Therapy Session

Moving from a prone (lying down) to an erect position decreases blood pressure and blood flow in the head and upper part of the body. The drop in pressure, however, is quickly counteracted by the *baroreceptor reflexes*. Sometimes these reflexes operate more slowly than normal, especially in older people. As a result, a person can faint due to reduced brain blood flow upon standing up too quickly. All manual therapy patients should be cautioned to **arise slowly from a massage therapy session** on a massage table. Certain persons, including elderly patients, females in their third trimester of pregnancy, and patients who are taking medications that lower blood pressure, should actually be assisted in arising. Pregnant females, especially if lying in the supine position for massage, may experience a drop in blood pressure as a result of the fetus compressing the inferior vena cava; the resultant decrease in blood return to the heart reduces cardiac output, which results in a drop in blood pressure.

Hypertension (high blood pressure) is becoming rampant. Once hypertension develops, it increases the risk for many diseases and typically lasts a lifetime. Medications to reduce hypertension commonly start with beta blockers and/or diuretics. In cases where these are not effective, vasodilators and/or angiotensin-converting enzyme (ACE) inhibitors are commonly prescribed. Manual therapists need to modify the massage for patients taking ACE inhibitors by limiting effleurage, increasing the rate of application of all strokes, using tapotement, and assisting the client's arise after the massage.

Figure 23.7 **Negative feedback regulation of blood pressure via baroreceptor reflexes.**

When blood pressure decreases, heart rate increases.

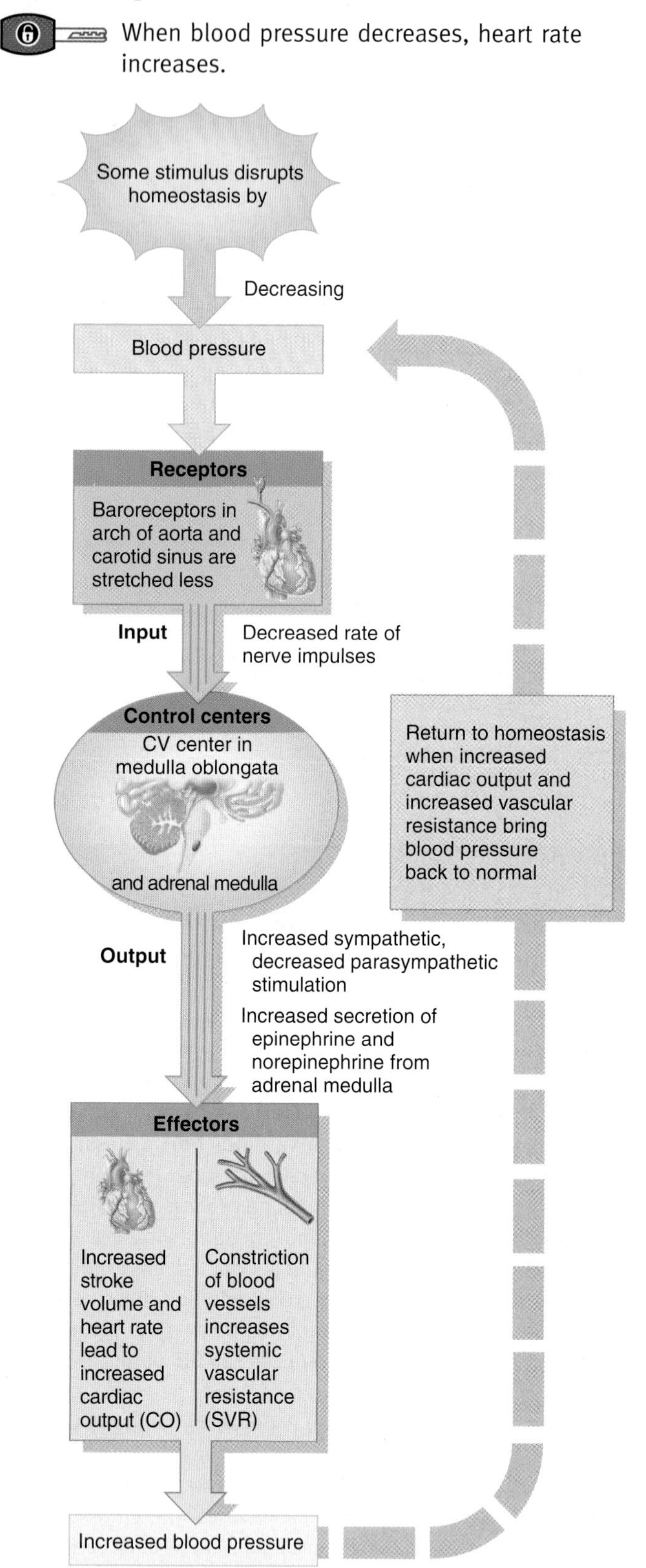

? Does this negative feedback cycle represent the changes that occur when you lie down or when you stand up?

Chemoreceptor reflexes *Chemoreceptors* (chemical receptors) that monitor blood levels of O_2, CO_2, and H^+ are located in the two ***carotid bodies*** in the common carotid arteries and in the ***aortic body*** in the arch of the aorta. *Hypoxia* (lowered O_2 availability), *acidosis* (an increase in H^+ concentration), or *hypercapnia* (excess CO_2) stimulates the chemoreceptors to send impulses to the cardiovascular center. In response, the CV center increases sympathetic stimulation of arterioles and veins, producing vasoconstriction and an increase in blood pressure.

Output Output from the cardiovascular center flows along sympathetic and parasympathetic fibers of the autonomic nervous system (see Figure 23.6, An increase in sympathetic stimulation increases heart rate and the forcefulness of contraction, whereas a decrease in sympathetic stimulation decreases heart rate and contraction force. The vasomotor region of the cardiovascular center also sends impulses to arterioles throughout the body. The result is a moderate state of vasoconstriction, called ***vasomotor tone,*** that sets the resting level of vascular resistance. Sympathetic stimulation of most veins results in movement of blood out of venous blood reservoirs, which increases blood pressure.

Hormonal Regulation of Blood Pressure and Blood Flow

Several hormones help regulate blood pressure and blood flow by altering cardiac output, changing vascular resistance, or adjusting the total blood volume.

1. **Renin–angiotensin–aldosterone (RAA) system.** When blood volume falls or blood flow to the kidneys decreases, certain cells in the kidneys secrete the enzyme renin into the bloodstream (see Figure 20.14, Section 20.7). Together, renin and angiotensin-converting enzyme (ACE) produce the active hormone *angiotensin II,* which raises blood pressure by causing vasoconstriction.

   Angiotensin II also stimulates secretion of *aldosterone,* which increases reabsorption of sodium ions (Na^+) and water by the kidneys. The water reabsorption increases total blood volume, which increases blood pressure.

2. **Epinephrine and norepinephrine.** In response to sympathetic stimulation, the adrenal medulla releases epinephrine and norepinephrine. These hormones increase cardiac output by increasing the rate and force of heart contractions; they also cause vasoconstriction of arterioles and veins in the skin and abdominal organs.

3. **Antidiuretic hormone (ADH).** ADH is produced by the hypothalamus and released from the posterior pituitary in response to dehydration or decreased blood volume. Among other actions, ADH causes vasoconstriction, which increases blood pressure. For this reason ADH is also called ***vasopressin.***

4. **Atrial natriuretic peptide (ANP).** Released by cells in the atria of the heart, ANP lowers blood pressure by causing vasodilation and by promoting the loss of salt and water in the urine, which reduces blood volume.

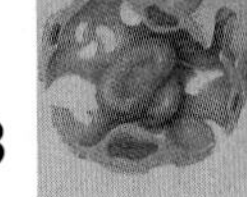

CHECKPOINT

4. What two factors influence cardiac output?
5. Describe how blood pressure decreases as distance from the left ventricle increases.
6. What factors determine vascular resistance?
7. Explain the roles of the cardiovascular center, reflexes, and hormones in regulating blood pressure.

23.3 CHECKING CIRCULATION

OBJECTIVE

- Explain how pulse and blood pressure are measured.

Pulse

The alternate expansion and elastic recoil of an artery after each contraction and relaxation of the left ventricle is called a ***pulse.*** The pulse is strongest in the arteries closest to the heart. It becomes weaker as it passes through the arterioles, and it disappears altogether in the capillaries. The radial artery at the wrist is most commonly used to palpate the pulse. Other sites where the pulse may be felt include the brachial artery along the medial side of the biceps brachii muscle; the common carotid artery, next to the voice box, which is usually monitored during cardiopulmonary resuscitation; the popliteal artery behind the knee; and the dorsal artery of the foot above the instep of the foot. All of these arteries are illustrated within the next few pages.

The pulse rate normally is the same as the heart rate, about 75 beats per minute at rest. ***Tachycardia*** (tak′-i-KAR-dē-a; *tachy-* = fast) is a resting heart or pulse rate over 100 beats/min. ***Bradycardia*** (brăd′-i-KAR-dē-a; *brady-* = slow) indicates a resting heart or pulse rate under 50 beats/min.

Measurement of Blood Pressure

In clinical use, the term ***blood pressure*** usually refers to the pressure in arteries generated by the left ventricle during systole and the pressure remaining in the arteries when the ventricle is in diastole. Blood pressure is usually measured in the brachial artery in the left arm (see Figure 23.10a). The device used to measure blood pressure is a ***sphygmomanometer*** (sfig-mō-ma-NOM-e-ter; *sphygmo-* = pulse; *-manometer* = instrument used to measure pressure). When the pressure cuff is inflated above the blood pressure attained during systole, the artery is compressed so that blood flow stops. The technician places a stethoscope below the cuff superficial to the brachial artery and then slowly deflates the cuff. When the cuff is deflated enough to allow the artery to open, a spurt of blood passes through, resulting in the first sound heard through the stethoscope. This sound corresponds to ***systolic blood pressure (SBP)***—the force with which blood is pushing against arterial walls during ventricular contraction. As the cuff is deflated further, the sounds suddenly become faint. This level, called the ***diastolic blood pressure (DBP),*** represents the force exerted by the blood remaining in arteries during ventricular relaxation.

The normal blood pressure of a young adult male is less than 120 mmHg systolic and less than 80 mmHg diastolic, reported, for example, as "110 over 70" and written as 110/70. In young adult females, both pressures are 8 to 10 mmHg less. People who exercise regularly and are in good physical condition may have even lower blood pressures.

CLINICAL CONNECTION | Hypertension

About 50 million Americans have **hypertension,** or persistently high blood pressure. It is the most common disorder affecting the heart and blood vessels and is the major cause of heart failure, kidney disease, and stroke. The current guidelines for blood pressure are as follows:

Category	Systolic (mmHg)		Diastolic (mmHg)
Normal	Less than 120	*and*	Less than 80
Prehypertension	120–139	*or*	80–89
Stage 1 hypertension	140–159	*or*	90–99
Stage 2 hypertension	Greater than 160	*or*	Greater than 100

Although several categories of drugs can reduce elevated blood pressure, the following lifestyle changes are also effective in managing hypertension: Lose weight; limit alcohol intake; exercise; reduce intake of sodium (salt); maintain recommended dietary intake of potassium, calcium, and magnesium; don't smoke; and manage stress. •

CHECKPOINT

8. What causes pulse?
9. Distinguish between systolic and diastolic blood pressure.

23.4 CIRCULATORY ROUTES

OBJECTIVE

- Compare the major routes that blood takes through various regions of the body.

Blood vessels are organized into ***circulatory routes*** that carry blood throughout the body (see Figure 23.8). As noted earlier, the two main circulatory routes are the systemic circulation and the pulmonary circulation.

Systemic Circulation

The ***systemic circulation*** includes the arteries and arterioles that carry blood containing oxygen and nutrients from the left ventricle to systemic capillaries throughout the body, plus the veins and venules that return blood containing carbon dioxide and wastes to the right atrium. Blood leaving the aorta and traveling through the systemic arteries is a bright red color. As blood flows through the capillaries, it loses some of its oxygen and takes on carbon dioxide, so that the blood of systemic veins is a

Figure 23.8 **Circulatory routes.** Red arrows indicate hepatic portal circulation. Details of the pulmonary circulation are shown here, and details of the hepatic portal circulation are shown in Figure 23.16.

Blood vessels are organized into routes that deliver blood to various tissues of the body.

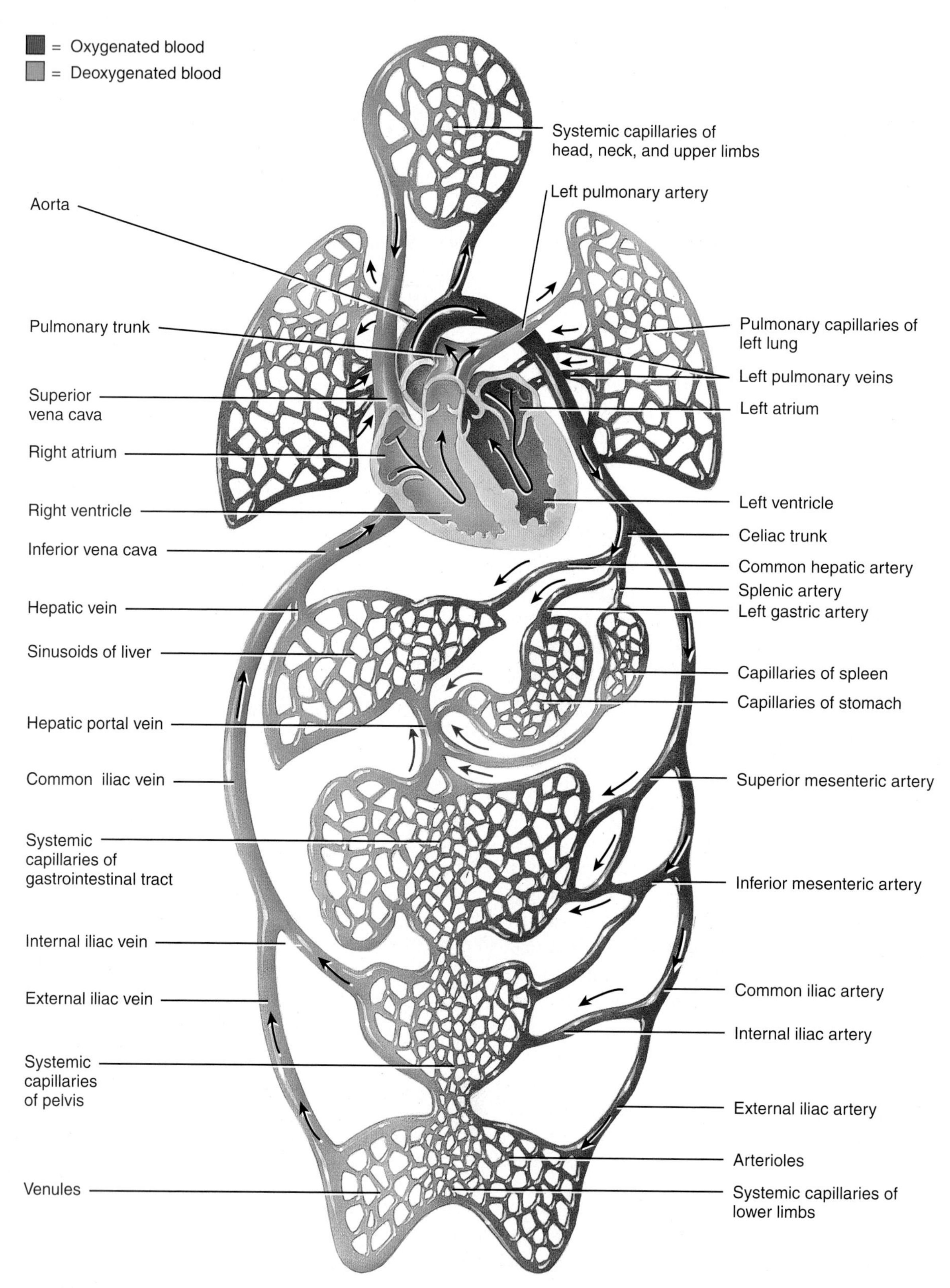

What are the two main circulatory routes?

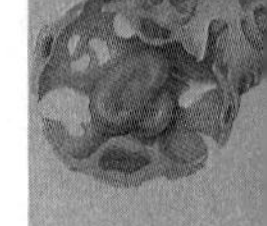

dark red color. This color, as viewed through the skin, appears blue. All systemic arteries *branch from the aorta,* which arises from the left ventricle of the heart (see Figure 23.9). Deoxygenated blood returns to the heart through the systemic veins. All the veins of the systemic circulation empty into the ***superior vena cava, inferior vena cava,*** or the ***coronary sinus,*** which, in turn, empty into the right atrium. The principal blood vessels of the systemic circulation are described and illustrated in Exhibits 23.1 through 23.7 and Figures 23.9 through 23.15.

• CLINICAL CONNECTION | Reversing Arterial Plaque Buildup

Not so long ago, scientists believed that once plaque formed in an artery, it never went away. Medical researchers thought that lifestyle changes and drugs could slow the process of atherosclerosis but could not undo damage already done. Researchers have discovered that the body's own healing process can help in **reversing arterial plaque buildup**. Lifestyle changes and drug treatments appear to stabilize the most dangerous atherosclerotic plaques and may even eliminate the need for surgical interventions, such as bypass surgery, in some people.

The health risk imposed by plaque that accumulates within the artery lining depends on several factors. Some plaque is fairly *stable*: It has a low lipid content, is not growing much in size, and has a strong fibrous cap that keeps it from rupturing when blood pressure rises. *Unstable plaque* is characterized by a large accumulation of lipid in its core and only a thin fibrous cap. In addition, unstable plaques contain a large number of macrophages. In a misguided attempt to heal endothelial damage, macrophages may ingest plaque lipids; the net result is increased arterial injury and lipid accumulation. An unstable plaque is apt to rupture, triggering formation of a life-threatening blood clot at the plaque site.

The first step in preventing, slowing, and possibly reversing artery disease is to control the risk factors associated with its progression. Recommendations for a heart-healthy and artery-healthy lifestyle include no smoking, regular exercise (at least 30 minutes of moderate-intensity exercise per day), stress management, and a heart-healthy diet. Diet recommendations include limiting fat intake and dramatically increasing consumption of plant foods, such as grains, fruits, and vegetables. These recommendations help prevent arterial disease by reducing obesity, blood lipids, platelet stickiness, and blood pressure, and by improving blood glucose control in people at risk for type 2 diabetes. •

Pulmonary Circulation

When deoxygenated blood returns to the heart from the systemic route, it is pumped out of the right ventricle into the lungs. In the lungs, it loses carbon dioxide and picks up oxygen. Now bright red again, the blood returns to the left atrium of the heart and is pumped again into the systemic circulation. The flow of deoxygenated blood from the right ventricle to the air sacs of the lungs and the return of oxygenated blood from the air sacs to the left atrium is called the ***pulmonary circulation*** (see Figure 23.8). The ***pulmonary trunk*** emerges from the right ventricle and then divides into two branches. The ***right pulmonary artery*** runs to the right lung; the ***left pulmonary artery*** goes to the left lung. After birth, the pulmonary arteries are the only arteries that carry deoxygenated blood. On entering the lungs, the branches divide and subdivide until ultimately they form capillaries around the air sacs in the lungs. Carbon dioxide passes from the blood into the air sacs and is exhaled, while inhaled oxygen passes from the air sacs into the blood (see Chapter 25). The capillaries unite, venules and veins are formed, and, eventually, two ***pulmonary veins*** from each lung transport the oxygenated blood to the left atrium. (After birth, the pulmonary veins are the only veins that carry oxygenated blood.) Contractions of the left ventricle then send the blood into the systemic circulation.

EXHIBIT 23.1 The Aorta and Its Branches

OBJECTIVE

- Identify the four principal divisions of the aorta and locate the major arterial branches arising from each.

The ***aorta*** (*aortae* = to lift up), the largest artery of the body, is 2 to 3 cm (about 1 in.) in diameter. Its four principal divisions are the ascending aorta, arch of the aorta, thoracic aorta, and abdominal aorta (Figure 23.9). The ***ascending aorta*** emerges from the left ventricle posterior to the pulmonary trunk. It gives off two coronary artery branches that supply the myocardium of the heart. Then it turns to the left, forming the ***arch of the aorta.*** Branches of the arch of the aorta are described in Exhibit 23.2. The part of the aorta between the arch of the aorta and the diaphragm, the ***thoracic aorta,*** is about 20 cm (8 in.) long. The part of the aorta between the diaphragm and the common iliac arteries is the ***abdominal aorta*** (ab-DOM-i-nal). The main branches of the abdominal aorta are the ***celiac trunk,*** the ***superior mesenteric artery,*** and the ***inferior mesenteric artery.*** The abdominal aorta divides at the level of the fourth lumbar vertebra into two common iliac arteries, which carry blood to the lower limbs.

CHECKPOINT

10. What general regions do each of the four principal divisions of the aorta supply?

DIVISION AND BRANCHES	REGION SUPPLIED
ASCENDING AORTA	
Right and left coronary arteries	Heart.
ARCH OF THE AORTA	
Brachiocephalic trunk (brā′-kē-ō-se-FAL-ik)	
Right common carotid artery (ka-ROT-id)	Right side of head and neck.
Right subclavian artery (sub-KLĀ-vē-an)	Right upper limb.
Left common carotid artery	Left side of head and neck.
Left subclavian artery	Left upper limb.
THORACIC AORTA (*thorac-* = chest)	
Pericardial arteries (per-i-KAR-dē-al)	Pericardium.
Bronchial arteries (BRONG-kē-al)	Bronchi of lungs.
Esophageal arteries (e-sof′-a-JĒ-al)	Esophagus.
Mediastinal arteries (mē′-dē-as-TĪ-nal)	Structures in mediastinum.
Posterior intercostal arteries (in′-ter-KOS-tal)	Intercostal and chest muscles.
Subcostal arteries (sub-KOS-tal)	Same as posterior intercostals.
Superior phrenic arteries (FREN-ik)	Superior and posterior surfaces of diaphragm.
ABDOMINAL AORTA	
Inferior phrenic arteries	Inferior surface of diaphragm.
Celiac trunk (SĒ-lē-ak)	
Common hepatic artery (he-PAT-ik)	Liver, stomach, duodenum, and pancreas.
Left gastric artery (GAS-trik)	Stomach and esophagus.
Splenic artery (SPLĒN-ik)	Spleen, pancreas, and stomach.
Superior mesenteric artery (MES-en-ter′-ik)	Small intestine, cecum, ascending and transverse colons, and pancreas.
Suprarenal arteries (soo-pra-RĒ-nal)	Adrenal (suprarenal) glands.
Renal arteries (RĒ-nal)	Kidneys.
Gonadal arteries (gō-NAD-al)	
Testicular arteries (tes-TIK-ū-lar)	Testes (male).
Ovarian arteries (ō-VAR-ē-an)	Ovaries (female).
Inferior mesenteric artery	Transverse, descending, and sigmoid colons; rectum.
Common iliac arteries (IL-ē-ak)	
External iliac arteries	Lower limbs.
Internal iliac arteries	Uterus (female), prostate (male), muscles of buttocks, and urinary bladder.

Figure 23.9 **Aorta and its principal branches.**

All systemic arteries branch from the aorta.

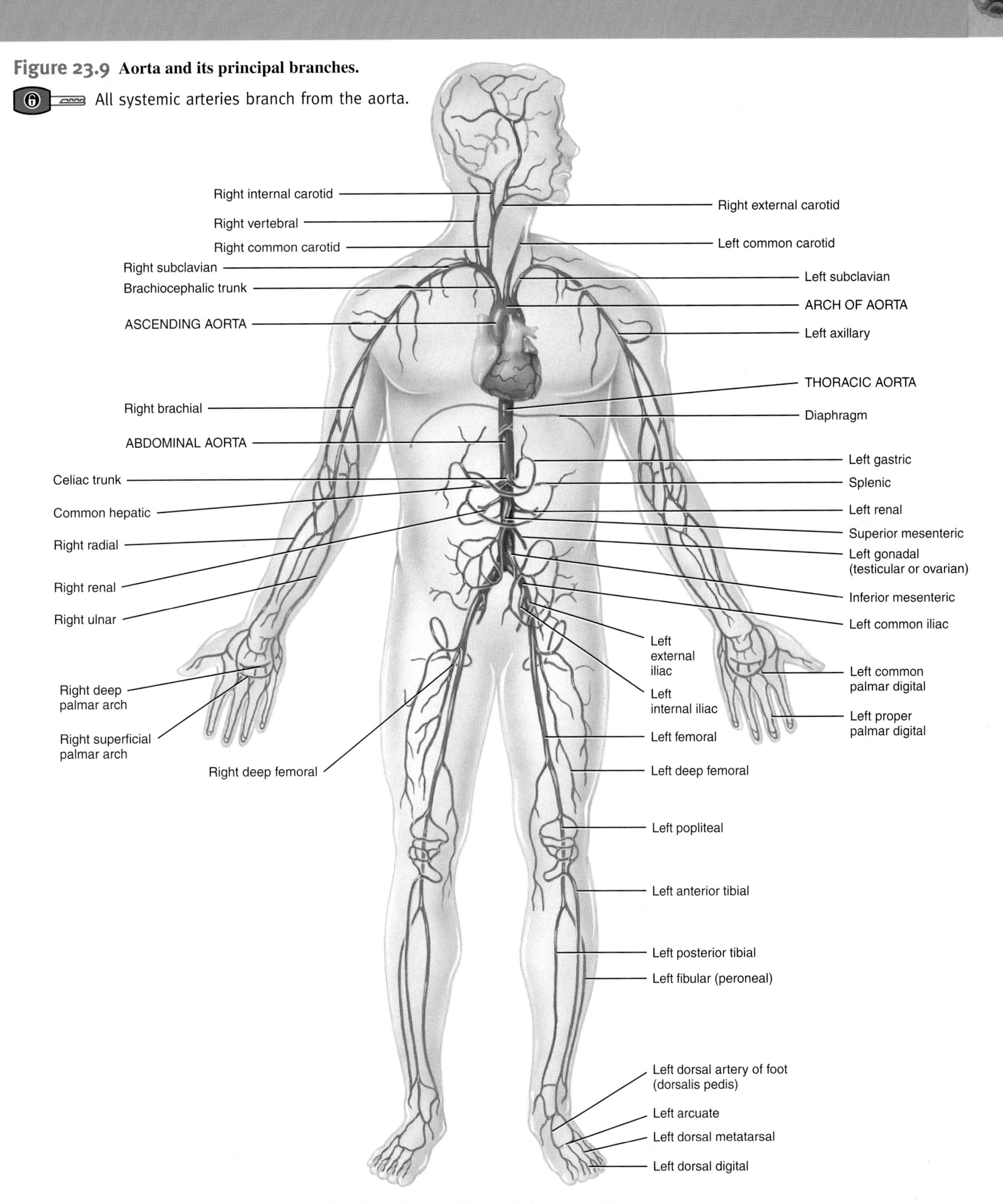

Overall anterior view of the principal branches of the aorta

What are the four subdivisions of the aorta?

EXHIBIT 23.2 The Arch of the Aorta

OBJECTIVE

- Identify the three arteries that branch from the arch of the aorta.

The ***arch of the aorta,*** the continuation of the ascending aorta, is 4 to 5 cm (almost 2 in.) in length. It has three branches. In order, as they emerge from the arch of the aorta, the three branches are the brachiocephalic trunk, the left common carotid artery, and the left subclavian artery (Figure 23.10).

CHECKPOINT

11. What general regions do the arteries that arise from the arch of the aorta supply?

BRANCH	DESCRIPTION AND REGION SUPPLIED
BRACHIOCEPHALIC TRUNK	The ***brachiocephalic trunk*** divides to form the right subclavian artery and right common carotid artery (Figure 23.10a).
Right subclavian artery (sub-KLĀ-vē-an)	The ***right subclavian artery*** extends from the brachiocephalic trunk to the first rib and then passes into the armpit (axilla). The general distribution of the artery is to the brain and spinal cord, neck, shoulder, and chest.
Internal thoracic or mammary artery (thor-AS-ik; *thorac-* = chest)	The ***internal thoracic (mammary) artery*** arises from the first part of the subclavian artery and descends posterior to the costal cartilages of the superior six ribs. It terminates at the sixth intercostal space. It supplies the anterior thoracic wall and structures in the mediastinum. In coronary artery bypass grafting, if only a single vessel is obstructed, the internal thoracic (usually the left) is used to create the bypass. The upper end of the artery is left attached to the subclavian artery and the cut end is connected to the coronary artery at a point distal to the blockage. The lower end of the internal thoracic artery is tied off. Artery grafts are preferred over vein grafts because arteries can withstand the greater pressure of blood flowing through coronary arteries and are less likely to become obstructed over time.
Vertebral artery (VER-te-bral)	Before passing into the axilla, the right subclavian artery gives off a major branch to the brain called the ***right vertebral artery*** (Figure 23.10b). The right vertebral artery passes through the foramina of the transverse processes of the sixth through first cervical vertebrae and enters the skull through the foramen magnum to reach the inferior surface of the brain. Here it unites with the left vertebral artery to form the ***basilar*** (BAS-i-lar) ***artery.*** The vertebral artery supplies the posterior portion of the brain with blood. The basilar artery passes along the midline of the anterior aspect of the brain stem. It gives off several branches (***posterior cerebral*** and ***cerebellar arteries***) that supply the cerebellum and pons of the brain and the inner ear.
Axillary artery (AK-sil-ār-ē = armpit)	The continuation of the right subclavian artery into the axilla is called the ***axillary artery.*** (Note that the right subclavian artery, which passes deep to the clavicle, is a good example of the practice of giving the same vessel different names as it passes through different regions.) Its general distribution is the shoulder, thoracic and scapular muscles, and humerus.
Brachial artery (BRĀ-kē-al = arm)	The ***brachial artery*** is the continuation of the axillary artery into the arm. The brachial artery provides the main blood supply to the arm and is superficial and palpable along its course. It begins at the tendon of the teres major muscle and ends just distal to the bend of the elbow. At first, the brachial artery is medial to the humerus, but as it descends it gradually curves laterally and passes through the cubital fossa, a triangular depression anterior to the elbow where you can easily detect the pulse of the brachial artery and listen to the various sounds when taking a person's blood pressure. Just distal to the bend in the elbow, the brachial artery divides into the radial artery and ulnar artery. Blood pressure is usually measured in the brachial artery. In order to control hemorrhage, the best place to compress the brachial artery is near the middle of the arm.
Radial artery (RĀ-dē-al = radius)	The ***radial artery*** is the smaller branch and is a direct continuation of the brachial artery. It passes along the lateral (radial) aspect of the forearm and then through the wrist and hand, supplying these structures with blood. At the wrist, the radial artery makes contact with the distal end of the radius, where it is covered only by fascia and skin. Because of its superficial location at this point, it is a common site for measuring the radial pulse.
Ulnar artery (UL-nar = ulna)	The ***ulnar artery,*** the larger branch of the brachial artery, passes along the medial (ulnar) aspect of the forearm and then into the wrist and hand, supplying these structures with blood. In the palm, branches of the radial and ulnar arteries anastomose to form the superficial palmar arch and the deep palmar arch.
Superficial palmar arch (*palma*= palm)	The ***superficial palmar arch*** is formed mainly by the ulnar artery, with a contribution from a branch of the radial artery. The arch is superficial to the long flexor tendons of the fingers and extends across the palm at the bases of the metacarpals. It gives rise to ***common palmar digital arteries,*** which supply the palm. Each divides into a pair of ***proper palmar digital arteries,*** which supply the fingers.
Deep palmar arch	Mainly the radial artery forms the ***deep palmar arch,*** with a contribution from a branch of the ulnar artery. The arch is deep to the long flexor tendons of the fingers and extends across the palm, just distal to the bases of the metacarpals. Arising from the deep palmar arch are ***palmar metacarpal arteries,*** which supply the palm and anastomose with the common palmar digital arteries of the superficial palmar arch.
Right common carotid artery	The ***right common carotid artery*** begins at the bifurcation (division into two branches) of the brachiocephalic trunk, posterior to the right sternoclavicular joint, and passes superiorly in the neck to supply structures in the head (Figure 23.10b). At the superior border of the larynx (voice box), it divides into the right external and right internal carotid arteries. Pulse may be detected in the common carotid artery, just lateral to the larynx. It is convenient to detect a carotid pulse when exercising or when administering cardiopulmonary resuscitation.
External carotid artery	The ***external carotid artery*** begins at the superior border of the larynx and terminates near the temporomandibular joint of the parotid gland, where it divides into two branches: the superficial temporal and maxillary arteries. The carotid pulse can be detected in the external carotid artery just anterior to the sternocleidomastoid muscle at the superior border of the larynx. The general distribution of the external carotid artery is to structures external to the skull.

BRANCH	DESCRIPTION AND REGION SUPPLIED
Internal carotid artery	The ***internal carotid artery*** supplies structures internal to the skull such as the eyeball, ear, most of the cerebrum of the brain, and pituitary gland. Inside the cranium, the internal carotid arteries along with the basilar artery form an arrangement of blood vessels at the base of the brain near the hypophyseal fossa called the ***cerebral arterial circle (circle of Willis)***. From this circle (Figure 23.10c) arise arteries supplying most of the brain. The cerebral arterial circle is formed by the union of the ***anterior cerebral arteries*** (branches of internal carotids) and ***posterior cerebral arteries*** (branches of basilar artery). The posterior cerebral arteries are connected with the internal carotid arteries by the ***posterior communicating arteries*** (kō-MŪ-nī-kā-tīng). The anterior cerebral arteries are connected by the ***anterior communicating artery.*** The internal carotid arteries are also considered part of the cerebral arterial circle. The functions of the cerebral arterial circle are to equalize blood pressure to the brain and provide alternate routes for blood flow to the brain, should the arteries become damaged.
LEFT COMMON CAROTID ARTERY	Divides into basically the same branches with the same names as the right common carotid artery.
LEFT SUBCLAVIAN ARTERY	Divides into basically the same branches with the same names as the right subclavian artery.

Figure 23.10 **Arch of the aorta and its branches.**

The arch of the aorta is the continuation of the ascending aorta.

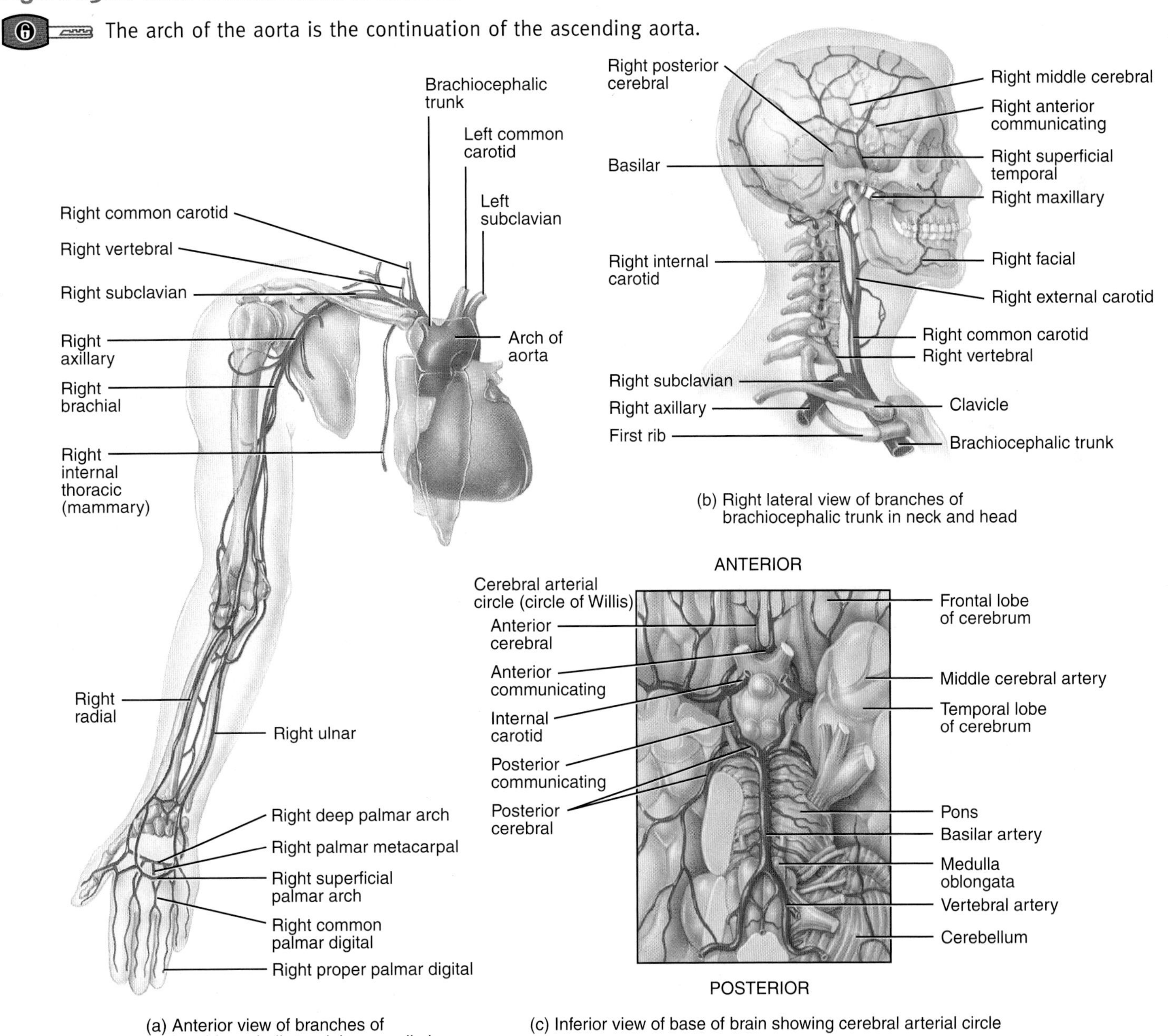

(a) Anterior view of branches of brachiocephalic trunk in upper limb

(b) Right lateral view of branches of brachiocephalic trunk in neck and head

(c) Inferior view of base of brain showing cerebral arterial circle

What is the functional significance of the cerebral arterial circle?

EXHIBIT 23.3 Arteries of the Pelvis and Lower Limbs

OBJECTIVE

- Identify the two major branches of the common iliac arteries.

The abdominal aorta ends by dividing into the right and left ***common iliac arteries*** (Figure 23.11a). These, in turn, divide into the ***internal*** and ***external iliac arteries.*** In sequence, the external iliacs become the ***femoral arteries*** in the thighs, the ***popliteal arteries*** posterior to the knee, and the ***anterior*** and ***posterior tibial arteries*** in the legs.

CHECKPOINT

12. What general regions do the internal and external iliac arteries supply?

ARTERY	DESCRIPTION AND REGION SUPPLIED
Common iliac arteries (IL-ē-ak = ilium)	At about the level of the fourth lumbar vertebra, the abdominal aorta divides into the right and left ***common iliac arteries*** (Figure 23.11a). Each gives rise to two branches: internal iliac and external iliac arteries. The general distribution of the common iliac arteries is to the pelvis, external genitals, and lower limbs.
Internal iliac arteries	The ***internal iliac arteries*** are the primary arteries of the pelvis. They supply the pelvis, buttocks, external genitals, and thigh.
External iliac arteries	The ***external iliac arteries*** supply the lower limbs.
Femoral arteries (FEM-o-ral = thigh)	The ***femoral arteries,*** continuations of the external illiacs, supply the lower abdominal wall, groin, external genitals, and muscles of the thigh.
Popliteal arteries (pop′-li-TĒ-al = posterior surface of the knee)	The ***popliteal arteries,*** continuations of the femoral arteries, supply muscles and skin on the posterior of the legs; muscles of the calf; knee joint; femur; patella; and fibula.
Anterior tibial arteries (TIB-ē-al = shin bone)	The ***anterior tibial arteries*** descend from the bifurcation of the popliteal arteries. They are smaller than the posterior tibial arteries. The anterior tibial arteries descend through the anterior muscular compartment of the leg. They pass through the interosseous membrane that connects the tibia and fibula, lateral to the tibia. The anterior tibial arteries supply the knee joints, anterior compartment muscles of the legs, skin over the anterior aspects of the legs, and ankle joints. At the ankles, the anterior tibial arteries become the ***dorsal arteries of the foot (dorsalis pedis arteries).*** A pulse in this artery may be taken to evaluate the peripheral vascular system. The dorsal arteries of the foot supply the muscles, skin, and joints on the dorsal aspects of the feet. On the dorsum of the feet, the dorsal arteries of the foot give off a transverse branch at the first (medial) cuneiform bone called the ***arcuate arteries*** (*arcuat-* = bowed) that run laterally over the bases of the metatarsals. From the arcuate arteries branch the ***dorsal metatarsal arteries,*** which supply the feet. The dorsal metatarsal arteries terminate by dividing into the ***dorsal digital arteries,*** which supply the toes.
Posterior tibial arteries	The ***posterior tibial arteries,*** the direct continuations of the popliteal arteries, descend from the bifurcation of the popliteal arteries. They pass down the posterior muscular compartment of the legs posterior to the medial malleolus of the tibia. They terminate by dividing into the medial and lateral plantar arteries. Their general distribution is to the muscles, bones, and joints of the leg and foot. Major branches of the posterior tibial arteries are the ***fibular (peroneal) arteries,*** which supply the fibularis, soleus, tibialis posterior, and flexor hallucis muscles. They also supply the fibula, tarsus, and lateral aspect of the heel. The bifurcation of the posterior tibial arteries into the medial and lateral plantar arteries occurs deep to the flexor retinaculum on the medial side of the feet. The ***medial plantar arteries*** (PLAN-tar = sole) supply the abductor hallucis and flexor digitorum brevis muscles and the toes. The ***lateral plantar arteries*** unite with a branch of the dorsal arteries of the foot to form the ***plantar arch.*** The arch begins at the base of the fifth metatarsal and extends medially across the metatarsals. As the arch crosses the foot, it gives off ***plantar metatarsal arteries,*** which supply the feet. These terminate by dividing into ***plantar digital arteries,*** which supply the toes.

Figure 23.11 Arteries of the pelvis and right lower limb.

The internal iliac arteries carry most of the blood supply to the pelvis, buttocks, external genitals, and thigh.

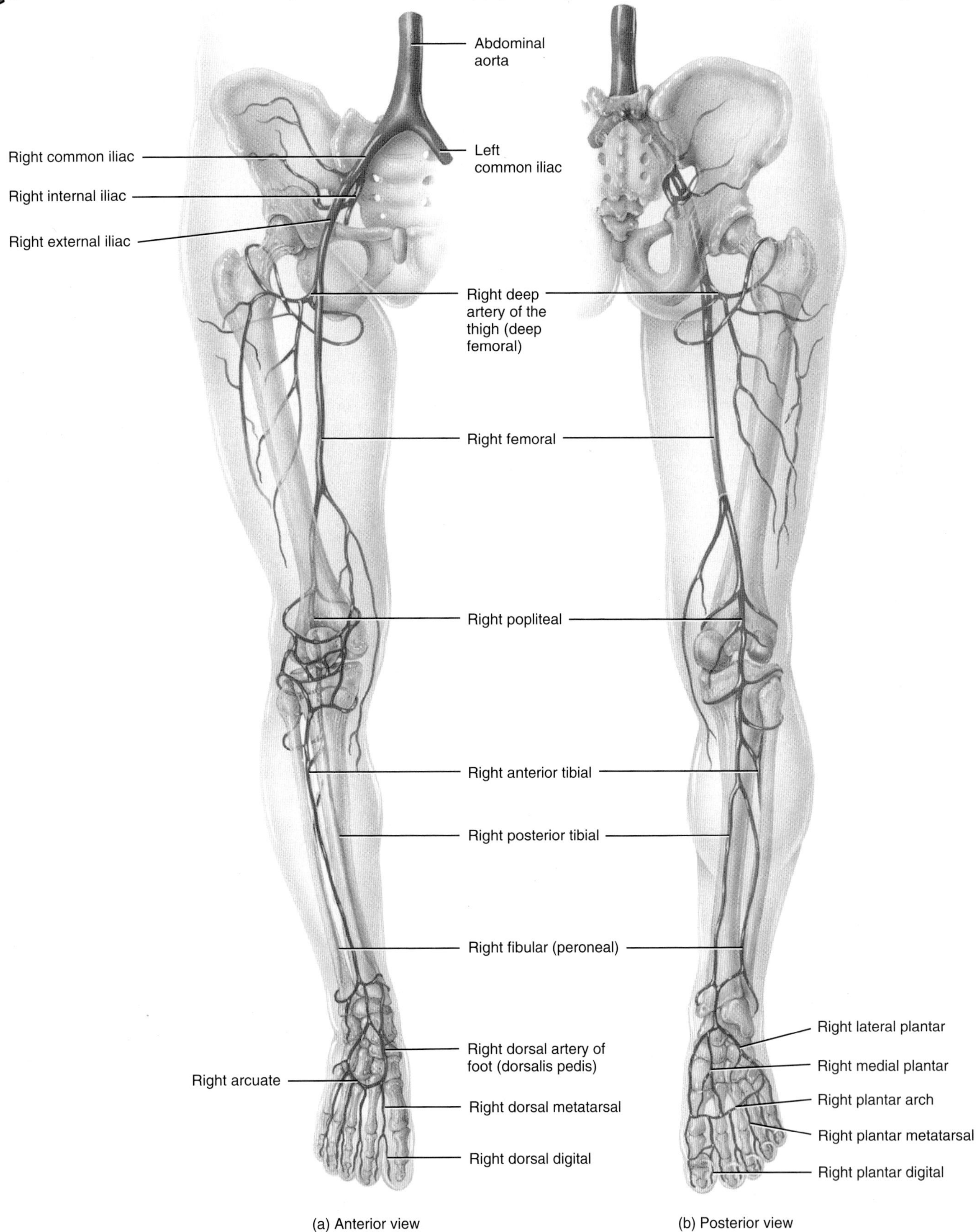

At what point does the abdominal aorta divide into common iliac arteries?

EXHIBIT 23.4 Veins of the Systemic Circulation

OBJECTIVE

- Identify the three systemic veins that return deoxygenated blood to the heart.

Arteries distribute blood to various parts of the body, and veins drain blood away from them. For the most part, arteries are deep. Veins may be ***superficial*** (located just beneath the skin) or ***deep.*** Deep veins generally travel alongside arteries and usually bear the same name. Because there are no large superficial arteries, the names of superficial veins do not correspond to those of arteries. Superficial veins are clinically important as sites for withdrawing blood or giving injections. Arteries usually follow definite pathways. Veins are more difficult to follow because they connect in irregular networks in which many smaller veins merge to form a larger vein. Although only one systemic artery, the aorta, takes oxygenated blood away from the heart (left ventricle), three systemic veins, the ***coronary sinus, superior vena cava,*** and ***inferior vena cava,*** deliver deoxygenated blood to the right atrium of the heart (Figure 23.12). The coronary sinus receives blood from the cardiac veins; the superior vena cava receives blood from other veins superior to the diaphragm, except the air sacs (alveoli) of the lungs; the inferior vena cava receives blood from veins inferior to the diaphragm.

CHECKPOINT

13. What are the three tributaries of the coronary sinus?

VEIN	DESCRIPTION AND REGION DRAINED
Coronary sinus (KOR-ō-nar-ē; *corona* = crown)	The ***coronary sinus*** is the main vein of the heart; it receives almost all venous blood from the myocardium. It is located in the coronary sulcus (see Figure 22.3b) and opens into the right atrium between the orifice of the inferior vena cava and the tricuspid valve. It is a wide venous channel into which three veins drain. It receives the ***great cardiac vein*** (in the anterior interventricular sulcus) into its left end, and the ***middle cardiac vein*** (in the posterior interventricular sulcus) and the ***small cardiac vein*** into its right end. Several ***anterior cardiac veins*** drain directly into the right atrium.
Superior vena cava (SVC) (VĒ-na CĀ-va; *vena* = vein; *cava* = cavelike)	The ***superior vena cava*** is about 7.5 cm (3 in.) long and 2 cm (1 in.) in diameter and empties its blood into the superior part of the right atrium. It begins posterior to the right first costal cartilage by the union of the right and left brachiocephalic veins and ends at the level of the right third costal cartilage, where it enters the right atrium. The SVC drains the head, neck, chest, and free upper limbs (Figure 23.12).
Inferior vena cava (IVC)	The ***inferior vena cava*** is the largest vein in the body, about 3.5 cm (1.4 in.) in diameter. It begins anterior to the fifth lumbar vertebra by the union of the common iliac veins, ascends behind the peritoneum to the right of the midline, pierces the caval opening of the diaphragm at the level of the eighth thoracic vertebra, and enters the inferior part of the right atrium. The IVC drains the abdomen, pelvis, and free lower limbs (Figure 23.12). The inferior vena cava is commonly compressed during the later stages of pregnancy by the enlarging uterus, producing edema of the ankles and feet and temporary varicose veins.

Figure 23.12 **Principal veins.**

Deoxygenated blood returns to the heart via the superior and inferior venae cavae and the coronary sinus.

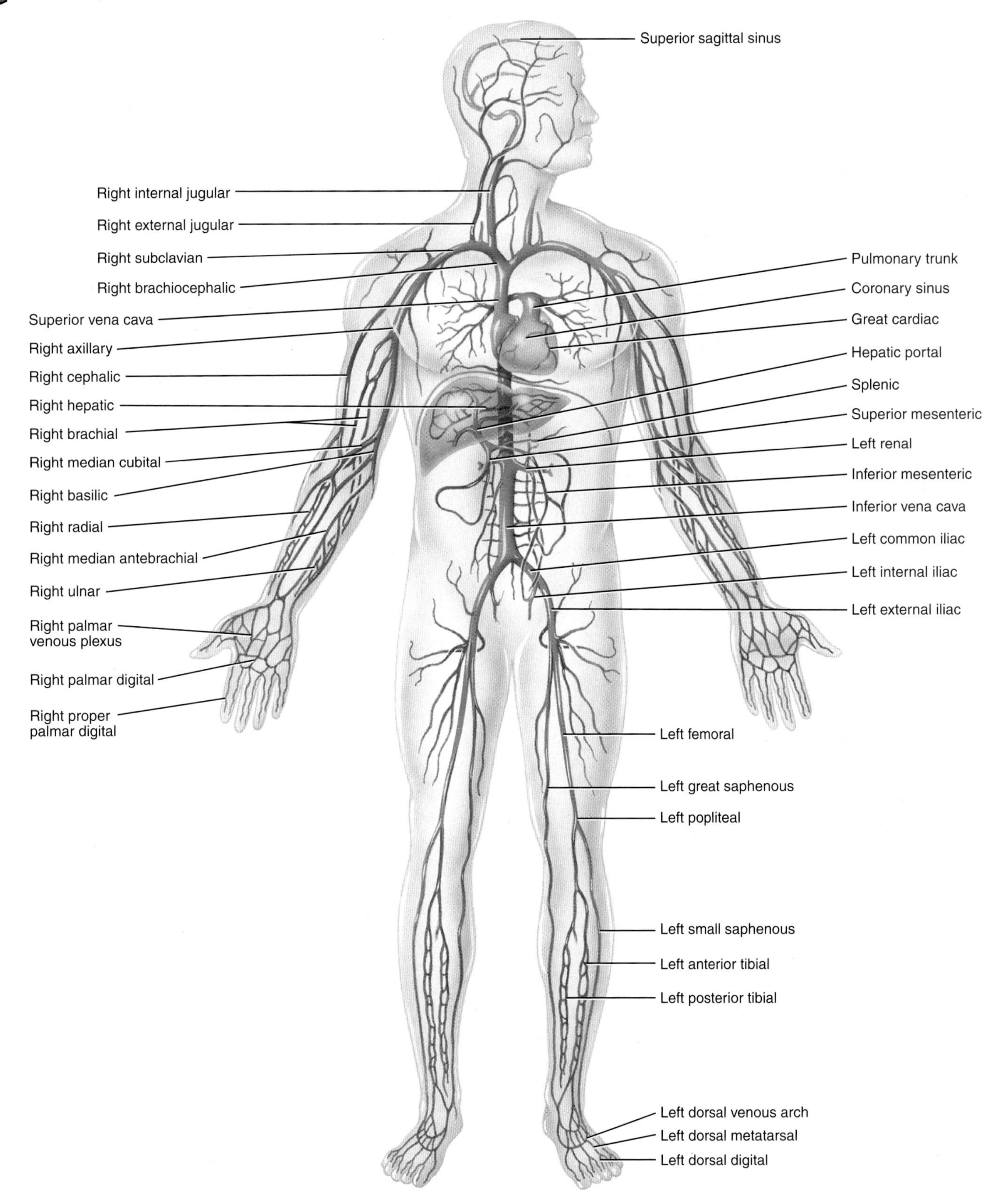

Overall anterior view of the principal veins

Which general regions of the body are drained by the superior vena cava and the inferior vena cava?

EXHIBIT 23.5 Veins of the Head and Neck

OBJECTIVE

- Identify the three major veins that drain blood from the head.

Most blood draining from the head passes into three pairs of veins: the ***internal jugular veins, external jugular veins,*** and ***vertebral veins*** (Figure 23.13). Within the brain, all veins drain into dural venous sinuses and then into the internal jugular veins. ***Dural venous sinuses*** are endothelium-lined venous channels between layers of the cranial dura mater.

CHECKPOINT

14. Which general areas are drained by the internal jugular, external jugular, and vertebral veins?

VEIN	DESCRIPTION AND REGION DRAINED
Internal jugular veins (JUG-ū-lar = throat)	The dural venous sinuses (the light blue vessels in Figure 23.13) drain blood from the cranial bones, meninges, and brain. The right and left ***internal jugular veins*** pass inferiorly on either side of the neck lateral to the internal carotid and common carotid arteries. They then unite with the subclavian veins to form the right and left ***brachiocephalic veins*** (brā-kē-ō-se-FAL-ik; *brachio-* = arm; *-cephalic* = head). From here blood flows into the superior vena cava. The general structures drained by the internal jugular veins are the brain (through the dural venous sinuses), face, and neck.
External jugular veins	The right and left ***external jugular veins*** empty into the subclavian veins. The general structures drained by the external jugular veins are external to the cranium, such as the scalp and superficial and deep regions of the face.
Vertebral veins (VER-te-bral = vertebrae)	The right and left ***vertebral veins*** empty into the brachiocephalic veins in the neck. They drain deep structures in the neck such as the cervical vertebrae, cervical spinal cord, and some neck muscles.

Figure 23.13 Principal veins of the head and neck.

Blood draining from the head passes into the internal jugular, external jugular, and vertebral veins.

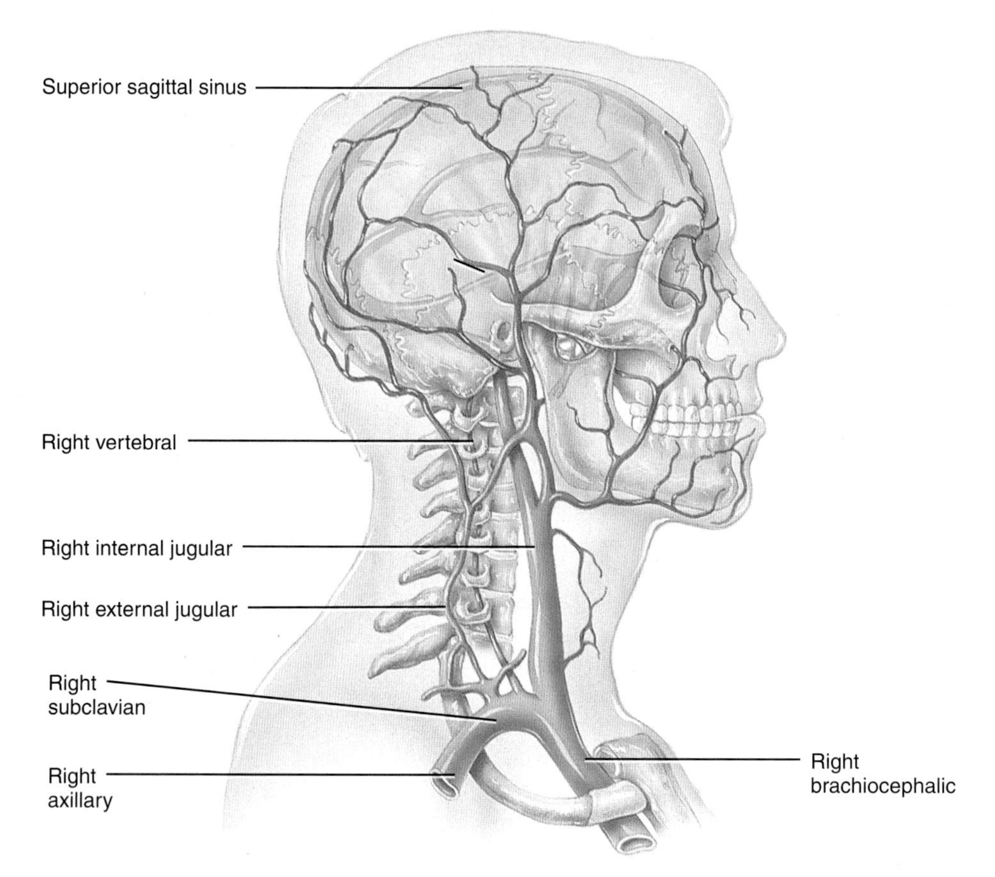

Into which veins in the neck does all the venous blood in the brain drain?

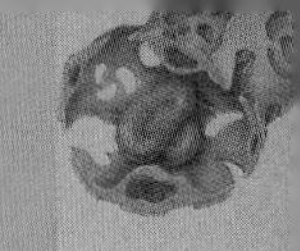

Exhibit 23.6 Veins of the Upper Limbs

OBJECTIVE

- Identify the principal veins that drain the upper limbs.

Blood from the upper limbs is returned to the heart by both superficial and deep veins (see Figure 23.14). Both sets of veins have valves, which are more numerous in the deep veins.

Superficial veins are larger than deep veins and return most of the blood from the upper limbs. ***Deep veins*** are located deep in the body. They usually accompany arteries and have the same names as the corresponding arteries. Both superficial and deep veins have valves, but valves are more numerous in the deep veins.

CHECKPOINT

15. Where do the cephalic, basilic, median antebrachial, radial, and ulnar veins originate?

VEIN	DESCRIPTION AND REGION DRAINED
SUPERFICIAL	
Cephalic veins (se-FAL-ik = pertaining to the head)	The principal superficial veins that drain the upper limbs are the cephalic and basilic vein. They originate in the hand and convey blood from the smaller superficial veins into the axillary veins. The ***cephalic veins*** begin on the lateral aspect of the ***dorsal venous networks of the hands (dorsal venous arches),*** networks of veins on the dorsum of the hands formed by the ***dorsal metacarpal veins*** (Figure 23.14a). These veins, in turn, drain the ***dorsal digital veins,*** which pass along the sides of the fingers. Following their formation from the dorsal venous networks of the hands, the cephalic veins arch around the radial side of the forearms to the anterior surface and ascend through the entire limbs along the anterolateral surface. The cephalic veins end where they join the axillary veins, just inferior to the clavicles. The cephalic veins drain blood from the lateral aspect of the upper limbs.
Basilic veins (ba-SIL-ik = royal, of prime importance)	The ***basilic veins*** begin on the medial aspects of the dorsal venous networks of the hands and ascend along the posteromedial surface of the forearm and anteromedial surface of the arm (Figure 23.14b). They drain blood from the medial aspects of the upper limbs. Anterior to the elbow, the basilic veins are connected to the cephalic veins by the ***median cubital veins*** (*cubital* = pertaining to the elbow), which drain the forearm. If veins must be punctured for an injection, transfusion, or removal of a blood sample, the median cubital veins are preferred. After receiving the median cubital veins, the basilic veins continue ascending until they reach the middle of the arm. There they penetrate the tissues deeply and run alongside the brachial arteries until they join the brachial veins. As the basilic and brachial veins merge in the axillary area, they form the axillary veins.
Median antebrachial veins (an′-tē-BRĀ-kē-al; *ante-* = before, in front of; *-brachi-* = arm)	The ***median antebrachial veins (median veins of the forearm)*** begin in the ***palmar venous plexuses,*** networks of veins on the palms. The plexuses drain the ***palmar digital veins*** in the fingers. The median antebrachial veins ascend anteriorly in the forearms to join the basilic or median cubital veins, sometimes both. They drain the palms and forearms.
DEEP	
Radial veins (RĀ-dē-al = pertaining to the radius)	The paired ***radial veins*** begin at the ***deep palmar venous arches*** (Figure 23.14c). These arches drain the ***palmar metacarpal veins*** in the palms. The radial veins drain the lateral aspects of the forearms and pass alongside the radial arteries. Just inferior to the elbow joint, the radial veins unite with the ulnar veins to form the brachial veins.
Ulnar veins (UL-nar = pertaining to the ulna)	The paired ***ulnar veins,*** which are larger than the radial veins, begin at the ***superficial palmar venous arches.*** These arches drain the ***common palmar digital veins*** and the ***proper palmar digital veins*** in the fingers. The ulnar veins drain the medial aspect of the forearms, pass alongside the ulnar arteries, and join with the radial veins to form the brachial veins.
Brachial veins (BRĀ-kē-al; *brachi-* = arm)	The paired ***brachial veins*** accompany the brachial arteries. They drain the forearms, elbow joints, arms, and humerus. They pass superiorly and join with the basilic veins to form the axillary veins.
Axillary veins (AK-sil-ār-ē; *axilla* = armpit)	The ***axillary veins*** ascend to the outer borders of the first ribs, where they become the subclavian veins. The axillary veins receive tributaries that correspond to the branches of the axillary arteries. The axillary veins drain the arms, axillas, and superolateral chest wall.
Subclavian veins (sub-KLĀ-vē-an; *sub-* = under; *-clavian* = pertaining to the clavicle)	The ***subclavian veins*** are continuations of the axillary veins that terminate at the sternal end of the clavicles, where they unite with the internal jugular veins to form the brachiocephalic veins. The subclavian veins drain the arms, neck, and thoracic wall. In a procedure called *central line placement,* the right subclavian vein is frequently used to administer nutrients and medication and measure venous pressure.

CONTINUES

Figure 23.14 Principal veins of the right upper limb.

Deep veins usually accompany arteries that have similar names.

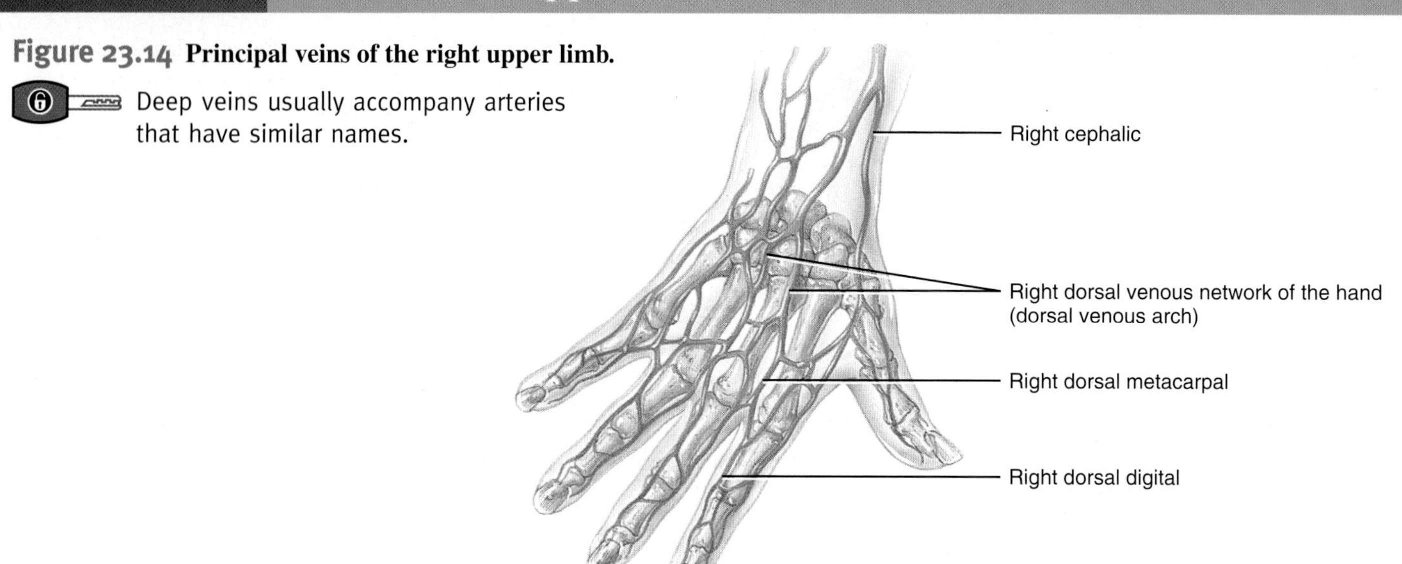

(a) Posterior view of superficial veins of the hand

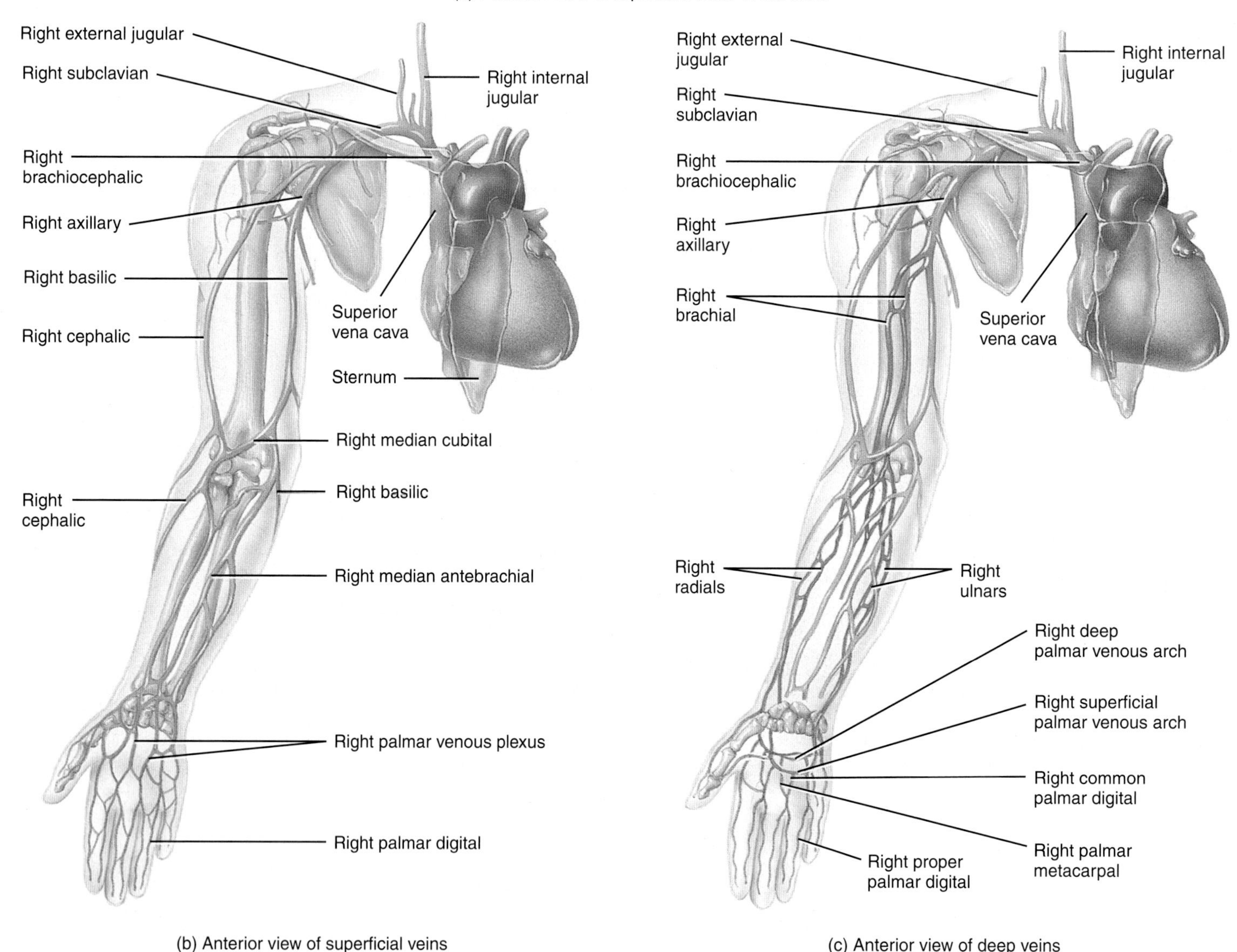

(b) Anterior view of superficial veins

(c) Anterior view of deep veins

? From which vein in the upper limb is a blood sample often taken?

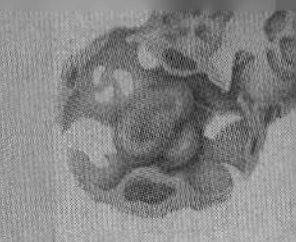

EXHIBIT 23.7 Veins of the Lower Limbs

OBJECTIVE

- Identify the principal veins that drain the lower limbs.

As with the upper limbs, blood from the lower limbs is drained by both superficial and deep veins (see Figure 23.15). The superficial veins often branch with each other and with deep veins along their length. All veins of the lower limbs have valves, which are more numerous than in veins of the upper limbs.

CHECKPOINT

16. Why are the great saphenous veins clinically important?

VEIN	DESCRIPTION AND REGION DRAINED
SUPERFICIAL VEINS	
Great saphenous veins (sa-FĒ-nus; *saphen-* = clearly visible)	The ***great (long) saphenous veins,*** the longest veins in the body, ascend from the foot to the groin in the subcutaneous layer. They begin at the medial end of the dorsal venous arches of the foot (Figure 23.15). The ***dorsal venous arches*** (VĒ-nus) are networks of veins on the dorsum of the foot formed by the ***dorsal digital veins,*** which collect blood from the toes, and then unite in pairs to form the ***dorsal metatarsal veins,*** which parallel the metatarsals. As the dorsal metatarsal veins approach the foot, they combine to form the dorsal venous arches. The great saphenous veins pass anterior to the medial malleolus of the tibia and then superiorly along the medial aspect of the leg and thigh just deep to the skin. They receive tributaries from superficial tissues and connect with the deep veins as well. They empty into the femoral veins at the groin. Generally, the great saphenous veins drain mainly the medial side of the leg and thigh, the groin, external genitals, and abdominal wall. Along their length, the great saphenous veins have from 10 to 20 valves, with more located in the leg than the thigh. These veins are more likely to be subject to varicosities than other veins in the lower limbs because they must support a long column of blood and are not well supported by skeletal muscles. The great saphenous veins are often used for prolonged administration of intravenous fluids. This is particularly important in very young children and in patients of any age who are in shock and whose veins are collapsed. In coronary artery bypass grafting, if multiple blood vessels need to be grafted, sections of the great saphenous vein are used along with at least one artery as a graft. After the great saphenous vein is removed and divided into sections, the sections are used to bypass the blockages. The vein grafts are reversed so that the valves do not obstruct the flow of blood.
Small saphenous veins	The ***small (short) saphenous veins*** begin at the lateral aspect of the dorsal venous arches of the foot (Figure 23.15). They pass posterior to the lateral malleolus of the fibula and ascend deep to the skin along the posterior aspect of the leg. They empty into the popliteal veins in the popliteal fossa, posterior to the knee. Along their length, the small saphenous veins have from 9 to 12 valves. The small saphenous veins drain the foot and posterior aspect of the leg. They may communicate with the great saphenous veins in the proximal thigh.
DEEP VEINS	
Posterior tibial veins (TIB-ē-al)	The ***plantar digital veins*** on the plantar surfaces of the toes unite to form the ***plantar metatarsal veins,*** which parallel the metatarsals. They in turn unite to form the ***deep plantar venous arches.*** From each arch emerges the ***medial*** and ***lateral plantar veins.*** The medial and lateral plantar veins, posterior to the medial malleolus of the tibia, form the paired ***posterior tibial veins*** (Figure 23.15), which sometimes merge into a single vessel. They accompany the posterior tibial artery through the leg. They ascend deep to the muscles in the posterior aspect of the leg and drain the foot and posterior compartment muscles. About two-thirds of the way up the leg, the posterior tibial veins drain blood from the ***fibular (peroneal) veins,*** which drain the lateral and posterior leg muscles. The posterior tibial veins unite with the anterior tibial veins just inferior to the popliteal fossa to form the popliteal veins.
Anterior tibial veins	The paired ***anterior tibial veins*** arise in the dorsal venous arch and accompany the anterior tibial artery (Figure 23.15). They ascend in the interosseous membrane between the tibia and fibula and unite with the posterior tibial veins to form the popliteal vein. The anterior tibial veins drain the ankle joint, knee joint, tibiofibular joint, and anterior portion of the leg.
Popliteal veins (pop′-li-TĒ-al = pertaining to the hollow behind knee)	The ***popliteal veins,*** formed by the union of the anterior and posterior tibial veins, also receive blood from the small saphenous veins and tributaries that correspond to branches of the popliteal artery (Figure 23.15). The popliteal veins drain the knee joint and the skin, muscles, and bones of portions of the calf and thigh around the knee joint.
Femoral veins (FEM-o-ral)	The ***femoral veins*** accompany the femoral arteries and are the continuations of the popliteal veins just superior to the knee (Figure 23.15). The femoral veins extend up the posterior surface of the thighs and drain the muscles of the thighs, femurs, external genitals, and superficial lymph nodes. The largest tributaries of the femoral veins are the ***deep veins of the thigh (deep femoral veins).*** Just before penetrating the abdominal wall, the femoral veins receive the deep femoral veins and the great saphenous veins. The veins formed from this union penetrate the body wall and enter the pelvic cavity. Here they are known as the ***external iliac veins.*** In order to take blood samples or pressure recordings from the right side of the heart, a catheter is inserted into the femoral vein as it passes through the femoral triangle. The catheter passes through the external and common iliac veins and inferior vena cava into the right atrium.

CONTINUES

Figure 23.15 Principal veins of the pelvis and lower limbs.

All veins of the lower limbs have valves.

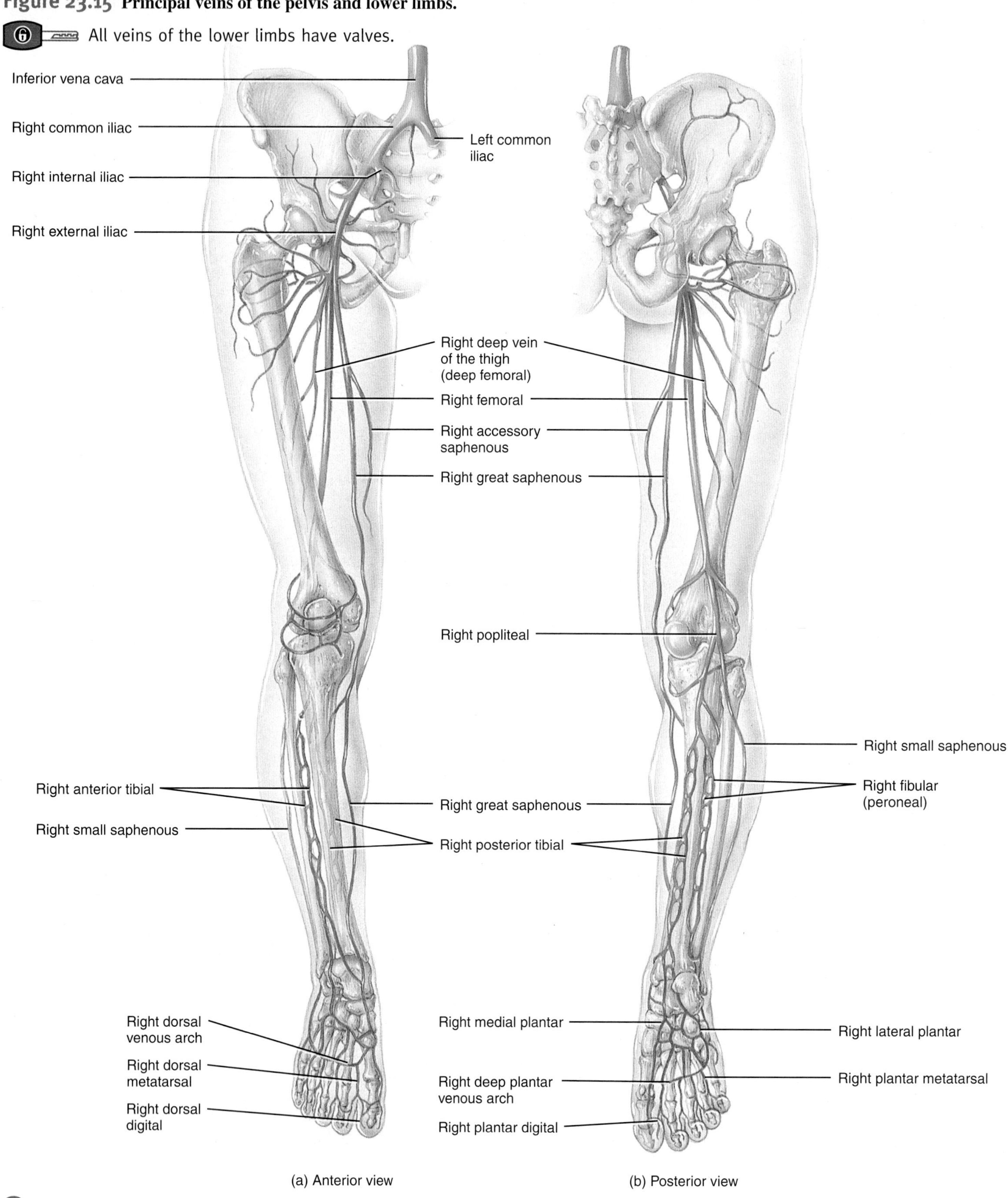

(a) Anterior view

(b) Posterior view

Which veins of the lower limb are superficial?

Hepatic Portal Circulation

Blood normally travels from arteries to arterioles to capillaries to venules to veins. A vein that carries blood between one capillary network and another is called a ***portal vein.*** The hepatic portal vein, formed by the union of the splenic and superior mesenteric veins (Figure 23.16), receives blood from capillaries of digestive organs and delivers it to capillary-like structures in the liver called sinusoids. In the ***hepatic portal circulation*** (*hepat-* = liver), venous blood from the gastrointestinal organs and spleen, rich with substances absorbed from the gastrointestinal tract, is delivered to the hepatic portal vein and enters the liver. The liver processes these substances before they pass into the general circulation. At the same time, the liver receives oxygenated blood

Figure 23.16 Hepatic portal circulation.

The hepatic portal circulation delivers venous blood from the gastrointestinal organs and spleen to the liver.

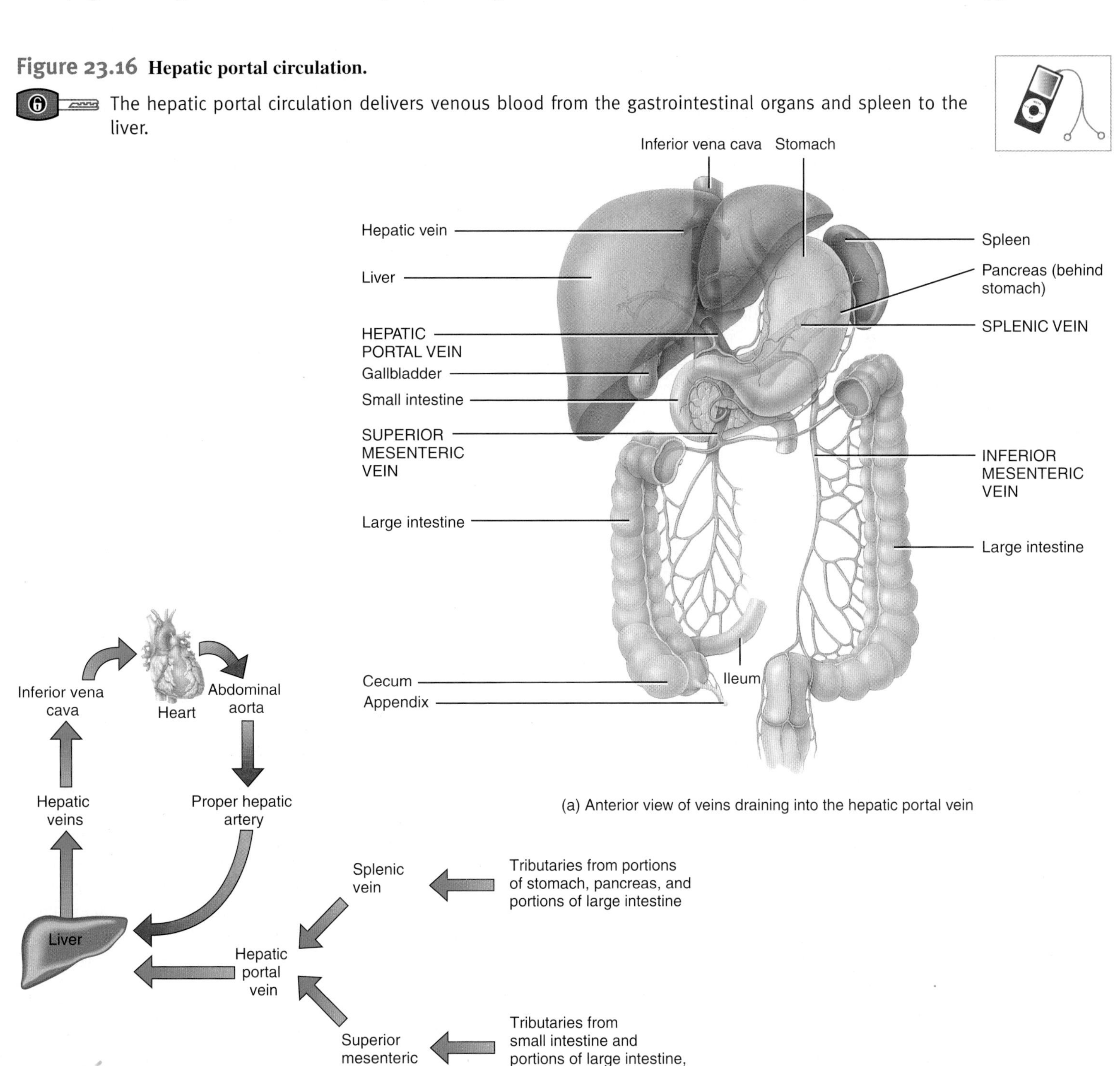

(a) Anterior view of veins draining into the hepatic portal vein

(b) Scheme of principal blood vessels of hepatic portal circulation and arterial supply and venous drainage of liver

Which veins carry blood away from the liver?

from the systemic circulation via the hepatic artery. The oxygenated blood mixes with the deoxygenated blood in sinusoids. ***Sinusoids*** (SĪ-nū-soyds) are large, thin-walled, and leaky types of capillaries, having large intercellular clefts. Ultimately, all blood leaves the sinusoids of the liver through the hepatic veins, which drain into the inferior vena cava. This pattern will be studied again in Chapter 26.

Fetal Circulation

The circulatory system of a fetus, called ***fetal circulation,*** contains special structures that allow the developing fetus to exchange materials with its mother (Figure 23.17). It differs from the postnatal (after birth) circulation because the lungs, kidneys, and gastrointestinal organs do not begin to function until birth. The fetus obtains O_2 and nutrients from and eliminates CO_2 and other wastes into the maternal blood.

The exchange of materials between fetal and maternal circulations occurs through the ***placenta*** (pla-SEN-ta), which forms inside the mother's uterus and attaches to the umbilicus (navel) of the fetus by the ***umbilical cord*** (um-BIL-i-kal). Blood does not normally mix between the mother and the fetus; capillaries of both individuals are close enough in the placenta for the diffusion of nutrients and waste products.

Blood passes from the fetus to the placenta via two ***umbilical arteries*** (Figure 23.17a). These branches of the internal iliac arteries are within the umbilical cord. At the placenta, fetal blood picks up O_2 and nutrients and eliminates CO_2 and wastes. The oxygenated blood returns from the placenta via a single ***umbilical vein.*** This vein ascends to the liver of the fetus, where it divides into two branches. Some blood flows through the branch that joins the hepatic portal vein and enters the liver, but most of the blood flows into the second branch, the ***ductus venosus*** (DUK-tus ve-NŌ-sus), which drains into the inferior vena cava.

Deoxygenated blood returning from lower body regions of the fetus mingles with oxygenated blood from the ductus venosus in the inferior vena cava. This mixed blood then enters the right atrium. Deoxygenated blood returning from upper body regions of the fetus enters the superior vena cava and also passes into the right atrium.

Most of the fetal blood does not pass from the right ventricle to the lungs, as it does in postnatal circulation, because an opening called the ***foramen ovale*** (fō-RĀ-men ō-VAL-ē) exists in the septum between the right and left atria. About one-third of the blood that enters the right atrium passes through the foramen ovale into the left atrium and joins the systemic circulation. The blood that does pass into the right ventricle is pumped into the pulmonary trunk, but little of this blood reaches the nonfunctioning fetal lungs. Instead, most is sent through the ***ductus arteriosus*** (ar-tē-rē-Ō-sus), a vessel that connects the pulmonary trunk with the aorta, so that most blood bypasses the fetal lungs. The blood in the aorta is carried to all fetal tissues through the systemic circulation. When the common iliac arteries branch into the external and internal iliacs, part of the blood flows into the internal iliacs, into the umbilical arteries, and back to the placenta for another exchange of materials.

After birth, when pulmonary (lung), renal (kidney), and digestive functions begin, the following vascular changes occur (Figure 23.17b):

1. When the umbilical cord is tied, blood no longer flows through the umbilical arteries, they fill with connective tissue, and the distal portions of the umbilical arteries become fibrous cords called ***medial umbilical ligaments.***

2. The umbilical vein collapses but remains as the ***ligamentum teres*** **(*round ligament*)**, a structure that attaches the umbilicus to the liver.

3. The ductus venosus collapses but remains as the ***ligamentum venosum,*** a fibrous cord on the inferior surface of the liver.

4. The placenta is expelled as the ***afterbirth.***

5. The foramen ovale normally closes shortly after birth to become the ***fossa ovalis,*** a depression in the interatrial septum. When an infant takes its first breath, the lungs expand and blood flow to the lungs increases. Blood returning from the lungs to the heart increases pressure in the left atrium. This closes the foramen ovale by pushing the valve that guards it against the interatrial septum. Permanent closure occurs in about a year.

6. The ductus arteriosus closes by vasoconstriction almost immediately after birth and becomes the ***ligamentum arteriosum*** (see Figure 22.3c, Section 22.1).

CHECKPOINT

17. What are the main functions of the systemic, pulmonary, hepatic portal, and fetal circulations?

23.5 AGING AND THE CARDIOVASCULAR SYSTEM

OBJECTIVE

- Describe the effects of aging on the cardiovascular system.

General changes in the cardiovascular system associated with aging include increased stiffness of the aorta, reduction in cardiac muscle fiber size, progressive loss of cardiac muscular strength, reduced cardiac output, a decline in maximum heart rate, and an increase in systolic blood pressure. As seen in cadavers of certain elderly persons, the aorta initially feels hard, but a little pressure on the aorta causes the internal contents to crack and break, somewhat analogous to punching through an uncooked chicken egg. Coronary artery disease (CAD) is the major cause of heart disease and death in older Americans. Congestive heart failure (CHF), a set of symptoms associated with impaired pumping of the heart, is also prevalent in older individuals. Changes in blood vessels that serve brain tissue—for example, atherosclerosis—reduce nourishment to the brain and result in the malfunction or death of brain cells. By age 80, blood flow to the brain is 20% less, and blood flow to the kidneys is 50% less, than it was in the same person at age 30.

To appreciate the many ways the cardiovascular system contributes to the homeostasis of other body systems, examine Exhibit 23.8. Next, in Chapter 24, we will examine the structure and function of the lymphatic system, and how it returns excess fluid filtered from capillaries to the cardiovascular system. We will also take a more detailed look at how some white blood cells function as defenders of the body by carrying out immune responses.

CHECKPOINT

18. What are some of the signs that the cardiovascular system is aging?

Figure 23.17 Fetal circulation and changes at birth.

The lungs and gastrointestinal organs do not begin to function until birth.

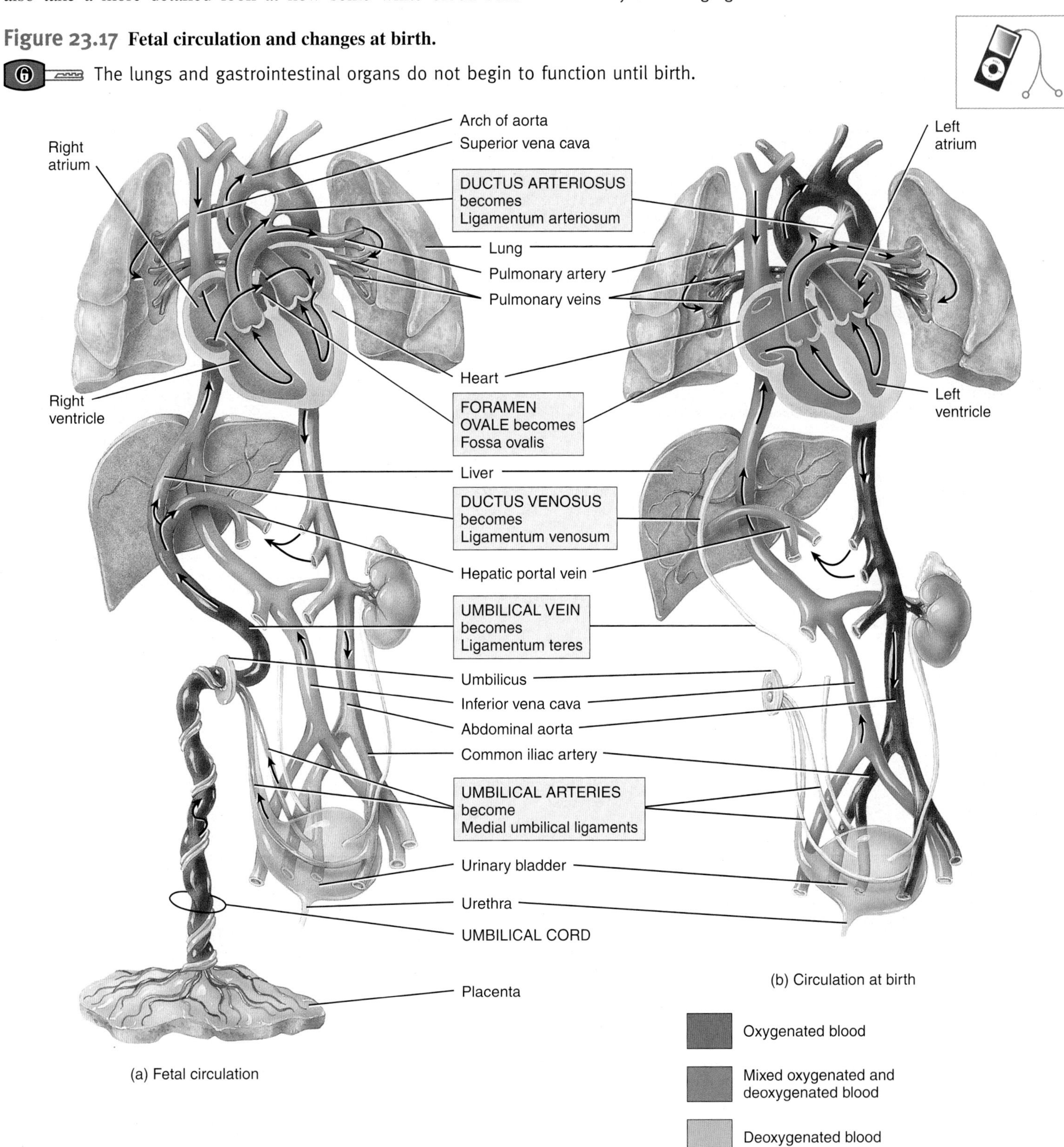

Which structure provides for exchange between mother and fetus?

EXHIBIT 23.8 Contributions of the Cardiovascular System to Homeostasis

BODY SYSTEM		CONTRIBUTION OF THE CARDIOVASCULAR SYSTEM
For all body systems		The heart pumps blood through blood vessels to body tissues, delivering oxygen and nutrients and removing wastes by means of capillary exchange. Circulating blood keeps body tissues at a proper temperature.
Integumentary system	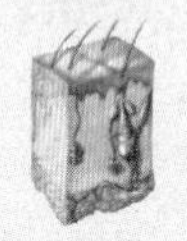	Blood delivers clotting factors and white blood cells that aid in hemostasis when skin is damaged and contribute to repair of injured skin. Changes in skin blood flow contribute to body temperature regulation by adjusting the amount of heat loss via the skin. Blood flowing in skin may give skin a pink hue.
Skeletal system	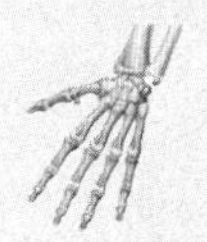	Blood delivers calcium and phosphate ions that are needed for building bone extracellular matrix, hormones that govern building and breakdown of bone extracellular matrix, and erythropoietin that stimulates production of red blood cells by red bone marrow.
Muscular system		Blood circulating through exercising muscles removes heat and lactic acid.
Nervous system	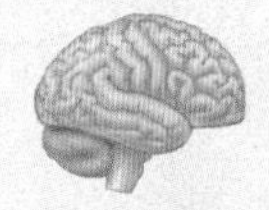	Endothelial cells lining choroid plexuses in brain ventricles help produce cerebrospinal fluid (CSF) and contribute to the blood–brain barrier.
Endocrine system		Circulating blood delivers most hormones to their target tissues. Atrial cells of the heart secrete atrial natriuretic peptide.
Lymphatic system and immunity		Circulating blood distributes lymphocytes, antibodies, and macrophages that carry out immune functions. Lymph forms from excess interstitial fluid, which filters from blood plasma due to blood pressure generated by the heart.
Respiratory system	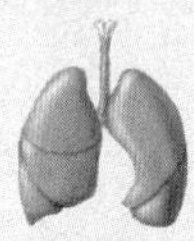	Circulating blood transports oxygen from the lungs to body tissues and carbon dioxide to the lungs for exhalation.
Digestive system	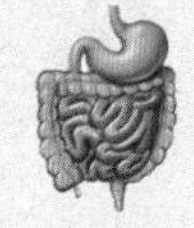	Blood carries newly absorbed nutrients and water to the liver. Blood distributes hormones that aid digestion.
Urinary system	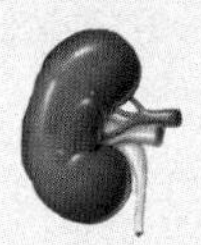	The heart and blood vessels deliver 20% of the resting cardiac output to the kidneys, where blood is filtered, needed substances are reabsorbed, and unneeded substances are eliminated as part of urine, which is excreted.
Reproductive systems	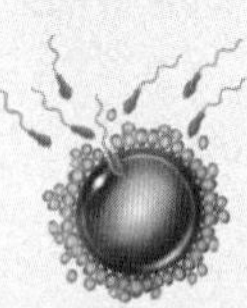	Vasodilation of arterioles in the penis and clitoris causes erection during sexual intercourse. Blood distributes hormones that regulate reproductive functions.

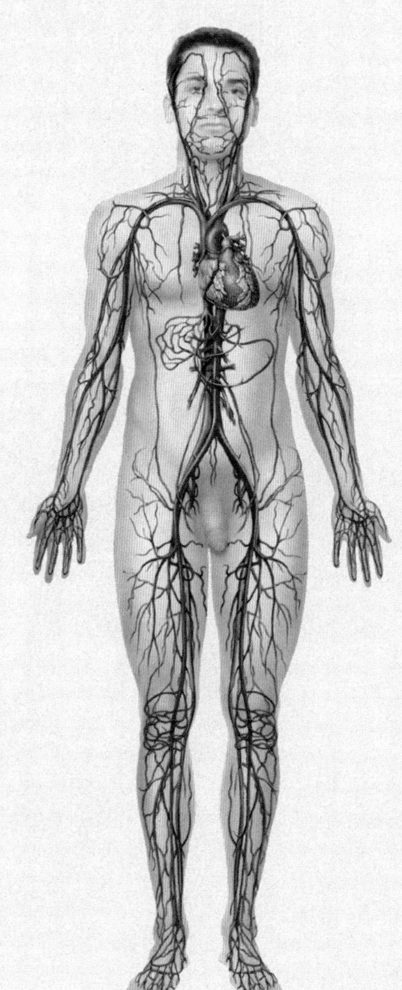

KEY MEDICAL TERMS ASSOCIATED WITH BLOOD VESSELS

Aneurysm (AN-ū-rizm) A thin, weakened section of the wall of an artery or a vein that bulges outward, forming a balloonlike sac. If untreated, the aneurysm enlarges and the blood vessel wall becomes so thin that it bursts. The result is massive hermorrhage with shock, severe pain, stroke, or death.

Angiogenesis (an′-jē-ō-JEN-e-sis) Formation of new blood vessels.

Aortography (ā-or-TOG-ra-fē) X-ray examination of the aorta and its main branches after injection of a dye.

Circulation time The time required for a drop of blood to pass from the right atrium, through the pulmonary circulation, back to the left atrium, through the systemic circulation, down to the foot, and back again to the right atrium; normally about 1 minute in a resting person.

Claudication (klaw-di-KĀ-shun) Pain and lameness or limping caused by defective circulation of the blood in the vessels of the limbs.

Deep-venous thrombosis (DVT) The presence of a thrombus (blood clot) in a deep vein of the lower limbs.

Hypotension (hī-pō-TEN-shun) Low blood pressure; most commonly used to describe an acute drop in blood pressure, as occurs during excessive blood loss.

Occlusion (ō-KLOO-zhun) The closure or obstruction of the lumen of a structure such as a blood vessel. An example is an atherosclerotic plaque in an artery.

Orthostatic hypotension (or-thō-STAT-ik; *ortho-* = straight; *-static* = causing to stand) An excessive lowering of systemic blood pressure when a person stands up; usually a sign of disease. May be caused by excessive fluid loss, certain drugs, and cardiovascular or neurogenic factors. Also called ***postural hypotension.***

Phlebitis (fle-BĪ-tis; *phleb-* = vein) Inflammation of a vein, often in a leg. The condition is often accompanied by pain and redness of the skin over the inflamed vein. It is frequently caused by trauma or bacterial infection.

Syncope (SIN-kō-pē) A temporary cessation of consciousness; a faint. One cause is insufficient blood supply to the brain.

Thrombophlebitis (throm-bō-fle-BĪ-tis) Inflammation of a vein involving clot formation. Superficial thrombophlebitis occurs in veins under the skin, especially in the calf.

White coat (office) hypertension A stress-induced syndrome found in patients who have elevated blood pressure when being examined by health-care personnel but otherwise have normal blood pressure.

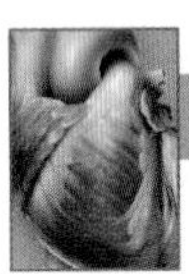

STUDY OUTLINE

Blood Vessel Structure and Function (Section 23.1)

1. Arteries carry blood away from the heart. Their walls consist of three layers.
2. The smooth muscle of the middle layer is responsible for vasoconstriction.
3. Arterioles are small arteries that deliver blood to capillaries.
4. Through constriction and dilation, arterioles play a key role in regulating blood flow from arteries into capillaries.
5. Capillaries are microscopic blood vessels through which materials are exchanged between blood and interstitial fluid.
6. Precapillary sphincters regulate blood flow through capillaries.
7. Capillary blood pressure "pushes" fluid out of capillaries into interstitial fluid (filtration).
8. Blood colloid osmotic pressure "pulls" fluid into capillaries from interstitial fluid (reabsorption).
9. Autoregulation refers to local adjustments of blood flow in response to physical and chemical changes in a tissue.
10. Venules are small vessels that emerge from capillaries and merge to form veins. They drain blood from capillaries into veins.
11. Veins consist of the same three layers as arteries but have less elastic tissue and smooth muscle. They contain valves that prevent backflow of blood.
12. Weak venous valves can lead to varicose veins.
13. Venous return, the volume of blood flowing back to the heart through systemic veins, occurs due to the pumping action of the heart, aided by skeletal muscle contractions (the skeletal muscle pump) and breathing (the respiratory pump).

Blood Flow through Blood Vessels (Section 23.2)

1. Blood flow is determined by blood pressure and vascular resistance.
2. Blood flows from regions of higher pressure to regions of lower pressure.
3. Blood pressure is highest in the aorta and large systemic arteries; it drops progressively as distance from the left ventricle increases. Blood pressure in the right atrium is close to 0 mmHg.
4. An increase in blood volume increases blood pressure, and a decrease in blood volume decreases it.
5. Vascular resistance is the opposition to blood flow mainly as a result of friction between blood and the walls of blood vessels.
6. Vascular resistance depends on size of the blood vessel lumen, blood viscosity, and total blood vessel length.
7. Blood pressure and blood flow are regulated by neural and hormonal negative feedback systems and by autoregulation.
8. The cardiovascular center in the medulla oblongata helps regulate heart rate, stroke volume, and size of blood vessel lumen.
9. Vasomotor nerves (sympathetic) control vasoconstriction and vasodilation.
10. Baroreceptors (pressure-sensitive receptors) send impulses to the cardiovascular center to regulate blood pressure.
11. Chemoreceptors (receptors sensitive to concentrations of oxygen, carbon dioxide, and hydrogen ions) also send impulses to the cardiovascular center to regulate blood pressure.
12. Hormones such as angiotensin II, aldosterone, epinephrine, norepinephrine, and antidiuretic hormone raise blood pressure; atrial natriuretic peptide lowers it.

Checking Circulation (Section 23.3)

1. Pulse is the alternate expansion and elastic recoil of an artery with each heartbeat. It may be felt in any artery that lies near the surface or over a hard tissue.
2. A normal pulse rate is about 75 beats per minute.

3. Blood pressure is the pressure exerted by blood on the wall of an artery when the left ventricle undergoes systole and then diastole. It is measured by a sphygmomanometer.
4. Systolic blood pressure (SBP) is the force of blood recorded during ventricular contraction. Diastolic blood pressure (DBP) is the force of blood recorded during ventricular relaxation. The normal blood pressure of a young adult male is less than 120/80 mmHg.

Circulatory Routes (Section 23.4)

1. The two major circulatory routes are the systemic circulation and the pulmonary circulation.
2. The systemic circulation takes oxygenated blood from the left ventricle through the aorta to all parts of the body and returns deoxygenated blood to the right atrium.
3. The parts of the aorta include the ascending aorta, the arch of the aorta, the thoracic aorta, and the abdominal aorta. Each part gives off arteries that branch to supply the whole body.
4. Deoxygenated blood is returned to the heart through the systemic veins. All the veins of systemic circulation flow into either the superior or inferior vena cava or the coronary sinus, which all empty into the right atrium.
5. The pulmonary circulation takes deoxygenated blood from the right ventricle to the air sacs of the lungs and returns oxygenated blood from the air sacs to the left atrium. It allows blood to be oxygenated for the systemic circulation.
6. The hepatic portal circulation collects deoxygenated blood from the veins of the gastrointestinal tract and spleen and directs it into the hepatic portal vein of the liver. This routing allows the liver to extract and modify nutrients and detoxify harmful substances in the blood. The liver also receives oxygenated blood from the hepatic artery.
7. Fetal circulation exists only in the fetus. It involves the exchange of materials between fetus and mother via the placenta. The fetus derives O_2 and nutrients from and eliminates CO_2 and wastes into maternal blood. At birth, when pulmonary (lung), digestive, and liver functions begin, the special structures of fetal circulation are no longer needed.

Aging and the Cardiovascular System (Section 23.5)

1. General changes associated with aging include reduced elasticity of blood vessels, reduction in cardiac muscle size, reduced cardiac output, and increased systolic blood pressure.
2. The incidence of coronary artery disease (CAD), congestive heart failure (CHF), and atherosclerosis increases with age.

SELF-QUIZ QUESTIONS

1. Sensory receptors that monitor changes in the blood pressure to the brain are
   a. chemoreceptors in the aorta
   b. baroreceptors in the carotid arteries
   c. the aortic bodies
   d. precapillary sphincters in the arterioles
   e. proprioceptors in the muscles.
2. The blood vessels that allow the exchange of nutrients, wastes, oxygen, and carbon dioxide between the blood and tissues are the
   a. capillaries **b.** arteries
   c. venules **d.** arterioles
   e. veins.
3. Substances undergo capillary exchange by means of
   a. simple diffusion and bulk flow
   b. endocytosis, exocytosis, and active transport
   c. simple diffusion and facilitated diffusion
   d. simple diffusion and active transport
   e. filtration, reabsorption, and secretion.
4. Blood flows through the blood vessels because of the
   a. establishment of a concentration gradient
   b. elastic recoil of the veins
   c. establishment of a pressure gradient
   d. viscosity (stickiness) of the blood
   e. thinness of the walls of capillaries.
5. Which of the following represents pulmonary circulation as the blood flows from the right ventricle?
   a. pulmonary trunk → pulmonary veins → pulmonary capillaries → pulmonary arteries
   b. pulmonary arteries → pulmonary capillaries → pulmonary trunk → pulmonary veins
   c. pulmonary capillaries → pulmonary trunk → pulmonary arteries → pulmonary veins
   d. pulmonary trunk → pulmonary arteries → pulmonary capillaries → pulmonary veins
   e. pulmonary veins → pulmonary capillaries → pulmonary arteries → pulmonary trunk
6. The tissue that allows arteries to stretch is
   a. endothelium **b.** collagen
   c. basement membrane **d.** cardiac muscle
   e. elastic lamina.
7. Match the following descriptions to the appropriate blood vessel:

____	**a.** composed of a single layer of endothelial cells and a basement membrane	**A.** arteries
		B. arterioles
		C. veins
____	**b.** formed by reuniting capillaries	**D.** venules
____	**c.** carry blood away from heart	**E.** capillaries
____	**d.** regulate blood flow to capillaries	
____	**e.** may contain valves	

8. Filtration of substances out of capillaries occurs when the capillary blood pressure
   a. is less than the blood colloid osmotic pressure
   b. and the blood colloid osmotic pressure are equal
   c. is high and the blood colloid osmotic pressure is high
   d. is higher than the blood colloid osmotic pressure
   e. is low and the blood colloid osmotic pressure is low.
9. Weakened leg muscles would slow the
   a. blood flow out of the heart
   b. respiratory pump
   c. venous return
   d. ability of arteries to vasodilate
   e. pulse.

10. Which of the following statements about blood vessels is true?
a. Capillaries contain valves.
b. Walls of arteries are generally thicker and contain more elastic tissue than walls of veins.
c. Veins carry blood away from the heart.
d. Blood flows most rapidly through veins.
e. Blood pressure in arteries is always lower than in veins.

11. Why is it important that blood flows slowly through the capillaries?
a. It allows time for the materials in the blood to pass through the thick capillary walls.
b. It prevents damage to the capillaries.
c. It permits the efficient exchange of nutrients and wastes between the blood and body cells.
d. It allows the heart time to rest.
e. It allows the blood pressure in capillaries to rise above the blood pressure in the veins.

12. Match the following:

____	**a.** source of all systemic arteries	**A.** hepatic portal vein
____	**b.** supplies a lower limb	**B.** pulmonary trunk
____	**c.** heart's blood system	**C.** pulmonary vein
____	**d.** returns blood to heart from lower limbs	**D.** common iliac artery
____	**e.** carries blood to liver	**E.** coronary circulation
____	**f.** leads to lungs	**F.** inferior vena cava
____	**g.** returns blood from lungs to heart	**G.** superior vena cava
____	**h.** supplies blood to brain	**H.** aorta
____	**i.** returns blood to heart from head and upper body	**I.** cerebral arterial circle

13. For each of the following factors, indicate if it increases (**A**) or decreases (**B**) blood pressure:
____ **a.** an increase in cardiac output
____ **b.** hemorrhage
____ **c.** vasodilation
____ **d.** vasoconstriction
____ **e.** stimulation of the heart by the sympathetic nervous system
____ **f.** hypoxia
____ **g.** epinephrine
____ **h.** increase in blood volume
____ **i.** bradycardia

14. Aldosterone affects blood pressure by
a. increasing heart rate
b. increasing vasoconstriction of arterioles
c. reducing blood volume
d. stimulating release of atrial natriuretic peptide by the heart
e. increasing reabsorption of sodium ions and water by the kidneys.

15. In a blood pressure reading of 110/70,
a. 110 represents the diastolic pressure
b. 70 represents the pressure of the blood against the arteries during ventricular relaxation
c. 110 represents the blood pressure and 70 represents the heart rate
d. 70 is the reading taken when the first sound is heard
e. the patient has a severe problem with hypertension.

16. Which of the following statements is NOT true?
a. Regulation of blood vessel diameter originates from the vasomotor region of the cerebral cortex.
b. The cerebral cortex may provide input to the CV center.
c. Baroreceptors may stimulate the cardiovascular center.
d. Activation of proprioceptors increases heart rate at the beginning of exercise.
e. Vasomotor tone is due to a moderate level of vasoconstriction.

17. Venous return to the heart is enhanced by all of the following EXCEPT
a. skeletal muscle "milking"
b. valves in veins
c. the pressure difference from venules to the right ventricle
d. vasodilation
e. inhalation during breathing.

CRITICAL THINKING QUESTIONS

1. The local anesthetic injected by a dentist often contains a small amount of epinephrine. What effect would epinephrine have on the blood vessels in the vicinity of the dental work? Why might this effect be desired?
2. In this chapter, you've read about varicose veins. Why didn't you read about varicose arteries?
3. Julie was flushed when she ran in late to her anatomy and physiology lab. She had spilled a cup of coffee on herself while she was weaving in and out of traffic. Then she missed her exit while she was changing the station of the radio, couldn't find a place to park, and missed the lab quiz. The lab today is learning to take blood pressures, and Julie's is high! (It's normally 110 over 70.) What is the physiological explanation for Julie's elevated BP?
4. Peter spent 10 minutes sharpening his favorite knife before carving the roast. Unfortunately, he sliced his finger along with the roast. His wife slapped a towel over the spurting cut and drove him to the emergency room. What type of vessel did Peter cut, and how do you know?

? ANSWERS TO FIGURE QUESTIONS

23.1 The femoral artery has the thicker wall; the femoral vein has the wider lumen.

23.2 Metabolically active tissues use O_2 and produce wastes more rapidly than inactive tissues, so they require more extensive capillary networks.

23.3 Excess filtered fluid and proteins that escape from plasma drain into lymphatic capillaries and are returned by the lymphatic system to the cardiovascular system.

23.4 The skeletal muscle pump and respiratory pump aid venous return.

23.5 Mean blood pressure in the aorta is closer to diastolic than to systolic pressure.

23.6 The effector tissues regulated by the cardiovascular center are cardiac muscle in the heart and smooth muscle in blood vessel walls.

23.7 It represents a change that occurs when you stand up because gravity causes pooling of blood in leg veins once you are upright, decreasing the blood pressure in your upper body.

23.8 The two main circulatory routes are the systemic circulation and the pulmonary circulation.

23.9 The subdivisions of the aorta are the ascending aorta, arch of the aorta, thoracic aorta, and abdominal aorta.

23.10 The cerebral arterial circle equalizes blood pressure to the brain and provides alternate routes for blood flow to the brain in case of damaged arteries.

23.11 The abdominal aorta divides into the common iliac arteries at about the level of L4.

23.12 The superior vena cava drains regions above the diaphragm, and the inferior vena cava drains regions below the diaphragm.

23.13 All venous blood in the brain drains into the internal jugular veins.

23.14 The median cubital vein of the free upper limb is often used for withdrawing blood.

23.15 Superficial veins of the free lower limbs are the dorsal venous arch and the great saphenous and small saphenous veins.

23.16 The hepatic veins carry blood away from the liver.

23.17 Exchange of materials between mother and fetus occurs across the placenta.

The Lymphatic System and Immunity

24

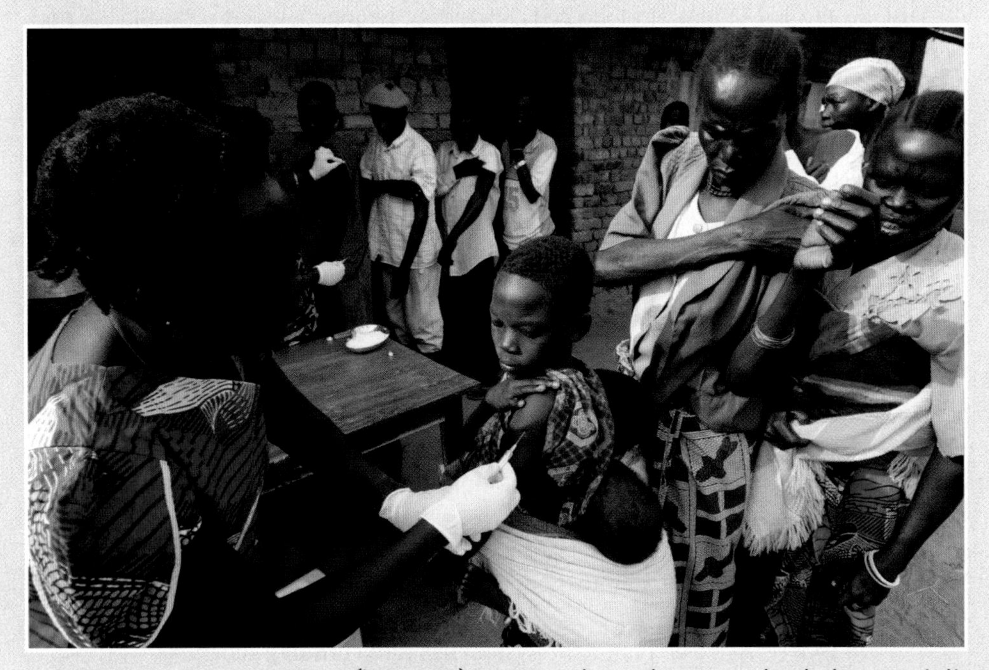

We often overlook the lymphatic system because we do not see evidence of its actions unless we are sick. But even as we sit here reading, the lymphatic system is constantly checking our body for foreign cells and materials. It looks at chemical markers on our own cells and determines what is us and what is not. Disease-causing agents such as viruses and bacteria are detected by the lymphatic (immune) system throughout our body by specialized cells and tissues. For example, when we eat food we scan it using collections of lymphatic tissues at the back of our throat called tonsils. The throat is a gateway to the rest of our body, so we must detect any potential danger right away. As food travels down to the stomach, then to the small intestine, we see special collections of tissue called Peyer's patches. They too monitor the food we are breaking down to prevent organisms from being absorbed into the bloodstream with the nutrition.

There are many organs that are specialized to assist immune function such as our thymus, spleen, and lymph nodes. The lymph nodes filter lymphatic fluid. These structures not only detect foreign agents but also assist in returning fluid to the cardiovascular system from the interstitial fluid. For manual therapists, this pattern of fluid return is an important component to treating inflammation and edema.

CONTENTS AT A GLANCE

24.1 OVERVIEW OF IMMUNITY

Maintaining homeostasis in the body requires continual combat against harmful agents in our internal and external environment. Despite constant exposure to a variety of ***pathogens*** (PATH-ō-jens), disease-producing microbes such as bacteria and viruses, most people remain healthy. The body surface also endures cuts and bumps, exposure to ultraviolet rays in sunlight, chemical toxins, and minor burns with an array of defensive ploys.

Immunity or ***resistance*** is the ability to ward off damage or disease through our defenses. Vulnerability or lack of resistance is termed ***susceptibility***. The two general types of resistance are (1) innate and (2) adaptive. ***Innate (nonspecific) immunity*** refers to defenses that are present at birth. They are always present and available to provide rapid responses to protect us against disease. Innate immunity does not involve specific recognition of a microbe and acts against all microbes in the same way. In addition, innate immunity does not have a memory component, that is, it cannot recall a previous contact with a foreign molecule. Among the components of innate immunity are the first line of defense (the physical and chemical barriers of the skin and mucous membranes) and the second line of defense (antimicrobial substances, phagocytes, natural killer cells, inflammation, and fever). Innate immune responses represent immunity's early warning system and are designed to prevent microbes from gaining access into the body and to help eliminate those that do gain access.

Adaptive (specific) immunity refers to defenses that involve specific recognition of a microbe once it has breached the innate immunity defenses. Adaptive immunity is based on a specific response to a specific microbe; that is, it adapts or adjusts to handle a specific microbe. Unlike innate immunity, adaptive immunity is slower to respond but it does have a memory component. Adaptive immunity involves lymphocytes (a type of white blood cell) called T lymphocytes (T cells) and B lymphocytes (B cells).

The body system responsible for adaptive immunity (and some aspects of innate immunity) is the lymphatic system. This system is closely allied with the cardiovascular system, and it also functions with the digestive system in the absorption of fatty foods. In this chapter, we explore the mechanisms that provide defenses against intruders and promote the repair of damaged body tissues.

24.2 LYMPHATIC SYSTEM STRUCTURE AND FUNCTION

OBJECTIVES

- Describe the components and major functions of the lymphatic system.
- Describe the organization of lymphatic vessels and the circulation of lymph.
- Compare the structure and functions of the primary and secondary lymphatic organs and tissues.

The body system responsible for adaptive immunity (and some aspects of innate immunity) is the ***lymphatic system*** (lim-FAT-ik), which consists of lymph, lymphatic vessels, a number of structures and organs containing lymphatic tissue (Figure 24.1), and red bone marrow, where stem cells develop into various types of blood cells, including lymphocytes (B cells and T cells). ***Lymphatic tissue*** is a specialized form of reticular connective tissue (see Table 4.4c, Section 4.4) that contains large numbers of lymphocytes.

Most components of blood plasma filter out of blood capillary walls to form *interstitial fluid,* the fluid that surrounds the cells of body tissues. After interstitial fluid passes into lymphatic vessels, it is called ***lymph*** (LIMF = clear fluid). Both fluids are chemically similar to blood plasma. The main difference is that interstitial fluid and lymph contain less protein than blood plasma because most plasma protein molecules are too large to filter through the capillary wall. Each day, about 20 liters of fluid filter from blood into tissue spaces. This fluid must be returned to the cardiovascular system to maintain normal blood volume. About 17 liters of the fluid filtered daily from the arterial end of blood capillaries return to the blood directly by reabsorption at the venous end of the capillaries. The remaining 3 liters per day pass first into lymphatic vessels and are then returned to the blood.

The lymphatic system has three primary functions:

1. **Draining excess interstitial fluid.** Lymphatic vessels drain excess interstitial fluid and leaked proteins from tissue spaces and return them to the blood. This activity helps maintain fluid balance in the body and prevents depletion of vital plasma proteins.
2. **Transporting dietary lipids.** Lymphatic vessels transport the lipids as well as lipid-soluble vitamins (A, D, E, and K) absorbed by the gastrointestinal tract into the blood.
3. **Carrying out immune responses.** Lymphatic tissue initiates highly specific responses directed against particular microbes or abnormal cells.

Lymphatic Vessels and Lymph Circulation

Lymphatic vessels begin as ***lymphatic capillaries.*** These tiny vessels are closed at one end and located in the spaces between cells (Figure 24.2 on page 630). Lymphatic capillaries are slightly larger than blood capillaries and have a unique structure that permits interstitial fluid to flow into them, but not out. The endothelial cells that make up the wall of a lymphatic capillary are not attached end to end; rather, the ends overlap (Figure 24.2b). When pressure is greater in interstitial fluid than in lymph, the cells separate slightly, like a one-way swinging door, and interstitial fluid enters the lymphatic capillary. When pressure is greater inside the lymphatic capillary, the cells adhere more closely and lymph cannot escape back into interstitial fluid.

Unlike blood capillaries, which link two larger blood vessels that form part of a circuit, lymphatic capillaries begin in the tissues and carry the lymph that forms there toward a larger

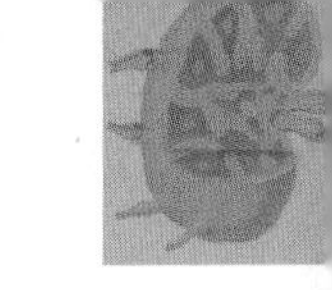

Figure 24.1 **Components of the lymphatic system.**

The lymphatic system consists of lymph, lymphatic vessels, lymphatic tissues, and red bone marrow.

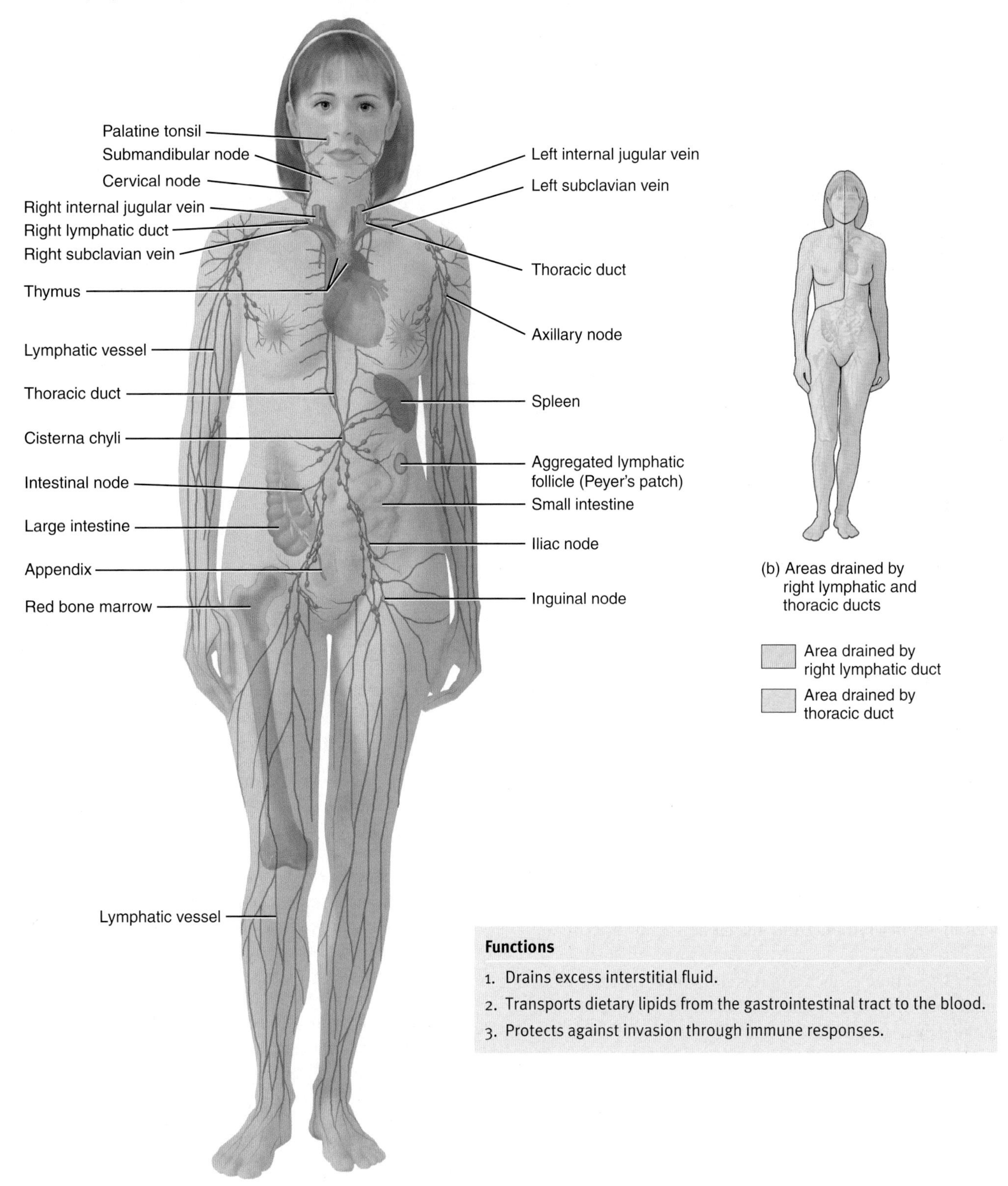

(a) Anterior view of principal components of lymphatic system

Functions

1. Drains excess interstitial fluid.
2. Transports dietary lipids from the gastrointestinal tract to the blood.
3. Protects against invasion through immune responses.

What tissue contains stem cells that develop into lymphocytes?

Figure 24.2 Lymphatic capillaries.

Lymphatic capillaries are found throughout the body except in avascular tissues, the central nervous system, portions of the spleen, and bone marrow.

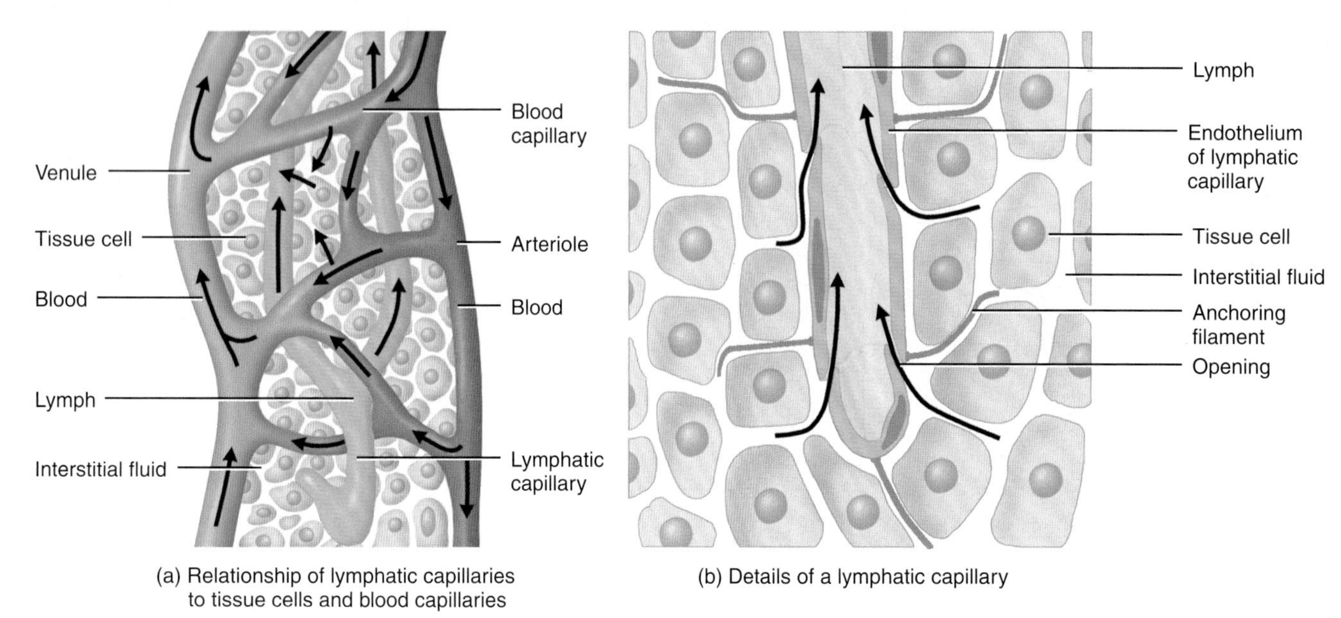

(a) Relationship of lymphatic capillaries to tissue cells and blood capillaries

(b) Details of a lymphatic capillary

Is lymph more similar to blood plasma or to interstitial fluid? Why?

lymphatic vessel. Just as blood capillaries unite to form venules and veins, lymphatic capillaries unite to form larger and larger ***lymphatic vessels*** (see Figure 24.1). Lymphatic vessels resemble veins in structure but have thinner walls and more valves. Located at intervals along lymphatic vessels are ***lymph nodes,*** masses of B cells and T cells that are surrounded by a capsule. Lymph flows through lymph nodes.

From the lymphatic vessels, lymph eventually passes into one of two main channels: the thoracic duct or the right lymphatic duct. The ***thoracic duct,*** the main lymph-collecting duct, receives lymph from the left side of the head, neck, and chest; the left upper limb; and the entire body below the ribs. The ***right lymphatic duct*** drains lymph from the upper right side of the body (see Figure 24.1b).

Ultimately, the thoracic duct empties its lymph into the junction of the left internal jugular and left subclavian veins, and the right lymphatic duct empties its lymph into the junction of the right internal jugular and right subclavian veins. Thus, lymph drains back into the blood (Figure 24.3).

The same two pumps that aid return of venous blood to the heart maintain the flow of lymph:

1. **Skeletal muscle pump.** The milking action of skeletal muscle contractions (see Figure 23.4) compresses lymphatic vessels (as well as veins) and forces lymph toward the subclavian veins.
2. **Respiratory pump.** Lymph flow is also maintained by pressure changes that occur during inhalation (breathing in). Lymph flows from the abdominal region, where the pressure is higher, toward the thoracic region, where it is lower. When the pressures reverse during exhalation (breathing out), the valves in lymphatic vessels prevent backflow of lymph.

MANUAL THERAPY APPLICATION

Edema and Lymphedema

Edema (e-DĒ-ma) is an excessive accumulation of interstitial fluid in tissue spaces that produces swelling or puffiness of parts of the body. Edema can be localized, for example, in a small region affected by a bee sting, or it can compromise an entire limb, the face, specific organs such as the lungs, or the entire body. When excess fluid accumulates within the peritoneal space, the condition is known as *ascites.* Edema may result from increased capillary blood pressure, which causes excess interstitial fluid to form faster than it can pass into lymphatic vessels or be reabsorbed back into the capillaries. Another cause is lack of skeletal muscle contractions, as in individuals who are paralyzed or who sit for too long without moving, as in persons on transcontinental flights. Eating food that contains too much salt can make the problem worse. Heart, kidney, liver, and thyroid diseases can cause edema. Pregnant women may also experience edema. Administration of a diuretic is a common treatment of edema because it helps to release the excess fluid. Elevation of the affected body part utilizes the effects of gravity to reduce symptoms. There is no direct cure for edema but the underlying cause(s) may be treated to reduce swelling.

Lymphedema is a condition that occurs when the pathways of the lymphatic system (see Figure 24.1) are blocked. This results in swelling of tissues as well as compromising the ability of immune cells to travel throughout the body. Causes of lymphatic obstruction include surgery, radiation therapy, tumors, parasitic infections, skin infections, or generalized

Figure 24.3 Schematic diagram showing the relationship of the lymphatic system to the cardiovascular system. Arrows show the direction of flow of lymph and blood.

The sequence of fluid is blood capillaries (blood plasma) → interstitial spaces (interstitial fluid) → lymphatic capillaries (lymph) → lymphatic vessels and lymph nodes (lymph) → lymphatic ducts (lymph) → junction of jugular and subclavian veins (blood).

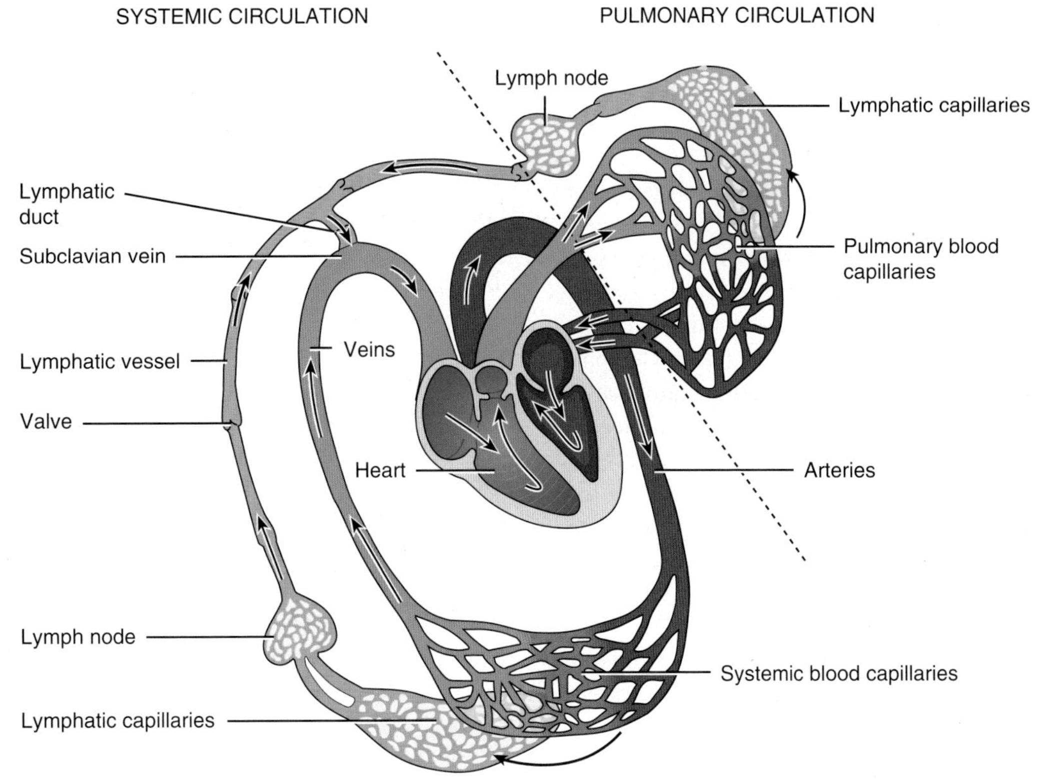

Does inhalation promote or hinder the flow of lymph?

trauma to a region. Lymphedema due to bacterial infections commonly resolves after administration of appropriate antibiotics. Lymphedema of an upper limb occurs in 12–15% of mastectomies with axillary dissection. Treatments include manual lymph drainage, compression, and range of motion exercises.

Manual lymph drainage is a light massage technique that moves the skin in directions that correspond to the vessels of the lymphatic system. This modality encourages movement of the fluids through the appropriate channels and is quite successful in the hands of knowledgeable manual therapists. Massage is often combined with a stretch wrap (such as an Ace bandage), used to exert pressure on the skin, which helps the body resorb excess fluids. For those persons with lymphedema of the lower limbs, special compression stockings may also be prescribed. For more severe lymphedema of the limbs, physicians may prescribe inflatable compression devices that alternately apply increased and reduced pressure to help reduce swelling.

Lymphatic Organs and Tissues

Lymphatic organs and tissues, which are widely distributed throughout the body, are classified into two groups based on their functions. ***Primary lymphatic organs*** and ***tissues,*** the sites where stem cells divide and develop into mature B cells and T cells, include the ***red bone marrow*** (in flat bones and the ends of most long bones of adults) and the ***thymus.*** The ***secondary lymphatic organs*** and ***tissues,*** the sites where most immune responses occur, include ***lymph nodes,*** the ***spleen,*** and ***lymphatic nodules.***

Thymus

The ***thymus*** is a two-lobed organ located posterior to the sternum, medial to the lungs, and superior to the heart (see Figure 24.1). It contains large numbers of T cells and scattered dendritic cells (so named for their long, branchlike projections), epithelial cells, and macrophages. Immature T cells migrate from red bone marrow to the thymus, where they multiply and begin to mature. Only about 2% of the immature T cells that arrive in the thymus achieve the proper "education" to "graduate" into mature T cells. The remaining cells die via apoptosis (programmed cell death). Thymic macrophages help clear out the debris of dead and dying cells. Mature T cells leave the thymus via the blood and are carried to lymph nodes, the spleen, and other lymphatic tissues where they populate parts of these organs and tissues.

Lymph Nodes

Located along lymphatic vessels are about 600 bean-shaped ***lymph nodes.*** They are scattered throughout the body, both superficially and deep, and usually occur in groups (see Figure 24.1). Lymph nodes are heavily concentrated near the mammary glands and in the axillae and groin. Each node is covered by a capsule of dense connective tissue (Figure 24.4). Internally, different regions of a lymph node may contain B cells that develop into plasma cells, as well as T cells, dendritic cells, and macrophages.

Lymph nodes filter lymph, which enters a node through one of several ***afferent lymphatic vessels*** (*af-* = toward; *-ferrent* = to carry). As lymph flows through the node, foreign substances are trapped by *reticular fibers* within the spaces between cells. Macrophages destroy some foreign substances by phagocytosis, and lymphocytes destroy others by a variety of immune responses. Filtered lymph leaves the other end of the node through one or two ***efferent lymphatic vessels*** (*ef-* = away). (The filtered lymph is somewhat analogous to clean air that leaves a furnace after passing through the furnace filter; substances collect on a furnace filter, and therefore it must be replaced every few months.) Plasma cells and T cells that have divided many times within a lymph node can also leave the node and circulate to other parts of the body. Valves direct the flow of lymph inward through the afferent lymphatic vessels and outward through the efferent lymphatic vessels (Figure 24.4).

MANUAL THERAPY APPLICATION

Metastasis

Metastasis (me-TAS-ta-sis; *meta-* = beyond; *-stasis* = to stand), the spread of a disease from one part of the body to another, can occur via lymphatic vessels. All malignant tumors eventually metastasize. Cancer cells may travel in the blood or lymph and establish new tumors where they lodge. When metastasis occurs via lymphatic vessels, secondary tumor sites can be predicted according to the direction of lymph flow from the primary tumor site. For example, malignant tumors of the breast, when they metastasize, are expected to form secondary tumor sites in the lymph nodes of the axilla. Cancerous lymph nodes feel enlarged, firm, nontender, and fixed to underlying structures. By contrast, most lymph nodes that are enlarged due to an infection are softer, tender, movable, and benign. Manual therapists may be the first to discover enlarged lymph nodes while working with patients and, with the generalized description just given, will know when it is appropriate to refer the patient to a physician.

• CLINICAL CONNECTION | Lymphomas

Lymphomas (lim-FŌ-mas; *lymph-* = clear water; *-oma* = tumor) are cancers of the lymphatic organs, especially the lymph nodes. Most have no known cause. The two main types of lymphomas are Hodgkin disease and non-Hodgkin lymphoma.

Hodgkin disease (HD) is characterized by painless, nontender enlargement of one or more lymph nodes, most commonly in the neck, chest, and axillae (armpits). If the disease has metastasized from these sites, fevers, night sweats, weight loss, and bone pain also occur. HD primarily affects individuals between ages 15 and 35 and those over 60; it is more common in males. If diagnosed early, HD has a 90–95% cure rate.

Non-Hodgkin lymphoma (NHL), which is more common than HD, occurs in all age groups. NHL may start the same way as HD but may also include an enlarged spleen, anemia, and general malaise. Up to half of all individuals with NHL are cured or survive for a lengthy period. Treatment options for both HD and NHL include radiation therapy, chemotherapy, and red bone marrow transplantation. •

Spleen

The ***spleen*** is the largest single mass of lymphatic tissue in the body (see Figure 24.1). It lies between the stomach and diaphragm and is covered by a capsule of dense connective tissue. The spleen contains two types of tissue called white pulp and red pulp. *White pulp* is lymphatic tissue, consisting mostly of lymphocytes and macrophages. *Red pulp* consists of blood-filled *venous sinuses* and cords of *splenic tissue* consisting of red blood cells, macrophages, lymphocytes, plasma cells, and granular leukocytes.

Blood flowing into the spleen through the splenic artery enters the white pulp. Within the white pulp, B cells and T cells carry out immune responses, while macrophages destroy pathogens by phagocytosis. Within the red pulp, the spleen performs three functions related to blood cells: (1) removal by macrophages of worn out or defective blood cells and platelets; (2) storage of platelets, perhaps up to one-third of the body's supply; and (3) production of blood cells (hemopoiesis) during fetal life.

The spleen is the organ most often damaged in cases of abdominal trauma. A ruptured spleen causes severe internal hemorrhage and shock. Prompt ***splenectomy,*** removal of the spleen, is needed to prevent bleeding to death. After a splenectomy, other structures, particularly red bone marrow and the liver, can take over functions normally carried out by the spleen.

Lymphatic Nodules

Lymphatic nodules are egg-shaped masses of lymphatic tissue that are not surrounded by a capsule. They are plentiful in the connective tissue of mucous membranes lining the gastrointestinal, urinary, and reproductive tracts and the respiratory airways. Although many lymphatic nodules are small and solitary, some occur as large aggregations in specific parts of the body. Among these are the ***tonsils*** in the pharyngeal region and the ***aggregated lymphatic follicles (Peyer's patches)*** in the ileum of the small intestine (see Figure 24.1). Aggregations of lymphatic nodules also occur in the appendix. The five ***tonsils,*** which form a ring at the junction of the oral cavity, nasal cavity, and throat, are strategically positioned to participate in immune responses against inhaled or ingested foreign substances. The single ***pharyngeal tonsil*** (fa-RIN-jē-al) or ***adenoid*** is embedded in the posterior wall of the upper part of the throat (see Figure 25.2, Section 25.2). The two ***palatine tonsils*** (PAL-a-tīn) lie at the back of the mouth, one on either side; these are the tonsils commonly removed in a tonsillectomy. The paired ***lingual tonsils*** (LIN-gwal), located at the base of the tongue, may also require removal during a tonsillectomy.

Figure 24.4 Structure of a lymph node. Arrows indicate direction of lymph flow through a lymph node.

Lymph nodes are present throughout the body, usually clustered in groups.

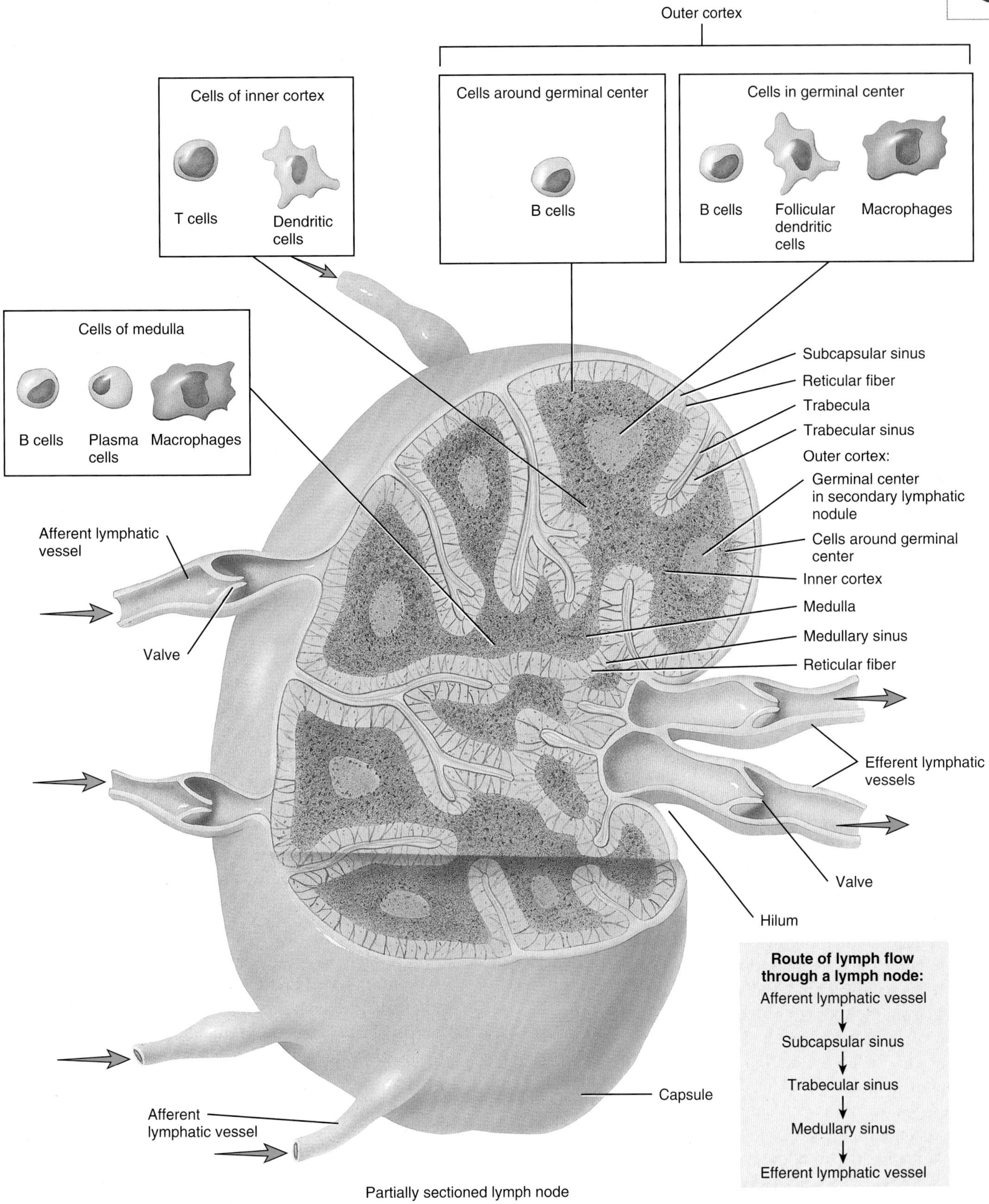

What happens to foreign substances in lymph that enter a lymph node?

CHECKPOINT

1. How are interstitial fluid and lymph similar, and how do they differ?
2. What are the roles of the thymus and the lymph nodes in immunity?
3. Describe the functions of the spleen and tonsils.

24.3 INNATE IMMUNITY

OBJECTIVE

- Describe the various components of innate immunity.

Innate immunity includes barriers provided by the skin and mucous membranes. It also includes various internal defenses, such as antimicrobial substances, natural killer cells, phagocytes, inflammation, and fever.

First Line of Defense: Skin and Mucous Membranes

Both ***physical barriers*** and ***chemical barriers*** to pathogens and foreign substances are found in the skin that covers the body and in mucous membranes that line body openings such as the mouth and breathing airways. With its many layers of closely packed, keratinized cells, the ***epidermis*** (the outer epithelial layer of the skin) provides a formidable physical barrier to the entrance of microbes (see Figure 5.1, Section 5.1). In addition, continual shedding of the top epidermal cells helps remove microbes at the skin's surface. Bacteria rarely penetrate an intact and healthy epidermis.

The epithelial layer of ***mucous membranes*** secretes a fluid called *mucus* that lubricates and moistens the surface of a body cavity. Because mucus is sticky, it traps many microbes and foreign substances. The mucous membrane of the nose has mucus-coated *hairs* that trap and filter microbes, dust, and pollutants from inhaled air. The mucous membrane of the upper airways contains *cilia,* microscopic hairlike projections on the surface of the epithelial cells, that propel inhaled dust and microbes trapped in mucus toward the throat.

Other fluids produced by various organs also help protect epithelial surfaces of the skin and mucous membranes. The ***lacrimal apparatus*** (LAK-ri-mal) of the eyes (see Figure 19.5b, Section 19.6) produces and drains away tears in response to irritants, diluting microbes and keeping them from settling on the surface of the eyes. ***Saliva,*** produced by the salivary glands, washes microbes from the surfaces of the teeth and from the mucous membrane of the mouth, much like tears wash the eyes. The cleansing of the urethra by the ***flow of urine*** retards microbial colonization of the urinary system. ***Vaginal secretions,*** likewise, move microbes out of the body in females. ***Defecation*** and ***vomiting*** also expel microbes; this action causes some persons to believe that it is often better to allow the harmful substances to exit the body as opposed to taking medications that slow vomiting and defecation.

Certain chemicals also contribute to the resistance of the skin and mucous membranes to microbial invasion. Sebaceous (oil) glands of the skin secrete an oily substance called ***sebum*** that forms a protective film over the surface of the skin. ***Perspiration*** helps flush microbes from the surface of the skin and contains ***lysozyme,*** an enzyme capable of breaking down the cell walls of certain bacteria. (Lysozyme is also found in tears, saliva, nasal secretions, and tissue fluids.) ***Gastric juice,*** a mixture of hydrochloric acid, enzymes, and mucus in the stomach, destroys many bacteria and most bacterial toxins. Vaginal secretions also are slightly acidic, which discourages bacterial growth.

Second Line of Defense: Internal Defenses

Although the skin and mucous membranes are very effective barriers in preventing invasion by pathogens, they may be compromised by injuries or everyday activities such as brushing the teeth or shaving. Any pathogens that get past the surface barriers encounter a second line of defense consisting of antimicrobial substances, phagocytes, natural killer cells, inflammation, and fever.

Antimicrobial Substances

Various body fluids contain four main types of ***antimicrobial substances*** that discourage microbial growth:

1. Lymphocytes, macrophages, and fibroblasts infected with viruses produce proteins called ***interferons*** (in-ter-FĒR-ons), or ***IFNs.*** After their release by virus-infected cells, IFNs diffuse to uninfected neighboring cells, where they stimulate synthesis of proteins that interfere with viral replication. Viruses can cause disease only if they can replicate within body cells.

2. A group of normally inactive proteins in blood plasma and on plasma membranes makes up the ***complement system.*** When activated, these proteins "complement" or enhance certain immune, allergic, and inflammatory reactions. One effect of complement proteins is to create holes in the plasma membrane of the microbe. As a result, extracellular fluid moves into the holes, causing the microbe to burst, a process called ***cytolysis***. Another effect of complement is to cause ***chemotaxis*** (kē-mō-TAK-sis), the chemical attraction of phagocytes to a site. Some complement proteins cause ***opsonization*** (op-son-i-ZĀ-shun), a process in which complement proteins bind to the surface of a microbe and promote phagocytosis.

3. ***Iron-binding proteins*** inhibit the growth of certain bacteria by reducing the amount of available iron. Examples include *transferrin* (found in blood and tissue fluids), *lactoferrin* (found in milk, saliva, and mucus), *ferritin* (found in the liver, spleen, and bone marrow), and *hemoglobin* (found in red blood cells).

4. ***Antimicrobial proteins*** (***AMPs***) are short peptides that have a broad spectrum of antimicrobial activity. Examples of AMPs are *dermicidin* (produced by sweat glands), *defensins* and *cathelicidins* (produced by neutrophils, macrophages, and epithelium) and *thrombocidin* (produced by platelets). Besides killing a wide range of microbes, AMPs can attract dendritic cells and mast cells, which participate in immune responses. Interestingly enough, microbes exposed to AMPs do not appear to develop resistance, as often happens with antibiotics.

Phagocytes and Natural Killer Cells

When microbes penetrate the skin and mucous membranes or bypass the antimicrobial proteins in blood, the next nonspecific defense consists of phagocytes and natural killer cells.

Phagocytes (*phago-* = eat; *-cytes* = cells) are specialized cells that perform ***phagocytosis*** (*-osis* = process), the ingestion of microbes or other particles such as cellular debris. The two main types of phagocytes are neutrophils and macrophages. When an infection occurs, neutrophils and monocytes migrate to the infected area. During this migration, the monocytes enlarge and develop into actively phagocytic cells called ***macrophages*** (MAK-rō-fā-jez) (see Figure 21.2, Section 21.2). Some are *wandering macrophages,* which migrate to infected areas. Others are *fixed macrophages,* which remain in certain locations, including the skin and subcutaneous layer, liver, lungs, brain, spleen, lymph nodes, and red bone marrow.

About 5–10% of lymphocytes in the blood are ***natural killer (NK) cells,*** which have the ability to kill a wide variety of microbes and certain tumor cells. NK cells also are present in the spleen, lymph nodes, and red bone marrow. Some cancer and AIDS patients have defective or decreased numbers of NK cells. They cause cellular destruction by releasing proteins that destroy the target cell's membrane.

Inflammation

Inflammation is a defensive response of the body to tissue damage. Because inflammation is one of the body's innate defenses, the response of a tissue to a cut is similar to the response to damage caused by burns, radiation, or invasion of bacteria or viruses. The events of inflammation dispose of microbes, toxins, or foreign material at the site of injury, prevent their spread to other tissues, and prepare the site for tissue repair. Thus, inflammation helps restore tissue homeostasis. The four signs and symptoms of inflammation are ***redness, heat, swelling,*** and ***pain.*** Inflammation can also cause the ***loss of function*** in the injured area, depending on the site and extent of the injury.

The stages of inflammation are as follows:

1. In a region of tissue injury, mast cells in connective tissue and basophils and platelets in blood release *histamine.* In response to histamine, two immediate changes occur in the blood vessels: *increased permeability* and *vasodilation,* an increase in the diameter of the blood vessels (Figure 24.5). Increased permeability means that substances normally retained in blood are permitted to pass out of the blood vessels. Vasodilation is an increase in the diameter of the blood vessels; it allows more blood to flow to the damaged area and helps remove microbial toxins and dead cells. Increased permeability permits defensive substances such as antibodies and clot-forming chemicals to enter the injured area from the blood.

From the events that occur during inflammation, it's easy to understand the signs and symptoms. Heat and redness result from the large amount of blood that accumulates in the damaged area. The area swells due to an increased amount of interstitial fluid that has leaked out of the capillaries (edema). Pain results from injury to neurons, from toxic chemicals released by microbes, and from the increased pressure of edema.

Figure 24.5 Inflammation.

6 The three stages of inflammation are as follows: (1) vasodilation and increased permeability of blood vessels, (2) phagocytic emigration, and (3) tissue repair.

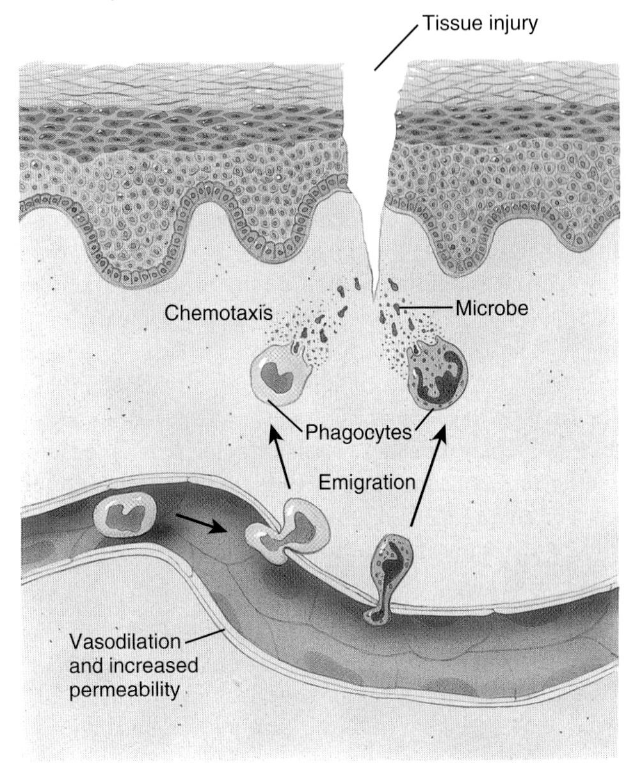

Phagocytes migrate from blood to site of tissue injury

? What causes each of the following signs and symptoms of inflammation: redness, pain, heat, and swelling?

2. The increased permeability of capillaries causes leakage of clotting proteins into tissues. Fibrinogen is converted to an insoluble, thick network of fibrin threads, which traps the invading organisms and prevents their spread. The resulting clot may isolate the invading microbes and their toxins.

3. Shortly after the inflammatory process starts, phagocytes are attracted to the site of injury by chemotaxis (Figure 24.5). Near the damaged area, neutrophils begin to squeeze through the wall of the blood vessel, a process called *emigration.* Neutrophils predominate in the early stages of infection, but they die off rapidly together with the microbes they have eaten. Within a few hours, monocytes arrive in the infected area. Once in the tissue, they turn into wandering macrophages that engulf damaged tissue, worn-out neutrophils, and invading microbes.

4. Eventually, macrophages also die. Within a few days, a pocket of dead phagocytes and damaged tissue forms; this collection of dead cells and fluid is called ***pus.*** At times, pus reaches the surface of the body or drains into an internal cavity and is dispersed; on other occasions the pus remains even after the infection is terminated. In this case, the pus is gradually destroyed over a period of days and is absorbed.

• CLINICAL CONNECTION | Ulcer

If pus cannot drain out of an inflamed region, the result is an *abscess*—an excessive accumulation of pus in a confined space. Common examples are pimples and boils. When superficial inflamed tissue sloughs off the surface of an organ or tissue, the resulting open sore is called an **ulcer.** People with poor circulation—for instance, diabetics with advanced atherosclerosis—are particularly susceptible to ulcers in the tissues of their legs. •

Fever

Fever is an abnormally high body temperature that occurs because the hypothalamic thermostat is reset. It commonly occurs during infection and inflammation. Many bacterial toxins elevate body temperature, sometimes by triggering release of fever-causing substances such as interleukin-1 from macrophages. Elevated body temperature intensifies the effects of interferons, inhibits the growth of some microbes, and speeds up body reactions that aid repair. For these reasons, some persons believe it is often better to permit the fever to rage (within limits) rather than take fever-reducing medications.

Table 24.1 summarizes the components of innate defenses.

CHECKPOINT

4. What physical and chemical factors provide protection from disease in the skin and mucous membranes?
5. What internal defenses provide protection against microbes that penetrate the skin and mucous membranes?
6. What are the main signs and symptoms of inflammation?

24.4 ADAPTIVE IMMUNITY

OBJECTIVES

- Define adaptive immunity and compare it with innate immunity.
- Explain the relationship between an antigen and an antibody.
- Compare the functions of cell-mediated immunity and antibody-mediated immunity.

The various aspects of innate immunity have one thing in common: they are not specifically directed against a particular type of invader. Adaptive (specific) immunity involves the production of specific types of cells or specific antibodies to destroy a particular antigen. An ***antigen*** is any substance—such as microbes, foods, drugs, pollen, or tissue—that the immune system recognizes as foreign (nonself). The branch of science that deals with the responses of the body to antigens is called ***immunology*** (im′-ū-NOL-ō-jē). The ***immune system*** includes the cells and tissues that carry out immune responses. Normally, a person's adaptive immune system cells recognize and do not attack their own tissues and chemicals. Such lack of reaction against self-tissues is called ***self-tolerance.***

At times, self-tolerance breaks down, which leads to an **autoimmune disease.** Sometimes tissues undergo changes that cause

TABLE 22.1

Summary of Innate Defenses

COMPONENT	FUNCTIONS
FIRST LINE OF DEFENSE: SKIN AND MUCOUS MEMBRANES	
Physical factors	
Epidermis of skin	Forms a physical barrier to the entrance of microbes.
Mucous membranes	Inhibit the entrance of many microbes, but not as effective as intact skin.
Mucus	Traps microbes in respiratory and gastrointestinal tracts.
Hairs	Filter out microbes and dust in nose.
Cilia	Together with mucus, trap and remove microbes and dust from respiratory tract.
Lacrimal apparatus	Tears dilute and wash away irritating substances and microbes.
Saliva	Washes microbes from surfaces of teeth and mucous membranes of mouth.
Urine	Washes microbes from urethra.
Defecation and vomiting	Expel microbes from body.
Chemical factors	
Sebum	Forms a protective acidic film over the skin surface that inhibits growth of many microbes.
Lysozyme	Antimicrobial substance in perspiration, tears, saliva, nasal secretions, and tissue fluids.
Gastric juice	Destroys bacteria and most toxins in stomach.
Vaginal secretions	Slight acidity discourages bacterial growth; flush microbes out of vagina.
SECOND LINE OF DEFENSE: INTERNAL DEFENSES	
Antimicrobial substances	
Interferons (IFNs)	Protect uninfected host cells from viral infection.
Complement system	Causes cytolysis of microbes, promotes phagocytosis, and contributes to inflammation.
Iron-binding proteins	Inhibit growth of certain bacteria by reducing amount of available iron.
Antimicrobial proteins (AMPs)	Have broad spectrum antimicrobial activities and attract dendritic cells and mast cells.
Natural killer (NK) cells	Kill infected target cells by releasing granules that contain perforin and granzymes. Phagocytes then kill the released microbes.
Phagocytes	Ingest foreign particulate matter.
Inflammation	Confines and destroys microbes and initiates tissue repair.
Fever	Intensifies the effects of interferons, inhibits growth of some microbes, and speeds up body reactions that aid repair.

the adaptive immune system to recognize them as foreign antigens and attack them. Among human autoimmune diseases are systemic lupus erythematosus (SLE), Addison's disease, Graves' disease, type 1 diabetes mellitus, myasthenia gravis, multiple sclerosis (MS), and ulcerative colitis.

Maturation of T Cells and B Cells

The cells that carry out adaptive immune responses are lymphocytes called B cells and T cells. Both develop from stem cells that originate in red bone marrow (see Figure 21.2, Section 21.2). B cells complete their development in red bone marrow; immature T cells migrate from red bone marrow to the thymus, where they mature. Before T cells leave the thymus or B cells leave red bone marrow, they begin to make several distinctive proteins that are inserted into their plasma membranes. Some of these proteins function as ***antigen receptors***—molecules capable of recognizing and binding to specific antigens.

Types of Adaptive Immune Responses

Adaptive immunity consists of two types of closely allied responses, both triggered by antigens. In ***cell-mediated immune responses,*** some T cells are like an army of soldiers that directly attack the invading antigen. In ***antibody-mediated immune responses,*** B cells change into plasma cells, which synthesize and secrete specific proteins called ***antibodies.*** A given antibody can bind to and inactivate a specific antigen. Other T cells aid both cell-mediated and antibody-mediated adaptive immune responses. Although each type of response is specialized to combat different aspects of an invasion, a given pathogen can provoke both types of adaptive immune responses.

Antigens and Antibodies

An ***antigen*** (meaning *anti*body *gen*erator) causes the body to produce specific antibodies and/or specific T cells that react with it. Entire microbes or parts of microbes may act as antigens. Chemical components of bacterial structures such as flagella, capsules, and cell walls are antigenic, as are bacterial toxins and viral proteins. Other examples of antigens include chemical components of pollen, egg white, incompatible blood cells, and transplanted tissues and organs. The huge variety of antigens in the environment provides myriad opportunities for provoking immune responses.

Located at the plasma membrane surface of most body cells are protein "self-antigens" known as ***major histocompatibility complex (MHC) antigens.*** Unless you have an identical twin, your MHC antigens are unique. Thousands to several hundred thousand MHC molecules mark the surface of each of your body cells except red blood cells. MHC antigens are the reason that tissues may be rejected when they are transplanted from one person to another, but their normal function is to help T cells recognize that an antigen is foreign, not self. This recognition is an important first step in any adaptive immune response.

The success of an organ or tissue transplant depends on ***histocompatibility*** (his′-tō-kom-pat-i-BIL-i-tē), the tissue compatibility between the donor and the recipient. The more similar the MHC antigens, the greater is the histocompatibility, and thus the greater is the chance that the transplant will not be rejected. In the United States, a nationwide computerized registry helps physicians select the most histocompatible and neediest organ transplant recipients whenever donor organs become available.

Antigens induce plasma cells to secrete proteins known as ***antibodies.*** Most antibodies contain four polypeptide chains (Figure 24.6a). At two tips of the chains are ***variable regions,*** so named because the sequence of amino acids there varies for each different antibody. The variable regions are the ***antigen-binding***

Figure 24.6 Structure of an antibody and relationship of an antigen to an antibody.

An antigen stimulates plasma cells to secrete specific antibodies that combine with the antigen.

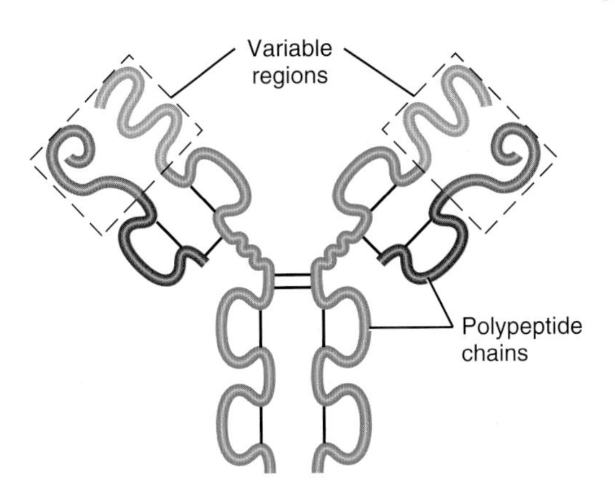

(a) Diagram of an antibody molecule

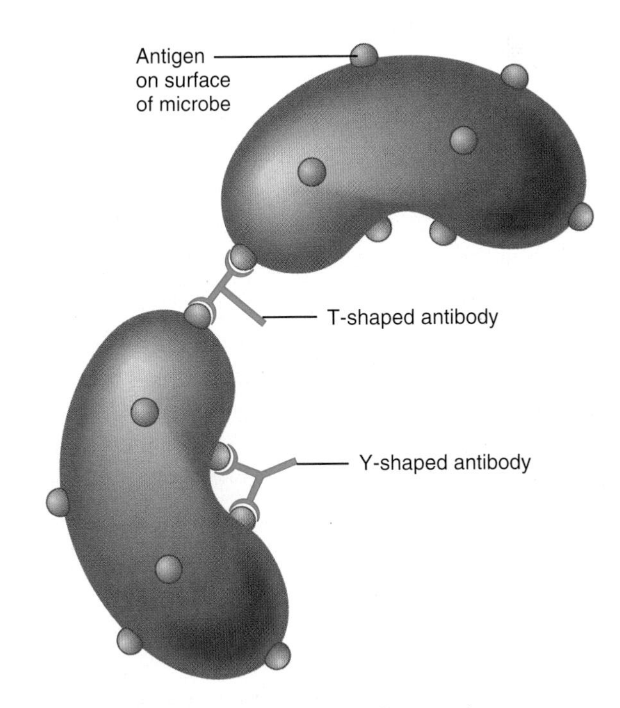

(b) Antibody molecules binding to antigens

What is the function of the variable regions of an antibody?

sites, the parts of an antibody that "fit" and bind to a particular antigen, much like your house key fits into its lock. Because the antibody "arms" can move somewhat, an antibody can assume either a T shape or a Y shape. This flexibility enhances the ability of the antibody to bind to two identical antigens at the same time—for example, on the surface of microbes (see Figure 24.6b).

Antibodies belong to a group of plasma proteins called globulins, and for this reason they are also known as ***immunoglobulins*** (im-ū-nō-GLOB-ū-lins). Immunoglobulins are grouped in five different classes, designated IgG, IgA, IgM, IgD, and IgE. Each class has a distinct chemical structure and different functions (Table 24.2). Because they appear first and are relatively short-lived, IgM antibodies indicate a recent invasion. In a sick patient, a high level of IgM against a particular pathogen helps identify the cause of the illness. Resistance of the fetus and newborn to infection stems mainly from maternal IgG antibodies that cross the placenta before birth and IgA antibodies in breast milk after birth.

Processing and Presenting Antigens

For an adaptive immune response to occur, B cells and T cells must recognize that a foreign antigen is present. B cells can recognize and bind to antigens in lymph, interstitial fluid, or blood plasma, but T cells only recognize fragments of antigens that are processed and presented in a certain way.

In ***antigen processing,*** antigenic proteins are broken down into peptide fragments and then combine with MHC molecules. Next the antigen–MHC complex is inserted into the plasma membrane of a body cell. The insertion of the complex into the plasma membrane is called ***antigen presentation.*** When an antigenic fragment comes from a *self-protein,* T cells ignore the antigen–MHC complex. However, if the fragment comes from a *foreign protein,* T cells recognize the antigen–MHC as an intruder, and an adaptive immune response takes place.

A special class of cells called ***antigen-presenting cells (APCs)*** process and present antigens. APCs include dendritic cells, macrophages, and B cells. They are strategically located in places where antigens are likely to penetrate innate defenses and enter the body, such as the epidermis and dermis of the skin (Langerhans cells are a type of dendritic cell); mucous membranes that line the respiratory, gastrointestinal, urinary, and reproductive tracts; and lymph nodes. After processing and presenting an antigen, APCs migrate from tissues via lymphatic vessels to lymph nodes.

The steps in the processing and presenting of an antigen by an APC occur as follows (Figure 24.7):

1. ***Ingestion of the antigen.*** Antigen-presenting cells ingest antigens by phagocytosis. Ingestion could occur almost anywhere in the body that invaders, such as microbes, have penetrated the nonspecific defenses.

TABLE 24.2

Classes of Immunoglobulins

NAME AND STRUCTURE	CHARACTERISTICS AND FUNCTIONS
IgG	Most abundant, about 80% of all antibodies in the blood; found in blood, lymph, and the intestines; monomer (one-unit) structure. Protects against bacteria and viruses by enhancing phagocytosis, neutralizing toxins, and triggering the complement system. It is the only class of antibody to cross the placenta from mother to fetus, conferring considerable immune protection in newborns.
IgA	Found mainly in sweat, tears, saliva, mucus, breast milk, and gastrointestinal secretions. Smaller quantities are present in blood and lymph. Makes up 10–15% of all antibodies in the blood; occurs as monomers and dimers (two units). Levels decrease during stress, lowering resistance to infection. Provides localized protection of mucous membranes against bacteria and viruses.
IgM	About 5–10% of all antibodies in the blood; also found in lymph. Occurs as pentamers (five units); first antibody class to be secreted by plasma cells after an initial exposure to any antigen. Activates complement and causes agglutination and lysis of microbes. Also present as monomers on the surfaces of B cells, where they serve as antigen receptors. In blood plasma, the anti-A and anti-B antibodies of the ABO blood group, which bind to A and B antigens during incompatible blood transfusions, are also IgM antibodies (see Figure 21.10).
IgD	Mainly found on the surfaces of B cells as antigen receptors, where it occurs as monomers; involved in activation of B cells. About 0.2% of all antibodies in the blood.
IgE	Less than 0.1% of all antibodies in the blood; occurs as monomers; located on mast cells and basophils. Involved in allergic and hypersensitivity reactions; provides protection against parasitic worms.

Figure 24.7 Processing and presenting of exogenous antigen by an antigen-presenting cell (APC).

Fragments of exogenous antigens are processed and then presented with MHC molecules on the surface of an antigen-processing cell (APC).

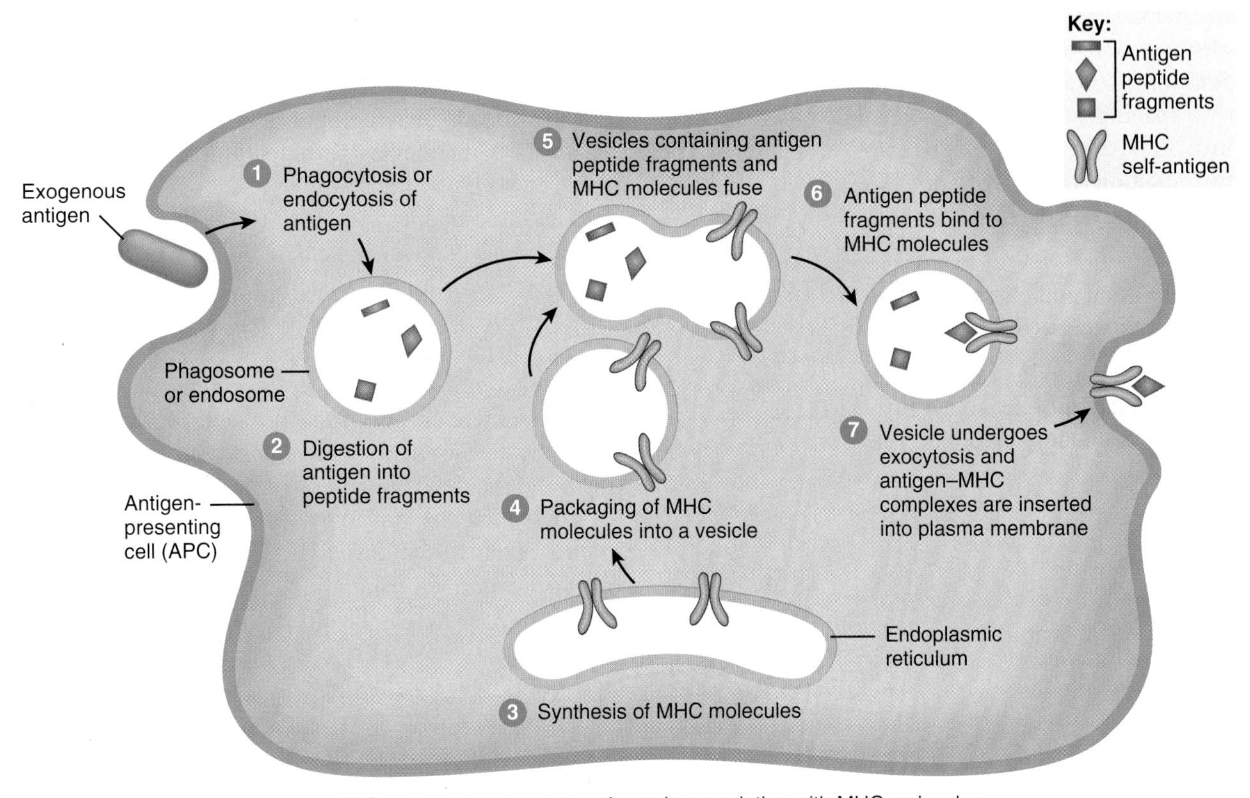

APCs present exogenous antigens in association with MHC molecules

? What types of cells are APCs, and where in the body are they found?

2 ***Digestion of antigen into peptide fragments.*** Within the APC, protein-digesting enzymes split large antigens into short peptide fragments.

3 ***Synthesis of MHC molecules.*** At the same time, the APC synthesizes MHC molecules at the endoplasmic reticulum (ER).

4 ***Packaging of MHC molecules.*** Once synthesized, the MHC molecules are packaged into vesicles.

5 ***Fusion of vesicles.*** The vesicles containing antigen fragments and MHC molecules merge and fuse.

6 ***Binding of fragments to MHC molecules.*** After fusion of the two vesicles, antigen fragments bind to MHC molecules.

7 ***Insertion of antigen–MHC complex into the plasma membrane.*** The combined vesicle that contains antigen–MHC complexes splits open and the antigen–MHC complexes are inserted into the plasma membrane.

• CLINICAL CONNECTION | Infectious Mononucleosis

Infectious mononucleosis or "mono" is a contagious disease caused by the *Epstein-Barr virus* (*EBV*). It occurs mainly in children and young adults, and more often in females than in males. The virus commonly enters the body through intimate oral contact such as kissing, which accounts for its being called the "kissing disease." EBV then multiplies in lymphatic tissues and spreads into the blood, where it infects and multiplies in B cells, the primary host cells. Because of this infection, the B cells become enlarged and abnormal in appearance so that they resemble monocytes, the primary reason for the term *mononucleosis*. Besides an elevated white blood cell count, with an abnormally high percentage of lymphocytes, signs and symptoms include fatigue, headache, dizziness, sore throat, enlarged and tender lymph nodes, and fever. There is no cure for infectious mononucleosis, but the disease usually runs its course in a few weeks. The EBV may lie dormant for years and then reappear. Adults with Epstein-Barr virus have symptoms, notably fatigue, similar to that of mononucleosis. It is typical for the EBV to reappear during times of stress when the person's immune system has been severely compromised. •

T Cells and Cell-Mediated Immunity

The presentation of an antigen together with MHC molecules by APCs informs T cells that intruders are present in the body and that combative action should begin. But a T cell becomes activated only if its antigen receptor (T cell receptor or TCR) binds to the foreign antigen (antigen recognition) and at the same time it receives a second stimulating signal, a process known as ***costimulation*** (Figure 24.8). A common costimulator is *interleukin-2* (*IL-2*). The need for two signals is a little like starting and driving a car. When you insert the correct key (antigen) in the ignition (T cell receptor) and turn it, the car starts (recognition of specific antigen), but it cannot move forward until you move the gear shift into drive (costimulation). The need for costimulation probably helps prevent immune responses from occurring accidentally.

Figure 24.8 Activation and clonal selection of a cytotoxic T cell.

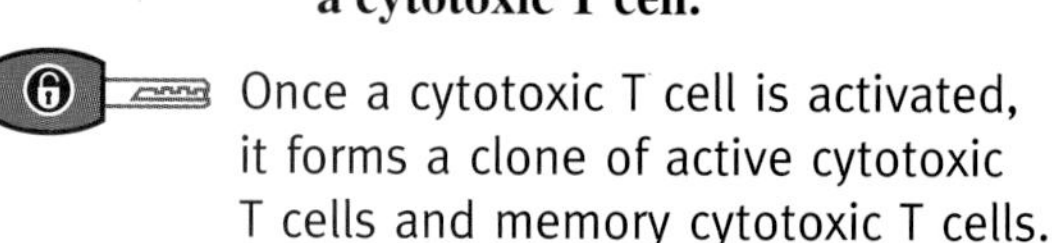

Once a cytotoxic T cell is activated, it forms a clone of active cytotoxic T cells and memory cytotoxic T cells.

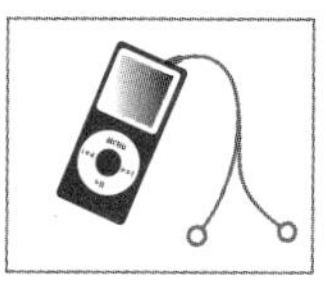

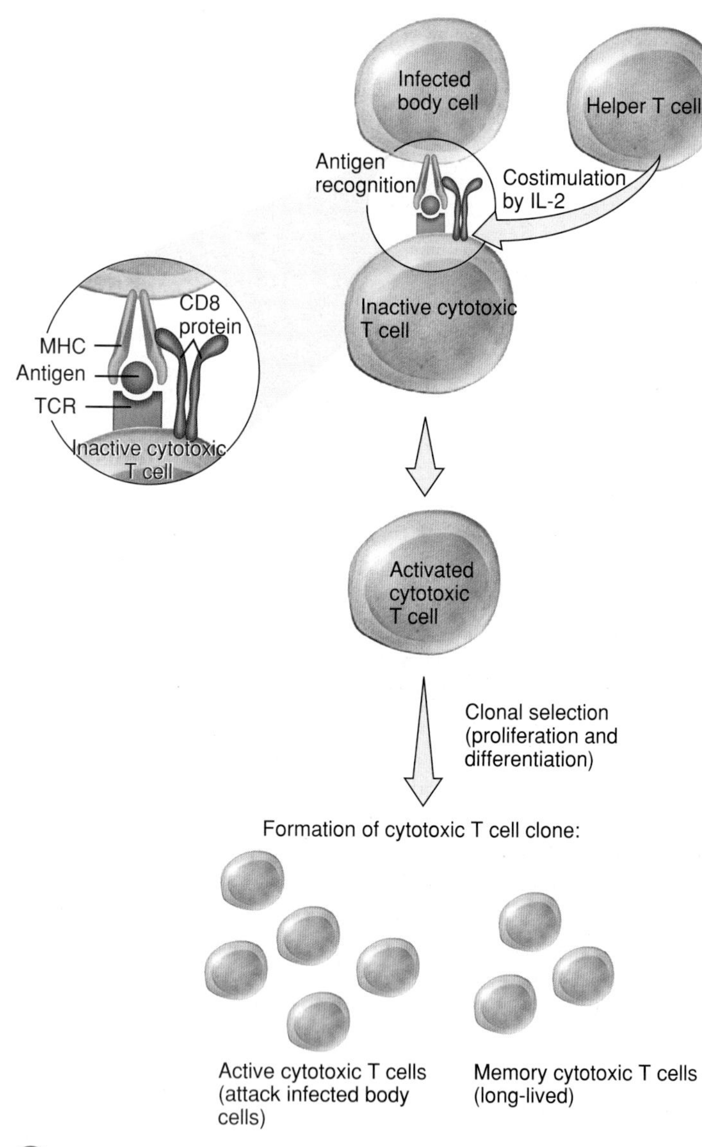

? **What is the function of the CD8 protein of a cytotoxic T cell?**

Once a T cell is activated, it enlarges and divides many times. The resulting population of identical cells, termed a ***clone,*** all recognize the same antigen. Before the first exposure to a given antigen, only a few T cells are able to recognize it. Once an adaptive immune response has begun, thousands of T cells can respond. Activation and division of T cells occur in the secondary lymphatic organs and tissues. If you have ever noticed swollen tonsils or lymph nodes in your neck, the continuous division of lymphocytes participating in an adaptive immune response was likely the cause.

The three major types of T cells are helper T cells, cytotoxic T cells, and memory T cells. ***Helper T cells*** help other cells of the adaptive immune system combat intruders. For instance, helper T cells release the costimulator protein interleukin-2, which enhances the activation and division of T cells. Other proteins released by helper T cells attract phagocytes and enhance the phagocytic ability of macrophages. Helper T cells also stimulate the development of B cells into antibody-producing plasma cells and the development of natural killer cells.

To reduce the risk of rejection, recipients of ***organ transplants*** receive immunosuppressive drugs. One such drug is *cyclosporine,* derived from a fungus, which inhibits secretion of interleukin-2 by helper T cells but has only a minimal effect on B cells. Thus, the risk of rejection is diminished while resistance to some diseases is maintained.

Cytotoxic T cells are the soldiers that march forth to do battle with foreign invaders in cell-mediated adaptive immune responses. The name "cytotoxic" reflects their function—killing cells. Cytotoxic T cells are especially effective against body cells infected by microbes, some tumor cells, and cells of a transplant. After they divide, cytotoxic T cells leave secondary lymphatic organs and tissues and migrate to sites of invasion, infection, and tumor formation. Cytotoxic T cells recognize and attach to cells bearing the antigen that stimulated their activation and division, operating in the following way (Figure 24.9):

1. Cytotoxic T cells, using receptors on their surfaces, recognize and bind to infected target cells that have microbial antigens displayed on their surface. The cytotoxic T cell then releases ***granzymes,*** protein-digesting enzymes that trigger apoptosis, the fragmentation of cellular contents (Figure 24.9a). Once the infected cell is destroyed, the released microbes are killed by phagocytes.

2. Cytotoxic T cells can also bind to infected body cells and release two proteins: perforin and granulysin. ***Perforin*** inserts into the plasma membrane of the target cell and creates channels in the membrane (Figure 24.9b). As a result, extracellular fluid flows into the target cell and cytolysis (cell bursting) occurs. ***Granulysin*** enters through the channels and destroys the microbes by creating holes in their plasma membranes. Cytotoxic T cells may also destroy target cells by releasing a toxic molecule called ***lymphotoxin,*** which activates enzymes in the target cell. These enzymes cause the target cell's DNA to fragment, and the cell dies. In addition, cytotoxic T cells secrete gamma interferon,

Figure 24.9 Activity of cytotoxic T cells. After delivering a "lethal hit," a cytotoxic T cell can detach and attack another infected target cell displaying the same antigen.

Cytotoxic T cells release granzymes that trigger apoptosis and perforin that triggers cytolysis of infected target cells.

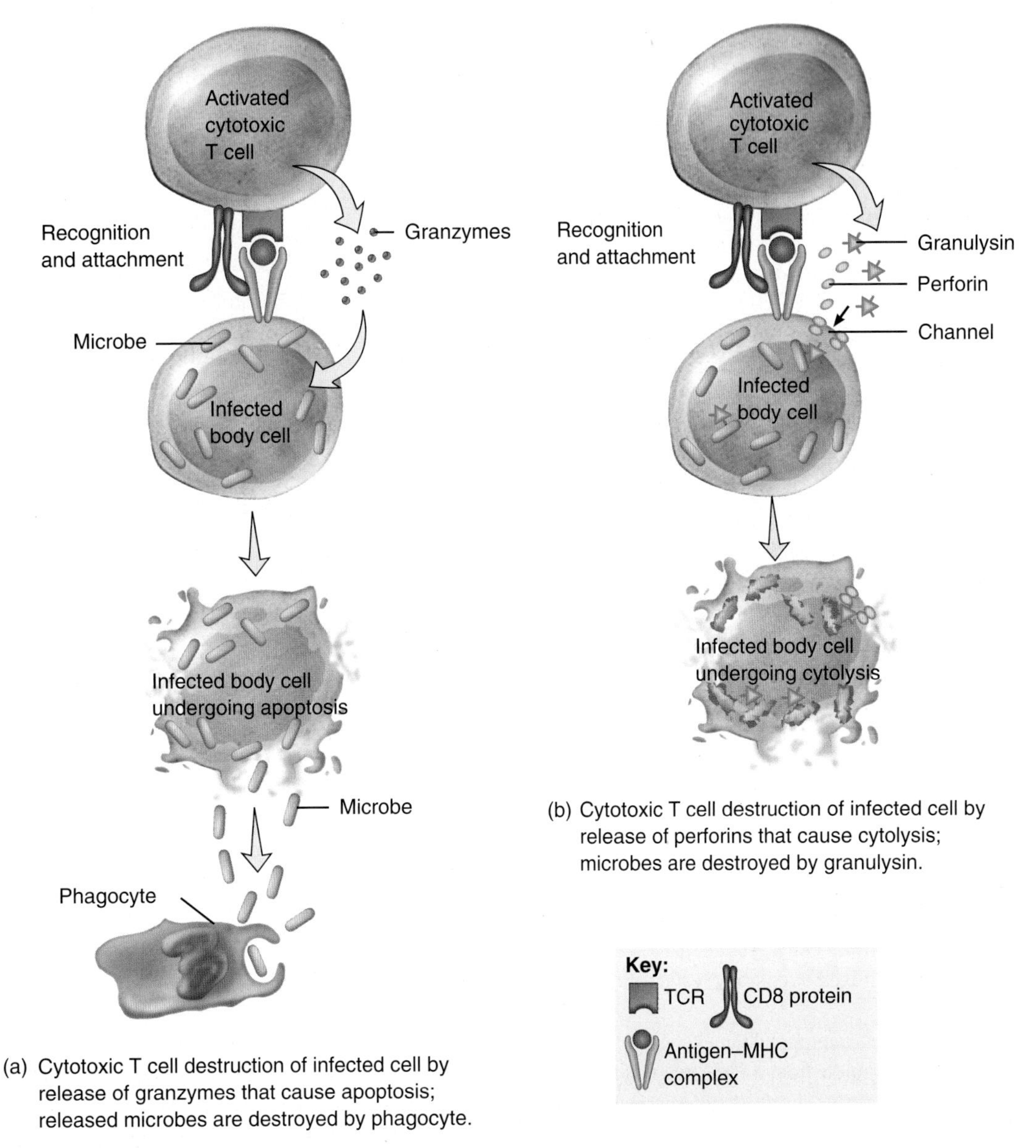

(a) Cytotoxic T cell destruction of infected cell by release of granzymes that cause apoptosis; released microbes are destroyed by phagocyte.

(b) Cytotoxic T cell destruction of infected cell by release of perforins that cause cytolysis; microbes are destroyed by granulysin.

? In addition to cells affected by microbes, what other types of target cells are attacked by cytotoxic T cells?

which attracts and activates phagocytic cells, and macrophage migration inhibition factor, which prevents migration of phagocytes from the infection site. After detaching from a target cell, a cytotoxic T cell can seek out and destroy another target cell.

Memory T cells remain in lymphatic tissue long after the original infection and are able to recognize the original invading antigen. Should the same antigen invade the body at a later date, the memory T cells initiate a faster reaction than occurred during the first invasion. The second response is so rapid that the pathogens are usually destroyed before any signs or symptoms of the disease occur. Memory T cells may provide immunity to a particular antigen for years. For instance, a person usually has the chickenpox only once because of memory T cells.

• CLINICAL CONNECTION | AIDS: Acquired Immunodeficiency Syndrome

Acquired immunodeficiency syndrome (AIDS) is a condition in which a person experiences an assortment of infections due to the progressive destruction of immune system cells by the *human immunodeficiency virus (HIV)*. AIDS represents the end stage of infection by HIV. A person who is infected with HIV may be symptom-free for many years, even while the virus is actively attacking the immune system. In the two decades after the first five cases were reported in 1981, 22 million people died of AIDS. Worldwide about 40 million people are currently infected with HIV.

Because HIV is present in the blood and some body fluids, it is most effectively transmitted by practices that involve the exchange of blood or body fluids. HIV is transmitted in semen or vaginal fluid during unprotected (without a condom) anal, vaginal, or oral sex. HIV also is transmitted by direct blood-to-blood contact, such as occurs in intravenous drug users who share hypodermic needles or in health-care professionals who may be accidentally stuck by HIV-contaminated hypodermic needles. In addition, HIV can be transmitted from an HIV-infected mother to her baby at birth or during breast-feeding.

HIV is a very fragile virus; it cannot survive for long outside the human body. The virus is not transmitted by insect bites. A person cannot become infected by casual physical contact with an HIV-infected person, such as by hugging or sharing household items. The virus can be eliminated from personal care items and medical equipment by exposing them to heat (135°F for 10 minutes) or by cleaning them with common disinfectants such as hydrogen peroxide, rubbing alcohol, household bleach, or germicidal cleansers such as Betadine or Hibiclens. Standard dish washing and clothes washing also kills HIV.

HIV consists of an inner core of ribonucleic acid (RNA) covered by a protein coat (capsid) surrounded by an outer layer, the envelope, composed of a lipid bilayer penetrated by proteins. Outside a living host cell, a virus is unable to replicate. However, when the virus infects and enters a host cell, its RNA uses the host cell's resources to make thousands of copies of the virus. New viruses eventually leave and then infect other cells.

HIV mainly damages helper T cells. Over 10 billion viral copies may be made each day. The viruses bud so rapidly from an infected cell's plasma membrane that the cell ruptures and dies. In most HIV-infected people, helper T cells are initially replaced as fast as they are destroyed. After several years, however, the body's ability to replace helper T cells is slowly exhausted, and the number of helper T cells in circulation gradually declines.

After a period of 2 to 10 years, the virus destroys enough helper T cells that most infected people begin to experience symptoms of immunodeficiency. HIV-infected people commonly have enlarged lymph nodes and experience persistent fatigue, involuntary weight loss, night sweats, skin rashes, diarrhea, and various lesions of the mouth and gums. In addition, the virus may begin to infect neurons in the brain, affecting the person's memory and producing visual disturbances.

As the immune system slowly collapses, an HIV-infected person becomes susceptible to a host of *opportunistic infections*. These are diseases caused by microorganisms that are normally held in check but now proliferate because of the defective immune system. In time, opportunistic infections usually are the cause of death.

At present, infection with HIV cannot be cured. Vaccines designed to block new HIV infections and to reduce the viral load (the number of copies of HIV RNA in a microliter of blood plasma) in those who are already infected are in clinical trials. Meanwhile, two categories of drugs have proved successful in extending the life of many HIV-infected people: reverse transcriptase inhibitors and protease inhibitors. Although HIV may virtually disappear from the blood with drug treatment, the virus typically still lurks in various lymphatic tissues. In such cases, the infected person can still transmit the virus to another person. •

CHECKPOINT

7. What is the normal function of major histocompatibility complex proteins (self-antigens)?
8. How do antigens arrive at lymphatic tissues?
9. How do antigen-presenting cells process antigens?
10. What are the functions of helper, cytotoxic, and memory T cells?
11. How do cytotoxic T cells kill their targets?

B Cells and Antibody-Mediated Immunity

The body contains not only millions of different T cells, but also millions of different B cells, each capable of responding to a specific antigen. Cytotoxic T cells leave lymphatic tissues to seek out and destroy a foreign antigen, but B cells stay put. In the presence of a foreign antigen, specific B cells in lymph nodes, the spleen, or lymphatic nodules become activated. They then divide and develop into plasma cells (see Figure 21.2, Section 21.2) that secrete specific antibodies, which in turn circulate in the lymph and blood to reach the sites of invasion.

During activation of a B cell, antigen receptors on the cell surface of a B cell bind to an antigen (Figure 24.10). B cell antigen receptors are chemically similar to the antibodies that eventually are secreted by the plasma cells. Although B cells can respond to an unprocessed antigen present in lymph or interstitial fluid, their response is much more intense when they process the antigen. Antigen processing in a B cell occurs in the following way: The antigen is taken into the B cell, broken into fragments and combined with MHC, and moved to the B cell surface. Helper T cells recognize the processed antigen–MHC complex and deliver the costimulation needed for B cell division and differentiation. The helper T cell releases interleukin-2 and other proteins that function as costimulators to activate B cells. Some of the activated B cells enlarge, divide, and differentiate into a clone of antibody-secreting ***plasma cells.*** A few days or weeks after exposure to an antigen, a plasma cell secretes hundreds of millions of antibodies daily, and secretion occurs for about 4 or 5 days, until the plasma cell dies. Most antibodies travel in lymph and blood to the invasion sites. Some activated B cells do not differentiate into plasma cells but rather remain as ***memory B cells*** that are ready to respond more rapidly and forcefully should the same antigen reappear at a future time. Although the functions of the five classes of antibodies differ somewhat, all attack antigens in several ways:

Figure 24.10 Activation and clonal selection of B cells. Plasma cells are actually much larger than B cells.

Plasma cells secrete antibodies.

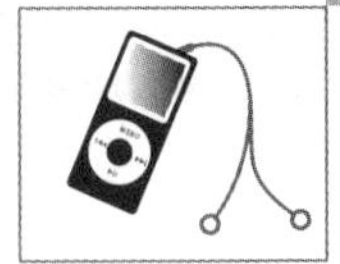

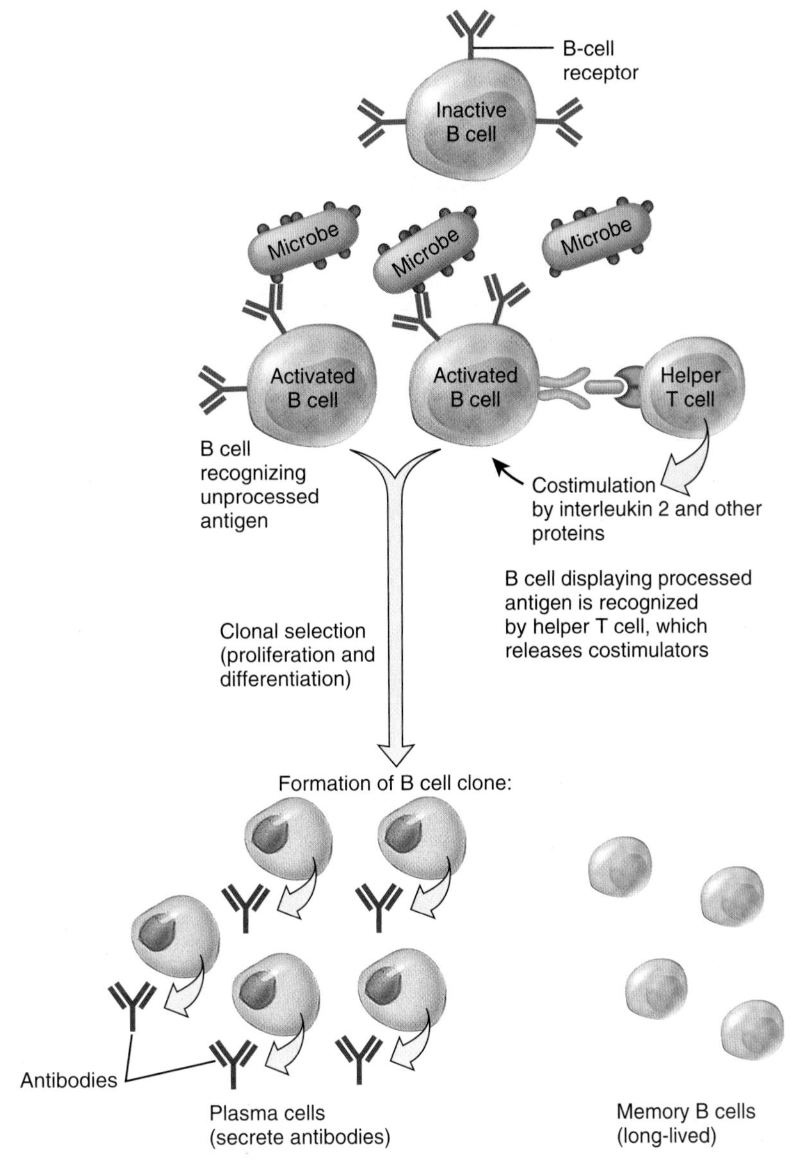

? How many different kinds of antibodies will be secreted by the plasma cells in the clone shown here?

1. **Neutralizing antigen.** The binding of an antibody to its antigen neutralizes some bacterial toxins and prevents attachment of some viruses to body cells.
2. **Immobilizing bacteria.** Some antibodies cause bacteria to lose their motility, which limits bacterial spread into nearby tissues.
3. **Agglutinating antigen.** Binding of antibodies to antigens may connect pathogens to one another, causing ***agglutination***, the clumping together of particles. Phagocytic cells ingest agglutinated microbes more readily.
4. **Activating complement.** Antigen–antibody complexes activate complement proteins, which then work to remove microbes through opsonization and cytolysis.
5. **Enhancing phagocytosis.** Once antigens have bound to an antibody's variable region, the antibody acts as a "flag" that attracts phagocytes. Antibodies enhance the activity of phagocytes by causing agglutination, by activating complement, and by coating microbes so that they are more susceptible to phagocytosis (opsonization).

Table 24.3 summarizes the functions of cells that participate in adaptive immune responses.

Immunological Memory

A hallmark of adaptive immune responses is memory for specific antigens that have triggered immune responses in the past.

TABLE 24.3

Summary of Cell Functions in Adaptive Immune Responses

CELL	FUNCTIONS
ANTIGEN-PRESENTING CELLs (APCs)	
Macrophage	Phagocytosis; processing and presentation of foreign antigens to T cells; secretion of interleukin-1, which stimulates secretion of interleukin-2 by helper T cells and induces proliferation of B cells; secretion of interferons that stimulate T cell growth.
Dendritic cell	Processes and presents antigen to T cells and B cells; found in mucous membranes, skin, and lymph nodes.
B cell	Processes and presents antigen to helper T cells.
LYMPHOCYTES	
Cytotoxic T cell	Kills host target cells by releasing granzymes that induce apoptosis, perforin that forms channels to cause cytolysis, granulysin that destroys microbes, lymphotoxin that destroys target cell DNA, gamma interferon that attracts macrophages and increases their phagocytic activity, and macrophage migration inhibition factor that prevents macrophage migration from site of infection.
Helper T cell	Cooperates with B cells to amplify antibody production by plasma cells and secretes interleukin-2, which stimulates proliferation of T cells and B cells. May secrete gamma interferon and tumor necrosis factor (TNF), which stimulate inflammatory response.
Memory T cell	Remains in lymphatic tissue and recognizes original invading antigens, even years after the first encounter.
B cell	Differentiates into antibody-producing plasma cell.
Plasma cell	Descendant of B cell that produces and secretes antibodies.
Memory B cell	Descendant of B cell that remains after an immune response and is ready to respond rapidly and forcefully should the same antigen enter the body in the future.

Immunological memory is due to the presence of long-lasting antibodies and very long-lived lymphocytes that arise during division and differentiation of antigen-stimulated B cells and T cells.

Primary and Secondary Responses

Adaptive immune responses, whether cell-mediated or antibody-mediated, are much quicker and more intense after a second or subsequent exposure to an antigen than after the first exposure. Initially, only a few cells have the correct antigen receptors to respond, and the immune response may take several days to build to maximum intensity. Because thousands of memory cells exist after an initial encounter with an antigen, they can divide and differentiate into helper T cells, cytotoxic T cells, or plasma cells within hours the next time the same antigen appears.

One measure of immunological memory is the amount of antibody in blood plasma. After an initial contract with an antigen, no antibodies are present for a few days. Then, the levels of antibodies slowly rise, first IgM and then IgG, followed by a gradual decline (Figure 24.11). This is the ***primary response.*** Memory cells may live for decades. Every new encounter with the same antigen causes a rapid division of memory cells. The antibody level after subsequent encounters is far greater than during a primary response and consists mainly of IgG antibodies. This accelerated, more intense response is called the ***secondary response.*** Antibodies produced during a secondary response are even more effective than those produced during a primary response. Thus, they are more successful in disposing of the invaders.

Primary and secondary responses occur during microbial infection. When you recover from an infection without taking antimicrobial drugs, it is usually because of the primary response. If the same microbe infects you later, the secondary response could be so swift that the microbes are destroyed before you exhibit any signs or symptoms of infection.

Figure 24.11 Production of antibodies in the primary (after first exposure) and secondary (after second exposure) responses to a given antigen.

Immunological memory is the basis for successful immunization by vaccination.

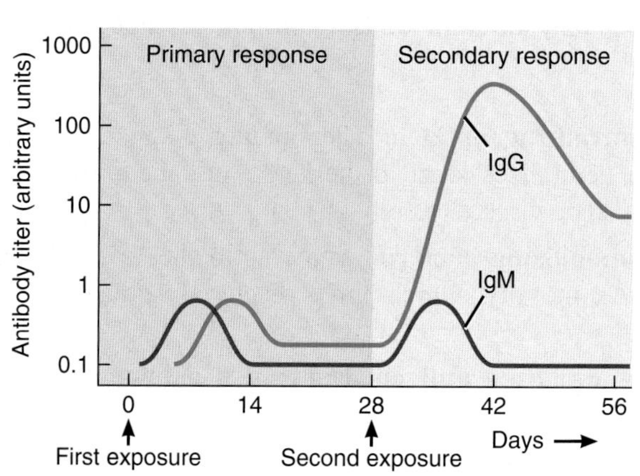

According to this graph, how much more IgG is circulating in the blood in the secondary response than in the primary response? (Hint: Notice that each mark on the antibody titer axis represents a 10-fold increase.)

• CLINICAL CONNECTION | Allergic Reactions

A person who is overly reactive to a substance that is tolerated by most other people is said to be *allergic* and the reactions that occur are called **allergic reactions** and result in some kind of tissue damage. The antigens that induce an allergic reaction are termed *allergens*. Common allergens include certain foods (milk, peanuts, shellfish, eggs), antibiotics (penicillin, tetracycline), vaccines (pertussis, typhoid), venoms (honeybee, wasp, snake), cosmetics, chemicals in plants such as poison ivy, pollen, dust, molds, iodine-containing dyes used in certain x-ray procedures, and even microbes.

Type I (anaphylactic) reactions are the most common and typically occur within a few minutes after a person who was previously sensitized to an allergen is reexposed to it. In response to certain allergens, some people produce IgE antibodies that bind to the surface of mast cells and basophils. The next time the same allergen enters the body, it attaches to the IgE antibodies already present. In response, both the mast cells and basophils release histamine, prostaglandins, and other chemicals. Collectively, these chemicals cause vasodilation, increased blood capillary permeability, increased smooth muscle contraction in the airways of the lungs, and increased mucus secretion. As a result, a person may experience inflammatory responses, difficulty in breathing through the narrowed airways, and a runny nose from excess mucus secretion. In *anaphylactic shock*, which may occur in a susceptible individual who has just received a triggering drug or been stung by an insect, wheezing and shortness of breath as airways constrict are usually accompanied by shock due to vasodilation and fluid loss from blood. Injecting epinephrine to dilate the airways and strengthen the heartbeat usually is effective in this life-threatening emergency. •

Naturally Acquired and Artificially Acquired Immunity

Immunological memory provides the basis for immunization by vaccination against certain diseases, for instance, polio. When you receive the *vaccine,* which may contain weakened or killed whole microbes or parts of microbes, your B cells and T cells are activated. Should you subsequently encounter the living pathogen as an infecting microbe, your body initiates a secondary response. However, booster doses of some immunizing agents must be given periodically to maintain adequate protection against the pathogen.

Table 24.4 summarizes the various types of antigen encounters that provide naturally and artificially acquired immunity.

CHECKPOINT

12. How are cell-mediated and antibody-mediated immune responses similar and different?
13. How is the secondary response to an antigen different from the primary response?

24.5 AGING AND THE IMMUNE SYSTEM

OBJECTIVE

- Describe the effects of aging on the immune system.

With advancing age, most people become more susceptible to all types of infections and malignancies. Their response to vaccines is decreased, and they tend to produce more autoantibodies (antibodies against their body's own molecules). In addition, the immune system exhibits lowered levels of function. For example, T cells become less responsive to antigens, and fewer T cells respond to infections. This may result from age-related atrophy of the thymus or decreased production of thymic hormones. Because the T cell population decreases with age, B cells are also less responsive. Consequently, antibody levels do not increase as rapidly in response to a challenge by an antigen, resulting in increased susceptibility to various infections. It is for this key reason that elderly individuals are encouraged to get influenza (flu) vaccinations each year.

CHECKPOINT

14. What are the consequences of decreases in the number of T cells and B cells with advancing age?

To appreciate the many ways that the lymphatic system and immunity contribute to homeostasis of other body systems, examine Exhibit 24.1.

TABLE 24.4

Types of Adaptive Immunity

METHOD	DESCRIPTION
Naturally acquired active immunity	Following exposure to a microbe, antigen recognition by B cells and T cells and costimulation lead to the formation of antibody-secreting plasma cells, cytotoxic T cells, and B and T memory cells.
Naturally acquired passive immunity	Transfer of IgG antibodies from mother to fetus across placenta, or of IgA antibodies from mother to baby in milk during breast-feeding.
Artificially acquired active immunity	Antigens introduced during a vaccination stimulate cell-mediated and antibody-mediated immune responses, leading to the production of memory cells. The antigens are pretreated to be immunogenic but not pathogenic; that is, they will trigger an immune response but not cause significant illness.
Artificially acquired passive immunity	Intravenous injection of immunoglobulins (antibodies).

EXHIBIT 24.1 Contributions of the Lymphatic System and Immunity to Homeostasis

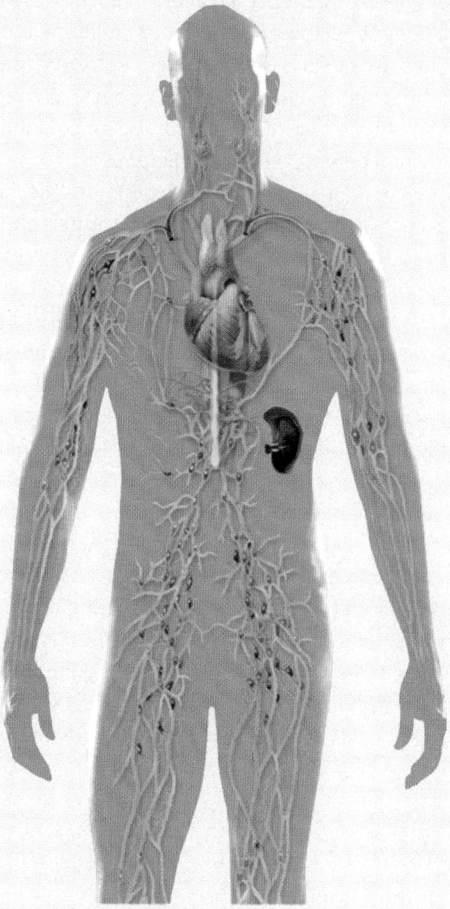

BODY SYSTEM		CONTRIBUTION OF THE LYMPHATIC SYSTEM AND IMMUNITY
For all body systems		B cells, T cells, and antibodies help protect all body systems from attack by harmful foreign microbes (pathogens), foreign cells, and cancer cells.
Integumentary system	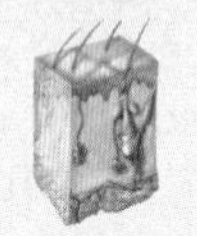	Lymphatic vessels drain excess interstitial fluid and leaked plasma proteins from the dermis of the skin. Immune system cells (Langerhans cells) in the skin help protect the skin. Lymphatic tissue provides IgA antibodies in sweat.
Skeletal system	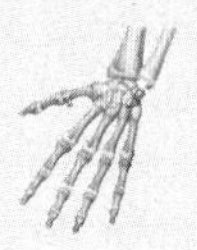	Lymphatic vessels drain excess interstitial fluid and leaked plasma proteins from connective tissue around bones.
Muscular system		Lymphatic vessels drain excess interstitial fluid and leaked plasma proteins from muscles.
Nervous system	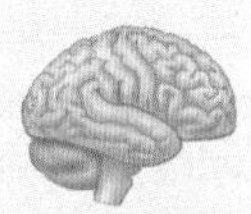	Lymphatic vessels drain excess interstitial fluid and leaked plasma proteins from the peripheral nervous system.
Endocrine system	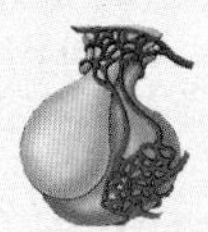	Flow of lymph distributes some hormones. Lymphatic vessels drain excess interstitial fluid and leaked plasma proteins from endocrine glands.
Cardiovascular system	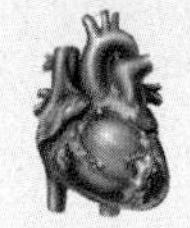	Lymph returns excess fluid filtered from blood capillaries and leaked plasma proteins to venous blood. Macrophages in spleen destroy aged red blood cells and remove debris in blood.
Respiratory system	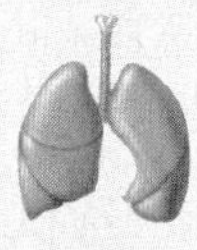	Tonsils, lymphatic nodules in the mucosa, and alveolar macrophages in the lungs help protect airways and lungs from pathogens. Lymphatic vessels drain excess interstitial fluid from the lungs.
Digestive system	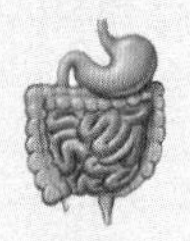	Tonsils and lymphatic nodules in the mucosa help defend against toxins and pathogens that penetrate the body from the gastrointestinal tract. Immune system provides IgA antibodies in saliva and gastrointestinal secretions. Lymphatic vessels pick up absorbed dietary lipids and fat-soluble vitamins from the small intestine and transport them to the blood. Lymphatic vessels drain excess interstitial fluid and leaked plasma proteins from organs of the digestive system.
Urinary system	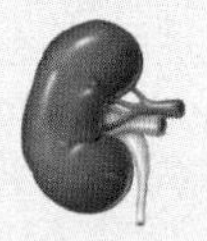	Lymphatic vessels drain excess interstitial fluid and leaked plasma proteins from organs of the urinary system. Lymphatic nodules in the mucosa help defend against toxins and pathogens that penetrate the body via the urethra.
Reproductive systems	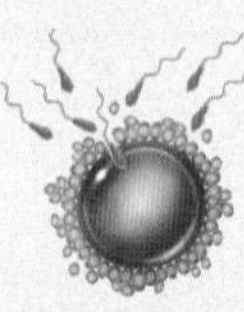	Lymphatic vessels drain excess interstitial fluid and leaked plasma proteins from organs of the reproductive systems. Lymphatic nodules in the mucosa help defend against toxins and pathogens that penetrate the body via the vagina and penis. In females, sperm deposited in the vagina are not attacked as foreign invaders due to components in the seminal fluid that inhibit immune responses. IgG antibodies can cross the placenta to provide protection to a developing fetus. Lymphatic tissue provides IgA antibodies in the milk of the nursing mother.

KEY MEDICAL TERMS ASSOCIATED WITH THE LYMPHATIC SYSTEM

Allograft (AL-ō-graft; *allo-* = other) A transplant between genetically different individuals of the same species. Skin transplants from other people and blood transfusions are allografts.

Autograft (AW-tō-graft; *auto-* = self) A transplant in which one's own tissue is grafted to another part of the body (such as skin grafts for burn treatment or plastic surgery).

Chronic fatigue syndrome (CFS) A disorder, usually occurring in young female adults, characterized by (1) extreme fatigue that impairs normal activities for at least 6 months and (2) the absence of other known diseases (cancer, infections, drug abuse, toxicity, or psychiatric disorders) that might produce similar symptoms.

Gamma globulin (GLOB-ū-lin) Suspension of immunoglobulins from blood consisting of antibodies that react with a specific pathogen. It is prepared by injecting the pathogen into animals, removing blood from the animals after antibodies have been produced, isolating the antibodies, and injecting them into a human to provide short-term immunity.

Graft Any tissue or organ used for transplantation or a transplant of such structures.

Lymphadenopathy (lim-fad-e-NOP-a-thē; *lymph-* = clear fluid; *-pathy* = disease) Enlarged, sometimes tender lymph glands as a response to infection, also called ***swollen glands.***

Splenomegaly (splē-nō-MEG-a-lē; *-mega-* = large) Enlarged spleen.

Tonsillectomy (ton-si-LEK-tō-mē; *-ectomy* = excision) Removal of a tonsil.

Xenograft (ZEN-ō-graft; *xeno-* = strange or foreign) A transplant between animals of different species. Xenografts from porcine (pig) or bovine (cow) tissue may be used in people as a physiological dressing for severe burns.

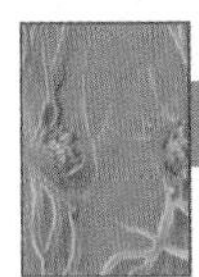

STUDY OUTLINE

Overview of Immunity (Section 24.1)

1. Despite constant exposure to a variety of pathogens (disease-producing microbes such as bacteria and viruses), most people remain healthy.
2. Immunity or resistance is the ability to ward off damage or disease. Innate immunity refers to defenses that are present at birth; they are always present and provide immediate but general protection against invasion by a wide range of pathogens. Adaptive immunity refers to defenses that respond to a particular invader; it involves activation of specific lymphocytes that can combat a specific invader.

Lymphatic System Structure and Function (Section 24.2)

1. The body system responsible for adaptive immunity (and some aspects of innate immunity) is the lymphatic system, which consists of lymph, lymphatic vessels, structures and organs that contain lymphatic tissue, and red bone marrow.
2. Components of blood plasma filter through blood capillary walls to form interstitial fluid, the fluid that bathes the cells of body tissues. After interstitial fluid passes into lymphatic vessels, it is called lymph. Interstitial fluid and lymph are chemically similar to blood plasma.
3. The lymphatic system drains tissue spaces of excess fluid and returns proteins that have escaped from blood to the cardiovascular system. It also transports lipids and lipid-soluble vitamins from the gastrointestinal tract to the blood, and it protects the body against invasion.
4. Lymphatic vessels begin as lymphatic capillaries in tissue spaces between cells. The lymphatic capillaries merge to form larger lymphatic vessels, which ultimately drain into the thoracic duct or right lymphatic duct. Located at intervals along lymphatic vessels are lymph nodes, masses of B cells and T cells surrounded by a capsule.
5. The passage of lymph is from interstitial fluid, to lymphatic capillaries, to lymphatic vessels and lymph nodes, to the thoracic duct or right lymphatic duct, to the junction of the internal jugular and subclavian veins.
6. Lymph flows due to the milking action of skeletal muscle contractions and pressure changes that occur during inhalation. Valves in the lymphatic vessels prevent backflow of lymph.
7. Primary lymphatic organs and tissues are the sites where stem cells divide and develop into mature B cells and T cells. They include the red bone marrow (in flat bones and the ends of the long bones of adults) and the thymus. Stem cells in red bone marrow give rise to mature B cells and to immature T cells that migrate to the thymus, where they mature into functional T cells.
8. The secondary lymphatic organs and tissues are the sites where most immune responses occur. They include lymph nodes, the spleen, and lymphatic nodules.
9. Lymph nodes contain B cells that develop into plasma cells, T cells, dendritic cells, and macrophages. Lymph enters nodes through afferent lymphatic vessels and exits through efferent lymphatic vessels.
10. Lymphatic nodules are oval-shaped concentrations of lymphatic tissue that are not surrounded by a capsule. They are scattered throughout the mucosa of the gastrointestinal, respiratory, urinary, and reproductive tracts.

Innate Immunity (Section 24.3)

1. Innate immunity defenses include barriers provided by the skin and mucous membranes (first line of defense). They also include various internal defenses (second line of defense): antimicrobial substances (interferons, complement, transferrin, and antimicrobial peptides), phagocytes (neutrophils and macrophages), natural killer cells (which have the ability to kill a wide variety of infectious microbes and certain tumor cells), inflammation, and fever.
2. Table 24.1 summarizes the components of innate immunity.

Adaptive Immunity (Section 24.4)

1. Adaptive immunity involves the production of specific types of cells or specific antibodies to destroy a particular antigen.
2. An antigen is any substance that the adaptive immune system recognizes as foreign (nonself). Normally, a person's immune system

cells exhibit self-tolerance: They recognize and do not attack their own tissues and cells.

3. B cells complete their development in red bone marrow, but mature T cells develop in the thymus from immature T cells that migrate from bone marrow.
4. The major histocompatibility complex (MHC) antigens are unique to each person's body cells. All cells except red blood cells display MHC molecules.
5. Antigens induce plasma cells to secrete antibodies, proteins that typically contain four polypeptide chains. The variable regions of an antibody are the antigen-binding sites, where the antibody can bind to a particular antigen.
6. Based on chemistry and structure, antibodies, also known as immunoglobulins (Igs), are grouped in five classes, each with specific functions: IgG, IgA, IgM, IgD, and IgE (see Table 24.2). Functionally, antibodies neutralize antigens, immobilize bacteria, agglutinate antigens, activate complement, and enhance phagocytosis.
7. Antigen-presenting cells (APCs) process and present antigens to activate T cells, and they secrete substances that stimulate division of T cells and B cells.
8. There are three main kinds of T cells: helper T cells, which stimulate growth and division of cytotoxic T cells, attract phagocytes, and stimulate development of B cells into antibody-producing plasma cells; cytotoxic T cells, which eliminate invaders by (1) releasing granzymes that cause target cell apoptosis (phagocytes then kill the microbes) and (2) releasing perforin, which causes cytolysis, and granulysin that destroys the microbes; and memory T cells, which recognize previously encountered antigens at a later date.
9. Antibody-mediated immunity refers to destruction of antigens by antibodies, which are produced by descendants of B cells called plasma cells.
10. B cells develop into antibody-producing plasma cells under the influence of chemicals secreted by antigen-presenting cells and helper T cells.
11. Table 24.3 summarizes the functions of cells that participate in adaptive immune responses.
12. Immunization against certain microbes is possible because memory B cells and memory T cells remain after a primary response to an antigen. The secondary response provides protection should the same microbe enter the body again. Table 24.4 summarizes the various types of antigen encounters that provide naturally and artificially acquired immunity.

Aging and the Immune System (Section 24.5)

1. With advancing age, individuals become more susceptible to infections and malignancies, respond less well to vaccines, and produce more autoantibodies.
2. T cell responses also diminish with age.

SELF-QUIZ QUESTIONS

1. Which of the following is NOT true concerning the lymphatic system?
   a. Lymphatic vessels transport lipids from the gastrointestinal tract to the blood.
   b. Lymph is more similar to interstitial fluid than to blood.
   c. Lymphatic tissue is present in only a few isolated organs in the body.
   d. The unique structure of lymphatic capillaries allows fluid to flow into them but not out of them.
   e. Lymphatic vessels resemble veins in structure.
2. Which of the following are produced by virus-infected cells to protect uninfected cells from viral invasion?
   a. complement molecules **b.** prostaglandins **c.** fibrins
   d. interferons **e.** histamines
3. A blockage in the right lymphatic duct would interfere with lymph drainage from the
   a. left arm **b.** right leg **c.** lower abdomen
   d. left leg **e.** right arm.
4. Which of the following best represents lymph flow from the interstitial spaces back to the blood?
   a. lymphatic capillaries → lymphatic ducts → lymphatic vessels → junction of internal jugular and subclavian veins
   b. junction of internal jugular and subclavian veins → lymphatic capillaries → lymphatic vessels → lymphatic ducts
   c. lymphatic capillaries → lymphatic vessels → lymphatic ducts → junction of internal jugular and subclavian veins
   d. lymphatic ducts → lymphatic vessels → lymphatic capillaries → junction of internal jugular and subclavian veins
   e. lymphatic capillaries → lymphatic vessels → junction of internal jugular and subclavian veins → lymphatic ducts
5. Lymph nodes
   a. filter lymph
   b. are another name for tonsils
   c. produce lymph
   d. are a primary storage site for blood
   e. produce a protective mucus.
6. Which of the following is NOT true about the role of skin in nonspecific immunity?
   a. Sebum inhibits the growth of certain bacteria.
   b. Epidermal cells produce interferons to destroy viruses.
   c. Shedding of epidermal cells helps remove microbes.
   d. Lysozyme in sweat destroys some bacteria.
   e. The skin forms a physical barrier to prevent entry of microbes.
7. Which of the following statements about B cells is true?
   a. They become functional while in the thymus.
   b. Some develop into plasma cells that secrete antibodies.
   c. Some B cells become natural killer cells.
   d. Cytotoxic B cells travel in lymph and blood to react with foreign antigens.
   e. They kill virus-infected cells by secreting perforin.
8. The cells that release granzymes, perforin, granulysin, and lymphotoxin are
   a. cytotoxic T cells **b.** plasma cells **c.** B cells
   d. natural killer cells **e.** helper T cells.
9. The secondary response in antibody-mediated immunity
   a. is characterized by a slow rise in antibody levels and then a gradual decline
   b. occurs when you first receive a vaccination against some disease
   c. produces fewer but more responsive antibodies than occur during the primary response

d. is an intense response by memory cells to produce antibodies when an antigen is contacted again
e. is rarely seen except in autoimmune disorders.

10. The ability of the body's immune system to recognize its own tissues is known as
a. immunological escape **b.** autoimmunity
c. nonspecific resistance **d.** hypersensitivity
e. self-tolerance

11. A disease that causes destruction of helper T cells would result in all of the following effects EXCEPT
a. inability to produce cytotoxic T cells
b. alteration of lymph flow
c. lack of development of plasma cells
d. decreased production of antibodies
e. increased risk of developing infections.

12. In which lymphatic organ do T cells mature?
a. thyroid gland **b.** spleen **c.** thymus
d. red bone marrow **e.** lymph node

13. Place the following steps involved in the process of inflammation in the correct order.
1. arrival of large numbers of neutrophils
2. vasodilation and increased permeability of blood vessels
3. formation of pus
4. increased migration of monocytes
5. formation of fibrin network to form a clot
6. release of histamine
a. 6, 2, 4, 1, 5, 3 **b.** 3, 6, 1, 4, 2, 5
c. 5, 1, 4, 2, 6, 3 **d.** 6, 2, 5, 1, 4, 3
e. 4, 6, 1, 3, 2, 5

14. Match the following:
____ **a.** destroy antigens by cytolysis
____ **b.** stimulate other cells of the adaptive immune response
____ **c.** are programmed to recognize the original invading antigen; allow immunity to last for years
____ **d.** function in innate immunity
____ **e.** develop into plasma cells

A. natural killer cells
B. helper T cells
C. B cells
D. memory T cells
E. cytotoxic T cells

15. What happens during opsonization?
a. engulfment of a microbe by a phagocyte
b. chemical attraction of a phagocyte
c. binding of complement to a microbe
d. attachment of a phagocyte to a microbe
e. breakdown of a microbe by enzymes

16. All of the following contribute to nonspecific immunity EXCEPT
a. complement **b.** immunoglobulins
c. natural killer cells **d.** lysozyme
e. interferons.

17. Inflammation produces
a. redness due to bleeding
b. heat due to fever-causing toxins
c. swelling due to increased permeability of capillaries
d. pain due to histamine release
e. mucus due to phagocytosis.

18. Place the phases of phagocytosis in the correct order.
1. adherence to foreign material
2. chemotaxis of phagocytes
3. exocytosis of indigestible materials
4. ingestion of foreign material
a. 1, 2, 3, 4 **b.** 2, 1, 4, 3
c. 1, 4, 3, 2 **d.** 4, 3, 2, 1
e. 2, 1, 3, 4

19. Antibodies attack antigens by all of the following methods EXCEPT
a. agglutination of antigens
b. activation of complement
c. opsonization to enhance phagocytosis
d. preventing attachment to body cells
e. producing acid secretions.

20. What is the importance of tonsils in the body's defenses?
a. They help destroy microbes that are inhaled.
b. They contain ciliated cells that move trapped pathogens from the breathing passages.
c. They are needed for T cell maturation.
d. They are needed for B cell maturation.
e. They filter lymph.

CRITICAL THINKING QUESTIONS

1. Nancy found a lump in her right breast during her monthly self-examination. The lump was found to be cancerous. The surgeon removed the breast lump, the surrounding tissue, and some lymph nodes. Which nodes were probably removed and why?

2. Years ago, a tonsillectomy was almost considered a "rite of passage" for children in elementary school. It seemed like all children were getting their tonsils removed. Why are tonsils frequently infected in young children?

3. Matt stepped on a rusty fishhook while walking along the beach. The emergency room nurse removed the fishhook and gave Matt a tetanus booster. Why?

4. You learned in Chapter 19 that the cornea and lens of the eye are completely lacking in capillaries. How is this fact related to the high success of corneal transplants?

? ANSWERS TO FIGURE QUESTIONS

24.1 Red bone marrow contains stem cells that develop into lymphocytes.

24.2 Lymph is more similar to interstitial fluid than to blood plasma because the protein content of lymph is low.

24.3 Inhalation promotes the movement of lymph from abdominal lymphatic vessels toward the thoracic region because the pressure in the vessels of the thoracic region is lower than the pressure in the abdominal region when you inhale.

24.4 Foreign substances in lymph that enter a lymph node may be phagocytized by macrophages or attacked by lymphocytes that mount immune responses.

24.5 Redness results from increased blood flow due to vasodilation; pain results from injury of nerve fibers, irritation by microbial toxins and pressure due to edema; heat results from increased blood flow and heat released by locally increased metabolic reactions; swelling results from leakage of fluid from capillaries due to increased permeability.

24.6 The variable regions of an antibody bind specifically to the antigen that triggered its production.

24.7 APCs include macrophages in tissues throughout the body, B cells in blood and lymphatic tissue, and dendritic cells in mucous membranes and the skin.

24.8 The CD8 protein of a cytotoxic T cell binds to the MHC molecule of an infected body cell to help anchor the T cell receptor (TCR)–antigen interaction so that antigen recognition can occur.

24.9 Cytotoxic T cells attack some tumor cells and transplanted tissue cells, as well as cells infected by microbes.

24.10 Because all of the plasma cells in this figure are part of the same clone, they secrete just one kind of antibody.

24.11 At peak secretion, approximately 1000 times more IgG is produced in the secondary response than in the primary response.

The Respiratory System

25

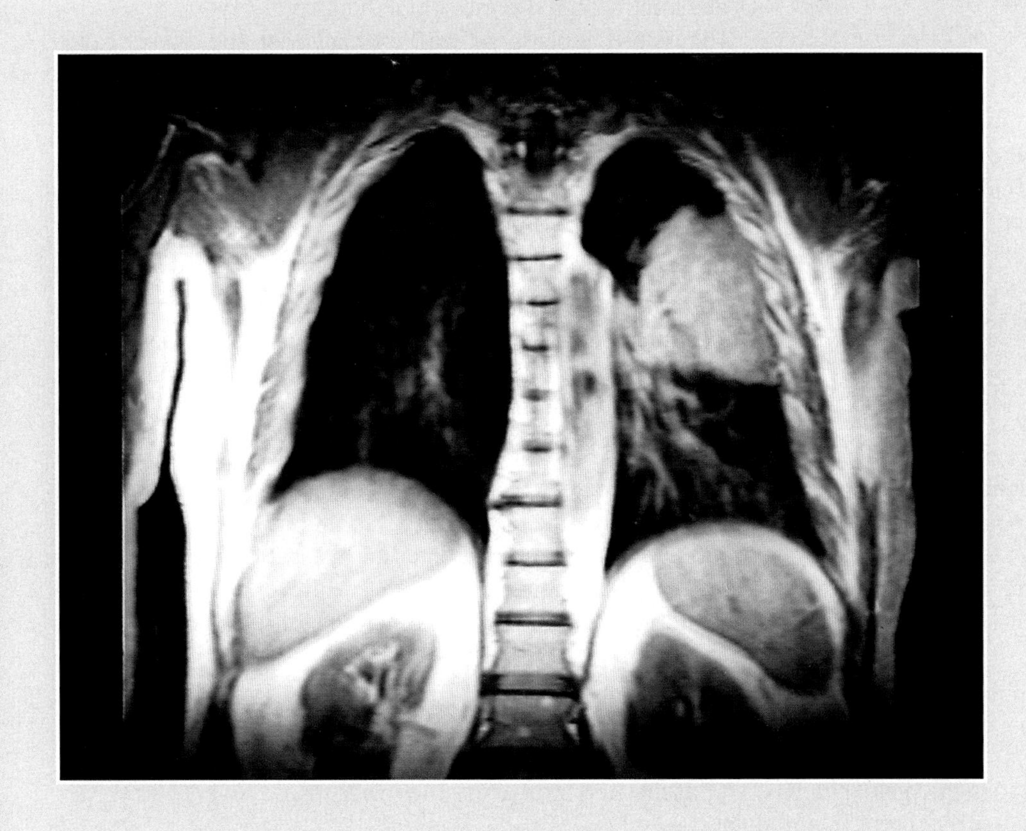

Every day we go about our activities of daily living and rarely think of the many functions that our body is automatically monitoring and maintaining such as breathing. Most do not consider their respiratory rate until they are under conditions of exertion or stress. The respiratory system quickly reacts to the various demands that we place on it, such as exercising, changing temperatures, changing altitudes, and other daily stresses. We begin inspiration by contracting our muscles to breathe in and relax to exhale. With every breath, molecules of oxygen are exchanged in the lungs across thin little cups called alveoli. Carbon dioxide as a waste product is exchanged for the oxygen. Oxygen is crucial to each and every cell in the body for survival. Without it cells become ischemic and cannot survive. The lungs are closely paired with the cardiovascular system in the delivery of oxygen to cells. They are also closely paired with the kidneys in balancing acid levels in our bodies, as we will see in our study of the urinary system (Chapter 28).

CONTENTS AT A GLANCE

25.1 OVERVIEW OF THE RESPIRATORY SYSTEM

OBJECTIVE

- Describe the steps of respiration.

Body cells continually use oxygen (O_2) for the metabolic reactions that release energy from nutrient molecules and produce ATP. These same reactions produce carbon dioxide (CO_2). Because an excessive amount of CO_2 produces acidity that can be toxic to cells, excess CO_2 must be eliminated quickly and efficiently. The ***respiratory system,*** which includes the nose, pharynx (throat), larynx (voice box), trachea (windpipe), bronchi, and lungs (Figure 25.1), provides for gas exchange, the intake of O_2, and the removal of CO_2. The respiratory system also helps regulate blood pH; contains receptors for the sense of smell; filters, warms, and moistens inspired air; produces sounds; and rids the body of some water and heat in exhaled air.

The branch of medicine that deals with the diagnosis and treatment of diseases of the ears, nose, and throat (ENT) is called ***otorhinolaryngology*** (ō′-tō-rī′-nō-lar′-in-GOL-ō-jē; *oto-* = ear; *-rhino-* = nose; *-laryngo-* = voice box; *-logy* = study of). A ***pulmonologist*** (*pulmon-* = lung) is a specialist in the diagnosis and treatment of diseases of the lungs.

The entire process of gas exchange in the body, called ***respiration,*** occurs in three basic steps:

1. ***Pulmonary ventilation,*** or ***breathing,*** is the flow of air into and out of the lungs.
2. ***External respiration*** is the exchange of gases between the air spaces (alveoli) of the lungs and the blood in pulmonary capillaries. In this process, pulmonary capillary blood gains O_2 and loses CO_2.
3. ***Internal respiration*** is the exchange of gases between blood in systemic capillaries and tissue cells. The blood loses O_2 and gains CO_2. Within cells the metabolic reactions that consume O_2 and give off CO_2 during the production of ATP are called ***cellular respiration*** (discussed in Chapter 21).

As you can see, two systems cooperate to supply O_2 and eliminate CO_2—the cardiovascular and respiratory systems. Pulmonary ventilation and external respiration are the responsibility of the respiratory system, while external respiration and internal respiration are functions of the cardiovascular system.

Figure 25.1 Organs of the respiratory system.

The upper respiratory system includes the nose, pharynx, and associated structures. The lower respiratory system includes the larynx, trachea, bronchi, and lungs.

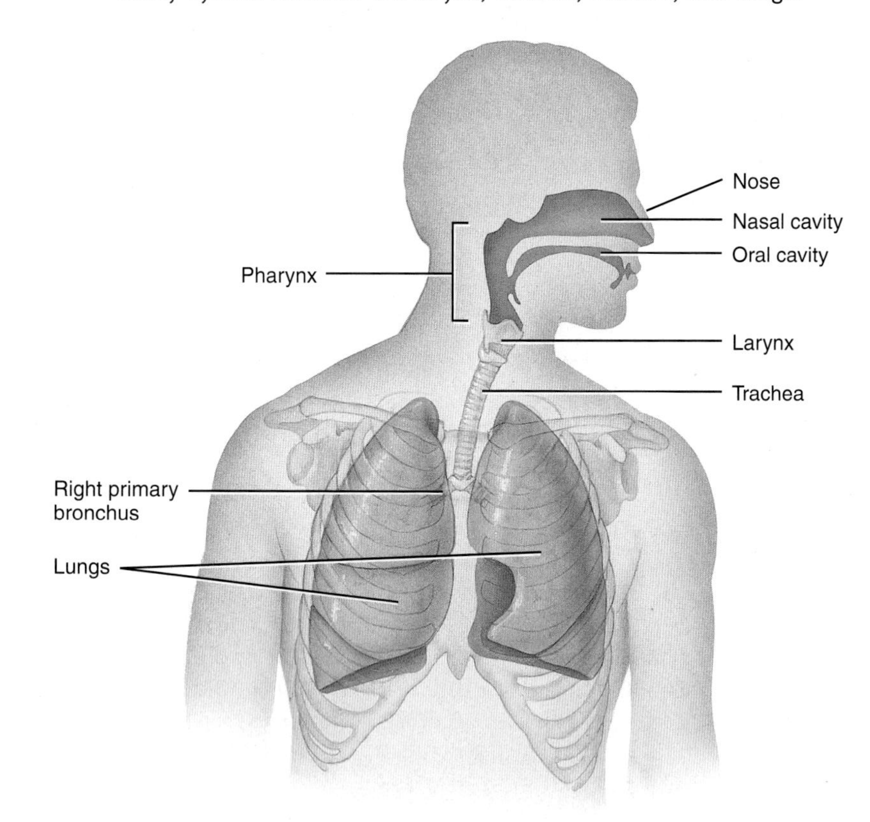

Anterior view showing organs of respiration

Functions

1. Provides for gas exchange—intake of O_2 for delivery to body cells and elimination of CO_2 produced by body cells.
2. Helps regulate blood pH.
3. Contains receptors for the sense of smell, filters inspired air, produces vocal sounds (phonation), and excretes small amounts of water and heat.

Which structures are part of the conducting zone of the respiratory system?

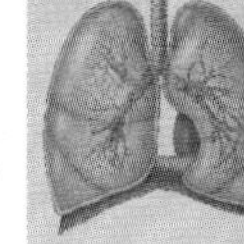

CHECKPOINT

1. What functions do the respiratory and cardiovascular systems have in common?

25.2 ORGANS OF THE RESPIRATORY SYSTEM

OBJECTIVE

- Describe the structure and functions of the nose, pharynx, larynx, trachea, bronchi, bronchioles, and lungs.

Structurally, the respiratory system consists of two parts: The ***upper respiratory system*** includes the nose, pharynx, and associated structures; the ***lower respiratory system*** consists of the larynx, trachea, bronchi, and lungs. The respiratory system can also be divided into two parts based on function. The ***conducting zone*** consists of a series of interconnecting cavities and tubes—nose, oral cavity, pharynx, larynx, trachea, bronchi, bronchioles, and terminal bronchioles—that conduct air into the lungs. The ***respiratory zone*** consists of tissues within the lungs where gas exchange occurs—the respiratory bronchioles, alveolar ducts, alveolar sacs, and alveoli.

Nose

The ***nose*** has a visible external portion and an internal portion inside the skull. The external nose consists of bone and cartilage covered with skin and lined with a mucous membrane. It has two openings called the ***external nares*** (NA-rēz; singular is *naris*) or ***nostrils*** (Figure 25.2).

The internal nose connects to the throat through two openings called the ***internal nares.*** Four paranasal sinuses (frontal, sphenoidal, maxillary, and ethmoidal) and the nasolacrimal ducts

Figure 25.2 Respiratory organs in the head and neck.

As air passes through the nose, it is warmed, filtered, and moistened.

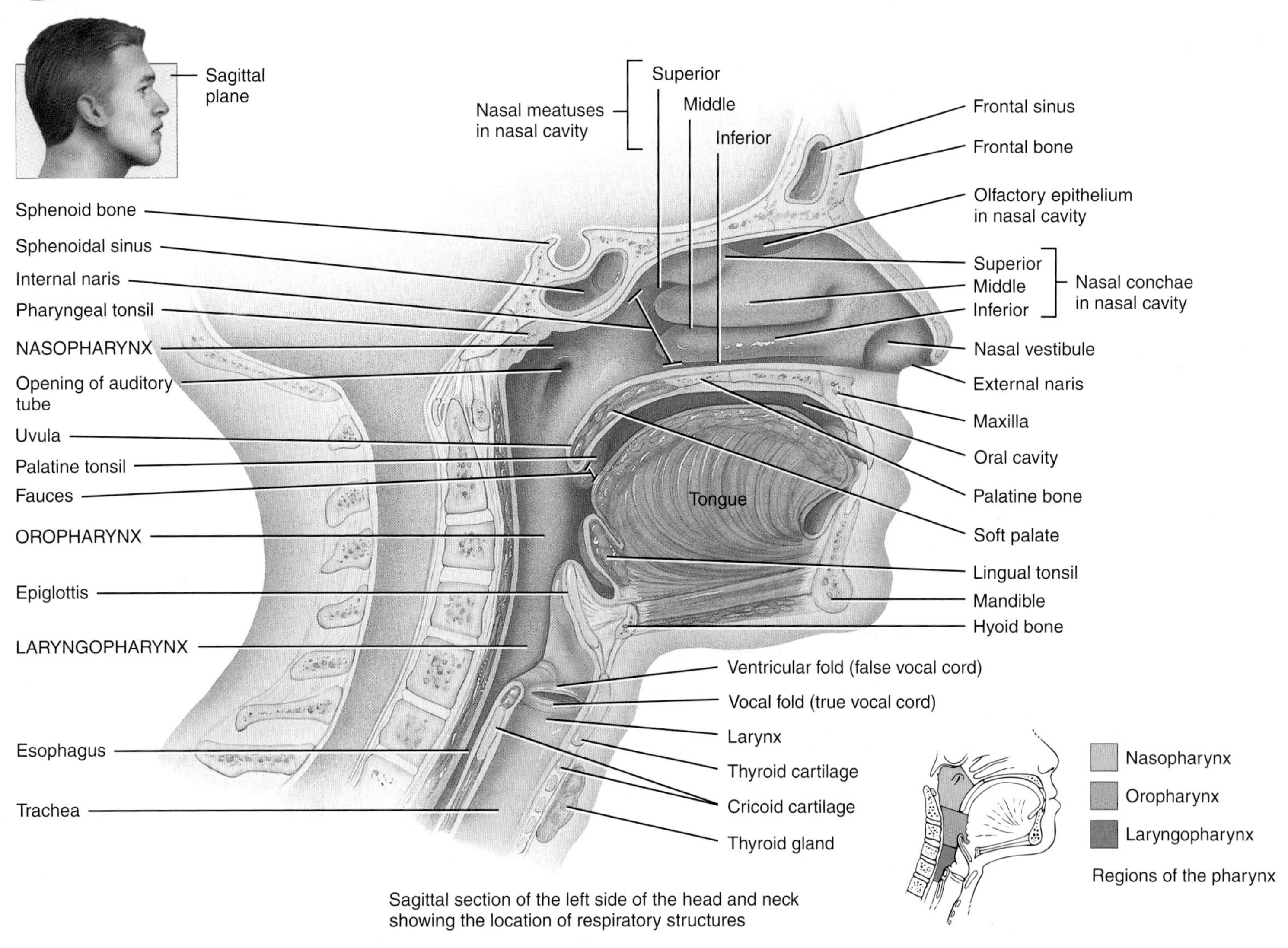

Sagittal section of the left side of the head and neck showing the location of respiratory structures

What is the path taken by air molecules into and through the nose?

also connect to the internal nose. The space inside the internal nose, called the ***nasal cavity,*** lies inferior to the cranium and superior to the mouth (oral cavity). The anterior portion of the nasal cavity, just inside the nostrils, is called the ***nasal vestibule.*** A vertical partition, the ***nasal septum,*** divides the nasal cavity into right and left sides. The septum consists of the perpendicular plate of the ethmoid bone, vomer, and cartilage (see Figure 7.11, Section 7.4).

The interior structures of the nose are specialized for three basic functions: (1) filtering, warming, and moistening of the incoming air; (2) detecting olfactory stimuli; and (3) modifying speech sounds. When air enters the nostrils, it passes coarse hairs in the nasal vestibule that trap some of the large dust particles. The air then flows over three shelves called the ***superior, middle,*** and ***inferior nasal conchae*** (KONG-kē) (see Figure 25.2). The conchae, almost reaching the nasal septum, subdivide each side of the nasal cavity into a series of groovelike passageways—the ***superior, middle,*** and ***inferior meatuses*** (mē-Ā-tus-ēz = openings or passages; singular is ***meatus***). A mucous membrane lines the nasal cavity and the three conchae. As inspired air whirls around the conchae, it is warmed by blood circulating in abundant capillaries. Anyone who experiences frequent nosebleeds is aware of the high degree of vascularity that is very superficial in the nasal cavity. The olfactory receptors also lie in the membrane lining the superior nasal conchae and adjacent septum. This region is called the ***olfactory epithelium.***

Pseudostratified ciliated columnar epithelial cells and goblet cells line the nasal cavity. Mucus secreted by goblet cells moistens the air and traps dust particles. Cilia move the dust-laden mucus toward the pharynx, at which point it can be swallowed or spit out, thus removing particles from the respiratory tract.

Pharynx

The ***pharynx*** (FAR-inks), or throat, is a funnel-shaped tube that starts at the internal nares and extends partway down the neck. It lies just posterior to the nasal and oral cavities and just anterior to the cervical (neck) vertebrae (see Figure 25.2). Its wall is composed of skeletal muscle and lined with a mucous membrane. The pharynx functions as a passageway for air and food, provides a resonating chamber for speech sounds, and houses the tonsils, which participate in immunological responses to foreign invaders.

The upper part of the pharynx, called the ***nasopharynx,*** connects with the two internal nares and has two openings that lead into the *auditory* (*eustachian*) *tubes.* The posterior wall contains the *pharyngeal tonsil.* The nasopharynx exchanges air with the nasal cavities and receives mucus–dust packages. The cilia of its pseudostratified ciliated columnar epithelium move the mucus–dust packages toward the mouth. The nasopharynx also exchanges small amounts of air with the auditory tubes to equalize air pressure between the pharynx and middle ear. The middle portion of the pharynx, the ***oropharynx,*** opens into the mouth and nasopharynx. Two pairs of tonsils, the *palatine tonsils* and *lingual tonsils*, are found in the oropharynx. The lowest portion of the pharynx, the ***laryngopharynx*** (la-rin′-gō-FAR-inks), connects with both the esophagus (food tube) and the larynx (voice box). Thus, the oropharynx and laryngopharynx both serve as passageways for air as well as for food and drink.

Larynx

The ***larynx*** (LAR-inks), or voice box, is a short tube of cartilage lined by mucous membrane that connects the pharynx with the trachea (Figure 25.3). It lies in the midline of the neck anterior to the fourth, fifth, and sixth cervical vertebrae (C4 to C6) (see Figure 25.2).

The ***thyroid cartilage,*** which consists of hyaline cartilage, forms the anterior wall of the larynx. Its common name (Adam's apple) reflects the fact that it is often larger in males than in females due to the influence of male sex hormones during puberty.

The ***epiglottis*** (*epi-* = over; *-glottis* = tongue) is a large, leaf-shaped piece of elastic cartilage that is covered with epithelium. The "stem" of the epiglottis is attached to the anterior rim of the thyroid cartilage and hyoid bone (Figure 25.3). The broad superior "leaf" portion of the epiglottis is unattached and is free to move up and down like a trap door. During swallowing, the pharynx and larynx rise. Elevation of the pharynx widens it to receive food or drink; elevation of the larynx causes the epiglottis to move down and form a lid over the larynx, closing it off. The closing of the larynx in this way during swallowing routes liquids and foods into the esophagus and keeps them out of the airways below. When anything but air passes into the larynx, a cough reflex attempts to expel the material.

• CLINICAL CONNECTION | Cough Reflex

The sensitivity of the mechanism just described can be illustrated when eating a powdered donut. If the donut is held close to the face and the person inhales (i.e., takes in air during the breathing process), the very small and light granules of powdered sugar may be drawn into the nose or mouth and proceed into the larynx. These small food particles, going "down the wrong tube," will initiate the **cough reflex.** •

The ***cricoid cartilage*** (KRĪ-koyd) is a ring of hyaline cartilage that forms the inferior wall of the larynx and is attached to the first tracheal cartilage. The paired ***arytenoid cartilages*** (ar′-i-TĒ-noyd) are located above the cricoid cartilage (Figure 25.3). They attach to the true vocal cords and pharyngeal muscles and function in voice production. The cricoid cartilage is the landmark for making an emergency airway called a *tracheotomy* in which an incision is made into the trachea, just inferior to the cricoid cartilage.

The Structures of Voice Production

The mucous membrane of the larynx forms two pairs of folds: an upper pair called the ***ventricular folds*** (***false vocal cords***) and a lower pair called the ***vocal folds*** (***true vocal cords***) (Figure 25.3c). The false vocal cords hold the breath against pressure in the thoracic cavity when you strain to lift a heavy object. They do not produce sound.

The true vocal cords produce sounds during speaking and singing. They contain elastic ligaments stretched between pieces

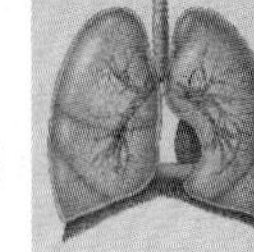

Figure 25.3 Larynx.

 The larynx is composed of cartilage.

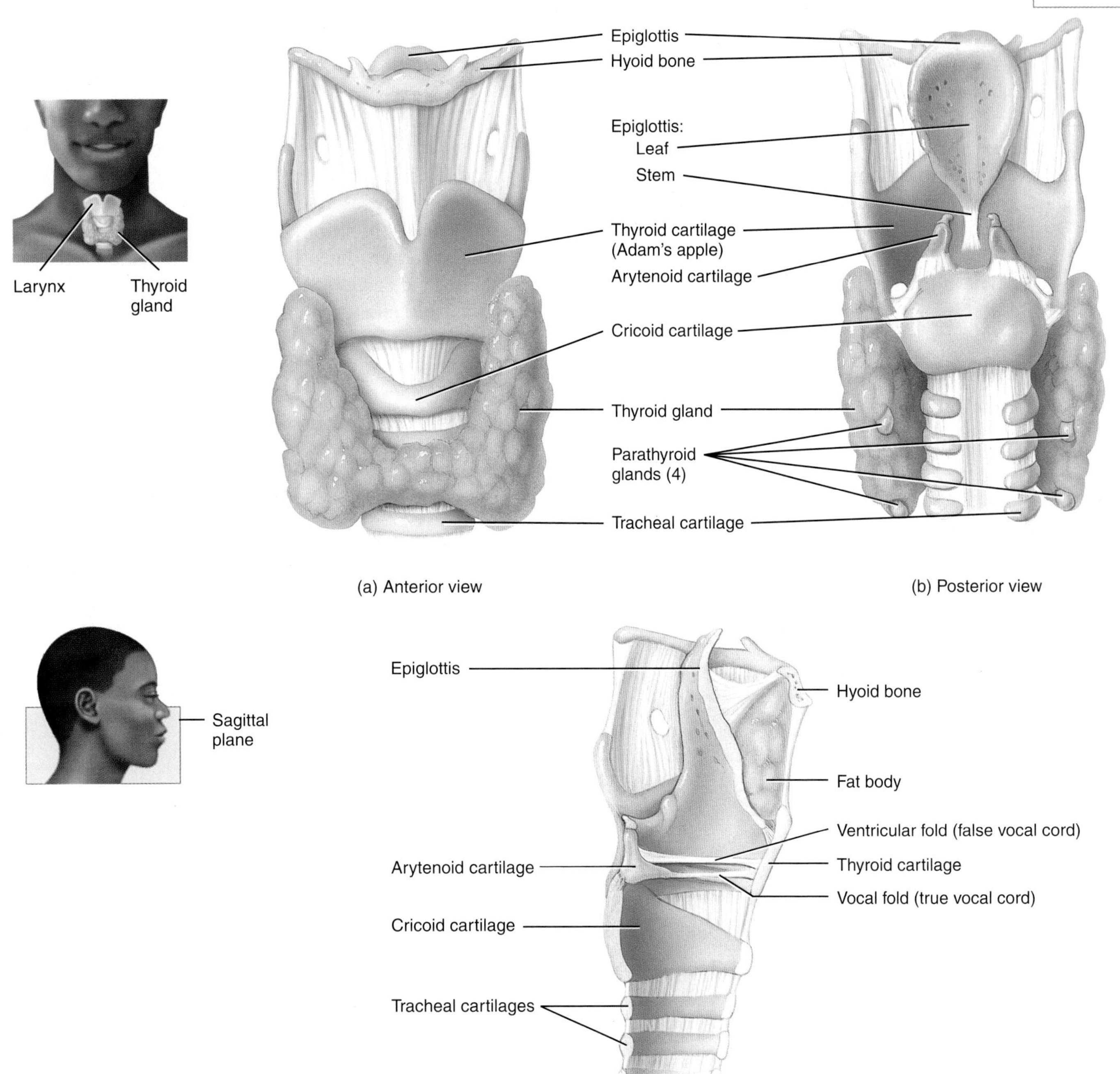

? How does the epiglottis prevent aspiration of foods and liquids?

of rigid cartilage like the strings on a guitar. Muscles attach both to the cartilage and to the true vocal cords. When the muscles contract, they pull the elastic ligaments tight, which moves the true vocal cords out into the air passageway. The air pushed against the true vocal cords causes them to vibrate and sets up sound waves in the air in the pharynx, nose, and mouth. The greater the air pressure, the louder is the sound.

Pitch is controlled by the tension of the true vocal cords. If they are pulled taut, they vibrate more rapidly and a higher pitch results. Lower sounds are produced by decreasing the muscular tension. Owing to the influence of male sex hormones, vocal cords are usually thicker and longer in males than in females. They therefore vibrate more slowly, giving men a lower range of pitch than most women.

Trachea

The ***trachea*** (TRĀ-kē-a), or ***windpipe,*** is a tubular passageway for air that is located anterior to the esophagus. It extends from the larynx to the upper part of the fifth thoracic vertebra (T5), where it divides into right and left primary bronchi (Figure 25.4).

The wall of the trachea is lined with a mucous membrane and is supported by cartilage. The mucous membrane is composed of pseudostratified ciliated columnar epithelium, consisting of ciliated columnar cells, goblet cells, and basal cells (see Table 4.1E), and provides the same protection against dust as the membrane lining the nasal cavity and larynx. The cilia in the upper respiratory tract move mucus and trapped particles *down* toward the pharynx, but the cilia in the lower respiratory tract move mucus and trapped particles *up* toward the pharynx. The cartilage layer consists of 16 to 20 C-shaped rings of hyaline cartilage stacked one on top of another. The open part of each C-shaped cartilage ring faces the esophagus and permits it to expand slightly into the trachea during swallowing. The solid parts of the C-shaped cartilage rings provide a rigid support so the tracheal wall does not collapse inward and obstruct the air passageway. The rings of cartilage may be felt deep to the skin below the larynx.

Bronchi and Bronchioles

The trachea divides into a ***right primary bronchus*** (BRON-kus = windpipe), which goes to the right lung, and a ***left primary bronchus,*** which goes to the left lung (Figure 25.4). Like the trachea, the primary bronchi (BRONG-kī) contain incomplete rings of cartilage and are lined by pseudostratified ciliated columnar epithelium. Pulmonary blood vessels, lymphatic vessels, and nerves enter and exit the lungs with the two bronchi.

Figure 25.4 Branching of airways from the trachea and lobes of the lungs.

The bronchial tree consists of airways that begin at the trachea and end at the terminal bronchioles.

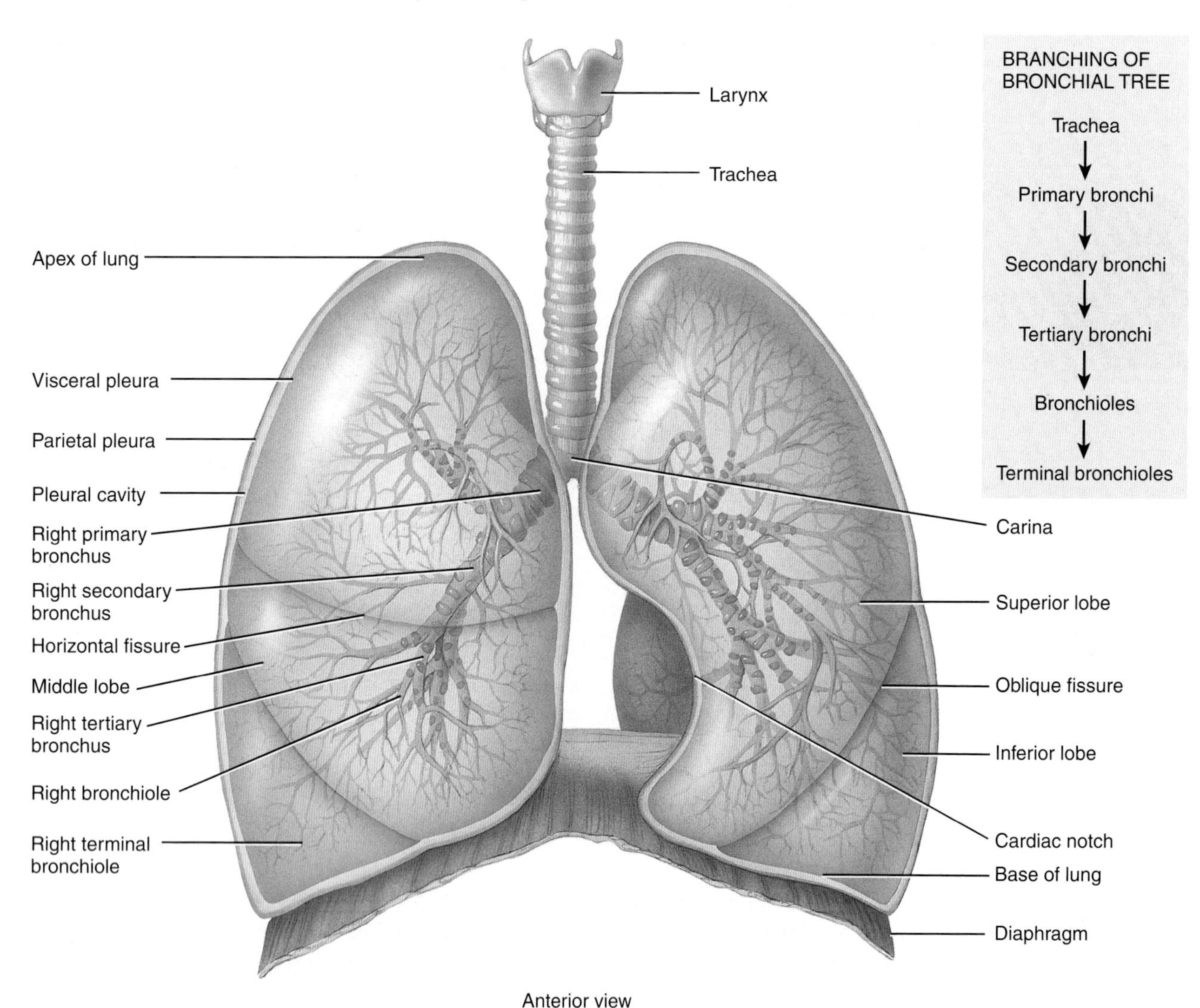

How many lobes and secondary bronchi are present in each lung?

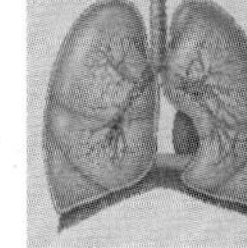

On entering the lungs, the primary bronchi divide to form the ***secondary bronchi,*** one for each lobe of the lung. (The right lung has three lobes; the left lung has two.) The secondary bronchi continue to branch, forming still smaller bronchi, called ***tertiary bronchi,*** that divide several times, ultimately giving rise to smaller ***bronchioles.*** Bronchioles, in turn, branch into even smaller tubes called ***terminal bronchioles.*** Because all of the airways resemble an upside-down tree with many branches, their arrangement is known as the ***bronchial tree.***

As the branching becomes more extensive in the bronchial tree, structural changes occur. First, plates of cartilage gradually replace the incomplete rings of cartilage in primary bronchi and finally disappear in the distal bronchioles. Second, as the amount of cartilage decreases, the amount of smooth muscle increases. Smooth muscle encircles the lumen in spiral bands. During exercise, activity in the sympathetic division of the autonomic nervous system (ANS) increases and causes the adrenal medullae to release the hormones epinephrine and norepinephrine (see Section 20.7). Both chemicals cause relaxation of smooth muscle in the bronchioles, which dilates (widens) the airways. The result is improved airflow, and air reaches the alveoli more quickly.

• CLINICAL CONNECTION | Asthma

During an **asthma** attack, bronchiolar smooth muscle goes into spasm. Because there is no supporting cartilage, the spasms can reduce the lumen or even close off the air passageways. Movement of air through constricted bronchioles causes breathing to be more labored. The parasympathetic division of the ANS and mediators of allergic reactions such as histamine also cause narrowing of bronchioles (bronchoconstriction) due to contraction of bronchiolar smooth muscle. Air moving through a restricted lumen causes a noise, and a true asthmatic can be heard breathing across the room. The principle is similar to that of a vacuum cleaner: It is so noisy because a large volume of air is moving through a small or restricted tube.

Asthmatics typically react to low concentrations of stimuli that do not normally cause symptoms in people without asthma. Sometimes the trigger is an allergen such as pollen, dust mites, molds, or a particular food. Other common triggers include emotional upset, aspirin, sulfiting agents (used in wine and beer and to keep greens fresh in salad bars), exercise, and breathing cold air or cigarette smoke. Symptoms include difficult breathing, coughing, wheezing, chest tightness, tachycardia, fatigue, moist skin, and anxiety. •

Lungs

The ***lungs*** (= lightweights, because they float) are two spongy, cone-shaped organs in the thoracic cavity. They are separated from each other by the heart and other structures in the mediastinum (see Figure 22.1, Section 22.1). The ***pleural membrane*** is a double-layered serous membrane that encloses and protects each lung (Figure 25.4). The outer layer is attached to the wall of the thoracic cavity and diaphragm and is called the ***parietal pleura.*** The inner layer, the ***visceral pleura,*** is attached to the lungs. Between the visceral and parietal pleurae is a narrow space, the ***pleural cavity,*** which contains a lubricating fluid secreted by the membranes. This fluid reduces friction between the membranes, allowing them to slide easily over one another during breathing.

The lungs extend from the diaphragm to slightly above the clavicles and lie against the ribs. The broad bottom portion of each lung is its ***base;*** the narrow top portion is the ***apex*** (Figure 25.4). The left lung has an indentation, the ***cardiac notch,*** in which the heart lies. Because of the space occupied by the heart, the left lung is about 10% smaller than the right lung.

Deep grooves called fissures divide each lung into lobes. The ***oblique fissure*** divides the left lung into ***superior*** and ***inferior lobes.*** The ***oblique*** and ***horizontal fissures*** divide the right lung into ***superior, middle,*** and ***inferior lobes*** (Figure 25.4). Each lobe receives its own secondary bronchus as well as its own vascular, lymphatic, and nerve supply.

MANUAL THERAPY APPLICATION

Postural Drainage of the Lungs

Normal volumes of mucus produced by cells of the trachea and bronchial tree are easily removed by healthy cilia and the *ciliary escalator* (movement of the cilia in unison toward the throat in which microbes and inhaled particles trapped in mucus can be swallowed or spit out). Diseases or the aftereffects of surgical anesthetics may, however, cause mucus to be retained, and the warm, damp, dark recesses of the lung become a prime site for infection.

Postural drainage of the lungs utilizes the effects of gravity in the removal of excess mucus. The patient is thus placed in sequential positions by the therapist. Each position is designed to facilitate the gravitational movement of mucus from a particular lobe of the lungs. A massage table or bed is tipped from 12 to 20 inches at one end such that the patient's head is higher or lower than the feet. The head, and thus the lungs, are lower than the feet in 7 of the 12 positions. Again, each position is designed to drain a particular lobe of each lung. Cystic fibrosis and other infections of the lungs of infants and children require postural drainage on a daily basis. In such instances, a parent is typically taught to hold the child on the lap and tip the position of the child's body by adjusting the height of the parent's thighs. Postural drainage at any age is accompanied by percussion and vibration techniques administered to appropriate parts of the patient's chest.

Blood Supply to the Lungs

The lungs receive blood via two sets of arteries: pulmonary arteries and bronchial arteries. As you learned in Chapter 22, deoxygenated blood passes through the pulmonary trunk, which divides into a left pulmonary artery that enters the left lung and a right pulmonary artery that enters the right lung. (The pulmonary arteries are the only arteries in the adult body that carry deoxygenated blood.) Return of the oxygenated blood to the heart occurs by way of the four pulmonary veins, which drain into the left atrium (see Figure 23.8, Section 23.4).

Bronchial arteries, which branch from the aorta, deliver oxygenated blood to the lungs. This blood mainly perfuses the walls of the bronchi and bronchioles. Connections exist between branches of the bronchial arteries and branches of the pulmonary arteries, however, and most blood returns to the heart via pulmonary veins. Some blood, however, drains into bronchial veins (see Figure 23.12, Exhibit 23.4) and returns to the heart via the superior vena cava (see Figure 23.3, Section 23.1).

Each lung lobe is divided into smaller segments that are supplied by a tertiary bronchus. The segments, in turn, are subdivided into many small compartments called ***lobules*** (Figure 25.5). Each lobule contains a lymphatic vessel, an arteriole, a venule, and a branch from a terminal bronchiole wrapped in elastic connective tissue. Terminal bronchioles subdivide into microscopic branches called ***respiratory bronchioles,*** which are lined by nonciliated simple cuboidal epithelium. Respiratory bronchioles, in turn, subdivide into several ***alveolar ducts.*** The two or more alveoli that share a common opening to the alveolar duct are called ***alveolar sacs*** (Figure 25.5).

Alveoli

An ***alveolus*** (al-VĒ-ō-lus; plural is *alveoli*) is a cup-shaped outpouching of an alveolar sac. Many alveoli and alveolar sacs surround each alveolar duct. The walls of alveoli consist mainly of thin ***type I alveolar cells,*** which are simple squamous epithelial cells (Figure 25.6). They are the main sites of gas exchange. Scattered among them are ***type II alveolar (septal) cells*** that secrete ***alveolar fluid,*** which keeps the surface between the cells and the air moist. Included in the alveolar fluid is ***surfactant*** (sur-FAK-tant), a mixture of phospholipids and lipoproteins that reduces the tendency of alveoli to collapse. Also present are ***alveolar macrophages (dust cells),*** wandering phagocytes that remove fine dust particles and other debris in the alveolar spaces. Underlying the layer of alveolar cells is an elastic basement membrane and a thin layer of connective tissue containing

Figure 25.5 Lobule of the lung.

Alveolar sacs are two or more alveoli that share a common opening into an alveolar duct.

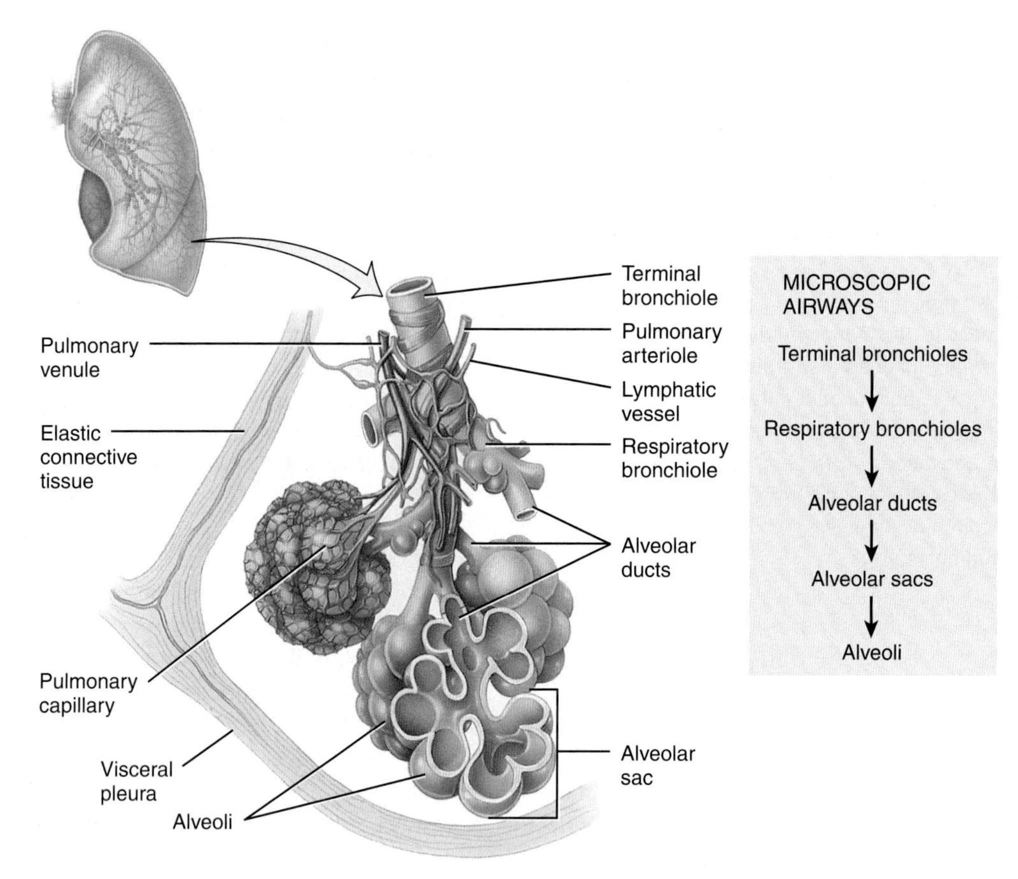

Diagram of a portion of a lobule of the lung

What structure receives air from a respiratory bronchiole?

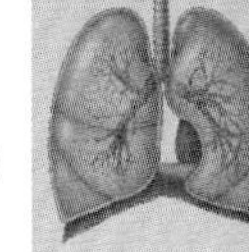

Figure 25.6 Structure of an alveolus.

6 The exchange of respiratory gases occurs by diffusion across the respiratory membrane.

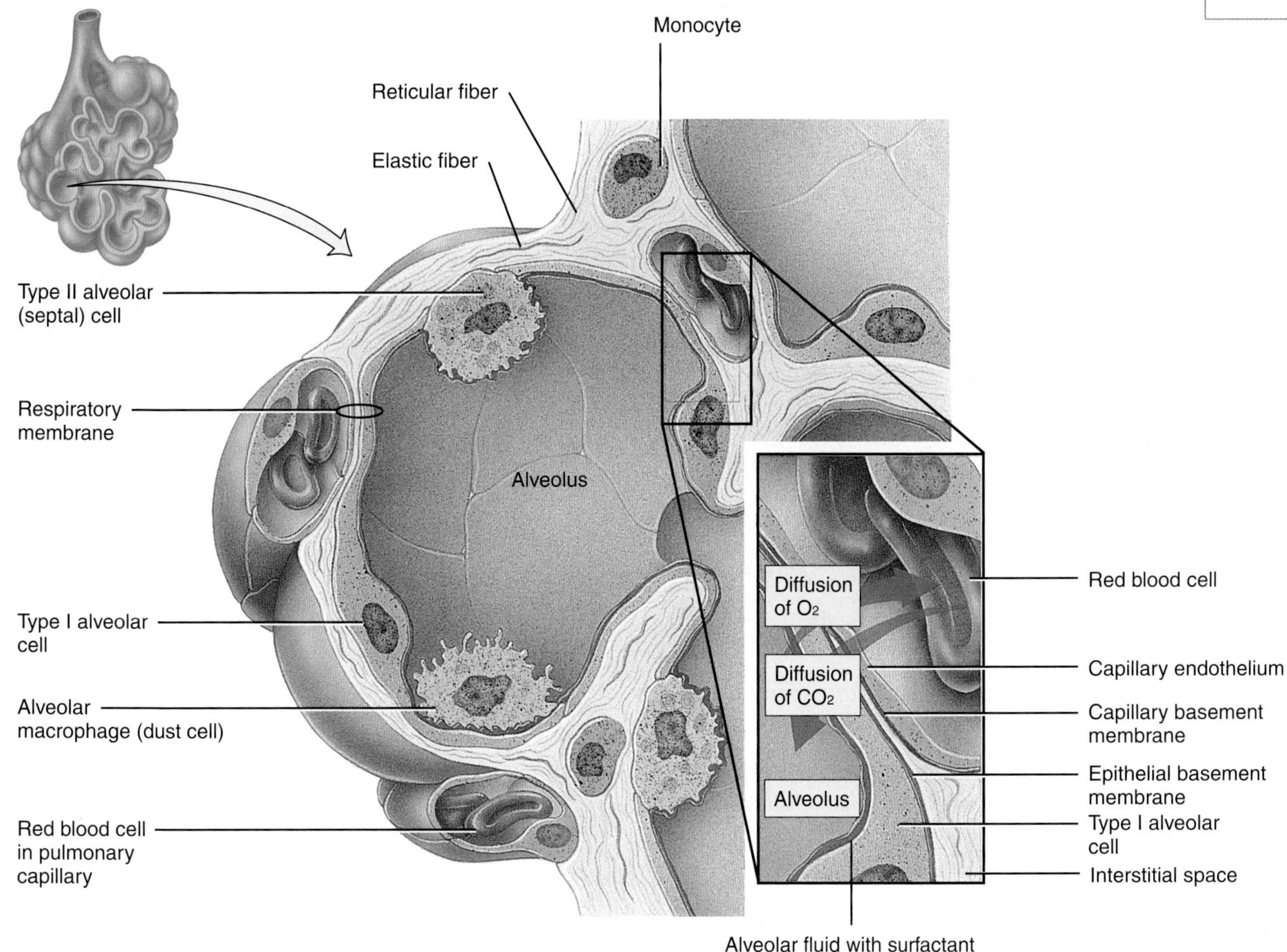

(a) Section through an alveolus showing its cellular components

(b) Details of respiratory membrane

How thick is the respiratory membrane?

plentiful elastic and reticular fibers (described shortly). Around the alveoli, the pulmonary arteriole and venule form lush networks of blood capillaries (Figure 25.5).

The exchange of O_2 and CO_2 between the air spaces in the lungs and the blood takes place by diffusion across the alveolar and capillary walls. Together these structures form the ***respiratory membrane.*** It consists of the following layers (see Figure 25.7):

1. The *alveolar cells* that form the wall of an alveolus.
2. An *epithelial basement membrane* underlying the alveolar cells.
3. A *capillary basement membrane* that is often fused to the epithelial basement membrane.
4. The *endothelial cells* of a capillary wall.

Despite having several layers, the respiratory membrane is only 0.5 μm* wide. This thin width, far less than the thickness of a sheet of tissue paper, permits O_2 and CO_2 to diffuse efficiently between the blood and alveolar air spaces. This exceedingly thin respiratory membrane is the only thing that separates the air that you breathe from your bloodstream. Moreover, the two lungs contain roughly 300 million alveoli. They provide a huge surface area for the exchange of O_2 and CO_2—about 30 to 40 times greater than the surface area of your skin or half the size of a tennis court!

*1 μm (micrometer) = 1/1,000,000 of a meter or 1/25,000 of an inch.

Figure 25.7 Respiratory membrane.

The respiratory membrane consists of two flat cells and a shared basement membrane.

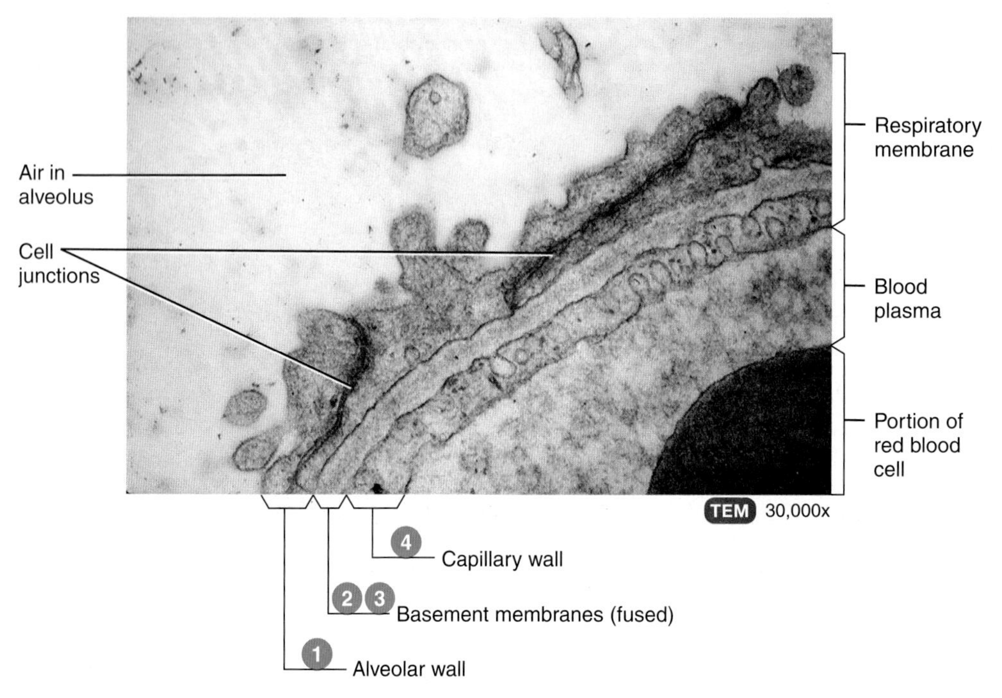

What three structures make up the respiratory membrane?

• CLINICAL CONNECTION | Cigarette Smoking

Cigarette smoking is the single most preventable cause of death and disability worldwide. All forms of tobacco use disrupt the body's ability to maintain homeostasis and health. Even in the short term, several factors decrease respiratory efficiency in smokers. (1) Nicotine in smoke constricts terminal bronchioles, which decreases airflow into and out of the lungs. (2) Carbon monoxide in smoke binds to hemoglobin and reduces its oxygen-carrying capability. (3) Irritants in smoke cause increased mucus secretion by the mucosa of the bronchial tree and swelling of the mucosal lining, both of which impede airflow into and out of the lungs. (4) Irritants in smoke also inhibit the cilia and destroy cilia in the lining of the airways. Thus, excess mucus and foreign debris are not easily removed, which further adds to the difficulty in breathing. Substances in cigarette smoke also inhibit movement of cilia; when the cilia are paralyzed, only coughing can remove mucus–dust packages from the airways. This is why smokers cough so much and are more prone to respiratory infections. (5) The endothelial lining of blood vessels is compromised, causing plaque formation and possible blood clots that could lead to pulmonary embolism, stroke, or heart attack.

Tobacco products contain numerous aromatic hydrocarbons that become tar when the tobacco is lit. The tar is thus inhaled, and much of it reaches alveoli. Alveolar macrophages eventually ingest the tar and encapsulate it. Enzymes of the macrophages, however, are not capable of destroying the tar, and these cells become engorged with tar-filled vesicles. This tar is similar to the tar that is used to manufacture asphalt for roads and similar to the tar that is used to seal a leaky roof. Visualize a miniature paintbrush used to coat the walls of more and more alveoli with thick, black tar. Each respiratory membrane thus coated has lost its ability to function effectively in the transport of gases. The situation is progressive, and each cigarette or exposure to secondhand smoke contributes to the deteriorating condition of the lungs.

Lung cancer was a rare disease in the early 1900s. Now it is the leading cancer killer for both men and women, due primarily to smoking. Cigarette smoke contains a number of known carcinogens, which may initiate and promote the cellular changes leading to lung cancer as well as cancers of the oral cavity, larynx, esophagus, stomach, kidney, pancreas, colon, urinary bladder, and other organs. •

CHECKPOINT

2. Compare the structure and functions of the external and internal nose.
3. How does the larynx function in respiration and voice production?
4. What is the bronchial tree? Describe its structure.
5. Where are the lungs located? Distinguish the parietal pleura from the visceral pleura.
6. Where in the lungs does the exchange of O_2 and CO_2 take place?

25.3 PULMONARY VENTILATION

OBJECTIVES

- Explain how inhalation and exhalation take place.
- Define the various lung volumes and capacities.

Pulmonary ventilation, the flow of air between the atmosphere and the lungs, occurs due to differences in air pressure. We in-

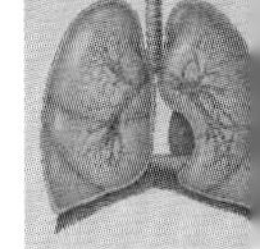

hale (breathe in) when the pressure inside the lungs is less than the atmospheric air pressure. We exhale (breathe out) when the pressure inside the lungs is greater than the atmospheric air pressure. Contraction and relaxation of skeletal muscles create the air pressure changes that power breathing.

Muscles of Inhalation and Exhalation

Breathing in is called ***inhalation*** or ***inspiration.*** The muscles of quiet (unforced) inhalation are the diaphragm, the dome-shaped skeletal muscle that forms the floor of the thoracic cavity, and the external intercostals, which extend between the ribs (Figure 25.8).

Figure 25.8 **Muscles of inhalation and exhalation and their actions.** A muscle of deep inhalation, the pectoralis minor muscle (not shown here), is illustrated in Figure 13.1, Exhibit 13.1.

During quiet inhalation, the diaphragm and/or external intercostals contract, the lungs expand, and air moves into the lungs. During exhalation, the diaphragm relaxes and the lungs recoil inward, forcing air out of the lungs.

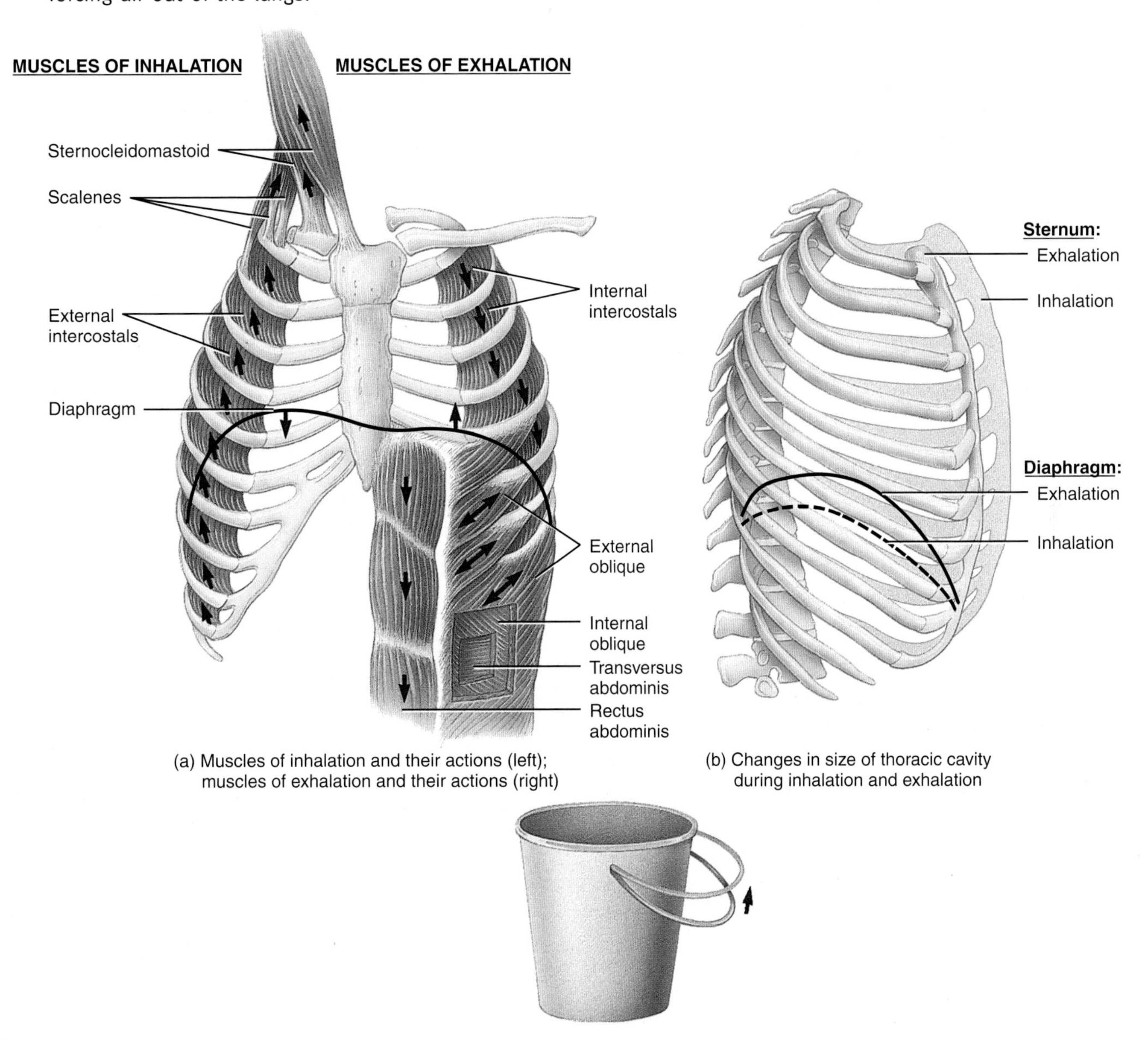

(a) Muscles of inhalation and their actions (left); muscles of exhalation and their actions (right)

(b) Changes in size of thoracic cavity during inhalation and exhalation

(c) During inhalation, the ribs move upward and outward like the handle on a bucket

? What is the main muscle that powers quiet breathing?

The diaphragm contracts when it receives nerve impulses from the phrenic nerves. As the dome-shaped diaphragm contracts, it descends and becomes flatter, which causes the volume of the attached lungs to expand. As the external intercostals contract, they pull the ribs upward and outward; the attached lungs follow, also increasing lung volume. Although persons vary in the manner in which they breathe, contraction of the diaphragm is typically responsible for about 75% of the air that enters the lungs during quiet breathing. Advanced pregnancy, obesity, stress, confining clothing, or increased size of the stomach after eating a large meal can impede descent of the diaphragm and may cause shortness of breath.

During deep, labored inhalations, the sternocleidomastoid muscles elevate the sternum, the scalene muscles elevate the two uppermost ribs, and the pectoralis minor muscles elevate the third through fifth ribs. As the ribs and sternum are elevated, the size of the lungs increases (Figure 25.8b). Movements of the pleural membrane aid expansion of the lungs. The parietal and visceral pleurae normally adhere tightly because of the surface tension created by their moist adjoining surfaces. Whenever the thoracic cavity expands, the parietal pleura lining the cavity follows, and the visceral pleura and lungs are pulled along with it.

Breathing out, called ***exhalation*** or ***expiration,*** begins when the diaphragm and external intercostals relax. Exhalation occurs due to *elastic recoil* of the chest wall and lungs, both of which have a natural tendency to spring back after they have been stretched. Although the alveoli and airways recoil, they don't completely collapse. Because surfactant in alveolar fluid *reduces* elastic recoil, a lack of surfactant causes breathing difficulty by increasing the chance of alveolar collapse.

Because no muscular contractions are involved, quiet exhalation, unlike quiet inhalation, is a *passive process.* Exhalation becomes *active* only during forceful breathing, such as in playing a wind instrument or during exercise. During these times, muscles of exhalation—the internal intercostals, external oblique, internal oblique, transversus abdominis, and rectus abdominis—contract to move the lower ribs downward and compress the abdominal viscera, thus forcing the diaphragm upward (Figure 25.8).

Pressure Changes during Ventilation

As the lungs expand, the air molecules inside occupy a larger *volume,* which causes the air *pressure* inside to decrease. The pressure of a gas in a closed container is inversely proportional to the volume of the container. This means that if the size of a closed container is increased, the pressure of the gas inside the container decreases, and that if the size of the container is decreased, then the pressure inside it increases. This inverse relationship between volume and pressure, called ***Boyle's law,*** may be demonstrated as follows (Figure 25.9): Suppose we place a gas in a cylinder that has a movable piston and a pressure gauge, and that the initial pressure created by the gas molecules striking the wall of the container is 1 atm. If the piston is pushed down, the gas is compressed into a smaller volume, so that the same number of gas molecules strike less wall area. The gauge shows that the pressure doubles as the gas is compressed to half its original volume. In other words, the same number of molecules in

Figure 25.9 Boyle's law.

6 The volume of a gas varies inversely with its pressure.

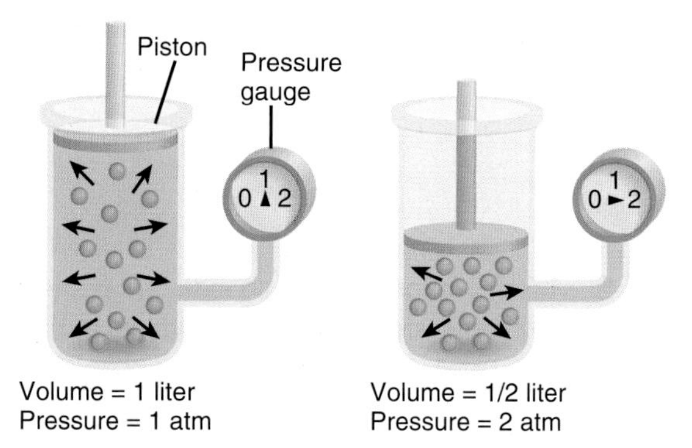

? If the volume is decreased from 1 liter to 1/4 liter, how would the pressure change?

half the volume produces twice the pressure. Conversely, if the piston is raised to increase the volume, the pressure decreases. Thus, the pressure of a gas varies inversely with volume.

Differences in pressure caused by changes in lung volume force air into our lungs when we inhale and out when we exhale. For inhalation to occur, the lungs must expand, which increases lung volume and thus decreases the pressure in the lungs to below atmospheric pressure. The first step in expanding the lungs during normal quiet inhalation involves contraction of the main muscles of inhalation, the diaphragm and/or external intercostals (see Figure 25.8).

Because the atmospheric air pressure is now higher than the ***alveolar pressure,*** the air pressure inside the lungs, air moves into the lungs. By contrast, when lung volume decreases, the alveolar pressure increases. Air then flows from the area of higher pressure in the alveoli to the area of lower pressure in the atmosphere. Figure 25.10 shows the sequence of pressure changes during quiet breathing.

1. At rest just before an inhalation, the air pressure inside the lungs is the same as the pressure of the atmosphere, which is about 760 mmHg (millimeters of mercury) at sea level.
2. As the diaphragm and external intercostals contract and the overall size of the thoracic cavity increases, the volume of the lungs increases and alveolar pressure decreases from 760 to 758 mmHg. Now there is a pressure difference between the atmosphere and the alveoli, and air flows from the atmosphere (higher pressure) into the lungs (lower pressure).
3. When the diaphragm and external intercostals relax, elastic recoil causes the lung volume to decrease, and alveolar pressure rises from 760 to 762 mmHg. Air then flows from the area of higher pressure in the alveoli to the area of lower pressure in the atmosphere.

Lung Volumes and Capacities

While at rest, a healthy adult breathes about 12 times a minute, with each inhalation and exhalation moving about 500 mL of air into and out of the lungs. The volume of one breath is called the

Figure 25.10 Pressure changes during pulmonary ventilation.

Air moves into the lungs when alveolar pressure is less than atmospheric pressure, and out of the lungs when alveolar pressure is greater than atmospheric pressure.

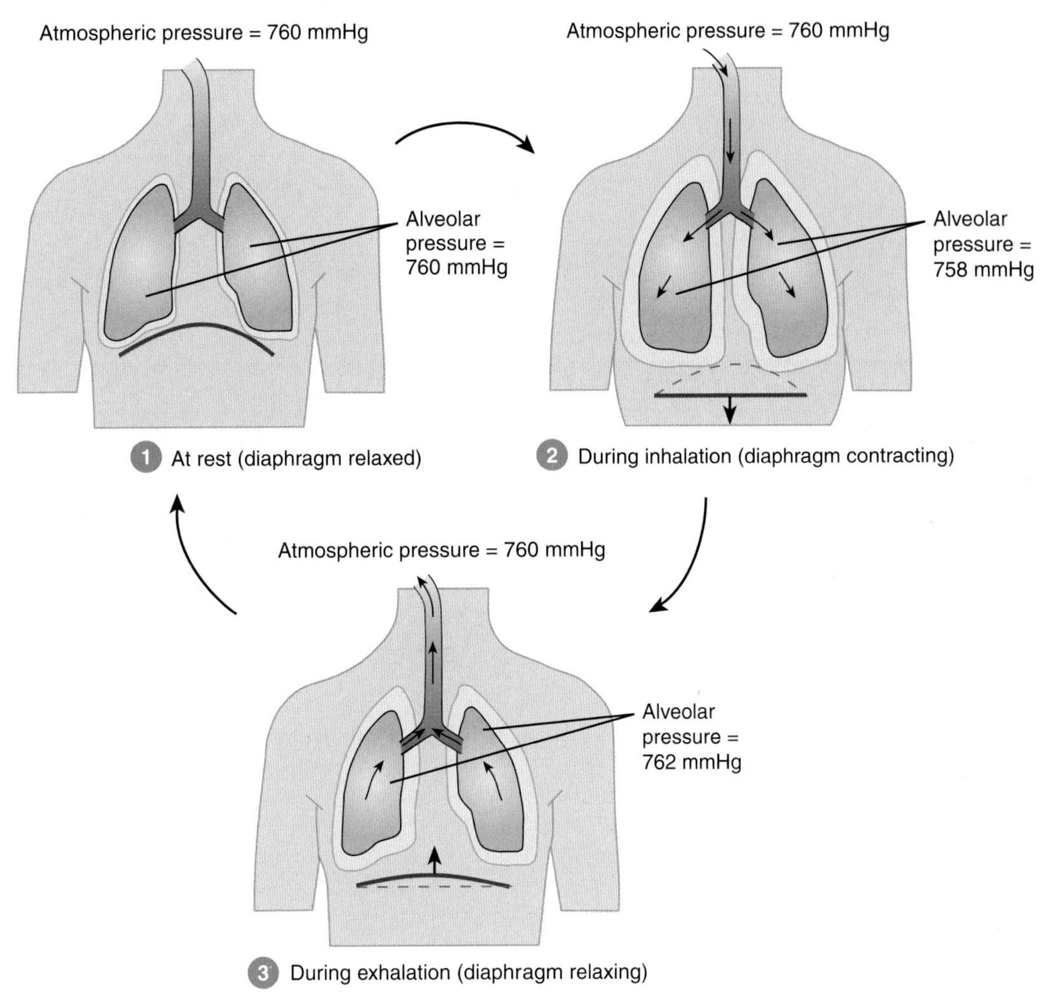

What is normal atmospheric pressure at sea level?

tidal volume. The ***minute ventilation (MV)***—the total volume of air inhaled and exhaled each minute—is equal to breathing rate multiplied by tidal volume:

$$\begin{aligned} \text{MV} &= 12 \text{ breaths/min} \times 500 \text{ mL/breath} \\ &= 6000 \text{ mL/min (6 liters/min)} \end{aligned}$$

Tidal volume varies considerably from one person to another and in the same person at different times. About 70% of the tidal volume (350 mL) actually reaches the respiratory bronchioles and alveolar sacs and thus participates in gas exchange. The other 30% (150 mL) does not participate in gas exchange because it remains in the conducting airways of the nose, pharynx, larynx, trachea, bronchi, bronchioles, and terminal bronchioles. Collectively, these conducting airways are known as the ***anatomic dead space.***

The apparatus commonly used to measure respiratory rate and the amount of air inhaled and exhaled during breathing is a ***spirometer*** (*spiro-* = breathe; *-meter* = measuring device). The record produced by a spirometer is called a ***spirogram.*** Inhalation is recorded as an upward deflection, and exhalation is recorded as a downward deflection (see Figure 25.11).

By taking a very deep breath, you can inhale a good deal more than 500 mL. This additional inhaled air is called the ***inspiratory reserve volume*** (see Figure 25.11). Even more air can be inhaled if inhalation follows forced exhalation. If you inhale normally and then exhale as forcibly as possible, you should be able to push out considerably more air in addition to the tidal volume. The additional volume is called the ***expiratory reserve volume.*** Even after the expiratory reserve volume is expelled, considerable air remains in the lungs and airways. This volume is called the ***residual volume.*** Lung *capacities* are combinations of specific lung *volumes* (see Figure 25.11). ***Inspiratory capacity*** is the sum of tidal volume and inspiratory reserve volume. ***Functional residual capacity*** is the sum of residual volume and

Figure 25.11 Spirogram showing lung volumes and capacities in milliliters (mL). The average values for a healthy adult male and female are indicated, with the values for a female in parentheses. Note that the spirogram is read from right (start of record) to left (end of record).

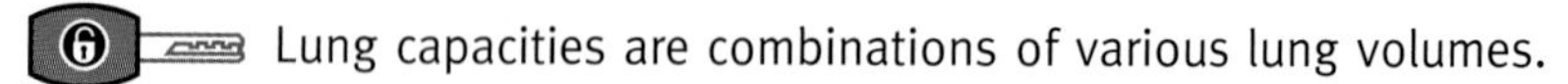

Lung capacities are combinations of various lung volumes.

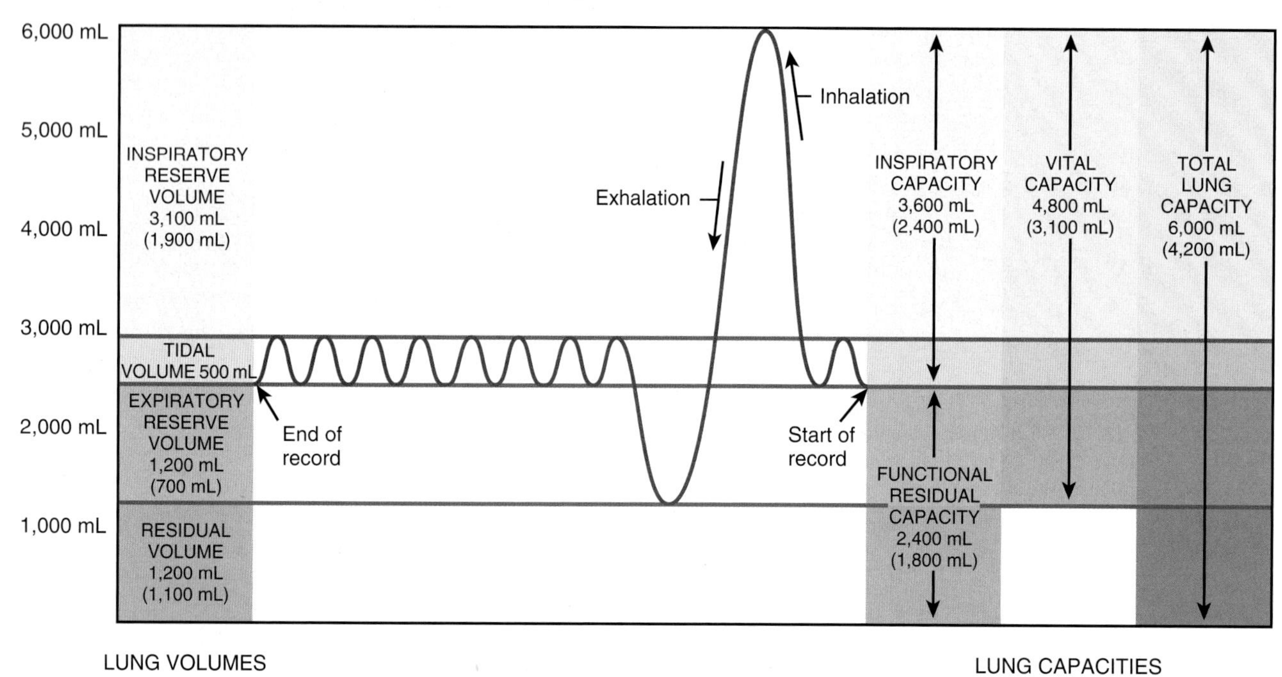

? If you breathe in as deeply as possible and then exhale as much air as you can, which lung capacity have you demonstrated?

expiratory reserve volume. ***Vital capacity*** is the sum of inspiratory reserve volume, tidal volume, and expiratory reserve volume. Finally, ***total lung capacity*** is the sum of vital capacity and residual volume. The values given in Figure 25.11 are typical for young adults. Lung volumes and capacities vary with age (smaller in older people), gender (generally smaller in females), and body size (smaller in shorter people). Lung volumes and capacities provide information about an individual's respiratory status since they are usually abnormal in people with pulmonary disorders.

Breathing Patterns and Modified Respiratory Movements

The term for the normal pattern of quiet breathing is ***eupnea*** (ūp-NĒ-a; *eu-* = good, easy, or normal; *-pnea* = breath). Eupnea can consist of shallow, deep, or combined shallow and deep breathing. A pattern of shallow (chest) breathing, called ***costal breathing,*** consists of an upward and outward movement of the chest due to contraction of the external intercostal muscles. A pattern of deep (abdominal) breathing, called ***diaphragmatic breathing,*** consists of the outward movement of the abdomen due to the contraction and descent of the diaphragm.

Respirations also provide humans with methods for expressing emotions such as laughing, sighing, and sobbing. Moreover, respiratory air can be used to expel foreign matter from the lower air passages through actions such as sneezing and coughing. Respiratory movements are also modified and controlled during talking and singing. Some of the modified respiratory movements that express emotion or clear the airways are listed in Table 25.1. All these movements are reflexes, but some of them also can be initiated voluntarily.

CHECKPOINT

7. Compare what happens during quiet versus labored ventilation.
8. What is the basic difference between a lung volume and a lung capacity?

25.4 EXCHANGE OF OXYGEN AND CARBON DIOXIDE

OBJECTIVE

- Describe the exchange of oxygen and carbon dioxide between alveolar air and blood (external respiration) and between blood and body cells (internal respiration).

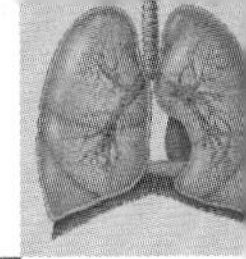

TABLE 25.1

Modified Respiratory Movements

MOVEMENT	DESCRIPTION
Coughing	A long-drawn and deep inhalation followed by a strong exhalation that suddenly sends a blast of air through the upper respiratory passages. Stimulus for this reflex act may be a foreign body lodged in the larynx, trachea, or epiglottis.
Sneezing	Spasmodic contraction of muscles of exhalation that forcefully expels air through the nose and mouth. Stimulus may be an irritation of the nasal mucosa.
Sighing	A long-drawn and deep inhalation immediately followed by a shorter but forceful exhalation.
Yawning	A deep inhalation through the widely opened mouth producing an exaggerated depression of the mandible. It may be stimulated by drowsiness, fatigue, or someone else's yawning, but precise cause is unknown.
Sobbing	A series of convulsive inhalations followed by a single prolonged exhalation.
Crying	An inhalation followed by many short convulsive exhalations, during which the vocal folds vibrate; accompanied by characteristic facial expressions and tears.
Laughing	The same basic movements as crying, but the rhythm of the movements and the facial expressions usually differ from those of crying.
Hiccupping	Spasmodic contraction of the diaphragm followed by a spasmodic closure of the larynx, which produces a sharp sound on inhalation. Stimulus is usually irritation of the sensory nerve endings of the gastrointestinal tract.

Air is a mixture of gases—nitrogen, oxygen, water vapor, carbon dioxide, and others—each of which contributes to the total air pressure. The pressure of a specific gas in a mixture is called its ***partial pressure*** and is denoted as P_X, where the subscript X denotes the chemical formula of the gas. The total pressure of air, the atmospheric pressure, is the sum of all the partial pressures:

$$\begin{aligned} &P_{N_2}\,(597.4 \text{ mmHg}) + P_{O_2}\,(158.8 \text{ mmHg}) \\ &+ P_{H_2O}\,(3.0 \text{ mmHg}) + P_{CO_2}\,(0.3 \text{ mmHg}) \\ &+ P_{\text{other gases}}\,(0.5 \text{ mmHg}) \\ &= \text{atmospheric pressure } (760 \text{ mmHg}) \end{aligned}$$

Partial pressures are important because each gas diffuses from areas where its partial pressure is higher to areas where its partial pressure is lower in the body.

External Respiration: Pulmonary Gas Exchange

External (pulmonary) respiration is the diffusion of O_2 from air in the alveoli of the lungs to blood in pulmonary capillaries and the diffusion of CO_2 in the opposite direction (see Figure 25.12a). External respiration in the lungs converts ***deoxygenated*** (low-oxygen) ***blood*** that comes from the right side of the heart to ***oxygenated*** (high-oxygen) ***blood*** that returns to the left side of the heart. As blood flows through the pulmonary capillaries, it picks up O_2 from alveolar air and unloads CO_2 into alveolar air. Although this process is commonly called an "exchange" of gases, each gas diffuses *independently* from an area where its partial pressure is higher to an area where its partial pressure is lower. An important factor that affects the rate of external respiration is the total surface area available for gas exchange. Any pulmonary disorder that decreases the functional surface area of the respiratory membrane, for example, emphysema (see Section 25.6), decreases the rate of gas exchange.

• CLINICAL CONNECTION | High Altitude Sickness

As a person ascends in altitude, the total atmospheric pressure decreases, with a parallel decrease in the partial pressure of oxygen. P_{O_2} decreases from 159 mmHg at sea level to 73 mmHg at 6000 meters (about 20,000 ft). Alveolar P_{O_2} decreases correspondingly, and less oxygen diffuses into the blood. The common symptoms of **high altitude sickness**—shortness of breath, nausea, and dizziness—are due to a lower level of oxygen in the blood. The 1968 Olympics were held in Mexico City at an altitude of nearly 2 miles above sea level. It is interesting to note that performances from athletes were very poor at that event. •

Internal Respiration: Systemic Gas Exchange

The left ventricle pumps oxygenated blood into the aorta and through the systemic arteries to systemic capillaries. The exchange of O_2 and CO_2 between systemic capillaries and tissue cells is called ***internal (tissue) respiration*** (see Figure 25.12b). As O_2 leaves the bloodstream, oxygenated blood is converted into deoxygenated blood. Unlike external respiration, which occurs only in the lungs, internal respiration occurs in tissues throughout the body.

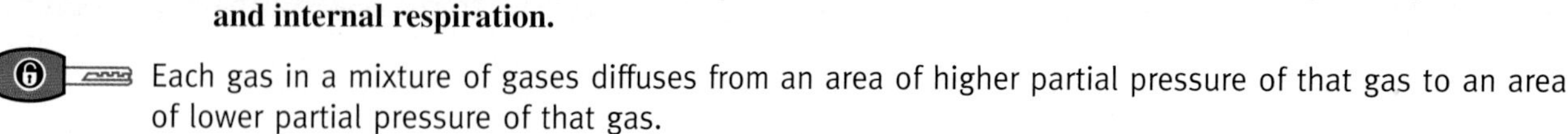

Figure 25.12 Changes in partial pressures of oxygen (O_2) and carbon dioxide (CO_2) in mmHg during external and internal respiration.

Each gas in a mixture of gases diffuses from an area of higher partial pressure of that gas to an area of lower partial pressure of that gas.

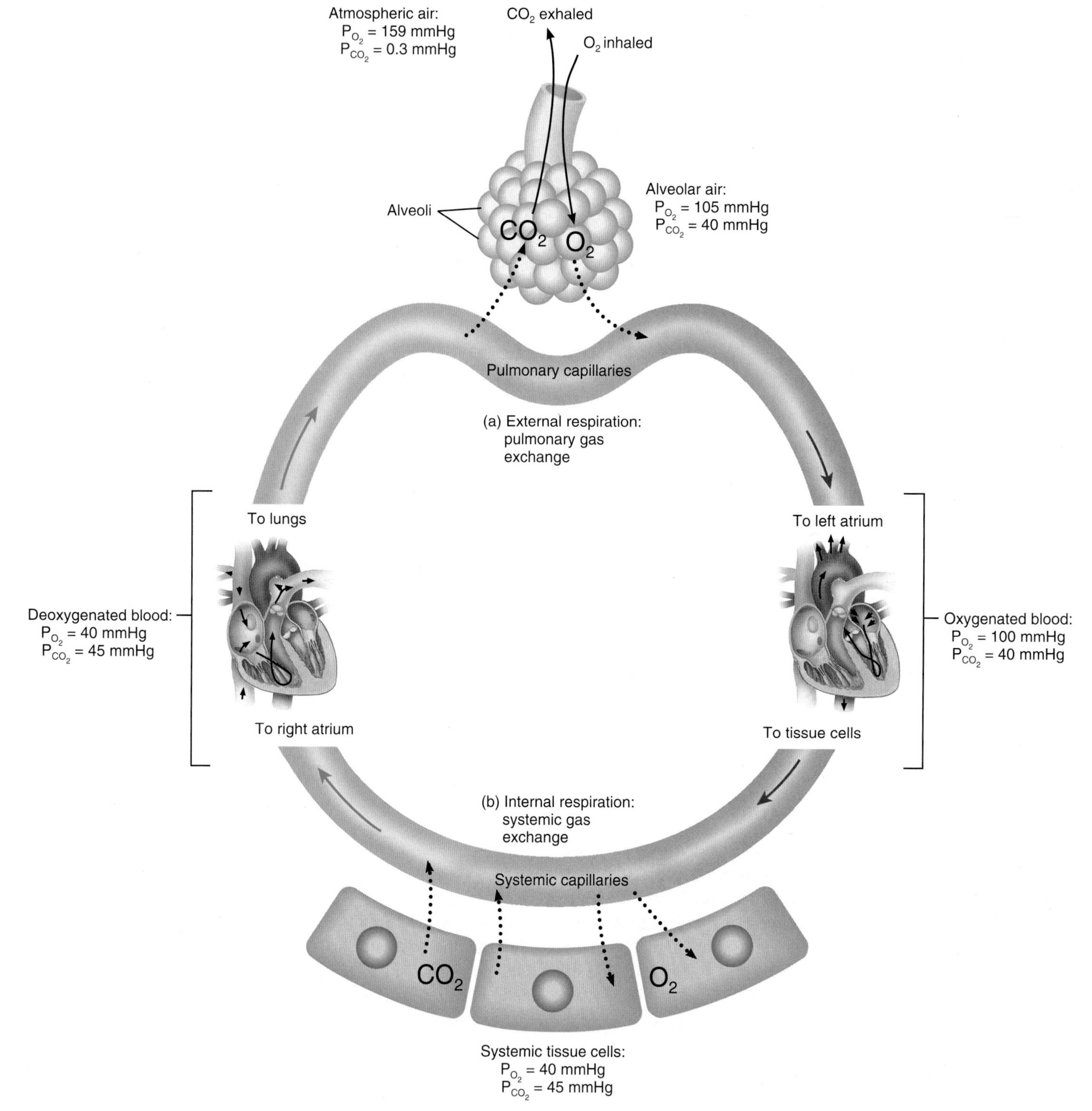

What causes oxygen to enter pulmonary capillaries from alveoli and to enter tissue cells from systemic capillaries?

CHECKPOINT

9. What are the basic differences among pulmonary ventilation, external respiration, and internal respiration?
10. In a person at rest, what is the partial pressure difference that drives diffusion of oxygen into the blood in pulmonary capillaries?

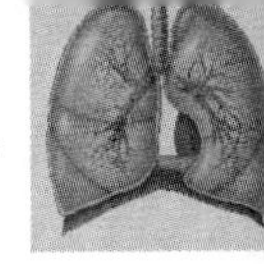

25.5 TRANSPORT OF RESPIRATORY GASES

OBJECTIVE

- Describe how the blood transports oxygen and carbon dioxide.

The blood transports gases between the lungs and body tissues. When O_2 and CO_2 enter the blood, certain physical and chemical changes occur that aid in gas transport and exchange.

Oxygen Transport

Oxygen does not dissolve easily in water, and for this reason only about 1.5% of the O_2 in blood is dissolved in blood plasma, which consists of mostly water. About 98.5% of blood O_2 is bound to hemoglobin in red blood cells (see Figure 25.13).

The heme part of hemoglobin contains four ions of iron, each capable of binding to a molecule of O_2. Oxygen and ***deoxyhemoglobin (Hb)*** bind in an easily reversible reaction to form ***oxyhemoglobin (Hb–O_2)***. When blood P_{O_2} is high, hemoglobin binds with large amounts of O_2 and is *fully saturated;* that is, every available iron atom has combined with a molecule of O_2. When blood P_{O_2} is low, hemoglobin releases O_2. Therefore, in systemic capillaries, where the P_{O_2} is lower, hemoglobin releases O_2, which then can diffuse from blood plasma into interstitial fluid and into tissue cells (see Figure 25.13b).

Besides P_{O_2}, several other factors influence the amount of O_2 released by hemoglobin:

1. **Carbon dioxide.** As the P_{CO_2} rises in any tissue, hemoglobin releases O_2 more readily. Thus, hemoglobin releases more O_2 as blood flows through active tissues that are producing more CO_2, such as muscular tissue during exercise.
2. **Acidity.** In an acidic environment, hemoglobin releases O_2 more readily. During exercise, muscles produce lactic acid, which promotes release of O_2 from hemoglobin.
3. **Temperature.** Within limits, as temperature increases, so does the amount of O_2 released from hemoglobin. Active tissues produce more heat, which elevates the local temperature and promotes release of O_2.

• CLINICAL CONNECTION | Carbon Monoxide Poisoning

Carbon monoxide (CO) is a colorless and odorless gas found in tobacco smoke and in exhaust fumes from automobiles, gas furnaces, and space heaters. CO binds to the heme group of hemoglobin, just as O_2 does, except that CO binds over 200 times more strongly. At a concentration as low as 0.1%, CO combines with half the available hemoglobin molecules and reduces the oxygen-carrying capacity of the blood by 50%. Elevated blood levels of CO cause **carbon monoxide poisoning,** which can cause the lips and oral mucosa to appear bright, cherry red (the color of hemoglobin with carbon monoxide bound to it). Administering pure oxygen, which speeds up the separation of carbon monoxide from hemoglobin, may rescue the person. •

Carbon Dioxide Transport

Carbon dioxide is transported in the blood in three main forms (see Figure 25.13):

1. **Dissolved CO_2.** The smallest percentage—about 7%—is dissolved in blood plasma. On reaching the lungs, it diffuses into alveolar air and is exhaled.
2. **Bound to amino acids.** A somewhat higher percentage, about 23%, combines with the amino groups of amino acids and proteins in blood. Because the most prevalent protein in blood is hemoglobin (inside red blood cells), most of the CO_2 transported in this manner is bound to hemoglobin. Hemoglobin that has bound CO_2 is termed ***carbaminohemoglobin (Hb–CO_2)***. In tissue capillaries P_{CO_2} is relatively high, which promotes formation of carbaminohemoglobin. But in pulmonary capillaries, P_{CO_2} is relatively low, and the CO_2 readily splits apart from hemoglobin and enters the alveoli by diffusion.
3. **Bicarbonate ions.** The greatest percentage of CO_2—about 70%—is transported in blood plasma as ***bicarbonate ions (HCO_3^-)***. As CO_2 diffuses into tissue capillaries and enters the red blood cells, it combines with water to form carbonic acid (H_2CO_3). The carbonic acid then breaks down into hydrogen ions (H^+) and HCO_3^-. Thus, as blood picks up CO_2, HCO_3^- accumulates inside RBCs. Some HCO_3^- moves out into the blood plasma, down its concentration gradient. In exchange, chloride ions (Cl^-) move from plasma into the RBCs. As a result of these chemical reactions, CO_2 is removed from tissue cells and transported in blood plasma as HCO_3^-.

As blood passes through pulmonary capillaries in the lungs, all these reactions reverse. The CO_2 that was dissolved in plasma diffuses into alveolar air. The CO_2 that was combined with hemoglobin splits and diffuses into the alveoli. The bicarbonate ions (HCO_3^-) reenter the red blood cells from the blood plasma and recombine with H^+ to form H_2CO_3, which splits into CO_2 and H_2O. This CO_2 leaves the red blood cells, diffuses into alveolar air, and is exhaled (see Figure 25.13).

CHECKPOINT

11. What is the relationship between hemoglobin and P_{O_2}?
12. What factors cause hemoglobin to unload more oxygen as blood flows through capillaries of metabolically active tissues, such as skeletal muscle during exercise?

25.6 CONTROL OF RESPIRATION

OBJECTIVE

- Explain how the nervous system controls breathing and list the factors that can alter the rate and depth of breathing.

At rest, body cells use about 200 mL of O_2 each minute. During strenuous exercise, however, O_2 use typically increases 15- to 20-fold in normal healthy adults, and as much as 30-fold in elite endurance-trained athletes. Several mechanisms help match respiratory effort to metabolic demand.

Figure 25.13 Transport of oxygen and carbon dioxide in the blood.

Most O_2 is transported by hemoglobin as oxyhemoglobin within red blood cells; most CO_2 is transported in blood plasma as bicarbonate ions.

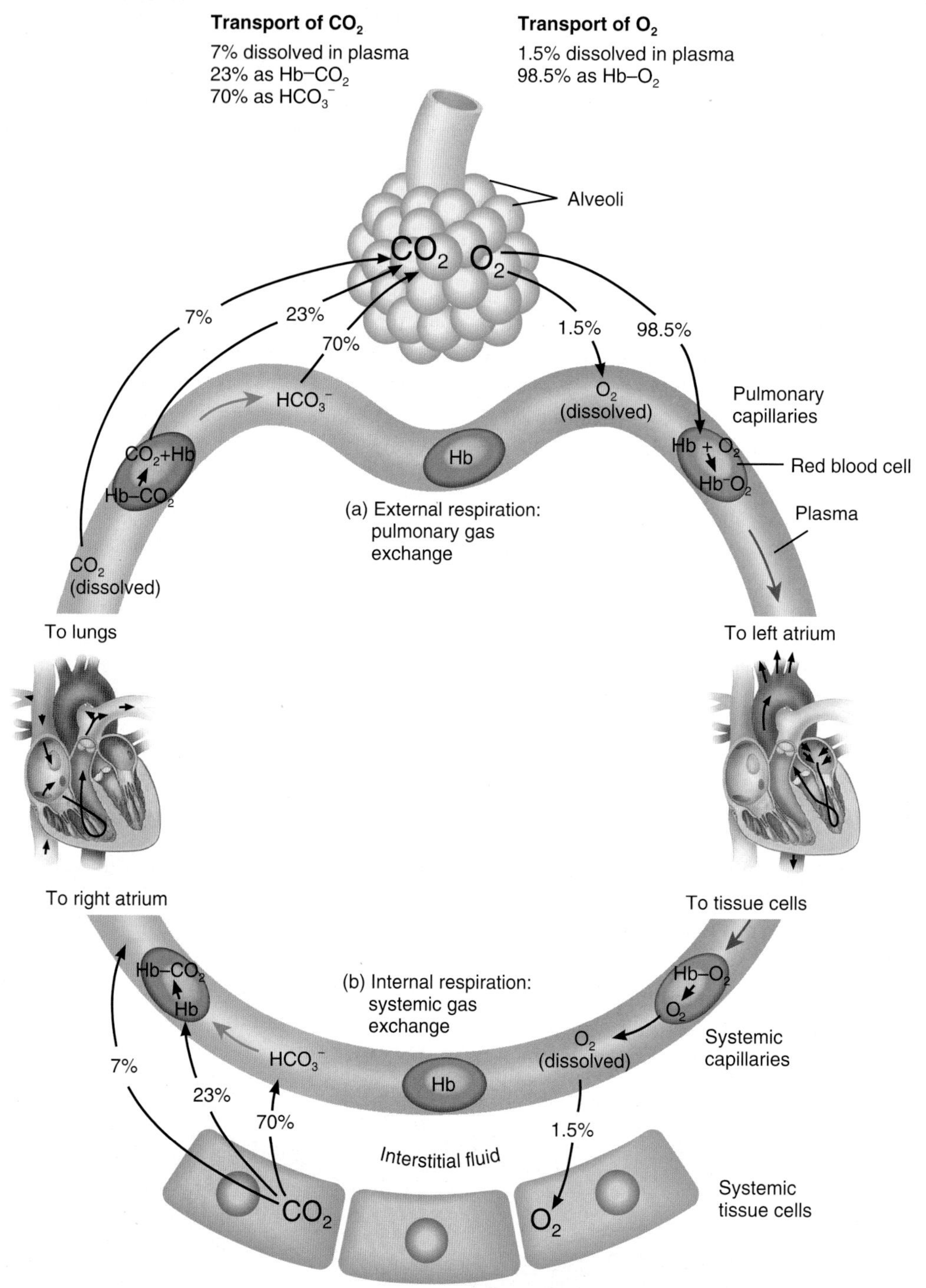

What is the most important factor that determines how much O_2 binds to hemoglobin?

Respiratory Center

The basic rhythm of respiration is controlled by groups of neurons in the brain stem. The area from which nerve impulses are sent to respiratory muscles is called the ***respiratory center*** and consists of groups of neurons in both the medulla oblongata and the pons.

The ***medullary rhythmicity area*** (rith-MIS-i-tē) in the medulla oblongata controls the basic rhythm of respiration. Within the medullary rhythmicity area are both inspiratory and expiratory areas. Figure 25.14 shows the relationships of the inspiratory and expiratory areas during normal quiet breathing and forceful breathing.

During quiet breathing, inhalation lasts for about 2 seconds and exhalation lasts for about 3 seconds. Nerve impulses generated in the ***inspiratory area*** establish the basic rhythm of breathing. While the inspiratory area is active, it generates nerve impulses for about 2 seconds (Figure 25.14a). The impulses propagate to the external intercostal muscles via intercostal nerves and/or to the diaphragm via the phrenic nerves. When the nerve impulses reach the diaphragm

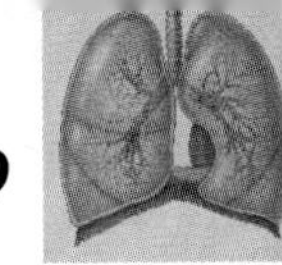

Figure 25.14 Roles of the medullary rhythmicity area in controlling (a) the basic rhythm of quiet respiration and (b) forceful breathing.

During normal, quiet breathing, the expiratory area is inactive. During forceful breathing, the inspiratory area activates the expiratory area.

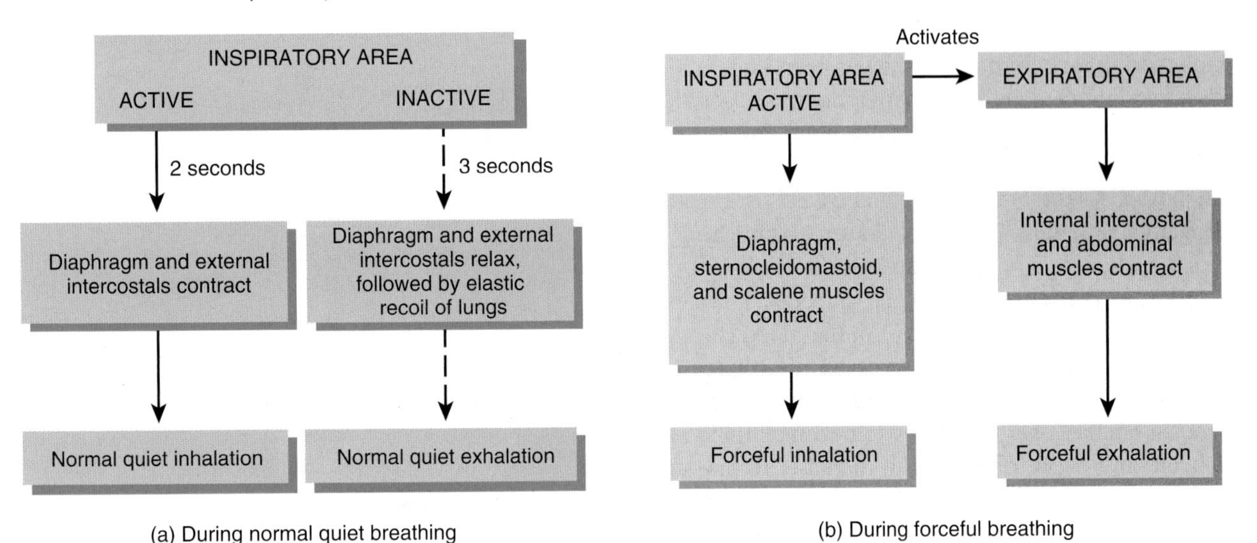

? **Which nerve conveys impulses from the respiratory center to the diaphragm?**

and/or external intercostal muscles, the muscles contract and inhalation occurs. Even when all incoming nerve connections to the inspiratory area are cut or blocked, neurons in this area still rhythmically discharge impulses that cause inhalation. At the end of 2 seconds, the inspiratory area becomes inactive and nerve impulses cease. With no impulses arriving, the diaphragm and/or external intercostal muscles relax for about 3 seconds, allowing passive elastic recoil of the lungs and thoracic wall. Then, the cycle repeats.

The neurons of the ***expiratory area*** remain inactive during quiet breathing. However, during forceful breathing, nerve impulses from the inspiratory area activate the expiratory area (Figure 25.14b). Impulses from the expiratory area then cause contraction of the internal intercostal and abdominal muscles, which decreases the size of the thoracic cavity and causes forceful exhalation.

The ***pneumotaxic area*** (noo-mō-TAK-sik; *pneumo-* = air or breath; *-taxic* = arrangement) in the upper pons helps turn off the inspiratory area to shorten the duration of inhalations and to increase breathing rate. The ***apneustic area*** (ap-NOO-stik) in the lower pons sends excitatory impulses to the inspiratory area that activate it and prolong inhalation. The result is a long, deep inhalation.

Regulation of the Respiratory Center

Although the basic rhythm of respiration is set and coordinated by the inspiratory area, the rhythm can be modified in response to inputs from other brain regions, receptors in the peripheral nervous system, and other factors.

Cortical Influences on Respiration

Because the cerebral cortex has connections with the respiratory center, we can voluntarily alter our pattern of breathing. We can even refuse to breathe at all for a short time. Voluntary control is protective because it enables us to prevent water or irritating gases from entering the lungs. The ability to not breathe, however, is limited by the buildup of CO_2 and H^+ in body fluids. When the P_{CO_2} and H^+ concentration reach a certain level, the inspiratory area is strongly stimulated and breathing resumes, whether the person wants it or not. It is impossible for people to kill themselves by voluntarily holding their breath. Even if the breath is held long enough to cause fainting, breathing resumes when consciousness is lost. Nerve impulses from the hypothalamus and limbic system also stimulate the respiratory center, allowing emotional stimuli to alter respirations as, for example, in laughing and crying.

Chemoreceptor Regulation of Respiration

Certain chemical stimuli determine how quickly and how deeply we breathe. The respiratory system functions to maintain proper levels of CO_2 and O_2 and is very responsive to changes in the levels of either in body fluids. Sensory neurons that are responsive to chemicals are termed ***chemoreceptors.*** *Central chemoreceptors,* located within the medulla oblongata, respond to changes in H^+ level or P_{CO_2}, or both, in cerebrospinal fluid. *Peripheral chemoreceptors,* located within the arch of the aorta and common carotid arteries, are especially sensitive to changes in P_{O_2}, H^+, and P_{CO_2} in the blood.

Because CO_2 is lipid-soluble, it easily diffuses through the plasma membrane into cells, where it combines with water (H_2O) to form carbonic acid (H_2CO_3). Carbonic acid quickly breaks down into H^+ and HCO_3^-. Any increase in CO_2 in the blood thus causes an increase in H^+ inside cells, and any decrease in CO_2 causes a decrease in H^+.

Normally, the P_{CO_2} in arterial blood is 40 mmHg. If even a slight increase in P_{CO_2} occurs—a condition called ***hypercapnia*** (*hypercarbia*)—the central chemoreceptors are stimulated and respond vigorously to the resulting increase in H^+ level. The peripheral chemoreceptors also are stimulated by both the high P_{CO_2} and the rise in H^+. In addition, the peripheral chemoreceptors (but not the central chemoreceptors) respond to a deficiency of O_2. When P_{O_2} in arterial blood falls from a normal level of 100 mmHg but is still above 50 mmHg, the peripheral chemoreceptors are stimulated. Severe deficiency of O_2 depresses activity of the central chemoreceptors and inspiratory area, which then do not respond well to any inputs and send fewer impulses to the muscles of inhalation. As the breathing rate decreases or breathing ceases altogether, P_{O_2} falls lower and lower, establishing a positive feedback cycle with a possibly fatal result.

The chemoreceptors participate in a negative feedback system that regulates the levels of CO_2, O_2, and H^+ in the blood (Figure 25.15). As a result of increased P_{CO_2}, decreased pH (increased H^+), or decreased P_{O_2}, input from the central and peripheral chemoreceptors causes the inspiratory area to become highly active. Then, the rate and depth of breathing increase. Rapid and deep breathing, called ***hyperventilation,*** allows the exhalation of more CO_2 until P_{CO_2} and H^+ are lowered to normal.

If the partial pressure of CO_2 in arterial blood is lower than 40 mmHg—a condition called ***hypocapnia***—the central and peripheral chemoreceptors are not stimulated, and stimulatory impulses are not sent to the inspiratory area. Then, the area sets its own moderate pace until CO_2 accumulates and the P_{CO_2} rises to 40 mmHg. People who hyperventilate voluntarily and cause hypocapnia can hold their breath for an unusually long time. Swimmers were once encouraged to hyperventilate just before a competition. However, this practice is risky because the O_2 level may fall dangerously low and cause fainting before the P_{CO_2} rises high enough to stimulate inhalation. A person who faints on land may suffer bumps and bruises, but one who faints in the water may drown.

Severe deficiency of O_2 depresses activity of the central chemoreceptors and inspiratory area, which then do not respond well to any inputs and send fewer impulses to the muscles of respiration. As the breathing rate decreases or breathing ceases altogether, P_{O_2} falls lower and lower, thereby establishing a positive feedback cycle with a possibly fatal result.

Figure 25.15 Negative feedback control of breathing in response to changes in blood P_{CO_2}, pH (H^+ level), and P_{O_2}.

6 An increase in blood P_{CO_2} stimulates the inspiratory area.

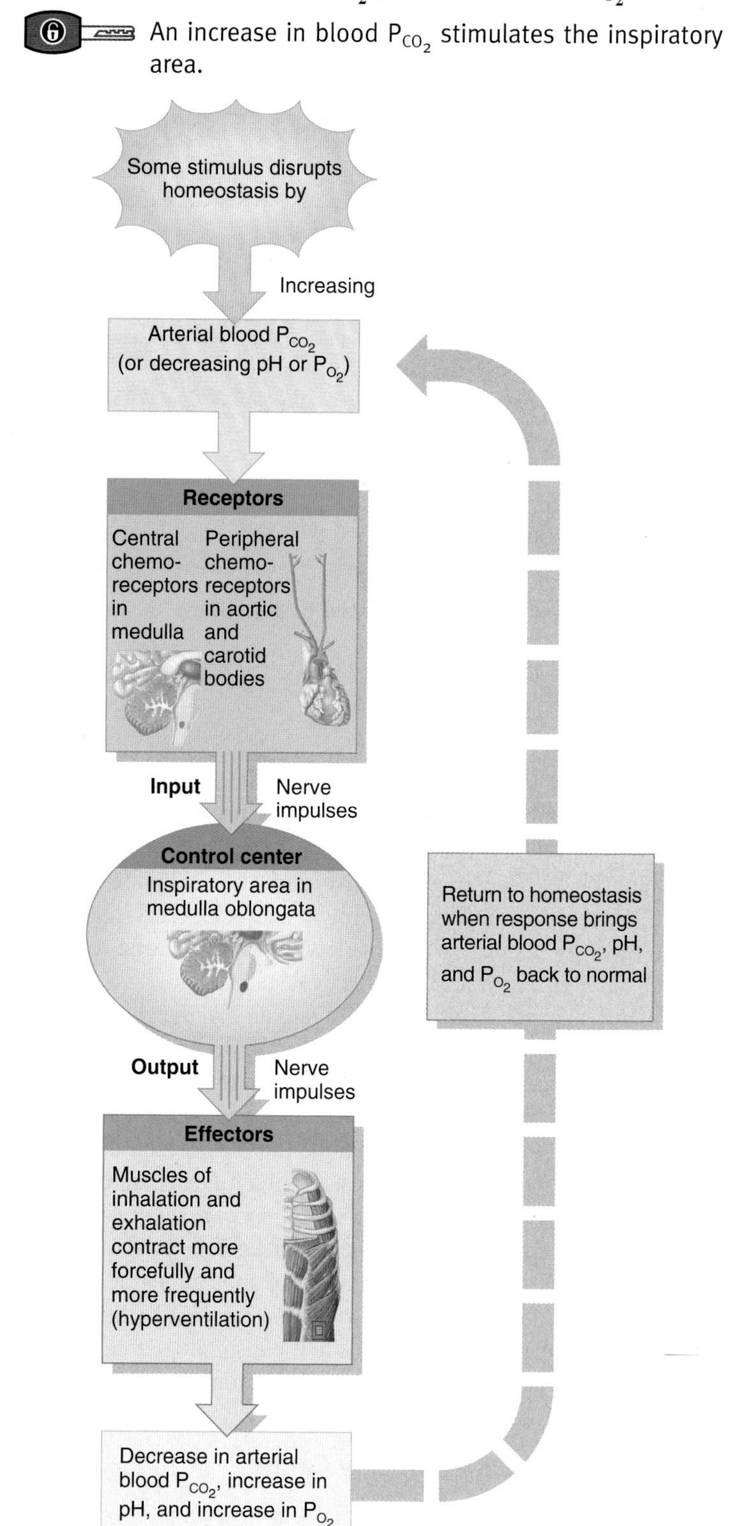

What is the normal arterial blood P_{CO_2}?

Other Influences on Respiration

Other factors that contribute to regulation of respiration include the following:

- **Limbic system stimulation.** Anticipation of activity or emotional anxiety may stimulate the limbic system, which then sends excitatory input to the inspiratory area, increasing the rate and depth of ventilation.
- **Proprioceptor stimulation of respiration.** As soon as you start exercising, your rate and depth of breathing increase, even before changes in P_{O_2}, P_{CO_2}, or H^+ level occur. The main stimulus for these quick changes in ventilation is input from proprioceptors, which monitor movement of joints and muscles. Nerve impulses from the proprioceptors stimulate the inspiratory area of the medulla oblongata.
- **Temperature.** An increase in body temperature, as occurs during a fever or vigorous muscular exercise, increases the rate of respiration; a decrease in body temperature decreases respiratory rate. A sudden cold stimulus (such as plunging into cold water) causes temporary ***apnea*** (AP-nē-a; *a-* = without; *-pnea* = breath), an absence of breathing.

- **Pain.** A sudden, severe pain brings about brief apnea, but a prolonged somatic pain increases respiratory rate. Visceral pain may slow the respiratory rate.
- **Irritation of airways.** Physical or chemical irritation of the pharynx or larynx brings about an immediate cessation of breathing followed by coughing or sneezing.
- **The inflation reflex.** Located in the walls of bronchi and bronchioles are pressure-sensitive ***stretch receptors.*** When these receptors become stretched during overinflation of the lungs, the inspiratory area is inhibited. As a result, exhalation begins. This reflex is mainly a protective mechanism for preventing excessive inflation of the lungs.

• CLINICAL CONNECTION | Chronic Obstructive Pulmonary Disease

Chronic obstructive pulmonary disease (COPD) is a respiratory disorder characterized by chronic obstruction of airflow. The principal types of COPD are emphysema and chronic bronchitis. In most cases, COPD is preventable because its most common cause is cigarette smoking or breathing secondhand smoke. Other causes include air pollution, pulmonary infection, occupational exposure to dusts and gases, and genetic factors.

Emphysema (em′-fi-SĒ-ma = blown up or full of air) is a disorder characterized by destruction of the walls of the alveoli, which produces abnormally large air spaces that remain filled with air during exhalation. With less surface area for gas exchange, O_2 diffusion across the respiratory membrane is reduced. Blood O_2 level is somewhat lowered, and any mild exercise that raises the O_2 requirements of the cells leaves the patient breathless. As increasing numbers of alveolar walls are damaged, lung elastic recoil decreases due to loss of elastic fibers, and an increasing amount of air becomes trapped in the lungs at the end of exhalation. Over several years, added respiratory exertion increases the size of the chest cage, resulting in a "barrel chest." Emphysema is a common precursor to the development of lung cancer.

Chronic bronchitis is a disorder characterized by excessive secretion of bronchial mucus accompanied by a cough. Inhaled pathogens become embedded in airway mucus and multiply rapidly. Besides a cough, symptoms of chronic bronchitis are shortness of breath, wheezing, cyanosis, and pulmonary hypertension. •

CHECKPOINT

13. How does the medullary rhythmicity area function in regulating respiration?
14. How do the cerebral cortex, levels of CO_2 and O_2, proprioceptors, inflation reflex, temperature changes, pain, and irritation of the airways modify respiration?

25.7 EXERCISE AND THE RESPIRATORY SYSTEM

OBJECTIVE

- Describe the effects of exercise on the respiratory system.

During exercise, the respiratory and cardiovascular systems make adjustments in response to both the intensity and duration of the exercise. The effects of exercise on the heart were discussed in Chapter 22; here we focus on how exercise affects the respiratory system.

Recall that the heart pumps the same amount of blood to the lungs as to all the rest of the body. Thus, as cardiac output rises, the rate of blood flow through the lungs also increases. If blood flows through the lungs twice as fast as at rest, it picks up twice as much oxygen per minute. In addition, the rate at which O_2 diffuses from alveolar air into the blood increases during maximal exercise because blood flows through a larger percentage of the pulmonary capillaries, providing a greater surface area for diffusion of O_2 into the blood.

When muscles contract during exercise, they consume large amounts of O_2 and produce large amounts of CO_2, forcing the respiratory system to work harder to maintain normal blood gas levels. During vigorous exercise, O_2 consumption and ventilation increase dramatically. At the onset of exercise, an abrupt increase in ventilation, due to activation of proprioceptors, is followed by a more gradual increase. With moderate exercise, the depth of ventilation rather than breathing rate is increased. When exercise is more strenuous, breathing rate also increases.

At the end of an exercise session, an abrupt decrease in ventilation rate is followed by a more gradual decline to the resting level. The initial decrease is due mainly to decreased stimulation of proprioceptors when movement stops or slows. The more gradual decrease reflects the slower return of blood chemistry and blood temperature to resting levels.

CHECKPOINT

15. How does exercise affect the inspiratory area?

25.8 AGING AND THE RESPIRATORY SYSTEM

OBJECTIVE

- Describe the effects of aging on the respiratory system.

With advancing age, the airways and tissues of the respiratory tract, including the alveoli, become less elastic and more rigid; the chest wall becomes more rigid as well. The result is a decrease in lung capacity. In fact, vital capacity (the maximum amount of air that can be expired after maximal inhalation) can decrease as much as 35% by age 70. A decrease in blood level of O_2, decreased activity of alveolar macrophages, and diminished ciliary action of the epithelium lining the respiratory tract occur. Owing to all these age-related factors, elderly people are more susceptible to pneumonia, bronchitis, emphysema, and other pulmonary disorders. Age-related changes in the structure and functions of the lung can also contribute to an older person's reduced ability to perform vigorous exercises, such as running.

CHECKPOINT

16. What accounts for the decrease in vital capacity with aging?

To appreciate the many ways that the respiratory system contributes to homeostasis of other body systems, examine Exhibit 25.1.

EXHIBIT 25.1 Contributions of the Respiratory System to Homeostasis

BODY SYSTEM		CONTRIBUTION OF THE RESPIRATORY SYSTEM
For all body systems		Provides oxygen and removes carbon dioxide. Helps adjust the pH of body fluids through exhalation of carbon dioxide.
Muscular system		Increased rate and depth of breathing support increased activity of skeletal muscles during exercise.
Nervous system	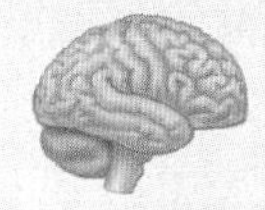	Nose contains receptors for the sense of smell (olfaction). Vibrations of air flowing across the vocal cords produce sounds for speech.
Endocrine system	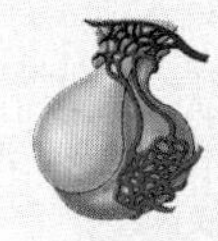	Angiotensin-converting enzyme (ACE) in the lungs promotes formation of the hormone angiotensin II, which in turn stimulates the adrenal gland to release the hormone aldosterone.
Cardiovascular system	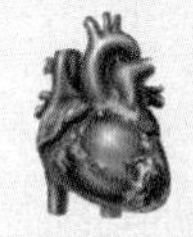	During inhalations, the respiratory pump aids the return of venous blood to the heart.
Lymphatic system and immunity		Hairs in the nose, cilia and mucus in the trachea, bronchi, and smaller airways, and alveolar macrophages contribute to nonspecific immunity to disease. The pharynx (throat) contains lymphatic tissue (tonsils). During inhalation, the respiratory pump promotes the flow of lymph.
Digestive system	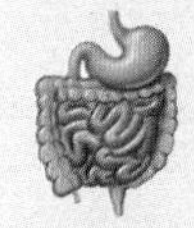	Forceful contraction of the respiratory muscles can assist in defecation.
Urinary system	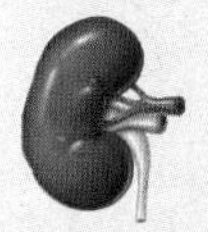	Together, the respiratory and urinary systems regulate the pH of body fluids.
Reproductive systems	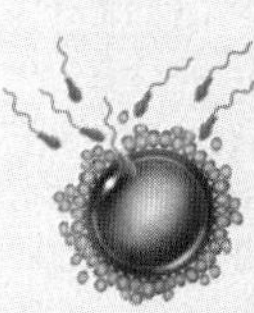	Increased rate and depth of breathing support activity during sexual intercourse. Internal respiration provides oxygen to the developing fetus.

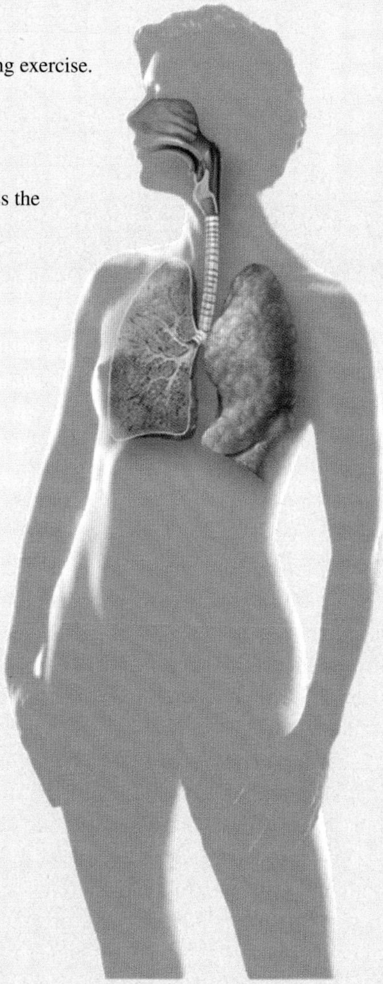

KEY MEDICAL TERMS ASSOCIATED WITH THE RESPIRATORY SYSTEM

Abdominal thrust maneuver (ATM) First-aid procedure to clear the airways of obstructing objects. It is performed by applying a quick upward thrust between the navel and lower ribs that causes sudden elevation of the diaphragm and forceful, rapid expulsion of air from the lungs, forcing air out of the trachea to eject the obstructing object. Also used to expel water from the lungs of near-drowning victims before resuscitation is begun. Previously known as the ***Heimlich maneuver*** (HĪM-lik ma-NOO-ver).

Asphyxia (as-FIK-sē-a; *-sphyxia* = pulse) Oxygen starvation due to low atmospheric oxygen or interference with ventilation, external respiration, or internal respiration.

Aspiration (as′-pi-RĀ-shun) Inhalation into the bronchial tree of a substance other than air, for instance, water, food, or a foreign body.

Bronchoscopy (brong-KOS-kō-pē) The visual examination of the bronchi through a ***bronchoscope,*** an illuminated, tubular instrument that is passed through the mouth (or nose), larynx, and trachea into the bronchi.

Coryza (ko-RĪ-za) Hundreds of viruses can cause coryza or the ***common cold.*** Typical symptoms include sneezing, excessive nasal secretion, dry cough, and congestion. The uncomplicated common cold is not usually accompanied by a fever. Complications may include sinusitis, asthma, bronchitis, ear infections, and laryngitis.

Cystic fibrosis (CF) An inherited disease of secretory epithelia that affects the airways, liver, pancreas, small intestine, and sweat glands. Clogging and infection of the airways leads to difficulty in breathing and eventual destruction of lung tissue.

Dyspnea (DISP-nē-a; *dys-* = painful, difficult) Painful or labored breathing.

Epistaxis (ep′-i-STAK-sis) Loss of blood from the nose due to trauma, infection, allergy, malignant growths, or bleeding disorders. It can be arrested by cautery with silver nitrate, electrocautery, or firm packing. Also called ***nosebleed.***

Hypoxia (hī-POK-sē-a; *hypo-* = below or under) A deficiency of O_2 at the tissue level that may be caused by a low P_{O_2} in arterial blood, as from high altitudes; too little functioning hemoglobin in the blood, as in anemia; inability of the blood to carry O_2 to tissues fast enough to sustain their needs, as in heart failure; or inability of tissues to use O_2 properly, as in cyanide poisoning.

Influenza (flu) Caused by a virus; symptoms include chills, fever (usually higher than 101°F, or 38°C), headache, and muscular aches. Coldlike symptoms appear as the fever subsides.

Mechanical ventilation The use of an automatically cycling device (ventilator or respirator) to assist breathing. A plastic tube is inserted into the nose or mouth and the tube is attached to a device that forces air into the lungs. Exhalation occurs passively due to the elastic recoil of the lungs.

Pleurisy (PLŪR-i-sē) Inflammation of the pleural membranes, which causes friction during breathing that can be quite painful when the swollen membranes rub against each other. Also known as pleuritis.

Pneumonia or ***pneumonitis*** (nū′-mō-NĪ-tis) Acute infection or inflammation of the alveoli. When certain microbes enter the lungs of susceptible individuals, they release harmful toxins that damage alveoli and bronchial mucous membranes. Inflammation and edema cause the alveoli to fill with debris and fluid, interfering with ventilation and gas exchange. The most common cause is the bacterium *Streptococcus pneumoniae,* but other bacteria, viruses, or fungi may also cause pneumonia.

Rales (RĀLS) Sounds sometimes heard in the lungs that resemble bubbling or rattling. Different types are due to the presence of an abnormal type or amount of fluid or mucus within the bronchi or alveoli, or to bronchoconstriction that causes turbulent airflow.

Respiratory distress syndrome (RDS) A breathing disorder of premature newborns in which the alveoli do not remain open due to a lack of surfactant. Surfactant reduces surface tension and is necessary to prevent the collapse of alveoli during exhalation.

Respiratory failure A condition in which the respiratory system either cannot supply enough O_2 to maintain metabolism or cannot eliminate enough CO_2 to prevent respiratory acidosis (a higher-than-normal H^+ level in interstitial fluid).

Rhinitis (rī-NĪ-tis; *rhin-* = nose) Chronic or acute inflammation of the mucous membrane of the nose.

Sudden infant death syndrome (SIDS) Death of infants between the ages of 1 week and 12 months thought to be due to hypoxia that occurs while sleeping in a prone position (on the stomach) and re-breathing exhaled air trapped in a depression of the mattress. It is now recommended that normal newborns be placed on their backs for sleeping (remember: "back to sleep").

Tachypnea (tak′-ip-NĒ-a; *tachy-* = rapid) Rapid breathing rate.

Tuberculosis *Mycobacterium tuberculosis* produces an infectious, communicable disease called tuberculosis (TB) that affects the lungs and the pleurae but may involve other parts of the body. Inflammation stimulates neutrophils and macrophages to engulf the bacteria to prevent their spread. If the immune system is not impaired, the bacteria may remain dormant for life. Impaired immunity may enable the bacteria to infect other organs.

Wheeze (HWĒZ) A whistling, squeaking, or musical high-pitched sound during breathing resulting from a partially obstructed airway.

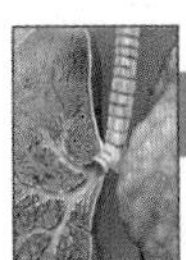

STUDY OUTLINE

Overview of the Respiratory System (Section 25.1)

1. Organs of the respiratory system include the nose, pharynx, larynx, trachea, bronchi, and lungs, and they act with the cardiovascular system to supply oxygen and remove carbon dioxide from the blood.
2. In addition to gas exchange, the respiratory system helps regulate blood pH and it filters, warms, and moistens inspired air.
3. The entire process of gas exchange (respiration) occurs in three steps: pulmonary ventilation (breathing), external respiration, and internal respiration (cellular respiration).

Organs of the Respiratory System (Section 25.2)

1. The external portion of the nose is made of cartilage and skin and is lined with mucous membrane. Openings to the exterior are the external nares.
2. The internal portion of the nose, divided from the external portion by the septum, communicates with the paranasal sinuses and nasopharynx through the internal nares.
3. The nose is adapted for warming, moistening, and filtering air; olfaction; and serving as a resonating chamber for special sounds.
4. The pharynx (throat), a muscular tube lined by a mucous membrane, is divided into the nasopharynx, oropharynx, and laryngopharynx.
5. The nasopharynx functions in respiration. The oropharynx and laryngopharynx function both in digestion and in respiration.
6. The larynx connects the pharynx and the trachea. It contains the thyroid cartilage (Adam's apple), the epiglottis, the cricoid cartilage, arytenoid cartilages, false vocal cords, and true vocal cords. Taut true vocal cords produce high pitches; relaxed ones produce low pitches.
7. The trachea (windpipe) extends from the larynx to the primary bronchi. It is composed of smooth muscle and C-shaped rings of cartilage and is lined with pseudostratified ciliated columnar epithelium.
8. The bronchial tree consists of the trachea, primary bronchi, secondary bronchi, tertiary bronchi, bronchioles, and terminal bronchioles.
9. Lungs are paired organs in the thoracic cavity enclosed by the pleural membrane. The parietal pleura is the outer layer; the visceral pleura is the inner layer.
10. The right lung has three lobes separated by two fissures; the left lung has two lobes separated by one fissure plus a depression, the cardiac notch.
11. Each lobe consists of lobules, which contain lymphatic vessels, arterioles, venules, terminal bronchioles, respiratory bronchioles, alveolar ducts, alveolar sacs, and alveoli.
12. Exchange of gases (oxygen and carbon dioxide) in the lungs occurs across the respiratory membrane, a thin "sandwich" consisting of alveolar cells, basement membrane, and endothelial cells of a capillary.

Pulmonary Ventilation (Section 25.3)

1. Pulmonary ventilation (breathing) consists of inhalation and exhalation, the movement of air into and out of the lungs. Air flows from higher to lower pressure.
2. Inhalation occurs when alveolar pressure falls below atmospheric pressure. Contraction of the diaphragm and external intercostals expands the volume of the lungs. Increased volume of the lungs decreases alveolar pressure, and air moves from higher to lower pressure, from the atmosphere into the lungs.
3. Exhalation occurs when alveolar pressure is higher than atmospheric pressure. Relaxation of the diaphragm and external intercostals decreases lung volume, and alveolar pressure increases so that air moves from the lungs to the atmosphere.
4. The sternocleidomastoids, scalenes, and pectoralis minors contribute to forced inhalation. Forced exhalation involves contraction of the internal intercostals, external oblique, internal oblique, transversus abdominis, and rectus abdominis.
5. The minute ventilation is the total air taken in during 1 minute (breathing rate per minute multiplied by tidal volume).
6. The lung volumes are tidal volume, inspiratory reserve volume, expiratory reserve volume, and residual volume.
7. Lung capacities, the sum of two or more lung volumes, include inspiratory, functional residual, vital, and total.

Exchange of Oxygen and Carbon Dioxide (Section 25.4)

1. The partial pressure of a gas (P) is the pressure exerted by that gas in a mixture of gases.
2. Each gas in a mixture of gases exerts its own pressure and behaves as if no other gases are present.
3. In external and internal respiration, O_2 and CO_2 move from areas of higher partial pressure to areas of lower partial pressure.
4. External respiration is the exchange of gases between alveolar air and pulmonary blood capillaries. It is aided by a thin respiratory membrane, a large alveolar surface area, and a rich blood supply.
5. Internal respiration is the exchange of gases between systemic tissue capillaries and systemic tissue cells.

Transport of Respiratory Gases (Section 25.5)

1. Most oxygen, 98.5%, is carried by the iron ions of the heme in hemoglobin; 1.5% is dissolved in plasma.
2. The association of O_2 and hemoglobin is affected by P_{O_2}, pH, temperature, and P_{CO_2}.
3. Hypoxia refers to O_2 deficiency at the tissue level.
4. Carbon dioxide is transported in three ways. About 7% is dissolved in plasma, 23% combines with the globin of hemoglobin, and 70% is converted to bicarbonate ions (HCO_3^-).

Control of Respiration (Section 25.6)

1. The respiratory center consists of a medullary rhythmicity area (inspiratory and expiratory areas) in the medulla oblongata and groups of neurons in the pons.
2. The inspiratory area sets the basic rhythm of respiration.
3. Respirations may be modified by several factors, including cortical influences; chemical stimuli, such as levels of O_2, CO_2, and H^+; limbic system stimulation; proprioceptor input; temperature; pain; the inflation reflex; and irritation to the airways.

Exercise and the Respiratory System (Section 25.7)

1. The rate and depth of ventilation change in response to both the intensity and duration of exercise.
2. The abrupt increase in ventilation at the start of exercise is due to neural changes that send excitatory impulses to the inspiratory area in the medulla oblongata. The more gradual increase in ventilation during moderate exercise is due to chemical and physical changes in the bloodstream.

Aging and the Respiratory System (Section 25.8)

1. Aging results in decreased vital capacity, decreased blood level of O_2, and diminished alveolar macrophage activity.
2. Elderly people are more susceptible to pneumonia, emphysema, bronchitis, and other pulmonary disorders.

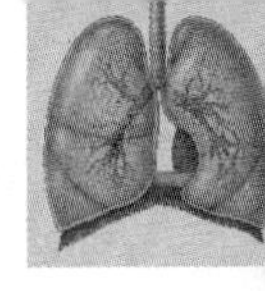

SELF-QUIZ QUESTIONS

1. Which of the following is NOT true concerning the pharynx?
   a. Food, drink, and air pass through the oropharynx and laryngopharynx.
   b. The auditory (eustachian) tubes have openings in the nasopharynx.
   c. The pseudostratified ciliated epithelium of the nasopharynx helps move dust-laden mucus toward the mouth.
   d. The palatine and lingual tonsils are located in the laryngopharynx.
   e. The wall of the pharynx is composed of skeletal muscle lined with mucous membranes.
2. During speaking, you raise your voice's pitch. This is possible because
   a. the epiglottis vibrates rapidly
   b. you have increased the air pressure pushing against the vocal cords
   c. you have increased the tension on the true vocal cords
   d. your true vocal cords have become thicker and longer
   e. the true vocal cords begin to vibrate more slowly.
3. Johnny is having an asthma attack and feels as if he cannot breathe. Why?
   a. His diaphragm is not contracting.
   b. Spasms in the bronchiole smooth muscle have blocked airflow to the alveoli.
   c. Excess mucus production is interfering with airflow into the lungs.
   d. The epiglottis has closed and air is not entering the lungs.
   e. Insufficient surfactant is being produced.
4. Which sequence of events best describes inhalation?
   a. contraction of diaphragm → increase in size of thoracic cavity → decrease in alveolar pressure
   b. relaxation of diaphragm → decrease in size of thoracic cavity → increase in alveolar pressure
   c. contraction of diaphragm → decrease in size of thoracic cavity → decrease in alveolar pressure
   d. relaxation of diaphragm → increase in size of thoracic cavity → increase in alveolar pressure
   e. contraction of diaphragm → decrease in size of thoracic cavity → increase in alveolar pressure
5. Which of the following does NOT help keep air passages clean?
   a. nostril hairs
   b. alveolar macrophages
   c. capillaries in the nasal cavities
   d. cilia in the upper and lower respiratory tracts
   e. mucus
6. If the total pressure of a mixture of gases is 760 mmHg and gas Z makes up 20% of the total mixture, then the partial pressure of gas Z would be
   a. 152 mmHg
   b. 175 mmHg
   c. 225 mmHg
   d. 608 mmHg
   e. 760 mmHg.
7. How does hypercapnia affect respiration?
   a. It increases the rate of respiration.
   b. It decreases the rate of respiration.
   c. It causes hypoventilation.
   d. It does not change the rate of respiration.
   e. It activates stretch receptors in the lungs.
8. Air would flow into the lungs along which route?
   1. bronchioles
   2. primary bronchi
   3. secondary bronchi
   4. terminal bronchioles
   5. tertiary bronchi
   6. trachea
   a. 6, 1, 2, 3, 5, 4
   b. 6, 5, 3, 4, 2, 1
   c. 6, 2, 3, 5, 4, 1
   d. 6, 2, 3, 5, 1, 4
   e. 6, 1, 4, 5, 3, 2
9. Match the following:
   ____ a. normally inactive; when activated, causes contraction of internal intercostals and abdominal muscles and forced exhalation
   ____ b. located in pons; stimulates inspiratory area to prolong inhalation
   ____ c. sets basic rhythm of respiration; located in medulla
   ____ d. transmits inhibitory impulses to inspiratory area; located in pons
   ____ e. allows voluntary alteration of breathing patterns

   A. inspiratory area
   B. expiratory area
   C. pneumotaxic area
   D. apneustic area
   E. cerebral cortex
10. Under normal body conditions, hemoglobin releases oxygen more readily when
    a. body temperature increases
    b. blood acidity decreases
    c. blood pH increases
    d. blood oxygen partial pressure is high
    e. blood CO_2 is low.
11. Match the following:
    ____ a. decreased carbon dioxide levels
    ____ b. normal, quiet breathing
    ____ c. rapid breathing
    ____ d. exchange of gases between the blood and lungs
    ____ e. inhalation and exhalation
    ____ f. increased carbon dioxide levels
    ____ g. exchange of gases between blood and tissue cells
    ____ h. absence of breathing

    A. external respiration
    B. apnea
    C. hypercapnia
    D. eupnea
    E. internal respiration
    F. hypocapnia
    G. pulmonary ventilation
    H. hyperventilation
12. Which of the following statements is NOT true concerning the lungs?
    a. The lungs contain about 300 million alveoli.
    b. The left lung is thicker and broader because the liver lies below it.
    c. The right lung is composed of three lobes.
    d. The top portion of the lung is the apex.
    e. The lungs are surrounded by a serous membrane.

13. Exhalation
    a. occurs when alveolar pressure reaches 758 mmHg
    b. is normally considered an active process requiring muscle contraction
    c. occurs when alveolar pressure is greater than atmospheric pressure
    d. involves the expansion of the pleural membranes
    e. occurs when the atmospheric pressure is equal to the pressure in the lungs.
14. In which structures would you find simple squamous epithelium?
    a. secondary bronchi
    b. larynx and pharynx
    c. tertiary bronchi
    d. primary bronchi
    e. alveoli
15. Overinflation of the lungs is prevented by
    a. the inflation reflex
    b. pain in the pleural membranes
    c. nerve impulses from proprioceptors
    d. control from the cerebral cortex
    e. controlling blood pressure.
16. The function of goblet cells in the nasal cavities is to
    a. warm the air entering the nose
    b. produce mucus to trap inhaled dust
    c. increase the surface area inside the nose
    d. help produce speech
    e. exchange O_2 and CO_2 within the nasal cavities.
17. Decreasing the surface area of the respiratory membrane would affect
    a. internal respiration
    b. inhalation
    c. speech
    d. external respiration
    e. mucus production.
18. In which form is carbon dioxide NOT carried in the blood?
    a. bicarbonate ion
    b. bound to globin
    c. oxyhemoglobin
    d. carbaminohemoglobin
    e. dissolved in plasma
19. Of the following, which would have the highest partial pressure of oxygen?
    a. alveolar air at the end of exhalation
    b. rapidly contracting skeletal muscle fibers
    c. alveolar air immediately after inhalation
    d. blood flowing into the lungs from the right side of the heart
    e. blood returning to the heart from the tissue cells
20. Match the following:

____	a. forceful exhalation of air	A. vital capacity
____	b. inspiratory reserve volume + tidal volume + expiratory reserve volume	B. inspiratory reserve volume
____	c. volume of air moved during normal quiet breathing	C. residual volume
____	d. air remaining after forced exhalation	D. expiratory reserve volume
____	e. forceful inhalation of air	E. tidal volume

CRITICAL THINKING QUESTIONS

1. Your 3-year-old nephew Levi likes to get his own way all the time! Right now, Levi wants to eat 20 chocolate kisses (one for each finger and toe), but you'll only give him one for each year of his age. He is at this moment "holding my breath until I turn blue and won't you be sorry!" Is he in danger of death?
2. Katie was diagnosed with exercise-induced asthma after she reported trouble catching her breath during a swim meet. Exercise-induced asthma is a particularly annoying condition for an athlete because the body's response to exercise is the exact opposite of the body's need. Explain this statement.
3. Brianna has a flare for being dramatic. "I can't come to work today," she whispered, "I've got laryngitis and a horrible case of coryza." What is wrong with Brianna?
4. The entire tour group was in fine health when they left the coast of China for their next stop—Tibet! After touring the mountainous area for a day, many of the group felt dizzy, nauseous, and exhausted. They were hyperventilating and could not catch their breath. The local physician had seen this condition many times before in people that did not take the time to acclimate to the mountains. What caused the tour group's symptoms?

? ANSWERS TO FIGURE QUESTIONS

25.1 The conducting zone of the respiratory system includes the nose, oral cavity, pharynx, larynx, trachea, bronchi, and bronchioles (except the respiratory bronchioles).
25.2 Air molecules flow through the external nares, the nasal cavity, and then the internal nares.
25.3 During swallowing, the epiglottis closes over the larynx to block food and liquids from entering.
25.4 There are two lobes and two secondary bronchi in the left lung and three lobes and three secondary bronchi in the right lung.
25.5 Alveolar ducts receive air from a respiratory bronchiole.
25.6 The respiratory membrane averages 0.5 μm in thickness.
25.7 The respiratory membrane consists of an alveolar wall, fused basement membranes, and a capillary wall.
25.8 The main muscle that powers quiet breathing is the diaphragm.
25.9 The pressure would increase fourfold, to 4 atm.
25.10 Normal atmospheric pressure at sea level is 760 mmHg.
25.11 You demonstrate vital capacity when you breathe in as deeply as possible and then exhale as much air as you can.
25.12 Oxygen enters pulmonary capillaries from alveolar air and enters tissue cells from systemic capillaries due to differences in P_{O_2}.
25.13 The most important factor that determines how much O_2 binds to hemoglobin is the P_{O_2}.
25.14 The phrenic nerves stimulate the diaphragm to contract.
25.15 Normal arterial blood P_{O_2} is 40 mmHg.

The Digestive System

26

The food we eat contains a variety of nutrients, which are used for building new body tissues and repairing damaged tissues. However, most of the food we eat consists of molecules that are too large to be used by body cells. Therefore, food must be broken down into molecules that are small enough to enter body cells, a process known as digestion. Collectively, the organs that perform these systems are known as the digestive system.

The digestive system is basically a hollow tube that runs through our body. It is a very muscular tube, noisily pushing materials along its length. Along the way, specialized organs assist in secreting fluids to maintain pH and turn on enzymes that catabolize (break down) food. Once those nutrients are broken down they can be absorbed through the stomach and the intestines, and allow nutrients to enter the blood. The digestive system is controlled by the parasympathetic nervous system, and therefore, when you are relaxed, you digest well. However, when you are under a lot of stress for prolonged periods it often leads to conditions such as high blood pressure or irritable bowel syndrome (IBS).

CONTENTS AT A GLANCE

26.1 OVERVIEW OF THE DIGESTIVE SYSTEM

OBJECTIVE

- Identify the organs of the digestive system and their basic functions.

Two groups of organs compose the digestive system (Figure 26.1): the gastrointestinal tract and the accessory digestive organs. The ***gastrointestinal (GI) tract*** or ***alimentary canal*** is a continuous tube that extends from the mouth to the anus. The GI tract contains food from the time it is eaten until it is digested and absorbed or eliminated from the body. Organs of the gastrointestinal tract include the mouth, pharynx, esophagus, stomach, small intestine, and large intestine. The teeth, tongue, salivary glands, liver, gallbladder, and pancreas serve as ***accessory digestive organs.*** Teeth aid in the physical breakdown of food, and the tongue assists in chewing and swallowing. The other accessory digestive organs never come into direct contact with food. The secretions that they produce or store flow into the GI tract through ducts and aid in the chemical breakdown of food.

Overall, the digestive system performs six basic processes:

1. **Ingestion.** This process involves taking foods and liquids into the mouth (eating).
2. **Secretion.** Each day, cells within the walls of the GI tract and accessory organs secrete a total of about 7 liters of water, acid, buffers, and enzymes into the lumen of the tract.
3. **Mixing and propulsion.** Alternating contraction and relaxation of smooth muscle in the walls of the GI tract mix food and secretions and propel them toward the anus. The ability of the GI tract to mix and move material along its length is termed ***motility***.
4. **Digestion.** Mechanical and chemical processes break down ingested food into small molecules. In ***mechanical digestion*** the teeth cut and grind food before it is swallowed, and then smooth muscles of the stomach and small intestine churn the food. As a result, food molecules become dissolved and thoroughly mixed with digestive enzymes. In ***chemical digestion*** the large carbohydrate, lipid, protein, and nucleic acid molecules in food are broken down into smaller molecules by digestive enzymes.
5. **Absorption.** The entrance of ingested and secreted fluids, ions, and the small molecules that are products of digestion into the epithelial cells lining the lumen of the GI tract is called ***absorption.*** The absorbed substances pass into interstitial fluid and then into blood or lymph and circulate to cells throughout the body.
6. **Defecation.** Wastes, indigestible substances, bacteria, cells shed from the lining of the GI tract, and digested materials that were not absorbed leave the body through the anus in a process called ***defecation.*** The eliminated material is termed ***feces (stool).***

The medical specialty that deals with the structure, function, diagnosis, and treatment of diseases of the stomach and intestines is ***gastroenterology*** (gas′-trō-en′-ter-OL-ō-jē; *gastro-* = stomach;

Figure 26.1 Organs of the digestive system and related structures.

Organs of the gastrointestinal (GI) tract include the mouth, pharynx, esophagus, stomach, small intestine, and large intestine. Accessory digestive organs are the teeth, tongue, salivary glands, liver, gallbladder, and pancreas.

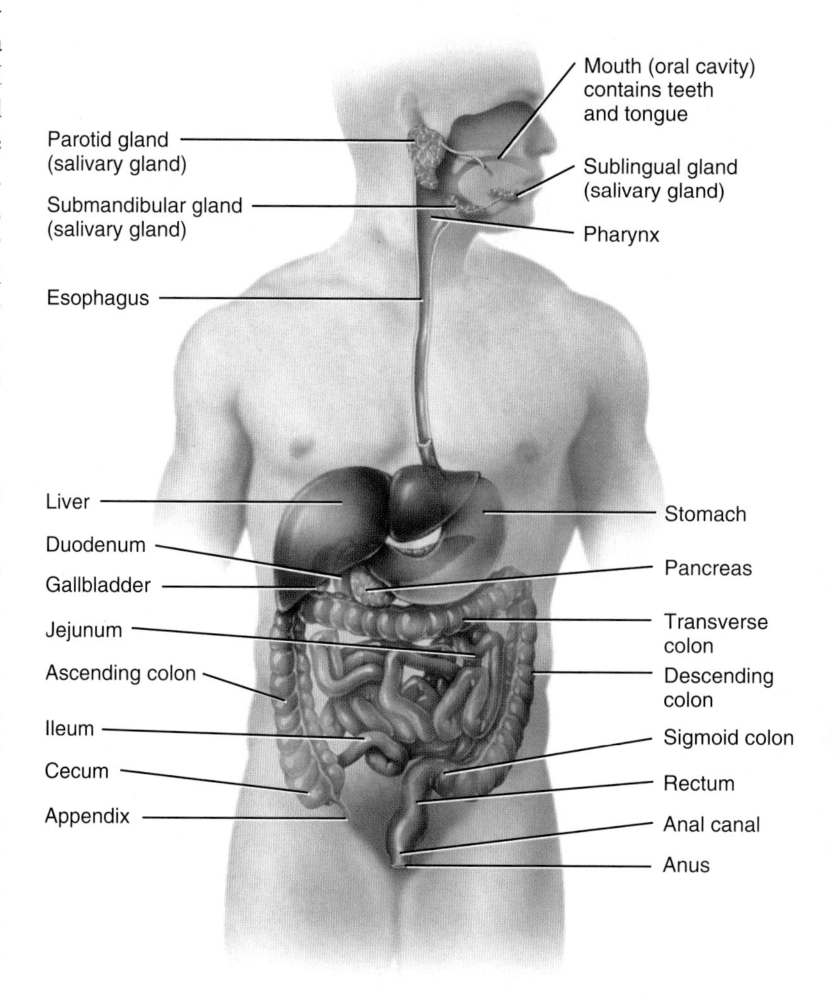

Right lateral view of head and neck and anterior view of trunk

Functions

1. Ingestion: taking food into the mouth.
2. Secretion: release of water, acid, buffers, and enzymes into the lumen of the GI tract.
3. Mixing and propulsion: churning and propulsion of the food through the GI tract.
4. Digestion: mechanical and chemical breakdown of food.
5. Absorption: passage of digested products from the GI tract into the blood and lymph.
6. Defecation: the elimination of feces from the GI tract.

? Which accessory digestive organs assist in the physical breakdown of food?

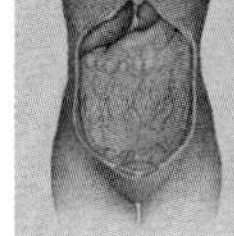

-entero- = intestines; *-logy* = study of). The medical specialty that deals with the diagnosis and treatment of disorders of the rectum and anus is ***proctology*** (prok-TOL-ō-jē; *proct-* = rectum).

CHECKPOINT

1. Which components of the digestive system are GI tract organs and which are accessory digestive organs?
2. Which organs of the digestive system come in contact with food, and what are some of their digestive functions?

26.2 LAYERS OF THE GI TRACT AND THE PERITONEUM

OBJECTIVE

- Describe the four layers that form the wall of the gastrointestinal tract.

The wall of the GI tract, from the lower esophagus to the anal canal, has the same basic, four-layered arrangement of tissues. The four layers of the tract, from the inside out, are the mucosa, submucosa, muscularis, and serosa (Figure 26.2).

1. Mucosa. The ***mucosa,*** or inner lining of the tract, is a mucous membrane. It is composed of a layer of epithelium in direct contact with the contents of the GI tract, a layer of areolar connective tissue called the ***lamina propria,*** and a thin layer of smooth muscle called the ***muscularis mucosae.*** Contractions of the muscularis mucosae create folds in the mucosa that increase the surface area for digestion and absorption. The mucosa also contains prominent lymphatic nodules that protect against the entry of pathogens through the GI tract.

2. Submucosa. The ***submucosa*** consists of areolar connective tissue that binds the mucosa to the muscularis. It contains many blood and lymphatic vessels that receive absorbed food molecules. Also located in the submucosa is a plexus of neurons which are subject to regulation by the autonomic nervous system (ANS) called the ***enteric nervous system (ENS),*** the "brain of the gut." This plexus, called the ***submucosal plexus,*** is found within the submucosa and controls the secretions of the organs of the GI tract.

3. Muscularis. As its name implies, the ***muscularis*** of the GI tract is a thick layer of muscle. In the mouth, pharynx, and up-

Figure 26.2 Layers of the gastrointestinal tract. Variations in this basic plan may be seen in the stomach (Figure 26.8), small intestine (Figure 26.13), and large intestine (Figure 26.16).

The four layers of the GI tract from inside to outside are the mucosa, submucosa, muscularis, and serosa.

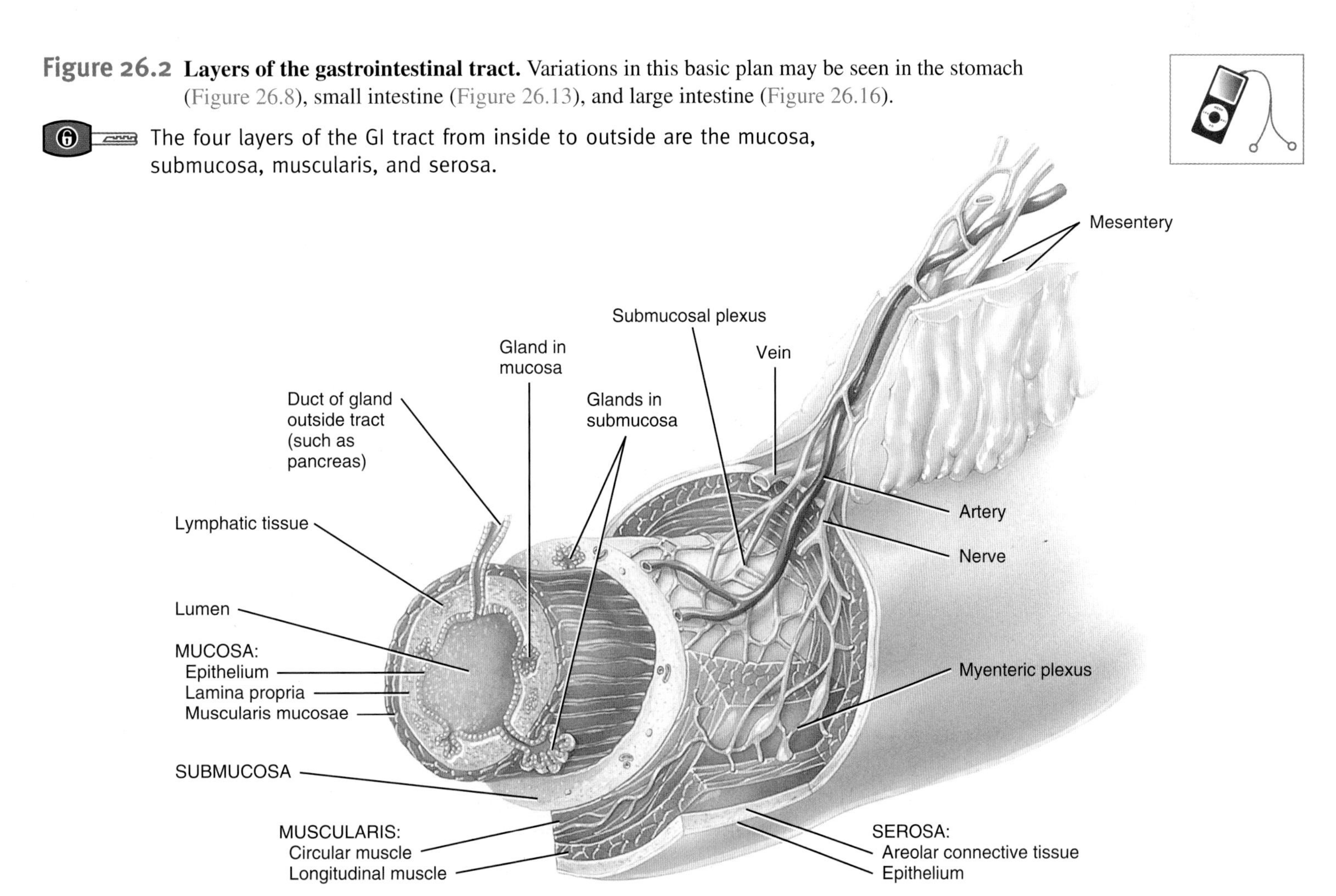

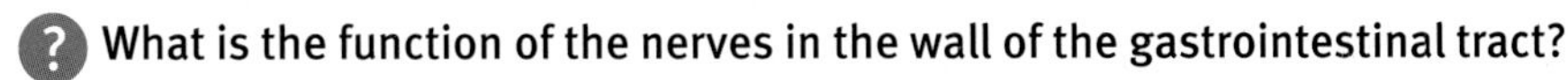

What is the function of the nerves in the wall of the gastrointestinal tract?

per esophagus, it consists in part of *skeletal muscle* that produces voluntary swallowing. Skeletal muscle also forms the external anal sphincter, which permits voluntary control of defecation. Recall that a sphincter is a thick circle of muscle around an opening. In the rest of the tract, the muscularis consists of *smooth muscle,* usually arranged as an inner sheet of circular fibers and an outer sheet of longitudinal fibers. Involuntary contractions of these smooth muscles help break down food physically, mix it with digestive secretions, and propel it along the tract. The ***myenteric plexus,*** ENS neurons located within the muscularis, controls the frequency and strength of its contractions.

4. Serosa and peritoneum. The ***serosa,*** the outermost layer around organs of the GI tract below the diaphragm, is a membrane composed of simple squamous epithelium and areolar connective tissue. The serosa secretes a slippery, watery fluid that allows the tract to glide easily against other organs. The serosa is also called the *visceral peritoneum* (per-i-tō-NĒ-um = to stretch over). Recall from Chapter 4 that the ***peritoneum*** is the largest serous membrane of the body. The *parietal peritoneum* lines the wall of the abdominal cavity; the visceral peritoneum covers organs in the cavity.

As you will see shortly, some organs lie on the posterior abdominal wall and are covered by peritoneum only on their anterior surface; they are not within the peritoneal cavity. Such organs, including the kidneys, ascending and descending colons of the large intestine, duodenum of the small intestine, and pancreas, are said to be ***retroperitoneal*** (*retro-* = behind).

In addition to binding the organs to each other and to the walls of the abdominal cavity, the peritoneal folds contain blood vessels, lymphatic vessels, and nerves that supply the abdominal organs. The ***greater omentum*** (ō-MEN-tum = fat skin) drapes over the transverse colon and small intestine like a "fatty apron" (Figure 26.3a,b). The many lymph nodes of the greater omentum contribute macrophages and antibody-producing plasma cells that help combat and contain infections of the GI tract. The

Figure 26.3 Views of the abdomen and pelvis. The relationship of the folds of the peritoneum (greater omentum, lesser omentum, mesentery, mesocolon, and falciform ligament) to each other and to organs of the digestive system is shown. The size of the peritoneal cavity in (a) is exaggerated for emphasis.

The peritoneum is the largest serous membrane in the body.

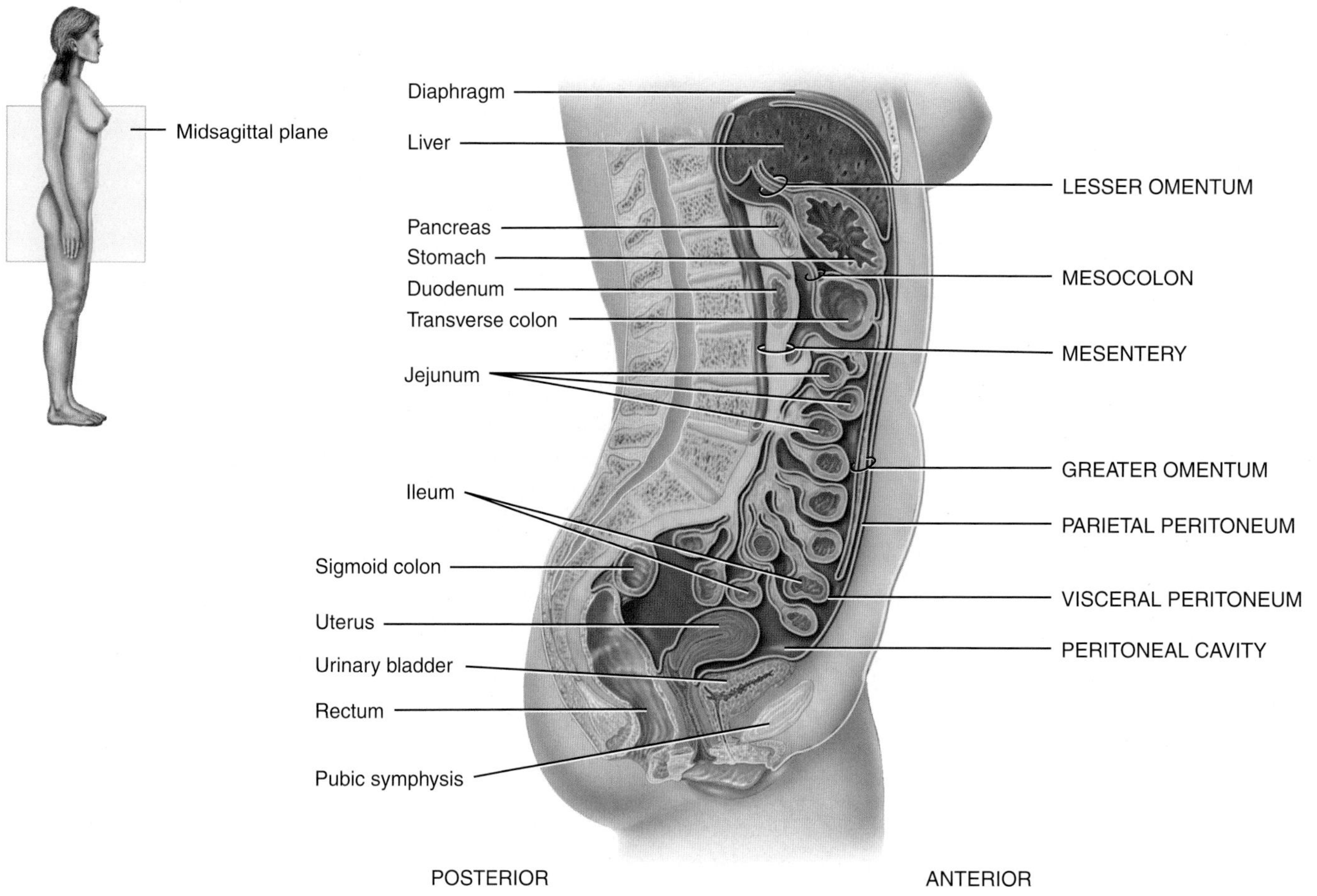

(a) Midsagittal section showing the peritoneal folds

greater omentum normally contains considerable adipose tissue. Its adipose tissue content can greatly expand with weight gain, giving rise to the characteristic "beer belly" seen in some overweight individuals. The ***falciform ligament*** (FAL-si-form; *falc-* = sickle-shaped) attaches the liver to the anterior abdominal wall and diaphragm. A part of the peritoneum, the ***mesentery*** (MEZ-en-ter′-ē; *mes-* = middle), binds the small intestine to the posterior abdominal wall (Figure 26.3a,c). The ***lesser omentum*** suspends the stomach and duodenum from the liver (Figure 26.3a). The ***mesocolon*** binds the transverse colon and sigmoid colon of the large intestine to the posterior abdominal wall.

CHECKPOINT

3. Where along the GI tract is the muscularis composed of skeletal muscle? Is control of this skeletal muscle voluntary or involuntary?
4. Where are the visceral peritoneum and parietal peritoneum located?

26.3 MOUTH

OBJECTIVES

- Identify the locations of the salivary glands, and describe the functions of their secretions.
- Describe the structure and functions of the tongue.
- Identify the parts of a typical tooth, and compare deciduous and permanent dentitions.

The ***mouth*** or ***oral cavity*** is formed by the cheeks, hard and soft palates, and tongue (Figure 26.4). The *cheeks* form the lateral walls of the oral cavity. The *lips* are fleshy folds around the opening of the mouth. Both the cheeks and lips are covered on the outside by skin and on the inside by a mucous membrane. During chewing, the lips and cheeks help keep food between the upper and lower teeth. They also assist in speech.

The *hard palate,* consisting of the maxillae and palatine bones, forms most of the roof of the mouth. The rest is formed by the muscular *soft palate.* Hanging from the soft palate is a projection called the ***uvula*** (Ū-vū-la). During swallowing, the uvula moves upward with the soft palate, which prevents entry of swallowed foods and liquids into the nasal cavity. At the back of the soft palate, the mouth opens into the oropharynx. The *palatine tonsils* are just posterior to the opening.

Tongue

The ***tongue*** forms the floor of the oral cavity. It is an accessory digestive organ composed of skeletal muscle covered with mucous membrane (Figure 26.4).

The muscles of the tongue maneuver food for chewing, shape the food into a rounded mass, force the food to the back of the mouth for swallowing, and alter the shape and size of the tongue for swallowing and speech. The ***lingual frenulum*** (LING-gwal FREN-ū-lum; *lingua* = tongue; *frenum* = bridle), a fold of mucous membrane in the midline of the undersurface of the tongue, limits the movement of the tongue posteriorly (Figure 26.4). The lingual tonsils lie at the base of the tongue (see Figure 19.4). The upper surface and sides of the tongue are covered with projections called *papillae* (pa-PIL-ē), some of which contain taste buds.

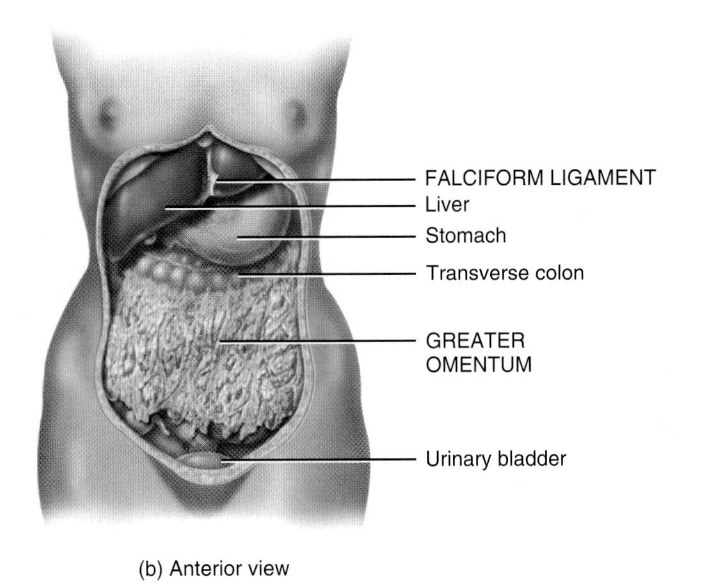

(b) Anterior view

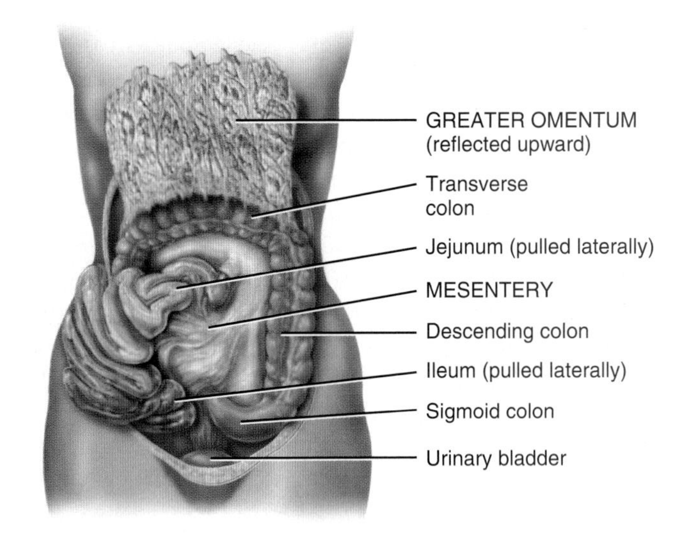

(c) Anterior view (greater omentum lifted and small intestine reflected to right side)

Which peritoneal fold binds the small intestine to the posterior abdominal wall?

Figure 26.4 Structures of the mouth (oral cavity).

The mouth is formed by the cheeks, hard and soft palates, and tongue.

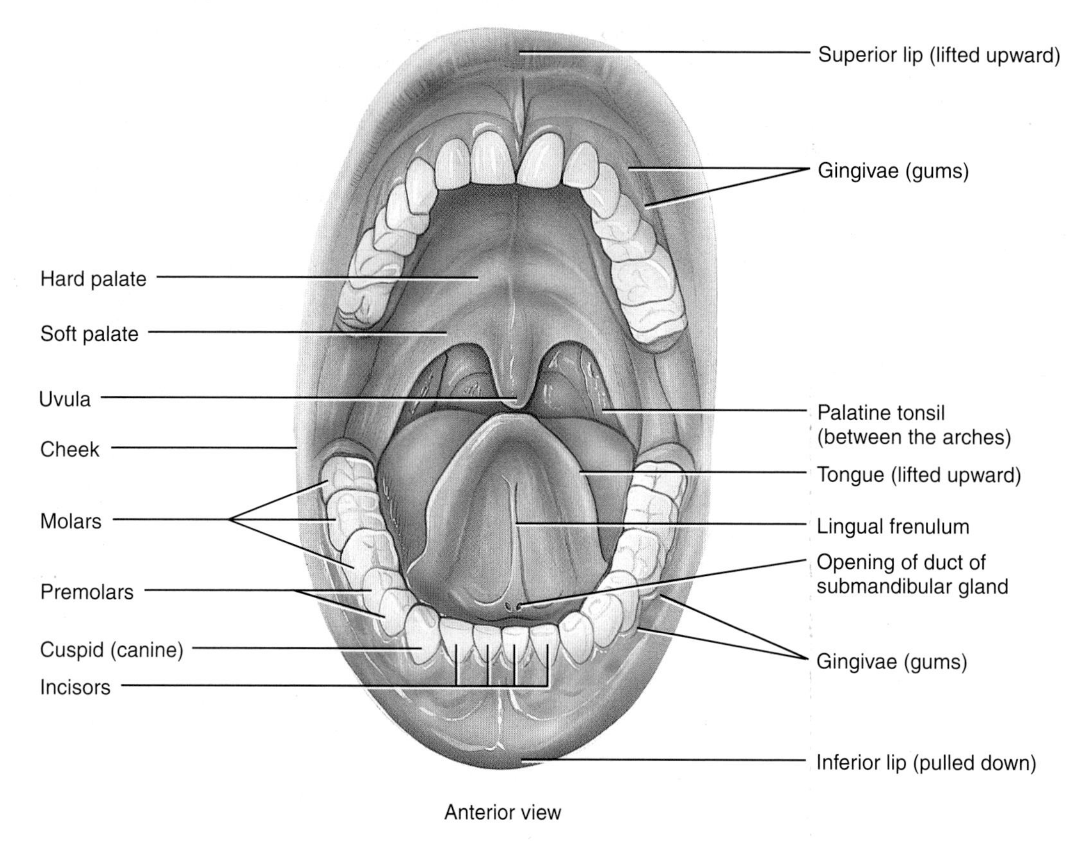

What is the function of the uvula?

Salivary Glands

The three pairs of ***salivary glands*** are accessory organs of digestion that lie outside the mouth and release their secretions into ducts emptying into the oral cavity (see Figure 26.1). The ***parotid glands*** are located inferior and anterior to the ears between the skin and the masseter muscle. The ***submandibular glands*** are found in the floor of the mouth; they are medial and partly inferior to the mandible. The ***sublingual glands*** are beneath the tongue and superior to the submandibular glands.

The fluid secreted by the salivary glands, called ***saliva,*** is composed of 99.5% water and 0.5% solutes. The water in saliva helps dissolve foods so they can be tasted and digestive reactions can begin. One of the solutes, the digestive enzyme ***salivary amylase,*** begins the digestion of starches in the mouth. Mucus in saliva lubricates food so it can easily be swallowed. The enzyme lysozyme kills bacteria, thereby protecting the mouth's mucous membrane from infection and the teeth from decay.

Secretion of saliva, called ***salivation*** (sal-i-VĀ-shun), is controlled by the autonomic nervous system. Normally, parasympathetic stimulation promotes continuous secretion of a moderate amount of saliva, which keeps the mucous membranes moist and lubricates the movements of the tongue and lips during speech. Sympathetic stimulation dominates during stress, resulting in dryness of the mouth.

Teeth

The ***teeth (dentes)*** are accessory digestive organs located in bony sockets of the mandible and maxillae. The sockets are covered by the ***gingivae*** (JIN-ji-vē; singular is *gingiva*) or *gums* and are lined with the ***periodontal ligament*** (*peri-* = around; *-odont-* = tooth). This dense fibrous connective tissue anchors the teeth to bone (Figure 26.5).

A typical tooth has three major external regions: the crown, root, and neck. The ***crown*** is the visible portion above the level of the gums. The ***root*** consists of one to three projections embedded in the socket. The ***neck*** is the junction line of the crown and root, near the gum line.

Internally, ***dentin*** forms the majority of the tooth. Dentin consists of a calcified connective tissue that gives the tooth its basic shape and rigidity. The dentin of the crown is covered by ***enamel*** that consists primarily of calcium phosphate and calcium carbonate. Enamel, the hardest substance in the body and the richest in calcium salts (about 95% of its dry weight), protects the tooth from the wear and tear of chewing. It is also a barrier against acids that easily dissolve the dentin. The dentin of the root is

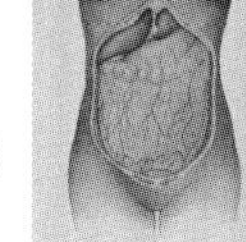

Figure 26.5 **A typical tooth and surrounding structures.**

There are 20 teeth in a complete deciduous set and 32 teeth in a complete permanent set.

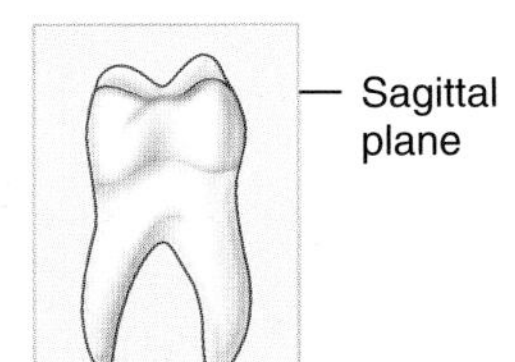

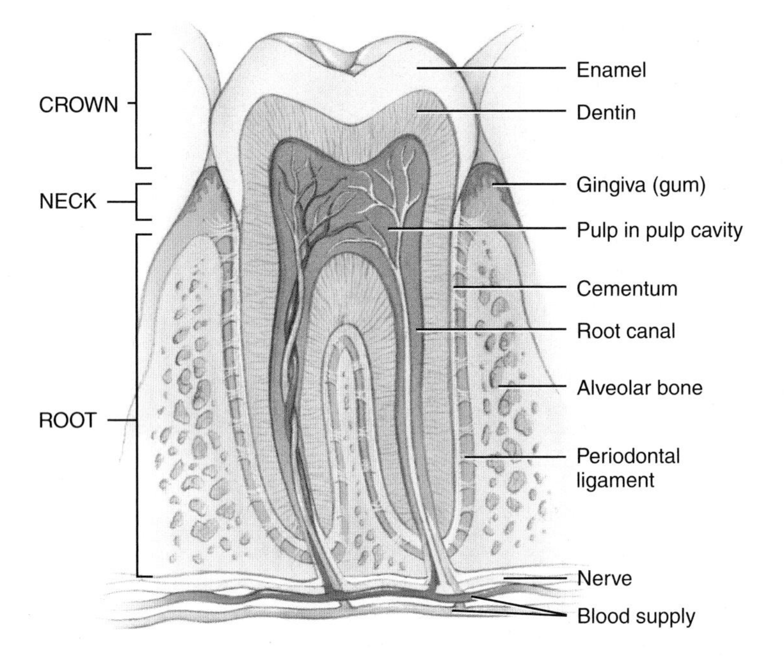

Sagittal section of a mandibular (lower) molar

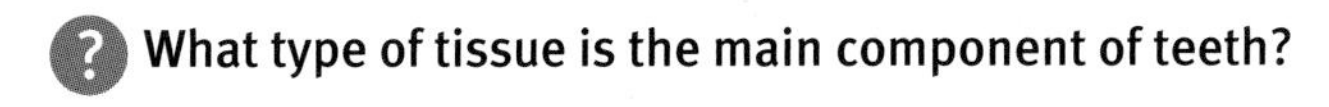

What type of tissue is the main component of teeth?

covered by ***cementum,*** a bonelike substance that attaches the root to the periodontal ligament. The dentin of a tooth encloses the ***pulp cavity,*** a space in the crown filled with ***pulp,*** a connective tissue containing blood vessels, nerves, and lymphatic vessels. Narrow extensions of the pulp cavity run through the root of the tooth and are called ***root canals.*** Each root canal has an opening at its base through which blood vessels bring nourishment, lymphatic vessels offer protection, and nerves provide sensation.

Humans have two sets of teeth. The ***deciduous teeth*** begin to erupt at about 6 months of age, and one pair appears about each month thereafter until all 20 are present. They are generally lost in the same sequence between 6 and 12 years of age. The ***permanent teeth*** appear between age 6 and adulthood. There are 32 teeth in a complete permanent set.

Humans have different teeth for different functions (see Figure 26.4). ***Incisors*** are closest to the midline, are chisel-shaped, and are adapted for cutting into food; ***cuspids (canines)*** are next to the incisors and have one pointed surface (cusp) to tear and shred food; ***premolars*** have two cusps to crush and grind food; and ***molars*** have three or more blunt cusps to crush and grind food.

Digestion in the Mouth

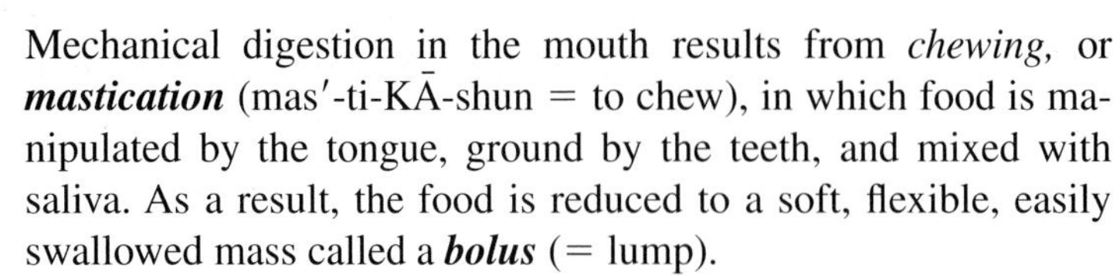

Mechanical digestion in the mouth results from *chewing,* or ***mastication*** (mas′-ti-KĀ-shun = to chew), in which food is manipulated by the tongue, ground by the teeth, and mixed with saliva. As a result, the food is reduced to a soft, flexible, easily swallowed mass called a ***bolus*** (= lump).

Dietary carbohydrates are either monosaccharide and disaccharide sugars or complex polysaccharides such as glycogen and starches (see Section 2.2). Most of the carbohydrates we eat are starches from plant sources, but only monosaccharides (glucose, fructose, and galactose) can be absorbed into the bloodstream. Thus, ingested starches must be broken down into monosaccharides. Salivary amylase begins the breakdown of starch by breaking particular chemical bonds between the glucose subunits. The resulting products include the disaccharide maltose (2 glucose subunits), the trisaccharide maltotriose (3 glucose subunits), and larger fragments called dextrins (5 to 10 glucose subunits). Salivary amylase in the swallowed food continues to act for about an hour until it is inactivated by stomach acids.

CHECKPOINT

5. What structures form the mouth (oral cavity)?
6. How is saliva secretion regulated?
7. What is a bolus? How is it formed?

26.4 PHARYNX AND ESOPHAGUS

OBJECTIVE

- Describe the location, structure, and functions of the pharynx and esophagus.

When food is swallowed, it passes from the mouth into the ***pharynx*** (FAR-inks), a funnel-shaped tube that is composed of skeletal muscle and lined by mucous membrane. It extends from the internal nares to the esophagus posteriorly and the larynx anteriorly (see Figure 26.6a). The nasopharynx is involved in respiration (see Figure 25.2); food that is swallowed passes from the mouth into the oropharynx and laryngopharynx before passing into the esophagus. Muscular contractions of the oropharynx and laryngopharynx help propel food into the esophagus.

The ***esophagus*** (e-SOF-a-gus = eating gullet) is a muscular tube lined with stratified squamous epithelium that lies posterior to the trachea. It begins at the end of the laryngopharynx, passes through the mediastinum and diaphragm, and connects to the superior aspect of the stomach. It transports food to the stomach and secretes mucus. At each end of the esophagus, the muscularis forms two sphincters—the ***upper esophageal sphincter (UES)*** (e-sof-a-JĒ-al), which consists of *skeletal muscle,* and the ***lower esophageal sphincter (LES),*** which consists of *smooth muscle.* The upper esophageal sphincter regulates the movement of food from the pharynx into the esophagus; the lower esophageal sphincter regulates the movement of food from the esophagus into the stomach.

Deglutition (dē′-glū-TISH-un), or *swallowing,* the movement of food from the mouth to the stomach, involves the mouth, pharynx,

Figure 26.6 **Deglutition (swallowing).** During the pharyngeal stage of swallowing (b), the tongue rises against the palate, the nasopharynx is closed off, the larynx rises, the epiglottis seals off the larynx, and the bolus passes into the esophagus. During the esophageal stage of swallowing (c), food moves through the esophagus into the stomach via peristalsis.

Deglutition (swallowing) is a mechanism that moves food from the mouth into the stomach.

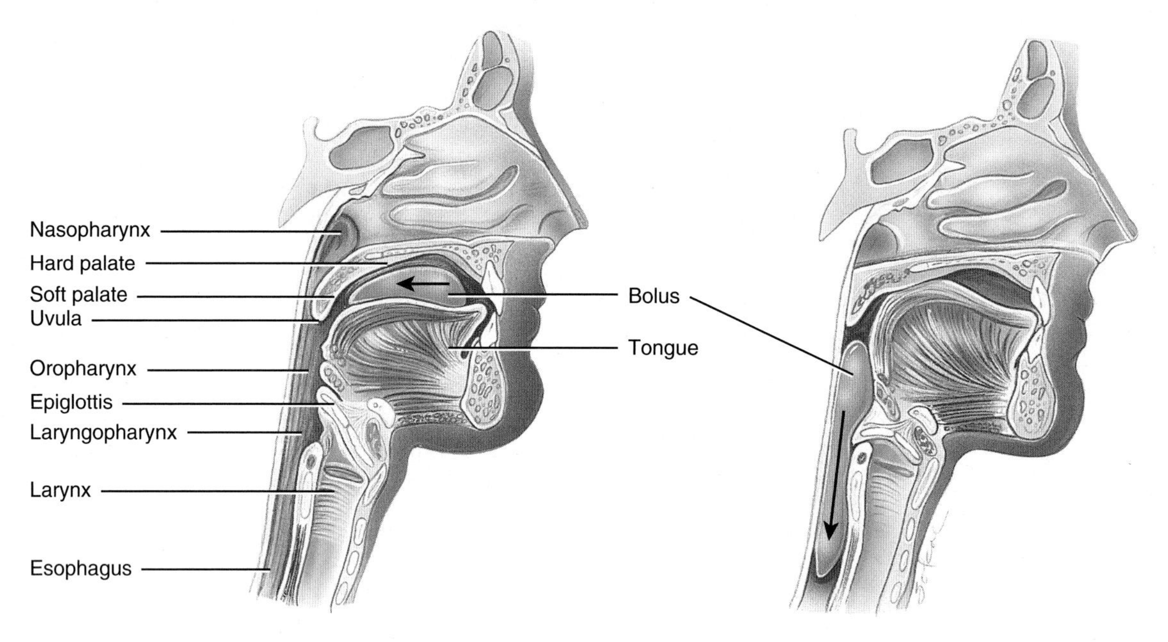

(a) Position of structures before swallowing

(b) During the pharyngeal stage of swallowing

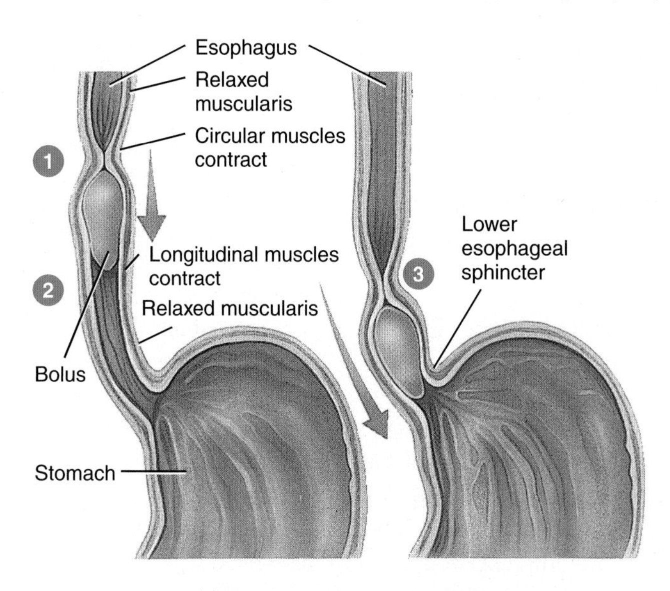

(c) Esophageal stage of swallowing

? Is swallowing a voluntary or an involuntary action?

and esophagus and is helped by saliva and mucus. Swallowing is divided into three stages: the voluntary, pharyngeal, and esophageal stages.

In the ***voluntary stage*** of swallowing, the bolus is forced to the back of the mouth cavity and into the oropharynx by the movement of the tongue upward and backward against the palate. With the passage of the bolus into the oropharynx, the involuntary ***pharyngeal stage*** of swallowing begins (Figure 26.6b). Breathing is temporarily interrupted when the soft palate and uvula move upward to close off the nasopharynx, the epiglottis seals off the larynx, and the vocal cords come together. After the bolus passes through the oropharynx, the respiratory passageways reopen and breathing resumes. Once the upper esophageal sphincter relaxes, the bolus moves into the esophagus.

In the ***esophageal stage,*** food is pushed through the esophagus by a process called ***peristalsis*** (Figure 26.6c):

1. The circular muscle fibers in the section of esophagus above the bolus contract, constricting the wall of the esophagus and squeezing the bolus downward.
2. Longitudinal muscle fibers around the bottom of the bolus contract, shortening the section of the esophagus below the bolus and pushing its walls outward.
3. After the bolus moves into the new section of the esophagus, the circular muscles above it contract, and the cycle repeats. The contractions move the bolus down the esophagus toward the stomach. As the bolus approaches the end of the esophagus, the lower esophageal sphincter relaxes and the bolus moves into the stomach.

• CLINICAL CONNECTION | Heartburn

Sometimes, after food has entered the stomach, the lower esophageal sphincter fails to close adequately and the stomach contents can back up (reflux) into the lower esophagus, a condition known as *gastroesophageal reflux disease (GERD).* Reflux of acid from the stomach can irritate the esophageal wall, causing a burning sensation known as **heartburn.** Although it is experienced in a region very near the heart, heartburn is unrelated to any cardiac problem. GERD also may increase the risk of esophageal cancer. •

CHECKPOINT

8. How does a bolus pass from the mouth into the stomach?

26.5 STOMACH

OBJECTIVE

- Describe the location, structure, and functions of the stomach.

The ***stomach*** is a J-shaped enlargement of the GI tract directly below the diaphragm. The stomach connects the esophagus to the duodenum, the first part of the small intestine (Figure 26.7). Because a meal can be eaten much more quickly than the intestines can digest and absorb it, one of the functions of the stomach is to serve as a mixing chamber and holding reservoir. At appropriate intervals after food is ingested, the stomach forces a small quantity of material into the duodenum. The position and size of the stomach vary continually; the diaphragm pushes it inferiorly with each inhalation and pulls it superiorly with each exhalation. The stomach is the most elastic part of the GI tract and accommodates a large quantity of food, up to about 6.4 liters (6 qt.).

Figure 26.7 External and internal anatomy of the stomach. The dashed lines indicate the approximate borders of the regions of the stomach.

The four regions of the stomach are the cardia, fundus, body, and pylorus.

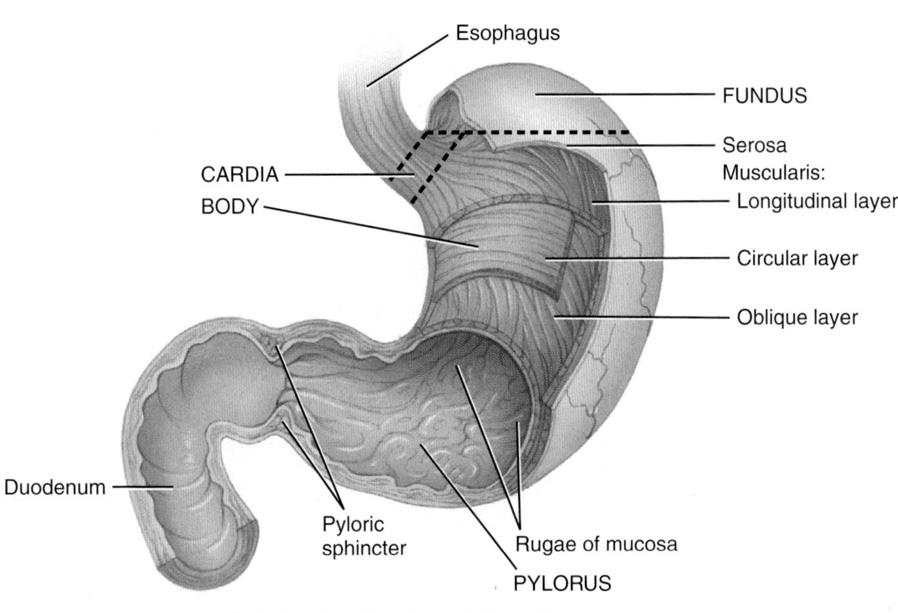

Functions of the Stomach

1. Mixes saliva, food, and gastric juice to form chyme.
2. Serves as a reservoir for food before release into small intestine.
3. Secretes gastric juice, which contains HCl (kills bacteria and denatures protein), pepsin (begins the digestion of proteins), intrinsic factor (aids absorption of vitamin B_{12}), and gastric lipase (aids digestion of triglycerides).
4. Secretes gastrin into blood.

? After a very large meal, does your stomach still have rugae?

Structure of the Stomach

The stomach has four main regions: cardia, fundus, body, and pylorus (see Figure 26.7). The ***cardia*** (CAR-dē-a) surrounds the superior opening of the stomach. The stomach then curves upward. The portion superior and to the left of the cardia is the ***fundus*** (FUN-dus). Inferior to the fundus is the large central portion of the stomach, called the ***body.*** The narrow, most inferior region is the ***pylorus*** (pī-LOR-us; *pyl-* = gate; *-orus* = guard). Between the pylorus and duodenum is the ***pyloric sphincter.***

The stomach wall is composed of the same four basic layers as the rest of the GI tract (mucosa, submucosa, muscularis, serosa), with certain differences (Figure 26.8a). When the stomach is empty, the mucosa lies in large folds, called ***rugae*** (ROO-jē = wrinkles). The surface of the mucosa is a layer of nonciliated simple columnar epithelial cells called ***surface mucous cells*** (Figure 26.8b). Epithelial cells also extend downward and form columns of secretory cells called ***gastric glands*** that line narrow channels called ***gastric pits.*** Secretions from the gastric glands flow into the gastric pits and then into the lumen of the stomach.

The gastric glands contain three types of *exocrine gland cells* that secrete their products into the stomach lumen: mucous neck cells, chief cells, and parietal cells (Figure 26.8b). Both surface mucous cells and ***mucous neck cells*** secrete mucus. The ***chief cells*** secrete an inactive gastric enzyme called *pepsinogen.* ***Parietal cells*** produce *hydrochloric acid,* which kills many microbes in food and helps convert pepsinogen to the active digestive enzyme ***pepsin.*** Parietal cells also secrete *intrinsic factor,* which is involved in the absorption of vitamin B_{12}. Inadequate production of intrinsic factor can result in pernicious anemia (see Chapter 21, Section 21.2) because vitamin B_{12} is needed for red blood cell production. The secretions of the mucous, chief, and parietal cells are collectively called ***gastric juice.*** The ***G cells,*** a fourth type of cell in the gastric glands, secrete the hormone *gastrin* into the bloodstream.

• CLINICAL CONNECTION | Peptic Ulcer Disease

Five to ten percent of the U.S. population develops **peptic ulcer disease (PUD)** each year. An *ulcer* is a craterlike lesion in a membrane; ulcers that develop in areas of the GI tract exposed to acidic gastric juice are called *peptic ulcers.* The most common complication of peptic ulcers is bleeding, which can lead to anemia. In acute cases, peptic ulcers can lead to shock and death. Three distinct causes of PUD are recognized: (1) the bacterium *Helicobacter pylori,* (2) nonsteroidal anti-inflammatory drugs (NSAIDs) such as aspirin, and (3) hypersecretion of HCl. •

Figure 26.8 Layers of the stomach.

Secretions from the gastric glands flow into the gastric pits and then into the lumen of the stomach.

Lumen of stomach
Gastric pits
Epithelium
Lamina propria
Gastric gland
Lymphatic nodule
Muscularis mucosae
Lymphatic vessel
Venule
Arteriole
Oblique layer of muscle
Circular layer of muscle
Myenteric plexus
Longitudinal layer of muscle
MUCOSA
SUBMUCOSA
MUSCULARIS
SEROSA

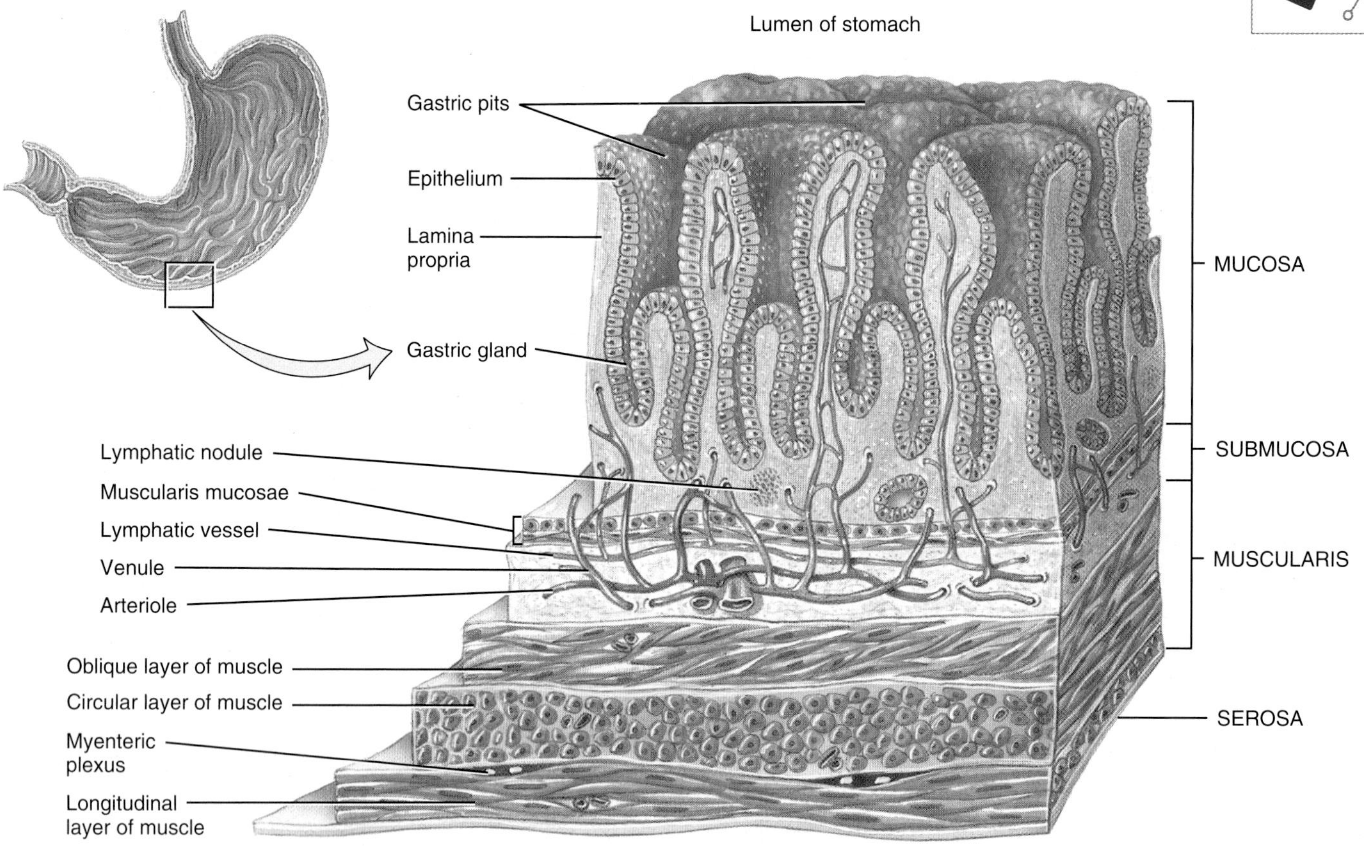

(a) Three-dimensional view of layers of the stomach

The submucosa of the stomach is composed of areolar connective tissue that connects the mucosa to the muscularis. The muscularis has three rather than two layers of smooth muscle: an outer longitudinal layer, a middle circular layer, and an inner oblique layer (see Figure 26.7). The serosa covering the stomach, composed of simple squamous epithelium and areolar connective tissue, is part of the visceral peritoneum.

Digestion and Absorption in the Stomach

Once food reaches the stomach, the stomach wall is stretched and the pH of the stomach contents increases because proteins in food have buffered some of the stomach acid. These changes in the stomach trigger nerve impulses that stimulate the flow of gastric juice and initiate ***mixing waves,*** gentle, rippling peristaltic movements of the muscularis. These waves macerate food and mix it with the secretions of the gastric glands, producing ***chyme*** (KĪM = juice), a thick liquid with the consistency of pea soup. Each mixing wave forces a small amount of chyme through the partially closed pyloric sphincter into the duodenum, a process called ***gastric emptying.*** Most of the chyme is forced back into the body of the stomach. The next mixing wave pushes chyme forward again and forces a little more into the duodenum. After the stomach has emptied some of its contents into the duodenum, reflexes begin to slow the exit of chyme from the stomach. This prevents overloading of the duodenum with more chyme than it can handle. Foods rich in carbohydrates spend the least time in the stomach; high-protein foods remain somewhat longer, and gastric emptying is slowest after a meal containing large amounts of fat.

The main event of chemical digestion in the stomach is the beginning of protein digestion by the enzyme pepsin, which breaks peptide bonds between the amino acids of proteins. As a result, the proteins become fragmented into *peptides,* smaller strings of amino acids. Pepsin is most effective in the very acidic environment of the stomach, which has a pH of 2. What keeps pepsin from digesting the protein in stomach cells along with the food? First, recall that chief cells secrete pepsin in an inactive form (pepsinogen). It is not converted into active pepsin until it contacts hydrochloric acid in gastric juice. Second, mucus secreted by mucous cells coats the mucosa, forming a thick barrier between the cells of the stomach lining and the gastric juice.

The epithelial cells of the stomach are impermeable to most materials, so little absorption occurs. However, mucous cells of the stomach absorb some water, ions, and short-chain fatty acids, as well as certain drugs (especially aspirin) and alcohol.

CHECKPOINT

9. What are the components of gastric juice?
10. What is the role of pepsin? Why is it secreted in an inactive form?
11. What substances are absorbed in the stomach?

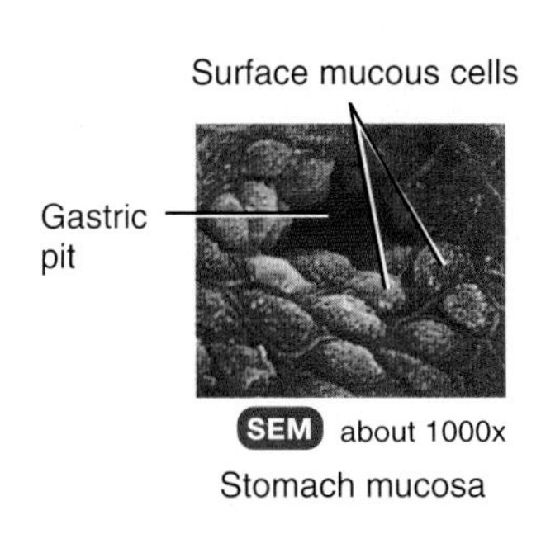

Stomach mucosa

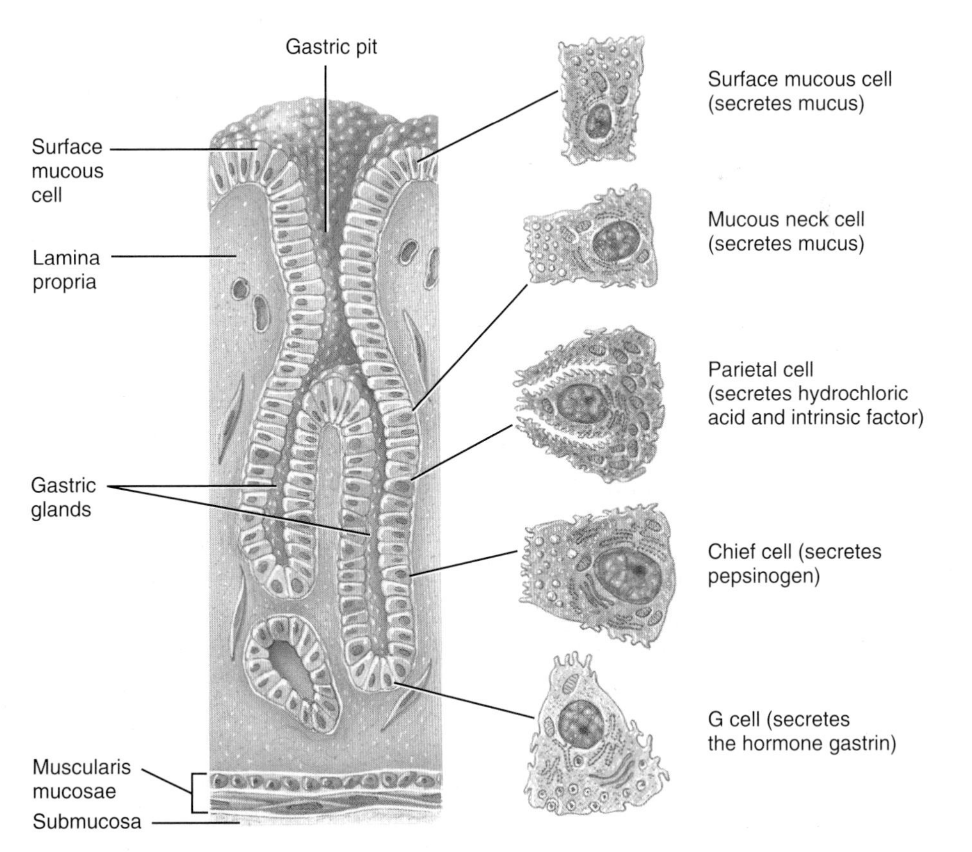

(b) Sectional view of the stomach mucosa showing gastric glands and cell types

? Which stomach layer is in contact with swallowed food?

26.6 PANCREAS

OBJECTIVE

- Describe the location, structure, and functions of the pancreas.

From the stomach, chyme passes into the small intestine. Because chemical digestion in the small intestine depends on activities of the pancreas, liver, and gallbladder, we first consider these accessory digestive organs and their contributions to digestion in the small intestine.

Structure of the Pancreas

The ***pancreas*** (*pan-* = all; *-creas* = flesh) lies posterior and inferior to the stomach (see Figure 26.1). Secretions pass from the pancreas to the duodenum via the ***pancreatic duct,*** which unites with the common bile duct from the liver and gallbladder and enters the duodenum as a dilated common duct called the ***hepatopancreatic ampulla.*** The passage of pancreatic juice and bile through the hepatopancreatic ampulla into the small intestine is regulated by a ring of smooth muscle known as the ***sphincter of the hepatopancreatic ampulla (sphincter of Oddi)*** (Figure 26.9a,b).

The pancreas is made up of small clusters of glandular epithelial cells, most of which are arranged in clusters called ***acini*** (AS-i-nī). The acini constitute the *exocrine* portion of the organ (see Figure 20.9, Section 20.6). The cells within acini secrete a mixture of fluid and digestive enzymes called ***pancreatic juice.*** The remaining 1% of the cells are organized into clusters called ***pancreatic islets (islets of Langerhans),*** the *endocrine* portion of the pancreas. These cells secrete the hormones glucagon, insulin, somatostatin, and pancreatic polypeptide, which are discussed in Chapter 20.

Pancreatic Juice

Pancreatic juice is a clear, colorless liquid that consists mostly of water, some salts, sodium bicarbonate, and enzymes. The bicarbonate ions give pancreatic juice a slightly alkaline pH (7.1 to 8.2), which inactivates pepsin from the stomach and creates the optimal environment for activity of enzymes in the small intestine. The enzymes in pancreatic juice include a starch-digesting enzyme called ***pancreatic amylase;*** several protein-digesting enzymes including ***trypsin*** (TRIP-sin), ***chymotrypsin*** (kī-mō-TRIP-sin), and ***carboxypeptidase*** (kar-bok-sē-PEP-ti-dās); the main triglyceride-digesting enzyme in adults, called ***pancreatic lipase;*** and nucleic acid–digesting enzymes called ***ribonuclease*** and ***deoxyribonuclease.*** The protein-digesting enzymes are produced in an inactive form, which prevents them from digesting the pancreas itself. Upon reaching the small intestine, the inactive form of trypsin is activated by an enzyme called ***enterokinase.*** In turn, trypsin activates the other protein-digesting pancreatic enzymes.

• CLINICAL CONNECTION | Pancreatic Cancer

Pancreatic cancer usually affects people over 50 years of age and occurs more frequently in males. Typically, there are few symptoms until the disorder reaches an advanced stage and often not until it has metastasized to other parts of the body such as the lymph nodes, liver, or lungs. The disease is nearly always fatal and is the fourth most common cause of death from cancer in the United States. Pancreatic cancer has been linked to fatty foods, high alcohol consumption, genetic factors, smoking, and chronic *pancreatitis* (inflammation of the pancreas). •

CHECKPOINT

12. What are the pancreatic acini? How do their functions differ from those of the pancreatic islets?
13. What is the role of pancreatic lipase?

26.7 LIVER AND GALLBLADDER

OBJECTIVE

- Describe the location, structure, and functions of the liver and gallbladder.

In an average adult, the ***liver*** weighs 1.4 kg (about 3 lb) and, after the skin, is the second largest organ of the body. It is located below the diaphragm, mostly on the right side of the body. It has two principal lobes, each divided into lobules. A connective tissue capsule covers the liver, which in turn is covered by peritoneum, the serous membrane that covers all the viscera. The ***gallbladder*** (*gall-* = bile) is a pear-shaped sac that hangs inferiorly from the posterior surface of the liver (Figure 26.9a).

Structure of the Liver and Gallbladder

Microscopically, the liver consists of several components (see Figure 26.10):

1. **Hepatocytes** (*hepat-* = liver; *-cytes* = cells). These are the major functional cells of the liver that perform metabolic, secretory, and endocrine functions. The hepatocytes are arranged in rows called ***hepatic laminae.***
2. **Bile canaliculi** (kan-a-LIK-ū-lī = small canals). These are small ducts between hepatocytes that collect bile produced by the hepatocytes. From bile canaliculi, bile passes into ***bile ducts.*** The bile ducts merge and eventually form the larger ***right*** and ***left hepatic ducts,*** which unite and exit the liver as the ***common hepatic duct*** (Figure 26.9). The common hepatic duct joins the ***cystic duct*** (*cystic* = bladder) from the gallbladder to form the ***common bile duct.*** From here, bile enters the small intestine to participate in digestion. When the small intestine is empty, the sphincter around the common duct at the entrance to the duodenum closes, and bile backs up into the cystic duct to the gallbladder for storage.
3. **Hepatic sinusoids.** These are highly permeable blood capillaries between hepatic laminae that receive oxygenated blood from branches of the hepatic artery and nutrient-rich deoxygenated blood from branches of the hepatic portal vein. Recall

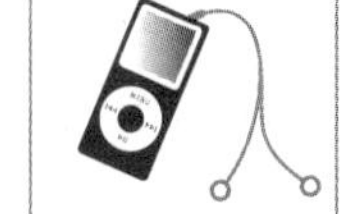

Figure 26.9 Relation of the pancreas to the liver, gallbladder, and duodenum. The inset (b) shows details of the common bile duct and pancreatic duct forming the hepatopancreatic ampulla and emptying into the duodenum.

Pancreatic enzymes digest starches (polysaccharides), proteins, triglycerides, and nucleic acids.

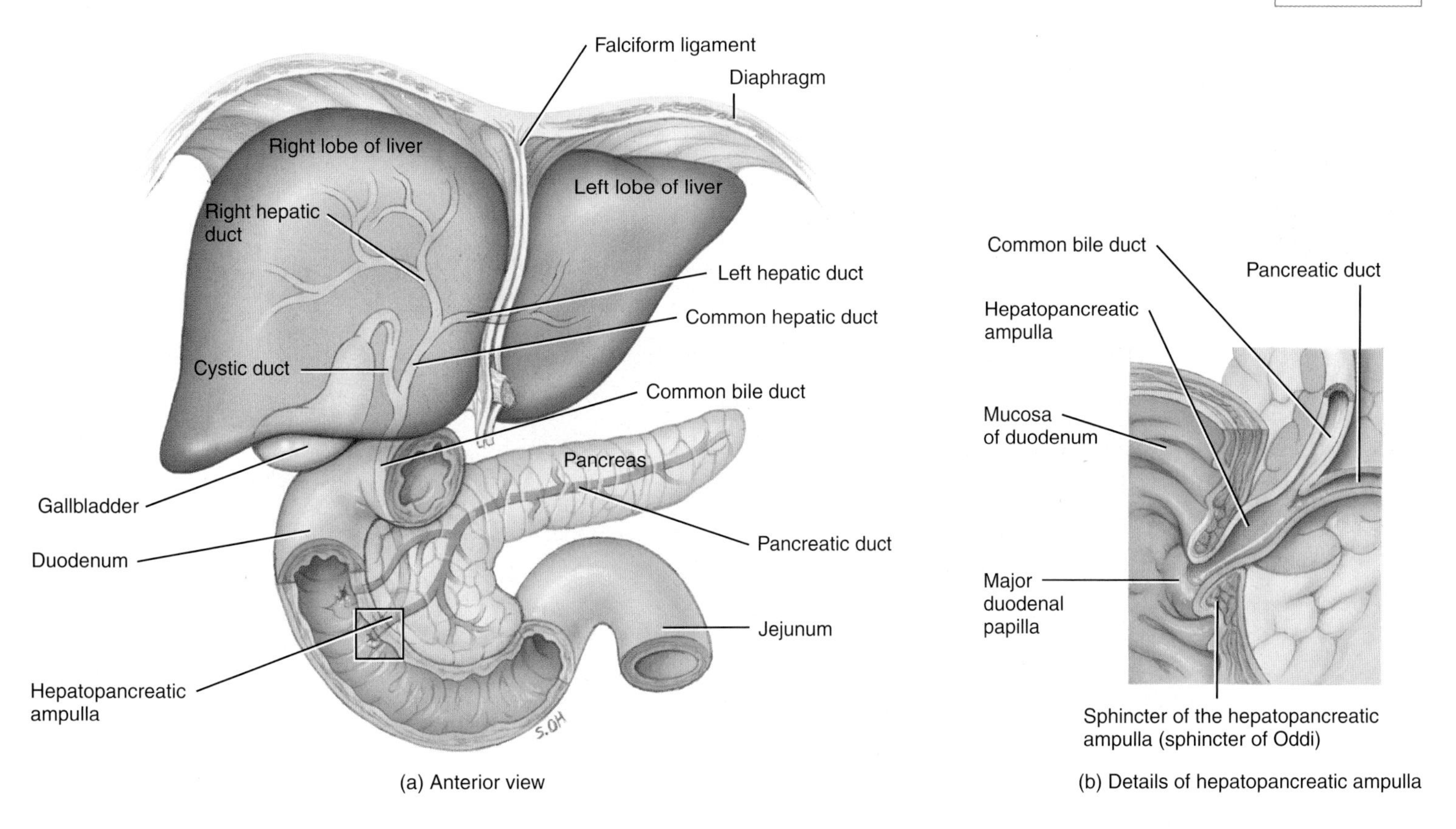

(a) Anterior view

(b) Details of hepatopancreatic ampulla

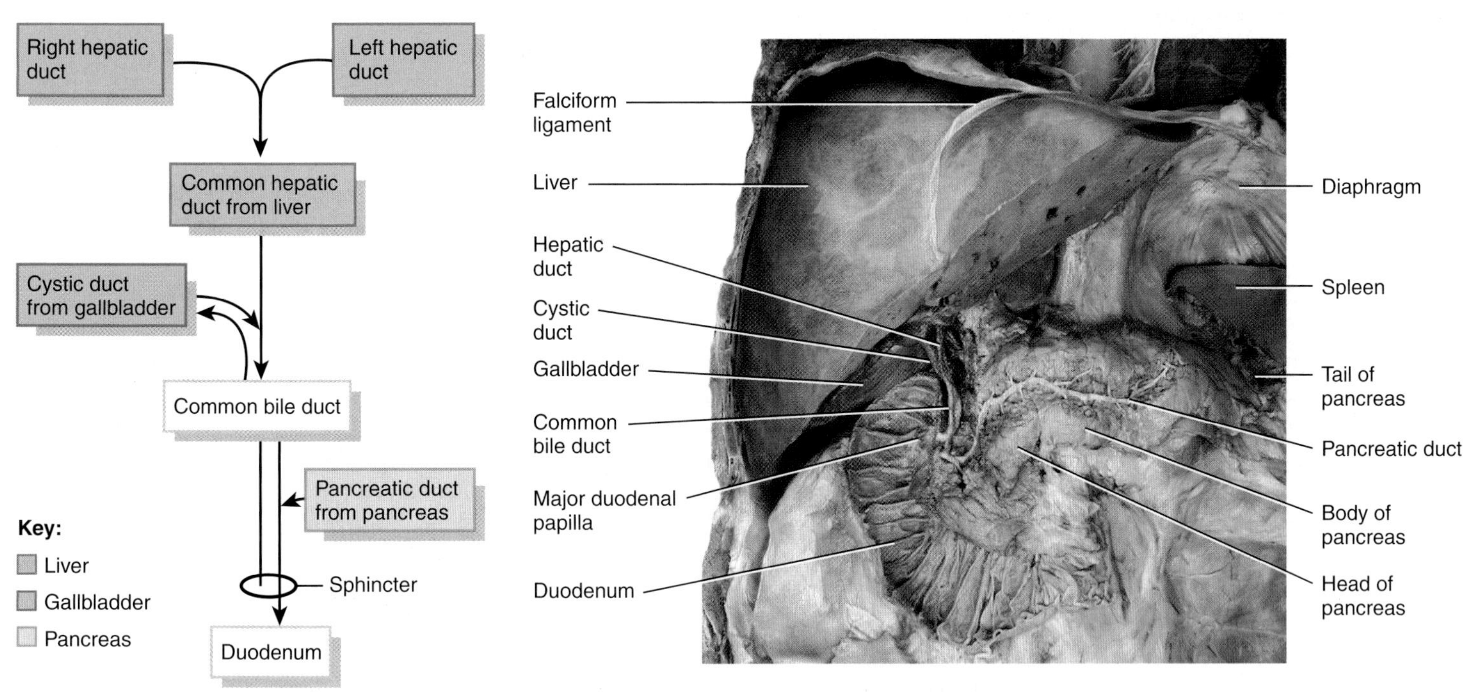

(c) Ducts carrying bile from liver and gallbladder and pancreatic juice from pancreas to the duodenum

(d) Anterior view

What substances are present in pancreatic juice?

Figure 26.10 Structure of the liver.

A liver lobule consists of hepatocytes arranged around a central vein.

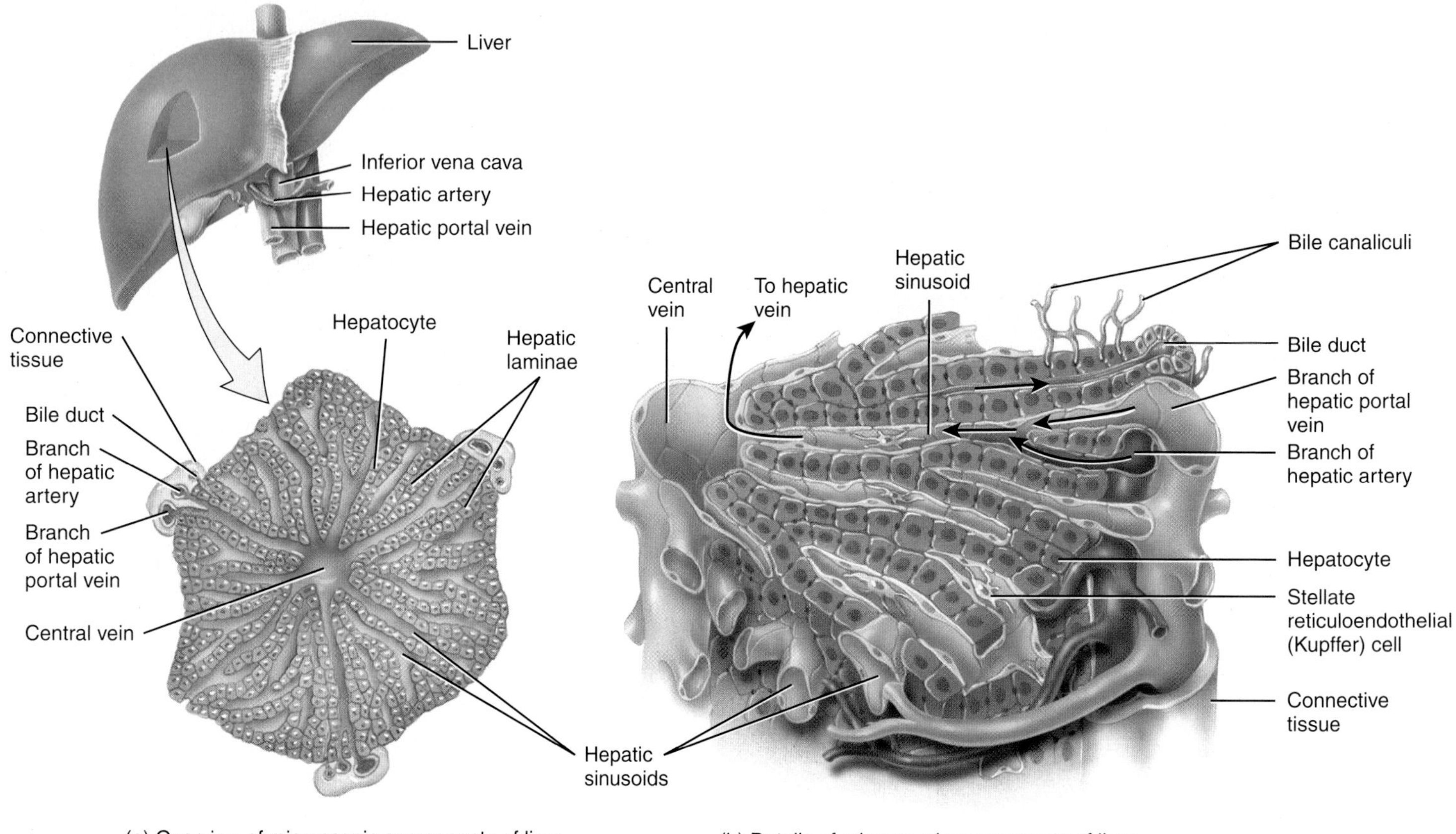

(a) Overview of microscopic components of liver

(b) Details of microscopic components of liver

Which cells in the liver are phagocytes?

that the hepatic portal vein brings venous blood from the gastrointestinal organs in to the liver. Hepatic sinusoids converge and deliver blood into a ***central vein.*** From central veins the blood flows into the ***hepatic veins,*** which drain into the inferior vena cava (see Figure 23.16, Section 23.4). Also present in the hepatic sinusoids are fixed phagocytes called ***stellate reticuloendothelial (Kupffer) cells,*** which destroy worn-out white and red blood cells, bacteria, and other foreign matter in the venous blood draining from the gastrointestinal tract.

Blood Supply of the Liver

The liver receives blood from two sources (Figure 26.11). From the hepatic artery it obtains oxygenated blood, and from the hepatic portal vein it receives deoxygenated blood containing newly absorbed nutrients, drugs, and possibly microbes and toxins from the gastrointestinal tract (see Figure 23.16, Section 23.4). Branches of both the hepatic artery and the hepatic portal vein carry blood into liver sinusoids, where oxygen, most of the nutrients, and certain toxic substances are taken up by the hepatocytes. Products manufactured by the hepatocytes and nutrients needed by other cells are secreted back into the blood, which then drains into the central vein and eventually passes into a hepatic vein. Because blood from the gastrointestinal tract passes through the liver as part of the hepatic portal circulation, the liver is often a site for metastasis of cancer that originates in the GI tract.

Bile

Bile salts in bile aid in ***emulsification,*** the breakdown of large lipid globules into a suspension of small lipid globules, and in absorption of lipids following their digestion. The small lipid globules formed as a result of emulsification present a very large surface area so that pancreatic lipase can digest them rapidly. The principal bile pigment is ***bilirubin,*** which is derived from heme. When worn-out red blood cells are broken down, iron, globin, and bilirubin are released. The iron and globin are recycled, but some of the bilirubin is excreted in bile. Bilirubin eventually is broken down in the intestine, and one of its breakdown products (stercobilin) gives feces their normal brown color (see Figure 21.4, Section 21.2). After they have served as emulsifying agents, most bile salts are reabsorbed by active transport in the final portion of the small intestine (ileum) and enter portal blood flowing toward the liver.

Figure 26.11 Hepatic blood flow: sources, path through the liver, and return to the heart.

The liver receives oxygenated blood via the hepatic artery and nutrient-rich deoxygenated blood via the hepatic portal vein.

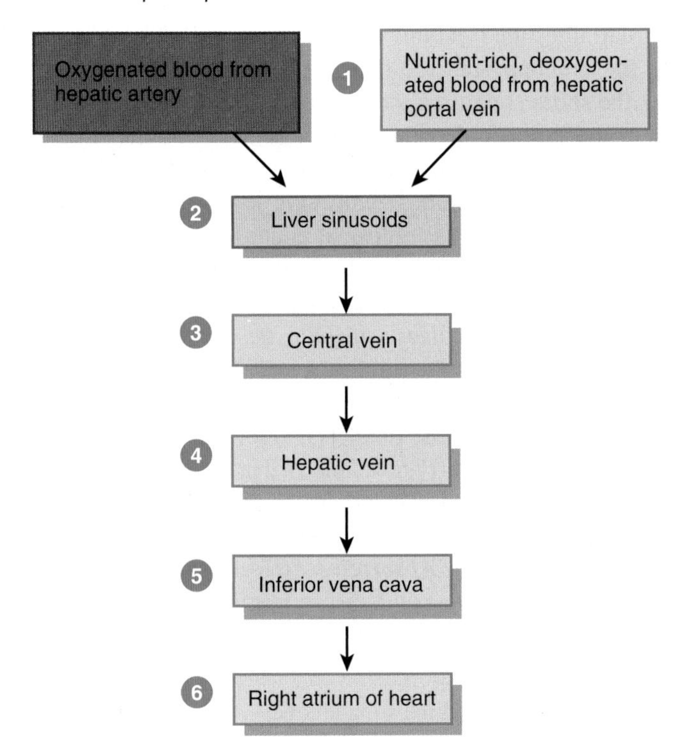

? During the first few hours after a meal, how does the chemical composition of blood change as it flows through the liver sinusoids?

• CLINICAL CONNECTION Gallstones

The components of bile sometimes crystallize and form **gallstones.** As they grow in size and number, gallstones may cause intermittent or complete obstruction to the flow of bile from the gallbladder into the duodenum. Treatment consists of using gallstone-dissolving drugs or *lithotripsy* (lith′-ō-TRIP-sē), a shock-wave therapy that smashes the gallstones into particles small enough to pass through the ducts. For people with recurrent gallstones or for whom drugs or lithotripsy is not indicated, *cholecystectomy* (ko′-lē-sis-TEK-tō-mē)—the removal of the gallbladder and its contents—is necessary. •

Functions of the Liver

The liver performs many other vital functions in addition to the secretion of bile and bile salts and the phagocytosis of bacteria and dead or foreign material by the stellate reticuloendothelial (Kupffer) cells. Many of these functions are related to metabolism and are discussed in Chapter 27. Briefly, however, other functions in the liver include the following:

- **Carbohydrate metabolism.** The liver is especially important in maintaining a normal blood glucose level. When blood glucose is low, the liver can break down glycogen to glucose and release glucose into the bloodstream. The liver can also convert amino acids and lactic acid to glucose, and it can convert other sugars, such as fructose and galactose, into glucose. When blood glucose is high, as occurs just after eating a meal, the liver converts glucose to glycogen and triglycerides for storage.
- **Lipid metabolism.** Hepatocytes store some triglycerides; break down fatty acids to generate ATP; synthesize lipoproteins, which transport fatty acids, triglycerides, and cholesterol to and from body cells; synthesize cholesterol; and use cholesterol to make bile salts.
- **Protein metabolism.** Hepatocytes remove the amino group (—NH_2) from amino acids so that the amino acids can be used for ATP production or converted to carbohydrates or fats. They also convert the resulting toxic ammonia (NH_3) into the much less toxic urea, which is excreted in urine. Hepatocytes also synthesize most plasma proteins, such as globulins, albumin, prothrombin, and fibrinogen.
- **Processing of drugs and hormones.** The liver can detoxify substances such as alcohol or secrete drugs such as penicillin, erythromycin, and sulfonamides into bile. It can also inactivate thyroid hormones and steroid hormones such as estrogens and aldosterone.
- **Excretion of bilirubin.** Bilirubin, derived from the heme of aged red blood cells, is absorbed by the liver from the blood and secreted into bile. Most of the bilirubin in bile is metabolized in the small intestine by bacteria and eliminated in feces.
- **Storage of vitamins and minerals.** In addition to storing glycogen, the liver stores certain vitamins (A, D, E, and K) and minerals (iron and copper), which are released from the liver when needed elsewhere in the body.
- **Activation of vitamin D.** The skin, liver, and kidneys participate in synthesizing the active form of vitamin D.

• CLINICAL CONNECTION Hepatitis

Hepatitis is an inflammation of the liver caused by viruses, drugs, and chemicals, including alcohol.

Hepatitis A (infectious hepatitis), caused by the hepatitis A virus, is spread by fecal contamination of food, clothing, toys, eating utensils, and so forth (fecal–oral route). It does not cause lasting liver damage.

Hepatitis B, caused by the hepatitis B virus, is spread primarily by sexual contact and contaminated syringes and transfusion equipment. It can also be spread by any secretion via saliva and tears. Hepatitis B can produce chronic liver inflammation. Vaccines are available for hepatitis B and are required for certain individuals, such as health-care providers.

Hepatitis C, caused by the hepatitis C virus, is clinically similar to hepatitis B. It is often spread by blood transfusions and can cause cirrhosis and liver cancer.

Hepatitis D is caused by the hepatitis D virus. It is transmitted like hepatitis B. A person must be infected with hepatitis B to contract hepatitis D. Hepatitis D results in severe liver damage and has a fatality rate higher than that due to infection with hepatitis B virus alone.

Hepatitis E is caused by the hepatitis E virus and is spread like hepatitis A. Although it does not cause chronic liver disease, the hepatitis E virus is responsible for a very high death rate in pregnant women. •

CHECKPOINT

14. How are the liver and gallbladder connected to the duodenum?
15. What is the function of bile?
16. List all of the major functions of the liver.

26.8 SMALL INTESTINE

OBJECTIVE

- Describe the location, structure, and functions of the small intestine.

Within 2 to 4 hours after eating a meal, the stomach has emptied its contents into the small intestine, where the major events of digestion and absorption occur. The ***small intestine*** averages 2.5 cm (1 in.) in diameter; its length is about 3 m (10 ft) in a living person and about 6.5 m (21 ft) in a cadaver due to the loss of smooth muscle tone after death.

Structure of the Small Intestine

The small intestine has three portions (Figure 26.12a): the duodenum, the jejunum, and the ileum. The ***duodenum*** (doo′-ō-DĒ-num), the shortest part (about 25 cm or 10 in.), attaches to the pylorus of the stomach. Duodenum means "twelve"; the structure is so named because it is about as long as the width of 12 fingers. The ***jejunum*** (je-Joo-num = empty) is about 1 m (3 ft) long and is so named because it is empty at death. It is mostly in the left upper quadrant. The final portion of the small intestine, the ***ileum*** (IL-ē-um = twisted), measures about 2 m (6 ft) and joins the large intestine at the ***ileocecal sphincter*** or ***valve*** (il-ē-ō-SĒ-kal). The ileum is mostly in the right lower quadrant.

The wall of the small intestine is composed of the same four layers that make up most of the GI tract: mucosa, submucosa, muscularis, and serosa (Figure 26.13). The epithelial layer of the small intestinal mucosa consists of simple columnar epithelium that contains several types of cells: absorptive, goblet, enteroendocrine, and paneth. ***Absorptive cells*** of the epithelium contain microvilli and digest and absorb nutrients in small intestinal

Figure 26.12 External and internal anatomy of the small intestine. (a) Regions of the small intestine are the duodenum, jejunum, and ileum. (b) Circular folds increase the surface area for digestion and absorption in the small intestine.

Most digestion and absorption occur in the small intestine.

Functions of the Small Intestine

1. Segmentations mix chyme with digestive juices and bring food into contact with the mucosa for absorption; peristalsis propels chyme through the small intestine.
2. Completes the digestion of carbohydrates, proteins, and lipids; begins and completes the digestion of nucleic acids.
3. Absorbs about 90% of nutrients and water that pass through the digestive system.

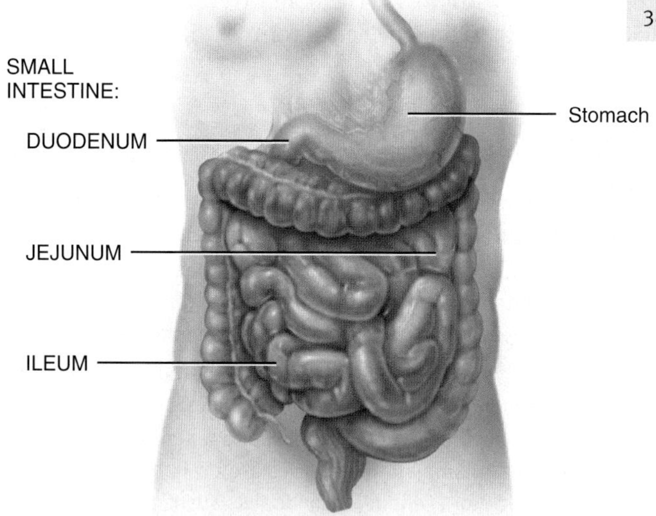

(a) Anterior view of external anatomy

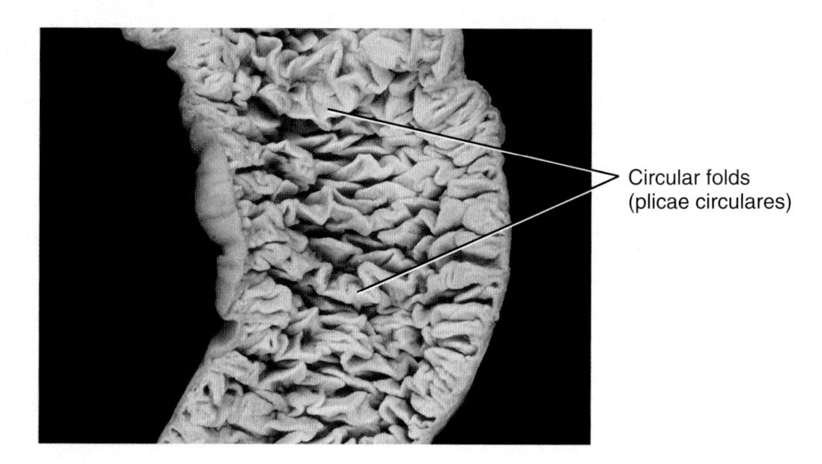

(b) Internal anatomy of jejunum

Which portion of the small intestine is the longest?

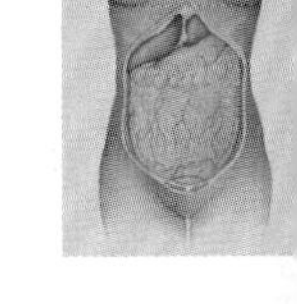

Figure 26.13 **Structure of the small intestine.**

Circular folds, villi, and microvilli increase the surface area of the small intestine for digestion and absorption in the small intestine.

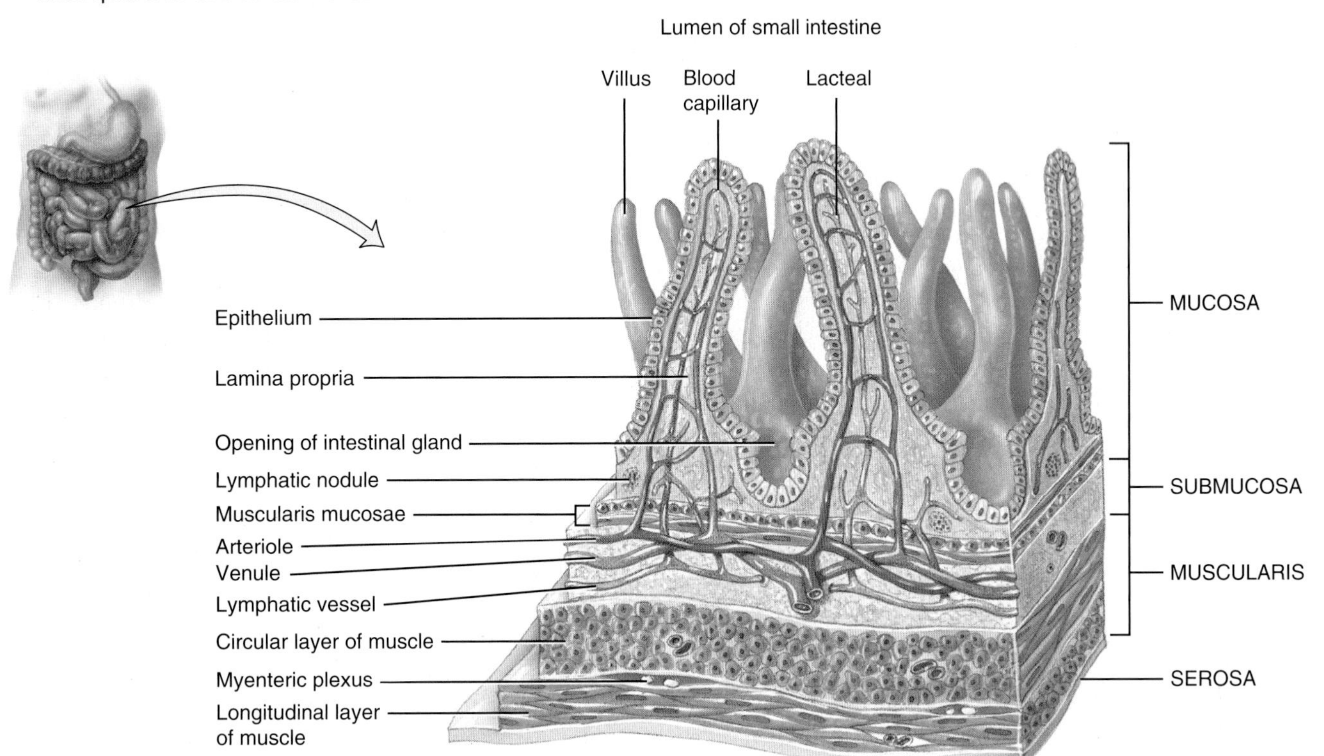

(a) Three-dimensional view of layers of the small intestine showing villi

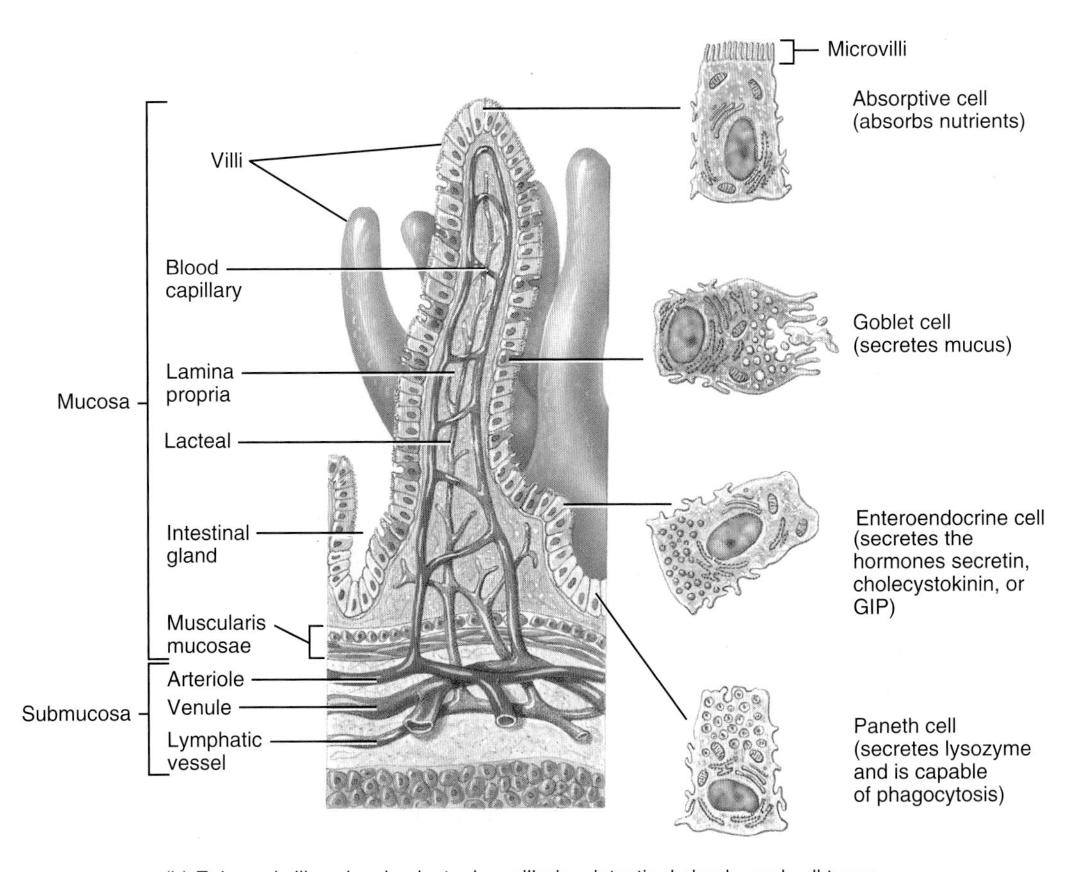

(b) Enlarged villus showing lacteal, capillaries, intestinal glands, and cell types

? **Where are the cells located that absorb dietary nutrients?**

chyme. *Goblet cells* secrete mucus. The small intestinal mucosa contains ***intestinal glands,*** which are deep crevices lined by epithelial cells that secrete intestinal juice. Besides absorptive cells and goblet cells, intestinal glands also contain ***enteroendocrine cells*** that secrete hormones into the bloodstream: ***secretin*** (se-KRĒ-tin), ***cholecystokinin*** (kō-le-sis-tō-KĪN-in) or ***CCK,*** and ***glucose-dependent insulinotropic peptide,*** or ***GIP*** (see Table 26.2 for secretin and CCK and Table 20.2 for GIP). ***Paneth cells*** secrete *lysozyme*, a bactericidal enzyme, and are also capable of phagocytosis. The lamina propria of the small intestinal mucosa contains areolar connective tissue that has an abundance of lymphatic tissue, which helps defend against pathogens in food. The submucosa of the duodenum contains ***duodenal glands*** that secrete alkaline mucus. It helps neutralize gastric acid in the chyme. The muscularis of the small intestine consists of two layers of smooth muscle—an outer longitudinal layer and an inner circular layer. The serosa is composed of simple squamous epithelium and areolar connective tissue.

Even though the wall of the small intestine is composed of the same four basic layers as the rest of the GI tract, special structural features of the small intestine facilitate the process of digestion and absorption. These structural features include circular folds, villi, and microvilli. ***Circular folds (plicae circulares)*** are permanent ridges of the mucosa and submucosa that enhance absorption by increasing surface area and causing the chyme to spiral, rather than move in a straight line, as it passes through the small intestine (see Figure 26.12b). Also present in the small intestine are numerous ***villi*** (= tufts of hair; singular is *villus*), finger-like projections of the mucosa that increase the surface area of the intestinal epithelium. Each villus consists of a layer of simple columnar epithelium surrounding a core of lamina propria. Within the core are an arteriole, a venule, a blood capillary network, and a ***lacteal*** (LAK-tē-al = milky), which is a lymphatic capillary. Nutrients absorbed by the epithelial cells covering the villus pass through the wall of a capillary or a lacteal to enter blood or lymph, respectively. Besides circular folds and villi, the small intestine also has ***microvilli*** (mī-krō-VIL-ī; *micro-* = small), tiny projections of the plasma membrane of absorptive cells that increase the surface area of these cells (see Figure 26.14). Thus, digested nutrients can move rapidly into absorptive cells.

Intestinal Juice

Intestinal juice, secreted by the intestinal glands, is a watery, clear yellow fluid with a slightly alkaline pH of 7.6 that contains some mucus. Together, pancreatic and intestinal juices provide a liquid medium that aids absorption of substances from chyme as they come in contact with the microvilli. Intestinal enzymes are synthesized in the absorptive cells that line the villi. Most digestion by enzymes of the small intestine occurs in or on the surface of these absorptive cells.

Mechanical Digestion in the Small Intestine

Two types of movements contribute to intestinal motility in the small intestine: segmentations and peristalsis. ***Segmentations*** are localized, mixing contractions that occur in portions of intestine distended by a large volume of chyme, mixing it with digestive juices and bringing food particles into contact with the mucosa for absorption. The movements are similar to alternately squeezing the middle and the ends of a capped tube of toothpaste. They do not push the intestinal contents along the tract.

After most of a meal has been absorbed, segmentation stops; ***peristalsis*** begins in the lower portion of the stomach and pushes chyme forward along a short stretch of small intestine. The peristaltic wave slowly migrates down the small intestine, reaching the end of the ileum in 90 to 120 minutes. Then another wave of peristalsis begins in the stomach. Altogether, chyme remains in the small intestine for 3 to 5 hours.

Chemical Digestion in the Small Intestine

The chyme entering the small intestine contains partially digested carbohydrates and proteins. The completion of digestion in the small intestine is a collective effort of pancreatic juice, bile, and intestinal juice. Once digestion is completed, the final products of digestion are ready for absorption.

Starches and dextrins not reduced to maltose by the time chyme leaves the stomach are broken down by ***pancreatic amylase,*** an enzyme in pancreatic juice that acts in the small intestine. Three enzymes located at the surface of small intestinal absorptive cells complete the digestion of disaccharides, breaking them down into monosaccharides, which are small enough to be absorbed. ***Maltase*** splits maltose into two molecules of glucose. ***Sucrase*** breaks sucrose into a molecule of glucose and a molecule of fructose. ***Lactase*** digests lactose into a molecule of glucose and a molecule of galactose.

Enzymes in pancreatic juice (trypsin, chymotrypsin, elastase, and carboxypeptidase) continue the digestion of proteins begun in the stomach, though their actions differ somewhat because each splits the peptide bond between different amino acids. Protein digestion is completed by ***peptidases,*** enzymes produced by absorptive cells that line the villi. The final products of protein digestion are amino acids, dipeptides, and tripeptides.

In an adult, most lipid digestion occurs in the small intestine. In the first step of lipid digestion, bile salts emulsify large globules of triglycerides and lipids into small lipid globules, giving pancreatic lipase easy access. Recall that triglycerides consist of a molecule of glycerol with three attached fatty acids (see Figure 2.10, Section 2.2). In the second step, ***pancreatic lipase,*** found in pancreatic juice, breaks down each triglyceride molecule by removing two of the three fatty acids from glycerol; the third remains attached to the glycerol. Thus, fatty acids and monoglycerides are the end products of triglyceride digestion.

Pancreatic juice contains two nucleases: ***ribonuclease,*** which digests RNA, and ***deoxyribonuclease,*** which digests DNA. The nucleotides that result from the action of the two nucleases are further digested by small intestinal enzymes into pentoses, phosphates, and nitrogenous bases.

Table 26.1 summarizes the enzymes that contribute to digestion.

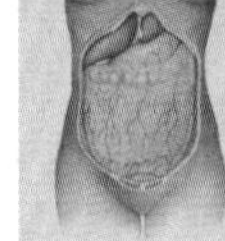

TABLE 26.1

Summary of Digestive Enzymes

ENZYME	SOURCE	SUBSTRATE	PRODUCT
CARBOHYDRATE DIGESTING			
Salivary amylase	Salivary glands.	Starches.	Maltose (disaccharide), maltotriose (trisaccharide), and dextrins.
Pancreatic amylase	Pancreas.	Starches.	Maltose, maltotriose, and dextrins.
Maltase	Small intestine.	Maltose.	Glucose.
Sucrase	Small intestine.	Sucrose.	Glucose and fructose.
Lactase	Small intestine.	Lactose.	Glucose and galactose.
PROTEIN DIGESTING			
Pepsin	Stomach (chief cells).	Proteins.	Peptides.
Trypsin	Pancreas.	Proteins.	Peptides.
Chymotrypsin	Pancreas.	Proteins.	Peptides.
Carboxypeptidase	Pancreas.	Amino acid at carboxyl (acid) end of peptides.	Peptides and amino acids.
Peptidases	Small intestine.	Amino acid at amino end of peptides and dipeptides.	Peptides and amino acids.
LIPID DIGESTING			
Pancreatic lipase	Pancreas.	Triglycerides (fats) that have been emulsified by bile salts.	Fatty acids and monoglycerides.
NUCLEASES			
Ribonuclease	Pancreas.	Ribonucleic acid.	Nucleotides.
Deoxyribonuclease	Pancreas.	Deoxyribonucleic acid.	Nucleotides.

Absorption in the Small Intestine

All the mechanical and chemical phases of digestion from the mouth down through the small intestine are directed toward changing food into molecules that can undergo ***absorption.*** Recall that absorption refers to the movement of small molecules through the absorptive epithelial cells of the mucosa into the underlying blood and lymphatic vessels. About 90% of all absorption takes place in the small intestine. The other 10% occurs in the stomach and large intestine. Absorption in the small intestine occurs by simple diffusion, facilitated diffusion, osmosis, and active transport. Any undigested or unabsorbed material left in the small intestine is passed on to the large intestine.

Absorption of Monosaccharides

All carbohydrates are absorbed as monosaccharides. Glucose and galactose are transported into absorptive cells of the villi by active transport. Fructose is transported by facilitated diffusion (Figure 26.14a). After absorption, monosaccharides are transported out of the epithelial cells by facilitated diffusion into the blood capillaries, which drain into venules of the villi. From here, monosaccharides are carried to the liver via the hepatic portal vein, then through the heart and to the general circulation (Figure 26.14b). Recall that the liver processes substances it receives from the hepatic portal vein before they pass into general circulation.

Absorption of Amino Acids

Enzymes break down dietary proteins into amino acids, dipeptides, and tripeptides, which are absorbed mainly in the duodenum and jejunum. About half of the absorbed amino acids are present in food, but half come from proteins in digestive juices and dead cells that slough off the mucosa. Amino acids, dipeptides, and tripeptides enter absorptive cells of the villi via active transport (Figure 26.14a). Inside the epithelial cells, peptides are digested into amino acids, which leave via diffusion and enter blood capillaries. Like monosaccharides, amino acids are carried in hepatic portal blood to the liver (Figure 26.14b). If not removed by liver cells, amino acids enter the general circulation. From there, body cells take up amino acids for use in protein synthesis and ATP production.

Figure 26.14 Absorption of digested nutrients in the small intestine. For simplicity, all digested foods are shown in the lumen of the small intestine, even though some nutrients are ingested at the surface of or in absorptive epithelial cells of the villi.

Long-chain fatty acids and monoglycerides are absorbed into lacteals; other products of digestion enter blood capillaries.

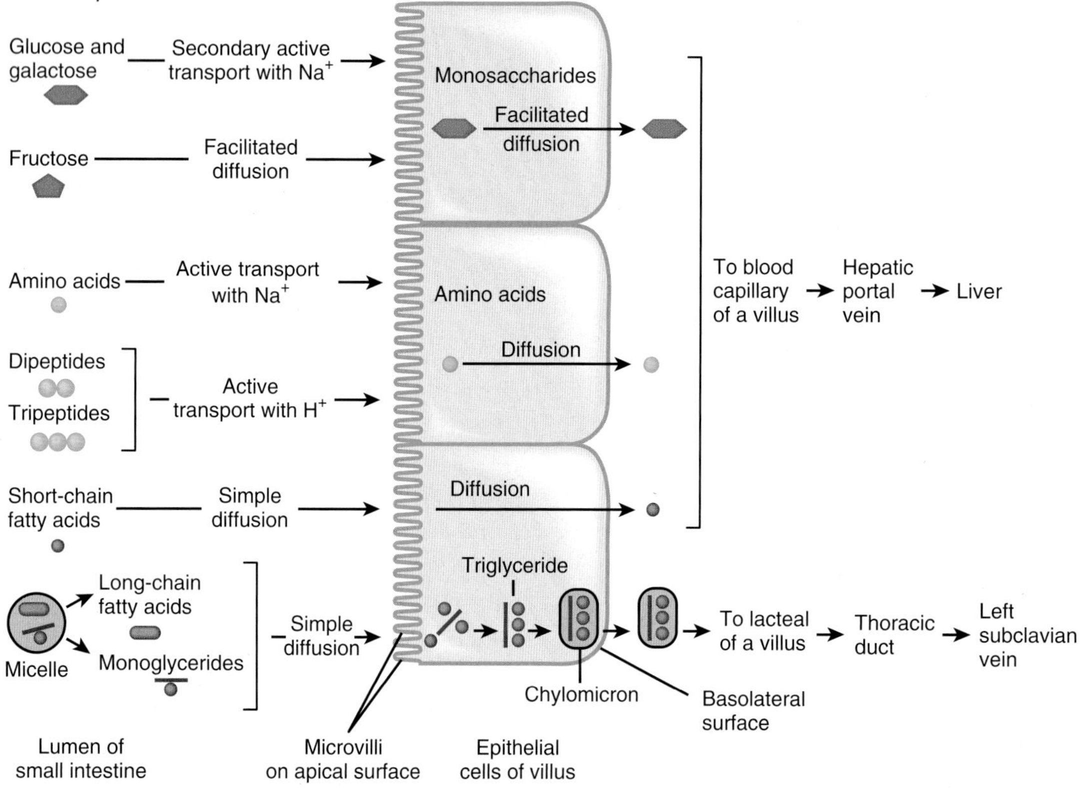

(a) Mechanisms for movement of nutrients through absorptive epithelial cells of the villi

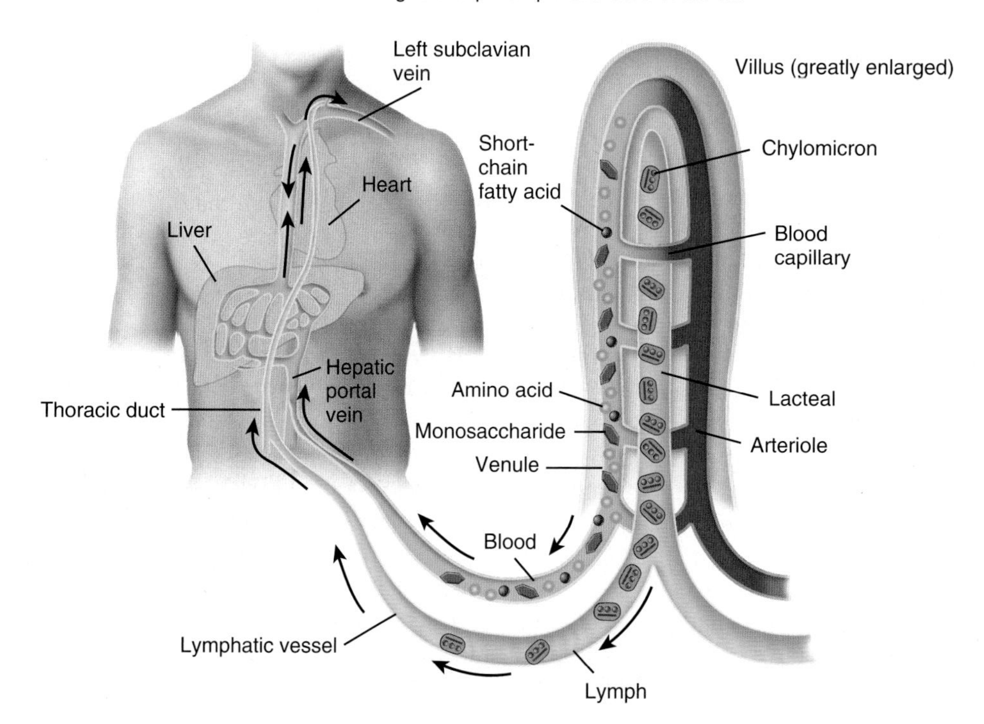

(b) Movement of absorbed nutrients into the blood and lymph

How are fat-soluble vitamins (A, D, E, and K) absorbed?

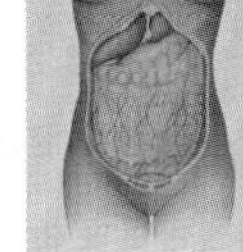

Absorption of Ions and Water

Absorptive cells lining the small intestine also absorb most of the ions and water that enter the GI tract in food, drink, and digestive secretions. Major ions absorbed in the small intestine include sodium, potassium, calcium, iron, magnesium, chloride, phosphate, nitrate, and iodide. All water absorption in the GI tract, about 9 liters (a little more than 2 gallons) daily, occurs via osmosis. When monosaccharides, amino acids, peptides, and ions are absorbed, they "pull" water along by osmosis.

Absorption of Lipids and Bile Salts

Lipases break down triglycerides into monoglycerides and fatty acids. The fatty acids can be either short-chain fatty acids (with fewer than 10–12 carbons) or long-chain fatty acids. The short-chain fatty acids are absorbed via simple diffusion into absorptive cells of the villi and then pass into blood capillaries along with monosaccharides and amino acids (Figure 26.14a). Bile salts emulsify the larger lipids, forming many ***micelles*** (mī-SELZ = small morsels), tiny droplets that include some bile salt molecules along with the long-chain fatty acids, monoglycerides, cholesterol, and other dietary lipids (Figure 26.14a). From micelles, these lipids diffuse into absorptive cells of the villi where they are packaged into ***chylomicrons,*** large spherical particles that are coated with proteins. Chylomicrons leave the epithelial cells via exocytosis and enter lymphatic fluid within a lacteal. Thus, most absorbed dietary lipids bypass the hepatic portal circulation because they enter lymphatic vessels instead of blood capillaries. Lymphatic fluid carrying chylomicrons from the small intestine passes into the thoracic duct and in due course empties into the left subclavian vein (Figure 26.14b). As blood passes through capillaries in adipose tissue and the liver, chylomicrons are removed and their lipids are stored for future use.

When chyme reaches the ileum, most of the bile salts are reabsorbed and returned by the blood to the liver for recycling. Insufficient bile salts, due to either obstruction of the bile ducts or liver disease, can result in the loss of up to 40% of dietary lipids in feces due to diminished lipid absorption.

Absorption of Vitamins

Fat-soluble vitamins (A, D, E, and K) are included along with ingested dietary lipids in micelles and are absorbed via simple diffusion. Most water-soluble vitamins, such as the B vitamins and vitamin C, are absorbed by simple diffusion. Vitamin B_{12} must be combined with intrinsic factor (produced by the stomach) for its absorption via active transport in the ileum.

CHECKPOINT

17. In what ways are the mucosa and submucosa of the small intestine adapted for digestion and absorption?
18. Explain the function of each digestive enzyme.
19. Define absorption and where does most of it occur? How are the end products of carbohydrate and protein digestion absorbed? How are the end products of lipid digestion absorbed?
20. By what routes do absorbed nutrients reach the liver?

26.9 LARGE INTESTINE

OBJECTIVE

- Describe the location, structure, and functions of the large intestine.

The large intestine is the last part of the GI tract. Its overall functions are the completion of absorption, the production of certain vitamins, the formation of feces, and the expulsion of feces from the body.

Structure of the Large Intestine

The ***large intestine*** averages about 6.5 cm (2.5 in.) in diameter and about 1.5 m (5 ft) in length. It extends from the ileum to the anus and is attached to the posterior abdominal wall by its mesentery (see Figure 26.3c). The large intestine has four principal regions: cecum, colon, rectum, and anal canal (see Figure 26.15).

At the opening of the ileum into the large intestine is a valve called the ***ileocecal sphincter.*** It allows materials from the small intestine to pass into the large intestine. Inferior to the ileocecal sphincter is the first segment of large intestine, called the ***cecum.*** Attached to the cecum is a twisted coiled tube called the ***vermiform appendix,*** or just simply, the ***appendix.*** This structure has highly concentrated lymphatic nodules which control the bacteria entering the large intestine by immune responses.

The open end of the cecum merges with the longest portion of the large intestine, called the ***colon*** (food passage). The colon is divided into ascending, transverse, descending, and sigmoid portions. The ***ascending colon*** ascends on the right side of the abdomen, reaches the undersurface of the liver, and turns to the left. The colon continues across the abdomen to the left side as the ***transverse colon.*** It curves beneath the lower border of the spleen on the left side and passes downward as the ***descending colon.*** The S-shaped ***sigmoid colon*** begins near the iliac crest of the left hip bone and ends as the ***rectum.***

MANUAL THERAPY APPLICATION

Abdominal Massage

Massage facilitates the flow of blood and therefore facilitates absorption of nutrients from the digestive system and distribution of nutrients around the body. Massage also shifts the balance of the autonomic nervous system such that the enteric nervous system (parasympathetic division) predominates, and innervation by the parasympathetic division enhances all activities of the digestive system. **Abdominal massage** reflexively stimulates peristalsis and therefore enhances elimination. Kneading of the abdomen also promotes production and utilization of digestive juices through activation of the parasympathetic system. Deep kneading of the colon, from the ileocecal valve to the ascending colon, transverse colon, and descending colon, provides mechanical assistance to remove fecal material in a more efficient manner and thus can have a very positive effect on constipation.

The last 2 to 3 cm (1 in.) of the rectum is called the ***anal canal*** (see Figure 26.15b). The opening of the anal canal to the exterior is called the ***anus.*** It has an internal sphincter of smooth

Figure 26.15 Anatomy of the large intestine.

The regions of the large intestine are the cecum, colon, rectum, and anal canal.

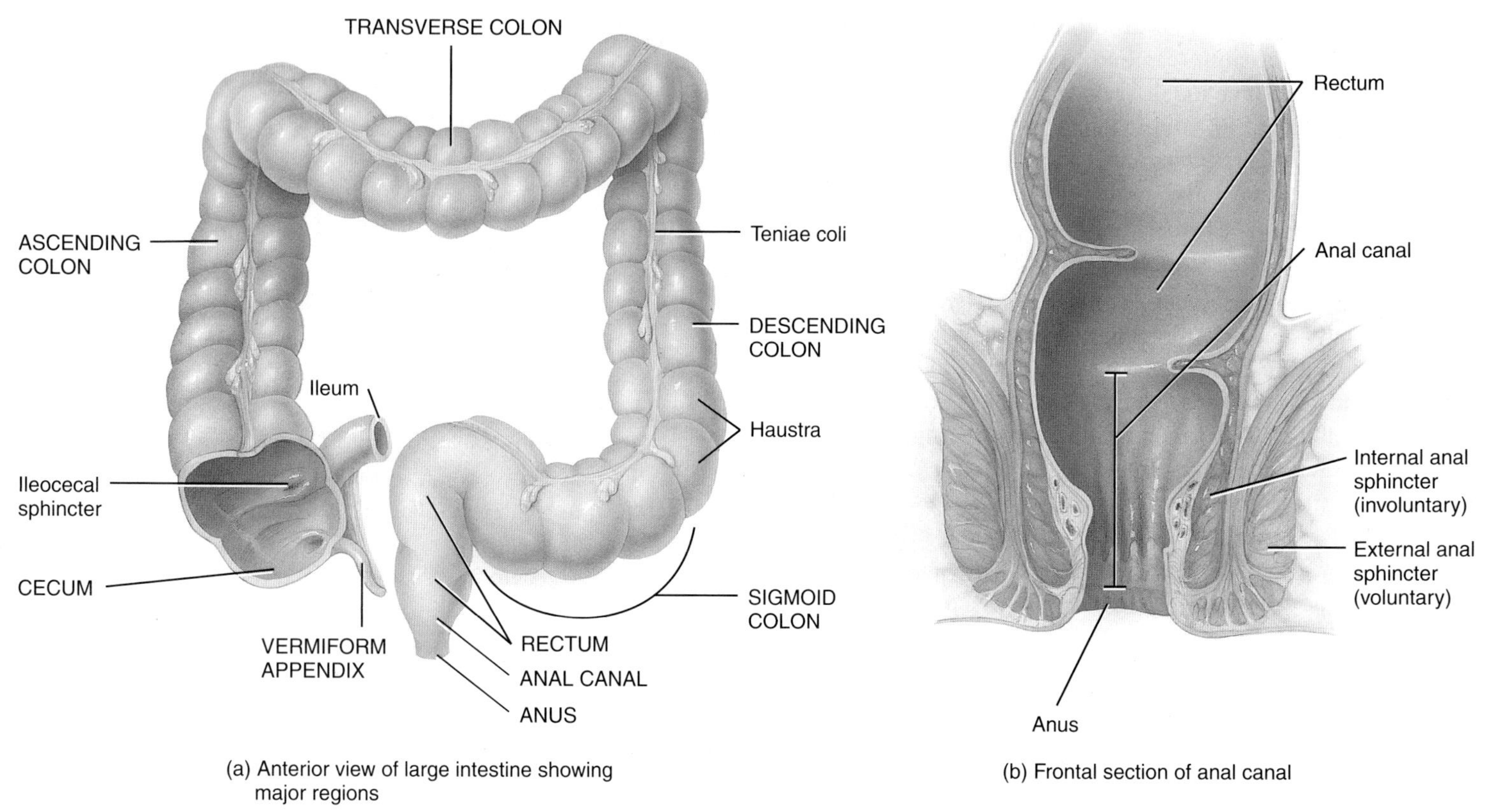

(a) Anterior view of large intestine showing major regions

(b) Frontal section of anal canal

Which portions of the colon are retroperitoneal?

(involuntary) muscle and an external sphincter of skeletal (voluntary) muscle. Normally, the anal sphincters are closed except during the elimination of feces.

The wall of the large intestine contains the typical four layers found in the rest of the GI tract: mucosa, submucosa, muscularis, and serosa. The epithelium of the mucosa is simple columnar epithelium that contains mostly absorptive cells and goblet cells (Figure 26.16). The cells form long tubes called ***intestinal glands.*** The absorptive cells function primarily in ion and water absorption. The goblet cells secrete mucus that lubricates the contents of the colon. Lymphatic nodules also are found in the mucosa. Compared to the small intestine, the mucosa of the large intestine does not have as many structural adaptations that increase surface area. There are no circular folds or villi; however, microvilli of the absorptive cells are present. Consequently, much more absorption occurs in the small intestine than in the large intestine. The muscularis consists of an external layer of longitudinal muscles and an internal layer of circular muscles. Unlike other parts of the gastrointestinal tract, the outer longitudinal layer of the muscularis is bundled into three longitudinal bands, called the ***teniae coli*** (TĒ-nē-ē KŌ-lī; *teniae* = flat bands), that run the length of most of the large intestine (see Figure 26.15a). Contractions of the bands gather the colon into a series of pouches called ***haustra*** (HAWS-tra = shaped like pouches; singular is *haustrum*), which give the colon a puckered appearance.

• CLINICAL CONNECTION Colon Pathologies

Polyps in the colon are generally slow-developing benign growths that arise from the mucosa of the large intestine. Often, they do not cause symptoms. If symptoms do occur, they include diarrhea, blood in the feces, and mucus discharged from the anus. The polyps are removed by colonoscopy or surgery because some of them may become cancerous.

Colorectal cancer is among the deadliest of malignancies. An inherited predisposition contributes to more than half of all cases of colorectal cancer. Intake of alcohol and diets high in animal fat and protein are associated with increased risk of colorectal cancer; dietary fiber, retinoids, calcium, and selenium may be protective. Signs and symptoms of colorectal cancer include diarrhea, constipation, cramping, abdominal pain, and rectal bleeding. Screening for colorectal cancer includes testing for blood in the feces, digital rectal examination, sigmoidoscopy, colonoscopy, and barium enema.

Diverticulosis is the development of *diverticula,* saclike outpouchings of the wall of the colon in places where the muscularis has become weak. Many people who develop diverticulosis have no symptoms and experience no complications. About 15% of people with diverticulosis eventually develop an inflammation known as **diverticulitis,** characterized by pain, either constipation or increased frequency of defecation, nausea, vomiting, and low-grade fever. Patients who change to high-fiber diets often show marked relief of symptoms. •

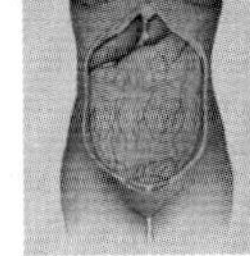

Figure 26.16 Structure of the large intestine.

Intestinal glands formed by absorptive cells and goblet cells extend the full thickness of the mucosa.

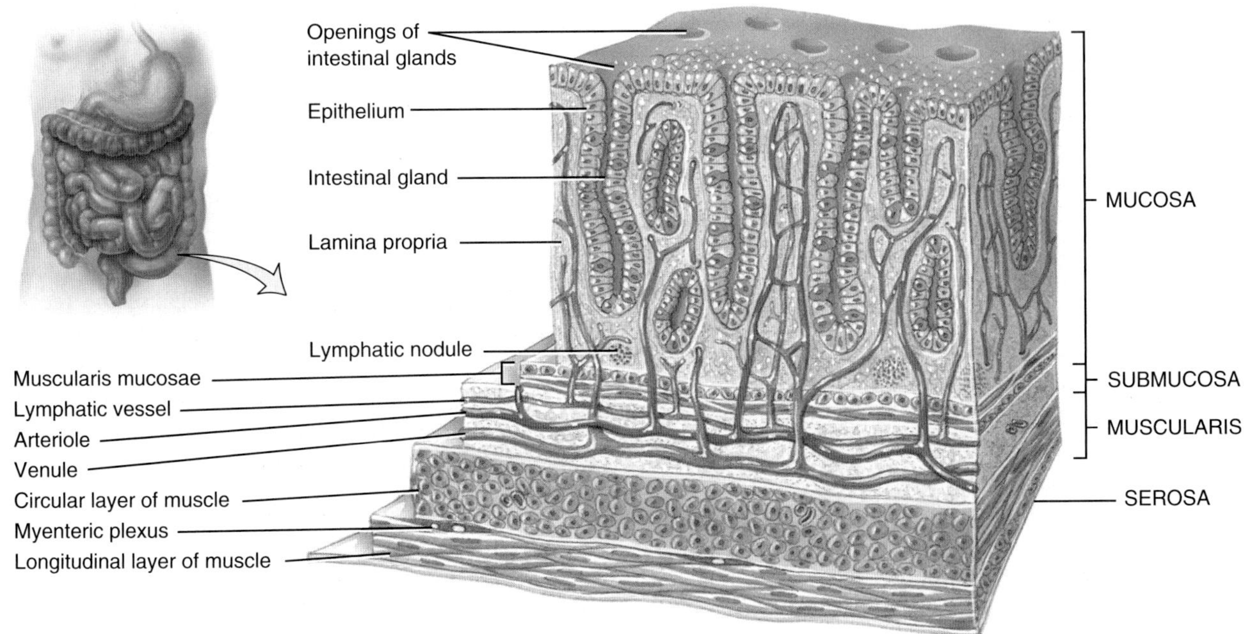

(a) Three-dimensional view of layers of the large intestine

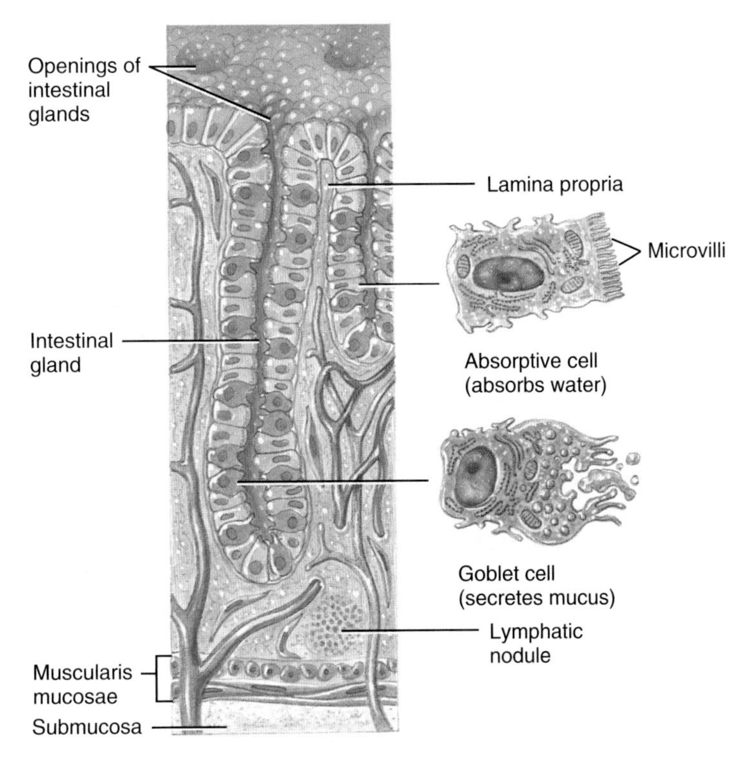

(b) Sectional view of intestinal glands and cell types

How does the muscularis of the large intestine differ from that of other parts of the GI tract?

Digestion and Absorption in the Large Intestine

The passage of chyme from the ileum into the cecum is regulated by the ileocecal sphincter. The sphincter normally remains slightly contracted so that the passage of chyme is usually a slow process. Immediately after a meal, a reflex intensifies peristalsis, forcing any chyme in the ileum into the cecum. ***Peristalsis*** occurs in the large intestine at a slower rate than in other portions of the GI tract. Characteristic of the large intestine is ***mass peristalsis,*** a strong peristaltic wave that begins in the middle of the colon and drives the colonic contents into the rectum. Food in the stomach initiates mass peristalsis, which usually takes place three or four times a day, during or immediately after a meal.

The final stage of digestion occurs in the colon through the activity of bacteria that normally inhabit the lumen. The glands of the large intestine secrete mucus but no enzymes. Bacteria ferment any remaining carbohydrates and release hydrogen, carbon dioxide, and methane gases. These gases contribute to flatus (gas) in the colon, termed *flatulence* when it is excessive. Bacteria also break down the remaining proteins to amino acids and decompose bilirubin to simpler pigments, including stercobilin, which give feces their brown color. Several vitamins needed for normal metabolism, including some B vitamins and vitamin K, are bacterial products that are absorbed in the colon.

Although most water absorption occurs in the small intestine, the large intestine absorbs a significant amount as well. The large intestine also absorbs ions, including sodium and chloride, and some dietary vitamins.

By the time chyme has remained in the large intestine 3 to 10 hours, it has become solid or semisolid as a result of water absorption and is now called ***feces.*** Chemically, feces consist of water, inorganic salts, sloughed-off epithelial cells from the mucosa of the gastrointestinal tract, bacteria, products of bacterial decomposition, unabsorbed digested materials, and indigestible parts of food.

The Defecation Reflex

Mass peristaltic movements push fecal material from the sigmoid colon into the rectum. The resulting distension of the rectal wall stimulates stretch receptors, which initiates a ***defecation reflex*** that empties the rectum. Impulses from the spinal cord travel along parasympathetic nerves to the descending colon, sigmoid colon, rectum, and anus. The resulting contraction of the longitudinal rectal muscles shortens the rectum, thereby increasing the pressure within it. This pressure plus parasympathetic stimulation opens the internal sphincter. The external sphincter is voluntarily controlled. If it is voluntarily relaxed, defecation occurs and the feces are expelled through the anus; if it is voluntarily constricted, defecation can be postponed. Voluntary contractions of the diaphragm and abdominal muscles aid defecation by increasing the pressure within the abdomen, which pushes the walls of the sigmoid colon and rectum inward. If defecation does not occur, the feces back up into the sigmoid colon until the next wave of mass peristalsis stimulates the stretch receptors. In infants, the defecation reflex causes automatic emptying of the rectum because voluntary control of the external anal sphincter has not yet developed.

Diarrhea (dī-a-RĒ-a; *dia-* = through; *-rrhea* = flow) is an increase in the frequency, volume, and fluid content of the feces caused by increased motility of and decreased absorption by the intestines. When chyme passes too quickly through the small intestine and feces pass too quickly through the large intestine, there is not enough time for absorption. Frequent diarrhea can result in dehydration and electrolyte imbalances. Excessive motility may be caused by lactose intolerance, stress, and microbes that irritate the gastrointestinal mucosa.

Constipation (kon-sti-PĀ-shun; *con-* = together; *-stip-* = to press) refers to infrequent or difficult defecation caused by decreased motility of the intestines. Because the feces remain in the colon for prolonged periods, excessive water absorption occurs, and the feces become dry and hard. Constipation may be caused by poor habits (delaying defecation), spasms of the colon, insufficient fiber in the diet, inadequate fluid intake, lack of exercise, emotional stress, or certain drugs.

CHECKPOINT

21. What activities occur in the large intestine to change its contents into feces?
22. What is defecation, and how does it occur?

26.10 PHASES OF DIGESTION

OBJECTIVES

- Describe the three phases of digestion.
- Describe the major hormones that regulate digestive activities.

Digestive activities occur in three overlapping phases: the cephalic phase, the gastric phasc, and the intestinal phase.

Cephalic Phase

During the ***cephalic phase*** of digestion, the smell, sight, sound, or thought of food activates neural centers in the brain. The brain then activates the facial (VII), glossopharyngeal (IX), and vagus (X) nerves. The facial and glossopharyngeal nerves stimulate the salivary glands to secrete saliva, whereas the vagus nerves stimulate the gastric glands to secrete gastric juice. The purpose of the cephalic phase of digestion is to prepare the mouth and stomach for food that is about to be eaten.

Gastric Phase

Once food reaches the stomach, the ***gastric phase*** of digestion begins. The purpose of this phase of digestion is to continue gastric secretion and to promote gastric motility. Gastric secretion during the gastric phase is regulated by the hormone ***gastrin.*** Gastrin is released from the G cells of the gastric glands in response to several stimuli: stretching of the stomach by chyme, partially digested proteins in chyme, caffeine in chyme, and the

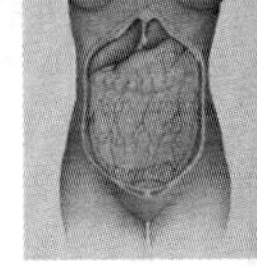

high pH of chyme due to the presence of food in the stomach. Gastrin stimulates gastric glands to secrete large amounts of gastric juice. It also strengthens the contraction of the lower esophageal sphincter to prevent reflux of acidic chyme into the esophagus, increases motility of the stomach, and relaxes the pyloric sphincter, which promotes gastric emptying.

Intestinal Phase

The ***intestinal phase*** of digestion begins once food enters the small intestine. In contrast to the activities initiated during the cephalic and gastric phases, which stimulate stomach secretory activity and motility, those occurring during the intestinal phase have inhibitory effects that slow the exit of chyme from the stomach and prevent overloading of the duodenum with more chyme than it can handle. In addition, responses occurring during the intestinal phase promote the continued digestion of foods that have reached the small intestine.

The activities of the intestinal phase are mediated by two major hormones secreted by the small intestine: cholecystokinin and secretin. ***Cholecystokinin (CCK)*** is secreted by intestinal glands of the small intestine in response to chyme containing amino acids from partially digested proteins and fatty acids from partially digested triglycerides. CCK stimulates secretion of pancreatic juice that is rich in digestive enzymes. It also causes contraction of the wall of the gallbladder, which squeezes stored bile out of the gallbladder into the cystic duct and through the common bile duct. In addition, CCK slows gastric emptying by promoting contraction of the pyloric sphincter, and it produces satiety (feeling full to satisfaction) by acting on the hypothalamus in the brain.

Acidic chyme entering the duodenum stimulates the release of ***secretin*** from glands of the small intestine. In turn, secretin stimulates the flow of pancreatic juice that is rich in bicarbonate (HCO_3^-) ions to buffer the acidic chyme that enters the duodenum from the stomach.

Table 26.2 summarizes the major hormones that control digestion.

CHECKPOINT

23. What are the stimuli that cause the cephalic phase of digestion?
24. Compare and contrast the activities that occur during the gastric phase of digestion with those that occur during the intestinal phase of digestion.

26.11 AGING AND THE DIGESTIVE SYSTEM

OBJECTIVE

- Describe the effects of aging on the digestive system.

Changes in the digestive system associated with aging include decreased secretory mechanisms, decreased motility of the digestive organs, loss of strength and tone of the muscular tissue and its supporting structures, changes in sensory feedback regarding enzyme and hormone release, and diminished response to pain and internal sensations. In the upper portion of the GI tract, common changes include reduced sensitivity to mouth irritations and sores, loss of taste, periodontal disease, difficulty in swallowing, hiatal hernia, gastritis, and peptic ulcer disease. Changes that may appear in the small intestine include duodenal ulcers, maldigestion, and malabsorption. Other pathologies that increase in incidence with age are appendicitis, gallbladder problems, jaundice, cirrhosis of the liver, and acute pancreatitis. Changes in the large intestine such as constipation, hemorrhoids, and diverticular disease may also occur. The incidence of cancer of the colon or rectum increases with age.

Now that our exploration of the digestive system is completed, you can appreciate the many ways that this system contributes to homeostasis of other body systems by examining Exhibit 26.1. Next, in Chapter 27, you will discover how the nutrients absorbed by the GI tract are utilized in metabolic reactions by the body tissues.

CHECKPOINT

25. List several changes in the upper and lower portions of the GI tract associated with aging.

TABLE 26.2

Major Hormones That Control Digestion

HORMONE	WHERE PRODUCED	STIMULANT	ACTION
Gastrin	Stomach mucosa (pyloric region).	Stretching of stomach, partially digested proteins and caffeine in stomach, and high pH of stomach chyme.	Stimulates secretion of gastric juice, increases motility of GI tract, and relaxes pyloric sphincter.
Secretin	Intestinal mucosa.	Acidic chyme that enters the small intestine.	Stimulates secretion of pancreatic juice rich in bicarbonate ions.
Cholecystokinin (CCK)	Intestinal mucosa.	Amino acids and fatty acids in chyme in small intestine.	Inhibits gastric emptying, stimulates secretion of pancreatic juice rich in digestive enzymes, causes ejection of bile from the gallbladder, and induces a feeling of satiety (feeling full to satisfaction).

EXHIBIT 26.1 The Digestive System and Homeostasis

BODY SYSTEM		CONTRIBUTION OF THE DIGESTIVE SYSTEM
For all body systems		The digestive system breaks down dietary nutrients into forms that can be absorbed and used by body cells for producing ATP and building body tissues; absorbs water, minerals, and vitamins needed for the growth and functions of body tissues; and eliminates wastes from body tissues in feces.
Integumentary system		The small intestine absorbs vitamin D, which the skin and kidneys modify to produce the hormone calcitriol. Excess dietary calories are stored as triglycerides in adipose cells in the dermis and subcutaneous layer.
Skeletal system	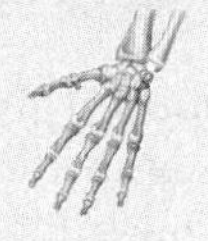	The small intestine absorbs dietary calcium and phosphorus salts needed to build bone extracellular matrix.
Muscular system	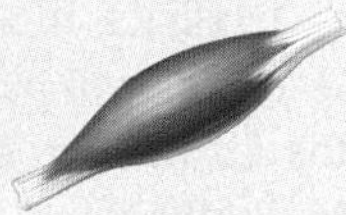	The liver can convert lactic acid produced by muscles during exercise to glucose.
Nervous system	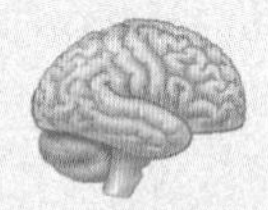	Gluconeogenesis (synthesis of new glucose molecules) in the liver plus digestion and absorption of dietary carbohydrates provide glucose, needed for ATP production by neurons.
Endocrine system	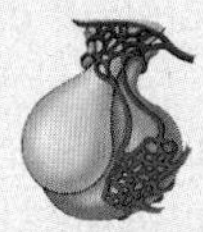	The liver inactivates some hormones, ending their activity. Pancreatic islets release insulin and glucagon. Cells in the mucosa of the stomach and small intestine release hormones that regulate digestive activities.
Cardiovascular system		The GI tract absorbs water, that helps maintain blood volume, and iron that is needed for the synthesis of hemoglobin in red blood cells. Bilirubin from hemoglobin breakdown is partially excreted in feces. The liver synthesizes most plasma proteins.
Lymphatic system and immunity		The acidity of gastric juice destroys bacteria and most toxins in the stomach.
Respiratory system	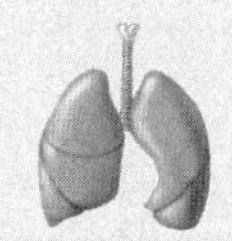	The pressure of abdominal organs against the diaphragm helps expel air quickly during a forced exhalation.
Urinary system	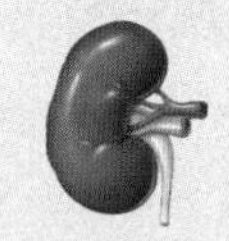	Absorption of water by the GI tract provides water needed to excrete waste products in urine.
Reproductive systems	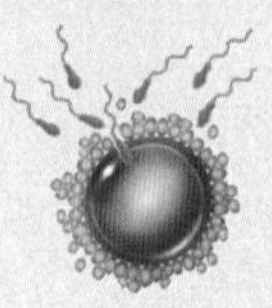	Digestion and absorption provide adequate nutrients, including fats, for normal development of reproductive structures, for the production of gametes (oocytes and sperm), and for fetal growth and development during pregnancy.

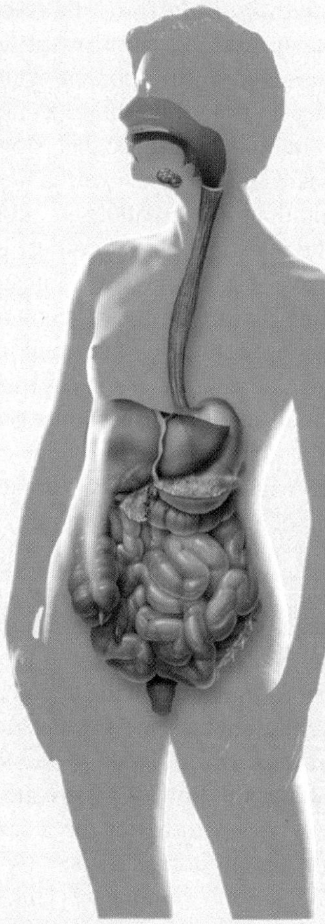

KEY MEDICAL TERMS ASSOCIATED WITH THE DIGESTIVE SYTEM

Anorexia nervosa A chronic disorder characterized by self-induced weight loss, negative perception of body image, and physiological changes that result from nutritional depletion. Patients have a fixation on weight control and often abuse laxatives, which worsens their fluid and electrolyte imbalances and nutrient deficiencies. The disorder is found predominantly in young, single females, and it may be inherited. Individuals may become emaciated and may ultimately die of starvation or one of its complications.

Bulimia (bū-LĒM-ē-a; *bu-* = ox; *-limia* = hunger) or ***binge–purge syndrome*** A disorder characterized by overeating at least twice a week followed by purging by self-induced vomiting, strict dieting or fasting, vigorous exercise, or use of laxatives or diuretics; it occurs in response to fears of being overweight, stress, depression, and physiological disorders such as hypothalamic tumors.

Canker sore (KANG-ker) Painful ulcer on the mucous membrane of the mouth that affects females more often than males, usually between ages 10 and 40; it may be an autoimmune reaction or result from a food allergy.

Cholecystitis (kō′-lē-sis-TĪ-tis; *chole-* = bile; *-cyst-* = bladder; *-itis* = inflammation of) In some cases, an autoimmune inflammation of the gallbladder; other cases are caused by obstruction of the cystic duct by gallstones.

Cirrhosis (si-RŌ-sis) Distorted or scarred liver as a result of chronic inflammation due to hepatitis, chemicals that destroy hepatocytes, parasites that infect the liver, or alcoholism; the hepatocytes are replaced by fibrous or adipose connective tissue. Symptoms include jaundice, edema in the legs, uncontrolled bleeding, and increased sensitivity to drugs.

Colostomy (kō-LOS-tō-mē; *-stomy* = provide an opening) The diversion of the fecal stream through an opening in the colon, creating a surgical "stoma" (artificial opening) that is affixed to the exterior of the abdominal wall. This opening serves as a substitute anus through which feces are eliminated into a bag worn on the abdomen.

Inflammatory bowel disease (in-FLAM-a-tō-rē BOW-el) Disorder that exists in two forms: (1) ***Crohn's disease,*** an inflammation of the gastrointestinal tract, especially the distal ileum and proximal colon, in which the inflammation may extend from the mucosa through the serosa, and (2) ***ulcerative colitis,*** an inflammation of the mucosa of the gastrointestinal tract, usually limited to the large intestine and usually accompanied by rectal bleeding.

Irritable bowel syndrome (IBS) Disease of the entire gastrointestinal tract in which a person reacts to stress by developing symptoms (such as cramping and abdominal pain) associated with alternating patterns of diarrhea and constipation. Excessive amounts of mucus may appear in feces; other symptoms include flatulence, nausea, and loss of appetite.

Malocclusion (mal-ō-KLOO-zhun; *mal-* = bad; *-occlusion* = to fit together) Condition in which the surfaces of the maxillary (upper) and mandibular (lower) teeth fit together poorly.

Nausea (NAW-sē-a = seasickness) Discomfort characterized by a loss of appetite and the sensation of impending vomiting. Its causes include local irritation of the gastrointestinal tract, a systemic disease, brain disease or injury, overexertion, or the effects of medication or drug overdose.

Traveler's diarrhea Infectious disease of the gastrointestinal tract that results in loose, urgent bowel movements; cramping; abdominal pain; malaise; nausea; and occasionally fever and dehydration. It is acquired through ingestion of food or water contaminated with fecal material typically containing bacteria (especially *Escherichia coli*); viruses or protozoan parasites are a less common cause.

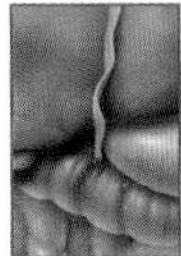

STUDY OUTLINE

Overview of the Digestive System (Section 26.1)

1. The breakdown of larger food molecules into smaller molecules is called digestion; the passage of these smaller molecules into blood and lymph is termed absorption.
2. The organs that collectively perform digestion and absorption constitute the digestive system.
3. The GI tract is a continuous tube extending from the mouth to the anus.
4. The accessory digestive organs include the teeth, tongue, salivary glands, liver, gallbladder, and pancreas.
5. Digestion includes six basic processes: ingestion, secretion, mixing and propulsion, mechanical and chemical digestion, absorption, and defecation.

Layers of the GI Tract and the Peritoneum (Section 26.2)

1. The basic arrangement of layers in most of the gastrointestinal tract, from the inside to the outside, is the mucosa, submucosa, muscularis, and serosa.
2. Folds of the peritoneum include the mesentery, greater omentum, lesser omentum, mesocolon, and the falciform ligament.
3. Structures not located within the peritoneal cavity are said to be retroperitoneal.

Mouth (Section 26.3)

1. The mouth is formed by the cheeks, hard and soft palates, lips, and tongue, which aid mechanical digestion.
2. The tongue forms the floor of the oral cavity. It is composed of skeletal muscle covered with mucous membrane. The superior surface and lateral areas of the tongue are covered with papillae. Some papillae contain taste buds.
3. Most saliva is secreted by the salivary glands, which lie outside the mouth and release their secretions into ducts that empty into the oral cavity. There are three pairs of salivary glands: parotid, submandibular, and sublingual. Saliva lubricates food and starts the chemical digestion of carbohydrates. Salivation is controlled by the autonomic nervous system.
4. The teeth, or dentes, project into the mouth and are adapted for mechanical digestion. A typical tooth consists of three principal portions: crown, root, and neck. Teeth are composed primarily of dentin and are covered by enamel, the hardest substance in the body. Humans have two sets of teeth: deciduous and permanent.

5. Through mastication, food is mixed with saliva and shaped into a bolus.
6. Salivary amylase begins the digestion of starches in the mouth.

Pharynx and Esophagus (Section 26.4)

1. Food that is swallowed passes from the mouth into the oropharynx.
2. From the oropharynx, food passes into the laryngopharynx.
3. The esophagus is a muscular tube that connects the pharynx to the stomach.
4. Swallowing moves a bolus from the mouth to the stomach by peristalsis. It consists of a voluntary stage, pharyngeal stage (involuntary), and esophageal stage (involuntary).

Stomach (Section 26.5)

1. The stomach connects the esophagus to the duodenum.
2. The main regions of the stomach are the cardia, fundus, body, and pylorus.
3. Adaptations of the stomach for digestion include rugae; glands that produce mucus, hydrochloric acid, a protein-digesting enzyme (pepsin), intrinsic factor, and gastrin; and a three-layered muscularis for efficient mechanical movement.
4. Mechanical digestion consists of mixing waves that macerate food and mix it with gastric juice, forming chyme.
5. Chemical digestion consists of the conversion of proteins into peptides by pepsin.
6. The stomach wall is impermeable to most substances. Among the substances the stomach can absorb are water, ions, short-chain fatty acids, some drugs, and alcohol.

Pancreas (Section 26.6)

1. Secretions pass from the pancreas to the duodenum via the pancreatic duct.
2. Pancreatic islets (islets of Langerhans) secrete hormones and constitute the endocrine portion of the pancreas.
3. Acinar cells, which secrete pancreatic juice, constitute the exocrine portion of the pancreas.
4. Pancreatic juice contains enzymes that digest starch (pancreatic amylase); proteins (trypsin, chymotrypsin, and carboxypeptidase); triglycerides (pancreatic lipase); and nucleic acids (nucleases).

Liver and Gallbladder (Section 26.7)

1. The liver has two principal lobes. The gallbladder is a sac located in a depression of the posterior surface of the liver that stores and concentrates bile produced by the liver.
2. The lobes of the liver are made up of lobules that contain hepatocytes (liver cells), sinusoids, stellate reticuloendothelial (Kupffer) cells, and a central vein.
3. Hepatocytes produce bile that is carried by a duct system to the gallbladder for concentration and temporary storage.
4. Bile's contribution to digestion is the emulsification of dietary lipids.
5. The liver also functions in carbohydrate, lipid, and protein metabolism; processing of drugs and hormones; excretion of bilirubin; synthesis of bile salts; storage of vitamins and minerals; phagocytosis; and activation of vitamin D.

Small Intestine (Section 26.8)

1. The small intestine extends from the pyloric sphincter to the ileocecal sphincter. It is divided into the duodenum, the jejunum, and the ileum.
2. The small intestine is highly adapted for digestion and absorption. Its glands produce enzymes and mucus, and the microvilli, villi, and circular folds of its wall provide a large surface area for digestion and absorption.
3. Mechanical digestion in the small intestine involves segmentations and migrating waves of peristalsis.
4. Enzymes in pancreatic juice and in the microvilli of the absorptive cells of the small intestine break down disaccharides to monosaccharides; protein digestion is completed by peptidase enzymes; triglycerides are broken down into fatty acids and monoglycerides by pancreatic lipase; and nucleases break down nucleic acids to pentoses and nitrogenous bases.
5. Absorption is the passage of nutrients from digested food in the gastrointestinal tract into the blood or lymph. Absorption, which occurs mostly in the small intestine, occurs by means of simple diffusion, facilitated diffusion, osmosis, and active transport.
6. Monosaccharides, amino acids, and short-chain fatty acids pass into the blood capillaries.
7. Long-chain fatty acids and monoglycerides are absorbed as part of micelles, resynthesized to triglycerides, and transported in chylomicrons to the lacteal of a villus.
8. The small intestine also absorbs water, electrolytes, and vitamins.

Large Intestine (Section 26.9)

1. The large intestine extends from the ileocecal sphincter to the anus. Its regions include the cecum, colon, rectum, and anal canal.
2. The mucosa contains numerous absorptive cells that absorb water and goblet cells that secrete mucus.
3. Mass peristalsis is a strong peristaltic wave that drives the contents of the colon into the rectum.
4. In the large intestine, substances are further broken down, and some vitamins are synthesized through bacterial action.
5. The large intestine absorbs water, electrolytes, and vitamins.
6. Feces consist of water, inorganic salts, epithelial cells, bacteria, and undigested foods.
7. The elimination of feces from the rectum is called defecation. Defecation is a reflex action aided by voluntary contractions of the diaphragm and abdominal muscles and relaxation of the external anal sphincter.

Phases of Digestion (Section 26.10)

1. Digestive activities occur in three overlapping phases; cephalic phase, gastric phase, and intestinal phase.
2. During the cephalic phase of digestion, salivary glands secrete saliva and gastric glands secrete gastric juice in order to prepare the mouth and stomach for food that is about to be eaten.
3. The presence of food in the stomach causes the gastric phase of digestion, which promotes gastric juice secretion and gastric motility.
4. During the intestinal phase of digestion, food is digested in the small intestine. In addition, gastric motility and gastric secretion decrease in order to slow the exit of chyme from the stomach, which prevents the small intestine from being overloaded with more chyme than it can handle.

5. The activities that occur during the various phases of digestion are coordinated by hormones. Table 26.2 summarizes the major hormones that control digestion.

Aging and the Digestive System (Section 26.11)

1. General changes with age include decreased secretory mechanisms, decreased motility, and loss of tone.
2. Specific changes may include loss of taste, hernias, peptic ulcer disease, constipation, hemorrhoids, and diverticular diseases.

SELF-QUIZ QUESTIONS

1. Which of the following is NOT an accessory digestive organ?
a. teeth **b.** salivary glands **c.** liver **d.** pancreas **e.** esophagus

2. Chewing food is an example of
a. absorption **b.** mechanical digestion **c.** secretion **d.** chemical digestion **e.** ingestion.

3. Which of the following is mismatched?
a. submucosa, enteric nervous system (ENS)
b. muscularis, lacteal **c.** serosa, greater omentum
d. mucosa, villi **e.** serosa, visceral peritoneum

4. Most chemical digestion occurs in the
a. liver **b.** stomach **c.** duodenum **d.** colon **e.** pancreas.

5. Absorption is defined as
a. the elimination of solid wastes from the digestive system
b. a reflex action controlled by the autonomic nervous system
c. the breakdown of foods by enzymes
d. the passage of nutrients from the gastrointestinal tract into the bloodstream
e. the mechanical breakdown of triglycerides.

6. The exposed portions of the teeth that you clean with a toothbrush are the
a. crowns **b.** periodontal ligaments **c.** roots **d.** pulp cavities **e.** gingivae.

7. The smell of your favorite food cooking makes "your mouth water"; this is due to
a. sympathetic stimulation of the salivary glands
b. mastication
c. parasympathetic stimulation of the salivary glands
d. increased mucus secretion by the pharynx
e. the enteric nervous system.

8. Match the following:

____ **a.** carries bile	**A.** pyloric sphincter
____ **b.** proteins combined with triglycerides and cholesterol	**B.** circular folds
____ **c.** surrounds the opening between the stomach and duodenum	**C.** micelles
____ **d.** secrete pancreatic juice	**D.** cystic duct
____ **e.** increase surface area in small intestine	**E.** ileocecal sphincter
____ **f.** bile salts combined with partially digested lipids	**F.** rugae
____ **g.** location between the opening of the small and large intestine	**G.** chylomicrons
____ **h.** large mucosal folds in stomach	**H.** acini

9. Which of the following correctly describes the esophagus?
a. Food enters the esophagus from the pyloric region of the stomach.
b. The movement of food through the entire esophagus is under voluntary control.
c. It allows the passage of chyme.
d. It produces several enzymes that aid in the digestion of food.
e. It is a muscular tube extending from the pharynx to the stomach.

10. If an incision were made into the stomach, the tissue layers would be cut in what order?
a. mucosa, muscularis, serosa, submucosa
b. mucosa, muscularis, submucosa, serosa
c. serosa, muscularis, mucosa, submucosa
d. muscularis, submucosa, mucosa, serosa
e. serosa, muscularis, submucosa, mucosa

11. Most water absorption in the digestive tract occurs in the
a. small intestine **b.** stomach **c.** mouth **d.** liver **e.** large intestine.

12. Which of the following would NOT result in secretion of gastric juices in the stomach?
a. secretion of gastrin
b. stimulation by the vagus nerves
c. the presence of partially digested proteins
d. stretching of the stomach
e. stimulation by the sympathetic nervous system

13. Bile
a. is produced in the gallbladder
b. is an enzyme that breaks down carbohydrates
c. emulsifies triglycerides
d. is required for the absorption of amino acids
e. enters the small intestine through the right hepatic duct.

14. Which of the following is NOT a function of the liver?
a. processing newly absorbed nutrients
b. producing enzymes that digest proteins
c. breaking down old red blood cells
d. detoxifying certain poisons
e. producing bile

15. The purpose of villi in the small intestine is to
a. aid in the movement of food through the small intestines
b. phagocytize microbes
c. produce digestive enzymes
d. increase the surface area for absorption of digested nutrients
e. produce acidic secretions.

16. Which of the following is NOT produced in the stomach?
a. sodium bicarbonate ($NaHCO_3$) **b.** gastrin **c.** pepsinogen **d.** mucus **e.** hydrochloric acid (HCl)

17. Which of the following is NOT correctly paired?
    a. esophagus, peristalsis
    b. mouth, mastication
    c. large intestine, mass peristalsis
    d. small intestine, segmentations
    e. stomach, emulsification
18. The enzyme pancreatic lipase digests triglycerides into
    a. glucose
    b. amino acids
    c. fatty acids and monoglycerides
    d. nucleic acids
    e. amylase.
19. Place the following in the correct order as food passes from the small intestine:
    1. sigmoid colon
    2. transverse colon
    3. ascending colon
    4. rectum
    5. cecum
    6. descending colon
    a. 1, 3, 2, 6, 5, 4
    b. 5, 1, 6, 2, 3, 4
    c. 4, 1, 6, 2, 3, 5
    d. 2, 3, 5, 6, 4, 1
    e. 5, 3, 2, 6, 1, 4
20. Lacteals function
    a. in the absorption of lipids in chylomicrons
    b. to produce bile in the liver
    c. in the absorption of electrolytes
    d. in the fermentation of carbohydrates in the large intestine
    e. to produce salivary amylase.

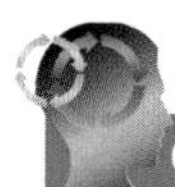

CRITICAL THINKING QUESTIONS

1. Antonio had dinner at his favorite Italian restaurant. His menu consisted of a salad, a large plate of spaghetti, garlic bread, and wine. For dessert, he consumed "death by chocolate" cake and a cup of coffee. He topped off his evening with a cigarette and brandy. He returned home and, while lying on his couch watching television, he experienced a pain in his chest. He called 911 because he was certain he was having a heart attack. Antonio was told his heart was fine, but he needed to watch his diet. What happened to Antonio?
2. Jared put a plastic spider in his sister's drink as a joke. Unfortunately, his mother doesn't think the joke is very funny because his sister swallowed it and now they're all at the ER (emergency room). The doctor suspects that the spider may have lodged at the junction of the stomach and the duodenum. Name the sphincter at this junction. Trace the path taken by the plastic spider on its journey to its new temporary home.
3. Jerry hadn't eaten all day when he bought a dried out, lukewarm hot dog from a street vendor for lunch. A few hours later, he was a victim of food poisoning and was desperately seeking a bathroom. After vomiting several times, Jerry noticed that he was expelling a greenish-yellow liquid. The hot dog may have been a bit shriveled, but it wasn't green! What is the source of this colored fluid?

? ANSWERS TO FIGURE QUESTIONS

26.1 The teeth cut and grind food.

26.2 Nerves in its wall help regulate secretions and contractions of the gastrointestinal tract.

26.3 Mesentery binds the small intestine to the posterior abdominal wall.

26.4 The uvula helps prevent food and liquids from entering the nasal cavity during swallowing.

26.5 The main component of teeth is connective tissue, specifically dentin.

26.6 Swallowing is both voluntary and involuntary. Initiation of swallowing, carried out by skeletal muscles, is voluntary. Completion of swallowing—moving a bolus along the esophagus and into the stomach—involves peristalsis of smooth muscle and is involuntary.

26.7 After a very large meal, the rugae stretch out and disappear as the stomach fills.

26.8 The simple columnar epithelial cells (surface mucous cells) of the mucosa are in contact with food in the stomach.

26.9 Pancreatic juice is a mixture of water, salts, bicarbonate ions, and digestive enzymes.

26.10 Stellate reticuloendothelial (Kupffer) cells in the liver are phagocytes.

26.11 While a meal is being absorbed, nutrients, O_2, and certain toxic substances are removed by hepatocytes from blood flowing through the sinusoids.

26.12 The ileum is the longest portion of the small intestine.

26.13 The absorptive cells cover the surface of the villi.

26.14 Fat-soluble vitamins are absorbed by simple diffusion from micelles.

26.15 The ascending and descending portions of the colon are retroperitoneal.

26.16 The muscularis of the large intestine forms three longitudinal bands (teniae coli) that gather the colon into a series of pouches called haustra.

Nutrition and Metabolism

27

Your body constantly rebuilds and remodels itself every day. What you choose to eat will determine how well your body can renew itself. To rebuild and remodel yourself appropriately, you need a variety of foods that contain water, carbohydrates, fats, proteins, vitamins, and minerals. Often in your busy life you do not eat regularly enough or properly. Good nutrition does not need to be difficult but does require some preparation. Many people find that within days of changing their diet to a well-balanced one, they feel more energetic and look more vibrant. If you reflect on diseases that have been mentioned in this text, how many of them could perhaps have been prevented or improved with good nutrition?

Each night when you go to sleep your body breaks down tissues. For example, a portion of the minerals in your bones (calcium and magnesium) are replaced with new minerals. This replacement keeps the main scaffold-like structure in your bones strong and it allows your bones to last for decades, even a century. With proper nutrition you can somewhat reduce the effects of aging. If you ignore your body's requirements for calcium, for example, osteoporosis may suddenly creep up on you. The keys to good nutrition are moderation (Calorie control), a balance of nutrients and minerals, and greater variety in the foods you eat. With a little effort these concepts can be balanced to create a healthy person of good weight, stature, and muscle tone.

CONTENTS AT A GLANCE

27.1 NUTRIENTS

OBJECTIVES

- Define a nutrient and identify the six main types of nutrients.
- List the guidelines for healthy eating.

Nutrients are chemical substances in food that body cells use for growth, maintenance, and repair. The six main types of nutrients are carbohydrates, lipids, proteins, water, minerals, and vitamins. ***Essential nutrients*** are specific nutrient molecules that the body cannot make in sufficient quantity to meet its needs and thus must be obtained from the diet. Some amino acids, some fatty acids, and vitamins and minerals are essential nutrients. The structures and functions of carbohydrates, proteins, lipids, and water were discussed in Chapter 2. In this chapter, we discuss some guidelines for healthy eating and the roles of minerals and vitamins in metabolism.

Guidelines for Healthy Eating

Each gram of protein or carbohydrate in food provides about 4 Calories; 1 gram of fat (lipids) provides about 9 Calories. We do not know with certainty what levels and types of carbohydrate, fat, and protein are optimal in the diet. Different populations around the world eat radically different diets that are adapted to their particular lifestyles. However, many experts recommend the following distribution of calories: 50–60% from carbohydrates, with less than 15% from simple sugars; less than 30% from fats (triglycerides are the main type of dietary fat), with no more than 10% as saturated fats; and about 12–15% from proteins.

The guidelines for healthy eating are to:

- Eat a variety of foods.
- Maintain a healthy weight.
- Choose foods low in fat, saturated fat, and cholesterol.
- Eat plenty of vegetables, fruits, and grain products.
- Use sugars in moderation only.

In 2005, the United States Department of Agriculture (USDA) introduced a food pyramid called ***My Pyramid,*** which represents a *personalized* approach to making healthy food choices and maintaining regular physical activity. By consulting a chart, it is possible to determine your desired Calorie intake based on your gender, age, and activity level. Once this is determined, you can choose the type and amount of food to be consumed.

If you carefully examine Figure 27.1, you will note that the six color bands represent the five basic food groups plus oils. Foods from all bands are needed each day. Also note that the overall size of the bands suggests the proportion of food a person should choose on a daily basis. The wider base of each band represents foods with little or no solid fats or added sugars, and these foods should be selected more often. The narrower top of each band represents foods with more added sugars and solid fats, which should be selected less frequently. The person climbing the steps is a reminder of the need for daily physical activity.

Figure 27.1 My Pyramid.

My Pyramid is a personalized approach to making healthy food choices and maintaining regular physical activity.

What does the wider base of each band mean?

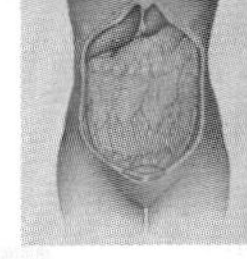

As an example of how My Pyramid works, let's assume based on consulting a chart that the calorie level of an 18-year-old moderately active female is 2000 Calories and that of an 18-year-old moderately active male is 2800 Calories. Accordingly, it is suggested that the following foods should be chosen in the following amounts:

CALORIE LEVEL	2000	2800
Fruits (includes all fresh, frozen, canned, and dried fruits and fruit juices)	2 cups	2.5 cups
Vegetables (includes all fresh, frozen, canned, and dried vegetables and vegetable juices)	2.5 cups	3.5 cups
Grains (includes all foods made from wheat, rice, oats, cornmeal, and barley such as bread, cereals, oatmeal, rice, pasta, crackers, tortillas, and grits)	6 oz	10 oz
Meats and beans (includes lean meat, poultry, fish, eggs, peanut butter, beans, nuts, and seeds)	5.5 oz	7 oz
Milk group (includes milk products and foods made from milk that retain their calcium content such as cheeses and yogurt)	3 cups	3 cups
Oils (choose mostly fats that contain monounsaturated and polyunsaturated fatty acids such as fish, nuts, seeds, and vegetable oils)	6 tsp	8 tsp

In addition, you should choose and prepare foods with little salt. In fact, sodium intake should be less than 2300 mg per day. If you choose to drink alcohol, it should be consumed in moderation (no more than 1 drink per day for women and 2 drinks per day for men). A drink is defined as 12 oz of regular beer, 5 oz of wine, or 1½ oz of 80 proof distilled spirits.

• CLINICAL CONNECTION | Diets

The carbohydrates, lipids, and proteins studied in Chapter 2 and again in this chapter may be produced by the body or ingested as food or drink. When intake exceeds the energy that the body uses, weight gain follows. Thirty-five hundred Calories equals one pound of weight. Thus, when intake of food and drink exceeds the energy that the body uses by 3500 Calories, a weight gain of one pound follows. Similarly, when the energy used exceeds the intake by 3500 Calories, a loss of one pound follows. The incidence of excessive weight and obesity has increased sharply for both children and adults in the United States during the past 30 years, and these increasing rates have important implications for health. Being overweight or obese increases the risk of many diseases and conditions, such as coronary heart disease, stroke, hypertension, diabetes mellitus, sleep apnea, certain cancers, gallbladder disease, osteoarthritis, and high levels of cholesterol or triglycerides.

Obesity is body weight more than 20% above a desirable standard due to an excessive accumulation of adipose tissue; it affects one-third of the adult population in the United States. In a few cases, obesity may result from trauma to or tumors in the food-regulating centers in the hypothalamus. In most cases of obesity, no specific cause can be identified. Contributing factors include genetic factors, eating habits taught early in life, overeating to relieve tension, and social customs.

The solution to the problem of excessive weight or obesity seems simple enough—either decrease the intake, increase the energy expended, or do a combination of the two. Weight-loss options, also called **diets,** sell a lot of books and tabloids. The fact that there are so many diets available indicates that none of them is the panacea claimed by the authors.

All diets involve one or more of the following weight-loss strategies:

- Low carbohydrate diets. A decrease in carbohydrate consumption may result in lower insulin levels, which causes the body to utilize stored fat for energy. Well-known examples are the Atkins diet, Protein Power, and the Zone diet.
- Glycemic-index diets. Similar to the low carbohydrate diet in theory, foods with a low-glycemic-index ranking are emphasized. The South Beach diet is an example. It should be noted, however, that many factors other than food influence the level of blood sugar. Age, weight, portion size, and the type of food preparation are variables.
- Low-fat diets. Lipids contain more calories per ounce than carbohydrates and proteins, and therefore dietary restriction or exclusion of lipids should be beneficial. Many people, however, find that fats are more satisfying than carbohydrates or proteins, and therefore they ingest more calories of these other sources to compensate for the hunger feelings.
- Meal providers. Jenny Craig, NutriSystem, and other companies provide members prepackaged portions of food for each meal and snack.
- Support groups. Many people will strive harder to succeed with their dietary goals when they know that weight will be checked on a weekly basis. Weight Watchers groups provide peer support as well as expertise on eating plans and exercise recommendations.
- Meal replacements. Slim-Fast and similar products provide relatively low-calorie meals that are nutritionally complete.

The sad experience of many dieters has been that weight loss, resulting from any diet, is temporary because of the dietary restrictions that they don't want to follow for very long. Successful weight loss requires the adoption of calorie intake and physical activity strategies that a person is willing to live with for the long term. •

Minerals

Minerals are inorganic elements that constitute about 4% of the total body weight and are concentrated most heavily in the skeleton. Minerals with known functions in the body include calcium, phosphorus, potassium, sulfur, sodium, chloride, magnesium, iron, iodide, manganese, copper, cobalt, zinc, fluoride, selenium, and chromium. Others—aluminum, boron, silicon, and molybdenum—are present but have no known functions. Typical diets supply adequate amounts of potassium, sodium, chloride, and magnesium. Some attention must be paid to eating foods that provide enough calcium, phosphorus, iron, and iodine. Excess amounts of most minerals are excreted in the urine and feces.

A major role of minerals is to help regulate enzymatic reactions. Calcium, iron, magnesium, and manganese are part of some coenzymes. Magnesium also serves as a catalyst for the conversion of ADP to ATP. Minerals such as sodium and phosphorus work in buffer systems, which help control the pH of body fluids. Sodium also helps regulate the osmosis of water and, with other ions, is involved in the generation of nerve impulses. Table 27.1 describes the roles of several minerals in various body functions.

Vitamins

Organic nutrients required in small amounts to maintain growth and normal metabolism are called ***vitamins.*** Unlike carbohydrates, lipids, or proteins, vitamins do not provide energy or serve as the body's building materials. Most vitamins with known functions serve as coenzymes.

Most vitamins cannot be synthesized by the body and must be ingested. Other vitamins, such as vitamin K, are produced by bacteria in the gastrointestinal (GI) tract and then absorbed. The body can assemble some vitamins if the raw materials, called ***provitamins,*** are provided. For example, vitamin A is produced by the body from the provitamin called beta-carotene, a chemical present in orange and yellow vegetables such as carrots and in dark green vegetables such as spinach. No single food contains all the vitamins required by the body—one of the best reasons to eat a varied diet.

Vitamins are divided into two main groups: fat-soluble and water-soluble. The ***fat-soluble vitamins*** are vitamins A, D, E, and K. They are absorbed along with dietary lipids in the small intestine and packaged into chylomicrons. They cannot be absorbed in adequate quantity unless they are ingested with other lipids. Fat-soluble vitamins may be stored in cells, particularly in the liver. The ***water-soluble vitamins*** include the B vitamins and vitamin C. They are dissolved in body fluids. Excess quantities of these vitamins are not stored but instead are excreted in the urine.

Besides their other functions, three vitamins—C, E, and beta-carotene (a provitamin for vitamin A)—are termed ***antioxidant vitamins*** because they inactivate oxygen free radicals. Recall from Chapter 2 that free radicals are highly reactive ions or molecules that carry an unpaired electron in their outermost electron shell. Free radicals damage cell membranes, DNA, and other cellular structures and contribute to the formation of atherosclerotic plaques. Some free radicals arise naturally in the body, and others come from environ-

TABLE 27.1

Minerals Vital to the Body

MINERAL	COMMENTS	IMPORTANCE
Calcium	Most abundant mineral in body. Appears in combination with phosphates. About 99% is stored in bone and teeth. Blood Ca^{2+} level is controlled by parathyroid hormone (PTH). Calcitriol promotes absorption of dietary calcium. Excess is excreted in feces and urine. Sources are milk, egg yolk, shellfish, and leafy green vegetables.	Formation of bones and teeth, blood clotting, normal muscle and nerve activity, endocytosis and exocytosis, cellular motility, chromosome movement during cell division, glycogen metabolism, and release of neurotransmitters and hormones.
Phosphorus	About 80% is found in bones and teeth as phosphate salts. Blood phosphate level is controlled by parathyroid hormone (PTH). Excess is excreted in urine; small amount is eliminated in feces. Sources are dairy products, meat, fish, poultry, and nuts.	Formation of bones and teeth. Phosphates constitute a major buffer system of blood. Plays important role in muscle contraction and nerve activity. Component of many enzymes. Involved in energy transfer (ATP). Component of DNA and RNA.
Potassium	Major cation (K^+) in intracellular fluid. Excess excreted in urine. Present in most foods (meats, fish, poultry, fruits, and nuts).	Needed for generation and conduction of action potentials in neurons and muscle fibers.
Sulfur	Component of many proteins (such as insulin), electron carriers in electron transport chain, and some vitamins (thiamine and biotin). Excreted in urine. Sources include beef, liver, lamb, fish, poultry, eggs, cheese, and beans.	As component of hormones and vitamins, regulates various body activities. Needed for ATP production by electron transport chain.
Sodium	Most abundant cation (Na^+) in extracellular fluids; some found in bones. Excreted in urine. Normal intake of NaCl (table salt) supplies more than the required amounts.	Strongly affects distribution of water through osmosis. Functions in nerve and muscle action potential conduction.
Chloride	Major anion (Cl^-) in extracellular fluid. Excess excreted in urine. Sources include table salt (NaCl), soy sauce, and processed foods.	Plays role in acid–base balance of blood, water balance, and formation of HCl in stomach.
Magnesium	Important cation (Mg^{2+}) in intracellular fluid. Excreted in urine and feces. Widespread in various foods, such as green leafy vegetables, seafood, and whole-grain cereals.	Required for normal functioning of muscle and nervous tissue. Participates in bone formation. Constituent of many coenzymes.

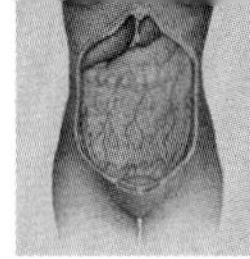

mental hazards such as tobacco smoke and radiation. Antioxidant vitamins are thought to play a role in protecting against some kinds of cancer, reducing the buildup of atherosclerotic plaque, delaying some effects of aging, and decreasing the chance of cataract formation in the lenses of the eyes. Table 27.2 lists the principal vitamins, their sources, their functions, and related deficiency disorders.

• CLINICAL CONNECTION | Supplements

Most nutritionists recommend eating a balanced diet that includes a variety of foods rather than taking vitamin or mineral **supplements,** except in special circumstances. Common examples of necessary supplementations include iron for women who have excessive menstrual bleeding; iron and calcium for women who are pregnant or breast-feeding; folic acid (folate) for all women who may become pregnant, to reduce the risk of fetal neural tube defects; calcium for most adults, because they do not receive the recommended amount in their diets; and vitamin B_{12} for strict vegetarians, who eat no meat. Because most North Americans do not ingest in their food the high levels of antioxidant vitamins thought to have beneficial effects, some experts recommend supplementing vitamins C and E. More is not always better; larger doses of vitamins or minerals can be very harmful. •

CHECKPOINT

1. Describe USDA's My Pyramid and give examples of foods from each food group.
2. Briefly describe the functions of the minerals calcium and sodium in the body.
3. Explain how vitamins are different from minerals, and distinguish between a fat-soluble vitamin and a water-soluble vitamin.

27.2 METABOLISM

OBJECTIVES

- Define metabolism and describe its importance in homeostasis.
- Explain how the body uses carbohydrates, lipids, and proteins.

Metabolism (me-TAB-ō-lizm; *metabol-* = change) refers to all the chemical reactions of the body. Recall from Chapter 2 that chemical reactions occur when chemical bonds between substances are formed or broken, and that ***enzymes*** serve as catalysts

MINERAL	COMMENTS	IMPORTANCE
Iron	About 66% found in hemoglobin of blood. Normal losses of iron occur by shedding of hair, epithelial cells, and mucosal cells, and in sweat, urine, feces, bile, and blood lost during menstruation. Sources are meat, liver, shellfish, egg yolk, beans, legumes, dried fruits, nuts, and cereals.	As component of hemoglobin, reversibly binds O_2. Present in electron transport chain.
Iodide	Essential component of thyroid hormones. Excreted in urine. Sources are seafood, iodized salt, and vegetables grown in iodine-rich soils.	Required by thyroid gland to synthesize thyroid hormones, which regulate metabolic rate.
Manganese	Some stored in liver and spleen.	Activates several enzymes. Needed for hemoglobin synthesis, urea formation, growth, reproduction, lactation, bone formation, and possibly production and release of insulin, and inhibition of cell damage.
Copper	Some stored in liver and spleen. Most excreted in feces. Sources include eggs, whole-wheat flour, beans, beets, liver, fish, spinach, and asparagus.	Required with iron for synthesis of hemoglobin. Component of coenzymes in electron transport chain and enzyme necessary for melanin formation.
Cobalt	Constituent of vitamin B_{12}.	As part of vitamin B_{12}, required for erythropoiesis.
Zinc	Important component of certain enzymes. Widespread in many foods, especially meats.	As a component of carbonic anhydrase, important in carbon dioxide metabolism. Necessary for normal growth and wound healing, normal taste sensations and appetite, and normal sperm counts in males. As a component of peptidases, it is involved in protein digestion.
Fluoride	Components of bones, teeth, other tissues.	Appears to improve tooth structure and inhibit tooth decay.
Selenium	Important component of certain enzymes. Found in seafood, meat, chicken, tomatoes, egg yolk, milk, mushrooms, and garlic, and cereal grains grown in selenium-rich soil.	Needed for synthesis of thyroid hormones, sperm motility, and proper functioning of the immune system. Also functions as an antioxidant. Prevents chromosome breakage and may play a role in preventing certain birth defects, miscarriage, prostate cancer, and coronary artery disease.
Chromium	Found in high concentrations in brewer's yeast. Also found in wine and some brands of beer.	Needed for normal activity of insulin in carbohydrate and lipid metabolism.

TABLE 27.2

The Principal Vitamins

VITAMIN	COMMENT AND SOURCE	FUNCTIONS	DEFICIENCY SYMPTOMS AND DISORDERS
FAT-SOLUBLE VITAMINS			
	All require bile salts and some dietary lipids for adequate absorption.		
A	Formed from provitamin beta-carotene (and other provitamins) in GI tract. Stored in liver. Sources of carotene and other provitamins include orange, yellow, and green vegetables; sources of vitamin A include liver and milk.	Maintains general health and vigor of epithelial cells. Beta-carotene acts as an antioxidant to inactivate free radicals. Essential for formation of light-sensitive pigments in photoreceptors of retina. Aids in growth of bones and teeth by helping to regulate activity of osteoblasts and osteoclasts.	Deficiency results in atrophy and keratinization of epithelium, leading to dry skin and hair; increased incidence of ear, sinus, respiratory, urinary, and digestive system infections; inability to gain weight; drying of cornea; and skin sores. ***Night blindness*** or decreased ability for dark adaptation. Slow and faulty development of bones and teeth.
D	In the presence of sunlight, the skin, liver, and kidneys produce active form of vitamin D (calcitriol). Stored in tissues to slight extent. Most is excreted in bile. Dietary sources include fish-liver oils, egg yolk, and fortified milk.	Essential for absorption of calcium and phosphorus from GI tract. Works with parathyroid hormone (PTH) to maintain Ca^{2+} homeostasis.	Defective utilization of calcium by bones leads to ***rickets*** in children and ***osteomalacia*** in adults. Possible loss of muscle tone.
E (tocopherols)	Stored in liver, adipose tissue, and muscles. Sources include fresh nuts and wheat germ, seed oils, and green leafy vegetables.	Inhibits catabolism of certain fatty acids that help form cell structures, especially membranes. Involved in formation of DNA, RNA, and red blood cells. May promote wound healing, contribute to the normal structure and functioning of the nervous system, and prevent scarring. May help protect liver from toxic chemicals such as carbon tetrachloride. Acts as an antioxidant to inactivate free radicals.	May cause oxidation of monounsaturated fats, resulting in abnormal structure and function of mitochondria, lysosomes, and plasma membranes. A possible consequence is ***hemolytic anemia.***
K	Produced by intestinal bacteria. Stored in liver and spleen. Dietary sources include spinach, cauliflower, cabbage, and liver.	Coenzyme essential for synthesis of several clotting factors by liver, including prothrombin.	Delayed clotting time results in excessive bleeding.
WATER-SOLUBLE VITAMINS			
	Dissolved in body fluids. Most are not stored in body. Excess intake is eliminated in urine.		
B_1 (thiamine)	Rapidly destroyed by heat. Sources include whole-grain products, eggs, pork, nuts, liver, and yeast.	Acts as a coenzyme for many different enzymes that break carbon-to-carbon bonds and are involved in carbohydrate metabolism of pyruvic acid to CO_2 and H_2O. Essential for synthesis of the neurotransmitter acetylcholine.	Improper carbohydrate metabolism leads to buildup of pyruvic and lactic acids and insufficient production of ATP for muscle and nerve cells. Deficiency leads to: (1) ***beriberi,*** partial paralysis of smooth muscle of GI tract, causing digestive disturbances; skeletal muscle paralysis; and atrophy of limbs; (2) ***polyneuritis,*** due to degeneration of myelin sheaths; impaired reflexes, impaired sense of touch, stunted growth in children, and poor appetite.

to speed up chemical reactions. Some enzymes require the presence of an ion such as calcium, iron, or zinc. Other enzymes work together with ***coenzymes,*** which function as temporary carriers of atoms being removed from or added to a substrate during a reaction. Many coenzymes are derived from vitamins. Examples include the coenzyme ***NAD*** $^{+}$***,*** derived from the B vitamin niacin, and the coenzyme ***FAD,*** derived from vitamin B_2 (riboflavin).

The body's metabolism may be thought of as an energy-balancing act between anabolic (synthesis) and catabolic

VITAMIN	COMMENT AND SOURCE	FUNCTIONS	DEFICIENCY SYMPTOMS AND DISORDERS
B_2 (riboflavin)	Small amounts supplied by bacteria of GI tract. Dietary sources include yeast, liver, beef, veal, lamb, eggs, whole-grain products, asparagus, peas, beets, and peanuts.	Component of certain coenzymes (for example, FAD) in carbohydrate and protein metabolism, especially in cells of the eye, integument, mucosa of intestine, and blood.	Deficiency may lead to improper utilization of oxygen resulting in blurred vision, cataracts, and corneal ulcerations. Also dermatitis and cracking of skin, lesions of intestinal mucosa, and one type of anemia.
Niacin (nicotinamide)	Derived from amino acid tryptophan. Sources include yeast, meats, liver, fish, whole-grain products, peas, beans, and nuts.	Essential component of NAD^+ and $NADH + H^+$, coenzymes in oxidation–reduction reactions. In lipid metabolism, inhibits production of cholesterol and assists in triglyceride breakdown.	Deficiency may lead to ***pellagra,*** characterized by dermatitis, diarrhea, and psychological disturbances.
B_6 (pyridoxine)	Synthesized by bacteria of GI tract. Stored in liver, muscles, and brain. Other sources include salmon, yeast, tomatoes, yellow corn, spinach, whole-grain products, liver, and yogurt.	Essential coenzyme for normal amino acid metabolism. Assists production of circulating antibodies. May function as coenzyme in triglyceride metabolism.	Most common deficiency symptoms are dermatitis of eyes, nose, and mouth. Other symptoms are retarded growth and nausea.
B_{12} (cyanocobalamin)	Only B vitamin not found in vegetables; only vitamin containing cobalt. Absorption from GI tract depends on intrinsic factor secreted by gastric mucosa. Sources include liver, kidneys, milk, eggs, cheese, and meat.	Coenzyme necessary for red blood cell formation, formation of the amino acid methionine, entrance of some amino acids into Krebs cycle, and manufacture of choline (used to synthesize acetylcholine).	Pernicious anemia, neuropsychiatric abnormalities (ataxia, memory loss, weakness, personality and mood changes, and abnormal sensations), and impaired activity of osteoblasts.
Pantothenic acid	Some produced by bacteria of GI tract. Stored primarily in liver and kidneys. Other sources include kidneys, liver, yeast, green vegetables, and cereal.	Constituent of coenzyme A, which is essential for transfer of acetyl group from pyruvic acid into Krebs cycle, conversion of lipids and amino acids into glucose, and synthesis of cholesterol and steroid hormones.	Fatigue, muscle spasms, insufficient production of adrenal steroid hormones, vomiting, and insomnia.
Folic acid (folate, folacin)	Synthesized by bacteria of GI tract. Dietary sources include green leafy vegetables, broccoli, asparagus, breads, dried beans, and citrus fruits.	Component of enzyme systems synthesizing nitrogenous bases of DNA and RNA. Essential for normal production of red and white blood cells.	Production of abnormally large red blood cells (macrocytic anemia). Higher risk of neural tube defects in babies born to folate-deficient mothers.
Biotin	Synthesized by bacteria of GI tract. Dietary sources include yeast, liver, egg yolk, and kidneys.	Essential coenzyme for conversion of pyruvic acid to oxaloacetic acid and synthesis of fatty acids and purines.	Mental depression, muscular pain, dermatitis, fatigue, and nausea.
C (ascorbic acid)	Rapidly destroyed by heat. Some stored in glandular tissue and plasma. Sources include citrus fruits, tomatoes, and green vegetables.	Promotes protein synthesis including laying down of collagen in the formation of connective tissue. As coenzyme, may combine with poisons, rendering them harmless until excreted. Works with antibodies, promotes wound healing, and functions as an antioxidant.	Scurvy; anemia; many symptoms related to poor collagen formation, including tender swollen gums, loosening of teeth (alveolar processes also deteriorate), poor wound healing, bleeding (vessel walls are fragile because of connective tissue degeneration), and retardation of growth.

(decomposition) reactions. Chemical reactions that combine simple substances into more complex molecules are collectively known as ***anabolism*** (a-NAB-ō-lizm; *ana-* = upward). Overall, anabolic reactions use more energy than they produce. The energy they use is supplied by catabolic reactions (Figure 27.2). One example of an anabolic process is the formation of peptide bonds between amino acids, combining them into proteins.

The chemical reactions that break down complex organic compounds into simple ones are collectively known as ***catabolism*** (ka-TAB-ō-lizm; *cata-* = downward). Catabolic reactions

Figure 27.2 Role of ATP in linking anabolic and catabolic reactions. When complex molecules are split apart (catabolism, at left), some of the energy is transferred to form ATP and the rest is given off as heat. When simple molecules are combined to form complex molecules (anabolism, at right), ATP provides the energy for synthesis, and again some energy is given off as heat.

The coupling of energy-releasing and energy-requiring reactions is achieved through ATP.

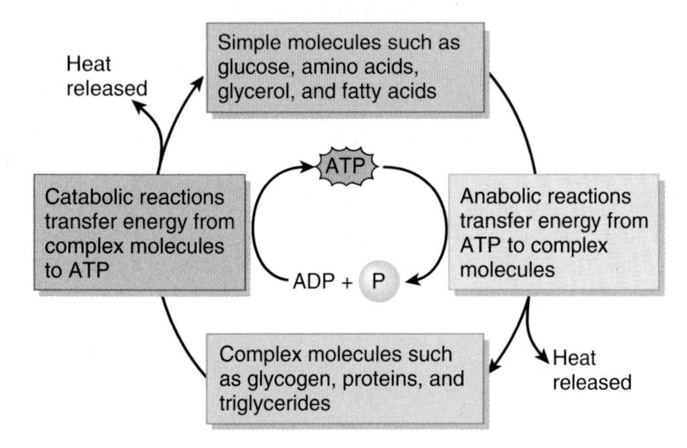

In a pancreatic cell that produces digestive enzymes, does anabolism or catabolism predominate?

release the energy stored in organic molecules. This energy is transferred to molecules of ATP and then used to power anabolic reactions. Important sets of catabolic reactions occur during glycolysis, the Krebs cycle, and the electron transport chain, which are discussed shortly.

About 40% of the energy released in catabolism is used for cellular functions; the rest is converted to heat, some of which helps maintain normal body temperature. Excess heat is lost to the environment. Compared with machines, which typically convert only 10–20% of energy into work, the 40% efficiency of the body's metabolism is impressive. Still, the body has a continuous need to take in and process external sources of energy so that cells can synthesize enough ATP to sustain life.

Carbohydrate Metabolism

During digestion, polysaccharide and disaccharide carbohydrates are catabolized to monosaccharides—glucose, fructose, and galactose—which are absorbed in the small intestine. Shortly after their absorption, however, fructose and galactose are converted to glucose. Thus, the story of carbohydrate metabolism is really the story of glucose metabolism.

Because glucose is the body's preferred source for synthesizing ATP, the fate of glucose absorbed from the diet depends on the needs of body cells. If the cells require ATP immediately, they oxidize the glucose. Glucose not needed for immediate ATP production may be converted to glycogen for storage in liver cells and skeletal muscle fibers. If these glycogen stores are full, the liver cells can transform the glucose to triglycerides for storage in adipose tissue. At a later time, when the cells need more ATP, the glycogen and triglycerides can be converted back to glucose. Cells throughout the body also can use glucose to make certain amino acids, the building blocks of proteins.

Before glucose can be used by body cells, it must pass through the plasma membrane by facilitated diffusion and enter the cytosol. Insulin increases the rate of facilitated diffusion of glucose.

Glucose Catabolism

The catabolism of glucose to produce ATP is known as ***cellular respiration.*** Overall, its many reactions can be summarized as follows.

$$\begin{aligned} &1 \text{ glucose} + 6 \text{ oxygen} \rightarrow 36\text{–}38 \text{ ATP} \\ &+ 6 \text{ carbon dioxide} + 6 \text{ water} \end{aligned}$$

Four interconnecting sets of chemical reactions contribute to cellular respiration (Figure 27.3):

1. During ***glycolysis*** (glī-KOL-i-sis; *glyco-* = sugar; *-lysis* = breakdown), reactions that take place in the cytosol convert one six-carbon glucose molecule into two three-carbon pyruvic acid molecules. The reactions of glycolysis directly produce two ATPs. They also transfer some chemical energy, in the form of high-energy electrons, from glucose to the coenzyme NAD^+, forming two NADH + H^+. Because glycolysis does not require oxygen, it is a way to produce ATP anaerobically (without oxygen) and is known as ***anaerobic cellular respiration.*** If oxygen is available, however, most cells next convert pyruvic acid to acetyl coenzyme A.
2. The formation of ***acetyl coenzyme A*** is a transition step that prepares pyruvic acid for entrance into the Krebs cycle. First, pyruvic acid enters a mitochondrion and is converted to a two-carbon fragment by removing a molecule of carbon dioxide (CO_2). Molecules of CO_2 produced during glucose catabolism diffuse into the blood and are eventually exhaled. Then, the coenzyme NAD^+ is converted to NADH + H^+. Finally, the remaining atoms, called an ***acetyl group,*** are attached to coenzyme A, to form acetyl coenzyme A.
3. The ***Krebs cycle*** is a series of reactions that transfer the chemical energy from acetyl coenzyme A to two other coenzymes—NAD^+ and FAD—thereby forming NADH + H^+ and $FADH_2$. Krebs cycle reactions also produce CO_2 and one ATP for each acetyl coenzyme A that enters the Krebs cycle. To harvest the energy in NADH and $FADH_2$, their high-energy electrons must first go through the electron transport chain.
4. Through the reactions of the ***electron transport chain,*** the energy in NADH + H^+ and $FADH_2$ is used to synthesize ATP. As the coenzymes pass their high-energy electrons through a series of "electron carriers," ATP is synthesized. Finally, lower-energy electrons are passed to oxygen in a reaction that produces water. Because the Krebs cycle and the electron transport chain together require oxygen to produce ATP, they are known as ***aerobic cellular respiration.***

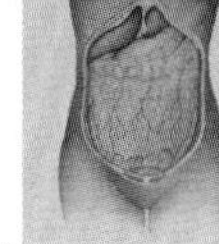

Figure 27.3 Cellular respiration.

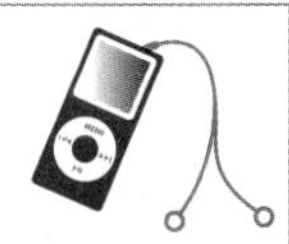

The catabolism of glucose to produce ATP involves glycolysis, the formation of acetyl coenzyme A, the Krebs cycle, and the electron transport chain.

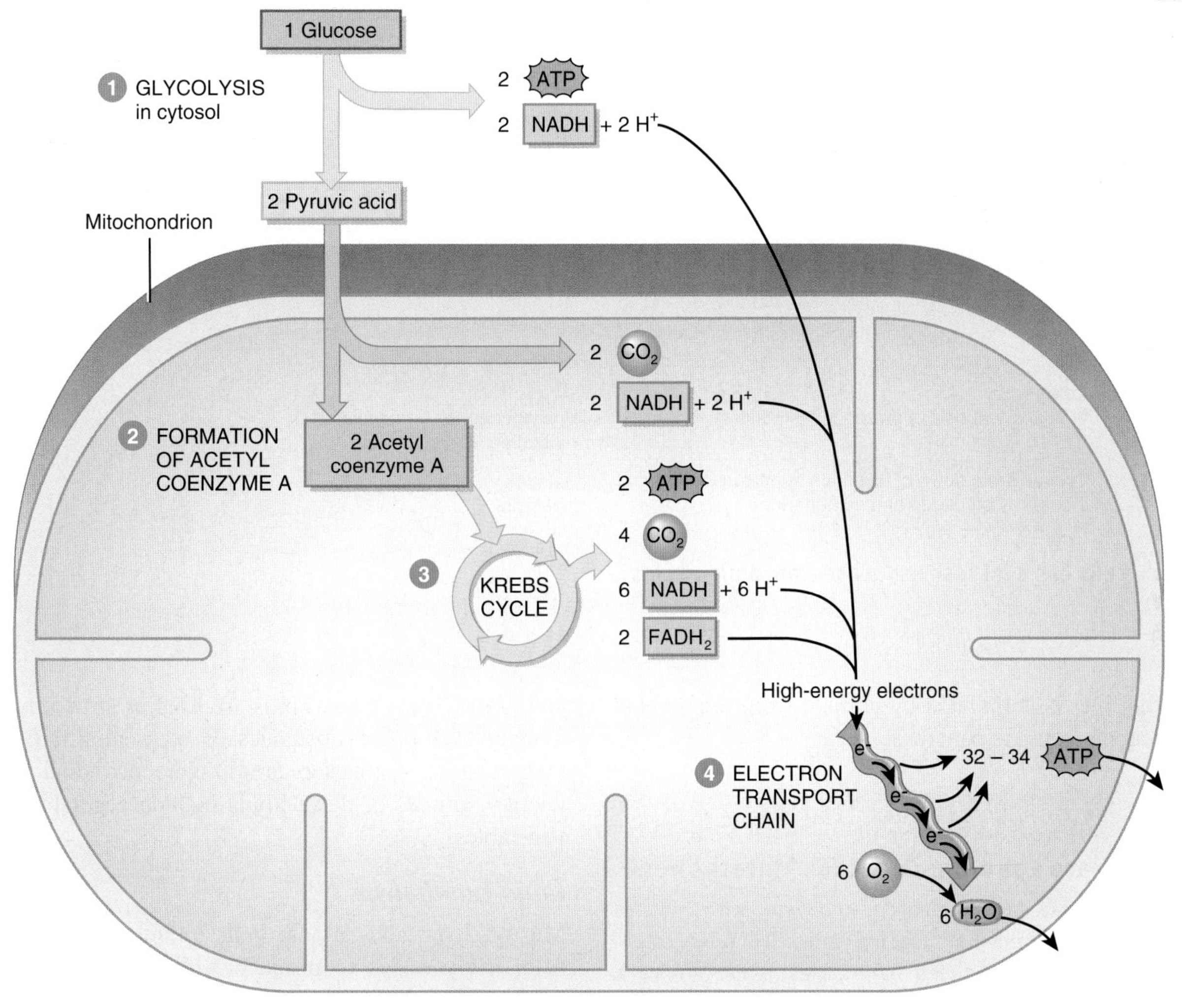

? How many molecules of ATP are produced during the complete catabolism of one molecule of glucose?

Glucose Anabolism

Even though most of the glucose in the body is catabolized to generate ATP, glucose may take part in or be formed via several anabolic reactions. One is the synthesis of glycogen; another is the synthesis of new glucose molecules from some of the products of protein and lipid breakdown.

If glucose is not needed immediately for ATP production, it combines with many other molecules of glucose to form a long-chain molecule called ***glycogen*** (see Figure 27.4). Synthesis of glycogen is stimulated by insulin. The body can store about 500 grams (about 1.1 lb) of glycogen, roughly 75% in skeletal muscle fibers and the rest in liver cells.

If blood glucose level falls below normal, glucagon is released from the pancreas and epinephrine is released from the adrenal medullae. These hormones stimulate breakdown of glycogen into its glucose subunits (see Figure 27.4). Liver cells release this glucose into the blood, and body cells pick it up to use for ATP production. Glycogen breakdown usually occurs between meals.

When your liver runs low on glycogen, it is time to eat. If you don't, your body starts catabolizing triglycerides (fats) and proteins. Actually, the body normally catabolizes some of its triglycerides and proteins, but large-scale triglyceride and protein catabolism does not happen unless you are starving, eating very few carbohydrates, or suffering from an endocrine disorder.

Liver cells can convert the glycerol part of triglycerides, lactic acid, and certain amino acids to glucose (see Figure 27.4). The series of reactions that form glucose from these noncarbohydrate sources is called ***gluconeogenesis*** (gloo′-kō-nē′-ō-JEN-e-sis; *neo-* = new). This process releases glucose into the blood, thereby keeping blood glucose level normal during the hours between meals when glucose is not being absorbed. Gluconeogenesis

Figure 27.4 Reactions of glucose anabolism: synthesis of glycogen, breakdown of glycogen, and synthesis of glucose from amino acids, lactic acid, or glycerol.

About 500 grams (1.1 lb) of glycogen are stored in skeletal muscles and the liver.

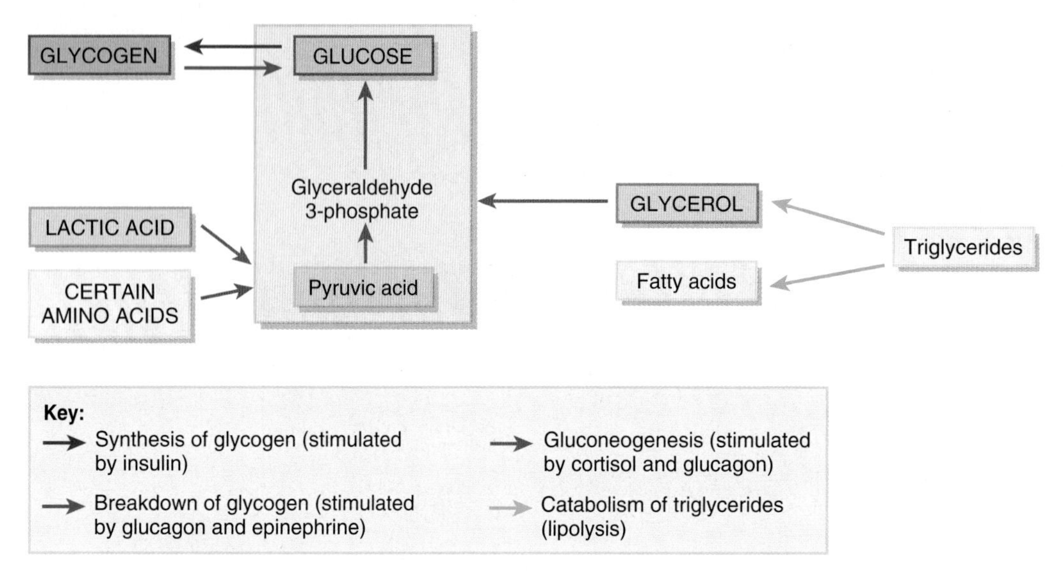

Which body cells can synthesize glucose from amino acids?

occurs when the liver is stimulated by cortisol from the adrenal cortex and glucagon from the pancreas.

MANUAL THERAPY APPLICATION

Carbohydrate Loading: Great for Marathoners

Physical therapists or massage therapists who work with athletes should understand how **carbohydrate loading** works. Two fuels, carbohydrates and fats, are primarily used by muscles while running a marathon race. As previously described, carbohydrate is stored in the liver and skeletal muscles as glycogen; fat is stored in adipose tissue. Typical marathon runners derive aboout 85% of their energy for the race from muscle glycogen.

Most people eating a normal diet can store no more than 2,000 Calories of glycogen in muscles. A mile of running burns approximately 100 Calories, and a marathon race is 26.2 miles. It is, therefore, common for less experienced marathoners to exhaust their glycogen near the 20-mile mark. Thus, many marathon runners and other endurance athletes follow a precise exercise and dietary regimen that includes eating large amounts of complex carbohydrates, such as pasta, rice, and potatoes, in the 3 days before an event. If last-minute practices are limited, endurance-trained muscles can store as much as another 900 Calories. Do the math!

Lipid Metabolism

Lipids, like carbohydrates, may be catabolized to produce ATP. If the body has no immediate need to use lipids in this way, they are stored as triglycerides in adipose tissue throughout the body, and in the liver. A few lipids are used as structural molecules or to synthesize other substances. Two essential fatty acids that the body cannot synthesize are linoleic acid and linolenic acid. Dietary sources of these lipids include vegetable oils and leafy vegetables.

Lipid Catabolism

Muscle, liver, and adipose cells routinely catabolize fatty acids from triglycerides to produce ATP. First, the triglycerides are split into glycerol and fatty acids—a process called ***lipolysis*** (li-POL-i-sis) (Figure 27.5). The hormones epinephrine, norepinephrine, and cortisol enhance lipolysis.

The glycerol and fatty acids that result from lipolysis are catabolized via different pathways. Glycerol is converted by many cells of the body to glyceraldehyde 3-phosphate. If the ATP supply in a cell is high, glyceraldehyde 3-phosphate is converted into glucose, an example of gluconeogenesis. If the ATP supply in a cell is low, glyceraldehyde 3-phosphate enters the catabolic pathway to pyruvic acid. Fatty acid catabolism begins as enzymes remove two carbon atoms at a time from the fatty acid and attach them to molecules of coenzyme A, forming acetyl coenzyme A (acetyl CoA). Then the acetyl CoA enters the Krebs cycle (Figure 27.5). A 16-carbon fatty acid such as palmitic acid can yield as many as 129 ATPs via the Krebs cycle and the electron transport chain.

As part of normal fatty acid catabolism, the liver converts some acetyl CoA molecules into substances known as ***ketone bodies*** (Figure 27.5). Ketone bodies then leave the liver to enter body cells, where they are broken down into acetyl CoA, which enters the Krebs cycle.

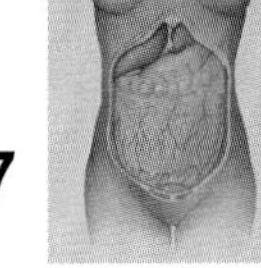

Figure 27.5 **Metabolism of lipids.** Lipolysis is the breakdown of triglycerides into glycerol and fatty acids. Glycerol may be converted to glyceraldehyde 3-phosphate, which can then be converted to glucose or enter the Krebs cycle. Fatty acid fragments enter the Krebs cycle as acetyl coenzyme A. Fatty acids also can be converted into ketone bodies.

Glycerol and fatty acids are catabolized in separate pathways.

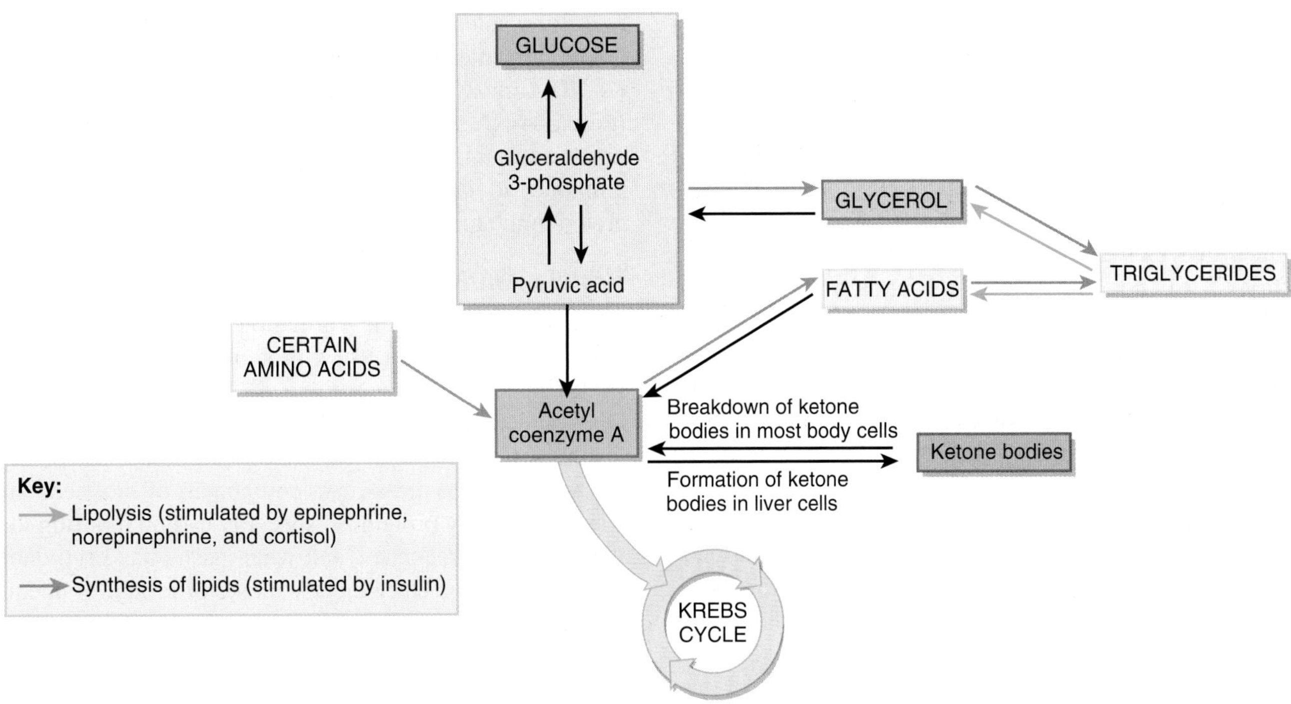

? Which cells form ketone bodies?

• CLINICAL CONNECTION | Ketosis

The level of ketone bodies in the blood normally is very low because other tissues use them for ATP production as fast as they are formed. When the concentration of ketone bodies in the blood rises above normal—a condition called **ketosis**—the ketone bodies, most of which are acids, must be buffered. If too many accumulate, blood pH falls. When a diabetic becomes seriously insulin deficient, one of the telltale signs is the sweet smell on the breath from the ketone body acetone. Prolonged ketosis can lead to *acidosis (ketoacidosis),* an abnormally low blood pH that can result in death. •

Lipid Anabolism

Insulin stimulates liver cells and adipose cells to synthesize triglycerides when more calories are consumed than are needed to satisfy ATP needs (Figure 27.5). Excess dietary carbohydrates, proteins, and fats all have the same fate—they are converted into triglycerides. Certain amino acids can undergo the following reactions: amino acid → acetyl CoA → fatty acid → triglycerides. The use of glucose to form lipids takes place via two pathways:

1. Glucose → glyceraldehyde 3-phosphate → glycerol; and
2. Glucose → glyceraldehyde 3-phosphate → acetyl CoA fatty acids.

The resulting glycerol and fatty acids can undergo anabolic reactions to become stored triglycerides, or they can go through a series of anabolic reactions to produce other lipids such as lipoproteins, phospholipids, and cholesterol.

Lipid Transport in Blood

Most lipids, such as triglycerides and cholesterol, are not water-soluble. For transport in watery blood, such molecules first are made more water-soluble by combining them with proteins. Such ***lipoproteins*** are spherical particles with an outer shell of proteins, phospholipids, and cholesterol molecules surrounding an inner core of triglycerides and other lipids. The proteins in the outer shell help the lipoprotein particles dissolve in body fluids and also have specific functions.

Lipoproteins are transport vehicles: They provide delivery and pickup services so that lipids can be available when cells need them or removed when they are not needed. Lipoproteins are categorized and named mainly according to their size and density. From largest and lightest to smallest and heaviest, the four major types of lipoproteins are chylomicrons, very low-density lipoproteins, low-density lipoproteins, and high-density lipoproteins.

1. ***Chylomicrons*** form in absorptive epithelial cells of the small intestine and transport dietary lipids to adipose tissue for storage.

2. ***Very low-density lipoproteins (VLDLs)*** transport triglycerides made in liver cells to adipose cells for storage. After depositing some of their triglycerides in adipose cells, VLDLs are converted to LDLs.

3. ***Low-density lipoproteins (LDLs)*** carry about 75% of the total cholesterol in blood and deliver it to cells throughout the body for use in repair of cell membranes and synthesis of steroid hormones and bile salts.

4. ***High-density lipoproteins (HDLs)*** remove excess cholesterol from body cells and transport it to the liver for elimination.

When present in excessive numbers, LDLs deposit cholesterol in and around smooth muscle fibers in arteries, forming fatty plaques that increase the risk of coronary artery disease (see Section 22.6). For this reason, the cholesterol in LDLs, called LDL-cholesterol, is known as "bad" cholesterol. Eating a high-fat diet increases the production of VLDLs, which elevates the LDL level and increases the formation of fatty plaques. Because HDLs prevent accumulation of cholesterol in the blood, a high HDL level is associated with decreased risk of coronary artery disease. For this reason, HDL-cholesterol is known as "good" cholesterol.

Desirable levels of blood cholesterol in adults are total cholesterol under 200 mg/dL, LDL under 130 mg/dL, and HDL over 40 mg/dL. The ratio of total cholesterol to HDL-cholesterol predicts the risk of developing coronary artery disease. A person with a total cholesterol of 180 mg/dL and HDL of 60 mg/dL has a risk ratio of 3. Ratios above 4 are considered undesirable; the higher the ratio, the greater the risk of developing coronary artery disease.

Protein Metabolism

During digestion, proteins are broken down into amino acids. Unlike carbohydrates and triglycerides, proteins are not warehoused for future use. Instead, their amino acids are either oxidized to produce ATP or used to synthesize new proteins for growth and repair of body tissues. Excess dietary amino acids are converted into glucose (gluconeogenesis) or triglycerides.

The active transport of amino acids into body cells is stimulated by insulin-like growth factors (IGFs) and insulin. Almost immediately after digestion, amino acids are reassembled into proteins. Many proteins function as enzymes; other proteins are involved in transportation (hemoglobin) or serve as antibodies, clotting factors (fibrinogen), hormones (insulin), or contractile elements in muscle fibers (actin and myosin). Several proteins serve as structural components of the body (collagen, elastin, and keratin).

Protein Catabolism

A certain amount of protein catabolism occurs in the body each day, stimulated mainly by cortisol from the adrenal cortex. Proteins from worn-out cells (such as red blood cells) are broken down into amino acids. Some amino acids are converted into other amino acids, peptide bonds are reformed, and new proteins are made as part of the recycling process. Liver cells convert some amino acids to fatty acids, ketone bodies, or glucose. Figure 27.4 shows the conversion of amino acids into glucose (gluconeogenesis). Figure 27.5 shows the conversion of amino acids into fatty acids or ketone bodies.

Amino acids also are oxidized to generate ATP. Before amino acids can enter the Krebs cycle, however, their amino group ($—NH_2$) must first be removed, a process called ***deamination*** (dē-am′-i-NĀ-shun). Deamination occurs in liver cells and produces ammonia (NH_3). Liver cells then convert the highly toxic ammonia to ***urea,*** a relatively harmless substance that is excreted in the urine.

Protein Anabolism

Protein anabolism, the formation of peptide bonds between amino acids to produce new proteins, is carried out on the ribosomes of almost every cell in the body, directed by the cells' DNA and RNA. Insulin-like growth factors, thyroid hormones, insulin, estrogens, and testosterone stimulate protein synthesis. Because proteins are a main component of most cell structures, adequate dietary protein is especially essential during the growth years, during pregnancy, and when tissue has been damaged by disease or injury. Once dietary intake of protein is adequate, eating more protein does not increase bone or muscle mass; only a regular program of forceful, weight-bearing muscular activity accomplishes that goal.

Of the 20 amino acids in the human body, 10 are ***essential amino acids:*** they must be present in the diet because they cannot be synthesized in the body in adequate amounts. ***Nonessential amino acids*** are those synthesized by the body. They are formed by the transfer of an amino group from an amino acid to pyruvic acid or to an acid in the Krebs cycle. Once the appropriate essential and nonessential amino acids are present in cells, protein synthesis occurs rapidly. Table 27.3 summarizes the processes occurring in both catabolism and anabolism of carbohydrates, lipids, and proteins.

CHECKPOINT

4. What happens during glycolysis?
5. What happens in the electron transport chain?
6. Which reactions produce ATP during the complete oxidation of a molecule of glucose?
7. What is gluconeogenesis, and why is it important?
8. What is the difference between anabolism and catabolism?
9. How does ATP provide a link between anabolism and catabolism?
10. What are the functions of the proteins in lipoproteins?
11. Which lipoprotein particles contain "good" and "bad" cholesterol, and why are these terms used?
12. Where are triglycerides stored in the body?
13. What are ketone bodies? What is ketosis?
14. What are the possible fates of the amino acids from protein catabolism?

TABLE 27.3

Summary of Metabolism

PROCESS	COMMENTS
CARBOHYDRATES	
Glucose catabolism	Complete oxidation of glucose (cellular respiration) is the chief source of ATP in cells and consists of glycolysis, the Krebs cycle, and the electron transport chain. Complete oxidation of one molecule of glucose yields a maximum of 36 or 38 molecules of ATP.
Glycolysis	Conversion of glucose into pyruvic acid results in the production of some ATP. Reactions do not require oxygen (anaerobic cellular respiration).
Krebs cycle	Cycle includes a series of oxidation–reduction reactions in which coenzymes (NAD^+ and FAD) pick up hydrogen ions and hydride ions from oxidized organic acids, and some ATP is produced. CO_2 and H_2O are by-products. Reactions are aerobic.
Electron transport chain	Third set of reactions in glucose catabolism is another series of oxidation–reduction reactions, in which electrons are passed from one carrier to the next, and most of the ATP is produced. Reactions require oxygen (aerobic cellular respiration).
Glucose anabolism	Some glucose is converted to glycogen (catabolized) for storage if not needed immediately for ATP production. Glycogen can be reconverted to glucose (glycogenolysis). The conversion of amino acids, glycerol, or lactic acid into glucose is called gluconeogenesis.
LIPIDS	
Triglyceride catabolism	Triglycerides are broken down into glycerol and fatty acids. Glycerol may be converted to glucose (gluconeogenesis) or catabolized via glycolysis. Fatty acids are catabolized into acetyl CoA that can enter the Krebs cycle for ATP production or be converted into ketone bodies (ketogenesis).
Triglyceride anabolism	The synthesis of triglycerides from glucose and fatty acids is called lipogenesis. Triglycerides are stored in adipose tissue.
PROTEINS	
Protein catabolism	Amino acids are oxidized via the Krebs cycle after deamination. Ammonia resulting from deamination is converted to urea in the liver, passed into blood, and excreted in the urine. Amino acids may be converted to glucose (gluconeogenesis), fatty acids, or ketone bodies.
Protein anabolism	Protein synthesis is directed by DNA and utilizes the cell's RNA and ribosomes.

27.3 METABOLISM AND BODY HEAT

OBJECTIVES

- Explain how body heat is produced and lost.
- Describe how body temperature is regulated.

We now consider the relationship of foods to body heat, heat production and loss, and the regulation of body temperature.

Measuring Heat

Heat is a form of energy that can be measured as ***temperature*** and expressed in units called calories. A ***calorie (cal)*** is defined as the amount of heat required to raise the temperature of 1 gram of water 1°C. Because the calorie is a relatively small unit, the ***kilocalorie (kcal)*** or ***Calorie (Cal)*** (always spelled with an uppercase C) is often used to measure the body's metabolic rate and to express the energy content of foods. A kilocalorie equals 1000 calories. Thus, when we say that a particular food item contains 500 Calories, we are actually referring to kilocalories. Knowing the caloric value of foods is important. If we know the amount of energy the body uses for various activities, we can adjust our food intake by taking in only enough kilocalories to sustain our activities.

Body Temperature Homeostasis

The body produces more or less heat depending on the rates of metabolic reactions. Homeostasis of body temperature can be maintained only if the rate of heat production by metabolism equals the rate of heat loss from the body. Thus, it is important to understand the ways in which heat can be produced and lost.

Body Heat Production

Most of the heat produced by the body comes from the catabolism of the food we eat. The rate at which this heat is produced, the ***metabolic rate,*** is measured in kilocalories. Because many factors affect metabolic rate, it is measured under standard conditions, with the body in a quiet, resting, and fasting condition called the ***basal state.*** The measurement obtained is the ***basal metabolic rate (BMR).*** BMR is 1200 to 1800 Calories per day in adults, which amounts to about 24 Calories per kilogram of body mass in adult males and 22 Calories per kilogram in adult females. The added Calories needed to support daily activities, such as digestion and walking, range from 500 Calories for a small, relatively sedentary person to over 3000 Calories for a person in training for Olympic-level competitions. The following factors affect metabolic rate:

1. **Exercise.** During strenuous exercise the metabolic rate increases by as much as 15 to 20 times the BMR.

2. **Hormones.** Thyroid hormones are the main regulators of BMR, which increases as the blood levels of thyroid hormones rise. Testosterone, insulin, and human growth hormone can increase the metabolic rate by 5–15%.

3. **Nervous system.** During exercise or in a stressful situation, the sympathetic division of the autonomic nervous system releases norepinephrine, and it stimulates release of the hormones epinephrine and norepinephrine by the adrenal medulla. Both epinephrine and norepinephrine increase the metabolic rate of body cells.

4. **Body temperature.** The higher the body temperature, the higher is the metabolic rate. As a result, metabolic rate is substantially increased during a fever.

5. **Ingestion of food.** The ingestion of food, especially proteins, can raise the metabolic rate by 10–20%.

6. **Age.** The metabolic rate of a child, in relation to its size, is about double that of an elderly person due to the high rates of growth-related reactions in children.

7. **Other factors.** Other factors that affect metabolic rate are gender (lower in females, except during pregnancy and lactation), climate (lower in tropical regions), sleep (lower), and malnutrition (lower).

Body Heat Loss

Because body heat is continuously produced by metabolic reactions, heat must also be removed continuously or body temperature would rise steadily. The principal routes of heat loss from the body to the environment are radiation, conduction, convection, and evaporation.

1. **Radiation** is the transfer of heat in the form of infrared rays between a warmer object and a cooler one without physical contact. Your body loses heat by radiating more infrared waves than it absorbs from cooler objects. If surrounding objects are warmer than you are, you absorb more heat by radiation than you lose.

2. **Conduction** is the heat exchange that occurs between two materials that are in direct contact. Body heat is lost by conduction to solid materials in contact with the body, such as your chair, clothing, and jewelry. Heat can also be gained by conduction, for example, while soaking in a hot tub.

3. **Convection** is the transfer of heat by the movement of a gas or a liquid between areas of different temperatures. The contact of air or water with your body results in heat transfer by both conduction and convection. When cool air makes contact with the body, it becomes warmed and is carried away by convection currents. The faster the air moves—for example, by a breeze or a fan—the faster the rate of convection.

4. **Evaporation** is the conversion of a liquid to a vapor. Under typical resting conditions, about 22% of heat loss occurs through evaporation of water—a daily loss of about 300 mL in exhaled air and 400 mL from the skin surface. Evaporation provides the main defense against overheating during exercise. Under extreme conditions, a maximum of about 3 liters of sweat can be produced each hour, removing more than 1700 kcal of heat if all of it evaporates. Sweat that drips off the body rather than evaporating removes very little heat.

Regulation of Body Temperature

If the amount of heat production equals the amount of heat loss, you maintain a nearly constant body temperature near 37°C (98.6°F). If your heat-producing mechanisms generate more heat than is lost by your heat-losing mechanisms, your body temperature rises. For example, strenuous exercise and some infections elevate body temperature. If you lose heat faster than you produce it, your body temperature falls. Immersion in cold water, certain diseases such as hypothyroidism, and some drugs such as alcohol and antidepressants can cause body temperature to fall. An elevated temperature may destroy body proteins, and a depressed temperature may cause cardiac arrhythmias; both can lead to death.

Fever

A ***fever*** is an elevation of body temperature that results from a resetting of the hypothalamic thermostat. The most common causes of fever are viral or bacterial infections and bacterial toxins; other causes are ovulation, excessive secretion of thyroid hormones, tumors, and reactions to vaccines. When phagocytes ingest certain bacteria, they are stimulated to secrete a ***pyrogen*** (PĪ-rō-gen; *pyro-* = fire; *-gen* = produce), a fever-producing substance. The pyrogen circulates to the hypothalamus and induces secretion of prostaglandins. Some prostaglandins can reset the hypothalamic thermostat at a higher temperature, and temperature-regulating reflex mechanisms then act to bring body temperature up to this new setting. *Antipyretics* are agents that relieve or reduce fever. Examples include aspirin, acetaminophen (Tylenol®), and ibuprofen (Advil®), all of which reduce fever by inhibiting synthesis of certain prostaglandins.

Although death results if core temperature rises above 44–46°C (112–114°F), up to a point, fever is beneficial. For example, a higher temperature intensifies the effect of interferon and the phagocytic activities of macrophages while hindering replication of some pathogens. Because fever increases heart rate, infection-fighting white blood cells are delivered to sites of infection more rapidly. In addition, antibody production and T cell proliferation increase.

The balance between heat production and heat loss is controlled by neurons in the hypothalamus. These neurons generate more nerve impulses when blood temperature increases and fewer impulses when blood temperature decreases. If body temperature falls, mechanisms that help conserve heat and increase heat production act by means of several negative feedback loops to raise the body temperature to normal (Figure 27.6). Thermoreceptors send nerve impulses to the hypothalamus, which produces a releasing hormone called thyrotropin-releasing hormone (TRH). TRH in turn stimulates the anterior pituitary to release thyroid-stimulating hormone (TSH). Nerve impulses from the hypothalamus and TSH then activate several effectors:

- Sympathetic nerves cause blood vessels of the skin to constrict (vasoconstriction). The decrease of blood flow slows the rate of heat loss from the skin. Because less heat is lost,

Figure 27.6 Negative feedback mechanisms that increase heat production.

When stimulated, the heat-promoting center in the hypothalamus raises body temperature.

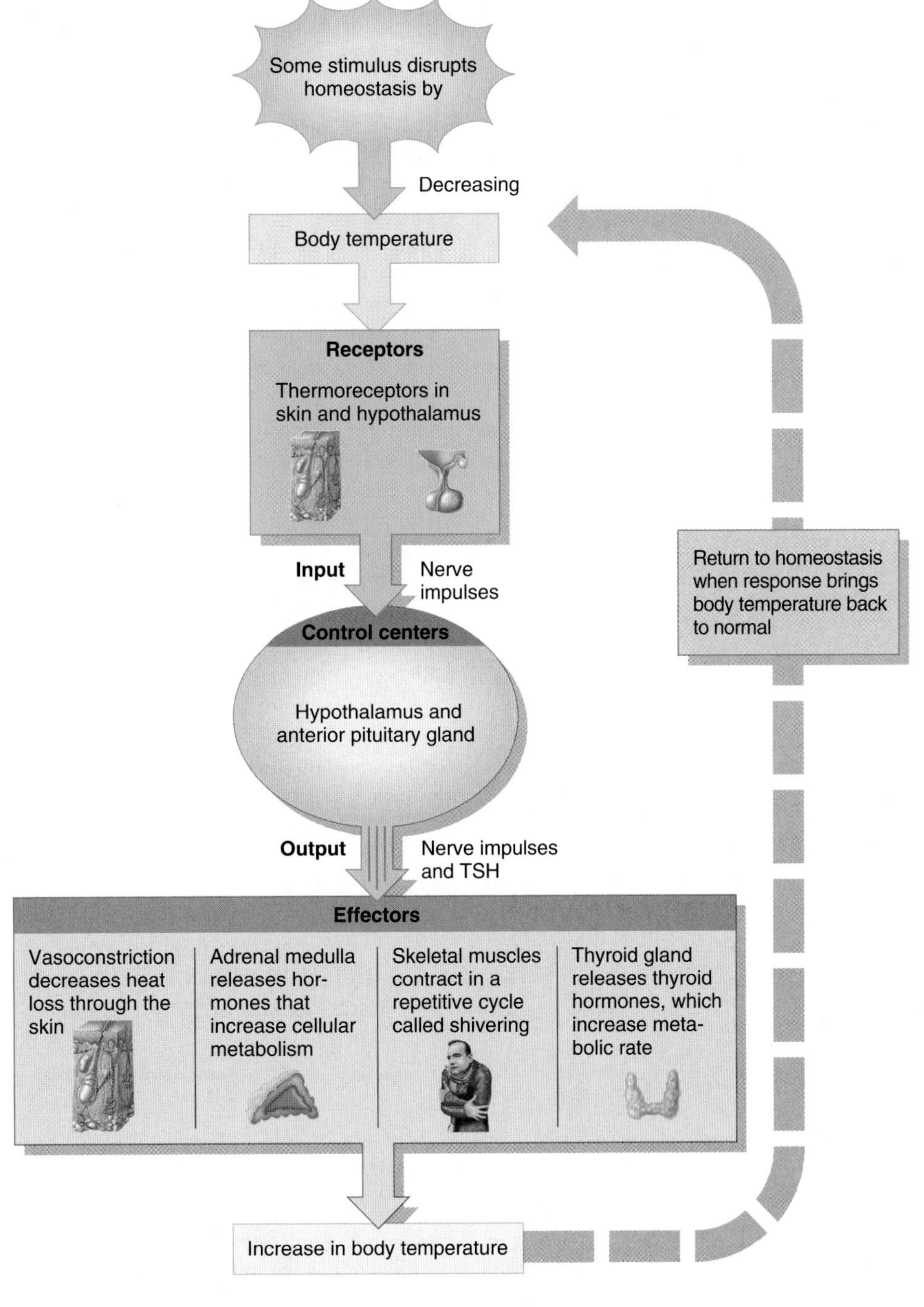

What factors can increase metabolic rate and thus increase the rate of heat production?

body temperature increases even if the metabolic rate remains the same.

- Sympathetic nerves stimulate the adrenal medulla to release epinephrine and norepinephrine into the blood. These hormones increase cellular metabolism, which increase heat production.
- The hypothalamus stimulates parts of the brain that increase muscle tone. As muscle tone increases in one muscle (the agonist), the small contractions stretch muscle spindles in its antagonist muscle, initiating a stretch reflex. The resulting contraction in the antagonist stretches muscle spindles in the agonist, and it too develops a stretch reflex. This repetitive cycle—called ***shivering***—greatly increases the rate of heat production. During maximal shivering, body heat production can rise to about four times the basal rate in just a few minutes.
- The thyroid gland responds to TSH by releasing more thyroid hormones into the blood, increasing the metabolic rate.

If body temperature rises above normal, a negative feedback system opposite to the one depicted in Figure 27.6 goes into action. The higher temperature of the blood stimulates the hypothalamus. Nerve impulses cause dilation of blood vessels in the skin. The skin becomes warm, and the excess heat is lost to the environment by radiation and conduction as an increased volume of blood flows from the warmer interior of the body into the cooler skin. At the same time, metabolic rate decreases, and the high temperature of the blood stimulates sweat glands of the skin by means of hypothalamic activation of sympathetic nerves. As the water in sweat evaporates from the surface of the skin, the skin is cooled. All these responses counteract heat-promoting effects and help return body temperature to normal.

Hypothermia is a lowering of core body temperature to 35°C (95°F) or below. Causes of hypothermia include an overwhelming cold stress (immersion in icy water), metabolic diseases (hypoglycemia, adrenal insufficiency, or hypothyroidism), drugs (alcohol, antidepressants, sedatives, or tranquilizers), burns, and malnutrition. Symptoms of hypothermia include sensation of cold, shivering, confusion, vasoconstriction, muscle rigidity, slow heart rate, loss of spontaneous movement, and coma. Death is usually caused by cardiac arrhythmias. Because the elderly have reduced metabolic protection against a cold environment coupled with a reduced perception of cold, they are at greater risk for developing hypothermia.

• CLINICAL CONNECTION | Exercise Training and Metabolism

Athletes spend hours a day training for their sports and there is a direct relationship between **exercise training and metabolism.** Many physiological changes occur as a result of all this training, including an increased ability to produce ATP for muscle contraction. These improvements are specific to the metabolic pathways that are used during the training. Athletes design their training programs to challenge the ATP production system or systems most vital to their sports.

Some sports require a short burst of power, a high energy output that lasts only a few seconds. Such events include the 100-meter sprint, the shot put, the discus throw, and the 25-meter swim. Muscle contraction for these events is supplied primarily by existing ATP and creatine phosphate. (Recall from Chapter 10 that creatine phosphate can donate a phosphate group to ADP to restore ATP.) Athletes improve their ability to generate power by practicing their events, over and over. Power lifting—forcefully lifting very heavy weights—also challenges the muscles to produce more ATP faster. In response to such training, the concentration of enzymes required for these ATP-production pathways as well as the levels of ATP and creatine phosphate increase in trained muscles.

Many athletic events require well-trained glycolytic pathways. Anaerobic glycolysis, combined with ATP and creatine phosphate, provides most of the energy for high-intensity exercise lasting up to 90 seconds, such as a 400-meter run or a 100-meter swim. Many sports, such as basketball, soccer, and tennis, require bursts of high-ATP production by anaerobic glycolysis interspersed with somewhat lower energy output. Athletes train the anaerobic glycolytic pathway by exercising at high intensities for periods of a minute or longer. Interval training includes both high-intensity work and periods of lower-intensity work or rest. In response to high-intensity exercise, the concentration of enzymes required for anaerobic glycolysis increases.

Most athletic events, including all events lasting longer than a few minutes, require aerobic cellular respiration. Athletes improve these ATP-production pathways by exercising at moderate to vigorous intensities for extended periods, with or without high-intensity intervals. Aerobic training increases the size and number of mitochondria as well as the concentration of enzymes required for ATP production by aerobic pathways. •

CHECKPOINT

15. In what ways can a person lose heat to or gain heat from the surroundings? How is it possible for a person to lose heat on a sunny beach when the temperature is 40°C (104°F) and the humidity is 85%?

KEY MEDICAL TERMS ASSOCIATED WITH NUTRITION AND METABOLISM

Heat cramps Cramps that result from profuse sweating. The salt lost in sweat causes painful contractions of muscles; such cramps tend to occur in muscles used while working but do not appear until the person relaxes once the work is done. Drinking salted liquids usually leads to rapid improvement.

Heatstroke (sunstroke) A severe and often fatal disorder caused by exposure to high temperatures. Blood flow to the skin is decreased, perspiration is greatly reduced, and body temperature rises sharply because of failure of the hypothalamic thermostat. Body temperature may reach 43°C (110°F). Treatment, which must be undertaken immediately, consists of cooling the body by immersing the victim in cool water and by administering fluids and electrolytes.

Kwashiorkor (kwash′-ē-OR-kor) A disorder in which protein intake is deficient despite normal or nearly normal caloric intake, characterized by edema of the abdomen, enlarged liver, decreased blood pressure, low pulse rate, lower than normal body temperature, and sometimes mental retardation. Because the main protein in corn lacks two essential amino acids, which are needed for growth and tissue repair, many African children whose diet consists largely of cornmeal develop kwashiorkor.

Malnutrition (*mal-* = bad) An imbalance of total caloric intake or intake of specific nutrients, which can be either inadequate or excessive.

Marasmus (mar-AZ-mus) A type of undernutrition that results from inadequate intake of both protein and calories. Its characteristics include retarded growth, low weight, muscle wasting, emaciation, dry skin, and thin, dry, dull hair.

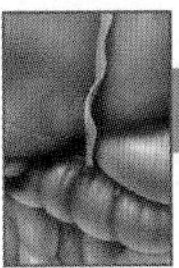

STUDY OUTLINE

Nutrients (Section 27.1)

1. Nutrients include carbohydrates, lipids, proteins, water, minerals, and vitamins.
2. Nutrition experts suggest dietary calories be 50–60% from carbohydrates, 30% or less from fats, and 12–15% from proteins.
3. The My Pyramid guide represents a personalized approach to making healthy food choices and maintaining regular physical activity.
4. Some minerals known to perform essential functions include calcium, phosphorus, potassium, sodium, chloride, magnesium, iron, manganese, copper, and zinc. Their functions are summarized in Table 27.1.
5. Vitamins are organic nutrients that maintain growth and normal metabolism. Many function as coenzymes.
6. Fat-soluble vitamins are absorbed with fats and include vitamins A, D, E, and K; water-soluble vitamins are absorbed with water and include the B vitamins and vitamin C.
7. The functions of the principal vitamins and their deficiency disorders are summarized in Table 27.2.

Metabolism (Section 27.2)

1. Metabolism refers to all chemical reactions of the body and has two phases: catabolism and anabolism. Anabolism consists of reactions that combine simple substances into more complex molecules. Catabolism consists of reactions that break down complex organic compounds into simple ones.
2. Metabolic reactions are catalyzed by enzymes, proteins that speed up chemical reactions without being changed.
3. Anabolic reactions require energy, which is supplied by catabolic reactions.
4. During digestion, polysaccharides and disaccharides are converted to glucose.
5. Glucose moves into cells by facilitated diffusion, which is stimulated by insulin. Some glucose is catabolized by cells to produce ATP. Excess glucose can be stored by the liver and skeletal muscles as glycogen or converted to fat.
6. Glucose catabolism is also called cellular respiration. The complete catabolism of glucose to produce ATP involves glycolysis, the Krebs cycle, and the electron transport chain. It can be represented as follows: 1 glucose + 6 oxygen $\longrightarrow$ 36 or 38 ATP + 6 carbon dioxide + 6 water.
7. Glycolysis is also called anaerobic respiration because it occurs without oxygen. During glycolysis, which occurs in the cytosol, one glucose molecule is broken down into two molecules of pyruvic acid. Glycolysis yields a net of two ATP and two NADH + H^+.
8. When oxygen is plentiful, most cells convert pyruvic acid to acetyl coenzyme A, which enters the Krebs cycle.
9. The Krebs cycle occurs in mitochondria. The chemical energy originally contained in glucose, pyruvic acid, and acetyl coenzyme A is transferred to the coenzymes NADH and $FADH_2$.
10. The electron transport chain is a series of reactions that occur in mitochondria in which the energy in the reduced coenzymes is transferred to ATP.
11. The conversion of glucose to glycogen for storage occurs extensively in liver and skeletal muscle fibers and is stimulated by insulin. The body can store about 500 g of glycogen.
12. The breakdown of glycogen to glucose occurs mainly between meals.
13. Gluconeogenesis is the conversion of glycerol, lactic acid, or amino acids to glucose.
14. Some triglycerides may be catabolized to produce ATP; others are stored in adipose tissue. Other lipids are used as structural molecules or to synthesize other substances.
15. Triglycerides must be split into fatty acids and glycerol before they can be catabolized. Glycerol can be transformed into glucose by conversion into glyceraldehyde 3-phosphate. Fatty acids are catabolized through formation of acetyl coenzyme A, which can enter the Krebs cycle.
16. The formation of ketone bodies by the liver is a normal phase of fatty acid catabolism, but an excess of ketone bodies, called ketosis, may cause acidosis.
17. The conversion of glucose or amino acids into lipids is stimulated by insulin.
18. Lipoproteins transport lipids in the bloodstream. Types of lipoproteins include chylomicrons, which carry dietary lipids to adipose tissue; very low-density lipoproteins (VLDLs), which carry triglycerides from the liver to adipose tissue; low-density lipoproteins (LDLs), which deliver cholesterol to body cells; and high-density lipoproteins (HDLs), which remove excess cholesterol from body cells and transport it to the liver for elimination.
19. Amino acids, under the influence of insulin-like growth factors and insulin, enter body cells by means of active transport. Inside cells, amino acids are reassembled into proteins that function as enzymes, hormones, structural elements, and so forth; stored as fat or glycogen; or used for ATP production.
20. Before amino acids can be catabolized, they must be deaminated. Liver cells convert the resulting ammonia to urea, which is excreted in urine.

21. Amino acids may also be converted into glucose, fatty acids, and ketone bodies.
22. Protein synthesis is stimulated by insulin-like growth factors, thyroid hormones, insulin, estrogen, and testosterone. It is directed by DNA and RNA and carried out on ribosomes.
23. Table 27.3 summarizes carbohydrate, lipid, and protein metabolism.

Metabolism and Body Heat (Section 27.3)

1. A calorie is the amount of energy required to raise the temperature of 1 gram of water 1°C.
2. The calorie is the unit of heat used to express the caloric value of foods and to measure the body's metabolic rate. One Calorie equals 1000 calories, or 1 kilocalorie.
3. Most body heat is a result of catabolism of the food we eat. The rate at which this heat is produced is known as the metabolic rate and is affected by exercise, hormones, the nervous system, body temperature, ingestion of food, age, gender, climate, sleep, and nutrition.
4. Measurement of the metabolic rate under basal conditions is called the basal metabolic rate (BMR).
5. Mechanisms of heat loss are radiation, conduction, convection, and evaporation.
6. Radiation is the transfer of heat from a warmer object to a cooler object without physical contact.
7. Conduction is the transfer of heat between two objects in contact with each other.
8. Convection is the transfer of heat by the movement of a liquid or gas between areas of different temperatures.
9. Evaporation is the conversion of a liquid to a vapor; in the process, heat is lost.
10. A normal body temperature is maintained by negative feedback loops that regulate heat-producing and heat-losing mechanisms.
11. Responses that produce or retain heat when body temperature falls are vasoconstriction; release of epinephrine, norepinephrine, and thyroid hormones; and shivering.
12. Responses that increase heat loss when body temperature rises include vasodilation, decreased metabolic rate, and evaporation of sweat.

SELF-QUIZ QUESTIONS

1. Creating a protein from amino acids is an example of
   a. deamination b. anabolism
   c. gluconeogenesis d. catabolism
   e. cellular respiration.
2. Free radicals
   a. are a type of provitamin
   b. are essential amino acids
   c. can cause damage to cellular structures
   d. help regulate enzymatic reactions
   e. are a form of energy.
3. Which of the following statements about vitamins is NOT true?
   a. Most vitamins are synthesized by body cells.
   b. Vitamins can act as coenzymes.
   c. Vitamin K is produced by bacteria in the GI tract.
   d. Lipid-soluble vitamins may be stored in the liver.
   e. Excess water-soluble vitamins are excreted in urine.
4. Match the following:
   ____ a. precursor for vitamin A
   ____ b. form in which lipids are transported in the blood plasma
   ____ c. needed to convert ADP to ATP
   ____ d. derived from vitamin B_2 (riboflavin)

   A. lipoproteins
   B. FAD
   C. beta-carotene
   D. magnesium
5. Body temperature is controlled by the
   a. pons b. thyroid gland
   c. hypothalamus d. adrenal medulla
   e. autonomic nervous system.
6. The removal of an amino group (—NH_2) from amino acids entering the Krebs cycle is known as
   a. deamination b. convection c. ketogenesis
   d. lipolysis e. aerobic respiration.
7. If your diet is low in carbohydrates, which compound(s) does your body begin to catabolize next for ATP production?
   a. vitamins b. lipids c. minerals
   d. cholesterol e. amino acids.
8. Cellular respiration includes the following steps in order:
   a. Krebs cycle, glycolysis, electron transport chain
   b. Krebs cycle, electron transport chain, glycolysis
   c. glycolysis, electron transport chain, Krebs cycle
   d. electron transport chain, Krebs cycle, glycolysis
   e. glycolysis, Krebs cycle, electron transport chain.
9. Which of the following is most often used to synthesize ATP?
   a. galactose b. triglycerides c. amino acids
   d. glucose e. glycerol.
10. How does glucose enter the cytosol of cells?
    a. facilitated diffusion b. simple diffusion
    c. active transport d. osmosis
    e. electron transport.
11. Sweat drying from a person's skin surface causes loss of body heat by
    a. radiation b. conduction c. convection
    d. evaporation e. conversion.
12. Glycolysis
    a. requires the presence of oxygen
    b. produces two ATP molecules per glucose molecule
    c. takes place in mitochondria
    d. is also known as the Krebs cycle
    e. is the conversion of glucose to glycogen.
13. Which of the following statements is NOT true?
    a. Triglycerides are stored in adipose tissue.
    b. Most lipids are water-soluble.
    c. Most of the body's cholesterol is carried in low-density lipoproteins.
    d. Lipids can be stored in the liver.
    e. High-density lipoproteins contribute to the formation of fatty plaques.
14. Which of the following equations summarizes the complete catabolism of a molecule of glucose?
    a. glucose + 6 water → 36 or 38 ATP + 6 CO_2 + 6 O_2
    b. glucose + 6 O_2 → 36 or 38 ATP + 6 CO_2 + 6 water
    c. glucose + ATP → 31 or 38 CO_2 + 6 water
    d. glucose + pyruvic acid → 36 or 38 ATP + 6 O_2
    e. glucose + citric acid → 31 or 38 ATP + 6 CO_2

15. Those amino acids that cannot by synthesized by the body and must be obtained from the diet are known as
    a. coenzymes b. ketones
    c. essential amino acids d. nonessential amino acids
    e. polypeptides.
16. Which of the following would NOT increase the metabolic rate?
    a. increased levels of thyroid hormones
    b. epinephrine
    c. old age
    d. fever
    e. exercise.
17. FAD and NAD^+ are examples of
    a. nutrients b. antioxidants c. pyrogens
    d. coenzymes e. minerals.
18. The process by which glucose is formed from amino acids is
    a. gluconeogenesis b. deamination
    c. anaerobic respiration d. ketogenesis
    e. glycolysis.
19. All of the following can contribute to an increase in body temperature EXCEPT
    a. shivering
    b. release of thyroid hormones
    c. sympathetic stimulation of the adrenal medulla
    d. vasodilation of blood vessels in the skin
    e. activation of the hypothalamus.
20. Match the following:

____	a. conversion of glucose to pyruvic acid	A. catabolism
____	b. the complete breakdown of glucose	B. anabolism
____	c. building simple molecules into more complex ones	C. glycolysis
____	d. NAD^+ and FAD pick up high-energy electrons	D. cellular respiration
____	e. the breakdown of organic compounds	E. Krebs cycle

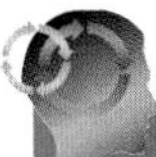

CRITICAL THINKING QUESTIONS

1. Carla and Ashley, members of their college's tennis team, ate lunch at McDonald's before their afternoon practice. Carla had a Quarter Pounder with cheese, small french fries, and a small chocolate shake; Ashley had a Chicken McGrill (plain, without mayonnaise), a garden salad with fat-free vinaigrette, and a glass of 1% low-fat milk. Critique their choices based on the recommended distributions of calories. How many calories are left for breakfast, dinner, and snacks?

QUARTER POUNDER MEAL (percentages of total calories)	GRILLED CHICKEN SANDWICH MEAL (percentages of total calories)
Total fats = 41%	Total fats = 24%
Saturated fats = 17%	Saturated fats = 9%
Total carbohydrates = 46%	Total carbohydrates = 47%
Simple sugars = 23%	Simple sugars = 15%
Proteins = 13%	Proteins = 29%
Total Calories = 1085	Total Calories = 592

2. It's noon on a hot summer day, the sun is directly overhead, and a group of sunbathers roasts on the beach. What mechanism causes their body temperature to increase? Several of the sunbathers jump into the cool water. What mechanisms decrease their body temperature?
3. Shannon is a morning person, but her roommate Darla is not. In fact, Shannon teases Darla for being a classic example of "BMR" (barely mentally responsive) during her 8 A.M. class. What does BMR really mean? How is metabolism measured?
4. Rob swallows a multivitamin tablet every morning and an antioxidant tablet containing beta-carotene, vitamin C, and vitamin E with his dinner every night. What are the functions of antioxidants in the body? What happens to the antioxidants if any exceed his daily requirements?

? ANSWERS TO FIGURE QUESTIONS

27.1 The wider base of each band represents foods with little or no solid fats or added sugars.

27.2 The formation of digestive enzymes in the pancreas is part of anabolism.

27.3 Complete catabolism of glucose yields 36 to 38 molecules of ATP.

27.4 Liver cells carry out gluconeogenesis.

27.5 Liver cells form ketone bodies.

27.6 Exercise, the sympathetic nervous system, hormones (epinephrine, norepinephrine, thyroid hormones, testosterone, and human growth hormone), elevated body temperature, and ingestion of food are factors that increase metabolic rate.

28 The Urinary System

As body cells carry out their metabolic functions, they consume oxygen and nutrients and produce substances, such as carbon dioxide, that have no useful functions and need to be eliminated from the body. While the respiratory system rids the body of carbon dioxide, the urinary system disposes of most other unneeded substances. As you will learn in this chapter, however, the urinary system is not merely concerned with waste disposal; it carries out a number of other important functions as well.

The kidneys allow you to conserve water by secreting hormones to reabsorb water that is normally lost to the urine. When you exercise you remove toxins in your body through increased sweating and therefore lose more water than when you are at rest. Not only do you lose water, but important ions such as sodium and potassium (important to electrical transmission in your body) are also lost. This is why athletes often drink sports drinks that contain water as well as sodium, potassium, and glucose (for quick energy supply). Most people think that the first sign of dehydration is thirst and are unaware that muscle fatigue and soreness occur first and thirst later. In order to ensure that you are getting enough fluids you should really never let yourself get to the point of thirst. When you partake in exercise or activities like massage that increase your metabolism, it is even more important to drink fluids or suffer the consequences of lactic acid buildup and muscle fatigue to the point of muscle spasm.

CONTENTS AT A GLANCE

28.1 OVERVIEW OF THE URINARY SYSTEM

OBJECTIVE

- List the components of the urinary system and their general functions.

The ***urinary system*** consists of two kidneys, two ureters, one urinary bladder, and one urethra (Figure 28.1). After the kidneys filter blood, they return most of the water and many of the solutes to the bloodstream. The remaining water and solutes constitute ***urine,*** which passes through the ureters and is stored in the urinary bladder until it is expelled from the body through the urethra. ***Nephrology*** (nef-ROL-ō-jē; *nephro-* = kidney; *-logy* = study of) is the scientific study of the anatomy, physiology, and disorders of the kidneys. The branch of medicine that deals with the male and female urinary systems and the male reproductive system is called ***urology*** (ū-ROL-ō-jē; *uro-* = urine). A physician who specializes in this branch of medicine is called a ***urologist*** (ū-ROL-ō-jist).

The kidneys do the major work of the urinary system. The other parts of the system are primarily passageways and temporary storage areas. The kidneys function to regulate the volume and chemistry of the blood. More specifically, functions of the kidneys include the following:

- **Regulation of ion levels in the blood.** The kidneys help regulate the blood levels of several ions, most importantly sodium ions (Na^+), potassium ions (K^+), calcium ions (Ca^{2+}), chloride ions (Cl^-), and phosphate ions (HPO_4^{2-}).
- **Regulation of blood volume and blood pressure.** The kidneys adjust the volume of blood in the body by returning water to the blood or eliminating it in the urine. They help regulate blood pressure by secreting the enzyme renin, which activates the renin–angiotensin–aldosterone pathway (see Figure 20.14, Section 20.7), by adjusting blood flow into and out of the kidneys, and by adjusting blood volume.
- **Regulation of blood pH.** The kidneys regulate the concentration of H^+ in the blood by excreting a variable amount of H^+ in the urine. They also conserve blood bicarbonate ions (HCO_3^-), an important buffer of H^+. Both activities help regulate blood pH.
- **Production of hormones.** The kidneys produce two hormones. *Calcitriol,* the active form of vitamin D, helps

Figure 28.1 **Organs of the urinary system in a female.**

Urine formed by the kidneys passes first into the ureters, then to the urinary bladder for storage, and finally through the urethra for elimination from the body.

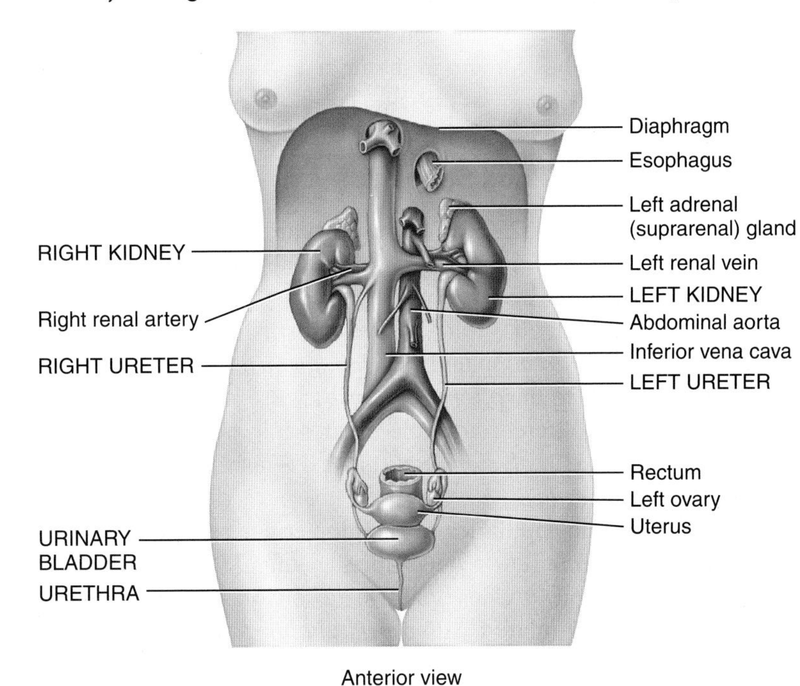

Functions of the Urinary System

1. The kidneys regulate blood volume and composition, help regulate blood pressure, synthesize glucose, release erythropoietin, participate in vitamin D synthesis, and excrete wastes by forming urine.
2. The ureters transport urine from the kidneys to the urinary bladder.
3. The urinary bladder stores urine.
4. The urethra discharges urine from the body.

 **Which organs constitute the urinary system?**

regulate calcium homeostasis (see Figure 20.10, Section 20.5), and *erythropoietin* stimulates the production of red blood cells (see Figure 21.3, Section 21.2).

- **Excretion of wastes.** By forming urine, the kidneys help excrete ***wastes***—substances that have no useful function in the body. Some wastes excreted in urine result from metabolic reactions in the body. These include ammonia and urea from the breakdown of amino acids; bilirubin from the breakdown of hemoglobin; creatinine from the breakdown of creatine phosphate in muscle fibers; and uric acid from the breakdown of nucleic acids. Other wastes excreted in urine are foreign substances from the diet, such as drugs and environmental toxins.

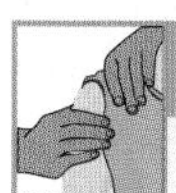

MANUAL THERAPY APPLICATION

Massage and the Kidney

The primary functions of the urinary system include regulation of ion levels in the *blood,* regulation of *blood* volume and *blood* pressure, regulation of *blood* pH, production of two hormones (one stimulates the development of red *blood* cells and the other regulates *blood* pressure), and the excretion of wastes from the *blood.* As discussed in chapters involving the cardiovascular system (Chapters 21–23), **massage** can clearly have a very positive effect on the movement of blood throughout the body and therefore very positively enhance the functions of the urinary system. Massage of the abdomen can also stimulate micturition.

CHECKPOINT

1. What are wastes, and how do the kidneys take part in their removal from the body?

28.2 STRUCTURE OF THE KIDNEYS

OBJECTIVE

- Describe the structure and blood supply of the kidneys.

The ***kidneys*** (KID-nēz) are a pair of reddish organs shaped like kidney beans (Figure 28.2). The kidneys are retroperitoneal, that is, they are between the peritoneum and the posterior wall of the abdominal cavity. The kidneys are located on either side of the vertebral column at the level of the 12th thoracic and first three lumbar vertebrae. The 11th and 12th pairs of ribs provide some protection for the superior parts of the kidneys. The right kidney is slightly lower than the left because the liver occupies a large area above the kidney on the right side.

External Anatomy of the Kidneys

An adult kidney is about the size of a bar of bath soap. Near the center of the medial border is an external indentation called the ***renal hilum*** (HĪ-lum), through which the ureter leaves the kidney and blood vessels, lymphatic vessels, and nerves enter and exit. Surrounding each kidney is the smooth, transparent ***renal capsule,*** a connective tissue sheath that helps maintain the shape of the kidney and serves as a barrier against trauma (Figure 28.2). Adipose (fatty) tissue surrounds the renal capsule and cushions the kidney. Along with a thin layer of dense irregular connective tissue, the adipose tissue anchors the kidney to the posterior abdominal wall.

Internal Anatomy of the Kidneys

Internally, the kidneys have two main regions: an outer, lighter red region called the ***renal cortex*** (*cortex* = rind or back) and an inner, darker red-brown region called the ***renal medulla*** (*medulla* = inner portion) (Figure 28.2). Within the renal medulla are several cone-shaped ***renal pyramids.*** The base (wider end) of each pyramid faces the renal cortex, and its apex (narrower end), called a ***renal papilla,*** points toward the renal hilum. Extensions of the renal cortex, called ***renal columns,*** fill the spaces between renal pyramids. A ***renal lobe*** consists of a renal pyramid, its overlying area of renal cortex, and one-half of each adjacent renal column.

Urine formed in the kidney drains into a large, funnel-shaped cavity called the ***renal pelvis*** (*pelv-* = basin). The rim of the renal pelvis contains cuplike structures called ***major*** and ***minor calyces*** (KĀL-i-sēz = cups; singular is *calyx*). Urine flows from several ducts within the kidney into a minor calyx and from there through a major calyx into the renal pelvis, which connects to a ureter. Water and solutes in the fluid that drains into the renal pelvis remain in the urine and are *excreted* (eliminated from the body).

Renal Blood Supply

About 20–25% of the resting cardiac output—1200 milliliters of blood per minute—flows into the kidneys through the right and left ***renal arteries*** (see Figure 28.3). Within each kidney, the renal artery divides into smaller and smaller vessels (*segmental, interlobar, arcuate,* and *interlobular arteries*) that eventually deliver blood to the ***afferent arterioles*** (*af-* = toward; *-ferre* = to carry). Each afferent arteriole divides into a tangled capillary network called a ***glomerulus*** (glō-MER-ū-lus = little ball; plural is *glomeruli*).

The capillaries of the glomerulus reunite to form an ***efferent arteriole*** (*ef-* = out). Upon leaving the glomerulus, each efferent arteriole divides to form a network of capillaries around the kidney tubules called ***peritubular capillaries*** (*peri-* = around). Extending from some efferent arterioles are long loop-shaped capillaries called ***vasa recta*** (VĀ-sa REK-ta; *vasa* = vessels; *recta* = straight) that supply tubular portions of the nephron in the renal medulla. The peritubular capillaries eventually reunite to form *peritubular veins,* which merge into *interlobular, arcuate,* and *interlobar veins.* Ultimately, all these smaller veins drain into the ***renal vein.***

• CLINICAL CONNECTION | Kidney Transplant

A **kidney transplant** is the transfer of a kidney from a donor to a recipient whose kidneys no longer function. In the procedure, the donor kidney is placed in the pelvis of the recipient through an abdominal incision. The renal artery and vein of the transplanted kidney are attached to a nearby artery or vein in the pelvis of the recipient and the ureter of the transplanted kidney is then attached to the urinary bladder. During a kidney transplant, the patient receives only one donor kidney, since only one kidney is needed to maintain sufficient renal function. The nonfunctioning diseased kidneys are usually left in place. As with all organ transplants, kidney transplant recipients must be ever vigilant for signs of infection or organ rejection. The transplant recipient will take immunosuppressive drugs for the rest of his or her life to avoid rejection of the "foreign" organ. •

Nephrons

The functional units of the kidney are the ***nephrons*** (NEF-ronz), numbering about a million in each kidney (see Figure 28.4). A nephron consists of two parts: a ***renal corpuscle*** (KOR-pus-el = tiny body), where blood plasma is filtered, and a ***renal tubule*** into which the filtered fluid, called ***glomerular filtrate,*** passes. As the fluid moves through the renal tubules, wastes and excess substances are added, and useful materials are returned to the blood in the peritubular capillaries.

The two parts that make up a renal corpuscle are the ***glomerulus*** and the ***glomerular (Bowman's) capsule,*** a double-walled cup of epithelial cells that surrounds the glomerular capillaries. Glomerular filtrate first enters the glomerular capsule and then passes into the renal tubule. In the order that fluid passes through them, the three main sections of the renal tubule are the ***proximal convoluted tubule,*** the ***loop of Henle,*** and the ***distal convoluted tubule.*** *Proximal* denotes the part of the tubule attached to the glomerular capsule, and *distal* denotes the part that is farther

Figure 28.2 Structure of the kidney.

 A renal capsule covers the kidney. Internally, the two main regions are the renal cortex and the renal medulla.

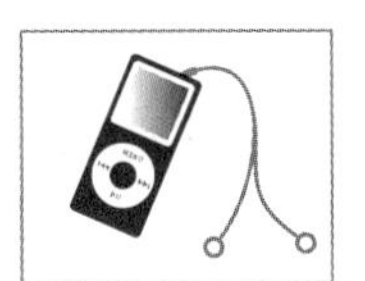

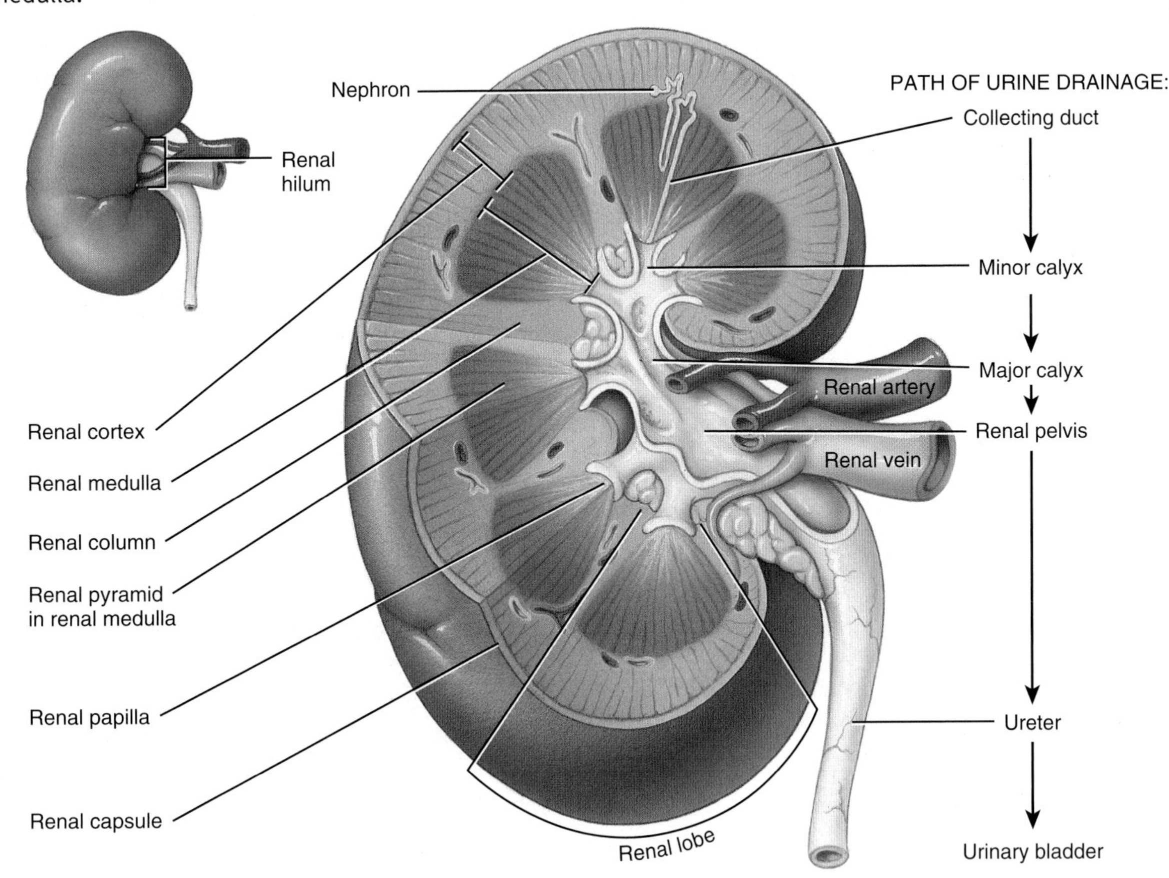

? What structures pass through the renal hilum?

Figure 28.3 Blood supply of the right kidney.

The renal arteries deliver about 20–25% of the resting cardiac output to the kidneys.

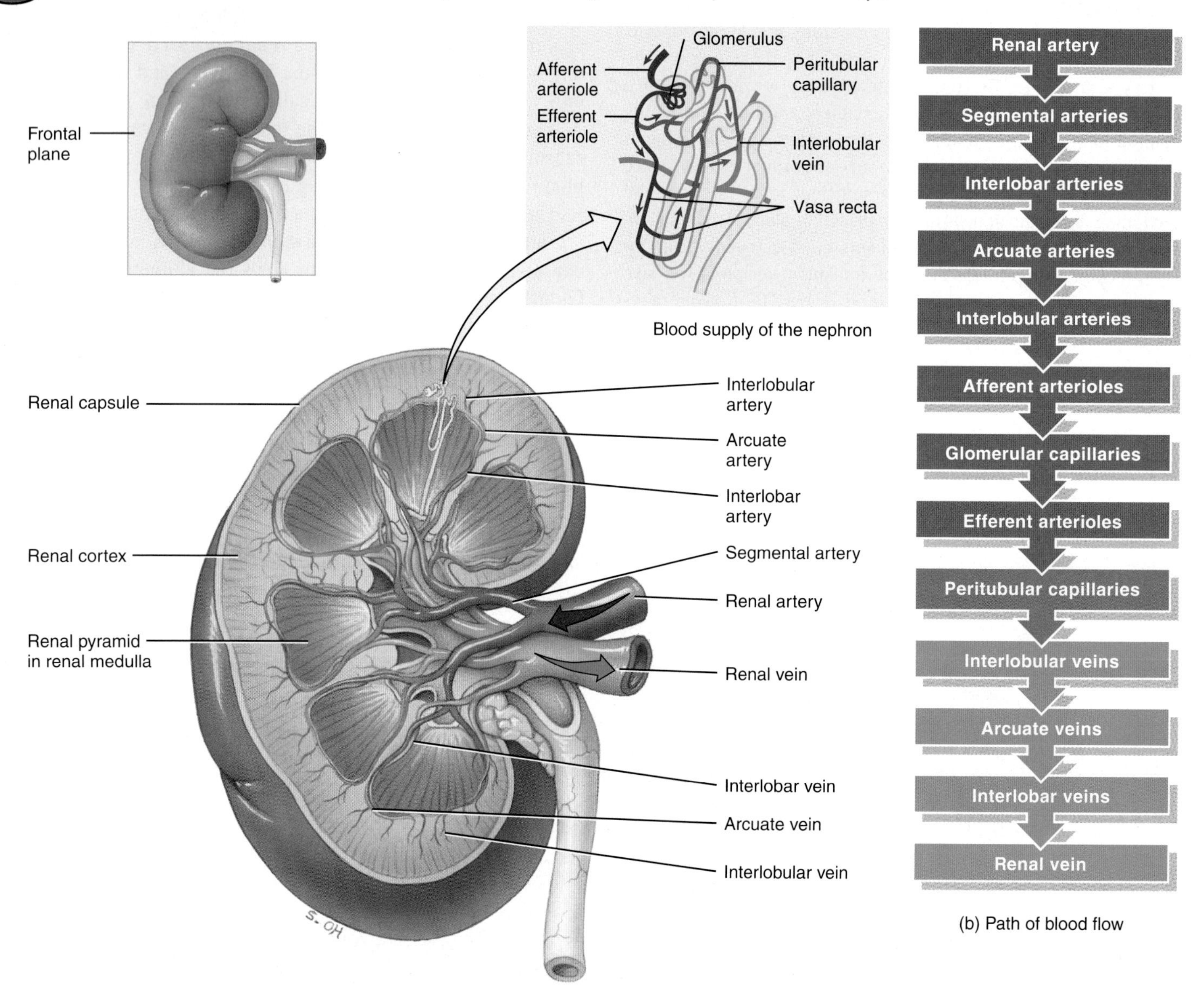

(a) Frontal section of right kidney

(b) Path of blood flow

What volume of blood enters the renal arteries per minute?

away. *Convoluted* means the tubule is tightly coiled rather than straight. The renal corpuscle and both convoluted tubules lie within the renal cortex; the loop of Henle extends into the renal medulla. The first part of the loop of Henle begins in the renal cortex and extends downward into the renal medulla, where it is called the ***descending limb of the loop of Henle*** (Figure 28.4). It then makes a hairpin turn and returns to the renal cortex as the ***ascending limb of the loop of Henle.*** The distal convoluted tubules of several nephrons empty into a common ***collecting duct.*** Collecting ducts then unite and converge into several hundred large ***papillary ducts,*** which drain into the minor calyces. The collecting ducts and papillary ducts extend from the renal cortex through the renal medulla to the renal pelvis.

• CLINICAL CONNECTION | Number of Nephrons

The **number of nephrons** is established at birth. New nephrons do not form to replace those that are injured or diseased. Signs of kidney damage often are not apparent, however, until the majority of nephrons are damaged because the remaining functional nephrons adapt to handle a larger-than-normal load. Surgical removal of one kidney, for example, stimulates enlargement of the remaining kidney, which eventually is able to filter blood at 80% of the rate of two normal kidneys. •

CHECKPOINT

2. Which structures help protect and cushion the kidneys?
3. What is the functional unit of the kidney? Describe its structure.

Figure 28.4 The structure of one type of nephron (cortical nephron) and associated blood vessels. Most nephrons are cortical nephrons; their renal corpuscles lie in the outer renal cortex and their short loops of Henle are mostly in the renal cortex.

Nephrons are the functional units of the kidneys.

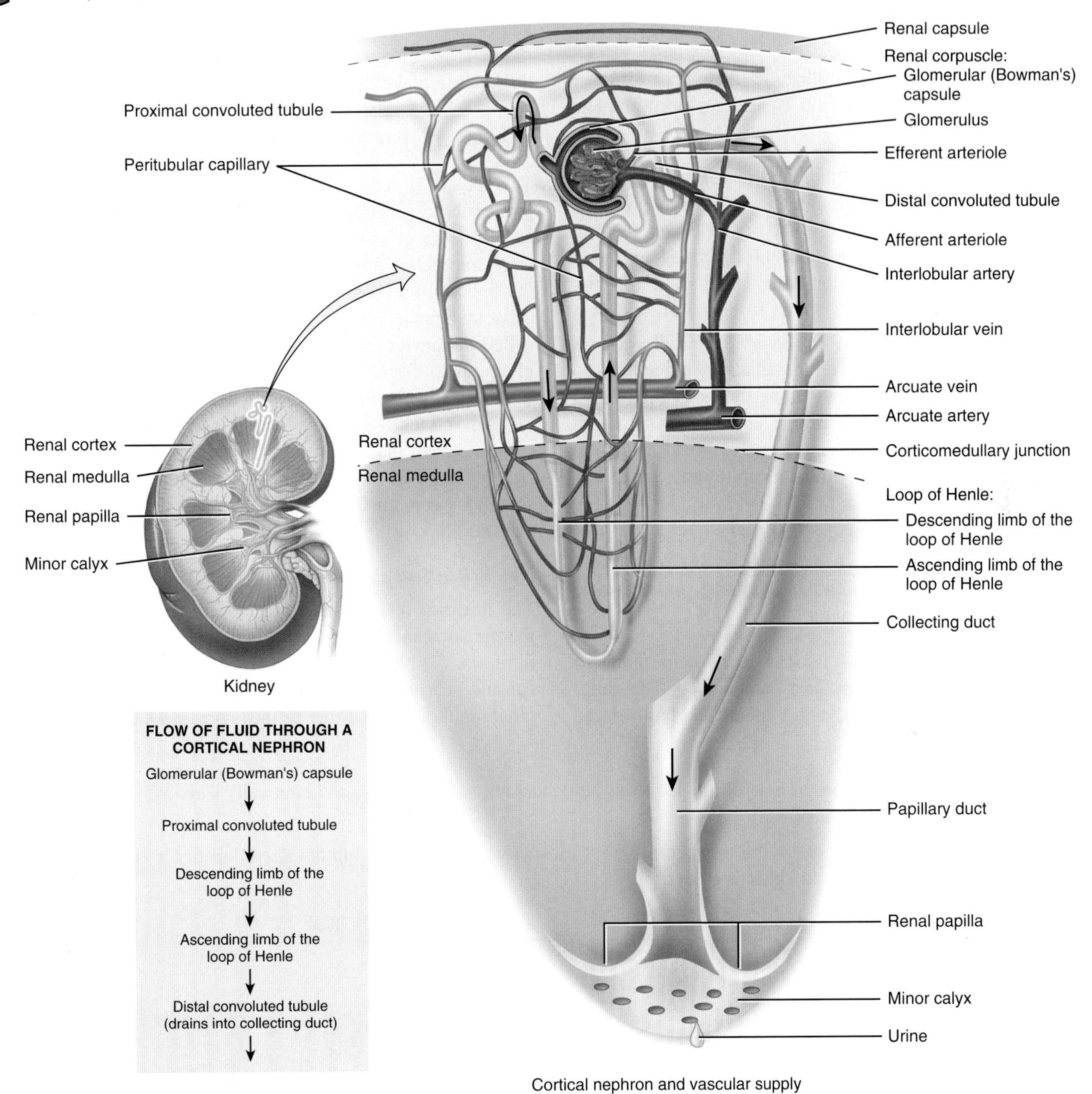

Cortical nephron and vascular supply

A water molecule has just entered the proximal convoluted tubule of a nephron. Which parts of the nephron will it travel through (in order) to reach the renal pelvis as part of a drop of urine?

28.3 FUNCTIONS OF THE NEPHRON

OBJECTIVE

- Identify the three basic functions performed by nephrons and collecting ducts and indicate where each occurs.

To produce urine, nephrons and collecting ducts perform three basic processes—glomerular filtration, tubular reabsorption, and tubular secretion (Figure 28.5):

1. *Filtration* is the forcing of fluids and dissolved substances smaller than a certain size through a membrane by pressure. ***Glomerular filtration*** is the first step of urine production: Blood pressure forces water and most solutes in blood plasma across the wall of glomerular capillaries, forming ***glomerular filtrate.*** Filtration occurs in glomeruli just as it occurs in other capillaries (see Figure 23.3, Section 23.1).
2. ***Tubular reabsorption*** is the return of water and solutes to the blood as filtered fluid flows along the renal tubule and through the collecting duct: Tubule and duct cells return about 99% of the filtered water and many useful solutes to the blood flowing through peritubular capillaries.
3. ***Tubular secretion*** also takes place as fluid flows along the tubule and through the collecting duct: The tubule and duct cells remove substances, such as wastes, drugs, and excess ions, from blood in the peritubular capillaries and transport them into the fluid in the renal tubules.

As nephrons perform their functions, they help maintain homeostasis of the blood's volume and composition. The situation is somewhat similar to a recycling center: Garbage trucks dump refuse into an input hopper, where the smaller refuse passes onto a conveyor belt (glomerular filtration of blood plasma). As the conveyor belt carries the garbage along, workers remove useful items, such as aluminum cans, plastics, and glass containers (reabsorption). Other workers place additional garbage and larger items onto the conveyor belt (secretion). At the end of the belt, all remaining garbage falls into a truck for transport to the landfill (excretion of wastes in urine).

Glomerular Filtration

Two layers of cells compose the capsule that surrounds the glomerular capillaries (Figure 28.6). Think of the renal corpuscle as a fist (the glomerular capillaries) pushed into a limp balloon (the glomerular capsule) until the fist is covered by two layers of the balloon with a space, the ***capsular space,*** in between. The cells that make up the inner wall of the glomerular capsule, called ***podocytes,*** adhere closely to the endothelial cells of the glomerulus. Together, the podocytes and glomerular endothelium form a ***filtration membrane*** that permits the passage of water and solutes from the blood into the capsular space. Blood cells and most plasma proteins remain in the blood because they are too large to pass through the filtration membrane. Simple squamous epithelial cells form the outer layer of the glomerular capsule.

Net Filtration Pressure

The pressure that causes filtration is the blood pressure in the glomerular capillaries. Two other pressures oppose glomerular filtration: (1) blood colloid osmotic pressure and (2) glomerular capsule pressure (due to fluid already in the capsular space and renal tubule). When either of these pressures increases, glomeru-

Figure 28.5 Overview of functions of a nephron. Excreted substances remain in the urine and eventually leave the body.

Glomerular filtration occurs in the renal corpuscle; tubular reabsorption and tubular secretion occur all along the renal tubule and collecting duct.

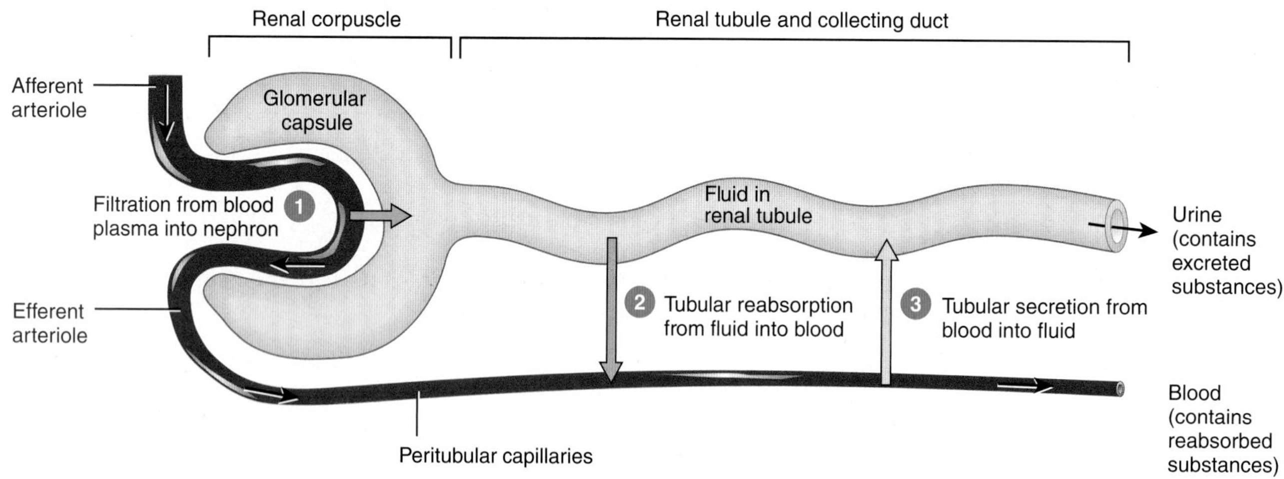

When cells of the renal tubules secrete the drug penicillin, is the drug being added to or removed from the bloodstream?

lar filtration decreases. Normally, blood pressure is greater than the two opposing pressures, producing a ***net filtration pressure*** of about 10 mmHg. Net filtration pressure forces a large volume of fluid into the capsular space, about 150 liters daily in average-sized females and 180 liters daily in average-sized males.

Because the efferent arteriole is smaller in diameter than the afferent arteriole, it helps raise the blood pressure in the glomerular capillaries. When blood pressure increases or decreases slightly, changes in the diameters of the afferent and efferent arterioles can actually keep net filtration pressure steady to maintain normal glomerular filtration. Constriction of the afferent arteriole decreases blood flow into the glomerulus, which decreases net filtration pressure. Constriction of the efferent arteriole slows outflow of blood and increases net filtration pressure.

• CLINICAL CONNECTION | Oliguria and Anuria

Conditions that greatly reduce blood pressure, for instance severe hemorrhage, may cause glomerular blood pressure to fall so low that net filtration pressure drops despite constriction of efferent arterioles. Then, glomerular filtration slows, or even stops entirely. The result is **oliguria** (*olig-* = scanty; *-uria* = urine production), a daily urine output between 50 and 250 mL, or **anuria,** a daily urine output of less than 50 mL. Obstructions, such as a kidney stone that blocks a ureter or an enlarged prostate that blocks the urethra in a male, can also decrease net filtration pressure and thereby reduce urine output. •

Glomerular Filtration Rate

The amount of filtrate that forms in both kidneys every minute is called the ***glomerular filtration rate (GFR).*** In adults, the GFR is about 105 mL/min in average-sized females and 125 mL/min in average-sized males. It is very important for the kidneys to maintain a constant GFR. If the GFR is too high, needed substances pass so quickly through the renal tubules that they are unable to be reabsorbed and pass out of the body as part of urine. On the other hand, if the GFR is too low, nearly all the filtrate is reabsorbed, and waste products are not adequately excreted.

Atrial natriuretic peptide (ANP) is a hormone that promotes loss of sodium ions and water in the urine, in part because it increases glomerular filtration rate. Cells in the atria of the heart secrete more ANP if the heart is stretched more, as occurs when blood volume increases. ANP then acts on the kidneys to increase loss of sodium ions and water in urine, which reduces the blood volume back to normal.

Like most blood vessels of the body, those of the kidneys are supplied by sympathetic neurons of the autonomic nervous system. When these neurons are active, they cause vasoconstriction. At rest, sympathetic stimulation is low and the afferent and efferent arterioles are relatively dilated. With greater sympathetic stimulation, as occurs during exercise or hemorrhage, the afferent arterioles are constricted more than the efferent arterioles. As a result, blood flow into glomerular capillaries is greatly

Figure 28.6 Glomerular filtration, the first step in urine formation.

Glomerular filtrate (red arrows) passes into the capsular space and then into the proximal convoluted tubule.

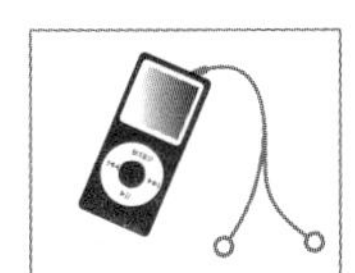

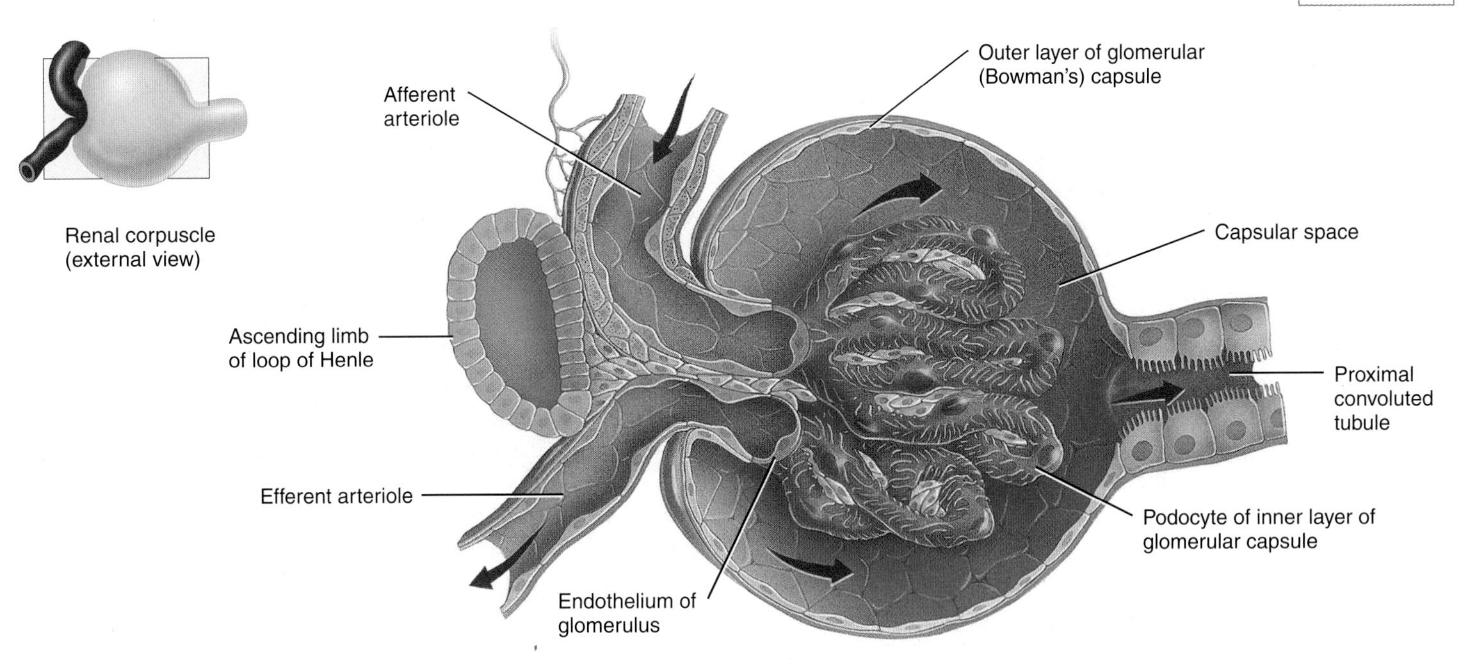

Which cells make up the filtration membrane in the renal corpuscle?

decreased, net filtration pressure decreases, and GFR drops. These changes reduce urine output, which helps conserve blood volume and permits greater blood flow to other body tissues.

Tubular Reabsorption and Secretion

Tubular reabsorption—returning most of the filtered water and many of the filtered solutes to the blood—is the second basic function of the nephrons and collecting ducts. The filtered fluid becomes ***tubular fluid*** once it enters the proximal convoluted tubule. Due to reabsorption and secretion, the composition of tubular fluid changes as it flows along the nephron tubule and through a collecting duct. Typically, about 99% of the filtered water is reabsorbed. Only 1% of the water in glomerular filtrate actually leaves the body as ***urine,*** the fluid that drains into the renal pelvis.

Epithelial cells all along the renal tubules and collecting ducts carry out tubular reabsorption (Figure 28.7). Some solutes are passively reabsorbed by diffusion; others are reabsorbed by active transport. Proximal convoluted tubule cells make the largest contribution, reabsorbing 65% of the filtered water, 100% of the filtered glucose and amino acids, and large quantities of various ions. Reabsorption of solutes also promotes reabsorption of water in the following way. The movement of solutes into peritubular capillaries decreases the solute concentration of the tubular fluid but increases the solute concentration in the peritubular capillaries. As a result, water moves by osmosis into peritubular capillaries. Cells located distal to the proximal convoluted tubule fine-tune reabsorption to maintain homeostatic balances of water and selected ions.

• CLINICAL CONNECTION | Glucosuria and Polyuria

When the blood concentration of glucose rises above normal, transporters in the proximal convoluted tubules may not be able to work fast enough to reabsorb all of the filtered glucose. As a result, some glucose remains in the urine, a condition called **glucosuria** (gloo′-kō-SOO-rē-a). The most common cause of glucosuria is diabetes mellitus, in which the blood glucose level may rise far above normal because insulin activity is deficient. Because "water follows solutes" as tubular reabsorption takes place, any condition that reduces reabsorption of filtered solutes also increases the amount of water lost in urine. **Polyuria** (pol′-ē-Ū-rē-a; *poly-* = too much), excessive excretion of urine, usually accompanies glucosuria and is a common symptom of diabetes. •

The third function of the nephrons and collecting ducts is ***tubular secretion,*** the transfer of materials from the blood through tubule cells and into tubular fluid. As is the case for tubular reabsorption, tubular secretion takes place all along the renal tubules and collecting ducts and occurs via both passive diffusion and active transport processes. Secreted substances include hydrogen ions (H^+), potassium ions (K^+), ammonia (NH_3), urea, creatinine (a waste from creatine in muscle cells), and certain drugs such as penicillin. Tubular secretion helps eliminate these substances from the body.

Ammonia is a poisonous waste product that is produced when amino groups are removed from amino acids. Liver cells convert most ammonia to urea, which is a less toxic compound. Although tiny amounts of urea and ammonia are present in sweat, most excretion of these nitrogen-containing waste products occurs in the urine. Urea and ammonia in blood are both filtered at the glomerulus and secreted by proximal convoluted tubule cells into the tubular fluid. Secretion of excess K^+ for elimination in the urine also is very important. Tubule cell secretion of K^+ varies with dietary intake of potassium to maintain a stable level of K^+ in body fluids.

Tubular secretion also helps control blood pH. A normal blood pH of 7.35 to 7.45 is maintained, even though the typical high-protein diet in North America provides more acid-producing foods than alkali-producing foods. To eliminate acids, the cells of the renal tubules secrete H^+ into the tubular fluid, which helps maintain the pH of blood in the normal range. Due to H^+ secretion, typical urine is slightly acidic.

• CLINICAL CONNECTION | Urinalysis

As a result of tubular secretion, certain substances pass from blood into urine and may be detected by a **urinalysis,** the analysis of various components of urine (discussed shortly). This is especially important in testing athletes for the presence of performance-enhancing drugs such as anabolic steroids, plasma expanders, erythropoietin, hCG, hGH, and amphetamines. Urine tests can also be used to detect the presence of alcohol or illegal drugs such as marijuana, cocaine, and heroin. •

Hormonal Regulation of Nephron Functions

Hormones affect the extent of Na^+, Cl^-, Ca^{2+}, and water reabsorption as well as K^+ secretion by the renal tubules. The most important hormonal regulators of ion reabsorption and secretion are ***angiotensin II*** and ***aldosterone.*** In the proximal convoluted tubules, angiotensin II enhances reabsorption of Na^+ and Cl^-. Angiotensin II also stimulates the adrenal cortex to release aldosterone, a hormone that in turn stimulates the tubule cells in the late part of the distal convoluted tubule and throughout the collecting ducts to reabsorb more Na^+ and Cl^- and secrete more K^+. When more Na^+ and Cl^- are reabsorbed, then more water is also reabsorbed by osmosis. Aldosterone-stimulated secretion of K^+ is the major regulator of blood K^+ level. An elevated level of K^+ in plasma causes serious disturbances in cardiac rhythm or even cardiac arrest. Besides increasing glomerular filtration rate, the hormone ***atrial natriuretic peptide*** **(*ANP*)** plays a minor role in inhibiting the reabsorption of Na^+ (and Cl^- and water) by the renal tubules. As GFR increases and Na^+, Cl^-, and water reabsorption decrease, more water and salt are lost in the urine. The final effect is to lower blood volume.

The major hormone that regulates water reabsorption is ***antidiuretic hormone*** **(*ADH*),** which operates via negative feedback

Figure 28.7 **Summary of filtration, reabsorption, and secretion in the nephron and collecting duct.** Percentages refer to the amounts initially filtered at the glomerulus.

Filtration occurs in the renal corpuscle; reabsorption occurs all along the renal tubule and collecting ducts.

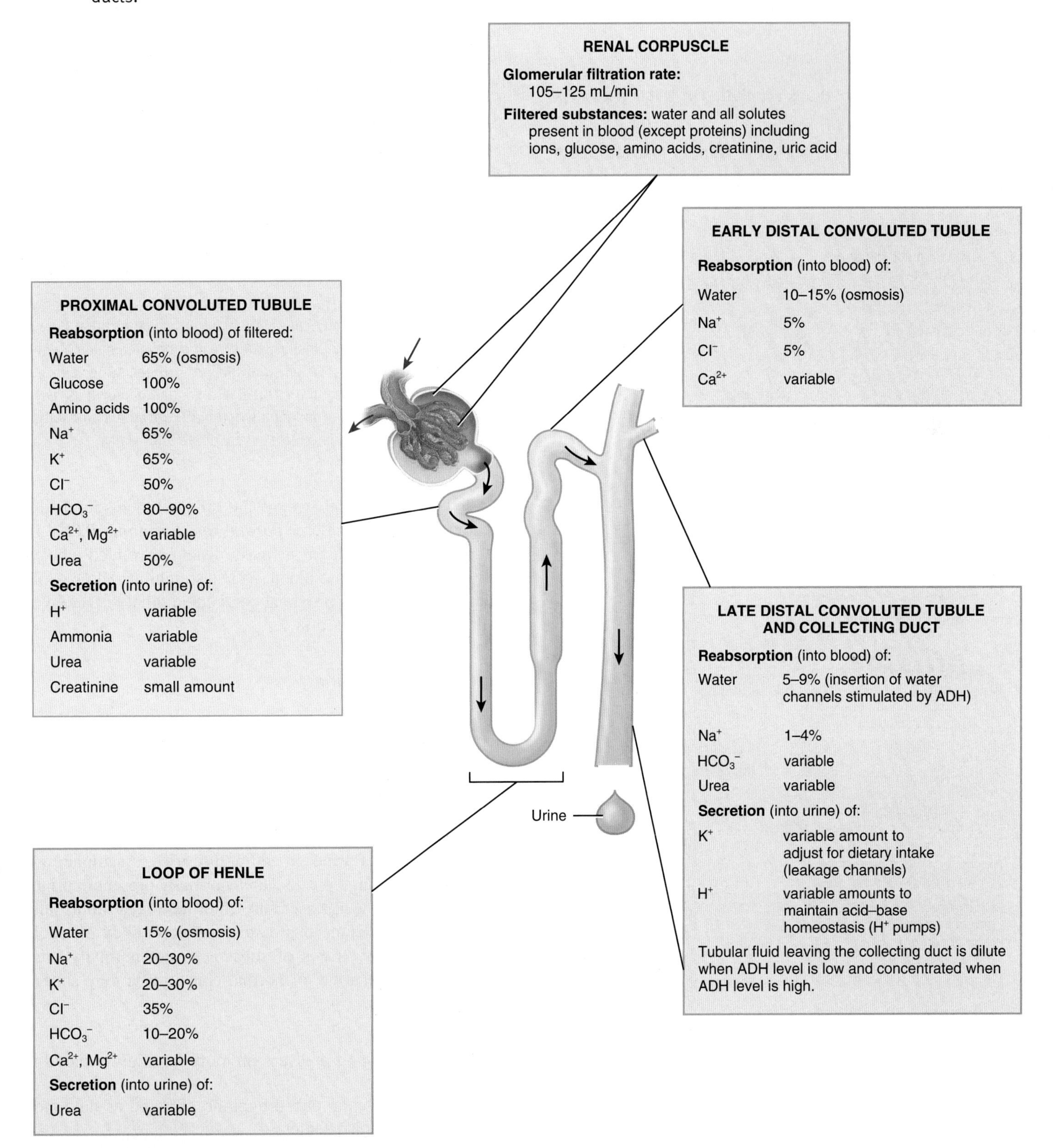

RENAL CORPUSCLE
Glomerular filtration rate: 105–125 mL/min
Filtered substances: water and all solutes present in blood (except proteins) including ions, glucose, amino acids, creatinine, uric acid
EARLY DISTAL CONVOLUTED TUBULE
Reabsorption (into blood) of:
Water 10–15% (osmosis)
Na^+ 5%
Cl^- 5%
Ca^{2+} variable
PROXIMAL CONVOLUTED TUBULE
Reabsorption (into blood) of filtered:
Water 65% (osmosis)
Glucose 100%
Amino acids 100%
Na^+ 65%
K^+ 65%
Cl^- 50%
HCO_3^- 80–90%
Ca^{2+}, Mg^{2+} variable
Urea 50%
Secretion (into urine) of:
H^+ variable
Ammonia variable
Urea variable
Creatinine small amount
LATE DISTAL CONVOLUTED TUBULE AND COLLECTING DUCT
Reabsorption (into blood) of:
Water 5–9% (insertion of water channels stimulated by ADH)
Na^+ 1–4%
HCO_3^- variable
Urea variable
Secretion (into urine) of:
K^+ variable amount to adjust for dietary intake (leakage channels)
H^+ variable amounts to maintain acid–base homeostasis (H^+ pumps)
Tubular fluid leaving the collecting duct is dilute when ADH level is low and concentrated when ADH level is high.
Urine
LOOP OF HENLE
Reabsorption (into blood) of:
Water 15% (osmosis)
Na^+ 20–30%
K^+ 20–30%
Cl^- 35%
HCO_3^- 10–20%
Ca^{2+}, Mg^{2+} variable
Secretion (into urine) of:
Urea variable

? **In which segments of the nephron and collecting duct does secretion occur?**

(Figure 28.8). When the concentration of water in the blood decreases by as little as 1%, osmoreceptors in the hypothalamus stimulate release of ADH from the posterior pituitary. A second powerful stimulus for ADH secretion is a decrease in blood volume, as occurs in hemorrhaging or severe dehydration. ADH acts on tubule cells in the last part of the distal convoluted tubules and throughout the collecting ducts. In the absence of ADH, these parts of the renal tubule have a very low permeability to water. ADH increases the water permeability of these tubule cells by causing insertion of proteins that function as water channels into their plasma membranes. When the water permeability of the tubule cells increases, water molecules move from the tubular fluid into the cells and then into the blood. The kidneys can produce as little as 400–500 mL of very concentrated urine each day when ADH concentration is maximal, for instance, during severe dehydration. When ADH level declines, the water channels are removed from the membranes. In contrast, the kidneys produce a large volume of dilute urine when ADH level is low.

Figure 28.8 Negative feedback regulation of water reabsorption by ADH.

When ADH level is high, the kidneys reabsorb more water.

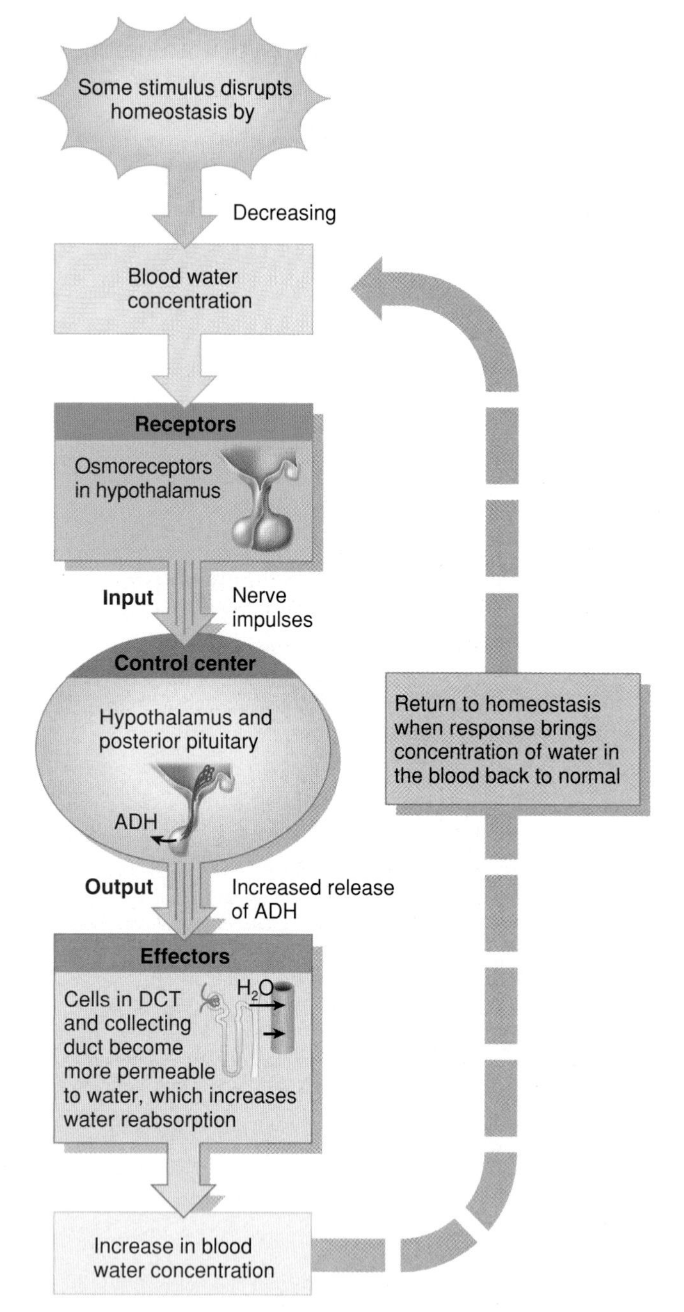

Would blood ADH level be higher or lower than normal in a person who has just completed a 5 km run without drinking any water?

• CLINICAL CONNECTION Diuretics

Diuretics (dī′-ū-RET-iks) are substances that slow reabsorption of water by the kidneys and thereby cause *diuresis,* an elevated urine flow rate. Naturally occurring diuretics include *caffeine* in coffee, tea, and cola sodas, which inhibits Na^+ reabsorption, and *alcohol* in beer, wine, and mixed drinks, which inhibits secretion of ADH. In a condition known as *diabetes insipidus*, ADH secretion is inadequate or the ADH receptors are faulty, and a person may excrete up to 20 liters of very dilute urine daily. •

A lower-than-normal level of Ca^{2+} in the blood stimulates the parathyroid glands to release ***parathyroid hormone (PTH)***. PTH in turn stimulates cells in the early distal convoluted tubules to reabsorb more Ca^{2+} into blood. PTH also inhibits HPO_4^{2-}(phosphate) reabsorption in proximal convoluted tubules, thereby promoting phosphate excretion.

Components of Urine

An analysis of the volume and physical, chemical, and microscopic properties of urine, called a ***urinalysis,*** tells us much about the state of the body. Table 28.1 summarizes the principal physical characteristics of urine.

The volume of urine eliminated per day in a normal adult is 1 to 2 liters (about 1 to 2 quarts). Water accounts for about 95% of the total volume of urine. In addition to urea, creatinine, potassium, and ammonia, typical solutes normally present in urine include uric acid as well as sodium, chloride, magnesium, sulfate, phosphate, and calcium ions. If disease alters body metabolism or kidney function, traces of substances not normally present may appear in the urine, or normal constituents may appear in abnormal amounts.

CHECKPOINT

4. How does blood pressure promote filtration of blood in the kidneys?
5. What solutes are reabsorbed and secreted as fluid moves along the renal tubules?
6. How do angiotensin II, aldosterone, and antidiuretic hormone regulate tubular reabsorption and secretion?
7. What are the characteristics of normal urine?

TABLE 28.1

Characteristics of Normal Urine

CHARACTERISTIC	DESCRIPTION
Volume	One to two liters (about 1 to 2 quarts) in 24 hours but varies considerably.
Color	Yellow or amber but varies with urine concentration and diet. Color is due to urochrome (pigment produced from breakdown of bile) and urobilin (from breakdown of hemoglobin). Concentrated urine is darker in color. Diet (reddish urine from beets), medications, and certain diseases affect color. Kidney stones may produce blood in urine.
Turbidity	Transparent when freshly voided but becomes turbid (cloudy) upon standing.
Odor	Mildly aromatic but becomes ammonia-like upon standing. Some people inherit the ability to form methylmercaptan from digested asparagus that gives urine a characteristic odor. Urine of diabetics has a fruity odor due to presence of ketone bodies.
pH	Ranges between pH 4.6 and 8.0, average 6.0. Varies considerably with diet. High-protein diets increase acidity; vegetarian diets increase alkalinity.

28.4 TRANSPORTATION, STORAGE, AND ELIMINATION OF URINE

OBJECTIVE

- Describe the structure and functions of the ureters, urinary bladder, and urethra.

As you learned earlier in the chapter, urine produced by the nephrons drains into the minor calyces, which join to become major calyces that unite to form the renal pelvis (see Figure 28.2). From the renal pelvis, urine drains first into the ureters and then into the urinary bladder; urine is then discharged from the body through the urethra (see Figure 28.1).

Ureters

Each of the two ***ureters*** (Ū-re-ters or ū-RĒ-ters) transports urine from the renal pelvis of one of the kidneys to the urinary bladder (see Figure 28.1). The ureters pass under the urinary bladder for several centimeters, causing the urinary bladder to compress the ureters and thus prevent backflow of urine when pressure builds up in the urinary bladder during urination. If this physiological valve is not operating, *cystitis* (urinary bladder inflammation) may develop into a kidney infection.

The wall of the ureter consists of three layers. The inner layer is the mucosa, containing *transitional epithelium* (see Table 4.1I, Section 4.3) with an underlying layer of areolar connective tissue. Transitional epithelium is able to stretch—a marked advantage for any organ that must accommodate a variable volume of fluid. Mucus secreted by the goblet cells of the mucosa prevents the cells from coming in contact with urine, the solute concentration and pH of which may differ drastically from the cytosol of cells that form the wall of the ureters. The middle layer consists of smooth muscle. Urine is transported from the renal pelvis to the urinary bladder by peristaltic contractions of this smooth muscle, but the fluid pressure of the urine and gravity also contribute. Unless contraindicated by the illness, most patients who are confined to bed are elevated at the head end to facilitate the effects of gravity on carrying urine from the kidney to the bladder. The outer layer of the ureter, called the *adventitia,* consists of areolar connective tissue containing blood vessels, lymphatic vessels, and nerves.

Urinary Bladder

The ***urinary bladder*** is a hollow muscular organ situated in the pelvic cavity behind the pubic symphysis (see Figure 28.9). In males, it is directly in front of the rectum (see Figure 29.1, Section 29.2). In females, it is in front of the vagina and below the uterus. Folds of the peritoneum hold the urinary bladder in position. The shape of the urinary bladder depends on how much urine it contains. When empty, it looks like a deflated balloon. It becomes spherical when slightly stretched and, as urine volume increases, becomes pear-shaped and rises into the abdominal cavity. Urinary bladder capacity averages 700–800 mL. It is smaller in females because the uterus occupies the space just superior to the urinary bladder. A pregnant female has an exceedingly small urinary bladder capacity due to an enlarging uterus. In the floor of the urinary bladder is a small triangular area called the ***trigone*** (TRĪ-gon = triangle). The two posterior corners of the trigone contain the two ***ureteral openings;*** the opening into the urethra, the ***internal urethral orifice,*** lies in the anterior corner. The ureters drain into the urinary bladder via the ureteral openings. Like the ureters, the mucosa of the urinary bladder contains transitional epithelium. The muscular layer of the urinary bladder wall consists of three layers of smooth muscle called the ***detrusor muscle*** (de-TROO-ser = to push down). The peritoneum, which covers the superior surface of the urinary bladder, forms a serous outer coat; the rest of the urinary bladder has a fibrous outer covering.

Urethra

The ***urethra*** (ū-RĒ-thra), the terminal portion of the urinary system, is a small tube leading from the floor of the urinary bladder to the exterior of the body (see Figure 28.9). In females, it lies directly behind the pubic symphysis and is embedded in the front wall of the vagina. The opening of the urethra to the exterior, the ***external urethral orifice,*** lies between the clitoris and vaginal opening. In males, the urethra passes vertically through the prostate, the deep perineal muscles, and finally the penis (see Figures 29.1 and 29.6, Section 29.2).

Figure 28.9 Ureters, urinary bladder, and urethra in a female.

Urine is stored in the urinary bladder before being expelled by micturition.

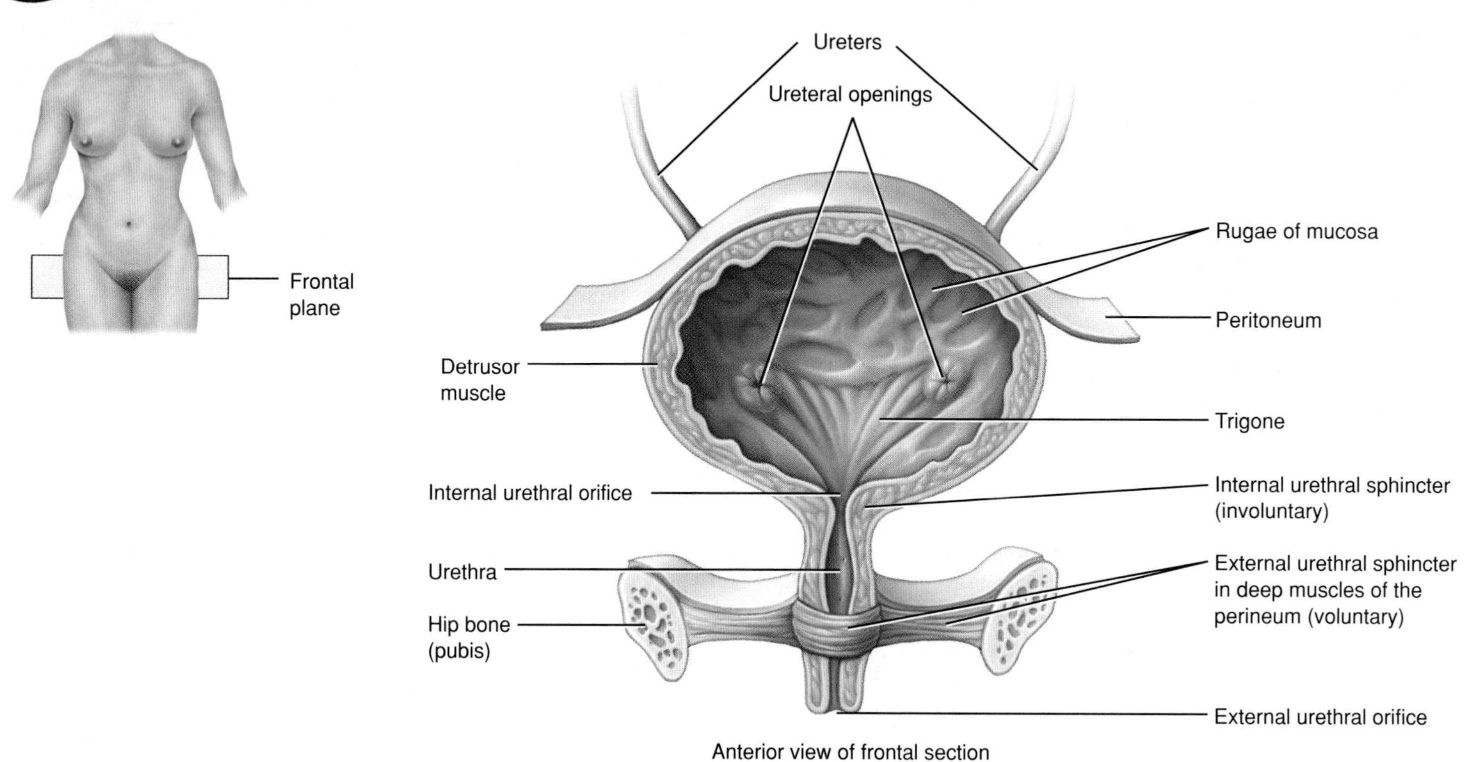

What is a lack of voluntary control over micturition called?

Around the opening to the urethra is an ***internal urethral sphincter*** composed of smooth muscle. The opening and closing of the internal urethral sphincter is involuntary. Inferior to the internal sphincter is the ***external urethral sphincter,*** which is composed of skeletal muscle and is under voluntary control. As discussed in Exhibit 12.3, the external urethral sphincter is one of the muscles of the deep perineum. In both males and females, the urethra is the passageway for discharging urine from the body. The male urethra also serves as the duct through which semen is ejaculated.

Micturition

The urinary bladder stores urine prior to its elimination and then expels urine into the urethra by an act called ***micturition*** (mik′-too-RI-shun = to urinate), commonly known as ***urination.*** Micturition requires a combination of involuntary and voluntary muscle contractions. When the volume of urine in the urinary bladder exceeds 200 to 400 mL, pressure within the bladder increases considerably, and stretch receptors in its wall transmit nerve impulses into the spinal cord. These impulses propagate to the lower part of the spinal cord and trigger a reflex called the ***micturition reflex.*** In this reflex, parasympathetic impulses from the spinal cord cause *contraction* of the detrusor muscle and *relaxation* of the internal urethral sphincter muscle. Simultaneously, the spinal cord inhibits somatic motor neurons, causing relaxation of skeletal muscle in the external urethral sphincter. Upon contraction of the urinary bladder wall and relaxation of the sphincters, urination takes place. Urinary bladder filling causes a sensation of fullness that initiates a conscious desire to urinate before the micturition reflex actually occurs. Although emptying of the urinary bladder is a reflex, in early childhood we learn to initiate it and stop it voluntarily. Through learned control of the external urethral sphincter muscle and certain muscles of the pelvic floor, the cerebral cortex can initiate micturition or delay it for a limited time.

A lack of voluntary control over micturition is termed ***urinary incontinence.*** Under about 2–3 years of age, urinary incontinence is normal because neurons to the external urethral sphincter muscle are not completely developed. Infants void whenever the urinary bladder is sufficiently distended to trigger the reflex. In *stress incontinence,* the most common type of urinary incontinence, physical stresses that increase abdominal pressure, such as coughing, sneezing, laughing, exercising, straining, lifting heavy objects, pregnancy, or simply walking, cause leakage of urine from the urinary bladder. See the Clinical Connection in Chapter 12, Exhibit 12.3, for more details on stress incontinence.

CHECKPOINT

8. What forces help propel urine from the renal pelvis to the urinary bladder?
9. What is micturition? How does the micturition reflex occur?
10. How does the location of the urethra compare in males and females?

28.5 AGING AND THE URINARY SYSTEM

OBJECTIVE

- Describe the effects of aging on the urinary system.

With aging, the kidneys shrink in size, have a decreased blood flow, and filter less blood. The mass of the two kidneys decreases from an average of 260 g in 20-year-olds to less than 200 g by age 80. Likewise, renal blood flow and filtration rate decline by 50% between ages 40 and 70. Kidney diseases that become more common with age include acute and chronic kidney inflammations and renal calculi (kidney stones). Because the sensation of thirst commonly diminishes with age, older individuals also are susceptible to dehydration. Urinary tract infections are more common among the elderly, as are polyuria, nocturia (excessive urination at night), increased frequency of urination, dysuria (painful urination), urinary retention or incontinence, and hematuria (blood in the urine).

To appreciate the many ways that the urinary system contributes to homeostasis of other body systems, examine Exhibit 28.1.

CHECKPOINT

11. Why are older individuals more susceptible to dehydration?

KEY MEDICAL TERMS ASSOCIATED WITH THE URINARY SYSTEM

Dialysis (dī-AL-i-sis; *dialyo* = to separate) is the separation of large solutes from smaller ones by diffusion through a selectively permeable membrane. It is used to cleanse a person's blood artificially when the kidneys are so impaired by disease or injury that they are unable to function adequately. One method of dialysis is ***hemodialysis*** (hē-mō-dī-AL-i-sis; *hemo-* = blood) which filters the patient's blood directly by removing wastes and excess electrolytes and fluid and then returning the cleansed blood to the patient. As a rule, most people on hemodialysis require about 6–12 hours a week, typically divided into three sessions.

Dysuria (dis-Ū-rē-a; *dys-* = painful; *-uria* = urine) Painful urination.

Enuresis (en′-ū-RĒ-sis = to void urine) Involuntary voiding of urine after the age at which voluntary control has typically been attained.

Glomerulonephritis is an inflammation of the glomeruli of the kidney. One of the most common causes is an allergic reaction to the toxins produced by streptococcal bacteria that have recently infected another part of the body, especially the throat.

Intravenous pyelogram (in′-tra-VĒ-nus PĪ-el-ō-gram′; *intra-* = within; -*veno-* = vein; *pyelo-* = pelvis of kidney; *-gram* = record) or ***IVP*** Radiograph (x-ray film) of the kidneys after venous injection of a dye.

Nocturnal enuresis (nok-TUR-nal en′-ū-RĒ-sis) Discharge of urine during sleep, resulting in bed-wetting; occurs in about 15% of 5-year-old children and generally resolves spontaneously, afflicting only about 1% of adults. Possible causes include smaller-than-normal urinary bladder capacity, failure to awaken in response to a full urinary bladder, and above-normal production of urine at night. Also termed ***nocturia.***

Polycystic kidney disease (PKD) is one of the most common inherited disorders. In PKD, the kidney tubules become riddled with hundreds or thousands of cysts (fluid-filled cavities). In addition, inappropriate apoptosis (programmed cell death) of cells in noncystic tubules leads to progressive impairment of renal function and eventually to end-stage renal failure.

Renal failure is a decrease or cessation of glomerular filtration. In ***acute renal failure (ARF)*** the kidneys abruptly stop working entirely (or almost entirely). The main feature of ARF is the gradual suppression or cessation of urine flow. ***Chronic renal failure (CRF)*** refers to a progressive and usually irreversible decline in glomerular filtration rate (GFR). CRF may result from chronic glomerulonephritis, pyelonephritis, polycystic kidney disease, or traumatic loss of kidney tissue. People with end-stage renal failure require dialysis therapy and are possible candidates for a kidney transplant operation.

Urinary retention A failure to completely or normally void urine; may be due to an obstruction in the urethra or neck of the urinary bladder, to nervous contraction of the urethra, or to lack of urge to urinate. In men, an enlarged prostate gland may constrict the urethra and cause urinary retention. If urinary retention is prolonged, a catheter (slender rubber drainage tube) must be placed into the urethra to drain the urine.

EXHIBIT 28.1 The Urinary System and Homeostasis

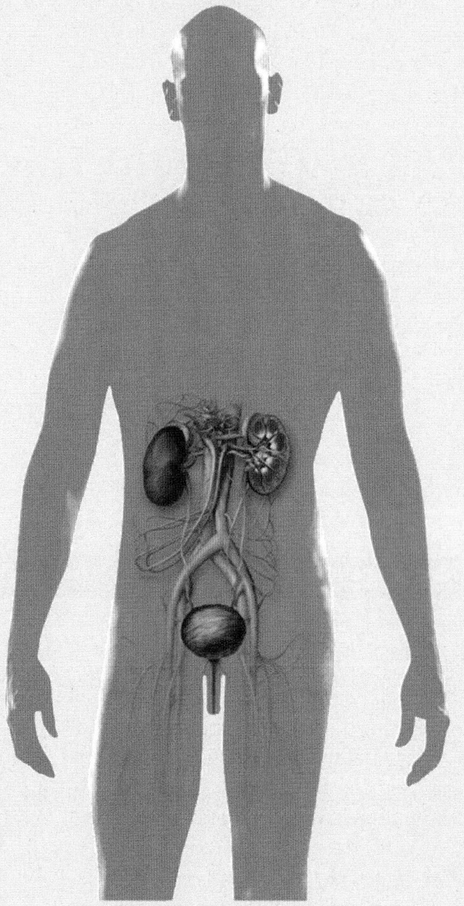

BODY SYSTEM		CONTRIBUTION OF THE URINARY SYSTEM
For all body systems		Kidneys regulate the volume, composition, and pH of body fluids by removing wastes and excess substances from blood and excreting them in the urine; the ureters transport urine from the kidneys to the urinary bladder, which stores urine until it is eliminated through the urethra.
Integumentary system		Kidneys and skin both contribute to the synthesis of calcitriol, the active form of vitamin D.
Skeletal system		Kidneys help adjust levels of blood calcium and phosphates, needed for building bone extracellular matrix.
Muscular system		Kidneys help adjust level of blood calcium, needed for contraction of muscle.
Nervous system	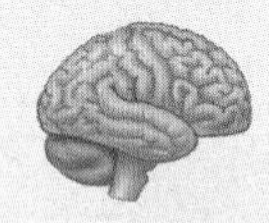	Kidneys perform gluconeogenesis, which provides glucose for ATP production in neurons, especially during fasting or starvation.
Endocrine system	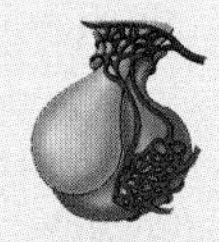	Kidneys participate in synthesis of calcitriol, the active form of vitamin D, and release erythropoietin, the hormone that stimulates production of red blood cells.
Cardiovascular system	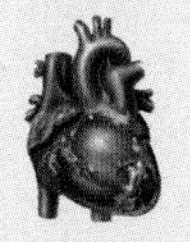	By increasing or decreasing their reabsorption of water filtered from the blood, the kidneys help adjust the blood volume and blood pressure; renin released by cells in the kidneys raises blood pressure; some bilirubin from hemoglobin breakdown is converted to a yellow pigment (urobilin), which is excreted in the urine.
Lymphatic system and immunity		By increasing or decreasing their reabsorption of water filtered from blood, the kidneys help adjust the volume of interstitial fluid and lymph; urine flushes microbes out of the urethra.
Respiratory system	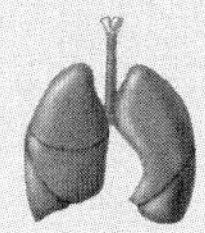	Kidneys and lungs cooperate in adjusting pH of body fluids.
Digestive system	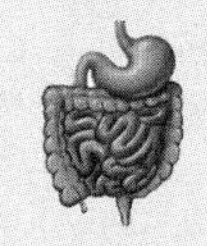	Kidneys help synthesize calcitriol, the active form of vitamin D, which is needed for absorption of dietary calcium.
Reproductive systems	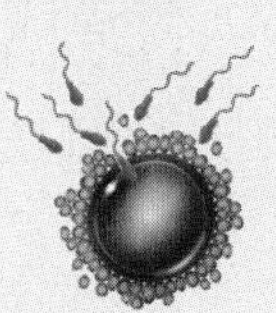	In males, the portion of the urethra that extends through the prostate and penis is a passageway for semen as well as urine.

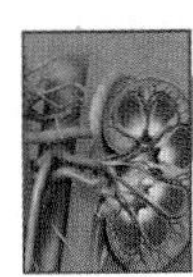

STUDY OUTLINE

Overview of the Urinary System (Section 28.1)

1. The organs of the urinary system include the kidneys, ureters, urinary bladder, and urethra.
2. After the kidneys filter blood and return most of the water and many solutes to the blood, the remaining water and solutes constitute urine.
3. The kidneys regulate blood ionic composition, blood volume, blood pressure, and blood pH.
4. The kidneys also release calcitriol and erythropoietin and excrete wastes and foreign substances.

Structure of the Kidneys (Section 28.2)

1. The kidneys lie on either side of the vertebral column between the peritoneum and the posterior wall of the abdominal cavity.
2. Each kidney is enclosed in a renal capsule, which is surrounded by adipose tissue.
3. Internally, the kidneys consist of a renal cortex, renal medulla, renal pyramids, renal columns, calyces, and a renal pelvis.
4. Blood enters the kidney through the renal artery and leaves through the renal vein.
5. The nephron is the functional unit of the kidney. A nephron consists of a renal corpuscle (glomerulus and glomerular or Bowman's capsule) and a renal tubule (proximal convoluted tubule, descending limb of the loop of Henle, ascending limb of the loop of Henle, and distal convoluted tubule). The distal convoluted tubules of several nephrons empty into a common collecting duct.

Functions of the Nephron (Section 28.3)

1. Nephrons perform three basic tasks: glomerular filtration, tubular reabsorption, and tubular secretion.
2. Together, the podocytes and glomerular endothelium form a leaky filtration membrane that permits the passage of water and solutes from the blood into the capsular space. Blood cells and most plasma proteins remain in the blood because they are too large to pass through the filtration membrane. The pressure that causes filtration is the blood pressure in the glomerular capillaries.
3. The amount of filtrate that forms in both kidneys every minute is the glomerular filtration rate (GFR). Atrial natriuretic peptide (ANP) increases GFR; sympathetic stimulation decreases GFR.
4. Epithelial cells all along the renal tubules and collecting ducts carry out tubular reabsorption and tubular secretion. Tubular reabsorption retains substances needed by the body, including water, glucose, amino acids, and ions.
5. Angiotensin II enhances reabsorption of Na^+ and Cl^-. Angiotensin II also stimulates the adrenal cortex to release aldosterone, which stimulates the collecting ducts to reabsorb more Na^+ and Cl^- and secrete more K^+. Atrial natriuretic peptide inhibits reabsorption of Na^+ (and Cl^- and water) by the renal tubules, which reduces blood volume.
6. Most water is reabsorbed by osmosis together with reabsorbed solutes, mainly in the proximal convoluted tubule. Reabsorption of the remaining water is regulated by antidiuretic hormone (ADH) in the last part of the distal convoluted tubule and collecting duct.
7. Tubular secretion discharges chemicals not needed by the body into the urine. Included are excess ions, nitrogenous wastes, hormones, and certain drugs. The kidneys help maintain blood pH by secreting H^+. Tubular secretion also helps maintain proper levels of K^+ in the blood.
8. Table 28.1 describes the physical characteristics of urine that are evaluated in a urinalysis: color, odor, turbidity, and pH.
9. Chemically, normal urine contains about 95% water and 5% solutes.

Transportation, Storage, and Elimination of Urine (Section 28.4)

1. The ureters transport urine from the renal pelves of the right and left kidneys to the urinary bladder and consist of a mucosa, muscularis, and adventitia.
2. The urinary bladder is posterior to the pubic symphysis. Its function is to store urine prior to micturition.
3. The mucosa of the urinary bladder contains stretchable transitional epithelium. The muscular layer of the wall consists of three layers of smooth muscle together referred to as the detrusor muscle.
4. The urethra is a tube leading from the floor of the urinary bladder to the exterior. Its function is to discharge urine from the body.
5. The micturition reflex discharges urine from the urinary bladder by means of parasympathetic impulses that cause contraction of the detrusor muscle and relaxation of the internal urethral sphincter muscle, and by inhibition of somatic motor neurons to the external urethral sphincter.

Aging and the Urinary System (Section 28.5)

1. With aging, the kidneys shrink in size, have lowered blood flow, and filter less blood.
2. Common problems related to aging include urinary tract infections, increased frequency of urination, urinary retention or incontinence, and renal calculi (kidney stones).

SELF-QUIZ QUESTIONS

1. Which of the following is NOT a function of the urinary system?
   a. regulation of blood volume and composition
   b. stimulation of red blood cell production
   c. regulation of body temperature
   d. regulation of blood pressure
   e. regulation of blood pH

2. Which of the following structures is located in the renal cortex?
   a. the renal pyramid **b.** the renal column
   c. the major calyx **d.** the minor calyx
   e. the renal corpuscle

3. Which of the following increases water reabsorption in the distal convoluted tubules and collecting ducts?
a. antidiuretic hormone (ADH)
b. angiotensin II
c. atrial natriuretic peptide (ANP)
d. diuretics
e. glucosuria

4. The major openings located in the base of the urinary bladder are orifices associated with the
a. renal artery, renal vein, urethra
b. renal artery, renal vein, ureter
c. ureter, urethra, collecting tubes
d. urethra and two ureters
e. external urethral sphincter and papillary ducts.

5. Which statement does NOT describe the kidneys?
a. They are protected by the 11th and 12th pairs of ribs.
b. The average adult kidney is 11 cm (4 inches) long and 6 cm (2 inches) wide.
c. The left kidney is lower than the right due to the presence of the large size of the liver.
d. Each kidney is surrounded by adipose and connective tissue.
e. The kidneys are surrounded by a renal capsule.

6. Place the following structures in the correct order for the flow of urine:
1. renal tubules **2.** minor calyx **3.** renal pelvis
4. major calyx **5.** collecting ducts **6.** ureters
a. 1, 2, 4, 3, 6, 5 **b.** 5, 1, 4, 2, 3, 6 **c.** 5, 1, 2, 4, 3, 6
d. 3, 5, 1, 2, 4, 6 **e.** 1, 5, 2, 4, 3, 6

7. The functional unit of the kidney where urine is produced is the
a. nephron **b.** pyramid **c.** pelvis
d. glomerulus **e.** calyx.

8. What causes filtration of plasma across the filtration membrane?
a. a full urinary bladder
b. control by the nervous system
c. water retention
d. the pressure of the blood
e. the pressure of urine in the glomerulus

9. Glomerular filtration rate (GFR) is the
a. rate of urinary bladder filling
b. amount of filtrate formed in both kidneys each minute
c. amount of filtrate reabsorbed at the collecting ducts
d. amount of blood delivered to the kidneys each minute
e. amount of urine formed per hour.

10. Which of the following is secreted into the urine from the blood?
a. hydrogen ions (H^+) **b.** amino acids
c. glucose **d.** water
e. white blood cells

11. In the nephron, tubular fluid that is reabsorbed from the renal tubules enters the
a. glomerulus **b.** peritubular capillaries
c. efferent arteriole **d.** afferent arteriole
e. renal artery.

12. Place the following structures in the correct order as they are involved in the formation of urine in the nephrons.
1. distal convoluted tubule
2. renal corpuscle
3. descending limb of loop of Henle
4. proximal convoluted tubule
5. collecting duct
6. ascending limb of loop of Henle
a. 4, 1, 6, 3, 2, 5 **b.** 2, 6, 3, 1, 5, 4
c. 2, 4, 3, 6, 5, 1 **d.** 5, 1, 4, 3, 6, 2
e. 2, 4, 3, 6, 1, 5

13. Blood is carried out of the glomerulus by the
a. renal artery **b.** afferent arteriole
c. peritubular venule **d.** segmental artery
e. efferent arteriole.

14. Which of the following increases glomerular filtration rate (GFR)?
a. atrial natriuretic peptide (ANP)
b. constriction of the afferent arterioles
c. increased sympathetic stimulation to the afferent arterioles
d. ADH **e.** angiotensin II

15. Which of the following statements concerning tubular reabsorption is NOT true?
a. Most reabsorption occurs in the proximal convoluted tubules.
b. Tubular reabsorption is a selective process.
c. Tubular reabsorption of excess potassium ions (K^+) maintains the correct blood level of K^+.
d. The reabsorption of water in the proximal convoluted tubules depends on sodium ion (Na^+) reabsorption.
e. Tubular reabsorption allows the body to retain most filtered nutrients.

16. The micturition reflex
a. is under the control of hormones
b. is activated by low pressure in the urinary bladder
c. depends on contraction of the internal urethral sphincter muscle
d. is an involuntary reflex over which normal adults have voluntary control
e. is also known as incontinence.

17. Which of the following is NOT normally present in glomerular filtrate?
a. blood cells
b. glucose
c. nitrogenous wastes such as urea
d. amino acids
e. water

18. Urine formation requires which of the following?
a. glomerular filtration and tubular secretion only
b. glomerular filtration and tubular reabsorption only
c. glomerular filtration, tubular reabsorption, and tubular secretion
d. tubular reabsorption, tubular filtration, and tubular secretion
e. tubular secretion and tubular reabsorption only

19. The transport of urine from the renal pelvis into the urinary bladder is the function of the
a. urethra **b.** efferent arteriole
c. afferent arteriole **d.** renal pyramids
e. ureters.

20. Incontinence is
a. failure of the urinary bladder to expel urine
b. a lack of voluntary control over the micturition reflex
c. an inability of the kidneys to produce urine
d. an ability to consciously control micturition
e. a form of kidney dialysis.

CRITICAL THINKING QUESTIONS

1. Yesterday, you attended a large, outdoor party where beer was the only beverage available. You remember having to urinate many, many times yesterday, and today you're very thirsty. What hormone is affected by alcohol, and how does this affect your kidney function?
2. Sarah is an "above average" 1-year-old whose parents would like her to be the first toilet-trained child in preschool. However, in this case at least, Sarah is average for her age and remains incontinent. Should her parents be concerned by this lack of success?
3. Kayla is a healthy, VERY active 4-year-old. She doesn't like to take the time to go to the bathroom because, as she says, "I might miss somethin'." Her mother is worried that Kayla's kidneys may stop working when her urinary bladder is full. Should her mother be concerned?

? ANSWERS TO FIGURE QUESTIONS

28.1 The kidneys, ureters, urinary bladder, and urethra are the components of the urinary system.

28.2 Blood vessels, lymphatic vessels, nerves, and a ureter pass through the renal hilum.

28.3 About 1200 mL of blood enters the kidneys each minute.

28.4 The water molecule will travel as follows: proximal convoluted tubule → descending limb of the loop of Henle → ascending limb of the loop of Henle → distal convoluted tubule → collecting duct → papillary duct → minor calyx → major calyx → renal pelvis.

28.5 Secreted penicillin is being removed from the blood.

28.6 Podocytes and the glomerular endothelium make up the filtration membrane.

28.7 Secretion occurs in the proximal convoluted tubule, the loop of Henle, the last part of the distal convoluted tubule, and the collecting duct.

28.8 The blood level of ADH would be higher than normal after a 5 km run, due to loss of body water in sweat.

28.9 A lack of voluntary control over micturition is termed incontinence.

29 The Reproductive Systems

Sexually transmitted diseases (STDs) are infectious conditions transmitted through sexual activity. The best way to avoid STD infection is to abstain from sex (vaginal, oral, or anal). Safe sex involves not allowing your partner's body fluids (such as semen, vaginal fluid, or blood) to enter or come in contact with your body.

For people who choose to have sex, the most reliable way to avoid STDs is to be in a long-term monogamous relationship with an uninfected partner. New sex partners should be checked by a physician. Practicing safe sex by using condoms can reduce the risk of transmission of HIV, human papillomavirus, gonorrhea, chlamydia, and other STD infections. The practice of safe sex and the prevention of STDs has become an important global issue.

Infection in both men and women may not show symptoms, so sometimes STDs can go undetected. Some STDs can be easily treated, but others are incurable. Unfortunately, the STDs that are most difficult to treat tend to be the most common and longest lasting, often having serious health consequences. Symptoms of these types of infections are quite varied and may include sores, ulcers, lumps, irritation, blisters, pain, itchiness, warts, or unusual discharge from the urogenital or oral regions. Be healthy. Be safe.

CONTENTS AT A GLANCE

29.1 INTRODUCTION TO THE REPRODUCTIVE SYSTEMS

OBJECTIVES

- Define gamete for the male and female.
- Define gonad for the male and female.

Sexual reproduction is the process by which organisms produce offspring by making germ cells called ***gametes*** (GAM-ēts = spouses). After ***fertilization,*** when the male gamete (sperm cell) unites with the female gamete (secondary oocyte), the resulting cell contains one set of chromosomes from each parent. The organs that make up the male and female reproductive systems can be grouped by function. The ***gonads***—testes in males and ovaries in females—produce gametes and secrete sex hormones. Various ***ducts*** then store and transport the gametes, and ***accessory sex glands*** produce substances that protect the gametes and facilitate their movement. Finally, ***supporting structures,*** such as the penis in males and the uterus in females, assist the delivery of gametes; in females, the uterus is also the site for the growth of the embryo and fetus during pregnancy.

Gynecology (gī′-ne-KOL-ō-jē; *gyneco-* = woman; *-logy* = study of) is the specialized branch of medicine concerned with the diagnosis and treatment of diseases of the female reproductive system. As noted in Chapter 28, ***urology*** (ū-ROL-ō-jē) is the study of the urinary system. Urologists also diagnose and treat diseases and disorders of the male reproductive system. The branch of medicine that deals with male disorders, especially infertility and sexual dysfunction, is called ***andrology*** (an-DROL-ō-jē; *andro-* = masculine).

CHECKPOINT

1. What are male gametes called and where are they produced?
2. What are female gametes called and where are they produced?

29.2 MALE REPRODUCTIVE SYSTEM

OBJECTIVES

- Describe the location, structure, and functions of the organs of the male reproductive system.
- Describe in general terms how sperm cells are produced.
- Explain the roles of hormones in regulating male reproductive functions.

The organs of the ***male reproductive system*** are the testes; a system of ducts (epididymis, ductus deferens, ejaculatory ducts, and urethra); accessory sex glands (seminal vesicles, prostate, and bulbourethral glands); and several supporting structures, including the scrotum and the penis (see Figure 29.1). The testes produce sperm and secrete hormones. Sperm are transported and stored, helped to mature, and conveyed to the exterior by a system of ducts. Semen contains sperm plus the secretions provided by the accessory sex glands.

Scrotum

The ***scrotum*** (SKRŌ-tum = bag) is a pouch that supports the testes; it consists of loose skin, subcutaneous tissue, and smooth muscle (see Figure 29.1). Internally, a septum divides the scrotum into two sacs, each containing a single testis.

The production and survival of sperm is optimal at a temperature that is about 2°C below normal body temperature. This lowered body temperature is maintained within the scrotum because it is outside the pelvic cavity. On exposure to cold, skeletal muscles contract to elevate the testes, moving them closer to the pelvic cavity, where they can absorb body heat. Exposure to warmth causes relaxation of the skeletal muscles and descent of the testes, increasing the surface area exposed to the air, so that the testes can give off excess heat to their surroundings.

Testes

The ***testes*** (TES-tēz; singular is *testis*), or *testicles,* are paired oval glands that develop on the embryo's posterior abdominal wall and usually begin their descent into the scrotum during the latter half of the seventh month of fetal development.

The testes are covered by a dense ***white fibrous capsule*** that extends inward and divides each testis into internal compartments called ***lobules*** (see Figure 29.2a). Each of the 200 to 300 lobules contains one to three tightly coiled ***seminiferous tubules*** (*semin-* = seed; *-fer-* = to carry) that produce sperm by a process called *spermatogenesis* (described shortly).

Seminiferous tubules are lined with ***spermatogenic cells,*** which are sperm-forming cells (see Figure 29.2b). Positioned against the basement membrane, toward the outside of the tubules, are the ***spermatogonia*** (sper-ma′-tō-GŌ-nē-a; *-gonia* = offspring), the stem cell precursors. Toward the lumen of the tubule are layers of cells in order of advancing maturity: primary spermatocytes, secondary spermatocytes, spermatids, and sperm cells. After a ***sperm cell*** or ***spermatozoon*** (sper′-ma-tō-ZŌ-on; *-zoon* = life) has formed, it is released into the lumen of the seminiferous tubule.

Large ***Sertoli cells,*** located between the developing sperm cells in the seminiferous tubules, support, protect, and nourish spermatogenic cells; phagocytize degenerating spermatogenic cells;

• CLINICAL CONNECTION | Cryptorchidism

The condition in which the testes do not descend into the scrotum is called **cryptorchidism** (krip-TOR-ki-dizm; *crypt-* = hidden; *orchid* = testis). It occurs in about 3% of full-term infants and about 30% of premature infants. Untreated bilateral cryptorchidism causes sterility due to the higher temperature of the pelvic cavity. The chance of testicular cancer is 30 to 50 times greater in cryptorchid testes, possibly due to abnormal division of germ cells caused by the higher temperature of the pelvic cavity. The testes of about 80% of boys with cryptorchidism will descend spontaneously during the first year of life. When the testes remain undescended, the condition can be corrected surgically, ideally before 18 months of age. •

Figure 29.1 Male organs of reproduction and surrounding structures.

Reproductive organs are adapted for producing new individuals and passing genetic material from one generation to the next.

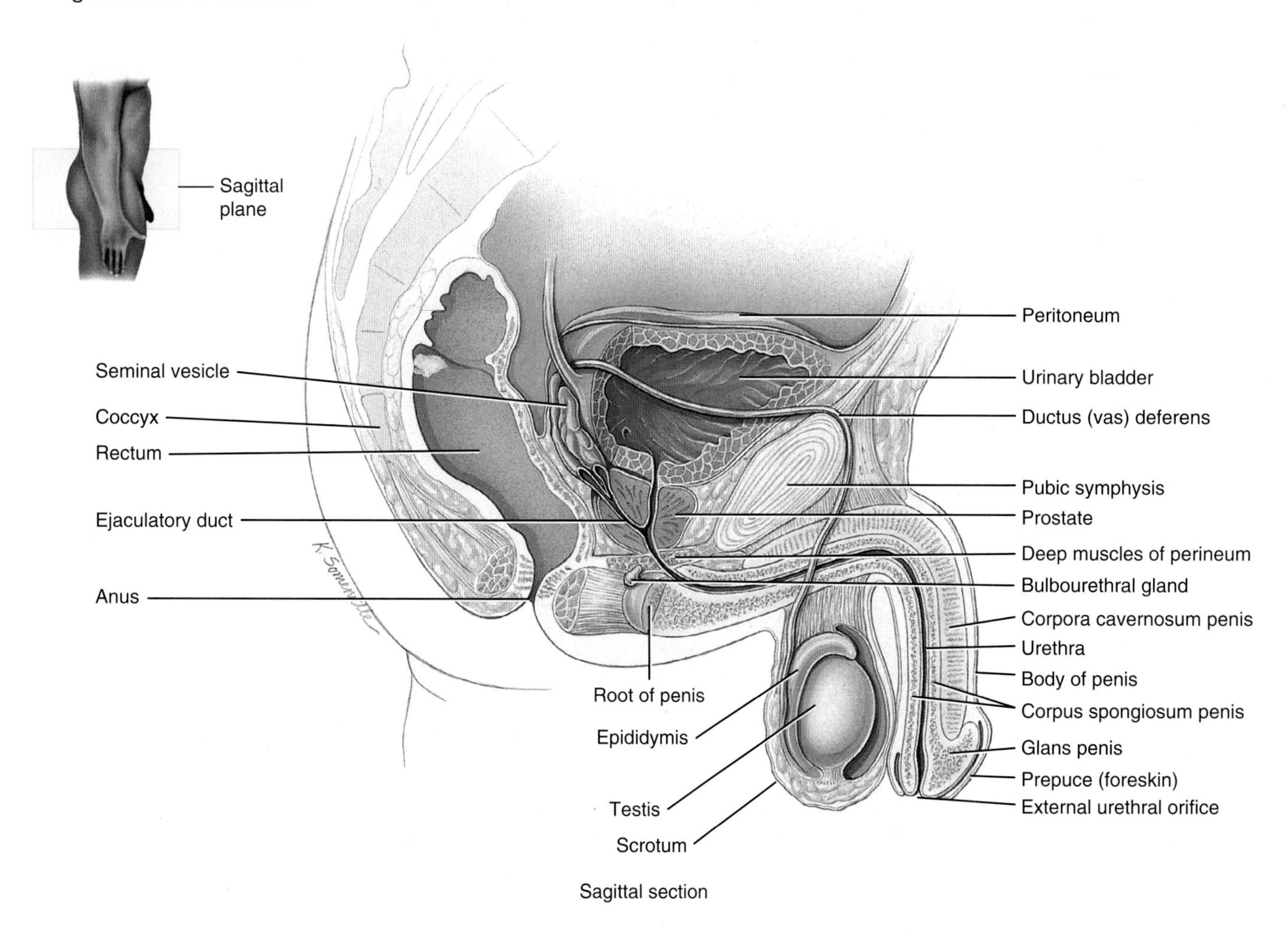

Functions of the Male Reproductive System
1. The testes produce sperm and the male sex hormone testosterone.
2. The ducts transport, store, and assist in maturation of sperm.
3. The accessory sex glands secrete most of the liquid portion of semen.
4. The penis contains the urethra, a passageway for ejaculation of semen and excretion of urine.

? What are the groups of reproductive organs in males, and what are the functions of each group?

secrete fluid for sperm transport; and release the hormone inhibin, which helps regulate sperm production (Figure 29.2b). Between the seminiferous tubules are clusters of ***Leydig cells.*** These cells secrete the hormone ***testosterone,*** the most important androgen. An ***androgen*** (AN-drō-jen) is a hormone that promotes the development of masculine characteristics. Testosterone also promotes a man's libido (sex drive).

Spermatogenesis

The process by which the seminiferous tubules of the testes produce sperm is called ***spermatogenesis*** (sper′-ma-tō-JEN-e-sis) (see Figure 29.3). As you learned in Chapter 3, most body cells (somatic cells), such as brain cells, stomach cells, kidney cells, and so forth, contain 23 pairs of chromosomes, or a total of 46 chromosomes, which is the ***diploid*** ($2n$) number. One member of each

Figure 29.2 Anatomy and histology of the testes. (a) Spermatogenesis occurs in the seminiferous tubules. (b) Stages of spermatogenesis. Arrows in (b) indicate the progression from least mature to most mature spermatogenic cells. The (*n*) and (2*n*) refer to haploid and diploid numbers of chromosomes, respectively.

6 The male gonads are the testes, which produce haploid sperm.

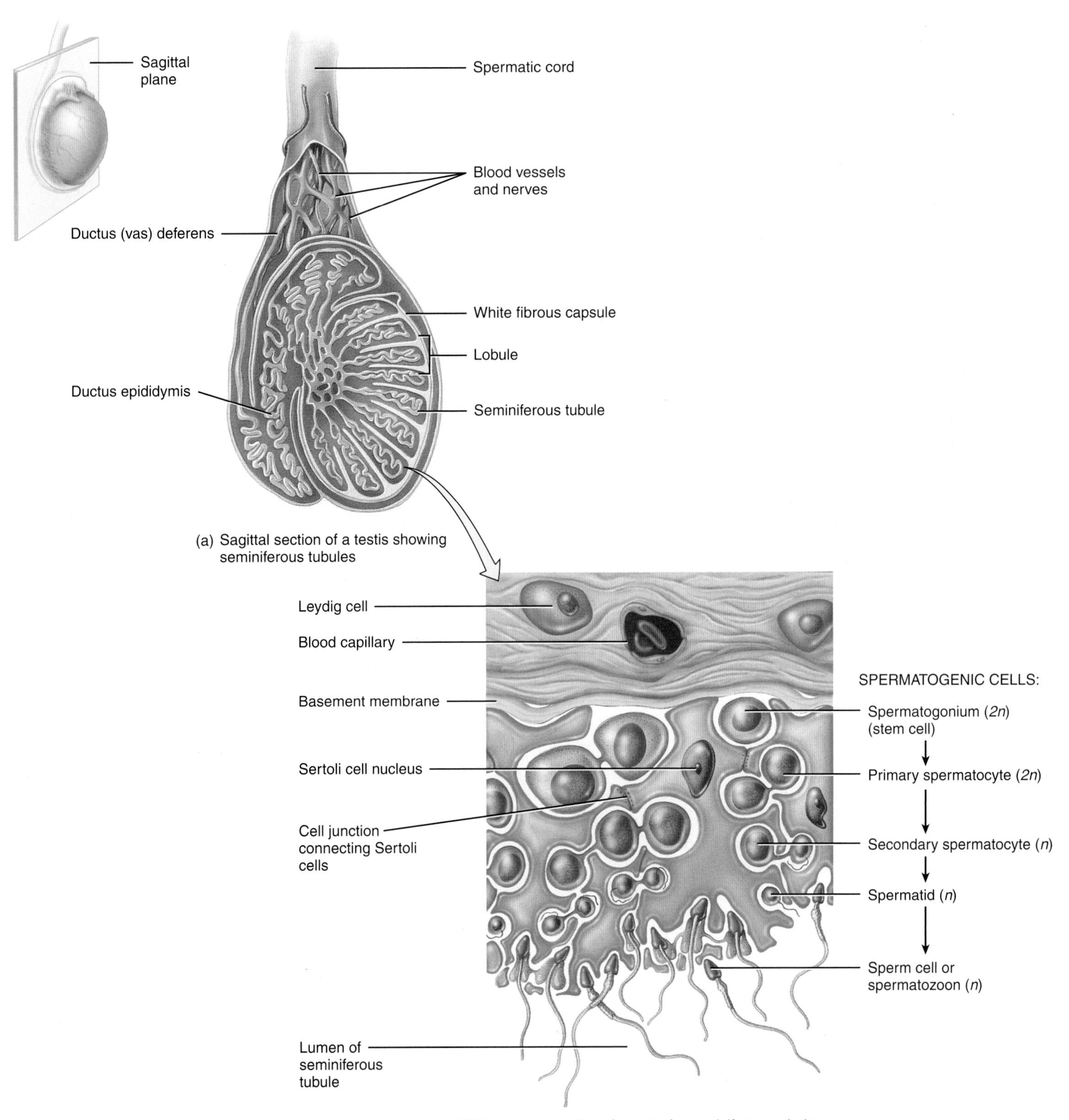

(a) Sagittal section of a testis showing seminiferous tubules

(b) Transverse section of a part of a seminiferous tubule

? Which cells secrete testosterone?

Figure 29.3 Events in spermatogenesis. Diploid cells (2*n*) have 46 chromosomes; haploid cells (*n*) have 23 chromosomes.

Spermiogenesis involves the maturation of spermatids into sperm.

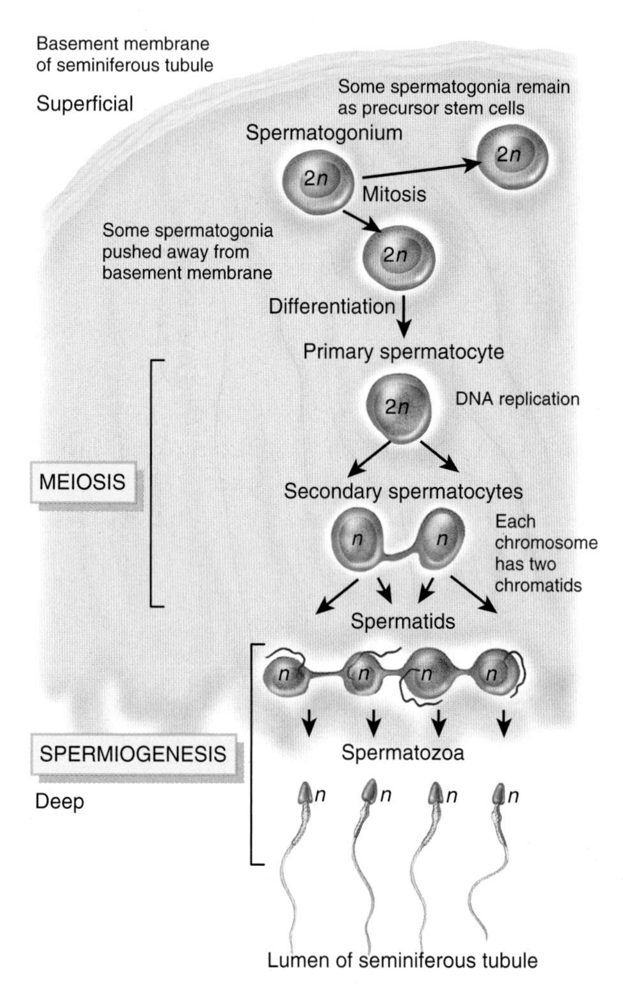

What is "reduced" during meiosis?

Figure 29.4 Parts of a sperm cell.

About 300 million sperm mature each day.

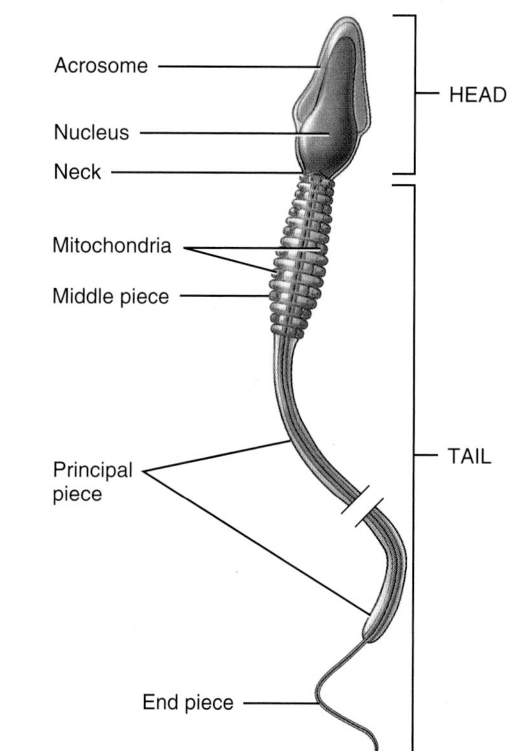

What is the function of the sperm middle piece?

pair is inherited from each parent. Gametes differ from somatic cells because they contain a single set of 23 chromosomes, symbolized as *n*; they are thus said to be ***haploid*** (HAP-loyd; *hapl-* = single). Spermatogenesis consists of meiosis and spermiogenesis (Figure 29.3). In sexual reproduction, an organism results from the fusion of two different gametes, one produced by each parent.

Sperm

Sperm are produced at the rate of about 300 million per day. Once ejaculated, most do not survive more than 24 hours in the female reproductive tract. The major parts of a sperm cell are the head and the tail (Figure 29.4). The ***head*** contains the nuclear material (DNA) and an ***acrosome*** (*acro-* = atop), a vesicle containing enzymes that aid penetration by the sperm cell into a secondary oocyte. The ***tail*** of a sperm cell is subdivided into four parts: neck, middle piece, principal piece, and end piece. The *neck* is the constricted region just behind the head. The *middle piece* contains mitochondria that provide ATP for locomotion. The *principal piece* is the longest portion of the tail, and the *end piece* is the terminal, tapering portion of the tail.

Hormonal Control of the Testes

At the onset of puberty, neurosecretory cells in the hypothalamus increase their secretion of ***gonadotropin-releasing hormone (GnRH).*** This hormone, in turn, stimulates the anterior pituitary to increase its secretion of ***luteinizing hormone*** (***LH***) and ***follicle-stimulating hormone*** (***FSH***)***.*** Figure 29.5 shows the hormones and negative feedback cycles that control the Leydig and Sertoli cells of the testes and stimulate spermatogenesis.

LH stimulates Leydig cells, which are located between seminiferous tubules, to secrete the hormone ***testosterone*** (tes-TOS-te-rōn). This steroid hormone is synthesized from cholesterol in the testes and is the principal androgen. Testosterone acts in a negative feedback manner to suppress secretion of LH by the anterior pituitary and to suppress secretion of GnRH by hypothalamic neurosecretory cells.

FSH and testosterone act together to stimulate spermatogenesis. Once the degree of spermatogenesis required for male reproductive functions has been achieved, Sertoli cells release ***inhibin,*** a hormone named for its inhibition of FSH secretion by the anterior pituitary (Figure 29.5). Inhibin thus inhibits the secretion of hormones needed for spermatogenesis. If spermatogenesis is

Figure 29.5 Hormonal control of spermatogenesis and actions of testosterone. Dashed red lines indicate negative feedback inhibition.

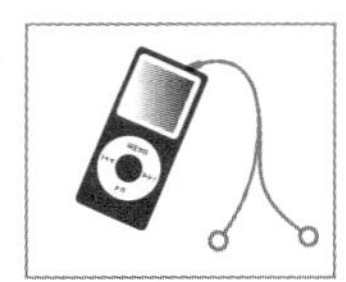

Release of FSH is stimulated by GnRH and inhibited by inhibin; release of LH is stimulated by GnRH and inhibited by testosterone.

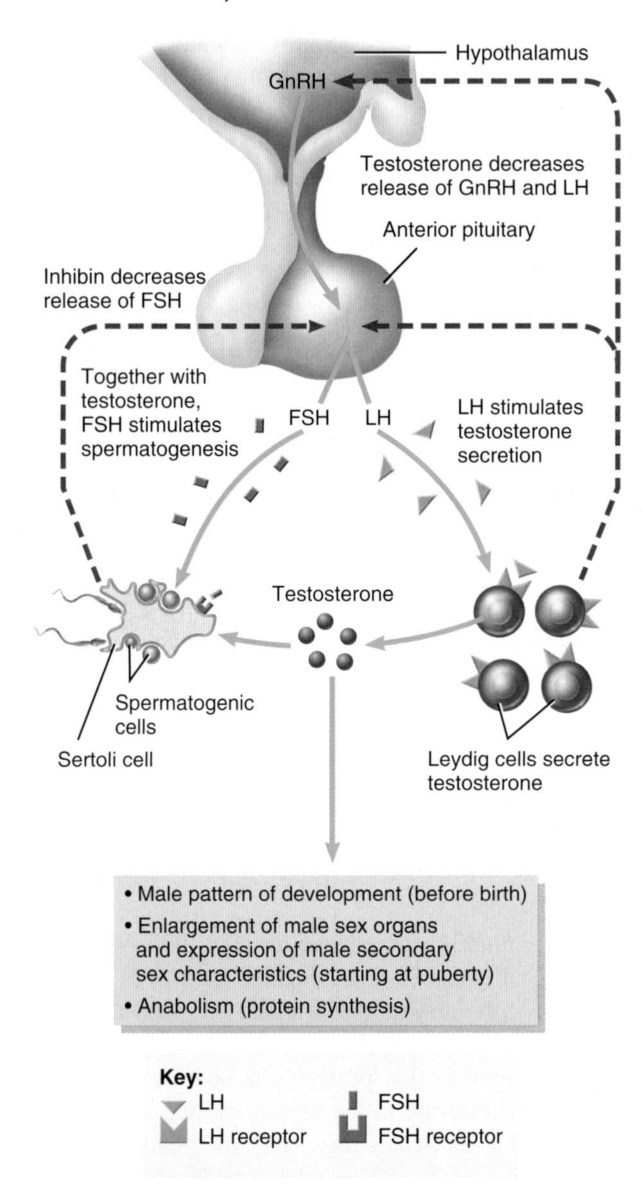

Which cells secrete inhibin?

proceeding too slowly, less inhibin is released, which permits more FSH secretion and an increased rate of spermatogenesis.

Testosterone binds to androgen receptors, producing several effects:

- **Prenatal development.** Before birth, testosterone (and its derivitives) stimulate the male pattern of development of reproductive system ducts, the descent of the testes, and development of the external genitals. Testosterone also is converted in the brain to estrogens (feminizing hormones), which may play a role in the development of certain regions of the brain in males.
- **Development of male sexual characteristics.** At puberty, testosterone (and derivatives) bring about development and enlargement of the male sex organs and the development of masculine secondary sexual characteristics. *Secondary sex characteristics* are traits that distinguish males and females but do not have a direct role in reproduction. These include muscular and skeletal growth that results in wide shoulders and narrow hips; pubic, facial, and chest hair (within hereditary limits), and more hair on other parts of the body; thickening of the skin; increased sebaceous (oil) gland secretion; and enlargement of the larynx and consequent deepening of the voice.
- **Development of sexual function.** Androgens contribute to male sexual behavior and spermatogenesis and to sex drive (libido) in both males and females. Recall that the adrenal cortex is the main source of androgens in females.
- **Stimulation of anabolism.** Androgens are anabolic hormones; that is, they stimulate protein synthesis. This effect is obvious in the heavier muscle and bone mass of most men as compared to women.

Ducts

Following spermatogenesis, pressure generated by the continual release of sperm and fluid secreted by Sertoli cells propels sperm and fluid through the seminiferous tubules and into the epididymis (see Figure 29.2a).

Epididymis

The ***epididymis*** (ep′-i-DID-i-mis; *epi-* = above or over; *-didymis* = testis; plural is *epididymides*) is a comma-shaped organ that lies along the posterior border of the testis (see Figures 29.1 and 29.2a). Each epididymis consists mostly of the tightly coiled ***ductus epididymis.*** Functionally, the ductus epididymis is the site of *sperm maturation,* the process by which sperm acquire motility and the ability to fertilize a secondary oocyte. This occurs over a 10- to 14-day period. The ductus epididymis also stores sperm and helps propel them during sexual arousal by peristaltic contraction of its smooth muscle into the ductus (vas) deferens. Sperm may remain in storage in the ductus epididymis for several months. Any stored sperm that are not ejaculated by that time are eventually phagocytized and reabsorbed.

Ductus (Vas) Deferens

At the end of the epididymis, the ductus epididymis becomes less convoluted, and its diameter increases. Beyond the epididymis, the duct is termed the ***ductus deferens*** or ***vas deferens*** (VAS DEF-er-enz; *vas* = vessel; *de-* = away). (See Figure 29.2a.) The ductus deferens ascends along the posterior border of the epididymis and penetrates the inguinal canal, a passageway in the front abdominal wall. Then, it enters the pelvic cavity, where it loops over the side and down the posterior surface of the urinary bladder (see Figures 29.1 and 29.6). The ductus deferens has a heavy coat of three layers of muscle.

Functionally, the ductus deferens stores sperm, which can remain viable here for up to several months. The ductus deferens also conveys sperm from the epididymis toward the urethra during sexual arousal by peristaltic contractions of the muscular coat.

Accompanying the ductus deferens as it ascends in the scrotum are blood vessels, autonomic nerves, and lymphatic vessels that together make up the ***spermatic cord,*** a supporting structure of the male reproductive system (see Figure 29.2a).

• CLINICAL CONNECTION | Vasectomy

The principal method for sterilization of males is a **vasectomy** (vas-EK-tō-mē; *-ectomy* = cut out), in which a portion of each ductus deferens is removed. An incision is made on either side of the scrotum, the ducts are located and cut, each is tied (ligated) in two places with stitches, and the portion between the ties is removed. Although sperm production continues in the testes, sperm can no longer reach the exterior. The sperm degenerate and are destroyed by phagocytosis. Because the blood vessels are not cut, testosterone levels in the blood remain normal, so vasectomy has no effect on sexual desire, performance, and ejaculation. If done correctly, it is close to 100% effective. The procedure can be reversed, but the chance of regaining fertility is only 30–40%. •

Ejaculatory Ducts

The ***ejaculatory ducts*** (e-JAK-ū-la-tō-rē; *ejacul-* = to expel) (Figure 29.6) are formed by the union of the duct from the ductus deferens and the seminal vesicles (to be described shortly). The short ejaculatory ducts carry sperm into the urethra.

Urethra

The ***urethra*** is the terminal duct of the male reproductive system, serving as a passageway for both sperm and urine. In the male, the urethra passes through the prostate, deep muscles of the perineum, and penis (see Figures 29.1 and 29.6). The opening of the urethra to the exterior is called the ***external urethral orifice.***

Accessory Sex Glands

The ducts of the male reproductive system store and transport sperm cells, but the ***accessory sex glands*** secrete most of the liquid portion of semen.

The paired ***seminal vesicles*** (VES-i-kuls) are pouchlike structures lying posterior to the base of the urinary bladder and anterior to the rectum (Figure 29.6). They secrete an alkaline, viscous fluid that contains fructose, prostaglandins, and clotting proteins (unlike those found in blood). The alkaline nature of the fluid helps to neutralize the acidic environment of the male urethra and female reproductive tract that otherwise would inactivate and kill sperm. The fructose is used for ATP production by sperm. Prostaglandins contribute to sperm motility and viability and may also stimulate muscular contraction within the female reproductive tract. Clotting proteins help semen coagulate after ejaculation. Fluid secreted by the seminal vesicles normally constitutes about 60% of the volume of semen.

The ***prostate*** (PROS-tāt) is a single, doughnut-shaped gland about the size of a golf ball (see Figures 29.1 and 29.6). It is inferior to the urinary bladder and surrounds the upper portion of the urethra. The prostate slowly increases in size from birth to puberty, and then it expands rapidly. The size attained by age 30 remains stable until about age 45, when further enlargement may occur. The prostate secretes a milky, slightly acidic fluid (pH about 6.5) that contains *citric acid,* which can be used by sperm for ATP production, and several protein-digesting enzymes, such as *prostate-specific antigen* (*PSA*), which eventually break down the clotting proteins from the seminal vesicles. Prostatic secretions make up about 25% of the volume of semen and contribute to sperm motility and viability.

• CLINICAL CONNECTION | Prostate Disorders

Because the prostate surrounds part of the urethra, any prostatic infection, enlargement, or tumor can obstruct the flow of urine. Acute and chronic infections of the prostate are common in adult males, often in association with inflammation of the urethra. In **acute prostatitis,** the prostate becomes swollen and tender. **Chronic prostatitis** is one of the most common chronic infections in men of the middle and later years; on examination, the prostate feels enlarged, soft, and very tender, and its surface outline is irregular.

Prostate cancer is the leading cause of death from cancer in men in the United States. A blood test can measure the level of prostate-specific antigen (PSA) in the blood. The amount of PSA, which is produced only by prostate epithelial cells, increases with enlargement of the prostate and may indicate infection, benign enlargement, or prostate cancer. Because many prostate cancers grow very slowly, some urologists recommend "watchful waiting" before treating small tumors in men over age 70. •

The paired ***bulbourethral glands*** (bul′-bō-ū-RĒ-thral) are about the size of peas. They are located inferior to the prostate on either side of the urethra (see Figures 29.1 and 29.6). During sexual arousal, the bulbourethral glands secrete an alkaline substance into the urethra that protects the passing sperm by neutralizing acids from urine in the urethra. At the same time, they secrete mucus that lubricates the end of the penis and the lining of the urethra, thereby decreasing the number of sperm damaged during ejaculation. Some men release a drop or two of this mucus upon each sexual arousal and erection. This fluid does not contain sperm cells.

Semen

Semen (= seed) is a mixture of sperm and the secretions of the seminal vesicles, prostate, and bulbourethral glands. The volume of semen in a typical ejaculation is 2.5 to 5 milliliters, with 50 to 150 million sperm per milliliter. When the number falls below 20 million per milliliter, the male is likely to be infertile. A very large number of sperm is required for fertilization because only a tiny fraction ever reaches the secondary oocyte.

Figure 29.6 Locations of several accessory reproductive organs in males. The prostate, urethra, and penis have been sectioned to show internal structures.

Ducts of the ductus (vas) deferens and the seminal vesicle join to form the ejaculatory duct within the prostate gland.

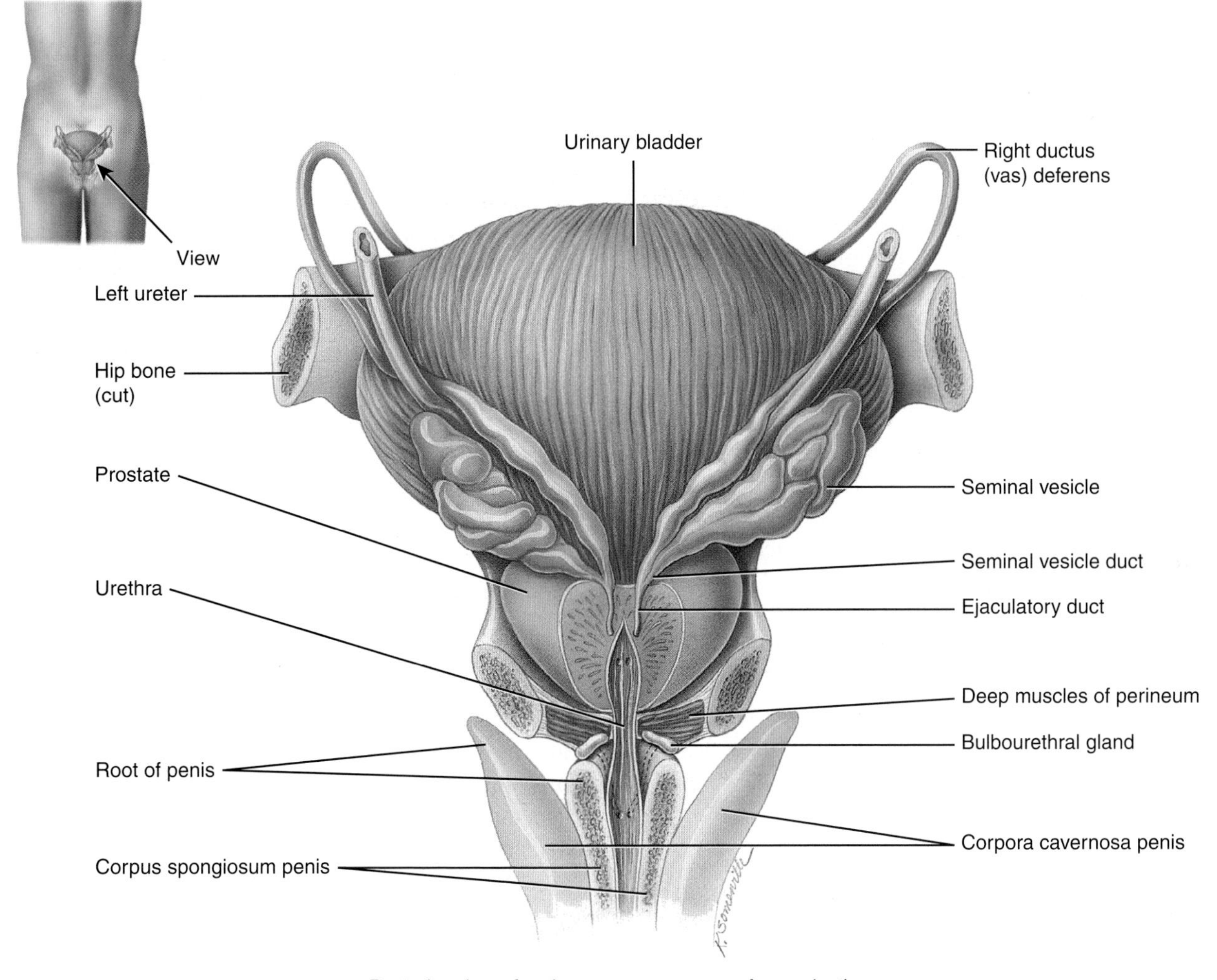

Posterior view of male accessory organs of reproduction

Functions of Accessory Sex Gland Secretions

1. The seminal vesicles secrete alkaline, viscous fluid that helps neutralize acid in the female reproductive tract, provides fructose for ATP production by sperm, contributes to sperm motility and viability, and helps semen coagulate after ejaculation.
2. The prostate secretes a milky, slightly acidic fluid that helps semen coagulate after ejaculation and subsequently breaks down the clot.
3. The bulbourethral (Cowper's) glands secrete alkaline fluid that neutralizes the acidic environment of the urethra and mucus that lubricates the lining of the urethra and the tip of the penis during sexual intercourse.

What accessory sex gland contributes the majority of the seminal fluid?

Despite the slight acidity of prostatic fluid, semen has a slightly alkaline pH of 7.2 to 7.7 due to the higher pH and larger volume of fluid from the seminal vesicles. The prostatic secretion gives semen a milky appearance, and fluids from the seminal vesicles and bulbourethral glands give it a sticky consistency.

Penis

The ***penis*** contains the urethra and is a passageway for the ejaculation of semen and the excretion of urine. It is cylindrical in shape and consists of a root, a body, and the glans penis (see Figure 29.1). The ***root of the penis*** is the attached portion (proximal portion). The ***body of the penis*** is composed of three cylindrical masses of tissue. The two dorsolateral masses are called the ***corpora cavernosa penis*** (*corpora* = main bodies; *cavernosa* = hollow). The smaller midventral mass, the ***corpus spongiosum penis,*** contains the urethra. Skin and a subcutaneous layer enclose all three masses, which consist of erectile tissue permeated by blood sinuses.

The distal end of the corpus spongiosum penis is a slightly enlarged region called the ***glans penis.*** In the glans penis is the opening of the urethra to the exterior, the ***external urethral orifice.*** Covering the glans in an uncircumcised penis is the loosely fitting ***prepuce*** (PRĒ-poos), or ***foreskin.***

• CLINICAL CONNECTION | Circumcision

Circumcision (= to cut around) is a surgical procedure in which part or the entire prepuce is removed. It is usually performed just after delivery, 3 to 4 days after birth, or on the eighth day as part of a Jewish religious rite. Although most health-care professionals find no medical justification for circumcision, some feel that it has benefits, such as a lower risk of urinary tract infections, protection against penile cancer, and possibly a lower risk for sexually transmitted diseases. Indeed, studies in several African villages have found lower rates of HIV infection among circumcised men. •

Most of the time, the penis is flaccid (limp) because its arteries are vasoconstricted, which limits blood flow. The first visible sign of sexual excitement is ***erection,*** the enlargement and stiffening of the penis. Parasympathetic impulses cause release of neurotransmitters and local hormones, including the gas nitric oxide, which relaxes vascular smooth muscle in the penile arteries. The arteries supplying the penis dilate, and large quantities of blood enter the blood sinuses. Expansion of these spaces compresses the veins draining the penis, so blood outflow is slowed.

Erectile dysfunction (ED), previously termed *impotence,* is the consistent inability of an adult male to ejaculate or to attain or hold an erection long enough for sexual intercourse. Many cases of impotence are caused by insufficient release of nitric oxide. The drug sildenafil (Viagra®) enhances the effect of nitric oxide.

Ejaculation (ē-jak-ū-LĀ-shun; *ejectus-* = to throw out), the powerful release of semen from the urethra to the exterior, is a sympathetic reflex coordinated by the lumbar portion of the spinal cord. As part of the reflex, the smooth muscle sphincter at the base of the urinary bladder closes. Thus, urine is not expelled during ejaculation, and semen does not enter the urinary bladder. Even before ejaculation occurs, peristaltic contractions in the ductus deferens, seminal vesicles, ejaculatory ducts, and prostate propel semen into the penile portion of the urethra. Typically, this leads to ***emission*** (ē-MISH-un), the discharge of a small volume of semen before ejaculation. Emission may also occur during sleep (nocturnal emission). The penis returns to its flaccid state when the arteries constrict and pressure on the veins is relieved.

CHECKPOINT

3. How does the scrotum protect the testes?
4. What are the principal events of spermatogenesis and where do they occur?
5. What are the roles of FSH, LH, testosterone, and inhibin in the male reproductive system? How is secretion of these hormones controlled?
6. Trace the course of sperm through the system of ducts from the seminiferous tubules through the urethra.

29.3 FEMALE REPRODUCTIVE SYSTEM

OBJECTIVES

- Describe the location, structure, and functions of the organs of the female reproductive system.
- Describe how oocytes are produced.

The organs of the ***female reproductive system*** (Figure 29.7) include the ovaries; the uterine (fallopian) tubes, or oviducts; the uterus; the vagina; and external organs, which are collectively called the vulva, or pudendum. The mammary glands also are considered part of the female reproductive system.

Ovaries

The ***ovaries*** (= egg receptacles) are paired organs that produce secondary oocytes (cells that develop into mature ova, or eggs, following fertilization) and hormones, such as progesterone and estrogens (the female sex hormones), inhibin, and relaxin. The ovaries arise from the same embryonic tissue as the testes, and they are the size and shape of unshelled almonds. One ovary lies on each side of the pelvic cavity, held in place by ligaments (see Figure 29.10). Figure 29.8 shows the histology of an ovary.

The ***germinal epithelium*** is a layer of simple epithelium that covers the surface of the ovary. Deep to the germinal epithelium is the ***ovarian cortex,*** a region of dense, irregular connective tissue that contains ovarian follicles. Each ***ovarian follicle*** (*folliculus* = little bag) consists of an ***oocyte*** and a variable number of surrounding cells that nourish the developing oocyte and begin to secrete estrogens as the follicle grows larger. The follicle enlarges until it is a ***mature (graafian) follicle,*** a large, fluid-filled follicle that will soon rupture and expel a secondary oocyte. The remnants of an ovulated follicle develop into a ***corpus luteum*** (= yellow body). The corpus luteum produces progesterone, estrogens, relaxin, and inhibin until it degenerates

Figure 29.7 Female organs of reproduction and surrounding structures.

The organs of reproduction in females include the ovaries, uterine (fallopian) tubes, uterus, vagina, vulva, and mammary glands.

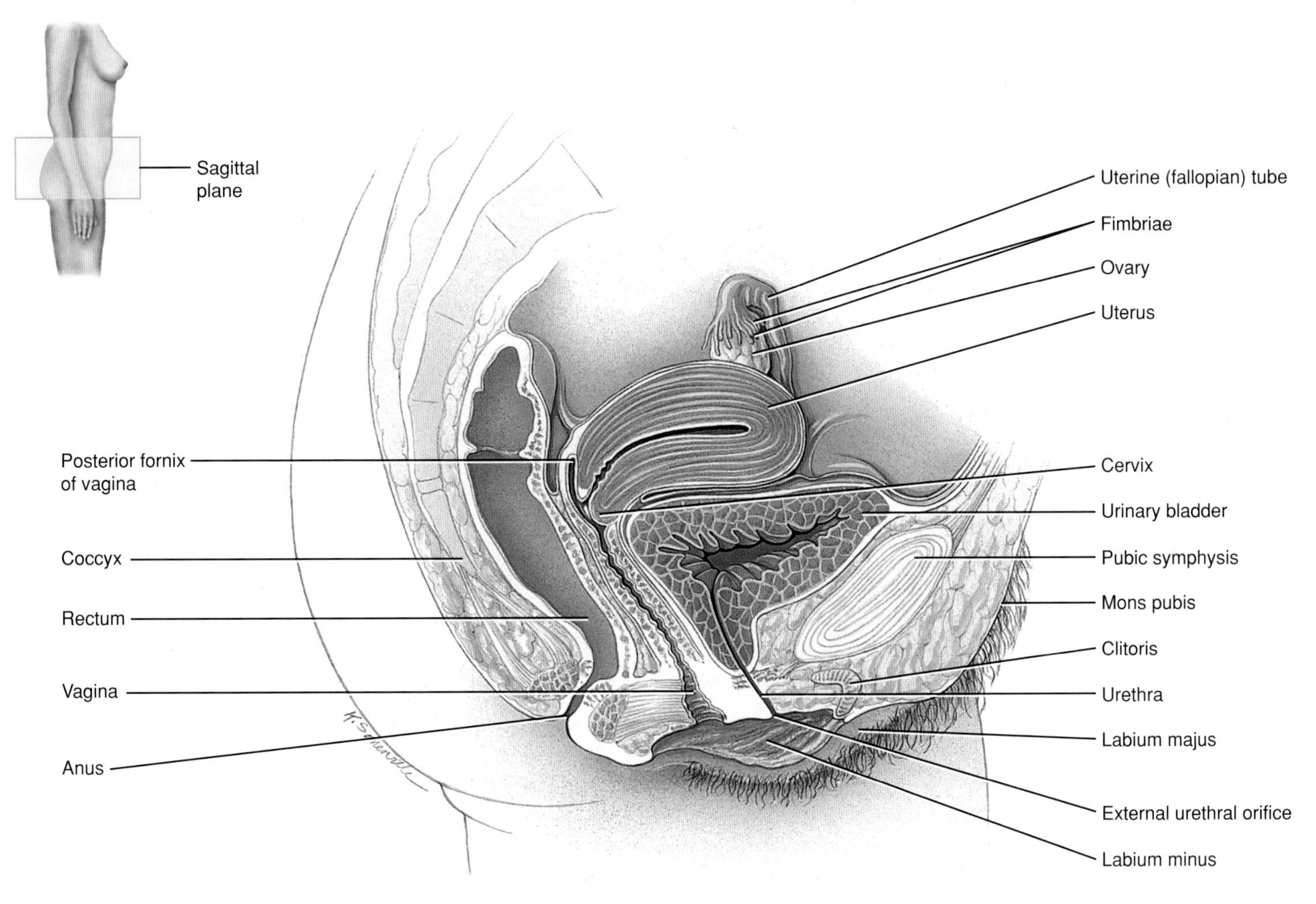

Sagittal section

Functions of the Female Reproductive System

1. The ovaries produce secondary oocytes and hormones, including progesterone and estrogens (female sex hormones), inhibin, and relaxin.
2. The uterine tubes transport a secondary oocyte to the uterus and normally are the sites where fertilization occurs.
3. The uterus is the site of implantation of a fertilized ovum, development of the fetus during pregnancy, and labor.
4. The vagina receives the penis during sexual intercourse and is a passageway for childbirth.
5. The mammary glands synthesize, secrete, and eject milk for nourishment of the newborn.

? What is the normal anatomical relationship of the vagina and uterus to the urethra and urinary bladder?

Figure 29.8 Histology of the ovary. The arrows indicate the sequence of developmental stages that occur as part of the maturation of an ovum during the ovarian cycle.

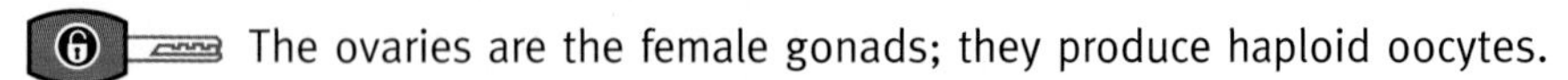
The ovaries are the female gonads; they produce haploid oocytes.

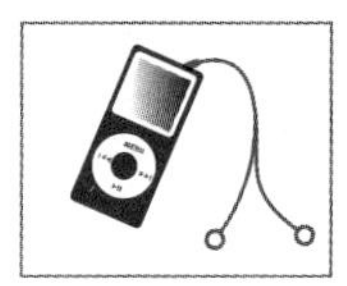

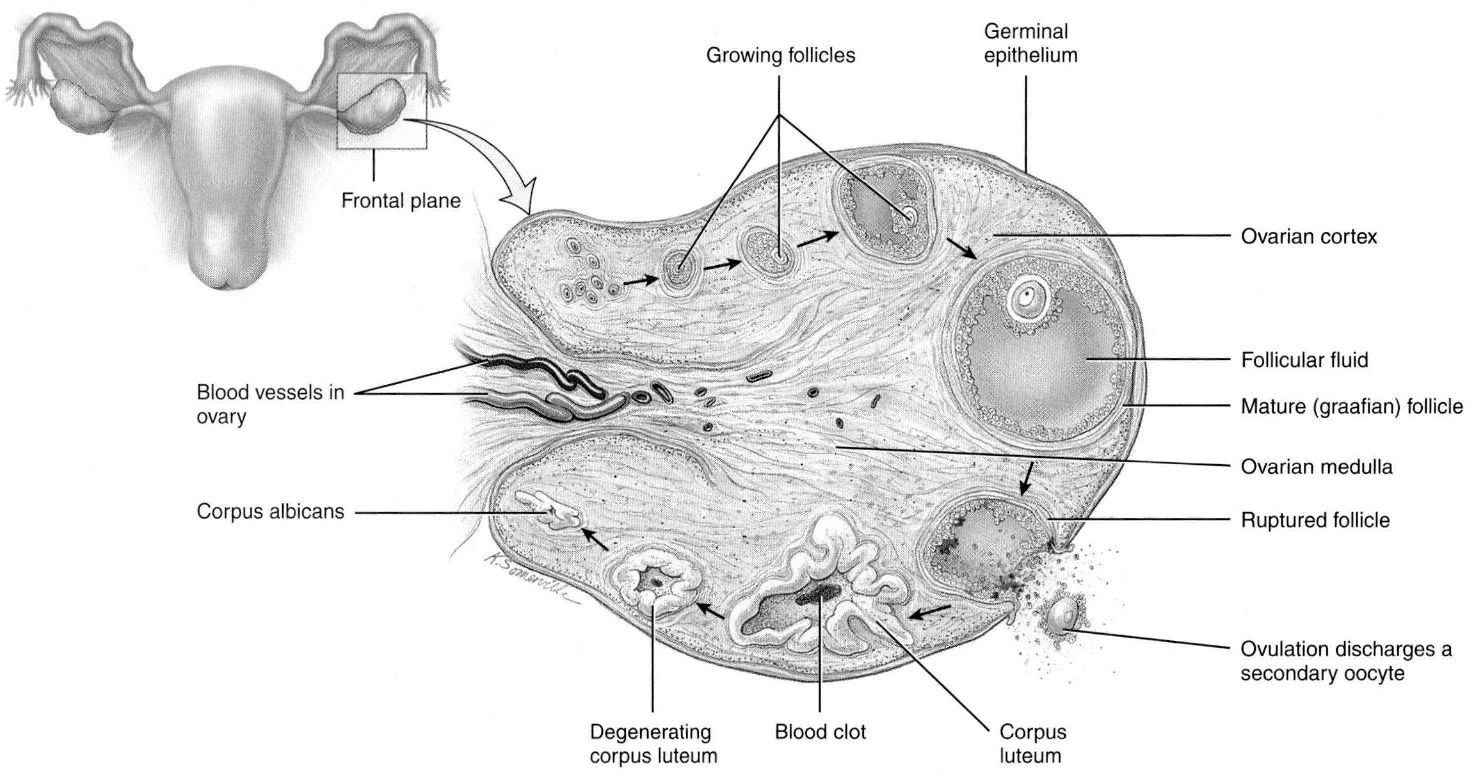

What structures in the ovary contain endocrine tissue, and what hormones do they secrete?

and turns into fibrous tissue called a ***corpus albicans*** (= white body). The ***ovarian medulla*** is a region deep to the ovarian cortex that consists of loose connective tissue and contains blood vessels, lymphatic vessels, and nerves.

• CLINICAL CONNECTION | Ovarian Cancer

Ovarian cancer is the sixth most common form of cancer in females, but it is the leading cause of death from all gynecological malignancies (excluding breast cancer) because it is difficult to detect before it metastasizes (spreads) beyond the ovaries. Risk factors associated with ovarian cancer include age (usually over age 50); race (whites are at highest risk); family history of ovarian cancer; more than 40 years of active ovulation; *nulliparity* (no pregnancies) or first pregnancy after age 30; a high-fat, low-fiber, vitamin A–deficient diet; and prolonged exposure to asbestos and talc. Early ovarian cancer may have no symptoms or mild ones such as abdominal discomfort, heartburn, nausea, loss of appetite, bloating, and flatulence. •

Oogenesis

Formation of gametes in the ovaries is termed ***oogenesis*** (ō′-ō-JEN-e-sis; *oo-* = egg). Unlike spermatogenesis, which begins in males at puberty, oogenesis begins in females before they are even born. Also, males produce new sperm throughout life, whereas females have all the eggs they will ever have by birth. Oogenesis occurs in essentially the same manner as spermatogenesis.

During early fetal development, cells in the ovaries differentiate into ***oogonia*** (ō′-ō-GŌ-nē-a), which can give rise to cells that develop into secondary oocytes (Figure 29.9). Before birth, most of these cells degenerate, but a few develop into larger cells called ***primary oocytes*** (Ō-ō-sīts). At birth, 200,000 to 2,000,000 primary oocytes remain in each ovary. Of these, about 40,000 remain at puberty, but only 400 go on to mature and ovulate during a woman's reproductive lifetime. The remainder degenerate.

After puberty, hormones secreted by the anterior pituitary stimulate the resumption of oogenesis each month. Development continues in several primary oocytes, although in each cycle only one follicle typically reaches the maturity needed for ovulation. The diploid primary oocyte results in two haploid cells of unequal size, both with 23 chromosomes (n) of two chromatids each. The smaller cell, called the *first polar body,* is essentially a packet of discarded nuclear material; the larger cell, known as the ***secondary oocyte,*** receives most of the cytoplasm. The follicle

Figure 29.9 Oogenesis. Diploid cells (2*n*) have 46 chromosomes; haploid cells (*n*) have 23 chromosomes.

In a secondary oocyte, meiosis is completed only if fertilization occurs.

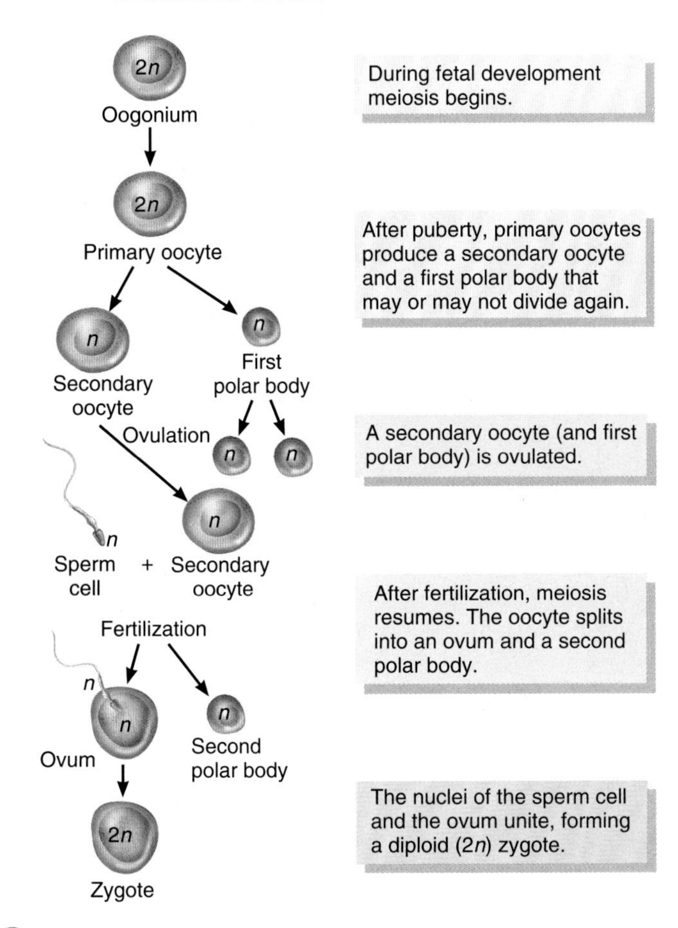

How does the age of a primary oocyte in a female compare with the age of a primary spermatocyte in a male?

in which these events are taking place—the mature (graafian) follicle—soon ruptures and releases its secondary oocyte, a process known as ***ovulation.***

At ovulation, usually a single secondary oocyte (with the first polar body) is expelled into the pelvic cavity and swept into the uterine (fallopian) tube. If a sperm penetrates the secondary oocyte (fertilization), the secondary oocyte splits into two haploid (*n*) cells of unequal size. The larger cell is the ***ovum,*** or mature egg; the smaller one is the *second polar body.* The nuclei of the sperm cell and the ovum then unite, forming a diploid (2*n*) ***zygote.***

Uterine Tubes

Females have two ***uterine (fallopian) tubes*** that extend laterally from the uterus and transport the secondary oocytes from the ovaries to the uterus (see Figure 29.10). The open, funnel-shaped end of each tube, the ***infundibulum,*** lies close to the ovary but is open to the pelvic cavity. It ends in a fringe of finger-like projections called ***fimbriae*** (FIM-brē-ē = fringe). From the infundibulum, the uterine tubes extend medially, attaching to the superior and lateral corners of the uterus.

After ovulation, local currents produced by movements of the fimbriae, which surround the surface of the mature follicle just before ovulation occurs, sweep the secondary oocyte into the uterine tube. The oocyte is then moved within the tube by cilia in the tube's mucous lining and peristaltic contractions of its smooth muscle layer. The usual site for fertilization of a secondary oocyte by a sperm cell is the uterine tube. (Fertilization may occur any time up to about 24 hours after ovulation, after which the unfertilized oocyte is no longer viable.) The fertilized ovum (zygote) descends into the uterus within 7 days. Unfertilized secondary oocytes disintegrate.

Uterus

The ***uterus (womb)*** serves as part of the pathway for sperm deposited in the vagina to reach the uterine tubes. It is also the site of implantation of a fertilized ovum, development of the fetus during pregnancy, and contraction during labor. During reproductive cycles when implantation does not occur, the uterus is the source of menstrual flow. The uterus is situated between the urinary bladder and the rectum and is shaped like an inverted pear.

Parts of the uterus include the dome-shaped portion superior to the uterine tubes called the ***fundus,*** the tapering central portion called the ***body,*** and the narrow portion opening into the vagina called the ***cervix.*** The interior of the body of the uterus is called the ***uterine cavity*** (see Figure 29.10).

The middle muscular layer of the uterus, the ***myometrium*** (*myo-* = muscle), consists of smooth muscle and forms the bulk of the uterine wall. During childbirth, coordinated contractions of uterine muscles help expel the fetus.

The innermost part of the uterine wall, the ***endometrium*** (*endo-* = within), is a mucous membrane. It nourishes a growing fetus or is shed each month during menstruation if fertilization does not occur. The endometrium contains many *endometrial glands* whose secretions nourish sperm and the zygote.

• CLINICAL CONNECTION Hysterectomy

Hysterectomy (hiss-ter-EK-tō-mē; *hyster-* = uterus), the surgical removal of the uterus, is the most common gynecological operation. It may be indicated in conditions such as fibroids, which are noncancerous tumors composed of muscular and fibrous tissue, endometriosis, pelvis inflammatory disease, recurrent ovarian cysts, excessive uterine bleeding, and cancer of the cervix, uterus, or ovaries. In a *partial (subtotal) hysterectomy,* the body of the uterus is removed but the cervix is left in place. A *complete hysterectomy* is the removal of both the body the body and cervix of the uterus. A *radical hysterectomy* includes removal of the body and cervix of the uterus, uterine tubes, possibly the ovaries, the superior portion of the vagina, pelvis lymph nodes, and supporting structures, such as ligaments. A hysterectomy can be performed either through an incision in the abdominal wall, or through the vagina. •

Figure 29.10 Uterus and associated structures. At left, the uterine tube and uterus have been sectioned to show internal structures.

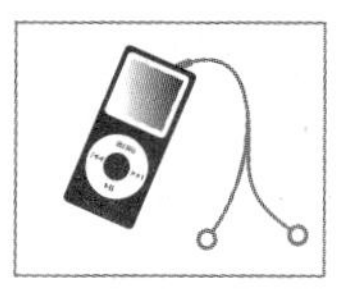

The uterus is the site of menstruation, implantation of a fertilized ovum, development of a fetus, and contraction during labor.

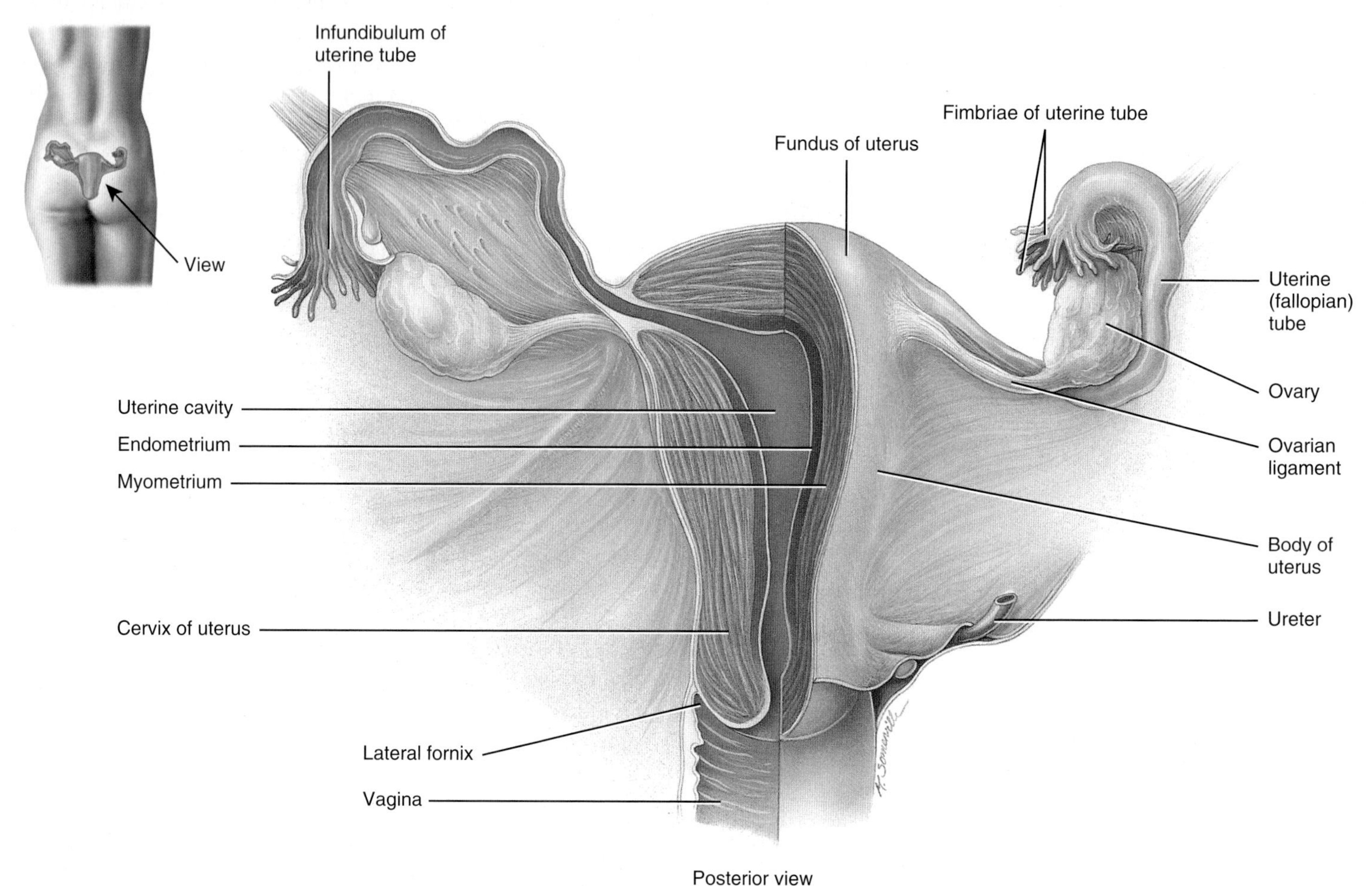

Which part of the uterine lining rebuilds after each menstruation?

Vagina

The ***vagina*** (va-JĪ-na = sheath) is a tubular canal that extends from the exterior of the body to the uterine cervix (Figure 29.10). It is the receptacle for the penis during sexual intercourse, the outlet for menstrual flow, and the passageway for childbirth. The vagina is situated between the urinary bladder and the rectum. A recess, called the ***fornix*** (= arch or vault), surrounds the cervix. When properly inserted, a contraceptive diaphragm rests in the fornix, where it is held in place as it covers the cervix.

The mucosa of the vagina contains large stores of glycogen, the decomposition of which produces organic acids. The resulting acidic environment retards microbial growth, but it also is harmful to sperm. Alkaline components of semen, mainly from the seminal vesicles, neutralize the acidity of the vagina and increase viability of sperm. The muscular layer of the vagina is composed of smooth muscle that can stretch to receive the penis during intercourse and allow for childbirth. There may be a thin fold of mucous membrane called the ***hymen*** (= membrane) partially covering the ***vaginal orifice,*** the vaginal opening (Figure 29.11). After its rupture, usually following the first sexual intercourse, only remnants of the hymen remain.

Perineum and Vulva

The ***perineum*** (per′-i-NĒ-um) is the diamond-shaped area between the thighs and buttocks of both males and females that contains the external genitals and anus (Figure 29.11).

The term ***vulva*** (VUL-va = to wrap around), or ***pudendum*** (pū-DEN-dum), refers to the external genitals of the female (Figure 29.11). The ***mons pubis*** (MONZ PŪ-bis; *mons* = mountain) is an elevation of adipose tissue covered by coarse pubic hair, which cushions the pubic symphysis. From the mons pubis, two longitudinal folds of skin, the ***labia majora*** (LĀ-bē-a ma-JŌ-ra; *labia* = lips; *majora* = larger), extend inferiorly and posteriorly (singular is *labium majus*). In females the labia majora develop from the same embryonic tissue that the scrotum develops from in males. The labia majora contain adipose tissue and

Figure 29.11 Components of the vulva (pudendum).

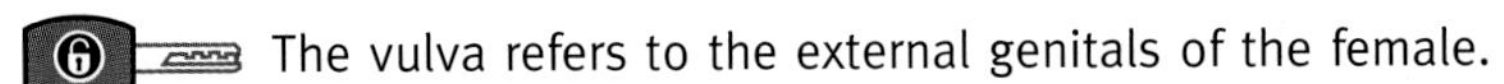

The vulva refers to the external genitals of the female.

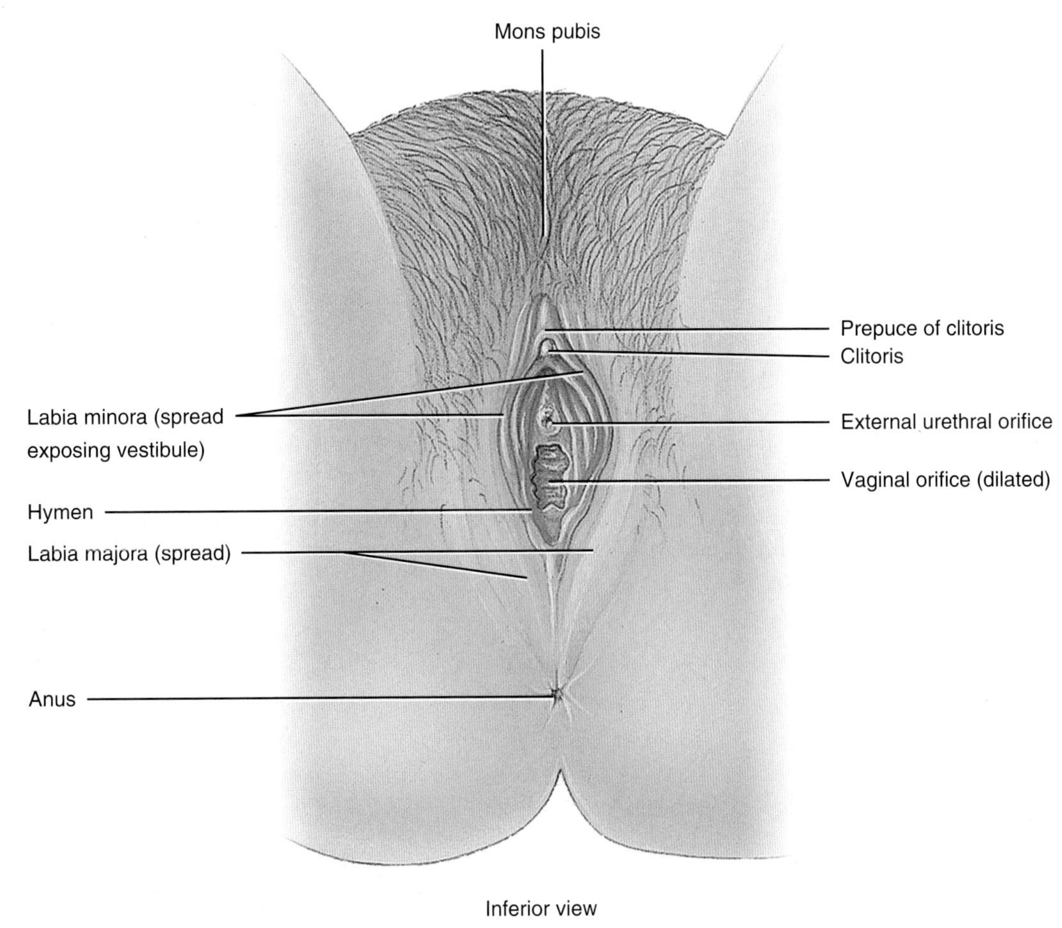

Inferior view

What surface structures are anterior to the vaginal opening? Lateral to it?

sebaceous (oil) and sudoriferous (sweat) glands. Like the mons pubis, they are covered by pubic hair. Medial to the labia majora are two folds of skin called the ***labia minora*** (mīn-ŌR-a = smaller; singular is *labium minus*). The labia minora do not contain pubic hair or fat and have few sudoriferous (sweat) glands; they do, however, contain numerous sebaceous (oil) glands.

The ***clitoris*** (KLIT-o-ris) is a small, cylindrical mass of erectile tissue and nerves. It is located at the anterior junction of the labia minora. A layer of skin called the ***prepuce*** (PRĒ-poos), also known as the ***foreskin,*** is formed at a point where the labia minora unite and cover the body of the clitoris. The exposed portion of the clitoris is the ***glans clitoris.*** Like the penis, the clitoris contains erectile tissue and is capable of enlargement upon sexual stimulation.

The region between the labia minora is called the ***vestibule.*** In the vestibule are the hymen (if present); ***vaginal orifice,*** the opening of the vagina to the exterior; ***external urethral orifice,*** the opening of the urethra to the exterior; and on either side of the external urethral orifice, the openings of the ducts of the ***paraurethral glands.*** These glands in the wall of the urethra secrete mucus. The male's prostate develops from the same embryonic tissue as the female's paraurethral glands. On either side of the vaginal orifice itself are the ***greater vestibular glands,*** which produce a small quantity of mucus during sexual arousal and intercourse that adds to cervical mucus and provides lubrication. In males, the bulbourethral glands are equivalent structures.

• CLINICAL CONNECTION Episiotomy

During childbirth, the emerging fetus stretches the perineal region. To prevent excessive stretching and even tearing of this region, a physician sometimes performs an **episiotomy** (e-piz-ē-OT-ō-mē; *episi-* = vulva or public region; *-otomy* = incision), a perineal cut made with surgical scissors. The cut may be made along the midline or at an angle of approximately 45 degrees to the midline. In effect, a straight, more easily sutured cut is substituted for the jagged tear that would otherwise be caused by passage of the fetus. The incision is closed in layers with sutures that are absorbed within a few weeks, so that the busy new mom does not have to worry about making time to have them removed. •

Mammary Glands

The ***mammary glands*** (*mamma* = breast), located in the breasts, are modified sudoriferous (sweat) glands that produce milk. The breasts lie over the pectoralis major and serratus anterior muscles and are attached to them by a layer of connective tissue (Figure 29.12). Each breast has one pigmented projection, the ***nipple,*** with a series of closely spaced openings of ducts where milk emerges. The circular pigmented area of skin surrounding the nipple is called the ***areola*** (a-RĒ-ō-la = small space). This region appears rough because it contains modified sebaceous (oil) glands. Internally, each mammary gland consists of 15 to 20 ***lobes*** arranged radially and separated by adipose tissue and strands of connective tissue called ***suspensory ligaments of the breast (Cooper's ligaments),*** which support the breast. In each lobe are smaller ***lobules,*** in which milk-secreting glands called ***alveoli*** (= small cavities) are found. When milk is being produced, it passes from the alveoli into a series of tubules that drain toward the nipple.

At birth, the mammary glands are undeveloped and appear as slight elevations on the chest. With the onset of puberty, under the influence of estrogens and progesterone, the female breasts begin to develop. The duct system matures and fat is deposited, which increases breast size. The difference between a larger and smaller breast in an adult is the amount of adipose tissue that is present and has no relationship to the ability of the breast to produce milk at the appropriate time. The functions of the mammary glands are the synthesis, secretion, and ejection of milk; these functions, called ***lactation,*** are associated with pregnancy and childbirth. Milk production is stimulated largely by the hormone prolactin from the anterior pituitary, with contributions from progesterone and estrogens. The ejection of milk is stimulated by oxytocin, which is released from the posterior pituitary in response to the sucking of an infant on the mother's nipple (suckling).

• CLINICAL CONNECTION | Breast Cancer

One in eight women in the United States faces the prospect of **breast cancer,** the second-leading cause of female deaths from cancer. Early detection by breast self-examination and mammograms is the best way to increase the chance of survival.

The most effective technique for detecting tumors less than 1 cm (0.4 in.) in diameter is ***mammography*** (mam-OG-ra-fē; *-graphy* = to

Figure 29.12 Mammary glands within the breasts.

The mammary glands function in the synthesis, secretion, and ejection of milk (lactation).

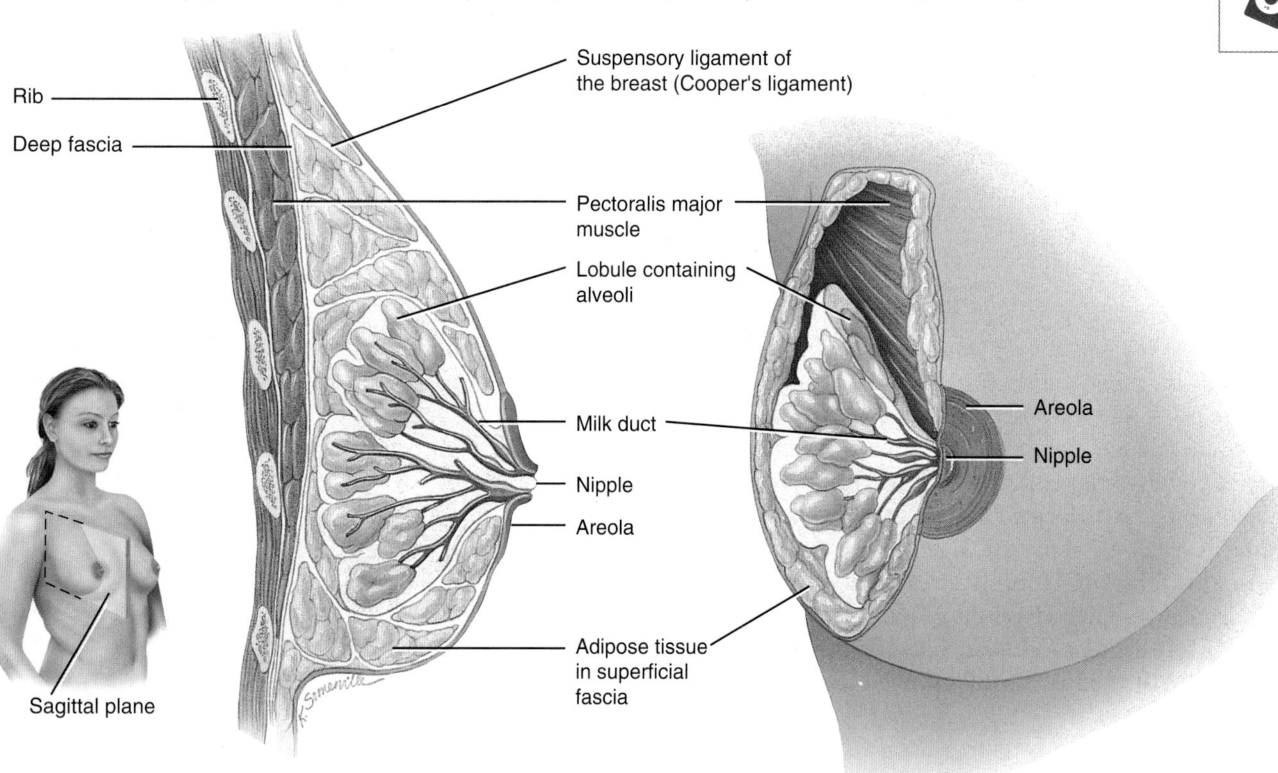

What hormone regulates the ejection of milk from the mammary glands?

record), a type of radiography using very sensitive x-ray film. The image of the breast, called a ***mammogram,*** is best obtained by compressing the breasts, one at a time, using flat plates. A supplementary procedure for evaluating breast abnormalities is ***ultrasonography.*** Although ultrasonography cannot detect tumors smaller than 1 cm in diameter, it can be used to determine whether a lump is a benign, fluid-filled cyst or a solid (and therefore possibly malignant) tumor.

Among the factors that increase the risk of developing breast cancer are (1) a family history of breast cancer, especially in a mother or sister; (2) never having borne a child or having a first child after age 35; (3) previous cancer in one breast; (4) exposure to ionizing radiation, such as x-rays; (5) excessive alcohol intake; and (6) cigarette smoking.

Treatment for breast cancer may involve hormone therapy, chemotherapy, radiation therapy, ***lumpectomy*** (removal of the tumor and the immediate surrounding tissue), a modified or radical mastectomy, or a combination of these approaches. A ***radical mastectomy*** (*mast-* = breast) involves removal of the affected breast along with the underlying pectoral muscles and the axillary lymph nodes. (Lymph nodes are removed because the spread of cancerous cells usually occurs through lymphatic or blood vessels.) Radiation treatment and chemotherapy may follow the surgery to ensure the destruction of any stray cancer cells. See Section 24.2 for a discussion of edema and lymphedema. •

CHECKPOINT

7. Describe the principal events of oogenesis.
8. Where are the uterine tubes located? What is their function?
9. Describe the histology of the uterus.
10. What is the function of the vagina? Describe its histology.
11. Describe the structure of the mammary glands. How are they supported?

29.4 FEMALE REPRODUCTIVE CYCLE

OBJECTIVE

- Describe the major events of the ovarian and uterine cycles.

During their reproductive years, nonpregnant females normally exhibit cyclical changes in the ovaries and uterus. Each cycle takes about a month and involves both oogenesis and preparation of the uterus to receive a fertilized ovum. Hormones secreted by the hypothalamus, anterior pituitary, and ovaries control the main events. You have already learned about the ***ovarian cycle,*** the series of events in the ovaries that occur during and after the maturation of an oocyte. Steroid hormones released by the ovaries control the ***uterine (menstrual) cycle,*** a concurrent series of changes in the endometrium of the uterus to prepare it for the arrival of a fertilized ovum that will develop there until birth. If fertilization does not occur, the levels of ovarian hormones decrease, which causes part of the endometrium to slough off. The general term ***female reproductive cycle*** encompasses the ovarian and uterine cycles, the hormonal changes that regulate them, and the related cyclical changes in the breasts and cervix.

Hormonal Regulation of the Female Reproductive Cycle

Gonadotropin-releasing hormone (GnRH) secreted by the hypothalamus controls the ovarian and uterine cycles (see Figure 29.13). GnRH stimulates the release of ***follicle-stimulating hormone (FSH)*** and ***luteinizing hormone (LH)*** from the anterior pituitary. FSH, in turn, initiates follicular growth and the secretion of estrogens by the growing follicles. LH stimulates the further development of ovarian follicles and their full secretion of estrogens. At midcycle, LH triggers ovulation and then promotes formation of the corpus luteum, the reason for the name luteinizing hormone. Stimulated by LH, the corpus luteum produces and secretes estrogens, progesterone, relaxin, and inhibin. ***Estrogens*** secreted by ovarian follicles have several important functions throughout the body:

- Estrogens promote the development and maintenance of female reproductive structures, feminine secondary sex characteristics, and the mammary glands. The secondary sex characteristics include distribution of adipose tissue in the breasts, abdomen, mons pubis, and hips; a broad pelvis; and the pattern of hair growth on the head and body.
- Estrogens stimulate protein synthesis, acting together with insulin-like growth factors, insulin, and thyroid hormones.
- Estrogens lower blood cholesterol level, which is probably the reason that women under age 50 have a much lower risk of coronary artery disease than do men of comparable age.

Progesterone, secreted mainly by cells of the corpus luteum, acts together with estrogens to prepare and then maintain the endometrium for implantation of a fertilized ovum and to prepare the mammary glands for milk secretion.

A small quantity of ***relaxin,*** produced by the corpus luteum during each monthly cycle, relaxes the uterus by inhibiting contractions of the myometrium. Presumably, implantation of a fertilized ovum occurs more readily in a "quiet" uterus. During pregnancy, the placenta produces much more relaxin, and it continues to relax uterine smooth muscle. At the end of pregnancy, relaxin also increases the flexibility of the pubic symphysis and helps dilate the uterine cervix, both of which ease delivery of the baby. ***Inhibin*** is secreted by growing follicles and by the corpus luteum after ovulation. It inhibits secretion of FSH and, to a lesser extent, LH.

Phases of the Female Reproductive Cycle

The duration of the female reproductive cycle typically varies from 24 to 35 days. As will be noted in the following discussion, the hypothalamus is greatly responsible for initiating various hormonal aspects of the female reproductive cycle. As you learned in Chapter 17, the hypothalamus is connected to many other areas of the brain, including the cerebral cortex. For this reason, stress from any source may influence the hypothalamus and thus the length of a particular reproductive cycle. For this discussion we assume a duration of 28 days and divide it into four phases: the menstrual phase, the preovulatory phase, ovulation, and the

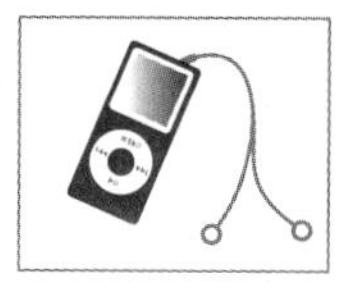

Figure 29.13 **The female reproductive cycle.** The length of the female reproductive cycle typically is 24 to 36 days; the preovulatory phase is more variable in length than the other phases. (a) Events in the ovarian and uterine cycles and the release of anterior pituitary hormones are correlated with the sequence of the cycle's four phases. In the cycle shown, fertilization and implantation have not occurred. (b) Relative concentrations of anterior pituitary hormones (FSH and LH) and ovarian hormones (estrogens and progesterone) during the phases of a normal female reproductive cycle.

6 Estrogens are secreted by the dominant follicle before ovulation; after ovulation, both progesterone and estrogens are secreted by the corpus luteum.

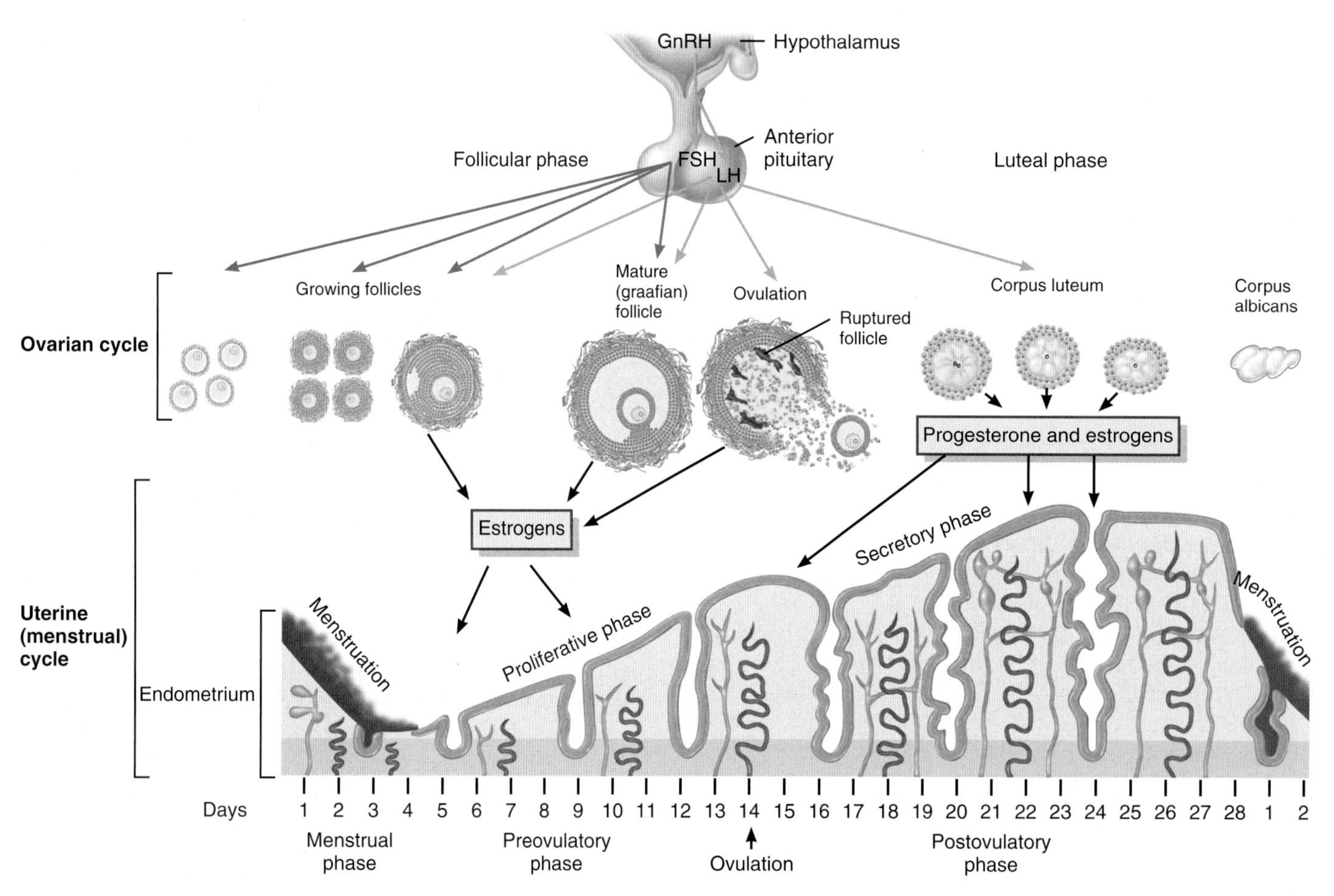

(a) Hormonal regulation of changes in the ovary and uterus

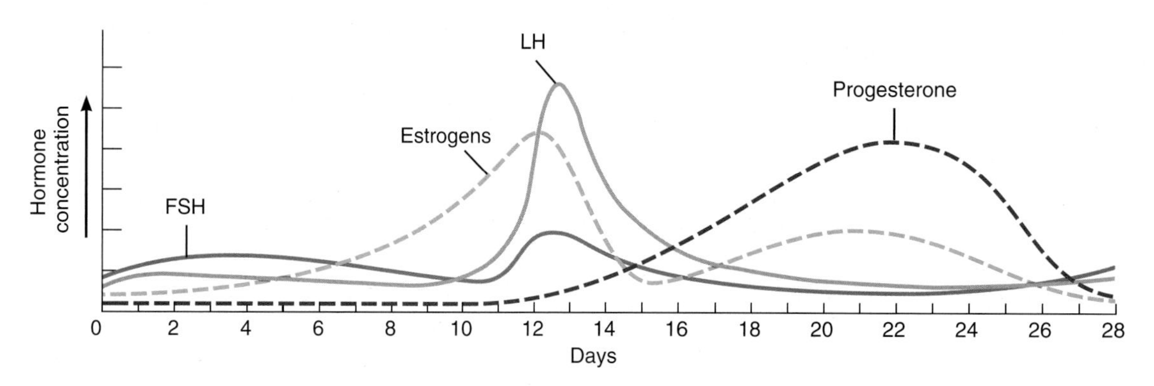

(b) Changes in concentration of anterior pituitary and ovarian hormones

? Which hormones are responsible for the proliferative phase of endometrial growth, for ovulation, for growth of the corpus luteum, and for the surge of LH at midcycle?

post-ovulatory phase (Figure 29.13a). Because they occur at the same time, the events of the ovarian cycle (events in the ovaries) and menstrual cycle (events in the uterus) are discussed together.

Menstrual Phase

The ***menstrual phase*** (MEN-stroo-al), also called ***menstruation*** (men′-stroo-Ā-shun) or ***menses*** (= month), lasts for roughly the first 5 days of the cycle. (By convention, the first day of menstruation marks the first day of a new cycle.)

Events in the Ovaries During the menstrual phase, several ovarian follicles grow and enlarge.

Events in the Uterus Menstrual flow from the uterus consists of 50 to 150 mL of blood, mucus, and tissue cells from the endometrium. This discharge occurs because the declining level of ovarian hormones (progesterone and estrogens) causes the uterine arteries to constrict. As a result, the cells they supply become oxygen-deprived and start to die. Eventually, part of the endometrium sloughs off. The menstrual flow passes from the uterine cavity to the cervix and through the vagina to the exterior.

Preovulatory Phase

The ***preovulatory phase*** is the time between the end of menstruation and ovulation. The preovulatory phase of the cycle accounts for most of the variation in cycle length. In a 28-day cycle, it lasts from days 6 to 13.

Events in the Ovaries Under the influence of FSH, several follicles continue to grow and begin to secrete estrogens and inhibin. By about day 6, a single follicle in one of the two ovaries has outgrown all the others to become the *dominant follicle.* Estrogens and inhibin secreted by the dominant follicle decrease the secretion of FSH (Figure 29.13b, see days 8 to 11), which causes other, less well-developed follicles to stop growing and die.

The one dominant follicle becomes the ***mature (graafian) follicle.*** The mature follicle continues to enlarge until it is ready for ovulation, forming a blister-like bulge on the surface of the ovary. During maturation, the follicle continues to increase its production of estrogens under the influence of an increasing level of LH. With reference to the ovarian cycle, the menstrual phase and preovulatory phase together are termed the ***follicular phase*** (fo-LIK-ū-lar) because ovarian follicles are growing and developing.

Events in the Uterus Estrogens liberated into the blood by growing ovarian follicles stimulate the repair of the endometrium. As the endometrium thickens, the short, straight endometrial glands develop, and the arterioles coil and lengthen. With reference to the uterine cycle, the preovulatory phase is also termed the ***proliferative phase*** because the endometrium is proliferating (growing rapidly).

Ovulation

Ovulation, the rupture of the mature (graafian) follicle and the release of the secondary oocyte into the pelvic cavity, usually occurs on day 14 in a 28-day cycle.

The high levels of estrogens during the last part of the preovulatory phase exert a *positive feedback* effect on both LH and GnRH. A high level of estrogens stimulates the hypothalamus to release more gonadotropin-releasing hormone (GnRH) and the anterior pituitary to produce more LH. GnRH then promotes the release of even more LH. The resulting surge of LH (Figure 29.13b) brings about rupture of the mature (graafian) follicle and expulsion of a secondary oocyte. An over-the-counter home test that detects the LH surge associated with ovulation can be used to predict ovulation a day in advance.

Postovulatory Phase

The ***postovulatory phase*** of the female reproductive cycle is the time between ovulation and onset of the next menstruation. This phase is the most constant in duration and lasts for 14 days, from days 15 to 28 in a 28-day cycle.

• CLINICAL CONNECTION | Premenstrual Syndrome

Premenstrual syndrome (PMS) is a cyclical disorder with a potential for severe physical and emotional distress. It appears during the postovulatory phase of the female reproductive cycle and dramatically disappears when menstruation begins. The signs and symptoms are highly variable from one woman to another. They may include edema, weight gain, breast swelling and tenderness, abdominal distension, backache, joint pain, constipation, skin eruptions, fatigue and lethargy, greater need for sleep, depression or anxiety, irritability, mood swings, headache, poor coordination and clumsiness, and cravings for sweet or salty foods. The cause of PMS is unknown. For some women, getting regular exercise; avoiding caffeine, salt, and alcohol; and eating a diet that is high in complex carbohydrates and lean proteins can bring considerable relief. •

Events in One Ovary After ovulation, the mature follicle collapses. Stimulated by LH, the remaining follicular cells enlarge and form the corpus luteum, which secretes progesterone, estrogens, relaxin, and inhibin. With reference to the ovarian cycle, this phase is also called the ***luteal phase.*** Subsequent events depend on whether or not the oocyte is fertilized. If the oocyte is not fertilized, the corpus luteum lasts for only 2 weeks, after which its secretory activity declines, and it degenerates into a corpus albicans (Figure 29.13a). As the levels of progesterone, estrogens, and inhibin decrease, release of GnRH, FSH, and LH rises due to loss of negative feedback suppression by the ovarian hormones. Then, follicular growth resumes and a new ovarian cycle begins.

If the secondary oocyte is fertilized and begins to divide, the corpus luteum persists past its normal 2-week lifespan. It is "rescued" from degeneration by ***human chorionic gonadotropin (hCG)*** (kō-rē′-ON-ik), a hormone produced by the embryo beginning about 8 days after fertilization. Like LH, hCG stimulates the secretory activity of the corpus luteum. The presence of hCG in maternal blood or urine is an indicator of pregnancy, and hCG is the hormone detected by home pregnancy tests.

Events in the Uterus Progesterone and estrogens produced by the corpus luteum promote growth of the endometrial glands, which begin to secrete glycogen, and vascularization and thickening of the endometrium. These preparatory changes peak about one week after ovulation, at the time a fertilized ovum might arrive at the uterus. Because of the secretory activity of the endometrial glands, this period is called the ***secretory phase*** of the uterine cycle.

Figure 29.14 summarizes the hormonal interactions and cyclical changes in the ovaries and uterus during the ovarian and menstrual cycles.

CHECKPOINT

12. Describe the function of each of the following hormones in the uterine and ovarian cycles: GnRH, FSH, LH, estrogens, progesterone, and inhibin.
13. Briefly outline the major events and hormonal changes of each phase of the uterine cycle, and correlate them with the events of the ovarian cycle.
14. Describe the major hormonal changes that occur during the uterine and ovarian cycles.

Figure 29.14 Summary of hormonal interactions in the ovarian and uterine (menstrual) cycles.

6 Hormones from the anterior pituitary regulate ovarian function, and hormones from the ovaries regulate the changes in the endometrial lining of the uterus.

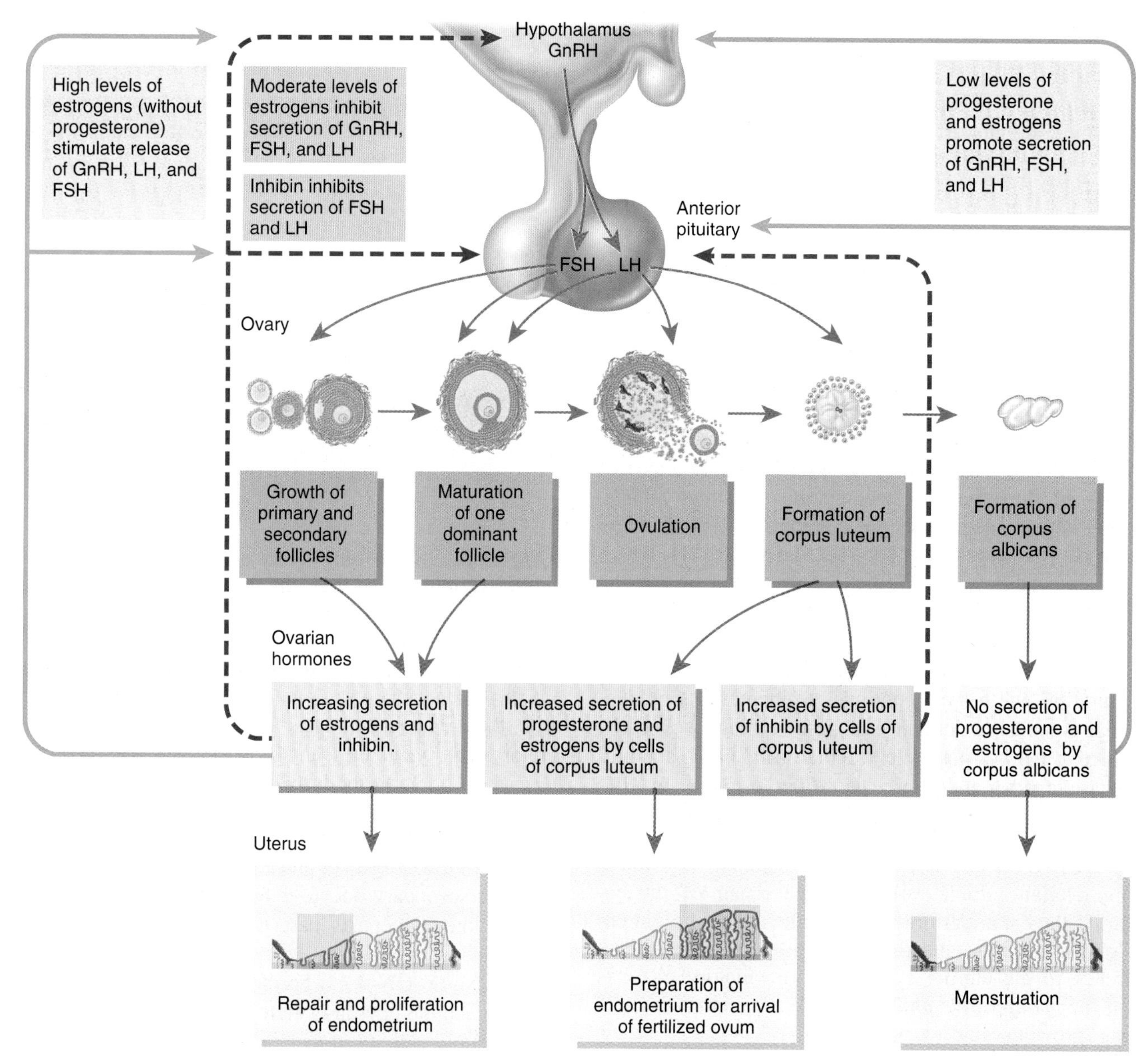

When declining levels of estrogens and progesterone stimulate secretion of GnRH, is this a positive or negative feedback effect? Why?

29.5 AGING AND THE REPRODUCTIVE SYSTEMS

OBJECTIVE

- Describe the effects of aging on the reproductive systems.

During the first decade of life, the reproductive system is in a juvenile state. At about age 10, hormone-directed changes start to occur in both sexes. ***Puberty*** (PŪ-ber-tē = a ripe age) is the period when secondary sexual characteristics begin to develop and the potential for sexual reproduction is reached. Onset of puberty is marked by bursts of LH and FSH secretion, each triggered by a burst of GnRH. The stimuli that cause the GnRH bursts are still unclear, but a role for the hormone leptin is starting to unfold. Just before puberty, leptin levels rise in proportion to adipose tissue mass. Leptin may signal the hypothalamus that long-term energy stores (triglycerides in adipose tissue) are adequate for reproductive functions to begin.

In females, the reproductive cycle normally occurs once each month from ***menarche*** (me-NAR-kē), the first menses, to ***menopause,*** the permanent cessation of menses. Thus, the female reproductive system has a time-limited span of fertility between menarche and menopause. Between the ages of 40 and 50 the pool of remaining ovarian follicles usually becomes exhausted. As a result, the ovaries become less responsive to hormonal stimulation. The production of estrogens declines, despite copious secretion of FSH and LH by the anterior pituitary. Many women experience hot flashes and heavy sweating, which coincide with bursts of GnRH release. Other symptoms of menopause are headache, hair loss, muscular pains, vaginal dryness, insomnia, depression, weight gain, and mood swings. Some atrophy of the ovaries, uterine tubes, uterus, vagina, external genitalia, and breasts occurs in postmenopausal women. Due to loss of estrogens, most women also experience a decline in bone mineral density after menopause. Sexual desire (libido) does not show a parallel decline; it may be maintained by adrenal androgens. The risk of having uterine cancer peaks at about 65 years of age, but cervical cancer is more common in younger women.

In males, declining reproductive function is much more subtle than in females. Healthy men often retain reproductive capacity into their eighties or nineties. At about age 55 a decline in testosterone synthesis leads to reduced muscle strength, fewer viable sperm, and decreased sexual desire. However, abundant sperm may be present even in old age.

Enlargement of the prostate to two to four times its normal size occurs in approximately one-third of all males over age 60. This condition, called ***benign prostatic hyperplasia (BPH),*** is characterized by frequent urination, nocturia (bedwetting), hesitancy in urination, decreased force of urinary stream, postvoiding dribbling, and a sensation of incomplete emptying.

CHECKPOINT

15. What changes occur in males and females at puberty?
16. What do the terms menarche and menopause mean?

To appreciate the many ways that the reproductive systems contribute to homeostasis of other body systems, examine Exhibit 29.1.

CLINICAL CONNECTION | The Female Athlete Triad—Disordered Eating, Amenorrhea, and Premature Osteoporosis

The female reproductive cycle can be disrupted by many factors, including weight loss, low body weight, disordered eating, and vigorous physical activity. The observation that three conditions—disordered eating, amenorrhea, and osteoporosis—often occur together in female athletes led researchers to coin the term **female athlete triad.** Many athletes experience intense pressure from coaches, parents, peers, and themselves to lose weight to improve performance. Consequently, many develop disordered eating behaviors and engage in other harmful weight-loss practices in a struggle to maintain a very low body weight.

Menstrual irregularity should never be ignored, because it may be caused by a serious underlying disorder for which the athlete should receive prompt medical treatment. Even when menstrual irregularity is apparently caused by disordered eating and physical training, and not associated with another physical disorder, it is still a cause for concern. One reason is that women with *amenorrhea,* the absence of menstrual cycles, are at increased risk for premature osteoporosis.

Why osteoporosis? Remember that the ovarian follicles produce estrogens when stimulated by FSH and LH. If ovulation is not occurring, then the ovarian follicles, and later the corpus luteum, are not producing estrogens. Chronically low levels of estrogens are associated with loss of bone minerals, as estrogens help bones retain calcium. The loss of the protective effect of estrogens explains why many women experience a decline in bone density after menopause, when levels of estrogens drop. Amenorrheic runners have been shown to experience a similar effect.

It is ironic that dedicated athletes should experience premature osteoporosis, because physical activity in general is associated with a *reduced* risk of osteoporosis. Exercise has been shown to increase bone density, especially if the exercise involves bone stress, such as running and aerobic dancing. However, in the presence of disordered eating and overtraining, exercise may simply add insult to injury. •

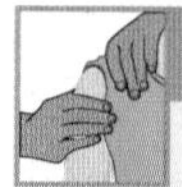

MANUAL THERAPY APPLICATION

Benefits of Massage during Pregnancy

Massage has been used by nearly all cultures for thousands of years as an effective therapy during pregnancy and childbirth. One of the primary **benefits of massage during pregnancy** is stress reduction, for both the pregnant woman and the fetus. Stress in the pregnant woman can increase the heart rate of the fetus. Stress can also retard fetal development and cause a higher risk of heart disease and diabetes later in life.

Here are some of the benefits of prenatal massage:

- Massage directly influences the amniotic environment of the developing fetus. Therefore, a pregnant mother receiving regular relaxation massage will lower her blood pressure and foster a healthier and calmer baby.
- Massage soothes and relaxes the nervous system by releasing endorphins into the mother's body to create a more relaxed and deeper sleep. Increased beta-endorphin released during pregnancy contributes to an easier childbirth for mother and baby. Endorphins modulate pain associated with anxiety and stress. When the mother's tolerance for pain is increased, the baby will undergo an easier passage.
- Progesterone supports the pregnancy but slows down the function of smooth muscle systems. Digestion, elimination, sinus drainage, and lymphatic drainage become sluggish. Massage during pregnancy promotes lymph drainage and reduces swelling.
- There is increased muscle fatigue during pregnancy because of increased cellular waste products. This fatigue can be reduced by the elimination of these wastes through the circulatory and lymphatic system by massage.
- Massage also promotes peristaltic activity in a sluggish digestive system. It stimulates glandular secretions which, in turn, help stabilize hormone levels. Improvement of the circulatory and lymphatic systems will help control the occurrence of varicose veins.
- The effects of massage on the nervous system may last for nearly 14 days. It is therefore suggested that a pregnant woman minimally receive massage every 2 weeks to maintain the balanced effects of massage.

As the pregnancy progresses, the pregnant mother is anything but comfortable; here are some of the ways that massage can relieve some of these discomforts:

- Massage eliminates or reduces many of the normal discomforts of pregnancy such as swollen feet and ankles, backache, stiff neck, leg cramps, uterine pain, headache, and sciatica.
- Muscle discomforts such as cramping, tension, stiffness, and tightening may be alleviated through massage, and massage promotes muscle flexibility, which is very beneficial.
- By relaxing tense muscles and toning loose ones, this flexibility may be enhanced. During the last trimester and the birth itself, increased flexibility can be the difference between a difficult labor and a good experience for mother and baby.
- Massage eases the burden placed on the pregnant woman's heart by increasing local and general blood circulation. Efficient circulation means more oxygen and nutrients are carried to the cells of both mother and fetus.

Massage can also help to relieve depression or anxiety caused by the hormonal changes the pregnant woman may be experiencing. A relaxed mother may have a less painful delivery and may deliver a baby that should gain weight more quickly, catch fewer colds because of a bolstered immune system, experience less colic, and have a better sleep pattern. Continued massage following childbirth will help the mother regain her strength more quickly and may be beneficial in reducing postpartum stress. Experience has shown that when the mother is recovering nicely and the baby is thriving, the benefits of massage during pregnancy are apparent.

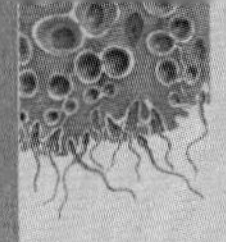

EXHIBIT 29.1 The Reproductive Systems and Homeostasis

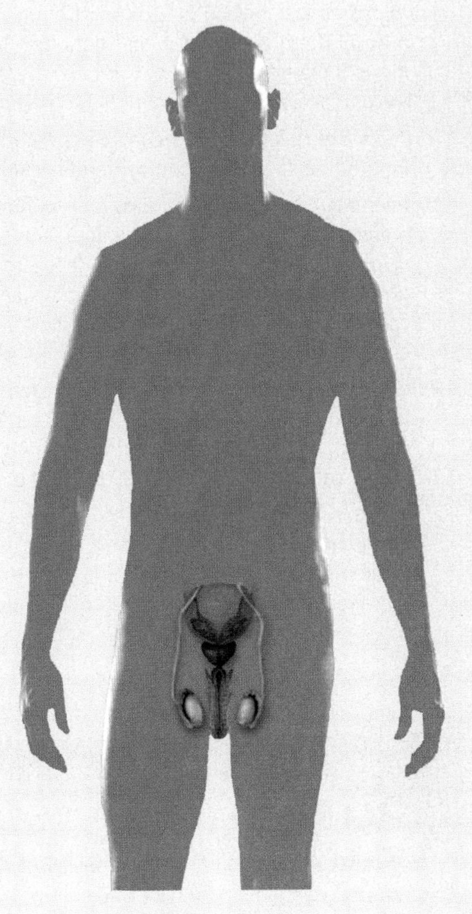

BODY SYSTEM		CONTRIBUTION OF THE REPRODUCTIVE SYSTEMS
For all body systems		The male and female reproductive systems produce gametes (oocytes and sperm) that unite to form embryos and fetuses, which contain cells that divide and differentiate to form all of the organ systems of the body.
Integumentary system		Androgens promote the growth of body hair. Estrogens stimulate the deposition of fat in the breasts, abdomen, and hips. Mammary glands produce milk. Skin stretches during pregnancy as the fetus enlarges.
Skeletal system	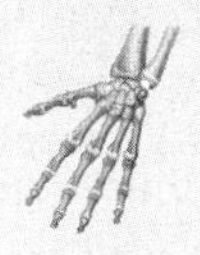	Androgens and estrogens stimulate the growth and maintenance of bones of the skeletal system.
Muscular system		Androgens stimulate the growth of skeletal muscles.
Nervous system	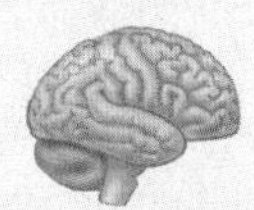	Androgens influence libido (sex drive). Estrogens may play a role in the development of certain regions of the brain in males.
Endocrine system	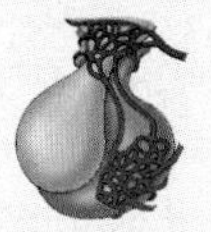	Testosterone and estrogens exert feedback effects on the hypothalamus and anterior pituitary gland.
Cardiovascular system	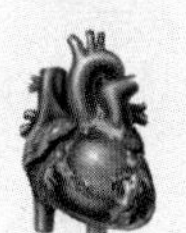	Estrogens lower blood cholesterol level and may reduce the risk of coronary artery disease in women under age 50.
Lymphatic system and immunity		The presence of an antibiotic-like chemical in semen and the acidic pH of vaginal fluid provide innate immunity against microbes in the reproductive tract.
Respiratory system	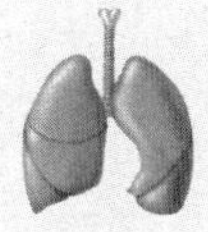	Sexual arousal increases the rate and depth of breathing.
Digestive system	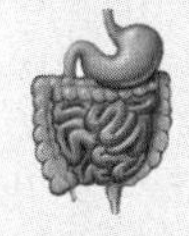	The presence of the fetus during pregnancy crowds the digestive organs, which leads to heartburn and constipation.
Urinary system	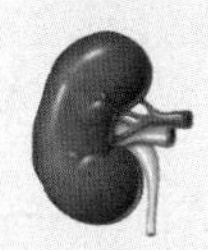	In males, the portion of the urethra that extends through the prostate and penis is a passageway for urine as well as semen.

KEY MEDICAL TERMS ASSOCIATED WITH THE REPRODUCTIVE SYSTEMS

Abortion (a-BOR-shun) The premature expulsion of the products of conception from the uterus, usually before the 20th week of pregnancy. An abortion may be spontaneous (naturally occurring; also called a *miscarriage*) or induced (intentionally performed). Induced abortions may be performed by vacuum aspiration (suction), infusion of a saline solution, or surgical evacuation (scraping).

Amenorrhea (ā-men′-ō-RĒ-a; *a-* = without; *-men-* = month; *-rrhea* = a flow) The absence of menstruation; it may be caused by a hormone imbalance, obesity, extreme weight loss, or very low body fat as may occur during rigorous athletic training.

Birth control Refers to restricting the number of children by various methods designed to control fertility and prevent conception.

Dysmenorrhea (dis′-men-ō-RĒ-a; *dys-* = difficult or painful) Painful menstruation; the term is usually reserved to describe menstrual symptoms that are severe enough to prevent a woman from functioning normally for one or more days each month. Some cases are caused by uterine tumors, ovarian cysts, pelvic inflammatory disease, or intrauterine devices.

Endocervical curettage (kū′-re-TAHZH; *curette* = scraper) A procedure in which the cervix is dilated and the endometrium of the uterus is scraped with a spoon-shaped instrument called a curette; commonly called a D and C (dilation and curettage).

Fibrocystic disease (fī-brō-SIS-tik) The most common cause of breast lumps in females in which one or more cysts (fluid-filled sacs) and thickenings of alveoli develop. It occurs mainly in females between the ages of 30 and 50 and is probably due to a relative excess of estrogens or a deficiency of progesterone in the postovulatory phase of the reproductive cycle. Fibrocystic disease usually causes one or both breasts to become lumpy, swollen, and tender a week or so before menstruation begins.

Menorrhagia (men-ō-RA-jē-a; *meno-* = menstruation; *-rhage* = to burst forth). Excessively prolonged or profuse menstrual period. May be due to a disturbance in hormonal regulation of the menstrual cycle, pelvic infection, medications (anticoagulants), fibroids, endometriosis, or intrauterine devices.

Ovarian cyst (ō-VAR-ē-an) A fluid-filled sac in or on an ovary. Such cysts are relatively common, are usually noncancerous, and frequently disappear on their own. Cancerous cysts are more likely to occur in women over 40. Ovarian cysts may cause pain, pressure, a dull ache, or fullness in the abdomen; pain during sexual intercourse; delayed, painful, or irregular menstrual periods; abrupt onset of sharp pain in the lower abdomen; and/or vaginal bleeding. Most ovarian cysts require no treatment, but larger ones (more than 5 cm = 2 in.) may be removed surgically.

Papanicolaou test (pa-pa-NI-kō-lō), or ***Pap smear*** A test to detect uterine cancer in which a few cells from the cervix and the part of the vagina surrounding the cervix are removed with a swab and examined microscopically. Malignant cells have a characteristic appearance that allows diagnosis even before symptoms occur.

Smegma (SMEG-ma) The secretion, consisting principally of sloughed off epithelial cells, found chiefly around the external genitals and especially under the foreskin of the male.

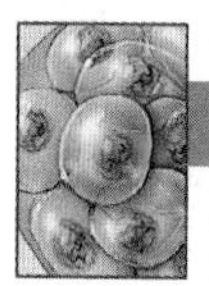

STUDY OUTLINE

Introduction to the Reproductive Systems (Section 29.1)

1. Sexual reproduction is the process of producing offspring by the union of gametes (oocytes and sperm).
2. The organs of reproduction are grouped as gonads (produce gametes), ducts (transport and store gametes), accessory sex glands (produce materials that support gametes), and supporting structures.

Male Reproductive System (Section 29.2)

1. The male reproductive system includes the testes, epididymis, ductus (vas) deferens, ejaculatory ducts, urethra, seminal vesicles, prostate, bulbourethral (Cowper's) glands, scrotum and penis.
2. The scrotum is a sac that supports and regulates the temperature of the testes.
3. The male gonads include the testes, oval-shaped organs in the scrotum that contain the seminiferous tubules, in which sperm cells develop; Sertoli cells, which nourish sperm cells and produce inhibin; and Leydig cells, which produce the male sex hormone testosterone.
4. Spermatogenesis occurs in the testes and results in the formation of haploid sperm cells.
5. Mature sperm consist of a head and a tail. Their function is to fertilize a secondary oocyte.
6. At puberty, gonadotropin-releasing hormone (GnRH) stimulates anterior pituitary secretion of LH and FSH. LH stimulates Leydig cells to produce testosterone. FSH and testosterone initiate spermatogenesis.
7. Testosterone controls the growth, development, and maintenance of sex organs; stimulates bone growth, protein anabolism, and sperm maturation; and stimulates development of male secondary sex characteristics.
8. Inhibin is produced by Sertoli cells; its inhibition of FSH helps regulate the rate of spermatogenesis.
9. Sperm are transported out of the testes into an adjacent organ, the epididymis, where their motility increases.
10. The ductus (vas) deferens stores sperm and propels them toward the urethra during ejaculation. Removing part of the vas deferens to prevent fertilization is called vasectomy.
11. The ejaculatory ducts are formed by the union of the ducts from the seminal vesicles and vas deferens, and they eject sperm into the urethra.
12. The male urethra passes through the prostate, deep perineal muscles, and penis.
13. The seminal vesicles secrete an alkaline, viscous fluid that constitutes about 60% of the volume of semen and contributes to sperm viability.
14. The prostate secretes a slightly acidic fluid that constitutes about 25% of the volume of semen and contributes to sperm motility.
15. The bulbourethral glands secrete mucus for lubrication and an alkaline substance that neutralizes acid.

16. Semen is a mixture of sperm and seminal fluid; it provides the fluid in which sperm are transported, supplies nutrients, and neutralizes the acidity of the male urethra and the vagina.
17. The penis consists of a root, a body, and a glans penis. It functions to introduce sperm into the vagina. Expansion of its blood sinuses under the influence of sexual excitation is called erection.

Female Reproductive System (Section 29.3)

1. The female organs of reproduction include the ovaries (gonads), uterine (fallopian) tubes, uterus, vagina, and vulva.
2. The mammary glands are also considered part of the reproductive system.
3. The female gonads are the ovaries, located in the upper pelvic cavity on either side of the uterus.
4. Ovaries produce secondary oocytes; discharge secondary oocytes (the process of ovulation); and secrete estrogens, progesterone, relaxin, and inhibin.
5. Oogenesis (production of haploid secondary oocytes) begins in the ovaries. The oogenesis sequence is completed only after an ovulated secondary oocyte is fertilized by a sperm cell.
6. The uterine (fallopian) tube, which transports a secondary oocyte from an ovary to the uterus, is the normal site of fertilization.
7. The uterus is an organ the size and shape of an inverted pear that functions in menstruation, implantation of a fertilized ovum, development of a fetus during pregnancy, and labor. It also is part of the pathway for sperm to reach a uterine tube to fertilize a secondary oocyte.
8. The innermost layer of the uterine wall is the endometrium, which undergoes marked changes during the menstrual cycle.
9. The vagina is a passageway for the menstrual flow, the receptacle for the penis during sexual intercourse, and the lower portion of the birth canal. The smooth muscle of the vaginal wall makes it capable of considerable stretching.
10. The vulva, a collective term for the external genitals of the female, consists of the mons pubis, labia majora, labia minora, clitoris, vestibule, vaginal and urethral orifices, paraurethral glands, and greater vestibular glands.
11. The mammary glands of the female breasts are modified sweat glands located over the pectoralis major muscles. Their function is to secrete and eject milk (lactation).
12. Mammary gland development depends on estrogens and progesterone.
13. Milk production is stimulated by prolactin, estrogens, and progesterone; milk ejection is stimulated by oxytocin.

Female Reproductive Cycle (Section 29.4)

1. The female reproductive cycle includes the ovarian and menstrual cycles. The function of the ovarian cycle is development of a secondary oocyte; that of the menstrual cycle is preparation of the endometrium each month to receive a fertilized egg.
2. The ovarian and menstrual cycles are controlled by GnRH from the hypothalamus, which stimulates the release of FSH and LH by the anterior pituitary.
3. FSH stimulates development of follicles and initiates secretion of estrogens by the follicles. LH stimulates further development of the follicles, secretion of estrogens by follicular cells, ovulation, formation of the corpus luteum, and the secretion of progesterone and estrogens by the corpus luteum.
4. Estrogens stimulate the growth, development, and maintenance of female reproductive structures; the development of secondary sex characteristics; and protein synthesis.
5. Progesterone works together with estrogens to prepare the endometrium for implantation and the mammary glands for milk synthesis.
6. Relaxin increases the flexibility of the pubic symphysis and helps dilate the uterine cervix to ease delivery of a baby.
7. During the menstrual phase, part of the endometrium is shed, discharging blood and tissue cells.
8. During the preovulatory phase, a group of follicles in the ovaries begins to undergo maturation. One follicle outgrows the others and becomes dominant while the others die. At the same time, endometrial repair occurs in the uterus. Estrogens are the dominant ovarian hormones during the preovulatory phase.
9. Ovulation is the rupture of the dominant mature (graafian) follicle and the release of a secondary oocyte into the pelvic cavity. It is brought about by a surge of LH.
10. During the postovulatory phase, both progesterone and estrogens are secreted in large quantity by the corpus luteum of the ovary, and the uterine endometrium thickens in readiness for implantation.
11. If fertilization and implantation do not occur, the corpus luteum degenerates, and the resulting low level of progesterone and estrogens allows discharge of the endometrium (menstruation) followed by the initiation of another reproductive cycle.
12. If fertilization and implantation occur, the corpus luteum is maintained by hCG.

Aging and the Reproductive Systems (Section 29.5)

1. Puberty is the period of time when secondary sex characteristics begin to develop and the potential for sexual reproduction arises. In older females, levels of progesterone and estrogens decrease, resulting in changes in menstruation and then menopause.
2. In older males, decreased levels of testosterone are associated with decreased muscle strength, waning sexual desire, and fewer viable sperm; prostate disorders are common.

SELF-QUIZ QUESTIONS

1. The testes are located in the scrotum because
   a. they must be separated from all other organs or sterility can occur
   b. sperm and hormone production and survival require a temperature lower than the normal body temperature
   c. the scrotum supplies the necessary hormones for sperm maturation
   d. sperm in the testes cannot survive without the nutrients supplied by the scrotum
   e. the scrotum produces alkaline fluids that neutralize the acids in the male urethra.

2. Match the following:

____ **a.** cell that supports, protects, and nourishes developing spermatogonia
____ **b.** contains developing oocytes
____ **c.** immature sperm cell
____ **d.** cell that secretes testosterone
____ **e.** produces progesterone and estrogens

A. corpus luteum
B. Leydig cell
C. Sertoli cell
D. follicle
E. spermatogonium

3. Removal of the prostate would
a. interfere with sperm production
b. inhibit testosterone release
c. decrease the volume of semen by about 25%
d. cause semen to become more acidic
e. affect semen clotting.

4. Which of the following is true?
a. Meiosis is the process by which somatic (body) cells divide
b. The haploid chromosome number is symbolized by $2n$
c. Meiosis results in diploid spermatocytes.
d. Gametes contain the haploid chromosome number.
e. Gametes contain 46 chromosomes in their nuclei.

5. The uterus is the site of all of the following EXCEPT
a. menstruation
b. implantation of a fertilized ovum
c. ovulation
d. labor
e. development of the fetus.

6. Menstruation is triggered by a
a. rapid rise in luteinizing hormone (LH)
b. rapid fall in luteinizing hormone (LH)
c. drop in estrogens and progesterone
d. rise in estrogens and progesterone
e. rise in inhibin.

7. An inflammation of the seminiferous tubules would interfere with the ability to
a. secrete testosterone **b.** produce sperm
c. void urine **d.** make semen alkaline
e. regulate the temperature in the scrotum.

8. Erection of the penis involves the release of which neurotransmitter?
a. norepinephrine **b.** serotonin
c. glycine **d.** dopamine
e. nitric oxide

9. Which of the following is NOT a function of semen?
a. transport sperm
b. lubricate the reproductive tract
c. provide an acidic environment needed for fertilization
d. provide nourishment for sperm
e. produce antibiotics to destroy some bacteria

10. Prior to ejaculation, sperm are stored in the
a. Leydig cells **b.** scrotum
c. Sertoli cells **d.** prostate
e. ductus (vas) deferens.

11. Place the following in the correct order for the passage of sperm from the testes to the outside of the body.
1. urethra
2. ductus (vas) deferens
3. seminiferous tubules
4. ejaculatory duct
5. external urethral orifice
6. epididymis
a. 6, 3, 2, 4, 1, 5 **b.** 3, 2, 6, 4, 1, 5 **c.** 3, 6, 2, 4, 1, 5
d. 3, 6, 2, 4, 5, 1 **e.** 2, 4, 6, 1, 3, 5

12. In males, the gland that surrounds the urethra at the base of the urinary bladder is the
a. glans penis **b.** prostate
c. seminal vesicle **d.** bulbourethral gland
e. greater vestibular gland.

13. An oocyte is moved toward the uterus by
a. peristaltic contractions of the uterine (Fallopian) tubes
b. contraction of the uterus
c. gravity
d. swimming
e. flagella.

14. Fertilization normally occurs in the
a. vagina **b.** cervix **c.** uterus
d. ovary **e.** uterine tube.

15. In the female reproductive system, lubricating mucus is produced by the
a. vulva **b.** clitoris
c. mons pubis **d.** greater vestibular glands
e. sudoriferous glands.

16. Ovarian follicles mature during
a. menstruation **b.** ovulation
c. the postovulatory phase **d.** the preovulatory phase
e. the secretory phase.

17. Match the following:

____ **a.** enlargement and stiffening of the penis
____ **b.** discharge of a small volume of semen before ejaculation
____ **c.** maturation of spermatids into sperm
____ **d.** powerful release of semen from the urethra to the exterior

A. emission
B. ejaculation
C. erection
D. spermiogenesis

18. The portion of the uterus responsible for its contraction is the
a. fundus **b.** infundibulum **c.** endometrium
d. myometrium **e.** perineum.

19. Match the following:

____ **a.** released by the hypothalamus to regulate the ovarian cycle
____ **b.** stimulates the initial secretion of estrogens by growing follicles
____ **c.** stimulates ovulation
____ **d.** stimulate growth, development, and maintenance of the female reproductive system
____ **e.** works with estrogens to prepare the uterus for implantation of a fertilized ovum
____ **f.** assists with labor by helping to dilate the cervix and increase flexibility of the pubic symphysis
____ **g.** inhibits release of FSH by the anterior pituitary

A. luteinizing hormone (LH)
B. gonadotropin releasing hormone (GnRH)
C. relaxin
D. progesterone
E. inhibin
F. follicle-stimulating hormone (FSH)
G. estrogens

20. Jody has a very regular menstrual cycle that lasts 31 days. She most likely ovulates on day
a. 1 **b.** 5
c. 14 **d.** 17
e. 31.

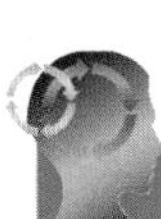

CRITICAL THINKING QUESTIONS

1. Janelle, who is 35 years old, has been advised to have a complete hysterectomy due to medical problems. She is worried that the procedure will cause menopause. Explain what is involved in the procedure and the likelihood that the procedure will result in menopause.
2. Phil has promised his wife that he will get a vasectomy after the birth of their next child. He is a little concerned, however, about the possible effects on his virility. What would you tell Phil about the procedure?
3. Julio and his wife have been trying unsuccessfully to become pregnant. The fertility clinic suggested that the problem may have something to do with Julio's habits of wearing very close-fitting briefs during the day and taking a long nightly soak in his hot tub. What effect could this have on fertility?

? ANSWERS TO FIGURE QUESTIONS

29.1 The gonads (testes) produces gametes (sperm) and hormones; the ducts transport, store, and receive gametes; the accessory sex glands secrete materials that support gametes; and the penis assists in the delivery and joining of gametes.

29.2 The Leydig cells of the testes secrete testosterone.

29.3 During meiosis the number of chromosomes in each cell is reduced by half.

29.4 The middle piece contains mitochondria which produce ATP that provides energy for locomotion of sperm.

29.5 The Sertoli cells secrete inhibin.

29.6 The seminal vesicles are the accessory sex glands that contribute the largest volume to seminal fluid.

29.7 The vagina is posterior to the urinary bladder and the urethra, and is also parallel to the urethra. The uterus is immediately superior to the urinary bladder and may move when the urinary bladder is extended.

29.8 Ovarian follicles secrete estrogens, and the corpus luteum secretes estrogens, progesterone, relaxin, and inhibin.

29.9 Primary oocytes are present in the ovary at birth, so they are as old as the woman is. In males, primary spermatocytes are continually being formed from spermatogonia and thus are only a few days old.

29.10 The endometrium is rebuilt after each menstruation.

29.11 Anterior to the vaginal opening are the mons pubis, clitoris, and prepuce. Lateral to the vaginal opening are the labia minora and labia majora.

29.12 Oxytocin regulates milk ejection from the mammary glands.

29.13 The hormones responsible for the proliferative phase of endometrial growth are estrogens; for ovulation, LH; for growth of the corpus luteum, LH; and for the midcycle surge of LH, estrogens.

29.14 This is negative feedback because the response is opposite to the stimulus. Decreasing levels of estrogens and progesterone stimulate release of GnRH, which, in turn, increases production and release of estrogens.

Appendix

Measurements

U.S. Customary System

PARAMETER	UNIT	RELATION TO OTHER U.S. UNITS	SI (METRIC) EQUIVALENT
Length	inch	1/12 foot	2.54 centimeters
foot	12 inches	0.305 meter	
yard	36 inches	9.144 meters	
mile	5,280 feet	1.609 kilometers	
Mass	grain	1/1000 pound	64.799 milligrams
dram	1/16 ounce	1.772 grams	
ounce	16 drams	28.350 grams	
pound	16 ounces	453.6 grams	
ton	2,000 pounds	907.18 kilograms	
Volume (Liquid)	ounce	1/16 pint	29.574 milliliters
pint	16 ounces	0.473 liter	
quart	2 pints	0.946 liter	
gallon	4 quarts	3.785 liters	
Volume (Dry)	pint	1/2 quart	0.551 liter
quart	2 pints	1.101 liters	
peck	8 quarts	8.810 liters	
bushel	4 pecks	35.239 liters	

International System (SI)

BASE UNITS		
UNIT	QUANTITY	SYMBOL
meter	length	m
kilogram	mass	kg
second	time	s
liter	volume	L
mole	amount of matter	mol

PREFIXES		
PREFIX	MULTIPLIER	SYMBOL
tera-	10^{12} = 1,000,000,000,000	T
giga-	10^{9} = 1,000,000,000	G
mega-	10^{6} = 1,000,000	M
kilo-	10^{3} = 1,000	k
hecto-	10^{2} = 100	h
deca-	10^{1} = 10	da
deci-	10^{-1} = 0.1	d
centi-	10^{-2} = 0.01	c
milli-	10^{-3} = 0.001	m
micro-	10^{-6} = 0.000,001	μ
nano-	10^{-9} = 0.000,000,001	n
pico-	10^{-12} = 0.000,000,000,001	p

Temperature Conversion

FAHRENHEIT (F) TO CELSIUS (C)

$$°C = (°F - 32) \div 1.8$$

CELSIUS (C) TO FAHRENHEIT (F)

$$°F = (°C \times 1.8) + 32$$

U.S To SI (Metric) Conversion

WHEN YOU KNOW	MULTIPLY BY	TO FIND
inches	2.54	centimeters
feet	30.48	centimeters
yards	0.91	meters
miles	1.61	kilometers
ounces	28.35	grams
pounds	0.45	kilograms
tons	0.91	metric tons
fluid ounces	29.57	milliliters
pints	0.47	liters
quarts	0.95	liters
gallons	3.79	liters

SI (Metric) To U.S. Conversion

WHEN YOU KNOW	MULTIPLY BY	TO FIND
millimeters	0.04	inches
centimeters	0.39	inches
meters	3.28	feet
kilometers	0.62	miles
liters	1.06	quarts
cubic meters	35.32	cubic feet
grams	0.035	ounces
kilograms	2.21	pounds

B Appendix

Periodic Table

The periodic table lists the known **chemical elements,** the basic units of matter. The elements in the table are arranged left-to-right in rows in order of their **atomic number,** the number of protons in the nucleus. Each horizontal row, numbered from 1 to 7, is a **period.** All elements in a given period have the same number of electron shells as their period number. For example, an atom of hydrogen or helium each has one electron shell, while an atom of potassium or calcium each has four electron shells. The elements in each column, or **group,** share chemical properties. For example, the elements in column IA are very chemically reactive, whereas the elements in column VIIIA have full electron shells and thus are chemically inert.

Scientists now recognize 117 different elements; 92 occur naturally on Earth, and the rest are produced from the natural elements using particle accelerators or nuclear reactors. Elements are designated by **chemical symbols,** which are the first one or two letters of the element's name in English, Latin, or another language.

Twenty-six of the 92 naturally occurring elements normally are present in your body. Of these, just four elements—oxygen (O), carbon (C), hydrogen (H), and nitrogen (N) (coded blue)—constitute about 96% of the body's mass. Eight others—calcium (Ca), phosphorus (P), potassium (K), sulfur (S), sodium (Na), chlorine (Cl), magnesium (Mg), and iron (Fe) (coded pink)—contribute 3.8% of the body's mass. An additional 14 elements, called **trace elements** because they are present in tiny amounts, account for the remaining 0.2% of the body's mass. The trace elements are aluminum, boron, chromium, cobalt, copper, fluorine, iodine, manganese, molybdenum, selenium, silicon, tin, vanadium, and zinc (coded yellow). Table 2.1 on page 29 provides information about the main chemical elements in the body.

23 ← Atomic number
V ← Chemical symbol
50.942 ← Atomic mass (weight)

Percentage of body mass
- 96% (4 elements)
- 3.8% (8 elements)
- 0.2% (14 elements)

	IA	IIA	IIIB	IVB	VB	VIB	VIIB	VIIIB	VIIIB	VIIIB	IB	IIB	IIIA	IVA	VA	VIA	VIIA	VIIIA
1	1 Hydrogen H 1.0079																	2 Helium He 4.003
2	3 Lithium Li 6.941	4 Beryllium Be 9.012											5 Boron B 10.811	6 Carbon C 12.011	7 Nitrogen N 14.007	8 Oxygen O 15.999	9 Fluorine F 18.998	10 Neon Ne 20.180
3	11 Sodium Na 22.989	12 Magnesium Mg 24.305											13 Aluminum Al 26.9815	14 Silicon Si 28.086	15 Phosphorus P 30.974	16 Sulfur S 32.066	17 Chlorine Cl 35.453	18 Argon Ar 39.948
4	19 Potassium K 39.098	20 Calcium Ca 40.08	21 Scandium Sc 44.956	22 Titanium Ti 47.87	23 Vanadium V 50.942	24 Chromium Cr 51.996	25 Manganese Mn 54.938	26 Iron Fe 55.845	27 Cobalt Co 58.933	28 Nickel Ni 58.69	29 Copper Cu 63.546	30 Zinc Zn 65.38	31 Gallium Ga 69.723	32 Germanium Ge 72.59	33 Arsenic As 74.992	34 Selenium Se 78.96	35 Bromine Br 79.904	36 Krypton Kr 83.80
5	37 Rubidium Rb 85.468	38 Strontium Sr 87.62	39 Yttrium Y 88.905	40 Zirconium Zr 91.22	41 Niobium Nb 92.906	42 Molybdenum Mo 95.94	43 Technetium Tc (99)	44 Ruthenium Ru 101.07	45 Rhodium Rh 102.905	46 Palladium Pd 106.42	47 Silver Ag 107.868	48 Cadmium Cd 112.40	49 Indium In 114.82	50 Tin Sn 118.69	51 Antimony Sb 121.75	52 Tellurium Te 127.60	53 Iodine I 126.904	54 Xenon Xe 131.30
6	55 Cesium Cs 132.905	56 Barium Ba 137.33		72 Hafnium Hf 178.49	73 Tantalum Ta 180.948	74 Tungsten W 183.85	75 Rhenium Re 186.2	76 Osmium Os 190.2	77 Iridium Ir 192.22	78 Platinum Pt 195.08	79 Gold Au 196.967	80 Mercury Hg 200.59	81 Thallium Tl 204.38	82 Lead Pb 207.19	83 Bismuth Bi 208.980	84 Polonium Po (209)	85 Astatine At (210)	86 Radon Rn (222)
7	87 Francium Fr (223)	88 Radium Ra (226)		104 Rutherfordium Rf (267)	105 Dubnium Db (268)	106 Seaborgium Sg (271)	107 Bohrium Bh (272)	108 Hassium Hs (270)	109 Meitnerium Mt (276)	110 Ds (281)	111 Rg (280)	112 Uub (285)	113 Uut (284)	114 Uuq (289)	115 Uup (288)	116 Uuh (293)		118 Uuo (294)

57–71, Lanthanides

57 Lanthanum La 138.91	58 Cerium Ce 140.12	59 Praseodymium Pr 140.907	60 Neodymium Nd 144.24	61 Promethium Pm 144.913	62 Samarium Sm 150.35	63 Europium Eu 151.96	64 Gadolinium Gd 157.25	65 Terbium Tb 158.925	66 Dysprosium Dy 162.50	67 Holmium Ho 164.930	68 Erbium Er 167.26	69 Thulium Tm 168.934	70 Ytterbium Yb 173.04	71 Lutetium Lu 174.97

89–103, Actinides

89 Actinium Ac (227)	90 Thorium Th 232.038	91 Protactinium Pa (231)	92 Uranium U 238.03	93 Neptunium Np (237)	94 Plutonium Pu 244.064	95 Americium Am (243)	96 Curium Cm (247)	97 Berkelium Bk (247)	98 Californium Cf 242.058	99 Einsteinium Es (254)	100 Fermium Fm 257.095	101 Mendelevium Md 258.10	102 Nobelium No 259.10	103 Lawrencium Lr 260.105

Appendix C

Answers

Answers to Self-Quiz Questions

Chapter 1 **1.** b **2.** d **3a.** nervous system **b.** brain, spinal cord, nerves, sense organs **c.** lymphatic system **d.** returns proteins and fluid to blood; sites of lymphatic maturation and proliferation to protect against disease; carries lipids from digestive system to blood **e.** respiratory system **f.** lungs, pharynx, larynx, trachea, bronchial tubes, **g.** testes, ovaries, vagina, uterine tubes, uterus, penis **h.** reproduces the organism and releases hormones **4.** d **5.** a **6.** c **7.** e **8.** a **9.** a **10.** d **11.** c **12.** e **13.** d **14.** c **15.** b **16.** e **17.** **(a)** D, **(b)** F, **(c)** H, **(d)** A, **(e)** I, **(f)** E, **(g)** B, **(h)** G, **(i)** C, **(j)** J **18.** **(a)** D, **(b)** A, **(c)** H, **(d)** F, **(e)** G, **(f)** E, **(g)** C, **(h)** B **19.** **(a)** G, **(b)** B, **(c)** E, **(d)** C, **(e)** F, **(f)** A, **(g)** D **20.** **(a)** C, **(b)** D, **(c)** A, **(d)** B

Chapter 2 **1.** d **2.** c **3.** a **4.** e **5.** d **6.** b **7.** b **8.** c **9.** a **10.** d **11.** e **12.** d **13.** c **14.** a **15.** b **16.** e **17.** a **18.** carbon, hydrogen, oxygen, nitrogen **19.** **(a)** D, **(b)** C, **(c)** A, **(d)** E, **(e)** B

Chapter 3 **1.** b **2.** e **3.** a **4.** a **5.** c **6.** c **7.** b **8.** d **9.** a **10.** **(a)** B, **(b)** F, **(c)** G, **(d)** H, **(e)** C, **(f)** E, **(g)** D, **(h)** A **11.** e **12.** c **13.** e **14.** d **15.** b **16.** b **17.** **(a)** C, **(b)** E, **(c)** F, **(d)** B, **(e)** D, **(f)** A **18.** e **19.** b **20.** b

Chapter 4 **1.** c **2.** d **3.** b **4.** d **5.** a **6.** **(a)** C, **(b)** G, **(c)** E, **(d)** D, **(e)** H, **(f)** F, **(g)** B, **(h)** A **7.** a **8.** e **9.** e **10.** b **11.** a **12.** c **13.** c **14.** d **15.** e **16.** d **17.** a **18.** c **19.** a **20.** b

Chapter 5 **1.** stratum lucidum **2.** eccrine, ceruminous, apocrine **3.** false **4.** true **5.** c **6.** e **7.** a **8.** c **9.** b **10.** e **11.** a **12.** c **13.** **(a)** C, **(b)** E, **(c)** D, **(d)** A, **(e)** F, **(f)** J, **(g)** B, **(h)** H, **(i)** I, **(j)** G **14.** **(a)** C, **(b)** D, **(c)** A, **(d)** B **15.** **(a)** D, **(b)** C, **(c)** B, **(d)** A, **(e)** inflammatory, migratory, proliferative, maturation

Chapter 6 **1.** interstitial, appositional **2.** hardness, tensile strength and flexibility **3.** true **4.** true **5.** true **6.** d **7.** a **8.** e **9.** c **10.** a **11.** **(a)** C, **(b)** I, **(c)** H, **(d)** A, **(e)** E, **(f)** D, **(g)** F, **(h)** G, **(i)** L, **(j)** B, **(k)** K, **(l)** J **12.** **(a)** B, **(b)** F, **(c)** D, **(d)** E, **(e)** G, **(f)** C, **(g)** A **13.** **(a)** L, **(b)** D, **(c)** H, **(d)** F, **(e)** C, **(f)** I, **(g)** M, **(h)** J, **(i)** G, **(j)** E, **(k)** B, **(l)** K, **(m)** A **14.** **(a)** A, **(b)** D, **(c)** C, **(d)** B **15.** **(a)** C, **(b)** G, **(c)** F, **(d)** A, **(e)** D, **(f)** B, **(g)** E, **(h)** I, **(i)** H, **(j)** J

Chapter 7 **1.** c **2.** d **3.** b **4.** a **5.** d **6.** a **7.** e **8.** a **9.** e **10.** e **11.** atlas, axis **12.** long, short, flat **13.** Wormian **14.** hyoid **15.** true **16.** true **17.** true **18.** false **19.** false **20.** **(a)** F, **(b)** K, **(c)** D, **(d)** H, **(e)** I, **(f)** K, **(g)** J, **(h)** F, **(i)** A, **(j)** B, **(k)** D, **(l)** K, **(m)** J, **(n)** F, **(o)** D, **(p)** E, **(q)** G, **(r)** C

Chapter 8 **1.** c **2.** e **3.** d **4.** e **5.** c **6.** b **7.** a **8.** b **9.** c **10.** b **11.** ischium **12.** patella **13.** posterior, femur **14.** false **15.** head, femur **16.** false **17.** true **18.** true **19.** false **20.** **(a)** C, **(b)** E, **(c)** G, **(d)** H, **(e)** D, **(f)** A, **(g)** I, **(h)** B, **(i)** F, **(j)** E, **(k)** C, **(l)** J, **(m)** E, **(n)** D, **(o)** F, K, **(p)** B, **(q)** G

Chapter 9 **1.** b **2.** c **3.** a **4.** e **5.** d **6.** e **7.** a **8.** c **9.** cartilaginous joint (synchondrosis), hyaline cartilage **10.** ligament **11.** diarthrosis **12.** hyaline cartilage (articular cartilage) **13.** bursae **14.** false **15.** true **16.** false **17.** false **18.** false **19.** **(a)** C, **(b)** E, **(c)** G, **(d)** B, **(e)** D, **(f)** H, **(g)** I, **(h)** A, **(i)** F **20.** **(a)** F, **(b)** D, **(c)** E, **(d)** C, **(e)** B, **(f)** G, **(g)** H, **(h)** A

Chapter 10 **1.** e **2.** **(a)** D, **(b)** E, **(c)** B, **(d)** A, **(e)** C **3.** d **4.** **(a)** C, **(b)** D, **(c)** A, **(d)** E, **(e)** B **5.** c **6.** a **7.** a **8.** **(a)** SM, CA, **(b)** SK, **(c)** SK, CA, **(d)** CA, **(e)** SK, **(f)** SK, **(g)** SM, **(h)** SM, **(i)** SM, **(j)** CA, SM **9.** b **10.** b **11.** e **12.** b **13.** d **14.** a **15.** a **16.** d **17.** **(a)** C, **(b)** F, **(c)** D, **(d)** B, **(e)** G, **(f)** H, **(g)** A, **(h)** E

Chapter 11 **1.** d **2.** d **3.** d **4.** b **5.** b **6.** a **7.** c **8.** c **9.** a **10.** c **11.** agonist, synergists **12.** origin, insertion **13.** bones, joints, skeletal muscles **14.** mastoid process **15.** sphincter **16.** false **17.** **(a)** C, **(b)** B, **(c)** A, **(d)** B, C

Chapter 12 **1.** central **2.** perineum **3.** sphincter **4.** d **5.** b **6.** a. **7.** c **8.** c **9.** a. **10.** c **11.** b **12.** c **13.** c **14.** b **15.** b **16.** a **17.** c **18** a **19.** d **20.** b

Chapter 13 **1.** d **2.** d **3.** c **4.** b **5.** c **6.** d **7.** a **8.** a **9.** c **10.** b **11.** b **12.** c **13.** c **14.** a **15.** a **16.** a **17.** b **18.** a **19.** c **20.** d

Chapter 14 **1.** b **2.** c **3.** d **4.** d **5.** d **6.** b **7.** a **8.** d **9.** e **10.** b **11.** d **12.** d **13.** c **14.** a **15.** d **16.** b **17.** a **18.** b **19.** a **20.** a

Chapter 15 **1.** c **2.** e **3.** a **4.** d **5.** a **6.** e **7.** c **8.** a **9.** b **10.** e **11.** c **12.** d **13.** b **14.** d **15.** **(a)** D, **(b)** E, **(c)** F, **(d)** C, **(e)** A, **(f)** B **16.** **(a)** I, **(b)** A, **(c)** D, **(d)** O, **(e)** E, **(f)** P, **(g)** H, **(h)** J, **(i)** G, **(j)** C, **(k)** B, **(l)** L, **(m)** N, **(n)** K, **(o)** F, **(p)** M

Chapter 16 **1.** mixed **2.** sensory receptor, sensory neuron, integrating center, motor neuron, effector **3.** true **4.** false **5.** c **6.** c **7.** a **8.** e **9.** c **10.** e **11.** a **12.** d **13.** **(a)** A, **(b)** H, **(c)** D, **(d)** B, **(e)** K, **(f)** A, **(g)** F, **(h)** E, **(i)** C, **(j)** I, **(k)** A, **(l)** L, **(m)** G, **(n)** B, **(o)** J **14.** **(a)** B, **(b)** A, **(c)** C, **(d)** D, **(e)** A, **(f)** E, **(g)** C, **(h)** B, **(i)** D, **(j)** A, **(k)** B, **(l)** D, **(m)** C, **(n)** E, **(o)** A **15.** **(a)** N, **(b)** L, **(c)** M, **(d)** A, **(e)** B, **(f)** E, **(g)** K, **(h)** H, **(i)** J, **(j)** I, **(k)** O, **(l)** D, **(m)** G, **(n)** C, **(o)** F

Chapter 17 **1.** e **2.** c **3.** a **4.** c **5.** a **6.** b **7.** e **8.** e **9.** a **10.** b **11.** **(a)** C, **(b)** D, **(c)** A, **(d)** B **12.** b **13.** a **14.** c, e **15.** **(a)** E, **(b)** F, **(c)** G, **(d)** A, **(e)** C, **(f)** B, **(g)** D

Chapter 18 **1.** c **2.** a **3.** c **4.** e **5.** d **6.** a **7.** d **8.** b **9.** d **10.** e **11.** b **12.** d **13.** b **14.** **(a)** C, **(b)** F, **(c)** E, **(d)** B, **(e)** A, **(f)** D **15.** **(a)** S, **(b)** P, **(c)** P, **(d)** S, **(e)** S, **(f)** P, **(g)** S, **(h)** S, **(i)** S

Chapter 19 **1.** b **2.** b **3.** a **4.** d **5.** e **6.** c **7.** **(a)** H, **(b)** E, **(c)** F, **(d)** C, **(e)** G, **(f)** A, **(g)** B, **(h)** D **8.** e **9.** d **10.** c **11.** a **12.** e **13.** e **14.** c **15.** **(a)** D, **(b)** F, **(c)** G, **(d)** B, **(e)** C, **(f)** A, **(g)** E **16.** a **17.** e **18.** c **19.** d **20.** a

Chapter 20 **1.** d **2.** c **3.** a **4.** d **5.** a **6.** e **7.** a **8.** b **9.** c **10.** c **11.** a **12.** b **13.** d **14.** b **15.** c **16.** b **17.** **(a)** C, **(b)** F, **(c)** G, **(d)** A, **(e)** D, **(f)** E, **(g)** B **18.** **(a)** B, **(b)** C, **(c)** E, **(d)** A, **(e)** F, **(f)** D **19.** **(a)** R, **(b)** F, **(c)** F, **(d)** R, **(e)** F, **(f)** E, **(g)** R, **(h)** E, **(i)** F

Chapter 21 **1.** c **2.** **(a)** D, **(b)** G, **(c)** C, **(d)** A, **(e)** E, **(f)** F, **(g)** B **3.** c **4.** d **5.** e **6.** e **7.** b **8.** b **9.** d **10.** a **11.** c **12.** a **13.** **(a)** D, **(b)** A, **(c)** C, **(d)** B **14.** d **15.** b **16.** a **17.** e **18.** a **19.** b **20.** c

Chapter 22 **1.** **(a)** D **(b)** H **(c)** F **(d)** G **(e)** C **(f)** B **(g)** A **(h)** E **2.** e **3.** e **4.** a **5.** c **6.** c **7.** a **8.** c **9.** b **10.** d **11.** d **12.** a **13.** c **14.** a **15.** d **16.** d **17.** b **18.** c **19.** **(a)** A, **(b)** A, **(c)** A, **(d)** B, **(e)** A, **(f)** B, **(g)** B, **(h)** A, **(i)** B, **(j)** A, **(k)** B, **(l)** A, **(m)** A **20.** **(a)** B, **(b)** G, **(c)** A, **(d)** F, **(e)** D, **(f)** C, **(g)** E

Chapter 23 **1.** b **2.** a **3.** a **4.** c **5.** d **6.** e **7.** **(a)** E, **(b)** D, **(c)** A, **(d)** B, **(e)** C **8.** d **9.** c **10.** b **11.** c **12.** **(a)** H, **(b)** D, **(c)** E, **(d)** F, **(e)** A, **(f)** B, **(g)** C, **(h)** I, **(i)** G **13.** **(a)** A, **(b)** B, **(c)** B, **(d)** A, **(e)** A, **(f)** A, **(g)** A, **(h)** A, **(i)** B **14.** e **15.** b **16.** a **17.** d

Chapter 24 **1.** c **2.** d **3.** e **4.** c **5.** a **6.** b **7.** b **8.** a **9.** d **10.** e **11.** b **12.** c **13.** d **14.** **(a)** E, **(b)** B, **(c)** D, **(d)** A, **(e)** C **15.** c **16.** b **17.** c **18.** b **19.** e **20.** a

Chapter 25 **1.** d **2.** c **3.** b **4.** a **5.** c **6.** a **7.** a **8.** d **9.** **(a)** B, **(b)** D, **(c)** A, **(d)** C, **(e)** E **10.** a **11.** **(a)** F, **(b)** D, **(c)** H, **(d)** A, **(e)** G, **(f)** C, **(g)** E, **(h)** B **12.** b **13.** c **14.** e **15.** a **16.** b **17.** d **18.** c **19.** c **20.** **(a)** D, **(b)** A, **(c)** E, **(d)** C, **(e)** B

Chapter 26 **1.** e **2.** b **3.** b **4.** c **5.** d **6.** a **7.** c **8.** **(a)** D, **(b)** G, **(c)** A, **(d)** H, **(e)** B, **(f)** C, **(g)** E, **(h)** F **9.** e **10.** e **11.** a **12.** e **13.** c **14.** b **15.** d **16.** a **17.** e **18.** c **19.** e **20.** a

Chapter 27 **1.** b **2.** c **3.** a **4.** **(a)** C, **(b)** A, **(c)** D, **(d)** B **5.** c **6.** a **7.** b **8.** e **9.** d **10.** a **11.** d **12.** b **13.** e **14.** b **15.** c **16.** c **17.** d **18.** a **19.** d **20.** **(a)** C, **(b)** D, **(c)** B, **(d)** E, **(e)** A

Chapter 28 **1.** c **2.** e **3.** a **4.** d **5.** c **6.** e **7.** a **8.** d **9.** b **10.** a **11.** b **12.** e **13.** e **14.** a **15.** c **16.** d **17.** a **18.** c **19.** e **20.** b

Chapter 29 **1.** b **2.** **(a)** C, **(b)** D, **(c)** E, **(d)** B, **(e)** A **3.** c **4.** d **5.** c **6.** c **7.** b **8.** e **9.** c **10.** e **11.** c **12.** b **13.** a **14.** e **15.** d **16.** d **17.** **(a)** C, **(b)** A, **(c)** D, **(d)** B **18.** d **19.** **(a)** B, **(b)** F, **(c)** A, **(d)** G, **(e)** D, **(f)** C, **(g)** E **20.** d

Answers to Critical Thinking Questions

Chapter 1

1. No. Computed tomography is used to look at differences in tissue density. To assess activity in an organ such as the brain, a positron emission tomography (PET) scan or a single-photo-emission computerized tomography (SPECT) scan would both provide a colorized visual assessment of brain activity.

2. Stem cells are undifferentiated cells. Research using stem cells has shown that these undifferentiated cells may be prompted to differentiate into the specific cells needed to replace those which are damaged or malfunctioning.

3. Homeostasis is the relative constancy of the body's internal environment. Homeostasis is maintained as the body changes in response to shifting external and internal conditions. Body temperature should vary within a narrow range around normal body temperature (37°C or 98° to 99°F), which is above normal room temperature (usually around 72°F (25°C).

Chapter 2

1. Neither butter nor margarine are particularly good choices for frying eggs. Butter contains saturated fats that are associated with heart disease. However, many margarines contain hydrogenated or partially hydrogenated trans-fatty acids that also increase the risk of heart disease. An alternative would be frying the eggs in any of the mono- or polyunsaturated fats such as olive oil, peanut oil, or corn oil.

2. High body temperatures can be life-threatening, especially in infants. The increased temperature can cause denaturing of structural proteins and vital enzymes. When this happens, the proteins become nonfunctional. If the denatured enzymes are required for reactions that are necessary for life, then the infant could die.

3. Simply adding water to the table sugar does not cause it to break apart into monosaccharides. The water acts as a solvent, dissolving the sucrose, and forming a sugar-water solution. To complete the breakdown of table sugar to glucose and fructose would require the presence of the enzyme sucrase.

Chapter 3

1. Synthesis of mucin by ribosomes on rough endoplasmic reticulum, to transport vesicle, to entry face of Golgi complex, to transfer vesicle, to medial cisternae where protein is modified, to transfer vesicle, to exit face, to secretory vesicle, to plasma membrane where it undergoes exocytosis.

2. Since smooth ER inactivates or detoxifies drugs, and peroxisomes also destroy harmful substances such as alcohol, we would expect to see increased numbers of these organelles in Sebastian's liver cells.

3. In order to restore water balance to the cells, the runners need to consume hypotonic solutions. The water in the hypotonic solution will move from the blood, into the interstitial fluid, and then into the cells. Plain water works well; sports drinks contain water and some electrolytes (which may have been lost due to sweating) but will still be hypotonic in relation to the body cells.

Chapter 4

1. The viscosity of the ground substance of connective tissue changes in response to pressure, passive motion, shearing, and heat. The ground substance becomes more fluid, and thus the exchange of nutrients and wastes between blood and cells is enhanced. Furthermore, connective tissue fascia can be manipulated and thus the patient's range of motion can be increased and pain is decreased.

2. There are many possible adaptations, including: more adipose tissue for insulation; thicker bones for support; more red blood cells for oxygen transport; increased thickness of skin to prevent water loss; etc.

3. Your bread-and-water diet is not providing you with the necessary nutrients to encourage tissue repair. You need proper amounts of many essential vitamins, especially vitamin C, which is required for repair of the matrix and blood vessels. Vitamin A is needed to help properly maintain epithelial tissue. Adequate protein is also needed in order to synthesize the structural proteins of the damaged tissue.

Chapter 5

1. The hair shaft is fused, dead keratinized cells. At the base of the hair, the hair bulb contains the living matrix where cell divison occurs. Hair root plexuses (nervous tissue) surround each hair follicle.

2. The epidermal layer is repaired by cell division of keratinocytes starting at the deepest layer of the epidermis, the stratum basale. The keratinocytes move toward the surface through the stratum spinosum, stratum granulosum, and stratum corneum. The cells become fully kertatinized, flat and dead as they travel to the surface. The process takes about 2–4 weeks.

3. A callus is an abnormal thickening of the stratum corneum due to constant friction/rubbing. Warts are caused by papilloma virus infection resulting in uncontrolled epithelial cell growth. Athlete's foot is a fungal infection.

4. Jeremy's blackheads are caused by oil (sebum) accumulation in the sebaceous glands. Sebum secretion often increases after puberty. The color is due to melanin and oxidized oil.

Chapter 6

1. Due to the strenuous, repetitive activity, Taryn has probably developed a stress fracture of her right tibia (leg bone). Stress fractures are due to repeated stress on a bone that causes microscopic breaks in the bone without any evidence of injury to other tissue. An x-ray would not reveal the stress fracture, but a bone scan would. Thus the bone scan would either confirm or negate the physician's diagnosis.

2. When Marcus broke his arm as a child, he injured one of his epiphyseal (growth) plates. Damage to the cartilage in the epiphyseal plate resulted in premature closure of the plate, which interfered with the lengthwise growth of the arm bone.

3. Exercise causes mechanical stress on bones, but because there is effectively zero gravity in space, the pull of gravity on bones is missing. The lack of stress from gravity results in bone demineralization and weakness.

Chapter 7

1. Inability to open mouth—damage to the mandible, probably at the temporomandibular joint; black eye—trauma to the ridge over the supraorbital margin; broken nose—probably damage to the nasal septum (includes the vomer, septal cartilage, and perpendicular plate of the ethmoid) and possibly the nasal bones; broken cheek—fracture of zygomatic bone; broken upper jaw—fracture of maxilla; damaged eye socket—fracture of parts of the sphenoid, frontal, ethmoid, palatine, zygomatic, lacrimal, and maxilla bones (all compose the eye socket); punctured lung—damage to the thoracic vertebrae, which have punctured the lung.

2. Due to the repeated and extensive tension on his bone surfaces by skeletal muscle contraction, Bubba would experience deposition of new bone tissue. His arm bones would be thicker and with increased raised areas (projections) where the tendons attach his very active muscles to bone.

3. The "soft area" being referred to is the anterior fontanel, located between the parietal and frontal bones. This is one of several areas of fibrous connective tissue in the skull that has not ossified; it should complete its ossification at 18–24 months after birth. Fontanels allow flexibility of the skull for childbirth and for brain growth after birth. The connective tissue will not allow passage of water; thus no brain damage will occur through simply washing the baby's hair.

Chapter 8

1. There are several characteristics of the bony pelves that can be used to differentiate male from female: (1) The pelvis in the female is wider and more shallow than the male's; (2) the pelvic brim of the female is larger and more oval; (3) the pubic arch of the female has an angle greater than 90º; (4) the female's pelvic outlet is wider than in a male's; (5) the female's iliac crest is less curved and the ilium less vertical. Table 8.1 provides additional differences between female and male pelves. Age of the skeleton can be determined by the size of the bones, the presence or absence of epiphyseal plates, the degree of demineralization of the bones, and the general appearance of the "bumps" and ridges of bones.

2. Infants do have "flat feet" because their arches have not yet developed. As they begin to stand and walk, the arches should begin to develop in order to accommodate and support their body weight. The arches are usually fully developed by age 12 or 13, so Dad doesn't need to worry yet!

3. There are 14 phalanges in each hand: two bones in the thumb and three in each of the other fingers. Farmer White has lost five phalanges on his left hand (two in his thumb and three in his index finger), so he has nine remaining on his left and 14 remaining on his right for a total of 23.

Chapter 9

1. Katie's vertebral column, head, thighs, lower legs, lower arms, and fingers are flexed. Her lower arms and shoulders are medially rotated. Her thighs and arms are adducted.

2. The knee joint is commonly injured, especially among athletes. The twisting of Jeremiah's leg could have resulted in a multitude of internal injuries to the knee joint but often football players suffer tearing of the anterior cruciate ligament and medial meniscus. The immediate swelling is due to blood from damaged blood vessels, damaged synovial membranes, and the torn meniscus. Continued swelling is a result of a buildup of synovial fluid, which can result in pain and decreased mobility. Jeremiah's doctor may aspirate some of the fluid ("draining the water off his knee") and might want to perform arthroscopy to check for the extent of the knee damage.

3. The condylar processes of the mandible passed anteriorly to the articular tubercles of the temporal bones and this dislocated Antonio's mandible. It could be corrected by pressing the thumbs downward on the lower molar teeth and pushing the mandible backward.

Chapter 10

1. Muscle cells lose their ability to undergo cell division after birth. Therefore, the increase in size is not due to an increase in the number of muscle cells but rather is due to enlargement of the existing muscle fibers (hypertrophy). This enlargement can occur from forceful, repetitive muscular activity. It will cause the muscle fibers to increase their production of internal structures such as mitochondria and myofibrils and produce an increase in the muscle fiber diameter.

2. The "dark meat" of both chickens and ducks is composed primarily of slow oxidative (SO) muscle fibers. These fibers contain large amounts of myoglobin and capillaries, which accounts for their dark color. In addition, these fibers contain large numbers of mitochondria and generate ATP by aerobic respiration. SO fibers are resistant to fatigue and can produce sustained contractions for many hours. The legs of chickens and ducks are used for support, walking, and swimming (in ducks), all activities in which endurance is needed. In addition, migrating ducks require SO fibers in their breasts to enable them to have enough energy to fly for extremely long distances while migrating. There may be some fast oxidative-glycolytic (FOG) fibers in the dark meat. FOG fibers also contain large amounts of myoglobin and capillaries, contributing to the dark color. They can use aerobic or anaerobic cellular respiration to generate ATP and have high-to-moderate resistance to fatigue. These fibers would be good for the occasional "sprint" that ducks and chickens undergo to escape dangerous situations. In contrast, the white meat of a chicken breast is composed primarily of fast glycolytic (FG) fibers. FG fibers have lower amounts of myoglobin and capillaries that give the meat its white color. There are also few mitochondria in FG fibers so these fibers generate ATP mainly by glycolysis. These fibers contract strongly and quickly and are adapted for intense anaerobic movements of short duration. Chickens occasionally use their breasts for flying extremely short distances, usually to escape prey or perceived danger, so FG fibers are appropriate for their breast muscle.

3. Destruction of the somatic motor neurons to skeletal muscle fibers will result in a loss of stimulation to the skeletal muscles. When not stimulated on a regular basis, a muscle begins to lose its muscle tone. Through lack of use, the muscle fibers will weaken, begin to decrease in size, and can be replaced by fibrous connective tissue, resulting in a type of denervation atrophy. A lack of stimulation of the breathing muscles (especially the diaphragm) from motor neurons can result in inability of the breathing muscles to contract, thus causing respiratory paralysis and possibly death of the individual from respiratory failure.

Chapter 11

1. All of the following could occur on the affected (right) side of the face: (1) drooping of eyelid—levator palpebrae superioris; (2) drooping of the mouth, drooling, keeping food in mouth—orbicularis oris, buccinator; (3) uneven smile—zygomaticus major, levator labii superioris, risorius; (4) unable to wrinkle forehead—occipitofrontalis; (5) trouble sucking through a straw—buccinator.

2. Refer to Exhibit 11.3 "MUSCLES THAT MOVE THE MANDIBLE AND ASSIST IN MASTICATION (CHEWING) AND SPEECH." You should have included temporalis (retract, elevate, i.e., closing the mouth); masseter (retract, elevate, i.e., closing the mouth); medial pterygoid (protract, elevate, i.e., closing the mouth); and lateral pterygoid (protract, depress, i.e., opening the mouth) Note that it is a suprahyoid group of muscles that aids the lateral pterygoid in depressing the mandible to opening the mouth (see Exhibit 11.4) but not with as much force as afforded by the temporalis and masseter muscles in closing the mouth; closing the mouth is a much more powerful movement.

Chapter 12

1. Bulbospongiosus, external urethral sphincter, and deep transverse perineal.

2. Wyman probably has an inguinal hernia. He should see a doctor, since constriction of the intestines pushing out through the opening can cause serious damage.

Chapter 13

1. The extensor muscles of the hand and wrist as well as the extensor muscles of the fingers originate on or near the lateral epicondyle of the humerus. The antagonistic flexor muscles originate on or near the medial epicondyle of the humerus.

2. The rotator cuff is formed by a combination of the tendons of four deep muscles of the shoulder—supscapularis, supraspinatus, infraspinatus, and teres minor. These muscles add strength and stability to the shoulder joint. Although any of the muscles' tendons can be injured, the supraspinatus is most often damaged. Dependent upon the injured muscle, Jose may have trouble medially rotating his arm (subscapularis), abducting his arm (supraspinatus), laterally rotating his arm (infraspinatus, teres minor), adducting his arm (infraspinatus, teres minor), or extending his arm (teres minor).

Chapter 14

1. The "sciatic nerve" is actually a misnomer. Two nerves, the tibial and common fibular nerves, are wrapped with common connective tissue fibers and the combined structure has been named the sciatic nerve. Impingement of the sciatic nerve could affect only the tibial nerve, only the common fibular nerve, or both. Muscles of the posterior thigh and posterior leg are innervated by the tibial nerve. The common fibular nerve and its branches innervate the anterior and lateral regions of the leg and the foot. All three men could have been properly diagnosed and yet have pain in different areas of the lower limb.

2. The soleus muscle lies deep to the gastrocnemius and is the more powerful of the two muscles. It is solely responsible for plantar flexion of the foot when the knee is bent, as when riding a bicycle.

Chapter 15

1. Smelling the coffee and hearing the alarm are somatic sensory, stretching and yawning are somatic motor, salivating is autonomic (parasympathetic) motor, stomach rumble is enteric motor.

2. Demyelinaton or destruction of the myelin sheath can lead to multiple problems, especially in infants and children whose myelin sheaths are still in the process of developing. The affected axons deteriorate, which will interfere with function in both the CNS and PNS. There will be lack of sensation and loss of motor control with less rapid and less coordinated body responses. Damage to the axons in the CNS can be permanent and Ming's brain development may be irreversibly affected.

Chapter 16

1. Karen's symptoms could be explained by compression of the brachial plexus, subclavian artery, subclavian vein, or a combination of these structures. These structures lie close together between the anterior and middle scalene muscles and again under the pectoralis minor muscle. Emotional stress, such as the death of Karen's cat, can increase the contraction of these and other muscles. When a muscle is contracted it becomes shorter and fatter and thus could compress nerves and blood vessels that have only a narrow point of exit through the thoracic outlet.

2. Bill has pain in his left hip, thigh, and leg. Only the sciatic nerve and its branches are sensory to all of these areas. Sitting on a large wallet for long periods could compress the sciatic nerve as it emerges near the piriformis muscle. Working the hip muscles, and especially the piriformis, could alleviate possible muscle spasms and could also help restore a normal blood supply to the area. Finally, Judy advised Bill to use a different type of wallet, perhaps one that is suspended by a short chain from his belt.

3. Kathy had not been taught the proper use of crutches. The weight of the body should be supported by the palms of the hands, not by the axilla. Major nerves of the brachial plexus could be compressed by improper use of the crutches. The radial nerve is most affected by "crutch palsy" and the numbness in Kathy's arms and hands can be traced to the sensory distribution of this nerve.

4. The pin in Lou's right foot evoked a crossed extensor reflex. The right hamstring muscles contract and the foot is typically withdrawn from the painful stimulus. In this case, the pin was stuck in the foot and the typical flexor reflex did not alleviate the pain. In response to lifting his right foot, the crossed extensor reflex was evoked and Lou's weight was supported by the left leg as he hopped around. Ascending interneurons carried the stimulus along the spinal cord to the cerebral cortex. Perception of the pain caused Lou to yell.

5. The needles will pierce the epidermis, dermis, and subcutaneous layer and then go between the vertebrae through the epidural space, the dura mater, the subdural space, the arachnoid mater, and into the CSF in the subarachnoid space. CSF is produced in the brain, and the spinal meninges are continuous with the cranial meninges.

6. The anterior gray horns contain cell bodies of somatic motor neurons and motor nuclei that are responsible for the nerve impulses for skeletal muscle contraction. Because the lower cervical region is affected (brachial plexus, C5–C8), you would expect that Sunil may have trouble with movement in his shoulder, arm, and hand on the affected side.

7. Allyson has damaged her posterior columns in the lower (lumbar) region of the spinal cord. The posterior columns are responsible for transmitting nerve impulses responsible for awareness of muscle position (proprioception) and discriminative touch—which are affected in Allyson—as well as other functions such as light pressure sensations, and vibration sensations.

Chapter 17

1. Movement of the right arm is controlled by the left hemisphere's primary motor area, located in the precentral gyrus. Speech is controlled by Broca's area in the left hemisphere's frontal lobe just superior to the lateral cerebral sulcus.

2. Nicki's right facial nerve has been affected; she is suffering from Bell's palsy due to the viral infection. The facial nerve (VII) controls contraction of skeletal muscles of the face, tear gland and salivary gland secretion, as well as conveying sensory impulses from many of the taste buds on the tongue.

Chapter 18

1. The parasympathetic division of the ANS directs rest-and-digestive activities. The organs of the digestive system will have increased activity to digest food, absorb nutrients, and defecate wastes. In the relaxed condition, the body will also exhibit slower heart rate, and bronchoconstriction.

2. The supermom effect was due to activation of the sympathetic nervous system resulting in a fight-or-flight response. The heart rate, force of contraction, and blood pressure increased, blood flow to muscles increased, and glucose and ATP production increased. The release of the hormones epinephrine and norepinephrine also increased.

3. The goose bumps are a sympathetic nervous system response. The cell bodies of sympathetic preganglionic neurons are in the thoracolumbar (T1–L2) segments of the spinal cord, their axons exit in the anterior roots of spinal nerves, and extend out to a sympathetic ganglion. From there, postganglionic neurons extend to hair follicle smooth muscles (arrector pili), which produce goose bumps when they contract.

4. Dual innervation refers to the innervation of most organs by both the sympathetic and parasympathetic divisions of the autonomic nervous system, not to the presence of more than one head.

Chapter 19

1. Pain may persist even after a pain-producing stimulus is removed because pain-causing chemicals linger and because nociceptors exhibit very little adaptation.

2. Some of the eyedrops placed in the eye may pass through the nasolacrimal duct into the nasal cavity where olfactory receptors are stimulated. Because most "tastes" are actually smells, the child will "taste" the medicine from her eye.

3. Gertrude has presbyopia, a loss of lens elasticity, which makes it difficult to read. She may also be experiencing age-related loss of sharpness of vision and depth perception. Gertrude's hearing difficulties could be a result of damage to hair cells in the organ of Corti or degeneration of the nerve pathway for hearing. The "buzzing" Gertrude hears may be tinnitus, which also occurs more frequently in the elderly.

Chapter 20

1. Patrick has Type 1 diabetes mellitus due to the destruction of the pancreatic beta cells. He must have injections of insulin to metabolize glucose. His aunt has Type 2 diabetes mellitus. She still produces insulin but her body cells have decreased sensitivity to the hormone.

2. Melatonin is released by the pineal gland during darkness and sleep. Melatonin helps set the biological clock, controlled by the hypothalamus, which sets sleep patterns. SAD may be caused by excess melatonin. Bright light inhibits melatonin secretion and is a treatment for SAD.

3. Dehydration will stimulate the release of ADH from the posterior pituitary. ADH will increase water retention by the kidneys, decrease sweating and constrict arterioles which will raise BP. Epinephrine and norepinephrine will be released by the adrenal medullae in response to stress.

Chapter 21

1. The broad spectrum antibiotics may have destroyed the bacteria that caused Shilpa's bladder infection but also destroyed the naturally occurring large intestine bacteria that produce vitamin K. Vitamin K is required for the synthesis of four clotting factors. Without these clotting factors present in normal amounts, Shilpa will experience clotting problems until the intestinal bacteria reach normal levels and produce additional vitamin K.

2. Mrs. Brown's kidney failure is interfering with her ability to produce erythropoietin (EPO). Her physician can prescribe Epoetin alfa, a manufactured EPO, which is very effective in treating the decline in RBC production with kidney failure.

3. A primary problem Thomas may experience is with clotting. Clotting time becomes longer because the liver is responsible for producing many of the clotting factors and clotting proteins such as fibrinogen. Thrombopoietin, which stimulates the formation of platelets, is also produced in the liver. In addition, the liver is responsible for eliminating bilirubin, produced from the breakdown of RBCs. With a malfunctioning liver, the bilirubin will accumulate, resulting in jaundice. In addition, there can be decreased concentrations of the plasma protein albumin, which can affect blood pressure.

Chapter 22

1. A pacemaker sends electrical impulses into the right side of the heart that stimulate contraction of the heart muscle. A pacemaker is used in conditions in which the heart rhythm is irregular to take over the function of the SA node.

2. The sudden appearance of the car activated the sympathetic division of the nervous system. The sympathetic signals from the cardiovascular (CV) center in the medulla oblongata travel down the spinal cord to the cardiac accelerator nerves which release norepinephrine that increases heart rate and forcefulness of contraction.

3. SV = CO ÷ HR (beats/min). Assuming an average CO of 5250 mL/min at rest, SV = 5250 mL/min ÷ 40 beats/min = 131.25 mL. Rearranging the equation, CO = SV × HR (beats/min). With exercise, Christopher's CO = 131.25 mL × 60 beats/min = 7875 mL/min.

4. Aunt Frieda's heart murmur may indicate a faulty heart valve. The peripheral edema in her feet may be related to the valve disorder or may be a sign of congestive heart failure of the right ventricle.

Chapter 23

1. One effect of epinephrine is to cause vasoconstriction of arterioles. If these vessels were constricted temporarily, this would reduce the blood flow locally at the site of dental work and reduce bleeding.

2. Varicose veins are caused by weak venous valves that allow the backflow of blood. With aging, venous walls may lose their elasticity and become stretched and distended with blood. Arteries rarely become distended because they have thicker inner and middle layers than veins and do not contain valves.

3. Julie's sympathetic nervous system is responding to the stress of her bad day. Impulses travel from the cardiovascular (CV) center to the heart which increases its rate and force of contraction. Impulses from the CV center also travel to the blood vessel walls resulting in vasoconstriction, which increases BP. Increased levels of epinephrine and NE due to stress amplify these effects.

4. Peter cut an artery. Blood flows from arteries in rapid spurts due to the high pressure generated by ventricular contraction.

Chapter 24

1. The right axillary lymph nodes were probably removed because lymph flowing in lymphatics away from the tumor (breast lump) was filtered by the axillary lymph nodes. Cancerous cells from the tumor may be carried in the lymph to the axillary node and spread the cancer by metastasis.

2. The five tonsils are positioned near the oral cavity, nasal cavity and throat—the ideal location to intercept microbes and other foreign invaders that enter the body through the mouth or nose. Young children will be exposed to many infectious agents and need to develop immunity to these invaders.

3. Initial tetanus immunization provided artificially acquired active immunity. The booster dose is needed to maintain immunity against the tetanus bacteria and toxin.

4. Without a blood supply, antibodies and T cells do not have easy access to the cornea. Therefore no immune response occurs to reject the transplanted foreign cornea.

Chapter 25

1. Holding the breath will cause blood levels of CO_2 and H^+ to increase and O_2 to decrease. These changes will strongly stimulate the inspiratory area, which will send impulses to resume breathing, whether Levi is still conscious or not.

2. Exercise normally induces the sympathetic nervous system to send signals to dilate the bronchioles, which increases air flow and oxygen supply. Asthma causes constriction of the bronchioles, making inhalation more difficult and reducing the airflow.

3. Brianna has a viral infection—the common cold (coryza). Laryngitis, an inflammation of the larynx, is a common complication from a cold. Inflammation of the true vocal cords can result in loss of voice.

4. The tour group members have altitude sickness. At high elevation, the partial pressure of oxygen is not sufficient for unadapted individuals to maintain an adequate level of oxygen in their blood.

Chapter 26

1. Antonio experienced gastroesophageal reflux. The stomach contents backed up (refluxed) into Antonio's esophagus due to a failure of the lower esophageal sphincter to fully close. The HCl from the stomach irritated the esophageal wall, which resulted in the burning sensation he felt; this is commonly known as "heartburn," even though it is not related to the heart. Antonio's recent meal worsened the problem. Alcohol and smoking both can cause the sphincter to relax, while certain foods such tomatoes, chocolate, and coffee can stimulate stomach acid secretion. In addition, lying down immediately after a meal can exacerbate the problem.

2. The pyloric sphincter is located at this junction. The spider traveled from the mouth through the oro- and laryngopharynx, then the esophagus and finally entered the stomach.

3. The repetitive vomiting had emptied Jerry's stomach of hot dog and gastric juice and now he was expelling fluid containing bile. Bile pigments such as bilirubin give bile its color. Bile is produced by the liver.

Chapter 27

1. The recommended percentages of Calories are: total fats = less than 30%, saturated fats = 10%, total carbohydrates = 50–60%, simple sugars = less than 15%, proteins = 12–15%. Total Calories for active women = 2000. The quarter pounder meal leaves 1115 Calories for the day. The grilled chicken leaves 1608 Calories.

2. Body temperature will increase due to radiation from the sun and hot surrounding sand, and possibly conduction from lying on the hot sand. Heat will be lost to the water by conduction and convection.

3. BMR means basal metabolic rate. BMR is the measurement of the rate at which heat is produced, i.e., the metabolic rate, under standard conditions is designed to be as close as possible to the basal state. Factors that would affect metabolic rate and should be avoided while measuring the BMR include exercise, stress, increased temperature, eating (all increase), and sleep (lower).

4. Antioxidants protect against free radical damage to cell membranes, DNA, and blood vessel walls. Vitamin C is water-soluble so excess quantities will be excreted in the urine. Vitamins A (from the beta-carotene) and E are lipid-soluble and might accumulate to toxic levels in tissues such as the liver.

Chapter 28

1. Alcohol inhibits the secretion of antidiuretic hormone (ADH). ADH is secreted when the hypothalamus detects a decrease in the amount of water in the blood. ADH makes the collecting ducts and distal portion of the distal convoluted tubules more permeable to water so that it can be reabsorbed.

2. Incontinence (lack of involuntary control over micturition) is normal in children of Sarah's age. Neurons to the external urethral sphincter are not completely developed until after about 2-3 years of age. The desire to control micturition voluntarily must also be present and initiated by the cerebral cortex.

3. No. Glomerular filtration is mainly driven by blood pressure and opposed by glomerular capsular pressure, not pressure from urine in the bladder. Under normal physiological conditions, urine remains in the bladder and does not back up into the kidneys.

Chapter 29

1. A complete hysterectomy is the surgical removal of the body and cervix of the uterus. The ovaries produce estrogens and progesterone, not the uterus, so menopause will not result since these organs will still be left in place.

2. Vasectomy cuts the ductus deferens so that sperm cannot be transported out of the body. The function of the testes is not affected. The Leydig cells secrete the hormone testosterone which maintains male sex characteristics and sex drive. Vasectomy will not affect production of the hormone or its transport to the rest of the body via the blood.

3. Sperm production is optimal at a temperature slightly below normal body temperature. The higher temperature that results from Julio wearing very close-fitting briefs and soaking in the hot tub inhibits sperm production and survival and therefore fertility.

GLOSSARY

Pronunciation Key

1. The most strongly accented syllable appears in capital letters, for example, bilateral (bī-LAT-er-al) and diagnosis (dī-ag-NŌ-sis).

2. If there is a secondary accent, it is noted by a prime (′), for example, constitution (kon′-sti-TOO-shun) and physiology (fiz′-ē-OL-ō-jē). Any additional secondary accents are also noted by a prime, for example, decarboxylation (dē′-kar-bok′-si-LĀ-shun).

3. Vowels marked by a line above the letter are pronounced with the long sound, as in the following common words:

ā as in *māke* | *ō* as in *pōle*
ē as in *bē* | *ū* as in *cūte*
ī as in *īvy*

4. Vowels not marked by a line above the letter are pronounced with the short sound, as in the following words

a as in *above* or *at* | *o* as in *not*
e as in *bet* | *u* as in *bud*
i as in *sip*

5. Other vowel sounds are indicated as follows:

oy as in *oil*
oo as in *root*

6. Consonant sounds are pronounced as in the following words:

b as in *bat* | *m* as in *mother*
ch as in *chair* | *n* as in *no*
d as in *dog* | *p* as in *pick*
f as in *father* | *r* as in *rib*
g as in *get* | *s* as in *so*
h as in *hat* | *t* as in *tea*
j as in *jump* | *v* as in *very*
k as in *can* | *w* as in *welcome*
ks as in *tax* | *z* as in *zero*
kw as in *quit* | *zh* as in *lesion*
l as in *let*

A

Abdomen (ab-DŌ-men *or* AB-dō-men) The area between the diaphragm and pelvis.

Abdominal (ab-DŌM-i-nal) **cavity** Superior portion of the abdominopelvic cavity that contains the stomach, spleen, liver, gallbladder, most of the small intestine, and part of the large intestine.

Abdominal thrust maneuver A first-aid procedure for choking. Employs a quick, upward thrust against the diaphragm that forces air out of the lungs with sufficient force to eject any lodged material.

Abdominopelvic (ab-dom′-i-nō-PEL-vik) **cavity** Inferior to the diaphragm; is subdivided into a superior abdominal cavity and an inferior pelvic cavity.

Abduction (ab-DUK-shun) Movement away from the midline of the body.

Abortion (a-BOR-shun) The premature loss (spontaneous) or removal (induced) of the embryo or nonviable fetus; miscarriage due to a failure in the normal process of developing or maturing.

Abscess (AB-ses) A localized collection of pus and liquefied tissue in a cavity.

Absorption (ab-SORP-shun) Intake of fluids or other substances by cells of the skin or mucous membranes; the passage of digested foods from the gastrointestinal tract into blood or lymph.

Accessory duct A duct of the pancreas that empties into the duodenum about 2.5 cm (1 in.) superior to the ampulla of Vater (hepatopancreatic ampulla). Also called the **duct of Santorini** (san′-tō-RĒ-nē).

Acetabulum (as′-e-TAB-ū-lum) The rounded cavity on the external surface of the hip bone that receives the head of the femur.

Acetylcholine (as′-ē-til-KŌ-lēn) **(ACh)** A neurotransmitter liberated by many peripheral nervous system neurons and some central nervous system neurons. It is excitatory at neuromuscular junctions but inhibitory at some other synapses (for example, it slows heart rate).

Achalasia (ak′-a-LĀ-zē-a) A condition, caused by malfunction of the myenteric plexus, in which the lower esophageal sphincter fails to relax normally as food approaches. A whole meal may become lodged in the esophagus and enter the stomach very slowly. Distension of the esophagus results in chest pain that is often confused with pain originating from the heart.

Achilles tendon *See* **Calcaneal tendon.**

Acini (AS-i-nē) Groups of cells in the pancreas that secrete digestive enzymes.

Acoustic (a-KOOS-tik) Pertaining to sound or the sense of hearing.

Acquired immunodeficiency syndrome (AIDS) A fatal disease caused by the human immunodeficiency virus (HIV). Characterized by a positive HIV-antibody test, low helper T-cell count, and certain indicator diseases (for example Kaposi's sarcoma, pneumocystis carinii pneumonia, tuberculosis, fungal diseases). Other symptoms include fever or night sweats, coughing, sore throat, fatigue, body aches, weight loss, and enlarged lymph nodes.

Acrosome (AK-rō-sōm) A lysosomelike organelle in the head of a sperm cell containing enzymes that facilitate the penetration of a sperm cell into a secondary oocyte.

Actin (AK-tin) A contractile protein that is part of thin filaments in muscle fibers.

Action potential (AP) An electrical signal that propagates along the membrane of a neuron or muscle fiber (cell); a rapid change in membrane potential that involves a depolarization followed by a repolarization. Also called a **nerve action potential** or **nerve impulse** as it relates to a neuron, and a **muscle action potential** as it relates to a muscle fiber.

Activation (ak′-ti-VĀ-shun) **energy** The minimum amount of energy required for a chemical reaction to occur.

Active transport The movement of substances across cell membranes against a concentration gradient, requiring the expenditure of cellular energy (ATP).

Acute (a-KŪT) Having rapid onset, severe symptoms, and a short course; not chronic.

Adaptation (ad′-ap-TĀ-shun) The adjustment of the pupil of the eye to changes in light intensity. The property by which a sensory neuron relays a decreased frequency of action potentials from a receptor, even though the strength of the stimulus remains constant; the decrease in perception of a sensation over time while the stimulus is still present.

Adduction (ad-DUK-shun) Movement toward the midline of the body.

Adenoid (AD-e-noyd) The pharyngeal tonsil.

Adenosine triphosphate (a-DEN-ō-sēn trī-FOS-fāt) **(ATP)** The main energy currency in living cells; used to transfer the chemical energy needed for metabolic reactions. ATP consists of the purine base *adenine* and the five-carbon sugar *ribose,* to which are added, in linear array, three *phosphate* groups.

Adhesion (ad-HĒ-zhun) Abnormal joining of parts to each other.

Adipocyte (AD-i-pō-sīt) Fat cell, derived from a fibroblast.

Adipose (AD-i-pōz) **tissue** Tissue composed of adipocytes specialized for triglyceride storage and present in the form of soft pads between various organs for support, protection, and insulation.

Adrenal cortex (a-DRĒ-nal KOR-teks) The outer portion of an adrenal gland, divided into three zones; the zona glomerulosa secretes mineralocorticoids, the zona fasciculata secretes glucocorticoids, and the zona reticularis secretes androgens.

Adrenal glands Two glands located superior to each kidney. Also called the **suprarenal** (soo′-pra-RĒ-nal) **glands.**

Adrenal medulla (me-DŪL-a) The inner part of an adrenal gland, consisting of cells that secrete epinephrine, norepinephrine, and a small amount of dopamine in response to stimulation by sympathetic preganglionic neurons.

Adrenergic (ad′-ren-ER-jik) **neuron** A neuron that releases epinephrine (adrenaline) or norepinephrine (noradrenaline) as its neurotransmitter.

Adrenocorticotropic (ad-rē′-nō-kor-ti-kō-TRŌP-ik) **hormone (ACTH)** A hormone produced by the anterior pituitary that influences the production and secretion of certain hormones of the adrenal cortex.

Adventitia (ad-ven-TISH-a) The outermost connective tissue covering of a structure or organ not covered by a serous coat.

Aerobic (air-Ō-bik) Requiring molecular oxygen.

Afferent arteriole (AF-er-ent ar-TĒ-rē-ōl) A blood vessel of a kidney that divides into the capillary network called a glomerulus; there is one afferent arteriole for each glomerulus.

Agglutination (a-gloo-ti-NĀ-shun) Clumping of microorganisms or blood cells, typically due to an antigen–antibody reaction.

Aggregated lymphatic follicles (AG-re-gā-ted lim-FAT-ik FOL-i-kas) Clusters of lymph nodules that are most numerous in the ileum. Also called **Peyer's** (PĪ-erz) **patches.**

Albinism (AL-bin-izm) Abnormal, nonpathological, partial, or total absence of pigment in skin, hair, and eyes.

Aldosterone (al-DOS-ter-ōn) A mineralocorticoid produced by the adrenal cortex that promotes sodium and water reabsorption by the kidneys and potassium excretion in urine.

Allantois (a-LAN-tō-is) A small, vascularized outpouching of the yolk sac that serves as an early site for blood formation and development of the urinary bladder.

Allergen (AL-er-jen) An antigen that evokes a hypersensitivity reaction.

Alopecia (al′-ō-PĒ-shē-a) The partial or complete lack of hair as a result of factors such as genetics, aging, endocrine disorders, chemotherapy, and skin diseases.

Alpha (AL-fa) **cell** A type of cell in the pancreatic islets (islets of Langerhans) in the pancreas that secretes the hormone glucagon. Also termed an **A cell.**

Alpha receptor A type of receptor for norepinephrine and epinephrine; present on visceral effectors innervated by sympathetic postganglionic neurons.

Alveolar (al-VĒ-ō-lar) **duct** Branch of a respiratory bronchiole around which alveoli and alveolar sacs are arranged.

Alveolar macrophage (MAK-rō-fāj) Highly phagocytic cell found in the alveolar walls of the lungs. Also called a **dust cell.**

Alveolar sac A cluster of alveoli that share a common opening.

Alveolus (al-VĒ-ō-lus) A small hollow or cavity; an air sac in the lungs; milk-secreting portion of a mammary gland. *Plural* is **alveoli** (al-VĒ-ōl-ī).

Alzheimer (ALTZ-hī-mer) **disease (AD)** Disabling neurological disorder characterized by dysfunction and death of specific cerebral neurons, resulting in widespread intellectual impairment, personality changes, and fluctuations in alertness.

Amenorrhea (ā-men-ō-RĒ-a) Absence of menstruation.

Amnesia (am-NĒ-zē-a) A lack or loss of memory.

Amnion (AM-nē-on) A thin, protective fetal membrane that develops from the epiblast; holds the fetus suspended in amniotic fluid. Also called the "**bag of waters.**"

Amniotic (am′-nē-OT-ik) **fluid** Fluid in the amniotic cavity, the space between the developing embryo (or fetus) and amnion; the fluid is initially produced as a filtrate from maternal blood and later includes fetal urine. It functions as a shock absorber, helps regulate fetal body temperature, and helps prevent desiccation.

Amphiarthrosis (am′-fē-ar-THRŌ-sis) A slightly movable joint, in which the articulating bony surfaces are separated by fibrous connective tissue or fibrocartilage to which both are attached; types are syndesmosis and symphysis.

Ampulla (am-PUL-la) A saclike dilation of a canal or duct.

Ampulla of Vater *See* **Hepatopancreatic ampulla.**

Anabolism (a-NAB-ō-lizm) Synthetic, energy-requiring reactions whereby small molecules are built up into larger ones.

Anaerobic (an-ar-Ō-bik) Not requiring oxygen.

Anal (Ā-nal) **canal** The last 2 or 3 cm (1 in.) of the rectum; opens to the exterior through the anus.

Anal column A longitudinal fold in the mucous membrane of the anal canal that contains a network of arteries and veins.

Anal triangle The subdivision of the female or male perineum that contains the anus.

Analgesia (an-al-JĒ-zē-a) Pain relief; absence of the sensation of pain.

Anaphase (AN-a-fāz) The third stage of mitosis in which the chromatids that have separated at the centromeres move to opposite poles of the cell.

Anaphylaxis (an′-a-fi-LAK-sis) A hypersensitivity (allergic) reaction in which IgE antibodies attach to mast cells and basophils, causing them to produce mediators of anaphylaxis (histamine, leukotrienes, kinins, and prostaglandins) that bring about increased blood permeability, increased smooth muscle contraction, and increased mucus production. Examples are hay fever, hives, and anaphylactic shock.

Anastomosis (a-nas-tō-MŌ-sis) An end-to-end union or joining of blood vessels, lymphatic vessels, or nerves.

Anatomic dead space Spaces of the nose, pharynx, larynx, trachea, bronchi, and bronchioles totaling about 150 mL of the 500 mL in a quiet breath (tidal volume); air in the anatomic dead space does not reach the alveoli to participate in gas exchange.

Anatomical (an′-a-TOM-i-kal) **position** A position of the body universally used in anatomical descriptions in which the body is erect, the head is level, the eyes face forward, the upper limbs are at the sides, the palms face forward, and the feet are flat on the floor.

Anatomy (a-NAT-ō-mē) The structure or study of the structure of the body and the relation of its parts to each other.

Androgens (AN-drō-jenz) Masculinizing sex hormones produced by the testes in males and the adrenal cortex in both sexes; also responsible for libido (sexual desire); the two main androgens are testosterone and dihydrotestosterone.

Anemia (a-NĒ-mē-a) Condition of the blood in which the number of functional red blood cells or their hemoglobin content is below normal.

Anesthesia (an′-es-THĒ-zē-a) A total or partial loss of feeling or sensation; may be general or local.

Aneurysm (AN-ū-rizm) A saclike enlargement of a blood vessel caused by a weakening of its wall.

Angina pectoris (an-JĪ-na *or* AN-ji-na PEK-tō-ris) A pain in the chest, upper limb, jaw, or shoulder related to reduced coronary circulation due to coronary artery disease (CAD) or spasms of vascular smooth muscle in coronary arteries.

Angiogenesis (an′-jē-ō-JEN-e-sis) The formation of blood vessels in the extraembryonic mesoderm of the yolk sac, connecting stalk, and chorion at the beginning of the third week of development.

Ankylosis (ang′-ki-LŌ-sis) Severe or complete loss of movement at a joint as the result of a disease process.

Antagonist (an-TAG-ō-nist) A muscle that has an action opposite that of the prime mover (agonist) and yields to the movement of the prime mover.

Antagonistic (an-tag-ō-NIST-ik) **effect** A hormonal interaction in which the effect of one hormone on a target cell is opposed by another hormone. For example, calcitonin (CT) lowers blood calcium level, whereas parathyroid hormone (PTH) raises it.

Anterior (an-TĒR-ē-or) Nearer to or at the front of the body. Equivalent to **ventral** in bipeds.

Anterior pituitary (pi-TOO-i-tār-ē) Anterior lobe of the pituitary gland. Also called the **adenohypophysis** (ad′-e-nō-hī-POF-i-sis).

Anterior root The structure composed of axons of motor (efferent) neurons that emerges from the anterior aspect of the spinal cord and extends laterally to join a posterior root, forming a spinal nerve. Also called a **ventral root.**

Anterolateral (an′-ter-ō-LAT-er-al) **pathway** Sensory pathway that conveys information related to pain, temperature, tickle, and itch.

Antibody (AN-ti-bod′-ē) A protein produced by plasma cells in response to a specific antigen; the antibody combines with that antigen to neutralize, inhibit, or destroy it. Also called an **immunoglobulin** (im-ū-nō-GLOB-ū-lin) or **Ig.**

Anticoagulant (an-tī-cō-AG-ū-lant) A substance that can delay, suppress, or prevent the clotting of blood.

Antidiuretic (an′-ti-dī-ū-RET-ik) Substance that inhibits urine formation.

Antidiuretic hormone (ADH) Hormone produced by neurosecretory cells in the paraventricular and supraoptic nuclei of the hypothalamus that stimulates water reabsorption from kidney tubule cells into the blood and vasoconstriction of arterioles. Also called **vasopressin** (vāz-ō-PRES-in).

Antigen (AN-ti-jen) A substance that has immunogenicity (the ability to provoke an immune response) and reactivity (the ability to react with the antibodies or cells that result from the immune response); contraction of *anti*body *gen*erator. Also termed a **complete antigen.**

Antigen-presenting cell (APC) Special class of migratory cell that processes and presents antigens to T cells during an immune response; APCs include macrophages, B cells, and dendritic cells, which are present in the skin, mucous membranes, and lymph nodes.

Antrum (AN-trum) Any nearly closed cavity or chamber, especially one within a bone, such as a sinus.

Anuria (an-Ū-rē-a) Absence of urine formation or daily urine output of less than 50 mL.

Anus (Ā-nus) The distal end and outlet of the rectum.

Aorta (ā-ŌR-ta) The main systemic trunk of the arterial system of the body that emerges from the left ventricle.

Aortic (ā-ŌR-tik) **body** Cluster of chemoreceptors on or near the arch of the aorta that respond to changes in blood levels of oxygen, carbon dioxide, and hydrogen ions (H^+).

Aortic reflex A reflex that helps maintain normal systemic blood pressure; initiated by baroreceptors in the wall of the ascending aorta and arch of the aorta. Nerve impulses from aortic baroreceptors reach the cardiovascular center via sensory axons of the vagus (X) nerves.

Apex (Ā-peks) The pointed end of a conical structure, such as the apex of the heart.

Aphasia (a-FĀ-zē-a) Loss of ability to express oneself properly through speech or loss of verbal comprehension.

Apnea (AP-nē-a) Temporary cessation of breathing.

Apneustic (ap-NOO-stik) **area** A part of the respiratory center in the pons that sends stimulatory nerve impulses to the inspiratory area that activate and prolong inhalation and inhibit exhalation.

Apocrine (AP-ō-krin) **gland** A type of gland in which the secretory products gather at the free end of the secreting cell and pinch off, along with some of the cytoplasm, to become the secretion, as in mammary glands.

Aponeurosis (ap′-ō-noo-RŌ-sis) A sheetlike tendon joining one muscle with another or with bone.

Apoptosis (ap-ō-TŌ-sis *or* ap′-ōp-TŌ-sis) Programmed cell death; a normal type of cell death that removes unneeded cells during embryological development, regulates the number of cells in tissues, and eliminates many potentially dangerous cells such as cancer cells. During apoptosis, the DNA fragments, the nucleus condenses, mitochondria cease to function, and the cytoplasm shrinks, but the plasma membrane remains intact. Phagocytes engulf and digest the apoptotic cells, and an inflammatory response does not occur.

Appositional (a-pō-ZISH-o-nal) **growth** Growth due to surface deposition of material, as in the growth in diameter of cartilage and bone. Also called **exogenous** (eks-OJ-e-nus) **growth.**

Aqueous humor (AK-wē-us HŪ-mer) The watery fluid, similar in composition to cerebrospinal fluid, that fills the anterior cavity of the eye.

Arachnoid mater (a-RAK-noyd MĀ-ter) The middle of the three meninges (coverings) of the brain and spinal cord. Also termed the **arachnoid.**

Arachnoid villus (VIL-us) Berrylike tuft of the arachnoid mater that protrudes into the superior sagittal sinus and through which cerebrospinal fluid is reabsorbed into the bloodstream.

Arbor vitae (AR-bor VĪ-tē) The white matter tracts of the cerebellum, which have a treelike appearance when seen in midsagittal section.

Arch of the aorta The most superior portion of the aorta, lying between the ascending and descending segments of the aorta.

Areola (a-RĒ-ō-la) Any tiny space in a tissue. The pigmented ring around the nipple of the breast.

Arm The part of the upper limb from the shoulder to the elbow.

Arousal (a-ROW-zal) Awakening from sleep, a response due to stimulation of the reticular activating system (RAS).

Arrector pili (a-REK-tor PĪ-lē) Smooth muscles attached to hairs; contraction pulls the hairs into a vertical position, resulting in "goose bumps."

Arrhythmia (a-RITH-mē-a) An irregular heart rhythm. Also called a **dysrhythmia.**

Arteriole (ar-TĒ-rē-ōl) A small, almost microscopic, artery that delivers blood to a capillary.

Arteriosclerosis (ar-tē-rē-ō-skle-RŌ-sis) Group of diseases characterized by thickening of the walls of arteries and loss of elasticity.

Artery (AR-ter-ē) A blood vessel that carries blood away from the heart.

Arthritis (ar-THRĪ-tis) Inflammation of a joint.

Arthrology (ar-THROL-ō-jē) The study or description of joints.

Arthroplasty (AR-thrō-plas′-tē) Surgical replacement of joints, for example, the hip and knee joints.

Arthroscopy (ar-THROS-kō-pē) A procedure for examining the interior of a joint, usually the knee, by inserting an arthroscope into a small incision; used to determine extent of damage, remove torn cartilage, repair cruciate ligaments, and obtain samples for analysis.

Arthrosis (ar-THRŌ-sis) A joint or articulation.

Articular (ar-TIK-ū-lar) **capsule** Sleevelike structure around a synovial joint composed of a fibrous capsule and a synovial membrane.

Articular cartilage (KAR-ti-lij) Hyaline cartilage attached to articular bone surfaces.

Articular disc Fibrocartilage pad between articular surfaces of bones of some synovial joints. Also called a **meniscus** (men-IS-kus).

Articulation (ar-tik-ū-LĀ-shun) A joint; a point of contact between bones, cartilage and bones, or teeth and bones.

Arytenoid (ar′-i-TĒ-noyd) **cartilages** A pair of small, pyramidal cartilages of the larynx that attach to the vocal folds and intrinsic pharyngeal muscles and can move the vocal folds.

Ascending colon (KŌ-lon) The part of the large intestine that passes superiorly from the cecum to the inferior border of the liver, where it bends at the right colic (hepatic) flexure to become the transverse colon.

Ascites (as-SĪ-tēz) Abnormal accumulation of serous fluid in the peritoneal cavity.

Association areas Large cortical regions on the lateral surfaces of the occipital, parietal, and temporal lobes and on the frontal lobes anterior to the motor areas, connected by many motor and sensory axons to other parts of the cortex. The association areas are concerned with motor patterns, memory, concepts of word-hearing and word-seeing, reasoning, will, judgment, and personality traits.

Asthma (AZ-ma) Usually allergic reaction characterized by smooth muscle spasms in bronchi resulting in wheezing and difficult breathing. Also called **bronchial asthma.**

Astigmatism (a-STIG-ma-tizm) An irregularity of the lens or cornea of the eye causing the image to be out of focus and producing faulty vision.

Astrocyte (AS-trō-sīt) A neuroglial cell having a star shape that participates in brain development and the metabolism of neurotransmitters, helps form the blood–brain barrier, helps maintain the proper balance of K^+ for generation of nerve impulses, and provides a link between neurons and blood vessels.

Ataxia (a-TAK-sē-a) A lack of muscular coordination; lack of precision.

Atherosclerotic plaque (ath′-er-ō-skle-RO-tic PLAK) A lesion that results from accumulated cholesterol and smooth muscle fibers (cells) of the tunica media of an artery; may become obstructive.

Atom Unit of matter that makes up a chemical element; consists of a nucleus (containing positively charged protons and uncharged neutrons) and negatively charged electrons that orbit the nucleus.

Atresia (a-TRĒ-zē-a) Degeneration and reabsorption of an ovarian follicle before it fully matures and ruptures; abnormal closure of a passage, or absence of a normal body opening.

Atrial fibrillation (Ā-trē-al fib-ri-LĀ-shun) Asynchronous contraction of cardiac muscle fibers in the atria that results in the cessation of atrial pumping.

Atrial natriuretic (na′-trē-ū-RET-ik) **peptide (ANP)** Peptide hormone, produced by the atria of the heart in response to stretching, that inhibits aldosterone production and thus lowers blood pressure; causes natriuresis, increased urinary excretion of sodium.

Atrioventricular (AV) (ā′-trē-ō-ven-TRIK-ū-lar) **bundle** The part of the conduction system of the heart that begins at the atrioventricular (AV) node, passes through the cardiac skeleton separating the atria and the ventricles, then extends a short distance down the interventricular septum before splitting into right and left bundle branches. Also called the **bundle of His** (HISS).

Atrioventricular (AV) node The part of the conduction system of the heart made up of a compact mass of conducting cells located in the septum between the two atria.

Atrioventricular (AV) valve A heart valve made up of membranous flaps or cusps that allows blood to flow in one direction only, from an atrium into a ventricle.

Atrium (Ā-trē-um) A superior chamber of the heart.

Atrophy (AT-rō-fē) Wasting away or decrease in size of a part, due to a failure, abnormality of nutrition, or lack of use.

Auditory ossicle (AW-di-tō-rē OS-si-kul) One of the three small bones of the middle ear called the malleus, incus, and stapes.

Auditory tube The tube that connects the middle ear with the nose and nasopharynx region of the throat. Also called the **eustachian** (ū-STĀ-shun *or* ū-STĀ-kē-an) **tube** or **pharyngotympanic tube.**

Auscultation (aws-kul-TĀ-shun) Examination by listening to sounds in the body.

Autoimmunity An immunological response against a person's own tissues.

Autolysis (aw-TOL-i-sis) Self-destruction of cells by their own lysosomal digestive enzymes after death or in a pathological process.

Autonomic ganglion (aw′-tō-NOM-ik GANG-lē-on) A cluster of cell bodies of sympathetic or parasympathetic neurons located outside the central nervous system.

Autonomic nervous system (ANS) Visceral sensory (afferent) and visceral motor (efferent) neurons. Autonomic motor neurons, both sympathetic and parasympathetic, conduct nerve impulses from the central nervous system to smooth muscle, cardiac muscle, and glands. So named because this part of the nervous system was thought to be self-governing or spontaneous.

Autonomic plexus (PLEK-sus) A network of sympathetic and parasympathetic axons; examples are the cardiac, celiac, and pelvic plexuses, which are located in the thorax, abdomen, and pelvis, respectively.

Autophagy (aw-TOF-a-jē) Process by which worn-out organelles are digested within lysosomes.

Autopsy (AW-top-sē) The examination of the body after death.

Autorhythmic cells (aw-tō-RITH-mik) Cardiac or smooth muscle fibers that are self-excitable (generate impulses without an external stimulus); act as the heart's pacemaker and conduct the pacing impulse through the conduction system of the heart; self-excitable neurons in the central nervous system, as in the inspiratory area of the brain stem.

Axilla (ak-SIL-a) The small hollow beneath the arm where it joins the body at the shoulders. Also called the **armpit.**

Axon (AK-son) The usually single, long process of a nerve cell that propagates a nerve impulse toward the axon terminals.

Axon terminal Terminal branch of an axon where synaptic vesicles undergo exocytosis to release neurotransmitter molecules.

B

B cell A lymphocyte that can develop into a clone of antibody-producing plasma cells or memory cells when properly stimulated by a specific antigen.

Babinski (ba-BIN-skē) **sign** Extension of the great toe, with or without fanning of the other toes, in response to stimulation of the outer margin of the sole; normal up to 18 months of age and indicative of damage to descending motor pathways such as the corticospinal tracts after that.

Back The posterior part of the body; the dorsum.

Ball-and-socket joint A synovial joint in which the rounded surface of one bone moves within a cup-shaped depression or socket of another bone, as in the shoulder or hip joint. Also called a **spheroid** (SFĒ-royd) **joint.**

Baroreceptor (bar′-ō-re-SEP-tor) Neuron capable of responding to changes in blood, air, or fluid pressure. Also called a **pressoreceptor.**

Basal ganglia (GANG-glē-a) Paired clusters of gray matter deep in each cerebral hemisphere including the globus pallidus, putamen, and caudate nucleus. Together, the caudate nucleus and putamen are known as the **corpus striatum.** Nearby structures that are functionally linked to the basal ganglia are the substantia nigra of the midbrain and the subthalamic nuclei of the diencephalon.

Basement membrane Thin, extracellular layer between epithelium and connective tissue consisting of a basal lamina and a reticular lamina.

Basilar (BĀS-i-lar) **membrane** A membrane in the cochlea of the internal ear that separates the cochlear duct from the scala tympani and on which the spiral organ (organ of Corti) rests.

Basophil (BĀ-sō-fil) A type of white blood cell characterized by a pale nucleus and large granules that stain blue-purple with basic dyes.

Belly The abdomen. The gaster or prominent, fleshy part of a skeletal muscle.

Beta (BĀ-ta) **cell** A type of cell in the pancreatic islets (islets of Langerhans) in the pancreas that secretes the hormone insulin. Also called a **B cell.**

Beta receptor A type of adrenergic receptor for epinephrine and norepinephrine; found on visceral effectors innervated by sympathetic postganglionic neurons.

Bicuspid (bī-KUS-pid) **valve** Atrioventricular (AV) valve on the left side of the heart. Also called the **mitral valve.**

Bilateral (bī-LAT-er-al) Pertaining to two sides of the body.

Bile (BĪL) A secretion of the liver consisting of water, bile salts, bile pigments, cholesterol, lecithin, and several ions; it emulsifies lipids prior to their digestion.

Bilirubin (bil-ē-ROO-bin) An orange pigment that is one of the end products of hemoglobin breakdown in the hepatocytes and is excreted as a waste material in bile.

Blind spot Area in the retina at the end of the optic (II) nerve in which there are no photoreceptors.

Blood The fluid that circulates through the heart, arteries, capillaries, and veins and that constitutes the chief means of transport within the body.

Blood–brain barrier (BBB) A barrier consisting of specialized brain capillaries and astrocytes that prevents the passage of materials from the blood to the cerebrospinal fluid and brain.

Blood island Isolated mass of mesoderm derived from angioblasts and from which blood vessels develop.

Blood pressure (BP) Force exerted by blood against the walls of blood vessels due to contraction of the heart and influenced by the elasticity of the vessel walls; clinically, a measure of the pressure in arteries during ventricular systole and ventricular diastole.

Blood reservoir (REZ-er-vwar) Systemic veins and venules that contain large amounts of blood that can be moved quickly to parts of the body requiring the blood.

Blood–testis barrier (BTB) A barrier formed by Sertoli cells that prevents an immune response against antigens produced by spermatogenic cells by isolating the cells from the blood.

Body cavity A space within the body that contains various internal organs.

Bolus (BŌ-lus) A soft, rounded mass, usually food, that is swallowed.

Bony labyrinth (LAB-i-rinth) A series of cavities within the petrous portion of the temporal bone forming the vestibule, cochlea, and semicircular canals of the inner ear.

Bowman's capsule *See* **Glomerular capsule.**

Brachial plexus (BRĀ-kē-al PLEK-sus) A network of nerve axons of the ventral rami of spinal nerves C5, C6, C7, C8, and T1. The nerves that emerge from the brachial plexus supply the upper limb.

Bradycardia (brād′-i-KAR-dē-a) A slow resting heart or pulse rate (under 50 beats per minute).

Brain The part of the central nervous system contained within the cranial cavity.

Brain stem The portion of the brain immediately superior to the spinal cord, made up of the medulla oblongata, pons, and midbrain.

Brain waves Electrical signals that can be recorded from the skin of the head due to electrical activity of brain neurons.

Broad ligament A double fold of parietal peritoneum attaching the uterus to the side of the pelvic cavity.

Broca's (BRŌ-kaz) **area** Motor area of the brain in the frontal lobe that translates thoughts into speech. Also called the **motor speech area.**

Bronchi (BRONG-kī) Branches of the respiratory passageway including primary bronchi (the two divisions of the trachea), secondary or lobar bronchi (divisions of the primary bronchi that are distributed to the lobes of the lung), and tertiary or segmental bronchi (divisions of the secondary bronchi that are distributed to bronchopulmonary segments of the lung). *Singular* is **bronchus.**

Bronchial tree The trachea, bronchi, and their branching structures up to and including the terminal bronchioles.

Bronchiole (BRONG-kē-ōl) Branch of a tertiary bronchus further dividing into terminal bronchioles (distributed to lobules of the lung), which divide into respiratory bronchioles (distributed to alveolar sacs).

Bronchitis (brong-KĪ-tis) Inflammation of the mucous membrane of the bronchial tree; characterized by hypertrophy and hyperplasia of seromucous glands and goblet cells that line the bronchi which results in a productive cough.

Bronchopulmonary (brong′-kō-PUL-mō-ner-ē) **segment** One of the smaller divisions of a lobe of a lung supplied by its own branches of a bronchus.

Brunner's gland *See* **Duodenal gland.**

Buccal (BUK-al) Pertaining to the cheek or mouth.

Bulb of penis Expanded portion of the base of the corpus spongiosum penis.

Bulbourethral (bul′-bō-ū-RĒ-thral) **gland** One of a pair of glands located inferior to the prostate on either side of the urethra that secretes an alkaline fluid into the cavernous urethra. Also called a **Cowper's** (KOW-perz) **gland.**

Bulimia (boo-LIM-ē-a *or* boo-LĒ-mē-a) A disorder characterized by overeating at least twice a week followed by purging by self-induced vomiting, strict dieting or fasting, vigorous exercise, or use of laxatives or diuretics. Also called **binge–purge syndrome.**

Bulk-phase endocytosis A process by which most body cells can ingest membrane-surrounded droplets of interstitial fluid. Also called **pinocytosis.**

Bundle branch One of the two branches of the atrioventricular (AV) bundle made up of specialized muscle fibers (cells) that transmit electrical impulses to the ventricles.

Bundle of His *See* **Atrioventricular (AV) bundle.**

Bursa (BUR-sa) A sac or pouch of synovial fluid located at friction points, especially about joints.

Bursitis (bur-SĪ-tis) Inflammation of a bursa.

Buttocks (BUT-oks) The two fleshy masses on the posterior aspect of the inferior trunk, formed by the gluteal muscles.

C

Calcaneal (kal-KĀ-nē-al) **tendon** The tendon of the soleus, gastrocnemius, and plantaris muscles at the back of the heel. Also called the **Achilles** (a-KIL-ēz) **tendon.**

Calcification (kal′-si-fi-KĀ-shun) Deposition of mineral salts, primarily hydroxyapatite, in a framework formed by collagen fibers in which the tissue hardens. Also called **mineralization** (min′-e-ral-i-ZĀ-shun).

Calcitonin (kal-si-TŌ-nin) **(CT)** A hormone produced by the parafollicular cells of the thyroid gland that can lower the amount of blood calcium and phosphates by inhibiting bone resorption (breakdown of bone extracellular matrix) and by accelerating uptake of calcium and phosphates into bone matrix.

Calculus (KAL-kū-lus) A stone, or insoluble mass of crystallized salts or other material, formed within the body, as in the gallbladder, kidney, or urinary bladder.

Callus (KAL-lus) A growth of new bone tissue in and around a fractured area, ultimately replaced by mature bone. An acquired, localized thickening.

Calyx (KĀL-iks) Any cuplike division of the kidney pelvis. *Plural* is **calyces** (KĀ-li-sēz).

Canal (ka-NAL) A narrow tube, channel, or passageway.

Canaliculus (kan′-a-LIK-ū-lus) A small channel or canal, as in bones, where they connect lacunae. *Plural* is **canaliculi** (kan′-a-LIK-ū-lī).

Canal of Schlemm *See* **Scleral venous sinus.**

Capacitation (ka′-pas-i-TĀ-shun) The functional changes that sperm undergo in the female reproductive tract that allow them to fertilize a secondary oocyte.

Capillary (KAP-i-lar′-ē) A microscopic blood vessel located between an arteriole and venule through which materials are exchanged between blood and interstitial fluid.

Carcinogen (car-SIN-ō-jen) A chemical substance or radiation that causes cancer.

Cardiac (KAR-dē-ak) **arrest** Cessation of an effective heartbeat in which the heart is completely stopped or in ventricular fibrillation.

Cardiac cycle A complete heartbeat consisting of systole (contraction) and diastole (relaxation) of both atria plus systole and diastole of both ventricles.

Cardiac muscle Striated muscle fibers (cells) that form the wall of the heart; stimulated by an intrinsic conduction system and regulated by autonomic motor neurons.

Cardiac notch An angular notch in the anterior border of the left lung into which part of the heart fits.

Cardinal ligament A ligament of the uterus, extending laterally from the cervix and vagina as a continuation of the broad ligament.

Cardiology (kar-dē-OL-ō-jē) The study of the heart and diseases associated with it.

Cardiovascular (kar-dē-ō-VAS-kū-lar) **center** Groups of neurons scattered within the medulla oblongata that regulate heart rate, force of contraction, and blood vessel diameter.

Carotene (KAR-ō-tēn) Antioxidant precursor of vitamin A, which is needed for synthesis of photopigments; yellow-orange pigment present in the stratum corneum of the epidermis. Accounts for the yellowish coloration of skin. Also termed **beta-carotene.**

Carotid (ka-ROT-id) **body** Cluster of chemoreceptors on or near the carotid sinus that respond to changes in blood levels of oxygen, carbon dioxide, and hydrogen ions.

Carotid sinus A dilated region of the internal carotid artery just superior to where it branches from the common carotid artery; it contains baroreceptors that monitor blood pressure.

Carpal bones The eight bones of the wrist. Also called **carpals.**

Carpus (KAR-pus) A collective term for the eight bones of the wrist.

Cartilage (KAR-ti-lij) A type of connective tissue consisting of chondrocytes in lacunae embedded in a dense network of collagen and elastic fibers and an extracellular matrix of chondroitin sulfate.

Cartilaginous (kar-ti-LAJ-i-nus) **joint** A joint without a synovial (joint) cavity where the articulating bones are held tightly together by cartilage, allowing little or no movement.

Catabolism (ka-TAB-ō-lizm) Chemical reactions that break down complex organic compounds into simple ones, with the net release of energy.

Cataract (KAT-a-rakt) Loss of transparency of the lens of the eye or its capsule or both.

Cauda equina (KAW-da ē-KWĪ-na) A tail-like array of roots of spinal nerves at the inferior end of the spinal cord.

Caudal (KAW-dal) Pertaining to any tail-like structure; inferior in position.

Cecum (SĒ-kum) A blind pouch at the proximal end of the large intestine that attaches to the ileum.

Celiac plexus (PLEK-sus) A large mass of autonomic ganglia and axons located at the level of the superior part of the first lumbar vertebra. Also called the **solar plexus.**

Cell The basic structural and functional unit of all organisms; the smallest structure capable of performing all the activities vital to life.

Cell cycle Growth and division of a single cell into two identical cells; consists of interphase and cell division.

Cell division Process by which a cell reproduces itself that consists of a nuclear division (mitosis) and a cytoplasmic division (cytokinesis); types include somatic and reproductive cell division.

Cell junction Point of contact between plasma membranes of tissue cells.

Cementum (se-MEN-tum) Calcified tissue covering the root of a tooth.

Central canal A microscopic tube running the length of the spinal cord in the gray commissure. A circular channel running longitudinally in the center of an osteon (haversian system) of mature compact bone, containing blood and lymphatic vessels and nerves. Also called an **haversian** (ha-VER-shun) **canal.**

Central fovea (FŌ-vē-a) A depression in the center of the macula lutea of the retina, containing cones only and lacking blood vessels; the area of highest visual acuity (sharpness of vision).

Central nervous system (CNS) That portion of the nervous system that consists of the brain and spinal cord.

Centrioles (SEN-trē-ōlz) Paired, cylindrical structures of a centrosome, each consisting of a ring of microtubules and arranged at right angles to each other.

Centromere (SEN-trō-mēr) The constricted portion of a chromosome where the two chromatids are joined; serves as the point of attachment for the microtubules that pull chromatids during anaphase of cell division.

Centrosome (SEN-trō-sōm) A dense network of small protein fibers near the nucleus of a cell, containing a pair of centrioles and pericentriolar material.

Cephalic (se-FAL-ik) Pertaining to the head; superior in position.

Cerebellar peduncle (ser-e-BEL-ar pe-DUNG-kul) A bundle of nerve axons connecting the cerebellum with the brain stem.

Cerebellum (ser′-e-BEL-um) The part of the brain lying posterior to the medulla oblongata and pons; governs balance and coordinates skilled movements.

Cerebral aqueduct (SER-ē-bral AK-we-dukt) A channel through the midbrain connecting the third and fourth ventricles and containing cerebrospinal fluid. Also termed the **aqueduct of Sylvius.**

Cerebral arterial circle A ring of arteries forming an anastomosis at the base of the brain between the internal carotid and basilar arteries and arteries supplying the cerebral cortex. Also called the **circle of Willis.**

Cerebral cortex The surface of the cerebral hemispheres, 2–4 mm thick, consisting of gray matter; arranged in six layers of neuronal cell bodies in most areas.

Cerebral peduncle (pe-DUNG-kel) One of a pair of nerve axon bundles located on the anterior surface of the midbrain, conducting nerve impulses between the pons and the cerebral hemispheres.

Cerebrospinal (se-rē′-brō-SPĪ-nal) **fluid (CSF)** A fluid produced by ependymal cells that cover choroid plexuses in the ventricles of the brain; the fluid circulates in the ventricles, the central canal, and the subarachnoid space around the brain and spinal cord.

Cerebrovascular (se rē′-brō-VAS-kū-lar) **accident (CVA)** Destruction of brain tissue (infarction) resulting from obstruction or rupture of blood vessels that supply the brain. Also called a **stroke** or **brain attack.**

Cerebrum (SER-e-brum *or* se-RĒ-brum) The two hemispheres of the forebrain (derived from the telencephalon), making up the largest part of the brain.

Cerumen (se-ROO-men) Waxlike secretion produced by ceruminous glands in the external auditory meatus (ear canal). Also termed **ear wax.**

Ceruminous (se-RŪ-mi-nus) **gland** A modified sudoriferous (sweat) gland in the external auditory meatus that secretes cerumen (ear wax).

Cervical ganglion (SER-vi-kul GANG-glē-on) A cluster of cell bodies of postganglionic sympathetic neurons located in the neck, near the vertebral column.

Cervical plexus (PLEK-sus) A network formed by nerve axons from the ventral rami of the first four cervical nerves and receiving gray rami communicantes from the superior cervical ganglion.

Cervix (SER-viks) Neck; any constricted portion of an organ, such as the inferior cylindrical part of the uterus.

Chemoreceptor (kē′-mō-rē-SEP-tor) Sensory receptor that detects the presence of a specific chemical.

Chiasm (KĪ-azm) A crossing; especially the crossing of axons in the optic (II) nerve.

Chief cell The secreting cell of a gastric gland that produces pepsinogen, the precursor of the enzyme pepsin, and the enzyme gastric lipase. Also called a **zymogenic** (zī′-mō-JEN-ik) **cell.** Cell in the parathyroid glands that secretes parathyroid hormone (PTH). Also called a **principal cell.**

Cholecystectomy (kō′-lē-sis-TEK-tō-mē) Surgical removal of the gallbladder.

Cholecystitis (kō′-lē-sis-TĪ-tis) Inflammation of the gallbladder.

Cholesterol (kō-LES-te-rol) Classified as a lipid, the most abundant steroid in animal tissues; located in cell membranes and used for the synthesis of steroid hormones and bile salts.

Cholinergic (kō′-lin-ER-jik) **neuron** A neuron that liberates acetylcholine as its neurotransmitter.

Chondrocyte (KON-drō-sīt) Cell of mature cartilage.

Chondroitin (kon-DROY-tin) **sulfate** An amorphous extracellular matrix material found outside connective tissue cells.

Chordae tendineae (KOR-dē TEN-di-nē-ē) Tendonlike, fibrous cords that connect atrioventricular valves of the heart with papillary muscles.

Chorionic villi sampling (CVS) The removal of a sample of chorionic villus tissue by means of a catheter to analyze the tissue for prenatal genetic defects.

Choroid (KŌ-royd) One of the vascular coats of the eyeball.

Choroid plexus (PLEK-sus) A network of capillaries located in the roof of each of the four ventricles of the brain; ependymal cells around choroid plexuses produce cerebrospinal fluid.

Chromaffin (KRŌ-maf-in) **cell** Cell that has an affinity for chrome salts, due in part to the presence of the precursors of the neurotransmitter epinephrine; found, among other places, in the adrenal medulla.

Chromatid (KRŌ-ma-tid) One of a pair of identical connected nucleoprotein strands that are joined at the centromere and separate during cell division, each becoming a chromosome of one of the two daughter cells.

Chromatin (KRŌ-ma-tin) The threadlike mass of genetic material, consisting of DNA and histone proteins, that is present in the nucleus of a nondividing or interphase cell.

Chromatolysis (krō′-ma-TOL-i-sis) The breakdown of Nissl bodies into finely granular masses in the cell body of a neuron whose axon has been damaged.

Chromosome (KRŌ-mō-sōm) One of the small, threadlike structures in the nucleus of a cell, normally 46 in a human diploid cell, that bears the genetic material; composed of DNA and proteins (histones) that form a delicate chromatin thread during interphase; becomes packaged into compact rodlike structures that are visible under the light microscope during cell division.

Chronic (KRON-ik) Long term or frequently recurring; applied to a disease that is not acute.

Chronic obstructive pulmonary disease (COPD) A disease, such as bronchitis or emphysema, in which there is some degree of obstruction of airways and consequent increase in airway resistance.

Chyle (KĪL) The milky-appearing fluid found in the lacteals of the small intestine after absorption of lipids in food.

Chyme (KĪM) The semifluid mixture of partly digested food and digestive secretions found in the stomach and small intestine during digestion of a meal.

Ciliary (SIL-ē-ar′-ē) **body** One of the three parts of the vascular tunic of the eyeball, the others being the choroid and the iris; includes the ciliary muscle and the ciliary processes.

Ciliary ganglion (GANG-glē-on) A very small parasympathetic ganglion whose preganglionic axons come from the oculomotor (III) nerve and whose postganglionic axons carry nerve impulses to the ciliary muscle and the sphincter muscle of the iris.

Cilium (SIL-ē-um) A hair or hairlike process projecting from a cell that may be used to move the entire cell or to move substances along the surface of the cell. *Plural* is **cilia.**

Circle of Willis *See* **Cerebral arterial circle.**

Circular folds Permanent, deep, transverse folds in the mucosa and submucosa of the small intestine that increase the surface area for absorption. Also called **plicae circulares** (PLĪ-kē SER-kū-lar-ēs).

Circumduction (ser-kum-DUK-shun) A movement at a synovial joint in which the distal end of a bone moves in a circle while the proximal end remains relatively stable.

Cirrhosis (si-RŌ-sis) A liver disorder in which the parenchymal cells are destroyed and replaced by connective tissue.

Cisterna chyli (sis-TER-na KĪ-lē) The origin of the thoracic duct.

Clitoris (KLI-tō-ris) An erectile organ of the female, located at the anterior junction of the labia minora, that is homologous to the male penis.

Clone (KLŌN) A population of identical cells.

Coarctation (kō′-ark-TĀ-shun) **of the aorta** A congenital heart defect in which a segment of the aorta is too narrow. As a result, the flow of oxygenated blood to the body is reduced, the left ventricle is forced to pump harder, and high blood pressure develops.

Coccyx (KOK-siks) The fused bones at the inferior end of the vertebral column.

Cochlea (KOK-lē-a) A winding, cone-shaped tube forming a portion of the inner ear and containing the spiral organ (organ of Corti).

Cochlear duct The membranous cochlea consisting of a spirally arranged tube enclosed in the bony cochlea and lying along its outer wall. Also called the **scala media** (SCĀ-la MĒ-dē-a).

Collagen (KOL-a-jen) A protein that is the main organic constituent of connective tissue.

Collateral circulation The alternate route taken by blood through an anastomosis.

Colliculus (ko-LIK-ū-lus) A small elevation.

Colon The portion of the large intestine consisting of ascending, transverse, descending, and sigmoid portions.

Colony-stimulating factor (CSF) One of a group of molecules that stimulates development of white blood cells. Examples are macrophage CSF and granulocyte CSF.

Colostrum (kō-LOS-trum) A thin, cloudy fluid secreted by the mammary glands a few days prior to or after delivery before true milk is produced.

Column (KOL-um) Group of white matter tracts in the spinal cord.

Common bile duct A tube formed by the union of the common hepatic duct and the cystic duct that empties bile into the duodenum at the hepatopancreatic ampulla (ampulla of Vater).

Compact (dense) bone tissue Bone tissue that contains few spaces between osteons (haversian systems); forms the external portion of all bones and the bulk of the diaphysis (shaft) of long bones; is found immediately deep to the periosteum and external to spongy bone.

Concha (KONG-ka) A scroll-like bone found in the nose. *Plural is* **conchae** (KONG-kē).

Concussion (kon-KUSH-un) Traumatic injury to the brain that produces no visible bruising but may result in abrupt, temporary loss of consciousness.

Conduction system A group of autorhythmic cardiac muscle fibers that generates and distributes electrical impulses to stimulate coordinated contraction of the heart chambers; includes the sinoatrial (SA) node, the atrioventricular (AV) node, the atrioventricular (AV) bundle, the right and left bundle branches, and the Purkinje fibers.

Condyloid (KON-di-loyd) **joint** A synovial joint structured so that an oval-shaped condyle of one bone fits into an elliptical cavity of another bone, permitting side-to-side and back-and-forth movements, such as the joint at the wrist between the radius and carpals. Also called an **ellipsoidal** (ē-lip-SOYD-al) **joint.**

Cone (KŌN) The type of photoreceptor in the retina that is specialized for highly acute color vision in bright light.

Congenital (kon-JEN-i-tal) Present at the time of birth.

Conjunctiva (kon′-junk-TĪ-va) The delicate membrane covering the eyeball and lining the eyelids.

Connective tissue The most abundant of the four basic tissue types in the body, performing the functions of binding and supporting; consists of relatively few cells in a generous extracellular matrix (the ground substance and fibers between the cells).

Consciousness (KON-shus-nes) A state of wakefulness in which an individual is fully alert, aware, and oriented, partly as a result of feedback between the cerebral cortex and reticular activating system.

Continuous conduction (kon-DUK-shun) Propagation of an action potential (nerve impulse) in a step-by-step depolarization of each adjacent area of an axon membrane.

Contraception (kon′-tra-SEP-shun) The prevention of fertilization or impregnation without destroying fertility.

Contractility (kon′-trak-TIL-i-tē) The ability of cells or parts of cells to actively generate force to undergo shortening for movements. Muscle fibers (cells) exhibit a high degree of contractility.

Contralateral (CON-tra-lat-er-al) On the opposite side; affecting the opposite side of the body.

Conus medullaris (KŌ-nus med-ū-LAR-is) The tapered portion of the spinal cord inferior to the lumbar enlargement.

Convergence (con-VER-jens) A synaptic arrangement in which the synaptic end bulbs of several presynaptic neurons terminate on one postsynaptic neuron. The medial movement of the two eyeballs so that both are directed toward a near object being viewed in order to produce a single image.

Cornea (KOR-nē-a) The nonvascular, transparent fibrous coat through which the iris of the eye can be seen.

Corona (kō-RŌ-na) Margin of the glans penis.

Corona radiata The innermost layer of granulosa cells that is firmly attached to the zona pellucida around a secondary oocyte.

Coronary artery disease (CAD) A condition such as atherosclerosis that causes narrowing of coronary arteries so that blood flow to the heart is reduced. The result is **coronary heart disease (CHD),** in which the heart muscle receives inadequate blood flow due to an interruption of its blood supply.

Coronary circulation The pathway followed by the blood from the ascending aorta through the blood vessels supplying the heart and returning to the right atrium. Also called **cardiac circulation.**

Coronary sinus (SĪ-nus) A wide venous channel on the posterior surface of the heart that collects the blood from the coronary circulation and returns it to the right atrium.

Corpus albicans (KOR-pus AL-bi-kanz) A white fibrous patch in the ovary that forms after the corpus luteum regresses.

Corpus callosum (kal-LŌ-sum) The great commissure of the brain between the cerebral hemispheres.

Corpuscle of touch See **Meissner corpuscle.**

Corpus luteum (LOO-tē-um) A yellowish body in the ovary formed when a follicle has discharged its secondary oocyte; secretes estrogens, progesterone, relaxin, and inhibin.

Corpus striatum (strī-Ā-tum) An area in the interior of each cerebral hemisphere composed of the caudate and putamen of the basal ganglia and white matter of the internal capsule, arranged in a striated manner.

Cortex (KOR-teks) An outer layer of an organ. The convoluted layer of gray matter covering each cerebral hemisphere.

Costal (KOS-tal) Pertaining to a rib.

Cramp A spasmodic, usually painful contraction of a muscle.

Cranial (KRĀ-ne-al) **cavity** A body cavity formed by the cranial bones and containing the brain.

Cranial nerve One of 12 pairs of nerves that leave the brain; pass through foramina in the skull; and supply sensory and motor neurons to the head, neck, part of the trunk, and viscera of the thorax and abdomen. Each is designated by a Roman numeral and a name.

Craniosacral (krā-nē-ō-SĀK-ral) **outflow** The axons of parasympathetic preganglionic neurons, which have their cell bodies located in nuclei in the brain stem and in the lateral gray matter of the sacral portion of the spinal cord.

Cranium (KRĀ-nē-um) The skeleton of the skull that protects the brain and the organs of sight, hearing, and balance; includes the frontal, parietal, temporal, occipital, sphenoid, and ethmoid bones.

Crista (KRIS-ta) A crest or ridged structure. A small elevation in the ampulla of each semicircular duct that contains receptors for dynamic equilibrium. *Plural* is **cristae.**

Crus (KRŪS) **of penis** Separated, tapered portion of the corpora cavernosa penis. *Plural* is **crura** (KROO-ra).

Crypt of Lieberkühn *See* **Intestinal gland.**

Cryptorchidism (krip-TŌR-ki-dizm) The condition of undescended testes.

Cuneate (KŪ-nē-āt) **nucleus** A group of neurons in the inferior part of the medulla oblongata in which axons of the cuneate fasciculus terminate.

Cupula (KU-pū-la) A mass of gelatinous material covering the hair cells of a crista; a sensory receptor in the ampulla of a semicircular canal stimulated when the head moves.

Cushing's (KUSH-ings) **syndrome** Condition caused by a hypersecretion of glucocorticoids characterized by spindly legs, "moon face," "buffalo hump," pendulous abdomen, flushed facial skin, poor wound healing, hyperglycemia, osteoporosis, hypertension, and increased susceptibility to disease.

Cutaneous (kū-TĀ-nē-us) Pertaining to the skin.

Cyanosis (sī-a-NŌ-sis) A blue or dark purple discoloration, most easily seen in nail beds and mucous membranes, that results from an increased concentration of deoxygenated (reduced) hemoglobin (more than 5 gm/dL).

Cyst (SIST) A sac with a distinct connective tissue wall, containing a fluid or other material.

Cystic (SIS-tik) **duct** The duct that carries bile between the gallbladder and the common bile duct.

Cystitis (sis-TĪ-tis) Inflammation of the urinary bladder.

Cytokinesis (sī′-tō-ki-NĒ-sis) Distribution of the cytoplasm into two separate cells during cell division; coordinated with nuclear division (mitosis).

Cytolysis (sī-TOL-i-sis) The rupture of living cells in which the contents leak out.

Cytoplasm (SĪ-tō-plasm) Cytosol plus all organelles except the nucleus.

Cytoskeleton (sī′-tō-SKEL-e-ton) Complex internal structure of cytoplasm consisting of microfilaments, microtubules, and intermediate filaments.

Cytosol (SĪ-tō-sol) Semifluid portion of cytoplasm in which organelles and inclusions are suspended and solutes are dissolved. Also called **intracellular fluid.**

D

Dartos (DAR-tōs) The contractile tissue deep to the skin of the scrotum.

Decidua (dē-SID-ū-a) That portion of the endometrium of the uterus (all but the deepest layer) that is modified during pregnancy and shed after childbirth.

Deciduous (dē-SID-ū-us) Falling off or being shed seasonally or at a particular stage of development. In the body, referring to the first set of teeth.

Decussation (dē′-ku-SĀ-shun) A crossing-over to the opposite (contralateral) side; an example is the crossing of 90% of the axons in the large motor tracts to opposite sides in the medullary pyramids.

Deep Away from the surface of the body or an organ.

Deep inguinal (IN-gwi-nal) **ring** A slitlike opening in the aponeurosis of the transversus abdominis muscle that represents the origin of the inguinal canal.

Deep-venous thrombosis (DVT) The presence of a thrombus in a vein, usually a deep vein of the lower limbs.

Defecation (def-e-KĀ-shun) The discharge of feces from the rectum.

Deglutition (dē-gloo-TISH-un) The act of swallowing.

Dehydration (dē-hī-DRĀ-shun) Excessive loss of water from the body or its parts.

Delta cell A cell in the pancreatic islets (islets of Langerhans) in the pancreas that secretes somatostatin. Also termed a **D cell.**

Demineralization (de-min′-er-al-i-ZĀ-shun) Loss of calcium and phosphorus from bones.

Dendrite (DEN-drīt) A neuronal process that carries electrical signals, usually graded potentials, toward the cell body.

Dendritic (den-DRIT-ik) **cell** One type of antigen-presenting cell with long branchlike projections that commonly is present in mucosal linings such as the vagina, in the skin (Langerhans cells in the epidermis), and in lymph nodes (follicular dendritic cells).

Dental caries (KA-rēz) Gradual demineralization of the enamel and dentin of a tooth that may invade the pulp and alveolar bone. Also called **tooth decay.**

Denticulate (den-TIK-ū-lāt) Finely toothed or serrated; characterized by a series of small, pointed projections.

Dentin (DEN-tin) The bony tissues of a tooth enclosing the pulp cavity.

Dentition (den-TI-shun) The eruption of teeth. The number, shape, and arrangement of teeth.

Deoxyribonucleic (dē-ok′-sē-rī-bō-nū-KLĒ-ik) **acid (DNA)** A nucleic acid constructed of nucleotides consisting of one of four bases (adenine, cytosine, guanine, or thymine), deoxyribose, and a phosphate group; encoded in the nucleotides is genetic information.

Depression (de-PRESH-un) Movement in which a part of the body moves inferiorly.

Dermal papilla (pa-PILL-a) Fingerlike projection of the papillary region of the dermis that may contain blood capillaries or corpuscles of touch (Meissner corpuscles).

Dermatology (der′-ma-TOL-ō-jē) The medical specialty dealing with diseases of the skin.

Dermatome (DER-ma-tōm) The cutaneous area developed from one embryonic spinal cord segment and receiving most of its sensory innervation from one spinal nerve. An instrument for incising the skin or cutting thin transplants of skin.

Dermis (DER-mis) A layer of dense irregular connective tissue lying deep to the epidermis.

Descending colon (KŌ-lon) The part of the large intestine descending from the left colic (splenic) flexure to the level of the left iliac crest.

Detrusor (de-TROO-ser) **muscle** Smooth muscle that forms the wall of the urinary bladder.

Diagnosis (dī′-ag-NŌ-sis) Distinguishing one disease from another or determining the nature of a disease from signs and symptoms by inspection, palpation, laboratory tests, and other means.

Dialysis (dī-AL-i-sis) The removal of waste products from blood by diffusion through a selectively permeable membrane.

Diaphragm (DĪ-a-fram) Any partition that separates one area from another, especially the dome-shaped skeletal muscle between the thoracic and abdominal cavities. Also a dome-shaped device that is placed over the cervix, usually with a spermicide, to prevent conception.

Diaphysis (dī-AF-i-sis) The shaft of a long bone.

Diarrhea (dī-a-RĒ-a) Frequent defecation of liquid feces caused by increased motility of the intestines.

Diarthrosis (dī-ar-THRŌ-sis) A freely movable joint; types are gliding, hinge, pivot, condyloid, saddle, and ball-and-socket.

Diastole (dī-AS-tō-lē) In the cardiac cycle, the phase of relaxation or dilation of the heart muscle, especially of the ventricles.

Diastolic (dī-as-TOL-ik) **blood pressure** The force exerted by blood on arterial walls during ventricular relaxation; the lowest blood pressure measured in the large arteries, normally about 80 mmHg in a young adult.

Diencephalon (dī′-en-SEF-a-lon) A part of the brain consisting of the thalamus, hypothalamus, and epithalamus.

Diffusion (di-FŪ-zhun) A passive process in which there is a net or greater movement of molecules or ions from a region of high concentration to a region of low concentration until equilibrium is reached.

Digestion (dī-JES-chun) The mechanical and chemical breakdown of food to simple molecules that can be absorbed and used by body cells.

Dilate (DĪ-lāt) To expand or swell.

Diploid (DIP-loid) Having the number of chromosomes characteristically found in the somatic cells of an organism; having two haploid sets of chromosomes, one each from the mother and father. Symbolized *2n.*

Direct motor pathways Collections of upper motor neurons with cell bodies in the motor cortex that project axons into the spinal cord, where they synapse with lower motor neurons or interneurons in the anterior horns. Also called the **pyramidal pathways** (pi-RAM-i-dal).

Disease Any change from a state of health.

Dislocation (dis′-lō-KĀ-shun) Displacement of a bone from a joint with tearing of ligaments, tendons, and articular capsules. Also called **luxation** (luks-Ā-shun).

Dissect (di-SEKT) To separate tissues and parts of a cadaver or an organ for anatomical study.

Distal (DIS-tal) Farther from the attachment of a limb to the trunk; farther from the point of origin or attachment.

Diuretic (dī-ū-RET-ik) A chemical that increases urine volume by decreasing reabsorption of water, usually by inhibiting sodium reabsorption.

Divergence (dī-VER-jens) A synaptic arrangement in which the synaptic end bulbs of one presynaptic neuron terminate on several postsynaptic neurons.

Diverticulum (dī-ver-TIK-ū-lum) A sac or pouch in the wall of a canal or organ, especially in the colon.

Dorsal ramus (RĀ-mus) A branch of a spinal nerve containing motor and sensory axons supplying the muscles, skin, and bones of the posterior part of the head, neck, and trunk.

Dorsiflexion (dor-si-FLEK-shun) Bending the foot in the direction of the dorsum (upper surface).

Down-regulation Phenomenon in which there is a decrease in the number of receptors in response to an excess of a hormone or neurotransmitter.

Duct of Santorini *See* **Accessory duct.**

Duct of Wirsung *See* **Pancreatic duct.**

Ductus arteriosus (DUK-tus ar-tē-rē-Ō-sus) A small vessel connecting the pulmonary trunk with the aorta; found only in the fetus.

Ductus (vas) deferens (DEF-er-ens) The duct that carries sperm from the epididymis to the ejaculatory duct. Also called the **seminal duct.**

Ductus epididymis (ep′-i-DID-i-mis) A tightly coiled tube inside the epididymis, distinguished into a head, body, and tail, in which sperm undergo maturation.

Ductus venosus (ve-NŌ-sus) A small vessel in the fetus that helps the circulation bypass the liver.

Duodenal (doo-ō-DĒ-nal) **gland** Gland in the submucosa of the duodenum that secretes an alkaline mucus to protect the lining of the small intestine from the action of enzymes and to help neutralize the acid in chyme. Also called a **Brunner's** (BRUN-erz) **gland.**

Duodenal papilla (pa-PILL-a) An elevation on the duodenal mucosa that receives the hepatopancreatic ampulla (ampulla of Vater).

Duodenum (doo′-ō-DĒ-num *or* doo-OD-e-num) The first 25cm (10 in.) of the small intestine, which connects the stomach and the jejunum.

Dura mater (DOO-ra MĀ-ter) The outermost of the three meninges (coverings) of the brain and spinal cord.

Dynamic equilibrium (ē-kwi-LIB-rē-um) The maintenance of body position, mainly the head, in response to sudden movements such as rotation.

Dysmenorrhea (dis′-men-ō-RĒ-a) Painful menstruation.

Dysplasia (dis-PLĀ-zē-a) Change in the size, shape, and organization of cells due to chronic irritation or inflammation; may either revert to normal if stress is removed or progress to neoplasia.

Dyspnea (DISP-nē-a) Shortness of breath; painful or labored breathing.

E

Ectopic (ek-TOP-ik) Out of the normal location, as in ectopic pregnancy.

Edema (e-DĒ-ma) An abnormal accumulation of interstitial fluid.

Effector (e-FEK-tor) An organ of the body, either a muscle or a gland, that is innervated by somatic or autonomic motor neurons.

Efferent arteriole (EF-er-ent ar-TĒ-rē-ōl) A vessel of the renal vascular system that carries blood from a glomerulus to a peritubular capillary.

Efferent (EF-er-ent) **ducts** A series of coiled tubes that transport sperm from the rete testis to the epididymis.

Ejaculation (e-jak-ū-LĀ-shun) The reflex ejection or expulsion of semen from the penis.

Ejaculatory (e-JAK-ū-la-tō-rē) **duct** A tube that transports sperm from the ductus (vas) deferens to the prostatic urethra.

Elasticity (e-las-TIS-i-tē) The ability of tissue to return to its original shape after contraction or extension.

Electrocardiogram (e-lek′-trō-KAR-dē-ō-gram) (**ECG** or **EKG**) A recording of the electrical changes that accompany the cardiac cycle that can be detected at the surface of the body; may be resting, stress, or ambulatory.

Elevation (el-e-VĀ-shun) Movement in which a part of the body moves superiorly.

Embolus (EM-bō-lus) A blood clot, bubble of air or fat from broken bones, mass of bacteria, or other debris or foreign material transported by the blood.

Embryo (EM-brē-ō) The young of any organism in an early stage of development; in humans, the developing organism from fertilization to the end of the eighth week of development.

Emesis (EM-e-sis) Vomiting.

Emigration (em′-i-GRĀ-shun) Process whereby white blood cells (WBCs) leave the bloodstream by rolling along the endothelium, sticking to it, and squeezing between the endothelial cells. Adhesion molecules help WBCs stick to the endothelium. Also known as **migration** or **extravasation.**

Emission (ē-MISH-un) Propulsion of sperm into the urethra due to peristaltic contractions of the ducts of the testes, epididymides, and ductus (vas) deferens as a result of sympathetic stimulation.

Emphysema (em-fi-SĒ-ma) A lung disorder in which alveolar walls disintegrate, producing abnormally large air spaces and loss of elasticity in the lungs; typically caused by exposure to cigarette smoke.

Emulsification (ē-mul′-si-fi-KĀ-shun) The dispersion of large lipid globules into smaller, uniformly distributed particles in the presence of bile.

Enamel (e-NAM-el) The hard, white substance covering the crown of a tooth.

Endocardium (en-dō-KAR-dē-um) The layer of the heart wall, composed of endothelium and connective tissue, that lines the inside of the heart and covers the valves and tendons that hold the valves open.

Endochondral ossification (en′-dō-KON-dral os′-i-fi-KĀ-shun) The replacement of cartilage by bone. Also called **intracartilaginous** (in′-tra-kar′-ti-LAJ-i-nus) **ossification.**

Endocrine (EN-dō-krin) **gland** A gland that secretes hormones into interstitial fluid and then the blood; a ductless gland.

Endocrinology (en′-dō-kri-NOL-ō-jē) The science concerned with the structure and functions of endocrine glands and the diagnosis and treatment of disorders of the endocrine system.

Endocytosis (en′-dō-sī-TŌ-sis) The uptake into a cell of large molecules and particles in which a segment of plasma membrane surrounds the substance, encloses it, and brings it in; includes phagocytosis, pinocytosis, and receptor-mediated endocytosis.

Endodontics (en′-dō-DON-tiks) The branch of dentistry concerned with the prevention, diagnosis, and treatment of diseases that affect the pulp, root, periodontal ligament, and alveolar bone.

Endolymph (EN-dō-limf′) The fluid within the membranous labyrinth of the internal ear.

Endometriosis (en′-dō-MĒ-trē-ō′-sis) The growth of endometrial tissue outside the uterus.

Endometrium (en′-dō-MĒ-trē-um) The mucous membrane lining the uterus.

Endomysium (en′-dō-MĪZ-ē-um) Areolar connective tissue separating each individual muscle fiber (cell).

Endoneurium (en′-dō-NOO-rē-um) Connective tissue wrapping around individual nerve axons.

Endoplasmic reticulum (en′-dō-PLAS-mik re-TIK-ū-lum) **(ER)** A network of channels running through the cytoplasm of a cell that serves in intracellular transportation, support, storage, synthesis, and packaging of molecules. Portions of ER where ribosomes are attached to the outer surface are called **rough ER;** portions that have no ribosomes are called **smooth ER.**

End organ of Ruffini *See* **Type II cutaneous mechanoreceptor.**

Endosteum (end-OS-tē-um) The membrane that lines the medullary (marrow) cavity of bones, consisting of osteogenic cells and scattered osteoclasts.

Endothelium (en′-dō-THĒ-lē-um) The layer of simple squamous epithelium that lines the cavities of the heart, blood vessels, and lymphatic vessels.

Enteric (EN-ter-ik) **nervous system** The part of the nervous system that is embedded in the submucosa and muscularis of the gastrointestinal (GI) tract; governs motility and secretions of the GI tract.

Enteroendocrine (en-ter-ō-EN-dō-krin) **cell** A cell of the mucosa of the gastrointestinal tract that secretes a hormone that governs function of the GI tract; hormones secreted include gastrin, cholecystokinin, glucose-dependent insulinotropic peptide (GIP), and secretin.

Enzyme (EN-zīm) A substance that accelerates chemical reactions; an organic catalyst, usually a protein.

Eosinophil (ē-ō-SIN-ō-fil) A type of white blood cell characterized by granules that stain red or pink with acid dyes.

Ependymal (ep-EN-de-mal) **cells** Neuroglial cells that cover choroid plexuses and produce cerebrospinal fluid (CSF); they also line the ventricles of the brain and assist in the circulation of CSF.

Epicardium (ep′-i-KAR-dē-um) The thin outer layer of the heart wall, composed of serous tissue and mesothelium. Also called the **visceral pericardium.**

Epidemiology (ep′-i-dē-mē-OL-ō-jē) Study of the occurrence and transmission of diseases and disorders in human populations.

Epidermis (ep′-i-DERM-is) The superficial, thinner layer of skin, composed of keratinized stratified squamous epithelium.

Epididymis (ep′-i-DID-i-mis) A comma-shaped organ that lies along the posterior border of the testis and contains the ductus epididymis, in which sperm undergo maturation. *Plural* is **epididymides** (ep′-i-di-DIM-i-dēz).

Epidural (ep′-i-DOO-ral) **space** A space between the spinal dura mater and the vertebral canal, containing areolar connective tissue and a plexus of veins.

Epiglottis (ep′-i-GLOT-is) A large, leaf-shaped piece of elastic cartilage lying on top of the larynx, attached to the thyroid cartilage; its unattached portion is free to move up and down to cover the glottis (vocal folds and rima glottidis) during swallowing.

Epimysium (ep-i-MĪZ-ē-um) Fibrous connective tissue around muscles.

Epinephrine (ep-ē-NEF-rin) Hormone secreted by the adrenal medulla that produces actions similar to those that result from sympathetic stimulation. Also called **adrenaline** (a-DREN-a-lin).

Epineurium (ep′-i-NOO-rē-um) The superficial connective tissue covering around an entire nerve.

Epiphyseal (ep′-i-FIZ-ē-al) **line** The remnant of the epiphyseal plate in the metaphysis of a long bone.

Epiphyseal plate The hyaline cartilage plate in the metaphysis of a long bone; site of lengthwise growth of long bones.

Epiphysis (e-PIF-i-sis) The end of a long bone, usually larger in diameter than the shaft (diaphysis).

Epiphysis cerebri (se-RĒ-brē) Pineal gland.

Episiotomy (e-piz′-ē-OT-ō-mē) A cut made with surgical scissors to avoid tearing of the perineum at the end of the second stage of labor.

Epistaxis (ep′-i-STAK-sis) Loss of blood from the nose due to trauma, infection, allergy, neoplasm, and bleeding disorders. Also called **nosebleed.**

Epithalamus (ep′-i-THAL-a-mus) Part of the diencephalon superior and posterior to the thalamus, comprising the pineal gland and associated structures.

Epithelial (ep-i-THĒ-lē-al) **tissue** The tissue that forms the innermost and outermost surfaces of body structures and forms glands.

Eponychium (ep′-o-NIK-ē-um) Narrow band of stratum corneum at the proximal border of a nail that extends from the margin of the nail wall. Also called the **cuticle.**

Erectile dysfunction Failure to maintain an erection long enough for sexual intercourse. Also known as **impotence** (IM-pō-tens).

Erection (ē-REK-shun) The enlarged and stiff state of the penis or clitoris resulting from the engorgement of the spongy erectile tissue with blood.

Eructation (e-ruk′-TĀ-shun) The forceful expulsion of gas from the stomach. Also called **belching.**

Erythema (er-e-THĒ-ma) Skin redness usually caused by dilation of the capillaries.

Erythrocyte (e-RITH-rō-sīt) A mature red blood cell.

Erythropoietin (e-rith′-rō-POY-e-tin) A hormone released by the juxtaglomerular cells of the kidneys that stimulates red blood cell production.

Esophagus (e-SOF-a-gus) The hollow muscular tube that connects the pharynx and the stomach.

Estrogens (ES-trō-jenz) Feminizing sex hormones produced by the ovaries; govern development of oocytes, maintenance of female reproductive structures, and appearance of secondary sex characteristics; also affect fluid and electrolyte balance, and protein anabolism. Examples are β-estradiol, estrone, and estriol.

Eupnea (ŪP-nē-a) Normal quiet breathing.

Eustachian tube *See* **Auditory tube.**

Eversion (ē-VER-zhun) The movement of the sole laterally at the ankle joint or of an atrioventricular valve into an atrium during ventricular contraction.

Excitability (ek-sīt′-a-BIL-i-tē) The ability of muscle fibers to receive and respond to stimuli; the ability of neurons to respond to stimuli and generate nerve impulses.

Excretion (eks-KRĒ-shun) The process of eliminating waste products from the body; also the products excreted.

Exocrine (EK-sō-krin) **gland** A gland that secretes its products into ducts that carry the secretions into body cavities, into the lumen of an organ, or to the outer surface of the body.

Exocytosis (ex′-ō-sī-TŌ-sis) A process in which membrane-enclosed secretory vesicles form inside the cell, fuse with the plasma membrane, and release their contents into the interstitial fluid; achieves secretion of materials from a cell.

Exhalation (eks-ha-LĀ-shun) Breathing out; expelling air from the lungs into the atmosphere. Also called **expiration.**

Extensibility (ek-sten′-si-BIL-i-tē) The ability of muscle tissue to stretch when it is pulled.

Extension (eks-TEN-shun) An increase in the angle between two bones; restoring a body part to its anatomical position after flexion.

External Located on or near the surface.

External auditory (AW-di-tōr-ē) **canal** or **meatus** (mē-Ā-tus) A curved tube in the temporal bone that leads to the middle ear.

External ear The outer ear, consisting of the pinna, external auditory canal, and tympanic membrane (eardrum).

External nares (NĀ-rez) The openings into the nasal cavity on the exterior of the body. Also called the **nostrils.**

External respiration The exchange of respiratory gases between the lungs and blood. Also called **pulmonary respiration.**

Exteroceptor (EKS-ter-ō-sep′-tor) A sensory receptor adapted for the reception of stimuli from outside the body.

Extracellular fluid (ECF) Fluid outside body cells, such as interstitial fluid and plasma.

Extracellular matrix (MĀ-triks) The ground substance and fibers between cells in a connective tissue.

Eyebrow The hairy ridge superior to the eye.

F

F cell A cell in the pancreatic islets (islets of Langerhans) that secretes pancreatic polypeptide.

Face The anterior aspect of the head.

Falciform ligament (FAL-si-form LIG-a-ment) A sheet of parietal peritoneum between the two principal lobes of the liver. The ligamentum teres, or remnant of the umbilical vein, lies within its fold.

Falx cerebelli (FALKS cer-e-BEL-li) A small triangular process of the dura mater attached to the occipital bone in the posterior cranial fossa and projecting inward between the two cerebellar hemispheres.

Falx cerebri (FALKS CER-e-brē) A fold of the dura mater extending deep into the longitudinal fissure between the two cerebral hemispheres.

Fascia (FASH-ē-a) Large connective tissue sheet that wraps around groups of muscles.

Fascicle (FAS-i-kul) A small bundle or cluster, especially of nerve or muscle fibers (cells). Also called a **fasciculus** (fa-SIK-ū-lus). *Plural* is **fasciculi** (fa-SIK-yoo-lī).

Fasciculation (fa-sik-ū-LĀ-shun) Abnormal, spontaneous twitch of all skeletal muscle fibers in one motor unit that is visible at the skin surface; not associated with movement of the affected muscle; present in progressive diseases of motor neurons, for example, poliomyelitis.

Fauces (FAW-sēs) The opening from the mouth into the pharynx.

Feces (FĒ-sēz) Material discharged from the rectum and made up of bacteria, excretions, and food residue. Also called **stool.**

Female reproductive cycle General term for the ovarian and uterine cycles, the hormonal changes that accompany them, and cyclic changes in the breasts and cervix; includes changes in the endometrium of a nonpregnant female that prepares the lining of the uterus to receive a fertilized ovum. Less correctly termed the **menstrual cycle.**

Fertilization (fer′-ti-li-ZĀ-shun) Penetration of a secondary oocyte by a sperm cell, meiotic division of secondary oocyte to form an ovum, and subsequent union of the nuclei of the gametes.

Fetal circulation The cardiovascular system of the fetus, including the placenta and special blood vessels involved in the exchange of materials between fetus and mother.

Fetus (FĒ-tus) In humans, the developing organism *in utero* from the beginning of the third month to birth.

Fever An elevation in body temperature above the normal temperature of 37 °C (98.6 °F) due to a resetting of the hypothalamic thermostat.

Fibroblast (FĪ-brō-blast) A large, flat cell that secretes most of the extracellular matrix of areolar and dense connective tissues.

Fibrous (FĪ-brus) **joint** A joint that allows little or no movement, such as a suture or a syndesmosis.

Fibrous tunic (TOO-nik) The superficial coat of the eyeball, made up of the posterior sclera and the anterior cornea.

Fight-or-flight response The effects produced upon stimulation of the sympathetic division of the autonomic nervous system.

Filiform papilla (FIL-i-form pa-PIL-a) One of the threadlike projections that are distributed in parallel rows over the anterior two-thirds of the tongue and lack taste buds.

Filtration (fil-TRĀ-shun) The flow of a liquid through a filter (or membrane that acts like a filter) due to a hydrostatic pressure; occurs in capillaries due to blood pressure.

Filum terminale (FĪ-lum ter-mi-NAL-ē) Nonnervous fibrous tissue of the spinal cord that extends inferiorly from the conus medullaris to the coccyx.

Fimbriae (FIM-brē-ē) Fingerlike structures, especially the lateral ends of the uterine (Fallopian) tubes.

Fissure (FISH-ur) A groove, fold, or slit that may be normal or abnormal.

Fixator A muscle that stabilizes the origin of the prime mover so that the prime mover can act more efficiently.

Fixed macrophage (MAK-rō-fāj) Stationary phagocytic cell found in the liver, lungs, brain, spleen, lymph nodes, subcutaneous tissue, and red bone marrow. Also called a **histiocyte** (HIS-tē-ō-sīt).

Flaccid (FLAS-sid) Relaxed, flabby, or soft; lacking muscle tone.

Flagellum (fla-JEL-um) A hairlike, motile process on the extremity of a bacterium, protozoan, or sperm cell. *Plural* is **flagella** (fla-JEL-a).

Flatus (FLĀ-tus) Gas in the stomach or intestines; commonly used to denote expulsion of gas through the anus.

Flexion (FLEK-shun) Movement in which there is a decrease in the angle between two bones.

Follicle (FOL-i-kul) A small secretory sac or cavity; the group of cells that contains a developing oocyte in the ovaries.

Follicle-stimulating hormone (FSH) Hormone secreted by the anterior pituitary; it initiates development of ova and stimulates the ovaries to secrete estrogens in females, and initiates sperm production in males.

Fontanel (fon-ta-NEL) A mesenchyme-filled space where bone formation is not yet complete, especially between the cranial bones of an infant's skull.

Foot The terminal part of the lower limb, from the ankle to the toes.

Foramen (fō-RĀ-men) A passage or opening; a communication between two cavities of an organ, or a hole in a bone for passage of vessels or nerves. *Plural* is **foramina** (fō-RAM-i-na).

Foramen ovale (fō-RĀ-men ō-VAL-ē) An opening in the fetal heart in the septum between the right and left atria. A hole in the greater wing of the sphenoid bone that transmits the mandibular branch of the trigeminal (V) nerve.

Forearm (FOR-arm) The part of the upper limb between the elbow and the wrist.

Fornix (FOR-niks) An arch or fold; a tract in the brain made up of association fibers, connecting the hippocampus with the mammillary bodies; a recess around the cervix of the uterus where it protrudes into the vagina.

Fossa (FOS-a) A furrow or shallow depression.

Fourth ventricle (VEN-tri-kul) A cavity filled with cerebrospinal fluid within the brain lying between the cerebellum and the medulla oblongata and pons.

Fracture (FRAK-choor) Any break in a bone.

Frontal plane A plane at a right angle to a midsagittal plane that divides the body or organs into anterior and posterior portions. Also called a **coronal** (kō-RŌ-nal) **plane.**

Fundus (FUN-dus) The part of a hollow organ farthest from the opening.

Fungiform papilla (FUN-ji-form pa-PIL-a) A mushroomlike elevation on the upper surface of the tongue appearing as a red dot; most contain taste buds.

Furuncle (FŪ-rung-kul) A boil; painful nodule caused by bacterial infection and inflammation of a hair follicle or sebaceous (oil) gland.

G

Gallbladder A small pouch, located inferior to the liver, that stores bile and empties by means of the cystic duct.

Gallstone A solid mass, usually containing cholesterol, in the gallbladder or a bile-containing duct; formed anywhere between bile canaliculi in the liver and the hepatopancreatic ampulla (ampulla of Vater), where bile enters the duodenum. Also called a **biliary calculus.**

Gamete (GAM-ēt) A male or female reproductive cell; a sperm cell or secondary oocyte.

Ganglion (GANG-glē-on) Usually, a group of neuronal cell bodies lying outside the central nervous system (CNS). *Plural* is **ganglia** (GANG-glē-a).

Gastric (GAS-trik) **glands** Glands in the mucosa of the stomach composed of cells that empty their secretions into narrow channels called gastric pits. Types of cells are chief cells (secrete pepsinogen), parietal cells (secrete hydrochloric acid and intrinsic factor), surface mucous and mucous neck cells (secrete mucus), and G cells (secrete gastrin).

Gastroenterology (gas′-trō-en′-ter-OL-ō-jē) The medical specialty that deals with the structure, function, diagnosis, and treatment of diseases of the stomach and intestines.

Gastrointestinal (gas-trō-in-TES-ti-nal) **(GI) tract** A continuous tube running through the ventral body cavity extending from the mouth to the anus. Also called the **alimentary** (al′-i-MEN-tar-ē) **canal.**

Gene (JĒN) Biological unit of heredity; a segment of DNA located in a definite position on a particular chromosome; a sequence of DNA that codes for a particular mRNA, rRNA, or tRNA.

Geriatrics (jer′-ē-AT-riks) The branch of medicine devoted to the medical problems and care of elderly persons.

Gestation (jes-TĀ-shun) The period of development from fertilization to birth.

Gingivae (jin-JI-vē) Gums. They cover the alveolar processes of the mandible and maxilla and extend slightly into each socket.

Gland Specialized epithelial cell or cells that secrete substances; may be exocrine or endocrine.

Glans penis (glanz PĒ-nis) The slightly enlarged region at the distal end of the penis.

Glaucoma (glaw-KŌ-ma) An eye disorder in which there is increased intraocular pressure due to an excess of aqueous humor.

Gliding joint A synovial joint having articulating surfaces that are usually flat, permitting only side-to-side and back-and-forth movements, as between carpal bones, tarsal bones, and the scapula and clavicle. Also called an **arthrodial** (ar-THRŌ-dē-al) **joint.**

Glomerular (glō-MER-ū-lar) **capsule** A double-walled globe at the proximal end of a nephron that encloses the glomerular capillaries. Also called **Bowman's** (BŌ-manz) **capsule.**

Glomerular filtrate (glō-MER-ū-lar FIL-trāt) The fluid produced when blood is filtered by the filtration membrane in the glomeruli of the kidneys.

Glomerular filtration The first step in urine formation in which substances in blood pass through the filtration membrane and the filtrate enters the proximal convoluted tubule of a nephron.

Glomerulus (glō-MER-ū-lus) A rounded mass of nerves or blood vessels, especially the microscopic tuft of capillaries that is surrounded by the glomerular (Bowman's) capsule of each kidney tubule. *Plural* is **glomeruli.**

Glottis (GLOT-is) The vocal folds (true vocal cords) in the larynx plus the space between them (rima glottidis).

Glucagon (GLOO-ka-gon) A hormone produced by the alpha cells of the pancreatic islets (islets of Langerhans) that increases blood glucose level.

Glucocorticoids (gloo′-kō-KOR-ti-koyds) Hormones secreted by the cortex of the adrenal gland, especially cortisol, that influence glucose metabolism.

Glucose (GLOO-kōs) A hexose (six-carbon sugar), $C_6H_{12}O_6$, that is a major energy source for the production of ATP by body cells.

Glucosuria (gloo′-kō-SOO-rē-a) The presence of glucose in the urine; may be temporary or pathological. Also called **glycosuria.**

Glycogen (GLĪ-kō-jen) A highly branched polymer of glucose containing thousands of subunits; functions as a compact store of glucose molecules in liver and muscle fibers (cells).

Goblet cell A goblet-shaped unicellular gland that secretes mucus; present in epithelium of the airways and intestines.

Goiter (GOY-ter) An enlarged thyroid gland.

Golgi (GOL-jē) **complex** An organelle in the cytoplasm of cells consisting of four to six flattened sacs (cisternae), stacked on one another, with expanded areas at their ends; functions in processing, sorting, packaging, and delivering proteins and lipids to the plasma membrane, lysosomes, and secretory vesicles.

Golgi tendon organ *See* **Tendon organ.**

Gomphosis (gom-FŌ-sis) A fibrous joint in which a cone-shaped peg fits into a socket.

Gonad (GŌ-nad) A gland that produces gametes and hormones; the ovary in the female and the testis in the male.

Gonadotropic hormone Anterior pituitary hormone that affects the gonads.

Gout (GOWT) Hereditary condition associated with excessive uric acid in the blood; the acid crystallizes and deposits in joints, kidneys, and soft tissue.

Graafian follicle *See* **Vesicular ovarian follicle.**

Gracile (GRAS-il) **nucleus** A group of nerve cells in the inferior part of the medulla oblongata in which axons of the gracile fasciculus terminate.

Gray commissure (KOM-mi-shur) A narrow strip of gray matter connecting the two lateral gray masses within the spinal cord.

Gray matter Areas in the central nervous system and ganglia containing neuronal cell bodies, dendrites, unmyelinated axons, axon terminals, and neuroglia; Nissl bodies impart a gray color and there is little or no myelin in gray matter.

Gray ramus communicans (RĀ-mus kō-MŪ-ni-kans) A short nerve containing axons of sympathetic postganglionic neurons; the cell bodies of the neurons are in a sympathetic chain ganglion, and the unmyelinated axons extend via the gray ramus to a spinal nerve and then to the periphery to supply smooth muscle in blood vessels, arrector pili muscles, and sweat glands. *Plural* is **rami communicantes** (RĀ-mē kō-mū-ni-KAN-tēz).

Greater omentum (ō-MEN-tum) A large fold in the serosa of the stomach that hangs down like an apron anterior to the intestines.

Greater vestibular (ves-TIB-ū-lar) **glands** A pair of glands on either side of the vaginal orifice that open by a duct into the space between the hymen and the labia minora. Also called **Bartholin's** (BAR-tō-linz) **glands.**

Groin (GROYN) The depression between the thigh and the trunk; the inguinal region.

Gross anatomy The branch of anatomy that deals with structures that can be studied without using a microscope. Also called **macroscopic anatomy.**

Growth An increase in size due to an increase in (1) the number of cells, (2) the size of existing cells as internal components increase in size, or (3) the size of intercellular substances.

Gustatory (GUS-ta-tō′-rē) Pertaining to taste.

Gynecology (gī′-ne-KOL-ō-jē) The branch of medicine dealing with the study and treatment of disorders of the female reproductive system.

Gynecomastia (gīn′e-kō-MAS-tē-a) Excessive growth (benign) of the male mammary glands due to secretion of estrogens by an adrenal gland tumor (feminizing adenoma).

Gyrus (JĪ-rus) One of the folds of the cerebral cortex of the brain. *Plural* is **gyri** (JĪ-rī). Also called a **convolution.**

H

Hair A threadlike structure produced by hair follicles that develops in the dermis. Also called a **pilus** (PĪ-lus).

Hair follicle (FOL-li-kul) Structure composed of epithelium and surrounding the root of a hair from which hair develops.

Hair root plexus (PLEK-sus) A network of dendrites arranged around the root of a hair as free or naked nerve endings that are stimulated when a hair shaft is moved.

Hand The terminal portion of an upper limb, including the carpus, metacarpus, and phalanges.

Haploid (HAP-loyd) **cell** Having half the number of chromosomes characteristically found in the somatic cells of an organism; characteristic of mature gametes. Symbolized *n*.

Hard palate (PAL-at) The anterior portion of the roof of the mouth, formed by the maxillae and palatine bones and lined by mucous membrane.

Haustra (HAWS-tra) A series of pouches that characterize the colon; caused by tonic contractions of the teniae coli. *Singular is* **haustrum.**

Haversian canal *See* **Central canal.**

Haversian system *See* **Osteon.**

Head The superior part of a human, cephalic to the neck. The superior or proximal part of a structure.

Heart A hollow muscular organ lying slightly to the left of the midline of the chest that pumps the blood through the cardiovascular system.

Heart block An arrhythmia (dysrhythmia) of the heart in which the atria and ventricles contract independently because of a blocking of electrical impulses through the heart at some point in the conduction system.

Heart murmur (MER-mer) An abnormal sound that consists of a flow noise that is heard before, between, or after the normal heart sounds, or that may mask normal heart sounds.

Hemangioblast (hē-MAN-jē-ō-blast) A precursor mesodermal cell that develops into blood and blood vessels.

Hematocrit (hē-MAT-ō-krit) **(Hct)** The percentage of blood made up of red blood cells. Usually measured by centrifuging a blood sample in a graduated tube and then reading the volume of red blood cells and dividing it by the total volume of blood in the sample.

Hematology (hēm-a-TOL-ō-jē) The study of blood.

Hematoma (hē′-ma-TŌ-ma) A tumor or swelling filled with blood.

Hemiplegia (hem-i-PLĒ-jē-a) Paralysis of the upper limb, trunk, and lower limb on one side of the body.

Hemoglobin (hē′-mō-GLŌ-bin) **(Hb)** A substance in red blood cells consisting of the protein globin and the iron-containing red pigment heme that transports most of the oxygen and some carbon dioxide in blood.

Hemolysis (hē-MOL-i-sis) The escape of hemoglobin from the interior of a red blood cell into the surrounding medium; results from disruption of the cell membrane by toxins or drugs, freezing or thawing, or hypotonic solutions.

Hemolytic disease of the newborn A hemolytic anemia of a newborn child that results from the destruction of the infant's erythrocytes (red blood cells) by antibodies produced by the mother; usually the antibodies are due to an Rh blood type incompatibility. Also called **erythroblastosis fetalis** (e-rith′-rō-blas-TŌ-sis fe-TAL-is).

Hemophilia (hē-mō-FIL-ē-a) A hereditary blood disorder where there is a deficient production of certain factors involved in blood clotting, resulting in excessive bleeding into joints, deep tissues, and elsewhere.

Hemopoiesis (hēm-ō-poy-Ē-sis) Blood cell production, which occurs in red bone marrow after birth. Also called **hematopoiesis** (hem′-a-tō-poy-Ē-sis).

Hemorrhage (HEM-o-rij) Bleeding; the escape of blood from blood vessels, especially when the loss is profuse.

Hemorrhoids (HEM-ō-royds) Dilated or varicosed blood vessels (usually veins) in the anal region. Also called **piles.**

Hepatic (he-PAT-ik) Refers to the liver.

Hepatic duct A duct that receives bile from the bile capillaries. Small hepatic ducts merge to form the larger right and left hepatic ducts that unite to leave the liver as the common hepatic duct.

Hepatic portal circulation The flow of blood from the gastrointestinal organs to the liver before returning to the heart.

Hepatocyte (he-PAT-ō-cyte) A liver cell.

Hepatopancreatic (hep′-a-tō-pan′-krē-A-tik) **ampulla** A small, raised area in the duodenum where the combined common bile duct and main pancreatic duct empty into the duodenum. Also called the **ampulla of Vater** (VA-ter).

Hernia (HER-nē-a) The protrusion or projection of an organ or part of an organ through a membrane or cavity wall, usually the abdominal cavity.

Herniated (HER-nē-ā′-ted) **disc** A rupture of an intervertebral disc so that the nucleus pulposus protrudes into the vertebral cavity. Also called a **slipped disc.**

Hiatus (hī-Ā-tus) An opening; a foramen.

Hilum (HĪ-lum) An area, depression, or pit where blood vessels and nerves enter or leave an organ. Also called a **hilus.**

Hinge joint A synovial joint in which a convex surface of one bone fits into a concave surface of another bone, such as the elbow, knee, ankle, and interphalangeal joints. Also called a **ginglymus** (JIN-gli-mus) **joint.**

Hirsutism (HER-soo-tizm) An excessive growth of hair in females and children, with a distribution similar to that in adult males, due to the conversion of vellus hairs into large terminal hairs in response to higher-than-normal levels of androgens.

Histamine (HISS-ta-mēn) Substance found in many cells, especially mast cells, basophils, and platelets, that is released when the cells are injured; results in vasodilation, increased permeability of blood vessels, and constriction of bronchioles.

Histology (hiss′-TOL-ō-jē) Microscopic study of the structure of tissues.

Holocrine (HŌ-lō-krin) **gland** A type of gland in which entire secretory cells, along with their accumulated secretions, make up the secretory product of the gland, as in the sebaceous (oil) glands.

Homeostasis (hō′-mē-ō-STĀ-sis) The condition in which the body's internal environment remains relatively constant within physiological limits.

Hormone (HŌR-mōn) A secretion of endocrine cells that alters the physiological activity of target cells of the body.

Horn An area of gray matter (anterior, lateral, or posterior) in the spinal cord.

Human chorionic gonadotropin (kō-rē-ON-ik gō-nad-ō-TRŌ-pin) **(hCG)** A hormone produced by the developing placenta that maintains the corpus luteum.

Human chorionic somatomammotropin (sō-mat-ō-mam-ō-TRŌ-pin) **(hCS)** Hormone produced by the chorion of the placenta that stimulates breast tissue for lactation, enhances body growth, and regulates metabolism. Also called **human placental lactogen (hPL).**

Human growth hormone (hGH) Hormone secreted by the anterior pituitary that stimulates growth of body tissues, especially skeletal and muscular tissues. Also known as **somatotropin** and **somatotropic hormone (STH).**

Hyaluronic (hī′-a-loo-RON-ik) **acid** A viscous, amorphous extracellular material that binds cells together, lubricates joints, and maintains the shape of the eyeballs.

Hymen (HĪ-men) A thin fold of vascularized mucous membrane at the vaginal orifice.

Hyperextension (hī′-per-ek-STEN-shun) Continuation of extension beyond the anatomical position, as in bending the head backward.

Hyperplasia (hī-per-PLĀ-zē-a) An abnormal increase in the number of normal cells in a tissue or organ, increasing its size.

Hypersecretion (hī′-per-se-KRĒ-shun) Overactivity of glands resulting in excessive secretion.

Hypersensitivity (hī′-per-sen-si-TI-vi-tē) Overreaction to an allergen that results in pathological changes in tissues. Also called **allergy.**

Hypertension (hī′-per-TEN-shun) High blood pressure.

Hyperthermia (hī′-per-THERM-ē-a) An elevated body temperature.

Hypertonia (hī′-per-TŌ-nē-a) Increased muscle tone that is expressed as spasticity or rigidity.

Hypertonic (hī′-per-TON-ik) Solution that causes cells to shrink due to loss of water by osmosis.

Hypertrophy (hī-PER-trō-fē) An excessive enlargement or overgrowth of tissue without cell division.

Hyperventilation (hī′-per-ven-ti-LĀ-shun) A rate of inhalation and exhalation higher than that required to maintain a normal partial pressure of carbon dioxide in the blood.

Hyponychium (hī′-pō-NIK-ē-um) Free edge of the fingernail.

Hypophyseal fossa (hī′-pō-FIZ-ē-al FOS-a) A depression on the superior surface of the sphenoid bone that houses the pituitary gland.

Hypophyseal (hī′-pō-FIZ-ē-al) **pouch** An outgrowth of ectoderm from the roof of the mouth from which the anterior pituitary develops.

Hypophysis (hī-POF-i-sis) Pituitary gland.

Hyposecretion (hī′-pō-se-KRĒ-shun) Underactivity of glands resulting in diminished secretion.

Hypothalamohypophyseal (hī′-pō-thal′-a-mō-hī-pō-FIZ-ē-al) **tract** A bundle of axons containing secretory vesicles filled with oxytocin or antidiuretic hormone that extend from the hypothalamus to the posterior pituitary.

Hypothalamus (hī′-pō-THAL-a-mus) A portion of the diencephalon, lying beneath the thalamus and forming the floor and part of the wall of the third ventricle.

Hypothermia (hī′-pō-THER-mē-a) Lowering of body temperature below 35 °C (95 °F); in surgical procedures, it refers to deliberate cooling of the body to slow down metabolism and reduce oxygen needs of tissues.

Hypotonia (hī′-pō-TŌ-nē-a) Decreased or lost muscle tone in which muscles appear flaccid.

Hypotonic (hī′-pō-TON-ik) Solution that causes cells to swell and perhaps rupture due to gain of water by osmosis.

Hypoventilation (hī-pō-ven-ti-LĀ-shun) A rate of inhalation and exhalation lower than that required to maintain a normal partial pressure of carbon dioxide in plasma.

Hypoxia (hī-POKS-ē-a) Lack of adequate oxygen at the tissue level.

Hysterectomy (hiss-te-REK-tō-mē) The surgical removal of the uterus.

I

Ileocecal (il-ē-ō-SĒ-kal) **sphincter** A fold of mucous membrane that guards the opening from the ileum into the large intestine. Also called the **ileocecal valve.**

Ileum (IL-ē-um) The terminal part of the small intestine.

Immunity (im-Ū-ni-tē) The state of being resistant to injury, particularly by poisons, foreign proteins, and invading pathogens.

Immunoglobulin (im-ū-nō-GLOB-ū-lin) **(Ig)** An antibody synthesized by plasma cells derived from B lymphocytes in response to the introduction of an antigen. Immunoglobulins are divided into five kinds (IgG, IgM, IgA, IgD, IgE).

Immunology (im′-ū-NOL-ō-jē) The study of the responses of the body when challenged by antigens.

Imperforate (im-PER-fō-rāt) Abnormally closed.

Implantation (im-plan-TĀ-shun) The insertion of a tissue or a part into the body. The attachment of the blastocyst to the stratum basalis of the endometrium about 6 days after fertilization.

Incontinence (in-KON-ti-nens) Inability to retain urine, semen, or feces through loss of sphincter control.

Indirect motor pathways Motor tracts that convey information from the brain down the spinal cord for automatic movements, coordination of body movements with visual stimuli, skeletal muscle tone and posture, and balance. Also known as **extrapyramidal pathways.**

Infarction (in-FARK-shun) A localized area of necrotic tissue, produced by inadequate oxygenation of the tissue.

Infection (in-FEK-shun) Invasion and multiplication of microorganisms in body tissues, which may be inapparent or characterized by cellular injury.

Inferior (in-FĒR-ē-or) Away from the head or toward the lower part of a structure. Also called **caudad** (KAW-dad).

Inferior vena cava (VĒ-na CĀ-va) **(IVC)** Large vein that collects blood from parts of the body inferior to the heart and returns it to the right atrium.

Infertility Inability to conceive or to cause conception. Also called **sterility.**

Inflammation (in′-fla-MĀ-shun) Localized, protective response to tissue injury designed to destroy, dilute, or wall off the infecting agent or injured tissue; characterized by redness, pain, heat, swelling, and sometimes loss of function.

Infundibulum (in-fun-DIB-ū-lum) The stalklike structure that attaches the pituitary gland to the hypothalamus of the brain. The funnel-shaped, open, distal end of the uterine (Fallopian) tube.

Ingestion (in-JES-chun) The taking in of food, liquids, or drugs, by mouth.

Inguinal (IN-gwi-nal) Pertaining to the groin.

Inguinal canal An oblique passageway in the anterior abdominal wall just superior and parallel to the medial half of the inguinal ligament that transmits the spermatic cord and ilioinguinal nerve in the male and round ligament of the uterus and ilioinguinal nerve in the female.

Inhalation (in-ha-LĀ-shun) The act of drawing air into the lungs. Also termed **inspiration.**

Inheritance The acquisition of body traits by transmission of genetic information from parents to offspring.

Inhibin (in-HIB-in) A hormone secreted by the gonads that inhibits release of follicle-stimulating hormone (FSH) by the anterior pituitary.

Inhibiting hormone Hormone secreted by the hypothalamus that can suppress secretion of hormones by the anterior pituitary.

Insertion (in-SER-shun) The attachment of a muscle tendon to a movable bone or the end opposite the origin.

Insula (IN-soo-la) A triangular area of the cerebral cortex that lies deep within the lateral cerebral fissue, under the parietal, frontal, and temporal lobes.

Insulin (IN-soo-lin) A hormone produced by the beta cells of a pancreatic islet (islet of Langerhans) that decreases the blood glucose level.

Integrins (IN-te-grinz) A family of transmembrane glycoproteins in plasma membranes that function in cell adhesion; they are present in hemidesmosomes, which anchor cells to a basement membrane, and they mediate adhesion of neutrophils to endothelial cells during emigration.

Integumentary (in-teg-ū-MEN-tar-ē) Relating to the skin.

Intercalated (in-TER-ka-lā t-ed) **disc** An irregular transverse thickening of sarcolemma that contains desmosomes, which hold cardiac muscle fibers (cells) together, and gap junctions, which aid in conduction of muscle action potentials from one fiber to the next.

Intercostal (in′-ter-KOS-tal) **nerve** A nerve supplying a muscle located between the ribs.

Intermediate (in-ter-MĒ-de-at) Between two structures, one of which is medial and one of which is lateral.

Intermediate filament Protein filament, ranging from 8 to 12 nm in diameter, that may provide structural reinforcement, hold organelles in place, and give shape to a cell.

Internal Away from the surface of the body.

Internal capsule A large tract of projection fibers lateral to the thalamus that is the major connection between the cerebral cortex and the brain stem and spinal cord; contains axons of sensory neurons carrying auditory, visual, and somatic sensory signals to the thalamus and cerebral cortex plus axons of motor neurons descending from the cerebral cortex to the thalamus, subthalamus, brain stem, and spinal cord.

Internal ear The inner ear or labyrinth, lying inside the temporal bone, containing the organs of hearing and balance.

Internal nares (NĀ-rez) The two openings posterior to the nasal cavities opening into the nasopharynx. Also called the **choanae** (kō-Ā-nē).

Internal respiration The exchange of respiratory gases between blood and body cells. Also called **tissue respiration.**

Interneurons (in′-ter-NOO-ronz) Neurons whose axons extend only for a short distance and contact nearby neurons in the brain, spinal cord, or a ganglion; they comprise the vast majority of neurons in the body.

Interoceptor (IN-ter-ō-sep′-tor) Sensory receptor located in blood vessels and viscera that provides information about the body's internal environment.

Interphase (IN-ter-fāz) The period of the cell cycle between cell divisions, consisting of the G_1-(gap or growth) phase, when the cell is engaged in growth, metabolism, and production of substances required for division; S-(synthesis) phase, during which chromosomes are replicated; and G_2-phase.

Interstitial cell of Leydig *See* **Interstitial endocrinocyte.**

Interstitial (in′-ter-STISH-al) **endocrinocyte** A cell that is located in the connective tissue between seminiferous tubules in a mature testis that secretes testosterone. Also called an **interstitial cell of Leydig** (LĪ-dig).

Interstitial (in′-ter-STISH-al) **fluid** The portion of extracellular fluid that fills the microscopic spaces between the cells of tissues; the internal environment of the body. Also called **intercellular** or **tissue fluid.**

Interstitial growth Growth from within, as in the growth of cartilage. Also called **endogenous** (en-DOJ-e-nus) **growth.**

Interventricular (in′-ter-ven-TRIK-ū-lar) **foramen** A narrow, oval opening through which the lateral ventricles of the brain communicate with the third ventricle. Also called the **foramen of Monro.**

Intervertebral (in′-ter-VER-te-bral) **disc** A pad of fibrocartilage located between the bodies of two vertebrae.

Intestinal gland A gland that opens onto the surface of the intestinal mucosa and secretes digestive enzymes. Also called a **crypt of Lieberkühn** (LĒ-ber-kūn).

Intracellular (in′-tra-SEL-yū-lar) **fluid (ICF)** Fluid located within cells.

Intrafusal (in′-tra-FŪ-sal) **fibers** Three to ten specialized muscle fibers (cells), partially enclosed in a spindle-shaped connective tissue capsule, that make up a muscle spindle.

Intramembranous ossification (in′-tra-MEM-bra-nus os′-i-fi-KĀ-shun) The method of bone formation in which the bone is formed directly in mesenchyme arranged sheet like layers that resemble membranes.

Intraocular (in′-tra-OK-ū-lar) **pressure (IOP)** Pressure in the eyeball, produced mainly by aqueous humor.

Intrinsic factor (IF) (in-TRIN-sik) A glycoprotein, synthesized and secreted by the parietal cells of the gastric mucosa, that facilitates vitamin B_{12} absorption in the small intestine.

Inversion (in-VER-zhun) The movement of the sole medially at the ankle joint.

In vitro (VĒ-trō) Literally, in glass; outside the living body and in an artificial environment such as a laboratory test tube.

Ipsilateral (ip-si-LAT-er-al) On the same side, affecting the same side of the body.

Iris The colored portion of the vascular tunic of the eyeball seen through the cornea that contains circular and radial smooth muscle; the hole in the center of the iris is the pupil.

Irritable bowel syndrome (IBS) Disease of the entire gastrointestinal tract in which a person reacts to stress by developing symptoms (such as cramping and abdominal pain) associated with alternating patterns of diarrhea and constipation. Excessive amounts of mucus may appear in feces, and other symptoms include flatulence, nausea, and loss of appetite. Also known as **irritable colon** or **spastic colitis.**

Ischemia (is-KĒ-mē-a) A lack of sufficient blood to a body part due to obstruction or constriction of a blood vessel.

Islet of Langerhans *See* **Pancreatic islet.**

Isometric (ī′-sō-MET-rik) **contraction** A muscle contraction in which tension in the muscle increases, but there is only minimal muscle shortening so that no visible movement is produced.

Isotonic (ī′-sō-TON-ik) Having equal tension or tone. A solution having the same concentration of impermeable solutes as cytosol.

Isotonic contraction Contraction in which the tension remains the same; occurs when a constant load is moved through the range of motions possible at a joint.

Isotopes (Ī-sō-tōps′) Chemical elements that have the same number of protons but different numbers of neutrons. Radioactive isotopes change into other elements with the emission of alpha or beta particles or gamma rays.

Isthmus (IS-mus) A narrow strip of tissue or narrow passage connecting two larger parts.

J

Jaundice (JON-dis) A condition characterized by yellowness of the skin, the white of the eyes, mucous membranes, and body fluids because of a buildup of bilirubin.

Jejunum (je-JOO-num) The middle part of the small intestine.

Joint kinesthetic (kin′-es-THET-ik) **receptor** A proprioceptive receptor located in a joint, stimulated by joint movement.

Juxtaglomerular (juks-ta-glō-MER-ū-lar) **apparatus (JGA)** Consists of the macula densa (cells of the distal convoluted tubule adjacent to the afferent and efferent arteriole) and juxtaglomerular cells (modified cells of the afferent and sometimes efferent arteriole); secretes renin when blood pressure starts to fall.

K

Keratin (KER-a-tin) An insoluble protein found in the hair, nails, and other keratinized tissues of the epidermis.

Keratinocyte (ker-a-TIN-ō-sīt) The most numerous of the epidermal cells; produces keratin.

Kidney (KID-nē) One of the paired reddish organs located in the lumbar region that regulates the composition, volume, and pressure of blood and produces urine.

Kidney stone A solid mass, usually consisting of calcium oxalate, uric acid, or calcium phosphate crystals, that may form in any portion of the urinary tract. Also called a **renal calculus** (KAL-kū-lus).

Kinesiology (ki-nē-sē′-OL-ō-jē) The study of the movement of body parts.

Kinesthesia (kin′-es-THĒ-zē-a) The perception of the extent and direction of movement of body parts; this sense is possible due to nerve impulses generated by proprioceptors.

Kinetochore (ki-NET-ō-kor) Protein complex attached to the outside of a centromere to which kinetochore microtubules attach.

Kupffer's cell *See* **Stellate reticuloendothelial cell.**

Kyphosis (kī-FŌ-sis) An exaggeration of the thoracic curve of the vertebral column, resulting in a "round-shouldered" appearance. Also called **hunchback.**

L

Labial frenulum (LĀ-bē-al FREN-ū-lum) A medial fold of mucous membrane between the inner surface of the lip and the gums.

Labia majora (LĀ-bē-a ma-JŌ-ra) Two longitudinal folds of skin extending downward and backward from the mons pubis of the female.

Labia minora (min-OR-a) Two small folds of mucous membrane lying medial to the labia majora of the female.

Labium (LĀ-bē-um) A lip. A liplike structure. *Plural* is **labia** (LĀ-bē-a).

Labor The process of giving birth in which a fetus is expelled from the uterus through the vagina.

Labyrinth (LAB-i-rinth) Intricate communicating passageway, especially in the internal ear.

Lacrimal canal (LAK-ri-mal) A duct, one on each eyelid, beginning at the punctum at the medial margin of an eyelid and conveying tears medially into the nasolacrimal sac.

Lacrimal gland Secretory cells, located at the superior anterolateral portion of each orbit, that secrete tears into excretory ducts that open onto the surface of the conjunctiva.

Lacrimal sac The superior expanded portion of the nasolacrimal duct that receives the tears from a lacrimal canal.

Lactation (lak-TĀ-shun) The secretion and ejection of milk by the mammary glands.

Lacteal (LAK-tē-al) One of many lymphatic vessels in villi of the intestines that absorb triglycerides and other lipids from digested food.

Lacuna (la-KOO-na) A small, hollow space, such as that found in bones in which the osteocytes lie. *Plural* is **lacunae** (la-KOO-nē).

Lambdoid (LAM-doyd) **suture** The joint in the skull between the parietal bones and the occipital bone; sometimes contains sutural (Wormian) bones.

Lamellae (la-MEL-ē) Concentric rings of hard, calcified extracellular matrix found in compact bone.

Lamellated corpuscle See **pacinian corpuscle.**

Lamina (LAM-i-na) A thin, flat layer or membrane, as the flattened part of either side of the arch of a vertebra. *Plural* is **laminae** (LAM-i-nē).

Lamina propria (PRŌ-prē-a) The connective tissue layer of a mucosa.

Langerhans (LANG-er-hans) **cell** Epidermal dendritic cell that functions as an antigen-presenting cell (APC) during an immune response.

Large intestine The portion of the gastrointestinal tract extending from the ileum of the small intestine to the anus, divided structurally into the cecum, colon, rectum, and anal canal.

Laryngopharynx (la-rin′-gō-FAR-inks) The inferior portion of the pharynx, extending downward from the level of the hyoid bone that divides posteriorly into the esophagus and anteriorly into the larynx. Also called the **hypopharynx.**

Larynx (LAR-inks) The voice box, a short passageway that connects the pharynx with the trachea.

Lateral (LAT-er-al) Farther from the midline of the body or a structure.

Lateral ventricle (VEN-tri-kul) A cavity within a cerebral hemisphere that communicates with the lateral ventricle in the other cerebral hemisphere and with the third ventricle by way of the interventricular foramen.

Leg The part of the lower limb between the knee and the ankle.

Lens A transparent organ constructed of proteins (crystallins) lying posterior to the pupil and iris of the eyeball and anterior to the vitreous body.

Lesion (LĒ-zhun) Any localized, abnormal change in a body tissue.

Lesser omentum (ō-MEN-tum) A fold of the peritoneum that extends from the liver to the lesser curvature of the stomach and the first part of the duodenum.

Lesser vestibular (ves-TIB-ū-lar) **gland** One of the paired mucus-secreting glands with ducts that open on either side of the urethral orifice in the vestibule of the female.

Leukemia (loo-KĒ-mē-a) A malignant disease of the blood-forming tissues characterized by either uncontrolled production and accumulation of immature leukocytes in which many cells fail to reach maturity (acute) or an accumulation of mature leukocytes in the blood because they do not die at the end of their normal life span (chronic).

Leukocyte (LOO-kō-sīt) A white blood cell.

Leydig (LĪ-dig) **cell** A type of cell that secretes testosterone; located in the connective tissue between seminiferous tubules in a mature testis. Also known as **interstitial cell of Leydig** or **interstitial endocrinocyte.**

Ligament (LIG-a-ment) Dense regular connective tissue that attaches bone to bone.

Ligand (LĪ-gand) A chemical substance that binds to a specific receptor.

Limbic system A part of the forebrain, sometimes termed the visceral brain, concerned with various aspects of emotion and behavior; includes the limbic lobe, dentate gyrus, amygdala, septal nuclei, mammillary bodies, anterior thalamic nucleus, olfactory bulbs, and bundles of myelinated axons.

Lingual frenulum (LIN-gwal FREN-ū-lum) A fold of mucous membrane that connects the tongue to the floor of the mouth.

Lipase An enzyme that splits fatty acids from triglycerides and phospholipids.

Lipid (LIP-id) An organic compound composed of carbon, hydrogen, and oxygen that is usually insoluble in water, but soluble in alcohol, ether, and chloroform; examples include triglycerides (fats and oils), phospholipids, steroids, and eicosanoids.

Lipid bilayer Arrangement of phospholipid, glycolipid, and cholesterol molecules in two parallel sheets in which the hydrophilic "heads" face outward and the hydrophobic "tails" face inward; found in cellular membranes.

Lipoprotein (lip′-ō-PRŌ-tēn) One of several types of particles containing lipids (cholesterol and triglycerides) and proteins that make it water soluble for transport in the blood; high levels of **low-density lipoproteins (LDLs)** are associated with increased risk of atherosclerosis, whereas high levels of **high-density lipoproteins (HDLs)** are associated with decreased risk of atherosclerosis.

Liver Large organ under the diaphragm that occupies most of the right hypochondriac region and part of the epigastric region. Functionally, it produces bile and synthesizes most plasma proteins; interconverts nutrients; detoxifies substances; stores glycogen, iron, and vitamins; carries on phagocytosis of worn-out blood cells and bacteria; and helps synthesize the active form of vitamin D.

Long-term potentiation (LTP) (PO-ten′-she′-Ā-shun) Prolonged, enhanced synaptic transmission that occurs at certain synapses within the hippocampus of the brain; believed to underlie some aspects of memory.

Lordosis (lor-DŌ-sis) An exaggeration of the lumbar curve of the vertebral column. Also called **hollow back.**

Lower limb The appendage attached at the pelvic (hip) girdle, consisting of the thigh, knee, leg, ankle, foot, and toes. Also called the **lower extremity.**

Lumbar (LUM-bar) Region of the back and side between the ribs and pelvis; loin.

Lumbar plexus (PLEK-sus) A network formed by the anterior (ventral) branches of spinal nerves L1 through L4.

Lumen (LOO-men) The space within an artery, vein, intestine, renal tubule, or other tubular structure.

Lungs Main organs of respiration that lie on either side of the heart in the thoracic cavity.

Lunula (LOO-noo-la) The moon-shaped white area at the base of a nail.

Luteinizing (LOO-tē-in′-īz-ing) **hormone (LH)** A hormone secreted by the anterior pituitary that stimulates ovulation, stimulates progesterone secretion by the corpus luteum, and readies the mammary glands for milk secretion in females; stimulates testosterone secretion by the testes in males.

Lymph (LIMF) Fluid confined in lymphatic vessels and flowing through the lymphatic system until it is returned to the blood.

Lymph node An oval or bean-shaped structure located along lymphatic vessels.

Lymphatic (lim-FAT-ik) **capillary** Closed-ended microscopic lymphatic vessel that begins in spaces between cells and converges with other lymphatic capillaries to form lymphatic vessels.

Lymphatic tissue A specialized form of reticular tissue that contains large numbers of lymphocytes.

Lymphatic vessel A large vessel that collects lymph from lymphatic capillaries and converges with other lymphatic vessels to form the thoracic or right lymphatic ducts.

Lymphocyte (LIM-fō-sīt) A type of white blood cell that helps carry out cell-mediated and antibody-mediated immune responses; found in blood and in lymphatic tissues.

Lysosome (LĪ-sō-sōm) An organelle in the cytoplasm of a cell, enclosed by a single membrane and containing powerful digestive enzymes.

Lysozyme (LĪ-sō-zīm) A bactericidal enzyme found in tears, saliva, and perspiration.

M

Macrophage (MAK-rō-fāj) Phagocytic cell derived from a monocyte; may be fixed or wandering.

Macula (MAK-ū-la) A discolored spot or a colored area. A small, thickened region on the wall of the utricle and saccule that contains receptors for static equilibrium.

Macula lutea (LOO-tē-a) The yellow spot in the center of the retina.

Major histocompatibility (MHC) antigens Surface proteins on white blood cells and other nucleated cells that are unique for each person (except for identical siblings); used to type tissues and help prevent rejection of transplanted tissues. Also known as **human leukocyte antigens (HLA).**

Malignant (ma-LIG-nant) Referring to diseases that tend to become worse and cause death, especially the invasion and spreading of cancer.

Mammary (MAM-ar-ē) **gland** Modified sudoriferous (sweat) gland of the female that produces milk for the nourishment of the young.

Mammillary (MAM-i-ler-ē) **bodies** Two small rounded bodies on the inferior aspect of the hypothalamus that are involved in reflexes related to the sense of smell.

Marrow (MAR-ō) Soft, spongelike material in the cavities of bone. Red bone marrow produces blood cells; yellow bone marrow contains adipose tissue that stores triglycerides.

Massage therapy The scientific application of soft tissue manipulation utilized in the treatment and prevention of structural and functional disorders. It can affect almost every system of the body, either directly or indirectly.

Mast cell A cell found in areolar connective tissue that releases histamine, a dilator of small blood vessels, during inflammation.

Mastication (mas′-ti-KĀ-shun) Chewing.

Mature follicle A large, fluid-filled follicle containing a secondary oocyte and surrounding granulosa cells that secrete estrogens. Also called a **graafian** (GRAF-ē-an) **follicle.**

Meatus (mē-Ā-tus) A passage or opening, especially the external portion of a canal.

Mechanoreceptor (me-KAN-ō-rē-sep-tor) Sensory receptor that detects mechanical deformation of the receptor itself or adjacent cells; stimuli so detected include those related to touch, pressure, vibration, proprioception, hearing, equilibrium, and blood pressure.

Medial (MĒ-dē-al) Nearer the midline of the body or a structure.

Medial lemniscus (lem-NIS-kus) A white matter tract that originates in the gracile and cuneate nuclei of the medulla oblongata and extends to the thalamus on the same side; sensory axons in this tract conduct nerve impulses for the sensations of proprioception, fine touch, vibration, hearing, and equilibrium.

Median aperture (AP-er-choor) One of the three openings in the roof of the fourth ventricle through which cerebrospinal fluid enters the subarachnoid space of the brain and cord. Also called the **foramen of Magendie** (ma-ghan-DĒ).

Median plane A vertical plane dividing the body into right and left halves. Situated in the middle.

Mediastinum (mē′-dē-as-TĪ-num) The anatomical region on the thoracic cavity between the pleurae of the lungs that extends from the sternum to the vertebral column and from the first rib to the diaphragm.

Medulla (me-DOOL-la) An inner portion of an organ, such as the medulla of the kidneys.

Medulla oblongata (me-DOOL-la ob′-long-GA-ta) The most inferior part of the brain stem. Also termed the **medulla.**

Medullary (MED-ū-lar′-ē) **cavity** The space within the diaphysis of a bone that contains yellow bone marrow. Also called the **marrow cavity.**

Medullary rhythmicity (rith-MIS-i-tē) **area** The neurons of the respiratory center in the medulla oblongata that control the basic rhythm of respiration.

Meibomian gland *See* **Tarsal gland.**

Meiosis (mī-Ō-sis) A type of cell division that occurs during production of gametes, involving two successive nuclear divisions that result in cells with the haploid (*n*) number of chromosomes.

Meissner (mīs-ner) **corpuscle** A sensory receptor for touch; found in dermal papillae, especially in the palms and soles. Also called a **corpuscle of touch.**

Melanin (MEL-a-nin) A dark black, brown, or yellow pigment found in some parts of the body such as the skin, hair, and pigmented layer of the retina.

Melanocyte (MEL-a-nō-sīt′) A pigmented cell, located between or beneath cells of the deepest layer of the epidermis, that synthesizes melanin.

Melanocyte-stimulating hormone (MSH) A hormone secreted by the anterior pituitary that stimulates the dispersion of melanin granules in melanocytes in amphibians; continued administration produces darkening of skin in humans.

Melatonin (mel-a-TŌN-in) A hormone secreted by the pineal gland that helps set the timing of the body's biological clock.

Membrane A thin, flexible sheet of tissue composed of an epithelial layer and an underlying connective tissue layer, as in an epithelial membrane, or of areolar connective tissue only, as in a synovial membrane.

Membranous labyrinth (mem-BRA-nus LAB-i-rinth) The part of the labyrinth of the internal ear that is located inside the bony labyrinth and separated from it by the perilymph; made up of the semicircular ducts, the saccule and utricle, and the cochlear duct.

Memory The ability to recall thoughts; commonly classifed as short-term (activated) and long-term.

Menarche (me-NAR-kē) The first menses (menstrual flow) and beginning of ovarian and uterine cycles.

Meninges (me-NIN-jēz) Three membranes covering the brain and spinal cord, called the dura mater, arachnoid mater, and pia mater. *Singular* is **meninx** (MEN-inks).

Menopause (MEN-ō-pawz) The termination of the menstrual cycles.

Menstrual (MEN-strū-al) **cycle** A series of changes in the endometrium of a nonpregnant female that prepares the lining of the uterus to receive a fertilized ovum.

Menstruation (men′-stroo-Ā-shun) Periodic discharge of blood, tissue fluid, mucus, and epithelial cells that usually lasts for 5 days; caused by a sudden reduction in estrogens and progesterone. Also called the **menstrual phase** or **menses.**

Merkel (MER-kel) **cell** Type of cell in the epidermis of hairless skin that makes contact with a Merkel (tactile) disc, which functions in touch.

Merkel Disc Saucer-shaped free nerve endings that make contact with Merkel cells in the epidermis and function as touch receptors. Also called a **tactile disc.**

Merocrine (MER-ō-krin) **gland** Gland made up of secretory cells that remain intact throughout the process of formation and discharge of the secretory product, as in the salivary and pancreatic glands.

Mesenchyme (MEZ-en-kīm) An embryonic connective tissue from which almost all other connective tissues arise.

Mesentery (MEZ-en-ter′-ē) A fold of peritoneum attaching the small intestine to the posterior abdominal wall.

Mesocolon (mez′-ō-KŌ-lon) A fold of peritoneum attaching the colon to the posterior abdominal wall.

Mesoderm The middle primary germ layer that gives rise to connective tissues, blood and blood vessels, and muscles.

Mesothelium (mez′-ō-THĒ-lē-um) The layer of simple squamous epithelium that lines serous membranes.

Mesovarium (mez′-ō-VAR-ē-um) A short fold of peritoneum that attaches an ovary to the broad ligament of the uterus.

Metabolism (me-TAB-ō-lizm) All the biochemical reactions that occur within an organism, including the synthetic (anabolic) reactions and decomposition (catabolic) reactions.

Metacarpus (met′-a-KAR-pus) A collective term for the five bones that make up the palm.

Metaphase (MET-a-phāz) The second stage of mitosis, in which chromatid pairs line up on the metaphase plate of the cell.

Metaphysis (me-TAF-i-sis) Region of a long bone between the diaphysis and epiphysis that contains the epiphyseal plate in a growing bone.

Metarteriole (met′-ar-TĒ-rē-ōl) A blood vessel that emerges from an arteriole, traverses a capillary network, and empties into a venule.

Metastasis (me-TAS-ta-sis) The spread of cancer to surrounding tissues (local) or to other body sites (distant).

Metatarsus (met′-a-TAR-sus) A collective term for the five bones located in the foot between the tarsals and the phalanges.

Microfilament (mī-krō-FIL-a-ment) Rodlike protein filament about 6 nm in diameter; constitutes contractile units in muscle fibers (cells) and provides support, shape, and movement in nonmuscle cells.

Microglia (mī-KROG-lē-a) Neuroglial cells that carry on phagocytosis.

Microtubule (mī-krō-TOO-būl′) Cylindrical protein filament, from 18 to 30 nm in diameter, consisting of the protein tubulin; provides support, structure, and transportation within a cell.

Microvilli (mī′-krō-VIL-ē) Microscopic, fingerlike projections of the plasma membranes of cells that increase surface area for absorption, especially in the small intestine and proximal convoluted tubules of the kidneys.

Micturition (mik′-choo-RISH-un) The act of expelling urine from the urinary bladder. Also called **urination** (ū-ri-NĀ-shun).

Midbrain The part of the brain between the pons and the diencephalon. Also called the **mesencephalon** (mes′-en-SEF-a-lon).

Middle ear A small, epithelial-lined cavity hollowed out of the temporal bone, separated from the external ear by the eardrum and from the internal ear by a thin bony partition containing the oval and round windows; extending across the middle ear are the three auditory ossicles. Also called the **tympanic** (tim-PAN-ik) **cavity.**

Midline An imaginary vertical line that divides the body into equal left and right sides.

Midsagittal plane A vertical plane through the midline of the body that divides the body or organs into *equal* right and left sides. Also called a **median plane.**

Mineralocorticoids (min′-er-al-ō-KOR-ti-koyds) A group of hormones of the adrenal cortex that help regulate sodium and potassium balance.

Mitochondrion (mī-tō-KON-drē-on) A double-membraned organelle that plays a central role in the production of ATP; known as the "powerhouse" of the cell. *Plural* is **mitochondria.**

Mitosis (mī-TŌ-sis) The orderly division of the nucleus of a cell that ensures that each new nucleus has the same number and kind of chromosomes as the original nucleus. The process includes the replication of chromosomes and the distribution of the two sets of chromosomes into two separate and equal nuclei.

Mitotic spindle Collective term for a football-shaped assembly of microtubules (nonkinetochore, kinetochore, and aster) that is responsible for the movement of chromosomes during cell division.

Modality (mō-DAL-i-tē) Any of the specific sensory entities, such as vision, smell, taste, or touch.

Monocyte (MON-ō-sit′) The largest type of white blood cell, characterized by agranular cytoplasm.

Monounsaturated fat A fatty acid that contains one double covalent bond between its carbon atoms; it is not completely saturated with hydrogen atoms. Plentiful in triglycerides of olive and peanut oils.

Mons pubis (MONZ PŪ-bis) The rounded, fatty prominence over the pubic symphysis, covered by coarse pubic hair.

Morula (MOR-ū-la) A solid sphere of cells produced by successive cleavages of a fertilized ovum about four days after fertilization.

Motor area The region of the cerebral cortex that governs muscular movement, particularly the precentral gyrus of the frontal lobe.

Motor end plate Region of the sarcolemma of a muscle fiber (cell) that includes acetylcholine (ACh) receptors, which bind ACh released by synaptic end bulbs of somatic motor neurons.

Motor neurons (NOO-ronz) Neurons that conduct impulses from the brain toward the spinal cord or out of the brain and spinal cord into cranial or spinal nerves to effectors that may be either muscles or glands. Also called **efferent neurons.**

Motor unit A motor neuron together with the muscle fibers (cells) it stimulates.

Mucosa-associated lymphatic tissue (MALT) Lymphatic nodules scattered throughout the lamina propria (connective tissue) of mucous membranes lining the gastrointestinal tract, respiratory airways, urinary tract, and reproductive tract.

Mucous (MŪ-kus) **cell** A unicellular gland that secretes mucus. Two types are mucous neck cells and surface mucous cells in the stomach.

Mucous membrane A membrane that lines a body cavity that opens to the exterior. Also called the **mucosa** (mū-KŌ-sa).

Mucus The thick fluid secretion of goblet cells, mucous cells, mucous glands, and mucous membranes.

Muscarinic (mus′-ka-RIN-ik) **receptor** Receptor for the neurotransmitter acetylcholine found on all effectors innervated by parasympathetic postganglionic axons and on sweat glands innervated by cholinergic sympathetic postganglionic axons; so named because muscarine activates these receptors but does not activate nicotinic receptors for acetylcholine.

Muscle An organ composed of one of three types of muscle tissue (skeletal, cardiac, or smooth), specialized for contraction to produce voluntary or involuntary movement of parts of the body.

Muscle action potential A stimulating impulse that propagates along the sarcolemma and transverse tubules; in skeletal muscle, it is generated by acetylcholine, which increases the permeability of the sarcolemma to cations, especially sodium ions (Na^+).

Muscle fatigue (fa-TĒG) Inability of a muscle to maintain its strength of contraction or tension; may be related to insufficient oxygen, depletion of glycogen, and/or lactic acid buildup.

Muscle spindle An encapsulated proprioceptor in a skeletal muscle, consisting of specialized intrafusal muscle fibers and nerve endings; stimulated by changes in length or tension of muscle fibers.

Muscle tone A sustained, partial contraction of portions of a skeletal or smooth muscle in response to activation of stretch receptors or a baseline level of action potentials in the innervating motor neurons.

Muscular dystrophies (DIS-trō-fēz′) Inherited muscle-destroying diseases, characterized by degeneration of muscle fibers (cells), which causes progressive atrophy of the skeletal muscle.

Muscularis (MUS-kū-la′-ris) A muscular layer (coat or tunic) of an organ.

Muscularis mucosae (mū-KŌ-sē) A thin layer of smooth muscle fibers that underlie the lamina propria of the mucosa of the gastrointestinal tract.

Muscular tissue A tissue specialized to produce motion in response to muscle action potentials by its qualities of contractility, extensibility, elasticity, and excitability; types include skeletal, cardiac, and smooth.

Myasthenia (mī-as-THĒ-nē-a) **gravis** Weakness and fatigue of skeletal muscles caused by antibodies directed against acetylcholine receptors.

Myelin (MĪ-e-lin) **sheath** Multilayered lipid and protein covering, formed by Schwann cells and oligodendrocytes, around axons of many peripheral and central nervous system neurons.

Myenteric plexus A network of autonomic axons and postganglionic cell bodies located in the muscularis of the gastrointestinal tract. Also called the **plexus of Auerbach** (OW-er-bak).

Myocardial infarction (mī′-ō-KAR-dē-al in-FARK-shun) **(MI)** Gross necrosis of myocardial tissue due to interrupted blood supply. Also called a **heart attack.**

Myocardium (mī′-ō-KAR-dē-um) The middle layer of the heart wall, made up of cardiac muscle tissue, lying between the epicardium and the endocardium and constituting the bulk of the heart.

Myofibril (mī-ō-FĪ-bril) A threadlike structure, extending longitudinally through a muscle fiber (cell) consisting mainly of thick filaments (myosin) and thin filaments (actin, troponin, and tropomyosin).

Myoglobin (mī-ō-GLŌB-in) The oxygen-binding, iron-containing protein present in the sarcoplasm of muscle fibers (cells); contributes the red color to muscle.

Myogram (MĪ-ō-gram) The record or tracing produced by a myograph, an apparatus that measures and records the force of muscular contractions.

Myology (mī-OL-ō-jē) The study of muscles.

Myometrium (mī′-ō-MĒ-trē-um) The smooth muscle layer of the uterus.

Myopathy (mī-OP-a-thē) Any abnormal condition or disease of muscle tissue.

Myopia (mī-Ō-pē-a) Defect in vision in which objects can be seen distinctly only when close to the eyes; nearsightedness.

Myosin (MĪ-ō-sin) The contractile protein that makes up the thick filaments of muscle fibers.

Myotome (MĪ-ō-tōm) A group of muscles innervated by the motor neurons of a single spinal segment. In an embryo, the portion of a somite that develops into some skeletal muscles.

N

Nail A hard plate, composed largely of keratin, that develops from the epidermis of the skin to form a protective covering on the dorsal surface of the distal phalanges of the fingers and toes.

Nail matrix (MĀ-triks) The part of the nail beneath the body and root from which the nail is produced.

Nasal (NĀ-zal) **cavity** A mucosa-lined cavity on either side of the nasal septum that opens onto the face at the external nares and into the nasopharynx at the internal nares.

Nasal septum (SEP-tum) A vertical partition composed of bone (perpendicular plate of ethmoid and vomer) and cartilage, covered with a mucous membrane, separating the nasal cavity into left and right sides.

Nasolacrimal (nā′-zō-LAK-ri-mal) **duct** A canal that transports the lacrimal secretion (tears) from the nasolacrimal sac into the nose.

Nasopharynx (nā′-zō-FAR-inks) The superior portion of the pharynx, lying posterior to the nose and extending inferiorly to the soft palate.

Neck The part of the body connecting the head and the trunk. A constricted portion of an organ, such as the neck of the femur or uterus.

Necrosis (ne-KRŌ-sis) A pathological type of cell death that results from disease, injury, or lack of blood supply in which many adjacent cells swell, burst, and spill their contents into the interstitial fluid, triggering an inflammatory response.

Neoplasm (NĒ-ō-plazm) A new growth that may be benign or malignant.

Nephron (NEF-ron) The functional unit of the kidney.

Nerve A cordlike bundle of neuronal axons and/or dendrites and associated connective tissue coursing together outside the central nervous system.

Nerve fiber General term for any process (axon or dendrite) projecting from the cell body of a neuron.

Nerve impulse A wave of depolarization and repolarization that self-propagates along the plasma membrane of a neuron; also called a **nerve action potential.**

Nervous tissue Tissue containing neurons that initiate and conduct nerve impulses to coordinate homeostasis, and neuroglia that provide support and nourishment to neurons.

Neuralgia (noo-RAL-jē-a) Attacks of pain along the entire course or branch of a peripheral sensory nerve.

Neuritis (noo-RĪ-tis) Inflammation of one or more nerves.

Neurofibral node *See* **Node of Ranvier.**

Neuroglia (noo-RŌG-lē-a) Cells of the nervous system that perform various supportive functions. The neuroglia of the central nervous system are the astrocytes, oligodendrocytes, microglia, and ependymal cells; neuroglia of the peripheral nervous system include Schwann cells and satellite cells. Also called **glial** (GLĒ-al) **cells.**

Neurohypophyseal (noo′-rō-hī′-pō-FIZ-ē-al) **bud** An outgrowth of ectoderm located on the floor of the hypothalamus that gives rise to the posterior pituitary.

Neurolemma (noo-rō-LEM-ma) The peripheral, nucleated cytoplasmic layer of the Schwann cell. Also called **sheath of Schwann** (SCHWON).

Neurology (noo-ROL-ō-jē) The study of the normal functioning and disorders of the nervous system.

Neuromuscular (noo-rō-MUS-kū-lar) **junction (NMJ)** A synapse between the axon terminals of a motor neuron and the sarcolemma of a muscle fiber (cell).

Neuron (NOO-ron) A nerve cell, consisting of a cell body, dendrites, and an axon.

Neurosecretory (noo-rō-SĒC-re-tō-rē) **cell** A neuron that secretes a hypothalamic releasing hormone or inhibiting hormone into blood capillaries of the hypothalmus; a neuron that secretes oxytocin or antidiuretic hormone into blood capillaries of the posterior pituitary.

Neurotransmitter (NOO′-rō-trans′-mit-er) One of a variety of molecules within axon terminals that are released into the synaptic cleft in response to a nerve impulse and that change the membrane potential of the postsynaptic neuron.

Neutrophil (NOO-trō-fil) A type of white blood cell characterized by granules that stain pale lilac with a combination of acidic and basic dyes.

Nicotinic (nik′-ō-TIN-ik) **receptor** Receptor for the neurotransmitter acetylcholine found on both sympathetic and parasympathetic postganglionic neurons and on skeletal muscle in the motor end plate; so named because nicotine activates these receptors but does not activate muscarinic receptors for acetylcholine.

Nipple A pigmented, wrinkled projection on the surface of the breast that is the location of the openings of the lactiferous ducts for milk release.

Nociceptor (nō′-sē-SEP-tor) A free (naked) nerve ending that detects painful stimuli.

Node of Ranvier (RON-vē-ā) A space along a myelinated axon between the individual Schwann cells that form the myelin sheath and the neurolemma. Also called a **neurofibral node.**

Norepinephrine (nor′-ep-ē-NEF-rin) **(NE)** A hormone secreted by the adrenal medulla that produces actions similar to those that result from sympathetic stimulation. Also called **noradrenaline** (nor-a-DREN-a-lin).

Nucleic (noo-KLĒ-ic) **acid** An organic compound that is a long polymer of nucleotides, with each nucleotide containing a pentose sugar, a phosphate group, and one of four possible nitrogenous bases (adenine, cytosine, guanine, and thymine or uracil).

Nucleolus (noo′-KLĒ-ō-lus) Spherical body within a cell nucleus composed of protein, DNA, and RNA that is the site of the assembly of small and large ribosomal subunits. *Plural* is **nucleoli.**

Nucleosome (NOO-klē-ō-sōm) Structural subunit of a chromosome consisting of histones and DNA.

Nucleus (NOO-klē-us) A spherical or oval organelle of a cell that contains the hereditary factors of the cell, called genes. A cluster of unmyelinated nerve cell bodies in the central nervous system. The central part of an atom made up of protons and neutrons.

Nucleus pulposus (pul-PŌ-sus) A soft, pulpy, highly elastic substance in the center of an intervertebral disc; a remnant of the notochord.

Nutrient (NOO-trē-ent) A chemical substance in food that provides energy, forms new body components, or assists in various body functions.

O

Obesity (ō-BĒS-i-tē) Body weight more than 20% above a desirable standard due to excessive accumulation of fat.

Oblique (ō-BLĒK) **plane** A plane that passes through the body or an organ at an angle between the transverse plane and either the midsagittal, parasagittal, or frontal plane.

Obstetrics (ob-STET-riks) The specialized branch of medicine that deals with pregnancy, labor, and the period of time immediately after delivery (about 6 weeks).

Olfactory (ōl-FAK-tō-rē) Pertaining to smell.

Olfactory bulb A mass of gray matter containing cell bodies of neurons that form synapses with neurons of the olfactory (I) nerve, lying inferior to the frontal lobe of the cerebrum on either side of the crista galli of the ethmoid bone.

Olfactory receptor A bipolar neuron with its cell body lying between supporting cells located in the mucous membrane lining the superior portion of each nasal cavity; transduces odors into neural signals.

Olfactory tract A bundle of axons that extends from the olfactory bulb posteriorly to olfactory regions of the cerebral cortex.

Oligodendrocyte (OL-i-gō-den′-drō-sīt) A neuroglial cell that supports neurons and produces a myelin sheath around axons of neurons of the central nervous system.

Oliguria (ol′-i-GŪ-rē-a) Daily urinary output usually less than 250 ml.

Olive A prominent oval mass on each lateral surface of the superior part of the medulla oblongata.

Oncogenes (ON-kō-jēnz) Cancer-causing genes; they derive from normal genes, termed proto-oncogenes, that encode proteins involved in cell growth or cell regulation but have the ability to transform a normal cell into a cancerous cell when they are mutated or inappropriately activated. One example is *p53.*

Oncology (on-KOL-ō-jē) The study of tumors.

Oogenesis (ō′-ō-JEN-e-sis) Formation and development of female gametes (oocytes).

Oophorectomy (ō′-of-ō-REK-tō-me) Surgical removal of the ovaries.

Ophthalmic (of-THAL-mik) Pertaining to the eye.

Ophthalmologist (of′-thal-MOL-ō-jist) A physician who specializes in the diagnosis and treatment of eye disorders using drugs, surgery, and corrective lenses.

Ophthalmology (of-thal-MOL-ō-jē) The study of the structure, function, and diseases of the eye.

Optic (OP-tik) Refers to the eye, vision, or properties of light.

Optic chiasm (kī-AZ-m) A crossing point of the two branches of the optic (II) nerve, anterior to the pituitary gland. Also called **optic chiasma** (kī-AZ-ma).

Optic disc A small area of the retina containing openings through which the axons of the ganglion cells emerge as the optic (II) nerve. Also called the **blind spot.**

Optic tract A bundle of axons that carry nerve impulses from the retina of the eye between the optic chiasm and the thalamus.

Ora serrata (Ō-ra ser-RĀ-ta) The irregular margin of the retina lying internal and slightly posterior to the junction of the choroid and ciliary body.

Orbit (OR-bit) The bony, pyramidal-shaped cavity of the skull that holds the eyeball.

Organ A structure composed of two or more different kinds of tissues with a specific function and usually a recognizable shape.

Organelle (or-gan-EL) A permanent structure within a cell with characteristic morphology that is specialized to serve a specific function in cellular activities.

Organism (OR-ga-nizm) A total living form; one individual.

Organogenesis (or′-ga-nō-JEN-e-sis) The formation of body organs and systems. By the end of the eighth week of development, all major body systems have begun to develop.

Orifice (OR-i-fis) Any aperture or opening.

Origin (OR-i-jin) The attachment of a muscle tendon to a stationary bone or the end opposite the insertion.

Oropharynx (or′-ō-FAR-inks) The intermediate portion of the pharynx, lying posterior to the mouth and extending from the soft palate to the hyoid bone.

Orthopedics (or′-thō-PĒ-diks) The branch of medicine that deals with the preservation and restoration of the skeletal system, articulations, and associated structures.

Osmoreceptor (oz′-mō-re-CEP-tor) Receptor in the hypothalamus that is sensitive to changes in blood osmolarity and, in response to high osmolarity (low water concentration), stimulates synthesis and release of antidiuretic hormone (ADH).

Osmosis (oz-MŌ-sis) The net movement of water molecules through a selectively permeable membrane from an area of higher water concentration to an area of lower water concentration until equilibrium is reached.

Osseous (OS-ē-us) Bony.

Ossicle (OS-si-kul) One of the small bones of the middle ear (malleus, incus, stapes).

Ossification (os′-i-fi-KĀ-shun) Formation of bone. Also called **osteogenesis.**

Ossification (os′-i-fi-KĀ-shun) **center** An area in the cartilage model of a future bone where the cartilage cells hypertrophy, secrete enzymes that calcify their extracellular matrix, and die, and the area they occupied is invaded by osteoblasts that then lay down bone.

Osteoblast (OS-tē-ō-blast′) Cell formed from an osteogenic cell that participates in bone formation by secreting some organic components and inorganic salts.

Osteoclast (OS-tē-ō-clast′) A large, multinuclear cell that resorbs (destroys) bone matrix.

Osteocyte (OS-tē-ō-sīt′) A mature bone cell that maintains the daily activities of bone tissue.

Osteogenic (os′-tē-ō-JEN-ik) **cell** Stem cell derived from mesenchyme that has mitotic potential and the ability to differentiate into an osteoblast.

Osteogenic layer The inner layer of the periosteum that contains cells responsible for forming new bone during growth and repair.

Osteology (os-tē-OL-ō-jē) The study of bones.

Osteon (OS-tē-on) The basic unit of structure in adult compact bone, consisting of a central (haversian) canal with its concentrically arranged lamellae, lacunae, osteocytes, and canaliculi. Also called a **haversian** (ha-VER-shan) **system.**

Osteoporosis (os′-tē-ō-pō-RŌ-sis) Age-related disorder characterized by decreased bone mass and increased susceptibility to fractures, often as a result of decreased levels of estrogens.

Otic (Ō-tik) Pertaining to the ear.

Otolith (Ō-tō-lith) A particle of calcium carbonate embedded in the otolithic membrane that functions in maintaining static equilibrium.

Otolithic (ō-tō-LITH-ik) **membrane** Thick, gelatinous, glycoprotein layer located directly over hair cells of the macula in the saccule and utricle of the internal ear.

Otorhinolaryngology (ō-tō-rī′-nō-lar-in-GOL-ō-jē) The branch of medicine that deals with the diagnosis and treatment of diseases of the ears, nose, and throat.

Oval window A small, membrane-covered opening between the middle ear and inner ear into which the footplate of the stapes fits.

Ovarian (ō-VAR-ē-an) **cycle** A monthly series of events in the ovary associated with the maturation of a secondary oocyte.

Ovarian follicle (FOL-i-kul) A general name for oocytes (immature ova) in any stage of development, along with their surrounding epithelial cells.

Ovarian ligament (LIG-a-ment) A rounded cord of connective tissue that attaches the ovary to the uterus.

Ovary (Ō-var-ē) Female gonad that produces oocytes and the estrogens, progesterone, inhibin, and relaxin hormones.

Ovulation (ov-ū-LĀ-shun) The rupture of a mature ovarian (Graafian) follicle with discharge of a secondary oocyte into the pelvic cavity.

Ovum (Ō-vum) The female reproductive or germ cell; an egg cell; arises through completion of meiosis in a secondary oocyte after penetration by a sperm.

Oxyhemoglobin (ok′-sē-HĒ-mō-glō-bin) **($Hb-O_2$)** Hemoglobin combined with oxygen.

Oxytocin (ok′-sē-TŌ-sin) **(OT)** A hormone secreted by neurosecretory cells in the paraventricular and supraoptic nuclei of the hypothalamus that stimulates contraction of smooth muscle in the pregnant uterus and myoepithelial cells around the ducts of mammary glands.

P

P wave The deflection wave of an electrocardiogram that signifies atrial depolarization.

Pacinian corpuscle (pa-SIN-ē-an) Oval-shaped pressure receptor located in the dermis or subcutaneous tissue and consisting of concentric layers of a connective tissue wrapped around the dendrites of a sensory neuron. Also called a **lamellated corpuscle.**

Palate (PAL-at) The horizontal structure separating the oral and the nasal cavities; the roof of the mouth.

Palpate (PAL-pāt) To examine by touch; to feel.

Pancreas (PAN-krē-as) A soft, oblong organ lying along the greater curvature of the stomach and connected by a duct to the duodenum. It is both an exocrine gland (secreting pancreatic juice) and an endocrine gland (secreting insulin, glucagon, somatostatin, and pancreatic polypeptide).

Pancreatic (pan′-krē-AT-ik) **duct** A single large tube that unites with the common bile duct from the liver and gallbladder and drains pancreatic juice into the duodenum at the hepatopancreatic ampulla (ampulla of Vater). Also called the **duct of Wirsung.**

Pancreatic islet (Ī-let) A cluster of endocrine gland cells in the pancreas that secretes insulin, glucagon, somatostatin, and pancreatic polypeptide. Also called an **islet of Langerhans** (LANG-er-hanz).

Papanicolaou (pa-pa-NI-kō-lō) **test** A cytological staining test for the detection and diagnosis of premalignant and malignant conditions of the female genital tract. Cells scraped from the epithelium of the cervix of the uterus are examined microscopically. Also called a **Pap test** or **Pap smear.**

Papilla (pa-PIL-a) A small nipple-shaped projection or elevation.

Paralysis (pa-RAL-a-sis) Loss or impairment of motor function due to a lesion of nervous or muscular origin.

Paranasal sinus (par′-a-NĀ-zal SĪ-nus) A mucus-lined air cavity in a skull bone that communicates with the nasal cavity. Paranasal sinuses are located in the frontal, maxillary, ethmoid, and sphenoid bones.

Paraplegia (par-a-PLĒ-jē-a) Paralysis of both lower limbs.

Parasagittal plane (par-a-SAJ-i-tal) A vertical plane that does not pass through the midline and that divides the body or organs into *unequal* left and right portions.

Parasympathetic (par′-a-sim-pa-THET-ik) **division** One of the two subdivisions of the autonomic nervous system, having cell bodies of preganglionic neurons in nuclei in the brain stem and in the lateral gray horn of the sacral portion of the spinal cord; primarily concerned with activities that conserve and restore body energy.

Parathyroid (par′-a-THĪ-royd) **gland** One of usually four small endocrine glands embedded in the posterior surfaces of the lateral lobes of the thyroid gland.

Parathyroid hormone (PTH) A hormone secreted by the chief (principal) cells of the parathyroid glands that increases blood calcium level and decreases blood phosphate level.

Paraurethral (par′-a-ū-RĒ-thral) **gland** Gland embedded in the wall of the urethra whose duct opens on either side of the urethral orifice and secretes mucus. Also called **Skene's** (SKĒNZ) **gland.**

Parenchyma (par-EN-ki-ma) The functional parts of any organ, as opposed to tissue that forms its stroma or framework.

Parietal (pa-RĪ-e-tal) Pertaining to or forming the outer wall of a body cavity.

Parietal cell A type of secretory cell in gastric glands that produces hydrochloric acid and intrinsic factor. Also called an **oxyntic cell.**

Parietal pleura (PLOO-ra) The outer layer of the serous pleural membrane that encloses and protects the lungs; the layer that is attached to the wall of the pleural cavity.

Parkinson disease (PD) Progressive degeneration of the basal ganglia and substantia nigra of the cerebrum resulting in decreased production of dopamine (DA) that leads to tremor, slowing of voluntary movements, and muscle weakness.

Parotid (pa-ROT-id) **gland** One of the paired salivary glands located inferior and anterior to the ears and connected to the oral cavity via a duct (Stensen's) that opens into the inside of the cheek opposite the maxillary (upper) second molar tooth.

Pars intermedia (in′-ter-MĒ-dē-a) A small avascular zone between the anterior and posterior pituitary glands.

Parturition (par′-too-RISH-un) Act of giving birth to young; childbirth, delivery.

Patent ductus arteriosus (PĀ-tent DUK-tus ar-tēr-ē-Ō-sus) A congenital heart defect in which the ductus arteriosus remains open. As a result, aortic blood flows into the lower-pressure pulmonary trunk, increasing pulmonary trunk pressure and overworking both ventricles.

Pathogen (PATH-ō-jen) A disease-producing microbe.

Pathological (path′-ō-LOJ-i-kal) **anatomy** The study of structural changes caused by disease.

Pectinate (PEK-ti-nāt) **muscles** Projecting muscle bundles of the anterior atrial walls and the lining of the auricles.

Pectoral (PEK-tō-ral) Pertaining to the chest or breast.

Pedicel (PED-i-sel) Footlike structure, as on podocytes of a glomerulus.

Pelvic (PEL-vik) **cavity** Inferior portion of the abdominopelvic cavity that contains the urinary bladder, sigmoid colon, rectum, and internal female and male reproductive structures.

Pelvic splanchnic (PEL-vik SPLANGK-nik) **nerves** Consist of preganglionic parasympathetic axons from the levels of S2, S3, and S4 that supply the urinary bladder, reproductive organs, and the descending and sigmoid colon and rectum.

Pelvis The basinlike structure formed by the two hip bones, the sacrum, and the coccyx. The expanded, proximal portion of the ureter, lying within the kidney and into which the major calyces open.

Penis (PĒ-nis) The organ of urination and copulation in males; used to deposit semen into the female vagina.

Pepsin Protein-digesting enzyme secreted by chief cells of the stomach in the inactive form pepsinogen, which is converted to active pepsin by hydrochloric acid.

Peptic ulcer An ulcer that develops in areas of the gastrointestinal tract exposed to hydrochloric acid; classified as a gastric ulcer if in the lesser curvature of the stomach and as a duodenal ulcer if in the first part of the duodenum.

Percussion (pur-KUSH-un) The act of striking (percussing) an underlying part of the body with short, sharp taps as an aid in diagnosing the part by the quality of the sound produced.

Perforating canal (PER-fō-rā′-ting) A minute passageway by means of which blood vessels and nerves from the periosteum penetrate into compact bone. Also called **Volkmann's** (FŌLK-mans) **canal.**

Pericardial (per′-i-KAR-dē-al) **cavity** Small potential space between the visceral and parietal layers of the serous pericardium that contains pericardial fluid.

Pericardium (per-i-KAR-dē-um) A loose-fitting membrane that encloses the heart, consisting of a superficial fibrous layer and a deep serous layer.

Perichondrium (per′-i-KON-drē-um) A covering of dense irregular connective tissue that surrounds the surface of most cartilage.

Perilymph (PER-i-limf) The fluid contained between the bony and membranous labyrinths of the inner ear.

Perimetrium (per′-i-MĒ-trē-um) The serosa of the uterus.

Perimysium (per-i-MĪZ-ē-um) Invagination of the epimysium that divides muscles into bundles.

Perineum (per′-i-NĒ-um) The pelvic floor; the space between the anus and the scrotum in the male and between the anus and the vulva in the female.

Perineurium (per′-i-NOO-rē-um) Connective tissue wrapping around fascicles in a nerve.

Periodontal (per-ē-ō-DON-tal) **disease** A collective term for conditions characterized by degeneration of gingivae, alveolar bone, periodontal ligament, and cementum.

Periodontal ligament The periosteum lining the alveoli (sockets) for the teeth in the alveolar processes of the mandible and maxillae.

Periosteum (per′-ē-OS-tē-um) The covering of a bone that consists of connective tissue, osteogenic cells, and osteoblasts; is essential for bone growth, repair, and nutrition.

Peripheral (pe-RIF-er-al) Located on the outer part or a surface of the body.

Peripheral nervous system (PNS) The part of the nervous system that lies outside the central nervous system, consisting of nerves and ganglia.

Peristalsis (per′-i-STAL-sis) Successive muscular contractions along the wall of a hollow muscular structure.

Peritoneum (per-i-tō-NĒ-um) The largest serous membrane of the body that lines the abdominal cavity and covers the viscera within it.

Peritonitis (per′-i-tō-NĪ-tis) Inflammation of the peritoneum.

Peroxisome (pe-ROKS-i-sōm) Organelle similar in structure to a lysosome that contains enzymes that use molecular oxygen to oxidize various organic compounds; such reactions produce hydrogen peroxide; abundant in liver cells.

Perspiration Sweat; produced by sudoriferous (sweat) glands and containing water, salts, urea, uric acid, amino acids, ammonia, sugar, lactic acid, and ascorbic acid. Helps maintain body temperature and eliminate wastes.

Peyer's patches (PĪ-erz) *See* **Aggregated lymphatic follicles.**

pH A measure of the concentration of hydrogen ions (H^+) in a solution. The pH scale extends from 0 to 14, with a value of 7 expressing neutrality, values lower than 7 expressing increasing acidity, and values higher than 7 expressing increasing alkalinity.

Phagocytosis (fag′-ō-sī-TŌ-sis) The process by which phagocytes ingest and destroy microbes, cell debris, and other foreign matter.

Phalanx (FĀ-lanks) The bone of a finger or toe. *Plural* is **phalanges** (fa-LAN-jēz).

Pharmacology (far′-ma-KOL-ō-jē) The science of the effects and uses of drugs in the treatment of disease.

Pharynx (FAR-inks) The throat; a tube that starts at the internal nares and runs partway down the neck, where it opens into the esophagus posteriorly and the larynx anteriorly.

Phlebitis (fle-BĪ-tis) Inflammation of a vein, usually in a lower limb.

Photopigment A substance that can absorb light and undergo structural changes that can lead to the development of a receptor potential. An example is rhodopsin. In the eye, also called **visual pigment.**

Photoreceptor Receptor that detects light shining on the retina of the eye.

Physical therapy One of the manual therapies. Physical therapy is concerned with the promotion of health, with prevention of physical disabilities, with the evaluation and rehabilitation of persons disabled by pain, disease, or injury and with treatment by physical therapeutic measures.

Physiology (fiz′-ē-OL-o-jē) Science that deals with the functions of an organism or its parts.

Pia mater (PĪ-a MĀ-ter *or* PĒ-a MA-ter) The innermost of the three meninges (coverings) of the brain and spinal cord.

Pineal (PĪN-ē-al) **gland** A cone-shaped gland located in the roof of the third ventricle that secretes melatonin. Also called the **epiphysis cerebri** (ē-PIF-i-sis se-RĒ-brē).

Pinealocyte (pin-ē-AL-ō-sīt) Secretory cell of the pineal gland that releases melatonin.

Pinna (PIN-na) The projecting part of the external ear composed of elastic cartilage and covered by skin and shaped like the flared end of a trumpet. Also called the **auricle** (OR-i-kul).

Pituicyte (pi-TOO-i-sīt) Supporting cell of the posterior pituitary.

Pituitary (pi-TOO-i-tā r-ē) **gland** A small endocrine gland occupying the hypophyseal fossa of the sphenoid bone and attached to the hypothalamus by the infundibulum. Also called the **hypophysis** (hī-POF-i-sis).

Pivot joint A synovial joint in which a rounded, pointed, or conical surface of one bone articulates with a ring formed partly by another bone and partly by a ligament, as in the joint between the atlas and axis and between the proximal ends of the radius and ulna. Also called a **trochoid** (TRŌ-koyd) **joint.**

Placenta (pla-SEN-ta) The special structure through which the exchange of materials between fetal and maternal circulations occurs. Also called the **afterbirth.**

Plantar flexion (PLAN-tar FLEK-shun) Bending the foot in the direction of the plantar surface (sole).

Plaque (PLAK) A layer of dense proteins on the inside of a plasma membrane in adherens junctions and desmosomes. A mass of bacterial cells, dextran (polysaccharide), and other debris that adheres to teeth (dental plaque). *See* also **Atherosclerotic plaque.**

Plasma (PLAZ-ma) The extracellular fluid found in blood vessels; blood minus the formed elements.

Plasma cell Cell that develops from a B cell (lymphocyte) and produces antibodies.

Plasma (cell) membrane Outer, limiting membrane that separates the cell's internal parts from extracellular fluid or the external environment.

Platelet (PLĀT-let) A fragment of cytoplasm enclosed in a cell membrane and lacking a nucleus; found in the circulating blood; plays a role in hemostasis. Also called a **thrombocyte** (THROM-bō-sīt).

Platelet plug Aggregation of platelets (thrombocytes) at a site where a blood vessel is damaged that helps stop or slow blood loss.

Pleura (PLOO-ra) The serous membrane that covers the lungs and lines the walls of the chest and the diaphragm.

Pleural cavity Small potential space between the visceral and parietal pleurae.

Plexus (PLEK-sus) A network of nerves, veins, or lymphatic vessels.

Plexus of Auerbach *See* **Myenteric plexus.**

Plexus of Meissner *See* **Submucosal plexus.**

Pluripotent stem cell (plu-RIP-ō-tent) Immature stem cell in red bone marrow that gives rise to precursors of all the different mature blood cells.

Pneumotaxic (noo-mō-TAK-sik) **area** A part of the respiratory center in the pons that continually sends inhibitory nerve impulses to the inspiratory area, limiting inhalation and facilitating exhalation.

Polycythemia (pol′-ē-sī-THĒ-mē-a) Disorder characterized by an above-normal hematocrit (above 55%) in which hypertension, thrombosis, and hemorrhage can occur.

Polyunsaturated fat (pol′-ē-un-SATCH-ū-rā′-ted) A fatty acid that contains more than one double covalent bond between its carbon atoms; abundant in triglycerides of corn oil, safflower oil, and cottonseed oil.

Polyuria (pol′-ē-Ū-rē-a) An excessive production of urine.

Pons (PONZ) The part of the brain stem that forms a "bridge" between the medulla oblongata and the midbrain, anterior to the cerebellum.

Portal system The circulation of blood from one capillary network into another through a vein.

Postcentral gyrus Gyrus of cerebral cortex located immediately posterior to the central sulcus; contains the primary somatosensory area.

Posterior (pos-TĒR-ē-or) Nearer to or at the back of the body. Equivalent to **dorsal** in bipeds.

Posterior column–medial lemniscus pathways (lem-NIS-kus) Sensory pathways that carry information related to proprioception, fine touch, two-point discrimination, pressure, and vibration. First-order neurons project from the spinal cord to the ipsilateral medulla in the posterior columns (gracile fasciculus and cuneate fasciculus). Second-order neurons project from the medulla to the contralateral thalamus in the medial lemniscus. Third-order neurons project from the thalamus to the somatosensory cortex (postcentral gyrus) on the same side.

Posterior pituitary (pi-TOO-i-tār-ē) Posterior lobe of the pituitary gland. Also called the **neurohypophysis** (noo-rō-hī-POF-i-sis).

Posterior root The structure composed of sensory axons lying between a spinal nerve and the dorsolateral aspect of the spinal cord. Also called the **dorsal (sensory) root.**

Posterior root ganglion (GANG-glē-on) A group of cell bodies of sensory neurons and their supporting cells located along the posterior root of a spinal nerve. Also called a **dorsal (sensory) root ganglion.**

Postganglionic neuron (pōst′-gang-lē-ON-ik NOO-ron) The second autonomic motor neuron in an autonomic pathway, having its cell body and dendrites located in an autonomic ganglion and its unmyelinated axon ending at cardiac muscle, smooth muscle, or a gland.

Postsynaptic (pōst-sin-AP-tik) **neuron** The nerve cell that is activated by the release of a neurotransmitter from another neuron and carries nerve impulses away from the synapse.

Pouch of Douglas *See* **Rectouterine pouch.**

Precapillary sphincter (SFINGK-ter) The distal most muscle fiber (cell) at the metarteriole-capillary junction that regulates blood flow into capillaries.

Precentral gyrus (JĪ-rus) Gyrus of cerebral cortex located immediately anterior to the central sulcus; contains the primary motor area.

Preganglionic (pre′-gang-lē-ON-ik) **neuron** The first autonomic motor neuron in an autonomic pathway, with its cell body and dendrites in the brain or spinal cord and its myelinated axon ending at an autonomic ganglion, where it synapses with a postganglionic neuron.

Pregnancy Sequence of events that normally includes fertilization, implantation, embryonic growth, and fetal growth and terminates in birth.

Premenstrual syndrome (PMS) Moderate to severe physical and emotional stress ocurring late in the postovulatory phase of the menstrual cycle and sometimes overlapping with menstruation.

Prepuce (PRĒ-poos) The loose-fitting skin covering the glans of the penis and clitoris. Also called the **foreskin.**

Presbyopia (prez-bē-Ō-pē-a) A loss of elasticity of the lens of the eye due to advancing age with resulting inability to focus clearly on near objects.

Presynaptic (prē-sin-AP-tik) **neuron** A neuron that propagates nerve impulses toward a synapse.

Prevertebral ganglion (prē-VER-te-bral GANG-glē-on) A cluster of cell bodies of postganglionic sympathetic neurons anterior to the spinal column and close to large abdominal arteries. Also called a **collateral ganglion.**

Primary motor area A region of the cerebral cortex in the precentral gyrus of the frontal lobe of the cerebrum that controls specific muscles or groups of muscles.

Primary somatosensory area (sō-ma-tō-SEN-sō-rē) A region of the cerebral cortex posterior to the central sulcus in the postcentral gyrus of the parietal lobe of the cerebrum that localizes exactly the points of the body where somatic sensations originate.

Prime mover The muscle directly responsible for producing a desired motion. Also called an **agonist** (AG-ō-nist).

Primordial (prī-MŌR-dē-al) Existing first; especially primordial egg cells in the ovary.

Principal cell Cell type in the distal convoluted tubules and collecting ducts of the kidneys that is stimulated by aldosterone and antidiuretic hormone.

Proctology (prok-TOL-ō-jē) The branch of medicine concerned with the rectum and its disorders.

Progeny (PROJ-e-nē) Offspring or descendants.

Progesterone (prō-JES-te-rōn) A female sex hormone produced by the ovaries that helps prepare the endometrium of the uterus for implantation of a fertilized ovum and the mammary glands for milk secretion.

Prognosis (prog-NŌ-sis) A forecast of the probable results of a disorder; the outlook for recovery.

Prolactin (prō-LAK-tin) **(PRL)** A hormone secreted by the anterior pituitary that initiates and maintains milk secretion by the mammary glands.

Prolapse (PRŌ-laps) A dropping or falling down of an organ, especially the uterus or rectum.

Proliferation (prō-lif′-er-Ā-shun) Rapid and repeated reproduction of new parts, especially cells.

Pronation (prō-NĀ-shun) A movement of the forearm in which the palm is turned posteriorly.

Prophase (PRŌ-fāz) The first stage of mitosis during which chromatid pairs are formed and aggregate around the metaphase plate of the cell.

Proprioception (prō-prē-ō-SEP-shun) The perception of the position of body parts, especially the limbs, independent of vision; this sense is possible due to nerve impulses generated by proprio-ceptors.

Proprioceptor (PRŌ-prē-ō-sep′-tor) A receptor located in muscles, tendons, joints, or the internal ear (muscle spindles, tendon organs, joint kinesthetic receptors, and hair cells of the vestibular apparatus) that provides information about body position and movements.

Prostaglandin (pros′-ta-GLAN-din) **(PG)** A membrane-associated lipid; released in small quantities and acts as a local hormone.

Prostate (PROS-tāt) A doughnut-shaped gland inferior to the urinary bladder that surrounds the superior portion of the male urethra and secretes a slightly acidic solution that contributes to sperm motility and viability.

Proteasome (PRŌ-tē-a-sōm) Tiny cellular organelle in cytosol and nucleus containing proteases that destroy unneeded, damaged, or faulty proteins.

Protein An organic compound consisting of carbon, hydrogen, oxygen, nitrogen, and sometimes sulfur and phosphorus; synthesized on ribosomes and made up of amino acids linked by peptide bonds.

Prothrombin (prō-THROM-bin) An inactive blood-clotting factor synthesized by the liver, released into the blood, and converted to active thrombin in the process of blood clotting by the activated enzyme prothrombinase.

Proto-oncogene (prō′-tō-ON-kō-jēn) Gene responsible for some aspect of normal growth and development; it may transform into an oncogene, a gene capable of causing cancer.

Protraction (prō-TRAK-shun) The movement of the mandible or shoulder girdle forward on a plane parallel with the ground.

Proximal (PROK-si-mal) Nearer the attachment of a limb to the trunk; nearer to the point of origin or attachment.

Pseudopods (SOO-dō-pods) Temporary protrusions of the leading edge of a migrating cell; cellular projections that surround a particle undergoing phagocytosis.

Pterygopalatine ganglion (ter′-i-gō-PAL-a-tīn GANG-glē-on) A cluster of cell bodies of parasympathetic post-ganglionic neurons ending at the lacrimal and nasal glands.

Ptosis (TŌ-sis) Drooping, as of the eyelid or the kidney.

Puberty (PŪ-ber-tē) The time of life during which the secondary sex characteristics begin to appear and the capability for sexual reproduction is possible; usually occurs between the ages of 10 and 17.

Pubic symphysis (SIM-fi-sis) A slightly movable cartilaginous joint between the anterior surfaces of the hip bones.

Puerperium (pū′-er-PER-ē-um) The period immediately after childbirth, usually 4–6 weeks.

Pulmonary (PUL-mo-ner′-ē) Concerning or affected by the lungs.

Pulmonary circulation The flow of deoxygenated blood from the right ventricle to the lungs and the return of oxygenated blood from the lungs to the left atrium.

Pulmonary edema (e-DĒ-ma) An abnormal accumulation of interstitial fluid in the tissue spaces and alveoli of the lungs due to increased pulmonary capillary permeability or increased pulmonary capillary pressure.

Pulmonary embolism (EM-bō-lizm) **(PE)** The presence of a blood clot or a foreign substance in a pulmonary arterial blood vessel that obstructs circulation to lung tissue.

Pulmonary ventilation (ven-ti-LĀ-shun) The inflow (inhalation) and outflow (exhalation) of air between the atmosphere and the lungs. Also called **breathing.**

Pulp cavity A cavity within the crown and neck of a tooth, which is filled with pulp, a connective tissue containing blood vessels, nerves, and lymphatic vessels.

Pulse (PULS) The rhythmic expansion and elastic recoil of a systemic artery after each contraction of the left ventricle.

Pupil The hole in the center of the iris, the area through which light enters the posterior cavity of the eyeball.

Purkinje (pur-KIN-jē) **fiber** Muscle fiber (cell) in the ventricular tissue of the heart specialized for conducting an action potential to the myocardium; part of the conduction system of the heart.

Pus The liquid product of inflammation containing leukocytes or their remains and debris of dead cells.

Pyloric (pī-LOR-ik) **sphincter** A thickened ring of smooth muscle through which the pylorus of the stomach communicates with the duodenum. Also called the **pyloric valve.**

Pyorrhea (pī-ō-RĒ-a) A discharge or flow of pus, especially in the alveoli (sockets) and the tissues of the gums.

Pyramid (PIR-a-mid) A pointed or cone-shaped structure. One of two roughly triangular structures on the anterior aspect of the medulla oblongata composed of the largest motor tracts that run from the cerebral cortex to the spinal cord. A triangular structure in the renal medulla.

Pyramidal (pi-RAM-i-dal) **tracts (pathways).** *See* **Direct motor pathways.**

Q

QRS wave The deflection waves of an electrocardiogram that represent onset of ventricular depolarization.

Quadrant (KWOD-rant) One of four parts.

Quadriplegia (kwod′-ri-PLĒ-jē-a) Paralysis of four limbs: two upper and two lower.

R

Radiographic (rā′-dē-ō-GRAF-ic) **anatomy** Diagnostic branch of anatomy that includes the use of x rays.

Rami communicantes (RĀ-mē kō-mū-ni-KAN-tēz) Branches of a spinal nerve. *Singular* is **ramus communicans** (RĀ-mus kō-MŪ-ni-kans).

Range of motion The range, measured in degrees of a circle, through which the bones of a joint can be moved.

Rathke's pouch (RATH-kē) *See* **Hypophyseal pouch.**

Receptor (rē-SEP-tor) A specialized cell or a distal portion of a neuron that responds to a specific sensory modality, such as touch, pressure, cold, light, or sound, and converts it to an electrical signal (generator or receptor potential). A specific molecule or cluster of molecules that recognizes and binds a particular ligand.

Receptor-mediated endocytosis (en′-dō-sī-TŌ-sis) A highly selective process whereby cells take up specific ligands, which usually are large molecules or particles, by enveloping them within a sac of plasma membrane. Ligands are eventually broken down by enzymes in lysosomes.

Rectouterine pouch (rek′-tō-Ū-ter-in) A pocket formed by the parietal peritoneum as it moves posteriorly from the surface of the uterus and is reflected onto the rectum; the most inferior point in the pelvic cavity. Also called the **pouch** or **cul de sac of Douglas.**

Rectum (REK-tum) The last 20 cm (8 in.) of the gastrointestinal tract, from the sigmoid colon to the anus.

Recumbent (re-KUM-bent) Lying down.

Red bone marrow A highly vascularized connective tissue located in microscopic spaces between trabeculae of spongy bone tissue.

Red nucleus A cluster of cell bodies in the midbrain, occupying a large part of the tectum from which axons extend into the rubroreticular and rubrospinal tracts.

Red pulp That portion of the spleen that consists of venous sinuses filled with blood and thin plates of splenic tissue called splenic (Billroth's) cords.

Referred pain Pain that is felt at a site remote from the place of origin.

Reflex Fast response to a change (stimulus) in the internal or external environment that attempts to restore homeostasis.

Reflex arc The most basic conduction pathway through the nervous system, connecting a receptor and an effector and consisting of a receptor, a sensory neuron, an integrating center in the central nervous system, a motor neuron, and an effector.

Regional anatomy The division of anatomy dealing with a specific region of the body, such as the head, neck, chest, or abdomen.

Regurgitation (rē-gur′-ji-TĀ-shun) Return of solids or fluids to the mouth from the stomach; backward flow of blood through incompletely closed heart valves.

Relaxin (RLX) A female hormone produced by the ovaries and placenta that increases flexibility of the pubic symphysis and helps dilate the uterine cervix to ease delivery of a baby.

Releasing hormone Hormone secreted by the hypothalamus that can stimulate secretion of hormones of the anterior pituitary.

Remodeling (rē-MOD-e-ling) Replacement of old bone by new bone tissue.

Renal (RĒ-nal) Pertaining to the kidneys.

Renal corpuscle (KOR-pus-l) A glomerular (Bowman's) capsule and its enclosed glomerulus.

Renal pelvis A cavity in the center of the kidney formed by the expanded, proximal portion of the ureter, lying within the kidney, and into which the major calyces open.

Renal pyramid (PIR-a-mid) A triangular structure in the renal medulla containing the straight segments of renal tubules and the vasa recta.

Reproduction (rē-prō-DUK-shun) The formation of new cells for growth, repair, or replacement; the production of a new individual.

Reproductive cell division Type of cell division in which gametes (sperm and oocytes) are produced; consists of meiosis and cytokinesis.

Respiration (res-pi-RĀ-shun) Overall exchange of gases between the atmosphere, blood, and body cells consisting of pulmonary ventilation, external respiration, and internal respiration.

Respiratory center Neurons in the pons and medulla oblongata of the brain stem that regulate the rate and depth of pulmonary ventilation.

Retention (rē-TEN-shun) A failure to void urine due to obstruction, nervous contraction of the urethra, or absence of sensation of desire to urinate.

Rete (RĒ-tē) **testis** The network of ducts in the testes.

Reticular (re-TIK-ū-lar) **activating system (RAS)** A portion of the reticular formation that has many ascending connections with the cerebral cortex; when this area of the brain stem is active, nerve impulses pass to the thalamus and widespread areas of the cerebral cortex, resulting in generalized alertness or arousal from sleep.

Reticular formation A network of small groups of neuronal cell bodies scattered among bundles of axons (mixed gray and white matter) beginning in the medulla oblongata and extending superiorly through the central part of the brain stem.

Reticulocyte (re-TIK-ū-lō-sīt) An immature red blood cell.

Reticulum (re-TIK-ū-lum) A network.

Retina (RET-i-na) The deep coat of the posterior portion of the eyeball consisting of nervous tissue (where the process of vision begins) and a pigmented layer of epithelial cells that contact the choroid.

Retinaculum (ret-i-NAK-ū-lum) A thickening of fascia that holds structures in place, for example, the superior and inferior retinacula of the ankle.

Retraction (rē-TRAK-shun) The movement of a protracted part of the body posteriorly on a plane parallel to the ground, as in pulling the lower jaw back in line with the upper jaw.

Retroperitoneal (re′-trō-per-i-tō-NĒ-al) External to the peritoneal lining of the abdominal cavity.

Reverse muscle action The "origin" becomes the "insertion." Muscles typically attach to bones that cross joints; the muscle attachment that most commonly moves has traditionally been called the insertion while the more stabile attachment site has been termed the origin. Many muscles are, however, capable of contracting whereby the attachment sites reverse and the other end of the muscle moves.

Rh factor An inherited antigen on the surface of red blood cells in Rh^+ individuals; not present in Rh^- individuals.

Rhinology (rī-NOL-ō-jē) The study of the nose and its disorders.

Ribonucleic (rī-bō-noo-KLĒ-ik) **acid (RNA)** A single-stranded nucleic acid made up of nucleotides, each consisting of a nitrogenous base (adenine, cytosine, guanine, or uracil), ribose, and a phosphate group; three types are messenger RNA (mRNA), transfer RNA (tRNA), and ribosomal RNA (rRNA), each of which has a specific role during protein synthesis.

Ribosome (RĪ-bō-sōm) A cellular structure in the cytoplasm of cells, composed of a small subunit and a large subunit that contain ribosomal RNA and ribosomal proteins; the site of protein synthesis.

Right lymphatic (lim-FAT-ik) **duct** A vessel of the lymphatic system that drains lymph from the upper right side of the body and empties it into the right subclavian vein.

Rigidity (ri-JID-i-tē) Hypertonia characterized by increased muscle tone, but reflexes are not affected.

Rigor mortis State of partial contraction of muscles after death due to lack of ATP; myosin heads (crossbridges) remain attached to actin, thus preventing relaxation.

Rod One of two types of photoreceptor in the retina of the eye; specialized for vision in dim light.

Root canal A narrow extension of the pulp cavity lying within the root of a tooth.

Root of penis Attached portion of penis that consists of the bulb and crura.

Rotation (rō-TĀ-shun) Moving a bone around its own axis, with no other movement.

Round ligament (LIG-a-ment) A band of fibrous connective tissue enclosed between the folds of the broad ligament of the uterus, emerging from the uterus just inferior to the uterine tube, extending laterally along the pelvic wall and through the deep inguinal ring to end in the labia majora.

Round window A small opening between the middle and internal ear, directly inferior to the oval window, covered by the secondary tympanic membrane.

Ruffini corpuscle (roo-FĒ-nē) A sensory receptor embedded deeply in the dermis and deeper tissues that detects the stretching of the skin.

Rugae (ROO-gē) Large folds in the mucosa of an empty hollow organ, such as the stomach and vagina.

S

Saccule (SAK-ūl) The inferior and smaller of the two chambers in the membranous labyrinth inside the vestibule of the internal ear containing a receptor organ for static equilibrium.

Sacral plexus (SĀ-kral PLEK-sus) A network formed by the ventral branches of spinal nerves L4 through S3.

Sacral promontory (PROM-on-tor′-ē) The superior surface of the body of the first sacral vertebra that projects anteriorly into the pelvic cavity; a line from the sacral promontory to the superior border of the pubic symphysis divides the abdominal and pelvic cavities.

Saddle joint A synovial joint in which the articular surface of one bone is saddle-shaped and the articular surface of the other bone is shaped like the legs of the rider sitting in the saddle, as in the joint between the trapezium and the metacarpal of the thumb.

Sagittal (SAJ-i-tal) **plane** A plane that divides the body or organs into left and right portions. Such a plane may be **midsagittal (median),** in which the divisions are equal, or **parasagittal,** in which the divisions are unequal.

Saliva (sa-LĪ-va) A clear, alkaline, somewhat viscous secretion produced mostly by the three pairs of salivary glands; contains various salts, mucin, lysozyme, salivary amylase, and lingual lipase (produced by glands in the tongue).

Salivary amylase (SAL-i-ver-ē AM-i-lās) An enzyme in saliva that initiates the chemical breakdown of starch.

Salivary gland One of three pairs of glands that lie external to the mouth and pour their secretory product (saliva) into ducts that empty into the oral cavity; the parotid, submandibular, and sublingual glands.

Sarcolemma (sar′-kō-LEM-ma) The cell membrane of a muscle fiber (cell), especially of a skeletal muscle fiber.

Sarcomere (SAR-kō-mēr) A contractile unit in a striated muscle fiber (cell) extending from one Z disc to the next Z disc.

Sarcoplasm (SAR-kō-plazm) The cytoplasm of a muscle fiber (cell).

Sarcoplasmic reticulum (sar′-kō-PLAZ-mik re-TIK-ū-lum) **(SR)** A network of saccules and tubes surrounding myofibrils of a muscle fiber (cell), comparable to endoplasmic reticulum; functions to reabsorb calcium ions during relaxation and to release them to cause contraction.

Satellite cell (SAT-i-līt) Flat neuroglial cells that surround cell bodies of peripheral nervous system ganglia to provide structural support and regulate the exchange of material between a neuronal cell body and interstitial fluid.

Saturated fat A fatty acid that contains only single bonds (no double bonds) between its carbon atoms; all carbon atoms are bonded to the maximum number of hydrogen atoms; prevalent in triglycerides of animal products such as meat, milk, milk products, and eggs.

Scala tympani (SKĀ-la TIM-pan-ē) The inferior spiral-shaped channel of the bony cochlea, filled with perilymph.

Scala vestibuli (ves-TIB-ū-lē) The superior spiral-shaped channel of the bony cochlea, filled with perilymph.

Schwann (SCHWON) **cell** A neuroglial cell of the peripheral nervous system that forms the myelin sheath and neurolemma around a nerve axon by wrapping around the axon in a jelly-roll fashion.

Sciatica (sī-AT-i-ka) Inflammation and pain along the sciatic nerve; felt along the posterior aspect of the thigh and may extend down the back and outside of the leg.

Sclera (SKLE-ra) The white coat of fibrous tissue that forms the superficial protective covering over the eyeball except in the most anterior portion; the posterior portion of the fibrous tunic.

Scleral venous sinus A circular venous sinus located at the junction of the sclera and the cornea through which aqueous humor drains from the anterior chamber of the eyeball into the blood. Also called the **canal of Schlemm** (SHLEM).

Sclerosis (skle-RŌ-sis) A hardening with loss of elasticity of tissues.

Scoliosis (skō-lē-Ō-sis) An abnormal lateral curvature from the normal vertical line of the backbone.

Scrotum (SKRŌ-tum) A skin-covered pouch that contains the testes and their accessory structures.

Sebaceous (se-BĀ-shus) **gland** An exocrine gland in the dermis of the skin, almost always associated with a hair follicle, that secretes sebum. Also called an **oil gland.**

Sebum (SĒ-bum) Secretion of sebaceous (oil) glands.

Secondary sex characteristic A characteristic of the male or female body that develops at puberty under the influence of sex hormones but is not directly involved in sexual reproduction; examples are distribution of body hair, voice pitch, body shape, and muscle development.

Secretion (se-KRĒ-shun) Production and release from a cell or a gland of a physiologically active substance.

Selective permeability (per′-mē-a-BIL-i-tē) The property of a membrane by which it permits the passage of certain substances but restricts the passage of others.

Semen (SĒ-men) A fluid discharged at ejaculation by a male that consists of a mixture of sperm and the secretions of the seminiferous tubules, seminal vesicles, prostate, and bulbourethral (Cowper's) glands.

Semicircular canals (sem-ī-SER-kū-lar) Three bony channels (anterior, posterior, lateral), filled with perilymph, in which lie the membranous semicircular canals filled with endolymph. They contain receptors for equilibrium.

Semicircular ducts The membranous semicircular canals filled with endolymph and floating in the perilymph of the bony semicircular canals; they contain cristae that are concerned with dynamic equilibrium.

Semilunar (sem′-ē-LOO-nar) **valve** A valve between the aorta or the pulmonary trunk and a ventricle of the heart.

Seminal vesicle (SEM-i-nal VES-i-kul) One of a pair of convoluted, pouchlike structures, lying posterior and inferior to the urinary bladder and anterior to the rectum, that secrete a component of semen into the ejaculatory ducts. Also termed **seminal gland.**

Seminiferous tubule (sem′-i-NI-fer-us TOO-būl) A tightly coiled duct, located in the testis, where sperm are produced.

Sensation A state of awareness of external or internal conditions of the body.

Sensory area A region of the cerebral cortex concerned with the interpretation of sensory impulses.

Sensory neurons (NOO-ronz) Neurons that carry sensory information from cranial and spinal nerves into the brain and spinal cord or from a lower to a higher level in the spinal cord and brain. Also called **afferent neurons** (AF-er-ent).

Septal defect An opening in the atrial septum (atrial septal defect) because the foramen ovale fails to close, or the ventricular septum (ventricular septal defect) due to incomplete development of the ventricular septum.

Septum (SEP-tum) A wall dividing two cavities.

Serous (SĒR-us) **membrane** A membrane that lines a body cavity that does not open to the exterior. The external layer of an organ formed by a serous membrane. The membrane that lines the pleural, pericardial, and peritoneal cavities. Also called a **serosa** (se-RŌ-sa).

Sertoli (ser-TŌ-lē) **cell** A supporting cell in the seminiferous tubules that secretes fluid for supplying nutrients to sperm and the hormone inhibin, removes excess cytoplasm from spermatogenic cells, and mediates the effects of FSH and testosterone on spermatogenesis. Also called a **sustentacular** (sus′-ten-TAK-ū-lar) **cell.**

Serum Blood plasma minus its clotting proteins.

Sesamoid (SES-a-moyd) **bones** Small bones usually found in tendons.

Sexual intercourse The insertion of the erect penis of a male into the vagina of a female. Also called **coitus** (KŌ-i-tus).

Sheath of Schwann *See* **Neurolemma.**

Shock Failure of the cardiovascular system to deliver adequate amounts of oxygen and nutrients to meet the metabolic needs of the body due to inadequate cardiac output. It is characterized by hypotension; clammy, cool, and pale skin; sweating; reduced urine formation; altered mental state; acidosis; tachycardia; weak, rapid pulse; and thirst. Types include hypovolemic, cardiogenic, vascular, and obstructive.

Shoulder joint A synovial joint where the humerus articulates with the scapula.

Sigmoid colon (SIG-moyd KŌ-lon) The S-shaped part of the large intestine that begins at the level of the left iliac crest, projects medially, and terminates at the rectum at about the level of the third sacral vertebra.

Sign Any objective evidence of disease that can be observed or measured, such as a lesion, swelling, or fever.

Sinoatrial (si-nō-Ā-trē-al) **(SA) node** A small mass of cardiac muscle fibers (cells) located in the right atrium inferior to the opening of the superior vena cava that spontaneously depolarize and generate a cardiac action potential about 100 times per minute. Also called the natural **pacemaker.**

Sinus (SĪ-nus) A hollow in a bone (paranasal sinus) or other tissue; a channel for blood (vascular sinus); any cavity having a narrow opening.

Sinusoid (SĪ-nū-soyd) A large, thin-walled, and leaky type of capillary, having large intercellular clefts that may allow proteins and blood cells to pass from a tissue into the bloodstream; present in the liver, spleen, anterior pituitary, parathyroid glands, and red bone marrow.

Skeletal muscle An organ specialized for contraction, composed of striated muscle fibers (cells), supported by connective tissue, attached to a bone by a tendon or an aponeurosis, and stimulated by somatic motor neurons.

Skene's gland *See* **Paraurethral gland.**

Skin The external covering of the body that consists of a superficial, thinner epidermis (epithelial tissue) and a deep, thicker dermis (connective tissue) that is anchored to the subcutaneous layer.

Skull The skeleton of the head consisting of the cranial and facial bones.

Sleep A state of partial unconsciousness from which a person can be aroused; associated with a low level of activity in the reticular activating system.

Small intestine A long tube of the gastrointestinal tract that begins at the pyloric sphincter of the stomach, coils through the central and inferior part of the abdominal cavity, and ends at the large intestine; divided into three segments: duodenum, jejunum, and ileum.

Smooth muscle A tissue specialized for contraction, composed of smooth muscle fibers (cells), located in the walls of hollow internal organs except for the heart, and innervated by autonomic motor neurons.

Sodium-potassium ATPase An active transport pump located in the plasma membrane that transports sodium ions out of the cell and potassium ions into the cell at the expense of cellular ATP. It functions to keep the ionic concentrations of these ions at physiological levels. Also called the **sodium-potassium pump.**

Soft palate (PAL-at) The posterior portion of the roof of the mouth, extending from the palatine bones to the uvula. It is a muscular partition lined with mucous membrane.

Somatic (sō-MAT-ik) **cell division** Type of cell division in which a single starting cell duplicates itself to produce two identical cells; consists of mitosis and cytokinesis.

Somatic nervous system (SNS) The portion of the peripheral nervous system consisting of somatic sensory (afferent) neurons and somatic motor (efferent) neurons.

Somite (SŌ-mīt) Block of mesodermal cells in a developing embryo that is differentiated into a myotome (which forms most of the skeletal muscles), dermatome (which forms connective tissues), and sclerotome (which forms the vertebrae).

Spasm (SPAZM) A sudden, involuntary contraction of skeletal muscles.

Spasticity (spas-TIS-i-tē) Hypertonia characterized by increased muscle tone, increased tendon reflexes, and pathological reflexes (Babinski sign).

Spermatic (sper-MAT-ik) **cord** A supporting structure of the male reproductive system, extending from a testis to the deep inguinal ring, that includes the ductus (vas) deferens, arteries, veins, lymphatic vessels, nerves, cremaster muscle, and connective tissue.

Spermatogenesis (sper′-ma-tō-JEN-e-sis) The formation and development of sperm in the seminiferous tubules of the testes.

Sperm cell A mature male gamete. Also termed **spermatozoon** (sper′-ma-tō-ZŌ-on).

Spermiogenesis (sper′-mē-ō-JEN-e-sis) The maturation of spermatids into sperm.

Sphincter (SFINGK-ter) A circular muscle that constricts an opening.

Sphincter of Oddi (OD-ē) *See* **Sphincter of the hepatopancreatic ampulla.**

Sphincter of the hepatopancreatic ampulla (HEP-a-tō-pan-crē-a′-tik am-PUL-a) A circular muscle at the opening of the common bile and main pancreatic ducts in the duodenum. Also called the **sphincter of Oddi** (OD-ē).

Spinal (SPĪ-nal) **cord** A mass of nerve tissue located in the vertebral canal from which 31 pairs of spinal nerves originate.

Spinal nerve One of the 31 pairs of nerves that originate on the spinal cord from posterior and anterior roots.

Spinal shock A period from several days to several weeks following transection of the spinal cord that is characterized by the abolition of all reflex activity.

Spinothalamic (spī-nō-tha-LAM-ik) **tract** Sensory (ascending) tract that conveys information up the spinal cord to the thalamus for sensations of pain, temperature, itch, and tickle.

Spinous (SPĪ-nus) **process** A sharp or thornlike process or projection. Also called a **spine.** A sharp ridge running diagonally across the posterior surface of the scapula.

Spiral organ The organ of hearing, consisting of supporting cells and hair cells that rest on the basilar membrane and extend into the endolymph of the cochlear duct. Also called the **organ of Corti** (KŌR-tē).

Splanchnic (SPLANK-nik) Pertaining to the viscera.

Spleen (SPLĒN) Large mass of lymphatic tissue between the fundus of the stomach and the diaphragm that functions in formation of blood cells during early fetal development, phagocytosis of ruptured blood cells, and proliferation of B cells during immune responses.

Spongy (cancellous) bone tissue Bone tissue that consists of an irregular latticework of thin plates of bone called trabeculae; spaces between trabeculae of some bones are filled with red bone marrow; found inside short, flat, and irregular bones and in the epiphyses (ends) of long bones.

Sprain Forcible wrenching or twisting of a joint with partial rupture or other injury to its attachments without dislocation.

Squamous (SKWĀ-mus) Flat or scalelike.

Starvation (star-VĀ-shun) The loss of energy stores in the form of glycogen, triglycerides, and proteins due to inadequate intake of nutrients or inability to digest, absorb, or metabolize ingested nutrients.

Static equilibrium (ē-kwi-LIB-rē-um) The maintenance of posture in response to changes in the orientation of the body, mainly the head, relative to the ground.

Stellate reticuloendothelial (STEL-āt re-tik′-ū-lō-en′-dō-THĒ-lē-al) **cell** Phagocytic cell within a sinusoid of the liver. Also called a **Kupffer** (KOOP-fer) **cell.**

Stem cell An unspecialized cell that has the ability to divide for indefinite periods and give rise to a specialized cell.

Stenosis (sten-Ō-sis) An abnormal narrowing or constriction of a duct or opening.

Stereocilia (ste′-rē-ō-SIL-ē-a) Groups of extremely long, slender, nonmotile microvilli projecting from epithelial cells lining the epididymis.

Sterile (STE-ril) Free from any living microorganisms. Unable to conceive or produce offspring.

Sterilization (ster′-i-li-ZĀ-shun) Elimination of all living microorganisms. Any procedure that renders an individual incapable of reproduction (for example, castration, vasectomy, hysterectomy, or oophorectomy).

Stimulus Any stress that changes a controlled condition; any change in the internal or external environment that excites a sensory receptor, a neuron, or a muscle fiber.

Stomach The J-shaped enlargement of the gastrointestinal tract directly inferior to the diaphragm in the epigastric, umbilical, and left hypochondriac regions of the abdomen, between the esophagus and small intestine.

Straight tubule (TOO-būl) A duct in a testis leading from a convoluted seminiferous tubule to the rete testis.

Strain Stretched or partially torn muscle.

Stratum (STRĀ-tum) A layer.

Stratum basalis (ba-SAL-is) The layer of the endometrium next to the myometrium that is maintained during menstruation and gestation and produces a new stratum functionalis following menstruation or parturition.

Stratum functionalis (funk′-shun-AL-is) The layer of the endometrium next to the uterine cavity that is shed during menstruation and that forms the maternal portion of the placenta during gestation.

Stretch receptor Receptor in the walls of blood vessels, airways, or organs that monitors the amount of stretching. Also termed **baroreceptor.**

Stroma (STRŌ-ma) The tissue that forms the ground substance, foundation, or framework of an organ, as opposed to its functional parts (parenchyma).

Subarachnoid (sub′-a-RAK-noyd) **space** A space between the arachnoid mater and the pia mater that surrounds the brain and spinal cord and through which cerebrospinal fluid circulates.

Subcutaneous (sub′-kū-TĀ-nē-us) Beneath the skin. Also called **hypodermic** (hi-pō-DER-mik).

Subcutaneous layer A continuous sheet of areolar connective tissue and adipose tissue between the dermis of the skin and the deep fascia of the muscles. Also called the **hypodermis.**

Subdural (sub-DOO-ral) **space** A space between the dura mater and the arachnoid mater of the brain and spinal cord that contains a small amount of fluid.

Sublingual (sub-LING-gwal) **gland** One of a pair of salivary glands situated in the floor of the mouth deep to the mucous membrane and to the side of the lingual frenulum, with a duct (Rivinus') that opens into the floor of the mouth.

Submandibular (sub′-man-DIB-ū-lar) **gland** One of a pair of salivary glands found inferior to the base of the tongue deep to the mucous membrane in the posterior part of the floor of the mouth, posterior to the sublingual glands, with a duct (Wharton's) situated to the side of the lingual frenulum. Also called the **submaxillary** (sub′-MAK-si-ler-ē) **gland.**

Submucosa (sub-mū-KŌ-sa) A layer of connective tissue located deep to a mucous membrane, as in the gastrointestinal tract or the urinary bladder; the submucosa connects the mucosa to the muscularis layer.

Submucosal plexus A network of autonomic nerve fibers located in the superficial part of the submucous layer of the small intestine. Also called the **plexus of Meissner** (MĪZ-ner).

Substrate A molecule upon which an enzyme acts.

Subthalamus (sub-THAL-a-mus) Part of the diencephalon inferior to the thalamus; the substantia nigra and red nucleus extend from the midbrain into the subthalamus.

Sudoriferous (soo′-dor-IF-er-us) **gland** An apocrine or eccrine exocrine gland in the dermis or subcutaneous layer that produces perspiration. Also called a **sweat gland.**

Sulcus (SUL-kus) A groove or depression between parts, especially between the convolutions of the brain. *Plural* is **sulci** (SUL-sī).

Superficial (soo′-per-FISH-al) Located on or near the surface of the body or an organ.

Superficial inguinal (IN-gwi-nal) **ring** A triangular opening in the aponeurosis of the external oblique muscle that represents the distal end of the inguinal canal.

Superior (soo-PĒR-ē-or) Toward the head or upper part of a structure.

Superior vena cava (VĒ-na CĀ-va) **(SVC)** Large vein that collects blood primarily from parts of the body superior to the heart and returns it to the right atrium.

Supination (soo-pi-NĀ-shun) A movement of the forearm in which the palm is turned anteriorly.

Surface anatomy The study of the structures that can be identified from the outside of the body.

Surfactant (sur-FAK-tant) Complex mixture of phospholipids and lipoproteins, produced by type II alveolar (septal) cells in the lungs, that decreases surface tension.

Suspensory ligament (sus-PEN-so-rē LIG-a-ment) A fold of peritoneum extending laterally from the surface of the ovary to the pelvic wall.

Sutural (SOO-chur-al) **bone** A small bone located within a suture between certain cranial bones. Also called **Wormian** (WER-mē-an) **bone.**

Suture (SOO-chur) An immovable or slightly movable fibrous joint that joins skull bones.

Sympathetic (sim′-pa-THET-ik) **division** One of the two subdivisions of the autonomic nervous system, having cell bodies of preganglionic neurons in the lateral gray columns of the thoracic segment and the first two or three lumbar segments of the spinal cord; primarily concerned with processes involving the expenditure of energy.

Sympathetic trunk ganglion (GANG-glē-on) A cluster of cell bodies of sympathetic postganglionic neurons lateral to the vertebral column, close to the body of a vertebra. These ganglia extend inferiorly through the neck, thorax, and abdomen to the coccyx on both sides of the vertebral column and are connected to one another to form a chain on each side of the vertebral column. Also called **sympathetic chain** or **vertebral chain ganglia.**

Symphysis (SIM-fi-sis) A line of union. A slightly movable cartilaginous joint such as the pubic symphysis.

Symptom (SIMP-tum) A subjective change in body function not apparent to an observer, such as pain or nausea, that indicates the presence of a disease or disorder of the body.

Synapse (SIN-aps) The functional junction between two neurons or between a neuron and an effector, such as a muscle or gland; may be electrical or chemical.

Synapsis (sin-AP-sis) The pairing of homologous chromosomes during prophase I of meiosis.

Synaptic (sin-AP-tik) **cleft** The narrow gap at a chemical synapse that separates the axon terminal of one neuron from another neuron or muscle fiber (cell) and across which a neurotransmitter diffuses to affect the postsynaptic cell.

Synaptic end bulb Expanded distal end of an axon terminal that contains synaptic vesicles. Also called a **synaptic knob.**

Synaptic vesicle Membrane-enclosed sac in a synaptic end bulb that stores neurotransmitters.

Synarthrosis (sin′-ar-THRŌ-sis) An immovable or slightly movable joint such as a suture, gomphosis, or synchondrosis.

Synchondrosis (sin′-kon-DRŌ-sis) A cartilaginous joint in which the connecting material is hyaline cartilage.

Syndesmosis (sin′-dez-MŌ-sis) A slightly movable or slightly movable joint in which articulating bones are united by fibrous connective tissue.

Synergist (SIN-er-gist) A muscle that assists the prime mover by reducing undesired action or unnecessary movement.

Synergistic (syn-er-JIS-tik) **effect** A hormonal interaction in which the effects of two or more hormones acting together is greater or more extensive than the sum of each hormone acting alone.

Synostosis (sin′-os-TŌ-sis) A joint in which the dense fibrous connective tissue that unites bones at a suture has been replaced by bone, resulting in a complete fusion across the suture line.

Synovial (sī-NŌ-vē-al) **cavity** The space between the articulating bones of a synovial joint, filled with synovial fluid. Also called a **joint cavity.**

Synovial fluid Secretion of synovial membranes that lubricates joints and nourishes articular cartilage.

Synovial joint A fully movable or diarthrotic joint in which a synovial (joint) cavity is present between the two articulating bones.

Synovial membrane The deeper of the two layers of the articular capsule of a synovial joint, composed of areolar connective tissue that secretes synovial fluid into the synovial (joint) cavity.

System An association of organs that have a common function.

Systemic (sis-TEM-ik) Affecting the whole body; generalized.

Systemic anatomy The anatomic study of particular systems of the body, such as the skeletal, muscular, nervous, cardiovascular, or urinary systems.

Systemic circulation The routes through which oxygenated blood flows from the left ventricle through the aorta to all the organs of the body except the lungs and deoxygenated blood returns to the right atrium.

Systole (SIS-tō-lē) In the cardiac cycle, the phase of contraction of the heart muscle, especially of the ventricles.

Systolic (sis-TOL-ik) **blood pressure** The force exerted by blood on arterial walls during ventricular contraction; the highest pressure measured in the large arteries, about 120 mmHg under normal conditions for a young adult.

T

T cell A lymphocyte that becomes immunocompetent in the thymus and can differentiate into a helper T cell or a cytotoxic T cell, both of which function in cell-mediated immunity.

T wave The deflection wave of an electrocardiogram that represents ventricular repolarization.

Tachycardia (tak′-i-KAR-dē-a) An abnormally rapid resting heartbeat or pulse rate (over 100 beats per minute).

Tactile (TAK-tīl) Pertaining to the sense of touch.

Tactile disc See **Merkel disc.**

Target cell A cell whose activity is affected by a particular hormone.

Tarsal bones The seven bones of the ankle. Also called **tarsals.**

Tarsal gland Sebaceous (oil) gland that opens on the edge of each eyelid. Also called a **Meibomian** (mī-BŌ-mē-an) **gland.**

Tarsal plate A thin, elongated sheet of connective tissue, one in each eyelid, giving the eyelid form and support. The aponeurosis of the levator palpebrae superioris is attached to the tarsal plate of the superior eyelid.

Tarsus (TAR-sus) A collective term for the seven bones of the ankle.

Tectorial (tek-TŌ-rē-al) **membrane** A gelatinous membrane projecting over and in contact with the hair cells of the spiral organ (organ of Corti) in the cochlear duct.

Teeth (TĒTH) Accessory structures of digestion, composed of calcified connective tissue and embedded in bony sockets of the mandible and maxilla, that cut, shred, crush, and grind food. Also called **dentes** (DEN-tēz).

Telophase (TEL-ō-fāz) The final stage of mitosis.

Tendon (TEN-don) A white fibrous cord of dense regular connective tissue that attaches muscle to bone.

Tendon organ A proprioceptive receptor, sensitive to changes in muscle tension and force of contraction, found chiefly near the junctions of tendons and muscles. Also called a **Golgi** (GOL-jē) **tendon organ.**

Tendon reflex A polysynaptic, ipsilateral reflex that protects tendons and their associated muscles from damage that might be brought about by excessive tension. The receptors involved are called tendon organs (Golgi tendon organs).

Teniae coli (TĒ-nē-ē KŌ-lī) The three flat bands of thickened, longitudinal smooth muscle running the length of the large intestine, except in the rectum. *Singular* is **tenia coli.**

Tentorium cerebelli (ten-TŌ-rē-um ser′-e-BEL-ē) A transverse shelf of dura mater that forms a partition between the occipital lobe of the cerebral hemispheres and the cerebellum and that covers the cerebellum.

Terminal ganglion (TER-min-al GANG-glē-on) A cluster of cell bodies of parasympathetic postganglionic neurons either lying very close to the visceral effectors or located within the walls of the visceral effectors supplied by the postganglionic neurons.

Testis (TES-tis) Male gonad that produces sperm and the hormones testosterone and inhibin. Also called a **testicle.**

Testosterone (tes-TOS-te-rōn) A male sex hormone (androgen) secreted by interstitial endocrinocytes (Leydig cells) of a mature testis; needed for development of sperm; together with a second androgen termed **dihydrotestosterone (DHT),** controls the growth and development of male reproductive organs, secondary sex characteristics, and body growth.

Tetralogy of Fallot (tet-RAL-ō-jē of fal-Ō) A combination of four congenital heart defects: (1) constricted pulmonary semilunar valve, (2) interventricular septal opening, (3) emergence of the aorta from both ventricles instead of from the left only, and (4) enlarged right ventricle.

Thalamus (THAL-a-mus) A large, oval structure located bilaterally on either side of the third ventricle, consisting of two masses of gray matter organized into nuclei; main relay center for sensory impulses ascending to the cerebral cortex.

Thermoreceptor (THER-mō-rē-sep-tor) Sensory receptor that detects changes in temperature.

Thigh The portion of the lower limb between the hip and the knee.

Third ventricle (VEN-tri-kul) A slitlike cavity between the right and left halves of the thalamus and between the lateral ventricles of the brain.

Thoracic (thor-AS-ik) **cavity** Cavity superior to the diaphragm that contains two pleural cavities, the mediastinum, and the pericardial cavity.

Thoracic duct A lymphatic vessel that begins as a dilation called the cisterna chyli, receives lymph from the left side of the head, neck, and chest, left arm, and the entire body below the ribs, and empties into the junction between the internal jugular and left subclavian veins. Also called the **left lymphatic** (lim-FAT-ik) **duct.**

Thoracolumbar (thōr′-a-kō-LUM-bar) **outflow** The axons of sympathetic preganglionic neurons, which have their cell bodies in the lateral gray columns of the thoracic segments and first two or three lumbar segments of the spinal cord.

Thorax (THŌ-raks) The chest.

Thrombosis (throm-BŌ-sis) The formation of a clot in an unbroken blood vessel, usually a vein.

Thrombus (THROM-bus) A stationary clot formed in an unbroken blood vessel, usually a vein.

Thymus (THĪ-mus) A bilobed organ, located in the superior mediastinum posterior to the sternum and between the lungs, in which T cells develop immunocompetence.

Thyroid cartilage (THĪ-royd KAR-ti-lij) The largest single cartilage of the larynx, consisting of two fused plates that form the anterior wall of the larynx.

Thyroid follicle (FOL-i-kul) Spherical sac that forms the parenchyma of the thyroid gland and consists of follicular cells that produce thyroxine (T_4) and triiodothyronine (T_3).

Thyroid gland An endocrine gland with right and left lateral lobes on either side of the trachea connected by an isthmus; located anterior to the trachea just inferior to the cricoid cartilage; secretes thyroxine (T_4), triiodothyronine (T_3), and calcitonin (CT).

Thyroid-stimulating hormone (TSH) A hormone secreted by the anterior pituitary that stimulates the synthesis and secretion of thyroxine (T_4) and triiodothyronine (T_3).

Thyroxine (thī-ROK-sēn) **(T_4)** A hormone secreted by the thyroid gland that regulates metabolism, growth and development, and the activity of the nervous system.

Tic Spasmodic, involuntary twitching of muscles that are normally under voluntary control.

Tissue A group of similar cells and their intercellular substance joined together to perform a specific function.

Tissue rejection Phenomenon by which the body recognizes the protein (HLA antigens) in transplanted tissues or organs as foreign and produces antibodies against them.

Tongue A large skeletal muscle covered by a mucous membrane located on the floor of the oral cavity.

Tonsil (TON-sil) An aggregation of large lymphatic nodules embedded in the mucous membrane of the throat.

Topical (TOP-i-kal) Applied to the surface rather than ingested or injected.

Torn cartilage A tearing of an articular disc (meniscus) in the knee.

Trabecula (tra-BEK-ū-la) Irregular latticework of thin plates of spongy bone tissue. Fibrous cord of connective tissue serving as supporting fiber by forming a septum extending into an organ from its wall or capsule. *Plural* is **trabeculae** (tra-BEK-ū-lē).

Trabeculae carneae (KAR-nē-ē) Ridges and folds of the myocardium in the ventricles.

Trachea (TRĀ-kē-a) Tubular air passageway extending from the larynx to the fifth thoracic vertebra. Also called the **windpipe.**

Tract A bundle of nerve axons in the central nervous system.

Transplantation (tranz-plan-TĀ-shun) The transfer of living cells, tissues, or organs from a donor to a recipient or from one part of the body to another in order to restore a lost function.

Transverse colon (trans-VERS KŌ-lon) The portion of the large intestine extending across the abdomen from the right colic (hepatic) flexure to the left colic (splenic) flexure.

Transverse fissure (FISH-er) The deep cleft that separates the cerebrum from the cerebellum.

Transverse plane A plane that divides the body or organs into superior and inferior portions. Also called a **cross-sectional** or **horizontal plane.**

Transverse tubules (TOO-būls) **(T tubules)** Small, cylindrical invaginations of the sarcolemma of striated muscle fibers (cells) that conduct muscle action potentials toward the center of the muscle fiber.

Tremor (TREM-or) Rhythmic, involuntary, purposeless contraction of opposing muscle groups.

Triad (TRĪ-ad) A complex of three units in a skeletal muscle fiber composed of a transverse tubule and the sarcoplasmic reticulum terminal cisterns on both sides of it.

Tricuspid (trī-KUS-pid) **valve** Atrioventricular (AV) valve on the right side of the heart.

Triglyceride (trī-GLI-cer-īd) A lipid formed from one molecule of glycerol and three molecules of fatty acids that may be either solid (fats) or liquid (oils) at room temperature; the body's most highly concentrated source of chemical potential energy. Found mainly within adipocytes. Also called a **neutral fat** or a **triacylglycerol.**

Trigone (TRĪ-gon) A triangular region at the base of the urinary bladder.

Triiodothyronine (trī-ī-ō-dō-THĪ-rō-nēn) **(T_3)** A hormone produced by the thyroid gland that regulates metabolism, growth and development, and the activity of the nervous system.

Tropic (TRŌ-pik) **hormone** A hormone whose target is another endocrine gland.

Trunk The part of the body to which the head and upper and lower limbs are attached.

Tubal ligation (lī-GĀ-shun) A sterilization procedure in which the uterine (fallopian) tubes are tied and cut.

Tubular reabsorption (TOO-bū-lar rē-ab-SORP-shun) The movement of filtrate from renal tubules back into blood in response to the body's specific needs.

Tubular secretion The movement of substances in blood into renal tubular fluid in response to the body's specific needs.

Tumor suppressor gene A gene coding for a protein that normally inhibits cell division; loss or alteration of a tumor suppressor gene called *p53* is the most common genetic change in a wide variety of cancer cells.

Tunica albuginea (TOO-ni-ka al′-bū-JIN-ē-a) A dense white fibrous capsule covering a testis or deep to the surface of an ovary.

Tunica externa (eks-TER-na) The superficial coat of an artery or vein, composed mostly of elastic and collagen fibers. Also called the **adventitia** (ad-ven-TISH-a).

Tunica interna (in-TER-na) The deep coat of an artery or vein, consisting of a lining of endothelium, basement membrane, and internal elastic lamina (in an artery). Also called the **tunica intima** (IN-ti-ma).

Tunica media (MĒ-dē-a) The intermediate coat of an artery or vein, composed of smooth muscle and elastic fibers.

Tympanic antrum (tim-PAN-ik AN-trum) An air space in the middle ear that leads into the mastoid air cells or sinus.

Tympanic (tim-PAN-ik) **membrane** A thin, semitransparent partition of fibrous connective tissue between the external auditory meatus and the middle ear. Also called the **eardrum.**

U

Umbilical cord (um-BIL-i-kal) The long, ropelike structure containing the umbilical arteries and vein that connect the fetus to the placenta.

Umbilicus (um-bi-LĪ-kus *or* um-bil-Ī-kus) A small scar on the abdomen that marks the former attachment of the umbilical cord to the fetus. Also called the **navel.**

Upper limb The appendage attached at the shoulder girdle, consisting of the arm, forearm, wrist, hand, and digits. Also called **upper extremity.**

Uremia (ū-RĒ-mē-a) Accumulation of toxic levels of urea and other nitrogenous waste products in the blood, usually resulting from severe kidney malfunction.

Ureter (Ū-rē-ter) One of two tubes that connect the kidney with the urinary bladder.

Urethra (ū-RĒ-thra) The duct from the urinary bladder to the exterior of the body that conveys urine in females and urine and semen in males.

Urinalysis (ū-ri-NAL-i-sis) An analysis of the volume and physical, chemical, and microscopic properties of urine.

Urinary (Ū-ri-ner-ē) **bladder** A hollow, muscular organ situated in the pelvic cavity posterior to the pubic symphysis; receives urine via two ureters and stores urine until it is excreted through the urethra.

Urine The fluid produced by the kidneys that contains wastes and excess materials; excreted from the body through the urethra.

Urogenital (ū′-rō-JEN-i-tal) **triangle** The region of the pelvic floor inferior to the pubic symphysis, bounded by the pubic symphysis and the ischial tuberosities, and containing the external genitalia.

Urology (ū-ROL-ō-jē) The specialized branch of medicine that deals with the structure, function, and diseases of the male and female urinary systems and the male reproductive system.

Uterine (Ū-ter-in) **tube** Duct that transports ova from the ovary to the uterus. Also called the **fallopian** (fal-LŌ-pē-an) **tube** or **oviduct.**

Uterosacral ligament (ū′-ter-ō-SĀ-kral LIG-a-ment) A fibrous band of tissue extending from the cervix of the uterus laterally to the sacrum.

Uterovesical (ū′-ter-ō-VES-i-kal) **pouch** A shallow pouch formed by the reflection of the peritoneum from the anterior surface of the uterus, at the junction of the cervix and the body, to the posterior surface of the urinary bladder.

Uterus (Ū-te-rus) The hollow, muscular organ in females that is the site of menstruation, implantation, development of the fetus, and labor. Also called the **womb.**

Utricle (Ū-tri-kul) The larger of the two divisions of the membranous labyrinth located inside the vestibule of the inner ear, containing a receptor organ for static equilibrium.

Uvea (Ū-vē-a) The three structures that together make up the vascular tunic of the eye.

Uvula (Ū-vū-la) A soft, fleshy mass, especially the V-shaped pendant part, descending from the soft palate.

V

Vagina (va-JĪ-na) A muscular, tubular organ that leads from the uterus to the vestibule, situated between the urinary bladder and the rectum of the female.

Vallate papilla (VAL-āt pa-PIL-a) One of the circular projections that is arranged in an inverted V-shaped row at the back of the tongue; the largest of the elevations on the upper surface of the tongue containing taste buds. Also called **circumvallate papilla.**

Varicocele (VAR-i-kō-sēl) A twisted vein; especially, the accumulation of blood in the veins of the spermatic cord.

Varicose (VAR-i-kōs) Pertaining to an unnatural swelling, as in the case of a varicose vein.

Vas A vessel or duct.

Vasa recta (VĀ-sa REK-ta) Extensions of the efferent arteriole of a juxtamedullary nephron that run alongside the loop of the nephron (Henle) in the medullary region of the kidney.

Vasa vasorum (va-SŌ-rum) Blood vessels that supply nutrients to the larger arteries and veins.

Vascular (VAS-kū-lar) Pertaining to or containing many blood vessels.

Vascular (venous) sinus A vein with a thin endothelial wall that lacks a tunica media and externa and is supported by surrounding tissue.

Vascular spasm Contraction of the smooth muscle in the wall of a damaged blood vessel to prevent blood loss.

Vascular tunic (TOO-nik) The middle layer of the eyeball, composed of the choroid, ciliary body, and iris. Also called the **uvea** (Ū-ve-a).

Vasectomy (va-SEK-tō-mē) A means of sterilization of males in which a portion of each ductus (vas) deferens is removed.

Vasoconstriction (vāz-ō-kon-STRIK-shun) A decrease in the size of the lumen of a blood vessel caused by contraction of the smooth muscle in the wall of the vessel.

Vasodilation (vāz′-ō-DĪ-lā-shun) An increase in the size of the lumen of a blood vessel caused by relaxation of the smooth muscle in the wall of the vessel.

Vein (VĀN) A blood vessel that conveys blood from tissues back to the heart.

Vena cava (VĒ-na KĀ-va) One of two large veins that open into the right atrium, returning to the heart all of the deoxygenated blood from the systemic circulation except from the coronary circulation.

Ventral (VEN-tral) Pertaining to the anterior or front side of the body; opposite of dorsal.

Ventral ramus (RĀ-mus) The anterior branch of a spinal nerve, containing sensory and motor fibers to the muscles and skin of the anterior surface of the head, neck, trunk, and the limbs.

Ventricle (VEN-tri-kul) A cavity in the brain filled with cerebrospinal fluid. An inferior chamber of the heart.

Ventricular fibrillation (ven-TRIK-ū-lar fib-ri-LĀ-shun) Asynchronous ventricular contractions; unless reversed by defibrillation, results in heart failure.

Venule (VEN-ūl) A small vein that collects blood from capillaries and delivers it to a vein.

Vermiform appendix (VER-mi-form a-PEN-diks) A twisted, coiled tube attached to the cecum.

Vermis (VER-mis) The central constricted area of the cerebellum that separates the two cerebellar hemispheres.

Vertebral (VER-te-bral) **canal** A cavity within the vertebral column formed by the vertebral foramina of all the vertebrae and containing the spinal cord. Also called the **spinal canal.**

Vertebral column The 26 vertebrae of an adult and 33 vertebrae of a child; encloses and protects the spinal cord and serves as a point of attachment for the ribs and back muscles. Also called the **backbone, spine,** or **spinal column.**

Vesicle (VES-i-kul) A small bladder or sac containing liquid.

Vesicouterine (ves′-ik-ō-Ū-ter-in) **pouch** A shallow pouch formed by the reflection of the peritoneum from the anterior surface of the uterus, at the junction of the cervix and the body, to the posterior surface of the urinary bladder.

Vestibular (ves-TIB-ū-lar) **apparatus** Collective term for the organs of equilibrium, which includes the saccule, utricle, and semicircular ducts.

Vestibular membrane The membrane that separates the cochlear duct from the scala vestibuli.

Vestibule (VES-ti-būl) A small space or cavity at the beginning of a canal, especially the inner ear, larynx, mouth, nose, and vagina.

Villus (VIL-lus) A projection of the intestinal mucosal cells containing connective tissue, blood vessels, and a lymphatic vessel; functions in the absorption of the end products of digestion. *Plural* is **villi** (VIL-ī).

Viscera (VIS-er-a) The organs inside the ventral body cavity. *Singular* is **viscus** (VIS-kus).

Visceral (VIS-er-al) Pertaining to the organs or to the covering of an organ.

Visceral effectors (e-FEK-torz) Organs of the ventral body cavity that respond to neural stimulation, including cardiac muscle, smooth muscle, and glands.

Vitamin An organic molecule necessary in trace amounts that acts as a catalyst in normal metabolic processes in the body.

Vitreous (VIT-rē-us) **body** A soft, jellylike substance that fills the vitreous chamber of the eyeball, lying between the lens and the retina.

Vocal folds Pair of mucous membrane folds below the ventricular folds that function in voice production. Also called **true vocal cords.**

Volkmann's canal *See* **Perforating canal.**

Vulva (VUL-va) Collective designation for the external genitalia of the female. Also called the **pudendum** (poo-DEN-dum).

W

Wallerian (wal-LE-rē-an) **degeneration** Degeneration of the portion of the axon and myelin sheath of a neuron distal to the site of injury.

Wandering macrophage (MAK-rō-fāj) Phagocytic cell that develops from a monocyte, leaves the blood, and migrates to infected tissues.

Whiplash injury A motion injury which may or may not involve direct impact of the head. The head moves forcibly into both hyperflexion and hyperextension with resulting damage to neck muscles, spinal ligaments, blood vessels, intervertebral discs, and/or the spinal cord.

White matter Aggregations or bundles of myelinated and unmyelinated axons located in the brain and spinal cord.

White pulp The regions of the spleen composed of lymphatic tissue, mostly B lymphocytes.

White ramus communicans (RĀ-mus kō-MŪ-ni-kans) The portion of a preganglionic sympathetic axon that branches from the anterior ramus of a spinal nerve to enter the nearest sympathetic trunk ganglion.

X

Xiphoid (ZĪ-foyd) Sword-shaped.

Xiphoid (ZĪ-foyd) **process** The inferior portion of the sternum is the **xiphoid process.**

Z

Zona fasciculata (ZŌ-na fa-sik′-ū-LA-ta) The middle zone of the adrenal cortex consisting of cells arranged in long, straight cords that secrete glucocorticoid hormones, mainly cortisol.

Zona glomerulosa (glo-mer′-ū-LŌ-sa) The outer zone of the adrenal cortex, directly under the connective tissue covering, consisting of cells arranged in arched loops or round balls that secrete mineralocorticoid hormones, mainly aldosterone.

Zona reticularis (ret-ik′-ū-LAR-is) The inner zone of the adrenal cortex, consisting of cords of branching cells that secrete sex hormones, chiefly androgens.

Zygote (ZĪ-got) The single cell resulting from the union of male and female gametes; the fertilized ovum.

CREDITS

Illustration Credits

Chapter 1 Figure 1.1: Kevin Somerville. Table 1.1: Kevin Somerville. 1.2: Imagineering. 1.3-1.4: Imagineering. 1.5: Imagineering. 1.6: Kevin Somerville. 1.7: Imagineering. 1.8: Imagineering. 1.9: Kevin Somerville. 1.10: Imagineering. 1.11: Kevin Somerville.

Chapter 2 Figure 2.1-2.15: Imagineering.

Chapter 3 Figure 3.1-3.22: Imagineering. Table 3.2: Imagineering.

Chapter 4 Figure 4.1-4.3: Imagineering. Table 4.1: Kevin Somerville. 4.4: Imagineering. 4.5: Imagineering. Table 4.3: Imagineering. 4.6: Imagineering. Table 4.5: Kevin Somerville/Imagineering.

Chapter 5 Figure 5.1: Kevin Somerville. 5.2: Imagineering. 5.4: Kevin Somerville. 5.5: Imagineering. 5.6: Kevin Somerville. 5.8: Imagineering. 5.10: Imagineering. 5.11: Imagineering.

Chapter 6 Figure 6.1: Imagineering. 6.2: Kevin Somerville/Imagineering. 6.3: Kevin Somerville. 6.4: Kevin Somerville. 6.5: Kevin Somerville. 6.6: Kevin Somerville. 6.7: Kevin Somerville. 6.8: Kevin Somerville. 6.9: Imagineering. 6.10: Kevin Somerville. 6.11: Imagineering.

Chapter 7 Table 7.1: John Gibb. Figure 7.1: John Gibb. 7.2: John Gibb. Table 7.3: John Gibb. 7.3-7.8: John Gibb. 7.9: John Gibb. 7.10-7.12: John Gibb. 7.13-7.15: John Gibb. 7.16: John Gibb/Imagineering. 7.17: John Gibb. 7.18-7.22: John Gibb. Table 7.5: John Gibb. 7.23-7.24: John Gibb. 7.25: Imagineering.

Chapter 8 Figure 8.1: John Gibb. 8.2-8.11: John Gibb. Table 8.1: John Gibb. 8.12-8.17: John Gibb. Exhibit 8.1: Keith Kasnot/Imagineering.

Chapter 9 Figure 9.1: John Gibb. 9.2: John Gibb. 9.3: Imagineering. 9.4: John Gibb/Imagineering. 9.11: John Gibb. 9.12: Imagineering. 9.13: Imagineering. 9.14: Imagineering. 9.15: Imagineering. 9.16: Imagineering. 9.17: John Gibb.

Chapter 10 Figure 10.1: Kevin Somerville. 10.2: Kevin Somerville/Imagineering. 10.3: Imagineering. 10.4: Imagineering. 10.5: Imagineering. 10.6: Imagineering. 10.7: Imagineering. 10.8: Imagineering. 10.9-10.10: Imagineering. 10.12: Kevin Somerville/ Imagineering. 10.13: Imagineering. Table 10.2: Imagineering.

Chapter 11 Figure 11.1: Kevin Somerville. 11.2: Kevin Somerville. Table 11.1: Kevin Somerville. 11.3: John Gibb. 11.4: John Gibb/Imagineering. 11.6: John Gibb. 11.7: John Gibb. 11.8: John Gibb. 11.10: John Gibb. 11.11: John Gibb/Imagineering. 11.12: Imagineering.

Chapter 12 Figure 12.1: John Gibb. 12.2: Imagineering. 12.3: John Gibb. 12.5: John Gibb. 12.6: John Gibb. 12.7: John Gibb. 12.8: John Gibb.

Chapter 13 Figure 13.1: John Gibb. 13.3: Imagineering. 13.4: John Gibb. 13.6: John Gibb/Imagineering. 13.8: John Gibb/Imagineering. 13.9: John Gibb. 13.11: John Gibb.

Chapter 14 Figure 14.1: John Gibb/Imagineering. 14.2: John Gibb. 14.4: John Gibb. 14.6: John Gibb/Imagineering. 14.8: John Gibb/Imagineering. 14.9: Imagineering. Exhibit 14.5: Imagineering/Keith Kasnot.

Chapter 15 Figure 15.1: Kevin Somerville/Imagineering. 15.2: Imagineering. 15.3: Kevin Somerville. 15.4: Imagineering. 15.5: Imagineering. 15.6: Imagineering. 15.7: Imagineering. Table 15.2: Imagineering. 15.8: Imagineering. 15.9: Imagineering. 15.10: Imagineering. 15.11: Imagineering. 15.12: Imagineering. 15.13: Imagineering. 15.14.

Chapter 16 Figure 16.1-16.3: Imagineering. 16.4: Kevin Somerville. 16.5-16.6: Imagineering. 16.7: Imagineering. 16.8-16.17: Kevin Somerville.

Chapter 17 Figure 17.1-17.4: Imagineering. 17.5: Imagineering. 17.6: Imagineering. 17.7: Imagineering. Table 17.2: Imagineering. 17.8: Imagineering. 17.9-17.13: Imagineering.

Chapter 18 Figure 18.1: Imagineering. 18.2: Kevin Somerville/Imagineering. 18.3-18.4: Imagineering. 18.5-18.6: Kevin Somerville/Imagineering. 18.7: Kevin Somerville. 18.8: Imagineering. 18.9: Imagineering. 18.10-18.13: Imagineering.

Chapter 19 Figure 19.1: Kevin Somerville. 19.2: Kevin Somerville. 19.3: Imagineering. 19.4: Imagineering. 19.5: Imagineering. 19.6-19.7: Imagineering. 19.8: Imagineering. Table 19.3: Imagineering. 19.9: Imagineering. 19.10: Imagineering. 19.11: Imagineering. 19.12: Imagineering. 19.13: Imagineering. 19.14: Imagineering. Table 19.4: Imagineering.

Chapter 20 Figure 20.1: Kevin Somerville. 20.2: Kevin Somerville/Imagineering. 20.4: Kevin Somerville/Imagineering. 20.5: Imagineering. 20.6: Imagineering. 20.7: Imagineering. 20.9: Imagineering. 20.10-20.14: Imagineering.

Chapter 21 Figure 21.1-21.2: Imagineering. 21.3-21.4: Imagineering. Table 21.2: Imagineering. 21.8: Imagineering. 21.9: Imagineering. 21.10: Imagineering.

Chapter 22 Figure 22.1: Imagineering. 22.2: Imagineering. 22.3: Kevin Somerville. 22.4: Kevin Somerville. 22.5: Imagineering. 22.6: Kevin Somerville. 22.7: Kevin Somerville/Imagineering. 22.8: Imagineering. 22.9: Imagineering. 22.11: Kevin Somerville. 22.12: Kevin Somerville.

Chapter 23 Figure 23.1: Kevin Somerville. 23.2: Imagineering. 23.3: Imagineering. 23.4: Kevin Somerville. 23.5: Imagineering. 23.6: Imagineering. 23.7: Imagineering. 23.8: Kevin Somerville. 23.9: Kevin Somerville. 23.10-23.16: Kevin Somerville. 23.17: Kevin Somerville.

Chapter 24 Figure 24.1-24.5: Imagineering. 24.6: Imagineering. Table 24.2: Imagineering. 24.7: Imagineering. 24.8-24.10: Imagineering. 24.11: Imagineering.

Chapter 25 Figure 25.1: Imagineering. 25.2: Kevin Somerville/Imagineering. 25.3: Imagineering. 25.4: Imagineering. 25.5-25.6: Kevin Somerville/Imagineering. 25.7: Kevin Somerville. 25.8: Kevin Somerville. 25.9-25.15: Imagineering.

Chapter 26 Figure 26.1: Kevin Somerville. 26.2: Kevin Somerville. 26.3: Kevin Somerville. 26.4: Imagineering. 26.5: Imagineering. 26.6: Imagineering. 26.7: Kevin Somerville. 26.8-26.9: Kevin Somerville/Imagineering. 26.10: Imagineering. 26.11: Imagineering. 26.12: Kevin Somerville. 26.13: Kevin Somerville. 26.14: Imagineering. 26.15: Kevin Somerville. 26.16: Kevin Somerville.

Chapter 27 Figure 27.1-27.6: Imagineering.

Chapter 28 Figure 28.1: Kevin Somerville. 28.2-28.5: Imagineering. 28.6: Kevin Somerville. 28.7: Imagineering. 28.8: Imagineering. 28.9: Imagineering.

Chapter 29 Figure 29.1-29.2: Kevin Somerville. 29.3: Imagineering. 29.4: Kevin Somerville. 29.5: Imagineering. 29.6: Kevin Somerville. 29.7: Kevin Somerville. 29.8: Kevin Somerville/Imagineering. 29.9: Imagineering. 29.10: Kevin Somerville/Imagineering. 29.11: Kevin Somerville. 29.12: Kevin Somerville/Imagineering. 29.13: Imagineering. 29.14: Imagineering.

Photo Credits

Chapter 1 Opener: Chris Cole/Iconica/Getty Images. Fig. 1.1: Rubberball Productions/ Getty Images. Fig. 1.8a-c: Dissection Shawn Miller; Photograph Mark Nielsen. Fig. 1.11a: Andy Washnik. Page 21 (center left): Biophoto Associates/Photo Researchers. Page 21 (center): Breast Cancer Unit, Kings College Hospital, London/Photo Researchers, Inc. Page 21 (center right): Zephyr/Photo Researchers, Inc. Page 21 (bottom left): Cardio-Thoracic Centre, Freeman Hospital, Newcastle-Upon-Tyne/Photo Researchers, Inc. Page 21 (bottom center): CNRI/Science Photo Library/Photo Researchers, Inc. Page 21 (bottom right): Science Photo Library/Photo Researchers, Inc. Page 22 (center left): Scott Camazine/ Photo Researchers, Inc. Page 22 (center right): Simon Fraser/Photo Researchers, Inc. Page 22 (bottom): Courtesy Andrew Joseph Tortora and Damaris Soler. Page 23 (center left): Howard Sochurek/Medical Images, Inc. Page 23 (top right): SIU/Visuals Unlimited. Page 23 (center right): Dept. of Nuclear Medicine, Charing Cross Hospital/Photo Researchers, Inc. Page 23 (bottom right): ©Camal/Phototake.

Chapter 2 Opener: Ernie Friedlander/Cole Group/PhotoDisc/Getty Images.

Chapter 3 Opener: ©Masterfile. Fig. 3.5: Andy Washnik. Fig. 3.10b: David Phillips/ Photo Researchers, Inc. Fig. 3.12b-c: Courtesy Abbott Laboratories. Fig. 3.25: Courtesy Michael Ross, University of Florida.

Chapter 4 Opener: Corbis Digital Stock. Table 4.1a (top): Biophoto Associates/ Photo Researchers, Inc. Table 4.1a: (center) Mark Nielsen. Table 4.1b-f: Mark Nielsen. Table 4.1g: Courtesy Michael Ross, University of Florida. Table 4.1h: Courtesy Michael Ross, University of Florida. Table 4.1i: Mark Nielsen. Table 4.2a: Mark Nielsen. Table 4.2b: Mark Nielsen. Table 4.3a: Courtesy Michael Ross, University of Florida. Table 4.3b: Mark Nielsen. Table 4.4a-j: Mark Nielsen. Table 4.4k: Courtesy Michael Ross, University of Florida. Table 4.5a: Courtesy Michael Ross, University of Florida. Table 4.5b-c: Mark Nielsen. Table 4.6: Mark Nielsen.

Chapter 5 Opener: Stella/Getty Images. Fig. 5.1b: Courtesy Michael Ross, University of Florida. Fig. 5.2b: Courtesy Andrew J. Kuntzman. Fig. 5.3a: Courtesy Andrew J. Kuntzman. Fig. 5.3b: David Becker/Photo Researchers, Inc. Fig. 5.4b: VVG/Science Photo Library/Photo Researchers, Inc. Fig. 5.5a: Mark Nielsen. Fig. 5.5b: Mark Nielsen. Fig. 5.5c: Mark Nielsen. Fig. 5.7a-b: Courtesy Andrew J. Kuntzman. Fig. 5.9a: Alain Dex/Photo Researchers, Inc. Fig. 5.9b: Biophoto Associates/Photo Researchers, Inc. Fig. 5.10a: Sheila Terry/Science Photo Library/Photo Researchers, Inc. Fig. 5.10b: St. Stephen's Hospital/Science Photo Library/Photo Researchers, Inc. Fig. 5.10c: St. Stephen's Hospital/Science Photo Library/Photo Researchers, Inc. Fig. 5.12 Dr. P. Marazzi/ Science Photo Library/Photo Researchers, Inc.

Chapter 6 Opener: Volker Steger/Photo Researchers, Inc. Fig. 6.1b: Mark Nielsen. Fig. 6.2 (left): CNRI/Photo Researchers, Inc. Fig. 6.2 (center and right): Dr. Richard Kessel and Randy Kardon/Tissues & Organs/Visuals Unlimited. Fig. 6.5: Courtesy Andrew J. Kuntzman. Fig. 6.7a: The Bergman Collection. Fig. 6.7b: Courtesy Andrew J. Kuntzman. Fig. 6.9a: Courtesy Department of Medical Illustration, University of Wisconsin Medical School. Fig. 6.9b: Courtesy Department of Medical Illustration, University of Wisconsin Medical School. Fig. 6.9c: Courtesy Department of Medical Illustration, University of Wisconsin Medical School. Fig. 6.9d: Dr. Andrew Schmidt/ The Bergman Collection. Fig. 6.9e: Courtesy Department of Medical Illustration, University of Wisconsin Medical School. Fig. 6.9f: Watney Collection/Phototake. Fig. 6.12a: P. Motta/Photo Researchers, Inc. Fig. 6.12b: P. Motta/Photo Researchers, Inc.

Chapter 7 Opener: Dr. T. Pichard/Photo Researchers, Inc. Fig. 7.25a: Princess Margaret Rose Orthopaedic Hospital/Photo Researchers, Inc. Fig. 7.25b: Dr. P. Marazzi/Photo Researchers, Inc. Fig. 7.25c: Custom Medical Stock Photo, Inc. Fig. 7.26: Center for Disease Control/Project Masters, Inc.

Chapter 8 Opener: Joao Canziani/Getty Images.

Chapter 9 Opener: Andersen Ross/Digital Vision/Getty Images. Fig. 9.5-9.10: John Wilson White. Fig. 9.12d: Dissection Shawn Miller, Photograph Mark Nielsen. Fig. 9.15f: Dissection Shawn Miller, Photograph Mark Nielsen. Fig. 9.17b: SIU/Visuals Unlimited. Fig. 9.17c: ISM/Phototake. Page 250: Carolyn A. McKeone/Photo Researchers, Inc.

Chapter 10 Opener: Corbis Digital Stock. Fig. 10.11: John Wiley & Sons. Table 10.1: Biophoto Associates/Photo Researchers, Inc.

Chapter 11 Opener: Simon Watson/Getty Images. Fig. 11.5a: Andy Washnik. Fig. 11.5b: John Wiley & Sons. Fig. 11.9: Andy Washnik.

Chapter 12 Opener: PhotoDisc, Inc./Getty Images. Fig. 12.4: John Wiley & Sons

Chapter 13 Opener: Corbis Digital Stock. Fig. 13.2: Andy Washnik. Fig. 13.5: John Wiley & Sons. Fig. 13.7a-c: John Wiley & Sons. Fig. 13.10a-b: John Wiley & Sons. Fig. 13.11g: Andy Washnik. Fig. 13.12a-d: John Wiley & Sons.

Chapter 14 Opener: Purestock. Fig. 14.3: John Wiley & Sons. Fig. 14.5a-c: John Wiley & Sons. Fig. 14.7a-d: John Wiley & Sons.

Chapter 15 Opener: Helene Rogers/Age Fotostock America, Inc.

Chapter 16 Opener: Will & Deni McIntyre/Photo Researchers, Inc. Fig. 16.5b: ©Dr. Richard Kessel and Dr. Randy Kardon/Visuals Unlimited.

Chapter 17 Opener: ©SCPhotos/Alamy. Fig. 17.1: Dissection Shawn Miller, Photograph Mark Nielsen. Table 17.1: Dissection Shawn Miller, Photograph Mark Nielsen.

Chapter 18 Opener: Thomas Northcut/Getty Images.

Chapter 19 Opener: David M. Grossman/Phototake. Fig. 19.15: John Wiley & Sons.

Chapter 20 Opener: Andersen Ross/Photodisc/Getty Images. Fig. 20.2c: Courtesy Andrew J. Kuntzman. Fig. 20.3a From New England Journal of Medicine, February 18, 1999, vol. 340, No. 7, page 524. Photo provided courtesy of Robert Gagel, Department of Internal Medicine, University of Texas M.D. Anderson Cancer Center, Houston Texas. Fig. 20.3b: The Bergman Collection/Project Masters, Inc. Fig. 20.6b: Mark Nielsen. Fig. 20.8a: Martin Rotker/Phototake. Fig. 20.8b: The Bergman Collection/Project Masters, Inc. Fig. 20.13c: Mark Nielsen. Fig. 20.15: Biophoto Associates/Photo Researchers, Inc.

Chapter 21 Opener: Yoav Levy/Phototake. Fig. 21.5: Stanley Fleger/Visuals Unlimited. Fig. 21.6a-e: Courtesy Michael Ross, University of Florida. Fig. 21.7a: Juergen Berger/Photo Researchers, Inc. Fig. 21.7b: Mark Nielsen. Fig. 21.9: Dennis Kunkel/ Phototake.

Chapter 22 Opener: Purestock. Fig. 22.10a: Vu/Cabisco/Visuals Unlimited. Fig. 22.10b: W. Ober/Visuals Unlimited. Fig. 22.11d: ©ISM/Phototake.

Chapter 23 Opener: Alex Bartel/Photo Researchers, Inc.

Chapter 24 Opener: Andrew Caballero-Reynolds/Getty Images.

Chapter 25 Opener: Zephyr/Photo Researchers, Inc. Fig. 25.7b: Courtesy Andrew J. Kuntzman.

Chapter 26 Opener: Asia Images Group/Getty Images. Fig. 26.8 (inset): ©Hessler/Vu/Visuals Unlimited. Fig. 26.12b: Dissection Shawn Miller, Photograph Mark Nielsen.

Chapter 27 Opener: Susie M. Elsing Food Photography/StockFood America.

Chapter 28 Opener: Scott Schubach/CSM/Landov LLC.

Chapter 29 Opener: REUTERS/Adnan Abidi /Landov LLC.

INDEX

B

H

T

COMMON EPONYMS

In the life sciences, an eponym is the name of a structure, drug, or disease that is based on the name of a person. For example, you may be more familiar with the Achilles tendon than you are with its more anatomically descriptive term, the calcaneal tendon. Because eponyms remain in frequent use, this listing correlates common eponyms with their anatomical terms.

EPONYM	ANATOMICAL TERM
Achilles tendon	calcaneal tendon
Adam's apple	thyroid cartilage
ampulla of Vater (VA-ter)	hepatopancreatic ampulla
Bartholin's (BAR-tō-linz) gland	greater vestibular gland
Billroth's (BIL-rōtz) cord	splenic cord
Bowman's (BŌ-manz) capsule	glomerular capsule
Bowman's (BŌ-manz) gland	olfactory gland
Broca's (BRŌ-kaz) area	motor speech area
Brunner's (BRUN-erz) gland	duodenal gland
bundle of His (HISS)	artrioventricular (AV) bundle
canal of Schlemm (SHLEM)	scleral venous sinus
circle of Willis (WIL-is)	cerebral arterial circle
Cooper's (KOO-perz) ligament	suspensory ligament of the breast
Cowper's (KOW-perz) gland	bulbourethral gland
crypt of Lieberkühn (LE-ber-kyūn)	intestinal gland
duct of Santorini (san′-tō-RĒ-nē)	accessory duct
duct of Wirsung (VĒR-sung)	pancreatic duct
Eustachian (yoo-STĀ-kē-an)	auditory tube
Fallopian (fal-LŌ-pē-an) tube	uterine tube
gland of Littré (LĒ-tra)	urethral gland
Golgi (GOL-jē) tendon organ	tendon organ
Graafian (GRAF-ē-an) follicle	mature ovarian follicle
Hassall's (HAS-alz) corpuscle	thymic corpuscle
Haversian (ha-VĒR-shun) canal	central canal
Haversian (ha-VĒR-shun) system	osteon
Heimlich (HĪM-lik) maneuver	abdomial thrust maneuver
islet of Langerhans (LANG-er-hanz)	pancreatic islet
Kupffer (KOOP-fer) cell	stellate reticuloendothelial cell
Leydig (LĪ-dig) cell	interstitial endocrinocyte
loop of Henle (HEN-lē)	loop of the nephron
Luschka's (LUSH-kaz) aperture	lateral aperture
Magendie's (ma-JEN-dēz) aperture	median aperture
Meibomian (mi-BŌ-mē-an) gland	tarsal gland
Meissner (MĪS-ner) corpuscle	corpuscle of touch
Merkel (MER-kel) disc	tactile disc
Müllerian (mil-E rē-an) duct	paramesonephric duct
organ of Corti (KOR-tē)	spiral organ
Pacinian (pa-SIN-ē-an) corpuscle	lamellated corpuscle
Peyer's (PĪ-erz) patch	aggregated lymphatic follicle
plexus of Auerbach (OW-er-bak)	myenteric plexus
plexus of Meissner (MĪS-ner)	submucosal plexus
pouch of Douglas	rectouterine pouch
Purkinje (pur-KIN-jē) fiber	conduction myofiber
Rathke's (rath-KĒZ) pouch	hypophyseal pouch
Ruffini (roo-FĒ-nē) corpuscle	type II cutaneous mechanoreceptor
Sertoli (ser-TŌ-lē) cell	sustentacular cell
Skene's (SKĒNZ) gland	paraurethral gland
sphincter of Oddi (OD-dē)	sphincter of the hepatopancreatic ampulla
Volkmann's (FŌLK-manz) canal	perforating canal
Wernicke's (VER-ni-kēz) area	auditory association area
Wharton's (HWAR-tunz) jelly	mucous connective tissue
Wolffian duct	mesonephric duct
Wormian (WER-mē-an) bone	sutural bone

COMBINING FORMS, WORD ROOTS, PREFIXES, AND SUFFIXES

Many of the terms used in anatomy and phsiology are compound words; that is, they are made up of word roots and one or more prefixes or suffixes. For example, *leukocyte* is formed from the word roots *leuk-* meaning "white," a connecting vowel (o), and *cyte* meaning "cell." Thus, a leukocyte is a white blood cell. The following list includes some of the most commonly used combining forms, word roots, prefixes, and suffixes used in the study of anatomy and physiology. Each entry includes a usage example. Learning the meanings of these fundamental word parts will help you remember terms that, at first glance, may seem long or complicated.

COMBINING FORMS AND WORD ROOTS

Acous-, Acu- hearing Acoustics.
Acr- extremity Acromegaly.
Aden- gland Adenoma.
Alg-, Algia- pain Neuralgia.
Angi- vessel Angiocardiography.
Anthr- joint Arthropathy.
Aut-, Auto- self Autolysis.
Audit- hearing Auditory canal.

Bio- life, living Biopsy.
Blast- germ, bud Blastula.
Blephar- eyelid Blepharitis.
Brachi- arm Brachial plexus.
Bronch- trachea, windpipe Bronchoscopy.
Bucc- cheek Buccal.

Capit- head Decapitate.
Carcin- cancer Carcinogenic.
Cardi-, Cardia-, Cardio- heart Cardiogram.
Cephal- head Hydrocephalus.
Cerebro- brain Cerebrospinal fluid.
Chole- bile, gall Cholecystogram.
Chondr-, cartilage Chondrocyte.
Cor-, Coron- heart Coronary.
Cost- rib Costal.
Crani- skull Craniotomy.
Cut- skin Subcutaneous.
Cyst- sac, bladder Cystoscope.

Derma-, Dermato- skin Dermatosis.
Dura- hard Dura mater.

Enter- intestine Enteritis.
Erythr- red Erythrocyte.

Gastr- stomach Gastrointestinal.
Gloss- tongue Hypoglossal.
Glyco- sugar Glycogen.
Gyn-, Gynec- female, woman Gynecology.

Hem-, Hemat- blood Hematoma.
Hepar-, Hepat- liver Hepatitis.
Hist-, Histio- tissue Histology.
Hydr- water Dehydration.
Hyster- uterus Hysterectomy.

Ischi- hip, hip joint Ischium.

Kines- motion Kinesiology.

Labi- lip Labial.
Lacri- tears Lacrimal glands.
Laparo- loin, flank, abdomen Laparoscopy.
Leuko- white Leukocyte.
Lingu- tongue Sublingual glands.
Lip- fat Lipid.
Lumb- lower back, loin Lumbar.

Macul- spot, blotch Macula.
Malign- bad, harmful Malignant.
Mamm-, Mast- breast Mammography, Mastitis.
Meningo- membrane Meningitis.
Myel- marrow, spinal cord Myeloblast.
My-, Myo- muscle Myocardium.

Necro- corpse, dead Necrosis.
Nephro- kidney Nephron.
Neuro- nerve Neurotransmitter.

Ocul- eye Binocular.
Odont- tooth Orthodontic
Onco- mass, tumor Oncology.
Oo- egg Oocyte.
Opthalm- eye Ophthalmology.
Or- mouth Oral.
Osm- odor, sense of small Anosmia.
Os-, Osseo-, Osteo- bone Osteocyte.
Ot- ear Otitus media.

Palpebr- eyelid Palpebra.
Patho- disease Pathogen.
Pelv- basin Renal pelvis.
Phag- to eat Phagocytosis.
Phleb- vein Phlebitis.
Phren- diaphragm Phrenic.
Pilo- hair Depilatory.
Pneumo- lung, air Pneumothorax.
Pod- foot Podocyte.
Procto- anus, rectum Proctology.
Pulmon- lung Pulmonary.

Ren- kidneys Renal artery.
Rhin- nose Rhinitis.

Scler-, Sclero- hard Atherosclerosis.
Sep-, Septic- toxic condition due to microrganisms Septicemia.
Soma-, Somato- body Somatotropin.
Sten- narrow Stenosis.
Stasis-, Stat- stand still Homeostasis.

Tegument- skin, covering Integumentary.
Therm- heat Thermogenesis.
Thromb- clot, lump Thrombus.

Vas- vessel, duct Vasoconstriction.

Zyg- joined Zygote.

PREFIXES

A-, An- without, lack of, deficient Anesthesia.
Ab- away from, from Abnormal.
Ad-, Af- to, toward Adduction, Afferent neuron.
Alb- white Albino.
Alveol- cavity, socket Alveolus.
Andro- male, masculine Androgen.
Ante- before Antebrachial vein.
Anti- against Anticoagulant.

Bas- base, foundation Basal ganglia.
Bi- two, double Biceps.
Brady- slow Bradycardia.

Cata- down, lower, under Catabolism.
Circum- around Circumduction.
Cirrh- yellow Cirrhosis of the liver.
Co-, Con-, Com- with, together Congenital.
Contra- against, opposite Contraception.
Crypt- hidden, concealed Cryptorchidism.
Cyano- blue Cyanosis.

De- down, from Deciduous.
Demi-, hemi- half Hemiplegia.
Di-, Diplo- two Diploid.
Dis- separation, apart, away from Dissection.
Dys- painful, difficult Dyspnea.

E-, Ec-, Ef- out from, out of Efferent neuron.
Ecto-, Exo- outside Ectopic pregnancy.
Em-, En- in, on Emmetropia.
End-, Endo- within, inside Endocardium.
Epi- upon, on, above Epidermis.
Eu- good, easy, normal Eupnea.
Ex-, Exo- outside, beyond Exocrine gland.
Extra- outside, beyond, in addition to Extracellular fluid.

Fore- before, in front of Forehead.

Gen- originate, produce, form Genitalia.
Gingiv- gum Gingivitis.

Hemi- half Hemiplegia.
Heter-, Hetero- other, different Heterozygous.
Homeo-, Homo- unchanging, the same, steady Homeostasis.
Hyper- over, above, excessive Hyperglycemia.
Hypo- under, beneath, deficient Hypothalamus.

In-, Im- in, inside, not Incontinent.
Infra- beneath Infraorbital.
Inter- among, between Intercostal.
Intra- within, inside Intracellular fluid.
Ipsi- same Ipsilateral.
Iso- equal, like Isotonic.

Juxta- near to Juxtaglomerular apparatus.

Later- side Lateral.

Macro- large, great Macrophage.
Mal- bad, abnormal Malnutrition.
Medi-, Meso- middle Medial.
Mega-, Megalo- great, large Megakaryocyte.
Melan- black Melanin.
Meta- after, beyond Metacarpus.
Micro- small Microfilament.
Mono- one Monounsaturated fat.

Neo- new Neonatal.